AF572500

Soil Mechanics in Highway Engineering

The photo on the front cover is by courtesy of
STRABAG BAU-AG, Köln, Federal Republic of Germany
showing the Qurum-Darsait-Street, Oman, under construction in December 1983

Soil Mechanics in Highway Engineering

Alfonso Rico Rodriguez

Professor of Soil Mechanics, National Autonomous University of Mexico
Head of the Geotechnical Department, Public Works Ministry, Mexico

Hermillo del Castillo

Civil Engineer, Geotechnical Department, Public Works Ministry, Mexico

George F. Sowers

Law Engineering Testing Company, USA

1988

TRANS TECH PUBLICATIONS

Prof. Alfonso Rico Rodriguez
Hermillo del Castillo
Dr. George F. Sowers

Soil Mechanics in Highway Engineering

TRANS TECH PUBLICATIONS
D-3392 Clausthal-Zellerfeld
Federal Republic of Germany

ISBN 0-87849-072-8

Library of Congress Cataloging-in-Publication Data

Rico Rodríguez, Alfonso.
(Ingeniería de suelos en las vías terrestres. English)
Soil mechanics in highway engineering / Alfonso Rico Rodriguez, Hermillo del Castillo, George F. Sowers.
p. cm. — (Series on rock and soil mechanics: 16)
Translation of: Ingeniería de suelos en las vías terrestres.
ISBN 0-87849-072-8
1. Highway engineering. 2. Soil mechanics. I. Castillo, Hermillo del. II. Sowers, George F. III. Title. IV. Series: Series on rock and soil mechanics: v. 16.
TE153.R4613 1987
625.7'32-dc19

Preface to the Completely Revised and Updated English Edition

This book, since its appearance in 1973 in the Spanish edition, has made many friends throughout the Spanish speaking world. The preparation of the totally revised and updated English edition has been a labour of love which has now reached fruition. The expertise of the authors, involved with the wide variety of soils and terrain found in Mexico, is now available to a much wider audience — in particular in the developing world where often significantly different weights are given to the priorities involved in road and railway construction. This book is excellently self-contained in as much as it includes large quantities of the most relevant information from other sources in the form of figures, graphs and tables which will avoid any unnecessary dependence on further reading. Where, however, a deeper insight is required ample references are provided to enable the reader to enquire further.

The authors and the publisher would like to thank Dr. G.F. Sowers for accepting the co-authorship and for critical reading of the English manuscript and Dr. P.I. Welch for preparing the manuscript for publication. Mrs. Ruth Welch was assiduous in her proof-reading.

I hope this book will provide as much pleasure to the reader as we have had in its preparation.

REINHARD H. WÖHLBIER
Editor-in-Chief
Trans Tech Publications

Clausthal-Zellerfeld 1988

Foreword

"According to the grace of God which is given unto me as a wise master builder, I have laid the foundation and another buildeth thereon".

(I: Corinthians. 3:10)

The above biblical quotation seems peculiarly appropriate for this book on soil mechanics as it is evident that soil or soil materials compose the ultimate foundation for most engineering structures. The reference seems especially appropriate as it is the expressed hope of the authors that many young engineers will *build upon the foundation* being laid by this book.

Soil mechanics, especially as it applies to road engineering, has had to progress through a maze of difficulties. Hard *practical* men were prone to regard all excavation for roads as being either *dirt* or *rock*. The wide variety of natural materials found on the surface of the earth have been studied, analyzed, classified and written about by farmers, agronomists, geologists, petrographers and mining engineers. As a result such simple terms as *good soil* or *poor soil* have meaning only for a particular specialist and what may be thought *good* for the agriculturist may be very poor for the engineer and vice versa.

It is especially true that the descriptive terms and classifications established by the geologist often have rather vague or misleading connotations for the engineer who is trying to determine just how a given soil will perform in the conditions of service.

It must be recognized that the successful utilization of soils as an engineering material is an activity that involves both art and science. Man has been shaping and utilizing soils in his structures since the dawn of civilization and before. Soils may or may not be the oldest engineering material but the Ancients had to learn a great deal in the art of making pottery and constructing buildings. It is one of the major developments of modern civilization that engineers have been able to introduce the scientific approach to this oldest of the construction arts.

The ultimate design and final plans for any large engineering project today must in effect be a compromise. For example, a major highway project is developed by the composite efforts of many individuals contributing detailed knowledge from many specialties.

As most engineering works rest upon the surface of the earth, the ability of the soil to support loads becomes a primary consideration. The expression "common as dirt" signifies that most people find little that is new or novel about soils. Nevertheless, soils have many special and peculiar properties which, through the ages, have led a few men to devote a lifetime to their study.

Today the competent highway materials engineer needs to be familiar with the work and findings of men who have worked in many different fields involving the use of soils. Present day knowledge of the materials composing the earth's crust has been slowly accumulated by such diverse technologists as: agricultural soil specialists, soil chemists, ceramicists, engineers, geologists, mining engineers, and several varieties of civil engineers and military engineers individually engaged in work on dams, canals, railroads, soil erosion control, airports, building foundations, etc.

Let us again consider the development of a large highway project. Here there is a need to correlate the efforts and knowledge of still another group of specialists. Those preparing the project plans and specifications must assimilate and include the preliminary information gleaned from the ground and aerial surveys followed by investigations and test data furnished by the materials engineer. Considered with the other basic data should be an estimate of the "load", i.e., the number and weight of vehicles which are expected to make up the traffic on the road. After the plans and specifications have been prepared they must be reviewed by the legal advisor in as much as a contract to build all, or a portion, of a project becomes a legal document and in the final analysis legal interpretations will usually take precedence over purely engineering considerations and opinions in the event of a dispute between the contractor and the engineer. In a well integrated highway department all plans and specifications should also be reviewed and examined by both the construction engineer and the maintenance engineer. It will be obvious that cooperation and team work are essential.

Any treatise on soil mechanics is an attempt to arrive at an orderly understanding of the factors responsible for the behavior of the soil and its ability to sustain loads. In order to realize the benefits from such a theoretical analysis it is further necessary that the essential requirements for handling and construction should be clearly set forth in the contract specifications. Then it is equally mandatory that the essential requirement will be fullfilled and the work properly carried out by the contractor or the actual construction forces. Again there is the need for team work and a sympathetic understanding on the part of all concerned for the purposes, aims and function of the other members of the team.

Mention has been made of the numerous individuals and specialists who usually have some part in the preparation of plans and the execution of a large engineering project. There is still another *specialist*, or group of them, — I refer to those who control the funds available. It is an old and trite saying that *given enough time and money any fool can build anything;* this saying might be debated but in any event it is the province of the engineer to construct a satisfactory project with limited funds and within a limited time.

This book comes from a land that offers enduring evidence of the skill of man in coping with the problems posed by the soils underlying both ancient and modern structures. The pyramids and temples built long ago in Mexico are as impressive and durable as any left by the builders of antiquity in any part of the world. Mexico presents a wide variety of soil types and great variations in rainfall and ground water conditions. Engineers ALFONSO RICO RODRIGUEZ and HERMILO DEL CASTILLO have had unique opportunities in their positions as practicing soil mechanics engineers for the Mexican Ministry of Public Works (SOP). They have had to deal with the problems of soils under highways, bridges, airports, and other forms of Public Works. In addition, the practice of teaching courses in soil mechanics at the National University of Mexico has kept them abreast of the theoretical developments in other countries of the world.

The authors are to be complimented upon their broad knowledge of both theoretical and practical aspects of soil mechanics and perhaps even more for the energy and devotion to their profession that has led to the writing of this book.

FRANCIS N. HVEEM

Consulting Engineer
Former Head of the Central Laboratory,
California Highway Department

Sacramento, CA
April, 1973

Preface to the First Edition

For the last 25 years, the authors of this book have devoted themselves to activities which basically could be described as the application of soil mechanics to the design and construction of roads in Mexico. They performed this work at the Mexican Ministry of Public Works which is the federal agency responsible for the planning, design, construction and maintenance of roads, among other functions.

In the course of their daily work, they have found that among all the fields where soil mechanics is likely to be of some assistance, road engineering is one of the most complete, fascinating and highly complicated ones. If it is simply remembered that roads are earth structures which are built on natural earth ground, it will be realized why soil mechanics inevitably plays such an important part at every stage of their design and construction. Sometimes the mechanical properties of the foundation ground are so critical that without the solutions provided by soil mechanics it would be impossible, or at least unreasonably risky, to confront the problems that may arise. At other times, the more favorable foundation conditions may apparently (as frequently used to happen in the past, but is fortunately becoming less frequent in Mexico) allow one to proceed without the guide of soil mechanics, but it becomes clear with just a minimum of experience in applying these techniques that even in this case an opportunity to optimize projects and reduce costs would be lost, which would be totally absurd at modern technological levels. Road construction techniques are now inconceivable without the extensive, continuous and detailed use of the principles of applied soil mechanics, just as they would be without a similar utilization of geology and rock mechanics.

Roads represent one of the most complete fields for the application of soil mechanics; this fact can be corroborated by a simple enumeration of some of the many problems involved, such as the stability of hillsides and artificial slopes, the construction of embankments on soft soils, the earth thrust against all kinds of retaining structures, the foundations for bridges and road structures, etc. Moreover, the action of water on roads occurs in the widest variety of ways that can be conceived in civil engineering, and it is common knowledge that this element can cause complications in soil mechanics whenever it seeps, flows or tries to surface. Further, road engineering involves two extremely important aspects which, despite the relatively intensive study which has been dedicated to them in the last few years, are nonetheless still difficult and little known. They are soil compaction and pavement design. In both cases it is felt that an approach based on soil mechanics can be of considerable value for clearing up ancient problems inherited from a more empirical, less scientific practice. Research on compaction is fairly recent and still very incomplete, but already opens up fascinating horizons to the observer. It falls into the category of problems relating to unsaturated soils, where soil mechanics reaches the highest degrees of complexity and unreliability. Pavement technology has made extraordinary progress, to the extent that it is now the object of a new specialty which at times seems to claim independence within the field of civil engineering. It should not, however, be forgotten that a pavement is a structure in which the soils and their combined properties will have a lot to say. The day is long off when sufficient progress has been made so that an optimist is able to say that he knows something sure and definitive about it.

The amounts invested by almost every country in the world and the variety and complexity of the problems involved readily justify the dedication of so many soil mechanics specialists to roads, railroads and runways.

In the course of their daily work, the authors have, however, observed another curious fact for which they can find no ready explanation. As far as they know, no book has yet been published on the subject of the application of soil mechanics to the road engineering field. Numerous books are available on soil mechanics together with excellent papers dealing with one of the many aspects of the theories involved. There are even many books on its application to other fields, such as earth dams or foundations, but the beauties of roads were never proclaimed by anyone. Different aspects of the subject are dealt with every year on tons of paper, but always in the form of articles, monographs or that curious new (most useful) phenomenon inexplicably known as *State of the Art Reports.* Although the authors of this book agree that these are the only publications that can suitably capture the different aspects of a constantly changing technical discipline, they are nonetheless strongly convinced of the value of a book, which will both lay down the already well established principles and inform of the most outstanding experiences. Such a book could be compared with a halt on the road, almost a still photograph of a scene in perpetual motion, which captures a mere instant, but which may avoid possible confusion owing to the incessant apparent disorder on the stage.

This is the goal the authors have set themselves, and it is only the absence of books by more competent men that has induced them to undertake this task.

In view of the fact that this book serves a didactic purpose, certain problems relating to earth thrust and slope stability are included to illustrate the different methods of analysis.

Many people have collaborated in this book. To all of them the authors wish to express their most grateful thanks.

Juan Manuel Orozco, Manuel Jara and Manuel Zárate read parts of the manuscript and made most useful comments. Eulalio Juarez Badillo, Jesús Alberro and Daniel Reséndiz discussed many delicate points with the authors.

Esteban Meneses is responsible for the figures and Ma. Esther Escoto, Ma. Antonieta Cárdenas and Graciela Reyes accomplished the most unrewarding task of typing the original with the greatest enthusiasm.

Grateful thanks are also extended to the Mexican Ministry of Public Works and the National Autonomous University of Mexico both of which gave warm encouragement and numerous facilities without which this book would have proved most difficult to write.

Alfonso Rico Rodriguez
Hermillo del Castillo

México, D.F.
November 1973.

CONTENTS

CHAPTER 1

FUNDAMENTAL CONCEPTS OF SOIL MECHANICS

1.1 Introduction

For the purposes of this book *roads* are understood to include the highways, railways and runways which constitute the basic elements of the infrastructure of a national transport grid. Within this context should be included both the most modern freeway, and the most humble rural road, likewise the runway of a great airport with facilities for jet aircraft, along with the simple landing strip for the operation of small planes. The roads thus defined are built fundamentally of soil and on soil. For some time now, modern technology has recognized the influence exerted by the supporting ground on a structure of this type, implying by this not only the soil or rock at the site, passively speaking, but also a whole series of conditions ranging from the mineralogical make-up, the structure of the soil, the quantity and state of pore-water and the manner in which it flows to a whole group of factors foreign to the traditional concept of soil, but which define its behavior in relation to time, such as climatic conditions, economic factors and those which refer to *soil usage* in activities which have little or nothing to do with the technology of road engineering. Nevertheless, only recently have engineers understood that the use of materials commonly found in a great variety in nature, within the framework of the structure, is neither indifferent nor arbitrary, but selective. Using the same materials, very different structural sections can be obtained, depending on the use made of the materials, the position of each material within the section and the placement procedures, including the mechanical and even chemical treatments to which they have been subjected.

Road engineering implies not only the use of soils, but a selective, judicious and, whenever possible, *scientific* use. Modern engineering has developed branches whose objectives are the rational handling, in the best possible manner from an engineering point of view, of the soil and rocks with which roads are built. These branches are soil and rock mechanics, closely assisted by applied geology. It is not, therefore, surprising that, leaving aside aspects of planning and land survey, as well as others of an economic and social character, the design and construction of roads is a matter of judiciously applying the concepts of soil and rock mechanics.

Today, soil and rock mechanics have diversified to such an extent that each constitutes an independent branch of knowledge, with different methodology and objectives, both within the framework of engineering specialties. Although the dividing line between the two branches is often very hard to distinguish, in the same way as it is often hard to distinguish between soils and rocks, they are gradually becoming two quite separate fields.

This book deals with the application of soil mechanics in road engineering. Rock mechanics is included only when the methods of the two disciplines overlap and the solutions are common or when differentiation is practically impossible.

The application of soil mechanics to any field requires a basic knowledge of that discipline, which is intentionally considered as being beyond the scope of this book. Fortunately, there are many books, some of them very good, to which the reader may refer for additional information. Nevertheless, for the purposes of unity of thought and uniform nomenclature this first chapter is devoted to the fundamental concepts of soil mechanics.

1.2 Nature and Origin of Soils

Soils are an assemblage of mineral particles, products of the physical disintegration or chemical decomposition of pre-existing rocks. The assemblage possesses two essential properties which must be appreciated by those who wish to understand its engineering behavior.

— It possesses a defined organization and properties which vary *vectorially*. In most soils the properties change more rapidly vertically than horizontally.

— The organization of the mineral particles is such that water, which is present in all soils will, if sufficient, have *continuity* of pressure distribution. The water does not occupy isolated voids without intercommunication; it can fill all the pores between the mineral particles and intercommunicate in such a manner that water forms a continuous mass, partially enveloping the minerals.

Soils are residual or transported, depending on whether they have remained where they developed or have moved elsewhere. Transportation by air, water or ice, followed by deposition, constitutes the usual process forming a transported soil.

The structure and *internal distribution* of the properties evinced by residual soils are completely different from those of transported soils. In the former, the mechanical attack and the chemical disintegration tend to produce a final result whose structure and disposition reflect, though distantly, those of the parent rock. Soils transported by air or water and deposited on land or in water, develop structures in accordance with the mechanisms of deposition and no longer reflect the characteristics and conditions of the original rock.

Unfortunately, early soil mechanics emphasized deposited soils and largely ignored residual soils. The real problems which led to systematic studies of the engineering behavior of soils were building settlements and landslides in urban areas. Most cities, situated in valleys or along the sea coast are underlain by deposited soils.

Residual soils predominate in more rugged terrain and particularly in warm humid regions where engineering construction has been less intense. Therefore, most site investigations, laboratory tests and research on fundamental properties have been made on transported, deposited soils. The developing methodologies of analysis, design and construction were based on the data and theories from work on deposited soils. If residual soils were encountered in more remote studies, scientists and engineers assumed them to behave in the same way as the transported, deposited soils.

Although deposited soils abound in nature, residual soils are no less abundant, and in highways and railroads that traverse all types of terrain, residual soils are encountered frequently. In the late 1960s engineers began to investigate residual soils, and learn about their composition, fabric and engineering behavior. Although there are only small differences in the behavior of residual and transported soils, such differences could lead to some changes in the mental attitudes in relation to residual soils and to experimental methodologies, including testing and equipment design.

The engineer who applies soil mechanics to road engineering should bear in mind the aforementioned ideas, and exercise a critical attitude towards the conclusions obtained from the current soil mechanics concepts when dealing with residual soils. This critical attitude, will allow the discovery of current deficiencies and new areas of interest for further study. Some countries are particularly rich in residual soils associated with construction problems. In the Union of South Africa, Brazil, India, Indonesia and in the south-eastern United States, concern has already been expressed about the lack of information specifically focused on residual soils. In Mexico, residual soils also abound. Doubtless they are more commonly found in regions with a tropical climate, where the chemical activity of waters, charged with soluble agents produced by rife vegetation, results in a decomposition and *in-situ* attack being carried out faster than the erosion capacity of natural agents.

1.3 Gravimetric and Volumetric Relationships

Three phases are distinguished in soils: solid (mineral particles), liquid (generally water), and gaseous (generally air). Among these phases it is necessary to define relationships between weights and volumes, that will establish the nomenclature and measurable parameters through which it is possible to describe the engineering processes affecting soils.

Figure 1-1 gives an idealized drawing of a soil sample, showing the three phases with weights and volumes.

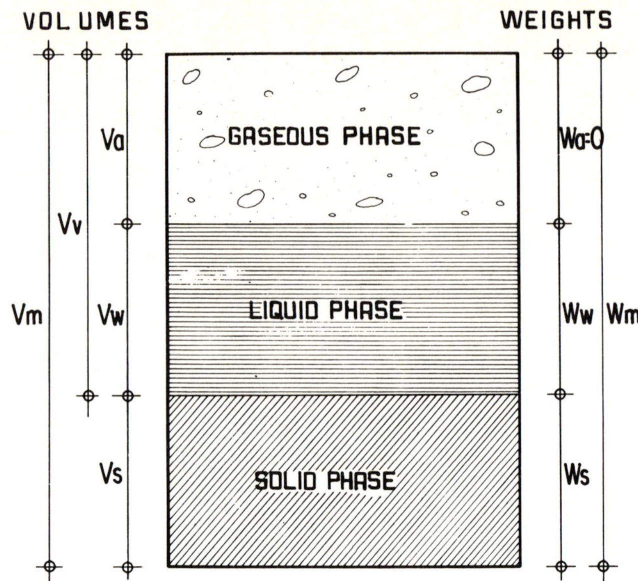

Fig. 1-1 Diagram showing a sample of soil, with the weights and volumes of components defined

The relationships between the weights and the volumes are established through the concept of unit weight (density) which is defined as the relationship between both quantities:

$$\gamma_m = \frac{W_m}{V_m} = \frac{W_s + W_w}{V_m} \tag{1-1}$$

In the expression γ_m is known as the bulk unit weight. The symbol γ_s is defined by:

$$\gamma_s = \frac{W_s}{V_s} \tag{1-2}$$

and is called the unit weight of the solid. In problems of compaction the term dry unit weight (dry density), is defined as the relationship between the weight of solids and the total volume of the soil:

$$\gamma_d = \frac{W_s}{V_m} \tag{1-3}$$

Note that Eq. (1-3) can be written:

$$\gamma_d = \frac{W_s\, W_m}{V_m W_m} = \frac{W_m/V_m}{W_m/W_s} = \frac{\gamma_m}{\dfrac{W_w + W_s}{W_s}} = \frac{\gamma_m}{1 + w} \tag{1-4}$$

Equation (1-4) is used in soil compaction. Also used is the relative unit weight or specific gravity of the solids, defined as:

$$S_s = \frac{\gamma_s}{\gamma_o} = \frac{W_s}{V_s\, \gamma_o} \tag{1-5}$$

Where γ_o is the unit weight of water (also written as γ_w). The following are relationships between weights and volumes which are also much used in practice, because they represent concepts that easily describe changes in soil structure.

The *void ratio, e,* is the quotient of the volume of voids and the volume of solids.

$$e = \frac{V_v}{V_s} \quad (1\text{-}6)$$

Theoretically, e, can vary from 0 to infinity (total vacuum), but in practice it varies from 0.25 for highly compacted sands with fines, to 15 for highly structured compressible clays, or even greater for peats.

The ***degree of saturation*** is the relationship between the volume of water and the volume of soil voids; mathematically:

$$G_w\,(\%) = 100\,\frac{V_w}{V_v} \quad (1\text{-}7)$$

The degree of saturation varies from 0% in a dry soil to 100% in a soil with all its voids full of water, known as a saturated soil.

The ***water content*** of a soil is the relationship between the weight of the water and the weight of the solid phase:

$$w(\%) = 100\,\frac{W_w}{W_s} \quad (1\text{-}8)$$

The water content varies theoretically from 0 to infinity but in practice it is rare to encounter values greater than 1,000%, which have been measured in clays from south-east Mexico; the well-known Mexico Valley clays have water contents between 400% and 600%.

The foregoing establish some useful relationships, which obviate the need to measure them all in the laboratory.For example, in a totally saturated soil it is sufficient to know two independent parameters to establish the others; in this case useful are:

$$e = wS_s \quad (1\text{-}9)$$

$$\gamma_s = \frac{S_s + e}{1 + e}\,\gamma_w = \frac{S_s\,(1 + w)}{1 + S_s\,w\,\gamma_w} \quad (1\text{-}10)$$

The derivation of these Equations, and of those which are mentioned below, referring to volumetric and gravimetric relationships are found in [1].

In the case of partially saturated soils (that is to say with part of their voids occupied by air) three independent quantities are required to define any other specific quantity. The more usual relationships which may be derived are:

$$eG_w = wS_s \quad (1\text{-}11)$$

and:

$$\gamma_m = \frac{1 + w}{1 + e}\,\gamma_s \quad (1\text{-}12)$$

Special attention must be given to the calculation of unit weights for soils located below the water table. In such a case, the hydrostatic buoyancy exerts an influence on weights, in accordance with the laws of buoyancy (ARCHIMEDEŚ Principle). The relative unit weight of submerged solid material is equal to:

$$S'_s = S_s - 1 \quad (1\text{-}13)$$

and the unit weight of submerged solids is:

$$\gamma'_s = \gamma_s - 1$$

$$\gamma'_s = \gamma_s - \gamma_o = \gamma_s - \gamma_w \quad (1\text{-}14)$$

That is to say, one cubic meter of solid soil displaces one cubic meter of water; therefore it is subject to a buoyancy of 1000 kg which is the weight of the cubic meter of water. The following equations are obtained for the submerged unit weight of the mass of soil [1]:

$$\gamma'_m = \frac{S_s - 1}{1 + S_s w}\,\gamma_w \quad (1\text{-}15)$$

and

$$\gamma'_m = \frac{S_s - 1}{S_s}\,\gamma_d \quad (1\text{-}16)$$

1.4 Characteristics and Structure of Mineral Particles

The shape of the mineral particles of a soil is of paramount importance in mechanical behavior. In coarse grained soils the characteristic shape is equidimensional; the three dimensions of the particle are of a comparable size. This is due to the action of mechanical break-down of rocks and only in exceptional cases corresponds to particles that have been affected by chemical agents. Since mechanical agents generally do not act with preference to a specific direction, the final product naturally tends to take a equidimensional form. An example of this is the rounded form characteristic of gravels and sands which have undergone mechanical abrasion in rivers and seas. If the rock particle or mineral exhibits planes of weakness, as does mica, then the broken fragment will not be equidimensional. In coarse grained soils, gravity forces are predominant over all other forces. Therefore, coarse particles all behave in a similar way.

In fine soils, which are generally the product of chemical decay, the shape of the mineral components tends to be flat (clay minerals usually take on a sheet-like shape, similar to mica flakes. In it, two of the dimensions are considerably greater than the third. As an exception, some clay minerals have an acicular or needle-like form, in which one dimension is much greater than the other two).

On account of their shape and very small size, the minerals of fine soil particles are affected by forces other than gravitational ones. In fine particles the ratio between the surface area and the weight (specific surface) reaches high values, and the electromagnetic forces on the surface of the minerals take on considerable importance. The internal structure of clays can be conceived in an elementary way. In [3&4], the reader will find some investigations that will enable him to take a closer look at the physico-chemical make-up of clays. This subject has gradually been gaining major importance in soil mechanics and has proved extremely useful in explaining the macroscopic behavior of earth masses that an engineer finds in his everyday activities.

1.4.1 Physico-chemical Make-up of Clays

The surface of each soil particle possesses a negative electric charge on its flat surfaces (contrarily, there seems to be evidence of a positive charge on the edges). The intensity of the charge depends on the composition of the clay. In this way, the particle attracts the positive ions of the water around it (H^+), along with the cations of the different elements in the water, such as Na^+, K^+, Ca^{++}, Mg^{++}, Al^{+++}, and Fe^{+++}. Thus each individual clay particle is surrounded by a layer of water molecules which are oriented in a specific direction and joined in a structure termed adsorbed water. When the clay particle attracts cations of other elements, these in turn attract other water molecules that become oriented, so the thickness of the absorbed water in a clay crystal is not only a function of the nature of the particle itself, but also of the cations attracted.

Due to the enormous specific surface of clay crystals, the electrical forces developed on its surface play a much more important role than gravity on the behavior of the particle. In the first place, this is reflected in the extremely open structures that fine soil particles can adopt when they sediment in the appropriate medium. Moreover, among the crystals of fine soils, the adsorbed layers provide a *bonded* contact which helps explain macrophysical properties, familiar to the engineer, such as plasticity and shear resistance.

The mechanical properties of a clay may change if there is a variation in the cations contained in the adsorbed complex. This presents possibilities for the physico-chemical treatment of many soils on an engineering level; unfortunately, such methods have not been sufficiently developed. In general, cations can be ranked according to their decreasingly beneficial effect on the resistance of clays as follows: $(NH_4)^+$, H^+, K^+, Fe^{+++}, Al^{+++}, Mg^{++}, Ba^{++}, Ca^{++}, Na^{++}, Li^+.

Briefly, one concludes that the shape of the mineral particles constituting the soil primarily determines the relative importance of the gravitational or electromagnetic forces among the crystals; and this in turn determines the general structure of the soil and the nature of the contact between the individual particles. In coarse and equidimensional particles the minimum area encloses the maximum weight of the particle (remembering that a sphere has the minimum area for a given volume), so in these soils gravitational activity is predominant. In fine soils, as a result of the flat or elongated form of their minerals, the particles contain a very large area with relatively small weight. The net electrical charge of the crystal acts on and depends on the surface area. Therefore, for very fine clay crystals the electrical activity on the surface prevails over the gravitational field. When the particles are small enough and the soils formed by deposition in a continuous medium, there are other effects, such as BROWNIAN motion, that contribute to minimizing the effect of gravity. This will be discussed later.

The arrangement or disposition adopted by the mineral particles is referred to as the structure of a soil. It is obvious that the structure of a given soil will play a fundamental role in its behavior, especially with reference to shear resistance, compressibility and permeability.

1.4.2 Physico-chemical Make-up of Coarse Soils

The structure of coarse soils is very different from that of fine soils. In coarse soils the agglomeration of particles is produced only by gravitational action; the particles of sand or gravel are arranged like marbles in a box. The structuration mechanism is easily visualized (men live in a gravitational world, in which these mechanisms are familiar) and, given the size of the particles in question, any hypothesis can immediately be verified by physical models. Any hypothesis of structuring in fine soils cannot be verified on sight, because of the small size of the crystals, therefore, the problem of fine soils structure is more difficult and controversial. The methods of investigation used to discover the structure of fine soils, such as electron microscopy and diffraction of X-rays, etc., are all of indirect type and subject to the interpretation of specialists; therefore it is not usually surprising to find a diversity of opinions on the subject.

The structure typical of a coarse soil is simple, like marbles in a box. Its mechanical behavior is primarily determined by its density. TERZAGHI has proposed a concept of relative density to measure this condition. The relative density is determined in the laboratory [5]:

$$C_r = 100 \frac{e_{max} - e_{nat}}{e_{max} - e_{min}} \tag{1-17}$$

where:

e_{max} = void ratio corresponding to the loosest type of arrangement, obtained by placing the sample in a container, without any compaction.

e_{min} = void ratio corresponding to the densest arrangement, obtained by tamping a soil within a container.

e_{nat} = void ratio corresponding to a soil in its natural condition.

The relative density is usually expressed as a percentage. Values higher than 50% are generally described as dense. This percentage is sometimes mentioned as a reasonable security limit for coarse soils involved in practical problems, such as bearing capacity of shallow foundations, and liquefaction (sudden loss of strength) of sands and non-plastic silts from shock.

Besides density, a coarse soil's mechanical behavior is influenced by the angularity of its particles (at equal density, greater angularity produces more interlocking and, consequently, greater shear resistance) and by the orientation of its particles, which admittedly mainly influences permeability, and isotropy of strength.

There are several types of structure of fine soils. TERZAGHI originally suggested the names of honeycomb and flocculent [6] which are shown in Figs. 1-2 and 1-3.

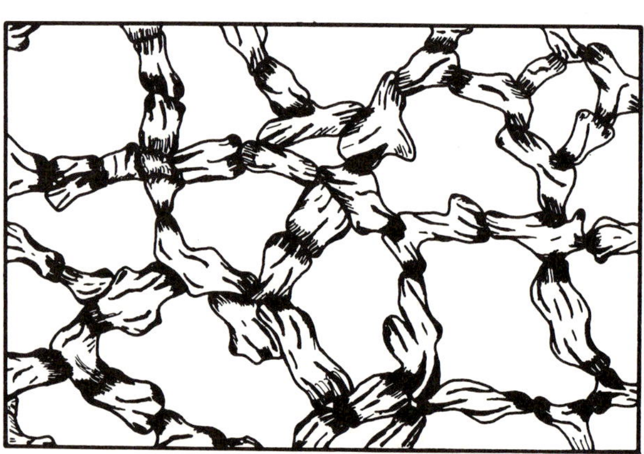

Fig. 1-2 Honeycombe structure

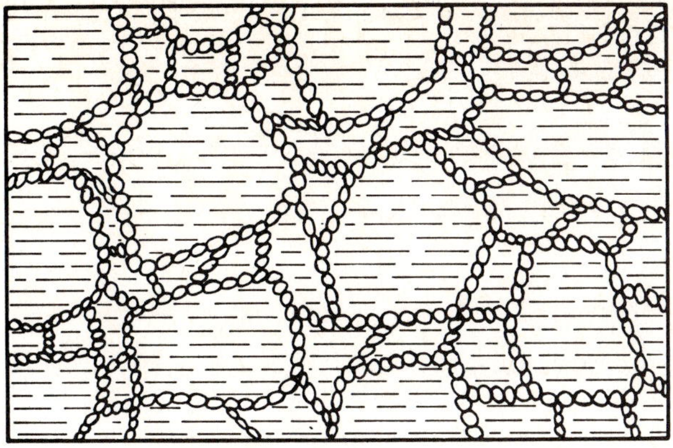

Fig. 1-3 Flocculent structure

The honeycomb structure is considered typical of grains of 0.02 mm (0.0008 in) or less, which are deposited in water or air; gravity exerts a certain influence, but electrical forces are of comparable magnitude. The particles wedge together, forming crude arches with large voids between.

The flocculent structure is typical of particles of a much smaller size, that by themselves would not settle because of the effect of the impact caused by molecular vibrations of the medium in which sedimentation occurs; the particles would move at random with a characteristic motion known as Brownian. TERZAGHI hypothesized that these particles link, forming an aggregate or floccules with a honeycomb structure. These would acquire enough weight to settle, producing a structure of honeycombs formed within honeycombs. If an electrolyte exists in the medium, this increases the capacity of the particles to flocculate into heavier aggregates. It was assumed that this structure was typical of very fine soils deposited in the sea or in salt-water lakes, susceptible of undergoing electrolytic dissociation.

CASAGRANDE [7] presented an extended hypothesis for fine soils, which appears in Fig. 1-4. It considers that not all particles of soil are of the same size. However, the most interesting idea is the concept of a structural skeleton, made up of the larger inert particles (silt in Fig. 1-4) and honeycombs of finer grains and floccules among them. Under the weight of the overlying soil or of a load acting on the surface, a mechanism of load transmission is formed in the interior of the soil, which functions like a skeleton of the whole. Among the coarse, more rigid, particles there is large amounts of fine material that is hardly, or not at all, compressed. The links between the coarse particles that form the skeleton, will have undergone a slow process of compression and adaptation to the load, this giving the soil its resistance. If this hypothesis is accepted, it helps to explain the different resistance that exists between an intact and a remodeled clay, in which the skeleton has been broken and the load is transmitted to the masses of floccules that are not precompressed.

Recently, the concepts of flocculation and dispersion, [8], have clarified the explanation of soil behavior. If the net effect of the attractive and repulsive forces among clay crystals is one of attraction, adjoining particles can link (possibly edge to face); it can then be said that they are flocculated. If the net effect is one of repulsion, they separate, producing a dispersed structure.

The alteration of the adsorbed layer can produce a tendency towards flocculation or dispersion in a system of clay crystals. The tendency towards flocculation increases principally when there is an electrolyte in the water that surrounds the clay

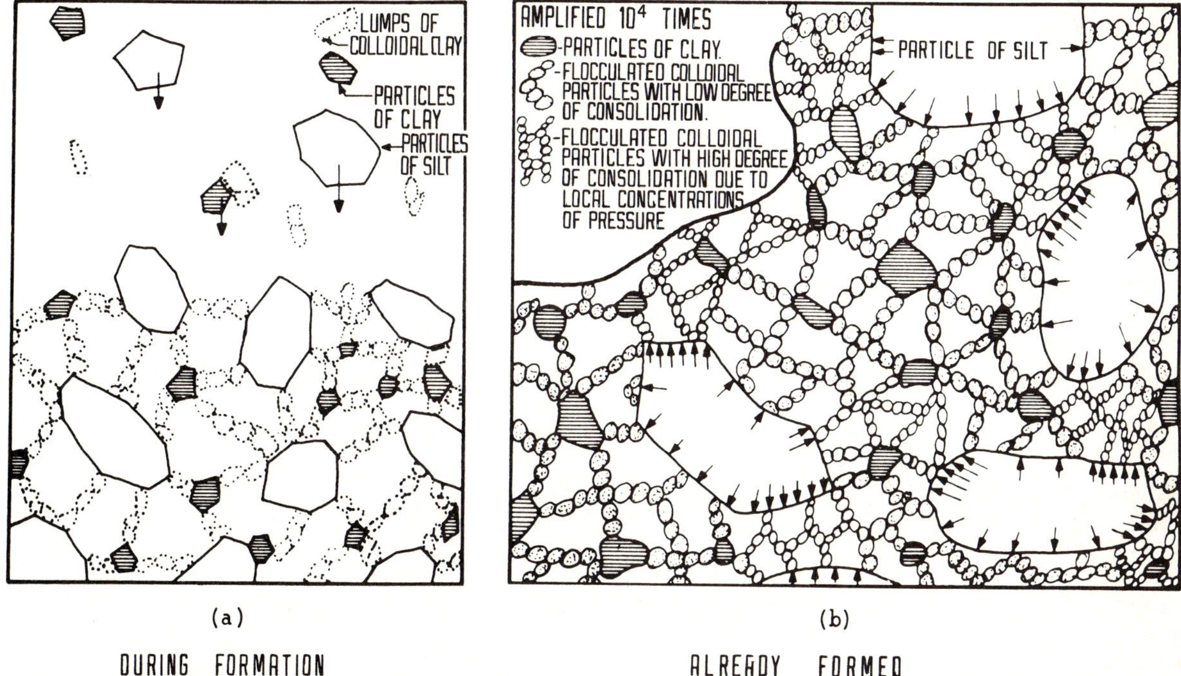

Fig. 1-4 Composite structure (from CASAGRANDE)

crystals or when the temperature is increased. Figures 1-5 and 1-6 show arrangements typical of flocculation and dispersion respectively. The group of structures for fine soils briefly described previously does not necessarily constitute a series of real possibilities found in nature, but are hypotheses of structuration being talked about today. Many researchers accept some of the previous explanations, but there is no complete agreement.

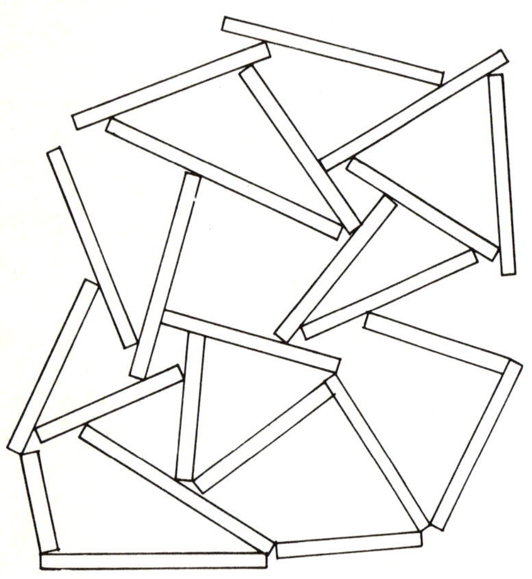

Fig. 1-5 Cardhouse structure

It should also be pointed out that there is the possibility of conjugating the aforementioned structures, thus coming up with a number of different combinations.

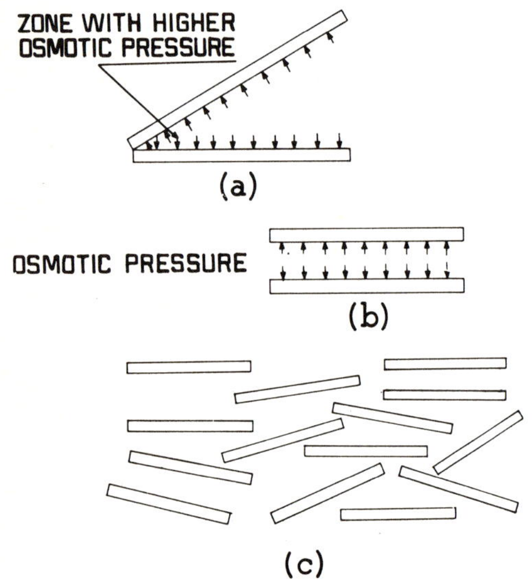

Fig. 1-6 Dispersed structure

1.5 Grading of Soils

When dividing a soil into different fractions, defined by particle size, the grain size distribution is obtained. The particle size of each fraction lies within maximum and minimum limits, in such a way that the maximum limit of a fraction is the minimum limit of the next coarser one. The division into fractions is made by sieving when possible. However, in very fine grain soils, that form clusters, more complicated methods must be employed to separate the individual particles. Later it will be seen that in order to determine different fractions in fine soils one has to employ unsatisfactory theories to obtain doubtful results.

The most important characteristics in assessing the resistance of coarse grained soils (non-plastic silts, gravels and sands) are denseness, angularity and orientation of the particles; although the latter is of minor importance. Obviously, no sieve analysis will give any information on these aspects. On the other hand, even though the compressibility of these soils depends partly on their structure and denseness, it is principally influenced by grading, as has been pointed out by recent research and as will be discussed later. All the efforts made to establish a correlation between grading and permeability [2] have been discouraging.

It has been said that coarse soils, with an ample range of sizes (well graded) can be more easily compacted than very uniform soils (poorly graded) with the same compactive effort. This is undoubtedly true because, especially with vibration, the small particles can accommodate themselves between the larger particles, thus producing greater compactness. However, the relation between grading and compactability is only qualitative, as vague as the one stated here. Consequently, in field compactation little or no benefit has been obtained from improving the grading of coarse soils. It is more difficult to correlate the mechanical properties of fine grained soils (traditionally called cohesive soils, clays and plastic silts) on the basis of grading, for they depend on a larger number of factors than those of coarse soils. All that needs to be said (as the reader will have opportunity to see later) is that none of the circumstances that define the mechanical properties of a fine soil is described by the grain size distribution. To a greater degree than in coarse soils, information on the grain size distribution of fine soils is useless.

Many specifications, used to accept or reject materials for pavements, are based to a greater or lesser degree on grading requirements (the influence of tradition and habit). This is an undesirable situation since the grain size distribution of a fine soil does not define its mechanical behavior. Any standard based on such a criterion risks accepting a bad soil and rejecting something that could be better. A kaolinitic clay for example, which is relatively stable in the presence of water and which could perfectly well be employed in various ways, can have a grain size distribution similar to a montmorillonitic clay (perhaps with extremely active organic material) which in most cases should be rejected in road construction.

One of the reasons for the popularity of grading techniques is that they somehow constitute a classification criterion. This is the origin of well-known terms such as clay, silt, sand and gravel. A soil is classified as a clay or a sand depending on its maximum size. The need for a system for classifying soils is undeniable, but the engineer must seek a useful one, where the soil categories are in accordance with their fundamental engineering properties and not with the size of their particles, which really does not matter.

The modern engineer, however, still makes frequent use of grading curves. Some details on the methods employed are given in the following paragraphs.

Whenever there is a sufficient number of points, a graphic representation of grain size distribution should always be considered preferable to numerical tables, or scaled sizes. Grain size distribution curves are usually drawn with the percentage passing as the ordinate and the particle size as the abscissa. The ordinate refers to the percentage passing, by weight, a certain sieve size. Semi-logarithmic representation of the abscissa is preferable to an arithmetic one, because it leaves greater space for the fine and very fine sizes than on a natural scale. The slope of the grading curve gives an idea of the grain size distribution of the soil. A soil made up of particles of only one size would be shown by a vertical line (because all of its particles are smaller than a certain sieve size and larger than the next one); a very flat curve indicates a great variety in size (a well graded soil). Some grading curves are shown in Fig. 1-7.

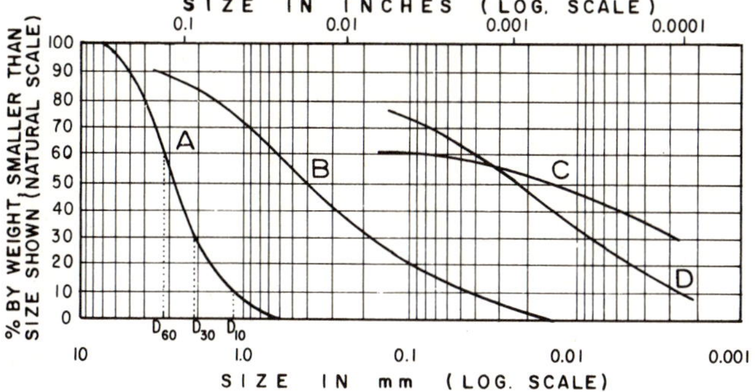

Fig. 1-7 Grain-size distribution curves for some soils: A) Very uniform sand from Cuauhtémo city, Mexico; B) Well-graded soil, Puebla, Mexico; C) Mexico Valley clay (curve obtained with hydrometer); D) Mexico Valley clay (curve obtained with hydrometer)

As a simple soil uniformity measure, HAZEN proposed the coefficient of uniformity, defined as:

$$C_u = \frac{D_{60}}{D_{10}} \tag{1-18}$$

where:

D_{60} : Size where 60%, by weight, of the soil is finer

D_{10} : Called by HAZEN the effective diameter; is the size where 10%, by weight, of the soil is finer.

In reality, Eq. (1-18) is a non-uniformity coefficient, since its numerical value decreases when the uniformity increases. Soils with $C_u < 3$ are considered very uniform; even very uniform natural sands seldom present $C_u < 2$.

As a complementary index to define uniformity, the coefficient of curvature of a soil is defined by the following expression:

$$C_c = \frac{D_{30}^2}{D_{60}\, D_{10}} \tag{1-19}$$

D_{30} is defined in the same way as were D_{10} and D_{60} previously. This relationship has a value between 1 and 3 in well graded soils with a wide range of particle sizes and appreciable quantities of each size.

From a smooth grading curve as described, it is possible to find the curve corresponding to the function: $y = d(p)/d(\log D)$ where p is the percentage, by weight, of the particles between any two selected limits, namely D and $10D$. The previous curve is usually referred to as the histogram of the soil and shows the frequency with which particles within certain size limits are present. The area under the histogram is 100%, since it represents all the particles in the soil. Figure 1-8 shows the histogram of a soil in which particles of approximately one mm predominate. The highest points of the histogram correspond to very steep zones of the grading curve, the lowest values correspond to flat zones. Nowadays, histograms are not extensively used in laboratories.

Grading curves have also been represented in a double logarithmic scale, with the advantage, for certain uses, that many natural soils are represented by straight line curves.

For separating soils into different sizes there are two methods which deserve special attention: sieve analysis and the hydrometer method. The first is used to obtain fractions corresponding to the largest sizes of the soils; generally larger than sieve No. 200 (0.074 mm, 0.003 in). The soil sample is passed through a set of sieves with decreasing openings down to sieve No. 200. The particles retained in each sieve are weighed and the percentage each represents of the total weight of the soil sample is added to the percentages retained in all the sieves of a larger size. The difference from 100% gives the percentage of the soil which is smaller than the size of the sieve in question. In this way a point in the grading curve can be obtained which corresponds to a certain size. The method presents difficulties when the sieve openings are small; for example, sieving through mesh No. 100 (0.149 mm, 0.006 in) and No. 200 (0.074 mm, 0.003 in) usually requires water.

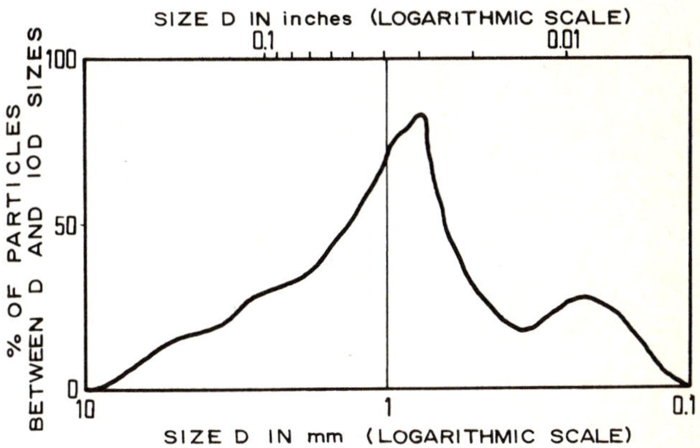

Fig. 1-8 Histogram of a soil

The smaller sizes of soil require testing techniques based on other principles. The hydrometer method is the one most extensively used and the only one that will be discussed in detail. Like all the methods within this group, it is based on the fact that the speed of sedimentation of particles in a fluid depends on their size. The method was proposed independently by GOLDSCHMIDT in Norway (1926) and by BOUYOUCOS in the United States (1927).

Due to the importance of the errors affecting the original tests, the methods did not satisfy many specialists. Therefore, the Public Road Administration of the United States commissioned DR. A. CASAGRANDE to investigate those errors and make the necessary corrections.

As a result of his studies, CASAGRANDE proposed the aerodynamic hydrometer, graduated in relative unit weights (instead of its original graduation in grams of a standardized soil per liter). He also recommended some radical changes in the test procedure in order to eliminate the principal errors, and obtained formulas to make the necessary corrections for certain steps which could not be remedied by changing the procedure.

The man responsible for the basic law used in the hydrometer method was STOKES. His law gives the relationship between the sedimentation velocity of soil particles in a liquid and the size of these particles. This relationship may be established empirically, making observations with a microscope, or theoretically. Following the latter, STOKES in 1850 obtained a relationship applicable to a sphere that falls through a homogeneous fluid of infinite extent. Even with this important limitation (since real soil particles are very different to the spherical shape) STOKES' Law is preferable to empirical observations. By applying this law, the equivalent diameter of a particle is obtained, that is, the diameter of a sphere of the same S_s as the soil that sediments with the same velocity as real particles. In equidimensional particles this equivalent diameter is equal approximately to half the real one, but for thin flat particles the real diameter can be several times the equivalent. It should be pointed out that in very fine particles this is the most frequent shape. This is another reason why two identical grading curves, corresponding to two different soils, do not necessarily indicate that the soils are equal. One can be a clay with a flocculent structure and the other a rock flour with a behaviour similar to that of a sand.

STOKES' Law has the following form:

$$V = \frac{2}{9} \frac{\gamma_s - \gamma_f}{\eta} \left(\frac{D}{2}\right)^2 \text{ (cm/s)} \tag{1-20}$$

$$V = 0.0874 \frac{\gamma_s - \gamma_f}{\eta} \left(\frac{D}{2}\right)^2 \text{ (in/s)}$$

where:
V = sedimentation velocity of solid, cm/sec (in/sec)
γ_s = specific gravity of solid, g/cm^3 (lb/in^3)
γ_f = specific gravity of solid, g/cm^3 (lb/in^3)
η = viscosity of fluid, g.sec/cm^2 (lb.sec/in^2)
D = diameter of solid, cm (in)

In the above Equation, if D is expressed in mm, one obtains:

$$D = \sqrt{\frac{1800\eta v}{\gamma_s - \gamma_f}} \tag{1-21}$$

Applied to real soil particles, that sediment in water, STOKES' Law is only valid for sizes smaller than approximately 0.2 mm. (For larger sizes, the law of sedimentation is altered considerably by the turbulence caused by movement of the particles). It applies also only to particles larger than 0.2 microns. Below this limit the particles are affected by BROWNIAN motion and do not sediment. Using sieve analysis, sizes of 0.074 mm (0.003 in) can be reached, which fall within the applicability of STOKES' Law. This fortunate fact allows an uninterrupted grading test.

The hydrometer method is based on the following assumptions.

— STOKES' Law is applicable to a soil suspension

— At the beginning of the test, the particle suspension is uniform and has a concentration low enough to avoid particle interference during sedimentation. (Generally an appropriate concentration is about 50 g/l)

— The cross-section of the hydrometer bulb is small in comparison to that of the tube where sedimentation takes place, so that the bulb does not interfere with the sedimentation of the particles at the moment of reading.

1.6 Plasticity

Plasticity and its extensive use by specialists in soil mechanics is one of the most difficult subjects to understand for an engineer foreign to the field; and yet, the concept underlying the use of plasticity is quite familiar in our everyday life. It is not unusual to find quantities which are impossible to measure or ones which involve difficult or costly direct measurement procedures; consequently the attempt to obtain indirect measurements is common in many fields of science. An attempt is made to find an auxiliary quantity, different from the one to be evaluated, that is easily measurable and whose correlation with the quantity in question is known and reliable. In this way, by measuring the changes in the auxiliary quantity and using the correlation, the changes in the quantity in question can be estimated. For example, it is very difficult to measure temperature directly, but it can be quite easily measured with a clinical thermometer, where what is in fact measured is the length of a mercury column; this is feasible because there is a known correlation between the increase in length of the mercury and the increase of temperature. This is an example of how a concept which in itself is difficult to measure can be easily and cheaply measured by indirect means.

The same applies to plasticity in soil mechanics. The engineer is really interested in the fundamental properties of soils, such as resistance, compressibility, permeability, and their relation to water content, etc. Today these properties can be measured directly with reasonable accuracy, as is testified by many engineering structures, but such methods take time and are expensive. Furthermore, the work carried out by ATTERBERG and CASAGRANDE [9] defined an auxiliary quantity in fine soils, which is easily measurable in the most rudimentary laboratories, using the simplest and cheapest soil samples imaginable. This auxiliary quantity is plasticity; its utility lies in that it has been possible to correlate it with the fundamental properties of the soil. These correlations are reliable enough to be utilized in the initial stages of a project, when the identification and classification of soils is so important. However, these correlations are not precise enough to rely on them for the quantitative detailed design data needed by a project in its advanced stages. The use of plasticity as an index to the corresponding engineering properties of the soils does not exempt the engineer from making the necessary tests of compressibility, elasticity and shear strength. They do, however, enable him to identify and classify soils on his first contact with them, so that he does not have to work blindly. From them he can obtain extremely valuable information for exploration programs, and for planning final sampling for more elaborate, and expensive, laboratory tests. Therefore, plasticity provides preliminary information which saves time and effort in subsequent stages of a project, and frequently helps to avoid serious errors.

Within the limits of its significance in the field of soil mechanics, the term plasticity can be defined as the property which enables a material to undergo rapid deformation, without appreciable volumetric change, without elastic rebound and without crumbling or cracking. This definition, as will be discussed later, ascribes the property to clayey soils, under certain circumstances.

ATTERBERG showed that plasticity is not a general property of all soils, because coarse soils do not exhibit it under any circumstances. Secondly, he showed that in fine soils it is not a permanent property, but is dependent on water content. A clay or silt which may be plastic can also be as hard as a brick when very dry. With a large water content both can exhibit the properties of a semi-liquid mud or even those of a liquid suspension. Between the two extremes there are certain water contents at which the soil shows plasticity.

With decreasing water contents a plastic soil can have any of the following consistencies, defined by ATTERBERG:

Liquid state, with the properties and appearance of a suspension.

Semi-liquid state, with the properties of a very viscous fluid.

Plastic state, in which the soil behaves plastically, according to the former definition.

Semi-solid state, in which the soil has the appearance of a solid, but reduces its volume if further dried.

Solid state, in which the volume does not vary if dried.

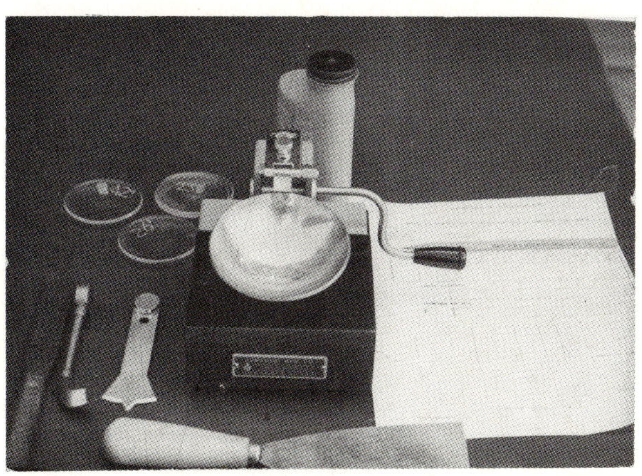

Plate 1-1 CASAGRANDE Cup

The preceding states describe the changing condition of a soil when it is dried. They are no strict criteria to define their limits, which are established arbitrarily. ATTERBERG established the original limits; CASAGRANDE subsequently refined them to become the ones which are known today, [10]. The boundary between the semi-liquid and the plastic states is called the liquid limit, which is defined in terms of a certain laboratory technique. The soil is placed in a CASAGRANDE cup, Plate 1-1, with a groove of specific dimensions, Fig. 1-9. If the groove closes in a certain way when the soil receives 25 standard blows, the water content of the soil is known to be at the *liquid limit,* W_l A higher water content makes the groove close with fewer blows, and the soil is in a semi-liquid state. If the soil had a lower water content, more blows are required to close the groove and it would be in a plastic state. Reference [10] provides details of this test, as well as of other tests mentioned

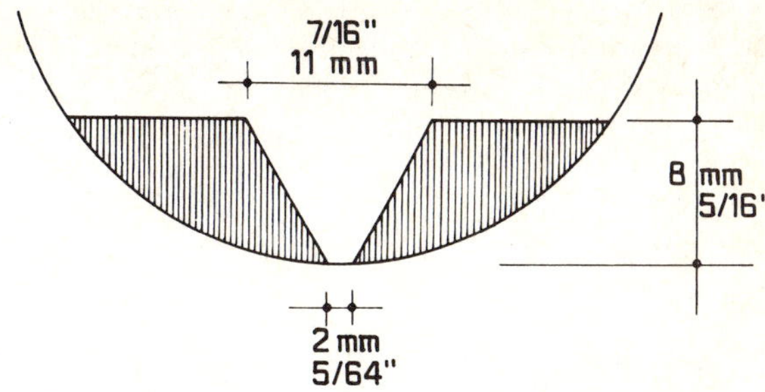

Fig. 1-9 Dimensions of the groove in CASAGRANDE cup

in this section. The boundary between the plastic and semi-solid states is known as the *plastic limit,* W_p. This is defined by the water content at which the soil sample breaks or crumbles when rolled into a cylinder 3mm (0.12 in) in diameter, Plate 1-2 [10]. Today, the *Plasticity Index,* is widely used:

$$I_p = W_l - W_p \tag{1-22}$$

This value measures the plastic interval. In order to locate it within the general scale of water contents, another value is needed, either the liquid or plastic limit. Thus in order to define the plasticity of a soil, two parameters are needed.

The third limit of practical interest between states of consistency is the shrinkage limit. It is defined as the water content below which the volume of the soil does not decrease when dried. This limit reveals itself during drying with a characteristic change of color, from a dark tone to a light one, produced by the withdrawal of the water towards the interior of the mass. This visible sign enables an approximate determination of the limit. Of all limits in use this is the only one that is not purely arbitrary, but is related to a significant physical phenomenon. With gradual drying, the shrinkage limit represents the moment at which capillary tension reaches its maximum value (the menisci reach their maximum curvature at the surface of the soil), and subsequent evaporation produces the withdrawal of water towards the interior of the soil. The soil is no longer saturated and the capillary tension is no longer uniformly distributed. As will be explained later, the fact that this occurs almost instantaneously on the entire surface of the sample indicates that, statistically speaking, the channels throughout the soil mass have a similar diameter.

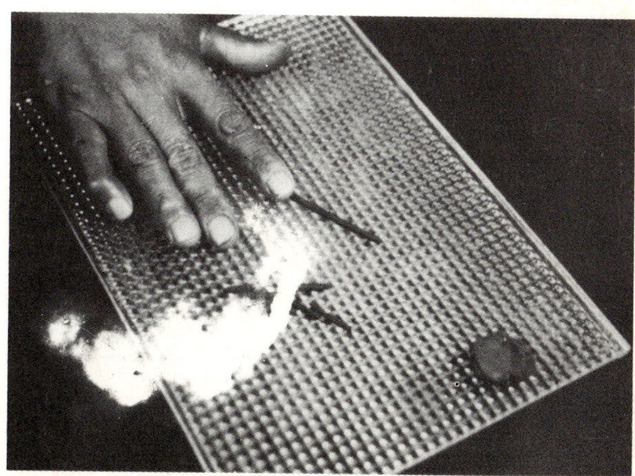

Plate 1-2 Plastic limit test

The adsorbed water around mineral crystals logically does not behave like a free liquid, subjected only to gravitational forces. For example, comparing two soils, *1* and *2* [11], if *1* has a greater tendency towards adsorption, it is to be expected that the water content as which both soils will start behaving like a liquid will be greater in *1* than in *2*. In other words, soil *1* will have a greater liquid limit than soil *2*, if its crystals have larger adsorption. A similar explanation would appear logical for the plastic limit and, therefore, the plasticity index. However, because the limits have been fixed in a completely arbitrary way, it is hard to believe that the size of either one of them, by itself, can be related in a quantitative way to the thickness of the adsorbed water.

Because of the large increase of specific surface that is associated with the decreasing size of the soil particles, the intensity of the adsorption phenomenon is greatly influenced by the clay content of the soil. SKEMPTON [12] defined a quantity known as the activity of a clay:

$$A = \frac{I_p}{\text{\% by weight of soil particles smaller than 0.002 mm}} \qquad (1\text{-}23)$$

The activity can be 0.38 in kaolinitic clays 0.90 in illitic clays and may reach values higher than 7 in montmorillonitic clays, which illustrates how the plasticity characteristics of clays vary with their mineralogical composition.

The plasticity limits have proved useful for classifying and identifying soils, as will be seen in Chapter 2. They are also used for establishing specifications to control the use of soils in earthworks. Some interesting correlations between the plasticity limits and certain fundamental properties of soils will be presented later in the chapter.

Plate 1-3 Sample of fine soil dried by evaporation

1.7 The Shrinkage Mechanism of Fine Soils upon Drying

When the surface of a liquid is in contact with a different material, stresses are produced on that surface, owing to the attraction between the neighboring molecules in the two different materials (Plate 1-3).

Highway engineers are particularly interested in the contact between water and the mineral particles of soils, and between water and air; the unequal forces in these phases cause surface tension stresses. The attraction between the neighboring molecules of the different substances in contact with each other can be measured by the surface tension coefficient, this is a characteristic property of each pair of substances. Reference [13] describes the physical concepts that enable the definition of this coefficient and a better understanding of the processes occurring on the contact of water and soils, which have repercussions in soil mechanics. Probably the most familiar evidence of surface phenomena is capillarity, which is the property that enables water to rise and remain above the level represented by atmospheric pressure within a glass capillary tube or within a channel between the mineral particles of a soil. Reference [13] shows that the maximum capillary height to which water can rise under such conditions is:

$$h_{cr} = \frac{2\,T_s \cos\alpha}{r\gamma_w} \qquad (1\text{-}24)$$

Where T_s is the surface tension coefficient of the water 0.074 g/cm at 20°C (0.005 lb/ft at 68°F); α is the angle of contact between the water and the wall of the channel [13] and r is the radius of the channel. In soil materials it is reasonable to consider the limiting value for $\alpha = 0$ where the spherical meniscus that is formed by the water is tangential to the walls of the soil channel or tube. In this case Eq. (1-24) can be written simply as:

$$h_{cr} = \frac{0.3}{D} \text{ (cm)} \qquad (1\text{-}25)$$

$$h_{cr} = \frac{0.046}{D} \text{ (in)}$$

Where D is the diameter of the channel in cm (in) and h is the height, also in cm (in).

Figure 1-10 shows the distribution of stresses in a channel between soil grains which has been idealized in the form of a real capillary tube, as is done in theoretical analysis. At any given point in the columm, the tension can be obtained by multiplying the vertical distance to the free surface by the specific weight of water:

$$u = h\gamma_w = \frac{2T_s \cos\alpha}{r} \tag{1-26}$$

hence:

$$u = \frac{2T_s}{R} \tag{1-27}$$

In the above equation u is the tension stress of water in g/cm^2 (or lb/in^2) and R is the meniscus radius of the water. The meniscus radius and the radius of the capillary tube are related as shown in Fig. 1-11.

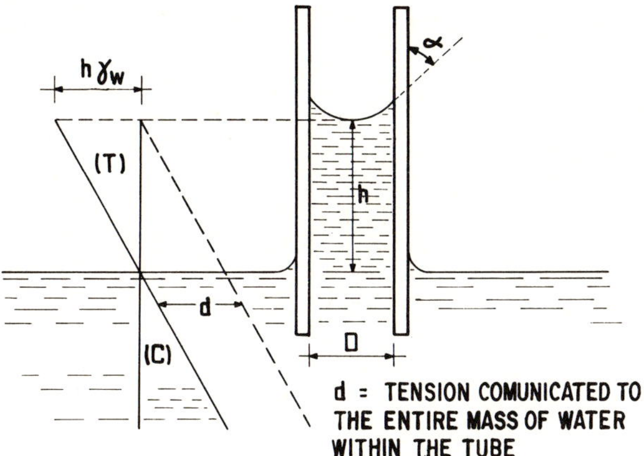

Fig. 1-10 Distribution of stress in a vertical capillary tube

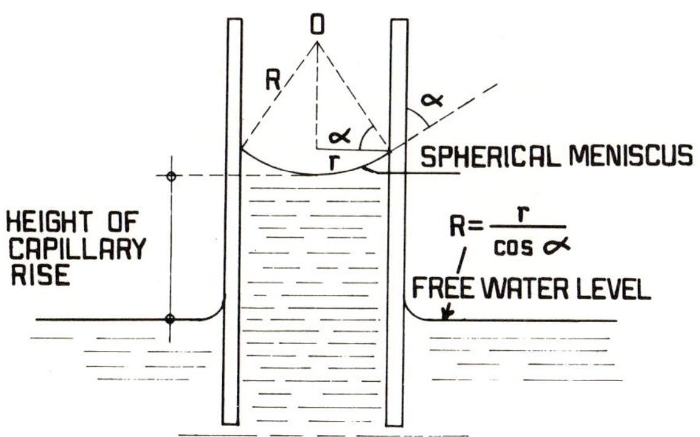

Fig. 1-11 Relationship between the radius of the meniscus and the radius of the capillary tube

Figure (1-27) establishes the important point that the tension to which the water is subjected is inversely proportional to the meniscus radius that develops in the channels. Naturally this depends on the diameter of the channel itself; the minimum meniscus radius (which will correspond to the maximum tension) is equal to half of the capillary channel diameter, which corresponds to a semi-spherical meniscus (completely developed). This means that the water will develop high tensions within the soil, when mineral particles are very close together; this occurs especially in fine soils.

A completely developed meniscus will be obtained, provided that the capillary channel is long enough to allow the columm of water to rise to the maximum capillary height. If the tube is shorter, capillary rise is restricted, and a meniscus will form with a radius such that there is hydraulic equilibrium between the capillary tension stress in the water (less than the possible maximum) and the column of water above the free surface.

If the capillary channel is horizontal, as is the case in Fig. 1-12 menisci will be formed gradually at the ends due to the evaporation of water. At each end the meniscus curvature will increase to the maximum, and, at the same time, the tension stress in the water will increase to its maximum value. If evaporation continues, the menisci will withdraw towards the interior of the channel, keeping their curvature and consequently maintaining constant tension in the water. Thus it can be seen that in a horizontal capillary channel the tension in the water is the same throughout its length, differing from the vertical channel, where the stresses vary hydrostatically in a triangular distribution law.

In the case of the channel in Fig. 1-12, when the menisci form, tension forces, F_T, will appear all along the outside caused by the attractions between the water molecules and the tube walls. As a reaction to these tension forces in the water there will be corresponding compression forces, F_R, in the tube wall, as shown. As a result of these forces, the capillary channel will tend to close and shorten. Throughout the mass of water, tensions exist between menisci; therefore compressions will exist in the channel walls. As a consequence of these stresses, a compressible mass containing capillary channels and subjected to evaporation will shrink volumetrically.

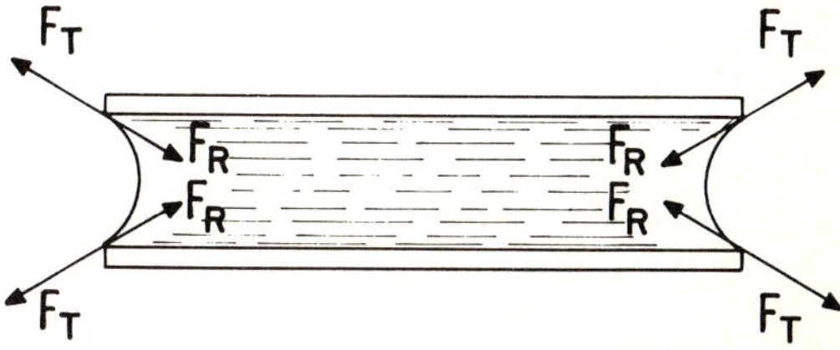

Fig. 1-12 Diagram illustrating the generation of capillary tension in a capillary tube

The shrinkage mechanism in fine soils can be understood in following way [13]: A saturated soil first shows a shiny surface, due to the presence of water. As evaporation takes place, menisci will form at the ends of the channels. As evaporation continues, the radius of the menisci will diminish and therefore the water tension, Eq. (1-27), will increase, and correspondingly the capillary compression acting on the solid structure of the soil will also increase.

Evaporation will continue to reduce the menisci radius and compress the soil structure to a point where the capillary tension is unable to produce any further deformation. At this point the menisci will start to withdraw towards the interior of the soil mass. Macrophysically this point is detected by a change in the appearance of the soil surface from wet to dry. This point is known as the *Shrinkage Limit,* because even though evaporation can continue, the soil volume will not diminish, because the water has reached its maximum tension. Note that at the shrinkage limit the soil is still as saturated, as at the beginning of the evaporation process, because even though evaporation has caused a loss of water, this has been compensated for by exactly the same reduction in voids. A loss of one gram of water will produce one cubic centimeter of volumetric shrinkage.

1.8 Permeability

Water generally flows through soils because of gravity augmented in some cases by capillary tension. The flow is said to be laminar when flow lines do not intercept each other, excluding the microscopic effect of molecular interchange. Flow is said to be turbulent when the lines intersect.

At a typical very low velocity through soils, the flow is laminar, but if the velocity increases beyond a certain limit, it becomes turbulent. In returning to laminar flow, it has been observed that the transition occurs at a lower speed than the one that was reached when the laminar flow became turbulent. This indicates an interval in the velocity at which the flow can either be laminar or turbulent. REYNOLDS [14] found that in all fluids there is a certain velocity below which, for a certain channel diameter and a given temperature, flow is always laminar. This is known as the critical flow velocity. Similarly there is also a velocity above which the flow is always turbulent; for water, this second velocity is about 6.5 times the critical flow velocity.

The earliest theory on flow through soils is based on the experimental work of DARCY [15]. While working with a model like the one shown in Fig. 1-13, he found that for low enough speeds, the discharge through the soil is expressed by:

$$Q = kiA \tag{1-28}$$

where: A is the cross-section area of soil through which the water seeps; i is the gradient or head loss over a given flow distance, given by:
$i = (h_1 - h_2)/L$; and k is the coefficient of permeability (DARCY)

Furthermore, the continuity equation for flow established that:

$$Q = Av \tag{1-29}$$

where v is the discharge velocity

If Eq. (1-29) is compared with Eq. (1-28), it can be established that:

$$v = ki \tag{1-30}$$

which is a common way to express DARCY's Law, even though it was originally proposed as seen in Eq. (1-28).

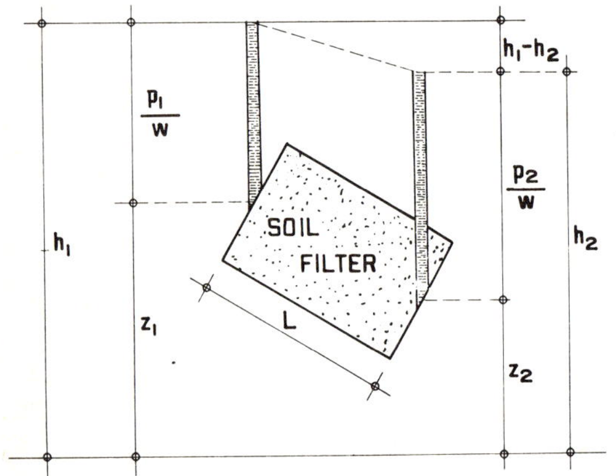

Fig. 1-13 Diagram showing experimental device for permeability coefficient by DARCY

An excellent definition for the permeability coefficient k, can be established by analysing Eq. (1-30), according to which k is the velocity at which a quantity of water passes through a soil when submitted to a unit hydraulic gradient. Naturally, k has units of velocity remembering that i is dimensionless. It is obvious that in the numerical value of k the physical properties of both soil and fluid are reflected.

In Eqs. (1-29) and (1-30), the velocity v does not represent the real velocity of the water that flows through the filter shown in Fig. 1-13. This represents the water discharge velocity, related to an area A of the soil, which does not correspond to the real area of the cross-section of voids or pores, A_v, through which the water has to flow. An approximate idea can be obtained of the real velocity by remembering that flow is only possible through the voids. From Fig. 1-14 it can be seen that if a corresponding seepage velocity (v_1) is defined, the flow continuity gives:

$$A_v v = Av,$$

hence: $v_1 = Av/A_v$

If a unit thickness is considered and the definition of void ratio is taken into account, then:

$$e = A_v/(A-A_v);$$

$$1/e = A/A_v-1;$$

and: $A/A_v = (1 + e)/e$

Based on this relationship the seepage velocity can be expressed in terms of the discharge velocity, as:

$$v_1 = \frac{1 + e}{e} v \tag{1-31}$$

Strictly speaking, this seepage velocity is not a *real* velocity because soil channels are not straight lines as shown in Fig. 1-14, but rather a network of irregular and sinuous conduits. The discharge and seepage velocities are simply calculation elements in the mathematical formulas.

DARCY's Law, as was said previously, was derived experimentally; therefore its applicability cannot go beyond the specific conditions under which the initial experiments that sustain it were developed. It is fortunate that DARCY experimented with groundwater flow using a great variety of soil

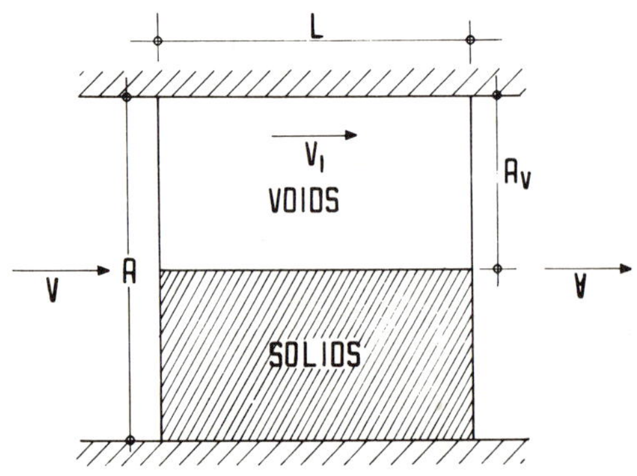

Fig. 1-14 Diagram to illustrate the difference between water discharge velocity and flow velocity

types and hydraulic gradients, for this makes his results applicable to the practical problems of soil mechanics. Reference [16] presents a justification of the use of DARCY's Law in soil mechanics which is more appropriate than simple intuition, and discusses the limits of its applicability on the basis of REYNOLD's Number. Reference [2] gives another analysis of the applicability limits of DARCY's Law using a different criterion. The conclusion in both cases is that DARCY's Law is applicable to groundwater flow through soils finer than medium or coarse sands, for almost any hydraulic gradient imaginable in a practical problem. It also applies to coarse sands and gravels at gradients less than 0.1.

Reference [2] discusses in detail the different methods used to measure the permeability coefficient of a soil. Soil permeability is one of those values which changes greatly depending on the type of material being dealt with. It varies between limiting values as great as 100 cm/s in clean gravels, and as little as 10^{-8} or 10^{-9} cm/s in homogeneous montmorillonitic clays. A typical permeability for clean sands can be of about 10^{-2} — 10^{-3} cm/s (28.34 — 2.834 ft/day) reaching values of 10^{-4} cm/s, (2.834×10^{-1} ft/day) in fine sands. Silts and moraine deposits can have permeabilities as low as 10^{-5} or 10^{-6} cm/s. (2.834×10^{-2} or 2.834×10^{-3} ft/day). Generally clays have permeabilities lower than 10^{-5} cm/s (2.834×10^{-2} ft/day). With permeabilities under 10^{-3} cm/s, (2.834 ft/day) a soil should be considered unsuitable as a draining material and with a permeability below 10^{-7} cm/s, (2.834×10^{-4} ft/day) a soil can be considered practically impervious.

The permeability of a soil is influenced by the following characteristics:
— The void ratio
— Particle size
— Mineralogical and physico-chemical composition
— Structure
— Degree of saturation
— Holes, fissures, etc.
— The temperature of the water.

In [2], a fairly complete discussion is given about the relationship between the permeability coefficient of a fine soil and its void ratio. The conclusion is reached that the first is in direct proportion to the square of the second.

No reliable relationship has been established between the permeability coefficient and the grain size distribution of soils. For fine sands, HAZEN obtained in 1892 his well-known relationship:

$$R = CD_{10}^2 \qquad (1\text{-}32)$$

where k is in cm/s and D_{10} is the effective diameter of the soil (10% by weight of the soil is smaller than this diameter), expressed in cm. Despite its popularity, Eq. (1-32) should be seen simply as a rough means of establishing an order of magnitude for the permeability coefficient in clean medium to coarse sands (which were the ones used by HAZEN to obtain his relationship) and, only if an approximation is required, never as a substitute for laboratory tests. The value of the constant C varied between 41 and 146 in HAZEN's tests and a value of 120 is usually mentioned as an accepted average for using the formula. In [2] other more complicated, formulas are mentioned, which relate the permeability coefficient to the size distribution of the soil particles, but they are of dubious applicability.

The mineral composition of clays greatly influences soil permeability, because of the absorption layers that form around, and adhere very strongly to, the mineral crystals, thus hindering ground-water flow.

The structure of the soil also affects its permeability. In very fine soils, with sheet-like minerals, it is important to know whether the structure is flocculent or oriented, since in the second case there are higher permeabilities parallel to the aligned faces of the particles, and marked anisotropy in the permeability. These phenomena are often found in compacted soils, in which the structure obtained is flocculent or dispersed, depending on the compaction procedures employed.

At degrees of saturation less than 100% this permeability is low. Cracks and fissures increase permeability greatly sometimes by orders of magnitude, their quantification is, however, difficult.

1.9 Effective and Neutral Stresses

A soil is composed of three phases: solid, liquid and gaseous. It is difficult to visualize mechanical components with such different behavior: a very rigid incompressible mineral crystal or fragment with high shear resistance; water, which is relatively incompressible under engineering pressures, but has insignificant static shear strength, and air, which is highly compressible, with negligible strength. The engineering properties of a soil such as compressibility and shear strength, depend on the three elements acting in close union. Their overall reaction to a load is a combined reaction of the three elements. Therefore in soils, the phenomena of stress transmission and resistance are subject to some of the most complicated and variable mechanisms of all engineering materials.

The strength, compressibility and deformability of the same soil may vary greatly according to loading conditions and their effect on each of the three components. The influence of time-rate of loading on soils depends on the variable response of water and air to slow or rapid loading. If the water acquires high pressures because of external loading, it will flow towards zones where the pressure is lower at a rate controlled by soil permeability; this will reflect on the rate of compressibility and the changing stress conditions in the loaded zones. In summary, the constant interaction of the three soil phases and their very different reactions to stresses, produce in each loading process a complex distribution of stresses among the three phases. This varies in accordance with the rate of load change, with each different load, and if a change occurs to any one of the three phases.

A load, P, is assumed to be evenly distributed on a plate of area A, which rests on irregularly shaped mineral particles with voids between them. It is clear that the uniform distribution of the load, which is admissible for the plate of area A, is no longer true for the soil particles. The irregular and variable shape of the particles makes it impossible to define exactly how the load is distributed among them and what the stresses are at any point. These stresses will be very high at the contact points and much lower at the intermediate points and inside the particles. Because it is impossible to work with the *real* stresses to which the particles are subjected, a fictitious stress is used in soil mechanics to describe the condition existing under the plate. This fictitious stress is obtained by relating the total acting load with the total area covered by the plate

($\sigma = P/A$). This is known as the total stress. It is of course lower than the mean stress in the solids under the plate and much lower than the stress at the contact point between the particles.

If the load P is applied to a soil with its voids filled with water, the distribution of the load will be even more complex (Fig. 1-15b). If u is the pressure of the water in the voids and A_v is the area of the voids measured on a plane that is parallel to the base of the plate, then uA_v will represent the part of the load P that is supported by the water in the soil voids. The rest of load P is supported by the solid structure of the soil and is transmitted through its particles. In Fig. 1-15b the solid structure has been represented by a coil. The load distribution is: $P = P' + uA_v$,

where P' represents the part of the load carried by the solid structure of the soil or the coil in Fig. 1-15b.

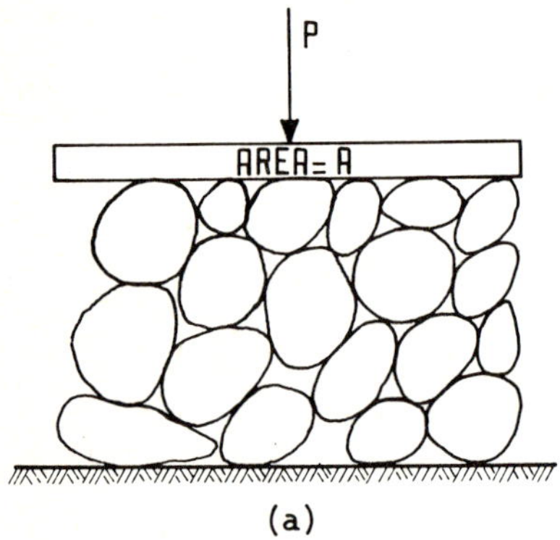

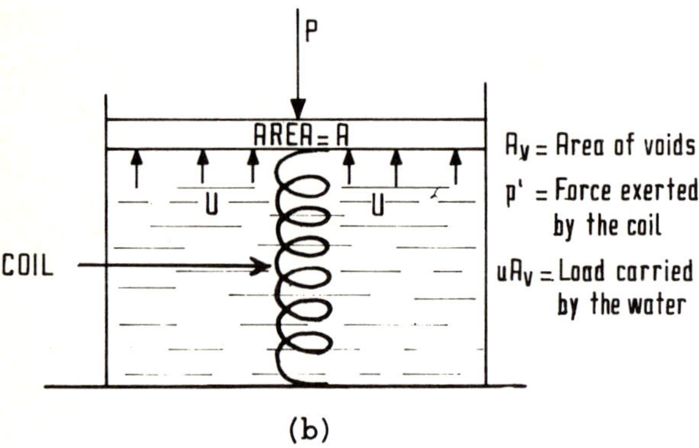

Fig. 1-15 Distribution of an external load in a soil mass

By dividing the two sides of the previous equation by A, which is the area of the plate, one obtains: $P/A = P'/A + uA_v/A$ or, using the stress notation:

$$\sigma = \sigma' + \frac{uA_v}{A} \tag{1-33}$$

Equation (1-33) is known as the effective stress equation and plays an important role in soil mechanics. Represented in this equation are the total stress, σ, already defined, and stresses σ' and u, called effective stress and pore pressure, respectively. The first represents the part of the total stress which affects the solid structure of the soil, and is transmitted from particle to particle. The second represents the pressure to which the water in the soil voids is subjected. Because of the water's incapacity to withstand shear stress, the pressure u is frequently referred to as neutral pressure, neutral stress or pore water pressure. Equation (1-33) also gives the relationship known as the neutral stress relation:

$$N = \frac{A_v}{A} \tag{1-34}$$

Since the contact area between the particles of soil on a given horizontal plane is much smaller than the total area covered by the plate with area A, the approximate value for relation N in soils is 1. Under these circumstances Eq. (1-33) can be written simply as:

$$\sigma = \sigma' + u \tag{1-35}$$

which was first proposed by TERZAGHI and opened the door to the development of soil mechanics.

In concrete or rocks, where solid grains interconnect through crystals, the value of N is appreciably less than 1, reaching values of about 0.5 in marble, granites and in concrete itself.

The concept of effective stress thus defined, describes the behavior of soils better than the concept of total stress. When effective stress increases, the solid particles are seen to press against each other, sliding and interlocking until they form more structures. On the other hand, the same increase in total stress and pore pressure, but with the effective stress remaining the same, will have no effect on the particles.

1.10 Stress-Strain Behavior

One of the most important characteristics of a material, in engineering problems, is its strain under stress: the stress-strain behavior.

When dealing with a construction material, the engineer is primarily concerned about two basic aspects of behavior, around which all the other aspects can be said to revolve. The first is the resistance of the material to stress and material failure, which will be briefly discussed later. The second is the deformability of the material under the stresses to which it is subjected including the intensity of these stresses, their distribution and the rate of application. This second aspect of behavior is what the engineer describes by a stress-strain relation. If soils were homogeneous, isotropic and linearly elastic, it would be possible to describe their stress-strain behavior by the YOUNG's modulus (E) (Elasticity modulus) and POISSON's ratio. These are obtained from a simple test, such as a simple tension test, where a bar of material is stretched, measuring the tension and the resulting longitudinal and transverse deformation. With the elastic constants it is possible in an ideal material to calculate the relationship between stresses and deformation for other types of loading than simple tension.

Soils are not ideal materials, even though in some specific cases it is expedient to use a constant elasticity modulus and a constant POISSON's ratio. In soils these are not constants, but variables which at best approximate the soil behavior for a given stress condition. They can change, sometimes radically, if the stress conditions change or if the stresses are applied in a different way. Therefore, when the above elastic constants are applied to soil behavior, they represent only the specific conditions for which measurements were taken or calculations made.

The amount of deformation caused by stresses depends on the composition of the soil, its void ratio, its previous stress history and the way in which the new stresses are applied. In the majority of practical problems, the best way to find out the stress-deformation characteristic is to measure it directly, in laboratory or field tests. The deformation produced by stresses should be as similar as possible to those that will occur in the soil mass under consideration.

In real situations, stress can be applied in many ways. All these cannot be represented in a single laboratory test, nor can one hope to design a special test for each case. The engineer tries, therefore, to provide reasonable data using several different laboratory tests that represent the conditions that are most frequently met in engineering.

The principal laboratory tests used to determine the stress-deformation of soils are the following:

Hydrostatic or isotropic compression test. This is useful only for the study of volumetric deformation. A hydrostatic stress is applied to the soil sample; equal compressive stress in all directions. This test is not often employed.

Confined compression or consolidation test. The test is carried out in an apparatus known as a consolidometer or oedometer. Normal vertical stresses are applied to the soil sample (in a cylinder of greater diameter than height), while preventing lateral deformation, by confining the specimen in a rigid ring. In this way, axial deformation determines the volumetric deformation. In this test the relation between the normal lateral and normal vertical stress is the value K_0, the stress coefficient or coefficient of earth pressure at rest, which plays an outstanding role in applied soil mechanics. In most consolidometers, only normal vertical stress and axial deformation (also vertical) are measured, but in [18] one is described that can also measure normal lateral stress.

Vertical deformation is measured with extensometers, while the normal vertical stress is found by controlling the loads that are applied to the apparatus, which are distributed over the area of the sample. This test was originally developed by TERZAGHI.

Triaxial test. This is the most common and versatile of all the tests that are used to find the stress-deformation relation in soils. It is also the most useful laboratory test for determining soil strength, and will be described in detail later. It measures the axial deformation of a cylindrical sample with a height double or triple its diameter, which is subjected to known normal vertical and lateral stresses (confining pressure). The later is usually equal in all horizontal directions.

The sample is first subjected to confining pressure, usually produced by water pressure in the triaxial chamber, and then the vertical stress (deviatoric stress) is increased until the sample fails.

A simple compression test is a variant of the triaxial test, in which the external confining pressure is zero. A closed triaxial chamber is not, therefore, necessary. It is similar to the compression test performed in concrete cylinders.

In the triaxial test the axial stress applied is measured by some form of loading machine in which the load is controlled and the deformation determined (controlled load or stress test), or in which a controlled rate of axial deformation is introduced and the load measured by a weighting system (controlled strain or deformation test). Extensometers are used to measure the axial deformation.

There are currently many other ways of causing failure in a sample. Apart from the method briefly described, the normal vertical stress is maintained constant and the confining pressure is increased (or decreased) until the sample fails by deforming upwards. This variant is known as the triaxial extension test and is used to simulate lateral thrust on a soil mass.

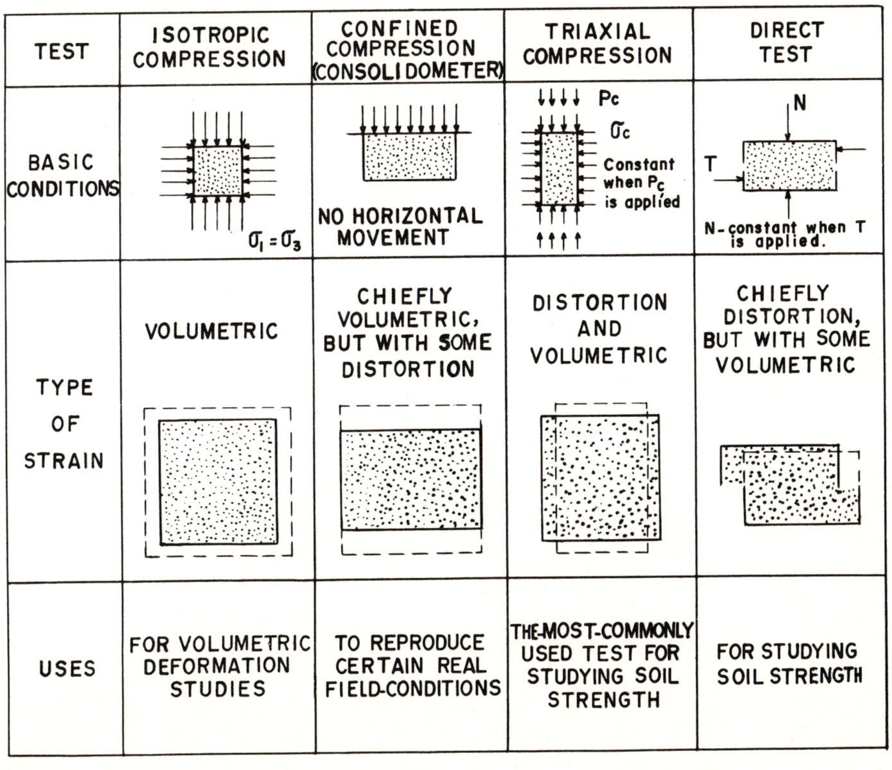

TEST	ISOTROPIC COMPRESSION	CONFINED COMPRESSION (CONSOLIDOMETER)	TRIAXIAL COMPRESSION	DIRECT TEST
BASIC CONDITIONS	$\sigma_1 = \sigma_3$	NO HORIZONTAL MOVEMENT	Pc, σ_c, Constant when P_c is applied	N, T, N-constant when T is applied.
TYPE OF STRAIN	VOLUMETRIC	CHIEFLY VOLUMETRIC, BUT WITH SOME DISTORTION	DISTORTION AND VOLUMETRIC	CHIEFLY DISTORTION, BUT WITH SOME VOLUMETRIC
USES	FOR VOLUMETRIC DEFORMATION STUDIES	TO REPRODUCE CERTAIN REAL FIELD-CONDITIONS	THE MOST COMMONLY USED TEST FOR STUDYING SOIL STRENGTH	FOR STUDYING SOIL STRENGTH

Fig. 1-16 Common types of stress-strain tests [18]

The direct shear test. A sample with a small height in relation to its transverse area is placed in a box with two sections. The lower section is fixed and the upper one can be moved horizontally. A vertical load is applied to the upper face of the device in order to produce a known normal vertical stress. Failure is produced by applying a force to the moveable upper frame, so that the sample is forced to fail on the plane that is defined by the gap between the fixed and mobile parts of the device.

Figure 1-16 shows the different conditions of stress and deformation and the use of the test loading [18].

The stress-strain curves obtained through the test mentioned briefly above, generally correspond to one of the two forms shown in Fig. 1-17.

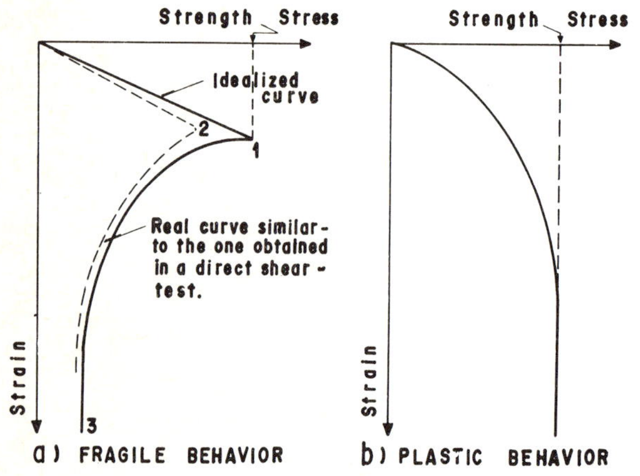

Fig. 1-17 Types of soil failure

The full curve in Part *a* of the figure is representative of so-called *fragile behavior*, whose stress-strain behavior is characterized by the fact that after the stress reaches a well-defined maximum, in an approximately linear manner, it descends rapidly as strain increases. Materials with this type of failure characteristics resist stresses with only minor deformation until the maximum stress (peak strength) is reached, and from there on their strength diminishes rapidly, while strain increases until rupture occurs. These materials are reliable only as long as their peak strength is not reached, for at that point they suffer complete collapse.

The stress-strain curve shown in Part *b* is typical of *plastic behavior* in which, once a limiting stress is reached, the material yields under constant stress. In these materials failure is not well defined, but what is important is that a plastic material will mobilize its strength in accordance with the increase of stress applied, until the yield stress (maximum strength) is reached, from there on the material is incapable of mobilizing any further strength and, in fact, starts to deform under constant stress, unless there is an external restriction impeding the deformation, as might be the case of a soil mass that has already reached the yield stress but is surrounded by other masses that have not, and consequently confines the yielding mass. What is important, from a practical point of view is that the *plastic behavior* material continues to mobilize its maximum strength even if it continues to deform under yielding stresses. This can have important repercussions on the structural behavior of the material, which continues to resist load even after it *fails*. *Fragile* materials, however, collapse, accompanied by a great loss of strength when they undergo any additional strain beyond that of peak strength.

The strain that can be absorbed by a plastic material at the yield stress before it breaks is variable. Lambe and Whitman present in [19] and [20] various stress-strain curves obtained in direct and triaxial tests. From these it is seen that there is a wide variety of curves, even though they can all be identified as one of the two types shown in Fig. 1-17.

The stress-strain relationship of a material is not a unique characteristic, but varies for different conditions. Plastic behavior is typical of loose sands and soft clays, with a relatively high water content, whereas fragile behavior is typical of compact sands and hard clays. There is no precise limit at which the behavior of sands changes from plastic to fragile; the limits vary for different sands. Skempton and Bishop [21] reported, for example, a case in which a sand with an initial porosity of 37.5% showed fragile behavior, which became plastic when porosity reached 45.6%. Lambe and Whitman [19] also presented a case in which a sand with a void ratio of 0.065 showed fragile behavior, whereas for a void ratio of 0.834 its behavior was plastic. Similar comments apply to clays, even though more factors are involved as will be discussed later.

1.11 Compressibility of Granular Soils

Until recently, little attention has been paid to the compressibility of granular soils. Many engineers thought that granular soils did not present many serious compressibility problems: any deformation was always minor and occurred almost instantaneously.

This simple attitude is probably acceptable if the stress applied to a granular soil is small. This may also be acceptable for foundations that transmit moderate loads to granular soils, especially if the quality of the soil is improved when its natural density is low.

Modern engineering, however, has imposed other uses on granular soils. They now provide the fills for huge dams or the large embankments required by modern highways. Granular soils, including broken rock are now subjected to stress levels which were previously unheard of. Rockfills higher than 150 m (492 ft) are quite common in embankment dams, and rockfills 50 to 60 m (164 to 197 ft) high are no longer unusual on highways and railways. Because the high embankments in road engineering are often built with soils in which rock fragments, gravels and sands predominate, the engineer has to be familiar with the behavior of granular materials under relatively high stress levels in which serious compressibility problems can arise. The deformation undergone by an element of granular soil is the result of deformation in the particles themselves, plus relative interparticle movement. The deformation at the interparticle contact may be quite large, and consists initially of distortion and eventually, local fracture and crumbling. Interparticle movement takes place by sliding or rolling. These movements are often aggravated by the distortion suffered by the particles, and the relative importance of these two causes of the total deformation changes as deformation takes place.

1.11.1 Behavior during Isotropic Compression

When a sand sample is subjected to isotropic compression, §1.10, extensive volumetric deformation can occur as a consequence of local changes in the grain structure. This involves sliding and rolling of the particles and results in large tangential forces exerted at the points of contact. However, the vectorial sum of these forces is zero on any plane that cuts through a large group of contact points; therefore the average shear stress on any plane can be zero.

1.11.2 Behavior during Confined Compression

The compressibility of granular soils and their stress-strain characteristics during confined compression are very important because this condition is commonly found under real circumstances, for example when a soil is subjected to vertical loads over large areas. LAMBE and WHITMAN [19] present data on the behavior of initially compact, uniform, medium and coarse quartz sands (quartz is the dominant mineral in most sands). When tested in a consolidometer, they showed yielding at stresses about 140 kg/cm^2 (1990lb/in^2). Beyond this level behavior was plastic, due to the fracturing of the individual particles, and subsequent interparticle movement, and the sand density increased, accompanied by significant vertical strain.

Figure 1-18 shows the results of consolidation tests on various typical sands, at high stress levels. It illustrates the great compressibility that is exhibited by granular soils, as a consequence of the sliding of the particles and their fracture. Although particle movement and fracture can start at low stress levels they greatly increase at high stress levels. The larger the particle size, the lower the stress which produces plastic behavior and consequently large deformation. The more angular, the looser, the more uniform the soil particles are and the weaker the individual particles, the larger the deformation will be.

Of course it is true that the stress levels referred to by LAMBE and WHITMAN are unusually high in relation to the ones commonly met with in engineering practice. They illustrate, however, the basic principle.

The deformation of frictional soils under confined compression is accompanied by the production of fines caused by particle breakage. There is considerable breakage when the grains are uniform, and far less if the grain size distribution curve is flat. The production of fines also increases with the angularity of the particles and with the effective pressure. It is also greater the looser the material.

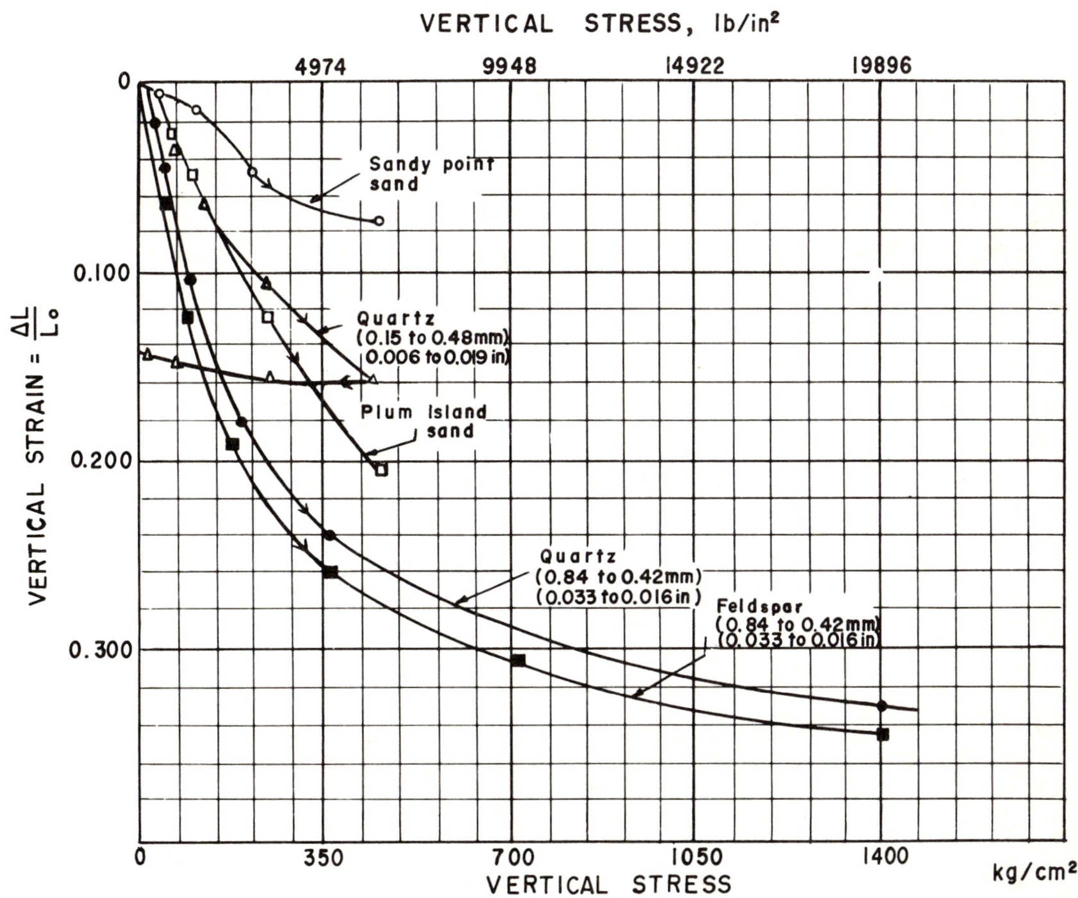

Fig. 1-18 Confined compression tests results showing the compressibility of several different sands subjected to very high stress levels [19]

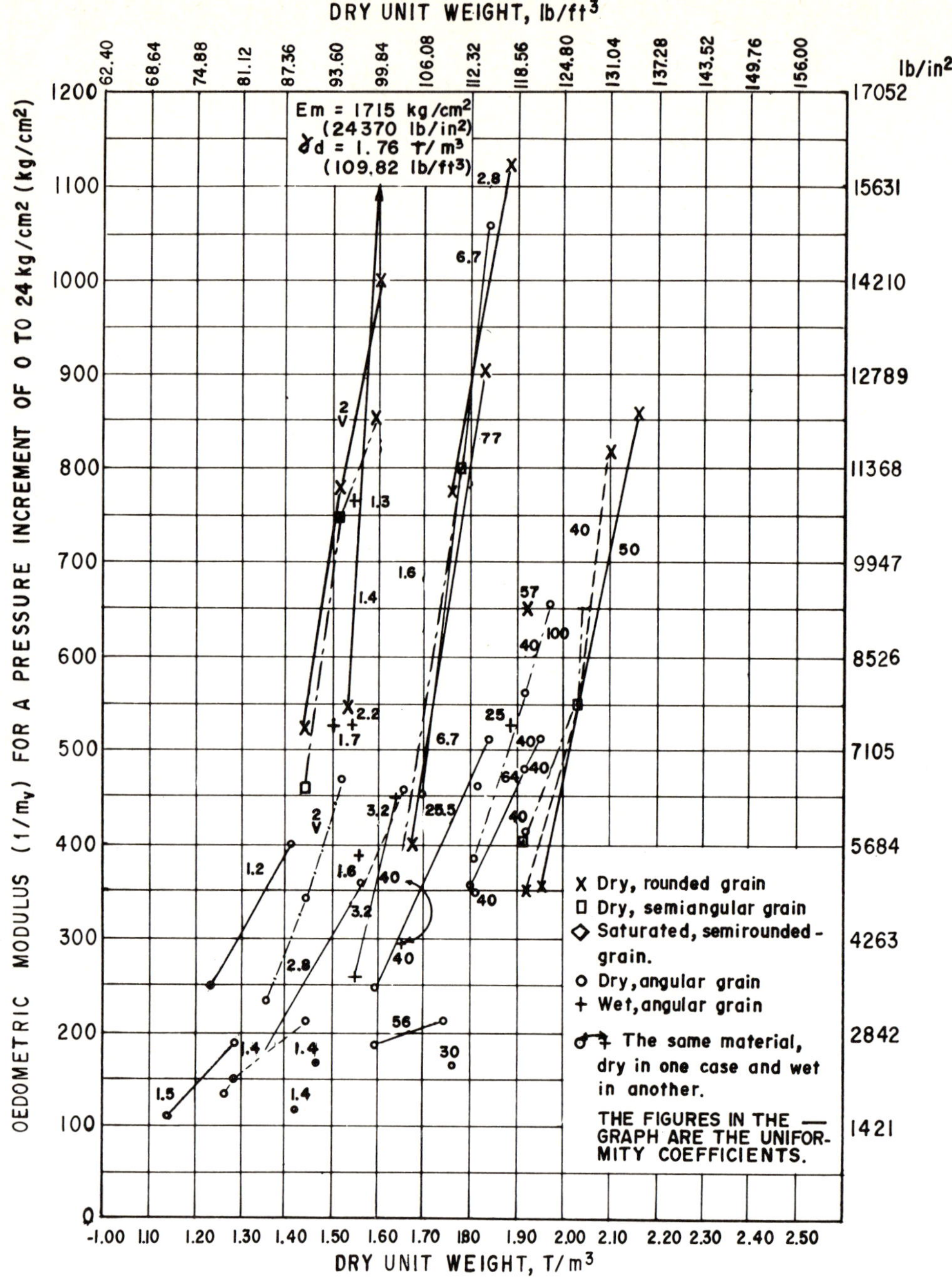

Fig. 1-19 Relationship between the oedometric modulus and the dry specific weight in different laboratory tests performed on granular materials [22]

Figure 1-19 [22] presents a relationship between the compressibility characteristics of various granular materials, represented by what the authors define as an *oedometric modulus* $E_m = \sigma/\epsilon$ (where σ is the stress and ϵ the corresponding strain; $E_m = 1/m_v$ where m_v is the modulus of volumetric change, as defined in [17] and used in the US) correlating it with the dry specific weight corresponding to different degrees of compaction.

The points that are joined in the graph correspond to the same material at different degrees of compaction; it can immediately be observed how the oedometric modulus increases when the material is compacted. It can also be seen how the same material is more compressible when wet than dry. Angular grains are more compressible than rounded ones, which is logical in the light of the ideas expressed above. Generally soils with a low uniformity coefficient are situated on the left of the graph and those with a high uniformity coefficient on the right. The rounder the particles and the greater the variety of sizes, the greater the specific weight that is reached with the same compactive effort, as will be seen in Chapter 4.

1.11.3 Behavior during Triaxial Compression

The compressibility characteristics of a wide variety of granular materials have been studied in detail, using triaxial tests. Lambe and Whitman made a general study [19] of the behavior of sands in triaxial tests. In the course of the test, two stages of deformation behavior are distinguished. The first stage is the beginning of the loading process, in which a small deformation is produced, and is generally accompanied by a reduction in the volume of the sample. This is caused by the particles seeking to adopt more compact structural arrangements. This is followed by the failure stage, during which a peak strength is exhibited if the sand shows fragile behavior. Large vertical deformation can occur if lateral particle movement is allowed; the final result is usually an increase in volume. This is termed dilatancy, which was first observed and investigated by O. Reynolds in 1885. After reaching peak strength, the sand shows a drop in strength as deformation continues. This drop in strength, which is more pronounced in sand that is initially compact can be explained by the rearrangement of individual particles. If a mass of individual particles is imagined on a horizontal surface, the contact planes between the grains will be inclined, not horizontal, so that in order to produce shear failure, it will not only be necessary to overcome the inter-particle friction, but also to force the particles to roll and slide over each other. Normal stress produces the frictional component of the shear strength that has traditionally been included in the angle of internal friction (which will be discussed later) but the movement between particles that is needed for failure, is an additional source of strength and deformation, which depends on the initial arrangement of the grains. If the initial arrangement is compact, the amount of stress and deformation needed to move the grains will be considerable. Gradually, as the grains move they acquire a more favourable position for sliding (the planes through their points of contact become more horizontal). Thus beyond peak strength, the material will exhibit less overall strength as deformation increases. This drop in strength has a lower limit, represented by the particle arrangement which allows sliding without any particle rearrangement. If the initial condition of the particles is loose, the material will have a stress-strain curve corresponding to plastic behavior and the strength component due to particle movement will be almost insignificant.

The initial void ratio of the sand will have a decisive influence on its stress-strain behavior; this is compatible with fragile or plastic behavior of a sand depending mainly on its initial compactness.

One of the most significant investigations on the compressibility and strength of granular materials is the one carried out by Marsal and his collaborators for the design of large dams. This investigation, sponsored by the Mexican Federal Electricity Commission and partially carried out at the Engineering Institute of the Mexican National Autonomous University, is reported in [23-27].

Marsal and his colleagues had a wide variety of sizes and types of laboratory equipment which allowed them to make representative investigations of the behavior of coarse materials at high stress levels. This equipment included a high pressure triaxial chamber up to 25 kg/cm^2 (355 lb/in^2) capable of testing samples with a diameter of 113 cm (44-½ in) and a height of 250 cm (98-½ in) with a maximum particle size of 20 cm (8 in) and one that could test samples with a maximum size of 15 cm (6 in) in plane-strain conditions, with a confinement pressure of up to 22 kg/cm^2 (313 lb/in^2). It also included compaction equipment for large specimens. Some of the conclusions of the studies on strength will be discussed later. In this section only some conclusions relating to compressibility are presented.

Figure 1-20 gives the results obtained when measuring the compressibility of three different materials [24]. Material *1* was a crushed basalt. The fragments were sound, with an unconfined compressive strength above 1,000 kg/cm^2 (14,210 lb/in^2). The dry unit weight of the sample was 2.14 t/m^3 (133.5 lb/ft^3). Material *2* was a granitic gneiss, excavated with explosives. The particles contained thin layers of schist. Its unconfined compressive strength was 740 kg/cm^2 (10,515 lb/in^2) and its dry unit weight was 1.98 t/m^3 (123.5 lb/ft^3). Material *3* was another granitic gneiss, with a more uniform grain size distribution than Material *2* and with a unit weight of 1.62 t/m^3 (101 lb/ft^3); there was no report on its unconfined strength. It can be seen that the void ratio-chamber pressure curves reflect the characteristics of preconsolidated soils, §1-12. The same figure also shows the compressibility coefficient a_v, for the three materials. The compressibility coefficients are sufficiently high to indicate large settlements in high embankments like the ones currently employed in road building.

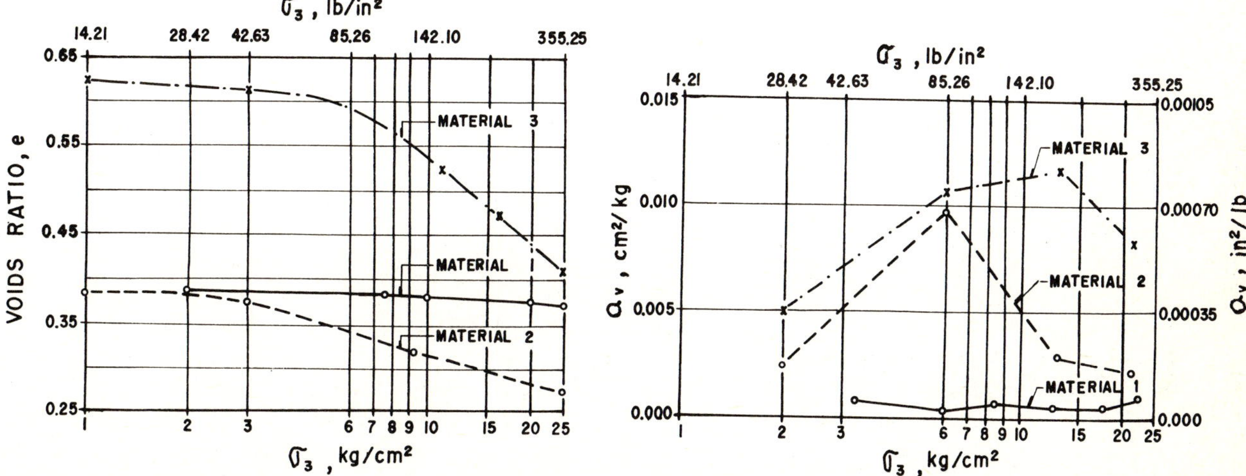

Fig. 1-20 Compressibility data for three rockfill materials

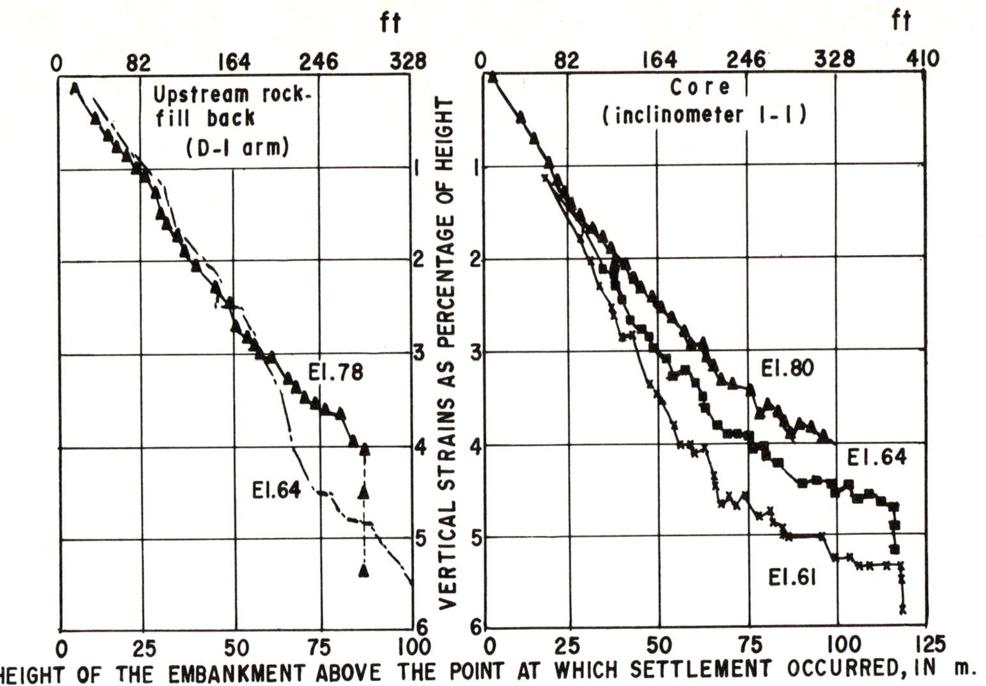

Fig. 1-21 Settlement in El Infiernillo dam [26]

Results like these are contrary to the traditional attitude in the field of highway construction, that rockfills behave *well*, regardless of their height and construction procedures. Instead Marsal and his co-workers [26] have found that the rockfill of the Infiernillo Dam, which is 148 m (485 ft) high (Fig. 1-21), has undergone settlement the same as those which occurred in the impermeable clayey core, constructed with materials traditionally considered compressible.

Particle breakage and the part it plays in overall deformation [24,25], is closely related to the compressibility of coarse materials under heavy loads, as has been shown by modern investigations. The phenomenon brings about changes in the grain size distribution and in the mechanical properties, especially in compressibility.

Figure 1-22 shows the grain size distribution curves of the three rockfill materials studied by the Marsal team, discussed previously [24], before and after testing in the giant triaxial chamber, where confining pressures of 25 kg/cm² (355 lb/in²) were reached. The uniform gneiss (material *3*) in particular, exhibited degradation, although the phenomenon is perceptible in all three materials. It appears that the greater the uniformity, the greater the particle breakage.

Marsal proposes a coefficient of particle breakage. It is represented by B, and is obtained as follows. The grain size distribution curves before and after the triaxial test are obtained, as well as the percentages retained on each sieve originally and after consolidation. When the original percentage is larger, the differences are considered positive. The sum of

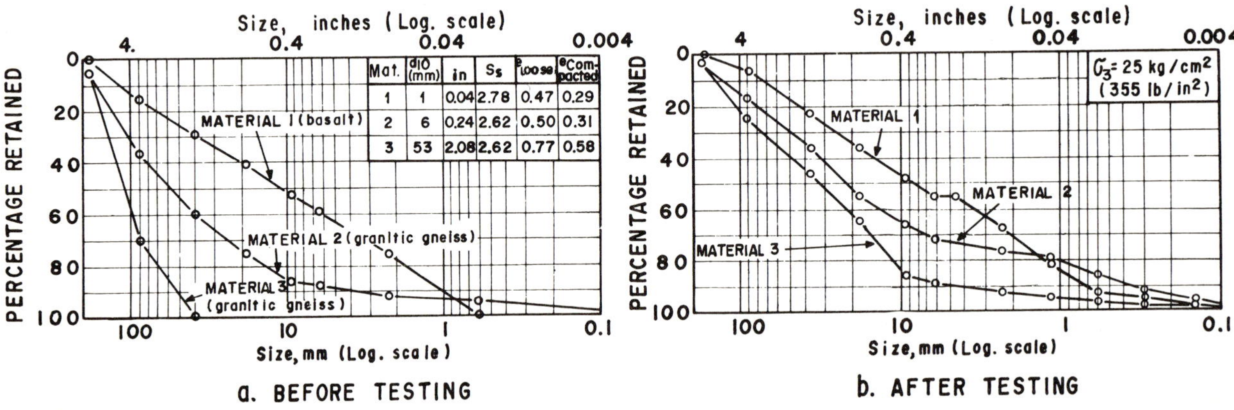

Mat.	d_{10} (mm)	in	S_s	e_{loose}	$e_{compacted}$
1	1	0.04	2.78	0.47	0.29
2	6	0.24	2.62	0.50	0.31
3	53	2.08	2.62	0.77	0.58

Fig. 1-22 Grain-size distribution curves for three rockfill materials

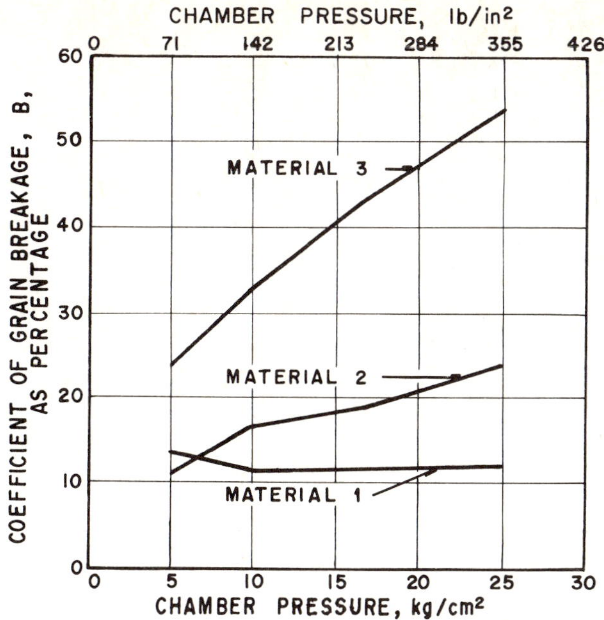

Fig. 1-23 Particle breakage in triaxial tests

the positive differences is the value B. The difference in the amount retained on each sieve represents the particle breakage which has taken place. In Fig. 1-23 the coefficient B is related to the value of the confining pressure used in each test.

1.12 Compressibility of Cohesive Soils

1.12.1 Consolidation

The deformation of cohesive soils, even under relatively small loads, has traditionally been recognized as a problem of fundamental interest, because it leads to serious settlement, especially of foundations of soft clays and plastic silts. Some of the earliest triumphs and initial fame of soil mechanics were due to the success of theories and techniques to predict settlement.

The deformation of clays under loading is of great significance not just because of the large settlement that can occur, but also because most of this settlement develops during a long period of time after the load is applied; as a result, a structure can suffer continuous large deformation years after its erection. The volume decrease which occurs in fine cohesive soils (clays and plastic silts, and also in micaceous and organic soils), as a result of loading, and which develops over a long period, is known as consolidation.

Frequently during the consolidation process, the relative horizontal position among particles remains essentially the same and the movement of the particles takes place in a vertical direction. This is known as one-dimensional consolidation. In embankments it occurs in a compressible formation whose width is larger than its thickness, and that is compressed under loads covering wide areas. It also occurs when a thick layer of compressible soil contains so many thin layers of sand that lateral strain is negligible.

In these and similar cases, the consolidation characteristics of clay strata can be quantitatively studied with a reasonable degree of accuracy in the confined compression or one dimensional consolidation test, see Plates 1-4, 1-5. The samples should be representative of the soil and as undisturbed as possible. In this way the magnitude and time-rate of settlement due to the load applied can be measured.

Plate 1-4b Detailed view of a consolidometer

Plate 1-4a Pneumatic consolidometers of the Geotec type

Plate 1-5 Consolidometers

In laboratory tests performed with small samples, consolidation is obtained in a short time compared to the time required for a real clay stratum to consolidate under a structure. When applying soil mechanics theories, it is assumed that all the consolidation constants determined in the short laboratory process are the same as in the slower field process. Whether this is the case has not yet been proven. This may be one of the reasons why predicted settlements are usually larger than the real ones.

The standard unidimensional consolidation test is performed on a cylindrical sample placed in a ring, usually of bronze or stainless steel which provides complete lateral confinement. The ring is placed between two circular porous stones, one on each face of the sample. The stones have a diameter that is slightly smaller than the interior diameter of the ring. All are placed in a consolidometer pan (Fig. 1-24). The consolidometer or oedometer shown is the *floating ring* type, one commonly used and thus named because the ring can move during consolidation.

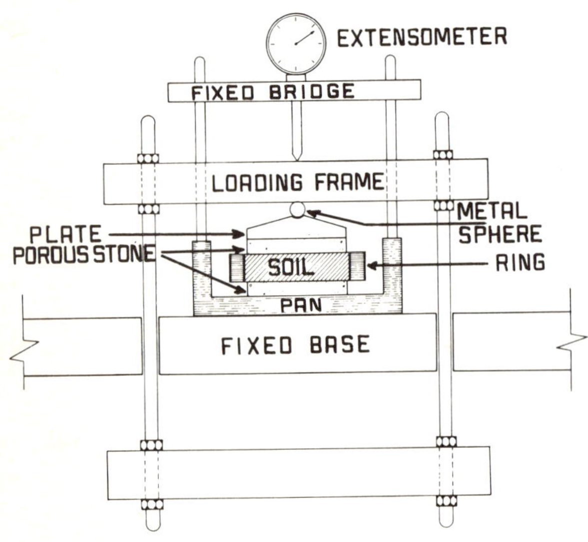

Fig. 1-24 Diagram showing in detail how the sample is placed in a floating ring consolidometer

By means of a loading frame or other testing machine that can apply a constant load, as shown in Fig. 1-24, different loads can be applied to the sample through a metal ball and a plate placed on the upper porous stone. An extensometer resting on the moveable loading frame and attached to the fixed base allows the soil deformation to be measured. The load is applied with increasing intensity, each being allowed to act for a long enough period of time, for the rate of deformation to be reduced to almost zero.

During each load increase, extensometer readings are taken at fixed times in order to characterize the deformation process. These readings are plotted on a graph with the time as the abscissa (logarithmic scale), and the corresponding extensometer readings on an arithmetic scale as the ordinate. These curves are known as consolidation curves and one is obtained for each increment in the applied load.

Figure 1-25 shows a typical curve (not to scale). As the soil reaches its maximum deformation under an increased load, the void ratio reaches a correspondingly low value. This can be determined from the initial data on the sample and the

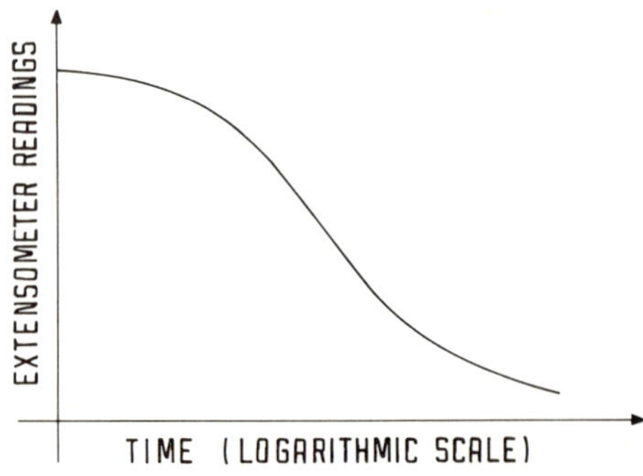

Fig. 1-25 Typical consolidation curve for clays (not to scale)

extensometer readings. Thus, for each load increase a value for the void ratio is obtained. Once all the load increments have been applied, a graph can be plotted using loads or stresses as abscissa and void ratios as ordinate. Loads can be plotted on an arithmetic or logarithmic scale, the void ratios are always plotted on an arithmetic scale. The resulting compressibility curve is obtained in each complete consolidation test. Figure 1-26 shows some typical curves (not to scale).

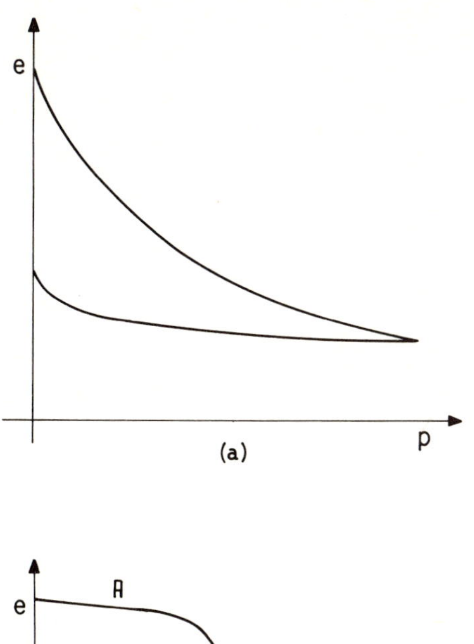

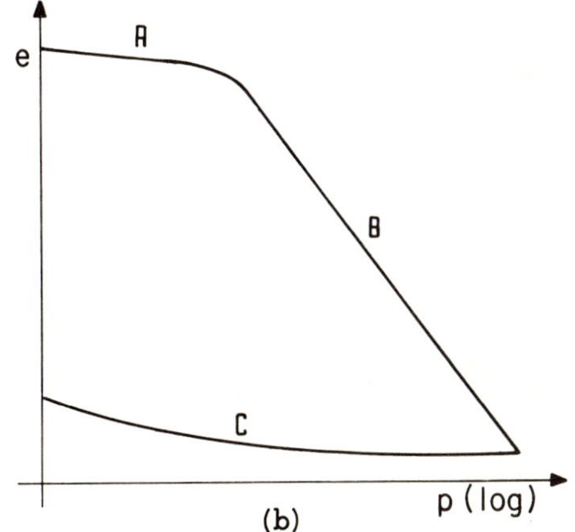

Fig. 1-26 Typical compressibility curve for compressible soils: a) Arithmetical representation b) Semilogarithmic representation

In a compressibility curve three different stages are usually defined. Stage A (Fig. 1-26b) starts almost horizontally and gradually becomes curved downward, reaching its maximum as it joins stage B, which is approximately straight for many soils. After reaching a maximum load, that is greater than the load anticipated in practice, the sample is unloaded, each unloading decrement remaining for enough time for the rate of deformation to be reduced practically to zero. In this stage the sample makes a certain recovery by expanding, although it may never again reach its initial void ratio. Stage C of Fig. 1-26b corresponds to this third stage where the specimen is unloaded to a final zero load.

Stage A of the compressibility curve is known as the *recompression section,* stage B as the *virgin stage* and stage C as the *unloading stage*. The reason for these names will be made clear in the following.

Consider a test in which a clay sample is subjected to a cycle of loading, and complete unloading, in a consolidation test. Immediately after it has been unloaded, it is loaded again, to a pressure significantly greater than the maximum reached in the first cycle. Finally, the sample is unloaded until it returns to the zero load condition.

Disregarding certain secondary behavior, the graphs obtained will be like the one which appears in Fig. 1-27.

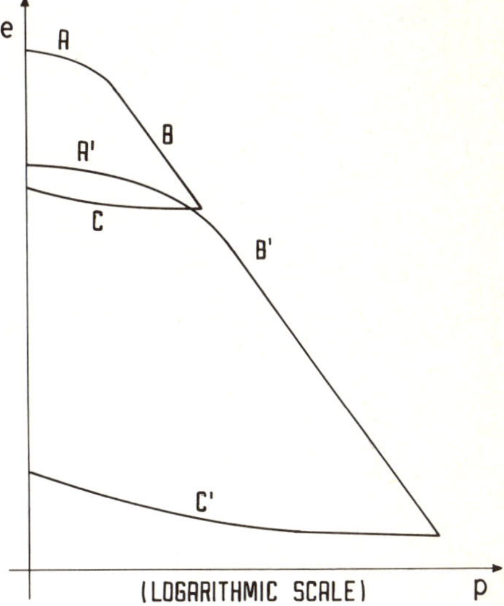

Fig. 1-27 Compressibility curves for two consecutive loading and unloading processes

In the graph, A', B', C', correspond to the second cycle. Section A', (recompression) extends to the maximum pressure at which the soil was loaded in the previous cycle. The new virgin section B', rapidly evolves as the prolongation of the virgin section corresponding to the first cycle. Section C' (unloading) is similar to section C, which was obtained initially.

From the relative position of sections A', B' and C' in the second loading and unloading cycle, one concludes that a recompression section is produced, like A', when the pressures being applied have previously been supported by the soil. A virgin section, like B', is the result of pressures being applied that have never been supported before. The names adopted for the different sections are therefore seen to be logical.

When a natural soil sample is subjected to only one loading and unloading cycle in the laboratory (as is usual in a normal unidimensional consolidation test) and a graph is obtained like the one which appears in Fig. 1-26b, there is sufficient experimental evidence to conclude that the pressures corresponding to section A have already been applied to the soil at some earlier date, while those corresponding to section B are greater than any previously supported.

In order to obtain an objective concept of the time rate of the unidimensional consolidation process in fine soils, TERZAGHI proposed the adoption of a mechanical model originally proposed by LORD KELVIN for other purposes.

A cylinder is considered with a cross-section, A, fitted with a non-frictional piston, and with a small perforation in it, as shown in Fig. 1-28.

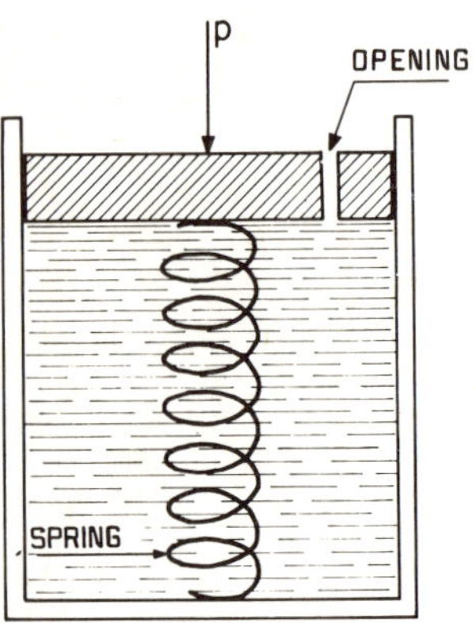

Fig. 1-28 Diagram showing TERZAGHI's mechanical model for the representation of the consolidation process in fine soils

The piston is supported by a spring which is attached to the bottom of the cylinder, and the cylinder is completely filled with a non-compressible fluid. If a load P is placed on top of the piston, and the orifice is closed, it is obvious that the spring cannot deform and, the entire load P will be supported by the fluid.

If the fluid is allowed to escape by opening the orifice, the load will be gradually transferred from the fluid to the spring. Indeed, between the interior and exterior of the cylinder, in the opening, there will initially be a difference in pressure, equal to P/A which will provide the necessary gradient for the fluid to flow out through the opening, allowing deformation of the spring which will take the load, according to HOOKE's Law. The rate of load transfer will depend on the size of the orifice and the viscosity of the fluid. Of course, if the spring is allowed to deform sufficiently, it will support the entire load P and the fluid will return to its condition prior to the application of P.

If instead of a single cylinder with its spring, one now considers a series of communicated cylinders as shown in Fig. 1-29, the initial pressure distribution in the water will be linear (line *1-2*). The fluid will not show any tendency towards movement if the weight of the pistons and springs is disregarded or if the device reached equilibrium before the beginning of the experiment. If the load P is suddenly applied to the first piston, the fluid should immediately support it. The fluid pressure will now become greater than hydrostatic, and the increase is transmitted to any depth with the same value. The new fluid pressure diagram will now be line *3-4*. There is still no hydraulic gradient to encourage movement of the fluid, if the upper orifice remains closed as previously described for the case of a single chamber. The difference in pressures, creates a hydraulic gradient which causes fluid to flow out of the first chamber, when the orifice is opened. As soon as the flow starts, the pressure in the fluid in the first chamber is reduced, and part of the load is simultaneously transferred to the spring. The reduction in the pressure of the fluid in the first chamber in relation to the fluid in the second chamber brings about a difference in the pressures across the second orifice, and consequently fluid will pass from the second chamber to the first. As a consequence, the pressure of the fluid in the second chamber also drops, and fluid flows from the chambers below. The end of the process will be the moment when the pressure in all the fluid returns to a hydrostatic condition, with the entire load being taken by the springs.

At any moment after the application of the load, the pressure distribution in the fluid and springs, u and p respectively, is indicated by the zig-zag line (Fig. 1-29). Note that in each chamber the pressure on the fluid is governed by a linear law and the discontinuities in pressure occur only at the openings. As time goes by, the zig-zag line moves continuously towards the left.

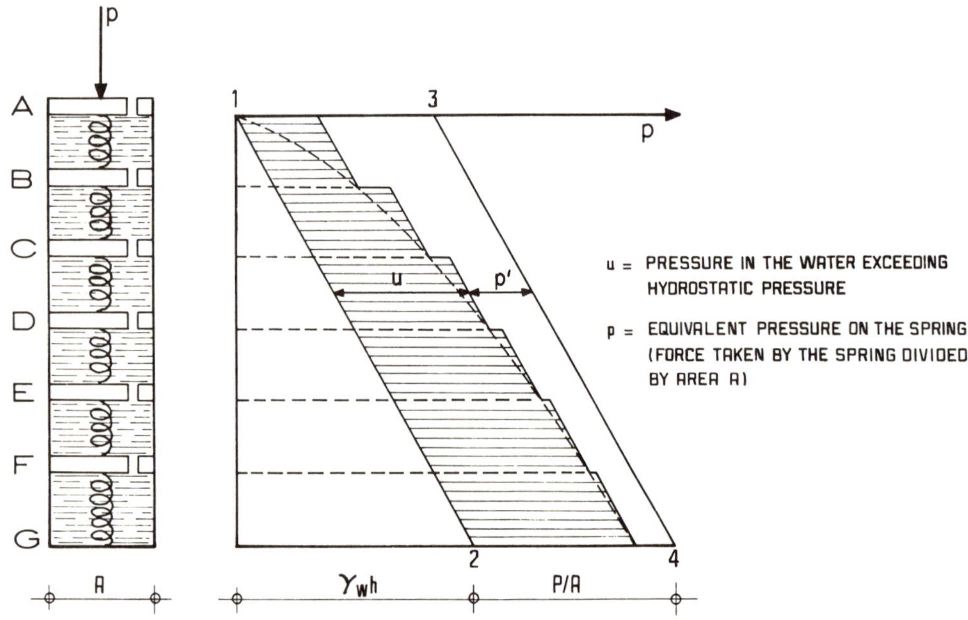

Fig. 1-29 TERZAGHI's model, including several different chambers

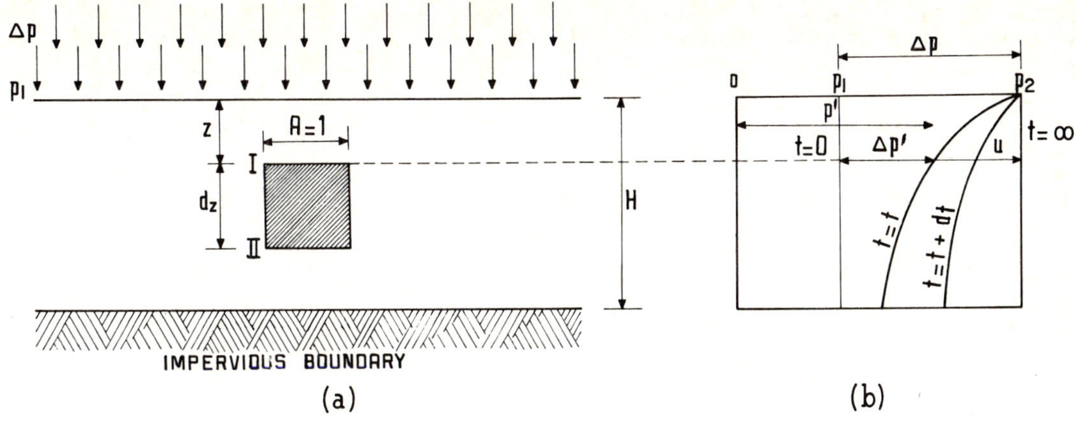

Fig. 1-30 Infinitely wide stratum of soil subjected to a unidimensional consolidation process

If the chambers are numerous, and each has a very small volume, the model will be closer to real soil conditions; the zig-zag line representing pressure distribution for a number of very small chambers will become a continuous curve.

The structure of a soil is represented by the springs in the model, the pore water by the non-compressible fluid, and the capillary channels by the openings in the pistons.

Now a soil stratum is considered of infinite width (in both horizontal directions) with a thickness, H, such that the stress due to the weight of the soil and its water can be ignored, when compared with the stresses produced by the loads applied.

It is assumed that water can only be drained through the upper face of the stratum, the lower boundary being impermeable. It is also assumed that the stratum has consolidated under a pressure p_1. Under these conditions, a pressure increment, Δp, is applied to the stratum. The total pressure acting on the stratum will be $p_2 = p_1 + \Delta p$. Immediately after the load increment is applied, it will be supported by the pore water, and its pressure will rise to a higher value than the hydrostatic one. This increment in pressure will be equal to Δp throughout the entire thickness, as shown in Fig. 1-30b.

After time t has elapsed, a certain amount of water will have drained through the upper face and consequently some of the excess pore pressure will have been transferred to the solids, $\Delta p'$ The new distribution of pressure among solids and water ($p' = p_1 + \Delta p'$ and u) is represented by the curve $t = t$, in Fig. 1-30b.

Obviously:

$$\Delta p = \Delta p' + u \tag{1-36}$$

This equation is valid at any instant, t, and at any depth z. An instant later, $t+dt$, the new pressure distribution is also shown in Fig. 1-30b. In this figure it can be seen that both the pressure Δp in the solid structure, and u in the pore water, are functions of depth z, and of time t. It can, thus, be written:

$$u = f(z,t) \tag{1-37}$$

and:

$$\Delta p' = \Delta p - u = \Delta p - f(z,t) \tag{1-38}$$

This equation expresses the evolution of unidimensional consolidation with vertical flow. It has a mathematical solution [17]:

$$\frac{k(1+e)}{a_v \gamma_w} \frac{d^2u}{dz^2} = \frac{du}{dt} \tag{1-39}$$

This is known as the unidimensional consolidation differential equation, with vertical flow, for it was obtained in accordance with these hypotheses. Here k, is the permeability coefficient of the soil, e is the void ratio (before the consolidation process starts), and a_v is the compressibility coefficient of the soil given by:

$$a_v = \frac{de}{dp'} \simeq \frac{\Delta e}{\Delta p'} \tag{1-40}$$

This coefficient expresses the change in the void ratio due to an increment of the effective pressure; it corresponds to the slope of the compressibility curve (Fig. 1-26). It the *coefficient of volumetric compressibility*, m_v is defined by

$$m_v = \frac{a_v}{1+e} \tag{1-41}$$

It expresses the soil's compressibility, relating it to the initial volume [17].

Finally, the expression

$$C_v = \frac{k(1+e)}{a_v \gamma_w} \tag{1-42}$$

defines the *consolidation coefficient*, C_v, of the soil. It is assumed to be constant during the consolidation process although it varies within limits.

To obtain a workable solution, Eq. (1-39) must be solved for the initial and boundary conditions of the particular case in hand.

The solution that is discussed, considers that the pressure, $\Delta p'$, that induces consolidation in a soil stratum of thickness H, is constant throughout the whole stratum (the solution is also applicable to a hydrostatic pressure distribution). The solution [17] is:

$$u = \Delta p' \sum_{n=0}^{n=\infty} \left\{ \frac{4}{(2n+1)\pi} \sin \left[\frac{(2n+1)\pi}{2} \cdot \frac{z}{H} \right] \cdot e^{-\frac{(2n+1)^2\pi^2 C_v}{4H^2} t} \right\} \quad (1\text{-}43)$$

Where: u is the excess water pressure, at depth z, and at instant t of the consolidation process; z is the depth at which u is calculated; H is the depth of the stratum under consolidation; t is any instant within the consolidation process; and e is the base number of natural logarithms. Equation (1-43) cannot be used for the solution of practical problems. To transform it into a workable expression, it is necessary to define the following.

The *Degree of Consolidation* of the stratum at an instant t of the consolidation process, is the relation between the consolidation that has already taken place and the total consolidation that will take place. It is represented by U. In [17] it is shown that the consolidation degree can be expressed as:

$$U\% = 100 \left[1 - \frac{\int_0^{2H} u dz}{\Delta p \, 2H} \right] \quad (1\text{-}44)$$

where u is given by Eq. (1-43)

The *Time Factor, T*, is a dimensionless parameter:

$$T = \frac{C_v}{H^2} t \quad (1\text{-}45)$$

Using the foregoing definitions and inserting Eq. (1-45) into (1-43) and the resulting one into Eq. (1-44), the following equation is obtained:

$$U\% = 100 \left[1 - \sum_{n=0}^{n=\infty} \frac{8}{(2n+1)^2\pi^2} \cdot e^{-\frac{(2n+1)^2\pi^2}{4} T} \right] \quad (1\text{-}46)$$

Equation (1-46) establishes the relationship between the degree of consolidation and the time factor, and is the final expression of TERZAGHI'S *Unidimensional Consolidation Theory*. Using Eq. (1-46), giving values to T and calculating the corresponding values of U, the values in Table 1-1 are obtained; these values are represented graphically in Fig. 1-31

Table 1-1
Theoretical relationship $U\%$ *vs.* T, calculated from Eq. (1-46)

$U(\%)$	T
0	0.000
10	0.008
15	0.018
20	0.031
25	0.049
30	0.071
35	0.096
40	0.126
45	0.159
50	0.197
55	0.238
60	0.287
65	0.342
70	0.405
75	0.477
80	0.565
85	0.684
90	0.848
95	1.127
100	∞

The unidimensional consolidation theory has been obtained under the following assumptions [17];

a) The soil consolidates in just one direction, (vertically for example)

b) Water flows only in the direction in which the soil consolidates

c) DARCY'S Law applies to the water flow

d) The soil is completely saturated

e) Water and mineral particles of the soil are non-compressible

f) The variation in the thickness of the stratum is small enough for the variable z to be regarded as constant throughout the consolidation process

g) $\Delta p'$ is constant

h) The consolidation coefficient, C_v, is constant throughout the consolidation process

i) When applying the consolidation theory to a settlement calculation, using soil parameters (for example, C_v) obtained from a consolidation test run in the laboratory, it is assumed that these parameters have the same value as in the field. The test represents field conditions and any scale effects between laboratory and field are disregarded

The above assumptions define the applicability of TERZAGHI'S theory. It has already been mentioned that a) and b) are reasonable in wide strata with a much smaller thickness; however the flow is vertical *and* horizontal if the soil mass under consolidation has three very similar dimensions or if the horizontal permeability is much greater than the vertical. (In [17] there is an extension of the consolidation theory to bidimensional and tridimensional flow cases). Assumption c) is reasonable in fine grained soils. Assumptions d) and e) will not cause large errors when the theory is applied to clayey soils below the water table, as is usually the case for transported soils deposited in lacustrine areas, river banks or shores.

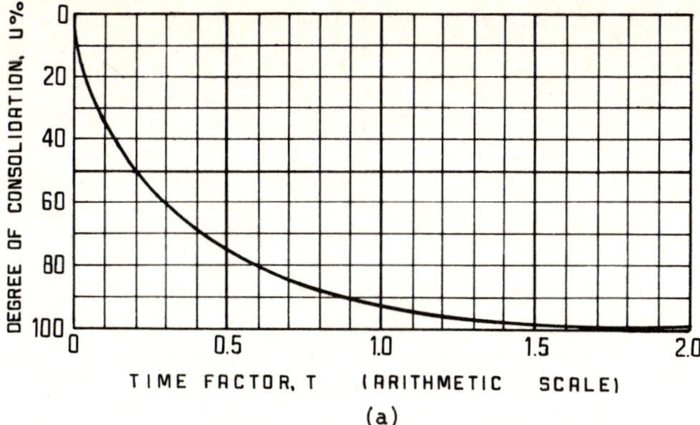

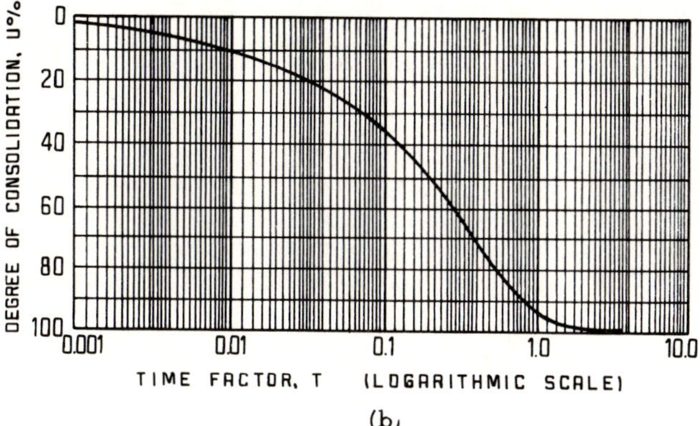

Fig. 1-31 Theoretical consolidation curves a) Plotted arithmetically b) Plotted semi-logarithmically

The importance of these assumptions can only be judged by comparing predictions based on the theory with the observation of real structures. Values obtained from the consolidation theory are often in accordance with the real behavior of clays, within an acceptable range from the engineering point of view.

The time factor was defined as:

$$T = \frac{k(1+e)}{a_v \gamma_w H^2} t \tag{1-47}$$

This expression can be rewritten as:

$$t = \frac{a_v \gamma_w H^2}{k(1+e)} T \tag{1-48}$$

From this expression some important concepts can be deduced: a) If everything else remains constant, the time necessary to reach a certain degree of consolidation (corresponding to a given time factor) is directly proportional to the square of the effective thickness of the stratum. This point deserves some comment. The thickness of the stratum that governs the evolution of an unidimensional consolidation process with vertical water flow, is the path that the water has to follow to drain out from the stratum. If the stratum has an impermeable boundary, the length of this path, known as the effective thickness, coincides with the real thickness of the stratum (Fig. 1-32 a). If the stratum can be drained through both upper and lower faces, the longest path water has to follow in order to drain out is half the stratum thickness (Fig. 1-32b). In the equations from the unidimensional consolidation theory, the value of H that appears is always the effective thickness, as far as the consolidation time is concerned.

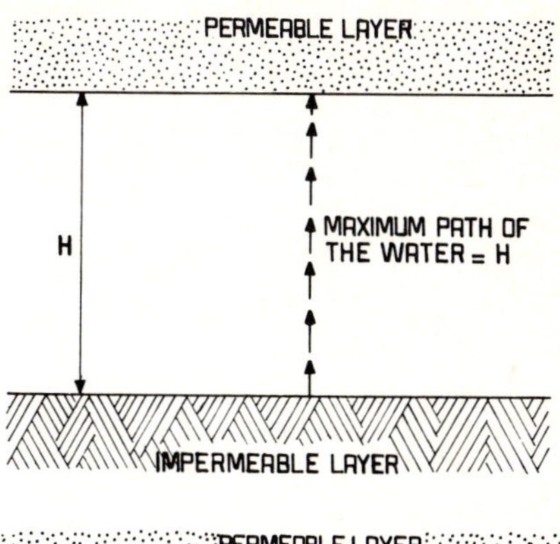

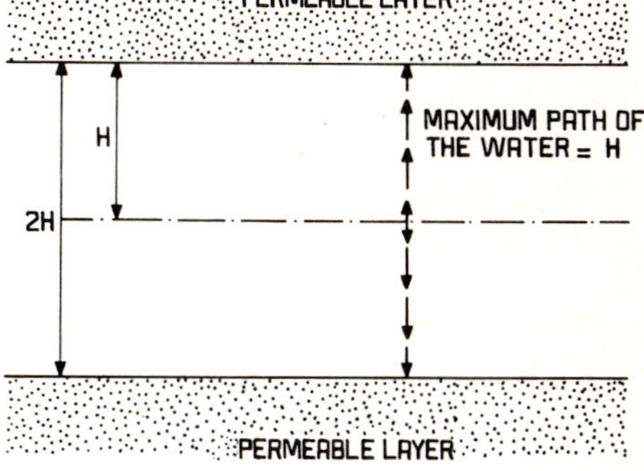

Fig. 1-32 Diagrams to illustrate the concept of effective thickness which determines consolidation time

If two layers of the same material have different effective thicknesses, H_1 and H_2 then the time t_1 and t_2 necessary for each layer to reach a certain degree of consolidation are related as follows:

$$\frac{t_1}{t_2} = \frac{H_1^2}{H_2^2} \tag{1-49}$$

b) If everything else remains constant, the time t, necessary for a soil to reach a certain degree of consolidation is inversely proportional to the permeability coefficient k. If, therefore, two layers with the same effective thickness have different permeabilities, k_1 and k_2, the times necessary for each layer to reach a certain degree of consolidation are related as follows:

$$\frac{t_1}{t_2} = \frac{k_2}{k_1} \tag{1-50}$$

c) If all other factors remain constant, the time necessary for a soil to reach a certain degree of consolidation is directly proportional to the compressibility coefficient, a_v. Therefore if both strata are considered as having the same effective thickness, but their compressibility coefficients are different, a_{v1}, and a_{v2}, the times t_1 and t_2 needed for each stratum to reach the same degree of consolidation, are related as follows:

$$\frac{t_1}{t_2} = \frac{a_{v1}}{a_{v2}} \tag{1-51}$$

When a consolidation test is performed on a sample of soil, consolidation curves are obtained for each of the load increments applied. It has already been seen that these curves relate the readings made by a micrometer at the corresponding times. By applying TERZAGHI's theory, a curve U (%) vs T was obtained, where T is the time factor which involves all the parameters that affect the consolidation process. T and t are, of course, directly proportional for a given sample under specific loading conditions.

If one imagines that the soil behavior strictly follows the theory, the degree of consolidation and the micrometric readings will be proportional, because under such conditions, at 50% consolidation, for example, half of the soil deformation will have occurred. Therefore, if a soil follows TERZAGHI's theory, the theoretical curve U (%) $-$ T and the laboratory consolidation curves should be similar in shape differing only in the scales employed.

In reality, no soil strictly follows the theoretical curve. To compare an observed curve with the theoretical one, it will be necessary to define at which point of the consolidation curve 0% and 100% consolidation will occur, so that the U (%) scale can be adjusted to the micrometric readings.

If the soil contains some air, or if it does not adjust perfectly to the test ring, an instantaneous deformation will occur immediately upon application of the load. By observing the micrometer readings, one cannot determine if the first deformation is due to those rapid adjustments or represents the initiation of the consolidation phenomenon. Fortunately, the consolidation curve before 50% consolidation is practically a parabola and a *theoretical* 0% can be determined by the application of a simple property of these curves.

It is more difficult to determine the theoretical point corresponding to 100% primary consolidation. Of the various methods proposed, that of CASAGRANDE is simple and applicable to most soils. A consolidation curve is plotted in a semi-logarithmic form (Fig. 1-33).

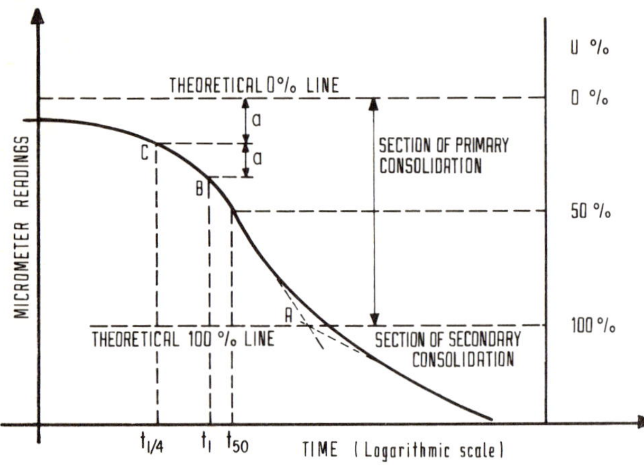

Fig. 1-33 Determination of 0% and 100% primary consolidation on a consolidation curve

In this form of the curve, the part where *secondary consolidation* (see later) becomes noticeable is defined by a straight section sloping downwards. This is usually evident, and can be determined accurately. Thus the time period where primary consolidation dies out can be determined at a glance. This corresponds to the transition between the steeply sloping curve and the final straight section (see Fig. 1-33). It has been empirically observed (CASAGRANDE) that the point A obtained as the intersection of the straight secondary compression section and the tangent to the curved part at its point of contraflexure, is a reasonable representation of the border point between primary and secondary consolidation; that is 100% primary consolidation.

Since secondary consolidation is present from the beginning of the test, it is not really possible to fix a precise point at which the primary effect ends and the secondary effect commences. Therefore, the previous definition of 100% consolidation is arbitrary. However in the first part of the consolidation curve, the secondary effect is not yet very noticeable and because of this the parabolic relationship already discussed, is found to be largely correct within a reasonable approximation. The 0% consolidation line can now be found as follows (Fig. 1-33):

Choose an arbitrary time t_1, such that the corresponding point B on the observed curve represents less than 50% consolidation. Obtain point C, corresponding to a time $t_{1/4}$ and determine the difference between the ordinates a at the top points. Since between these two points there is an abscissa ratio of 4 and since they are points on a parabola, it follows that the ordinate ratio must be $\sqrt{4}=2$. Thus, the origin of the parabola is at a distance $2a$ above C. It is advisable to repeat this simple construction several times, starting at different points and representing 0% consolidation as the average ordinate thus obtained. In Fig. 1-33, the U (%) scale drawn from the limits found, can be seen on the right.

Observe that all the previous construction presumes that the U (%) scale can be applied to different consolidation curves, presuming that it is possible to identify 0% and 100% primary consolidation. This depends on the shape of the actual consolidation curve and on how well it agrees with the theoretical curve. Unfortunately this does not always happen in practice, sometimes the shape of the curve obtained in the laboratory is inappropriate. TAYLOR developed an alternative method for computing the consolidation limits that can be successfully applied where the previously described method fails.

For this method the theoretical curve is drawn using the U% values as ordinate and the $\sqrt{T}$ values as abscissa (Fig. 1-34b). The theoretical curve is approximately a straight line to near 60% consolidation, as should occur considering that it is nearly parabolic up to that point.

From the table of values for $U\% - T$, already obtained, the abscissa corresponding to 90% consolidation is 1.15 times the abscissa corresponding to the continuation of the straight section. This peculiarity is used to find 90% consolidation on the curve obtained in the laboratory. In Fig. 1-34a, the typical shape of a real curve is shown with the micrometric readings as a function of $\sqrt{t}$. Enough precision will be obtained by simply extending the straight section. Another straight line should be drawn, with its abscissa values equal to the values of the original curve multiplied by 1.15. This second line crosses the consolidation curve at the point corresponding to 90% primary consolidation. Notice that the extension of the straight section of the laboratory curve cuts the ordinate axis at a point that should be considered as 0% primary consolidation, and it is at this point that the second straight line should commence.

With this method, C_v is computed using the expression:

$$C_v = \frac{T_{90}}{t_{90}} H^2 = \frac{0.848\, H^2}{t_{90}} \qquad (1\text{-}52)$$

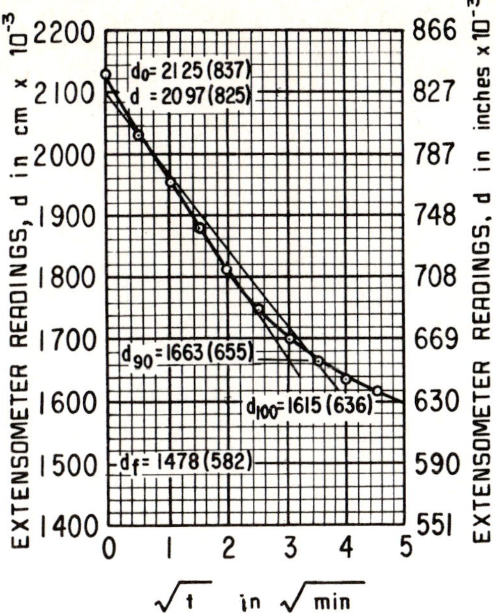

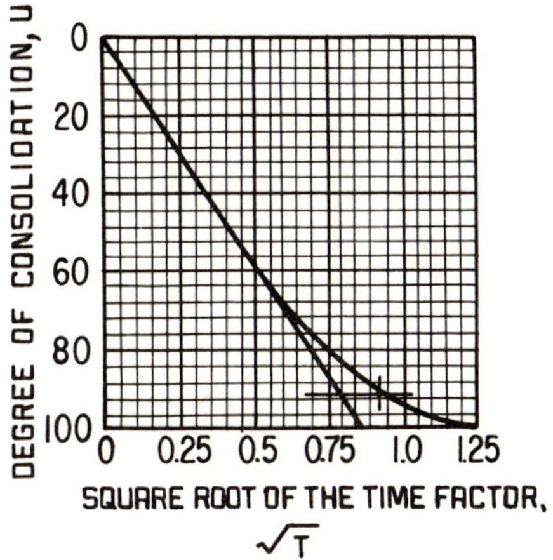

Fig. 1-34 TAYLOR's Method for computing the values of C_V

From the ideas expressed and from the similarity of the curves obtained in the succesive loading cycles (Fig. 1-27) it is deduced that the maximum pressure that the soil withstood before the loading cycle must be in the vicinity of the transition from the recompression to the virgin sections. This pressure, which represents the maximum pressure that the soil has withstood in its geological history, before it was subjected to the consolidation test, is known as the *preconsolidation load.* It plays a very important role in the soil's behavior. However, the transition from the recompression to the virgin section is not abrupt but gradual, and cannot be determined at first sight. CASAGRANDE developed an empirical method for determining the preconsolidation load (p_c), that is effective for practical purposes. This is illustrated in Fig. 1-35:

Once the compressibility curve has been obtained, determine the point of maximum curvature (T) in the transition region between the recompression (II) and the virgin (I) section. Through T draw a horizontal line (h) and a line (t) tangential to the curve. Draw a line bisecting the angle formed by the straight lines, h and t. Prolong the virgin section upwards until it crosses this last line. The abscissa of this intersection point (C), can be approximately taken as the preconsolidation load (p_c) of the soil.

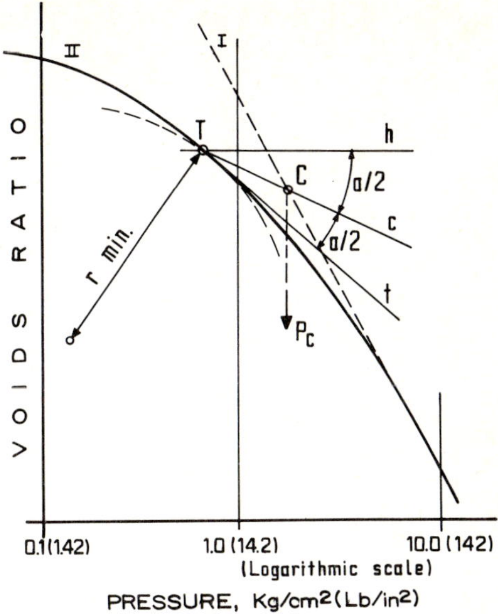

Fig. 1-35 Determination of the preconsolidation load

The most important practical application of the preconsolidation load concept is found in settlement analysis: it may also be of importance in geological investigations.

The virgin section is not noticeably affected by expansion and other minor deformation of the sample, which is fortunate. From this it follows that if a soil has been completely consolidated under a pressure p_1 (usually the weight of the overlying material), any additional consolidation under a load increment Δp can be estimated by the simple expression:

$$\Delta H = \frac{e_1 - e}{1 + e} H$$

where H is the total thickness of the soil stratum. Figure 1-36 shows that the total settlement under a given load increment Δp is smaller, the larger the initial effective pressure (p_1).

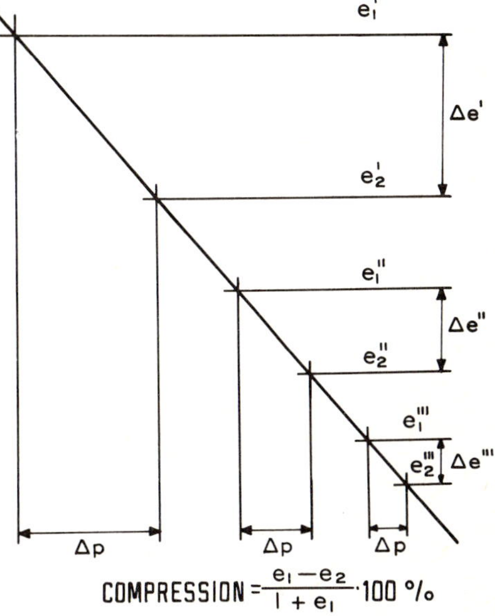

Fig. 1-36 Diagram showing the reduction in settlement at higher initial p_1

If the maximum thickness of soil over a certain point has at some time been partially eroded, the settlement due to the load increment will be much smaller, even though the virgin section of the compressibility curve remains unaltered. For example, in Fig. 1-37, if a clay layer has been subjected to pressures of 3 kg/cm^2 (42.63 lb/in^2) from overlying soil, this pressure being subsequently reduced to 1 kg/cm^2 (14.21 lb/in^2) by erosion and afterwards increased to 2 kg/cm^2 (28.42 lb/in^2) by the construction of a structure, the compression under that structure will follow the line between points B and C of the compressibility curve and will produce Δ_1 settlement. On the contrary, if the soil had been consolidated only under a pressure of 1 kg/cm^2, (14.21 lb/in^2), the appropriate consolidation would be represented between points D and E, leading to Δ_2 settlement, which is larger. This illustrates the importance of the preconsolidation pressure in settlement analysis.

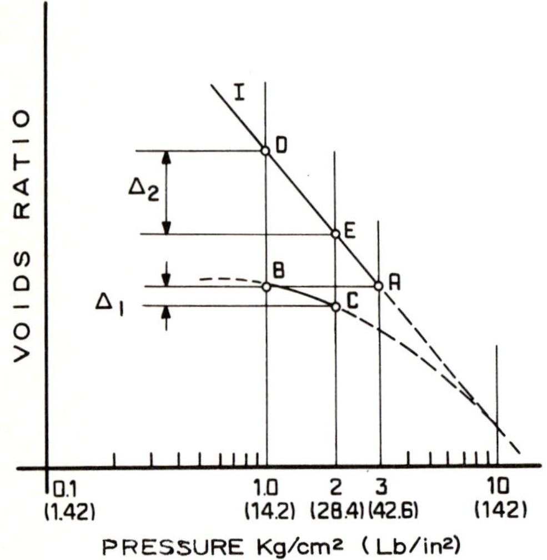

Fig. 1-37 Influence of the preconsolidation load on settlement calculations

1.12.2 Settlement and Expansion

The most useful application of the unidimensional consolidation theory, as well as of the foregoing ideas on the compressibility of cohesive soils, is the prediction of the total settlement that a layer of soil will undergo when subjected to an external load and the evolution of that settlement with time, both of which are equally important for road engineering. The importance of the total amount of settlement is obvious. It can show how much settlement will occur in an embankment built on a soft clay, or how much differential settlement will occur between a bridge and an embankment, depending on the type of bridge foundation used.

The evolution of settlement with time is another essential factor for the engineer handling settlement problems. The effect of a 30 cm (12 in) settlement on a rigid structure, such as a bridge, is completely different if it takes place rapidly or over a period of several years. In the example of the bridge and its access embankment, it will not be enough for the engineer to know the total settlement of both structures in order to understand their interaction. He will also need to know how the movement of both structures occurs with time, because only in this way will he be able to have a clear idea of which is the best foundation and how to prepare for levelling in some parts of the bridge. Often the knowledge that most of the settlement of an access embankment will take place in a short time, such as during construction, will lead to simple and safe solutions for an acceptable interaction between the embankment and the bridge. Building the embankment first, to allow it to settle and then designing a bridge foundation that will not settle is one such solution.

The total settlement of a clay layer of thickness H, due to a unidimensional consolidation process and induced by an overload Δp applied on the upper surface of the layer, can be determined from the data obtained in a consolidation test and using Fig. 1-38.

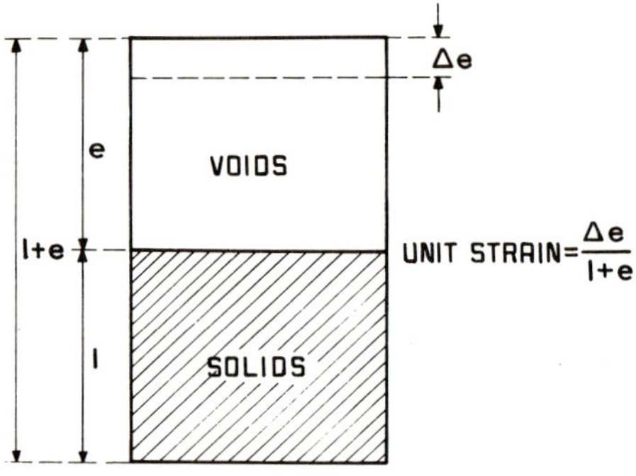

Fig. 1-38 Diagram to illustrate how the total settlement of a soil stratum is obtained

If Δe represents the decrease in the thickness of a soil sample whose original thickness was $dz = 1 + e_o$, (e_o being the initial void ratio), the change in height of the element can be expressed as:

$$\Delta dz = \frac{\Delta e}{1 + e_o} dz \tag{1-53}$$

By integrating this equation over the total thickness of the compressible stratum, H, one obtains:

$$\Delta H = \int_o^H \frac{\Delta e}{1 + e_o} dz \tag{1-54}$$

if the upper boundary of the layer is taken as the origin of z. Equation (1-54) is the general equation for computing the total settlement due to primary consolidation, unidimensional consolidation being assumed, and suggests a simple and practical method for assessing settlement in real situations (Fig. 1-39).

If consolidation tests are performed on unaltered samples taken from different depths in a compressible stratum, a compressibility curve can be plotted for each test which will be representative of the behavior of the soil at that specific depth (Part a of Fig. 1-39). First, the value of p'_o is found, which is the real effective pressure of the soil at a given depth. From this value it will be possible to obtain the corresponding e_o from the graph. Next, to p_o, Δp is added, to obtain the new effective stress that must be withstood by the solid structure of the soil when it has been totally consolidated under the new external loading conditions. The ordinate of the value $p' = p'_o + \Delta p'$ will give the final e which will theoretically be reached by the soil at the depth being dealt with. In this way, it is possible to determine $\Delta e = e - e_o$ and consequently, $\Delta e/(1 + e_o)$.

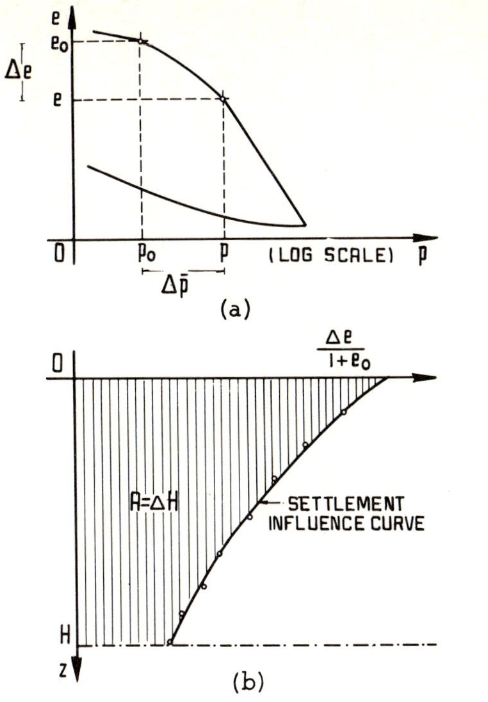

Fig. 1-39 Methods for obtaining settlement influence curves

Part *b* of Fig. 1-39 shows the graph $\Delta e/(1 + e_o)$vs.z which is plotted once the points have been established for different depths in accordance with the procedure just described. Equation (1-54) shows that the area between O and H, known as the *settlement influence curve,* gives the value of ΔH.

In some special cases, the settlements can be assessed using simplified versions of the above method. For example, in the case of a homogeneous, not very thick, compressible stratum, in which the coefficient m_v can be regarded as constant for the pressure interval being handled, the following equation is derived:

$$\Delta H = \int_0^H \frac{\Delta e}{1 + e_o} dz = \int_0^H m_v \Delta p' dz = m_v \int_0^H \Delta p' dz \qquad (1\text{-}55)$$

The integral represents the pressure increase area between depths O and H and can be calculated graphically.

If $\Delta p'$ can be regarded as constant, Eq. (1-55) is simply reduced to:

$$\Delta H = m_v \Delta p' H \qquad (1\text{-}56)$$

The popularity of Eq. (1-56) is perhaps unjustified, for it has limitations which are not always taken into consideration.

In order to be able to calculate the evolution of ΔH with time, which is essential in numerous practical engineering problems, it is first necessary to determine the consolidation coefficient, C_v of the soil, which appears in Eq. (1-45). This equation can be applied to the consolidation test sample, taking into consideration the time corresponding to 50% consolidation. Indeed, $T_{50} = 0.197$, as is deduced from the theoretical consolidation curve. The time t_{50} can be found once the $U(\%)$ scale has been established in the consolidation curve (see Fig. 1-33), and H is the effective thickness of the sample used at the moment when it reached 50% consolidation under the load increment. If, as is usually the case, the sample is drained through both face, half of the thickness of the specimen should be used, and this is calculated as an average of the initial and ultimate semi-thicknesses of the sample corresponding to that load increment. Thus:

$$C_v = \frac{T_{50}}{t_{50}} H^2 = \frac{H^2}{5t_{50}} \qquad (1\text{-}57)$$

Note, however, that for each load increment applied during the consolidation test, Eq. (1-57) can be used. Thus, a value of C_v is obtained for each load increment. In this way it is possible to plot a graph of C_v versus the mean pressure applied during that increment, (the arithmetical average of the initial and ultimate pressures). For an existing stratum subjected to a load increment Δp, the mean value for the C_v corresponding to the zone of the curve of Δp will be used for calculations.

Once the mean C_v of the soil has been obtained, Eq. (1-45) can be applied as follows:

$$t = \frac{H^2}{C_v} T \qquad (1\text{-}58)$$

H is the effective thickness of the soil stratum, computed according to the drainage conditions, described previously. C_v is the consolidation coefficient of the soil, just calculated, within the pressure interval represented by the load increment applied to the stratum. Thus, given values of T, for example the ones in Table 1-1, the times in which the stratum reaches the degrees of consolidation corresponding to those time factors are computed. Because settlement is in direct proportion to the degree of consolidation, final values can be tabulated for settlement corresponding to different times, depending on the evolution of the consolidation process.

The results can be depicted on an arithmetic or logarithmic scale with time as the abscissa. In this way, a curve is obtained for the predicted settlement and its evolution with time.

In many practical problems, especially in cases where a soil is unloaded, as in an excavation, it is of interest to determine the expansion which takes place as a result of unloading.

The problem is basically comparable to settlement assessment, and a similar procedure would, to a certain extent, be acceptable. Expansion, however, has certain peculiarities which deserve attention, and must be considered in conducting a reasonable analysis of an existing situation.

The first case considered is that of a homogeneous clay soil with a surface horizontal before unloading commences. For simplicity, it is assumed that the water table coincides with the surface of the ground. Neutral, effective and total stress conditions correspond to the dashed lines in Fig. 1-40. It is now assumed that a rapid excavation is made with a depth h and infinite width. The total pressure removed will be $\gamma_m h$ and, consequently the total pressure diagram will be reduced by this amount. Because the effective stress condition of the soil mass cannot change instantaneously, the water saturating the soil will bear the unloading and the neutral stress will be reduced by $\gamma_m h$. Because the original pressure of the water at depth h was $\gamma_w h$, the new pressure at that depth, after the rapid excavation,

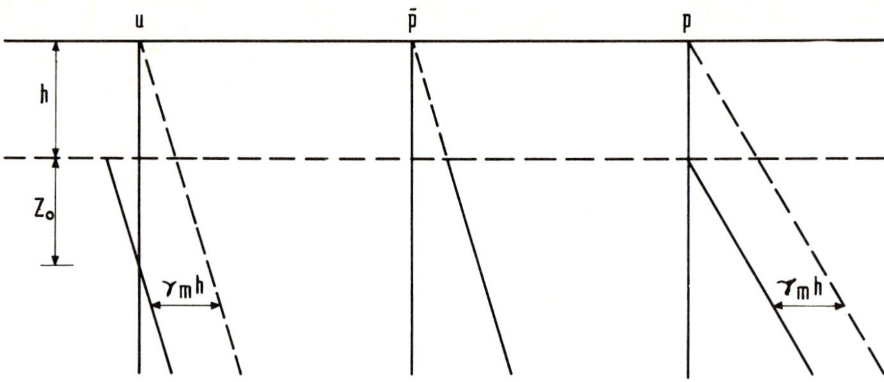

Fig. 1-40 Distribution of vertical stress underneath an infinitely wide excavation

will be $\gamma_w h - \gamma_m h = -\gamma'_m h$. In other words, tension appears in the water equal to the effective pressure at depth h, which, in this case, is the submerged specific weight of the soil for that depth.

It should be noted that, because the excavation is infinitely wide, the new pressure distribution in the water is linear and parallel to the original one. The new pressure distribution will be hydrostatic and in equilibrium; therefore the water will not flow in any direction. The previous neutral, effective and total pressure conditions will therefore be maintained and will correspond both to the moment at which excavation commenced and any subsequent time. The effective pressures which are maintained in the soil will not in this case allow any expansion.

If the water pressure diagram after excavation is observed (the solid lines in Fig. 1-40), it is noticed that the level at which the neutral pressure is zero (water table) corresponds to the depth

$$z_o = \frac{\gamma'_m}{\gamma_w} h \qquad (1\text{-}59)$$

This lowering of the water table, theoretically occurs immediately after the removal of the excavated material. Thus, by simply excavating the soil the depth h (infinitely wide), the water table is lowered to $h + z_o$, that is to say a depth of z_o from the bottom of the excavation.

It is now assumed (Fig. 1-41) that the subsoil in the previous case contains a water-bearing sandy layer, in which the water pressure is maintained. If a rapid excavation of infinite width is made to depth h, the pressure diagrams immediately after excavation will be identical to the ones in the previous analysis, except in the vicinity of the water-bearing layer, where the neutral pressure does not change, but the effective pressure will be reduced by $\gamma_m h$. If d is the depth at which the water-bearing layer is located, the new effective pressure on the upper boundary of that layer immediately after excavation ($t = 0$), will be:

$$p' = \gamma'_m d - \gamma_m h$$

The minimum value which can be reached by effective pressure in the sand is obviously zero. In this case, there will be a maximum depth (h) to which excavation can be taken without neutral pressure in the water-bearing layer (uplift pressure) heaving up the bottom, thus causing failure. This depth will be:

$$h_{crit} = \frac{\gamma'_m}{\gamma_m} d \qquad (1\text{-}60)$$

In Fig. 1-41, it is assumed that $h < h_{crit}$. In this case from the moment of excavation ($t = 0$), an expansion process commences both in the clayey layer above the water-bearing layer and in the underlying clay mass. It is triggered by the water which seeps into the clay from the water-bearing layer. This expansion process increases the neutral pressures in the clayey strata, thus reducing the effective pressures. In Fig. 1-41, isochrones have been drawn for $t = t$, this being an intermediate moment in the process. The final pressure condition in the upper stratum of clay will depend on the boundary conditions in the bottom of the excavation. If it is

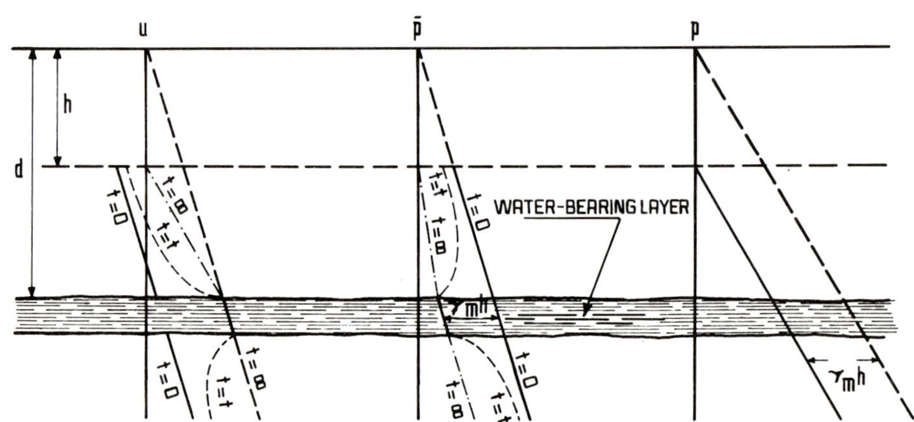

Fig. 1-41 Distribution of vertical stress underneath an infinitely wide excavation with a water-bearing stratum

assumed that all the water which surfaces on the bottom of the excavation is immediately drained, the final condition will be given by the lines $t = \infty$. In the lower stratum, since it is very thick (semi-infinite), the expansion process will continue indefinitely, although at a decreasing rate, and the final pressure condition is the one given by the lines $t = \infty$, as is shown in that zone of the same figure. The expansion process analyzed is uni-dimensional and the flow is vertical. The data obtained from the unloading section of a consolidation test should therefore, in principle, apply. In a case like the one just analyzed, the heave in the bottom of the excavation in time t has two components: the expansion occurring in the clay stratum of finite thickness which overlies the water-bearing layer, and the expansion of the semi-infinite mass located below. First we will discuss the expansion process in the finite stratum.

Before unloading commences, an element of soil at depth z is subjected to an effective pressure $p'_1 = \gamma'_m z$; when expansion terminates it is subjected to p_2, which can be determined as was discussed previously. If a consolidation test is performed on a representative sample of the soil at depth z, a maximum load of p_1 is imposed, followed by unloading to a minimum value of p_2. In the unloading section of the compressibility curve thus obtained, it will be possible to determine the variation Δe corresponding to the soil expansion during unloading. Adopting a similar procedure for other depths, it will be possible to plot the heave influence curve, $\Delta e/(1 + e_o) - z$. The area under the curve corresponds to the total heave in the finite stratum. The heave during time t can be determined by studying the evolution of the expansion with time, in the same way as the evolution of primary settlement was previously studied.

Parameters a_v, m_v and C_v of the unidimensional consolidation theory have corresponding similar concepts a_{vs}, m_{vs} and C_{vs} for unloading, and these can be used in the same way.

For the semi-infinite mass which is placed below the water-bearing layer, its total heave will theoretically be infinite, and so it will only be practical to calculate heave for a finite time t.

Note that for expansion to take place, the water-bearing layer must maintain its neutral pressure. If, by some artificial means, this pressure is lowered to $\gamma_m h$ (Fig. 1-41), the expansion process will not take place. In practice, this is done by wells, from which a suitable amount of water is pumped out of the water-bearing layer. The situation now becomes similar to the one first dealt with in this section, where there was no water-bearing layer.

If, in the case now being analyzed, the water-bearing layer were a hydraulically closed system, with no water supply, (for example, the case of a sandy lens of finite width), the neutral pressure in the sandy stratum would drop as soon as the water drained, and the expansion would not occur. (Theoretically, because water is non-compressible, the neutral pressure will be relieved if only the smallest amount of water drains from the sandy stratum). The case thus becomes similar to the one dealt with in the first part of this section, where the soil mass was clayey and homogeneous.

In real projects, infinitely wide excavations do not exist. The foregoing ideas, nevertheless, provide the basis for discussing idealized finite excavations. Figure 1-42 shows the case of a finite excavation made in a homogeneous clayey medium. The water table is at a depth of h_o below the surface. In this case, the effect of the excavation will not be uniform throughout the entire layer. This reduction in total pressure will have to be estimated at the different points using for example the BOUSSINESQ theory. In an initial approximation, the reduction in neutral pressure at each point in the mass is assumed to be equivalent to the reduction in total pressure (remembering the first of the two cases of infinitely wide excavations above). For this reason, the neutral pressure will decrease more in the central areas of the excavation and in the levels close to the bottom; these reductions will become gradually smaller towards the edges of the excavation and deeper into the homogeneous clay mass. This causes a flow of water from the outside towards the center and from the deep-seated zones towards the bottom of the excavation (Fig. 1-42).

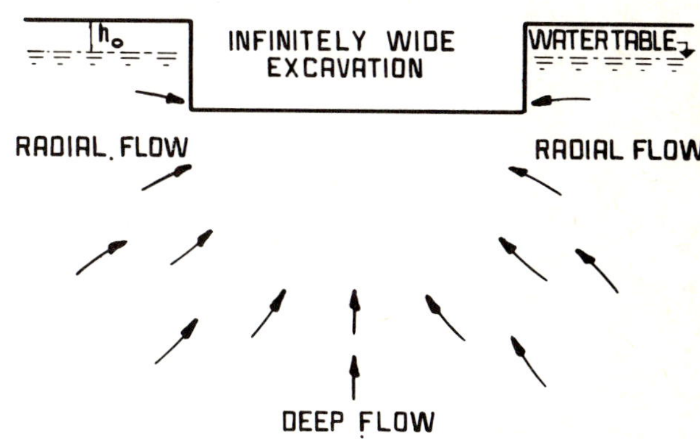

Fig. 1-42 Water flowing towards finite excavation

Expansion of the soil mass underneath the excavation will therefore be greater at the center of the bottom of the excavation, gradually decreasing towards the edges. In natural clay deposits, permeability is generally higher in the horizontal direction than in the vertical direction, which is why the radial flow towards the excavation affects expansion more than vertical flow from deep-down zones. It is striking how an excavation in the clayey mass can cause a reduction in the neutral pressures underneath it, and if the geometrical position of the points at which the neutral pressure is zero is called the water table (atmospheric pressure being the origin), this level drops of its own accord, even lower than the bottom of the excavation.

If, underneath the bottom of the excavation, there are extensive permeable strata providing a water supply, the expansion process will be far more rapid (see the second case of an infinitely wide excavation already discussed). In order to reduce the velocity of expansion to a minimum in the bottom of an excavation, measures are adopted, to resist the forces involved. First deeply driven sheet-piles are driven around the edges of the excavation. This impedes radial flow and allows only vertical flow, which is far slower. Second, pumping wells and other methods (electro-osmosis, for example) are used to lower the neutral pressures at specific points and zones around them. In this way, a low pressure region is built around the excavation which intercepts horizontal flow. Because these excavations are temporary and are built to last for a relatively short time, expansion does not have the opportunity to reach very high values.

The fact that in permeable soils, such as sands and gravels, the water table has to be lowered for a dry excavation, has often led engineers to think that this should also be the same with clays, forgetting that in these materials the water table drops of its own accord when excavation takes place.

Real excavations are not instantaneous, but take a certain period of time. This does not invalidate the above reasoning, but implies that the reductions in neutral pressure will take place gradually, as unloading progresses.

The amount of expansion of soils can be obtained by calculating their expansion index, defined by the expression

$$C_e = -\frac{\Delta e}{\Delta(\log p')} \quad (1\text{-}61)$$

determined in the consolidation test carried out in an oedometer (consolidometer). Thus defined, the expansion index is a measure of how steep the compressibility curve is in the unloading interval, during which the soil expands. Sets of expansion curves can be obtained in the consolidometer by loading a series of specimens at different effective vertical pressures, and subsequently unloading them. These curves tend to be parallel, showing that the expansion coefficient varies very little with the effective pressure to which the soil was subjected prior to expansion. Figure 1-43 [28] shows how the expansion index varies with the liquid limit of a clay. C_e can be seen to increase with an increase in the liquid limit, but the nature of the relationship is such that it can be regarded as merely qualitative.

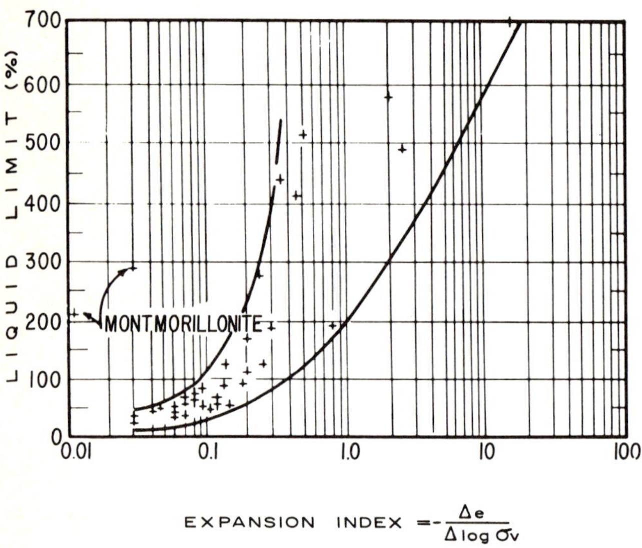

Fig. 1-43 Correlation between the expansion index and the liquid limit in fine soils [28]

For sodium montmorillonite, with a liquid limit of 500%, expansion indices may be as high as 2.5. The values for natural soils are, however, far lower (for example, 0.09 for Boston blue clay, during unloading from 1 to 0.1 kg/cm^2 (14.21 to 1.42 lb/in^2).

1.12.3 Secondary Consolidation

Consolidation consists of two superimposed and combined phenomena. The first one was described in some detail earlier in this section, and consists of the transmission of the external load, originally withstood by the pore-water, to the solid structure of the soil. This is accompanied by a reduction in soil volume and a corresponding loss of pore water, which drains out through the permeable boundaries of the layer. This is primary consolidation. However, because the shrinkage process is accompanied by an increase in effective pressure, it leads to another source of strain, this time due to the effects of the rearrangement of the mineral particles as they adapt themselves to the new, more closed structure. This process is given the name of secondary consolidation, and is not taken into account in TERZAGHI'S unidimensional consolidation theory.

During the initial stages of primary consolidation, almost the entire increased load is supported by the pore-water, and little volume change takes place in the solid structure. Therefore, the effects of strain as a result of particle rearrangement are not very noticeable, and may consist only of minor slides, slippage and rotation of some particles in relation to others. For this reason, secondary consolidation is scarcely perceptible during the early stages of primary consolidation. In the final stages of the primary consolidation process, much of the external pressure has already been transmitted to the mineral particles in the form of effective pressure, and a large part of the volume change that is expected has already taken place. Consequently, the strain due to the rearrangement of mineral particles will be far more important when adapting to the new, more closed fabric. Secondary consolidation will become more and more important, as primary consolidation progresses. In the final stages of a primary consolidation process, secondary consolidation may be of capital importance; and in some soils continues long after the primary process has terminated.

There is no theory permitting calculation of the strain that may be suffered by a soil due to secondary consolidation, with the degree of reliability with which TERZAGHI'S theory enables primary settlement to be assessed. Some very important laboratory investigations have been carried out, and some attempts have been made to achieve a mathematical model for this behavior. References [29, 30] are among the ones available on this subject.

There is experimental evidence that secondary consolidation is approximated by a straight line in a graph showing the strain undergone by a sample in the consolidometer versus testing time, on a logarithmical scale (consolidation curve). This fact explains the difference in shape between the theoretical consolidation curve (Fig. 1-31) and the one obtained in the laboratory (Fig. 1-25), which adopts a straight line in the final stages of the primary process, when secondary consolidation becomes predominant.

Secondary consolidation is more important wherever primary consolidation is brief, as occurs in organic soils, thin strata or strata with numerous sand lenses providing drainage. Secondary consolidation is especially important in peat deposits, where primary consolidation sometimes occurs in a manner almost simultaneous to loading. Therefore, for an embankment built on a peat deposit, where it is of interest to establish the progress of the settlement which takes place after construction, special attention must be paid to secondary consolidation, for it will be responsible for almost all subsequent settlement.

1.13 Introduction to the Problem of Shear Strength in Soils

1.13.1 General Comments and the Failure Theory

In soil mechanics, shear strength is the fundamental characteristic which determines the ability of soils to resist loading without failing.

There are several reasons. First, the resistance of soils to some stresses other than shear, such as tension, is so low that it is not of great importance to the engineer. The structures in which the engineer uses soil are generally of such a nature that the shear stress is the basic one, and whether the structure fails or not depends on shear strength. However, in these structures, those other stresses which are not pure shear, often have a stronger influence than the engineer would wish. For example, tension stresses sometimes play an important part in the cracking of earthworks. It is sometimes felt that tension has been unjustly disregarded as a stress worthy of investigation in soils. The essential fact remains, however, that the engineer uses shear strength more than any other, and logically therefore gives it preference.

Second, the resistance of soils to other types of stress, such as pure compression is so high that it is not of a practical interest either. Soils that were subjected to compression in a real situation would suffer shear failure before their resistance to compression alone was exhausted.

Third, the almost exclusive interest shown by soil engineers in shear strength is encouraged because the failure theory that is most widely used in soil mechanics is a shear theory.

There is not yet a universally accepted definition for failure. The word *failure* may imply the commencement of inelastic behavior in a material, or the moment at which the material breaks, to mention just two very common interpretations. Often the failure is associated with economic or aesthetic factors. Failure can involve personal preference to such an extent that its meaning varies from one specialist to another, from one engineering field to another or from one country to another, depending on their respective resources or wealth. For example, what exactly is implied by pavement failure and how could one define it?

Even under favorable circumstances the question will arise as to whether the criteria that are adopted for the design or protection of a certain structure will ensure that that structure will not fail. This question leads to another: What causes failure in the material? If it is not established why materials fail, it will not be possible to predict whether a material will fail or not in a given situation.

In current soil mechanics, the *failure theory* most often used is what might be considered a combination of two somewhat similar-classic theories. The first one, proposed in 1776 by Coulomb [33], says that a material fails when the shear stress acting on a flat element reaches a critical value.

$$\tau_f = c + \sigma \tan \varnothing \qquad (1\text{-}62)$$

where: τ_f is the shear strength; c the cohesion of the soil, assumed constant by Coulomb. It is the strength of the soil under zero normal pressure; σ the normal stress acting on the failure plane; and $\varnothing$ the angle of internal friction of the soil, also assumed constant by Coulomb.

The other failure theory is by Mohr [34] and establishes that, generally speaking, failure by sliding will occur along the surface on which the relationship of the shear or tangential stress to the normal stress reaches a certain maximum value. This maximum value was proposed by Mohr as a function both of the arrangement and shape of the soil particles, and of the coefficient of friction between them. Mathematically, the failure condition can be represented as:

$$\tau_f = \sigma \tan \varnothing \qquad (1\text{-}63)$$

Originally Mohr established his theory with special relation to granular soils, whereas Coulomb proposed Eq. (1-62) as a failure criterion for cohesive soils (including granular soils as a special case where the resistance to shear is zero for a zero acting normal stress). This is equivalent to $c = 0$ in Eq. (1-62). Strictly speaking, the essential difference between Mohr's theory and Coulomb's lies in the fact that in Mohr's the value of $\varnothing$ is not necessarily constant. In a graph with normal stress as abscissa and shear stress as ordinate, Eq. (1-62) of Coulomb will be represented by a straight line, and Eq. (1-63) will be represented by a curve, which in special cases can also be a straight line.

The failure criterion generally used in current soil mechanics is commonly referred to as the Mohr-Coulomb Theory. In this, Eq. (1-62) is used as a mathematical representation, but Coulomb's original idea that c and $\varnothing$ are constants is abandoned, and these values are regarded as variable. It is observed, therefore that the failure theory most used in modern soil mechanics, attributes the failure of soils to shear stress. Consequently, it is logical that shear strength becomes the fundamental parameter to be defined for strength and failure problems.

For practical problems, the Mohr-Coulomb failure theory leads to satisfactory results but it is not a perfect theory because it does not make it possible to predict all failures, nor does it explain all experimental evidence. Perhaps the deficiencies of this theory can be explained by the omission of one stress component if it is accepted that a material fails as a consequence of the stress condition prevailing within it. This stress condition can be described by three independent parameters, for example the three principal stresses, σ_1, σ_2, σ_3. A stress condition cannot be fully described with fewer than three independent parameters. The Mohr-Coulomb theory relates failure to the acting shear stress, which is in turn related to the difference between the principal major and minor stresses, $\tau_f = f(\sigma_1 - \sigma_3)$, but it does not take into account the intermediate principal stress, σ_2. Consequently, the failure theory cannot hope to cover fully all real failure cases.

Current experimental work indicates that the value of the stress σ_2 during failure, does affect the strength parameters, C, and $\varnothing$ which can be obtained in the laboratory, although this influence is probably minor. It is also becoming evident that the failure of the real material is influenced by how σ_2 varies during the loading process which leads to failure. A more detailed discussion on this subject is beyond the scope of this book, but more specialized publications, such as [32, 35, 36] can be consulted.

1.13.2 Nature of Shear Strength in Granular and Cohesive Soils

Here the factors influencing the shear strength of frictional and cohesive soils will be briefly analyzed. It is generally accepted that the shear strength of soils is partly due to the friction which develops between the particles, when they show a tendency to slide. The concept of friction is used in the sense familiar in mechanics (Fig. 1-44).

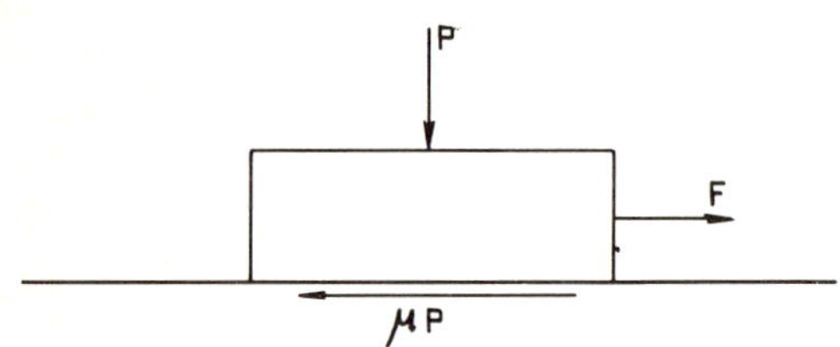

Fig. 1-44 Mechanical concept of friction

The force that is required to trigger the sliding of the body in the figure is: $F = \mu P$, where μ is a coefficient of friction between the contact surfaces.

Similarly, frictional strengths develop between the soil particles, in such a way that if a potential slip surface is considered and σ' is the normal pressure acting on that surface, the shear strength required to produce sliding, τ_f, is related to σ' by

$$s = \tau_f = \sigma' \tan \varnothing \qquad (1\text{-}64)$$

Thus the frictional strength *(s)* must be determined by the normal effective stress. In the above equation, $\tan \varnothing$ represents the coefficient of friction, and at the same time defines the angle of internal friction of the soil.

Equation (1-64) was first proposed by Coulomb, in a rather stricter sense than the one that is attributed to it today. For Coulomb, $\varnothing$ was an absolute constant typical of the soil being dealt with, whereas it later became necessary to consider the possibility of certain variations in the angle of internal friction. Similarly, Coulomb, historically established the concept of cohesion when he observed that some materials (clays) show shear resistance under zero external normal pressure. He proposed the following strength law for these materials.

$$s = \tau_f = c \qquad (1\text{-}65)$$

where c is the cohesion of the soil (which was also considered constant by Coulomb, whereas today it is regarded as variable). These were called purely cohesive materials, and in them $\varnothing$ was regarded as being equal to zero.

For a more general case, Coulomb attributed the strength of the soils to both causes, in accordance with an expression which combines the above two equations for a cohesive, frictional soil.

$$s = \tau_f = c + \sigma' \tan \varnothing \qquad (1\text{-}66)$$

Friction is currently considered the fundamental source of strength in granular soils, although it is not the only one, as has already been said, §1.11. Consequently, the shear strength of granular soils depends basically on the normal pressure between the particles and the value of the angle of internal friction, $\varnothing$. This in turn depends on the compactness of the material and the shape of its particles which, when their edges are sharper or less rounded, will develop greater friction.

Reference [37] describes a study conducted on the influence of water on the angle of friction which develops between equidimensional quartz particles. According to this study, whether or not there is any water between the particles has little effect on the angle of friction between them.

If the behaviour of granular soils were purely frictional, as was suggested by Coulomb, Eq. (1-64), their strength law when represented on axes $\tau - \sigma$ (as obtained in a triaxial test) would be a straight line passing through the origin, and the angle, $\varnothing$, would be constant, exactly as established by Coulomb. However what normally happens is that the strength law shows a curved line (although usually nearly straight). This is due to the effect of particle arrangement on the strength of the material. The particles have to deform and roll over one another for failure to occur. The effect of particle arrangement lessens as confining stress increases, for the particles become smoother at their contact points as a result of flattening and breakage. As a consequence, the sample of granular soil is compacted, but even like this it will fail more easily due to particle arrangement. Thus as σ' becomes greater, $\varnothing$ becomes smaller, and the strength law gradually becomes more horizontal.

Curvature appears to be more distinct the greater the size of the particles (see [23], where the case of rockfill is mentioned). This seems to be related to particle breakage. This is confirmed by tests on some sands with relatively small grain-size but particles that are weak and crumbly (for example, shell-bearing sands), which also show very curved strength envelopes. The curvature also seems to be greater in plane strain than in triaxial compression.

To summarize, granular soils are frictional materials, but with certain deviations from purely frictional behavior due to the effects of particle shape and arrangement. As a result, the material suffers distortion, breaks at the contact points and rolls and slides over itself. If the shear stress is sufficiently high, the effect of overcoming friction, plus the effects of particle arrangement, is continuous movement or distortion of the mass which is in shear failure. The phenomenon is not affected by the water in the voids of the granular soil. The angle of internal friction includes both the coefficient of particle-to-particle friction, and all the effects of particle arrangement. It is striking how little the coefficient of particle-to-particle friction (which under natural circumstances is rather variable) affects the angle of internal friction [38]. This can be explained if one considers that particles always move in the manner that proves easiest for them. If the coefficient of material friction is low, they slide, and if it is high, they roll.

Shear strength mechanisms are somewhat different in fine laminar soils, which are traditionally referred to as cohesive soils. Like granular soils, cohesive soils are discrete accumulations of particles which must slide over one another or roll for shear failure to occur. There are, however, certain important differences. First, when an external load is applied to a saturated clay, it is initially withstood by the water in the soil, in the form of the neutral pressure, u. This is a consequence of the compressibility of the solid structure of the soil in relation to the water. Second, the permeability of the soil is so low that the neutral pressure produced requires time to dissipate, assuming that drainage conditions are appropriate. Third, there are very significant forces acting between the soil particles, due to attractive and repulsive electrical effects.

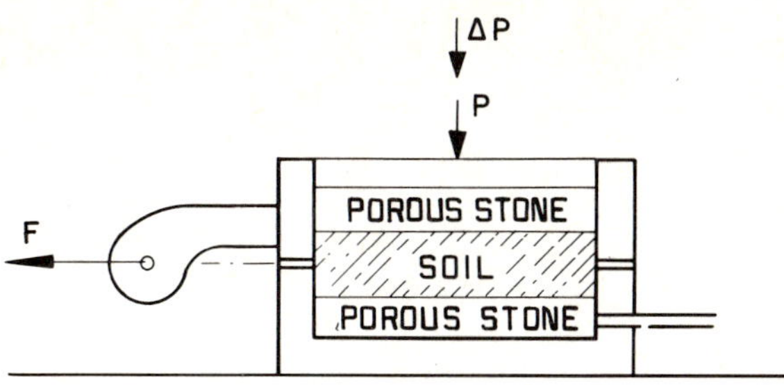

Fig. 1-45 Diagram to illustrate the influence of various factors on the shear strength of a cohesive soil

There is much evidence that the strength of fine cohesive soils is also fundamentally an effect of friction, but simple mechanical friction is disguised by many secondary effects, which may complicate the picture considerably. For example, the laminated flat particles of clay, although very close to one another throughout almost their entire areas, at no point make actual contact. It is thought that the ions and adsorbed water between the mineral surfaces, are not removed by normal pressures (that tend to join surfaces) smaller than 5,000 kg/cm^2 (71 Kips/in^2), therefore, these ions and adsorbed water may participate in the transmission of normal shear stresses. Perhaps the frictional effect between the clay crystals is more similar to the case of frictional soils, where there is contact between the edge and flat face of two particles, which occurs in some structural arrangements.

In clays an increase in pore water volume brings about a reduction in shear strength. A possible explanation [37] lies in the fact that in a very dry clay, the surface ions of its crystals are not completely hydrated, which permits closer particle spacing and stronger bonds between the crystals. With the appearance of water, the ions are hydrated and the bonds between the crystals are substantially weakened.

From an engineering viewpoint, the principal factors which influence the shear strength of *saturated cohesive* soils (and which should be carefully considered in each individual case) are: consolidation history of the soil, drainage conditions, loading rate and sensitivity of the soil structure.

In order to visualize the mechanism by which each of these factors exerts its influence, the case is considered of a totally saturated clay which is subjected to a direct shear test.

The sample is assumed to have been previously consolidated under normal pressure, σ_1', due to a load, P. It is also assumed that in the course of its geological history the sample was never subjected to a stress greater than this. In other words, the sample is normally consolidated. Under the circumstances, the neutral pressure in the water must be $u = 0$.

If the normal pressure is now rapidly increased by $\Delta\sigma_1$, with a load increment of ΔP a total pressure of $\sigma_f = \sigma_1 + \Delta\sigma_1$ will act on the sample. This load increment may have extremely varied effects on the shear strength of the sample, depending on the time it is allowed to act before the application of the shear force F which causes failure, on the drainage of the sample and on the rate at which F is applied. In a direct shear test the sample has very good drainage with water escaping easily through the porous stones. At first, $\Delta\sigma_2$ will be supported by the water in the sample, but if sufficient time elapses, the clay will consolidate under the new stress condition and $\Delta\sigma_1$ will also become an effective stress. If the sample is now brought to failure, by applying F in small increments and allowing sufficient time between each application for any neutral pressure originating in the zone close to the failure surface to dissipate, the strength of the clay can be expressed as follows:

$$s = (\sigma_1 + \Delta\sigma_1) \tan \varnothing = \sigma_f \tan \varnothing$$

for σ_1 and $\Delta\sigma_1$ are effective at all times and there are no excess pore pressures in the water.

If, on the other hand, F is applied rapidly, neutral pressures will appear in the vicinity of the failure surface as a result of a volume change under tangential stress. In normally consolidated clays, this is always towards a reduction, which is why the stresses in the water are now pressures which lower the effective stresses. If u represents the neutral pressures at the moment of failure, the strength of the clay will be:

$$s = (\sigma_1 + \Delta\sigma_1 - u) \tan \varnothing = (\sigma_f - u) \tan \varnothing$$

The shear strength has varied simply because of a change in the rate at which F was applied. The value of u depends largely on the sensitivity of the structure of the soil. Under the strain which occurs in the test, a sensitive structure breaks down, tending to shrink further, and consequently u becomes larger than in the case of a clay which is not very sensitive to strain.

If, on the other hand, water is not allowed to drain out through the porous stones during testing, the stress, $\Delta\sigma_1$, will never become effective, for the clay cannot consolidate, and will continue to be neutral, $\Delta\sigma_1 = u_1$. When F is applied the added neutral pressures which may be generated by the tangential stress will not dissipate either, even if F is applied slowly. (Although it is assumed that water is not allowed to escape, this is very difficult, if not impossible to achieve in a direct shear apparatus). Assuming that the neutral pressure originated by tangential strain is also u, in reality it is a little lower, and the shear strength of the clay will now be:

$$s = (\sigma_1 + \Delta\sigma_1 - u_1 - u) \tan \varnothing = (\sigma_f - u) \tan \varnothing$$

which is different from the two previous equations, because of the change in the drainage conditions of the sample.

This same strength would have been obtained if $\Delta\sigma_1$ and F had been applied rapidly, one after the other, even with free drainage, for in this case there would be no time for any neutral pressure to dissipate in the pores of the soil.

These arguments are applicable to a normally consolidated soil under natural conditions. If the soil is preconsolidated, similar arguments apply. The same sample as above is now considered to be consolidated by a pressure, σ_1 of great magnitude. If the sample is now rapidly unloaded, removing force P which was producing σ_1 the clay will tend to expand. Since the sample cannot instantaneously take the water required, even assuming that there is water available, the pore water will develop a state of tension that will confine the mineral particles, with sufficient pressure to retain the same total volume. Obviously, this pressure must be the same as the one which acted previously on the clay from the outside, $u_2 = -\sigma_1$.

If, immediately after the load P is withdrawn, the sample is brought to failure, by applying F rapidly, the tangential stress on the failure plane will bring about disturbance in the solid structure. The resulting pore water pressure, u, will reduce the existing tension u_2, as above. In this case, the shear strength can be written thus:

$$s = (-u_2 - u) \tan \varnothing = (\sigma_1 - u) \tan \varnothing$$

This takes into account that the total pressure is zero since P has been withdrawn.

This is the strength which has traditionally been interpreted as *cohesion* in clays, because it occurs at zero external stress. As can be seen, it is friction resulting from preconsolidation (previous consolidation history) acquired by the clay owing to the action of σ_1. If there is no water available on the outside, it does not matter how much time is allowed to elapse after the removal of load P until failure of the sample, as a result of the rapid application of F the strength will remain the same. It should be observed that if drainage facilities are non-existent, that is, if there is no possibility of the sample gaining or losing water, whatever the decrease or increase in external pressure, all the additional pressure will be taken by the water. Therefore when force F is applied rapidly, the material will have exactly the same strength owing to the preconsolidation under σ_1. Thus the material would behave as though purely cohesive. On the other hand, if the soil has the possibility of absorbing water, and time is allowed for this to occur after removing P, the sample will expand and the tension in the water will gradually dissipate. Finally the effective stress will be practically zero and the strength of the material (cohesion) will have been reduced to almost zero.

Obviously the above arguments can be applied to natural clay deposits, where the strength of the material will increase or decrease as the compression or tension originated in the water by loading gradually dissipates with time.

Therefore it is concluded that friction is by far the most important factor in explaining the shear strength of all types of soils. This is a somewhat simple image, because in flat clay particles, at the contacts between edges and flat face, bonds may develop that are so strong that the expression *real cohesion* is justified. This is, however, beyond the scope of this book. Friction explains strength clearly enough for the application of soil mechanics to road engineering. Those ideas, which have scarcely been touched on here, are dealt with in greater detail in [39].

To bring to a conclusion these ideas on shear strength mechanisms it will be necessary to establish the concept of residual strength, which occupies an important place in soil stability and problems in road building. In Fig. 1-17a, it was seen that in the materials showing fragile behavior, the stress-strain curve reaches a condition where the soil shows increasing strain for a practically constant stress. This effect can be observed in all soils (sands or clays) with peak strength; it becomes more marked the more pre-consolidated the clay or the more compact the sand. It is perceptible in normally consolidated clays and in relatively loose sands. This strength, referred to as ultimate or residual, was studied by SKEMPTON [40] in clays. For sands, this strength occurs with a void ratio that is independent of the initial one which prevailed before the shear strain process commenced; eventually deformation takes place at a constant volume. The influence of the initial particle arrangement is minimal, although it still plays a certain part, in spite of the large deformation that has taken place. In clays, the residual strength is independent of the previous stress history, as is shown by the fact that it has the same value for natural soils and remolded soils. The drop in strength after peak strength is due both to a progressive breaking of the bonds between the particles and to their rearrangement in such a way that their faces are parallel.

The shear strength of partially saturated cohesive soils (so important for the road engineer owing to the wide use that is made of compacted soils, which generally belong to the above category), involve the same concepts as saturated soils. Nevertheless, because there is both air and water in the voids of the soil, the neutral pressure generation mechanisms are far more complicated. They involve capillary tension and pressure of gases, which in turn depend on the degree of saturation and the size of the voids. With the data currently available, it is practically impossible to determine the effective stresses which really act between the grains of a soil.

1.13.3 Tests to Determine the Shear Strength of Soils

In Section 1-10, the principal laboratory tests employed today to measure the shear strength of soils were briefly described. Now they will be discussed in slightly more detail. A description will also be given of the apparatus used, for it is felt that it would otherwise not be possible to reach an understanding of the conclusions that will be drawn in the next two sections. Plate 1-6 shows the preparation of a sample for testing.

Plate 1-6 Preparation of a sample for testing

.1 Direct Shear Test

The direct shear apparatus is the most intuitive response to the problem of measuring the strength of soils. It is shown in Fig. 1-46. The apparatus consists of two frames which contain the sample of soil, one fixed and another movable.

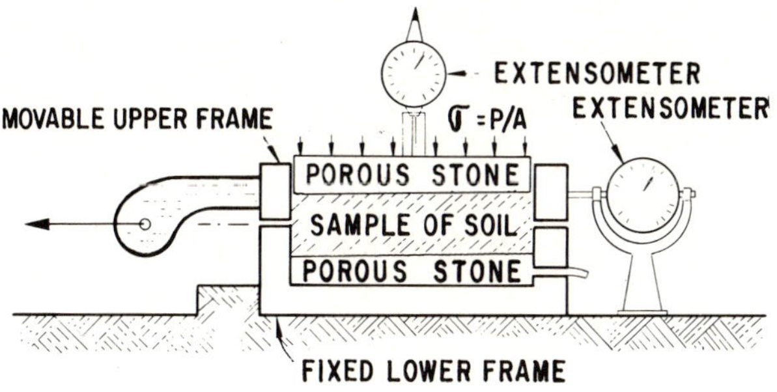

Fig. 1-46 Diagram of direct shear apparatus

Two porous stones, one above and the other below, provide free drainage to saturated samples, when required, and are replaced by confining plates when dry samples are to be tested.

The movable part has an attachment to which a shear force can be applied to cause failure of the specimen along a plane which, due to the way in which the apparatus is designed, is predefined. Loads, which can be varied as desired, provide a normal pressure on the failure plane, σ. They are applied to the upper face. Strain is measured with extensometers, both in horizontal and the vertical directions.

Depending on the drainage conditions established for the sample there are three types of tests:

Undrained: in which no drainage is permitted in the sample, either during application of normal stress, or during application of shear stresses.

Consolidated undrained: where the sample is allowed to consolidate during the application of vertical normal stress, until all pore pressure has dissipated. No additional drainage is permitted during the application of the shear stress.

Drained: where the sample is allowed to consolidate during both stages of the test, so that the neutral pressures dissipate both when normal stress is applied and when the shear stress is applied.

.2 Triaxial Compression Test

The tests most commonly used to determine the strength of soils are triaxial ones. Triaxial compression tests are more refined than direct shear tests, and they are currently used in most laboratories for determining the stress-strain and strength characteristics of soils. Theoretically, the pressures could be varied at will in three orthogonal directions on a specimen of soil with detailed measurements of its mechanical response. For the sake of simplicity the same stresses are used in two lateral directions. The specimens are cylindrical and are subjected to lateral pressures from a liquid, usually water, from which they are protected by an impermeable membrane. In order to achieve the appropriate confinement, the sample is placed inside a cylindrical, sealed chamber made of clear plastic with metal bases (Fig. 1-47). Porous stones are placed on the ends of the sample, which can be made to communicate at will with a buret on the outside by plastic tubing.

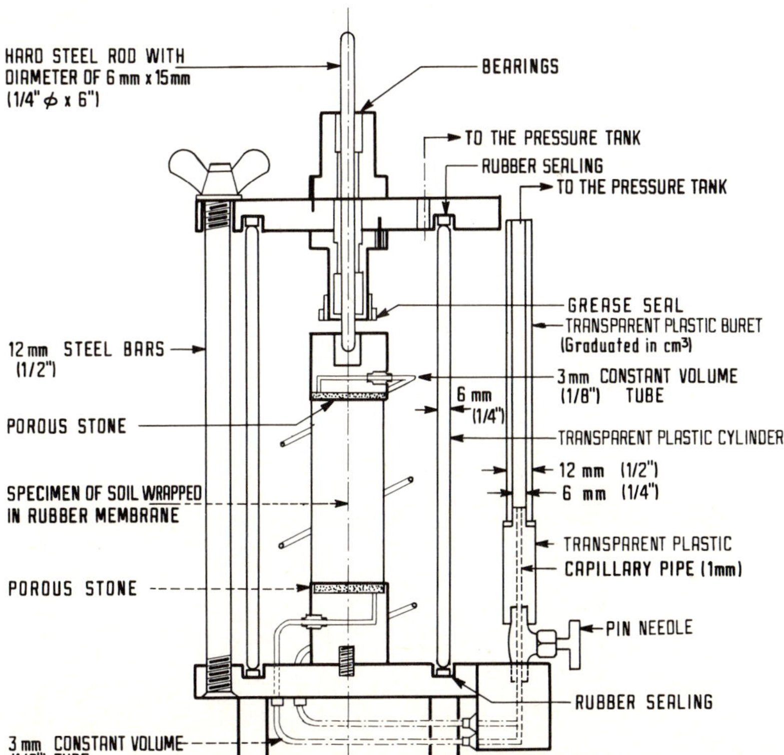

Fig. 1-47 Triaxial compression chamber

The water in the chamber can achieve any pressure that is desired by an attached compressor. Axial loads are transmitted to the specimen by a rod which goes through the top of the chamber, or cables drawn through the lower part of the chamber.

The lateral pressure produced by the water in the chamber is only normal, for it is hydraulic and therefore produces principal stresses on the specimen of: $\sigma_2 = \sigma_3$. Naturally, this same pressure, σ_3 acts on the ends of the specimen too, but these sections are also subjected to the effect of the load transmitted from the outside by the rod, which exerts a net pressure p on the specimen. In soil mechanics, this pressure is usually referred to as the *deviator stress*. The total pressure acting in the axial direction is σ_1, which is also a principal stress and is equivalent to $\sigma_1 = \sigma_3 + p$. At a given moment, the stress condition is assumed uniform throughout the entire sample and is analyzed with the aid of MOHR's graphical solutions, with σ_1 and σ_3 as principal major and minor stresses, respectively. The soil in a triaxial chamber is subjected to a tridimensional stress condition and should be dealt with using MOHR's general solution, which involves three different circles. However, because in the test two of the principal stresses are identical, the minor stress and the intermediate stress, the three circles become just one and the treatment is simplified.

It has already been seen that shear strength is variable, especially in *cohesive* soils, and depends on various environmental factors. When attempting to reproduce field conditions in the laboratory each of these factors must be taken into consideration. Therefore no single test can reflect all the conditions that may arise in natural circumstances. The correct approach might appear to be a special test for each individual case, but this is seldom practical. What is done is to reproduce the most characteristic and influential circumstances in standardized tests. These tests cover extreme types of behavior and conditions. The results have to be adapted to the situation being handled, which is generally somewhere between the two extremes. The results must be interpreted with common sense, constantly bearing in mind all those norms that have been established from previous experience.

The types of triaxial compression test most commonly performed today in soil mechanics laboratories are the ones briefly described below:

Drained (S). The chief characteristic of this test, sometimes termed the *slow* test, is that the stresses applied to the specimen are effective. First, the soil is subjected to a hydrostatic pressure σ_3, maintaining the valve that communicates with the buret open, and allowing sufficient time to elapse for complete consolidation to take place under the acting pressure. When the internal static equilibrium has been re-established, all the external forces will be acting on the solid phase of the soil; they produce effective stresses, whereas the neutral stresses in the water are hydrostatic. The sample is then brought to failure by applying an axial load in small increments, each of which is maintained for sufficient time for the pressure in the water, exceeding hydrostatic pressure, to be reduced to zero.

Consolidated — Undrained test (R_c). In this test, sometimes termed the *consolidated-quick* test, the specimen is first consolidated under the hydrostatic pressure, σ_3, as in the first stage of a drained or slow test. In this way, the stress σ_3 becomes the effective stress σ_3', acting on the solid phase of the soil. Next, the sample is brought to failure by a rapid increase in axial loading, so that no volume change is allowed. No additional consolidation is permitted during the failure period. This is easily achieved in a triaxial chamber by closing the outlet valve from the porous stones to the buret. Once this is done, the requirement is fulfilled independently of the axial loading rate. However, this rate does influence the strength of the soil, even if drainage is totally restricted.

In the second stage of a consolidated-undrained test, it might appear that all the deviator stress is supported by the water in the voids of the soil in the form of neutral pressure. This is not, however, so. Part of the axial pressure is supported by the solid phase of the soil, although it has not yet been possible to discover either the stress distribution or its causes. There is no reason why the deviator stress should be supported entirely by the water in the form of neutral pressure. If the sample were laterally confined, as in the case of a consolidation test, then this simple distribution of the deviator stress would occur. However in a triaxial test the sample can deform laterally and its structure can therefore withstand shear stresses from the very beginning.

Undrained test (Q). In this type of test, sometimes termed the *quick* test, at no stage is consolidation of the sample allowed. The value connecting the specimen and the buret remains closed, thus impeding drainage. First a hydrostatic pressure is applied to the specimen exterior, and immediately afterwards the soil is brought to failure by rapidly applying an axial load. The effective stresses in this test are not clearly determined, nor is their distribution at any given moment, before or during application of the axial load.

Unconfined compression test (q_u). This test is not really triaxial and is not usually classified as such, but from many points of view it is similar to an undrained test. At the beginning of the test, the external stresses are zero, but in the soil structure there are certain (but not very clearly defined) effective stresses due to capillary tensions in the pore water.

The triaxial tests to which reference has been made, where the deviator stress is applied by compression from the rod is the traditional test historically, and is still the one most frequently used. In recent years other versions of the triaxial test have been developed. In one of them, which is used for special studies, the stress transmitted by the rod is smaller than the confining stress, and therefore becomes the minor principal stress acting on the sample during the test. In another the lateral pressure is varied, by changing the chamber pressure, but maintaining the axial pressure constant. In order to do this, the necessary adjustments have to be made in the load transmitted by the rod. Lastly especially for research, tests are run in which both the axial stress and the lateral stress are made to vary. Such, triaxial tests are currently classified in two large categories, compression and extension tests. In the first, the axial load is larger and in the second it is lower than the confining stress.

Both compression tests and extension tests can undergo variations in the laboratory. For example, the axial length of the specimen can be reduced and the axial stress increased, by increasing the load transmitted by the rod, or by maintaining the axial stress constant, but making the lateral chamber stress diminish; or lastly, by increasing axial pressure and reducing lateral pressure simultaneously. The most common of these last tests is where each increment in axial pressure on the sample is double the decrease in lateral pressure, in such a way that the arithmetical average of the principal normal stresses is maintained constant.

Similarly, there are corresponding variations for extension tests.

In a compression test, the axial pressure is always the major principal stress, σ_1. In an extension test, however, the axial pressure is always the minor principal stress, σ_3.

Triaxial equipment has also been developed for the application of three different principal stresses [41]. Plane strain apparatus is also available [42, 43], in which strains can vary axially and in one lateral direction, the dimension of the specimen remaining fixed in the other lateral direction.

In order to be able to measure the dynamic properties of soils, the pulsating triaxial test has been developed, in which σ_3 is applied as in the standard test, but σ_1 is applied in cycles.

.3 Other Tests

The ***annular shear test*** [44] is performed with an apparatus which is almost identical to the direct shear device, the only difference being that the shear stress is produced by applying torsion around an axis which is normal to the plane of the sample. Because the area of the sample undergoes no change, the test is appropriate for the determination of the residual strength of soils.

In the ***simple shear apparatus,*** the specimen deforms in a manner similar to that in the direct shear test, but in such a way that during strain all the horizontal sections of the sample remain equal. The two principal types of apparatus for this purpose are described in detail in [45, 46]. The simple shear apparatus is more appropriate than the direct shear apparatus for studying strain in soils, for almost the entire volume of the specimen is sheared, instead of just a narrow strip, with its uncertainties in strain analysis [47]. Simple shear apparatus produces plane strain conditions, which have often been regarded as representative of the situation prevailing in many real problems.

The ***vane test*** is a relatively modern contribution to the study of shear strength in soils. In principle, the test has a clear advantage; it is performed on soils in place, in the field, instead of on samples that have been removed from the ground and which may be disturbed. Thus the test is on materials in the place where they were actually deposited under natural circumstances. Disturbance of soils subjected to this test is, however, significant, for the vane has to be sunk into the stratum being studied, and this operation can disrupt the soil structure. The interpretation of results is somewhat similar to that of the direct shear test, and is affected by some of the same limitations. In addition the cylindrical shear surface is different from the shear surface in real problems.

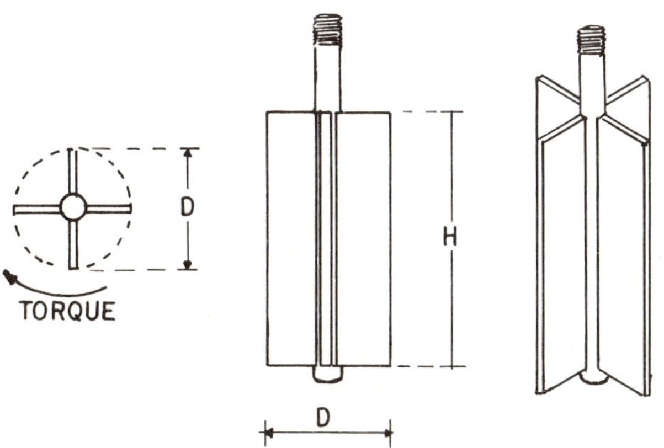

Fig. 1-48 Vane apparatus for shear strength determinations

The apparatus consists of a rod, which can be dismantled part by part, to the bottom of which is attached the vane. It, usually consists of four blades firmly secured to an axle, which is a prolongation of the rod (Fig. 1-48). Once the vane has been sunk to the desired depth, a moment is applied to the upper end of the rod, where there is a mechanism permitting torque measurement and circular displacement. The sinking process is usually made easier by boring to a depth slightly less than the level at which the test is to be performed. The upper part of the vane must be well below the bottom of the hole. As the torque is gradually applied, the vane tends to rotate, shearing a cylinder of soil. A laboratory apparatus is shown in Plate 1-7.

Plate 1-7 Laboratory vane apparatus

Referring to the shear strength of the soil as s, the maximum shear withstood by the soil will be measured by means of the resisting moments that are generated, both at the ends of the cylinder and in the surface of the cylinder. The resisting moment that develops in this last area will be:

$$M_{RL} = \pi DHs \frac{D}{2} = \frac{\pi sHD^2}{2}$$

and disregarding the effect of the rod, the moment generated at each end will be:

$$M_{RB} = \frac{\pi D^2}{4} \frac{2}{3} \frac{D}{2} = \frac{\pi s D^3}{12}$$

Note that, at the base, the lever arm of the resisting force is taken as 2/3-D/2, which is equivalent to considering resisting elements in the form of circular sectors.

The total resisting moment, at the instant failure commences will be the same as the torque that is applied $M_{max} = M_{RL} + 2M_{RB}$:

$$M_{max} = \pi D^2 s \left(\frac{H}{2} + \frac{D}{6}\right) \tag{1-67}$$

Thus:

$$s = \frac{M_{max}}{\pi D^2 \left(\frac{H}{2} + \frac{D}{6}\right)} = \frac{M_{max}}{C} \tag{1-68}$$

The value of C is a constant of the apparatus and can be calculated once and for all. Frequently $H = 2D$ and so:

$$C = \frac{7}{6} \pi D^3 \tag{1-69}$$

It can be readily observed that the type of failure that is produced by the vane is progressive, with maximum deformation at the blade end and minimum deformation on the bisecting planes of the blades, and it can therefore be concluded that the vane is most reliable in materials with a plastic behavior, of the soft clay type.

In sands, even loose ones, introduction of the vane modifies the compactness of the strata and, what is even more important, the general condition of the mass. Therefore any results that are obtained will be hard to interpret.

In finely stratified clays, in which thin layers of clay alternate with others of fine sand which provide ready drainage, the stresses due to rotation induce some consolidation in the clay. As a consequence, strengths are obtained which are higher than the real ones.

A vane that is suitable for measuring high strengths has been operated by MARSAL [48]. The same reference describes other test equipment currently being developed and used for measuring the strength of soils in the field.

1.14 Shear Strength in Granular Soils

As was seen in the previous section, the factors affecting the shear strength of granular soils fall into two categories. The first includes those which affect the shear strength of a given soil, the most important factors being compactness (often with reference to the initial void ratio or the initial relative density) and the confining stress (under natural circumstances or in the triaxial chamber), but the loading rate also plays a part. The second category includes those factors which make the strength of one granular soil different from that of another granular soil with the same confining stress and the same degree of compactness. Among these factors are the size, shape, texture and grain-size distribution of the particles and their degree of soundness and hardness, these last two conditions determining the particle breakage phenomenon which affects strength so basically.

Here are some conclusions that may be considered of interest, they are taken from the results of laboratory tests and field experiences.

First, for practical purposes, the shear strength of granular soils can be expressed as in Eq. (1-64): $s = \sigma \tan \varnothing$ where s represents the strength of the soil, the maximum shear stress to which the soil can be subjected without failing, τ_{max}.

Figure 1-49 shows the failure envelopes obtained in conventional triaxial tests, conducted at relatively low levels of stress for three sands, one loose, one compact and the third cemented. The points at which failure occurred due to a particular combination of normal stress and maximum shear stress, are marked for each test. In the case of the loose sand, the failure envelope is practically a straight line passing through the origin. The material obeys a law similar to Eq. (1-64), and the angle of internal friction of the sand ($\varnothing_s$) can be obtained from the set of tests.

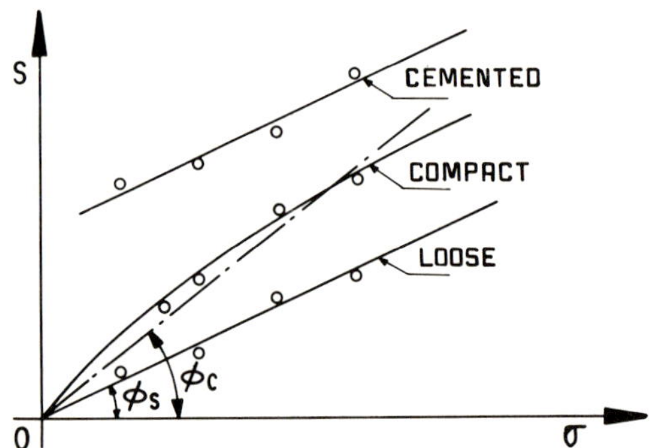

Fig. 1-49 Strength envelopes for a sand in loose, compact and cemented conditions

In the case of the compact sand, the resulting points give a very slightly curved line, passing through the origin, with the average angle of slope $\varnothing_c$. For practical purposes, the curve can be regarded as a straight line obeying the law in Eq. (1-64), and the angle $\varnothing_c$ (compact condition) can be found from the tests. This angle is required if Eq. (1-64) is to be applied to field problems.

The law for cemented sands may be like the previous ones, depending on whether the sands are loose or compact. The difference lies in the strength that is shown by the sand under a zero external normal pressure, owing to the effect of cementation (ordinate at the origin). It is better to express the strength by a law of the Eq. (1-66) type, for c can be calculated from the triaxial tests that are conducted, bearing in mind that it represents an effect of cementation.

From the above, it is possible to obtain manageable expressions for the shear strength of sands at relatively low stress levels. When these levels increase, the situation becomes more complicated, as will be discussed later.

It is the effective stress that must be taken into account when applying the above strength law to sands. If the sand is saturated, pressures, u, may appear in the water owing to external loading or seepage. If the normal pressure which is to be included in Eq. (1-64) is calculated like a total stress, (starting from the specific weight of the saturated soil, γ_m, which involves the weight of the soil and of the water contained by it) Eq. (1-64) must be written in either one of the two forms,

$$s = \sigma' \tan \varnothing = (\sigma - u) \tan \varnothing \qquad (1\text{-}70)$$

where σ' represents the effective stress and σ the total stress as defined previously. Laboratory experience has shown that the value of $\varnothing$ changes relatively little between the dry sand and the saturated sand, at the same void ratio. The real change in the strength of the sand lies in the neutral pore pressure, u, which when large, reduces the strength substantially. If the sand were *dry*, for strength purposes there would be a normal pressure at depth z within the mass of $\sigma' = \gamma_d z$.

If the water table rises to the surface of the sand, the value γ_d increases to γ_m which is higher. If, however, neutral pressure u develops in the water, the stress available for strength will be:

$$\sigma' = (\sigma - u) = \gamma_m z - u$$

If u is sufficiently large, the strength will be reduced to an insignificant value. Then the influence of the water and the pressures that may develop on earth stability problems can be clearly seen. Fluctuations in the water table or groundwater flow are common causes of the development of neutral pressure.

If the neutral pressure increases sufficiently, the difference $\sigma - u$ may become zero, and the sand will have lost all its strength, now behaving like a heavy fluid. This condition is associated not only with the cause of u, as might be the case for groundwater flow, but also with the characteristics of the sand itself. In fine and uniform sands or in cohesionless silts, the permeability is relatively low and any neutral pressure that may develop will have difficulty in dissipating. These are the soils which run the greatest risk of suffering a drop in strength or total loss of strength under these circumstances. Coarse sands and gravels may reach the zero strength condition if the flow is sufficiently large, but the effect is usually short-lived.

When sands deform under a shear stress, they undergo a volume change. If the sand is saturated, this change is accompanied by re-distribution of the water in the voids. If the permeability of the soil is high or the changes occur very slowly, only very small neutral pressures will appear and they will have no major influence on the strength of the soil. If, however, the changes are very rapid or the permeability is relatively low, high neutral pressures can accumulate, and the strength is consequently strongly influenced.

Compact soils expand or dilate when they deform, which tends to generate pore tensions with a limiting value equal to the maximum capillary tension of the soil. This effect leads to a temporary increase in the strength of the soil.

In loose sands, deformation due to shear causes a volume reduction and the water generates neutral pressure. The limiting value of u is now the confining pressure of the soil (σ_3) and the minimum effective stress that can be reached is:
$\sigma = \sigma - U = 0$.

When sand deforms as a result of shear, the neutral pressures develop at first only in the deformation zone. Whether the neutral pressure is maintained or spreads throughout the mass of sand depends on permeability and internal flow conditions. As a result of this weakening of the soil beyond the initially deformed zone, the failure zone enlarges and more neutral pressures are generated in the water, in such a way that a progressive failure may be triggered. These phenomena are responsible for many large-scale flow-type landslides.

Relatively small loads can trigger failure, due to the development of neutral pressure, when the load acts repeatedly in a more or less cyclical manner. Each load application produces an increment in the neutral pressure. If the grain-size distribution and permeability conditions do not allow this neutral pressure to dissipate before the following application, conditions will be appropriate for failure. This is the situation which may arise underneath a foundation supporting vibrating equipment. It is also caused by blasting and earthquakes, during, or after which, zero strength may occur with disastrous consequences (liquefaction).

Capillary tension may cause variations in the shear strength of a sand as compared to the dry condition. In wet sands, menisci may develop between the grains and cause high tension stresses in the water, accompanied by strong compressions among the grains. This is equivalent to an increase in effective pressure, and consequently in strength. This effect of apparent cohesion due to capillarity, is responsible for the nearly vertical slopes that many faces of partially saturated sand can hold without collapsing. This is obviously not a permanent effect. If the engineer relies on it, he will almost certainly be confronted with a failure when the sand loses water by evaporation, or when it becomes saturated during intense rain; it will then produce a continuous sand fall.

As has already been mentioned, the shear strength law for granular soils is close to a straight line Eq. (1-64), as long as the normal stresses acting on the failure plane at the instant of failure are maintained at a low level. There is no specific boundary for defining high or low stress levels. In [49], LAMBE and WHITMAN mention experiments in which this limit was found to be between 5 and 10 kg/cm^2 (71 and 142 lb/in^2), depending largely on the compactness of the granular soil. When the level of normal stress on the failure surface is combined with the compactness of the soil, in such a way that curved strength envelopes are obtained (to the extent that the approximation to a straight line is carried out with an undesirable lack of precision), one of the following three procedures may be adopted. First, the curved envelope obtained in the tests can be used, which complicates any calculation that may have to be made using that envelope. Second, it is possible that the only part of the curved envelope that is almost straight will be the section between the extreme values of the normal pressure on the failure plane which it is thought will apply in the specific problem under analysis. This leads to a strength law of the Eq. (1-66) type, because the prolongation of the straight line approximation cuts the axis τ above the origin. The value of c thus obtained will, of course, have little to do with the cohesion concept already discussed and should be regarded as a calculation parameter. Third, Eq. (1-64) can be used, but considering $\varnothing$ as variable and dependent on the confining pressure at failure; $\varnothing = f(\sigma_3)$. This method is not very convenient for practical calculations.

This brief analysis of the shear strength of frictional soils concludes with some observations about the influence on test results of some of the factors affecting that strength, which have been mentioned previously.

The first factor to be considered will be the effect of the confining stress σ_3, used in the test. As has already been said, this stress is fundamental for defining the additional strength of the granular soil resulting from particle arrangement. When the confining stress is increased, the strength component due to particle arrangement undergoes a reduction because the particles become abraded at the contact points and may even break. This tendency is shown in Fig. 1-50, from MARSAL [24].

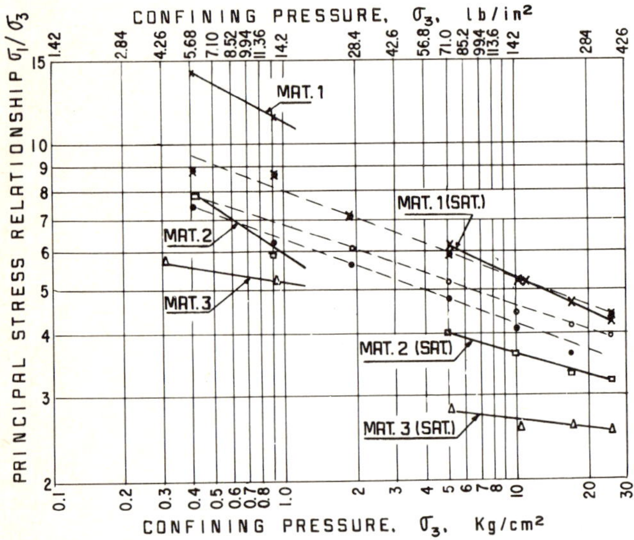

Fig. 1-50 Relationship between the quotient σ_1/σ_3 during failure and σ_3 for three granular materials

The figure gives results for the three rockfill materials already mentioned in Section 1.11. There are two series of tests: on the left, with relatively low (up to 1 kg/cm², 14.21 lb/in²) confining pressures, carried out in triaxial apparatus on samples 113 cm (44-1/2 in) in diameter and 250 cm (98 -1/2 in) in height; and on the right, tests carried out in giant triaxial apparatus (see Plates 1-8 and 1-9), with confining pressures of up to 25 kg/cm² (355 lb/in²). In both cases, there is a tendency towards a reduction in the effect of particle arrangement with an increment of σ_3.

The tests on the left were conducted with dry specimens, and those on the right with saturated specimens. The change in slope and trend of the lines obtained indicates the effect of saturation on the shear strength of granular soils (the scale used to draw σ_3 is logarithmic). The initial void ratio or initial compactness have a decisive influence on the shear strength. With an initially smaller void ratio or an initially greater relative compactness, the shear strength is greater; Fig. 1-51 [49] illustrates this tendency for a particular sand. The figure also shows the values of $\varnothing_\mu$, angle of particle-to-particle friction of the material in the mechanical sense of the term, which is of course independent of the initial compactness.

On the other hand, the initial void ratio of a given soil appears to have no influence on the value of the angle of friction corresponding to the residual or ultimate strength of the soil, nor does it have any effect on the void ratio when the residual condition is reached (where the soil deforms at a constant volume and with a deviator stress that is constant). This angle

Plate 1-8 Triaxial chamber used to study coarse granular soils and gravels

Plate 1-9 Triaxial chamber used to study coarse granular soils and gravels (inside view)

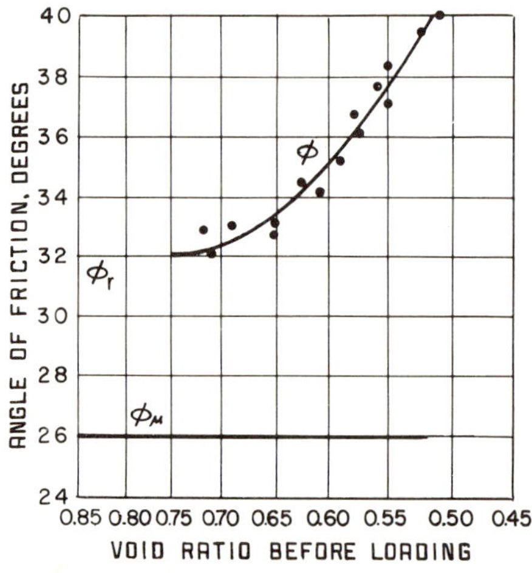

Fig. 1-51 Angle of internal friction as a function of the initial void ratio of a medium to fine sand [49]

of residual strength is greater than $\emptyset_{\mu}$ and is shown in Fig. 1-51 for the sand in question; Fig. 1-52 [49] shows the relationship between the angle of internal friction, $\emptyset$, and the initial void ratio in several different granular soils.

Since the values of $\emptyset_{\mu}$, which determine the effect of particle-to-particle friction alone, vary relatively little among the mineral particles of different sizes of which granular soils are composed, it follows that the wide differences that are observed in $\emptyset$ for a given initial void ratio must be due to the effect of particle arrangement.

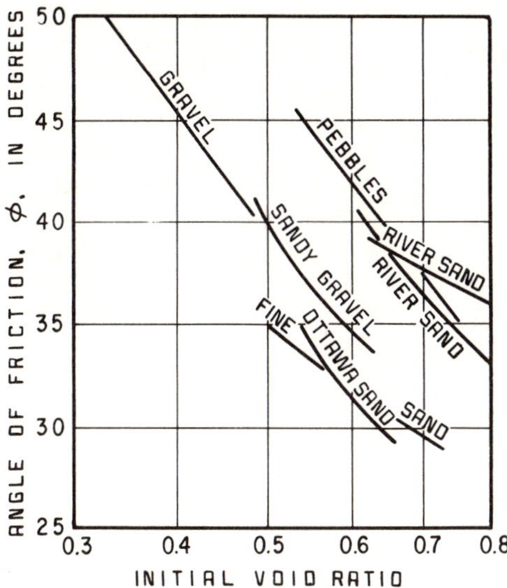

Fig. 1-52 Values for the angle $\emptyset$ versus initial void ratio in various granular soils [49]

The grain-size distribution of a granular soil affects its angle of internal friction in two ways. First, it affects the void ratio that is reached with a given compactive effort, if the soil is compacted as is so often the case; second as can be seen in Fig. 1-52, it affects the value of $\emptyset$ that is reached with a given initial void ratio. For a specific practical problem (for example, the construction of an embankment), the effect of the grain-size composition of the soil can be studied by means of a series of triaxial tests that are run to determine $\emptyset$ for several different grain-size distributions, subjecting the sand to the same compactive effort in each case.

The most common procedure for determining $\emptyset$ in the field is by means of correlations with results from boring tests. The study of these correlations is therefore essential. This important aspect will be considered in greater detail later.

Last, the influence of the particle breakage on the shear strength of granular soils is significant. In all the materials investigated by Marsal [24], there is a reduction in strength with an increase in breakage coefficient B. The date in Fig. 1-53 depict this. It also shows that with an increase in confining pressure, σ_3, there is a corresponding increase in particle breakage. Among the phenomena affecting breakage, Marsal mentions confining pressure, grain-size distribution, average size and shape of the particles, void ratio, and of course the nature and soundness of the grains.

The reason why breakage occurs more readily as the confining stress σ_3 increases is thought to lie in the high local stresses at the points of contact between the particles. These stresses increase with the average size and the coefficient of uniformity. Marsal [50] compared these inter-particle stresses for a typical sand and a rockfill, both subjected to a confining pressure of 1 kg/cm^2 (14.21 lb/in^2). He concluded that they are about two million times greater in the rockfill than in the common sand which explains many of the differences in behavior found between these materials. This should not be forgotten by engineers working with rockfills in relation to both strength and compressibility.

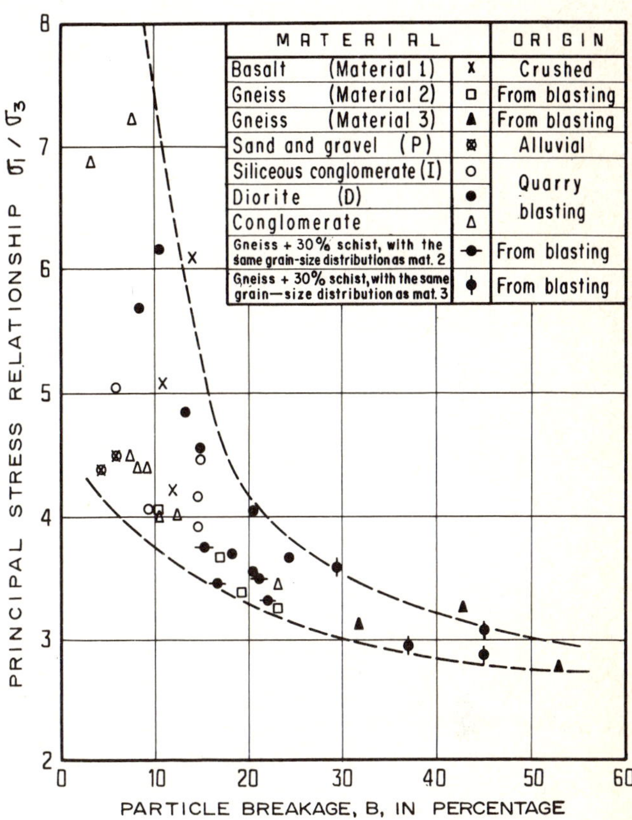

Fig. 1-53 Correlation of the σ_1 / σ_3 ratio during failure and particle breakage [24]

1.15 Shear Strength of Cohesive Soils

1.15.1 Saturated Soils

The following is an analysis of the basic conclusions reached from results for different triaxial tests on saturated soils. As was indicated in Section 1.13, each triaxial test represents specific load and environmental circumstances, principally where consolidation and drainage are concerned, and is not a random division or one based on simple work methodology. Below, the results of each traditional test are analyzed individually. Plates 1-10 and 1-11 show typical apparatus used in such tests.

Plate 1-10 Triaxial chambers

.1 Drained Test (slow test)

As already stated, the stresses acting on the specimen in this test are effective at every stage. This is achieved by permitting free drainage of the sample and consequently the complete consolidation of the soil under the different stress conditions to which it is subjected. During the first stage, the specimen is subjected to an external confining pressure (σ_3) acting in all directions, and in the second stage it is brought to failure with increments of axial load, p, (deviator stress). Figure 1-54 [47] is a diagram showing the distribution of total and effective stresses during the test.

In this test, there are no changes in the neutral stresses, and any increase in the total stress brings about a corresponding increase in the effective stress. During testing, the soil consolidates, its void ratio and water content decreasing. Although the mechanism of this consolidation is essentially the same as the one described when discussing the compressibility of cohesive soils, the compressibility curve is different, for the field of the acting stresses is different. The effect of the confining ring during a conventional consolidation test imposes the condition that the strains in the two horizontal directions are zero ($\epsilon_2 = \epsilon_3 = 0$), the principal stresses in those directions are equal to one another and equal to a fraction, k, of the principal vertical normal stress, σ_1, $\sigma_2 = \sigma_3 = K\sigma_1$. Thus, if successive conventional consolidation tests were conducted for increasing vertical loads, the Mohr circles shown in Fig. 1-55 [51] would be obtained.

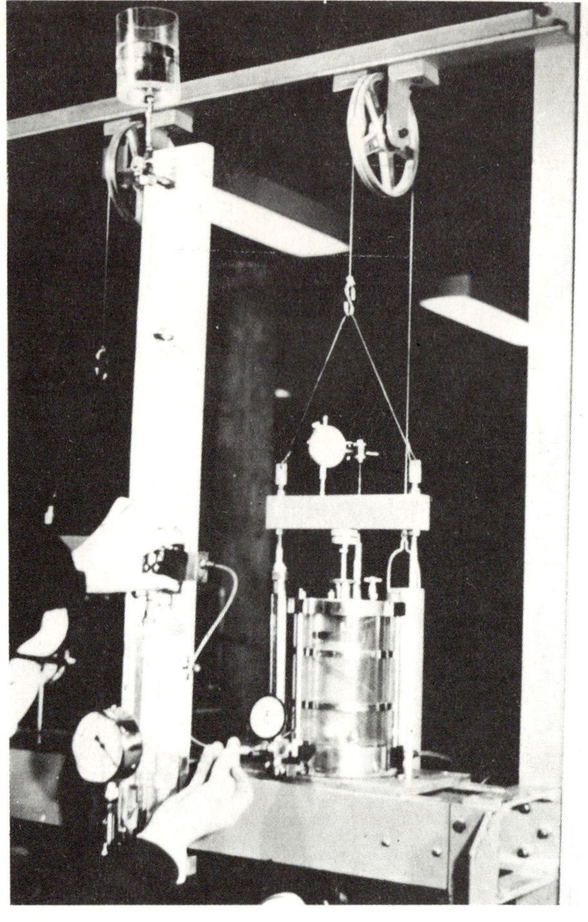

Plate 1-11 Bench with triaxial chamber and apparatus for measuring pore pressure

The geometrical location of a point on the successive Mohr circles obtained from a series of tests is called the stress path acting on a certain plane; it represents a combination of normal and shear stresses acting on that plane in each test. In Fig. 1-55, the stress path was drawn for three successive unidimensional consolidation tests, selecting the plane of the maximum shear stress (line *1-2-3*). The stress path is seen to be a straight line.

The drained test is different from the conventional unidimensional consolidation test. Consolidation of the specimen during the first stage of the drained test is usually isotropic ($\sigma_1 = \sigma_2 = \sigma_3$). After consolidation, the deviator stress is usually

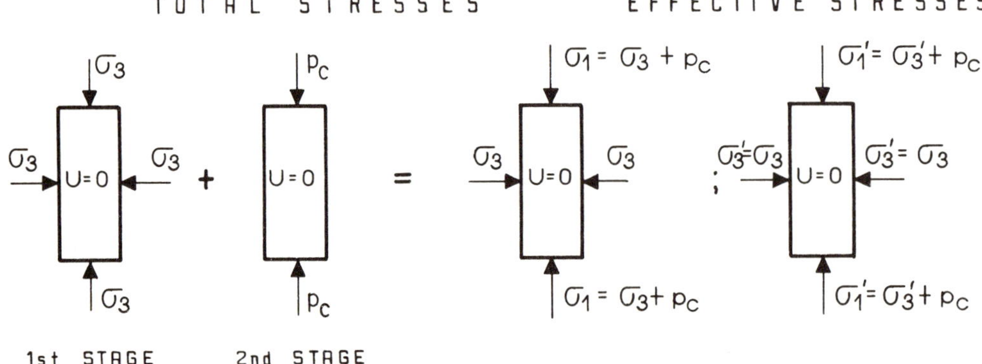

Fig. 1-54 Distribution of total and effective stresses in drained triaxial compression test

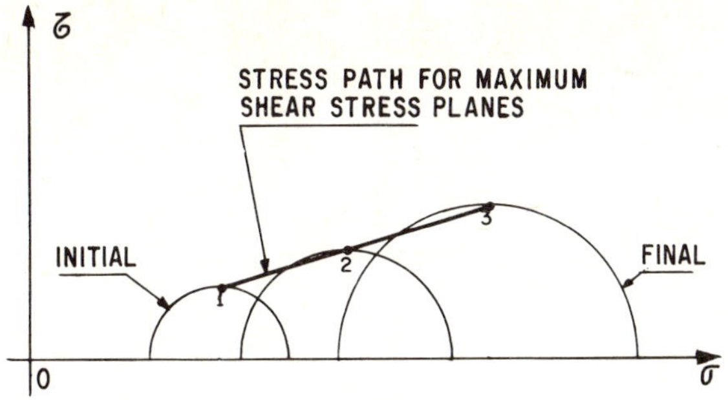

Fig. 1-55 MOHR circles and stress path in unidimensional consolidation test

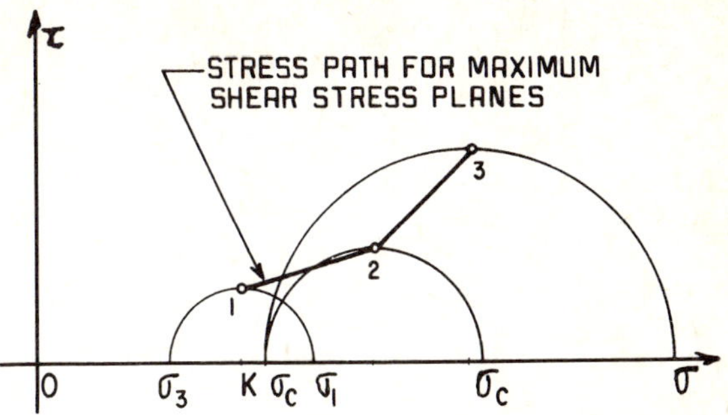

Fig. 1-56 Stress path in a drained triaxial test

increased, the lateral stress, σ_3 induced by the external confining water pressure being maintained constant. Figure 1-56 [51] shows a typical stress path on the maximum shear plane (for data comparable to Fig. 1-55).

The tests in Fig. 1-56 were conducted as follows. A confining stress, σ_3, was applied to the specimen with external water, together with a normal stress, σ_1, greater than σ_3, which is equivalent to producing an anisotropic consolidation ($\sigma_1 > \sigma_2 = \sigma_3$) in the first stage of the test. This is a common practice in laboratories. Next, a deviator stress σ_c, equal to the preconsolidation load of the soil, is applied, and the confining chamber pressure varied to a value $K\sigma_c$. Consolidation of the specimen under these stresses takes place. The effective stress condition represented by Circle *2* was thus obtained. Immediately afterwards, without any further variations in chamber pressure $\sigma_3 = K\sigma_c$, the second stage of the test commenced, applying a vertical stress, σ, to the soil by means of a rod in order to obtain Circle *3*.

The stress path for the maximum shear plane is now (line *1-2-3*,) different from the one shown in Fig. 1-55 for the case of the consolidation test; this is to be expected because in the consolidation test there is rigid lateral confinement, which is not true of the triaxial test.

The results of triaxial consolidation which are usually expressed in terms of vertical stress versus vertical strain (or settlement) graphs, are becoming more and more widely used. Some engineers think that results of triaxial consolidation may be more appropriate for describing the settlement of thick strata of clays or plastic silts, but conventional consolidation is still widely used to determine the compressibility of all types of cohesive soils.

As a result of triaxial consolidation, during a drained or slow test there is a reduction both in the spacing between the particles and in the water content of the soil. Inter-particle bonds become stronger in proportion to the confining stress. This means that strength increases proportionally to the effective confining stress. A strength envelope can be obtained for this situation by a series of tests with increasing stresses. The results are a straight line passing through the origin (Fig. 1-57).

The angle $\varnothing$ is called the angle of strength, or angle of internal friction of a cohesive soil, and usually varies between 20° and 30°. The highest values usually correspond to clays with a plasticity index between 5 and 10 and the lowest ones to clays with indices higher than 50 or 100. This reflects the effect of the repulsion among particles and of the absorbed water at the inter-crystal bonds, because the inter-particle repulsive forces are greater for high plasticity indices.

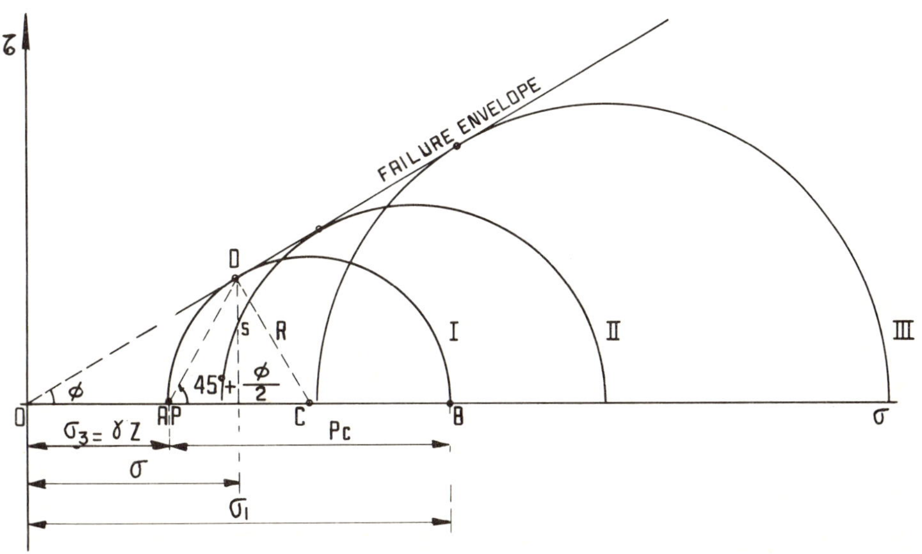

Fig. 1-57 Failure envelope of saturated, normally consolidated clays

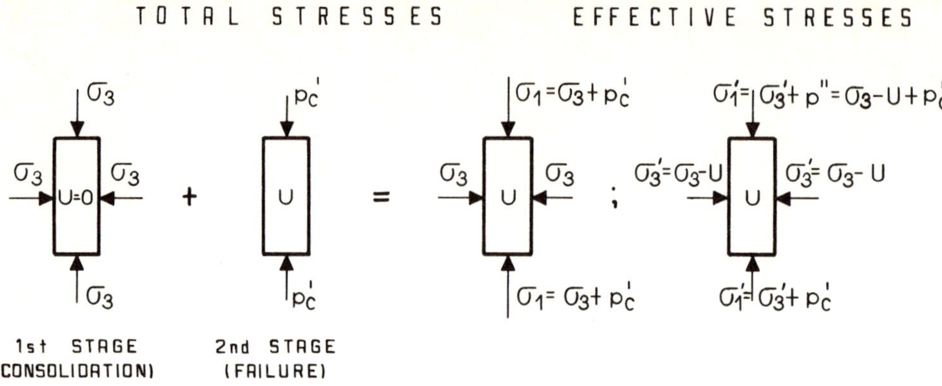

Fig. 1-58 Distribution of total and effective stresses in a drained-undrained triaxial compression test

When a clay is loaded in the triaxial chamber with stresses lower than those of preconsolidation ($\sigma_1<\sigma_c$), even if there is a tendency towards expansion due to the absorption of water, the particles do not recover their original spacing and the void ratio does not reach its original value prior to consolidation under σ_c. Consequently, the attractive forces between the particles are not reduced as much as they might be. Their resistance to stresses smaller than the preconsolidation load is no longer proportional to the effective confining stress, but somewhat larger. This makes the strength envelope (Fig. 1-57) deviate from the straight line, above it for stresses lower than σ_c. This part of the envelope which is not straight, represents the behavior of the soil in a drained strength test. Thus, the strength of a clay in a drained test may be represented by Eq. (1-64): $s = \sigma' \tan \varnothing$, for values above the preconsolidation load (condition of normally consolidated soil) and by Eq. (1-66): $s = c + \sigma' \tan \varnothing_A$ for values below the preconsolidation load (condition of a preconsolidated soil). In this last case, c and $\varnothing_A$ are obtained by making a straight line approximation to the curved envelope, which is why they cannot be regarded as signifying anything more than calculation parameters without any theoretical significance.

The drained strength of a cohesive soil, as obtained in a slow or drained test, represents the strength that is developed by the soil when it is subjected to changes in stress in such a way that the soil becomes completely consolidated under the new stresses. This implies appropriate drainage conditions and an adequate lapse of time. It represents the strength that will be reached in the long term, under conditions where there is no impediment to consolidation of the soil under the stresses that are applied to it. The drained strength is also used to solve practical problems by the effective stress method, which will be described in detail later and in which the failure conditions are determined using total stresses and neutral pressure. It is especially useful in problems where complicated changes occur in the loading conditions and in the movements of water in the subsoil.

.2 Drained Undrained Test (consolidated-quick or R_c)

In the first stage, the soil consolidation under the confining stresses, is usually applied in a hydrostatic condition ($\sigma_1 = \sigma_2 = \sigma_3$), but sometimes in an anisotropic condition. In the second, or failure stage, the specimen is loaded with a deviator stress applied without permitting drainage, and therefore without additional consolidation. The distinction between the two stages is established more clearly in this test than in the drained one. As the deviator stress is applied, neutral pressure gradually develops in the pore water; therefore during the entire second stage of the test the effective stresses will no longer be equal to the total stresses, but will be reduced vertically and laterally by the neutral pressure.

Figure 1-58 shows the distribution of total and effective stresses in this test. The major principal total stress at failure is $\sigma_1 = \sigma_3+p'_c$ and the minor total stress is σ_3. The value reached by the neutral pressure, u, which develops in the axial loading stage, is fundamental for an adequate comprehension of the test. In normally consolidated clays, the value of u depends mainly on the sensitivity of the structure, that is, on the readiness with which it becomes degraded with shear strain. If the behavior of the soil were perfectly elastic, $u = p'_c/3$, as is pointed out in [47]. In actual fact, there are plasticity effects in the soil that move it far away from its purely elastic behavior. As a result of structural changes, the structure transmits to the water what it can no longer support as effective pressure. In low and medium sensitivity soils, neutral pressures between $p'_c/2$ and p'_c have been measured during failure, at the end of the loading stage of an undrained-drained test. In highly sensitive soils values as high as $1.5p'_c$ may be reached. At first sight, it appears paradoxical that $u>p'_c$, that is, during the second stage of the test the water develops pressures higher than the total vertical stress applied during failure. The paradox disappears if one recalls the partial collapse of the solid structure due to strain which takes place in very sensitive clays and which even affects their capacity to withstand the confining pressures in the chamber corresponding to the first stage of the test (which are effective in the second stage). In this way, the pore water not only has to support all the deviator stress, but is obliged to help withstand the confining pressure too.

A general equation representing neutral pressure is:

$$\Delta u = A\,(\Delta\sigma_1 - \Delta\sigma_3) \qquad (1\text{-}71)$$

In this relationship, A is a coefficient of pore pressure which describes the effect of the change in the difference between the principal stresses [47, 52, 53]. For many unconsolidated saturated clays, A is approximately 1. For strongly overconsolidated clays or dense mixtures of sand and clay, the increase in shear stress described by the difference $\Delta\sigma_1-\Delta\sigma_3$ brings about an increase in volume similar to the one which occurs in compact sands when subjected to shear strain. For soils of this type, $A<0$. In slightly overconsolidated clays, A varies between

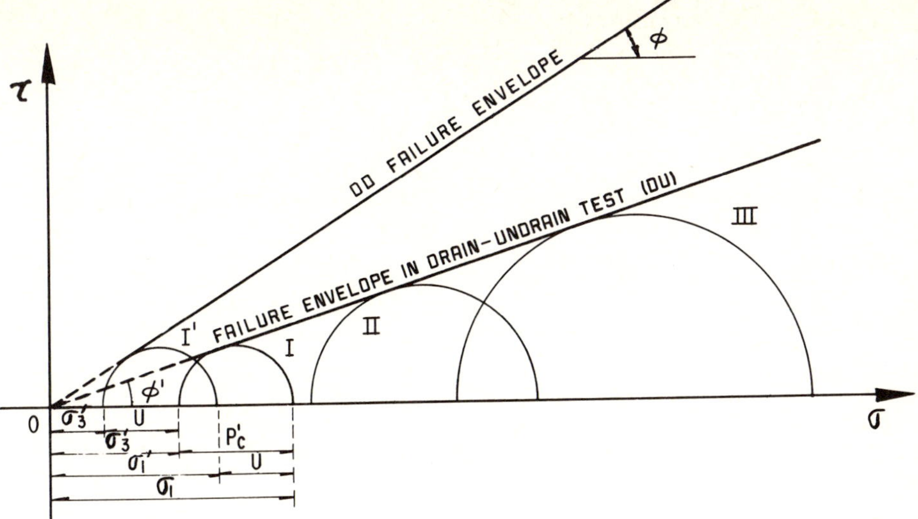

Fig. 1-59 Failure line for saturated, normally consolidated soils in drained-undrained test

0.25 and 0.75. In sensitive clays, as has been seen, A may exceed 1. In each case, the correct value for A will have to be determined in tests in which the neutral pressure at the moment of failure commences is measured.

If several different drained-undrained tests are run with increasing stresses applied to various specimens of the same soil, it will be possible to draw MOHR circles in a $\tau - \sigma$ diagram and obtain the strength envelope of the soil. This can be done now in two different ways: (1) using total stresses, which are known to the operator at every moment of the test and especially during failure, and (2) using effective stresses, for which it will be necessary to know the neutral pressure, at least at the moment failure commences. This requires tests in which the neutral pressure is measured directly in the triaxial chamber [52]. Figure 1-59 shows the typical envelopes obtained in both cases. With the same argument as for the drained test, it is easy to understand why envelopes are straight above the preconsolidation load, σ_c. Below the preconsolidation load the soil shows a strength somewhat higher than the one corresponding to the straight envelope.

When running tests with neutral pressure measurements, the effective stress circles are approximately tangential to the failure line obtained in drained tests.

If the total stress criterion is adopted, the strength law of the soil above the preconsolidation load can be expressed as in Eq. (1-63): $s = \sigma \tan \varnothing_u$ where $\varnothing_u$ is given the name of angle of apparent or undrained strength of the soil. This is strictly only a calculation parameter, the real theoretical meaning of which is very hard to establish.

In terms of effective stresses, the strength for the normally consolidated interval can be established in the drained-undrained test by Eq. (1-64): $s=(\sigma-u) \tan \varnothing=\sigma' \tan \varnothing$ using the angle of strength, $\varnothing$, obtained from the effective stress envelope, as it would be obtained in drained tests. The angle $\varnothing_u$ usually has a value of about $\varnothing/2$.

The drained-undrained test represents the conditions of a soil which is initially consolidated under the weight of a structure and is subsequently subjected to a rapid increase in stress by the construction of another structure or by a large sudden live load. It is sometimes used to represent the conditions of embankment foundations in which construction takes longer than the time required by the soil to reach a significant degree of consolidation.

.3 Undrained Test (quick)

In this test, both the confining stress, produced by the pressure of the water in the chamber, and the deviator stress, are applied in such a way that the specimen is not allowed to consolidate at all. This is achieved by closing the outlet valve between the chamber and the buret and/or applying the stresses with sufficient speed so that no drainage occurs. The void ratio of the sample and its water content remain constant and neutral pressures develop within the soil.

If the sample comes from depth z and γ is its specific weight, it represents a soil which was consolidated at pressure γz. If the sample is subjected to this pressure inside the chamber in the first stage of the test, the solid structure of the soil will theoretically carry the entire load and the pore water in the sample develop zero pressure. If, on the other hand, the pressure exerted by the water is greater than the natural pressure of the soil, the entire excess load will in theory be carried by the water in the sample, without any change in the degree of consolidation of the specimen, in void ratio or in the magnitude of the effective stresses. The interparticle spacing and the strength of the soil remain the same whatever the value of the pressure applied in the chamber. Consequently, since the effective stresses do not vary, the strength shown by the soil, p''_c, is constant, whatever the confining pressure of the chamber water during the initial stage. This is reflected in all the MOHR circles corresponding to total stresses having identical diameters, and the strength envelope corresponding to these total stresses is a horizontal line. Figure 1-60 shows the distribution of stresses within the specimen during the undrained test.

In the first stage, the hydrostatic pressure in the chamber typically is the γz of the natural soil, plus a certain arbitrary value, Δ. Consequently, a neutral pressure $u_1 = \Delta$ will develop in the water in the sample. In the second stage, the deviator stress, p''_c, is applied with the chamber rod, and at the end of this stage an additional neutral pressure, u_2, will have developed in the water.

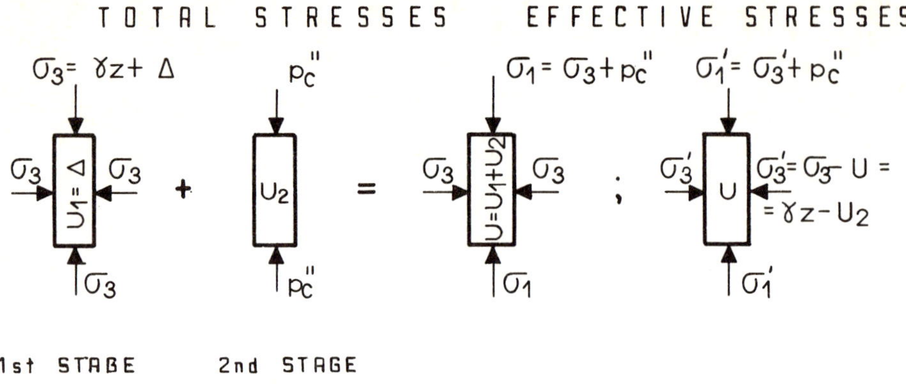

Fig. 1-60 Distribution of total and effective stresses in undrained triaxial compression test

By adding the two stages, a total neutral pressure of $u=u_1+u_2$ is obtained. The effective stresses will be the total stresses minus this value of u:

$$\sigma_3'=\sigma_3-u=\sigma_3-(u_1+u_2)=(\gamma z+\triangle)-(\triangle+u_2)=\gamma z-u_2$$

$$\sigma_1=\sigma_3+p'{}_c'=\gamma z-u_2+p'{}_c'$$

The value of the effective stresses turns out to be independent of $\triangle$ and all the total stress circles obtained in a series of tests using increasing total stresses, are only one circle for the corresponding effective stresses. Therefore all the total stress circles must have the same diameter and the strength envelope for total stresses must be a horizontal line, as has already been established. Figure 1-61 shows this strength envelope and relates it to the ones corresponding to drained and quick drained-undrained tests.

The ordinate at the origin of the strength envelope, is very similar to the shear strength of the soil in its original condition, consolidated under the load applied by the overlying soil. This ordinate at the origin is referred to as the cohesion of the soil, and any soil to which a horizontal strength envelope is applicable is referred to as a purely cohesive soil. When undrained test conditions (undrained-unconsolidated) are applicable, the strength of the soil will be simply as in Eq. (1-65): $s=c$ and the apparent angle of friction is zero in this case. This angle is no more than a calculation parameter, which will be used when the total stress method is used in a practical problem, in which the conditions of the undrained test are representative of the ones to which the soil is really subjected. In the real test, however, the inclination of the failure plane in the sample is not 45°, as it would be if the angle of apparent friction were really representative of the frictional strength of the soil (instead this is $\varnothing$, associated with the acting effective stresses, which can be measured in a drained or in a drained-undrained test with neutral pressure determination).

The undrained strength represents the strength of the soil under natural conditions. Because most construction is carried out in briefer times than those required by a clay to consolidate, the undrained strength, c, with total stresses is used in the majority of problems.. Even in those cases where construction is so slow that a significant increase occurs in the strength due to consolidation, the undrained strength is generally used for obtaining design data, for it represents a minimum, and consequently conservative value. When utilization of the undrained

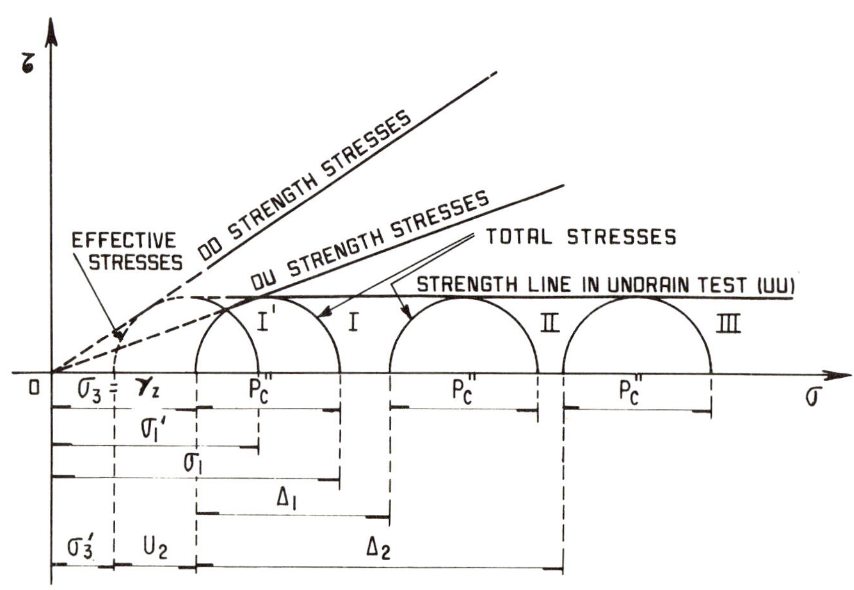

Fig. 1-61 Strength envelope in undrained triaxial test

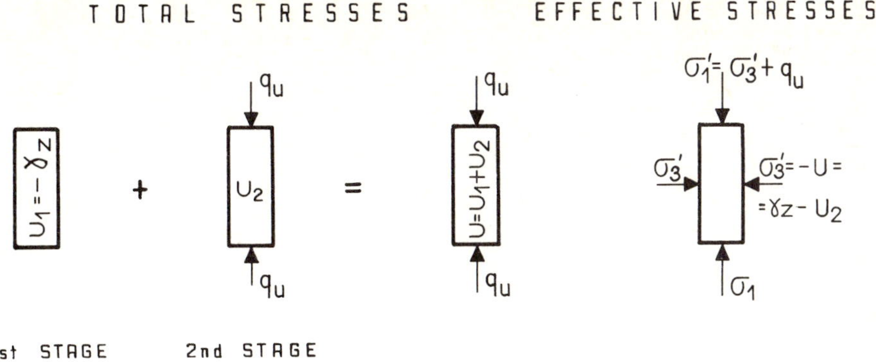

Fig. 1-62 Distribution of total and effective stresses in unconfined compression test

strength is considered for obtaining design values, it is important to recognize those cases in which the final stresses applied to the soil may be smaller than the initial load to which it was subjected. This is frequently the situation in excavations and slope stability problems. In these cases, for short term design conditions, when the soil does not have sufficient time to expand, undrained strength conditions may be applicable. In the long term, however, the soil becomes weakened and the use of the undrained test will prove unsafe.

The undrained strength depends on the initial stress to which the soil is subjected in its original location, its preconsolidation load and the MOHR's failure envelope corresponding to drained conditions. In compressible soils, the pressure withstood by the soil at its natural location is related to the void ratio by the compressibility curve. As a result, the undrained strength of a saturated clay increases when the void ratio and/or the water content are reduced. In normally consolidated soils, a graph of the void ratio or the water content versus the undrained strength, is approximately a straight line.

.4 Unconfined Compression Test.

As has been mentioned, this test, see Plate 1-12, is performed by applying an axial stress to the specimen, without the initial hydrostatic pressure stage. This test includes only the loading stage which brings the soil to failure. However, the initial condition of the sample reflecting the load in the ground without external stresses, could be regarded as the first stage of the test. As the test begins in the laboratory (Fig. 1-62), the total stresses are zero and the water acquires tension equivalent to the preconsolidation stress (γz) of the natural soil. This tension transmits to the solid structure the effective stresses that maintain the samples' volume.

In the second stage, the sample is brought to failure by the application of the axial stress (q_u), which measures the strength in this type of test, at the same time producing an additional neutral pressure u_2. The effective stresses which appear at the end of the test, at the moment of failure, are also shown in Fig. 1-62 and are

$$\sigma_3 = -(u_1 + u_2) = -(-\gamma z + u_2) = \gamma z - u_2$$

$$\sigma_1 = \sigma_3' + q_u = \gamma z - u_2 + q_u$$

Observe that the minor effective principal stress is theoretically the same as for the undrained triaxial test.

Consequently, the maximum deviator stress required for the sample to be brought to failure in this test, q_u, referred to as the resistance of soil to unconfined compression, is the same as in the undrained test. The unconfined compression test is not, however, an undrained triaxial test. The test method is basically different because it relies on capillary tension. It is not permissible to use the data from that test to complete envelopes obtained with undrained tests. It is very common for q_u to be a little smaller than p''_c but for practical purposes in saturated clays it can be regarded as the same. In fissured soils or those containing sand seams, there may be

Plate 1-12 Unconfined compression test

no relation of the unconfined strength to the undrained. Figure 1-63 shows the total (I) and effective (I') stress circles corresponding to the moment failure commences in this type of test and their position in relation to the strength envelopes in triaxial tests. It should be observed that the figure has been drawn under the assumption that the preconsolidation load of the soil is γz.

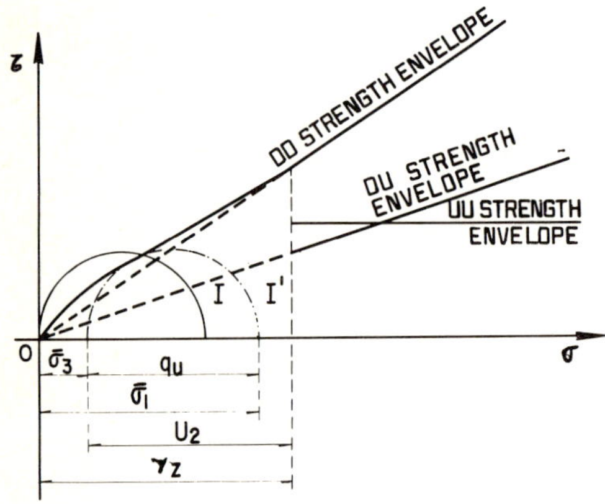

Fig. 1-63 Total and effective stress circles in unconfined compression test

The resistance of the soil to unconfined compression has been used as a means of measuring the sensitivity of the structure of the soil to strain, comparing the value of q_u in the same soil in undisturbed and remolded conditions. The sensitivity of a soil is defined as

$$S_t = \frac{q_u \text{ (unaltered)}}{q_u \text{ (remolded)}} \qquad (1\text{-}72)$$

1.15.2 Non-Saturated Soils

Basically, the shear strength of non-saturated soils involves the same concepts as that of saturated soils. There are, however, certain very significant differences between the two. In non-saturated soils, the pores are only partially filled with water and they contain air in accordance with the degree of saturation. The vast difference in the mechanical behavior of the two fluids (air and water) leads to behavior characteristics which are very complex. As the shear strength of soils is currently conceived, it is still true that the effective stress controls the frictional component of that strength. The shear stresses are withstood only by the solid particles of the non-saturated soil (skeleton), except at very high levels of strain. The total normal stress on any plane can be broken down into two parts, one corresponding to the effective stress transmitted to the mineral skeleton and the other supported by the pressure of the fluid in the pores of the soil. The neutral pressure is, however, a very complicated combination of pressure and capillary tension in the water and pressure in the air, which depends on the degree of saturation and the size of the pores.

If there is just one fluid in the pores, either air or water, the effective normal stress measured by the already established equation is: $\sigma' = \sigma - u$, where σ' is the effective stress, σ the total stress and u the neutral pressure. In partially saturated soils, there are two fluids in the pores, which are in equilibrium at pressures that are considerably different in each because of surface tension. For representing the effective stress in this case, Bishop [54] has proposed an expression:

$$\sigma' = \sigma - u_a + X(u_a - u_w) \qquad (1\text{-}73)$$

where u_a represents the pressure in the gaseous phase (gas or vapor) and u_w the pressure in the liquid phase. The parameter X is 1 for saturated soils and zero for dry soils. The intermediate values depend mainly on the degree of saturation, but are influenced by other factors such as the structure of the soil, the cycles of wetting and drying to which it is subjected and the changes in stresses which may occur in reaching the degree of saturation. In [54], determinations of X are shown for some specific soils, the value increases with the degree of saturation.

The values of u_a and u_w which are available when the soil is subjected to a change in stress $\Delta\sigma$ have been studied by Bishop and Eldin [55] and by Skempton [56]. According to these authors, when a hydrostatic increase in stress, $\Delta\sigma_3$ is applied to a partially saturated soil, an increase occurs both in the pressure of the water, and in the pressure of the air, in accordance with the relationships

$$\Delta u_a = B_a \, \Delta\sigma_3$$

$$\Delta u_w = B_w \, \Delta\sigma_3 \qquad (1\text{-}74)$$

The above equations define the coefficients of neutral pressure, B_a and B_w. In [56], typical values of B_w are given for partially saturated soils, with variations between 0.10 to 0.89, indicating in each case which part of the stress applied is carried by the water.

A similar approach can be adopted for expressing the increase in the pressure in the water and in the air when there is an increment in the deviator stress acting on the soil sample. Now:

$$\Delta u_a = A_a \, (\Delta\sigma_1 - \Delta\sigma_3)$$

$$\Delta u_w = A_a \, (\Delta\sigma_1 - \Delta\sigma_3) \qquad (1\text{-}75)$$

Typical values for A_w during failure have been reported by Bishop and Henkel [57] as being located between – 0.28 and + 0.27 for samples of partially saturated, compacted soils.

In undrained triaxial tests on partially saturated soils, the shear strength increases with the external normal pressure, for the compression of the air permits the development of an effective stress. However, the increase in strength becomes gradually smaller, due to the effect of dissolution of the air in the water in the pores, which becomes easier as the pressure in the air increases. When the stress levels are sufficiently high, the low compressibility of the water-dissolved air combination and the reduction in volume of voids due to strain, produce in the specimen a behavior similar to that of saturated soils, with an angle $\varnothing$ in the total stress failure envelope close to zero. The total stress envelope is not, therefore, a straight line, but a curve which tends to the horizontal. The strength parameters c and $\varnothing$ can only be defined if that part of the curve which includes the interval of normal stress governing the specific problem is approximated by a straight line. If a problem is to be solved using the total stress criterion, (and this is the most common case in non-saturated soils) it is of the greatest importance to reproduce, in the laboratory, test conditions. Figure 1-64 shows a typical envelope for non-saturated soils in undrained triaxial tests.

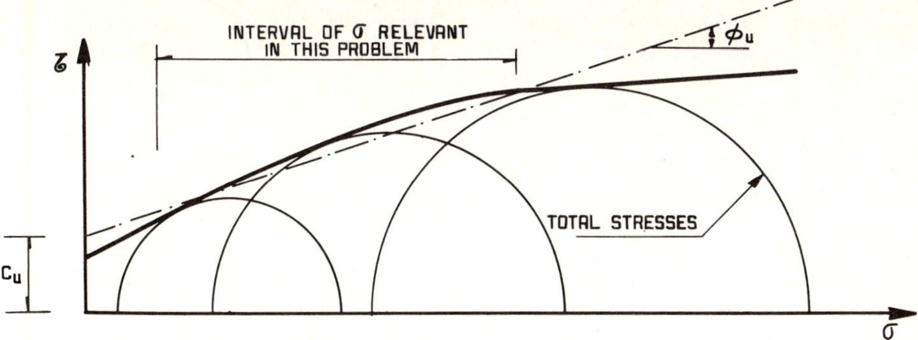

Fig. 1-64 Undrained triaxial test on a partially saturated soil

It is not possible to carry out drained tests on partially saturated soils, with the same significance and reliability as normal drained tests (that is, tests in which the neutral pressure is zero at every stage of any significance), for that would imply destroying the capillary tensions. In order to achieve this, it is necessary to saturate the sample. If effective stresses are to be used, what is done to obtain the corresponding envelope is to saturate the sample and assume that this process does not bring about any significant changes in the value of ∅. This is a conservative criterion for practical analyses, for strength is usually reduced with saturation (except in cemented soils).

In non-saturated soils, the drained test is commonly used, but at constant water contents. The sample is maintained without modification of the water content, and the pressure of the air is controlled as far as possible in order to achieve this. In this type of test, it is customary to measure the neutral pressure in the pore-water in order to find the pore-water pressure.

The strength envelopes of non-saturated soils in undrained tests become more like the shape corresponding to saturated soils, as the degree of saturation increases. In [47], some laboratory results confirming this can be seen.

A case of fundamental importance in non-saturated soils, which is of great interest for the road engineer, is that of compacted soils. There is already considerable information on this subject, but it will be discussed in the chapter on compacted soils, later in this book.

1.15.3 Application of the Results of Triaxial Tests to Practical Problems

When an engineer needs to find out the stress-strain and strength characteristics of a soil, with a view to obtaining data for design, he usually turns to triaxial compression tests. The question immediately arises as to which of these tests should be performed in order to solve the specific problem and what interpretation should be given to the results obtained. In each case, the test or tests to be run will be the ones which best reflect the circumstances to which the soil will be exposed in the field.

The engineer must first analyze the different stages through which the soil will pass during the life of the project from the very beginning of the construction. Only in this way will he be able to judge correctly the critical conditions for which the design must be established. It should be recalled that it is common for these critical conditions to occur in the soil mass a long time after the construction. It is also essential for the engineer to have an intimate knowledge of the soil profile, its basic properties and the drainage conditions that will occur, over long periods of time, and in a changing environment. Preconsolidation conditions merit special investigation, for they will have an important influence on overall behavior.

At the present time, there are two criteria for the practical determination of the shear strength of soils.

.1 The Effective Stress Method.

This is the stress which really determines the shear strength of the soil. Once the effective stress, which will act between the particles of the soil at a given point in the mass, has been found, it will be sufficient to multiply this value by the tangent of the angle of internal friction obtained in a drained test in order to obtain the real shear strength available at that point in the soil. This criterion poses few difficulties of a theoretical nature. It is the logical conclusion to all that has been studied in this chapter until now, in relation to the shear strength of soil. Figure 1-65 is an example of the effective stress criterion for interpreting the shear strength of soils from the results of triaxial tests.

The first requirement for the application of the method is to know the strength envelope of the soil obtained in relation to the effective stresses, such as for example a series of drained tests, drawing the failure circle for each one and then plotting the *DD* strength envelope accordingly, tangential to all of them. The *DD* strength envelope could theoretically be defined with a single circle drawn in the normally consolidated interval, but given the mistakes inherent in laboratory work, it is advisable to obtain at least two or three failure circles and to draw the *DD* strength envelope as a straight line which is closest to the common tangent.

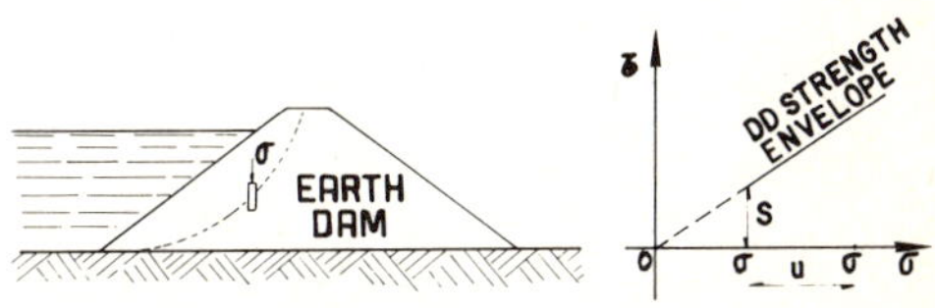

Fig. 1-65 How to obtain the shear strength of a soil using effective stress

In the dam in the figure, the strength of the soil is to be computed for the purposes of studying the stability of the upstream slope. The *DD* envelope, which is assumed to have been obtained already, is also shown. The material of which the dam is made is assumed to be saturated and normally consolidated. If σ is the total pressure on the element and u the neutral pressure in it at the amount of failure, the effective stress prevailing in the structure of the soil will be $\sigma' = \sigma - u$ and the strength of the element will simply be the ordinate of the *DD* strength envelope corresponding to σ'.

The above criterion, which is apparently so simple, presents serious practical drawbacks. Among these are the ones involving the need to obtain the *DD* strength envelope in the laboratory, not to mention the ones which arise subsequently, at later stages of the practical application of the method.

In order to obtain the *DD* strength envelope in the laboratory, drained tests could be performed, and this would apparently be a simple and satisfactory solution to the problem. This is not, however, true. Drained tests are time consuming, and therefore the most expensive ones. The solution based exclusively on them cannot be considered devoid of practical difficulties. Apart from this economic reason, and the time they involve, drained tests present difficulties inherent to their very nature of which only two will be discussed here. First there is a problem in the laboratory which has not been completely solved in relation to the impermeable membrane which isolates the specimens from water in the triaxial chambers. Very thin membranes, which are flexible enough not to influence the stress conditions of the specimen, with time, allow small quantities of water to seep into the soils. This can be sufficient to introduce considerable errors into the results, especially when relatively high pressures are involved. Membranes that are sufficiently thick to ensure complete impermeability, owing to their greater rigidity have a significant influence on the results of triaxial tests. This effect is notable in drained tests, although it is usually insignificant in other triaxial tests, for in drained tests the chamber water may be subjected to higher pressures and the times of exposure of the membrane to the water itself are also far longer. A second practical difficulty which arises when performing drained tests in the laboratory, comes from the notably greater strains that the sample undergoes than in other triaxial tests, under rod pressures that are also greater. These strains tend to make the specimen shorten in length and increase in diameter. A restriction is established due to friction between the ends of the specimen, as the soil tends to displace laterally, and the porous stones, which remain fixed, stop this movement. This restriction due to friction produces shear stresses on the end of the specimen which are then no longer principal stresses, with the consequent error in the interpretation of the test using MOHR's theory, which considers them principal stresses.

It is clear, therefore, that efficiency and accuracy will not always be ensured if the *DD* strength envelope is obtained by means of drained tests, which are both lengthy and costly.

Currently it is possible to obtain the *DD* strength envelope in the laboratory using triaxial tests other than the drained ones, for example drained-undrained tests. There is a variety of equipment available for measuring the pore-pressure that develops in the specimen at the time of failure. Once the total deviator stress is found, it is easy to obtain the effective stress acting at that moment. Nevertheless, the equipment for measuring pore-pressure is expensive and requires relatively delicate handling, which is why it is not commonly found in many soil mechanics laboratories, especially those that are set up at a job location.

Last, there are theoretical methods for estimating pore-pressure at the instant of failure in a specimen subjected to a drained-undrained test. SKEMPTON, HENKEL and JUÁREZ-BADILLO have established methods to meet this purpose [47]. To summarize, it can be said that there are now some reasonably reliable methods for obtaining the *DD* strength envelope, either in the laboratory or with the aid of methods which cannot yet be considered within the reach of many. This arouses the hope that in the near future it will be possible to apply the effective stress method more readily than at present, at least in so far as this first requirement is concerned.

Once the *DD* strength envelope has been obtained, there still remains an important problem for applying the effective stress method to practical situations. The situation shown in Fig. 1-65 is now considered. In order to carry out an analysis it will be necessary to find the effective stress condition at all points of interest within the soil mass, which, in this specific case, will be the points on the assumed slip surface. This is a problem which until now has not been solved, for it is clear that if it has not been possible to discover completely the effective stress condition inside the specimen within a triaxial chamber subjected to a test control, it will be even more difficult to discover in detail this stress condition in the immense masses of soil that are involved in a real project. Thus, even if the drained strength envelope has been established, there will in practice be the additional difficulty that the effective stresses acting at the different points of the soil mass are unknown. Certain organizations that specialize in the construction of earth dams overcome this difficulty by designing their projects in accordance with the effective stress method, basing them on a prediction of the effective stresses that are likely to develop during construction. By installing piezometers to measure the pore-pressure as construction progresses, it can be seen whether the predictions are correct or whether modifications must be made in the design in the light of the readings taken. This method is practical only for organizations possessing sufficient field experience, backed by extensive records for similar dams that have been built previously.

Despite all these difficulties, which should not be underestimated, especially in projects smaller than earth dams, it is safe to say that, with the constant progress in the field of soil mechanics, the effective stress method will eventually be the most widely used. It is the most rational method where the basic ideas governing shear strength are concerned.

.2 Total Stress Method

In this method, the total stresses used in triaxial tests are directly applied; envelopes *DD* or *DU* are used, according to the specific problem. Because the strength values for the same soil are different for each tests, owing to the different circumstances in which the tests are run, it is logical that the test will only be representative if the stress conditions are as close as possible to those to which the soil in the prototype will be subjected. Consequently, for this method it is essential that the engineer be experienced when selecting the type of test or tests to be run.

There is no rigid rule to determine which tests should be performed, and it is the common sense and experience of the design engineer which will decide. In order to help the reader establish his own criterion, here are some general comments.

A structure must be designed bearing in mind which will eventually be the most critical stages in its lifetime. In structures erected on soil or made of soil, it is very common for the critical stages to occur either during the earliest moments in their lifetime or else in the long term. It is interesting to analyze what will be the critical moments in the lifetime of a given structure, and consequently what laboratory tests will be required.

For example, a building is to be erected on a flat clayey ground. As the soil consolidation that is induced by the building weight progresses, the strength of the soil increases accordingly. The critical condition therefore will correspond to the initial stages in the lifetime of the structure. Because the clay is very impermeable, the consolidation processes will be slow and the construction time by comparison will be almost negligible. As a result, the critical moment will probably be when the load exerted by the building is complete. Here field conditions will be represented by a test in which the deviator stress is applied rapidly. An undrained test will therefore be appropriate.

If, on the other hand, the building is erected on a clay similar to the one in the previous case, but with numerous sand strata providing quick and efficient drainage, the soil may consolidate at the same rate as the construction progresses. A drained test will therefore be the most appropriate one for determining shear strength. If the structure to be erected is an embankment (Fig. 1-66), for a road or a dike, and the strength of the bearing ground must be established, it should be remembered that the weight of the embankment will cause consolidation in the soil, and the foundation shear strength will therefore increase as time goes by. If the embankment is erected quickly and drainage is difficult in the clayey ground, the most critical moment will be the early stages in the lifetime of the project, before the soil consolidates, as was true for the building. If the soil consolidates at the same rate as construction progresses, a drained test will be the correct one for obtaining design data.

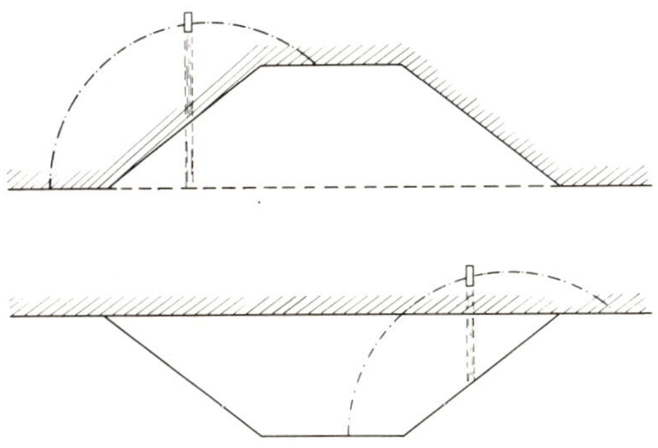

Fig. 1-66 Variation in the shear strength of the same clayey soil in a cutting and an embankment

The situation will change radically if an excavation is made in the same soil for the foundations of a structure. In this case, expansion will develop in the soil mass due to unloading and the shear strength will drop as time goes by. The critical condition of the soil will be during the final stages of the expansion process, that is, during the final stages in the lifetime of the excavation A drained test or a drained-undrained test will be the most appropriate ones for representing this situation.

Once the type or types of triaxial test have been selected, several tests of the type chosen are performed, obtaining the MOHR failure circle for each test and drawing a straight line (the normally consolidated section) which approximates the envelope of these circles. In the preconsolidated section, a second envelope is drawn, following the procedures already discussed in this chapter, at a tangent to the circles. Once the approximate envelope for the soil has been obtained in this type of test, it is good practice to select the portion of that envelope that corresponds to the stress range that will develop in the soil during the project life and approximate the envelope with a straight line. This straight line, especially in preconsolidated or non-saturated soils, will seldom go through the origin of the coordinates, and the mathematical equation will be:

$$s = a + \sigma \tan \alpha \qquad (1\text{-}76)$$

with a and α as parameters defining the strength of the soil in a specific test and within a specific pressure interval (a is the ordinate at the origin and α the angle of slope in relation to the horizontal part of the straight line in question). Note that Eq. (1-76) is like the classic COULOMB law. However, it is no longer worthwhile discussing the basic differences in interpretation between the two. The parameters a and α no longer have a characteristic physical meaning as inherent soil properties, but are only elements for calculation purposes. Partly by tradition and partly by habit certain authors have referred to a as *apparent cohesion of soil* in its natural conditions and α as the *angle of apparent friction.* It is even common practice in literature on the subject to continue to use the symbols c and $\varnothing$ for strength parameters, but subjecting them to modern interpretation. Symbols c and $\varnothing$ should be interpreted thus when appearing in subsequent pages of this book. Since currently used triaxial tests represent extreme soil conditions, some specialists prefer to draw their own envelopes between the two extremes. When this procedure is backed up by wide experience, far more realistic data can be obtained than with any other test.

1.15.4 Peak and Residual Strengths in Clays

Consider a preconsolidated clay that is subjected to a simple or direct shear test with free drainage throughout (which are the characteristics of a drained test). Presume that it is a controlled strain test, slow enough for pore-pressures to dissipate. In it, the stresses corresponding to the induced strains are duly measured. As displacement increases and the sample of preconsolidated clay deforms angularly, the shear load increases, and therefore the shear stress. For a given effective normal pressure applied to the sample there is a limit to the shear stress that can be withstood by the sample. This limit, which up to now has been referred to as the shear strength of the clay, will now be called peak strength. If the test continues, and greater displacements take place, the shear load that is applied will drop, together with the shear stress. In practice, the test is often suspended once the peak strength has been determined. Should the test continue, however, a reduction is observed in the strength of the clay as the displacement and strain increase. This reduction also has a limit though, and once it has been reached the shear stress remains constant even if displacement and strain continue to increase. There is field evidence that this strength is maintained by the clay even for displacements running into meters. If different tests are carried out in this manner, using a different effective

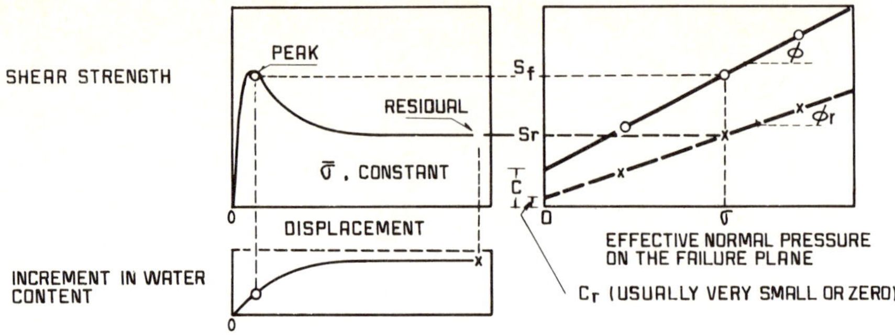

Fig. 1-67 Peak and residual strengths and shear strength characteristic of a preconsolidated clay

normal pressure in each, results similar to the above ones will be obtained, although the ultimate value shown for the strength of the clay will be different in each case. This ultimate strength, beyond peak strength, is referred to as residual strength [40]. Part (*a*) of Fig. 1-67 shows the shear stress-displacement relationship, that is usually obtained in a test like the one described above. The change in the water content that is undergone by the specimen during the test can also be observed.

Part (*b*) of the same figure shows the failure envelopes obtained by plotting the effective normal stress, versus the peak and residual strengths obtained in these tests. It can be observed could therefore be written as in Eq.(1-66): $s_f = c + \sigma' \tan \varnothing$ and residual strength, as:

$$s_r \quad = \quad c_r + \sigma' \tan \varnothing_r. \qquad (1\text{-}77)$$

Test results show that c_r is usually very small, and negligible. Consequently, for residual strength the following equation can be written:

$$s_r \quad = \quad \sigma' \tan \varnothing_r \qquad (1\text{-}78)$$

It has also been observed that $\varnothing_r$ is smaller than the angle $\varnothing$. In some clays this difference is of just 1 or 2 degrees, but clays have been reported in which this difference is as high as 10°.

According to Skempton [40], the reasons for these differences may be the following: first, when strongly preconsolidated clays deform under shear stress, beyond their peak strength, expansion occurs. Part of the reduction in strength can therefore be blamed on the increase in the void ratio and water content in the drained test which accompanies large strains. Second, there is the development of thin fringes within the clay mass where the flat particles are orientated in the direction of the displacement. It is reasonable to deduce that the strength of a group of these particles with random orientation may be greater than that of a parallel arrangement.

Independently of the reasons that might be given to explain the drop in strength suffered by clays beyond peak strength, there is especially evidence of this reduction when the clays are preconsolidated. If for any reason strains progress beyond those of peak strength at any point in the clay mass, the strength at that point will drop. This brings about a redistribution of the stresses as a consequence of which the neighbouring zones become overloaded. As a result the peak strength may also be passed at other nearby points. In this way a progressive failure develops. This occurs in landslides, and when the limit is reached, the strength along the entire failure surface will be reduced to the value for residual strength. However, the amounts of strain required for residual strength to develop are so large that this condition should only be considered for design, or calculation purposes, when the clay has suffered a slide over an already existing failure surface or when there are generalized creep conditions.

Skempton also points out that the presence of a large number of tiny fissures, little cracks and other similar defects in the clay mass represents another case in which the residual strength should be used during design in order to achieve a more realistic analysis.

There is no standard test for determining the residual strength of clays in the laboratory. Skempton [40] describes one that was performed for a case in which a direct shear apparatus was used. After making the specimen deform about one centimeter in a certain direction, the sliding part of the shear box was returned to its original position. The shear and return procedure was repeated until the strength of the clay reached a final constant value, which was taken as the residual strength. The disadvantage of the test was the time it took, 6 days, because the pore pressures were allowed to dissipate throughout. Skempton comments that this technique is not perfect and suggests that a better test would be one causing a continuous shear in just one direction without the return to the original position. He also points out that an annular shear apparatus might prove appropriate. Other authors have suggested the suitability of torsion shear tests.

The reduction from peak strength to residual strength occurs not only in preconsolidated clays, but also in normally consolidated clays, although the difference between the two strengths is smaller. In normally consolidated clays, the reduction in the angle of internal friction is chiefly attributed to the effect of particle orientation, when the strain along a given failure surface is of an appreciable magnitude. The results available so far seem to indicate that the residual strength of a clay is the same, under a certain effective normal stress, whether the clay is preconsolidated or normally consolidated. In other words, $\varnothing_r$ is constant for a certain clay, independent of its consolidation history, and its magnitude depends on the nature of the mineral particles, the value diminishing when the percentage of particles smaller than two microns increases. Skempton reports a value of about 10°, when the percentage by weight of particles smaller than two microns is between 60% and 80%.

The important matter, from a practical point of view, is to decide which strength to use for designing the stability of a given slope. Skempton applied most of his ideas on the subject of residual strength to slope stability. For this purpose, he defined the *Residual Factor R* by means of the expression:

$$R = \frac{s_f - s'}{s_f - s_r} \quad (1\text{-}79)$$

where: s_f is the peak strength of the clay, s_r the residual strength of the clay, and s' the average shear stress acting on the failure surface being studied.

SKEMPTON analyzed the stability of several failed slopes and found the average effective normal stress and average shear stress on the failure surface. Since the slides were real ones, s' can be obtained by back calculation if the safety factor is regarded as being equal to unity. SKEMPTON later compared this s' with the peak and residual strengths of the clay corresponding to the effective normal stress existing on the failure surface. In this way, he was able to compute the residual factor for each case. If for a given situation the strength at which the slope failed was the peak strength, $R = 0$, and if equal to the residual strength, R will be equal to 1.

Another alternative interpretation for the residual factor is obtained by writing Eq. (1-79) as follows:

$$s' = R s_r + (1 - R) s_f \quad (1\text{-}80)$$

In this expression, R can be interpreted as a number which indicates the proportion of the total failure surface along which the strength has been reduced to its residual value.

SKEMPTON's objective was to relate as far as possible the value of R with the type of clay of which the slope was composed. If residual strength can develop, he recommends its use for practical analyses. In clays with no fissures and cracks, he finds that the reduction in strength during failure is negligible in relation to peak strength, and consequently peak strength could be used. He also considers that the stability of embankments made of compacted clay can be analyzed using peak strength. Lastly, if a slide has taken place, any subsequent movement on the failure surface that is formed will occur with residual strength, independently of the type of clay.

REFERENCES

1. JUÁREZ BADILLO, E. and RICO, A., *Mecánica de Suelos, Vol. I: Fundamentos de la Mecánica de Suelos,* Limusa: México, 1977, 3rd Edition, Chap. III.
2. ibid., Chap. IX.
3. ibid., Chap. II.
4. LAMBE, T. W. and WHITMAN, R. V., *Soil Mechanics,* John Wiley and Sons, 1971, Chap. 4.
5. Reference [1], Chap. IV.
6. TERZAGHI, KARL, "Modern conceptions concerning Foundation Engineering", *Journal of the Boston Society of Civil Engineers,* Contributions to Soil Mechanics 1925-1940, 1959.
7. CASAGRANDE, A., "The Structure of Clay and its Importance in Foundation Engineering", *Journal of the Boston Society of Civil Engineers,* Contributions to Soil Mechanics 1925-1940, 1959.
8. Reference[4], Chap. 5.
9. CASAGRANDE, A., "Classification and Identification of Soils", *Trans. ASCE,* Vol. 113, 1948, p. 901.
10. Reference [1], Chap. VI.
11. Reference [4], Chap. 3.
12. SKEMPTON, A. W., "The Colloidal Activity of Clays", *Procs. III. International Conference on Soil Mechanics and Foundations Engineering,* Zurich, 1953, Vol. I.
13. Reference [1], Chap VIII.
14. REYNOLDS, O., "An Experimental Investigation of the Circumstances which determine whether the Motion of Water shall be Direct or Sinuous and the Law of Resistance in Parallel Channels", *Phil. Transactions Royal Society*, Vol. 174, London, 1883.
15. DARCY, H., *Les Fontaines publiques de la Ville de Dijon,* Paris, 1856.
16. JUÁREZ BADILLO, E. and RICO, A., *Mecánica de Suelos, Vol. III: Flujo de Agua en Suelos,* Limusa: México, 1977, Chap. I.
17. Reference [1], Chap. X.
18. Reference [4], Chap. 9.
19. ibid., Chap. 10.
20. ibid., Chap. 20.
21. SKEMPTON, A. W. and BISHOP, A.W., *Soil Building Materials, their Elasticity and Inelasticity,* North Holland Publ. Co.: Amsterdam, 1954, Chap. X.
22. JIMÉNEZ SALAS, J. A. and DE JUSTO ALPAÑES, J. L., *Geotecnia y Cimientos, Vol. I: Propiedades de los Suelos y de las Rocas,* Rueda: Madrid, 1971, Chap. 6.
23. MARSAL, R. J., MORENO, E., NÚÑEZ, A., CUÉLLAR, R. and MORENO, R., "Investigación sobre el comportamiento de suelos granulares y muestras de enrocamiento", Federal Electricity Commission, México, 1965.
24. MARSAL, R. J., "Large scale testing of Rockfill Materials", *Journal of the Soil Mechanics and Foundations Division, ASCE,* March, 1967.
25. MARSAL, R. J., "Contributions and Discussions on Mechanical Properties of Rockfill and Gravel Materials", *VII. International Conference on SMFE,* México, August, 1969, Special Session N° 13
26. MARSAL, R. J. and RAMÍREZ DE ARELLANO, L., "Performance of El Infiernillo Dam", *Journal of the Soil Mechanics and Foundations Division, ASCE,* July, 1967.
27. MARSAL, R. J., RAMÍREZ DE ARELLANO, L. and NÚÑEZ, A., "Plane strain of Rockfill Materials", *III. Panamerican Conference on SMFE,* Caracas, 1967.
28. Reference [4], Chap. 22.
29. LEONARDS, G. A. and GIRAULT, P., "A Study of the One-Dimensional Consolidation Test", *V. International Conference on SMFE,* Paris, 1961.
30. BARDEN, L., "Primary and Secondary Consolidation of Clay and Peat", *Geotechnique,* Vol. 18, 1968.
31. Reference [1], Chap. XI.
32. SCOTT, R. F., *Principles of Soil Mechanics,* Addision Wesley Publ. Co., Inc., 1963, Chaps. 7 & 8.

33. Coulomb, Ch. A., "Essai sur une application des regles des maximes et minimes a quelques problems de statique relatifs a l'architecture", *Memoir to the Royal French Academy,* 5, 7, Paris, 1776.

34. Mohr, O., *Abhandlungen aus dem Gebiete der Technischen Mechanik*, W. Ernst: Berlin, 1914, 2nd Edition.

35. Newmark, N. M., "Failure hypotheses for Soils," *ASCE Research Conference on Shear Strength of Cohesive Soils*, Colorado, 1960.

36. Scott, R. F. and Hon-Yim Ko., "Stress-Deformation and Strength Characteristics," *VII. International Conference on SMFE,* Mexico, 1969, Vol. III.

37. Reference [4], Chap. 6.

38. Skinner, A. E., "A note on the influence of Interparticle Friction on the Shearing Strength of a Random Assembly of Spherical Particles," *Geotechnique*, Vol. 19, 1969.

39. Scott, R. F. and Schoustra, J. J., *Soil Mechanics and Engineering*, McGraw-Hill Book, Co., 1968, Chap. 5.

40. Skempton, A. W., "Long-Term Stability of Clay Slopes," IV. Rankine Lecture, *Geotechnique,* Vol. 14, 1964.

41. Hambly, E. C., "A new Triaxial Apparatus," *Geotechnique,* Vol. 19, 1969.

42. Bishop, A. W., "The Strength of Soils as Engineering Materials," VI. Rankine Lecture, *Geotechnique,* Vol. 16, 1966.

43. Cornforth, D. H., "Some experiments on the Influence of Strain conditions on the Strength of Sand," *Geotechnique,* Vol. 16, 1964.

44. Hvorslev, M. J. and Kaufman, R. I., "Torsion Shear Apparatus and Testing Procedures," *Bulletin N° 38, Waterways Experiment Station*, Vicksburg, Miss., 1952.

45. Roscoe, K. H., *Procs. International Conference on SMFE*, Paris, 1961, Volume 3: Discussion pp. 105–107.

46. Bjerrum, L. and Landva, A., "Direct Simple Shear Tests on a Norwegian quick clay," *Geotechnique*, Vol. 16, 1966.

47. Reference [1], Chap. XII.

48. Wilson, S. D. and Squier, R., "Earth and Rockfill Dams," *VII. International Conference on SMFE*, México, 1969, Vol. III.

49. Reference [4], Chap. 11.

50. Marsal, R. J. and Ramírez de Arellano, L., "Field measurements in Rockfill Dams," *II. Panamerican Conference on SMFE*, Sao Paulo: Brazil, 1963, Vol. II.

51. Sowers, G. B. and Sowers, G. F., *Introductory Soil Mechanics and Foundations,* McMillan, 1972, Chap. 3.

52. Reference [4], Chap. 26.

53. Skempton, A. W., "The Pore Pressure Coefficients A and B," *Geotechnique*, Vol. 4, 1954.

54. Bishop, A. W., Alpan, I., Blight, G. E. and Donald, I. B., "Factors controlling the Strength of Partly Saturated Cohesive Soils," *Research Conference on Shear Strength of Cohesive Soils*, ASCE, Boulder, Colorado, 1960.

55. Bishop, A. W. and Eldin, G., "Undrained triaxial Tests on Saturated Sands and their significance in the General Theory of Shear Strength," *Geotechnique*, Vol. 2, 1950.

56. Bishop, A. W. and Henkel, D. J., *The Measurement of Soil Properties in the Triaxial Test*, Edward Arnold, Ltd.: London, 1957.

CHAPTER 2

CLASSIFICATION OF SOILS FOR ROAD ENGINEERING PURPOSES

2.1 Introduction

In the particular field of road engineering there is an unending variety and complexity of soils. Before making any attempt, therefore, at scientific systematization, with the corresponding tendency to generalize, soils should be classified as carefully as possible.

Soil classification systems are as ancient as soil mechanics itself, but so little was known about soils that the first systems to appear were based on not very relevant properties (odor, color, texture, etc.) or on characteristics which were very difficult to correlate with the fundamental ones. Today these systems have been greatly improved and deserve attention.

Gradation provides a simple method of soil classification. The soil is divided into fractions according to size. Each fraction is defined by an equivalent particle diameter. Each is given a common name that is compatible with the layman's concept of that size.

Grading classification systems, which were so popular in the past, started in this simple manner. The terms gravel, sand, silt and clay for civil engineers, geologists, and laymen alike bear a significance related to the size of the particles.

For engineering purposes, a classification system must group soils according to the important basic mechanical properties. However, the criteria of classifying are qualitative, because a system which includes quantitative relationships would be too difficult and complicated to handle. The least an engineer can expect from a classification system is that it serves as a guide to a soil's usual behavior, before he acquires more detailed specific knowledge of its properties. By using the system he will, amongst other things, be able to decide in which direction his investigations should be oriented.

In spite of their simplicity, purely grain-size criteria are not very appropriate, because the correlations between grain size distribution and the fundamental engineering properties (strength, compressibility, stress-strain ratios, permeability, etc.) are poor.

The most efficient system of soil classification today is the one proposed by CASAGRANDE [1], slightly modified by the U.S. Corps of Engineers and the Bureau of Reclamation, known as the Unified Soil Classification System (U.S.C.S.), and is the most widely used system in the world today. In [2] there is a detailed description of this system and the experimental work carried out by CASAGRANDE in order to formulate it.

The system classifies fine soils primarily on the basis of their plasticity characteristics, which have a reasonable correlation with some of the basic mechanical properties, as was mentioned in Chapter 1. In classifying coarse soils (with 50% of the particles larger than 0.075 mm, N° 200 sieve, or 0.003 in), greatest weight is given to the grain size distribution. In addition it reflects the plasticity characteristics of the fine particles. The maximum particle size of the soils included in this system is not clearly defined; because the test requires sieving many users arbitrarily select 7.6 cm (3 in) as the size limit. For the road engineer this system has the disadvantage that rock fragments of different sizes are not included in the original system.

2.2 "Soils" Classification System Used by the Mexican Ministry of Public Works (SOP)

For many years, Mexican civil engineers have used the Unified Soil Classification System, with satisfactory results: their familiarity with the system has induced them to make minor modifications to fit their practical needs. As a complement to the system, they have established a procedure for classifying rock fragments larger than 7.6 cm. The Ministry of Public Works (Secretaría de Obras Públicas), the official organization which is responsible for road engineering in Mexico, has created its own soils, rock fragments and rocks classfication system, [3].

For classification purposes, the materials which form the earth's crust fall into 3 groups: *soils*, *rock fragments* and *rocks*.

The term *soil* is applicable to all those particles smaller than 7.6 cm (3 in), the term *rock fragments* appies to fragments larger than 7.6 cm (3 in) which do not form part of a massive rock formation, and the term *rock* is used for reasonably continuous massive rock formations.

Soils are subdivided into fine-grained or *fines* and coarse-grained or *sands and gravels*. *Fines* are soils with particles smaller than 0.075 mm (N° 200 sieve) and *sands and gravels* are those with particles larger than 0.075 mm (N° 200 sieve) but smaller than the 7.6 cm (3 in) sieve. Fines include organic soils, silts and clays. Organic soils are those which contain a considerable amount of organic matter, a fine organic soil being silt or clay, depending on the plasticity characteristics, as will be described later. Soils in which organic matter predominates fall into the category of *peat.*

Coarse soils include the groups called sand and gravel. The dividing line between sand and gravel is 4.76 mm (N° 4 sieve).

Rock fragments are subdivided into *small*, *medium* and *large*. Small fragments are those which are larger than a 7.6 cm (3 in) sieve and with a maximum dimension smaller than 30 cm (1 ft). Medium fragments are those with a maximum dimension between 30 cm (1 ft) and 100 cm (3.28 ft). Large fragments are those with a maximum dimension bigger than 100 cm (3.28 ft).

Each of these large groups has a group symbol which can be one or more letters [2]. Table 2-1 gives a summary of the groups which together form the classification system of the Mexican Ministry of Public Works (SOP).

2.2.1 Unified Soil Classification System (SOP Version)

The basis of the Unified Soil Classification System is the original plasticity chart which is the result of laboratory research carried out by CASAGRANDE [1,4]. This research revealed that if the plasticity characteristics of soils are placed in a coordinate system, with the liquid limit on the abscissa and the plastic index on the ordinate, soils having similar hydraulic and mechanical properties form definite elongated groups. The soils of neighbouring groups have similar properties; those which are far apart have different ones. On the basis of this observation, CASAGRANDE subdivided the fine materials into groups with similar properties (by the *A* and *B*, lines). This graph or *plasticity chart*, as it is known, appears in Fig. 2-1 in the form in which it is used by the Mexican Ministry of Public Works. The original CASAGRANDE chart, also utilized in the Unified Classification System includes an additional subdivision, shown as a dotted line in Fig. 2-1.

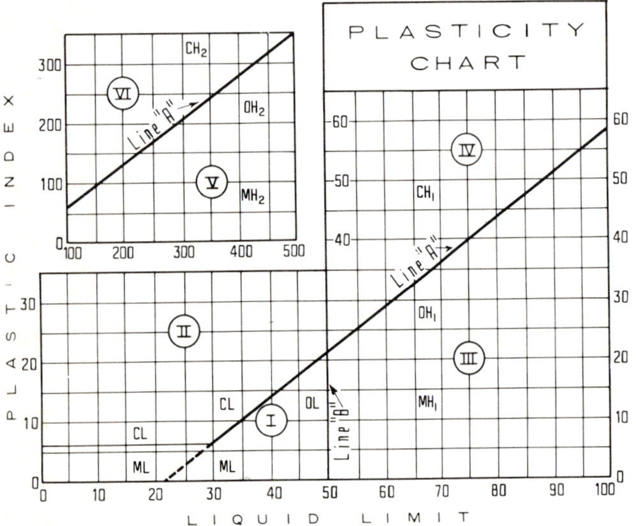

Fig. 2-1 Plasticity chart as used by the Mexican Ministry of Public Works

The Unified System in its original form and Mexican modification covers both coarse and fine soils. A soil is considered coarse if more than 50% of its particles by weight are coarser than 0.075 mm (N° 200 sieve), and fine if more than half its weight in particles are finer than 0.075 mm, (see Tables 2-1, 2-2, and 2-3). The different groups of coarse soils will be described first.

.1 Coarse Soils

The symbol belonging to each category is made up of two capital letters, which are the initials of the English names of the most typical soils in the group, as follows:

— Gravels and soils in which gravel by weight exceeds sand
— Sands and sandy soils in which sand by weight exceeds gravel

Gravels and sands are separated at 4.76 mm (N° 4 sieve) so that a soil belongs to the *G* group if more than 50% of the coarse particles (larger than N° 200 sieve) are coarser than N° 4 sieve, and to the *S* group if the contrary is true.

Gravels and sands are subdivided into four types:

— For a well-graded material which is practically free of fines, the symbol is *W* (well-graded). In combination with the group symbols, groups *GW* and *SW* are obtained.
— For a poorly-graded material which is practically free of fines, the symbol is *P* (poorly-graded). In combination with the group symbols, groups *GP* and *SP* are obtained.
— For a material containing a considerable quantity of non-plastic fines, the symbol is *M* (from the Swedish *mo* and *mjala*). In combination with the group symbols, groups *GM* and *SM* are obtained.
— For a material containing a considerable quantity of plastic fines, the symbol is *C* (clay). In combination with the group symbols, groups *GC* and *SC* are obtained.

The following description of the groups is intended to give more detailed identification criteria for field and laboratory.

Groups GW and SW: These are well-graded soils with few or no fines. The fines these groups may contain should neither produce noticeable changes in the friction and interlocking of the coarse fraction, nor reduce permeability. In practice, the above requirements are usually met if the fines content does not exceed 5% by weight. In the laboratory, grading is judged by means of the uniformity (C_u) and curvature (C_c) coefficients. For a well-graded gravel, the uniformity coefficient must be higher than 4, and the curvature coefficient must lie between 1 and 3. For well-graded sands, the uniformity coefficient must be higher than 6, and the curvature coefficient must also lie between 1 and 3.

Groups GP and SP: These soils are poorly-graded, either they are uniform or one or two size ranges predominate, with intermediate sizes being absent. In the laboratory tests they contain less than 5% fines, but they do not meet the grading requirements as well-graded. These groups include uniform gravels, such as those deposited on river-beds, and uniform sands from dunes and beaches which are found at different levels during excavation. They also include gap gravel; sands and gravels when they are mixtures of two or more uniform soils.

Groups GM and SM: In these groups the fines affect the strength, rigidity and permeability of the coarse particles. It has been observed in practice that this occurs when the fines are greater than 12% by weight, (this quantity has been adopted as the

Table 2-1

Classification of Rock Fragments and Soils

TYPE OF MATERIAL	GROUP	SUB-GROUP	SYMBOL	DIMENSIONS OF PARTICLES OR FRAGMENTS
SOILS	HIGHLY ORGANIC	PEAT	P_t	--------------------------------
	FINE	ORGANIC	*O*	< N° 200 sieve
		SILTS	*M*	< N° 200 sieve
		CLAYS	*C*	< N° 200 sieve
	COARSE	SANDS	*S*	> N° 200 and < N° 4
		GRAVELS	*G*	> N° 4 and < 7.6 cm (3 in)
ROCK FRAGMENTS		SMALL	*Fs*	> 7.6 cm (3 in) and < 30 cm (1 ft)
		MEDIUM	*Fm*	> 30 cm (1 ft) and < 100 cm (3.28 ft)
		LARGE	*Fl*	> 100 cm (3.28 ft)
ROCKS	IGNEOUS	EXTRUSIVE	*Rie*	--------------------------------
		INTRUSIVE	*Rii*	--------------------------------
	SEDIMENTARY	CLASTIC	*Rsc*	--------------------------------
		CHEMICAL	*Rsq*	--------------------------------
		ORGANIC	*Rso*	--------------------------------
	METAMORPHIC	FOLIATED	*Rmf*	--------------------------------
		NON-FOLIATED	*Rmn*	--------------------------------

lower dividing line for the fines content). The plasticity of the fines in these groups varies from *nil* to *moderate*. Therefore, on the plasticity chart, the fraction of the soil finer than N° 40 sieve either occurs below the *A* line, or the plasticity index must be lower than 6%. In his system, CASAGRANDE fixed this last figure at 4%. When the percentage of fines is between 5 and 12, a double symbol must be used. For example *GW-GM* indicates a well-graded gravel with non-plastic fines between 5 and 12% by weight.

Groups GC and SC: As for groups *GM* and *SM* the fines content of these groups must be greater than 12% by weight. The plasticity of the fines is moderate to high. The fraction which passes N° 40 sieve plots above the *A* Line on the plasticity chart, and in addition the plastic index is higher than 6% (7% in CASAGRANDE's original system).

If a material does not fall clearly into a specific group, double symbols are used. For example, the symbol *GW-SW* is used for a well-graded material with less than 5% fines and with a coarse fraction made up of equal proportions of gravel and sand.

.2 Fine Soils

The group symbols for fine grained soils are made up of two capital letters. The first letters denote three principal types:

— Inorganic silts, group symbol *M*
— Inorganic clays, group symbol *C* (clay)
— Organic silts and clays, group symbol *O* (organic)

Each of these three soil-types is subdivided into two groups, depending on its liquid limit. If this limit is less than 50%, (in other words if they are soils likely to have low or medium compressibility), the letter *L* (low compressibility) is added to the group symbol, creating groups *ML, CL* and *OL*. Fine soils with a liquid limit more than 50%, (high compressibility) take letter *H* (high compressibility) after the group symbol, thus creating groups *MH*, *CH* and *OH*.

It should be noted that letters *L* and *H* do not refer to low or high plasticity; this property is described by reference to two parameters (w_L and I_p), as indicated on the CASAGRANDE chart whereas the classification symbol depends on the liquid limit alone. Further, it has already been shown that the higher the liquid limit, the more compressible the soil is likely to be. It should also be recalled that compressibility, as used here, refers to the slope of the virgin portion of the compressibility curve and not to the preconsolidated range due to preloading or drying.

Highly organic, fibrous soils, such as highly compressible peats, fall into an independent category which takes the symbol P_t (from the English *peat*).

The groups of predominately fine grained soils are described in more detail, as follows:

Groups CL and CH: Inorganic clays fall into these categories. Group *CL* includes the area above the *A* line, defined by $w_L < 50$ and $I_p > 6$ ($I_p > 7$ in the system originally proposed by CASAGRANDE).

Group *CH* corresponds to the area above the *A* line, defined by $w_L > 50\%$. Into this category fall those clays which have been formed by the chemical decomposition of volcanic ashes, such as bentonite or the clay found in the Valley of Mexico, with liquid limits as high as 500.

Groups ML and MH: Group *ML* includes the area below the *A* line defined by $w_L < 50\%$ and the portion above the *A* line, with $I_p < 6\%$ ($I_p < 4\%$ in the original system). Group *MH* corresponds to the area below the *A* line defined by $w_L > 50\%$. These groups include typical inorganic silts and clayey silts. Common types of inorganic silts and rock dust with $w_L < 30\%$ fall into the *ML* category. Aeolian deposits such as loess with $25\% < w_L < 35\%$ usually appear in this group too.

An interesting type of fines in this category are the kaolin clays, which are derived from the feldspars in granitic rocks. They are an interesting contradiction because although the term clay is appropriate mineralogically, some of their engineering characteristics correspond to inorganic silts. For example, in a dry state their strength is relatively low and in a wet state they show a moderate to low reaction in the dilatancy test. However, their grain size and compressibility are comparable to those other typical clays that plot above the *A* line. Some kaolins are borderline, *ML-CL* and *MH-CH*, (because they plot very near this line, see Table 2-2). Nearly pure diatomaceous earths are seldom plastic, even though their liquid limits may exceed 100% *MH*. When mixed with other fine soils they fall into categories *ML* or *MH*.

Groups OL and OH: The areas of the plasticity chart corresponding to these two groups are the same as for groups *ML* and *MH* respectively. However organic soils are usually near the *A* line. Adding a little colloidal organic matter to an inorganic clay will increase its liquid limit without causing a noticeable change in the plastic index. This makes the soil move towards the right on the plasticity chart, to a position further away from the *A* line.

Group P_t: For most peaty soils ATTERBERG limit tests can be carried out after a complete remoulding. The liquid limit of these soils usually exceeds 300% and sometimes exceeds 500%, occupying a position well below the *A* Line on the plasticity chart. The plastic index typically lies between 100 and 200%. As with granular soils, when a fine material does not fall clearly into one of the categories, double borderline symbols are used. For example, *MH-CH* represents a fine soil with $w_L > 50\%$ and a plastic index which places the material practically on the *A* line.

The Unified Soil Classification System is not limited to placing the material within one of the above groups. It also includes its description, both remoulded and undisturbed. This description plays a leading part in the development of a healthy technical criterion and can sometimes be of vital importance in showing up certain characteristics which escape the mechanical performance tests being carried out. A typical example of this is density or compaction.

Generally speaking, in the case of coarse-grained soils, the following data must be provided: typical name that reflects approximate percentages of gravel and sand, maximum dimension, flatness, angularity, slabbiness and hardness of the particles, local and geological names, together with any further information that might be of interest, depending on the engineering purposes for which the material is to be used.

For undisturbed soils, added terms describe structure, fabric and stratification, consistency (strength) in undisturbed and remoulded conditions, moisture and drainage characteristics.

.3 Identification of Soils

The problem, of soil identification is of fundamental importance in engineering. To identify a soil is strictly speaking to locate it in a classification system such as one of the categories of the Unified Soil Classification System on the basis of its index characteristics. Identification of a soil makes it possible to estimate the mechanical and hydraulic properties based on the group to which it belongs. The usefulness of such classification depends on experience and the correlation of the mechanical and hydraulic properties with the classification. One of the advantages of the system is that it also provides criteria for field identification by simple hand examination when there is no laboratory apparatus available for precise identification tests.

Field Identification of Granular Soils: Those materials which are made up of large particles can be identified in the field on a practically visual basis. By spreading a dry sample of the soil on a flat surface, it is possible to make an approximate estimation of particle-size distribution, maximum particle size, shape and mineralogical composition. In order to distinguish between gravels and sands, the 0.5 cm (0.2 in) dimension can be used instead of N° 4 sieve, and to estimate the fines content it is sufficient to remember that particles corresponding in size to N° 200 sieve are the smallest ones which are visible to the naked eye.

Considerable experience is required to be able to differentiate between well and poorly-graded soils on sight. This experience can be obtained by comparing estimated gradings with those measured in the laboratory whenever the opportunity arises. In order to examine the fines of a soil, field identification tests, which will be described later, must be carried out on the part finer than 0.425 mm (N° 40 sieve). If this sieve is not available, particles can be separated by hand instead.

It is sometimes important to judge the physical integrity of the particles. Particles from sound igneous rocks can be easily identified. Weathered or decomposed particles can be recognized by their discoloration, loss of luster and the relative ease with which they crumble.

Field Identification of Fine Soils: One of the great advantages of the Unified Soil Classification System is the criteria proposed for the field identification of fine soils if the engineer has some experience. There is still no better way of acquiring this experience than to work beside an experienced person. If this is not possible, then it is advisable to make a systematic comparison of field and laboratory identification results whenever possible.

The principal basis for the identification of fine soils in the field is the dilatancy, toughness and dry strength crushing characteristics. The color and odor of the soil may help, especially in organic soils (see Identification Procedures in Table 2-2).

Dilatancy: clean, very fine sands show the fastest and most distinctive reaction, plastic clays show no reaction at all. Inorganic silts such as a typical rock flour show a rapid but moderate reaction. The speed with which the pat changes its consistency and the water appears and disappears defines the intensity of the reaction and denotes the nature of the fines content of the soil. A quick and distinctive reaction is typical of uniform, non-plastic, fine sands (*SP* and *SM*) and some inorganic silts (*ML*), especially of the rock flour type. It also occurs in the case of diatomaceous earth (*MH*). For less uniform soils, the reaction becomes less rapid. Small proportions of colloidal clay give the soil a certain degree of plasticity, one reason the reaction becomes much slower in these materials. The slow reaction is typical in inorganic and slightly plastic organic silts (*ML, OL*), very silty clays (*CL-ML*) and many kaolin-type clays (*ML, ML-CL, MH* and *MH-CH*). An extremely slow reaction or none at all is typical of clays located above the *A* line (*CL, CH*) and highly plastic organic clays.

The phenomenon of the appearance of water on the surface of the sample is due to the densification of silty soils, and even more so of sandy soils subjected to the dynamic impacts of one hand against the other. The reduction of the void ratio is accompanied by expulsion of the water. The subsequent squeezing process increases the void ratio and causing the water to disappear from the surface. Clayey soils do not react under dynamic loading and consequently show no reaction.

Toughness: the plasticity of the clay fraction is identified by the toughness of a small thread of soil as it approaches the plastic limit and by the stiffness of the sample when it finally crumbles in the fingers. A weak thread at the plastic limit and rapid crumbling of the sample just drier than the limit denote low-plasticity inorganic clay or materials such as kaolin-type clays. Organic clays are very weak and spongy at their plastic limit.

The higher a soil's position in relation to the *A* line (*CL, CH*), the stiffer and tougher the thread will be when nearing the plastic limit and the stiffer will be the sample when it is squeezed by the fingers just drier than the plastic limit. In soils which are slightly above the *A* line, such as glacial clays (*CL, CH*), the threads are moderately tough when nearing their plastic limit and the sample begins to crumble when squeezed just drier than the plastic limit. Almost all soils below the *A* line (*ML, MH, OL* and *OH*) produce threads which are not very tough near the plastic limit. For organic and micaceous soils which are well below the *A* line, the threads are very weak and spongy; and for all soils below the *A* line, (except the *OH* ones which are near it) the soil is loose and crumbly when kneaded with the fingers if the moisture content is slightly lower than the plastic limit.

If the atmospheric humidity is almost constant, the time required to reach the plastic limit by rolling the soil into a thread is a rough measurement of the plastic index of the soil. For example, a *CH* clay with $w_L = 70\%$ and $I_p = 50\%$ or an *OH* with $w_L = 100\%$ and $I_p = 50\%$ will have to be kneaded and rolled for a longer time than a glacial clay of the *CL* type before the plastic limit is reached. For silts of the *ML* group, the plastic limit is reached very rapidly. (All the tests will obviously have to start with the soils at approximately the same consistency, preferably near the liquid limit, for consistent meaning).

Dry strength (crushing characteristics): great strength when dry is a characteristic of *CH* group clays. An inorganic silt shows very little strength, but is recognizable to the touch when the dry specimen is powdered. Fine sand feels granular, whereas a typical silt has a soft feel, like flour.

Non-plastic *ML* or *MH* silts show practically no strength when dry. Samples crumble at the slightest pressure of the fingers. Rock dust and diatomaceous earth are typical examples. Little strength when dry, is characteristic of all low-plasticity soils below the *A* line and of some very silty inorganic clays that are slightly above the *A* line, *CL*. Moderate strength generally denotes *CL* clays and sometimes those belonging to groups *CH, MH* (kaolin-type clays) or *OH* which are very near the *A* line. Most *CH* clavs, however, exhibit great dry strength — likewise those *CL* .ays which are well above the *A* line. *OH* materials which have high liquid limits and are near the *A* line also show great strength. Inorganic *CH* clays very high above the *A* line are known for their great strength when dry.

Color: in field studies the color of the soil is usually useful for distinguishing between different layers and when local experience is available, for identifying specific strata. In addition, there are also more specific characteristics that can be identified, for example, black and other dark colors suggest colloidal organic matter. Light, bright colors are more characteristic of inorganic soils. Reds and yellows indicate oxidation of iron minerals and possibly iron cementing.

Odor: organic soils (*OH* and *OL*) usually have a distinctive odor which helps identification. The odor is particularly strong if the soil is damp, becoming even stronger if the damp sample is heated, and less noticeable on exposure to air.

2.2.2 Classification of Rock Fragments

Rock fragments are all those rock pieces bigger than 7.6 cm (3 in) which do not belong to a rock formation.

.1 Division of the Fragments

Rock fragments are subdivided as follows:

Small fragments *Fs*. Those with two dimensions larger than 7.6 cm (3 in) and one maximum dimension of 30 cm (1 ft).

Medium-sized fragments *Fm*. Those with a maximum dimension between 30 cm (1 ft) and 100 cm (3.28 ft).

Large fragments *Fl*. Those with a maximum dimension greater than 100 cm (3.28 ft).

.2 General Characteristics

When describing these materials, the following characteristics must be mentioned: mineralogical or petrographic classification, fragment size distribution, maximum size of the fragments, shape including flatness and elongation surface characteristics, degree of weathering and any other descriptive information which may be relevant.

When referring to *in-situ* materials, information must be added about their structure, stratification, denseness, cementation, water content and drainage characteristics. A description of these characteristics together with the adjectives which should be used to specify each one of them follows.

Petrographic Classification: When possible, petrographic classification must be carried out. This involves identifying the mineral present, particle size and internal arrangement or fabric (see Section 2 ... 3).

Fragment Size Distribution: When establishing the fragment size distribution, it must be made clear whether the material in question has a uniform size or if there are different sizes, i.e., whether the material should be considered *poorly-graded* or *well-graded*. Similar criteria apply to those used for field soil identification procedures described previously for coarse-grained soils. Maximum fragment size must also be given.

Shape: The shape of the fragments will be referred to as *elongated* if its length exceeds 3 times its thickness, *slab-like* if its length and width exceed 3 times its thickness and *equidimensional* when the three dimensions have the same order of magnitude. This shape should be further described as *angular* when the fragments have sharp edges; *subangular* when the edges are slightly rounded, *subrounded* when the edges are moderately rounded and *rounded* when they are practically spherical.

Surface Characteristics: Surface characteristics are described as follows: *smooth, slightly rough, moderately rough* and *very rough*.

Degree of Weathering: The degree of weathering is described in the following terms: *intact*, *slightly weathered*, *moderately weathered* and *very weathered*. The degree of weathering can be judged by the following characteristics: loss of luster, local changes in color and the dull sound when hit with a hammer. Some fragments which appear to be sound when they have just been extracted should be exposed to weathering for a time. Disintegration may occur if they are not absolutely intact. However, intact shales often split apart in layers, although the undisturbed rock is unweathered.

Fabric (or Structure): The term structure as used here refers to the way in which the different fragments are arranged. Structure is important from the point of view of the mechanical behavior of any rock fragment deposit. Either the fragments are in close contact or they are separated by soil. In the latter case, the mechanical behavior of the whole is largely determined by the properties of the soil which separates the fragments. In a deposit made up of rock fragments and soil, extreme cases which may arise are: 1) a deposit where all the fragments are sound and in contact with one another, thus forming a simple structure, where the fine soil only partially fills the spaces in this simple structure, or 2) a deposit which is predominantly composed of fragments and where the individual pieces are separated by small amounts of soil. Obviously when loaded in undrained conditions, material 1) will behave as though it were *purely frictional*, whereas the mechanical behavior of material 2) will be closer to that of a fine soil.

Stratification: is described by stating the thickness of the individual strata, the type of material of which they are made and their dip and strike. When a material is not layered, this must be clearly stated.

Denseness: is judged in the following terms: *very loose*, *loose*, *fairly dense*, *dense* and *very dense*.

Cementation: between fragments should be referred to as: *non-existent*, *slight*, *moderate* and *high*, according to the force required to separate the fragments. Where possible, it must be indicated whether cementation is due to carbonates, sulfates, silicates, aluminates or iron oxides, or dried clay. It goes without saying that high cementation borders on what might be called a sedimentary rock. Estimation of the degree of cementation should be made both on a dry sample and one which has been submerged in water for a day.

Water Content: is described in the following terms: *dry*, *damp*, *wet* and *saturated*.

Drainage Characteristics: of a deposit refers to the ease with which a deposit of a given material can drain itself should it become saturated. Adjectives used to describe drainage characteristics are: *very poor*, *poor*, *fair* and *good*. These characteristics depend not only on the hydraulic and capillary relations of the component materials, but also on the topography and the nature of the surrounding geological formations.

Table 2-2

Unified Soil Classification System (S.O.P. Version)
(for nomenclature and procedural notes see Table 2-2b)

<table>
<tr><th colspan="4">SOIL CHARACTERISTICS</th><th>LABORATORY CLASSIFICATION
(Based on grain-size curve to quantify fractions of soil)</th><th>GROUP SYMBOL</th><th>TYPICAL DESCRIPTIVE NAMES</th></tr>
<tr><td rowspan="8">COARSE — GRAINED SOILS
(More than 50% retained by sieve No. 200)</td><td rowspan="4">GRAVELS
(> 50% of coarse fraction retained by sieve No. 4)</td><td rowspan="2">Clean Gravel
< 5% passing sieve No. 200</td><td rowspan="8">Borderline cases between 5% and 12% require double symbols</td><td>C_u greater than 4; C_u between 1 and 3</td><td>GW</td><td>Well graded gravels, gravel and sand mixtures, little or no fines</td></tr>
<tr><td>Not satisfying all requirements for GW</td><td>GP</td><td>Poorly graded gravels, gravel-sand mixtures, little or no fines</td></tr>
<tr><td rowspan="2">Gravels with Fines
> 12% passing sieve No. 200</td><td>ATTERBERG limit below "A" Line or I_p less than 6</td><td>GM</td><td>Silty gravels, poorly graded gravel-sand-silt mixtures</td></tr>
<tr><td>ATTERBERG limit above "A" Line with I_p greater than 6</td><td>GC</td><td>Clayey gravels, poorly graded gravel-sand-clay mixtures</td></tr>
<tr><td rowspan="4">SANDS
(> 50% of coarse fraction passing sieve No. 4)</td><td rowspan="2">Clean Sand
< 5% passing sieve No. 200</td><td>C_u greater than 6; C_c between 1 and 3</td><td>SW</td><td>Well graded sands, gravelly sands, little or no fines</td></tr>
<tr><td>Not satisfying all requirements for SW</td><td>SP</td><td>Poorly graded sands, gravelly sands, little or no fines</td></tr>
<tr><td rowspan="2">Sand with Fines
> 12% passing sieve No. 200</td><td>ATTERBERG limit below "A" Line or I_p less than 6</td><td>SM</td><td>Silty sands, poorly graded sand-silt mixtures</td></tr>
<tr><td>ATTERBERG limit above "A" Line with I_p greater than 6</td><td>SC</td><td>Clayey sands, poorly graded sand-clay mixtures</td></tr>
<tr><td rowspan="6">FINE — GRAINED SOILS
(More than 50% passing sieve No. 200)</td><td rowspan="6">SILTS AND CLAYS</td><td rowspan="3" colspan="2">Liquid limit < 50</td><td rowspan="6">Classification of soils on the basis of liquid limit and plasticity index
IN SOILS WITH THE SAME LIQUID LIMIT, TENACITY AND DRY STRENGTH INCREASE WITH PLASTIC INDEX
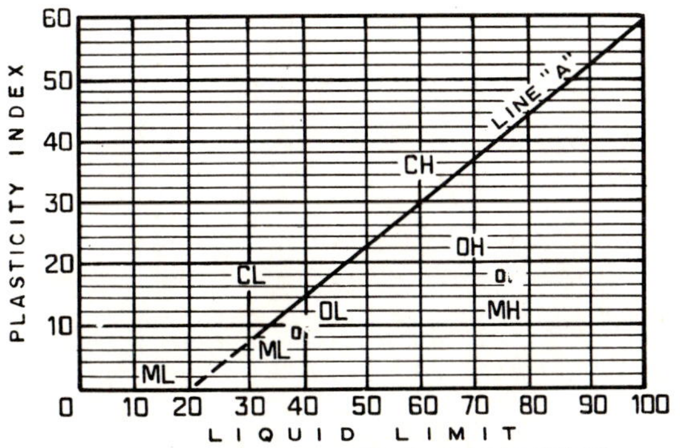

PLASTICITY CHART FOR CLASSIFICATION OF FINE SOILS IN LABORATORY</td><td>ML</td><td>Inorganic silts, rock flour, sandy silts and clayey silts with slight plasticity</td></tr>
<tr><td>CL</td><td>Inorganic clays of low to medium plasticity, gravelly clays, sandy clays, silty clays, lean clays</td></tr>
<tr><td>OL</td><td>Organic silts and silty clays of low plasticity</td></tr>
<tr><td rowspan="3" colspan="2">Liquid limit > 50</td><td>MH</td><td>Inorganic silts, micaceous or diatomaceous, elastic silts</td></tr>
<tr><td>CH</td><td>Inorganic clays of high plasticity, fat clays</td></tr>
<tr><td>OH</td><td>Organic clays of medium to high plasticity, organic silts of medium plasticity</td></tr>
<tr><td colspan="5">Highly Organic Soils</td><td>P_t</td><td>Peat and other highly organic soils</td></tr>
</table>

Table 2-2 (Cont.)

INFORMATION REQUIRED FOR DESCRIBING SOILS	FIELD IDENTIFICATION PROCEDURES (basing fractions on estimated weights)			GROUP SYMBOL
Give typical name; indicate approximate percentages of sand and gravel; maximum size, angularity; surface condition and hardness of the coarse grains; local or geological name and other pertinent descriptive information; symbols in parentheses.	Wide range in grain size and substantial amounts of all intermediate particle sizes			*G W*
	Predominantly one size or range of sizes with some intermediate sizes missing			*G P*
For undisturbed soils add information on stratification, degree of compactness, cementation, moisture conditions and drainage characteristics.	Little or non-plastic fines (for identification procedures see *ML* below)			*G M*
	Plastic fines (for identification procedures see *CL* below)			*G C*
Example: Silty sand, gravelly; about 20% hard, angular gravel particles, 1.5 cm (0.6 in) maximum size; rounded and subangular sand, grains coarse to fine; about 15% non-plastic fines, with low dry strength; well compacted and moist in place; alluvial sand; (*SM*).	Wide range in grain sizes and substantial amounts of all intermediate particle sizes			*S W*
	Predominantly one size or range of sizes with some intermediate sizes missing			*S P*
	Little, or non-plastic, fines (for identification procedures see *ML* below)			*S M*
	Plastic fines (for identification procedures see *CL* below)			*S C*
	DRY STRENGTH	DILATANCY	TOUGHNESS	
Give typical name; indicate degree and character of plasticity; amount and size of coarse grains; color in wet condition, odor if any; local or geological name and other pertinent descriptive information; symbol in parentheses.	None to slight	Quick to slow	None	*M L*
	Medium to high	None to very slow	Medium	*C L*
For undisturbed soils add information on structure, stratification, consistency in undisturbed and remolded states, water content and drainage conditions.	Slight to medium	Slow	Slight	*O L*
	Slight to medium	Slow to none	Slight to medium	*M H*
Example: Clayey silt, brown, slightly plastic; small percentage of fine sand; numerous vertical root holes; firm and dry in place; loess; (*ML*).	High to very high	None	High	*C H*
	Medium to high	None to very slow	Slight to medium	*O H*
	Readily identified by color, odor, spongy feel and frequently by fibrous texture			P_t

Table 2-2b

Nomenclature and Procedural Notes for Soil Classification (U.S.C.S.)

Meaning of group symbols

G = Gravel S = Sand M = Silt C = Clay O = Organic P_t = Peat
W = Well graded P = Poorly graded
L = Low compressibility H = High compressibility

Borderline classification: Soils possessing characteristics of two groups are designated by combinations of group symbols, e.g.: $GW - GC$, well graded gravel-sand mixture with clay binder.

Laboratory identification

Coefficient of uniformity = $C_u = \dfrac{D_{60}}{D_{10}}$, Eq. (1-18)

Coefficient of curvature = $C_c = \dfrac{(D_{30})^2}{D_{10} \cdot D_{60}}$, Eq. (1-19)

Field identification

Visual estimation of grain-size fractions:
- — Remove all particles larger than 7.6 cm (3 in)
- — Grains of 0.074 mm dia (200 mesh) are approximately the smallest visible to the naked eye
- — Sieve No. 4 corresponds to approximately 0.5 cm (0.2 in)
- — Sieve No. 40 corresponds to approximately 0.425 mm (0.016 in)

Field identification for fine-grained soils or fractions — carried out on the minus No. 40 sieve size:

— Dilatancy (reaction to shaking): after removing particles larger than No. 40 sieve size, prepare a pat of moist soil with a volume of about 10 cm^3 (0.6 in^3); add enough water to make the soil soft but not sticky. Place the pat in the open palm of one hand and shake horizontally, striking vigorously against the other hand several times. A positive reaction consists of the appearance of water on the surface of the pat which changes to a livery consistency and becomes glossy. When the sample is squeezed between the fingers the water and gloss disappear from the surface, the pat stiffens and finally it cracks or crumbles. The rapidity of appearance of water during shaking and its disappearance during squeezing assist in identifying the character of the fines in a soil.

— Dry strength (crushing characteristics): after removing particles larger than No. 40 sieve size, mold a pat of soil to the consistency of putty adding water if necessary. Allow the pat to dry completely by oven, sun or air drying and then test its strength by breaking and crumbling between the fingers. This strength is a measure of the character and quantity of the colloidal fraction contained in the soil. The dry strength increases with increasing plasticity.

— Toughness (consistency near the plastic limit): After removing particles larger than the No. 40 sieve size, a specimen of soil of about 10 cm^3 (0.6in^3) in size, is molded to the consistency of putty. If too dry, water is added and if too sticky, the specimen is spread out in a thin layer and allowed to lose moisture by evaporation. The specimen is rolled out by hand on a smooth surface or between the palms into a thread of about 3 mm (0.12 in) in diameter. The thread is then folded and re-rolled repeatedly. During this manipulation the moisture content is gradually reduced and the specimen stiffens, finally loses its plasticity and crumbles when the plastic limit is reached. After the thread crumbles, the pieces should be lumped together and a slight kneading action continued until the lump crumbles.

All sieve sizes given are U.S. Standard

Table 2-3

Flow diagram for classification of soils according to the Unified Soil Classification System (S.O.P. version)

Auxiliary procedure for laboratory identification of soils
Unified Soil Classification System (SOP version)

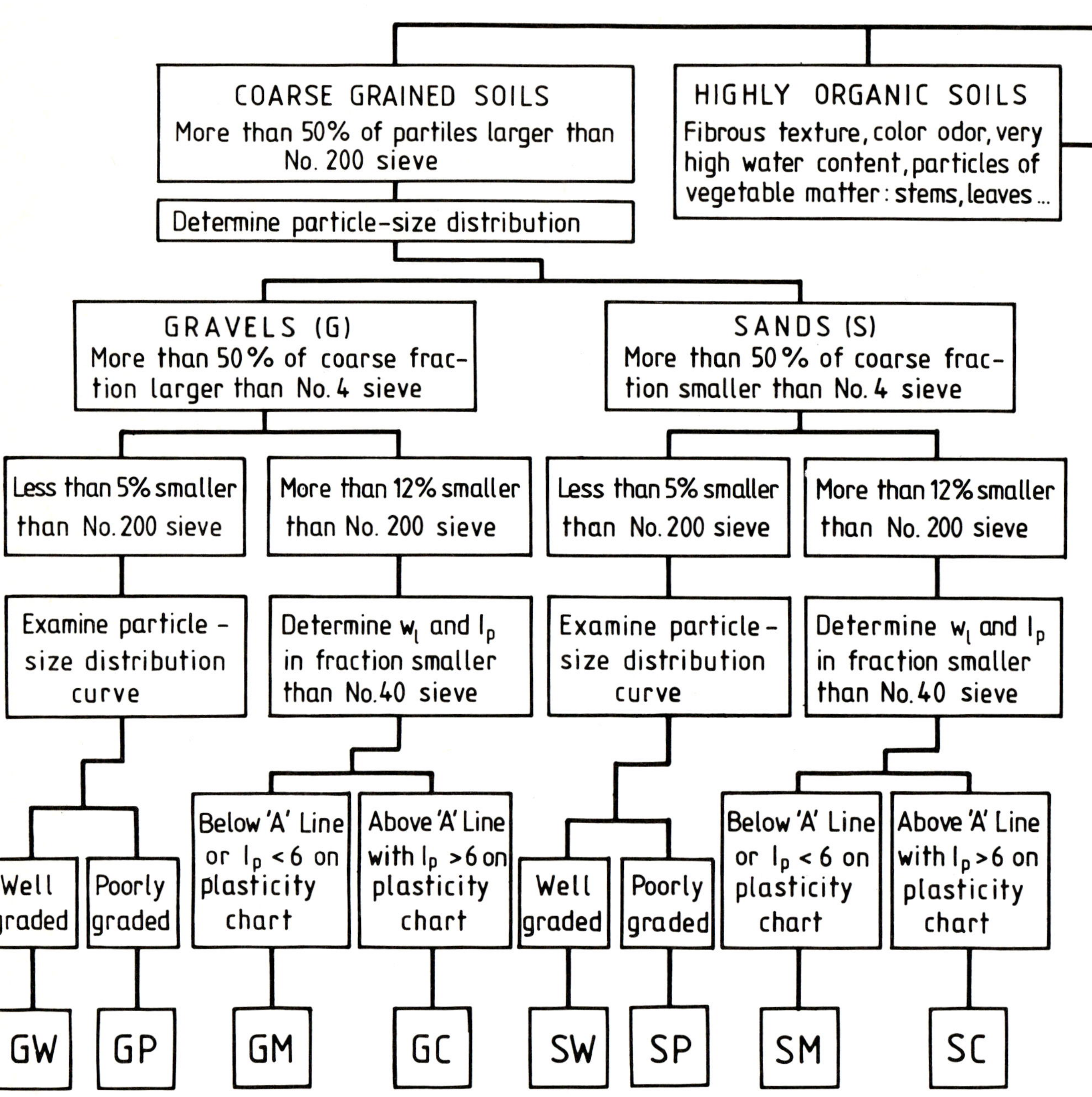

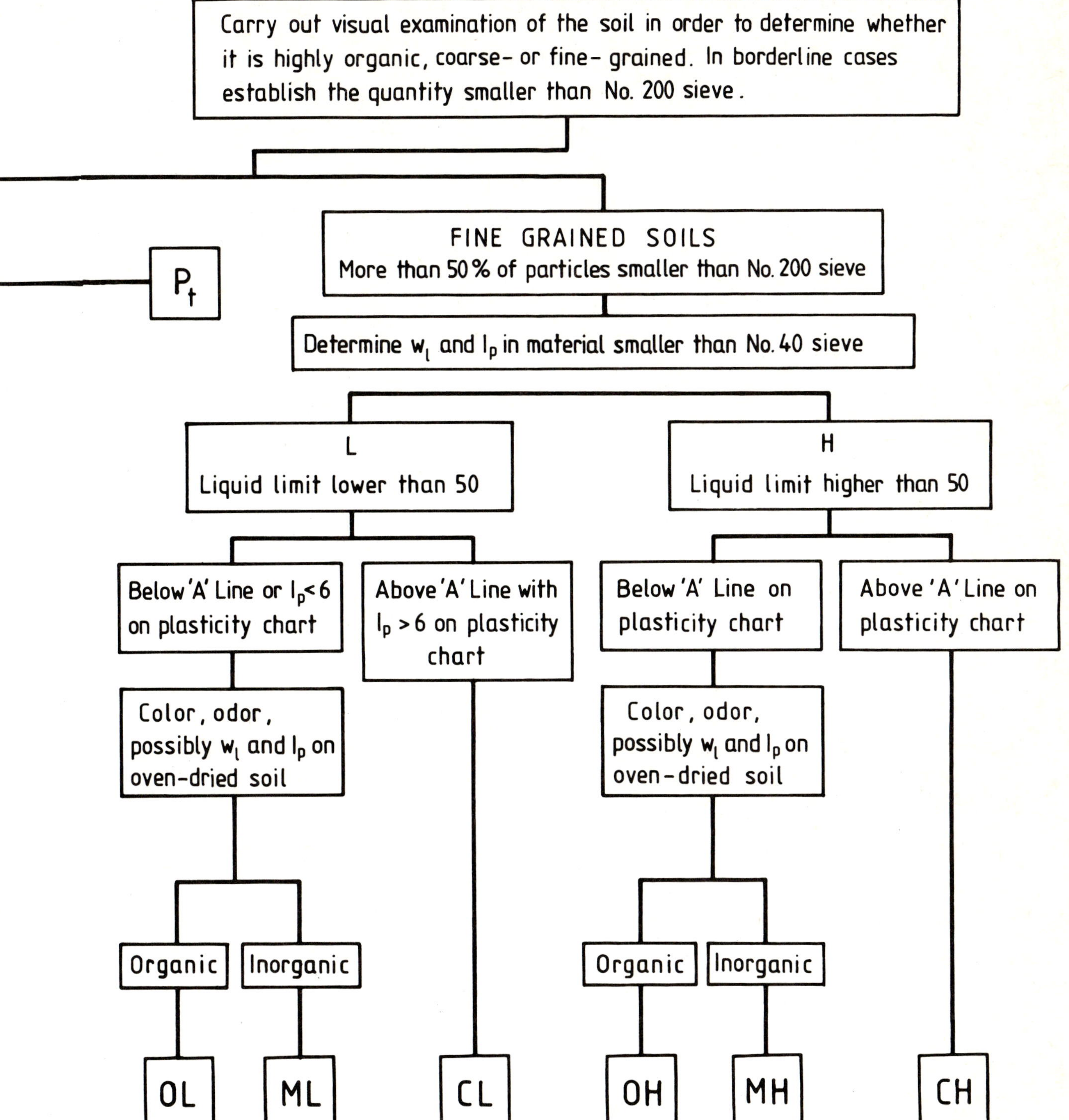
Carry out visual examination of the soil in order to determine whether it is highly organic, coarse- or fine-grained. In borderline cases establish the quantity smaller than No. 200 sieve.
FINE GRAINED SOILS
More than 50% of particles smaller than No. 200 sieve
P_t
Determine w_l and I_p in material smaller than No. 40 sieve
L
Liquid limit lower than 50
H
Liquid limit higher than 50
Below 'A' Line or I_p<6 on plasticity chart
Above 'A' Line with I_p >6 on plasticity chart
Below 'A' Line on plasticity chart
Above 'A' Line on plasticity chart
Color, odor, possibly w_l and I_p on oven-dried soil
Color, odor, possibly w_l and I_p on oven-dried soil
Organic
Inorganic
Organic
Inorganic
OL
ML
CL
OH
MH
CH

2.3 Classification of Rocks

2.3.1 General Guidelines for the Classification of Rocks

The classification of rocks which appears in this chapter (Tables 2-4, 2-5 and 2-6) is based on those characteristics which can be observed in the field without the help of a microscope. Consequently, in order to classify a rock, the principal factors to be considered are individual mineralogical composition of the rock and its texture.

In Table 2-7, a list is given of the principal mineral components of rocks and their most important physical characteristics, which will help identify them. A description is also given in the text of the most common textures and types of rock.

The following procedure is recommended when classifying a rock:

— Once a sample of the rock has been carefully examined, the following three fundamental aspects must be defined in this order: texture, mineral components, and group to which it belongs (igneous, sedimentary or metamorphic).
— Once these facts have been established, Table 2-4, 2-5, or 2-6 should be consulted in order to determine the kind of rock being dealt with.
— Once the name of the rock has been determined, the corresponding description is consulted in order to prove its identification.

2.3.2 Texture of Rocks

The texture of a rock depends on the arrangement, orientation, size, shape and interlocking of its component minerals, which can be seen with the naked eye or with the help of a hand-lens. This excludes textures that can only be seen under a microscope through a thin section of rock.

.1 The Most Common Textures of Igneous Rocks

The differences in the degree of crystallization and the size of the crystals determine the texture of an igneous rock. Both factors depend principally on the cooling rate, although the chemical constitution of the magma, particularly the volatile materials, is also involved.

Pyroclastic: These are composed of particles of volcanic glass, small pieces of pumice and fragments of rock cemented or fused together. The particles of glass and the pumice may be altered partially into clays. Pyroclastic rocks are the product of violent volcanic explosions.

Vitreous: These rocks are essentially volcanic glass, and are non-crystalline. Small phenocrysts of feldspar or other minerals may be scattered throughout the glass.

Aphanitic: Composed principally of minute crystals, smaller than 0.5 mm (0.02 in), with or without vitreous residue between the crystals. Even if the crystals can be seen with the naked eye, they cannot be identified without the aid of a microscope. They give the rock a stony or opaque luster in contrast to the glassy luster of vitreous-textured rocks. Most lava flows have an aphanitic texture. In some cases, the flow has aligned the small mineral grains giving the rock a banded or laminated appearance.

Phaneritic: Composed of crystals which are big enough to be seen and identified without the help of a lens or microscope. The range in average grain-size is between 0.5 mm (0.02 in) and several centimeters. Common phaneritic rocks, such as granite, have grains with an average size of 3 to 5 mm (0.12 to 0.2 in).

Porphyritic: These are composed of minerals of two greatly different sizes, which give the rock a mottled appearance. The texture has been attributed to a change in the cooling rate while the magma was crystallizing. The process of formation can be explained by imagining a large body of magma which slowly cools until it reaches a temperature at which one or more of the mineral components crystallize. Because the cooling rate is slow, these specific mineral crystals grow to a considerable size. Then, when the magma is partially crystallized, it is expelled with its suspended crystals forming a lava flow on the ground surface. The fluid part of the magma cools rapidly on the surface surrounding the big crystals (known as phenocrysts) leaving them embedded in an aphanitic matrix. The phenocrysts are formed underground; the aphanitic matrix on the surface. Such a lava has a *porphyritic* texture. Rocks with a prophyritic texture are common amongst intrusive bodies near the surface. Vitreous porphyritic texture appears in some lava flows and in the pumice fragments of pyroclastic rocks. Conditions other than a change in the cooling rate can seldom produce porphyritic rocks.

.2 Common Textures of Sedimentary Rocks

The differences between the nature of the component particles and the way in which they are joined, determines the texture of a sedimentary rock.

Clastic: Composed of rock fragments and mineral particles or shells which have been cemented together into a whole. Further distinctions are made as to the size of the particles and the degree of roundness of the individual fragments.

Organic: Composed of accumulated organic detritus (shells, residues, bones, etc.) in which the individual organic particles are so well preserved (neither broken nor notably worn) that the organic features predominate in the texture of the rock.

Crystalline: Composed of crystals which have precipitated from solutions and grown with a mutual interpenetration. The rock owes its cohesiveness to this interlocking of crystals and not to a cementing agent, as is the case in clastic and organic textures.

.3 Common Textures of Metamorphic Rocks

The differences in the orientation and alignment of the crystals and their sizes determine the texture of a metamorphic rock. There are two general texture groups:

— *Foliated*, in which the laminar minerals or those in sheet form, such as mica and clorite, are almost all in parallel lines in such a way that the rock can be easily split along nearly parallel, well oriented, cleavage planes of its mineral particle components.
— *Non-foliated,* composed either of equidimensional minerals or laminar minerals with random orientation, which causes the rock to break into angular particles.

Gneissic. Roughly foliated; the thickness of the individual sheets is 1 mm (0.04 in) or more, sometimes reaching several centimeters. The sheets can be straight flat, undulated or serrated. They usually differ in composition. For example, feldspars may alternate with dark minerals. The individual mineral grains are not easy to identify.

Table 2-4a
Composition and Classification of Igneous Rocks

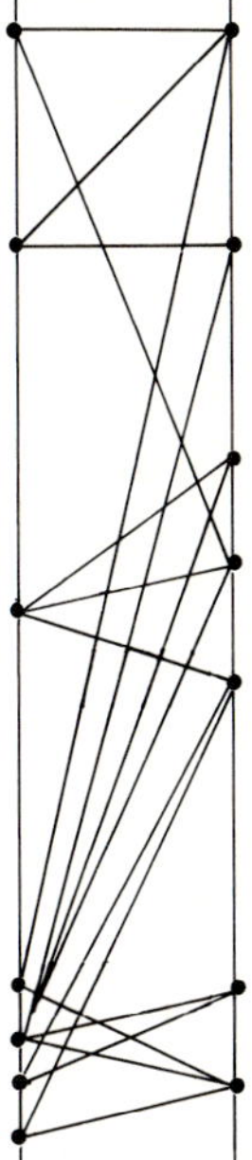

PRINCIPAL ROCK FORMING MINERALS	ROCKS: EXTRUSIVE fine-grained or porphyritic)	ROCKS: INTRUSIVE (coarse-grained)	
QUARTZ SiO_2	RHYOLITE	GRANITE	LIGHT ROCKS
POTASSIUM FELDSPAR (Orthoclase) $K.AlSi_3O_8$	TRACHYTE	SYENITE	LIGHT ROCKS
	ANDESITE	DIORITE	DARK ROCKS**
SODIUM-CALCIUM FELDSPAR (Plagioclases) $Ca.Al_2Si_2O_8$ $K.AlSi_3O_8$	DACITE	GRANODIORITE	DARK ROCKS**
	BASALT	GABBRO	DARK ROCKS**
Ferro-magnesian Silicates:			DARK ROCKS**
Micas (Muscovite Biotite*) Hornblende	AUGITE	PYROXENITE	DARK ROCKS**
Pyroxene Olivine	LIMBURGITE	PERIDOTITE	DARK ROCKS**

*Included here only because vitreous.
**Basic rocks

Table 2-4b
Composition and Classification of Igneous Rocks

ORIGIN	NATURE	ROCK	
QUIET ERUPTIONS	VITREOUS	OBSIDIAN PERLITE PUMICE STONE* RETINITE (Fish stone)	
EXPLOSIVE ERUPTIONS	PYROCLASTIC (FRAGMENTARY)	PUMICE STONE BLOCKS BOMBS	
		Loose	Consolidated
		GRAVELS	BRECCIAS (AGGLOMERATES)
		PUMICEOUS SANDS	TUFFS (AGGLOMERATES)
		SANDS	SANDSTONES (AGGLOMERATES)
		VOLCANIC ASHES	TUFFS
		VOLCANIC DUSTS	TUFFS

Table 2-5a
Composition and Classification of Sedimentary Rocks

ORIGIN	TRANSPORTING AGENTS	LOOSE SEDIMENTS	CONSOLIDATED SEDIMENTS
MECHANICAL	WATER	GRAVEL (ROUNDED	CONGLOMERATE
		GRAVEL (ANGULAR)	BRECCIA
		SAND	SANDSTONE
		SILT	SILTSTONE
		CLAY	CLAYSTONE
	WIND	DUNES	SANDSTONE
		LOESS	
	ICE	ANGULAR GRAVELS	
		SAND	
		SILT	TILLITE
		CLAY	
	GRAVITY	ANGULAR GRAVEL	TALUS BRECCIA

Schistose. Very fine flakes which form thin parallel bands along which the rock can be easily split. The individual minerals look different. They are mainly planar or elongated, especially mica, chlorite and amphibole. Some equidimensional minerals such as feldspar, garnet and pyroxene may be present, but not in large quantities.

Slaty. Very fine foliation in flat, almost invariably parallel, flakes which can be easily separated because of the almost perfect parallelism of microscopic and ultramicroscopic crystals of laminar minerals, principally mica.

Granoblastic. Non-foliated or very slightly foliated. Composed of mineral grains which mutually interpenetrate one another and which crystallized simultaneously. The minerals are large enough to be easily identified without the use of a microscope and are equidimensional, such as feldspar, quartz, garnet and pyroxene.

Felsitic Corneous. Non-foliated. The mineral grains are usually microscopic or ultramicroscopic, although a few may be visible to the naked eye. They break into very sharp-edged fragments with conchoidal fracture (shell-shaped).

2.3.3 Rock Formations

Reference to the morphological and megascopical features of rocks caused by cavities, deformities or discontinuities, as well as their bedding and texture are required to fully identify rock formations in place.

When the minerals which form an igneous rock appear in the form of large crystals visible to the naked eye, as in the case of a granite or a diorite, then it is almost certainly intrusive.

When the igneous rock has a glassy, aphanitic or porphyritic texture, it is almost certainly extrusive. These rocks sometimes present a rhyolitic structure in the field. Basalts often present a rough or ropy structure; rhyolites have the fluid-like appearance which has given them their name, and this can be observed even in the hand sample, for the crystals (especially the quartz ones) are aligned in the direction of the flow. At other times, rhyolite, andesite and dacite occur in different colored bands, which proves that they originated from a single lava flow.

Table 2-5b

Composition and Classification of Sedimentary Rocks

ORIGIN	NATURE	CONSOLIDATED SEDIMENTS
CHEMICAL	CALCAREOUS	LIMESTONE
		DOLOMITE
		ARAGONITE
		TRAVERTINE
	CLAYEY CALCAREOUS	MARL
	SILICEOUS	FLINT
		GEYSERITE
	SALINE	EVAPORITES:
		ROCK SALT
		GYPSUM
		BORAX
ORGANIC	CALCAREOUS	LIMESTONE
		CORAL
		COQUINA
		CRETE
	SILICEOUS	DIATOMITE
	CARBONACEOUS	PEAT
		LIGNITE
		COAL
		ANTHRACITE

Table 2-6

Composition and Classification of Metamorphic Rocks

ORIGINAL ROCK	METAMORPHIC PRODUCT
SANDSTONE	QUARTZITE
LIMESTONE	MARBLE
CLAY SHALE	SLATE
BASIC ROCKS	SCHISTS, SERPENTINE, ETC.
GRANITE, DIORITE AND CONGLOMERATE	GNEISS

Sedimentary rocks with a clastic texture are not difficult to identify when their grains are coarse, apart from sandstones. A very coarse-grained sandstone can sometimes be confused with a fine conglomerate; (in this case, it is referred to as a conglomerate-sandstone).

Identification of fine-grained sedimentary rocks with a clastic texture is difficult. Of course, the majority of this type of rocks are deposited in thin layers (or strata) but there are some, such as marly limestones, which appear in thick beds. When subjected to large stresses, both undergo changes in their original position, which should be horizontal, and may adopt inclined positions and simple or complex folds which are sometimes observable even in hand samples. In this last case, not only the texture of the rock should be taken into consideration, but also the structure. This is important as there may be cases where the size of the hand samples cannot give an idea of the thickness of the layer of rock nor its variation in structure in the field.

Some limestones of organic origin occur in thick banks or large masses (corals) and this cannot be appreciated in a hand sample.

Diatomite is made up of light colored shells so small that they can only be seen under a powerful microscope, but when this powder is rubbed between the fingers very close to the ear, it produces a slight sound which distinguishes it from kaolin; it is also different to the touch. A little water immediately shows that diatomite is not plastic.

Clay shales (hardened muds) include not only clayey and silty formations, but also marls. When exposed to weathering, these rocks disintegrate rapidly; first they crack, then they break and disintegrate with the consequent formation of highly plastic soils. The behavior of *choy* in the north of the Mexican Republic should be recalled.

It should be specified that marls are included here amongst clay shales in general, as their appearance is the same when they occur in not very thick layers. They can be easily distinguished from one another with hydrochloric acid, as it causes effervescence only in the marls because of their calcium carbonate content.

When dealing with foliated metamorphic rocks, structure plays an important part in identification. Gneiss occurs in the form of belts or bands which separate the component minerals; there are some bands of quartz, some of mica, some of feldspars, etc. If the sample under examination is broken parallel to one of these bands, it will have the corresponding texture. This is especially so in schists, whose most noticeable feature is the predominating mineral component (chlorite, mica, etc.), whereas their structure is almost always very folded, more so than in gneiss, and may pass unnoticed.

The property slates have, of splitting into sometimes quite broad leaves or sheets in a direction which is independent of the stratification planes, is called *fissility*.

One of the non-foliated metamorphic rocks that is worthy of mention is marble, where the calcium carbonate of the limestones from which it originates is turned into calcite with crystals of all different sizes, from very big ones to very small ones. Large-crystal marble is crushed and used in the manufacture of artificial granite. The fine-grained species, which can be either white or colored, is used for ornamental purposes. Black marble contains graphite and comes from limestones which originally contained finely divided coal. Hand samples of these black limestones have sometimes been mistaken for basalt, although they are easily scratched by a knife.

When handling calcium carbonates, or rocks containing them, the application of hydrochloric acid causes them to give off the carbon dioxide that forms them; this occurs as an effervescence; the stronger the concentration of acid, and of course the concentration of carbonates in the rock, the greater the effervescence. A 10% concentration is sufficient, and it has the advantage that when dealing with a double carbonate of calcium and magnesium (Dolomite) or dolomitic limestones, these rocks do not effervesce at all, or only very slightly, with diluted acid. When a little bit of powder is scraped from the sample with a pocket knife, effervescence is obtained, and consequently identification is made.

2.4 Common Rocks

2.4.1 Common Igneous Rocks

.1 Extrusive Pyroclastic Rocks

Tuff: is a fine-grained pyroclastic rock made up of pieces smaller than 5 mm (0.2 in). The majority of the fragments are broken pieces of volcanic glass and fragments of solidified, often porous, lava. Other common components are crystals, either microscopic splinters called scales, or frothy fragments known as pumice. Particles of the basal rock on which the volcano rests may also be present. They are bonded either by microscopic fusion or by chemical cementing into a weak sandstone. Tuffs are usually not very strong rocks, although some are consolidated well enough to support bridge or building foundation loads or to be excavated in vertical cuts. Many tuffs are very susceptible to landsliding. Montmorillonite clay is a component of bentonite, a mineral which is commonly produced by weathering of tuffs and which must always be considered a danger sign.

Tuff occurs in large quantities in the whole Sierra Madre Occidental, usually interbedded with rhyolitic lavas, in much of Central America, and in the western United States.

Volcanic Breccia: is chiefly composed of fragments measuring more than 5 mm (0.2 in). Lava fragments are generally more numerous than in tuff; glass and pumice particles may be few and far between. Slag, *tezontle*, is found in large quantities in some breccias. It may form large angular blocks or acicular bombs 2 to 15 cm (0.8 to 6 in) long, in the shape of a spindle or teardrop from being thrown into the air when still molten.

Some volcanic breccias are formed like tuffs, but others from streams of volcanic mud. It has been observed that when heavy rains have fallen on the steep slopes of a volcanic cone, there have been land-slides consisting of unconsolidated volcanic debris. Other breccias have formed from hot clouds produced by eruptive explosions through side craters. Volcanic debris may travel several kilometers along valleys in the form of clouds. Very coarse volcanic breccias are sometimes termed agglomerates, a sort of angular conglomerate.

Lapilli (Pumitic Sand): is a fragmentary material formed by loose particles measuring about 2 cm (0.8 in) made up of frothy lava, which were expelled by volcanoes. This material is often found in volcanic cones and is almost always excavated for an abrasive.

Obsidian: is a natural glass principally formed by magmas of rhyolitic, dacitic or andesitic composition. It is shiny and has a conchoidal fracture. The majority of obsidians are black because they are not well crystallized; they can be red or grey on account of iron oxidization caused by the hot magmatic gases. Thin fragments of obsidian are practically transparent. They are found along the edges of intrusions and on rare occasions form small intrusive masses. Most intrusive obsidians have an opaque luster similar to that of fish; hence this variety is called fishstone.

Pumice: is glass froth typified by a whitish-grey color and full of tiny bubbles. The bubbles are so numerous that pumice can float on water. Fragments of pumice are found in large quantities in tuffs and breccias. It also forms different streams or, more commonly, tops streams of obsidian and rhyolite and gradually becomes lost in the non-frothy lava underneath.

.2 Extrusive Rocks (Lavas)

Rhyolite: has an aphanitic mass splashed with phenocrysts of quartz or potassium feldspar. The color of rhyolitie varies enormously, but it is usually white, pale yellow, grey or red. The structure of some rhyolites is banded, that is to say they have a series of apparent layers which formed when the magma flowed as a paste, before it became solid. A large display of rhyolites and their tuffs can be found on the summits of the Sierra Madre Occidental, which includes areas belonging to the states of Nayarit, Zacatezas, Sinaloa, Durango and Chihuahua. Large quantities can also be found in the centre of the Mexican Republic.

Dacite: is similar to rhyolite, with the exception that the predominant mineral is plagioclase instead of potassium feldspar. Its relationship to rhyolite is the same as that of granodiorite to granite (see later). Rhyolite and dacite appear in the form of lava flows and small intrusions.

Andesite: is an aphanitic and often porphyritic rock, which resembles dacite, but contains no quartz. Plagioclase feldspar is the most common phenocryst, but pyroxene, amphibole and biotite are sometimes present. The structure of most andesites is banded, but not so noticeably as in rhyolites. The color of andesites varies from white to black, although the majority are a dark grey or greenish-grey. Andesite occurs in large quantities in lava flows and also in fragments of volcanic breccia, especially in mountain ranges capped by volcanoes, such as the Andes (where the name came from), the Cascadas in the NW United States and the Carpathians. Andesite can also form small intrusive masses.

Two thirds of the base of the Pachuca Range of Mexico is made up of andesites. The summit, is composed of dacites. Mineral seams in the mining districts of Pachuca and Real del Monte are full of andesites. A large part of the Valley of Mexico basin is made up of andesites; the Sierra Nevada, the Ajusco, the Sierras de Las Cruces and Monte Alto are all andesitic. The great fan of fluvio-glacial origin which covers the slopes of the mountain ranges to the South of Mexico City is composed of retransported pyroclastic andesitic material (sands and gravels). This same material can be found to the east of Texcoco, likewise in the *lomas* (hills) of Cuernavaca and Morelos and in the Ocuilan, Malinalco and Malinaltenango mountains in the State of Mexico.

Basalt: is a grey or black aphanitic rock. Most basalts are not porphyritic, but some contain phenocrysts of plagioclase and olivine. Basalt is the lava which exists in largest quantities throughout the world and it is widely spread out in the form of huge plateaus which cover thousands of square kilometers. Although it usually takes the form of lava flows, basalt is also found in small intrusive masses. There is a lot of basalt in the Mexican Republic, especially in the Valley of Mexico and surrounding areas. It can be found in almost every state of Mexico and in much of the western United States.

.3 Intrusive Rocks

Granite: which is characterized by its granular texture, contains the minerals quartz and feldspar in abundance; therefore most of its grains are light in color. Most granites also contain biotite and hornblende or one of the two.

Large quantities are found on the Pacific Ocean coasts of Mexico, in the Sierra de Chiapas, the Acapulco batholiths, in Guerrero, Michoacan and Jalisco, at the two extremes of the

Table 2-7
List of Common Minerals

MINERAL	FORM	CLEAVAGE	HARDNESS	SPECIFIC GRAVITY	OTHER PROPERTIES
COMMON CARBONATES, SULPHATES, CHLORIDES AND OXIDES					
Calcite: calcium carbonate-$CaCo_3$	Dogtooth or flattened crystals with excellent cleavage; granular with cleavage; also in masses with grain too fine to see cleavage clearly	Three prominent cleavages at oblique angles which give the fragments rhombohedral shapes	3	2.72	Usually colorless, white or yellow or any color owing to impurities. Transparent or opaque: transparent variety shows very strong double refraction (1 point viewed through **Calcite** looks like two). Luster: vitreous or opaque. Boils easily with cold dilute hydrochloric acid
Dolomite: carbonate of calcium and magnesium-$CaMg(CO_3)_2$	The rhombic-faced crystals show good cleavage. Also occurs in fine-grained masses	Three perfect cleavages at oblique angles, like **Calcite**	3.5 – 4	2.9	Color variable but usually white. Transparent or translucid. Luster: vitreous or pearly. When in powder form it boils slowly with dilute hydrochloric acid, not as big crystals
Gypsum: hydrated calcium sulphate-$CaSO_4.2H_2O$	Tabular crystals and earthy fibrous granular masses, even with cleavage	One perfect cleavage which produces thin flexible sheets. Two others are less perfect	2	2.2 – 2.4	Colorless or white, other colors because of impurities. Transparent or opaque. Luster: vitreous, pearly or silky. The cleaved sheets are flexible but not as elastic as **Mica**
Halite (rock salt): sodium chloride-NaCl	Cubic crystals. Granular masses	Excellent cubic cleavage, three cleavages at right angles	2 – 2.5	2.1	Colorless or white but other colors from impurities. Color may be non-uniform. Transparent or translucid. Luster: vitreous. Taste salty
Opal: hydrated silica with 3-12% water $SiO_2.nH_2O$. No defined internal structure — a mineraloid	Amorphous. Usually in veins or irregular masses with banded structure. May be earthy	None — Conchoidal fracture	5 – 6.5	2.1 – 2.3	Color extremely variable, often in belts or wavy. Translucid or opaque. Luster similar to that of wax
Chalcedony (cryptocrystalline quartz): siliceous dioxide-SiO_2	Crystals too small to be visible. Sometimes clearly marked bands, also in masses	None — Conchoidal fracture	6 – 6.5	2.6	Color is usually white or pale gray, any color from impurities. Can be distinguished from **Opal** by opaque and cloudy luster
Quartz (rock crystal): siliceous dioxide-SiO_2	Prismatic, six-faced crystals, also massive	None or scarcely noticable — Conchoidal fracture	7	2.65	Usually colorless or white, but may be yellow, pink, a smoky translucent greyish-brown and even black. Transparent or opaque. Luster: vitreous or greasy

Magnetite: combined ferrous and ferric oxides-Fe_3O_4	Well formed, eight-faced crystals usually in compact aggregates or scattered grains or loose in sand	None. Conchoidal or uneven fracture. Can be split in such a way that it looks like cleavage	5.5 – 6.5	5 – 5.2	Black. Opaque. Luster: metallic or submetallic. Streak: black. Strongly magnetic. Important mineral of iron ore.
Haematite: ferric oxide-Fe_2O_3	Very varied; compact, granular, fibrous or earthy — micaceous; seldom in well formed crystals	None but some fibrous or micaceous specimens split as though they had cleavage. Uneven, splintery fracture	5 – 6.5	4.9 – 5.3	Color steel-grey, reddish brown, red or iron black. Luster metallic or earthy. Characteristic streak brownish red. Most important mineral of iron ore

IMPORTANT METALLIC MINERALS — ORES (See Fe Oxides above)

Galena: lead sulphide-PbS	Cubic crystals common but generally coarse to fine grained granular masses predominate	Three perfect 90° cleavages	2.5	7.3 – 7.6	Silvery grey, metallic luster. Streak: silvery grey or greyish black. Principal lead mineral
Sphalerite: zinc sulphide-ZnS, almost always with iron as impurity	Crystals common but mainly granular masses	Six perfect 60° cleavages	3.5 – 4	3.9 – 4.2	Varies from white to black, commonly yellowish brown. Translucid or opaque. Luster: resinous or adamantine. Streak: white, pale yellow or brown. Principal zinc mineral
Pyrite (Fools' gold): iron [illegible] Fragile.	Well formed, commonly cubic crystals with striated faces, also in granular masses	None, uneven fracture	6 – 6.5	4.9 – 5.2	Pale, brassy yellow. Opaque. Luster: metallic. Streak: greenish or brownish black. Fragile. Not an iron ore. Previously used in sulphuric acid manufacture. Usually accompanies minerals of different metals
Chalcopyrite: mixed iron-copper sulphide-$CuFeS_2$	In compact or scattered masses, occasionally wedge shaped crystals	None, uneven fracture	3.5 – 4	4.2 – 4.3	Golden or bronzy yellow. As result of oxidation purplish blue or irridescent red. Streak: greenish black. Distinguished from **Pyrite** because brighter yellow and not so hard. Common copper mineral
Chalcocite (shiny copper): copper sulphide-Cu_2S	Massive, occasionally roughly hexagonal crystals	Can seldom be observed	2.5 – 3	5.5 – 5.8	Blackish grey or steel grey. Turns blue or green as result of oxidation. Streak: dark grey. Very heavy. Luster: metallic. Important copper ore
Copper (native copper): chemical element-Cu	Folded and twisted leaves and wire-shaped, flattened and rounded grains	None	2.5 – 3	8.8 – 8.9	Characteristic copper color but almost always stained with green. Extremely ductile, excellent conductor of heat and electricity. Very heavy

Table 2-7 (cont.)
List of Common Minerals

MINERAL	FORM	CLEAVAGE	HARDNESS	SPECIFIC GRAVITY	OTHER PROPERTIES
Gold: chemical element-Au	Massive or in thin sheets, also in flattened grains or flakes. Separate crystals very rare	None	2.5 – 3	15.6 – 19.3	Characteristic gold yellow, also in streak. Extremely heavy. Very malleable and ductile
Silver: chemical element-Ag	In flattened grains and scales, occasionally wire-shaped or in irregular needle-shaped crystals	None	2.5 – 3	10 – 11	Color and streak silver white but may present grey or black coloration on surface. Very ductile and malleable. Very heavy. Luster: mirror like on clean surface
Cassiterite: dioxide of tin-SnO_2	Well-shaped 4-faced prismatic crystals ending in pyramid. May be interwoven, also in kidney-shaped masses like river pebbles	None, curved or irregular fracture	6 – 7	7	Brown or black. Luster: adamantine. Streak: white or pale yellow. Principal tin mineral
Uraninite (pitchblende): uranium oxide $-UO_2$ or U_3O_8	Regular 8-faced or cubic crystals; massive	None, conchoidal or uneven fracture	5 – 6	6.5 – 10	Black or brownish black. Luster: submetallic, resinous or opaque. Principal mineral producing uranium, radium etc.
Carnotite: potassium and uranium vanadate $K_2(UO_2)_2\ (VO_4)_2.8H_2O$	Earthy dust	None apparent	Very soft	~ 4.1	Bright canary yellow. Vanadium and uranium mineral
Limonite: the majority of limonites are very fine crystals of Goethite containing absorbed water. Hydrated ferric oxide with minor contents of other elements-$Fe_2O_3.H_2O$	Compact or earthy masses, may present radial fibrous structure	None. Conchoidal or earthy fracture	1 – 5.5	3.4 – 4	Color yellow, greyish-brown or black. Luster: earthy, opaque, which distinguishes it from **Haematite.** Characteristic streak: yellowish brown. Mineral common in iron ore
Ice: hydrogen oxide-H_2O	Irregular grains, irregular straws in the form of fringes with hexagonal symmetry, massive	None. Conchoidal fracture	1.5	0.9	Colorless, white or blue. Luster: vitreous. Melts at 0°C therefore is liquid at room temperature. Low specific gravity

COMMON SILICATES IN ROCK FORMATION

Potassium Feldspar (orthoclase, microcline and sanidine): silicate of aluminium and potassium-$KAlSi_3O_8$	Box-shaped crystals, massive with excellent cleavage	One perfect and one good at right angles	6	2.5 – 2.6	Usually white, grey, pink or pale yellow; seldom colorless. Opaque, may be transparent in volcanic rocks. Vitreous. Luster: pearly at best cleavage. Distinguished from **Plagioclase** by absence of striae
Plagioclase Feldspar (sodium and calcium feldspars: group of solid solutions of silicates of aluminium, sodium and calcium — $CaAl_2Si_2O_8$ or $NaAlSi_3O_8$	In well-formed crystals and in granular masses or with cleavage	Two good cleavages almost at right angles (86°). Not clear in same volcanic rocks	6 – 6.5	2.6 – 2.7	Usually white or grey, may present other colors. Some grey varieties display color-play called opalescence. Transparent in some volcanic rocks. Luster: vitreous or pearly. Striae (fine parallel lines) on best cleavage face unlike **Orthoclase**
Muscovite (white mica, fish tail): a complex silicate of potassium and aluminium-app. $KAl_2\ Si_3O_{10}(OH)_2$ but varies	Thin flake-like crystals or in foliated, scaley aggregates	Perfect in one direction which gives thin flexible flakes	2 – 3	2.8 – 3.1	Colorless but can be grey, green or pale greyish brown in thick pieces. Transparent or translucid. Luster: pearly or vitreous
Biotite (black mica): complex silicate of potassium, iron aluminium and magnesium of variable composition-app. $K\ (Mg, Fe)_3AlSi_3O_{10}(OH)_2$	Thin scale-like crystals, usually 6-sided and in foliated, scaley masses	Perfect in one direction which gives thin flexible flakes	2.5 – 3	2.7 – 3.2	Black or dark-brown. Translucid or opaque. Luster: pearly or vitreous. Streak: white or greenish
Pyroxene: group of solid solutions of silicates of Ca, Mg and Fe with variable quantities of other elements. Most common varieties augite, hiperstine	Usually short 8-faced prismatic crystals. Angle between faces app. 90°. Also in compact masses and scattered grains	Two cleavages almost at right angles, not always well developed. In some specimens fracture is uneven or conchoidal	5 – 6	3.2 – 3.6	Usually greenish or black. Luster: opaque or vitreous. Streak: greenish grey. Distinguished from **Amphibole** by 90° cleavage, 8-faced crystals, and most crystals are short and very hard rather than long, thin prisms as in **Amphibole**
Amphibole: group of complex silicates in solid solution, chiefly Ca, Mg, Fe and Al. Similar to Pyroxene but contains a little hydroxide ion (OH). Most common variety is hornblende	Long 6-faced prismatic crystals, also in irregular or fibrous masses with interwoven crystals, also in scattered grains	Two good cleavages at angles of 56° and 124°	5 – 6	2.9 – 3.2	Black or pale green or even colorless. Opaque. Luster: intensely vitreous on cleavage plane. Distinguished from **Pyroxene** by cleavage angle and crystal shape. Has far better cleavage and greater luster than **Pyroxene**.
Olivine: silicate of iron and magnesium-$(Fe,Mg)_2SiO_4$	Usually in glass-like grains and in granular aggregates	So poor that it is rarely seen. Conchoidal fracture	6.5 – 7	3.2 – 3.6	Various shades of green, also opalescent yellowish and brownish when slightly weathered. Transparent or translucid. Luster: vitreous like small fragments of quartz but has a characteristic greenish color except when weathered

Table 2 – 7 (Cont)

List of common minerals

MINERAL	FORM	CLEAVAGE	HARDNESS	SPECIFIC GRAVITY	OTHER PROPERTIES
COMMON SILICATES (Cont.)					
Garnet: group of complex silicates in solid solution with variable proportions of different metallic elements, most commonly Ca, Fe, and Al	Generally in well-formed equidimensional crystals but also massive or granular	None. Conchoidal or uneven fracture	6.5 – 7.5	3.4 – 4.3	Generally red, brown or yellow but may present other colors. Transparent or opaque. Luster: resinous or vitreous
Sillimanite (Fibrolite): silicate of aluminium-Al_2SiO_5	In long, thin crystals, or fibrous	Parallel lengthwise but seldom noticeable	6 – 7	3.2	Grey, white, greenish grey or colorless; thin prismatic crystals or in velvety fibre-masses. Streak: white or colorless
Kyanite or **Cyanite** (Distene): silicate of aluminium-Al_2SiO_5	Long crystals in the shape of a knife blade	One perfect, other scarcely noticable. Parallel to length of crystal. Very rough split through crystal	4 – 7	3.5 – 3.7	Colorless, white or a distinctive pale blue. Can be scraped with a knife parallel to cleavage but across the grain is harder than steel
Staurolite: silicate of iron and aluminium-$Fe(OH)_2(Al_2SiO_5)_2$	Strong prismatic crystals and cross-shaped pairs	Scarcely noticeable	7 – 7.5	3.7	Reddish brown, yellowish brown or brownish black. Usually in well-formed crystals, larger than the matrix they are in
Epidote: complex group of silicates of Ca, Al and Fe-$Ca_2(Al,Fe)_3(SiO_4)_3\,(OH)$	Short 6-faced crystals or groups of radiant crystals, also in compact granular masses	One good cleavage, in some specimens second, weak, cleavage at 115° to first	6 – 7	3.4	Characteristic yellowish green (pistachio green). Luster: vitreous
Chlorite: complex hydrated silicates of Mg and Al, containing Fe etc. as trace elements	Commonly foliated or scaly masses; may occur in tabular 6-faced crystals resembling **Mica**	One perfect, giving thin, flexible but inelastic flakes	1 – 2.5	2.6 – 3	Grass green or blackish green. Translucid or opaque. Streak: greenish. Luster: vitreous. Very crumbly
Serpentine: complex hydrated silicates of Mg-app. $H_4Mg_3Si_2O_9$	Foliated or fibrous, mainly massive	Commonly only one cleavage but may be in prisms. Fracture usually conchoidal or splintery	2.5 – 4	2.5 – 2.65	Soft to the touch, sometimes greasy; leek green or blackish green varying to brownish red, yellow etc. Luster: resinous or greasy. Translucid or opaque. Streak: white
Talc: hydrated silicate of Mg-$Mg_3(OH)_2Si_4O_{10}$	In tiny scales and soft, compact masses	One perfect cleavage which forms scales and small pieces	1	2.8	White, silvery white or apple green. Very soft, greasy to the touch. Luster: pearly on cleavage surfaces
Kaolinite: hydrated silicate of Al, representative of 3 or 4 similar clay minerals-$H_4Al_2Si_2O_9$	Generally in soft, compact, earthy masses	Cleavage only visible microscopically as crystals so small	1 – 2	2.2 – 2.6	White but may be stained with impurities. Greasy to touch. Sticks to tongue and becomes plastic when wet. When breathed on smells of clay

Baja California peninsular, to the south in the El Cabo region and to the north in the Sierra de Kukapas and Rumorosa region. In some areas of the states of Sonora and Sinaloa, where it is very weathered it is known by the name of Tucuruguay and looks like a poorly-cemented coarse sand. In the United States granites are found in localized areas of the Appalachians, Texas, the Rocky mountains and Sierra Nevada mountains.

Technically speaking, the term granite is reserved for those granular quartziferous igneous rocks whose predominant mineral is potassium feldspar. Rock with a predominance of plagioclase is called granodiorite (see *rhyolite* and *dacite*, above). Granodiorite can be distinguished from granite by the fine striae which can be seen on the cleavage surfaces of the plagioclase. It is frequently found in the same areas as granite.

Geological studies have shown that granite and granodiorite occur in large quantities in the earth's crust. They form large intrusive masses along the core of several mountain ranges, as well as in areas which have suffered deep erosion. Typically, they are continental rocks and they have never been found in isolated oceanic islands or far from continental masses. Some granites are of a metamorphic, and not igneous, origin.

Diorite: is a granular rock made up of plagioclase and small quantities of ferro-magnesian minerals (and sometimes a little quartz). The most common ferro-magnesian minerals are hornblende, biotite and pyroxene. Diorite masses are usually smaller than granite or granodiorite ones. Diorite occurs in huge masses in Mexico, south of Zitacuaro (Cerro de la Coyota) and many parts of the Sierra Madre Occidental and the Pacific Coasts. In the big Acapulco batholith, diorite dikes crossing the granite mass are seen at frequent intervals alongside the highway. At the River Aguacatillo crossing, a granite with diorite contact can be seen. The little peninsular called La Quebrada is a diorite intrusion in granite.

Gabbro: is a granular rock, composed chiefly of plagioclase and pyroxene and usually containing small quantities of other ferro-magnesian minerals, especially olivine. If the ferro-magnesian minerals predominate over the plagioclase in such a way that the rock is dark in color, it is likely to be gabbro, although its distinction from diorite usually depends on the microscopic identification of the plagioclase.

Gabbro is found in the form of both large and small masses. It is especially common in dikes and thin, fine-granted intrusive sheets. In most of these small intrusions the mineral grains are so small that they are hard to recognize without the aid of a microscope. Those gabbros with a grain size between that of basalt and normal gabbro are called dolerites or diabases.

Sound gabbro occurs in several places in Mexico: in the State of San Luis Potosi, at the La Ventilla reservoir; in the State of Hidalgo, near Tialchinol; on the road from Pachuca to Huejutla; in the State of Sinaloa and in some cuttings of the Chihuahua-Pacific Railroad. During work on certain cuttings on the road from Tijuana to Ensenada, in the State of Baja California, a gabbro has been discovered which is sometimes full of caves, but is nevertheless strong, despite severe weathering caused by a long period spent under the sea.

Peridotite, Pyroxenite and Serpentine: In some areas, rocks with a granular texture and composed almost entirely of ferro-magnesian minerals, without feldspar, are common. If the chief component of the rock is olivine, it is called peridotite; if it is made up entirely of pyroxenes, it is called pyroxenite. Peridotites and pyroxenites altered by weathering (and metamorphism) become serpentines. Because serpentine is almost entirely composed of secondary minerals which did not solidify directly from the magma, it is often classified as a metamorphic, not igneous, rock. Serpentine forms intrusive sheets, dikes and other small intrusive masses.

Porphyry: This ancient term is used here to describe grain texture. It is commonly applied to fine-grained intrusive igneous rocks with a porphyritic texture, with 25% or more of their volume made up of phenocrysts. The ground mass can be either coarse-grained aphanitic or fine-grained phaneritic. The word *porphyry* is preceded by the name of the rock whose composition and texture correspond to the ground mass, such as diorite porphyry. If a rock contains only scattered phenocrysts it is described as *porphyritic* such as a porphyritic andesite. Granite porphyry, granodiorite porphyry and diorite porphyry usually are found as dikes near masses of granite and granodiorite. Rhyolite porphyry, dacite porphyry and andesite porphyry are common in volcano necks and other small intrusive masses.

2.4.2 Common Sedimentary Rocks

.1 Conglomerate

Conglomerate is cemented gravel. Gravel is an unconsolidated deposit principally made up of fragments transported by rivers. These fragments may be of any kind of rock or mineral and of any size. Most conglomerates, especially those deposited by rivers, contain a large quantity of sand and other fine materials which fill the spaces between the gravels. Some beach conglomerates, when carefully washed, contain little sand.

In the Chilapa and Tiapa areas in the State of Guerrero, Mexico, conglomerate is found in the form of large masses considerably altered by weathering. In the north, large extensions of it are often found. It usually occurs everywhere in small masses, often associated with larger deposits of sandstone.

.2 Breccia

Sedimentary breccias are similar to conglomerates, except that most of the fragments are angular instead of rounded, (although there is no clearly defined dividing line between the two rocks). In view of the angular form of the component fragments, it is obvious that the components of breccia underwent relatively little wear and transportation before being deposited. There are other types of breccias which are not sedimentary, like volcanic and fault breccias. Massive breccias of volcanic origin are sometimes termed "agglomerates". Breccias are found in mountainous areas which have been subjected to intense tectonic movements, and volcanism.

.3 Sandstone

Sandstone is cemented sand. By definition, sand consists of particles with a diameter between 4.76 mm (3/16 in) (N° 4 sieve) and 0.074 mm (0.003 in) (N° 200 sieve).

Sands accumulate in different ways. Some are deposited by rivers, others are accumulated by winds and form dunes, others are spread out by the waves and currents along beaches or in shallow waters on the continental shelves. Others are carried by turbid currents along submarine slopes to the bottom of the sea.

Three main varieties of sandstones are recognized:

Quartz Sandstone: is composed mainly of grains of quartz mineral, although it may contain small quantities of other minerals.

Arkose: is a sandstone rich in feldspar. It may contain almost as many particles of partially weathered feldspars as of quartz, or even more. Most arkoses have been formed by the rapid erosion of rocks rich in feldspar grains, such as granites and gneisses, and by the rapid deposition of the eroded debris, without the feldspar having time to be converted into clay by weathering. Their grains are often more angular than quartz sandstones.

Graywacke: is a cemented *dirty sand* which contains large quantities of clay and rock fragments, in addition to quartz and feldspar. Some graywackes contain large quantities of pyroclastic debris at different stages of weathering. Others contain a large number of small fragments of slate, greenstones and other metamorphic rocks. They are often rocks with a high content of ferro-magnesian minerals.

All graywackes contain considerable quantities of clay. They are commonly dark grey, dark green or black. Like arkoses, they were formed in an environment of rapid erosion and deposition without much chemical weathering.

Sandstones form large lithological units which occupy considerable areas. In Mexico, there are outcrops of sandstones in the Ciudad Altamirano area in Guerrero; in the Sierra Madre Occidental, interstratified with clay shales; on the coastal plain of the Gulf of Mexico, in the northern area of the Isthmus and, in smaller masses, throughout the Mexican Republic. They are widespread in much of the United States such as the Appalachian Ridge and Valley Province and the Appalachian Plateau.

.4 Mudstones (shales)

Strictly speaking, the term mudstone refers to a laminated or finely stratified rock with a fissile structure together with a basically clayey composition, although it may contain large quantities of silt, sand, organic matter and calcium carbonate. Mudstones accumulate in very different environments. Since the principal load carried by big rivers to the sea is made up of silt, clay and fine sand, it is not surprising that mudstones are commonly-found in sedimentary marine rock. In addition the mud which is deposited in deltas, at the bottom of lakes and on plains which once bordered on ancient diverging rivers hardens and turns into mudstone or shale.

In addition high consolidation stresses can produce some mineral changes and a direct bonding of some clay minerals. Some mudstones are also indurated by cementing with carbonates, sulfates, and iron oxides. Mudstones deposited from alternating high and low stream flows from alternating thin strata of silt and clay or fine sand, silt, and clay. Repeating alternating thin strata are termed laminated. Fissility develops if the micas and clay mineral sheets are reoriented parallel to one another by the consolidating stresses. Such mudstones are shales. If the consolidating stress is perpendicular to the stratification or lamination, the fissility exaggarates the laminations. If it is in a different direction, the fissility will cross the laminations producing two different directions of weakness in the shale.

.5 Marl

Marl is a term for rocks composed of clay, silt (or even sand) and calcium carbonate. Its properties are between those of a well indurated mudstone and soft limestone. If the clay content predominates, they are sometimes termed calcareous mudstones; if calcium carbonate predominates, they may be referred to as clayey limestones.

.6 Limestone

Limestone is composed of calcium carbonate ($CaCO_3$) often with impurities which may be clay, silt, sand, chert or flint organic matter and iron oxides. Limestones of organic origin are common rocks and there is a great variety of them owing to the many different types of debris from which they originate. The most common ones are: coral limestone, which contains a network of coral deposits, but also contains shells from other animals, especially foraminiferes, molluscs and gasteropods; algae limestone composed mainly of calcite precipitated from algae and bacteria; foraminiferes limestone chiefly made up of tiny shells from foraminiferes; coquina composed mainly of large shells, molluscs and gasteropods and chalk which is chiefly composed of microscopic sheets and spines of calcite. Clastic limestones are made up of broken and worn shell fragments or calcite crystals.

Their engineering properties are extremely variable depending on their void ratio and induration. Young, poorly indurated, algae or precipitated lime muds have void ratios as great as firm soils with similar strengths and compressibilities. Poorly consolidated coral or coquina limestones may have void ratios of 1 to 3, and are as strong as poor concrete but as compressible as very loose sand. Well indurated limestones can be as strong as granite and are thinly stratified. Deformed and fractured they may cause serious problems in building foundations, being very unfavourable on account of their low shear resistance and the seepage they allow. In other respects, they are considered a magnificent building material.

Limestone is the rock which forms the general mass of the Sierra Madre occidental in Mexico, and there are outcrops of it all along this range. It occurs in the form of large masses in the south of Mexico, the whole of the Yucatan Peninsular is made of soft porous limestone and it is found in almost every state of the Republic. Soft porous limestones are found in the Atlantic and Gulf Coastal plains; well indurated limestones are found in much of the remainder of the United States.

.7 Travertine

Travertine is formed by impure calcium carbonate which is deposited where thermal springs come to the surface. It is heavily crystalline and is usually full of small irregular holes through which the water that formed it once flowed. Travertine is found in large quantities in Mexico in the area between Valsequillo and Tehuacan in the State of Puebla. In Viesca, Coahuila, there are also large deposits of it. In the State of Nuevo Leon, it is used for construction work.

.8 Dolomite

Dolomite consists largely of the mineral of the same name: dolomite (double carbonate of calcium and magnesium); it resembles limestone and it may gradually change into limestone as a result of changes in the calcite content. Microscopic and chemical tests are usually necessary in order to determine the relative quantities of calcite and dolomite in the rock. Dolomite effervesces gently in HCl and to make this effect more noticeable it should be scraped with a pocket knife and the acid applied to the resulting powder. Limestones and dolomites have very similar engineering properties. Dolomites are not very common in the Mexican Republic. However, dolomitic limestone does exist in quarryable quantities in Teapa, Tabasco; it is also found in the Petaquillas Canyon in Guerrero. They occur frequently in the United States in the same areas as limestones.

.9 Fine-Grained Siliceous Rocks

Rocks composed almost entirely of fine-grained silica are frequently found, but not often in the form of large masses; the most common ones are:

Flint and Chert: Flint is a hard rock so finely grained that the fracture plane has a uniform and glossy appearance. It occurs as nodules and lenses in limestones and dolomites. Flint is very hard and is usually black, yellowish and white. Chert is similar in composition and occurrence. It is less glassy than flint. Both minerals can react with alkali and in Portland cement swell. Limestones containing flint and chert should be tested for this reaction.

Diatomite: is a soft white rock composed almost entirely of siliceous shells belonging to microscopic plants called diatoms.

Not all fine-grained siliceous rocks are of organic origin. Some are thought to have been precipitated by submarine thermal springs of siliceous waters. Others have formed as a result of the replacement of wood, limestone, clay shale or other materials by siliceous solutions. Petrified wood is a common example.

.10 Carbonaceous Rocks (Peat and Coal)

Peat is a mixture of slightly decomposed plant remains. It can be found in accumulation processes of swamps and shallow lakes in temperate climates and even on steep hill-slopes in damp regions. Lignite and Coal are the result of the compression and a more advanced decomposition of plant matter in ancient peat bogs which were buried by later sediments.

.11 Evaporites or Salt Deposits

Evaporites vary greatly in composition and texture. Even at the present time, they are being formed by the evaporation of waters saturated in dissolved minerals.

Halite: When sea water evaporates completely, several salts are precipitated; the largest quantities are of rock salt (NaCl), *halite*.

Gypsum and Anhydrite: Calcium sulphate, both in its hydrated form, *gypsum* ($CaSO_42H_2O$) and in its anhydrous mineral form, called *anhydrite* ($CaSO_4$) is far more plentiful in a natural form than rock salt.

Caliche: is a deposit resulting from the evaporation of incrusting waters which appears in the form of surface crusts or layers intercalated with soils, especially in semi-arid regions. Its composition is $CaCO_3$ mixed with the soil components and in some cases it serves as a cementing agent for soils.

2.4.3 Metamorphic Rocks

.1 Quartzite

It is a very hard granoblastic rock with a sugary texture, composed mainly of grains of quartz either strongly cemented or intergrown by fusion under pressure. Quartzite differs from sandstones in that it breaks across the grains instead of around them. Color varies from white to black, going through cream, pink, red and grey, but most quartzites are light in color. Quartzite forms as a result of the metamorphism of quartz sandstone. It is a widespread metamorphic rock.

Sandstone with silicate cement (ortho-quartzite) is hard to distinguish from metamorphic quartzite, for both break across the grains. They can be distinguished under a petrographic microscope, because the cement crystallizes differently than the original grains of sand. Quartzite associates with other metamorphic rocks and orthoquartzite with other sedimentary rocks. They are among the strongest, toughest, rocks.

.2 Marble

It is a fine or coarse-grained granoblastic rock, predominantly calcite, dolomite or a mixture of them. Several marbles have alternately dark and light bands; others have breccia structures healed by calcite cementing veins. Marble forms as a result of the metamorphism of limestone and dolomite. If it originates from dolomite, it usually contains magnesiferous silicates such as pyroxene, serpentine and amphibole. Like limestones, marbles dissolves readily; joints and cracks become solution channels.

.3 Slate and Phyllite

These are very fine-grained, exceptionally well-foliated rocks. Because of their excellent foliation, they split into thin plates. The mineral grains are so small that they can only be identified with a microscope or X-rays. Slate is opaque on its cleavage surfaces, phyllite is shiny and coarse-grained and contains some mineral grains large enough to be identified with the naked eye. Slate, and less so phyllite, often show remains of sedimentary features such as stratification, gravels and fossils.

Slates and phyllites are found in large quantities. The majority formed as a result of the metamorphism of clay shales, but others are derived from tuffs and fine-grained rocks. Because of their foliation they are very anisotropic, shearing far more easily parallel to foliation than perpendicular to it.

.4 Chlorite Schist

This is a very fine-grained greenstone that is schistose and slaty. It is a soft rock, greasy to the touch and easily powdered, made up of chlorite, plagioclase and epidote: all, with the exception of chlorite, may be present in the form of grains too small for identification. They may have remains of the original volcanic structures such as phenocrysts and slag. Chlorite schists are common. They are often called green schists or, if their foliation is very poor, greenstone, because of the color of the chlorite. Most were formed as a result or the metamorphism of basalt of andesite and their corresponding tuffs, but some are derived from dolomitic shale, gabbro and other ferro-magnesian rocks.

.5 Mica Schist

This is a schistose rock chiefly composed of muscovite, quartz and biotite in variable proportions; any of these minerals may predominate. Mica schist is one of the most plentiful metamorphic rocks. Like slate, most have their origin in clay shales and tuffs, though some derive from arkose, clayey sandstone, rhyolite and other rocks. The metamorphism of schist is more intense than that of slate.

.6 Amphibole Schist

This is a schistose rock composed chiefly of amphiobole and plagioclase with variable quantities of garnet, quartz and biotite. It is a metamorphic derivation of basalt, gabbro, chlorite schist and related rocks. Slates and sericite schists, with imperceptible differences in grading, can be found in Mexico in the States of Hidalgo, Puebla and Veracruz, in the deep gorges near the borders between these states (Huayacocotla and Vinazco gorges), on the road from Vizarron to Jalpan (Queretaro State) where, for paleontological reasons in the first case, and stratigraphical in the second, they have been attributed to a Jurassic period. In the States of Mexico and Michoacan (Tlalpujahua, Janapeo, and in most of the region where hydroelectric plants are being developed such as Miguel Aleman, Ixtapan de la Sal) large outcrops of sericitic slate have been found. In many places they have the appearance of schists.

.7 Gneiss

Gneiss is a coarse-grained gneissic rock with various lenses or layers of different materials. It has a variable mineral composition, but with especially large quantities of feldspar. Other minerals common in gneiss are quartz, amphibole, garnet and mica. Gneisses are found among the most common metamorphic rocks. They may derive from a variety of rocks such as granite, granodiorite, clay shale, rhyolite, diorite, slate and schist. They are often associated with schists, one grading into the other. Gneisses occur in large quantities in the areas of the Mexican Republic in the same areas as large masses of granite. In such cases, the gneiss has originated from the granite as in the States of Oaxaca, Guerrero and Michoacan. They have also been found with no apparent relation to granites in the Tomellin Canyon and the Sierra de Ixtlan in the States of Oaxaca. In the United States gneiss is the predominant rock of the Predmont region of the eastern end of the Rocky Mountains.

REFERENCES

1. Casagrande, A., "Classification and Identification of Soils", *Transactions American Society of Civil Engineers*, Vol. 113, 1948.
2. Juárez Badillo, E. and Rico, A., *Mecánica de Suelos, Vol. I: Fundamentos de la Mecanica de Suelos,* Limusa: Mexico, 1977, 3rd Edition, Chap. 7.
3. "Sistema de Clasificación de Materiales Pétreos Suelos", Technical publication of the Public Works Ministry, Mexico, 1970.
4. Reference [2], Chap. 6.

CHAPTER 3

THE EARTH FOUNDATION AND SOIL EXPLORATION

3.1 Introduction

The earthworks that are necessary for road construction transmit stresses to the natural ground beneath, and these stresses in turn cause deformation which reflects on the structural behaviour of the earthworks. Thus it is necessary to study the earth foundation. Other factors, such as water, for example, affect the mechanical behavior of the earth foundation and although this is independent of the superstructure of the road, it is sometimes affected by it. Such factors must also be studied, for they produce effects which reflect on the behavior of the road. Lastly, the interaction of the earth foundation and any superstructure has such an important influence on overall behavior that it is vital for the engineer to study every available method for improving any unfavorable conditions existing in the foundation.

The earth foundation or natural ground is understood as that part of the earth crust which must bear the road structure, and will consequently be affected by it. Its function is to support the road under conditions involving a reasonable degree of resistance to failure and deformation.

3.2 General Comments Concerning the Foundation Ground

The earth foundation, the natural ground, may be composed of rock or soils. Rock generally poses no problems, because the stresses to which it is subjected by the road are usually of very low intensity in comparison with the strength of the rock itself. The susceptibility of the rock formation to weathering by the action of mechanical or chemical agents also does not greatly affect its ability to bear the road structure.

Owing to their hardness, igneous rocks may produce very high excavation costs. When reasonably sound, they generally allow vertical or near vertical slopes. If they are to support a pavement, an intermediate layer of soil must be placed in the cutting to eliminate any irregulatities that may remain after grading.

Sedimentary rocks are extremely variable in hardness, often being not nearly so hard as igneous rocks, which makes excavation easier. This category comprises a large number of crumbly rocks, especially ones with an agglomerated structure. Special mention is deserved by limestones. These show a wide range of behavior. The fine-grained well indurated ones are hard and durable but the young, coarse grained or porous limestones are soft and crumbly. Excavation in clay-shales and marls is usually fairly easy. They are often unstable in the presence of water; like gypsums and other similar rocks, they are inclined to absorb water and expand, which makes them dangerous materials for both cutting beds and retaining-wall fills. Lastly, it should be emphasized that water that has flowed through marly rocks, gypsums or anhydrites can be highly dangerous, for it collects calcium salts that can cause decomposition of the cement content of the concretes that are used in the different structures of the road. Relatively sound sedimentary rocks, however, also allow almost vertical stable slopes.

Gneisses, schists and slates are perhaps the metamorphic rocks most commonly used in road engineering. They are excavated by mechanical methods if thinly laminated, and explosives are often unnecessary. Since they can usually be split along their very well marked foliation planes, their dip is of vital importance in the case of cuts. Schists and slates are rather crumbly rocks, and as a result of weathering eventually produce highly unstable clays, sometimes during the lifetime of the road.

Like rock, foundation ground composed of soils usually has sufficient load bearing capacity for roads, although there are certain conditions that pose serious problems of design and construction. Some of these will be described in detail in the pages that follow; indeed, they may well represent the most difficult contingencies that are confronted by the road engineer. The best solution to these problems is invariably to avoid them by relocating the projected road. Although it is most unusual for a natural foundation composed of soils to present really difficult problems involving costly solutions, it is nevertheless essential to detect such problems during the initial investigation stage so that they can be avoided where possible. Otherwise they must be subjected to special and often very extensive (expensive!) studies when the engineer finds himself *obliged* to include them in his project.

Frictional soils (gravels, sands and non-plastic silts, or mixtures where they predominate) usually have an adequate bearing capacity, and compressibility characteristics that do not give rise to important settlement problems.

Very loose sands and silts may lead to erosion and sudden settlement because of a rapid collapse in their simple structure when subjected to fairly heavy loads. As pointed out in Chapter 1, these collapses are usually associated with movements of the subsoil water, either saturation due to the seepage of surface water or a rise in the water table. Embankments are not, however, seriously affected by this, for they absorb much of the resulting movement without difficulty; it is far more dangerous, of course, when the ground has to support one of the rigid structures that are generally necessary for a road.

Sometimes when the hydrodynamic forces produced by rising water flow overcome the weight of the particles, buoyancy occurs, with the result that the soil loses all or almost all its bearing capacity, with the inevitable consequences for the road under construction. This problem does not often arise and need not be feared once the embankments that are constructed on the ground have reached a certain height, but it may be of some importance in the case of certain cut road-beds. The solution is always to interrupt the flow or to reduce its gradient to suitable levels. Fortunately it is a situation which the engineer can assess using theoretical methods.

Another effect of flowing water in foundation ground is *piping,* which occurs when the hydraulic gradient of the water seeping through a foundation soil is greater than a critical value, and particles are consequently transported by the flow [1]. Piping is not usually a very dangerous condition in the foundations for earth structures; it is more likely to affect embankments and is therefore a factor to be considered when assessing stability (and as such it will be discussed in a later chapter). It may occasionally occur, for example when a flow breaks through the side of an embankment owing to the presence of accumulated water on the other side. Most susceptible to piping are fine, cohesionless, permeable, uncemented soils, with a plasticity index under 10%. Soils which, apart from possessing the foregoing characteristics, are also light (pumitic sands, for example) are especially sensitive to the effects of flowing water. Table 3-1 [1] describes the susceptibility of different soils to piping.

Piping is most common in natural grounds with erratic stratification and permeable mantles which both tend to accelerate the phenomenon. Graded filters, which will be described later, provide the best means of avoiding piping in foundation soils, although, for reasons of cost, their use in road engineering is limited to those places where there is an undeniable risk and where the phenomenon might have very serious consequences.

Liquefaction has been the cause of highly dramatic and spectacular slides, owing to the magnitude of the mass of soil it involves. In soils such as relatively loose saturated sands, a rapid dynamic stress, of the kind that may occur during an earthquake, can generate high water pressures, which increase faster than the water is able to escape through the pores of the material. As the pressures in the water within the structure increase, the contact between the grains of sand weakens, with a consequent reduction in shear strength to zero or very close to zero. Under these conditions, the mass of sand behaves like a viscous liquid and is obliged to flow by the loads which are responsible for the phenomenon. This explains the origin of some of the most impressive slides on record, where the superstructure of a road has suffered displacements measuring tens and even hundreds of meters. In Mexico, examples are the slides which occurred as a result of the earthquakes that affected Coatzacoalcos and Jáltipan in 1957 (Plate 3-1).

Table 3-1
Susceptibility of soils to piping

Large resistance to piping	1. Very plastic $I_p > 15\%$, well compacted clays 2. Very plastic $I_p > 15\%$, poorly compacted clays
Medium resistance to piping	3. Well graded, well compacted sands or mixtures of sand and gravel, containing moderately plastic clay $I_p > 6\%$ 4. Well graded, poorly compacted sands or mixtures of sand and gravel, containing moderately plastic clay $I_p > 6\%$ 5. Well graded, well compacted non-plastic mixtures of gravel, sand and silt with $I_p < 6\%$
Low resistance to piping	6. Well graded, poorly compacted non-plastic mixtures of gravel, sand and silt with $I_p < 6\%$ 7. Well compacted fine uniform clean sands with $I_p < 6\%$ 8. Poorly compacted fine uniform clean sand with $I_p < 6\%$

Plate 3-1 Liquefaction of foundation ground under road embankments as a result of an earthquake (Jáltipan, 1957)

The soils most susceptible to liquefaction are loose sands (deformation tends to compact their structure, and pressures are transmitted to the water which trigger the phenomenon), these sands must at the same time be uniform, fine and saturated. For such fine sands the permeability is low, and dissipation of the pressures in the water is slow. Deposits of loose, non-plastic silts are also dangerous (Plate 3-2).

According to experience, the only reliable means of preventing liquefaction appears to be compaction of the susceptible soils by means of any of the currently used procedures. In the case of highway construction, compaction of large masses of foundation is difficult and, what is most important, expensive. In earthquake zones where important roads are to be built, it may well prove advisable and economical, especially if the road in question must cross dangerous areas of limited proportions.

In foundations composed of plastic silts and clays, there are two different cases: one, when the compressibility of the soils is relatively low (*CL, ML* and *OL*) and the other, when the materials are very compressible (*CH, MH, OH* and P_t).

Soils with relatively low compressibility present no special problems in relation to the pavement of a road. Any small settlements that may occur are easily absorbed by the flexibility of the pavement itself and the bearing capacity of the ground is usually sufficient to support whatever embankments are to be constructed. For special, more rigid structures, such as bridges and culverts, the problems can be adequately solved by applying the available theories, which consider the load bearing capacity of the ground and the movements that can be tolerated by the structure.

Plate 3-2 View showing typical problems of road building on soft soils (construction of the Villahermosa-Escárcega road)

Owing to their content of organic matter, *OL* materials may not, in certain extreme cases, be appropriate for construction purposes.

As already mentioned, the prospect is different when the ground is made up of highly compressible silts or clays (*OH*, MH, CH and P_t). This case will be dealt with in greater detail later in this chapter, but a few general points can be made here.

Firstly, no fixed relationship exists between the unfavorable characteristics of the ground in relation to strength and compressibility, and its geographical or topographical situation, although unfavorable foundation soils are usually more common in river, lake or sea deposits. The most reliable methods of detecting problem areas requiring detailed investigation are photo-interpretation and surface geological investigations.

The strength of the foundation soil is of special importance when high road embankments are required. This is usually the case for bridge approaches, flood plains adjacent to rivers and lakes, and areas where there is a certain depth of water. A 3.0 m height limit has been proposed for embankment construction after which special investigations would be required, including soil exploration and detailed laboratory tests of the characteristics of the soil in question as a basis for the stability analyses are required in these special cases. However, it is hard to establish limits like these, for the seriousness of a specific case depends not only on the height of the embankments, but also on the nature of the materials at the location, and the importance of the consequences of a failure.

Low strength in the foundation ground beneath an embankment can bring about bearing capacity failure accompanied by a sudden destructive settlement in the embankment, and heave in the soil on either side of it, not far from the embankment. Failure may occur without warning, but it is sometimes preceded by deformation at the crest of the road, settlement along the center line and the appearance of cracks in the virgin ground, parallel to the embankment, and at a distance which depends on the height and width of the embankment. These cracks are usually accompanied by a perceptible swelling of the ground. As soon as the engineer detects these signs, he should immediately proceed to take action, either adding berms or by reducing the weights induced by the embankment. A reasonable means of preventing a slide is to construct the embankment in stages. This is done by first building only part of the embankment, and then, little by little adding height as the foundation ground consolidates under the load applied, and gradually develops sufficient bearing capacity.

The procedure for estimating the increase in shear strength that takes place in a soil during a consolidation process is based on concepts that have already been discussed in Chapter 1. Let us suppose, for instance, that consolidation is produced by an embankment which is built on a normally consolidated compressible soil, with an initial strength that does not ensure stability. It is decided to commence by building the embankment with only half of its proposed height; completion will not be effected until the soil is partially consolidated, and, as a consequence, the initial strength of the ground sufficiently increased.

With rapid loading, that is if the embankment is built in a short period of time in relation to the time required for the soil to achieve a significant degree of consolidation, the strength of the foundation soil will be represented by the envelope obtained in a consolidated undrained test using total stresses. If this envelope (Fig. 3-1) is analyzed, it can be seen that the shear strength is proportional to the load which consolidated the material.

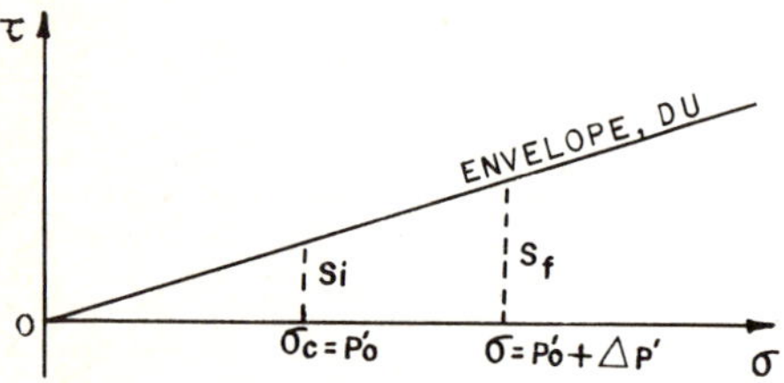

Fig. 3-1 Increase in the undrained strength with consolidation load

In a normally consolidated compressible layer the undrained strength will therefore be proportional to the depth. By constructing half of the embankment, consolidation will be induced, which will bring about an increase in the effective stress in the foundation ground. Once 100% consolidation has been achieved under the new load, the new strength at any point in the foundation soil can be determined from the new effective pressures existing at the end of the process. This can be computed by applying the BOUSSINESQ theory, as shown in Section 3.3. Thus, if s_i is the initial strength at one point in the consolidated mass under the effective pressure of its own weight (p'_o), the final undrained strength, s_f, will be the one corresponding to the new consolidation pressure, $p'_o + \Delta p'$, where $\Delta p'$ represents the increment in effective stress produced by the first half of the embankment that was built. In this way, s_f will be the strength that can be expected when construction of the second half of the embankment commences, if the ground has reached 100% consolidation under the first half. The strength corresponding to any percentage of consolidation between 0% and 100% will also have a value between s_i and s_f; this can be obtained by linear interpolation, as shown in Fig. 3-1. If the foundation soil were preconsolidated, the problem could be dealt with as above, but including the interval of consolidation in the envelope *CD*.

Another problem related to very low strength foundation grounds is concerned with failures in the embankment itself; these may be either rotational or translational and occur on totally or partially developed slide surfaces that extend into the foundation soil. However, this type of failure will be discussed in the chapter on Slope Stability.

3.3 Settlement of the Foundation

The most serious problem posed by a fine, compressible foundation soil are the settlements that may occur under the load to which it is subjected by overlying embankments. These settlements cause (See Plates 3-3 to 3-7):

— Loss of pavement camber, for the settlement induced by the embankment is greater in the center than at the shoulders.
— Appearance of differential settlements in the longitudinal direction, owing to heterogeneities in the yield characteristics of the foundation ground. These are detrimental to the performance of the road, the pavement and surface drainage.
— Reduction in the height of the embankment, which is of great importance when the road has to cross flood zones.
— Damage to drainage systems, which acquire hydraulically unsuitable slopes and crack as the center sinks more than the shoulders.
— Cracks in the crest of the embankment, specially when it is very wide, and when the embankment has berms.
— Loss of the appropriate transition between access embankments and structures, when these, for example on pointed pile foundations, do not participate in the general settlement.

Plate 3-3 Effect of total settlement, sinking of a drain

In Mexico it is not unusual to find areas where settlement in the foundation play such an important part that the whole design of the road, including the possibility of altering its location, is conditioned by them. If only the traditional concepts of road design were taken into account, projects would be developed which are not optimal.

In Section 1.12.2, the methods for assessing settlement in a compressible foundation soil under an embankment were discussed in considerable detail. It was stated that a basic requirement for computations is the knowledge of the variation of $\Delta p'$ with depth, $\Delta p'$ being the increase in stress that is transmitted to the foundation by the embankment, assuming that before construction of the embankment the foundation ground was consolidated only by its own weight p'_0.

In [2], a detailed description is given of the application of the theory of elasticity, in particular the BOUSSINESQ theory, to the calculation of the distribution of stresses in soil mass, when a certain load is applied to the horizontal surface. Now attention is drawn to the specific case where this load corresponds to an embankment. The problem can be dealt with as an extension of the case of a concentrated load with a magnitude p, as originally proposed and solved by BOUSSINESQ. In the solution, the loaded medium, which represents the real soil, is regarded as a semi-infinite continuum, which is homogeneous, isotropic and linearly elastic. This hypothesis is so far removed from the con-

Plate 3-4 Another effect of embankment settlement on transversal drainage, observe the sunken pipe culvert

ditions of the real soil, that it is rather surprising that settlements can be assessed using the BOUSSINESQ Theory (in combination with the Theory of Consolidation) with an approximation that is regarded by the engineer as adequate.

The problem of computing the stresses transmitted by a very long embankment to a semi-infinite mass of soil was solved by CAROTHERS [2,5] for the cross-section in Fig. 3-2

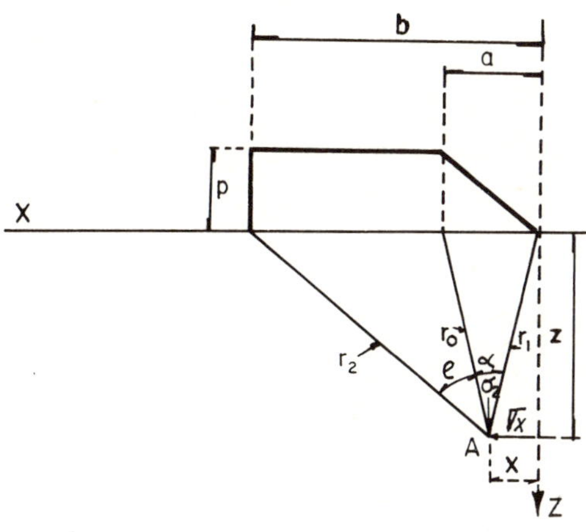

Fig. 3-2 Distribution of stress under a trapezoidal load of infinite length (rectangle — triangle)

The stresses in the directions which are shown are computed as:

$$\sigma_z = \frac{p}{\pi}\left[\beta + \frac{x}{a}\alpha - \frac{z}{r_2^2}(x-b)\right]$$

$$\sigma_x = \frac{p}{\pi}\left[\beta + \frac{x}{a}\alpha + \frac{2x}{a}\ell_n\frac{r_o}{r_1} + \frac{z}{r_2^2}(x-b)\right]$$

$$\tau_{xz} = \frac{p}{\pi_3}\left[\frac{z}{a}\alpha - \frac{z^2}{r_2^2}\right] \qquad (3\text{-}1)$$

Figure 3-3 is a graphical depiction of the solution for σ_z in Eqs. (3-1) carried out by OSTERBERG, which permits the stress, σ_z, to be computed at the points shown.

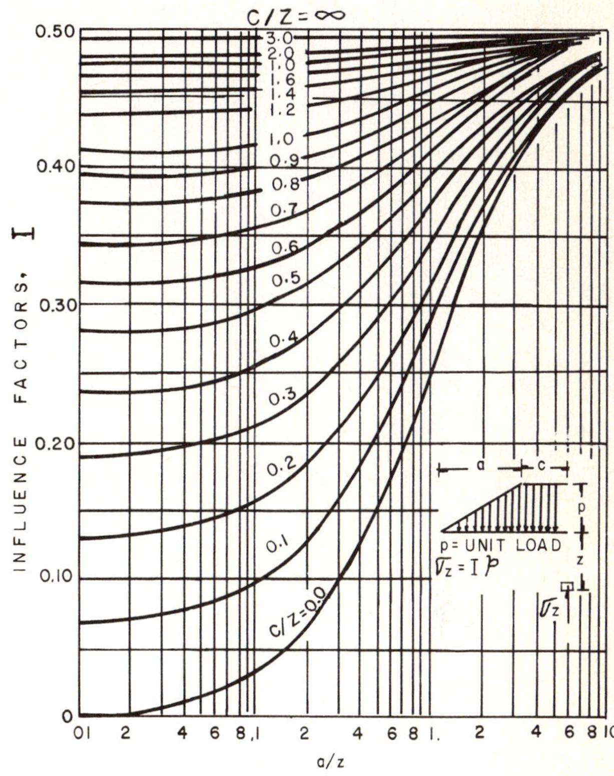

Fig. 3-3 Influence factors for vertical stresses due to a trapezoidal load of infinite length, according to J.O. OSTERBERG.

Plate 3-5 Movements due to settlement of the access embankment caused the problem that is shown in the abutment of the pile-driven bridge

Plate 3-6 Effect of settlement on the access embankments of a structure

Plate 3-7 Another example of the effect of settlement on the access embankments of an overpass. The access slopes were very steep so as to reduce the height of the embankments.

In order to calculate the values of σ_z under the center of the embankment, which is assumed to be infinite in length, it is necessary to multiply by two the value of σ_z obtained for each depth, z, because only half of the embankment is considered in this graph and the principle of superposition is employed. To compute the stresses under the center of the end of an embankment, it is sufficient to use half of the value of σ_z obtained for the entire embankment of infinite length.

In Figs. 3-4 and 3-5, a graphical solution is given for another case which is useful to the road engineer; it was originally proposed by GRAY [2,6]. This is a triangular load, of width B, and finite length, L, useful for the assessment of the stresses induced by embankment slopes. On the basis of the principle of superposition, the solutions in Figs. 3-3, 3-4 and 3-5 can, of course, be combined to provide a better reproduction of the geometry of real cases.

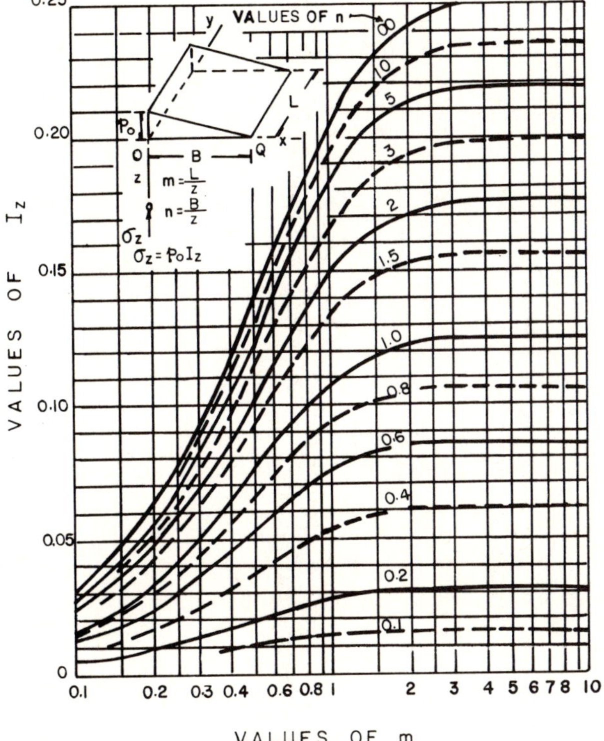

Fig. 3-4 Vertical stresses induced beneath point O by a triangular load of width B and of finite length L

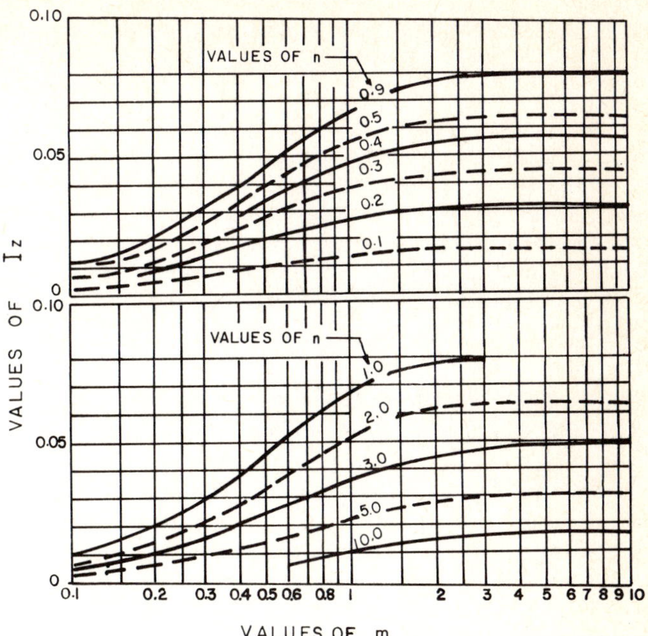

Fig. 3-5 Vertical stresses induced beneath point Q by a triangular load of finite length L. See Fig. 3-4 for position of Q

With a known distribution of the stresses induced by embankments on a soil mass, the settlement of foundations can be computed on the basis of TERZAGHI'S theory of consolidation, following the steps indicated in § 1.12.2.

It is very hard to define the permissible settlement for the construction of a road on soft soils. First, it must be remembered that the total settlement of an embankment may not be of great importance (except embankments joining rigid structures that do not settle, or embankments in flood zones), compared with differential settlements within a significant distance. Owing to the nature of its traffic, a highway usually tolerates differential settlements better than a railroad. On the other hand, with modern equipment it is possible to level a railway track with ease and speed, by laying more ballast until the original alignment is recovered. However in the case of a highway, levelling is generally done with asphalt or concrete which is a very expensive procedure. In airports, differential settlement limits, are usually very strict, for the undulations caused in an aeroplane by differential settlements may interfere with aircraft operation. Furthermore, differential settlements in runways encourage the formation of dangerous pools of water when it rains. This can, of course, occur in roads too, but on a smaller scale.

In short, no hard-and-fast rule can be given for definition design policies with regard to settlement in road building. In each particular case, the engineer must establish permissible values according to the importance of the problem and any other considerations. In Mexico, there have been cases of settlement of more than 1.0 m (3.5 ft) which have occurred without causing undue damage to the highway. An example of this is the direct Mexico-Puebla highway, which for several kilometers crosses an ancient lake basin. The subsoil formation is very homogeneous and the large total settlements do not, therefore, produce serious differential effects. Cases like this can be solved if the road is provided with an adequate freeboard in the first place. But even in these favorable cases, excessive settlement can lead to serious problems with access embankments, drains, and pile supported structures.

It is a fact well known to specialists in soil mechanics that the time rate of settlement is far harder to determine accurately than the magnitude of settlement. This is due to various factors, the principal one being that it is very hard to determine correctly the draining layers (and, consequently, the thickness of the compressible layers), which is so important if the results of the computation are to be correct (see §1.12). This is an unfortunate circumstance, for many important decisions that are reached by the road engineer are based on the determination of the time required for settlements to take place. As a consequence, there is often dangerous uncertainty about recommendations to construct embankments on soft soils before the rest of the road is built, with the intention that when the road is completed, any detrimental settlements will have occurred already.

3.4 Improvement of the Soil Foundation

Usually the soil foundation is capable of supporting roads under normal conditions, because the stresses to which it is subjected are relatively small, and an embankment usually adapts to any movements that may occur. The problems that have been mentioned so far, and the methods of improvement which will be described here, usually apply to restricted areas. They are not applied to all embankments owing to their high cost.

The principal methods for improving either the strength or reducing the compressibility of the ground, are discussed in the following.

.1 Use of Lightweight Materials

The aim here is to find sources of materials with a low unit weight within an economic haulage distance. These materials will be used to construct embankments, so that both the load on the ground and the geometry of the cross-section under construction can be minimized. Because settlement is usually associated with low strength, the slopes of an embankment built with heavy materials will have to be very flat, and perhaps with berms. The use of lightweight materials may reduce these problems. Since the settlement is less if the width of the embankment is less, this reflects favorably on the final settlement.

.2 Elevation of the Surface

The top of the embankment is raised initially, so that it reaches the required level after settlement. Whether this solution is effective or not depends on the capacity of the ground to support the higher embankment.

.3 Advance Construction of Embankments

In this case, the embankment is built long before paving, so that most settlement can occur during that period. The pavement is built afterwards, when the embankment no longer deforms significantly. Sometimes owing to the low strength of the foundation ground, it will be necessary to complete construction using successive surcharges, taking advantage of the strength that is generated by consolidation. There will, of course, have to be several surcharges, the final one producing insignificant settlement. The solution is highly successful, especially for bridges and bridge accesses; but it is limited by the availability of time.

.4 The Use of Vertical Drains

Because settlement is a consolidation process, any procedure that encourages consolidation will speed up settlement too. If settlement takes place during construction, the structure will be virtually unaffected during its active lifetime. Moreover, if consolidation is accelerated, there will be a corresponding increase in the speed at which shear strength is generated. Vertical drains have proved to be useful in speeding up consolidation processes, and their influence on these processes can be theoretically established [7]. They consist of vertical drill holes filled with permeable material. They are small in diameter and long enough to effect the entire body of the compressible mantle, or at least the thickness that will produce the greater part of the settlement.

Their function is to reduce the distance that water must travel in order to drain from the compressible strata that consolidate. This is achieved by promoting horizontal flow as well as the customary vertical flow. Because most fine clayey soils are stratified to some degree, their horizontal permeability is greater than their vertical permeability. Therefore the radial flow towards the vertical sand drains is, in principle, very efficient.

There are two basic systems, with several variants of each. The older system is the vertical sand drain or sand pile (Plates 3-8 to 3-10). A pipe is driven through the compressible stratum. In one variant the soil is kept out of the pipe by a precast concrete plug. Simultaneously, sand is placed inside the pipe and the pipe withdrawn, leaving the plug and a column of sand in the ground. In another variant the pipe is open-ended, and the soil which enters the pipe is jetted or drilled out at intervals during driving. Sand is again placed in the pipe as it is withdrawn, leaving the sand column. The second variant is more effective than the first because the compressible soil and any thin sand seams are not seriously disturbed by the pipe displacement.

Plate 3-8 Preparation of the work platform and drainage apron for the installation of vertical sand drains

The newer systems employ a plastic drain about 10 cm wide and 4 mm thick encased in a fabric or porous paper filter. The plastic drain is punched into the compressible soil, protected by a steel sleeve, about 15 cm wide and 2 cm thick. The sleeve is withdrawn leaving the filter in contact with the soil. This system is about 1/10 of the cost of sand drains. The flat installation sleeve minimizes soil displacement.

Installation of the vertical drains does cause some soil disturbance and a temporary loss of strength. It is more critical for sand drains installed by procedures that displace the soil, suitable field exploration and precise knowledge of the subsoil must be involved. They are, perhaps, most effective in cases where the drains cross lenses of more permeable materials. In homogeneous clayey soils, vertical drains sometimes do not cause enough acceleration of consolidation to justify the high costs. The cheaper plastic drains permit closer spacing, for a given cost.

For the efficient performance of vertical sandrains, it is important that the material used to fill them be very permeable. Experience indicates that it is advisable to surpass the usual standards for drainage or filter materials (discussed later). Special care must be taken that the amount of particles not retained by sieve No. 100 be very small. These fines represent a serious danger to overall permeability; variations of just 1% or

Plate 3-9 Installation of vertical sand drains

Plate 3-10 Machine for installing vertical sand drains

2% can reduce the permeability by 3 or 4 times [10]. Likewise, care should be taken to avoid segregation of the filter material when placing it in the drain.

The installation of vertical drains should be complimented by a horizontal drain blanket 30 cm to 1 m thick, which will cover the entire area under treatment. This will ensure adequate drainage of the water that collects in the vertical drains, and also some vertical flow in addition to the radial flow. The greater thickness is necessary to support construction equipment if the foundation soil is weak as well as compressible.

The spacing of vertical sand drains has a great effect on acceleration of the consolidation process. The rate of drainage increases inversely to the square of the spacing. The cost increases also at the same rate. The engineer has to find a happy medium. The diameter or width of the drains is one factor, though not critical. Typical sand drain diameter ratios are 30 to 40 cm (12 to 16 in) with spacings of 2.50 to 3.50 m (8 to 12 ft); typical plastic drain spacings are 1.50 to 3.00 m (5 to 10 ft).

The correct computation of the rate of consolidation in a vertical drain system depends largely on the accuracy with which vertical and radial permeabilities or vertical and horizontal coefficients of consolidation have been determined [11]. Vertical permeability can be measured in the laboratory using the methods for stratified soils; but radial permeability is best measured with field tests. The wells for installing piezometers, which are needed in any sizeable vertical sand drain system, can be used. Lack of precision in measuring permeability has been cited as an important cause of failure in predicting the benefits of sand drains.

Sufficient study has not yet been devoted to the effect that vertical sand drains may have on the shear resistance of the compressible stratum, by acting as genuine piles made of sand, that reinforce the soil.

Vertical sand drains are generally expensive, especially in countries where specialized machinery for their construction is not generally available. The plastic drains are far cheaper but also require special machinery. They cannot, therefore, be recommended without first carrying out a study of their effectiveness and a complete economic consideration of other alternatives.

.5 Compensation of the Load Represented by the Embankment

If the embankment material is added by excavation or displacement accompanied by a lateral displacement of the soft foundation soil, the net load on the foundation soil will only be the difference between the weight of the embankment and that of the soil excavated or displaced. The easier the lateral displacement of the foundation soil or its excavation, the more feasible the method. Best results are therefore obtained in organic clayey soils and peats. Displacement of natural soil is sometimes aided by surcharges and explosives. For airport runways, in which lengths are more limited, a procedure involving total compensation has been used. A box-like excavation is made and a thin slab of weak concrete placed on the bottom. It is then filled with lightweight but strong materials, so as to achieve total compensation. Foam plastic and porous volcanic ash have been used as fill. An example of this is the construction of extensions for some of the runways and taxiways at Mexico City Airport. The cost is usually prohibitive for highways.

.6 Removal of Compressible Material

This case is based on the following simple idea: if the foundation soil is poor and compressible, remove it and put another one of better quality in its place. The highways department of the State of California considers that this is the best solution in very soft and compressible soils occurring in layers no thicker than 4 to 5 m (13 to 16 ft). The substitute material must be granular where good drainage is not ensured. If the construction budget is low, even this may be impractical. In Mexico, for example, the substitution of stable soils for poor materials under embankments has been little used. Experience has shown that when the natural foundation soil is less than 4 to 5 m (13 to 16 ft) thick, favorable behavior can be achieved more economically by using one of the other methods described. When the layer of poor soil is thicker than 4 to 5 m (13 to 16 ft) the cost of material substitution is usually prohibitive. In short, material substitution should be regarded as yet another alternative available to the engineer, to be used only when it turns out to be the most economical or convenient method.

.7 Physico-Chemical Treatment of Compressible Ground

Even though these techniques are still in their early stages, it is known that if certain substances are added to a soil, ionic interchanges occur between the mineral particles in that soil and the materials dissolved in the interstitial water. The structural bonds can be altered to increase the strength of the soil and reduce its compressibility. A physico-chemical analysis of the soil will be necessary in order to define the substance or substances which might produce the favorable effects. These can then be made to diffuse through the soil by injecting them in water solution. In Mexico, various studies have been carried out to determine the applicability of these techniques, but they have never actually been used because of their high cost.

.8 Soil Calcination

This method consists of burning or calcining the soil fabric, using high temperatures produced by the combustion of gas. In some cases, notable reductions in compressibility, and consequently in settlement, have been reported. The method should be regarded as being at the experimental stage. It appears only to be cost effective in areas where natural gas is otherwise wasted.

.9 Networks of Branches, Palm Leaves, Etc. Placed Underneath the Embankment

This method consists of making a genuine raft underneath the embankment, to distribute the load and to enable the embankment to *float*. Excellent results have been achieved with this method for narrow embankments in jungle swamps. In Mexico there has been no conclusive experience.

.10 The Use of Berms or Very Flat Slopes

The pressures transmitted to the ground underneath the embankment can be more evenly distributed by flat slopes or by composite slopes that flatten toward the toe and by berms. Settlements are made more uniform, and differential settlements are reduced. On the other hand, it should not be forgotten that the greater the width of the loaded area, the greater the total settlement, and the methods dealt with under this heading, therefore, can increase these total settlements. The effectiveness of these methods will depend on the balance of these contradictory factors.

.11 Terracing of Natural Slopes

In natural ground with a steep tranversal slope, there is the danger that the embankments may slip downhill, on the plane of contact with the virgin ground. Terracing (benching or stepping) the ground in a manner that is in keeping with the geometry of the embankment and the natural topography is the method most often used in Mexico to combat these problems. The terraces or steps with their horizontal tread and vertical rise, provide the embankment with horizontal support, eliminating the component of its weight along the contact surface with the ground. The horizontal width of the bench or step tread is sufficient to operate the construction equipment. A detailed description of the shape and dimensions of the terraces must be included in the design. The benching or terracing must be in firm enough ground such that the embankment cannot slide by shearing the foundation soil.

.12 Construction of Fills on Irregular Rock Support

When excavating cuttings in rock, a common consequence of blasting is that the bed of the road is left rough, full of sharp, irregular edges. In this case, a sufficiently firm thick layer of soil must be placed between the rock and the pavement, so that these irregularities will not affect the pavement. This is a case which illustrates how improvement of the ground can consist of replacing hard support by a material of apparently lower quality. The lesson is that problems involving the interaction of the superstructure and the foundation of a road are so complex that the improvement measures selected sometimes appear to be self-defeating. The important thing in these cases is to weigh all the benefits and shortcomings of the method that is adopted.

.13 Compaction

The uppermost portion of the foundation is often improved by compaction after clearing the land of vegetation and roots. This treatment termed proof-ralling is frequently used, especially in runways. The density required is not high, 85% to 90% of what would be achieved by any usual standard.

.14 Anchorage of Blocks of Fractured Rock

On steep rocky hillsides, and in cases where the fracture planes are not favorable to road building, blocks of rock are anchored by introducing steel rods into drill holes, which are afterwards sealed with concrete or cement; the fragments that are hazardous are literally laced or stitched together.

.15 Crack Filling

The surface of the foundation ground often presents cracks. When this occurs, the cause of the cracks must always be investigated, for the phenomenon may be a sign of pending failure that is still relatively simple to correct for example in a sloping hillside. It also can be a generalized surface slide condition or the result of high tension stresses of the type described by JUÁREZ BADILLO, in [13].

In order to correct these cracks, it is essential to eliminate the cause. In some cases, the best solution may be a change of location, for, as it has been said, the cracks may be part of large-scale phenomena that are extremely difficult and costly to correct. Nevertheless, once the cause has been eliminated, it may prove helpful to fill the already formed cracks with clay, asphalt, soil or some other similar material of a plastic nature. Open cracks may be dangerous, because when they become filled with water they generate hydrostatic thrusts that can aggravate any already existing tendency towards instability.

As can be appreciated, none of the methods proposed [12] with a view to improving the strength or compressibility characteristics of the foundation ground of an embankment provide a universal solution. Therefore wherever improvement is absolutely essential, each circumstance must be carefully analyzed, so that the most suitable solution or, combination of solutions, can be chosen. Some of the methods proposed are contradictory, in that although they satisfy one particular aspect of the problem, they may prove unfavorable in other aspects. The choice in each case is not subject to hard-and-fast rules, but depends on the judgment of the design engineer. Fortunately, the theoretical methods provided by soil mechanics (see Chapter 1) allow computation of both the magnitude of the settlement and their evolution with time, though the latter not quite so accurately.

This computation demands a far more detailed knowledge of the properties of the subsoil than can be achieved by following the usual procedures for exploring foundation soils for highways and runways. Thus, when areas of soft clayey soils posing special problems are involved, exploration must also be of a special nature, and undisturbed soil specimens will be necessary. Consequently, the laboratory test program cannot be a routine one either, and must include unconfined and triaxial compression tests to determine the shear strength, and consolidation tests to define compressibility characteristics.

3.5 Water in the Foundation

Of the water that falls on the ground where a road is to be built, part runs along the surface, part infiltrates the soil and part evaporates. The relationship between the water that runs off and the total precipitation is described by the *runoff coefficient* of the soil. It varies in accordance with type of soil, inclination, type of vegetation and other factors, such as previous rainfall.

The water that runs over the surface of the ground often erodes it, polluting the runoff and damaging the ground surface.

The water which filters into the soil penetrates it until it reaches an impermeable layer. It then saturates the area overlying that layer, thus forming the groundwater table. The groundwater maintains a fairly constant level so long as there is no substantial change in climate, ground cover or groundwater flow. When the groundwater table is near the surface, or not very deep, the result is swampy or marshy terrain. When the water table is relatively deep, but the overlying soil is fine, with a high capillary potential, moisture can rise to great heights and affect earthworks and pavements. The following is a description of the principal effects produced by changes in phreatic and capillary water in the foundation ground.

— A variation in the water content of a soil brings about changes in important mechanical properties, such as shear strength. In clayey soils, for example, or ones containing appreciable quantities of fines, an increase in water content will lead to a notable reduction in the shear strength, as well as an increase in compressibility. In sandy soils, especially those that are cemented with soluble substances, the addition of water can bring about drastic changes in structure and, consequently, a loss in strength due to the buoyant forces and to loss of cementing. All the foregoing considerations influence the settlement that may occur in an embankment, the possible failure of that embankment, and deformation that may be suffered by the subgrade and pavement.

— The movements and variations in the groundwater and their effects are never uniform, which explains the existence of zones with different types of behavior within the ground.

— Changes in water content bring about changes in volume, which are detrimental in expansive soils.

— In soils that are susceptible to frost [14], the presence of water is particularly dangerous, owing to the variations in volume and strength that occur with periodic freezing and thawing.

— The action of water on a pavement can have destructive effects that are different from the ones associated with changes in volume or strength. These may be pumping of the base and subgrade or separation of the asphalt film from the particles of aggregate in asphalt base courses or surfaces. Many of the most successful methods of treatment which can be applied to foundation soils, apart from the previously listed ones, are related to the removal of water from these soils. These are subdrainage techniques which play a fundamental role in road engineering. They will become increasingly important in the future. When an embankment is built, the watertable of the foundation soil undergoes a change: evaporation of water no longer occurs in the area which was previously exposed. This is why the level of the groundwater table will rise in foundation soils beneath embankments. The embankment often becomes a dam, obstructing surface water drainage, and increasing infiltration and the groundwater level.

Subdrainage techniques are basic for the treatment and improvement of foundation soils. Drainage will be considered in a separate chapter, because it is common to most aspects of road engineering.

3.6 Foundations of Clean Sands

Roads in desert areas and near the sea coast must cross sand dunes. Although this should be avoided by relocation whenever possible, there are times when it is unavoidable. The resulting problems are very costly and hard to solve. There are, however, certain practical rules and remedial methods which can lead to successful results. In [15], an analysis is made of a real case, which summarizes the chief precautions to be taken in such cases.

In general, there is an inverse relationship between the movement of dunes (which is the principal cause of road construction problems) and their weight. A large dune, 100 m (320 ft) high for example, can advance as little as two centimeters per year, whereas dunes 2 or 3 m (7 or 10 ft) high can move tens of centimeters per hour during a violent storm. A dune 10 m (33 ft) high can easily advance as much as a meter (3 ft) per year.

A dune represents a mound of sand that has been dumped by wind action so that the slope of the advancing front will be very close to the angle of limiting equilibrium, or $\varnothing$ minimum. The windward part of the dune has a flatter slope, because of the way in which it advances under the action of dominating winds. The standard cut slope that is made in the advancing front will sometimes be steeper than the angle of limiting equilibrium of sand; it will not, therefore, be stable and will encourage sand to slide to the base of the cut. Sometimes as a result of capillary tension, the sand has apparent cohesion and occurrence of the sliding may be delayed.

The general stability of the dune is not basically affected by the cut, which means that the volumes of sliding material are not usually very large, but the fact that the phenomenon is continuous implies a danger to traffic and makes any cutting in dunes inadvisable. The road is,therefore, best located on an embankment, or at least on a level with the highest dunes. This can lead to prohibitive costs, and the engineer will sometimes have to be satisfied with a location corresponding to the level of the most mobile dunes, taking care not to cut into the higher ones. It should not be thought that by avoiding cuttings and locating the road on embankments all the problems involved in crossing a duny area are automatically solved. The embankment is a barrier against winds and general movement of the sand. Without a detailed study, it would be very hard to say what the effects of this barrier are; they may sometimes be quite unfavorable.

It has always been common practice to protect roads located in dune areas by planting appropriate vegetation on the side from which the sand advances. This can be carried out either on a massive scale or else by planting in rows parallel to the road. A complementary effect can be provided by curbs, fences and groups of trees and shrubs. Plant species are selected with the aid of an agronomical study (which is beyond the scope of this book), but the engineer will find that knowledge of the most appropriate plants for the region is an indispensable basis for designs of this kind.

The mechanical properties of dune and beach sands have some interesting features that have been the object of specific studies. In [16] for example, some important data and conclusions are given, showing the influence of stress history on current characteristics.

3.7 Foundations of Very Soft Clays and Peats

3.7.1 Introduction

Deposits of soft soils and peats generally have three conditions in common: they are flat areas, they have poor surface drainage and they are made up of very fine clayey or organic soils. To overcome these problems, it is essential to detect them early in the design stage, before they cause costly damage to the road and while the engineer still has freedom of action, including the ability to study a change of location to avoid any zone that turns out to be critical. For this, the interpretation of aerial photographs is of great assistance.

Once the decision is taken, for whatever reason, to confront the dangers and high costs of crossing a zone of soft or organic soils, the engineer must obtain thorough knowledge of the compressibility and strength characteristics of both the soils on which the road is to be built, and those that will be used in its construction. This case should be regarded as special as far as soil exploration and laboratory tests are concerned, and therefore justifies the use of advanced methods to obtain undisturbed specimens, and complete laboratory programs including consolidation tests and triaxial tests.

Exploration must be carried out with a very clear distinction between the two traditional stages [17]: first, preliminary sampling using simple, economical procedures that will produce intact but disturbed specimens for soil classification; then the final investigation using careful, far more demanding (and expensive) methods capable of providing undisturbed samples. The data collected in the first stage, should provide reasonably reliable soil profiles that are fundamental for planning the second stage at optimum time and cost. The information that is obtained concerning the ground should be sufficient to study the two major problems of embankment stability and settlement of the embankment, in detail [18].

It is desirable for any significant settlement to occur *during construction;* however this is not usually possible without the use of methods to accelerate the consolidation process, such as sand drains and surcharges. Settlement time is not greatly affected by loading, but the magnitude of the settlement is; a surcharge will therefore enable the embankment to reach its ultimate settlement in less time than it would otherwise. If these methods are uneconomical, temporary pavements with elevated grades should be considered so that settlement which occurs during the operation stage can be tolerated. The magnitude and nature of the problem is in each case strongly influenced by the characteristics listed below:

Embankment dimensions. The height and width of the embankment have an important influence on the choice of a suitable solution. A high, narrow embankment is far more likely to settle as a result of shear displacement than a lower wider one. For the first kind a construction procedure based on the displacement of the weak foundation material may prove effective.

Characteristics of the foundation ground. The most important factors here are the strength profile of the soft soil, its thickness, and its compressibility.

Construction materials. The engineer's decision will be greatly influenced by the availability and cost of the materials he will use to build his embankment. If, for example, there is no granular material within a reasonable distance, underwater dumping will be out of the question. The use of lightweight materials such as volcanic ash, will be possible only when they are within economic haulage distance. The use of such materials gives, however, opportunities for adopting numerous solutions which would otherwise be unthinkable.

The construction program. Program requirements have an important influence on possible design methods. The date by which the final pavement is to be laid determines whether sufficient time is available for construction by stages, use of surcharges, or if vertical drains are needed.

Location. The topographical conditions of the site: whether natural, or man-made as a consequence of other construction, also have an important influence on the methods that can be selected. For example, the presence of people imposes severe restrictions on the use of explosives; likewise the existence of a narrow right of way will restrict the use of berms or the formation of mud waves during displacement.

3.7.2 Mechanical Properties of Soft Soils

Table 3-2 gives a brief summary of the methods for building embankments on very soft soil foundations. Before proceeding to a rapid description of the principal methods, let us recall briefly the main properties of very soft foundation soils, together with the most reliable methods of obtaining the required information [19,20].

From this particular point of view, the significant properties of peats and very soft soils are density, water content, permea-

Table 3-2

Methods for building embankments on very soft foundation ground

I. Removal By:

a) Excavation
   1. Complete
   2. Partial

b) Displacement
   1. By the weight of the embankment, with or without surcharges
   2. With explosives

II. Improvement of the Ground

a) To improve stability
   1. Construction in advance or in stages
   2. Use of lightweight materials
   3. Stabilising Berms
   4. Interceptor drainage

b) Primarily to reduce settlement
   1. Construction by stages
   2. Surcharges
   3. Compaction with heavy equipment

c) To improve the stability *and* to reduce settlement
   1. Construction by stages or with surcharger
   2. Vertical drains
   3. Combination of the foregoing methods

bility, shear strength and compressibility. Owing to their high water content and the difficulty in obtaining and working with specimens, special care must be taken in handling these soils for sampling and testing, so that results will be obtained that agree statistically.

The water content of peats and highly compressible clays can vary from 400% to 1500%. Occasionally values of over 2000% have been reported. The uppermost meter of soil commonly has a far lower water content, even in areas where there is a greater predominance of peats. Estimation of the voids ratio is usually based on the water content, and the relative density can also be assessed, with values as low as 1.5 or 1.6 for very pure peats. The air and gases contained in peats may be of interest. Since for this there is no standard test of recognized value, estimation is by means of consolidation tests. Values as high as 10% are not unusual in peats.

Peats show a marked reduction in permeability when subjected to increasing loads. In virgin formations, the permeability coefficient is usually between 10^{-2} and 10^{-4} cm/s (4×10^{-3} and 4×10^{-5} in/s), but may fall to 10^{-9} cm/s (4×10^{-10} in/s) in consolidated peats underneath an embankment one or two meters high. Figure 3-6 shows the correlation between values of the voids ratio and the permeability of British Columbia peats, which is dealt with in [19].

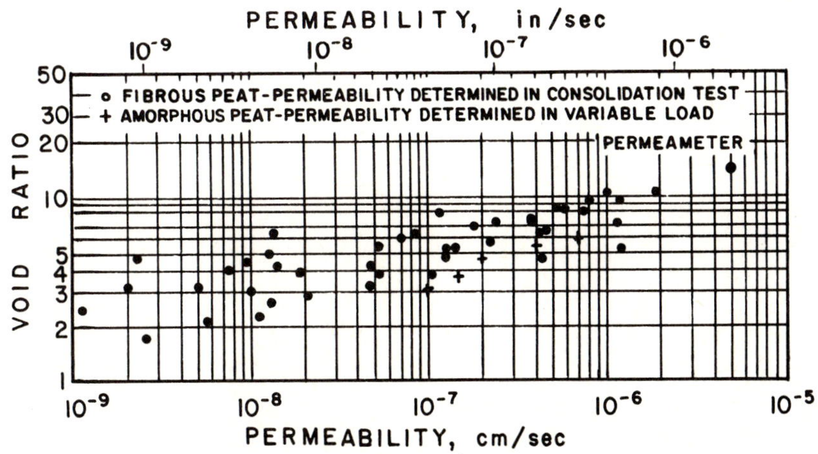

Fig. 3-6 Void ratio vs permeability in peats [19]

Table 3-3
Shear strength in peats

Reference	Location	Shear Strength		Natural Water Content
		t/m²	lb/in²	
ANDERSON and HEMPSTOCK, [21]	Canada (Alberta)	0.50-1.25	0.71-1.78	700-1400
CASAGRANDE, A and L, [22]	U.S.A. (Mass.)	0.50-1.85	0.71-2.63	230-750
		0.35-2.90	0.50-4.12	400-800
		1.35	1.92	400-550
		0.35-0.95	0.50-1.35	250-380
		5.00(1)	7.1	110
DÜCKER, [23]	Germany (Holstein)	0.10-5.00(1)	0.14-7.1	400-800
FRASER, [24]	Northern Ireland	1.40-2.80	1.99-3.98	680-1450
HARDY and THOMSON, [25]	Canada (N.O.)	0.50-3.00	0.71-4.26	470-760
LEA and BRAWNER, [19]	Canada (Alberta)	0.55-1.50	0.78-2.13	No Data
MARGASON and FRASER, [26]	Northern Ireland	1.70	2.41	790
MOOS and SCHNELLER, [27]	Switzerland	0.50-1.50	0.71-2.13	220-1460
RIPLEY and LEONOFF, [28]	Canada	1.00-2.25	1.42-3.20	100-2100
SMITH, [29]	England	0.35-1.80	0.50-2.56	No Data
TRESIDDER and FRASER, [30]	Scotland	0.35-9.35	0.50-13.28	400-1600
WARD, [31]	England, Wales	0.65	0.92	800-1000

Shear strength can be determined by means of simple compression tests or triaxial tests, although the more organic soils may present increasing difficulties as regards specimen handling, which is why vane tests are often used to determine and estimate strength, by applying the calculations to sections where a failure may already have occurred, or where one may be induced. Table 3-3, taken from [20], compares the shear strength obtained for various different peats and their water content. The majority of the strengths in the table correspond to in-situ vane readings. The table also gives an interesting list of publications on the subject, compiled by L. CASAGRANDE.

Figure 3-7, [20], shows mean results obtained by different researchers at different locations, and correlates shear strength with depth. In it can be observed the important drying effect which is characteristic of peaty soils. The majority of vane test results are concentrated in the middle section of the figure (shaded) and show substantial preconsolidation by drying, even at the greatest depths.

Because of the relatively high degree of permeability in some peats, primary consolidation can take place very rapidly [19,32-38]. However, some also consolidate very slowly as suggested by the low permeabilities of Fig. 3-6.

Secondary consolidation predominates when primary consolidation is concluded, and generally follows a linear law when settlement vs. logarithms of time is plotted. Secondary consolidation of peats can be far more important than primary consolidation and lasts for many years; it is affected by the decay of organic matter during the lifetime of the structure, [36,38,39]. Prediction of settlement in peats by normal soil mechanics methods is unreliable because of the foregoing; even more unreliable is prediction of the evolution of settlements with time. Simple one-dimensional consolidation is not sufficient. Most horizontal drainage is very rapid in peats because of their anisotropic permeability. There is little point in determining 100% primary consolidation in peats [19], because of the continuing secondary consolidation.

If the soft soil is inorganic, and the clays are reasonably homogeneous, TERZAGHI's theory can be applied to calculate settlement, and even the evolution of that settlement with time, although the latter will be determined far less accurately.

For these reasons, the results of the calculation of settlement and its time-rate, will not be sufficiently reliable to provide a basis for an important construction project, especially in the case of peats. In such situations the instrumentation of test embankments, preferably full scale, can give very good results [40].

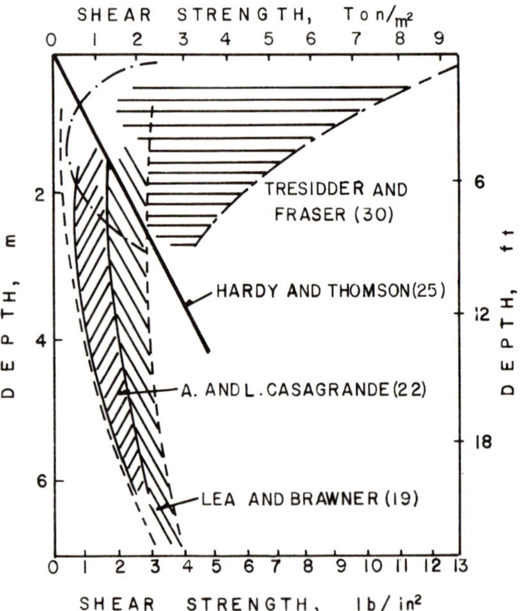

Fig. 3-7 Shear strength vs depth in peats (in situ vane readings) [20]

3.7.3 Construction Methods

Some brief comments follow in relation to the construction methods that are listed in Table 3-2.

.1 Excavation

Total excavation: is economical when the layers of peat or soft soil are not very thick, and when very rapid stabilization of the embankments is desired. It usually proves economical to dispose of the waste material right beside the excavation site so that no transportation costs are involved. The finished excavation is usually full of water and often has flat slopes. The soil fill will have to be granular.

Generally speaking, the narrower the embankment, the more effective is total excavation. This is because with a wide embankment and mudwaving there is a greater risk of masses of peat or soft soil being trapped underneath and subsequently causing problems.

The depth at which total removal of the foundation soil should be regarded as possible is variable and depends on the project. It has been done even with a layer 10 m (33 ft) thick.

Partial excavation: is recommended if the strength of the soft soil increases with depth and the compressibility decreases. It is also used as an aid to other construction methods, such as displacement.

.2 Displacement

When the stress which is transmitted by the embankment to the foundation ground is greater than the strength of the latter, (plus any restrictive force that may exist) a displacement will occur in the foundation soil in the direction offering least resistance. The intensity of the displacement depends on the relationship between the thickness of the soft stratum, the height and width of the embankment and the magnitude of the previously mentioned lack of balance. The displacement causes waves of mud at the sides of the embankment, which act as a restriction to future displacements. Generally, in embankments on uniform soft soils, once subsidence of the structure and displacement of the soft soil have commenced, the process will continue if the same elevation is maintained at the crest of the embankment by continued filling, or if the mud waves that form are removed. This is because the embankment material has a greater unit weight than the soft soil, so that the total superimposed weight increases gradually with an increment in fill height. If the height of the embankment is maintained constant, the displacement will lead to partial compensation of the embankment weight. If the mud waves that form at the sides are not removed, they can generate sufficient resistance to prevent any further displacements. Displacement is encouraged by the drop in strength suffered by a sensitive soft soil as a result of remolding.

Displacement due to the weight of the embankment, with or without surcharges: It is sometimes possible to use displacement to remove the soft soil if there is a great enough difference between the weight of the soil displaced and that of the fill, so that movement will continue despite the balancing effect of the mud waves. If the fill is sufficiently heavy, any of the soft clay that is trapped under the fill will subsequently consolidate. In order to accelerate the process, excess fill (a surcharge) can be added. Surcharges have two favorable effects; they induce a larger displacement and make settlements occur more rapidly by consolidation. Figure 3-8 illustrates the final situation reached by an embankment that is built using the method of displacement under a surcharge.

In the case of embankments on very soft soils, it is good practice to place a surcharge of earth on top of the embankment. The weight limit of this surcharge is imposed by the strength of the soft soil, because a violent failure, as a result of a slope slide or lack of load bearing capacity, can have very dangerous consequences because of the inevitable remolding and loss of strength that occurs in the soft materials.

The effect of a surcharge on future settlements can be estimated by the methods already described. The accuracy is poor because simple one-dimensional consolidation may not be applicable and because the amount of soft clay trapped under the embankment cannot be predicted.

Displacements by blasting: Here the objective is to cause an instantaneous increase in neutral pressure, with a consequent reduction in the strength of the soil, followed by gas pressure.

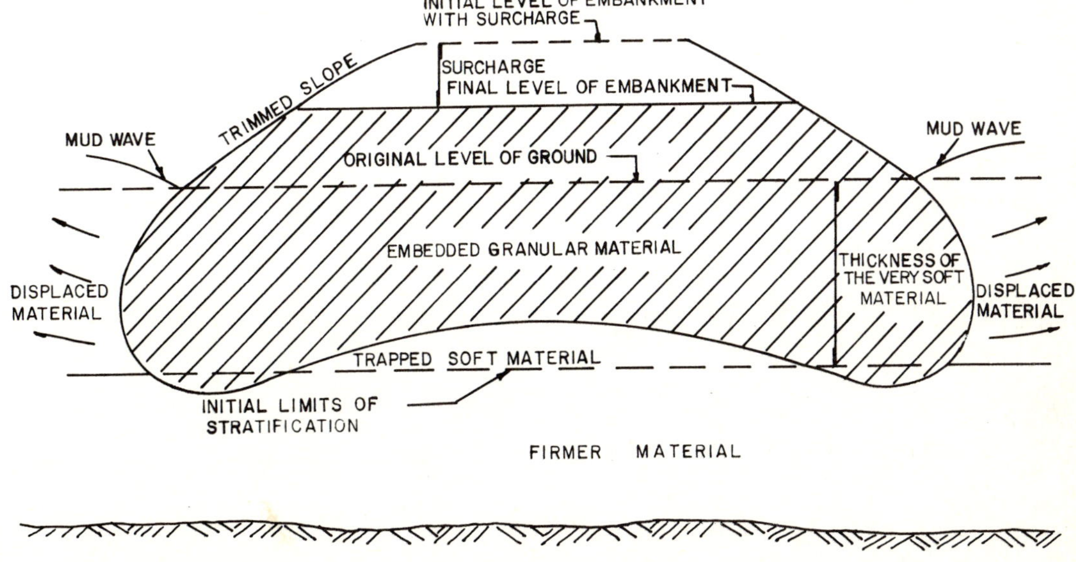

Fig. 3-8 Placing an embankment by displacement of a soft soil, using a surcharge

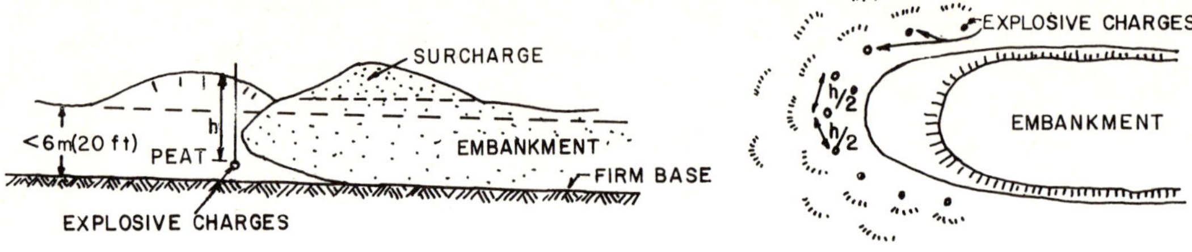

Fig. 3-9 Displacement of peats by blasting the advancing front [20]

This method is becoming more and more popular for stabilizing embankments on very soft soils, and it will undoubtedly be even more widely used in the future. Reference [20] is an important source of information on this subject, for it not only offers a detailed description of the method, but also gives a very complete list of references. The description that follows is based on this publication.

The following methods are the ones that have been most widely used for displacing very soft soils with the aid of explosives.

Blasting the advancing front: The purpose of this method is to modify and shift deposits of very soft soils by means of explosives which are placed in bore holes situated around the advancing end of the embankment under construction, at a distance of 8 to 10 m (26 to 33 ft) (Fig. 3-9).

It is best to set off a whole row of charges each time. The explosive charge in each hole must be so small that it will not t damage the adjoining embankment. It has to be determined with the aid of experiment, and often turns out to be about $\frac{h}{4}$ in kg (ft/6 in pounds), h referring to Fig. 3-9.

It is a slow method, and sometimes leaves large amounts of soft soil trapped underneath the embankment. German practice has established an operational sequence which can be regarded as a variation on the traditional method. It consists of the following steps.

1) A work platform is made of sand ahead of the advancing end of the embankment (Fig. 3-10). The thickness of this platform can be between 30 and 60 cm (1 and 2 ft).
2) On this platform, holes with a diameter of 20 to 30 cm (8 to 12 in) are bored, until firm ground is reached. Spacing between the holes can vary from 2 to 5 cm (7 to 16 ft), depending on the thickness of the soil layer to be displaced.
3) Explosive charges are laid at the bottom of the holes, in quantities between 8 and 40 kg (18 and 88 lb).
4) The necessary electrical connections are made for the explosion, and the wires are suitably protected.
5) The embankment is then built up over the boreholes until the desired height is reached, including any surcharge that is to be used.
6) The explosion takes place.
7) The fill subsides into the gap.

Blasting underneath the body of the embankment. After clearance of the ground surface, the embankment is built and 4 to 12 cm (1.5 to 5 in) holes are then bored in it, following any suitable procedure. When thick layers of peat or very soft soil are to be displaced, blasting is recommended in stages, each time affecting a layer of peat 4 to 5 m (13 to 16 ft) thick in embankments from 30 to 50 m (98 to 164 ft) long. About 25 kg (55lb) of explosive can be used in each borehole [41-43].

The New Hampshire Method. The New Hampshire Department of Highways has developed an economical method of shifting soft soils with a layer thickness between 3 and 15 m (10 and 50 ft). It enables the constructed embankment to rest on the firm underlying strata by alternating blasting. Once the vegetative covering has been cleared away, construction of the embankment commences, starting with the two end segments.

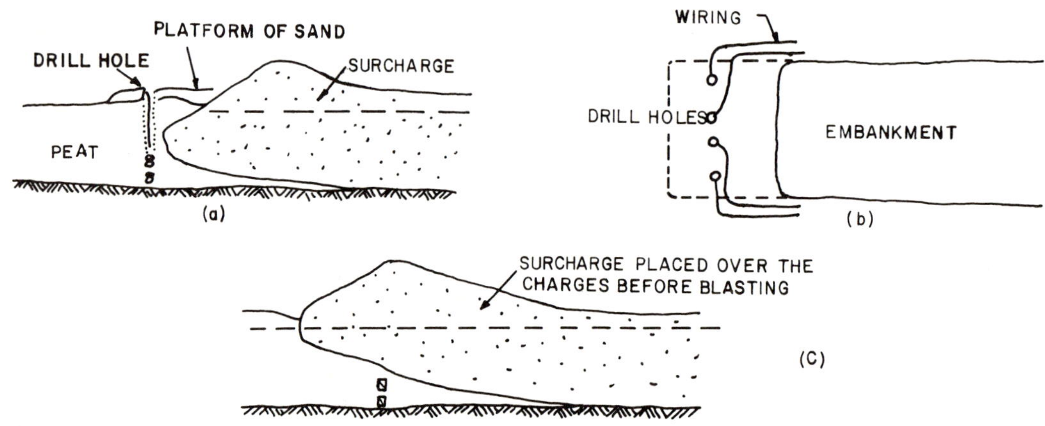

Fig. 3-10 German method of blasting the advancing front [20]

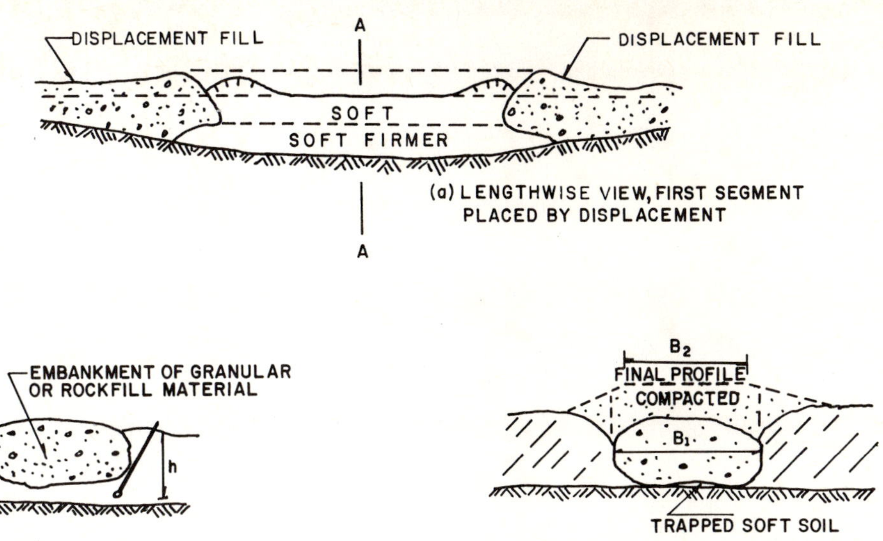

Fig. 3-11 New Hampshire Method [20]

These are constructed using only surcharges and displacement until these sections of the embankment finally rest on the underlying firm strata. The two end segments are then joined, by dumping material onto the soft soil between so that this is trapped underneath. The trapped soft soil is removed by blasting and construction of the embankment can be completed (Fig. 3-11).

The soft material that is trapped underneath the embankment thus constructed is removed by means of the following procedure. Holes are bored on both sides of the fill, about 3 m (10 ft) apart, as shown in part (*b*) of Fig. 3-11. The holes are from 4 to 5 cm (1.5 to 2 in) in diameter and must contain a quantity of explosive in kg corresponding to about one third of the thickness of the soft material in m (ft/4.5 in pounds).

If construction of the embankment causes large-scale mud waves at the sides, a second row of boreholes can be made about 3 m (10 ft) from the previous ones, as shown in Fig. 3-11. This second row of charges must be set off a fraction of a second after the first, for experience has proved that maximum efficiency is reached if when the principal rows explode they meet with resistance on both sides. Part (*c*) of Fig. 3-11 shows the final position of the embankment.

This method has proved useful when the body of the embankment is made of very coarse granular material or even rock fill for its settles more uniformly than when it is made of finer materials, such as sands. Moreover, coarser soils form a better arch over the small pockets of soft material which might in any case become trapped.

The material (Fig. 3-11c) that is used to give the embankment its ultimate shape, can be any suitable one, and should be compacted in the usual way.

The German Method. This method was developed for the network of highways which were built in Germany between 1934 and 1940 [44-47]. Once the vegetation has been cleared away, the entire length of the embankment is constructed on soft soil, as shown in Fig. 3-12. A large quantity of charges are then placed underneath the embankment, and made to explode simultaneously throughout the entire length and width. This large-scale explosion is highly effective for destroying the strength of the soft soil, so that the embankment settles easily until it reaches the final position shown in Fig. 3-12c.

The explosives can be laid out in 4 rows, or more, underneath the embankment, and in each borehole up to 100 kg (220 lb) of explosive can be placed when the layers of soft soil are thick, or 25 kg (55 lb) when the layers are not so thick.

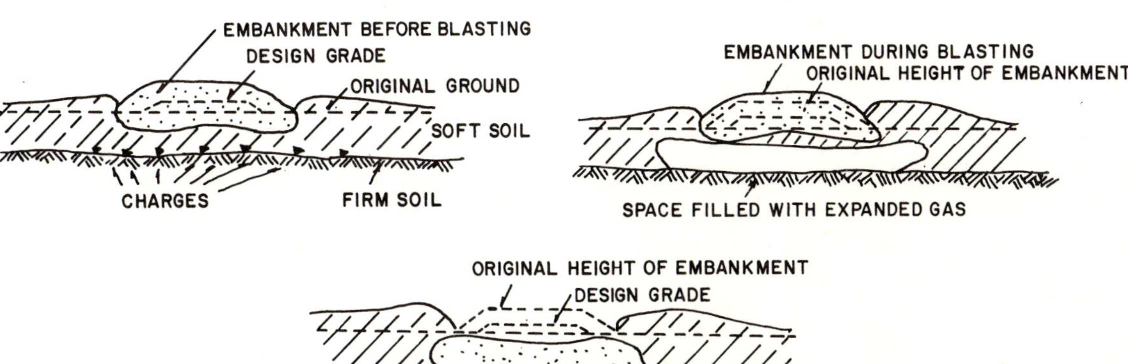

Fig. 3-12 German Method [20]

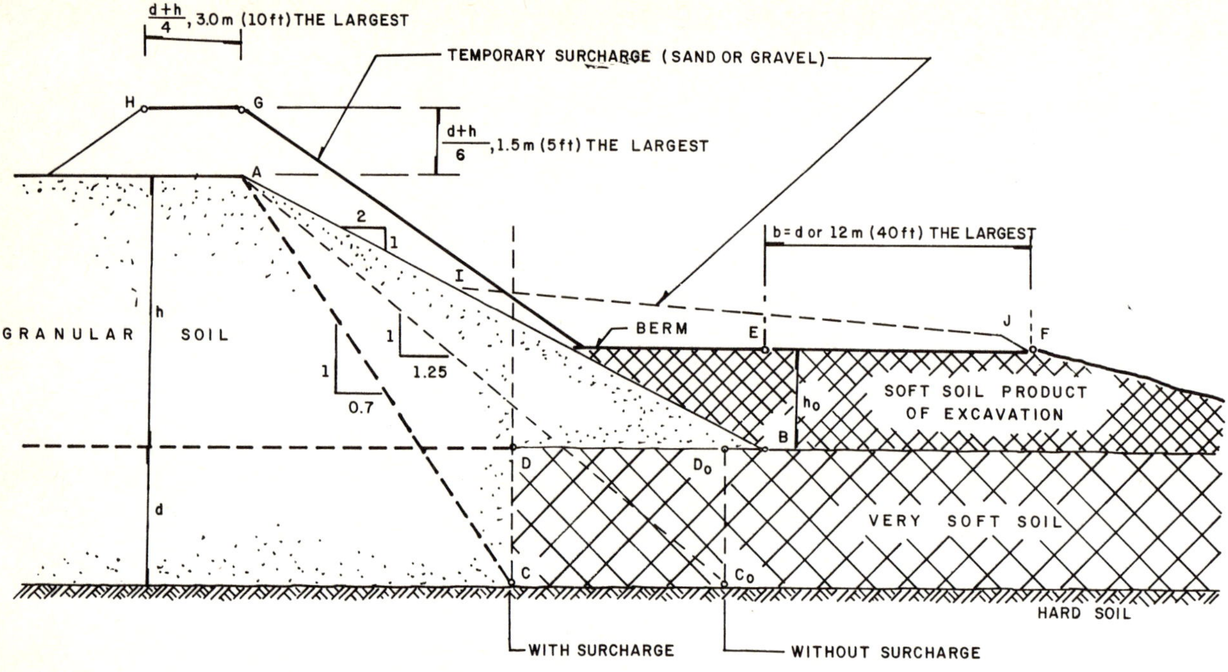

Fig. 3-13 Determination of minimum width of fill for small *h/d* [20]

Determination of Width of Fill: Apart from the decision as to what procedure should be followed to embed the embankment into the soft ground, the question also arises of the width of the fill to be used in order to avoid subsequent deformation, and cracks, especially at the edges of the embankment. In [20] L. CASAGRANDE gives semi-empirical rules for establishing this width, which are referred to in Figs. 3-13 and 3-14.

When the embankment becomes embedded in the foundation ground *without* the help of a temporary surcharge of soil, the method used to determine the width of the excavation is that illustrated in Fig. 3-13:

Starting at A, on the embankment shoulder, a line is drawn with a slope of 1.25: until it intersects at C_0 with the firm stratum which is underneath the soft soil that is to be displaced. The width of the excavation in soft soil is determined by the vertical line C_oD_o. The slope of the final embankment is shown in Fig. 3-13, with an slope of 2:1, and its toe B coincides approximately with D_0, but is located slightly outside the fill zone, so that there is no danger of substantial deformation taking place in the toe of the slope. If, however, h/d is very large, which is the case illustrated in Fig. 3-14, the previously described structure leads to B, far to the right of D_0. As a consequence a large part of the final embankment would still be on very soft soil, susceptible to deformation. This situation can be handled by increasing the width of the excavation, so that D_o coincides approximately with B, constructing a berm as illustrated in the figure or making the embankment slope steeper, until B is near D_o.

When a temporary surcharge of soil is used to help the embankment become embedded in the foundation ground, the method proposed by L. CASAGRANDE for assessing the width of the excavated and filled area is determined according to the following rule, which also relates to Figs. 3-13 and 3-14.

A line is drawn with a slope of 0.7:1, thus determining C on the firm soil. The vertical line CD defines the width of the excavation in soft soil. The relative position of B and D, is the same as discussed previously, so long as a substantial part of the embankment is not resting on soft soil.

When h/d is small, the above procedure may lead to an excavation width that is much wider than the embankment. In this case the position of B determines the width. For very thick layers of soft soil, it is advisable to use stabilizing berms on both sides of the embankment. Any kind of material including excavated soft soil can be used for the construction of these berms, for their only function is to apply weight.

.3 Treatment of the Natural Foundation Soil to Improve Stability

Removal of the foundation soil by excavation or displacement, which involves the substitution of a poor soil by another one of better quality, often leads to excessive earth hauling, unpractical construction procedures and high costs. Criteria which require substitution methods for all soft foundations do not, moreover, allow correct discrimination of those cases where the foundation soil is genuinely unsuitable. Simple treatment of the natural ground will often make it possible to use soils which, if a stricter criteria were followed, would be removed or displaced at great expense. When considering methods for improving the condition of the natural soil, it is necessary to consider both strength and settlement.

It is sometimes possible to improve stability simply by using berms that are calculated as shown in Chapter 6 on Slope Stability. At other times, the solution may be to flatten the grade, consequently reducing the height of the embankment. Likewise the angle of slope can be dealt with by giving the smallest possible inclination to the stretch of road where maximum settlement is expected, so that when this settlement occurs the differential effect will be a minimum.

The advantages of advance construction are obvious, and it is not felt necessary to go into details. The only requirement is

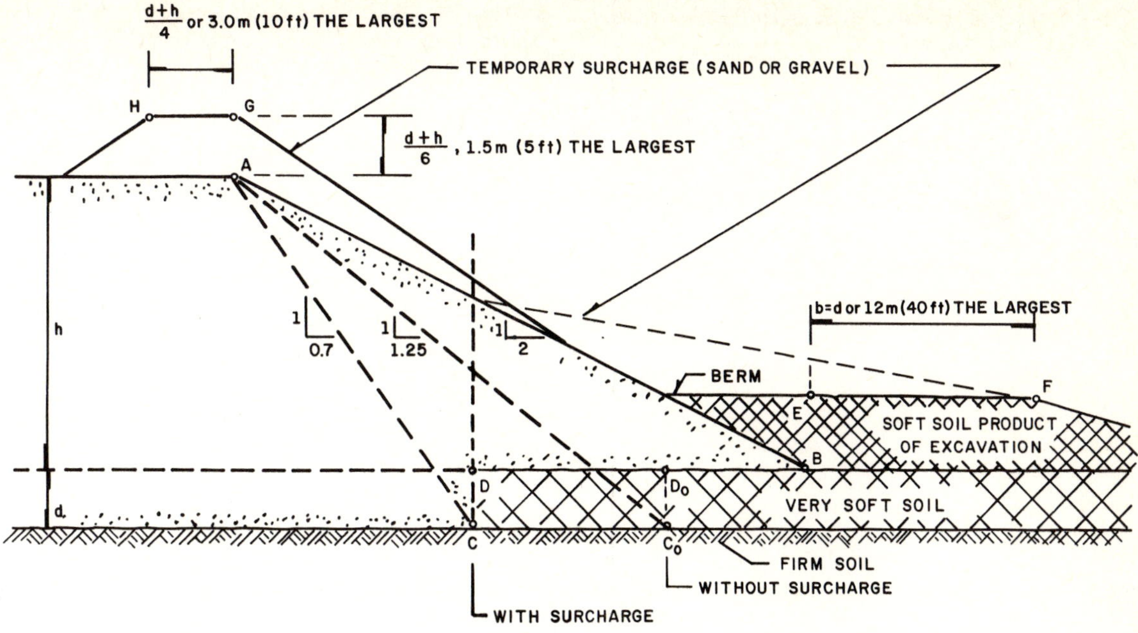

Fig. 3-14 Determination of minimum width of fill for large h/d [20]

compatibility with the project schedule. The advantage of construction by stages, and the use of lightweight materials in the embankment, depending on their availability, have also already been mentioned.

Sometimes the condition of the soil foundation can be greatly improved by using interceptor drainage, uphill from the embankments. The methods most often used for this are underdrainage ditches, transversal drains and even pumped wells. As water flows through the subsoil, the seepage forces and pressures cause a reduction in shear strength. Wherever such conditions exist, careful exploration is indispensable in order to determine the conditions of the ground water and its movements.

.4 Treatment of the Natural Foundation Soil to Promote Settlement

The choice of a suitable criterion for reducing the effects of settlements depends on the magnitude and rate of the settlements, the uniformity and continuity of the embankment and foundation soil, the existence of any special features such as access to bridges, and, lastly, the characteristics of the pavement and the traffic which will be handled.

It should always be remembered that uniform settlement, even when relatively large, in most cases does not seriously affect the embankment. Construction by stages is a simple solution in cases where serious problems of differential settlement arise, and where surcharges cannot be handled economically. Construction by stages usually entails postponing the final paving until the road has attained its ultimate equilibrium.

When the foundation ground is sufficiently firm, good results may be achieved by using surcharges to add height to the embankment. The weight of the surcharge required depends on the time-settlement relationship, the thickness of the compressible stratum, the height of the embankment and the time available for the construction program. The surcharge brings about an increase in settlement rate; it can be removed after the settlement corresponding to the ultimate weight of the embankment has taken place.

When the layer of compressible soil is very thick or the embankment very high, a surcharge capable of producing a significant effect will require such quantities of earth that it turns out to be uneconomical. The cost of the material used for the surcharge will be double, because it is handled twice, unless it can be used in other sections of the road and the cost partially recovered. Part of the surcharge can be used, so that once settlement has occurred, the embankment will have the required height.

Sometimes, when the layer of soft soil is rather thin, it can be compacted with very heavy equipment so that subsequent settlement resulting from the weight of the embankment can be minimized. This method can also be used if there are thin layers of loose sand in the foundation soil. The method is not very effective in the case of very soft clays; it becomes more so with higher contents of particles of the size of sand or gravel in the natural soil. Peats sometimes can be partially densified by high impact compaction.

.5 Choice of Method

At this point, a few further comments will be made to complete the section on construction of embankments on very soft soils or peats. First, it must be emphasized how vital it is to detect any problems during the initial stages of the project. The best solution may be a change in the location of the road. Other economical solutions are sometimes acceptable when sufficient time is available for them to take effect.

The different methods that have been briefly described in the foregoing paragraphs cannot be applied indiscriminately; they will have to be analyzed for each particular situation, so that the one which is most economical and suitable can be selected. The best solution sometimes turns out to be a combination of several different methods.

There is no hard-and-fast rule for establishing the safety factor for each method; this is a point which must be determined in each individual situation. For reasons of cost, the most serious problems associated with construction on soft soils do not usually allow very high safety factors. Moreover failure in a stretch of embankment built on very soft material or peat may have very serious consequences because with the internal remolding which accompanies the movement, there is such a reduction in the already deficient strengths of the materials, that after failure, the strengths can make it impossible for them to support significant loads. An exhaustive laboratory investigation of the structural breakdown caused by remolding becomes indispensable for establishing a margin of safety for the solutions to be adopted.

Lastly, when high costs are involved, it is vital to establish for each case how suitable any of the foregoing methods will be. The authors have the impression that road engineers have exaggerated the importance of these problems and that some expensive and complicated projects could perhaps have been carried out with greater simplicity and economy by building embankments on soft soils by conventional methods, taking care to avoid any shear failures and loss of strength by remolding. So long as a suitable remedy can be found for differential settlement, total settlement is not necessarily damaging. If the correct amount of initial over-building of the embankment is chosen, the need for a more costly method can be eliminated. A good example of the application of this approach is the 10 km (6 mile) stretch of the direct Mexico-Puebla road which crosses the region formerly occupied by Lake Chalco. It was constructed using partial displacement of the natural foundation soil. A temporary pavement, followed by relevelling, and then laying of the final pavement once the embankment-foundation system had stabilized completely, provided a simple solution to a serious problem.

3.8 Embankments on Sloping Hillsides

The construction of embankments on sloping hillsides can prove difficult. There are usually two adverse geological circumstances. Firstly, the boundary between the more weathered zone and the sounder materials is weak and tends to follow the inclination of the hillside; therefore there is a consequent tendency to sliding along that boundary. Second, the embankment blocks the natural movement of surface and subsoil water. Water which collects at the uphill base of the embankment causes an increase in the unit weight of the materials and a reduction in shear strength. This, in turn, leads to increased slide risks. Even when the water is not visible on the surface of the embankment, it nevertheless moistens the potential slide surfaces.

For this reason, seepage water must be carefully controlled when embankments are to be constructed on sloping hillsides. Also the varying ground water flow conditions from one time of the year to another must be considered because there may be no sign of the seepage at the time of the investigation.

It is usually not possible to avoid sloping hillsides when selecting the location for a road or a railroad. The inherent problems must be confronted and solved whenever they arise. Apart from the precautions relating to drainage and subdrainage, (considered in the corresponding chapters), benches or steps, like the ones which appear in Fig. 3-15, have proved a very practical and generally indispensable solution in hillsides with a slope steeper than 4:1.

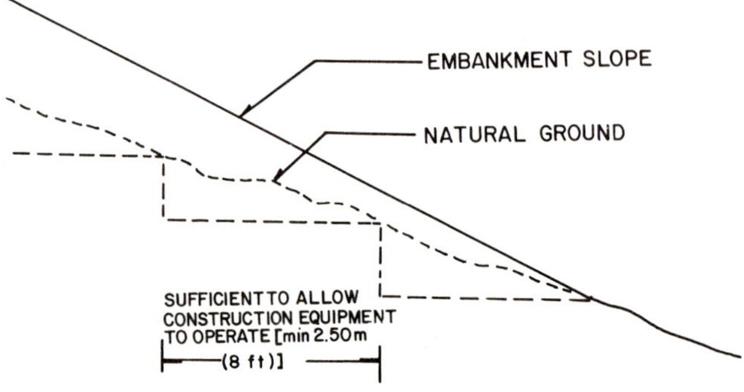

Fig. 3-15 Tie benches or steps in hillsides

The dimensions of the bench steps must be established for each particular case, but the width, or tread must be sufficient to enable construction equipment to operate, which is usually about 2.50 m (8 ft). With benches or steps, the load on the embankment is transmitted to horizontal planes. The value of these benches or steps is further enhanced if they are located on relatively firm ground, beneath the most severely weathered layers.

3.9 Clearing and Grubbing of the Foundation

Before the actual building operations for an earthwork can commence, the foundation ground must first be completely cleared of vegetation, including trees.. Stumps and roots are then removed and the topmost layer of soil removed and often stockpiled for later landscaping; this operation is termed grubbing. The objectives of clearing and grubbing the land are:

— To make it possible for the road-building equipment to operate in woody areas.
— To enable the earthworks and the foundation ground to be suitably joined.
— To do away with unwanted materials, such as weeds, shrubs and trees in cuts and borrow pits.
— To avoid subsequent falls of branches or trees onto the road owing to their proximity to the slopes. In these cases the trees will be felled, but, the areas not grubbed so that the covering of soil at the crest of the cut will not be loosened.
— To improve visibility at curves, especially where the ground is flat and the vegetation thick.
— To prevent the growth of roots which may afterwards affect the road surface, especially in very low embankments or road sections that are practically on a level with the surrounding land.
— To avoid subsequent behavior problems in the embankment due to rotting of the tree trunks or large roots trapped in or underneath them.

It is good practice to prepare the foundation ground before construction by removing a certain amount of surface soil. Usually a maximum thickness of 30 cm (12 in) is removed, and sometimes less. The objectives are:

- To avoid movement in the embankment; since the vegetal covering and root zone is generally spongy and compressible, it can affect low embankments.
- To remove soils that are unsuitable for construction in borrow pits or cut used for borrow.
- To remove organic vegetative matter likely to cause problems underneath low embankments owing to future growth.

The volume removed during clearing and grubbing must be included in the calculation of soil volumes because the waste represented by these operations can reflect on the total volume of soil displaced.

3.10 Soil Exploration for Road Building Purposes

The methods of soil exploration and sampling that are used in road building are basically the same as the ones that are commonly applied in all the other fields of soil mechanics. Therefore, it is unnecessary to discuss these methods in this book; traditional literature on soil mechanics can be consulted for information on this subject.

For general information on these points, [17,48,49] can be consulted, and with specific reference to road engineering, [8,9,50]. Methods of boring and sampling will not be discussed here. There are, however, certain ideas for establishing criteria which should be mentioned.

3.10.1 Geological Investigation and Photo-Interpretation

Geological maps provide an invaluable basis for road engineering projects. In some parts of the world, there has already been a sufficiently detailed geological mapping for other purposes so that only a limited field examination will be necessary to verify the accuracy of the existing data. In other areas, it will be necessary to supplement the existing maps or develop new ones for the specific project being considered. By means of field gelogic reconnaissance, aided by interpreting stereoscopic pairs of aerial photographs, the types of soil and rock formations can be observed, together with their limits and sequences. These will give a basic idea of the mechanical properties of the soils in the area in which the road is to be built and some fundamental information with regard to stability problems. In addition, the geological structures of interest, such as faults, the paths of joints and fissures, and previous landslides can be defined. An investigation including a sufficiently detailed geological map is, therefore, the first important step to be taken when designing a road. These investigations must always be regarded as economical and indispensable.

.1 Photo-Interpretation Techniques

These [51] have become an increasingly essential part of design. They speed up the production of information and save a great many inspections in the field. The principal data that can be obtained from photo-interpretation are:

- Social and economical characteristics of the area where the road is to be built, including towns, industries, crops, mines, and a survey of the engineering projects already existing in the region.
- Topography of the region, including access facilities.
- Data related to climate, vegetation, humidity, etc.
- Hydrological factors, such as large flows, length and location of bridges and exact configuration of gullies, creeks and rivers.
- General nature of rocks and soils.
- Identification of land forms: features of geological interest, such as lacustrine or swampy formations, flood plains, unstable formations, and details such as places where erosion is pronounced, shortage or abundance of building materials, possible excavation difficulties, and seepage zones.
- Definition of the use of the land, including types of crops.

Photo-interpretation work is generally best divided into two, and possibly three, successive stages for which photographs of increasing scale are used (the scale 1:50,000 is appropriate for the first stage, and 1:25,000 to 1:10,000 are almost certainly suitable for further refinement. The choice of scale is a question of experience but also depends on the photogrammetric equipment that is available for use in combination with photo-interpretation). Once photo-interpretation work is concluded at any stage of the design, the conclusions drawn from it will have to be verified in the field; this is done by inspecting the zone through which the road will run and checking the features identified in the study. The previous photo-interpretation work is then revised as necessary, confirming or correcting the conclusions obtained.

.2 Geological Investigation

A geological report for any stage of the design should include data on the following points on a scale giving the amount of detail appropriate for associated design decisions.

Rocks. Petrographic classification; morphological description; degree of weathering; classification and description of fractures, cracks, and shears; thickness of overburden materials that cannot be used for construction; general recommendations for cut stability, and classification from the point of view of ease of construction for estimating costs.

Soils. Origin, layer thickness, compactness, plasticity, organic content and water content. Also useful will be any information that can be given on variations in the vertical and horizontal directions, and general recommendations for using soils as embankment materials.

Crossings. A note must be made of all the features of interest where the road under design crosses any watercourse, indicating what materials are deposited there, the estimated water velocity and the run-off discharge, elevation during floods, erosion and sedimentation, and the stability of the river bed itself.

Finally the geologic report should suggest what additional data should be obtained by more detailed exploration in order to analyze soil behavior for final design of embankments, road beds, and structures. The degree of precision with which the geological engineer can supply this data is of course variable. It will be necessary to carry out more direct exploration or detailed geotechnical investigations wherever broader or more precise geological knowledge is required. A carefully conducted geological investigation provides an invaluable basis for analyzing possible alternatives, detecting important problems, planning more precise future studies, and in interpreting those studies.

Geological investigation in general, and photo-interpretation in particular, are of fundamental importance in road location, the analysis of alternatives and at each stage of road design.

3.10.2 Direct Exploration of Soils and Rocks

Soil exploration for designing and constructing roads is a subject for which there are few criteria to guide the engineer who must plan and execute the direct investigation. Exploration must be planned differently in each case, not only because of the differences between one road and another, but because each stretch of road is different from the rest.

There are five types of fundamental problems involved with road building which make soil exploration necessary. They are:

1. Stability analysis of cuts and embankments
2. Investigation of sources for borrow material, both soil and rocks
3. Foundation studies for bridges and other structures
4. Exploration for purposes of quality control
5. Groundwater control

Apart from the above problems, there are other less common ones which also require direct exploration. An example is the need for subdrainage in an airport, or when water must be obtained to compact a road crossing a desert region. The comments here will be confined to the first four cases mentioned above, especially 1 and 3. The other points will be dealt with in the respective chapters.

.1 Direct Exploration to Determine Stability Conditions in Cuts and Embankments

In theory, an adequate exploration is one which leads to a complete knowledge of the mechanical properties of the soils involved, with the aid of the laboratory. It is, however, practically impossible to find out in detail the mechanical properties of the soil at every point of a highway or railroad, with sufficient accuracy to allow a design based on theoretical methods in every cut and embankment. The reasons are largely economic, because it is impossible to conceive of a road-building organization with the finances, staff, and administration capable of undertaking such an intricate, extensive task. The sheer length of highways and railroads makes any uniformly detailed investigation, such as those employed for buildings, utterly impractical. Instead, the exploration focusses on critical features such as deep cuts and high embankments, structures such as bridges and stretches of the route that the geology indicates will be troublesome (Plate 3-11).

On the other hand, it is also worth stressing that an over-theoretical treatment of the stability problems of a road project can easily turn sterile: a large increase in the size of the investigation may not result in a proportionate increase in the knowledge and reliability achieved. There are two reasons for this. Firstly, there are the inherent uncertainties of any theory that is used, which are relatively independent of the knowledge that is obtained of the materials to which the theory is applied. Secondly, any field exploration, however thorough, will always produce scatter in the data obtained, chiefly because of the variability between one point and another (however close the points may be) and because of difficulties in obtaining samples and interpreting results. In embankments, the material used is controlled both with regard to excavation and construction, (or can be controlled in theory). In cuts, however work has to be carried out with an infinitely complex natural material. It is obvious, therefore, that any theoretical method will have far greater possibilities of success in an embankment with its controlled material than in a cut where the probability of obtaining representative data is far smaller. This should be taken into consideration when planning exploration work and establishing the extent of a program.

As a consequence of the foregoing, as far as the authors' know, in almost every country of the world the general selection of slopes for cuts and embankments relies on the recommendations of the soils engineers in charge of a road design who are guided by their previous experience and general knowledge of the materials involved.

The foregoing apply to the road project as whole, however certain cuts and embankments require detailed theoretical studies, supported by direct exploration and laboratory testing. In field investigations, the information supplied by geology must be used to full advantage, for it has a relatively low cost and can aid in interpreting data to the road engineer.

Indirect exploration methods (geophysical methods), which will be discussed later, are also very useful. At a cost which is also relatively low, very extensive and detailed general information can be obtained, For example, the Mexican Ministry of Public Works, which is responsible for the building of roads throughout Mexico, insists on a geophysical study wherever a cut deeper than 7 m (23 ft) is to be made. The information

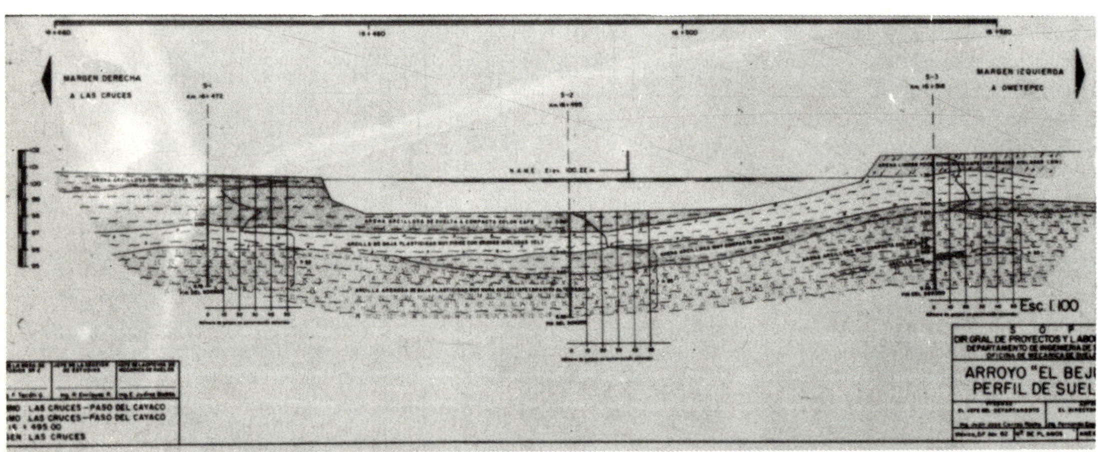

Plate 3-11 A soil profile for bridge foundation design

obtained is used both to help establish the slope of the cut and to determine the rippability of the materials, so that the most appropriate method can then be selected.

However, despite the invaluable aid provided by the foregoing information, the ultimate recommendation for the slope of all cut and embankments that do not merit special study, should be left to the judgment of the soils engineer, aided by simple methods of exploration and quick, preliminary investigations. The engineer who is responsible for these decisions must, therefore, be an expert in this field. Road-building enterprises must be careful to entrust this work to staff who are technically capable, selecting them from among engineers with post-graduate studies in soil mechanics and rock mechanics. The engineers who conduct geotechnical investigations will provide design criteria for routine cuts and embankments based on general information resulting from preliminary exploration. These criteria are not infallible, and will require constant field checks during construction, with modifications wherever necessary.

The preliminary exploration referred to will include open pits, hand auger and power auger borings as well as geophysical methods. The exploration must extend deeper than the upper weathered soil and should terminate in the materials on which the stability will depend. This requirement is usually met by prospecting to depths of about 1 to 1.5 m (3 to 5 ft.), although costwise, greater depths could be covered. The open pit is the best available method because it provides a three-dimensional view of the soils encountered. The effectiveness of these methods can be enhanced by examining old road cuts, natural gullies and ditches.

The distance between the points to be explored must be established in each case by the engineer in charge of geotechnical investigations, with aid from the geologist. Many organizations specify a maximum distance between prospecting points of 500 m, (1650 ft). These are arbitrary norms, which follow tradition rather than logic.

For cuts or embankments demanding special detailed exploration on account of their height, risks implied by failure and materials involved, the exploration methods are those commonly used in all fields of soil or rock mechanics, and will be briefly discussed below.

Tunnels and embankments on soft compressible soils will always be regarded as special cases. First, very complete information must be obtained concerning the materials in which the tunnel is to be made. These data are usually obtained from a combination of geophysical exploration (generally seismic methods for the analysis of geological formations,) and rock core drilling. In soft soils, the exploration methods are the ones used in foundation engineering. This includes obtaining undisturbed samples for triaxial and consolidation tests. In this case, increasing importance is being given to methods for determining the in-situ shear strength, such as the vane test, mentioned in Chapter 1. Because these devices are easy to operate and relatively economical, they are used to find out the variations in shear strength over a period of time, by using periodical investigations, and to determine the variations in strength from one place to another on the road under construction.

.2 Investigation of Quarries

As has already been mentioned, the methods of exploration used are common practice, and will be dealt with in the chapter on material quarries.

.3 Exploration for Design of Bridges and Other Structures

As will be discussed in detail in the chapter on this subject, the design of foundation for bridges and other structures requires detailed exploration, for it is based on the same data that are used in all fields of soil mechanics. The methods of boring and sampling are described in detail in the references mentioned at the beginning of this section.

For bridges, exploration methods based on the use of penetrometers are very popular. The standard penetration test [17,49, 52] is a good example. Static or dynamic cone penetrometers are also very often used.

It is not easy to establish rigid rules for the optimum distance between boring in river beds. If, at the time of exploration, the layout of the bridge to be built is known, the soundings can be made at the foundation locations. But if, as is often the case, no preliminary design of the bridge is available at the time of exploration, the boreholes will have to be done in such a way that they will enable a reasonable accurate soil profile to be drawn up. Here the engineer will have to be aware that a great deal of information will have to be obtained for this profile, because river beds are usually extremely heterogeneous and show great changes in the soils over very small distances. The amount of exploration that will be required to produce a *reliable* soil profile will not eliminate the risks of error. However, when the geotechnical engineer is also involved in the construction he can correct such errors before they become disasters.

A practical rule is to make borings in the cross-section of the river bed 20 to 25 m (65 to 82 ft) apart; this is adequate if there are no special circumstances. In very wide river beds or ones where very homogeneous conditions are known to exist, the distance between the boreholes can be slightly increased.

For bridges, deep foundations with piles or cylinders are very common; therefore the borings must be relatively deep. Drilling decisions should be entrusted to responsible engineers, so that exploration depths can be reliably established.

The cheapest type of foundation is usually the shallowest one. For this reason if the first few meters of explored and sampled soils turn out to have acceptable strength and compressibility, some engineers believe that it will not be necessary to continue exploration at greater depths. However shallow foundations for bridges, have three risks: undermining, the possibility of flooding off the excavations below the water-table and weak or

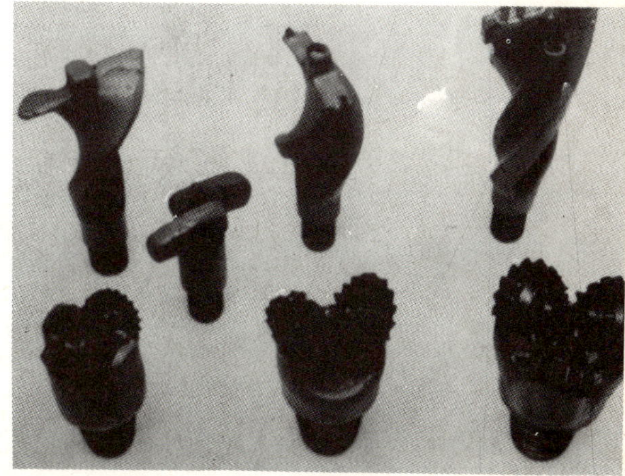

Plate 3-12 Common types of drills

compressible materials at greater depths. Any of these can make shallow foundations impossible, despite shallow soils with suitable strength and low compressibility. When deep foundations are essential, the entire depth must be explored, including a sufficient thickness of the firm stratum, below end-bearing piles or cylinders, or possible compressible strata underneath friction piles.

The significant depth, (that affected by the foundation below the level of applied loads), can be approximated as the level where the stresses applied to the soil by the foundation are reduced to 10% of the original pre-construction stress at the same level. The ground must be systematically explored to this depth, except for sound rock formations.

Soft compressible soils require undisturbed specimens for consolidation and shear tests. Drilling methods with thin-walled tube samplers, pressure driven, which are available in a wide variety of types, are used to obtain these samples, Plates 3-12 to 3-15.

Plate 3-13 A more elaborate type of soil exploration device

Investigations for determining the foundation of culverts and other small structures are usually so numerous that cost prohibits a detailed investigation for each. These designs usually rely on the judgment of a specialized engineer, aided by preliminary exploration techniques. Here too education and the experience of the man who conducts these studies will be of vital importance.

For retaining walls, the investigation will vary from the simplest to the most elaborate. There is a tendency to disregard the foundation of these most important and complex structures. This attitude has led to many mistakes, which could otherwise have been avoided.

Plate 3-14 Soil exploration device mounted on a barge

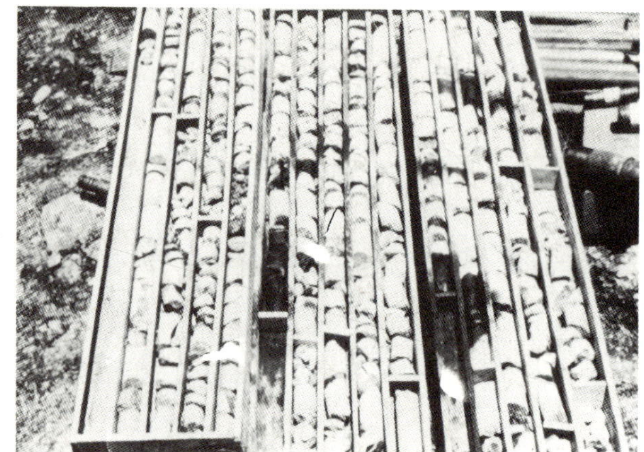

Plate 3-15 Samples from rock exploration

.4 Exploration for Quality Control

This will be briefly discussed in the chapter on the general problems of quality control in road engineering.

3.10.3 Sampling Requirements

The nature of the samples which must be obtained is determined by the method of exploration that will be used. Table 3-4 [48] illustrates this important relationship.

3.10.4 Indirect Exploration Methods: Geophysical Prospecting

The use of geophysical methods in the field of engineering explorations, and road building in particular, is becoming increasingly popular. For road building, these methods enable long stretches of route to be explored at a relatively low cost and with a degree of precision which is often adequate.

Geophysics [53] is a science which allows physical parameters of the ground to be correlated with stratification. It also enables identification of some geological features such as fractures, and sometimes even estimates of engineering properties. It also provides explanation and proof of certain theories about the constitution of the earth.

Table 3-4
Sampling of soil and rocks for road engineering

EXPLORATION STAGE	TYPE OF TEST TO BE PERFORMED	TYPE OF SAMPLE	QUANTITY OR SIZE OF SAMPLE
Reconnaissance	Classification on sight Water content Plasticity limits	Representative	Samples from hand-auger drill holes or standard penetration test. Also from open pits or ditches
Detailed exploration minor tests	Plasticity limits Grain size Relative density	Representative	About 1l (60 in^3) About 50kg (110 lb) About 1l (60 in^3)
	Water Density content	Representative, well sealed	Samples with a diameter of 5cm (2 in) are usually adequate, but are often somewhat larger. Samples measuring about 30 cm (12 in) are usually taken from open pits.
	Simple compression Direct shear test	Undisturbed	For shear tests a sample of 10 cm (4 in) in diameter is desirable.
Detailed exploration major tests	Permeability Consolidation Triaxial compression	Undisturbed	Occasionally samples of 5 cm (2 in) in diameter but diameters of 7.5 cm (3 in) to 15 cm (6 in) are more appropriate.
	Multiple compression; direct tests; special shear tests	Undisturbed	Samples with a minimum diameter of 10 cm (4 in), preferably 15 cm (6 in). In open pits, cubic samples measuring 25-40 cm (10-16 in).
Building materials-exploration	Grain size Compaction and *CBR* Triaxial compression Tests on concrete aggregates	Naturally representative or laboratory prepared to simulate field	50-100 kg (110-220 lb) but sometimes the whole series of tests on one sample will demand 250 kg (550 lb)
Building materials-quality control	Dry unit weight Water content *CBR* Triaxial compression	Undisturbed	Samples 5-10 cm (2-4 in) diameter. In open pits, cubic samples measuring 30 cm (12 in) at least. Samples from the *CBR* mold
Water	Chemical analysis Bacteriological analysis	Representative	10 l (600 in^3)
Rock cores	Visual inspection Mineralogical composition Compression, shear tests Porosity, permeability to air	Undisturbed	2.2-2.9 cm samples (7/8 in and 1-1/8 in: *EX* and *AX* barrels). Preferably 4.13-5.40 cm (1-5/8 in—2-1/8 in, *BX* and *NX* barrels). In soft or very fractured rock it is advisable to secure 10-15 cm (4-6 in) samples.

A phenomenon which can be measured on the ground surface and correlated with the underground structure can provide the basis for a geophysical method of prospecting. The following are the methods most often used at the present time:

— Magnetic
— Gravimetric
— Nuclear
— Geothermic
— Seismic
— Electrical resistivity

The last two are the methods most often used in road design and construction technology, therefore more emphasis will be placed on them in the brief description which follows. The reader will find further details in [17,48,49,53] with lists of bibliographical references for specialized study.

.1 Magnetic Method

This is the oldest of all the geophysical methods. It is used to determine the strength of the earth's magnetic field at different points and correlate this with geological formations. It gives good results only in large-scale explorations, which are not required by road engineering.

.2 Gravimetric Method

This method has been used throughout the world for oil prospecting for locating anomalies in deep seated structures. Only in fairly recent years has it been used for shallow investigations and prospecting.

The method measures the effect of the different densities in the ground. It is especially useful for detecting large shallow natural cavities, which form where gypsum-bearing or calcareous rocks have dissolved, or large artifical cavities such as sand mines, wells, galleries, and tunnels etc. Such cavities represent large contrasts in ground density and therefore in both cases changes can be detected in the earth's gravitational field. Values of gravity slightly higher than normal indicate masses of dense rock. The contrary will be a sign of the presence of light masses or soils.

The size of the contrasting feature must be larger than its depth below the ground surface to make a gravity anomaly that can be detected by available instruments. Wide shears or cracks near the surface can be detected; deep-seated anomalies are, however, still hard to interpret.

Insufficient experience has been acquired in applying this method to road engineering problems; the development of high precision, easily portable, instruments makes it far more attractive today than it was in the past, and justifies its use, at least experimentally.

.3 Nuclear Method

This method consists of recording the radiation which reaches the atmosphere from subsoil formations. A variation in these radiations can indicate the nature of, and certain features of, the geological formations of the local soil profile. Rocks rich in active matter emit measurable amounts of radiation. Radiactive sources owe their properties largely to the weathering of granite and other potassium-rich rocks by atmospherical agents. Methods of surface prospecting are currently being tested based on the emission of radiation on the soil within a depth of about 2 m (7 ft).

.4 Geothermic Method

This is based on measuring the temperature of the ground at different depths, and comparing it with the mean geothermic gradient for the earth. At the surface, it has proved useful in detecting springs, caverns and fractures, but its value in subsoil investigation remains to be demonstrated.

.5 Seismic Method

This method is based on the differences in the velocity of propagation of elastic waves in different media. Although in general, different soils and rocks have rather similar densities and unit weights, their moduli of elasticity are very different. The velocity of propagation of the elastic waves depends largely on the modulus of elasticity. Thus there is a fairly reliable correlation between changes in wave propagation, velocities and underground structure such as stratification.

With the seismic method, elastic waves are provoked by artificial means, such as impact or explosions. The vibrations transmitted by the ground are picked up by geophones and recorded on sensitive instruments called seismographs. If several geophones are placed at different distances from the impact or explosion, the different wave arrival times can be measured. The geophones are placed at distances of 15 to 30 m (50 to 100 ft) from one another and the velocities of propagation computed from the times required for the wave to travel from the source to them, see Plates 3-16 and 3-17.

The seismic method is applied in two different ways, by reflection and by refraction.

Plate 3-16 A geoseismic study in progress

The seismic method by *reflection* measures the time taken by a wave to travel from the origin through the ground to some contrasting object such as a different soil or rock layer and then reflect back to the surface. In theory, the procedure is very simple but very precise instruments are required, and the origin of the shock must be located at a certain depth. The complexity of the instruments is due to recording the wave on its return, while the surface of the ground is still in motion. It gives accurate results for great depths and is therefore frequently used in oil prospecting. Owing to its complexity it is, however, little used for civil engineering purposes.

The seismic *refraction* is based on an elastic wave crossing the boundary between different materials and being refracted towards the plane of this boundary when it enters a material

Plate 3-17 Recording device used in the seismic method by refraction

which transmits the wave at a greater velocity than the one it had in its first material. It is refracted towards a plane which is perpendicular to the boundary when the velocity of propagation is smaller in the material it enters than its velocity in the medium in which it propagated. Geophones are placed at varying distances from the wave origin and generally in line with it. The distance from the origin point to the furthest geophone is from 3 to 12 times the depth that is to be explored. Fig. 3-16 is a diagram showing the typical layout of the geophones and the type of graph that is obtained. This must subsequently be interpreted.

Only the time taken by the initial impulse to reach a geophone is used. A graph like the one shown in Fig. 3-16a is obtained if the successively deeper strata transmit waves at increasing velocities.

The geophones that are nearest to the origin receive waves transmitted only through the layer of surface soil; the ones in the middle receive them refracted through the upper boundary of the clay and returned to the surface, and the geophones that are furthest away receive waves which have been refracted on the lower boundary of the clay with the rock. From the curve in Fig. 3-16a, the velocities in each stratum can be deduced, and from these the depths at which the different borderlines appear can be determined.

The seismic refraction method can only be used when the velocity of wave propagation increases in the successively deeper strata. The presence of a stratum where the waves propagate at a lower velocity than in the overlying strata cannot be determined. Complications sometimes arise in loose deposits where the velocity of transmission increases gradually with depth. The path of the first impulses and the time-distance graphs are then curves, which makes determination of the velocities of propagation and the thicknesses of the non-uniform strata difficult. In sloping strata only the average thicknesses can be determined, but adjustments can be made by inverting the positions of the geophones and the explosion point. Table 3-5 is a chart showing the ranges of velocity of propagation of elastic waves corresponding to different types of soils and rocks.

The table also shows the rippability characteristics which have been experienced with some materials, although these must be treated with caution, because it is very hard to estimate

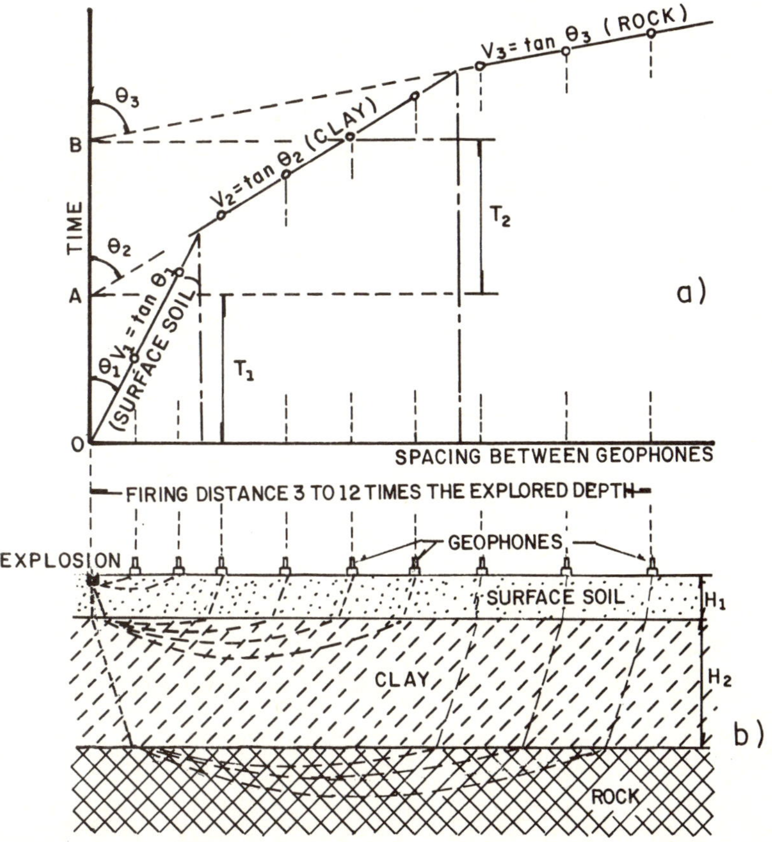

Fig. 3-16 Seismic refraction method

Table 3-5

Propagation velocity of elastic waves in soils and rocks

Material	Condition	Velocity		Rippability
		km/s	mi/s	
Soils				
Silts and sands	Compact Loose	0.4-0.7 0.2-0.4	0.25-0.43 0.12-0.25	Rippable Rippable
Clays	Hard Soft	0.6-1.0 0.2-0.3	0.37-0.62 0.12-0.19	Rippable Rippable
Boulders and gravels	—	0.2-0.4	0.12-0.25	Rippable
Igneous Rocks				
Granite	Sound Slightly fractured Highly fractured Weathered	4.5-6.0 1.5-4.5 0.7-1.8 0.4-1.0	2.80-3.73 0.93-2.80 0.43-1.12 0.25-0.62	Explosives Explosives Explosives Explosives or possibly rippable
Rhyolite and andesite	Slightly fractured Highly fractured Weathered	3.8-5.0 1.5-3.8 0.9-1.5	2.36-3.10 0.93-2.36 0.56-0.93	Explosives Explosives Rippable
Basalts	Sound Slightly fractured Highly fractured Weathered	5.0-6.0 1.4-5.0 0.7-1.4 0.5-0.7	3.10-3.73 0.87-3.10 0.43-0.87 0.31-0.43	Explosives Explosives Explosives or possibly rippable
Tuffs	Sound Slightly fractured Highly fractured Weathered	1.4-1.8 1.2-1.6 0.4-1.2 0.3-0.7	0.87-1.12 0.74-1.00 0.25-0.74 0.19-0.43	Possibly rippable Possibly rippable Rippable Rippable
Sedimentary and Metamorphic Rocks				
Limestones		1.5-4.0	0.93-2.48	Explosives
Sandstone		0.6-2.5	0.37-1.55	Rippable up to 8 km/s (0.5 mi/s) afterwards explosives
Agglomerates		0.2-0.9	0.12-0.56	Rippable
Conglomerates		1.0-3.0	0.62-1.86	Generally explosives
Clay shale	Hard Soft	1.2-4.0 0.6-1.4	0.74-2.48 0.37-0.87	Explosives Rippable

the state of the formations on the basis of just the velocity of propagation of elastic waves. Other relationships have been published, some more detailed than the one shown here, all of which must be applied in a real situation using careful judgment and experience. Rippability is specified by three words. The term *rippable* means that the material can be tackled with only a power shovel, ripper, tractor-drawn ripper, etc. The term *possibly rippable* refers to those cases where a limited use is made of explosives, with a view to breaking up or loosening a layer which afterwards becomes rippable, or for breaking up fragments which are too large. The term *explosives* is applied to those cases where the entire material must be loosened by this method.

For road engineering purposes, velocities below 800 m/s (2625 ft/s) correspond to rippable materials and those higher than 1500 m/s (4920 ft/s) indicate the need for explosives. Values between these limits correspond to the possibly rippable cases referred to, and these are in fact the hardest ones to define, because a velocity of 1000 m/s (3280 ft/s) for example, may correspond to rippable material if the structure is very granular, or may require more elaborate methods of approach if the material is a highly fractured granite.

.6 Electrical Resistivity

The electrical method is based on the differences in electrical conductivity in subsoil materials, which can be correlated with other geological and mechanical characteristics. The resistivity of sound igneous rocks is far higher than that of loose saturated soils. However, some dry sedimentary deposits can have fairly high resistivities. Resistivity depends mainly on the quantity and ionization of the water contained in the subsoil, and to a lesser degree, on the mineralogical composition of the soils and rocks (Plate 3-18)

Plate 3-18 A geoelectrical study being carried out

There are two main varieties of geophysical electrical methods: the method of *resistivity* and the *potential drop* method.

The method of resistivity is based on the production of an electrical field in the soil by means of two current electrodes (Fig. 3-17).

By measuring the current and the difference in voltage between the two potential electrodes, the resistivity at a point located between the potential electrodes and at a depth equal to the distance between these can be measured. If resistivity is plotted against the spacing between the potential electrodes,

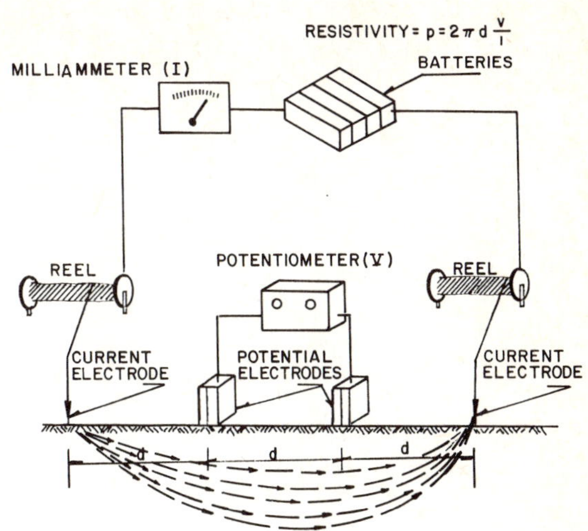

Fig. 3-17 Diagram showing the equipment set-up for geophysical exploration by electrical resistivity

preliminary indications as to the nature of the subsoil can be acquired. Detailed evaluation is far more complicated and must be carried out by specialists in the method. In the simple diagram referred to, there are generally abrupt changes in curvature when the spacing between the electrodes reaches a value equal to the depth at which there is a deposit with different resistivity from that of the overlying material.

In the method involving a potential drop, the current electrodes are placed wide apart (5 to 10 times the depth that is to be explored) and measurements are taken close to one of these electrodes. A diagram is given in Fig. 3-18.

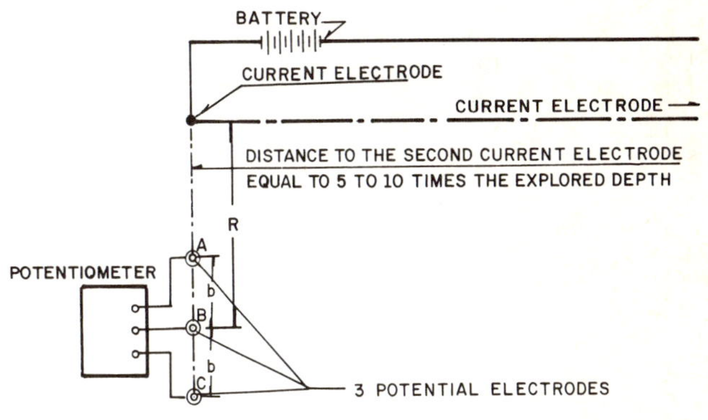

Fig. 3-18 Electrical method of potential drop [48]

Three potential electrodes are used in line with the current electrode and the drop in potential between *A-B* and *B-C* is measured. The distance *R* varies, while the value *b* remains constant, the distance between the potential electrodes generally being about *R*/3.

The relationship between the potential drops and the value of *R* is plotted. An abrupt change in curvature indicates the presence of a soil with a resistivity that is different from that of the overlying material. The results must, of course, be interpreted by a specialist.

Table 3-6
Electrical resistivity of some common rocks and soils

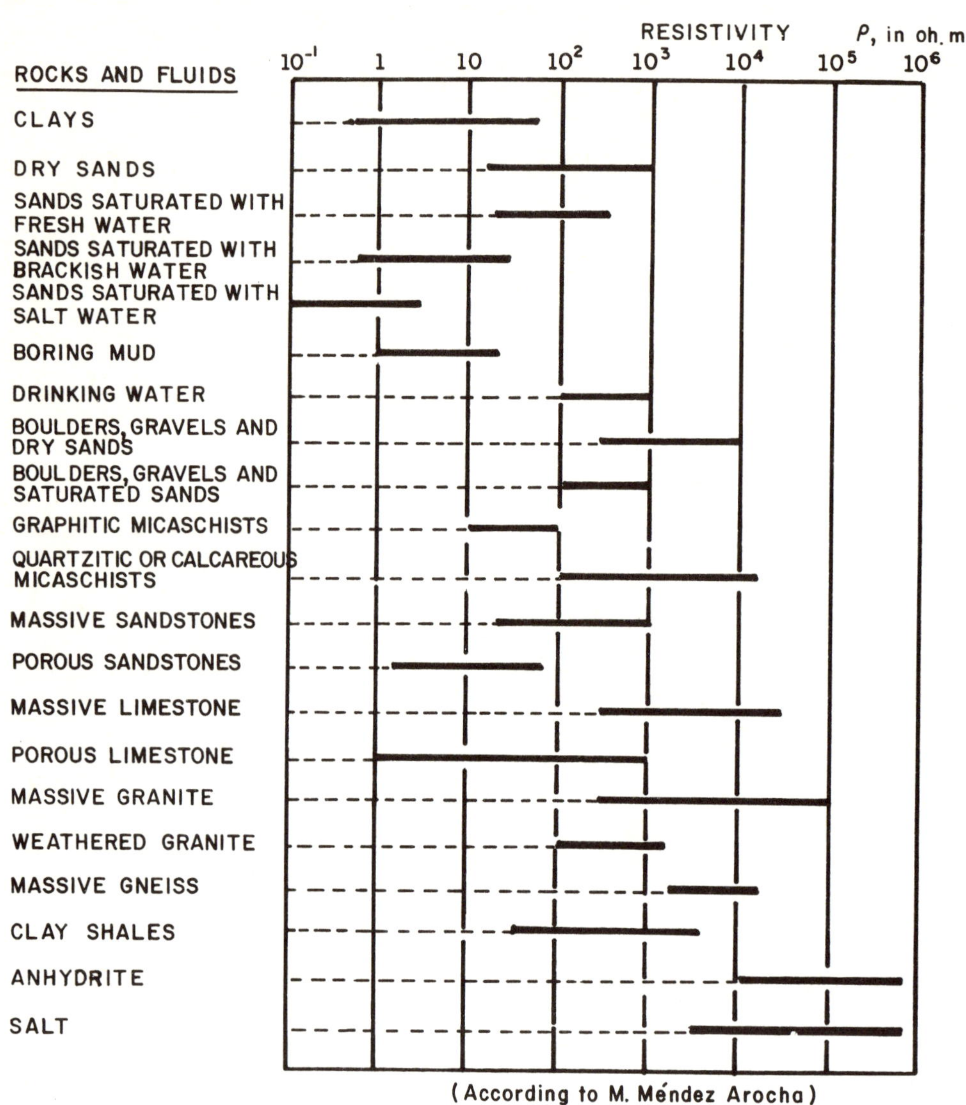

(According to M. Méndez Arocha)

This method supplies fairly acceptable indications as to sloping strata and, in certain cases, is more precise than the electrical resistivity method, but the information is not so well differentiated for horizontal stratification (which is the most frequent kind). Therefore, the method of resistivity is usually preferred for purposes of civil engineering and exploration in relation to roads. Table 3-6 gives a list of the resistivities corresponding to the most common types of soils and rocks.

None of the geophysical methods that have been so briefly described here can be used alone, and in most of cases they must be checked and their results compared with those of direct exploration by means of boring. Used like this, though, they do give excellent results, for they enable long distances to be covered at relatively low costs, thus avoiding a great deal of direct exploration which would otherwise have to be done.

In road engineering, they are used to best advantage for tunnels and large cuts, and for prospecting for water.

The optimum field of application of each geophysical method is obvious from its characteristics. Since resistivity is highly sensitive to the water contained by voids, this method will be appropriate for studying porosity or permeability or for prospecting for water. The structural conditions of a formation are better described with relation to its modulus of elasticity which can be obtained using the seismic method based on the propagation velocities of longitudinal and transversal waves [53]; this is why seismic methods are the best ones for investigating the mechanical properties of a formation. The seismic and electrical methods identify the boundaries between different strata fairly accurately, although the electrical method is more precise. The electrical method does not have the problems which arise when soft strata are located beneath hard strata. It has sometimes been said that the best geophysical exploration would be one where the stratification boundaries would be determined by the electrical method and the mechanical properties by the seismic method. Naturally, in the case of road engineering, for the sake of simplicity and economy, it is common practice to use just one method for each investigation.

REFERENCES

1. JUÁREZ BADILLO, E. and RICO, A., *Mecánica de Suelos, Vol. II: Teoria y Aplicaciones de la Mecánica de Suelos,* Limusa: México, 1977, 3rd Edition, Chap. XI
2. ibid., Chap. II
3. CAPPER, P. L. and CASSIE, W. F., *The Mechanics of Engineering Soils,* E. and F.N. Spon, 1960, Chap. 6.
4. RUTLEDGE, P. C., "Results on the investigation on triaxial compression", Original Publication of the Waterways Experiment Station, trans. R. J. MARSAL and M. MAZARI, in *Contributions of Soil Mechanics to Design and Construction of Earth Dams,* Secretaria de Recursos Hidráulicos, México, 1961.
5. JURGENSON, L., "The Application of Theories of Elasticity and Plasticity to Foundation Problems", *Contributions to Soils Mechanics,* Boston Society of Civil Engineers 1925-1940, 1959.
6. GRAY, H., "Charts to Facilitate the Determination of Stresses under Loaded Areas", *Civil Engineering,* June, 1948.
7. JUÁREZ BADILLO, E. and RICO, A., *Mecánica de Suelos, Vol. I; Fundamentos de la Mecánica de Suelos,* Limusa: México, 1977, 3rd Edition, Chap. X (Annex X-g).
8. "Slope Stability and Foundation Investigation", Institute of Transportation and Traffic Engineering, University of California at Berkeley, 1967.
9. "Notes on Procedures, Testing Methods and Use of Material for Highway Purposes", California Highways Department, Foundations Section, Sacramento, 1968.
10. SMITH, T. W., CEDERGREN, H. R. and REYNER JR., C. A., "Permeable Materials for Highway Drainage", California Highways Department, Sacramento, California, 1964.
11. WEBER, JR., W. G., "Experimental Sand Drain Fill at Napa River", California Highways Department, Sacramento, California, 1966.
12. RICO, A., MORENO, G., MEDINA, J. and DE LA TORRE, F.", Terreno de cimentación", *Seminar on Pavements,* Public Works Ministry, México, 1970.
13. JUÁREZ BADILLO, E. and RICO, A., *Mecánica de Suelos, Vol. III; Flujo de agua en suelos,* Limusa: México, 1977, Appendix I.
14. Ref. [1], Chap. I
15. SULLIVAN, E. Q., "Sand Dune Road 20 years old", California Highways Department, San Bernardino, California, 1946.
16. ZOLKOV, E. and WISEMAN, G., "Engineering Properties of Dune and Beach Sands and the Influence of Stress History", *VI. International Conference on SMFE,* Montreal, Canada, 1965.
17. Ref. [7], Appendix.
18. SINACORI, M. N., HOFMANN, W. P. and EMERY, A. H., "Treatments of Soft Foundation for Highway Embankments", *Proc. Highway Research Board,* Vol. 31, 1952.
19. LEA, N. D. and BRAWNER, C. O., "Highway Design and Constructions over Peat Deposits in Lower British Columbia", *Highway Research Board Record* N° 7, Publication 1103, Washington, 1963.
20. CASAGRANDE, L., "Construction of Embankments Across Peaty Soils", *Journal of the Boston Society of Civil Engineers,* Vol. 53, N° 3, 1966.
21. ANDERSON, K. O. and HEMPSTOCK, R. A., "Relating the Engineering Properties of Muskeg to some Problems of Fill Construction", *Procs. of V. Organic Soils Conference,* Technical Report N° 61, National Research Council, Canada, 1959.
22. CASAGRANDE, A. and CASAGRANDE, L., "Investigation for Neponset Swamp Crossing and River Bridge", Unpublished Report, mentioned in Ref. [19], 1953.
23. DUCKER, A., "Properties of Peat Deposits in Schleswig — Holstein, Germany", Unpublished Report, Geologisches Landesamt, Kiel, mentioned in Ref. [19], 1954.
24. FRASER, C. K., "Observations during Bog Blasting in Co. Armagh, Northern Ireland", Dept. Scient. and Ind. Research, Road Research Laboratory, note N° RN/2881/CKF, U.K. 1956.
25. HARDY, R. M. and THOMSON, S., "Measurement of the Shearing Strength of Muskeg", *Proc. Eastern Muskey Research Meeting,* Technical Report N° 42, National Research Council, Canada, 1956.
26. MARGASON, G. and FRASER, C. K., "Bog Blasting in Northern Ireland', Dept. of Scient. and Ind. Res., Road Research Laboratory, note N° LN/328/G.M. CKF, U.K. 1963.
27. MOOS, A. VON and SCHNELLER, A., "Slide of Highway Embankment on Peat near Sargans, St. Gallen", *Strassen und Verkehr* N° 10, 1961.
28. RIPLEY, C. F. and LEONOFF, C. E., "Embankment Settlement Behavior on Deep Peat", *Proc. VII. Muskeg Res. Conf,* Technical Report N° 71 National Research Council, Canada, 1961.
29. SMITH, A. H. V., "A Survey of Some British Peats and Their Strength Characteristics", Army Operational Research Group, Report N° 32/49, U.K. 1950.
30. TRESIDDER, J. O. and FRASER, C. K., "The Construction of an Experimental Road over Peat at Hammar, Shetland", Dept. Scient, and Ind. Res., Road Research Laboratory, note N° RN/2394/J.O.T. CKF, U.K. 1955.
31. WARD, W. H., "A slip in a Flood Defense Bank Constructed on a Peat Bog", *II. International Conference on SMFE,* Rotterdam, 1948, Vol. II.
32. BRAWNER, C. O., "The Principle of Preconsolidation in Highway Construction over Muskeg", Technical Report N° 61, National Research Council, Canada, 1959.
33. JOHNSON, A. W., "Resumé of Peat Excavation Procedure", *XIII. Meeting of NJ, NY and New England Testing Engineers Association,* Boston, Mass, 1954.
34. MOOS, A. VON, "Results of Several Test Embankments on Soft Ground in Switzerland", *Strassen und Verkehr,* N° 9, 1962.

35. Ringeling, J. C. N., "Measuring Groundwater Pressures in a Layer of Peat caused by an imposed Load", *I. International Conference on SMFE,* Cambridge, Mass, 1936, Vol. I.

36. Casagrande, A., "Notes on Soil Mechanics", Harvard University, 1938, (Unpublished, mentioned in Ref. [20]).

37. Hanrahan, E. T., "An Investigation of Some Physical Properties of Peat", *Geotechnique,* Vol. 4, 1954.

38. Thomson, J. B. and Palmer, L. A., "Report of Consolidation Tests with Peat", Special Tech. Publ. N° 126, ASTM, 1952.

39. Root, A. W., "California Experience in Construction of Highways Across Marsh Deposits", Highway Res. Board Bull. N° 173, Nat. Res. Council Publication N° 533, Washington, 1958.

40. Rico, A., Moreno Pecero, G. and Garcia Altamirano, G., "Test Embankments on Texcoco Lake", *VII. International Conference on SMFE,* México, 1969, Vol. II.

41. Curtis, A. R., "Blowing the Muck, Example of Swamp Filling Practice", *Roads and Streets,* Vol. 93, 1950.

42. Jeffries, J. M., "Building Roads through Unstable Formations", *Civil Engineering,* Vol. 6, 1936.

43. Parsons, A. W., "Accelerated Settlement of Embankments by Blasting", *Public Roads,* Vol. 20, N° 10, 1939.

44. Casagrande, L. *Blasting of Peat near Saarmund,* Volk und Reich Verlag: Berlin, 1938.

45. Casagrande, L., "The Method of Blasting Peat", *Die Strasse,* 1939.

46. Casagrande, L., "The Construction of Embankments on Soft Ground", Note N° CII. Dept. of Scient. and Indust. Research, Watford, U.K. 1947.

47. Casagrande, L. and Siedek, P., "Recent Experiences of Roads Construction over Marshes", *Die Bautechnik,* Vol. 15, N° 48, 1937.

48. Hvorslev, J., "Subsurface Exploration and Sampling of Soils for Civil Engineering Purposes", The Waterways Experiment Station, U.S. Corps of Engineers, U.S. Army, Vicksburg, Miss, 1962.

49. Terzaghi, K. and Peck, R. B., *Soil Mechanics in Engineering Practice,* John Wiley and Sons, Inc: New York, 1967, 2nd Edition

50. "La Secretaria de Obras Públicas y los Métodos Modernos de Investigación", Obras Públicas, Public Works Ministry Review, Supplement 14, Mexico, August, 1969.

51. Puig, J. B., *Geologiá Aplicada a la Ingeniería Civil y Fotointerpretación,* Juventud: México, 1970.

52. De Mello, U. F. B., "The Standard Penetration Test", *IV. Panamerican Conference on SMFE,* San Juan, Puerto Rico, 1971.

53. Vignaud, R., "Nociones sobre la geofísica empleada para reconocimiento de aguas subterráneas", unpublished notes, Mexico, 1970.

OTHER RELATED REFERENCES

Amar, S., Baguelin, F. and Jezequel, J.F., "Le Pressio-Penetrometre pour la Reconnaissance des Sols a Terre et en Mer", Colloque International sur la Gestion des Ouvrager, Paris, 1981.

Assonyi, Cs. and Richter, R., *The Continuum Theory of Rock Mechanics,* Trans Tech Publications, Clausthal-Zellerfeld, F.R.G., 1979.

Giroud, J., and Noiray, L., "Geotextile-Reinforced Unpaved Road Design", *Journal of the Geotechnical Engineering* Division, Proc. ASCE., Vol. 107, No. GT 9, September, 1981.

Gregory, C.E., *Explosives for North American Engineers — Third Edition,* Trans Tech Publications, Clausthal-Zellerfeld, F.R.G., 1983

Heraud, H. and Livet, M., "Reconnaissance des Massifs Rocheux. Prise D'empreintes dans un Forage", Colloque International sur la Gestion des Ouvrager, Paris, 1981.

Jumikis, A.R., *Rock Mechanics — Second Edition,* Trans Tech Publications, Clausthal-Zellerfeld, F.R.G., 1983

Karafiath, L.L. and Nowatzki, E.A., *Soil Mechanics for Off-Road Vehicle Engineering,* Trans Tech Publications, Clausthal-Zellerfeld, F.R.G., 1978.

Proceedings of the Tenth International Conference on Soil Mechanics and Foundation Engineering, Stockholm 1981, A.A. Balkema, Rotterdam, Netherlands.

Schmertmann, J.H., "Use the SPT to Measure Dynamic Soil Properties? — Yes, but . . !", Special Technical Publication 654, American Society for Testing and Materials, Philadelphia, Pa., 1978.

Sherif, M.A., Ishibashi, I., and Ding, W.W., "Heave of Silty Sands", *Journal of the Geotechnical Engineering Division,* Proc. ASCE.,Vol. 103, No. GT3, March, 1977.

Singh, R. D., Dobry, R. Doyle, E.H., and Idriss, I.M. "Nonlinear Seismic Response of Soft Clay Sites", *Journal of the Geotechnical Engineering Division,* Proc. ASCE. Vol. 107, No. GT9, September, 1981.

Waschkowsky, E., "Le Penetrometre Dynamique, La Reconnaissance des Sols, Ses Applications," Colloque International sur la Gestion de Ouvrager, Paris, 1981.

CHAPTER 4

SOIL COMPACTION

4.1 Introduction

Soil compaction is mechanical densification to improve the strength, reduce the compressibility and enhance the rigidity of soils. It usually entails a fairly rapid reduction in the void volume and a corresponding reduction in the volume of the soil. These changes are usually equal to the loss in volume of air, as water is seldom driven out of the voids during the process (except in coarse sands and gravels). All the air is not expelled, so the soil remains only partially saturated.

The principal aim of compaction is to obtain a soil which is structured in such a way that it will maintain its appropriate mechanical behavior throughout the effective life of the project. The properties sought vary from one case to another. Strength, incompressibility and rigidity are among those where improvement is usually desired. Compaction is sometimes employed to reduce permeability or control flexibility. Compaction usually helps an earth structure to withstand erosion, but can increase the tendency of some soils to swell.

The multiple effects of compaction complicate its application in improving soils in civil engineering applications. For example a stronger, less compressible soil produced by compaction is also more rigid and likely to crack when used as a subgrade layer over a foundation of more yielding soil. Further, the effects can be contradictory in some applications. A compacted plastic clay will shrink less on drying, but will swell more upon wetting. These possible contradictions become still more complicated if one bears in mind that compacted soils are expected to have an extended life and to maintain these properties throughout the whole of that lifespan under such effects as changing water, or loading. For example, a subgrade that is compacted to obtain strength will also develop greater potential capillary tension (or suction) and can absorb water and thus become weaker. The same soil with less compaction would have less suction. Although initially weaker than if highly compacted, it might retain that strength better in an environment in which it could absorb water from below.

Soil compaction is directly related to the quality control of construction. Therefore, after compacting, it is always necessary to verify whether the objectives have been met. Since roads are usually built under contract, this verification must precede contract payment. The vast number of problems involved in soil compaction, so often beyond the technical details, require intensive liaison and cooperation between the engineer and the constructor. They share the responsibility for producing the required qualities in the completed embankment.

The required qualities of incompressibility, rigidity, and permeability are measured by tests that are specialized and expensive. Moreover they require too much time for such tests to be used to verify that compaction has produced the desired result. However, investigations by Proctor and others, [1,2] carried out during the early years of modern compaction techniques, prove that there is a correlation between the basic properties specified above and the dry unit weight of the compacted material: the higher the dry unit weight of a compacted soil, the better. There are some exceptions and limitations to this simple rule. Because, however, the unit weight test is simple and easy to carry out, it is customary to control compaction by determining the dry unit weight of the compacted materials.

The correlations between basic engineering properties and the dry unit weight are neither simple, nor always reliable. Therefore, using density as the sole criterion for the required compaction without due care and attention to the peculiarities of the material and the particular quality desired, as well as the reliability of the correlation, is perhaps a common cause of poor performance of compacted soil, as well as controversies between the engineer and the constructor.

The increase in unit weight is simply a means to an end. Where the improvement of the basic properties is directly related to the increase in unit weight, it is a simple, easily evaluated, measure of contract performance. In some situations and particularly with fine grained soil, other criteria must be considered [3].

Compaction has been included in construction techniques ever since mankind can recall, although in ancient times its use was neither general nor systematic. Methods whereby the ground was trampled by people or animals were used in the very distant past, as for example in the construction of dikes and dams in Asia. According to the historian Silvanus G. Morley, a compaction roller was used by the ancient Mayans

Plate 4-1 Stone roller which appears to have been used by the ancient Mayans for compacting their roads

Plate 4-3 View of an ancient Mayan road

Plate 4-2 Another view of the Mayan compaction roller

Plate 4-4 Another view of the same Mayan road. Note the heavy stones bordering the finer material in the center

for the construction of the huge network of roads that served to link the principal ceremonial centers of what today are the States of Yucatan and Quintana Roo in Mexico, see Plates 4-1 and 4-2. Plates 4-3 and 4-4 show these roads today as restored by the Public Works Ministry of the Mexico Federal government. The roller was found on the road between Cobá and Yaxuná. The roller was originally 4 m long (13 ft 1 in) (although now broken) with a diameter of 65 cm (2 ft 1 in) and an approximate weight of 5 tons (11 kips). It was probably pushed by 15 men on the basis of some recent tests with it.

The development of modern compaction techniques took place during the last few years of the nineteenth century and the first few years of the twentieth century, especially in the United States. In 1906, Fitzgerald's 2000 kg (4.4 kip) sheepsfoot roller made its appearance; it was the first of these devices [4].

In 1928 and 1929, O. J. Porter, of the California Division of Highways, developed the basic laboratory research which made it possible to make a rational use of compaction techniques for highway construction [5]; Purcell [6] was largely responsible for the popularization of these techniques. In 1933, Proctor began to carry out the important work which paved the way for many of the techniques which are in use today [1,2,7]. Since then, the development of compaction equipment has made great progress and considerable research has since been carried out on the properties of compacted soils as well as field and laboratory techniques for measuring these properties. They will be briefly described later in this chapter.

Compaction is the most efficient and widely used method available today for improving the condition of a soil to be used for construction purposes. Other methods are, however, also available. Table 4-1 charts the various methods of soil improvement which are recognized today.

Compaction is usually required for fills, such as cores for earth dams, dikes, embankments for roads and railroads, airfields and revetments, harbors, and base courses for pavements. In addition, it is sometimes necessary to compact natural ground, as in the case of foundations on loose sands or soft clay. Soil compaction is therefore a field construction technique to solve an engineering design problem.

The effectiveness of any compaction process depends on several factors. Testing is necessary in order to evaluate the particular influence of each. Field tests using the actual construction are most meaningful, and are employed on large projects as well as for research. Because such tests are very

Table 4-1
Soil improvement methods

Methods	In-situ Mechanical	Confinement (frictional soils) Preconsolidation (fine clayey soils) Vibroflotation
	Chemical (Stabilization)	With salt With cement With asphalt With lime With other substances
	Mechanical	Compaction Mixtures of soils

expensive and time consuming, most projects rely on laboratory testing. The laboratory tests are designed to reproduce field conditions as closely as possible; in addition they have been standardized to help compare work by different laboratories on a variety of soils. Laboratory compaction tests are usually the basis for quality control, because they can be made quickly and without interfering with the project compaction. Finally, research into the soil properties which can be obtained by means of compaction, is done most conveniently in the laboratory.

In this chapter, the three concepts of the compaction problem are discussed: (1) Obtaining the best compaction, (2) quality control, and (3) the relation of soil qualities to compaction.

4.2 Variables which Affect the Soil Compaction Process

A soil can be compacted in various ways; in each case the result obtained will be different. Furthermore, the same compaction method applied to different soils will produce different results. Lastly, if the same compaction method is applied to one specific soil, very different results will be obtained if in each case there are variations in the conditions prevailing in that soil.

Thus, compaction depends on various factors: those related to the type of soil; those related to the compaction method used; and those which depend on the circumstances prevailing at the particular moment at which the soil is being compacted. These factors are usually referred to as the *variables* which govern compaction. The principal variables are given below.

.1 The Nature of the Soil

The type of soil obviously plays a decisive role in the compaction process. Throughout this chapter the different techniques used and the results obtained will be specified in accordance with the type of soil. The usual gross distinction between fine grained and coarse grained (or cohesive and cohesionless, or frictional) soils applies. However, it is helpful to refine this distinction in the analysis of compaction processes by classifying the soils in accordance with the standards discussed in Chapter 2.

.2 The Method of Compaction

In the laboratory the currently used compaction methods can be classified into three types: dynamic or impact, kneading, static and vibratory compaction. Each produces different results both in the structure acquired by the soil and in the properties of the material being compacted.

It is more difficult to classify field compaction methods in the same way. These are commonly described by the mechanical equipment used in the process; tamper, steel-wheeled roller, pneumatic-tired roller, vibratory roller. Although the laboratory methods are supposed to reproduce field conditions, they do not always do so faithfully.

.3 Specific Energy

Specific energy is the energy which is applied to the soil per unit of volume during a given mechanical compaction process. Specific energy E_e is very easily measured in a laboratory test when the soil is compacted by blows with a tamper. It is expressed by the following

$$E_e = \frac{Nn\,Wh}{V} \tag{4-1}$$

where:

E_e = specific energy
N = number of blows with a tamper for each of the layers of soil placed in the compaction mold
n = number of layers required to fill the mold
W = weight of the tamper
h = height from which tamper falls when applying blows to the soil
V = total volume of soil compacted (usually the mold volume)

In laboratory tests where soil is subjected to static compaction, specific energy can be evaluated theoretically in terms of the size of the mold, the number of layers of soil, the pressure applied to each layer and the length of application. However, evaluation is not so easy in this case as the specific energy is affected by the deformability of the soil and pressure application time.

In the case of tests where the kneading compaction method is used, evaluation of specific energy is even more complex, for each layer of soil in the mold must be compacted with a certain number of blows with a tamper that produces pressures which vary gradually from zero to a maximum value, the process being inverted for unloading. There is no simple method of measuring the compaction effort, but it can be made to vary at will by introducing changes in the tamping pressure, number of layers, number of applications of the tamper per layer, area of the tamper or size of the mold.

The specific energy concept maintains its full intrinsic value when related to field compaction procedures. When rollers are used, it depends mainly on the pressure and the contact area between roller and soil, the thickness of the layers being compacted and the number of passes. Simply evaluating the compaction *energy* in absolute terms for a field compaction case is not, however, sufficient since various combinations of the above mentioned factors can produce the same specific energy but differing compaction degrees.

.4 The Water Content of the Soil

In PROCTOR's very earliest studies, it was shown that water content is another fundamental variable of the process. PROCTOR observed that starting with low values and gradually increasing water contents, higher dry unit weights were obtained for the material compacted, using the same compactive efforts. He also observed that this tendency does not hold indefinitely; once the moisture passes a certain value, there is a reduction in the dry unit weights achieved. He demonstrated that for a given soil and using a particular compaction procedure, there is an optimum water content which produces the maximum dry unit weight obtainable with that procedure. For field compaction processes, this water content is the optimum for the corresponding equipment and effort.

Optimum moisture can be explained by the effect of moisture on the soil grains. In fine grained soils at low water contents, the water occurs in capillary form, producing compression between the particles and lumps that cannot be easily broken. This makes compaction difficult. An increase in water content reduces the capillary tension, the lumps are not so hard, and the soil compacts more readily. If, however, the water content is so great that there is free water, so that the voids in the soil are nearly filled, the soil cannot be compacted because water cannot be instantaneously squeezed out. The optimum moisture is a compromise between additional moisture enhancing soil grain mobility and added moisture interfering with void reduction.

.5 The Direction of the Water Content Scale During Compaction

This aspect affects mainly laboratory compaction tests, where results are often presented in graphs of $\gamma_d - w$ (dry unit weight vs. water content). The curves are different if the tests are carried out starting with a relatively dry soil to which water is gradually added, or a damp soil which gradually dries out as the test progresses. Experimental research proves that in the first case higher dry specific weights are achieved than in the second, using the same soil and with the same water contents; this effect is particularly noticeable in fine plastic soils and at water contents below optimum. An explanation for this phenomenon may be that when a soil is dry and water is added, this water tends to remain on the exterior of the lumps, penetrating them only after some time; on the other hand, when water evaporates as a damp soil dries, the surface moisture of the lumps becomes less than the internal moisture. Like this, lumps of soil with identical water contents have different conditions; in the first case, where water was added, the capillary pressure among the lumps is less because of the excess water, compared to the second case where the menisci are more highly developed as a result of evaporation. In the first case, therefore, the lumps will not be so strongly bound together and the same effort will compact the soil more effectively than in the second case.

The above reasonings are of course influenced by the time that is allowed to elapse between the incorporation of the water and the moment when the compactive effort is applied, for if the lapse of time is long enough, the water will be incorporated uniformly into the lumps of soil, with the consequent reduction in surface moisture and increase in capillary pressures. The salt content and the nature of the clay also play a part.

In the laboratory, the process is often started with a relatively dry soil; water is incorporated and sufficient time is allowed to go by (24 hours or so) so that the water will be evenly distributed.

.6 The Original Water Content of the Soil

This concept concerns the natural water content of the soil before any moisture is added or taken away, in order to obtain the optimum content or any other which may have been chosen for compaction.

In field processes, the original water content not only has a pronounced effect on the response of the soil to the compaction equipment, but is also largely responsible for the future behavior of the compacted mass. Although only relatively small changes can be achieved by wetting or drying the soil at the location, it is highly recommendable to always try and find natural moisture conditions as close as possible to the optimum for the compaction process to be used.

In laboratory processes, the natural water content of the soil has a particularly strong influence on the compactions achieved with a certain effort at lower than optimum moisture levels, especially when the soil is compacted immediately after incorporation of the water. This phenomenon can be understood if one bears in mind the explanation given in the previous Point 5, for in a soil which is fairly dry originally, the water added will produce a greater immediate difference between the internal and external moisture conditions of the lumps than in another which originally is moister; this is why the dry unit weights obtained can be expected to be higher when the original water contents of the soil are lower.

.7 Re-use of Samples

In many laboratories, it is common practice to use the same soil sample for successive steps of the compaction tests; this implies continued recompaction of the same soil. This practice has been proved *totally inadvisable* as experiments have shown, without any doubt, that when working with recompacted soils the unit weights obtained are higher than those achieved with virgin samples in identical circumstances, so with recompacted soils the test will fail to be representative. A simple explanation of this effect is apparently the volumetric strain of a plastic nature caused by successive compactions [8,9].

.8 Temperature

Temperature, and temperature changes, have an influence on field compaction processes. First, high temperature enhances evaporation of the water incorporated in a soil. Temperature drops cause condensation of the natural water content.

Lowering the temperature of the soil surface (such as produced by radiation at night) induces capillary flow of water upward and an increase in moisture at the soil surface. Finally, low temperatures increase capillary tension and water viscosity and correspondingly reduce soil workability and ease of compaction.

.9 Other Variables

Other variables affect laboratory and field compaction tests, including the number of layers into which the soil is placed and their depth, the number of passes of the compaction equipment over each point (or the number of blows with the tamper on each layer). The effects of these will be discussed in the description of field compaction processes and the different laboratory tests.

4.3 The Compaction Curve

As we have seen before, according to history, compaction processes developed in the field as construction techniques. It was not until the effects of these techniques were studied more carefully that quality control procedures were established, field results verified and laboratory compaction tests came into being. At first they were based on PROCTOR's original test; afterwards many different tests, including variations on the first, were developed to gradually bring laboratory work as close as possible to field processes. At the same time, field processes were improved and expanded owing to the introduction of new equipment produced by ever more knowledge and demanding technology.

As previously mentioned, PROCTOR formalized the correlation between compaction and the increase in dry unit weight of the material compacted. He established the practice, which is still observed today, of judging a compaction process by the dry unit weight produced. He also established the fundamental role played by the water content in producing compaction by a certain procedure. By combining these two principles, he established the practice of representing the progress of a compaction process by a graph showing the dry unit weight produced during the compaction as a function of the water content. Because different compaction processes produce varying degrees of compaction in the same soil, one soil may have diverse compaction curves, each corresponding to a different method of compaction, either in the field or the laboratory.

A representation like this γ_d vs. w is usually termed the *compaction curve,* (or PROCTOR curve). However, it is not the only graphical means of representing the results of a compaction process. Other graphical ways of representing or analyzing certain interesting conditions in compaction processes will be given later in this chapter. In [10] a triangular representation is shown, though actually this is not in common use.

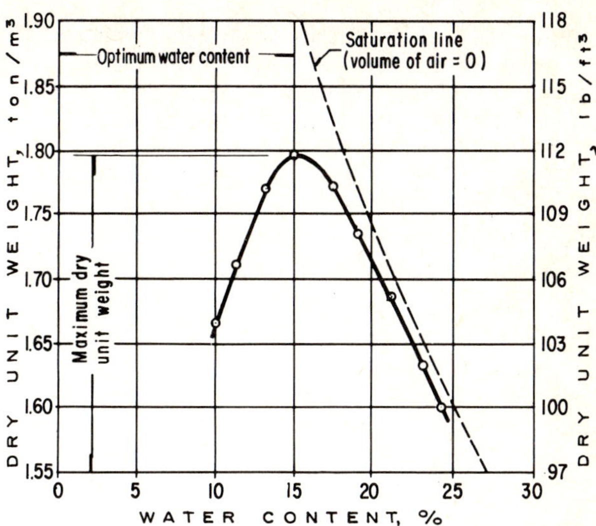

Fig. 4-1 Typical compaction curve

Whatever the compaction procedure, the curve will be similar to the one shown in Fig. 4-1. The curve shows an absolute maximum, sometimes accompanied by another secondary one of lesser value. The dry unit weight corresponding to the absolute maximum is called the *maximum dry density* or more correctly the *maximum dry unit weight*; the water content which corresponds to this maximum is known as the *optimum moisture*. It represents the water content at which the compaction procedure being used produces maximum density. It also reflects maximum efficiency, if efficiency is judged by the dry unit weight achieved.

Curves similar to Fig. 4-1 can be established either in the laboratory or in the field, starting with pairs of values which can be obtained if the same compaction procedure is applied to several specimens of the same soil with different water contents. The curve can then be drawn starting with the above values and using Eq. (1-4):

$$\gamma_d = \frac{\gamma_m}{1 + w} \qquad (4\text{-}2)$$

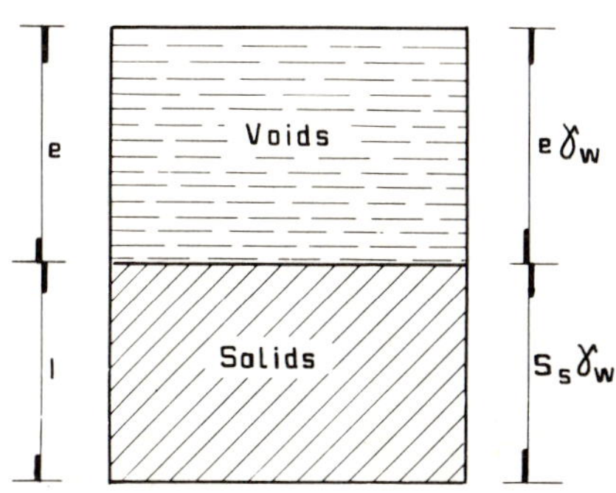

By definition:

$$\gamma_d = \frac{W_s}{V_m} = \frac{S_s \gamma_w}{1 + e}$$

According to formula (1-9)

$$e = w\, S_s \quad \text{(Saturated soil)}$$

$$\therefore \; \gamma_d = \frac{S_s}{1 + w\, S_s} \gamma_w$$

Fig. 4-2 Diagram of a saturated soil and deduction of Eq. (4-3)

In Fig. 4-1 the curve corresponding to 100% saturation of the soil also appears. As previously discussed, the state of a soil compacted under normal circumstances is unsaturated. Therefore, the compaction curve lies below the saturation curve. By comparing both, it is possible to find the water content which would be necessary to saturate a sample compacted to any dry unit weight. The saturation curve can be obtained by calculating the dry unit weights which would correspond to saturation of a given water content. The expression

$$\gamma_d = \frac{S_s}{1 + wS_s}\gamma_w \tag{4-3}$$

for the dry unit weight in the saturated state is obtained by referring to Fig. 4-2.

The moisture-dry unit weight curve for a clean cohesionless sand produced by tamping in the laboratory is shown in Fig. 4-3 [4]. The soil is sufficiently permeable so that pore-water pressures do not build up during the compaction process. When the water content is very low, particle movement during tamping is resisted largely by the grain friction generated by the tamping forces. When some moisture is added, the frictional resistance of the grains is augmented by capillary tension. Therefore the compacted density will be less. With more water the capillary forces decrease and the compacted dry unit weight becomes equal to or slightly greater than that of dry soil. The optimum moisture approximately corresponds to saturation at the maximum dry unit weight; if more water is added it is expelled readily from the sand voids.

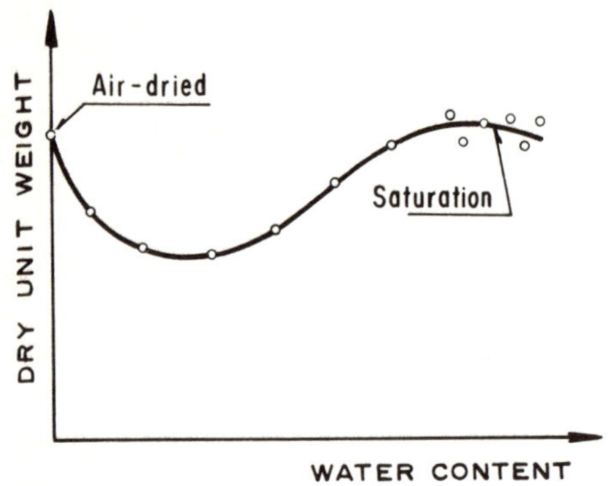

Fig. 4-3 Compaction curve obtained with dynamic test on clean sands and sandy gravels [4]

4.4 Field Compaction Processes

The energy that is required to compact soils in the field can be applied by means of any of the four methods listed below, which are differentiated by the nature of the forces applied and the duration of these forces. The methods are: Kneading compaction, Static compaction, Dynamic or impact compaction, and Vibratory compaction. All have been used for several decades; however the vibratory types have undergone the greatest improvement during recent years. These do not represent every possible method of soil compaction, but are the commercial and industrial solutions which are in common use.

4.4.1 Kneading Compactors; Sheepsfoot Rollers

The weight of these compactors is centered on the relatively small surface of a series of differently shaped projections or feet, Fig. 4-4, Plates 4-5 to 4-7. A concentrated pressure that changes in both angle and magnitude is thus applied at the points where the feet penetrate the soil. As the ground is rolled and the material compacted, the feet penetrate less into the soil. Eventually there comes a moment when additional compaction is no longer being produced and the roller is said to *walk out*. The surface invariably remains distorted to a depth of about 2 to 5 cm (1 to 2 in), but it is compacted under the next layer.

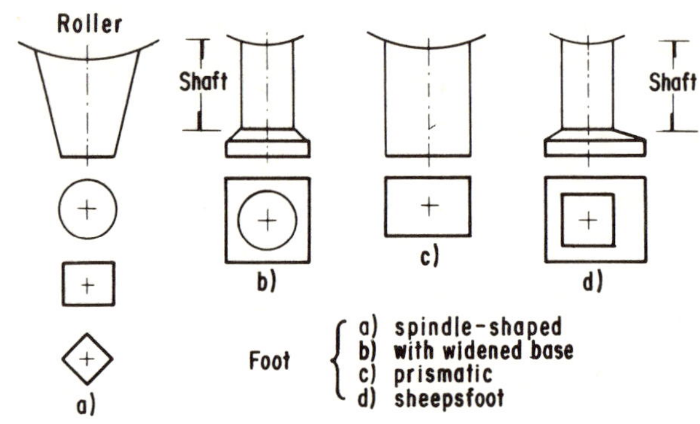

Fig. 4-4 Usual types of sheepsfoot roller shafts [3]

Figure 4-5 [11] shows the effect on compaction of the end of the foot. The contours depict the dry unit weights obtained for each of three shapes, all 15 cm by 15 cm (6 in by 6 in) in cross-section. In all the cases shown, the foot applied the same average pressure and in the same manner throughout. The greater efficiency of the flat foot and the wide wedge shape is revealed by the more regular shape of the better compacted area (1300 kg/m^3 or more) and the greater soil volume that this area represents.

Figure 4-6 [11] shows the results of tests on the effect of the shape of the cross-section of the foot and its area. The greater effectiveness of the larger foot can be seen, as well as the superiority of the square section compared with the rounded one. This will be discussed further in this chapter (see Fig. 4-4).

The pressure exerted by the sheepsfoot roller as it goes over the soil with its feet is not constant timewise; the feet penetrate the soil, exerting ever-increasing pressures which reach a maximum at the moment the foot is vertical and thus at its maximum penetration. From that moment on, the pressure reduces until the foot is withdrawn. The action of the roller is such that compaction of the soil takes place from the bottom upwards. In the first few passes, the feet and part of the drum penetrate the soil deeply which causes large pressures to be exerted on the bottom part of the layer being compacted. For best results, however, the depth of the uncompacted layer should be less than about 4/3 the length of the foot. This variable compaction force is referred to as *kneading*. In recent years it has been incorporated into laboratory compaction devices, so as to achieve greater similarity in laboratory tests to field compaction with sheepsfoot rollers.

The most common rollers have feet 20 to 25 cm (8 to 10 in) long and are used to compact layers of loose soil about 30 cm (12 in) deep. As already mentioned, as the number of passes increases, the strength attained by the lower parts of the layer increases which prevents deep penetration of the feet, so that they eventually compact the upper soil. The process reaches a limit where the roller *walks out* on the surface. This limit is sometimes specified to control layer compaction. However when the

Plate 4-5 Common sheepsfoot roller drawn by a caterpillar tractor. It is equipped with a blade to level off the strip of ground being compacted

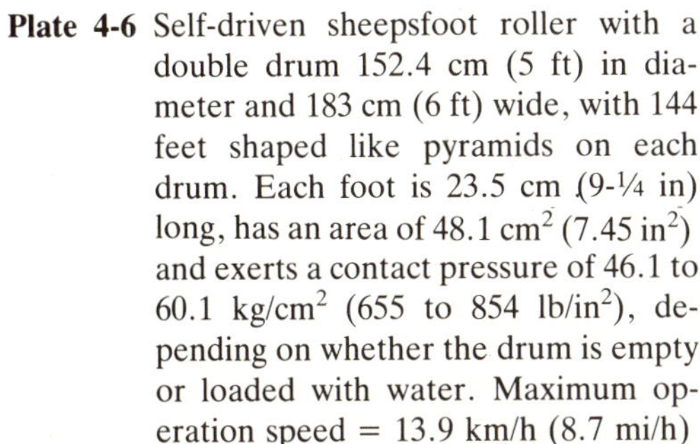

Plate 4-6 Self-driven sheepsfoot roller with a double drum 152.4 cm (5 ft) in diameter and 183 cm (6 ft) wide, with 144 feet shaped like pyramids on each drum. Each foot is 23.5 cm (9-¼ in) long, has an area of 48.1 cm^2 (7.45 in^2) and exerts a contact pressure of 46.1 to 60.1 kg/cm^2 (655 to 854 lb/in^2), depending on whether the drum is empty or loaded with water. Maximum operation speed = 13.9 km/h (8.7 mi/h)

Plate 4-7 Heavyweight high capacity self-driven sheepsfoot roller, equipped with 4 identical drums with diameters and widths of 152.4 cm (5 ft). It has 120 cone-shaped feet per drum; each foot is 23 cm (9 in) long and has a contact area of 64.4 cm^2 (10 in^2). Maximum operation speed = 8 km/h (5 mi/h)

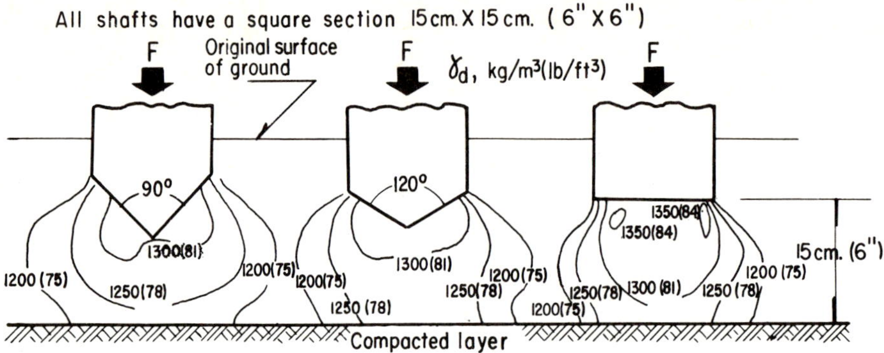

Fig. 4-5 Effect of the shape of the shaft tip in sheepsfoot rollers

soils have a relatively high water content, or when very heavy rollers are used, the roller may not walk out and the drum may make continuous contact with the soil. This can occur regardless of the number of passes [12,13].

The compaction has generally reached its limit of effectiveness when 1/5 to 1/3 of the length of the foot still penetrates the soil, depending on the plasticity of the soil material. For a soft clay smaller penetrations are necessary, to minimize soil sticking to the shaft and reducing the efficiency of the equipment, than for a low plasticity sandy clay. In all cases, the roller feet always penetrate the soil to some degree.

The sheepsfoot roller produces, therefore, two very desirable results in earthfills composed of compacted fine soils, namely a reasonably uniform distribution of the compactive effort within each layer and an irregular dimpled surface from the remaining foot penetration which prevents a smooth weak surface between successive layers.

Figure 4-7 depicts families of moisture-dry unit weight curves for a single soil and single type of roller. Each different curve represents the variation of dry unit weight with soil moisture using a single level of compacted effort. Each successive curve from right to left represents a larger compactive effort, achieved by increasing the number of passes or decreasing the layer thickness. If the same equipment is used, any increase in the compactive effort will cause an increase in maximum dry unit weight and a decrease in optimum water content.

Figure 4-8 [13,14] depicts the effect of the number of passes of a typical roller on different types of soil. For all soils the rate of dry unit weight increase becomes less as the number of passes increases. Moreover, the rate of unit weight increase is less for a highly plastic or heavy clay than for a sandy clay or silty clay.

Table 4-2 [13,15] gives a summary of information from different sources on the effect of the contact pressure under the feet of a roller; the calculation consisted of dividing the total weight of the roller by the product of the number of feet in one row by the contact area of each foot. It can be seen that the pressures and the number of passes specified cause scarcely any change in the compaction of the soils, even with increase of more than three times in the contact pressure.

The above seems to indicate that the intensity of the pressure of the foot is of no importance in the process, but there must, of course, be a minimum value for this concept if effective compaction is to be achieved. With the information available today, it is not possible to say what this value is, but Table 4-2 leads us to believe that it should not be lower than 8 kg/cm^2 (114 lb/in^2) if the contact area of the foot is not larger than 75 or 90 cm^2 (11.6 or 14.0 in^2). On the other hand, although maximum dry unit weight does not depend on contact pressure, the water content with which this maximum is obtained does increase with the contact pressure.

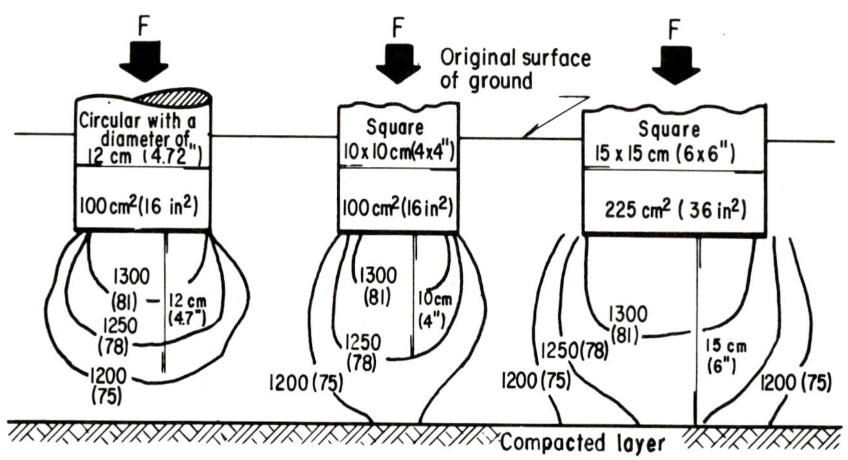

Fig. 4-6 Effect of the size and area of the cross-section of the shafts of sheepsfoot rollers

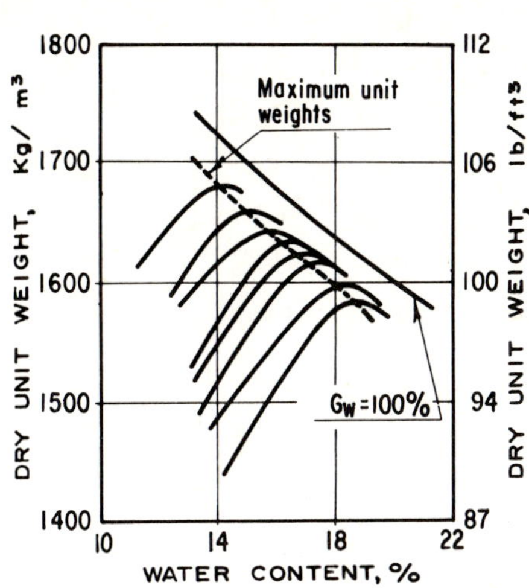

Fig. 4-7 Qualitative diagram of γ vs *w* for various compactive efforts [13]

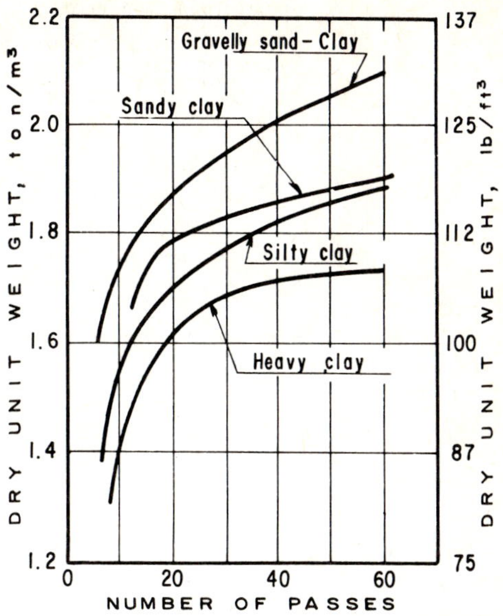

Fig. 4-8 Compaction with sheepsfoot roller. Effect of the number of passes on the degree of compaction of various soils [13]

Table 4-2

Sheepsfoot rollers: effect of contact pressure† on the maximum dry unit weight

Type of Soil	Contact pressure		Contact area		Number of passes	Compaction — standard PROCTOR %
	kg/cm^2	lb/in^2	cm^2	in^2		
Clayey sand	17.5	248.67	43.75	6.78	9	99
	31.5	447.61	43.75	6.78	9	99
Silty clay I	17.5	248.67	43.75	6.78	8	102
	35.0	497.35	43.75	6.78	8	101
	52.5	746.02	43.75	6.78	8	101
Lean clay	8.7	123.63	87.5	13.56	12	101
	26.2	372.3	87.5	13.56	12	101
Heavy clay	8.0	113.68	75.25	11.66	64	108
	17.5	248.67	31.5	4.88	64	108
Silty clay II	8.0	113.68	75.25	11.66	64	112
	17.5	248.67	31.5	4.88	64	111
Sandy clay	8.0	113.68	75.25	11.66	64	104
	17.5	248.67	31.5	4.88	64	104
Mixture of gravel sand and clay	8.0	113.68	75.25	11.66	64	100
	17.5	248.67	31.5	4.88	64	99

† In all cases the depth of the compacted layer was approximately 15 cm (6 in)

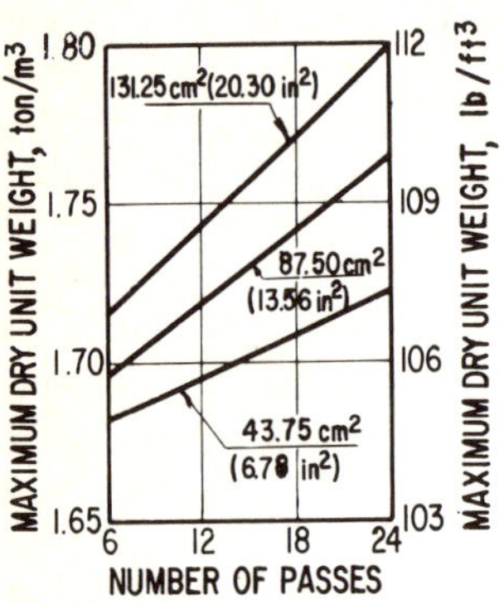

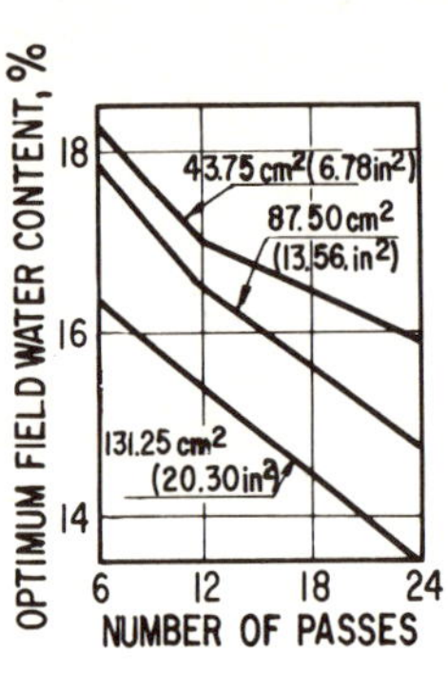

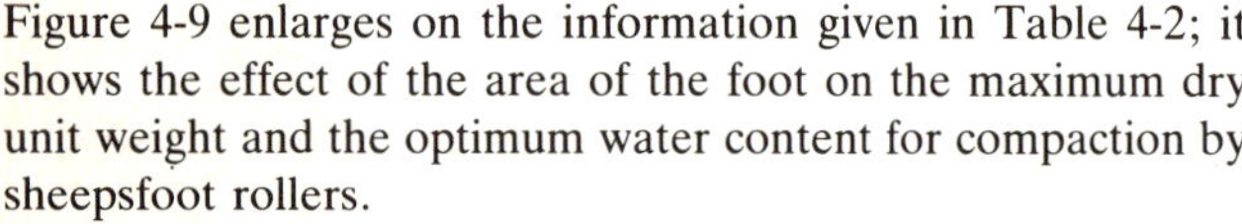

Fig. 4-9 Compaction with sheepsfoot roller. Effect of the contact area of the shafts on the dry unit weight and optimum field water content [13,15]

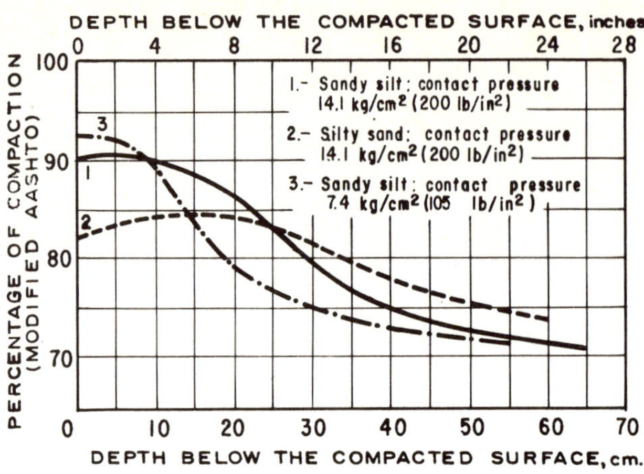

Fig. 4-10 Effect of compaction with sheepsfoot roller in relation to the depth of the compacted layer [16]

Figure 4-9 enlarges on the information given in Table 4-2; it shows the effect of the area of the foot on the maximum dry unit weight and the optimum water content for compaction by sheepsfoot rollers.

For a given number of passes, an increase in the contact area brings about an increase in the dry unit weight obtained (see also Fig. 4-6); further, an increase in the contact area permits a reduction in the number of passes required to achieve a specific result. It is, therefore advisable to make the area of the cross-section of the feet as big as possible (most modern rollers tend to be like this) so as to obtain maximum efficiency, though the required pressure must be above the minimum.

Figure 4-10 [16] illustrates the way in which the compactive effort produced by sheepsfoot rollers affect different depths of soil, measured from the surface of the layer. The graph corresponds to a sheepsfoot roller with a cone-shaped foot, a contact area per foot of 32.2 cm^2 (5 in^2) and a contact pressure equivalent to 7.4 kg/cm^2 (105 lb/in^2) with the roller drum empty and 14.1 kg/cm^2 (200 lb/in^2) with the drum ballasted. With the drum empty, tests were carried out only on a sandy silt. The compaction percentage refers to the maximum dry unit weight obtained with the modified AASHTO test. By examining the curves in Fig. 4-10, it can be seen that in these sandy soils, the compaction percentage is relatively constant above a depth of about 25 cm (10 in) when the ballasted roller is used. However, when the empty roller is used there is a marked decrease in the dry unit weight below a depth of 10 cm (4 in).

Most sheepsfoot rollers in use are within the limits of the specifications given in Table 4-3 [16]. The percentage of area covered per pass of the sheepsfoot roller (two successive applications on the same point) is usually between 4% and 12%, which is considerably smaller than for other types of compaction equipment. If the number of feet per drum is increased, the percentage of coverage increases correspondingly, but the contact pressure is reduced. Therefore, the number of feet for commercial rollers must be determined by careful evaluation of these factors. Moreover there must be a minimum space between the feet to allow for cleaning forks; otherwise the roller efficiency will drop.

Sheepsfoot rollers leave a larger percentage of holes or surface indentations than other types of compaction equipment such as smooth steel-wheel rollers, pneumatic-tired rollers, grid rollers and segmented steel-wheeled rollers. This unfavorable condition can be avoided if a combination of different types of compaction equipment is used for cohesive soils. If the feet penetrate the same indentations several successive times, the efficiency of the rolling is less. To avoid this, the operator should make slight changes in the roller's route. A better procedure is to alternate rolling with two rollers having different foot spacings.

Table 4-3
Sheepsfoot rollers : common specifications

Width of drum	1.22 to 1.98 m (4 to 6.5 ft)
Diameter of drum	1.02 to 1.83 m (3.35 to 6.0 ft)
Number of feet or shafts	64 to 144
Area of straight section of feet	33 to 135 cm^2 (5.12 to 20.93 in^2)
Length of feet	18 to 46 cm (7.08 to 18.11 in)
Weight of empty roller	1.6 to 7.0 t (3.52 to 15.40 kips)
Weight of roller filled with water	2.5 to 11.5 t (5.50 to 25.30 kips)
Contact pressure, empty	5.2 to 30.00 kg/cm^2 (73.89 to 426 lb/in^2)
Contact pressure filled with water	8.0 to 55.0 kg/cm^2 (113.7 to 781 lb/in^2)

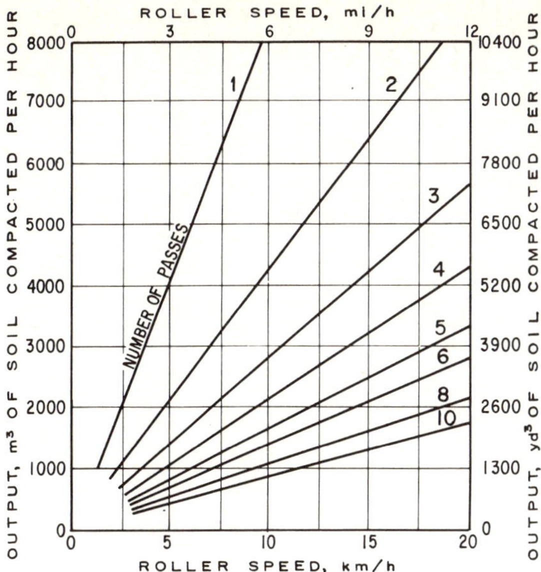

Fig. 4-11 Relationship between the output of a sheepsfoot roller, operation speed and number of passes [16]

For equipment with specific characteristics, the maximum possible operation efficiency can be roughly calculated by the following equation:

$$E = \frac{ahv}{10n} \ (\text{m}^3/\text{h})$$

$$E = \frac{1.36\, ahv}{n} \ (\text{yd}^3/\text{h}) \qquad (4\text{-}4)$$

in which: E is the efficiency of the compactor, m^3/h (yd^3/h), a the width of roller in cm (in), h the depth of the layer compacted in cm (in), v the speed of compactor in km/h (mi/h), n the number of passes of roller over the same point.

The theoretical compaction efficiency of sheepsfoot rollers always increases with speed; this relationship is even stronger with a reduction in the number of passes. However, at speeds exceeding 10 to 15 km/h, there appears to be insufficient time for the soil to densify, and greater speed reduces efficiency.

Figure 4-11 is a graph which shows how the efficiency of the roller increases when the required results can be achieved with fewer passes for a constant operation speed. The graph refers to a self-driven sheepsfoot roller with two pairs of drums each 183 cm (6 ft) wide; it also corresponds to a layer thickness of 23 cm (9 in).

Sheepsfoot rollers produce best results in fine soils. The concentrated pressure of the feet has shown itself to be very useful in breaking up the lumps which form in homogeneous clays as a result of the action of capillary-type forces among the particles. In non-homogeneous fine soils, with different particle-size ranges sheepsfoot rollers are also very good at breaking up and separating the different particles. Sheepsfoot rollers aid is bonding together the layers of compacted material because the distorted surface of each layer is compacted together with the following layer, eliminating the tendency to slip.

In soft heavy clays, sheepsfoot rollers are useful not only because they get rid of the lumps, but also on account of the kneading action already described. Recently sheepsfoot rollers have even been used in combination with vibratory equipment to increase the concentration of effort on small areas. There is little evidence of the effectiveness of the combination, except for breaking down hard lumps.

Two types of compactor have been developed which can be considered variations on the traditional sheepsfoot: the grid roller, Plate 4-8, and the segmented roller, Plate 4-9. The grid roller has been successfully used on materials that have to be broken up and then compacted; it has also produced good compaction in a great variety of soils, including homogeneous clays or mixtures of sands, silts and clays. The surface of the drum consists of a grid or mesh made of steel bars which form squares. They are usually ballasted with blocks of concrete or wet sand. Generally they are made very heavy (over 14 tons (31 kips) when ballasted) and exert high contact pressure (more than 20 kg/cm^2 (284 lb/in^2)).

The segmented roller is also used with materials that have to be broken up; however its use is being extended to various soils including low plasticity clays. Each drum usually consists of three disc wheels side by side each with a broken or segmented rim to which the roller owes its name.

Plate 4-8 Double-drum grid roller, with a free square of 8.9 × 8.9 cm (3½ in × 3½ in) between bars; its overall weight varies between 2.82 t when empty and 6.87 (15.13 kips) when fully ballasted. This compactor is equipped with 4 metal deposits for ballast and is drawn by a caterpillar tractor

Plate 4-9 Self-driven compactor with segmented metal wheels, capable of speeds of up to 9.65 km/h (6 mi/h)

4.4.2 Smooth Steel-Wheeled and Pneumatic-Tired Rollers

.1 Smooth Steel-Wheeled Rollers

These are divided into two groups: tractor-drawn and self-driven. The former generally consists of two drums mounted on a frame to which the axles are attached; they usually weigh between 14 and 20 t (31 and 44 kips) and can be ballasted by filling the box, which is situated above the frame, with water or wet sand. The self-driven type consists of one front wheel and one or two rear wheels and weigh between 3 and 13 t (6.6 and 29 kips) Plate 4-10 [17]. They run on a petrol or diesel engine and can go both backwards and forwards. The use of smooth steel-wheeled rollers is restricted to materials that do not require great pressures (either because the soils do not form lumps or else because they do not need to be broken, such as relatively clean sands and gravels). They are also used to finish the upper surface of the compacted layers (subgrade, base and asphalt surfaces).

The compaction efficiency of smooth rollers is impaired if they sink into the layer being compacted, and compaction takes place from top to bottom. Figure 4-12 illustrates how pressure and compacted density diminish with depth below a smooth 3-wheeled roller, having an overall weight of 9.5 t (21 kips) compacting a clayey-sandy soil with a water content of 13.5% [16].

When a smooth drum roller is used on clays and plastic silts, cracks often appear in the upper part of the layer after a large number of passes. The well compacted rigid surface breaks from the distortion of the deeper, poorly compacted lower part of the same layer.

Smooth rollers are characterized by diameter (which can greatly improve their efficiency), width and overall weight. The depth of layers of loose material that can be compacted with a smooth roller varies between 10 and 20 cm (4 and 8 in). Table 4-4 shows the typical characteristics of self-driven smooth rollers with three wheels [16].

An approximate calculation of the efficiency of a smooth roller can be made with Eq. (4-4). For three-wheeled rollers value a should be taken as the width of the compacted layer equal to the sum of the width of the three wheels minus the amount the rear wheels overlap the front one.

Figure 4-13 shows the theoretical efficiency of a three-wheeled smooth roller weighing 10 t (22 kips) with driving wheels 51 cm (1 ft 8 in) wide and with a space of 91.5 cm (3 ft) between them. Efficiency was calculated taking into account only the effect of the driving wheels as they worked on a layer 15 cm (6 in) deep. At speeds exceeding 15 km/h or 10 m/h, the real efficiency drops.

.2 Pneumatic-Tired Rollers

The compacting action of pneumatic-tired rollers (with air-filled tires) is basically a result of the pressure which is exerted on the layer of soil. In addition, these rollers produce some kneading which causes large angular deformations and more effective densification. The kneading is caused by the tire tread and the tire flexing. The kneading apparently occurs on a smaller scale than with sheepsfoot rollers, but is nevertheless significant, especially in the topmost part of the layer being compacted. From the first pass the roller exerts a changing

Plate 4-10 Smooth steel-wheeled roller with 3 wheels, compacting a layer of crushed rock. Behind it is attached a compactor with 3 vibrating plates

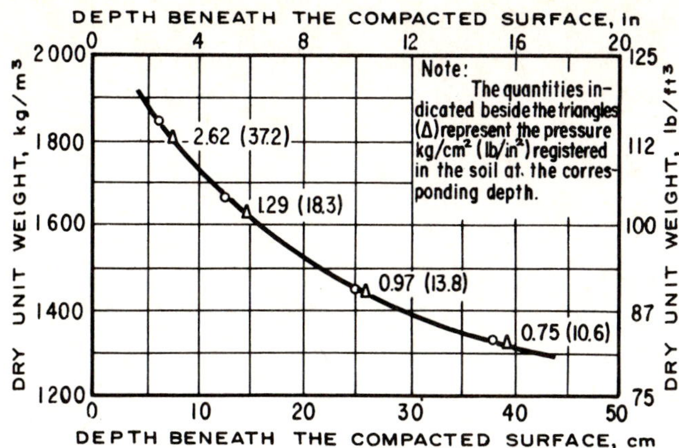

Fig 4-12 Disappearance with depth of contact pressure of a smooth steel-wheeled roller

pressure on the surface of the layer. The pressure exerted is not uniform but the average is equal to or slightly less than the air pressure.

The contact surface of the tire depends on the weight of the roller and the tire pressure. The contact area is more or less elliptical. In order to achieve more uniform compaction below the surface, the front and rear wheels of the roller are slightly offset. The compactive effort increases with the tire pressure; however if the pressure is too high a bearing capacity failure will occur and the soil layer will be rutted without compaction. The surface finish of layers compacted with pneumatic-tired rollers is usually sufficiently rough to assure a strong bond with the layer above.

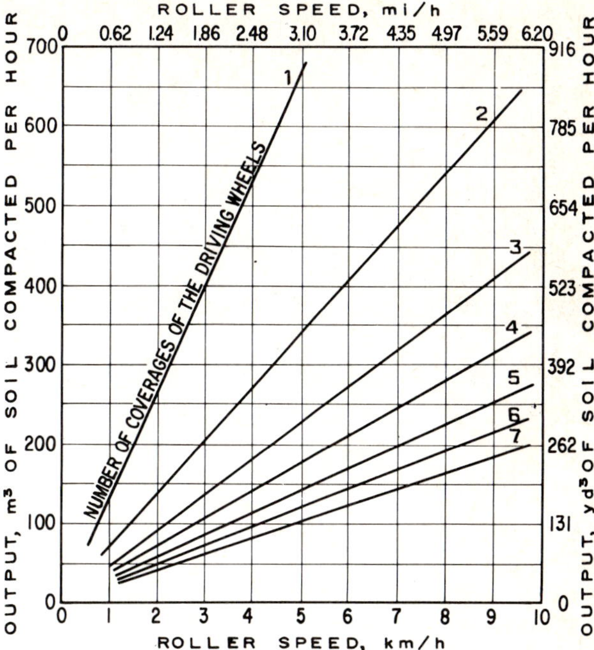

Fig. 4-13 Relationship between the output of a smooth steel-wheeled roller, operation speed and number of passes of the wheels over the same point [16]

Figure 4-14 illustrates the effect of the number of passes and tire pressure on the dry unit weight obtained in several soils [13,18]. The shape of the curves is the same for the three soils. In all cases the increase in dry unit weight after 16 passes is insignificant. This would not occur if the water content were below optimum, for it has been observed that in such cases the dry unit weight increases even with a far greater number of passes. Examination of the figure also shows the vital importance of the tire pressure on the compaction process.

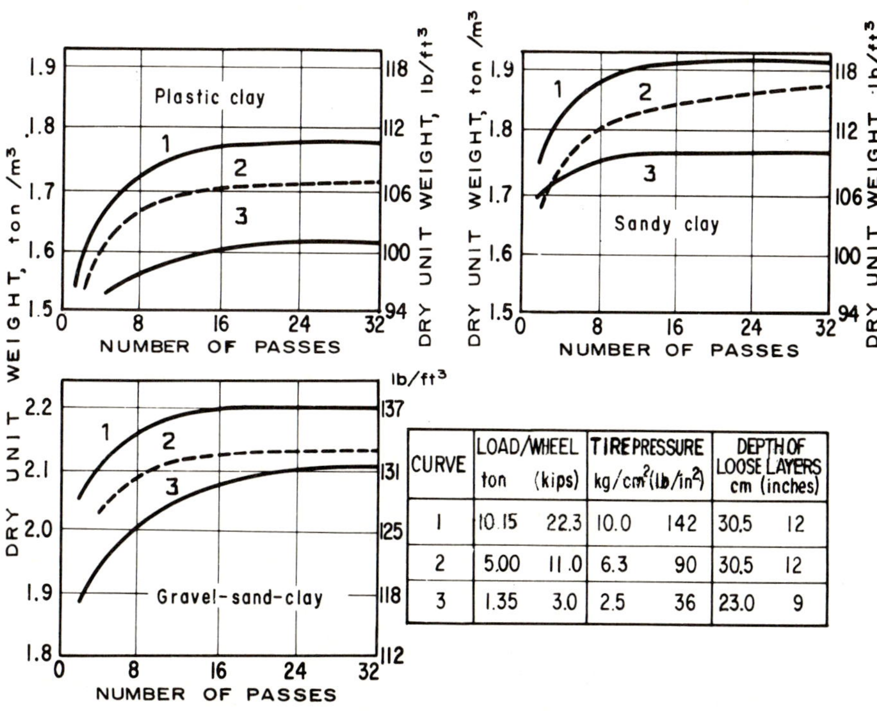

CURVE	LOAD/WHEEL ton	(kips)	TIRE PRESSURE kg/cm²	(lb/in²)	DEPTH OF LOOSE LAYERS cm	(inches)
1	10.15	22.3	10.0	142	30.5	12
2	5.00	11.0	6.3	90	30.5	12
3	1.35	3.0	2.5	36	23.0	9

Fig. 4-14 Compaction with pneumatic-tyred roller. Effect of the number of passes and tyre pressure on the dry unit weight of various soils [13,18]

Table 4-4
Smooth steel-wheeled self-driven rollers with 3 wheels : common specifications

Overall weight	3 to 13 t (6.62 to 28.7 kips)
Diameter of front roller	86 to 120 cm (2.82 to 3.94 ft)
Diameter of rear rollers	94 to 160 cm (3.08 to 5.25 ft)
Width of front roller	61 to 122 cm (2.00 to 4.00 ft)
Width of rear rollers	38 to 58 cm (1.25 to 1.90 ft)
Load per unit width of front roller	14 to 43 kg/cm^2 (78.3 to 240.6 lb/in^2)
Load per unit width of rear rollers	25 to 80 kg/cm^2 (140 to 448 lb/in^2)

Figure 4-15 [4,19] shows the dry unit weights obtained with a pneumatic-tired roller in relation to tire pressure, the number of passes and the water content of the soil, for a low compressibility silt (*ML*). It will be observed that for the highest water content in the test (18% above optimum) the increase in the number of passes from 4 to 16 had little effect. There is also little increase in density for increases in tire pressure above a limit (the bearing capacity). With less compaction moisture, an increase in tire pressure produces greater compaction. Moreover, increasing the number of passes also produces greater compaction. Also for a given compaction water content, an increase in pressure of the roller tire makes it possible to reduce the number of passes required to obtain a particular dry unit weight.

In any type of soil an increase in tire pressure brings about an increase in the maximum dry unit weight, as can be appreciated from Fig. 4-16. This increase is accompanied by a decrease in the optimum water content. However, it is not advisable to

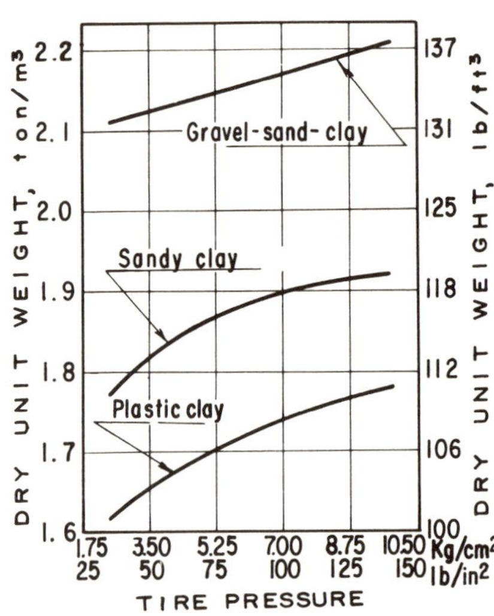

Fig. 4-16 Compaction with pneumatic-tired roller, relationship between tire pressure and maximum dry unit weight [13,18]

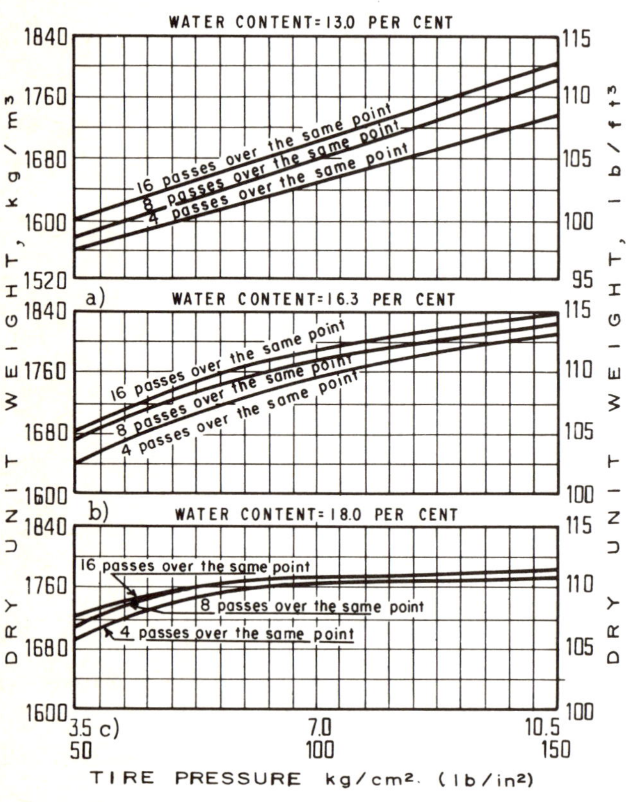

Fig. 4-15 Effect of the tire pressure, number of passes and water content, Pneumatic-tired roller [4]

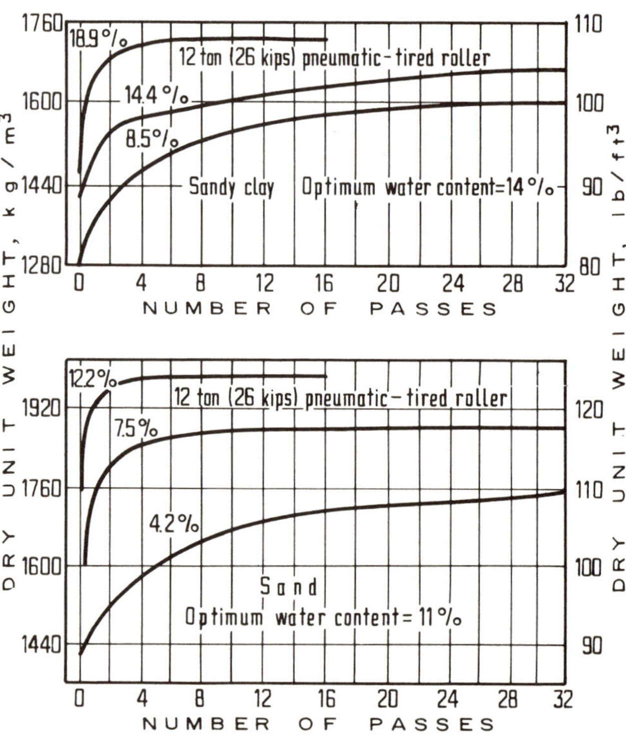

Fig. 4-17 Graph showing the effect on compaction of the water content and the number of passes of the equipment

increase tire pressure unless the load per wheel is increased in the same proportion, because this reduces the contact area and horizontal confinement, and tends to cause reduced compaction with increasing depth.

Figure 4-17 depicts data from research by the Road Research Laboratory in London, England. It shows the results of compaction of two soils, a sand and a clayey sand. Compaction was by a light pneumatic-tired roller with multiple wheels. The optimum water content given is the one corresponding to the British Standard Test which is very similar to the AASHTO Standard Test (described later in this chapter). The figure shows curves of the dry unit weight* as the function of the number of passes over the soil for different water contents. The great influence of the water content on the efficiency of the equipment should be noted. For each soil there is a moisture content that produces a higher density with 6 passes than can be obtained by a lower moisture regardless of the number of passes. In both cases that moisture was slightly greater than the british laboratory's optimum. Again, the appropriate water content for compacting soil with a particular type of equipment is not the same as the standard laboratory optimum moisture that is used to control compaction work. The reason, of course, is that the compactive efforts are different.

Figure 4-17 also shows how the increase in density per pass (or the efficiency of compaction) drops after a certain number of passes, depending on the soil and its water content; Fig. 4-18 [4,19] shows the effectiveness of compacting by a heavy pneumatic-tired roller with varying depth of the layers. Three different depths were used: 15, 30 and 60 cm (6 in, 1 ft and 2 ft). The dry unit weights were obtained at different depths within the layer with three different water contents of a clayey soil. There are a number of practical lessons to be learned from this graph: (1) for field compaction the constructor is required to produce a certain minimum unit weight throughout the depth of the layer; (2) in order to achieve this (and to avoid quality control problems) he should use equipment which produces higher unit weights than are required in the upper part of a thick layer so as to produce the minimum required in the lower part; (3) the choice of layer depth is not arbitrary, but is related to the equipment available, and compaction water content, etc. For an effective balance, all these factors usually require the use of test earthfills where the variables can be tested before production compaction begins.

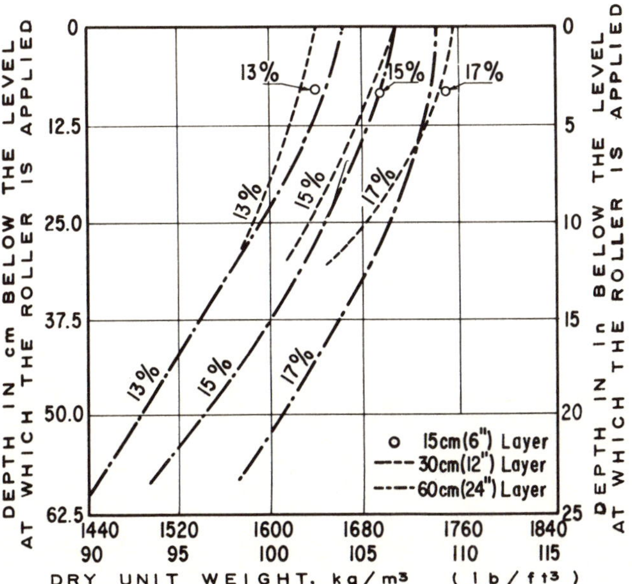

Fig. 4-18 Effect of depth of layer and water content. Pneumatic-tired rollers [4,19]

Figure 4-19 [16] complements the information given in the previous figure. It shows the compacted density as a function of depth for different soils. A light pneumatic-tired roller was used, weighing 14 t (31 kips) and with 13 tires on two axles, with a load per tire of approximately one t (2.2 kips). The contact area was 19 x 38 cm (7-1/2 x 15 in) and the tire pressure 2.5 kg/cm^2 (36 lb/in^2). The three soils were placed in loose layers 75 cm (2.5 ft) deep and were compacted by 6 passes. It can be observed that the effectiveness of compaction drops fairly rapidly with depth, although not as rapidly as with smooth steel-wheeled rollers.

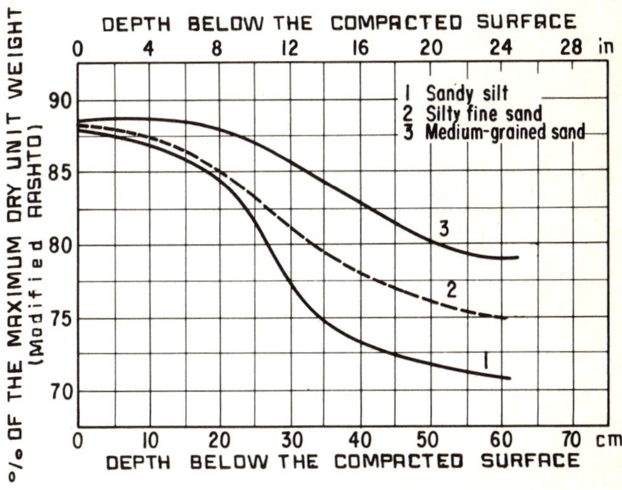

Fig. 4-19 Effect of a pneumatic-tired roller throughout the depth of the compacted layer [16]

Pneumatic-tired rollers are usually built with one or two axles, on which there is a platform or deposit for ballast; they can either be tractor-drawn or self-propelled. Lightweight rollers are usually self-propelled, weigh less than 13 t (29 kips) and have 9 to 13 wheels on two axles. Medium-weight rollers vary between 13 and 25 t (29 and 55 kips) and usually have 4 to 11 wheels on one or two axles. Heavy rollers weigh from 25 to 110 t (55 to 242 kips) and generally consist of 7 wheels on two axles or 4 on one, Plates 4-11, 4-12.

There is a pneumatic-tired compactor, called a wobble wheeled roller, which has the wheels of one of its axles mounted at an oblique angle to the axle, causing an increase in the kneading effect; this slightly increases the effectiveness of the equipment in fine soils and produces a smoother surface. Table 4-5 shows the most common characteristics of pneumatic-tired rollers.

The effectiveness of pneumatic-tired rollers is affected by the load per wheel, tire pressure, width of the roller, percentage of coverage per pass, overlapping of passes and speed of the compactor. Although each case is different, Table 4-6, by way of illustration, gives the average output of several pneumatic-tired rollers, obtained during compaction of a clayey sand until a dry unit weight of 95% of a Standard Proctor Test was reached.

Plate 4-11 Tractor-drawn pneumatic-tired roller with 4 tires, each of which carries an oscillating box. This 4-section unit is built with an overall weight of 13.5 to 91 t (30 to 200 kips) and tire pressures of 5.6 to 10.6 kg/cm^2 (80 to 151 lb/in^2)

Table 4-5
Pneumatic-tired rollers : common specifications

Total width of equipment	152 to 305 cm (5 to 10 ft)
Size of tire	7.5 × 15 to 30 × 40 in
Space between wheels (center to center)	46 to 76 cm (1.5 to 2.5 ft)
Overall weight of roller	6 to 110 t (13.2 to 242 kips)
Load per wheel	0.6 to 27 t (1.33 to 59.4 kips)
Tire pressure	1.76 to 10.6 kg/cm^2 (25 to 150 lb/in^2)
Contact pressure	1.5 to 8.5 kg/cm^2 (21.3 to 120.8 lb/in^2)
Contact area	480 to 3730 cm^2 (75 to 578 in^2)

Figure 4-20 shows the theoretical maximum output of a heavy pneumatic-tired roller with a load per wheel of 11.4 t (25 kips), tire pressure of 10.6 kg/cm^2 (150 lb/in^2) and a width of 3.05 m (10 ft) compacting a layer of material 23 cm (9 in) deep, complete coverage per pass is considered with a width equal to that of the roller.

As compaction progresses in each layer, the soil gradually becomes more and more resistant to compaction. Therefore it is sometimes advisable to begin each layer with equipment that exerts relatively low contact pressure. As the layer becomes more dense and stronger, a roller with greater pressures can be used for the final passes. Some self-propelled pneumatic-tired rollers are equipped with a special device which enables the driver to vary tire pressure within limits without interrupting the compaction process, Plate 4-14. This is done by a compressor which is connected to the tires. Such equipment is available with tire pressures varying from 2.1 to 7 kg/cm^2 (30 to 100 lb/in^2). The change improves the efficiency and probably reduces costs. These rollers are mainly used for compacting subgrades for pavements.

Pneumatic-tired rollers are principally for sandy soils and clays with low to medium plasticity; such soils have no hard lumps to be broken and consequently do not require such concentrated pressures as those produced by sheepsfoot rollers. Pneumatic-tired rollers also produce good results in low to medium plasticity silts.

Plate 4-12 Self-driven pneumatic-tired roller, with a maximum weight of 30 (66 kips) and 7 wheels

Table 4-6
Average efficiency of pneumatic-tired rollers

Weight of roller		Load per wheel		Tire pressure		Width of compacted strip		Speed of roller		Number of passes	Depth of compacted layer		Output of compacted soil	
t	kips	t	kips	kg/cm^2	lb/in^2	m	ft	km/h	mi/h	—	cm	in	m^3/h	yd^3/h
13.44	29.60	1.35	2.97	2.54	36.09	2.08	6.82	3.65	2.27	4	12.7	5	199	260
22.4	49.34	2.26	4.98	5.64	80.14	2.13	6.99	3.65	2.27	4	15.2	6	245	320
50.4	111.01	5.09	11.21	6.34	90.09	2.35	7.71	3.65	2.27	4	17.7	7	321	420
50.4	111.01	5.09	11.21	9.86	140.11	2.35	7.71	3.65	2.27	4	20.3	8	367	480
50.4	111.01	10.18	22.42	6.34	90.09	2.35	7.71	3.65	2.27	4	22.8	9	550	720
50.4	111.01	10.18	22.42	9.86	140.11	2.35	7.71	3.65	2.27	4	25.4	10	611	800

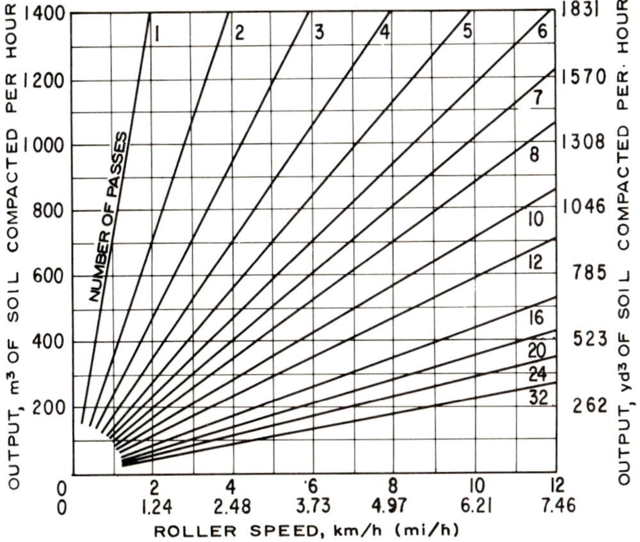

Fig. 4-20 Relationship between the output of a pneumatic-tired roller, operation speed and number of passes of the wheels over the same point [16]

.3 Comparison of Pneumatic-Tired and Sheepsfoot Rollers

As has been established already, it is often impossible to distinguish between the practical uses of pneumatic-tired and sheepsfoot rollers, so that in many cases the choice of equipment is a matter of individual preference or equipment availability. Some constructors maintain that more uniform compaction can be expected with a sheepsfoot roller than with a pneumatic-tired one. Since uniformity is a highly desirable characteristic, many constructors prefer to use sheepsfoot rollers. The quantitative data available do not, however, support this opinion. On the contrary, the test data are in favor of the pneumatic-tired roller. These data are illustrated in Table 4-7, which is taken from [13]. Here it can be seen that with the same tire pressure the degree of uniformity obtained with pneumatic-tired rollers is in direct relation to the increase of load per wheel; in the case of sheepsfoot rollers the prism-shaped shaft (type *c* in Fig. 4-4) appears to hold a slight advantage.

Plate 4-13 Tractor-drawn pneumatic-tired roller with 4 tires, a maximum weight of 100 t (220 kips) and tire pressure of 3.5 to 10.5 kg/cm^2 (50 to 150 lb/in^2)

Table 4-7

Variation of the dry unit weight within the depth of the compacted layer : ratio of the average dry unit weight in the lower third of the layer to that in the remainder

Sheepsfoot roller : depth of compacted layer 15 cm (6 in)		
Type of soil	Type *a* foot as in Fig. 4-4, 17.5 cm (6.9 in) long	Type *c* foot as in Fig. 4-4, 19.5 cm (7.68 in) long
Plastic clay	0.88	0.88
Silty clay	0.82	0.88
Sandy clay	0.80	0.90
Mixture of gravel sand and clay	1.00	0.86

Pneumatic-tired roller : depth of compacted layer 30 cm (12 in)				
Type of soil	10,300 kg (22,687 lb) per wheel 9.9 kg/cm^2 (140 lb/in^2)	10,300 kg (22,687 lb) per wheel 9.9 kg/in^2 (140 lb/in^2)	5,150 kg (11343 lb) per wheel 6.35 kg/cm^2 (90 lb/in^2)	5,150 kg (11343 lb) per wheel 6.35 kg/cm^2 (90 lb/in^2)
Plastic clay	0.93	0.93	0.88	0.88
Sandy clay	0.95	0.94	0.90	0.89
Mixture of gravel sand and clay	0.95	0.95	0.95	0.96

There are other recognized advantages of one type of roller over another:

— In residual soils, the sheepsfoot roller achieves a higher degree of uniformity and is more efficient than the pneumatic-tired one because the concentrated pressure produced by its feet enables it to break up weathered rock fragments.

— Because of the indented layer surface, sheepsfoot rollers produce a better bond between successive layers than the pneumatic-tired variety.

— Pneumatic-tired rollers can compact thicker layers at higher speeds than sheepsfoots. Apart from the economical advantage this implies, the greater layer thickness makes it possible to include coarser, bigger particles.

— In soils containing large pebbles, pneumatic tires allow a more even distribution of the compactive effort, whereas the rigid drum of the sheepsfoot roller with its shafts usually bridges the gaps between these pebbles, leaving the soil between virtually uncompacted.

4.4.3 Impact Compaction

In impact compaction processes, the application of load is very brief. Equipment to be included in this category are the different types of tampers, which are used only for small areas and certain types of tamping rollers which in many ways are similar to sheepsfoot rollers, but can operate at far higher speeds, producing an impact or dynamic effect on the layer of soil being compacted.

Tampers go from the most elementary free-fall, hand-operated type to considerably more complicated devices driven by compressed air or internal combustion (Plates 4-14, 4-15). Chiefly because of expense their use is limited to small areas such as ditches, foundation excavation, areas adjacent to drains, drain covers or bridge abutments, where large-scale compaction equipment cannot be used because of lack of space or for fear of overweight.

Free-fall tampers range from straightforward hammers with a handle attached and operated by one man, to 2 or 3 t (4.4 or 6.6 kips) hammers which are raised by cables and allowed to fall from a height of one or two meters. These heavy machine-operated models have been successfully used for compacting large rock fragments.

Pneumatic or gasoline-operated tampers rise to a height of 15 or 20 cms (6 or 8 in) above the ground with the energy which they themselves produce reacting on the soil itself. They are considered suitable for compacting cohesive soils, but are also useful for other types of soil.

Tampers weighing 30 to 1,000 kg (66 lb to 2.2 kips) are manufactured today. Excellent results have been achieved with 5 or 6 coverages by half-ton (1.1 kips) tampers on layers 20 to 25 cm (8 to 10 in) thick. Outputs have been reported of about 200 to 250 m^3/hour (260 to 327 yd^3/h).

Plate 4-14 Self-driven tamping roller which is shaped like a tamping foot, specially designed to work at speeds of up to 24.1 km/h (15 mi/h)

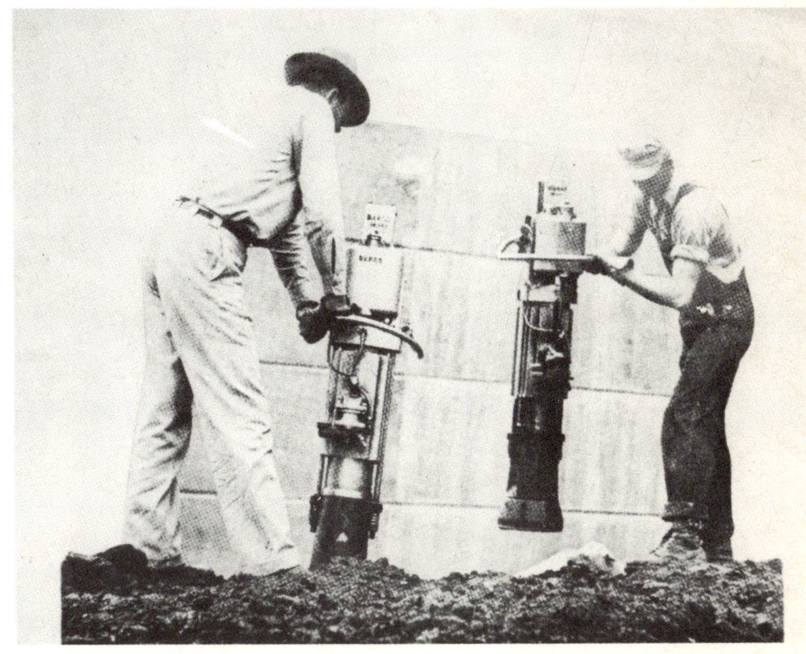

Plate 4-15 Gasoline engine tamping compactors working on a small area

High energy impact compaction, dropping weights of several tons from heights exceeding 1 m, was first used in European road construction during the 1930's. The technique has recently been revived, using weights as great as 10 t (44,000 lb) with end areas of 2 to 4 m^2 (22 to 43 ft^2) dropped 15 m (50 ft) produce both impact forces and some vibration. Sands have been densified to depth of 5 to 10 m (16 to 33 ft). Partially saturated cohesive soils have been densified to depths of more than 5 m (16 ft). The uppermost 1 m (3 ft) is loosened and must be compacted by rolling.

4.4.4 Vibratory Compaction

For vibratory compaction, the mechanism used is either a rotating or reciprocating mass, actuated by a hydraulic motor. The vibrator is attached to a flat plate or to a roller (typically the smooth drum type). The frequency of the vibration is of extraordinary importance in a compaction process. There is evidence that the optimum lies between 0.5 and 1.5 times the natural frequency of the soil-vibrator system. Thus the required compaction frequencies are between 1,500 to 2,000 cycles per minute. Some types of equipment available commercially have frequencies as high as 5,000 cycles per minute.

Several factors have a considerable influence on the results produced by the equipment. The principal ones are:

— The *frequency,* that is the number of revolutions per minute of the oscillator.
— The *amplitude,* which is usually measured by a vertical distance in almost all commercial types of equipment.
— The *dynamic force* generated with every impulse of the oscillator. This is computed theoretically by assuming that the vibrator rests on a non-yielding surface. On soil, the force is much less.
— The *dead load,* that is, the weight of the compaction equipment, excluding the vibrator.
— The *shape and size of the contact area* of the vibrator in contact with the soil.
— The *water content* and nature of the soil.

Maximum effectiveness is usually obtained at a water content considerably below the optimum required for any other compaction method (or found by the laboratory standards). The chief advantage of vibration is that thicker layers of cohesionless soils can be compacted than with other types of equipment; this increases the efficiency of the process and reduces operation costs. For example, with soils of the *GW* or *GP* type, vibratory compaction can easily achieve the same result on 60 cm (2 ft) layers as heavy pneumatic-tired rollers would produce on 20 or 30 cm (8 or 12 in) layers. The U.S. practice of compacting 1.20 m (4 ft) layers using extremely heavy vibratory rollers, has already been mentioned.

In granular soils, vibration momentarily reduces the internal friction between the particles. This enables the particles to re-orient in a more dense arrangement. Vibration alone gives poor results. Pressure is necessary to force the particles together after their resistance to movement has been reduced by the vibration. However, static force such as produced during rolling is less effective because the same force that must overcome the particle resistance by creating interparticle shear also increases the interparticle normal forces and the frictional resistance to compaction.

Several tests [20] have led to quantitative evaluations of the reduction in internal friction achieved during a vibratory process. This has proved as great as 1/15 for sands and 1/40 in gravels. Added to this effect of reducing friction is the pressure of the compactor which forces the particles together in a more dense arrangement. This also counteracts any capillary tension forces between the grains of sand. This apparent cohesion by capillarity has also been quantified in experimental form [20]. The pressures required to overcome it are about 0.5 to 1 kg/cm^2 (7.1 to 14.2 lb/in^2) in gravels and sands and 4 to 7 kg/cm^2 (57 to 100 lb/in^2) in clays compacted at 90% of the maximum dry unit weight as found by the modified PROCTOR test.

The larger the predominant particles, the smaller the apparent cohesion forces, so that in gravels and broken rock they are insignificant. In spite of this, however, it has been observed that the water content of the material being compacted also plays an important part in these materials. This will be discussed in greater detail when discussing rockfills later in this chapter. However, when a very coarse soil is compacted by vibration there is a rapid loss of excessive pore-water during the process. This observation leads to the practical conclusion that gravels and rock fragments can be successfully compacted at very low water contents.

If the granular soil (sand or gravel) contains an appreciable amount of fines and its water content is high, vibratory compaction is less effective. For vibratory compaction the fines content should not exceed about 10% [21].

The more uniform the sand or gravel, the more difficult it will be to compact the upper part of the layer. A layer only 10 cm (4 in) thick will be less easily compacted than thicker ones. However, a thin layer will be densified by vibrating a second layer above it. In road building, the last layer of a base is compacted together with the first layer of surface or pavement. The vibratory compaction of sands and fine gravels can be improved by adding water and then making the final coverages at high speed. It also helps if small-amplitude vibrations are used during the final roller passes. The part played by the water in these cases has not yet been explained. It is possible that wetting the edges of the coarse particles makes them more easily crushed.

The part played by the water is, however, quite clear when granular soils with high capillary tension between the particles are compacted by vibratory methods. Adding water reduces capillary tension, which enables the particles to move readily into denser arrangements.

The layer to be compacted is given the first one or two coverages with a low water content because the capillary tension in the loose cohesionless soil prevents the equipment from sinking into the surface. Water is gradually added in the course of subsequent passes to eliminate the effects of capillarity after the partially compacted layer is strong enough to support the roller without the temporary cohesion of capillarity.

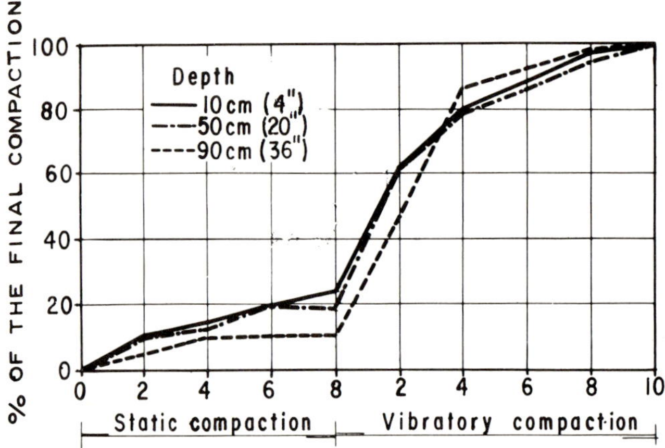

Fig. 4-21 Illustration of the efficiency of vibratory compaction [20]

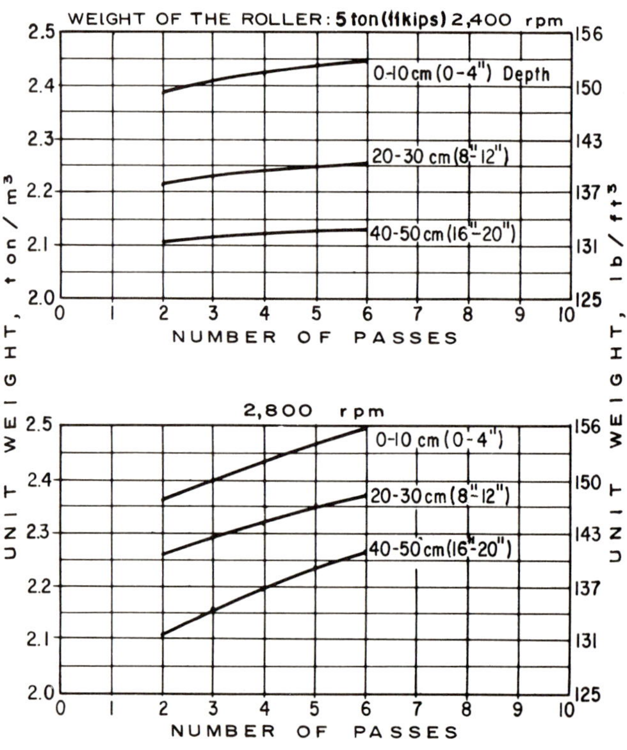

Fig. 4-22 Effect of frequency on a compaction process [20]

Figure 4-21 illustrates the favorable effects of vibratory compaction methods. It refers to the Lancashire-Yorkshire highway in England. Rockfills with a maximum size of 60 cm (2 ft) were compacted in 90 cm (3 ft) layers, using pneumatic-tired rollers weighing 50 t (110 kips) and grid rollers weighing 13.3 t (30 kips). These were followed by vibratory rollers weighing 8 and 5 t (18 and 11 kips). The amount of densification by vibration was 5 times that of the static rollers.

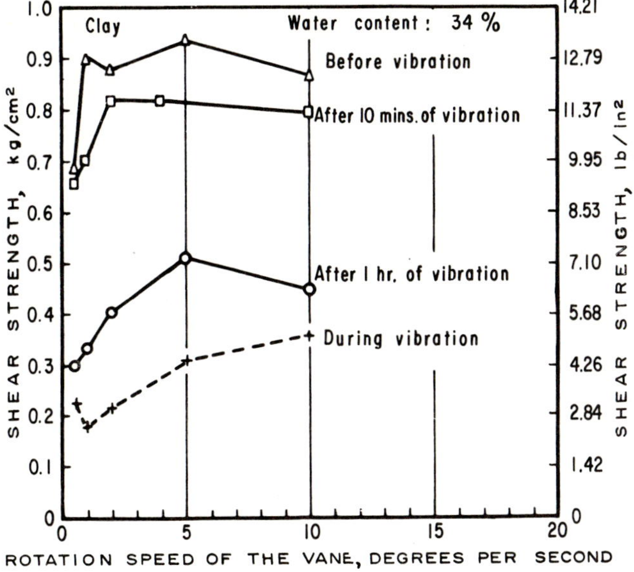

Fig. 4-23 Variation in the shear strength of a clay with vibration [22]

Figure 4-22 [20] provided limited data on the influence of vibration frequency on compaction processes for a 5 t vibratory roller compacting a layer of gravelly-sandy soil 60 cm (2 ft) deep. The same figure also shows that compaction is less with increasing depth.

The influence of the water content is apparent when fine clayey soils are subjected to vibratory compaction. Clays much drier than the standard AASHTO optimum require great compactive efforts and equipment exerting very high pressures. The compactor has to overcome the interparticle forces of the clay minerals. As mentioned previously, additional (vibratory) stresses of about 8 kg/cm^2 (114 lb/in^2) are required for this. As a result, the depth of the layers which can be compacted is considerably less than in the case of sands and other frictional soils. Compaction has to be carried out with heavy sheepsfoot or pneumatic-tired rollers capable of providing the required additional pressure.

The mechanisms by which vibration acts on moist plastic clays have not yet been clearly defined. Some favorable effect seems to be produced by a reduction in the apparent viscosity of the adsorbed water. This phenomenon has been observed in some clay masses during vibratory compaction. Clays that are wetter than optimum can be compacted with smaller efforts than the drier clays and with equipment capable of exerting considerably lower pressure. On the other hand, [21], it has been observed that in soft wet homogeneous clays, where it is relatively easy to achieve some compaction, vibration may be of little benefit. The limited tests in [21] suggest that in the field it may be impossible to make a soft homogeneous clay reach dry unit weights higher than 90% with any type of vibratory equipment, whereas in the laboratory the same material can be compacted to 100% of the dry unit weights in a Modified AASHTO test (described later in this chapter).

It has been observed that vibration can reduce the shear strength of some clays, probably by producing gradual, permanent structural degradation (Chapter 1). The more sensitive the clay, the more noticeable the effect. Fig 4-23 [22] shows these effects. The shear strength was measured by means of vane tests.

Silts and silty soils can be adequately compacted by vibratory equipment when their water content is near optimum and when the layers are 5 to 10 cm (2 to 4 in).

One of the most widely used types of vibratory equipment is the hand-operated plate, maneuvered by a handle or lever, Plate 4-16. When operated efficiently, it can cover about 10 m (33 ft) per minute. Vibratory plates can also be gang-mounted on a chassis, Plate 4-17, which is drawn by a tractor. Table 4-8 gives the most common characteristics of vibratory plate compactors.

Vibrating drum rollers are the most effective of the vibratory methods, Fig. 4-24 [22] shows the range of the dynamic pressures exerted at different depths by some of the types of vibratory equipment in use today. (Plates 4-18 to 4-20). The benefit of vibration can be seen in the pressure interval 0.5-1.0 kg/cm (7.1-14.2 lb/in^2) which is necessary to break the capillary tension in frictional soils, and in the range of highest pressures such as required by clays. In the case of smooth steel-wheeled rollers, results are given with and without vibration which identifies the effectiveness of vibration.

Plate 4-16 One-plate vibratory compactor operating on an electric engine and equipped with two handles

Table 4-8
Vibrating plate compactors : common specifications

Overall weight of compactor	70 to 6000 kg (154 to 13,216 lb)
Weight of each vibratory unit	70 to 204 kg (154 to 449 lb)
Contact area of plate	1540 to 13,900 cm^2 (239 to 2155 in^2)
Contact pressure	0.04 to 0.43 kg/cm^2 (0.6 to 6.0 lb/in^2)
Amplitude of vibration	2.03 to 12.7 mm (0.08 to 0.5 in)
Frequency	420 to 2800 /min
Width of compacted strip	38 to 380 cm (15 to 150 in)
Operation speed	0.05 to 26.0 km/h (0.03 to 16.2 mi/h)

Plate 4-17 Multiple-plate vibratory compactor

Figure 4-25 [22] illustrates the effects of vibrational frequency and amplitude on the pressures exerted by a vibratory smooth steel-wheeled roller, equipped with rotating eccentrical masses. Note the increase in pressure with vibration, and also the rapid increase in pressure with frequency up to about 1,500 rpm (or slightly higher). With higher frequencies the benefit of increasing frequency diminishes rapidly. Note also the influence of the vibration amplitude.

Many of the concepts that apply to other types of compactors also fit vibratory methods in the field. For example, the basic information contained in Fig. 4-7 is valid. Likewise, the rate of density increase per pass of the vibrator is greatest for the first few passes, and then diminishes.

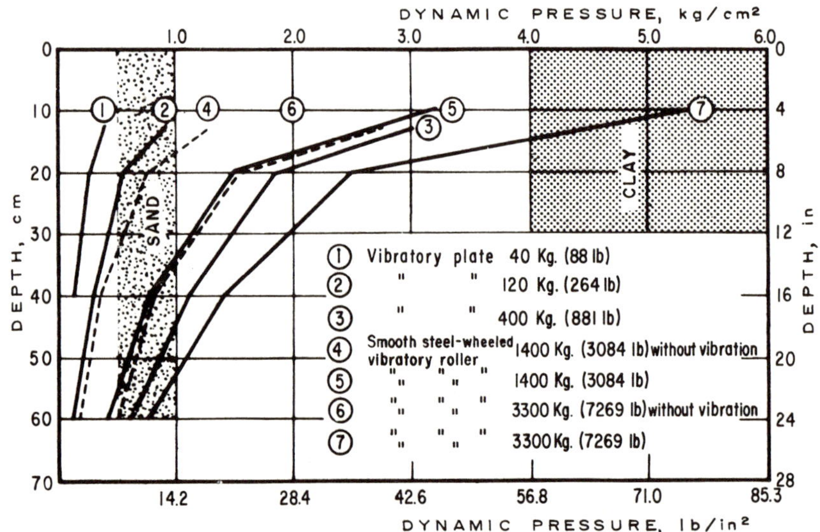

Fig. 4-24 Dynamic pressure exerted at different depths by various vibratory equipment [22]

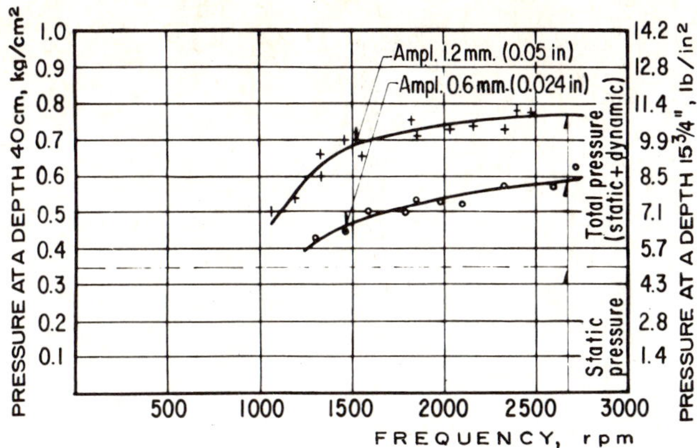

Fig. 4-25 Effect of the frequency and the amplitude on the pressures exerted by a smooth steel-wheeled vibratory roller [22]

4.4.5 Compaction Using Combined Methods

Present-day technology is developing numerous types of equipment where the effects of two or more of the traditional systems are combined to achieve the best effects of each and to optimize results. Naturally the use of many of these types of equipment by contractors and public organizations on small jobs cannot be justified. Moreover, there is insufficient knowledge about the compactors themselves.

The vibration unit of the smooth steel-wheeled vibratory roller, is within the drum powered by an external engine. These rollers, Plates 4-18 to 4-20, are both tractor-drawn and self-propelled. Their effectiveness is greatest in granular soils. They can combine the effects of vibration and pressure, which have already been discussed, in layers much thicker than could be compacted by the smooth-wheeled roller alone. They are also very effective for compacting asphaltic concrete.

Plate 4-18 Smooth steel-wheeled hand-operated vibratory roller, weighing 203 kg (447 lb), with a diameter of 53 cm (1 ft 9 in) and a width of 61 cm (2 ft). The vibratory mechanism operates on a gasoline engine

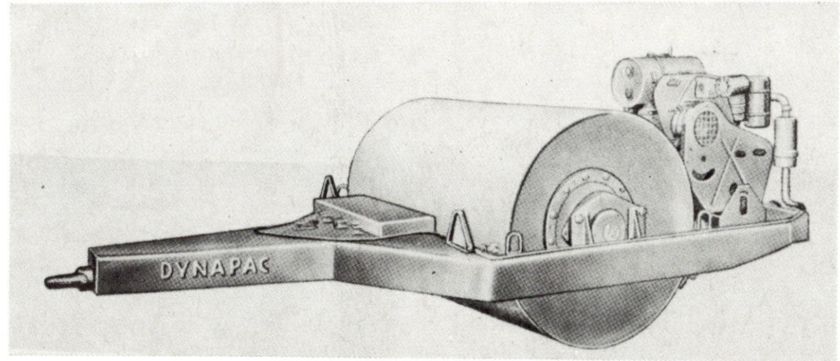

Plate 4-19 Smooth steel-wheeled tractor-drawn vibratory roller, weighing 3.9 (8.6 kips) and with a drum 1.9 m (6.23 ft) wide. It has a dynamic force of 8 t (18 kips) and produces 1,400 to 1,600 vibrations per minute. Operation speed varies between 3 (1.9) and 5 km/h (3.1 mi/h)

Plate 4-20 Combination of 3 smooth steel-wheeled vibratory rollers drawn by just one tractor

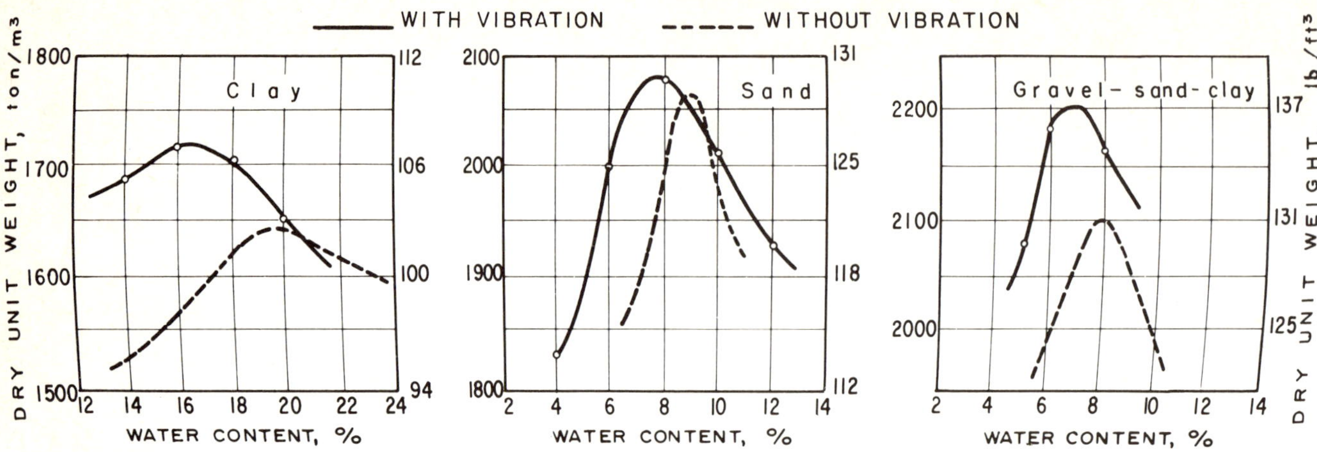

Fig. 4-26 Optimum water contents for compaction with a smooth steel-wheeled roller, with and without vibration [16]

Figure 4-26 [16] gives dry unit weight-water content graphs for three types of soil which were compacted in layers 23 cm (9 in) deep with 32 passes of a double smooth-wheeled roller with vibration in the front drum; the overall weight of the roller was 2.2 t (5 kips) producing pressures of 12 and 17 kg/cm^2 (67 and 95 lb/in^2) per unit width of the front and rear rollers, respectively. The solid lines correspond to the vibratory rolling and the broken lines to the equipment without vibration. Observe again that when vibration is used the water content required is less than when the same system is used without vibration. Of these data, the operation speed of the roller is of special importance, for as it is independent of frequency, it consequently has a great effect on the compactive effort.

Table 4-9 shows the most common characteristic of smooth steel-wheeled vibratory rollers. Small hand-operated types of equipment are available.

Pneumatic-tired vibratory compactors are usually tractor-drawn (Plate 4-21). They are best used for well-graded sandy soils, silty sands and even clayey sands. They are more efficient than smooth-wheeled rollers when there is a greater fines content in a frictional soil because their efforts can be transmitted to greater depths. Unfortunately, the rubber tires absorb some of the vibration energy.

Sheepsfoot rollers with vibratory attachments, Plate 4-22, are usually tractor-drawn. They are most effective for compacting fine clayey soils. Apart from other advantages already mentioned, they allow somewhat greater layer depths. It has been one author's experience that their added effectiveness seldom justifies their additional cost. Combined smooth-wheeled and pneumatic-tired rollers usually have smooth drums at the front with tires at the rear for propulsion. Some are equipped with an added rear vibrating drum attachment. This enables either of the two types of drums to be raised, so that they can be operated in three different ways, including vibration. This explains their popularity with some construction firms.

The smooth-wheeled roller can also be combined with vibrating plates or platforms. This makes it effective for compacting small rock fragments, gravels and mixtures of these soils with sand, and allows thicker layers to be compacted than would be possible with just a smooth-wheeled roller.

Smooth-wheeled rollers with vibratory rollers, Plate 4-23 and sometimes an additional axle with segmented rollers (Plate 4-24) are also available. They usually have mechanisms for raising any one of the rollers, which makes them even more versatile. However, they are expensive and seldom as effective in any one mode of operation as the more specialized roller.

Plate 4-21 Heavyweight pneumatic-tired vibratory roller with only one axle and two tires, drawn by a caterpillar tractor

Table 4-9
Smooth vibratory rollers : common specifications

Diameter of roller	53 to 122 cm (21 to 48 in)
Width of roller	61 to 183 cm (24 to 72 in)
Overall weight of roller	0.2 to 13† (441 to 28,634 lb)
Operation speed	0.5 to 6 km/h (0.31 to 3.73 mi/h)
Vibration frequency	1050 to 5000 /min
Amplitude of vibration	about 1 mm (0.04 in)

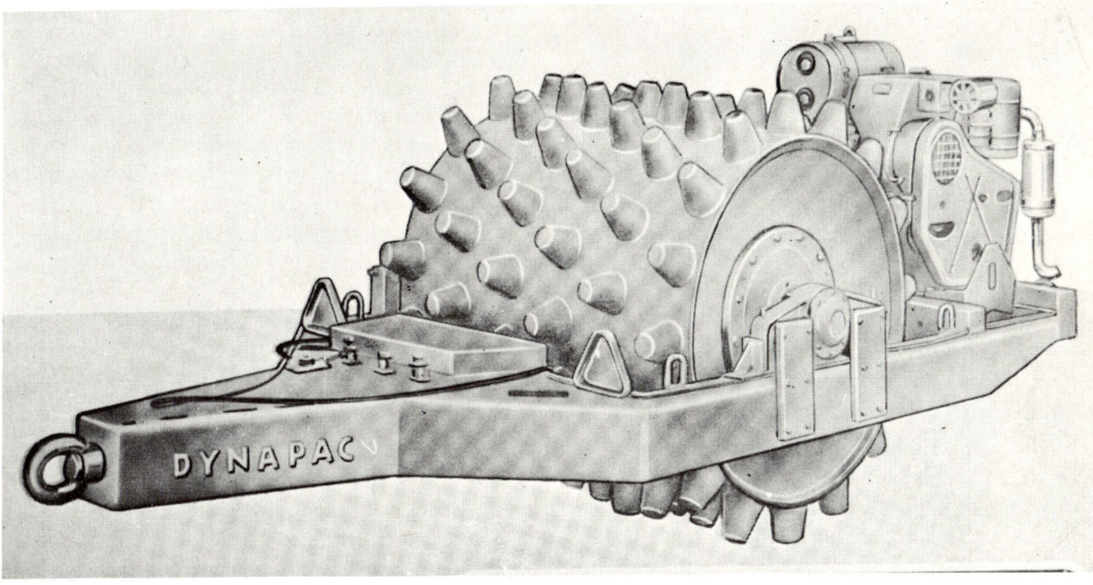

Plate 4-22 Tractor-drawn vibratory sheepsfoot roller, with an overall weight of 3.9 t (8.6 kips), dynamic force of 10 t (22 kips) and frequency of 1,400 to 1,600 cycles per minute. The drum is equipped with 98 cone-shaped feet

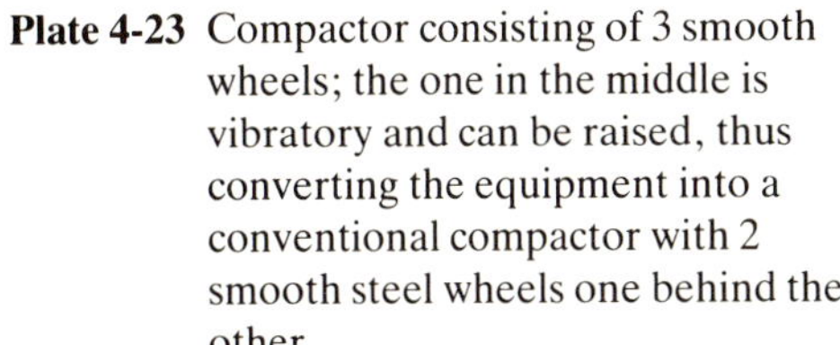

Plate 4-23 Compactor consisting of 3 smooth wheels; the one in the middle is vibratory and can be raised, thus converting the equipment into a conventional compactor with 2 smooth steel wheels one behind the other

Plate 4-24 Compactor consisting of a segmented steel roller in front, a smooth vibratory roller, and a smooth roller at the back

4.5 Some Useful Ideas for Carrying Out Compaction Jobs in the Field: Degree of Compaction

The first requirement for a good compaction job is for the engineers and construction supervisors to be as familiar as possible with the soils that are going to be compacted. This can be achieved by studying the deposit from which the materials are to be extracted. For this purpose, representative samples must be obtained (40 to 50 kg) (88 to 110 lb) so that the necessary laboratory tests can be performed. In addition, smaller sealed samples from different levels in the deposit are obtained to determine the natural water content of the soil. Compaction curves following the laboratory procedure which is considered most representative of field conditions are made of the large samples, as will be discussed later in this chapter.

The expansion and shrinkage characteristics of the soil by wetting and drying should be studied also, so as to establish the percentage of volume change the soil may undergo during operation of the road. Expansion should be studied in compacted specimens, letting them saturate and then shrink by drying the compacted soil. The saturation and drying establish the limits of potential volume change. The actual volume change will be less, depending on the actual moisture content changes induced by the environment.

The choice of compaction equipment is of course very important. Before making a decision, not only should the characteristics of the soils be carefully considered, but also the structural conditions desired in accordance with the requirements of the road to be built and the position of the soil to be compacted in relation to the embankment cross-section. The most important points to ponder before selecting the type of equipment are the following:

— Type of soil.
— Variations of the soil at the location.
— Size and importance of the job to be carried out.
— Compaction specifications (degree of compaction required).
— Time available for the job.
— Equipment already available before commencing work.

The ultimate choice of compaction equipment depends on cost. The reader who has carefully followed all that has been said about the characteristics and possible uses of the different types of equipment available will have noticed how the uses of different compactors overlap one another. In fact the development of the different types of equipment has been empirical, ruled by market trends and specific contractor requirements; only very rarely has it been the result of a strictly scientific investigation process. Consequently, the engineer usually has several alternatives from which he must choose which ones provide the required performance at least cost.

In Section 4-4 the uses of the different types of equipment were discussed together with the characteristics that make them more efficient and effective. Table 4-10 [11] gives a summary of these criteria based on the authors' opinions derived from their personal experiences; the conclusions may not be shared by all specialists.

There are a number of similar tables like 4-10, each of them reflecting the experience which has been acquired by different institutions and technical teams in varying environments.

Table 4-10

Guide to the choice of compaction equipment : [11]
Number indicate preference (1 is best)

Usage	Group symbol U.S.C.S.	Material	Self-driven tamper	Tractor-drawn tamper	Self-driven sheepsfoot	Tractor-drawn sheepsfoot	Small smooth vib. roller	Heavy smooth vib. roller	Small vib. sheepsfoot	Heavy vib. sheepsfoot	Light tired roller	Heavy tired roller
Base		Clean granular					1	1			3	2
Sub base		Granular-few fines	1	1			1	1	2	2		2
Earth fill		Rock	2	2				1		2		
	GW,SP,SW	Sands, gravels	2	2			1	1	2	2		2
	SP	Uniform sand					1	1	2	2		3
	SM,GM	Silty sands or gravels	1	1	4	4	3	3	2	2		2
	ML,MH	Silts	1	1	2	2			3	3		2
	GC,SC	Clayey sands or gravels	1	1	2	2			3	3		2
	CH,CL	Clays	1	1	2	2				3		3

Table 4-11
Soil utilization characteristics grouped according to U.S.C.S.

Symbol	Compaction characteristics	Typical maximum dry unit weight (standard PROCTOR test)		Compressibility and expansion	Permeability and drainage characteristics	Characteristics as earth fill material	Characteristics as sub-grade	Characteristics as pavement base	Characteristics as temporary surfacing	
		t/m³	lb/ft³						With thin surface treatment	With asphalt surface treatment
GW	Good. Smooth vib. rollers, pneumatic-tired roller	1.9-2.1	119-131	Practically non-existent	Permeable. Very good	Very stable	Excellent	Very good	Fair to good	Excellent
GP	Good. Smooth vib. rollers, pneumatic-tired roller	1.8-2.0	112-125	Practically non-existent	Permeable. Very good	Stable	Good to Excellent	Fair	Poor	Fair
GM	Good. Lightweight pneumatic-tired or sheepsfoot roller	1.9-2.2	119-137	Slight	Semi-permeable Poor drainage	Stable	Good to excellent	Fair to bad	Poor	Fair to poor
GC	Good or fair. Pneumatic-tired or sheepsfoot roller	1.8-2.1	112-131	Slight	Impermeable Poor drainage	Stable	Good	Fair to good	Excellent	Excellent
SW	Good. Pneumatic-tired or vibratory roller	1.7-2.0	106-125	Practically non-existent	Permeable. Good drainage	Very stable	Good	Fair to bad	Fair to bad	Good
SP	Good. Pneumatic-tired or vibratory roller	1.6-1.9	100-119	Practically non-existent	Permeable. Good drainage	Reasonably stable when compacted	Fair to good	Bad	Bad	Fair to bad
SM	Good. Pneumatic-tired or sheepsfoot roller	1.7-2.0	106-125	Slight	Impermeable Bad drainage	Reasonably stable when compacted	Fair to good	Bad	Bad	Fair to bad
SC	Good or fair. Pneumatic-tired or sheepsfoot roller	1.6-2.0	100-125	Slight to medium	Impermeable Bad drainage	Resonably stable	Fair to good	Fair to bad	Excellent	Excellent
ML	Good to bad. Pneumatic-tired or sheepsfoot roller	1.5-1.9	94-119	Slight to medium	Impermeable Bad drainage	Bad stability if not very well compacted	Fair to bad	Must not be used	Bad	Bad
CL	Fair to good. Pneumatic-tired or sheepsfoot roller	1.5-1.9	94-119	Medium	Impermeable Bad drainage	Good	Fair to bad	Must not be used	Bad	Bad
OL	Fair to bad. Sheepsfoot or pneumatic-tired roller	1.3-1.6	81-100	Medium to high	Impermeable Bad drainage	Unstable-avoid using	Bad	Must not be used	Must not be used	Must not be used
MH	Fair to bad. Sheepsfoot or pneumatic-tired roller	1.1-1.6	69-100	High	Impermeable Bad drainage	Unstable-avoid using	Bad	Must not be used	Very bad	Very bad
CH	Fair to bad. Sheepsfoot roller	1.3-1.7	81-106	Very high	Impermeable Bad drainage	Fair. Watch expansion	Bad or very bad	Must not be used	Very bad	Must not be used
OH	Fair to bad. Sheepsfoot roller	1.0-1.6	62-100	High	Impermeable Bad drainage	Unstable-avoid using	Very bad	Must not be used	Must not be used	Must not be used
P_t	Must not be used			Very high	Fair or bad drainage	Must not be used	Must not be used	Must not be used	Must not be used	Must not be used

Obviously it would be impossible to reproduce even the most important and complete ones. Table 4-11 [21] refers to soil utilization characteristics, not only in relation to compaction problems, but also to various others. The reader will have to be careful in using such generalized and condensed information. A table such as 4-11 is an excellent guide, but it does not exempt the engineer from carrying out all the studies which are necessary to define the conditions with which he will be confronted.

As discussed previously, the requirements for a specific compaction job are usually defined by a certain dry unit weight to be achieved with the equipment. This weight is generally the result of laboratory studies of that compacted soil. From among all the different tests available, the one is chosen which is most representative of the field compaction process and will produce a degree of compaction sufficiently high to produce the desired behavior of the material in the field. Unfortunately, design engineers often forget the need for representativity in the laboratory test. They choose, relying on past experience, the one which is sufficiently energetic to produce compaction that almost invariably produces a well-behaved material in the field; only occasionally do they evaluate the final engineering properties of that material. The authors of this book trust that when the reader finishes this chapter, especially the part which deals with the properties of the materials compacted, he will understand that to ensure an adequate job it is not enough to specify a high compaction based on a laboratory compaction test with high specific energy. The test might not be representative of the type of compaction to be used in the field. Therefore even if the soils compacted in the laboratory by these methods have suitable properties the equipment in the field might, for the same compacted density, produce a soil with different (and even unsuitable) properties.

In any case, the unit weight for the field is specified on the basis of a laboratory test. As a result of the fundamental difference existing between the two compaction processes, and also all the problems which may arise in the field, the unit weight which is obtained during the job is almost always different from the maximum dry unit weight produced by the laboratory test on which the study was based. The difference between the two values is defined by the *degree of compaction* or *percentage of compaction.* The percent relation between the dry unit weight obtained by the field equipment and the maximum corresponding to the laboratory test on which the study was based is defined as the percent or degree of compaction:

$$G_c(\%) = 100 \frac{\gamma_d}{\gamma_{d\,max}} \% \tag{4-5}$$

Despite the widespread use that is currently made of the percent or degree of compaction, it is in fact a far from perfect measure of compaction. Moreover, in some materials whose particles fracture with impact it can be misleading. A material such as a dry silty clay in a completely loose state, just as deposited at the site, may have a degree of compaction of about 80% according to Eq. (4-5) before undergoing any sort of compaction; another material such as volcanic ash or tephra in the same circumstances, may have a degree of compaction of 60%. If the ash is compacted until it reaches the same 80% as the silty clay, some would argue that both soils have the same degree of compaction and thus similar engineering qualities. This is false, for the silty clay is in a loose uncompacted state, with all that this implies as regards its mechanical behavior, whereas the ash has already been partially compacted. The strength of the ash therefore may be greater, and its compressibility lower. This difference in significance of percentage of compaction is especially significant in cohesionless soils. For such materials, a different relation for measuring the compaction achieved by a soil in the field has evolved. It is *relative compaction* or relative density, C_r. It is defined by:

$$C_r(\%) = 100 \frac{\gamma_d - \gamma_{d\,min}}{\gamma_{d\,max} - \gamma_{d\,min}} \tag{4-6}$$

where: $\gamma_{d\,max}$ is the maximum dry unit weight obtainable in the laboratory, $\gamma_{d\,min}$ is the minimum dry unit weight obtainable in the laboratory of the same material, and γ_d is the dry unit weight of the material in the field conditions.

This second relation has the advantage that it does not imply that 70% compaction represents a 70% effort although 70% is produced by no compaction. The equivalent relative density would be about 0%. There is, however, the disadvantage that there is no consistent reliable procedure to determine the minimum dry unit weight. Even the maximum weight is not easily defined. Therefore, field procedures sometimes produce computed relative densities less than zero and greater than 100%.

The degree or percentage of compaction continues to be the most commonly used method for defining the compaction requirement. It is usually expressed in the following terms: *This material is to be compacted to 95% of the maximum dry unit weight obtained by the standard laboratory compaction test.* Consequently the field work of a compaction equipment is planned to obtain the desired percentage or degree of compaction in the most economical way.

The percentage of compaction established for a given job should be realistic in the sense that it should not make excessive demands, either in relation to the properties that are to be obtained or the equipment available and size of the job to be undertaken. It must be compatible with the environmental restrictions of rain and temperature. Otherwise there will be continual adjustment problems in the field which will hinder the job's progress. For example, it would be futile to specify compaction to 100% of the modified Proctor maximum if the soil moisture in the borrow areas is 6% greater than the modified optimum moisture and there is rain throughout the year. Instead, the best compaction that could be expected might be 95% of the standard Proctor maximum. The lack of high density and the high strength associated with the 100% compaction must be offset by a design that compensates for the poorer engineering qualities of the obtainable density. Some large organizations establish preliminary recommended percentages of compaction, based on experience within a particular area.

The Mexican Ministry of Public Works, for instance, requires that earthfills shall be compacted no less than 90%, with 95% for the upper third or half meter. For moisture tolerant subgrades 100% is required. (These degrees of compaction correspond to the specific laboratory compaction tests used by the Ministry of Public Works). Table 4-12 [23] is a guide to the degrees of compaction rather than fixed figures that can be indiscriminately applied.

The compaction requirement is established by finding a balance between the following properties [13]:

1. Homogeneity
2. Favorable permeability characteristics
3. Low compressibility, to avoid the development of excessive pore pressures or unacceptable deformations. The higher the earthfill, the more important is this requirement.
4. Shear strength
5. Permanence of the mechanical properties when saturated.
6. Flexibility, so as to withstand flexure or distortion without cracking.

Fulfilling condition *1* depends on the placing and compaction equipment available and careful control of the process. Requirements *3* and *4* together are in conflict with *6* and often with *2* and *5*.

For most soils, the best compromise is to compact with a water content very close to the field optimum. When one of the conflicting requirements is considered more important than the other, the specification of the compaction moisture should be modified appropriately. If, for example, conditions *3* and *4* are considered of greater interest than *5* and *6*, a lower-than-optimum water content should be specified, and, if the contrary

Table 4-12
Tentative values for suitable degrees of compaction

Type of soil: U.S.C.S. Symbol	Degrees of compaction depending on type and importance of job (standard PROCTOR test)		
	Type 1	Type 2	Type 3
GW	97	94	90
GP	97	94	90
GM	98	94	90
GC	98	94	90
SW	97	95	91
SP	98	95	91
SM	98	95	91
SC	99	96	92
ML	100	96	92
CL	100	96	92
OL	—	96	93
MH	—	97	93
CH	—	—	93
OH	—	97	93

Type 1: Earth fills more than 30 m (33 yd) high. Subgrade beneath final surfaces, not thicker than 30 cm (1 ft). The upper 2 m (6.58 ft) underneath foundations for buildings of two or more stories or bridges and overpasses

Type 2: The lower parts of fills under buildings. Upper layer of common earth fills, under subgrades with a minimum depth of 30 cm (1 ft). Earth fills less than 30 m (33 yd) high

Type 3: Other soils requiring compaction, without great strength and compressibility requirements

is true, vice-versa. (Of course the job moisture environment (rainfall and evaporation) will limit drier-than-optimum compaction).

Condition *5* can be investigated by means of consolidation tests in which the sample is subjected to saturation under different loads; usually an acceptable minimum value can be reached for the compaction water content.

In order to estimate the maximum compaction water content acceptable for conditions *4* and *5*, triaxial tests are suggested, either without consolidation or drainage or consolidated undrained with measurement of pore pressure coefficients *A* and *B*. The minimum water content required to fulfill condition *6* can only be estimated qualitatively, for there are only limited correlations available between prototype behaviors and the stress-strain characteristics of the soils. Finally, experience from similar jobs is always a useful guide.

The compaction requirements are eventually established in terms of the equipment to be used and the results expected. There must be detailed knowledge of the sensitivity of the soil to all the variables of importance in the compaction process. Of these, the water content is the most important. Very often the compaction requirements contain no reference to the water content; specifications imply that a wide range of water contents are acceptable by varying the equipment and the way it is used. However, at varying moistures the soil may also behave in a variety of ways whether or not the same dry unit weight is reached. Figure 4-27 [3,28] shows a study of laboratory specimens (clayey sand) which were compacted by the kneading

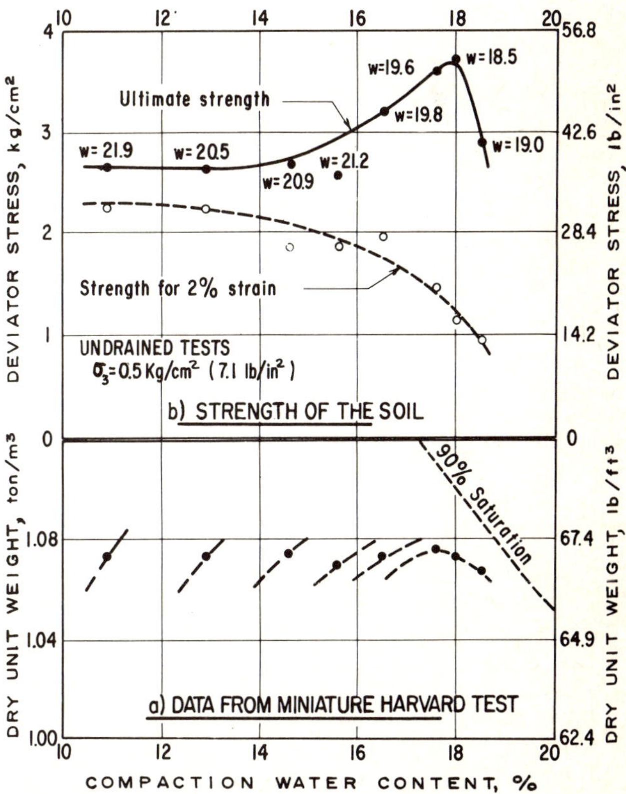

Fig. 4-27 Variation in the strength of a soil for various compaction water contents [3,28]

method, using various different compactive efforts, but to the same dry unit weight (Part *a* of Fig. 4-27). Afterwards, the soils were saturated under low confinement pressure comparable to those immediately below a pavement. Finally they were tested in the triaxial chamber without drainage. Figure 4-27a shows the big difference in the compaction moisture necessary to reach the same unit weight using different efforts. Figure 4-27b shows the variation in the final strength after saturation with different compaction moistures. At large strains the strength of the saturated soil is greater when it has been compacted with the optimum water content corresponding to the compactive effort used. If confining pressure is increased during saturation, the final strength of the soil will increase also. High compactive efforts are required to achieve a specified unit weight at low water contents. As can be seen in Fig. 4-27, the soil strength can be less than that when compaction is carried out at the appropriate optimum water content.

Reference [15] contains several examples of the variation in dry unit weight with compaction moisture and the way this is reflected in the characteristics of the compacted soil.

As has been seen already, compaction moisture is a basic variable in any field process and there is an optimum water content at which compaction efficiency is greatest for combinations of equipment and fill lift thickness. Unfortunately, many engineers treat the optimum moisture as though it were a basic constant of the soil and not a variable which also changes in accordance with the conditions of compaction. The compactive effort is usually the most important variable. (It has already been said that with an increase in specific energy there is invariably a reduction in the optimum water content). In the field, therefore, the optimum water content varies with the type of compaction equipment and the way in which it is used.

The water content at which the soil should be compacted in each case and with a specific equipment can be best determined by building test earthfills where the soil is compacted by the same equipment and with the same fill layers as in the field. All the alternatives that might prove necessary for the projected job are tried. From such tests the appropriate water content, the thickness of layers, the number of passes of the equipment, and all the other variables that may be of importance are quantified. The optimum water content corresponding to the laboratory test utilized for design will not be the same as the optimum moisture for the field. However, it serves as a guide or reference value for finding moisture contents in the field.

This is where one of the most delicate problems of road compaction arises. In the case of dams, very large volumes of soil from a few sources have to be compacted, which provides a large economic incentive to use test earthfills. For road building, however, the materials vary, along the line of the road over relatively short distances, so that many test fills would be required. Too many contractors feel that would be time-consuming and not cost-effective. However, careful control of production compaction can serve as a test fill. For larger fills the small test fills usually save more time and money than they cost. Unfortunately, the highway or railroad builder would like to settle all job details in advance before bringing equipment to a job and therefore he does not obtain the benefits provided by a test earthfill. In such cases the engineer has to use his own common sense and experience. The engineer should be prepared to modify specifications whenever necessary, paying careful attention to the variables that affect compaction results. Intensive use should be made of the field laboratories that are set up at the job site; they allow continued study of performance providing the basis for on-the-job experience and the basis for changes in the compaction procedures.

Soils often have to be moistened or dried at the source (borrow pit) or on the fill. It is difficult to add water to the soil on an earthfill, because the mixing delays compaction. Adding water is much easier at the source but requires planning. It is even more difficult to dry the soil on the job site, particularly when the spread layer is exposed to rain in a damp climate. Drying is usually done by airing, sometimes with the help of mechanical mixers. Wherever the original water content of the soil has to be modified, a homogeneous distribution of the new water content is essential. When modification of the water content becomes so difficult that the conditions required by the job cannot be achieved, then the project requirements and design will have to be changed accordingly.

The compactive effort for a given equipment is represented by the number of passes of that equipment over the same point. We have already seen that the increase in unit weight is not proportional to the number of passes. Instead, as illustrated by Figs. 4-8 and 4-14, there is a number of passes beyond which the results obtained show only slight improvement. When the critical number of passes is reached without the requirement for the field being met, the procedure has to be studied again so as to find out which other factors must be modified.

The operating speed of the equipment is important, though its effect has not yet been carefully studied. Within the speeds recommended by the different manufacturers, there does not seem to be any fundamental difference in the way the equipment works. It has sometimes been observed that when compaction equipment operates at high speed (greater than 6 or 10 km/h), which economically speaking is desirable, the topmost surface of the layer tends to be slightly irregular. However, this is not always detrimental and could be beneficial in the construction of massive earthfills. It is not difficult to deal with when compacting layers of subgrade. Test fills or correlating compacted density with equipment speed, on carefully observed layers, is helpful, job by job.

What is of great importance in a compaction process is the layer thickness. For a particular soil, compaction equipment and objective, the number of passes required increases rapidly with the increase in layer thickness. However, as a general rule, the thicker the layer, the more economical it proves to meet compaction requirements. For this reason builders prefer the greatest possible thickness. However, this is limited by the fact that the thicker the layer, the smaller the compactive effect in the lower portion, as pointed out in Fig. 4-18, for example, and the less uniform the compaction. The higher the degree of compaction required, the thinner the layers will have to be. Figure 4-28 illustrates the variation of the chief factors on which the choice of layer thickness depends.

Figure 4-28a, also, shows the great increase in the number of passes required when the layer thickness goes beyond a certain value, making it almost impossible to meet compaction requirements. An optimum layer thickness would be somewhat less than this value. This can be determined if a test earthfill is used. On the other hand, a very thin layer requires almost the same number of passes as a moderately thick one with a consequent increase in cost per unit volume of compacted soil. If the layer thickness is greater than the *optimum*, then either a disproportionately large number of passes will be necessary or it may not be possible to meet the requirements. If the cost of

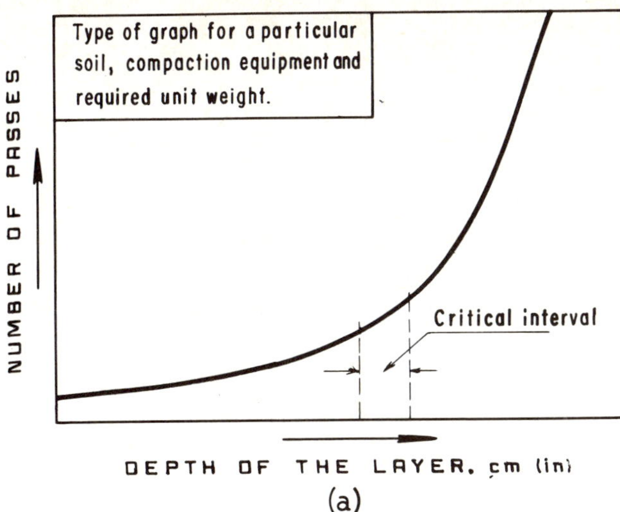

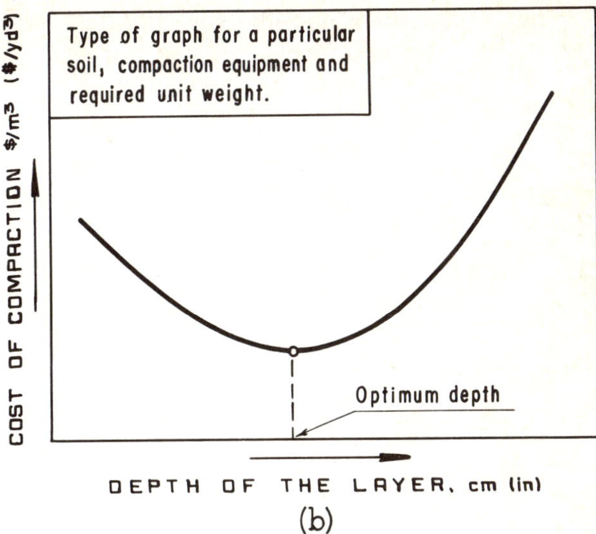

Fig. 4-28 Determination of appropriate layer depth

compaction is represented as a function of layer thickness for a given compaction machine, a graph will be obtained like the one in Fig. 4-28b, which clearly defines *optimum thickness.*

For earthfills and heavier road building equipment, the optimum layer thickness often varies between 20 and 30 cm (8 and 12 in). Vibratory equipment in cohesionless materials usually permit thicker layers.

Successive layers must bond together. More or less horizontal layers are recommended, especially where there is a very marked transversal slope, so as to minimize any sliding tendency. Whenever there is a doubt as to whether the finished surface of a compacted layer will make a successful bond with the following one, it should be slightly scarified before the next layer is added. The pockmarked surface left by sheepsfoot rollers aids bonding.

4.6 Some Special Problems Regarding Compaction in the Field

There are certain problems of compaction in the field that require special consideration.

.1 Compaction of Difficult Areas Inaccessible to Conventional Types of Equipment

These conditions appear quite frequently and may demand the inefficient use of equipment, or the use of small low-output equipment, thin layers or other special procedures. This problem is often very serious at the bottom of deep, narrow gorges, where special access roads are not justified because of the small volumes requiring compaction. In such cases, it is common practice to fill in the bottom of the gorge by dumping only, until a level is reached at which compaction can start. If care is taken with the filling operation, and the height of material dumped is only a fraction of the total fill height, this result may not necessarily be detrimental. It does, however, become more critical when there is a drainage structure at the bottom of the gorge (as is often the case). The structure system demands low compressibility of the material around and above it. It is also common, in the case of highways and railroads, to place coarse rock at the bottom of gorges and hollows. The compaction techniques required for these materials are those corresponding to rockfills. The least difficulties arise when any dumped fill is composed of crushed rock which densifies somewhat upon dumping and which can be compacted in thick lifts by vibratory rollers.

If large rocks are used in the bottom of fills, there will be slow but continuing ravelling of fines from the fill above into the large voids between the rocks. This produces delayed settlement and occasionally even sink holes in humid climates. Unfortunately the delayed reaction of a decade or two encourages the practice.

.2 Areas Adjoining Bridges, Culverts, Retaining Walls, etc.

The good results produced by hand-operated compaction equipment using thin fill lifts have already been mentioned. The cushions used to protect drainage structures are a special case, as they have to be built to specifications which depend on the type of job; they will be described in detail in a later chapter.

.3 Compaction at the Edge of Earthfills

As construction of an earthfill progresses in height, the equipment cannot go too near the edge during the operation because there is no lateral confinement. The edge can collapse and cause the equipment to roll down the slope. The problem is usually solved by giving extra width to both sides of the earthfill (30 or 40 cm) or (12 to 16 in) may be sufficient on each side. This is eventually cut back and trimmed on completion of the job. A curved or crowned fill surface helps in minimizing overfill (as well as improving surface drainage).

For very low earthfills, the embankments can be made with slopes gentle enough for the compaction equipment to move about on them. Complementary surface drainage facilities and protection against erosion by means of vegetation help combat the effects of poor compaction of the embankment surface.

.4 Compaction of the First Layers of an Earthfill on Soft Ground

When an earthfill is to be built on very soft ground, there is usually a lack of adequate support for the compaction equipment to be able to work efficiently on the first layers. Clearing and cleaning the natural ground, followed by exposure to air and sunlight, when possible, may help solve the problem. If not, then a working blanket of clean sand, river gravel or crusher run stone, 20 to 30 cm (8 to 12 in) thick is placed over the whole area. Compaction of this layer also improves the uppermost part of the natural soil. When there is water, the working blanket will have to be as thick as 1 m (3 ft). If the water is very deep, then it will be more economical to consider dumped sand fill, rockfills, or other procedures including muck removal.

Soft clayey soils often have a relatively hard crust, as a result of surface drying by evaporation. If this layer is either removed or weakened by excessive stripping, it will make working conditions more difficult.

.5 Frictional Soils Which Become "Quick"

Experience has proved that some soils, such as lean silts, very fine sands and rock flour, when compacted above a high water table can absorb water by capillarity and become *quick,* with an almost total loss of strength. The same effect occurs if they are compacted with an excessive water content. When this problem arises, they can be dried by harrowing and airing if the source of the water can be eliminated. In small areas, the rise of water can be stopped by placing a layer of coarse-grained material on the fill bottom to break the capillarity. In other cases, the water table must be lowered by lateral trench subdrains. If drainage is not possible, a clean sand, gravel or crushed stone blanket as previously described can be helpful.

.6 Problems Caused by Overcompaction

The common belief that any increase in the dry unit weight of a soil as a result of compaction is accompanied by a general improvement in that soil's engineering behavior is contradicted in some materials, where unfavorable properties are produced by compacting soils beyond a certain limit. Some of the most common ones are described below.

Soil in which compaction produces particle breakdown: Examples are *tezontle* (basalt foam) which is used as a lightweight earthfill on compressible soft soils, porous volcanic ash and some porous industrial wastes. Overcompaction breaks the porous fragments, producing a wider range of grain sizes with large quantities of fines. Thus the maximum dry density is greater than found by laboratory tests and the advantage of light weight is lost. Moreover moisture trapped in the porous grains can be converted to excess pore pressure and a loss of strength upon compaction. The excess fines also reduce the material's permeability and increase capillary retention; the material is no longer free-draining and it becomes frost susceptible.

Expansive materials or ones with elastic resilience: Expansive materials can cause very serious problems, especially where the rainfall and evaporation variations during the year lead to significant changes in the water content. This is especially serious when construction work takes place during the dry season and the soil absorbs moisture during the subsequent rainy season. The greater the degree of compaction, the greater their subsequent expansion with wetting, accompanied by very high expansion pressures. If, however, they are compacted with the appropriate water content and to modest densities, expansion can be minimized. For this purpose, the compaction requirements are not established by the maximum dry unit weight or the optimum moisture as found by the usual laboratory tests. Instead, the appropriate water content and unit weight for compaction must be found by studying expansion as a function of soil moisture and density. Control of compaction moisture throughout the process is of fundamental importance. The same can be said of soils exhibiting elastic resilience under transient loads. This *rubbery* effect can lead to the rapid deterioration of a pavement. It is far more serious when the soil is compacted wetter than optimum. The best moisture is determined from field tests, accompanied by laboratory tests of the engineering properties of the compacted soil.

4.7 Compaction of Rockfills [29,30]

In other parts of this text, it has been mentioned how the development of roads and railways with flatter grades and gentle curvature has made it necessary to build ever higher embankments. As a consequence of the practical need for using the materials available at the location, it is also very common for these embankments to be made of blasted rock. Therefore the construction of very high rockfills is becoming more and more common. As described in Chapter 1, coarse soils and rock fragments lead to very serious compressibility problems when subjected to the high stress levels imposed by high rockfills. These problems are most critical in structures higher than 30 m or about 100 ft.

In recent years there has been great progress in the technology of rockfill construction. This is largely due to the experience achieved by building large dams. Although valuable for highway building, this knowledge must not be applied blindly, for there are differences that cannot be ignored. For example, rockfill dams are usually made of selected, very clean materials; for roads it is essential, economically speaking, to use more deeply weathered materials containing considerable percentages of fine soils, or rocks that weather after placement in the fill.

Obviously, rock fragments have always been used for road building. Engineers tend to consider them inert material that cannot be expected to cause serious behavior problems. This is true of many rocks, but not of all rocks. Particularly shales, mudstones, some sandstones and tuffs break down both physically and chemically, particularly at the stress levels in today's higher fills. Therefore, past experience is not always a good guide to today's conditions.

Common road building practice allows the base of the rockfill to be formed by filling in the bottom of the gorge, with little or no treatment of the foundation. Rockfill is dumped until a working surface is achieved, which is large enough to allow construction equipment to operate. It is far better to clear the foundation and muck out unstable materials into which the rock might squeeze. Under water, pre-excavation or displacement of the soft foundation is sometimes required (see [31] for a very interesting case of construction under these conditions).

The coarse rock fill at the bottom often receives no compaction. Instead, compaction is reserved for the occasional layer of finer material on top and, of course, for the subgrade and layers of pavement. However, rock fill embankments require as thorough a study and as careful a construction as other embankments if they are to behave properly.

Dumping rock fragments produces a loose, highly compressible mass. Observations based on test fills [32] and the results of recent investigations ([33,34], Sowers *Settlement of Rockfill,* and also the paragraphs on the compressibility of granular soils in Chapter 1) have brought about a fundamental change in the attitude of engineers towards rockfill embankments for highways.

Research priority should be given to the classification of rockfill materials and simple index tests which enable one to distinguish between those that are likely to perform adequately and those which are likely to settle and deteriorate structurally. In Mexico a fill consisting of rock fragments and small quantities of fines that do not pass through No. 4 sieve (4.75 mm) is considered clean. Tentatively a rockfill containing more than 5% of material passing through No. 4 sieve is considered contaminated. Material larger than 6 mm (1/4 in) is coarse and smaller fine. There are no standard tests to classify the coarse fraction of rockfills (larger than 6 mm). In Mexico the traditional tests for checking the quality of concrete aggregates have been adopted, in combination with the type of rock, shape of fragments, and weathering characteristics. The fine fraction is judged in Mexico by the Unified Soil Classification System.

As we saw in Chapter 1, particle size distribution is an important factor in the behavior of rockfills. In a uniform material ($C_u < 10$) the particle contact points are few, and stresses are very high at these points. This causes particle breakage and plastic flow within the rockfill mass with consequent deformation. A well-graded rockfill should, therefore, prove less deformable.

Particle breakage starts at relatively small stress levels, as has been seen in triaxial tests of confining pressures lower than 5 kg/cm^2 (71 lb/in^2). In Chapter 1 emphasis was laid on the significance of this breakage with regard to the behavior of granular soils, both in terms of compressibility and strength.

The way in which rockfills are compacted has a very marked effect on their compressibility and strength. This is true both for clean rockfills and more contaminated ones. Particle size distribution has a notable influence on the compaction results for these materials. For the same compactive effort, a well-graded material achieves greater density than a uniform one. However, it is not always easy to obtain an appropriate particle size distribution, especially when the rock has been broken by explosives (except when a well-graded material can be obtained as a result of closely spaced natural fractures of the rock. Mixtures of gravel and river sand are usually well-graded and if their particles are sound they are excellent rockfill materials. When well-graded materials are used it is essential to avoid size segregation during transportation and placement operations. Fill should therefore be dropped from the lowest possible height and special precautions should be taken when spreading it.

It is not easy to measure the density of rockfill materials once they have been compacted. Test pits of 1 to 2 m^3 are excavated in the fill and the materials weighed. The volume is measured by lining the pit with sheet rubber or plastic and filling the pit with water. When the fragments are not very large, the relative density concept is used, Eq. (1-17), for determining if the fill is loose or dense.

Currently, rockfills that are relatively clean and are not made up of fragments bigger than about 50 cm (20 in) are compacted with vibratory rollers. The heavier vibratory rollers have been reasonably effective in compacting rockfill particles up to 1 m in diameter. Larger particles are usually rolled to the outside of the fill by a bulldozer. Well-graded and even *contaminated* rockfills are sometimes best compacted by alternating the heaviest vibratory rollers with 50 to 100 t rubber-tired rollers: 3 to 4 passes of each. An old method was to dump coarse rockfill down a slope at least 10 m (33 ft) high accompanied by sluicing with water jets 4 cm diameter at pressures of 15 kg/cm^2 (200 lb/in^2). The densities produced did not, however, approach those of vibratory rolling.

The layer thickness of the fill depends on the maximum size of the rock fragments. Those smaller than 30 cm (1 ft) are usually spread in 50 cm (20 in) layers in a loose state. In the case of the large fragments, layer thicknesses are increased to one meter.

When constructing large rockfills, it is helpful to select material from the cut or borrow pit, sorting it into two different types using the loading machinery; one with fragments smaller than 30 cm (1 ft) and the other with larger fragments. Separation also can be carried out on the embankment. The finer material should be placed in the center of the rockfill, leaving the large fragments for the edges.

Experience has shown that the compressibility of rockfills can be reduced by adding water during placement. This has been verified by laboratory research, where it has been seen that during compressibility tests wetting causes a drastic and rapid increase in deformation. Adding water to the rockfill similarly causes deformation during construction, thus reducing its occurrence later. The reason for this rapid increase in deformability as a result of adding water is not clear; it has been explained by local softening of the edges and sharp points of the rock fragments. According to currently available information, water should be incorporated at a rate of 300 or 400 litres/m^3 (25 gallon/yd^3). The behavior of a large rockfill should always be monitored by instruments to gain experience for future projects.

4.8 Laboratory Compaction Tests

Full-scale test-fill compaction is too slow and costly to be repeated whenever any particular variable is to be studied. It is not a practical tool for analysis, study and research, as required by the problem of soil compaction with its numerous complications and complexities. Therefore simple and economical simulation of these processes by laboratory tests has proved to be useful for detailed research.

For the same reasons, laboratory tests have become the basis for project studies and for planning efficient field processes. The alternative would be either to evaluate these processes on the basis of past experience, or else to develop them from full-size models, but at great expense.

Laboratory compaction tests are only justified in terms of their simulation of the field processes they are intended to represent. The greatest care must be taken that the field behavior is reproduced in the laboratory; otherwise the laboratory results will have little significance in the field compaction process which is supposedly being studied. This could lead to serious consequences regarding the practical conclusions and would mislead engineers who relied principally on the results of the laboratory tests for making decisions concerning the field compaction process.

Laboratory standard compaction tests have two main uses. First, soils are compacted to obtain data for earth structure projects; the compacted soils are tested for such properties as strength, deformability, permeability, and tendency to crack. The representativity of the test is essential, in that the soil produced in the laboratory must have the same mechanical properties as those which will be obtained when the materials have been compacted in the field. The second use of compaction tests is for field quality control. Here the test acts as a comparative index of the laboratory and field unit weights; the similarity of the mechanical properties is less important. The essential quality of a comparative index test is that it should always be reproducible.

Table 4-13
Characteristics of the most widely used dynamic compaction tests

Test	Processing	Mold dimensions				Hammer		Height of fall		No. of layers	Blows per layer	Soil re-use	Compactive energy	
		Diameter		Height										
		cm	in	cm	in	kg	lb	cm	in				kg-cm/cm³	lb-in/in³
Standard Proctor	Sieve through 1/4 in mesh	10.16	4	12.70	5	2.49	5-1/2	30.48	12	3	25	yes	4.02	57
E-10 Test, U.S.B.R.	Sieve through No. 4 mesh after air-drying and breaking up the lumps	10.80	4-1/4	15.25	6	2.49	5-1/2	45.72	18	3	25	yes	6.05	86
Standard AASHTO (Variant A)	Sieve through No. 4 mesh after air-drying	10.16	4	11.43	4-1/2	2.49	5-1/2	30.48	12	3	25	yes	6.05	86
Modified AASHTO (Variant D*)	After air-drying break up the lumps and sieve through 3/4 in mesh replacing material retained with material between 3/4 in and No. 4 mesh	15.24	6	17.78	7	4.53	10	45.72	18	5	55	no	27.31	389
California (Variant A)	Sieve dry soil using 3/4 in mesh	7.30	2-7/8	91.44	36	4.53	10	45.72	18	5	20	no	17.70	252
California (Variant B)	Sieve moist soil through 3/4 in mesh	7.30	2-7/8	91.44	36	4.53	10	45.72	18	5	20	no	35.40	504
British Standard Test	Oven or air drying, sieve through 3/4 in mesh	10.16	4	11.68	4.6	2.49	5-1/2	30.48	12	3	25	yes	6.05	86
Proctor (SOP)	Air-drying and sieve through No. 4 mesh	10.16	4	11.68	4.6	2.49	5-1/2	30.48	12	3	30	yes	6.65	95

* This test has been replaced by test AASHTO T180-70 with a compactive energy of 17 kg-cm/cm³ (242 in-lb/in³)

We have already discussed in considerable detail the factors that affect a compaction process; obviously they must all be taken into consideration when designing a laboratory test. Because there are so many different methods available for compacting soils in the field, one may reasonably conclude that it will be impossible to create a single test that will represent all of them. Therefore, there should be various different types of compaction tests. As discussed previously, the compactive effort also has an important effect on process results; because the equipment available today is capable of applying effort in so many different ways. This too leads to variations in the corresponding test requirements. Although other variables affect a compaction process, only the above differentiate between the common laboratory tests.

Since 1933, when PROCTOR developed his, historically first, test, many others have appeared; all fall into one of the following categories: Dynamic or impact tests, Static load tests, Kneading tests, Vibratory tests and Special tests or those under development.

4.8.1 Dynamic Tests

All the dynamic tests in use today have the following characteristics:

— A soil is compacted by layers within a rigid cylindrical mold; the size of the mold and thickness of the layer vary from one test to another.
— Compaction is achieved by applying to each layer within the mold a certain number of evenly distributed blows from a tamper. The weight, dimensions and height from which this tamper falls vary from one test to another. The number of blows applied to each layer also varies.
— The specific energy is accurately calculated by using Eq. (4-1). It is defined by the number of blows per layer, number of layers, weight of the tamper, height from which it falls and total volume of the mold. It is assumed that each layer is entirely compacted by the blows falling directly on that layer, and not on the layers above it.
— In all cases, a maximum soil particle size is specified, the larger sizes being eliminated by sieving before the test. A specification is often also made with reference to the re-use of material during the test.

The value of each one of the test variables can be changed as necessary, so that field compaction conditions can be simulated; however, it has become a custom for those who use compacted soils to establish a *master test*, in accordance with their own experience, on the results of which field compaction specifications are based. Naturally, because the mechanical properties of compacted soils depend on compaction conditions and the properties which are desirable for one particular structure are not necessarily so for another, one laboratory master test would not fit all the possible field situations [13].

Some of the standarized dynamic tests which are now most widely used are the Standard PROCTOR Test (which is the one originally proposed by PROCTOR), the Standard (AASHTO and ASTM) PROCTOR Test (with four variants), the U.S. Bureau of Reclamation's E-10 test, the California Compaction Test (two variants) and the British Standard Test (BS-1377, 1948). The main characteristics of some of these tests are given in Table 4-13 [4].

At the foot of Table 4-13 the PROCTOR S.O.P. (Ministry of Public Works of Mexico) variant has been included. This is the one generally used by the Ministry to control compaction on earth structures of fine materials. Also worth mentioning is the variant of the dynamic compaction test stipulated by the Texas Highway Department (USA) which is similar to the Modified (AASHTO) Test.

Details of the AASHTO Standard and Modified Test procedures are given in Appendix 4-a in this chapter. Both have four variants based on the following criteria. In the first place, two sizes of mold are used, one with a diameter of 10.16 cm (4 in) and the other with a diameter of 15.24 cm (6 in). The first is the traditional mold used by PROCTOR (1/30 ft^3) and which is now used by force of habit; the second, an enlargement of the first, was introduced later to measure soil strength in the mold (*CBR* tests). The *CBR* test, used in pavement technology will be described later. For it, the 4 in mold is too small, so the 6 in one is used (techniques of the US Army Engineers, [41]). Two different particle limits are imposed, one with a maximum size of No. 4 sieve (4.75 mm) and the other with a maximum size of 3/4 in (19.1 mm). The larger size limit permits the use of the test with a wider variety of materials.

The California Compaction Method, which is described in Appendix 4-b, was used before the original PROCTOR Test as a method of controlling field compaction. Basically it is similar to the AASHTO Standard tests, although the specific energy is different as shown in Table 4-13. The California Division of Highways has been using it for many years for field compaction control, although other state transportation departments favor AASHTO standard or modified PROCTOR tests. Possibly their unstated reason is not to lose the experience of many field engineers who have *gotten the feel of* a certain test. This is also the reason why many organizations continue to use other similar but non-standard tests which really are not very different from others and which do not substantially contribute to the technology being used.

The same is true of the Mexican Ministry of Public Works test for fine soils; their PROCTOR-type test only differs from the Standard AASHTO in that 30 blows are applied per layer instead of 25. This change was introduced more than 35 years ago when it was thought to produce a more even distribution of blows per layer. For the last 35 years the Ministry of Public Works has acquired all its experience on the basis of this test and it is still used instead of a worldwide standard.

The Texas Highway Department has developed another dynamic test which, as we have already mentioned, is of special interst [36]. The essential part of the test is its mechanization; otherwise it is similar to the Modified (AASHTO) PROCTOR Test. The purpose of this mechanization is to eliminate the variations due to the operator. The differences in particle size and specimen size are minor, and compaction is carried out with automatic equipment, which uses automatic tampers (Plate 4-25). The material must not be re-used. Four different energies are specified for different types of soils. The specific energy is reduced in fine soils according to their tendency to expand or crack. There is a special compaction procedure for clean sands.

One of the most serious objections to dynamic compaction tests is that their simulation of field conditions is suspect because of the very strict confinement the rigid mold imposes on the soil. Confinement interferes with the free movement of the soil particles, thus making compaction different from that in the field where there is far less lateral confinement. To reduce this confinement, FRANCIS HVEEM, one of the most brilliant and original men to have studied these problems, proposed that these tests be performed in conventional molds, but using

Plate 4-25 Mechanical dynamic compactor, in Texas

specimens in the shape of a hollow cylinder, inside which a rubber cylinder would be placed to give the particles greater freedom of movement. The authors of this book were personally informed of some of the preliminary results of this experiment, which show that in some soils, higher degrees of compaction are obtained with smaller compactive efforts compared to the standardized tests. However, these interesting investigations were cut short before definite conclusions could be reached.

Figure 4-7 illustrates the basic principle of dynamic compaction processes in the laboratory. With an increasing effort, a higher maximum dry unit weight is achieved at a decreasing optimum water content. By comparing the different compaction curves in Fig. 4-7, it can also be seen that above optimum moisture a great increase in compaction energy has very little effect on the dry unit weight obtained, whereas below the optimum the effect of an increase in compactive effort is significant.

Figure 4-29 [37] shows the effect of the type of soil (in this case the particle size distribution) on the compaction results achieved for two different sands. In both cases the British Standard Test was used. Note the advantage of the well-graded sand, where the fine particles can fit into the spaces between the coarse particles.

The influence of the content of coarse particles on the soil sample was investigated by MADDISON [38], who found that mixing 25% of any aggregate of one single size, up to 2.5 cm (1 in) has little effect on the compaction of the soil as a whole. Adding larger percentages of that same size can cause a rapid decrease in the dry unit weights obtained; when this percentage reaches 70%, the soil's behavior is comparable to that of a mass of coarse particles of the size selected.

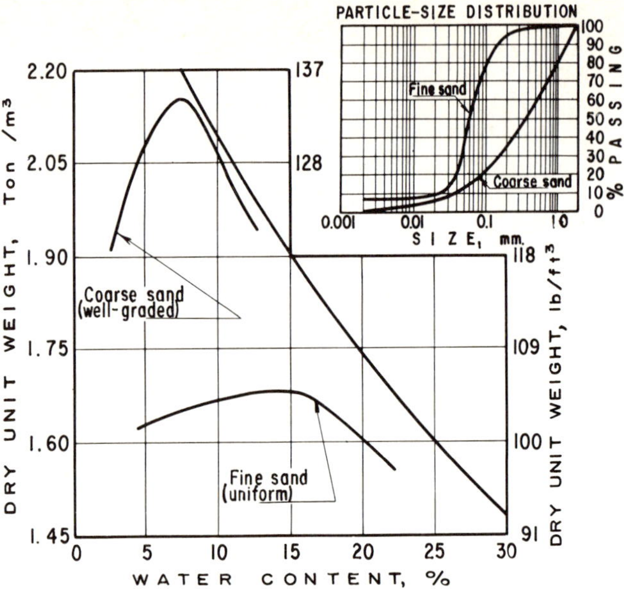

Fig. 4-29 Compaction curves for two differently graded sands and the same compaction energy [37]

Further information concerning the effect of the soil type can be obtained by analyzing Fig. 4-30 [15], where compaction curves are given for 8 different soils compacted according to the standard PROCTOR (AASHTO) test.

TYPES OF SOIL

No	DESCRIPTION	SAND %	SILT %	CLAY %	W_L	I_p
1	WELL-GRADED SAND	88	10	2	16	–
2	WELL-GRADED SANDY MARL	72	15	13	16	–
3	MEDIUM SANDY MARL	73	9	18	22	4
4	SANDY CLAY	32	33	35	28	9
5	SILTY CLAY	5	64	31	36	15
6	LOESS SILT	5	85	10	26	2
7	CLAY	6	22	72	67	40
8	POORLY–GRADED SAND	94	6	–	–	–

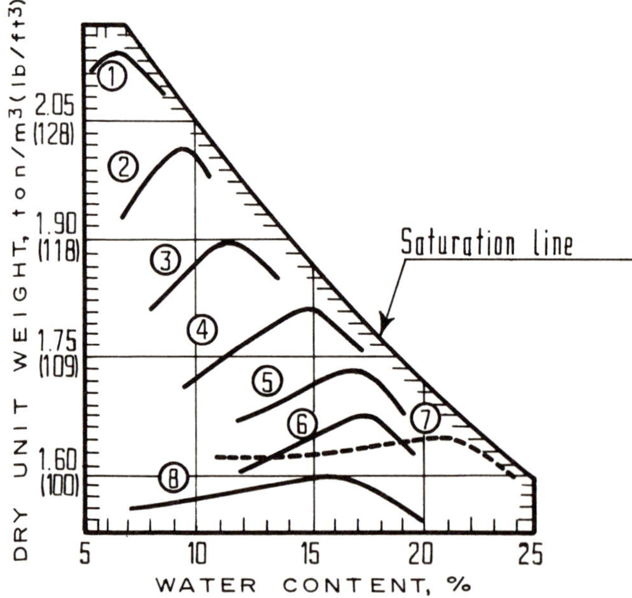

Fig. 4-30 Compaction curves for eight different soils using the standard (AASHTO) PROCTOR test [15]

Figure 4-31 [37] gives the results of drying and adding water to a clayey sand compacted according to the British Standard Test. The corresponding compaction curve is also shown. The soil was dried and water was added starting at different points of the compaction curve. The figure shows the evolution of the water content and dry unit weight as a consequence of these operations. Maximum volume changes occur when the soil is compacted at higher densities. The ability to absorb water and expand is reduced when the soil is compacted near saturation. The smallest changes in volume occur when the soil is compacted to low densities. These soils reach the lowest unit weights and highest water contents when saturated.

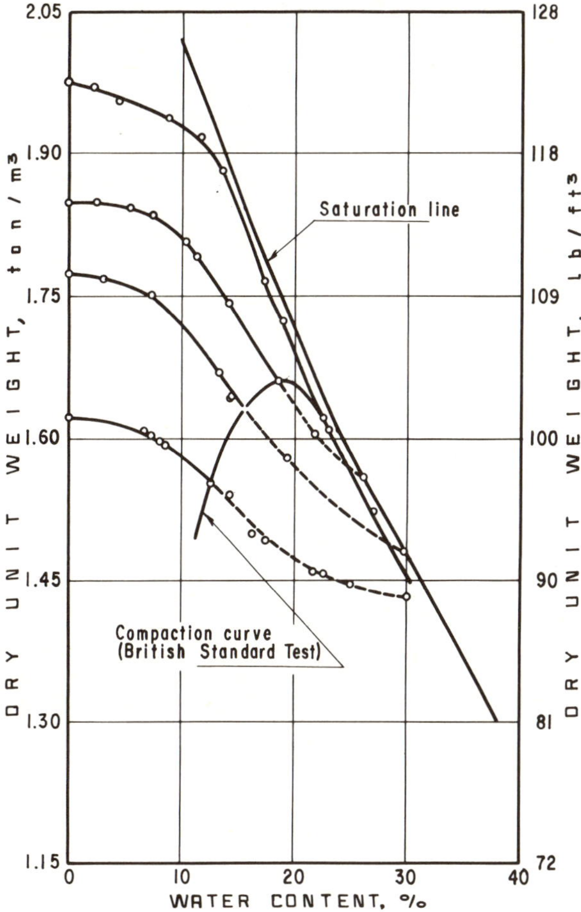

Fig. 4-31 Relationship between the dry unit weight and the water content of a sandy clay when slowly wet or dried after compaction [37]

The shape of the compaction curves obtained is considered "regular" when somewhat parabolic, as in many of the figures. Many lateritic soils, uniform sands (see Fig. 4-3) and some highly plastic clays often have very irregular compaction curves. The shape of the curves is related to the compaction effort; for example, in highly plastic clays, the irregular shape obtained in the standard PROCTOR (AASHTO) test usually becomes regular when the modified PROCTOR (AASHTO) test is used with its higher specific energy.

A semilogarithmical representation, such as given in Fig. 4-32 [39], enables us to appreciate the variation in maximum dry unit weights for 17 different soils compacted with different energy. This is obtained by plotting the maximum dry unit weights achieved in the laboratory for each soil as a function of

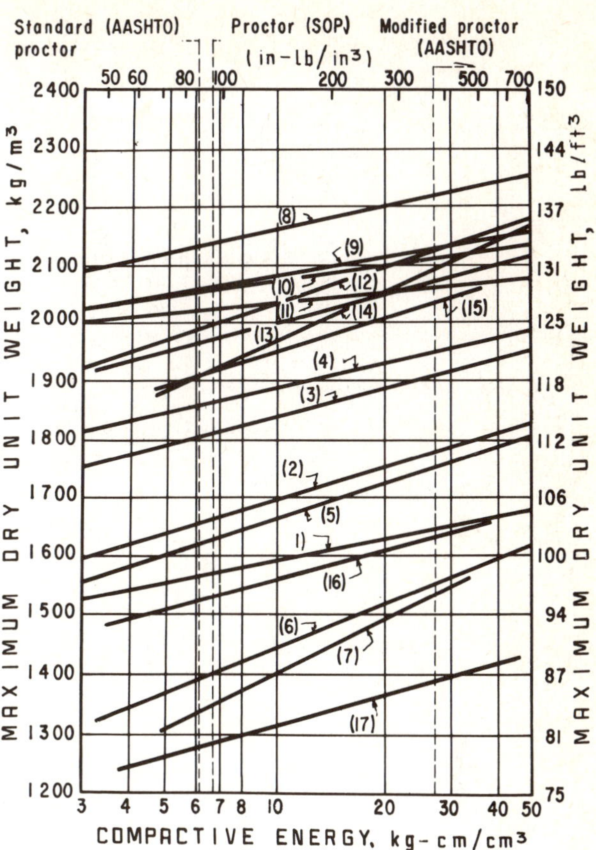

Fig. 4-32 Variation of maximum unit weight with the compactive effort for different types of soil [39]

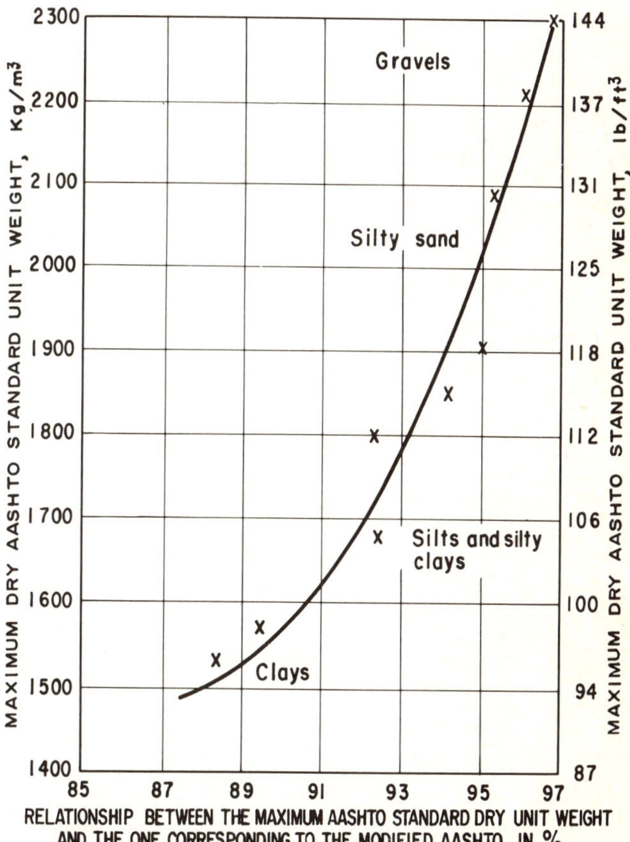

Fig. 4-33 Relationship between standard and modified maximum unit weights (AASHTO)

Table 4-14
Comparison of maximum dry unit weights and optimum water contents obtained with various laboratory tests and field compaction equipment

Type of test	Heavy clay		Silty clay		Sandy clay		Sand		Mixture of sand gravel and clay	
	γ_d	w	γ_d	w	γ_d	w	γ_d	w	γ_d	w
British Standard	1.56 (97)	26	1.67 (104)	21	1.85 (115)	14	1.94 (121)	11	2.08 (130)	9
Modified AASHTO	1.81 (113)	17	1.93 (122)	14	2.05 (128)	11	2.08 (130)	9	2.22 (139)	7
2.5t (5.5 kips) smooth roller	1.52 (95)	21	1.77 (110)	17	1.83 (114)	16	2.06 (129)	10	2.15 (134)	8
7.5t (16.5 kips) smooth roller	1.67 (104)	20	1.78 (111)	16	1.86 (116)	14	2.12 (132)	8	2.22 (139)	7
Pneumatic-tired roller	1.57 (98)	25	1.67 (104)	20	1.78 (111)	19	2.04 (127)	11	2.02 (126)	7
Sheepsfoot roller (foot *b* in Fig. 4-4)	1.72 (107)	16	1.86 (116)	14	1.91 (119)	12	—	—	2.08 (130)	6
Sheepsfoot roller (foot *a* in Fig. 4-4)	1.72 (107)	15	1.85 (115)	14	1.92 (120)	12	—	—	2.06 (129)	5
Hand-operated 450 kg (990 lb) vibratory plate	1.72 (107)	17	1.76 (110)	15	1.86 (116)	13	2.05 (128)	10	2.18 (136)	7

Maximum dry unit weight, γ_d, in t/m^3 (lb/ft^3); Optimum water content, w, in%.

the logarithm of the specific energy; the result is approximately linear. In this graph, the influence of the type of soil on maximum density is again seen, together with the decreasing benefit of an increase in the compactive energy of the test.

Field experience indicates that with many soils it is very difficult to achieve densities greater than 100% of the modified PROCTOR maximum test; with others, particularly in some sandy soils and some highly plastic clays it is easier. This is reflected in the lines in the graph.

Figure 4-33 provides a comparison of standard PROCTOR (AASHTO) and modified PROCTOR (AASHTO) maximums for 43 different soils. The maximum dry unit weight for the standard PROCTOR (AASHTO) varies between 85% and 97% of the maximum corresponding to the modified test depending on the type of soil. It is seen that the results of the two tests are almost identical in the case of granular materials.

To complete the information concerning dynamic tests, Table 4-14 [37] shows compaction results for several types of soils, which were subjected to both laboratory tests and various kinds of field compaction equipment; and the great difference that can be obtained, both in maximum dry unit weight and in optimum water content, by applying different compaction methods to the same soil can be clearly seen. The relativity of the two previously mentioned concepts is once again obvious. Although the information is limited to a specific investigation, it nevertheless illustrates the general tendencies.

4.8.2 Static Compaction Tests

With reference to Fig. 4-3, we have already shown that in frictional soils it is very common for dynamic or impact tests to produce a compaction curve that does not define a maximum dry unit weight or optimum water content. We also said that for these soils there are other tests where the compaction curve has a more typical shape, and therefore may be appropriate for these purposes.

One of these is the static load compaction test, which was introduced by O. J. PORTER and reached its ultimate form about 1935. A soil is placed inside a cylindrical mold 15.24 cm (6 in) in diameter. It is deposited in three layers and is first lightly rammed 25 times by a rod with a bullet-shaped end; this does not apply intensive compaction because the rod is light and the height from which it strikes, though not specified, is the minimum that can be used by the operator to allow comfortable handling. Actual compaction is achieved by applying a pressure of 140.6 kg/cm^2 (2 kips/in^2), which is maintained for a whole minute, to all three layers together. Details of this test are included in Appendix 4-c. There, reference is also made to the version adopted by the Mexican Ministry of Public Works, which is often used to control field compaction with soils that are predominantly frictional (PORTER SOP test).

As can be seen, the static load compaction test is as old as dynamic tests. Although it is not as universally recognized as these, it has the same advantages. The original static-load test was originally used with the California Bearing Ratio, (*CBR*), test which is widely used for pavement design; perhaps this is another reason for its survival.

It is doubtful (though relatively little research has been carried out on the subject) that a static-load compaction simulates any field compaction process. The application of pressure is not an efficient method for compacting frictional soils (which are the ones to which the static load test is most often applied). It does not involve vibration or any of the modern compaction methods used for these soils in the field. Besides, there is evidence that the application of confined static pressure to granular materials can produce important changes in particle size distribution during the test.

There was a time when it was thought that a static-load compaction test simulated a smooth drum roller when used in granular soils. This led to the idea that whereas dynamic tests better simulated compaction processes on clays, static tests were more appropriate for sands and gravels. This was based on intuition and was never demonstrated by evidence as far as the authors of this book are aware. Because smooth drum rollers are seldom used to compact granular soils, the idea has been forgotten.

In [39] AGUIRRE presented the results of a comparative study of the achievements of the static-load test in relation to those of dynamic tests. Seventeen soils were included, from gravels to highly plastic clays. Table 4-15 gives the chief characteristics of these 17 soils. Table 4-16 shows the maximum dry unit weights obtained and the optimum water contents corresponding to each one of the tests performed on the different soils. Note that for fine soils the miniature Harvard test mold (2 in diameter) was used, but the soil was subjected to blows from a hammer.

The main conclusions of the study were that in coarse sands and gravels, either clean or with non-plastic fines, the results of the PORTER SOP test are similar to those obtained for the same soils with the PROCTOR (AASHTO) standard test (see Fig. 4-34). In moderately plastic clays, fine sands containing any type of fines, coarse sands with plastic fines and gravels with plastic fines, the results of the static load test are comparable to those of the PROCTOR (AASHTO) modified test, as can be seen in Fig. 4-35. Finally, in highly plastic clays, the maximum densities obtained with the PORTER test are considerably greater (by 10%) than those of the PROCTOR (AASHTO) modified test, as shown by Fig. 4-36.

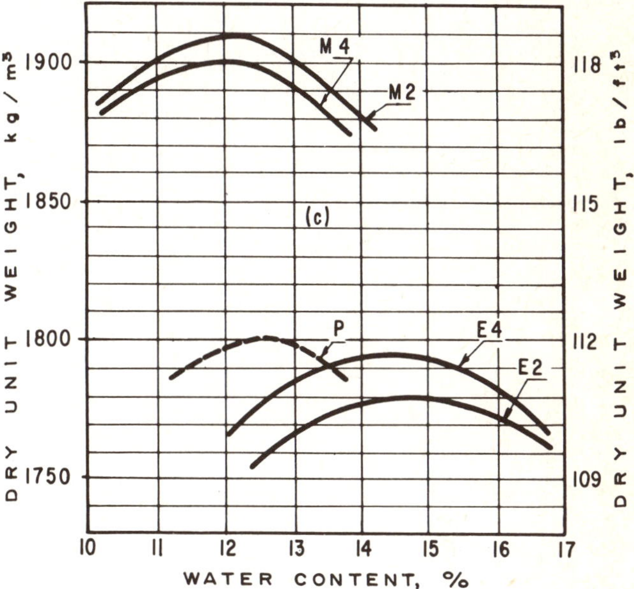

Fig. 4-34 Comparison between the static test and dynamic test results for a coarse-grained sand with non-plastic fines [39]

Table 4-15

Comparison of results from static load and dynamic tests: Part 1: list of soils used [39]

Soils	Description	Classification (U.S.C.S)	Plasticity indices		Percentage passing sieve			
			W_l	I_p	3/4 in	No. 4	No. 40	No. 200
1	Uniform fine sand	SP	22	n.a.	100	100	100	5
2	Medium silty sand, well graded	SW-SM	25	n.a.	100	100	40	10
3	Angular coarse sand	SW-SM	20	n.a.	100	100	30	12
4	Silty clayey sand	SM-SC	21	7	100	100	55	25
5	Silty clay	CL-ML	34	12	100	100	70	45
6	Highly plastic clay	CH	62	36	100	100	96	90
7	Highly plastic clay	CH	71	35	100	100	96	92
8	Angular gravel with 9% non-plastic fines	GW-GM	21	6	100	49	21	9
9	Rounded gravel with 9% non-plastic fines	GW-GM	21	6	100	47	21	9
10	Angular gravel with 18% non-plastic fines	GM	21	6	100	55	25	18
11	Rounded gravel with 18% non-plastic fines	GM	21	6	100	55	25	18
12	Angular gravel with 9% plastic fines	GW-GC	49	29	100	42	19	9
13	Rounded gravel with 9% plastic fines	GW-GC	49	29	100	42	19	9
14	Angular gravel with 18% plastic fines	GC	49	29	100	54	28	18
15	Rounded gravel with 18% plastic fines	GC	49	29	100	54	28	18
16	Clayey sand with app. 30% gravel	SC	38	12	100	72	55	37
17	Clayey sand	SC	38	12	100	100	80	45

Table 4-16

Comparison of results from static and dynamic tests: Part 2 : Summary of maximum dry unit weights and optimum water contents for the soils listed in Table 4-15 and various testing procedures

SANDS AND FINE SOILS											
	Test	E–2		E–4		P		M–2		M–4	
	Soil	γ_d	w	γ_d	w	γ_d	w	γ_d	w	γ_d	w
1	SP	1.55 (97)	17.1	1.56 (97)	17.0	1.64 (103)	16.8	1.63 (102)	15.3	1.64 (102)	15.4
2	SW-SM	1.64 (102)	16.8	1.64 (103)	15.0	1.72 (107)	14.0	1.73 (108)	15.5	1.76 (110)	12.2
3	SW-SM	1.78 (111)	14.7	1.79 (112)	14.3	1.80 (112)	12.3	1.91 (119)	12.2	1.90 (119)	12.0
4	SM-SC	1.83 (114)	14.0	1.85 (115)	13.7	1.90 (119)	11.5	1.91 (119)	12.0	1.94 (121)	11.6
5	CL-ML	1.56 (97)	22.0	1.61 (100)	19.8	1.74 (109)	15.3	1.67 (105)	17.3	1.74 (109)	16.7
6	CH	1.31 (82)	31.8	1.38 (86)	30.4	1.71 (107)	18.9	1.51 (94)	25.4	1.54 (96)	24.6
7	CH	1.29 (80)	32.2	1.32 (82)	32.4	1.63 (102)	21.8	1.45 (90)	26.1	1.51 (95)	23.8
GRAVELS											
	Test	E–4		E–6		P		M–4		M–6	
	Soil	γ_d	w	γ_d	w	γ_d	w	γ_d	w	γ_d	w
8	GW-GM	2.12 (133)	9.3	2.12 (132)	9.5	2.09 (131)	7.5	2.21 (138)	7.6	2.17 (136)	7.9
9	GW-GM	2.05 (128)	8.8	2.06 (129)	9.1	2.03 (127)	8.1	2.12 (132)	7.7	2.09 (131)	8.0
10	GM	2.04 (128)	10.1	2.05 (128)	9.1	2.05 (128)	9.9	2.10 (131)	8.5	2.10 (131)	8.6
11	GM	2.01 (126)	10.0	1.99 (124)	10.0	2.01 (126)	10.9	2.06 (128)	9.0	2.04 (127)	9.3
12	GW-GC	1.97 (123)	11.9	1.98 (124)	10.9	2.08 (130)	10.6	2.11 (132)	8.7	2.11 (131)	8.4
13	GW-GC	1.95 (122)	11.5	1.96 (122)	10.2	2.02 (126)	10.7	2.06 (128)	8.3	2.06 (128)	8.3
14	GC	1.89 (118)	12.6	1.92 (120)	12.0	2.10 (131)	10.4	2.08 (130)	8.3	2.07 (129)	8.8
15	GC	1.89 (118)	11.0	1.89 (118)	11.6	2.05 (128)	10.1	2.02 (126)	9.6	2.02 (126)	9.3
16	SC	1.51 (94)	22.9	—	—	1.63 (112)	21.4	1.62 (101)	19.4	—	—
17		No data									

The values for maximum dry unit weight, γ_d in t/m^3 (lb/in^3) and optimum water content, w, in % are the average of 5 tests.

Symbols: E–2 : Standard (AASHTO) PROCTOR, in 5.08 cm (2 in) mold — miniature
E–4 : Standard (AASHTO) PROCTOR, in 10.16 cm (4 in) mold
E–6 : Standard (AASHTO) PROCTOR, in 15.24 cm (6 in) mold
M–2 : Modified (AASHTO) PROCTOR, in 5.08 cm (2 in) mold
M–4 : Modified (AASHTO) PROCTOR, in 10.16 cm (4 in) mold
M–6 : Modified (AASHTO) PROCTOR, in 15.24 cm (6 in) mold
P : Static load test (PORTER SOP) 15.24 cm (6 in) mold

Figure 4-37 [59] gives the results of two correlations between the PROCTOR SOP and PORTER SOP compaction tests. The first is between the quotient of the γ_d PROCTOR divided by γ_d PORTER and the product of the plastic index of the soil with the percentage of particles smaller than No. 200 sieve size (clay activity). The second is the same relationship of unit weights and with the sand equivalent of the soils (a test used in pavement technology, which will be described in detail in the corresponding chapter).

For the first correlation it will be observed that the γ_d (PROCTOR) is less than γ_d (PORTER) in plastic soils when the activity is greater than 100. The more frictional the material and the lower the activity, the smaller the maximum dry unit weights produced in the PORTER test, and the more plastic the material, the greater the unit weights. These results agree with those obtained by AGUIRRE [39] for the same 17 soils.

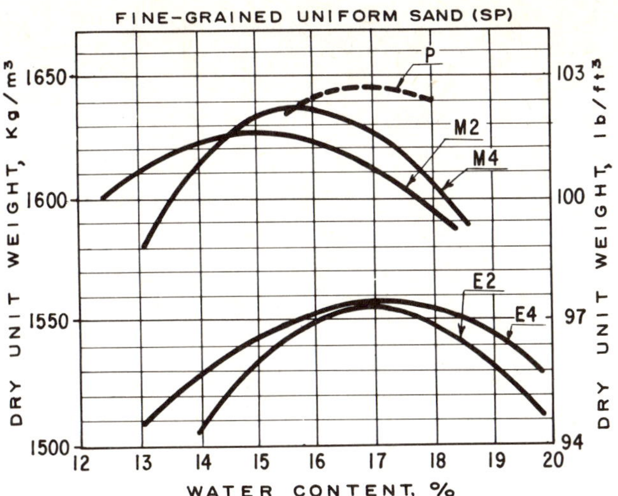

Fig. 4-35 Comparison between the static test and dynamic test results for a fine sand [39]

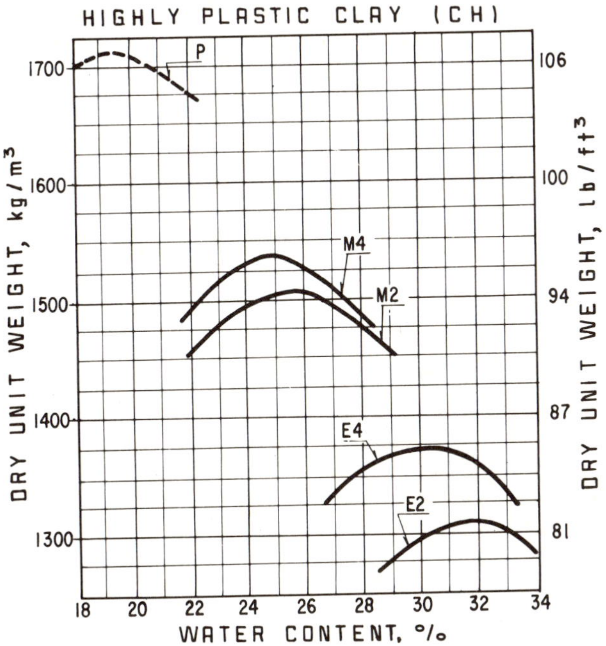

Fig. 4-36 Comparison between the static test and dynamic test results for a highly plastic clay [39]

The second correlation restates the same relation. For sand equivalents less than 20 (plastic soils), the maximum dry unit weight by PROCTOR is smaller than the maximum dry unit weight by PORTER. The more frictional the soil, the greater the PROCTOR maximum dry unit weight compared to the PORTER maximum.

A few organizations use the PORTER test as a standard compaction test for frictional soils, the PROCTOR-type tests for fine soils. The data of Figs. 4-36 and 4-37 lead us to reconsider the advisability of two different types of compaction control tests. One standard sometimes implies extremely high requirements with regard to the other, whereas on other occasions the reverse is true.

Finally, it is difficult to measure or compute the specific energy employed in laboratory static-pressure compaction. Based on these studies it is our opinion that the best policy is to relate

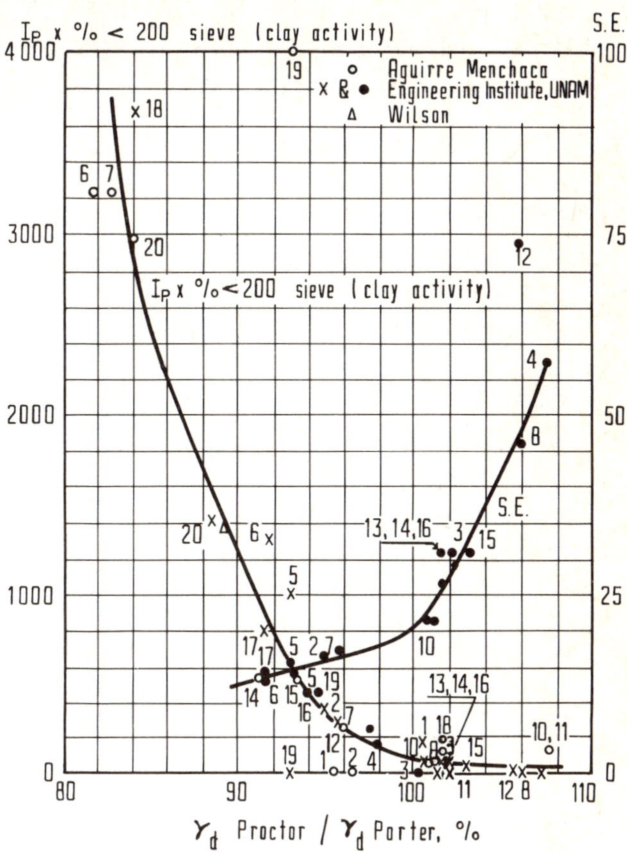

Fig. 4-37 Correlation between the maximum dry unit weights obtained in PORTER and PROCTOR tests

field compaction by just one standard, taking into consideration all its limitations. Each particular discrepancy should be considered on the basis of a thorough understanding of field compaction mechanisms and project needs.

4.8.3 Kneading Tests

With just one exception, kneading compaction methods are relatively recent in laboratory research technology. The exception is the so-called *Harvard miniature test* which was developed by S. D. WILSON at Harvard University (USA). This is an attempt to reproduce in the laboratory the typical mechanical effect produced by many sheepsfoot and pneumatic-tired rollers so as to achieve in the specimen the same internal grain structure that is acquired by the soil in the field.

In the Harvard test, the kneading effect is achieved as follows. A piston with a specific area exerts pressure on the surface of the different layers of soil sample within a mold which has the dimensions for producing a specimen suitable for conventional triaxial tests. A more or less constant final pressure is transmitted to the piston during each application or stroke by means of a spring gage. In Appendix 4-d, there is a detailed description of this test, which can only be performed on soils with a maximum particle size of 2 mm (0.08 in); this however, is not a serious disadvantage, for the test is limited to clayey soils.

HVEEM [42] has developed a mechanical compactor for laboratory use, which requires no hand operation and produces specimens by means of a kneading process (Plate 4-26). Even though not enough has been published about the results and conclusions for us to be able to define how well it simulates field compaction, our engineering judgement leads us to believe that such a method should produce laboratory specimens that best represent the soils compacted in the field by sheepsfoot or pneumatic-tired rollers. In Appendix 4-d, details are also given of HVEEM's kneading compaction test. The major drawback of kneading tests is that the specific energy cannot be determined readily.

Plate 4-26 Mechanical kneading compactor, by HVEEM

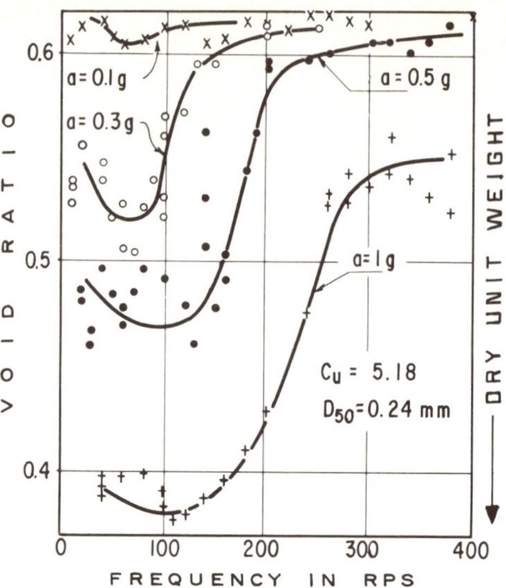

Fig. 4-38 Compaction of sands by vibration. Effect of acceleration and frequency [4]

4.8.4 Vibratory Tests

Numerous researchers have in recent years taken a keen interest in vibratory compaction tests [43]. Many use a PROCTOR mold mounted on a vibratory table. The effects of the frequency, amplitude and acceleration of the vibratory table are being studied, as well as the influence of confining loads, the particle size distribution of the soil and the water content. SCHAFFNER [44], also mentioned in [43], studied the compaction of dry sands on vibratory tables. Figure 4-38 shows typical results. It can be seen how the void ratio decreases with acceleration and how the greatest dry unit weights were obtained at frequencies of about 6000 rpm.

Similar results were reported by SELIG [44]. He found that maximum dry unit weights are obtained with accelerations between 1g and 2g. In addition, when the confining pressure exerted on the sand is increased, greater acceleration is required in order to reach a certain dry unit weight.

ORTIGOSA and WHITMAN [46] found that with accelerations greater than 2g, the dry unit weight drops again, but if the sand is saturated or moist, the dry unit weight continues to rise even with accelerations greater than 3g [47].

Swedish technology [21] has developed a different laboratory vibratory test in which a specimen is placed in the lower part of a cylinder which is attached to a massive block of concrete. Above the specimen, covering its entire surface, is a plate attached by a shaft to a vertical impulse vibrator.

Laboratory vibratory methods have become standardized by using a vibratory table in combination with a confining load, or a vibratory tamper. These are described in [48-50].

Figure 4-39 compares the efficiency of compacting a sand in the field with vibratory methods with that of a dynamic test in the laboratory. This shows the great effect of the size of the vibratory plate and the high degree of efficiency that can be achieved with vibratory compaction.

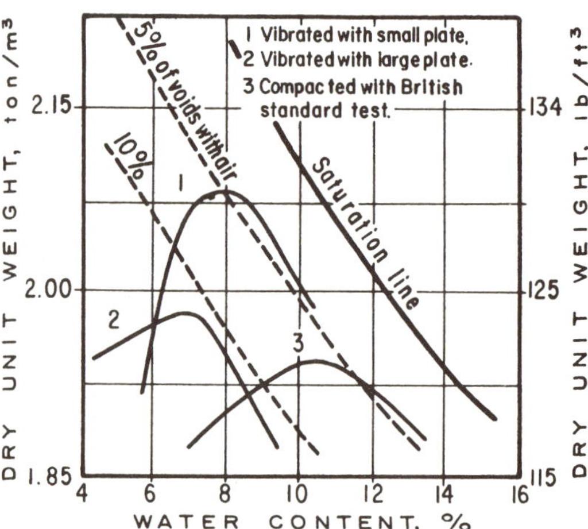

Fig. 4-39 Compaction curves for a sandy soil compacted with two types of vibrator and with the British Standard Test [15]

Owing to the importance of natural frequency when specifying the range of frequencies at which compactors should be used in the field or vibrations applied in the laboratory, Table 4-17 below gives the natural frequencies of some soils and rocks using a vibrator. The table is taken from [15] and refers to a specific vibrator.

Table 4-17
Natural self-vibration frequencies for several different soils and rocks

Type of soil or rock	Natural frequency/min
2m (6.56 ft) of peat on top of sand	750
2m (6.56 ft) of fill with sands and fine soils	1145
Sand and gravel with clay lenses	1165
Earth fill compacted by the action of traffic	1280
Moist clay	1430
Very uniform medium sand	1445
Uniform coarse sand	1570
Almost dry clay	1650
Limestone	1800
Sandstone	2040

An increase in vibration amplitude brings about a corresponding increase in efficiency of the vibration and depth of compaction at all frequencies. A high amplitude is especially favorable for coarser gravel and crushed stone. When very high amplitudes are used, the vibration frequencies of the equipment can be reduced, which usually leads to more economical compaction.

Laboratory research [21] has also shown that the greater the pressure exerted on the compacted soil, the more useful are the resonant frequencies of the soil-vibrator system. In practice, this has led to the use of higher frequencies for lighter types of compaction equipment.

4.8.5 Special Tests or Those Under Development

Of these, the *revolving compaction machine* deserves special mention [51,52]. It was constructed with the intention of reproducing in the laboratory specimen the grain structure and other characteristics that are acquired by a soil when compacted in the field with the rolling equipment. In fact this machine is a kneading compactor. Figure 4-40 shows how pressure combined with a rocking effect is transmitted to the specimen.

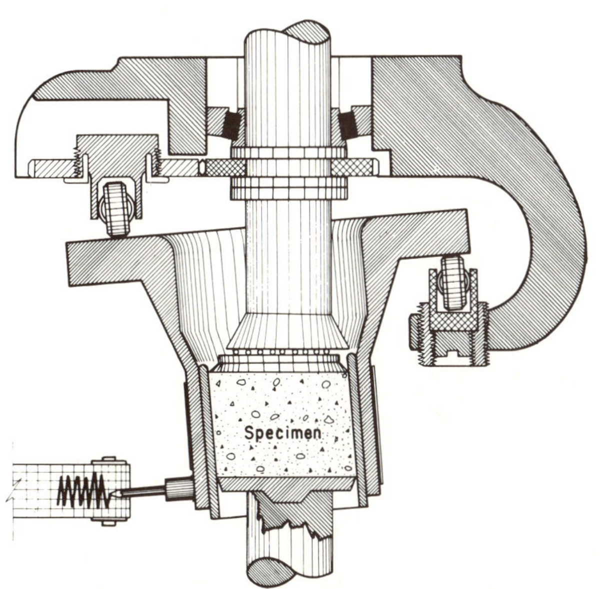

Fig. 4-40 Diagram showing the gyratory compaction machine [52]

Up to now, many of the tests that have been carried out with this machine are linked with pavement technology; different ways of applying the compactive energy at varying intensities are beginning to be developed with a view to distinguishing between larger or smaller traffic volumes. The machine has also proved useful for detecting the influence of the plasticity of the fine fraction of the soils and studying the structural degradation of the materials under dynamic loads.

It is hoped that in a not too distant future there will be more detailed information concerning this compaction system.

4.9 Criteria for the Selection of Laboratory Tests. Comparison of Results Obtained in Laboratory and Field

A standard laboratory compaction test serves several purposes:
- A norm or standard of comparison for quality control of field compaction.
- A basis for comparing the relative compaction qualities of different soils.
- A standard for comparing research by different organizations with different soils.

A test must be chosen that by reproducing the relationship existing between unit weights and water contents, and the structure of the soil in the field, will make it possible to study the effect of compaction variables on the specific soils to be used. In this way the engineer can define the conditions that will govern the compaction process in the field [13]. In the case of fine soils, the compaction should reproduce the structure of the soil produced by field compaction. This has been proven by comparing mechanical properties [57]. Dynamic compaction is less appropriate and less representative, but is probably acceptable in road building technology, especially for quality control purposes. The major differences surely fall within the range of the deviations inherent to practical construction processes [13].

Finally, the laboratory test should include the compactive energy that will best reproduce the relationship between the dry unit weight and water contents expected in the field.

4.9.1 A Comparison of Laboratory Tests with Sheepsfoot Rollers

Figure 4-41 [13,58] shows that the field moisture density curve with a sheepsfoot roller corresponds to slightly higher degrees of saturation than the ones for a laboratory dynamic compaction test (standard PROCTOR or AASHTO). In the same figure can be seen a curve corresponding to a kneading compaction test (miniature Harvard) fairly near the field curve. The field compaction was carried out using 12 coverages of a heavy roller over a clayey soil placed in a loose state in layers 23 cm (9 in) thick. Figure 4-42 [12,13] gives similar information corresponding to fine materials from the three dams. The field process was carried out with a heavy sheepsfoot roller and the material was a clay which was spread loosely in layers 20 cm (8 in) thick. The data from Figs. 4-41 and 4-42 should be compared with those from Fig. 4-43 [13,15], showing similar results but with a light sheepsfoot roller. Note that in the latter case the relative positions of the optimum field and laboratory curves are inverted. A roller pressure of about 25 kg/cm² (355 lb/in²) provides the distinction between the lightweight and heavyweight equipment for the purposes discussed here.

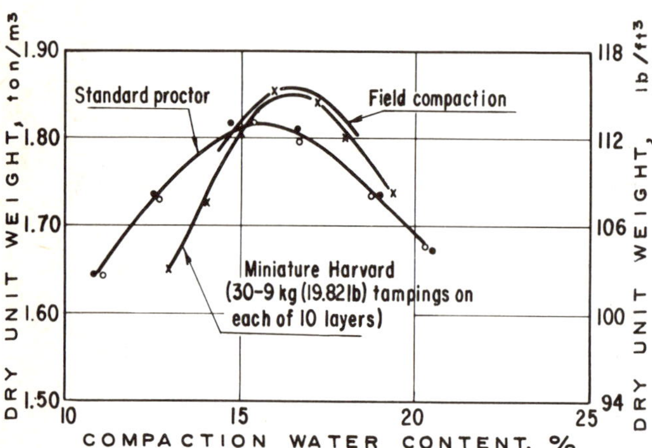

Fig. 4-41 Comparison of field compaction curves (very heavy sheepsfoot roller) and laboratory compaction curves (Standard PROCTOR and miniature Harvard [13,58]

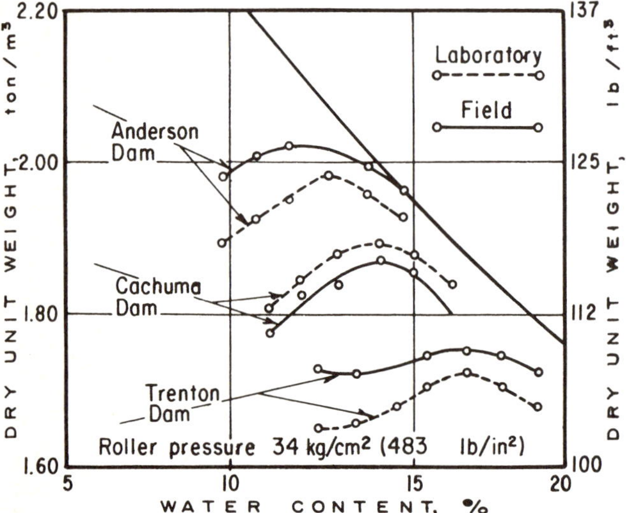

Fig. 4-42 Comparison of field compaction curves (very heavy sheepsfoot roller) and laboratory curves (Standard PROCTOR for material passing sieve No 4) [12,13]

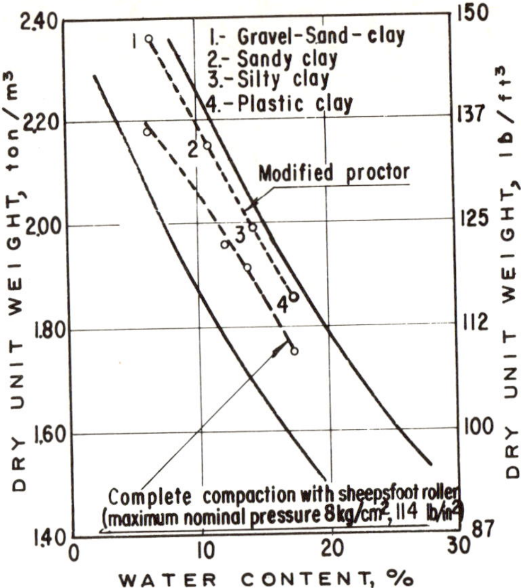

Fig. 4-43 Comparison of field (lightweight sheepsfoot roller) and laboratory (blows) optimum water content paths [13,14,15]

The data from [12-15] are summarized in [13] with the following conclusions:

— For pressures between 8 and 18 kg/cm² (114 and 256 lb/in²), the field compaction line of optimum is slightly to the left of the one corresponding to dynamic standard PROCTOR test. The differences are greater between the field curves and those for kneading compaction of the miniature Harvard type, for the latter are to the right of the ones which are obtained in dynamic tests.

— For pressures of 18 to 35 kg/cm² (256 to 497 lb/in²), the field compaction optimum line almost coincides with the one obtained in a dynamic test, the standard PROCTOR, and is slightly to the right of it. The kneading curves produced by the miniature Harvard type tests practically coincide with the field curves. Unfortunately there is not sufficient information regarding the specimens produced with other types of kneading compactors, which are more widely used today.

In Fig. 4-44 [14,15], the field compaction curves for the four soils shown are compared with the values obtained in the laboratory by means of the British Standard Test (very similar to the standard PROCTOR, AASHTO test). The field compaction used a lightweight sheepsfoot roller, 8 kg/cm² (114 lb/in²) contact pressure, with 64 passes and foot contact area of 75 cm² (12 in²). It can be seen that the relationship between the maximum unit weights obtained in the field and in the laboratory for the four soils is not consistent. Observe that Fig. 4-44 refers to the same investigation as Fig. 4-43.

The comparative information between the results of field compaction processes using a sheepsfoot roller and laboratory tests is completed by the data provided by Fig. 4-45 [15]. The dynamic tests compared are the modified PROCTOR (AASHTO) (1), the standard PROCTOR (AASHTO) (3) and a dynamic test with intermediate energy (2). The line of optimums obtained from the three tests represents the range that can be expected when using such tests in the laboratory (with the efforts indicated).

The soil that was tested was a clay with $w_L = 38\%$ and $I_p = 18\%$ and was spread in the field in compact layers 15 cm (6 in) thick. There are three field curves. *A* corresponds to 6 passes of a

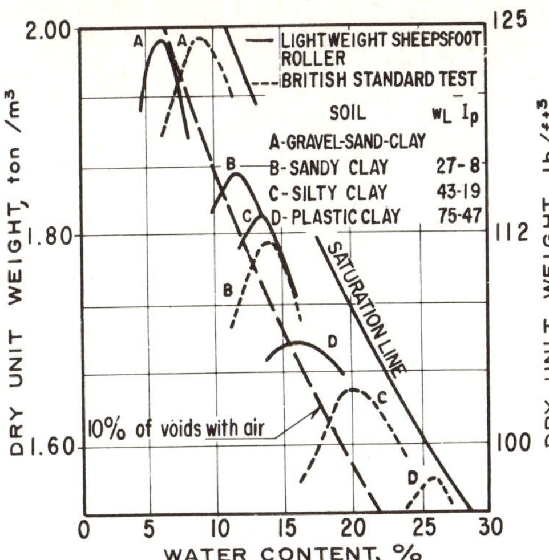

Fig. 4-44 Comparison of field compaction curves (lightweight sheepsfoot roller) and laboratory compaction curves (British Standard Test) [14,15]

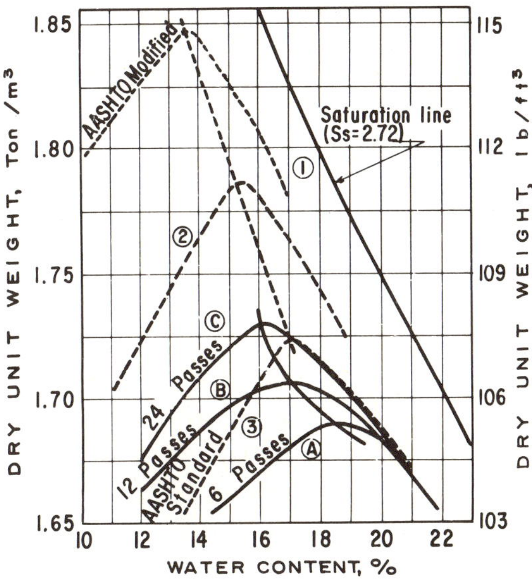

Fig. 4-45 Comparison between the results of a field compaction process with sheepsfoot roller and dynamic tests [15]

sheepsfoot roller with a foot area of 44 cm^2 (6.8 in^2) and contact pressure of 18 kg/cm^2 (256 lb/in^2); *B* to 12 passes of the same equipment, and finally, *C* to 24 passes. Note that the optimum moisture for the roller is below the laboratory optimum for fewer than approximately 22 passes. Comparisons like this one show the dangers of fixing the degree of field compaction arbitrarily on the basis of a laboratory test chosen without performing a study to determine the relationship between the methods of laboratory compaction and field results.

4.9.2 Comparison of Laboratory Tests with Pneumatic-Tired Rollers

Figures 4-46 and 4-47 [13] show that for a given soil the line of optimums corresponding to field processes using pneumatic-tired rollers with tire pressures between 2.80 and 10.50 kg/cm^2 (40 and 150 lb/in^2) is to the right of the line of optimums for dynamic laboratory tests. Figure 4-48 [13,15] shows that the difference in the position of the two curves of optimums diminishes as the compactive effort is increased. In Fig. 4-49, the effect of the type of soil is shown.

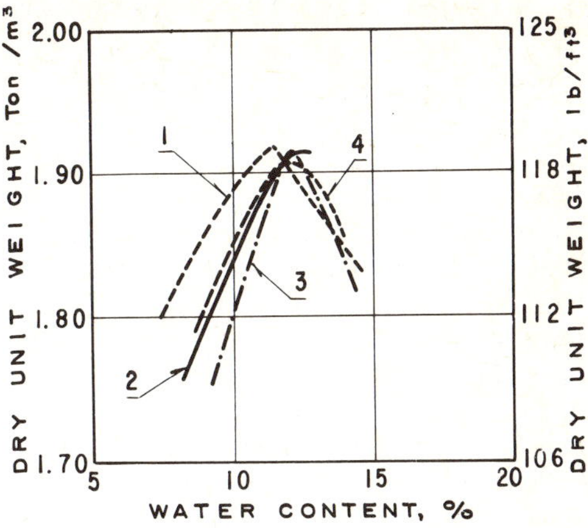

Fig. 4-46 Comparison of field compaction (pneumatic-tired roller) and laboratory compaction (Standard PROCTOR) curves for a sandy clay (W_l = 18, I_p = 16) [13]

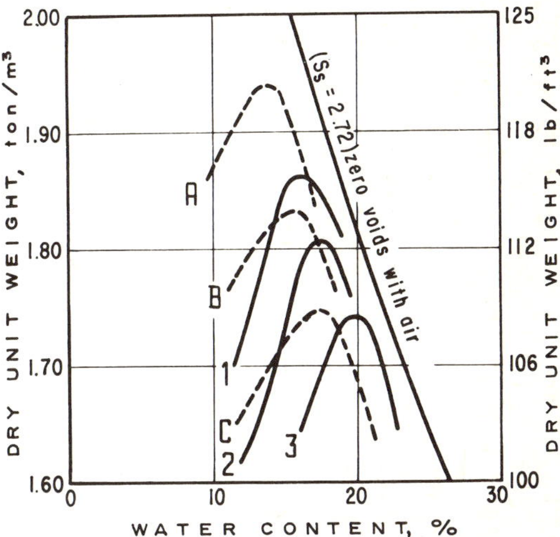

Fig. 4-47 Comparison of field compaction (pneumatic-tired roller) and laboratory compaction (Standard PROCTOR) curves [13]

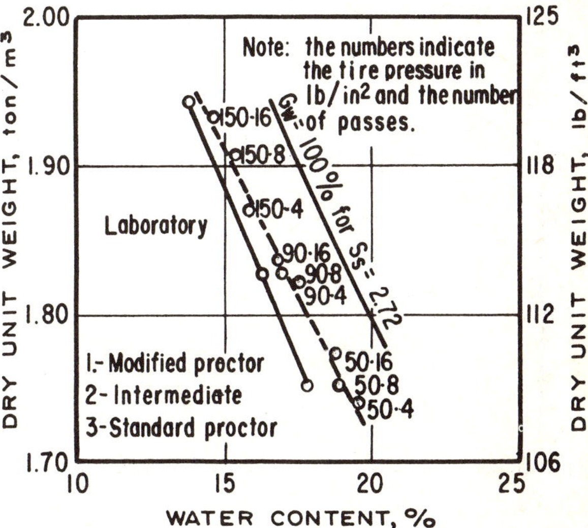

Fig. 4-48 Comparison of field (pneumatic-tired roller) and laboratory (PROCTOR type blows) optimum water content paths [13,15]

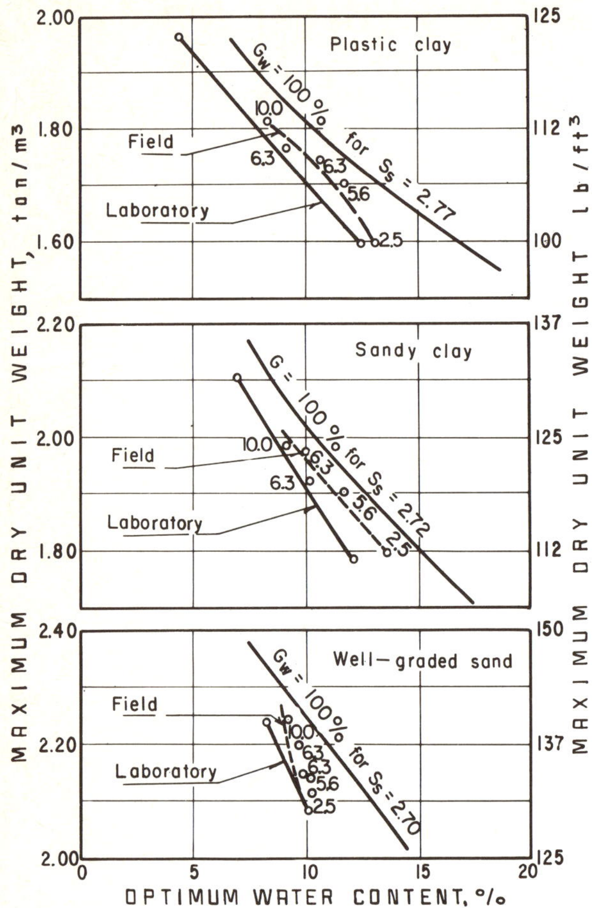

Fig. 4-49 Comparison of field (pneumatic-tired roller) and laboratory (PROCTOR type blows) optimum water content path [13]

In compaction processes using pneumatic-tired rollers, the curve of optimums for a wide range of tire pressures is located to the right of the one obtained for dynamic tests. It appears that kneading compaction tests reproduce the field curves of pneumatic-tired rollers better than dynamic tests, but the difference between the two types of tests is not great.

4.10 Mechanical Properties of Compacted Sands

If, in a direct shear apparatus, a loose sand is tested to obtain its shear resistance, a *plastic type* stress-strain curve will be obtained like the one shown in Fig. 4-50 [10]. In order to produce increasing strains, increasing shear stresses are necessary. In the same test, a very dense sand will produce a stress-strain curve like the one with the broken line shown in this figure. At first, increasing stress is required in order to increase the strain, but once a maximum stress value has been reached, the stress drops without an increase in strain. It has also been shown already that this difference in behavior can be explained by grain structure changes. In dense sand it is essential not only to overcome the friction between the particles, but also to force them to turn and move, rolling over one another. Once the initial compact structure has been broken and the density drops, relative motion becomes much easier. In loose sand, however, the initial structure is loose and unstable; it is easy to begin particle motion producing a greater density (Chapter 1). During shear of a loose sand the resistance of the sand gradually increases until it reaches a limit. This behavior is almost identical to that obtained when compacting sand; the ultimate or residual strength is the same for a loose sand as for a dense one.

The lower part of Fig. 4-50 shows the variations in volume of the sample under strain. The volume of the loose sand decreases from the start because of the reorientation of grains and the destruction of the initially unstable structure. In the dense sand, at first there is a slight reduction in volume from the general increase in stress level, but the continuing deformation process causes an increase in volume with a maximum rate corresponding to the maximum stress.

The figure shows that the maximum resistance that can be developed by a dense sand is far greater than the resistance of the same sand in a loose state. This high peak resistance in dense sands is not always more favorable. For example, in loose sand an increasing resistance is always developed with increasing strain. In dense sand there is a loss of strength with strain. Although stronger dense sands have a tendency to suffer progressive failure once failure begins.

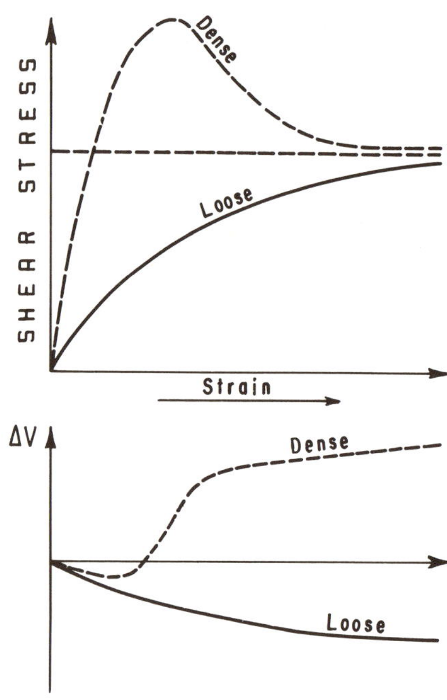

Fig. 4-50 Differences in mechanical behaviour between a loose and a dense sand [10]

Therefore, when a sand is compacted in the field, its maximum resistance may increase greatly, but only part of the increase obtained can be taken into account in a project. Beyond a certain point progressive failure can develop, with continuing strains at stresses well below the maximum. The compressibility of compacted sands is also smaller than that of loose sands. Here the reader should consult Chapter 1 so as to review the compressibility problems of sands and coarse soils at high pressures.

An effect which deserves more attention than it has received is the structural degradation suffered by many coarse soils as a result of compaction. This causes important changes in particle size distribution, so that the one obtained in the field is not the same as the one found previously in the laboratory. Particularly the permeability is reduced and the capillary tension increased, material. The strength and the compressibility are changed slightly, and for bituminous base-courses there can be an alteration of the optimum asphalt content. Obviously, the weaker the particles of the material being compacted, the more notable will be the effect. In [39], AGUIRRE MENCHACA gives particle size distribution curves for coarse materials compacted in the laboratory by means of various procedures. In some cases the effects of structural degradation are considerable (for example, from 9% of the material smaller than No. 200 sieve before compaction to 18% after).

4.11 Mechanical Properties of Compacted Fine Soils

4.11.1 Introduction

In order to study the properties of compacted fine soils, an analysis must first be made of the compaction variables which govern the mechanical properties of these soils. These are the void ratio (or dry unit weight), degree of saturation and structure acquired by the solid particles. Much of the information that follows has been taken from [13], an excellent summary. Further information can be found in [53-56], and particularly in [60].

In Chapter 1 of this book we discussed the interaction of the solid and liquid phases of a water air clay system. We saw that each clay crystal appears to behave as though it had a negative electrical charge, attracting positive ions (double electrical layer).

Besides this interaction, there is also particle-to-particle interaction, as a result of longer distance action forces. These consist of an electromagnetic attraction (VAN DER WAALS forces) and a repulsion between the positive layers of the double electrical layers of each particle. The repulsion forces are a function of interparticle spacing and increase with a reduction in the concentration of electrolytes. The VAN DER WAALS forces are independent of the concentration of electrolytes.

Figure 4-51 [13] shows two extreme solid particle arrangements between which a pure clay soil may vary. The structure is determined by two principal factors, these are the relative magnitude of the interparticle attractive and repulsive forces and the degree of angular deformation and shear the soil may have suffered during compaction. The greater the repulsion and angular deformation, then the greater will be the degree of orientation of the particles. The water content, specific energy, compaction effort, compaction procedure, method of soil preparation and the proportion and characteristics of the non-clayey fraction, and the clay minerals all influence the soil structure and the resulting physical properties, as discussed below.

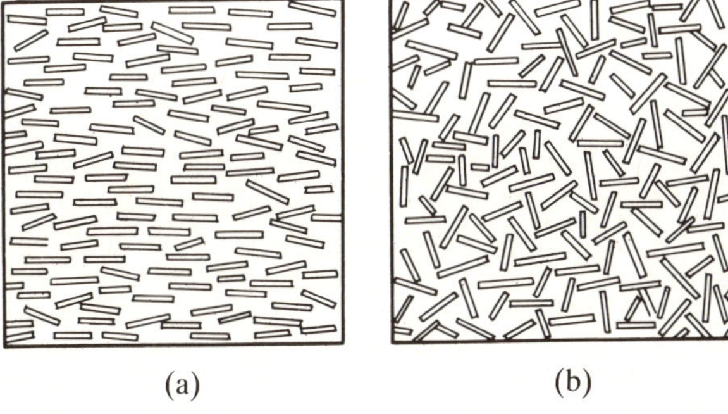

Fig. 4-51 Extreme structures of a clayey soil: (a) high degree of particle orientation; (b) low degree of particle orientation [13]

.1 Effect of Water Content

On the dry unit weight: As established earlier in this chapter, maximum dry unit weight is reached at optimum water content (Fig. 4-1).

On the degree of saturation: This effect can also be seen in the compaction curve. For any pair of values $\gamma_d - w$, G_w can be calculated using the equation:

$$G_w = \frac{w\gamma_d S_s}{S_s\gamma_w - \gamma_d} \tag{4-7}$$

where G_w and w are in % and S_s is the specific gravity of the solids. Using Eq. (4-7), the curve corresponding to any degree of saturation can be plotted on the compaction graph.

The degree of saturation diminishes very rapidly for water contents below optimum and remains almost constant for water contents above optimum (see, for example, Fig. 4-7).

On the structure: When the water content is low, development of the double electrical layer is limited and the concentration of ions is very high. This causes limited interparticle repulsion and high capillary tension. The result is a soil with great shear resistance that resists particle movement and thus compaction produces a soil with a low degree of particle orientation.

If there is an increase in the water content, the repulsive forces become stronger and the capillary forces weaker, with a consequent reduction in the soil's resistance to deformation. Compacted with the same method and energy, a soil with a higher water content will undergo greater angular deformations and develop a structure with a higher degree of orientation. Thus the degree of particle orientation increases with higher water contents. If the compactive energy is increased, the tendency towards particle orientation is also enhanced.

.2 Effect of the Specific Energy or Compactive Effort

On the dry unit weight: In Fig. 4-7 is seen how the compaction curve changes with variations in specific energy: as the energy increases, at low water contents, the greater will be the unit weight. Any increment in the energy applied to a soil with above-optimum water content is used to produce angular deformation, but not a reduction in volume. Although a soil with a high water content is more deformable, the water in the voids is incompressible and resists densification.

On the degree of saturation: During a compaction process, the water content of fine soils remains nearly constant; therefore the degree of saturation increases if there is an increase in the compactive energy and a higher unit weight is produced. When the water content of a soil is above optimum, an increase in compactive energy has little effect because the voids contain less compressible air and more incompressible water.

On the structure: The energy applied to a soil is used to reduce the volume and cause angular deformations. An increment in compactive energy produces, therefore, further orientation of the clay particles. If the water content of the soil is above optimum, an increase in compactive energy brings the soil closer to the extreme dispersive condition shown in Fig. 4-51a.

.3 Effect of the Compaction Method

Unfortunately it is not possible to compare the different methods of compaction that are used at the same compactive energy because this energy cannot always be accurately quantified. Moreover it is affected by certain imponderable factors that influence process efficiency. So, instead, one compares the procedures that bring a soil to the same dry unit weight at the same water content. Therefore the difference in the properties obtained by a soil are largely due to a variation in structure, which reflect differences in the angular deformations caused by different compaction methods.

In the laboratory, at the same unit weight and the same water content the maximum degree of particle orientation is reached by kneading and the minimum by static load compaction. Reference [63], from which Fig. 4-52 is taken, shows an interesting investigation where it is shown that under static load compaction a clay maintains a completely random or flocculated structure throughout a wide range of moistures. A clay which at low water contents had a flocculated structure with kneading compaction, developed a dispersed structure (with maximum particle orientation) when compacted at water contents corresponding to the optimum water content for that same procedure. This dispersed structure is maintained at water contents greater than optimum.

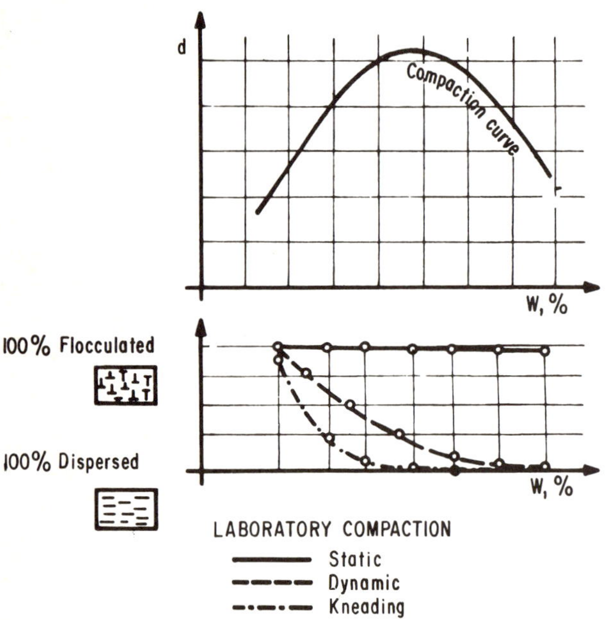

Fig. 4-52 Effect of the type of compaction on the structure acquired by a compacted soil [63]

Dynamic compaction produces a flocculated structure at very low water contents. The structure gradually becomes dispersed with increasing water contents. A totally oriented structure is reached at above-optimum water contents (in the wet section of the curve).

A number of conclusions can be drawn from this study. First, different soil structures correspond to different engineering properties (this will be discussed in the following pages). Second, almost all the most widely used field compaction processes, excluding vibration (which unfortunately was not included in the above-mentioned investigation) include some degree of kneading effects, whereas static load laboratory compaction cannot be considered representative of any current field method. Therefore, static-load compaction tests cannot be considered to reproduce the same soil behavior as field compaction process. The use of a static-load laboratory test to evaluate soils with a view to designing an earth structure design is questionable.

In the field, the sheepsfoot roller produces a higher degree of particle orientation than the pneumatic-tired one.

A well-known difference between laboratory kneading and dynamic tests is that the optimums correspond to higher degrees of saturation in the first case than in the second. In Fig. 4-53 [13], the order of magnitude of this difference is given for a sandy clay of low plasticity. In more plastic soils, the difference may be greater.

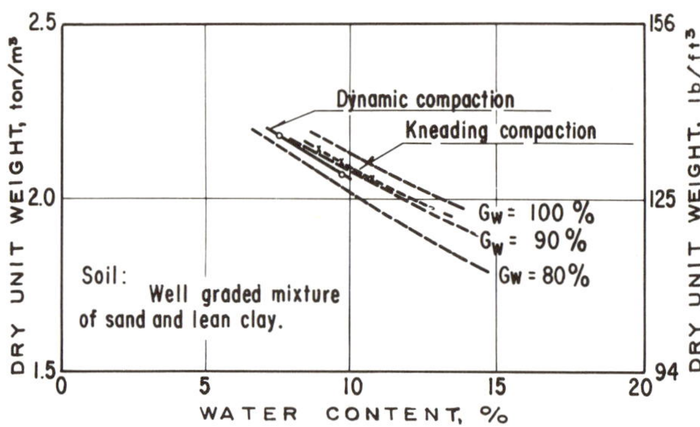

Fig. 4-53 Optimum water content paths for dynamic compaction (PROCTOR type) and kneading compaction (miniature Harvard type) of the same soil [13]

.4 Effect of the Coarse Fraction

The influence of the coarse fraction on the unit weight that is obtained during laboratory dynamic tests was discussed earlier in this chapter. The dry unit weight increases with an increase in the percentage of coarse particles to a certain limit, beyond which it then drops. If the percentage of coarse particles is constant, but there is a change in the grain size distribution of the coarse fraction, then the maximum dry unit weight increases as grading improves. This is why it is not advisable to use a laboratory compaction process in which the fraction coarser than a specified size (often the 3/4 in size or 1.90 cm) is replaced by the same weight of material that passes through that sieve, but is retained by the No. 4 size. This method can lead to results very different from the ones obtained in the field.

4.11.2 Permeability

The permeability of a compacted soil, like the other mechanical properties, depends on the void ratio (or dry unit weight), structure and degree of saturation.

As stated in Section 1-8, the coefficient of permeability varies approximately with the square of the void ratio. The typical variation of the permeability coefficient of a soil with the compaction water content, at a constant specific energy of compaction, is shown in Fig. 4-54 [61]. Figure 4-55 [13,62] shows the effect of the degree of saturation on the permeability of compacted clays; the permeability always increases with an increasing degree of saturation. The figure also illustrates the changes in structure due to thixotropy (an increase in adsorbed water orientation), when specimens are stored for 21 days at a constant water content.

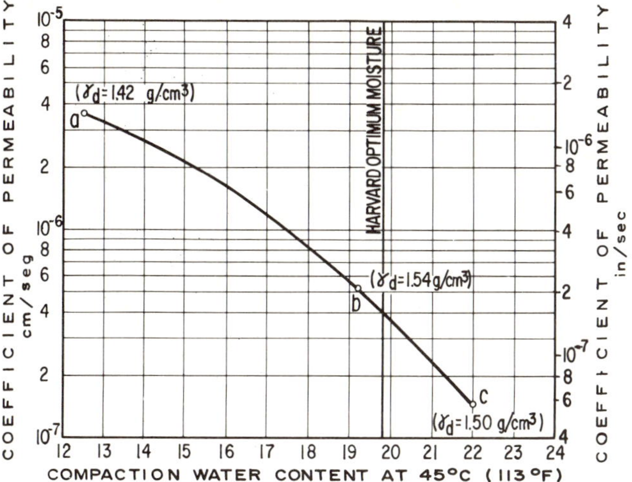

Fig. 4-54 Relationship between the compaction moisture and the permeability in a gypsum-containing silt saturated under a backpressure of 6kg/cm² (85 lb/in²) [61]

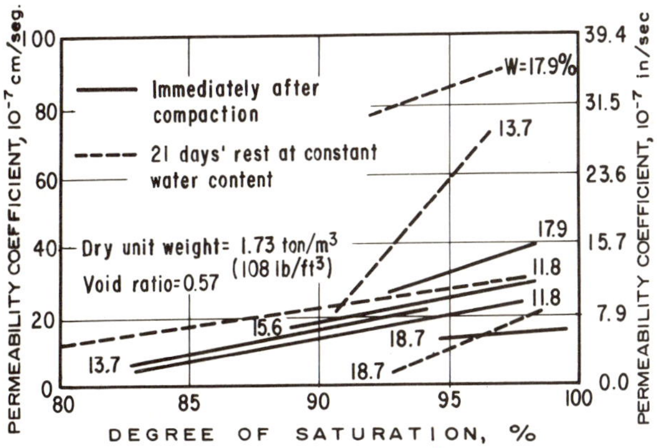

Fig. 4-55 Effect of the degree of saturation on the permeability of a silty clay compacted by kneading [13,62]

Structure is the most important factor that affects the permeability of a compacted soil, Fig. 4-56, [13,62]. Note that at a higher water content and greater strain produced by compaction (with a corresponding increase in the degree of particle orientation), lower permeability is obtained for flow perpendicular to the long axis of the orientation, than parallel. The permeability coefficient perpendicular to the axis is less than that of the flocculated soil.

Figure 4-57 [13,57] gives the differences in permeability obtained when compacting the same soil in the field with a sheepsfoot roller and in the laboratory with a kneading compactor, at the same dry unit weight and the same water content. Note the differences caused by the compaction method, and also those obtained in the field between vertical and horizontal permeabilities, which are much higher than the ones for the laboratory specimen.

The permeability of a compacted clayey soil may vary greatly with compaction conditions, especially those affecting the structure of the soil.

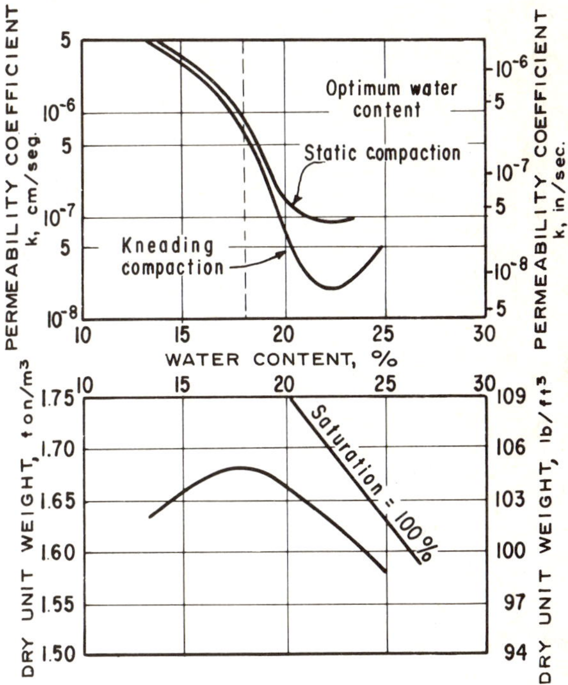

Fig. 4-56 Effect of the structure on the permeability of a silty clay [13,62]

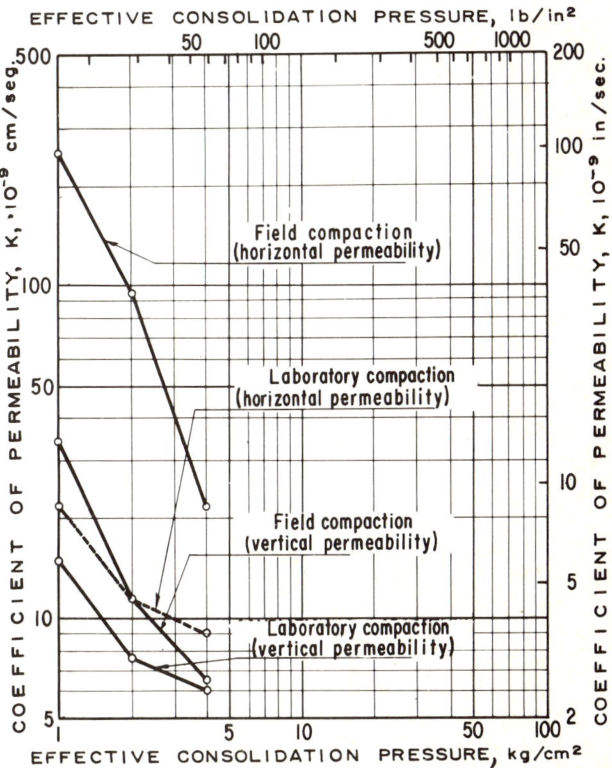

Fig. 4-57 Differences in vertical and horizontal permeability of samples compacted in field and laboratory (13,57)

4.11.3 Compressibility and Expansion

When there is a change of volume in a clayey soil, there are two structural components of deformation. The first corresponds to variations in the distance between the particles, with a constant degree of orientation; the second occurs as a result of reduction in the average distance between particles, without a change in the minimum distance, due to an increase in the degree of orientation. The second component operates only when compression occurs in the soil, and is irreversible. Therefore expansion can be attributed almost exclusively to the first component [13].

If two samples of a clayey soil are compacted following the same method and energy to the same dry unit weight, but using in one case a below-optimum water content and in the other an above-optimum one, and then both samples are submitted to a load process in which volumetric deformation can be measured, a behavior like that shown in Fig. 4-58 is obtained. The sample compacted wet of optimum (sample *2*) shows a regular compressibility curve, typical of the type of soil being tested (clayey), whereas sample *1*, compacted dry of optimum, shows a peaked compressibility curve.

Under low stresses, the coefficient of compressibility of the soil compacted on the dry side is lower than for the soil that was compacted on the wet side, but under high pressures the situation is reversed. This is due to the fact that under low pressure there are insignificant changes in the degree of particle orientation of the two samples. As the minimum distance between particles is greater in sample *2* (the denser soil compacted wet of optimum) the particles are more easily forced together than the particles in the sample compacted dry of optimum. Under high pressures, volumetric deformation occurs in sample *1*, compacted dry of optimum, as a result of the increase in the degree of particle orientation, but not so in sample *2*, compacted wet of optimum. Under very high pressures, the two samples reach the same void ratio because a similar oriented structure is produced in both.

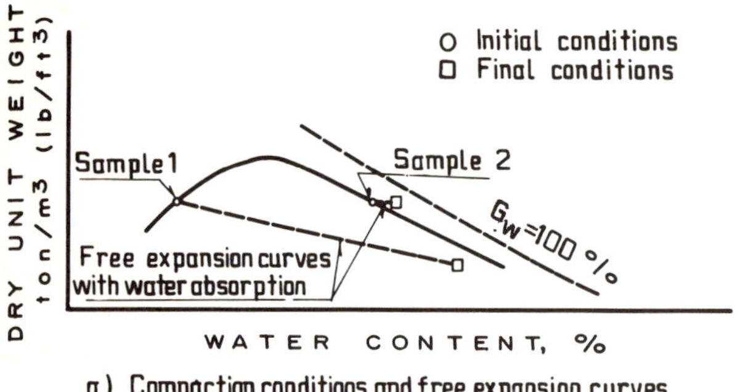

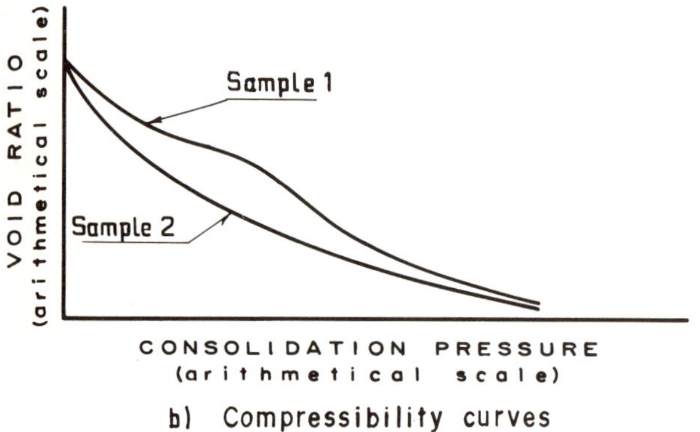

Fig. 4-58 Compressibility and expansion characteristics of a clayey soil, with the same void ratio and compacted with water contents on both sides of the optimum [13]

In soils compacted by kneading, it has been observed that the stress at which the abrupt variation in slope occurs is slightly below the compactive pressure [64].

In the soil compacted on the dry side, the free expansion that takes place when the soils are allowed to absorb water is far greater (Fig. 4-58a). This is due to the fact that the minimum distance between particles is far less in the soil compacted on the dry side; the potential net repulsions are, therefore, much greater. Expansibility increases with the compactive energy.

It is difficult to decide whether to compact the soil on the dry side of optimum to reduce compressibility or on the wet side where the compressibility will be offset in many soils by less rigidity. Therefore the more compressible, wet, soil is better able to adapt to differential settlement, particularly where there are great differences in fill height over small distances. Soils compacted on the dry side are usually more brittle and crack more readily. Figure 4-59 [28] shows the compressibility curves obtained for two samples of a clayey sand; one compacted wet of optimum and the other dry of optimum. This confirms the effect of moisture on compressibility.

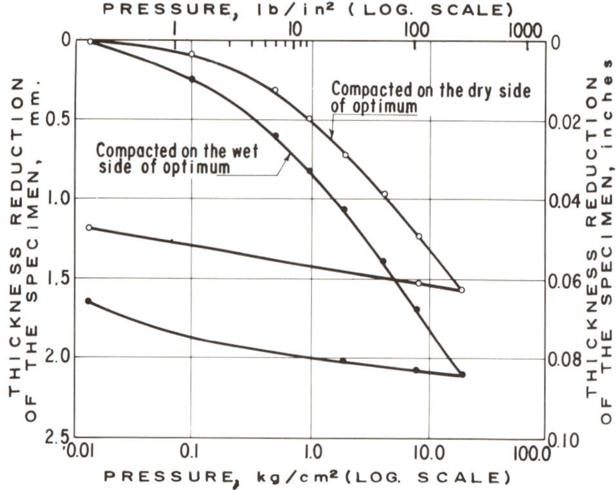

Fig. 4-59 Effect of compaction moisture on the compressibility of a clayey sand [28]

Once a soil has been compacted as part of an earth fill, changes occur in the water content and saturation; these produce changes in volume and stress that may or may not be transitory and about which little quantitative information is available. These changes are very hard to reproduce in the laboratory, even qualitatively.

The circumstances that influence these changes most are [3]: the increase in stress and strain due to additional layers of soil being placed on top, the change in water content produced by weather, and the corresponding compression or expansion (also depending on the compaction water content and confining pressures). In [65] Bishop and Henkel point out that some expansion should be expected in clayey soils compacted approximately at their optimum water contents and under loads of 10 m (32 ft 10 in) of earth fill. Finally, there is shrinkage caused by a

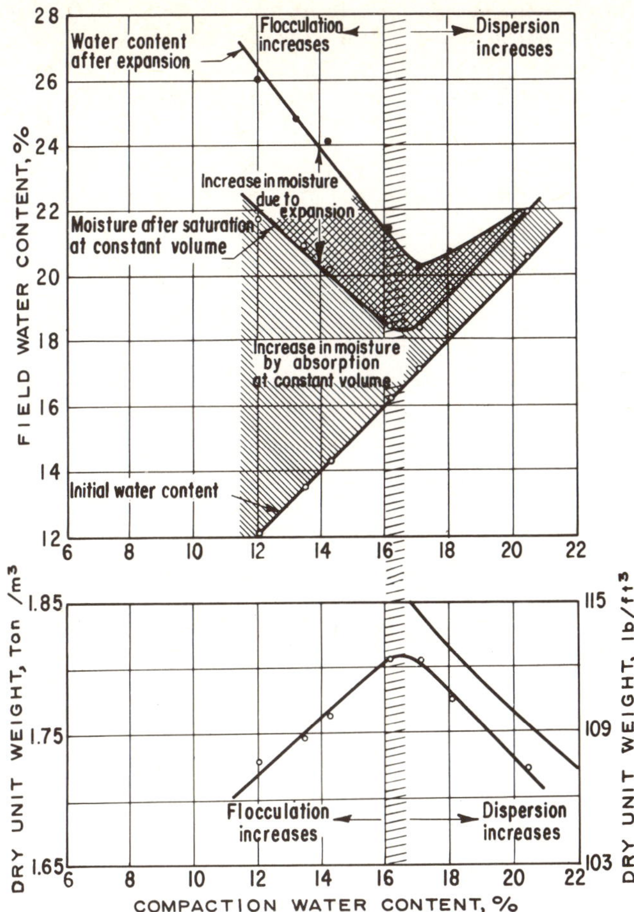

Fig. 4-60 Effect of compaction moisture and soil structure on the expansion characteristics of a clayey sand [3,60]

reduction in the water content during dry weather. If such is the case, Fig. 4-60 [3,60] shows the changes that can occur in the water content of a compacted clayey soil. Expansion is related to the degree of orientation of the structure of the clay. The greatest dry unit weight and minimum expansion is obtained when the soil is compacted near the optimum water content corresponding to the method and specific energy of the compactive effort used.

The expansion potential also varies in accordance with the method of compaction used. In Fig. 4-61 [3,60] expansion and shrinkage are given for a sandy clay subjected to both kneading and static compaction in the laboratory. Measurements were taken using samples with the same dry unit weight, on samples compacted both dry and wet of the optimum.

Figure 4-62 [3,28] shows that the expansion potential is also a function of the compactive or specific energy and increases with it. Data are presented for a clay subjected to laboratory static compaction under different pressures. Note that for this soil, expansion is proportional to the compaction pressure.

The expansion potential also depends a lot on the method of compaction. Generally speaking, it is higher with static load methods than with kneading methods; the higher the compactive effort used and dry unit weight obtained, the greater the difference.

Finally, expansion potential depends on which clay minerals are present, as was discussed in Chapter 1.

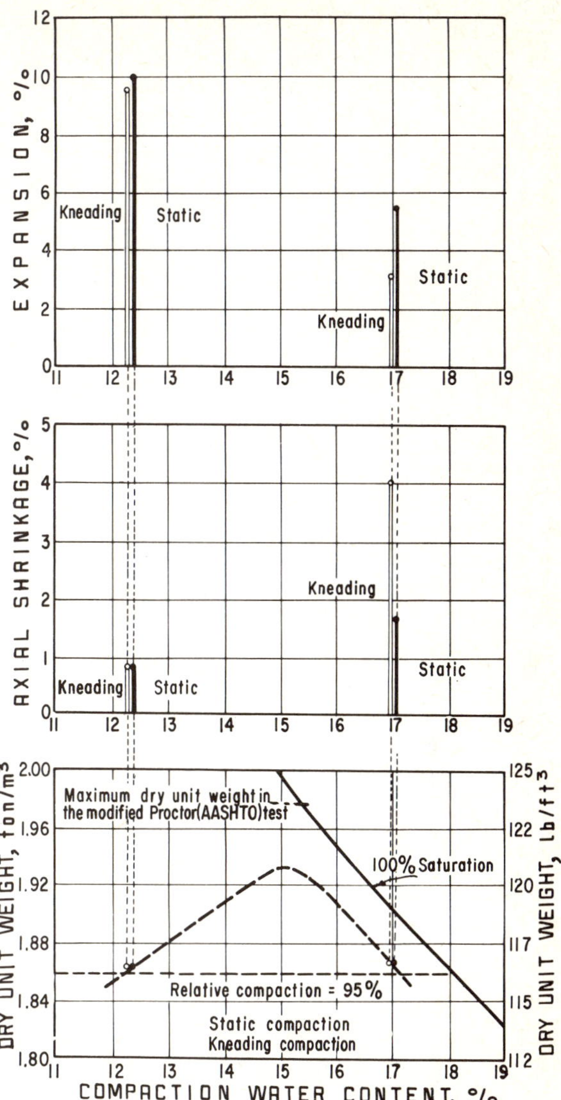

Fig. 4-61 Expansion and shrinkage of a sandy clay subjected to static or kneading compaction [3,60]

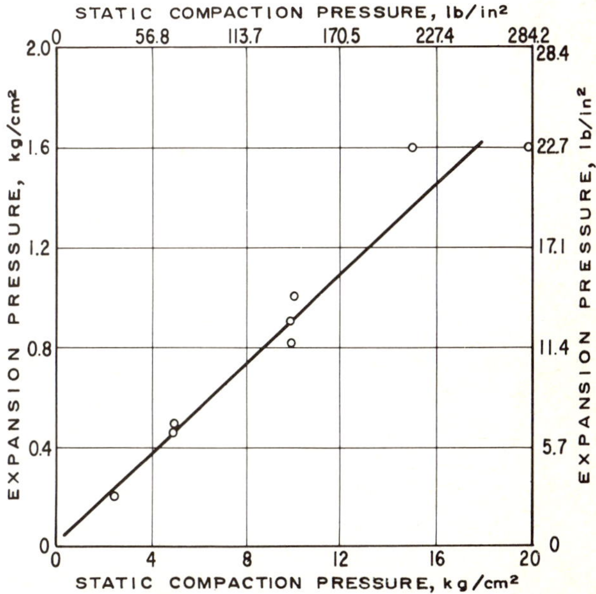

Fig. 4-62 Relationship between the static compaction pressure and the expansion pressure of a clay [3,28]

4.11.4 Shear Resistance

The shear resistance that clayey soil exhibits depends on the normal effective stress and the average value of the minimum distance between the particles. Resistance increases with a reduction in this average value, as was discussed in Chapter 1.

The minimum distance between particles depends on the void ratio and the degree of particle orientation. The normal effective stress that develops under a certain external stress depends on the water pressure that, in turn, is a function of the void ratio, degree of saturation and type of structure in the soil. Consequently these last three parameters govern the stress-strain and strength characteristics of compacted clays.

In compacted fine soils, negative water pressures usually develop once compaction is completed, as the soil tries to rebound. These negative pressures depend chiefly on the degree of saturation of the soil [63]. Laboratory research suggests that the stress state of water changes very rapidly immediately after compaction. A state of negative pore pressure or water tension is reached quickly, and is relatively constant thereafter so long as there is no change in external environmental conditions, such as rainfall or severe drought. The lower the initial degree of saturation, the higher are the tensions that develop in the water, and the greater its strength and the less its deformability.

It should also be taken into consideration that the greater the compactive effort, the higher will be the water tensions that develop in a compacted soil immediately after compaction. Experiments [63] give reason to believe that the higher the rate of strain to which a compacted soil is subjected, the lower is its resistance and also its deformability. This effect of the rate of strain becomes less noticeable at decreasing degrees of saturation; it is evidently due to the surface tension developed in the water.

As layers of fill are added above, the soil recompresses and the negative stress or water tension is reduced. If the degree of saturation is high, the compression can produce positive pore-water pressures, a loss of effective stress, and corresponding loss of strength. These effects are most serious in embankments higher than 30 m and sometimes have been responsible for sliding in the fill during construction.

.1 Behavior in Unconsolidated-Undrained Test (*UU* or *Q* Test)

Since the degree of saturation greatly influences the properties of compacted soils, the response of the same sample in a *Q* triaxial test will depend on whether it is tested with the degree of saturation it reaches when compacted, or whether it is saturated at a constant volume before testing. In the first case, strength is a function of the confining pressure in the chamber, for the compressibility of the air lets the void ratio decrease under this pressure. In the second case, the behavior of the soil is largely independent of the confining pressure (see Chapter 1).

Figures 4-63 and 4-64 [13] show the typical behavior of compacted clayey soils tested with the degree of saturation achieved during compaction. Both figures show lines of equal resistance to compression in a *Q* triaxial test. Numerous samples of the same soil were prepared, each with a different water content and different dry unit weight. Each sample was subjected to a *Q* test until failure occurred. The maximum deviator stress obtained is shown at the point defined by its water content-unit weight coordinates. Then equal resistance curves were drawn, which appear in the figures. The samples were tested as-compacted and using a confining pressure of 4 kg/cm² (57 lb/in²), sufficient to dissolve the air that was left in the sample

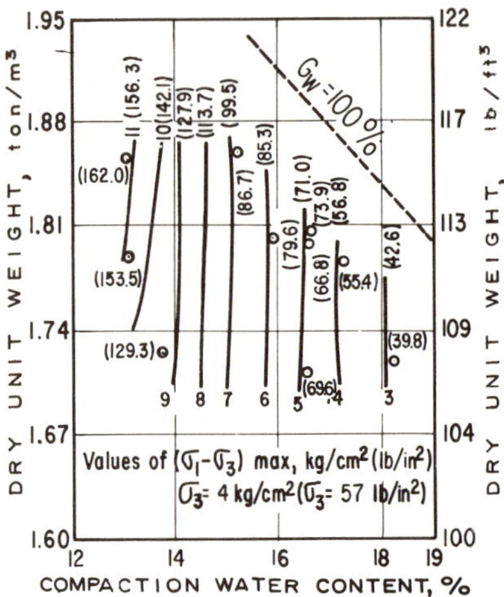

Fig. 4-63 Lines of equal resistance to compression in quick test, without pre-saturation and with confined compression of 4 kg/cm² (57 lb/in²) [13]

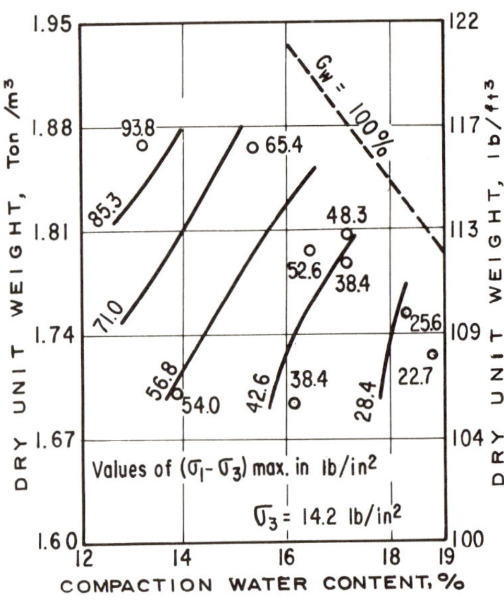

Fig. 4-64 Line of equal resistance to compression in quick test, without pre-saturation and with confined compression of 1 kg/cm² (14.2 lb/in²) [13]

after the compaction process. Note that strength decreases appreciably with an increase in the compaction water content and is practically independent of the dry unit weight. The reason for this relative independence of strength with the unit weight is that when all the air in the sample is dissolved as chamber pressure is applied, all samples having the same initial moisture reach new unit weights that are about the same. Therefore their strengths are the same for the same initial water content.

Figure 4-64 gives the strengths of the same soil but tested under a confining pressure of 1 kg/cm^2 (14.2 lb/in^2), at which the air in the sample is not completely dissolved. Now there is a loss of strength with an increase in compaction moisture, but also with a decrease in the as-compacted dry unit weight of the sample.

Figures 4-65 and 4-66 [13,60] show strength results for the silty clay referred to in the two previous figures; this is also a Q triaxial test, but with the specimen saturated before testing. Volume changes are not permitted during saturation, and the samples were compacted by kneading, using three different compactive energies.

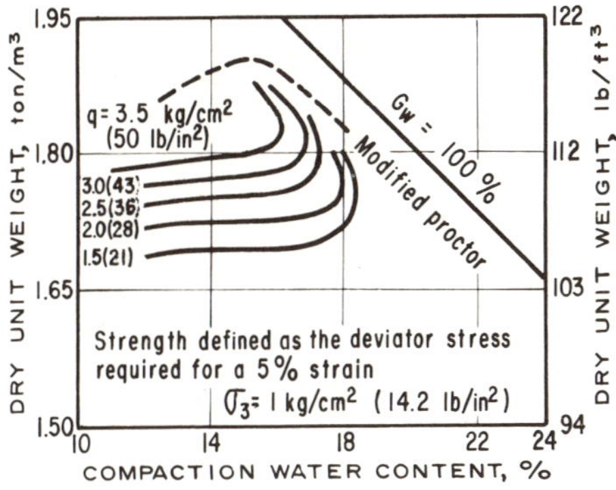

Fig. 4-65 Lines of equal resistance to compression in quick test, with saturation (resistance for small strain) [13,60]

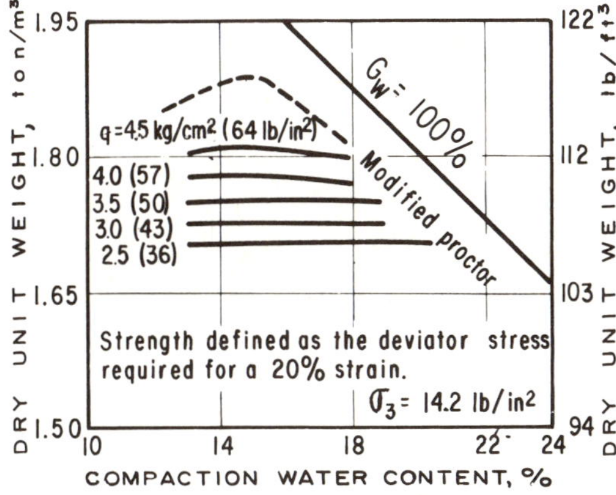

Fig. 4-66 Lines of equal resistance to compression in quick test, with saturation (resistance for large strain) [13,60]

In Fig. 4-65, strength is defined as the deviator stress which produces a strain of 5%. In this case, strength is greater at a lower compaction water content. This is because at higher water contents there is an increase in particle orientation as well as in the water pressure induced by the deviator stress; both factors cause lower strengths. At a constant dry unit weight, strength diminishes with increasing compaction water contents, particularly wet of optimum.

Figure 4-66 presents similar equal resistance curves, but now strength is defined as the stress which produces a strain of 20% in the specimen. These large strains produce practically identical grain structures in all the samples with the same dry density or void ratio. Therefore, the strengths are similar in all the specimens, independently of the compaction water content.

The behavior of the presaturated samples can also be appreciated by comparing stress-strain curves of specimens with the same void ratio, but one compacted dry of optimum and the other wet of optimum.

Figure 4-67 [13] shows two curves of this kind. It can be seen that the less oriented structure, dry of optimum, is stiffer. Also the strengths tend to approach the same limit at high strains because the grain orientations become similar.

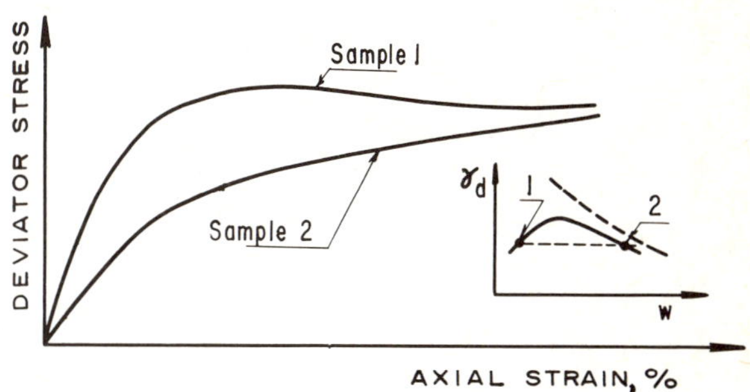

Fig. 4-67 Stress-strain relations in quick test (with pre-saturation at constant volume) of samples compacted at the same unit weight with different water contents [13]

Structure also has the following effect on strength. Samples compacted by different methods, but to the same dry unit weight with the same water content, have different strengths under the Q test conditions, especially when the compaction water content is above optimum and strength is defined at low strains. Nevertheless, in different soils the effects of the compaction method are very different as is illustrated by Fig. 4-68 [60]. This figure shows the difference obtained by static and kneading compaction on the strength of 3 different soils, defined by the stress which produces in one case a strain of 5%, and in the other 20%. At high strains the strengths are almost identical. At low strains the soil's susceptibility to structural alteration is less and it varies considerably between different soils. This may be due to the different interparticle forces which develop in the clayey fraction. When these are highly attractive, there is a tendency towards very flocculated structures, and when they are repulsive to dispersed structures.

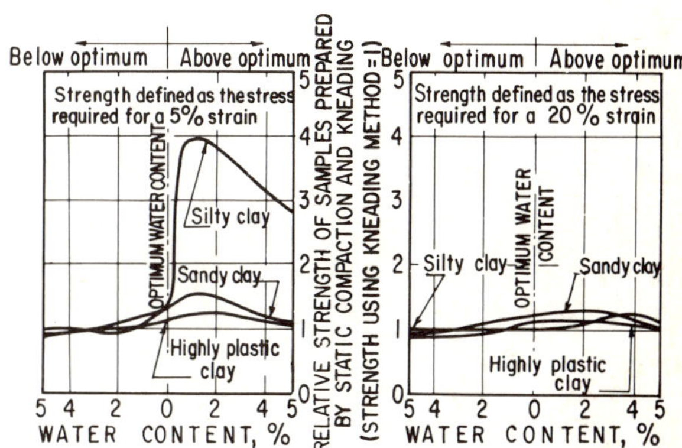

Fig. 4-68 Effects of compaction method on the resistance obtained in quick test (pre-saturation of samples compacted at same γ_d and w) using static and kneading methods [60]

The most important conclusion that can be drawn from the above is that the undrained strength of a fine grained clayey soil is not always linked with unit weight. Therefore it may prove dangerous to rely solely on the principle, *the higher the unit weight obtained, the better the soil.* Strength can be practically independent of the unit weight, and in these cases any attempt to improve the unit weight will prove useless (or even detrimental). There are some situations where an increase in unit weight produces improvement of strength. Further, even though higher unit weights produce an increase in undrained strength, it should be noted that when degrees of saturation approach 100%, the rate of strength increase becomes less and sometimes the strength even decreases.

The fundamental importance of the compaction method on the undrained strength is also evident. Note in Fig. 4-68 where a soil is compacted by a static load method until it reaches a certain unit weight and at a specific water content, the strength obtained may (depending on the soils) be several times greater than if the soil is compacted by kneading to the same unit weight and with the same water content.

.2 Behavior in Consolidated-Undrained Test (*R*)

As far as the stability of earth fills is concerned, this strength is only of interest in the case of conditions of presaturation in the compacted soil, and is seldom of great importance in road engineering problems.

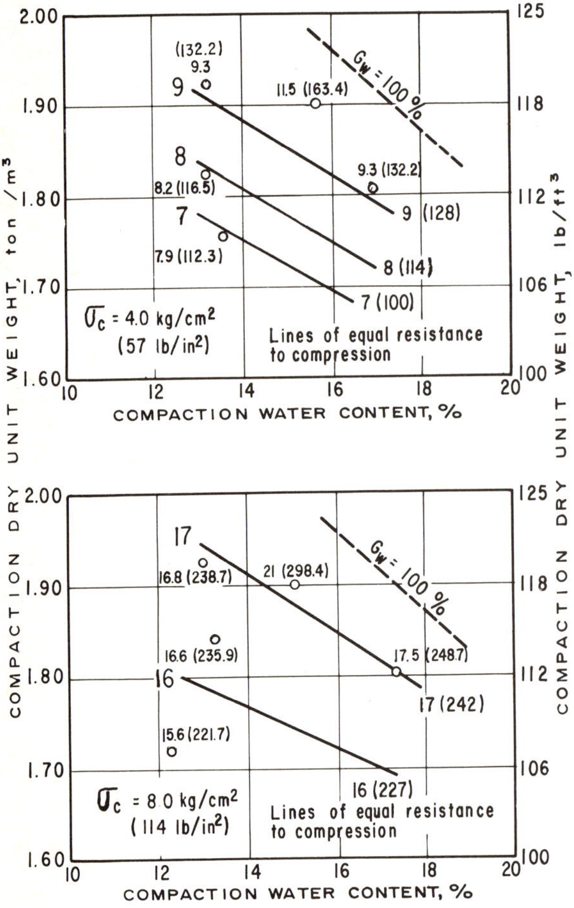

Fig. 4-69 Effects of compaction conditions and effective pressure of the triaxial test on the drained strength, without pre-saturation [13,67]

Some investigations show that with the same compaction water content, strength increases with unit weight. The higher the unit weight obtained during compaction, the greater will the density be after consolidation, when the deviator stress is applied.

At the same dry unit weight from compaction, strength increases with compaction water content; the higher this content the more compressible is the soil and the higher the unit weight obtained after consolidation and immediately before the deviator stress is applied. This usually produces higher strengths.

.3 Behavior in Consolidated-Drained Test (*S*)

Here too, for practical purposes, strength depends on the conditions of saturation. The little experience available indicates that for consolidation pressures greater than 1 kg/cm^2 (14.2 lb/in^2), the *S* strength is similar with and without presaturation.

Fig. 4-69 [13,67] shows the combined effects of the compaction conditions and confining pressure on the drained strength of a compacted clayey soil. The tests were carried out without presaturation.

In a drained test, at a constant compaction water content, strength increases with the dry unit weight, because of the smaller interparticle spacing achieved at higher densities. At a constant dry unit weight soil compressibility increases with the compaction water content. Therefore, the strength increases with increasing moisture at the same compacted dry unit weight. After consolidation the dry unit weight is higher with a corresponding increase in strength.

4.11.5 Resistance to Piping

The resistance to piping depends on the interlocking and bonding of the particles, which is determined by the geometry of the structure and magnitude of the interparticle electromagnetic forces. In clays there are usually particles that are so small that they can migrate through the pores if they become detached and suspended in water. Thus, for a clayey soil, the stronger the repulsive forces between the particles, the more susceptible it will be to piping. This is confirmed by analyzing the piping failures of many dams [13].

Compaction on the dry side of the optimum produces a low degree of orientation and high permeability. If, in such a situation, water containing a low concentration of salts flows through the soil, the repulsive forces between the particles will increase. Consequently some particles will be carried away. If compaction is carried out on the wet side, with above-optimum water contents, the opposite effect will be true and there will be less susceptibility to piping.

The erosion of clay particles can obviously not be stopped with filters. Therefore, the susceptibility of a clay to erosion should be tested by a dispersion test or pinhole erosion test [73].

4.11.6 California Bearing Ratio (*CBR*)

The California Bearing Ratio test is still widely used as a basis for pavement design. In some cases it also serves as a quality control test. The variation of the CBR under different compaction conditions is, therefore, important. In the chapter dealing with flexible pavements this will be discussed at greater length.

Figure 4-70 [68] shows how the *CBR* value of a silty clay varies according to compaction conditions; naturally, the *CBR* value depends not only on the water content, but also on the dry unit

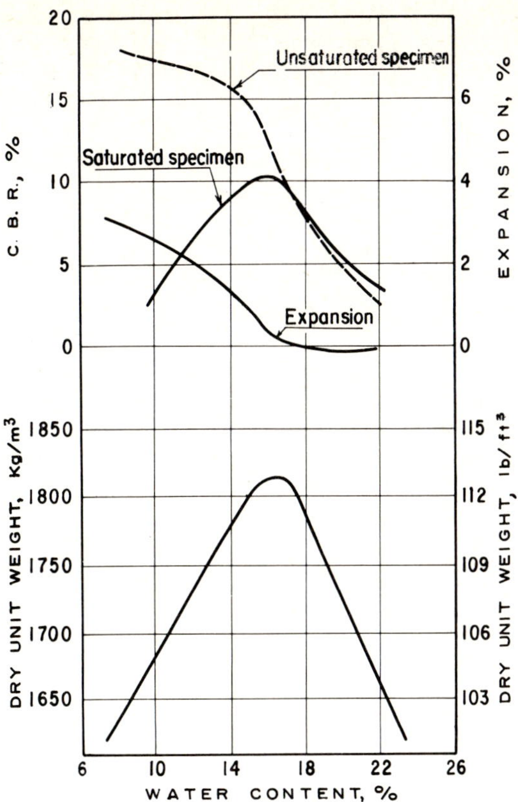

Fig. 4-70 Variation of the *CBR* of a silty clay with compaction [68]

weight that is reached. Curves are given for specimens that were tested at compaction water content and for specimens saturated after four days of inundation in water in the laboratory.

For the specimens tested after saturation, a curve similar to the compaction curve is obtained, because of the water absorbed and expansion undergone by the specimen during saturation. The figure also shows the expansion suffered by the specimens as a function of the water content at which they were compacted. Those samples compacted wet of optimum did not expand or

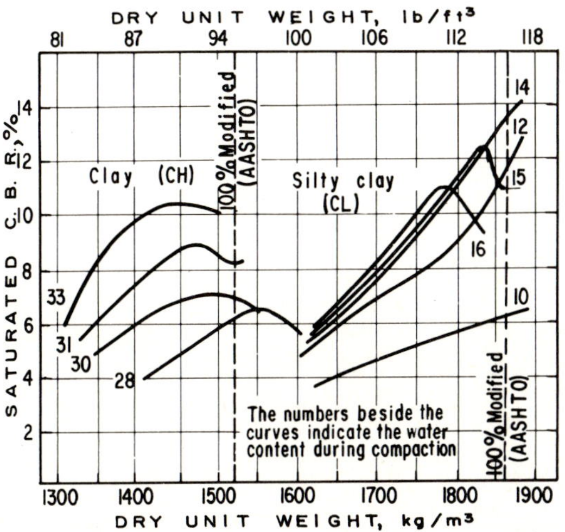

Fig. 4-71 Variation of the *CBR* with the dry unit weight of different soils [68]

change *CBR* significantly after inundation. This again illustrates the advisability of compacting expansive soils on the wet side, regardless of whether a lower *CBR* value is reached under the circumstances.

Figure 4-71 [68] illustates the variation of the saturated *CBR* of two soils (a *CH* and a *CL*) in relation to the dry unit weight. In both cases, the specimens tested were saturated in the laboratory for four days, after compaction at the water contents shown. In the *CH* material, the saturated *CBR* increases with an increase in compaction water content, for the same compacted dry unit weight. The *CBR* also increases if the dry unit weight increases. There is a limit to this increase beyond which the *CBR* decreases with increasing unit weight. This phenomenon is the result of an increment in the water pressure within the soil when it is compacted beyond a certain limit. The *CL* soil shows similar tendencies. The curves in the figure correspond to dynamic laboratory compaction, and might be different for static-load or kneading compaction methods.

4.11.7 Effects of Time

Investigations on this subject have shown that time after compaction, has a significant effect on the strength of compacted clays. References [60,69-72] deal with this aspect of the properties of compacted soils in considerable detail, although it has been almost completely ignored by embankment designers.

A typical example of the way in which time can affect strength is shown in Fig. 4-72 [60]. A silty clay with thixotropical properties was tested until failure in a simple unconfined compression test at load application rates which varied from 5 minutes to 10 days per test. There were two series of tests: in one case they were performed immediately after compaction and in the other after a storage period of 18 days at a constant water content and unit weight. Strength was defined as the deviator stress required to cause a strain of 10%.

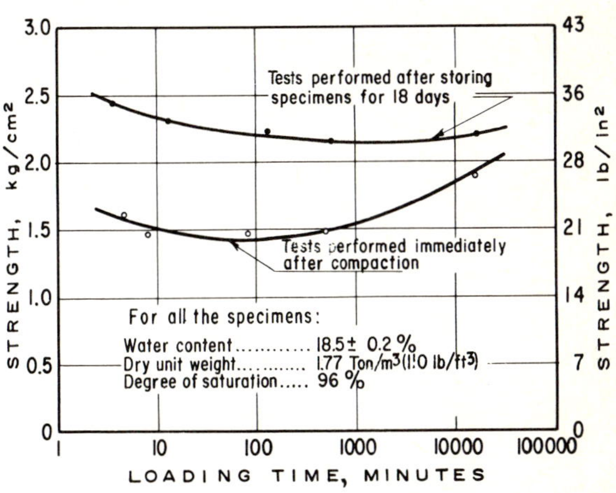

Fig. 4-72 Effects of time on the strength of a silty clay with high degree of saturation [60]

In the specimens that were tested immediately after compaction, there was a slight loss of strength between loading times of 5 and 100 minutes and a continuous increase for tests lasting more than 100 minutes, so that strength was 30% higher in a 10 day long test than in the standard test which lasts about 10 minutes.

In the specimens that were tested after 18 days' storage, there was a slight loss of strength for the loading times lasting up to a day, and afterwards a slight increase for up to 10 days, but in this case the differences produced were not more than 4% of the standard strength.

The considerable increase in strength of the specimens on which long tests were carried out immediately after compaction is probably due to the thixotropy, which occurs rapidly. In the samples stored for days, all the strengths were higher than for those tested immediately, but the slower tests did not gain as much strength as in those tested immediately. The thixotropic strength gain occurred before testing; therefore less took place during testing.

Most clayey soils become stronger with time. The data presented in [60] for the same silty clay tested after 9 months' storage, show that the strength continued to increase, reaching values of 3 kg/cm^2 (42.6 lb/in^2) or 20 to 50% stronger than for the samples stored 18 days and up to twice the strength of the samples tested immediately.

It should be noted that these strengths are in terms of total stresses. The associated effective stresses are unknown, and could be a subject for future research.

APPENDIX 4A

DYNAMIC TESTS PROCTOR TYPE

4a.1 Standard (AASHTO) Proctor Test [35] (ASTM D-698 is identical)

.1 Test Objective

The objective of the test is to determine the relationship between the dry unit weight and water content of soils compacted by the standardized methods described below.

There are four test alternatives:
Method A: In 10.16 cm (4 in) mold, with soil passing Sieve No. 4.
Method B: In 15.24 cm (6 in) mold, with soil passing Sieve No. 4.
Method C: In 10.16 cm (4 in) mold, with soil passing 3/4 in sieve.
Method D: In 15.24 cm (6 in) mold, with soil passing 3/4 in sieve.
When the method is not specified, Method A will automatically be used.

.2 Apparatus and Supplies (Plate 4a-1)

For this test, the following equipment is necessary:
- A standard compaction mold with collar. For the 4 in mold a volume of 1/30 ft^3 is established, with a tolerance of $\pm$ 0.0003 ft^3, and for the 6 in one a volume of 1/13.33 ft^3 with $\pm$ 0.00075 ft^3 tolerance.
- Hammer, 5.08 cm (2.0 $\pm$ 0.005 in) in diameter, and weighing 2.49 kg (5.5 $\pm$ 0.02 lb) and means for controlling its drop to 30.5 cm (12 in)
- Balance with a capacity of 15 kg and sensitivity of 5 g.
- Laboratory balance with a capacity of 200 g and sensitivity of 0.1 g.
- Drying oven
- Straight edge and knife
- 2 in, 3/4 in and No. 4 sieves
- Various items of supplementary equipment such as spatulas, watch glasses, and sample containers.

.3 Test Procedure

Method A

1) If necessary, dry the sample when received in the laboratory until it can easily be handled. Drying can be by air or by oven with a temperature not exceeding about 60°C.* Afterwards, break all soil lumps without breaking the particles.
2) Sieve the soil using No. 4 mesh and eliminate the intact particles that are retained.
3) Select a representative sample of about 3 kg.
4) Add to the sample sufficient water to place it 4 to 6% below the estimated optimum water content.

Plate 4a-1 Equipment for standard (AASHTO) Proctor test

1. Compaction mold
2. and 3. Tampers with means for controlling drop
4. Straight-edge
5. Glass jar
6. Drying cans
7. Balances
8. Spatula
9. Large pan and scoop
10. Metal can
11. Rubber-tipped pestle
12. Solid

5) Divide the sample into three portions according to the layers in which it is to be deposited in the mold with a diameter of 10.16 cm (4 in). Install the collar of the mold; the total thickness of the three layers of sample when compacted will be about 13 cm (5-1/8 in). Compact each layer with 25 blows from the hammer, distributing them evenly and allowing the hammer to fall from a height of 30.5 cm (12 in). During the operation, the mold will have to stand on a rigid base such as a concrete floor or a concrete block weighing 100 kg. After compaction, remove the collar from the mold and level off the compacted soil using the straight-edge. Weigh the mold with soil and subtract the tare weight of the mold in order to obtain the wet weight of the compacted soil. Divide by the volume of the mold to obtain the unit weight of the soil mass (γ_m)
6) Remove the material from the mold, without allowing it to crumble, and divide the specimen into two portions by a vertical cut through the center of the cross-section. Take a representative sample from one of the cut faces (about 100 g) and determine the water content of the soil.
7) Crumble the remaining soil material until it once again passes through sieve No. 4. Add sufficient water to increase the water content by 1 or 2% and repeat the whole procedure. Repeat the procedures until the wet weight of the compacted soil decreases or ceases to change. This procedure works satisfactorily in many cases. If there is a change in the particle size distribution because of recompaction, or in highly plastic clays in which it is very difficult to incorporate water, re-use of the material should be avoided and a fresh uncompacted sample prepared for each compaction. The water content should vary from one specimen to another by approximately 2%. The samples should be placed in closed containers after adding moisture where they remain for 12 hours before testing.

Method B

The sample is selected as in Method *A*, but about 7 kg is required. Test procedure will be the same as for Method *A*, except that a 15.24 cm (6 in) mold with collar is used and the soil will be placed in 3 identical layers, so that a total sample thickness of about 13 cm (5-1/8 in) is reached when compacted. The surface of each layer will receive 56 evenly distributed blows from a height of 30.5 cm (12 in).

Method C

If the soil sample received in the laboratory has excessive moisture, dry until it crumbles and can be easily handled. Drying can be by air or in an oven at a temperature not exceeding about 60°C (140°F).* Afterwards, break all the material lumps and sieve using a 3/4 in mesh, eliminating the material that is retained. If it is desirable to maintain in the sample the same percentage of coarse material (material between the 2 in size and No. 2) as in the original field material, the particles which are retained by the 3/4 in sieve should be replaced as follows:

Sieve a sufficient quantity of soil using 2 in and 3/4 in meshes and eliminate the material retained by the 2 in size. Withdraw the material which passed through the 2 in sieve and was retained by the 3/4 in sieve, and replace it by an equal weight of material passing through the 3/4 in sieve, but retained by the No. 4 sieve. Take the replacement material from a portion of the original field sample that is not going to be used. For Method *C*, a sample of soil weighing about 5 kg will be needed.

Test procedure:
1) Mix the soil with sufficient water to produce a moisture content 4 to 6% below the estimated optimum.
2) Make a specimen by compacting the soil in a 10.16 cm (4 in) mold in 3 equal layers, until a total compacted thickness of about 13 cm (5-1/8 in) is obtained. Follow the compaction procedure described for Method *A* until the wet unit weight and water content of the sample are determined.
3) Break up the remaining material until it all passes through the 3/4 in sieve and 90% through No. 4. The criterion for this can be established visually. Add to the sample the water it needs in order to increase the water content one or two %, and repeat the whole procedure of the test to obtain another point in the compaction curve. Repeat the procedure enough times until the wet unit weight of the soil shows no change or drops. The same comments apply as for Method *A* as regards re-use.

* For clays containing halloysite, the drying temperature should not exceed the air temperature of the construction site and the moisture should not drop below the lowest moisture obtainable at the site.

Method D

The sample is prepared in the same way for the other methods, but it is advisable to have about 12 kg (27 lb) of sample after preparation.

The test procedure is the same as for Method *C*, but a 15.24 cm (6 in) mold is used, with 3 layers and 56 blows per layer.

.4 Calculations

To complete the test, calculations must be made to determine the water content and dry unit weights required. Also, the compaction curve must be plotted and in it must be determined the maximum dry unit weight and optimum water contents.

4a.2 Modified (AASHTO) Proctor Test [40] (ASTM D-1557 is identical)

The test is identical to the Standard (AASHTO) Proctor Test except for hammer weight, drop, and number of layers (Plate 4a-2). Figure 4a-1 shows a drawing of the setup.

There are 4 variants (*A, B, C* and *D*) and their descriptions are identical to those of the corresponding AASHTO Standard variants. In the modified test, the greatest compactive effort is achieved with the weight of the hammer, which is now 4.530 kg (10 ± 0.02 lb) and is dropped from a height of 45.71 cm (18 in ± 1/16 in).

In Method *A* the soil is placed in 5 layers and each receives 25 blows. In *B* there are 5 layers too and each receives 56 blows. In *C* there are 5 layers and 25 blows per layer, and finally, in *D* there are 5 layers and 56 blows per layer.

Recently (1970) the same organization, AASHTO, established an intermediate test, which is also dynamic, with an energy between the standard and modified tests (about 17 kg-cm/cm^3) (240 lb-in/in^3). There are 4 variants of this test also, *A* and *C*

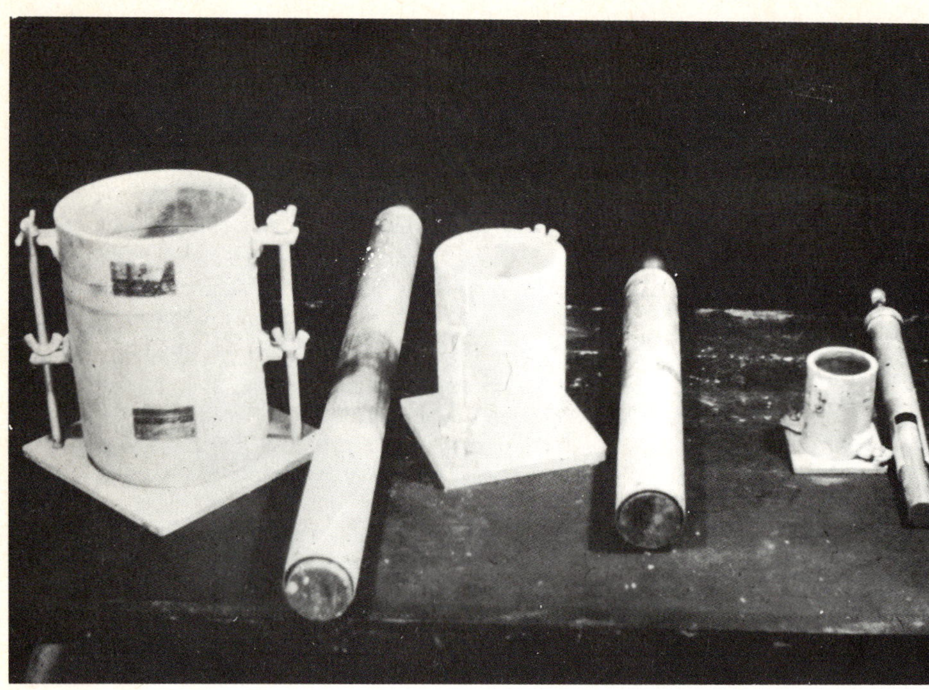

Plate 4a-2 Molds and hammers for use in modified and standard AASHTO tests, together with the miniature test apparatus

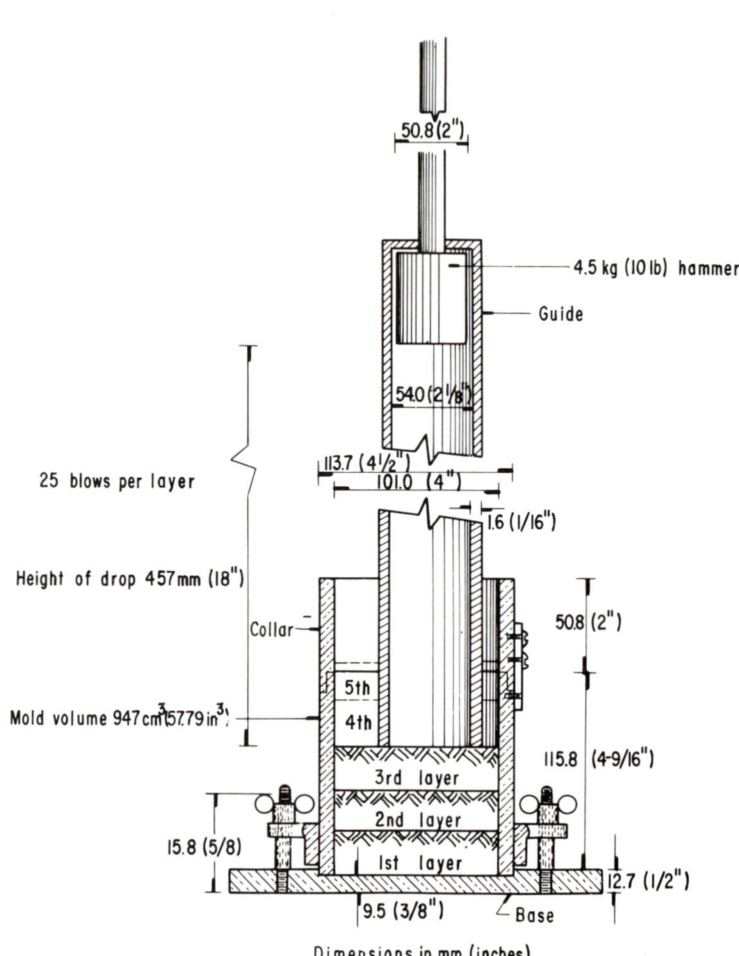

Fig. 4a-1 Dynamic compaction (Modified AASHTO)

with a 10.16 cm (4 in) mold, and *B* and *D* with a 15.24 cm (6 in) mold. In versions *A* and *C* there are 3 layers and 25 blows per layer from a 4.54 kg (10 lb) hammer which is dropped from a height of 45.7 cm (18 in). In versions *B* and *D* there are also 3 layers, but the number of blows per layer is 56.

This new test has been created in an attempt to reproduce more realistically in the laboratory the compaction conditions that actually occur in the field.

APPENDIX 4B

DYNAMIC COMPACTION TEST: CALIFORNIA METHOD

This test method makes it possible to determine the maximum dry unit weight and optimum water content for unstabilized soils or ones that have been stabilized with non-asphalt products, which are used for the construction of earth fills. The method consists of splitting an initial sample into smaller portions from which test specimens are prepared with different water contents and compacted by means of blows in order to determine the maximum dry unit weight and optimum water content. This method has the advantage that, thanks to a corrective factor, it takes into consideration the sizes larger than 3/4 in contained by the materials to which it is applied.

4b.1 Method *A*

.1 Equipment

The equipment required for this test is the following:

- Standard California-type dynamic compaction apparatus, consisting of a cylindrical mold, a hammer weighing 4.53 kg (10 lb) and the means for controlling its drop.
- A concrete base consisting of a 30 cm (1 ft) side cubic block.
- Balances with a minimum capacity of 3 kg and 1 gram sensitivity.
- An oven with a thermostat that will maintain a constant temperature between 105 and 110°C (221 and 230°F).
- A US Standard 3/4 in sieve.
- Large pans, scoops, spatulas, and sample containers for weighing.

.2 Sample Preparation

Preparation of the sample is as follows:

1) With the product of the sampling that is carried out to determine the unit weight at the location, make a sample of soil weighing between 15 and 20 kg (30 and 45 lb), completing it with material obtained from the walls of the boring.
2) Break up the lumps by hand and sieve using 3/4 in size; dry the particles retained until the weight is constant and determine the dry weight, W_s.
3) Determine the relative density S_s of the fraction retained by the 3/4 in sieve.
4) Split the material passing through the 3/4 in sieve into four or five representative portions with equal weights. Each portion or test sample must be a sufficient quantity for specimens to be obtained between 25.4 and 30.48 cm (10 and 12 in) in height, compacted in the standard mold. For each specimen approximately 2.7 kg of moist soil is required. If necessary, this weight can be found more accurately by making up a preliminary specimen.

.3 Test Procedure

Test procedure is as follows:

1) Adjust the moisture in the different portions, so that from one sample to the next there is an increase in water content of approximately 2%. In order to obtain these intervals, water will have to be added (or removed by means of drying) but the two operations must never be performed on the same portion, nor must test portions be dried out completely. When choosing moisture percentages for the test, it is best to have two portions with below-optimum and two with above-optimum water contents. The approximate test optimum is usually the maximum moisture at which the soil, squeezed in the palm of the hand, will not stick to the hand nor leave the hand moist. The same compressed material can be taken between two fingers without disintegrating. Once the quantity of water required by each portion has been added, mix thoroughly and cover with canvas or place in sealed containers to minimize moisture loss.

2) Divide one of the test samples into five almost equal parts, either by weight or volume. Place one of these parts in a test mold and compact with twenty blows from the hammer, which should fall freely from a height of 45.7 cm (18 in) measured from the surface of the material being compacted. Repeat the operation with each of the four remaining parts. After compacting the fifth, place the hammer guide and piston in the mold and level off the upper face of the compacted specimen with five blows from the hammer with a free fall of 45.7 cm (18 in) measured from the upper surface of the piston. During the compaction operation, the mold should be stood either on the standard block of concrete or on something equally firm. If, when compaction of the specimen is completed, water is observed on the bottom of the mold, the compaction water content is above optimum. If, on the other hand, the bottom of the mold is dry or dusty, the water content is below optimum.

3) While the hammer is over the piston, read the graduated rod of the hammer at the point where it coincides with the edge of the mold and make a note of this value in centimeters, to the nearest tenth, in column *a* of the test report sheet in Table 4b-1.

4) Remove the specimen from the mold, taking care that no material is lost. Determine the wet weight, W_1, in kg, to the nearest gram and write this value down in column *c* of the record sheet.

5) Cut the specimen lengthwise to obtain a representative portion of approximately 1000 grams. Determine the weight W_m of this portion to the nearest gram, and write the value down in column *k* of the record sheet.

6) Dry the above-mentioned portion until a constant weight is reached and weigh to the nearest gram, writing down its value (W_s) in column *l* of the record sheet.

7) Repeat the procedure with the remaining test samples.

Table 4b-1
Test report sheet — California Method

DYNAMIC COMPACTION TEST; CALIFORNIA METHOD

PROJECT______ LOCATION______ BORING______

SAMPLE______ DEPTH______ DATE______ SAMPLING______

TEST					DATA					
TAMPER READING	FACTOR	WEIGHT OF SPECIMEN (kg)		WET UNIT WEIGHT	DRY UNIT WEIGHT	WATER CONTENT %	RELATIVE DENSITY	CORRECTED RELATIVE DENSITY	CORRECTED UNIT WEIGHT
		MOIST	DRY						
a	b	c	d	e	f	g	h	i	j

WATER CONTENT			
WET WEIGHT IN GRAMS	DRY WEIGHT IN GRAMS	WATER LOST IN GRAMS	WATER CONTENT %
k	l	m	n

GRADING AND RELATIVE DENSITY		
W	OVERALL WEIGHT OF THE SAMPLE, IN GRAMS	
X	WEIGHT OF THE MATERIAL > 3/4 in	
Y	WEIGHT OF THE MATERIAL < 3/4 in	
Z	RELATIVE DENSITY OF MATERIAL>3/4 in	
R	COEFFICIENT	

CALCULATIONS:

$f = b\,d \qquad g = \dfrac{c-d}{d} \qquad i = \dfrac{100}{\dfrac{\%X}{Z} + \dfrac{\%Y}{Rh}}$

$h = f/1000$ kg/m^3 ($f/62.4$ lb/ft^3)

$j = 1000\,i$ kg/m^3 ($62.4\,i$ lb/ft^3)

DESCRIPTION OF THE SAMPLE______

OPERATOR______

DATE OF THE TEST______

.4 Calculations

In the case of each specimen, calculate and make a note of the following:

1) The water content, by means of the equation given below, making a note of its value in columns *n* and *g* of the record sheet.

$$w = \frac{W_m - W_s}{W_s} 100$$

Where

w = water content, %.
W_m = weight of the fraction of moist soil, in grams.
W_s = weight of the fraction of dry soil, in grams.

2) The dry weight, by means of the following formula, making a note of its value in column *d* of the record sheet.

$$W_2 = \frac{100 W_1}{100 + w}$$

Where

W_2 = dry weight of the specimen, in kg.
W_1 = wet weight of the specimen, in kg.
w = water content, percentage.

3) The dry unit weight, using the following formula and writing down its value in column *f* of the record sheet.

$$\gamma_d = W_2 C$$

Where

γ_d = dry unit weight of the specimen, in kg/m^3 (lb/ft^3).
W_2 = dry weight of the specimen, in kg.
C = factor obtained from Table 4b-2, which corresponds to the shaft reading.

Table 4b-2
C factor for the calculation of unit weights

Shaft reading		C factor		Shaft reading		C factor	
cm	in	for t/m³	for lb/ft³	cm	in	for t/m³	for lb/ft³
25.4	10.00	.9479	59.18	28.0	11.02	.8595	53.66
.5	.04	.9438	58.92	.1	.06	.8567	53.48
.6	.08	.9398	58.67	.2	.10	.8539	53.31
.7	.12	.9361	58.44	.3	.14	.8506	53.10
.8	.16	.9326	58.22	.4	.18	.8474	52.90
.9	.20	.9292	58.01	.5	.22	.8443	52.71
26.0	.24	.9252	57.76	.6	.26	.8414	52.53
.1	.28	.9215	57.53	.7	.30	.8389	52.37
.2	.32	.9180	57.31	.8	.34	.8355	52.16
.3	.35	.9146	57.10	.9	.38	.8325	51.97
.4	.39	.9114	56.90	29.0	.42	.8296	51.79
.5	.43	.9076	56.66	.1	.46	.8269	51.62
.6	.47	.9039	56.43	.2	.50	.8243	51.46
.7	.51	.9004	56.21	.3	.54	.8211	51.26
.8	.55	.8972	56.01	.4	.57	.8181	51.07
.9	.59	.8941	55.82	.5	.61	.8152	50.89
27.0	.63	.8905	55.59	.6	.65	.8126	50.73
.1	.67	.8869	55.37	.7	.69	.8102	50.58
.2	.71	.8836	55.16	.8	.73	.8072	50.39
.3	.75	.8805	54.97	.9	.77	.8043	50.21
.4	.79	.8776	54.79	30.0	.81	.8016	50.04
.5	.83	.8748	54.61	.1	.85	.7990	49.88
.6	.87	.8720	54.44	.2	.89	.7966	49.73
.7	.91	.8696	54.29	.3	.93	.7942	49.58
.8	.94	.8661	54.07	.4	.97	.7919	49.44
27.9	10.98	.8627	53.86	30.5	12.00	.7899	49.31

To obtain values of C for units of kg/m³, multiply by 1000

4) When the soil sample contains more than 10% in weight of particles larger than 3/4 in, obtain the corrected maximum dry unit weight by means of the following equations:

$$(\gamma_{dm})_c = \frac{100}{\dfrac{X}{S_s} + \dfrac{Y}{R\gamma_{dm}/1000}} (1000) \_\_\_\_\_\_ \text{kg/m}^3$$

$$(\gamma_{dm})_c = \frac{100}{\dfrac{X}{S_s} + \dfrac{Y}{R\gamma_{dm}/62.4}} (62.4) \_\_\_\_\_\_ \text{lb/ft}^3$$

Where

$(\gamma_{dm})_c$ = corrected maximum dry unit weight of the specimen, in kg/m³ (lb/ft³)
X = material retained by the 3/4 in sieve, %
Y = material that passes the 3/4 in sieve, %
S_s = relative density of the material retained by the 3/4 in sieve.
γ_{dm} = maximum dry unit weight of the specimen, in kg/m³ (lb/ft³).
R = coefficient, the value of which is given according to the values of O, in Table 4b-3.

.5 How to Obtain the Compaction Curve

Obtain the dry unit weight-water content curve as follows:

1) In a system of coordinate axes, plot the point corresponding to each specimen, taking as ordinate the dry unit weight and as abscissa the corresponding water content.

2) Join the dots corresponding to each of the specimens to form a curve. The maximum of the curve represents the maximum dry unit weight, and the corresponding optimum water content.

.6 Precautions to be Taken During the Test

When performing this test, the following precautions should be taken:

1) Do not use material that has been subjected to any laboratory compaction procedure.

2) The sample of soil that is used to determine the water content must always be obtained by cutting the specimen lengthwise, because when some soils are subjected to dynamic compaction, the moisture tends to concentrate in the lower part of the specimen.

3) The layers that are compacted to form the specimen must be practically identical so as to ensure uniform compaction.

Table 4b-3
Values of the coefficient R

X, in percentage	R
20 or less	1.00
21–25	0.99
26–30	0.98
31–35	0.97
36–40	0.96
41–45	0.95
46–50	0.94
51–55	0.92
56–60	0.89
61–65	0.86
66–70	0.83

4) The wing nuts must not be tightened with a spanner or the section of the mold will lose its shape. The spanner should only be used to loosen these nuts should they become tight on account of the expansion of soil within the mold.

4b.2 Method *B*

This version of the method is used to determine the maximum wet unit weight of soils where the fraction retained by the 3/4 in sieve is less than 10% in weight.

1) The equipment used is the same as for Method *A* of the test.
2) The sample is prepared in the usual way, except that the fraction retained by the 3/4 in sieve must be eliminated, and consequently the relative density (S_s) is not determined.
3) Test procedure is the same as for Method *A*.
4) In this method, calculate and record the following values:
   a) The water content, by means of the formula given below, writing down its value in columns *n* and *g* of the record sheet:

$$w = \frac{W_m - W_s}{W_s} 100$$

Where

w = water content, percentage.
W_m = weight of the fraction of moist soil.
W_s = weight of the fraction of dry soil.

   b) the wet unit weight, by means of the following formula, writing down its value in column *e* of the record sheet.

$$\gamma_h = W_1 C$$

Where

γ_h = wet unit weight of the speciment, in kg/m^3 (lb/ft^3).
W_1 = wet weight of the specimen, in kg.
C = corrective factor obtained from Table 4b.3, corresponding to the shaft reading.

5) Obtain the wet unit weight - water content curve as follows:
   a) In a system of coordinated axes, plot the point corresponding to each specimen, taking as ordinate the wet unit weight and as abscissa the corresponding water content.
   b) Join the dots corresponding to each of the specimens to form a curve. The maximum of the curve represents the maximum wet unit weight of the material, and the corresponding optimum water content.
6) During this test the same precautions should be taken as for Method *A*.

APPENDIX 4C

STATIC LOAD COMPACTION TESTS

4c.1 SOP Porter Test

.1 Test Objective

This test method is used to determine the maximum dry unit weight and optimum water content of soils containing coarse particles. It can also be performed on sands and fine materials with a plasticity index lower than 6. The method consists of preparing specimens from material which passes the 25.4 mm (1 in) sieve, adding different quantities of water and subjecting them to static load compaction.

.2 Equipment Required (Fig. 4c-1)

- A cylindrical compaction mold with an interior diameter of 15.24 cm (6 in) and a height of 22.86 cm (9 in) including the collar, on a base equipped with a mechanism for holding the cylinder in place.
- Compression machine with a minimum capacity of 30 t to the nearest 100 kg.

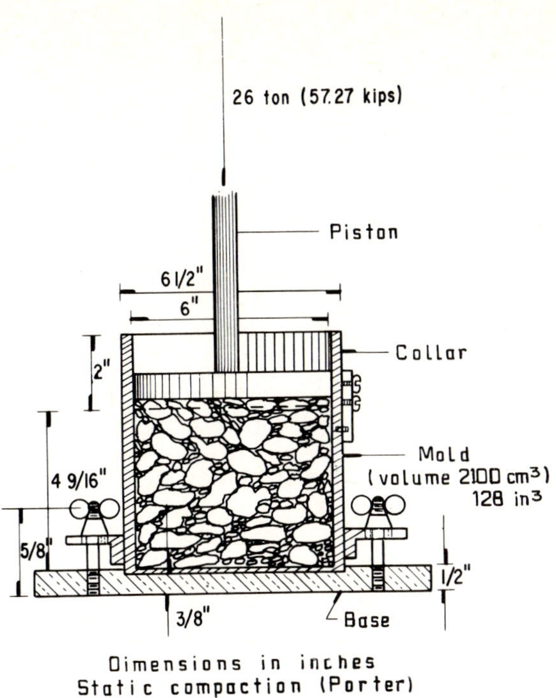

Fig. 4c-1 Static load compaction (PORTER)

Plate 4c-1 Static compaction test

- Metal rod 1.9 cm (3/4 in) in diameter and 30 cm (1 ft) long, with a bullet shaped end.
- Circular compacting plate, slightly smaller than the interior diameter of the cylinder, with a diameter of 15 cm (5.9 in) which can be fixed to the load-applying head.
- U.S. Standard sieve with square 25.4 mm (1 in) mesh.
- U.S. Standard sieve with square 4.76 mm (No. 4) mesh.
- Balances with a minimum capacity of 10 kg to the nearest gram
- Pans
- The usual accessory equipment.

.3 Sample Preparation

The preparation of the sample is as follows:

1) From a field sample which has been prepared taking care to dry the material only enough for the lumps to be more easily broken, take and sieve a sufficient quantity to obtain a 16 kg (35 lb) portion of material which passes the 25.4 mm (1 in) sieve.
2) Split this portion into four equal portions.

.4 Test Procedure

The test procedure is as follows (Plates 4c-1, 4c-2):

1) Take one of the representative portions of material and mix with it sufficient water so that, when squeezed, moisture is not left in the palm of the hand, and at the same time the compressed material will cohere in lumps. In some cases, to achieve uniform moisture distribution it will be necessary to store the moist material for several hours, in a sealed container or covered with a damp piece of canvas.

Plate 4c-2 Static compaction, note the water expelled from the specimen

2) In 3 different layers, place the moistened material in the mold; ram each layer 25 times with the tip of the rod in a uniformly distributed pattern.
3) When the last layer is in place, place the mold containing the material, in the compression machine. Compact the soil by slowly applying a uniform load, so that in five minutes a pressure of 140.6 kg/cm^2 (2 kips/in^2) is achieved, equivalent to a 26.5 t (60 kips) load, approximately. Maintain this load for one minute. When the maximum load is reached, examine the bottom of the mold; if it is slightly moist, the material will have reached optimum water content and its maximum dry unit weight.
4) If, when the maximum load is reached the bottom of the mold is not moist, the water content with which the sample was prepared is below optimum. Take, therefore, another representative portion of the material, add to it the same amount of water as in the previous specimen, plus 80 cm^3, (4.9 in^3) mix thoroughly and repeat the steps explained in paragraphs 2) and 3). Prepare as many specimens as necessary following the steps given in this paragraph, until with one of them the bottom of the mold begins to become moist at the maximum load (Plate 4c-2). This is usually achieved with less than four specimens.
5) If the bottom of the mold shows signs of moisture before the maximum load is reached, because water is expelled, the water content with which the sample was prepared is above optimum. In this case, proceed as explained in paragraph 4), but instead of adding 80 cm^3 (4.9 in^3) of water, reduce the amount in each new representative portion of material until in one of them moisture just begins to show on the bottom of the mold at the maximum load.
6) On completion of compaction of the specimen prepared with optimum moisture, take the mold out of the compression machine and determine the height (h_e) by subtracting the height from the upper face of the specimen to the upper edge of the mold, from the total height of the mold. Record this value in cm (in) to the nearest 1/10 mm (0.004 in).
7) Weigh the mold containing the compacted specimen and record this weight W_i in kg to the nearest gram.
8) Remove the specimen from the cylinder, cut it lengthwise and take from the center a representative sample; determine water content, and record its value (w).

.5 Calculations

In this test, calculate and record the following facts:

1) Using the specimen compacted at optimum water content, calculate:

a) The volume, by means of the following formula:

$$V = \frac{A_m h_e}{1000} \ \ell$$

$$V = \frac{A_m h_e}{1728} \ \text{ft}^3$$

Where

V = volume of the specimen in ℓ (ft^3)
h_e = height of the specimen, in cm (in).
A_m = area of the cross-section of the compaction cylinder, in cm^2 (in^2).

b) The wet unit weight, by means of the following formula:

$$\gamma_m = \frac{W_i - W_t}{V} \times 1000 \quad (\text{kg/m}^3)$$

$$\gamma_m = \frac{W_i - W_t}{V} \times 8 \quad (\text{lb/ft}^3)$$

Where

γ_m = wet unit weight of the specimen in kg/m^3 (lb/ft^3).
W_i = weight of the wet specimen plus the weight of the compaction mold, in kg (lb).
W_t = weight of the compaction mold, in kg (lb).
V = volume of the specimen, in ℓ (ft^3).

c) The maximum dry unit weight γ_{dm}, by means of the following formula:

$$\gamma_{dm} = \frac{\gamma_m}{100 + w} \times 100$$

Where

γ_{dm} = maximum dry unit weight of the specimen, in kg/m^3 (lb/ft^3).
γ_m = wet unit weight of the specimen, in kg/m^3 (lb/ft^3)
w = optimum water content of the specimen, %.

2) Record the maximum dry unit weight (γ_{dm}) and optimum water content as values corresponding to the material tested.

.6 Common Errors

The most common errors are the following:
— The water is not sufficiently well mixed with the material.
— The loading rate is not the one specified.
— The material is not sufficiently well mixed before it is placed in the test cylinder.

APPENDIX 4D

KNEADING COMPACTION TESTS

4d.1 Miniature Harvard Compaction Test

.1 Test Objective

The objective of this test method is to determine the maximum dry unit weight and optimum water content for fine plastic soils with particles smaller than 2 mm. Specimens of material that pass No. 10 sieve (US Bureau of Standards 2.00 mm) are prepared, and different amounts of water added to them. The specimens are placed in a metal mold and compacted by means of a piston which applies pressure that is transmitted by the action of a gaged spring.

.2 Equipment Required for the Test (Plates 4d-1, 4d-2)

For this test, the following equipment is used:
— A cylindrical metal compaction mold, with metal collar and base plate. The dimensions of the mold are 3.3 cm (1-5/16 in) interior diameter and 7.2 cm (2-13/16 in) height: its volume is 62 cm^3 (1/454 ft^3); the extension is 3.5 cm (1-3/8 in) high.

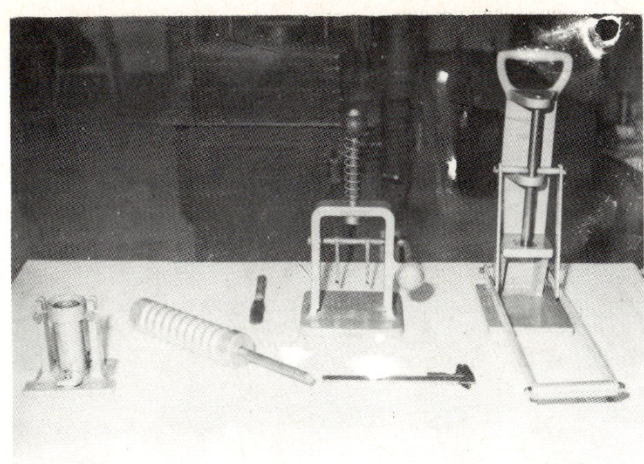

Plate 4d-1 Equipment required for the miniature Harvard test

Plate 4d-2 Miniature Harvard test, Specimen Press

— A metal tamper with a piston at the lower end, which can apply pressure through a spring (the pressure applied can be made to vary widely by using springs with different elasticity). The pressure-applying piston is a metal bar with a diameter of 1.3 cm (1/2 in), and a wooden handle; inside this handle is the spring.
— A mechanism for removing the mold collar, equipped with a foot which holds the soil in place during removal.
— An extractor for removing the compacted sample from the mold with minimum disturbance.
— Laboratory balances to the nearest 0.1 g.
— A straight-edge, an oven, a No. 10 sieve and other items of equipment such as spatulas, and sample containers.

.3 Sample Preparation

1) For this test, a duly quartered sample of material is required, weighing between 1 and 1.5 kg. It is oven-dried to make it easier to break up the lumps.

2) Once it has been broken up by hand the sample is sieved using No. 10 mesh.

3) Because the dry unit weight-water content curve must have 6 or 8 values, prepare in the containers 6 to 8 portions of soil each with the desired water content, and leave in a sealed container overnight; this enables the water to diffuse into fine soils. When working with soils that absorb water rapidly, such as silts, the soil and water can be mixed just before the test.

.4 Test Procedure

1) In a mold which is fixed to its base and equipped with an extension, place the required amount of loose soil.

2) The soil is placed in the mold in the number of layers that are desired (usually five): level off each layer with slight pressure from a rubber tamper.

3) After adjusting the tamper spring, insert the tamper piston in the soil and press until the spring begins to be compressed (Plate 4d-2). Withdraw the pressure, slightly changing the position of the piston and repeat the operation, applying the pressure evenly to each layer until completing as many applications as desired.

4) Repeat this procedure for each layer. Be sure that the top layer is at least 1 cm (3/8 in) above the edge of the mold (entering the metal extension).

5) Transfer the mold as a whole to the appliance for removing the collar. Press firmly the foot of this appliance while operating the extractor mechanism.

6) Remove the mold from its base and level off the soil at the upper case fully with a straight-edge. With the same straight-edge, check that the soil at the lower end of the mold is levelled.

7) Weigh the mold containing the compacted soil to the nearest 0.1 g. Find the soil weight by subtracting the tare weight of the mold.

8) Remove the sample from the mold using the extractor and place it in a container for oven drying and moisture content determination. If the material is used again for other points of the compaction curve, the water content will be determined using the excess material from the upper end of the mold.

9) Compact further specimens with increasing water contents, until the wet weight of the sample begins to decrease, which is an indication that one has gone beyond the optimum water content.

10) Calculate the dry unit weight corresponding to each water content by means of the following formula:

$$\gamma_d = 100 \frac{\gamma_m}{100 + w}$$

and plot the compaction curve so as to obtain the maximum dry unit weight and optimum water content.

11) If desired, change the compaction energy by varying the number of applications of the tamper per layer, the pressure applied or the number of layers.

4d.2 Hveem's Kneading Compaction Test

.1 Test Objective

To represent in the laboratory the kneading conditions produced by field compaction equipment. This procedure was developed by F. N. Hveem from the California Division of Highways (U.S.A.) with the intention of preparing specimens to measure stability in the method of pavement design he developed.

.2 Test Equipment

Apparatus

1) A mechanical kneading compactor.
2) Compactor accessories:
   - High resistance steel mold with an interior diameter of 10.16 cm (4 in ± 0.002 in) and an exterior diameter of 11.40 cm (4.49 in ± 0.005 in). The interior of the cylinder must be smooth, with a maximum roughness of 250 in. 10^{-6}. The height of the mold is 12.70 cm (5 in ± 0.008 in).
   - Mold extension with handle and funnel.
   - 50.8 cm (20 in) feeder and spatula.
   - Rubber discs 10.00 cm (3-15/16 in) in diameter and 0.32 cm (1/8 in) thick.
3) Basket making equipment:
   - Wooden cylinder 9.84 cm (3-7/8 in) diameter and a 1.27 cm (1/2 in) masking tape dispenser.
   - Phosphor-bronze perforated discs, for exudation pressure, of 10.08 cm (3-31/32 in) diameter.
4) Metal collar for shaping the stainless steel baskets.

Materials

- Discs made of paper 10.00 cm (3-15/16 in) in diameter.
- For making the baskets, strips of notched 60-lb brown paper 6.35 cm (2-1/2 in) wide by 34.29 cm (13-1/2 in) long, with 3.42 cm (1-7/8 in) slots, all 1.91 cm (3/4 in) apart.
- 1.27 cm (1/2 in) masking tape.

.3 Compaction Procedure

1) In paragraphs 2 to 10 a description is given of the normal procedure for making up specimens from fine soils and aggregates with sufficient natural cohesion for the specimens to remain intact during testing. For cohesionless materials, such as base aggregates, paper baskets are used so that they can be handled without suffering disturbance. When baskets are used, the special procedure is described in paragraphs 11 to 19.
2) Place the mold in the base with handle, which has a rubber disc 10.00 cm (3-15/16 in) in diameter and 0.32 cm (1/8 in) thick and is stuck to the plate. Adjust the mold in such a way that 0.32 cm (1/8 in) are left between the lower edge of the mold and the base of the mold which has a handle. Hold in place like this. Place a 10.00 cm (3-15/16 in) cardboard disc in the mold over the rubber disc. Fit the funnel extension and the mold on the revolving plate of the compactor and screw into place.
3) Place a well-mixed sample in the feeder; the material should be loose and evenly distributed throughout the extension.
4) Start the compactor and adjust the air pressure until it is 1.05 kg/cm^2 (15 lb/in^2), which is equivalent to a tamper pressure of approximately 16.85 kg/cm^2 (240 lb/in^2). Wait until the tamper is in its lowest position before placing the material in the mold.
5) With the spatula, gradually transfer the material from the feeder to the mold until the bottom is covered. The remainder of the sample will be emptied in 20 equal parts, one with each application of the tamper. Then apply the tamper ten more times so as to settle and level off all the material. Raise the tamper, clean it and place a 10 cm (3-15/16 in) rubber disc on the top of the specimen. If, during all the previous operations, the 1.05 kg/cm^2 (15 lb/in^2) pressure proved too great and caused the material to raise up around the tamper, it can be reduced.
6) Loosen the mold within the collar with the handle by means of the screws, lower the tamper and increase the air pressure until a tamper pressure of 24.6 kg/cm^2 (350 lb/in^2) is obtained, which is normally achieved when the manometer which measures the air pressure shows a reading of 1.48 kg/cm^2 (21 lb/in^2).
7) Lower compaction pressures may be required for clays, because they are easily penetrated by the tamper. In these cases, the factor to be watched is tamper penetration; care should be taken that it is not greater than 0.64 cm (1/4 in).
8) Apply the tamper to the specimen 100 times.
9) If before 100 applications free water appears on the bottom of the mold, interrupt the process immediately and make a note of the number of applications.
10) If the surface of the specimen is left rough after compaction, smooth it off.

Procedure when baskets are required

11) Make the baskets in accordance with the following steps.
    a) Take a piece of notched paper and place it round the cylindrical wooden block with the ends touching.
    b) With the masking tape stick the phosphor-bronze disc to the paper, in such a way that the holes in the disc are not covered.
12) Place the mold in the outer collar with the handle, after sticking a rubber disc to the plate and place a paper disc inside the mold over the rubber one.
13) Slip a basket into the mold in such a way that its upper edge stands out from the upper edge of the mold by about 2.54 cm (1 in). In the same way, place the metal collar in the basket until its lower edge is about 2.54 cm (1 in) below the upper edge of the basket. Now slide the basket and metal collar simultaneously until the phosphor-bronze perforated disc (stuck to the bottom of the basket) rests on the cardboard disc and the upper edge of the metal collar coincides with the upper edge of the mold.
14) Adjust the mold so that about 3 mm (1/8 in) are left free between the lower edge of the mold and the base of the outside mold with the handle. This is done by sliding the mold and the collar it contains. Screw up the device.
15) Fit into place the funnel, put the whole device on the revolving table and fix.
16) Follow step 3.
17) Start the compactor and adjust air pressure to 0.7 kg/cm^2 (10 lb/in^2), which is equivalent to a tamper pressure of about 11.25 kg/cm^2 (160 lb/in^2).
18) Using the spatula, transfer half the material from the feeder to the mold, spreading it out. Lower the tamper and apply 10 times to the material. Raise the tamper, transfer the other half of the material to the mold, and apply the tamper another 10 times, maintaining the air pressure at 0.7 kg/cm^2 (10 lb/in^2).
19) Raise the compactor and clean it. Renew the metal collar and place a rubber disc over the upper part of the specimen, which is now ready for the actual compaction process described in paragraphs 6 to 9.

.4 Precautions

— The material must be evenly placed in the mold. Failure to observe this point leads to an immediate increase in the effort required to produce exudation.
— The coarse particles must be evenly distributed throughout the length of the feeder in order to avoid segregation.
— It is very important to place the basket carefully by hand before starting compaction operations. If compaction is commenced without the whole device being perfectly set on the base of the collar which holds the mold, the strips of masking tape which hold the basket and the bronze disc together may be torn.

REFERENCES

1. Proctor, R. R., "Fundamental Principles of Soil Compaction", *Engineering News Record,* Vol. III, August and September, 1933.
2. Proctor, R. R., "Design and Construction of Rolled Earth Dams", *Engineering News Record* , Vol. III, 1933.
3. Altschaefel, A. G., and Lowell, Jr. C. W., "Compaction Variables and Compaction Specification", *Purdue University Engineering Reprints,* June, 1969.
4. Foster, C. R. "Field Problems: Compaction," Chap. 12 of *Foundation Engineering,* Ed. G. A. Leonards, McGraw-Hill Book Co., Inc, 1962.
5. Porter, O. J., "Method of Determining Relative Compaction and Shinkage of Soil Materials", Research Department, California Division of Highways, August, 1930.
6. Purcell, C. H. "Grading Methods and Grading Equipment", *Road Builders Association Bulletin,* No 17, 1931.
7. Proctor, R. R. "Description of Field and Laboratory Methods", *Engineering News Record,* Vol. III, No 10, September, 1933.
8. Tamez, E., "Algunos factores que afectan a la prueba de compactación dinámica", *Conference on Soils for Engineering Purposes,* Committee D-18 A.S.C.E., Mexican Society of Soil Mechanics Meeting, México, 1957.
9. Holtz, W. G., and Lowitz, C. A., "Compaction Characteristics of Gravelly Soils", ibid.
10. Jiménez Salas, J. A., "Suelos y rocas compactadas como materiales de construcción", Chap. I of *Compactación de Terrenos, Terraplenes y Pedraplenes,* Técnicos Asociados S. A: Barcelona, 1970.
11. Road and Streets, Publication of Reuben H. Donnelley, Co., June, July, and August, 1970.
12. Hilf, J. W., "Compacting Earth Dams with Heavy Tamping Rollers", *Transactions A.S.C.E.,* Vol. CXXIV, 1959.
13. Marsal, R. J., and Reséndiz, D., "Compactación de Suelos Arcillosos compactados", Publication of the Engineering Institute, National University of México, 1968.
14. Williams, F. H. P., and McLean, D. F., "The Compaction of Soil: A Study of the Performance of Plant", Road Research Laboratory, Technical Bulletin No 17, London, 1950.
15. Johnson, A. W., and Sallberg, J. R., "Factors that influence Field Compaction of Soils", Highway Research Board, Bulletin No 272, 1960.
16. Ramos Medina. J. E., "Métodos de compactación en el campo", Note for a paper in the Seminar on Pavements, Public Works Ministry, México, 1971.
17. Highway Research Board, "Compaction of Embankments, Subgrades and Bases", Bulletin No 58, 1952.
18. Lewis, W. A., "Investigation to the Performance of Pneumatic Tyred Rollers in the Compaction of Soil", Bulletin No 45, Road Research Laboratory, London, 1959.
19. Juárez Badillo, E., and Rico, A., *Mecánica de Suelos, Vol. II: Teoria y Aplicaciones de la Mecanica de Suelos,* Limusa: México, 1977, Chap. XI (Appendix XI-i).
20. Liamazares Gómez, O., "Compactación por vibración", Chap. III of *Compactación de Terrenos, Terraplenes y Pedraplenes,* Técnicos Asociados: Barcelona, 1970.
21. AB Vibro-Verken, "Manual on Vibratory compaction of soil and rock fill", Research Department, Solna, Sweden, 1972.
22. Forssblad, L., "Vibratory Soil Compaction, Research Results and Practical Applications", Lecture AB Vibro-Verken, Solna, Sweden, 1972.
23. Sowers, G. B., and Sowers, G. F., *Introductory Soil Mechanics,* The McMillan Book, Co., 1972.
24. Hveem, F. N., "Maximum Density and Optimum Moisture of Soils. What do these terms mean?", 36th Annual Meeting of the H. R. B., Washington D. C., 1957.
25. Juárez Badillo, E., and Rico, A., *Mecánica de Suelos, Vol. I: Fundamentos de la Mecánica de Suelos,* Limusa: México, 1979, 2nd Ed., Chap. XIII.
26. Skempton, A. W., "The Pore Pressure Coefficients A and B", *Geotechnique,* Vol. IV, 1954.
27. Reference [25], Chap. XII (appendix f).
28. Wilson, S. D., "Effect of Compaction on Soil Properties", *Proceedings of the Conference on Soil Stabilization,* Massachussets Institute of Technology, Boston, 1952.
29. Marsal, R. J., "Resistencia y Compresibilidad de enrocamientos y gravas", Publication of the Engineering Institute, National University México, 1971.
30. Marsal, R. J., "Pedraplenes", Unpublished Lecture, México, 1974.
31. Casagrande, A., "An Unsolved Problem of Embankment Stability of Highly Plastic Clays", *Procs. I. Panamerican Conference on SMFE,* México, 1959.
32. Bertram, G. E., "Rockfill Compaction by Vibratory Rollers", *Procs. II. Panamerican Conference on SMFE,* Brazil, 1963.

33. MARSAL, R. J., MORENO, E., NÚÑEZ, A., CUÉLLAR, R., and MORENO, R., "Investigación sobre el comportamiento de los suelos granulares y muestras de enrocamiento", Publication of the Federal Electricity Commission, México, 1965.

34. MARACHI, N. D., "Strength and Deformation Characteristics of Rockfill Materials", Report No. TE-69-5, Department of Civil Engineering, University of California, Berkeley, 1969.

35. *A.S.T.M.,* Designation D-698-64T, 1964.

36. Texas Highway Department, Materials and Test Division, Test Method Tex-113-E (revised April 1970), Austin, Texas, 1970.

37. Road Research Laboratory, *Soil Mechanics for Road Engineers,* Her Majesty's Stationery Office: London, 1961, Chap. 9.

38. MADDISON, L., "Laboratory Test on the Effect of Stone Content on the Compaction of Soil Mortar", *Roads and Road Construction,* No 22, 1944.

39. AGUIRRE M., L. M., "Correlación entre las pruebas estáticas y dinámicas de compactación de suelos en el laboratorio", Master Degree Thesis, National University of México, 1964.

40. *A.S.T.M.,* Designation D1557-64-T, 1964.

41. U.S. Army Engineers Waterways Experiment Station, Soil Compaction Investigation, Waterways Experiment Station Technical Memorand (3-271), Vicksburg, USA, 1957.

42. Specifications of the California Highway Department, Test California 301-E, Sacramento, Cal., 1964.

43. BROMS, B. B., and FORSSBLAD, L., "Vibratory Compaction of Cohesionless Soils", Soil Dynamics Speciality Conference, *VII. International Conference on SMFE,* México, 1969.

44. SCHAFFNER, H. J., "Unlagerung Rolliger Erdstoffe durch Vibration", *Mitteilungen der Forschungsanstalt für Schiffahrt, Wasser und Grundbau,* Berlin, No 6, 1962.

45. SELIG, E. T., "Effect of Vibration on Density of Sand", *Procs. II. Panamerican Conference on SMFE,* Brazil, 1963, Vol. I.

46. ORTIGOSA, P., and WHITMAN, R. V., "Densification of Sand by Vertical Vibrations with almost Constant Stresses", Dept. of Civil Engineering, M.I.T., Research Report No 206, Boston, 1968.

47. FORSSBLAD, L., "Investigations of Soil Compaction by Vibration", *Scandinavian Technical Act,* C i 34, 1965.

48. JOHNSON, A. W., and SALLBERG, J. R., "Factors Influencing Compaction. Test Results", Highway Research Board, Bulletin No 319, 1962.

49. PETTIBONE, H. C., and HARDIN, J., "Research on Vibratory Maximum Density Test for Cohesionless Soils", *A.S.T.M. Special Technical Publication,* No 377, 1964.

50. A.S.T.M., Designation D-2049-69, 1969.

51. Texas Highway Department, Materials and Test Division, Test Method Tex-114-E (revised in April, 1970), Austin, Texas, 1970.

52. Engineering Developments Co, Inc., Test Catalogue for Machine Model 4C, Vicksburg, Miss. 1970.

53. RESÉNDIZ, D., "Considerations on the Solid Liquid interaction in Clay Water Systems", *Procs. VI. International Conference on SMFE,* Montreal, Canada, 1964.

54. Lambe, T. W., "Compacted Clay: Structure", *Transactions ASCE* Vol. C-XXV, Part I, 1960.

55. SEED, H. B., and CHAN, C. K., "Structure and Strength Characteristics of Compacted Clays", *Journal of Soil Mechanics and Foundations Division, ASCE,* Vol. LXXXV, SM5, 1959.

56. RESÉNDIZ, D., "On the Strength of Claying Soils: A Study of the Shearing Resistance Mechanism at the Structural Level", Institute of Engineering, Publication 126, National University of México, 1965.

57. CASAGRANDE, A., HIRSCHFELD, R. C., and POULOS, S. J., "Investigation of Stress-Deformation and Strength Characteristics of Compacted Clays", Harvard Soil Mechanics Series, Vol. LXX, Harvard, Mass. 1963.

58. CASAGRANDE, A., and HIRSCHFELD, R. C., "Investigation of Stress-Deformation and Strength Characteristics of Compacted Clays", Harvard Soils Mechanics Series, Vol. LXI, Harvard, Mass. 1960.

59. ALBERRO, J., "Estudio de una correlación entre pruebas de compactación estática y dinámica", Report from the Institute of Engineering, National University of México, 1966.

60. SEED, H. B., MITCHEL, J. K., and CHAN, C. K., "The Strength of Compacted Cohesive Soils", *Research Conference on Shear Strength of Cohesive Soils: ASCE,* University of Colorado, USA, 1960.

61. JIMÉNEZ SALAS, J. A., and DE JUSTO, J. L., *Geotecnia y Cimientos,* Rueda: Madrid, 1971, Vol. I. Chap. 7.

62. MITCHEL, J. K., HOOPER, D. R., and CAMPANELLA, R. G., "Permeability of Compacted Clay", *Journal of Soil Mechanics and Foundations Division: ASCE,* Vol. 91, SM4, 1965.

63. MARANHA DAS NEVES, E., "Influencia das Tensoes Neutras Negativas nas Caracteristicas Estructurais dos Solos Compactados", National Laboratory of Civil Engineering, Report No 386, Public Works Ministry, Lisbon, 1971.

64. YOSHIMI, Y., and OSTERBERG, J. O., "Compressibility of Partially Saturated Cohesive Soils", *Journal of Soil Mechanics and Foundations Division: ASCE,* Vol. 89, SM4, 1963.

65. BISHOP, A. W., and HENKEL, D. J., *The Triaxial Test,* Edward Arnold: London, 1962.

66. SEED, H. B., and CHAN, C. K., "A symposium on Compacted Clays", *Transactions ASCE,* Vol. 126, 1961.

67. CASAGRANDE, A., and HIRSCHFELD, R. C., "Investigation of Stress-Deformation and Strength Characteristics of Compacted Clays: 2nd Part", Harvard Soil Mechanics Series No 65, Harvard, Mass, 1962.

68. YODER, E. J., *Principles of Pavement Design,* John Wiley and Sons, Inc, 1967.

69. CASAGRANDE, A., and SHANNON, W. L., "Research on Stress-Deformation and Strength Characteristics of Soil and Soft Rocks under Transient Loadings", Harvard Soil Mechanics Series No 31, Harvard University, Mass., 1948.

70. WHITMAN, R. V., "The Behavior of Soil under Transient Loadings," *Procs. IV. International Conference on SMFE,* London, 1957, Vol. I.

71. HAMPTON, D., "Effect of Rate of Strain on the Strength of Remolded Soils", Purdue Joint Highway Research Project, Purdue University, Lafayette, Indiana, 1958.

72. CASAGRANDE, A., and WILSON, S. D., "Effect of Rate of Loading on Strength of Clays and Shales at Constant Water Content", *Geotechnique,* Vol. II, 1951.

73. SHERARD, J. L., "Identification and Nature of Dispersive Clays", *Journal of the Geotechnical Engineering Division, ASCE,* Vol. 102, No GT4, April 1976, pp. 287-301.

OTHER RELATED REFERENCES

LEDOUX, J. L. MÉNARD, J. and SOULARD, P., "Le Penetro-Gammadensimetre", Colloque International sur la Gestion des Ouvrager, Paris, 1981.

Proc. of the International Conference on Compaction, 3 Volumes, Ecole Nationale des Ponts et Chaussées, Paris, April, 1980.

PUIATTI, D., and GESTIN, G., "Tassement Propre des Remblais, Influence de la Compacite, du Degre de Saturation et du Mode de Compactage", Colloque International sur la Gestion des Ouvrager, Paris, 1981.

RESÉNDIZ, D., *Compaction Conditions, Stage Variables and Engineering Properties of Compacted Clay,* Mexico, 1980.

RESÉNDIZ, D., *Full-scale Observations, Numerical Modeling, and Generalization in Soil Mechanics,* Mexico, 1979.

YODER, E. J. and WILLIAMSON, T. G., "Techniques for Compaction Control", CE249: Civil Engineering, Purdue University Engineering Reprints, January, 1969.

CHAPTER 5

EARTH RETAINING STRUCTURES

5.1 Introduction

Earth retaining structures frequently provide the solution to numerous road engineering problems. Situations where two neighbouring masses of earth must be maintained at different levels constantly arise. A typical solution to this problem is careful slope construction, but retaining structures are often necessary, too. In highways and railroads, most of these structures are rigid concrete or masonry walls. They are usually not very high (generally less than 8 to 10 m (26 to 33 ft), so that their design implies a generous degree of safety, independent of the complications and theoretical uncertainties regarding earth pressure on retaining structures (which become increasingly important for higher structures). Although large-scale retaining structures are not common practice in road engineering (especially for reasons of cost), they should not, however, be excluded, and they can present major design problems owing to the numerous uncertainties that still prevail with regard to earth thrust theories.

Currently available knowledge enables a reasonable degree of safety to be achieved in small structures. For this, not only a *simple and reasonable* earth thrust theory is necessary, but also knowledge of practical design rules, which are not usually considered by theories, such as the need for adequate drainage of the backfill.

A design based on retaining walls is often adopted in a particular case, after preliminary comparison with other alternative designs which turn out to be more expensive. However, on detailed analysis of the drawn design and cost estimates, it is observed that the construction of the retaining wall has been given only trivial consideration, without paying attention to details. In many cases, the installation of filters in the back of the wall, and drainage of the backfill, have not been taken into account, and no precautions have been taken for special foundation conditions which may prove necessary. With omissions like these, the retaining wall will often cost more than the other solutions and, more seriously, it is also very likely to fail. It is very common in the case of retaining structures to ignore all those design and construction details, without which earth thrust theories are often unable to ensure success. This is sometimes because the details involve circumstances not included by the theories, and at other times because they compensate for deficiencies of the theories that are not understood by designers and builders. If the retaining structure and all the details that ensure its efficient performance are included in any economic comparison, the result will often be different and the cost of the structure excessive.

If the foregoing is true, given the large number of walls that can be seen in roads and railroads, it is reasonable to believe that many of them were built without the necessary design and construction precautions. This, unfortunately, is the experience of the authors of this book. Retaining walls are among the structures which usually receive insufficient attention from design construction engineers, in relation to the dangers they involve such as drainage to eliminate hydrostatic thrusts and the detrimental effects of water. Consequently collapses still do occur in retaining structures, despite the good intentions of the engineer.

Serious flaws exist in the theories currently available for the calculation of pressures against earth retaining structures. Moreover it is often very hard to know whether the conditions of the theories are being correctly met in actual projects. There is no universal standard for the application of these theories. Their usefulness in a given situation always depends on the interactive conditions of the structure and the fill which cannot easily be predicted, such as vertical and horizontal deformability of the structure and saturation of the fill or the evolution of its shear strength with time. Thus, several different thrust theories and types of retaining structures have to be considered. These will be summarized in the pages that follow, with special emphasis on the structures which are most commonly used in the road engineering field.

The situations where retaining structures are most frequently used are:

— Confinement of embankments, either because there is not enough space for their slopes, as often occurs in urban areas, or because their slopes would be too long, narrow, and difficult to build, as usually occurs in sections of roads which are located on steep hillsides, Plates 5-1 to 5-4.
— Confinement of embankments that give access to bridges, culverts, and other structures. Embankments with a large slope are avoided, either owing to lack of space, or so as to avoid invading river beds and unsuitable zones, or else to save earth removal. (Plates 5-5, 5-6, 5-7).
— Retention of masses of earth that are in themselves unstable. In this case, the retaining structure is used as a solution to the instability of the slope.

Plate 5-1 Retaining wall on a highway

Plate 5-3 Cracks in a masonry wall

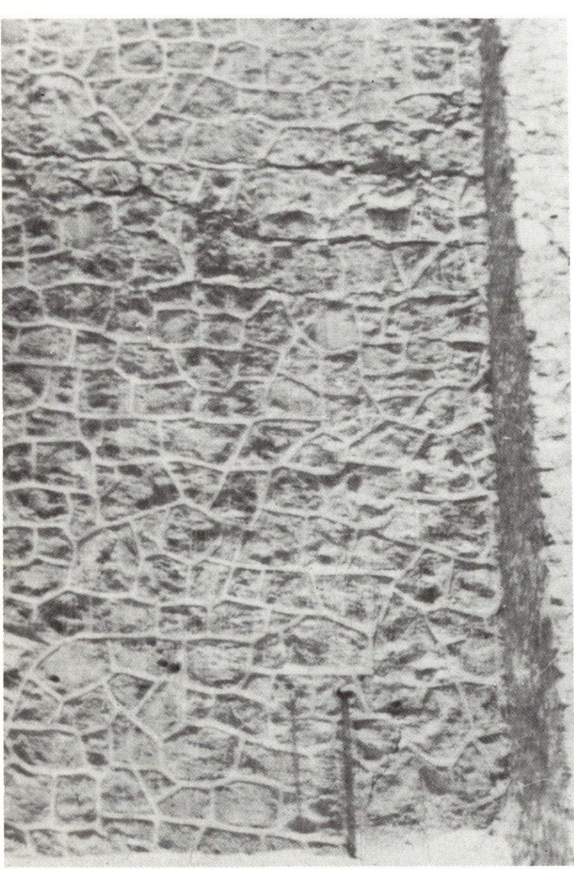

Plate 5-2 Crack in a retaining wall

Plate 5-4 Crack in a retaining wall with no through drainage

Plate 5-5 Failure in the wing of a bridge abutment

Plate 5-6 Crack in the wing of a bridge with no through drainage

Retaining structures are classified in accordance with two criteria, which in turn depend largely on their mechanical behavior and on the applicability of available earth thrust theories. According to the first criterion, structures can be *rigid* or *flexible,* depending on their tendency to deform under the pressures exerted by the fill. Structures usually belong to one category or the other, depending on the material of which they are made. Concrete and masonry give rigid structures, when used for fairly thick walls. The term "rigid" clearly applies only to structures which do not easily give way under pressure. Later there will be occasion to discuss in detail the limitations of this expression and its highly important significance in mechanisms of earth pressure generation.

Plate 5-7 Horizontal displacement in a retaining wall

Flexible structures are ones which, on account of their cross-section and component materials, are highly susceptible to deformation. The most representative varieties are sheet-piles made of wood, steel or concrete, which are not considered in this book because they are not often used in road engineering technology.

Following the second criterion, retaining structures are classified in accordance with their expected lifespan, and can be permanent or temporary. The permanent retaining structure characteristic of roads is the concrete or masonry wall (Plates 5-2, 5-3). Temporary structures are called bracing or struts and are usually made of wood or steel when they are obliged to withstand strong thrusts. In road engineering, they are chiefly used to support unstable excavation walls, for bridge foundations, in tunnels and, not so frequently, to support the walls of excavations for special large-scale drainage projects associated with slope stability problems. Temporary structures have a brief life, that is the time it takes to build the principal project. This reflects on their design and construction, but without implying lack of interest in their problems, for their good behavior is often of the utmost importance. Special attention will be paid in this chapter to rigid permanent retaining walls, which are the ones most frequently used in road engineering, but some of the most common types of bracing will also be mentioned.

The first point to be considered will be the classic earth thrust theories and certain empirical methods for assessing this concept. After this, norms will be given for defining the possibilities for applying the methods to different practical problems. Details will also be given of a few tricks of the trade which, as has already been said, are so often responsible for the success or the failure of a particular solution. Bracing will be dealt with separately, but tunnels will be excluded from this chapter and will later be the object of a separate chapter.

5.2 Classical Earth Thrust Theories

5.2.1 RANKINE's Theory [1,2]

.1 Limiting Equilibrium Conditions. RANKINE's Theory Applied to Frictional Soils.

An element of soil is considered with a height *dz,* located at a depth *z* within a semi-infinite mass of soil at rest (that is, without allowing any displacement from a natural state, which is what will hereafter be implied by the term *at rest*). Let the boundary of the mass be horizontal (Fig. 5-1). Under these conditions, the effective vertical stress acting on the structure of the element is:

$$P_v = \gamma z \qquad (5\text{-}1)$$

where γ is the unit weight.

Under the vertical acting stress, the element of soil exerts a lateral stress, so a horizontal pressure, P_h is thus created which, according to experience, is accepted as being in direct proportion to P_v

$$P_h = K_0 \gamma z \qquad (5\text{-}2)$$

The constant K_0 relating P_v and P_h is called the coefficient of earth pressure at rest. Values for it have been obtained experimentally in the laboratory and the field, and it has been observed that for granular soils without fines, it is between 0.4 and 0.8. The first value corresponds to loose sands and the second one to highly compacted sands. A dense natural sand usually has a K_0 of about 0.5

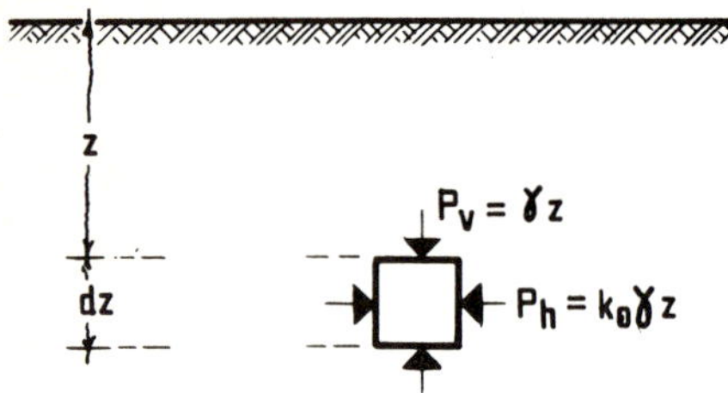

Fig. 5-1 Forces acting on an element of soil at rest

In the case of clays that have become overconsolidated as a result of drying, K_0 may come close to 1 [3]. During wetting, the value of K_0 will depend on already existing cracks and expansion properties [4], but may increase considerably in certain cases. Also affecting K_0 are precompression phenomena in sands, which depend on the stress history of the deposit and on the geological history of the material in situ. It is very hard to determine the value of K_0 in a particular place, by means of the appropriate tests [5,6], so that any attempt at this is usually abandoned, even if only for reasons of cost.

If a MOHR's circle is drawn corresponding to the stress conditions described for the element mentioned (Fig. 5-2), a circle like *N°1* will be obtained; this is obviously not a rupture circle.

Starting with these *at rest* stress conditions, rupture can be reached by two interesting stress paths. In the first case, the horizontal stress is reduced, and the vertical stress maintained constant. In this way rupture circle *N°2* is reached, with a minor

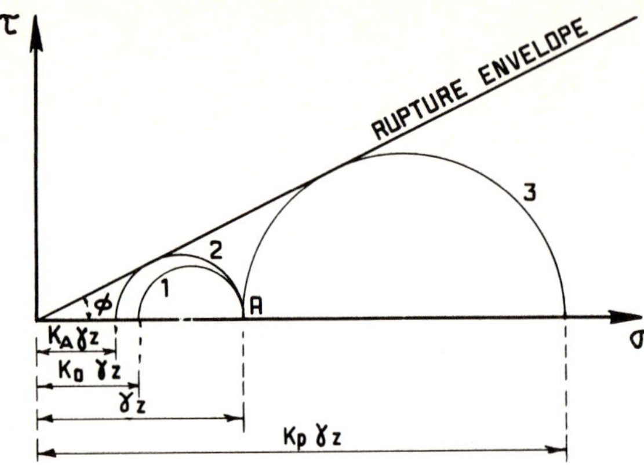

Fig. 5-2 Rupture conditions represented as MOHR Circles (Frictional soils)

principal stress $\sigma_3 = K_a \gamma z$, where K_a is referred to as the active earth pressure coefficient. Observe that in this circle the stress σ_3 corresponds to a horizontal pressure, since the corresponding major principal stress is by hypothesis γz, the vertical pressure due to the weight of the overlying soil. The second way of reaching rupture is to allow the stress γz to be the minor principal stress, increasing the horizontal stress until a value $K_p\ \gamma z$ is reached with the resulting circle at a tangent to the rupture envelope. The value K_p is termed the coefficient of passive earth pressure.

These are the limiting possibilities of practical interest for reaching rupture conditions starting from the at rest condition. They include the value γz for the vertical pressure for a level backfill in this simplified analysis.

In accordance with RANKINE, a soil is said to be in a limiting plastic equilibrium state when a generalized failure condition is commencing. There are two limiting equilibrium conditions: the one when horizontal stress reaches a minimum value of $K_p \gamma z$ and the one which occurs when this pressure reaches a maximum value of $K_p \gamma z$. These are known as the active and passive states, respectively.

In the active condition (Fig. 5-2), the following is true:

$$\frac{P_h}{P_v} = \frac{\sigma_3}{\sigma_1} = \frac{1}{N_\varnothing} \qquad (5\text{-}3)$$

It can be seen from the geometry of the MOHR's circle that:

$$K_a = \frac{1}{N_\varnothing} = \tan^2\left(45° - \frac{\varnothing}{2}\right) \qquad (5\text{-}4)$$

which gives the value of the coefficient for the active condition. Likewise, in the passive condition:

$$\frac{P_h}{P_v} = \frac{\sigma_1}{\sigma_3} = N_\varnothing \qquad (5\text{-}5)$$

and the result will be:

$$K_p = N_\varnothing = \tan^2\left(45° + \frac{\varnothing}{2}\right) \quad (5\text{-}6)$$

for the passive state.

These two limiting conditions have a relationship with actual engineering which lends them a practical interest.

Take the case of a wall with a fill that is originally assumed to be at rest. Physically, the soil retained by this wall can be brought to failure in two different ways: one by expansion of the fill, with the structure tilting away from the soil and the other, by some external thrust, that forces the wall against the fill and rotates to the back.

RANKINE reasoned that with thrust from the fill and the wall moving outwards, the pressure of the fill dropped to values below those corresponding to at rest conditions. The shear stresses thus generated enable the soil mass to partially support itself. If the wall moves enough, the horizontal pressure can become the active one. This is a minimum value which cannot be reduced, even if the wall continues to move.

In this way, it is reasoned that by designing a wall to withstand active pressure, stability will be ensured, so long as the structure can move enough for the active state to develop. A similar argument could be used for the condition where the wall shifts backwards owing to an external force that is large enough to cause the passive state to develop.

If, following RANKINE's Theory, the pressures obtained for active and passive states for a depth z are integrated along the height H of a retaining wall, the corresponding total thrusts can be obtained. For this procedure, the respective limiting conditions are assumed to be fully developed throughout the fill; in other words, the wall has deformed as much as necessary to produce active or passive pressures.

Thus, on the basis of Eq. (5-3), the active state can be written:

$$P_h = \frac{P_v}{N_\varnothing} = \frac{\gamma z}{N_\varnothing} \quad (5\text{-}7)$$

This expression gives the horizontal pressure acting on the wall at the depth z, in the case of a fill with a horizontal surface.

In an element dz of the back of the wall, at depth z, the thrust is

$$dE_a = \frac{1}{N_\varnothing}\gamma z \, dz$$

per unit of wall length. At height H, therefore, the total thrust per unit of wall length will be:

$$E_a = \frac{\gamma}{N_\varnothing}\int_0^H z\,dz = \frac{1}{2N_\varnothing}\gamma H^2 = \frac{1}{2} K_a \gamma H^2 \quad (5\text{-}8)$$

Eq. (5-8) gives the total active thrust which is applied by a fill with a horizontal surface to a wall with a vertical back, in a purely frictional soil.

Similarly, for the passive state, using Eq. (5-6), the following value is obtained for the total passive thrust:

$$E_p = \frac{1}{2} N_\varnothing \gamma H^2 = \frac{1}{2} K_p \gamma H^2 \quad (5\text{-}9)$$

It is also valid for a wall with a vertical back and a fill with a horizontal surface in frictional soil.

For the purposes of assessing the stability of the wall, which is considered a rigid element, pressure distribution can be regarded as being replaced by an equal number of concentrated forces, with magnitudes that are given by E_a and E_p. The triangular distribution existing for both conditions in RANKINE's Theory, means that the point at which the resulting forces are applied is located at one third of the height of the wall above the base. In this case both forces are horizontal.

In a situation where the surface of the fill slopes at an angle β to the horizontal, it is possible to derive the following expressions for the active and passive thrusts by means of an integration process, similar to the one previously described [1]:

$$E_a = \frac{1}{2}\gamma H^2 \left[\cos\beta \frac{\cos\beta - \sqrt{\cos^2\beta - \cos^2\varnothing}}{\cos\beta + \sqrt{\cos^2\beta - \cos^2\varnothing}}\right] \quad (5\text{-}10)$$

$$E_p = \frac{1}{2}\gamma H^2 \left[\cos\beta \frac{\cos\beta + \sqrt{\cos^2\beta - \cos^2\varnothing}}{\cos\beta - \sqrt{\cos^2\beta - \cos^2\varnothing}}\right] \quad (5\text{-}11)$$

Seeing that the pressure distributions are also linear and they are parallel to the surface of the fill, the resultants will also be parallel to the surface of the fill and will be applied to a third of the height of the wall from the base.

Note that for $\beta = 0$, Eqs. (5-10) and (5-11) become (5-8) and (5-9).

An interesting practical condition is the one that arises when the surface of the fill, assumed to be horizontal, is subjected to a uniform surcharge, with a value q per unit of area. For an active stress, this situation can be analyzed as follows:

It has already been seen that: $\sigma_3/\sigma_1 = 1/N_\varnothing = K_a$. Under the surcharge q, the vertical stress becomes: $\sigma^*_1 = \sigma_1 + q$ and the horizontal stress: $\sigma^*_3 = \sigma_3 + \Delta\sigma_3$. Therefore, it can be said that:

$$\frac{1}{N_\varnothing} = \frac{\sigma_3 + \Delta\sigma_3}{\sigma_1 + q}$$

Consequently:

$$\sigma_3 + \Delta\sigma_3 = \frac{\sigma_1}{N_\varnothing} + \frac{q}{N_\varnothing}$$

By comparison with the zero surcharge condition, the following equation can be deduced:

$$\Delta P_h = \Delta\sigma_3 \frac{q}{N_\varnothing} = K_a q \quad (5\text{-}12)$$

In other words, for the active state the effect of the uniform surcharge is simply a uniform increase in the pressure acting against the wall by the value given for Eq. (5-12).

Similarly, for the passive state the effect of the uniform surcharge is to increase this pressure by:

$$\Delta P_h = \Delta\sigma_1 = qN_\varnothing = K_p q \quad (5\text{-}13)$$

Note that the application of Eqs. (5-12) and (5-13) is restricted to fills with a horizontal surface. For fills with a sloping surface similar expressions can be obtained [1].

Another interesting case is the one where part of the horizontal sandy backfill is submerged. If H is the total height of the wall and H_1 is the height of the unsubmerged sand, measuring from the crest (Fig. 5-3), the vertical pressure of the fill at a point below the water-table will be:

$$P_v = \gamma H_1 + z'\gamma' \quad (5\text{-}14)$$

The horizontal pressure of the sand below the water-table will be:

$$P_h = \frac{P_v}{N_\varnothing} = \frac{1}{N_\varnothing}\left(\gamma H_1 + z'\gamma'\right) \quad (5\text{-}15)$$

Additionally, there will be a hydrostatic pressure of:

$$P_w = \gamma_w z' \quad (5\text{-}16)$$

on the wall below the water-table.

Consequently, the total active thrust is given by:

$$E_a = \frac{1}{2N_\varnothing}\gamma H_1 + \frac{1}{N_\varnothing}\gamma H_1 H_2 + \frac{1}{2N_\varnothing}\gamma' H_2^2 + \frac{1}{2}\gamma_w H_2^2 \quad (5\text{-}17)$$

Note that in this case, despite the sand being submerged and the values of γ reduced to γ' the total thrust against the wall increases considerably, because the hydrostatic effect of the water is not reduced since it has no shear resistance. Similar equations to (5-14) to (5-17) can be obtained for the passive state and for cases where the fill is not horizontal.

If, apart from the effects now under consideration, there is a uniform surcharge, q, its influence must be superimposed. This is illustrated in Fig. 5-3.

All these equations are frequently used for the design of retaining walls made of masonry, or of masonry and reinforced concrete, which is why it is vital to define the conditions under which they can be applied. These are the assumptions which affect RANKINE's Theory:

— Limiting stress conditions, both active and passive, must develop completely throughout the entire mass of soil adjacent to the wall. It has already been mentioned that this is a reasonable hypothesis in real walls, which can undergo a sufficient degree of deformation so long as the design or construction does not restrict the movements of the structure to that of a rigid body. The type of movement that enables a limiting state to develop is a slight tilt of the wall about its base.

— When the surface of the fill is horizontal and the back of the wall vertical, which have been the implicit conditions up to now, the wall should be smooth, the coefficient of friction between the wall and the fill must be zero. When the surface of the fill is a sloping plane at an angle of β to the horizontal, the wall acts as if it were rough, with a coefficient of friction with the soil such that the resulting pressures on the vertical back of the wall act at the same angle.

.2 RANKINE's Theory Applied to Cohesive Soils

In purely cohesive soils, for the formulas that follow to be put to practical use, it is essential to remember that cohesion is not an intrinsic property of clays. It is a circumstantial property, susceptible to change with time, as the clay consolidates or as it expands owing to absorption of water. It is therefore vital to ensure in each case that time will bring no change in the cohesion used in the design formulas. This is very hard to guarantee.

Take the case of an element of purely cohesive soil at a depth z. As in the case of frictional soils, if the horizontal mass of soil is at rest, the horizontal stress on the element, when it is submitted to vertical stress γz will be $K_0\gamma z$. As before the value of K_0 depends on the material and its previous stress history.

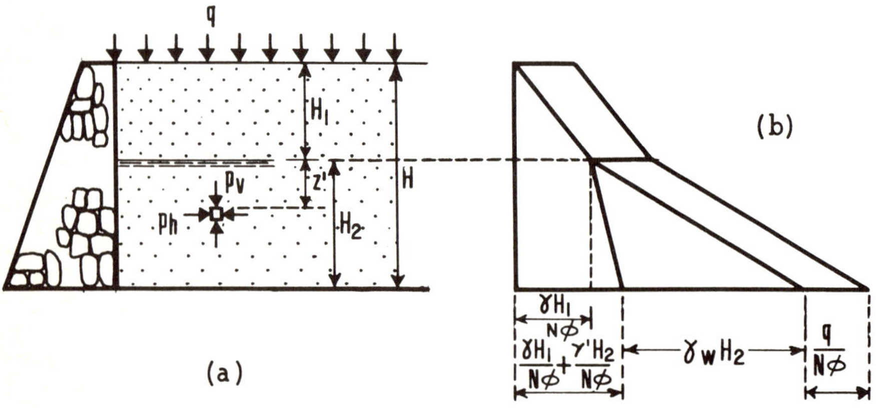

Fig. 5-3 Active pressure of a sandy fill that is partially submerged and subjected to a uniform surcharge

MOHR's circle *N°1* of Fig. 5-4 represents the stress conditions for the previously mentioned element. Here too, if lateral deformation is permitted, the material can reach rupture in two different ways. In the first case, the element is allowed to deform outward laterally, by reduction of the horizontal stress to the minimum value that is compatible with equilibrium. This new stress condition is represented by Circle *N°2* and corresponds to the active limiting condition or state (see Fig. 5-4), in which the pressures are:

Horizontal: $P_a = \gamma z - 2c$ and Vertical: $P_v = \gamma z$ (5-18)

where P_v is the major principal stress and P_a the minor principal stress in failure circle *N°2* that is a tangent to the rupture envelope $s = c$, (which was obtained in an undrained test).

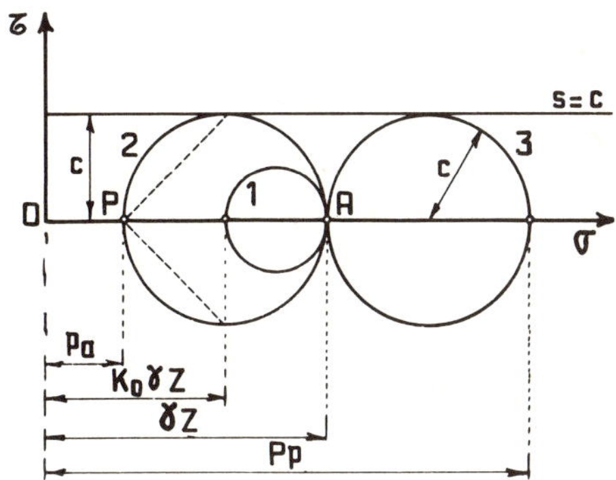

Fig. 5-4 Rupture conditions in the MOHR Circle (Cohesive soils)

The other way to bring the element located at a depth z to rupture is to increase the horizontal stress beyond γz until a stress is reached such that the new MOHR's circle *(N°3)* is also at a tangent to the horizontal rupture line. At this point, the limiting stress condition or state is passive and the following values are attained:

Horizontal: $P_p = \gamma z + 2c$; Vertical: $P_v = \gamma z$ (5-19)

and p_p is equal to the major principal stress.

Here too the same practical interpretation applies to the generation of limiting pressure conditions when designing retaining walls. The formulas for active pressure can be related to the thrust of soils against walls, whereas passive stress is related to the thrust of walls toward their backfills.

As in the case of frictional soils, formulas can be obtained for total active and passive thrust by integrating the respective horizontal pressures to the height H of the wall. The procedure is the one which has already been described, and the results obtained are as follows:

$$E_a = \frac{1}{2}\gamma H^2 - 2cH \qquad (5\text{-}20)$$

$$E_p = \frac{1}{2}\gamma H^2 + 2cH \qquad (5\text{-}21)$$

These thrusts are horizontal and go through the centroid of the pressure distribution area.

Note that Eqs. (5-20) and (5-21) apply only if the surface of the backfill is horizontal and if the corresponding limiting equilibrium condition develops completely throughout the fill. Note also Eq. (5-20) presumes a tensile connection between the upper part of the wall and the soil.

Eq. (5-20) shows a simple procedure for calculating the maximum height which can be reached for a vertical cutting in cohesive material, without using any support. A vertical cutting without support will hold up without failing if $E_a = 0$, which leads to:

$$H_c = \frac{4c}{\gamma} \qquad (5\text{-}22)$$

The critical height of a cohesive material is usually referred to as H_c. Eq. (5-22) gives values for the actual stable height that are rather high; if it is to be used in practice, there should be a minimum safety factor of 2.

When applied to cohesive soils, RANKINE's Theory should be subjected to careful consideration. It has already been mentioned that cohesion is not a reliable shear strength parameter, but a parameter which undergoes important changes with time that are hard to predict and which usually cause a reduction in the initial value. As a general rule, the designer should not rely on it for retaining structures, however attractive it may appear at first sight. There is always the possibility that a fill may become saturated with water. In an excavation, flow is always induced towards the edges. Thus, a cohesive material does tend to lose strength, and a design based on the cohesive strength of a material will eventually becomes unsafe.

From Eq. (5-18) it can be seen that the distribution of the stresses in the fill is theoretically linear, with an upper zone under tension and a lower zone under compression. The value of the tension on the surface of the fill is $2c$ and the depth to which the tensile zone extends, characterized by $P_a = 0$ is:

$$Z_0 = \frac{2c}{\gamma} \qquad (5\text{-}23)$$

Figure 5-5a shows the distribution of active stress in the case under consideration, together with the depth to which the tension zone extends. Part *(b)* shows the theoretical distribution of passive stress.

Because the soil does not have the capacity to exist under tension for extended periods, vertical cracks will develop in the active state; their depth is given in Eq. (5-23).

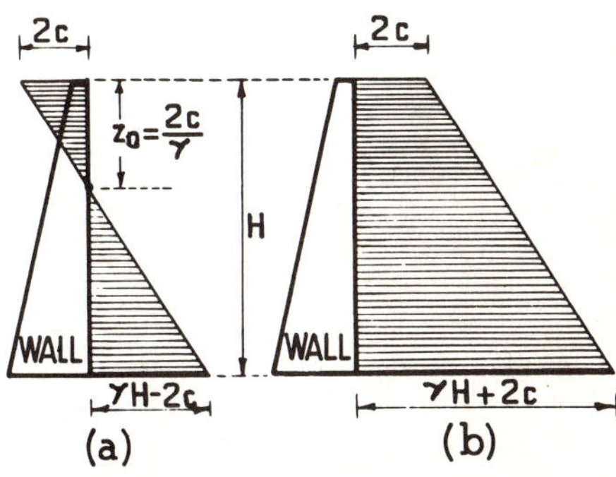

Fig. 5-5 Theoretical distribution of active and passive pressures in purely cohesive soils

In view of the foregoing comments relating to cohesion and how it changes with time, it is not considered necessary to extend the analysis of limiting stress conditions to include cases of sloping fills and non-vertical wall backs. These can be found in [7].

.3 Rankine Theory Applied to Cohesive-Frictional Soils

If the upper surface of the fill is horizontal, the same reasoning applies as for purely frictional material (see Section 1). In Fig. 5-6, an element of soil at a depth z, which is assumed to be at rest, is subjected to a stress condition represented by Mohr's circle *N°1*. Once again, rupture can be reached either by reducing the lateral pressure or by increasing it from $K_0 \gamma z$ upwards. Two Mohr's circles are obtained which are representative of the active (circle *N°2*) and passive (circle *N°3*) stress conditions. In this case, the relationship between the major principal stress and the minor principal stress is given by: $\sigma_1 = \sigma_3 N_\varnothing + 2c\sqrt{N_\varnothing}$ In the active state, $P_a = \sigma_3$ and $\sigma_1 = \gamma z$, so that:

$$P_a = \frac{\gamma z}{N_\varnothing} - \frac{2c}{\sqrt{N_\varnothing}} \qquad (5\text{-}24)$$

On the other hand, in the passive state, $\sigma_1 = P_p$ and $\sigma_3 = \gamma z$; therefore:

$$P_p = \gamma z N_\varnothing + 2c\sqrt{N_\varnothing} \qquad (5\text{-}25)$$

Equations (5-24) and (5-25) give the horizontal stresses for both conditions. As usual, the corresponding thrusts are obtained by integrating the pressures along the height H of the wall, thus:

$$E_a = \frac{1}{2N_\varnothing}\gamma H^2 - \frac{2c}{\sqrt{N_\varnothing}} H \qquad (5\text{-}26)$$

$$E_p = \frac{1}{2} N_\varnothing \gamma H^2 + 2c\sqrt{N_\varnothing} \qquad (5\text{-}27)$$

The directions of these forces are theoretically horizontal through the centroid of the entire pressure diagram. The active case includes tension; the expression for thrust is not valid unless there is a tensile connection between wall and soil.

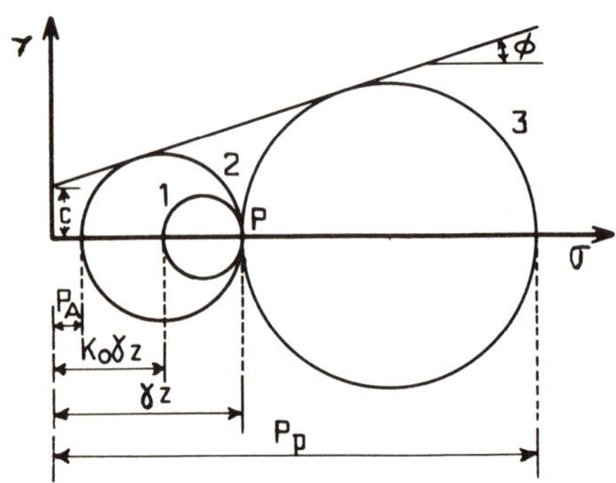

Fig. 5-6 Rupture conditions in the Mohr Diagram (Cohesive and frictional soils)

As in the case of purely cohesive soils, for the active stress there is now a zone in the pressure diagram which corresponds to tension. The depth to which this zone extends, measuring from the crest of the wall, can be obtained by assuming that $p_a = 0$ at that point. If $p_a = 0$ then:

$$\frac{\gamma z}{N_\varnothing} = \frac{2c}{\sqrt{N_\varnothing}} \quad \text{and: } z_o = \frac{2c}{\gamma}\sqrt{N_\varnothing} \qquad (5\text{-}28)$$

If, as a result of this tension, the crack opens, the tensile stress disappears, therefore the point at which the thrust is applied can be calculated using only the lower compression triangle. As before, Eq. (5-28) gives an idea of how to calculate the depth of the crack that forms.

The critical height at which the soil in a vertical cutting can be maintained stable without support, can also be calculated by assuming that $E_a = 0$. Here:

$$\frac{1}{2N_\varnothing}\gamma H^2 = \frac{2c}{\sqrt{N_\varnothing}} H \quad \text{and: } H_c = \frac{4c}{\gamma}\sqrt{N_\varnothing} \qquad (5\text{-}29)$$

Reference [1] gives appropriate equations for these cases where the surface of the fill is not horizontal.

5.2.2 Coulomb's Method

.1 Cases Where Friction at the Back of Retaining Walls Can Develop: Coulomb's Method for Frictional Soils

Except for the already mentioned possibility of an extension to include fills with a sloping surface, Rankine's Theory basically considers cases where there are no frictional forces between the back of the wall (which is assumed to be vertical) and the fill. Usually, however, shear stresses develop between the backfill of a retaining wall and the adjacent soil, owing to the relative movement of the two materials and the resulting friction.

In the active state, displacement away from the backfill of the wall leads to a downwards movement of the soil relative to the wall. This movement produces a tangential frictional stress which acts downwards on the wall. In the passive state, the horizontal thrust of the wall against the soil is usually accompanied by an upward movement of the soil in relation to the wall. This produces an ascending tangential stress, which acts on the back of the wall. In the passive state, however, the direction of this force may sometimes become inverted; cases like these must be determined by means of a specific analysis.

In order to analyze the cases where there is friction between the back of the wall and the soil, the wedge method proposed by Coulomb in 1776 can be used. This was the first rational attempt at calculating thrusts on retaining walls. This method considers that the thrust on a wall is due to the action of a wedge of soil limited by the face of the wall, the surface of the fill and an assumed plane slide surface that develops within the fill (Fig. 5-7).

Wedge *OAB* tends to slip under the effect of its own weight and as a consequence, frictional stresses are produced both in the back of the wall and along plane *OB*. Since the frictional resistances are fully developed, forces E_a and F turn out to be sloping, in relation to the corresponding normal directions, angles δ and $\varnothing$ which correspond to the angles of friction between wall and fill and between soil and soil respectively.

The numerical value of the angle δ is obviously limited by:

$$0 \leqslant \delta \leqslant \varnothing$$

In effect, $\delta = 0$ corresponds to perfectly smooth walls, a smaller value for the angle of friction is impossible. If, on the other hand, $\delta > \varnothing$, which is theoretically possible, failure will occur immediately behind the wall, within the soil. This is the same condition as if the sliding should occur between the wall and the soil. Therefore the maximum value that can be considered for δ is actually $\varnothing$. Following TERZAGHI's observations, the value of δ can in practice be taken as:

$$\frac{\varnothing}{2} \leqslant \delta \leqslant \frac{2}{3}\varnothing \tag{5-30}$$

In [8] it is recommended that δ be taken as equal to the angle of residual friction of the frictional soil of the backfill. This is the angle of friction after considerable deformation of the material, when the void ratio reaches a constant value, independently of the subsequent deformation. The value of the angle of residual friction is greater than that of simple mechanical friction between the particles of sand, and in a continuous deformation process reflects both this property and the degree of interlocking which occurs between the grains.

Taking into consideration the force equilibrium of the wedge, it is observed that the force polygon formed by W, F and E must close. Because the direction and magnitude of W are known, and the direction of E and F were known in advance, this polygon can be constructed for a given wedge. The magnitude of the thrust on the wall can thus be determined. Obviously there is no reason why the wedge selected should be the one producing the biggest thrust. The method will entail trial and error work, where various different wedges are assumed, and the thrust corresponding to each one is calculated. The maximum value produced by the *critical* wedge can be assessed. It is the thrust for which the wall is designed.

It should be noted that if the slip plane selected coincides with the back of the wall, the thrust corresponding to that wedge will obviously be zero, and if the slip plane selected forms an angle $\varnothing$ with the horizontal, the thrust will also be zero. In this condition (see Fig. 5-7), stress F is vertical in an upward direction. Since W is vertical in a downward direction, the only possibility of equilibrium will be $W = F$ and $E = 0$. For wedges with a plane situated between these two extremes, the thrust on the wall is not zero. Between, there is a maximum which must be assessed by the above mentioned trial and error procedure. In [1] some methods are described which enable a value of maximum thrust to be reached that is suitable for practical purposes, and which does away with the trial and error procedures.

In the case of a frictional backfill with a sloping plane, and a wall with a plane back, a mathematical treatment can be used, in combination with COULOMB's hypotheses in order to arrive at the following equation for maximum thrust:

$$E_a = \frac{1}{2}\gamma H^2 \bullet \frac{\cos^2(\varnothing - \omega)}{\cos^2\omega \cos(\delta + \omega)\left[1 + \sqrt{\dfrac{\sin(\delta + \varnothing)\sin(\varnothing - \beta)}{\cos(\delta + \omega)\cos(\omega - \beta)}}\right]^2} \tag{5-31}$$

where:

E_a = maximum active thrust, according to COULOMB's Theory
$\varnothing$ = angle of internal friction of the sand
ω = angle formed between the back of the wall and the vertical
δ = angle formed between the smooth surface of the fill and the horizontal

The other letters have the usual significance throughout this chapter.

If the back of the wall is vertical, $\omega = 0$ and Eq. (5.31) becomes:

$$E_a = \frac{1}{2}\gamma H^2 \bullet \frac{\cos^2(\varnothing)}{\cos\delta\left[1 + \sqrt{\dfrac{\sin(\delta + \varnothing)\sin(\varnothing - \beta)}{\cos\delta\cos\beta}}\right]^2} \tag{5-32}$$

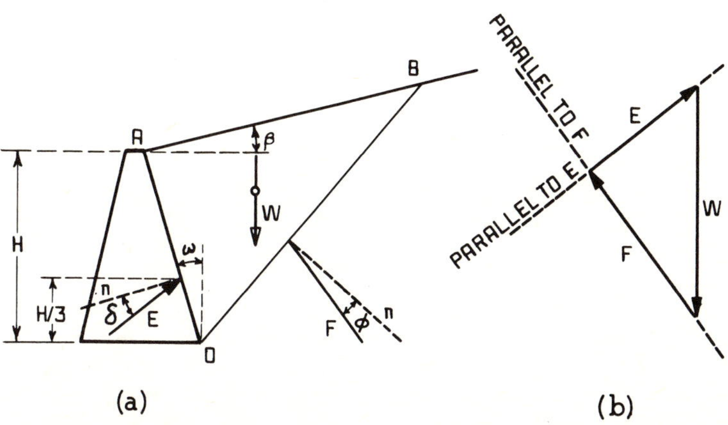

Fig. 5-7 Thrust mechanism of frictional soils against walls, according to COULOMB

If, moreover, the fill is horizontal, and $\beta = 0$, Eq. (5-32) simplifies to:

$$E_a = \frac{1}{2}\gamma H^2 \bullet \frac{\cos^2(\varnothing)}{\cos\delta\left[1+\sqrt{\dfrac{\sin(\varnothing+\delta)\sin\varnothing}{\cos\delta}}\right]^2} \tag{5-33}$$

It should be noted that if $\delta = 0$, meaning no friction between the wall and the fill, Eq. (5-33) leads to the following:

$$E_a = \frac{1}{2}\gamma H^2 \frac{1-\sin\varnothing}{1+\sin\varnothing} = \frac{1}{2N_\varnothing}\gamma H^2 \tag{5-34}$$

Here RANKINE's and COULOMB's theories are identical.

It is also important to point out that if in Eq. (5-32) $\delta = \beta$, Eq. (5-10) of RANKINE's Theory is obtained: that is to say, COULOMB's Theory agrees with RANKINE's if the thrust is regarded as parallel to the surface of the fill.

Coulomb did not himself consider the passive stress state, but his hypotheses have nevertheless been applied to this case and equations can be obtained that are similar to the ones presented for the active state. The equation for the passive state is Eq. (5-31) again, but replacing $\varnothing$ by $-\varnothing$, δ by $-\delta$ and changing the radical sign of the denominator. The equation thus becomes:

$$E_p = \frac{1}{2}\gamma H^2 \bullet \frac{\cos^2(\varnothing+\omega)}{\cos^2\omega\cos(\omega-\delta)\left[1-\sqrt{\dfrac{\sin(\delta+\varnothing)\sin(\varnothing+\beta)}{\cos(\omega-\delta)\cos(\omega-\beta)}}\right]^2} \tag{5-35}$$

Justification of the change is illustrated in Fig. 5-8.

If the angle δ is large, the actual slip surface is very different from the plane assumed by COULOMB's Theory. This leads to serious errors in the computed passive thrust which are sometimes greater than appropriate for safe design. TERZAGHI and PECK assess this error as being as high as 30% if $\delta = \varnothing$, with lower values for smaller δ angles. In the case of active thrust, the influence of the value of angle δ is far less and is usually disregarded.

COULOMB's Theory does not enable the distribution of stresses on the wall to be computed, for the wedge of earth producing the thrust is regarded as a rigid body with the boundary stress distribution not specified. For this reason, the Theory says nothing about the point at which active thrust is applied. To overcome this difficulty, COULOMB assumed that every point at the back of the wall represents the toe of a potential slip surface. The thrust produced against the wall by the soil above that point can be calculated. If a small increase is now considered in the height of the wall segment a new thrust can be calculated, then the difference $\triangle E$ is obtained. If this process is repeated, it is possible to find out the approximate distribution of pressures on the wall throughout the entire height; the resulting thrust will pass through the centroid of the pressure distribution diagram. This leads to hydrostatic stress distribution, with a thrust at height $H/3$ in a flatbacked wall with a uniformly sloping backfill. For situations where these conditions do not pervail, this method is laborious. TERZAGHI proposed a construction which, although approximate, nevertheless shows accurately enough the point at which thrust is applied. A line parallel to the slip surface is simply drawn through the center of gravity of the critical wedge. The point at which it intersects the back of the wall will be the point at which thrust is applied.

Here it is worth making general comments about COULOMB's Theory. As described, the method appears to take into consideration the stability equations in two directions (this explains essentially why the force polygon is closed), with two unknown factors, E and F of which only one is of real interest. It should be noted, however, that it would have been possible to work with just one stability equation and one unknown factor, *(E)*, had the stresses been projected on a line perpendicular to the direction known for F. COULOMB's Theory can thus be said to use just one equation to establish the equilibrium of the rigid wedge, which according to statics, is insufficient. Another way of viewing this deficiency is to examine equilibrium of moments. The three forces in vertical and horizontal equilibrium do not act through a common point; therefore the solution is not statically valid.

2. COULOMB's Method Applied to Cohesive-Frictional Soils

In practice, it is usually sufficient to regard the assumed slip surface as a plane which extends from the base of the wall to the crack zone, as shown in Fig. 5-9. COULOMB's Theory can then be applied as follows. (A more detailed description appears in [1]).

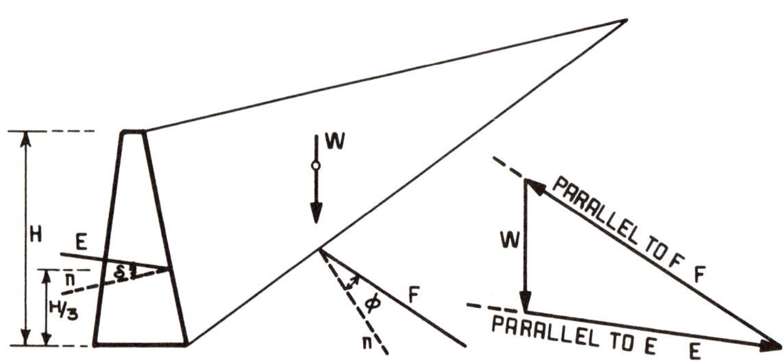

Fig. 5-8 Passive thrust of frictional soils, according to COULOMB

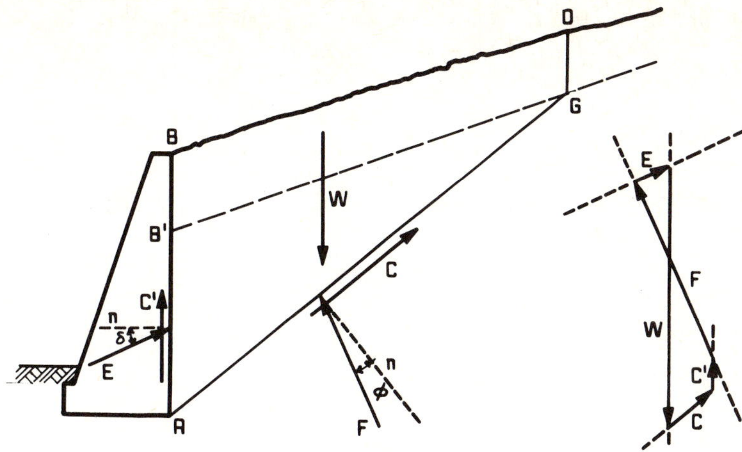

Fig. 5-9 Application of COULOMB's Theory to cohesive and frictional soils

If a sliding wedge is assumed, equilibrium involves the following forces; its total weight, W, calculated as a product of the area of the wedge multiplied by the specific weight of the soil; the reaction between the wedge and the soil with two components, F, due to normal reaction and friction, and C due to cohesion; adhesion, C', between the soil and the wall, and lastly, the active thrust, E.

These forces form the closed polygon which appears in Fig. 5-9, where it is possible to calculate the value of E corresponding to an assumed slip surface. Note that the magnitude of the forces C and C', can be found out by multiplying parameter c of the soil by lengths AG and AB' respectively.

This method of calculation leads to a trial and error procedure to determine the maximum possible E. The wall will, of course, have to be computed to withstand the combination of forces C' and E_{max}.

In the case of a passive thrust, COULOMB's Theory can also be applied if the shape of the failure surface is simplified by making it straight instead of curved. Under these circumstances, the design thrust can also be found by means of a trial and error procedure like that described for active thrust. However for passive thrusts, COULOMB's Theory does not give close enough approximations, and it is considered unsafe. Its use is not, therefore, recommended.

5.2.3 Other Calculation Methods Based on the Classical Theories

In engineering literature, various methods of calculation are described, which are based on the application of the classical RANKINE and COULOMB Theories. In [1], for example, graphical methods can be seen for applying COULOMB's Method to frictional backfills, which makes calculation work much easier. Also included in [1] are the friction circle and logarithmical spiral methods; the latter is useful for assessing passive thrust in cohesive-frictional fills. In these methods, the failure wedge of the backfill against the wall is no longer defined by a plane. A discussion on this important subject, is also published in [1].

5.2.4 Some Comments on the Classical Theories

The principal hypotheses involved in the use of RANKINE's and COULOMB's theories have already been briefly discussed. It is worth adding here that not one of them considers the effects of water that might accumulate in the fill. Such effects have to be considered independently, adding them to the thrusts of the dry soil, as in the analysis which was presented in §5.2.1.1. It is not advisable to design a retaining wall to withstand hydrostatic thrusts, because this leads to a very large structure which is expensive in comparison with one that will withstand just dry soil. Therefore it is better to minimize such hydrostatic thrusts. This can generally be achieved by means of drainage, as will be described later in this chapter. If adequate drainage is used, a wall design to withstand only soil thrusts is justified. If, however water should accumulate behind a wall designed for dry soil, the structure will probably fail; the hydrostatic pressure plus submerged soil produces about three times the thrust of dry soil (Plates 5-4, 5-6).

Knowledge of boundary conditions is essential for assessing the pressures on retaining structures. These conditions should be representative of the construction procedure and the interaction between soil and structure, and should not be based on simple analytical assumptions, such as a *rigid wall, smooth wall* or *rough wall.* For example, the movement of the wall in relation to its backfill is of vital importance. This movement is hard to observe and even harder to predict. On it depends the development of the low thrust values, of the active state, or the far higher values corresponding to the *at rest* condition if wall deflection is effectively prevented.

In all the classical theories, the entire wedge of soil behind the wall defined by the failure surface must be at limiting equilibrium and the strength of the soil mobilized as expressed in terms of the MOHR-COULOMB strength law (Chapter 1). Different hypotheses exist in relation to the shape of the slip surface. In [4], other methods and earth thrust theories are mentioned which are of interest in certain specific cases.

A reasonable degree of accuracy in the calculation of earth pressures is possible, when the classical theories are applied under suitable stress conditions; however, in practice there are several factors which impose limitations on the precision of calculations. One of these is the assessment of the angle of friction between wall and backfill, which is not only a property of the material but also depends on the nature and direction of the relative movements. It may vary all along the back of the wall. The value of this angle has been studied by ROWE and PEAKER [10] and by SCHOFIELD [11]. In all the classical theories, it is implicitly assumed that shear strength is fully mobilized in a uniform manner along the entire slip surface. In the field, this hypothesis is only partially met; this can bring about important differences in certain materials; progressive failure may invalidate the application of limiting equilibrium theories to dense sands and very stiff clays [10].

The most important condition in the application of the classical theories to the design of a retaining wall is the deflection of the wall: not only its magnitude, but also the way in which it occurs, translation or rotation, [12,13]. This condition becomes even more important in the case of excavation bracing.

5.2.5 Application of the Classical Theories to Practical Problems Posed by Retaining Walls.

In 1934, TERZAGHI [9] reported the results of a study, today a classic, in which the lateral pressures on a rigid wall were measured, as a function of the movement of the wall due to pressure from the backfill. TERZAGHI made the wall tilt about its base, in the direction of the active and passive thrusts, using backfills of loose and dense sand. The results reported are summarized in Fig. 5-10 [4,9]. Note the small displacement that is required in sand to produce pressures very close to those for the active state. For example, in the dense sand, a wall 10 m (33 ft) high would have to move approximately one centimeter (0.4 in) in order to reach minimum thrust. In loose sands, this tilting deflection may be 2 to 3 times greater. In [16] studies are presented which are basically in agreement with the foregoing

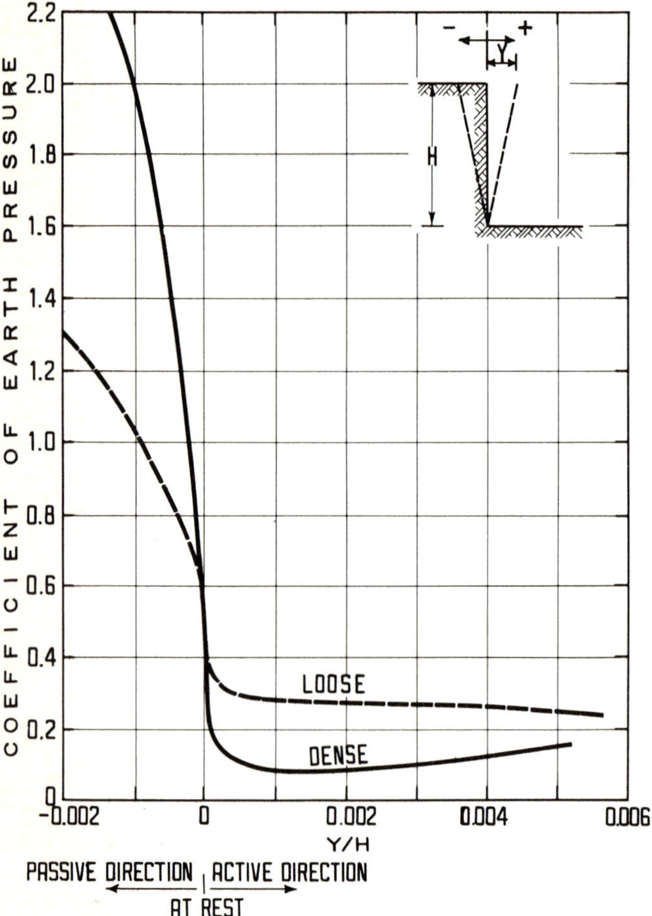

Fig. 5-10 Relationship between the movement of a wall and the earth pressure generated for loose and dense sands [4,9]

conclusions, but extended to soils other than clean sands. A far greater movement is required to mobilize passive strength. Reference [2] gives the results of various studies, which show that a deformation of 0.5% of the height of the wall is necessary to achieve half of RANKINE's passive resistance. Tilting of 2% of the height of the wall may be necessary for the passive strength to develop fully in dense sands, and up to 15% in loose sands.

From the foregoing, it is deduced that in masonry or concrete walls the necessary movements for achieving pressures close to RANKINE's active thrust can develop readily unless movement is in some way restricted or the walls are very rigid and directly founded on sound rock.

Furthermore, if the thrust acting on the wall were greater than the active thrust the wall would not run any serious risk in most designs. It would simply mean that the supporting soil at the base of the wall was resisting far more than would be necessary, with minimum movement. Long before breaking, the wall will move enough for the shear resistance of the soil backfill to be mobilized and the minimum, active pressure or thrust reached (Plate 5-7). The resistance of the backfill of a retaining wall will usually be mobilized long before the foundation soil reaches its peak strength. It will thus be reasonable to design the wall to withstand the active thrust alone, but at the same time adopt a sensible safety factor for the bearing capacity of the foundation soil. The bearing capacity of the foundation is the most difficult of the factors to assess.

Care must be taken not to restrict the wall's ability to move, because if the wall cannot move, the soil thrust will increase considerably. For example, in the wings of bridge abutments designed for active thrust and where the backfill is placed afterwards, sufficient space must be left between the wall and the beams and other rigid parts of the bridge for the necessary movement to take place.

Figure 5-11 [14] shows two typical situations where a restriction is imposed on the movement of the walls. TSCHEBOTARIOFF [14] comments on both cases and recommends the use of an earth thrust coefficient close to that of at rest. A value of $K = 0.5$ is appropriate for a sandy fill. Very high degrees of density are not recommended, at least in the immediate neighbourhood of the back of the wall because of the pressures generated by compaction, although care must be taken to avoid a condition so loose that it may cause settlement of the sand under the dynamic effects of traffic.

Another important factor is the parameter defining the shear strength of the fill. The material for the fill must first be selected and its properties then analyzed in the laboratory. A series of requirements will also have to be met if the conditions in the fill material are to be faithfully reproduced in the laboratory. If, for example, the fill is simply dumped behind the wall by a truck, a heterogeneous structure will be obtained in which the average strength cannot be reliably predicted. Because of exposure to heat and cold there will be changes in the strength with time that are difficult to anticipate. The classical theories do not take into consideration the changes which may occur in the condition of the fill as a result of seasonal climate, such as saturation or drying. Although the classical theories do not consider any effect of hydrostatic pressure that may develop in the water contained by the fill, methods for calculating water pressure have been presented earlier in this section.

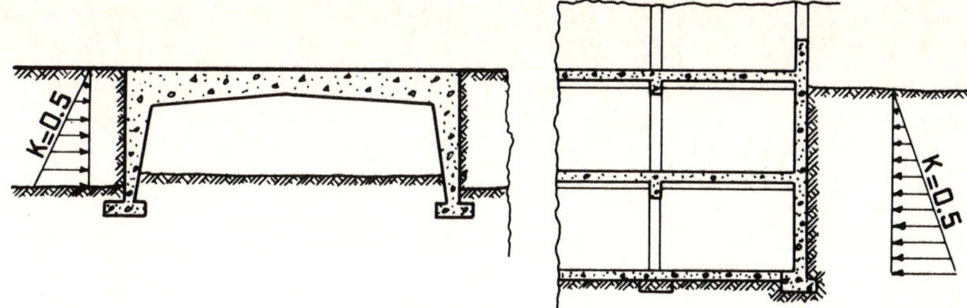

Fig. 5-11 Movement restrictions imposed on a retaining wall [14]

5.3 TERZAGHI's Empirical Method for Calculating Thrusts Against Retaining Walls

This method [1,15] was developed by TERZAGHI to provide a simple design approach for routine work, such as the smaller walls which are built alongside roads. It would be uneconomical and impractical to carry out all the preliminary studies to eliminate the uncertainties arising from seasonal variations and the other sources which have already been mentioned. The method is based on stability analysis for existing walls, of which only a few failed. It undoubtedly provides a conservative estimate of thrusts. It is therefore an appropriate method within the scope of the road engineer, so long as the walls under consideration are not very high, about 7 m (23 ft) maximum.

The first step is to classify the backfill material. There are five categories:

I. Coarse granular soil, without fines

II. Coarse granular soil, with silty fines

III. Residual soil, with pebbles, blocks of stone, gravels, fine sands and appreciable quantities of clayey fines.

IV. Soft plastic clays, organic silts or silty clays

V. Fragments of hard or moderately hard clay, protected in such a way that no water from any source can penetrate between the fragments.

Generally, categories *IV* and *V* are undesirable for fills, and should be discarded whenever possible. Category *V* in particular should be turned down if there is a risk of water entering the spaces between the fragments of clay, for this would cause expansion and a corresponding increase in the pressures on the wall. If, for any reason (which should be avoided whenever possible) the wall is designed before it is known what material will be used for the fill, the design must be drawn up on the most unfavorable basis. The method proposed is applied in four frequently occurring practical situations, in so far as the geometry of the fill and loading conditions are concerned.

1. The surface of the fill is flat, either sloping or not and without any surcharge.
2. The surface of the fill is sloping from the crest of the wall to a certain level where it becomes horizontal.
3. The surface of the fill is horizontal and subjected to a uniform surcharge.
4. The surface of the fill is horizontal and subjected to a uniform linear surcharge, parallel to the crest of the wall.

In case *1*, the problem can be solved by applying the following equations:

$$E_H = \frac{1}{2} K_H H^2; \quad E_v = \frac{1}{2} K_v H^2 \tag{5-35}$$

which give the horizontal and vertical components of the thrust acting on the vertical plane which goes through the extreme lower point of the wall, on the fill side. Figure 5-12 consists of graphs to obtain values of K_H and K_v, which are necessary for the application of these equations, as a function of the inclination of the surface of the fill and the type of material that is used. The criterion used to measure the height H must be taken from the figures. The equations and graphs give the value of thrust per linear meter of wall. The thrust must be applied to height $H/3$, measuring from the lower face of the wall.

Where a category *V* fill is used, the value of H considered in the calculations must be reduced by 1.20 m (4 ft) in relation to that shown in the figures, and the thrust that is obtained is applied at a distance of d where:

$$d' = \frac{1}{3}(H - 1.20), \text{ in m}; \quad d' = \frac{1}{3}(H - 4.0), \text{ in ft} \tag{5-36}$$

measuring from the lower level of the wall.

In case *2*, where the fill has a sloping surface to a certain height and then becomes horizontal, the values of K_H and K_v must be obtained from the graphs in Fig. 5-13. The same figure shows the rules which must be observed for the height measurements that are used and the points and planes where thrust is applied. When a category *V* fill is used, the height at which thrust is applied will also be given by Eq. (5-36).

When the fill is horizontal, and is subjected to a uniform surcharge, case *3*, the horizontal pressure on the vertical plane on which the thrust is assumed to act, must undergo a uniform increase of:

$$p = Cq \tag{5-37}$$

Where q is the value of the uniform surcharge in the appropriate units. The value of C of the previous equation will be selected from Table 5-1.

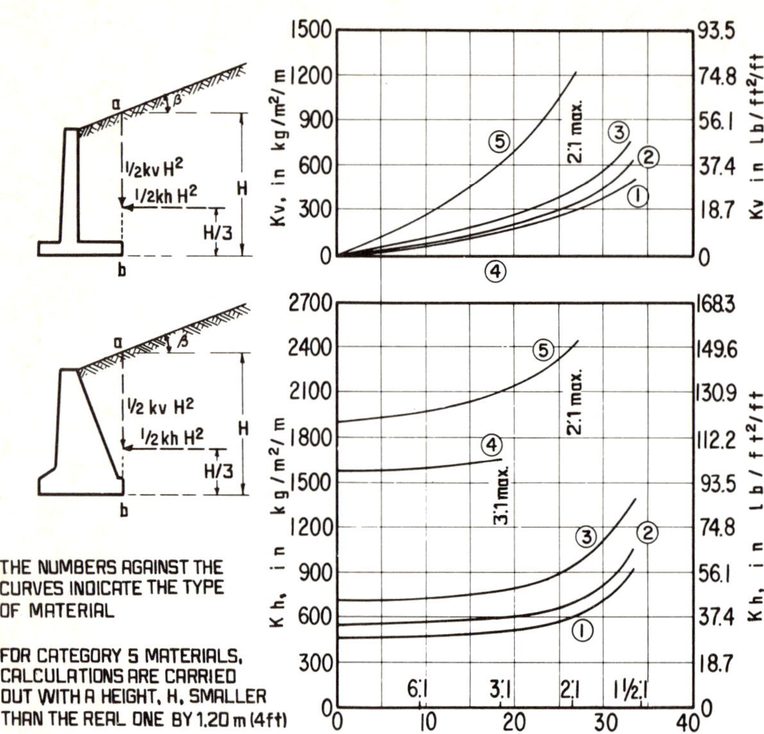

Fig. 5-12 Graphs to determine the thrust of flat-surfaced fills, according to Terzaghi [1,15]

Table 5-1
Values of C for various types of fill

Type of fill	C
I	0.27
II	0.30
III	0.39
IV	1.00
V	1.00

If the surface of the horizontal fill is subjected to a uniform linear load that is parallel to the crest, case *4,* it will act on a vertical plane, where the thrusts apply a concentrated linear load of:

$$p = C\,q'$$

Where q' is the value of the uniform linear surface load and C is obtained, as before, from Table 5-1. The application point P can be obtained from the construction which is shown in Fig. 5-14. If, by drawing the line down at 40°, P is below the base of the wall, the effect of q' can be disregarded. Load q' also produces on the foundation slab a vertical pressure that can be calculated (Fig. 5-14) by considering an influence at 60° from q', which is uniform throughout ab and with an intensity of q'/ab, considering for calculation purposes only the part of the pressure affecting the foundation slab ($a'b'$).

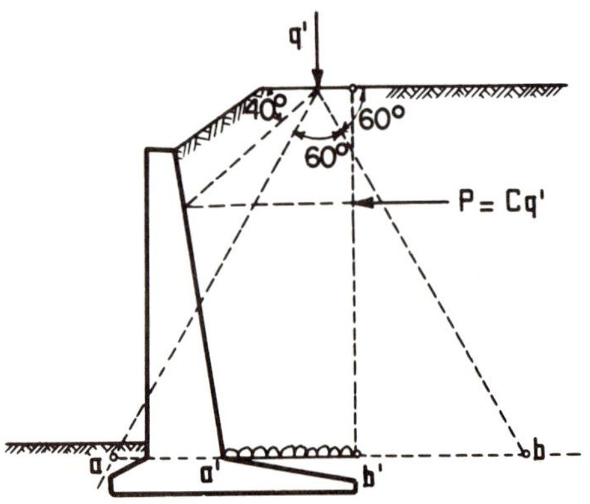

Fig. 5-14 Method of calculating the effect of a linear surcharge, Terzaghi's Method [1,15]

The methods described above refer to walls with firm foundations, where the friction and adhesion between soil and wall are directed downwards, with a stabilizing effect that tends to reduce thrust. If the wall rests on soft ground, its settlement may cause an inversion of the vertical thrust component. This leads to a considerable increase in thrust. In this case Terzaghi recommends that the values obtained for the thrust in the previous graphs be systematically increased by 50%.

In walls designed by Terzaghi's semi-empirical method, a good drainage system should be designed to ensure that there will be no hydrostatic pressures against the wall that were not considered in the previous graphs.

5.4 Drainage in Retaining Walls

As has already been said, the classic earth thrust theories do not include any effect caused by the pressure of water accumulating in the fill. Therefore for a safe and economical design they must always be used in combination with adequate drainage facilities. Terzaghi's empirical method takes into consideration the effect of seepage pressures and foresees that the characteristics of the fill may change with time as a result of accumulated water. He nevertheless recommends that drainage precautions be taken to avoid the accumulation of water behind the wall and the effects of freezing and thawing. The drainage of retaining walls should be regarded as compulsory, for it never proves economical to design a structure to withstand hydrostatic thrusts as well as earth thrust (Plates 5-4, 5-6).

The first precaution will be to provide outlet pipes or weep holes for the water that may accumulate within the fill. These go through the structure and have a diameter large enough to ensure that they will not become accidentally obstructed. A diameter of about 10 cm (4 in) is commonly used. The pipes are placed in parallel rows throughout the entire front of the wall.

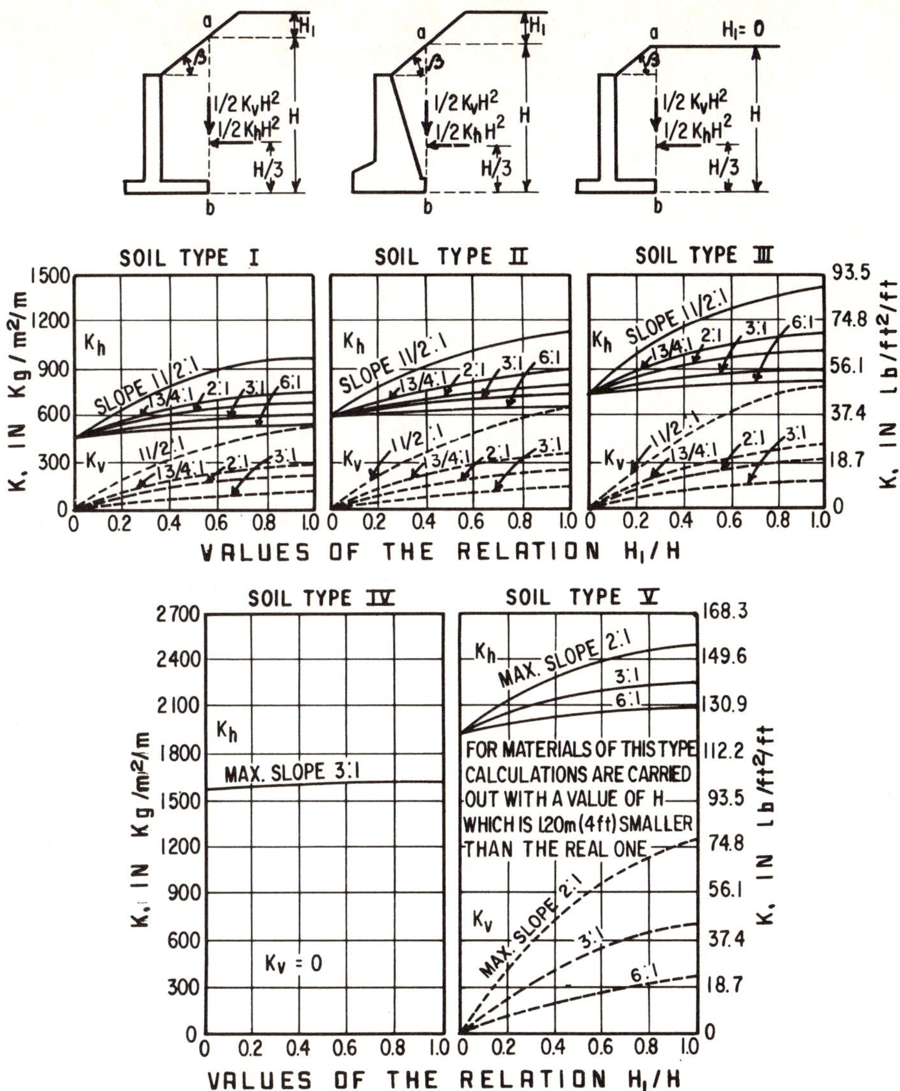

Fig. 5-13 Graphs to determine the thrust of embankment fills with a terraced backfill [1,15]

The vertical spacing between these rows is typically less than 2 m (6.5 ft), even in cases where fill perviousness allows water to accumulate at the base of the wall. Efficient drainage of the fill does not depend only on the outlet pipes which enable water to be drained. Only in exceptional cases, where the fill is highly permeable, (composed of rock pieces or gravel) can the pipes alone be considered sufficient. It will generally be necessary to provide the fill with a filter material, as will be described later. Let us consider the extreme situation of a clayey fill with very low permeability. In it, saturation and all the other harmful effects of water will continue to occur despite the fact that the water seeps towards the outlet pipes. This is due to the very slow rate of water movement towards the pipes, on account of the low permeability of the fill.

The horizontal spacing between the outlet pipes depends on the precautions that are taken to direct the flow towards the pipes. The engineer's influence is restricted to walls that are built in embankments, where the fill is placed after construction of the wall; here the fill material can be selected, at least within certain limitations imposed by availability cost. This is not true, however, in walls, that support cuts where the *fill* is determined by nature with all her inherent complexities.

The cheapest, but least effective, system for draining the fill is simply to place highly permeable granular material (about 50 kg (110 lb)) at the mouth of each of the pipes in the back of the wall. Under these circumstances, the horizontal spacing between the pipes should not be more than 1.5 m (5 ft). This system has the disadvantage that the fines in the fill can be eroded into the voids in the highly permeable material, thus contaminating it and making it useless. The system can only be used, therefore, when the backfill material contains little fines. Further, the water flowing from the outlet pipes falls on to the base of the wall, wetting the soil where the foundation load is highest, which should obviously be kept dry. This condition can be corrected if, instead of each of the rows of outlet pipes, drains of permeable material are installed throughout the entire length of the wall. These drains are made to discharge beyond the wall, where the water can have no harmful effects. Among the most elaborate systems of fill drainage is one with continuous layers of permeable material throughout the entire back of the wall or even inside of the fill. In this way the flow can be suitably directed and the effect of seepage forces becomes minimal or disappears [1,17]. Figure 5-15 depicts different possibilities for drainage installations. These become gradually more complicated, until systems are reached which modify the shape of the flow net through the fill and neutralize the effect of the seepage forces, as is described in [1, 17].

System *(a)* (Fig. 5-15) will only be useful in fills composed of highly permeable granular material with no fines. The system shown in diagram *(b)* cannot be used when the fill contains fines that are likely to contaminate the pockets of the permeable material. It will only be appropriate in permeable fills where the water can seep towards the pockets of more permeable material and into the outlet pipes.. Part *(c)* illustrates the drain which unites the mouths of the outlet pipes (or which takes the place of these pipes); they discharge laterally beyond the wall. Their effect is sometimes complemented by a discontinuous system of vertical drains placed against the back of the wall between the outlet pipes. Part *(d)* illustrates the drainage system most commonly used in the backfill of retaining walls. The drains are installed in a continuous blanket against the wall for the entire backfill height, preferably in a single layer (for reasons of cost and ease of construction) or in two or three layers, when necessary. This will be discussed later. Diagrams *(e)*, *(f)* and *(g)* show more complicated, and consequently more expensive drainage systems which are used when it is vital to alter the direction of the flow or to prevent water from being absorbed within the backfill.

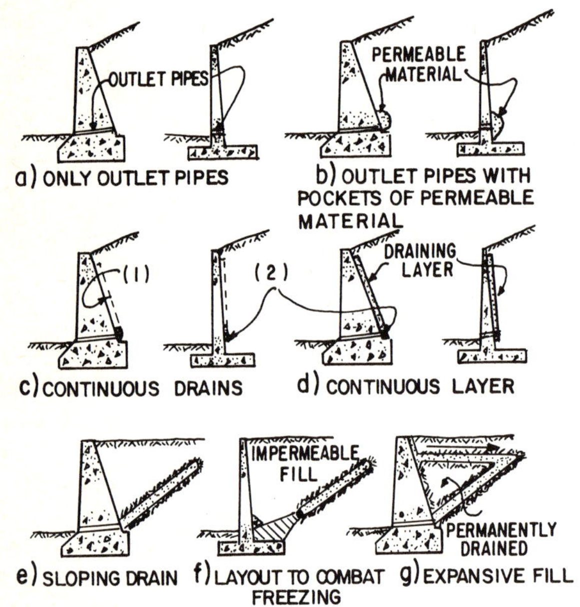

Fig. 5-15 Drainage systems at the back of a retaining wall and in the backfill [7]

All the above systems are designed to prevent the accumulation of water behind the wall and the consequent development of hydrostatic pressures. Only the last three systems are effective in redirecting the flow and minimizing seepage forces in fine soil backfills. The installation of drains also reduces the uplift under the wall that must be taken into consideration when calculating thrust in steady flow conditions [1,17]. This effect will have to be analyzed in each case, so that the most appropriate drainage system can be adopted.

The detailed design of a drainage system for a retaining wall belongs to the general problem of filter design and will be discussed in the corresponding chapter. The reader should refer to that chapter for the materials that are used for drainage as well as drain calculations.

The requirements that have to be met by a drainage design for a retaining wall, or for any other filter, are contradictory. What is required is a material that is sufficiently permeable for water to seep through it, and be easily eliminated without accumulating pressure. However, a material with these characteristics will be so porous that the water that seeps through it will carry away any fines in suspension. From this standpoint, a drain material should be fine grained and with a low permeability and low seepage velocity. It is hard to fulfill both requirements in a single layer of material. This is why a filter is used in two or three layers, with strict grading requirements, the material being placed according to increasing grain size, from backfill to drain pipe. The draining layer next to the fill (assumed to contain abundant fines) must be sufficiently fine to filter the fines suspension. It will not therefore be permeable enough to ensure free drainage. Therefore the water goes through to a second layer, with larger grains, which will act as a filter to the previous one but is more pervious. In some cases, a third more pervious layer will be necessary.

The above system implies a technology that is too rigid and delicate to be regarded as practical for routine road engineering purposes. It will be justified for some walls which, because of their height or some other special reason, involve unusual dangers. However it is always preferable to design draining layers consisting of a single layer, with just one type of material that can be produced and placed by one process. Moreover it is best if the drain material is as similar as possible to some other material which has to be manufactured for other purposes on the project, such as, a subbase or base material.

The chapter dealing with subdrainage gives the general criteria, (almost all based on grading) which have become established for the design of materials for draining layers. These criteria generally lead to drain layers or blankets made up of two or three different layers. In the same chapter, some Mexican experience is mentioned where a single draining layer can be used with a reasonable degree of reliability and, of course, great savings in work and cost.

When the drainage system includes a blanket which is placed against the back of the retaining wall, the thickness of this blanket must be such that it can be constructed as a continuous layer without undue expense. This is generally achieved with a minimum thickness of 30 cm (12 in) in low walls, increasing to 50 cm (20 in) in higher walls. The draining blanket discharges preferably by means of outlet pipes through the wall, or a perforated collector pipe which is installed at the base of the blanket and provides an easy exit for the water at both ends of the wall. The characteristics of the perforated pipe are described in the chapter on subdrainage.

5.5 Some Considerations Regarding the Design of Retaining Walls

It is not the authors' intention here to describe in detail the steps that are usually followed to design a retaining wall. Reference [7] describes design involving the application of the classical theories, and gives numerous details and considerations, [18] gives many practical design criteria and is very useful for the engineer who is faced with a specific problem. The aim of this section is to comment on those aspects which should be considered for a successful design for retaining walls. It is especially based on [16].

Usually the analyses, which are based on the classical theories, consider that the backfill develops peak shear strength. They apply a substantial safety factor to the proportions of the wall to avoid the possibility of sliding or overturning in the active state. More elaborate, but not more realistic attempts have been made using KOTTER's equations [19]. COULOMB's Wedge Method is valuable, because it makes it easy to consider factors such as the friction between the wall and the fill, the effect of seepage within the fill [20] and the influence of varying surcharges.

In the case of frictional fills made up of coarse permeable materials, in which hardly any pore pressures occur in undrained conditions, the angle of peak strength in drained conditions is usually used for the limiting analysis of the wall. If the movement of the wall is somewhat restricted, slightly smaller angles can be selected. The soil shear conditions with plane strain prevailing in the prototype will probably imply an increase of several degrees compared to the angle obtained in the laboratory with drained tests. It does not appear important whether the triaxial tests are carried out with an increasing vertical stress or a decreasing lateral stress.

In clays, the selection of the angle of strength for limiting analysis calculations is far more complicated. In an active state, when the average total normal stresses decrease, overconsolidated clays tend to expand and pore pressures drop temporarily. For non-sensitive clays, with an overconsolidation ratio between 1 and 3, the volume change is not very important, so that the drained and undrained strengths with decreasing chamber pressure are similar and there is little difference between the short-term and long-term strengths. If the movement of the wall is restricted, a strength smaller than maximum should be considered: that at which the stress remains constant under gradual deformation conditions (creep). In more highly overconsolidated clays, the movements that are required to reach the active state cause considerable expansion. The strength parameters reduce until they are the same as those corresponding to the drained strength under similar expansion conditions. In order to assess the effect of expansion on loss of strength, triaxial tests will be necessary with a decreasing chamber pressure; and to assess the long-term strength which should be used when computing permanent structures, cohesion values that are likely to drop with time should not be relied on.

It has been seen that the overconsolidation of a clay is of vital importance in the values of the earth thrust coefficients [16, 21]. For overconsolidation ratios between 3 and 5, the horizontal stresses are equal to the vertical stresses present ($K = 1$), and for values between 15 and 20 the horizontal stresses may be double the present vertical stresses ($K = 2$). In sands, with initial values of 0.35 for K_0, the residual lateral stresses which appear during the unloading process corresponding to the active state, may be equivalent to a value of K_0 between 0.6 and 0.9, if the sand has been previously subjected to loads equivalent to overconsolidation ratios between 2 and 5.

In many overconsolidated clays and clayshales, effective lateral pressures exist which are greater than the vertical pressures present. Later, the influence of compaction on the development of strong horizontal residual pressures will be discussed. From this point of view, compaction is a form of overconsolidation.

In highly consolidated or compacted clays, it is hard to distinguish the lateral pressure that is generated during swelling, from the pressure produced by overconsolidation or compaction. Experienced engineers recommend that clay backfills should not be used; but sometimes they have to be used. The way in which fills are usually placed sometimes enables them to develop pore tensions, which compress the mineral skeleton and make the lateral pressure on the wall temporarily smaller than that corresponding to effective stresses in the fill. Even with drainage, however, the seepage and capillary rise of water into the fill will change this state of pressures. Pore pressure will develop which will eliminate the compression on the mineral skeleton and the pressure on the wall will become equal to, or greater than, that equivalent to effective pressure. Then, if the wall moves, the pressure on it will be temporarily relieved, but with seasonal variations, especially rain, the clay can expand applying increasing pressures on the wall which may damage it.

If the wall is constructed against the face of a cut in overconsolidated clay, the above mechanisms can be expected, and water seeping into the fill, or rising by capillarity independently of drainage, will eliminate pore tensions. As a result, expansion will develop accompanied by large pressures, even if the wall is capable of moderate subsidence. This is particularly serious in expansive clays at low initial moisture contents.

The materials that are most susceptible to expansion are clayshales with a high plasticity index and with no cementation of their minerals. In these materials, walls must be designed to withstand pressures even higher than the at-rest state.

Most engineers with experience in designing retaining walls agree that assessment of the lateral pressure is less important than designing an effective drainage system for the fill or determining foundation bearing. In [22], numerous cases of failures are analyzed; the conclusion is that two thirds occurred in walls built on clay foundations. The other third were walls with a fill consisting of clayey soil or a material whose nature was not reported.

Active thrust is usually recommended as a basis for design of granular backfills behind gravity retaining walls, or cantilever walls founded on soil. For cantilever walls founded on rock or any kind of wall on piles, the use of higher than active pressures is usually recommended. The pressure corresponding to the at-rest condition is used for designing gravity retaining walls founded on rock or piles. When the fill consists of overconsolidated clay it is not considered appropriate to base the design on the active thrust calculated with the maximum undrained strength. The strength is usually selected from among the following values: undrained strength parameters close to the limiting or residual value; drained strength parameters, preferably without apparent cohesion; or a value of the angle of strength between the peak drained strength and the residual strength.

The wing walls of bridge abutments and similar structures should be computed with a thrust that is larger than active. This depends on the degree of restriction to rotation or translation. The design basis that is recommended for walls with total restriction of movement and a sandy fill has already been mentioned. References [16, 23] mention some cases where the conditions of the interaction between soil and structure (relating here to restrictions on the movement of the wall) cause higher than active pressures.

In walls above sandy ground, either no perceptible settlements occur, or else they occur during the construction period. When the wall is founded on clay, large settlements may continue to occur over a long period of time as a result of consolidation. In this case, it is recommended that the resultant of the external forces acting on the wall pass about halfway along the base, so that the settlements will be more uniform and risks of overturning avoided. Walls have relatively little resistance to differential settlements in the longitudinal direction. Therefore if there are important variations in the foundation ground along the wall, they must be considered when designing the different wall sections. Settlements sometimes cause fracture of the wall structure or total collapses because of unexpected variations in the pressures of the fill related to the soil-structure interaction.

Figure 5-16 shows the most common types of retaining walls that are used in road engineering, Fig. 5-17 illustrates the most common types of failure likely to occur in retaining walls.

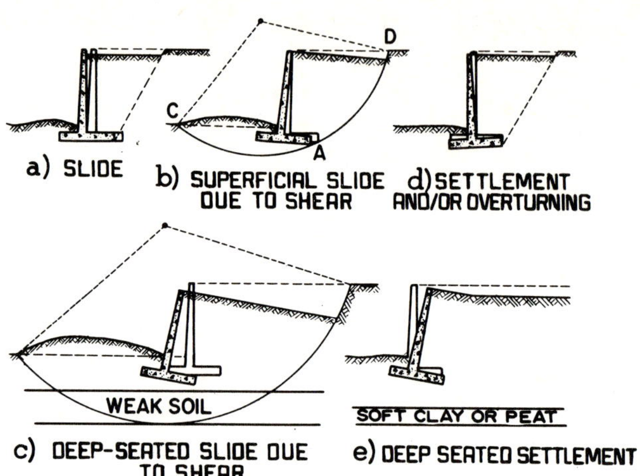

Fig. 5-17 Common failures in walls

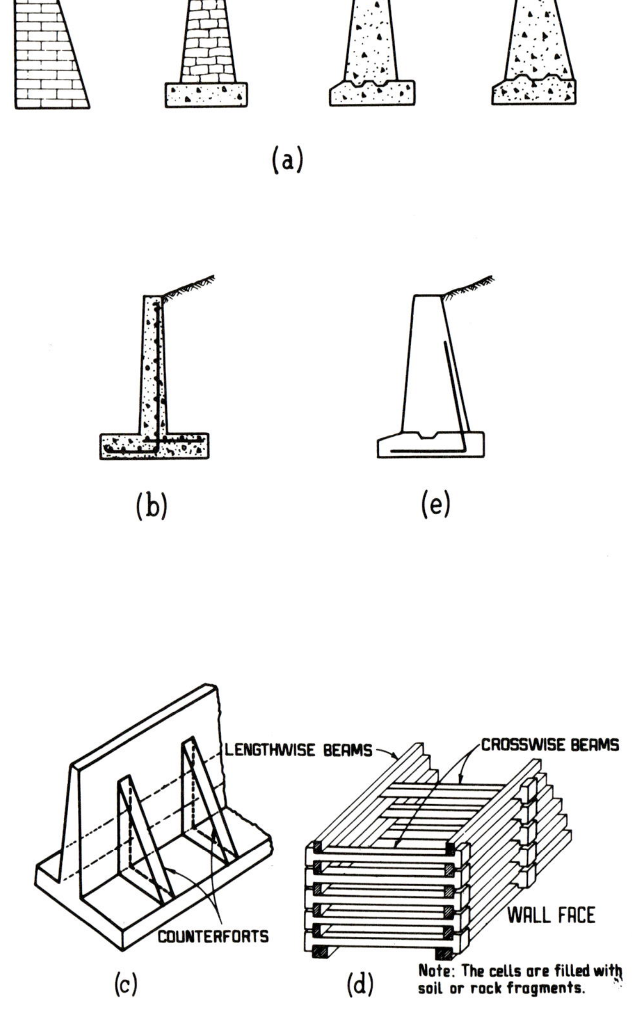

Fig. 5-16 Different types of retaining walls [18] a) Gravity retaining walls made of masonry or plain concrete, b) Cantilever wall, c) Counterfort wall, d) Crib wall, e) Semi-gravity retaining wall (with small amount of steel reinforcement)

5.6 Design of Deadman Anchors

This section deals with design criteria for deadman anchors, either in plate or massive form, which are placed well below the ground surface. The principal source of the recommendations given is [14]. Fig. 5-18 shows a system of anchor plates in sand.

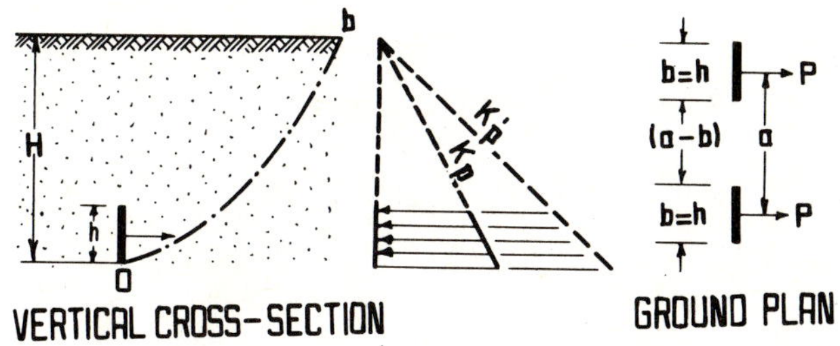

Fig. 5-18 Resistance of deadman anchors (massive or plates) in sand [14]

In the past, the only factor considered used to be the passive strength that developed in height, *h,* of the plate, as is shown with arrows in the conventional diagram given in Fig. 5-18. In addition, low values of K_p were selected for the sand and a value $\delta = 0$ was adopted for the friction between plate and sand. Consequently, the values obtained for the passive thrust were too conservative. Figure 5-18 also depicts the form of experiments conducted at the Franzius Hannover Institute [14], with results which allow far more realistic values for the passive thrust. The test refer to the resistance of a very long (continuous) plate and to that of a series of square plates with a width $b = h$ at different depths, *H*. The results of the experiments are given in Fig. 5-19 [14].

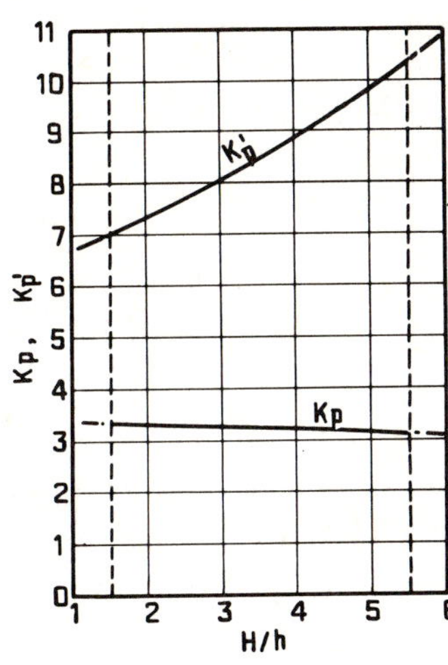

Fig. 5-19 Coefficient of passive thrust for deadman anchors (plates) in sand [14]

The resistance of a continuous plate per unit of width can be computed by means of the equation:

$$E_p = \frac{1}{2} \gamma H^2 K_p \qquad (5\text{-}38)$$

where the values of the coefficient K_p must be taken from the graph in Fig. 5-19 as a function of the relation $\frac{H}{h}$. The graph is considerable to be reliable for $\frac{H}{h}$ between 1.5 and 5.5. When applying these results, it should also be remembered that the Hamburg experiments were carried out on medium compact sands with an angle of friction $\varnothing = 32.5°$.

The resistance of square plates with a width $b = h$ can be computed using the equation:

$$E_p = \frac{1}{2} \gamma H^2 K'_p b \qquad (5\text{-}39)$$

The values of K'_p must also be taken from the graph in Fig. 5-19, within the same limits. Equations (5-38) and (5-39) can be used with a small safety factor (1.25) for temporary structures or with 1.5 to 2 for large-scale structures of a permanent nature.

For the case of deadman anchors in clay, the solution proposed by BRINCH HANSEN [24], quoted in [14], can be considered. Figure 5-20 shows a deadman anchor with a rectangular cross-section installed in a plastic clay. The clay does not undergo a change in volume and moves around the deadman, offering an undrained shear resistance of $s = c$ to that movement in the flow lines. The dot-and-dash lines indicate the slip surfaces.

The total strength per unit width of the deadman obtained theoretically by HANSEN was:

$$p = 11.4ch \qquad (5\text{-}40)$$

where c is the cohesion of the soil.

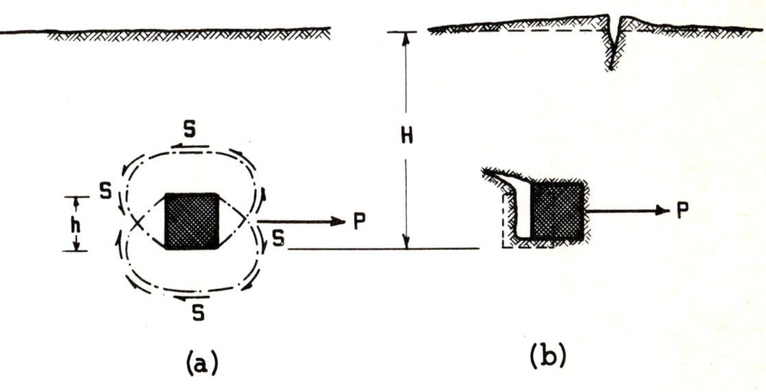

Fig. 5-20 Slip surface around deadman anchors in clay.
a) Theoretical considerations by J. B. HANSEN [24]
b) Experimental results by MACKENZIE [25]

In [25], details are given of the investigations carried out by MACKENZIE on scale models of deadmen placed in two different types of clay. In these tests, the value of γh was very small in comparison with the value Eq. (5-18), which is possibly why the deformations around the deadman were like the ones shown in Part (*b*) of Fig. 5-20. The total strength measured by MACKENZIE was smaller than the one obtained theoretically by HANSEN. MACKENZIE obtained as a maximum:

$$p = 8.5\, ch \qquad (5\text{-}41)$$

The graphs in Fig. 5-21 show that the total strength ($2ch$) of the wedge opposite the deadman was effective only for a relation $\frac{H}{h} = 1$, that is when the deadman was touching the surface of the ground. For increasing values of the $\frac{H}{h}$ relation, the total strength is smaller than the one which can be computed using Eqs. (5-18) or (5-20), and its value is given by Eq. (5-41), as a lower limit. The experimental values which are shown in Fig. 5-21 can be used for design purposes, with a reasonable safety factor (3 according to [14]).

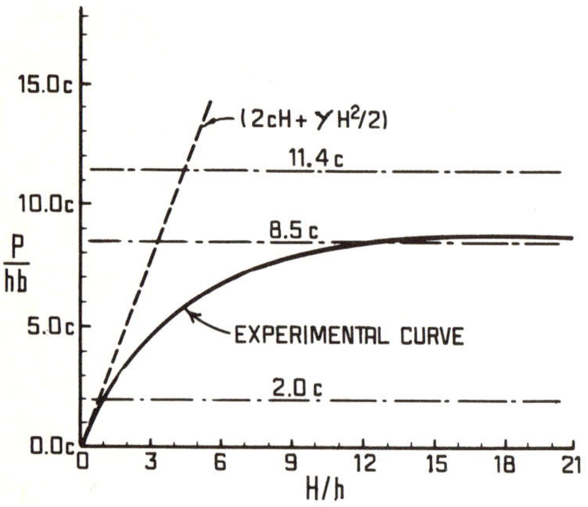

Fig. 5-21 Resistance of deadman anchors in clay [24,25]

5.7 Crib Walls

The characteristic type of crib wall is illustrated in Fig. 5-16. It is built with timber prefabricated reinforced concrete or special steel parts. The parts are placed in such a way that they form parallelepipedic cells which are afterwards filled with soil or rock. Figure 5-22 [14] shows a detail of these structures; it includes a ground plan and a profile (see Plates 5-8, 5-9).

The width b of the wall is determined by analyzing the whole structure in the same way as a common gravity retaining wall, under the effect of its own weight, thrust and reactions of the ground. The dimensions of the prefabricated parts are selected so that $f \geqslant 2e$ (Fig. 5-22). Otherwise, a granular fill material would leak out. The lengthwise beams B must be designed to withstand bending, as a uniformly loaded beam between two supports. The force that causes bending, on account of the earth thrust, will be $0.5\ ab\gamma\ (d + e)$, γ being the unit weight of the fill. All the other quantities are defined in Fig. 5-22. The beams marked B must also withstand half of the total vertical load acting on the cross-beams A, which will be described in detail later. This lateral pressure, which corresponds to the thrust in a silo, develops when soil is placed in the crib before the back fill is place. This fill does not produce flexural moments of significance. When A and B are made of concrete, they must be reinforced symmetrically on the sides of tension and compression. The crosswise beams A must also be reinforced to the total lateral stress which is exerted against B as tension. The head of A must be capable of resisting the same stress as shear. A must be designed like a beam, with a span b and two supports. On them will be considered a total vertical stress equal to: $0.58.05.\ ab\gamma\ (d + e)$, where 0.58 is the coefficient of friction between beam and fill (equivalent to $\delta = 30°$). This stress is transmitted by friction in both sides of the element.

The chief virtue of crib walls is that they are capable of resisting considerable differential settlements without suffering undue damage. They can also withstand tilt around the base and horizontal movements which, in any other type of structure, would be destructive.

The material that is used to fill the cells of the wall must be both frictional and permeable. This last condition will be a valuable contribution to the general drainage of the backfill. Adequate compaction should be used when placing the material in the cells, so as to avoid subsequent grain rearrangement and settlement; otherwise there will be the risk that the earth thrust, which should normally act on the wall in a downward direction, will become inverted and act on the wall in an upward direction, thus producing far greater thrusts, as can be seen in the analysis [26]. A similar effect may occur when the wall settles in relation to the backfill. Because these walls are commonly built on soft ground, this condition should always be kept in mind at the time of analysis.

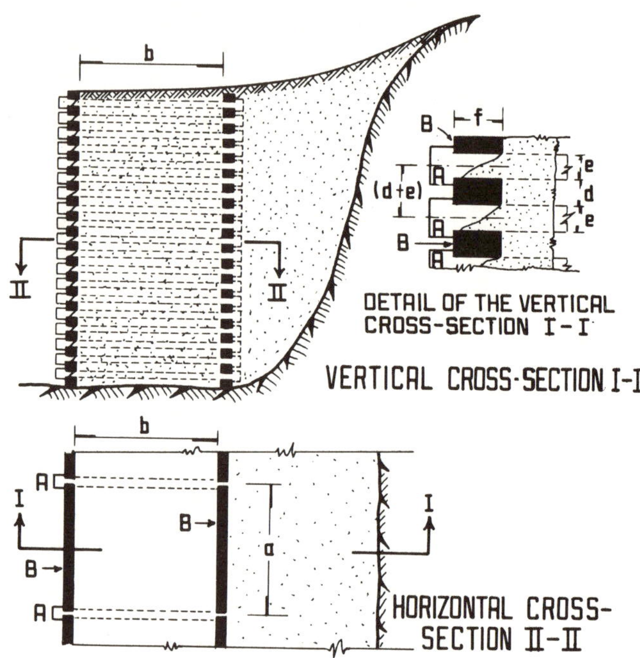

Fig. 5-22 Crib wall [14]

Plate 5-8 Crib wall

Plate 5-9 Another view of a crib wall

5.8 Backfills for Retaining Walls

The backfill which plays such a vital role in the behaviour of the foundation soil and structure as a whole, is responsible for the success of a retaining wall. The principal factors to consider are: the nature of the fill materials; the environmental conditions; the methods used; the compaction procedure and its intensity, and the drainage systems that are installed.

The type of material used for the fill, and the placement method, greatly influence earth pressure and settlement of the surface of the fill, (which is frequently a problem in road engineering). The classification of fill materials, which was included in the description of TERZAGHI's empirical method, should be regarded as a general guide. *Fat* (highly plastic) clays should be avoided, and if possible the design should use non-cohesive permeable soils. Expansive clays are undesirable materials, but they can be used with due regard to the pressures they generate. In cases where settlement of the surface of the fill is an important factor, the use of materials containing large quantities of shells is not recommended, although the general behavior of such materials is frictional. Also to be avoided for the same reasons are mixed fills with materials that are likely to penetrate one another; for example, alternate layers of rock fragments and fine sand, or rock fragments placed directly on top of soft clays. In both examples, a layer of coarse sandy material would be necessary in-between to act as a filter or cushion to prevent embedding. Fill materials should also have low susceptibility to freezing [27].

The ideal materials are sands and gravel or graded crushed stone. These are stable soils with high shear resistances; they do not settle when correctly placed and their good qualities are not affected by water. The material that is used should fit this description as closely as possible. The design engineers can seldom obtain data on the precise mechanical properties of the fills that will be used during construction; these are arbitrarily established for calculation purposes. This practice is admissible for low walls with good quality fill, where serious errors in design are unlikely.

If material that is very clayey has to be used for a fill, it can be expected to lose its cohesion, and subsequently act like a fluid having the same unit weight as the saturated soil.

When a backfill is placed after the construction of a retaining wall (as is usually the case), it is usually constructed in horizontal layers with a slight downward slope away from the wall. A layer thickness (loose) of about 20 to 30 cm (8 to 12 in) is acceptable, although thinner layers of about 10 cm (4 in) may prove necessary in cases where intensive compaction must be avoided; this will be discussed later. Compaction increases the shear strength of a fill. The earth pressure is thus reduced, and settlement is minimized. On the other hand, permeability drops, and this may restrict drainage. An uncompacted fill will settle behind the wall, and this can affect the direction of the earth thrust and cause a notable increase in its magnitude.

From the foregoing, it may appear that compaction has favorable effects on a fill. Unfortunately, however, this is not always the case. When a backfill is subjected to intensive compaction, excessive deformation usually occurs in the structure of the wall, even if the precaution is taken of not allowing heavy compaction equipment to work very close to the back of the wall. It has also been observed that after compaction at least part of the horizontal pressures which developed during the process remain in the fill soil. G. F. SOWERS et al. [28] have made a study of the intensity and effect of these residual pressures. They report the results of an investigation involving different sands and a sandy-silty clay (CL-ML) with $W_\ell = 41\%$ and $I_p = 14\%$, part of which took place in the laboratory and part in the field. The laboratory research included static and dynamic compaction in a cylinder with a diameter of 10 cm (4 in) provided with a device for measuring lateral pressure. For each test, several different compaction water contents were used between the dry condition and the optimum water content. Field tests consisted of in-situ readings taken in a low wall.

In the laboratory it was seen that the residual pressures in sands scarcely developed, but in clays they were found to be high. Figure 5-23 shows the residual pressures that developed as a function of the compactive effort in a dynamic or impact compaction test. Results of the static and dynamic compaction tests indicated that residual pressures are far greater when the material is compacted on the dry side, decreasing considerably when the compaction water content is close to optimum. In the clay, however, even within these limits the pressures were still high.

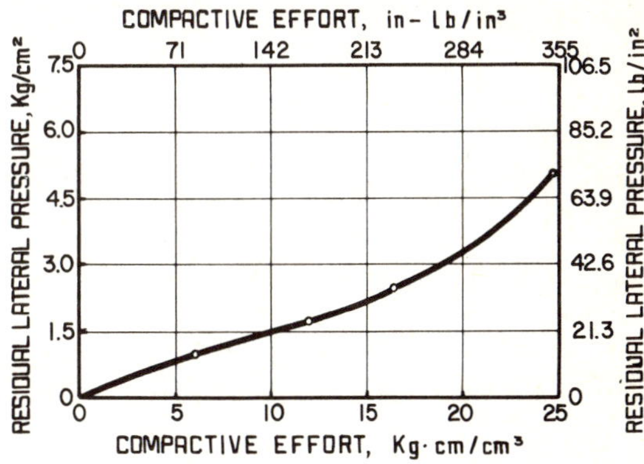

Fig. 5-23 Relationship between the compactive effort and the residual lateral pressure for a silty clay with $w = 14\%$, Dynamic test [28]

Figures 5-24 and 5-25 show the results of field measurements taken on a river sand and a silty clay, respectively. It can again be seen how the residual lateral pressures are far smaller in the sands.

In view of the foregoing, it can be said that compaction of the backfill of a retaining wall must be handled with care. It is

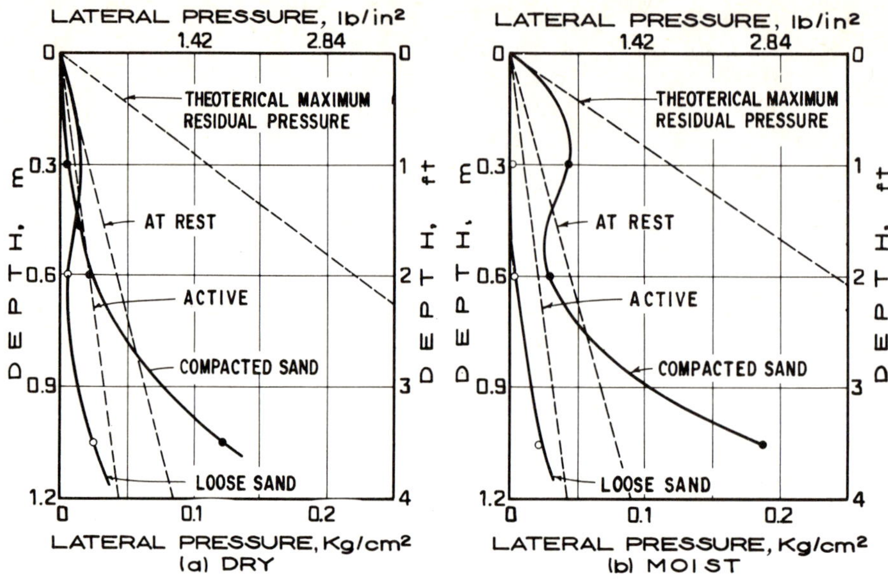

Fig. 5-24 Residual lateral pressures as a function of the depth of the backfill [28], River bed material a) dry, b) with $w = 14\%$

necessary to use sufficient compaction to avoid settlement of the fill due to its own weight or any other effect, but it should always be kept in mind that the development of residual lateral pressures that may cause a large increment in earth thrusts must be avoided. If the fill is built before the retaining wall, it seems likely that the residual pressures will be far smaller than if the construction order is inverted. This of course depends on the degree of lateral confinement with which the fill is compacted, and the construction procedure for the wall.

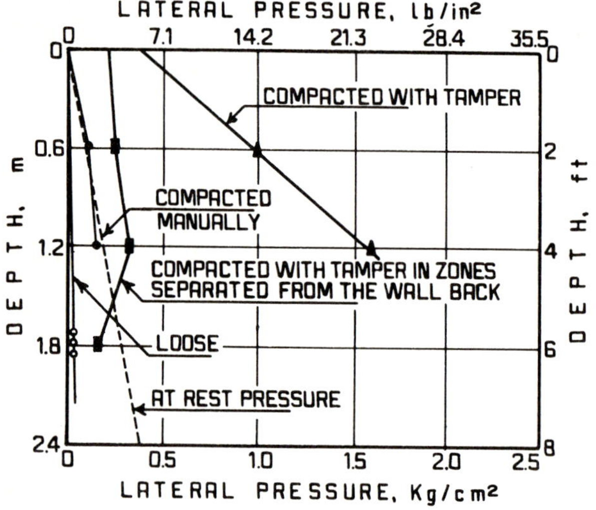

Fig. 5-25 Residual lateral pressures as a function of the depth of the backfill [28], Silty clay fill with $w = 18\%$

5.9 Bracing

Temporary excavations or cuts can be made without any support. The sides of the excavation have the steepest slope that is compatible with the stability of the soils during the period of excavation. Often the risk imparted by a slope that is too steep is severe, but the alternative solution of flattening the slopes would be too expensive or not possible owing to lack of space. In situations like these, a temporary support is used for the excavation faces, which are then generally vertical. This temporary solution is termed excavation bracing or earth cofferdam.

The depth of the excavation and the nature of the soil or rock determine the magnitude of the problem, the need for bracing and the type that should be used. TERZAGHI and PECK [29] have mentioned the arbitrary figure of 6 m (20 ft) for distinguishing shallow excavations (where special precautions are not usually required for the design and installation of bracing) from deep excavation where large-scale investigations, a careful determination of earth pressures and engineering design are necessary. This criterion may prove correct in the light of practical experience; if so, most braced excavations in road engineering will be simplified. The excavation bracing theory will be described in subsequent sections including the experimental design data that have been attained through recent research, but without recommending detailed analysis of the conditions of each specific problem. In certain cases that will be described later, some of the procedures that are recommended for specific road engineering problems are proposed (by their authors) for excavations that are considerably deeper than 6 m (20 ft). To follow these recommendations for shallower excavations may appear conservative, but it is wise in view of the uncertainties and limitations arising in road engineering.

Excavations for relatively shallow foundations for bridges, or retaining walls or structures for subdrainage, which are built to stabilize large masses of earth, are the situations where bracing is most commonly used in road engineering. This does not include tunnels, which are a special case of bracing that will not be discussed in this section.

Both theory and experience show that for temporary bracing it is impossible to support the hypothesis of a triangular distribution of earth pressure of the type derived by the classical theories. TERZAGHI's work on models of retaining structures with sand fills has already been mentioned. The object of his investigation was to discover how pressure is affected by movement of the structure. It was seen that when the wall tilts about its base, the hypothesis of a linear distribution of pressure agrees reasonably with the pressures measured, and the thrust can be represented by a force that is applied at the lower third of the height of the triangular diagram. This is the type of movement which normally occurs in a retaining wall. TAYLOR [3] carried these studies further by making the wall turn about its top. He found that the pressure measured is very different from the triangular linear distribution: a parabolic distribution with a maximum that is approximately at the center of the height of the retaining structure. The resulting thrust is notably larger than the one which was found using the classical theories.

The movement is similar to that which can be expected in a braced excavation. The usual procedure is to add struts from the top down as the excavation progresses. Therefore the ground's ability to move becomes more restricted in the upper part of the excavation than in the lower part. This is one fundamental difference between a retaining wall and excavation bracing. Another difference is that a retaining wall is a structural unit and fails as such, whereas bracing can fail strut by strut. Each piece that breaks causes an overload that affects adjacent struts, encouraging progressive failure. In retaining walls, it is reasonable to talk about the shear strength of the fill as a whole. In bracing, however, the excavation procedure, local variations in the soil, small alignment errors, time, temperature and many other factors have such a marked influence on the acting pressures that it is difficult to think of a comprehensive theory that would enable realistic values to be found for these pressures ([1, 29] describe some theoretical attempts at assessing lateral pressures in braced excavations).

Consequently, the practical methods of computing braced excavations that are in use today are of a more empirical nature. They are based on general conclusions from major projects, where instrumentation and measurement of the acting pressures have been carried out [31]. Pressure distribution curves have been obtained, and the corresponding envelopes plotted, which give a conservative design pressure for a specific situation. The application of these envelopes to situations different from the specific one from which they were obtained is problematical. The possibility of application will depend not only on the similarity of the situation under consideration, but also on the validity of the data and the criteria for plotting the envelope.

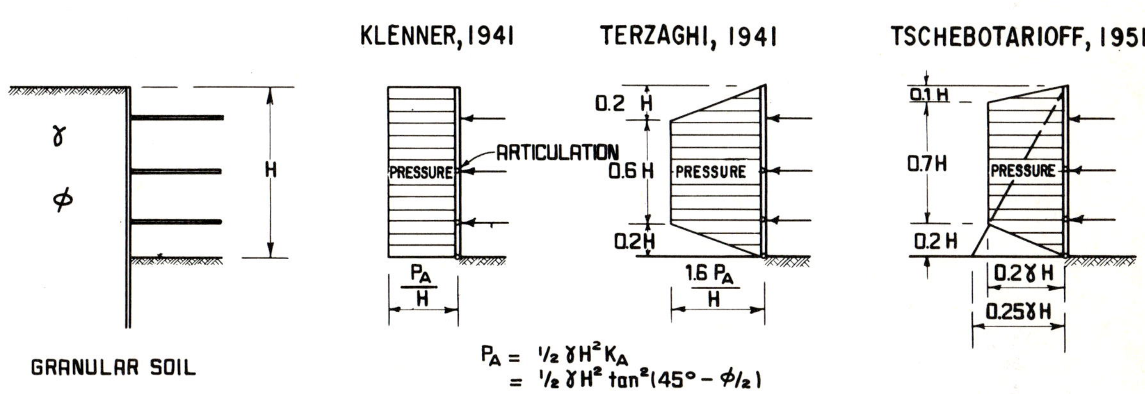

Fig. 5-26 Earth pressure envelope in sand-retaining bracing [31]

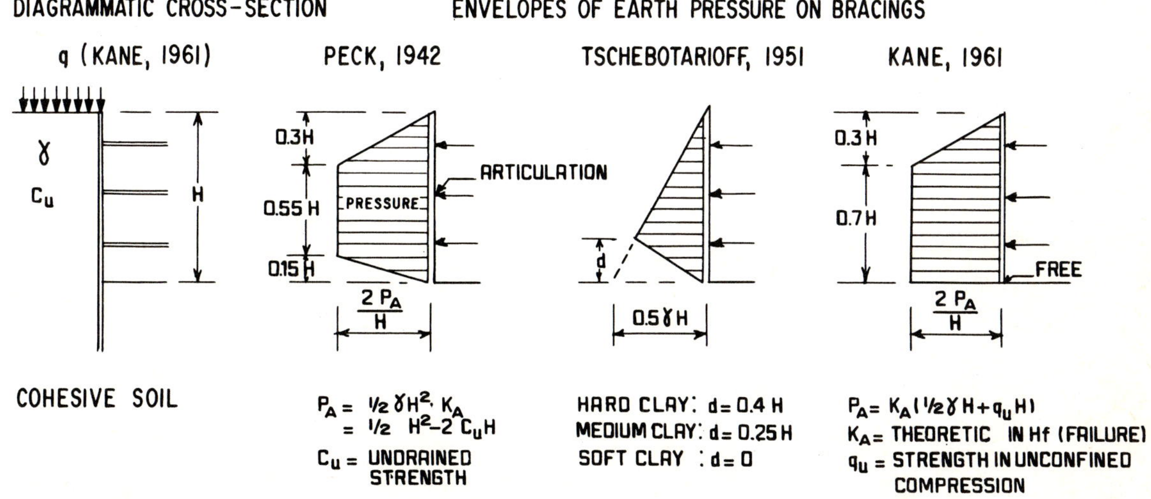

Fig. 5-27 Earth pressure envelopes in bracing retaining cohesive soil [31]

The first large-scale measurements in sands were carried out in the Berlin underground railway in 1936 by SPILKER [31] and in New York in the same year by WHITE and PRENTIS [32]. These investigations were responsible for the hypothesis that the envelope of the distribution of the pressures is uniform or trapezoidal. KLENNER reached similar conclusions for sands in 1941 [31], Fig. 5-26, as a result of studies conducted in Berlin and Munich. On the basis of the investigations he carried out on sandy soils in the Berlin underground railway, TERZAGHI developed the design envelope which appears in Fig. 5-26 and which has been widely used in professional practice [1,29].

The first measurements in clayey soils were made by BRUGGEN in Rotterdam [31]. PECK later reported a very complete set of data from the Chicago underground [33], Fig. 5-27. From these observations not only was the design pressure envelope obtained, but also an analytical method for calculating overall stability, based on the hypothesis of a failure circle which extends to the bottom of the cut. A theoretical analysis of PECK's data was carried out by WU [34]. TSCHEBOTARIOFF made yet another proposal for pressure envelopes, both for sands and clays [35], Figs. 5-26, 5-27. On the basis of a practical case, SKEMPTON and WARD [36,37] discussed the possibility of redistributing pressures in a parabolic form. They concluded that theoretically a parabolic distribution exists in sands, but may not exist in clays.

Reference [31] contains two diagrams proposed by the author for the special soils that are mentioned. They are largely based on field data. Both distributions are shown in Figs. 5-28 and 5-29.

BJERRUM and EIDE [38] studied the problem of base failure, which has been the cause of collapse in many braced excavations. In [39], a similar detailed study of this type of failure can also be found. The procedure proposed by BJERRUM and EIDE is shown in Fig. 5-30.

References [40-45] hold a special interest for the reader who wishes to go deeper into earth pressures against bracing, [31] is also of particular interest. Reference [1] deals briefly with the soil arching concept, mentioned by TERZAGHI, which is very useful for understanding the redistribution of stresses which occurs in braced excavations, for example between a tightly braced cut and its less restrained neighbours.

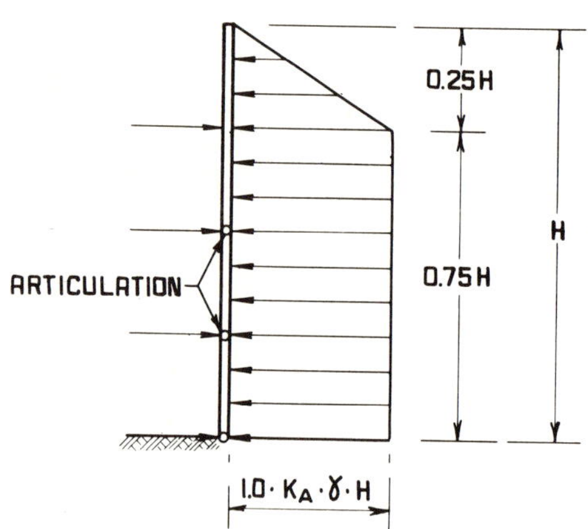

Fig. 5-28 Earth pressure envelopes in bracing retaining soft and medium clays [31]

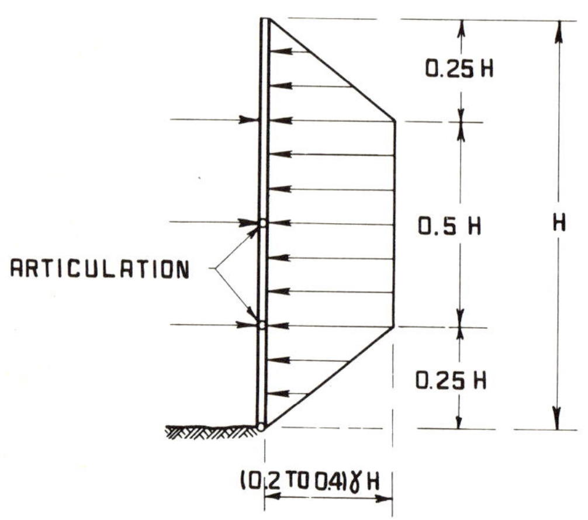

Fig. 5-29 Earth pressure envelope in bracing retaining hard clay [31]

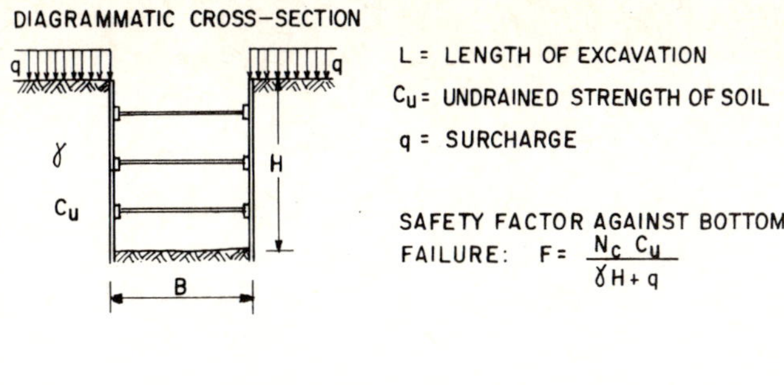

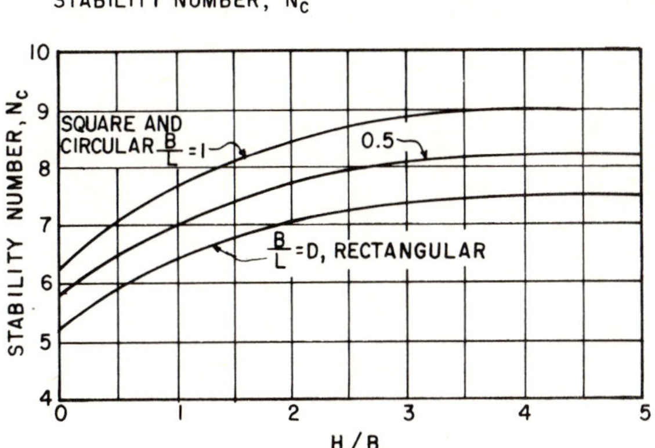

Fig. 5-30 Analysis of bottom failure [31,38,39]

5.10 Reinforced Earth

A new system for using earthy materials for construction made its appearance fairly recently. Owing to its similarity with the coupling of concrete and its steel reinforcement, the new system has been given the highly suggestive name of *reinforced earth*.

References [46,47] by the inventors of the new system, give details of how and when it can be used, as well as design techniques.

Reinforced earth is a combination of earth and linear reinforcing strips that are capable of bearing large tensile stresses. The strips, or linear elements, are usually of metal or plastic. The reinforcement provided by these strips enables the mass to resist tension in a way which the earth alone could not, with the added advantage that reinforcement can be placed in the most convenient directions. The source of this resistance to tension is the internal friction of the soil, because the stresses that are created within the mass are transferred from the soil to the reinforcement strips by friction.

The diagram in Fig. 5-31 shows a retaining wall made of reinforced earth. It also illustrates some of the applications of this material (Plate 5-10).

The stability of a retaining wall built with reinforced earth should include two types of analysis: first, a general analysis of the structure as a whole, which is no different from that of a conventional gravity retaining wall; and second, an internal stability analysis, the chief purpose of which will be to determine the length of the reinforcement strips and the horizontal and vertical spacing between them, so as to ensure that no sliding will occur between the earthy material and the strips. It will also be necessary to evaluate other aspects of design, such as the risk of corrosion to metal strips, and the type of facing which covers the exposed surface of the front of the retaining wall to prevent earth from escaping between the reinforcement strips. Drainage must be planned as for conventional walls.

Three different types of studies have been carried out on the subject of reinforced earth:

— Studies to determine methods for design. Available methodology has been applied in utilizing traditional earth thrust theories

— Laboratory studies using two-dimensional analog models in which granular earth is represented by metal bars which are long in relation to their diameter. For the reinforcement strips the same material is used as in the prototypes. The models are basically qualitative. They are used to study the modes of failure that are likely to occur

— Measurements in prototypes constructed to solve specific road engineering problems. Most of these studies have been conducted in France, Great Britain and the U.S.A.

Previous analyses and studies indicate that there is a risk of any one of the following three types of failure:

— Failure involving a collapse of the entire structure of reinforced earth, but without important deformation occurring within it. This failure can occur by sliding or overturning and is similar to that of a conventional retaining wall that falls for the same reasons

— Failure where the earth slips, but the reinforcement strips do not. It is accompanied by disruption within the body of the reinforced earth

— Failure due to breakage of the reinforcement strips, which appears to be associated with progressive failure

The comments made by the inventors of these techniques [46, 47] maintain that there is a reasonable agreement between the results of the theoretical analysis, behavior of the models, and performance of the prototypes that are built. Their conclusions are based on full-scale walls.

A major consideration is the selection of the earthy material to be used. So far, it has always been a free draining material with a frictional nature. It is felt that the use of purely cohesive materials requires further research. Structures have nevertheless been built using natural materials with a fines content of about 10 to 20% finer than sieve N° 200, without the need for any special processing process.

Plate 5-10 Example of situation where reinforced earth is used

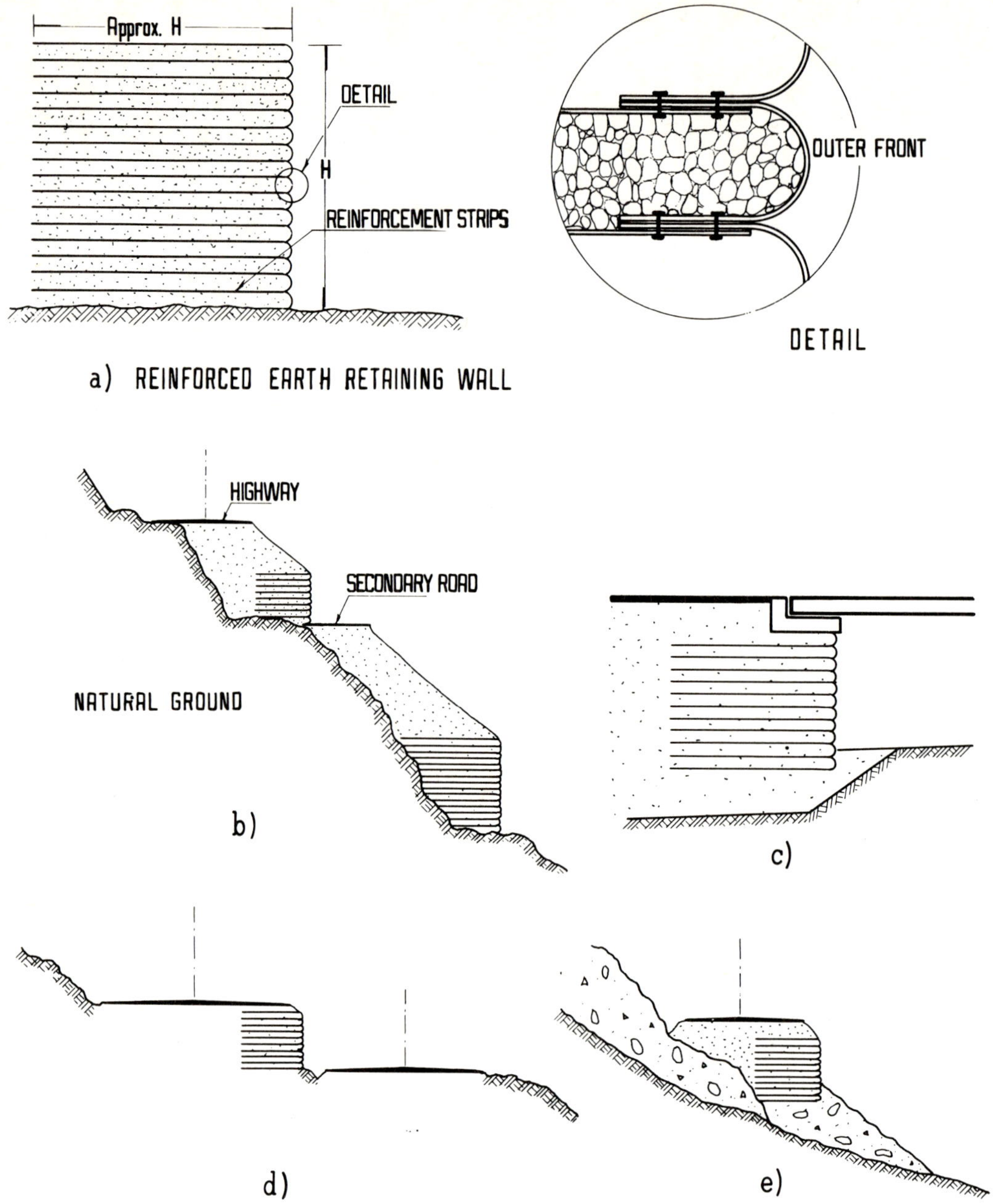

Fig. 5-31 Some examples of how reinforced earth is used [46,47]

Figure 5-32 shows a procedure proposed by SCHLOSSER and VIDAL [46] for computing the internal stability of the mass of reinforced earth. For the mass of reinforced earth, a nearly rectangular cross-section is recommended, where the width is approximately equivalent to the height of the wall. The surface *A-C* limits a wedge of reinforced earth; its equilibrium can be analyzed by expanding on the ideas governing COULOMB's Method for conventional cases of earth thrust. The equilibrium of each wedge depends on the following forces: weight *W*, whose magnitude and position are known; reaction *R*, resulting from the normal and frictional effects along *A-C* of which only the direction is known; and force *T*, resulting from all the tension stresses in the reinforcement strips, which will be horizontal. With these factors, the force polygon for the wedge can be constructed, which will give the value of *T*. A trial-and-error analysis using the necessary number of potential sliding surfaces like *A-C*, will allow assessment of the T_{max} that is likely to occur.

To continue with the analysis and distribute force T_{max} among the reinforcement strips, it will be necessary to formulate a hypothesis regarding the distribution of the tension forces in these strips. In [46], a triangular distribution hypothesis is pro-

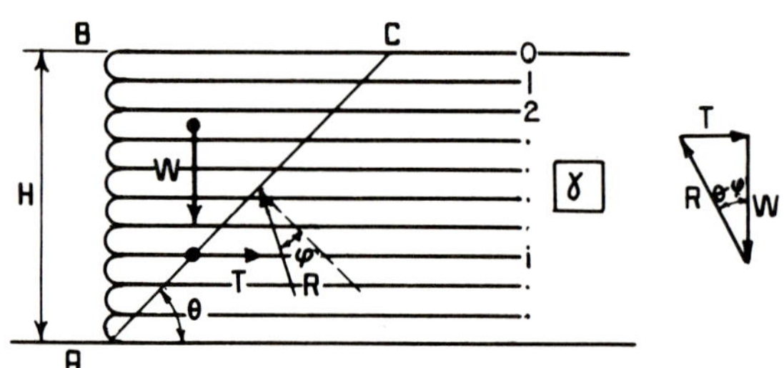

Fig. 5-32 Internal stability analysis of a mass of reinforced earth [46]

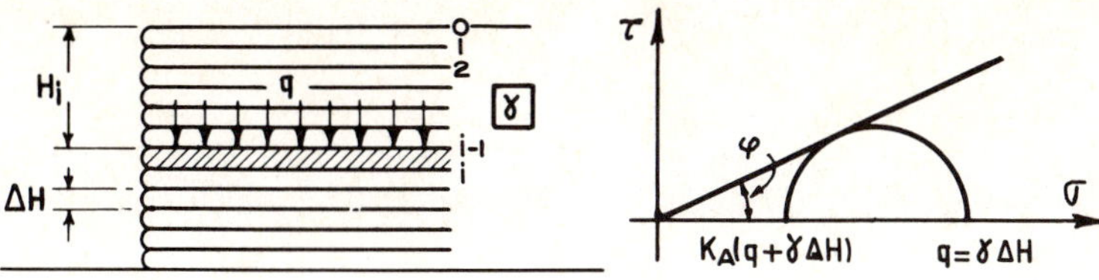

Fig. 5-33 Another method of analyzing the internal stability of reinforced earth [46]

posed with maximum tension in the lowest strip and zero tension in the highest one. In this way, the tensile forces acting in each row of reinforcement strips can be found; this, multiplied by the vertical spacing between the rows of strips, will give the tension force for each row (per meter of length of wall).

Each row must now be considered individually, taking into account the width of the strips. The normal pressure acting in the strip is known. This, multiplied by the area of the strips, gives the total normal force acting on it, and this, multiplied by twice the coefficient of friction between the strip and the soil, will (because friction acts on the top and bottom faces) give the tensile force that can be resisted by the strips without them slipping. This is the value which must be compared with the tension force acting in each strip. The previous procedure makes it possible to find the most suitable vertical and horizontal spacing for the reinforcement strips.

The internal stability of the mass of reinforced earth can also be analyzed using RANKINE's Theory, just as it is applied to conventional earth thrust problems. Figure 5-33 shows the position of the i'th row, at depth H_i, $\triangle H$ being the thickness of fill corresponding to each row of reinforcement strips.

The weight of the reinforced earth at depth H_i is regarded as a surcharge acting at that level. If K is the coefficient of earth pressure, from RANKINE's theory, this surcharge produces the thrust

$$E_q = K_a q \triangle H$$

Furthermore, at depth H_i and in the thickness $\triangle H$, in accordance with RANKINE's Theory and considering the soil as purely frictional, there will be the following thrust:

$$E_a = \frac{1}{2} K_a \gamma \triangle H^2$$

The total thrust at the level H_i will therefore be the sum of the two previous values. It is regarded as being applied on the reinforcement strip and represents the value which must be resisted by the friction developing between the soil and the reinforcement strip. This is assumed to develop only beyond the wedge $A - C$ of Fig. 5-32.

Reinforced earth still requires a great deal of research, especially in the application of the classical earth thrust theories to design calculations. For example, there has been some discussion regarding the value of the coefficient of earth pressure to be used in the last method of calculation described, because the reinforcement strips and the protective covering on the front of the wall are sufficiently rigid to produce a severe restriction to lateral deformation of the structure as a whole. Thus it is questionable that the coefficient of earth pressure will reach the minimum value. The use of a value of K_0 corresponding to the at-rest condition has been proposed in the upper part of the wall, because in the lower part there

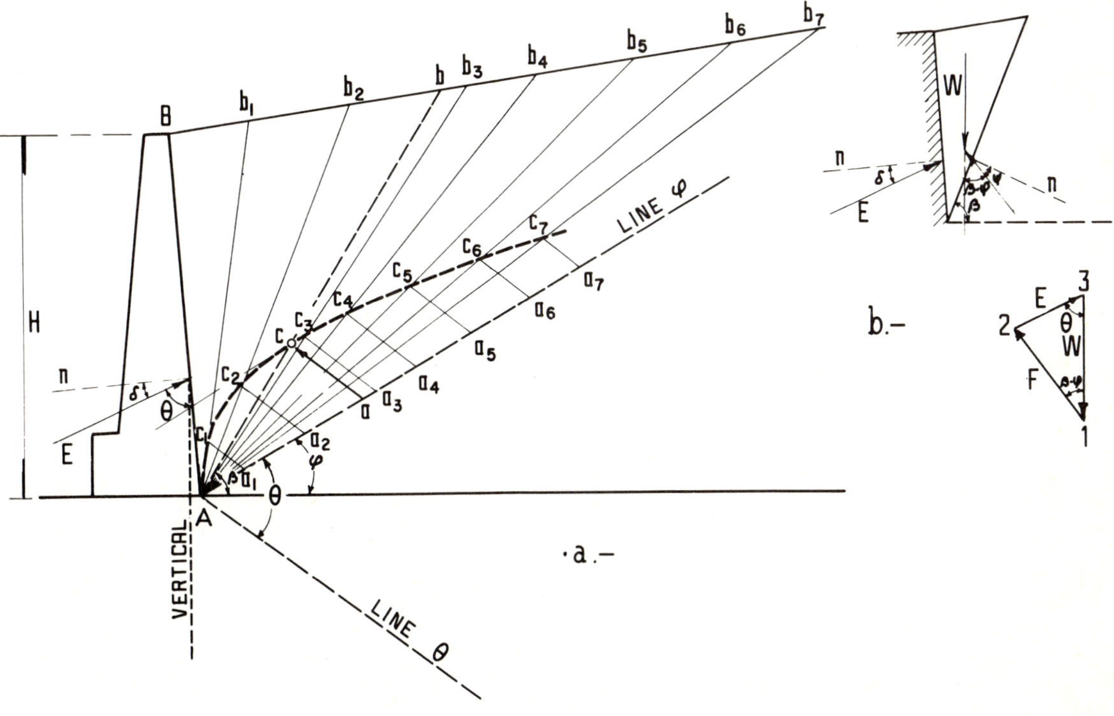

Fig. 5a-1 Graphical method by CULMANN

are greater horizontal shear stresses which allow the soil to deform more freely.

It is important to determine the angle of friction between the reinforcement strips and the frictional material of which the soil is composed. To encourage the development of maximum friction, corrugated metal and perforated plastic strips have been used, which slide along surfaces that are located slightly outside the strip, in the frictional fill material. In this way, it is reasonable to consider an angle of friction equal to that of the frictional fill material. If the strip is smooth, the value of the angle of friction between the strip and the frictional material may be as low as $\frac{\varnothing}{2}$

Another factor requiring determination is a suitable compaction process when building a mass of reinforced earth. If the soil is well compacted, the internal friction of the fill will be increased and the risk of disruption as a result of settlement will be reduced. However, excessive compaction can lead to even more serious risks, because residual compaction pressures can cause important deformations in the wall facing or a serious reduction in the efficiency of the reinforcement strips, which are then required to counteract these parasitic pressures.

APPENDIX 5A

EXERCISES

5a.1 CULMANN's Graphical Method Applied to Frictional Fills [1]

This example will be used to give a brief description of CULMANN's graphical method, since it was not described in the text. The method is a graphical application of COULOMB's theory and enables the problem to be solved in sandy fills without trial-and-error procedures.

Data for the wall and the sandy fill are as follows (Fig. 5a-1):

H = height of wall = 3.00 m (10 ft)

$\varnothing$ = angle of internal friction of the sandy material = 30°

δ = slope angle of the thrust, in relation to the normal one $= \frac{2}{3}\varnothing = 20°$

γ = unit weight of fill = 1.8 t/m³ (112 lb/ft³)

θ = the first thing that can be calculated is the slope angle of the thrust in relation to the vertical line = 65° 30', obtained graphically in Fig. 5a.1

By this method, the maximum value of the thrust applied to a wall by a sandy fill can be determined graphically. The method can be applied to a fill of any shape. Through A, at the base of the wall, lines $\varnothing$ and θ are drawn, the first forming an angle of 30° with the horizontal and the second an angle of 65°30′ with the previous one.

Next, various different assumed sliding planes are selected, Ab_1, Ab_2... etc. The weight of the sliding wedges is calculated by multiplying their area by the specific weight, 1.8 t/m³ (112 lb/ft³), of the fill sand (per unit of length).

$$W_1 = \frac{0.7 \times 3}{2} \times 1.8 = 1.8 \text{ t/m}$$

$$\left(\frac{2.3 \times 10}{2} \times 112 = 1288 \text{ lb/ft}\right)$$

$$W_2 = \frac{1.55 \times 3}{2} \times 1.8 = 4.19 \text{ t/m}$$

$$\left(\frac{5.08 \times 10}{2} \times 112 = 2845 \text{ lb/ft}\right)$$

$$W_3 = \frac{2.55 \times 3}{2} \times 1.8 = 6.89 \text{ t/m}$$

$$\left(\frac{8.36 \times 10}{2} \times 112 = 4682 \text{ lb/ft}\right)$$

$$W_4 = \frac{3.15 \times 3}{2} \times 1.8 = 8.50 \text{ t/m}$$

$$\left(\frac{10.33 \times 10}{2} \times 112 = 5785 \text{ lb/ft}\right)$$

$$W_5 = \frac{4.00 \times 3}{2} \times 1.8 = 10.80 \text{ t/m}$$

$$\left(\frac{13.12 \times 10}{2} \times 112 = 7347 \text{ lb/ft}\right)$$

$$W_6 = \frac{4.84 \times 3}{2} \times 1.8 = 13.07 \text{ t/m}$$

$$\left(\frac{15.88 \times 10}{2} \times 112 = 8893 \text{ lb/ft}\right)$$

$$W_7 = \frac{5.60 \times 3}{2} \times 1.8 = 15.12 \text{ t/m}$$

$$\left(\frac{18.37 \times 10}{2} \times 112 = 10{,}287 \text{ lb/ft}\right)$$

On a suitable scale of forces, these weights are plotted from A on the "$\varnothing$ line"; points a_1, a_2... etc. are thus obtained.

Through these last points, lines parallel to the "θ line" are drawn until they intersect the respective failure planes of the wedges at c_1, c_2... etc. Segments a_1c_1, a_2c_2... etc, represent, on the scale of forces that is used, the thrusts that are produced by each one of the wedges arbitrarily chosen. Part *(b)* of Fig. 5a-1 shows a force triangle corresponding to wedge ABb_2. Thrust E and weight W form the angle θ, since this, by definition, is the angle formed by E and the vertical. Between the reaction along the failure plane, F, and W, angle β–$\varnothing$ is formed, being the one formed by the sliding plane and the horizontal.

Now, triangle Aa_2c_2 is regarded as connected to the same sliding wedge. Aa_2 is in proportion to the weight of the wedge, W, by geometry. The angle at A_2 is θ since a_2c_2 is parallel to the "θ line". Obviously in triangle Aa_2c_2 the angle at A is $\beta-\varnothing$, β being the angle that is formed by the sliding plane Ab_2 and the horizontal. The triangle Aa_2c_2 is therefore similar to *123* of Part *(b)* of Fig. 5a-1. By comparing these triangles, it can be seen that the side a_2c_2 is the homologue of E in the force triangle. These two magnitudes are therefore proportional and c_2a_2 represents E on the scale of forces chosen.

A curve can now be plotted through all the points c. This is the *thrust line* or CULMANN's line. A line parallel to the "$\varnothing$ line", the tangent to CULMANN's line, makes it possible to calculate the maximum thrust, segment ac, using the same scale of forces for interpretation, c being the resulting tangential point on CULMANN's line. In this case, the thrust E was 3 t/m (2016 lb/ft). Line Ac, prolonged as far as b, gives the most critical sliding plane associated with maximum thrust.

CULMANN's method also leads to the maximum thrust produced by the combination of a frictional fill and a linear surcharge with an intensity of q force units per unit of length. In this case $q = 2$ t/m (1344 lb/ft) (Fig. 5a-2).

The same procedure is followed, with the difference that to the right of plane $A\,b'_1$ defined by the position of q not only must the weight of the sliding wedge be carried on the "$\varnothing$ line", but also the value of q must be added on the same scale of forces used. To be precise, a discontinuity due to the surcharge must appear in Ab'_1 of CULMANN's line.

$W_2+q=4.19+2.00=6.19$ t/m (2845 + 1344 = 4189 lb/ft)

$W_3+q=6.89+2.00=8.89$ t/m (4682 + 1344 = 6026 lb/ft)

$W_4+q=8.50+2.00=10.50$ t/m (5785 + 1344 = 7129 lb/ft)

$W_5+q=10.80+2.00=12.80$ t/m (7347 + 1344 = 8691 lb/ft)

$W_6+q=13.05+2.00=15.05$ t/m (8893 + 1344 = 10,237 lb/ft)

$W_7+q=15.15+2.00=17.15$ t/m (10,287 + 1344 = 11,631 lb/ft)

Thrust E', given by segment $a'c'$ is the maximum considering the surcharges, whereas segment $a\ c$ would be the maximum thrust if there were no surcharge. In the example thrust E' was 4 t/m (2688 lb/ft). If the surcharge were situated to the right of b'', it would no longer have any effect, for the thrust would then be equal to the maximum obtained with CULMANN's dashed line. The line cc'' has been drawn parallel to the "$\varnothing$ line". A fair approximation for the point at which maximum thrust is applied is obtained by means of a graph, which is shown in Fig. 5a-3. If there is no linear surcharge, a parallel line to the critical sliding surface $A\ b$ through G, (center of gravity of the sliding wedge) cuts the wall at a point where the thrust E can be regarded as applied.

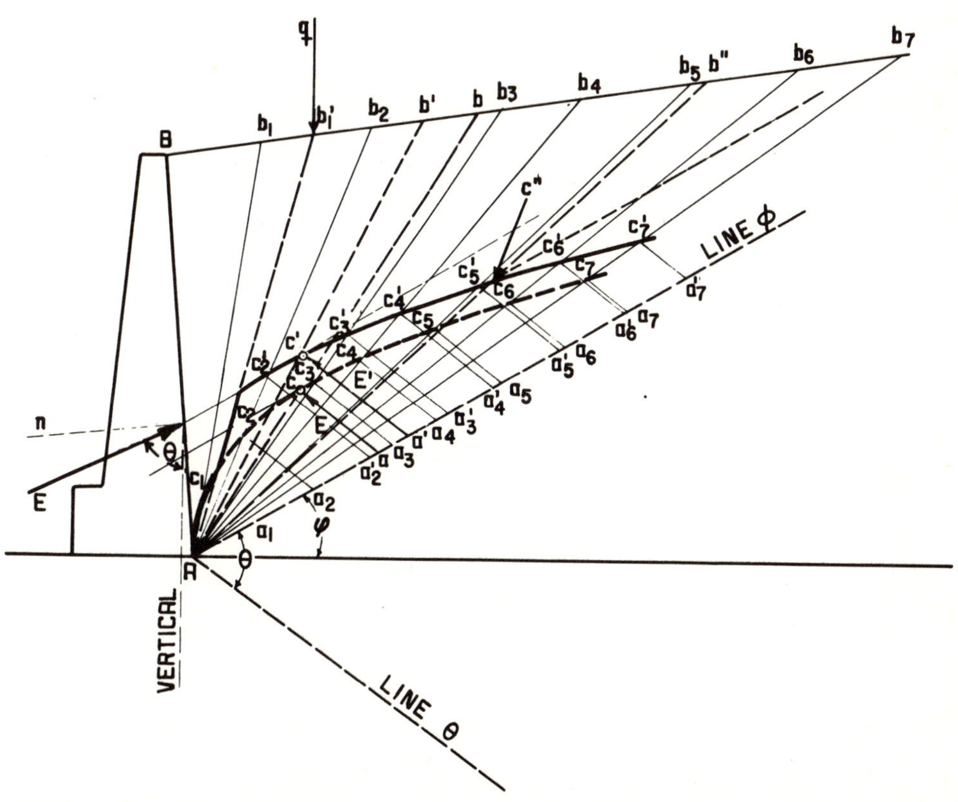

Fig. 5a-2 CULMANN's method for a linear surcharge

If there is a surcharge, to the previous force will be added for design purposes another, ΔE, calculated by subtracting $E' - E$, in this case equal to 1 t/m (672 lb/ft) obtained as shown in Fig. 5a-2, and applied in the upper third of the segment ff', where f is the intersection of a line parallel to the "∅ line", drawn through q, with the back of the wall. Here f' is the intersection with the same plane of a parallel to the critical sliding surface, also drawn through q Fig. 5a-3.

CULMANN's method can be used to calculate the passive thrust that is applied on a sandy fill. The procedure is identical, with the difference that the "∅ line" must now be drawn downwards to form the angle ∅ with the horizontal.

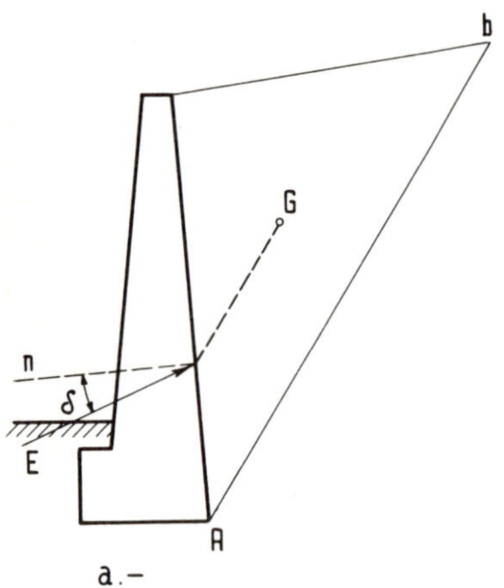

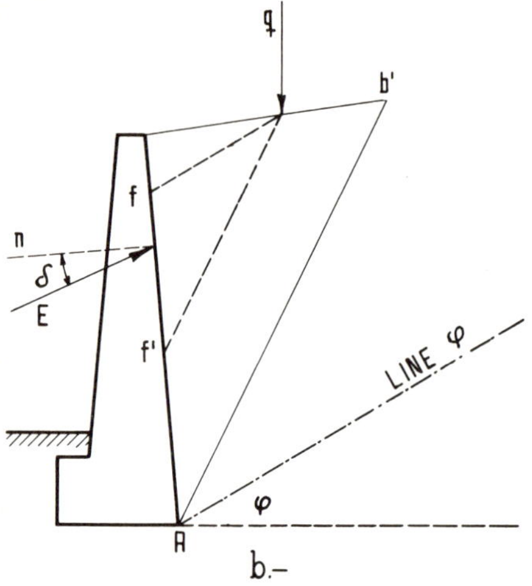

Fig. 5a-3 Point at which thrust is applied according to CULMANN's method

5a.2 Design of a Retaining Wall with Several Variants

A concrete wall 10 m (33 ft) high is to be constructed to withstand a 12 m (40 ft) embankment set on firm flat ground. The embankment will be built with a compacted silty sand, with the following characteristics: c = 1 t/m², (205 lb/ft²), ∅ = 30°, γ_d = 1 800 kg/m³ (112 lb/ft³), γ_m = 2 000 kg/m³ (125 lb/ft³) and S_s = 2.4 (Fig. 5a-4).

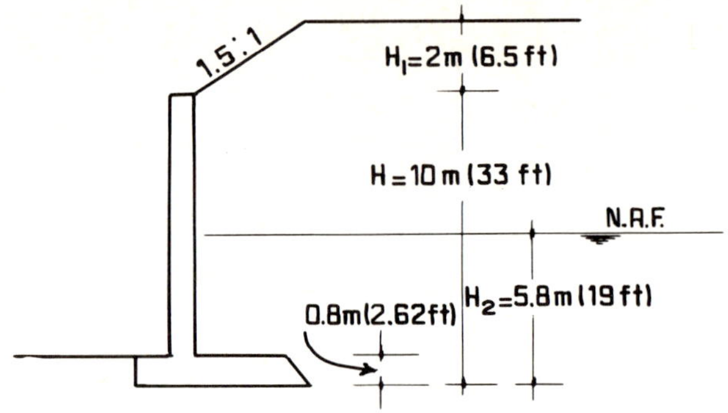

Fig. 5a-4 Wall to be computed

1. Calculate the thrust of the fill against the wall using COULOMB's, RANKINE's and TERZAGHI's Methods, and regarding the fill as dry.
2. Calculate the thrust of the fill on the wall considering a water-table at 5.8 m (19 ft) above the base of the wall, using RANKINE's Method and regarding the wall as impermeable.
3. Calculate the thrust on the wall, considering the installation of a filter 0.40 m (1.31 ft) thick, directly adjacent to the back of the wall.

The characteristics of the filter material are as follows: $c = 0$; ∅ = 32°; γ_d = 1 700 kg/m³ (106 lb/ft³); γ_m = 1 900 kg/m³ (118 lb/ft³); S_s = 2.6.

Solution:

An approximate geometry of the wall foundation is assumed, with dimensions affecting thrust calculations.

.1 Regarding the Soil as Dry

COULOMB's Method (Fig. 5a-5) The equilibrium of the sliding wedge is analyzed, assuming several different values of angle β. The weight of wedge W and the values of c and c' for a one meter strip in the direction parallel to the wall are calculated.

$$W = \frac{b\,(H + H_1)}{2}\gamma_d;\; C = cl;\; C' = (H + H_1)\,c$$

For β = 52° 30′

$$W = \frac{9.20 \times (10 + 2)}{2} \times 1.8 = 99.4 \text{ t}$$

$$\left(\frac{30 \times (33 + 6.5)}{2} \times 112 = 66{,}360 \text{ lb}\right)$$

$$C = 1 \times 15 = 15 \text{ t } (205 \times 49 = 10{,}045 \text{ lb})$$

$$C' = (10 + 2) \times 1 = 12 \text{ t}$$

$$((33 + 6.05) \times 205 = 8097 \text{ lb})$$

With these values and the directions of forces R and E the force polygon is constructed and the magnitude of E_a determined.

$$E_a = 20 \text{ t/m } (13{,}440 \text{ lb/ft})$$

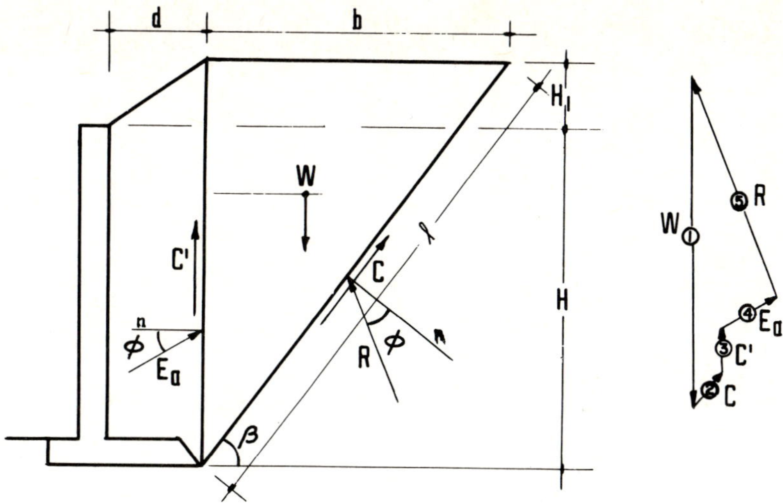

Fig. 5a-5 COULOMB's method, dry fill

The following values can be obtained from other trials:

β	E_a
45°	18 t/m (12,096 lb/ft)
50°	19 t/m (12,768 lb/ft)
52° 30′	20 t/m (13,440 lb/ft)
55°	20 t/m (13,440 lb/ft)

and the *maximum* value of the thrust is found to be $E_a = 20$ t/m (13,440 lb/ft)

RANKINE's Method According to RANKINE theory, the active thrust is:

$$E_a = \frac{1}{2N_\varnothing}\gamma_d (H + H_1)^2 - \frac{2c}{\sqrt{N_\varnothing}}(H + H_1)$$

where:

$$N_\varnothing = \tan^2\left(45° + \frac{\varnothing}{2}\right) = \tan^2 60° = 3$$

$$E_a = \frac{1}{2 \times 3} \times 1.8 \times (10 + 2)^2 - \frac{2 \times 1}{\sqrt{3}} \times (10 + 2)$$

$$= 43.2 - 13.8$$

$$E_a = 29.4 \text{ t/m } (19{,}773 \text{ lb/ft})$$

TERZAGHI's Semi-empirical Method The fill soil is a silty sand corresponding to category II.

The thrust will be:

$$E_H = \frac{1}{2} K_H (H + H_1)^2$$

The value of the relation H_1/H, which in this case is equal to zero since $H_1 = 0$, is calculated. This value is used in the corresponding graph and the following value is obtained:

$$K_H = 600 \text{ kg/m}^3 \ (37.4 \text{ lb/ft}^3)$$

Hence:

$$E_H = \frac{1}{2} \times 600 \times (10 + 2)^2 = 43{,}200 \text{ kg/m}$$

$$E_H = 43.2 \text{ t/m } (29{,}176 \text{ lb/ft})$$

For this case $H_1/H = 0$, $K_v = 0$, therefore $E_v = 0$.

For the final design of the wall, the value obtained for the thrust must be combined with the weight of the fill on the shoe of the wall foundation.

.2 Considering a Water-table at 5.80 m (19 ft) Above the Base of the Wall

RANKINE's Method: According to RANKINE's theory, the thrust in these conditions will be:

$$E_a = \frac{1}{2N_\varnothing}\gamma_m (H + H_1 - H_2)^2 + \frac{1}{N_\varnothing}\gamma_m (H + H_1 - H_2) H_2 + \frac{1}{2N_\varnothing}\gamma'_m H_2^2 + \frac{1}{2}\gamma_w H_2^2$$

where:

$\gamma_m = 2$ t/m³ (125 lb/ft³); $\gamma_d = 1.8$ t/m³ (112 lb/ft³); $S_s = 2.4$;

$H = 10$m (33 ft); $H_1 = 2$m (6.5 ft); $H_2 = 5.8$m (19 ft);

$N_\varnothing = 3$; and $\gamma_w = 1$ t/m³ (62.4 lb/ft³)

γ'_m is given by Eq. (1-16) $= \dfrac{S_s - 1}{S_s}\gamma_d = \dfrac{2.4 - 1.0}{2.4} \times 1.8$

$$\gamma'_m = 1.05 \text{ t/m}^3 \ (65.2 \text{ lb/ft}^3)$$

The thrust will therefore be:

$$E_a = \frac{1}{2 \times 3} \times 2 \times (10 + 2 - 5.8)^2 + \frac{1}{3} \times 2 \times$$

$$(10 + 2 - 5.8) \times 5.8 + \frac{1}{2 \times 3} \times 1.05 \times 5.8^2 + \frac{1}{2} \times 1 \times 5.8^2$$

$$E_a = 12.8 + 24.0 + 5.88 + 16.8$$

$$E_a = 59.48 \text{ t/m } (40{,}170 \text{ lb/ft})$$

.3 Assuming Steady Flow Towards a Vertical Filter Drain Placed Against the Back of the Wall

The flow net is drawn as shown in Fig. 5a-6 and the values of the uplift calculated:

$$U = \frac{5.3 \times 2.6}{2} = 6.9 \text{ t/m}$$

$$\left(\frac{17.4 \times 3.7 \times 144}{2} = 4635 \text{ lb/ft}\right)$$

$$U' = \frac{2.6 \times 3}{2} = 3.9 \text{ t/m} \left(\frac{8.53 \times 613.7}{2} = 2617 \text{ lb/ft}\right)$$

The weight of the wedge is calculated

$$W = \frac{9.12 \times 12 \times 2}{2} = 109.5 \text{ t/m}$$

$$\left(\frac{30 \times 39}{2} \times 125 = 73{,}125 \text{ lb/ft}\right)$$

together with the values of cohesion on the failure surface and on the vertical surface on which the thrust is being calculated. In both cases, a zero cohesion condition is assumed below the water-table:

$$L = 15.1 - 5.3 = 9.8;\ cL = 1 \times 9.8 = 9.8 \text{ t/m } (6569 \text{ lb/ft})$$

$$L' = 12.0 - 3.0 = 9.0;\ cL' = 1 \times 9.0 = 9.0 \text{ t/m } (6055 \text{ lb/ft})$$

With the values of U, U', W, cL, cL' and the directions of reaction R and thrust E_a, the force polygon shown in Part *(b)* of the figure is constructed, the magnitudes of reaction R and thrust E thus being determined. The result is:

$$R = 80 \text{ t/m } (53{,}760 \text{ lb/ft})$$

$$E_a = 28 \text{ t/m } (18{,}816 \text{ lb/ft})$$

.4 Comments

- As can be seen, if the effect of water is not considered, Coulomb's and Rankine's Methods lead to thrusts of the same order of magnitude, although in this case Rankine's gave a result 50% higher, because it regards the surface on which the thrust is calculated as being ideally smooth.
- If the unneutralized effect of the water is considered, the thrust increases greatly (by about 100%).
- If the wall has an adequate drainage system, the thrust drops again to values that are about the same as the ones for the dry soil.
- Terzaghi's semi-empirical method provides values that are between the dry soil ones, and the ones with the unneutralized effect of water. This appears to imply that Terzaghi considered deficient drainage when drawing his graphs.

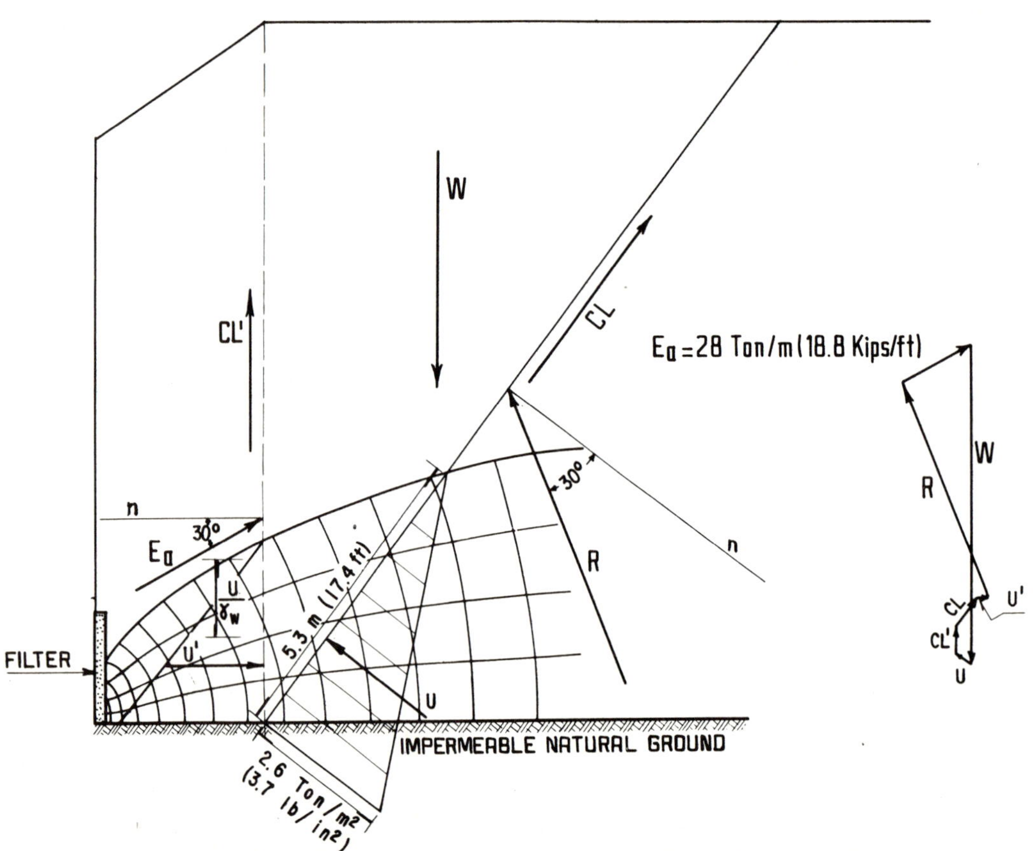

Fig. 5a-6 Flow net (regarded as established) in the backfill of the wall towards the filter

5a.3 Case Study Reinforced Earth

A reinforced earth structure is to be built as shown in Fig. 5a-7. Calculate the stability conditions of the wall.

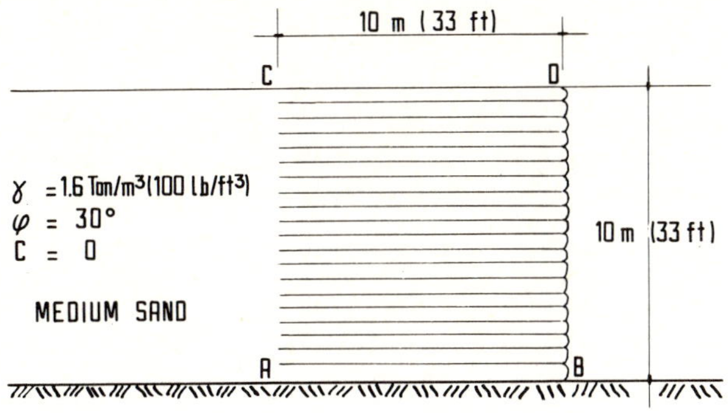

Fig. 5a-7 Conditions and data of a reinforced earth structure

.1 Analysis of the Collapse of the Reinforced Earth Structure as a Whole

The force equilibrium and moment equations can be written as follows:

$$W = \frac{L}{2}(a + b) \qquad Pe = (b - a)\frac{L^2}{12}$$

Thus

$$b = \frac{1}{L}\left(W + \frac{6\,Pe}{L}\right) \qquad a = \frac{1}{L}\left(W - \frac{6\,Pe}{L}\right)$$

For the problem proposed: $P = \frac{1}{2} K_a \gamma H^2$ and $e = \frac{H}{3}$
Therefore:

$$b = \gamma H\left[1 + K_a\left(\frac{H}{L}\right)^2\right] = 1.6 \times 10 \times$$

$$\left[1 + 0.33\left(\frac{10}{10}\right)^2\right] = 21.28 \text{ t/m}^2$$

$$a = \gamma H\left[1 + K_a\left(\frac{H}{L}\right)^2\right] = 1.6 \times 10$$

$$\left[1 - 0.33\left(\frac{10}{10}\right)^2\right] = 10.72 \text{ t/m}^2$$

$$b = 21.28 \text{ t/m}^2 \text{ (4389 lb/ft}^2\text{)}$$

$$a = 10.72 \text{ t/m}^2 \text{ (2211 lb/ft}^2\text{)}$$

Since $a > 0$, the wall will not overturn.

The condition where the base $A\ B$ does not slide will be considered later.

.2 Analysis of Failure due to Breaking of the Reinforcement Strips (Fig. 5a-8)

First trial-and-error procedure. There is assumed to be a spacing of $\Delta H = 0.25$ m (10 in) between the strips. The tension forces in the strips will therefore be:

$$T_i = K_a\, q\, \Delta H + \frac{1}{2} K_a\, \gamma\, \Delta H^2$$

$$q = \gamma H_i$$

$$T_i = 0.132\, H_i + 0.0165 \text{ t/m } (27\, H_i + 11 \text{ lb/ft})$$

For different values of i, the following values are obtained:

H_i in m (ft)	T_i in t/m (lb/ft)
1 (3.28)	0.148 (99.45)
3 (9.84)	0.412 (276.86)
5 (16.40)	0.676 (454.27)
8 (26.25)	1.072 (720.38)
10 (32.81)	1.336 (897.79)

Trial-and-error procedures are conducted to select the most suitable design values for the tension in the strips.

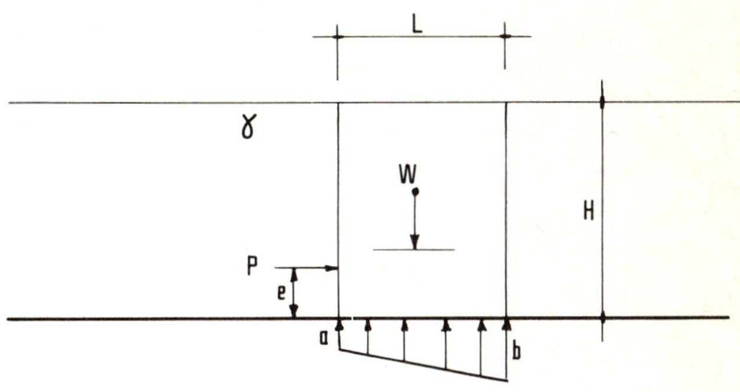

Fig. 5a-8 Force analysis

.3 Analysis of Failure due to the Earth Sliding in Relation to the Reinforcement Strips, Accompanied by Disruption in the Body of Reinforced Earth

The safety factor, F_s, against any horizontal failure by shear between the reinforcement strips and the soil will be:

$$F_s = \frac{2 \tan \varnothing}{K_a}\left(\frac{L}{H}\right)$$

For the structure under consideration ($L = H$)

$$F_s = \frac{2 \tan \varnothing}{K_a} = 3.5$$

In order to obtain these equations, it was assumed that the shear strength develops on horizontal planes halfway between the strips like a system of rigid blocks as shown in Fig. 5a-9.

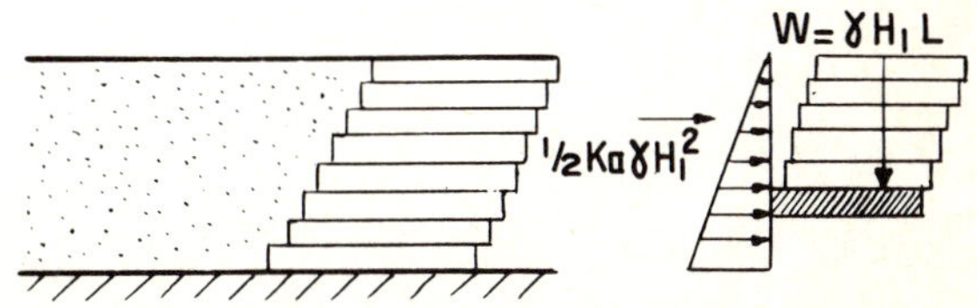

Fig. 5a-9 Analysis to prevent distortion

REFERENCES

1. Juárez Badillo, E. and Rico, A., *Mecánica de Suelos, Vol. II: Teoría y Aplicaciones de la Mecánica de Suelos,* Limusa: México, 1977, Chap. IV.
2. Lambe, T. W. and Whitman, R. V., *Soil Mechanics,* John Wiley and Sons, 1971, Chap. 13.
3. Moretto, O., Contribution to Discussion, *Procs. of VII. International Conference on SMFE,* México, 1969, Vol. III, pp. 357-359.
4. Morgenstern, N. R., and Eisenstein, Z., "Methods of Estimating Lateral Loads and Deformations," *ASCE Specialty Conference on Lateral Stress in the Ground and the Design of Earth Retaining Structures,* Cornell University, 1970.
5. Kenney, T. C., "Field Measurements of in situ Stresses in Quick Clays," *Proc. of the Oslo Geotechnical Conference,* 1967, Vol. I.
6. Peterson, R., "Rebound in the Bearpaw Shale, Western Canada," *Bulletin Geological Society of America,* No 69, 1958.
7. Huntington, W.C., *Earth Pressures and Retaining Walls,* John Wiley and Sons, 1967.
8. Reference [2], Chap. 2.
9. Terzaghi, K., "Large Retaining Wall Test, I: Pressure of Dry Sand," *Engineering News Record,* No 112, 1934.
10. Rowe, P. W., and Peaker, K., "Passive Earth Pressure Measurements," *Geotechnique,* Vol. 15, 1965.
11. Schofield, A. N., "The Development of Lateral Force of Sand against the Vertical Face of a Rotating Model Foundation," *Procs. V. International Conference on SMFE,* Paris, 1961.
12. James, R. G., and Bransby, P. L., "Experimental and Theoretical Investigations of a Passive Earth Pressure Problem," *Geotechnique,* Vol. 20, 1970.
13. Finn, W. D. L., "Boundary Value Problems of Soil Mechanics," *Journal Soil Mechanics Division ASCE,* Vol. 89, SM-5, 1963.
14. Tschebotarioff, G. P., "Retaining Structures," Chap. 5 of *Foundation Engineering,* Ed. G. A. Leonards, McGraw-Hill Book Co, 1962.
15. Terzaghi, K., and Peck, R. B., *Soil Mechanics in Engineering Practice,* John Wiley and Sons, 1967, Article 46.
16. Gould, J. P., "Lateral Pressures on Rigid Permanent Structures," *ASCE Specialty Conference on Lateral Stress in the Ground and the Design of Earth Retaining Structures,* Cornell University, 1970.
17. Terzaghi, K., *Theoretical Soil Mechanics,* John Wiley and Sons, 1956, Art. 91.
18. Bowles, J. E., *Foundation Analysis and Design,* McGraw-Hill Book Co, 1968, Chaps. 6 & 7.
19. Reference [1], Appendix V-C.
20. Juárez Badillo, E., and Rico, A., *Mecánica de Suelos, Vol. III: Flujo de agua en suelos,* Limusa: México, 1979, Chap. II.
21. Brooker, W. E., and Ireland, H. O., "Earth Pressures at Rest Related to Stress History," *Canadian Geotechnical Journal,* Vol. II, No 1, 1965.
22. Ireland, H. O., "Design and Construction of Retaining Walls," Proceedings of Lectures Series, ASCE Illinois Section, Chicago, 1964.
23. Tschebotarioff, G. P., "Bridge Abutments on Piles through Plastic Clay," Conference on Design and Installation of Pile Foundations and Cellular Structures, Lehigh University, 1970.
24. Hansen, J. B., "The Stabilizing Effect of Piles in Clay," Christiani and Nielsen Post, 1953.
25. Mackenzie, T. R., "Strength of Deadman Anchors in Clay," Masters Thesis, Princeton University, 1955.
26. Tschebotarioff, G. P., "Analysis of a High Crib Wall Failure," *Procs. of VI. International Conference on SMFE,* Montreal, Canada, 1965.
27. Reference [1], Chap. 1.
28. Sowers, G. F., Robb, A. D., Mullis, C. H., and Glenn, A. J., "The Residual Lateral Pressures produced by Compacting Soils," *Procs. of IV. International Conference on SMFE,* London, 1957, Vol. II.
29. Reference [15], Art. 48.
30. Taylor, D. W., "Abstracts of Selected Thesis on Soil Mechanics," Massachussetts Institute of Technology, Department of Civil Engineering, Tech. Publ. No 79, 1941.
31. Kaare, S. F., "Stresses and Movements in Connection with Braced Cuts in Sand and Clay," Doctor Thesis, University of Illinois, Urbana, Ill. 1966.
32. White, L., and Prentis, E. A., *Cofferdams,* Columbia University Press, New York, 1940.
33. Peck, R. B., "Soil Studies for the Initial System of Chicago Subways," Unpublished Report, Department of Subways and Traction, City of Chicago, March, 1943, Mentioned in [31].
34. Wu, T. H., "An Analytical Study of Earth Pressure Measurements in Open Cuts in Chicago Subway," Doctor Thesis, University of Illinois, Urbana, Ill. 1951.
35. Tschebotarioff, G. P., *Soil Mechanics, Foundations and Earth Structures,* McGraw-Hill Book Co. 1951.
36. Skempton, A. W. and Ward, W. H., "Investigations Concerning a Deep Cofferdam in the Thames Estuary Clay at Shellhaven," *Geotechnique,* Vol. 3, No 3, 1951.
37. Ward, W. H., "Some Comparisons between Measured and Calculated Earth Pressures," Lecture for the Institution of Civil Engineers, London, 1955.
38. Bjerrum, L., and Eide, O., "Stability of Strutted Excavations in Clay," *Geotechnique,* Vol. 6, No 1, 1956.
39. Reference [1], Chap. 8.

40. DiBiagio,, E., and Bjerrum, L., "Earth Pressures Measurements in a Trench Excavated in Stiff Marine Clay," *Procs. of IV. International Conference on SMFE,* Vol. II, London, 1957.

41. Kotoda, K. A., Minokuma, A., Endo, M., and Kawasaki, T., "Measurement of Earth Pressure Acting on the Struts of Cofferdam," (In Japanese), Journal Society Geotechnique of Japan, Earth and Foundations, 1959. Mentioned in [31]

42. Endo, M., "Earth Pressure in the Excavation Work of Alluvial Clay Stratum," Procs. of the International Conference on SM organized by the Hungarian Academy of Sciences, Budapest, 1963.

43. Kane, H., "Earth Pressures on Braced Excavations in Soft Clay in Oslo Subway," Doctor Thesis, University of Illinois, Urbana, Ill., 1961.

44. Kjaernsli, B., "Review of Measurements of Strutted Excavations in Clay in Oslo (1957-1962)," Norwegian Soc. of Professional Engineers, Oslo, 1962.

45. Humphers, J. D., "The Measurement of Loads on Timber Supports in a Deep Trench," *Geotechnique,* Vol. 12, No 1, 1962.

46. Scholosser, F., and Vidal, H., "La Terre Armée," *Bulletin de Liaison des Laboratoires Routiers, Ponts et Chaussées,* No 41, 1969.

47. Vidal, H., "La Terre Armée," Annales de L'Institut Technique du Batiment et des Travaux Publiques, No 259-260, 1969.

OTHER RELATED REFERENCES

Bustamante, M., Gianselli, L., and Cambier, J., "Calcul d'un Pieu Visse Moule dans une Argile Plastique," Colloque International sur la Gestion des Ouvrager, Paris, 1981.

Mexican Society for Soil Mechanics, *Reinforced Earth,* International Symposium, Mexico City, July, 1980.

CHAPTER 6

SLOPE STABILITY

6.1 Introduction

The term *slope* is applied to any surface which is at an angle to the permanently horizontal surface of an earth mass. When a slope forms naturally, without any human interference, it is also referred to as a hillside; when man-made, it is called a cut or artificial slope. The slopes of cuts are the result of excavations made in a natural formation, whereas artificial slopes are the sides of embankments. From these terms come *cut slopes, embankment slopes*, and *excavation slopes.*

Slopes are the most complex structures that exist in road engineering, in so far as soil and rock mechanics are concerned. The most complicated soil and rock mechanics problems are associated with slope stability, not forgetting the vital part played by applied geology in evaluating their behavior. Firstly it is necessary to define slope stability criteria. These entail such simple things as the appropriate angle for a cut or an embankment. This angle is the steepest one at which the slope will remain in place as long as necessary. This is the essential point of the problem and the reason why it must be carefully studied. For flatter cut slopes, larger masses of material have to be moved with the consequent increase in cost. Occasionally, to provide a large borrow volume, the desirable slope will be flat; in this case *slope stability problems* will not be posed. Usually the best design is when the minimum amount of earth has to be moved; in other words, the steepest slope.

Slopes are, therefore, structures which should be designed and built with the objective of minimum cost consistent with safety. The amounts of money which are saved by using technically sound slope stability criteria represent a very large part of the total investment which is made in the construction of a road. In Mexico, for example, statistics show that 50% of all future highways will be built in very mountainous areas, 30% on hilly ground and 20% on level ground. In the mountainous areas, 70% of the total cost of the highway will be moving earth; thus any change in the slope will affect the total cost enormously.

Figure 6-1 shows a specific case of the volume moved in relation to the angle of the slope. For purely illustrative purposes, a cut is presumed to be made in a natural slope with angles of 60° and 30°. The cut has an angle of slope δ_3 (variable) and the resulting volumes corresponding to several different heights are given. On flat ground, the percentage of highway cost corresponding to earth moving can be as low as 40%. The slope angles of shallow cuts and low embankments will have less relative effect.

Many of the difficulties that are associated with slope stability problems are due to the many, sometimes radically different factors that must be considered. A study of only one factor can only lead to considerable confusion. The complex nature of slopes requires study of many factors involving many engineering and scientific disciplines. These must be identified and their impact evaluated to develop total criteria for slope design.

Many of the problems of the stability of *natural* slopes are radically different from those of *man-made* slopes. Cuts and embankments should, therefore, be dealt with as a totally different issue. The main differences are, firstly the nature of the materials involved, and, secondly, the environment, which depends on the geologic history, the climatic conditions that prevailed and will prevail in the future and the influence of man. The history of the formation of hillsides and slopes, including the stresses to which they have been subjected and the influence of environmental conditions, define the nature and structure of soils and rocks and the flow of underground water through the materials forming the hillside, which have a decisive influence on stability conditions.

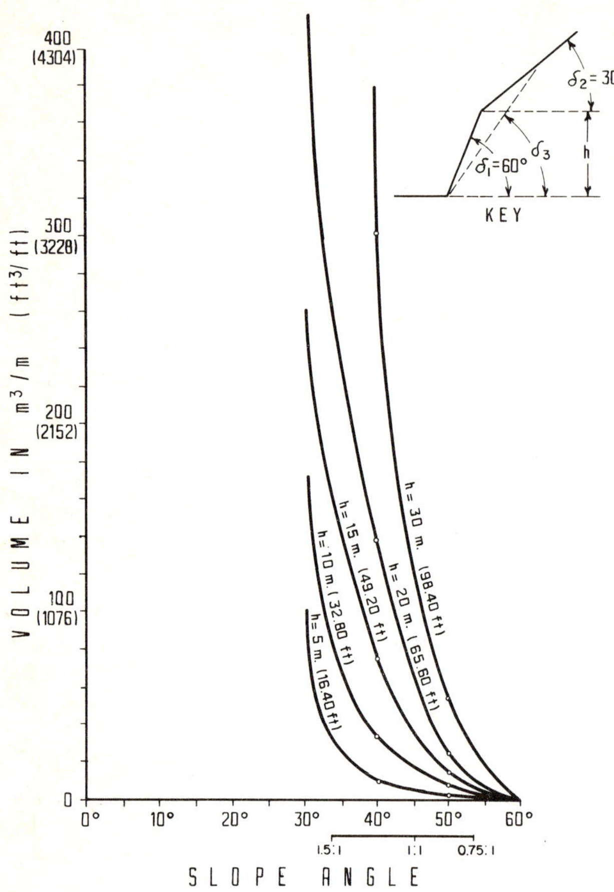

Fig. 6-1 Variation of the volume moved in relation to the angle of slope

In artificial slopes, there are also essential differences between cuts and embankments. The latter are structures which are (or at least can be) built with relatively well-controlled material. In cuts, however, (see Chapter 3), this possibility does not exist. These different formation conditions obviously impose variations in the homogeneity and structure of the materials, which must be reflected in the final structure and all the aspects of its behavior.

Another cause of confusion in slope stability concepts is the extraordinary complexity and multiplicity of slope failure. There is no universal consensus as to what is implied by this term. Most of these failures are described as *slides* or *collapses*, by which it is understood that something is seriously wrong with the structure of the slope: its displacements are so large as to be incompatible with the engineer's conception of the behavior of the slope and the purpose for which it was built. Although there are many elements in this *failure* that can be subjectively interpreted (border-line cases, which may be very hard to define) the authors feel that this is not the real source of confusion in the *slope failure* concept. This confusion is more likely due to the great variety of phenomena that are generally involved; a rotational slide affecting the entire body of soil comprising the slope and the soil beyond, may endanger the integrity of that slope to the same extent as a more limited translational slide involving a deep narrow part of the mass or as a slow shallow superficial sloughing in a hillside. In all these cases, a failure or problem has occurred, according to the usual terminology. In order to solve the problem or correct that failure the engineer may consult standard manuals where the corrective measures are described indiscriminately, with no regard for the different mechanism of rotational slide from a translational movement, and where the corrective measures in many cases should be handled in a different way. It is necessary, therefore, that the many different ways in which a slope may fail to fulfill the purpose for which it was intended should be studied, each one being regarded as a separate problem with respect to its origin, manifestation and correction.

Knowledge of the nature and homogeneity of the component materials is essential when defining the aspects of the stability of a slope. Engineers obtain sufficient general information to enable them to make a mathematical model in which stability can be analyzed by applying some mathematical procedure. The first objective is to find a satisfactory method of calculation. In this way a design procedure can be established with input of the individual characteristics of each case. Thus, with a reliable mathematical method of analysis, even a beginner can make calculations as accurately as those made by an experienced engineer, who relies only on his engineering instinct. If this can be achieved, slope engineering will cease to be an *art* and will be transformed into a routine technique. In the application of soil and rock mechanics to these problems, some very commendable advances have been made, which are of great utility after many years of general application.

Unfortunately however, there is no general method of analysis that can be applied to all slopes. First, it must be realized that the traditional and most widely used method of structural analysis cannot be applied to slopes. In designing a beam, for example, the external loads acting on the structure must be known in order to determine the internal stresses. These are then compared with the strength of the material, defined by a previously accepted failure theory. This simple approach cannot be applied to slopes, because there is no practical procedure that can determine the state of internal stress in the soil mass, corresponding to the external loads. A solution to this problem has not yet been found even by utilizing the simplifying assumptions of continuum mechanics. Therefore, other methods must be used for assessing the stability of slopes. Almost all those in use today are limit analysis methods. These start by establishing a kinematic failure mechanism, which is chosen from experience. The forces that produce motion (active forces) are then analyzed, and are subsequently compared with the forces that can develop to resist failure (resisting forces). Most methods of calculation are associated with a kinematic failure mechanism; therefore they can only be applied to slope failures of this type.

There is another reason why there cannot be a general method of analysis that can be applied in all cases. The application of any theoretical method of analysis requires that the strength parameters of the soil be presented in a simple, mathematical form. Consequently, some simple geometry such as homogeneity or well-defined stratification is required. This condition is usually met in road embankments with fill of select materials and uniform compaction, but this is rarely the case in cuts and natural slopes owing to the heterogeneity of the materials and the variations in properties that can occur in just a small area. Moreover a very thorough exploration of the entire cut or hillside is not possible. Therefore, in the great majority of cuts and natural slopes a mathemetical method of calculation cannot be applied. There will be certain cases where, owing to a combination of their special importance (and reasonable conditions of homogeneity) the necessary explorations, sampling and laboratory tests can be made to define the strength parameters, so that an appropriate method of theoretical analysis can be applied. It is, nevertheless, still true that it is practically impossible to be familar with all the mechanical properties of the soils at every point of the road to be built, with sufficient precision to allow the design of every cut and embankment to be based on theoretical methods. As was pointed out in §3.10, the slope angle for the majority of cuts and embankments will be recommended by specialists on the basis of reconnaissance explorations and simple laboratory techniques. These recommendations are based on previous experience, knowledge of the materials and the general policy established by the organization that is responsible for the project design.

In economically developed countries, these policies are usually very conservative for the volumes of traffic and are so high that high construction cost are justifiable so long as the road will not at some future date have to be closed on account of failures or collapses. Besides, in those countries, construction is so mechanized that large masses of earth can be shifted fairly economically and fast; whereas in order to clean and level slopes that fail after construction, expensive hand labor would be required.

In developing countries, however, the recommendations given by field engineers are usually less conservative in an attempt to minimize construction costs. This criterion generally leads to some failures, both during construction and afterwards. Because the interruptions in traffic that may be caused are not so serious in under-developed countries, a less conservative, more daring policy may be advisable in these areas. This daring must be carefully controlled, because it has been observed that many supposedly low-cost roads that are designed with very steep slopes, after all their problems have been corrected, are more expensive than if they had been originally designed using somewhat more conservative criteria. Furthermore, in many developing countries (Mexico is a good example) there are now some very important roads where the volumes of traffic are so high that non conservative design criteria are no longer advisable. Policies involving slope angle standards should, therefore, be sufficiently flexible to take into consideration all these details. Railroads, because of their limitations in diverting traffic during failure repairs, and the potential damage to traffic, should be designed following more conservative criteria (in terms of slope stability) than are applied to roads with low to moderate volumes of traffic.

So, once again it is emphasized that the angle of slope for most road cuts and embankments should be left to the judgment of field engineers, whose recommendations will be based on quick, superficial studies. The great importance of extensive exploration methods thus becomes obvious, for in this way the conditions prevailing in large areas can be identified at a low cost. Photointerpretation and geophysics are particularly useful tools when applying routine standards to slopes. It is important that the professional level of the field engineers in charge of these studies be improved; they should work in close collaboration with specialists in soil and rock mechanics and engineering geology. Also, work should be organized in such a way that the specialist who was responsible for the initial recommendations can have the opportunity to check them during construction, and if necessary make adjustments.

All this applies to roads and railroads, but runways, where the investment is far greater in relation to the area occupied, must be the object of far more detailed studies. On the other hand, airfield projects do not usually involve serious cutting and embankment problems.

Although we have explained why mathematical methods of slope stability analysis are not usually applied in road engineering, either because of lack of homogeneity in the construction materials or the large number of structures involved, there are those cases which require special study. These include cases of slopes which have failed and must be reconstructed.

6.2 Slope Failures Commonly Encountered in Road Engineering.

The following is a description of the types of slope failure that most frequently occur in road engineering. First of all, we will distinguish between the types which are most likely on hillsides or natural slopes and those which are most likely in artificial slopes.

It is not our intention to include rock mechanics, and we will only occasionally deal with rock slopes, for such study is beyond the scope of this book. It requires very detailed knowledge and specialized methodology which are available within that speciality.

It is not easy to differentiate between all the types of landslides worthy of attention, and it is even more difficult to classify them rationally. An attempt is made in the pages that follow, but the specialists who feel the categories are not altogether satisfactory should complete them using their own valuable experience. The factors responsible for the stability of earth masses can be grouped as shown in Table 6-1, based on [1].

Table 6-1

Factors affecting the stability of soil slopes

- a) Geomorphological factors
  - a.1 Topography of the surroundings and geometry of the slope
  - a.2 Distribution of discontinuities and stratification
- b) Internal factors
  - b.1 Mechanical properties of the soil
  - b.2 State of stress
  - b.3 Climatic factors — specifically surface and underground water

When considering the different types of failure that can occur in natural and artificial slopes, it will also be necessary to distinguish between the ones which occur in residual soils, transported soils and materials that have been compacted. Residual soils will receive special attention in a later paragraph.

6.2.1 Failures in Natural Hillsides

In this category are included movements that are typically found in hillsides, even though they do also occur occasionally in artificial slopes.

.1 Imperceptible Down-Slope Movement or Creep

Creep is the more or less continuous, and usually slow, downhill movement which occurs near the surface of some hillsides. Creep usually affects large areas and takes place without a definite slide or shear surface between the upper mobile part and the deeper immobile masses. Creep is usually caused by a combination of gravitational forces and various other factors. The downhill movement of a typical creep is very slow and rarely exceeds a few centimeters per year [2]; Plates 6-1, 6-2.

Plate 6-1 View of creep conditions in a natural slope near the Huixtla-Motozintla highway

Plate 6-2 Signs of creep in a natural slope

According to Terzaghi, there are two types of creep: seasonal creep, which affects only the upper crust of a hillside subjected to the effects of climatic changes with freezing and thawing, or wetting and drying, and massive creep, which occurs in deeper layers of earth which are not exposed to atmospheric conditions and consequently can be attributed largely to gravitational effects. The first kind, which is always present in some degree, produces movements which can vary from one season to another; the second shows almost constant movements. The surface layer affected by seasonal creep is extremely thin, with a thickness of one or two meters [3].

It has not yet been discovered why a massive creep process commences in a hillside, causing a surface crust, which may in this case be several meters thick, to start moving slowly downhill. The possibility of an *intrinsic strength* limit has been discussed [2,4]. It appears that if the shear stresses are below this limit, the surface of the hillside will remain at rest, but if the stresses are above it, continous slow shear and massive creep will occur. In addition, the hillside material also exhibits a peak shear strength; if the stresses exceed this, there will be a rapid movement: a landslide.

Although the intrinsic strength concept and the other possible causes of creep have not yet been clearly defined, this movement occurs at very much lower stress levels than the peak shear strength of the soil. This has been established by Griggs [4] and Bishop [5], who, while performing drained triaxial tests on clays, discovered that the stresses required to produce very slow long-term deformations were only a fraction of the peak shear strength. It also seems reasonable that the mechanism of these slides is related to the low shear strength of the surface materials, produced by low normal effective stresses.

In the third Terzaghi Lecture [46] Bjerrum proposed another mechanism which may in some cases contribute to the surface movements of hillsides composed of over-consolidated clays or shales. According to his theory, with the weathering of these materials and the consequent weakening of their interparticle physicochemical bonds, energy stored as residual strain is released. As a result the materials expand and weaken, while at the same time additional forces are generated which act down-hill on the outer layers of the slope.

Another fundamental mechanism which undoubtedly plays an important part in triggering creep is illustrated by Fig. 6-2, which was pointed out by Goldstein and Ter-Stepanian [6] some years ago.

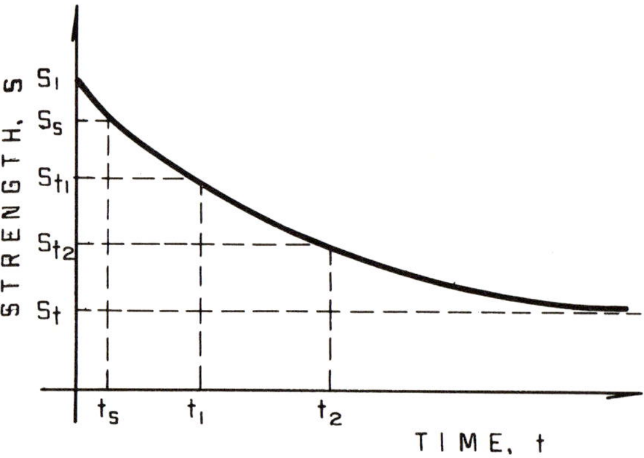

Fig. 6-2 Evolution of the strength of a clay subjected to continuous stress equal to or less than required for failure [6]

This figure shows the evolution of the strength of a clay when subjected to continous stress equal to or less than failure. The S_s value in the figure corresponds to the conventional peak strength, as obtained in a triaxial test performed in t_s time. The reduction in strength by long term stress can be explained in terms of the progressive rupture of the interparticle bonds which occurs in the clay as a result of deformation under the acting shear forces. In a natural slope, both conditions are present: the existence of shear stress, and that shear acting over a long period of time. This is consistent with the lowering in strength of the slope material, even at low stress levels, of the type reported by GRIGGS and BISHOP.

As we have already mentioned, creep often affects wide expanses of sloping ground. No reliable method has been found for deterring the process once started; hence the importance of locating it in advance, either in the initial field studies that are carried out before commencing a project or at the pre-project stage. That is why it is vital to consider the external signs of the phenomenon that can be noticed by the engineer or geologist undertaking site reconnaissance. (Fig. 6-3 and Plate 6-2).

Typically the displacement rate is fastest on the surface of the slope, gradually decreasing with depth, where the movement is more restricted. There is ample experimental evidence of this velocity gradient. It can be recognized in the tilting of the trees, posts and other similar elements, which become perpendicular to the hillside instead of their natural vertical position. This is a bad omen, which engineers should always be on the look-out for when exploring a natural hillside. Also, all the heterogeneities of the surface area of the hillside are reflected in the varying rates of movement which causes cracks, terraces, fractures and disruption of walls and fences and other structures.

All these external signs help to identify creep. A most useful aid is systematic and careful photointerpretation. By studying photographs, creep can be easily detected even by the beginner, for a creeping hillside acquires a strange wavey configuration, similar in appearance to a moving viscous liquid. When aerial photos are studied with the aid of a stereoscope, the evidence of creep can be seen even in vegetation which might hide the ground surface.

Once creep is discovered, there should be no hesitation in altering the location of the road so as to avoid problems, because no reliable remedy has yet been found for this type of failure. If the problem is not avoided, the road cuttings and embankments will be in continuous movement, with all the consequent disadvantages to service capacity and appearance, and with high maintenance costs and the ever-imminent risk of all more profound shear slides that might be triggered by the superficial slide itself.

.2 Landslides Associated with Accumulated Deformation (Unfavorable geological profiles).

Landslides can be generated in natural slopes where deformation is accumulated by the tendency of large soil masses to creep or slide downhill. This type of slide is typical of natural slopes in hillside deposits made up of rather heterogeneous unconsolidated materials, such as old landslide debris or colluvium which are affected almost exclusively by gravitational forces. Movement occurs on the surface where these deposits come into contact with other firmer underlying ones. The hillside is formed with slope angles that are locally slightly greater than that of limiting equilibrium. Inside the mass there must be strong, local sliding tendencies, which produce significant deformations in the affected soils. During the prolonged period of time during which the gravitational forces and their associated shear stresses act on the materials within the hillside, the shear strength is reduced by the accumulating strains (Fig. 6-2). In certain areas inside the hillside creep occurs in the sense described by GOLDSTEIN and TER-STEPANIAN in [6]. According to these authors, very slow continuous deformation develop within the hillside where there are local concentrations of shear stresses.

Under these conditions, a hillside deforms over a long period of time. Eventually, as a result of accumulated strain and reducing strength, the soil fails and a failure surface forms inside the hillside. The loss of strength through deformation (Fig. 6-2) plays an important part in the development of both the failure surface and the progressive failure (see §6.4), because the soil begins to slide first in areas with the highest concentration of shear stresses. Failure causes redistribution of the stresses and extension of the failure surface. The residual strength of the soil should be the one considered available, to resist sliding because of the advanced degrees of deformation required to produce these failures (Chapter 1), are well beyond the strain associated with peak strength.

Once the failure surface has developed, a rapid sliding of the affected masses may occur; if not, the mass will remain in place, not at stresses close to limiting equilibrium. This will depend primarily on the angle of the failure surface formed and, to a lesser extent, on the shear restraints created by the heterogeneities in materials and irregularities in shape along the failure surface.

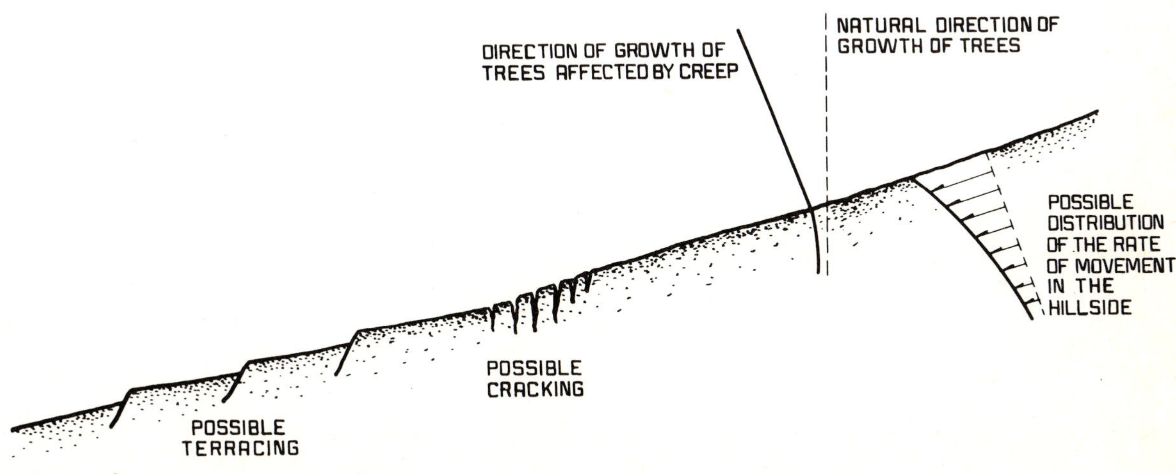

Fig. 6-3 Signs of a surficial slide or creep

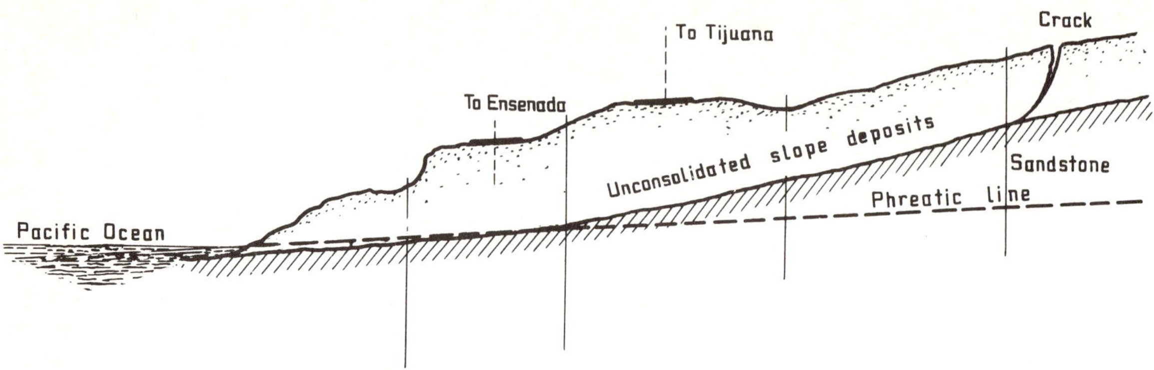

Fig. 6-4 Failure surface formed in a natural slope

In theory, this case can be compared with the equilibrium of a body on an inclined plane. The observed failure surface typical of a process where deformation is accumulated, is almost plane (Fig. 6-4). Several factors may contribute to this; first and most important is the pre-failure, slow deformation processes which are more likely to mobilize strength mechanisms of a purely frictional type, that are conducive to plane flat sliding planes. If the angle of a bedding plane is steeper than the angle of friction (which will have a value about equal to or slightly greater than the angle of residual strength of the soil), the mass above will move very slowly downhill along the bedding plane; if the angle is smaller, then the failure plane will develop within the soil mass along some surface of weakness.

The authors believe that mechanisms like these are very common in natural slopes. Wherever a weak surface occurs with an appropriate slope angle, owing to local geology (especially the distribution of the contacts between hillside deposits or weathered residual materials with much firmer underlying layers), the mass may stay in place, even if a localized sliding surface has already formed. This situation may last for a long time, until one day an engineer makes a cut in that slope, destroying the already precarious equilibrium. The engineer finds himself faced with a problem which usually has serious consequences, with huge masses of earth that begin to move for no apparent reason. The authors are also of the opinion that many of the large landslides (probably far more than are generally thought) recorded on natural slopes alongside of roads, occur on localized or ancient failure surfaces which formed long before the engineer disturbed the precarious equilibrium with his project.

Figure 6-4, [7], shows a case of the type described. It is a huge slide of unconsolidated hillside deposits located above unweathered sandstone. The average angle of the sliding plane is about 15°; a value of perhaps 13° can be attributed to the angle of residual shear strength of the hillside deposits. The effects of this failure are shown in Plates 6-3 to 6-6.

The groundwater levels, in the vicinity of the failure surface, have a fundamental influence on stability. Water complicates and aggravates the slope failure mechanism that has been described.

.3 Flows

This type of earth movement refers to more or less rapid displacement in a natural slope in which the mass of soil loses its shape and internal continuity. The movement and the apparent distribution of speeds and displacements are similar to the behavior of a viscous liquid. The sliding surface can either not be distinguished or else disappears in a relatively brief lapse of time. Also the contact between the moving soil and the immobile remainder of the slope is often a zone of plastic flow, rather than a discrete surface.

The flowing material is usually an unconsolidated formation, and so the phenomenon may present itself in rock fragments, hillside deposits, loose silts or plastic clays; flows commonly occur in saturated landslide debris as mudslides.

Plate 6-3 Typical formation of slope deposits leaning against a platform of massive igneous rock (failure zone on the Tijuana-Ensenada highway)

Plate 6-4 Panoramic view of a slow slide in slope deposit over a predeveloped failure surface

Plate 6-5 Another view of the effect of slow deformation on a predeveloped failure surface (Tijuana-Ensenada highway)

Plate 6-6 Aspect of the failure surface at the toe of one of the largest slides on the Tijuana-Ensenada highway

Following [8] flows will be divided in this book into two large categories, depending on whether or not the water contained by the materials is responsible for the phenomenon. Thus, a distinction will be made between flows in relatively dry materials, flows in wet materials, or the extreme case of mud flows.

Flows in Relatively Dry Materials: Into this category fall flows of rock fragments, from the very fast ones to those which occur so slowly that some geologists describe them as creep. These movements can be explained in terms of the localized plastic flow of the deep-seated contacts between the rock fragments; they usually affect large masses of fragments, often with catastrophic consequences. It has been said [8] that the air trapped between the fragments, when heavily compressed, may play an important part in triggering the flow, through mechanisms similar to those by which pore pressure shows its influence in water. Possibly all real fragment flows originate with a conventional rock slide or by a large fall of rock masses from uphill formations. In any case, for fragments to flow, a large deposit of considerable volume is necessary.

Second, flows in relatively dry soils have occurred in loess, often associated with earthquakes. Apparently the effect of the earthquake has been to cause very rapid destruction of the bonds between the silt particles, bringing about genuine liquefaction, but with pore air pressure playing the part which, in these phenomena, usually corresponds to pore water pressure. Similar phenomena have been recorded in dry sands. Figure 6-5 [8], shows the typical shape adopted by these movements (Plate 6-7).

Plate 6-7 Flow in relatively dry soil

Flows in Wet Materials Including Mud Flows: For these flows, the soils must contain a considerable proportion of water which determines the nature of the slide. The amount of water that can be contained by the materials is very variable;

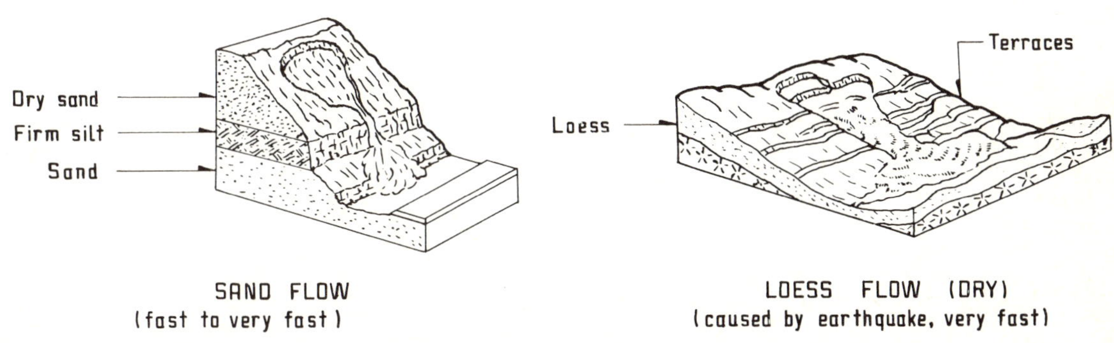

Fig. 6-5 Flows in dry soils [8]

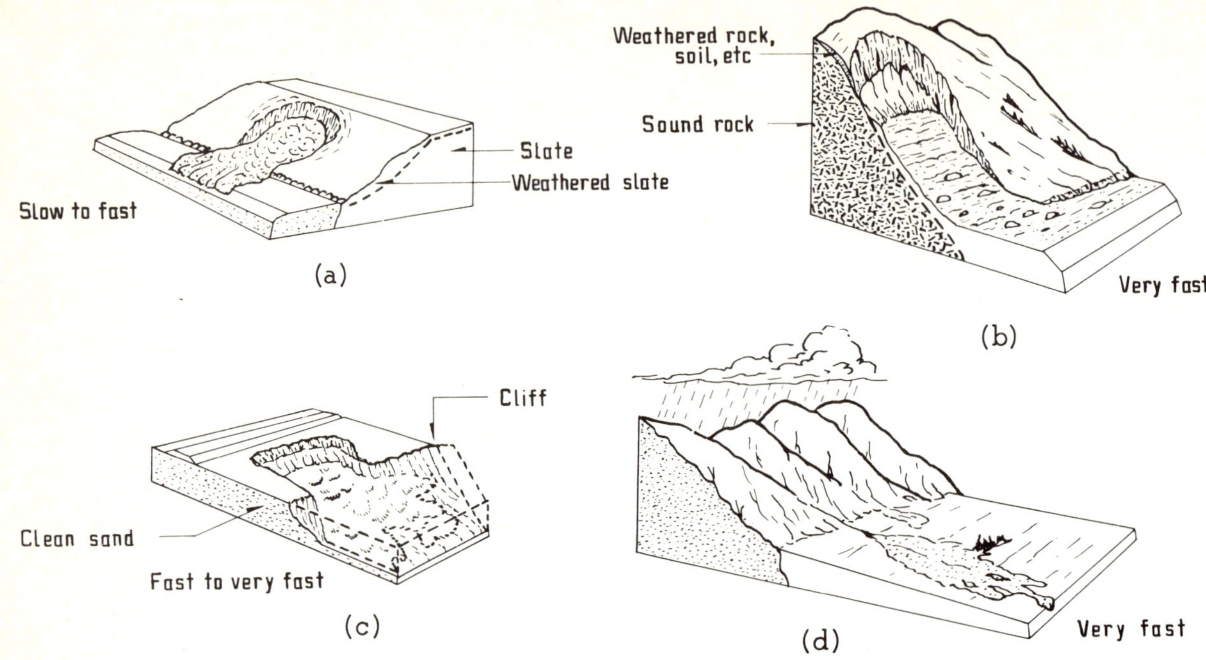

Fig. 6-6 Flows in moist materials [8,9].

likewise the role it may play in the development of the slide. Figure 6-6, [8,9] shows some drawings of typical slides of this variety (Plate 6-8).

Plate 6-8 A flow in moist materials

Flows in wet materials are referred to as mud flows when the water content is very high, but naturally there is no clear difference between *earth flows* and *mud flows*. *Detritus flows* are sometimes mentioned, when the flow material contains a large amount (at least 50%) of gravels, boulders or rock fragments, embedded in finer soil, as commonly occurs in old landslide deposits or many natural slopes made up of residual soil with inclusions of unweathered rock.

Earth flows (in earth materials which are not too wet) often develop at the foot of rotational-type slides, which will be described later. They sometimes occur with extraordinary speed as a secondary movement of the initial slide [10]. These earth flows usually retain large quantities of the original vegetation, together with jumbled stratification but the general appearance of the formation in which the initial slide occurred.

On other occasions, earth flows occur independently of previous slides. In this case they are movements with speeds which can vary greatly and occur especially in wet clayey soils and very fine frictional soils. Movements generally follow significant increases in the water contents of the materials and at high water pressures. In clayey soils, the flow often begins rapidly but may continue slowly for a long time. In highly sensitive clays, flows can occur at constant water contents, because of the great reduction in shear strength owing to structural breakdown from some initial shear displacement [3].

Earth flows in fine-grained soils are typical of coastal formations. They are generally associated with marine erosion and repeated fluctuations in pore pressure as a result of changes in the water level because of tides [11]. They originate from processes similar to liquefaction.

Mud flows are slides in fine materials with a very high water content. The movement causes complete structural disorder. Slides typically resemble an advancing glacier and the displacement rate may vary between a few centimeters per year [9] and meters per second, corresponding to catastrophic slides [8]. In slow-moving flows, seasonal changes in climate commonly influence the rate of movement, whereas rapid flows usually follow periods of heavy rains or ice and snow melting. Slow movements usually occur in clayey materials which are either fissured or finely interstratified with layers of sand with high water contents [12]. They occur along shallow surfaces with a moderate slope angle, which cannot be far from the value of the angle of residual strength of the soil.

Very rapid mud flows are often encountered on hillsides from which the overlying vegetation has been removed by fire or stripping for construction. They usually begin with very modest sizes. They rapidly enlarge as they collect and transport the soil over which they pass until they are out of all proportion to their initial size. In this way, genuine rivers of mud, capable of causing real catastrophes, are formed. The soil liquefaction phenomenon is probably among the factors at their origin.

Detritus flows occur when there is a drop in the shear strength of the fine-grained matrix of such formations. The mobile mass breaks into ever smaller fragments as it advances downhill.

6.2.2 Failures in Artificial Slopes

.1 Rotational Slides

Rotational slides (Plate 6-9) are rapid or almost instantaneous movements that occur in slopes, with movement along a curved surface which develops well within the body of the slope. Sometimes, it extends into the foundation soil and is termed a base slide. Sometimes it only extends to the toe of the slope, a toe slide. The sliding surface forms when shear stresses greater than the strength of the material act along a curved, continuous surface that can permit the mass above it to move. The strength to be considered in each particular case is vital in analyzing the potential for failure. This will be dealt with separately later in this Chapter. For the purpose of discussion the strength which is overcome when a rotational slide occurs is generally the peak strength, in the sense that is used in Chapter 1. Thus, at failure within the slope there are shear stresses which fairly rapidly overcome the shear strength of the soil. Consequently the soil fails, and a sliding surface forms. These movements are typical of road cuts, embankments, and weak soil foundation for embankments.

Plate 6-9 Effect of a rotational slide in a road

Plate 6-10 Typical view of the head of a rotational slide

Creep can occur [6] in the initial stages of a slide of this kind, but its practical importance is now less. Usually even the fastest landslides are preceded by movements, cracks and other signs that the relation between stress and strength is becoming critical within the slope. The formation of curved cracks at the head of the slope is typical of this condition (Plate 6-10). On the other hand, the progressive failure mechanisms that may occur along the future sliding surface probably plays a very important part in its development.

Rotational-type slides may occur along cylindrical or conchoidal surfaces which in cross-section parallel to the direction of movement is an arc of a circle (or a reasonable

Plate 6-11 Rotational slide by the toe of the slope (Puebla-Orizaba highway)

Plate 6-12 Rotational bottom failure. Note the failure surface day lighting at the toe of the slope

approximation). This shape proves very convenient when establishing a mathematical model of the failure. The circle is often elongated or flattened, usually depending on local geological sequence, stratigraphical profile, and the nature of the materials. Figure 6-7 shows the profile of some typical examples of rotational slides, see Plates 6-9 to 6-13.

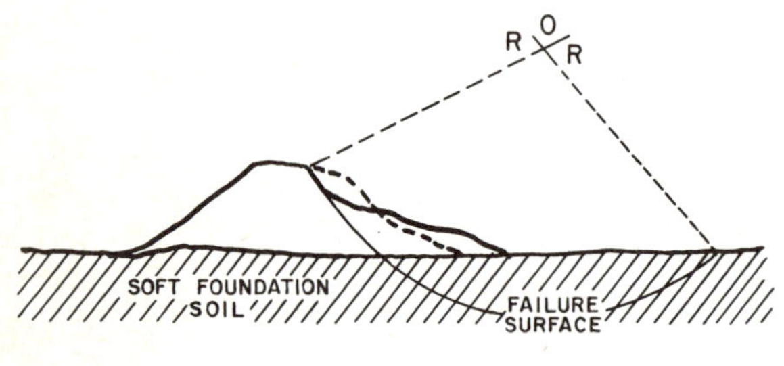

a) Bottom failure

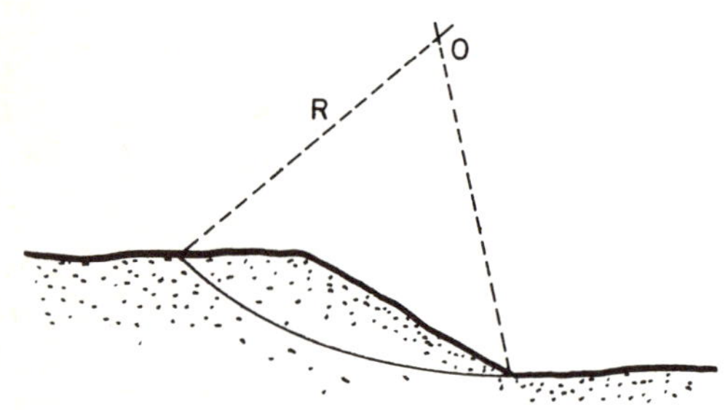

b) Failure through the toe of the slope

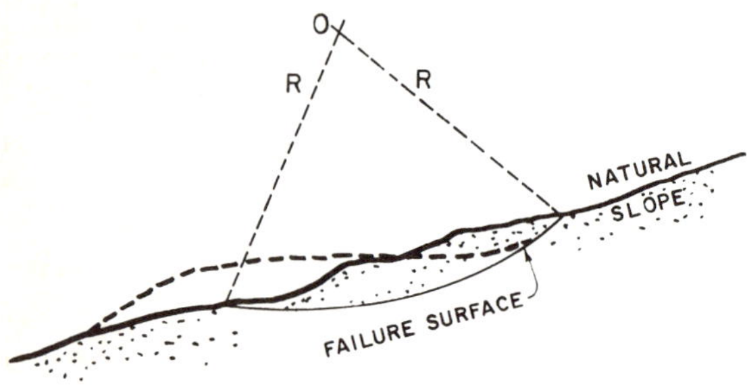

c) Superficial failure

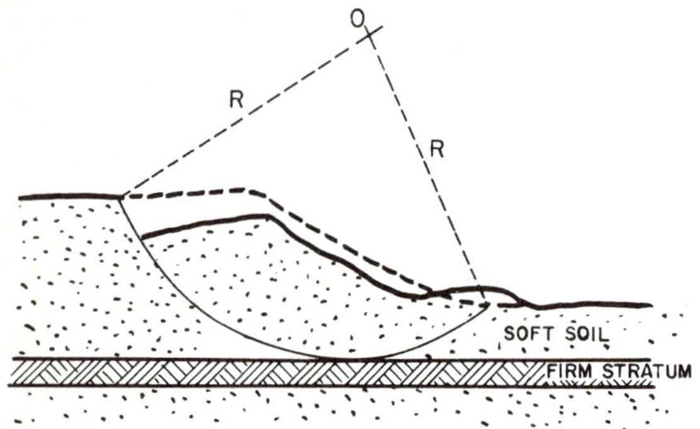

d) Failure limited by a firm stratum

Fig. 6-7 Rotational slides. Typical cross-sections

Plate 6-13 Example of a large-scale rotational slide (by courtesy of T. SMITH, Laboratory of the California Department of Highways, U.S.A.)

Circular rotational slides usually occur, of course, in homogeneous clayey materials or soils whose mechanical behavior is governed basically by their clay fraction. The areas affected are usually rather deep; the steeper the slope, the deeper (talking about the actual slope alone and not taking the foundation ground into consideration). Although engineers tend to associate circular rotational slides with man-made cuts and embankments, these failures are in fact also common in natural slopes composed of homogeneous fine materials (often overconsolidated clays). When the slopes of the hillside are not at all steep, failure surfaces may develop at not very great depths (Part *c* of Fig. 6-7). Circular rotational slides may be found in the body of the slope or at the base. The first kind do not affect the foundation soil, but the second kind do, because part of their development takes place in it.

The masses affected by a circular slide may be very long in relation to the general dimensions of the slide itself, leading to cylindrical slides, or be very short of a conchoidal shape, the width being then much smaller than the length (Fig. 6-8). A typical slide is illustrated in this figure showing the usual terminology for the features.

Rotational slides that are not of the typical circular type are sometimes associated with overconsolidated clays. Owing to different degrees of weathering, stratification and other causes, such as discontinuities or structural disorder, the slopes formed by these clays are not homogeneous. They are, therefore, typical of cuts. The materials involved always suffer severe disintegration [8].

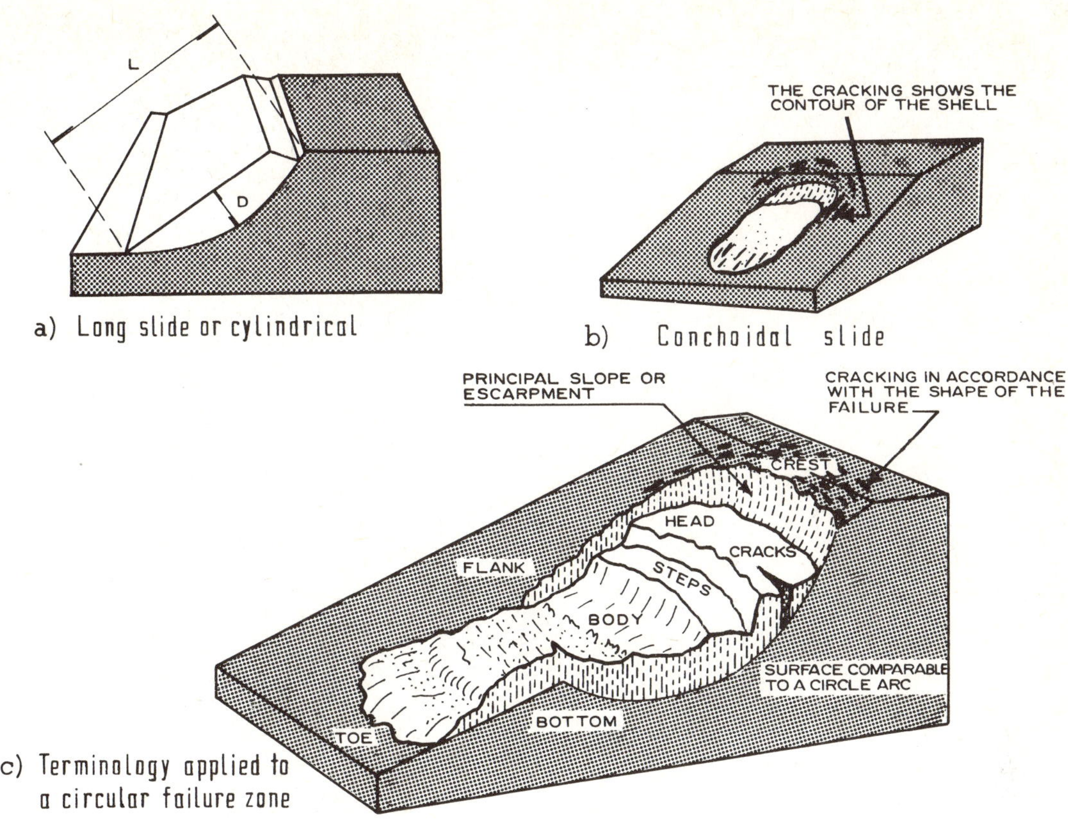

Fig. 6-8 Type of rotational slides

The shape of the curved sliding surface, is idealistically considered circular or partially made up of circle arcs. This is a convenient approximation, especially where calculations are concerned. The circular shape is altered by joints, contacts and other discontinuities in the materials, particularly in residual soils.

Once a slide has taken place, the scarp at the head is almost vertical. This tends to encourage new slides if the engineer does not correct it. The same effects can occasionally occur in the sides of the slide.

Plate 6-14 Effect of a translational slide

.2 Translational Slides

These slides usually consist of linear or translational movements of the body of the slope over essentially flat sliding surfaces, that are associated with weak layers, located usually relatively at shallow depths (Plates 6-14, 6-15). The sliding surface develops parallel to the weak layer and terminates in two steep faces, which usually form as a result

Plate 6-15 Development of a translational slide. (By courtesy of T. SMITH, California Department of Highways, U.S.A.)

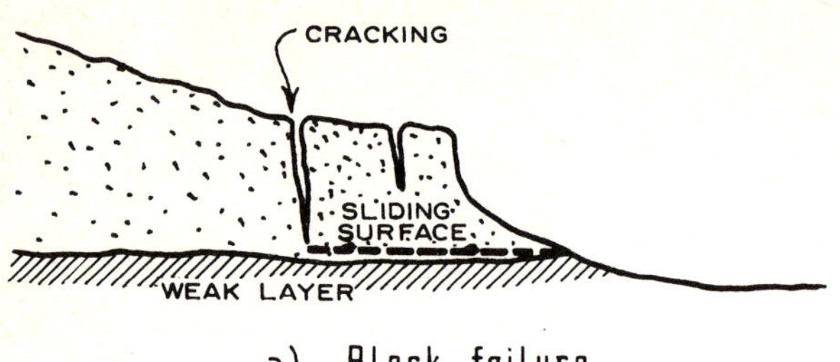

a) Block failure

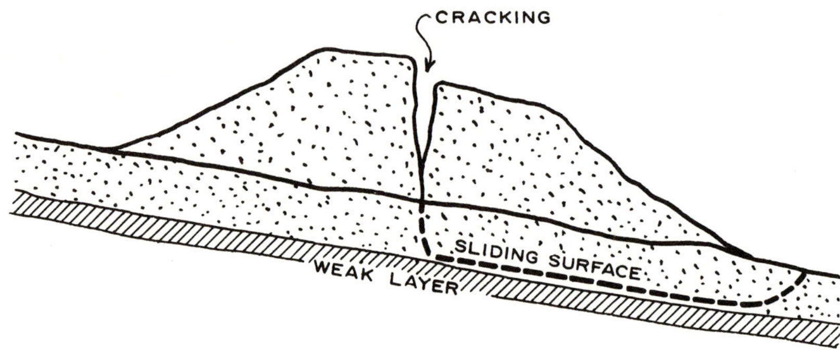

b) Block failure encouraged by the stratification of the natural ground

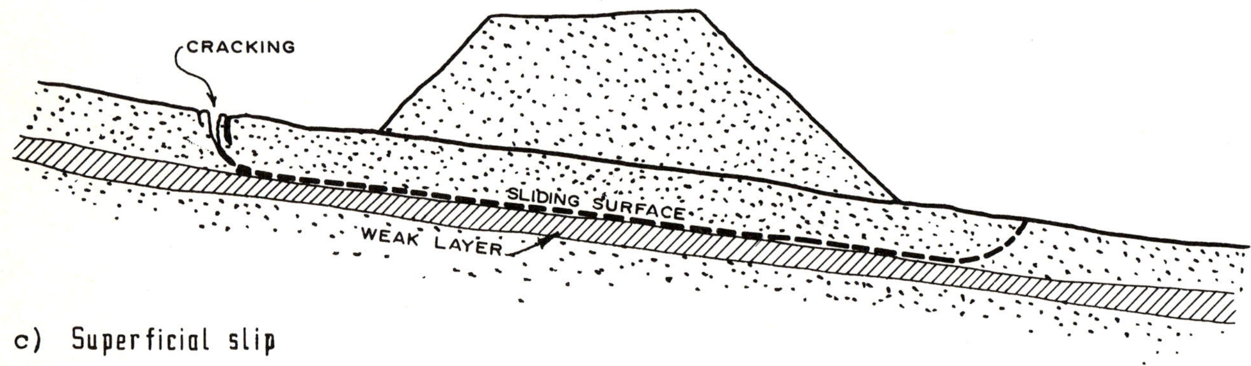

c) Superficial slip

Fig. 6-9 Translational slides

of cracking. The weak layers which encourage these slides are usually made of soft clays, fine sands or loose non-plastic silts. The weakness of the layer is very often related to high pore water pressures in both the silt or sand layers. Most of such slides occur during or after the rainy season.

Block slides (Part *a* of Fig. 6-9) are often associated with pre-existing discontinuities and fractures in the materials that help define the moving mass, but always as an effect which complements that of the weak underlying layer. However, the movements also produce similar fractures.

Slides in a surficial layer (Part *c* of Fig. 6-9) are typical of natural slopes made of clayey materials which have formed as a result of in-place weathering. They are usually brought about by overloading from an earth structure built on the hillside. In these slides, the movement occurs almost without distortion [13].

.3 Slides with Compound Sliding Surfaces

In this type of slide, rotation and translation are combined producing compound sliding surfaces on which flat zones and curved stretches develop, resembling circular arcs. These surfaces are usually predetermined by surfaces of weakness or discontinuity within the slope. The slide shown in Part *d* of Fig. 6-7 can be classified as either compound or circular. Classification of the slide as rotational or translational generally depends on whether it is the circular or the flat parts that predominate. Compound slide is the category used when both surfaces are more or less equal.

Usually the nearer the weakness or discontinuity (failures, joints or a weak layer) appears to the surface, the greater will be the translational component of the slide. Figure 6-10 shows a typical compound slide. These slides usually cause distortion in the materials which is typical of circular failures [14,15].

.4 Multiple Slides

These are movements with several different slip surfaces that occur either simultaneously or in rapid succession. It is advisable to distinguish between progressive and retrogressive slides (Fig. 6-11). Both are common in natural slopes where a cut is made.

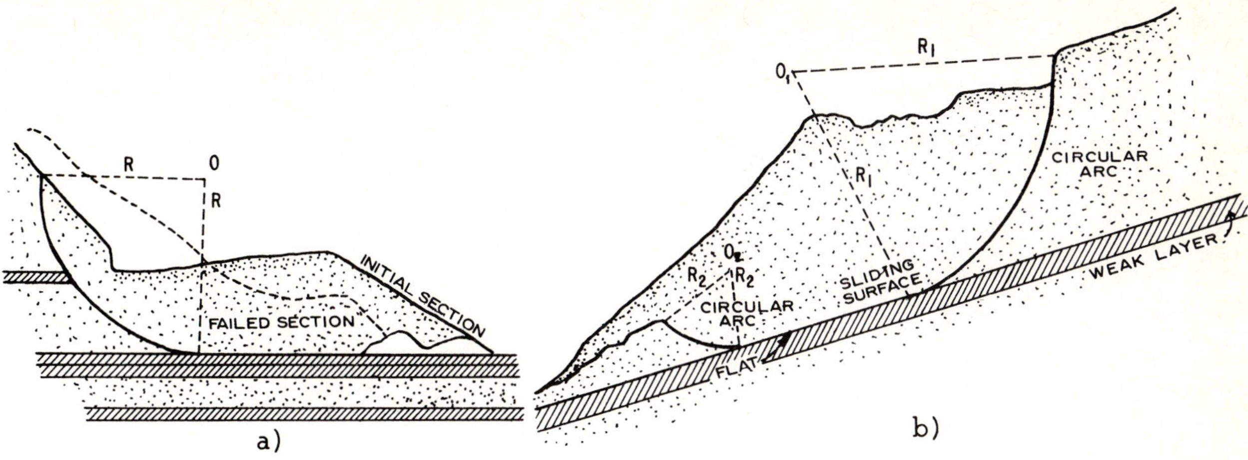

Fig. 6-10 Compound slides

Retrogressive slides originate in an initial failure situated furthest downhill with successive slides uphill. They are usually attributed to the instability of the soil at the head of each new slide. All slip surfaces usually converge in one extended surface. They may be rotational or translational (Parts *a* and *b* of Fig. 6-11). Retrogressive rotational slides often occur in regions where the topography changes abruptly such as where steps have formed in which important erosion features [9] take place, especially if there are thick layers of overconsolidated fissured clays or shales, with thick overlying layers of rock or firm soils (see Plate 6-16). Retrogressive translational slides occur in surface layers, often associated with fissured clays and shales. It appears that the more cohesive the material, the fewer the independent sliding units that form the sliding mass [9,16].

Progressive slides (Part *c* of Fig. 6-11) usually consist of a series of rotational surface slides that progress downhill. They are a characteristic of the final stages of erosion in hillsides made of overconsolidated or fissured clay (like London clay, [9], in which they form slopes with an average angle of up to 8°). These failures sometimes form extremely regular small steps.

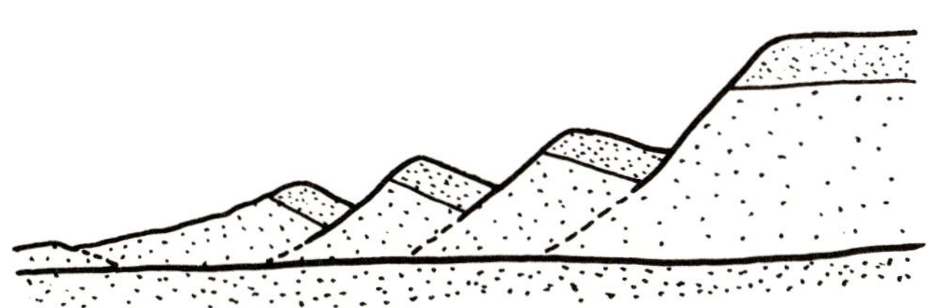

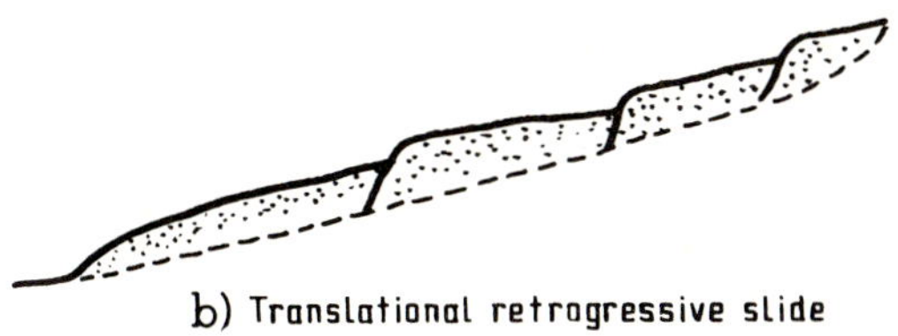

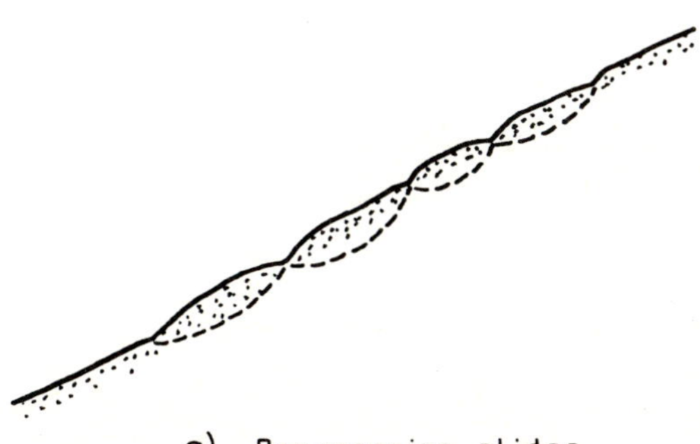

Fig. 6-11 Multiple slides

Plate 6-16 A retrogressive rotational slide in fractured tuff

6.2.3 Collapses and Falls

These failures are typical both of steep natural slopes and the cuts made in them. They do not necessarily happen alongside roads; for example, a collapse can occur in a river bank as a result of erosion by the current.

Plate 6-17 Limestone formation that encourages falls and collapses (Puebla-Orizaba highway)

Plate 6-18 A large collapse in a cut for a railway. (Viborillas-Villa de Reyes)

They generally consist of local slides, usually of small masses of soil or rock which tumble or fall freely down the steep face. However they can take the form of great masses of fragments which are classified as collapses. In these slides, one cannot refer to a sliding surface, although the fall is usually predetermined by already existing discontinuities and fissures. The latter usually open when the cut is being excavated. The blocks of soil or rock find themselves deprived of the lateral confinement they enjoyed previously, with the result that the fragments come loose. Falls are aggravated by frost action and hydrostatic pressure in cracks.

Collapses and falls are always associated with cliffs or steep-sided cuts. They often occur in highly overconsolidated clays and badly fractured rocks [17,18], Plates 6-17 to 6-20.

Plate 6-19 Fragmented dolomite formation that encourages falls

Plate 6-20 A fall of rock fragments

6.2.4 Damage Not Related to Shear Strength of Soil

Here we will discuss three special types of damage, whose mechanism does not depend, either directly or entirely, on the shear strength of the soil. However this does not mean that this extremely important property plays no part in the development of the damage.

First let us deal with the damage caused by erosion, which so often occurs in road embankments and cuts. It is the result of the action of erosive agents on the slope materials. Wind, rain and rainfall run-off are the agents most often responsible for these harmful effects which have to be controlled by road engineers. They cause caverns and gullies and irregular surfaces in the slope, which was originally regular. If these defects are not deterred in accordance with the standards which will be described later, they may progress until an embankment slope is eventually destroyed or a cut face very badly damaged. In this last case, it is very difficult to distinguish between the damage caused by erosion and the attack on slope materials by weathering or chemical decay, which sometimes play an important part in the stability of rock slopes with soil-filled joints or in slopes in heavily overconsolidated clays, shales, and slates. (Plates 6-21, 6-22).

Plate 6-21 Effect of erosion by a river

Plate 6-22 Effect of erosion. Note that as material is lost in the body of the slope and at the toe, a rotational slide starts to develop

Piping or seepage erosion within the soil or rock [18,19], although not very common in road engineering, may be responsible for more problems than engineers realize. Piping typically occurs in an embankment when water accumulates on one side for quite a long time, generating seepage through the embankment. Water will be on both sides of an embankment crossing swampy areas, reservoir basins, areas that are flooded by rivers, and inlets. However, it is unusual for a road embankment to have very different water heads on opposites sides, like a dam, creating a sufficiently large hydraulic head for flows' velocities sufficient to cause piping. (In Appendix 7a.11 the order of magnitude is given for the hydraulic gradient at which piping can occur. Also standards are given for calculating this gradient from a flow net which is drawn across the embankment).

In Chapter 3, great emphasis was laid on the importance the type of material has in relation to the risk of piping through the embankment, Table 3-1 giving a summary of U.S. experience on the subject. Piping starts when soil particles are eroded from the downstream slope by the flow of water. Once these particles begin to move, small channels develop within the soil mass, along which the water flows at higher speeds, with increased erosive power. Once started, the piping tends to grow continuously and the diameters of the channel get bigger and bigger. Another curious characteristic of the phenomenon is that it begins at the downstream slope, and grows backwards, towards the opposite face of the embankment. The phenomenon ceases when the holes in it are large enough that the soil above collapses into the holes or when enough soil has been removed that the soil mass can slide.

A factor which greatly contributes to piping is insufficient compaction of the embankment, when susceptible soils are involved (Table 3-1). Insufficient compaction is a common fault in the vicinity of walls or rigid surfaces such as ducts or culverts. Because culverts are places where water is present and around which compaction is difficult, they are always critical points where piping problems are concerned. Very careful attention should, therefore, be paid to the nature of the materials used in the areas surrounding them, and to the compaction.

Damage by cracking is of special importance in embankments. The cracks that occur in road embankments can be both transversal or longitudinal. The transversal type are induced by differential settlement along the axis of the road, and are of importance with embankments built on soft soils. They are likely in transition areas between poor and good foundation soils or in places where differential settlements are especially large. However, it is hard to conceive that this type of cracking is only dangerous when there are unbalanced water levels on the opposite sides of the embankment and when the soils are erodible. Cracks along the axis of the road are far more usual, or at least far more often perceptible. They are generally caused by differential movements between the shoulders and center of the embankment. They are identified by the two groups of cracks parallel to and usually symmetrical about the axis of the road. They are usually located on the shoulders, and on the edge of areas which are usually paved. These cracks extend almost without interruption for tens or hundreds of meters. This type of failure often presents a serious problem for the engineer, both because of the size of the cracks and their continuous enlarging, until eventually the whole embankment is destroyed. The causes of the largest longitudinal cracks that have been reported are still under discussion. The differential movements caused by difference in drying in the materials in the shoulders and slopes of the embankment, compared with the central part which is far less exposed to solar evaporation, seem to be largely responsible.

The above theory is illustrated by Fig. 6-12. Consider a foundation soil of soft, compressible clay, located in a flat area or a dip, with a groundwater level close to the surface, on which an embankment is built. Let us suppose that this embankment is made of fine materials, with a large clay fraction susceptible to volume changes with variations in the water content. Evaporation is restricted in the area underneath the embankment; so water accumulates there. In some cases, the groundwater level under the embankment may rise slightly.

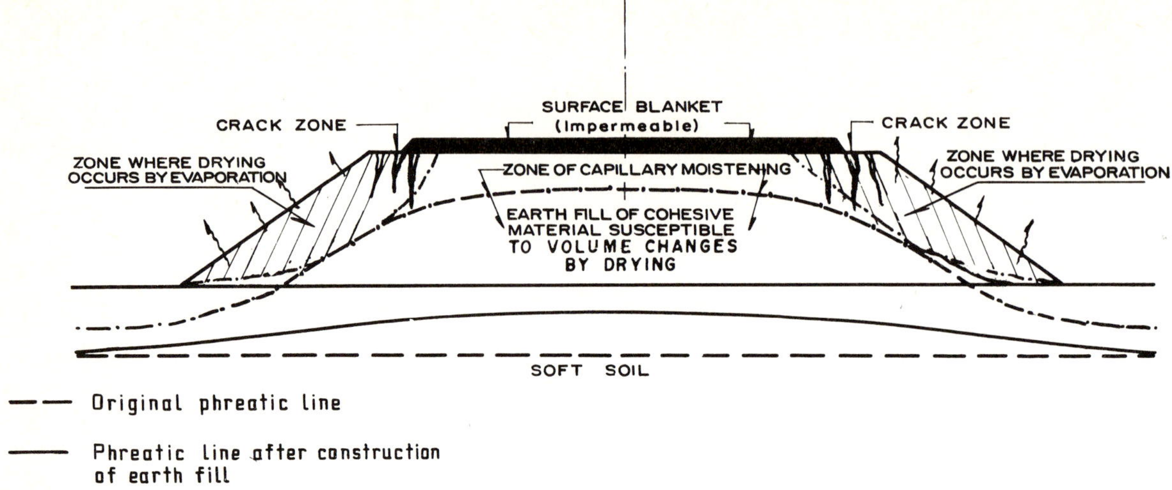

Fig. 6-12 Longitudinal crack mechanism in embankments [20]

If the local climate is extreme, with a prolonged and severe rainy season followed by drought, both the foundation soil and the embankment material will accumulate high water contents as a result of a combination of the rain and capillary rise from the rising groundwater. When the dry season starts, there will be intense evaporation in the exposed soils in the shoulders and slopes of the embankment (striped zones of Fig. 6-12). As a result of this evaporation, these soils will suffer severe shrinkage. It is hard to define the border-lines between these areas of intense evaporation and the central zone, where the soils are not affected. In Fig. 6-12 they have been tentatively drawn on the basis of the data given in [20], where an initial (though maybe not conclusive) study of the problem is presented. As a consequence of these differential volumetric deformations, two symmetrical crack zones are formed, like the ones shown. The central parts of the embankment will be protected from solar and wind evaporation and will remain moist as a result of capillarity. This explains how the effect of the differential changes in volume is apparent even in embankments that are not protected by an impervious paving material. Pavement will only intensify the phenomenon.

Plate 6-23 Longitudinal crack before the Escárcega-Chetumal road was surfaced

Plate 6-24 Longitudinal cracks in an earth fill. (Apaseo-Irapuato highway)

The conditions just described are usually associated with longitudinal cracking problems (Plates 6-23,6-24). Furthermore, longitudinal cracking usually appears where the foundation soil is predominantly clayey, soft and compressible. The sequence of rains and droughts causes violent volumetrical changes in this soil, which can be identified by rises and falls in the level of the ground. The part that is played by this phenomenon in the formation of longitudinal cracks is not yet clear and this requires additional research. However, longitudinal cracks have been observed in embankments made of material that is susceptible to volume changes by drying, but on firm foundations. In these areas there is also longitudinal cracking, but less pronounced.

The nature of the embankment materials and the way in which they have been compacted have a strong influence on the development of the cracks. Figure 6-13, [21], shows the range in grading curves for soils that have been found to be more susceptible to cracking. Although the evidence available is far from complete, it appears that inorganic clays with plasticity indexes below 15 and with grading within the area marked in Fig. 6-13 are more susceptible to cracking than other finer or coarser soils (this appears to be especially when the soils are compacted on the dry side of optimum). Clays with a plasticity index above 20, although they may be finer than the previous kind, can be subjected to greater stress without cracking.

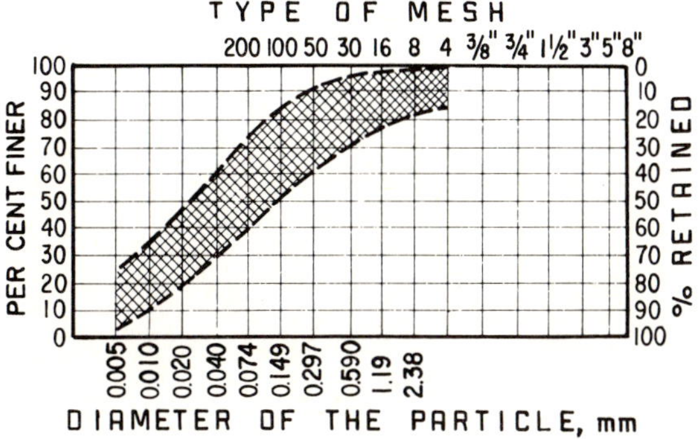

Fig. 6-13 Particle size distribution of the soils most susceptible to cracking

Residual soils with coarse particles of soft rock that are powdered during compaction are particularly prone to cracking. These soils are often compacted on the dry side, owing to the difficulty of adding water to them. Cracked embankments are often found to have been compacted at water contents considerably drier than the field optimum.

Narain [22] reported that if clays are compacted at the field optimum their flexibility is greatly increased, compared to compaction performed at water contents 2% or 3% below that optimum. Subsequent increases in the water content seem to have little effect. It is also stated in the same study that there is no good correlation between the volumetric changes that cause cracks in an embankment and those that are obtained by performing a typical laboratory expansion test. Therefore, the laboratory test is not a reliable index for estimating cracking susceptibility. In Chapter 4, some interesting data appear for establishing the influence of compaction on the flexibility of the embankment that is obtained.

6.2.5 Failures Caused by Liquefaction

Liquefaction phenomena consist of the rapid total loss of shear strength, which can be either temporary or semi-permanent [18]. This loss leads to collapse in any road structure that is built on or composed of a material that liquefies.

The two factors to which this loss of strength can be attributed were discussed in §1.14. They are an increase in the shear stresses and accompanied by a corresponding development in pore pressure, or (and this is the case that interests us here) the rapid development of high pressures in the pore water, as a result of a localized shear earthquake, or an explosion. This second cause is associated with a rapid collapse in the structure of the soil, whose voids, when the soil is saturated, become smaller; there is a consequent increase in pore water pressure.

Almost instantaneous liquefaction occurs in very sensitive saturated clays and loose fine sands, when saturated or nearly saturated.

As was explained in §3.2, the granular soils that are most susceptible to liquefaction are the fine-grained, loose-structured ones, when saturated. These characteristics are typical of uniform fine sands and non-plastic silts, or mixtures of the two. The most dangerous materials are loose sands, with $D_{10} < 0.1$ mm (0.004 in) and a coefficient of uniformity, $C_u < 5$, and silts with a plasticity index below 6, either when forming part of the body of the embankment or in the foundation soil.

In clays, liquefaction failures [23,24] are always associated with marine clays which have emerged as a result of the isostatic uplift of continents and have afterwards been leached so that the salt water originally contained by the pores is gradually replaced by fresh water. This leads to cationic exchanges (loss of sodium ions), some loss of peak strength, a great loss of the residual shear strength and a large increase in sensitivity. Local shear causes a structural breakdown and a drop in strength from the peak to the very low residual strength. The clay is remolded to such an extent that it takes on the consistency of a liquid (its residual strength) which becomes a semipermanent condition, since the absence of ions in the water impedes restructuring.

Earlier in this Chapter it was mentioned how liquefaction phenomena can occur in sands and dry granular soils through the development of pore air pressure. This is an interesting type of liquefaction, which has so far been studied very little.

6.2.6 Failure by Permanent Settlement of Embankment Shoulders

This failure consists of a progressive, cumulative deformation under the embankment shoulders, which suffer a vertical downward displacement, producing a rounded or terraced section at the crest. Between the settled zone and the rest of the embankment, a crack or scarp sometimes appears parallel to the axis of the road, which in turn can be the beginning of slope failure.

In roads suffering this condition, 30 to 40 cm (12 to 16 in) vertical displacements have been found, sometimes without any evidence of rupture in the affected materials. The failure appears to be associated with embankments which have either been inadequately compacted or else are made of fine plastic soils, in areas with poor surface drainage. However, it sometimes occurs in areas where the general stability cannot be blamed. Although further research into this type of failure is necessary, it often seems to be a process of accumulated deformations linked with stresses acting on the boundary of the embankment. The embankment deforms downwards in the upper part and outward in the lower part.

This type of failure has sometimes been blamed on vehicle wheels on the shoulder from improper passing where the outer wheels cause the ground to give way because there is no confinement. However, such slides at the edges of roads that have never been subjected to the effects of traffic, together with recent research into embankment failure using the finite element techniques suggest that the problem is probably caused by the internal state of stresses of the structure and not by any external load, aggravated by poor compaction at the edge of the embankment.

6.3 Theories on Stability of Slopes in Residual Soils

Residual soils have certain characteristics that deserve individual attention when evaluating the stability of both natural and artifical slopes. A recent publication, [25], has provided the principal basis for these comments. There are three concepts which play an important part in the stability of slopes in residual soils. They are the weathering profile, relic structure inherited from the rock and the effects of underground water.

The weathering profile is the sequence of layers with different properties which have formed in place, where they are found directly above the unweathered rock. Certain other soil profiles that are not residual, but have undergone limited transportation, such as colluvial profiles and hillside deposits are so similar to those of residual soils that they can be included here for discussion.

The weathering profile forms from mechanical fracturing and chemical decay. It can be very different from one place to another, especially because of local variations in the rock type and structure, topography, erosion conditions, underground water and climate, particularly rainfall and temperature.

In almost all metamorphic and intrusive igneous rocks, the weathering profile consists of a layer of residual soil underlain by weathered rock, and then the sound parent rock which is scarcely weathered. Sketches of such profiles are shown in Fig. 6-14.

Many of the engineering problems in connection with roads on residual soils are caused by the transition layer of weathered rock, between the upper stratum of soil and the lower stratum of sound rock. The limits between the different areas of a weathering profile are indefinite. There are several arbitrary definitions. For example, the boundary between residual soil and weathered rock is based on a 10% core recovery whereas 75% distinguishes the weathered rock layer from the reasonably sound bedrock.

Figure 6-15 shows some typical weathering profiles of common sedimentary rocks. Substantial differences can be observed between these and those of igneous and metamorphic rocks. For example, in Part *a*, a typical profile of carbonate rocks (limestones, dolomites and marbles) shows how the residual soil can vary greatly in thickness and quality. This soil, derived from the original rock by its dissolution, consists of the insoluble residue. It is often clayey, but can be sandy and gravelly. It usually represents only a small percentage of the original rock. The carbonate dissolved was subsequently leached away. (By way of contrast in igneous and metamorphic rocks, the residual soil contains almost all the components of the original rock). The weathering profile of many sedimentary rocks of the limestone type is often extremely irregular, with cavities that may or may not be filled with clay [28], Parts *a* and *b* of Fig. (6-15), and its transition from the residual soil is usually not abrupt but irregular.

In clay shales, the layer of residual soil (clayey) is often thin. This is attributed to the low permeability of the rock and to the resistance to weathering that has been developed by many of the minerals which are already a result of previous weathering of other rocks. In such materials, there are sometimes systems of small cracks and fissures which can easily be opened by stress relief, thus triggering mechanical weathering processes that can be very rapid. The fact that clay shales are generally more prone to mechanical disintegration than to chemical decay makes them different from most other rocks.

Most of the problems related to the stability of residual soils resulting from the weathering of metamorphic and igneous rocks occur in the upper layer of residual soil due to phenomena connected with increases in pore pressure (rain flows), or in the intermediate layer of weathered rock due to discontinuities and fractures inherited from the original rock. In these profiles, there are often strong seasonal fluctuations in the water-table.

The most common stability problems encountered in limestones and other carbonate rocks are when sink holes, intense fracturing and frequent interbedding of soft clays appear (Part *b* of Fig. 6-15). Highly concentrated water flow zones are often found in these formations.

The most typical kind of failure in clay shales is a shallow slide marked *A* in Part *c* of Fig. 6-15. These slides are usually preceded by an abnormally high groundwater level in the underlying fissured clay shale. Slides of the type marked *B* in this figure are associated with small weak or very permeable layers interstratified with shale, which is a common situation. Slides of the *A* type are typically associated with a layer of very soft weak clay. *B* and *C* type slides often become retrogressive. Eventually deep slides accompanied by small movements develop, the permeability of the failed mass increases, with the consequent possibility of accumulating seepage pressures; however, flow paths for the water behind and beneath the sliding mass generally remain the same. If the sliding mass does not collapse, steps taken to improve drainage will greatly benefit overall stability.

Skempton [29,30] has shown that hillsides in overconsolidated shales and clays will be stable with slope angles somewhat greater than half of the angle of residual strength but no higher than the residual. The limit can vary greatly owing to the pore pressure distributions within the mass. The limit is theoretical; in practice it is relatively common to find stable natural slopes with inclinations of the order of the residual angle of friction (or even slightly greater). In terms of peak strength, the maximum stable slope, in the sense used by Skempton, will be about half of the peak strength effective stress angle of friction (as obtained in a drained triaxial test).

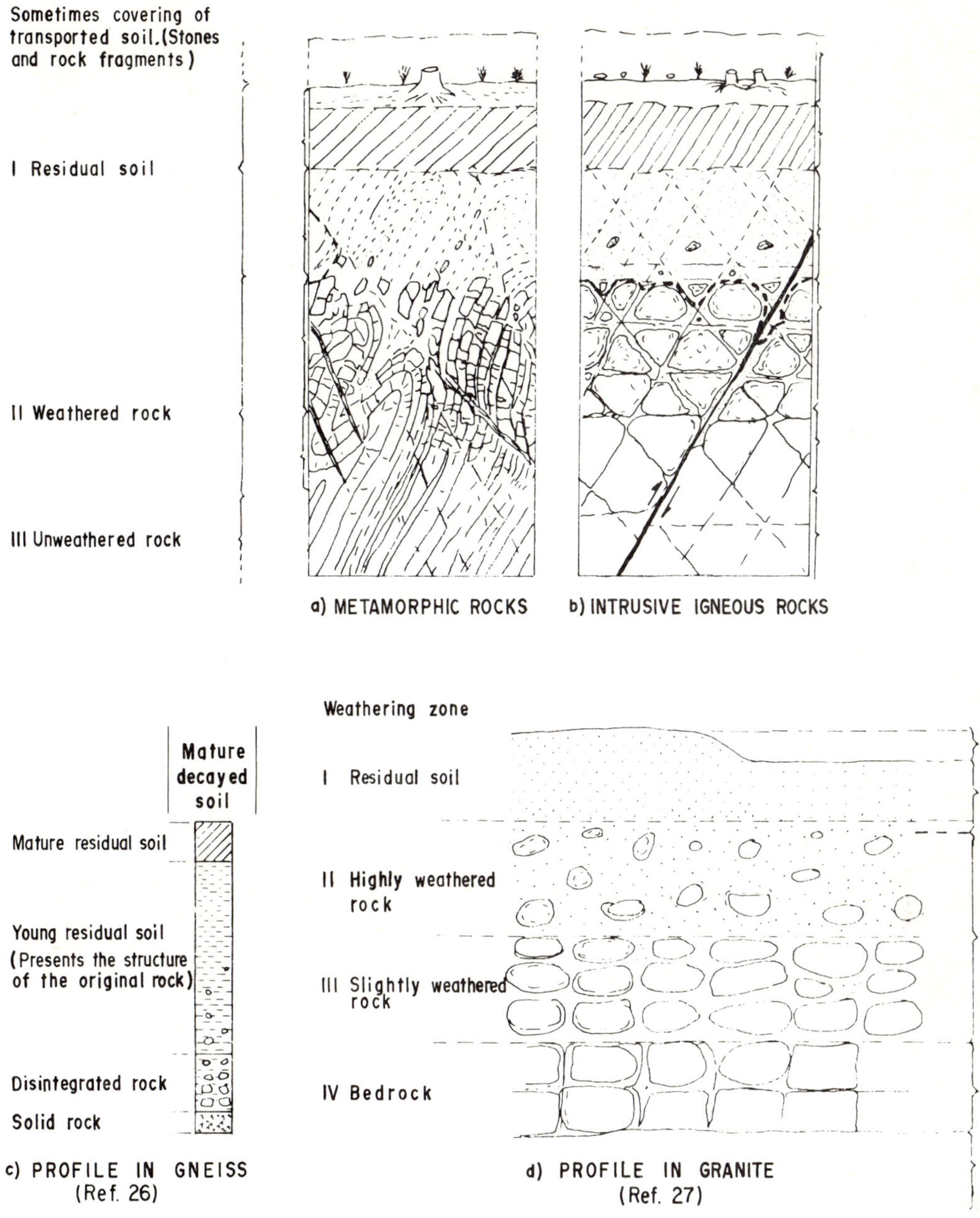

Fig. 6-14 Weathering profiles characteristic of igneous and metamorphic rocks

A special case of geological sequence which is of interest in many problems of residual soils is where shales are interstratified with sandstones. Sandstones are often more rigid than shales, but are also far more permeable and allow the seepage water to diffuse freely. The profiles produced by interbedded sandstone and clay shale can vary greatly as a result of folding and the different degrees of cracking and fracturing it may cause.

HENKEL [31] has made a theoretical study of the slope angle these profiles may exhibit under stable conditions, with conclusions similar to those that have been mentioned for clay shales alone. The critical slope angle is between half of the angle of friction that can be attributed to the material and the full angle. Note that it is often the residual angle of friction that should be considered for shales interstratified with sandstones.

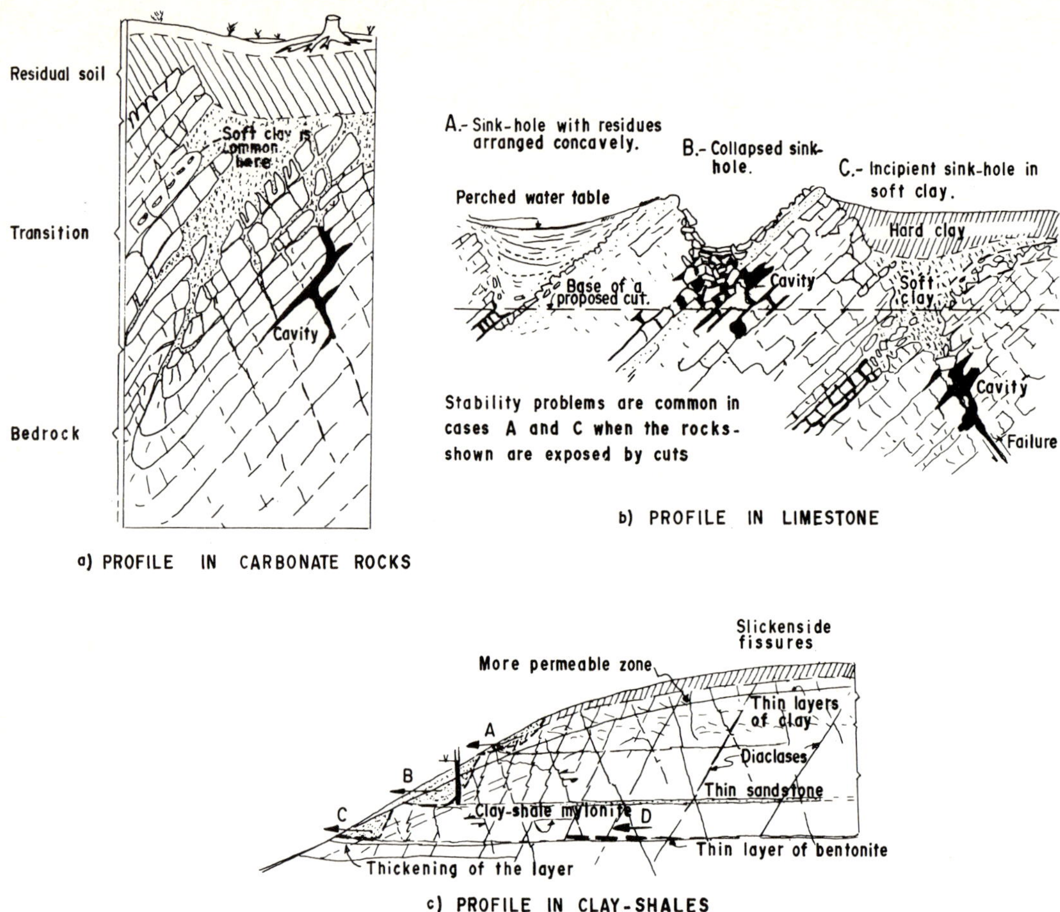

Fig. 6-15 Weathering profiles characteristic of some sedimentary rocks

Figure 6-16 shows some typical profiles of clay shales and sandstones, and the slope stability problems that are most often encountered in them. The figure illustrates the types of slides (*A* and *B*) most common in these cases, that are always associated with inherited structures, openings in the sandstone because of increases in pore pressure, bedding surfaces weakened by the expansion of the shales, or extrusion and erosion underneath the blocks of sandstone.

As mentioned previously, underground water, its flow and pressures, play a fundamental part in the stability of residual soils. An excellent summary of these flow systems can be found in [32].

Underground flow systems usually resemble the above-ground ones, which drain chiefly toward large valleys and rivers, but when important variations occur in the permeability of relatively deep formations, these systems can become very complex and hard for engineers and geologists to define. A critical condition is that found in large-scale cuts where a covering of highly weathered impermeable soil coexists with the natural tendency of water to drain at the surface of the cut. This case is particularly detrimental when the residual soils or weathered rocks under the impermeable covering retain inherited fractures with adverse orientation. High hydrostatic pressures often develop in the partially weathered rock, where it is not unusual for the piezometric level to rise above the ground level.

Inherited structures consist of exfoliations, joints, cracks, shear zones and other structural defects inherited by a soil from the original rock. Their influence is so great that the

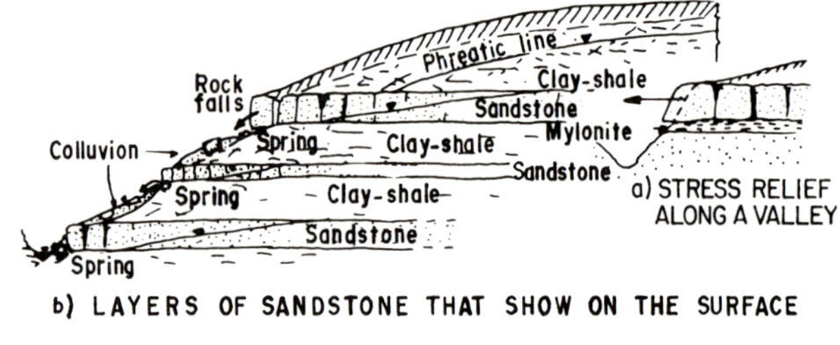

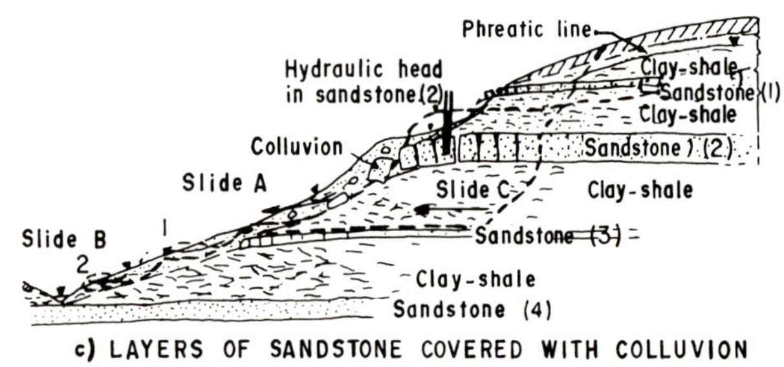

Fig. 6-16 Typical weathering profiles and stability problems of clay-shales interstratified with sandstone [25]

shear strength of the *intact* material is in no way representative of that of the structure as a whole. Almost all deep slides are to some extent related to structural weak planes. References [33-36] mention some interesting cases of this type.

The danger from these discontinuities increases when they are filled with clayey soils. Shear strength usually drops with an increase in the degree of weathering of the material. Figure 6-17 [25] shows this phenomenon in granite with increasing degrees of weathering (index values related to the weathering degree are indicated on the curves). The specimens correspond to fragments with no discontinuities, which means that they do not represent the strength of the whole mass affected by inherited structures.

The real problem lies in the assessment of the strength of the mass as a whole, considering the structure and condition of the in-situ materials. Information on the subject is not very consistent, for it is obtained by different methods, such as laboratory tests on samples containing defects (sometimes in terms of total stresses and sometimes in terms of effective stresses), direct field tests or computations from slides that have already occurred. The local direction of the discontinuity plays the most important part, especially if the slide follows it. The degree of weathering is important too, as well as the presence of water and its pressure in the discontinuity. Since the strength envelopes are usually curves that approach straight lines (Chapter 1), the type of envelope that is obtained will depend on the stress intervals with which the tests were carried out. This leads to serious confusion when estimating equivalent values of c and $\emptyset$, especially when results obtained by different researchers are compared so that conclusions of a general nature can be reached. The values of the angle of residual strength are particularly useful in that they help avoid the above-mentioned differences in results. With due regard for all these limitations, Table 6-2 gives some general information, based on [25], which will prove useful for routine design but which will not abolish the need to study special cases where the stability of a given cut or specific natural slope have to be analyzed.

Probably the best way to design slopes in residual soils is to follow previous experience as presented in Table 6-2, taking into consideration the weathering profile, the nature and position of inherited structures and local underground water conditions. It is difficult to envisage an exploration program sufficiently extensive or a test program sufficiently rational to allow a design based on calculations alone.

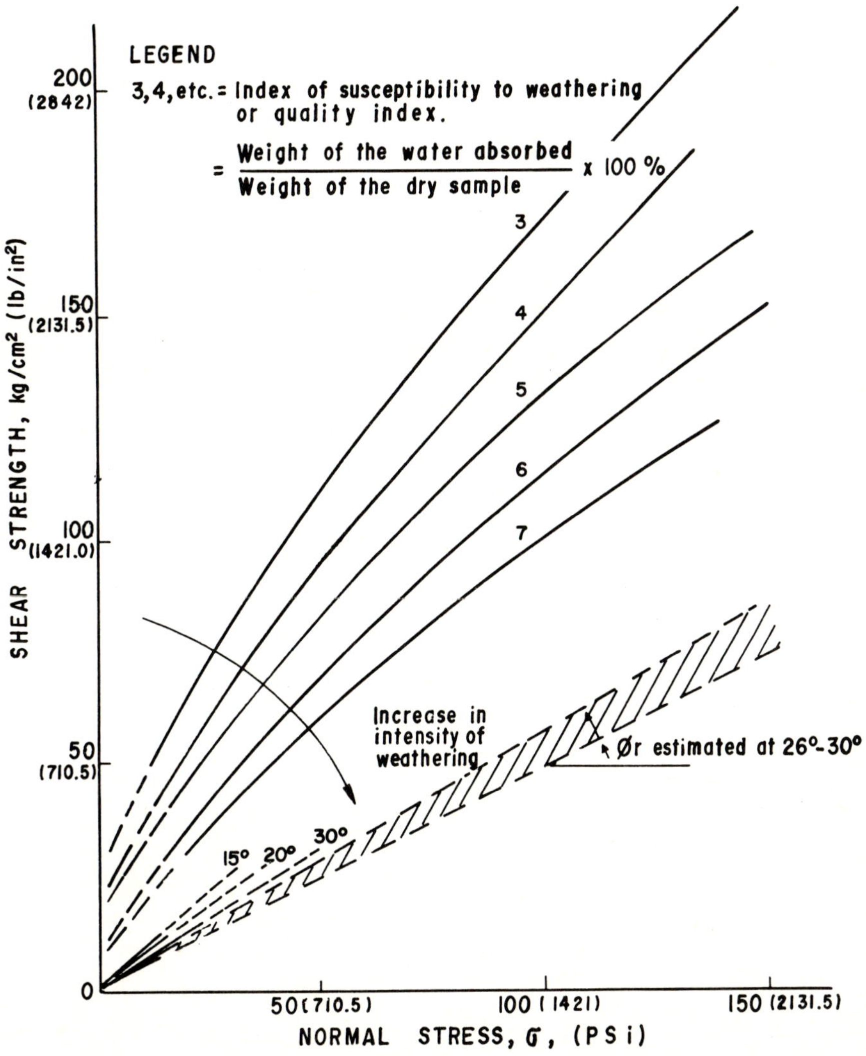

Fig. 6-17 Shear strength envelopes of weathered *intact* granite

Table 6-2
Typical shear strength parameters characteristic of residual soils and partially weathered rocks, based on [25]

Type of rock or soil	Degree of weathering	Strength parameters					Type of shear test	Reference
		C_u Total		$\varnothing_u$ Stress	$\varnothing$ Effective stress	$\varnothing_r$ Residual		
METAMORPHIC ROCKS		kg/cm²	lb/in²					
Gneiss	Sound	12.5	177.6	60°			Direct shear tests with rock-concrete contacts	[37]
	Moderately decomposed	8	113.7	35°				
	Very decomposed	4	56.8	29°				
	Very decomposed (Failure zone)	1.5	21.3	27°				
	Decomposed	—	—	18.5°			Consolidated-undrained triaxial tests	[38]
Schists	Partially weathered	0.7	10.0	35°			Sheared normal to schistosity	[33]
	Weathered	—	—	24.5°				
	Moderately weathered	—	—	15°			Consolidated-undrained triaxial test with degree of saturation at 50% and 100%	[39]
		—	—	21°				
	Weathered	—	—	26-30°			Direct shear tests on compacted rockfill	[40]
Phyllites	Residual soil	0	0	24°			Sheared normal to schistosity	[33]
		0	0	18°			Sheared parallel to schistosity	
IGNEOUS ROCKS								
Granite	Quality Index (Fig. 6-17)						Direct in-situ shear tests	[25]
	3	6-13	85-185	62-63°				
	5	5	71.0	57°				
	7	3	42.6	49-52°				
	10	2	28.4	45°				
	15	1	14.2	41°				
	Relatively sound	—	—			29-32°	Direct shear tests in the laboratory	[25]
	Partially weathered	—	—			27-31°		
	Weathered	—	—			26-33°		
	Very decomposed	0	0	25-34°	35°			
	Residual soil	—	—		28°			
Diorite	Decomposed	0.1	1.4	30°			Consolidated-undrained triaxial tests	[36] [38]
	Partially weathered	0.3	4.2	22°				
Rhyolite	Decomposed	—	—		30°			[25]
SEDIMENTARY ROCKS								
Marl	Sound	—	—		> 40°	23-32°	Drained and consolidated-undrained triaxial tests	[42]
	Moderately weathered	—	—		32-42°	22-29°		
	Highly weathered	—	—		25-32°	18-24°		
London Clay	Weathered	—	—		19-22°	14°		[29]
	Unweathered	—	—		23-30°	15°		
Black clay	Fissured	—	—			10.5°	Consolidated-undrained triaxial tests	[38]
	Unfissured	—	—			14.5°		
SOILS AND MINERALS								
Quartz sand		—	—			30-35°		[25]
Kaolinite		—	—			12°		
Illite		—	—			6.5°		[43]
Montmorillonite		—	—			4-11°		[44]
Muscovite		—	—			17-24°		
Hydrated Mica		—	—			16-26°		

6.4 Shear Strength Parameters for Numerical Calculations of the Stability of Slopes

Engineers have to assess the stability of a natural or artificial slope, both for design purposes and in order to review the stability of an existing slope. In the case of design, the slope exists only on paper, whereas in the case of review, the slope is already there and the need for approximate knowledge of its stability conditions by means of numerical calculations may be of dramatic urgency. Most failures that can be mathematically analyzed are associated with insufficient data on shear strength of the soil mass versus the shear stresses. Therefore there is great need to determine parameters to express this strength. In Chapter 1 it was seen how there are various laboratory and field tests for assessing these parameters. It was also seen how shear strength, far from being a constant that uniquely characterizes the behavior of soils, is a circumstantial variable. The conclusion is reached that before applying any particular mathematical method of analysis, attention must be paid to the conditions in which the strength parameters are measured. The following must be considered: how to obtain this strength, which laboratory tests should be performed, what use can be made of the results obtained and to what degree do these results represent the conditions to which the project will be subjected throughout its lifetime.

In very few natural slopes are the materials sufficiently homogeneous for a single set of strength parameters to represent the soil involved. About the only exception are hillsides of soft clays because even hard clays that are apparently homogeneous, in the natural state have a pronounced secondary structure with fissures and cracks, which upsets any attempt to reduce the complexities of nature to one set of numbers.

As mentioned, most of the doubts concerning cuts in artificial slopes are similar to those for hillsides; therefore it is usually more realistic to talk about homogeneity, mathematical models and numerical calculations in connection with embankment slopes.

In recent years, the importance of the progressive failure has been recognized in problems related to the stability of slopes in general and hillsides in particular [45]. Studies of this mechanism have led to a better understanding of the behavior of natural slopes of medium to hard cohesive soils, weak clay shales and other similar materials. There is evidence that the relation between the field strength of a soil compared to that obtained in the laboratory using undrained samples, is significantly less the firmer the soil. This has led many researchers to propose an empirical factor for reducing the laboratory undrained strength of soils when required for slope stability analysis.

In 1963, SKEMPTON [30] introduced the concept of the difference between the peak strength and the residual strength of soils. He showed that the strength that develops in a sliding mass may not be as high as the conventional peak strength along the entire length of the failure surface. In most cases, however, that strength is not reduced to such an extent that it reaches the residual value at all points of the failure surface, although this might provide a good boundary for limiting stability conditions. SKEMPTON did not suggest a satisfactory method for predicting the average strength that is mobilized between peak and residual strength values, and these values are often very far apart.

Considerable emphasis was laid by BJERRUM [46] on the significance of physico-chemical structural phenomena in progressive failure mechanisms. The soils that are most difficult to evaluate in stability of a natural or artificial slope seem to be highly overconsolidated clays with very strong structural bonds after they have been subjected to weathering. During this process a great deal of energy is released from these bonds, with the consequent development of strong tendencies to expand. BJERRUM pointed out that paradoxically the same highly overconsolidated clays are the most reliable materials when they have not been subjected to weathering.

In 1966, BISHOP [5] demonstrated that the actual strength of a natural slope is far better represented by the results of a large-scale field test than by those of laboratory tests on small samples. BISHOP assumed general validity for his conclusions, although they were obtained using London clay, which has a marked secondary structure as a result of cracking. Sufficient evidence is available to corroborate a generalization for many materials.

A very interesting series of experiments on synthetic rocks with geometrically controlled cracks and fissures was carried out by PATTON [47] in 1966. One of the conclusions of these experiments was that the deformation at which peak strength occurs in a given material depends on normal stress. At low normal stresses, little deformation is required. The value increases for moderate pressures and decreases again for even higher ones. Figure 6-18 [45] is an experimental confirmation of the results previously achieved by CONLON for Canadian clays, on which he performed direct shear tests using specimens obtained with a thin-wall sampler, 12.5 cm (5 in) in diameter. During the tests, the direction of the deformation was reversed several times, in an attempt to reach residual strength.

Theory suggests [45] that in a rotational failure the degree of deformation in the upper part of the sliding mass is usually sufficient (according to information of the type given in Fig. 6-18) to pass the peak strength of the material; values close

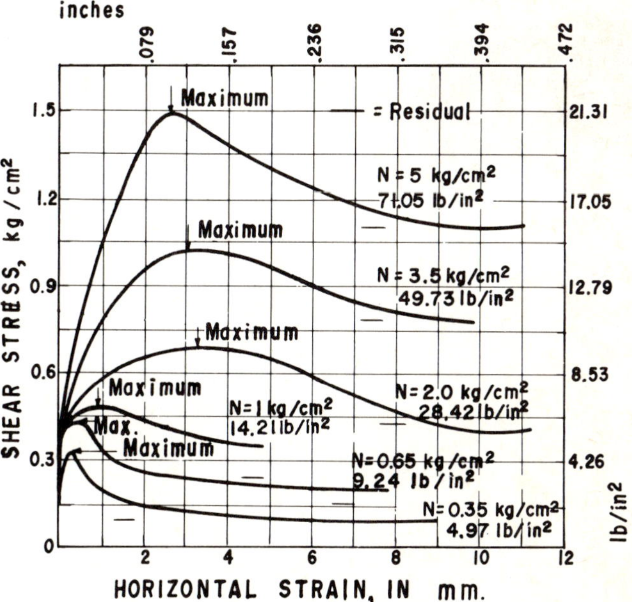

Fig. 6-18 Relationship between the shear strength and deformation in undisturbed samples of clay, according to CONLON [45]

to the residual strength are reached. In the center portions of the slip surface, where the normal stress is maximum, and consequently large deformations are necessary for peak strength to develop, the peak strength will probably be available. In the lower portion of the sliding surface, the deformations will probably cause the strength that develops to be intermediate between the peak and residual strengths.

The foregoing ideas led CONLON to make a suggestion for analysis [45]. The slip surface is divided into three portions. To the upper one he assigns the value of the residual strength, to the intermediate portion the peak strength, and to the lower portion a strength from the stress-strain relation of the soil obtained under average normal pressure, but corresponding to the degree of deformation required for peak strength to develop in the center portion.

These suggestions probably represent a level of design refinement hardly compatible with road engineering technology. Their chief purpose here is just to acquaint the reader with these criteria.

It is not easy to establish the exact mechanism that triggers a process of progressive failure in a natural or artificial slope. These mechanisms are not yet clearly comprehended. In [46], however, BJERRUM suggests an analysis of the conditions in which a progressive failure may occur in a hillside composed of overconsolidated clays or shales. Today it is generally accepted that progressive failure processes are very common in hillsides and artificial slopes. Because these processes cause radical changes in strength in the traditional model of a slope with a simple circular failure surface along which the peak strength of the soil acts, it is felt appropriate to present the concepts of BJERRUM's analysis, even though they can scarcely provide an element of quantitative analysis for use in an actual project. It is the authors' intention to supply the reader with ideas that will enable him to form his own judgement, rather than to give detailed methods of calculation.

The example in Fig. 6-19 is a portion of stable hillside, with slope angle α in relation to the horizontal plane. If the equilibrium of portion $OA'A'O'$ is considered, it can be concluded that the forces acting will be two equal side pressures of earth E on each side, and a shear stress due to gravitational forces, acting in the plane OA, equal to:

$$\tau = \gamma z \sin \alpha \cos \alpha \tag{6-1}$$

which is obtained by dividing the force in the direction OA ($\gamma z \sin \alpha$) by the length of the slice base, which, if it is unity will be $1/\cos \alpha$. If τ is smaller than the peak strength of the material, the slice will be stable. Now suppose that a vertical cut is made to a depth z in the section $O'O$. This or any other similar disturbance will cause a redistribution of

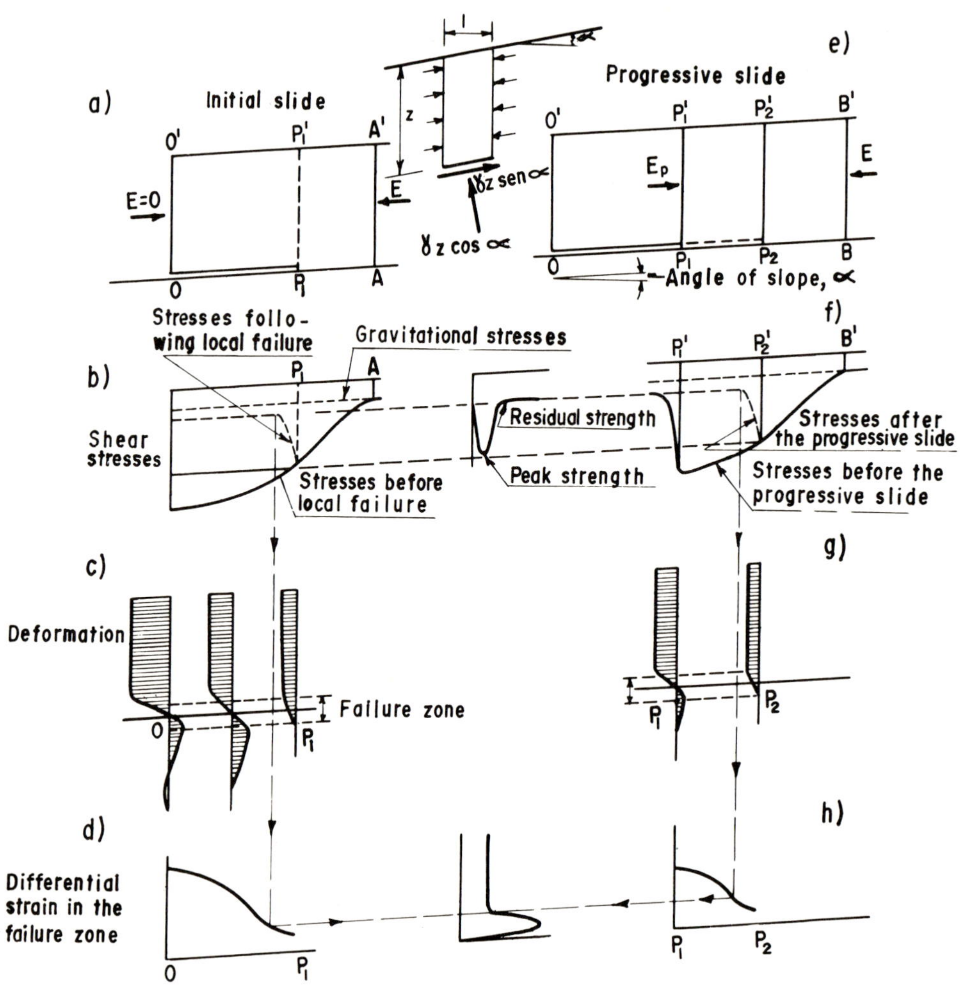

Fig. 6-19 Progressive failure mechanism [46]

stresses in the slice $OAA'O'$. If the AA' section is sufficiently far away that the side forces acting on it do not change, the equilibrium of the whole slice can only be maintained if the shear strength is sufficient to counteract E, as it is distributed along surface OA.

This additional shear stress produced by the unbalanced E will not be evenly distributed in OA: Part b of the figure suggests a possibility for its distribution. Other causes than an excavation can bring about a redistribution of stresses like the one shown here. The new acting redistributed shear stress from the unbalanced E may already be greater than the peak strength of the material. In this case a local shear failure will occur, starting at O. It will extend to a point where the shear stresses are once again below the peak strength of the material, at P_1.

The local failure in the slice $OP_1P'_1\ O'$ will cause the shear stresses along OP_1 to slacken; consequently the internal forces in the slice $OP_1P'_1\ O'$ will be reduced. The clay within this slice will tend to expand towards the excavation because of the reduced stress (Part c of the figure), sliding along the newly formed failure surface OP_1. As a result of this process, deformation will be produced, reducing the available strength in OP_1, from peak to residual. From the foregoing it can be deduced that, if equilibrium is to be maintained, there will be a large increase in the shear stresses acting in the plane OA, uphill from P_1.

The next stage of the process is finding out the equilibrium of slice $P_1BB'P'_1$. The shear stresses along plane P_1B will be the original gravitational ones, plus whatever increase may have occurred as a consequence of the foregoing mechanism. If this new value of τ is greater than the peak strength of the material, the progressive failure will continue to develop. This will depend on the difference E–E_p (Part e of the figure) where E_p represents the lateral thrust of the slice after suffering the progressive failure. E_p will depend on the value of the residual strength of the soil, the angle of the failure surface that gradually forms, and especially on the decrease in the internal stresses as a consequence of the change of the shear stresses acting in the failure plane that may form.

These conditions encourage the development of a failure surface almost parallel to the ground surface, with the phenomenon progressing uphill. If the residual strength is very great or the angle of slope small, there will soon come a point when E_p will be sufficiently large for equilibrium to be reached.

According to this mechanism, progressive failure can only develop if there is a discontinuity in the clay mass of the hillside, such as the excavation shown here, which will cause the initial local loss of equilibrium and start the progressive deformations. This discontinuity may be a cut, as in this example, erosion at the toe of the hillside or a far softer formation somewhere embedded in the slope.

The risk of progressive failure generally becomes greater when there is an increase in the ratio between the internal stresses and the peak strength of the material, or between the lateral deformation and the deformation corresponding to peak strength. These relations might well provide a laboratory criterion for assessing the risks of progressive failure.

For a progressive failure to occur in clay, deformation must be accompanied by a significant and abrupt drop in strength after peak strength has been reached, so that the strength available in the failed portion will not be sufficient to restrict the uphill deformations that will be required to move the zone of concentrated shear stresses towards the clay that has not yet failed. The ratio between the peak strength and the residual strength will thus be a good index for estimating the possibility of progressive failure, which is more likely to occur in materials with fragile stress-strain characteristics.

Two different strength conditions usually arise in roads, one in cuts and one in embankments (Fig. 6-20). Case a corresponds to an embankment. During construction there is an increase in the major and minor principal stresses. Case b corresponds to an excavation or cut in a homogeneous soil. Here there is a large decrease in the minor principal stress (σ_3), associated with some reduction in the major principal stress during construction.

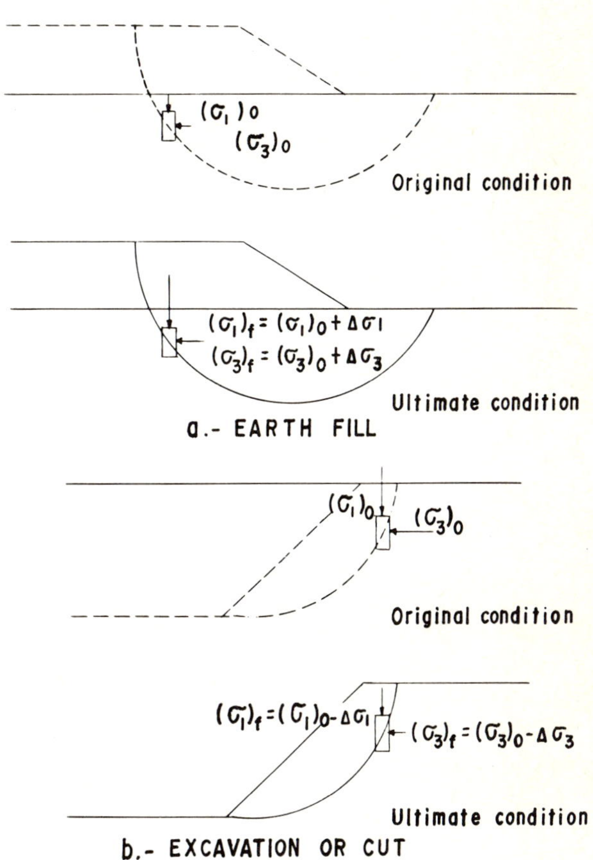

Fig. 6-20 Typical cases of an increase or decrease in stress with construction

Shear strength parameters are usually obtained from one of the triaxial tests. The use of triaxial tests for this purpose today is routine. In all such tests the laboratory should reproduce whatever conditions prevail in the field as nearly as practicable. Analysis of the stress paths and the type of drainage that will be found in the field is therefore of vital importance. This same criterion of representativity should be used to define whether the type of load applied in the test is axial compression, extension or some other type.

The results of any laboratory test are most simply presented as in-situ strength corresponding to the natural ground or to an embankment. This is expressed either in terms of the total stresses or the effective stresses acting at the time of failure. It can also be expressed in terms of the maximum effective stresses resisted in the failure plane after a period of consolidation [48].

The in-situ strength is frequently obtained with a vane test, which is from many points of view equivalent to an undrained test. It can also be obtained with simple compression tests, and with triaxial tests. In this last case, in order to reproduce field conditions in the chamber, the specimen is consolidated at the same major and minor principal stresses as those that were present in the field, but a chamber where the vertical and horizontal stresses are initially equal is generally used. A chamber pressure of 75% of the normal vertical field stress is often considered to be an acceptable representation of actual conditions. If the specimen is tested without drainage in the triaxial chamber, the undrained strength of the soil under the pressure acting in the field will be obtained. If a drained test is used, allowing sufficient time for drainage, the strength associated with the effective stresses will be obtained.

In Chapter 1 the most important characteristics of shear strength tests were discussed, together with the way results are represented in the Mohr circle (axes $\sigma - \tau$).

In [9] there are some interesting observations regarding the variation of strength with effects of sampling, sample size and anisotropy. This last item refers to changes in the strength of the specimen that depend on the orientation of its vertical axis in relation to the soil stratification (Fig. 6-21).

Stability problems can be handled either in terms of the total or the effective stress. The question is, therefore, which one to use in a specific case. This will depend on the type of problem that is posed and the stage at which stability conditions are to be assessed (long or short term).

For roads, the slopes are frequently treated as though they were above the groundwater level, either because they really are (which is often the case), or simply because the effect of water or its flow is not considered important. As will be seen when dealing with subdrainage, the chief reason for this is economical, because neutral pressures impose harsher conditions, especially as regards the shear strength of soils. In these cases, the shear strength of clayey soils is usually obtained by vane tests (which can be used for heavy and soft clays, but not for clayey soils containing sands or silts), simple unconfined compression tests or undrained triaxial tests. Thus, the stability analysis in routine road engineering is based on total stresses.

Situations frequently arise, however, where a road cut or embankment has to be analyzed under seepage conditions. In these circumstances, the effective stresses will be different from total stresses; pore pressure will exist and play an active part in stability. Therefore, it will be necessary to use the effective stresses stability analysis criteria. When a cut is made or an embankment built, variations of the pore water pressures occur with time. For example, when the cut is being made, the pore pressures in the soil vary; the reduction in the total principal stresses imposed by the cutting, leads to a momentary drop in the pore pressures of the neighbouring material [9]. The pore pressure adjusts with time until it eventually reaches values that are in keeping with seepage conditions and the new ground surface-profile. This final

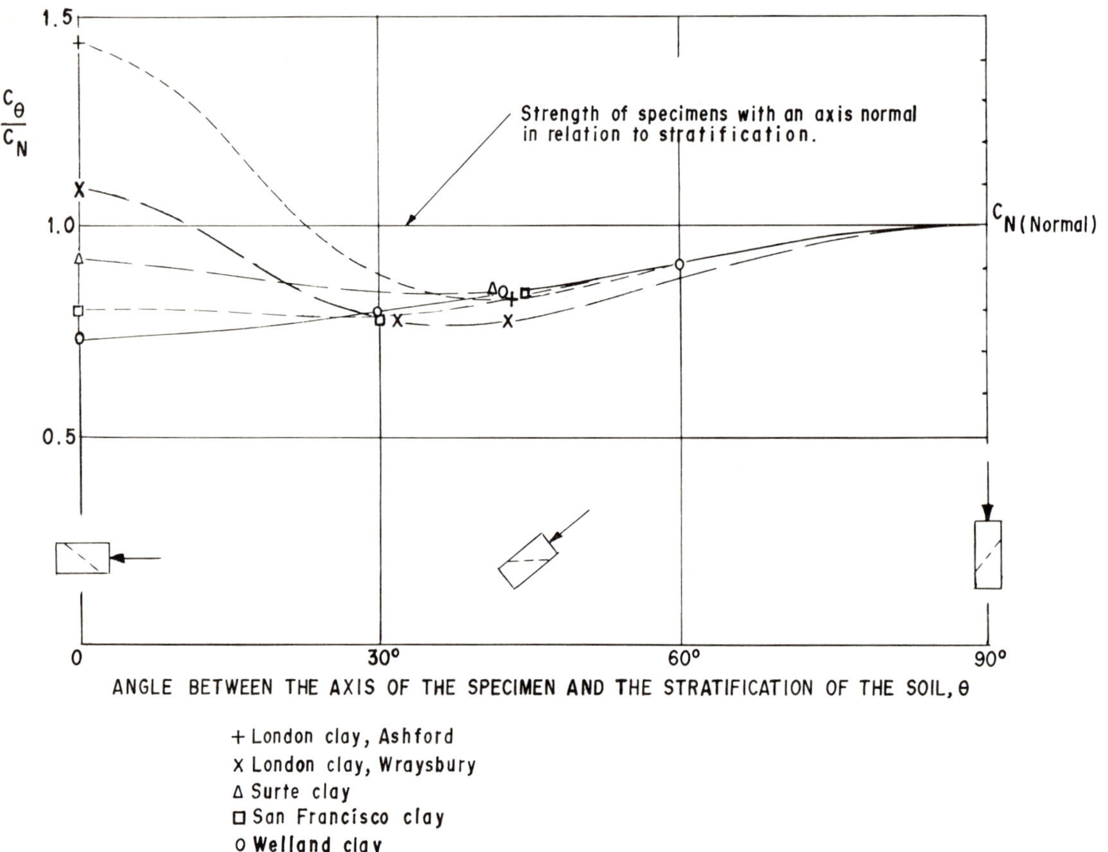

Fig. 6-21 Effect of orientation on the undrained strength of the specimen [9]

condition is referred to as the long-term condition. It is different from the initial condition and corresponds to the long period required for the changes of pressure to occur. In embankments too, there is usually a difference between the initial condition and the long-term condition. In embankments built with clayey materials or on normally or slightly overconsolidated clays, the initial condition is most critical, because time consolidation increases the strength.

Between the initial condition and the long-term condition there is usually a pore pressure adaptation stage. Figure 6-22 [9,49] gives a qualitative illustration of the changes that may occur in the pore pressures and stability conditions of an excavation slope in homogeneous clay. The case is shown where $A = 0$ and $A = 1$ for this clay (see §1.15.1). In the lower part of the figure, the line representing stability conditions is often different, with an absolute minimum starting from which there is a continuous improvement asymptotic to the line that is shown.

In previous soils, such as sands and gravels, the pore pressure adjustment period is minimal and generally the stability problems are of the long-term type, but in clays the transition period may last for years. In addition to the foregoing considerations, the engineer in each case will have to define whether the most critical conditions of the project correspond to the initial stage or to the long-term situation, so that the appropriate laboratory tests and methods of stability analysis can be chosen.

6.4.1 Parameters Used in Common Conditions

In the paragraphs that follow, a brief description of the usual types of analysis for the most common conditions in road cuttings and embankments is given.

Slopes in normally consolidated, saturated clays. Embankments in homogeneous clay hillsides, soft soils and clay embankments in a saturated state: In these cases, the initial condition will be the critical stage, for any additional consolidation which occurs will bring about an increase in strength. Under these circumstances, the adequate parameters will be those that are determined in an unconsolidated-undrained triaxial test (Q test) and used for an analysis in terms of total stresses where $C_u = 0$, $\varnothing_u = 0$.

Embankments in partially saturated soils. End-of-construction condition: In order to obtain laboratory results representative of the field condition, the following procedure is recommended. First the specimen is compacted at the unit weight and water content that are to be obtained in the field, using a similar compaction procedure, (Chapter 4). In this unsaturated state, the specimen is subjected to a chamber pressure similar to the pressure expected in the field, depending on the position the material will occupy in the embankment. At this stage of the test, no drainage will be allowed. Next, the deviator stress is applied without drainage

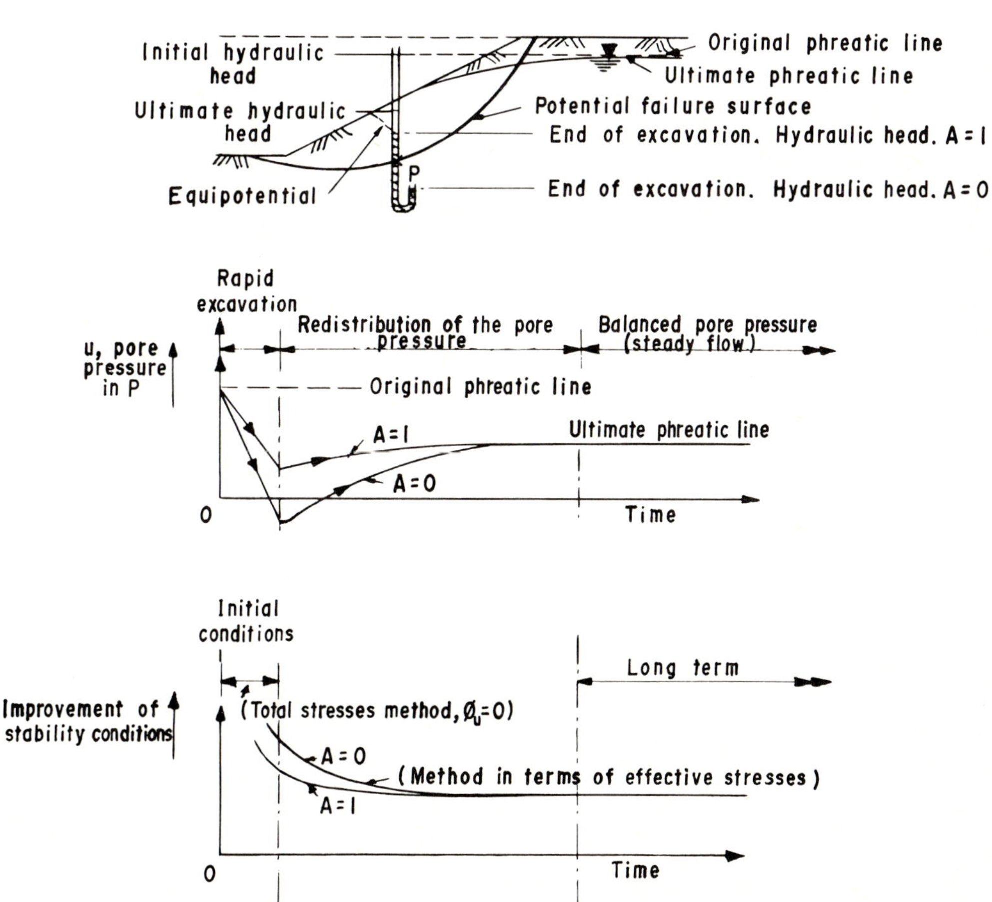

Fig. 6-22 Changes in pore pressure and stability conditions in an excavation in clay [9,49]

until failure. These are the conditions of an unconsolidated-undrained triaxial test. The stability analysis can be carried out in terms of total stresses. The effects of the pore pressures that may develop during the test and at failure are automatically taken into account by making the analysis in terms of total stresses [48], so long as compaction conditions in the laboratory specimen are a fair reproduction of real conditions.

Problems related to unloading processes. Cuts in homogeneous clays, with possible presence of water: As has been mentioned in §1.15, the critical conditions that correspond to this case are the long-term ones, which is why a method of analysis based on effective stresses is necessary. The neutral pressures are assessed in accordance with the real conditions of the water at the location, for example by means of a flow net.

If the excavations are temporary, as in the case of the slopes for foundation excavations, the unconsolidated-undrained strength of the soil could be considered, and the design based on total stresses for short-term conditions. However, it must be remembered in this case that the time required for the detrimental evolution of the strength may be brief.

Steady flow conditions: This is the case of a cut or embankment that is exposed to internal seepage for long enough for steady flow to be established (see Appendix 7a for concepts related to seepage or internal water flows). Under these circumstances, the appropriate flow net can generally be relied on, and from it can be obtained the pore pressure at any point. In steady flow conditions, analysis can be performed by two methods which, although different in appearance, agree in the long run. Stability can be analysed if the total weights of soil and the effect of the seepage forces are taken into consideration, or if the submerged weight of the soil plus the weight of the water in the slice, plus the pressures of the water (which can be obtained from the flow net) are considered. In either case, it will be necessary to obtain the strength parameters in a consolidated-drained triaxial test.

The reason why the results for both methods of analysis are the same is that the strength that is used is the same in both cases; the system of the total weight of the earth plus the seepage forces is statically equivalent to the system of submerged weights plus all the forces of water and its weight.

Steady flow conditions are not commonly encountered in road engineering, but some cuts and embankments, because of their special importance or consequences of their failure, will have to be analyzed for this condition, particularly if the pore water pressures can be predicted by a combination of flow net and geologic study.

Drawdown condition [50]: This is an even more rare condition in road slope engineering and is almost never considered. However, it may be necessary to analyze for this condition some large embankments that are to be built across flood plains near rivers, reservoir basins, river banks and lakes subject to sudden changes in level.

In order to reproduce drawdown conditions in the laboratory, these steps are followed [48]. First the specimen is compacted to the same unit weight, water content and by the same compaction procedure that will be used in the field. Next it is saturated and placed in the triaxial chamber. Then a confining pressure is applied to it equal to the field pressure

Table 6-3

Strength parameters for stability problems in clay slopes and hillsides, [9]

Type of failure	Condition of Clay	Cuts		Natural Slopes	
		Initial	Long-term	After 100 yrs	After 1000 yrs
Conventional Slide	Soft, normally consolidated, intact	$x;c_u$		$c;\varnothing$	
	Slightly overconsolidated, intact		$c;\varnothing$	$c;\varnothing$	
	Stiff, intact			$c;\varnothing$	
	Stiff, fissured	$f;\ x;\ c_u$	$r;\ c;\ \varnothing$	$c \simeq 0;\ \varnothing$	$c \simeq 0;\ \varnothing \rightarrow \varnothing_r$
	Very fissured and cracked	$c \simeq 0;\ \varnothing$			
Predeveloped failure surface	All cases	$c_r;\ \varnothing_r$	$c_r;\varnothing_r$	$c_r;\varnothing_r$	

c_u – Peak strength parameter, undrained test
$c;\varnothing$ – Peak strength parameters, drained test
$c_r;\varnothing_r$ – Residual strength parameters
x – Reduction factor for test conditions, anisotropy, etc.
f – Reduction factor for fissuring
r – Reduction factor for time

that will develop when the embankment is under a maximum depth of water. At the same time, a vertical stress is applied equal to twice that value. The objective is to represent the ultimate consolidation conditions of the material in its lifetime prior to drawdown. Thus, during the initial stage of the triaxial test the specimen is consolidated somewhat differently from the usual hydrostratic conditions ($\sigma_1 = \sigma_3$), using a stress condition in which $\sigma_1 = 2\sigma_3$. The next step is to bring the specimen to failure without allowing any further drainage.

These conditions are similar to the performance of a consolidated-undrained triaxial test. Reference [50] includes a discussion to justify the use of this procedure. Another procedure, possibly more refined, appears in [48].

Slides following predetermined failure surfaces: In these cases, it should be considered that deformations have occurred or are occurring to such a degree that the strength available will always be the residual strength which provide the basis for computations.

Although the foregoing conditions are analyzed from time to time in road engineering, a majority of slope analyses are made without considering seepage, using the strength parameters obtained from an unconsolidated-undrained triaxial test and following the total stresses method.

Table 6-3 is a personal communication from SKEMPTON and HUTCHINSON to complement their work in [9]. It refers to cuts and natural slopes in clay, and completes the information given in the foregoing paragraphs. The specific values of the reduction factors which are given in this table are tentative and their selection should be left to the design engineer. They originated in the analyses that have been mentioned at different points under this heading. Today there is some discrepancy between the stability calculations that are made on the basis of laboratory test data and those that are performed by reviewing the conditions of failed slopes. Some of these prove to have very reasonable safety conditions whereas other stable slopes, when analyzed on the basis of laboratory data, should have failed. This shows the role that is played by irregularities in the soil, the difficulties involved in obtaining good undisturbed samples, the drawbacks that still exist in laboratory testing and the mistakes that are made when establishing the effects of seepage.

6.5 Slope Stability Analysis

Here are some methods that can be used for assessing whether a given slope will be stable or not at the design stage, to review the stability condition of an already constructed slope, and to evaluate the influence of a stabilizing procedure that may prove necessary.

First, it should be emphasized that all the mathematical models for these methods are based on homogeneity, simple stratification, layout, simple environmental conditions and predictable behavior of the natural agents, which the road engineer will very seldom find in his projects.

Once again the enormous difference that exists between cuts and embankments must be remembered. In cuts it will be far more difficult to find conditions that agree with those of the mathematical models. In embankments, however, it will be easier, so long as they are built following a known procedure with a certain degree of uniformity in the use and treatment of soils, and the necessary field and laboratory studies are carried out.

Not all the types of failure that were described in §6.2 can be represented by a mathematical model that will serve as a basis for a stability analysis. Some of the more common and dangerous varieties, such as flows, cannot be numerically analyzed either because knowledge of their mechanisms is not satisfactory, or simply because the mechanisms are so complex that they challenge any attempt at analysis.

The most widely used stability analysis methods, with a description of the type of failure to which each one can be applied, are described below.

6.5.1 Slopes in Clean Sands

A slope in clean, dry sand will be stable, independently of its height, so long as its inclination, β, is smaller than the angle of internal friction of the sand. In this case, the risk of failure can be expressed by a factor of safety, F_s, which is simply defined as

$$F_s = \frac{\tan \varnothing}{\tan \beta} \tag{6-2}$$

Here the stability of a sand grain at the surface of the slope, or at any point within the mass, can be compared with the equilibrium of a body on a sloping plane. Since the resistance to shear as expressed by downhill sliding is purely frictional, a grain will only slide if it finds itself on a plane that is steeper than the friction angle. If the grain at the very surface of the slope does not slide, none of those within the mass will do so either. Thus, the flatter the slope, the more stable will it be (Fig. 6-23). Note that even in the extreme condition ($\beta = \varnothing$), the body of the mass is still stable, because the angle of any potential slide plane within the mass would be flatter and thus more stable. For this reason, it should be possible for slopes in clean sands to be designed with a $F_s = 1$, as in Eq. (6-2). However, this is not advisable, since the sand grains close to the slope surface would be in a precarious position. Wind, rain or any other agent might make them move easily, causing erosion and small spills of sand that will go into the ditches of a cut slope. This explains why the angle of slope should be a little smaller than the angle $\varnothing$; a few degrees will probably be sufficient.

If the clean sand slope is submerged in water or if the sand is moist, the foregoing reasonings will be valid only if the effective angle of internal friction of the sand is considered (in the case of the dry sand, the angle of internal friction is effective, but in that case the distinction between total and effective stresses was irrelevant, because it was a dry sand). The factor of safety is expressed in the same way, Eq. (6-2).

Tension may well appear in slopes in fine wet sands, especially in the part nearest the slope surface, where the effects of evaporation are most noticeable. This capillary tension produces an increase in effective stresses between sand grains which can make steeper slopes temporarily stable, corresponding to an apparently larger angle of internal friction than the real one. The engineer should always be prepared for this situation, of which he will never be able to take advantage, because the capillary tension may suddenly disappear (for example, if the sand gradually dries out as a result of evaporation or is saturated because of rain or seepage). In

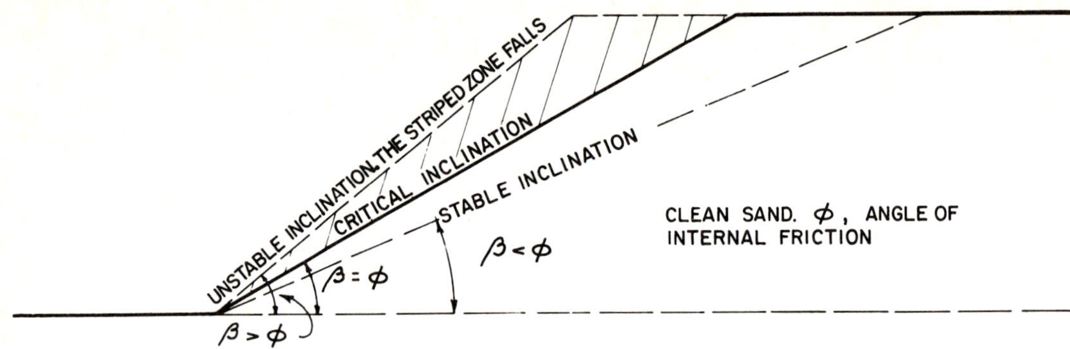

Fig. 6-23 Slopes in clean sands

this case, the excess of effective stress produced by capillary tension will disappear and the sand will fall if the angle of slope is steeper than the real angle of internal friction (see Chapter 1).

6.5.2 Rotational Slides. Swedish Method

The methods of limiting analysis available for assessing the possibility of a rotational slide in the body of a slope, like almost all stability analysis methods, follow three basic steps:

1. A hypothesis is established regarding the slide mechanism that may occur. This includes the shape of the failure surface, a complete kinematic description of the movements that will occur on it and a detailed analysis of the driving forces.
2. A strength criterion is adopted for the soil. Those currently in use have already been discussed in sufficient detail in previous Chapters. The available resisting forces can be computed on the basis of the criteria adopted.
3. A mathematical procedure is established to determine whether it will be possible for the proposed slide mechanism to occur under the action of the driving forces, overcoming the effect of the resisting forces.

The reason the above method is so widely used is that no satisfactory method has yet been developed to determine the stress distribution within the slope mass. No promising solution to this basic problem has yet been found, which is why the more traditional kinds of calculation cannot be used in engineering problems of the type described in the introduction to this Chapter.

On the basis of his own work and that of his collaborators (PETTERSON et al.), FELLENIUS [51] proposed the circular surface as a reasonable approximation of the shape of failure surface for many slides in the body of a slope. The failure surface is presumed to be a cylinder which, in cross-section appears as an arc of a circle. Adoption of this hypothesis defines the slide which in this Chapter has been described as rotational. Proposed by FELLENIUS and his group at the Royal Swedish Geotechnical Institute, this method has become extraordinarily useful. It covers Step 1 (above) very simply. Furthermore, progress in the field of soil mechanics has made it possible for Step 2 to be approached in a reasonable way. A large number of procedures soon made their appearance to cover Step 3, beginning with the one originally proposed by FELLENIUS, until the circular failure hypothesis took its place in applied soil mechanics. Today most procedures for assessing slope stability that make use of the circular failure hypothesis are known as the "Swedish Method". However, a circular or cylindrical method can be handled in several different ways (especially varying Steps 2 and 3). It is not the intention of the authors to describe all the procedures that are in use today (most of which are very similar), but just the basic ones that are applied when handling different types of soils in the most commonly encountered conditions.

.1 The Swedish Method: Strength Law $s = c_u$.

Cases will be analyzed here in which the shear strength of the soils is expressed in accordance with the results of an unconsolidated-undrained shear test in terms of total stresses.

The case of a slope with a height h, excavated in clay with perfect homogeneity of material in both the slope and the foundation ground, will be studied first. The procedure that is proposed for this case was established by A. CASAGRANDE, and can be used to analyze failures both at the toe and base of the slope. It is illustrated by Fig. 6-24.

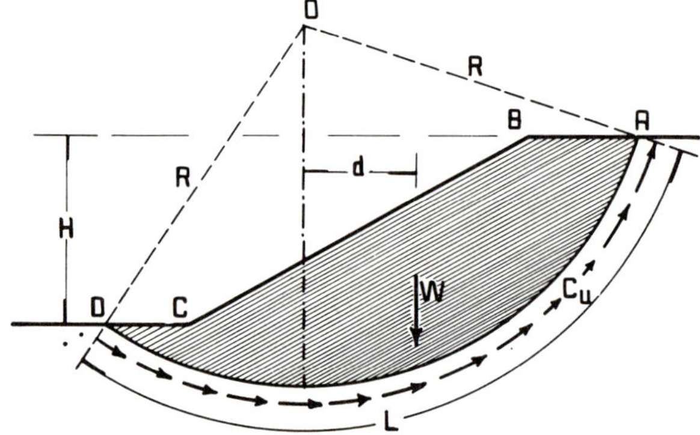

Fig. 6-24 A. CASAGRANDE's procedure for applying the Swedish Method to a purely cohesive slope

Let us consider that the arc with radius R and center O represents an assumed failure surface, defining the zone of movement (shaded). The forces that tend to cause sliding will be the weight (W) of the area $ABCDA$, plus any surcharges that act on the crest of the slope. The weight W is computed considering a unit thickness perpendicular to the cross-section.

The moment of the driving forces can be expressed as

$$M_m = \sum Wd \tag{6-3}$$

which includes the weight of soil plus any surcharges.

The resisting forces will depend on the shear strength along the entire length of the assumed failure surface. Their moment about the center O will be

$$M_r = c_u LR \quad (6\text{-}4)$$

In the limiting conditions $M_m = M_r$ and thus:

$$\sum Wd = c_u LR \quad (6\text{-}5)$$

If a factor of safety, F_s, is defined as

$$F_s = \frac{M_r}{M_m} = \frac{c_u LR}{\sum Wd} \quad (6\text{-}6)$$

the safety of the slope can be expressed in terms of F_s. At the limiting equilibrium condition $F_s = 1$.

There is no assurance that the circle chosen for this analysis is the one which will lead to the smallest factor of safety; this means that the foregoing procedure will require a series of trial-and-error calculations, by which the stability of a large number of circles will be computed, until the one which will produce the smallest possible factor of safety has been found (the critical circle). In this analysis, circles through both the toe and the base of the slope will be considered, until the smallest factor of safety is found.

It is not easy to define the value of F_s that should be accepted for design in a particular analysis. It will depend on the dimensions of the potential slide, on those of the slope, on the characteristics of the soil, on how detailed and reliable are the load analyses, on the evolution of the strength of the material with time and on the consequences of a possible failure. However, a few general comments can be made in order to norm the decision criteria.

1. The type of analysis under consideration applies particularly to slopes and hillsides of soft clays, in which the effects of consolidation tend to increase strength with time, with a corresponding increase in the safety factor. This increase can be estimated [52]. In many cases, this will allow initially low factors of safety to be accepted. Section 6.6 of this Chapter will deal with a possible drop in the strength of many soils (at least in the short term) when a slope is built on them. This should not be disregarded.

2. The slope of road cuts and embankments is largely a matter of general design standards, which explains why the majority of slopes are neither studied nor calculated. If just one is analyzed, it will require the adoption of a policy in keeping with the rest of the road. Sometimes a very audacious policy is seen to prevail in non-calculated slopes, while for analyzed slopes a very conservative one is applied. The analyzed slopes are often the most important or problematical ones, or those where failure would have the most serious consequences. Therefore it is natural that the required safety factors of the analyzed slopes are greater than those for non-analyzed slopes. However, in our opinion, the general criteria should be the same for the entire road.

3. The natural tendency to accept safety factors that are initially low, which was mentioned in Point 1, and which is reasonable, should be changed in special cases, some of which will be easily recognized by the engineer. It is, however, of vital importance to bear in mind the awful consequences of a failure in very sensitive soft clayey soils, where the remolding caused by a slide can lead to an abrupt drop in shear strength (this, moreover, can be recovered only very slowly and with time). In such situations it is advisable to require greater safety. This is also true of embankments founded on very soft clayey soils or peats.

In the literature on this subject, 1.5 is usually the value given for the minimum initial safety factor. This has been firmly established for permanent slopes as a result of experience. However, in many practical cases it is possible to use smaller factors, depending on the specific case. Permanent slopes with an initial safety factor of 1.1 or 1.2 have shown excellent behavior in soils where there are continuing increases in strength due to loading. Such small safety factors have been established taking into consideration the transient nature of live loads and the infrequency of seismic effects. In non-permanent slopes, safety factors of 1.1 and 1.2 are usual.

TERZAGHI [53] proposed procedures to include the effect of the tension cracks that open at the crest of a cohesive slope before failure. Many engineers include these recommendations in their analysis for defining the most critical condition of the slope. These are illustrated in Fig. 6-25, according to which the appearance of such cracks generally has three independent effects.

a) A drop in the resisting forces, as the available length of the sliding surface is reduced.
b) A small decrease in the driving forces when eliminating the weight of the wedge, e_1fe.
c) The development of hydrostatic thrusts caused by the rainwater that collects in the crack; these thrusts are always dangerous.

TERZAGHI pointed out that the last two effects tend to counteract one another, so that their net influence sometimes can be ignored and only the first effect should be taken into consideration. TERZAGHI himself suggests that the *cohesion* of the soil (c_u) be replaced by a corrected value c_c in accordance with the relation:

$$c_c = \frac{be_1}{be} c_u \quad (6\text{-}7)$$

The position of point e_1 depends on that of the crack and is usually determined as shown in Fig. 6–25. For a critical circle at the toe of the slope, the crack develops vertically from the point at the crest that is half the distance from the edge of the slope to the failure surface. In base failure circles, the crack is in the vertical segment and goes from the failure surface to the ground surface a distance of $H/2$. This analysis is based on the critical circle.

There is a whole series of papers of a theoretical nature showing cumulative and repetitive calculations intended to supply the engineer designing slopes in purely cohesive soils with charts that will minimize guesswork and take him straight to the results of the calculations described in Fig. 6-2. Reference [52] (Supplement V-a) deals with those papers which lead to the most practical conclusions. Many of these are mentioned in [3] and [54]. Here reference will be made to only the most important conclusions taken from TAYLOR's work [55,56].

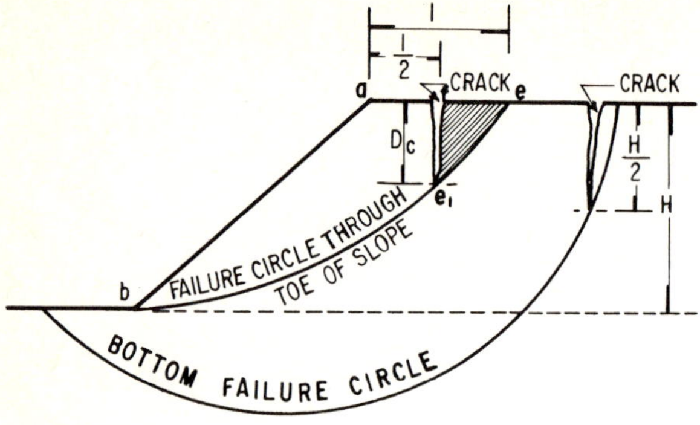

Fig. 6-25 Tension cracks at the crest of a slope

TAYLOR related the stability of a cohesive, homogeneous slope together with the foundation ground, to a number which is called the Stability Number and is defined by the expression:

$$N_e = \frac{c_u}{\gamma_m h} \tag{6-8}$$

He demonstrated theoretically that in a graph where the ordinates are values of N_e and the abscissa values of the angle of slope, β (Fig. 6-26), the value $\beta = 53°$ is of special importance.

All slope angles smaller than 53° have the same stability conditions (the same $N_e = 0.181$). Under these circumstances, the most critical circle possible always corresponds to base failure. If the angle of slope is greater than 53°, the stability number is variable, with a relationship that is approximately linear between $N_e = 0.181$ for $\beta = 53°$ and $N_e = 0.26$ for $\beta = 90°$. In this case, the most critical circle corresponds to failure through the toe of the slope.

The graph in Fig. 6-26 spares the design engineer calculations by providing him with N_e in relation to each slope angle, from which he will be able to find the value of the required c_u in the critical condition. This can be compared with the cohesion available in the soil. Now the safety factor can be defined as

$$F_s = \frac{c_u \text{ (available)}}{c_u \text{ (required)}} \tag{6-9}$$

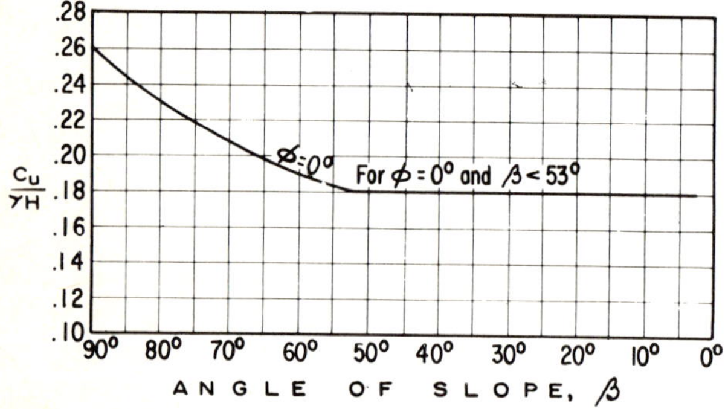

Fig. 6-26 TAYLOR's graph for determining the stability numbers for slopes composed of cohesive materials that are homogeneous with the foundation ground [55,56]

Taylor also studied the practical case where at a certain depth, in a purely cohesive foundation soil, there is a horizontal firm stratum. This is illustrated graphically in Fig. 6–27.

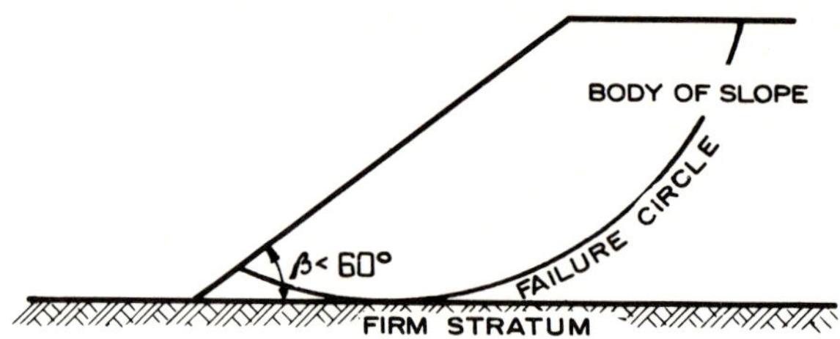

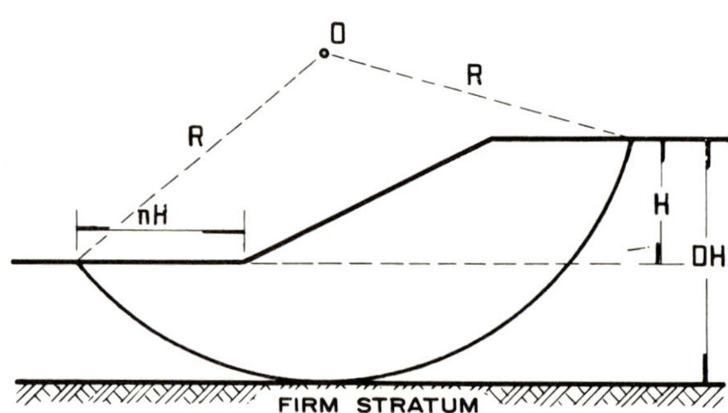

Fig. 6-27 Failure circle for a slope made of cohesive material when there is a firm stratum in the foundation ground [55,56]

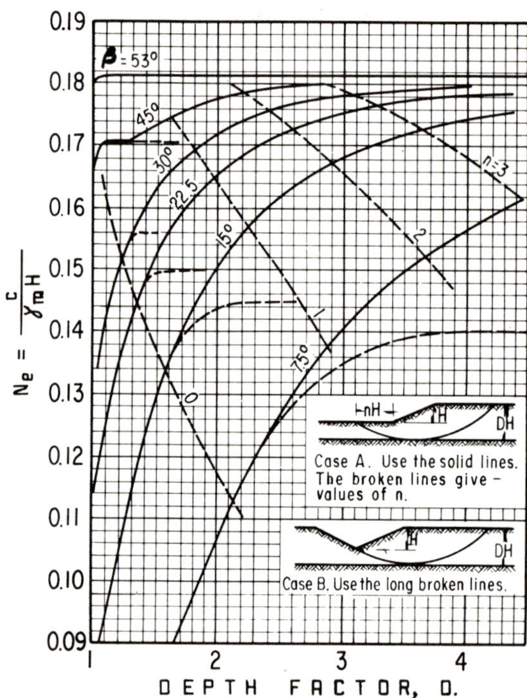

Fig. 6-28 TAYLOR's graphs for determining the stability number and the distance factor in circles at a tangent to a firm layer [55,56]

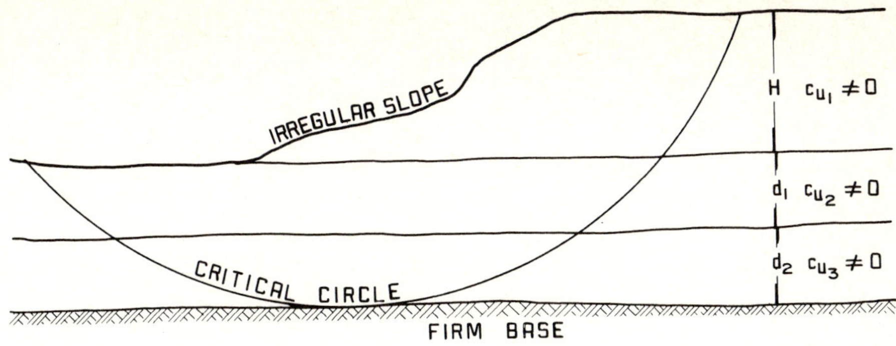

Fig. 6-29 Case of an irregular slope or stratified soil in the foundation ground

Now the most critical circle will be tangential to the firm stratum (so long as this stratum is at a maximum depth of four times the height of the slope. At a greater depth it scarcely has any effect and the case becomes the same as that of a homogeneous foundation soil). If the firm stratum is very close to the surface, the critical circle gradually becomes more and more similar to a bottom failure circle.

Figure 6-27 illustrates the depth factor and distance factor concepts which are used in the chart in Fig. 6-28 and which make it possible to solve these problems without circular failure computations by giving the stability number for each geometrical condition. Operation of this chart requires no explanation.

The foregoing simplified methods cannot be used when the slope has an irregular geometrical shape or when the soil involved is stratified with several layers of soft clayey soil, each with different c_u values. These cases, which have to be solved using trial-and-error procedures, are shown in Fig. 6-29.

Naturally, trial-and-error procedures can be guided. For example, if one of the strata is notably weaker than the rest, the circle that develops most within that stratum is usually the critical one. If there is a very firm stratum at a significant depth, the most critical circle is usually tangential to that stratum.

.2 The Swedish Method: Strength Law $s = c_u + \sigma \tan \varnothing_u$

This is the case of an analysis in terms of total stresses for soils located above the phreatic level. In these cases, the strength parameters are obtained in an unconsolidated-undrained triaxial test, or vane test. The analysis method described here is the slices method proposed by FELLENIUS [51], which has been very widely used. It is illustrated in Fig. 6-30.

First, a slip circle is assumed and the sliding mass is divided into slices as illustrated in the figure. Part *b* shows the stresses acting in a slice, when the sliding mass is located above the phreatic level and the water forces are not taken into consideration. The forces in each slice, like the forces acting throughout the entire sliding mass, must be in equilibrium. However, forces *E* and *S*, acting on the sides of the slices, depend on the stress-strain characteristics of the material and cannot be rationally evaluated. In order to handle them, a reasonable estimate at their value must first be made.

The simplest hypothesis is that the combined effect of the four side forces is nil and these forces consequently play no part in the analysis. This assumption was used by FELLENIUS in his original work. It is equivalent to regarding each slice as independent of the others and considering that components N_i and T_i balance the weight W_i of the i-th slice (Fig. 6-30).

For each slice, the ratio N_i/L_i can be computed, which is considered a close approximation to the value of σ_i, total average normal stress acting in the base of the slice. With this value of σ_i, and using the shear strength along the base of the slice which is found from the MOHR's envelope or the COULOMB parameters from the appropriate shear tests (in this case it is generally expressed in terms of total stress) the value of s_i, and from it the average available shear resistance in the arc L_i is found.

Now a driving moment can be calculated about *O*, which is the center of the circle chosen for the analysis, corresponding to the weight of the slices; this movement will be:

$$M_m = R \sum |T_i| \qquad (6\text{-}10)$$

Note that the normal component of the weight of the slice, N_i, does not give a moment about *O* because the surface being circular, its line of action goes through *O*. If there were

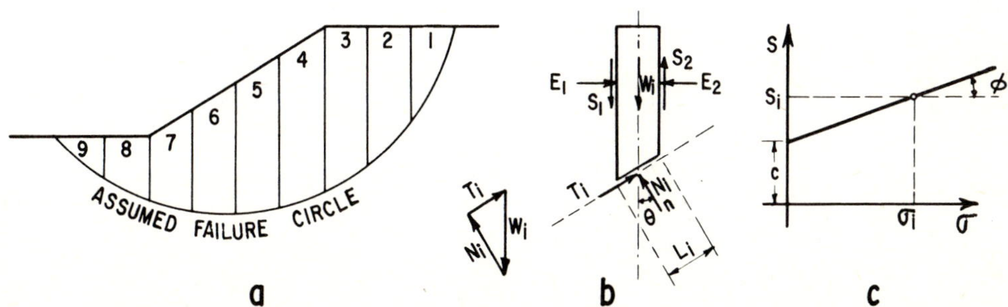

Fig. 6-30 The method of slices or FELLENIUS' method

surcharges at the crest of the slope, the effect would be included in the sum of Eq. (6-10). Note also that the sum in Eq. (6-10) is algebraic; for the slices located beyond the vertical through O, the component of the weight acts in the opposite way, tending to balance the mass.

The resisting moment depends on the shear strength s_i which develops in the base of the slices.

Thus:

$$M_r = R \sum s_i L_i \tag{6-11}$$

which is an arithmetical sum, for the resistance always acts in the same direction.

Once M_m and M_r have been calculated, the factor of safety can be defined as:

$$F_s = \frac{M_r}{M_m} = \frac{\sum s_i L_i}{|T_i|} \tag{6-12}$$

This method of calculation once again requires trial-and-error work, because the critical circle with the smallest safety factor must be found. Both toe failure circles and base failure circles will have to be analyzed. Table 6-4 shows one of several different ways in which computations can be laid out.

In determining the appropriate safety factor, similar observations can be made to the ones mentioned previously, bearing in mind that the type of analysis described here is usually applied to soils where consolidation adds very little to the shear strength of the material. In road engineering, safety factors of 1.2 or 1.3 are often accepted for design in normal conditions, and 1.5 when greater safety is required. This last value is the one most often recommended for slopes in general.

Figure 6-31 is a graph by Taylor similar to those described in Section 6.5.2.1 [55,56]. Note that this figure includes the information presented in Fig. 6-26 as a special case ($\varnothing$ = O). The graph eliminates trial-and-error work; it introduces the angle of slope and the value of $\varnothing$ available in the soil, so that the c required for the slope can be calculated, and then compared with the available c. Obviously the available c can be given together with the angle of slope, so that the required $\varnothing$ can be calculated. The graph (in Fig. 6-31) corresponds to failure circles at the toe of the slope only. It has been demonstrated [3] that in this case there is no base failure unless $\varnothing$ is less than approximately 3°. Therefore if base failure occurs in a homogeneous soil, the value of $\varnothing$ at the moment of failure must have been practically zero in terms of total stresses.

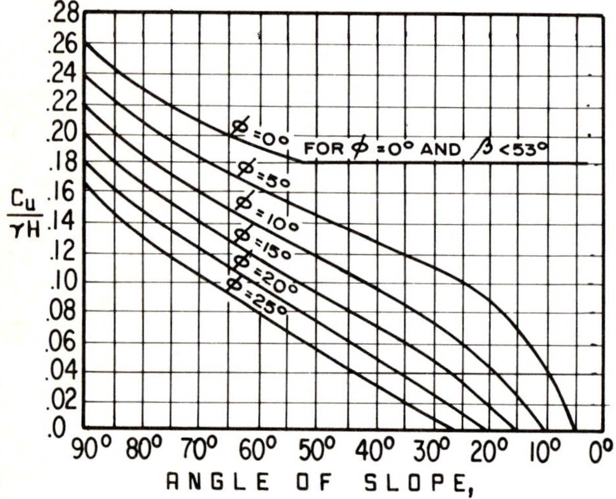

Fig. 6-31 TAYLOR's graph for determining the stability numbers for cohesive frictional materials [55,56]

In [52] and [57] graphs can be seen which were established by N. JAMBU from fairly refined theoretical analyses to provide the stability number for conditions of failure at the toe of simple slopes made up of frictional-cohesive soils.

In practice, slopes composed of stratified soils often occur, as illustrated in Fig. 6-32. Although it refers to a specific case, it provides a fairly good general description of the method.

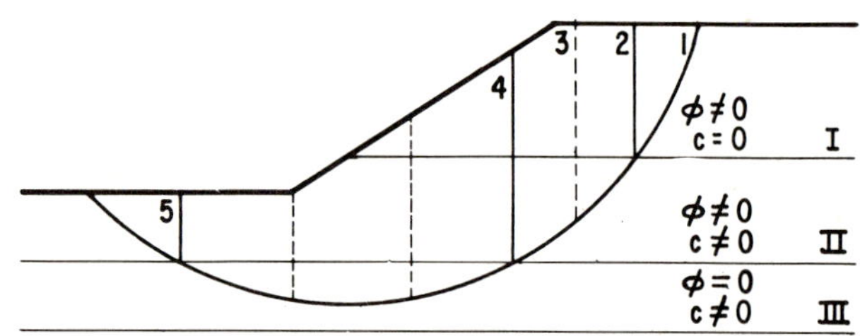

Fig. 6-32 Application of the Swedish Method to slopes in stratified soils

The sliding mass can be regarded as divided into slices, which are selected in such a way that the base of each one coincides with just one material and never comes between two strata. The weight of the slice is calculated as the sum of partial weights obtained by multiplying the corresponding area of each stratum by its own unit weight. The problem can be solved with a chart like the one given in Table 6-4, using for each slice the appropriate shear strength in accordance with the nature of the material.

The rest of the method is the same as the one for homogeneous slopes. The problem will have to be solved by trial-and-error procedures, for in this case there are no charts or graphs available for everyday use. The search for the critical

Table 6-4

Calculation layout for FELLENIUS' method

Slice No.	W_i	N_i	T_i	$\frac{N_i}{\Delta L_i} = \sigma_i$	S_i	$S_i L_i$

$\sum$ = Driving moment (algebraic) $\sum$ = Resisting moment (arithmetic)

circle is simplified considerably is some strata are either not as firm or much firmer than the others. In the first case, the critical circle will probably be the one that develops more in the weak stratum. In the second case, the circle will probably be tangential to the firm stratum, for circles deeper into this stratum will have large increase in average strength.

.3 The Swedish Method: Strength Law $s = c + \sigma \tan \varnothing$

This is an analysis in terms of effective stress applied to slopes located totally or partially below the phreatic level or subjected to seepage. This type of analysis has to be carried out in terms of effective stresses obtained from a consolidated-drained or consolidated-undrained triaxial test, with pore pressure measurement.

The method of slices described for slopes above the phreatic level is still valid; the only change concerns the forces acting in the slices. Figure 6-33 illustrates the computation procedure using the submerged weights of the soil (when in that condition), the total weights of the soils above the phreatic level and the water pressures acting in the slice. The figure shows a general sketch of the slope, with an assumed circular failure surface as one of the trial-and-error solutions to be found. An analysis is made of the forces acting in a typical slice (Part *b* of the figure). Finally the force polygons corresponding to the — equilibrium of that slice are given. Part *c* shows the whole set of forces acting in the slice, while Part *d* gives the force polygon on the basis that forces *E* and *S* are nil on the vertical faces of the slice, as is generally accepted in the original version of the Swedish Method established by FELLENIUS.

The piezometer shown in Part *b* of the figure indicates that in addition to the partial submersion of the material there is a neutral pressure *u* as a result of seepage, at O_i.

The acting forces will be the weight of the slice, which can be calculated using the expression:

$$W = W_l + \overline{W} + zb\gamma_w \tag{6-13}$$

W_l corresponds to the part of the slice that is located above the phreatic level and must be calculated with the γ_m of the material. $\overline{W}$ corresponds to the submerged part and must be calculated with γ'_m. The component $zb\gamma_w$ represents the weight of the water included in the submerged part of the slice. If the entire slice were under water, like slice *j* in Part *a* of the figure, the weight of all the water on it should be considered in the last part of Eq. (6-13).

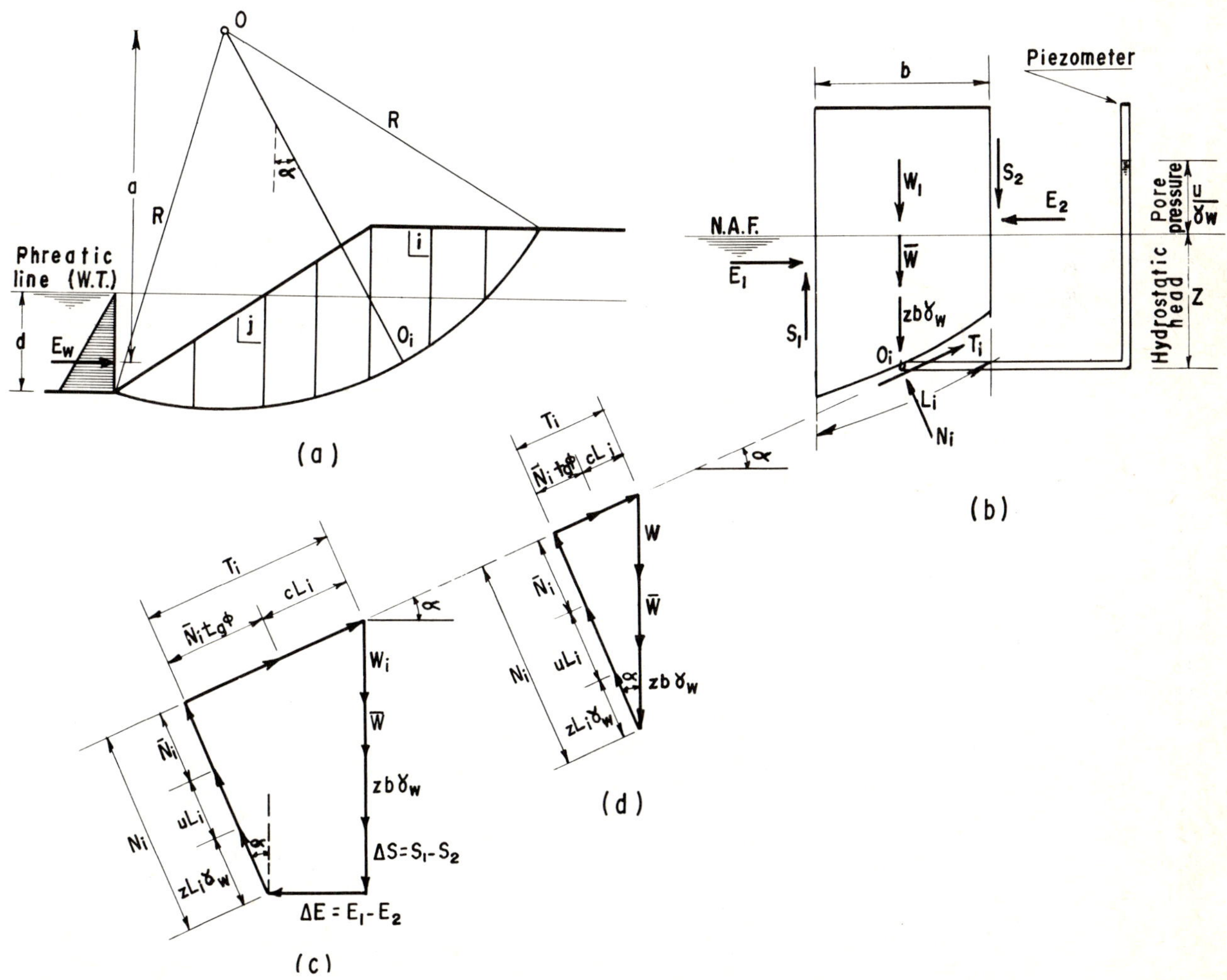

Fig. 6-33 Application of the Swedish Method to the case of a slope with flow and partially below the phreatic line. Analysis with pore pressures and in terms of effective stresses

The total pressure of the water at O_i is given by the piezometer and is equivalent to:

$$u_t = z\gamma_w + u \quad (6\text{-}14)$$

where $z\gamma_w$ is the hydrostatic pressure corresponding to the position of the phreatic level and u is the pore pressure in excess of the hydrostatic head, caused, for example, by seepage. For analysis, this excess of pressure must be found, either from a flow net or field measurements. Details of the first case are given in Appendix 7a (see text related to Fig. 7a-10); the second was briefly dealt with in Chapter 1, and the third will be described in Chapter 13 on field instrumentation. Triaxial test data supply other pore pressure changes produced by shear as described in Chapter 1. If the phreatic level is located below O_i, the pore pressure at O_i is $h\gamma_w$, h being the height to which the water would rise in a piezometer placed at O_i. If the pore pressure is due to capillarity (tension in the water), it must be considered negative in the following analyses.

The driving moment is:

$$M_m = \sum (W_1 + \overline{W} + zb\gamma_w) R \sin \alpha \quad (6\text{-}15)$$

but since below the phreatic level the water should be in equilibrium, the following should be true:

$$\sum zb\gamma_w R \sin \alpha = 1/2\ \gamma_w d^2 a \quad (6\text{-}16)$$

where the right side of Eq. (6-16) represents the effect of the hydrostatic thrust of the water at the toe of the slope. In this way, the driving moment is:

$$M_m = \sum (W_1 + \overline{W}) R \sin \alpha = R\sum \overline{T}_i \quad (6\text{-}17)$$

so that the driving moment depends on the effective weight of the slice whose component in the direction of the slide is referred to as $\overline{T}_i$.

The resisting moment will depend on the shear strength that actually develops in the base of the slice. This strength can be calculated if the total weight of the slice $(W_1 + \overline{W} + zb\gamma_w)$ is multiplied by cos α, which will give the total normal force N_i. The value of N_i divided by L_i, gives the total normal stress on the base of the slice, σ_i. The effective normal pressure, $\bar{\sigma}_i$, will be: $\bar{\sigma}_i = \sigma_i - z\gamma_w - u = \sigma_i - u_t$. This value is used in the shear strength envelope of effective stresses, in order to obtain $\bar{s}_i$, which is the shear strength in the base of the slice.

The resisting moment will therefore be:

$$M_r = \sum \bar{s}_i L_i R \quad (6\text{-}18)$$

The safety factor of the circle will be:

$$F_s = \frac{\sum \bar{s}_i L_i}{\sum T_i} \quad (6\text{-}19)$$

A trial-and-error procedure will be necessary to determine the critical circle that is associated with the smallest safety factor. With regard to the choice of the smallest safety factor to be accepted in the design, the previous observations are valid, bearing in mind that the loading condition now considered is more realistic and a smaller degree of uncertainly can be expected. Calculations should be written down as in Table 6-4.

As was shown in §6.4, the foregoing procedure is not the only possible one for performing the stability analysis in this case. It can also be done using the total weights of the soil and the seepage forces of the water against the sides of the slices. In this case, the shear strength of the soil must again be taken from an effective stress envelope of the type obtained from consolidated-drained triaxial tests. Figure 6-34 shows which acting stresses should be considered in each slice when using this procedure.

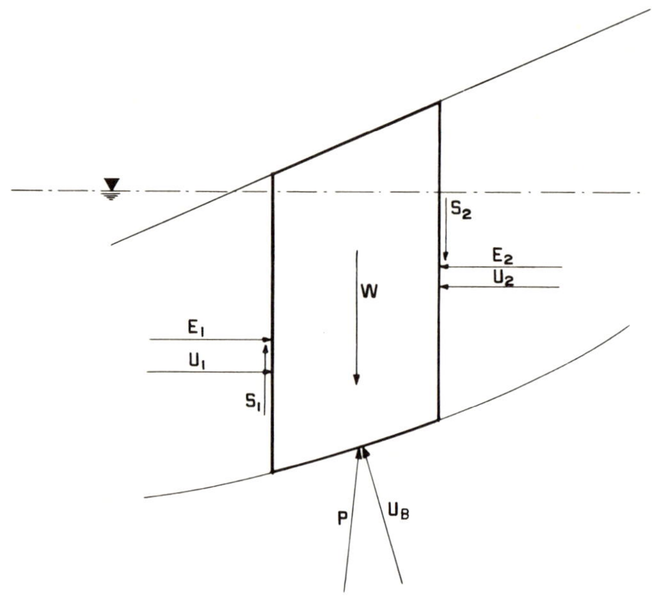

Fig. 6-34 Forces acting in a slice. Analysis using seepage forces

First, the weight W of the slice is considered as obtained from γ_m of the soil. Second, come forces E and S, on both faces of the slice. If the simplified version of the Swedish Method is applied, these forces are considered nil. Also to be taken into account are the water forces U_1 and U_2 on the sides of the slice and U_B on the base.

If there is no seeping water, (the water is in a hydrostatic condition) the only water forces will be those resulting from the hydrostatic thrusts against the sides, and the hydrostatic pressure on the base; but if there is flowing water, these forces must be obtained from the flow net, using the methods explained in Appendix 7a.

Once the forces in each slice have been established, the resisting and driving moments can be assessed in the usual way for each circle and the method of calculation continues as described previously.

.4 Swedish Method: More Sophisticated Procedures

If the lateral earth forces in the slices are taken into account, the results obtained may be more accurate. In [3], TERZAGHI and PECK mention that for circular surfaces the error with the original method may not be more than 10% or 15%, and is on the conservative side. Reference [48] states that in the case of earth dams with large rock backfills, the increase in the safety factor by taking into consideration the side forces, may be as much as 30%, which justifies a more sophisticated analysis for reasons of cost. Lastly, in [58] this difference is said to reach as much as 60% in some cases. The difference becomes greater with values of ∅ exceeding 20°.

Despite these observations, only occasionally can the use of more sophisticated procedures be justified. The ones previously described do not take into account the effects of lateral earth forces on the slices. By way of illustration, and without going into a detailed explanation of its theoretical development, a description will be given here of one method which was originally proposed by BISHOP. It was later simplified, in that only horizontal side forces are considered. Details of this method can be found in [59] and [60]. The information given here is taken from [61].
The safety factor of the slope is expressed by:

$$F_s = \frac{\sum [cb_i + (W_i - b_i u_i) \tan \varnothing] \dfrac{1}{M_i(\alpha)}}{\sum (W \sin \alpha)_i} \qquad (6\text{-}20)$$

where: b_i is the width of the i-th slice, measured horizontally; c, $\varnothing$ are the shear strength parameters in terms of effective stresses; W_i is the total weight of the i-th slice; u_i is the average neutral or water pressure in the base of the slice, and:

$$M_i(\alpha) = \cos \alpha_i \left(1 + \frac{\tan \alpha_i \tan \varnothing}{F_s}\right) \qquad (6\text{-}21)$$

Note that Eq. (6-20) requires a trial-and-error solution, for F_s is included in both sides. Fortunately this is very quick, Fig. 6-35 helps by giving the value $M_i(\alpha)$ corresponding to each slice.

The BISHOP Method does not lead to theoretically correct values of the safety factor, but it does provide greater accuracy. In [62] a procedure is suggested for applying the BISHOP Method in graphical form, [63] and [64] describe a way of applying it with the aid of a computer. More sophisticated versions of the method do exist, where certain inclinations are considered for the lateral earth forces [19,48,65]. The angle usually given to these forces is the slope inclination. Lastly in [60] and [66], procedures are described for taking into account the effect of the lateral earth forces in the slices, in cases where circular slip surfaces are not used. These methods will be dealt with in §6.5.3.

.5 Swedish Method: Further Comments

The main hypotheses used in the Swedish Method are the following:

1. A failure surface that is circular in cross-section.
2. A bi-dimensional analysis is made corresponding to a plain deformation field.
3. The MOHR-COULOMB strength law is considered valid.
4. The shear strength is mobilized completely at all points of the slip surface. This has already been seen to be contradictory to some present-day observations and schools of thought.
5. In analyses under seepage, 100% consolidation of the soil in a steady flow condition is supposed, the pore pressure from the flow net being the only one considered.

The Swedish Method poses the problem that in theory there are more unknown quantities than the three equations of static equilibrium provided for a given system of forces [61]. Figure 6-36 illustrates this statement.

Weight W is a force with a known magnitude and location. Reactions to normal and tangential forces due to friction (N and R in the figure) are unknown in magnitude and location, although they must be perpendicular to each other. It should also be understood that:

$$R_\varnothing = \frac{\bar{N} \tan \varnothing}{F_s}$$

where F_s is the safety factor associated with the circle, and is also unknown. The cohesive reaction R_c is determined entirely by the variation of c along the failure surface; its magnitude could also be found in terms of c and F_s. The four unknown quantities of the analysis are F_s, the magnitude and location of $\bar{N}$ and the magnitude of $R_\varnothing$. Three equations

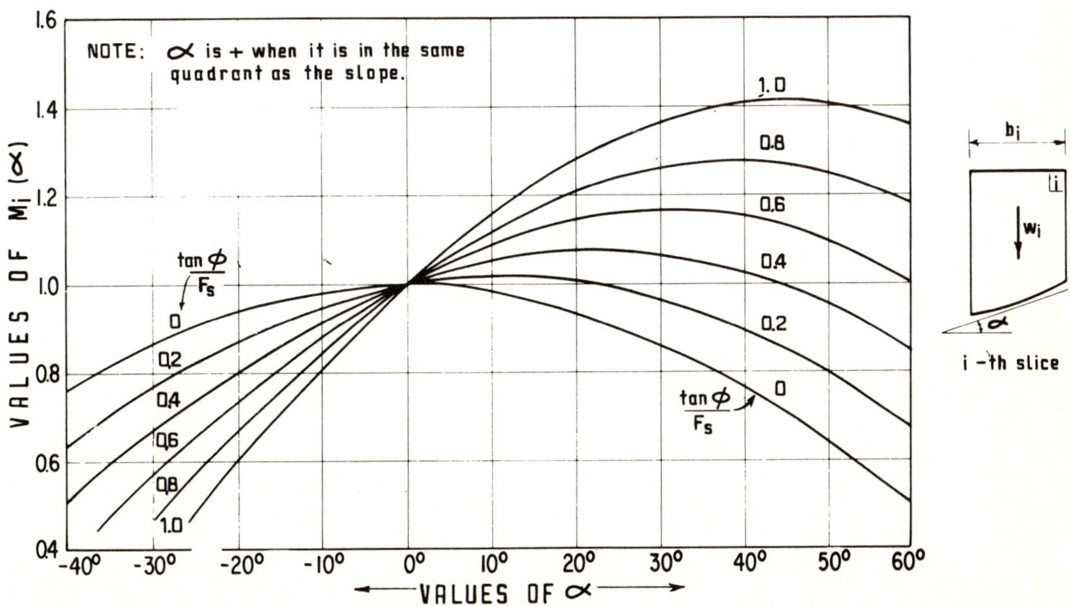

Fig. 6-35 Graph to determine $M_1(\alpha)$

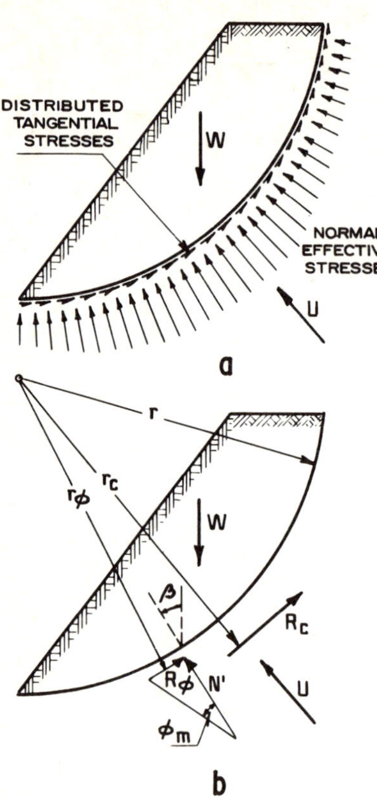

Fig. 6-36 Forces acting on a sliding mass with a circular end a) Normal and shear forces distributed over the surface b) Resulting forces

of equilibrium are provided by statics, which means that the problem cannot be solved unless the deformation characteristics of the soil are considered.

This is chiefly why the Swedish Method requires simplifying assumptions to solve the problem. A discussion of the best way to establish these assumptions and of those included in the different methods that are in use today is presented in detail in [58] and [67].

6.5.3 Stability Analysis of Noncircular Failure Surfaces

Under this heading are the failure surfaces of natural or artificial slopes which are almost flat or made up of a combination of shapes that are so far from circular, that the Swedish Method does not prove very satisfactory, [60,66,67] describe methods of calculation for such cases. A useful variety of available methods is given in [68]. Parts of the following description are taken from [3], where the problem receives more analytical treatment than has been used in this book. This treatment, which is very common in modern literature and can also be used for the traditional forms of the Swedish Method, makes it possible to find a safety factor by trial-and-error procedures. This was the procedure followed in Eqs. (6-20) and (6-21).

Figure 6-37 shows a noncircular failure surface. Part *a* is a general sketch of the hillside and position of the *i*-th slice. Part *b* shows the stresses acting in each slice, and Part *c* the force polygon corresponding to equilibrium.

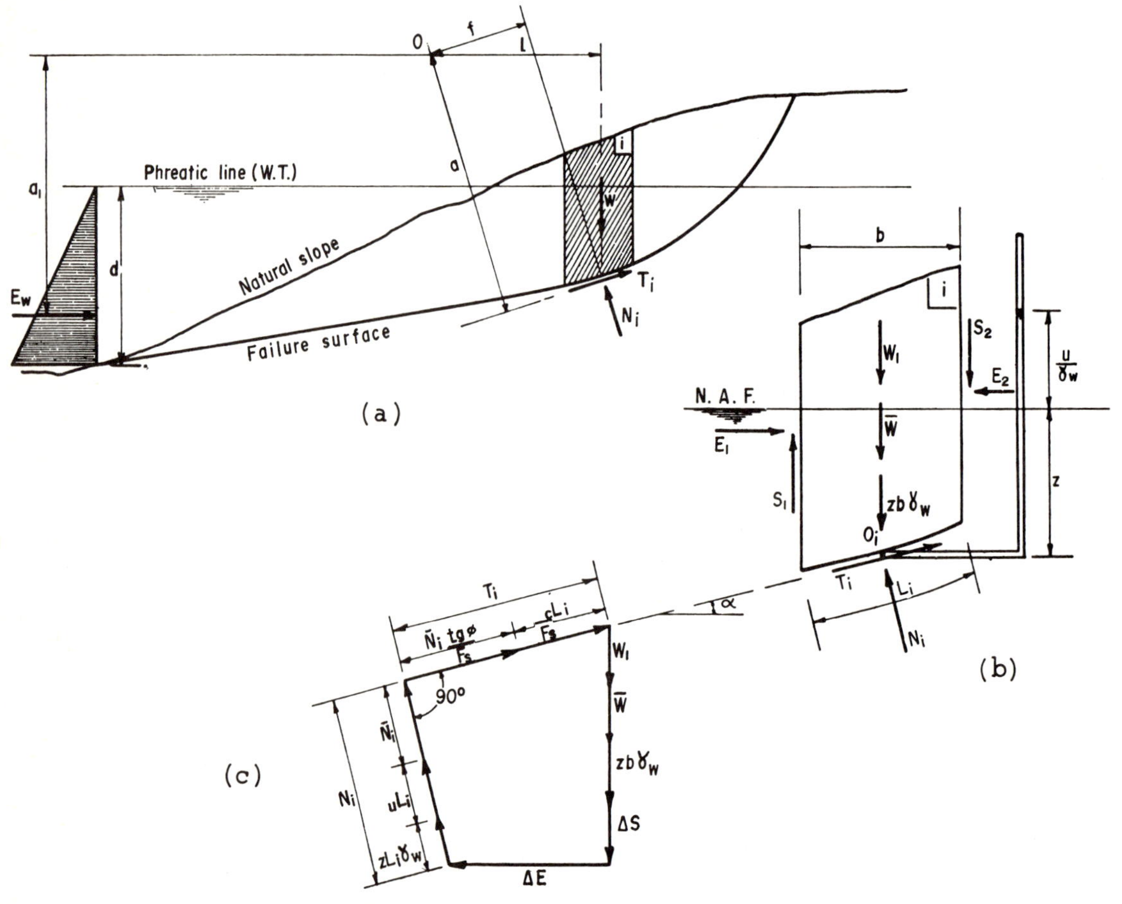

Fig. 6-37 Stability analysis with a non-circular failure surface [3]

If an arbitrary moment center, O, is adopted, the equilibrium of the whole sliding mass will demand that:

$$\Sigma Wl = \Sigma (T_i a + N_i f) + 1/2\, \gamma_w d^2 a_1 \quad (6\text{-}22)$$

But:

$$W = W_1 + \bar{W} + zb\, \gamma_w \quad (6\text{-}23)$$

and from the force polygon:

$$T_i = \frac{c}{F_s} L_i + \bar{N}_i \frac{\tan \varnothing}{F_s} \quad (6\text{-}24)$$

T_i is the force acting in the tangential direction, neutralized by the fraction of soil strength that is mobilized. This is why this last value is affected by the safety factor (if the slice were at limiting equilibrium, maximum strength would be mobilized, but if the conditions do not represent failure, a fraction of the strength in inverse proportion to the safety factor is mobilized).

Bearing in mind Eq. (6-24), Eq. (6-22) can be written as:

$$\Sigma(W_1 + W + zb\, \gamma_w)l = \Sigma \left(\frac{c}{F_s} L_i + \bar{N}_i \frac{\tan \varnothing}{F_s} \right) + \Sigma N_i f + \frac{1}{2} \gamma_2 d^2 a_1 \quad (6\text{-}25)$$

The safety factor is computed:

$$F_s = \frac{\Sigma (cL_i + N_i \tan \varnothing)\, a}{\Sigma(W_1 + \bar{W} + zb\gamma_w)\, l - \Sigma N_i f - 1/2\, \gamma_w d^2 a_1} \quad (6\text{-}26)$$

Below the water level, its mass should be in equilibrium, which means that:

$$\Sigma\, zb\gamma_w l - 1/2 \gamma_w d^2 a_1 = \Sigma z \gamma_w L_i f \quad (6\text{-}27)$$

In other words, the weight of the water, the hydrostatic thrust at the toe of the slope, and the force due to the hydrostatic pressure of the water on the base of the slice must be in moment equilibrium about O.

If the results of Eq. (6-27) are introduced in the denominator of Eq. (6-26), what remains will be:

$$\Sigma (W_1 + \bar{W})\, l - \Sigma (N_i - z\gamma_w L_i)\, f \quad (6\text{-}28)$$

By defining

$$\bar{N}_i = N_i - (z\gamma_w + u)\, L_i \quad (6\text{-}29)$$

as the normal effective stress on the base of the slice (the existence of a neutral pressure, u, for example through seepage, has been considered), the denominator of Eq. (6-26) can be written as follows:

$$\Sigma (W_1 + \bar{W})\, l - \Sigma (\bar{N}_i + uL_i)\, f \quad (6\text{-}30)$$

and Eq. (6-26) becomes:

$$F_s = \frac{\Sigma (cL_i + N_i \tan \varnothing)\, a}{\Sigma (W_1 + \bar{W})\, l - \Sigma (\bar{N}_i + uL_i)\, f} \quad (6\text{-}31)$$

Equation (6-31) would give the value of the safety factor for the failure surface being studied if the shear strength parameters of the soil in terms of effective stresses and the pore pressures on the base of the slice are known, but ignoring the effect of the lateral earth forces E and S.

If the effect of these forces is to be considered, the sum of the forces in the vertical direction can be included in the force polygon as in Fig. 6-37c:

$$W_1 + \bar{W} + zb\gamma_w + \Delta s = (z\gamma_w L_i + uL_i + \bar{N}_i) \cos \alpha$$
$$\frac{1}{F_s}(cL_i + \bar{N}_i \tan \varnothing) \sin \alpha \quad (6\text{-}32)$$

From this it can be deduced that:

$$\bar{N}_i = \frac{W_1 + \bar{W} + \Delta s - ub \frac{c}{F_s} b \tan \alpha}{M_i(\alpha)} \quad (6\text{-}33)$$

To obtain the above expression, it must be taken into account that $L_i \cos \alpha = b$ and that the function $M_i(\alpha)$ has already been defined in Eq. (6-21).

By introducing the values from Eq. (6-33) into (6-31), the following equation can finally be obtained:

$$F_s = \frac{\Sigma \left[cb + (W_1 + \bar{W} + \Delta s - ub) \tan \varnothing \right] \frac{\alpha}{M_i(\alpha)}}{\Sigma (W_1 + \bar{W})\, l - \Sigma \left[W_1 + \bar{W} + \Delta s + (ub \tan \varnothing - cb) \bullet \frac{\tan \alpha}{F_s} \right] \frac{f}{M_i(\alpha)}} \quad (6\text{-}34)$$

The solution of Eq (6-34) requires successive approximations for it contains F_s on both sides. Calculation is made easier by the graph in Fig. 6-35 for the determination of $M_i(\alpha)$. Equation (6-34) gives F_s in connection with a given failure surface. Many others will have to be analyzed in order to find the smallest F_s.

The value of F_s depends on Δs and it must be introduced in Eq. (6-34) with one of the values that are given in the different methods referred to on previous pages. In the majority of practical problems, application of Eq. (6-34) with $\Delta s = O$ will be sufficient. Convergence of trials for Eq. (6-34) is very rapid.

6.5.4 Translational Slides

The mathematical model for this type of failure is illustrated in Fig. 6-38. The weak stratum shown is usually made of soft clays or fine sands and silts that are subjected to water pressures which reduce the effective stresses and the shear strength. The risk of this type of failure is particularly serious in hillsides where the weak stratum has the same inclination as the hillside itself.

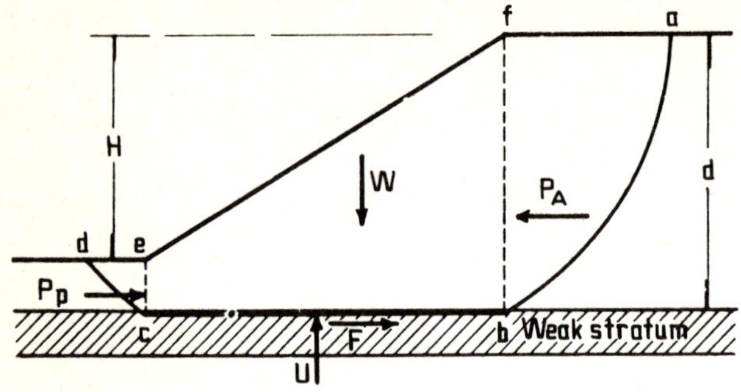

Fig. 6-38 Failure surface corresponding to a translational slide

If the weak stratum is below an embankment on a hillside or on a soft foundation the most critical condition will be during or immediately after construction if the weak stratum is made of clay. In this case, strength parameters will be obtained in an unconsolidated-undrained triaxial test, and analysis can be made in terms of total stresses. In this same case, but with a weak stratum made of sand below the phreatic level, and subjected to excessive pore pressures, the analysis will have to be based on effective stresses, with the force U, which is the total uplift from the water pressure. If the slope corresponds to a cut in a hillside, the critical condition will be the long-term one, and analysis using effective stresses will be necessary.

The calculation procedure is carried out as follows.

Wedge *bfec* will move towards the left due to the earth thrust in plane *bf*. This thrust may be the active pressure. The resisting forces are the effect of a passive thrust in plane *ec* and the shear strength along the slip surface *cb* (F). Earth thrusts can be evaluated using the methods described in Chapter 5.

In an analysis using total stresses (embankment built on ground containing a weak clayey stratum), force F will simply be equal to $c_u \cdot \overline{cb}$. In an analysis using effective stresses, force F will be:

$$F = c \cdot \overline{cb} + (W - U) \tan \varnothing \qquad (6\text{-}35)$$

where c and $\varnothing$ must be expressed in terms of effective stresses. Force U, or total water pressure uplift, will have to be obtained as the area of the seepage pressure or uplift diagram in plane *cb*, which in turn can be obtained from a flow net.

The safety factor which indicates the risk of failure can be written thus:

$$F_s = \frac{F + P_P}{P_A} \qquad (6\text{-}36)$$

In practical designs it will probably not be wise to accept a safety factor smaller than 1.5.

It should be noted that in this case location of planes *fb* and *ec* for calculation of the earth thrusts and force F lead to the smallest safety factor in the case of Fig. 6-38, for any displacement of point b towards the right or of point c towards the left force F increases with the same earth thrusts. If b or c move towards the inclined part of the slope, F decreases as a linear function of H, but the active thrust decreases as a function of d^2 (if b moves to the left) the passive thrust also increases as a function of d^2 (if c moves to the right), from which a safety factor greater than the one corresponding to the case shown in this figure can easily be expected. However, in some situations with irregular slope geometry the positions of the critical planes *ec* and *fb* must be found by trial.

6.5.5 The Wedge Method

This is a method for analyzing the stability of the body of a slope, and it can be applied to the same cases as the Swedish Method by means of the circular failure hypothesis. However, because of the nature of the failure surfaces now being dealt with (flat surfaces), in practical calculations the wedge method is limited to translational slides. A typical case for its application is that of an embankment built on foundation ground containing a very soft stratum close to the surface (or actually on the surface, as might be the case of intensely weathered surface zones underlain by harder residual soils at a greater depth) or that of a poorly compacted earth embankment built on a hard foundation ground.

In this method, the potential or actual slip surface is represented by two or more straight segments, as shown in Fig. 6-39. Wedges are thus defined within the sliding mass (*I* and *II* in the case shown in the figure). The shear strength along the slip surface must be expressed as a function of the applicable strength parameters.

In the equilibrium of the two wedges, there are four unknown mechanical elements (E, $\overline{N}_1$, $\overline{N}_2$ and α) and a fifth unknown factor which is the safety factor corresponding to the failure surface chosen. For a given geometry and specific strength parameters, stability conditions must be defined for the sliding mass, which will be reflected in the safety factor that is determined.

To solve this problem there are two equations of force equilibrium in each wedge, thus this force equilibrium is indeterminate. When the diagram of the free body of wedge *I* or *II* is drawn, the following forces appear on the wedge:

— A force, $c_1 = \dfrac{c}{F_s} \cdot \overline{AB}$ (6-37)

— A force, $\overline{T}_1$, depending on $\overline{N}_1$, the strength parameters and, F_s

— The force $\overline{N}_1$

— The weight of the wedge W_1

— The earth thrust E produced by wedge *II*, on wedge *I*.

— A force, $C_3 = \dfrac{c}{F_s} \cdot \overline{BC}$ (6-38)

From this it can be deduced than an assumption must be made to eliminate one or other of the unknown factors in order to solve the problem. This assumption usually refers to the direction of force E. It is generally assumed that E is parallel to the slope plane, or that it makes an angle $\varnothing_E$, with the contact surface between the wedges, defined by the expression:

$$\varnothing_E = \tan^{-1}\left(\frac{\tan \varnothing}{F_s}\right)$$

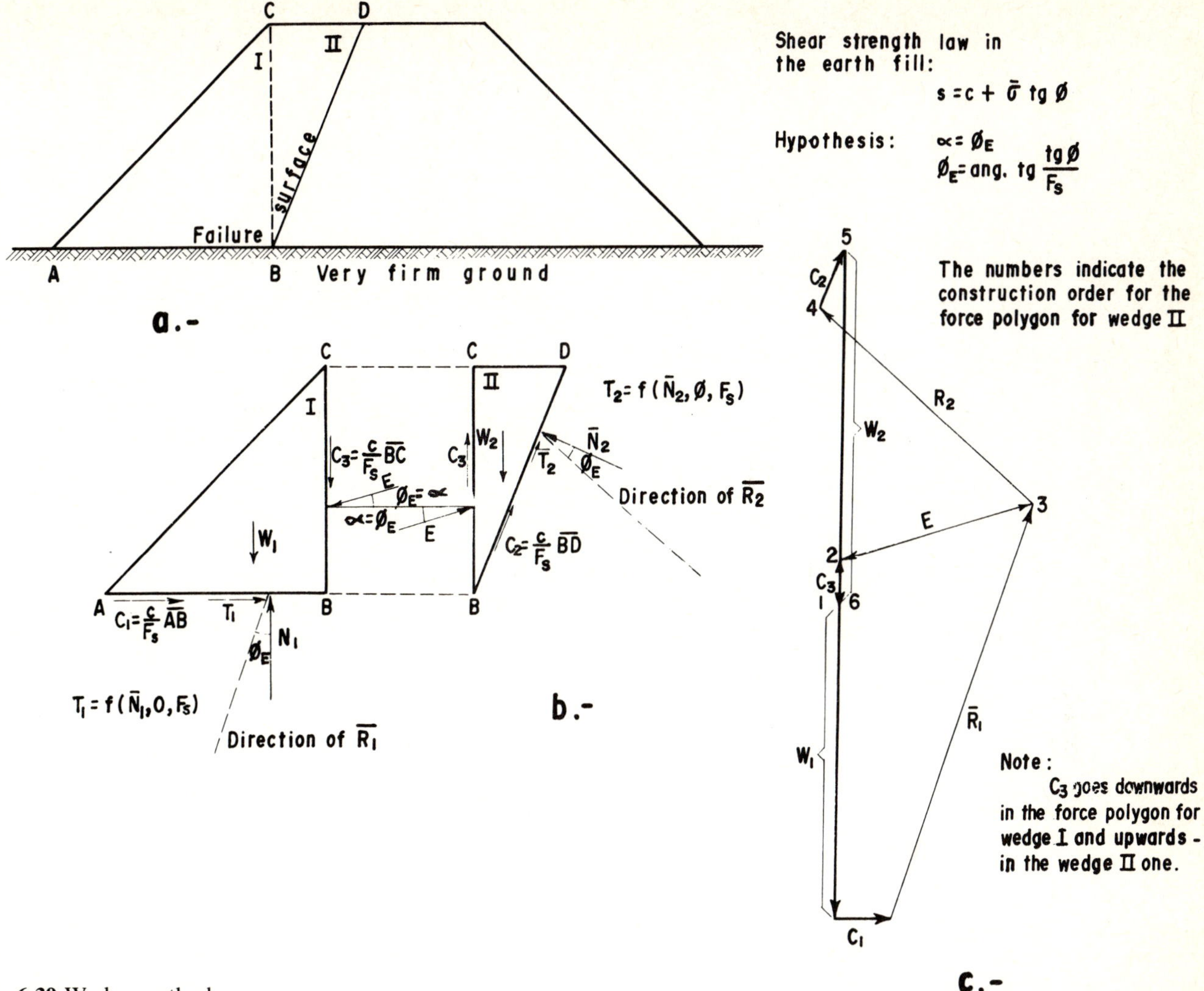

Fig. 6-39 Wedge method

This last hypothesis was the one accepted for construction of Fig. 6-39.

Force $\overline{T}_i$ and $\overline{N}_i$ are of unknown magnitude, but not direction. The direction of their resultant, $\overline{R}_i$, will therefore be known. It must form an angle, $\varnothing_E$, which represents the angle of friction with force N; taking into account the effect of the safety factor. It is these forces $\overline{R}_1$ and $\overline{R}_2$ that are used for constructing the force polygon that appears in Part *c* of the figure, instead of components $\overline{T}_i$ and $\overline{N}_i$.

Construction of the force polygon of wedge *I* starts at W_1, which has a known magnitude and position. Next a safety factor will have to be assumed for the combination of both wedges. On the basis of this hypothesis and with Eq. (6-37), the magnitude and position of forces C_1 and C_3 are found and duly included in the polygon. In the figure, the force polygon was started at C_3, but this was for ease of drawing. A line with the same direction as $\overline{R}_1$ (Part *b* of the figure) can be drawn through the C_1 end, and through the origin of C_3 a line with the same direction as E. In this way, forces $\overline{R}_1$ and E are determined for the assumed safety factor.

The force polygon for wedge *II* can be constructed on the one for wedge *I*, adding W_2 to C_3 and E, which are known; (see Part *c* of the figure). By applying Eq. (6-37) to wedge *II*, C_2 can be computed for the assumed safety factor. Through the C_2 end, a line can be drawn with the direction of $\overline{R}_2$.

If the value of the safety factor assumed happens to be correct, the force polygon thus constructed will close, with the line of action of $\overline{R}_2$ passing through the origin of C_2. However, this will probably not be the case, which will mean that a safety factor was chosen that does not reflect the actual conditions of the problem. In this case, trial-and-error procedures must be followed until the correct safety factor is assumed and the polygon closes. This factor will, of course, be associated with the assumed failure surface. Calculations will have to be repeated for other possible surfaces, until the surface with the minimum safety factor is found.

6.6 Embankments on Soft Soils

Much of what can be said here about the important problem of embankments built on very soft soils or peats has already been discussed in Chapter 3 in relation to soil foundations. There are, however, a few comments which are not out of place in this Chapter. The first problem to consider is the assessment of the stability of the embankment in combination with its soil foundation, which in these cases usually constitutes a vital element.

In many soft soils, it is essential to avoid a catastrophic slide in the embankment because of a drop in shear strength usually suffered as a result of the intense remolding which accompanies a total collapse of the soil structure. Afterwards the shear strength is so slowly recovered that insoluble problems can arise. Figure 6-40a [69] shows the type of analysis that should be done in these cases, which has already been discussed under the previous heading. Part *b* of the same figure shows the loading rate, and Part *c* the evolution of the pore pressures within the natural soil foundation that can be expected. Part *d* expresses the variation of the safety factor with time. Analysis generally uses the undrained strength of the soil foundation (c_u) and total stresses. There are, however, two points that are worthy of mention here.

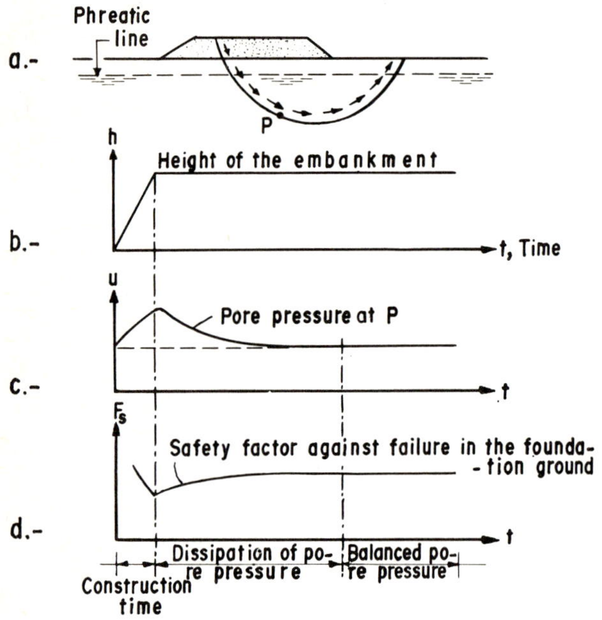

Fig. 6-40 Variation in the conditions of an earth fill built on a soft soil as time passes [69]

The first refers to the concepts mentioned in relation to Fig. 6-2 [6]. If the strength of the soil foundation gradually decreases with strain or time, until values far below peak strength are reached, a design based on peak strength may be inappropriate. In this case, there will be no alternative but to determine in the laboratory the response of the clay to long-term tests, so as to establish an appropriate value for the design strength.

In 1960, A. Casagrande [70] reported a very interesting practical construction case where an important variation in the undrained strength of a clay (obtained with uniaxial compression tests) was observed in the long-term tests (up to 2 weeks) in relation to the peak strength in a standard test (about 5 min). The stress-strain curves varied from typical fragile failure to plastic failure. The most important practical fact is that the c_u decreased to 30% of its value in the standard unconsolidated-undrained test. Many engineers believe that a drop in strength like this occurs with time in a very soft sensitive soil foundation on which an embankment has been built, as a result of the effect of the increased shear strains which disrupt the structure of the clay, even when the stresses are far from failure, as defined by the peak strength obtained in a conventional test (c_u). Consolidation will contribute to an increase in the strength offsetting the shear strain effect, but this minimum slow strain strength represents a critical condition that is felt by many designers to be worthy of attention. There has not been sufficient investigation to judge what percentage reduction in the conventional peak strength should be considered safe. In many design offices, this conventional peak strength is arbitrarily reduced 25% or 30% in order to obtain the value of the design strength.

The strength of the undisturbed soil foundation can probably be obtained approximately and economically by means of vane tests [69], whose main features were discussed in Chapter 1. A rotation of less than 10° is usually necessary to obtain the peak strength of an *intact* soil, whereas several turns are usually required to reach the residual strength. Figure 6-41 [71] shows the typical curve obtained from a vane test on a soft clay. The strength of the intact soil divided by the residual strength is usually taken as a measure of the sensitivity of the clay. The vane test is of course no longer considered representative if the foundation soil is hard or non homogeneous.

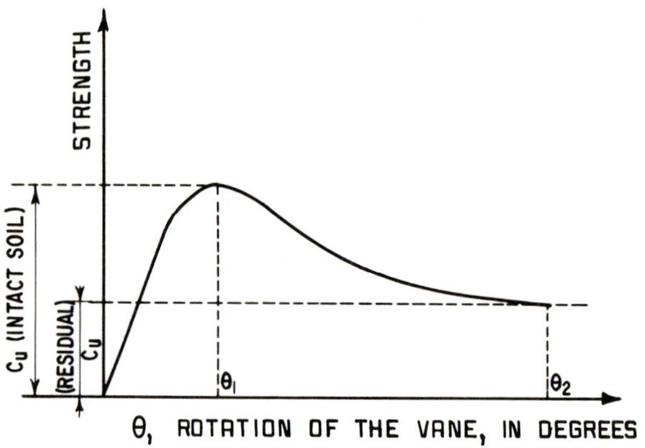

Fig. 6-41 Typical strength curve of a soft clay in a vane test [71]

To take into consideration the lower strength from triaxial tests compared to vane tests, Bjerrum gives in [69] a graph of a corrective factor μ which, when multiplied by the vane strength provided by the test, gives the strength that should be used in the design (Fig. 6-42).

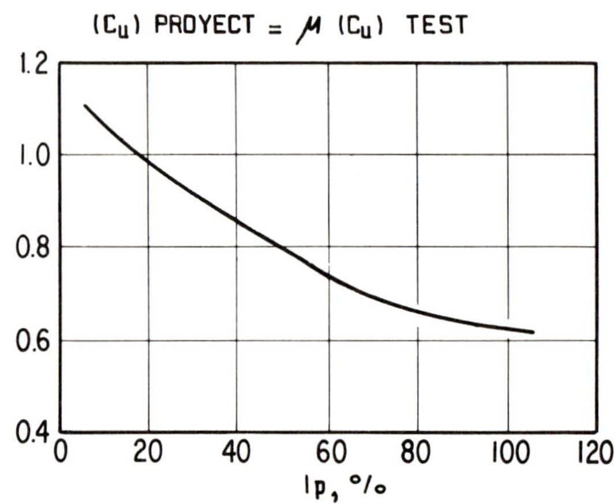

Fig. 6-42 Reduction factor for obtaining design strength based on vane test results [69]

In terms of the plasticity index of the clay, BJERRUM'S correlation is purely statistical and has been obtained on the basis of the relation observed between the plasticity index of 14 foundation soils that failed under embankments and the safety factor computed retrospectively for these failures. This safety factor was systematically somewhat larger than one, which is a sign that the strength of the soil was overestimated in the field vane tests.

The doubts involving the value of the soil strength that should be used in stability analyses have led many designers to obtain strengths from scale test embankments, [72-78] give descriptions of this type of research and contain some interesting results.

Many engineers have simplified and added conservatism to establish design methods by ignoring the contribution of the embankment soil strength to general stability. This procedure is most often followed when an embankment is low and the clay crust of the soil foundation hardened by drying is thin, for it has been seen that in cases like these failure of the embankment is usually preceded by cracking throughout almost the entire embankment.

Reference [69] also includes some interesting discussions about the reliability of settlement assessments for embankments on very soft soils. This is where important discrepancies can usually be observed between theory and reality, further justifying the use of test embankments. What is far more difficult to predict is the evolution of these settlements with time. This is a problem which well deserves the use of test embankments, so long as sufficient time is available for observations.

The information that can be obtained from a test embankment is extremely varied and instructive. The embankment is a full scale model of the structure under study. Figures 6-43 and 6-44 [77] are an example of the data that

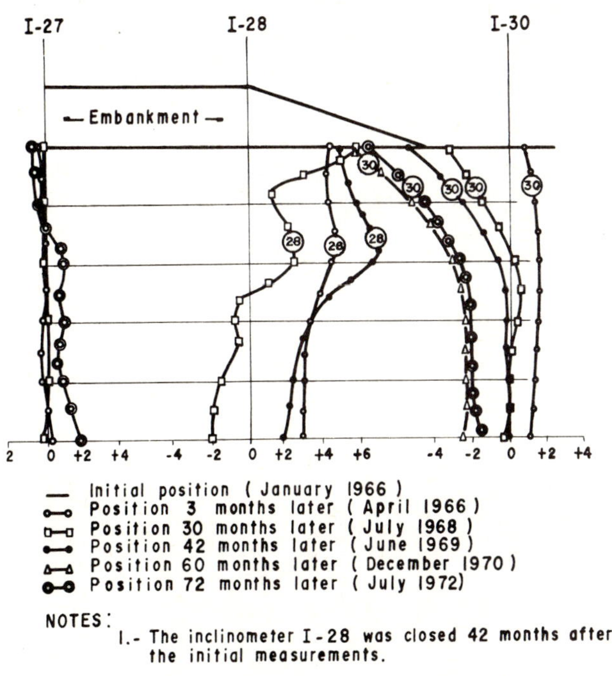

Fig. 6-43 Inclinometer measurements in a section of a test embankment built in the Texcoco Lake, Mexico

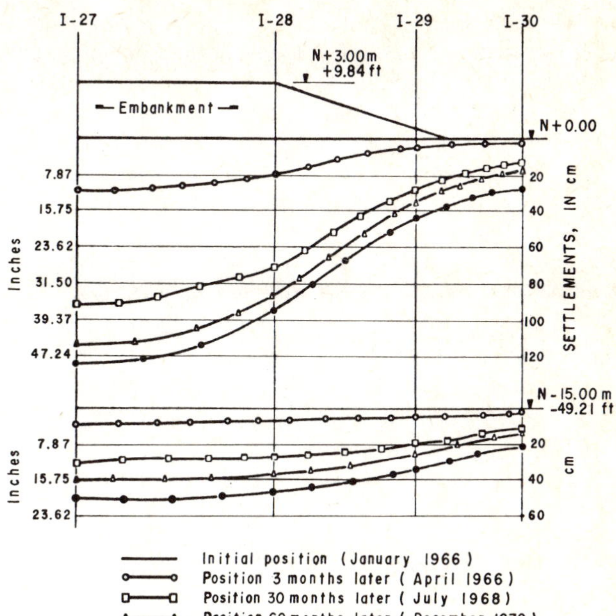

Fig. 6-44 Results of measurements with a settlement torpedo in a section of the same test embankment in Fig. 6-43.

can be obtained, Fig. 6-43 gives data from inclinometers located in an instrumented cross-section of a test embankment. These instruments will be described in greater detail in a later Chapter, which is devoted to field instrumentation. The data given here are taken from [77] and have been slightly modified to take into account the time that has passed since publication. The gradual lateral deformation of the foundation soil under the weight of the embankment follows an interesting course. First, as would be expected, outward lateral displacement occurred; but after a time the lateral deformations were reversed and started to move towards the center underneath the embankment. It seems that the decrease in volume as a result of consolidation, which was at its maximum under the center of the structure, was sufficient to reverse the direction of the lateral deformation. Figure 6-44 gives data regarding magnitude, distribution and evolution of the settlements in the same test embankment. The data were obtained by using the settlement torpedo developed by WILSON, which will also be described in a later Chapter.

6.7 Establishing the Inclination of Uncomputed Road Cuttings

As previously stated at the beginning of this Chapter, the majority of road cuts are designed without any field study (including sampling) and without laboratory tests; therefore a detailed mathematical analysis cannot be envisaged. It was also mentioned that in many cases the heterogeneity of the formations involved makes any type of analysis that might be attempted of limited value. This means that a high percentage of road and railway cuts have to be designed on the basis of the experience of the engineer in charge, guided by the behavior of similar cuts in the same area (when there are any), the conditions of local natural slopes and possibly some field reconnaissance that may have been carried out within the general framework of the geotechnical study of the road. No hard-and-fast rules can be laid down here. Each individual case is different and should be treated as such.

Previous experience provides valuable help in the task of establishing the angle of cut slopes. With this idea in mind the authors wish to justify the information that is given in the following pages. Little success can be expected by the engineer who applies this information blindly. It should be regarded as a general reference or simply as the personal opinion of engineers who have come up against the same problems in the past.

Figure 6-45 [79] presents a collection of experiences of a group of engineers of the California Department of Highways, USA. The angle of the cut slopes is given as a function of the height for a range of *c* and ∅. These parameters are estimated by the engineer on the basis of a general knowledge of the materials involved. The graph includes a *reasonable* safety factor.

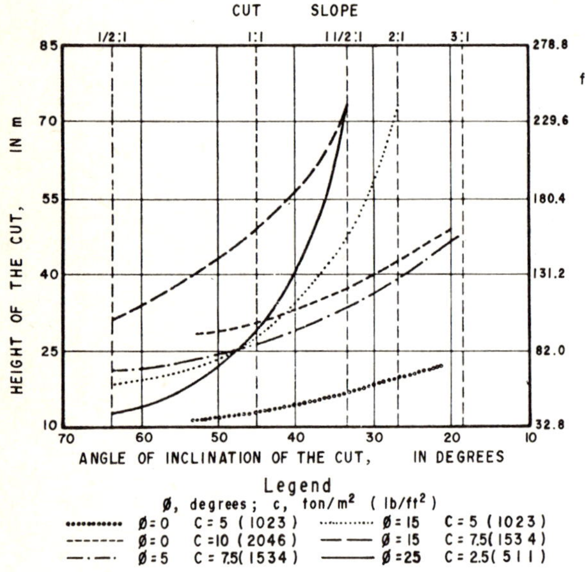

Fig. 6-45 Height of a cut as a function of values of *c* and ∅ [79]

A graph like the one in Fig. 6-45 is useful for rapid visualization of the influence of the different strength parameters on general stability conditions, using different sets of values.

Table 6-5 is a complete summary of recommended angles for cuts in very different materials, including many types of rocks as well as soils. It summarizes the experience of the Geotechnical Department of the Mexican Ministry of Public Works.

6.8 Factors Causing Stability Failures in Natural and Artificial Slopes

It is sometimes hard to establish the relative significance of the causes of slides in hillsides or in artificial slopes. The outstanding influence of internal water flows (seepage) and the pressures they may provoke in the soil masses, is emphasized by the majority of important slides that occur in the period following the start of the rainy season and whose rates of movement are closely connected with seepage conditions. Figure 6-46 shows the type of relation that can be established between the rainfall and the incidence of stability problems in an area.

In this case the information handled was collected over a period of more than 2 years for three failures on the Tijuana-Ensenada highway [7]. The three were of the type of failure surface which formed prior to construction and in all three, movements of enormous masses of earth were registered on sloping planes.

The rainy season and dry periods in the region have been well defined. The rains commence at about the beginning of November and last until the second half of February. The dry periods are correspondingly from mid-February until the beginning of November. The increase in the displacements beginning in January is remarkable. This indicates that from the first rains a period of about two and a half months is necessary for internal flows or seepage to be established. Similarly, from the end of April there is a marked decrease in the movements. This shows that the effects of seepage do not disappear for about two months after the last rains. Figure 6-46 makes the relation between rainfall and failure movements clear.

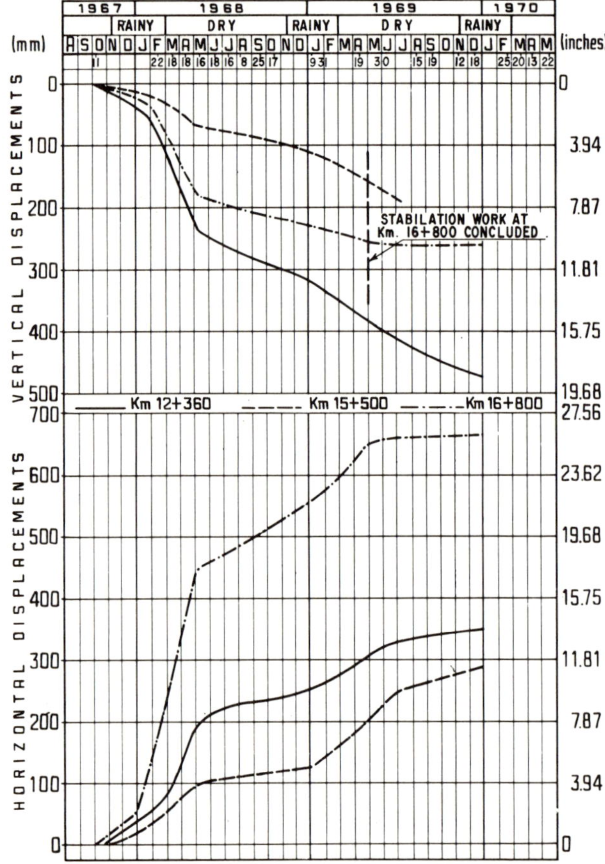

Fig. 6-46 Relationship between regional rainfall and the magnitude of movements in several slides. Tijuana-Ensenada highway

Table 6-5
Slopes recommended for cuts

TYPE OF MATERIAL	RECOMENDED SLOPE (HORIZONTAL DISTANCE: VERTICAL DISTANCE)				OBSERVATIONS
	Up to 5m (16ft)	From 5 to 10m (16 to 33ft)	From 10 to 15m (33 to 50ft)	Greater than 15m (50ft)	
Sound massive granite	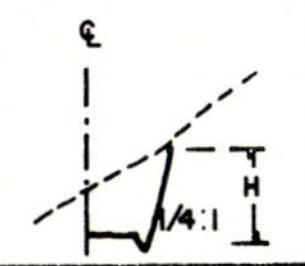 1/4:1	 1/4:1	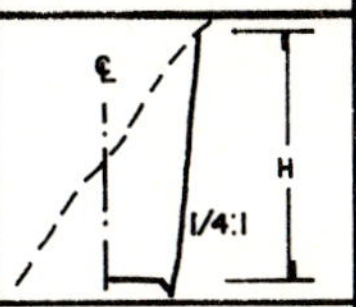 1/4:1	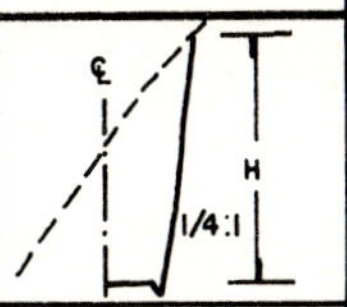 1/4:1	Remove the weathered part at the crest at 1/2:1 (if there is any).
Blocky sound fissured granite	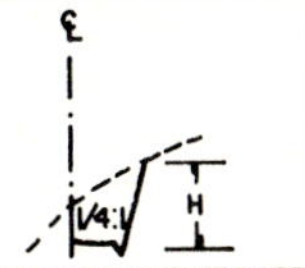 1/4:1	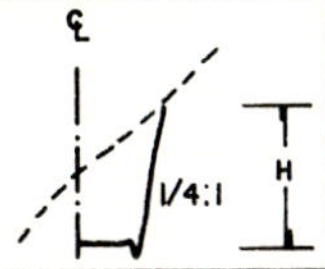 1/4:1	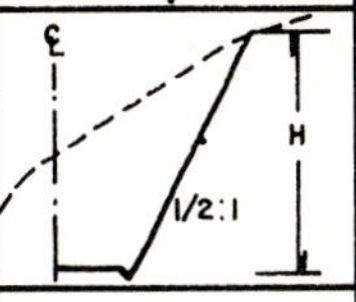 1/2:1	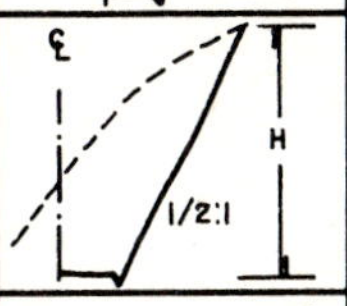 1/2:1	Remove loose blocks according to the layout of the fissures
Exfoliated granite; large blocks packed in sand	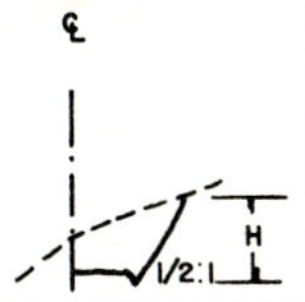 1/2:1	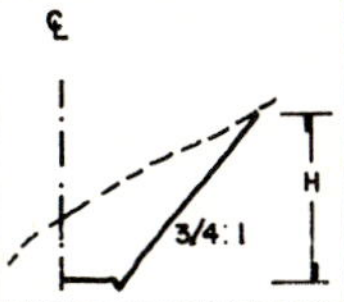 3/4:1	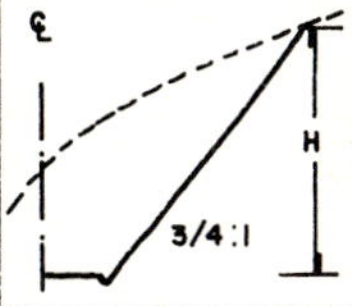 3/4:1	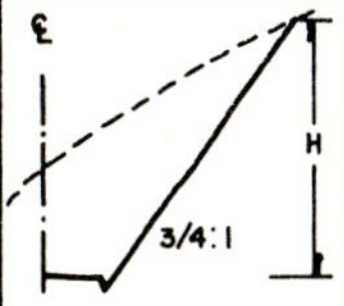 3/4:1	Construction of berm at the change of the slope not considered advisable if the weathered part at the crest is removed at 1/2:1.
Exfoliated granite; large blocks in a matrix of sandy clay	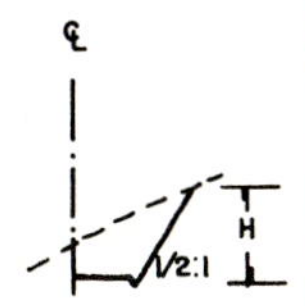 1/2:1	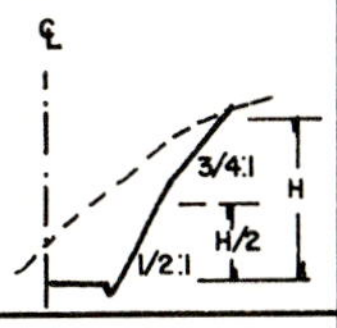 3/4:1; H/2; 1/2:1	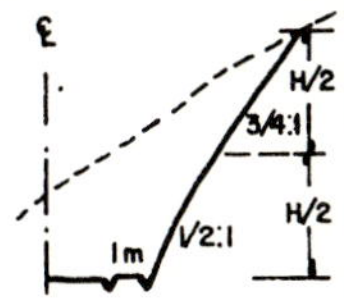 H/2; 3/4:1; H/2; 1m; 1/2:1	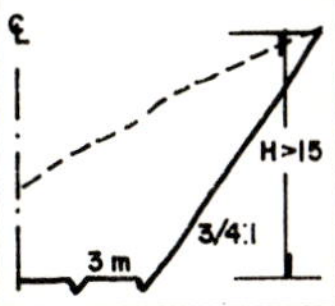 H>15; 3m; 3/4:1	Construction of toe berm recommended to catch the small surficial slides that usually occur.
Fully weathered granite	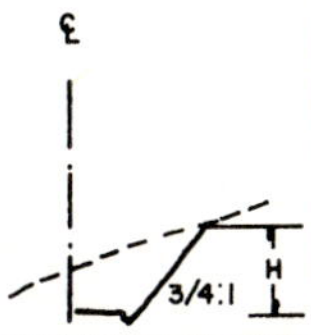 3/4:1	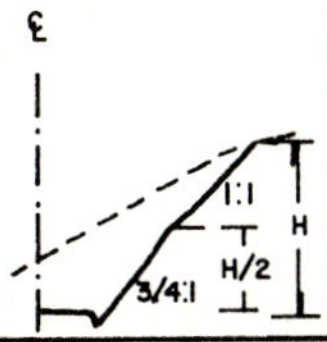 1:1; H/2; 3/4:1	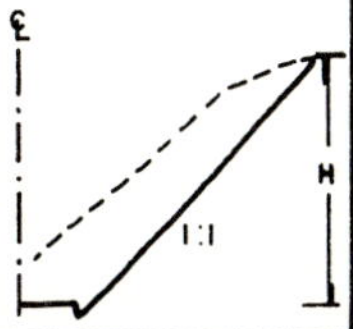 1:1	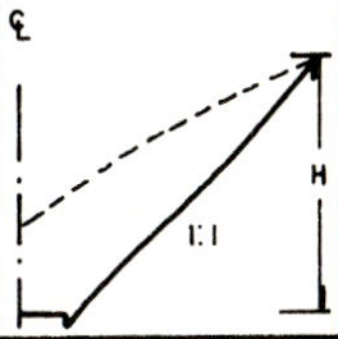 1:1	If the product of the weathered granite is fine silty or clayey sand, a 1m (3.3ft) foot berm is recommended for cuts up to 15m (50ft) and a 3m (10ft) foot berm for larger cuts.
Diorities	The same observations as for granites, depending on the degree of weathering of the rock.				
Unweathered fissured andesite	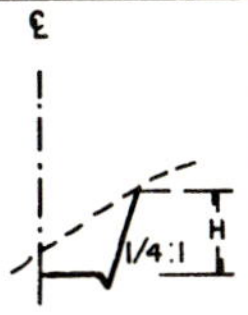 1/4:1	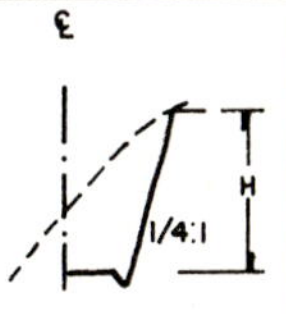 1/4:1	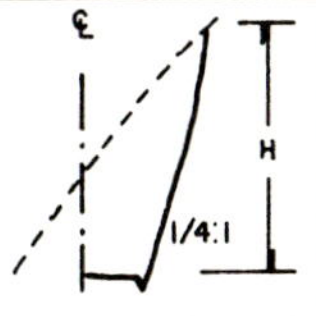 1/4:1	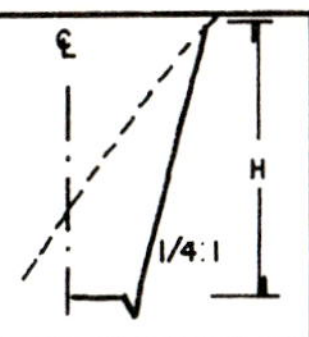 1/4:1	Loose block removal following the fissure planes is recommended
Fractured, slightly weathered andesite	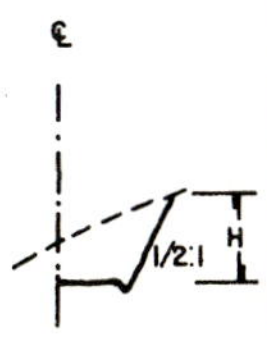 1/2:1	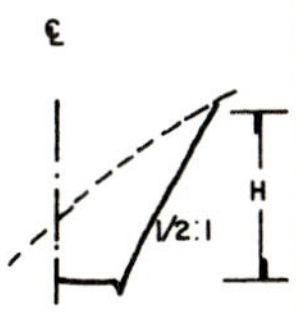 1/2:1	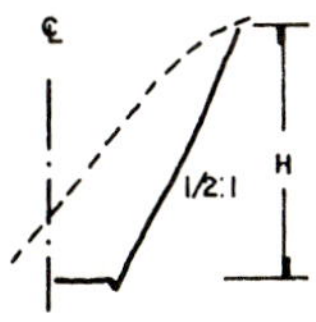 1/2:1	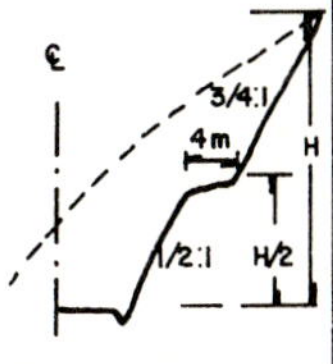 3/4:1; 4m; 1/2:1; H/2	A 4m (13ft) berm can be built at the slope, change if the lower part of the cut does not contain clay in the fractures and the fractures are closed.
Fractured, weathered andesite	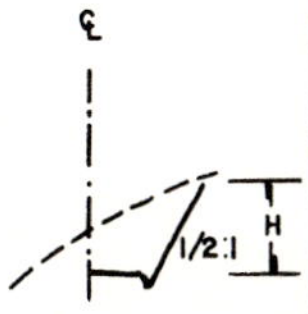 1/2:1	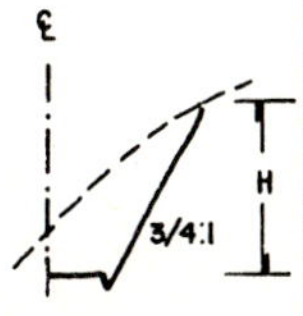 3/4:1	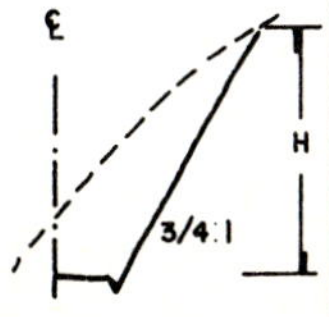 3/4:1	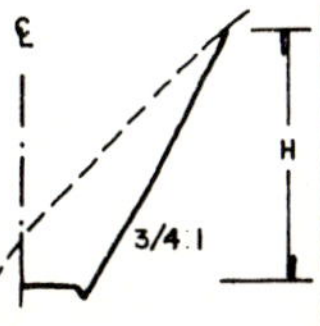 3/4:1	It is advisable to remove the crest at a slope of 1:1, (the most highly weathered part). If there is seepage, adequate subdrainage must be planned.

Table 6-5
(continued)

TYPE OF MATERIAL	RECOMENDED SLOPE (HORIZONTAL DISTANCE: VERTICAL DISTANCE)				OBSERVATIONS
	Up to 5m (16ft)	From 5 to 10m (16 to 33ft)	From 10 to 15m (33 to 50ft)	Greater than 15m (50ft)	
Sound or fractured rhyolites in large blocks with fracture systems at 90°, horizontally and vertically.	1/4:1 H	1/4:1 H	1/4:1 H	1/4:1 H	Loose block removal following the fracture planes is recommended; also removal of the weathered part at the crest at 1:1.
Sound, slightly fractured diabase	1/4:1 H	1/4:1 H	1/4:1 H	1/4:1 H	Loose block removal is recommended
Sound fractured basalt	1/4:1 H	1/4:1 H	1/4:1 H	1/4:1 H	Remove the crest of the cut at 1/2:1 if fracturing is very intense. If there is a weathered layer, remove at 1:1.
Fractured basalt in blocks of all sizes	3/4:1 H	3/4:1 H	3/4:1 H	1:1 H	If the fragments are loose and without soil, or packed in clay or soft silt with water flow.
Fractured basalt in blocks of all sizes	1/2:1 H	1/2:1 H	1:1 H/2; 1/2:1 H-2	1:1 H/2; 1/2:1 H/2	If the fragments are packed in firm clay with no seepage
Very fractured, highly weathered basalt	1/2:1 H	1/2:1 H	3/4:1 H	3/4:1 H	In very rainy areas, construction of a 1m (3.3ft) foot berm for cuts up to 15m (50ft) and 3.0m (10ft) foot berm for cuts larger than 15m (50ft) is recommended at the toe of the slope.
Basaltic flows interbeded with pyroclastic rocks	Basalt; 1:1 Pyroclastics	Definition of the contact between the basalt and the pyroclastic rocks is needed, so the coresponding slope can be given for each. Pyroclastic rocks require a 1:1 slope when loose or 3/4:1 when compact or for very coarse materials.			
Massive basalt Pyroclastic Rocks	1/2:1 H	3/4:1 H	3/4:1 H	1:1 H/2; 3/4:1 H/2	If the basalt pyroclastic rock is fine-grained and loose, the same recommendations apply as for the other pyroclastics.
Sound or slightly fissured tuffs and brecciated, andesitic, rhyolitic or basaltic tuffs.	1/8:1 H	1/4:1 H	1/4:1 H	1/4:1 H	If they are weathered in the upper portion of the cut, it is advisable to remove the crest at 1/2:1.
Sound or slightly fissured tuffs and brecciated, andesitic, rhyolitic or basaltic tuffs.	1/4:1 H	1/4:1 H	1/4:1 H	1/4:1 H	If there is a large water flow, construction of a 4m (13ft) waterproofed berm half way up is recommended.

Table 6-5
(continued)

TYPE OF MATERIAL	RECOMENDED SLOPE (HORIZONTAL DISTANCE: VERTICAL DISTANCE)				OBSERVATIONS
	Up to 5m (16ft)	From 5 to 10m (16 to 33ft)	From 10 to 15m (33 to 50ft)	Greater than 15m (50ft)	
Slightly weathered tuffs and brecciated, rhyolitic, andesitic or basaltic tuffs.	1/4:1 H	1/2:1 H	1/2:1 H	1/2:1 H	Removal of the upper part of the crest at 3/4:1 is recommended if there is intense fracturing or weathering.
Highly weathered tuffs and brecciated, rhyolitic, basaltic or andesitic tuffs.	1/2:1 H	1/2:1 H	3/4:1 H	H/2 1:1, 3/4:1 H/2	Change in slope half way up cuts deeper than 15m (50ft).
Hard, firm, slightly fractured clay-shale with almost horizontal dip.	1/4:1 H	1/4:1 H	1/4:1 H	1/2:1 H/2, 1/4:1 H/2	Do not excavate crest ditches if not thoroughly impermeable. Remove the topmost weathered portion of the crest at 3/4:1.
Soft, medium-strength, highly fractured clay-shale	1/2:1 H	3/4:1 H	3/4:1 H	1:1 H/2, 3/4:1 H/2	Do not excavate crest ditches if not thoroughly impermeable. Remove the most weathered part of the crest at 1:1.
Strongly cemented sound sandstones, poorly defined stratification, horizontal or dipping to the cut.	1/4:1 H	1/4:1 H	1/4:1 H	1/4:1 H>15	Remove the weathered portion of the crest at 3/4:1.
Poorly cemented, highly weathered sandstone, with seepage.	1/4:1 H	1/4:1 H	1/21 H	3/4:1 H/2, 1/2:1 H/2	Remove the weathered portion of the crest at 1:1
Well-cemented brecciated conglomerate with siliceous or calcareous matrix.	1/8:1 H	1/4:1 H	1/4:1 H	1/4:1 H	Removal of all loose fragments is recommended
Poorly cemented conglomerate with clayey matrix	1/2:1 H	3/4:1 H	H/2 1:1, 3/4:1 H/2	H/2 1:1, 3/4:1 H/2	If the clayey matrix is saturated or subjected to marked changes in moisture, construction of a 1m (3.3ft) foot berm is recommended for cuts deeper than 10m (33ft), with 4m (13ft) berms half way up.
Fractured limestone with thick or poorly defined stratification dipping toward the cut.	1/8:1 H	1/4:1 H	1/4:1 H	1/4:1 H	Removal of the weathered or very fractured upper portion of the crest at 1:1 is recommended.
Sound limestones with thin horizontal stratification dipping toward the cut.	1/4:1 H	1/2:1 H	1/2:1 H	3/4:1 H/2, 1/2:1 H/2	Remove to 1:1 the upper portion

Table 6-5
(continued)

TYPE OF MATERIAL	RECOMENDED SLOPE (HORIZONTAL DISTANCE: VERTICAL DISTANCE)				OBSERVATIONS
	Up to 5m (16ft)	From 5 to 10m (16 to 33ft)	From 10 to 15m (33 to 50ft)	Greater than 15m (50ft)	
Weathered limestone with seepage	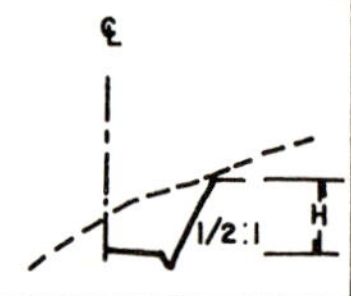 1/2:1 H	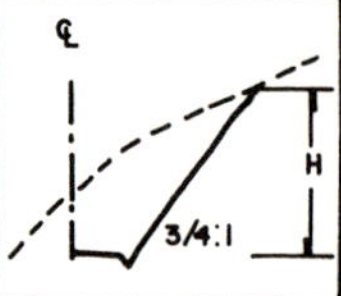 3/4:1 H	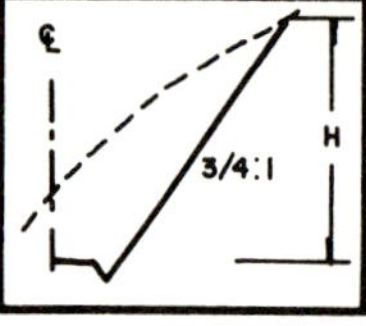 3/4:1 H	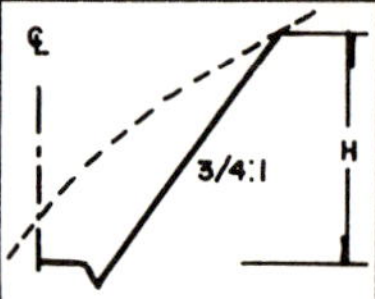 3/4:1 H	Plan for subdrainage and impermeable crest ditches
Unweathered limestone with dip between 90° and 45° to the outside of the cut, with clay between strata.	Give the slope corresponding to the dip. If the rock is highly fractured, design waterproofed 4m (13ft) berm half way up. Impermeable crest ditches.				
Very fractured weathered limestone	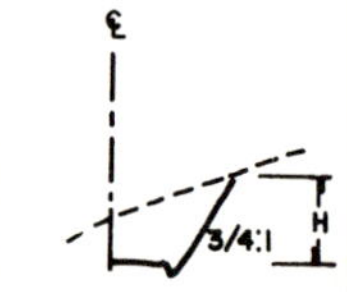 3/4:1 H	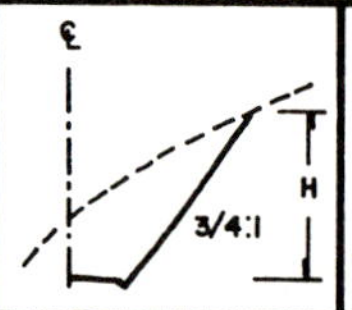 3/4:1 H	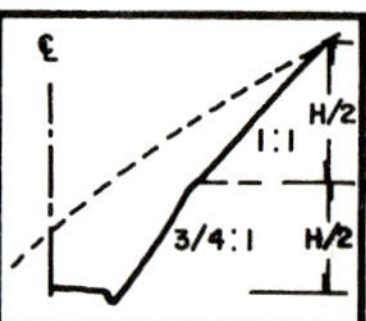 1:1 H/2; 3/4:1 H/2	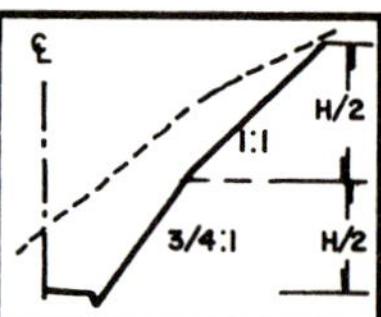 1:1 H/2; 3/4:1 H/2	Impermeable crest ditch
Slightly fractured unweathered limestone, with dip between 30° and 45° to the outside of the cut.	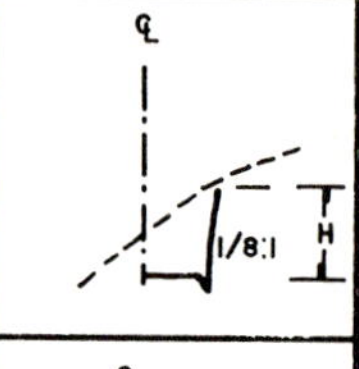 1/8:1 H	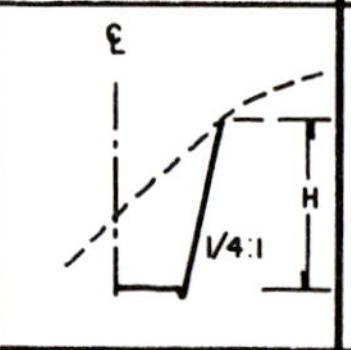 1/4:1 H	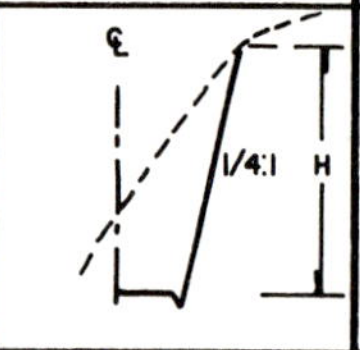 1/4:1 H	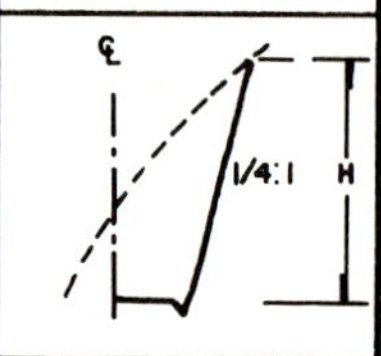 1/4:1 H	Can be regarded as though the dip were horizontal
Very slightly weathered and fractured limestone with dip between 45° and 30° to the outside of the cut.	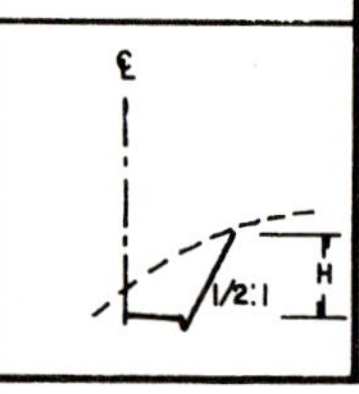 1/2:1 H	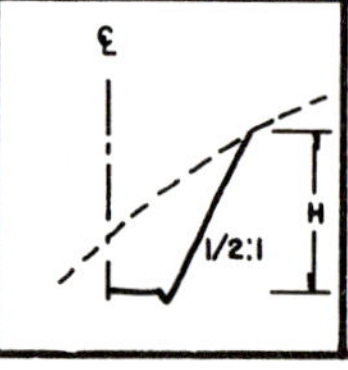 1/2:1 H	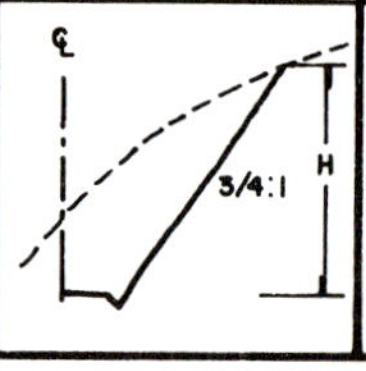 3/4:1 H	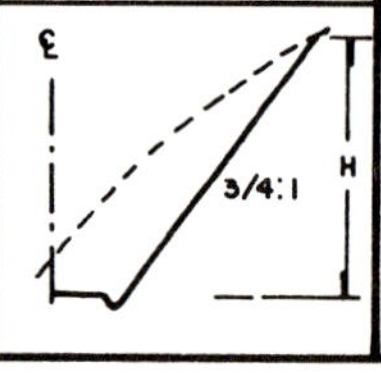 3/4:1 H	Remove the most fractured portion at 1:1. Waterproofed crest ditch.
Slates	Same recommendations as for limestones				
Moderately compact aglomerate with non-plastic fines	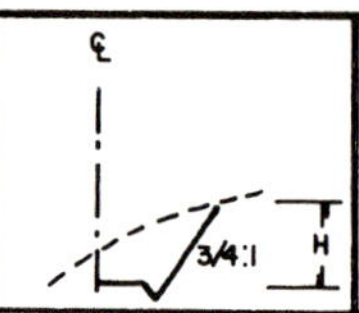 3/4:1 H	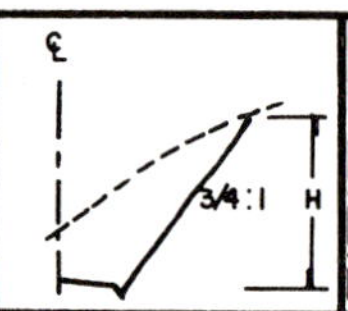 3/4:1 H	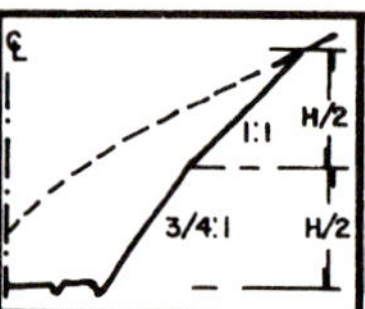 1:1 H/2; 3/4:1 H/2	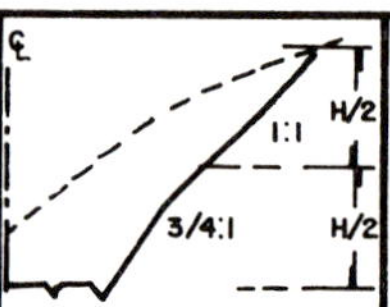 1:1 H/2; 3/4:1 H/2	Waterproofed crest ditch. For cuts deeper than 10 m (33 ft), construct 1m (3.3 ft) berm at toe of slope.
Moderately compact aglomerate with plastic fines	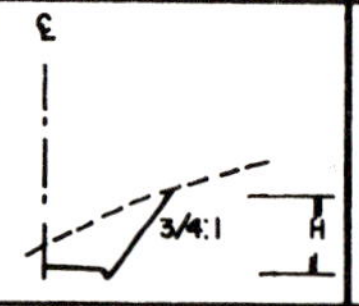 3/4:1 H	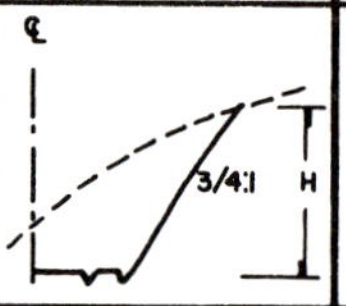 3/4:1 H	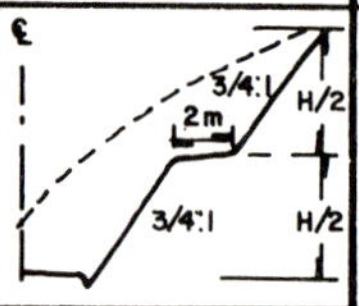 3/4:1 H/2; 2m; 3/4:1 H/2	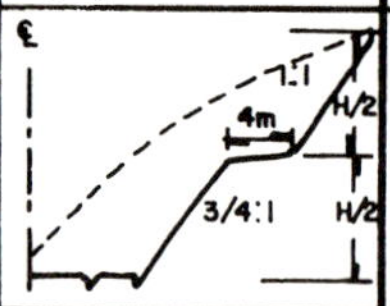 1:1 H/2; 4m; 3/4:1 H/2	Waterproofed crest ditches. For cuts deeper than 10m (33ft), design a 2m (6.6ft) berm half way up and for cut deeper than 15m (50ft) increase the width to 4m (13ft).
Silty sands and compact silts	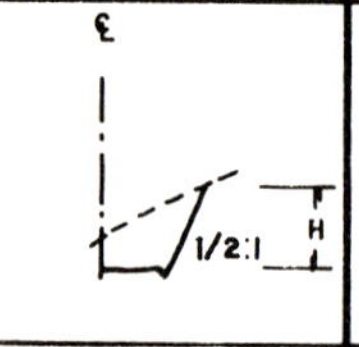 1/2:1 H	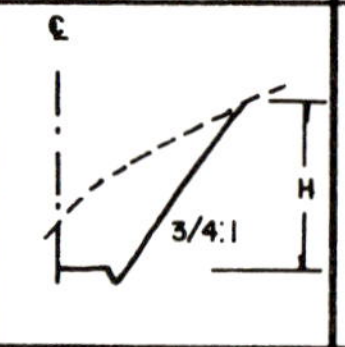 3/4:1 H	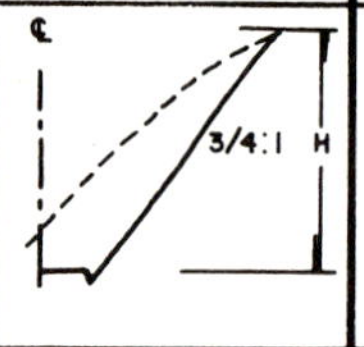 3/4:1 H	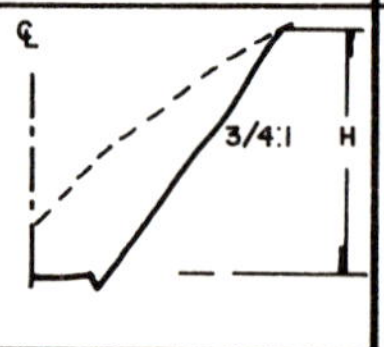 3/4:1 H	Remove the upper more weathered portion at 1:1. If the materials are susceptible to erosion, a slope of 1:1 should be designed and protected with grass.
Silty sands and not very compact silts	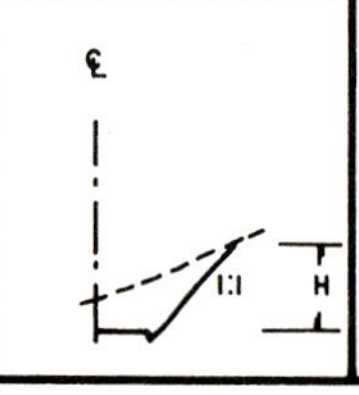 1:1 H	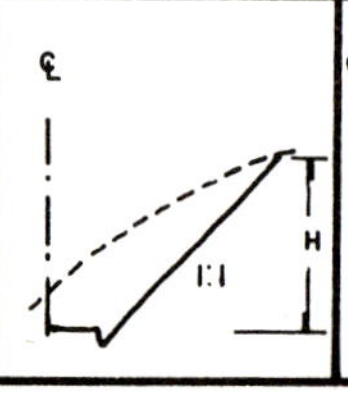 1:1 H	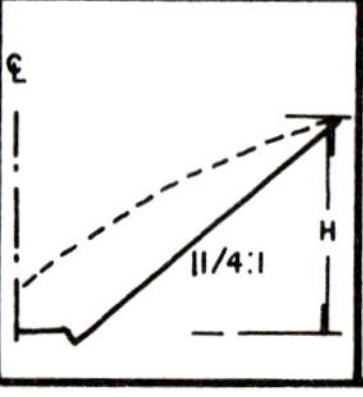 11/4:1 H	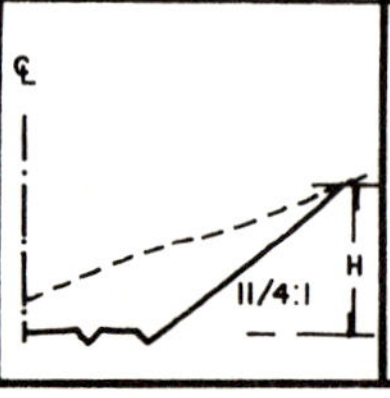 11/4:1 H	Impermeable crest ditch. Remove the most weathered part at 1.5:1. For cuts greater than 15m (50ft), design a 3m (10ft) berm at the toe of the slope.

Table 6-5
(continued)

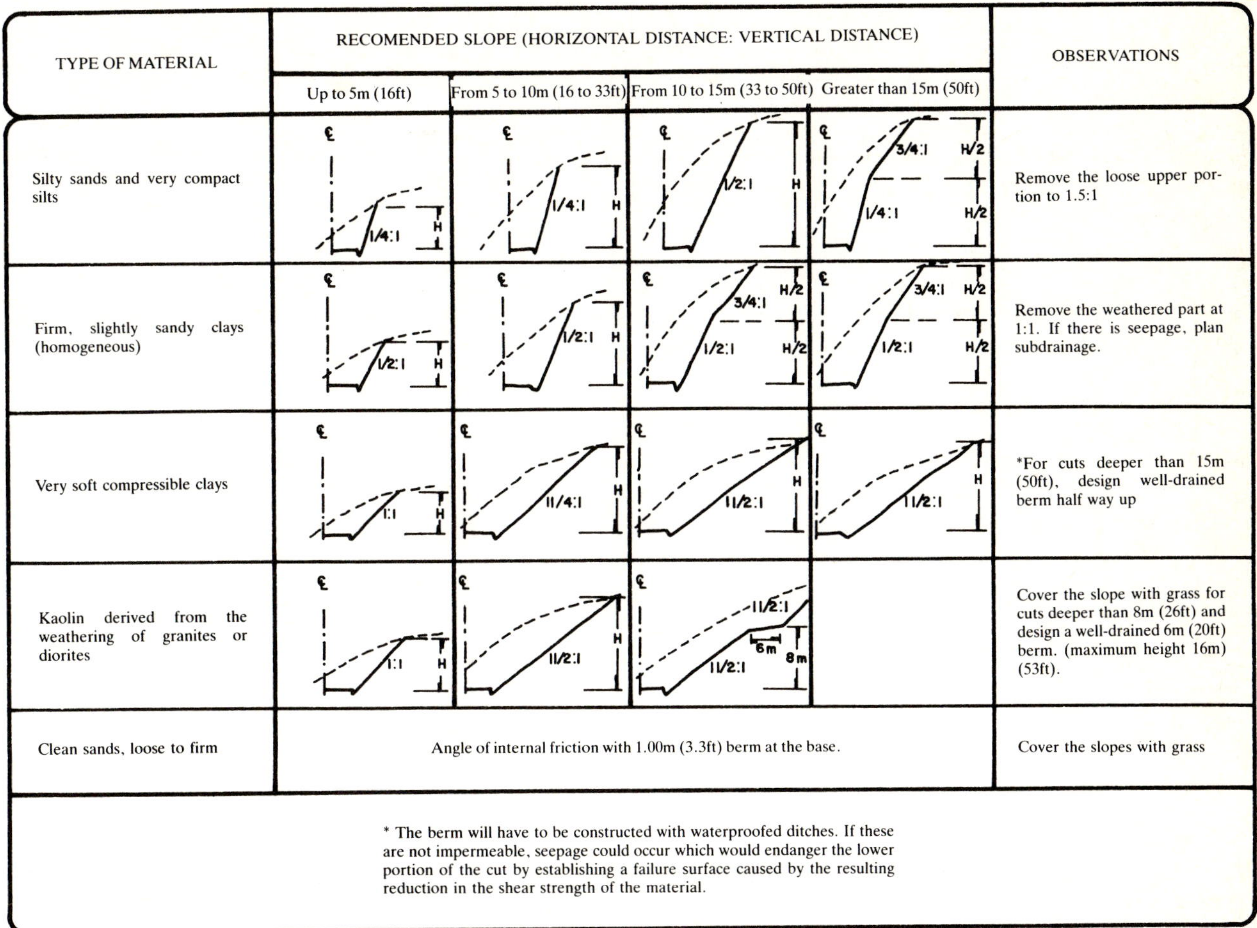

TYPE OF MATERIAL	RECOMENDED SLOPE (HORIZONTAL DISTANCE: VERTICAL DISTANCE)				OBSERVATIONS
	Up to 5m (16ft)	From 5 to 10m (16 to 33ft)	From 10 to 15m (33 to 50ft)	Greater than 15m (50ft)	
Silty sands and very compact silts	1/4:1, H	1/4:1, H	1/2:1, H	3/4:1, H/2; 1/4:1, H/2	Remove the loose upper portion to 1.5:1
Firm, slightly sandy clays (homogeneous)	1/2:1, H	1/2:1, H	3/4:1, H/2; 1/2:1, H/2	3/4:1, H/2; 1/2:1, H/2	Remove the weathered part at 1:1. If there is seepage, plan subdrainage.
Very soft compressible clays	1:1, H	1 1/4:1, H	1 1/2:1, H	1 1/2:1, H	*For cuts deeper than 15m (50ft), design well-drained berm half way up
Kaolin derived from the weathering of granites or diorites	1:1, H	1 1/2:1, H	1 1/2:1; 6m; 8m; 1 1/2:1		Cover the slope with grass for cuts deeper than 8m (26ft) and design a well-drained 6m (20ft) berm. (maximum height 16m) (53ft).
Clean sands, loose to firm	Angle of internal friction with 1.00m (3.3ft) berm at the base.				Cover the slopes with grass

* The berm will have to be constructed with waterproofed ditches. If these are not impermeable, seepage could occur which would endanger the lower portion of the cut by establishing a failure surface caused by the resulting reduction in the shear strength of the material.

Table 6-6, [2], is a comprehensive summary of the factors causing slides and their mechanisms.

Work done by the engineer and the constructors can often be the cause of serious slope stability problems. The following [8] is a list of the construction processes that most often cause instability problems:

1. Modification of the natural conditions of seepage due to fills, ditches or excavations
2. Overloading of weak strata due to fill, and sometimes waste
3. Overloading of soils with weak stratification planes due to fill
4. Removal, by cutting, of a thin stratum of permeable material which acts as a natural draining blanket of the soft clay
5. Detrimental increase in seepage pressures or orientation of seepage forces when changes occur in the direction of seepage, as a result of cuts or fills
6. Exposure of hard fissured clays to air and water, due to cuts
7. Removal of surface layers of soil due to stripping or excavation, which may cause layers of the same stratum further uphill to slide over the underlying layers of harder soil or rock
8. Increase in hydrostatic loads or hydraulic heads below the surface of a cut when its bed is covered with an impermeable layer

Generally speaking, the causes of slides can be external or internal. External causes bring about an increase in the acting shear stresses, without altering the shear strength of the material. Causes of this type are an increase in the height or steepness of the slope, any structural load or embankment that is placed on the crest of the slope and earthquakes. Internal causes are those which occur without any change in the external conditions of the slope. They are always associated with a loss of shear strength of the soil. An increase in pore pressure or dissipation of cohesion by weathering are causes of this type.

Table 6-7 [8] gives the factors that most commonly lead to an increase in acting shear stresses in a natural or artificial slope. Table 6-8 [8] gives the factors that most often cause a reduction in the shear strength of the materials of natural and artificial slopes.

Table 6-6
Factors causing slides [2]

AGENT	ACTIVATING PROCESS	WAY IN WHICH AGENT ACTS	MOST SUSCEPTIBLE MATERIALS	PHYSICAL NATURE OF ACTION	EFFECTS ON STABILITY
Erosion and transport	Construction processes or erosion	Increases height or steepness of slope	All Materials	Changes in state of stress	Increase in shear stresses
			Stiff or fissured clays, clay-shales	Changes in state of stress and opening of fissures	Increase in shear stresses. Process § is triggered
Tectonic forces	Tectonic movement	Large deformations in the earth's crust	All materials	Increase in angle of slope	Increase in shear stresses
Tectonic forces or the use of explosives	Earthquakes or blasting	High frequency vibrations	All materials	Transient loading	Increase in shear stresses
			Loess, slightly cemented sands and gravels	Alteration of interparticle bonds	Reduction in cohesion and increase in shear stresses
			Fine or medium grained sand, loose and saturated	Rearrangement of particles	Liquefaction
Weight of the slope material	Construction of the slope	Surface slide	Hard or fissured clay, clay-shale, remains of old slides	Opening of closed fissures and creation of new fissures	Reduction in cohesion. Process § is accelerated
		Slide in weak strata at toe of slope	Hard materials on soft strata		
Water	Rain or thaw	Removal of air from the voids	Moist sand	Increase in pore water pressure	Drop in strength
		Removal of air from open joints	Jointed rock, clay-shales		
		§) Reduction in capillary tension associated with expansion	Hard and fissured clays, some clay-shales	Expansion	Reduction in cohesion
		Chemical decay	Any rock	Weakening of interparticle bonds	
	Freezing of the ground	Expansion of water by freezing	Jointed rock	Opening of closed fissures and creation of new fissures	Reduction in cohesion
		Formation of ice lenses in the soil	Silts and sandy silts	Increase in water content of the frozen soil	Reduction in frictional strength
	Period of drought	Shrinkage	Clay	Cracking by shrinkage	Reduction in cohesion
	Drawdown	Flow towards the toe of the slope	Silts and fine sands	Increase in pore water pressure	Reduction in frictional strength
	Fluctuations in the phreatic level	Rearrangement of particles	Medium to fine grained sands, loose, saturated	Increase in pore water pressure	Liquefaction
	Rise in the phreatic level of a distant aquifer	Rise in the hydraulic head of the slope material	Strata of sand or silt between or below strata of clay	Increase in pore water pressure	Reduction in frictional strength
	Internal water flow or seepage	Seepage toward the slope	Saturated silt	Increase in pore water pressure	Reduction in frictional strength
		Removal of air from the voids	Moist fine sand	Dissipation of surface tension	Reduction in cohesion
		Removal of soluble cementing agents	Loess	Weakening of the interparticle bonds	
		Internal erosion	Silt or fine sand	Piping	Increase in shear stresses

Table 6-7

Factors that most commonly cause an increase in the shear stresses in a slope

1. Removal of support, which includes:
   - 1-a Erosion
     - 1-a.1 By streams and rivers
     - 1-a.2 By glaciers
     - 1-a.3 By action of waves or marine currents
     - 1-a.4 By successive wetting and drying (winds, freezing, etc.)
   - 1-b Modification of the initial slope by falls, slides, settlements or other causes.
   - 1-c Human activity
     - 1-c.1 Cuts and excavations
     - 1-c.2 Removal of retaining walls or sheet piles
     - 1-c.3 Drawdown of lakes, lagoons or bodies of water
2. Overloading
   - 2-a Natural causes
     - 2-a.1 Weight of rains, snow, etc.
     - 2-a.2 Accumulation of materials due to falls, slides or other causes
   - 2-b By human activity
     - 2-b.1 Construction of fills
     - 2-b.2 Buildings and other overloads at the crest
     - 2-b.3 Possible leakage of water in pipes and sewers
3. Transitory effects, such as earthquakes
4. Removal of underlying materials that provided support
   - 4-a By rivers or sea
   - 4-b By weathering
   - 4-c By underground erosion due to seepage (piping) solvent agents, etc.
   - 4-d By human activity. Excavation or mining
   - 4-e By loss of strength of the underlying material
5. Increase in lateral pressure
   - 5-a By water in cracks and fissures
   - 5-b By freezing of the water in the cracks
   - 5-c By expansion of clays

Table 6-8

Factors that most commonly cause a reduction in the shear strength of natural and artificial slopes [8].

1. Factors inherent in the nature of the materials
   - 1-a Composition
   - 1-b Structure
   - 1-c Secondary or inherited structures
   - 1-d Stratification
2. Changes due to weathering and physico-chemical activity
   - 2-a Wetting and drying processes
   - 2-b Hydration
   - 2-c Removal of cementing agents
3. Effect of pore pressures, including those due to seepage
4. Changes in structure, including fissuration due to stress release and structural degradation under the acting shear stresses

6.9 Identification of Slope Stability Problems in the Field

Field recognition and identification of possible failures in natural and artificial slopes, that provide the engineer in charge with useful classifications or standards have traditionally been based on the interpretation of local signs in accordance with past experience. The most common signs are deformations, and irregular topography, cracks and indications of seepage. Such a system leads to a qualitative and subjective interpretation of the stability of a slope and its risk of future failure. In the following pages, some practical ideas will be given to help put this method for assessing risk in the right perspective.

It would be best if some safe and reliable theoretical method could be established for assessing if a specific hillside or slope is in a critical condition and evaluating the risk of a catastrophic landslide. Many inexperienced engineers believe that if a slope is *analyzed* and a safety factor computed (especially by a computer) a quantitative, reliable assessment of its condition should be available immediately. Apart from the fact that in road engineering many slopes *cannot* be analyzed, confidence in computed safety factors is hardly acceptable in the light of the doubts that have already been expressed in relation to analysis methods in general. The safety factor must be verified or gauged in accordance with data provided by already constructed slopes. Means of achieving these data are largely missing today. Encouraging attempts have been made, some of which will be mentioned here.

The objective is to find a theoretical relation between the condition of the slope and laboratory test results that can be easily obtained and interpreted. The condition of the slope is expressed in terms of its behavior. This is obtained from continuous, detailed sets of field measurements. These are the result of an appropriate and careful program of field instrumentation. This aspect will be discussed later in a special Chapter.

References [81] and [82] relate the stability of a slope to the uniaxial and triaxial test parameters, measuring the deformation velocity of the soils under loads smaller than those corresponding to failure. In this way, the deformation velocity under different test conditions can be determined. In these references, a method is established that permits assessment of the risk of failure and even the moment when this failure will occur, as a function of the deformation velocity and the time necessary for the laboratory specimens to fail. Although these ideas may at the present time appear esoteric to engineers who are used to designing and constructing roads, they provide very valuable directions for research that may prove very useful in the future.

Slope failure is not necessarily associated with a catastrophic landslide, for an excessive deformation may lead to a functional failure. There are no conventionally accepted or commonly used methods for assessing the magnitude of the deformations that may occur in an embankment built with a compacted material. An interesting method for this purpose appears in [83].

The stability of a slope generally depends on features of the materials, such as mineralogy, grain size and shape, structure, stratigraphy, weathering, and external circumstances that are not governed by the slope itself, including the environment and adjoining topography, climate, and vegetation. Surface flow

Table 6-9
Aids to recognition of active or recently active slides
(Consult terminology in Fig. 6-8)

TYPE OF MOVEMENT	TYPE OF MATERIAL	STABLE ZONES AROUND THE SLIDE			PARTS THAT HAVE MOVED			
		CREST OR HEAD OF FAILED ZONE	PRINCIPAL SLOPE (behind failed zone)	FLANKS	HEAD	BODY	BASE	TOE
Falls and collapses								
Rock falls	Rock	Loose rock, probably cracks behind the failure line, irregular appearance characterized by joints	Usually almost vertical, irregular, smooth, appearance of fresh rock. Jointed rock	Generally clean rock edges	Generally not clearly defined. Fallen material forms loose pile of rocks near the escarpment	Irregular surface with rock fragments. If big with trees or contrasting materials may indicate-direction of radial movement from escarpment. May contain depressions	Base is usually buried if visible, generally shows causes of failure such as weak underlying rock or strata undermined by water	If slide is small, has an irregular slope of debris. If rock fall is large the toe may be rounded
Soil falls (Collapse)	Soils	Cracks behind the failure line	Almost vertical. Moist soil. Very cracked on the surface	Often almost vertical	Usually unclear. The fallen material forms a pile of rocks near escarpment	Irregular	As Above	Irregular
Slides								
Circular	Soils	Numerous cracks, the majority concave towards the slide	Inclined, clean, concave towards slide, often high. May have striae and grooves on surface, from crest to head. Upper portion of slope behind failure may be vertical	Striae in flanks of the escarpment have large vertical components near the head and notable horizontal components near base. Height of flanks reduced towards base. Flanks can be higher than original surfaces of ground between toe and base. In initial stages slides surrounded by terraced cracks	Upper part of failure conserves parts of the natural ground before failure. Pools form at toe of principal slope. Entire head of failure furrowed by cracks. Trees in fallen zone point uphill	The moving part of the soil breaks up and distintegrates. Longitudinal cracks, heave. Generally pools develop just above the base	Transversal heave and cracks normally on the base. Upheaval zone, absence of large individual blocks. The trees lean downhill	Often a lobular earth flow zone, material that has been rolled over and buried. The trees are flattened or at different angles among the material at the toe
	Rocks	Cracks tend to follow the fractures in the original rock	As Above	As Above	As Above	As above but material not so broken nor deformed	As Above	Little earth flow. Toe often almost straight and near base. May have a sharp face.
Translational	Rock or soil	Most cracks almost vertical, tend to follow outline of slope	Almost vertical in the upper portion, in lower almost flat with gradual transition	Lateral flanks very low vertical cracks. Downhill the cracks generally diverge	Relatively intact. No rotation	Generally intact units except for tension cracks. Little or no vertical displacement in cracks	No base and no upheaval zone	Sliding over the surface of the ground
Rock slide	Rock	Loose rock, cracks between the blocks	Generally terraced after joints or strata. Irregular surface in upper portion, sloping in lower. Can be flat or composed of rock flows	Irregular	Many blocks of rock	Rough surface with many blocks which may be in original position but lower if movement was slow translational translational	Generally there is no real base	Accumulation of rock fragments
Flow of dry material								
Flow of rock fragments	Rock	Same as in rock falls	Same as in rock falls	Same as in rock falls	No head	Irregular surface of mixed rock fragments spreading fan-shaped downhill. Valleys, lobular transversal hills	No base	Composed of tongues. May slide following natural channels
Sand flow	Soil	No cracks	Funnel-shaped when reaches angle of repose	Form continuous curve starting at crest	Generally no head	Cone-shaped mound of sand, same volume as portion emptied from head	No base	No toe or a wide, hardly perceptible fan
Flow of moist material								
Mud flows	Soil	Few cracks	Upper part notched or V-shaped. Long and narrow, smooth, often striated	Sloping, upper portion irregular. Heaps of material in the lower portion of the flanks	May be no head	Moist, very moist. May contain large blocks in fine matrix. Flow lines. Follows drainage lines, may have sharp bends. Long and narrow	No base or buried in debris	Extends sideways in lobes. When toe dries, may have a low step at the front
Earth flow	Soil	May be some cracks	Concave to slide, sometimes almost circular, slide occurs through a narrowed portion	Curves, steep sides	Usually consists of a sunken block	Broken into numerous small pieces. Moist, shows flow structure	No base	Extends in lobes
Sand or silt flow	Soil	Few cracks	Sloping, concave to slide, variety of shape (almost straight, arc, bottle-shaped)	Flanks often converge in the direction of movement	Generally under water	The body extends like a fluid	No base	Extends in lobes

and subsurface seepage are of vital importance. The overall influence of so many factors leads to such a vast number of variables that the problem becomes one of the most complex ones that face road engineers. Photointerpretation, which cannot be described in detail in this book, should be regarded as an invaluable help in identifying problems and interpreting field data.

The problem of recognizing and identifying slides has two important aspects [8]. The first concerns the identification of an actual slide to estimate movements that have occurred or may occur in the future. The second aspect, which is just as important, refers to the identification and classification of the type of slide that is actually occurring or may at some time occur.

In the case of a newly built road, once a general idea of the stability of an area has been established by means of geological maps and photointerpretation, it will be necessary to visit the area and inspect the site. This inspection should always start with a broad view and then move on to the more specific details. Special attention must be given to the different slopes of the hillsides, relating them to variations in the materials indicated by surface geology. Of particular interest will be specific signs, such as springs, pools, cracks, and irregular, hummocky topography. Such an estimate is vital in the initial stages of the project.

In many cases, however, it will be very hard to detect the existence of slides and failures or predict future ones, and the engineer can but take extreme precautions in those places where the stratigraphy arouses suspicion. Some such stratigraphies are:

1. Any kind of rock formation or hard soil overlying very fractured rocks, shales, soft soils or materials highly susceptible to weathering
2. Hillsides made of soft clay or fissured shales, especially if old failures are detected in other parts of the hillside
3. Hillside deposits resting against or on beds of firm rock
4. Hillsides at whose toe erosion may occur, such as by the sea or river banks
5. Rock or residual soil formations with a dip or inherited structures that will prove detrimental to the excavation undertaken for the road

When an existing slide is detected, it is important to classify it, for the correction methods to be used will depend on the type of slide. Field instrumentation is almost the only effective, reliable means of achieving full knowledge of the slide. Its use has increased considerably over the last few years, and will continue to increase in the future. Table 6-9 [8] lists the most usual external signs of the different types of failures for purposes of recognition and classification.

The ability to spot small cracks and interpret their significance is one of the greatest gifts that can be possessed by any engineer who constantly has to handle this type of problem. It is a gift that has to be cultivated, gauged and developed with care. It can be a most useful guide to discovering the kinematic mechanism of a failure where detailed instrumentation is not available and it will always be an invaluable help in designing the instrumentation program.

The direction of the cracks is often perpendicular to the movements, but this is not a hard-and-fast rule. For example, the cracks in the sides of a failure can be practically parallel to the movement. In rotational slides, the cracks are usually curved, limiting the failure. Cracks showing relative movement between their sides are sometimes the first signs of instability (Plate 6-25). A complete survey of such cracks usually provides a magnificent description of the failure to come. Fig. 6-47 [8] shows a typical example of the cracking that accompanies a landslide.

In a translational slide, cracks are commonly slightly curved, with similar openings at the head or at the toe of the failure.

Plate 6-25 Crack at the crest of a cut

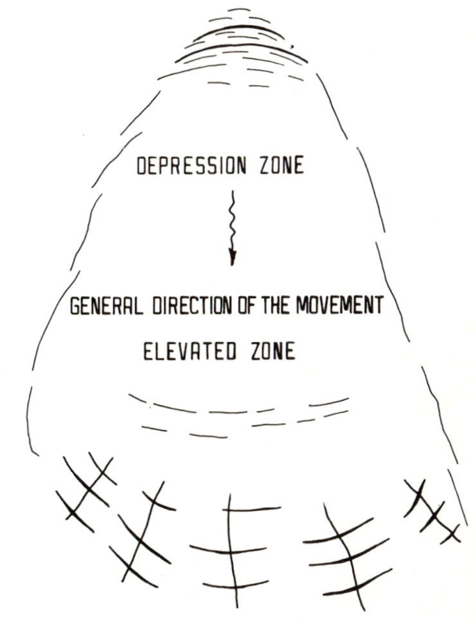

Fig. 6-47 Typical cracking in a landslide [8]

6.10 Slide Prevention

The best way to avoid problems from landslides is to avoid their location by changing the project alignment. This is a golden rule which should always be kept in mind by the road designer and no alternative criterion is more reliable or economical. Although risks sometimes have to be confronted because it is impossible to locate the project anywhere else this does not make this rule any less valid.

The road designer usually has considerable freedom of choice as to avoiding locations where future failure is predictable. The characteristics of these locations should be described, if only briefly, since they are one of the factors that will help to define the final project alignment.

First, failures will be more easily prevented by a more accurate design (by computation or experience). This greatly depends on exploration (not only geological exploration but also boring, sampling and testing that is commonly associated with soil mechanics). Today, the routine exploration for roads is usually minimal and does not allow for very sophisticated designs. When danger is suspected in a location, exploration should be more thorough. This is done already in extreme cases, such as swampy areas or where the soil is very soft, but the transition from the routine exploration to the specialized is usually abrupt; it is either cursory or very extensive and detailed. Serious consideration should be given to establishing intermediate degrees of exploration for different places and formations, to be adopted depending on the magnitude of the risk involved.

Many of the factors governing failure are very hard to detect with conventional exploration methods. This is true of features such as cracks, fissures, small discontinuities, pre-existing slide surfaces, and small flows. Indeed, many of these factors can be more easily determined using field instrumentation techniques, which should be regarded as invaluable tools for detecting and even preventing failures.

Failure prevention in road engineering is chiefly a matter of making changes in the design of the road. Even if the horizontal alignment is not modified, prevention can be achieved through efficient teamwork on the part of surveyors and geotechnical engineers by adjusting the grades of the road. This is easier in roads than in railways. Today the same road design rules apply for a wide variety of types of terrain, considering problems of grades and curvature and earthworks only from the point of view of achieving an adequate material balance and (though often rather illusorily) low transportation costs. However, there are places where any embankment would be problematical and others where any cut will be unsafe. Coordination of careful surveying and geotechnical considerations is therefore desirable in all cases. Plates 6-26 and 6-27 show unfavourable geological formations mentioned earlier.

Some important failures can be prevented by simply bearing in mind basic geotechnical considerations. In cuts, for example, an attempt should always be made to maintain at least the toe of the slope as *loaded* as possible. Unloading usually causes a drop in the shear resistance of soils and rocks and releases the residual horizontal stresses, which tends to cause expansion of the slope.

The smaller the volume of soil excavated for a cut and the steeper the slope, the smaller will be the amount of water received by the cut surface when it rains. In large cuts, this water may in itself be a good reason to require a berm, with ditches on the uphill side, so that any water collected can be quickly

Plate 6-26 An example of a poor geological formation, limestone formation on the road from Tula to Ciudad Victoria

Plate 6-27 Another example of a poor geological formation, slates on the road from Iguala to Ciudad Altamirano

eliminated. Often a simple, steep slope, will prove successful (if the soil is strong) where a very complicated one with berms might fail. Cuts in loess, where the rain dissolves the natural cementing agents, are an example.

The construction of embankments with steep or flat slopes has already been discussed. In the case of steep slopes, there is a concentration of stresses at the toe. In very flat slopes, settlement is encouraged as a consequence of the increase in the area of contact between the embankment and the soil foundation.

In construction procedures, there are many other possibilities for increasing or reducing the risk of failure. The problems caused by the unskillful use of explosives when making cuts in rocks will not be discussed here, though many failures in rock can be blamed on this factor.

There are several construction rules that help maintain stability. For example, a cut should preferably be excavated in the uphill direction, so that rain or spring water can be easily drained. Good results are also obtained by excavating the cut stratum by stratum, throughout almost the entire length; in this way, the water-table can be evenly lowered, avoiding the development of large localized zones of water pressure in the undrained slopes.

The omission of steps (Chapter 3) to support embankments on sloping hillsides has frequently led to problems, not only caused by catastrophical landslides, but also by demands of excessive maintenance for slowly moving structures.

It is usual to excavate cuts with slopes that are initially steeper than the design ones, trimming them after mass excavation. This practice should always be considered inappropriate, for the stability of the cut is reduced during the interim. This not only encourages failures, but if soils are subjected to excessive stress, especially near the crest of the slope, cracks and fissures will open which can lead to a degradation of the slope which is very detrimental to its future performance.

6.11 Mechanical Methods for Correcting Failures in Slopes

Here the chief methods available to the engineer for correcting stability problems in hillsides or artificial slopes or for reconstructing failed zones will be briefly described. Methods involving drainage and subdrainage techniques will not be dealt with under this heading; they will be the object of a later special Chapter. Many of the corrections in failed zones will nevertheless require drainage for, as has been constantly reiterated, the action of surface and underground water has profound influence on the stability of earth masses. For classification purposes, however, the remedial methods that are based on drainage and subdrainage will be mentioned in the corresponding charts. For want of a better term, these remedial methods are referred to as *mechanical*.

All remedial methods take one or more of the following lines of action.

1. Avoid the failure zone
2. Reduce the driving forces
3. Increase the resisting forces

Avoiding the failure zone is usually associated with changes in either the horizontal or vertical alignment of the road, totally removing the unstable materials or building structures such as bridges or viaducts founded on stable ground.

Reducing the driving forces can generally be achieved by two methods: removing material from the appropriate part of the moving mass, and subdrainage which reduces the hydrostatic thrusts and the weight of the earth masses by removing water.

The most varied line of action is usually that which increases the resisting forces. Some of its variants are: subdrainage, which increases the shear strength of the soil; eliminating weak strata or other potential failure zones; building retaining structures or other supports, and treatment, generally chemical, to increase the shearing resistance of the soils.

Table 6-10 [8] gives a summary of the principal methods of failure correction. The boundary between methods of correction and prevention is rather flexible. The title includes both concepts and the table can therefore, be considered complementary to §6.11.

A fairly detailed description follows of the principal mechanical methods of natural slope correction.

6.11.1 Methods That Avoid Unstable Zones

These are the safest methods for preventing problems caused by slides and failures, but they cannot always be applied. At other times, they can be used only partially, in the sense that an unstable zone cannot be avoided altogether, but often by a slight change in alignment the road can avoid the worst part; sometimes a long stretch of road can be changed to bypass the zone.

One of the conditions that best responds to these methods is road building on sloping formations of soil or rock, with an unsuitable dip. Small changes in the horizontal alignment can make the zones far less dangerous, or even harmless. The problems also can be greatly reduced by raising the grade of the road. If one side of a valley has a poor dip, the other side will very probably have a favorable dip.

Where it is not possible to avoid a potential slip zone or where a slide has already occurred, the problem can be sometimes be avoided by constructing a viaduct founded on the firm soil on both sides of the problem area. The cost of this solution is usually very high and, if it is adopted, a structural design will have to be developed that will accommodate moderate movements, for total immobility cannot be guaranteed. The building of such bridge structures is often complemented by totally removing the failed material so as to protect the structure's foundations from the risks of a sudden sliding of the earth mass and possible thrusts against its supports.

In many cases of road construction on sloping hillsides with difficult stability conditions, the best rule is to alter the natural formations as little as possible. This idea favors the *half viaduct* solution, where the supports of the inner side of road structure are fixed to the hillside, leaving the outer side overhanging, with support columns founded in stable formations. Conditions for the success of this solution are, first, the hillside is at present stable, second, that the columns be well supported, and third, that the hillside be altered only very slightly when placing the supports corresponding to the inner side of the road.

Methods of avoiding failures do not help stabilize them. This, plus generally high cost, are the principal limitations. The influence of cost is usually decisive in small failures. Construction cost becomes considerably less important when the unstable zone is very extensive, for the costs of any remedial method tend to increase and are usually many times greater than those of the original construction under such circumstances.

6.11.2 Excavation Methods

A wide range of these methods are mentioned in Table 6-10, from minor excavations made only at the head of the failure to the complete removal of unstable material. Slope flattening and the use of berms are methods requiring excavation in the case of cut, and fills in the case of embankments. Both methods will be discussed separately under this heading.

Removal of material at the head of the failure (or throughout the entire body up to total removal), is a method which in practice is applied only to existing failures. Very seldom is there sufficiently detailed knowledge of future failures in a potentially unstable zone for it to be wise to proceed to large-scale removal of materials. Removal work at the head reduces the driving forces and tends to balance the failure. Total removal eliminates the cause radically, although under these circumstances a problem of instability in the slopes of the new excavation may arise. These slopes should always be carefully studied, together with the new drainage conditions. Material removal usually leads to fairly permanent solutions, so long as the drainage of the excavation receives careful consideration.

Table 6-10

Summary of methods for preventing and correcting slides [8]

METHOD of treatment and its effect on stability of slide	TREATMENT	GENERAL USE		FREQUENCY OF SUCCESS (1)			POSITION of treatement related to the actual or potential sliding mass	POSSIBILITIES AND LIMITATIONS
		Pre-ven-tion	Cor-rec-tion	Col-lapse	Slide	Flow		
I. AVOIDANCE - No effect	A. Relocation	X	X	2	2	2	Outside slide limits	The best method if economical
	B. Construction of viaduct	X	X	3	3	3	Outside slide limits	Applicable to short stretches of sloping hillsides
II. MOVEMENT OF EARTH	A. Removal from the head	X	X	N	1	N	Upper part and head	Large masses of cohesive material
	B. Slope flattening	X	X	1	1	1	In the cut or embankment slopes	More effective in earth fills on frictional soils
Reduction in slopes Shear Stress	C. Terracing in slopes	X	X	1	1	1	In the cut or embankment slopes	—
	D. Removal of all unstable material	X	X	2	2	2	Throughout the slide	In relatively small superficial masses of moving material
III. DRAINAGE - Reduction in Shear Stresses and Increase in Shear Strength of the Soil	A. Surface							
	1) Ditches	X	X	1	1	1	Above the crown	Essential in all types
	2) Slope treatment	X	X	3	3	3	On surface of sliding mass	Rock covering or permeable apron to control the flow
	3) Subgrade trimming	X	X	1	1	1	On surface of sliding mass	Beneficial in all types
	4) Sealing of cracks	X	X	2	2	2	Throughout, crest to toe	Beneficial in all types
	5) Sealing of joint and fissure planes	X	X	3	3	N	Throughout, crest to toe	Applicable to rocky formations
	B. Subdrainage							
	1) Horizontal drains	X	X	N	2	2	Located to	Large masses of soil with underground flow
	2) Stabilising trenches	X	X	N	1	3	intercept	Relatively superficial masses of soil with underground flow
	3) Drainage galleries	X	X	N	3	3	and divert	Deep-seated and large masses of soil with significant permeability
	4) Vertical drainage wells	X	X	N	3	3	underground	Deep-seated sliding masses, underground water in strata or lenses
	5) Continuous siphon	X	X	N	2	3	water	Chiefly used as a ditch or drainage well opening
IV. RETAINING STRUCTURES	A. Support at the Base							
	1) Rock fill	X	X	N	1	1	Base and toe	Sound rock or firm soil at a reasonable depth
- Sliding Resistance Increases	2) Earth fill	X	X	N	1	1	Base and toe	As counterweight in the toe gives additional strength
	B. Common or crib retaining walls	X	X	3	3	3	Base	Relatively small moving masses
	C. Piles							
	1) Fixed in the slip surface	-	X	N	3	N	Base	The strength of the failure surface is increased by the amount of stress required to make the piles fail
	2) Not fixed to slip surface	-	X	N	3	N	Base	
	D. Anchorages in rock	X	X	3	3	N	Uphill from the highway or structure (cuts)	Stratified rock
	E. Short anchorages in slopes	X	X	3	3	N	Uphill from highway or structure	Eroding slope protected by screen anchored to a solid underlying formation
V. OTHER METHODS - Chiefly an Increase in Shear Strength	A. Hardening of the sliding mass							
	1) Cementing or chemical treatment							
	a) At the base	-	X	3	3	3	Base and toe	Cohesionless soils
	b) Throughout the mass	-	X	N	3	N	Throughout sliding mass	Cohesionless soils
	2) Freezing	X	-	N	3	3	Throughout sliding mass	To prevent temporary movement in large masses
	3) Electro-osmosis	X	-	N	3	3	Throughout sliding mass	Hardens the soil by reducing the water content
	B. Use of explosives	-	X	N	3	N	In the lower part of the slide	Relatively superficial cohesive mass overlying a mass of rock. Fragmented sliding surface. Explosives may also enable water to drain from the sliding mass

KEY to (1) 1: Frequent; 2: Occasional; 3: Rare; N: Not considered applicable

These methods are better for preventing rather than correcting failures, since the relatively high costs involved in moving earth are lower for new construction than for remedial jobs.

Careful removal of material should improve the drainage of the zone. This method can be used in practically all types of slides, but is especially effective in the rotational variety. Apart from cost, which can be high in the case of large failures, the principal disadvantage of this method lies in the fact that the material that is excavated is wasted. This may prove difficult and dangerous in some cases, not to mention the cost. Another factor that also raises costs is that excavation frequently has to start at the highest part of the slope and continue downhill. Another possible disadvantage is that by removing material and reducing the driving forces, reductions may also occur in the resisting forces.

Figure 6-48 shows in diagram form the procedure that was used to stabilize the slide at km 16+800 (sta. 551+18.1) of the Tijuana-Ensenada highway. 40,000 m^3 (52,300 yd^3) of material were removed from the head.

A solution which combines material removal and relocation is that of lowering the road grade in order to reduce the weight of the embankment on areas composed of weak soils or with predeveloped failure surfaces.

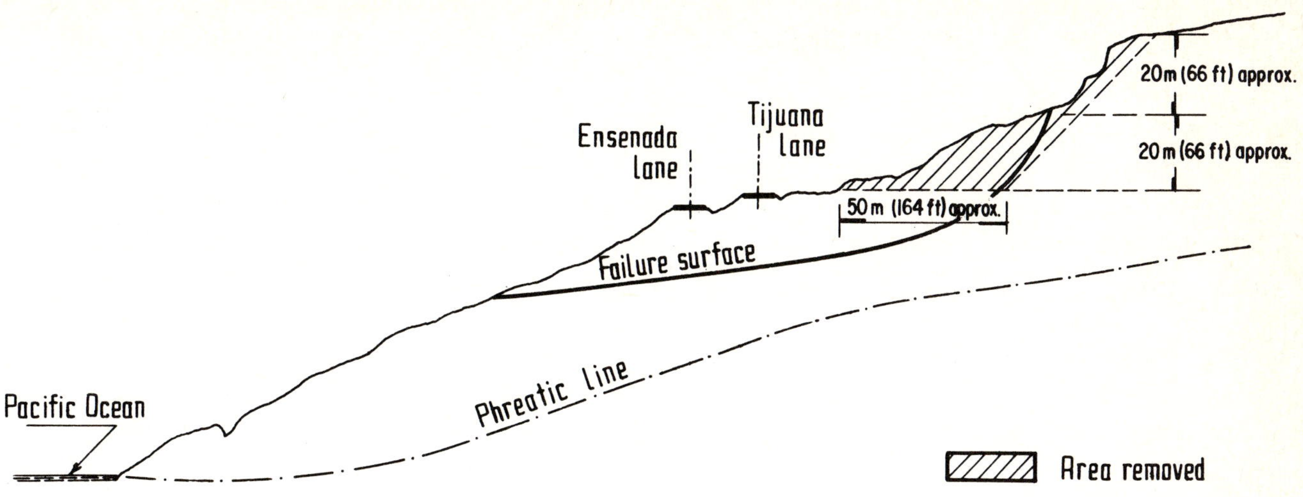

Fig. 6-48 Stabilization of a slide by removal of material from the head; km 16+800 (551+18) of the Tijuana-Ensenada highway

6.11.3 Slope Flattening

This is one of the most widely applied methods for improving slope stability. It is a remedial method that is used to correct slides in the body of the slope. Usually this is the first point to be considered when stabilizing a slope. Like all the other methods, it is not universal, and its degree of effectiveness can vary greatly from one case to another.

The flattened slope will be different from the original one, with all the corresponding implications. For example, if a critical circle for the original slope has been determined by the procedures discussed earlier in this Chapter, the critical circle for the flattened slope will be different, and consequently the safety factor for the new slope will be different. Therefore, a new stability analysis will have to be made to calculate the safety factor corresponding to the new critical circle.

When describing slope stability analysis, it was seen that in purely frictional soils the stability depends only on the angle of the slope, whereas in purely homogeneous cohesive soils, that extend into the foundation, stability depends more on the height of the slope (for inclinations smaller than 53° and base failure, stability is independent of the angle of slope, and for inclinations from 53° to 90°, stability varies only a little with the angle of slope). Most slopes for road engineering purposes are built in soils that are regarded as having two strength components, frictional and cohesive, but the extreme conditions mentioned here are still valid as design criteria. In soils where the frictional component has greater importance than the cohesive component, stability will be determined by the angle of slope, whereas in soils of a more cohesive nature the height of the slope, rather than the inclination, will be the decisive factor.

The foregoing general comments give an idea of the different factors that should be taken into account when choosing a method for correcting slope failure. In soils with a large frictional component, slope flattening will prove effective; in more cohesive soils, other methods that will be discussed later may prove more successful. These methods are in a sense equivalent to working with lower slopes (terracing for example).

Independently of these general considerations, Fig. 6-49 illustrates the consequences of flattening an embankment slope. The discussion is very general, for the sketches in this figure do not represent all the possibilities that can occur in practice. Some of the conclusions that can be drawn from this figure may even be inverted or their relative importance may vary greatly owing to changes in the position of the critical circles with the soil stratification. Thus, both the sketches in Fig. 6-49 and the corresponding discussion should be regarded as a guide but not as a rule as to what occurs when slopes are flattened. If consequences are analyzed for each individual case, it will be seen how from one slope to another there may be great variations in the mechanisms involved, the effectiveness of the solution, and the causes of possible changes in stability.

Part *a* of the figure shows an embankment where the critical circle originally corresponded to base failure (L_1). By flattening the embankment slope a new critical circle (L_2) is obtained. In this case, the change will probably lengthen the failure surface, causing an increase in the resisting forces due to the action of the shear strength of the soil over a wider area. Also, the new critical circle will tend to be deeper than the original one. This will increase the shear strength of the soil if it depends on normal pressure, but will not affect the shear strength if it is of a cohesive nature. Therefore, the solution will be more effective in frictional soils than in cohesive soils if these conditions prevail. The tendency of the failure surface to go deeper will also generally encourage stability in frictional soils, because their strength normally increases with depth owing to greater overburden weight and to less exposure to weathering. In clays, however, this effect can be detrimental, for these soils frequently have a stronger crust on the surface as a result of preconsolidation by evaporation; at greater depths a normally consolidated clay is softer.

Still referring to Fig. 6-49a, the wedge of fill that serves to flatten the slope causes an increment in the driving moment and a corresponding reduction in the stability. Evolution of the

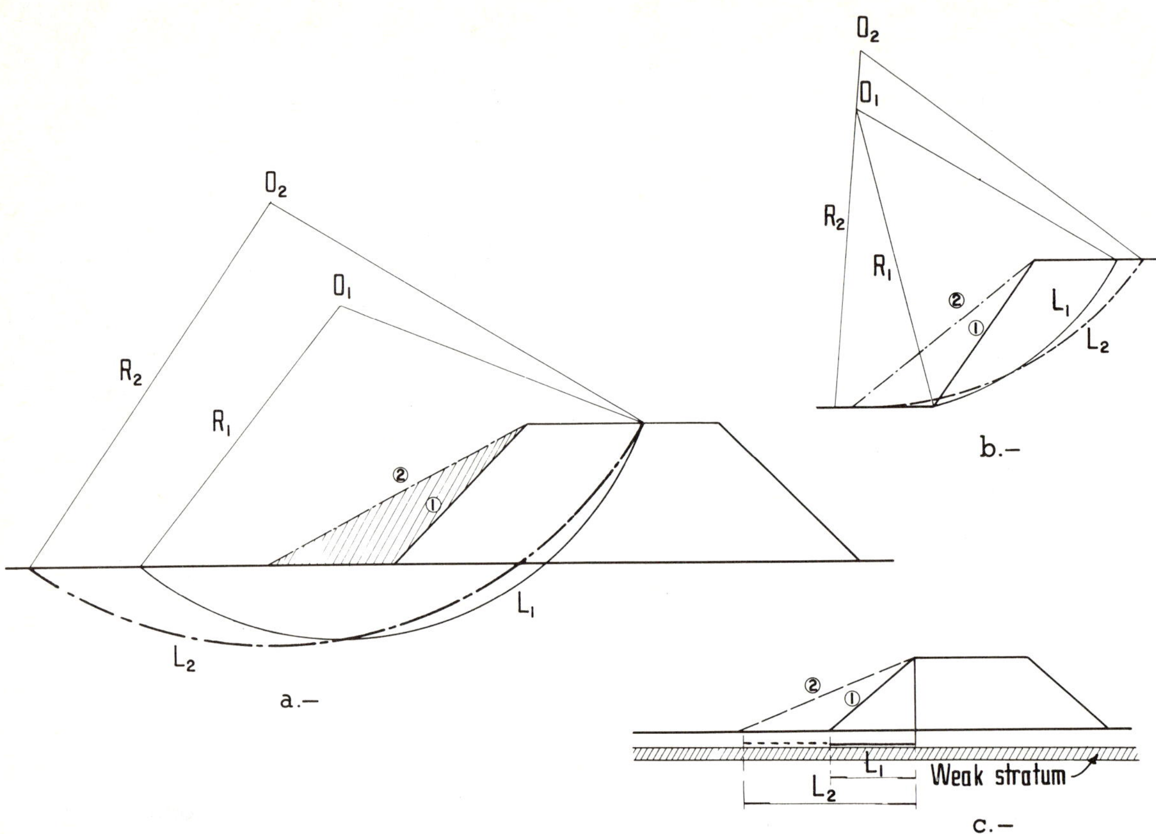

Fig. 6-49 Slope flattening in earth fills

safety factor cannot be predicted, for this depends on the relation between the resisting and driving moments. In each case, the corresponding calculation will have to be made to assess what improvement in safety is gained compared to the cost of the work.

Part *b* of Fig. 6-49 shows the flattening of an embankment slope where the critical circle passes through the toe of the slope. Since this type of failure generally occurs in soils where the prevailing strength component is frictional, the added weight increases the soil strength, thus the solution is likely to be more effective. The failure surface tends to become longer too.

In Part *c* of the same figure, the effect of slope flattening in a translational failure can be appreciated. This increases the length of the failure surface within the weak stratum. Moreover if that stratum is of a frictional nature, the weight of the fill will cause an increase in shear strength.

Figure 6-50 is similar, but this time refers to slope flattening in a cut. This case is different from the embankment, because flattening is achieved by excavating instead of by filling. This appears to be an advantage, for it may bring about a decrease in the driving forces. Slope flattening in this case tends to force the failure surface deeper into the soil mass of the cut. This is probably an advantage, because the soil there is likely to be firmer due to less weathering, less dissipation of residual stresses through expansion, and a greater normal effective pressure. (This last condition will affect only the frictional component). It should also be emphasized that the degree to which flattening a cut will benefit stability cannot be presumed. It must be assessed for each case; the effectiveness may vary greatly.

Slope flattening demands careful construction procedures; otherwise any mechanical benefit that might be obtained will be lost. In the case of embankments, slope flattening requires appropriate computation of the improvement in stability. Construction must be performed from the bottom upwards, compacting the fill and linking the new part of the slope with the original, so that there will be no problems of continuity. For this, the original section usually has to be terraced or benched and the new fill compacted layer by layer, working on benches with dimensions suited to the compaction equipment. In the case of cuts, it will also be necessary to make a design of the slope flattening. Here work will be done preferably from the top downwards. Here the major problem is safety instead of keying new constructions into old dangerous ones. If explosives are used during the slope flattening, which will be the case of cuts in rock, care must be taken to avoid incorrect use so that the faces obtained will not be shattered and weakened.

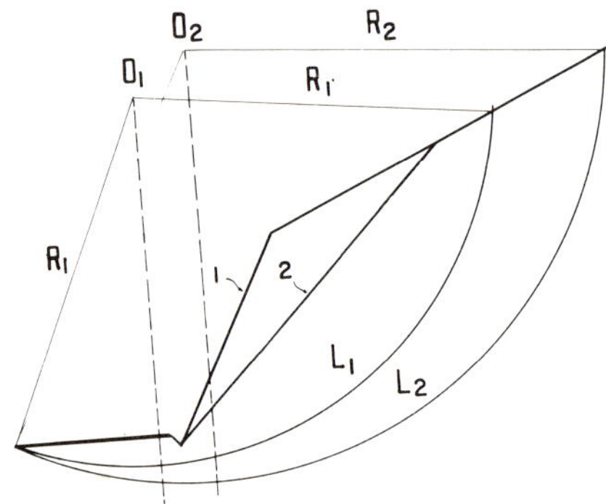

Fig. 6-50 Mechanism by which flattening the slopes of a cut often leads to better stability conditions

6.11.4 Use of Berms and Terracing

Berms is the name given to a terrace of new fill placed against an existing slope, and constructed of a similar material. These berms are placed against the slope to enhance its stability (see Fig. 6-51). Berms increase the stability of the slope for reasons similar to the ones mentioned for slope flattening (Plate 6-28). In many cases the construction of berms is equivalent to slope flattening. Thus, many of the comments made with reference to slope flattening can also be applied to the berms. A berm will change the critical circle. Therefore, a new safety factor will be determined by analyzing the new cross-section.

Berms increase the length and depth of the failure surface, and increase the resisting moment as discussed for slope flattening. Furthermore, the weight of the berm material increases the shear strength of the frictional part of the foundation soil. The most interesting effect of berms in cohesive soils is that the slope is divided into two parts, the second lower than the first, which has an important influence on general stability.

There is no rule regarding the appropriate dimensions for a berm in a given case. The design dimensions will have to be calculated by means of successive approximations, the designer having previously selected a safety factor in keeping with the slope in question. A good basis for commencing trial-and-error work is usually to give the berm half of the height of the embankment that is to be stabilized and a width roughly corresponding to that of the top width of the embankment. In bridge access embankments, front berms are sometimes used; these are built parallel to the road axis.

The ideal way to stabilize the embankment in Fig. 6-51 would appear to be with another parallel embankment. This will have all the advantages of the berm in the sketch, but the driving moment would not be increased by the weight of the material to the right of the vertical that passes through the center of the critical circle (O_2). Although this idea is in principle correct, it is often not practical, because drainage of the space between the two embankments may prove difficult if the berm is long; moreover, the resulting structure does not look nice and may endanger passing vehicles.

Terracing is a solution similar to berms. Figure 6-52 shows two typical terraces (often termed benches), one in purely cohesive soils and the other in soils with cohesive and frictional strength. It can be appreciated that in clay one objective of terracing is to transform the high slope into several lower ones, because in these soils, height is the decisive factor in stability. For this reason, the terraces must be sufficiently wide to enable them to fulfil the functions of independent slopes. For slopes in soils with cohesion and friction, the chief purpose of terracing is to flatten the slope. Other purposes, that are sometimes of great importance, are to collect falls of material and to divert surface water.

Plate 6-28 A stabilizer berm on the Mexico-Puebla highway

Terracing is defined by the width of the terraces, the vertical distance and inclination of the slopes between them. Whether the slopes of the different terraces will be parallel or built with variable slopes will greatly depend on the condition of the cut material. Variable slope terraces like the ones in Part *b* of Fig. 6-52 are recommended for material with a weathered upper layer, with the soils increasing in strength with increasing depth.

An important function of terraces is to protect the cut against erosion by surface water, by diverting the downhill flow of that water. For this, the terraces must be carefully designed. In soils that are highly susceptible to erosion, it may be advisable to tilt the terrace surface towards the hillside and to construct an impermeable apron and ditch on the inside to ensure rapid elimination of water. If there is great fear of rainwater infiltrating the terrace, the extreme precaution of water-proofing the surface can be taken.

As mentioned, another purpose of terracing is to retain any small surficial falls. This (together with cost) sometimes determines the width.

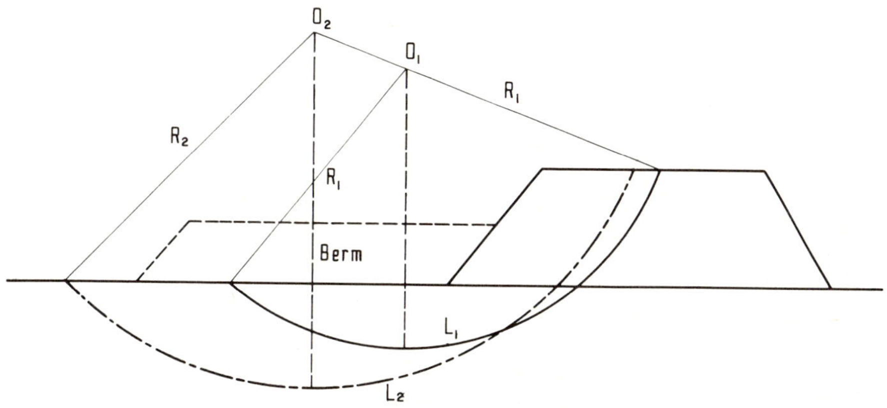

Fig. 6-51 Effect of a berm

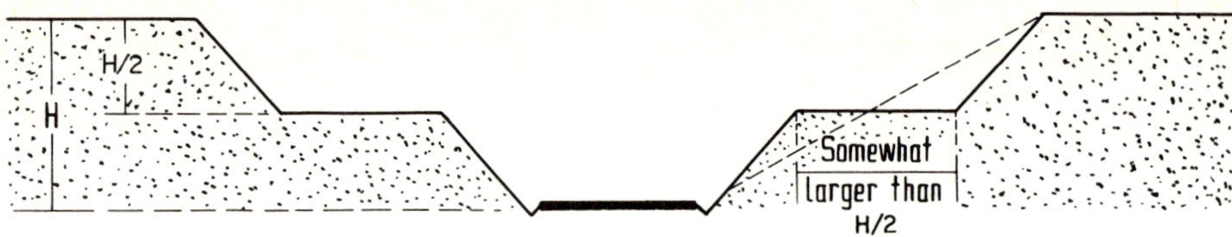

a). Terracing in cohesive materials

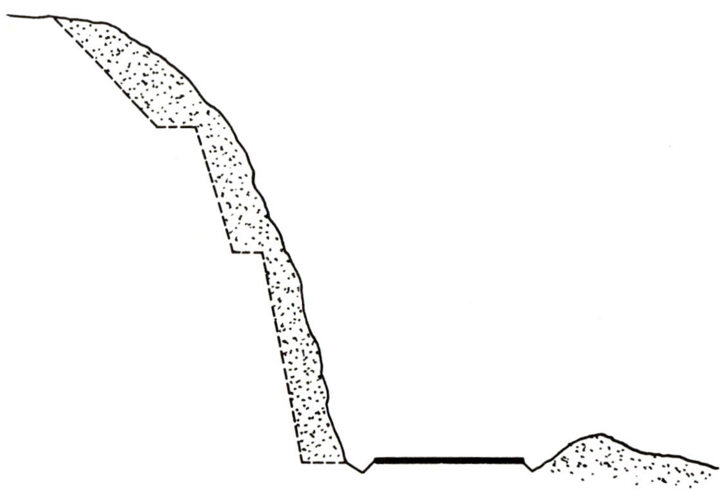

b). Terracing in soils with frictional strength

Fig. 6-52 Terraces in slopes

There are no rigid rules regarding the design of terraces for a cut. Each case must receive individual attention. In the case of cohesive and frictional soils, the terracing profile should be such that a reasonable average slope is reached for the entire cut; one simple slope as the equivalent of all the terraces. The height and width of the terraces, should also consider the previously mentioned requirements of preventing erosion by surface water and retaining of falls. The tread and rise of the terraces are frequently smaller towards the top of the slope because there is less water and falling material towards the top.

6.11.5 Use of Lightweight Materials

This method is applicable only to embankments. It is most effective in purely cohesive foundation soils, such as soft clays or peats, because in frictional foundation soils the advantage of the lightness is offset by smaller normal pressure and less frictional strength. This method was also mentioned in Chapter 3; Therefore it is not considered necessary to describe it in detail here. The driving forces are reduced by using materials of low unit weight in the body of the embankment. *Tezontle,* which is volcanic basalt foam or pumice, with a unit weight generally between 0.8 and 1.2 t/m^3 (50 and 75 lb/ft^3) is widely used in Mexico for this purpose. Other materials, almost always of volcanic origin, can also be used; among them, many pumitic sands. Foamed plastics have been used in local, critical locations.

It has also been mentioned (Chapter 4) that the behavior of lightweight materials must be clearly understood when compacting embankments, because the porous materials can be structurally degraded by high compaction forces so they are no longer lightweight.

Other solutions of this type, such as the replacement of part of the embankment by empty concrete pipes or empty concrete boxes, are usually extremely expensive, and their use is therefore limited.

6.11.6 Preconsolidation of Compressible Soils

This solution, which is based on preloading, has been discussed in §3.2. Preconsolidation of foundation ground can also be achieved by some of the other methods mentioned in §3.4. Chapter 3 in general deals with a series of solutions for improving foundation soils, all of which can be regarded as methods of improving the stability of any embankment that may be constructed on those foundation soils.

6.11.7 Use of Stabilizing Materials

One aspect of this solution is to add to the soil some substance that will improve its strength. This type of solution is generally more feasible in embankments. The substances most commonly used for this purpose are cement, portland cement, chemical cements, asphalts and chemical salts. However, these procedures are very expensive, and their use is therefore limited.

The usual procedure is to add artificial cementing to the grains of soil. Most chemical grouting processes that are in current use employ chemical mixtures of sodium silicate. This becomes a siliceous gel that will fill cracks, gaps and voids in the soil. Experience has demonstrated these methods can only be applied to sandy soils with a minimum effective diameter of 0.1mm (0.004 in). The majority of the reports that can be found in literature on the subject of these techniques refer to temporary treatments.

As an exception, thermal treatment has been used to stabilize landslides. The method was developed by LITVINOV [84]. Essentially it is a calcination method, where gases are injected into the soil at a temperature of over 1,000°C (1,832°F) to harden it. Radii of action of 2-3m (7-10 ft) can be achieved around the injection hole (Fig. 6-53). In [85] a means of applying this method to slope stabilization problems is described.

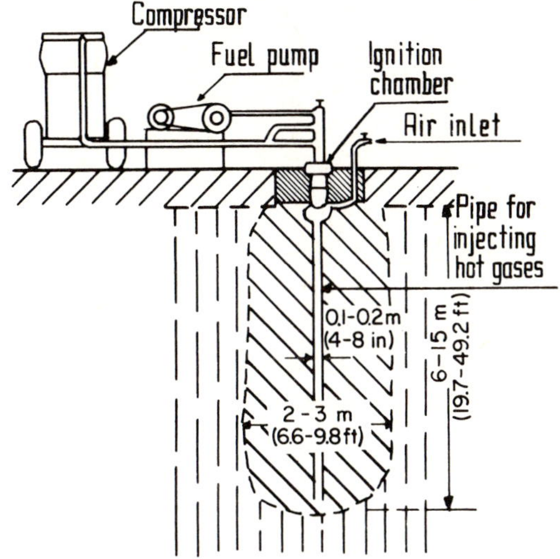

Fig. 6-53 Thermal treatment of soils, diagram showing the appliance [84]

Another method of hardening soils is cement grouting. It has been especially used in Europe for railroad construction. In England it is a fairly popular method, and is even used for cuts and embankments in clay [84]. Best results have been reported for the treatment of existing failure surfaces, in hard materials such as clay shales, argilites and stiff fissured clays. It does not give good results in soft clay.

Grouting forces the water from the fissures and fills them with cement mortar, which forms a good joint between the blocks. It does not modify the intrinsic characteristics of the soil mass, for the cement does not penetrate it. In order for the cement to penetrate the fissures and the existing failure surface, grouting must begin at higher pressures that the one already existing at the points under consideration. AYRES [86] reports a case where a continuous layer of cement mortar 6-12 cm (2-3/8in–4-3/4in) thick was applied to an entire failure surface which proved highly successful in stabilizing a large failure.

A grouting program demands detailed knowledge of a failure surface, so that the grouting holes can be appropriately placed. The injection holes are usually 3–5m (10–16 ft) apart. Grouting operations usually start at the foot of the slope and work upwards.

Asphaltic emulsions have also been used as grouting materials. Better penetration is achieved with these than with cement on account of their lower degree of viscosity. Cost is comparable to or somewhat higher than cement grouting, although costs depend on the availability of asphalt or cement in the locality where the method is to be applied. Costs also depend on experience in handling one or other of these products. The use of asphalt grouting is very limited because of seepage, which can easily remove the asphaltic material before the emulsion breaks into insoluble asphalt.

Another method of treating soils for these purposes is freezing. It is a slow, very expensive method, which is only applied as a temporary emergency solution.

Electro-osmosis (electrically induced seepage) is another method that can be used to improve the characteristics of slope materials.

6.11.8 Retaining Structures

The use of crib walls, sheet piles and other retaining structures is very common for correcting existing slides or preventing slides in zones where they are feared. In our experience, prevention is their principle field of application. (Plates 6-29, 6-31). The mechanics of this method require no further description. However, results have been disappointing in many cases, and the authors feel that the engineer should understand why retaining structures are often ineffective.

First, the object of the retaining structure is to support the soil mass above the failure surface that has either already formed

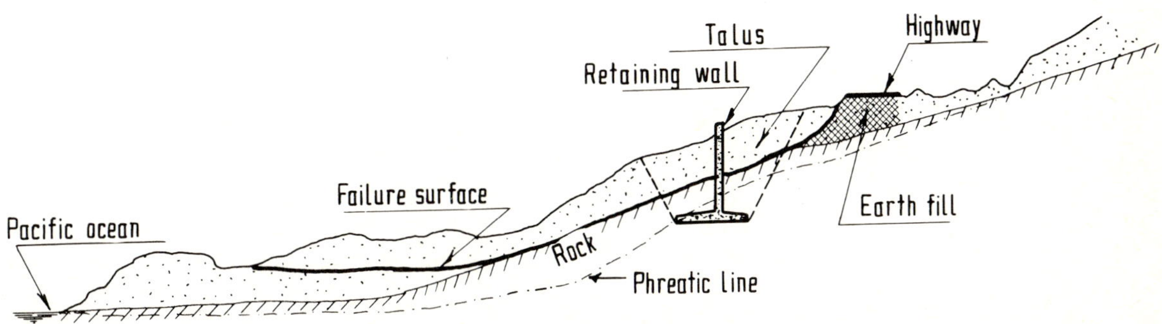

Fig. 6-54 Sketch showing a slide in a natural slope with a retaining wall; Km 16+000 (Sta. 524+93) of the Tijuana-Ensenada highway

Plate 6-29 Protection against falls in the form of *dry* masonry walls, Rio Verde-Valles Road

or is about to form. If this failure surface extends below the wall instead, then the effect of this wall on general stability will be non-existent. To prevent wall failure, very high walls, which require very deep footings are necessary. Because wall cost increases almost as the square of its height, wall costs are usually prohibitive.

Another reason why this method sometimes falls is that care is not taken to assure adequate drainage of the retaining structure. Drainage, which is always of fundamental importance for walls, becomes more critical when the wall must play a part in solving slope stability problems, many of which are caused by water.

A reasonably priced retaining structure does not greatly increase overall sliding resistance. If the forces tending to produce failure are not much greater than the resisting forces, construction of the wall may be worthwhile, but if there is considerable unbalance, then a wall will probably not provide the degree of safety that is required without exhorbitant costs.

Retaining structures are sometimes built at the toe of embankment slopes on steep hillsides where the embankment slope is not steep enough to join the natural ground slope. They are also built at the toe of cuts to provide visibility or to reduce the height of cuts in materials with a strength that is predominantly or purely cohesive, (where stability is largely dependent on height) this is perhaps one of their most effective uses.

The advantage of retaining structures is that they require little space. The volume of excavation for their foundations depends on the nature of the existing soil. This is one of the circumstances that should be most carefully considered before deciding to use this method, because a weak foundation soil can allow very undesirable movements in the wall. Such movements, in combination with the movements of the hillside, may well lead to uncontrollable situations.

One of the most common uses of retaining walls for slope stabilization is when there is not sufficient space available for slope flattening; this is quite a usual situation in embankments. Retaining walls are also successfully used for confining the toe of failures in clays and clay shales, by impeding the opening of cracks and fissures from free expansion.

Long high retaining walls are very expensive, which makes it hard for them to compete with other alternative methods. They involve several auxiliary construction tasks, such as subdrainage, bracing, and surface drainage, which are time consuming

Plate 6-30 A gunite covering comes away owing to pressure from accumulated water; Viborillas-Villa de Reyes railway

Plate 6-31 Protection against erosion; Masonry steps

and raise the total cost considerably. Crib walls have the advantage compared to common retaining walls, in that they can be built rapidly with common labor and that they can absorb settlements.

Surfacing methods increase slope stability by protecting the materials against the effects of erosion and weathering. They include dry masonry, masonry, gunites, shotcretes, thin concrete slabs (often anchored), asphaltic coating, and surface paving with stones or concrete block.

It should be mentioned that continuous coatings such as shotcrete have often failed because no attention has been paid to water flowing from the slope. If water collects behind a coating, a hydraulic pressure will develop until it is large enough to tear down the coating. Thus, when seepage is suspected, corresponding subdrainage precautions should be taken (see Plate 6-30).

Steel or plastic mesh is used to hold back falling materials on steep slopes. The mesh is anchored at the crest of the cut and attached to the slope with clamps or anchors.

Many stability problems have been solved by false tunnels or roofs. These are roofs of reinforced concrete covered with a sufficiently large mass of soil, so that no further collapses can damage the structure or road. The false tunnel must not, however, be placed in a deep seated failure zone. Therefore they are not well suited to major landslides, but are more appropriate in zones where there are local falls, collapses, or soil flows. False tunnels are expensive, but provide a radical solution when the material at the slope surface is unstable.

6.11.9 Use of Piles

Pile driving is the most controversial method of mechanical stabilization of slides in hillsides and artificial slopes. Nevertheless, some spectacular successes have been reported that have been achieved at a comparatively low cost. In almost all the successful cases, two or three rows of piles were driven. In other cases one or two rows of piles were driven to bring a displacement to a temporary standstill. Additional rows were driven as the material adapted to the restriction and movements

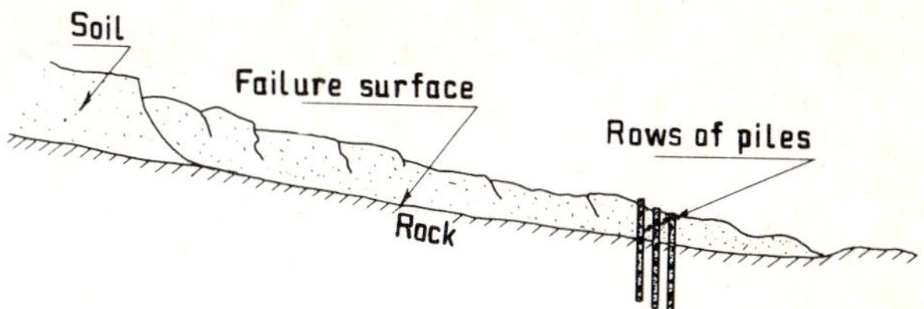

Fig. 6-55 Sketch showing how piles are used to stabilize a slide

recommended. Under such circumstances, there are continuous failures where pile driving has been going on for 20 years.

The method is appropriate only in the case of surface slides. Deep-seated slides generate very strong forces, which cannot easily be resisted by the piles. Moreover, even if these piles did resist, the soil would move between them.

Anchorage is vital. Poorly embedded or anchored piles will be uprooted and overturned, and this movement will bring about changes in the failure surface with detrimental results. There are no rules as to the length of embedment or anchorage, which should be decided in each individual case.

This method should only be attempted in rock or rigid materials, for soft soils are inclined to flow around the pile, thus reducing greatly the effectiveness. When there is considerable friction along the potential failure surface, pile driving will prove to be a worth-while preventive measure because it can bring about an important increase in friction. The diagram in Fig. 6-55 illustrates this approach.

The action of the piles in sometimes complemented by placing slabs of reinforced concrete between them, forming a wall.

6.11.10 Use of Counterweights at the Toe of Failure

There are two objectives of this method: first, to balance the effect of the driving forces at the head of the failure, in a way similar to berms (counterweights are similar to berms in several aspects); and second, to increase the shear strength of the underlying material, when frictional.

An appropriately shaped failure surface (it should rise under the counterweight) and soil in the placement zone that is sufficiently strong to bear the weight to which it is subjected, are the two conditions that must be fulfilled before application of this method can be considered.

The method consists of placing a sufficient weight of soil or rock in the appropriate zone at the toe of the failure. Figure 6-56 is a sketch which shows how this method was applied on the Tijuana-Ensenada highway, for the dual purpose of stabilizing the slope and preventing marine erosion. It is a large rockfill counterweight.

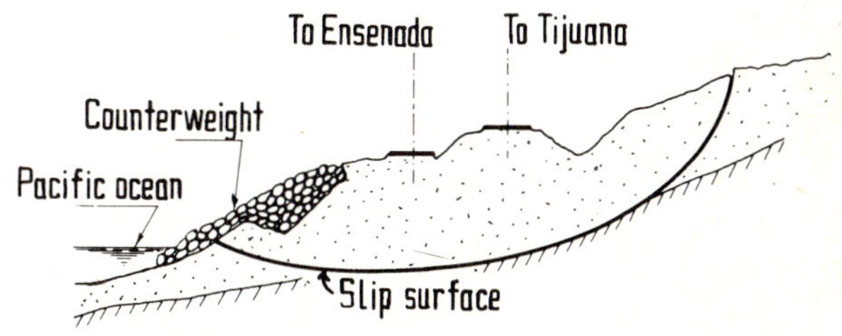

Fig. 6-56 Rockfill counterweight; Km 12+360 (Sta.405+51) of the Tijuana-Ensenada highway

Several methods are available, combining the effect of the counterweight with other desirable effects. For example, failures have been stabilized by using rock backfills [87], where the effect of the counterweight is added to the replacement of poor materials by rock having far higher strength and much better drainage.

Figure 6-57 refers to an actual landslide. A large mass slipped partially along the contact surface made of very fissured and fragmented rock, where there were large quantities of water. The rock backfill, apart from holding back the embankment, provided drainage and introduced a large quantity of material which was very useful for preventing any potential slip surface that might have formed after stabilization. A slide like this one is described in detail in [87].

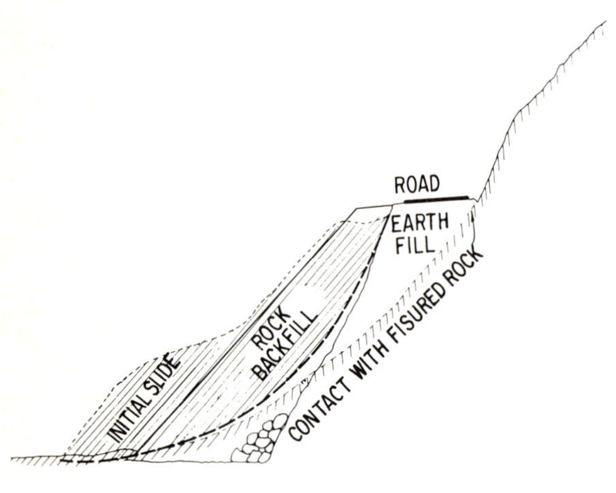

Fig. 6-57 Rock backfill where the effect of the counterweight is combined with those of replacement and subdrainage

6.11.11 Anchorages

Anchors are well known and widely used for stabilizing rock slopes. In the last few years these techniques have been extended spectacularly to both cohesive and frictional soils [88]. This subject will not be dealt with here, instead a more detailed description will be given in the Chapter devoted to special problems. Here we will simply say that some very important slope stability problems have been solved using anchorages in soils at very reasonable costs. This method should therefore be included among the ones that have already been considered.

A variant of anchorage techniques that has been little used in road engineering (but which should perhaps receive more attention from designers), is the use of anchor ties in retaining structures. These are especially effective when the structures are to be built on weak soils with contact pressures close to the bearing capacity. Very successful results can be achieved by anchoring retaining walls with pile foundations when the soil is not firm enough to prevent lateral deflection of the piles.

Anchorages usually consist of steel cables attached to deadmen and well secured to the retaining structure. They are more easily used in embankments than in cuttings.

6.11.12 Use of Explosives

The failure surface on which a slide moves is often smooth and polished. These occur typically in masses of cohesive soil which slide over mantles or rock or much harder soils. Contacts of this type also represent a potential slip surface.

In situations like this, explosives have sometimes been used to break and roughen the contact surface. The contact surface between the two materials thus becomes frictional. Efficiency of the method is increased if at a certain depth below the failure surface there are previous strata, which as a result of the explosion are made to communicate with the failure surface, thus providing it with the necessary drainage. When using this method, care should be taken in handling the explosives; otherwise, there is the risk that the explosion will accelerate the slide over any predeveloped surface or trigger it on a potential failure surface.

Whether correction by means of explosives is permanent is frequently discussed, but no definite conclusion has so far been reached. The opinion of the majority of specialists is apparently that for the method to be successful there must be a hard formation underneath the failure surface. Also, in all the cases where the method has been used, large-scale settlements have been reported during the following months. Any drainage, that may be obtained through the use of explosives is questionable. The resulting fragments do not form a filter, so the small spaces between them are sealed by very fine material deposited by the water flow. Very deep-seated slides are not within the scope of this method, an account of the violence of the explosions required.

Explosives can also be used in collapses and falls, not as a remedial method, but to remove fallen material.

The greatest advantage of explosives is cost. This is usually so very much lower than that of other methods, that a program of several successive applications of the procedure over a period of years is usually still economically advantageous.

6.11.13 Use of Vegetation

Vegetation both prevents and corrects failures by erosion (Plates 6-32, 6-33). The removal of earth in the construction of cuts and embankments inevitably causes very undesirable destruction of the vegetal covering, and the soils are left exposed to the onslaught of surface water and wind. Vegetation fulfils two important purposes; first, it reaches the water content on the surface, and second, it stabilizes the surface by the intertwining of its roots. Since plants and grass take their water from the soil in which they grow, there are several different criteria for the selection of the most appropriate species. It is advisable to use local plants, thus avoiding problems of adaptation to climate. These are hard for the civil engineer to predict and require the aid of horticulture experts. There are species that take much water from the soil and others that take little, leading to very different reductions in the surface water content. In clayey soils, the first species are usually the best, because they ensure a drier, stronger soil crust; however, intense drying of sandy surface soils makes them more susceptible to erosion, and this is not good.

When vegetation consists of trees planted at the crest of cuts or as barriers to hold back sand, the foregoing considerations are

Plate 6-32 Stabilization of a cut slope using vegetation; Villa Cardel-Veracruz highway; isolated cactus plants were used

Plate 6-33 Stabilization of the slope of a large earth fill using vegetation

important; it will probably be best to choose local species that adapt comfortably to the specific place where they are to be planted.

Experience has shown that slopes can be more effectively protected by planting grasses and herbaceous plants over the entire slope rather than by planting brush in isolated areas. Because the two methods involve different costs, the engineer's decision should be guided by how fast the plants grow in the region. There are zones where revegetation occurs quickly and naturally, and others where plant growth proves very difficult. Isolated planting greatly increases the risk of local seepage and scour. However, for very high embankments, good results have been achieved by planting rows of bushes so as to reduce surface runoff erosion. Grass plays a very important part in avoiding shrinkage cracks in soils that would otherwise be exposed.

Figure 6-58 [84] shows the hydraulic equilibrium which tends to become established with time in clayey soils with a covering of different species of plants. Note how there is a reduction in the water content to depths of about 2–2.50 m (7–8 ft), and as much as 3.00 m (10 ft) where there is a covering of shrubs.

A detailed study of the different plant species that can be used in each place and region is beyond the scope of this book and should be left to a specialist. Likewise, it is not felt necessary to analyze all the techniques that have recently been developed for improving the growth of plant species in inhospitable terrains such as arid zones, where the vegetation dries up and dies almost as soon as it is planted.

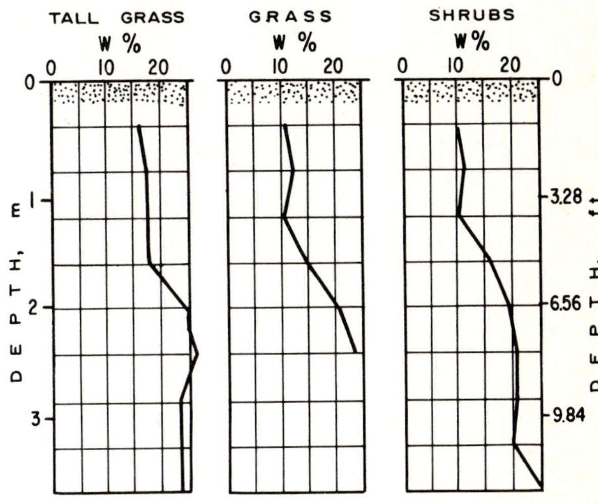

Fig. 6-58 Effect of vegetal covering on the surface water content of clayey soils

Irrigation of plantings, when abundant and prolonged, should be considered undesirable because of cost and conservation. To avoid it, *crusts* of clay and humus are placed on slopes built in soils that offer no encouragement to plant life. These crusts conserve moisture, which favors vegetation. In pursuit of similar objectives, a thin spraying of asphalt, asphalt and straw and other substances has sometimes been applied to slopes. By impeding evaporation and erosion, at least initially, the growth of a dense covering of vegetation is encouraged, which, in the years to come, will be able to fend for itself.

6.11.14 Correction of Other Types of Failure

It is not felt necessary to discuss methods of preventing or correcting other types of failures, such as piping or liquefaction. These measures are almost always related to the compaction conditions of the slope material or soil foundation.

Special mention is deserved by the methods for preventing and correcting longitudinal cracks. Since these cracks occur in the embankment zones that are most susceptible to variations in moisture, the shoulders, the simplest solution is to make the embankment wider than strictly necessary for geometrical reasons. In this way, the variations in moisture below the paved zone will be kept to a minimum and the cracks will appear outside that zone. The same criterion leads to the widening of existing embankments in which cracks appear. The solution is only a palliative (which has given excellent practical results), but it does not go to the root of the problem. Moreover it has a very high cost.

On other occasions, with a view to economizing on earth removal, embankments have been widened by constructing berms two or three meters wide and approximately half the height of the embankment, often with satisfactory results

Plate 6-34 Prevention of longitudinal cracking using berms; note how cracks develop in the berms themselves

(Plate 6-34). Even better results have been obtained by making the slopes of the embankment very flat where cracks are to be avoided (2.5:1, 3:1, or more) without widening the crest. Experience has shown that the cracks are kept well away from the paved zone even when the widening of the slopes (in the case of the correction of existing embankments) is done carelessly, without adequate compaction. Better results have been achieved by widening slopes with fine clayey soils, with the added advantage that vegetation is thus encouraged. The same technique has also given good results in cases of failed embankment shoulders.

6.11.15 Other Remedial Methods

All the above-mentioned methods refer to solutions that have been tested with varying degrees of success, but the engineer who is confronted with a stability problem must not allow his mind and his imagination to be overruled by solutions with which he is already familiar. He has the detailed knowledge, information and vision of the case he is handling, that cannot be achieved by any of those people who hand out advice by systems of remote control. Similarly standard reference books cannot substitute for local understanding. The engineer must, therefore, put this knowledge to good use, reaching for the best solution for the particular case in which he is interested. This solution may be original, and at other times a combination of several recognized methods.

Figure 6-59 illustrates a solution of the type suggested. It represents a very wide embankment built for a highway where the foundation soil was very soft and compressible. At this location there was a horizontal curve in the highway, requiring the corresponding superelevation. The engineer in charge modified a less imaginative design and constructed the cross-section that is shown in the diagram. Although it is not a new solution to such a problem, and information on such solutions might have been available, the man who modified the project had not seen the method previously.

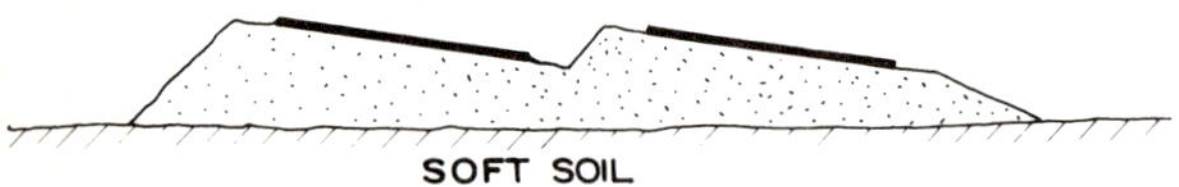

Fig. 6-59 Serrated embankment, useful for solving problems of superelvation by curvature on very soft soils

This is a good example of what can be achieved by a little open-minded thinking when one is faced with a specific problem.

Special reference should be made to all the methods that reduce the harmful effects of settlements in existing soft soil embankments. These were considered in Chapter 3 as measures for correcting foundation soils. Methods of preconsolidation are most often used together with the use of lightweight materials in the body of the embankment. Imaginative solutions may well exist for this case too. Here the solution originally conceived by L. M. Aguirre [89] should be mentioned. Its aim was to minimize the large settlements that occur in land once belonging to Texcoco Lake [90] during the construction of new runways or additions to already-existing ones, as at the Mexico City International Airport. In the area, there are enormously thick layers of highly compressible materials (Fig. 6-60). The Airport is old now and its runways were built before the introduction of modern techniques. The results can be observed in Plate 6-35. When major settlements occur in the center of an earth fill, the transverse surface slope is lost, and drainage is impossible. The formation of pools is thus encouraged and the runways become so dangerous that they cannot be used. The solution to this problem was for many years based on costly overlaying with asphalt concrete, until the extreme was reached that can be appreciated in the photograph.

Plate 6-35 Deformations in one of the runways at Mexico City International Airport; note how asphaltic concrete has been used to relevel

The fundamental aim of design for the new runway is to bring the increase in pressures produced by the earthfill on the foundation soil as close to zero as possible. In this case, a compensation principle was followed: the pavement was built in an excavated section, so that the weight of the material removed was the same as that of the runway above. For the runway (see the cross-section in Fig. 6-61), light materials were used [pumice gravel, locally known as *tezontle*, with a unit weight of 0.8 t/m^3 (50 lb/ft^3)].

The lower bed of the cross-section is a thin slab of plain concrete resting on a layer of sand at the bottom of the excavation. The purpose of this slab is to give uniform support to the cross-section, distribute the transferred forces evenly and improve compensation throughout. Note that on both sides of the runway there are sections of crushed stone with a normal unit weight, so that settlement will be uniform and not differential throughout the cross-section.

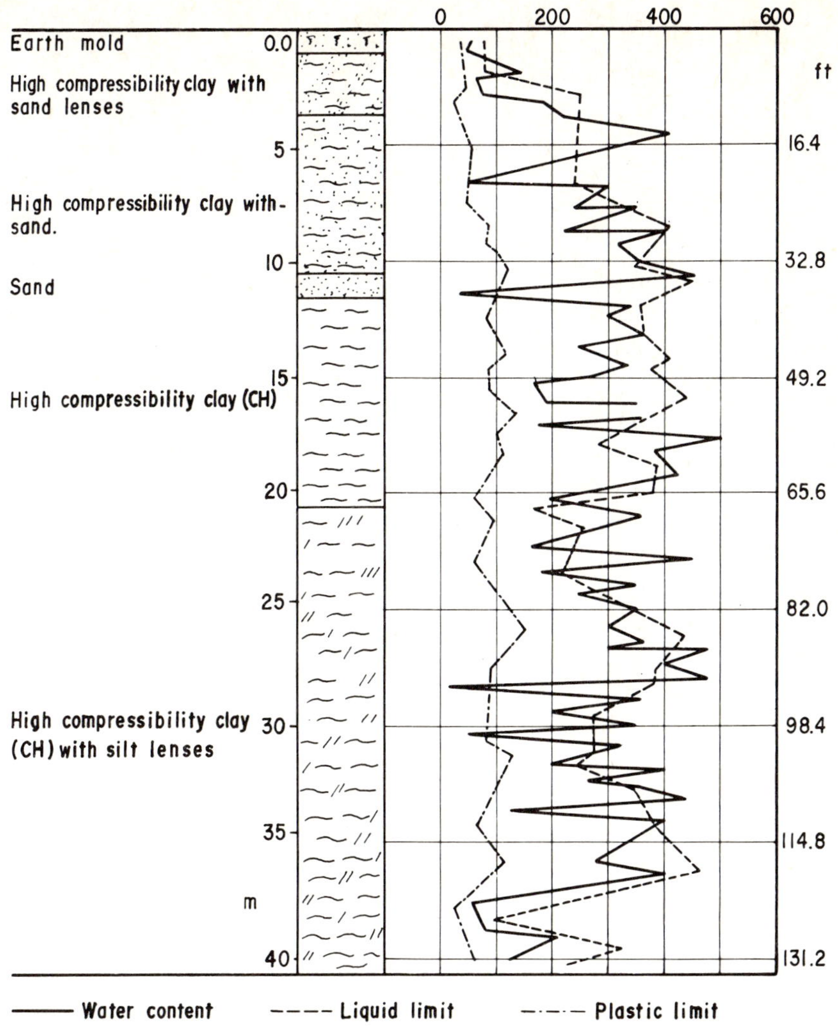

Fig. 6-60 Profile of soils in the Mexico City airport zone [89]

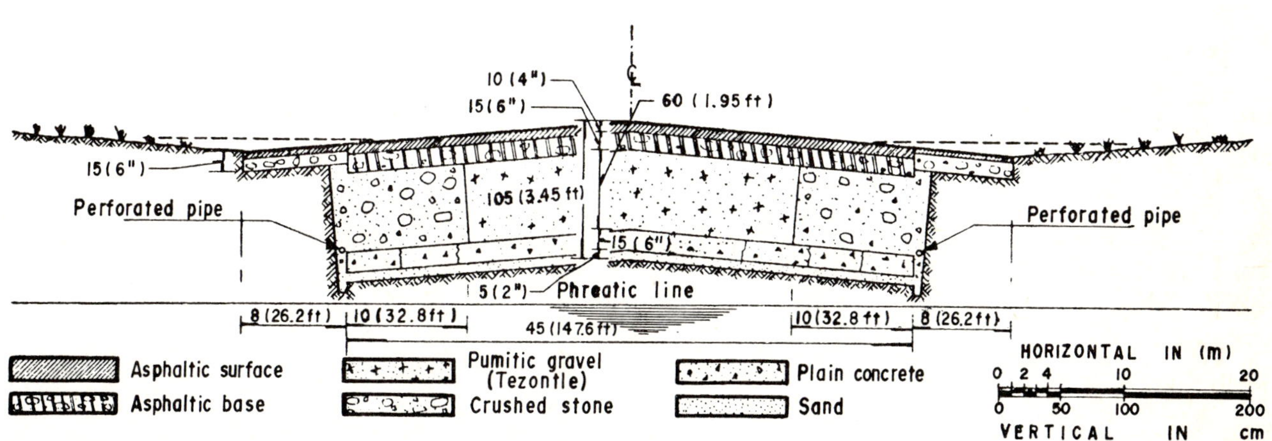

Fig. 6-61 Structural cross-section of the extensions added to the runways at Mexico City Airport [89]

Performance of this cross-section in preventing differential settlement can be seen in Fig. 6-62, where levelling data are given for the first 4 years of life of the structure. It should be mentioned that from that time until 1972, the movements observed have been as small as the accuracy of the instruments used to measure them.

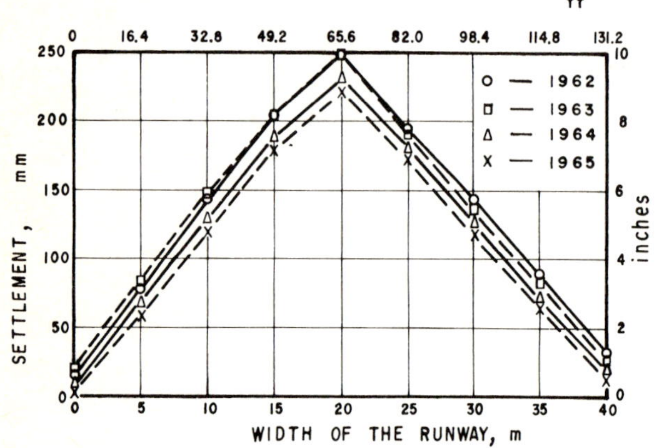

Fig. 6-62 Settlements observed in a cross-section of the extensions added to the runways at Mexico City Airport [89]

It has already been seen that there is a large variety of methods for preventing or correcting slope failure. All the methods involving drainage or subdrainage (which provide the most important solutions to these types of problems) will be discussed in a future Chapter. For each situation, the engineer will generally be able to choose between several different alternatives, all of which may appear attractive. Some of the possible solutions will, however, prove unsuitable. It will thus be necessary to carry out a selective investigation, in the course of which the alternatives will be compared and some eliminated. The investigation will be of a distinctly technical nature; it may lead to several possible solutions, all technically sound. The final selection is based on cost, speed of execution and aesthetics. There are many cases where the choice of the solution is determined by the need to complete remedial work by a specific date, for example when a road has to be opened to traffic, or finished before the start of the rainy season. However, the decisive factor is usually cost, in the widest sense of the word.

6.11.16 Applicability in Practice

Not all the foregoing methods are appropriate for all types of failures in hillsides and artificial slopes. Although it is very hard to generalize, a few comments follow concerning the different remedial methods that have been mentioned, and the types of failure where they have shown the best results.

.1 Soil and Rockfalls

Here, stabilizing measures usually correspond to one of the following: relocation, slope flattening, terracing and surface drainage. Retaining methods have been used on a smaller scale, to supply a protective covering to materials that are prone to weathering. They include masonry and thin concrete slabs. In the case of minor falls, protection with gunite, shot-cretes, wire meshes, etc. has given excellent results (see Plate 6-36).

Plate 6-36 Protection against falls using mesh and a retaining structure in the background

Anchorages are becoming a more and more popular solution to this type of problem.

In the case of those cuts where the falls occur at the crest, the periodical removal of loose material is useful.

.2 Landslides

The following are the most common methods for solving landslide problems.

a. Relocation
b. Slope flattening
c. Berms
d. Removal of material at the head of the failure
e. Surface drainage and sealing of cracks
f. Modification of grade
g. Use of counterweights
h. Retaining walls
i. Piles
j. Use of explosives

Counterweights, walls and explosives should be used only for small slides; very seldom have they proved effective in large-scale slides. Rockfills and walls have been successfully used to prevent erosion, for example of a river, at the toe of very large landslides.

Subdrainage has once again been omitted from the list, although it is one of the most effective and often most rapid, economical and elegant solutions to landslide problems.

.3 Mud Flows

The following are the most commonly used solutions to mud flows:

a. Relocation
b. Slope flattening
c. Terracing
d. Partial or total removal of failed material
e. Surface drainage, including crack sealing

In this case too, subdrainage provides a whole set of often successful solutions that should always be taken into consideration.

Retaining structures can only be used for very small mud flows. Bridges to pass over the failure zone have been used more often in flows than in other types of failure, because the failure zone is often narrow.

APPENDIX 6A
PRACTICAL EXERCISES

6a.1 Computation of the Safety Factor for a Slope on Cohesive Soil Founded on the Same Material and Limited by an Underlying Horizontal Stratum of Firm Soil

To illustrate how Taylor's graphs can be applied to this particular case, let us solve the following problem.

.1 Data:

$$c = 2 \text{ t/m}^2 \ (410 \text{ lb/ft}^2)$$
$$\gamma_m = 1.8 \text{ t/m}^3 \ (112 \text{ lb/ft}^3)$$
$$H = 3.0 \text{ m} \ (9.84 \text{ ft})$$
$$DH = 4.5 \text{ m} \ (14.76 \text{ ft})$$
$$\beta = 30°$$

.2 Solution:

a) *The critical circle* must be tangent to the firm stratum, with its center on a vertical that goes through the middle of the slope.

b) To determine the required stability number (N_e) and the *position of the critical circle* (value of n), values D and B of Taylor's graph (Fig. 6-28) are used.

For $D = 1.5$ and $\beta = 30°$; $N_e = 0.1625$ and $n = 0.55$ are obtained

c) To calculate the stability number using the available strength in the embankment, the following expression is applied:

$$N_e = \frac{c}{\gamma_m H} = \frac{2}{1.8 \times 3} = 0.370$$

d) Finally, the safety factor of the slope is calculated by dividing the value of N_e available by the value of N_e required.

$$F_s = \frac{N_e \text{ (available)}}{N_e \text{ (required)}} = \frac{0.370}{0.1625} = 2.27$$

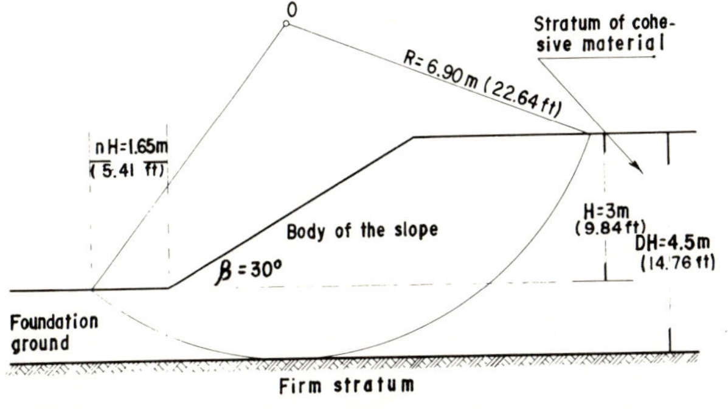

Fig. 6a-1 Critical circle corresponding to the example proposed

6a.2 Example on an Analysis Using Trial and Error Procedures

It is required to calculate the stability of a retaining wall, assuming a circular slide surface.

.1 Data:

Width of footing $2b = 4$ m (13.12 ft); height of wall $H = 8$ m (26.24 ft); width at foundation level = 3.4 m (11.15 ft); width at crest 1.4 m (4.59 ft); foundation depth $h = 2$ m (6.56 ft); unit weight of the masonry $\gamma = 2.0$ t/m³ (125 lb/ft³).

Mechanical Characteristics of the fill and foundation soil: Unit weight $\gamma_m = 2$ t/m³ (125 lb/ft³); angle of internal friction $\varnothing_u = 12°$; cohesion $c_u = 1$ t/m² (205 lb/ft²).

.2 Solution:

The problem is solved by finding the smallest factor of safety. It is usually advisable to move the slide surface center along a vertical until a minimum factor of safety is determined. Then, at this level, the center is moved along a horizontal line.

1. From an arbitrarily selected center, O_1, an arc is drawn through the heel E of the retaining wall (Fig. 6a-2).

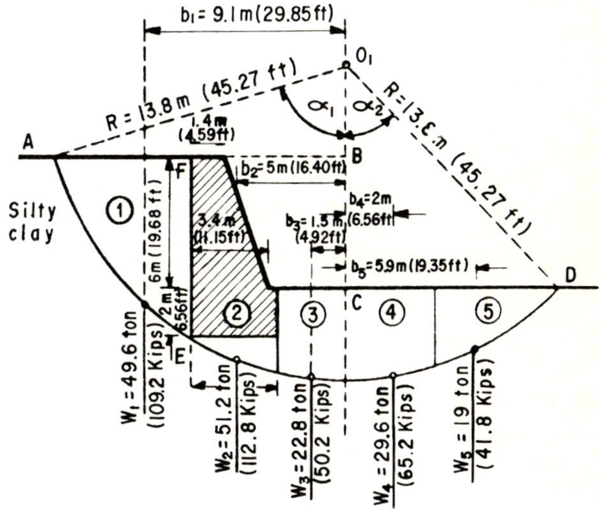

Fig. 6a-2 Presentation of the problem and first trial

2. The radius is determined graphically; in this example, it will be $R_1 = 13.8$ m (45.27 ft).

3. For triangles ABO_1 and CDO_1 the values of the angles α_1 and α_2 can be determined by measuring the values of the sides O_1B and O_1C.

$$\cos\alpha_1 = \frac{O_1B}{R_1} = \frac{4}{13.8} = 0.29; \ \alpha_1 = 73°8'$$

$$\cos\alpha_2 = \frac{O_1C}{R_1} = \frac{10}{13.8} = 0.725; \ \alpha_2 = 43°28'$$

4. The length of the sliding arc L_1 along AED is determined

$$L_1 = (\alpha_1 + \alpha_2) R_1 = \frac{116°36'}{180°} \pi \; 13.8 \text{ m} = 28.22 \text{ m}$$

$$= \frac{116°36'}{180°} \pi \times 45.27 \text{ ft} = 92.6 \text{ ft}$$

(the sum of the angles α_1 and α_2 is expressed in radians).

5. The moment of the cohesive resistant forces along the sliding arc AED is computed:

$$L_1 c_u R_1 = 28.22 \times 1 \times 13.8 = 390 \frac{\text{t-m}}{\text{m}}$$

$$= 92.6 \times 0.205 \times 45.27 = 858 \frac{\text{ft-kips}}{\text{ft}}$$

6. The sliding mass is divided into 5 slices

To simplify calculation, the width of the slices is selected according to the location of the strata (when the mass is stratified) and bearing in mind that the base of the slice will be taken as a straight line.

7. The geometrical elements for Slice *1* are determined. The area of triangle AEF is:

$$A_1 = \frac{AF \times FE}{2} = \frac{6.2 \times 8.0}{2} = 24.8 \text{ m}^2$$

$$= \frac{20.34 \times 26.25}{2} = 267 \text{ ft}^2$$

The weight of Slice *1* will be:

$$W_1 = A_1 \gamma_m = 24.8 \times 2 = 49.6 \text{ t/m}$$

$$= 267 \times 125 \times \frac{1}{1000} = 33.375 \text{ kips/ft}$$

Its distance from the center of gravity to the center O_1 is:

$$b_1 = BF + \frac{AF}{3} = 7 + \frac{6.2}{3} = 9.1 \text{ m}$$

$$= 23 + 20.34/3 = 29.85 \text{ ft}$$

The driving moment generated by the weight of the slice about O_1 is:

$$W_1 b_1 = 49.6 \times 9.1 = 452 \text{ t-m}$$

$$= 33.375 \times 29.8 = 996 \text{ ft-kips}$$

The moment generated by the frictional forces along failure arc AE is:

$$W_1 \times \cos\theta_1 \times \tan \varnothing_u \times R_1 = 49.6 \times 0.7547 \times 0.213 \times 13.8$$

$$= 110 \text{ t-m}$$

$$= 33.375 \times 0.7547 \times 0.213 \times 45.27 = 242 \text{ ft-kips}$$

Computations for the rest of the slices are conducted in the same way.

8. Since the retaining wall is subjected to no additional vertical load, apart from the weight of the masonry itself, which has a unit weight equal to that of the soil (2 t/m³) (125 lb/ft³), its weight might be included in that of Slice *2*. To simplify calculations, it is assumed that the weight of the wall is evenly distributed over its base, and that the center of gravity of Slices *2, 3, 4* and *5* is located at its geometrical center or centroid.

All data corresponding to the first trial are given in Table 6a-1.

The safety factor is determined by the equation:

$$F_s = \frac{\sum W_i \cos\theta_i \tan \varnothing_u R_1 + \sum c_u L_i R}{Pa + \sum W_i b_i}$$

$$F_s = \frac{453.8 + 390}{588.9} = \frac{998 \times 858}{1294.5} = 1.42$$

Product Pa of the equation is not taken into consideration, since the weight of the retaining wall P is included in the weight of Slice *2*.

9. As a second trial, choose a center, O_2, for the second failure arc at 2 m (6.56 ft) above, on the same vertical line (Fig. 6a-3).

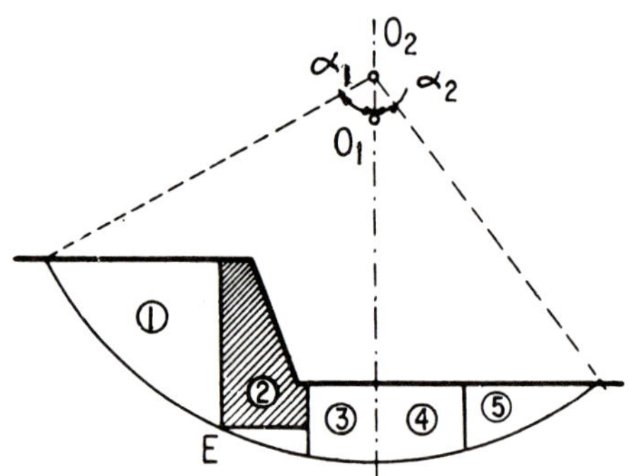

Fig. 6a-3 Second trial

As in the first case, draw the failure arc with radius O_2E, determining graphically the length of the radius and all the data necessary for the computation of the corresponding safety factor. Calculations are conducted as previously, and the data are given in Table 6a-2. Value of radius R_2 = 15.7 m (51.51 ft); length of failure arc L_2 = 29.6 m (97.11 ft).

The safety factor for the second failure arc is:

$$F_s = \frac{550.3 + 29.6 \times 1 \times 15.7}{708.7}$$

$$= \frac{1212 + 97.11 \times 0.205 \times 51.51}{1560} = 1.43$$

10. Now a new center is placed 2 m (6.56 ft) higher on the same vertical line (Fig. 6a-4), at point O_3. Calculation data for the third trial are given in Table 6a-3.

The value of radius R_3 = 17.6 m (57.74 ft); the length of the sliding arc is L_3 = 30.41 m (99.77 ft).

Table 6a-1
Stability analysis by trial and error procedure
First trial

No of slice	Area of slice (A_i)		Weight of the slice (W_i)		Moment distance (b_i)		W_ib_i		θ_i	$W_i \cos \theta_i \tan \varnothing_u R_1$	
	m^2	ft^2	t/m	kips/ft	m	ft	t-m/m	ft-kips/ft	deg	t-m/m	ft-kips/ft
1	24.8	267	49.6	33.2	9.1	29.8	452.0	993.0	41.0	110.0	242.0
2	25.8	279	51.6	34.6	5.0	16.4	258.0	568.3	21.5	141.1	310.1
3	11.4	123	22.8	15.3	1.5	4.9	34.2	75.1	6.3	66.6	146.2
4	14.4	155	28.8	19.3	−1.8	− 5.9	− 51.8	− 114.2	7.0	84.0	185.2
5	9.6	104	19.2	12.9	−5.4	−17.7	−103.5	− 227.7	22.5	52.1	114.5
					Total:		588.9	1294.5		453.8	998.0

Table 6a-2
Second trial

No of slice	Area of slice (A_i)		Weight of slice (W_i)		Moment distance (b_i)		W_ib_i		θ_i	$W_i \cos \theta_i \tan \varnothing_u R_2$	
	m^2	ft^2	t/m	kips/ft	m	ft	t-m/m	ft-kips/ft	deg	t-m/m	ft-kips/ft
1	29.60	320	59.2	39.7	9.6	31.4	568.3	1250.7	36.8	158.8	350.2
2	25.40	275	50.8	34.0	5.3	17.3	270.0	593.5	14.4	164.8	363.1
3	10.80	117	21.6	14.4	1.5	4.9	32.4	71.8	5.0	72.1	158.2
4	14.4	155	28.8	19.3	−1.9	− 6.2	− 54.8	− 120.0	7.0	95.8	211.0
5	9.4	101	18.8	12.6	−5.7	−18.7	−107.2	− 236.0	21.0	58.8	129.5
					Total:		708.7	1560.0		550.3	1212.0

Table 6a-3
Third trial

No of slice	Area of slice (A_i)		Weight of the slice (W_i)		Moment distance (b_i)		W_ib_i		θ_i	$W_i \cos \theta_i \tan \varnothing_u R_3$	
	m^2	ft^2	t/m	kips/ft	m	ft	t-m/m	ft-kips/ft	deg	t-m/m	ft-kips/ft
1	33.6	363	67.2	45.0	9.4	30.8	631.7	1385.0	32.3	210.7	462.9
2	24.8	267	49.6	33.3	5.1	16.7	253.0	558.0	17.0	175.9	387.0
3	10.2	110	20.4	13.7	1.4	4.6	28.6	64.9	4.8	75.4	165.6
4	13.4	144	26.8	18.0	−2.0	− 6.5	− 57.6	− 126.6	6.1	98.8	218.1
5	8.7	94	17.4	11.7	−5.8	−19.0	−100.9	− 221.2	19.0	61.0	134.0
					Total:		754.8	1660.1		621.8	1367.6

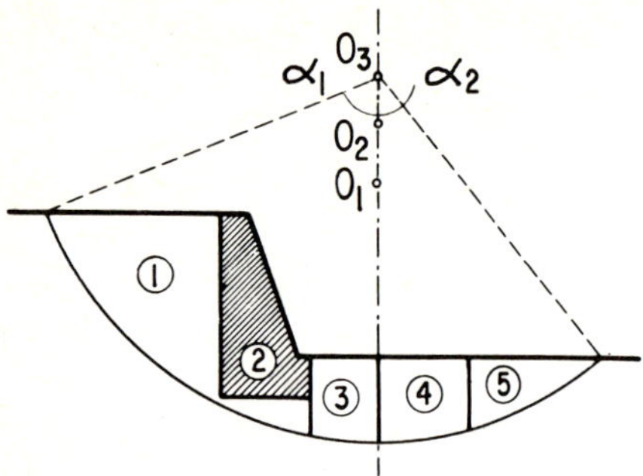

Fig. 6a-4 Third trial

The safety factor for the third trial is:

$$F_s = \frac{621.8 + 30.41 \times 1 \times 17.6}{754.8}$$

$$= \frac{1367.6 + 99.77 \times 0.205 \times 57.74}{1660.1} = 1.53$$

11. The safety factor increases as point O is located higher on the vertical. Values of the safety factor for centers below the previous ones must therefore be found. Thus, O_4 is placed 2m (6.56 ft) below O_1 on the same vertical and similarly the fourth failure arc is obtained (Fig. 6a-5)

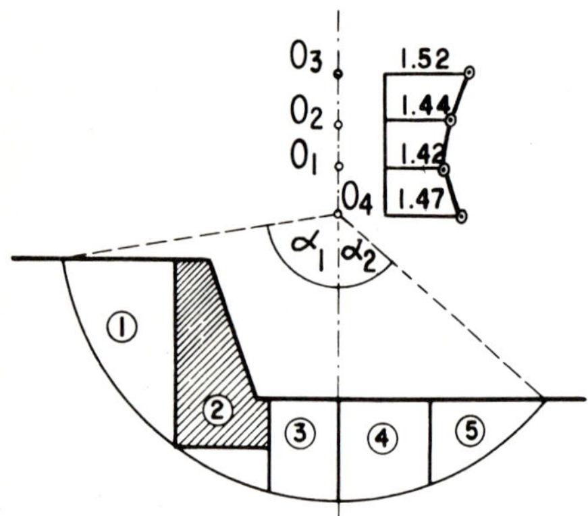

Fig. 6a-5 Fourth trial

The corresponding safety factor is computed as previously (Table 6a-4).

The value of the radius is R_4 = 12.4 m (40.68 ft); the length of the failure arc is L_4 = 27.31 m (89.60 ft).

The safety factor for the fourth trial is:

$$F_s = \frac{387.1 + 27.31 \times 1 \times 12.4}{495.7}$$

$$= \frac{851.2 + 89.6 \times 0.205 \times 40.68}{1092.0} = 1.46$$

Consequently the cylindrical sliding surface for the smaller safety factor corresponds to O_1, the safety factor here being F_s = 1.42.

12. After finding the centre at the vertical where the minimum value of the safety factor is located, the safety factor is investigated in the horizontal direction. For this, sliding surfaces must be chosen with centers located on a horizontal line passing through O_1.

O_5 is located to the left of O_1 (Fig. 6a-6) and the safety factor computed for this fifth trial. The data are presented in Table 6a-5. The value of the radius is R_5 = 13.1 m (42.98 ft); the length of the failure arc is L_5 = 25.7 m (84.32 ft).

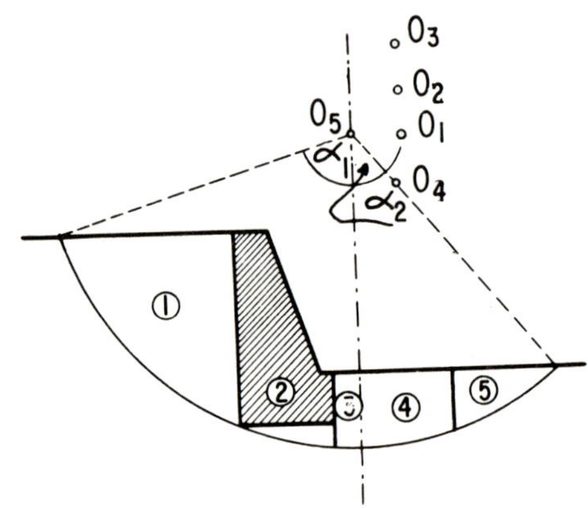

Fig. 6a-6 Fifth trial

The safety factor for this fifth trial is:

$$F_s = \frac{362.8 + 25.7 \times 1 \times 13.1}{509.8}$$

$$= \frac{800 + 84.32 \times 0.205 \times 42.98}{1120.3} = 1.37$$

The safety factor decreases towards the left, so the values to the left of O_5 will be investigated.

O_6 is placed 2 m (6.56 ft) to the left of O_5 (Fig. 6a-7) and the same computations conducted as previously (Table 6a-6). The value of the radius is R_6 = 12.4 m (40.68 ft) the length of the sliding surface is L_6 = 23.24 m (76.24 ft).

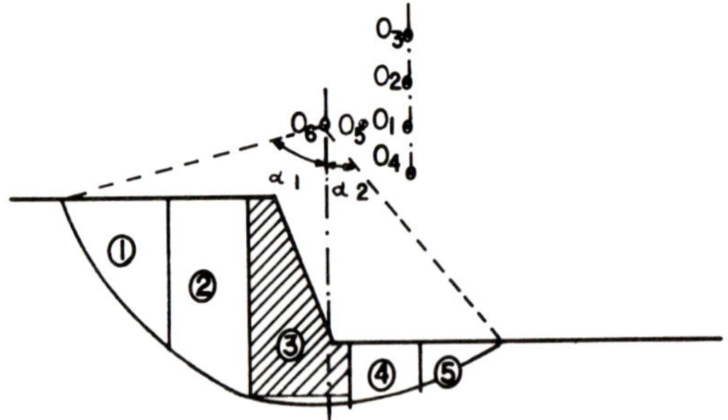

Fig. 6a-7 Sixth trial

The safety factor for the sixth failure arc is:

$$F_s = \frac{352.5 + 23.24 \times 1 \times 12.4}{525.9}$$

$$= \frac{776.1 + 76.24 \times 0.205 \times 40.68}{1160.0} = 1.22$$

Table 6a-4
Fourth trial

No of slice	Area of slice (A_i)		Weight of slice (W_i)		Moment distance (b_i)		W_ib_i		θ_i	$W_i \cos \theta_i \tan \varnothing_u R_4$	
	m²	ft²	t/m	kips/ft	m	ft	t-m/m	ft-kips/ft	deg	t-m/m	ft-kips/ft
1	19.6	212	39.2	26.3	8.8	28.8	344.3	758.5	46.0	71.2	156.4
2	26.2	283	52.4	35.1	5.3	17.3	278.0	612.5	25.5	123.5	272.0
3	12.6	136	25.2	16.9	1.4	4.6	35.2	77.5	6.8	65.2	143.3
4	15.1	163	30.2	20.2	−1.7	− 5.6	− 51.3	− 113.0	8.1	78.0	171.4
5	10.4	112	20.8	13.9	−5.3	−17.3	−110.5	− 243.5	25.0	49.2	108.1
					Total:		495.7	1092.0		387.1	851.2

Table 6a-5
Fifth trial

No of slice	Area of slice (A_i)		Weight of slice (W_i)		Moment distance (b_i)		W_ib_i		θ_i	$W_i \cos \theta_i \tan \varnothing_u R_5$	
	m²	ft²	t/m	kips/ft	m	ft	t-m/m	ft-kips/ft	deg	t-m/m	ft-kips/ft
1	29.2	316	58.4	39.2	7.2	23.6	421.0	925.8	33.6	134.4	296.5
2	24.3	262	48.6	32.6	3.6	11.8	175.0	385.0	16.0	129.0	284.6
3	2.9	31	5.8	3.9	0.5	1.64	2.9	6.4	2.0	15.9	35.2
4	9.5	102	19.0	12.7	−1.7	− 5.6	− 32.3	− 71.4	7.0	52.0	114.5
5	6.1	66	12.2	8.2	−4.65	−15.2	− 56.8	− 125.5	20.4	31.5	69.2
					Total:		509.8	1120.3		362.8	798.7

Table 6a-6
Sixth trial

No of slice	Area of slice (A_i)		Weight of slice (W_i)		Moment distance (b_i)		W_ib_i		θ_i	$W_i \cos \theta_i \tan \varnothing_u R_6$	
	m²	ft²	t/m	kips/ft	m	ft	t-m/m	ft-kips/ft	deg	t-m/m	ft-kips/ft
1	16.0	173	32.0	21.4	8.2	26.9	262.0	578.0	42.6	62.0	136.6
2	25.2	272	50.4	33.8	4.9	16.0	248.0	547.0	23.5	122.0	268.5
3	23.8	257	47.6	31.9	1.5	4.9	71.2	156.5	7.0	124.5	274.0
4	6.0	65	12.0	8.0	−2.4	− 7.8	− 28.8	− 63.3	11.4	31.0	68.3
5	2.7	29	5.4	3.6	−4.9	−16.0	− 26.5	− 58.2	23.4	13.0	28.7
					Total:		525.9	1160.0		352.5	776.1

As the safety factors continue to decrease towards the left, O_7 is located (Fig. 6a-8) and the safety factor computed for this seventh trial. The data are presented in Table 6a-7. The value of the radius is R_7 = 12.0 m (39.37 ft); the length of the failure arc is L_7 = 21.8 m (71.52 ft). The safety factor for this seventh trial is:

$$F_s = \frac{373.6 + 21.8 \times 1 \times 12.0}{480.6}$$

$$= \frac{822.0 + 71.52 \times 0.205 \times 39.37}{1056.6} = 1.32$$

In this way, the critical circle turns to be the one with center in point O_6, and the associated safety factor is 1.22.

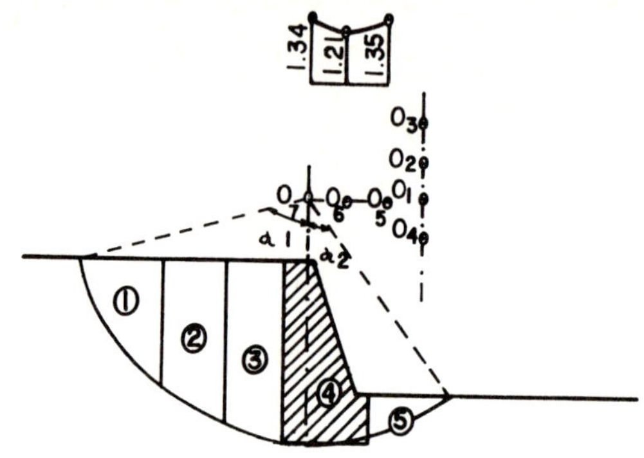

Fig. 6a-8 Seventh trial

Table 6a-7
Seventh trial

No of slice	Area of slice (A_i)		Weight of slice (W_i)		Moment distance (b_i)		W_ib_i		θ_i	$W_i \cos \theta_i \tan \varnothing_u R_7$	
	m²	ft²	t/m	kips/ft	m	ft	t-m/m	ft-kips/ft	deg	t-m/m	ft-kips/ft
1	12.2	132	24.4	16.4	8.3	27.2	199.0	438.0	44.3	44.5	98.0
2	20.5	221	41.0	27.4	5.5	18.0	226.0	496.0	27.0	93.0	204.3
3	22.6	244	45.2	30.3	2.3	7.5	104.0	229.0	11.5	113.0	248.7
4	21.5	232	43.0	28.8	−0.6	− 1.96	− 25.8	− 56.8	3.0	109.0	239.7
5	2.9	31	5.8	3.9	−3.9	−12.8	− 22.6	− 49.6	19.2	14.1	31.3
					Total:		480.6	1056.6		373.6	822.0

6a.3 Analysis in Terms of Total Stresses

Now the stability of a homogeneous embankment, located above the groundwater table, will be analyzed. From an unconsolidated-undrained triaxial test run on the embankment material, the following strength parameters were determined:

$\varnothing_u$ = 4°, c_u = 4 t/m² (5.68 lb/in²) and a unit weight of approximately 1.6 t/m³ (100 lb/ft³). The geometry of the slope is shown in Fig. 6a-9.

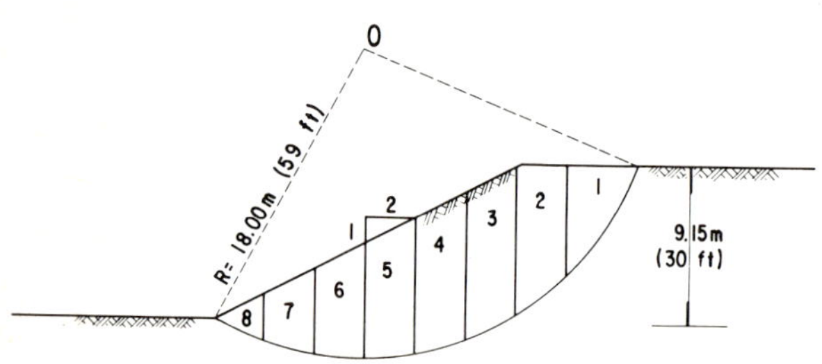

Fig. 6a-9 Geometry of the cross-section

In this case, the total stress criterion and the method proposed by Fellenius are used. The analysis for the critical sliding surface is given here, corresponding in this case to the one passing through the toe of the slope. Table 6a-8 can be used to mechanize the calculation procedure.

After dividing the soil mass into slices, in accordance with the assumed sliding surface, the volume of soil in each slice can be estimated. This volume will be numerically equal to the area, considering a unitary thickness. The weight of each slice can then be estimated and entered under heading (W_i).

The reaction to W_i must be broken down into directions that are normal and tangential to the sliding surface at the base of each one of the slices. These two components are entered under headings (N_i) and (T_i) respectively.

Component (N_i) generates a normal stress which, as already mentioned, can be approximated as:

$$\sigma_i = \frac{N_i}{L_i}$$

This stress is entered under heading (σ_i). Now the shear resistance that develops at the sliding surface for each slice can be found, as:

Table 6a-8
Calculation of stability in terms of total stresses

Slice No.	W_i		N_i		T_i		$\sigma_i = \frac{N_i}{L_i}$		s_i		s_iL_i	
	t	Kips	t	Kips	t	Kips	t/m²	KSf	t/m²	KSf	t	Kips
1	26.64	58.73	16.97	37.41	21.67	47.77	2.09	0.43	4.14	0.85	33.53	73.92
2	37.08	81.75	28.98	63.89	23.29	51.34	7.60	1.55	4.53	0.93	17.26	38.05
3	43.20	95.24	39.73	87.59	20.37	44.91	12.04	2.46	4.84	0.99	15.97	35.21
4	40.32	88.89	37.44	82.54	11.26	24.82	11.55	2.36	4.80	0.98	15.55	34.28
5	35.28	77.78	34.22	75.44	3.94	8.68	11.40	2.33	4.79	0.98	14.37	31.68
6	28.08	61.90	27.94	61.59	− 1.58	− 3.48	9.31	1.90	4.65	0.95	13.95	30.75
7	18.75	41.27	18.22	40.17	− 4.29	− .46	5.62	1.15	4.39	0.90	14.22	31.35
8	6.48	14.28	5.91	13.03	− 2.39	− 5.27	1.82	0.37	4.13	0.84	13.38	29.50
				$\sum =$	72.27	159.31				$\sum =$	138.23	304.74

$$s_i = 4 + \sigma_i \tan 4° \ [\text{t/m}^2]$$

$$s_i = 5.68 + \sigma_i \tan 4° \ [\text{lb/in}^2]$$

This value is entered under heading (s_i).

The resisting force can be calculated for each slice as the product s_iL_i.

Finally the safety factor is computed as:

$$F_s = \frac{\sum s_iL_i}{\sum T_i} = \frac{138.23}{72.27} = \frac{304.74}{159.31} = 1.91$$

The proposed slope can be regarded as stable.

6a.4 Analysis with Circular Failure Surface and in Terms of Effective Stresses

The example here corresponds to a section of road which, like the one shown in Fig. 6a-10, is built partially between embankments and partially through a cut. This section is subjected to seepage from uphill. The material found after field exploration and laboratory testing had been carried out was a stiff clay, with a wet unit weight of 2.0 t/m³ (125 lb/ft³) and drained shear strength parameters of c = 0.5 t/m² (102 lb/ft²) and $\varnothing$ = 30°, which are shown graphically in Fig. 6a-11. Stability analysis is carried out assuming a cylindrical sliding surface through the toe of the slope, and following the two procedures below:

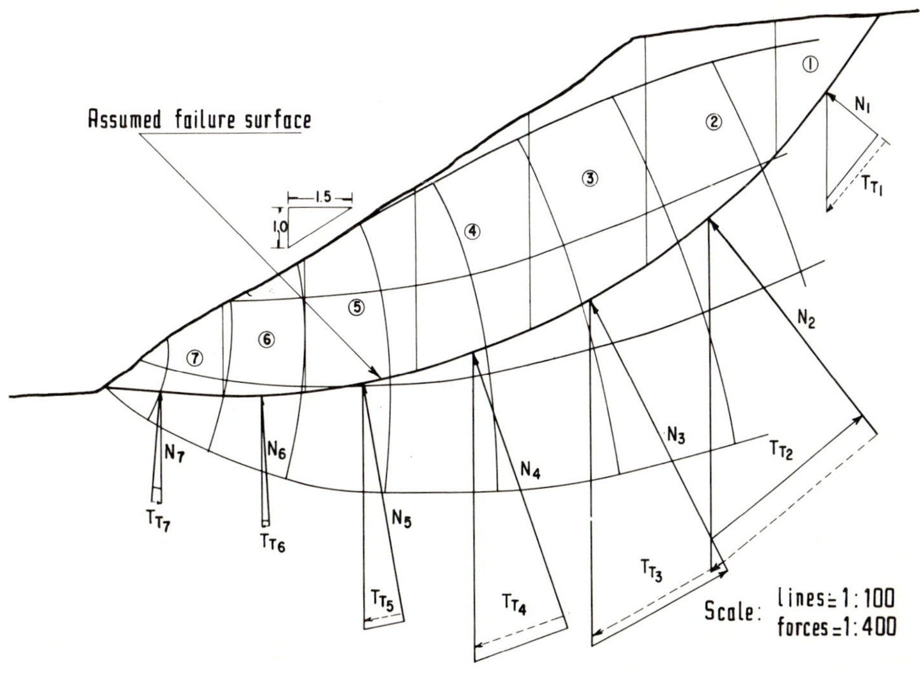

Fig. 6a-10 Profile of the hillside, flow net for steady flow conditions and stability analysis

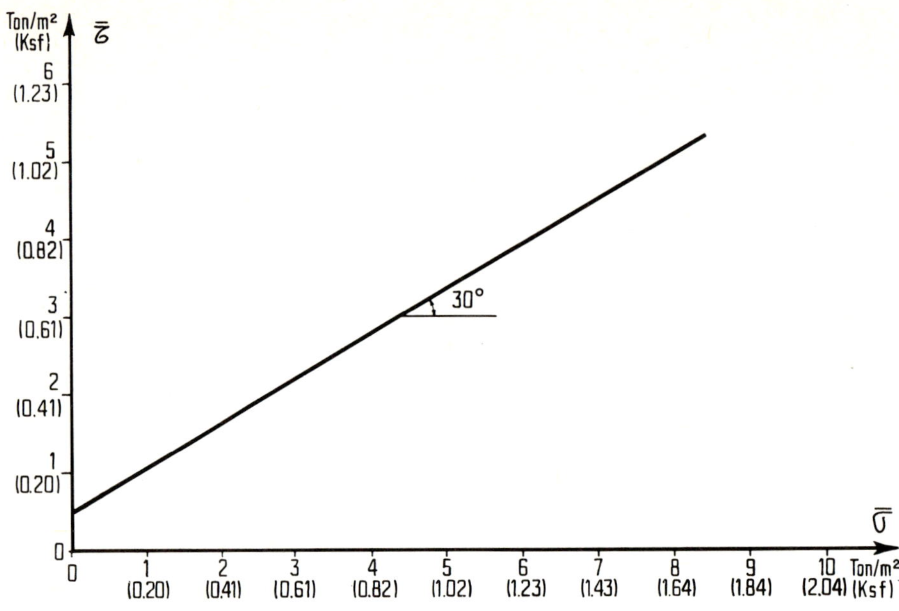

Fig. 6a-11 Strength law in terms of effective stress

Since the slope is subjected to steady flow conditions, the flow net must be drawn as illustrated in Fig. 6a-10.

Next, draw the sliding surface through the toe of the slope.

Analysis with Pore Pressures

Divide the sliding mass into slices. Here there are seven. Draw up a chart to include all the numerical figures, as in Table 6a-9.

Analysis Considering Seepage Forces

Table 6a-10 summarizes the computations. The seepage forces cause an increase in the driving forces, which tends to make the sliding mass turn about the center of the sliding surface.

Table 6a-9

Analysis with circular failure surface in terms of effective stresses — with pore pressures

Slice	W_i		N_i		L_i		$\frac{N_i}{L_i} = \sigma_i$		u_i		$\sigma_i\text{-}u_i = \bar{\sigma}_i$		$\bar{s}_i$		$\bar{s}_i\ L_i$		W_{Total}		T_{Total}	
	t/m	Kips/ft	t/m	Kips/ft	m	ft	t/m²	Ksf	t/m²	Ksf	t/m²	Ksf	t/m²	Ksf	t/m	Kips/ft	t/m	Kips/ft	t/m	Kips/ft
1	10.92	7.34	7.0	4.70	4.2	13.78	1.67	0.34	1.5	0.31	0.17	0.03	0.6	0.12	2.52	1.69	12.65	8.43	10.0	6.72
2	33.80	22.71	26.3	17.67	4.5	14.76	5.85	1.20	3.6	0.74	2.25	0.46	1.75	0.36	7.88	5.29	37.2	25.00	24.0	16.13
3	36.00	24.19	31.8	21.37	3.5	11.48	9.10	1.86	4.3	0.88	4.80	0.98	3.22	0.66	10.61	7.13	34.8	23.38	17.0	11.42
4	32.10	21.57	30.2	20.29	3.1	10.17	9.75	1.99	4.5	0.92	5.25	1.07	3.50	0.72	10.85	7.29	30.5	20.49	10.0	6.72
5	25.90	17.40	25.2	16.93	3.1	10.17	8.36	1.71	4.1	0.84	4.26	0.87	2.91	0.59	11.95	8.03	25.3	17.00	4.8	3.22
6	13.10	8.80	13.0	8.73	2.3	7.54	5.65	1.15	3.0	0.61	2.65	0.54	2.00	0.41	4.60	3.09	13.5	9.04	1.0	0.67
7	9.00	6.05	8.9	5.98	3.6	11.81	2.48	0.51	1.5	0.31	0.98	0.20	1.05	0.21	3.78	2.54	9.9	6.65	−0.8	−0.54
—	—	—	—	—	—	—	—	—	—	—	—	—	—	—	52.19	35.06	—	—	66.0	44.34

$$F_s = \frac{\sum_1^7 \bar{s}_i L_i R}{\sum_1^7 T_{Total} R} = \frac{52.19}{66.00} = \frac{35.06}{44.34} = 0.79$$

Table 6a-10

Analysis with circular failure surface in terms of effective stresses — considering seepage forces (symbols as in Table 6a-9)

Slice	W_i		T_i		N_i		L_i		$\frac{N_i}{L_i} = \sigma_i$		u_i		$\sigma_i - u_i = \bar{\sigma}_i$		$\bar{s}_i$		$\bar{s}_i L_i$	
	t/m	Kips/ft	t/m	Kips/ft	t/m	Kips/ft	m	ft	t/m²	Ksf	t/m²	Ksf	t/m²	Ksf	t/m²	Ksf	t/m	Kips/ft
1	10.9	7.34	8.6	5.78	7.0	4.70	4.2	13.78	1.66	0.34	1.5	0.31	0.17	0.03	0.60	0.12	2.52	1.69
2	33.8	22.71	21.0	14.11	26.3	17.67	4.5	14.76	5.85	1.20	3.6	0.74	2.25	0.46	1.75	0.36	7.88	5.29
3	36.0	24.19	17.0	11.42	31.8	21.37	3.5	11.48	9.10	1.86	4.3	0.88	4.80	0.98	3.22	0.66	10.61	7.13
4	32.1	21.57	10.5	7.05	30.2	20.29	3.1	10.17	9.75	1.99	4.5	0.92	5.25	1.07	3.50	0.72	10.89	7.29
5	25.9	17.40	5.0	3.36	25.2	16.93	3.1	10.17	8.36	1.71	4.1	0.84	4.26	0.87	2.91	0.59	11.95	8.03
6	13.1	8.80	1.0	0.67	13.0	8.73	2.3	7.54	9.69	1.97	3.0	0.61	2.65	0.54	2.00	0.41	4.60	3.09
7	9.0	6.05	−0.9	−0.60	8.9	5.98	3.6	11.81	2.48	0.51	1.5	0.31	0.98	0.20	1.05	0.21	3.78	2.54
	—	—	62.2	41.79	—	—	—	—	—	—	—	—	—	—	—	—	52.19	35.06

$$F_s = \frac{\sum_1^7 \bar{s}_i L_i R}{\sum_1^7 T_i R + \sum_1^7 J R} = \frac{52.19 \times 20.6}{62.2 \times 20.6 + 507.7}$$

$$= \frac{35.06 \times 67.58}{41.79 \times 67.58 + 1118.78} = 0.60$$

Key:

W_i — Weight of material of slice, regarding it as totally saturated
N_i — Normal component of W_i obtained graphically in Fig. 6a-10
L_i — Length of base of slice
σ_i — Total normal stress
u_i — Pressure exceeding hydrostatic, obtained from flow net
$\bar{\sigma}_i$ — Effective normal stress
$\bar{s}_i$ — Shear strength obtained from graph in Fig. 6a-11
$\bar{s}_i L_i$ — Resisting tangential or shear force
W_{Total} — Weight of saturated slice material above the saturation line, plus the submerged weight of the slice below the saturation line, plus the uplift pressure (which is the excess hydrostatic pressure obtained from the flow net for that slice multiplied by the length of the base of the slice) (algebraic sum).
T_{Total} — Acting tangential force, component of W_{Total} obtained graphically from Fig. 6a-10
R — Radius of failure circle under analysis. In this case, 20.6 m (67.6 ft).
F_s — Safety factor

The increment in the driving forces can be computed as follows. In each square of the flow net, if L_i is the average side of the square, the seepage force (J) in the square under consideration is equal to the unit weight of water times the drop in potential Δh times the average side L_i of that square, that is

$$J = j \times L^2 = \gamma_w i L^2 = \gamma_w \frac{\Delta h}{L} L^2 = \gamma_w \Delta h L$$

The direction of this force will be that of the flow line passing through the center of the square, it being possible then to define the normal distance to the center of the circle and the product of J times that distance will be the increase in the driving moment corresponding to the square under analysis. The sum of the moments of all the squares will give the total increment of the driving moment due to the seepage forces.

Calculation of the seepage forces moment is illustrated in Table 6a-11.

6a.5 Stability of a Natural Slope with a Non-Circular Sliding Surface, including Seepage. Analysis in terms of Effective Stresses.

Suppose that in a natural slope there is seepage, as shown in Fig. 6a-12. The slope is composed of a slightly preconsolidated clay. Consolidated drained tests performed on undisturbed specimens gave the following results: $c = 0.5$ t/m² (102 lb/ft²); $\varnothing = 32°$; $\gamma_{sat} = 2$ t/m³ (125 lb/ft³). Calculate the safety factor corresponding to the sliding surface shown in the figure.

Table 6a-11

Calculation of the seepage forces moment

Number of square	$\triangle h$		L		R		$M_j = \triangle h\, L\, R$	
	m	ft	m	ft	m	ft		
I	0.9	2.95	2.0	6.56	17.25	56.59	31.00	68.31
II	1.2	3.94	2.4	7.87	17.25	56.59	49.60	109.30
III	1.1	3.61	2.4	7.87	17.25	56.59	45.50	100.26
IV	1.0	3.28	2.9	9.51	17.70	58.07	51.10	112.60
V	0.9	2.95	2.9	9.51	18.60	61.02	48.50	106.88
VI	0.7	2.29	2.6x0.5	4.25	19.00	62.33	17.30	38.12
VII	1.0	3.28	1.6	5.25	19.00	62.33	30.40	66.99
VIII	0.9	2.95	1.9	6.23	19.00	62.33	32.50	71.62
IX	0.9	2.95	2.2	7.22	19.00	62.33	37.50	82.63
X	1.2	3.94	2.6x0.8	6.82	19.30	63.32	48.10	105.99
XI	1.1	3.61	2.5x0.7	5.74	19.00	62.33	36.60	80.65
XII	1.0	3.28	3.1x0.5	5.08	19.70	64.63	30.50	67.21
XIII	0.9	2.95	3.0x0.3	2.95	20.20	66.27	16.30	35.92
XIV	0.5	1.64	1.2x0.6	2.36	20.10	65.94	7.20	15.87
XV	1.0	3.28	1.6x0.4	2.10	20.15	66.11	13.55	29.86
XVI	0.9	2.95	2.1x0.2	1.38	20.30	66.60	7.65	16.86
XVII	0.9	2.95	2.4x0.1	0.79	20.40	66.93	4.40	9.69
Sum							507.70	1118.78

Where:

$\triangle h$ = pressure head at point under consideration
L = average side of square
R = Lever arm in relation to center of circle

Solution:

This problem can be considered a case of non-circular failure surfaces. To analyze it, Eq. (6-34) from §6.5.3 will be used [3].

The meaning of the symbols that appear in this equation can be seen in Fig. 6-37.

The sliding mass was divided into 6 slices and Table 6a-12 was used to solve Eq. (6-34). It should be mentioned that the term $\triangle s$ is ignored in the solution of this problem.

The way in which quantities appearing in Table 6a-12 are obtained is shown graphically in Fig. 6a-12 for Slice 5.

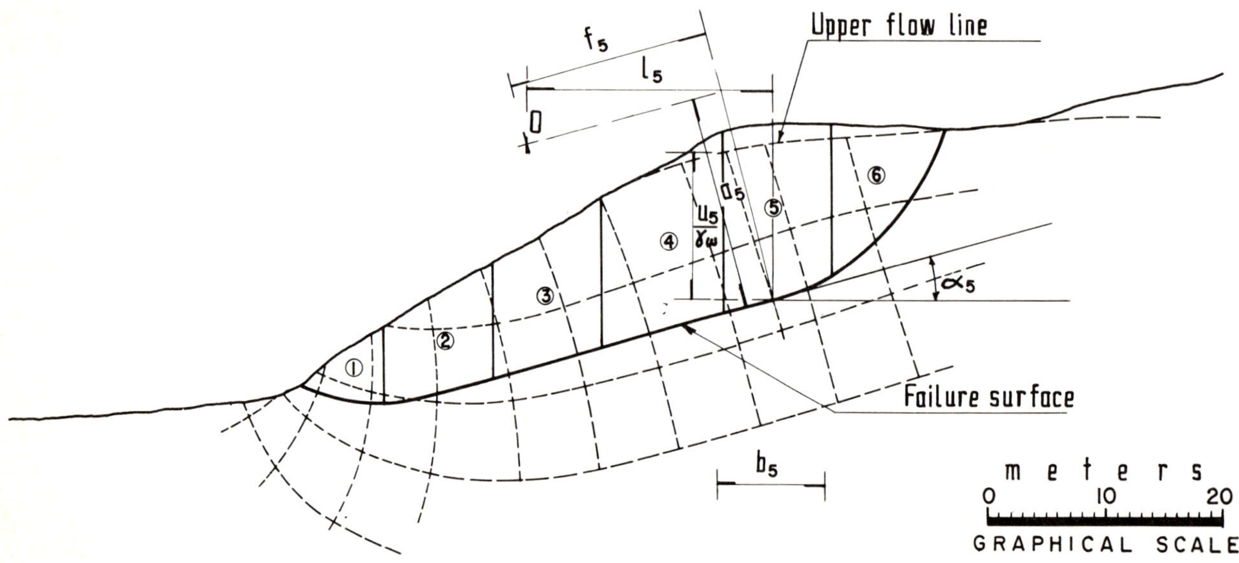

Fig. 6a-12 Stability analysis

Table 6a-12

Calculation of the stability of a natural slope with a non-circular sliding surface, including seepage. Analysis in terms of effective stresses.

Col.	1		2		3		4		5	6	7		8	9		10		11		12	
	b		a		l		f		α	tan α	c		tan $\varnothing$	(7).(1)		u		(10).(1)		W_1	
Slice	m	ft	m	ft	m	ft	m	ft	deg		t/m^2	Ksf		t/m	Kips/ft	t/m^2	Ksf	t/m	Kips/ft	t/m	Kips/ft
1	1.0	22.4	24.5	80.4	−14.8	−48.5	− 9.1	−29.8	−14	−0.249	0.5	0.10	0.625	3.5	2.4	4.6	0.94	32.2	21.6	0.0	0.0
2	9.4	30.8	17.9	58.7	− 7.5	−24.6	−12.9	−42.3	11	0.194	0.5	0.10	0.625	4.7	3.2	7.9	1.62	74.2	49.9	0.0	0.0
3	9.4	30.8	17.9	58.7	1.9	6.2	− 3.1	−10.2	11	0.194	0.5	0.10	0.625	4.7	3.2	8.9	1.82	83.6	56.2	0.0	0.0
4	10.0	32.8	17.9	58.7	11.6	38.1	6.9	22.6	11	0.194	0.5	0.10	0.625	5.0	3.4	11.4	2.33	114.0	76.6	20.7	13.9
5	9.2	30.2	17.9	58.7	21.3	69.9	17.0	55.8	11	0.194	0.5	0.10	0.625	4.6	3.1	11.9	2.43	109.4	73.5	33.6	22.6
6	10.0	32.8	25.1	82.4	29.2	95.8	17.0	55.8	40	0.839	0.5	0.10	0.625	5.0	3.4	8.6	1.76	86.0	57.8	14.0	9.4

Col.	13		14		15		16		17		18		19		20		21		22	
	$\overline{W}$		(12)+(13)		(14)−(11)		(15).(8)		(9)+(16)		(17).(2)		(14).(3)		(11).(8)		(20)−(9)		(21).(6)	
Slice	t/m	Kips/ft	t/m	Kips/ft	t/m	Kips/ft	t/m	Kips/ft	t/m	Kips/ft	$\frac{\text{t-m}}{\text{m}}$	$\frac{\text{ft-Kips}}{\text{ft}}$	$\frac{\text{t-m}}{\text{m}}$	$\frac{\text{ft-Kips}}{\text{ft}}$	t/m	Kips/ft	t/m	Kips/ft	t/m	Kips/ft
1	21.6	14.5	21.6	14.5	−10.6	− 7.2	− 6.63	− 4.45	− 3.13	− 2.10	− 76.8	− 169	− 320	− 705	20.1	13.7	16.6	11.3	− 4.14	− 2.81
2	73.0	49.0	73.0	49.0	− 1.2	− 0.8	− 0.75	− 0.50	− 3.95	2.65	70.7	− 156	− 547	−1206	46.4	31.5	41.7	28.3	− 8.10	5.50
3	100.1	67.3	100.1	67.3	16.5	11.1	10.30	6.92	15.00	10.08	268.5	592	190	419	52.2	35.5	47.5	32.3	9.21	6.26
4	125.0	84.0	145.7	97.9	31.7	21.3	19.80	13.30	24.80	16.66	444.0	979	1690	3726	71.3	48.5	66.3	45.1	12.90	8.77
5	110.3	74.1	143.9	96.7	34.5	23.2	21.60	14.51	26.20	17.60	469.0	1034	3070	6768	68.4	46.5	63.8	43.4	12.40	8.43
6	55.0	36.9	69.0	46.4	17.0	11.4	−10.60	− 7.12	− 5.60	− 3.76	−140.6	− 310	2020	4453	53.7	36.5	48.7	33.1	40.90	27.80
											Sum		6103	13455						

FIRST TRIAL

Col.	23	24		25		26		27	28		29	
	F_{s1}	(22)/(23)		(14)+(24)		(25).(4)		M_i (α)	(18)/(27)		(26)/(27)	
Slice		t/m	Kips/ft	t/m^2	Ksf	$\frac{\text{t-m}}{\text{m}}$	$\frac{\text{ft-Kips}}{\text{ft}}$		$\frac{\text{t-m}}{\text{m}}$	$\frac{\text{ft-Kips}}{\text{ft}}$	$\frac{\text{t-m}}{\text{m}}$	$\frac{\text{ft-Kips}}{\text{ft}}$
1	1.2	− 3.45	− 2.34	18.15	3.71	− 165	− 364	0.84	− 91.4	− 201	− 196	− 432
2	1.2	6.75	4.59	79.75	16.31	−1030	−2271	1.07	66.1	146	− 963	−2123
3	1.2	7.67	5.21	107.77	22.04	334	736	1.07	251.0	553	− 312	− 668
4	1.2	10.74	7.30	156.44	32.00	1080	2381	1.07	415.0	915	−1010	−2227
5	1.2	10.32	7.01	154.22	31.55	2620	5776	1.07	438.0	966	2450	5401
6	1.2	34.10	23.18	103.10	21.09	1753	3865	1.09	−129.0	− 284	1610	3549
						Sum			949.7	2094	1579	3481

SECOND TRIAL

Slice	23	24		25		26		27	28		29	
1	0.7	− 5.91	− 4.02	15.70	10.67	− 143	− 315	0.77	− 99.5	− 219	− 186	− 410
2	0.7	11.50	7.82	84.50	57.43	−1090	−2403	1.15	61.5	136	− 948	−2090
3	0.7	13.10	8.90	113.20	76.44	− 351	− 773	1.15	234.0	516	− 305	− 672
4	0.7	18.40	12.51	164.10	111.54	1130	2491	1.15	386.0	851	983	2167
5	0.7	17.70	12.03	161.60	109.84	2750	6063	1.15	408.0	899	2390	5269
6	0.7	58.40	39.69	127.40	86.59	2170	4784	1.34	−105.0	− 231	1620	3571
						Sum			885.0	1951	3554	7835

CALCULATION

$$F_{s2} = \frac{\Sigma(28)}{\Sigma(19) - \Sigma(29)}$$

First Trial: $F_{s1} = 1.2$;

$$F_{s2} = \frac{949.7}{6103 - 1579} = \frac{2094}{13455 - 3481} = 0.21$$

Second Trial: $F_{s1} = 0.7$;

$$F_{s2} = \frac{885.0}{6103 - 3554} = \frac{1951}{13455 - 7835} = 0.34$$

Weights W_1 and $\overline{W}$ were determined by dividing the slices into simple geometrical shapes in order to calculate their area.

For this particular stability analysis, the values of f and l in the table are positive when they are to the right of O. To calculate M_i (α), Fig. 6-35 was used.

Because to find the F_s in Eq. (6-34) several trials have to be done, two of them were performed in the table. In the first one, a $F_{s1} = 1.2$ was assumed, and a $F_{s2} = 0.2$ was obtained. In the second, a $F_{s1} = 0.7$ was used and a $F_{s2} = 0.5$ was found. It is very likely that in a third trial a $F_s = 0.6$ would be obtained; that is to say that the slope is unstable in the conditions given.

Table 6a-13

Embankment on soft soil: strength parameters from vane test

Depth		Shear Strength-S		W_L	W_p	I_p
m	ft	t/m²	psi	%	%	
2	6.56	2.74	3.89	225	130	95
4	13.12	2.25	3.20	242	129	113
6	19.68	1.80	2.56	251	146	105
8	26.24	1.92	2.73	248	138	110
10	32.81	2.05	2.91	301	190	111
12	39.37	2.12	3.01	272	147	125
14	45.93	1.87	2.66	290	179	111
16	52.49	1.93	2.74	248	126	122
18	59.05	1.95	2.77	253	146	107
20	65.62	2.11	3.00	274	168	106

6a.6 Embankment on Soft Soil

An embankment is to be built with the cross-section shown in Fig. 6a-13. It will lie on a soft soil with strength parameters determined by a vane test, appearing in Table 6a-13, together with the ATTERBERG limits. The embankment will have a height of 3 m (9.84 ft), a slope angle $\beta = 18° 24'$ (3:1) and it will be built with a carefully compacted clayey material, with $\gamma_m = 1.67$ t/m³ (104 lb/ft³) and $c = 4$ t/m² (818 lb/ft²).

The γ_{sat} of the foundation soil is 1.2 t/m³ (75 lb/ft³).

Taking 2 t/m² (409 lb/ft²) as the average value of the cohesion, in accordance with the values of the I_p (mean $I_p = 110$), the strength obtained in the vane test must be multiplied by a correction factor of 0.61 (Fig. 6-42).

For purposes of stability analysis, the embankment will be regarded as cracked, which means that there will be no shear strength along the sliding surface within the embankment.

Here is a description of the procedure used to calculate the safety factor of sliding surface No. 1 (Fig. 6a-13). The driving moment will be given by the weight of the mass tending to cause the slide, times its distance from the vertical axis passing through the center of the circle under consideration.

The moment of the forces that are opposed to the slide, that is the resisting moment, will be given by the cohesion along the entire assumed sliding surface, times the radius of the assumed circle.

To simplify calculation of the driving moment, the sliding mass will be divided into slices, as shown in Fig. 6a-13. The

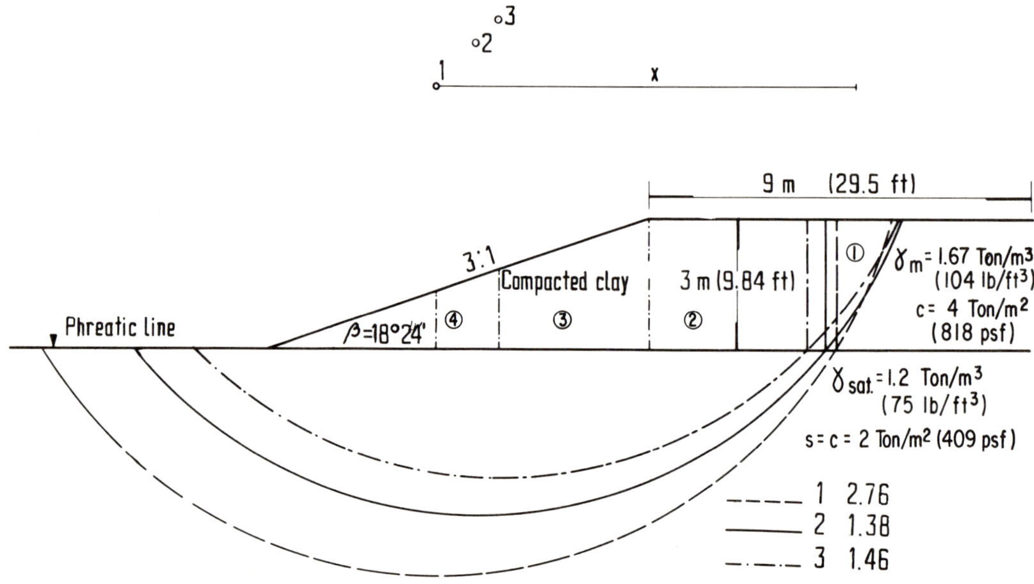

Fig. 6a-13 Geometry of the cross-section and stability analysis

Table 6a-14
Embankment on soft soil: Calculation of driving moment

Slice	W		x		Wx	
	t	Kips	m	ft	t-m	ft-kips
2	13.50	29.76	7.25	23.79	97.9	215.83
3	7.25	15.98	2.50	8.20	18.1	39.90
4	4.16	9.17	−1.33	− 4.36	− 5.5	− 12.12

$$M_m = \sum W \cdot x = 110.5\left(\frac{\text{t-m}}{\text{m}}\right) = 243.61\left(\frac{\text{ft-Kips}}{\text{ft}}\right)$$

natural ground is in equilibrium by itself. The weight of Slice *1* is omitted due to the presence of a crack between Slices *1* and *2*.

Table 6a-14 summarizes calculations of the driving moment.

In order to calculate the resisting moment, the shear strength s will be determined by a vane test and modified by the correction factor of 0.61 (Fig. 6-42), so that $M_r = s.L.r = 0.61 \times 2.00 \times 22.66 \times 11.2 = 309.6$ t-m/m ($= 0.61 \times 0.41 \times 74.34 \times 36.74 = 683.1$ ft-kips/ft) where L = length of the sliding surface, where shear strength (s) can develop, r = radius of the circle. The corresponding factor of safety will be $F_s = M_r/M_m$ and for Circle *1* it is equal to 2.76.

With a similar analysis for sliding surfaces *2* and *3*, the following safety factors were obtained: $F_{s2} = 1.38$; $F_{s3} = 1.46$.

6a.7 The Wedge Method

The wedge method is used here to determine the safety factor for an embankment 10 m (33 ft) high and with a slope of 1.5:1, built on firm rock (Fig. 6a-14). The embankment is built with clayey sand and its strength parameters, as determined in a drained test, are: $\varnothing = 27°$, $c = 0.6$ t/m^2 (123 lb/ft^2). The material was compacted until a unit weight $\gamma_m = 1.8$ t/m^3 (112 lb/ft^3) was reached.

DATA:

h = 10 m (33 ft) Slope = 1.5:1
γ_m = 1.8 t/m^3 (112 lb/ft^3)
$\varnothing$ = 27°
c = 0.6 t/m^2 (123 lb/ft^2)

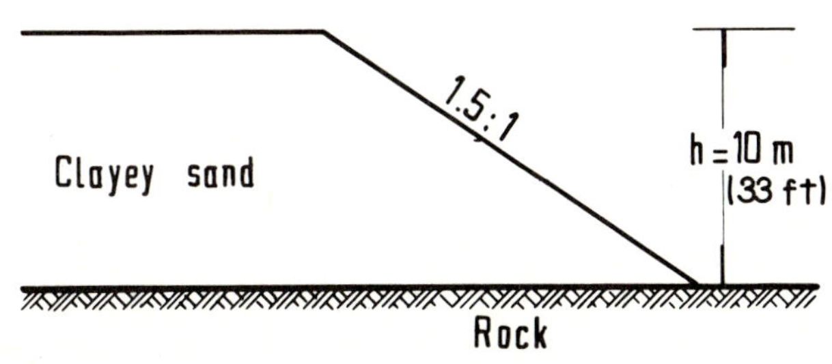

Fig. 6a-14 Presentation of the problem

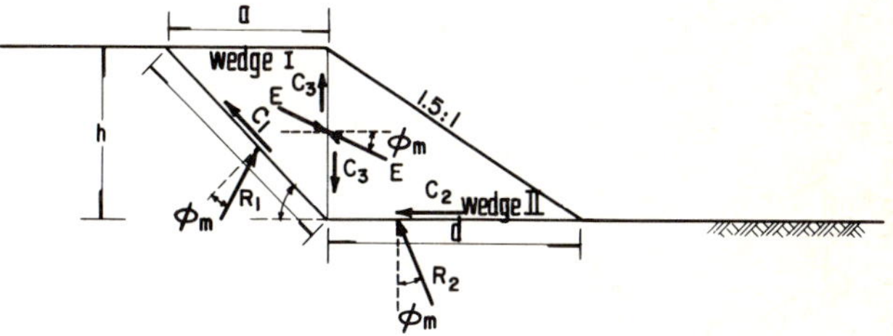

Fig. 6a-15 Stability analysis

SOLUTION:

It is assumed that the inclination of the thrust E between the wedges is:

$$\varnothing_m = \tan^{-1}\frac{\tan\varnothing}{F_s}$$

A trial and error procedure is followed, dividing the embankment into the two wedges shown in Fig. 6a-15 and making the angle α vary until the minimum F_s at which the force polygon can be closed is determined. With the forces that intervene in the equilibrium of the two wedges, for a certain value of α and an assumed F_s, the force polygon in Fig. 6a-16 is drawn.

For α = 42° 30′ and $F_s = 1.9$

$$\varnothing_m = \tan^{-1}\frac{\tan 27°}{1.9} = 15°$$

$$C_m = \frac{c}{F_s} = \frac{0.6}{1.9} = 0.315 \text{ t/m}^2;\quad c_m = \frac{0.123}{1.9} = 0.0647 \text{ kips/ft}^2$$

$$C_1 = bc_m = 14.8 \times 0.315 = 4.66 \text{ t/m};\ 48.55 \times 0.0647 = 3.14 \text{ kips/ft}$$

$$C_2 = dc_m = 15.0 \times 0.315 = 4.74 \text{ t/m}; 49.21 \times 0.0647 = 3.18 \text{ kips/ft}$$

$$C_3 = hc_m = 10.0 \times 0.315 = 3.15 \text{ t/m}; 32.81 \times 0.0647 = 2.12 \text{ kips/ft}$$

$$W_I = \frac{ha}{2} \gamma_m \frac{10 \times 10.9}{2} \times 1800 = 98 \text{ t/m};$$

$$\frac{32.81 \times 35.76}{2} \times 0.11232 = 65.9 \text{ kips/ft}$$

$$W_{II} = \frac{hd}{2} \gamma_m = \frac{10 \times 15}{2} \times 1800 = 135 \text{ t/m};$$

$$\frac{32.81 \times 49.21}{2} \times 0.11232 = 90.7 \text{ kips/ft}$$

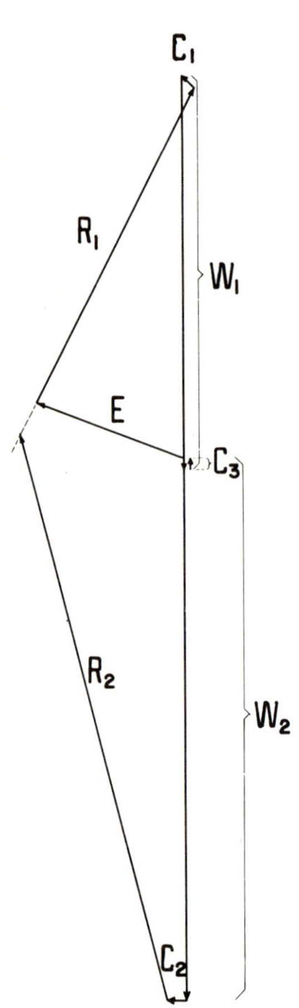

Fig. 6a-16 Force polygon

Making an approximation, the safety factor required for the force polygon to close is 1.95. Using similar trial and error procedures for different values of α, Table 6a-15 is obtained, in which it is shown that the safety factor for the embankment which corresponds to the wedge geometry selected is 1.95.

If the geometry of Wedge *II* is also made to vary, different values for F_s will be found. It is suggested that the reader calculate these values. The problem is solved when the smallest safety factor is found.

Table 6a-15

Trial and error calculation of F_s in Wedge Method

α	F_s
40°	2.10
42.5°	1.95
45°	2.00
50°	2.20

6a.8 Translational Failure

The stability of the embankment slope in Fig. 6a-17 is to be analyzed. The uppermost stratum and embankment are composed of a stiff clay with vertical cracks. The characteristics of the materials are given in the figure. In the region where the embankment is located there are plentiful autumn rains.

A translational failure will be analyzed. Since the cracks are open, active pressure will be zero. Nevertheless, the water that seeps into the cracks may cause a hydraulic thrust of:

$$\frac{1}{2} \cdot \gamma w \cdot H^2 = \frac{1}{2} \times 1.00 \times 6.5^2 = 21.10 \text{ t/m}$$

$$= \frac{1}{2} \times 62.4 \times 21.32^2 = 14.18 \text{ kips/ft}$$

The value of force F will be:

$$F = cL = 2 \times 7.5 = 15 \text{ t/m}$$
$$= 0.20457 \times 2 \times 24.6 = 10.08 \text{ kips/ft}$$

The passive thrust will be:

$$P_p = \frac{1}{2} \cdot \gamma_w \cdot H_1^2 \cdot K_p + 2cH_1 \sqrt{K_p} \therefore K_p = 1.52 \quad .52$$

$$P_p = 3.00 + 14.80 = 17.80 \text{ t/m}$$

$$= 2.01 + 9.95 = 11.96 \text{ kips/ft}$$

The safety factor is:

$$F_s = \frac{17.80 + 15}{21.10} \left(= \frac{11.96 + 10.08}{14.18} \right) = 1.55$$

and the slope is stable.

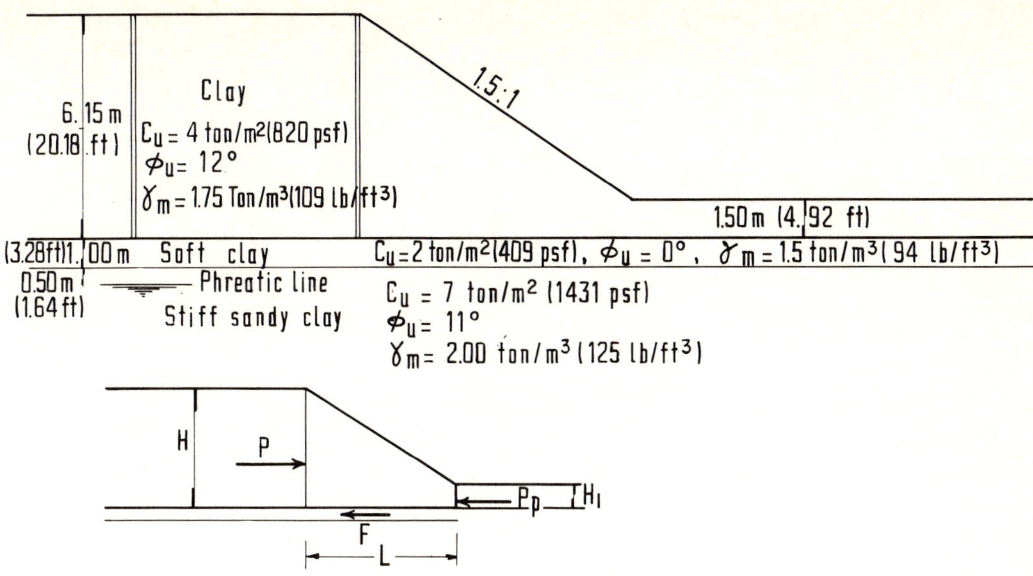

Fig. 6a-17 Presentation of the problem and stability analysis

REFERENCES

1. Da Costa Nunes, A. J., "Factores Geomorfológicos e climáticos na estabilidade de Taludes de Estradas," *Revista Latinoamericana de Geotecnia.* Vol. I, No 3., October-December, 1971.

2. Terzaghi, K., "Mechanism of Landslides," *From Theory to Practice in Soil Mechanics* (Selections of the Papers of K. Terzaghi) L. Bjerrum, A. Casagrande, R. B. Peck and A. W. Skempton), John Wiley and Sons, 1960, Part III.

3. Terzaghi, K., and Peck, R. B., *Soil Mechanics in Engineering Practice,* John Wiley and Sons, 1967, 2nd Edition.

4. Griggs, D. T., "Deformations of Rocks under High Confining Pressures," *Journal of Geology,* Vol. 44, 1936.

5. Bishop, A. W., "The Strength of Soils as Engineering Materials," *Geotechnique,* Vol. 16, No 2, 1966.

6. Goldstein, M., and Ter Stepanian, G., "The Long Term Strength of Clays and Depth Creep of Slopes," *Procs. of the IV. ICSMFE,* London, 1957, Vol. II.

7. Rico, A., Springall, G., and Springall, J., "Failures at the Tijuana-Ensenada Highway," Contribution of the Mexican Ministry of Public Works to the VII. ICSMFE. Publication of the Ministry of Public Works, Mexico, 1969.

8. Highway Research Board, "Landslides and Engineering Practice," Committee on Landslide Investigation, E. B. Eckel Ed., Special Report No 29, Washington D. C., 1958.

9. Skempton, A. W., and Hutchinson, J., "Stability of Natural Slopes and Embankment Foundations." *Procs. VII. ICSMFE,* Mexico, 1969, Session 5.

10. Sharpe, C. F. S., *Landslides and Related Phenomena: A Study of Mass-movements of Soil and Rock,* Columbia University Press, New York, 1938.

11. Koppejan, A. W., Van Wamelon, B. M., and Weinberg, L. J. H., "Coastal Flow Slides in the Dutch Province of Zeeland," *Procs. II. ICSMFE,* Rotterdam, 1948, Vol. 5.

12. Hutchinson, J. N., "The Stability of Cliffs Composed of Soft Rocks, With Particular Reference to the Coasts of South East England," Ph. D. Dissertation, Cambridge University, 1965, mentioned in [9]

13. Hutchinson, J. N., "The Free Fragmentation of London Clay Cliffs," *Proc. Geotechnical Conference,* Oslo, 1967, Vol. I.

14. Hutchinson, J. N., and Hughes, M. J., "The Application of Micropaleontology to the Location of a Deep Seated Slip Surface in London Clay," *Geotechnique,* Vol. 18, 1968.

15. EIDE, O., and BJERRUM, L., "The Slide at Bekkelaget," *Proc. European Conference on Stability of Earth Slopes,* Stockholm, 1954.

16. HENKEL, D. J., and SKEMPTON, A. W., "A Landslide at Jackfield, Shropshire, in an Over-consolidated clay," *Proc. European Conference on Stability of Earth Slopes,* Stockholm, 1954, Vol. I.

17. BAZETT, D. J., ADAMS, J. L., and MATYAS. E. L., "An Investigation of a Slide in a Test Trench Excavated in Fissured Sensitive Clay," *Proc. V. ICSMFE,* Paris, 1961, Vol. I.

17'. SKEMPTON, A. W., and LA ROCHELLE, P., "The Bradwell Slip, A Short Term Failure in London Clay," *Geotechnique,* Vol. 15, 1965.

18. JUÁREZ-BADILLO, E., and RICO, A., *Mecánica de Suelos, Vol. II: Teoría y Aplicaciones de la Mecánica de Suelos,* Limusa: Mexico, 1977, Chapter XI.

19. SHERARD, J.L., WOODWARD, R. J., GIZIENKI, S. F., and CLAVENGER, W. A., *Earth and Earth-Rock Dams,* John Wiley and Sons, 1963.

20. RICO, A., and OROZCO, R. V., "Formación de Grietas Longitudinales en Caminos," *Proc. XI. Panamerican Road Congress,* Quito, 1971.

21. SHERARD, J. L., "Influence of Soil Properties and Construction Methods on Performance of Homogeneous Earth Dams," US Bureau of Reclamation Technical Memorandum No 645, Denver, U.S.A., 1953.

22. NARAIN, J., "Flexibility of Compacted Clays," Ph. D. Thesis, Purdue University, 1962.

23. MEYERHOF, G. G., "The Mechanism of Flow Slides in Cohesive Soil," *Geotechnique,* Vol. 5, 1957.

24. HOLMSEN, P., "Landslips in Norwegian Quick Clays," *Geotechnique,* Vol. 3, 1953.

25. DEERE, D. U., and PATTON, F. D., "Estabilidad de Taludes en Suelos Residuales," (Translation by ALBERTO S. NIETO-PESCETTO) *Proc. IV. Panamerican Conference on SMFE,* San Juan, Puerto Rico, 1971, Vol. I.

26. VARGAS, M., "Some Engineering Properties of Residual Clay Soils occurring in Southern Brazil," *Proc. of the III. ICSMFE,* Zurich, 1953, Vol. I.

27. RUXTON, B. P., and BERRY, L. "Weathering of Granite and Associated Erosional Features in Hong Kong," *Bulletin of the American Geological Society,* Vol. 68, 1957.

28. JENNINGS, J. E., "Building on Dolomites in the Transvaal," *Trans. South African Inst. of Civil Engineers,* Vol. 8, No 2, 1966.

29. SKEMPTON, A. W., and DELÓRY, F. A., "Stability of Natural Slopes in London Clay," *Proc. IV. ICSMFE,* London, 1957, Vol. II.

30. SKEMPTON, A. W., "Long Term Stability of Clay Slopes," *Geotechnique,* Vol. 14, No 2, 1964.

31. HENKEL, D. J., "Local Geology and the Stability of Slopes," *A.S.C.E. Jour. Soil Mechanics,* Vol. 93, S.M. 4, 1967.

32. DEERE, D. U., and PATTON, F.D., "Effect of Pore Pressures on the Stability of Slopes," *GSA-ASCE Symposium,* New Orleans, 1967.

33. DE FRIES, C. K., and STOLK, E. P., "High Fills on Residual Soils", *IV. Panamerican Conference on SMFE,* San Juan, Puerto Rico, 1971, Vol. II.

34. NETO, N. A. F., "Discusión sobre Suelos Metamórficos y Residuales," *Proc. III. Panamerican Conference on SMFE,* Caracas, 1967, Vol. II.

35. SOWERS, G. F., "Discussion," *Proc. III. Panamerican Conference on SMFE,* Caracas, 1967, Vol. III.

36. LI, C. Y., and MEJÍA, O., "Building Earth Dams in a Region of Residual Soil in Colombia," *Proc. III. Panamerican Conference on SMFE,* Caracas, 1967, Vol. II.

37. EVDOKIMOV, P. D., and CHIRIAEV, R. A., "Quelques Lois de la Resistance au Cisaillement des Ouvrages de retenue en Beton sur Fundations Rocheuses," *Proc. I. International Conference on Rock Mechanics,* Lisbon, 1966, Vol. II.

38. ST. JOHN, B. J., SOWERS, G. F., and WEAVER, CH.E., "Slickensides in Residual Soils and their Engineering Significance," *Proc. VII. ICSMFE,* Mexico, 1969, Vol. II.

39. SOWERS, G. F., "Engineering Properties of Residual Soils Derived from Igneous and Metamorphic Rocks," *Proc. II. Panamerican Conference on SMFE,* Brazil, 1963, Vol. I.

40. WILSON, S. D., and MARANO, D., "Performance of Muddy River Embankment," *A.S.C.E., Jour. Soil Mech,* Vol. 94, SM-4, 1968.

41. ROCHA, M., "Mechanical Behavior of Rock Foundations in Concrete Dams," *Proc. VIII. International Conference on Large Dams (VIII. ICOLD),* Edinburgh, 1964, Vol. I.

42. CHANDLER, R. J., "The Effect of Weathering on the Shear Strength Properties of Keuper Marl," *Geotechnique,* Vol. 19, No 3, 1969.

43. KANJI, M. A., "Shear Strength of Soil-Rock Interfaces," M.Sc. Thesis, Geology Department, University of Illinois, Urbana, 1970.

44. KENNEY, T. C., "The Influence of Mineral Composition of the Residual Strength of Natural Soils," *Proc. Geotechnical Conference,* The Norwegian Geotechnical Institute, Oslo, 1967.

45. PECK, R. B., "Stability of Natural Slopes," A.S.C.E. Special Meeting on Stability and Performance of Slopes and Embankments, Berkeley, California, 1966.

46. BJERRUM, L., "Mechanism of Progressive Failure in Slopes of Over-consolidated Plastic Clay and Clay-Shales," *III. Terzaghi Lecture,* A.S.C.E. Meeting, Miami, 1966. (Also: *A.S.C.E. Jour. Soil Mech,* Vol. 93, SM-5. 1967).

47. PATTON. F. D., "Multiple Modes of Shear Failure in Rock and Related Materials," Ph. D. Thesis, University of Illinois, Urbana, 1966.

48. LOWE III, J., "Stability Analysis of Embankments", A.S.C.E. Special Meeting, Berkeley, California, 1966.

49. BISHOP, A. W., and BJERRUM, L., "The Relevance of the Triaxial Test to the Solution of Stability Problems," *Proc. Research Conference on Shear Strength of Cohesive Soils,* Boulder, Colorado, 1960.

50. JUÁREZ-BADILLO, E., and RICO, A., Reference [18], Vol. III, Chap. III.

51. Fellenius, W., "Calculation of the Stability of Earth Dams," *Proc. II. ICOLD,* Washington, 1936, Vol. IV.

52. Juárez-Badillo, E., and Rico, A., Reference [18], Vol. II, Chap. V.

53. Terzaghi, K., *Theoretical Soil Mechanics,* John Wiley and Sons, 1956, Art. 62.

54. Lambe, T. W., and Whitman, R. V., *Soil Mechanics,* John Wiley and Sons, 1971.

55. Taylor, D. W., "Stability of Earth Slopes," Contributions to Soil Mechanics, Boston Society of Civil Engineers, 1925-1940.

56. Taylor, D. W., *Fundamentals of Soil Mechanics,* John Wiley and Sons, 1956, Chap. 16.

57. Janbu, N., "Stability Analysis of Slopes with Dimensionless Parameters," Harvard Soil Mechanics Series No 46, Harvard University, 1954.

58. Whitman, R. V., and Bailey, W. A., "Use of Computers for Slope Stability Analysis," *Proc. A.S.C.E.,* Vol. 93, SM-4, 1967.

59. Bishop, A. W., "The Use of the Slip Circle in the Stability Analysis of Slopes," *Geotechnique,* Vol. 5, 1955.

60. Janbu, N., "Application of Composite Slip Surfaces for Stability Analysis," *Proc. European Conference on Stability of Earth Slopes,* Sweden, 1954.

61. Lambe, T. W., and Whitman, R. V., Reference [54], Chap. 24.

62. Arnold, M., "Slope Stability Analysis by a New Graphical Method," *Journal A.S.C.E.,* SM-5, 1961.

63. Martins, J. B., Maranha das Neves, E., and Guedes de Melo, F., "A Flexible Program for Automatic Analysis of Stability of Slopes," Proc. No 385, National Laboratory of Civil Engineering, Lisbon, 1971.

64. Carter, R. K., Lovell Jr., C. W., and Harr, M. E., "Computer Oriented Stability Analysis of Reservoir Slopes," Tech. Rep. No 17, Purdue University, 1971.

65. Lowe III, J., and Karafiath, L., "Stability of Earth Dams upon Drawdown," *Proc. I. Panamerican Conference on SMFE,* Mexico, 1960, Vol. II.

66. Nonveiller, E., "The Stability Analysis of Slopes with Slip Surfaces of General Shape," *Proc. VI. ICSMFE,* Montreal, 1965.

67. Morgenstern, N. R., and Price, V. E., "The Analysis of the Stability of General Slip Surfaces," *Geotechnique,* Vol. 15, 1965.

68. Escario, V., "Estabilidad de presas de tierra y escollera," Monograph No 1, Laboratorio del Transporte y Mecánica del Suelo, Madrid, 1966.

69. Bjerrum, L., "Embankments on Soft Ground," *Proc. of the II. A.S.C.E. Specialty Conference* on Performance of Earth and Earth-Supported Structures, Purdue University, Lafayette, 1972.

70. Casagrande, A., "An Unsolved Problem of Embankment Stability on Soft Ground," *Proc. of the I. Panamerican Conference on SMFE,* Mexico, 1960, Vol. II.

71. Dreyfus, G., "Etude des remblais Sur Sols compressibles," Recommendations des Laboratoires des Ponts et Chaussees, Dunod, Paris, 1971.

72. Pilot, G., "Study of Five Embankment Failures on Soft Soils," Reference [69] ibid.

73. Dascal, O., Tournier, J. P., Tavenas, F., and La Rochelle, P., "Failure of a Test Embankment on Sensitive Clay," ibid.

74. Wilkes, P. F., "An Induced Failure at a Trial Embankment at King's Lynn, Norfolk, England," ibid.

75. Ladd, Ch.C., "Test Embankment on Sensitive Clay," ibid.

76. Eide, O. and Holmerg, S., "Test Fills to Failure on the Soft Bangkok Clay," ibid.

77. Rico, A., Moreno, G. and Garcia, G., "Test Embankments on Texcoco Lake," *Proc. of VII. ICSMFE,* Mexico, 1969, Vol. II.

78. Holtz, R. D., and Lindskog, G., "Soil Movements below a Test Embankment," Reference [69] ibid.

79. "Slope Stability and Foundation Investigation," Prepared by Personnel of the Department of Investigations and Materials of the California Division of Highways, Transport and Traffic Engineering Institute, University of California at Berkeley, 1967.

80. Juárez-Badillo, E., and Rico, A. Reference [18], Vol. II, Chap. I.

81. Saito, M., and Uezawa, H., "Failure of Soil due to Creep," *Proc. of the V. ICSMFE,* Paris, 1961, Vol. I.

82. Saito, M., "Forecasting the Time of Occurrence of a Slope Failure," *Proc. VI. ICSMFE,* Montreal, 1965, Vol. II.

83. Resendiz, D., and Romo, M., "Analysis of Embankment Deformations," Reference [69] ibid.

84. Zaruba, Q., and Mencl, V., *Landslides and Their Control,* Czechoslovak Academy of Sciences – Elsevier, Prague and Amsterdam, 1969.

85. Beles, A. A., "Le Traitment thermique des Sols," *Proc. of the IV. ICSMFE,* London, 1957, Vol. III.

86. Ayres, D. J., "The Treatment of unstable Slopes and Railway track formation," *Journal Soc. of Engineering,* No 52, 1961.

87. Juárez-Badillo, E., and Rico, A., "Estabilización de un deslizamiento de tierras en el camino Toluca-Ixtapan de la Sal," *Journal of Construction,* August-September, 1962.

88. Habid, P., "Anchorages, especially in Soft ground," *Proc. VII ICSMFE,* Mexico, 1969, special session No. 15. Also: Mechanics of Solids Laboratory of the Ecole Polytecnique of Paris, Technical Publication.

89. Aguirre, L. M., Sánchez, D., and Zárate, M., "Performance Studies of the Mexico City International Airport," *Proc. II. International Conference on the Structural Design of Asphalt Pavements,* Michigan University, Ann Arbor, Mich., 1967.

90. Marsal, R. J., and Mazari, M., "El Subsuelo de la Ciudad de Mexico," Publication of the Institute of Engineering of the National Autonomous University of Mexico (UNAM), Mexico, 1959.

OTHER RELATED REFERENCES

CLOUGH, G. W., SITAR, N., BACHUS, R. C., AND SHAFII RAD, N., "Cemented Sands under Static Loading," *Journal of the Geotechnical Engineering Division,* Proc. ASCE, Vol. 107, No. GT6, June, 1981.

DEEN, R. C., HOPKINS, T. C. AND ALLEN, D. L., "Some Uncertainties of Slope Stability Analyses," Division of Research, Kentucky Department of Transportation, Frankfurt.

DODD, J. S. AND ANDERSON, W., "Tectonic Stresses and Rock Slope Stability," *13th Symposium on Rock Mechanics,* Urbana, Ill., August, 1971.

JAEGER, J. C., "Friction of Rocks and Stability of Rock Slopes," Eleventh Rankine Lecture, 1971.

LEWIS, W. A., MURRAY, R. T., AND SYMONS, I. F., "Settlement and Stability of Embankments constructed on Soft Alluvial Soils," Proceedings Inst. Civ. Engrs., Part 2, 1975, Vol. 59, pp. 571-593.

MENEROUD, J. P., "Relations Entre la Pluviosité et le Declenchement des Mouvements de Terrain," Colloque International sur la Gestion des Ouvrager, Paris, 1981.

MURPHY, D. J., "High Pressure Experiments on Soil and Rock," *13th Symposium on Rock Mechanics,* Urbana, Ill., 1971.

Proceedings of the International Symposium on Landslides, New Delhi, India, April, 1980.

Transportation Research Board, "Evaluating Strength Parameters of Simple Clays: Geotechnical Consideration of Residual Soils," Transportation Research Record 919, Washington, D. C., 1984.

CHAPTER 7

SUBSOIL DRAINAGE IN ROAD ENGINEERING

7.1 Introduction

Engineers who work with earth masses in roads, railroads or airfields observe that the collapses, slides and flows they are so often confronted with are largely due to the effects of water. The correlation between the intensity of the rainfall and earth movements is nearly infallible. The pattern repeats year after year and any engineer can observe from the obvious signs left by water after failure that unless precautions are taken to control this element, it can become one of his worst enemies. The correlation between water and failure is so marked that the search for water as a cause becomes instinctive.

Paradoxically, however, the mechanisms by which water affects stability are not always clearly understood. Engineers who feel sure that water has a distinct influence and who design routine measures to counteract its harmful effects, nevertheless frequently either do not know the ways in which water acts or else misinterpret them.

TERZAGHI [1] pointed out that many engineers in charge of large-scale projects, when asked to explain the influence of water on the stability of earth masses, talk about its lubricating effect. As indicated by TERZAGHI, this explanation is unacceptable for two reasons. First, water acts not as a lubricant, but as an antilubricant on many of the contacts between the minerals of which soils are most commonly composed. For example, the coefficient of friction between two surfaces of dry quartz varies between 0.17 and 0.20; if the quartz is moistened, the coefficient of friction rises to between 0.36 and 0.41. Second, the amount of water required to produce thorough "lubrication" between the particles of soil is surprisingly small. Sufficient moisture is present in most soils (except in outstandingly dry zones) that the mechanical interaction between the grains is not modified by adding more water. Furthermore, it is observed that the relation between rainfall and landslides holds for both wet regions (where the water contents of the soils are relatively high) as well for very dry regions (where a moisturizing effect could be expected).

It appears, therefore, that the mechanisms by which water acts must be due to other types of phenomena. These are numerous. First, if soil voids are partially filled with air and the water content of that soil increases substantially, part of the surface tension (which provided the soil with an apparent cohesion that aided stability: see Chapter 1), is eliminated. Second, an increase in water content brings about an increase in the weight of the soil, which may have repercussions on the overall stability of the mass. Third, groundwater flow may affect the stability of a mass of soil by dissolving any natural cementing agents. This is well known in loess soils, where the grains are often cemented by soluble calcium carbonates.

Added to the above three effects, the water that penetrates a mass of soil and subsequently flows through it has yet a fourth effect which is usually the most influential on stability. This is the raising of the water-table or piezometric level, which in turn brings about an increase in the neutral pressures of the water in the soil and a corresponding reduction in shear strength.

The water-table or piezometric level of a mass of soil is the elevation that would be reached by the water in piezometers installed in the mass. If h is the water-table or piezometric height above a specific point in the mass, the pore water pressure at that point will be:

$$u = h\gamma_w \tag{7-1}$$

The effective shear strength of the soil at that point will be (Chapter 1):

$$s = c + (\sigma - h\gamma_w)\tan\varnothing = c + (\sigma - u)\tan\varnothing \tag{7-2}$$

Equation (7-2) defines the drop in strength which occurs with any increase in h.

Each rainy season usually brings with it a rise in h and a corresponding reduction of the safety factor for hillsides and slopes. These periodical changes may not be sufficient to directly impair the stability during the lifetime of the structure. However, the repeating increases and decreases in the forces tending to cause sliding and the corresponding decreases and increases in strength sometimes cause a gradual and cumulative weakening of the soil not defined by Eq. (7-2). Gradual reductions in strength were mentioned in Chapter 6. SKEMPTON [2] made a singular quantitative application of these ideas to the problems in the London clay. In accordance with a statistical analysis and calculation, a vertical slope in this clay 6 or 7 meters (20 or 23 ft) high will hold up for several weeks, a 1:2 slope of the same height may fail after 10 to 20 years and a 1:3 slope may fail after 50 years. SKEMPTON concluded that in London clay it is hard to find a natural slope with an inclination steeper than 1:6. The effect of progressive structural changes and loss of strength of a hillside or slope material is often very important, but always extremely hard to predict. The part played in these effects by changes in the water content, water pressure or seepage is one of the most difficult factors to establish when analyzing the effects of water. Nevertheless an important one.

Lastly, in this brief review of the reasons why water may affect the stability of an earth mass, rapid drawdown [3], liquefaction and the various phenomena which are described as piping and internal erosion should be mentioned. These phenomena have already been described in the corresponding Chapters of this book.

Design engineers routinely consider one or both of the following designs to control water which affects their projects.

1. Keep the water away from zones where it may cause damage
2. Control any water that enters the danger zones by diversion and elimination. These are known as subsoil drainage to distinguish them from those classified as drainage, which usually refers to the control of surface water.

Both designs are often used. Cutoff walls, impermeable aprons, grouting curtains, etc. are examples of the first. This Chapter emphasizes rather the second: subdrainage. Any water that has entered the subsoil must be diverted and eliminated so that it can cause no damage; either from harmful pressures or increased water content and weight. Most problems caused by subsoil water in road engineering are associated with uncontrolled saturation and the unexpected development of pore pressures. The capillary action of water often plays an important part in these processes.

In the field of road engineering, designs based on subsoil drainage are generally more promising. It is doubtful that an engineer can all prevent water from going where it is driven by natural forces. At the most, he will be able to make its course more difficult. However, this may well have results that are contrary to his intentions since water that is constrained and hindered can develop excessive pore pressures. Only at a cost that would be completely unreasonable for road building purposes could an engineer consider impeding all groundwater flow.

A better design is to accept the presence of water and provide earth structures with systems of internal channels, through which water can flow more freely in its gravitational form, at the lowest possible pressures. This should be kept in mind when considering all the types of subsoil drainage systems which will be described later. The theory of Flow Nets used in this chapter is dealt with in Appendix 7a.

7.2 Subterranean Water

The water that is found in the subsoil usually has one of 3 origins. First, there is the water precipitated from the atmosphere in the form of rain or snow. Second, there is connate water, which occupies the spaces between the sediments that were left at the bottom of oceans and lakes. This is generally salt water, because the largest masses of sediments that can be found today are ones which formed in sea water. Lastly, there is magmatic or juvenile water, which is a product either of magmatic volcanic activity, or the condensation of vapors derived from deep-seated magmas. There is probably more of this water than one might suspect. By simply considering that 90% of the total product erupted by volcanoes is steam, and that much of that steam must have been supplied to the volcano by various subterranean sources, the immense amounts of magmatic water can be readily imagined.

The quantity of water percolating into the earth is determined by several factors [4]:

1. Intensity and amount of precipitation.
2. Rhythm of precipitation: The faster the rain falls, the less water infiltrates, because the surface of the ground becomes saturated.
3. Surface slope. Infiltration is greater in flat ground, where the velocity of surface runoff is lower.
4. Porosity of the surface soils and rocks.
5. Permeability of the surface soils and rocks. A very porous formation is not necessarily very permeable. Clay, for example, is very porous but not very permeable.
6. Structure of the soils and rocks, especially fracturing, stratigraphy and the sequence of permeable and impermeable strata.
7. Quantity and type of vegetation, and its effect on surface permeability.
8. Atmospheric humidity. If humidity is low, a large part of the water that falls evaporates before it can penetrate the ground.

Subterranean water can be stored in several different ways. Most is found in the voids between particles of soil or in the cavities, fractures and faults in rocks. A smaller percentage may go to form subterranean rivers or lakes. Sometimes conditions are actually modified as a result of storage of subterranean water over a period of time. For example, cavities, pores or fractures may become sealed by soluble minerals that precipitate in the subterranean water.

When studying flow conditions in subterranean water (its storage, movements and possible surfacing) a fundamental role is played by the geology of the formations, referring both to near-surface and more deep-seated formations. First, the types of soils and rocks that are present must be considered. Unconsolidated pervious sediments, such as gravels, sand or those composed of mixtures of these materials are very important because of their large pores, these are capable of storing and releasing large quantities of water. Water-bearing formations are common in alluvial deposits along existing streams, abandoned or buried river valleys, alluvial fans, and mountain outwash deposits, glacial outwash and marine deltas and beaches.

Plate 7-1 Typical talus deposit

There are other types of formations where appreciable volumes of water can be found. Limestones, for example, are very variable water-bearing formations; their porosity is largely the cavities left as materials dissolved inside them. Limestone terranes often include numerous springs, caves and even subterranean rivers. Volcanic rocks usually form springs. Sometimes the rocks are highly porous, but their pores do not necessarily communicate with one another. Water flows through the interconnected pores and especially through the cracks that have formed by cooling, the fractures caused by deformation and in the open spaces and rough surfaces between successive lava flows.

Crystalline igneous rocks and metamorphic rocks usually contain little water; any that is found in them is in their fracture.

Clays and clayey soils are capable of storing enormous quantities of water, as has frequently been mentioned in this book, but they release water very slowly because of their low permeabilities.

Figure 7-1 [5] shows the different types of subsoil water. Near the surface is the aeration or discontinuous moisture zone, where the soil pores contain both air and water. This is known as vadose water. It is held in the soil by capillarity and constitutes the soil moisture. The thickness of the aeration zone may vary from zero to hundreds of meters in particularly arid zones and in those with an abrupt relief. Underneath the aeration zone is the saturation zone, where the pores of soil are virtually filled with water. The boundary between the two zones is above the water-table, by the amount of capillary rise. The water-table is the level at which the water pressure is atmospheric. Below the water-table, the saturation zone may extend for hundreds of meters. Its water content usually becomes smaller with depth, as the voids gradually become smaller due to the weight of the overlying masses.

Local saturation zones occasionally form above impermeable strata, producing a perched water-table (Fig. 7-1). Sometimes the water-table underlies an impermeable stratum, forming a confined stratum. In this, confined aquifer artesian water may develop due to the weight of the overlying soil and any hydrostatic head which exists from tilted strata. If a piezometer is installed in the confined water-bearing stratum, the water will rise to the piezometric level, which, in this case, is similar to the water-table. When this piezometric level line is above the surface of the ground, then any hole drilled to the aquifer produces a flowing artesian well.

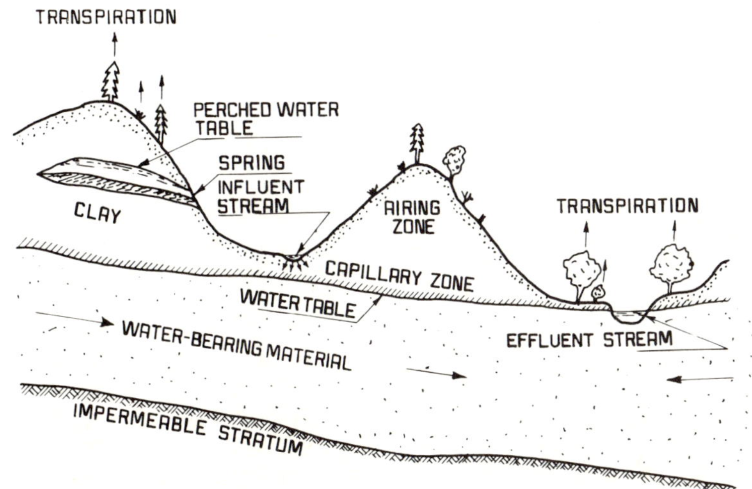

Fig. 7-1 Types of subsoil water [5]

The water in the aeration zone may be free, capillary or hygroscopic. Of the three, capillary water is susceptible to the greatest variations. Most of the water in aeration and saturation zones comes from atmospheric precipitation. The ways in which this water penetrates the subsoil through the aeration zone are extremely varied and are subject to highly complex laws. An important part is played by the gravity, surface tension and molecular attractions and osmotic forces. Infiltration is usually rapid in the aeration zone, but in the saturation zone it is slow. In Chapter 1, the basic laws governing the movement of water in the saturation zone were discussed.

The configuration of the water-table depends largely on the shape of the ground surface topography, which it reflects although with less abrupt contours. It also depends on the permeability of the soil and the amount of water in the ground. It is generally further from the surface of the ground underneath hills, and closer to the surface in valleys, and especially near rivers and lakes.

The water-table usually drops considerably during periods of drought and rises after periods of heavy rains. These fluctuations are generally more significant in permeable, granular soils. There are occasions when the water-table is lowered in a dry region so much that rivers and lakes lose water due to downward seepage (influent rivers). Under normal conditions, however, the water-table supplies them with water. The phreatic level of the ground water joins the free surface of the effluent rivers and lakes. The common layman's belief that subterranean water consumes large quantities of water from rivers and lakes is usually false.

The height the water-table rises after rain is usually less than the depth of precipitation. However, in some fine sandy soils that are partially saturated by capillary saturation the rise can exceed the precipitation depth following long steady rains because the void volume to be filled is only a fraction of the soil volume.

Figure 7-2 [6] shows the regional differences which can be found in river valleys, depending on whether the rivers gain or lose water from the water-table. In the case of effluent or draining rivers, (see Fig. 7-2a), the water-table is relatively high in the adjacent hillsides, sloping towards the valley and finally joining the river. On the other hand, in influent rivers the water table is located far deeper beneath the hillsides, with all the corresponding implications for the construction of roads. The problems relating to each different type of valley will obviously be quite different when cuts are to be made in the hillsides. Therefore, it is always wise to define the ground water regime of the valley under consideration so that appropriate design and construction criteria can be selected.

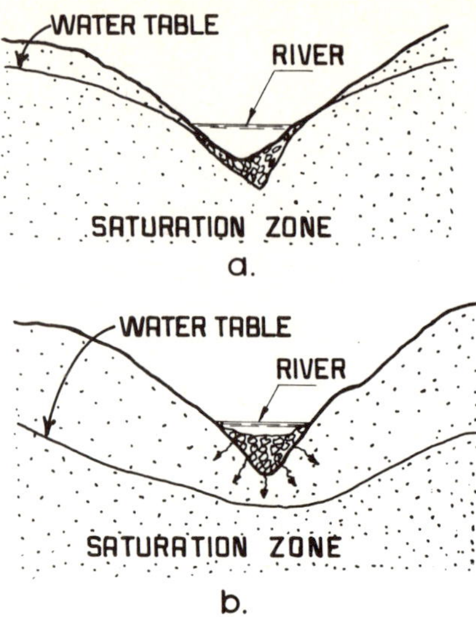

Fig. 7-2 Infiltrating and draining rivers. Difference between regional water-tables

Figure 7-3 shows a geological profile that encourages the formation of a perched water-table. This often means large quantities of water in hillsides where it would otherwise not be suspected. A timely geological exploration that uncovers this situation may avoid many problems during the construction or the maintenance of a road built on such a hillside.

The probability of finding large quantities of water in natural slopes composed of porous materials and in the flat zones of valleys, is greatest in undrained subterranean basins like the one shown in Fig. 7-4 (See Plate 7-1). This condition is far more frequent than might at first be supposed. It appears in almost all the large natural amphitheaters: flat zones surrounded by mountains of impermeable crystalline rocks. These basins sometimes cover wide areas and are excellent sources of water for human consumption.

Figure 7-5 shows typical cases of geological profiles which are favourable for shallow water or springs in natural slopes. It is not the authors' intention to give examples of all possible profiles, but simply to illustrate the type of geological sequence which is commonly associated with subsoil drainage problems in the road engineering.

A typical profile leading to artesianism is shown in Fig. 7-6. A well bored in a water-bearing stratum like this one would be an artesian well and the water would burst forth of its own accord

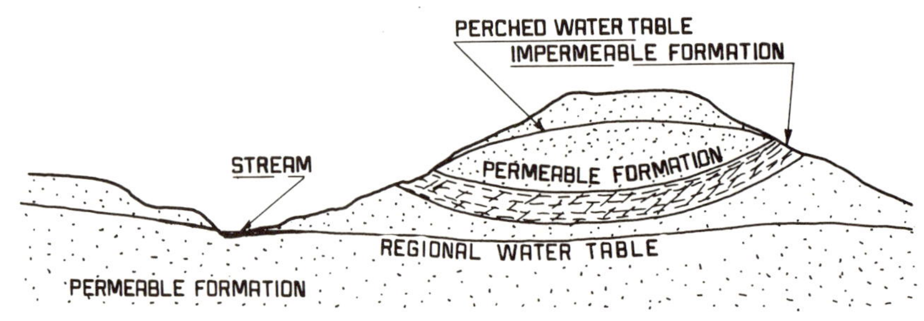

Fig. 7-3 A geological profile which encourages the formation of a perched water-table

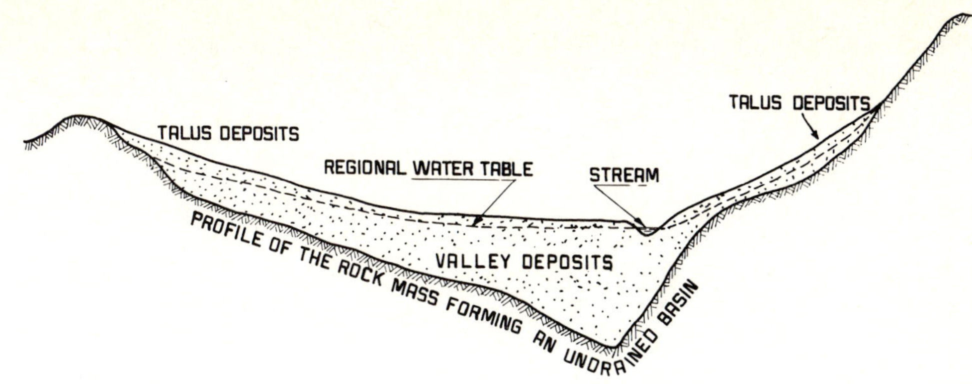

Fig. 7-4 Undrained subterranean basin

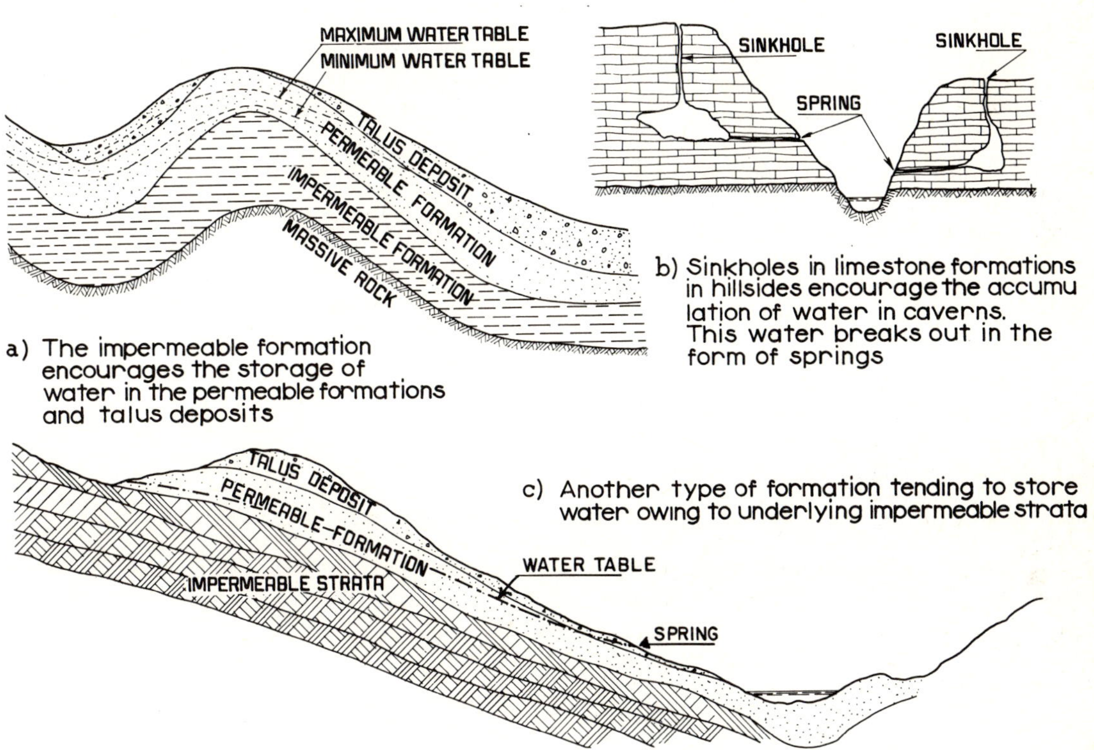

Fig. 7-5 Examples of typical formations which encourage the storage of water in natural slopes

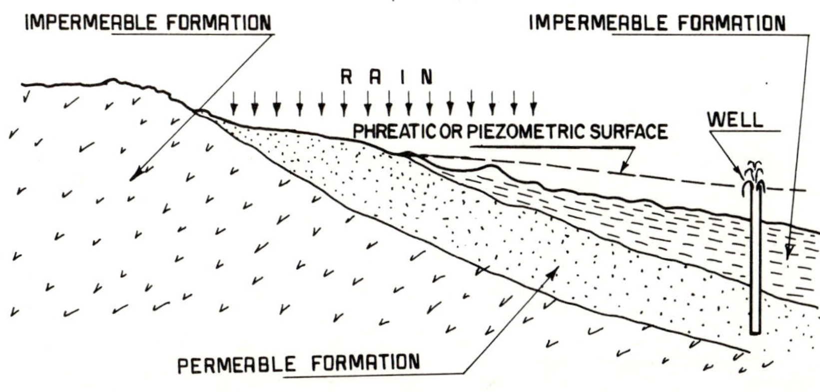

Fig. 7-6 An artesian formation

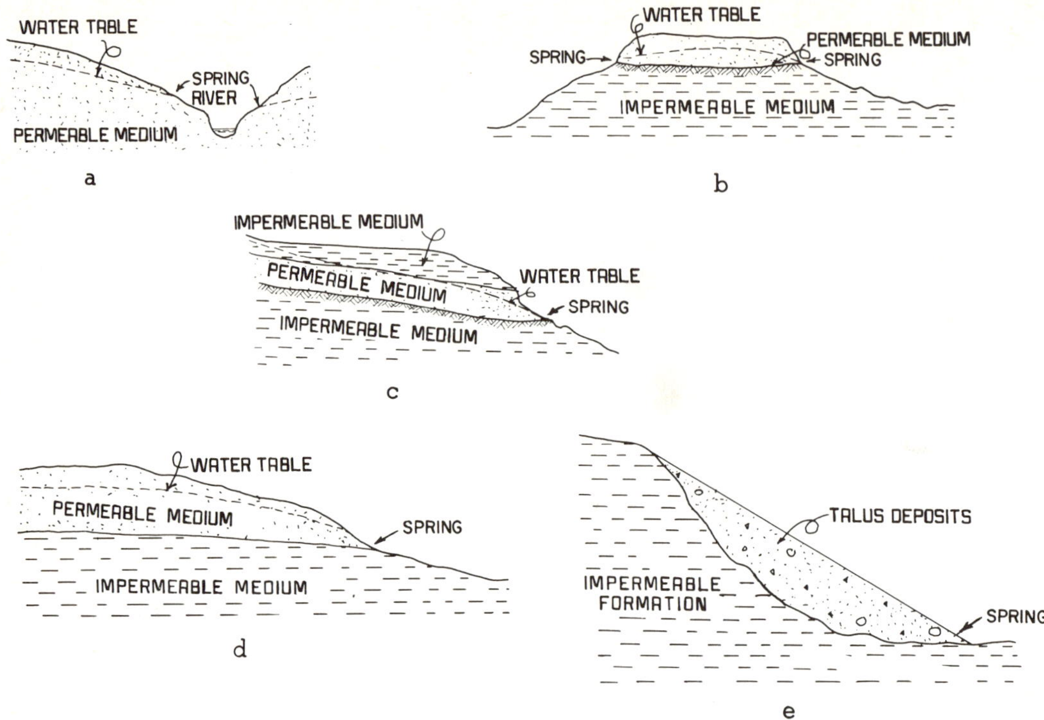

Fig. 7-7 Example of formations which encourage the appearance of springs in hillsides

Plate 7-2 Superficial water-table at the toe of a cut

Plate 7-3 Water surfacing at the toe of a failed hillside

(Plates 7-2 and 7-3). Where the water-table intercepts the ground surface or where artesian water surfaces, a spring occurs, Fig. 7-7 illustrates some of the conditions which most commonly lead to springs in natural slopes. The arrows mark the place where water is likely to appear. Plants and trees, that are water-loving, usually indicate the presence of a seep or spring.

7.3 Filter Design [7]

Construction work for roads usually involves soils or rocks which contain water. Sound rock formations can be drained by simply allowing the water to flow freely from the water bearing cracks or porous zones out into conduits such as ditches, wells or tunnels. The cohesion or cementing of these materials is sufficient for water to be able to flow through them without causing harmful erosion. Poorly cemented rocks and uncemented soils may, however, be easily eroded due to the traction forces produced by the water as it flows through them. If no restrictions are imposed on these processes, they will lead to serious problems of internal erosion and piping. Any surface where water comes out into the open must therefore be protected in such a way that the water can find its way out easily, but at the same time ensuring that the soil or rock particles stay in place (See Plates 7-4 and 7-5).

The system responsible for allowing water to surface freely and at the same time preventing particles from being carried away are called reverse filters or simply filters. Numerous materials are used today for this purpose. Paper, various woven and non woven fabrics of synthetic fiber termed geotextiles and fiberglass are examples, and are used in many situations where their high cost is offset by their ease of installation.

Plate 7-4 Failure of a retaining wall owing to lack of draining material in the backfill

Natural aggregates of sand and gravel sizes are the filter materials most widely used because of their availability and durability. When these natural materials are of good quality, they are virtually indestructible and everlasting, compared to the lifetime of the project. When correctly placed, their performance is excellent, both as filters and as support materials that require strength and low compressibility. Lastly, they are usually plentiful, which means they are usually relatively economical to obtain and handle. They are generally used in combination with perforated or porous pipes, which serve to collect and eliminate the water.

Plate 7-5 Failure in a wall with no drainage pipes or filter material at the back

In order for natural granular materials to play their protective role as a filter efficiently and justify the investment that is made in them, they must fulfill certain requirements which are based on theory and experiment. The filter will fail if these requirements are not strictly observed. Seldom is there such a small margin for carelessness or error in road building as in the case of the selection, treatment and placement of granular filter materials.

Many of the requirements for filter materials refer to grain-size distribution of grading. Other very important ones refer to careful handling so as to avoid contamination and size segregation. Compaction to reduce the possibility of changes in grading due to invasion by fines from the soil that is to be protected is also important. As was indicated already in Chapter 5, filters must here satisfy two contradictory conditions:

1. The spaces between the particles of the filter material where it is in contact with the soil to be protected must be small enough to prevent fines from that soil from penetrating the filter.
2. The spaces between the particles of the filter material must be large enough that there is sufficient permeability for the water to flow freely through the filter towards the collecting drain and discharge point without generating undesirable pore pressures.

If these criteria are to be satisfied, more than one filter layer may be necessary, with each successive layer filtering and draining the adjacent one. Although some designs have required four such layers, they are unnecessarily difficult to construct. With good design a two layer filter can be developed for almost any situation.

The chief concepts and conclusions relating to filter design will now be considered; the above two requirements are handled separately.

7.3.1 Prevention of Internal Erosion and Piping

The first rule for avoiding piping and internal erosion is that the soil particles must not be exposed to open spaces that are larger than themselves. In [8] there is a very interesting analysis of the influence of cracks and fissures in the deep rocky subsoils on the surface topography, resulting from the migration of appreciable quantities of finer overlying soils (silts, sands, and gravels into the deeper openings). The second rule is that it is essential to seal any cracks, construction joints and contacts between different materials that may exist in the drain conduits, such as pipes and culverts. Serious seepage may occur through a poorly sealed crack under high water pressures, forming large cavities in the filter or adjoining soil, which make the presence of any filter useless.

A few criteria for grain-size distributions will now be given. These are based on experience.

If the voids in the filters are small enough to prevent a particle representing D_{85} of the size of the soil to be protected from going through them, experience has demonstrated that the soil that is to be protected will be held in place. In other words, it is regarded as tolerable that the size of 15% in weight of the soil to be protected be smaller than the voids that are left between the particles of the filter.

As a result of investigations directed by TERZAGHI, CASAGRANDE and BERTRAM [9], the following rule has been established for relating filter material to the soil to be protected:

$$\frac{D_{15} \text{ of the filter}}{D_{85} \text{ of the soil}} < 4 \text{ or } 5 < \frac{D_{15} \text{ of the filter}}{D_{15} \text{ of the soil}} \quad (7\text{-}3)$$

By satisfying the first of the above inequalities, the migration of fine particles from the material to be protected to the spaces in the filter material is avoided. Experience has shown that the second inequality included in Eq. (7-3) ensures sufficient permeability in the filter for large seepage forces or undesirable pore pressures not to develop.

The investigations mentioned in [9] were verified in experimental work carried out by the U.S. Army Corps of Engineers [10] and the U.S. Bureau of Reclamation [11]. As a result of these, some additional criteria were established to prevent the migration of particles from the soil to the filter. The U.S. Army Corps of Engineers selected 5 as the figure to be used in Eq. (7-3), and also established the following additional requirement:

$$\frac{D_{50} \text{ of the filter}}{D_{50} \text{ of the soil}} \leq 25 \quad (7\text{-}4)$$

The above criteria cannot ensure protection in high plasticity clays, whose grain size distribution would require filters made up of several layers of material. For these soils, D_{15} of the filter would be 0.4 mm (0.016 in) and Eq. (7-4) is not considered applicable. Ignoring Eq. (7-4) may lead to a single-layer filter. However, broadly graded single layer filters often segregate and thus can fail. A single layer filter requires a coefficient of uniformity smaller than 20.

As a result of their findings, the U.S. Bureau of Reclamation recommended that the grain-size distribution curves of the filter and the soil should be noticeably parallel; this is the general effect of Eq. (7-4).

SHERARD and his collaborators [12] add the precaution that when the soil to be protected contains large quantities of gravel, the grain-size distribution curve to be considered in the application of the above rules should correspond to a material smaller than 2.5 cm (1in), eliminating the larger size materials.

The results of all the studies indicate that a combination of the rules given in Eqs. (7-3) and (7-4) will prevent internal erosion and piping even in the harshest conditions. This also applies to horizontal layers of granular soil overlying layers of finer material, which is a common situation in pavements.

7.3.2 Prevention of Obstruction of Pipe Perforations or Migration of Fine Particles

Subsoil drainage systems often include a slotted or a perforated pipe with circular holes inside the filters for collecting and eliminating water rapidly. As a consequence, the filter material must be sufficiently coarse so as not to escape through these perforations or block them. The U.S. Corps of Engineers [13] make the following recommendations:

For slots:

$$\frac{D_{85} \text{ of the filter}}{\text{Width of the slot}} > 1.2 \quad (7\text{-}5)$$

For circular perforations:

$$\frac{D_{85} \text{ of the filter}}{\text{Diameter of the hole}} > 1.0 \quad (7\text{-}6)$$

The U.S. Bureau of Reclamation [14] gives the following rule:

$$\frac{D_{85} \text{ of the filter (adjacent to the perforations)}}{\text{Maximum perforation of the pipe}} \geq 2.0 \quad (7\text{-}7)$$

In [15] the following requirement is given:

$$\frac{D_{85} \text{ of the filter}}{\text{Maximum perforation of the pipe}} \geq 1.5 \quad (7\text{-}8)$$

As it can be seen, the different criteria recommended by those who have carried out experimental work are not in total agreement. The discrepancies are not, however, significant. Any of the above listed rules will lead to adequate performance, with no loss of filter material through the perforations in the pipe.

7.3.3 Permeability Requirements in the Filter Material

This section will deal with the selection of filter material. Care must be taken that there is sufficient discharge capacity to ensure the rapid and efficient elimination of the water which accumulates without seepage forces or harmful pressures being generated. As indicated, the second inequality Eq. (7-3) establishes the necessary permeability.

As a general rule, filters should be at least 20 to 25 times as permeable as the soil that is to be protected. The effective permeability improves if the water in the filter can discharge freely; unfortunately it is relatively common for the efficiency of very costly filter installations to be partially annulled by defective outlets. This problem deserves special attention in draining layers underneath pavements or in capillarity breaking layers, which will be described later. In these cases, discharge problems may be serious.

Sometimes the hydraulic capacity of a drain can be considerably increased by using a filter consisting of more than one layer. In [16] some cost estimates are given which show that in some cases a multiple layer filter may prove more economical than a single layer one. Because the benefits and costs of multi-layer filters vary, the discharge capacity of several different filter alternatives should be analyzed according to the flow theory methods as given in Appendix 7a or in [15] which also describes the research carried out by A. CASAGRANDE and W. L. SHANNON (originally contained in [17]) on the drainage of a horizontal layer, such as the base course for a highway or runway. In the case they considered, flow is unsteady. A calculation method is given which may provide useful for design.

The dewatering capacity of a drain system may be established by applying DARCY's Law to simple geometric approximations or performing an analysis by means of flow nets. Some engineers use a combination of both. In order to apply DARCY's Law the permeability of the soils that are to be drained (k) must be found, either by means of laboratory tests or field tests. This is usually the most difficult step involved, for in many cases these values have not been measured during site studies and it may not be possible to obtain in time for design in many practical road engineering situations. In such cases permeability must be estimated, which is difficult and may involve serious errors. The guides given in [18] may prove useful for these purposes; they provide methods (though not very reliable ones) for estimating the coefficient of permeability on the basis of grain-size distribution. At best such estimates can vary from the actual permeability by one or two decimal places.

The mean hydraulic gradients in the soil to be protected and in the filter material, i, must also be estimated in order to be able to calculate the volume of flow to be removed by the drain. The discharge area, A, in which both materials are in contact with one another must also be available. Once these data have been estimated, the permeability of the filter can be obtained by simply applying DARCY's Law:

$$k = \frac{Q}{iA} \tag{7-9}$$

Several trials usually have to be performed with different values for A, until a satisfactory combination of the permeability and thickness of the filter is obtained.

Equation (7-9) may also be handled by starting with permeability values corresponding to the filter materials available and seeing what thicknesses of A must be used for the filter. However, a minimum thickness of 10 to 20 D_{85}, but at least 8 cm should be used. It is very hard to estimate the hydraulic gradients existing in the filter, for in the general expression $i = \frac{h}{L}$, the value of h is usually unknown, although L can be found (it is the length of the entire discharge path of the water in the drain).

Figure 7-8 shows two typical flow nets for vertical upward groundwater flow, from a permeable stratum with a slight artesian head to the base of an embankment where a horizontal draining blanket has been laid. Dewatering is by means of two lateral underdrains. The case presents a groundwater flow through soils with different permeabilities. It is analyzed in detail in [19]. If the coefficient of permeability of the filter material is represented by the symbol k_f and that of the soil by k_s, according to the above-mentioned reference the solution must be:

$$\frac{k_s}{k_f} = \frac{b}{c} \tag{7-10}$$

In this example b is the width and c the length of the rectangles forming the flow net in the filter material, assuming that the flow net in the embankment material (with a thickness of 0.5 D in Fig. 7-8) which in this case is the soil to be protected, is made up of squares measuring a.

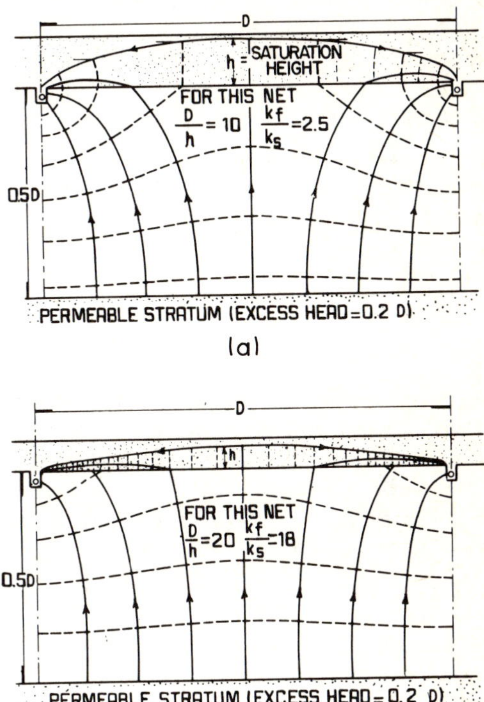

Fig. 7-8 Vertical flow nets in horizontal permeable blankets [7]

Construction of the flow net in these cases can be complicated or even impossible if the ratio between the permeabilities of the two materials is large. If the net can be drawn, Eq. (7-10) can be used to calculate the permeability necessary in the filter material, either by assuming certain dimensions for the filter and drawing several trial nets with different $\frac{b}{c}$ relations until a suitable permeability is found for the filter, or by starting out with a suitable k_f/k_s relation in accordance with the materials available and seeing what thickness this gives for the draining layer, after establishing in it the corresponding flow net. Once the net and the permeabilities for the two materials are obtained, the volume of flow which the layer is capable of draining can then be calculated.

7.3.4 Segregation Requirements

An ever-present danger in the construction of filters for subsoil drainage is a possible change in the grain-size distribution of any mixture as a result of segregation during placement. To avoid this danger, [15] gives the following rule:

$$\frac{D_{60} \text{ of the filter}}{D_{10} \text{ of the filter}} \leqslant 20 \qquad (7\text{-}11)$$

Also, the grain-size distribution curve corresponding to the draining material should be smooth, without any discontinuities denoting the shortage of some intermediate size. For the same reason, some moisture is recommended when placing the filter material, although care must be taken not to jeopardize compaction efficiency.

7.3.5 Layout of Pipe Perforations

As has already been said, perforated pipes are frequently installed in the filter material in subsoil drainage systems. The purpose of the pipe is to discharge the water easily and rapidly and the object of the perforations is to offer the water access to the inside of the pipe, Fig. 7-9 shows the layout which is generally recommended for the perforations.

It is not advisable to perforate the upper portion of the pipe for this would encourage the entry of fine particles of filter material. Some engineers prefer not to perforate the lower portion of the pipe, for this might encourage the water that is captured to run out, either with a decrease in flow velocity or a low volume of flow. Sometimes the joints between the sections of an unperforated pipeline are left open as an alternative to perforations. This practice should be regarded as unsuitable, for it exaggerates the two above mentioned disadvantages.

In most subsoil drainage systems, pipes with a diameter of 10 to 20 cm (4 to 8 in) are used. These are often made of concrete, but other materials can be used should they prove more economical and sufficiently durable. In installations where there is a particularly large volume of flow or the permeability is not known accurately, pipes with a far greater diameter may prove necessary. According to experience, the diameter of perforations is usually somewhere between 5 and 10 mm (0.2 and 0.4 in).

7.3.6 Summary

All the above-mentioned criteria usually define just one type of filter material and do not include composite filters consisting of several layers. Three or more are generally avoided in road building technology because of cost and ease of construction. A material which can be used as sand for the manufacture of concrete will usually satisfy all the requirements for the first filter layer reasonably well.

Although these rules provide a good criterion for the design of filters, there is the disadvantage that in order to apply them the permeability of the soil to be protected must be known. This requirement usually proves impractical for roads, where there are enormous variations in the soils, even between points that are not far apart. It would be expensive and time consuming to perform the number of permeability tests which would be necessary in order to fulfill these requirements. It is impractical to keep changing the filter material to fit the varying soils. For roads, the ideal solution would be to have a standard or universal filter material independent of the characteristics of the soil to be protected. In other words, a filter material would be produced which could be used for all cases. Theoretically this is impossible, but some organizations have found it possible to use a single material, with carefully selected grain-size distribution characteristics that enable it to filter a wide range of soils. It is hard to specify the characteristics of such a filter material, for the organizations which use them modify them slightly from time to time, on the basis of the results of field experience. Thus any recommendation that may be established at a given moment runs the risk of rapidly becoming outdated. The two grain-size distributions which follow are recommended by two organizations specializing in road building. The first one [20] appears in Table 7-1. The second grain-size distribution recommended is used by the Mexican Ministry of Public Works and appears in Table 7-2. Part *a* of Fig. 7-10 shows this second grain-size distribution in a graph which indicates three zones within the limits of which base-course materials should be confined in accordance with Mexican engineering practice. As a guide, Part *b* of Fig. 7-10 gives some grain-size distribution curves typical of the ones used in road construction for different drainage purposes, and the approximate permeabilities to which they correspond. If there is any montmorillonite present in the finer curves, the permeabilities will be much lower.

Table 7-1

Grain size distribution for filter materials recommended by the California Division of Highways

Sieve	Opening in mm	Percentage passing
1 in	25.4	100
3/4 in	19.1	90-100
3/8 in	9.52	40-100
No. 4	4.76	25-40
No. 8	2.38	18-33
No. 30	0.59	5-15
No. 50	0.297	0-7
No. 200	0.074	0-3

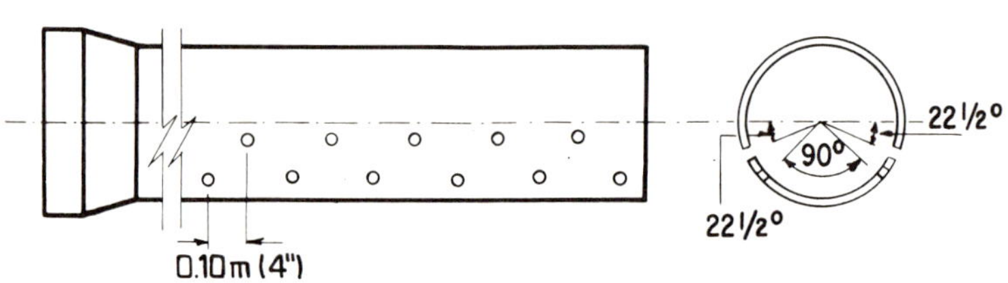

Fig. 7-9 Layout recommended for the perforations in subsoil drainage pipes

Table 7-2

Grain size distribution for filter materials recommended by the Mexican Ministry of Public Works

Sieve	Opening in mm	Percentage passing
1-1/2 in	38.1	100
1 in	25.4	80-100
3/4 in	19.1	65-100
3/8 in	9.52	40-80
No. 4	4.76	20-55
No. 10	2.00	0-35
No. 20	0.84	0-20
No. 40	0.42	0-12
No. 100	0.149	0-7
No. 200	0.074	0-5

7.4 Methods of Subsoil Drainage for Road Building

The following describes the principal methods that are used for subsoil drainage in road building. The most frequent uses of the different methods will be discussed later, for in road building technology there are some important distinctions which must be made in relation to subsoil drainage, compared to runways.

7.4.1 Permeable Blankets in Pavements

Large quantities of water often appear in cut subgrades for roads and railroads. In these cases it is useful to place a permeable blanket underneath the pavement in order to protect it. These blankets are reasonably thick and are laid underneath the road surface or the paved surface. They are made of filter material, so that with the aid of a suitable transverse slope and efficient disposal of water, they can drain any water which filters through from the pavement or shoulders of the road or which rises from lower levels due to uplift pressure.

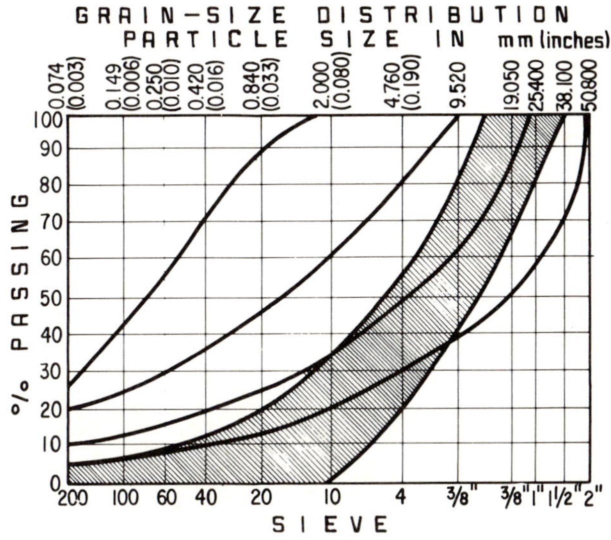

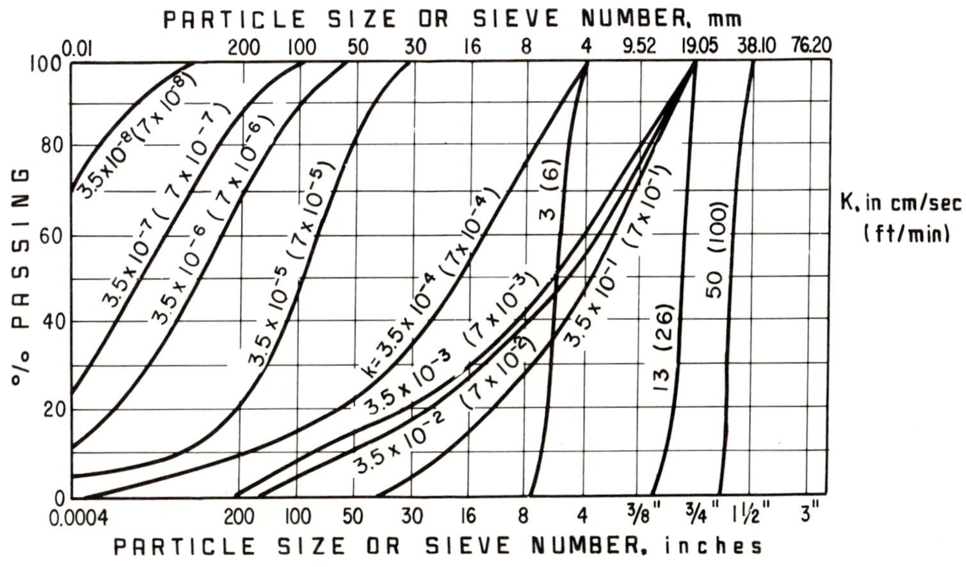

Fig. 7-10 Grain-size distributions for subsoil drainage

These draining layers often form part of the pavement, utilizing the granular nature of filter materials which makes them also highly suitable for structural support. Another important function of filter layers is that they protect the fine soil subgrade or embankment materials from a layer of coarse crushed base course material which is placed on the top, preventing the coarse fragments from becoming embedded in the finer matrix.

Figure 7-11 is a diagram showing the utilization of permeable blankets to control seepage from the upper part of the pavement and at the same time water rising from lower layers by an uplift pressure. In Part *a* of the figure (downward flow), the sub-base is made up of suitable materials and is used as a draining layer. Since it is accepted that there will be no upward flow, the cut bed can consist of a subgrade course with no special qualities. Part *b* of the figure considers an upward flow due to uplift pressures in a fractured rock. Here the draining layer is the subgrade which is the cut bed. The relatively common practice of converting the sub-base into a draining layer too by placing it on top of a conventional sub-grade (cut bed), should be regarded as unwise, for any soil that is placed underneath the draining layer will tend to become saturated, consequently losing strength and becoming more susceptible to deformation. As a result the pavement and the draining layer together may have a poor performance despite their quality, for they will in fact *float* on a saturated subgrade layer.

Sometimes a permeable blanket of coarse material is placed in the lower part of a pavement or even in the body of the embankment, to interrupt an upward capillary flow which would otherwise damage the subgrade. These are capillarity breaking layers, whose function it is to prevent water from entering the subgrade but not necessarily to drain it. They are not true draining layers. The fine embankment material is brought into contact with the air in the large spaces between the coarse particles, providing an opportunity for the formation of discontinuous menisci which prevent the water from continuing its upward journey. Thus any layers of soil that are placed on the top will be free of water.

There are important differences in the design of the permeable blanket which depend on whether its function is to intercept (and eliminate) flow, or to act as a capillarity breaking layer to combat water rising from lower levels. First, the blanket will be subjected to a groundwater flow when laid. This requires suitable transverse slopes, and perforated collector pipes. The most important requirement is that the material which is used for the blanket must be a genuine filter, in keeping with the criteria given in §7.3 of this Chapter. If, on the other hand, the purpose of the blanket is to break the capillary potential of rising water, the material should be coarse, granular and very permeable. In this way the material with capillary potential will come into contact with large pores with little capillarity. The

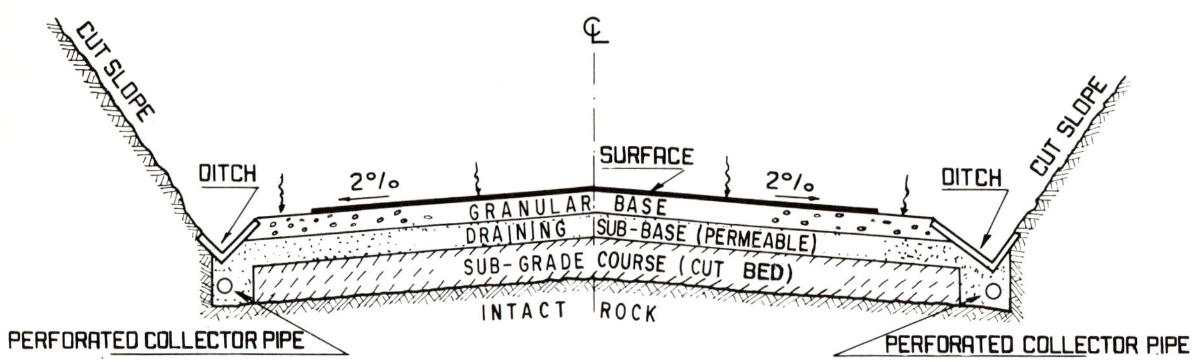

a.- Sub-base use as a permeable layer to intercept water coming from the pavement

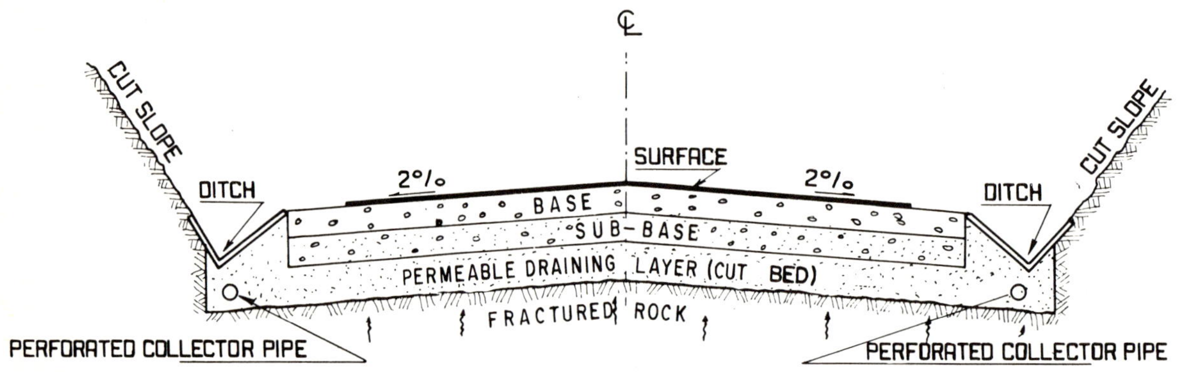

b.- Sub-grade course used as a permeable layer to intercept upward flow due to uplift pressure

Fig. 7-11 Permeable layers

water will not be able to rise through the capillarity breaking layer and will be confined to the lower layers. The necessary discontinuous menisci will develop in this zone where the capillary channels come into contact with the air. Under these circumstances, it is not advisable for the capillarity breaking layer to meet filter criteria because the capillary height in filters may be appreciable. This makes it appear that the ideal material for a capillarity-breaking layer would be relatively large, uniform-sized fragments of stone, but it would be unwise that any material that is placed in the structural section of a road has just one objective. A very coarse material would be subject to clogging by fines under flow conditions. Therefore, the materials selected for capillarity breaking layers are usually graded.

The permeable blankets that will be considered here are those intended to capture either water flowing downwards from the pavement surface and shoulders, or from the sides of a cut, or an upward flow due to a water head. Some of the aspects involved will be similar to those for a capillarity-breaking layer.

Figure 7-8 shows two flow nets for a real situation. It illustrates two different relations between the coefficient of permeability of the filter and that of the soil, as well as two layer thicknesses. They correspond to an upward flow coming from lower levels as a result of high water head. The flow is vertical until it reaches the permeable blanket, becoming nearly horizontal within that layer. The head within the permeable blanket is usually somewhat smaller than the thickness of the layer itself. This requires relatively low hydraulic gradients. This is a very common condition in drain systems. Water that is collected from large masses of soil, and which in the neighbourhood of the filter may have very high hydraulic gradients, must be drained through small aquifers by means of flow that is subjected to similarly small hydraulic gradients.

Figure 7-12 [7] shows design curves for horizontal permeable blankets. They are the result of analyses performed with a flow net. Part *a* of the figure gives the relation $\frac{D}{h}$ (see Fig. 7-8) as a function of the ratio of the permeability of the filter to that of the soil $\frac{k_f}{k_s}$, and in Part *b* the same concepts are related, but with individual values for k_f and k_s. The curves may be used to design capillarity breaking layers where the geometry of the problem corresponds to that of the figure (for other geometries, similar graphs would have to be plotted). This is done by first assuming a thickness for the capillarity breaking layer (usually between 20 and 40 cm (8 and 16 in), 30 cm (12 in) being a very typical value). Then, one estimates, or measures, the permeability of the soil underneath the capillarity breaking layer, through

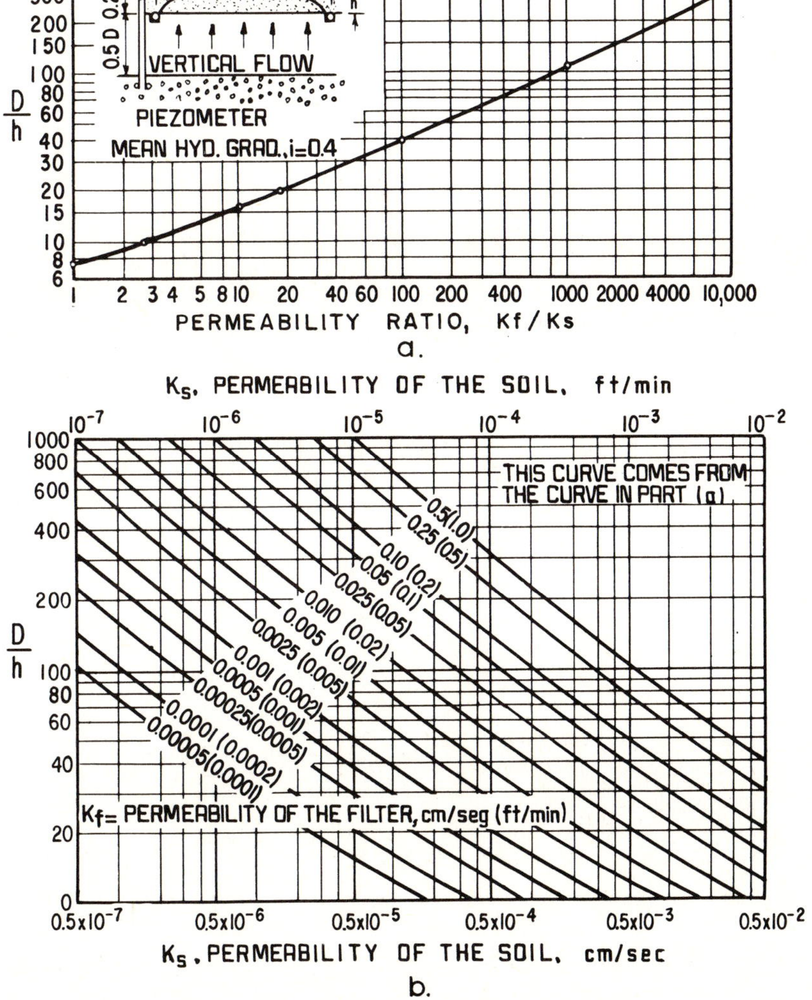

Fig. 7-12 Design graphs for horizontal isotropic permeable layers [7]

which the water rises due to uplift pressures in this case. Next a suitable value for h (value of the saturation height within the permeable blanket) must be chosen, generally 4 or 5 cm (1.6 to 2 in) below the upper level of the layer. In this way the relation $\frac{D}{h}$ can be computed, and the necessary permeability for the filter material obtained.

As has already been said, a permeable blanket that is designed in accordance with the previous procedures is often rather inefficient for removing large volumes of water. On airfields, this dewatering capacity is usually considerably less than that needed to drain the amounts of rainfall that are anticipated. Therefore, exceptionally permeable materials are frequently used [21]. Although pavements are usually made relatively watertight, so that most rainfall is eliminated as surface runoff enhanced by the crown or camber, no pavement is totally impermeable. Some rainwater seeps through the surface. To this infiltration must be added the water that penetrates through the shoulders, and any water from groundwater flow through neighbouring soils, particularly at the base of cuts. The frequent case of upward flow due to capillarity or uplift pressure has also been mentioned previously. Assessment of the quantity of water that may have to be drained in a given period of time is very complex. Specialists in pavement design or laboratory researchers may be able to throw some light on the discharge relation through a road surface. At least a rough assessment of surface infiltration can thus be made, but the other origins of the water that collects may be harder to quantify. References [16] and [22] describe some investigations relating the length, width and thickness of the permeable layer with the discharge and the flow volume that can be drained. Figure 7-13 shows the basic conclusions of these investigations. The ordinate on the right gives the thicknesses and characteristics of permeable blankets that are capable of draining the volumes of flow produced by the discharge relations and the lengths of the

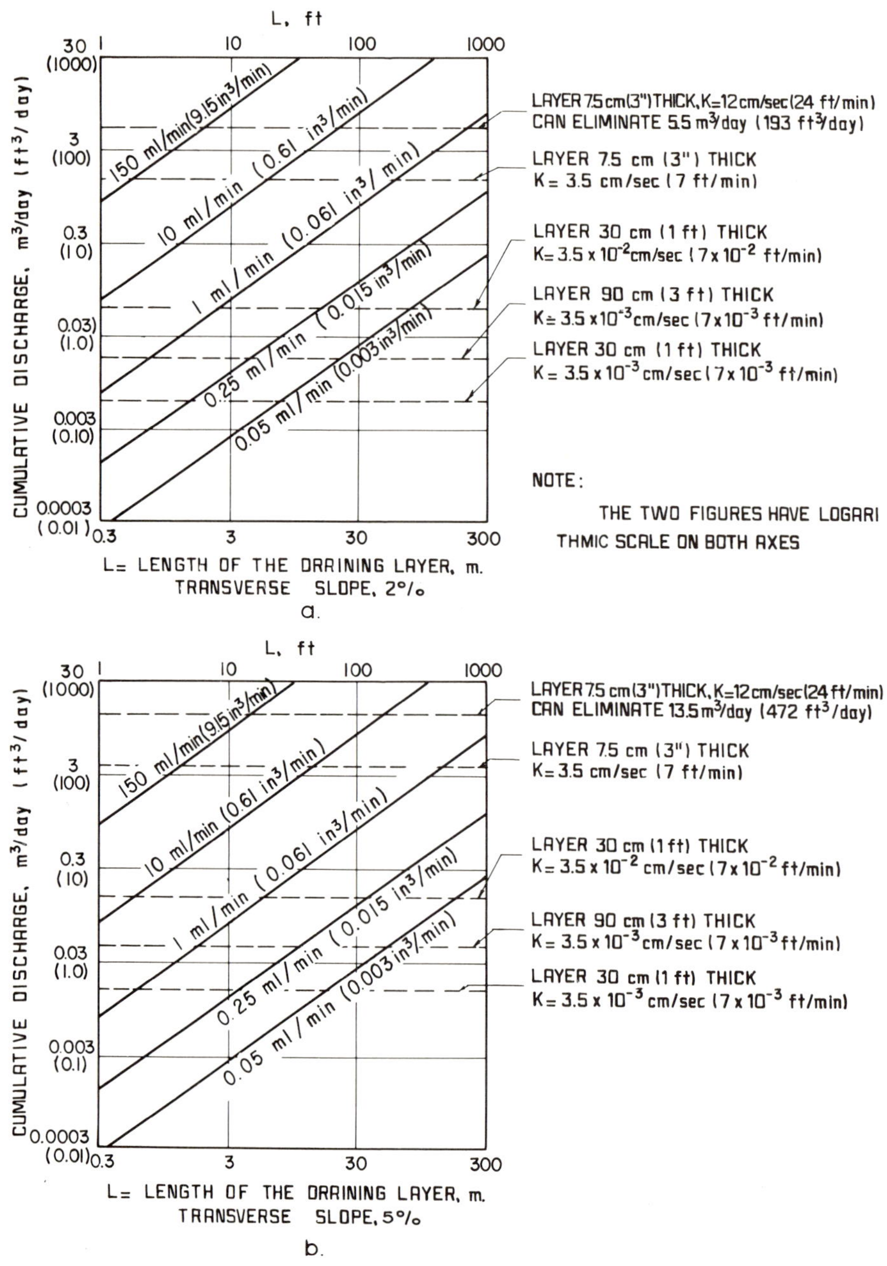

Fig. 7-13 Discharge capacity of drain layers [16,22]

permeable blanket (for a unit width) which are shown by the diagonal straight lines and on the abscissa. Part *a* refers to pavements constructed with a 2% transverse slope, and Part *b* to pavements with a 5% slope. The discharge relations of the pavements studied are given in milliliters per minute (cubic inches per minute) and refer to the water which penetrates through a circle on the surface with a diameter of 15 cm (6 in).

The figures indicate, for example, that in order to drain the water that penetrates a strip of pavement 5 m (16.4 ft) long and 1 m (3.3 ft) wide, with a discharge relation of 0.25 ml/min (0.015 in^3/min), a permeable blanket with a thickness of 90 cm (3 ft), a permeability of 3.5×10^{-3} cm/sec (7×10^{-3} ft/min) and a 2% transverse slope will be necessary. On the other hand, to remove the water that penetrates a band of pavement 300 m (328 yd) long and 1 m (3.3 ft) wide with the same discharge relation, the thickness required for the permeable blanket will be only 7.5 cm (3 in) if the permeability is 3.5 cm/sec (7 ft/min). In both figures, the ordinates axis on the left shows the cumulative discharge in the pavement in m^3/day (ft^3/day).

Table 7-3 [22] shows the discharges through the surface of a pavement which can be removed, depending on whether or not an efficient permeable blanket is laid under the conventional sub-base course. Discharge is given in cm of rainfall per hour (in per hour).

The table refers to a layout which, although frequent, is not considered ideal by the authors of this book, who prefer the permeable blanket to form part of the structure of the pavement so that it acts as a sub-base when the water is the result of infiltration from overlying layers (Fig. 7-11a). Nevertheless it provides a guide and illustrates the enormous draining power of horizontal permeable blankets. An installation like this can increase the dewatering capacity of a road or runway by hundreds to thousands of times. These permeable blankets which will be subjected to flow *must,* it is stressed, be designed with the filter requirements, which were described earlier in this Chapter.

An important complement to a filter layer is a system for collecting or eliminating water, this is installed on both sides. Figure 7-14 shows a detail of one such installation for a runway.

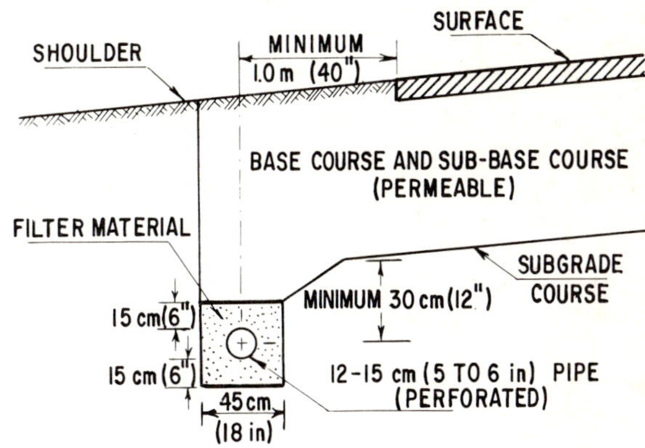

Fig. 7-14 Detail of a set-up for eliminating water from a permeable layer

7.4.2 Longitudinal Underdrains

Cost is an important aspect of the design of permeable blankets; it is usually high. Therefore, any reduction in the thickness of the layer that can be achieved without causing an excessive decrease in the dewatering capacity is desirable. Draining layers that are too thin, however, can complicate construction to such a point that all economical advantage is lost. Layers under 15 cm (6 in) thick should prehaps not be used, 20 to 30 cm (8 to 12 in) being the most common dimensions. Greater layer thicknesses will probably be excessively expensive, especially in highways; in runways the margins are generally wider.

Table 7-3

Theoretical discharge relations which can be drained from the layers of a pavement (calculated using flow nets [22])

SUB-BASE MATERIAL OVER THE PERMEABLE LAYER	PERMEABILITY		Theoretical discharge relation			
			WITHOUT PERMEABLE LAYER		WITH PERMEABLE LAYER	
	cm/sec	ft/min	cm/h	in/h	cm/h	in/h
Conventional, well-graded sub-base material	3×10^{-4}	6×10^{-4}	0.0015	0.0006	1.25	0.50
Sand with high content of silty and clayey fines	6×10^{-4}	1.2×10^{-3}	0.0030	0.0012	2.50	1.00
Sand for concrete with very low content of silty and clayey fines	6×10^{-3}	1.2×10^{-2}	0.030	0.0118	25.00	10.00

In sloping hillsides or undulating and mountainous ground, ground-water commonly slopes downward in the same direction as the ground surface; the water table topography is similar to that of the ground, although usually less irregular. When a deep excavation is made for a road, as in the case of cuts, flow will occur towards the excavation (Fig. 7-15). This flow can be intercepted by a longitudinal underdrain like the one in Fig. 7-15. This shows flow directions before and after installation of the drain. The effect of the underdrain in this case is to intercept and eliminate flow towards the cut bed and, on a smaller scale, to reduce the saturation zone in the slope. These are the objectives of most longitudinal underdrains that are installed in roads and railroads.

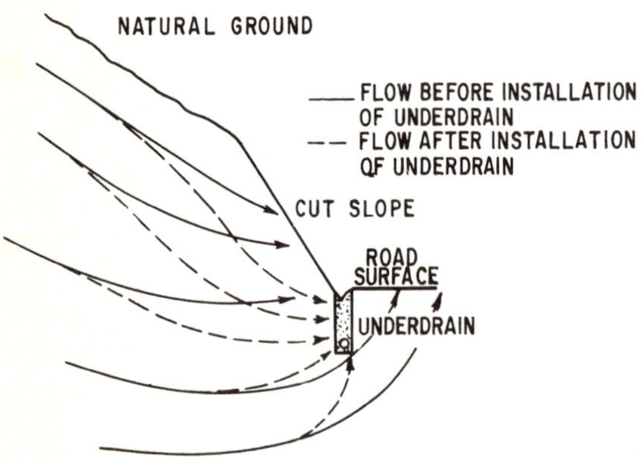

Fig. 7-15 Flow towards the slope and the bed of a cut

Another common use of longitudinal underdrains is shown in Fig. 7-16. It depicts three cases where drainage makes it possible to maintain a pavement in a situation where ground water would otherwise flood it. Here, rather than to intercept flow, the purpose of the drain is to lower the water-table. This is frequently required in flat ground where the water-table is very close to the surface. This is typical of runways, although in the figure the three cases refer to highways. The underdrain consists of a ditch 1 to 1.5 m (3 to 5 ft) deep although sometimes as deep as 4.0 m (13 ft) with a perforated pipe in the bottom. It is filled with filter material. Water is discharged from the pipe by gravity to some low-lying area or canyon where it can cause no damage. A sketch of such an underdrain can be seen in Fig. 7-17 (Plates 7-6 and 7-7). The filter material and the perforations in the pipe must meet the requirements discussed in §7.3.

Plate 7-6 Construction of a lateral drainage ditch

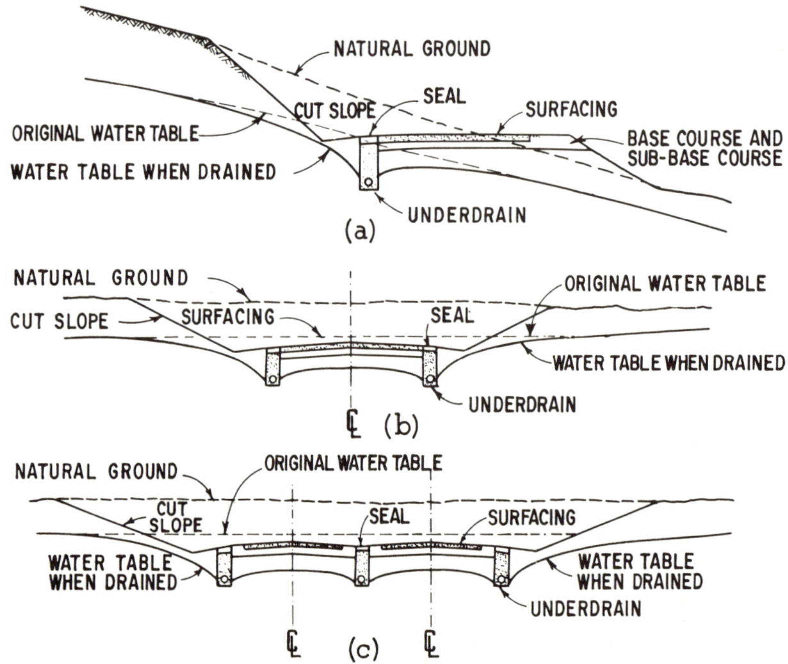

Fig. 7-16 Cases where longitudinal underdrains are used to lower the water-table

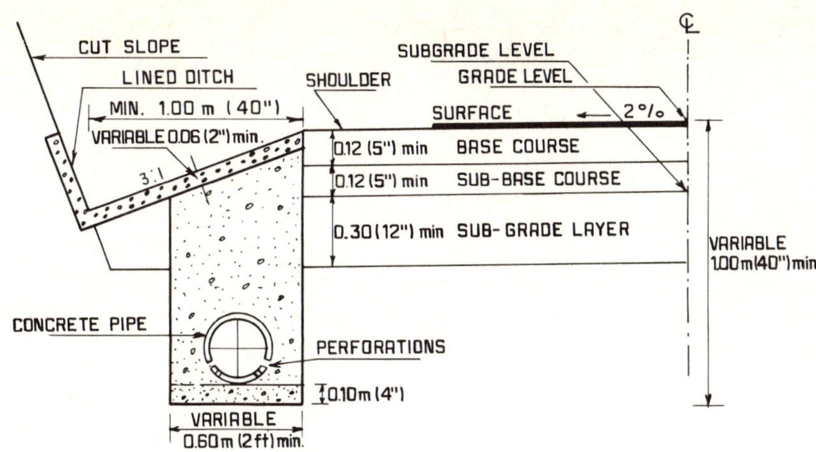

Fig. 7-17 Cross-section of an underdrain, according to Mexican practice (Mexican Ministry of Public Works)

Figure 7-16 also illustrates cases where drains have to be installed on one side of the highway, on both, or even where three or more drains are needed, as may be the case in wide modern motorways or runways. In runways, the effect of these longitudinal underdrains is often combined with transverse drains which will be discussed later. The ditches must be deep and close together if there is a high water head under the road, and the permeable fill must be suitably compacted. When the ground in which the drains are to be installed is soft and moist, care must be taken to place a sufficiently thick layer of filter material in the bottom of the ditch in order to ensure a stable footing for the drain pipe. In some cases, porous weak concrete is used for the footing.

Plate 7-7 Underdrain in action. Pátzcuaro-Uruapan highway

Figure 7-18 [22] shows typical flow nets towards longitudinal underdrains. Figure 7-19 [22] gives a relation between the depth of the drains and the maximum hydrostatic head which can be generated by water between the two parallel drains, underneath a road. The graph was plotted on the basis of flow nets and an isotropic permeability. A case is illustrated where pavement is underlain by a thickness of soil d, possibly embankment material, and underneath that a source of water. The water has an uplift pressure equivalent to the head h. As a result, an upward flow develops. For the geometry that is given, Part *b* of Fig. 7-19 makes it possible to relate the maximum head which affects the water between the two drainage ditches with the depth of these underdrains.

When the uplift pressure is high, considerations like the ones in Fig. 7-19 lead to underdrains that are either very close to one another or very deep. In such cases, the combination of longitudinal underdrains with permeable blankets forming part of the pavement, gives optimum results. These permeable blankets greatly contribute to the dissipation of any pressures that may develop in the water.

There are differences in opinion among engineers specializing in drainage as to the comparative virtues of permeable blankets and longitudinal underdrains. Some favor the systematic use of permeable blankets, even for capturing flows coming from cut slopes. Their argument is based on the total protection which is given by these blankets to the entire supporting of the pavement. They also argue that if the layer is part of the pavement it serves both structural and drainage functions, saving money. The excavations for longitudinal underdrains are costly even if the ground where they are to be excavated is not very hard. Those engineers who favor longitudinal underdrains maintain that they operate reliably and are backed by greater tradition and experience.

The authors feel that each solution has its application, which greatly depends on the geometry of the road section, the flow pattern, the water pressures and the permeabilities, the materials that are present, and construction costs. In general terms, they feel that the draining blanket is most useful for capturing upward flows and pavement infiltration (both virtually vertical). Drainage ditches give best results when flows coming from cut slopes must be intercepted. In the design of subsoil drainage, personal preference plays an important role, as well as cost.

Figure 7-20 [22] shows some typical layouts for longitudinal drainage ditches in runways. Part *a* of the figure shows a

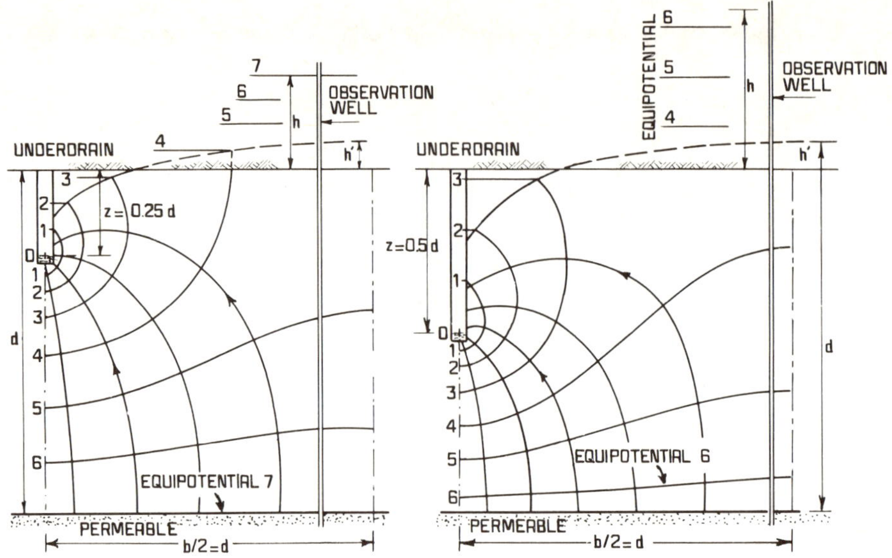

Fig. 7-18 Typical flow nets towards underdrains [22]

combination of the effect of a permeable draining sub-base with a system of longitudinal underdrains, and an appropriate transverse drainage system based on trenches and channels. Part *b* of the figure shows an underdrain upstream from the runway, replacing subsoil drainage in the runway itself. Part *c* shows an installation which includes under-drains for general protection and a complete system of pluvial drainage on both sides of the structure.

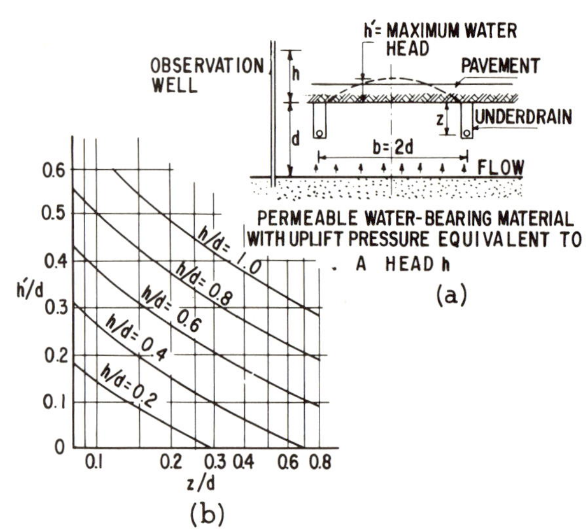

Fig. 7-19 Graph relating the depth of the underdrain and the maximum hydrostatic head that may be generated between two underdrains [22]

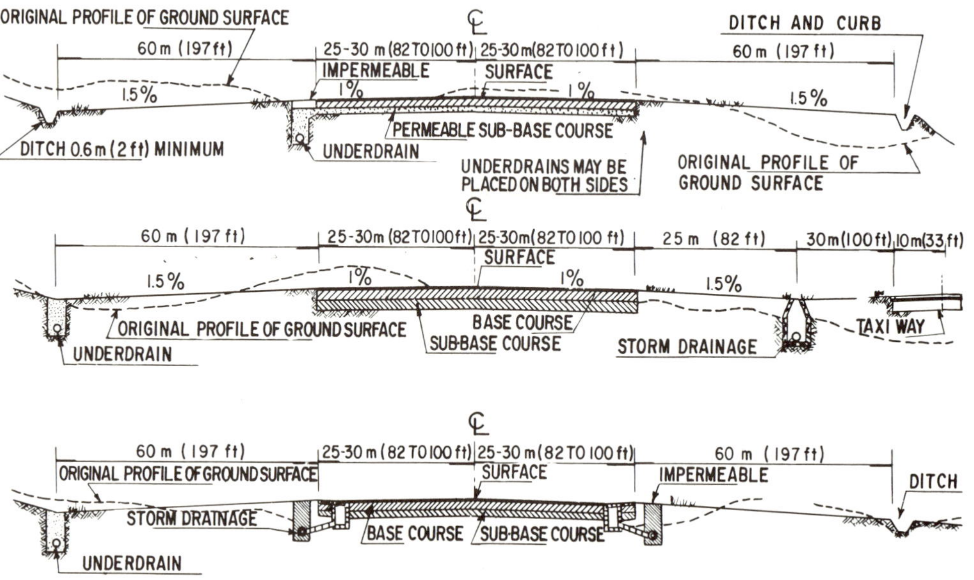

Fig. 7-20 Typical installations for subsoil drainage in runways [22]

7.4.3 Transverse Underdrains

Transverse drains are similar to ditches; the major difference is that they are more or less perpendicular to the road centerline. A typical case of the installation of these underdrains in highways is illustrated in Fig. 7-21, which shows the transition from a cut section to an embankment section. Without the transverse underdrain, the flow coming from the cut might penetrate the

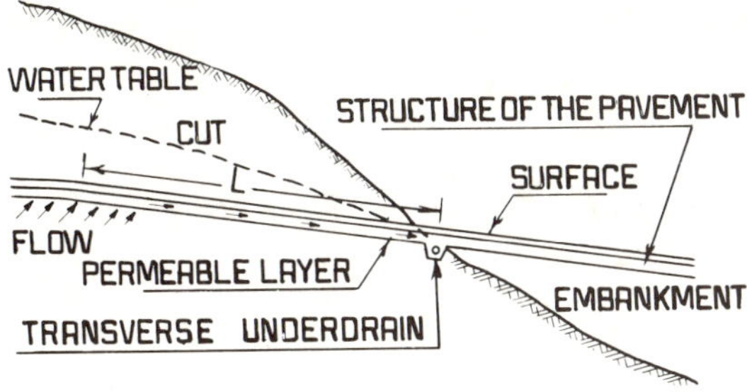

Fig. 7-21 Transverse underdrain [22]

embankment, causing settlement or sliding. The effect of a transverse drain is greatly increased in cases like the one shown in Fig. 7-21 if a short section of permeable drain blanket is installed on both sides of it. The advisability of installing transverse underdrains in combination with longitudinal ones, when the latter are far apart or the water-table must be lowered over wide areas, has already been discussed.

Transverse underdrains must be capable of quickly eliminating any water that may reach them; therefore their permeability requirements are of special importance.

7.4.4 Horizontal Drains

The harmful effects of water seeping out of a cut face, including erosion, as well as the effect of water in the mass on slope stability have already been discussed. The same mechanisms can jeopardize the stability of a natural slope subject to water pressure and seepage. It has also been seen how most slides that are water related occur hours, days, or sometimes weeks after periods of heavy rains. The delay represents the time required for groundwater flow to become established, accompanied by accumulating water in slopes and hillsides. Ground water in cut slopes should not be considered a rare or fortuitous phenomenon. On the contrary, it should be expected in any ground where the water-table is not considerably deeper than the grade of the road or where rainfall is plentiful. When a cut is made, the water-table is lowered to its base, creating a deep zone at atmospheric pressure, towards which the water in the neighbouring soil will seep. In other words a cut acts as a drain. A cut may be stable at one ground water level (and storage of water in the ground). But if an additional quantity of water flows towards it, neutral pressures will increase in the soil and sometimes cause failure. For this reason, a cut made many years ago may suddenly fail after a period of extraordinarily heavy rain.

Horizontal drains [15,23-28] are subsoil drainage installations specifically designed to lower the pressures generated by water in cut slopes and prevent failure. They were first used by the California Department of Highways (U.S.A.) in the late 1930's They consist of pipes with perforations throughout, placed in holes drilled into the ground more or less perpendicular to the road centerline. They capture groundwater and lower its neutral pressures. First, a hole is bored with a diameter of 7.5 to 10 cm (3 to 4 in). For this operation there is special automatic machinery which can work backwards and forwards to make manouvering easier. The holes are installed so as to slope downward toward the road. Slopes of about 5 to 10% help to remove accumulating silt. Steeper slopes are sometimes necessary to reach suspected aquifers. In order to minimize setting up the drilling machine at many locations, several holes are drilled from one point in a fan pattern.

There is always a tendency for the drilling process to cause a substantial change in the inclination that is adopted, generally reducing it due to gravity. This of course also depends on the nature and uniformity of the soils that are drilled, for there is a tendency for the boring tools to follow the weakest paths, such as fissures, cracks or soft strata. Boulders, open cracks and cavities are usually the most serious problems when boring for horizontal drains. Since for reasons of cost the initial drill hole is not usually cased, the collapse of soil or fractured rock is another common problem. When drilling in zones where sliding has already occurred and there are still movements, this danger is of special significance.

The drilling equipment consists of a rotating tool, preferably mounted on a self-driven machine (like a small tractor; See Plate 7-8). Metal rods 1.52 m (5 ft) long are used. Triconic

Plate 7-8 Boring of a horizontal drain

drills have proved versatile for many types of soil, so long as there is not an excessive content of pebbles and rock fragments.

The hole is permanently lined with protective casing. Galvanized or asphalt steel are used when impact-resistance is necessary. Most installations today are rigid plastic pipe, 5 to 7 cm (2 to 2-1/2 in). The pipe is perforated for the entire length. Saw cuts perpendicular to the pipe are economical, and if narrow enough, act as filters. Pipe with drilled holes should be wrapped with a geotextile filter to prevent seepage erosion.

Water captured by the drains is often discharged on the road shoulder ditch or, in large-scale installations, into collector pipes about 20 cm (8 in) in diameter. The one or two meters of drain pipe immediately next to the outlet, should not be perforated, so as to avoid the invasion of vegetation and consequent obstruction. The length of horizontal drains largely depends on the geometry of the zone in which they are to be installed, as will be briefly discussed later, but they can be easily made from 50 to 70 m (164 to 230 ft) long and have often been used more than 100 m (328 ft) long. The nature of the ground in which they are installed often controls the length.

As has been discussed, the purpose of horizontal drains is to drain water and lower its neutral pressures deeper within the soil mass than other subsoil drainage methods. Also, they usually cause a favorable modification in the direction of seepage forces. For these reasons their usual application is in cut slopes and hillsides, especially where an embankment must be supported. For effective performance, a large number of drains are required. In impermeable ground or masses of cracked rock with no ready internal communication, the effect of any one drain may be relatively small, so that they must be placed very close to one another. They are often installed as close as 2 to 5 meters (7 to 17 ft) from each other in two or more rows that are separated vertically 5 to 10 m (17 to 33 ft). Figure 7-22 is a sketch showing how they are installed and their effects in the case of a cut and embankment section.

Like all other methods of subsoil drainage, horizontal drains must not be regarded as a universal panacea. They should be installed only after careful study to ensure their efficiency and economy. This study may consist of soundings, a field inspection of the zone, geological investigations or stability analyses of a slope before and after failure, or a combination of several different investigations. The geological sequence of the materials, existence and level of water must be determined, and the harmful influence of this water assessed, if only qualitatively. Very often the first drains that are installed must play the role of genuine trial drains, especially in zones where no previous soundings have been made. Often only one drain in five produces water. Thus many must be installed. Performance depends on whether the principal cause of stability problems is in fact water, and whether this water is located in such a way that the water-table or the upper flow line can be intercepted by the drains. The one producing drain in five may intercept the critical aquifer. These should not be installed blindly, simply because of a dangerous stability condition, even if subterranean water is seen to play a significant part.

Horizontal drains should be installed in such a way that they can be readily maintained during the normal lifetime of the road. This maintenance consists of cleaning them on the inside, which includes freeing their perforations of obstructions. Special machinery is available for this work, generally equipped with wire brushes and water jets mounted on tractors. For this reason, it is sometimes necessary to construct a tunnel or large pipe to provide access to the drain exits. The performance of a set of horizontal drains can be measured by recording variations in the height of the water-table in observation wells that are strategically placed throughout the drained zone.

Length is one of the hardest factors to decide, especially in areas where there is not sufficient information from detailed sounding programs, as often occurs in road building. In the case of a recently built cut slope, a fair idea of the length required can be obtained by drawing a cross-section of the slope with its probable slip circles superimposed on a geologic cross-section depicting aquifers. The drains must be installed in such a way that they will give thorough protection to any soil mass that is likely to slide. Similar studies are made when horizontal drains are used to stabilize an existing slope that is sliding or shows signs of instability. When a slip surface is either recognized or suspected, its actual shape based on instrumentation or deduced from its scarp or bulge will help decide both the layout and length of the drains. When a hillside is to be stabilized underneath an embankment, the drains should be made long enough to cover the entire zone that is likely to be affected. Drains installed in an active slide can be broken by the moving ground. However, depending on the rate of movement they often help drain the soil and retard the movement before they are damaged. It is sometimes necessary to re-install drains several times before a large slide can be stopped. However, such multiple installations may be the cheapest solution to slide control. Figure 7-23 is a graph which provides a general illustration of these criteria.

When horizontal drains are installed without specialized equipment, which is commonly the case in developing countries, conventional drilling tools are often used which require fairly large quantities of drilling water. Diffusion of this water throughout the soil mass due to pressure may cause instability

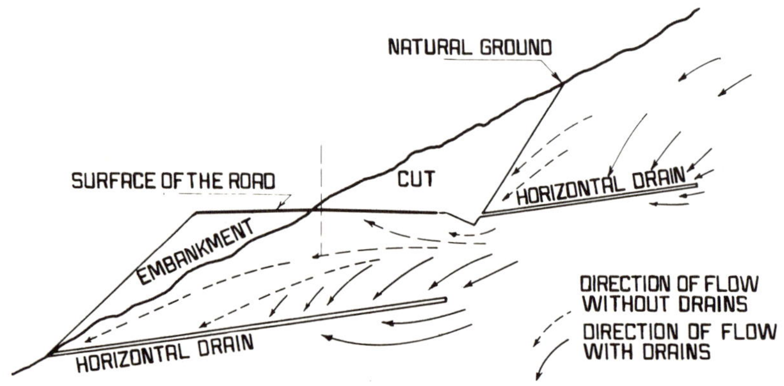

Fig. 7-22 Diagram showing the effect of horizontal drains in a cut and embankment section

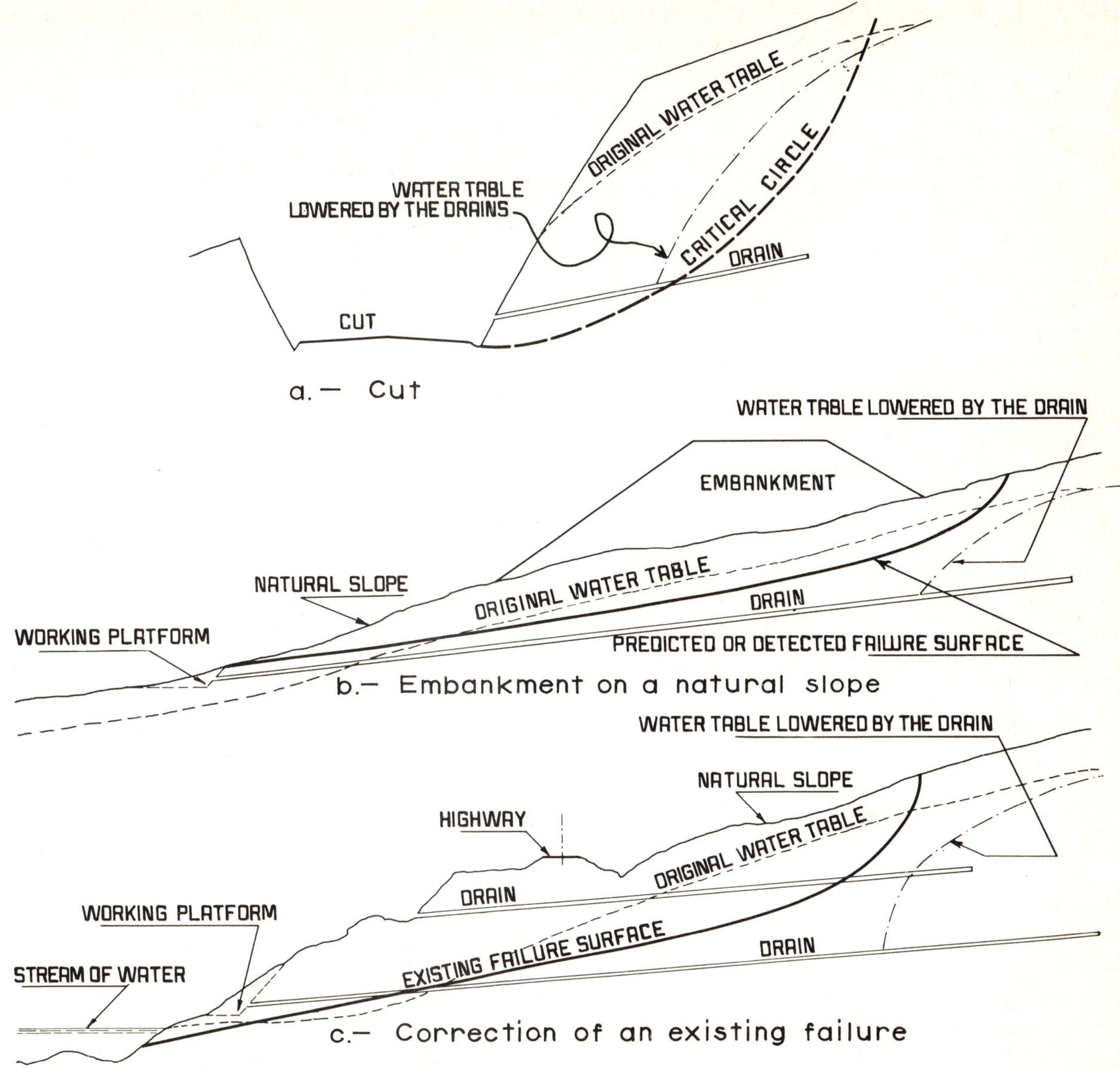

Fig. 7-23 Sketch showing layout for horizontal drains

conditions which though only transitory, are nevertheless very critical. In [28] a detailed description is given of the drilling equipment used by the California Department of Highways, which does not require water. In Mexico and other States horizontal drains have several times been installed using conventional equipment. Experience demonstrates that if the operations and use of water are carefully controlled, the unfavorable consequences can be overcome; however, it is always preferable not to use water, if possible.

The two formations in which it is probably most difficult to install horizontal drains are fine silty sands and soils containing large pebbles and rock fragments, the former owing to their tendency to collapse and form cavities during boring, and the latter owing to the difficulties caused by their hardness and heterogeneity, which are slow and costly to drill.

The basic objective of the drains is to lower the neutral pressures in soil masses where stability is critical. Success cannot necessarily be measured in terms of the quantity of water that is collected by the drains. A very permeable water-bearing stratum with free water may be intercepted, in which case the volume of flow that is drained may be impressive. On the other hand, drains may be installed in impermeable clayey formation, where they can very efficiently lower pore pressure and contribute greatly to stability, although the quantities of water they collect are small. The horizontal drain transfers atmospheric pressure throughout its length and establishes a zone of decreased pore pressure. Within this zone of influence, the water flows towards the drain, but the quantity that reaches it depends first and foremost on the permeability of the formation. In many situations the soils in which a reduction in pore pressure is of the greatest interest may be clayey, which cannot be expected to have large volumes of flow. The amount of flow that is collected usually also varies seasonally.

7.4.5 Drainage Wells

Although these are not very commonly used in road building technology, drainage wells provide a useful solution to certain specific problems. They are vertical holes with a diameter of about 0.40 to 0.60 m (16 to 24 in), inside which a perforated pipe with a 10 to 15 cm (4 to 6 in) diameter is placed. The annular space which is left between the hole and pipe is filled with filter material. Drainage wells with a depth of up to 50 m (165 ft) have been built. They are installed in such a position that they will capture harmful flows toward the zone that is to

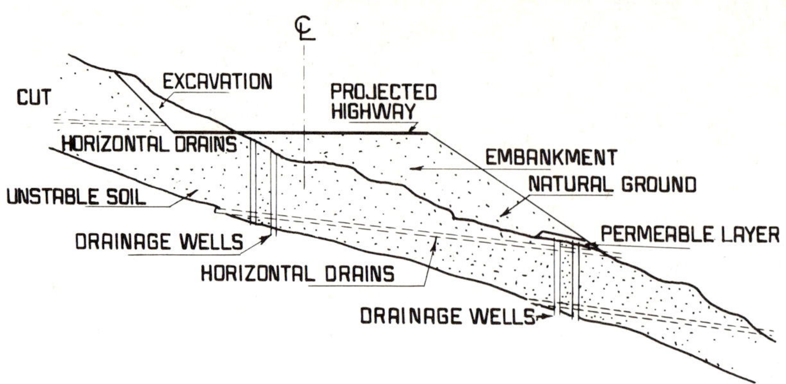

Fig. 7-24 Drainage wells combined with horizontal drains

be protected. Their principal function is to lower the pressures in the water existing in layers deep down in the subsoil, which cannot be reached by means of open excavation or horizontal drains, either because of cost, or because of construction difficulties. They are not usually very effective in eliminating all the water in the soil, except when very pervious aquifers are to be drained.

Wells should incorporate a water disposal system to eliminate the water they drain, otherwise the relief in the pressure they provide will be only temporary or localized in time, without continuous pumping they become filled with water, the levels and pressure conditions that prevailed before their installation will be restored. Pumping is expensive and greatly complicates normal maintenance of the road. There is the risk that pumps will become poorly maintained or even abandoned. On some occasions, a gallery or small tunnel has been constructed to intercept the bottoms of several wells in mountainous regions. This procedure is also costly, and it may complicate the diversion of water to somewhere where it will do no damage. Therefore such galleries are used only in special cases. Figure 7-24 illustrates what is probably the best system of eliminating water collected by drainage wells. This is by horizontal drains. It is of course very hard to achieve a physical connection between the well and the horizontal drain owing to the inexactness of the drilling process, but this limitation is not always critical in pervious strata for the horizontal drains can capture a great deal of the water which accumulates in the wells, and thus reduce neutral pressures. In Fig. 7-24, the wells have been placed in two tiers, with two rows in each tier. These are intercepted by a double row of horizontal drains.

Figure 7-25 [26] shows an ingenious procedure developed by the Washington State Department of Highways to provide drainage to the wells. It is a siphon which, like all other siphons, is limited by the depth at which it can operate. Two design alternatives are shown in the figure; in one the siphon is made of metal and in the other it consists of a tube within a perforated concrete pipe. The system can also be used in other drainage structures, such as stabilizing trenches or drainage galleries, which will be discussed later.

The spacing between drainage wells is very important because it affects both the performance and cost of the system. Spacings between 5 and 10 m (16 and 33 ft) are common; many systems consist of two closely overlapping rows (Plate 7-9). The method of drainage wells usually has the disadvantage of high costs. Economically it is seldom justified where drilling is difficult or where the open well has to be supported before it is filled.

The collection capacity can be computed from Fig. 7-26, given the flow that must be drained. This is based on calculations from flow nets. Part *a* of the figure refers to a horizontal flow which is directed towards the well from ground around it, which is assumed to be saturated. D_p is the diameter of the well and D_f the diameter of the perforated pipe, half of the difference is the thickness of the filter. The graph makes it possible to

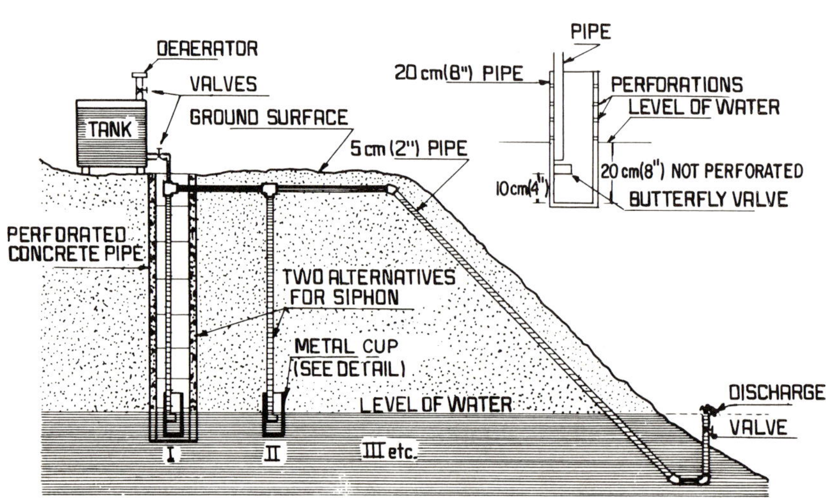

Fig. 7-25 Discharge siphon for drainage wells [26]

Plate 7-9 Drainage wells.

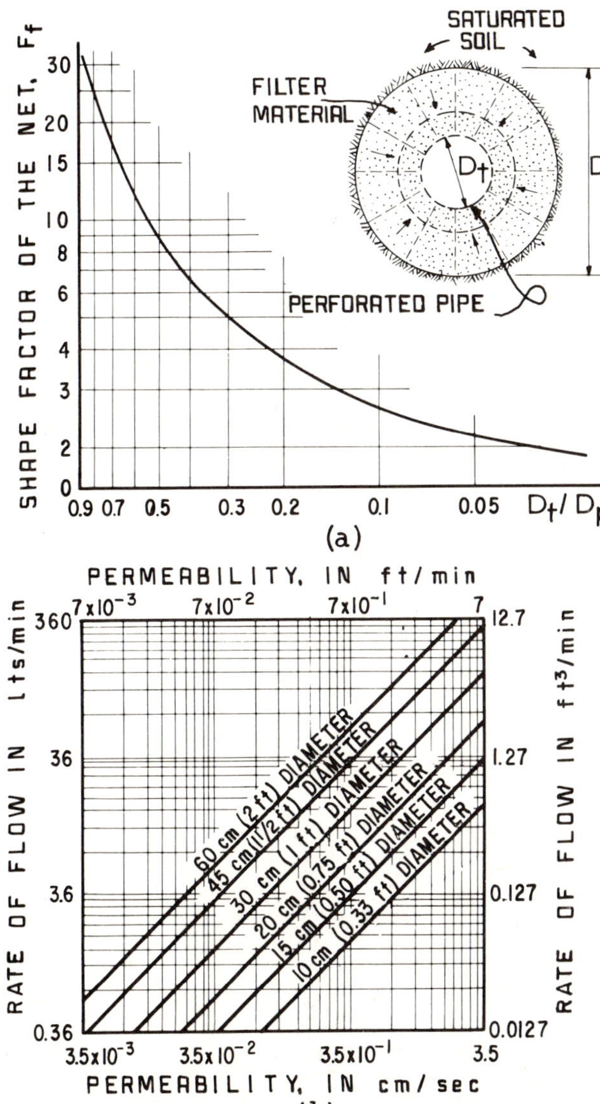

Fig. 7-26 Volumes of flow for drainage wells [23]

determine the shape factor of the flow net around the perforated pipe as a function of the relation of the diameters of the pipe and the well. Once this shape factor has been found, the volume of flow to be eliminated will be:

$$Q = khF_fL \tag{7-12}$$

where k is the coefficient of permeability of the filter material, h the hydraulic head inducing the flow (which generally has to be estimated) and L the depth of the well.

Part b of Fig. 7-26 refers to the rate or flow at the bottom of a drainage well, when drainage takes place through a gallery or horizontal drain connected to the well bottom. The flow rate is given in terms of the permeability of the filter and the diameter of the well.

7.4.6 Drain Blanket

When there is a thin layer (not more than 3 or 5 m (10 or 13 ft)) of poor quality saturated soil underneath an embankment and below that layer there are materials of far better quality, it is often practical to totally remove as much of the poor soil, as it necessary. Figure 7-27 illustrates this procedure. The excavation that is made in order to remove the poor soil can be covered with a 1/2 to 1 m (1.64 to 3.28 ft) layer of filter material, with a perforated pipe to capture flow and a drainage system. Later, the excavation will be filled with a good quality, adequately compacted material. The purpose of the drain blanket is to prevent the fill from suffering any future damage from water. Also, the system enables the embankment to rest on firm ground, which is why this solution should be regarded as a combination of improvement of the foundation ground and subsoil drainage. This solution is limited by the thickness of the layer of poor material; it can be economically prohibitive when the layer is very thick. Exploration is important to determine whether the material being removed is the only one that is affected by flow. There are situations where underneath the firm stratum there is another weak saturated stratum, so that removal of the overlying material leads to little improvement in the stability conditions.

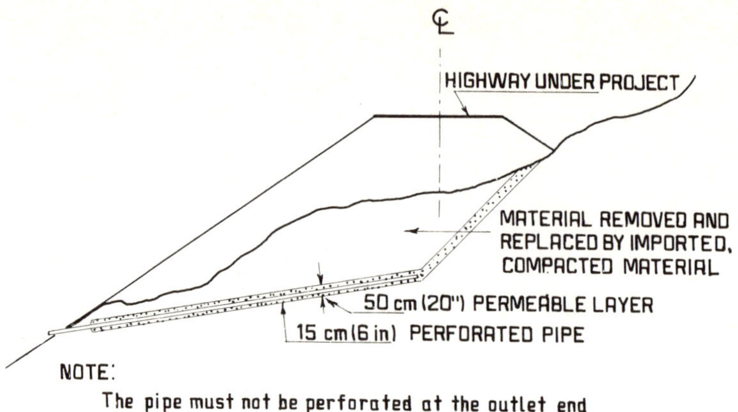

Fig. 7-27 Removal of poor material and placement of a drain blanket underneath embankments

7.4.7 Stabilizing Trenches

When the stability of a hillside is threatened by groundwater flow and on that hillside an embankment is to be built, the replacement of all the poor materials is both difficult and uneconomical. In such cases, local stability can be improved by capturing and disposing of the water from below the embankment. This can be achieved by draining the water from the entire zone in which a failure circle might develop, benefiting the embankment and foundation soil as a whole.

Figure 7-28 shows four different possibilities for stabilizing trenches adapted to specific situations [29]. A stabilizing trench is an excavation which has a blanket of filter material between 0.50 m and 1.00 m (1.64 and 3.28 ft) thick against its upstream slope and at the bottom a system for collecting and eliminating water. The collecting system usually consists of another blanket of filter material of the same thickness, in which a perforated pipe is installed to lead the captured water away quickly. A typical diameter is between 15 and 20 cm (6 to 8 in), or even larger if an important volume of water is expected. This pipe is connected to a drain pipe which carries the water away. Drainage may be a difficult problem if the excavation is deep and the topography unfavorable. Sometimes the problem can be solved by extending the longitudinal pipe to some appropriate low-lying land or canyon. In other situations, it is necessary to provide the trench with horizontal drain pipes installed in a ditch or tunnel. Lastly, there are times when the drainage problems are so complicated and lead to such expensive solutions that the water captured by the trench is eliminated by pumping. It is possible to drain the bottom of a stabilizing trench by horizontal drains, to lower disposal areas. The horizontal drain casing is not necessarily perforated. It is helpful for the bottom of the trench to be constructed with suitable slopes and on a firm impermeable stratum to direct water to the horizontal pipes.

The bottom of the trench should have sufficient width to allow the efficient operation of construction equipment; this can be achieved with about 3 to 4 m (10 to 13 ft). The excavation side slopes need to be stable only during construction, because the fill that is subsequently used eliminates the long-term instability. Since trenches are commonly located where there is poor material, problems of stability in the excavation slopes are common. Rapid construction usually helps overcome these problems. In extreme cases construction is performed in sections of suitable length with each one being filled before the next one is excavated. The material used to fill the trench is generally

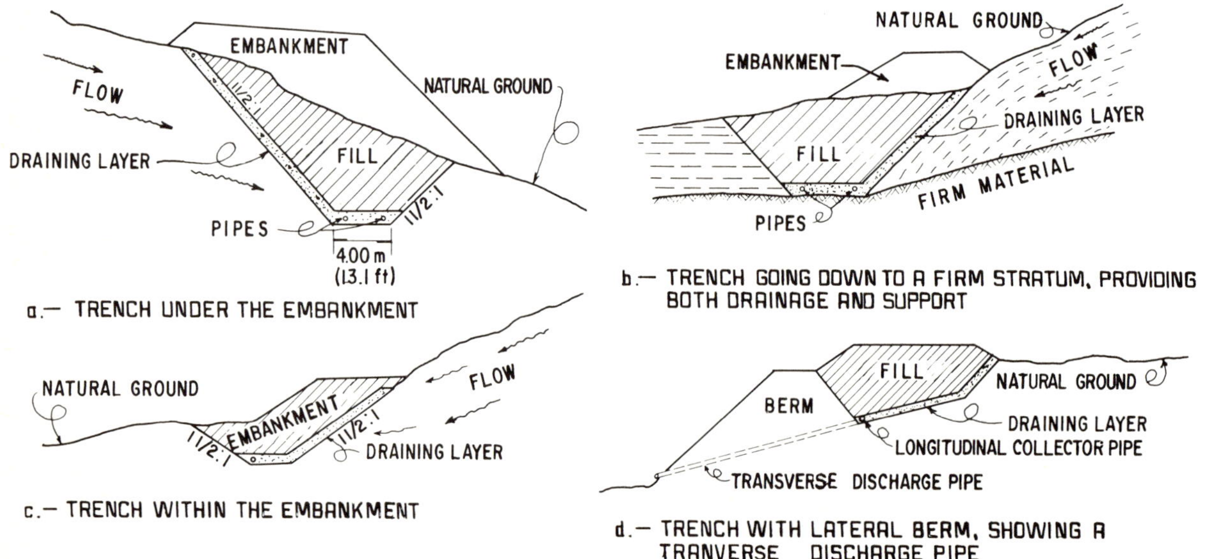

Fig. 7-28 Different types of stabilizing trenches

taken from a borrow pit. It is preferably pervious and incompressible when suitably compacted layer by layer.

A stabilizing trench improves the stability of an embankment or its soil foundation in several different ways.

1. It provides drainage, as has already been described
2. It replaces a weaker material with a better material as a result of which the embankment and the trench together rest on a firmer soil (Part *b* of Fig. 7-28). The potential slip surface is made so long and deep that failure becomes unlikely (Part *a* of Fig. 7-28). The quality of the material used to fill the trench must be so good that no slip surface can develop through the trench.

A stabilizing trench therefore has a dual function. Downslope, the subsoil drainage it provides improves the mechanical properties of the soil by physically interrupting the flow, and up-slope, it lowers the pressures in the water over an extensive area. In addition, the improvement of the mechanical properties of the soil that is replaced creates a mechanical restriction to failure, which may be of vital importance in many cases.

An example of these two actions is provided by the large-scale trench that was built to stabilize the slide at km 15+050 of the Tijuana-Ensenada highway in Northwest Mexico [30] (Plate 7-10). A sketch showing the general profile of the stabilized zone and a cross-section of the stabilizing trench is given in Fig. 7-29. At the location there was an existing slip surface which developed at the contact between a talus deposit and a layer of clay-shales. Part *a* of the figure shows this situation, as well as the original water table and the location of both road and trench. It was decided to fill the trench with high strength rocks so as to take advantage of both the draining and the mechanical reinforcement. The success of the solution, which stabilized a large slide, is somewhat surprising if the amount of corrective work is compared to the size of the unstable zone. This demonstrates the great efficiency of these solutions, which often exceed the limits of theoretical assessment.

Plate 7-10 Construction of a large stabilizing trench on the Tijuana-Ensenada highway

In cases like this, the mechanical effect of the trench on the safety factor of the entire system can be calculated. The draining effect is, however, far harder to assess.

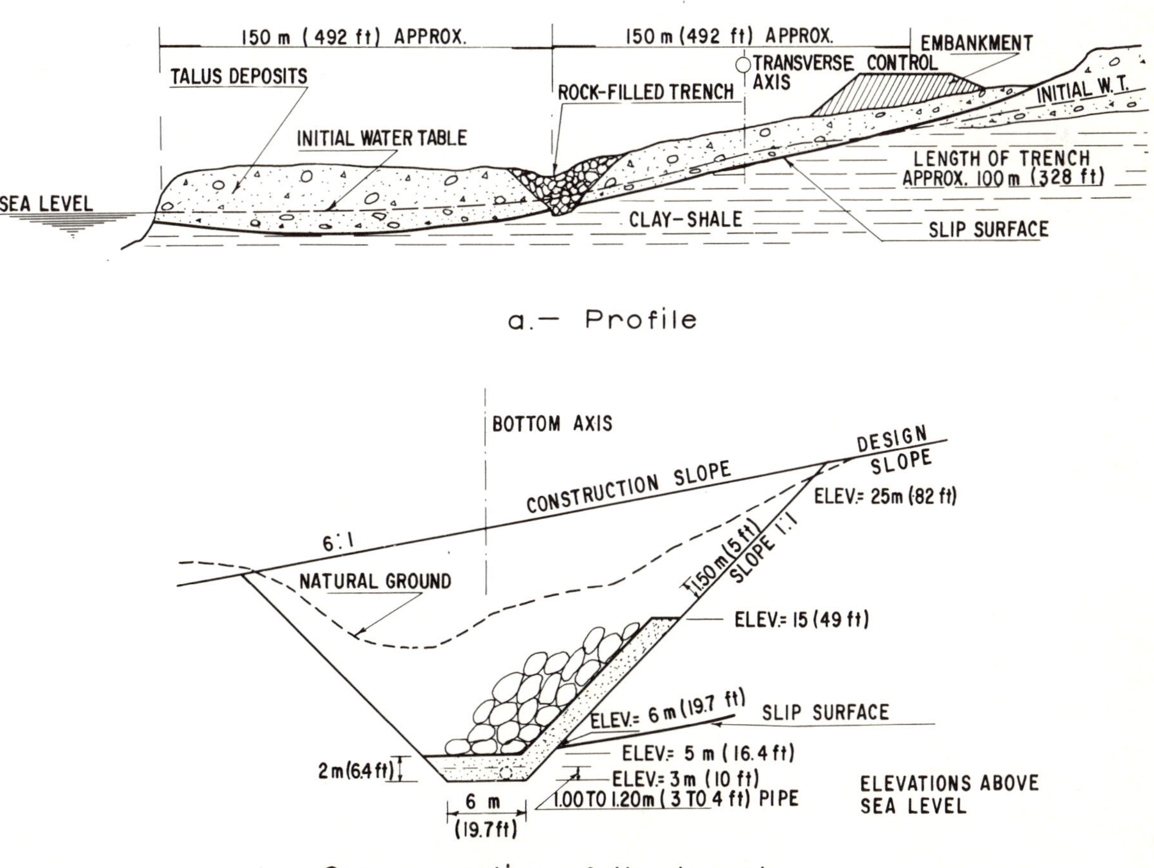

Fig. 7-29 Rock-filled trench at km 15+050 (9+623) of the Tijuana-Ensenada highway, Mexico [30]

Figure 7-30 shows another solution which was used with success to stabilize another slide that appeared in the same zone of the Tijuana-Ensenada highway [30,31]. Because of the topography of the zone (Part *a* of the figure) a hybrid solution was used; a combination of a rock backfill, whose stabilizing effect lies in its weight, and a stabilizing trench that is a drain. A large excavation was made which was filled with high quality rockfill (Plate 7-11). Appropriate filter sections were installed to prevent the fill from becoming clogged. Figure 7-31 refers to the above two cases, and shows the behavior of the trench at km 15+050 (9+623) during and after construction. Note the large movements which took place during the construction period, despite excavation in sections no more than 10 m (33 ft) wide, each of which was carefully filled before the next one was opened. Note too the rapid response of the zone to the stabilizing trench, which has continued from 1971 until today. Behavior is described by the horizontal and vertical displacement at three representative points at the crown of the highway. By observing the displacements that occurred, the importance of rapid construction in these jobs can be appreciated.

Stabilizing trenches can be constructed not only with their axis parallel to the road, but also transverse perpendicular to the road. The choice depends on the topography of the zone.

As a result of experience in California [29] stabilizing trenches are recommended in those places where, owing to already existing or predictable stability problems and groundwater flow, the water-table is 10 to 15 m (33 to 50ft) below the natural ground. This recommendation is rather restrictive, in view of the versatile role attributed to the stabilizing trench in this book. In our opinion it does much more than just eliminate water, which is the only function usually considered by the Californian engineers.

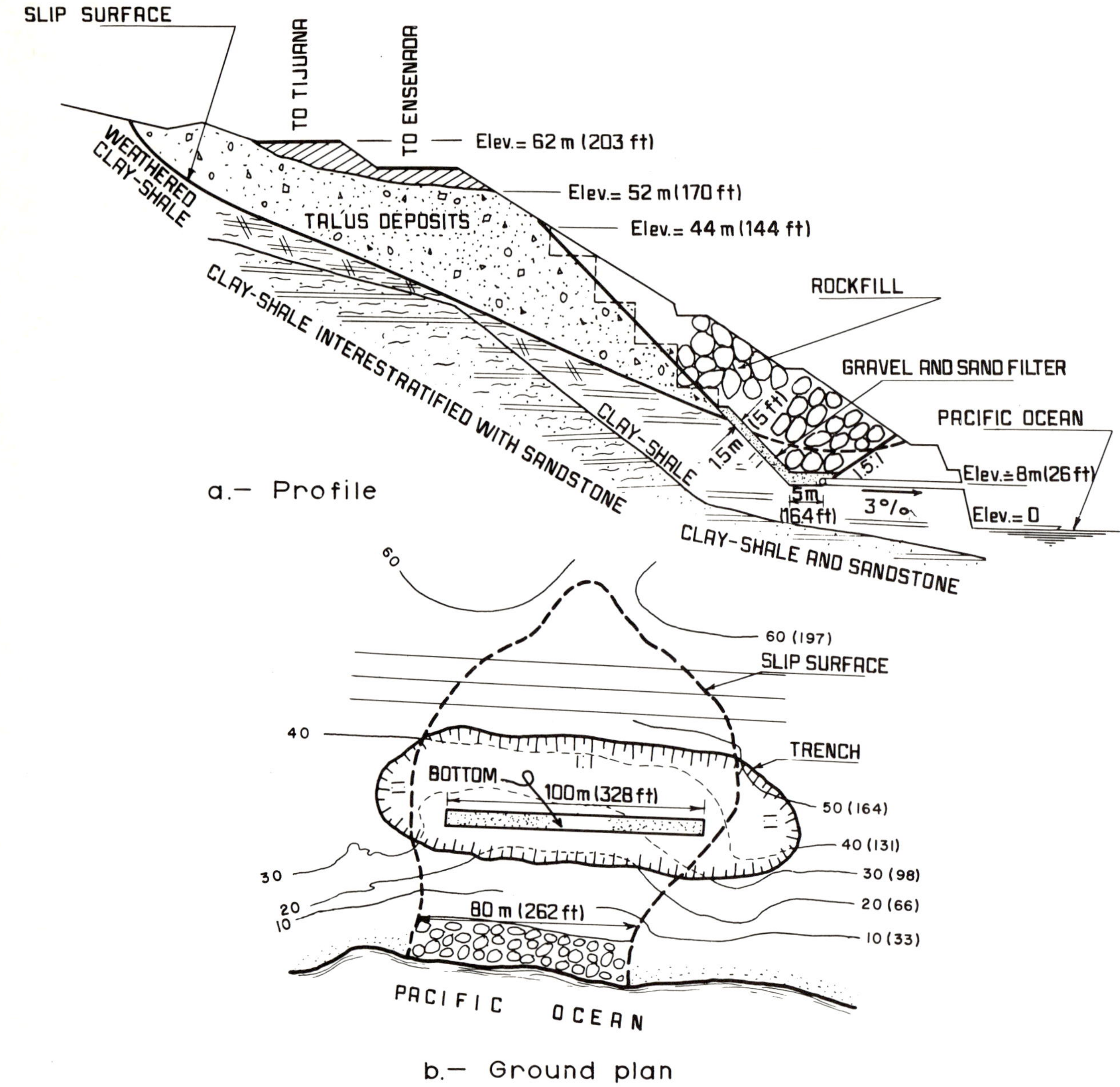

Fig. 7-30 Rockfill acting as a stabilizing trench in the landslide at km 20+400 (12+1190) on the Tijuana-Ensenada highway, Mexico [30]

Plate 7-11 Construction of the stabilizing trench in Fig. 7-30

By drawing the appropriate flow nets, it is usually possible to obtain an idea of the flow of water and the permeability requirements of the filter material which is placed against the upstream excavation slope. Figure 7-32 [7] summarizes the results of flow net studies. The relation between the permeabilities of the filter and soil, $\frac{k_f}{k_s}$ is needed for practical design, H is the height of the filter which remains below the upper flow line. The figure also makes it possible to relate the permeabilities required with design filter thicknesses.

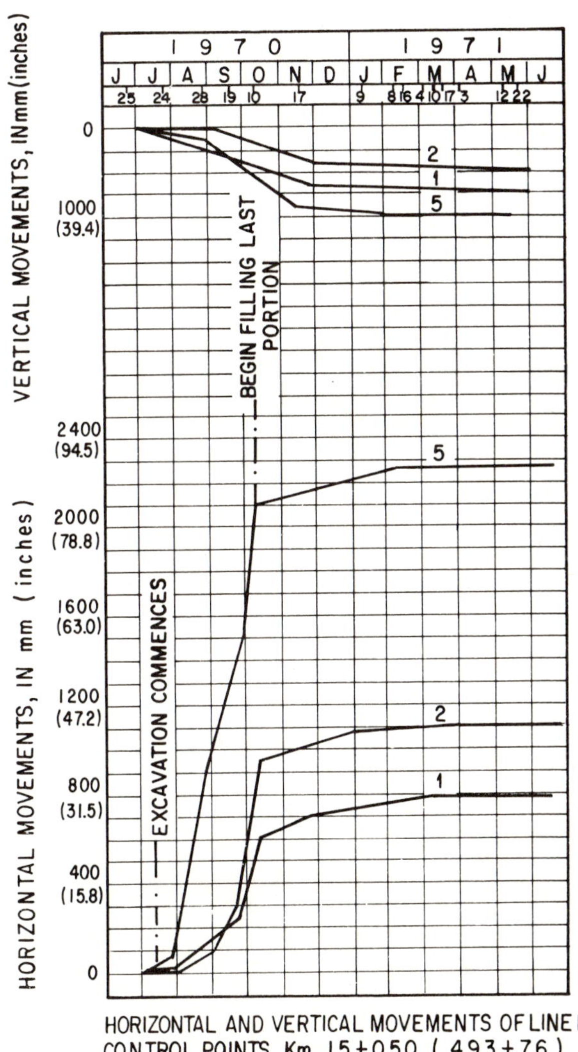

Fig. 7-31 Behavior of the trench at km 15+050 (9+623) of the Tijuana-Ensenada highway during and after construction [30]

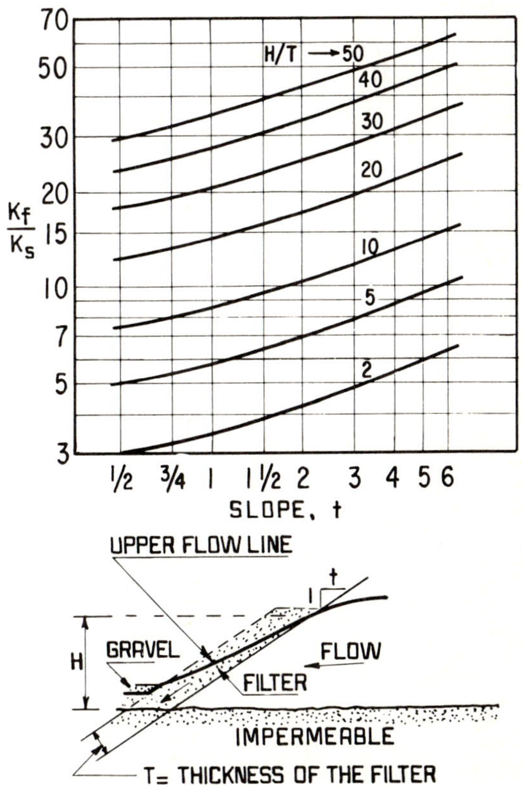

Fig. 7-32 Requirements for sloping filters [7]

7.4.8 Drainage Galleries

When subterranean water is located at such a depth that it is impossible to reach it by open excavation methods and the topography makes horizontal drains impractical, drainage galleries are sometimes used. This technique has been widely used in dam foundations, but galleries are now being installed for subsoil drainage in roads. They are especially effective for correcting unstable zones of large proportions. They are, however, seldom used as a preventive measure.

A drainage gallery is a tunnel with a suitable cross-section for excavation purposes. It is located where it can most efficiently capture and eliminate water affecting the stability of a slope or hillside that is used as a foundation for a road. The construction techniques are same as for a tunnel made for other purposes. They will not be given attention here, but will be briefly discussed in later pages. If lining proves necessary, it should be designed so that it will not interfere with the gallery's efficient performance as a drain. A large perforated metal pipe embedded in filter material has been used for a permanent lining. Because of material cost and durability a conventional lining made of concrete, masonry or both (masonry walls and concrete crown) can also be used. The lining incorporates numerous holes to permit drainage, without impairing structural support. Drainage galleries are often used underneath an existing failure surface, in which case drainage is greatly improved by installing a fan of perforated pipes as far as the failure zone.

The most important design consideration for drainage galleries is their location in relation to the bottom of the zone where stability is to be improved. Here a detailed exploration using borings will be necessary to define the shape of the slip surface in three dimensions. Instrumentation with inclinometers (see the section dealing with field instrumentation) to determine the shape of the failure surface within the subsoil usually provides the required data. Once this failure surface has been defined, the drainage gallery can be built through the lowest zone, so it collects water from the part that is hardest to drain. It has already been mentioned that the usual procedure is to build the gallery just below the failure surface; this avoids damage to the gallery from any movements that may occur, and improves drainage possibilities. Dewatering of the drainage gallery is very simple when the entrance can be drained by gravity. When this is not possible, pumping is sometimes the only solution.

Figures 7-33 to 7-35 depict the first drainage gallery which was built in Mexico (1965). They give a general ground plan of that particular zone of the Tijuana-Ensenada highway, the geologic cross-section of the failure zone and a structural cross-section of the gallery. It is about 200 m (660 ft) long and has additional horizontal drains installed to form a fan at the upper end and a drainage pipe for the last 100 m (330 ft) (Fig. 7-35). This system stabilized the most dramatic of all the slides which occurred on this particular stretch of the Tijuana-Ensenada highway. This was the first slide which occurred during the construction period, with maximum cumulative movements of 2 m (6.56 ft) in the vertical direction on the crown of the road and 1.80 m (6 ft) towards the sea in the horizontal direction in the highest part of the failure surface, at the toe of the high cliff of sound clay-shale (Fig. 7-33).

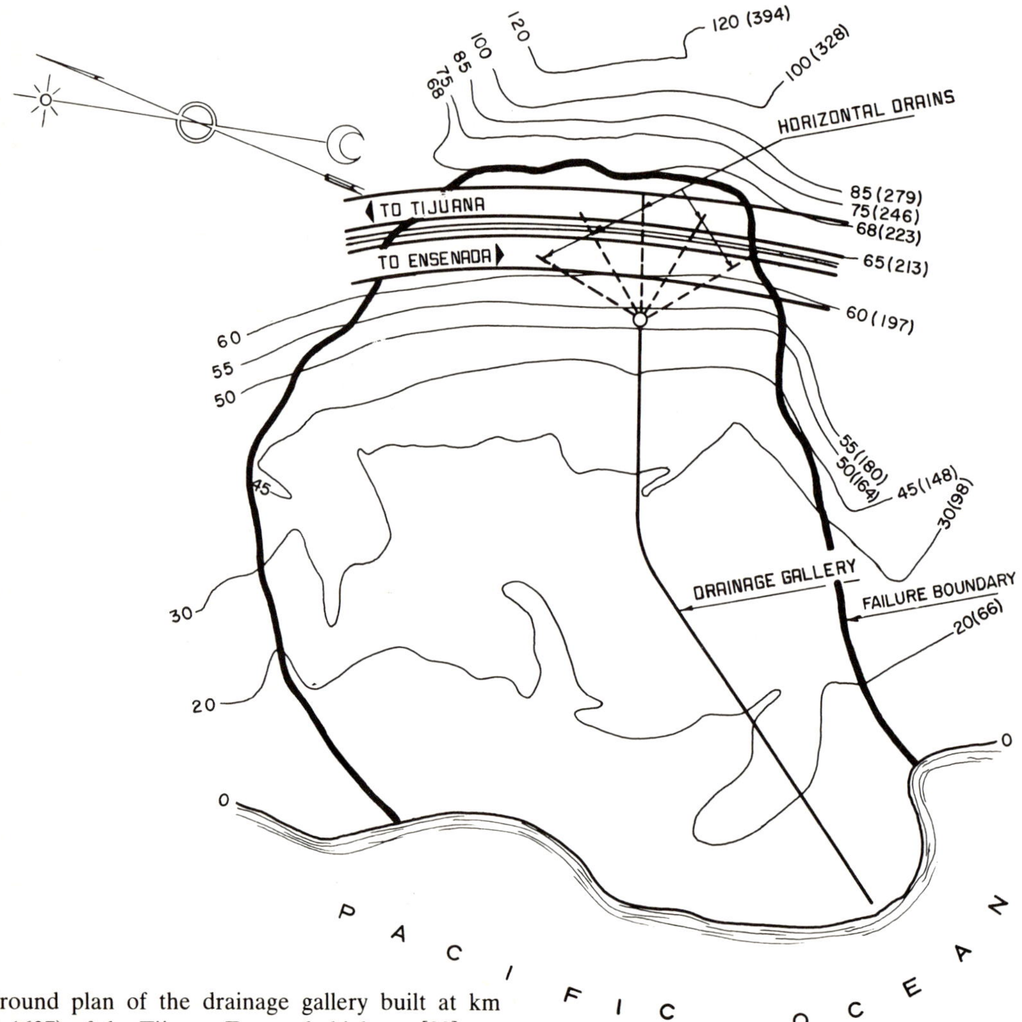

Fig. 7-33 Ground plan of the drainage gallery built at km 19+200 (11+1637) of the Tijuana-Ensenada highway [30]

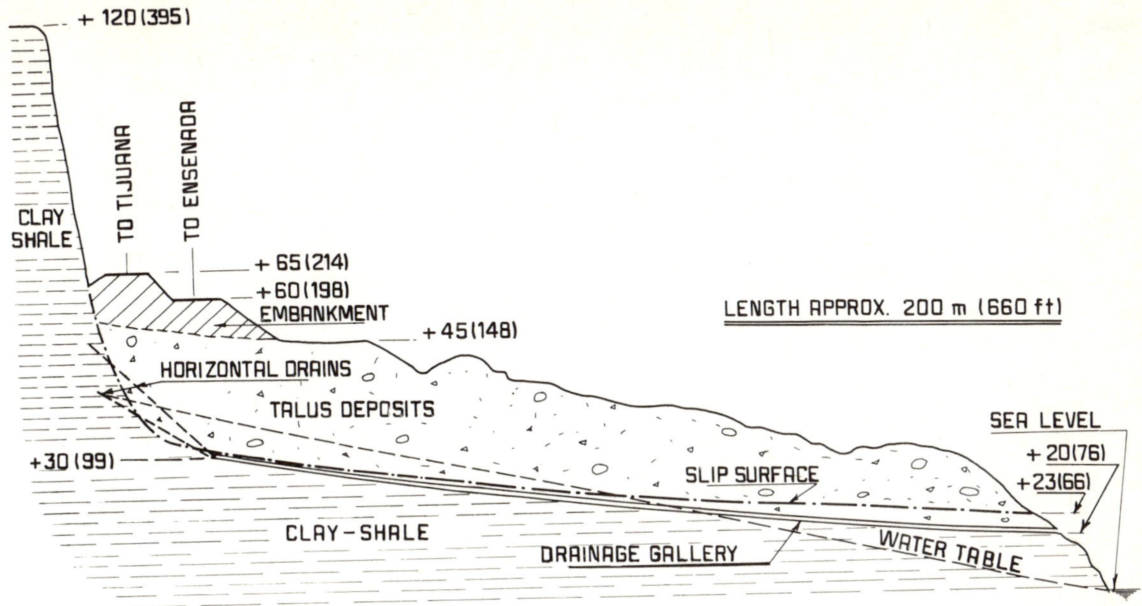

Fig. 7-34 Geologic cross-section of the drainage gallery at km 19+200 (11+1637) of the Tijuana-Ensenada highway [30]

During geological investigation it was possible to accurately define the internal configuration of the failure surface on which the movements occurred. Once the drain structure had been built, the response of the moving mass to its effect was no less dramatic than the slide itself. Movements were brought to a complete halt and no further slides have been recorded.

Construction of the gallery commenced with installing a large corrugated metal pipe (1.80 m (6 ft) in diameter) (Plate 7-12) embedded in filter material within the tunnel of about 2.50 m (8 ft) in diameter. This section changed to the one in Fig. 7-35 after the first 30 m (98 ft) for reasons of cost. Water disposal is by pumping at the lower end of the gallery.

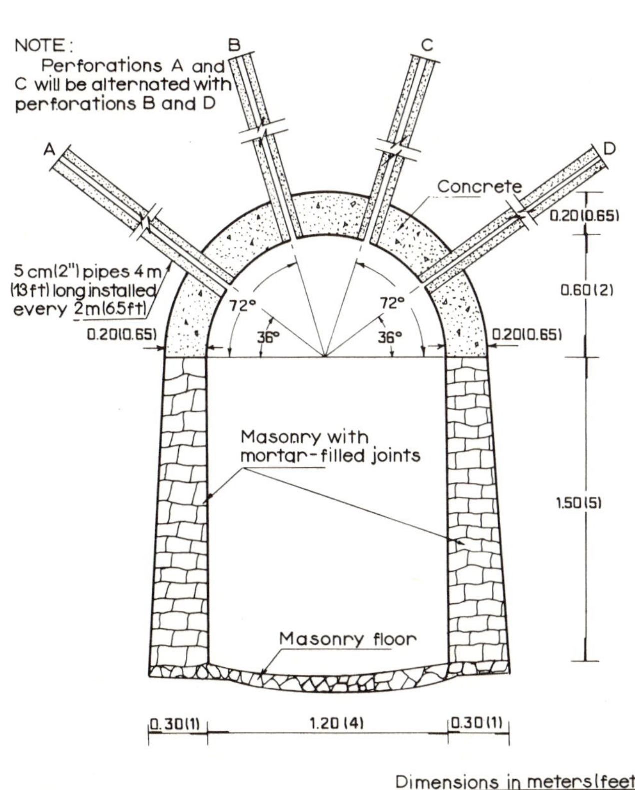

Fig. 7-35 Cross-section of the drainage gallery at km 19+200 (11+1637) of the Tijuana-Ensenada highway [30]

Plate 7-12 Inside view of the drainage gallery at km 19+200 (11+1637) of the Tijuana-Ensenada highway

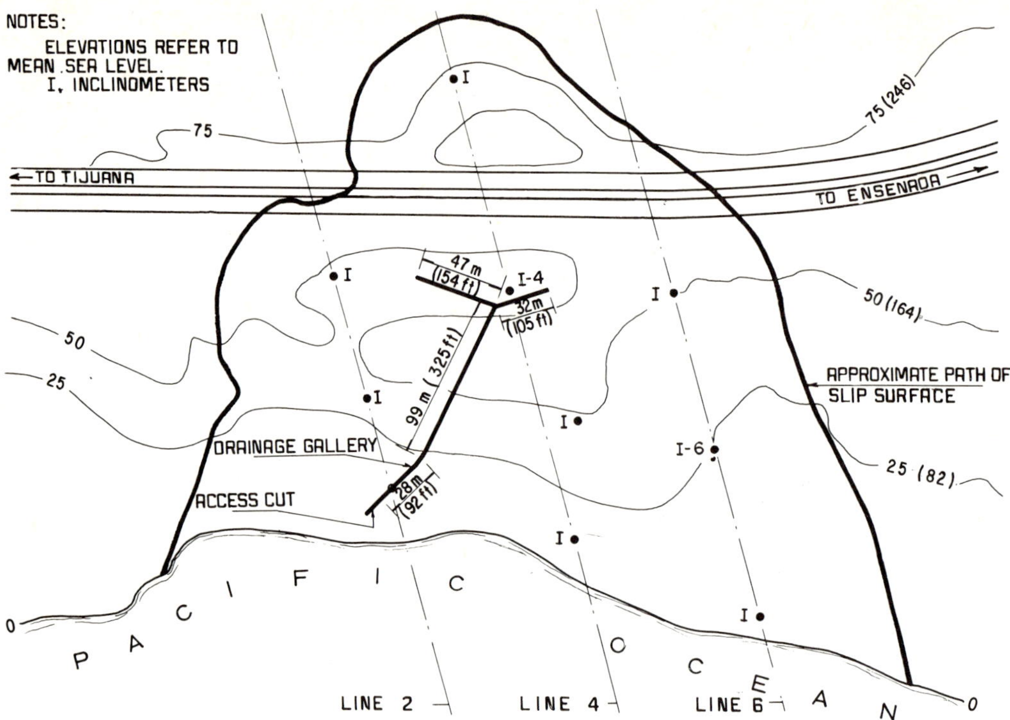

Fig. 7-36 Ground plan of the drainage gallery at km 15+500 (9+1115) of the Tijuana-Ensenada highway [30]

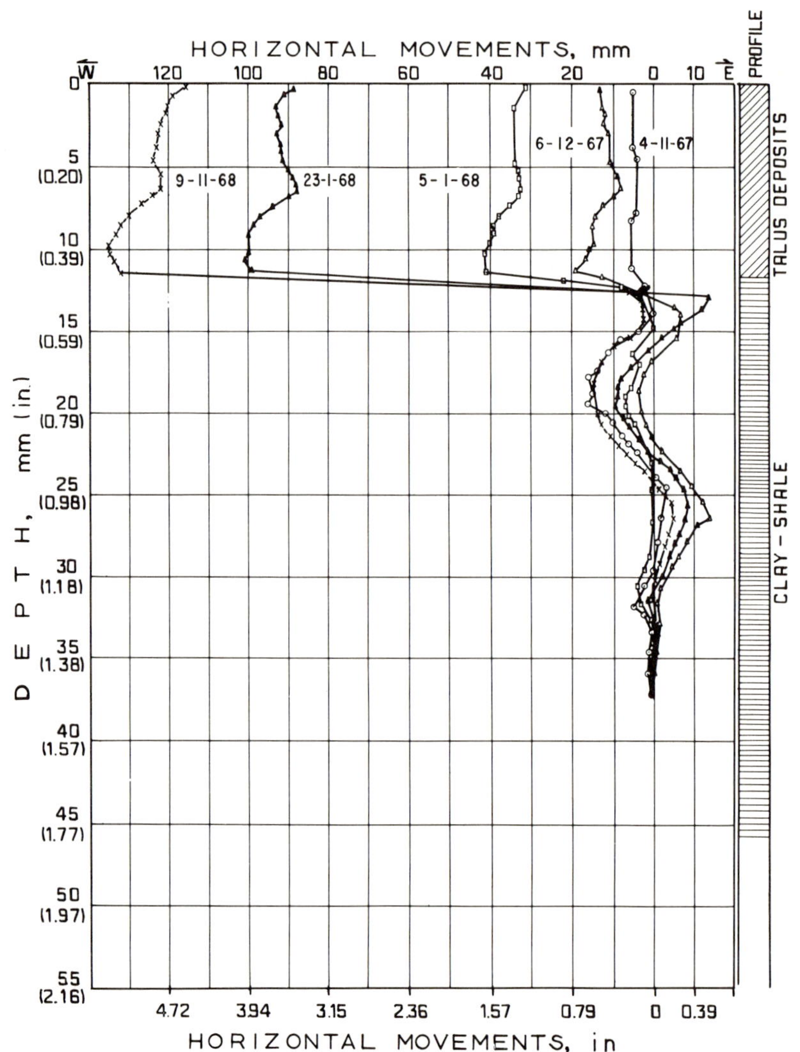

Fig. 7-37 Depth of the slip surface as given by an inclinometer, at km 15+500 (9+1115) of the Tijuana-Ensenada highway [30]

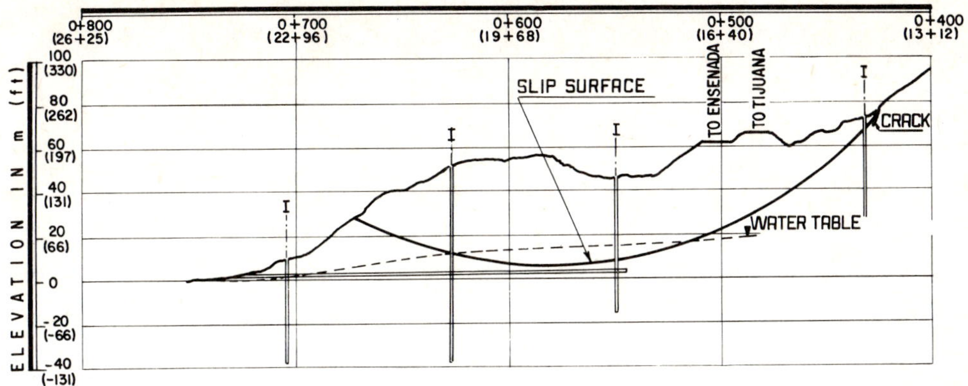

Fig. 7-38 Cross-section of the slide at km 15+500 (9+1115) of the Tijuana-Ensenada highway, showing the drainage gallery [30]

Figure 7-36 shows a plan of the slide zone at km 15+500 (9+1115) of the Tijuana-Ensenada highway, where another large drainage gallery was built, in the shape of a T (Plate 7-13). The same figure shows the location of 10 inclinometers which made it possible to determine the configuration of the failure surface, Fig. 7-37 shows typical horizontal displacement data recorded by one of the inclinometers. Figure 7-38 illustrates the approximate cross-section of the failure surface in one of the sections where it was determined. Zero is the elevation of the pacific Ocean. This failure surface basically developed at a contact between colluvial deposits and a very thick layer of clay-shale. An embankment 17 m (56 ft) high was built there for the highway. A ground plan of the gallery which was constructed to drain the zone is shown in Fig. 7-36 and a cross-section in Fig. 7-38 in which the gallery (not quite in the plane of the cross-section) has been projected onto it. Figure 7-39

Plate 7-13 Interior of the drainage gallery at km 19+200 (11+1637) of the Tijuana-Ensenada highway

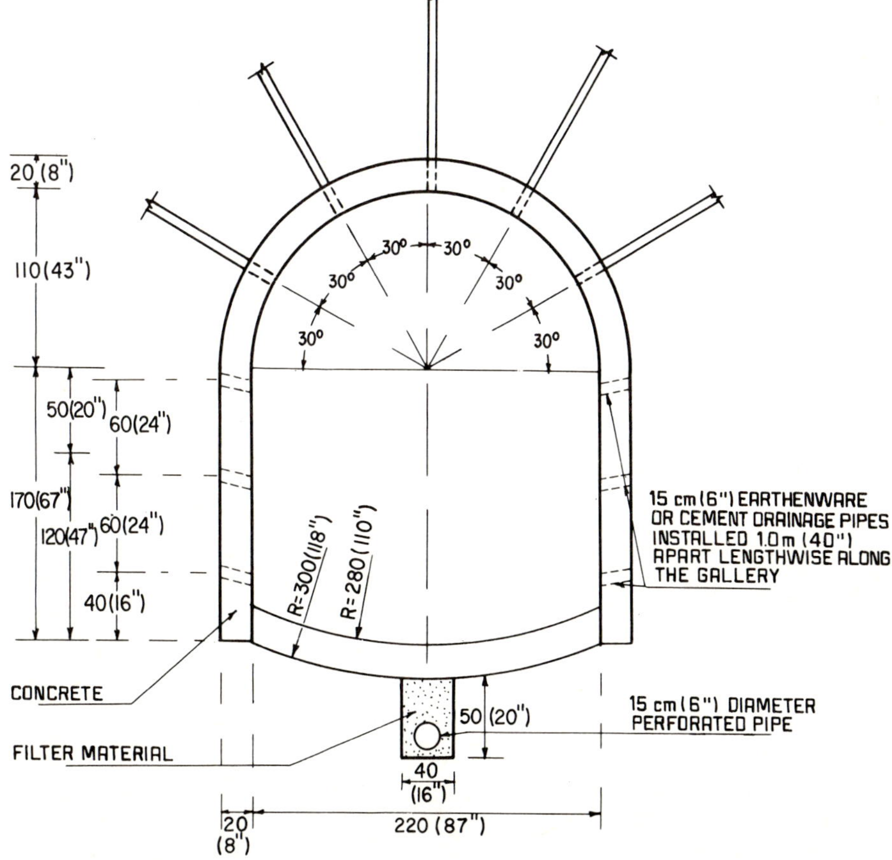

Fig. 7-39 Cross-section of the drainage gallery at km 15+500 (9+1115) of the Tijuana-Ensenada highway [30]

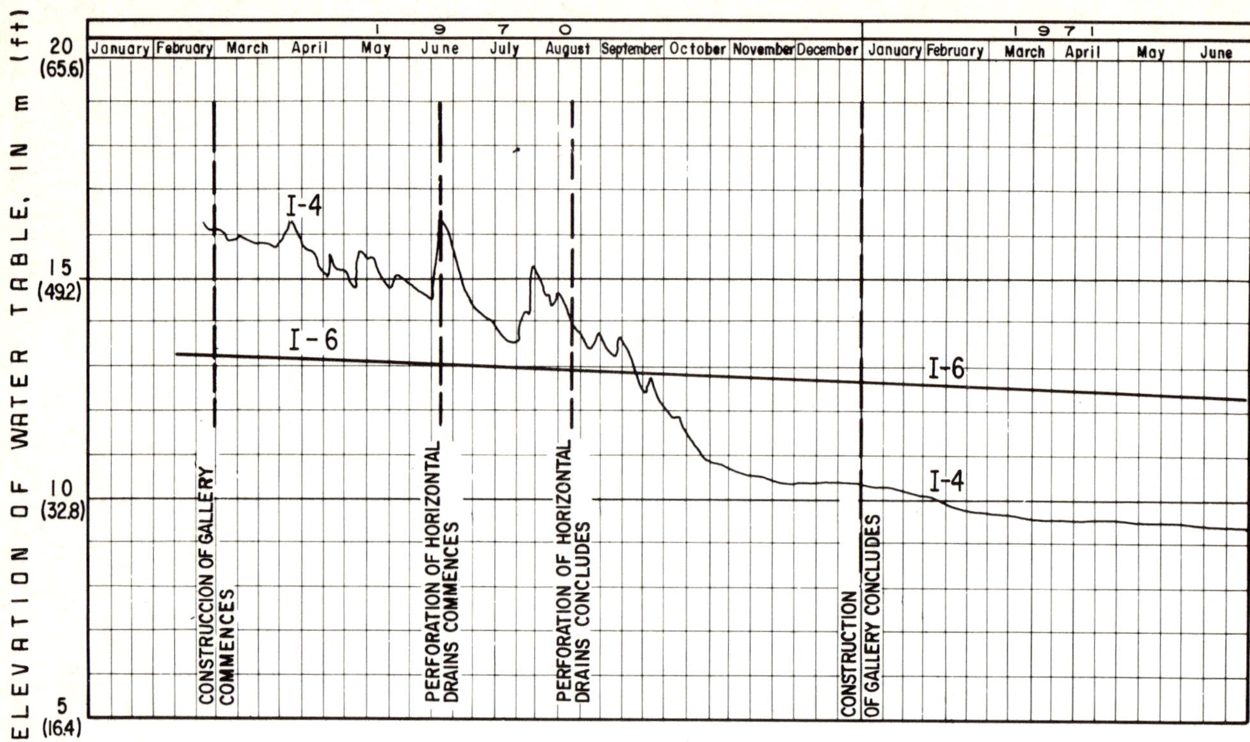

Fig. 7-40 Variation of the water-table in the slide zone at km 15+500 (9+1115) of the Tijuana-Ensenada highway [30]

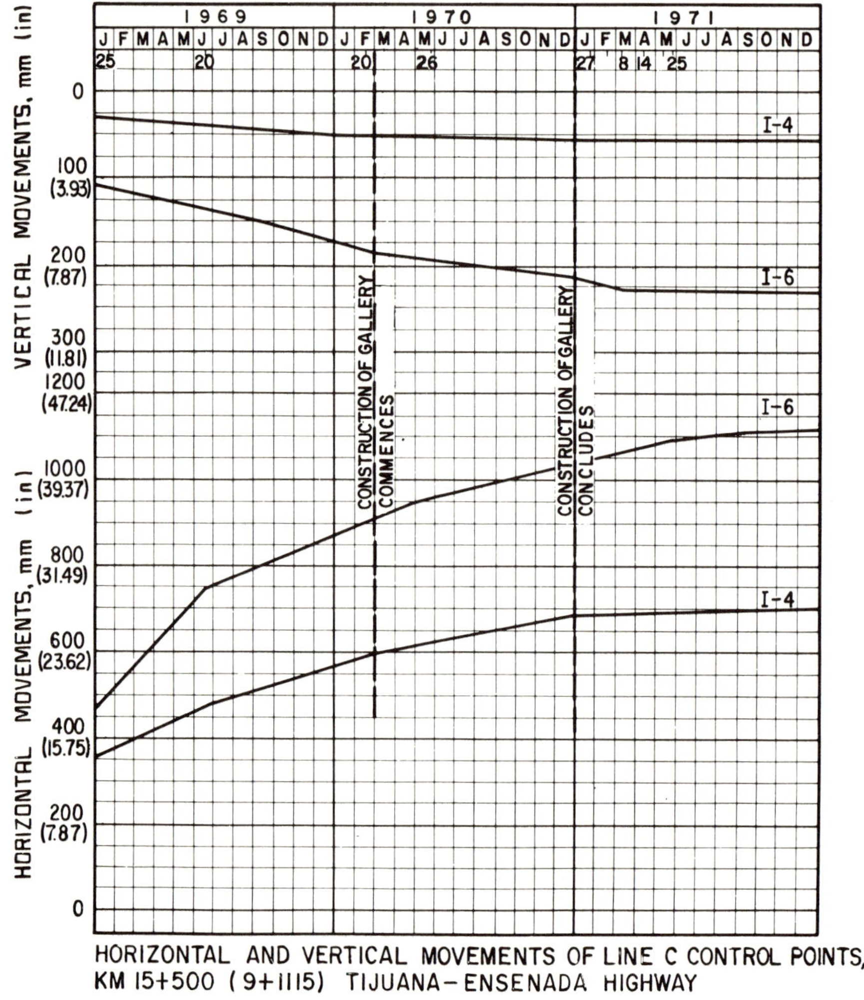

Fig. 7-41 Movements of two control points of the slide zone at km 15+500 (9+1115) of the Tijuana-Ensenada highway [30]

shows a cross-section of the gallery. It can be seen that its drainage capacity was enhanced by a fan of drilled drains with perforated pipes extending upward to intercept the failure surface. Figure 7-40 shows variations in the water-table at two points in the zone affected by the drainage gallery. Note that this was very marked in well *I-4* (Fig. 7-36) very close to the *T* of the gallery, and less marked, but nevertheless consistent, in well *I-6* much further away. The effect of the fans of drilled drains is significant, especially in the zones nearest to the gallery. By 1973 the water-table appeared to be nearing a final stabilization. The movements of the failed zone were halted, as can be seen in Fig. 7-41, where the records for *I-4* and *I-6* are shown. The relatively slow response of the movements to the effects of the gallery should be noted, especially at the points that are furthest away from it.

7.5 Capillary Effects in Subsoil Drainage

Explanation of the presence of water and its effects and movement above the water-table is somewhat complex. The soil is virtually saturated above the water-table to the height of capillary rise, but above that level it has a lower degree of saturation. Even though the forces of gravity and viscosity continue to play an important role, capillary forces dominate above the water-table. These forces were briefly analyzed in Chapter 1 and include effects due to surface tension and to physico-chemical interaction between the water and the mineral walls of the conduits of the soil. References [32,33] consider some of the theoretical aspects of the mechanisms for studying these phenomena.

The stresses resulting from surface tension are tensions in the water; negative pore pressures. These tensions increase when the degree of saturation is reduced and increase with decreasing temperature. In the zone of partial saturation, above the capillary height of the soil, there is also vaporized water. The vapor pressure drops with a drop in temperature.

Figure 7-42 [34] shows the state of equilibrium of water above the water-table. As has already been said, within the capillary height, h_c, the soil is saturated. There is continuity in the water and the pressure in it obeys the hydrostatic law. Above this zone, there is a zone of partial saturation, where the degree of saturation rapidly decreases with height and where the water no longer fills all the voids, but nevertheless still maintains its continuity at the contacts between the grains of soil.

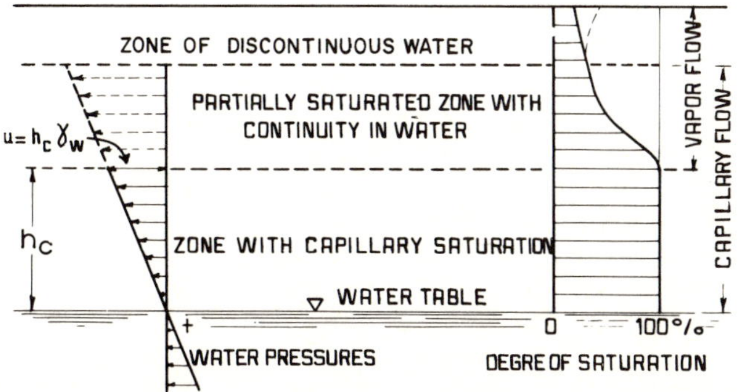

Fig. 7-42 Equilibrium condition of the water above the water-table [34]

In this zone, the effective stress is no longer the total stress minus the pressure in the water, since the water pressure does not act on the entire area of the voids.

Above the partial saturation zone with continuity in the water, there is a second with partial saturation with an even lower degree of saturation which decreases with height; here the continuity in the water is interrupted. The hydrostatic pressure law is no longer valid. In this zone, menisci develop in the water at the contacts between the soil grains, and the stress in the water depends on the radius of these menisci.

Throughout the entire partial saturation zone, both where there is continuity in the water and where there is not, there is vaporized water. This also can move. Whether it moves and how it moves depends on the gradient of vapor pressure. For example, surficial evaporation reduces the pressure of vapor in the upper layers of soil. This causes an upward flow of moisture in vapor form. If the ground surface temperature drops, the reduced vapor pressure at the surface, also causes an upward flow of moisture in vapor form (the cause of dew). A warmer ground surface temperature induces downward flow.

Flow of capillary water can only occur in the zone of partial saturation where there is continuity in the water. If there is equilibrium, the capillary tension must be equal to the hydrostatic pressure γ_w and there will be no movement. If the capillary tension changes, either increasing or decreasing, a corresponding flow will occur. Evaporation in the upper layers of the partial saturation zone reduces the degree of saturation, the radii of the menisci of the water in the voids of the soil decreases, which leads to an increase in capillary tension. At the same time, as a result of the loss of water, the thickness of the saturation zone is reduced to the value h_c' (Fig. 7-43, [34]). At the new level, h_c', which is the new boundary between the saturated and unsaturated zones, the capillary tension remains the same as it was at the previous level, h_c, the menisci being modified accordingly. The tension is now greater than the hydrostatic pressure at the same level, $\gamma_w h_c'$ (see Part *a* of Fig. 7-43). Consequently an upward flow will occur with the following gradient:

$$i = \frac{\Delta u/\gamma_w}{h_c'} \tag{7-13}$$

This is the reason why in arid regions with intense solar activity and surface evaporation there is a continuous groundwater flow towards the surface. The rising water carries salts which are deposited as a residue in the partial saturation zone when the water evaporates. The increase in the concentration of salts in the upper layers of soil may contribute to cementation and make the soil impermeable.

An effect similar to that of evaporation in arid regions can be caused in humid regions by the loss of surface water through intense evapo-transpiration in the layer of vegetal covering. As the surface of the soil is heated by solar action, evaporation increases and there is a corresponding acceleration in the rising flow. In compressible clayey soils, increases in capillary tension cause shrinkage and cracking. As vegetation absorbs or transpires water, an effect similar to evaporation occurs; zones where considerable volumetric shrinkage occurs are created around many species of trees.

When a structure is built on natural ground, evaporation is retarded in the covered area. An embankment may practically prevent all evaporation underneath it. Static equilibrium is

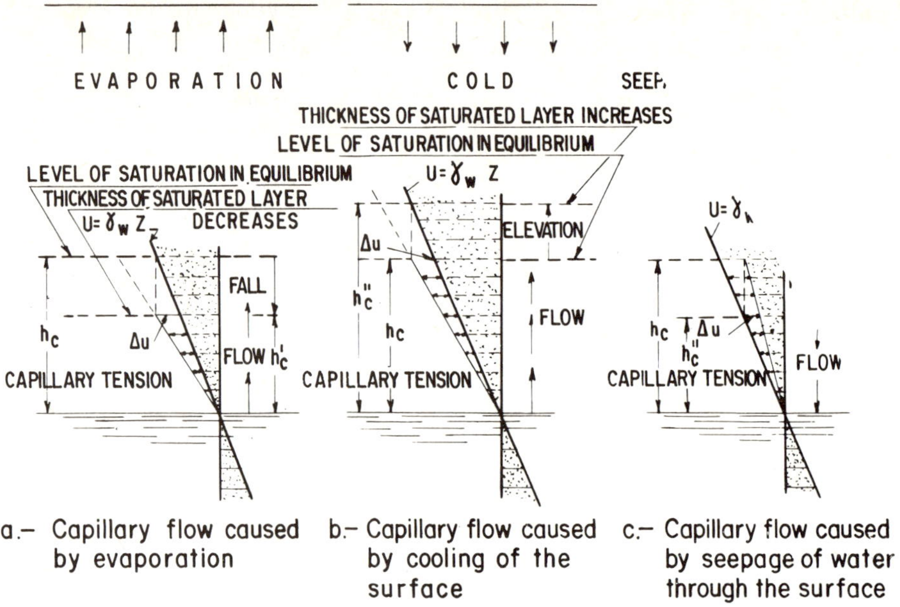

Fig. 7-43 Cases of capillary flow [34]

reached here with a rise in the capillary saturation line, which usually causes loss of strength and swelling of clay soils as a result of the increase in water content (Part *b* of Fig. 7-43). This effect will be more marked in zones subjected to intense sunshine, where the soils are normally desiccated. An increase in the water content of the normally desiccated pervious surface produces severe expansion and usually distortion of the structures that retard the evaporation.

If there is an abrupt drop in temperature of the surface of the ground, the mechanisms shown in Part *b* of Fig. 7-43 may also be triggered. In this case, capillary tension increases by a value Δu to level h_c, and there is a corresponding capillary rise to the new level, h''_c. Part *c* of Fig. 7-43 shows another case which is important in road engineering. The increase in the degree of saturation of the soil that is close to the surface as a result of seepage or rain infiltration for example causes an increase in the radius of the menisci. There is a corresponding reduction in capillary tension and reduction in capillary rise, causing downward seepage that feeds the groundwater.

In sands, variations in capillary tension are comparatively small because the voids of the soil are so large that the magnitude of these forces is small. The effects are far more notorious in clays and some silts. The proximity of the water-table to the surface of the ground is also significant in the magnitude of these effects. Very deep-seated water-tables do not encourage important changes in the surface water contents, even in areas where solar action is more intense or where large areas of the surface are watertight.

The different subsoil drainage systems which have been discussed in this Chapter will provide an outlet for the water contained in the soil, so long as in the immediate vicinity of the drain the water pressures are equal to or higher than the atmospheric pressure which prevails inside that drain. A drain will be unable to dewater zones where the water is at a lower than atmospheric pressure; in other words, where the water is under tension (zones of capillary water). During the construction of a subsoil drainage system, such as a horizontal drain, for example, atmospheric pressure is introduced into the soil. As a result the drain introduces a boundary of zero hydrostatic pressure within the groundwater. If the water around the drain has higher-than-atmospheric pressure, a gradient is created towards the drain and the water flows towards this drain. Therefore subsoil drainage can only be effective in, and should therefore be built only in, zones that are below the water-table, where the pressure of the water is higher than the atmospheric pressure. Above the water-table, the water is under capillary tension. These tension stresses generate compressions between the particles of the solid structure with a corresponding increase in the effective stresses between them. This increases the stability of the soils. Removing the capillary water would be detrimental even if it were possible. An objective of an under-drain, apart from eliminating gravitational water, is to reduce the neutral stresses in a saturated soil by lowering the water pressures and converting them into tension stresses.

Figure 7-44 [35] shows how this objective can be achieved with a horizontal drain. In the figure it is assumed that a cut has been made in a clayey soil where the water-table originally occupied position *I*. Just by making the cut there has been a change in the water-table which is now *II*. The open cut, where atmospheric pressure now prevails, will attract groundwater flow from neighbouring soils. This effect should be remembered by engineers working on road projects. Any cut that is made below the original water-table is a drain, in the sense that it attracts water from the neighbouring soil masses which is at a higher-than-atmospheric pressure. This is why water surfaces naturally in cuts. The water content of the soils adjacent to the cut face tends to increase, with the corresponding consequences for the general stability of the face.

Still referring to Fig. 7-44, a transverse drain is installed as shown in Part *c*. The water-table will adopt a profile similar to *III*, and throughout the entire cross hatched zone, neutral pressures that were originally higher than the atmospheric pressure will be replaced by tensions in the water. The water tension increases the effective stresses in the soil mass causing compression in the mineral structure of the soil. This effect, which increases the stability of the cut, is independent of the favorable reorientation of the seepage forces for which the under-drain is also responsible.

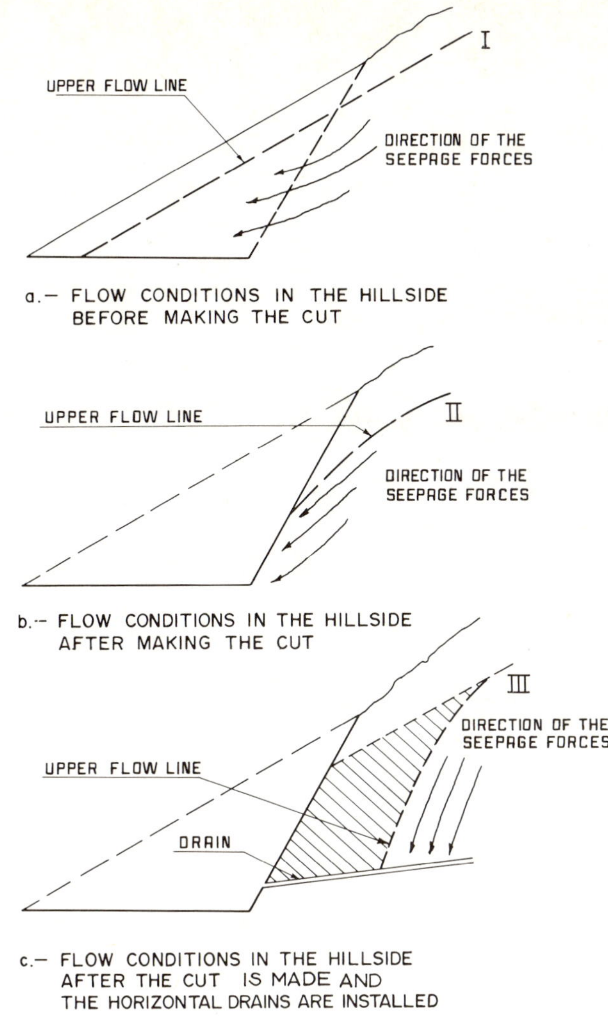

Fig. 7-44 Effect of a horizontal drain on the pressure condition of water in a cut [35]

To conclude these comments regarding the influence of surface physics on subsoil drainage processes, some attention should be given to the influence of the relative humidity of the atmosphere and the possibility that atmospheric moisture condenses. Experience shows that there is a specific relation between the relative atmospheric humidity and the stress in the water in the soil close to the surface. If the relative atmospheric humidity were everywhere 100% and the soil were clayey, the water-table would tend to be very close to, if not coinciding with, the surface. Similarly, a clayey subsoil with a water-table very near to the ground surface usually reflects atmospheric humidity conditions that are very close to saturation. Likewise, a very low relative humidity, implies a deep water-table.

The importance of the relation of humidity to soil moisture is greatest in the layers of soil in contact or near the natural ground surface such as the typical case of pavements, which may become saturated. If relative humidity is high and the subsoil clayey, the base and sub-base tend to have low capillary tension. They tend toward saturation, as a result of long term capillarity, although they may have had low water contents at the time of construction. In such cases, it is desirable for there to be a layer to isolate the capillary effect. This is known as a capillarity breaking layer. According to Terzaghi [36,37], this is equivalent to converting an open system into a closed system, where the mean water content of the isolated zone virtually does not change because water is prevented from migrating from neighbouring soil masses. Fluctuations in the water-table as a consequence of atmospheric humidity will in turn cause shrinkage or expansions of those clayey soils which are susceptible to volume changes.

7.6 Subsoil Drainage in Highways

The importance given to the control of subterranean water has varied greatly from one period to another in the history of road building, including relatively modern times. It is evaluated differently from one country to another and from one group of engineers to another. Most engineers agree that subsoil drainage is advisable and beneficial. However, its cost is often very high and opinions start to diverge when costs are compared with benefits. The differences in opinion are so marked that some engineers regard subsoil drainage as an essential part of the design and construction routine for a highway, in the same way as surface drainage or compaction, while there are other engineers, organizations and even countries who seldom consider drainage. A common viewpoint, especially in the so-called developing nations, is that subsoil drainage is a luxury for the wealthy. This argument is unfounded. If subsoil drainage is beneficial and its function is of vital importance, then it is more a necessity for the poor. Those nations with limited income should take special care to secure the vast investment that is represented by the construction of a highway. If subsoil drainage is not important, then it is not a luxury for the wealthy but an absurd expenditure. It appears, therefore, that the problem should not be considered in terms of relative wealth, but instead, looking at the technological root of the problem, and considering the benefits which the highway will gain from the subsoil drainage. Those benefits must be compared with the amounts invested as well as the money which will be spent for repair if there is no subsoil drainage.

Difficulties usually arise in establishing the benefit-cost relations and behavior alternatives which were mentioned in the preceding paragraph. For example, despite the numerous theoretical and computational aids which are available today, an accurate evaluation of the benefit of a specific subsoil drainage project continues to be difficult. In many cases it is hard to establish the change in safety factor in a certain part of the road. It is even more difficult if the safety factor concept is extended to include the numerous concepts which determine the behavior of a whole stretch of that road. The above is true, despite the considerable improvement in theory of groundwater flow and exploration methods during the last few years.

It is even harder to develop reliable alternatives to improve the behavior of a certain stretch of highway when no subsoil drainage system is used, or when subsoil drainage is constructed on the basis of mediocre systems, to which an equally mediocre efficiency can be attributed. It is evident, therefore, that any subsoil drainage design, whether for a specific case or for general use, must be largely subjective. As always, the analyses will depend fundamentally on the abilities and integrity of the engineer as a human being and a member of society.

The authors believe that many of the uncertainties and controversies concerning subsoil drainage can be attributed to insufficient consideration being given to its functions and role. It is common belief that the principal (or exclusive) object of subsoil drainage is to *eliminate* water. Therefore, drainage should be considered only in places where large quantities of water can be seen on the surface. At the beginning of this Chapter and in many other parts of this book, it was established that the ultimate objective of a subsoil drainage system is to reduce neutral pressures. The fundamentals of soil and rock mechanics demonstrate that reducing neutral stress is very beneficial to the stability of a mass of soil or rock. It also induces a change in the direction of the seepage forces to eliminate subsurface erosion and local softening and sloughing. With this improvement in strength conditions and reorientation of seepage forces, cross-sections can be designed for each individual case which are not only more reliable, but also more economical in that there will be smaller earth movements, and savings in investment and maintenance.

The authors feel that subsoil drainage is too costly to adopt indisciminately in a routine manner; but there is no doubt that it is technically and economically necessary in many cases. Subsoil drainage is invaluable both for improving the stability of slopes, hillsides and embankments and protecting pavements, and it should always be considered in design, weighing costs and total benefits. The major questions that must be given careful consideration are where, when and what system should be used. Because highway engineers so often have to work with the guidance of only rather superficial information, obtained by hasty exploration and sampling, and with limited aid from the laboratory, it again becomes essential that the geotechnical investigations that are carried out during the design stage must be entrusted to specialists in soil and rock mechanics, geology and pavement technology. They will be able to make the necessary recommendations wherever subsoil drainage may prove essential. Whenever problems are detected that are too complex to evaluate from the routine information, these specialists will immediately order more detailed exploration. In existing roads, the capacity of the field geotechnical engineers and geologists for diagnostics must be good enough to enable them to analyze both the places where soil and rock behavior is likely to be bad including any slides that may occur. They can detect causes and make recommendations for appropriate subsoil drainage systems for those extremely common cases where subterranean water is a major factor.

The extreme alternative, if subsoil drainage is not used, may be a landslide, or the destruction of a pavement. The implications of these occurrences depend largely on the economico-social importance of the road in question. Thus, the frequency and intensity with which subsoil drainage is applied are partially governed by the importance of the road including the goods and services that use it. It should not be concluded, however, that in roads with little traffic or a low economic level of services (which are so common in developing countries) subsoil drainage should not be used. The importance of drainage is often so vital that the existence of the road depends on it. What is stressed here is that the importance of a landslide that endangers traffic is economically different for a much-used highway than it is for a minor road with little traffic. Nevertheless, the authors re-emphasize that subsoil drainage often plays a part which is quite independent of the relative importance of the road: the very existence of the road depends on drainage.

The objectives of subsoil drainage vary somewhat for cuts, embankments and road surfaces.

When making a cut, water tends to surface in the face of the slope. This is because construction modifies the external boundary of zero water stresses. The natural ground is unloaded which causes a local decrease in the normal stresses and an increase in the shear stresses in the soil or rock immediately behind and below the excavation, both of which affect the stability of the slope. As was stated in Chapter 6, a consequence of all this is that the critical condition of the cut develops some time after excavation, generally following periods of intense rainfall. The use of subsoil drainage in cuts helps to control groundwater flow within the slope, not allowing it to come to the surface. As a result, volume changes in the soil are restricted and seepage forces suitably redirected. The change in the state of hydrostatic stresses in the water in the slope zone is of great advantage, even if the soils are not dried by the subsoil drainage system. This is readily understandable if it is recalled that the strength of soils depends on the effective stresses to which they are subjected, and not the total stresses.

In the case of embankments, the need for subsoil drainage in the foundation soil on which they are built can be explained in terms of similar mechanisms. Construction of an embankment on a hillside causes an increase in the actuating shear stresses. As the slope of the embankment is steeper than that of the natural ground, the increase in the shear resistance of the ground, which is due to the increase in normal stresses, may not adequately compensate for the increase in shear stresses. The stability of the hillside is consequently reduced. A rational and efficient way of improving it is to bring about an increase in the effective normal stresses, which can be achieved by lowering the neutral stresses in the water in the voids of the hillside soil or rock.

When water tends to rise underneath a road surface, as occurs at the bottom of a cut, the layers of material comprising that road surface may be damaged. This is not the only reason for a possible increase in the water content of these layers, but it may well be one of the most serious ones. As the traffic load acts on the pavement, normal and shear stresses are transmitted to the base course, sub-base and subgrade. If these layers are dry or their water content is low, the normal stresses will be supported by the granular structure and the shear strength will increase as required. However, if the base course and the other layers are saturated, part of the traffic load will be transmitted to the pore water and will not contribute to the creation of shear strength. The traffic load will have to be supported just by the strength generated by the already low normal stresses caused by the weight of the pavement itself. In this case, a major aim of a subsoil drainage system should be to achieve low degrees of saturation in the base course, sub-base and subgrade.

In order to establish an adequate system of subsoil drainage, information must be available concerning the geology and structure of the soil and rock that are involved. This may be obtained from field inspection, geological investigations or borings with sampling, followed by laboratory tests. Since the information from these sources is usually incomplete and unreliable, a subsoil drainage design should never be considered finalized, but should always be kept open to any changes and adaptations that may prove necessary either at the construction stage or during the lifetime of the road.

Field inspection should commence with route selection, analyzing all possibilities and carefully weighing the importance of subsoil drainage in every case. Once the general route for the future road has been selected, a more detailed examination will be necessary to detect hazardous zones and determine corrective measures. The location of springs and seeps, natural or artificial seepage barriers and geological formations that are aquifers is essential. A clear picture must be acquired of groundwater flow conditions and groundwater changes. Observations during periods of drought and periods of rain will be helpful and correlations of groundwater level with rainfall established. All this information should be as complete as possible during the construction stage when cuts are opened and embankments built, so that there will be time to correct or improve the drain systems.

Geological investigations are of a fundamental value. Existing formations and their sequence must be determined, together with all kinds of discontinuities, such as folds, fissures, fractures and faults. Cracks and fissures must be examined to find out whether they are open, closed, or filled with impervious materials. Permeability of strata and formations measured in the field will provide a valuable guide to both understanding the ground water flow and controlling it by drainage.

Much of the information for planning subsoil drainage is obtained from borings, starting with minimum borings for the geotechnical investigation of the road. Fluctuations in the water table, especially due to rainfall, are valuable indicators of potential problems, together with all other information about seepage that can be obtained during boring operations, such as loss of drilling water or gain of water from artesian flow. For detailed investigations, piezometers are installed in slide zones, together with a network of observation wells so that the fluctuations in the water-table and possible artesian aquifers can be recorded.

Table 7-4 and Fig. 7-45, both taken from [23], together give an idea of the need for subsoil drainage in several different conditions, and illustrate the stability problems which often arise in cut slopes and hillsides. The relation N/T is shown between normal stresses, N, and tangential (or shear) stresses, T. The most favorable conditions are Cases *1* and *4*, where there is either no flow or the flow occurs vertically downwards due to subsoil drainage. The effectiveness of the different subsoil drainage systems depend mainly on the geology and

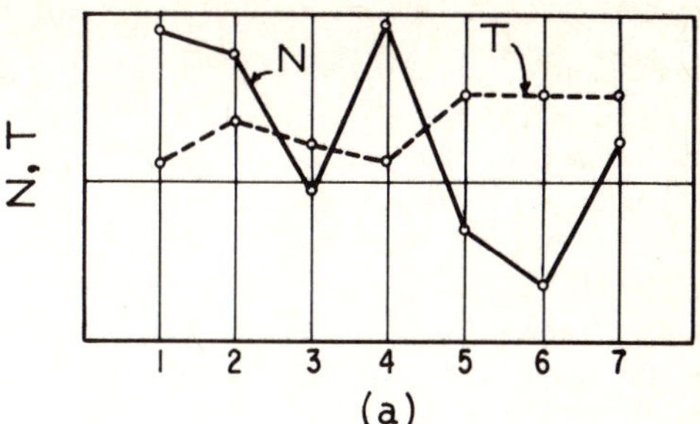

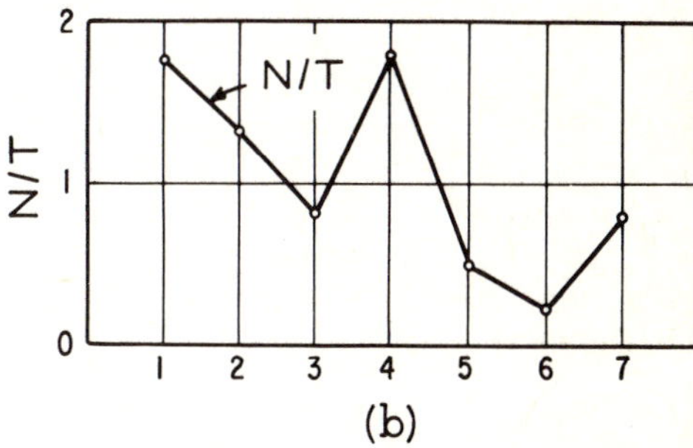

Fig. 7-45 Comparison of the stability conditions of slopes under several conditions [23]

Table 7-4

Some common causes of slope stability problems from seepage and earthquakes [23]

Description of slope or hillside	Observations on stability
Naturally dry or well-drained, with little seepage	Favourable conditions: the aim of every subsoil drainage system
— Subjected to a significant earthquake	Serious slides seldom occur
— Subjected to a normal flow that is uncontrolled and harmful, due to rainfall infiltration	The flow is generally parallel to the slope. Pore pressures are produced and stability is reduced
— Subjected to a favourable vertical downward flow, encouraged by draining layers in the lower portion of the slope	Vertical downward flow which greatly reduces pore pressures
Saturated, with no volume changes in the soils which are subjected to a severe earthquake	It is a common design condition in seismic regions
— Composed of soils or weathered rocks, with a tendency to liquefy, subjected to a severe earthquake	This condition must always be avoided
— Composed of dense soils or rock formations with a tendency to expand, subjected to a severe earthquake	The stronger the materials, the more able they will be to withstand an earthquake without damage

climatic conditions of the zone. An important factor, which is sometimes disregarded, is local practice; engineers in a specific region may favor one type of solution to others.

If there is a weak stratum of soil where an embankment is to be built and that stratum is close to the surface, replacement will usually be the most economical solution. If the soft stratum is located at a considerable depth, stabilizing trenches can be considered. If the unstable area is in a natural depression, the stabilizing trench can be continued through the depression with its axis perpendicular to the road. If, on the other hand, large areas are to be drained, multiple stabilizing trenches will be necessary. The drainage gallery finds its most effective use in those cases where the depth of the subterranean water is such that replacement cannot be considered and stabilizing trenches would prove uneconomical. The roles played by horizontal drains, longitudinal underdrains and permeable blankets have already been discussed. Here attention will be drawn to the possibilities of achieving versatility by combining different solutions. It is stressed that the effectiveness of both permeable blankets and longitudinal underdrains is closely related to their hydraulic capacity and appropriate location. The need for installing subsoil drainage in the transition areas between cuts and embankments should be emphasized.

When groundwater flow is heavy, a continuous drain layer will probably be more economical than a system of longitudinal underdrains and will also be more effective.

7.7 Subsoil Drainage in Runways

The chief purpose of subsoil drainage in runways is to protect the pavement, the subgrade and the upper layers of embankments and cut virgin soil. The fundamental systems of subsoil drainage will, therefore, be permeable blankets, longitudinal underdrains and interceptor underdrains. The water to be drained is either rainwater which seeps through the pavement, groundwater flow or, on a smaller scale, capillary rise and condensation of atmospheric humidity.

Where possible, airfields are built on ground that is either flat or very slightly undulating, where large areas are exposed to seepage and there is generally not enough slope to encourage natural surface drainage. They are therefore critical zones from the point of view of subsoil drainage. Despite this, the subsoil drainage systems required by most airfields are comparatively modest, especially when surface drainage is well taken care of.

Whenever a runway is to be built, exploration should be carried out to determine the presence, origin and aquifers transmitting subterranean water. One of the most reliable signs of this problem is a high water-table in part of the structure that is to be built or throughout the entire area. The investigation which is referred to will make it possible to determine whether subsoil water is:

a) Confined to permeable strata overlying impermeable strata
b) In low-lying portions of an undulating permeable stratum
c) Confined to a permeable stratum underlying other impermeable ones: typical of artesian conditions
d) In zones flood of potential flooding of a lake, river or sea.

Cases *a* and *b* can generally be handled by subsoil drainage in areas with a high water-table. The system may be a ditch filled with filter material and a perforated pipe. In Cases *c* and *d* drainage ditches will usually be necessary to protect the pavement plus interceptor underdrains to eliminate the flow. The following is an analysis of five typical soil profiles, with comments on conditions where subsoil drainage systems are usually necessary owing to the presence of subterranean water.

Uniform permeable soil: In this case, probably no subsoil drainage will be required, for these soils are self-draining. Problems are usually caused by erosion as a result of surface flow. This must be combatted with the help of flat slopes and by protecting shoulders with soil-cement or even hydraulic concrete surfacing.

Uniform impermeable soil: Such soils do not usually respond to subsoil interceptor drainage either, because seepage of any importance is unlikely in them owing to their low permeability. If the water-table is very close to the surface, subsoil drainage may be necessary to protect the pavement.

Stratum of permeable soil overlying another impermeable stratum: Here, any water that seeps through the upper stratum is trapped at the boundary with the impermeable stratum; seepage will follow the natural slope of that boundary. Drainage ditches are installed as deep as the boundary. If the boundary is very deep, the ditches can be made just deep enough that deeper seepage will not cause any damage.

Impermeable stratum overlying a permeable stratum: This case can be regarded in the same way as the Uniform Permeable Soil and generally requires no subsoil interceptor drainage (Fig. 7-46). Subsoil drainage to protect the pavement will only be necessary if the water-table reaches the impermeable stratum and comes close to the pavement.

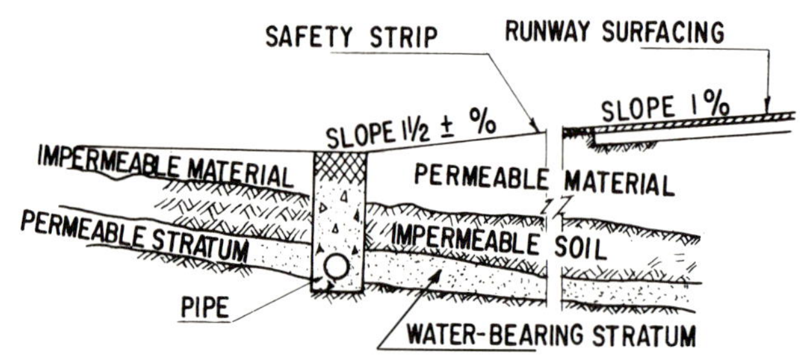

Fig. 7-46 Interceptor underdrain

Erratic strata with alternatively permeable and impermeable layers: Subsoil drainage is generally required here, although there are rigid rules. The design will depend on the geometry of the strata in each particular case. Unnecessary expenditure on subsoil drainage systems can often be avoided by efficient surface drainage.

According to the U.S. Corps of Engineers [22], subsoil drainage systems must be installed in the base course and sub-base of all pavements where the subgrade is exposed to frost, where the water-table reaches the upper layer of subgrade, or where the road or runway surface is likely to become flooded, especially if the subgrade is not very permeable. They also requires subsoil drainage in the subgrade in zones where the water-table is likely to rise to as much as 30 cm (12 in) under the lower layer of the base course.

7.8 Subsoil Drainage in Railroads

The rails on which the train runs transmit loads through the sleepers or cross ties to a layer of crushed stone called ballast. In normal construction practice, the ballast is placed directly on top of the embankment or virgin ground. Sometimes, however, a layer of subgrade at least 30 cm (12 in) thick, consisting of high-quality, well-compacted materials, is placed on top of the embankment. This practice is to be recommended. If the quality of the embankment is sufficiently good, it is usually enough to compact the upper part of the subgrade somewhat better than the rest so that it will simply play the part of an improvement layer in this case. It is becoming more and more common practice in Mexico to use a different layer in railroads which, for want of a better name, is frequently called sub-ballast. The quality of this material is similar to that of the sub-base for a highway and its functions are the same. Sub-ballast is particularly useful when the subgrade or the uppermost layer of embankment (if there is no such sub-grade) is made up of fine materials. The crushed ballast materials can become very easily embedded in the fine soil, eventually disappearing, so that the ties have to be frequently retamped. This is not only a nuisance, but also leads to high maintenance costs. Ballast does not usually pose any subsoil drainage problems. Because its capillary potential is zero, it is not liable to be subjected to a vertical upward flow and because its permeability is very high, it readily eliminates seepage water due to rainfall.

The subsoil drainage problems related to the protection of what might by extension be called the *surface* of the railway includes the sub-ballast, subgrade and body of the embankment. From this point of view, the subsoil drainage for a railroad is no different from that for a road. The draining layers or the longitudinal underdrains and the interceptor underdrains play the same part as the one they play in roads. Figure 7-47 shows two typical cross-sections of a railroad with subsoil drainage. Part *a* shows a railroad constructed in accordance with the procedures that are still frequently used today. There is neither a layer of subgrade nor of sub-ballast, which means that with the passage of time the cross-section of the ballast will become similar to that shown. Part *b* of the figure gives a more rational design for the same situation. In Part *a*, the ballast becomes embedded, forming irregular pockets in the lengthwise direction of the railroad, which trap water. Underdrains must be installed and the principal pockets are connected with the underdrain by means of horizontal underdrains. None of these problems are expected in Part *b*, where the interceptor underdrain was originally part of the design to protect the crown of the embankment from the very start.

Stabilizing hillsides and artificial slopes for railroads (the other objective of subsoil drainage) requires the same drains as for highways. For railroads, the more rigid requirements of grade and curvature, usually lead to deeper cuts and higher embankments which makes subsoil drainage more difficult. Another problem which frequently arises in railroads is the draining of tunnels; this will not be dealt with here.

7.9 Special Aspects of Subsoil Drainage

All the aspects of subsoil drainage are special in the sense that they are all different and require solutions in keeping with their individual characteristics. The title of this section therefore demands an explanation. Certain, sometimes minor, cases are dealt with which do not fit into previous parts of this Chapter, but which nevertheless deserve attention. In subsoil drainage, it is very common to disregard initially some minor cases which could have been solved with a minimum of cost and effort, but which eventually lead to catastrophic slides of vast proportions.

The first special case to be discussed refers to the potential draining function of the layers or blankets of frictional sandy material that are placed on top of the natural ground when embankments are built on soft tuffs, swampy zones and very soft clayey soils. These sand layers can provide an outlet for water over wide areas. The thinner the layer of soft soils in relation to the area covered, the faster the consolidation process. The collector layers which are placed on the surface of the ground in installations for vertical sand drains serve as such blanket drains, their purpose being to accelerate consolidation, as described in Chapters 1 and 3. In earlier pages of this Chapter, some attention was paid to the draining capacity of these blanket layers, which are similar to the drain layers that have already been discussed.

Another case of special interest is that posed by springs and water surfacing within the road area. It is essential to capture and eliminate any such water. This can be achieved with drain or blanket layers, small stabilizing trenches or suitably orientated drainage ditches.

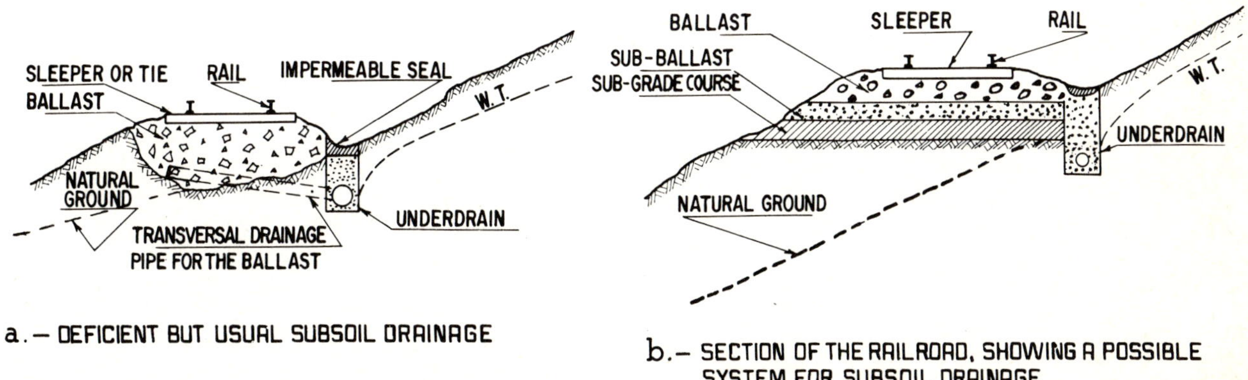

Fig. 7-47 Protection of the railroad bed by means of underdrains

With springs the difficulty usually lies in the need to eliminate large volumes of water by drains that are small to minimize cost. It is essential for the filter and drain materials to have a high permeability; special care must be taken to see that the amount of fines passing sieve N°40 is as small as possible. If this is impossible, several layers of filters will be used. Engineers should not have undue confidence in the draining capacity of the blankets made of frictional materials, even if they are very permeable. On the basis of observations of slope, flow velocity, and discharge, CEDERGREN [38] has calculated the discharge capacity of a pipe with a diameter of 15 cm (6 in) (180 cm^2 (28 in^2) in area), and has compared it with some typical filter blankets under the same conditions. The same draining capacity is obtained for a section of 3 m^2 (32 ft^2) of crushed stone with a size of 2 cm (3/4 in), 13 m^2 (140 ft^2) of crushed stone with a size of 0.75 cm (5/16 in) and lastly in a section of 400 m^2 (4305 ft^2) of gravel-sand with a coefficient of permeability of 3×10^{-1} cm/sec (6×10^{-1} ft/min) (which is considered satisfactory by many specialists). Since the volume of water from a spring is continuous and may be considerable a restrictive drain blanket is likely to be damaged (though perhaps locally). If the water is not removed sufficiently fast, then the need for a high coefficient of permeability in the filters can be appreciated.

Figure 7-48 is a sketch showing ground plans of typical installations for draining isolated springs. Part *b* of the figure illustrates a case where the spring is received in an open pit or well which is cut in the embankment and filled with highly permeable filter material. Outlet pipes must be installed to take the water to a place where it can be discharged. Part *c* shows the case of several springs, each controlled by an underdrain, and wet zones are shown which are drained by means of combs of ditches filled with filter material and equipped with perforated pipes.

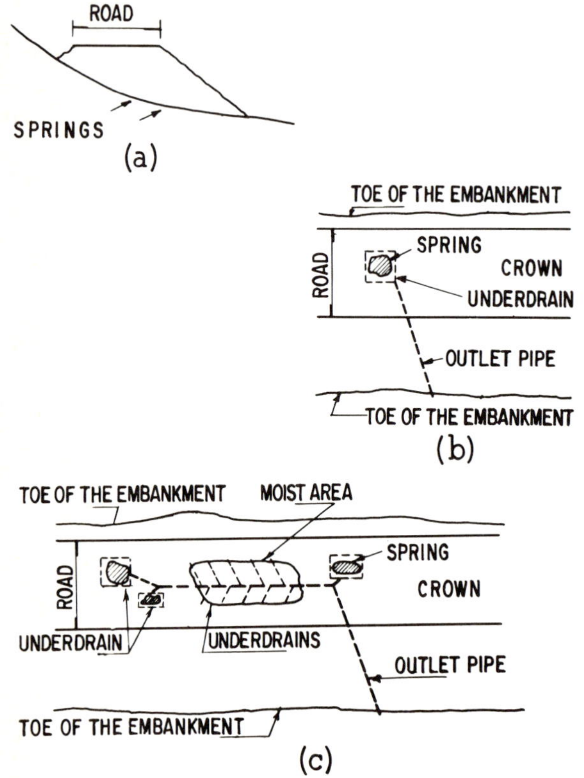

Fig. 7-48 Typical installations for draining springs [38]

Sometimes it has been preferable to intercept water rising inside the embankments by placing a sufficient depth of very permeable rockfill at the bottom of the embankments. On top of this a section is laid with a transitional grain-size distribution as a filter, on which a conventional embankment can be built. This solution can be economical for large areas, when the embankment is sufficiently high.

A third case is posed by the cracks which open in natural hillsides and unstable slopes as a result of movements. When water penetrates these cracks and fills them, it causes hydrostatic thrusts which may very seriously reduce the general stability of the entire zone. It is a vital precaution to fill these cracks and calk them. This can be done with clay or impermeable materials such as soil-asphalt. In extremely serious cases it may be necessary to cut slots in the upper part of the crack, which are then filled with the calking materials. Since filling the cracks cannot be expected to permanently remedy an instability which is already in the process of development, the cracks will continue to open up. Therefore, the cracks will have to be recalked periodically.

APPENDIX 7A

A THEORETICAL APPROACH TO THE PROBLEM OF FLOW IN SOILS — FLOW NETS

7a.1 Introduction

The purpose of this Appendix is to provide basic ideas to aid comprehension of the theoretical approach that is currently suggested for seepage problems. A brief description will also be given of the method most frequently used for handling the conclusions to which this theory leads. The bibliographical source consulted throughout is [1].

Any of the Chapters of this book will make it clear to the reader that the problems relating to the seepage of water through soils are of great importance in road engineering technology. Of special importance is the influence of seepage on the general stability of soil masses and to a lesser degree the possibilities that seepage water may cause piping, internal erosion, etc.

Water flowing through a mass of soil causes a hydrodynamic pressure that is greater than the hydrostatic pressure corresponding to static equilibrium. This has several important consequences. First, depending on the direction of the flow, the hydrodynamic pressure may alter the submerged unit weight of the soil. For example, if the flow is vertical upwards, it has a buoyant effect on the soil particles which is equivalent to a reduction in unit weight. Second, in accordance with COULOMB equation

$$s = (\sigma - u) \tan \varnothing$$

the increase in the pressure of the water brings about a

corresponding reduction in effective pressure, and consequently in the shear resistance of the mass subjected to seepage, so that a slope that under no-flow conditions is stable may not be stable if flow occurs. Moreover, water seeping through a soil may carry solid particles with it, and if due attention is not paid to this, the stability of any earth structure may be seriously jeopardized by the tunnels and cavities that form as a result of erosion.

The problem of flow in soils can be established on a theoretical basis, so long as the geometry of the flow is relatively uniform and the soils are fairly homogeneous. These conditions are rare in road engineering problems, and that is why the conclusions obtained from a theoretical analysis are seldom fully applicable. What is needed is a great deal of improvisation, based on the sensible handling of uncertainties, which is related to other fields of engineering and other kinds of problems. The role played by a theoretical approach to seepage problems and their solution (even if the assumed conditions are somewhat different from the real one) is of vital importance in that it helps the practical decisions that will have to be made. Even if they provide only an intellectual basis for thought and action, the theoretical solutions to seepage problems are useful to the engineer. Besides, the degree of representativity of the solutions obtained by means of careful investigations is sometimes remarkable.

Groundwater falls into three different categories, depending on its mobility within the soil. First, there is absorbed water which becomes attached to the soil particles due to forces of electrical nature. Since this water does not move about inside the porous mass, it does not participate in seepage. Second, there is capillary water, which is of great importance in certain aspects of soil mechanics, such as a pavement, which is moistened by upward flow. In the majority of seepage problems, however, the effect of flow in the capillary zone is relatively unimportant and is usually disregarded because of the complications that would arise if it were taken into account theoretically. Third and last there is the free water which, influenced by the earth's gravity, can circulate within the soil mass without any restriction other than its own viscosity and the solid structure of the soil. To avoid confusion, the authors specify that the flow theory that is described here refers only to free water.

In a soil mass, the free water is separated from the capillary water by a surface which is called the water-table (or phreatic line). It is not always easy to determine or locate the water-table. During excavation in a soil that is sufficiently fine, the water surface that appears in the open hole after a time defines the water-table. In the adjacent soil, however, there is no such distinctive surface, for the soil above this level may be totally saturated due to capillarity, and consequently the water-table in that soil does not really exist.

The authors have found it hard to agree on a definition for the water-table, for the purposes of this book, however, the water-table will be regarded as the surface which represents the geometric location of the points at which the pressure of the groundwater is equal to atmospheric pressure; in seepage conditions, where measurements are taken with a manometer, this pressure is regarded as zero. Consequently this is the pressure at all points throughout the water surface that appears in an excavation, and in the soil adjacent to it where water pressure is also zero.

In static conditions, the water-table for a given soil will be a horizontal surface. If, however, there is seepage through the soil, the water-table is no longer horizontal.

7a.2 Hydrodynamic Equations Governing Flow in Soils

By means of the mathematical process that follows, the basic equations that are used today for the theoretical approach to seepage problems can be derived.

Let us consider a flow region (that is, an area of soil through which water flows) which includes a rectilinear element with dimensions, dx, dy and dz, as shown in Fig. 7a-1.

It is assumed that the velocity v at which the water flows through the element has three components, v_x, v_y and v_z. These are functions only of x, y and z respectively, but not of time (since steady flow conditions are assumed) or of any other variable. It is also assumed that these components are continuous functions which allow for any order of derivation that may prove necessary for the above argument.

Under these circumstances, if on faces I (see Fig. 7a-1) the velocity components of the water are v_x, v_y, and v_z, on faces II the same components will be respectively,

$$v_x + \frac{\partial v_x}{\partial x} dx;$$

$$v_y + \frac{\partial v_y}{\partial y} dy;$$

$$v_z + \frac{\partial v_z}{\partial z} dz$$

It is now assumed that the voids of the soil through which the flow occurs are saturated with water and that both this water and the solid particles forming the soil structure are incompressible. Under steady flow conditions, the quantity of water that enters the element must be the same as the quantity that leaves it. Considering that the volume of flow passing through a particular section can be expressed as the product of the area of the section times flow velocity, the following expression will be true:

$$v_x\, dy\, dz + v_y\, dx\, dz + v_z\, dx\, dy =$$

$$= \left(v_x + \frac{\partial v_x}{\partial x} dx\right) dy\, dz + \left(v_y + \frac{\partial v_y}{\partial y} dy\right) dx\, dz +$$

$$+ \left(v_z + \frac{\partial v_z}{\partial z} dz\right) dx\, dy$$

Here, the first part of the equation represents the volume of flow that enters the element and the second the volume of flow that leaves it.

In similar terms,

$$\frac{\partial v_x}{\partial x} dx\, dy\, dz + \frac{\partial v_y}{\partial y} dx\, dy\, dz + \frac{\partial v_z}{\partial z} dx\, dy\, dz = 0$$

Hence

$$\frac{\partial v_x}{\partial x} + \frac{\partial v_y}{\partial y} + \frac{\partial v_z}{\partial z} = 0 \qquad (7a\text{-}1)$$

The preceding equation plays an important part in the flow theory and is known as the *Continuity Equation.*

A brief summary now follows of the hypotheses that are implied by acceptance of the Continuity Equation. They are:

1. There are steady flow conditions
2. The soil is saturated
3. The water and the solid particles are incompressible
4. The flow in no way modifies the soil structure

If DARCY'S Law is now assumed valid, the discharge velocity of the water through the element can be written as follows:

$$v = -k\frac{\partial h}{\partial l}$$

Thus, if the gradient is expressed in terms of its three components:

$$v_x = -k_x\frac{\partial h}{\partial x}; \quad v_y = -k_y\frac{\partial h}{\partial y}; \quad v_z = -k_z\frac{\partial h}{\partial z} \tag{7a-2}$$

Equations (7a-2) assume the general situation, where the soil is anisotropic insofar as permeability is concerned, with a permeability k_x in the direction of the axis $X-X'$, another with a value k_y in the direction of the axis $Y-Y'$ and, lastly another with a value k_z in the direction of the axis $Z-Z'$.

If Eqs. (7a-2) are introduced into Eq. (7a-1) we obtain:

$$k_x\frac{\partial^2 h}{\partial x^2} + k_y\frac{\partial^2 h}{\partial y^2} + k_z\frac{\partial^2 h}{\partial z^2} = 0 \tag{7a-3}$$

Equation (7a-3) is a mathematical description of flow in the region under consideration and implies all the above-mentioned hypotheses, plus the applicability of DARCY'S Law.

For the purposes of practical soil mechanics, the seepage in a given section, which is transversal in relation to the longitudinal axis, is frequently identical to the seepage in any other section. An example of this is an earth embankment which has a long axis in comparison with its height. This means that the effects at the edges of the flow zone can be ignored, and the problem can be studied two-dimensionally, as though the entire seepage zone were represented by the plane $X-Y$. Under these circumstances, Eq. (7a-3) can be simplified thus:

$$k_x\frac{\partial^2 h}{\partial x^2} + k_y\frac{\partial^2 h}{\partial y^2} = 0 \tag{7a-4}$$

which is the basic equation used to analyze a two-dimensional seepage in a given flow region.

If the soil in which the seepage occurs is also isotropic, then:

$$k_x = k_y = k$$

and Eq. (7a-4) can be simplified even further, leading to

$$\frac{\partial^2 h}{\partial x^2} + \frac{\partial^2 h}{\partial y^2} = \nabla^2 h = 0 \tag{7a-5}$$

Equation (7a-5) is a very well known differential equation which describes mathematically many physical phenomena of great practical importance besides flow. It is known as LAPLACE'S equation. A function that satisfies LAPLACE'S equation, like h in Eq. (7a-5), is said to be harmonious.

Since LAPLACE'S equation has been very carefully studied, and in view of its individual and general solutions, it is extremely fortunate that it describes seepage problems from the engineering point of view. Equation (7a-5), however, represents a specific situation, where the soil is isotropic in permeability (the particularity of two-dimensional flow is also implied, but this assumption is a reasonable approximation for the majority of practical cases. Therefore, its limitations can usually be disregarded). Anisotropy is a common condition in soils. Many earth structures on which a seepage study is to be performed, are compacted by layers during the construction process, and this procedure leads to horizontal permeabilities that are considerably higher than those obtained for vertical seepage. Consequently, an awkward situation arises, and it appears that Eq. (7a-4) will have to be applied instead of the simpler Eq. (7a-5). Fortunately there is a mathematical trick of the trade which makes it possible to handle anisotropic problems as though they occurred in isotropic soils. This is known as the *Transformed Section Theory* and is described further on in this Appendix. It enables any anisotropic soil to be studied as though it were isotropic. Thanks to this theory, the practical importance of Eq. (7a-5) becomes immense and it is regarded as the basic equation for solving problems of ground-water seepage in soils.

The general solution to LAPLACE'S equation is given by two sets of functions. These in turn have a very useful geometrical interpretation, both sets of functions can be represented within the flow zone and be studied as two sets of orthogonal lines. A solution that fulfills the boundary conditions of a specific flow zone gives the individual solution for LAPLACE'S equation for that zone.

On the basis of Fig. 7a-1, an expression is now obtained to give the volume of flow passing through the element in dt time. Remembering that the volume of flow can be expressed as the product of the area of the section times the flow velocity:

$$dq = k_x\frac{\partial h}{\partial x}dy\,dz + k_y\frac{\partial h}{\partial y}dx\,dz + k_z\frac{\partial h}{\partial z}dx\,dy \tag{7a-6}$$

If the soil is isotropic in permeability Eq. (7a-6) becomes:

$$dq = k\left(\frac{\partial h}{\partial x}dy\,dz + \frac{\partial h}{\partial y}dx\,dz + \frac{\partial h}{\partial z}dy\,dx\right) \tag{7a-7}$$

With two-dimensional flow:

$$dq = k\left(\frac{\partial h}{\partial x}dy + \frac{\partial h}{\partial y}dx\right) \tag{7a-8}$$

In Eq. (7a-8) the element in Fig. 7a-1 is regarded as flat and contained entirely by the plane $X-Y$. It is assumed to have a normal unit thickness so that the areas that are normal in relation to the directions of the flow are dx and dy.

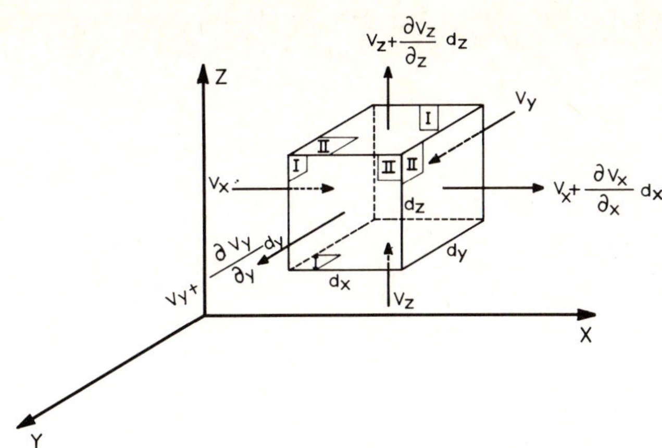

Fig. 7a-1 Element of a region with a three-dimensional flow

Equation (7a-8) gives a differential expression for the volume of a two-dimensional flow in an isotropic soil. As was specified previously, this is a common practical situation.

7a.3 The Solution of LAPLACE's Equation

Let's suppose two-dimensional flow, and LAPLACE's equation Eq. (7a-5). Consider a function: $\varnothing = -kh + c$; which is known as the potential velocities function. When differentiated, this function leads to:

$$\frac{\partial^2 \varnothing}{\partial x^2} + \frac{\partial^2 \varnothing}{\partial y^2} = 0 \qquad (7a\text{-}9)$$

In this way, the function $\varnothing(x, y)$ = constant, is one solution to LAPLACE's equation. This solution represents an infinite number of functions, depending on the value of the constant c. A geometrical interpretation can be given to this solution: $\varnothing(x, y)$ = constant, may represent a set of curves which develop in the flat zone where flow occurs, a specific curve being obtained for each value that is chosen for the constant.

Let us now consider a function: $\psi(x, y)$ = constant, known as a function of flow and defined in such a way that:

$$v_x = \frac{\partial \psi}{\partial y} \qquad v_y = \frac{\partial \psi}{\partial x} \qquad (7a\text{-}10)$$

It can be shown that a function ψ thus defined agrees with LAPLACE's equation too, so that

$$\frac{\partial^2 \psi}{\partial x^2} + \frac{\partial^2 \psi}{\partial y^2} = 0 \qquad (7a\text{-}11)$$

Indeed, the function of flow will be: $\psi(x, y)$ = constant, defined at each point of the zone by Eq. (7a-10). Bearing in mind that $\varnothing = -kh + c$ and noting Eq. (7a-2) it follows that:

$$v_x = \frac{\partial \varnothing}{\partial x} \; ; \quad v_y = \frac{\partial \varnothing}{\partial y} \qquad (7a\text{-}12)$$

By comparing Eqs. (7a-10) and (7a-11), the well-known CAUCHY-RIEMANN conditions are obtained, which are familiar in the theory of complex variable functions. First differentiating the (7a-10) Equations with respect to y, and second with respect to x:

$$\frac{\partial^2 \varnothing}{\partial x\, \partial y} = \frac{\partial^2 \psi}{\partial y^2} \; ; \quad -\frac{\partial^2 \varnothing}{\partial x\, \partial y} = \frac{\partial^2 \psi}{\partial x^2}$$

By adding, the following is obtained:

$$\frac{\partial^2 \psi}{\partial x^2} + \frac{\partial^2 \psi}{\partial y^2} = \nabla^2 \psi = 0$$

In other words, the function ψ complies with LAPLACE's equation (Eq.(7a-11)) and therefore provides a solution to it.

Moreover, if a geometrical interpretation is given to the set of functions $\psi(x, y)$ = constant, so that these functions are also represented by a set of curves (ψ = constant) in the flow region, the set ψ = constant is orthogonal to the set $\varnothing$ = constant, so that the curves of the two different sets intersect at right angles.

The total derivates along each of these curves will thus be:

$$d\varnothing = \frac{\partial \varnothing}{\partial x} dx + \frac{\partial \varnothing}{\partial y} dy$$

$$d\psi = \frac{\partial \psi}{\partial x} dx + \frac{\partial \psi}{\partial y} dy \qquad (7a\text{-}13)$$

On the basis of the above equation, the slopes, dy/dx, can be obtained for each set:

$$\left(\frac{dy}{dx_\varnothing}\right) = -\frac{\dfrac{\partial \varnothing}{\partial x}}{\dfrac{\partial \varnothing}{\partial y}}$$

$$\left(\frac{dy}{dx_\psi}\right) = -\frac{\dfrac{\partial \psi}{\partial x}}{\dfrac{\partial \psi}{\partial y}}$$

Now, by applying the CAUCHY-RIEMANN conditions (which fulfill the functions ψ and $\varnothing$, as has been seen) to the second of the above expressions, leaving the first one unchanged, the following equation is obtained:

$$\left(\frac{dy}{dx_\varnothing}\right) = -\frac{\dfrac{\partial \varnothing}{\partial x}}{\dfrac{\partial \varnothing}{\partial y}}$$

$$\left(\frac{dy}{dx_\psi}\right) = \frac{\dfrac{\partial \varnothing}{\partial y}}{\dfrac{\partial \varnothing}{\partial x}} \qquad (7a\text{-}14)$$

The slopes for the two sets are therefore reciprocal and have opposite signs, which means that the curves ψ = constant and $\varnothing$ = constant are orthogonal.

Books specializing on this subject show that for a given situation with fixed boundary conditions, the solution of LAPLACE's equation by two sets of curves ψ = constant and $\varnothing$ = constant, that also agree with existing boundary conditions, leads to a unique solution for each problem. This should be kept in mind when reading the paragraphs that follow.

A general solution has been found for LAPLACE's equation and a geometrical interpretation has been given which later will be seen to be very useful. However, because seepage is really a physical problem, it is important that a physical interpretation also be found for the two sets of curves. This interpretation does exist and is essential if the solutions to flow problems are to be fully understood from an engineering point of view. This physical interpretation is described in the paragraphs that follow.

Function $\varnothing$ is defined by the expression $\varnothing = -kh + c$. If a curve unites points at which $\varnothing$ is constant, h will be constant too at these points. In other words the water head will be the same at all points on the curve $\varnothing$ = constant. The physical significance of the curves $\varnothing$ = constant is thus made clear. These curves are sets of points with the same water head in the flat flow zone, which is why they are called equipotential lines.

An analysis follows of the physical significance of the curves ψ = constant (see Fig. 7a-2). Let us consider the path of the water passing through P (x, y). At this point the water has a velocity, v, which will naturally be at a tangent to its path. Now the mathematical equation must be found for this path. The following is true along the curve:

$$\tan \theta = \frac{v_y}{v_x} = \frac{dy}{dx}$$

Thus,

$$v_y \, dx - v_x \, dy = 0$$

but, according to Eqs. (7a-10), this can be written as follows:

$$\frac{\partial \psi}{\partial x} dx + \frac{\partial \psi}{\partial y} dy = 0$$

The above expression is the total differential of function $\varnothing$, so that all along the path of the water $d\psi = 0$ and therefore ψ = constant. The equation for the path of the water is therefore ψ = constant; in other words, the set of curves ψ = constant is made up of the real physical paths of the water through the seepage zone. This is why the curves ψ = constant are known as flow lines.

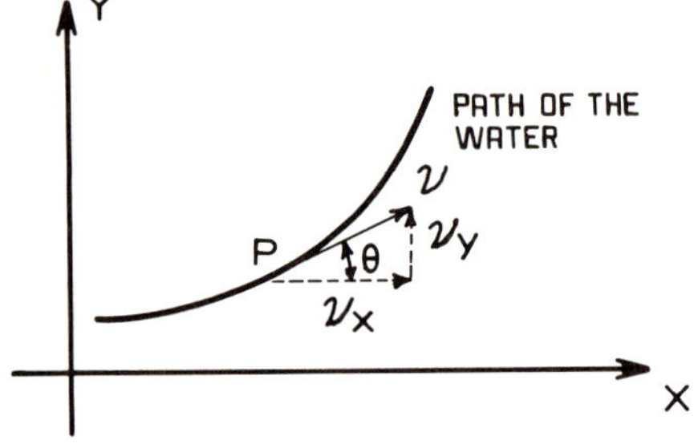

Fig. 7a-2 Physical interpretation of the curve ψ = constant

The first and most important property of flow lines is that the volume of water passing between any two of them is constant in any section that is chosen. This space between two flow lines is usually called a flow channel. In fact:

$$q = \int_{\psi_2}^{\psi_1} v_x dy = \int_{\psi_2}^{\psi_1} d\psi = \psi_1 - \psi_2 = cte$$

where q represents the rate of flow through the channel per unit of length (Fig. 7a-3). A second important property of flow lines is that they cannot cross each other. Indeed, if the two flow lines converge at the point of contact, no area is left for the flow, and the continuity of the flow volume is no longer possible, which according to the hypotheses of the theory being studied cannot occur. A third important property of these lines is that equipotentials cannot intersect, for the water at that point would have two different hydraulic heads at the same time.

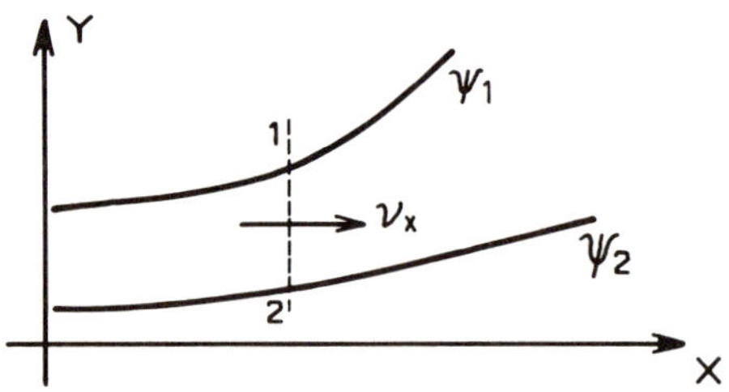

Fig. 7a-3 An important property of flow lines

7a.4 The Transformed Section Theory

The Transformed Section Theory, which has already been mentioned, enables a soil with different permeabilities for vertical (k_y) and horizontal (k_x) seepage to be reduced to the case of a homogeneous isotropic soil. As a result of this reduction, LAPLACE's equation and its solutions can be applied to describe flow through an anisotropic medium. The Transformed Section Theory is really a simple calculation device involving a simple transformation of coordinates, which modifies the dimensions of the flow region on paper, so that the seepage pattern for the new section obtained, which is isotropic with $k_x = k_y$, will represent that prevailing in the real section where $k_x \neq k_y$.

An example is the flow region shown in Fig. 7a-4. The permeabilities here are $k_x \neq k_y$. The flow region is subjected to a transformation of coordinates where all coordinate y are transformed into y', so that

$$y' = \sqrt{\frac{k_x}{k_y}} \, y \qquad \text{(7a-15)}$$

Equation (7a-4) describes two-dimensional flow in an anisotropic medium. It can be rewritten as follows:

$$\frac{k_x}{k_y} \frac{\partial^2 h}{\partial x^2} + \frac{\partial^2 h}{\partial y^2} = 0$$

Differentiating Eq.(7a-15) and incorporating it we can now write Eq.(7a-4) as:

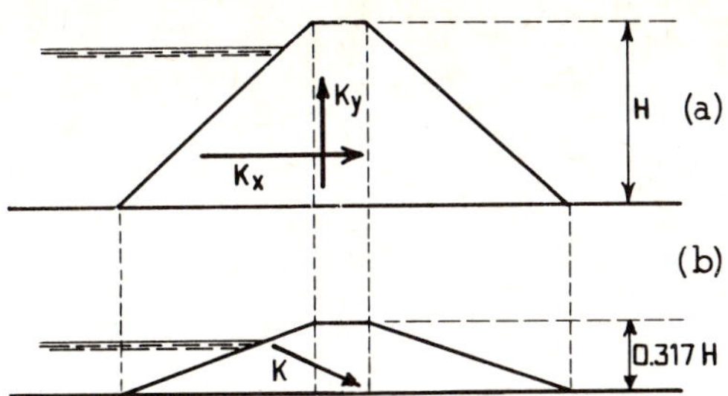

Fig. 7a-4 The theory of the Transformed Section

$$\frac{\partial h}{\partial y} = \frac{\partial h}{\partial y'}\frac{\partial y'}{\partial y} = \sqrt{\frac{k_x}{k_y}}\frac{\partial h}{\partial y'} \qquad \text{(7a-16)}$$

and also

$$\frac{\partial^2 h}{\partial y^2} = \frac{k_x}{k_y}\frac{\partial^2 h}{\partial y'^2} \qquad \text{(7a-17)}$$

If these relations are incorporated into Eq.(7a-4):

$$\frac{k_x}{k_y}\frac{\partial^2 h}{\partial x^2} + \frac{k_x}{k_y}\frac{\partial^2 h}{\partial y'^2} = 0$$

which reduces to:

$$\frac{\partial^2 h}{\partial x^2} + \frac{\partial^2 h}{\partial y'^2} = 0 = \nabla^2 h \qquad \text{(7a-18)}$$

In this way, the transformation of coordinates of Eq.(7a-15) has enabled Eq.(7a-4) to be reduced to the form that is shown in Eq.(7a-18), which is LAPLACE's equation for isotropic soils. The transformation of coordinates is performed both in the equations, and also physically in the section being studied. In this way, the initial seepage zone in Fig. 7a-4a is transformed for all subsequent computations into the transformed zone shown in Fig. 7a-4b. (In Fig. 7a-4, k_x/k_y is assumed to be equal to 10^{-1}). The vertical dimensions are all modified in accordance with Eq.(7a-15) but the horizontal dimensions remain unchanged. With the transformation

$$x' = \sqrt{\frac{k_y}{k_x}}\, x$$

another isotropic section could have been achieved where the horizontal dimensions would have been modified instead of the vertical ones. This simple exercise will be left to the reader.

Now the volume of flow given by Eq.(7a-6) is considered. In the two-dimensional case, it can be visualized that this above equation is reduced to:

$$dq = k_x \frac{\partial h}{\partial x} dy + k_y \frac{\partial h}{\partial y} dx \qquad \text{(7a-19)}$$

If the transformation of Eq.(7a-15) is applied here, bearing in mind the relation in Eq.(7a-16), the following is obtained:

$$dq = k_x \frac{\partial h}{\partial x}\sqrt{\frac{dy'}{k_x/k_y}} + k_y \frac{\partial h}{\partial y'}\sqrt{\frac{k_x}{k_y}}\, dx$$

for

$$dy' = \sqrt{\frac{k_x}{k_y}}\, dy$$

Rearranging the terms we obtain:

$$dq = \sqrt{k_x k_y}\left(\frac{\partial h}{\partial x} dy' + \frac{\partial h}{\partial y'} dx\right) \qquad \text{(7a-20)}$$

This equation must now be compared with Eq.(7a-8) which gave the volume of flow in an isotropic medium.

Equations (7a-20) and (7a-8) refer to the same rate of seepage, that is really passing through the section where it occurs. By comparing the two equations, it is seen that the permeability in the transformed section, equivalent to the combined permeabilities in the real section, is:

$$k = \sqrt{k_x k_y} \qquad \text{(7a-21)}$$

In other words, since the transformed section is considered isotropic, a value must be used for the permeability equal to the geometrical average for the real permeabilities. In this way, any calculation referring to the rate of seepage can be performed in the transformed section and the same result will be obtained as if the section were anisotropic, but the process is far less complicated.

Thanks to the Transformed Section Theory, anisotropic soils can be treated in a simple way instead of with their troublesome and complicated flow theory. When an anisotropic soil appears in a real situation, it is transformed and the theory for isotropic soils is then applied.

7a.5 The Flow Net

In §7a.3, it was shown how LAPLACE's equation is satisfied by two sets of orthogonal curves, which are the flow lines and the equipotential lines. It was also mentioned that those two sets of lines, which are orthogonal and must comply with the boundary conditions of the flow zone, are the only solution to LAPLACE's equation and consequently to the seepage it describes. The method of flow nets provides a simple graphical solution to the problem. The boundary conditions are defined for each individual case, and the two sets of orthogonal curves are drawn accordingly, so that a graphical image is obtained.

By plotting the two sets of curves in a hand drawing, where the boundary conditions are correct and the lines are made orthogonal, an approximation will be obtained for the only solution to the problem. If the drawing is carefully done, the approximation will be good enough for engineering purposes and will provide solutions which will be just as valuable as the ones which would be obtained by strictly mathematical methods which, although mathematically more exact, are far more complicated.

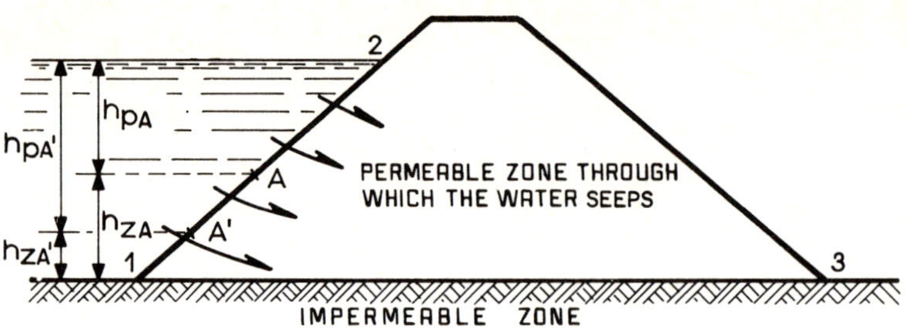

Fig. 7a-5 Analysis of some boundary conditions in flow nets

The steps to follow when drawing a flow net are:

1. Define the flow region that is to be studied, analyzing the specific boundary conditions, in two dimensions.
2. Plot two sets of orthogonal curves that comply with the boundary conditions. These are the only solution to LAPLACE's equation. There are not many general rules available for defining the boundaries of a given flow zone, but a few common situations will be mentioned here which may serve as a guide to the engineer.

First let us consider the case illustrated by Line *1-2* in Fig. 7a-5. This is clearly a boundary of the seepage region through the dam. By analyzing what occurs at Points A and A', it can be seen that the pressure heads along that line are different (represented by the heights of water, measuring from point to surface). The elevation heads are also different if Plane *1-3* is used for comparison purposes, but the sum of both, that is the total head is the same at all the points and is represented by the distance between the horizontal line *1-3* and the level of the water. Line *1-2* is therefore an equipotential line. This is the usual situation as the contact between free water and a permeable medium through which the water seeps is invariably an equipotential line.*

Now let us consider the case of Boundary *1-3*. Any water which succeeds in making contact with this line is obliged to follow right along it, for the impermeable rock does not allow it to seep through. *1-3* is therefore a flow line. Thus, the general rule can be established that the contact between an impermeable medium and a permeable medium through which seepage is occurring is a flow line. Following the above procedure it is possible to define an equipotential line or flow line for each of the boundaries of the flow region. For the time being, all these boundaries are assumed to be known in advance: the flow region is clearly defined. There are certain cases where the boundaries of the flow region are not known in advance and must be studied before the flow net can be drawn.

Once the boundaries are defined, drawing of the flow net can commence. For this, two sets of orthogonal curves are sketched which must agree with the boundary conditions. This simply means that the theoretical requirements for the net must be satisfied. Thus, if the boundary is a flow line, it must be orthogonally cut by a set of equipotential lines.

*Strictly speaking, the total head is the sum of the elevation head, pressure head and velocity head. The velocity head was not considered in the previous argument because, owing to the low velocities at which the water circulates through the soil, it is very small and can be disregarded.

7a.6 Drawing the Flow Net: Calculation of Flow Volume

When plotting the set of equipotential lines and flow lines the problem arises that a flow line and an equipotential line should in principle pass through each point of the flow region, because the water at each of these points has a velocity and a water head. If every possible line were drawn, it would be impossible to identify the pattern. This would be a procedure with no practical value. To achieve a discriminatory solution, which will differentiate one flow condition from another, only a few carefully selected regularly spaced lines will be drawn. This procedure will be recognized by those readers who are familiar with the graphical representation of other vectorial fields of a scalar function, such as the electrical field for example, or the representation of topography. The solution for seepage problems is similar to the one recommended for the other cases; that is, only some of the infinity of possible lines will be drawn.

The following recommendations are made:

a) Draw the flow lines in such a way that the volume of flow, Δq, passing through each channel that is formed between every adjacent two of them will be the same.

b) Draw the equipotential lines in such a way that the drop in head, Δh, between every adjacent two of them will be the same.

Let us assume that a flow net has been drawn in accordance with the above two requirements, so that a fragment of it, limited by the flow lines ψ_i and ψ_j and the equipotentials $\varnothing_i$ and $\varnothing_j$ is like the one shown in Fig. 7a-6. According to DARCY's Law, the rate of seepage, Δq, passing through the channel is

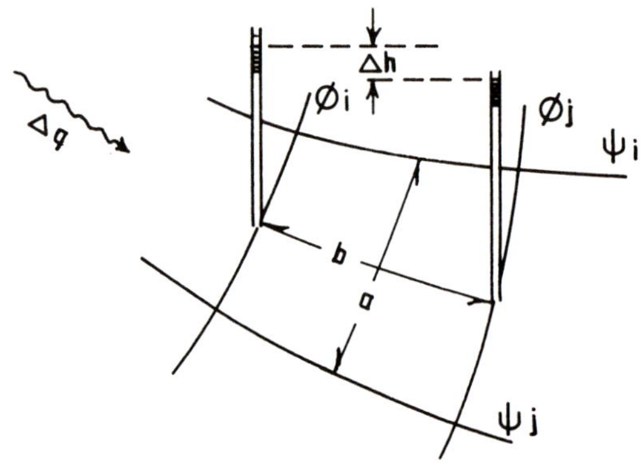

Fig. 7a-7 Portion of a flow net. How to obtain the formula for calculating the flow volume

$$\Delta q = ka \frac{\Delta h}{b} \tag{7a-22}$$

for the average area of the curvilinear rectangle through which seepage occurs is a (considering a unit thickness). The constant fall in hydraulic potential, Δh, between $\varnothing_i$ and $\varnothing_j$ and b is the average distance covered by the water.

If n_f is the total number of flow channels in the net and n_e the number of drops in potential throughout the entire flow zone, bearing in mind the two rules that were observed when drawing the flow net, the following will be true:

$$\Delta q = \frac{q}{n_f}; \; \Delta h = \frac{h}{n_e} \tag{7a-23}$$

where q is the total rate of seepage and h the total head lost throughout the entire flow zone.

Consequently Eq.(7a-22) can be written as:

$$q = kh \frac{n_f}{n_e} \frac{a}{b} \tag{7a-24}$$

In Eq.(7a-24), it will be observed that since q, k, h, n_f and n_e are constants for a given flow net, the relation a/b must also be constant. In order for the two conditions to be fulfilled, therefore, the relation between the width and the length of all the curvilinear rectangles of a flow net, a/b, must be the same. In other words, all the curvilinear rectangles must be similar, and if this is so, then the two conditions for the net that were mentioned at the beginning of this section will automatically be fulfilled. It should also be noted that the only requirement which must be met by relation a/b if it is to fulfill the two conditions that establish the pattern of the flow lines and equipotentials, is that it must be constant throughout; it may be any constant. For the sake of simplicity and elegance, a simple value for a/b is 1. In this way, the curvilinear rectangles are transformed into curvilinear squares. The resulting net will fulfill the conditions that the rate of flow passing through each channel will be the same and the drop in potential between every pair of equipotential lines will be the same. The square is the simplest and most suitable figure, with the added advantage that one can see at a glance how well the net has been drawn. This is not the case with rectangles since they vary in size and it is difficult to tell without measuring whether their proportions are correctly depicted or whether they have been drawn differently with the corresponding error.

If the flow net is square, Eq.(7a-24) can be written thus:

$$q = kh \frac{n_f}{n_e} \tag{7a-25}$$

The term n_f/n_e depends only on the shape of the seepage element defined by the equipotential and flow lines. It is called the *Shape Factor* and is represented as:

$$F_f = \frac{n_f}{n_e} \tag{7a-26}$$

and Eq.(7a-24) can consequently be rewritten as:

$$q = khF_f \tag{7a-27}$$

which is a simple formula that permits calculation of the rate of seepage occurring per unit of length in a flow region for which the corresponding net has been drawn.

Before describing other important concepts which can be computed by means of a flow net, a little more detail will describe how these nets should be drawn. In [2] Casagrande gives the following advice to young students and engineers with little or no experience in this field:

1. Take every available opportunity to study carefully drawn flow nets, afterwards trying to repeat them without the model, until satisfactory drawings are obtained.
2. It is usually sufficient to draw the net with four to five flow channels. If more channels are used, drawing becomes considerably more difficult and attention is diverted from the essential aspects.
3. Always observe the net as a whole, without trying to correct details until the entire net is fairly well sketched.
4. There are often parts of the net where the flow lines are approximately straight and parallel. Here the channels are more or less of the same width and the squares should be very similar. Drawing can be made easier by starting with this zone.
5. Flow nets in confined regions limited by parallel boundaries (especially the upper and the lower ones), are often symmetrical and the flow lines and equipotentials are then almost elliptical in shape.
6. A mistake commonly made by beginners is to make a very abrupt transition from the straight portion to the curved portion of the different lines. These transitions must always be very gentle and either parabolical or elliptical. The size of the different squares also changes very gradually.
7. Generally, the first attempt does not provide a flow net made up of squares throughout the entire seepage zone. The drop in potential, corresponding to a certain whole number of channels with which the solution was attempted does not usually provide a whole number of equipotential drops, Δh. Thus, when the net is finished, a last row of rectangles is usually left between two equipotential lines where the drop in head is a fraction of the Δh that prevailed in the rest of the net. This is generally not detrimental; this last row can be considered for the calculation of n_e, assessing the resulting fraction of head loss, Δh. If for the sake of appearance all the rows of squares must have the same Δh, the net can be corrected, changing the number of flow channels either by interpolation or by starting again. No attempt should be made to convert the incomplete row into one made up of squares by means of purely graphical local corrections, unless the space that is either lacking or in excess is very small.
8. Boundary conditions may give rise to special circumstances in the net. These will be discussed in further detail in the paragraphs that follow.
9. In the net, an exit surface which is in contact with the air is neither a flow line nor an equipotential one if it is not horizontal. Therefore, the squares that are limited by that surface cannot be complete. Nevertheless, as will be shown further on, the drops in elevation between the points on these surfaces that are cut by equipotential lines must be equal.

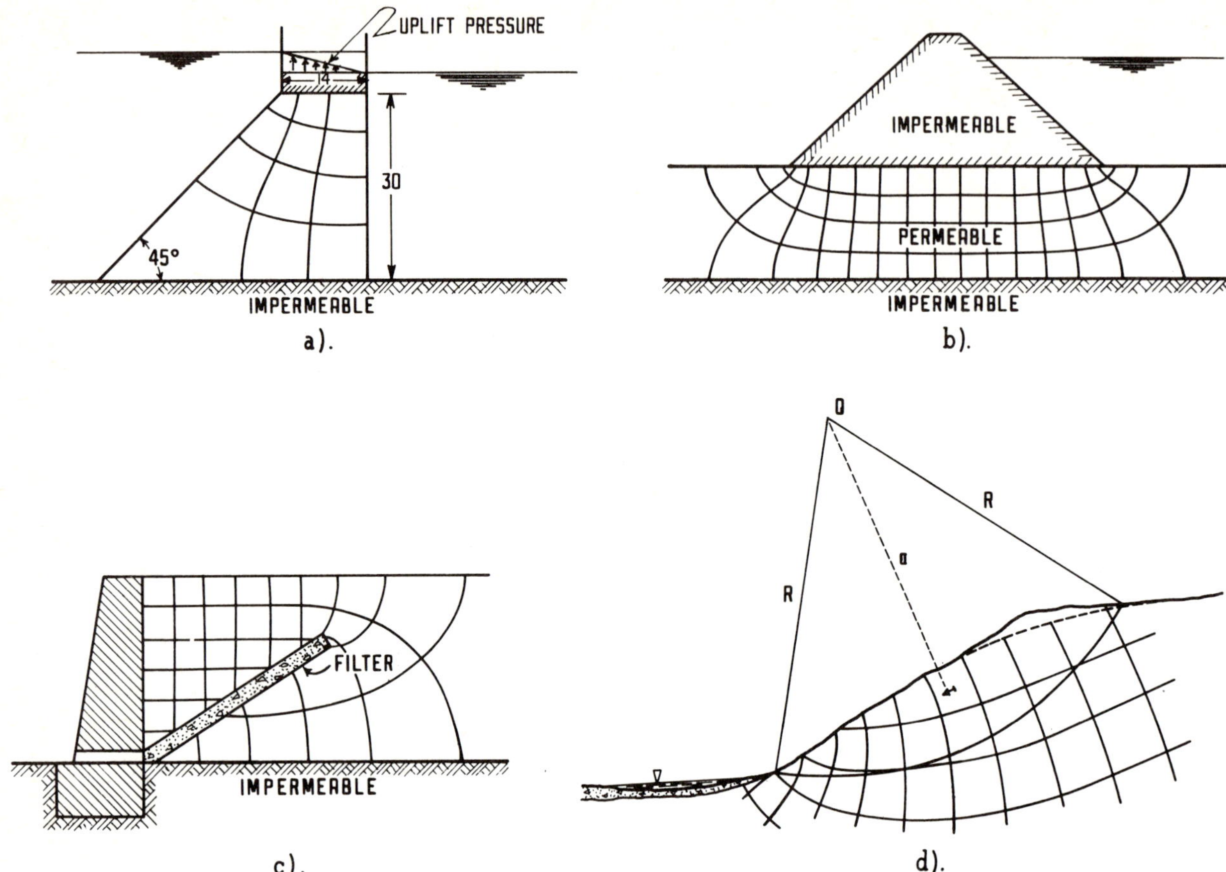

Fig. 7a-7 Examples of flow nets

Besides the above, it is recommended that flow lines and equipotentials always be drawn complete. Beginners make numerous conceptual errors when drawing flow nets, many of which could be avoided by leaving no lines incomplete. Fig. 7a-7 show four different flow nets for study.

7a.7 Surfaces that are Exposed to Atmospheric Pressure

A very common boundary in flow nets is the one consisting of a surface that is exposed to the air or, in more general terms, a surface on which all the points are at atmospheric pressure. In the case of these surfaces there is a theoretical condition which must be fulfilled. This is done graphically and can be easily checked.

AB is assumed to be a surface exposed to the air, on which all the points have the same pressure head, corresponding to atmospheric pressure (Fig. 7a-8). Two points of that surface cut by two successive equipotentials are separated vertically by a distance Δh which is equal to the drop in head between these two equipotentials, for since the pressure head is the same, zero, the difference in head can only be a change of position. Between all the equipotentials that cut the free surface there is the same loss in head; therefore, it follows that between all the points at which these equipotentials cut the free surface there must be the same difference in elevation equal to Δh. This is illustrated graphically in Fig. 7a-8.

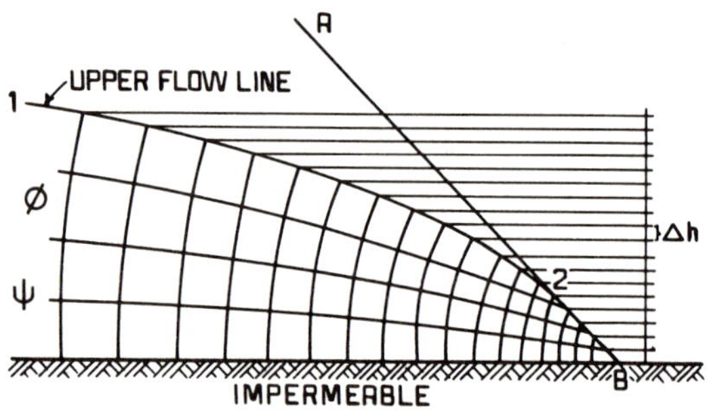

Fig. 7a-8 Free (unconfined) upper boundary of seepage

7a.8 Special Squares

In flow nets, there are occasions when the geometrical circumstances of the flow zone affect in such a way that a special case is produced, giving rise to squares which are geometrically different from the usual ones. Part *a* of Fig. 7a-9 shows a very common case which has already been shown in the nets in Fig. 7a-7.

The upper boundary of the fragment of the flow zone, which is reproduced, is an equipotential line, whereas the lower boundary is a flow line. Both lines are parallel; therefore, the last square, from a_1 b_1 to the left, is an unusually shaped open square. Observe that the same volume of flow Δq passes to the left of the flow line which starts at a_1 — as through the remaining flow channels to the right. If this special square is subdivided into halves (lines through points a_2 and b_2 of the figure), the volume of flow $\Delta q/2$ will pass through each subdivision. If the subdivisions are continued towards the left it will be possible to obtain the channels through which a quarter, an eighth, etc. of the volume of flow passes. It can be seen that these channels tend to be similar towards the left, whereas the volume of flow passing through them becomes rapidly smaller. From this it can be deduced that the seepage velocity of the water in the permeable zone becomes steadily lower towards the left, so that it becomes asymptotically close to zero. This is a general rule; it can therefore be said that when a flow line and an equipotential line are parallel owing to special circumstances in a net, the seepage velocity of the water drops to zero at their intersection (point ∞).

Part *b* of Fig. 7a-9 illustrates another special case which is fairly common in flow nets. A flow line and an equipotential meet at *A*. They are collinear; that is to say between them they make an angle of 180° instead of the usual 90°. Here again, if the original channel with a volume of flow of Δq is subdivided, two channels are obtained with a volume of $\Delta q/2$ in each. By further subdivision, channels can be obtained for a quarter of the flow volume, an eighth, etc. But the situation is now different from the one in Case *a)*. By observing Fig. 7a-9b, it can be seen that the cross-section of each channel gradually becomes considerably smaller than half of the preceding one, while the volume of flow passing through it is exactly half of the volume which passed through the channel prior to subdivision. Consequently, as *A* is approached, the seepage velocity of the water in the soil must increase. At point *A* the velocity is theoretically infinite. This too can be regarded as a general rule, so it can be said that if a flow line and an equipotential meet at an angle greater than 90° (180° is no more than a specific case), the seepage velocity of the water at the intersection will be infinite.

When considering the theoretical fact that the velocity at *A* is infinite, the following points should be kept in mind. The theory which led to the conclusion under consideration was drawn up according to the laminar flow hypothesis and assuming DARCY's Law to be valid. Since, as it has already been emphasized, this hypothesis demands low flow velocities, the theory does not apply to a point at which the velocities undergo a considerable increase. The conclusion that the velocity becomes infinite should not, therefore, be taken literally. The real conclusion that can be drawn is that in the neighbourhood of *A* the velocities of the water increase greatly and the flow becomes more concentrated, which is why zones like these will be critical with a change to turbulent flow or aggravated seepage erosion, whenever they are located at the net exit where the material has no confinement.

Part *c* of the figure shows another special case which frequently occurs in flow nets. Here an equipotential line and a flow line intersect at an angle α smaller than 90°. In this case, it can be seen that each time a subdivision is made, the volume of flow is equivalent to half of the preceding one, and passes through a section which is larger than half of the previous one. In this way, the seepage velocity steadily decreases as it approaches *A*, so that when this point is reached the velocity is zero. This can also be regarded as a general rule; that is to say, when, due to some special circumstance in a flow net, an equipotential and a flow line intersect at an angle $\alpha < 90°$, the seepage velocity of the water at the intersection will be zero. The value $\alpha < 90°$ includes zero, as was seen when discussing Fig. 7a-9a, which is therefore a special case of the type now being dealt with.

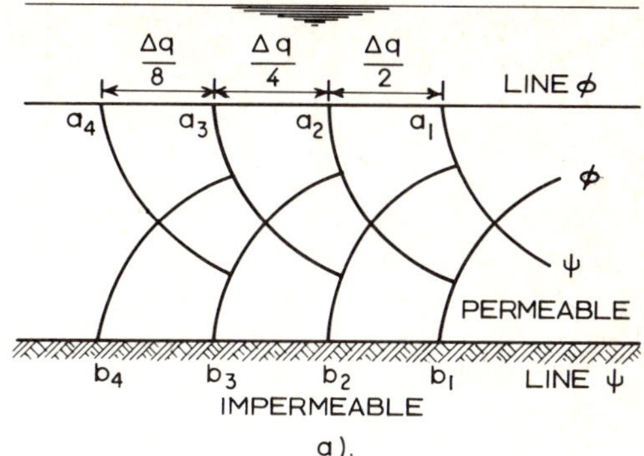

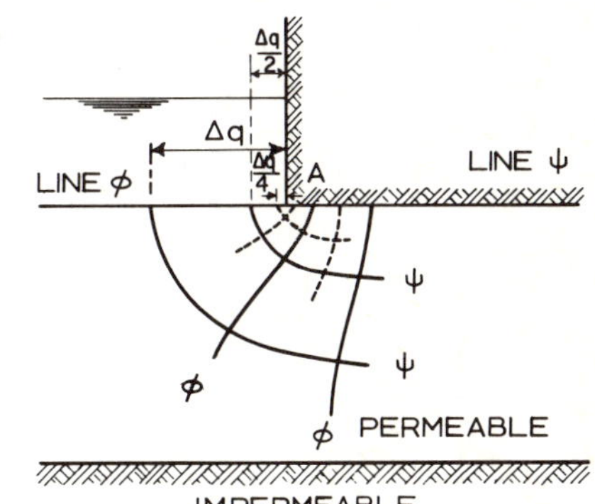

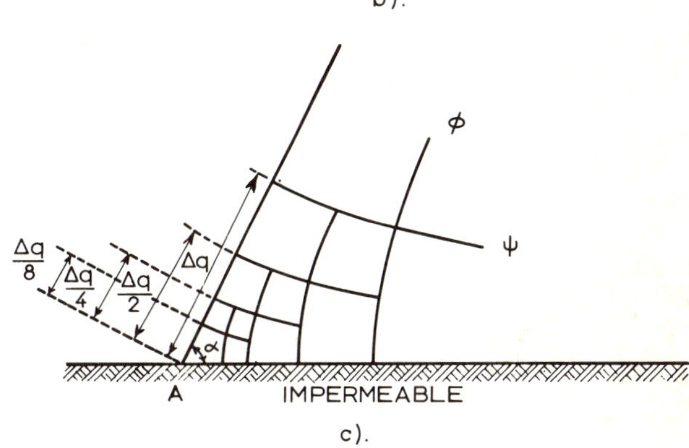

Fig. 7a-9 Special squares

7a.9 Computation of Hydrodynamic Pressures in a Flow Net

Here one of the most useful applications of a flow net will be shown. It enables the hydrodynamic pressures of the water seeping through a flow region to be calculated. This computation is applied to the design of structures that are subjected to seepage, such as slopes, retaining walls and foundations. By way of illustration, the paragraphs that follow give an analysis of the calculation of pressures in the water in two cases of practical interest. The first is a slope with a flow net which is partially shown in Fig. 7a-10. The pressures in the water within the slope are to be computed.

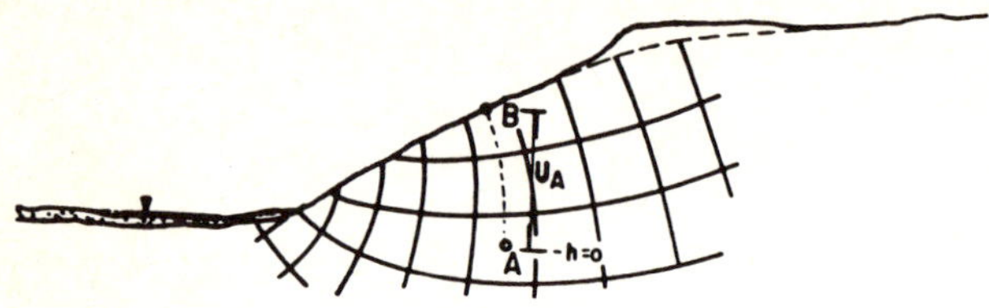

Fig. 7a-10 Computation of the pressures in the water inside a slope

Let us suppose that the hydrodynamic pressure at a point like A is to be calculated. If the corresponding equipotential is drawn through this point, the line comes out into the open at B. A and B must have the same head, since they belong to the same equipotential. If a horizontal plane ($h = 0$) is made to pass through A for the sake of reference, the elevation head will be zero and the total head will be a pressure head and will correspond to the pressure of the water at that point. The pressure head at B is zero, as it is in contact with the atmosphere and its total head is therefore an elevation head. The following must be true.

$$(\text{Elevation head})_B = (\text{Pressure head})_A$$

The pressure at A can then be calculated by drawing a horizontal line through the exit point B and measuring the distance between A and this reference line, which is the pressure head required.

Now let us consider the case illustrated in Fig. 7a-11, where water seeps into a permeable zone underneath an impermeable structure. What is to be calculated now are the pressures of the water at the points directly below the structure (which are known as uplift pressures and play an important role in the design of the stability of the structure as a whole) and in any other part of the permeable zone.

First let us consider the case of the point marked *1* in the foundation of the structure. Since the initial head of the water is h, the head at *1* will be $h - \triangle h$, for this point is located on the next equipotential, with a drop in head $\triangle h$ in comparison with the initial value. But *1* also has an elevation head, which will be the distance between it and AB, the plane that is used for comparison purposes ($h = 0$). If the head h is divided into n_e equal parts (11 in this case), since there are 11 drops in potential in the net) and horizontal reference lines are drawn through these divisions, the distance from AB to the corresponding division will give the hydrostatic head at any point. In the case of *1*, this head is graphically the vertical distance between plane AB and the level of the first division. From this head must be subtracted the elevation head represented by the vertical distance from *1* to AB, which in this case is negative. In this way, the pressure head ($h_{\text{pressure}} = h - h_{\text{elevation}}$) at *1* (that is the value of the uplift pressure) is the distance u_1.

In the case of the point marked *2* which can be located anywhere in the permeable mass, the pressure head can be computed similarly. Observe that *2* is at a distance of one and a half drops in potential $\triangle h$ from the initial head. The head at this point will therefore be the vertical distance between AB and a horizontal line drawn one and a half divisions below the level h. Moreover, the distance between *2* and AB gives the elevation head at that point, which is also negative, so that the pressure head at *2* is the segment u_2 which is obtained by subtracting the elevation head (negative) from the head. The heads at *3* and *4* are calculated graphically. Explanation of the procedure is left to the reader. Once the pressure of the water at all points underneath the structure has been computed (uplift pressures) a diagram can be drawn to include them on a suitable scale. The total uplift force will be the area of this figure and will go through the centroid.

7a.10 Computation of Velocities and Hydraulic Gradients in a Flow Net

At the points of a flow region which are represented in a flow net, both the critical gradient and the velocity of the water can be found. This is done by simply drawing the segment of the flow line passing through a given point within the corresponding square. Then the fall between equipotentials, $\triangle h$, divided by the length of the flow line where this fall occurs, will give the mean gradient in that portion which includes the point in question. A closer approximation can be found for the specific gradient at the point if the square is subdivided into other smaller ones, gradually nearer and nearer the point.

Once the gradient has been found at that point, it is simply multiplied by the coefficient of permeability of the soil to obtain the magnitude of the velocity of the water, according to DARCY's Law. At the point under consideration, this velocity will be at a tangent to the flow line passing through it and will have the same direction as the flow. This is not the true water velocity but that of water passing through a conduit whose area is equal to the total area of the soil; solids and voids.

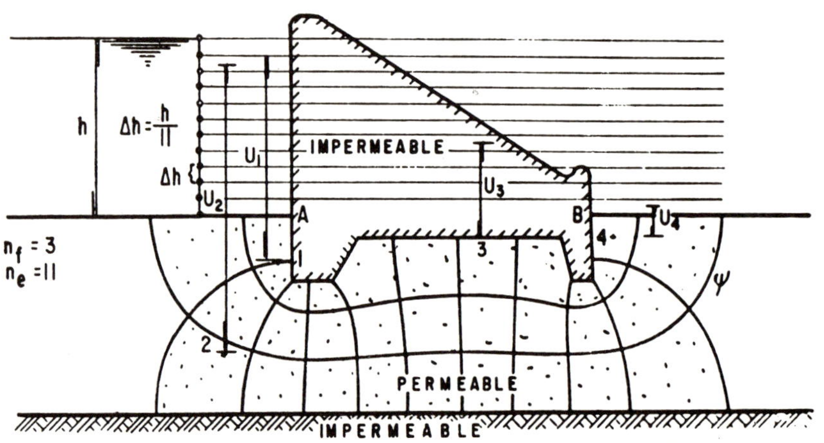

Fig. 7a-11 Computation of the pressures in the water underneath an impermeable structure

7a.11 Seepage Forces — Critical Liquefaction Gradient

When water flows through a mass of soil, its effect is not limited to the hydrostatic pressure occurring in the water in equilibrium; a hydrodynamic pressure is also applied on the particles of soil, in the direction of the flow, and this effect can be represented by hydrodynamic thrusts in the direction of the flow and at a tangent to the respective flow lines. The magnitude of these pressures or hydrodynamic thrusts depends especially on the prevailing hydraulic gradient.

Let us consider a square in a flow net, like the one that is shown in Fig. 7a-12. The hydrodynamic pressure applied by the water on the soil particles in section $\triangle A$ of the square (assuming that the square has a unit thickness), is

$$P_D = \triangle h \, \gamma_w$$

because the loss in head $\triangle h$ has been transmitted to the soil particles due to viscosity. This pressure produces a hydrodynamic thrust of:

$$J = \triangle h \gamma_w \triangle A \tag{7a-28}$$

This force is commonly expressed per unit of volume, the value for the square under consideration being:

$$j = \frac{J}{\triangle A \triangle L} = \frac{\triangle h \gamma_w \triangle A}{\triangle A \triangle L}$$

In other words,

$$j = \gamma_w i \tag{7a-29}$$

Using Eq.(7a-29), any seepage force corresponding to a square in a flow net can be calculated. Once the volume of flow is known, which is its area multiplied by normal unit thickness, the total force can be computed. This will act in the same direction as the flow, at the controid of the volume of the square and tangent to the flow line passing through that point.

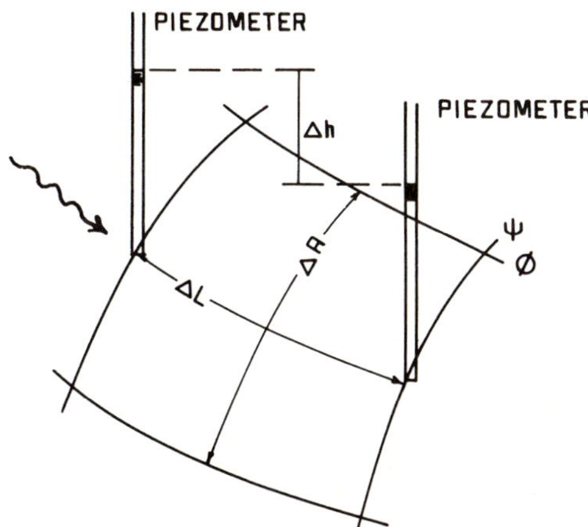

Fig. 7a-12 Seepage forces

Note that the seepage force depends on the specific weight of the water and the hydraulic gradient prevailing in the square under consideration, but it is independent of the flow velocity and the coefficient of permeability of the soil. Therefore, it is the same for cohesive soils as for frictional soils, although the flow velocities in both are very different. The seepage force is due to the viscous resistance that is generated in the fluid by the solid structure of the soil; owing to this, the water consumes energy in the form of hydrodynamic pressure loss and is capable of overcoming it, as is shown in Eq.(7a-28), where it can be seen that the hydrodynamic thrust is due to the head $\triangle h$ lost by the water during its course $\triangle L$ through the square.

Another phenomenon in direct connection with water flowing through soils is a localized *quick* condition in sands, which is really a form of piping. TERZAGHI [2] presented an interesting analysis on the subject, which will be described in the following.

The flow net considered is the one corresponding to the sheet pile appearing in Fig. 7a-13. The equilibrium of the water exit zone underneath the sheet pile will be studied in this net. As a result of tests on models and experience gleaned from structures that have been built, it is known that the sand in the zone under consideration stays in equilibrium so long as the head h is smaller than a certain limiting value h_p. As soon as this critical value is passed, there is a marked increase in discharge at the exit, the permeability of the sand suddenly increases, and the water starts to carry sand with it. Local liquefaction (a quick condition) occurs, followed often by piping. Experience has shown that the maximum concentration of flow occurs within a distance $D/2$ from the sheet pile.

The quick condition commences when the hydrodynamic pressure of the upward flow overcomes the submerged weight of the sand deposited in the zone where the phenomenon first appears. It can be said with a fair amount of accuracy that the sand moved by the water has the shape of a prism with a width $D/2$ and a height D_3. The tendency of this prism to be carried away is counteracted by its own weight. The moment the sand starts to be dragged away; the effective pressure on the sides of the prism, and consequently the frictional resistance, is zero. The prism moves upwards when the upward hydrodynamic pressure produced by the water overcomes the downward pressure produced by the submerged weight of the material. The head, h_p, that is produced by this unstable situation is the critical head. The level of the base of the prism to be analyzed will be determined for a minimum h_p, for the particles will

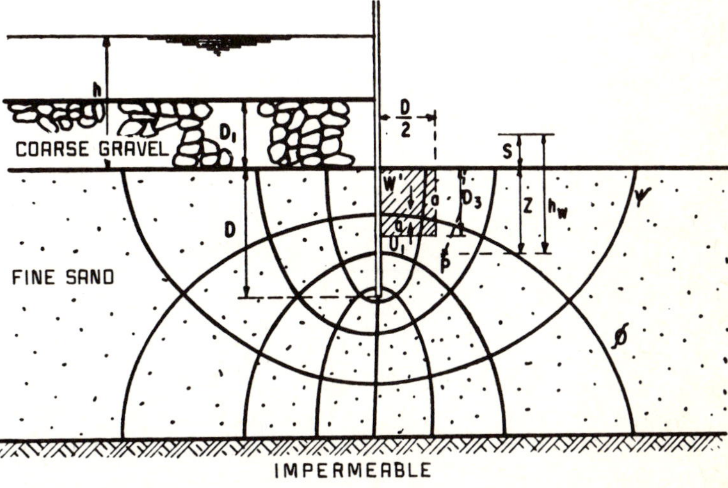

Fig. 7a-13 Physical gradient for quicksand (local liquefaction)

naturally respond to the pull of the smallest head capable of producing the phenomenon. In the figure, this level is represented by the dimension D_3.

In order to find the hydrodynamic pressure at this level, the pressure of the water at that depth will have to be known. This is done by first studying what this pressure will be at any point in the net, such as P. The pressure at P is given by the value h_w, which is the height to which the water rises within a piezometer installed at P, multiplied by the specific weight γ_w. The height h_w is composed of two parts, z and s, so that the neutral stress at P is

$$u_p = z\gamma_w + s\gamma_w \qquad (7a\text{-}30)$$

The first part of this equation represents the hydrostatic pressure at the depth of P. Its effect is to reduce the specific weight of the sand from γ_m to γ'_m, corresponding to the submerged condition. The second part, $s\gamma_w$, is the pressure existing in the water at P in excess of the hydrostatic pressure (hydrodynamic pressure). Thus, for the prism under consideration the condition that is required for it to be carried away is that the pressure in excess of the hydrostatic pressure on its base must not be higher than its submerged weight, which is (1/2) $DD_3\gamma'_m$. The excess pressure at P can be computed from the flow net and has already been seen to be

$$s\gamma_w = n_d \Delta h \gamma_w \qquad (7a\text{-}31)$$

where n_d is the number or fraction of drops in potential from P to the net exit. According to the above, the distribution of hydrodynamic pressures on the base of the prism can be drawn. The mean pressure on this base is called $h_a\gamma_w$ and the upward hydrodynamic thrust, U, in the same zone will be:

$$U = \frac{1}{2} D h_a \gamma_w \qquad (7a\text{-}32)$$

*

The value of s can be expressed as follows:

$$s = \frac{h}{n_e} n_d = hx \text{ (constant)} \qquad (7a\text{-}33)$$

where n_e = total number of drops in potential in the net and the constant shown has a value which depends only on the position of P within the net.

The hydrodynamic heads on the base of the prism under consideration can therefore be expressed thus:

$$h_a = mh \qquad (7a\text{-}34)$$

where m is a constant. The values of h_a and h are found either from the approach to the problem or the flow net, and the value of m in Eq.(7a-34) can then be computed. (For this, it will in fact be necessary to know D_3). The prism of sand under consideration will be lifted by the water when the hydrodynamic pressure exceeds the value which satisfies the equation.

$$\frac{1}{2} D h_a \gamma_w = \frac{1}{2} D D_3 \gamma'_m$$

Consequently,

$$h_a = D_3 \frac{\gamma'_m}{\gamma_w} \qquad (7a\text{-}35)$$

is the value of the hydrodynamic head at the base of the prism at the moment it becomes suspended. At that moment, the head h by definition has the critical value h_p and, in accordance with Eq.(7a-34).

$$h_a = mh_p \qquad (7a\text{-}36)$$

By substituting this value in Eq.(7a-35), the following are obtained:

$$mh_p = D_3 \frac{\gamma'_m}{\gamma_w} \qquad (7a\text{-}37)$$

$$h_p = \frac{D_3}{m} \frac{\gamma'_m}{\gamma_w} \qquad (7a\text{-}38)$$

Equation (7a-38) can be applied for different values of D_3, so long as a flow net to enable calculation of m has been drawn, Eq.(7a-34). Different values of h_p are thus obtained for different values of D_3. The minimum h_p is obviously the most critical value of the head and is the one which governs, and the corresponding level D_3 is the critical section, where uplift, a quick condition, and piping commence. The phenomeon may appear in this section if the prevailing head is greater than the value that was found for h_p. In the case of a simple sheet pile, like the one shown here the calculations lead to: $D_3 = D$, in the critical section. This result could have been directly deduced by observing the flow net, for as D_3 increases the value of the hydrodynamic pressures rises more quickly than the submerged weight of the sand. Note that, in accordance with Eq.(7a-38), the value of the critical height does not depend on the angle of internal friction of the sand but is proportional to the submerged weight of the sand. Also worth mentioning is the agreement between the theoretical prediction based on the preceding calculations and the results of experiments that have been reported as very satisfactory [8].

For a real water head, h, the factor of safety against piping can be readily computed using the expression

$$F_s = \frac{h_p}{h} \qquad (7a\text{-}39)$$

Values of about 3 or 4 are usually considered appropriate for the F_s.

If Eq.(7a-35) is observed, the mean value for the critical hydraulic gradient can be obtained, that is the value of the mean hydraulic gradient acting at the critical level, when heave and the quick condition commence. This value is:

$$i_c = \frac{h_a}{D_3} = \frac{\gamma'_m}{\gamma_w} \qquad (7a\text{-}40)$$

Considering a typical value of γ'_m, it is deduced that for piping to occur at D_3, which is assumed to be the critical level, the following must be true:

$$i_c = 1 \tag{7a-41}$$

Since the gradient at that level can be readily calculated from the flow net, if it is compared with the critical value equal to one, a simple criterion is available for finding out what the risks of piping are in a given problem.

REFERENCES

1. Terzaghi, K., "Mechanism of Landslides", *From Theory to Practice in Soil Mechanics. Part III.*, L. Bjerrum, A. Casagrande, R. B. Peck and A. W. Skempton, John Wiley, 1960.
2. Skempton, A. W., "The Rate of Softening of Stiff, Fissured Clays", *Proc. III. ICSMFE,* Rotterdam, 1948, Vol. II.
3. Juárez Badillo, E., and Rico, A., *Mecánica de Suelos. Vol. III: Flujo de agua en suelos,* Limusa: Mexico, 1977, (3rd Edn.) Chap. III.
4. Emmons, W. H., Allison, I. S., Stauffer, C. R., and Thiel, G. A., *Geologia: Principios y procesos,* (Translation by Francisco Alvarez Ros). Castilla Editions: Madrid, 1965, Chap. 14.
5. Linsley, R. K., Kohler, M. A. and Paulhus, J. L. H., *Hydrology for Engineers,* McGraw Hill Book Co., Inc. 1958, Chap. 6.
6. Spencer E. W., *Geology: A survey of Earth Science,* Thomas Y. Crowell Co: New York, 1966.
7. Cedergren, H. R., *Seepage, Drainage and Flow Nets.* John Wiley, 1967, Chap. 5.
8. Parker, G. G., "Piping, a Geomorphic Agent in Landform Development of the Drylands", *International Assoc. of Scientific Hydrology,* Berkeley, Cal., 1963. Pub. No. 65.
9. Bertram, G. E., "An Experimental Investigation of Protective Filters." *Harvard Soil Mechanics Series,* Harvard University, 1940, No. 267.
10. U.S. Army Corps of Engineers, "Investigation of Filter Requirements for Underdrains", *Waterways Experiment Station,* Vicksburg, Pa., 1941, Technical Memorandum 183-1.
11. Karpoff, K. P., "The Use of Laboratory Tests to Develop Design Criteria for Protective Filters", *Proc. A.S.T.M.,* 1955, Vol. 55.
12. Sherard, J. L., Woodward, R. J., Gizienski, S. F., and Clevenger, W. A., *Earth and Earth-Rock Dams,* John Wiley, 1963.
13. U.S. Army Corps of Engineers, "Drainage and Erosion Control-Subsurface Drainage Facilities for Airfields", Part XIII of Chap. 2 in *Engineering Manual, Military Construction,* Washington, D.C., 1955.
14. U.S. Bureau of Reclamation. *Design of Small Dams,* U.S. Government Printing Office: Washington, D.C., 1965.
15. Reference [3], Chap. VI.
16. Lovering, W. R. and Cedergren, H. R. "Structural Section Drainage", *Proc. I. Conference on Structural Design of Asphalt Pavements,* University of Michigan, Ann Arbor, Mich., 1963.
17. Casagrande A., and Shannon, W. L, "Base Course Drainage for Airport Pavements", *Trans. A.S.C.E.* 1952, Vol. 117.
18. Juárez Badillo, E., and Rico, A., *Mecánica de Suelos. Vol. I: Fundamentos de la Mecánica de Suelos,* Limusa: Mexico, 1977, 3rd. Edn. Chap IX.
19. Reference [3], Chap. II.
20. Smith, T. W., Cedergren, H. R. and Reyner, Jr. C. A., "Permeable Materials for Highway Drainage", *California Division of Highways. Materials and Research Department,* 1964. (Also Presented to the 43 Annual Meeting of the H.R.B. at Washington, D.C.)
21. Barber, E. S., "Discussion to the paper: Seepage requirements of Filters and Pervious Bases by H. R. Cedergren", *Trans. ASCE.*, 1962, Vol. 127, Part 1.
22. Reference [7], Chap. 9.
23. Reference [7], Chap. 8.
24. Smith, T. W., Notes for a course presented at the Laboratory of the California Department of Highways (Procedures, Testing Methods and use of Materials for Highway Purposes. Highway Subdrainage). Sacramento, Cal., 1964.
25. California Division of Highways. Materials and Research Department, Slope Stability and Foundation Investigation, Ed. University of California of Berkeley, 1967.
26. Highway Research Board, "Landslides and Engineering Practice", Special Report No 29, Washington, D. C., 1958.
27. Highway Research Board, "Construction of Embankments", N.C.H.R.P. Synthesis No 8., 1971.
28. Smith, T. W., and Stafford, G. V. "Horizontal Drains on California Highways". *Journal ASCE. Soil Mechanics and Foundations Division,* 1957, Vol. 83, SM 3.
29. Smith, T. W., "Groundwater Control for Highways", *California Division of Highways, Materials and Research Department,* Sacramento, Cal., 1964.
30. Rico, A., Springall, G., Springall, J., Moreno, G., and Mendoza, J. A., "Estabilización de fallas en la Autopista Tijuana-Ensenada", Technical Proc. in preparation. Related publication between the Ministry of Public Works and Federal Highways and Bridges of Profits and Connected Services, Mexico, 1973.
31. Rico, A., Springall, G., and Springall, J., "Deslizamientos en la Autopista Tijuana-Ensenada", Contribution of the Mexican Ministry of Public Works to the VII. ICSMFE, Mexico, D. F., 1969.

32. Reference [18], Chap. VIII.

33. PERUCCA, E., *Física General y Experimental. Vol. I.* Trans: J. Mañas B., Labor, 1948.

34. SOWERS, G. B., and SOWERS, G. F., *Introduction to Soil Mechanics,* John Wiley, 1972, Chap. 4.

35. JUÁREZ BADILLO, E., OROZCO, R. V., and BARRAGÁN, S., "Subdrenaje para carreteras, vías férreas y aeropistas", Paper presented to the Seminary of Embankments and Pavements of the Ministry of Public Works of Mexico, Mexico, D.F., 1968.

36. JUÁREZ BADILLO, E., and RICO, A., *Mecánica de Suelos. Vol. II: Teoría y aplicaciones de Mecánica de Suelos,* Limusa: Mexico, 1977, 3rd Edn. Chap. I.

37. TERZAGHI, K., "Frost and Permafrost", *Soil Mechanics Series,* Harvard University, 1952, No. 37.

38. Reference [7], Chap. 7.

OTHER RELATED REFERENCES

Departments of the Army, Navy, and Air Force. *Dewatering and Groundwater Control for Deep Excavations,* Technical Manual, Washington, D.C., 1971, No. 5-818-5.

HOARE, DAVID J., "Synthetic Fabrics as Soil Filters: A Review", *Journal of the Geotechnical Engineering Division. Proc. ASCE.,* October, 1982, Vol. 103, No. GT. 10.

REFERENCES — APPENDIX 7A

1. JUÁREZ BADILLO, E., and RICO, A., *Mecánica de Suelos. Vol. III: Flujo de Agua en Suelos*. Limusa: Mexico, 1977, 3rd. Edn. Chaps. I & II.

2. TERZAGHI, K., *Theoretical Soil Mechanics,* John Wiley, 1956, Art. 94

3. CEDERGREN, H. R., *Seepage, Drainage and Flow Nets,* John Wiley, 1967.

4. CASAGRANDE, A., "Seepage through Dams", Contributions to *Soil Mechanics,* Boston Society of Civil Engineers, 1940.

5. MUSKAT, M., *The Flow of Homogeneous Fluids through Porous Media,* McGraw Hill, 1937.

6. HARR, M. E., *Groundwater and Seepage,* McGraw Hill, 1962.

7. SCOTT, R. F., *Principles of Soil Mechanics,* Addison Wesley, 1963.

8. TERZAGHI, K., *Theoretical Soil Mechanics,* John Wiley, 1956.

9. SHERARD, J. L., WOODWARD, R. J., GIZIENSKI, S. F., and CLEVENGER, S. A., *Earth and Earthrock Dams,* John Wiley, 1963.

10. TAMEZ, E., *Principios del diseño y construcción de presas de tierra,* Ministry of Hydraulic Resources, Mexico, 1963.

11. MANSUR, C. I., and KAUFMAN, R. I., "Dewatering", Chapter 3 of *Foundation Engineering,* Ed. G. A. Leonards, McGraw Hill, 1962.

12. TODD, D. K., *Groundwater Hydrology,* John Wiley, 1960.

CHAPTER 8

FOUNDATIONS FOR HIGHWAY STRUCTURES

8.1 Introduction

This chapter is devoted to those concepts of applied soil mechanics used for the design and construction of foundations for highway structures. Here the word *structure* is used in a somewhat restricted sense, though a common, one, which includes the masonry, concrete or steel structures: bridges and drains. It also refers to retaining walls and similar structures. The fundamental principles of soil mechanics are the same for all these cases; likewise the methods of applying them in analysis and design. Consequently, what might be considered the theoretical content of this chapter is the same for the different types of structures involved. Differences do occur in the information obtained, either from the behavior of existing foundations, the study of models or the analysis of field tests. Since test data play an important part in current foundation techniques, the discussions that follow differentiate between the criteria that should be applied in accordance with the characteristics of each particular structure. Also, there are different types of foundations best adapted to different types of terrain, so that it will be necessary to differentiate between their special applications in considering the many types of foundations.

Lastly, as has been so often said, current foundation techniques involve considerable *art*, in the sense that many of the criteria, and design rules were not developed from theory. The success of a foundation will depend then largely on previous experience, intuition and ingenuity as well as on other qualities which are harder to define, such as audacity or prudence.

This chapter is not intended to compete with the numerous texts, papers, or state of the art reports, that are available on the subject of foundations; therefore, the reader is asked to approach it bearing in mind the focus of this book on transportation. For this reason, detailed analyses of bearing capacity theories or of theoretical considerations are not included; any reference is of a merely informative nature, and the corresponding justification available in other sources. An attempt will, however, be made to explain certain problems of a practical nature which are often a cause of anxiety at the time of construction.

From the point of view of soil mechanics, the design of a foundation is really the result of the superposition of two different questions. The first question has to do with the stress that can be transmitted by the foundation to the underlying ground without causing soil or rock failure. The second has to do with the equally-important aspect of what deformations the soil will undergo. In soil mechanics terminology, the first question is answered by what is known as bearing capacity theory and the second question by a settlement analysis. Bearing capacity theories and settlement (or expansion) analysis methods are vital contributions of soil mechanics to the problem of foundation design. However, a bearing capacity theory and a settlement analysis do not provide a solution to all the problems faced when designing and constructing a foundation: They merely tell what the problems are. For example, the ingenuity and experience of construction engineers have led to the development of various different methods of constructing a structure on a specific type of ground. Several of these may satisfy the requirements imposed by the bearing capacity and the settlement analyses. The choice of the type of foundation to be used is therefore based on other considerations, among which cost and ease of construction play a leading role. The engineer will be guided by soil mechanics in the correct assessment of these considerations. The optimum choice of the type of foundation will depend partly on the nature of the soil or rock and partly on the interaction between the ground soil and structure. There are situations where the conditions of the soil may be decisive owing to factors which are independent of the bearing capacity or the deformability of the ground. An example is the influence of the permeability of the soil and groundwater flow. A major design choice is whether a shallow or deep foundation is required. Details of the foundation design results from a careful consideration of the specific characteristics of the soil structure combination, costs and local circumstances, such as the availability of materials construction equipment and skilled labor. In both parts of the analysis, the role played by soil mechanics is of undeniable importance.

Once the design of the foundation has been made, the engineer will be faced with problems of construction. Here soil mechanics will give him guidance in the appropriate excavation of the soil or rock: that part of the foundation which has already been constructed. Bearing capacity theories usually originate from mathematical solutions to problems of continuum mechanics. After borrowing them from these sources, soil mechanics has adapted them for use in real earth materials. Almost all of these theories are based on PRANDTL's solution, to the problem of the indentation of a rigid solid in a continuous, semi-infinite, homogeneous and isotropic medium under plane strain conditions [1,2] in 1921. This solution, within the framework of the plasticity theory, assumes a perfect rigid-plastic medium. Figure 8-1 depicts the general mechanism of indentation, giving the solution for a purely cohesive, weightless, medium; Fig. 8-2 illustrates PRANDTL's solution applied to the more usual case in which the indented medium is a rigid plastic material which is weightless, but has cohesive and frictional strength components. It is also considered here that the body being indented is perfectly smooth, uniformly loaded and of infinite length, perpendicular to the page.

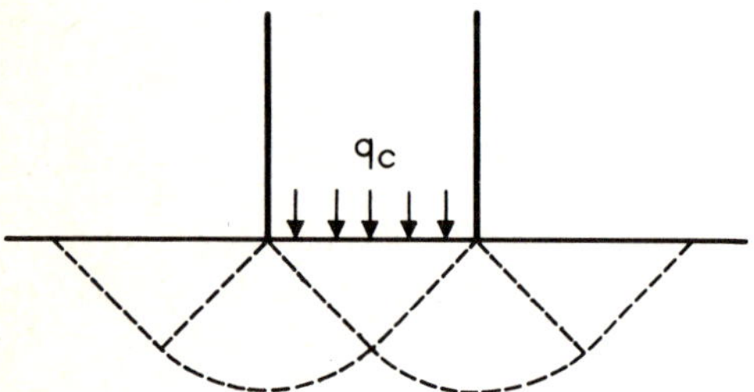

Fig. 8-1 Problem of indentation and solution by PRANDTL for a weightless medium with $c \neq 0$ and $\varnothing = 0$ [1,2]

In 1924, REISSNER extended PRANDTL's theory to include cases where the body is indented within the medium and no longer on the surface. For the case illustrated in Fig. 8-1, the maximum stress q_c that can be applied to the solid without indentation in the medium is:

$$q_c = (\pi + 2)\, c \qquad (8\text{-}1)$$

For settlement analysis soil mechanics has emphasized solutions for loads on clayey soil foundations, although the methods can be adapted to all soils. These are based on TERZAGHI's Consolidation Theory, the general principles of which were described in previous chapters. At the same time, comments were made on the most usual methods of applying the theory to real settlements. More recently some attention has been given to the compressibility of granular soils, which was also discussed previously. However, the methods available have not been as well verified by field measurements in comparison with the solutions for clay soils.

Foundations are generally regarded as belonging to one of two large categories: deep foundations and shallow foundations. Shallow foundations are those where the depth of the excavation does not exceed two to three times the width of the foundation. A more precise criterion for differentiation is impossible because no strict dividing line can be drawn. The most common types of shallow foundations are footings, strip footings and foundation slabs or mats (sometimes termed rafts). Of these, footings are the simplest and most widespread for bridges and similar structures. Shallow foundations are given preference, usually for reasons of economy and ease of construction.

Footings are usually square or rectangular and almost always made of concrete or masonry. Stone masonry is still much used for small projects, particularly in countries where labor is cheap or where for social reasons the use of labor-demanding methods is appropriate. The purpose of a footing is to widen the supporting area of a structural element such as a wall or column so as to transmit to the underlying ground stresses in keeping with its strength. When the strength of the ground is poor or the loads transmitted to the foundation are heavy, the areas supporting the foundation must be increased, by means of continuous strip footings which support several load concentrations, or foundation slabs or mats which are continuous in two directions throughout the entire loaded area. There is no rigid rule for distinguishing between the three types of shallow foundations, the only decisive factor being practice. Neither is it unusual to see combined foundations, in which the three basic types are joined to suit the design engineer.

If, even with a slab or mat, the pressure transmitted to the subsoil is still higher than its bearing capacity, or if analyses show that excessive settlements may occur, the structure must be placed on firmer supporting strata, at greater depths; hence the necessity for deep foundations. The search for firm strata is the origin of deep foundations. Sometimes, however, strata that are sufficiently firm are not found within economical depths, which leads to another type of deep foundation, in which the foundation elements distribute their loads by means of friction or adhesion into the thick soil in such a way that the

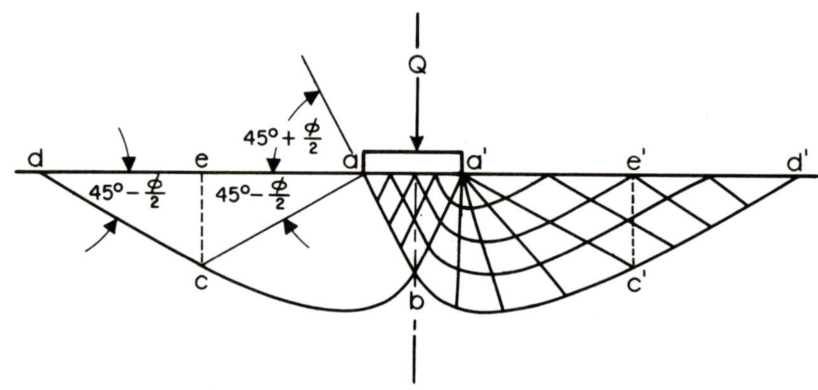

Fig. 8-2 Failure mechanism according to PRANDTL, Cohesive- frictional soil [2]

stress does not exceed the soil bearing capacity nor cause excessive settlements. The elements composing the most frequently used types of deep foundations are arbitrarily distinguished from one another by their diameter or width (depending on whether they are round or rectangular). The slimmest elements are piles, with a diameter or width up to 1.5 m (5 ft). Most have diameters or widths between 0.30 m and 0.60 m (1 and 2 ft). They may be of wood, concrete or steel in various cross-sections and structural varieties. Elements with a width greater than 1.5 m (5 ft) but not more than two times that value, are usually referred to as piers. There is no strict dividing line between piers and piles. Some engineers use the term pier for those deep foundations constructed by excavation as contrasted to piles which are usually driven. For others, a pier is an element which, working in exactly the same manner as a footing, transmits loads to greater depths.

Lastly, elements larger than the above ones are often required. These are called cylinders when round or caissons. However, the term caisson usually refers to a prefabricated hollow structure. The diameter of the cylinder usually is between 3 and 6 m (10 and 20 ft). They are usually built hollow in order to economize materials. They are almost always made of reinforced concrete. Figure 8-3 shows the different types of deep foundations.

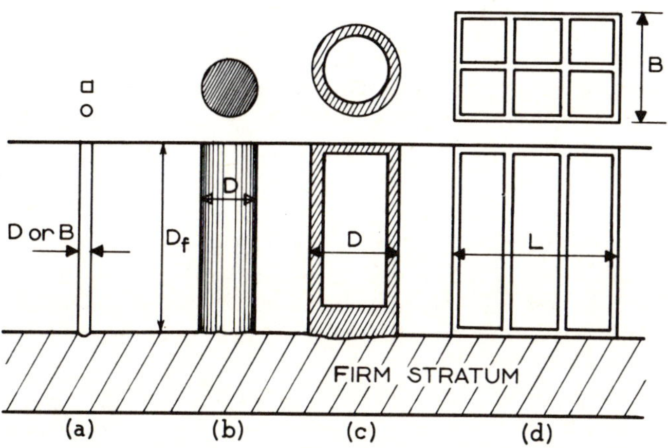

Fig. 8-3 Types of deep foundations

There are no very strict rules for determining the relative suitability of shallow or deep foundations. Economical considerations are usually the decisive factor in the final choice. As a general rule, shallow foundations are cheaper than deep ones. Therefore, in highway structures there is a marked tendency towards footings. This economic advantage however, gradually loses significance as the excavation becomes deeper, for the cost of open excavations increases very rapidly with depth. It has often been said that a shallow foundation on footings is suitable for depths of up to 5 to 6 m (17 to 20 ft), so long as there are no special problems involving water. Indeed, the use of shallow foundations in highways is frequently limited by groundwater. In addition to groundwater flow in the walls of the excavation there are the problems reducing slope stability making bracing necessary. There is the further disadvantage that the bottom of the excavation becomes flooded, making it very difficult (or even impossible) to work with simple open excavation methods. Controlling this water may require complicated and very costly pumping which may make deep foundations cheaper.

8.2 Bearing Capacity Theories

The following is a brief review of the more important bearing capacity theories. A full description of them is given in [3].

8.2.1 Terzaghi's Theory

In 1943, Terzaghi extended the Prandtl-Reissner theory to include practical soil mechanics problems [1-6]. His theory covers the general case of soils whose shear strength can be approximated by $s = c + \sigma \tan \varnothing$. Terzaghi ignored the shear strength (and consequently its contribution to bearing capacity) of the soil located above the depth of the excavation, D_f. According to this theory, this material only exerts a downward pressure as a surcharge acting at the level of the foundation (Fig. 8-4). Figure 8-5 shows the failure mechanism proposed for a foundation of infinite length, with a contact with the soil that is rough and uniformly loaded. The figure is divided into two parts. The part on the left shows conditions prior to soil failure, whereas on the right, failure has already occurred, and the foundation has moved downward, indenting the soil.

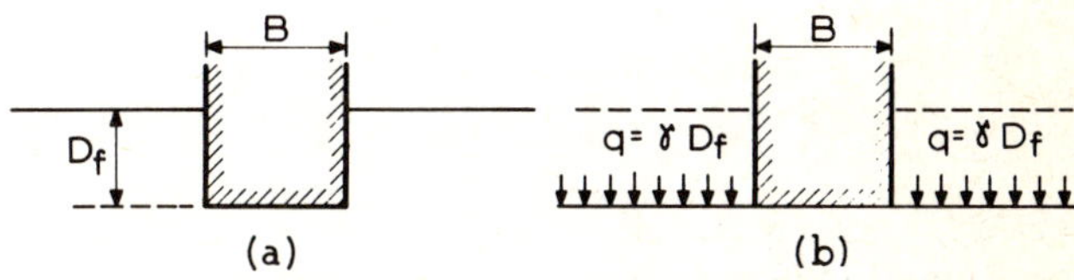

Fig. 8-4 Equivalence of the soil above the level of the excavation for a foundation, with an overload due to its weight

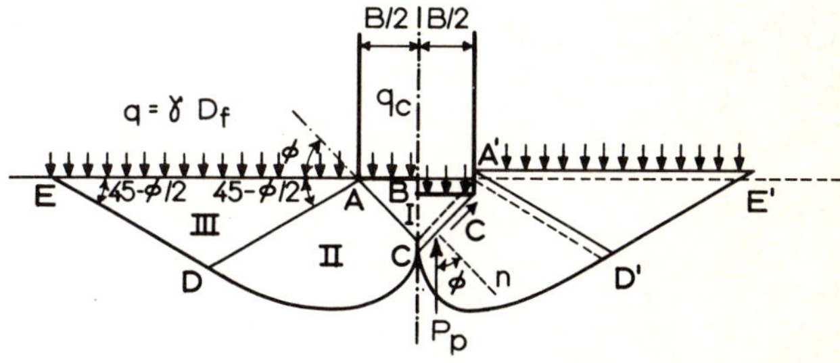

Fig. 8-5 Failure mechanism of a shallow strip foundation, according to Terzaghi

The chief assumptions made by Terzaghi (besides the surcharge) are the shape of the lines limiting Zones *II*, which appears as a logarithmic spiral, and the stress conditions in Zones *III* which correspond to Rankine's passive plastic state. He also assumed that the shear strength is simultaneously mobilized along the entire failure surface.

With a mathematical analysis of his failure model, which is described in [1] and [6], TERZAGHI obtained the following expression for the critical or failure load that can be transmitted by the foundation:

$$q_c = cN_c + \gamma D_f N_q + \frac{1}{2} \gamma B N_\gamma \tag{8-2}$$

In the above expression, c is the cohesion (shear strength) with no normal stress on the soil, B is the width of the foundation of infinite length and γD_f is the value of the surcharge at the foundation level. The terms N_c, N_q and N_γ are the bearing capacity factors of TERZAGHI's theory. In this theory, it is shown that they depend only on the angle of internal friction, $\varnothing$, of the soil and are non dimensional coefficients which characterize the bearing capacity of the soil. N_c is related to the cohesion of the soil, N_q to the surcharge existing at excavation level and N_γ to the unit weight of the soil supporting the foundation. All are proposed for both shallow and deep foundations.

The condition required for application of Eq.(8-2) to a practical case is knowledge of the values of N_c, N_q and N_γ. The theory enables the use of algebraic expressions to calculate these factors as a function of the angle $\varnothing$. The values obtained are illustrated in Fig. 8-6 [1,6].

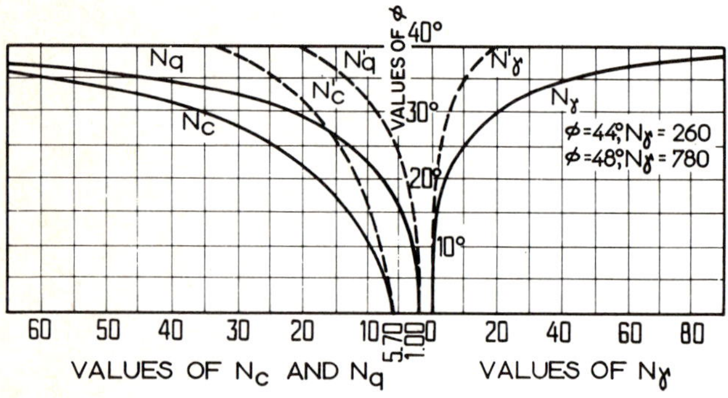

Fig. 8-6 Bearing capacity factors for application of TERZAGHI's Theory

Figure 8-6 shows three curves which give the values of N_c, N_q and N_γ as a function of the angle $\varnothing$. Another three curves (broken lines) which give modified values for these factors, $N'c$, $N'q$ and N'_γ, according to TERZAGHI, these last factors should be applied when a shear failure of the type termed as *local* is likely to occur in the foundation, in contrast to the mechanism presented in Fig. 8-5, which he termed *general shear failure*. TERZAGHI pointed out that as the foundation penetrates the soil, lateral displacements occur, and the plastified zones reach the extreme points E and E' (Fig. 8-5). Thus at the moment of soil failure the entire length of the failure surface is working at the limiting stress. This *general* mechanism does not, however, develop in all types of soil. In Fig. 8-7, a failure situation is depicted in loose sands or soft clays with stress-strain curves like c_2, where strains become very large. The plastic zones of Fig. 8-5 cannot develop fully, because the shear movement is absorbed in strain without failure at its outer limits. However the foundation has moved into the soil so far that the soil immediately around it has failed which, for practical purposes, is equivalent to failure. This is termed local shear failure and consequently lower corrected bearing capacity factors must be used.

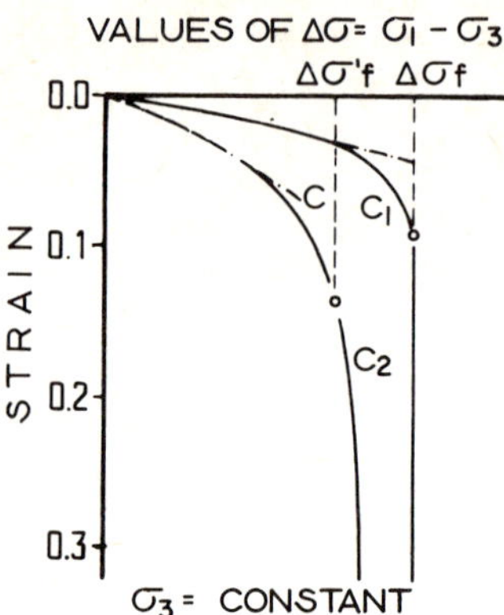

Fig. 8-7 Typical stress-strain curves for general *(1)* and local *(2)* shear failure mechanisms, according to TERZAGHI

TERZAGHI proposed that values for local shear failure be obtained by correcting the strength parameters c and $\varnothing$ of the soil whenever there is a possibility of this type of failure. The new values proposed are:

$$c' = \frac{2}{3} c$$

$$\tan \varnothing' = \frac{2}{3} \tan \varnothing \tag{8-3}$$

Given an angle $\varnothing$ in a soil in which local shear failure may occur, the corresponding value of $\varnothing'$ can be calculated with the second of the Eqs.(8-3). If this value of $\varnothing'$ is applied to the solid curves in Fig. 8-6, the same factors are obtained as by using the original value of $\varnothing$ and following the broken curves. In this manner, the repeated application of the expression can be avoided.

Consequently the ultimate bearing capacity for local shear failure is:

$$q_c = \frac{2}{3} cN'_c + \gamma D_f N'_q + \frac{1}{2} \gamma B N'_\gamma \tag{8-4}$$

The above theory refers to strip footings. For square or circular footings, which are so often used, no solution was provided. Nevertheless, TERZAGHI proposed the following.

Square footing: $q_c = 1.3cN_c + \gamma D_f N_q + 0.4\gamma B N_\gamma$ (8-5)

Circular footing: $q_c = 1.3cN_c + \gamma D_f N_q + 0.6\gamma R N_\gamma$ (8-6)

where R is the radius of the circular footing.

The bearing capacity factors in the above expressions are obtained from Fig. 8-6, both those corresponding to local shear and general shear failure. The above equations are valid for foundations subjected to vertical non-eccentric loading.

Many design engineers feel that the conditions under which the bearing capacity factors for local shear failure should be used are not clearly defined. Not enough reliable information is available regarding the stress-strain characteristics of soils, and their relation to either the development of either local or general shear. The design engineer cannot rely on any rigid rule; his own experience and judgement will have the last word. Some engineers relate general and local shear to the strain information supplied by triaxial tests. If strain at failure is below 5%, they apply the N factors; if it is higher than 15%, they apply the N' factors. For in-between cases they apply bearing capacity values which are linearly interpolated between the N and N' factors. Other authors have proposed the use of the N' factors for sands with relative densities lower than 30% and clays with sensitivities higher than 10.

More recently another mode of failure in shallow foundations has been proposed: *punching failure* [3]. As the load on the foundation increases, the soil beneath becomes compressed. The foundation may penetrate the ground if a vertical rupture occurs around the foundation perimeter. The equilibrium of the foundation is maintained, both vertically and horizontally, and the soil outside the loaded area is forced outward with very little change. To maintain the vertical movement of the foundation, a continuous increment in the vertical load would be required. This type of local failure is currently under investigation, but no analytical treatment has yet been established. Although experimentally it has been shown to develop, its limits have not been well defined.

For the case of purely cohesive soils (analysis based on an undrained triaxial test, for example), Fig. 8-6 gives the bearing capacity factors for general shear ($\varnothing_u = 0$) as: $N_c = 5.7$, $N_q = 1.0$, $N_\gamma = 0$. With these values, Eq.(8-2) becomes:

$$q_c = 5.7c_u + \gamma D_f \quad (8\text{-}7)$$

which is usually written in terms of unconfined compression strength ($q_u = 2c_u$). Thus:

$$q_c = 2.85q_u + \gamma D_f \quad (8\text{-}8)$$

Equation (8-8) is valid when the soil is purely cohesive, is insensitive and the foundation is of infinite length. An equivalent equation is obtained for square foundations from Eq.(8-5) in the same manner. Thus:

$$q_c = 1.3 \times 2.85q_u + \gamma Df \quad (8\text{-}9)$$

In practice, the following expression is frequently used for rectangular foundations with width B and length L, in purely cohesive soils, which is no more than an arbitrary superposition of Eqs. (8-8) and (8-9), both representing extreme cases [7]:

$$q_c = 2.85q_u \left(1 + 0.3\frac{B}{L}\right) + \gamma D_f \quad (8\text{-}10)$$

8.2.2 MEYERHOF's Theory

Since 1951 G. G. MEYERHOF [8-10] has made important contributions to the bearing capacity of soils. In addition to these references, [1] and [4] may prove useful for following the course of these contributions. MEYERHOF's theory proposed the added consideration of the shear stresses that may develop within the foundation ground above the level of the excavation. This effect was disregarded by TERZAGHI's theory, except as an overload. In MEYERHOF's theory the soil around the foundation above the level of the excavation is a medium in which shear surfaces develop, the failure mechanism proposed for foundations of infinite length appears in Fig. 8-8. According to this, the zone ABB' has uniform stresses and can be regarded as being in an active RANKINE state. The wedge ABC, limited by the arc of a logarithmic spiral, has radial shear stresses. The wedge $BCDE$ is a transition zone in which the stresses vary from a radial shear state to a passive plastic state.

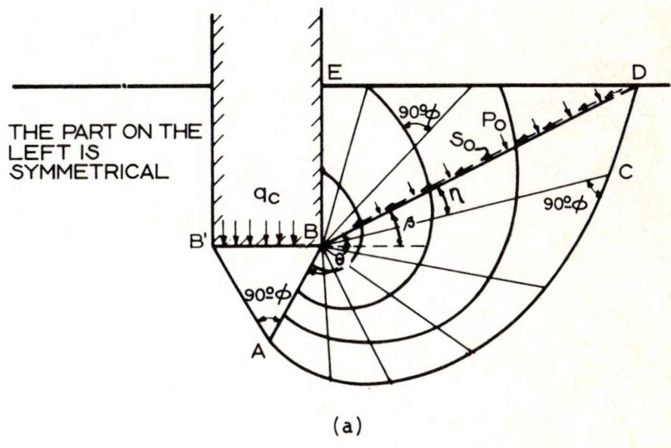

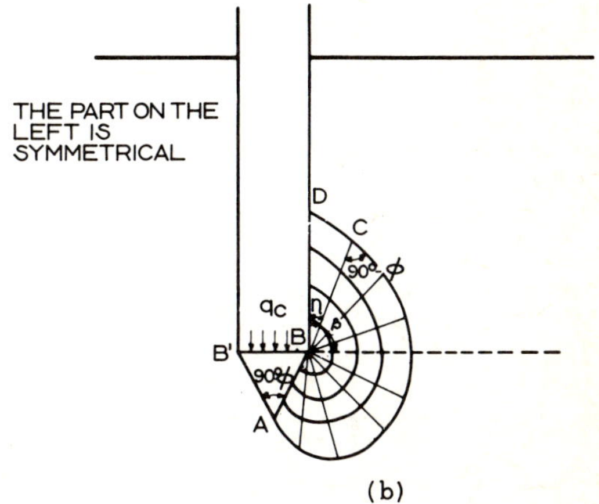

Fig. 8-8 Failure mechanisms proposed by MEYERHOF; *a*-shallow, *b*-deep

The equation that is derived in the MEYERHOF theory for expressing the bearing capacity of shallow foundations is of the same type as the one proposed in Eq.(8-2). For deep foundations, the following expression was reached:

$$q_c = cN_c' + \gamma D_f N_q' \quad (8\text{-}11)$$

which refers only to the capacity at the tip of the pile, it does not take into consideration the lateral friction on the pile shaft. The expression is valid only if the piles penetrate the firm stratum at as deep as $D = 4\sqrt{N\varnothing}\,B$. Figure 8-9 shows the values of the bearing capacity factors N_c, N_q and N_γ for shallow foundations, together with those for the N_c' and N_q' factors for piles, this figure originates from a later contribution by MEYERHOF [11] and was presented with Eq.(8-11) as a simplification of previous concepts. For shallow foundations, the figure gives factors for an element of infinite length and for a square one. The case of a rectangular foundation with a certain B/L relation is not solved, but the factors can be reasonably estimated [11] by linear interpolation between the extremes given by the graph (infinitely long $B/L = 0$; square $B/L = 1$).

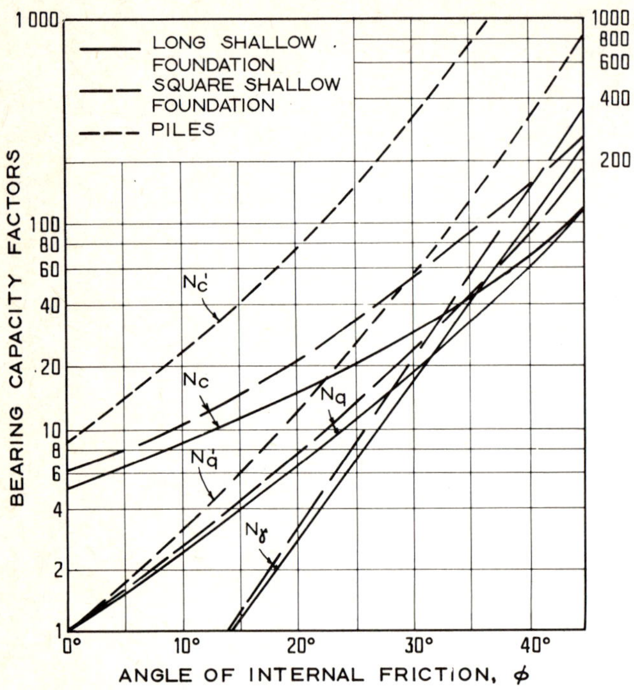

Fig. 8-9 Bearing capacity factors according to MEYERHOF [11]

For rectangular foundations, MEYERHOF recommends the use of Fig. 8-9, rectifying the angle of friction ($\varnothing$ rec) in such a way that:

$$\varnothing_{rec} = \left(1.1 - 0.1\frac{B}{L}\right)\varnothing \tag{8-12}$$

$\varnothing$ being the angle of internal friction as obtained in a triaxial test. The reason for this rectification is that a value of the angle of internal friction must be applied that is between the ones which are obtained in plane strain and triaxial tests [11]. In accordance with the type of analysis selected, total strength parameters (c_u, $\varnothing_u$) or effective strength parameters (c, $\varnothing$) are used. In the case of shallow foundations with an excavation depth equal to or smaller than their width, the factors in Fig. 8-9 can be increased due to the resistance which develops in the soil above the foundation level [11]. In order to do this, they are multiplied by the following factors:

$$d_c = 1 + 0.2\sqrt{N\varnothing}\ \frac{D}{B}\ f$$

$$d_q = d_\gamma = 1 \ ;\text{if} \quad \varnothing = 0$$

$$d_q = d_\gamma = 1 + 0.1\sqrt{N\varnothing}\ \frac{D_f}{B};\ \text{if} \quad \varnothing > 10° \tag{8-13}$$

where $N_\varnothing$ has the usual meaning with reference to earth thrust (Chapter 5) and the other letters have the usual meanings employed in this chapter. For foundation depths greater than the width, these *depth factors* become smaller and should no longer be taken into account.

8.2.3 SKEMPTON's Theory [12]

For purely cohesive soils when determining the value of N_c, TERZAGHI does not take into account the depth D to which the foundation penetrates the bearing stratum. Thus, in Fig. 8-10, the two foundations shown would have the same capacity in terms of the influence of cohesion (the value of N_c). According to TERZAGHI the total bearing capacity would not be the same for the two foundations owing to the different value of the term γD_f. However, our intuition tells us that the value N_c should be different in both cases. In terms of failure surfaces, the deeper foundation will have a larger shear surface where the total effect of cohesion will be greater, and should therefore deserve a higher value of N_c. SKEMPTON carried out experiments to quantify these ideas and found that N_c is not independent of the depth of the excavation, but increases with it.

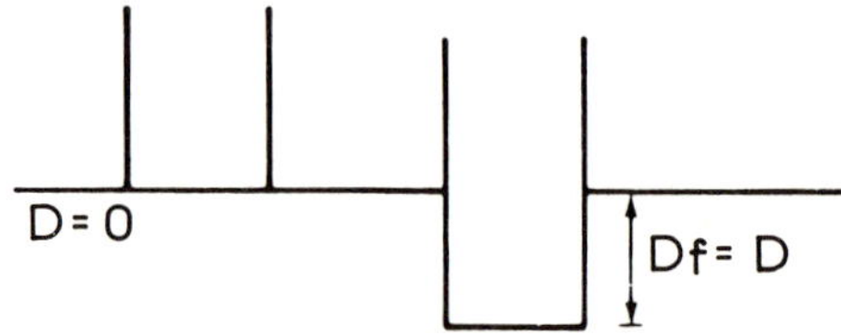

Fig. 8-10 Influence of the foundation depth on the N_c' value for purely cohesive soils

For the bearing capacity of cohesive soils, he proposed an expression similar to TERZAGHI's:

$$q_c = cN_c + \gamma D_f \tag{8-14}$$

The difference is that here N_c is not always 5.7, but varies with the D/B relation, where D is the depth at which the foundation is embedded in the firm stratum and B is the width of the foundation. Figure 8-11 shows the values obtained by SKEMPTON for the coefficient N_c to be applied in Eq.(8-14). Values are given for strip, square and circular footings. Both formula and coefficients can in principle be applied to shallow foundations and deep foundations in insensitive clay. In strati-

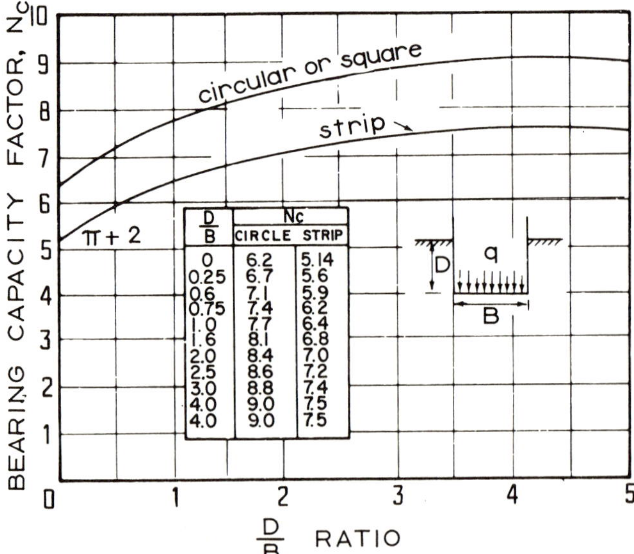

D/B	Nc CIRCLE	Nc STRIP
0	6.2	5.14
0.25	6.7	5.6
0.6	7.1	5.9
0.75	7.4	6.2
1.0	7.7	6.4
1.6	8.1	6.8
2.0	8.4	7.0
2.5	8.6	7.2
3.0	8.8	7.4
4.0	9.0	7.5
4.0	9.0	7.5

Fig. 8-11 N_c values for purely cohesive soils, according to SKEMPTON

fied heterogeneous soils, the term γD_f of Eq.(8-14), which represents the pressure of the overburden at the foundation level, should be calculated by the different thicknesses of the layers; with their respective specific weights and considering the condition in which the soil is found (saturated, dry, partially saturated or submerged). The distinction between D_f and D in SKEMPTON's theory, can be seen in Fig. 8-12.

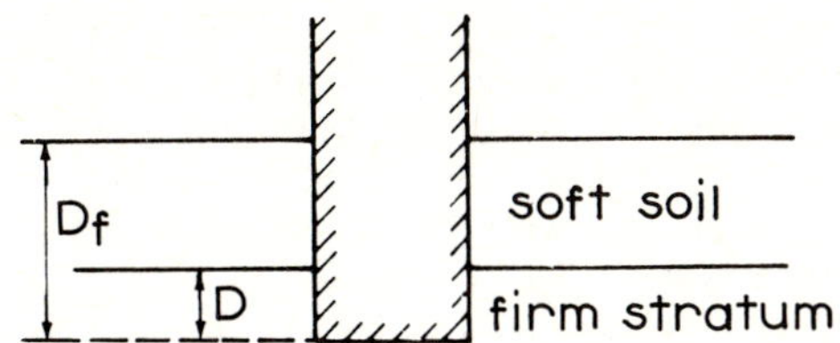

Fig. 8-12 Distinction between D and D_f for application of SKEMPTON's Theory

8.2.4 Other Bearing Capacity Theories

There are other bearing capacity theories developed by different investigators, in which Eq.(8-2) is usually maintained. The only difference between the theories is the values of the bearing capacity factors N_c, N_q and N_γ. This should be considered fortunate, for it makes it possible to compare the different theories in a very objective and simple manner.

HANSEN [13-15] gives the following for rectangular shallow or deep foundations excavated in any type of soil:

$$q_c = cN_c\left(1 + 0.2\frac{B}{L}\right)\left(1 + 0.35\frac{D_f}{B}\right) + \gamma D_f N_q\left(1 + 0.2\frac{B}{L}\right)\left(1 + 0.35\frac{D_f}{B}\right) + \frac{1}{2}\gamma B N_\gamma\left(1 - 0.4\frac{B}{L}\right) \qquad (8\text{-}15)$$

The second parenthesis of the second term should be taken as 1 for $\varnothing = 0°$. The bearing capacity factors with which Eq.(8-15) is applied appear in Table 8-1.

Another theory that is expressed in the same form as TERZAGHI's Eq.(8-2) is that of BELL [16]. Its values for the bearing capacity factors are shown in Fig. 8-13. This expression utilizes a simplified version of the shear zone of TERZAGHI's theory, the values that are shown refer only to very long

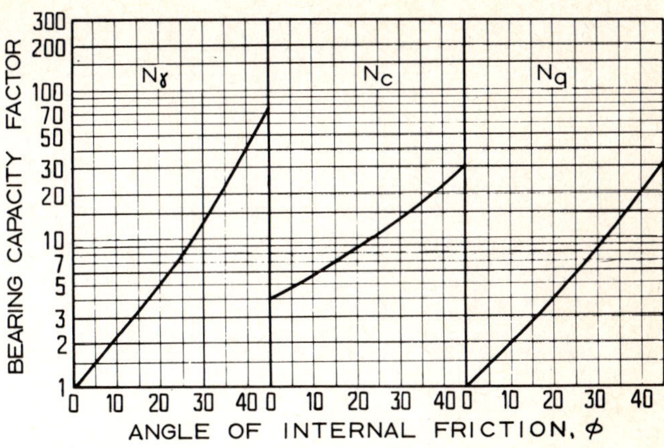

Fig. 8-13 Bearing capacity factors, according to BELL [16]

foundations. For square or circular footings, the values in Fig. 8-13 should be corrected with the coefficients in Table 8-2 [16].

Table 8-2
Correction coefficients for bearing capacity factors according to BELL [16]

Footing	Correction factor for N_c	Correction factor for N_γ
Square	1.25	0.85
Rectangular, $\frac{L}{B} = 2$	1.12	0.90
Rectangular, $\frac{L}{B} = 5$	1.05	0.95
Circular	1.20	0.70

Figure 8-14 shows the bearing capacity factors proposed by BEREZANTZEV [17] for deep foundations. The bearing capacity that is obtained with these factors is reported to be in close agreement with the results of pile tests performed on large-scale models and on actual deep foundations.

Another solution to the problem of bearing capacity is proposed by BALLA [18,19]. This theory has been said by some engineers to offer by-far the closest agreement with the few measurements that have been taken in relation to the behavior of real foundations. BALLA's theory refers to foundations in

Table 8-1
Bearing capacity factors according to Hansen [13,14,15]

Factor	Value of the angle of friction ∅ (degrees)										
	0	5	10	15	20	25	30	35	40	45	50
N_c	5.1	6.5	8.3	11.0	14.8	20.7	30.1	46.1	75.3	134	267
N_q	1.0	1.6	2.5	3.9	6.4	10.7	18.4	33.3	64.2	135	319
N_γ	.0	0.1	0.5	1.4	3.5	8.1	18.1	40.7	95.4	241	682

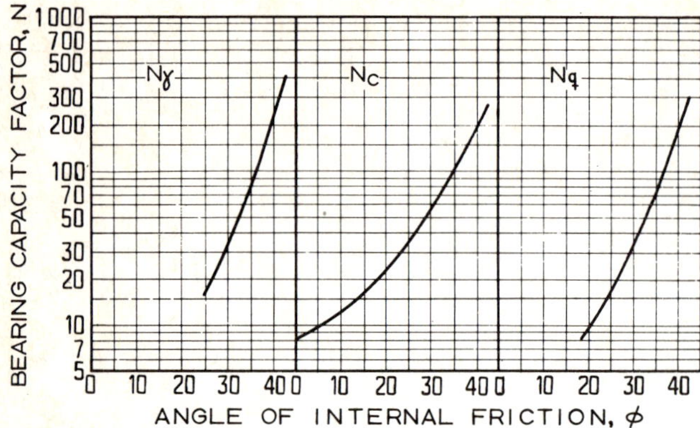

Fig. 8-14 Bearing capacity factors, according to BEREZANTZEV [17]

cohesionless soils or ones with little cohesion. It was originally intended for long foundations, and correction factors are proposed for its application to square or circular foundations, but recently FEDA [20] pointed out that it is not a serious error to take these correction factors as being equal to 1. A limitation of this theory is that it can only be used in shallow foundations where $D_f \leqslant 1.5\ B$. The bearing capacity is given by the expression:

$$q_c = c\,(\tan \varnothing + \rho F_6) + q\,(1 + \rho F_5) + \frac{1}{2}\,\gamma B\,(\rho F_4 + F_5 \tan \varnothing)\,\rho \quad (8\text{-}16)$$

which could be expressed thus:

$$q_c = cN_c + \gamma D_f\,N_q + \frac{1}{2}\,\gamma B N_\gamma \quad (8\text{-}17)$$

The term ρ is defined as:

$$\rho = \frac{2R}{B} \quad (8\text{-}18)$$

where B is the total width of the foundation and R the radius of the curved part of the failure surface. With the exception of the F factors, all the symbols of Eq.(8-16) have the usual meanings. The F factors have a mathematical form that can only be discussed by reviewing BALLA's theory in detail, which is beyond the scope of this book, but which can be done with the aid of the above-mentioned references. The factor ρ can be computed as a function of the D_f/B relation, using the curves in Fig. 8-15 [19]. The value of ρ is also a function of the $2c/B\gamma$ relation, these letters having the usual meanings. Once the value of ρ has been computed, the coefficients N_c, N_q and N_γ of BALLA's theory can be computed using the graphs in Fig. 8-16, and with them the bearing capacity of the shallow foundation can be computed using Eq.(8-17).

Other researchers have studied the effects of the anisotropy or the heterogeneity of soils, especially stratification. REDDY and SRINIVASAN [21] obtained solutions for the bearing capacity of long foundations in layered soils, considering variations in the properties, relating to both friction and cohesion. DAVIS and CHRISTIAN [22] obtained solutions for a long foundation in cohesive material, considering anisotropy in the cohesion. MEYERHOF and BROWN [23] carried out an experimental investigation on the bearing capacity of layered clayey soils, and proposed equations for such cases.

8.2.5 Comparison of the Different Theories

Table 8-3 presents a comparison of the different bearing capacity factors that can be obtained using the theories that have just been mentioned. Some of them, such as TERZAGHI's, BELL's or HANSEN's, do not distinguish between values for

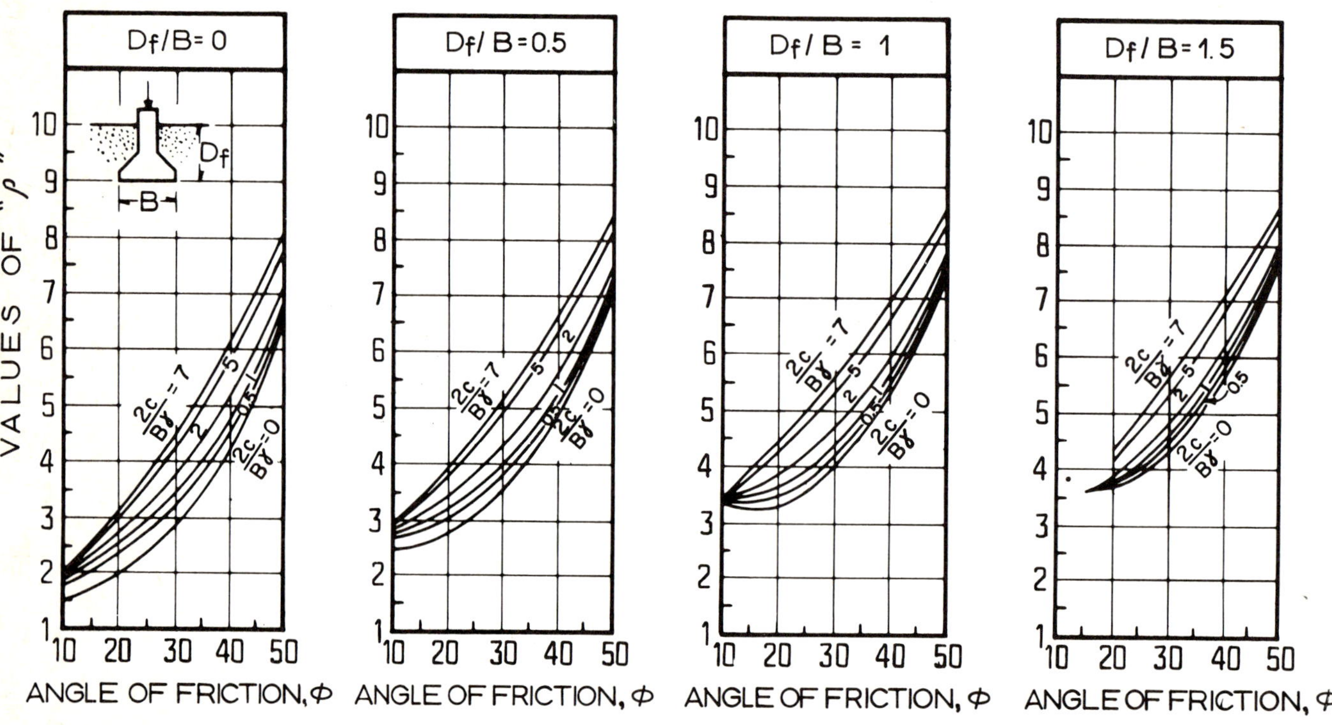

Fig. 8-15 Values of ρ in BALLA's Theory for several different $\frac{D_f}{B}$ ratios [19]

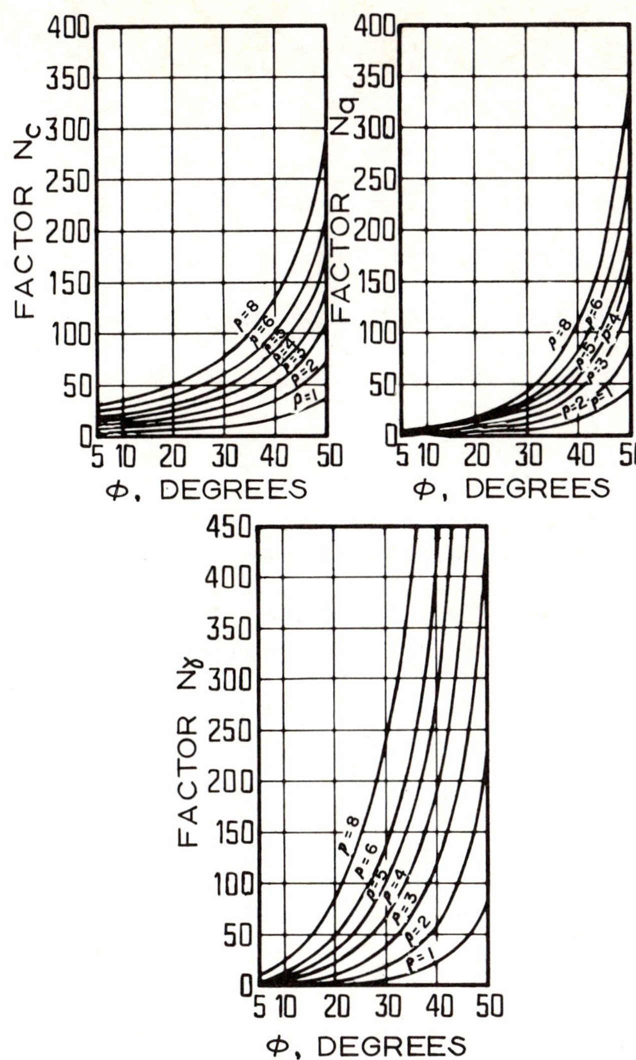

Fig. 8-16 Bearing capacity factors, according to BALLA [19]

shallow and deep foundations: the values are the same for both cases. In the TERZAGHI and BELL theories there is no difference between them in calculation, except in the usual consideration of D_f. HANSEN's Eq.(8-15) does differentiate between calculations by means of a set of factors relating to shape and depth, which correspond to the parentheses appearing in the mathematical expression. Table 8-3 also includes the values given by MEYERHOF for long shallow foundations and deep foundations. Lastly, the values given by BEREZANTZEV for piles are included. The coefficients from SKEMPTON's and BALLA's theories cannot be compared with the previous ones in such a table.

It is hard to say exactly which theory should be used in a specific case. The comparison of values in Table 8-3 shows that differences between them may be very large, so that the choice is not trivial. It has already been said that within its field of application, BALLA's theory has been demonstrated to be in close agreement with field observations. TERZAGHI's theory has often been used to designing shallow foundations, and has many supporters. However, very few designs have been verified with field measurements. For clays, SKEMPTON's theory has been more and more frequently used in shallow foundations. Of course the differences between SKEMPTON's and TERZAGHI's theories are not substantial (especially if the usual high safety factors are adopted). For shallow foundations, MEYERHOF's theory leads to results very similar to TERZAGHI's.

The choice becomes more difficult for deep foundations. When these rest on very hard strata topped by weak formations, it appears logical, for calculation purposes, to use the same theories as for shallow foundations. Note that in Table 8-3 theories developed specifically for deep foundations, like MEYERHOF's and BEREZANTZEV's, lead to bearing capacity values which are very high in the firmest soils. It thus becomes difficult to advise the use of such high bearing capacities except in the hardest clays or the densest sands, with the tip of the pile penetrating at least 10 diameters. The values given by MEYERHOF enjoy general acceptance, but the design engineer should be duly cautious with regard to utilizing high values for the angle ∅. SKEMPTON's formulation has many supporters for calculating piles in clay.

Table 8-4 [19,24] presents a comparison of the bearing capacities of eight shallow foundations, calculated by different methods, with results of experimental measurements in each one of the eight cases. Attention is drawn to the wide discrepancies both between the different theories and between the results of the theories and the experimental values. Note that the two cases in which $\varnothing = 38.5°$ give slightly different values for the bearing capacities computed and measured, which can only be explained if in the two cases foundations of a different width were used. Unfortunately the source of this information does not specify the real reason. Also to be noted is the close agreement of BALLA's results in cases where soil cohesion is small, and how the theory becomes more erratic as the values of this parameter increase. HANSEN's equation gives good results for higher cohesion values.

It should also be noted that TERZAGHI's and MEYERHOF's values are extraordinarily alike and rather conservative, especially for almost cohesionless soils. When cohesion increases, it gives bearing capacities that are very much in agreement with the values observed. If the trends shown by the table are extrapolated (of which practice the authors strongly disapprove), there is an indication that with increasing values for cohesion, TERZAGHI's and MEYERHOF's theories might overestimate bearing capacity. In any case, the confidence placed in TERZAGHI's theory for shallow foundations in sands seems to be justified by the conservative nature of the results obtained.

8.2.6 Bearing Capacity of Shallow Foundations Subjected to Eccentric or Inclined Loads

In the case of eccentric loads, acting at a distance e from the longitudinal axis of the foundation (eccentricity), MEYERHOF [25] recommends that the problems be approached the same as for axial loads, modifying the width of the foundation for calculation purposes:

$$B' = B - 2e \tag{8-19}$$

This is essentially equivalent to regarding the load as being centered on a width that is smaller than the real one, considering that a belt of foundation with a width $2e$ does not contribute to the bearing capacity. This reduced width B' should be used

Table 8-3

Values for bearing capacity factors according to several authors

Factor	Theory	Type of foundation	Angle of internal friction (∅°)									
			0	5	10	15	20	25	30	35	40	45
N_c	TERZAGHI: General shear	Shallow and Deep	5.7	7.3	9.6	12.9	17.7	25.1	37.2	57.8	95.7	172.3
	Local shear		5.7	6.7	8.0	9.7	11.8	14.8	19.0	25.2	34.9	51.2
	MEYERHOF	Continuous footing	6.2	7.1	8.7	10.5	16.0	20.0	36	55	75	110
		Deep (N_c')	9.0	14	23	50	83	150	400	850	—	—
	HANSEN	Shallow and Deep	5.1	6.5	8.3	11	14.8	20.7	30.1	46.1	75.3	139.9
	BEREZANTZEV	Deep	8	9.5	12.5	16	22	34	56	100	190	400
	BELL	Shallow and Deep	4	4.7	5.7	7	8.8	11	14.5	16.5	22	32
N_q	TERZAGHI: General shear	Shallow and Deep	1.0	1.6	2.7	4.4	7.4	12.7	22.5	41.4	81.3	173.3
	Local shear		1.0	1.4	1.9	2.7	3.9	5.6	8.3	12.6	20.5	35.1
	MEYERHOF	Continuous footing	1.0	1.7	2.8	5	7.3	11	19	40	70	115
		Deep (N_c')	1.0	1.9	4	7	11	30	68	120	400	875
	HANSEN	Shallow and Deep	1.0	1.6	2.5	3.9	6.4	10.7	18.4	33.3	64.2	134.9
	BEREZANTZEV	Deep	—	—	—	6	8.5	17	32	70	200	600
	BELL	Shallow and Deep	1	1.5	2	3	4	5.5	9	14	21	32
N_γ	TERZAGHI: General shear	Shallow and Deep	0	0.5	1.2	2.5	5	9.7	19.7	42.4	100.4	297.5
	Local shear		0	0.2	0.5	0.9	1.7	3.2	5.7	10.1	18.8	37.7
	MEYERHOF	Continuous footing	—	—	—	1.1	3	7.5	18	50	100	270
	HANSEN	Shallow and Deep	0	0.1	0.5	1.4	3.5	8.1	18.1	40.7	95.4	240.9
	BEREZANTZEV	Deep	—	—	—	—	—	17	32	80	200	700
	BELL	Shallow and Deep	1	1.5	2.2	3.4	5	8	13	23	42	80

Table 8-4

Comparison of theoretical bearing capacities and experimental results [19,24]

	Practically cohesionless soils				Cohesive soils			
D_f—m (ft)	0.0 (0.0)	0.5 (1.64)	0.5 (1.64)	0.5 (1.64)	0.4 (1.3)	0.5 (1.64)	0.0 (0.0)	0.3 (1.0)
Ø —degrees	37.0°	35.5°	38.5°	38.5°	22.0°	25.0°	20.0°	20.0°
c — t/m² (psi)	0.60 (0.85)	0.34 (0.48)	0.74 (1.05)	0.74 (1.05)	1.20 (1.70)	1.40 (1.98)	0.93 (1.32)	0.93 (1.32)
Method for determining bearing capacity	Bearing capacity — t/m² (lb/in²)							
TERZAGHI	7.62 (10.83)	7.80 (11.09)	15.23 (21.65)	18.55 (26.37)	4.47 (6.36)	5.77 (8.20)	2.51 (3.57)	2.90 (4.12)
MEYERHOF	6.68 (9.50)	7.60 (10.80)	15.50 (22.04)	19.00 (27.01)	4.40 (6.26)	5.60 (7.96)	2.30 (3.27)	2.80 (3.98)
B. HANSEN	6.23 (8.86)	8.80 (12.51)	17.53 (24.92)	22.52 (32.02)	3.98 (5.66)	5.74 (8.16)	1.98 (2.81)	2.57 (3.65)
BALLA	10.34 (14.70)	14.11 (20.06)	25.18 (35.80)	32.50 (46.20)	6.74 (9.58)	10.18 (14.47)	2.93 (4.17)	4.40 (6.26)
Exptl: MUHS	10.80 (15.35)	12.00 (17.06)	24.20 (34.40)	33.00 (46.92)	—	—	—	—
Exptl: MILOVIC	—	—	—	—	4.10 (5.83)	5.50 (7.82)	2.20 (3.13)	2.57 (3.65)

instead of B in the usual equations. In the case of a rectangular foundation with width B and length L, if the load is eccentric in relation to the two symmetry axes of the rectangle, two modified dimensions will be obtained according to (Fig. 8-17):

$$L' = L - 2e_y : \quad B' = B - 2e_z \qquad (8\text{-}20)$$

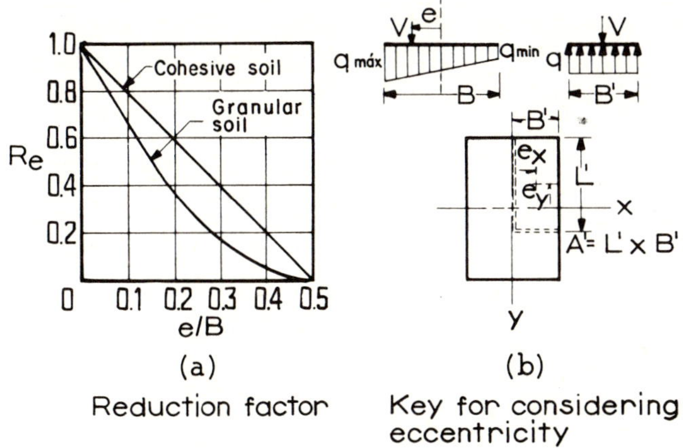

Fig. 8-17 Factor for reducing the bearing capacity due to eccentric loading, according to MEYERHOF [25]

Both modified factors define the reduced area A' which should be used to calculate the total load that can be supported by the foundation. Consequently, in the case of a circular foundation, the total load will be:

$$Q_{Tot} = \frac{\pi D D'}{4} q_c \qquad (8\text{-}21)$$

MEYERHOF [25] also proposes an alternative solution for taking into account the eccentricity of the load acting on a foundation. The alternative corrected bearing capacity is equal to the bearing capacity calculated with a centered load, multiplied by a factor of reduction R_e. The values of R_e can be obtained from Fig. 8-17.

$$q'_c = q_c R_e \qquad (8\text{-}22)$$

The figure was constructed assuming that for an eccentricity relation $e/B = 0.5$, the bearing capacity is zero ($R_e = 0$). Note that the reduction is linear for cohesive soils and is more or less parabolic for frictional soils. If the foundation is square and there is eccentricity in relation to the two symmetry axes, the correction should be applied twice, once in relation to each axis.

For loads which slope in relation to the surface of the foundation, there are also several different schools of thought. MEYERHOF [11] recommends that the bearing capacity factors supplied by his theory be multiplied by the following reduction factors for a load with an inclination of α degrees in relation to a horizontal line through the bottom of the foundation:

$$i_c = i_q = \left(1 - \frac{\alpha}{90°}\right)^2 ; \quad i_\gamma = \left(1 - \frac{\alpha}{\varnothing}\right)^2 \qquad (8\text{-}23)$$

As the inclination of the load increases, the bearing capacity of a square foundation gradually resembles that of a long foundation, until failure occurs by sliding, at which point the two capacities are the same.

HANSEN [13,14] gives the following, different, factors of slope to be used with Eq.(8-15):

$$i_c = i_q - \frac{1 - i_q}{N_q - 1}$$

$$i_q = 1 - \frac{H}{V + cB'L \cot \varnothing}$$

$$i_\gamma = (i_q)^2 \tag{8-24}$$

with the limitation: $H \leq V \tan \delta + cBL$, where H is the horizontal component of the inclined load and V the vertical component. The term $\tan \delta$ is the coefficient of friction between the foundation and the soil.The meaning of the other letters has already been discussed.

An interesting case of an inclined load is presented by MEYERHOF [25] in Fig. 8-18. This is a less favorable case (all the remaining conditions being the same) than that of a foundation with a horizontal base and a vertical load, but it is nontheless preferable to a foundation with a horizontal base and a load inclined at the same angle as the one shown for the entire foundation in the figure. The figure gives a correction factor, R_i, by which the bearing capacity obtained for a foundation with a horizontal base and a vertical load should be multiplied in order to obtain the bearing capacity of the inclined foundation with an inclined load and the same width and minimum depth as shown.

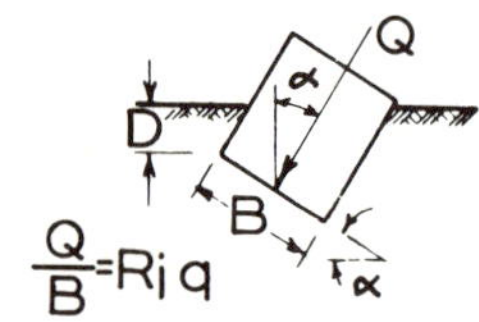

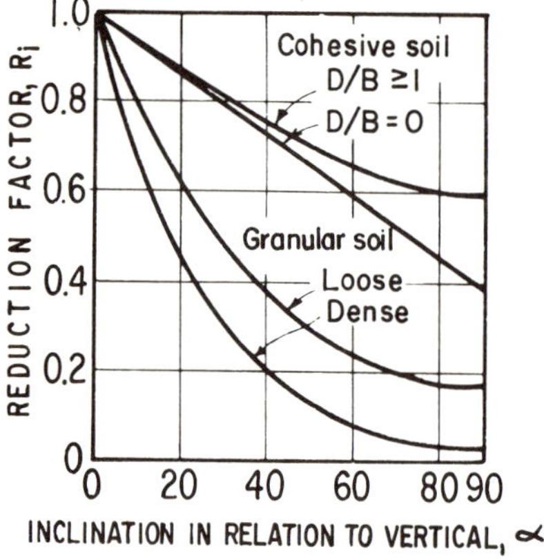

Fig. 8-18 A special case with inclined load [25]

8.2.7 Bearing Capacity in Layered Soils

All the bearing capacity theories mentioned in previous pages apply to homogeneous soils. Stratification poses a problem of lack of homogeneity which has been solved only for a few special cases.

For the case of two layers of purely cohesive clay, some approximate solutions have been proposed by BUTTON [26] and REDDY and SRINIVASAN [21] based on the consideration of cylindrical failure surfaces.

Figure 8-19 illustrates BUTTON's solution for a system of two purely cohesive strata with cohesions c_1 and c_2. The solution covers both the case where the firmer stratum is uppermost, and the opposite case, which is perhaps less frequently met in practice. In the figure it can be seen that the effect of the weak stratum, when underneath, is to reduce the bearing capacity of the firm stratum, and this reduction depends both on the relation between the cohesions in the two soils and on the d/B relation. If, on the other hand, the weak stratum is uppermost, its bearing capacity increases when there is a firm stratum below it. If the lower stratum is far firmer than the upper one, the failure surface will be tangential to its boundary and the strength of the lower stratum will not influence the bearing capacity of the foundation. This becomes apparent if one observes the way in which the curves become horizontal once a certain value is reached for the c_2/c_1 relation, at each ratio of d/B.

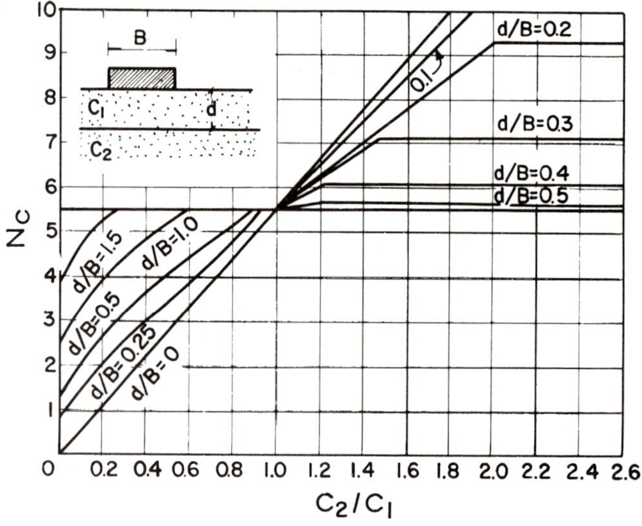

Fig. 8-19 BUTTON's solution for a system of two cohesive strata

Figure 8-20 [19,21] shows the solution to the same problem proposed by REDDY and SRINIVASAN. Here, too, the solutions are approximated as circular slide surfaces. The solution refers to two uniform level purely cohesive strata. The case is presented in which the cohesions of the soils, c_1 and c_2, are constant in the respective strata, but somewhat anisotropic in the upper one. The authors of the paper (which served as a reference) also solved the case in which both values, c_1 and c_2, increase linearly with the depth of these strata. The solutions given here will prove somewhat conservative in this last situation. The expression for the bearing capacity of a shallow foundation now becomes:

$$q_c = \left(1 + 0.2\frac{B}{L}\right)\left(1 + 0.35\frac{D_f}{B}\right) cN_c + \gamma D_f \tag{8-25}$$

In the above, the values of N_c are obtained from Fig. 8-20. N_c turns out to be a function of the parameter K, where $K = q_1/q_2$ and q_1 is the shear strength in the vertical direction in the stratum of clay that is immediately below the foundation and q_3 is the shear strength in the same stratum, but in the horizontal direction.

If the strata are not purely cohesive, no solutions of the above type are available. In this case it is common practice to ignore stratification, computing the bearing capacity of the foundation in a fictitious homogeneous soil obtained by averaging the strength parameters of the strata proportionally. However, this is possible only if the parameters do not differ too much from one another. In [16], SOWERS recommends that if the variations are greater than 20%, safety factors which are higher than usual should be used, which will be discussed later. In [26] it is recommended that this method should not go beyond variations of 50%.

When the situation involves a firm stratum with cohesion and friction, overlying a weak stratum, (where conditions are such that the above proportional average cannot be taken) a system has been adopted which makes use of BOUSSINESQ's theory. According to this, a comparison is made between the bearing capacity of the weak stratum (computed on the assumption that the foundation rests on its upper boundary and the overlying stratum is no more than a surcharge), and the maximum stress that is transmitted to it by the foundation from its actual base, computed according to BOUSSINESQ's theory. In this way, the capacity of the weak stratum can limit the allowable stress for foundation design. Computation of the maximum stress transmitted by the foundation must include the stresses from adjacent foundations that are close enough for their presence to be felt. If the weak stratum is located above a firmer stratum, the wisest solution is to limit the bearing capacity of the foundation to the capacity of the weak stratum. The bearing values are somewhat conservative, because the actual zone in which the failure surface will have an average strength that exceeds the one taken into consideration during computation.

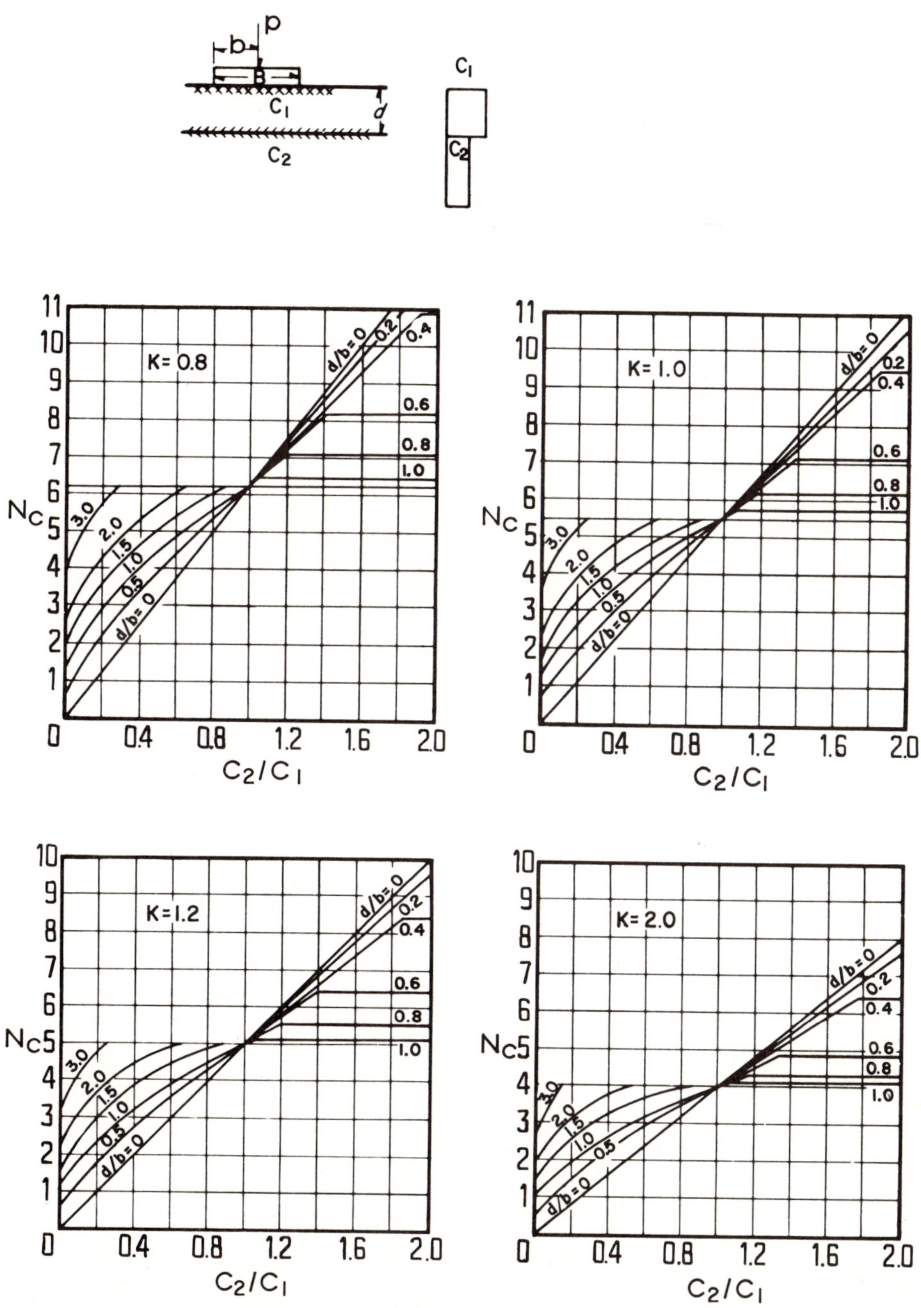

Fig. 8-20 Bearing capacity factors for statified cohesive soils [19,21]

8.2.8 Bearing Capacity of Shallow Foundations on Slopes

For highway design, it is especially important to consider shallow foundations that are placed at the crest of an embankment, or in the slope material. Both are cases which occur frequently in foundations for bridges and fly-overs (over passes). MEYERHOF [27] studied both cases and proposed the following expressions for bearing capacity:

For Cohesive soils:

$$q_c = cN_{cq}\left(1 + 0.2\frac{B}{L}\right)$$

(8-26)

For Frictional soils:

$$q_c = \frac{1}{2}\gamma BN_{\gamma q}\left(1 - 0.4\frac{B}{L}\right)$$

where B is the width of foundation and L is the length of the foundation.

Equations (8-26) correspond to a conception of bearing capacity which MEYERHOF superceded for foundations on conventional terrains (see §8.2.2). Design engineers continue to use Eqs.(8-26) because his more recent expressions do not include foundations on slopes. Figure 8-21 is a graph showing the values of N_{cq} for foundations built in purely cohesive slope material. It is regarded as a strip foundation. For rectangular foundations a shape factor like the one appearing in the first Eq.(8-26) is included. The N_{cq} factor is a function of the stability number of the slope, N_e (see Chapter 6). It also depends on β, which is the angle of inclination of the slope and the D/B relation between the smallest foundation depth and the width of the footing. The figure includes another graph which gives the value of the $N_{\gamma q}$ factor governing the bearing capacity of a strip foundation built in a slope of purely frictional material. Here, too, a shape factor must be used to assess the bearing capacity of a rectangular foundation, second Eq.(8-26). The $N_{\gamma q}$ factor depends on the angle of internal friction of the soil, $\varnothing$, the inclination of the embankment and once again the D/B relation.

It can be observed that in the case of cohesive materials, for $N_e = 5.53$ there is a critical condition in the stability of the slope which is independent of the foundation. Consequently, the bearing capacity of the foundation is zero in this case. Similarly, if $N_e = 0$ and $\beta = 0$, there is a horizontal surface and the N_{cq} factor will be equal to 5.2, which is the value given by PRANDTL for a very long common foundation in cohesive material. It can be observed that for a given value of N_e, the bearing capacity drops with an increase in the steepness of the slope, and for an increase in N_e corresponding to an increase in the height of the embankment, there is a rapid decrease in the bearing capacity. In purely frictional materials, the $N_{\gamma q}$ factor drops with a reduction in $\varnothing$, and also drops when β increases. Even for $D/B = 0$, where the foundation is excavated in an embankment with a critical slope ($\beta = \varnothing$), a bearing capacity is maintained.

Figure 8-22 shows similar graphs for foundations at the crest of the slope, but relatively close to the edge of it. Once again there are two graphs, one giving N_{cq} for strip foundations on embankments of purely cohesive materials, and the other giving the $N_{\gamma q}$ factor for long foundations on embankments of purely frictional soils. In both cases the corresponding shape factors can be used to obtain the bearing capacity for rectangular foundations. In the case of cohesive slopes, the value of N_{cq} depends on the stability number, N_e, of the embankment, the inclination β, and D/B ratio and the distance to the edge of the embankment, b, expressed by the relation b/B or b/H, as shown in the figure. The $N_{\gamma q}$ factor depends on the angle of internal friction of the soil, the inclination of the slope, the D/B relation and the b/B ratio. In both cases a value exists for distance b at which the bearing capacity of the foundation ceases to be influenced by the presence of the slope and the bearing capacity corresponds to a foundation on horizontal ground. This value, which is of great practical importance, lies between 2 and 6 times the width of the foundation, and depends on the D/B relation and the angle of internal friction of the soil.

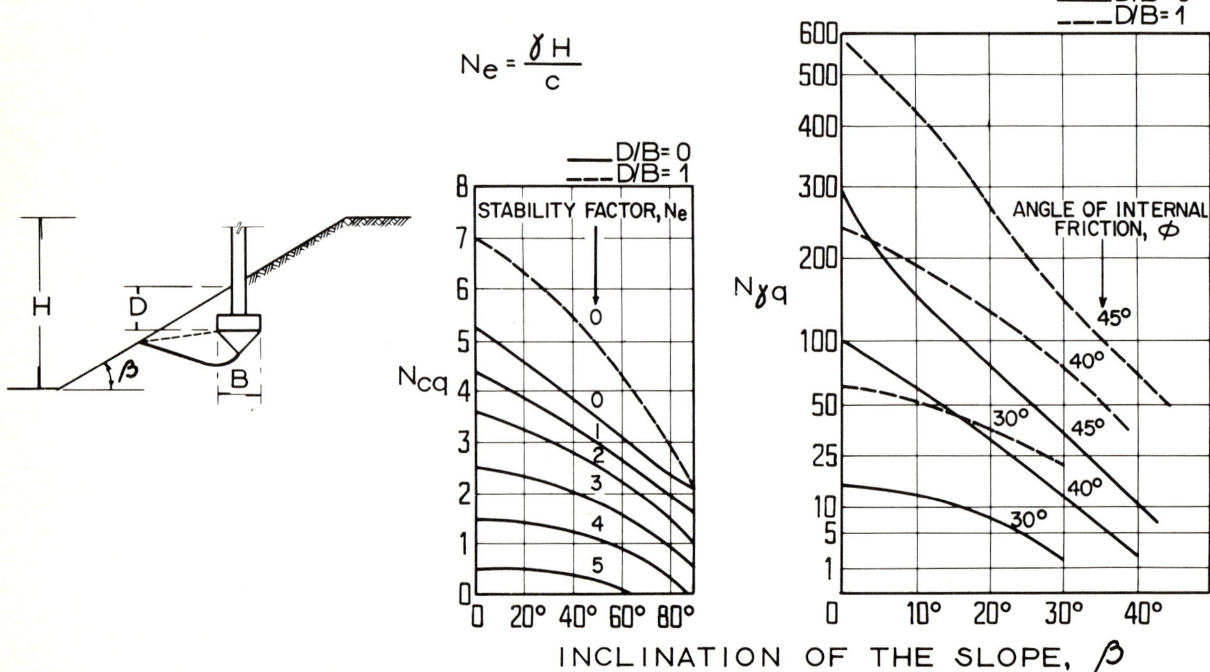

Fig. 8-21 Bearing capacity factors for a foundation on a slope

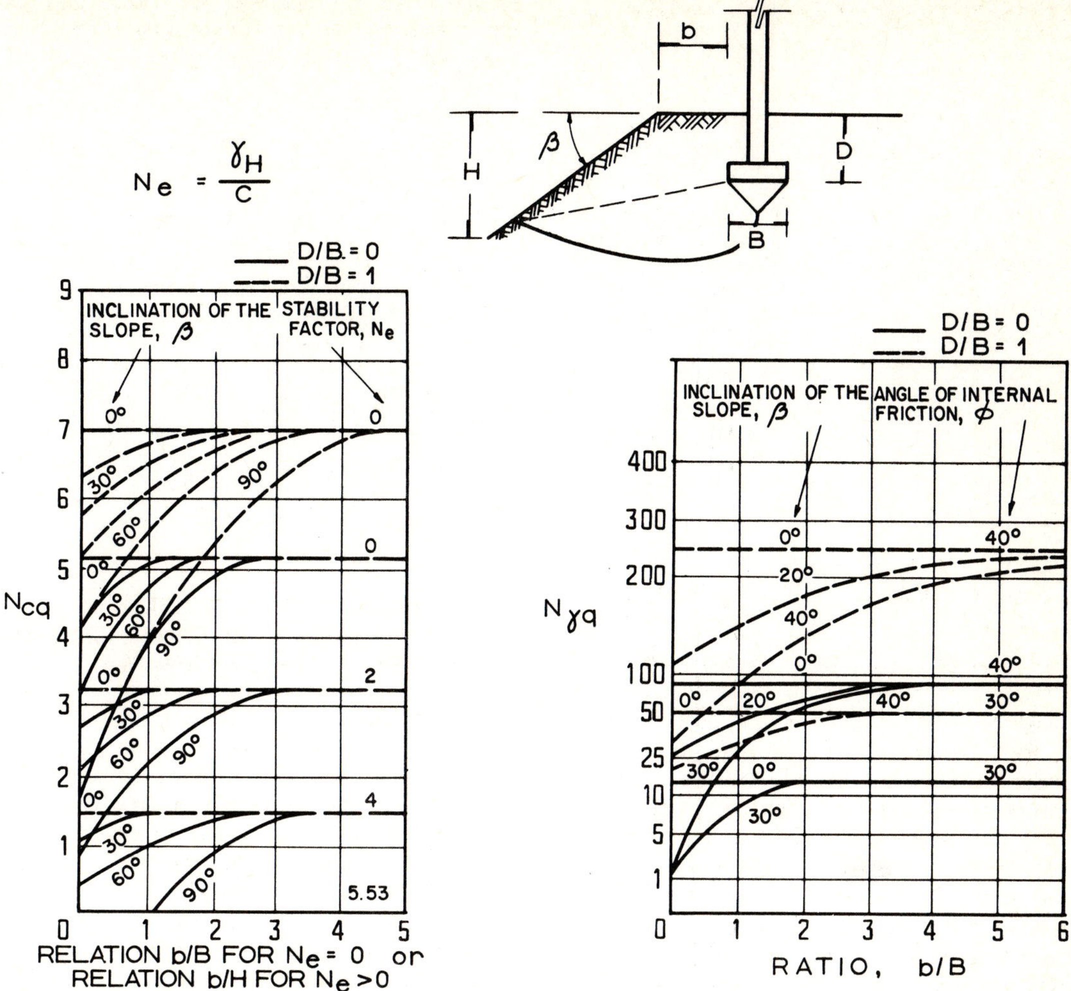

Fig. 8-22 Bearing capacity factors for a foundation at the crest of a slope

When a foundation is built on a slope, whatever its position, the stability of the slope will probably be affected. The slope should always be checked by the usual methods which were described in Chapter 6, taking into consideration the surcharge that is represented by the foundation.

8.2.9 Correction of the Bearing Capacity for the Position of the Water-Table

The unit weight of the soil that is included in the bearing capacity equations should be the one which gives the effective pressure at the level in question. Thus, below the water-table the specific submerged weight γ'_m should be applied, whereas above the water-table the unit weight of the soil mass γ_m should be used. If the water-table is at the level of the excavation for the foundation, or higher, it is very easy to handle the above weights in computating the surcharge, γD_f. If the water-table is below the excavation level, part of the soil within the shear zone will be in a submerged condition and another part will not be. A precise evaluation of each part is impossible.

In order to take this into account, TERZAGHI and PECK [28] proposed an empirical rule which is reasonably precise. The term of the bearing capacity which depends on the N_γ factor is reduced by half if the water-table is located exactly on a level with the bottom of the foundation. The factor 0.5 accounts for the specific submerged weight being approximately half of γ_m. If the water-table is at a depth which is the same as or greater than B (the width of the foundation) below the level of the excavation, the bearing capacity is no longer affected by the water-table since the shear zone does not reach so deep. For this position of the water-table, therefore, no correction will be necessary. For positions of the water-table between *0* and B, below the level of the excavation, linear interpolation is recommended between the correction factor 0.5 and 1.0. Figure 8-23 is a graph showing this correction factor for the term in N_γ, W'. The correction factor W is also shown for positions of the water-table that are above the level of the excavation and which should be used only for computing the surcharge γD_f.

8.2.10 Some Ideas on the Bearing Capacity of Piles and Other Deep Foundations

Several of the bearing capacity theories that have been briefly described in the preceding pages consider piles or deep foundations. TERZAGHI's theory makes no special provision regarding these, and the equation he proposes is applicable to any kind of foundation. The same can be said for SKEMPTON's equations

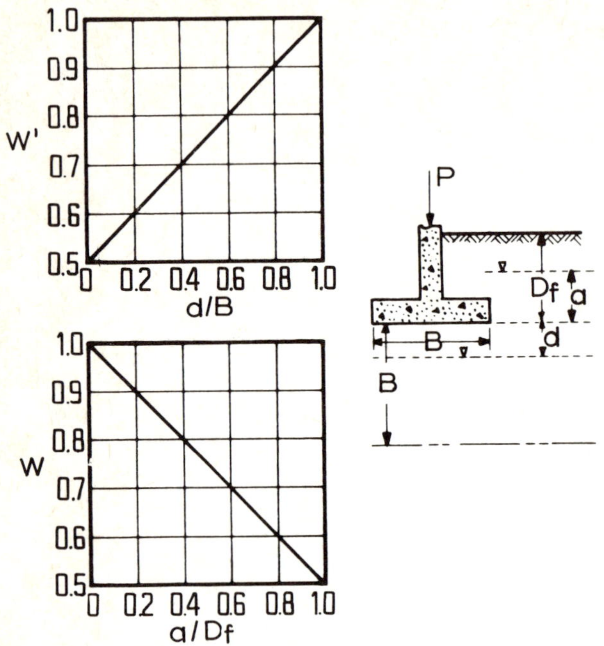

Fig. 8-23 Factors for correcting the bearing capacity due to the position of the water-table [19]

for foundations resting on clayey strata. In his more recent works, MEYERHOF distinguishes shallow foundations from deep ones, proposing Eq.(8-11) which has already been analyzed for the latter. As in the case of HANSEN'S, other bearing capacity theories offer the depth factors that have already been mentioned, for use with deeper foundations. These factors are becoming more frequently used by design engineers, regardless of the bearing capacity formula that is used. Lastly, some of the theories mentioned were intended to apply only to shallow foundations (BALLA) and others only to deep foundations (BEREZANTZEV).

Therefore, the problem of the bearing capacity of deep foundations could be regarded as having been dealt with already. There are, however, certain observations that should be made with specific reference to deep foundations. Dynamic equations [1,29] are used gradually less and less for computing the bearing capacity of piles for highway construction. In the opinion of the authors, this is a fortunate tendency and probably nothing should be done to modify it. The reasons can be seen in [3] or [30]. Another means of predicting the bearing capacity of piles and other types of deep foundations is load tests. These will not be discussed in detail at this point, but will be given some attention later in this chapter.

Whatever the theory adopted, the greatest difficulty to overcome when applying bearing capacity formulas to piles is that it is possible to calculate the bearing capacity of an individual pile, whereas in real practice piles are always placed in groups. Both theory and experience indicate that the behavior of a group of piles is significantly different from that of an isolated pile. There is yet a great deal of research to be carried out on the bearing capacity mechanisms of groups of point-bearing piles and the ways in which they are similar to or differ from those of individual piles. Owing to the high cost of bearing capacity tests on groups of piles, especially under conditions enabling the different design factors to be compared, most of the experimental results that are available are from research carried out on laboratory models. References [31-36] deal with some such investigations and offer some very useful information to the reader who is specifically interested in these matters. Much more unusual are bearing capacity tests on full scale pile groups, [37] and [38] include reports of two of the most complete tests of this type. The theories of pile group action are not fully developed nor adequately verified by full-scale tests. There are, however, certain rules which should be observed for pile group design.

For point-bearing piles resting on firm rock with no underlying compressible stratum, the bearing capacity of the group is the sum of those of the individual piles, so long as they are spaced in such a way that driving one will not interfere with the others that have already been installed. This interference may be where a pile loses its bearing and is raised by soil displacement during driving subsequent piles. It can be the direct impact of one pile against another, due to misalignment during driving. In both cases a minimum spacing of 2 to 3 diameters measured from the center of one pile to the center of its neighbor is a reasonable precaution. Interference problems can be minimized by driving piles in pre-bored holes.

For point-bearing piles resting on firm non-rocky strata, but with no underlying compressible strata, the bearing capacity of the group becomes gradually more similar to the above case the firmer and more rigid the support. In most practical situations arising in bridge design, it can be said that in this case, too, the bearing capacity of the group is very similar to that of an individual pile, multiplied by the number of piles. There are, however, still uncertanties that have not yet been resolved, especially if the supporting stratum is frictional. For example, many engineers and researchers are of the opinion that in this case the bearing capacity of a group of piles is *greater* than the capacity that is deduced from the sum of the capacities of the individual piles [39]. The reason is that the pile group is assumed to work like a pier with the same dimensions. Moreover, the piles densify the soil, increasing $\varnothing$. Many engineers have reached the conclusion that the capacity of a group driven in clay can be somewhat smaller than the one obtained by multiplying that of one pile by the total number of piles in the group, because the clay is not densified by pile driving.

If, under the supporting stratum, there are soft compressible layers, determination of the bearing capacity of the group becomes more complicated. It is limited by the bearing capacity of the soft stratum, and also by settlement problems. Settlement in this case may become far more serious because the influence of the group increases with depth, in comparison with that of an individual pile.

Some comments are necessary on piles without any rigid support at their tip: those which rest on strata that are liable to suffer important settlements. The behavior of these piles is mixed in that they act like point-bearing piles but also provide support by lateral friction. If the piles are very close together, the group acts as a block, whereas if they are very wide apart, each pile fails individually, and consequently they penetrate the soil as separate entities. There is a critical spacing to distinguish between the two types of failure [32,40]. Figure 8-24 shows not only the two types of pile failure, but also failure of the slab if there is one.

In cases of non-rigid support, it is currently common to express the bearing capacity of the group in terms of the capacity of one individual pile, as follows:

$$Q_G = ENQ_i \qquad (8\text{-}27)$$

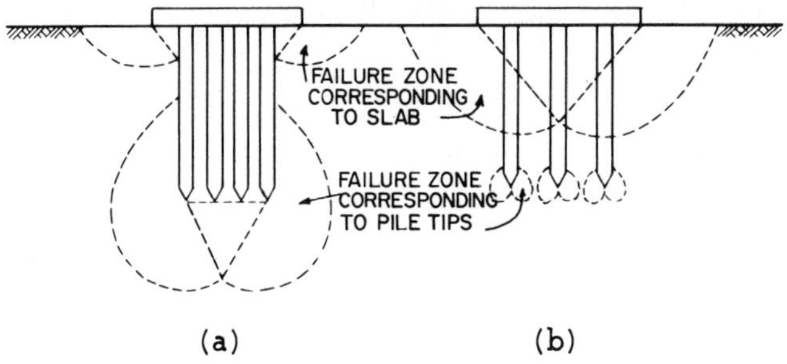

Fig. 8-24 Types of failure in the pile groups [32 and 40]; *a)* Block failure, *b)* Individual failure

where: Q_G is the bearing capacity of the group, in tons; Q_i is the bearing capacity of an individual pile, also in tons; N is the number of piles in the group; E is a group efficiency factor, often expressed as a percentage. Attempts have been made to assess E by means of simple empirical formulas, of which a typical example is the one by CONVERSE-LABARRE. According to this:

$$E = 1 - \varnothing \frac{(n-1)\, m + (m-1)\, n}{90mn} \quad (8\text{-}28)$$

where: m is the number of rows of piles in the group, n is the number of piles in each row, $\varnothing$ is the relation d/s in which d is the diameter of the piles and s the spacing from center to center.

Equations like (8-28) do not, however, take into account the type of soil, the length of the piles, the stratigraphical sequence and other important factors, and should therefore not be fully trusted [28,41].

Model tests conducted by WHITAKER [42] have indicated that for piles driven in clay, with little support at the tip, block failure occurs at very low efficiencies when the piles are closer than 1.5 to 2.5 diameters (center to center). For wider spacings individual pile failures occurred, and the efficiency factor increases to that for spacings of 8 diameters (Fig. 8-25). The

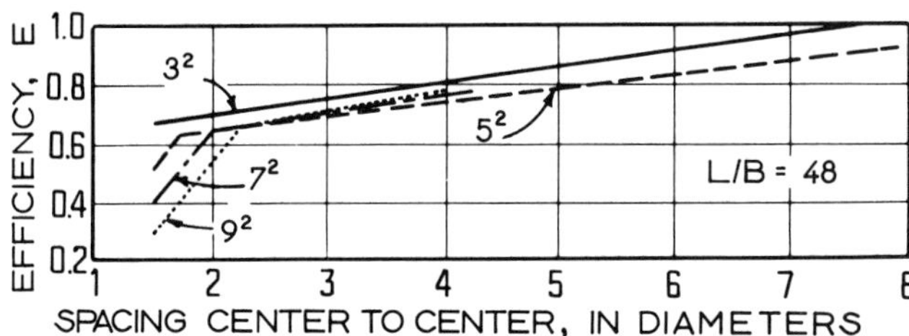

Fig. 8-25 Efficiencies of pile groups in clays, Model tests [40,42]

numerical symbols in the figure indicate square pile groupings. The piles used in the test did not have a footing bearing on the soil. When this was added [43], block failure occurred at wider spacings than those shown in the figure, but the trend towards an increase in efficiency with a corresponding increase in spacing persisted.

The tests reported by KOIZUMO and ITO [38] refer to groups of 9 piles installed in clay. The piles were 5.5 m (18 ft) long, 30 cm (1 ft) in diameter and were placed 3 diameters apart. An efficiency between 0.75 and 0.80 was obtained for the group without taking into account any effect of a footing on the soil surface. For spacings smaller than 3 diameters, block failure occurred and the bearing capacity of the group could be assessed by:

$$Q_G = 2L(A + B)f + 1.3c_u N_c AB \quad (8\text{-}29)$$

where: Q_G is the bearing capacity of the group of piles; L is the distance penetrated by the pile into a consistent clay soil, from which significant frictional support develops; A is the width of the group of piles; B is the length of the group of piles. AB = area covered by the group of piles; c_u is the shear strength of the clay in which the piles are driven, measured in an undrained test; N_c is the bearing capacity factor for a rectangular foundation and f is the value for the adhesion that can be regarded as acting on the walls of the block that fails.

Under these conditions, the first term of the second part of Eq.(8-29) represents the bearing capacity of the group due to friction or side shear which will be discussed later. The second term of the second part of the same equation is the tip resistance.

Groups of piles in sand have shown efficiencies greater than 1 in all the model tests, with one exception [44]: a test performed on very dense sand. The tests carried out by VESIC with piles 10 cm (4 in) in diameter [36] with and without a cap, gave maximum efficiency values for spacings of 3 to 4 diameters, but in all cases efficiency was greater than 1. The results of all the tests indicate an average value of E of 1.7 for groups of piles with a cap and 1.3 for capless piles [40].

COYLE and SULAIMAN [40,45] have summarized all the information available with the design criterion shown in Fig. 8-26, where groups of piles driven in both sands and clays are considered. The nomenclature and criteria expressed in the figure have already been discussed.

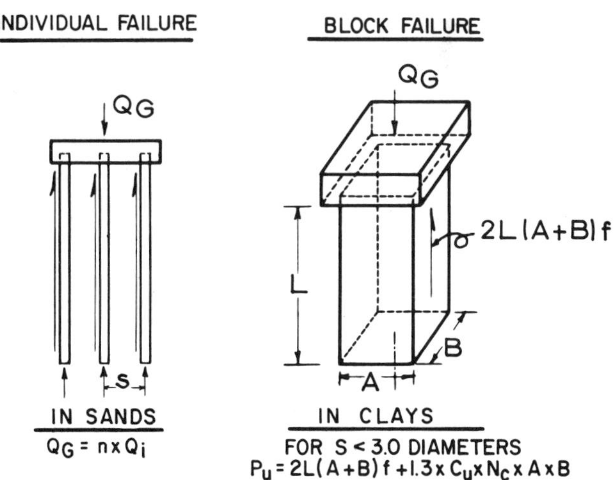

Fig. 8-26 COYLE's and SULAIMAN's criterion for assessing the bearing capacity of pile groups [40,45]

8.3 Bearing Capacity of Friction Piles

As has already been mentioned, not all piles are driven to a firm stratum to which they transmit the loads of the structure from the surface (point-bearing piles). Friction piles are totally embedded in strata that are not strong enough to tolerate the load exerted at the tip. They transmit their load to these strata by the frictional or side wall shear stresses that develop between the soil and the surface of the pile (shaft). The shear which develops between the pile shaft and a clayey soil is commonly termed *adhesion*; the term *friction* describes the same concept in sandy soils. In any pile, friction (shear) stresses develop between its shaft and the surrounding soil. Thus, in all piles some bearing capacity will develop as a result of side friction. If the tip of the pile rests on a soil that is sufficiently firm, there will also be point resistance and the total capacity of the pile will be the sum of tip or end bearing and side friction. If the pile has no point support, the only bearing capacity to be considered will be that resulting from lateral friction.

The following is a brief analysis to assess the bearing capacity of friction piles.

8.3.1 Piles in Clay

When a friction pile is hammer-driven, the clay around it is disturbed and softened during the driving operations. Pore pressures undergo enormous increases, dissipating slowly depending on soil permeability. New equilibrium conditions are established in the surrounding soil, sometimes in a short time in partially saturated soils with little pore pressure, but frequently in weeks or months in saturated clays. The new equilibrium conditions feature a smaller water content in the soil around the pile, greater shear strength and different earth pressure conditions. Figure 8-27 shows the typical conditions which are established for water content and strength. The data refer to research carried out by FLAATE [46] in the 7 upper meters (23 ft) of a mass of soil between two piles, 5.5 years after driving. The figure gives a clear indication of the increase in strength and decrease in water content in the area directly adjacent to the pile, which is offset by an increase in water content and decrease in strength in the soil between the piles.

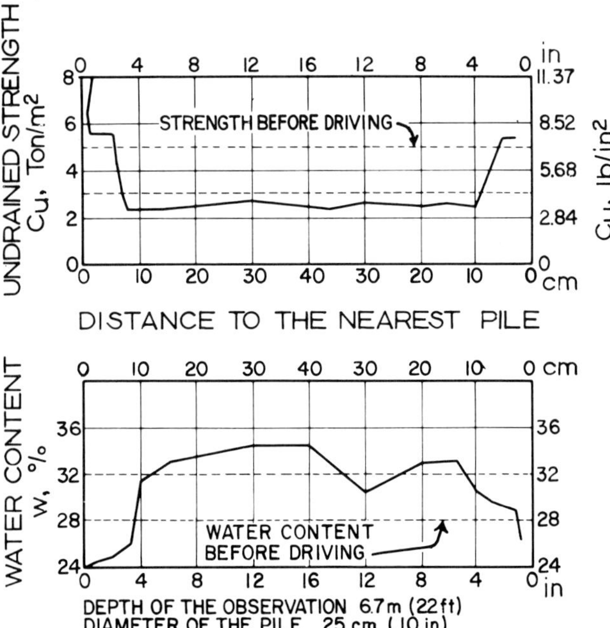

Fig. 8-27 Undrained strength c_u and water content between two hammer-driven piles, 5.5 years after driving [46]

Independently of the theoretical implications of the information given in Fig. 8-27, the most common method for assessing the bearing capacity due to pile friction is based on a total stress analysis. The adhesion, a, between pile and clay is evaluated by means of the expression.

$$a = \alpha C_u \tag{8-30}$$

A great deal of research has been carried out for the assessment of α. Reference [4] summarizes and quotes many of the principal papers on the subject, [40] completes the panorama of what has been discovered up till now. Results reported by PECK [47] and corroborated by other researchers appear to indicate that α is equivalent to 1 for values of c_u smaller than 4.5 t/m² (0.92 kips/ft²). Even though values of α higher than 1 have been reported [48], for practical purposes 1 is usually regarded as an upper limit of the value of the coefficient. Figure 8-28 [49] summarizes almost all of the experience that has been compiled on the subject for establishing the value of α as a function of c_u. The graph includes TOMLINSON's chart [50] which is widely used by design engineers throughout the world. It can be seen here how the value of α decreases as c_u increases. MCCLELLAND et al. [51] point out that almost all the pile tests that have served to establish the value of α have been performed on relatively shallow over-consolidated soils and suggest that higher values of this coefficient should be expected in deep normally consolidated clays, 30m (98 ft) or more. Recent investigation data compiled by VIJAYVERGIYA and FOCHT, however, show the contrary. They found the value of α continued to drop in relation to c_u to depths of about 60 m (197 ft) in normally consolidated clays.

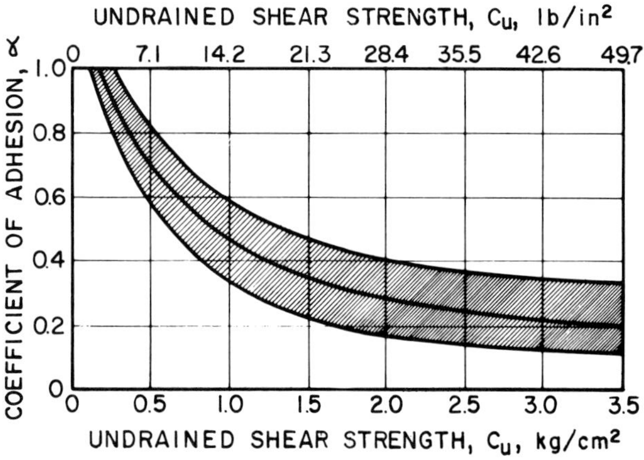

Fig. 8-28 Relation between the adhesion of friction piles and the strength of the soil, c_u [49]

All the above information is applicable to hammer-driven piles in clays, but if the piles are placed in bore-holes, the need is recognized for somewhat different criteria. This illustrates the important influence exerted by the construction procedure on the soils in the neighborhood of the piles. In [52] a case is mentioned where the adhesion of 11 bored piles scarcely reached half of the value it attained in two piles of the same dimensions hammer-driven into the same ground. This difference is attributed to softening of the surface of the soil in

Table 8-5
Values for adhesion between clay and hammer-driven piles, after TOMLINSON [50]

Pile material	Consistency of the clay	Cohesion c_u		Adhesion a	
		t/m²	lb/in²	t/m²	lb/in²
Concrete and wood	soft	0 — 4	0 — 5.7	0 — 3.5	0 — 5.0
	firm	4 — 8	5.7 — 11.4	3.5 — 4.5	5.0 — 6.4
	hard	8 — 15	11.4 — 21.3	4.5 — 7.0	6.4 — 10.0
Steel	soft	0 — 4	0 — 5.7	0 — 3	0 — 4.3
	firm	4 — 8	5.7 — 11.4	3 — 4	4.3 — 5.7
	hard	8 — 15	11.4 — 21.3	?	?

contact with the pile, due to expansion of the soil followed by the migration of water and the opening of fissures. In [53] SKEMPTON analyzed the data on 10 foundations with bored piles and found α (Fig. 8-28) between 0.3 and 0.6, depending on the construction procedures, the typical value being 0.45. In [54] the following equation is given for assessing adhesion in cast-in-place bored piles:

$$\alpha = \alpha_{11}\,\alpha_{12}\,\alpha_{13}\,\psi \qquad (8\text{-}31)$$

in which α_{11}, α_{12}, α_{13} represent the effects of the alteration brought about in the soil during installation of the pile, of the migration of water towards the drill hole and of the shrinkage occurring at the surface of the drill hole respectively. ψ represents the effect of the drill mud or fluid used. For the specific case that was studied, the authors give the following typical values:

$\alpha_{11} = 0.65$ (round pile) $\alpha_{12} = \alpha_{13} = 1 - 0.75/L$, where L is the length of the pile in meters and $\psi = 1.0$ for a dry hole, $\psi = 0.6$ for a mud-drilled hole.

If the friction piles are driven in silty soils or in loess, some further considerations can be made. In the case of plastic silts [55], the pile can be evaluated as for clays, but if there is no cohesion or little cohesion, formulas corresponding to piles driven in frictional soils must be used, paying attention to the aspect of soil liquefaction, especially in hammer-driven piles. For loess, [56] recommends the use of friction piles only if the material is sufficiently firm and if the structure or cementation of the material is such that it does not collapse or crumble. If the soils are crumbly, their use as a supporting stratum is not advisable.

8.3.2 Piles in Sands

The bearing capacity due to friction of piles embedded in sandy layers is traditionally expressed by the product of the lateral area of the pile multiplied by a coefficient of friction, f_r, as follows:

$$f_r = K\gamma_m\, z \tan\delta \qquad (8\text{-}32)$$

where: K is the coefficient of lateral earth pressure; γ_m is the unit weight of the sand, under the appropriate conditions (wet, submerged, etc.); z is the depth at which friction is measured within the sandy layer; and $\tan\delta$ is the coefficient of friction developing between the pile and the sand.

Equation (8-32) shows that the friction varies along the shaft. The normal pressure follows a law that is linear in relation to depth and is a fraction of the vertical pressure from the soils own weight at the level under consideration. It has been proposed that coefficient K be the value K_o, which corresponds to the condition of the soil *at rest* (Chapter 5). The values of K_o most often mentioned ranged between 0.4 for loose sands and 0.6 for denser sands. However, in some dense sands K sometimes exceeds 1 (Fig. 8-29). The value of δ most recommended was 2/3 $\varnothing$. Current research, however, casts more and more doubts on the validity of an equation like (8-32). In [40] several papers are mentioned which show that the average friction increases only to limited depths, deeper it remains practically constant. On the basis of this, [57] proposes that the lateral friction at depth z be expressed as:

$$f_r = \alpha K\gamma_m\, z \tan\varnothing \qquad (8\text{-}33)$$

where α should be taken as equal to 2 for depths of up to 6 m (20 ft) and equal to 0.5 for greater depths. Calculation is also proposed with the value $K = 1$. In [58] VESIC concluded that lateral friction increases to a depth of 10 diameters in very loose sands and 20 in very dense layers, and remains constant at greater depths.

Figure 8-29 [55-59] presents the values of the earth pressure coefficient K with which NORDLUND recommends the use of an equation of the Eq.(8-33) type. He considers that the coefficient of friction between pile and soil can be $\tan\varnothing$ but also $\tan\delta$, with $\delta \neq \varnothing$. A correction factor (α) is given to take this circumstance into account.

In sand, the bearing capacity of the piles due to friction is substantially modified by any jetting operation carried out during pile driving. In [60] the case is mentioned in which the bearing capacity was reduced to 2/3 for this reason. In [40], MCCLELLAND recommends adoption of this ratio for design purposes. It has also been found from tension pile tests that the bearing capacity due to friction is 30% of the total capacity of the same pile working under compression [40].

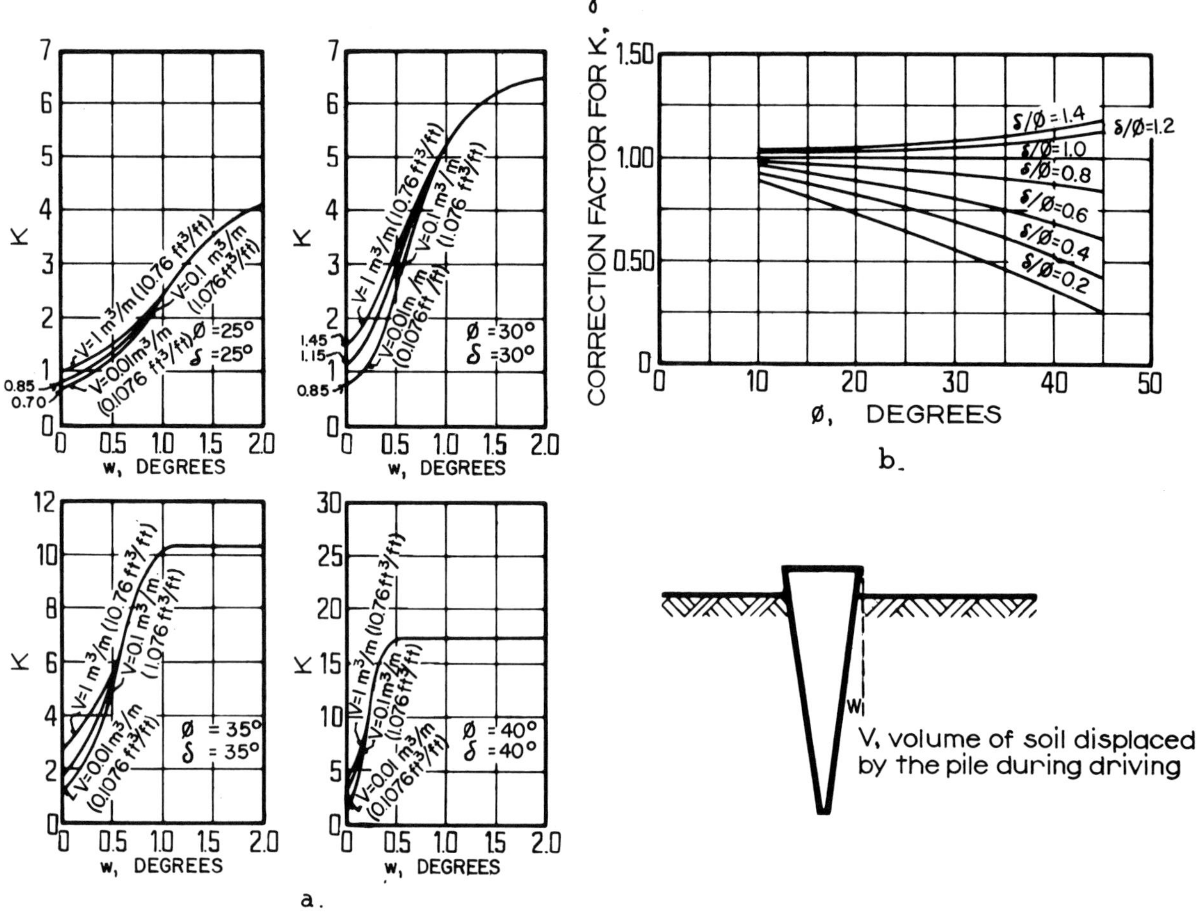

Fig. 8-29 Values of K for computing lateral friction in piles in sands [55,59]

The bearing capacity of piles due to friction in sands has not been clearly determined. All the above rules have been reported as reasonable for calculation purposes. Wide discrepancies exist between the different design methods and it is hard to sustain any one of them on a solid theoretical basis. The distribution of lateral friction along the pile shaft is not linear, so far as it can be determined on the basis of currently available information. Moreover, the coefficient of lateral friction depends on many factors that are hard to quantify, such as the relative density of the sand, the position of the water-table and the disturbances that may be suffered by the soil due to construction operations.

The principles above refer to hammer-driven piles. There is very little information regarding cast-in-place piles. It is, however, felt that there must be a difference between the two, even if this difference may be smaller than in the case of piles in clay. Friction piles in sands offer a good example of the case where load tests are justified. A design based exclusively on calculations must include high safety factors, dividing by 3 or 4 the bearing capacities that are obtained in accordance with the above procedures.

8.3.3 Groups of Piles

For friction piles, the most important group-effect involving bearing capacity is the possibility of a failure in the foundation as a whole (Fig. 8-30). The possibility of this failure is that the

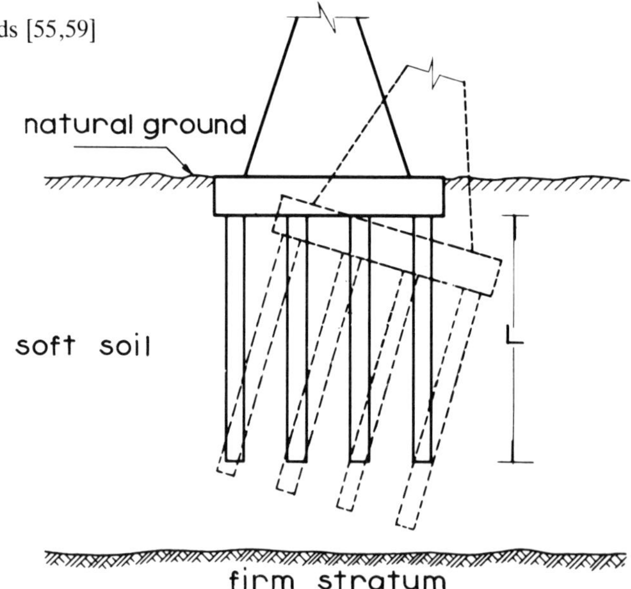

Fig. 8-30 Failure of an entire friction pile foundation

resistance by adhesion and friction in the lateral area of the prism with height L, the pile envelope, is smaller than the sum of the resistance of all the piles considered individually. If n is the number of piles in the foundation, p the perimeter of each one, P the perimeter of the group and a the adhesion between pile and soil, the danger of failure of the group as a whole will no longer be critical if the following inequality is true:

$$anpL \leqslant aPL \qquad (8\text{-}34)$$

i.e. $np \leqslant P$. In other words, the above gives rise to the well-known practical rule that there will be no risk of overall failure so long as the perimeter of the area occupied by the piles is greater than or equal to the sum of the perimeters of the individual piles. Equation (8-34) is conservative considering the practice of some engineers of including the resistance that is offered to failure by the lower base of the prism formed by the group of piles. Further, for groups of friction piles the same comments are applicable which were made about piles which, although they were basically conceived as point-bearing piles, do not receive a firm support at the tip, so that they in fact behave like combined (friction and point-bearing) piles.

8.3.4 Negative Skin Friction

When a foundation is made up of point-bearing piles within a stratigraphy including a compressible layer, if the thickness of this layer is reduced as a result of some induced consolidation process, a very common problem known as negative skin friction is originated. In the field of highway engineering, this problem occurs at the supports of structures that are installed on a firm footing by means of point-bearing piles, where the overlying compressible layer is consolidated under the weight of the access embankment. Since the piles remain fixed, the consolidated soil around them moves down the sides of the shaft, inducing downward friction stresses which can overload the piles. If these overloads are not taken into account in the design, the piles may collapse and penetrate the firm layer.

At the best (that is when the piles tolerate the overload induced by this negative skin friction without failing) the effect is harmful, for an appreciable part of the bearing capacity of the piles is wasted on carrying loads that serve no useful purpose. Furthermore, although the soil around the piles sinks, the piles themselves do not move, thus the structure they support *emerges* or appears to rise up and this usually leads to problems of access.

It is easy to see that in a structure supported by point-bearing piles, where the effects of negative skin friction may be felt, a pile on the inner part of the foundation may be overloaded at the most with a weight equal to the volume of clay that is tributary to that pile. In the case of a pile on the edge of the group, however, the overload may be greater for a larger clay volume is involved. The effect is accentuated in corner piles. In the case of the latter, the only limit to the overload from negative skin friction is imposed by the adhesion between the soil and the pile throughout the length of the pile shaft. If the piles are supported by a stratum that consolidates, the corner pile will settle the most, followed by the edge piles. Those of the inner zones of the foundation area will suffer the smallest settlements.

The overload that can be tolerated by a pile as a result of skin friction is not easy to compute. There has been little research on the subject [4,61-63]. In practice calculation is done by multiplying the lateral area of the pile by the adhesion between pile and soil which, owing to uncertainty, is usually considered equal to the shear strength of the soil (c_u). This is somewhat conservative, especially for the inner piles where the overload due to negative skin friction cannot exceed the weight of the tributary clay. For these piles the negative skin friction is the weight of the tributary volume, which can be evaluated by simple geometrical considerations. In the case of a group of friction piles, the soil on the inside of the group will almost certainly remain firmly attached to the piles and will not participate in any relative downward movement like the soil outside the pile group. Here, the total overload due to negative skin friction throughout the whole group is equal to the weight of the volume of soil located between all the piles in the foundation, plus the effect of the negative skin friction all along the perimeter of that block. The negative skin friction for the entire group can be computed as follows:

$$F_n = LPc_u + \gamma_m LA \qquad (8\text{-}35)$$

where: F_n is the total overload due to negative skin friction acting on the group of piles; L is the pile length that is subject to negative skin friction; P is the perimeter of the pile-enclosed area; c_u is the shear strength of the soil, obtained in an undrained unconsolidated test; γ_m is the unit weight of the soil between the piles, under whatever conditions may prevail; and A is the area of the pile foundation. The first term of the second part of Eq. (8-35) represents the negative skin friction that develops in the walls of the block, whereas the second term represents the weight of the block. The value of the overload due to the negative skin friction suffered by the group, given by Eq. (8-35), should be compared with the sum of the overloads induced by the phenomenon in the individual piles:

$$F_n = nc_u Lp \qquad (8\text{-}36)$$

where p is the perimeter of an individual pile and n the number of piles. The design value chosen will be the highest one produced by Eqs.(8-35) and (8-36).

ZEEVAERT [64] has pointed out a further consequence of the effects of negative skin friction in point-bearing pile foundations, the practical importance of which is perhaps greater than might at first appear. When the soil starts to sink around the pile, clinging to it due to adhesion, part of the soil weight, which was formerly supported by the zone of the pile tip on the firm layer, has been relieved. If the firm layer is frictional, this reduction in effective pressure causes a drop in the shear strength and a loss of the bearing capacity of that firm layer, thus encouraging the pile to penetrate the supporting layer.

8.4 Allowable Bearing Capacity — Safety Factor

All the bearing capacities that have been mentioned thus far correspond to maximum values; in other words, such stresses transmitted by the foundation to the soil correspond to an incipient failure. These are not the values which the designer assigns to real foundations. This is the origin of the concept of the allowable bearing capacity which is the basis for the foundation design. This bearing capacity will always have a lower value than the maximum one, and should be sufficiently far removed from it to provide the necessary margins of safety to cover all the uncertainties relating to the soil properties, the magnitudes of the acting loads, the uncertainties in the bearing capacity theory that is used, and construction problems. It has become the custom to express the allowable bearing capacity as a fraction of the maximum or failure bearing capacity. The maximum capacity is divided by a number greater than 1, which is referred to as the factor of safety (F_s). However, in cohesive soils, the above criterion is hard to defend, both from the conceptual standpoint and from the point of view of the numerical value of the bearing capacity that is thus obtained. This concept is discussed in [26].

In general terms, the ultimate bearing capacity of a foundation in a purely cohesive soil is given by Eq. (8-14) and the maximum safety condition is: $q_c = \gamma D_f$ for that is where the total strength of the soil is in reserve. If a safety factor is applied, it should act only on the part q_c that exceeds γD_f, that is above cN_c [26]. In this manner, the following expression is obtained for the allowable bearing capacity:

$$q_{ad} = \frac{cN_c}{F_s} + \gamma D_f \tag{8-37}$$

Of course this F_s value does not cover the unknowns in the load on the pile.

In the case of purely frictional soils, the bearing capacity is far greater than the pressure acting at the excavation level. This is why dividing the maximum bearing capacity by a factor of safety leads to an error which, although conceptually speaking is similar to the one that would be committed in the case of cohesive soils, is however numerically very small. For this reason, the allowable bearing capacity of a frictional soil is usually simply given by:

$$q_{ad} = \frac{q_c}{F_s} \tag{8-38}$$

The values of F_s to be used in a given case should vary slightly in accordance with the importance of the project and the magnitude of the uncertainties involved. Each case should be given individual attention and the appropriate F_s selected. For the sake of simplicity, however, there are certain typical values that have come to be regarded as acceptable. In the case of shallow foundations, if the analysis of the acting loads takes into account only the permanent sustained loads, a minimum F_s of 3 is recommended.

If sustained loads and eventual live loads are considered, the above value can be reduced to 2 or 2.5. If a very detailed load analysis is conducted, including earthquake effects, in regions that are prone to these phenomena, the safety factor may descend to values as low as 1.5.

For deep foundations, the uncertainties that are involved are usually greater, owing to the heterogeneity of subsoil and construction methods. A safety factor of 3 is customary when the foundation is designed for dead load and permanent live load (which is very common for highway structures). More detailed load analyses might make somewhat lower safety factors permissible.

8.5 Bearing Capacity in Rocks

As has already been frequently repeated, rocks and their important and varied problems are beyond the scope and intentions of this book. However, foundations on rock are so common in highways and receive so little attention, even from the most meticulous design engineers, that a few comments appear necessary. They are not made with the intention of exhausting the subject, but rather of pointing out that rocks tend to pose problems which are sometimes very critical, and that it is advisable for engineers confronted with these problems to probe deeper by studying the ever-increasing number of papers specializing in rock mechanics.

Plate 8-1 Installation of a support for the *Fernando Espinosa* bridge, near Guadalajara (Mexico)

The resistance of rock materials by definition is usually sufficiently great to inspire confidence in the face of bearing capacity and foundation problems. Plate 8-1 shows a typical foundation in rock. Often the resistance to compression of a sound limestone, sandstone, granite or basalt, is greater than that of a good concrete. It is also true that the rigidity of these materials is such that settlement is not usually a limitation to design. The problems here are due to two factors: 1) defects in the rock, such as cracks and fissures, and 2) the high stresses that must be withstood by the structure for which the foundation is being built, owing to the high contact pressures that are tolerated.

The resistance of a rock is usually obtained from a simple compression test or, as is very frequent in highway design, it is estimated. Given the difficulty involved in carrying out elaborate tests (triaxial, for example), it is common practice to resort to unconfined compression tests, from which a strength parameter is obtained, assuming that the rock is a purely cohesive material ($\varnothing = 0$). The result is an expression of the following type: $c = q_u/2$, where q_u is the resistance to simple compression that has been measured. With this value for the strength, the

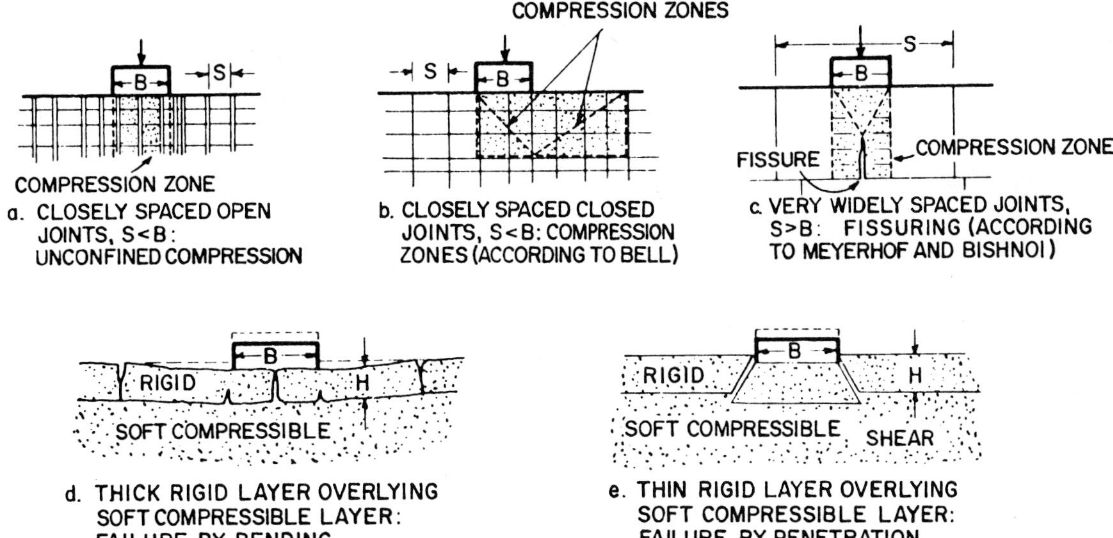

Fig. 8-31 Different ways in which rock can fail due to bearing capacity

bearing capacity of a homogeneous rock can be computed using one of the theories that are applicable to cohesive soils. In this maximum bearing capacity, a minimum safety factor of about 3 is usually included in order to obtain an allowable or working value. If more elaborate laboratory tests have been performed and the strength of a homogeneous rock formation is expressed in terms of two parameters, c and $\varnothing$, a formula of the Eq.(8-2) type can be used to compute the bearing capacity. It has become a custom to use the bearing capacity factors corresponding to local shear failure, so as to take into account the brittle behavior which is characteristic of rock.

If there are joints in the rocky mass, the failure process is different from that of a homogeneous, continuous mass, for it is influenced by the spacing between the joints and how wide open they are, together with the location of the load. Figure 8-31 [16] shows the different modes in which a rocky layer can fail as a result of insufficient bearing capacity. Part *a* of the figure shows the case where the spacing between joints, S, is a fraction of the width of the foundation, B, and the joints are open. Here the foundation is supported by columns of rock, and the bearing capacity can be estimated on the basis of the resistance of the columns to simple compression. The total bearing capacity should always be taken as being smaller than the sum of the resistances of all the columns involved, for not all of them have the same degree of rigidity and some will fail before others reach their maximum capacity. If the joints are closed (Part *b* of the figure) and the pressure is transmitted through them without any movement, the problem can basically be tackled as though the rock mass were sound. Part *c* of the figure shows the case where the spacing between the joints is far greater than the width of the foundation. This situation has been studied by MEYERHOF [65] and BISHNOI [66]. Underneath the foundation a conical failure zone forms. Capacity can be expressed with an equation of the type:

$$q_c = JcN_{cr} \tag{8-39}$$

Parts *d* and *e* refer to formations of stratified rock overlying soft compressible soil. Two types of failure may occur in this case, depending on the values of the relations H/S and S/B and the resistance of the rock to bending. Part *d* of the figure shows the case where the resistance to bending is low and relation H/B is large, whereas Part *e* illustrates the case of a failure by penetration typical of a small H/B.

Figure 8-32 [16,66] gives values of the N_{cr} factor obtained from models. Values are given for circular foundations, values for square foundations are 85% of the ones shown. Also included in the figure is a graph which gives the values of J, as a function of the relation H/B. $J = 1$ if $H/B > 8$ and $J = 0.5$, approximately, for $H/B = 1$. In very cracked rocks, it will be the weak zones which limit the design load, and the safety factor to be used in these cases should not be lower than 5.

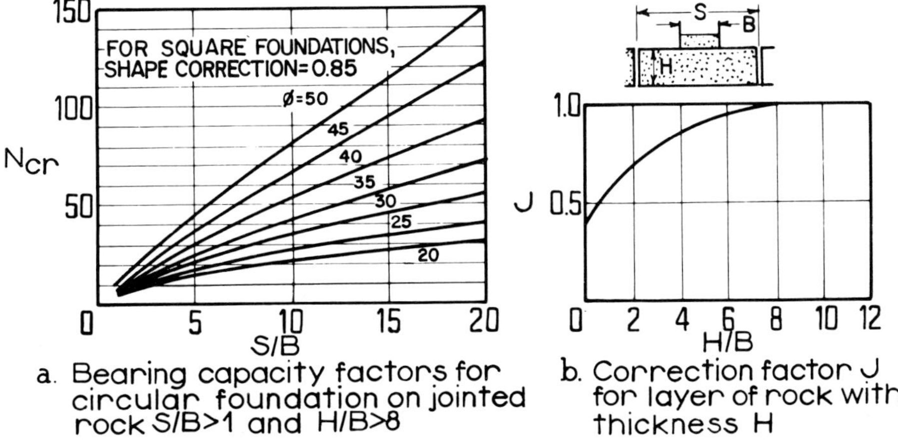

Fig. 8-32 Bearing capacity factors for rock with widely spaced fissures, according to BISHNOI [16,66]

Plate 8-2 Foundations of the Emperador bridge on the direct Mexico-Puebla highway. Caverns in the hillside had to be filled

Plate 8-3 Foundation work on the *Mariano García Sela* bridge. The presence of hollows in a fissured travertine formation made complementary grouting necessary in the supporting area

In certain rocky formations, of which limestones are a typical example, there may be caverns or hollows, and this risk must always be considered during exploration. If cavern ceilings are likely to jeopardize the safety of a foundation, it will be necessary to fill these caverns or extend the foundation down to their floor (see Plates 8-2 and 8-3). If the rock layer is tilted, the foundation will run the risk of sliding, especially if the strata have a dip greater than 30°. In the case of foundations covering a wide area, anchorages or steps are used under these circumstances. Foundations on rock slopes may be very delicate if they are jointed, especially if their dip is towards the cut or slope. Here the nature of the material that fill the cracks and joints takes on great importance. In situations like this, anchorages have proved useful.

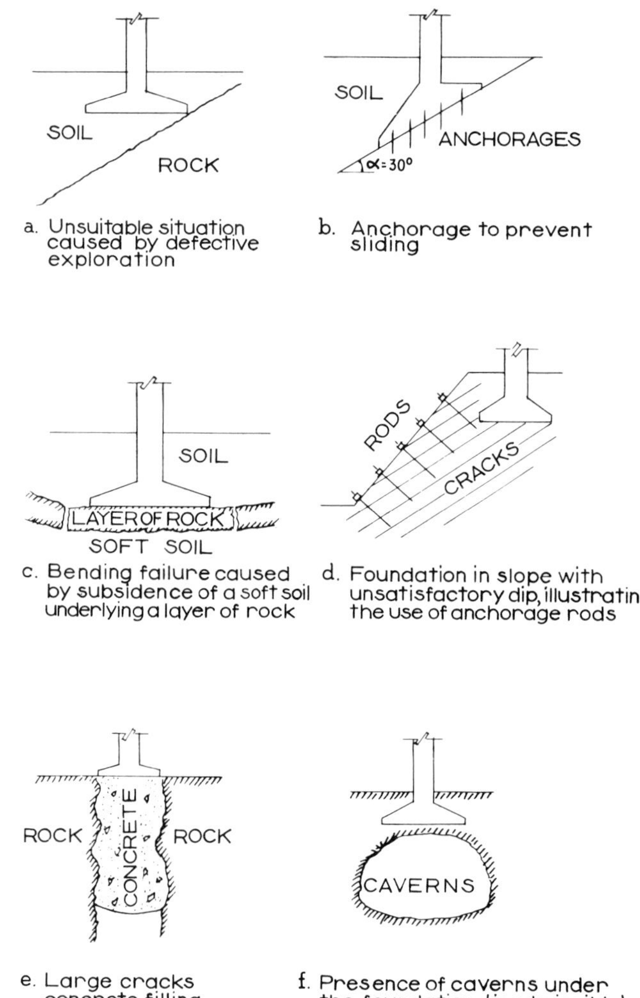

Fig. 8-33 Problems relating to foundations in rock

Figure 8-33 [26] illustrates some of the problems related to foundations on rock, and gives some of the solutions that are typically used to combat them. A case of special interest with regard to construction on rocks is the one represented by Karstic zones, of which a large part of the Yucatan Peninsular in Mexico is an example. The foundation conditions in this zone have recently been the object of a detailed investigation, which can be found in [67].

8.6 Laterally Loaded Piles

A vertically loaded pile bends like a cantilever beam that is partially embedded in a wall. If the loads are small, the behavior of the soil is fairly elastic. This can be represented by a model consisting of a series of horizontal springs, the rigidity of which can be expressed as a reaction modulus (relation between the applied normal stress and settlement under this stress. The result is expressed in force units for unit of area per unit length of displacement: F/L^3). An early attempt at evaluation of the horizontal reaction moduli in a deep foundation was made by TERZAGHI [78], who gave the following expressions:

Frictional soils:

$$K_h = 0.2768\,K_2\,\frac{z}{B};\ \left(144\,K_2\,\frac{z}{B}\right) \tag{8-40}$$

Cohesive soils:

$$K_h = 0.0562\,\frac{K_3}{B};\ \left(96\,\frac{K_3}{B}\right) \tag{8-41}$$

The reaction moduli obtained are in t/m^2.cm (lbs/ft^2.in), B is the diameter or width of the foundation in meters (ft), and z the depth at which the modulus is measured, also in meters (ft). The coefficients K_2 and K_3 necessary for applying Eqs.(8-40) and (8-41) can be obtained from the graph in Fig. 8-34, originally proposed by TERZAGHI, where they appear in lbs/in^3. Equations (8-40) and (8-41) without parenthesis already include these mixed units in order to give the values of K_h in t/m^2.cm. The horizontal reaction moduli, multiplied by the horizontal strain of the deep foundation, assumed to be a rigid body, give the pressure in t/m^2 (lbs/ft^2) that is developed in the element (without causing failure). In order to obtain the total horizontal load that developed, the formulas should be used successively after dividing the foundation (pile or cylinder) into spans by means of imaginary horizontal planes, obtaining the thrust that acts in each span for a given strain. For the case of frictional soils, Eq.(8-40) refers to dry or wet soils, but not submerged ones. The horizontal reaction modulus in this last case is 60% of the one computed.

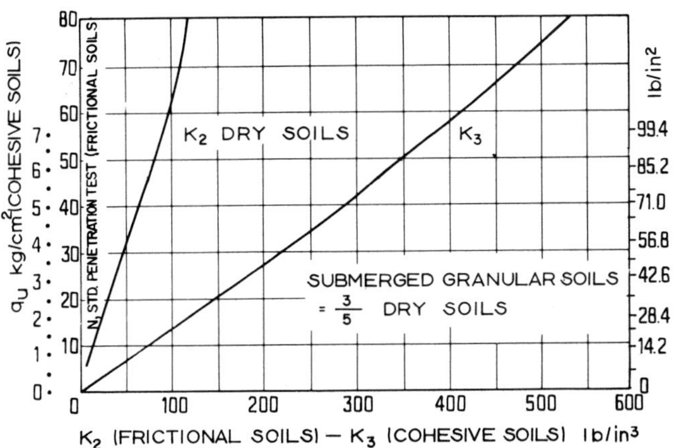

Fig. 8-34 Graph for computing the horizontal reaction modulus, according to TERZAGHI [78]

Reference [79] is a paper which is somewhat more varied and complete than TERZAGHI's in which the conditions of laterally loaded piles are analyzed in three different situations and is illustrated in Fig. 8-35. Part *a* of the figure refers to a pile cap that is allowed to rotate freely as a horizontal force is applied to it. Part *b* of the figure corresponds to a rigid cap, that resists rotation of the upper end of the pile on a level with the natural ground. The lateral load is applied to this upper end and the pile deforms, maintaining a vertical tangent at that point. Part *c* refers to piles fitted with a rigid cap above the level of the ground. The possibility of rotation of the upper end of the pile is now related to a combination of the effects of rigidity of the superstructure and resistance of the ground. Reference [79] also gives a calculation sequence for each of the three loading conditions, which makes it possible to obtain a value of the total allowable lateral load, P_t for a desired displacement. With this value of P_t, pile design should follow the conventional structural procedures which are beyond the scope of this book.

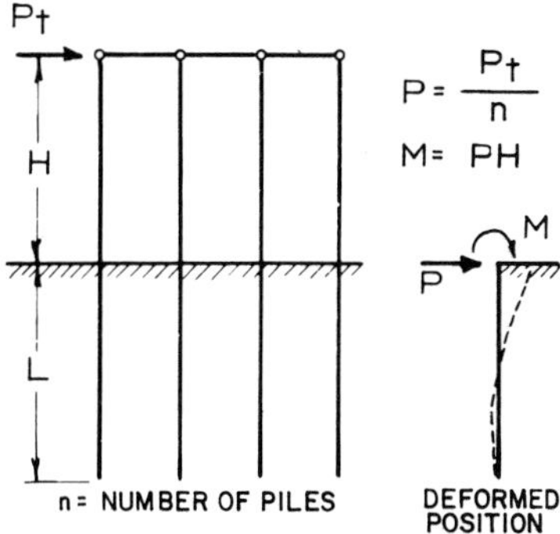

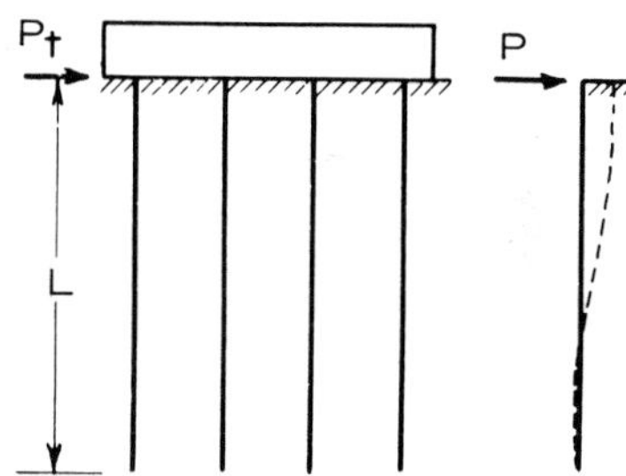

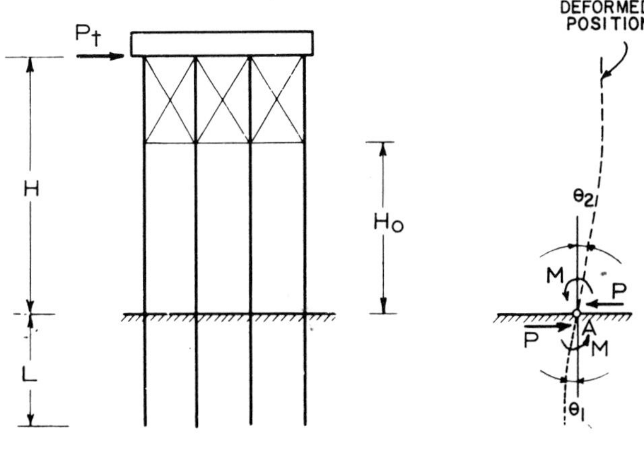

Fig. 8-35 Loading conditions for the study of lateral loads in piles [79]

The solutions proposed in [79] assume that the modulus of elasticity increases linearly with depth, which is usually regarded as a useful working hypothesis in granular soils and soft normally consolidated clays. For hard, strongly preconsolidated clays, it is more common to assume that the coefficient of elasticity is constant with depth. Later it will be seen that the work under consideration provides a means of converting the constant modulus into another equivalent one, which varies linearly with depth. The calculation sequences are as follows:

1) For case *a* in Fig. 8-35. The length of the pile, L, is assumed to have been determined by means of a conventional vertical local analysis. The first step consists of calculating the factor of relative rigidity.

$$T = 0.3\left(\frac{EI}{f}\right)^{1/5}; \quad \left(T = \left(\frac{EI}{f}\right)^{1/5}\right) \tag{8-42}$$

where E is the modulus of elasticity of the material of which the pile is made, in t/m^2 (Kips/ft^2); I is the moment of inertia of the cross-section of the pile in relation to an axis which is perpendicular to the direction of the force, in m^4 (in^4); f is the coefficient of variation of the modulus of the soil with depth, in t/m^3 (Kips/ft^3). Figure 8-36 [80] gives a graph from which the value of the coefficient f can be obtained for sands or for soft clays as a function of the relative density of the sand or the clays unconfined compressive strength.

With the value of T the following relation can be computed.

$$Z_{max} = \frac{I}{T} \tag{8-43}$$

The effect of the load on the upper end of each pile is evaluated at the level of the ground surface where in a similar manner there will be a force $P = \frac{P_t}{n}$ (n is the number of piles in the row) and a moment $M = PH$. The value of Z_{max} will make it possible to select the corresponding curves from among the ones appearing in Fig. 8-37. From the appropriate curves, a deflection factor (F_δ), a moment coefficient (F_M) and a shear stress factor (F_v), are found. These coefficients should be obtained at several different depths z (or several different depth factors).

$$Z = \frac{z}{T} \tag{8-44}$$

The respective mechanical responses (deflection, bending moment and shearing force) may now be obtained for the different depths selected by means of the following equations:

For the moment $M = PH$:

$$\delta_m = F_\delta \frac{MT^2}{EI}; \quad M_M = F_M M; \quad V_M = F_v \frac{M}{T} \tag{8-45a}$$

For the shearing force, P:

$$\delta_P = F_\delta \frac{PT^3}{EI}; \quad M_P = F_M PT; \quad V_P = F_v P \tag{8-45b}$$

Total diagrams for deflection, moment and shear can be obtained through the algebraical superposition of the effects of M and P at each of the depths that are indicated.

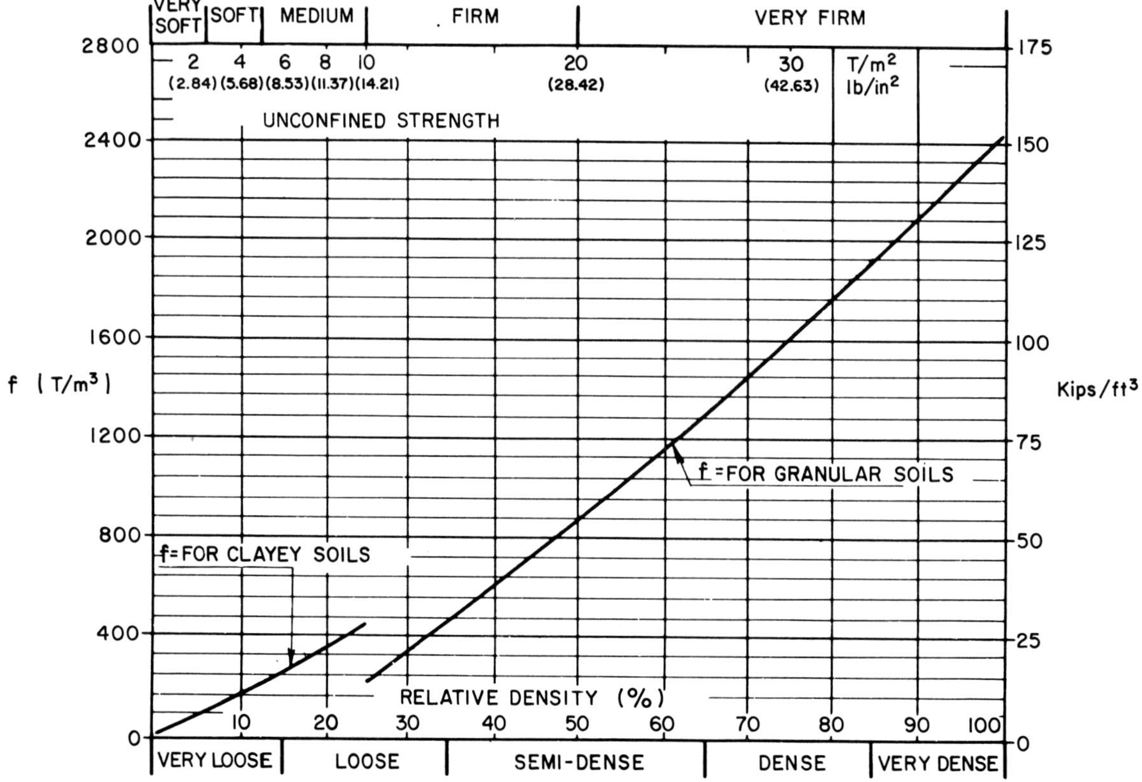

Fig. 8-36 Coefficient of variation of the elastic modulus with depth for analysis of piles subjected to lateral loading

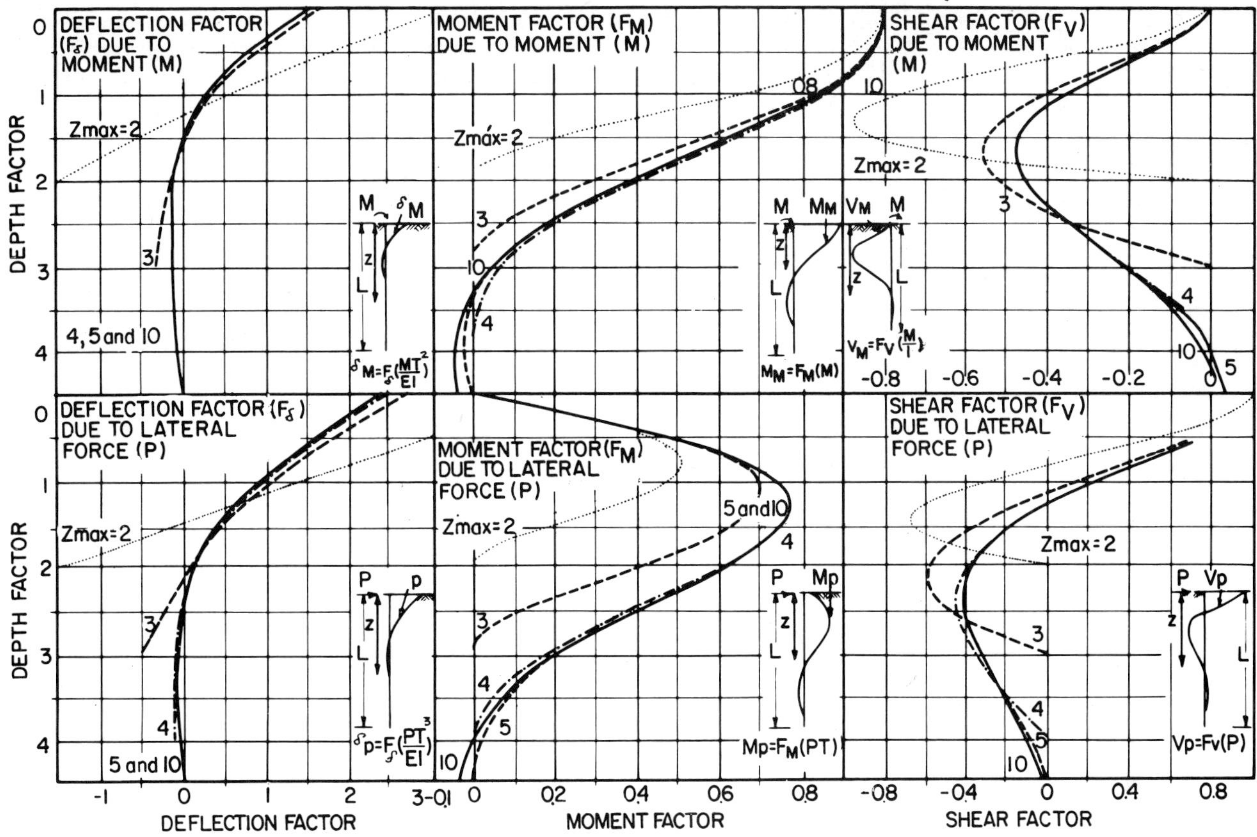

Fig. 8-37 Influence factors for mechanical elements of piles subjected to lateral loading. Loading condition *a* from Fig. 8-35

For the loading condition that is shown in Part *b* of Fig. 8-35 the procedure is the same as in the previous case as far as the determination of the deflection (F_δ) and moment (F_M) factors from Fig. 8-38. The deflection and moment for the depths under consideration can be computed with the following equations:

$$\delta_p = F_\delta \frac{PT^3}{EI}; \quad M_P = F_M PT \tag{8-46}$$

Maximum shear occurs in the upper part of the pile: a shearing force equal to $P = P_T/n$.

For the loading condition shown in Part *c* of Fig. 8-35, a hinge should be assumed at point *A* with an equivalent moment *M* applied at that point. The moment *M* should be computed by equalizing the rotation suffered by the pile and by the corresponding columns of the superstructure. The following apply:

Pile:

$$\theta_1 = F_{\theta P} \frac{PT^2}{EI} + F_{\theta M} \frac{MT}{EI}$$

Column:

$$\theta_2 = \frac{H}{3.5EI} M \tag{8-47}$$

In the above expressions, *P* is the lateral load applied to each pile. The expressions should be solved by regarding *M* as unknown and equating them, which makes calculation of *M* possible. The factors F_{θ_M} and F_{θ_P} should be obtained from the graphs in Fig. 8-39.

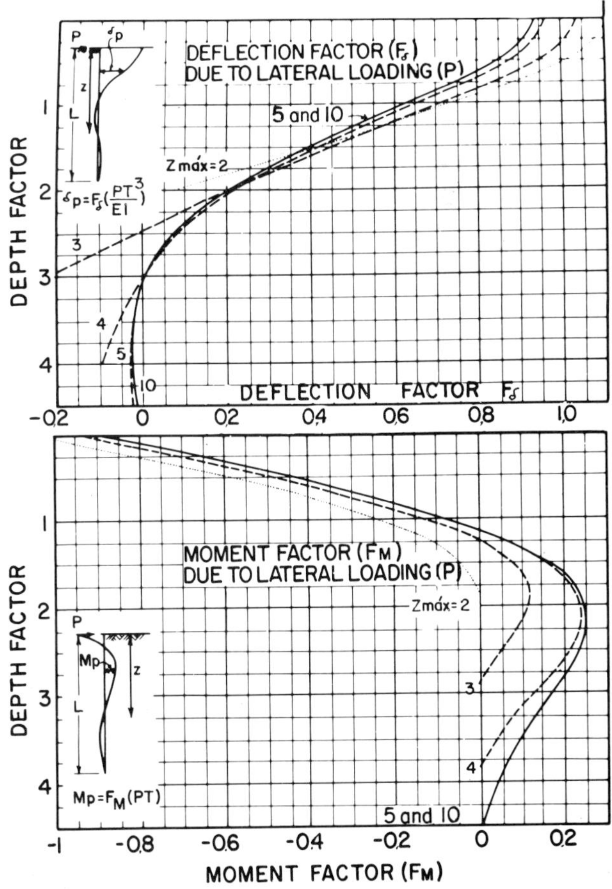

Fig. 8-38 Influence factors for mechanical elements of piles subjected to lateral loading. Loading condition *d* from Fig. 8-35

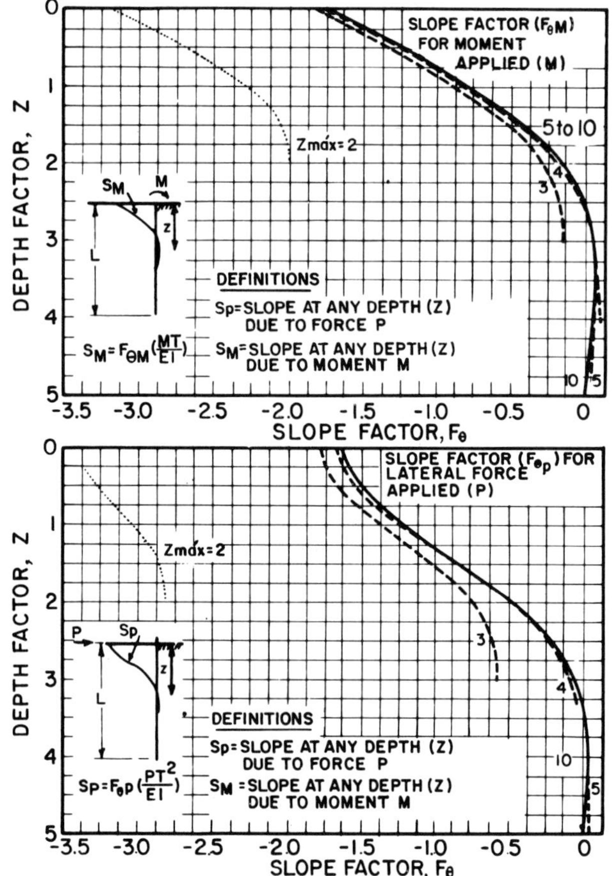

Fig. 8-39 Influence factors for mechanical elements of piles subjected to lateral loading. Loading condition *c* from Fig. 8-35

It has already been said that for the case of preconsolidated clays the constant modulus of elasticity of the soil must be converted into an equivalent modulus which increases linearly with depth.

The procedure for this is as follows:

a) Given the constant value E for the soil under consideration (stiff to hard clay), a value must be assumed for f_s, which is the variation factor of the modulus of reaction with depth, Eq.(8-42).

b) The steps described above should be followed in order to calculate the depth z at which deflection of the pile is zero, where the coefficient of deflection is zero (Figs. 8-37 or 8-38, depending on the loading conditions).

c) At this depth z, the following must be true:

$$f_s z = 2E \tag{8-48}$$

If this is not so, the procedure should be recommenced from *a)*, assuming a different value for f_s.

8.7 Settlements of Foundations

A foundation design is not complete when only the bearing capacity has been established (on the basis of any of the theories that have already been described) for although the strength of the ground is of vital importance, it is only part of the foundation analysis. The deformation that may be suffered by the ground as a result of the loads transmitted to it must be considered, independently of whether or not the loads may trigger sudden failure. Probably far more foundation disasters are the result of slow sustained deformation or minor rapid deformation rather than bearing capacity deficiencies involving violent failure.

The basic theories available for assessing settlement, which is the term traditionally applied to deformations and displacements occurring in the ground, were discussed in Chapter 1, where the compressibility characteristics of soils were analyzed. Chapter 3 included a brief discussion of the problem of the assessment of the stresses that are transmitted to the soil mass by superimposed loads. Although graphs were included which referred to the analysis of stresses produced by embankments, the fundamental ideas are the same for other structures erected on soil. References [68] and [69] give similar graphs for other types of loads which are a reasonably good representation of the ones which are developed by bridges and other highway structures.

Nevertheless, it is advisable to discuss briefly some of the specific criteria for analyzing settlements in foundations. First, shallow foundations will be considered, followed by deep foundations. The analyses applicable to cohesive soils and frictional soils are distinguished in both cases.

8.7.1 Shallow Foundations

.1 Cohesive Soils

Total settlement comprises three different deformation mechanisms. These are settlement due to primary consolidation, settlement due to secondary consolidation and, depending on restriction of lateral strain, instantaneous settlement due to shear strains at constant volume as well as compression of air in the soil voids.

Settlement due to primary consolidation is computed on the basis of TERZAGHI's theory, which was briefly described in Chapter 1 using the one-dimensional consolidation test with vertical water seepage with which the reader is already familiar.

Settlements due to secondary consolidation are far more difficult to assess. This is generally done by observing that there is a linear relation between the amount of any such deformation and the logarithm of time since loading began. A coefficient of secondary consolidation is determined, obtaining it as the increment in vertical strain (or void ratio change) corresponding to one cycle on the logarithmic time scale. On the basis of this coefficient, it is possible to estimate the total secondary settlement corresponding to a given time from the beginning of loading. Difficulties arise from omitting secondary consolidation in routine consolidation tests such as those conducted for highway design. Reference [70] may be of interest to the reader who wishes to obtain further information about the coefficient of secondary consolidation and the factors that influence it. In all matters associated with foundations for highway structures instantaneous settlements at a constant volume are usually disregarded, but [70] provides a reasonable method for assessing them.

When the compressible soil is partially saturated, some change in void ratio occurs rapidly upon loading as the air in the voids compresses. This is determined in the routine consolidation test as the percentage on the settlement from a load increment that occurs between applying the load and making the first settlement reading (usually within a few seconds). For some partially saturated clays from 1/4 to 1/2 the primary consolidation can occur virtually instantaneously.

References [71] and [72] give other methods for assessing the magnitude of settlements in shallow foundations on cohesive soils. These are based on the use of stress paths. The method consists of making the changes in stress which will take place in a prototype foundation act on representative specimens of soils. The vertical strain measured during these tests is assumed to represent in some way the one which will take place in the real foundation. This vertical strain includes the shear strain settlement as well as the change in void ratio accompanying compression of the air in the soil voids.

Skempton and Bjerrum [73] have presented data from the observation of full scale projects which suggest that with Terzaghi's theory it is possible to calculate total primary settlement with an error of 10 to 15% in normally consolidated clays, but this precision is greatly reduced for over-consolidated clays. The real settlements are smaller than the ones predicted: 1/5 to 1/2 the computed amounts. Reference [5] presents a panorama of current ideas on settlement prediction in shallow foundations. From this the conclusion can also be drawn that the customary computation of one-dimensional settlement, based on Terzaghi's theory may lead to excellent results in normally consolidated clays, but to large overestimates of settlement in overconsolidated clays. At various points in this book, emphasis has been laid on how difficult it is to predict the evolution of settlements with time, as well as the chief reasons for the difficulties. In principle, the evolution of total primary settlement with time can be computed on the basis of Terzaghi's theory, using the methods which were described in Chapter 1.

.2 Frictional Soils

In relation to frictional soils, there is no generally accepted theory for settlement assessment. Most theoretical attempts are based on some application of the elastic theory. The book [71] by T. W. Lambe and R. V. Whitman includes a methodology. The authors are sceptical about such a direct application of the elastic theory to foundations and real soils. Moreover these methods are virtually excluded from highway engineering technology owing to the practical difficulties in assessing the stress-strain characteristics of granular soils either in situ or in the laboratory. These difficulties are associated both with sampling and the difficulties in performing the necessary tests. Therefore, such methods are almost always impractical.

Some attempts have also been made to determine settlements empirically on the basis of the results of the standard penetration test, cone penetration tests, or plate load tests. Some of these methods will be mentioned later in this book, when the practical problems related to foundation construction are briefly discussed. Reference [5] specifies some significant papers that have been published which provide a basis for methods of this type. French engineers have developed another method for predicting settlements in footings in sands, using a pressurometer [74] and [75].

8.7.2 Deep Foundations

.1 Cohesive Soils

Settlement of point-bearing piles is commonly of little significance, for the supporting stratum, even when cohesive, must be so rigid and firm that settlement is negligible. More frequently the firm supporting stratum is underlain by a soft compressible layer, which may cause settlements owing to the stresses that are transmitted to it from the level of the pile tips. There is no satisfactory method for evaluating settlements in this case. The one generally applied [1] considers the total load of the structure acting at the level of the pile tips is uniformly distributed throughout the entire loaded area (Fig. 8-40). The difficulty lies in the evaluation of the stresses which reach the compressible layer from the supporting stratum. The most conservative procedure would be to assume a stress distribution according to the Boussinesq theory [69], which was discussed in Chapter 3. The vertical distribution that would be obtained is shown in Fig. 8-40. This is equivalent to disregarding the slab or load spreading effect of the firm stratum. If the mechanical properties of all the strata involved are known, it is possible to obtain a more realistic stress distribution by using Burmister's theory of elastic layers [69].

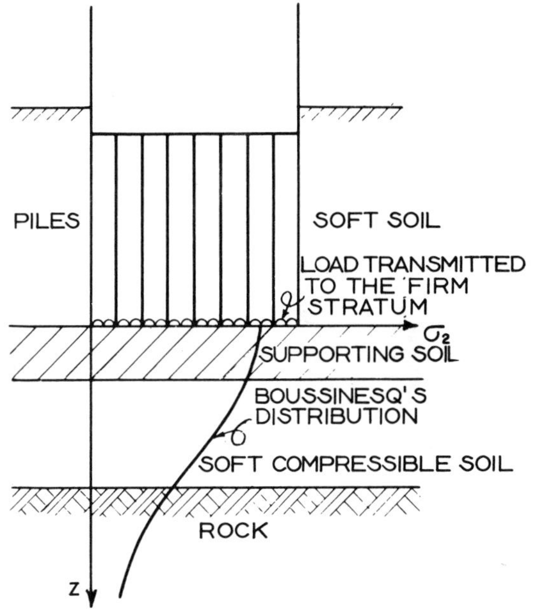

Fig. 8-40 Method for evaluating settlements underneath point-bearing piles

Where friction piles or combined piles, the tip of which rest in clay, are concerned, the problem is more critical because the clay may suffer large settlements. Although it has been seen [40] that a settlement of 2 mm (0.08 in) or a little more at the tip of a pile is sometimes sufficient for lateral friction to develop totally on its shaft, it can be concluded that virtually any pile that is affected by a settlement problem must be a friction pile or, at least a combined friction end-bearing pile. When computing settlements in pile foundations in clays, there is no sense in finding the settlement of a single pile and assuming that this will be representative of group settlement. This is shown in Fig. 8-41. The zone of soil influenced by a group of piles is far wider and deeper than that of a single pile. This makes a computation based on a single pile completely inadequate.

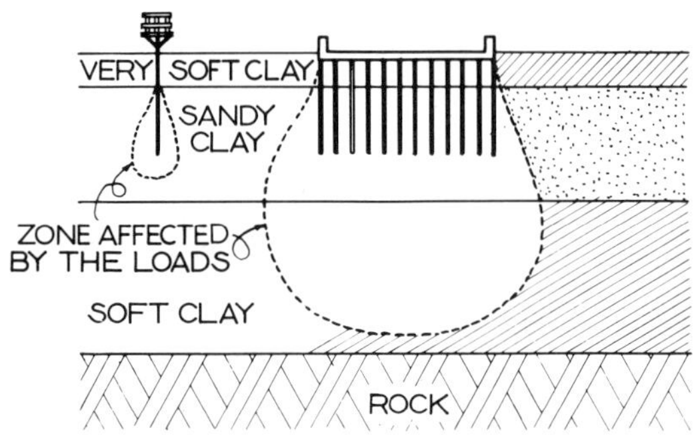

Fig. 8-41 Difference between the influence of a pile and a pile group on settlement

Settlement evaluation in groups of friction piles in clay demands the computation of the distribution of the stresses transmitted to the soil by the piles. This estimate can then be followed by a conventional computation based on TERZAGHI's consolidation theory. There is no rigid rule for finding the stress distribution. It is common practice to assume that the total load of the foundation acts on an imaginary flexible slab, located at some selected level within the length of the piles, and to calculate, by means of BOUSSINESQ's theory, the downward stress distribution that would be produced by that imaginary slab. The success of this assumption will depend on how the level for the imaginary slab is selected, and how the designer succeeds in evaluating the soil-structure interrelation.

The most common calculation technique is shown in Part *a* of Fig. 8-42 [40]. Originally presented by TERZAGHI and PECK [28], this solution consists of assuming the imaginary slab to be located at depth $\frac{2}{3}D$, where D is the total length of the piles within the soil. SOWERS [76] recommends that even with the imaginary flexible slab in this position, the compressible thicknesses to be considered for the calculation of the settlement should be measured from the tip of the piles. Parts *b* and *c* show other hypotheses used for the selection of a position for the imaginary slab. In Part *b*, it is placed at the tip of the piles, this position being regarded as suitable when the piles are embedded in soft clay, but rest on a firmer stratum of clay. In Part *c* the imaginary slab is placed on a level with the ground surface which is regarded as representing better the settlements which occur when the piles are embedded in a layer of sand, underlain by a stratum of soft clay. This is a rare situation in a pile foundation.

In any case, the problem of calculating settlement of pile groups in clays is far from being solved. All the comparisons between predicted and observed settlements reported in literature on the subject are retrospective, and at the best the calculation methods used cannot be extrapolated for different situations.

.2 Frictional Soils

Point-bearing piles resting on sands, should have a sufficiently rigid support to avoid any settlement that might generate lateral skin friction. Under these circumstances, settlements will be negligible. Consequently, they are only studied for friction piles or at least combined friction and end bearing piles.

Again it should be emphasized that the study cannot be conducted on a single pile; for each case the entire group will have to be considered. Group settlements will be larger than those for a single pile, because of the greater loaded areas. There is no satisfactory theoretical method for calculating settlements in pile groups in sands. If the piles are point-bearing, resting on a firm sandy stratum, under which there is a layer of soft clay, the problem can be solved approximately by assuming the imaginary slab already discussed for piles in clay, placed at the tip of the piles. For other stratigraphies, it will not be so easy to find a suitable line of action. Most of the systems for predicting settlements in pile groups in sands are based on the extrapolation of results of load tests, generally carried out on individual piles.

Figure 8-43 presents an empirical relation proposed by SKEMPTON [40] on the basis of field investigations on eight real cases. On the ordinate of the graph a *settlement relation* appears which is obtained by dividing the settlement expected for the group by the settlement measured for a single pile in a load test conducted in-situ or by a representative plate load test. The abscissa represents the total width of the pile foundation. For comparison, a similar curve which was proposed by TERZAGHI and PECK for individual footings in sands is shown. Note should be taken of the far greater settlement which has been measured in groups of piles compared to a single pile. Similar curves have been developed by VESIC [36].

Fortunately, settlement assessment of pile groups in sand is not usually very important, especially for highway design. This is because settlements are generally minor and usually occur abruptly during construction. The possibility of a failure as a result of liquefaction is excepted, but this particular type of settlement belongs rather within the category of catastrophic failures.

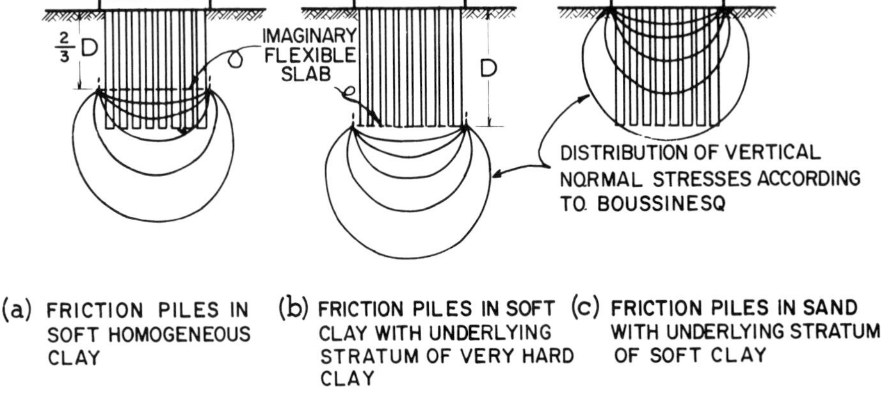

Fig. 8-42 Hypotheses for computing stress distribution under groups of friction piles [40]

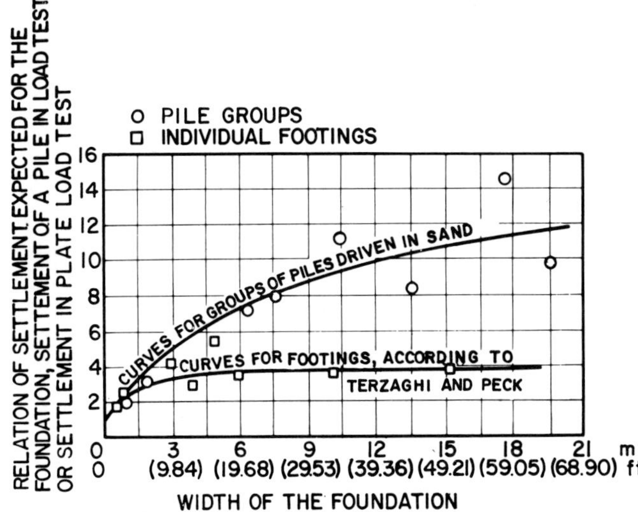

Fig. 8-43 Empirical curve for calculating settlements in groups of piles in sands [40]

8.8 Strength Parameters used for Computing Foundations in Clays

The use of the different shear strength parameters that can be obtained in the laboratory for foundation analysis has been discussed in Chapter 1. The strength parameters obtained in consolidated-drained (slow) tests represent the condition which develops in the soil after all the stress changes that are imposed by the foundation are supported by effective stress. This represents the long term strength. The strength parameters that are obtained in unconsolidated-undrained (quick) tests represent the condition of the undisturbed soil, before the foundation causes any change either in the stress conditions or in the water content of the soil. Since in most cases construction is completed before consolidation is completed, and because consolidation leads to an increase in shear strength, the parameters obtained in a quick test are those used in most designs for foundations in soft clays. Even if construction were so slow that a significant increase in strength were attained as a result of consolidation, the use of the strength parameters obtained from an unconsolidated-undrained test would be conservative; conservatism is essential in designs that are constructed beyond the control of the design engineer.

The strength parameters from a consolidated-undrained test represent the condition of a clay which initially undergoes total consolidation owing to the weight of a structure, and is subsequently subjected to a rapid increase in stress by an additional construction or by a sudden live load. These parameters are frequently used in foundation analysis for embankments, where construction takes longer than the period that is required for the soil to consolidate significantly.

For bridges and other highway structures which must be founded on saturated clays, the usual procedure is to carry out an anlysis in terms of total stresses ($c_u \neq 0$, $\varnothing_u = 0$), with strength parameters obtained in unconsolidated-undrained tests. The unconfined compression test is often used, using a correlation which has been little discussed; the shear strength, c_u is half of the unconfined compressive strength obtained in that test. The use of in-situ vane tests is also becoming increasingly popular in very soft clays. Reference [4] includes a complete analysis of these important matters, on which the success of any foundation study depends so fundamentally.

8.9 Load Tests

What is generally regarded as being one of the best methods for assessing the bearing capacity of a foundation element is to carry out a load test at the exact spot where the foundation is to be built, using a model which is as closely representative as possible of the real foundation. This model is usually a plate of small dimensions (generally between 30 cm and 50 cm (12 and 20 in) wide) in tests for shallow foundations, or a pile identical to the ones which will eventually be used for deep foundations. Reference has already been made to the cost and time required by load tests. This is their chief disadvantage, especially for piles, where these are of such great importance. Few tests on pile groups are described in literature on the subject. The importance of load tests is very different in shallow foundations and in piles. The uncertainties relating to pile behavior are far greater, at least from a practical standpoint, and the cost of such foundations is also high. Both aspects indicate the need for further investigation. In shallow foundations plate load tests are useful, especially in soils like fissured clays where it is practically impossible to shape the specimens, without which the corresponding laboratory research cannot be conducted.

In projects of considerable importance it is unwise to avoid load tests in order to save time or cost. Furthermore, they should be carried out in places where different conditions prevail, for variations in soil conditions are usual in highway routes. The appropriate location for a load test is not always the one where the soil conditions are most critical, because of the magnitude of the loads, the number and shapes of the foundations to be installed, and the consequences of a possible failure in each location.

For shallow foundations, plate load tests can be used for verifying bearing capacity and short term settlement calculations. The bearing capacity measured at the moment the plate fails, is included in a relation of the Eq.(8-14) type, so that by back-calculation a value can be obtained for c. The revised c is then used to evaluate the prototype foundation. This procedure is useful in cases where the value of c cannot be measured in laboratory tests, as occurs with fissured clays or gravelly clays where it is difficult and sometimes impossible to shape specimens for triaxial or unconfined compression tests.

When a plate load test is conducted for settlement analysis, the settlements of the plate are determined. The prototype foundation settlement is found from a correlation formula of the following type:

$$\Delta h = 4\Delta h_P \left(\frac{B}{B + 30} \right)^2 \qquad (8\text{-}49)$$

where: Δh is the settlement of a prototype foundation; Δh_P is the settlement of a 30 cm (12 in) square plate; and B is the width of the prototype foundation. Eq.(8-49) is only applicable for individual footings on homogeneous sand whose settlement is not influenced by any adjacent footing. It does not provide any information on clay consolidation. The inter-

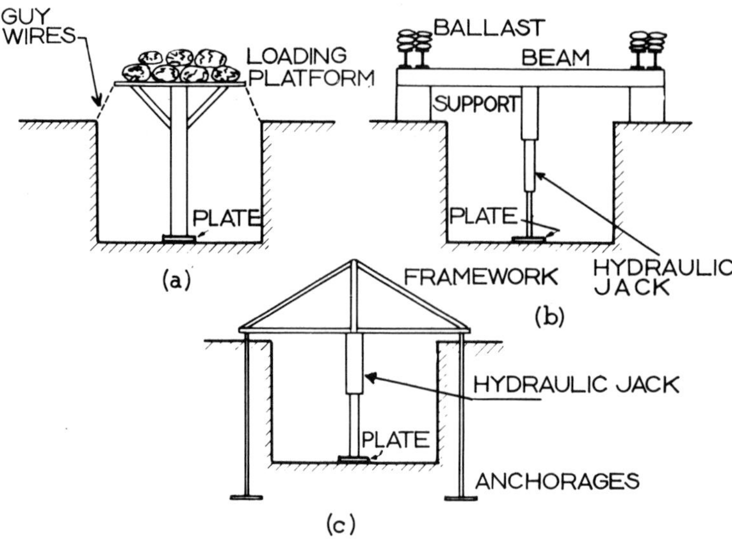

Fig. 8-44 Diagrams showing devices for plate load tests; *a)* — With platform, *b)* — With ballasted beam, *c)* — With anchored structure

pretation of a plate load test, for both settlement and bearing capacity, should take into careful consideration ideas of the type expressed in Fig. 8-41 which (although it illustrates a pile foundation) expresses the relation between the behavior of a small loaded area to that of a large loaded area.

In order to carry out a plate load test, the plate must be centered in the bottom of an excavation with a width of about 4 to 5 times the diameter or width of the plate, and at a depth corresponding to the level at which bearing capacity is to be assessed. For complete tests it will be necessary to carry out the test at several different levels to evaluate the larger potential shear zone of the prototype. The width of the excavation is large enough so that there is no significant surcharge load (γD_f), in the bearing capacity formula that may be used for back calculation. The size or diameter of the plate depends on the spacing between the fissures in the soil, the size of the prototype foundation and the degree of uniformity of the soil's strength. Experience has established the use of plates with widths between 30 and 50 cm (12 and 20 in), although much larger plates would provide more realistic results. The plate must bear uniformly on the soil. Therefore it is placed on a thin layer of fine uniform sand or wet plaster sufficiently thick to fill in the irregularities in the bottom of the excavation. Figure 8-44 shows three different types of loading devices, which go from the simple ballasted platform to an anchored device which generates the load by hydraulic jacks. The use of hydraulic jacks makes it possible to control the speed of the test and the loading process very effectively, but requires the constant presence of an operator.

The test consists of applying load increments to the plate and measuring the settlements it undergoes. These measurements can be taken with a level instrument or better still with a micrometer mounted on an independent structure supported outside the zone that is affected by the test. Load increments of about one tenth of the load estimated for failure or of about one fifth of the working load should be applied. Each increment is maintained constant until the settlement rate of the plate is less than 0.005 cm/h (0.013 in/h) and formation readings should be taken at increasing intervals, such as 1, 2, 5, 10, 30 min., 1 hour, 2 hours. The test is continued preferably until total failure of the soil under the plate occurs or until a load is applied which is at least double the working load. Test results can be presented in two types of graphs, examples of which are given in Fig. 8-45. At the end of each load increment, the settlement-log time curve is drawn and on it the rate of settlement can be measured, Fig. 8-45b. At the end of the test, a graph is plotted which relates the total and ultimate settlement of each load increment with the total load. In this graph the failure load can usually be distinguished by an abrupt change in direction between two straight tangents to the curve, Fig. 8-45a.

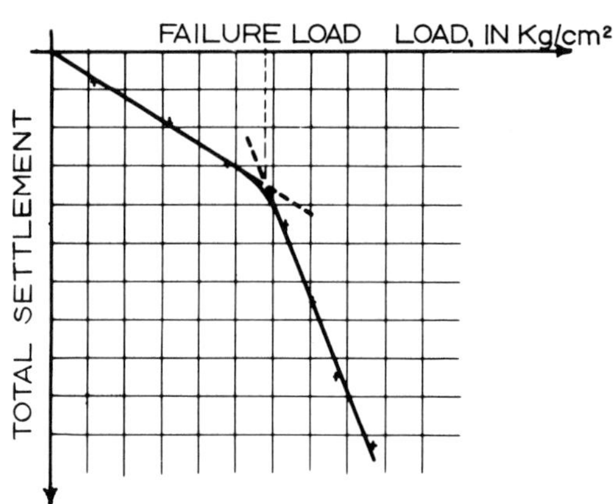

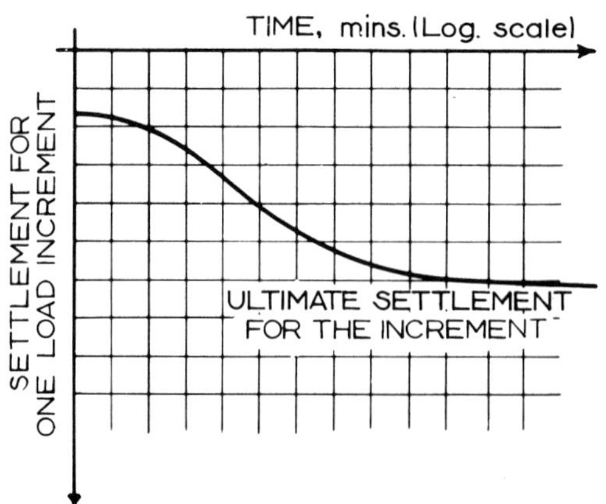

Fig. 8-45 Settlement-time and load-settlement curves in a plate load test

The results of a plate load test are not representative of the long-term settlement or bearing capacity of a foundation. The test is too short for this. Also, the conclusions of the test must be carefully compared with information from a complete exploration; otherwise interpretation is uncertain. The results that can be obtained from a careful plate load test are usually more reliable for bearing capacity than for settlement (where interpretation difficulties become more numerous). The test is not so reliable for design calculations as an analysis based on exploration, sampling and shear strength tests. Its use should, therefore, be restricted to cases like fissured clays and soils that are difficult to sample and test, or to verify uncertain results. It should be used only where previous exploration has provided complete information regarding the stratigraphy.

As has already been said, pile load tests have all the advantages of a full scale direct investigation. The disadvantages are the time and money that must be invested. An important limitation of these tests is the difference between the behavior of a single pile and a group of piles, which has already been discussed both from the bearing capacity and settlement points of view. (Fig. 8-41). A pile load test, see Plate 8-4, provides information on the following:

The ultimate point resistance of a pile: A carefully performed load test gives satisfactory values for the point resistance of piles driven to dense sands or hard clays. In order to determine the value of the point resistance, that part of the bearing capacity of the pile due to lateral or skin friction must either be eliminated or measured. The usual practice is to adopt the first solution, placing the pile within a hollow pipe from which only its tip emerges. Another approach is to mount a load measuring cell on the pile tip.

Plate 8-4 Supporting beam, jacks and gages in a pile load test

The bearing capacity of a pile due to lateral or skin friction: The result of a load test gives the bearing capacity due to friction when the point resistance is negligible. This can be produced using a hydraulically collapsible pile tip. In combined piles, either point-bearing or friction piles, a mechanism can be established using load measuring cells at the pile tips, which enable the point resistance to be evaluated independently of the total bearing capacity [77].

Total settlement of the pile under loading: This information is fairly reliable when the tip of the pile rests on non-compressible soils. However, in piles supported by cohesive soils susceptible to consolidation settlements, or in friction piles embedded in soft clays, the settlements obtained in the test are not representative of the ones that will develop during a long period of loading. The reason is that soil compressibility requires a very long time and the duration of the tests by no means represents the actual lifespan of the prototype piles. Moreover there are enormous differences between the settlement of a single pile and that of a group of piles. The paper by MURAYAMA and SHIBATA [81] maintains that the settlement of a friction pile in clay is proportional to the logarithm of time when the load applied is constant.

Once the pile is in position for testing, the load is applied by any of the following methods. Direct application of the load, by placing weights or ballast on a platform which rests directly on top of the pile cap. Application of pressure from a hydraulic jack, the reaction being absorbed by a weighted platform, the weight of an existing structure, or from a steel beam anchored to the ground by tension piles, or uplift anchors. Application of a load by leverage, using a beam which has tension piles at one end and is loaded at the other end. Figure 8-46 shows some typical arrangements for carrying out pile load tests. The weight or ballast usually consists of rails, ingots, blocks of concrete, tanks of water, or earth. Of the loading methods used, there are operational difficulties posed by the direct weights, especially if (as is usual) the pile behavior during unloading is measured. Unloading can be very tedious with the ballast system, but very rapid if jacks are used.

The sequence for conducting a pile load test is as follows. Load increments are applied to the piles until the maximum value for the test is reached, at least double the anticipated design load (or failure). The corresponding settlements at the cap of the pile are measured at regular time intervals, as for the plate load test. Each load increment must be left for a sufficient time for settlement to practically cease. Settlement at the pile cap is due to elastic deformations (which can be recovered on removal of the load) both in the soil and in the pile itself, and to plastic deformations of the soil (which continue after the load is removed). These deformations are the ones which generally cause excessive settlements in structures, and should therefore be minimized in design. In a pile load test the distinction should be made between the two types of deformation, because the ones which should really be defined in the test are of the plastic type. In order to do this, cyclic loading and unloading are carried out, using ever-increasing loads.

Figure 8-47 is a graph showing typical results of a load test. Part *a* of the figure illustrates the load increment process. It gives details of the duration of load applications and marks the settlements which occurred. Each increment was applied to the pile for 6 hours, which was assumed to be sufficient time for settlements to become negligible. The first unloading was carried out after the load had reached 35 t (77 kips). During unloading of the pile the previous settlement was fully recovered, which shows that it was of an elastic nature. After

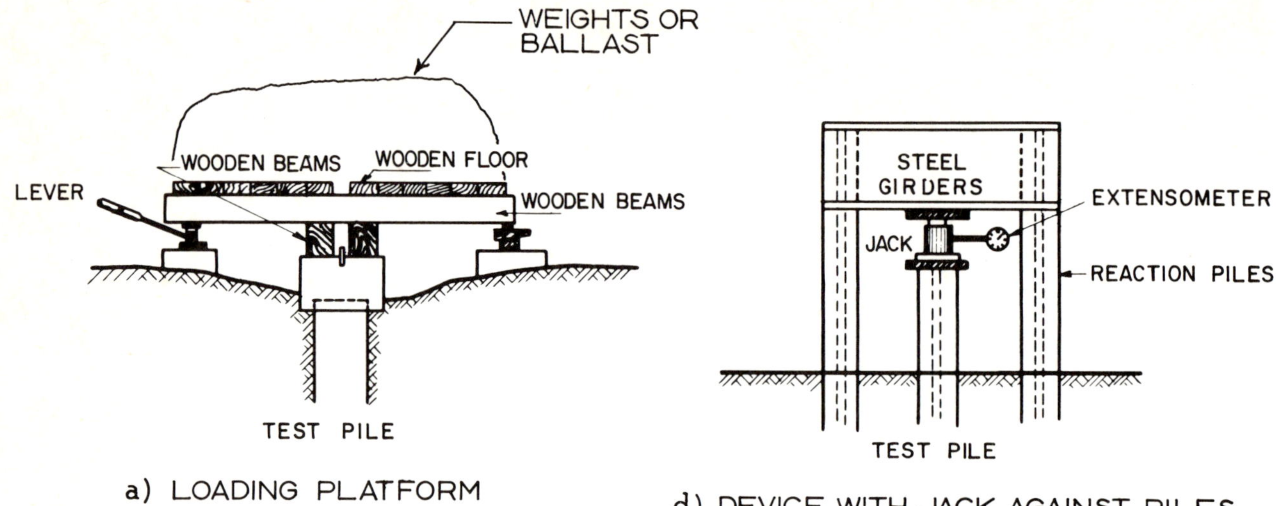

WEIGHTS OR WATER-FILLED TANK
C
GIRDERS
WOOD
C'
C-C'
TEST PILE
TEST PILE
REACTION PILES
TARE OF WATER
GIRDERS
TEST PILE
WATER LEVEL
b) LOADING PLATFORM
e) CANTILEVER WITH TARE OF WATER

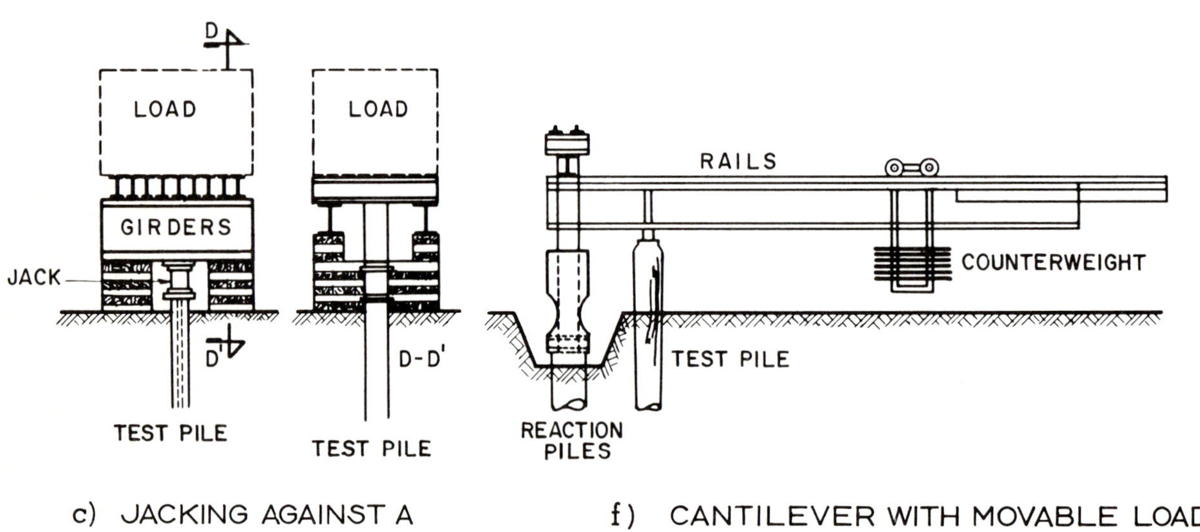

Fig. 8-46 Typical pile load test devices (according to R. D. CHELIS)

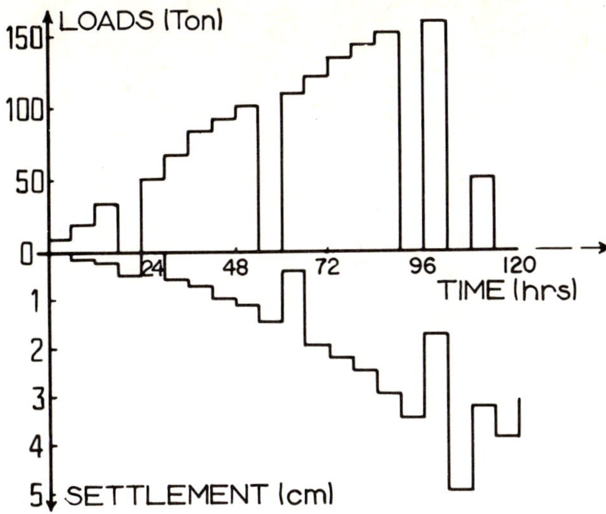

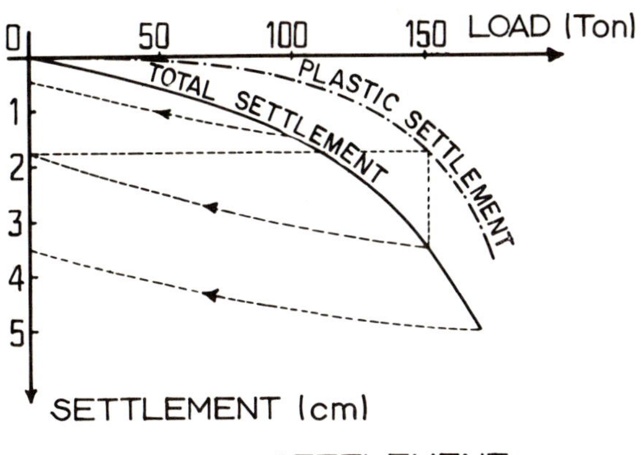

Fig. 8-47 Diagram of typical correlation between load, settlement and time in a pile load test

reloading to 100 t (220 kips), the pile was again unloaded; this time a settlement of 0.4 cm (0.16 in) remained. The third unloading followed a load of 150 t (330 kips) and there remained settlements of 1.75 cm (0.69 in). In Part *b* of the figure, the load-total settlement curve has been plotted with a solid line. The residual, no load or plastic settlements are shown with a broken line. This graph was obtained from the results of Part *a* of the figure, which enable plotting of the unloaded settlements (of which only the first and last points are known). With the remaining deformation under zero loading, as a function of the load before unloading commenced, points are obtained on the plastic settlement curve. Part *b* of the figure illustrates how the point corresponding to a load of 150 t (330 kips) is obtained. The curves for total and plastic settlement versus load depict both the response of the pile and the soil to load. An abrupt change in the slope of the curve will leave no doubt as to the load required for failure. In this case, the working load of the pile is found by choosing a suitable factor of safety by which to divide the failure load. This safety factor is frequently about 2.

In many tests it is not easy to determine the failure load, because the change in the slope of the settlement-load curves is very gradual. In this case, it is essential to define the ultimate load of the pile by means of some suitable, arbitrary, criterion. There are several different criteria: some apply to the total settlement curve; most refer to the settlement curve. Some that are widely used follow:

— Determine the load that produces a plastic settlement not exceeding 0.5 cm (0.20 in) in 48 hours. Divide this by a safety factor of 2, thus obtaining the design load. (Department of Highways of the State of Luisiana, U.S.A., and Department of Public Works of New York State, U.S.A.)

— Continue the test until a load is applied which is double the one which is intended to be supported by the pile. The test will be regarded as satisfactory if this load does not produce a plastic (net) total settlement greater than 0.025 cm for each ton (0.004 in/kip) applied. The plastic settlement is measured 24 hours after removal of the load, (Building Code of New York City, U.S.A.)

— Once the load-plastic settlement curve has been obtained, draw tangents to its initial and final spans. The load corresponding to the intersection of the two lines, divided by a factor of safety of 1.5 or 2 is the design load.

— Obtain the point at which total settlement exceeds 0.125 cm for each additional ton (0.022 in per kip), or where plastic settlement exceeds 0.075 cm for each ton (0.013 in per kip) of load. The load corresponding to either of these points is regarded as the ultimate load of the pile. To obtain the design load, this value is divided by 2 if the loads to which the pile is subjected are static, or by 3 if they are dynamic. (Dr. R L. Nordlung, Consulting Engineer).

8.10 Cylinders and Foundation Caissons

As has already been said, cylinders and caissons are used for deep foundations. The purpose is to find a firm stratum at an economical depth underneath a layer of poor quality soil. Given their complicated nature, size and other characteristics, they are economically efficient only when heavy loads can be transmitted to them, which is why the stratum which they penetrate must be firm enough to give them genuine support. There is no real difference between cylinders and caissons, except in size and shape. In road and railroad bridges, cylinders are far more frequently used, Plate 8-5, caissons are usually an alter-

Plate 8-5 Alvarado Bridge on foundation cylinders

native foundation for the largest structures. The bearing capacity and settlement problems are similar in both cases; the following discussions emphasize cylinders. Caissons will be mentioned only when a distinction must be made between them.

Cases in which these types of foundations are economically useful are as follows:

— When there are heavy concentrations of load on foundations, as occurs in widespan bridges
— In heavily loaded foundations where there is a large permanent depth of water
— When there are very serious problems of control of water or groundwater in any alternative excavation that might have to be made
— When the foundation will be subjected to severe horizontal forces
— When a deep foundation is necessary, but boulders or any other obstacle makes pile driving difficult

In deep foundations, the alternatives to the cylinder are piles and piers (little used for road bridges). Piles are often cheaper and can be built with less expert supervision.

Figure 8-48 [82] shows typical types of foundation cylinders and caissons. As it can be seen, the distinction between piers, piles, cylinders and caissons is not very clear, geometrically speaking. As will be discussed, it is easier to define by construction procedure. Part *a* of the figure shows an open shaft or well foundation, where the excavation is progressively braced. It is really a pier. Parts *c, d* and *e* show procedures for constructing cylinders which are seldom used today. In one case, a steel pipe is driven, into which concrete is subsequently poured. It is gradually withdrawn as it is filled. It can be a large diameter pile because it was installed by driving. The drawing in Part *e* shows a cylinder (or pier, because in this case the distinction is not clear) in which the excavation has been braced by means of pre-driven sheet-piles. Part *b* of the figure shows a cylinder which is typical of highway engineering technology, see Plates 8-5 and 8-6, excavated by a clamshell bucket or alternatively by a dredge. Large diameter augers are also used. Parts *f* and *g* illustrate variations on this element, the cylinder with piles being rarely used, at least in Mexico. Parts *h* and *j* distinguish between the closed or box caisson and the open, bottomless one, which is used for excavation, *k* shows a pneumatic caisson.

The bearing capacity of a cylinder or a caisson should be calculated on the basis of the theories which were discussed at the beginning of this chapter. The nature of the material in the firm stratum is the major consideration for determining which is the most suitable theory. The bearing capacity comes both from the support at the base (bearing in mind that in open cylinders and caissons a concrete seal or plug is usually placed after installation), and from the lateral or skin friction. It is very common to neglect the bearing capacity due to friction, because the end bearing support provided by the cylinders is so rigid that there is no opportunity for the necessary deformations to develop in such a way that the friction can develop. However, test of fullsize cylinders show that skin friction is always significant in firm soils. The strength of the supporting ground will determine whether skin friction should be included in any design situation. In rock supports, friction will play only a minor role unless fresh concrete is poured against the vertical rock surface. If the end of the cylinder settles, even a small amount, friction can be taken into account.

Lateral friction cannot be taken into account in the bearing capacity if: the length of cylinder below the foundation level is smaller than its width; the ground around the cylinder is susceptible to erosion or undermining; a compressible fill is placed between the cylinder and the excavation walls; as the cylinder sinks into the soil arches gradually form at its sides which may afterwards become filled with loose soft materials by erosion and local sliding; the caisson is embedded in artificial fills; or if the soil around the caisson is susceptible to shrinkage as a result of drying.

Plate 8-6 Construction of cylinders

Lateral friction has another effect which cannot be ignored. This is the restriction it offers to driving or sinking when the hand-driven or augered well procedure is followed in which the soil is excavated from within the hollow cylinder allowing it to sink as the material underneath it is removed. The weight of the cylinder will offset the friction forces that may develop in its walls. It is advisable to have a previous idea of any friction forces that may be present, in order to determine the need for using weights and jetting to aid sinking. Table 8-6 [83] shows some typical values for adhesion and lateral friction that have been observed in engineering practice, independently of those which may be computed by the semi-theoretical criteria which were described earlier in this Chapter.

Table 8-6

Typical values for lateral friction and adhesion when driving or sinking cylinders and caissons

Type of soil	Friction or adhesion t/m^2	Friction or adhesion lb/in^2
Silt and soft clay	0.75 — 3.0	1.06 — 4.3
Firm clay	5.0 — 19.5	7.1 — 27.7
Loose sand	1.2 — 3.7	1.7 — 5.3
Dense sand	3.5 — 7.0	5.0 — 10.0
Dense gravel	5.0 — 10.0	7.1 — 14.2

Reference [83] indicates that the friction force measured in several of the cases mentioned increases linearly with depth, which is in agreement with the theory. In the same article the values shown in Table 8-6 are recommended for computing negative skin friction.

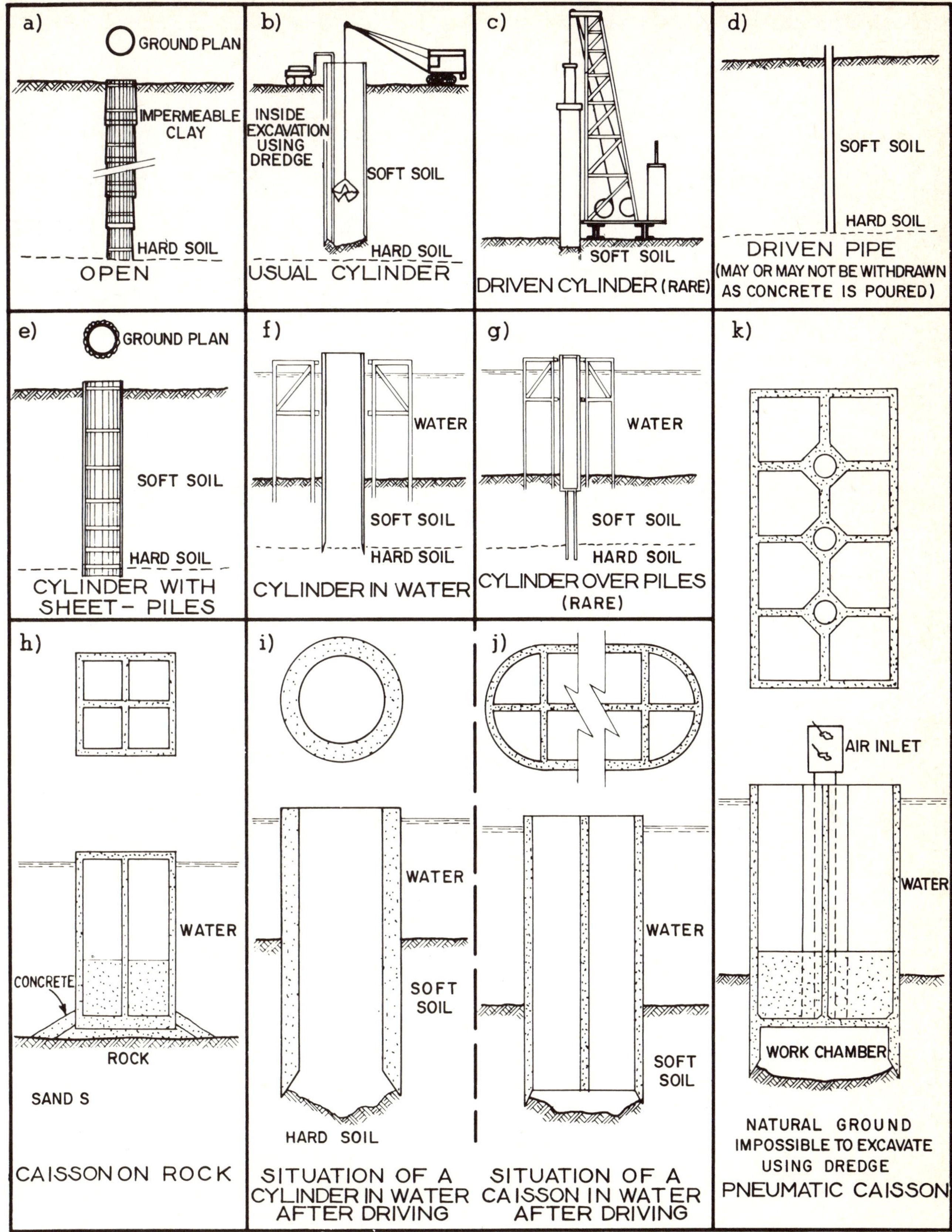

Fig. 8-48 Types of cylinders and caissons [82]

The best measure to overcome lateral friction during sinking or driving is for the cylinder to have thick walls to ensure sufficient weight. Walls with a thickness of 1.5 to 2.0 m (5 to 6.6 ft) of concrete provide sufficient weight to withstand large lateral friction. However, such a simple design may often be uneconomical. Ballast provides an economical solution to many problems. Jetting with water or even drilling-mud has been helpful. Good results have also been obtained with a projecting lateral blade that forms a chamber about 10 cm (4 in) wide around the cylinder. This is filled with a mixture of clay and water with a suitable amount of a dispersing agent, which reduces friction or adhesion.

In Mexico the procedure most often used for sinking large bottomless cylinders and caissons (the ones most widely used) is to remove the material from within the cylinder by a clamshell dredge or similar equipment. It operates inside the cylinder and must be small enough such that there is ample clearance. When the caisson is sunk in water, the inside level is maintained the same as the outside level. Excavation must be carried out in such a way that the cylinder sinks due to its own weight or with the aid of ballast. Sometimes it is necessary to excavate well below the caisson bottom but care is necessary to prevent collapse of the excavation sides. In multiple cell caissons, verticality is frequently maintained by excavating in the different cells alternately. When the cylinder crosses very soft soils, by means of pumping soil can be sucked from the inside. If needed, water is injected to keep the outside and inside water levels the same. Once the cylinder is in its ultimate design position, which has been pre-established by the appro-

Plate 8-7 Excavation inside a cylinder

priate foundation study, a concrete plug is cast at the lower extreme, using an elephant trunk if the bottom is dry and using the *treamic* procedure under water. There are occasions when it is possible to remove any water from inside the caisson so that concrete casting can be done dry, but the dangers this practice involves (such as bottom failure and sand boiling) make it unadvisable. The bottom of the cylinder is usually fitted with metal blade cutting edges to make penetration easier. Figure 8-49 [82] shows some typical cutting edge shapes.

A general diagram of the excavation process by the excavated well method is shown in Fig. 8-50 [84]. In this reference the engineer will find some useful practical material.

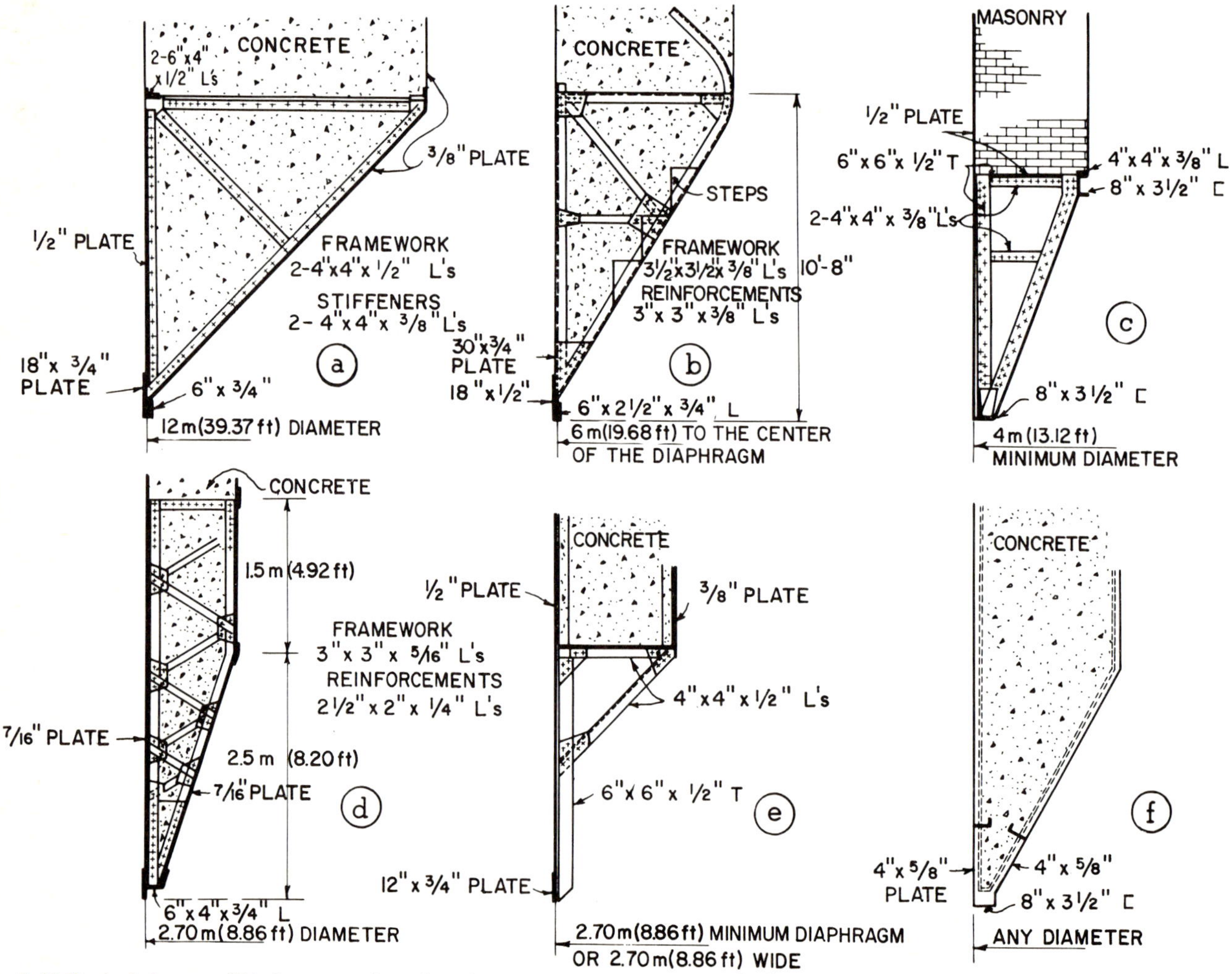

Fig. 8-49 Typical shapes of blades or cutting edges for cylinder driving or sink [82]

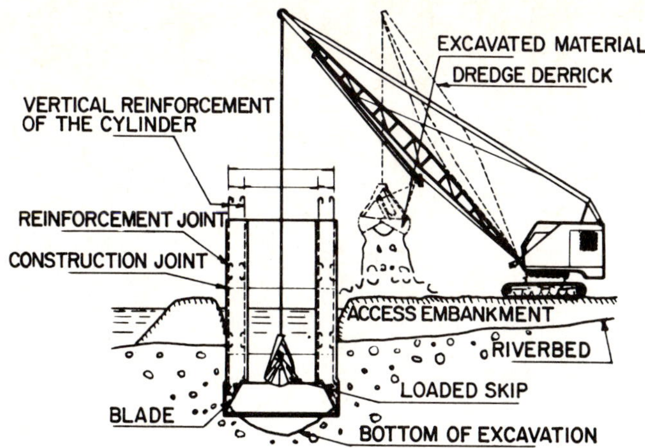

Fig. 8-50 Excavation for foundation cylinders and caissons using the excavated well method [84]

During the first stages of sinking, the cylinder must be guided so that it stays in place. Afterwards additional sections are cast to lengthen the top of the caisson as sinking progresses. Excavation must always be ahead of the penetration of the cutting edges; it should, however, be neither excessively far ahead nor uneven, so as to prevent blow-ins. Verticality can be controlled by shifting weights on the top of the caisson, differential dredging, forcing by means of cables or jacks, and driving aided by external guide piles. In the more complicated caissons which are driven-in, water control is aided, by balancing the effect of suitably placed chambers of air. Jets outside the caisson may also serve this purpose, so long as the pressure of the water is sufficient to eliminate sudden collapse of the local quick zones around the caisson wall. This usually requires pressures of 8 to 10 kg/cm^2 (114 to 142 lbs/in^2).

Obstacles to driving, such as large boulders, are usually eliminated with explosives. A frequent hazard during driving is local streams of water flowing upwards within the hollow cylinders. This makes it very hard to establish reasonable slopes below and to the sides of the cutting edge. Water-flows even cause

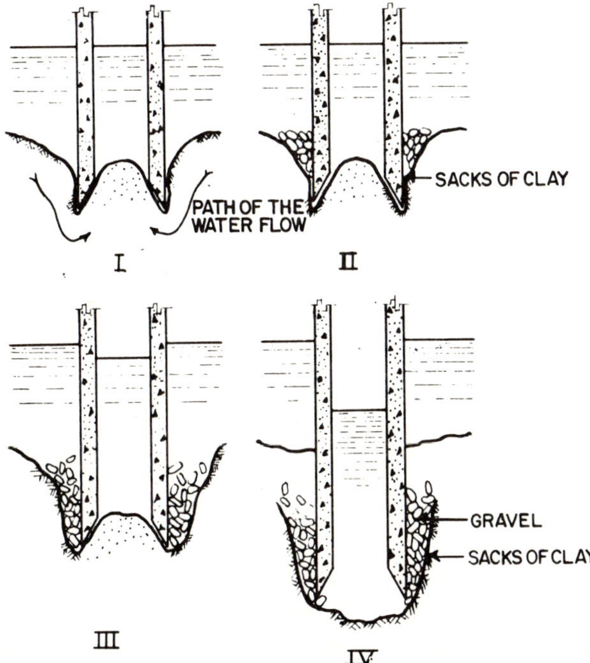

Fig. 8-51 Method for controlling water flowing towards the bottom of cylinders during driving [84]

bottom failure, Fig. 8-51 [84] shows a simple but very effective procedure for solving these problems, by dumping crushed stone or gravel on sacks of clay around the perimeter. The sacks of clay that are shown in the figure stabilize the excavation slopes and restrict lateral flow, which is usually the most harmful and abundant type of flow.

Pneumatic caissons (Fig. 8-52 [83]) are used when the conventional excavated well procedure can cause a loss in the ground around the foundation. They are also used when vertical driving is impeded by obstacles or when, in the case of underwater foundations, there is a risk that clayey materials or fine sands may flow underneath the cutting edges towards the inside. In these caissons, the air inside the work chamber is compressed to such an extent that it balances or is slightly higher than the hydrostatic pressure on the outside. In this way no water or soil is allowed to flow into the caisson. Care must be taken to keep the cutting edges far enough below the inner surface of the soil to avoid serious loss of air. Since it must be ensured that there is no water within the caisson, great care must be taken when making the joins that they are properly sealed on the inside. With pneumatic caissons, excavation work can be carried out by hand inside a dry work chamber. This permits a direct attack on all the obstacles to driving with first-hand knowledge of the supporting ground. These dry conditions are ideal for casting the concrete plug. The advantages are counterbalanced by extremely important disadvantages. The sinking process is slow and the men working inside the chamber cannot be exposed to the high air pressures that are required for very long. The maximum pressure which can be used is about 3.5 kg/cm^2 (50 lbs/in^2), which limits the driving depth to about 30 m (100 ft). Generally speaking, pneumatic caissons are expensive and their use for road building purposes is limited to the most critical conditions.

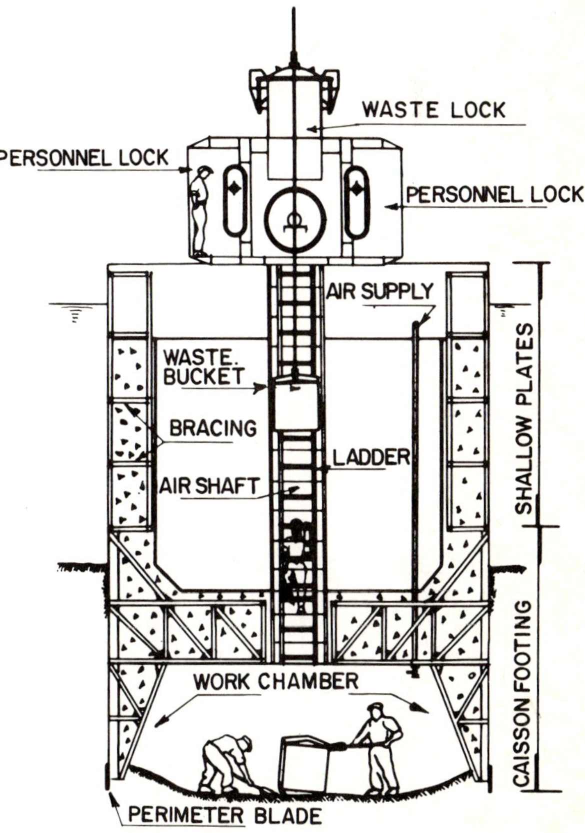

Fig. 8-52 Diagram of a typical pneumatic caisson [83]

8.11 Some Practical Aspects of Foundation Design

The following are some general considerations in the design and construction of foundations for road structures such as bridges, flyovers or overpasses and some particularly important or problematical drainage structures. The purpose of this chapter is not to describe the entire foundation field in detail, but to emphasize certain rules and reference papers which cannot be included in detail in the framework of bearing capacity theories or settlement analysis methods. However, because of their general validity or their experimental verification, these should be kept in mind as a background for engineering design when facing these types of problems.

Following the previous sequence, comments will be made first on shallow foundations and then on deep foundations, the distinction being made between piles, cylinders and caissons.

8.11.1 Shallow Foundations

.1 Foundations on Sands and Gravels

Bearing capacity, assessed in accordance with the theories mentioned earlier in this Chapter depends on the following concepts:

The relative density of the supporting layer: which reflects directly on the value of ∅ and consequently on the bearing capacity factors, N_q and N_γ. These increase greatly as density reaches high values. The relative density is usually estimated by penetration tests, of which the standard penetration test is by far the most widely used one in Mexico. The usefulness and importance of the standard penetration test [85,86] lies in the correlations made between field and laboratory. These correlations have proved to be reasonably reliable in sands and predominantly sandy soils. They are less reliable in clays and plastic soils. Figure 8-53 [7] shows a correlation between the number of blows for the 30 cm (1 ft) specified for penetration and the angle of internal friction in sands. It will be observed that in clean medium or coarse sands, for the same number of blows a higher value is obtained for the angle than in clean, fine sands or silty sands. The relations illustrated in Fig. 8-53 do not consider the influence of the vertical pressure on the number of blows, which becomes greater as greater depths are reached, and which thus incorrectly indicate excessive relative densities [86-88]. Figure 8-54 [87] presents experimental results obtained by the U.S. Bureau of Reclamation, which show that for the same number of blows in the standard penetration test the relative densities are less, with increasing vertical pressure acting on the sand, which in turn is a function of the depth to which the test is taken. This information enables corrections to be made in the calculations when the results of the standard penetration test are used.

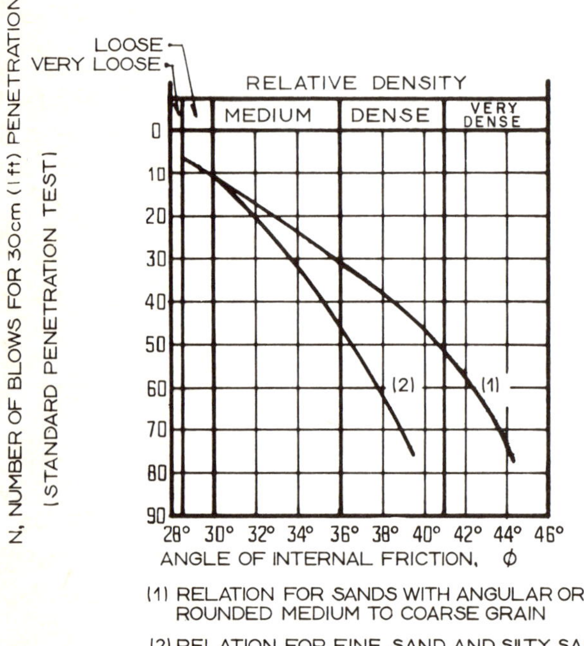

Fig. 8-53 Correlation between the number of blows for 30 cm (1 ft) of standard penetration and the angle of internal friction of sands

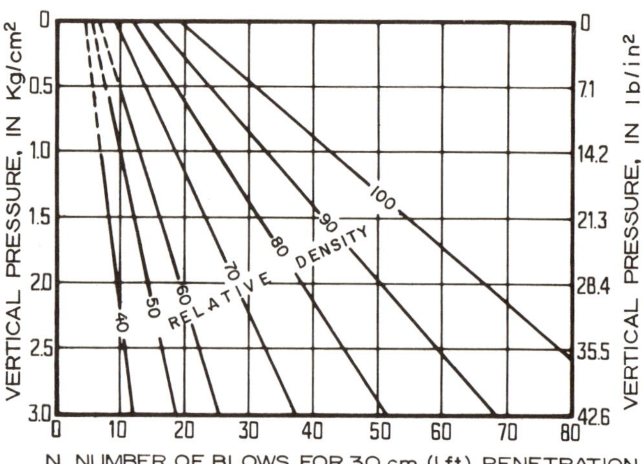

Fig. 8-54 Relation between the standard penetration resistance, the vertical pressure and the relative density for sands

When the sands being handled are very fine, have relative densities exceeding 50% and are located below the water-table, the value of N given by the standard penetration test is higher than the one that would be obtained for dry sand. The low permeability of the sand, does not allow the pore water to emigrate through the pores at the time of impact. Reference [7] gives an expression for correcting the values obtained in the test, N, under these circumstances. According to this expression, the corrected values, N, to be used are:

$$N = 15 + \frac{1}{2}(N' - 15) = \frac{N' + 15}{2} \qquad (8\text{-}50)$$

The above correction is recommended only when $N' > 15$. For clays and clayey soils, Terzaghi and Peck [28] give the correlation which is presented in Table 8-7. The fact is stressed that for these types of soils the correlations established on the basis of the standard penetration test should be used conservatively. It can be observed that the value of q_u given by the table is obtained by dividing the number of blows per foot by 8 (some authors recommend 6).

Despite the extensive use of the standard penetration test, especially in highway engineering, there is nevertheless a serious danger that its results may be misinterpreted. It has already been said that the reliability of the proposed correlations is very different in clays and in sands, but even in the latter, relatively low contents of coarse gravels and boulders render

Table 8-7
Correlation between the unconfined compressive strength and the number of blows in a standard penetration test on clays

Consistency	No. of blows, N	Resistance to unconfined compression, q_u kg/cm^2	lb/in^2
Very soft	< 2	< 0.25	< 3.55
Soft	2—4	0.25—0.50	3.55— 7.10
Medium	4—8	0.50—1.00	7.10—14.20
Firm	8—15	1.00—2.00	14.20—28.40
Very firm	15—30	2.00—4.00	28.40—56.80
Hard	> 30	> 4.00	> 56.80

the results of the test totally invalid. In these cases, especially below the water-table, the problem of determining the density characteristics by means of a simple test should be regarded as unsolved: experience and common sense are the engineer's best guides.

In [89] an interesting study of the correlation between the results of the standard penetration test and those of a dynamic cone penetration test [85] is given. The cone test was performed with a 60° pointed cone 5 cm (2 in) in diameter with the same striking energy and test method as in the standard test. The study is of a practical interest because the dynamic cone test is far more rapid and economical than the standard test. Therefore if a good correlation is established between the number of blows in both cases for a given location, a limited number of standard penetration tests, necessary for providing the soil samples and a basis for the correlation, can be very economically complemented by dynamic tests with the cone. It is concluded that $N_c = N_e$ (N_c, number of blows in the cone test and N_e the number of blows in the standard test) to a depth of about 1.2 times the depth of the water-table. From this level down, $N_c > N_e$ because of the lateral friction of the soil against the boring rods. The difference $N_c - N_e$ has been found to be proportional to the total force of friction. Lastly, it is recommended that a site-by-site local correlation be established as the safest method for design. This is not always practical for highway bridges, for the number of tests that are conducted is seldom large enough to justify an extensive study. It has also been said [85] that the relation does not exceed $2\,N_e = N_c$. If this rather rough guide were accepted, it could be used in combination with the previously described one. Whatever the case, only with well-established correlations and the judicious use of past experience will it be possible to use the dynamic cone test in combination with the standard test, [89] presents a guide for combining the above data, but it is unfortunately not supported by much experimental verification.

The position of the water-table: the influence of this was discussed in §8.2.9.

The width of the foundation: has a linear influence on the part of the bearing capacity related to the weight of the soil located below the foundation level, as can be seen in any of the equations applicable to gravels and sands.

The foundation depth, D_f: must be reduced by the depth of a future erosion or excavation. There is no rigid rule for establishing the depth that is suitable for shallow foundations for highway structures. In the case of bridges, when there are firm soils which would permit very shallow depths, the excavation depth is usually determined rather by the possible erosion by water along the riverbanks and deepening of the river bottom during floods as well as man-made excavation in the future, (assuming that the firm soils in question cannot be undermined). From this point of view, depths smaller than 2 m (6.6 ft) are probably not advisable, even in the most favorable cases. For foundations resting on sound rock the above limit can be reduced considerably.

In cold regions the foundation must be below the depth of frost action. In the arctic, it must be deeper than the thaw depth during the summer (depth of active permafrost zone). Moreover the foundation must be insulated from the ground so that it does not transmit heat to the frozen soil and thereby increase the depth of thawing.

In soils that are not particularly firm, or in which undermining may occur, this last condition is frequently the one which determines the minimum depth at which the foundation supports for a bridge should be placed. Possible future undermining must be investigated systematically in each case. If bridges rest on soils where excavation must be taken to great depths in order to find a suitable bearing capacity, or on account of undermining, seepage problems during excavation often become decisive. In riverbeds, the water-table is shallow and if excavation is taken below that level, construction procedures must prevent uplift and boiling. Some experts in the structural design of bridge foundations maintain that excavations deeper than 6 or 7 m (20 to 23 ft) lead to uneconomical foundations: in the sense that they are more expensive than some other alternatives for deep foundations. This economic depth is less if excavation goes below the water-table and seepage occurs on a large scale.

Another consideration of interest when establishing the excavation level for a shallow bridge foundation is the possibility of caverns or hollows in the subsoil or man-made underground structures such as sewer ducts, collectors and cables. The latter may be rare for road and rail bridges in rural areas but are far more often found in cities, especially in the case of flyovers or overpasses. In all these cases, only appropriate soil exploration and thorough study of past construction can avoid some very conflicting situations. Whenever the presence of any of these factors is suspected, a careful exploration must be conducted at the exact place where the support is to be constructed, a boring or pile striking a buried electrical conduct can be both spectacular and lethal.

Flyovers and other crossing structures are similar to bridges with regard to foundation conditions, especially in the case of sands and gravels under consideration.

Culverts and very small drainage bridges pose some very special foundation problems. Generally speaking, these can be solved with shallow foundations; the relative cost of a deep foundation is rarely tolerable. Moreover, culverts usually require only shallow excavations, thus avoiding large-scale excavations and heavy expenses. In sands and gravels there is usually no problem involved in finding the necessary bearing capacity, which is usually about 10 to 15 t/m^2 (2 to 3 kips/ft^2), at a reasonable depth such as 2 m (6.6 ft). Furthermore, owing to their number, culverts rarely allow a detailed foundation study, and the corresponding design recommendations have to be established in accordance with the scant information that is made available by the general geotechnical study of that road or railway. At most, exploration consists of a preliminary boring only a few meters deep.

The design for a shallow foundation on frictional soils is complicated by uncertain settlement of the structure. As mentioned in earlier chapters, the problem of assessing settlements in sands and gravels is far from being solved. Settlement underneath footings in sand will naturally depend on the stress-strain characteristics of the sand, especially its resistance to shear stresses, which depends on the confinement of the material and its relative density. The shear strength and rigidity increases in a roughly linear fashion with depth in the sand; and consequently the bearing capacity will follow a more or less similar law. However, the specific weight of submerged sand is about half of the unsubmerged weight. Therefore, the settlement of a footing on submerged sand will be approximately double the value in the same unsubmerged sand and its bearing capacity half, due to the fact that the confining pressure in the first case depends on γ'_m instead of γ_m and consequently the resistance of the material to shear stress is reduced by practically half.

For the same contact pressure of a footing in sand, settlement increases with width, although not in proportion, Fig. 8-43. The reason for this is that as the footing becomes wider, deeper zones in the sand are affected, and the strength and rigidity become steadily greater. Figure 8-55 [7] gives empirical graphs for obtaining the contact pressure that produces a maximum settlement of 2.5 cm (1 in) for footings in unsubmerged sands or in sands where the water-table is at a depth B (width of footing) or greater, below the foundation base. The information is given for different footing widths and different densities of the sand, as measured by the standard penetration test. The values of N number of blows in the standard penetration test are subject to all the previously mentioned corrections. For submerged sands, the allowable pressures can initially be regarded as half of the ones given in Fig. 8-55 for the same settlement, based on the analysis conducted earlier in this same section.

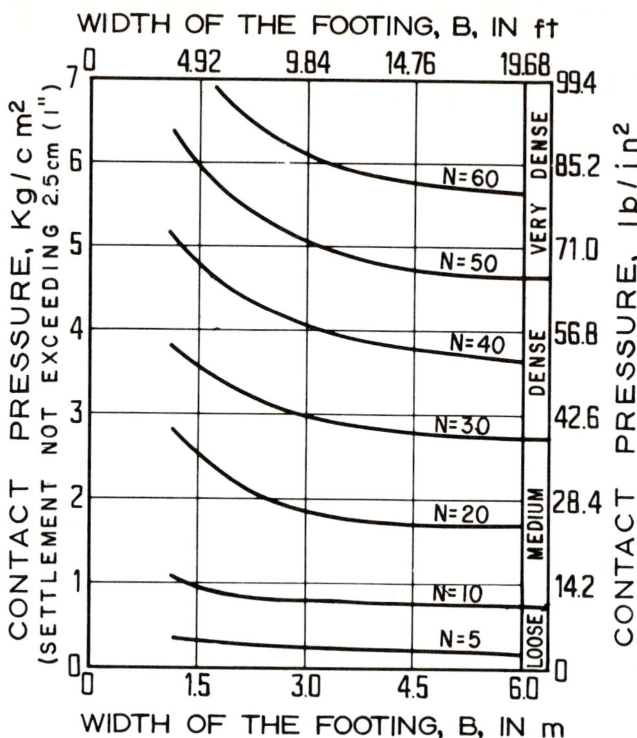

Fig. 8-55 Contact pressure corresponding to settlements of 2.5 cm (1 in) for footings in sand [7]

.2 Foundations in Homogeneous Clays

Earlier in this Chapter comments were made on Equations to be used for computing the bearing capacity in homogeneous clays, which depends on the unconsolidated-undrained strength of the soil (c_u) and the pressure acting as an overload or surcharge above the base of the foundation (γD_f). The value of c_u can be obtained in unconsolidated-undrained triaxial tests, as was mentioned, but also by the unconfined compression test and the vane test. The unconfined compression test is frequently used because it is easier and quicker to perform. The value of c_u that is obtained in this test is often lower than the one that is obtained with the triaxial test, because the lack of lateral support in unconfined compression, makes any fissure or minor structural irregularity reduce the unconfined strength. Further, under natural circumstances, clay does have some degree of confinement, which is why the results obtained in a triaxial chamber are considered rather more representative. In Chapter 6 a correlation was presented between the results of an unconsolidated-undrained triaxial test and normal vane tests. The correlation was developed for cold climates in a limited range of geologic environments and has not been verified worldwide.

In shallow foundations resting in homogeneous clays, there is an important difference in the calculation sequence to be followed when the level of the excavation is below the water-table, depending on whether the soil is impermeable or not. If it is impervious, the ground at the excavation level will have been relieved of a pressure which is the total pressure corresponding to that depth. In a permeable foundation, however, the excavation fills with water to a height equal to that of the water-table; the unloading due to excavation does not include the pressure of the water, and the term γD_f should therefore represent only effective pressure and should be computed accordingly. This means using the submerged specific weight of the soil that is below the water-table, or subtracting the pressure of the water corresponding to the water-table from the total pressure at that same level.

In truly clayey soils, the level of the excavation below the water-table is no longer such a serious problem as it is in sands and gravels. Thanks to the low permeability of clays, excavations can be maintained dry by means of limited, low cost pumping. Exceptions are excavations covering extensive areas or deep excavations for large bridges, where, as a result of intensive pumping, the water that flows towards the excavation walls and bottom may produce slope stability problems and heave in the bottom which will subsequently lead to settlements. In these cases (which rarely occur in road technology) the solution will be excavation by sections, or methods to reduce, intercept or control the water flowing towards the excavation.

In shallow foundations in clay, the problem of settlement is usually the most important consideration. Therefore the allowable bearing pressure is usually limited by the value of the maximum tolerable settlement. Settlements as a result of consolidation can be computed using the one-dimensional consolidation theory with vertical flow, described in Chapter 1. By calculating the settlements suffered by each support of a bridge, differential settlements can be found. These, in the long run are the most important to the design engineer. When assessing the settlement occurring for each footing, the pressures transmitted into the soil by any adjacent footings must be included. When calculating settlements due to consolidation, only the permanent dead and sustained live loads should be considered, because the duration of accidental loads including wind, is very brief in comparison with the time required to influence a consolidation process. The differential settlements that can be tolerated by a highway structure depend on the function it is to fulfill and its structural characteristics. This is a point where the foundations specialist must comply with the needs of the structural design engineer. Total settlements have intrinsic importance when there are adjacent structures such as approach fills and spans which may suffer damage as a result of movement of the structure under consideration. They are also important when there are drains and ducts, that cannot tolerate the settlements that may occur without damage.

The application of the traditional consolidation theory to roads is often problematical, especially for bridges, because riverbeds are commonly so heterogeneously stratified, that it is hard to meet some of the assumptions required by the application of TERZAGHI's theory. Use of the three-dimensional seepage theory [90] may lead to more reasonable solutions in some soil profiles, but the above statement remains valid. There is no simple and practical calculation method that takes all these problems into consideration, and the theory must, therefore, always be applied with a large measure of wise human judgment based on past experience.

The time required for settlement is even more uncertain. As has been pointed out, its assessment is of great practical interest, owing to the relation between the evolution of settlements undergone by the structure compared to its access embankments, on which so many design decisions are based. This is another case where the results of any computation must be tempered by all the past experience that is available.

.3 Foundations in Fissured Clays

Frequently, owing to desiccation and tectonic movements suffered by deposited clays in the course of their geological history, or as a consequence of natural heritage in the case of residual clays, the massive structure of these materials may be broken by a multitude of fissures, which may run in several different directions. When the fissures are very closely spaced, the specimens necessary for conducting shear strength tests are very difficult to trim. Furthermore, even if a suitable sample is dressed, interpretation of test results is dubious and the strengths obtained are usually smaller than the actual in-situ strength. In unconfined compression tests, the error is greatest, but even in triaxial tests the effect is significant, because the fissures are weakening planes which influence test results unless very high chamber pressures are used. Vane tests are preferable, so long as the clays were fairly soft, and without gravel seams, but fissured clays, especially residual ones, often exhibit fragile or brittle behavior, which would make test results inconclusive.

Perhaps the best means of assessing the shear strength of fissured clays for the bearing capacity of a shallow foundation is to conduct plate load tests on the soil at the location. The method consists of loading the soil with a 50 cm (20 in) square or circular plate until failure occurs. If the maximum load is regarded as being the ultimate bearing capacity, the value of the parameter c_u can be obtained using an expression of the Eq.(8-14) type. Once this value has been determined, fissured clays can be treated like homogeneous ones. The strength thus obtained is not, however, that which should be used in the stability analyses for the excavation slopes.

.4 Foundations in Silts and Loess

Silts belong to two clearly differentiated categories: plastic and non-plastic, for computing the bearing capacity of shallow foundations. The mechanical behavior of plastic silts is similar to that of clays, their degree of plasticity being generally low to medium, whereas the behavior of non-plastic silts is similar to that of the fine sands. Plastic silts owe their plasticity to a percentage of clay or clay-like flaky particles or to their content of fine-grained organic material. Rock flour is the typical example of a non-plastic silt. Interestingly, pure kaolin and bauxite often behave like non-plastic silts. In developing design data for road engineering, it is common practice to use the standard penetration test to determine the consistency of silts. If the number of blows in the test, N, is smaller than 10, the silts are regarded as loose or soft and usually unsuitable for bearing foundations. If N is higher than 10, then the material can often be used for foundation purposes and its bearing capacity can be computed with the methods indicated for sands or clays, as required. In normally consolidated plastic silts, located below the water-table, the problem of settlements is comparable to that in clays. Here, too, they can be analyzed in accordance with TERZAGHI's one-dimensional consolidation theory, with the corresponding laboratory tests. In loose or soft silts unsuitable for bearing foundations, deep foundations can be used.

Apart from the friction between the particles, the shear strength of many silts is due to a certain amount of *cohesion*, produced by a cementing agent. Triaxial tests are the best means of determining shear strength. Vane tests usually give reliable results in plastic silts below the water-table. The unconfined compression test sometimes gives highly exaggerated values for cohesion, because of the compression between the particles due to capillary tension in the water. This is equivalent to considerable confinement and is consequently a resistance due to friction. Some specialists like to repeat this test with specimens submerged in water. If the difference in strength proves to be very large in relation to the natural condition, or if the specimen collapses, it will have been established that what originally appeared to be cohesion is either frictional strength due to capillary tension or water-soluble cementing.

Loess is a silty material which forms as a result of Aeolian deposition; its particles are bonded to one another by a cementing agent. Deposition commonly gives rise to a rather open structure, similar to the honeycomb structure, and a relatively high voids ratio. Typically the voids are more highly interconnected vertically causing anisotropic high vertical permeability and vertical cleavage. The fundamental characteristic of loess deposits, from the point of view of their foundation bearing capacity, is their lack of uniformity: their strength may vary greatly over small distances or depths. The standard penetration test is a useful means of checking this uniformity. However, the strength estimated from N is too low, because the peculiar loose cemented structure breaks down, causing low resistance to the sampling spoon. Loess is a material in which load tests are very useful, so long as they are used sensibly, according to the uniformity of the deposit. Loess soils are seldom saturated; when they become saturated their cementing agent softens or dissolves and they lose most of or all their cohesion. Under these conditions, the soil structure may suffer a collapse accompanied by abrupt settlement. A rise in the water-table, irrigation, leaks in pipelines and drains or exposure to heavy rains are common causes of saturation that should be avoided.

8.11.2 Pile Foundations

.1 Hammer-Driven Piles

These are prefabricated piles that are driven into the ground by blows from a hammer or pile-driver. In Mexican and U.S. engineering practice, these are by far the most widely used piles for highway construction purposes. They are usually made of concrete and have a square cross-section, from 30 cm × 30 cm (12 × 12 in) to 50 cm × 50 cm (20 × 20 in). Steel piles are also used, generally tubular or of the H-beam shapes.

Sufficient emphasis has already been laid on the expressions available for calculating pile bearing capacity. The correct application of them in practice is usually very closely linked with the quality of the data obtained from the previous soils exploration program. The care and attention that must be exercised when conducing such an exploration can never be strongly enough emphasized. Point-bearing piles in particular rest on a firm layer whose thickness and characteristics must be carefully investigated, local variations in these qualities have been the cause of numerous failures. It has sometimes been incorrectly stated that a point-bearing pile is a structural element which works like a column, transferring the load from its cap to its tip, which is placed on firm soil or rock. On this basis, piles have been designed with stresses no greater than those for a column of the same material, with the same dimensions and subject to the same axial load based on slenderness ratio of an unbraced column. Experience has, however, shown that structural collapses and buckling are so rare in piles [96] that for design purposes they should not be considered likely events. Both theory and experience have proved that buckling cannot occur due to lack of lateral confinement, even in piles driven in the softest of soils. Stresses due to handling prior to driving may, however, play an important part in pile design, especially when lifting pre-stressed concrete piles, and placing them for driving. The bearing capacity of a point-bearing pile, therefore, depends almost entirely on the characteristics of the soil on which it rests and the area of its cross-section.

In olden days it was customary to drive the piles until they could be driven no further. This criterion was known as pile-driving refusal. Refusal should not be used as a basis for determining the acceptability of piles in a group, it is not adequate and may lead to serious errors. As an example, consider Fig. 8-56 [1] which illustrates a situation where refusal

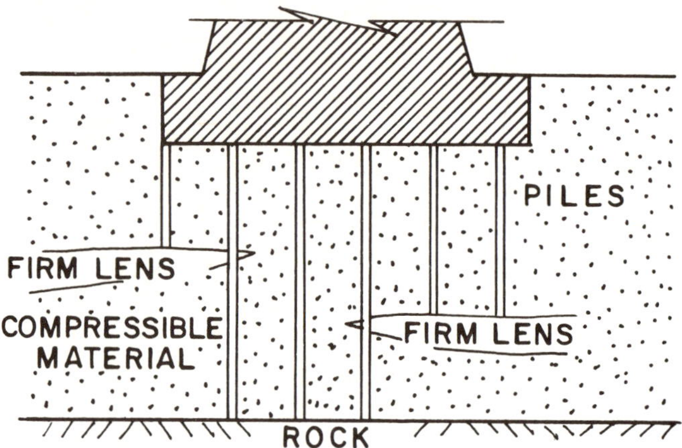

Fig. 8-56 Diagram to illustrate the dangers of driving piles until they are rejected

neither produces uniform bearing nor settlement. In the figure, lenses of firm materials, which have the capacity to stop the piles driven into them, are shown with horizontal lines. Since the distribution of these lenses is irregular, the piles in the group will be at different depths. As soil consolidation develops with time, those resting on rock will remain securely in place. The piles resting on the firm lenses will all suffer different settlements, owing to the difference in the thickness of the compressible material underneath each lens. The structure in the figure is sure to suffer damage due to differential settlements.

The device most commonly used for driving piles is the pile-driver shown in Fig. 8-57 [76]. It consists of a crane, generally mounted on tracks, with steel leads attached to the derrick,

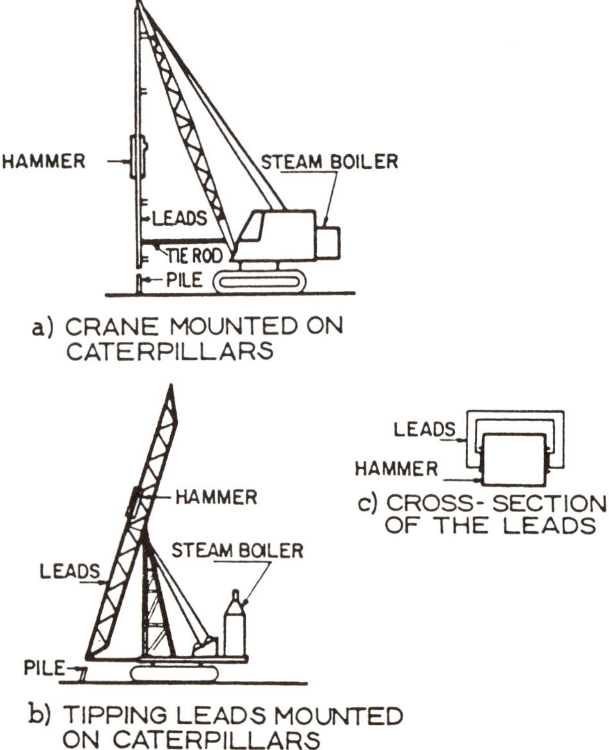

Fig. 8-57 Principal parts of a pile-driver

along which a hammer runs, striking the pile and driving it into the ground, Plate 8-8. The hammer is generally self-driven; in this illustration the system needs a source of energy, such as a steam boiler or air compressor. The pile is placed between the leads and is struck by the hammer directly on its cap, which is often protected by a piece of steel with a shock-absorber. Sliding parts known as followers are often used too, providing support about half or three-quarters of the way along the pile.

Plate 8-8 Pile-driving operations (Courtesy of CIESA)

The simplest kind of pile-driver is a free-fall drop hammer, weighing from 250 to 1,000 kg (550 to 2,200 lbs), which is allowed to fall from a height between 1.50 and 3.00 m (5 and 10 ft). The single-acting hammer [91] is fitted with a ram and a cylinder. Steam or compressed air enters the cylinder and lifts the hammer 60 to 90 cm (2 to 3 ft). Then the steam is released rapidly and the hammer falls on the pile cap. These hammers are easy to operate and are rugged. They are slow (60 to 75 blows per minute), but they strike the pile with a constant energy equal to the weight of the hammer multiplied by the height of fall. Figure 8-58a [76] is a diagram showing this device. The chief problems as regards operation may be due to mal-adjustments in the steam or air admission valve, and result in a drop in the driving speed and height of fall or an increase in loss by friction.

Plate 8-9 Driving a pile for a bridge buttress using a double-acting diesel pile hammer (Courtesy of CIESA)

In double-acting hammers, steam or compressed air is not only used to raise the hammer, but also to increase its downward momentum. A far quicker action is achieved, with 100 to 200 blows per minute, Plate 8-9. Driving time is reduced and the operation is made easier in some materials, such as loose sand. The valves require good adjustment. The striking energy may vary greatly if the hammer is not constantly subjected to careful inspections. Part *b* of Fig. 8-58 shows a double-acting hammer. These devices are referred to as differential-acting hammers when the steam or air cycle that is used to raise the striking parts is different from the one used to drive them downwards.

The differential-acting hammer combines the advantages and disadvantages of single- and double-acting ones. An ideal hammer would be one with large striking parts and a rapid operation speed, requirements which in practice are contradictory, not only because of the power that would be required to operate a large hammer quickly, but also because the increase in the speed of the blows requires greater system weight, when the cycle of a double-acting hammer is used. Consequently, a single-acting hammer has heavy striking parts, but operates at a relatively slow speed, whereas a double-acting one is rapid, but has lightweight striking parts for the same hammer weight. The use of independent steam cycles allows the hammer to be heavy without greatly affecting its striking speed [91].

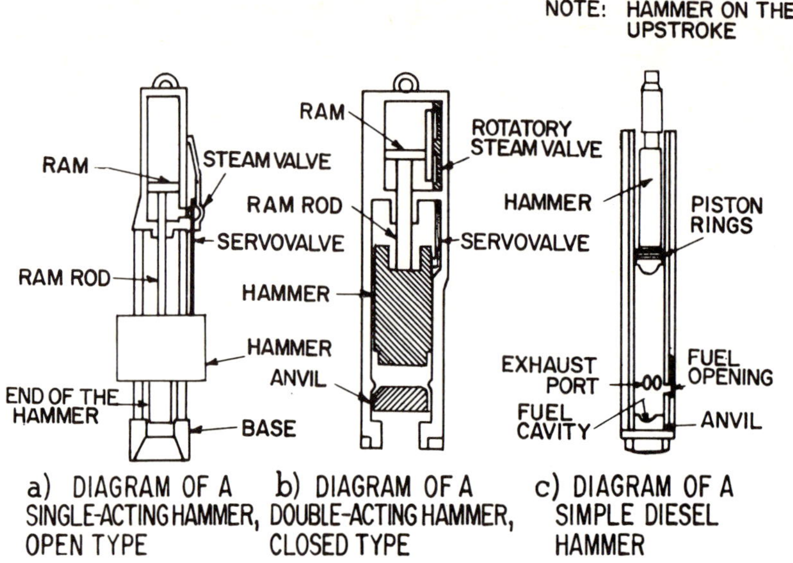

Fig. 8-58 Diagram showing the construction of steam hammers for pile-driving

Diesel pile-drivers consist of a cylinder fitted with an enourmous ram, which is the hammer. At first, the hammer is raised mechanically and is allowed to fall while the ignition mixture is injected into the cylinder. At this point the normal diesel operation commences, the hammer now being raised by the explosion, which at the same time drives the cylinder downwards onto the pile cap. Diesel pile-drivers are being built increasingly large, but even the largest models are manageable thanks to the fact that they have their own energy-producing source. Striking energy is high in relation to weight, but as in any hammer, the energy varies inversely with the operation speed. Moreover, energy increases with the resistance that is offered by the pile and is hard to evaluate at the work location. Table 8-8 [55,76] summarizes the chief characteristics of the some commonly used pile-drivers.

Table 8-8
Chief characteristics of some common types of pile-driver

Pile-driver	Type	Striking parts		Height of fall		Energy/blow		Blows per Min
		lb	kg	in	cm	lb-ft	kg-m	
Vulcan 2	Single-acting	3,000	1,360	29	74	7,620	1,055	70
Vulcan 1	Single-acting	5,000	2,270	36	91	15,000	2,075	60
Vulcan 0	Single-acting	7,500	3,400	39	99	24,375	3,370	50
McKiernan-Terry S-3	Single-acting	3,000	1,360	36	91	9,000	1,245	65
McKiernan-Terry S-5	Single-acting	5,000	2,270	39	99	16,250	2,250	60
McKiernan-Terry 9B3	Double-acting	1,600	725	—	—	8,700	1,205	145
McKiernan-Terry 11B3	Double-acting	5,000	2,270	—	—	19,150	2,650	95
McKiernan-Terry C-5	Differential	5,000	2,270	—	—	16,000	2,210	100
McKiernan-Terry S-14	Single-acting	14,000	6,360	32	81	37,500	5,185	60
Vulcan 50C	Differential	5,000	2,270	—	—	15,100	2,015	120
Vulcan 65C	Differential	6,500	3,630	—	—	19,200	2,655	117
Vulcan 80C	Differential	8,000	3,630	—	—	24,450	3,380	111
McKiernan-Terry DE20	Diesel	2,000	910	113*	287	18,800*	2,600	48*
McKiernan-Terry DE40	Diesel	4,000	1,820	129*	328	43,000*	5,945	48*
Link Belt 520	Diesel (DA)	5,000	2,270	—	—	30,000*	4,150	80*
Raymond 65 CH	Hydraulic	6,500	2,960	—	—	19,500	2,695	130*
Delmag D5	Diesel	1,100	500	—	—	9,100	2,600	42-60
Delmag D12	Diesel	2,750	1,250	—	—	22,500	3,100	42-60
Delmag D22	Diesel	4,850	2,200	—	—	39,700	5,480	42-60
Delmag D44	Diesel	8,819	4,000	—	—	72,300	9,980	40-60

* Maximum energy for maximum fall and minimum blows per minute

Table 8-9

Tentative guide for selection of the striking energy of a pile-driver

Length of pile		Depth of penetration	Metal sheet pile						Wooden pile				Concrete pile			
			Light		Medium		Heavy		Light		Heavy		Light		Heavy	
m	ft	%	kg-m/blow (ft-lb/blow)						kg-m/blow (ft-lb/blow)				kg-m/blow (ft-lb/blow)			
1. Driving in sand, loose gravel and not very consistent soils.																
8	26	50	140–250	1012–1806	140–250	1012–1806	250–350	1806–2529	500–590	3613–4263	500–1000	3613–7226	1000–1250	7226–9032	1250–2100	9032–15175
		100	140–500	1012–3613	250–500	1806–3613	250–500	1806–3613	500–1000	3613–7226	500–1250	3613–9032	1000–1250	7226–9032	1850–2100	13368–15175
16	52	50	250–500	1806-3613	250–500	1806–3613	500–590	3613–4263	500–1250	3613–9032	1000–1250	7226–9032	1250–2100	9032–15175	1850–3500	13368–25291
		100	500–590	3613–4263	500–590	3613–4263	500–1050	3613–7587	1000–1250	7226–9032	1000–2000	7226–14452	1850–2100	13368–15175	2100–3500	15175–25291
25	82	50	—	—	500–1050	3613–7587	500–1250	3613–9032	—	—	1850–2000	13368–14452	—	—	2700–5000	19510–36130
		100	—	—	—	—	500–1250	3613–9032	—	—	2000–2700	14452–19510	—	—	2500–5000	18065–36130
2. Driving in hard clays or very consistent soils																
8	26	50	250–350	1806–2529	250–350	1806–2529	250–600	1806–4336	1000–1250	7226–9032	1000–1250	7226–9032	1000–1250	7226–9032	1250–2100	9032–15175
		100	250–500	1806–3613	250–500	1806–3613	250–600	1806–4336	1000–1250	7226–9032	1000–1250	7226–9032	1000–2100	7226–15175	1850–2100	13368–15175
16	52	50	250–600	1806–4336	500–600	3613–4336	500–1250	3613–9032	1000–2100	7226–15175	1000–2100	7226–15175	1850–2100	13368–15175	1850–3500	13368–25291
		100	—	—	500–1250	3613–9032	500–1850	3613–13368	—	—	1850–2100	13368–15175	—	—	2700–5000	19510–36130
25	82	50	—	—	500–1250	3613–9032	500–1850	3613–13368	—	—	1850–2100	13668–15175	—	—	2700–5000	19510–36130
		100	—	—	—	—	1050–2700	7587–19510	—	—	2100–3500	15175–25291	—	—	2700–5000	19510–36310
Weight of pile kg/m (lbs per ft)			30	20	45	30	60	40	45	30	90	60	220	150	600	400

Table 8-9 [61] is useful as a tentative guide for the selection of the energy that must be applied by the pile-driver in a given case. This choice depends on many factors and specific conditions, and generalizations cannot therefore be made. Experience has shown that large hammers, weighing not less than half of the total weight of the pile being driven and with an energy of at least 0.30 kgm for each kilogram of weight (0.09 ft-lb/lb) of the pile, are the most effective ones. The table gives the striking energy with which the pile-driver should be selected, depending on the material of which the pile is made, the soil to be penetrated and the general driving conditions. Energies are given for driving piles of different weights.

In the last 30 years experiments have been carried out with vibratory pile driving [91], using low-frequency (5 to 35 cycles per second) vibrators. Vibration tends to reduce the friction between pile and soil. In some soils, the efficiency of the system is very high, but further research is necessary to obtain optimum results. Experience does, however, indicate that heavier, more powerful vibrators than the ones now in use will probably have to be developed.

The use of high-frequency (40 to 150 cycles per second) vibrators is even more recent [91]. These vibrators tend to make the pile vibrate at its natural frequency, which reduces shock and greatly increases the energy transmitted by the pile to the ground. It is clear that this system will in the future make it possible to greatly increase the driving efficiency and the work rate. The drawback is the design of a rugged variable frequency vibrator.

Hammer-driven piles have the outstanding advantage that they can transmit the design load with reasonable reliability, but they demand special reinforcement against handling, controlled manufacturing, and the use of heavy expensive driving equipment. Furthermore, even if the soil exploration is conducted with all due care and attention, surprises are not unusual during driving and can complicate operations severely. The main causes of problems are unpredicted changes in the consistency of the soils, boulders, large formations of very dense gravel or other obstacles. Many of the problems posed by hammer-driven piles can be solved by using cast-in-place piles, which will be briefly described later.

Most of the hammer-driven piles that are used in highway construction are made of concrete, are either circular or square and vary in width from 30 to 50 cm (12 to 20 in). Currently it is not unusual to see much larger piles in wide-span bridges, but these usually require special pile-drivers and pre-excavation.

With pre-drilling and large hammers, cross-sections of 1 or 1.20 m (3.3 or 4 ft) are sometimes used. Steel piles are used when a high bearing capacity is required, when they have to be driven through hard or dense strata or ones with boulders, or when only a small volume of soil is to be displaced during driving. They have the further advantage that they can take rough handling prior to driving. Their chief disadvantages are that they may suffer damage due to corrosion or electrolitic processes, their effectiveness as friction piles is comparatively smaller, and their cost relatively somewhat higher, and they bend easily when they strike the side of a boulder or a sloping rock surface.

Hammer-driven piles made of concrete usually have a uniform cross-section or are slightly cone-shape. Prestressing concrete piles has become a fairly popular technique, for it enables the necessary strength to be obtained with thin walls and a hollow core. These lightweight piles increase driving efficiency and make penetration easier with normal pile-drivers [92].

Hammer-driven piles provide a safe, permanent and very reliable foundation when installed in accordance with the rules of Soil Mechanics and the *rules of the art*. In practice, most of the problems they pose are related to driving difficulties. Excluding boulders or other obstacles, which would make driving impossible, but which can be identified in most (but not all) thorough investigations, driving becomes difficult when the ground becomes gradually harder (clays) or denser (sands) with increasing depth. When the consistency is sufficiently firm, driving becomes so difficult that it is no longer advisable to use these types of piles for foundations. This occurs in sandy formations requiring more than 40 blows per foot in a standard penetration test, especially if, as is common, density increases with depth. In clays, the limit varies between 30 blows and 50 blows per foot.

The possibility of driving depends largely on the interaction of three fundamental factors: the pile, the pile-driver and the soil [93]. The efficiency of the pile-driver is essential. It increases with the striking energy, when the weight of the pile is reduced and when the mass of soil that is mobilized with the pile is reduced. Reference [93] shows how by simply changing from concrete piles to steel piles there is an increase in the efficiency of a given pile-driver of 20 to 25%. This explains the success of steel piles in difficult ground. A reduction in the straight section of a concrete pile increases driving efficiency by reducing the weight of the pile, but in [93] it is pointed out that this process is of interest only in the case of piles not exceeding 15 m (50 ft) in length. For greater lengths, relatively important decreases in the weight of the pile have a gradually smaller influence on driving efficiency. For these great lengths, when a driving problem is predicted, the best solution is sometimes to use steel piles or ones made of a combination of concrete and steel. However, precast concrete piles as long as 40 m are not unusual.

Pre-boring is helpful in reducing the volume of soil mobilized by driving, thus increasing driving efficiency. For concrete piles, pre-boring should be carried out with a diameter or width that is slightly smaller than that of the pile, especially if important lateral friction is expected. Pre-boring is sometimes difficult because of the nature of the ground. The advantage of using combined concrete and steel piles is emphasized. Excellent results have often been achieved in difficult ground by using either a steel H-beam tip 1 to 2 m (3.3 to 6.6 ft) in length, or rail, on a precast concrete pile.

Jetting is often a valuable aid to pile-driving. The soil at the tip of the pile is loosened and lateral friction in the walls is reduced [94,95]. Figure 8-59 shows some typical arrangements for outlets for the jets in the pile caps. Jets consists of a pipe 5 to 7 cm (2 to 3 in) in diameter, which narrows to about half at the outlets for the jets in the pile caps. Jets consist of a pipe 5 to minute (264 gal/min) per pile should be expected, with water pressures of 10 to 20 kg/cm^2 (142 to 284 lbs/in^2). In coarser materials, both figures may increase considerably. Arrangements with just one injector at the tip of the pile are not recommended [1] for they become blocked, and also they tend to form a compact plug underneath the pile, making driving difficult. In the best arrangement, the jets should be placed symmetrically around the pile tip, so as to avoid any displacement of the pile away from the vertical, and they should be directed slightly upwards. Pipes on the outside of the pile become easily freed and shifted. Prejetting with a jet pipe extending to within 2 or 3 m (7 or 10 ft) of the proposed pile tip will loosen fine sands, so the pile can be driven after removing the jet pipe.

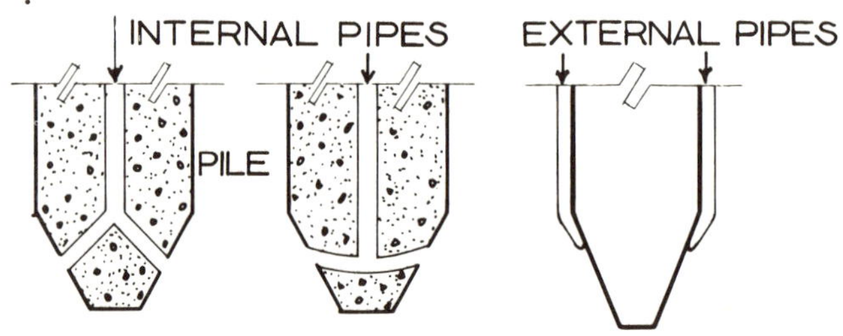

Fig. 8-59 Typical arrangements for injecting water to make pile-driving easier

Jetting requires a special high pressure, high volume pump with corresponding connections, pipes and nozzles. Table 8-10 [95] is useful for selecting a suitable type of pump. It gives the flow that is necessary in liters/min (gal/min) in accordance with the pressure required and the diameter of the nozzle of the jet. Table 8-11 gives a useful estimate of the losses by friction in the system of hoses and pipes between the pump outlet and nozzle of the jet. These values are essentially the result of experience and are somewhat smaller than are calculated by theoretical methods [98]. With the diameter of the nozzle in Table 8-10, it is possible to estimate an appropriate combination of flow and pressure for the pump that is to be used. The table gives these data for a jetting pipe with a specific diameter, but this is not a serious practical limitation, for the diameters included in the table are very similar to the ones used in real situations. Also the data it provides are for a jetting pipe 15 m (50 ft) long. For other lengths, adjustments will be necessary. With the combination of pressure and flow obtained from Table 8-10, Table 8-11 will enable adjustment of the estimate so that despite the losses due to friction the pressure at the nozzle of the jet will be adequate. Losses by friction increase when the diameter of the jetting pipe is reduced, and become excessive when this diameter reaches 2.54 cm (1 in).

Jetting can be used both in sands and in clays, it is not so effective in the latter. In clays, the danger of the tip of the jet becoming stopped up is greater, and dewatering is more difficult in the area of the pile tip. The operation is not effective in coarse gravel or rock. Injection must cease at least one meter above the ultimate tip level for the pile; otherwise the risk is run of the supporting stratum being loosened, with a consequent reduction in bearing.

Table 8-10

Volume of flow required for jetting operations with a 6.25 cm (2-½ in) pipe 15 m (50 ft) long

Pressure in the pump		Diameter of the nozzle; cm (in)				
		1.90 (¾)	2.54 (1)	3.18 (1-¼)	3.50 (1-⅜)	3.81 (1-½)
kg/cm²	lbs/in²	Volume of flow; l/min (gal/min)				
7	100	605 (160)	1040 (275)	1510 (400)	1740 (460)	1950 (516)
10	142	740 (196)	1290 (342)	1870 (494)	2160 (570)	2420 (640)
14	200	830 (220)	1500 (396)	2200 (581)	2540 (670)	2800 (740)

Table 8-11

Losses in pressure due to friction in pipes and hoses

Diameter of the pipe or hose		Jetting volumes; l/min (gal/min)								
		380 (100)	570 (150)	760 (200)	950 (250)	1140 (300)	1320 (350)	1510 (400)	1700 (450)	1890 (500)
cm	in	Loss by friction; kg/cm²/m (lbs/in²/ft)								
5.08	2	0.011 (0.048)	0.022 (0.095)	0.040 (0.173)	0.062 (0.268)	0.088 (0.38)	—	—	—	—
6.35	2-½	0.002 (0.009)	0.005 (0.022)	0.008 (0.035)	0.013 (0.056)	0.019 (0.082)	0.025 (0.108)	0.034 (0.147)	0.042 (0.182)	0.053 (0.23)
7.62	3	—	0.001 (0.004)	0.002 (0.009)	0.004 (0.017)	0.005 (0.022)	0.007 (0.030)	0.010 (0.043)	0.013 (0.056)	0.015 (0.065)
8.89	3-½	—	—	—	0.001 (0.004)	0.002 (0.009)	0.002 (0.009)	0.003 (0.013)	0.004 (0.017)	0.005 (0.022)
10.16	4	—	—	—	—	0.001 (0.004)	0.001 (0.004)	0.001 (0.004)	0.002 (0.004)	0.002 (0.009)

.2 Cast-in-Place Piles

There are many types of piles that are built in the ground. Their use in highway construction has not been too widespread in the past, but is becoming increasingly popular, Plate 8-10. References [1,76,92,95] describe the chief types of cast-in-place piles. Many have been patented, but such patents soon expire. All of them fall into two large categories: 1) those consisting of a metal case that is driven into the ground and emptied. The metal case is used as a mold, but may or may not be recoverable; and 2) those that are built by pouring the concrete directly into a previously bored hole, where the concrete is in direct contact with the soil. Piles with a metal case are sometimes preferred because a detailed inspection of their length, diameter and straightness can be conducted before the concrete is poured into them. Moreover they permit a more careful control of the placement of the concrete. Uncased piles are usually cheaper, but their use is limited to formations of soil where the excavation walls are strong enough, and ground water does not weaken the plastic concrete.

.3 Pre-Stressed Piles

Piles made of pre-stressed concrete are becoming increasingly popular because of their lighter weight, with the corresponding advantage in driving efficiency (when this is the installation process), their greater strength during handling and their resistance to breaking during driving and to the fact that they eliminate the problem of the cracks usually suffered by conventional concrete, which may be an important advantage in the case of piles installed in water.

Plate 8-10 Equipment for the construction of wide-diameter piles

8.11.3 Cylinders and Foundation Caissons

One of the chief problems of cylinder or caisson construction is to maintain the cylinder in position during the sinking operation. The different methods for doing this fall into one of the following four general procedures: a temporary guide structure is built with piles; a series of tie posts is built with piles, or chambers are formed with sheet-piles; an artificial island is built, using sand; or the cylinder is held in position with the aid of a set of anchored steel cables. The choice of method depends in each case on the size of the cylinder or caisson on whether construction is to take place on land or in water, and in the case of water, on the stability of the underlying material.

Excavation consists of removing the material from inside the cylinder, using a clamshell dredge or some similar mechanical device. The volume that is extracted always exceeds the volume of the embedded part of the cylinder, due to the way in which the surrounding soil inevitably flows inward towards the bottom of the cylinder. In granular soils, there may be a 100% difference in volumes for this reason. It is always advisable for divers to be on hand during sinking. They can efficiently lay explosives at the bottom of the excavation when hard or cemented materials are met. They may also be a great help in removing boulders or other obstructions.

Cylinders often tilt from the vertical or move from their position on the bottom during sinking. When these movements exceed the structurally permissible value (it is sometimes said that the geometrical center of the bottom section may move up to 20 to 30 cm (8 to 12 in) without causing problems in the structural performance of a deep foundation), one of the following remedial methods can be used, Plate 8-11. Excavate within the cylinder, advancing slightly more on the higher side than on the lower side, but without interrupting excavation on the lower side; excavate the higher side outside the cylinder; use jets on the exterior of the higher side; pull the cylinder in the appropriate direction by means of cables and deadmen; add weight on the high side. Sometimes it is sufficient to advance construction of the cylinder on the higher side, adding concrete to provide more weight. Underwater concrete casting for making the final plug should be performed with special care to avoid contamination of the concrete with soil or water or segregation of its aggregates.

Plate 8-11 Lateral excavation in a cylinder in order to restore verticality

Plate 8-12 Metal frame for a bridge pier, an effective element for helping to control sinking in water

Large caissons are built in two situations for which different techniques must be adopted: on land or in deep water. If there is no water, the excavation procedure is similar to that for an open cylinder. If, however, there is water present, either a steel shell or form, Plate 8-12, which constitutes the lower section of the caisson, can be floated to the right position, or else an island can be built, generally of sand. The construction area is isolated beforehand by a ring or box of sheet-piles that is filled with sand. This method has the advantage of being relatively little affected by fluctuations in the water level. It is illustrated in Fig. 8-60 [97]. The island should be large enough to provide a suitable working area outside the caisson and allow the movement of the necessary construction equipment.

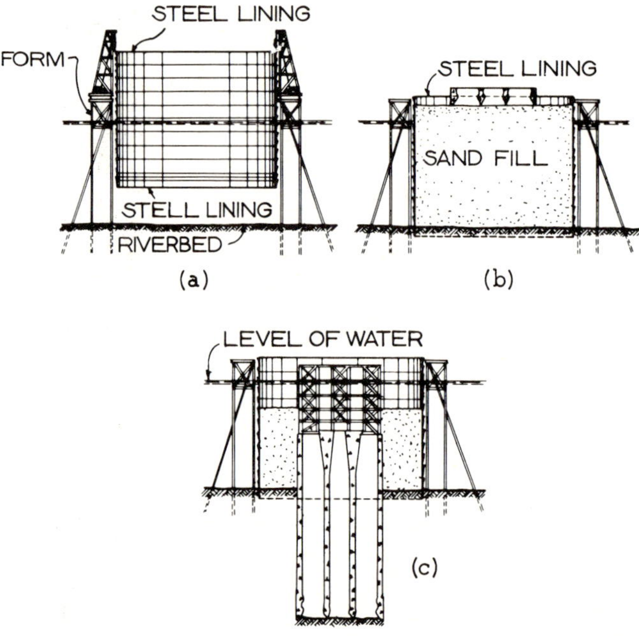

Fig. 8-60 The *sand island* method for driving caissons [97]

8.12 Control of Water in Excavations

One of the key problems in the construction of a shallow foundation is the control of water that may seep into the excavation below the water-table. It is also one of the problems that has been least investigated; consequently it often gives rise to difficult, unpredictable situations at the time of construction. Many highway bridges, on account of their magnitude or subsoil conditions, do not justify the detailed site investigations which are necessary for determining seepage problems in deep foundations. In such cases, the judicious use of past experience, making suitable adjustments as construction progresses, is the best way to find the correct approach. The neglectful attitude and lack of interest in these problems has, however, gone too far and often leads to excessive costs in excavations for large bridges or even in much smaller bridges when excavation work extends well below the water-table.

When special investigations have not been conducted, water is controlled by means of pumping, usually with centrifugal pumps. A minimum of pumps is used, using only one that produces the same volume of flow as a combination of several smaller pumps. This is justified by the cost per cubic meter of water removed being inversely proportional to the capacity of the pump. The suction lift of the pump is watched so that it always remains within operational limits. In very deep excavations, ramps or platforms are required with the pumps being lowered as depth increases. To mobilize pumps with a discharge diameter greater than 15 cm (6 in), special efforts will be required. Pumps are usually placed downstream of the excavation so that the water that is removed is less likely to seep back again. Emergency pumping equipment for use in case of a breakdown should be on hand to avoid expensive setbacks. In bridges with several different spans, several adjacent excavations should be made simultaneously, so as to increase the output of the equipment and the effectiveness of the operation in each excavation. This technique incorporates the effect of adjacent drains, hydraulically enhancing the drainage efficiency. A collector channel is often built around the perimeter of the bottom of the excavation with a pit in the lowest part where the pump intake is placed.

It is a widespread construction practice to cover any excavation slopes that are subjected to heavy pumping with stone, sandbags or some other heavy materials. This practice has been justified by the argument that it prevents fines in the soil mass from being carried away by seepage. Anyone who understands the design of filters and their effectiveness (as described in the Chapter on Subdrainage) will realize how dubious this argument is. If there is any advantage in covering the walls of an excavation with heavy materials, it is of a purely mechanical nature for protecting the geometry of the excavation. It is not, however, easy to believe that a measure like this can have any outstanding influence on the stability of an excavation. If the heavy material is underlain by a properly designed filter. The seepage will be stabilized.

When special seepage studies are not carried out and very serious pumping problems are expected, other solutions are often adopted, such as sheet-piles, coffer dams made of different materials and catchment wells. However, if these solutions are not based on good geotechnical data and sound analyses, then any attempt at using them will be irrational. Experience has shown only too clearly the disadvantages of attacking seepage problems in deep excavations without first bringing them into proper focus. Imperfect pumping schemes are the result, and their inefficiency may prove dangerous, causing upset programs and excessive costs. A rational approach to the problem is usually to determine the permeability of the soil as well as its relation to the less pervious strata. Following is a list of the concepts to be analyzed if a rational seepage control program is to be formulated for a deep excavation [99]:

- Lowering of the water-table and control of flow towards the excavation
- Pumping of seepage water
- Possibilities of piping
- Stability analysis of the bottom of the excavation
- Expansion or heave of the bottom of the excavation
- Settlements in zones adjacent to the excavation, from lowering the water-table
- Effect of rain
- Interaction between stability problems and construction schedule (analysis of the periods of time during which the excavation will remain open)
- Influence of construction procedures in relation to the construction equipment available
- Costs analysis

Of the above, attention will be given here only to the first two. The others have been dealt with in previous chapters or are beyond the scope of this book, (as is the case for costs analysis or construction equipment).

The rate of seepage through any soil towards a deep excavation can be determined using the general theories of groundwater flow [100,101] and the permeability of the formation; hence the importance of this determination. In practice, however, it is more complex. The heterogeneity of riverbed soils usually makes a simple application of this theory dubious, because it implies a homogeneous soil. In any case, an assessment of the permeability conditions is necessary for the formulation of a suitable study. Permeability tests in the laboratory are not truly representative of the permeability of large masses of soil for these problems. Therefore, tests capable of enabling assessment of the seepage conditions in large masses of in-situ soil must be conducted in the field. Field permeability tests are usually carried out in three different ways: 1) simple ditches or open wells are used, 2) pumping tests are performed, sometimes with the aid of observation wells, 3) other, special techniques are used, although not often for highway construction purposes. These special techniques trace seepage paths and measure velocities with materials such as radioisotopes, fluorescent or radioactive materials.

Tests in ditches [102] or open wells consist of the excavation of a drain or sump with defined geometrical characteristics (Fig. 8-61). A known volume of flow is introduced into the

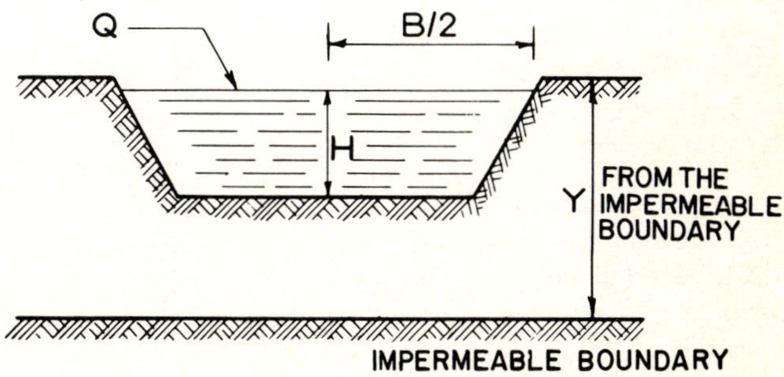

Fig. 8-61 Permeability tests in simple ditches

ditch, and the level of the water must be maintained constant throughout the duration of the test. This volume of flow must be expressed in units of volume per unit of length of the ditch. The equations applicable in this case are:

$$k = \frac{Q}{B + 2H}; \text{ if } y < \frac{3}{2}(H + 2B) \tag{8-51}$$

$$k = \frac{Q}{B - 2H}; \text{ if } y > \frac{3}{2}(H + 2B) \tag{8-52}$$

This determination assumes homogeneity and isotropy of the soils and steady flow. In practice none of these are completely satisfied. Reference has already been made that homogeneity is rare in riverbed foundations, and the same can be said about the isotropy of soils in general and of these formations in particular. As a consequence, together with the uncertainties introduced by test techniques and the lack of representativity of measurements taken in shallow ditches, the results of such a test are only very approximate.

Sometimes a similar test is performed, but using two parallel ditches (Fig. 8-62), of known geometry. From one a volume of water is withdrawn such that the difference in levels becomes constant. These levels can be measured by any procedure. When the levels of the water in both ditches stabilize, it is possible to assess k using the flow net that is drawn between the two excavations. The same comments apply to this test as to the above-mentioned tests involving a single ditch. The values obtained for permeability are only approximately representative of the mass of soil between the two ditches.

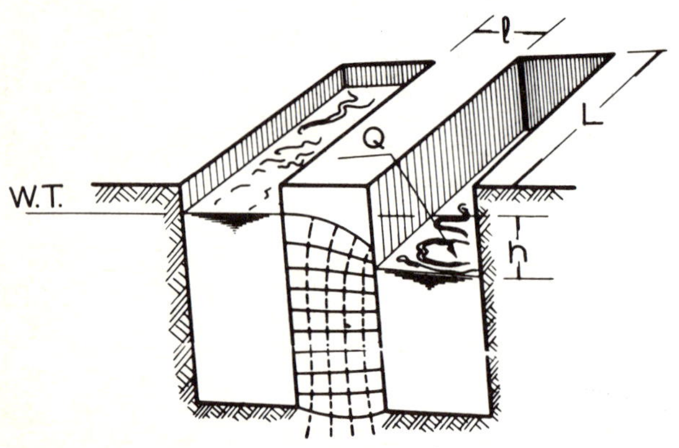

Fig. 8-62 Permeability test with two parallel ditches

Among the test for determining permeability with a single well, by pumping or injection, the most popular ones are of the LEFRANC type, where a cased well is built. The lower end is left free or surrounded by a slotted pipe and is filled with gravel over a length l (Fig. 8-63). A flow of water is then injected into (or withdrawn from) the inside of the well, in such a way that the level in the well is maintained constant throughout the test. If H is the difference between the height of the water in the well and the water-table corresponding to a flow Q that is either injected or withdrawn, the coefficient of permeability can be computed using an expression of the following type:

$$k = c\,\frac{Q}{H} \tag{8-53}$$

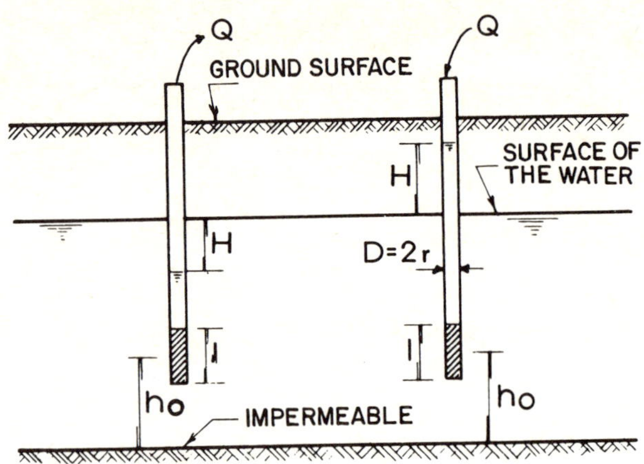

Fig. 8-63 Diagram showing a LEFRANC type test

The constant c, whose units are the reciprocal of a length, can be expressed as:

$$c = 0.366\,\frac{\log\left[\dfrac{l + \sqrt{l^2 + D^2}}{D}\right]}{l} \tag{8-54}$$

the symbols having the same meaning as in Fig. 8-63. Fig. 8-64 is a graph which enables c to be calculated in two dimensions, which are very usual for the well. This graph was supplied by the Advisory-Technical Department of the Mexican Federal Electricity Commission. A test of the LEFRANC type assumes homogeneity, isotropy and steady flow conditions; therefore it should be used with due care.

The tests with one pumping well and several observation wells are more elaborate and can be more reliable. Chapter 7 of [100] contains the theory to be applied, together with a description of the way to conduct the test (Supplement VII-a). In [103], a detailed presentation of the supporting theory can be found. Such tests can be carried out using two different methods, supported by two different theories, depending on whether flow conditions are steady or not. The above-mentioned references enable both cases to be handled. For steady flow conditions, the equation to be used is:

$$k = \frac{2.3q \log \dfrac{r_2}{r_1}}{\pi(h_2^2 - h_1^2)} \tag{8-55}$$

where h_1 and r_1 are the height of the water in the pumping well (above an impermeable stratum) and the radius of the well, respectively. Value h_2 is the height of water in the observation and r_2 its distance from the pumped well. The measurements are made after pumping and flow are stabilized. If two observation wells are used at different distances from the pumped well, both pairs of r and h refer to them. A test like this assumes homogeneity and isotropy, and assumes that the wells are of the type where total penetration occurs in a free water-bearing layer [100].

The determination of the coefficient of permeability based on pumping wells is expensive, and is reasonably safe when the assumptions mentioned are satisfactorily met. Strictly speaking, the theory that supports them assumes that DUPUIT's

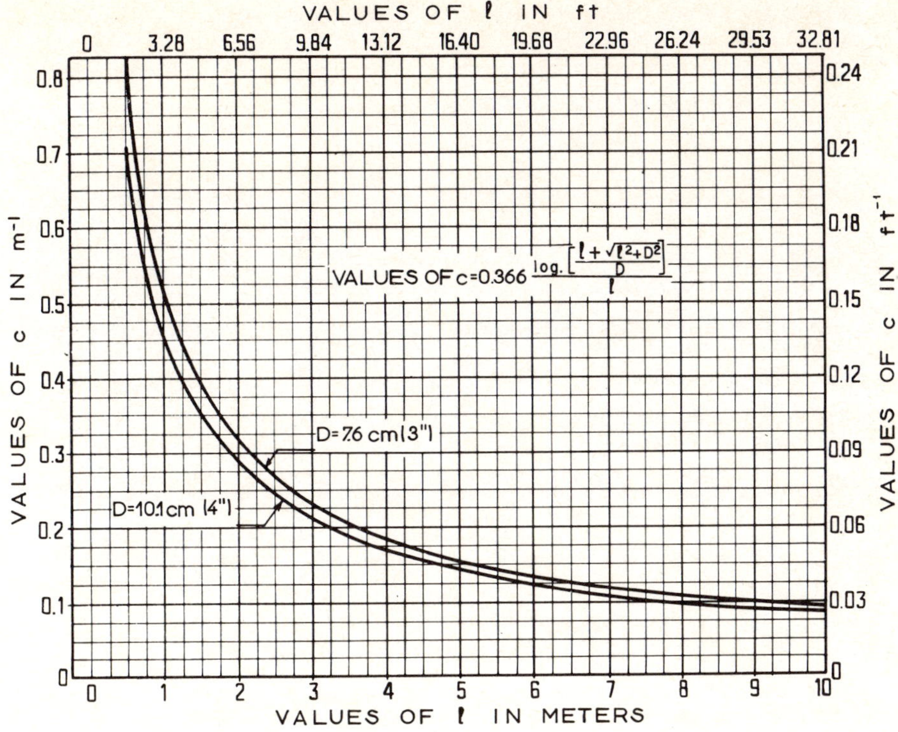

Fig. 8-64 Graph for determination of constant c in a LEFRANC type test

General Theory is valid and that the water-table spreads indefinitely in all directions beyond the pumping zone.

When the unsteady flow analysis is adopted, the theory supporting the test becomes complicated [100,103], but the test procedure is similar, and consequently there can be no special problems if tables and charts are available to avoid complicated calculations. However, it is doubtful whether the needs of highway engineering technology justifies such refinements. The main reason that is usually given for using tests with unsteady flow conditions, is the saving in the pumping time that is required to achieve steady flow conditions. This urgency is rare in studies for bridges. For these reasons, in highway technology, tests using the steady flow hypothesis are far more popular. The test is more accurate than the steady flow test (because steady flow never really develops). It still assumes isotropy and homogeneity. Reference [103] also includes a complete analysis of a test by the U.S. Bureau of Reclamation (USA), which is well supported by experience and which has the advantage that it does not require any type of observation well. Instead it is performed using a single well with a small diameter. References [100, 102] contain observations regarding the utilization of catchment wells and electro-osmosis. Both techniques are generally too refined and expensive for highway engineering. Their use is therefore very limited for controlling water in excavations for bridge foundations, where the soils have low permeabilities. Table 8-12 [102] gives a list of the typical permeabilities of various different granular soils, which usually pose serious excavation problems. The table should be used exclusively as a rough guide not as a basis for final design.

Table 8-12

Characteristic coefficients of permeability (k) for several different granular soils

Soil	$k \times 10^{-4}$	
	cm/sec	ft/min
Very fine sand	50	100
Fine sand	200	400
Fine to medium sand	500	1000
Medium sand	1000	2000
Medium to coarse sand	1500	3000
Coarse sand and gravel	3000	6000

APPENDIX 8A

FOUNDATION PROBLEMS WORKED EXAMPLES

8a.1 Foundation with Shallow Footings

Determine the bearing capacity of the foundations of the bridge shown in Fig. 8a-1. The stratigraphy of the subsoil, elevation of the maximum high water level and probable erosion undermining are shown in the same figure.

Solution

Since at a depth of 4.5 m (15 ft) there is a deposit with favorable mechanical properties (Stratum *3*), it is advisable for the foundation of the bridge supports to be composed of slab foot-

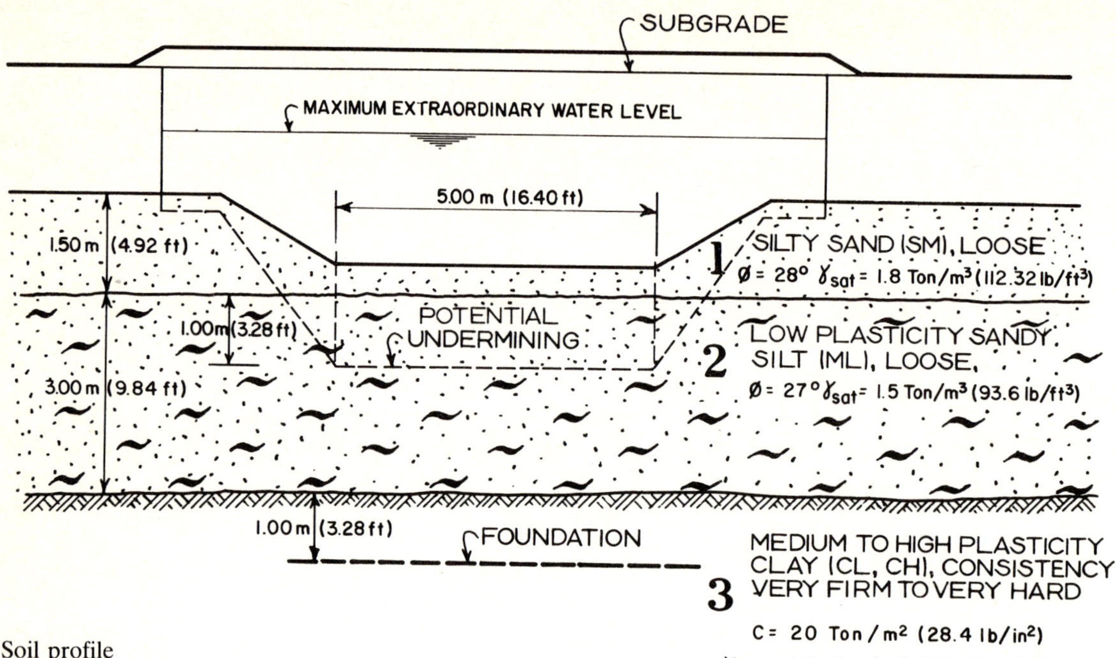

Fig. 8a-1 Soil profile

ings excavated at a depth of 5 m (16.5 ft). Since the soil is a clayey material, the bearing capacity will be obtained using SKEMPTON's criterion:

$$q_c = cN_c + \gamma D_f \tag{8-14}$$

where:

q_c = ultimate bearing capacity, in t/m² (kips/ft²)

γD_f = effective pressure at foundation level $0.5 \times 2 + 0.8 \times 1 = 1.8$ t/m² (0.37 kips/ft²). Note that for surcharge purposes only the soil below the potential undermining or erosion level is considered.

γ' = submerged unit weight of the soil, in t/m³ (kips/ft³)

D_f = depth of excavation for the foundation = $5 - 2.50 = 2.5$ m (8.25 ft) (not taking into account probable undermining).

N_c = a bearing capacity factor which depends on the D/B ratio

D = 1 m (3.28 ft) (Fig. 8-12)

B = 3 m (10 ft)

$D/B = \frac{1}{3} = 0.33 \therefore N_c = 5.8$ (Fig. 8-11)

Applying Eq.(8-14) and a factor of safety, $F_s = 3$, the following is obtained:

$$q_a = cN_c/F_s + \gamma D_f = \frac{20 \times 5.8}{3} + 1.8$$

$$= 40 \text{ t/m}^2 \text{ (8.21 kips/ft}^2\text{)}$$

where q_a is the allowable bearing capacity of the foundation soil.

8a.2 Foundation with Cylinders

It is wished to select the suitable type of foundation and the bearing capacity of the soil for the bridge shown in Fig. 8a-2. The profile of the potential general undermining is shown in the same figure, and a possible local undermining of the piles of about 2.4 m (8 ft) was also computed in the high plasticity clay stratum. Piles and cylinders must be embedded to a minimum depth of 5 m (16.5 ft) if the bridge is to be stable in the face of lateral stresses.

Solution

Strata *1* and *2* do not have favorable supporting conditions, for the fine sand is of low density and is affected by erosion in the riverbed, and the high plasticity clay deposit is soft and is also affected by erosion. Under the circumstances, the foundation should consist of cylinders resting on the layer made up of boulders, gravels and sands, Stratum *3*. In order to compute the bearing capacity, the theories by TERZAGHI, MEYERHOF and BALLA will be used. Cylinders with a diameter of 4.0 m (13 ft) will be tested.

a) TERZAGHI's method:

$$q_c = 1.3\, cN_c + \gamma D_f N_q + 0.6\, \gamma R N_\gamma \tag{8-6}$$

where:

q_c = ultimate bearing capacity, in t/m² (kips/ft²)

c = soil cohesion = 0

γ = submerged unit weight of the soil, in t/m³ (kips/ft³)

D_f = depth of foundation = $12.8 - 5.6 = 7.2$ m (23.6 ft) (allowing for a probable total undermining of 5.6 m (18.4 ft)).

R = radius of the cylinder = diameter /2 = 4/2 = 2 m (6.6 ft)

γD_f = effective pressure at depth of foundation = $0.6 \times 0.5 + 5.0 \times 1.0 = 5.3$ t/m² (1.08 kips/ft²).

The total undermining or erosion is subtracted N_c, N_q and N_γ are non-dimensional factors which depend on the angle of internal friction of the foundation soil.

General shear failure for $\varnothing = 36°$, $N_q = 52$, $N_\gamma = 50$ (Fig. 8-6)

Substituting values in Eq.(8-6):

$q_c = 5.3 \times 52 + 0.6 \times 1.0 \times 2 \times 50 = 335.6$ t/m² (68.7 kips/ft²)

q_a = allowable bearing capacity = q_c/F_s, F_s being the factor of safety = 3.

$\therefore q = 335.6/3 = 112$ t/m² (22.9 kips/ft²)

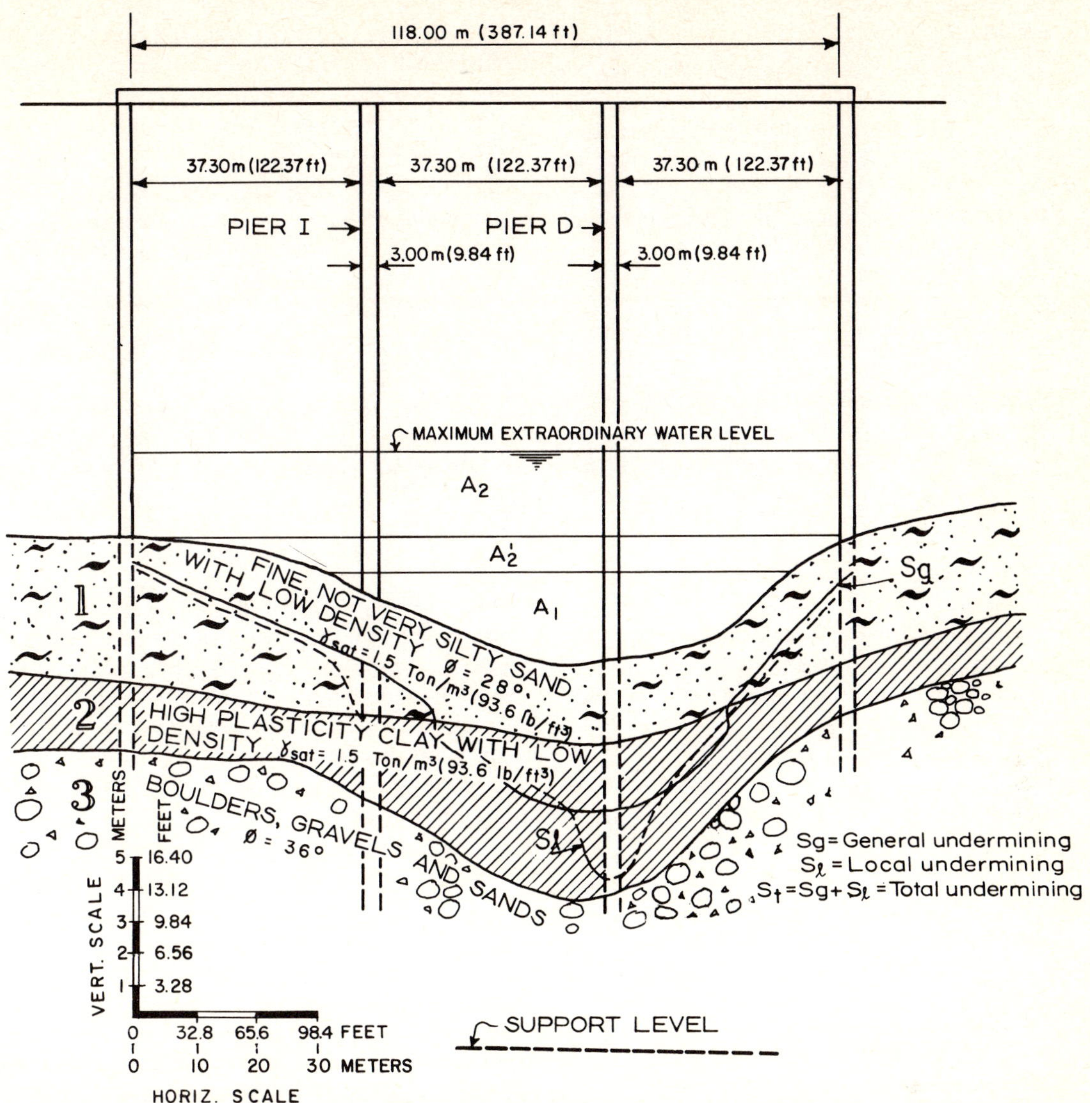

Fig. 8a-2 Profile of the river crossing

b) Meyerhof's method:

$$q_c = cN_c + \gamma D_f N_q + \frac{1}{2} \gamma B N_\gamma \qquad (8\text{-}2)$$

In order to obtain the values of N_q and N_γ, it is first necessary to compute the length of pile that must be embedded so that the failure surface of a deep foundation can develop completely. This is given by:

$$D = 4\sqrt{N_\varnothing}\, B$$

where:

$$N_\varnothing = \tan^2(45° + \varnothing/2) = \tan^2(45° + 36°/2) = 3.852$$

$$B = \text{width of the foundation} = 4 \text{ m (13 ft)}$$

$$\therefore D = 4\sqrt{3.852} \times 4 = 31.41 \text{ m (103 ft)}$$

Interpolating proportionally between the values of N_q and N_γ for shallow foundations and for piles (Fig. 8-9), bearing capacity factors are obtained for an embedment length of 5 m (16.5 ft). This results in N_q = 66 and N_γ = 68. Applying Eq.(8-2):

$$q_c = 5.3 \times 66 + \frac{1}{2}\, 1.0 \times 4 \times 68 = 485.8 \text{ t/m}^2 \text{ (99.36 kips/ft}^2\text{)}$$

$$q_a = q_c/F_s = 485.8/3 = 162 \text{ t/m}^2 \text{ (33.12 kips/ft}^2\text{)}$$

c) Balla's method:

$$q_c = cN_c + \gamma D_f N_q + \frac{1}{2} \gamma B N_\gamma \qquad (8\text{-}17)$$

D_f = length of pile embedded in the supporting stratum = 5 m (16.5 ft); B = 4 m (13 ft); D_f/B = 5/4 = 1.25; ρ = 4.7 (Fig. 8-15).

N_q = 45; N_γ = 165 (Fig. 8-16)

Applying Eq.(8-17):

$$q_c = 5.3 \times 45 + \frac{1}{2} \times 1 \times 4 \times 165 = 568.5 \text{ t/m}^2 \text{ (116.25 kips/ft}^2\text{)}$$

$$q_a = 190 \text{ t/m}^2 \text{ (38.88 kips/ft}^2\text{)}$$

(Problems 1 and 2 are by courtesy of A.D. Colina, M. Eng.)

8a.3 Pile Foundation

A flyover or overpass is to be built at a location with a stratigraphy as shown in Fig. 8a-3. The total load that is transmitted by the structure to the foundation level in the intermediate supporting layer is 645 t (1421 kips). The dimensions of the supporting footing are tentatively to be 4.00 m × 13.00 m (13 × 43 ft); 52 m² (560 ft²). The foundation of the structure is to be analyzed using: a) Friction piles in the first stratum, b) Point-bearing piles resting on the intermediate layer of dense silty sand.

Friction piles

The bearing capacity of a single pile will be computed

$$q_{ad} = \frac{a}{F_s} p L \qquad (8\text{-}30)$$

$$a = \alpha c_u$$

where:

q_{ad} = working bearing capacity, in t (kips)
a = adhesion between piles and soil
p = perimeter of the straight section of the pile, in m (ft)
L = effective length of the pile, in m (ft)
α = coefficient which depends on the roughness of the pile walls (according to TOMLINSON [50]).
c_u = cohesion of the soil, in t/m² (kips/ft²)
F_s = factor of safety

In Fig. 8a-3, c = 0.4 kg/cm² (0.82 kips/ft²); according to Table 8-5, a = 3.5 t/m² (0.72 kips/ft²). If a factor of safety F_s = 2 is selected, the result will be:

$$\frac{a}{F_s} = \frac{3.5}{2} = 1.75 \text{ t/m}^2 \text{ (0.36 kips/ft}^2\text{)}$$

The thickness of compressible soil, according to Fig. 8a-3, is 30 m (100 ft); 15% of the total thickness, that is 4.5 m (14.8 ft), will remain as a compressible layer underneath the tip of the pile. Taking the above into account, the length of the piles will be 25.5 m (83.7 ft).

In Table 8a-1, values are given for the bearing capacity by adhesion for different perimeters of circular and square cross-sections, calculated with the expression for q_{ad} which was given above. A pile is considered with a square cross-section, 30 cm × 30 cm (1.0 × 1.0 ft). The number of piles will be:

$$N_p = \frac{\text{Total load imposed by the structure}}{\text{Bearing capacity of the pile}} = \frac{645}{53} = 13$$

Analyzing the bearing capacity of the group of piles, according to Eq.(8-29):

$$Q_g = 2L(A + B)f + 1.3c_u N_c A - B \qquad (8\text{-}29)$$

Since they are adhesion piles, the second term of the second part of the equation does not have to be considered in the analysis. In the above expression: L is the length of the pile; A is the width of the area formed by the group of piles; B is the length of the area formed by the group of piles; AB is the area covered by the group of piles = 52 m² (560 ft²); f is the value of adhesion in the walls of the block of soil formed by the group of piles. It is taken as equal to c_u in t/m² (kips/ft²).

Now $q_g = Q_g/F_s$ and taking a value for F_s of 1.5:

$$q_g = \frac{2 \times 25.5\ (4 + 13)}{1.5} \times 4$$

$$= 2{,}312 \text{ t (5092 kips)}$$

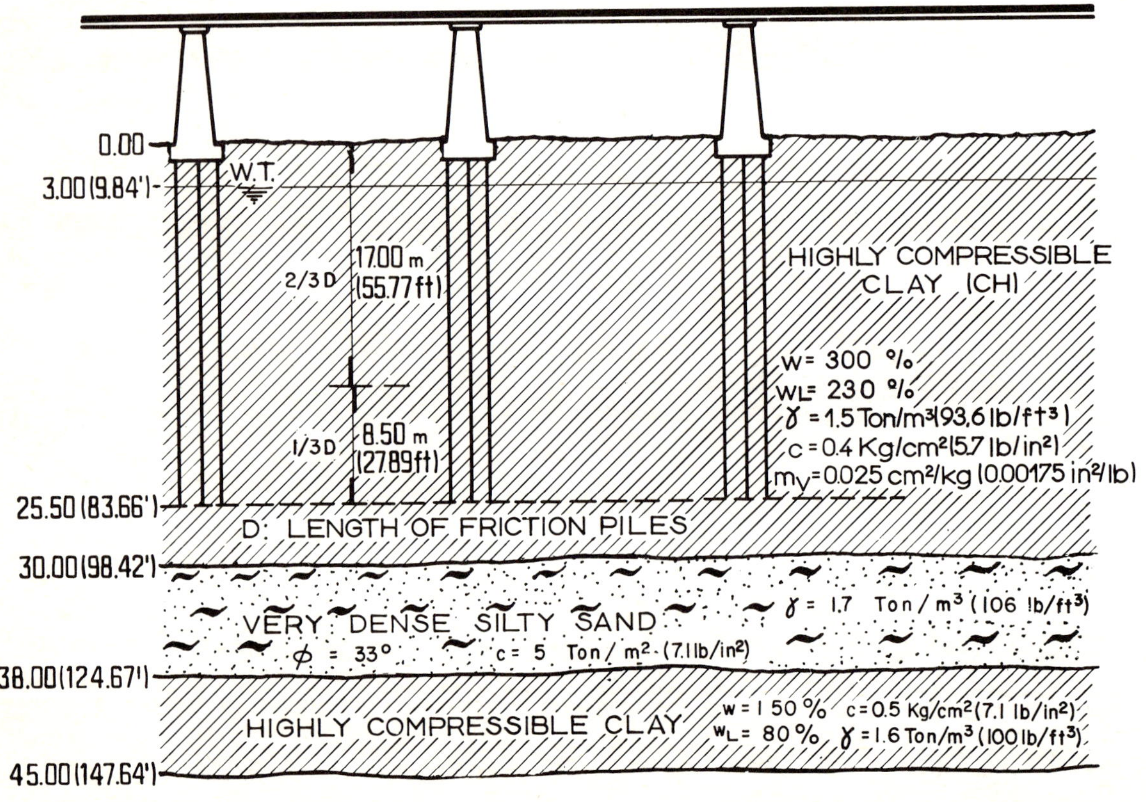

Fig. 8a-3 Structural and soil profile

Table 8a-1

Values of Qad calculated from Eq.(8-30) For Pile Foundation

L = 25.2 m (83 ft) Square section							
Side		Perimeter		Lateral area		Qad	
m	ft	m	ft	m^2	ft^2	t	kips
0.25	0.82	1.00	3.28	25.5	274	44.6	98.5
0.30	1.00	1.20	4.00	30.6	330	53.6	118.0
0.35	1.15	1.40	4.64	35.7	385	62.5	137.5
0.40	1.32	1.60	5.32	40.8	440	71.4	157.5
Circular section							
Diameter		Perimeter		Lateral area		Qad	
m	ft	m	ft	m^2	ft^2	t	kips
0.25	0.82	0.7854	2.58	20.0	216	35.0	77.0
0.30	1.00	0.94	3.09	24.0	258	42.1	92.7
0.35	1.15	1.10	3.62	28.0	302	49.1	108.0
0.40	1.32	1.26	4.14	32.0	344	56.1	123.6

According to Eq.(8-34): $apnL \leqslant aPL$; $3.5 \times 1.20 \times 25.5 \times 13 = 1392$ t and thus $1392 < 2312$ and there will be no risk of failure in the group of piles.

The *Settlement analysis* will now be carried out. For calculation purposes, the technique presented by TERZAGHI and PECK will be adopted, assuming the presence of an imaginary slab at $\frac{2}{3}D$, where D is the depth of the piles within the soil. The stress at 6.5 m (21.32 ft) below that level (center of the soil that is regarded as compressible), will be computed according to the following expression, assuming that the influence of the overload drops to 60°, as is usual:

$$\Delta p' = \sigma_z = \frac{wBL}{(B + 2z \tan 30°)(L + 2z \tan 30°)}$$

where: B and L are the dimensions of the area where the piles are placed; w is the surcharge imposed on the surface by the structure (645 t/52m^2 = 12.4 t/m^2 (2.54 kips/ft^2)); and σ_z is the stress at the depth z = 6.5 m (21.32 ft).

$$\Delta p' = \sigma_z = \frac{12.4 \times 4 \times 13}{(4 + 2 \times 6.5 \times 0.576)(13 + 2 \times 6.5 \times 0.576)}$$

$$= 2.73 \text{ t/m}^2 \text{ (0.56 kips/ft}^2\text{)}$$

$$\Delta H = m_v \Delta p'H = 0.025 \times 2.73 \times 13 = 0.89 \text{ m (2.9 ft)}$$

Point-Bearing Piles

It is assumed that the piles are resting on a firm stratum which starts at a depth of 30 m (98 ft). MEYERHOF's method (1963) is used to calculate bearing capacity:

$$q_c = cN'_c + \gamma D_f N'_q \qquad (8\text{-}11)$$

where: q_c is the ultimate bearing capacity of the pile, t/m^2 (kips/ft^2); c is the cohesion of the material below the foundation level; D_f is the depth of the pile, in m (ft); and N'_c, N'_q are coefficients of bearing capacity that depend on the angle of internal friction, $\varnothing$, of the material and of the type and dimensions of the foundation.

In Fig. 8-9: $N'_c = 600$, $N'_q = 100$. In order to apply these factors, the base of the pile must penetrate the stratum under consideration by at least:

$$D = 4\sqrt{N_\varnothing}\, B$$

where: D is the length of pile that must penetrate the supporting layer; $N_\varnothing$ is the $\tan^2(45° + \varnothing/2)$; and B is the width of the pile. (Square section 30 × 30 cm (1.0 × 1.0 ft)).

$$N_\varnothing = \tan^2\left(45° + \frac{33°}{2}\right) = \tan^2 61.5° = 3.4$$

$$D = 4\sqrt{3.4} \times 0.30 = 2.2 \text{ m (7.0 ft)}$$

Therefore the piles must be placed at a depth of 32 m (105 ft). Calculation of γD_f must take the water-table into account:

$$\gamma D_f = 1.5 \times 3 + 0.5 \times 27 + 2 \times 0.7 = 19.4 \text{ t/m}^2 \text{ (4 kips/ft}^2\text{)}$$

$$q_c = 5 \times 600 + 19.4 \times 100 = 4940 \text{ t/m}^2 \text{ (1011 kips/ft}^2\text{)}$$

and the ultimate bearing capacity of the pile will be: 444.6 t (978 kips).

Considering $F_s = 3$:

$$Q_{ad} = \frac{444.6}{3} = 148 \text{ t (326 kips)}$$

and the number of piles will be:

$$n = \frac{645}{148} = 5 \text{ piles in the central support}$$

The calculation is repeated using TERZAGHI's theory (N coefficients):

$$q_c = cN_c + \gamma D_f N_q + \frac{1}{2}\gamma B N_\gamma \qquad (8\text{-}2)$$

$$q_c = 5 \times 50 + 19.4 \times 30 + 0.35 \times 0.30 \times 30$$

$$q_c = 250 + 582 + 3.15 = 835 \text{ t/m}^2 \text{ (171 kips/ft}^2\text{)}$$

The result is: $Q_c = 835 \times 0.09 = 75$ t/pile (165 kips/pile) (ultimate load).

The difference between TERZAGHI's and MEYERHOF's results is significant.

(This problem is by courtesy of J. A. MENDOZA-MÁRQUEZ)

REFERENCES

1. JUÁREZ BADILLO, E. and RICO, A., *Mecánica de Suelos. Vol. II: Teoría y Aplicaciones de la Mecánica de Suelos,* Limusa: Mexico, 1977, 3rd. Edn., Chaps. VII and IX.
2. WU, T. H., *Soil Mechanics,* Allyn and Bacon Inc., 1966, Chap. 8.
3. VESIC, A. S., "Análisis de la capacidad de carga de cimentaciones superficiales," Journal Engineering, Faculty of Engineering, U.N.A.M. México, 1973.
4. DE MELLO, V. F. B., "Foundations of Buildings in Clay. Summary about the State-of-the-Art." *VII. ICSMFE,* Mexico, 1969.
5. MOORHOUSE, D. C., "Shallow Foundations. State-of-Art Paper," Proc. of the Specialty Conference on Performance of Earth and Earth-Supported Structures, Purdue University, Indiana, 1972.
6. TERZAGHI, K., *Theoretical Soil Mechanics,* John Wiley and Sons, 1956, Chap. VIII.
7. PECK, R. B., HANSON, W. E. and THORNBURN, T. H., *Foundation Engineering,* John Wiley and Sons, 1957.
8. MEYERHOF, G. G., "The Ultimate Bearing Capacity of Foundations," *Geotechnique,* December, 1951, Vol. II.
9. MEYERHOF, G. G., "Recherches sur la force portante des pieux." Suplements des Annaux du Institute du Batiment et Travaux Publiques, Paris, March-April, 1953.
10. MEYERHOF, G. G., "Influence of Roughness of Base and Groundwater Conditions on the Ultimate Bearing Capacity of Foundations," *Geotechnique,* 1955, Vol. V.
11. MEYERHOF, G. G., "Some Recent Research on the Bearing Capacity of Foundations," *Canadian Geotechnical Journal,* 1963, Vol. I, No. 1.
12. SKEMPTON, A. W., "The Bearing Capacity of Clays," Building Research Congress, The Institution of Civil Engineers, Div. I. London, 1951.
13. HANSEN, B., "A General Formula for Bearing Capacity," The Danish Geotechnical Institute, 1961, Tech. Bull. No. 11.
14. HANSEN, B., "Note Concerning Geotechnical Institute Bulletin No 1," The Danish Geotechnical Institute, 1966, Tech. Bull. No. 21.
15. HANSEN, B., "A Revised and Extended Formula for Bearing Capacity," The Danish Geotechnical Institute, 1968, Tech. Bull. No. 28.
16. SOWERS, G. B. and SOWERS, G. F., *Introductory Soil Mechanics,* John Wiley and Sons, Chap. 9.
17. BEREZANTZEV, V. G., KHRISTOFOROV, V. S. and GOLUBKOV, V. V., "Load Bearing Capacity and Deformation of Pile Foundations," *Proc. V. ICSMFE,* Paris, 1961, Vol. II.
18. BALLA, A., "Bearing Capacity of Foundations," *Journal of Soil Mechanics and Foundations Division, ASCE.* October 1962, Vol. SM5-89.
19. BOWLES, J. E., *Foundation Analysis and Design,* The McGraw Hill Book Co., 1968, Chap. 2.
20. FEDA, J., "Discussion of Balla's Bearing Capacity of Foundations," *Journal of Soil Mechanics and Foundations Division, ASCE,* May, 1963, Vol. SM3-89.
21. REDDY, A. S and SRINIVASAN, R. J., "Bearing Capacity of Footings on Layered Clays," *Journal of the Soil Mechanics and Foundations Division, ASCE,* March, 1967, Vol. SM2-93.
22. DAVIS, E. H. and CHRISTIAN, J. T., "Bearing Capacity of Anisotropic Cohesive Soils," *Journal of the Soil Mechanics and Foundations Division, ASCE,* May, 1971, Vol. SM5-97.
23. MEYERHOF, G. G. and BROWN, J. D., "Experimental Study of Bearing Capacity of Layered Clays," *Proc. VII. ICSMFE,* Mexico, 1969, Vol. II.
24. MILOVIC, D. M., "Comparison between the Calculated and Experimental Values of the Ultimate Bearing Capacity," *Proc. VI. ICSMFE.* Montreal, Canada, 1965, Vol. II.
25. MEYERHOF, G. G., "The Bearing Capacity of Foundations under Eccentric and Inclined Loads," *Proc. III. ICSMFE, Zurich, 1953, Vol. I.*
26. Reference [1], Chap. VIII
27. MEYERHOF, G. G., "The Ultimate Bearing Capacity of Foundations on Slopes," *Proc. IV. ICSMFE,* London, 1957, Vol. I.
28. TERZAGHI, K. and PECK, R. B., *Soil Mechanics in Engineering Practice,* John Wiley and Sons, Inc., 1948.
29. OLSON, R. E. and FLAATE, K. S., "Pile Driving Formulas for Friction Piles in Sand," *Journal of the Soil Mechanics and Foundations Division. ASCE,* 1967, Vol. SM6-93.
30. TERZAGHI, K., "Discussion of Pile Driving Formulas Progress Report of the Committee on the Bearing Value of Pile Foundations." *Proc. ASCE,* 1942, Vol. 68 No. 2.
31. KEZDI, A., "Bearing Capacity of Piles and Pile Groups," *Proc. IV. ICSMFE,* London, 1957, Vol. II.
32. KISHIDA, H. and MEYERHOF, G. G., "Bearing Capacity of Pile Groups under Eccentric Loads in Sand," *Proc. VI. ICSMFE,* Montreal, 1965, Vol. II.

33. SAFFERY, M. and TATE, A. P. K., "Model Test on Pile Groups in a Clay Soil with Particular Reference to the Behavior on the Group when it is Loaded Eccentrically," *Proc. V. ICSMFE,* Paris, 1961.

34. SOWERS, G. F., MARTIN, C. B., WILSON, L. L. and FAUSOLD, M., "The Bearing Capacity of Friction Pile Groups in Homogeneous Clay from Model Studies," ibid.

35. CORREA, J. J., RICO, A., MORENO G. and ESQUIVEL, R., "Pruebas de carga en modelos de cimientos profundos en arenas," Conference on Deep Foundations, Mexico, December, 1964.

36. VESIC, A. S., "Experiments with Instrumented Pile Groups in Sand. Performance of Deep Foundations," *ASTM* 1969, STP No. 444.

37. SCHLITT, H. G., "Group Pile Loads in Plastic Soils," *Proc. H.R.B.,* 1952, Vol. 31.

38. KOIZUMO, Y. and ITO, K., "Field Tests with regard to Pile Driving and Bearing Capacity of Pile Foundations," *Soils and Foundation,* 1967, Vol. 7, No. 3.

39. MEYERHOF, G. G., *General Report,* presented to Session I. Conference on Deep Foundations, Mexico, 1964.

40. McCLELLAND, B., "Design and Performance of Deep Foundations," Specialty Conference on Performance of Earth and Earth-Supported Structures, ASCE, Purdue University, Lafayette, Indiana, 1972.

41. MOORHOUSE, D. C. AND SHEEHAN, J. B., "Predicting Safe Capacity of Pile Groups," *Civil Engineering,* 1968, Vol. 38, No. 10.

42. WHITAKER, T., "Experiments with Model Piles in Groups," *Geotechnique,* 1957, Vol. VII.

43. WHITAKER, T., "Some Experiments on Model Pile Foundations in Clay," Symposium on Design of Pile Foundations, Part I, Stockholm, 1960.

44. HANNA, T. H., "Model Studies of Foundation Groups in Sand," *Geotechnique,* 1963, Vol. XIII.

45. COYLE, H. M. and SULAIMAN, I. H., "Bearing Capacity of Foundation Piles: State-of-the-Art," *HRB,* 1970, Rec. No. 333.

46. FLAATE, K., "Effects of Pile Driving in Clays," *Canadian Geotechnical Journal,* 1972, Vol. 9, No. 1.

47. PECK, R. B., "A Study of the Comparative Behavior of Friction Piles," *HRB,* 1958, Special Report No. 36.

48. SEED, H. B. and REESE, L. C., "The Action of Soft Clay along Friction Piles," *Trans. ASCE,* 1957, Vol. 122.

49. KERISEL, J. L., "Vertical and Horizontal Bearing Capacity of Deep Foundations in Clays," Procs., Symposium on Bearing Capacity and Settlement of Foundations, Duke University, 1965.

50. TOMLINSON, M. J., "The Adhesion of Piles Driven in Clay Soil," *Proc. IV. ICSMFE,* London, 1957, Vol. II.

51. McCLELLAND, B., FOCHT, J. A. and EMRICH, W. J., "Problems in Design and Installation of Offshore Piles," *Journal of the Soil Mechanics and Foundations Division, ASCE.,* 1969, Vol. SM6-95.

52. MEYERHOF, G. G. and MURDOCK, L. J., "An Investigation of the Bearing Capacity of some Bored and Driven Piles in London Clay," *Geotechnique, 1953, Vol. III.*

53. SKEMPTON, A. W., "Cast in-Situ Bored Piles in London Clay," *Geotechnique, 1959, Vol. IX.*

54. O'NEILL, M. W. and REESE, L. C., "Behavior of Bored Piles in Beaumont Clay," *Journal of the Soil Mechanics and Foundations Division, ASCE,* 1972, Vol. SM2-98.

55. BOWLES, J. E., *Foundation Analysis and Design,* McGraw Hill Book Co., 1968, Chap. 9.

56. HOLTZ, W. G. and GIBBS, H. J., "Field Test to Determine the Behavior of Piles in Loess," *Proc. III. ICSMFE,* Zurich, 1953, Vol. II.

57. COYLE, H. M. and SULAIMAN, I. H., "Skin Friction for Steel Piles in Sand," *Journal of Soil Mechanics and Foundations Division, ASCE,* 1967, Vol. SM6-93.

58. VESIC, A. S., "Tests on Instrumented Piles: Ogeechee River site," *Journal of Soil Mechanics and Foundations Division, ASCE,* 1970, Vol. SM2-96.

59. NORDLUND, R. L., "Bearing Capacity on Piles in Cohesionless Soils," *Journal of Soil Mechanics and Foundations Division, ASCE,* 1963, Vol. SM3-89.

60. HUNTER, A. H. and DAVISSON, M. T., "Measurements of Pile Load Transfer," *Performance of Deep Foundations,* ASTM, 1969, STP 444.

61. TENG, W. C., *Foundation Design,* Prentice Hall, Inc. 1962, Chap. 8.

62. HANSEN, B., "A Theory for Skin Friction on Piles," Danish Geotechnical Institute, 1968, Bull. No. 25.

63. MAZURKIEWICZ, B. K., "Skin Friction on Model Piles in Sand," Danish Geotechnical Institute, ibid.

64. ZEEVAERT, L., "Reducción de la capacidad de carga en pilotes apoyados de punta, debido a la fricción negativa," I. Panamerican Conference on SMFE, Mexico, 1959, Vol. I.

65. MEYERHOF, G. G., "Bearing Capacity of Rock," *Magazine of Concrete Research,* April, 1953.

66. BISHNOI, B. W., "Bearing Capacity of Jointed Rock," Ph. D. Thesis Georgia Institute of Technology, Atlanta, Ga., 1968.

67. SPRINGALL, G. and ESPINOSA, L., "El subsuelo de la Península de Yucatán. Present State of Knowledge," VI. National Meeting on Soil Mechanics, Foundations in Mexican Urban Areas, (Acapulco, Morelia, Tampico and Yucatán). Publication of the Mexican Society of Soil Mechanics, Mexico, 1972, Vol. I.

68. JÜRGENSON, L., "The Application of Elasticity and Plasticity to Foundation Problems," Contribution to Soil Mechanics, Boston Society of Civil Engineers, 1925-1940.

69. Reference [1], Chap. II.

70. LADD, C. C., "Settlement Analysis of Cohesive Soils," M.I.T. Special Summer Program 1.34S, M.I.T., Boston, Mass., 1971.

71. LAMBE, T. W., "Methods of Estimating Settlements," *Journal of the Soil Mechanics and Foundations Division, ASCE.,* 1964, Vol. SM5-90, Also mentioned in the book *Soil Mechanics,* T. W. LAMBE AND R. V. WHITMAN.

72. DAVIS, E. H. and POULOS, B. E., "The Use of Elastic Theory for Settlement Prediction under Three-dimensional Conditions," *Geotechnique,* 1968, Vol. 18.

73. Skempton, A. W. and Bjerrum, L., "A Contribution to the Settlement Analysis of Foundations on Clay," *Geotechnique,* 1957, Vol. 7.

74. Komornik, A., Wiseman, G. and Frydman, S., "A Study of In-Situ Testing with the Pressurometer," Conference on In-Situ Investigations in Soils and Rocks, British Geotechnical Society, Session III, 1969.

75. Thorley, A., "Borehole Instruments for Economical Strength and Deformation In Situ Testing," ibid.

76. Reference [16], Chap. 10.

77. Correa, J. J., Quintero, J. and Aztegui, E., "Pruebas de carga en pilotes para la cimentación del Puente Alvarado," Conference on Deep Foundations, Mexico, 1964.

78. Terzaghi, K., "Evaluation of Coefficients of Subgrade Reaction," *Geotechnique,* 1955.

79. Reese, L. C. and Matlock, H., "Non-dimensional Solutions for Laterally Loaded Piles with Soil Modulus Assumed Proportional to Depth," *Proc. VIII. Texas Conference on Soil Mechanics and Foundations Engineering,* The University of Texas, Austin, 1956.

80. Department of the Navy. "Bureau of Yards and Docks," *Design Manual Soil Mechanics: Foundations and Earth Structures,* (DM7), Washington D.C., 1962.

81. Murayama, S. and Shibata, T. "The Bearing Capacity of a Pile Driven into Soil and its New Measuring Method," *Soil Foundations,* Japan, 1960, Vol. 1, No. 2.

82. Lee, D. H., *An Introduction to Deep Foundations and Sheetpiling,* Concrete Pub. Ltd.: London, 1961, Chap. IX.

83. Tomlinson, M. J., *Foundation Design and Construction,* Pitman, 1975, 3rd. Edn, Chap. 6.

84. Lasso Herrera, R., *Procedimientos de construcción para puentes,* Author's Edition, Mexico, 1964.

85. Juárez Badillo, E. and Rico, A., *Mecánica de Suelos, Vol. I: Fundamentos de la Mecánica de Suelos.* Limusa: Mexico, 1977, Appendix.

86. De Mello, V. F. B., "El ensayo de penetración estándar. State-of-the-Art Paper." IV. Panamerican Conference on SMFE, S. Juan, Puerto Rico, 1971, Vol. I.

87. Holtz, W. G. and Gibbs, H. J., "Research on Determining Density of Sand by Spoon Penetration Testing," *Proc. IV. ICSMFE,* London, 1957.

88. Coffman, B. S., "Estimating the Relative Density of Sands," *Civil Engineering,* October, 1960.

89. Dobry, R., "Cono dinámico y prueba estándar en arenas limosas," *IV. Panamerican Conference on SMFE,* S. Juan, Puerto Rico, 1971, Vol. II.

90. Reference [85], Chap. X.

91. Gendron, G. J., "Pile Driving: Hammers and Driving Methods," Highway Research Board, Washington D.C., 1970, Record No 333 (Pile Foundations).

92. Grand, B. A., "Types of Piles: Their Characteristics and General Use," ibid.

93. Moreno Pecero, G., "Hincado de pilotes," Unpublished notes for internal use of the Ministry of Public Works, Mexico, 1971.

94. Gerwick, B. C., "Current Construction Practices in the Installation of High-Capacity Piling," Highway Research Board, Washington D. C., 1970, Record No. 333 (Pile Foundations).

95. Chellis, R. D., *Pile Foundations,* McGraw Hill Book Co., Inc., 1951.

96. York, D. L., "Structural Behavior of Driven Piling," Highway Research Board, Washington D. C., 1970, Record N° 333 (Pile Foundations).

97. Reference [82], Chap. X.

98. Demeneghi, A., "Chiflonado en pilotes," Unpublished notes, for internal use of the Ministry of Public Works, Mexico, 1971.

99. De la Fuente, E., "Estudio de permeabilidad," Unpublished notes. Personal Communication with the Author, Mexico, 1972.

100. Juárez Badillo, E. and Rico, A., *Mecánica de Suelos, Vol. III: Flujo de agua en suelos,* Limusa: Mexico, 1969.

101. Harr, M. E., *Groundwater and seepage,* McGraw Hill Book, Co., 1962.

102. Mansur, Ch. I. and Kaufman, R. I., "Dewatering," Chapter 3 of *Foundation Engineering,* G. A. Leonards, McGraw Hill Book, Co. Inc., 1962.

103. Cedergren, H. R., *Seepage, Drainage and Flow Nets,* John Wiley and Sons, Inc., 1967, Chap. 2.

OTHER RELATED REFERENCES

1. Aguirre, M., "Dispositivo para controlar hundimientos de estructuras piloteadas," Institute of Engineering, UNAM, May, 1981, No. 439.

2. Akinmusuru, J. O., and Akinbolade, J. A., "Stability of Loaded Footings on Reinforced Soil," *Journal of the Geotechnical Engineering Division Proc. ASCE,* June, 1981, Vol. 107, No. GT6

3. Lobdell, G. T., "Hydroconsolidation Potential of Palouse Loess," *Journal of the Geotechnical Engineering Division, Proc. ASCE,* June, 1981, Vol. 107, No. GT6.

4. Martin, R. E., "Estimating Foundations Settlements in Residual Soils," *Journal of the Geotechnical Engineering Division, Proc. ASCE,* March, 1977, Vol. 103, No. GT 3.

5. Meyerhof, G. G., "Ultimate Bearing Capacity of Footings on Sand Layer overlying Clay," *Canadian Geotechnical Journal,* 1974, Vol. II.

6. Oda, M., "Anisotropic Strength of Cohesionless Sands," *Journal of the Geotechnical Engineering Division, Proc. ASCE,* September, 1981, Vol. 107, No. GT 9.

7. O'Neill, M. W. and Heydinger, A. G., "Design Inferences for Pile Groups," *Public Roads,* September, 1981, Vol. 45, No. 2, pp. 53-60.

8. Reimbert, M. & A., *Retaining Walls. Vol. I: Anchorages and Sheet Piling. Vol. II: Study of Passive Resistance in Foundation Structures,* Trans Tech Publications, Clausthal-Zellerfeld, 1974/1978.

9. Sogge, R. L., "Laterally Loaded Pile Design," *Journal of the Geotechnical Engineering Division, Proc. ASCE,* September, 1981, Vol. 107, No. GT 9.

CHAPTER 9

FLEXIBLE PAVEMENTS

9.1 Introduction

A pavement is a structural system composed of a layer or layers of appropriate materials comprising the upper level of an embankment or compacted original ground and the wearing surface of a road. The chief functions of a pavement are to provide a smooth wearing surface with a suitable color and texture to withstand traffic, weathering and other harmful agents, and efficiently transmit the stresses produced by traffic loads to the underlying ground. In other words, a pavement is the superstructure of a road. On it depends the speed and ease with which vehicles can travel from one point to another with the degree of safety, comfort and economy prescribed by the design. The structure of the components of a pavement, including the characteristics of the materials used to construct it, provide a wide variety of designs. It may consist of just one layer or, what is more commonly the case, several layers, Plate 9-1, each of which may be of selected natural materials each with a different function. The wearing surface may be an asphalt overlay, a slab of Portland cement concrete, or duly compacted stony materials. Current technology includes an extensive range of different structural sections. Selection of the most appropriate one for the specific conditions under consideration is a demanding task that requires a pavement specialist.

In a somewhat arbitrary fashion (and for practical purposes) pavements are divided into two categories: flexible and rigid. However, the rigidity or flexibility of a pavement is not so easily defined as to permit a sharp differentiation between the two categories. The decision as to just how rigid a flexible pavement may be, or how flexible a rigid pavement may be, is governed by available materials, possible settlement and ultimately the pavement specialist's wise judgement. Although it may appear that the terms employed to distinguish one type of pavement from another are not altogether appropriate, they are so widely accepted that it is advisable to continue to use them. Also, there is seldom a cause of any important confusion where practical communication is concerned. The pavements are differentiated and defined more in terms of the materials of which they are made and how these materials are structured, and less by the way in which they transmit the stresses and deformations produced by traffic to the underlying layers. For the purposes of this book, a rigid pavement will be regarded as one made of slabs of portland cement concrete. Any other pavement will be regarded as flexible. This somewhat arbitrary classification is in agreement with worldwide practice.

Plate 9-1 Cut through a flexible pavement

Clearly the strength and smoothness of the natural ground surface can rarely meet the demands of modern transportation systems. This is true even if the adjective *modern* is applied well into the past. With the gradual evolution of vehicle weight, speed, comfort and autonomy, improvements became more and more necessary in the curvature, slope, visibility, cross-section, uniformity and texture of roads. These demands led to the evolution and subsequent construction of embankments. The surface of these embankments had to provide conditions appropriate to the increasing volumes of faster and heavier vehicles. For economy, embankments are built with materials found in the immediate vicinity; therefore this led to the utilization of soils and rock fragments. The natural road surface, represented by the uppermost layer of an embankment which is made only of natural stone materials, can only be used if the traffic volume and weight is very small. Even if the earth materials or rock fragments are carefully selected and mechanically treated (compaction), a suitable surface cannot be obtained if the traffic volume is moderate and the loads high. Natural materials, when used for surfacing, may provide suitable conditions for a short time, but it has not been possible to assure the permanence of these conditions when traffic exceeds some minimum limit, although these minimums do exist in many countries where industrial development is limited.

Consequently, an initial distinction must be established for the wearing surface. For roads with little traffic (typically, those with less than 200 vehicles per day), low-cost wearing surfaces made of carefully selected and compacted rock fragments, or mixtures of rock fragments and soils are often used for economy. A low-cost surface can be obtained, capable of providing suitable wearing conditions for a limited period of time. However, the recognized susceptibility of these materials to the effects of water must be given due attention in other aspects of the design, such as longitudinal and transverse slope, curvature and surface drainage. It is worth repeating that in many developing countries, roads with low traffic volumes are the rule rather than the exception; the use of solutions just described should be kept in mind by the design engineer. These solutions may lead to the creation of a transport network capable of satisfying the genuine social and economic needs of a country, so long as the essential requirement is fulfilled: the level of technology with which the solutions are applied must be the highest economically possible. *Cheap* solutions are rarely as durable as those of very expensive projects. Unfortunately the engineer devotes more technical attention to important highway projects than to modest roads, with a consequent loss of performance for many simple solutions which, under suitable circumstances, could lead to enormous savings.

Independently of the above arguments, which are of great importance in highway engineering, when the level of traffic reaches a certain degree of importance, it becomes essential to add to the surface embankments a layer of material that meets the following requirements:

— Stability in the face of weathering: Rain, sun, freezing.

— Resistance to traffic loads

— Suitable texture for smooth driving

— Durability (in addition to weathering resistance)

— Appropriate permeability conditions

— Economy

— Constructability

The above requirements are satisfied by a layer of high-quality granular material which cannot be obtained under entirely natural circumstances, because the component particles must be bonded together by artificial means. Natural cohesive soils are rarely able to withstand the direct, prolonged action of traffic; and granular materials, in their natural form, despite their greater potential strength, provide an unstable surface owing to a lack of cohesion.

The layer under consideration are therefore more expensive than fill materials, which means that the economics must play an important role. The economic problem might be solved by using an expensive, but very thin pavement. This layer would also satisfy the requirements of stability, durability, texture and permeability; however, because it is so thin, very high stress levels would be transmitted to the fill below. The corresponding deflection would soon damage the pavement because there would not be sufficient support. Conflicting requirements are, therefore, involved, and must be reconciled. Two different lines of action have been followed to achieve this purpose.

1) The pavement system is made sufficiently thick and of a quality which enables the stresses transmitted to the embankment to be compatible with its resistance. This line of action leads to rigid pavements made of slabs of Portland cement concrete. Any deflections occurring in the underlying soils are absorbed by the resistance of the slab to tension.

2) The upper or wearing surface is a relatively thin bituminous layer, which is expensive and of high quality. Between this layer and the embankment there is a system of carefully selected layered materials; the strength and rigidity of each is less with increasing depth, in the same way the traffic stress decreases with increasing strength. The problem of layer design consists of varying the thickness, strength and rigidity of the materials of each layer in such a way that the material properties fit the stresses. These are the fundamental concepts involved in the design of flexible pavements.

Ultimately the pavement system thickness depends fundamentally on the materials of the embankment that provides the foundation.

As a further refinement to the design philosophy described above, it is sometimes advisable to include in the pavement, a system of layers with considerable rigidity and resistance to tension; this is done by adding a cementing material, such as cement, asphalt or lime. Layers that are subjected to this treatment show a corresponding improvement in their capacity to spread the stress from traffic, which makes it possible to reduce the total pavement thickness. Those systems that include semi-rigid layers such as soil-cement or soil-asphalt, constitute a third type of pavement, which is becoming increasingly important. Engineers regard these as semi-rigid pavements, (within the category of flexible pavements).

The above description implies that the problem of the layering of a pavement is simple; this is indeed true for the basic design, however, when these criteria are applied to a specific case, a vast number of uncertainties arise, so that detailed optimum design becomes very difficult. The difficulties are of several types. First, there are no rigid rules for the design of a pavement. The distribution of stresses and deformations cannot be easily computed in a multi-layer system made up of earth materials subject to dynamic traffic loads. Some theoretical solutions to this problem exist; reference will be made here to some of them. All, however, are based on simplifying assumptions that do not accurately depict material and pavement behavior. An example is the solutions in which the system is assumed to consist of homogeneous, isotropic, linearly elastic layes. Even if such assumptions were valid their use in pavement design becomes very complicated mathematically, to the extent that it goes beyond the capabilities of many experienced engineers. More important, pavement design cannot be carried out with the refinement that would be demanded by a detailed application of such theories. It is not efficient to use very detailed and complex theories for the design, when the construction processes are such that there is no assurance that the sophisticated design conditions will be fulfilled in the actual project.

Second, there are the insuperable difficulties involved in a reasonable evaluation of the natural climatic agents, to which all pavements are exposed.

Finally, the design is complicated by the many possible variants in the criteria adopted. In any situation, the engineer is presented with a variety of possible materials, some distant and some nearby, with different properties, each with advantages and disadvantages. The design is complicated by the possible combinations of thicknesses of the different layers. A greater thickness of cheap, poor quality material may prove to be a successful substitute for a smaller thickness of a better but more expensive material. The already varied and complex rules of design are further complicated when considering the limits for the quality of the materials, which vary from one layer to another, one climate to another, and one region to another.

Traffic constitutes the pavement load, and the effects of this load, plus climate, must be maintained at non-destructive levels. Unfortunately little is known of traffic induced stresses. Traffic variations include the number of vehicles, their weight and load distribution. Furthermore, traffic moves, causing repeated loads which causes short term, changing stresses, temporary and permanent strains, and other effects that are relatively ill understood such as fatigue and elastic distortion, all of which complicate any attempt at defining a single *external loading condition*. Such a definition for a pavement is illusory compared to a similar definition for other structures.

Another consideration, which greatly complicates the choice of pavement design, is the enormous variety of circumstances involved in such a project. Pavement design for a large highway imposes criteria which are substantially different from those of a country road. These variations in criteria are of major importance, although they are sometimes neglected.

The economic factors of cost, useful lifespan, definition of acceptable service conditions (including damage or deterioration that may lead to repairs or reconstruction) add to the complex background to all decisions relating to pavement design and construction. All possible variants and criteria must be examined within an economic framework which goes beyond the simple consideration of what is cheaper or more expensive to build. It involves the analysis of the entire range of political-social factors connected with public investment and all considerations relating to the degree and quality of service the users expect.

The following factors can be regarded as the essential features of a flexible pavement, considered as a whole: Structural strength; Deformability; Durability; Cost; Maintenance requirements; Comfort: Riding qualities.

9.1.1 Structural Strength

The first condition that must be met by a pavement is to withstand traffic stresses within the limits of deterioration and gradual destruction predicated by the design. Traffic loads produce normal and shear stresses at all points throughout the pavement structure and the embankment below. The theoretical methodology analyzing pavement strength is provided by soil mechanics, and in this field the failure theories now most widely accepted are those relating to combined shear and normal stresses. Consequently, when studying flexible pavements, shear stresses are the main cause of structural failure; consequently, the shear strength of the soils is a fundamental property. The bearing capacity theories of soil mechanics usually refer to homogeneous, isotropic media. Consequently the heterogeneity of the structure of flexible pavements, plus their anisotropy, complicate simple theoretical strength considerations. Fortunately modern soil mechanics is developing solutions that are more realistic for pavements. Some praise-worthy theories have been developed by BURMISTER, IVANOV, BACHALEZ, ACUM, HOGG, etc.

Pavements are not only subjected to shear stresses from vertical loads but also more complex shear caused by vehicle acceleration and braking and to tension stresses which develop in the lower levels of each layer in the pavement structure, both beneath the wheel load and at the outer limits of vertical downward deformation. The problem of strength is closely related to the structure of the pavement materials, for even when the foundation materials are of poor quality, it is the protective thickness provided by the pavement that reduces the stresses to the safe bearing capacity of the foundation soils. This is a very general condition. The authors of this book have observed that many of the failures that are currently regarded as occurring in pavements, actually commence in the underlying soil foundation. The attitude of some designers that such layers only play a passive role, should be revised. This will give rise to greater demands for the quality of fill materials, the treatment to which they are subjected, their protection from water, and to the rejection of a large number of materials that are currently regarded as suitable for highway construction. The application of such stringent criteria will doubtless lead to a general rise in construction costs, which means that the extent of these demands should be limited to cases where they are truly justifiable. Acceptable roads have been built all over the world in accordance with current standards (and low traffic). When it is stated that a lack of care is observed in the quality of fill materials and construction methods, it is not the authors' intention to introduce unrealistic requirements indiscriminately, but rather to create a better understanding of the joint behavior of a pavement system and its corresponding fill, which at present is sometimes disregarded.

As has already been outlined in previous pages, determining the strength of the materials in a pavement is a difficult problem that has not yet been satisfactorily solved. It is affected not only by the type of soil and its treatment, but also by the soil's reaction to weathering, of which the variations in the water content are most important. The engineer can only rarely predict the most detrimental water content that may be reached by the materials he handles. However, knowledge of this is necessary for design, because it controls the strength in that critical condition. This is another of the major uncertainties of design, that has been solved on the basis of arbitrary assumptions which, to some degree, have been justified by experience. For example in a humid region it is reasonable to assume the soil will reach saturation. In other regions the soil will reach an *equilibrium moisture* or maintain the optimum compaction water content.

Another factor with a considerable influence on material strength is the type of loads that are applied and the loading rate. Pavements are subject to *moving* loads, and the effects of these are different from those of static loads (and not so well known). This is a further source of uncertainty. It is usually solved by means of theoretical analyses (BOUSSINESQ, BURMISTER, etc.), assuming that the loads are static. In laboratory tests and the design methods based on them, the situation is a little more realistic, for although the tests are conducted using static loads or very slow loading rates, the results are modified using empirical correlations that are based on the real behavior of the pavements under moving loads.

Repetitive loads, in the long term, affect the strength of the layers of pavements that are relatively rigid. This is why this effect is most evident in the surfacings and stabilized base courses of flexible pavements where fatigue phenomena occur. These are very hard to test and to quantify. In resilient soils, the repeated loading may eventually lead to collapse, a phenomenon which it has not been possible to introduce into the design by means of sufficiently reliable readings from laboratory or field tests. Moreover, repeated loading leads to particle breakage (Chapter 1) in some granular soils, which modifies the strength of the layer in a manner which cannot be easily quantified. Repeated loading also brings about penetration of granular particles into the layers of finer soil.

The strength of pavement materials is considered from two points of view.

1) The bearing capacity develops in the pavement layers, enabling them to withstand traffic stress in an appropriate manner.
2) The bearing capacity of the subgrade (which is the layer that unites the pavement and the embankment) which enables it to withstand the stresses it receives and in turn to transmit suitable stress levels to the embankment below.

Both points are vital when selecting materials for the different layers of a pavement. The closer the layer to the surface of the road, the greater the importance. The satisfactory fulfilment of bearing capacity requirements by a certain layer is to an extent independent of the thickness of that layer; it is more related to the spreading and transmission of stresses to underlying layers. A thin layer capable of withstanding the loads to which it is subjected, may transmit high stresses to the layers below it; whereas a thick layer, which is not necessarily stronger, will transmit much lower stress levels to the underlying layers. Thickness is particularly important when the layer has tensile strength. If it has tensile strength there is an important increase in the capacity to distribute stresses to greater underlying areas. In such a case the capacity to transmit low stress levels depends rather more on the tensile strength of the layer than on its thickness.

The importance of the subgrade on the general equilibrium of a pavement can never be emphasized enough. A strong subgrade layer will be capable of tolerating relatively high stresses, therefore smaller pavement thicknesses can be used on top of it without endangering total stability. This will lead to considerable economies in investment cost, for the costs of the different layers of a flexible pavement generally increase the nearer they are to the surface.

9.1.2 Deformability

In some ways, the problem of the deformability of pavement layers is the reverse of that of strength. Deformability of the materials usually increases greatly with depth layer by layer. The natural ground and even a compacted embankment are more susceptible to deformation than the pavement. Within the pavement, the subgrade, or lowest layer, is more deformable than the overlying layers. Therefore, deformability is of particular concern at relatively deep levels, because the stronger, upper layers are likely to be relatively less deformable even considering the higher stresses to which they are subjected. Deformations in pavements are of concern from two points of view. First, because excessive deformations are associated with failure, and second because a badly deformed pavement ceases to fulfil its function of providing smooth support for traffic regardless of whether the deformation is the result of structural collapse or not.

Traffic loads cause several types of deformations in pavements. Elastic deformations are instantly recovered. Any deformations which remain in the pavement after the cause of deformation has ceased are referred to as plastic deformations. Under repeated, moving loads, plastic deformation becomes cumulative and may reach inadmissible values. Paradoxically, this process is often accompanied by an increase in density of the deformable materials, in such a way that they become stronger and more resistant to deformation after they deform than before. However if deformation is so great that a layer in the system shears, that layer becomes looser. Repeated elastic deformation is of special importance in materials that have significant tensile strength, usually those in the upper part of the pavement system. Such layers may fail progressively by fatigue if the magnitude of the strain is considerable. Materials that undergo marked elastic deformation under loading are the most dangerous ones in this respect, and are often of volcanic origin.

Today there is a strong belief that the deformability of flexible pavements is the most important factor to be considered. A large number of *fashionable* design methods are focused on maintaining deformation within tolerable limits. Identification of these limits is a more complex task than might at first be supposed. They are usually defined on the basis of the experience of groups of engineers. A part of the problem is measuring the deformation that is suffered by a pavement under loading. In design this problem is considered in two phases. First, estimate the elastic deformation. This can be carried out with reasonable precision once the pavement materials are known, obtaining their deformation modulus by means of one of the different field tests available, which can be performed on test pavement systems under critical conditions of load and environment. These may be plate-bearing tests, or Benkelman beam tests, or dynamic sensors such as the Dynaflect, sonic or electric deflectometers. Some organizations make observations on constructed pavements, developing correlations for design between the elastic deformations and climate, traffic and characteristics of the materials. Measurements of the deformation modulus in the laboratory have also been made (for example, the Kansas triaxial test). However it is hard to reproduce critical field conditions and overcome scale problems in the laboratory. Once the deformation modulus is obtained for the different layers, elastic deformation can be computed in accordance with one of the theories that have been mentioned in earlier chapters.

The second problem involved in measuring strain is posed by plastic deformation, which is a cumulative effect of repeated loading. This has been approached with purely empirical data, which for design purposes require experimental extrapolations. For example, the many varieties of traffic loads are represented by a single value known as the *standard load*. This is the result of statistical studies on experimental sections or on highways subjected to the effects of real and classified traffic. The standard load takes into account the effects of repeated loading, for in order to determine it, its destructive effect has been correlated with those effects caused by the real repetitive loads. Once the *design* traffic has been established, a maximum permanent ultimate deformation is established from experimental evidence. The pavement is designed in such a way that this deformation develops at the end of the predicted useful lifespan.

There are two criteria for establishing maximum permissible deformation. One is the deformation which causes failure of a road. Failure is the condition at which the pavement loses the charactertistics of service for which it was designed (AASHTO criterion or serviceability index). The second is the deformation which demands reconstruction work of a certain economical importance (British criterion).

9.1.3 Durability

The practical uncertainties that are associated with the durability of a flexible pavement are not easy to describe, even in a general manner. It is hard to define how much durability will be best in each case. This is linked with the economic and social factors relating to the use of the road. In the case of a minor project, the life of the pavement may be far shorter than that of the road. If the series of reconstruction jobs that subsequently prove necessary are not as expensive as the initial cost of a far more durable pavement, plus the cost represented by the interruptions in service caused by the reconstruction work, limited durability can prove cheaper. On the other hand, for projects with heavy volumes of traffic and great social importance, very durable pavements are necessary to avoid costly interruptions in service. In both cases the availability of maintenance money compared to initial construction money must be considered, as well as the political implications of both.

Once the criterion for the desired duration of a pavement has been established, many practical uncertainties arise before this duration or life can be achieved. As has already been stated, the effects of climate and traffic on the life of a pavement cannot be determined with any degree of accuracy. During their useful lifetime, pavements may be exposed to extraordinary conditions, such as torrential rains, floods and earthquakes. It is questionable to establish a suitable resistance to these special circumstances, or the design criteria for these unpredictable events to achieve a specific pavement life. As a consequence of the above, the authors do not know of any design method that takes durability requirements into consideration in a quantitative, rational manner, unaffected by personal feelings.

9.1.4 Cost

Like all engineering structures, a pavement represents a successful balance between general strength and stability requirements on the one hand, and cost on the other. A correct design will be one which can meet service requirements satisfactorily and at a minimum ultimate expense. In order to achieve this equilibrium, there are a vast number of possible lines of action. This is the most uncertain aspects of design; one which demands the highest degree of wise judgment.

The first alternative that arises is the choice of the type of pavement to be used. Rigid, flexible or semi-rigid pavements have their advantages and disadvantages, which vary from one situation to another. Rigid pavements usually involve low maintenance costs because they do not deteriorate easily, but the cost of their construction is high, and they are limited by the availability of the necessary materials and specialized construction equipment. Flexible pavements require a smaller initial investment, but usually require higher maintenance costs. Semi-rigid pavements may provide, very economical, compromise solutions when suitable materials are available, because they permit appreciable reductions in thicknesses. There are no rigid rules for establishing the type of pavement that is appropriate in each case. Each specific situation must be evaluated in detail. Based on maintenance compared to first cost, rigid pavements are particularly desirable in urban areas, streets and avenues and in highways with heavy volumes of traffic, where any interruption or deterioration in service is costly (and politically troublesome). Pilots have a marked preference for rigid pavements in runways, because they permit smoother operations, when well built, and they are far more permanent than flexible pavements. Because of the color and nature of an asphalt pavement, the layers of air closest to the ground become very hot in bright sun. As a result, the air loses density and aircraft landing and especially take-off operations become more difficult. For these reasons (amongst others), pavements made of portland cement concrete are favoured for airports for high performance aircraft throughout the world. However, in many areas such as Mexico, a flexible pavement may be only half to one third the cost of a rigid one. This fact favors asphalt runways for rather less critical airfields, where the smaller volume of traffic or less demanding aircraft permit rougher landings and lower speeds, and where interruptions in service due to periodical maintenance work do not cause important upsets in transportation.

Once the type of pavement has been chosen, the materials that are to be used must be selected. These may be available in abundance, and the design adapted to the cheapest, but they may be so scarce that the pavement design will have to adapt to the materials available. When choosing borrow pits for the construction of a pavement, many problems arise from the homogeneity of the materials, the methods of excavation to be used, the treatments to be given to the different materials to improve them, the volume of waste and the volume of usable materials. All have an influence on cost.

Another decisive factor in the cost of a pavement which cannot be determined by rigid rules, concerns the construction procedures to which the different materials must be subjected if they are to meet the requirements of the project. Compaction, for example, includes a large number of important uncertainties many of which must be resolved during construction, relying on the experience and judgment of the design and construction engineers and the cooperation of the constructor.

9.1.5 Maintenance Requirements

Many of the uncertainties in pavement design are related to maintenance. Climatic factors have a decisive influence on the life of pavements. Therefore they must be considered in the design, so that the cost of maintenance will be reasonable. However, such factors involve many elements that are hard to assess. However they should be evaluated, in so far as possible, using previous experience and reliable information about local conditions. The intensity of traffic also reflects on maintenance. Future traffic must be predicted: number and type of vehicles.

Another factor to be considered is the future behavior of the embankments such as deformation, possible collapse and local saturation. These could cause excessive maintenance or even reconstruction. Embankment behavior frequently has a decisive influence on pavement life. A typical case is that of pavements that are placed on embankments which are likely to suffer deformations because of the soft, compressible foundation grounds on which they are located. Because the pavement suffers from embankment settlement, a low-cost temporary pavement is appropriate. The surface and soil drainage are among the most important factors for determining both the lifespan of a pavement and the need to maintain it. For design purposes the drainage systems are regarded as essential parts of the pavement; they form an inseparable whole. All the uncertainties relating to drainage in roads and runways consequently affect pavement design.

As already mentioned, the structural degradation of the component materials from repeated loading is another important aspect of maintenance. Although there are some helpful tests available to help define the behavior of materials under these conditions, many doubts arise in specific cases. It is essential that these uncertainties be resolved with wise judgment and experience, for any mistakes in this field are reflected in costly maintenance and even reconstruction work.

Pavements often suffer from a lack of systematic maintenance; as a result their service life is greatly reduced. This is particularly true when resources are insufficient. A network of highways and runways is a costly asset on which much of the economy of the region depends. They cannot be allowed to deteriorate without deterioration of the region. Regions that feel the social and political obligation to devote much of their assets and energy to the construction of new roads must therefore have an equal obligation to conserve the ones they already possess. Consequently, administrators should establish a very realistic level of serviceability (which can be as modest as is convenient) but once established it must be respected, and maintenance work performed regularly.

9.1.6 Comfort

For important highways and long distance roads in particular, pavement design should be influenced by the comfort of the user when travelling at the appropriate speed. This requirement includes related ones, the most important of which is safety. Aesthetics and their effect on the psychological reactions of the user also deserve consideration.

From a strictly mechanical viewpoint, longitudinal deformations in a pavement represent no more than a very slight structural deficiency or risk of structural failure. They are, however, very uncomfortable and tiring to the driver. In important roads, therefore, the designer must raise his demands to include considerations of this type. These are less important in more modest roads, where lower speeds or less traffic are involved. This is not a treatise on pavements. The vast developments in this field during the last few years has made it a discipline in its own right. Today there are pavement specialists in the same way that there are specialists in applied soil mechanics, hydrology, planning and the many other disciplines that should be represented in a team devoted to road engineering. Pavement technology has its own research institutes, congresses and information media, that can in no way be fully detailed in a text on applied Soil Mechanics. The authors have always believed that because pavements are composed of soils and soil-like materials the mechanics of pavement systems cannot be separated. A good grounding in soil mechanics is the best foundation for someone who specializes in the highly difficult field of pavement analysis and design.

It is questionable whether pavement technology can be developed without an intimate knowledge of material technology including the strength, deformability and stress-strain relations of soils, all of which are provided by soil mechanics. The authors question whether a pavement should be regarded as a layered structural system that forms the upper part of a road (which is the common conception today). Instead they believe that it is more rational to regard the pavement as a part of a structural system of the road, considering the entire set of layers, (the original ground, fill, subgrade, sub-base, base course and surface course) as a complete, indivisible whole. It is very difficult for successful pavement designs to be developed so long as the designer's attention is centered only on the upper layers, because the part played by the lower layers is always important and often decisive. In this Chapter little will be said about the theories that apply only to the specialized field of pavements, or the detailed methods of design that have been developed within this field. These matters are dealt with in specialized literature on the subject.

9.2 The Structure of Flexible Pavements

This section deals with the system that is currently used for designing most flexible pavements. A nomenclature will be established, and the role that is assigned to each of the different layers will be discussed.

Figure 9-1 (Plate 9-1) shows a typical structure for a cut and embankment section. First is a bituminous surface layer, typically consisting of a mixture of stony aggregate and an asphaltic cementing material. This constitutes the wearing surface. Immediately below are at least two very different layers: first,

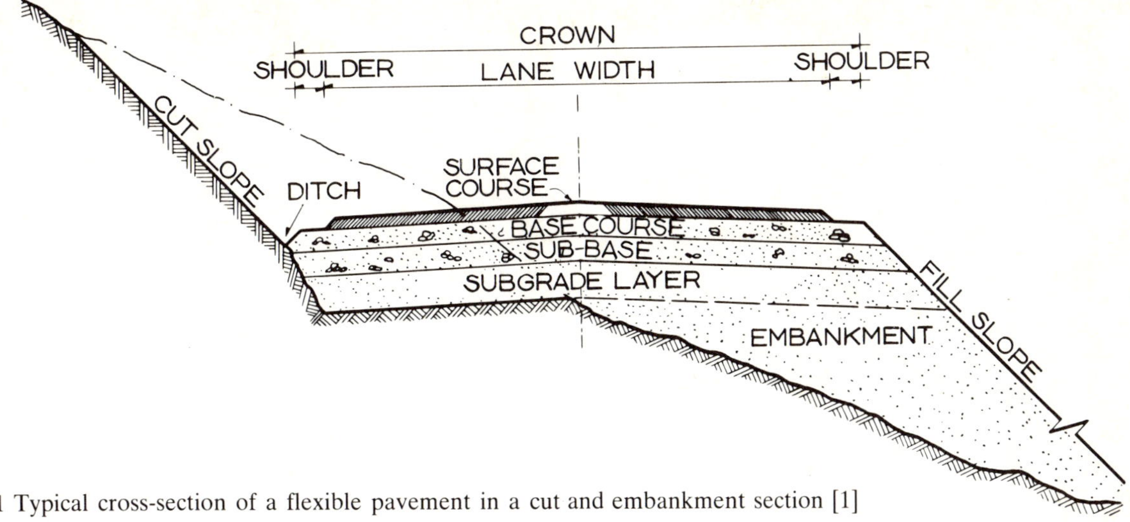

Fig. 9-1 Typical cross-section of a flexible pavement in a cut and embankment section [1]

a base course, made of granular material, and second, a sub-base, usually of a granular material too (although this is not so important as in the base course and lower quality materials with a larger fines content and poorer grading are sometimes permitted). The sub-base can be of lower quality because it is located at a greater distance from the wearing surface, and the stresses it receives are therefore less intense. Beneath the sub-base, the upper surface of fill, or virgin ground is usually considered to be another layer, known as the subgrade. This has even lower minimum quality requirements than the sub-base, because of lower stresses. However its mechanical and economic role is becoming gradually more widely acknowledged. Underneath the subgrade is the conventional fill material, which is subjected to compaction or possibly some additional form of improvement.

9.2.1 Fill Material

Little attention has been paid in the past to the interaction of the fill and the pavement structure; little has been written on the subject. From certain experimental facts, which will be mentioned later in this section, and the opinion of the authors, the following conclusions can be drawn.

.1 Shear Strength

It seems reasonable to believe that the shear strength of the soils in the embankment is not a fundamental requirement for pavement support. The levels of stress that reach the embankment after traversing the entire protective thicknesses of the pavement system, are always lower than the maximum bearing capacity of the fill material, in accordance with other requirements that will be mentioned below. (However the fill must be strong enough that its slopes do not shear from the fill weight).

.2 Deformability

Deformability is the basic requirement for the acceptance or rejection of a fill material; on it depends the successful performance of the fill as an adequate support for a pavement. All factors contributing to the absence of deformability in a fill material are of importance. Among these, the quality of the materials plays an important role, especially in two extreme cases; (1) materials containing appreciable amounts of large fragments, and (2) materials in which the smallest particle sizes predominate.

Materials in which large and medium fragments predominate are deformable, because they are hard to place and compact suitably, leading to errors during construction which have detrimental repercussions that become more serious as the embankment becomes higher. In Chapter 4, the construction procedures for erecting rockfills and embankments with a predominance of rock fragments were discussed. Another special deformability problem is the one which occurs when one end of an embankment made of large rock fragments is covered only by thin layers of soil. In this case, the soil layers are not usually uniform, being thicker at the sides and in between the fragments and thinner above them. It is difficult to compact such an embankment correctly, and it may pose serious deformability problems. A minimum thickness of soil is specified to cover the rock fragments; the greater the thickness, the smaller will be the importance of this problem.

The second important problem relating to embankment materials arises when these consist of compressible soils. Many *MH* and *CH* soils are so deformable that their use should be prohibited. The deformability is aggravated even further if the soils are organic or micaceous. The Mexican Ministry of Public Works, for example, forbids the use of *MH, OH* and *CH* materials in the body of an embankment when their liquid limits are higher than 100 [2]. It also prohibits the use of materials referred to as P_t in the Unified Soils Classification System. It cannot be determined to what extent a strict classification limit may prevent solve such problems, for the behavior of a soil varies according to climate, drainage and subdrainage, geometry of the embankment in which it is to be used, and local topography. Compaction also has a fundamental influence on the ultimate deformability of a material. Such requirements therefore, are regarded as only guides and, as always occurs with specifications, they are not complete substitutes for the rational consideration of each specific situation.

A deformable fill will require a very thick pavement that will reduce stresses to sufficiently low levels. Such a pavement will be costly and uneconomical for, as has already been said, pavement material is more expensive than fill material. If deformability is not given sufficient consideration in the design phase, (as so often is the case) a pavement with suitable behavior will never be achieved on that fill, no matter how much maintenance and reconstruction work is expended on it.

.3 Climatic Conditions

Careful attention to climatic effects has frequently been mentioned as fundamental, if a well-behaved embankment is to be built that is capable of successfully supporting a pavement. However, this statement deserves elaboration.

Reference [3] presents a study that has very seldom been conducted on such a large scale. In order to evaluate the combined strength of pavement and fill, including its seasonal variations, the influence of climatic effects, two sets of surface deflection measurements were made at 52 road sections throughout Mexico. The first reading was taken in October, at the close of the rainy season, and the second in March and April, at the close of the dry season. The deflection shown by the pavement surface is a measurement of the structural condition prevailing throughout the entire pavement thickness beneath that point. Therefore the condition of a pavement with high deflections is regarded as worse than that of a pavement with low deflections. Fig. 9-2 shows the results obtained.

The information given in Fig. 9-2 is surprising in that it contradicts the general belief of many specialists. Statistically there is no appreciable difference in the structural condition of the road sections, even at two such drastically different times at which readings were taken. One of two conclusions can be drawn: either (1) deflection of the pavement surface is not an adequate criterion for judging the structural condition of a pavement-fill system, (which is contradictory to all modern experience, and especially to other evaluation studies conducted on existing pavements), or (2) past considerations of the effects of climate on the structural behavior of pavement fill systems need to be reviewed.

It is possible that this conclusion is limited to the particular environment in Mexico, where seasonal variations are not very pronounced. The seasonal effect may be far more noticeable in other latitudes, especially in countries where climatic effects include freezing and thawing. Here it is worth mentioning that in many cases it is in such regions where much of the practical methodology involved in pavement evaluation and design has evolved. The question arises as to whether these methods are not excessively conservative for other regions like Mexico.

Some of the data points in Fig. 9-2 exhibit far more differences in deflection than the general trend. In a personal communication to the authors of this book, the author of [3] pointed out that many of these points reflect a borrow pit close to the section of road under construction. The relatively deep adjacent excavation made that section far more vulnerable to climatic effects.

The conclusions drawn from Fig. 9-2 are so interesting that the authors are conducting a general review of all the information obtained by studying these same sections plus some newly selected sections. All the results obtained until now in similar studies in Mexico produced the same results, and similar

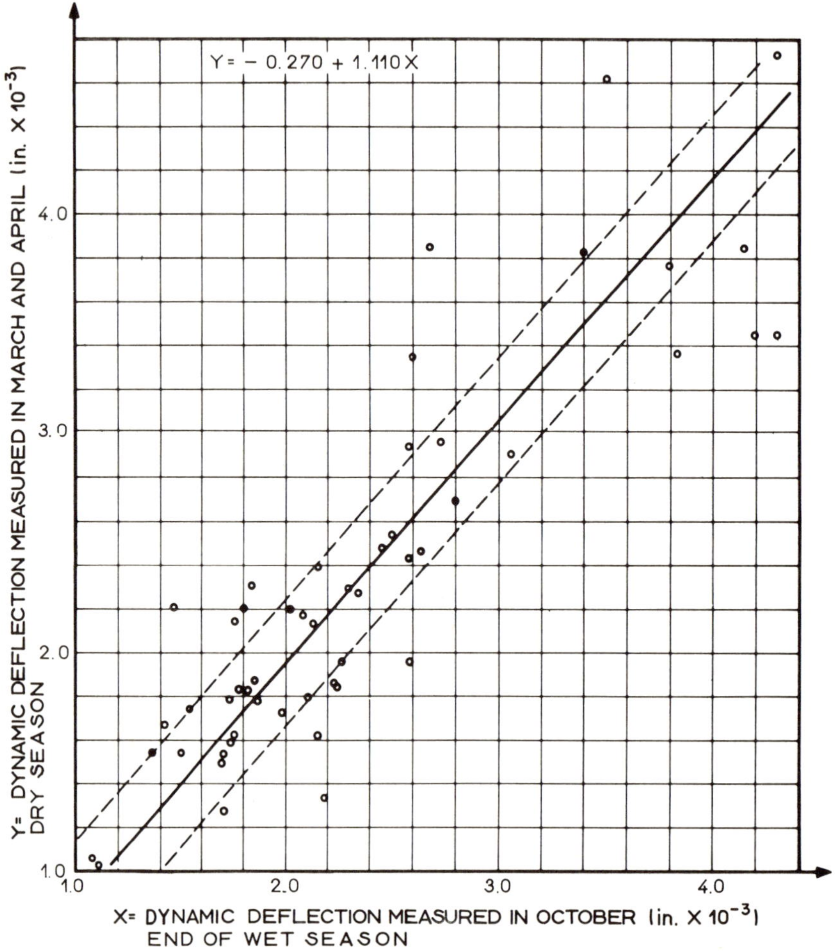

Fig. 9-2 Correlation between two sets of readings with Dynaflect equipment in 52 selected sections of Mexican Highways [3]

investigations conducted in 1972 in two important test sections of the network controlled by the Ministry of Public Works have given similar results. Reference [4] presents a similar investigation conducted in Australia, where deflections were measured with a BENKELMAN Beam. The final results corroborate Fig. 9-2. *Climatic effects* means the effect of seasonal variations in rainfall, evaporation and temperature. We agree that this should have very pronounced, strong repercussions on the life of the road. The Mexican and Australian data suggest that this is not so. This criterion must, therefore, be revised in regions with similar environments. It seems that some time after the construction of a road is completed, an equilibrium condition is reached which is relatively independent of seasonal changes in regions with similar environments to Mexico. The equilibrium condition reached will depend on the climate and climatic changes prevailing in the zone. However, the topographic and geologic conditions and the location of the road in relation to groundwater and drainage are also important. The general conditions of a road may be very good even in a place where the climate is detrimental. When this is the case, the equilibrium condition that is achieved will also be favorable. Similarly, in a zone where the climate is apparently kind, a poorly located road may attain equilibrium conditions that imply serious dangers to its future existence. Climatic effects cannot be regarded as factors independent of equally important considerations. The important consideration in design is not to define the climate where the road is to be constructed, but what the effect of that climate and climatic changes will be on the embankment and pavement system. We believe that the system approach will make design more independent of local seasonal changes alone than would have been possible in the past.

Encouraged by the results of the above investigation and its experimental nature, the authors suggest that the expression *climatic effects* is perhaps too broad for describing the important changing effects of water content on the structural strength of pavements. This effect includes the water built into the fill and the subgrade layer as well as that added by rainfall and capillary rise. Consequently, one should not give so much thought to general climatic effects, but to the specific effects of adequate subdrainage in the strength of road in which special saturation conditions such as a high water-table or poor natural drainage are anticipated. Of course, climatic effects may influence other problems relating to pavement behavior, such as longitudinal cracks due to repeated evaporation and wetting processes already described (Chapter 4), in which solar activity plays an important role, or the aging of a pavement surface due to sunlight. Reference [18] describes field investigations which give support to the statement regarding the great importance of subdrainage, relating variations in the water content of the subgrade with fluctuations in the water-table. The similarity between the curves in Fig. 9-3 illustrates the practical importance of fluctuations in the water-table in the subgrade beneath the pavement crown. It infers how subdrainage is capable of controlling such fluctuations by maintaining the water content in the subgrade at a steady value.

Another result of climatic effects are the potholes that form on account of local rainwater seepage, and as a result of frost action in colder regions.

9.2.2 Subgrades

The subgrade layer plays an important role in the interaction of a pavement and its supporting embankment, [5] is an initial report of an experiment carried out at the Engineering Institute of the Mexican National Autonomous University, sponsored by the Ministry of Public Works. It studied the effect of the thickness and quality of the subgrade layer on the performance of flexible pavements. The investigation was carried out on a circular test road at the Fernando Espinosa Laboratory of the Engineering Institute. Photographs of the test road are shown in Plates 9-2 and 9-3.

Scale models with a thickness of soil of 1.5 m (5 ft) and the length which corresponds to a diameter of 13 m (42.7 ft) can be studied. A system of 3 dual wheels operating in a circle with a diameter of 10 m (33 ft), with a load of 10 t (22 kips) per axle, runs along the top of the model embankments. The eccentric system of the 3-axle assembly has a planetary movement (Plate 9-4) and enables the reproduction of a histogram of load applications corresponding to the cross-section of a real road. Most of the tests were performed with wheel loads concentrated within a path 0.80 m (2.63 ft) wide, which represents a heavy traffic lane. The system can operate between 4 and 40 km/h (2.5 and 25 mi/h); it is usually maintained between 10 and 20 km/hour (6.2 and 12.4 mi/h). The trough in which the embank-

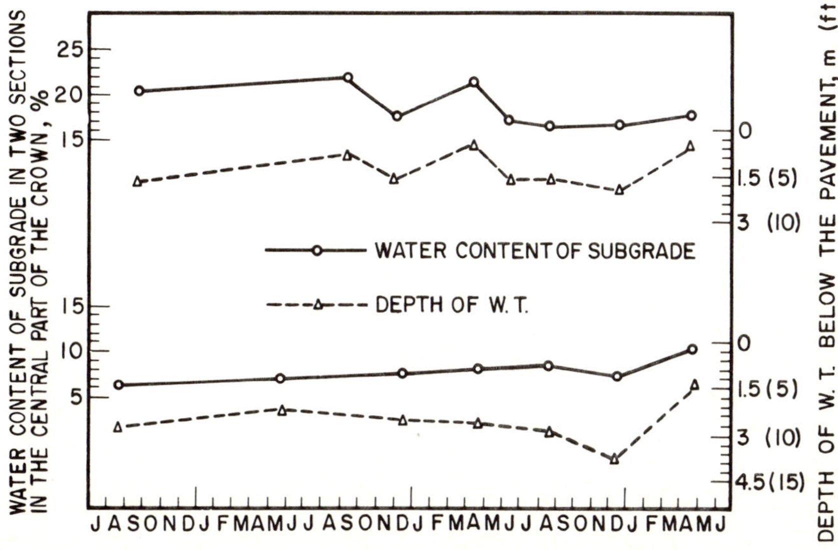

Fig. 9-3 Variations in the water content of the subgrade layer with changes in the position of the Water-Table [18]

Plate 9-2 General view of the test road. *Fernando Espinosa* Laboratory, Engineering Institute of the UNAM, Mexico

Plate 9-3 Circular test road in the *Fernando Espinosa* Laboratory. A pavement is being tested

Plate 9-4 Detail of the mechanical device, Circular test road

ment is constructed is made of concrete and is fitted with all the installations necessary for complete instrumentation. For the purposes of the investigation, it was assumed that there was a linear relationship between the log. of the strength required by a pavement and the log. of the number of load repetitions to which it may be subjected without failing. This correlation is commonly used in the field of pavements [3,7].

The first stage of the investigation was performed with 6 series of test rings each consisting of 3 different pavement sections, a total of 18 test sections. The pavements under consideration were not of a very high quality, with no surface course worthy of consideration, and only one with a prime coat. The embankment material was always the same and its quality very close to the minimum established by the Mexican Ministry of Public Works [2]. The California Bearing Ratio *CBR,* in a saturated condition was about 8.

Rings *1, 3* and *4* were constructed using subgrade layers with thicknesses of 20, 30 and 40 cm (8, 12 and 16 in). The subgrade material used was the same as in the body of the embankment, the only difference being in compaction, which was 90 to 95% (PROCTOR, Ministry of Public Works) in the subgrade and 85 to 90% in the embankment. Rings *1, 3* and *4* each utilized three thicknesses of the crushed gravel base course: 15, 20 and 25 cm (6, 8 and 10 in). These thicknesses, combined with the 20, 30 and 40 cm (8, 12 and 16 in) of subgrade, provided 9 different models. The moisture condition selected for the body of the embankment and the subgrade was intended to reproduce a likely construction water content. Under these conditions, the rings failed with a number of load repetitions shown in Table 9-1.

Ring *2* was built with the same materials and given the same treatments as *1, 3* and *4,* using a base course 20 cm (8 in) thick and subgrade thicknesses of 20, 30 and 40 cm, (8, 12 and 16 in), thus providing 3 models. The difference was that these pavements were exposed to saturation for a month, with a depth of water around them rising to 60 cm (23.6 in) below the lower

Table 9-1

Number of load repetitions required for failure in 5 test rings on the circular test road at the Engineering Institute of the Mexican Autonomous University [5]

Ring	Geometry: Base course (cm)	(in)	Subgrade (cm)	(in)	Number of load repetitions required for failure
1	15	6	20	8	25,000
	15	6	30	12	102,000
	15	6	40	16	155,000
2 (saturated)	20	8	20	8	105
	20	8	30	12	10
	20	8	40	16	85
3	20	8	20	8	790,000
	20	8	30	12	1.160,000
	20	8	40	16	2.940,000
4	25	10	20	8	800,000
	25	10	30	12	1.800,000
	25	10	40	16	13.000,000
5 (saturated)	20	8	20	8	5
	20	8	30	12	10
	20	8	40	16	20

bed of the base course. Table 9-1 shows the drastic reduction in the number of load repetitions that were required to bring this saturated subgrade model to failure.

Ring *5*, also with 3 sections, was designed to establish how much the results might be affected by a very good subgrade. For this purpose, 20, 30 and 40 cm (8, 12 and 16 in) layers of a sandy-silty material with 95% compaction and a saturated California Bearing Ratio of 40, were placed on top of the same embankment material as was used in the previous rings. On top of this a gravel base course 20 cm (8 in) thick was laid. This ring was saturated for a month too. Table 9-1 indicates that the number of load repetitions required for failure was the same as in Ring *2*, which was also saturated.

Ring *6* was also prepared with 3 different sections, to investigate whether a poor-quality saturated subgrade would cause failure in an overlying pavement, regardless of pavement quality and thickness. The aim of this test was also to find out whether saturation is as destructive as it appeared in the previous results, independent of the thickness and degree of compaction of the subgrade. Subgrade thicknesses of 0.80, 0.90 and 1.00 m (2.63, 2.96 and 3.28 ft) were therefore laid of the same poor-quality material that was used in the fills for the 5 previous rings, but compacted uniformly to 100% (Proctor standard, Ministry of Public Works). On top of these subgrades, pavement thicknesses of 70, 60 and 50 cm (27.6, 23.6 and 19.7 in) were placed, using a very well compacted high-quality crushed basalt. The whole structure was then subjected to the same saturation condition as in Rings series *2* and *5*. Results were highly satisfactory, with millions of applications in all cases without failure.

The above results suggest that the contribution of the subgrade to the stability of fill and pavement system is fundamental. Ring series number *6* illustrates the advisability of using a carefully compacted fill material with no sudden changes in relation to the quality and condition of the subgrade. In all cases better performance is observed from the thicker subgrades. The tests also show the enormous importance of saturation in fills and subgrades. Considering the behavior of many minor roads throughout the world, saturation does not appear to be a realistic design consideration for many geographical situations and climatic factors. If it were, it seems that there would be no roads.

The test results support the efforts of highway engineers throughout the world who are fighting for better construction quality. The role of compaction is decisive. The same material that immediately collapsed in the embankment and subgrade in Rings *2* and *5* under saturated conditions, showed excellent strength and deformability in Ring series *6*, under the same severe conditions. The only factor to which this difference can be attributed is the better compaction that was given to the material in Ring series *6*. Compaction appears to be even more significant than material quality, because the subgrade in Ring series *5* was made of a material with far better characteristics than the one used in Ring series *6*. Finally this investigation illustrates the importance of the extensive use of subdrainage. It is not, however, felt necessary to comment further on this point; the results were only too obvious.

The foregoing considerations strongly demonstrate the importance of the subgrade. Economic considerations can be added in favor of its importance (although it has already been mentioned). A thick enough layer of good quality well compacted subgrade will permit appreciable economies in the thickness of overlying pavement, without affecting the combined structural function. It will absorb relatively high levels of stress coming from the surface and will reduce them sufficiently before transmitting them to the embankment below. From the economic point of view, quality and thickness are of equal importance. The quality of the materials that are used in the subgrade layer is seldom very good; therefore the contribution of the layer usually depends on its thickness rather than on its quality. However if the subgrade is of a high quality (maintaining reasonable proportions in relation to the qualities of the sub-base and base course), enormous economies will be obtained in the thickness of the pavement layers.

In many cases the subgrade is made of the same material as the fill, the only difference being that it is compacted better. This is acceptable when the quality of the fill material is sufficiently good. However if the fill quality is poor, Mexican engineering experience indicates that it is economical and advisable to look for a better material and transport it from a convenient borrow area.

Few general rules can be given regarding the thickness that is advisable for the subgrade layer, whatever the construction process. The Mexican Ministry of Public Works has established a minimum of 30 cm (12 in) for its highways. This thickness may reach 50 cm (20 in) in roads with high volumes of traffic or in places where the fill material is poor. For runways the same 50 cm (20 in) is ordinarily used. However it is sometimes increased, as in the case of Mexico City Airport. The Ministry of Public Works further requires that for highways the subgrade material must not contain particles larger than 7.6 cm (3 in). It eliminates fine soils (*MH, CH*) with liquid limits greater than 100%, and all organic soils with liquid limits greater than 50% (*OH*). It specifies minimum degrees of compaction of 95% in relation to the standards in use at the Ministry, which were described in Chapter 4. Lastly, for runways, a minimum *CBR* [7] of 5% is demanded with the material in a saturated condition; a degree of compaction of 100% is required; and the use of any *MH, CH* or *OH* materials is prohibited. These regulations are intended as guides rather than strict rules, for we have already made it clear that many circumstantial factors may influence the behavior of a material in special situations. Traffic intensity and subdrainage conditions must also be taken into account.

9.2.3 Sub-bases

For many designs the chief function of the sub-base of a flexible pavement is to reduce cost [7]. The objective is to achieve required pavement thickness with the cheapest possible materials. The entire thickness could be constructed with a high-quality material such as that used in the base course. However it is generally preferred to make the base thinner and substitute a sub base layer of poorer quality material, although the total pavement thickness may have to be increased. The poorer the quality of the material that is used, the greater will be the thickness required to tolerate and transmit the stresses.

Another function of the sub-base is to provide a transition layer between the material of the base course, which is generally coarse-grained, and that of the subgrade, which is usually much finer. The sub-base also serves to absorb detrimental deformations in the subgrade such as volume changes associated with variations in water content, which might eventually be reflected in the pavement surface. Another function of the sub-base is to drain off any water that infiltrates from the surface and to prevent water from the fill from rising towards the base course due to capillarity. Of these factors, the structural and economic ones are common to all sub-bases; the others depend partly on specific circumstances and partly on the quality of the material that is used in the sub-base.

Sub-base materials are expected to meet requirements relating to maximum particle size, grading, plasticity, sand equivalent and *CBR* [7] when; as it is still common, this last concept is used to determine the thickness of pavement layers. Minimum compaction requirements are usually established too. Figure 9-4 [2] shows the range within which the grain-size distribution curve of a sub-base material should lie, according to the Mexican Ministry of Public Works. Apart from keeping within Zones *1*, *2* or *3*, the grain size curve must also follow the general direction of the lines limiting these zones, without any abrupt changes in curvature. The ratio of the percentage by weight passing No. 200 sieve to that passing No. 40 sieve must not exceed 0.65. The maximum size of the material is limited to 51 mm (2 in).

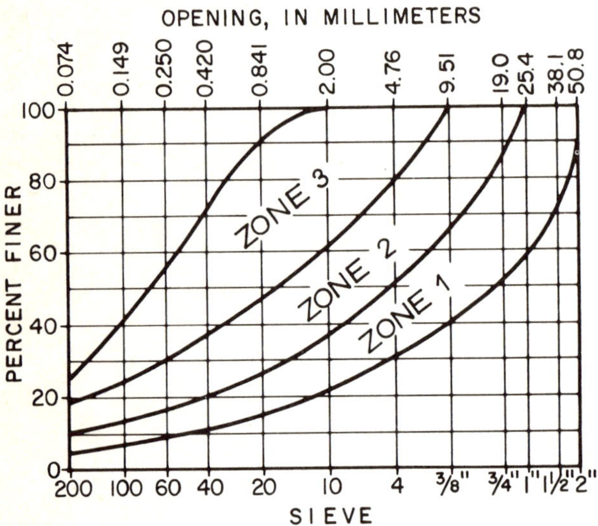

Fig. 9-4 Grading requirements for a sub-base material [2]

A minimum of 20 is tentatively established for the sand equivalent of the material. This important test is mentioned further on in this volume. The Mexican Ministry of Public Works uses the California Bearing Ratio method very extensively for designing its pavements. A minimum *CBR* value of 50% is given for sub-base materials, in a saturated condition. The degree of compaction stipulated by the Mexican Ministry of Public Works in accordance with the laboratory methods described in Chapter 4 is 95%. The Ministry also recommends methods for stabilizing those soils which do not satisfy the general requirements of plasticity, (measured with the sand equivalent test and plasticity limits). Possible methods include portland cement, lime, and asphalt.

The above requirements are merely guides; the same comments apply as for similar information given in previous pages. Basically, two main qualities should be sought in a sub-base material; frictional strength and draining capacity. The first will enhance overall strength, and at the same time will reduce deformability because most materials with frictional strength are rigid when well compacted. The draining capacity reflects a dual function: drainage enables the pavement to eliminate water filtering through its surface and at the same time blocks water rising by capillarity. Sub-base thicknesses are very variable depending on each specific design; 12 to 15 cm (4.7 to 5.9 in) are usually regarded as typical minimums.

9.2.4 Base Courses

The base course, which is the layer above the sub-base, has an economic function similar to that of the sub-base. It enables a reduction in the thickness of the more expensive surface course. The fundamental function of the base course of a flexible pavement is structural. It is resistant to traffic loads and capable of spreading those loads to the layers below at suitable intensities. The base course also has an important draining function. It must be capable of eliminating quickly and easily any water that may filter through the surface course, and also impede radically any capillary flow rising from below.

The base course of a flexible pavement must, therefore, be made of a material that is frictional and with sufficiently large voids. Friction will ensure adequate structural strength, along with permanence of that strength under variations in environment, such as water content. Only a granular or frictional material is sufficiently reliable from these points of view. A fine, cohesive clayey soil would hardly be able to develop sufficient strength to tolerate the intense application of traffic stresses, and this strength would, moreover, be extremely variable in the face of the environmental changes already mentioned. Of course, adequate strength or suitable deformability characteristics are not ensured only by the use of a frictional material; it is essential that the potentially good characteristics of this material are developed by adequate compaction. Once the frictional material has the compactness and interlocking that are acquired from good compaction, a suitable base course is achieved. Furthermore, it is usually necessary to produce the stony coarse aggregates or rock fragments by crushing. This crushing enhances strength and low deformability. The sharp-edged particles, which interlock structurally are one of the sources of strength (Chapter 1) at the same time favoring low deformability. Another process to which base course materials are frequently subjected is screening, by means of which a pre-established grading can be met. In Chapter 1 it was also shown that a suitable grain-size distribution improves the strength of a material, allows greater densities to be achieved with compaction and reduces particle breakage. Unfortunately little importance is given to breakage in pavements, despite repeated dynamic loading with high levels of stress to which a base course is subjected. In Chapter 4 studies were presented which show the change in grain-size distribution that may be suffered by base courses under traffic stresses.

The benefits of crushing and screening have not been sufficiently well evaluated by many engineers; often the two operations are valued only on account of the part they play in achieving a suitable grain-size distribution (which is included in standard specifications for highway construction). Washing is another operation that is often specified in designs where the materials to be used for a base course come from natural deposits. The beneficial effects of this operation are obvious, because it eliminates fines that would otherwise have a harmful effect on the material's strength and increase its deformability, as well as their notoriously detrimental effect on drainability; although detrimental regardless of their nature their effects are far more pronounced as the fines become more clayey and active.

The same gradation ranges or zones shown in Fig. 9-4 for sub-base layers are used by the Mexican Ministry of Public Works for defining the grain-size distribution curves of its base courses. However, a preference is shown for those in Zones *1* or *2*. Grain-size distribution curves must be similar in shape to those shown, with no abrupt changes in curvature. The ratio of the percentage by weight passing the No. 200 sieve to that passing No. 40 must not be larger than 0.65. The maximum size of the stony aggregate is 51 mm (2 in) for natural materials requiring no treatment, and 38 mm (1-1/2 in) for materials that have to be sieved or crushed.

The Mexican Ministry of Public Works [2] specifies a maximum liquid limit of 30% for base course materials, a minimum sand equivalent of 30 for roads with fewer than 1,000 heavy vehicles per day, and 50 for those with greater volumes of traffic. In runways, the minimum sand equivalent is 50 for commercial aircraft. The *CBR* is specified a minimum of 80% for roads with fewer than 1,000 heavy vehicles per day, and 100% both for roads with more than 1,000 heavy vehicles per day and for commercial aviation runways. In Mexico, the minimum degree of compaction established for base courses is 95%, in accordance with the standards usually observed in that country (Chapter 4) but 100% compaction is very common. The general specifications of the Mexican Ministry of Public Works also include stabilizing with asphalt, cement or lime for natural materials which would not otherwise be satisfactory. Base course thicknesses vary greatly in accordance with design, but 12 to 15 cm (4.72 to 5.9 in) is usually regarded as the minimum thickness.

A point on which not all specialists agree is whether it is advisable for the base course of a flexible pavement to contain fine materials smaller than No. 200 sieve, what the quality of these fines should be, and what advantages or disadvantages can be expected from their presence. Figure 9-4 shows Mexican practice for limited quantities of fines smaller than No. 200 sieve to be accepted in the grain-size distribution for base courses. However the Mexican Ministry of Public Works almost always requires gradations defined by Zones *1* and *2* of the graphs, Zone *3* is accepted only in very special cases. There is a minimum requirement of 5% of fine material, with practical maximum amounts of about 18 to 20%. Amounts of fines larger than these limits could lead to an unstable base, which makes it reasonable not to use Zone *3* gradations. The minimum requirement may be questionable, as will be seen in the paragraphs that follow. Under some conditions there is no reason why a base course with a fines content lower than the lower limit should not be used, especially in the case of natural materials that do not contain any such fines. Likewise, the maximum permissible fines content in Zone *2* of Fig. 9-4 (18%) often proves very high. A specification including smaller permissible percentages might be advisable. This is particularly true if one considers how difficult it is to use only inactive, non-plastic fines. However some material in the finest zone of the grain-size distribution curve may possibly prove beneficial if all the factors that influence the behavior of a base course are considered. The evaluation that must be made in each situation deserves a few general observations.

First, in order to withstand traffic stresses, the base course requires frictional strength. This is affected by fines. The larger the fines content, the more this strength suffers, aggravated by the activity and plasticity of the fines. The same applies to the resistance to deformability of the base course and its drainage characteristics. From this point of view, the base course should be made up of pure frictional material, without any fines. However, when purely frictional materials without fines, whether natural or crushed, are compacted for a base course, the upper part of the layer rapidly loses compaction under construction traffic. It becomes unable to resist lateral displacement of the particles, from the construction vehicles that must place the wearing surface on it. Experience has demonstrated all over the world that if a small percentage of fines is incorporated into the base course, the foregoing disadvantages are greatly relieved; and the material becomes easier to work with because it has a more stable surface when exposed.

For a road with heavy traffic, which requires a very thick surface course, the above setbacks can be easily overcome. Compaction of the first layer of surfacing restores the upper part of the base course to its original density and strength. It may even be further protected by a prime coat. Moreover, the thick surface course will provide the granular material with sufficient confinement to develop the good strength and resistance to deformability of a well compacted granular material. In this respect, the reader may recall that the surcharge thickness required to develop the bearing capacity of a dense granular material is small. For example a plank or a light mattress of branches is sufficient to make a loose sand fit for traffic. Therefore, it has been the authors' experience that the conditions in roads with heavy traffic are suitable for base course materials with little or no fines (sufficiently small amounts that they do not affect the strength and resistance to deformation requirements). The percentage of fines will usually be less than 5%.

In roads with light traffic, at the other end of the scale, conditions are different. Here a minimum thickness of surface course is used, possibly only a prime coat, which would not produce sufficient confinement to maintain the strength in a purely granular base course. The incorporation of a certain quantity of fines in the base course is fundamentaly a problem of mechanical stabilization here. The purpose is to give the layer a certain cohesive strength without interfering too much with the frictional strength, resistance to deformability and ability to drain. From this point of view, a grain-size distribution that includes limited quantities of fines passing the No. 200 sieve usually improves these base courses. The quantity to be included, (within the limits established above), depends largely on the nature of the fines that are used and the environmental conditions to which the road will be subjected. Owing to their low cost, roads with low volumes of traffic cannot be expected to have good surface drainage or subdrainage systems. Therefore, the fines content that is tolerated must consider total pavement environment. The presence of excessively active fractions is never advisable. When materials suitable for a reliable base course are not available, the use of soils stabilized with lime, asphalt or cement should be considered, for even if they are more expensive, they can be used in smaller thicknesses and may prove an appropriate solution when the materials available are not of reliable quality.

In roads with moderate to light traffic (which represent a very large part of the road networks in many countries) the situation is intermediate between the two extreme cases that have been discussed. The strength, resistance to deformability and drainage characteristics of the base courses require the use of granular materials with a minimum content of fines or without any fines at all. On the other hand, the workability and maintenance of these qualities under small surcharges (in this case surface courses about 5 cm (2 in)) demand some cohesion in the layer. The precautions that must be observed for the construction of a base course make it possible to use percentages of fines that do not exceed perhaps 8 to 10%, so long as this fraction is relatively inactive in the presence of water and is not highly plastic. Variations in the handling of the above criteria will depend largely on the environmental conditions to which the road is exposed, the drainage and subdrainage that are adopted and the possible introduction of capillary water. The grain-size distribution of the coarse fraction must be considered for if the voids are relatively small a small percentage of fines may be very harmful. The use of certain percentages of fines for the purposes of mechanical stabilization may lead to important economies in construction and acceptable designs; however, the overuse of fines is one of the surest causes of pavement failure. The designer must balance these two contradictory concepts.

When a natural material with very little or no fines passing No. 200 sieve is available for the construction of a base course, it is common practice in many countries to find another natural material (generally passing No. 4 sieve) which, when mixed in proper proportions with the previous material, will produce a grain-size distribution curve with a larger fraction of smaller sizes that meets the need for particles passing No. 200 sieve. This must be very carefully done, because a relatively small gain in the workability of the material (which is a short-term advantage), may cause a reduction in the mechanical behavior of the base course throughout its future life. The importance of the quality as well as the quantity of the fines incorporated is again stressed. Too often an excellent granular base obtained at great expense (such as by crushing) is mixed with materials containing fines to produce a material that is inferior. The loss of quality is even greater if the coarser particles of the material containing fines are weaker than those of the high-quality base. The loss of quality can be aggravated by degregation of weaker particles producing even more fines. The loss of quality of the combined material, plus the cost of hauling and mixing even small amounts of materials containing fines from some distance, is seldom worth the small gain in workability during construction.

Where the base course is laid directly on top of the subgrade, some contamination of the base course material with fines from the subgrade will be inevitable. Thus final percentages of fines are higher than originally specified. Finally where the surcharge weight of pavement will be sufficient to ensure the permanence of the properties of the compacted granular soils without fines, careful precautions will have to be taken to protect the exposed surface of the base course during pavement construction. Protective prime coats will be necessary, together with suitable compaction; however the risk of a future more costly failure of the pavement, will be avoided.

9.2.5 Surfacing Coats

The surfacing material of the flexible pavement must provide a stable wearing course, capable of withstanding direct application of loads, tire friction, shear stresses caused by braking, centrifugal forces and impact. It must have a texture that allows safe, comfortable driving and adequate braking. The nature of the surfacing material must be such that it will withstand the action of weathering. Its color should discourage reflections from the sun during the day and from artificial lights during the night. A very thick asphalt concrete wearing course is becoming increasingly popular, because in addition to the foregoing traditional functions, it also has an important structural function, greatly influencing the stress-strain behavior of the entire section of a road by its great stiffness.

The direct exposure to traffic stresses and the lack of deformability that are necessary if a road is to give good service, imply that the surfacing must be made of a material that will offer sufficient strength with virtually zero normal external pressure (which is the case in the upper boundary of a pavement). In other words, a material is required with some *cohesion*. The asphaltic product which cements the stony aggregates together provides this cohesion in bituminous surfacings. In surfacings without asphalt, that are placed on roads with very low cost, the materials used are usually coarse and granular, with a maximum particle size of 75 mm (3 in). The weight of these particles and the structural bond that is achieved with their interlocking by compaction provide the necessary resistance to traffic stress (which is small). It is also common practice to add small percentages of silty or even clayey fines to these granular materials to cement them together. The result of this practice is, at best, questionable, for the strength of the surfacing material is reduced, encouraging the formation of troublesome ruts due to the larger wheel loads. Also, some of the fines thus added are quickly washed away by rains. The best practice is to begin with a material in which the particles are protected from abrasion by their weight and by a strong interlocking. Such good materials can be improved by a relatively broad grain-size distribution, where the particles up to the size of sands fill the voids between the larger particles, thus *toughening* the whole. Here the addition of fines does not have the harmful effect that is so pronounced in base courses for such surface courses are subjected to low traffic loads and their strength and deformability requirements are therefore smaller.

The grain-size distributions that are stipulated by the Mexican Ministry of Public Works for surface course materials are similar to those for sub-base, but with a maximum grain size of 7.6 cm (3 in). Materials with a curve within Zone *1* should preferably not be used.

9.3 Traffic Stresses — Systems of Representation

In order to design a flexible pavement, all traffic variables are combined in a convenient equivalent load which can be conveniently handled in the mathematical formulas or design criteria. Hveem [8] has distinguished the following group of factors as the ones which play a part in the overall concept referred to as *traffic stress*. He considers that four of these factors have a leading influence, and three secondary effects. The Principal factors are: Load transmitted by the wheel; Area of load influence; Number of load repetitions; Speed of vehicle. The Secondary factors are: Contact area of the tire, which determines contact pressure; Number of tires in the assembly; Spacing between axles. Many of these factors are difficult or impossible to reproduce in the laboratory for investigation purposes.

The magnitude of the load that is applied to flexible pavements varies between very wide limits. For trucks it can be as much as 12 t (26,000 lb) per axle. It may reach 355 t (780 kips), which is the weight of a recent version of the Boeing 747. If load applications are to be used in comparative analyses, they must be reduced to simple, idealized loads, such as the standard wheel load and the equivalent load. No current design method takes the variability of traffic into consideration in a complete manner. Flexible pavements for roads are normally designed to tolerate the load transmitted by a single idealized wheel; runways are designed by an equivalent tire assembly. In order to determine these equivalents in a representative manner, vehicles must first be selected which suitably represent the heavier types of traffic anticipated. In airports, this is commonly the aircraft whose landing gear transmits the heaviest load. In roads, the most frequent or the heaviest truck that is legally permitted is usually chosen. In order to achieve a design load which will represent the overall effect, it will also be necessary to establish a relation or equivalence between the load transmitted by the tire assembly of the chosen vehicles and the ideal load. Two criteria have been followed for roads. Either a single wheel is sought that produces the same vertical stresses as the tire system of the vehicle at a certain depth, or else the single wheel which produces the same strains.

Figure 9-5 shows a widely used idealization of the effect of a dual system, by which an equivalent wheel load is derived that produces the same intensity in the stresses transmitted.

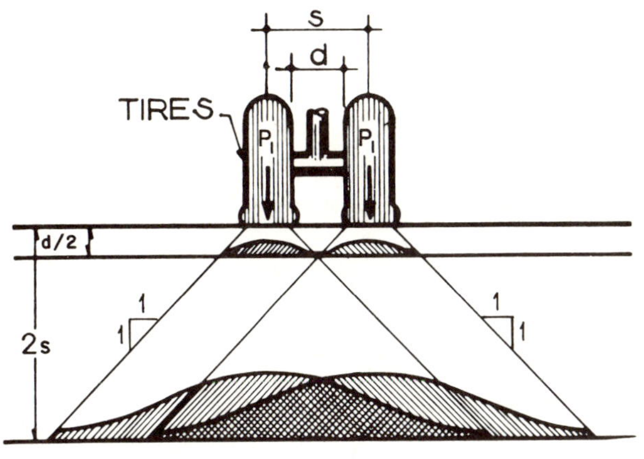

Fig. 9-5 Influence of a dual wheel load on stresses in the pavement system

Both theory and experimental readings show that the cumulative effect of the two tires becomes significant at a depth $d/2$ below the wearing course. They also show that the superposition of the stresses from the two tires is practically complete at depth $2S$, that is, at any depth below that level the stresses produced by two wheels is the same as if the center of the loading area at the surf were subjected to a single force, $2P_1$. On the same basis if it is assumed that between depths $d/2$ and $2S$ the variation in load that produces a given stress at a given depth is linear, a simple criterion can be adopted for obtaining the single load that is equivalent to a dual system. Indeed, a point between the pavement surface and depth $d/2$ is subjected to a stress that is due only to load P_1. A point deeper than $2S$ suffers a stress due to a single load $2P_1$. An intermediate point between $d/2$ and $2S$, will have a stress equivalent to a load between P_1 and $2P_1$. A linear relation between load and depth is, however, not correct, linearity is more nearly achieved [7] when the logarithm of equivalence is plotted against logarithm of depth. From the above considerations the method for finding the load that is equivalent to a dual system is shown in Fig. 9-6.

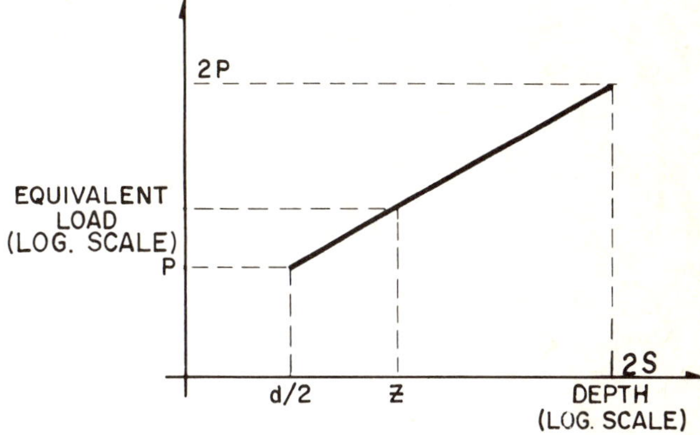

Fig. 9-6 Graphical method of determining equivalent single wheel for any dual-wheel load

Reference [7] presents another means of defining the single load equivalent to a multiple system, using equal strain criteria. It also presents an expression for defining the equivalent load in accordance with Burmister's theory, which regards the soil as being made up of two homogeneous, isotropic, linearly elastic layers. This enjoys popularity for theoretical pavement analyses, and is regarded by many as being more representative than Boussinesq's theory, which is based on a homogeneous, isotropic semi-infinite elastic continuum. Burmister's theory is described in some detail in [9-12] and brings the pavement problem into focus. However given the hypotheses involved and the real nature of the layers that constitute a flexible pavement, it is hard to determine how appropriate or precise it really is. It is even harder to justify using this more elaborate method of analysis to others, including Boussinesq's which give more conservative results.

Traffic stress applications usually include load repetitions. A repetition is said to occur in a road or runway when the same tire passes twice over the same point. In roads; it is frequently considered that two vehicles of the same type must pass over the same point for there to be a load repetition in the pavement. For runways, the relation between the number of aircraft operations and the number of repetitions is given in Table 9-2 [1]. In the case of runways, the concept of coverage rather than repetition is more commonly used today. A coverage is the number of passes of a wheel or wheel assembly that are required in order to cover the central third

Table 9-2

Number of operations to produce one repetition for several aircraft types

Aircraft	Runway	Taxiway
DC-3	3.3	7.0
DC-4	9.6	27.0
DC-6	10.4	30.0
DC-8	7.6	21.0

of the runway with contiguous parallel paths. Coverages are determined from the number of passes of the load, relating the width of the strip under consideration, the arrangement and disposition of the wheels, the width of the contact area of each tire and the distribution of traffic. Table 9-3 [1] gives the number of coverages that are sometimes used for design of runways, including traffic intensity and channelization.

Table 9-3
Number of coverages for the purpose of runway design

Airport conditions	Number of coverages to consider
Intensely channelized traffic	25,000
Normal operation capacity	5,000
Limited operation	3,000
Minimum operation	200
Emergency operation	40

During its service life, a road is subjected to millions of load repetitions, and a runway to thousands. The effect of the repetitions is such that the thickness of the pavement is sometimes the same in both cases, for although the wheel loads applied to runways are far heavier, roads are subjected to a far greater number of repetitions. Figure 9-7 [8] shows one of the interesting aspects of the practical considerations of the effect of moving loads on a pavement. It shows that the deterioration that is suffered by a pavement, expressed in this case by the thickness necessary to ensure a suitable performance (on log. scale), is a linear function of the logarithm of the number of load repetitions, [13] gives several experimental graphs from which similar conclusions can be drawn.

Among the procedures that are currently popular for homogenizing the extremely variable traffic that uses roads today, are those employed by the California Division of Highways [14]. This is one of the more rational and complete attempts that have been made at achieving such a homogenization. However because it is based on local statistical data, it cannot be used for other locations without adaptation. It is mentioned here chiefly to illustrate the way in which these types of problems are generally solved.

Traffic is expressed by the traffic index, described by:

$$TI = 6.7 \left(\frac{EWL}{10^6}\right)^{0.119} \tag{9-1}$$

where: EWL is the number of 2,270 kg (5,000 lb) wheel loads that are equivalent to the real traffic on a rod during the design period. It is calculated using the equivalency factors for several different vehicle axles, as shown in Table 9-4.

The engineers in the California Division of Highways consider that the equivalent wheel has a tire pressure of 4.9 kg-cm^2 (70 psi), similar to truck tires. First an estimate must be made of the number of vehicles per day, grouped in accordance with the number of axles. The reduction to the standard load is done using the factors in Table 9-4. The factor, multiplied by the number of vehicles of each type per day, gives the equivalent number of vehicles per year with the uniform design load that will produce the same effect on the pavement as caused by real vehicles using it for a whole year. By applying the rule to the different types of real vehicles and adding up the final results, the number of standard design load repetitions during the year is reached. The factors appearing in Table 9-4 refer to the average number of vehicles travelling along the road each day in just one direction. The value of a study like this depends on just how representative of local traffic conditions the factors may be.

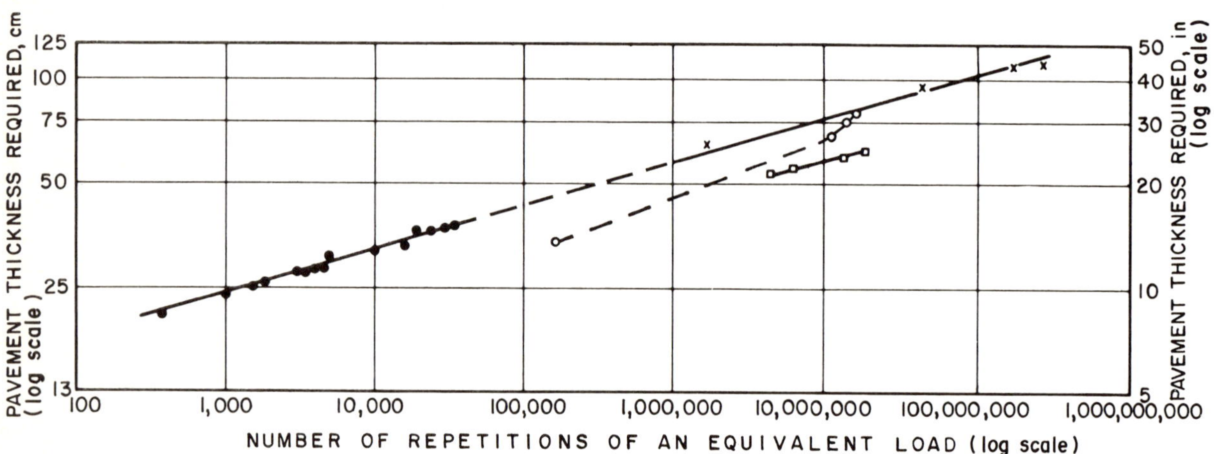

TEST	WHEEL LOAD lbx10³	WHEEL LOAD kg	TYPE OF SOIL (Subgrade) CBR
□ WASHO	—	—	17
○ Brighton	6	2800	4
● Stockton	25	11300	5
x Stockton	40	18100	5

Fig. 9-7 Experimental graphs showing the relationship between the deterioration of a pavement and traffic load [8]

This depends on how careful the statistical analyses are. Section 9-7 of this chapter describes another currently popular method of homogenizing traffic.

Table 9-4

Equivalency factors for multiple axle vehicles with dual tire assemblies in relation to the standard 2,270 kg (5,000 lb) wheel

Number of axles	Value of equivalent wheel load, *EWL*, for one year	
	Primary roads	Secondary roads
2	280	200
3	930	690
4	1320	1070
5	3190	1700
6	1950	1050

In Mexico traffic volumes are usually assessed on a far more simple basis. For its pavement designs, the Mexican Ministry of Public Works counts the number of vehicles with loads equal to or greater than 3 t (6.6 kips) running in one direction. It divides pavements into 4 categories: those with under 500 vehicles, those with 500 to 1,000, 1,000 to 2,000, and over 2,000 per day. This means of assessing traffic has the disadvantage [15] that it does not consider time as being associated with the service life of the road and the distribution of the traffic. Note that with the above criterion the design is reached regardless of how the number of vehicles is distributed which is against all logic. The upper limit of 2,000 vehicles or more proves inadequate today, and may lead to pavements that are unable to meet the requirements of traffic volumes much greater than the ones anticipated in the design, there are many such roads in Mexico. The 3 t (6.6 kips) limit is also insufficient now that there are so many different types of trucks including many models with far heavier loads. For example, the same pavement design would be required for 2,000 heavy vehicles as for 20,000; likewise the same design would be required for 500 3 t (6.6 kips) trucks as for 500 30 t (66 kips) trucks. It should be clear from these considerations that the characterization of traffic for the purpose of simplifying such a complicated variable cannot be reduced to merely counting vehicles. Vehicle caracteristics must be taken into account too.

The structural damage caused by loading increases rapidly with increases in the axle load. The information in Fig. 9-8 [16] illustrates this. It presents results of the AASHTO test showing the load equivalency factor for single axles and tandem axles. It can be seen that one application of an 8.2 t (18000 lb), single axle causes the same damage as 100 2.7 t (5950 lb) single axles. Likewise there is a great difference between the damage produced by single or tandem axles of the same weight; a single axle causes the same deterioration in a pavement as 12 tandem axles of the same total load.

Another interesting aspect of the repeated application of traffic loads is their lateral distribution within the pavement, Figure 9-9. [13] illustrates this lateral distribution, obtained from readings taken across the width of an active runway. A normal statistical distribution of the load application is seen. In 2-lane roads, heavy traffic is channeled into the lane adjacent to the shoulder. In 4-lane highways, this same channelization occurs in the two outside lanes. The effect of the outer tires of the vehicles is always more noticeable than that of the inner ones [13], possibly because of the pavement slope away from the crown.

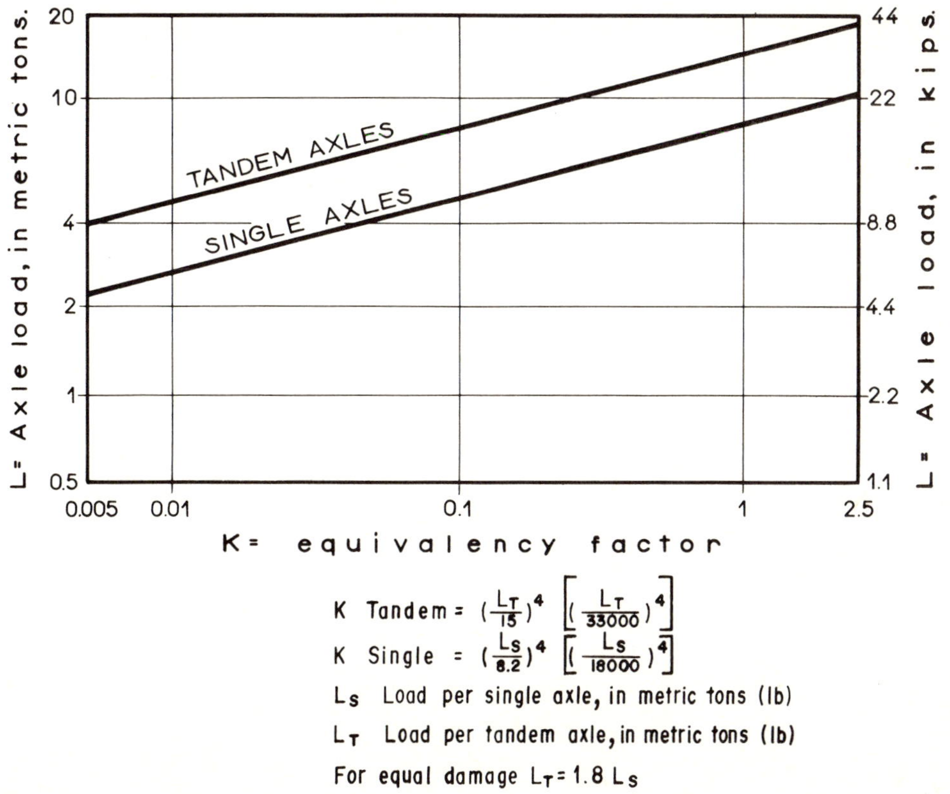

$$K\ \text{Tandem} = \left(\frac{L_T}{15}\right)^4 \left[\left(\frac{L_T}{33000}\right)^4\right]$$

$$K\ \text{Single} = \left(\frac{L_S}{8.2}\right)^4 \left[\left(\frac{L_S}{18000}\right)^4\right]$$

L_S Load per single axle, in metric tons (lb)

L_T Load per tandem axle, in metric tons (lb)

For equal damage $L_T = 1.8\ L_S$

Fig. 9-8 Coefficients of load equivalence [16]

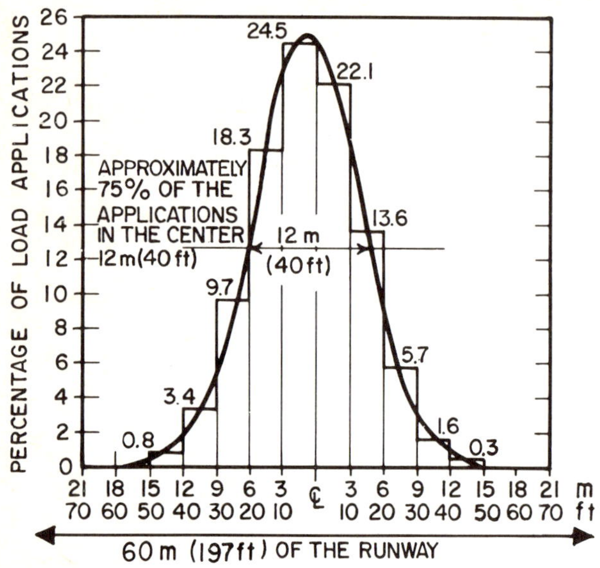

Fig. 9-9 Distribution of load application in the cross-section of a runway [13]

The loading rate has an influence on the pavement. Static or slow loads produce greater distress than more rapid ones. Consequently, uphill lanes are often more badly damaged than downhill ones. For the same reason the pavements in the taxiway, warm-up pad and apron of an airfield must be stronger than at the center of the runways.

It has been observed [1] that a moving load causes a mean permanent deformation, in each load application, of 0.1% of the total strain that is observed for that application. This shows that deformation under loading is 99.9% elastic. This is true even in pavements that have been inappropriately supported on plastic soils. The above refers to readings taken directly under passing loads on the surface of deteriorated pavements. Despite the predominance of elastic phenomena, however, pavement failure is the result of the accumulation of small plastic deformations. Underneath a moving load, the plastic flow of the material is more pronounced in the direction of the traffic. The stress model corresponding to this flow can theoretically be approximated to that produced by successive, closely spaced circular loads, along the wheel path.

Figure 9-10 [15] shows the types of vehicles that are usually considered in road design. Type *1* has single wheels, Types *2*, *3* and *4* have dual tire rear axle assemblies, and Types *5*, *7* and *8* have dual-tandem assemblies. Type *6* is a less common tractor and trailer with only dual tires on the rearmost axle. The figure does not include triple axle arrangements, which are very common at the moment, and very unfavorable ones. The figure also shows load equivalency factors for the different vehicles in use, which were obtained from AASHTO tests. These factors, which are given for empty and loaded vehicles, are obtained directly from the graphs in Fig. 9-8, by entering the axle load.

The tire assembly influences the superposition of the stresses induced. The contact area of the tires depends on the tire pressure and wheel load. The depth to which stresses are transmitted is partially determined by the contact area, and increases with it for a constant tire pressure. At the same time, tire assemblies with a large contact area tend to produce more uniform stress states for a given load than those with more concentrated loads. In practice, when load applications with large contact areas are anticipated, materials with uniform quality should be used throughout the profile of the pavement, whereas when the contact areas are small, a far higher quality will be required in the upper layers of pavement than in the lower ones. Table 9-5 [1] shows the data corresponding to aircraft landing gear.

Table 9-5
Load characteristics of common aircraft [1]

Type of aircraft		BOEING 747		DC-8		BOEING 707		COMET		BOEING 727		DC-9		DC-6		DC-4		DC-3	
Maximum weight at take-off, ton (kips)		352.9	777	162.4	357	152.5	336	73.5	162	78.5	173	52.2	115	48	105	33	73	12.2	27
Distribution of the load, ton (kips)	Nose gear	22.9	50.4	8.0	17.6	10.7	23.6	3.2	7.0	6.3	13.8	2.9	6.4	2.1	4.6	1.6	3.5	0.6	1.3
	Main gear	330.0	726	154.4	340	141.8	312	70.3	154	72.2	158	49.3	108	45.9	101	31.4	69	11.6	25
Tire pressure, kg/cm² (lb/in²)		13.01	185	13.7	195	12.7	180	11.8	168	11.8	168	9.1	129	7.45	106	5.5	78	3.5	50
Tire contact area, cm² (in²)		1585	245	1409	218	1396	216	745	115	1530	237	1354	210	1540	238	1427	221	1657	256
Geometry of the principal landing gear, cm (in)	S	112	44	55	22	88	35	49	19	86	34	66	26	78	31	76	30	—	—
	S_t	147	58	140	55	142	56	114	45	—	—	—	—	—	—	—	—	—	—
	L	55	22	52	20	52	20	38	15	54	21	51	20	54	21	52	20	56	22
	W	33	13	31	12	31	12	22.5	9	32.5	13	30	12	33	13	31	12	34	13
Tire arrangement		W, L, S_t, S								W, L, S								W, L	

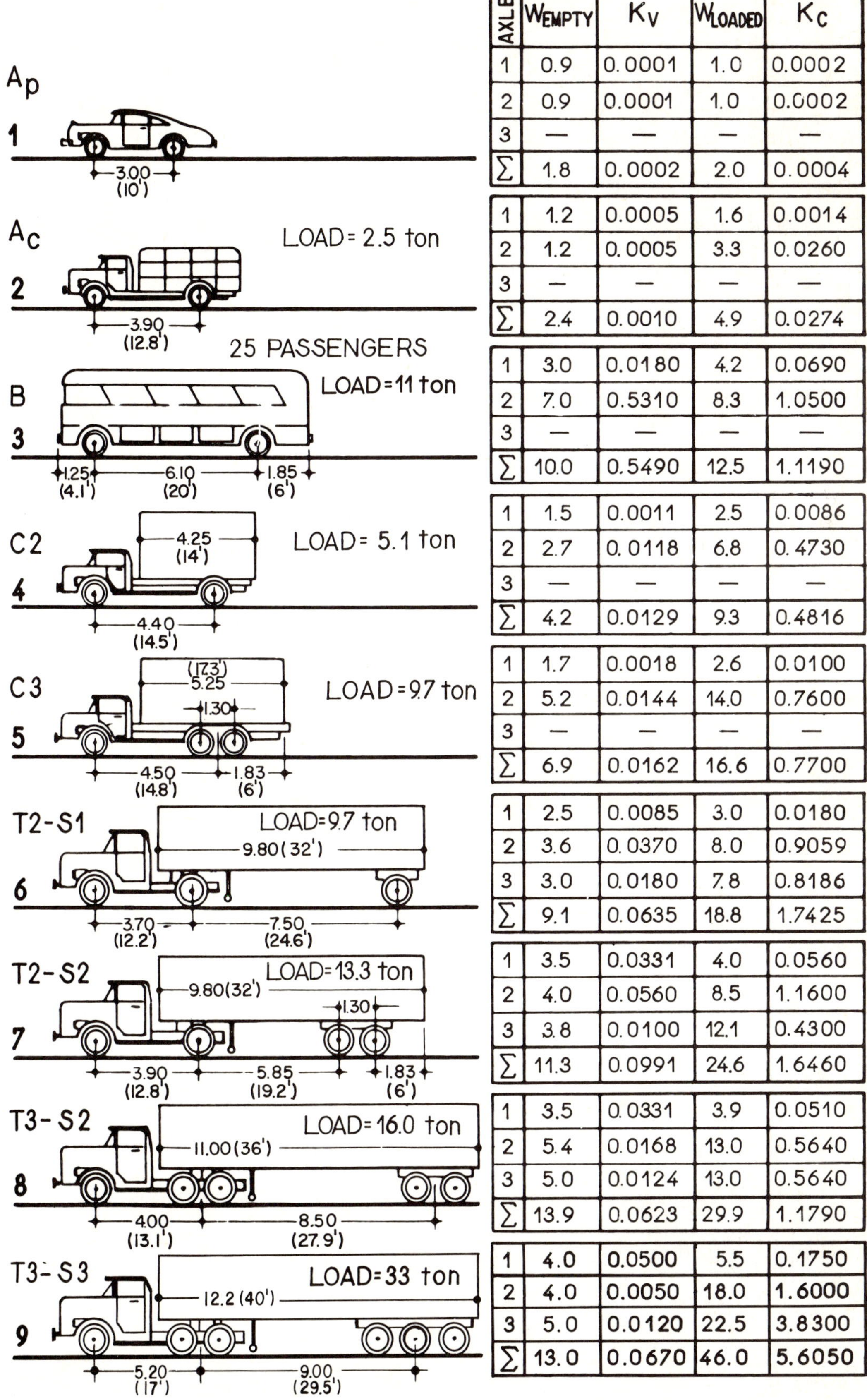

AXLE	W EMPTY	Kv	W LOADED	Kc
1	0.9	0.0001	1.0	0.0002
2	0.9	0.0001	1.0	0.0002
3	—	—	—	—
Σ	1.8	0.0002	2.0	0.0004
1	1.2	0.0005	1.6	0.0014
2	1.2	0.0005	3.3	0.0260
3	—	—	—	—
Σ	2.4	0.0010	4.9	0.0274
1	3.0	0.0180	4.2	0.0690
2	7.0	0.5310	8.3	1.0500
3	—	—	—	—
Σ	10.0	0.5490	12.5	1.1190
1	1.5	0.0011	2.5	0.0086
2	2.7	0.0118	6.8	0.4730
3	—	—	—	—
Σ	4.2	0.0129	9.3	0.4816
1	1.7	0.0018	2.6	0.0100
2	5.2	0.0144	14.0	0.7600
3	—	—	—	—
Σ	6.9	0.0162	16.6	0.7700
1	2.5	0.0085	3.0	0.0180
2	3.6	0.0370	8.0	0.9059
3	3.0	0.0180	7.8	0.8186
Σ	9.1	0.0635	18.8	1.7425
1	3.5	0.0331	4.0	0.0560
2	4.0	0.0560	8.5	1.1600
3	3.8	0.0100	12.1	0.4300
Σ	11.3	0.0991	24.6	1.6460
1	3.5	0.0331	3.9	0.0510
2	5.4	0.0168	13.0	0.5640
3	5.0	0.0124	13.0	0.5640
Σ	13.9	0.0623	29.9	1.1790
1	4.0	0.0500	5.5	0.1750
2	4.0	0.0050	18.0	1.6000
3	5.0	0.0120	22.5	3.8300
Σ	13.0	0.0670	46.0	5.6050

Fig. 9-10 Conversion of vehicles to equivalent axles

9.4 Types of Failure in Flexible Pavements

Most of the technology that has gradually been developed by pavement engineers is aimed at preventing the appearance of a wide range of forms of deterioration and failure. These have been characterized and described in great detail to develop a cause-effect relationship. From this has been developed a whole set of criteria for design and maintenance, which are a vital part of the knowledge and experience of the engineer. Before applying soil mechanics principles to pavement technology, it is necessary to review the most important and common types of distress suffered by pavements, relating them to the factors that cause them. In this way it will be possible to discuss these applications in a practical context, from the engineering viewpoint.

It is not a simple task to describe and discuss pavement failures, because of their great variety as well as other difficulties that are not even connected with engineering facts. To this last category belong the problems posed by the lack of a universal nomenclature. This is aggravated by the subjective, personal nature of describing a failure which varies from one place to another, one country to another and one organization to another. Here again there are ambiguities in the use of the word *failure*. In pavements, the word is commonly used both for total collapse, local distress as well as to describe simple deterioration or places where detrimental future evolution may be anticipated from large deflections [17]. The matter becomes even more complicated if the concept of distress or failure is associated with the serviceability level. This is basically subjective, depending on the level of requirements (or the affluence of whoever establishes it). A behavior that is quite simply different from what would have been regarded as *perfect* is sometimes described as a failure. An engineer with experience on pavements in several different regions or countries will have heard of wearing courses referred to as excellent that would never have passed inspection in more demanding areas. In a place without paved roads, a pavement that offers bad weather trafficability, even if only for special vehicles, may be regarded as excellent.

Pavement failures can be divided into three categories, of very different nature.

1) Failures due to structural insufficiency (Plate 9-5). These occur in pavements constructed with materials with inadequate strength, or with an insufficient thickness of good-quality materials. This is the failure which occurs when the combinations of the shear strength of each layer and the respective thicknesses do not develop an appropriate strength mechanism for the system.

Plate 9-5 Typical structural failure in a flexible pavement

2) Failures due to defects in construction. This occurs when pavements may be well designed and made of sufficiently resistant materials, but built with errors of dimensions or defects in workmanship that endanger the overall behavior.

3) Failures due to fatigue. This is the case of pavements which initially may have had suitable strength, but on account of the continual repetition of traffic stresses have suffered the effects of fatigue, structural degradation and, in general, loss of strength and accumulated strain. Although these phenomena are closely linked with the number of load repetitions, failures due to fatigue are clearly influenced by the duration of service (exposure to weather). They are typical in pavements which have operated for a long time without any problems (Plate 9-6).

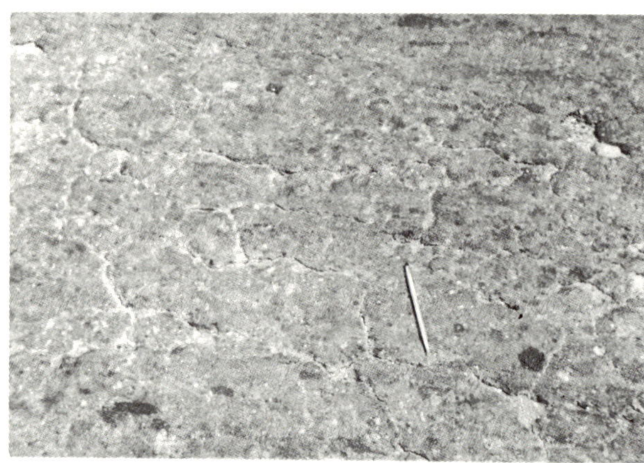

Plate 9-6 Reticulated cracking due to failure of the pavement structure by fatigue

In addition to the previous classification, based on their origin, failures in flexible pavements can also be grouped in accordance with the manner in which they occur and manifest themselves. Table 9-6 [19] presents such a classification. All the failures are divided into three types: (1) fractures (2) deformation and (3) break-up or disintegration. The table also presents a further subdivision into other categories according to the seriousness of the defect, their links with the most common mechanical causes, and in some cases the origins. The causes are the same for the three main types of failure (Table 9-6) and are related to the effect of traffic, the characteristics and structure of the pavement system and the nature of the support provided by the fill. However the specific principal variables with the greatest influence on each type of failure may be different. Table 9-7 groups the characteristics of the three causes that have the greatest influence in each of the three principal types of failure.

In conclusion, it can be seen that failures due to structural insufficiency, defects in construction, or fatigue are manifested in three ways: fractures, deformation and break-up or disintegration. Whether a specific deficiency originates one type of failure or another will depend on the combination of all the variables that are generically grouped under the headings 1) effect of traffic, 2) mechanical characteristics and structuration of the pavement materials, 3) support from

Table 9-6
Types of failures in flexible pavements and their principal characteristics [19]

TYPE	CHARACTERISTICS	CAUSES
Breaking	Cracking	Overloading (structural insufficiency) Repeated loading (fatigue) Changes in temperature Changes in moisture content (construction defect) Corrugation due to horizontal stresses (structural deficiency or construction defect) Shrinkage
	Destruction by Cracking	Overloading (structural insufficiency) Repeated loading (fatigue) Changes in temperature Changes in moisture content (construction defect)
Deformation	Permanent deformation	Overloading (structural insufficiency) Viscous deformation processes (fatigue, structural insufficiency and construction defect) Increase in density (construction defect. Particle breakage) Consolidation Expansion
	Failure	Overloading (structural insufficiency) Increase in density (construction defect. Particle breakage) Consolidation Expansion
Break-up (failure of surface course)	Removal	Loss of adhesion in surface course Chemical reactivity Abrasion caused by traffic
	Detatchment	Loss of adhesion in surface course Chemical reactivity Abrasion caused by traffic Degradation of the aggregates

underlying layers from the fill or, 4) from the soil or rock foundation. Recently attempts have been made to introduce all these variables into mathematical distress models with loading and environmental effects being represented by stochastic methods. [19] and [20] are examples of such analyses. From a mechanical point of view, pavement failures are usually the result of deformation due to shear stresses, or a deformation due to consolidation (an increase in density). These processes may take place in *any* of the pavement layers, including in the embankment.

A brief description follows of some of the most common forms of failures in flexible pavements:

Alligator cracking (or map cracking): This is a cracking which spreads over the entire wearing course, or at least a very substantial part of it, breaking the surface into irregular figures like the back of an alligator or the boundaries on a map (Plate 9-7). This condition indicates widespread localized movement in one or more of the pavement layers including fatigue. Although often limited to the wearing course, alligator cracking is common in flexible pavements constructed on resilient fills or subgrades. It is also typical of weak or poorly compacted base courses. The phenomenon is sometimes progressive. When it is, the local distress in the pavement starts with the detachment of parts of the surfacing and the rapid but local disappearance of the exposed granular

Plate 9-7 Alligator or map cracks on a flexible surface

Table 9-7
Principal factors affecting the three basic types of failure in flexible pavements

Type of failure	Ultimate cause		
	Traffic	Pavement	Foundation (support)
Breaking	Wheel load (magnitude) Repetitions Area of load influence Speed Arrangement and layout of wheels and axles	Rigidity of the various layers Flexibility (adaptability to fatigue) Durability Plastic deformation Elastic deformation	Rigidity of base course and sub-base Plastic deformation Elastic deformation
Deformation	Wheel load (magnitude) Repetitions Area of load influence Speed Arrangement and layout of wheels and axles	Thickness Strength Compressibility Susceptibility to volume change Plastic deformation Elastic deformation	Susceptibility to volume change Plastic deformation Elastic deformation
Break-up (failure in surface courses)	Tire pressure Repetitions Speed	Asphalt characteristics Aggregate characteristics (porosity, lack of adhesion to the asphalt)	Resilience of pavement layers Water seepage Changes in temperature

materials below. When the phenomenon reaches such destructive levels, it is almost certainly associated with structural deficiencies in the base course.

In countries where freezing weather occurs regularly, cracking is frequent when frost susceptible materials are used [21]. It may also be indicative of places where subdrainage is required. When this type of cracking is first observed it is essential to determine whether it is likely to evolve further. Processes associated with aging and fatigue in the wearing course progress very slowly. However, evolution of the phenomenon is very rapid when associated with serious structural deficiencies or excess water. Consequently, in order to predict the future evolution when detected, detailed pavement studies are usually necessary.

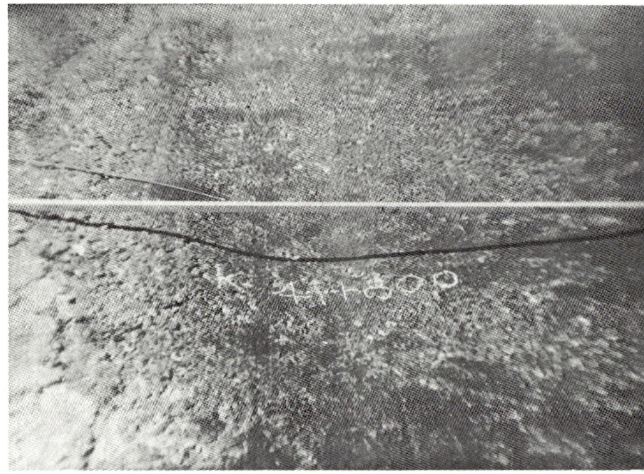

Plate 9-8 Permanent deformation at the surface of a flexible pavement

Permanent deformation in the pavement surface, Ruts: This is frequently associated with an increase in the density of the granular layers of the base course or sub-base. The causes can be over-loading, repeated loading (increase in density due to vibration) or particle breakage. It may also be due to consolidation of the subgrade or even the body of the fill (Plate 9-8). The width of the rut exceeds that of the tire. It becomes larger the deeper the layer causing the subsidence. The deformation that is referred to should be distinguished from the rut that occurs as a result of simple lateral displacement of a defective surface course, Plate 9-9. The distinctive sign is that in this last case the surface bulges upward on both sides of the rut, whereas in a rut that originates deep down, these bulges do not occur.

Plate 9-9 Deformation of a flexible surface due to lack of stability in the asphalt mixture

Shear failures: These are typically associated with a lack of shear strength in the base course or sub-base of the pavement, and more rarely in the subgrade. They generally consist of deep, clean, clearly marked ruts, which do not greatly exceed the width of the tire. In this case, the material on either side of the rut usually bulges upward, but the failure is easily distinguished from a simple displacement of the wearing course by the greater depth that is affected (Plate 9-10)

Plate 9-10 Shear failure in the base of a flexible pavement

Longitudinal cracking: This consists of the appearance of not very open longitudinal cracks (about 0.5 cm (0.2 in) wide) throughout the entire area corresponding to the heaviest traffic loads. Cracks of this type are due to movements in the pavement layers, which occur predominantly in the horizontal direction. The movement may occur in the base course, the sub-base, or, with reasonable frequency, in the subgrade. They are indicative of freezing and thawing or of volume changes due to variations in the water content, especially in the subgrade.

Consolidation of the original foundation soil: The consolidation of soft foundation soils may cause distortion of the pavement, regardless of pavement layer thicknesses or its structural condition. Deformations in the cross-section may lead to longitudinal cracks. When the stability of the embankment is jeopardized by deficient strength in the underlying soil, typical semi-circular cracks occur, marking what may become the head of an eventual slide. These cracks, of course, damage the pavement.

9.5 Theoretical Approaches to Flexible Pavement Design

More than 50 years have gone by since the development of the California Bearing Ratio method (*CBR*) by O. J. PORTER [22]. This was the first rational design for flexible pavements, based on an empirical relationship between the performance of a pavement and that of its component material determined by a laboratory test. More than 30 years have passed since BURMISTER [10] presented the first analysis of the stress distribution in a pavement using a model of several different elastic layers. These two approaches to the problem of designing flexible pavements through explaining their structural behavior continue to serve as examples of most current methodology used for these problems. However no fully satisfactory solution has yet been found, although over the years certain improvements have been made to both original methods.

The *CBR* method [7], which is semi-empirical, has taken the lead in practice almost everywhere. It has been subjected to improvements by the U.S. Army Corps of Engineers [23], some of which will be briefly mentioned later in this chapter. Although conceived as a purely empirical correlation, so much attention, experience, research and analysis has been devoted to this method that it is now becoming more complex and rational. It is still far from being a *scientific* method, and rarely provides a suitable cause-effect relation and does not lead to a better knowledge of the structural behavior of a flexible pavement.

The theories that view pavements as a system of elastic layers have developed slowly, on account of their mathematical complexity (see [11] for example). Substantial progress was made with the development of the charts and tables of influence factors for stresses and strains, which are now calculated with electronic computers [10,24,25]. Nowadays, such stress influence factors are computed for three-layer systems and for a wide variety of circumstances [26]. Strain influence factors are more difficult to obtain, but [27] provides an approximate solution to the problem.

The numerical computation of influence factors made it possible to verify experimentally many of the ideas generated by the theories in which flexible pavements are represented as a system of elastic layers. In this way the stresses and strains measured are compared with those computed theoretically [28]. Reasonable agreements are found only in pavements with high tensile strength (cement-treated base courses). Subsequent comparisons of theory and field readings have given results which are often erratic and deceptive.

The theory of elastic layers has been extended more recently to visco-elastic systems, both for stress analysis and strain analysis [29-32]. Because the representation of a layer of pavement as a linearly elastic solid with fixed constants is not altogether satisfactory, attempts have also been made to compute the stresses and strains in systems using *variable elastic constants* [33], by analyses based on the non-linear elastic theory [34] or and on stochastic analyses [35].

Fairly complete summaries of the applications and conclusions of elastic theories to the problem of pavement design are included in [36] and [37].

A certain disenchantment is perceived in the attitude of investigators towards the application of the pure elastic theory, when taken to its ultimate consequences for the solution of pavement design problems. Many theoretical investigators have abandoned this line of action, preferring to devote their attention to general systematic approaches, in which computers are used to evaluate the relative contribution of the factors that influence design. When fed with specific information relating to a problem obtained from the actual field locations or from road structures with similar conditions, the computers help to develop schemes for optimum lines of action. These systems-engineering approaches can hardly be expected to bring about fundamental improvements in pavement design and construction. The result that is obtained by following a certain line of optimization of decisions will depend on how correct the data fed into the computer are and how correct the basic conceptions of the problem are. These approaches appear to be more promising in other aspects of the pavement design, such as the evaluation of investment or the distribution of expenditure.

It seems clear that the approaches to the analysis of the structural behavior of pavements should be reviewed and accordingly modified. Much is unknown about the behavior of the different layers, the role of each and their interrelation, plus the relationships of the pavement to the underlying fill. A highly promising future lies in the study of these relations. Consequently, the measurement of material properties and their variations in existing pavements, the use of laboratory test roads of the type mentioned earlier in this chapter and the realistic determination of the effects of the different factors that influence pavement performance will be some of the most rewarding tasks of the future.

Once the part played by each layer and the interaction of all the layers is understood, and once the requirements that must be met by each soil or aggregate layer have been established, including suitable or unsuitable soil properties, and the role played by water and climate in the different soils, it will be possible to undertake a generalization leading to a rational calculation method in accordance with the laws of soil mechanics and the structural properties of pavements. Until such methods are available, this deficiency must be duly acknowledged, abandoning any attempts at theorization that may include hypotheses that cannot be verified in practical construction and by keeping as closely as possible to sound past experience the performance of whatever projects are undertaken must be measured, in order to detect failures and limitations and correct them in time. The need for a design method that will act as a gauge of common experiences and as a guide for those who do not have sufficient personal experience, can be satisfied by simple methods of a semi-empirical nature (*CBR*, or the use of HVEEM's stabilometer). Tests should be made in accordance with the conditions likely to prevail in the field: compaction, the water content and its variations, and including the effects of traffic.

In highways and even runways, a reasonable hypothesis for the analysis of load-bearing capacities appears to be that the weight of the pavement material can be disregarded. Moreover so long as the tire pressures are approximately constant and the tire imprints are the same shape, the thickness of the flexible pavement for the same traffic will be proportional to the width of the tires, or even better, to the square root of the contact area of those tires [7]. The relationship between the thickness required and the contact pressure, when the area is not constant, is not determined. However, the comparisons that have been made between the effects of a single wheel and those of dual wheel assemblies seem to indicate that the thickness of pavement required depends on the correct pressure of the tire. Likewise, the analysis of deflection readings in existing pavements indicates that the deflections caused by loading are approximately proportional to the thickness of pavement required to suitably tolerate that loading. Theoretically and experimentally, the deflections, when measured in thicknesses of pavement that are proportional to the square root of the area of the tire, turn out to be proportional to the product of the contact pressure of that tire multiplied by the square root of the bearing area.

The desirable relation between traffic volume and pavement thickness has been determined on the basis of experimental results obtained from test sections of road. As has been said, these results show that there is a linear relation between the logarithm of the thickness required and the logarithm of the number of load applications.

The relation between the quality of the subgrade and the thickness of the pavement has not been established quantitatively such that it can be utilized in design calculations. The important contribution a prototype test road can make to clarifying these problems has already been mentioned. Representative tests are now required to determine the *quality* of the soils, as applied to pavement design.

Failures in the subgrade often resemble shear failures under specific confinement conditions. Therefore many engineers have proposed traditional shear tests for evaluating the risk of such failures. In traditional laboratory interpretation, however, the shear strength of materials is expressed in terms of two parameters, one related to internal friction and another which expresses the shear strength of a soil under zero normal external pressure (see Chapter 1). Combination of these two parameters for defining total strength is problematical, depending on the magnitude and variation in the principal pressures involved, as well as other factors that are not well known. The magnitude and direction of the principal stresses applied to the soil during the passing of a moving load vary greatly. When confinement pressures are small, the frictional component will be by far the more critical one. All these considerations make it hard to determine a simple shear test that will express the capacity of a soil to tolerate traffic stresses.

9.6 Special Laboratory Tests used in Pavement Technology

Within the field of pavement technology, some special tests have been developed which are worth mentioning on account of their increasing popularity, or because they provide a basis for design methods with very wide applications. With these ideas in mind, a brief description follows of plate-bearing tests, *CBR* tests and some triaxial tests.

9.6.1 Plate-Bearing Tests

The purpose of these is to evaluate the bearing capacity of subgrades (and sometimes sub-bases and base courses). They are currently used for both rigid and flexible pavements.

The test [7] consists of loading a circular plate, that is in close contact with the layer to be tested, and measuring the ultimate deflections corresponding to the different loads applied. The plates used for runways are usually 76.2 cm (30 in) in diameter, but for roads they are usually smaller, with a diameter of 30.5 cm (12 in). Both areas are similar to those of the loaded area of the subgrade produced by a tire at the surface (Plate 9-11). In order to prevent bending of the plate, others with decreasing diameters are stacked on top of it, thus providing the necessary rigidity. The load is transmitted to the plates by hydraulic jacks, acting against loaded trucks (Plate 9-12). The deflections produced are usually measured at four points of the plate, at right angles to one another, by means of dial gages attached to a horizontal beam with its supports placed far enough away from the plate for it to be regarded as fixed. Figure 9-11 gives a diagram of the arrangement.

Plate 9-11 Equipment for conducting plate-bearing tests in runways

Plate 9-12 Special equipment for loading plates for bearing tests in runways

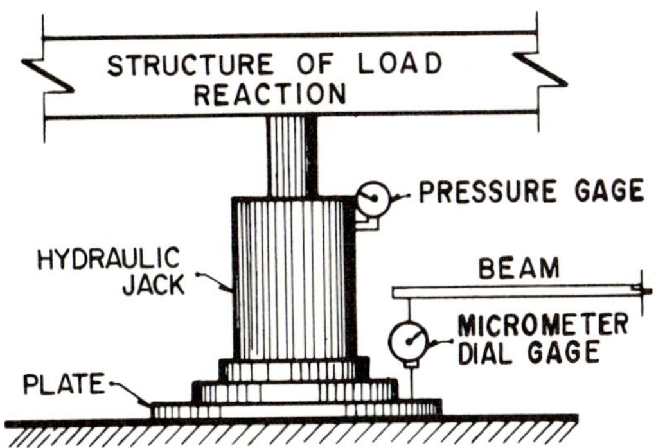

Fig. 9-11 Plate-bearing test set-up

The stress that is transmitted by the plate to the layer depends on the relation between its perimeter and its area, and also on the strength of the soil. The unit load (normal pressure) that is transmitted by the plate for a given deflection, corresponds to the following equation [39]:

$$\sigma = n + m \frac{p}{A} \tag{9-2}$$

where: σ is the normal pressure transmitted by the plate; n, m are empirical coefficients obtained experimentally; and p/A is the relation between the perimeter and the area of the plate.

The equation is based on empirical relations and does not presuppose any special soil behavior (for example, elastic). Values n and m must be determined by conducting at least two tests with different size plates, with the same deflection, measuring the pressure in each one. Fig 9-12 [38] shows the type of information that can be obtained from a plate-bearing test, together with the typical shapes of the resulting curves.

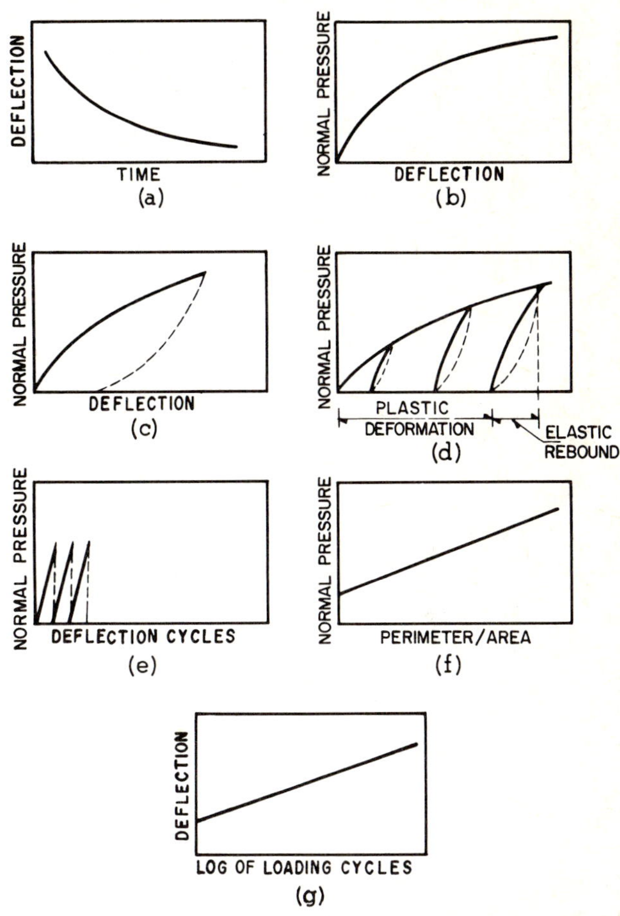

Fig. 9-12 Typical information that can be obtained from a plate-bearing test [38]
a) Deflection-time curve
b) Pressure-deflection curve
c) Pressure-deflection curve with unloading
d) Curve with cyclic application of pressure
e) Repeated loading curve
f) Effect of the plate size
g) Repeated loading-deflection curve

Plate-bearing tests enable computation of the coefficient of subgrade reaction, k. This is defined as the pressure that must be transmitted to the plate in order to produce a predetermined deflection in the soil.

$$k = \frac{p}{\Delta}, \left[\frac{\text{force}}{\text{length}^3}\right] \tag{9-3}$$

A direct application of this concept is in the design of rigid pavements (Chapter 10). It is also used in some design methods for flexible pavements, such as the one by McLeod [7], in which several design methods are described.

The modulus of subgrade reaction thus defined depends on the diameter of the plate that is used to compute it, for at a constant pressure the settlement of the circular plate increases with its diameter. Consequently if a given settlement is required, the pressure required to produce it will become gradually smaller as the diameter of the plate increases. This is one of the reasons why the diameter of the plates should be standardized.

In spite of the wide use that has been made of the modulus of subgrade reaction concept in pavement technology, its lack of intrinsic significance as a measurement of any fundamental soil property should be pointed out. Its value lies in its usefulness as a calculation parameter when comparing the coefficients obtained in the same manner for different soils.

The coefficient of subgrade reaction, like any other parameter of subgrade behavior, depends on the water content of the soil. In laboratory or field tests, the water content that will be attained by the soil in the pavement should be used. This is referred to as the equilibrium moisture (usually different from the compaction water content), but it is not initially known. What is done is to work with a water content that is regarded as critical. Some organizations use that corresponding to saturation. Others, like the State of Texas (U.S.A.) use that resulting from a pre-specified curing process, which is described later in the Texas triaxial test. It is a sensitive matter and one where the judgement of the engineer is decisive. Further, it does not apply exclusively to the type of tests that are described here, but affects all the field and laboratory tests that have to be carried out in order to design future pavements.

When the water content is representative of the average future conditions of the soil, the results of the plate-bearing tests conducted on soils with a water content that is different must be corrected. The correction factor depends on the relation of the resistance to simple compression shown by two soil specimens, one tested in the natural condition and the other with the water content considered critical for design (or representative of future equilibrium conditions). Use of the simple compression test in this case, although it may appear reasonable, is nonetheless arbitrary.

During the test, the load is applied in increments. A new increment is applied to the plate when the rate of deflection under the previous increment is of about 0.001 cm/min (0.0004 in/min). In some cases, it is advisable to determine the relative amounts of elastic deformation and permanent (plastic) deformation. To accomplish this, loads varying in intensity are applied and are maintained until all deflection ceases. At this moment the load is removed and a greater increment is applied, so that a graph like the one in Fig. 9-12 can be plotted. After the final load has been applied, an unloading decrement process is followed with the same loads as in the loading process, so as to obtain soil recovery graphs. The difference between the two graphs is regarded as elastic.

Appendix 9a describes in some detail the test sequence proposed by McLeod [1,39], which was adopted by the Asphalt Institute. McLeod carried out the test with 10 load repetitions and established deflections of 0.508 cm (0.2 in) for roads and 1.27 cm (0.5 in) for runways, using the diameters that were specified for each case in the initial paragraphs of this section.

9.6.2 California Bearing Ratio Test

This test was originally developed by the California Division of Highways. Currently it is used very widely. The design method based on it is perhaps responsible for more than half of all the pavements now under construction throughout the world. The California Bearing Ratio (*CBR*) is obtained from a penetration test, in which a piston with an area of 19.4 cm^2 (3 in^2) is made to penetrate a sample of soil at a rate of 0.127 cm/min (0.05 in/min). The load applied for penetrations varying by 0.25 cm (0.1 in) is measured. The California Bearing Ratio of the soil is defined as the ratio (expressed as a percentage) between the pressure that is necessary to penetrate the first 0.25 cm (0.1 in) and the pressure required to obtain the same penetration in an arbitrarily selected material, taken as a reference. The reference is a crushed stone in which the piston pressures that are shown in Table 9-8 are produced.

Table 9-8

Pressures for different piston penetrations of the sample material in the *CBR* test

Penetration		Piston pressure	
cm	in	kg/cm^2	lb/in^2
0.25	0.1	70	1,000
0.50	0.2	105	1,500
0.75	0.3	133	1,900
1.00	0.4	161	2,300
1.25	0.5	182	2,600

As stated previously, the penetration that is used to compute the *CBR* is that corresponding to the first 0.25 cm (0.1 in). As a general rule, the *CBR* diminished when the penetration for which it is computed is greater, however sometimes, when it is computed with the 0.5 cm (0.2 in) penetration, it turns out to be greater than the one obtained for the first penetration. If this is the case, then the *CBR* adopted is the one obtained from the second penetration (0.5 cm) (0.2 in).

The soil specimen on which the test is conducted is confined in a mold 15.2 cm (6 in) in diameter and 20.3 cm (8 in) high. In the original California test method, the specimen was prepared in three rodded layers until the mold was filled. The material was then subjected to a pressure of 140 kg/cm^2 (2000 lb/in^2), applied uniformly to its upper surface. Specimens with different water contents were prepared thus, until one was found which, when subjected to this pressure of 140 kg/cm^2 (2000 lb/in^2), sent exuded water into the lower part of the mold. After a saturation period of 4 days, this specimen was assumed to be representative of the worst conditions that might prevail in the future pavement.

In more recent years, the U.S. Army Corps of Engineers [39] has developed a test method which differs from the traditional one in the specimen preparation procedures. This method is described in detail in Appendix 9b. The specimens are compacted using a dynamic method, the AASHTO standard and modified tests procedures (see Chapter 4) as well as with an intermediate compactive effort. The objective is a reasonable reproduction of the compaction conditions achieved with field equipment. The overlying pavement confines the soil (or deeper pavement layers) by a surcharge pressure which may enhance the soil's resistance. In order to simulate this a plate is placed on top of the soil tested which transmits to the specimen a pressure equivalent to the surcharge anticipated for the finished pavement. The plate has a hole in the center for the piston.

The factors with the strongest influence on the values obtained in the *CBR* test are the texture of the soil, its water content and its compacted density. In frictional soils, there is practically no expansion during saturation, and the surcharge produced by the perforated plate is therefore not significant during this stage of the test. However, the value of the surcharge does have a considerable influence during the penetration stage, for the strength of frictional soils is greatly affected by confinement. In clayey soils exactly the opposite occurs.

Generally the pressure-penetration curve obtained from a *CBR* test is linear for small penetrations, and tends to become slightly curved concave downwards for greater penetrations. Sometimes, however, the curve is concave upwards for a small span corresponding to the initial penetrations. This happens most often when the piston is not exactly at right angles to the surface of the sample at the start of the test or the surface is not smooth. When this occurs, the results of the test have to be corrected by moving the graph towards the left, so that the straight section, which is duly lengthened, passes through the origin, disregarding the small initial curvature. The *CBR* values thus obtained are referred to as the *corrected CBR.*

The results of a complete test to determine the *CBR* are used in a combination of three graphs. These are referred to in Fig. 9-13. Part *(a)* of the figure shows graphs resulting from the compaction tests that were used to prepare the specimens on which the *CBR* tests were conducted. According to the norm established by the Corps of Engineers, compaction tests must be of the dynamic type, with decreasing compactive efforts. Part *(b)* of the figure gives the results characteristic of *CBR* tests for the same specimens referred to in Part *(a)*. Note that the *CBR* is not a constant characteristic of the soil, but a circumstantial one, which depends on the water content and the compaction condition (compactive effort and method of compaction). There is a maximum *CBR* corresponding to a near-optimum compacting water content. Note also that for soils with a high water content, the *CBR* of the soil compacted with a greater effort may be smaller than the one that is obtained with a smaller effort, so long as the water content is lower. However, the maximum *CBR* that can be obtained is higher the greater the effort used to compact the specimen.

In Part *(c)* of Fig. 9-13, a graph is shown from which conclusions of practical value can be drawn and which illustrates the procedure that is recommended by the Corps of Engineers [39] for selecting the design *CBR*. Values for the corrected *CBR* have been plotted against the dry densities of the specimens tested. Each curve drawn corresponds to penetration tests in which the soil had the same compacting water content, but was compacted with different efforts. The water content, for example 14%, is represented by the curve labeled 14. Part *(a)* enables the three densities to be obtained, which in this particular case corresponded to the water content of 14%, with the different compactive efforts used, *I, II,* and *III.* In Part *b* the corrected *CBR* values for these three compactive efforts can be obtained. Three densities and three *CBR* values are thus obtained for three specimens compacted with a water content of 14%, using the three compactive efforts described. With these three pairs of values, Curve *14* of Part *(c)* of Fig. 9-13 is plotted.

The curves indicate that higher *CBR* values do not always correspond to a greater density. For example, Curve *20* shows conditions which deterioriate as density increases because of increasing saturation. It all depends on the water content of the soil. The above provides a practical working approach. A water content between 14% and 18% is assumed to prevail in the field. It is also assumed that a dry density between 95 and 99% of the maximum corresponding to effort *I* is to be obtained in the field. These values determine the range in the water contents and densities that should be stipulated for the field (shaded zone in Fig. 9-13a). Now, in Part *(c)* it can be seen that for water contents between 14 and 18% and the selected densities, the *CBR* ranges between 11% and 25% approximately. It can also be seen how dangerous it would be if the water

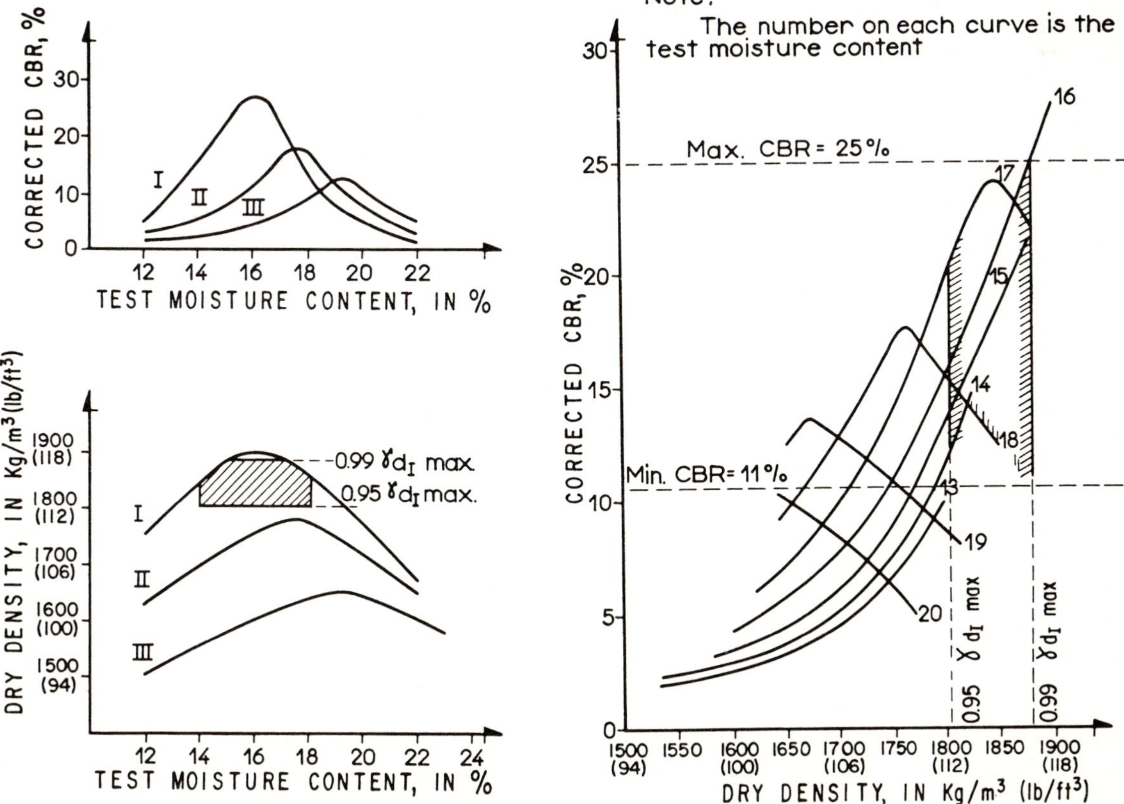

Fig. 9-13 Criterion of the Corps of Engineers for selecting the *CBR* for design [39]

content in the field were to rise above 18%, for in that case the *CBR* value of the soil would drop appreciably. On the basis of the above, a minimum *CBR* for design could be fixed at about 12%, for example. With graphs similar to these, the designer will be able to adopt a logical *CBR* value for design, set the density that should be stipulated for the field, and get an idea of the seriousness of an error due to the deficient or excessive field water content.

Given the extensive, almost universal use that is made of the *CBR* test as a basis or guide for designing flexible pavement thicknesses, one wonders what its value would be as a test for measuring soil strength and what its situation would be among the laboratory and field techniques that have been developed and recommended by soil mechanics for measuring this fundamental characteristic. Such a test is a penetration or punching test, where a piston is forced into a soil specimen in a mold, by a pressure applied at a controlled rate. The pressure is measured for a pre-established series of penetrations. At the end of the test a pressure-penetration graph is plotted in which the pressure values are pre-selected and the penetration values may vary widely from one test to another, depending on the nature of the soil, its water content, and compaction conditions. Thus, in addition to the practical aspect of the magnitude of the pressure required to bring about a certain pre-established penetration, compared to the pressure that is required to achieve the same penetration in an ideal material (Table 9-8), the *CBR* test is a type of stress-strain test.

From this point of view, the question arises as to its representativity. The same material may have very different stress-strain behaviors, depending on the circumstances in which it is loaded. Thus, a specific stress-strain behavior obtained in the laboratory has an engineering value only when it refers to conditions that correspond to the structural behavior common in engineering works. Even then the conclusions that can be drawn from such a laboratory analysis will only be applicable to those practical cases in which the same work conditions or environment prevail as the ones in the laboratory test.

Punching is a peculiar way of working with earth materials, and only remotely resembles the loading of the soil beneath for a pavement that is subjected to traffic loads and climatic variations. Consequently, what might be regarded as representativity as a model only remotely exists in a *CBR* test. In principle, little that is implied by the results of the test can be considered applicable to the structural behavior of a pavement. Furthermore, the test has characteristics which make it hard to interpret. Its boundary conditions are both kinematically (walls and bottom of the mold) and dynamically (surcharge from the plates and pressure from the piston) different from the pavement and cannot be readily reconciled to provide a clear interpretation. The viscous characteristics of the soil play an important part in pavement behavior and are not easy to characterize in the *CBR* either. Moreover the design of a layer of pavement is related to the strength of the soils involved as well as with their deformability. The test emphasizes this second characteristic. So far as the authors know, the effect of scale in the *CBR* test has not been investigated, but it is probably enormously important in many soils; consequently the test has indeterminate theoretical significance.

The result of a *CBR* test depends on the resistance of the soil to penetration of the piston; therefore, although it may not be a strength test, the shear strength of the soil nevertheless is indirectly involved through the punching or bearing capacity mechanism. However no unique quantitative relationship with that strength can be established. The shear strength of soils depends on many factors (Chapter 1), none of which is fully controlled or modeled in a *CBR* test. Under the circumstances, measuring strength realistically by *CBR* cannot be considered.

The differences in the results of the test are so great when the water content of the soil or its compaction conditions are made to vary, that almost any value can be obtained for the *CBR* by simply modifying these variables at will. Consequently, what the water content should be that will represent the condition of the finished pavement, and what the compaction of the soil and its evolution with time should be, become fundamental aspects of the analysis, more important even than a single *CBR* value. The test is, therefore, not a reliable predictor of the future behavior of a soil below a pavement. This explains why in different organizations and different countries the use of the *CBR* may lead to practical results with discrepancies between design and performance.

Another problem is the variability of test results, which is shown by the great differences obtained by different operators, working correctly on the same soil and using the same technique. Table 9-9 [40] is significant. It summarizes the results of identical tests, conducted on subgrade, sub-base and base course materials. In all cases, operators were competent and techniques appropriate. The information contained in the table, not only shows the variability of *CBR* test results, but also demonstrates that the engineers must have a healthy skepticism of laboratory data, which are so often blindly trusted. As far as could be determined by a team of experts, test errors are excluded from the investigation reported in Table 9-9. Indeed it probably represents more careful work conditions than is characteristic of most road engineering jobs.

Table 9-9
Variability in the *CBR* test [40]

Material	No. of tests	Max. value	Min. value
Subgrade	11	10	2
Sub-base course	6	142	67
Base course	6	172	85

It has been said that it is sometimes possible to evaluate a laboratory technology, by considering an extreme, almost grotesque situation. For example consider a *CBR* test carried out on water in a mold and a piston without any leaks. The result would theoretically be an infinite *CBR*, because water is non-compressible at the stresses involved in civil engineering. Clays or soils in general are not water but when saturated are incompressible; but the anxiety that is provoked by this extreme nevertheless remains. It has been suggested that a *CBR* test produces a type of failure in the soil that is similar to that considered by Prandtl, Terzaghi and others in their bearing capacity theories. The foregoing statement is not correct. The indentation studied by Prandtl considers the complete development of plastic shear but with the very small penetration in the *CBR* test penetration is finite and shear incomplete. The concepts are similar but also different. However the same criticism applies to using bearing capacity theory to determine the safety of foundations against shear failure.

The merits of the *CBR* test do not therefore lie in its theoretical value, its representativity as a model, or its measure of soil strength as a reliable means of correlating the behavior of a soil in the laboratory with field experience. What, then, is the reason for its popularity? The authors feel that historical reasons and an ever-increasing daily use have contributed to this popularity. The test and the design method based on it were an early quantitative tool for designing pavements, that superseded personal whim or fancy. From the time of its development highway engineers have grown accustomed to referring their daily experiences to the test and its results, often without realizing that the great variability of the test makes it a limited basis for correlation. Many experienced engineers have never used any other design method, and are consequently mentally gauged to the *CBR* alone, no matter how difficult it may be to achieve an adequate calibration. Moreover the design method based on the *CBR* test is remarkably simple, requires little previous field information and is quick and economical. It is also unfortunately true that current engineering research has not developed another design method that does not also have serious limitations. In the light of all these considerations, the design method based on the *CBR* test must be regarded as empirical, lacking in scientific universality and difficult to interpret from a theoretical viewpoint. As such, it permits soil behavior in the laboratory to be correlated with the behavior of pavements in the field. How well such a correlation is corroborated will depend largely on the personal observation and attention to detail of the engineer applying the method. We stress that the basis for correlation is vague.

In our opinion the time has arrived when the methods based on the *CBR* test will have to be reinforced or replaced by others for routine design. Other, more rational methods are being developed and it is hoped that the situation will improve even more in the future. However when a pavement construction or design organization introduces a new design method, it must be aware that pavement design and construction is, to a great extent, empirical. It is largely what is referred to as an *engineering art.* Any method that may supersede the *CBR* will also require large amounts of experience and as well as verification of laboratory data with field behavior. Most or all the experience of the engineers of such an organization will refer to the CBR test, so that the engineers will find themselves unprepared for the drastic introduction of a new design method within another experimental framework. The introduction of new design methods, which is advisable, will have to be done parallel to using the *CBR*. In this way, the organization will use both the *CBR* test and the new method to be introduced until adequate comparisons have been established between the two, and the personal experience of the engineers can be transferred to the new method.

9.6.3 Triaxial Tests

Pavement technology has developed a series of special triaxial test procedures, on which various pavement design methods are based [7]. The tests have been applied to determine the properties of subgrades and as well as pavement layers, including wearing courses. In general, they are similar to the common quick or undrained test in soil mechanics practice (see Chapter 1) and add little to it from a theoretical point of view.

.1 Kansas Triaxial Test [41]

The triaxial test developed by the State of Kansas (U.S.A.) is the basis of a complete method of pavement design in use in that State [7]. The test measures the deformation modulus of a soil, defined as the slope of the stress-strain curve obtained. To avoid problems relating to representativity of the sample, the specimen used in the triaxial chamber is large (about 10 cm (4 in) in diameter). The moisture conditions most detrimental to the existence of the pavement are reproduced by saturating the specimen. However it is acknowledged that this condition may prove excessively conservative, and a correction factor, *n,* is included which is a function of the rainfall in the construction zone (Table 9-10). The deformation modulus is determined in

Table 9-10

Correction factors for saturation conditions in the Kansas triaxial test

Correction factor n	mean rainfall	
	cm/year	in/year
0.5	38-50	15-20
0.6	51-63	20-25
0.7	64-76	25-30
0.8	77-89	30-35
0.9	90-101	35-40
1.0	102-127	40-50

the test using a deviator stress ($\sigma_1 - \sigma_3$) – strain graph and marking on it the deviator stress that is presumed will act in the finished pavement. The deformation modulus is the secant corresponding to that point (Fig. 9-14).

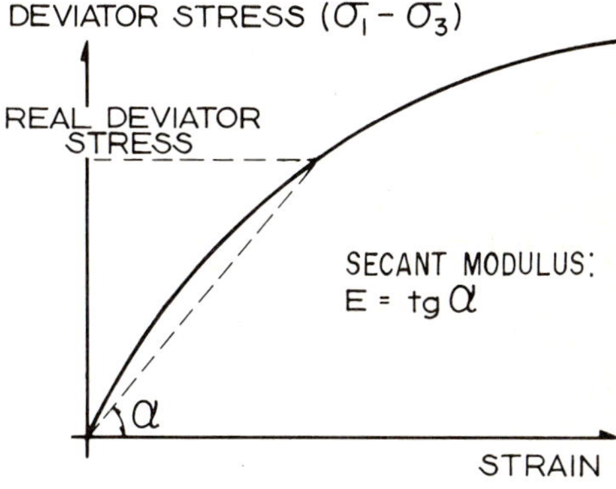

Fig. 9-14 Criterion for obtaining the deformation modulus in the Kansas test [7]

.2 Texas Triaxial Test

The Texas Highway Department (Dept. of Transportation) developed a triaxial test procedure for obtaining the strength envelopes of soils [43]. These envelopes are used in a specific design method that was also developed. The triaxial chamber used is a stainless steel cylinder with a rubber membrane inside. Air is forced in between the membrane and the chamber to produce the confining stress. The chamber is relatively large (30.5 cm (12 in) high, with an internal diameter of 17.2 cm (6.77 in)), so as to be able to test materials with abundant coarse particles, which are common in paving technology.

The material is compacted in four layers by blows from an automatic mechanical compactor which was described in Chapter 4. The material is confined in a mold similar to the one that is used in the *CBR* tests. After compaction, it is oven-dried at a temperature of 60°C for 8 hours and is subsequently left in contact with a source of water for a minimum of 10 days or for a number of days that is numerically equal to the plasticity index for the soil if larger. During this period of capillary absorption, the soil is subjected to a surcharge of 0.07 kg/cm^2 (1 psi). In this way, the conditions most detrimental to the existence of a pavement are supposedly reproduced in the laboratory.

.3 HVEEM's Stabilometer [42]

HVEEM's stabilometer is another triaxial device which is the basis for a specific method for designing flexible pavements. The apparatus enables a test to be conducted which measures the mechanical behavior of materials under stress combinations below those that produce shear failure. For the preparation and compaction of the specimens to be tested in the stabilometer, HVEEM developed the mechanical kneading compactor and the compaction method that were described in some detail in Chapter 4. The stabilometer is basically a triaxial chamber (Fig. 9-15) consisting of a metal cylinder in which there is a rubber membrane. Between the metal cylinder and the rubber membrane the annular space is filled with oil that transmits lateral pressure to the specimen.

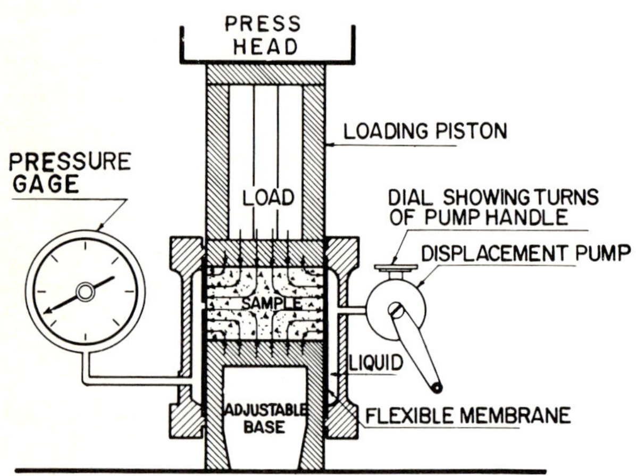

Fig. 9-15 Diagram showing HVEEM's stabilometer [42]

When the specimen is vertically loaded, a horizontal pressure is transmitted to the oil, which can be measured on the pressure gage. The vertical pressures applied are of 5.6 and 11.2 kg/cm^2 (80 and 160 psi), comparable to very high tire pressures. The results of the test are interpreted by means of a value called the stability number, also proposed by HVEEM in accordance with the following equation

$$R = 100 - \frac{100}{\frac{2.5}{D}\left(\frac{p_v}{p_h} - 1\right) + 1} \qquad (9\text{-}4)$$

where: R is Hveem's stability number, without dimensions; p_v is the vertical pressure applied. The value of R is generally measured for 11.2 kg/cm^2 (160 psi); p_h is the horizontal pressure on the walls of the specimen measured with a pressure gage; D is the horizontal displacement of the specimen, corresponding to a horizontal pressure of 7 kg/cm^2 (100 psi). Displacement is measured by the number of turns that are recorded by the dial on the pump handle, and which were necessary for the lateral pressure measured on the manometer to vary from the value that was recorded in the test on application of vertical pressure to the value indicated of 7 kg/cm^2 (100 psi). The procedure described for the test with HVEEM's stabilometer is included in Appendix 9c.

9.6.4 Other Types of Test

A description follows of some special tests, developed by the California Division of Highways, U.S.A., which can be used for the application of HVEEM's method of design for flexible pavements [42].

.1 Exudation Pressure

After compacting specimens using the method recommended by the California Division of Highways, which is described in Appendix 9c, the test to determine exudation pressure for soil moisture must be carried out. This consists of measuring the compressive stress that is necessary for the specimen compacted with a certain water content to expel some of the water used for molding. For this purpose, there is a device which consists of a base with seven photoelectric cells and a recording instrument. The specimen in the mold is placed on the base and is subjected to continuous load increments, exudation pressure being recorded as the one which makes the expelled water close the circuit of at least five of the perimeter cells. The central cell indicates the contact between the specimen and the base of the apparatus and must therefore be continuously recorded. The California Division of Highways specifies that the stability number obtained with HVEEM's stabilometer, Eq. (9-4), which is used to compute the necessary pavement thickness, must correspond to a water content that requires an exudation pressure of 21 kg/cm^2 (300 psi). For this reason, it is common practice to measure exudation pressure in specimens prepared with water contents that make this vary between 7 and 56 kg/cm^2 (100 to 800 psi).

.2 Swell Pressure

This test measures the pressure that is developed under confined conditions by a specimen of soil which is allowed to absorb water freely. The test is performed in such a way that significant changes in the density of the soil are not permitted during testing. The device is shown in Fig. 9-16. It consists of a calibrated metal beam on which the vertical force exerted by a 10.1 cm (4 in) circular plate resting on the specimen can be read, pressure being transferred to the beam by a rod. Swell pressure is measured by saturating the specimen as shown in the figure. In the design procedure currently used in the state of California, U.S.A., one of the conditions is that the weight of the pavement must be sufficient to neutralize the swell pressure that is measured. The test is described in some detail in Appendix 9c.

.3 Cohesiometer Value

The cohesiometer measures the resistance of a specimen of soil to tension produced by flexure (Fig. 9-17). This value is presumed to be related to the shear strength that is developed by the specimen under a confinement representative of that in the finished pavement. The test is applied particularly to materials that will be used in the upper layers of the pavement system. The specimen is placed inside two

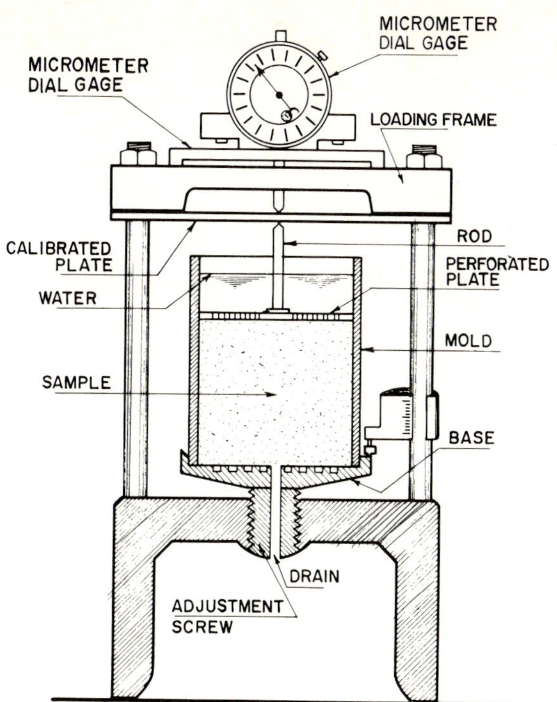

Fig. 9-16 Diagram showing HVEEM's expansiometer [42]

articulated clamps, one movable and the other fixed. The first one is attached to a bar (loading arm) to the end of which an increasing stress is applied until failure of the sample takes place. The cohesiometer value is given by the formula:

$$C = \frac{W}{D\,(0.20\,H + 0.044\,H^2)} \qquad (9\text{-}5)$$

where, using the (mixed) dimensions provided by the California Division of Highways, U.S.A. [43]: C is the cohesiometer value, in g/in^2; W is the weight of the shot in the end deposit, in g; D is the diameter or width of the specimen, in in; H is the height of the specimen in in. The procedure for obtaining the cohesiometer value is described in Appendix 9c.

9.7 Design Methods for Flexible Pavements

It is not intended here to describe all the different design methods that have been developed or that are used for flexible pavements. References [7,40,41,43,44] contain most of the design methods that are used, particularly the ones which will not be described in detail in the paragraphs that follow. Such methods are used in different parts of the world with local variations which usually add little to the basic method. Reference has already been made to the great difficulty that is currently experienced in establishing a theoretical approach to pavement analysis and design. This difficulty or the lack of a satisfactory theoretical solution, is reflected in the existing design methods. If there is no satisfactory theory, the design methods will have to be based either on incomplete theories, approximations or on considerations that do not relate to theory.

Most thickness design methods are based on a laboratory test or series of tests, which provide a guide for representing the real behavior of the pavements by a more or less reliable correlation or set of correlations between the behavior of the materials in the laboratory and in the finished pavement. The reader who has followed the brief discussion that has been made in this chapter of the mechanics of flexible pavements and particularly the factors that affect their performance, will have visualized the extraordinary complexity and diversity of the problem. He will be prepared to understand how difficult it is to establish a simple and manageable correlation between a laboratory test, which must be easy to standardize and simple to interpret, and extremely complicated, varied, real behavior. The design methods that are based on a laboratory test and on its empirical correlation with structural behavior can be expected to have the corresponding limitations and deficiencies. It is also clear that the more representative the test of real soil conditions, the more possibilities will there be of making good use of it. Bearing in mind these complexities, it is logical to expect that there can be no design method that can be applied with absolute reliability or even with a tranquillizing degree of confidence. Therefore, the design method that is applied should be regarded not only as guide and a basis for calculation, but also as something requiring *art* and experience.

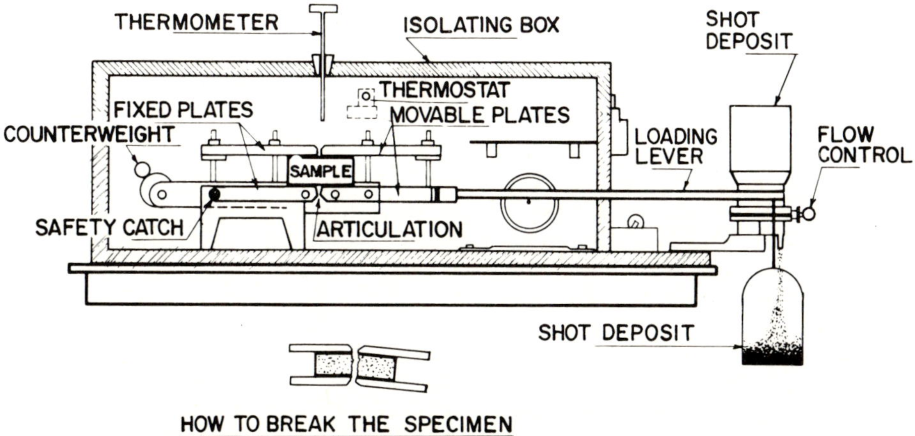

Fig. 9-17 Diagram showing HVEEM's cohesiometer [42]

Now it can be understood why pavement technology is so full of factors that have nothing to do with the specific methodology of design. For example, it is common practice to establish fairly strict quality control standards for the materials to be used. These materials are often improved by the addition of substantial quantities of cement, lime, and asphalt. This technique, referred to as soil stabilization, will be discussed later. The mechanical treatment of paving materials (compaction, crushing, washing, etc.) is also governed by the same requirements.

Of the methods that are listed in [7], which will not be described here, two deserve some brief comments. The design method based on the use of the triaxial test that was developed by the State of Texas, U.S.A., has two effective advantages. First, it is based on a triaxial test, which has a stress similar to real pavement conditions. The results of a triaxial test can be readily correlated with a great deal of valuable non-pavement experience for the engineer with experience in soil mechanics. A greater advantage is the large triaxial chamber that is used which permits a large, more representative sample to be tested. Second, the Texas method is simple to apply. The Texas triaxial test is of interest regardless of whether the design method that is based on it is followed. The test has an independent value for detecting the quality of a layer once in place and compacted. In Mexican road building the doubt often arises as to the ultimate condition of a compacted layer in which the design requirements were not fully satisfied. The ultimate question is whether the layer should be removed. If specimens of this layer are tested in the triaxial chamber, it is possible to obtain a strength envelope that shows what level of performance can be expected of it. In this way, the triaxial test provides an independent criterion for assessing the quality of an existing pavement layer, which has proved useful in many cases. Another of the recent methods of flexible pavement design which deserves attention is McLeod's, which uses the plate-bearing tests that were discussed in §9.6.1. It is a simple method that is based on more rational concepts than others. McLeod's method has the advantage that the plate-bearing tests on which it is based measure the behavior of the layers of the pavement system as constructed and thus are also used in reconstruction work and evaluation of the capacity of existing pavements.

9.7.1 The *CBR* Method

.1 Application to Runways

The method is based on the performance and results of the test of the same name, as described in §9.6.2. A description was also given of the version of the test developed by the U.S. Army Corps of Engineers. The design method that is described below is the one based on this version of the test as proposed by the Corps of Engineers. Of all the numerous versions of the *CBR* method applied throughout the world (all of which are similar) this appears to be one of the most complete. The method can be applied to pavements for both highways and runways.

On the basis of observation of the performance of pavements more than 20 years old and comparisons between that performance and the *CBR* values shown for the different layers of those pavements, the U.S. Army Corps of Engineers proposed the following expression for determining pavement thickness for runways [45-47]:

$$e = 2.5\,F \sqrt{\frac{P}{8.1\,CBR} - \frac{A}{\pi}} \quad \text{(cm)} \qquad \text{(9-6)}$$

$$e = F \sqrt{\frac{P}{17.85\,CBR} - \frac{6.45\,A}{\pi}} \quad \text{(in)}$$

where: e is the total thickness of material that must be placed on top of the soil, for which the *CBR* is given in cm (in) in Eq. (9-6); $F = 0.23 \log C + 0.15$; C is the volume of traffic (The number of coverages anticipated in runway design); P is the simple load equivalent to the multiple tire assembly of the design aircraft, in kg (lb) obtained as indicated in §9.3 for two-tire assemblies. For more complicated assemblies, there are conceptually similar methods of obtaining it [47]. They are not included here, because relating thickness to load equivalent of different assemblies is included later. Eq. (9-6) then becomes unnecessary; A is the contact area, in cm^2 (in^2); *CBR* is the California Bearing Ratio of the subgrade layer.

The above equation is only valid for runways and for *CBR* values below 10 or 12%. However these cover the *CBR* range for subgrade that is most frequently met with in practice. For these values, the equation represents the shape or trend of the design curves, which were plotted on the basis of purely empirical data. For higher *CBR* values, the equation no longer represents the design curves, and these should therefore be consulted for each specific case. It should be made clear that the design curves that are included later cover any *CBR* value, including the interval covered by the equation, therefore in practice it is sufficient to utilize the curves. Figures 9-19 to 9-25 are design curves for pavement thicknesses as a function of the *CBR* of the soil for some common types of civil aircraft.

The DC-3 is the design aircraft most often used in Mexico for small rural airfields. It is a good example of the larger light aircraft, although the aircraft is obsolete. The different versions of the DC-9 and Boeing 727 have been used in Mexico as design aircraft for medium-load runways, whereas the DC-8 has been used for those catering for international traffic. Design graphs are also included for the Boeing 747, although it is not necessarily more critical than the DC-8. The reason for this is that it is not only the weight of the aircraft that determines the effects on the pavement, but as has already been mentioned, the tire spacing, tire pressure, number of tires, and the assembly arrangements. Figure 9-24 gives the design curves that have been plotted for the once proposed supersonic Boeing SST aircraft. It will be observed that an aircraft of this type would be only a little more critical than one of those currently in use. Naturally the word *critical* refers only to the destructive effect on the pavement. Figure 9-25 refers to the Franco-British supersonic Concorde.

It can be seen that some of the aircraft that are frequently seen in airports all over the world are not included. This is because they are similar to some of the ones presented, from the operational point of view, but have less notorious effects on pavements. A typical example is the very common Boeing 707, which is similar to the DC-8. There are [44] also many design curves in which the thickness of the pavement is given for certain specific tire assemblies, specifying the geometry of the assembly and the corresponding spacings for different tires, and their contact pressures. These graphs are useful when aircraft types are to be handled for which no

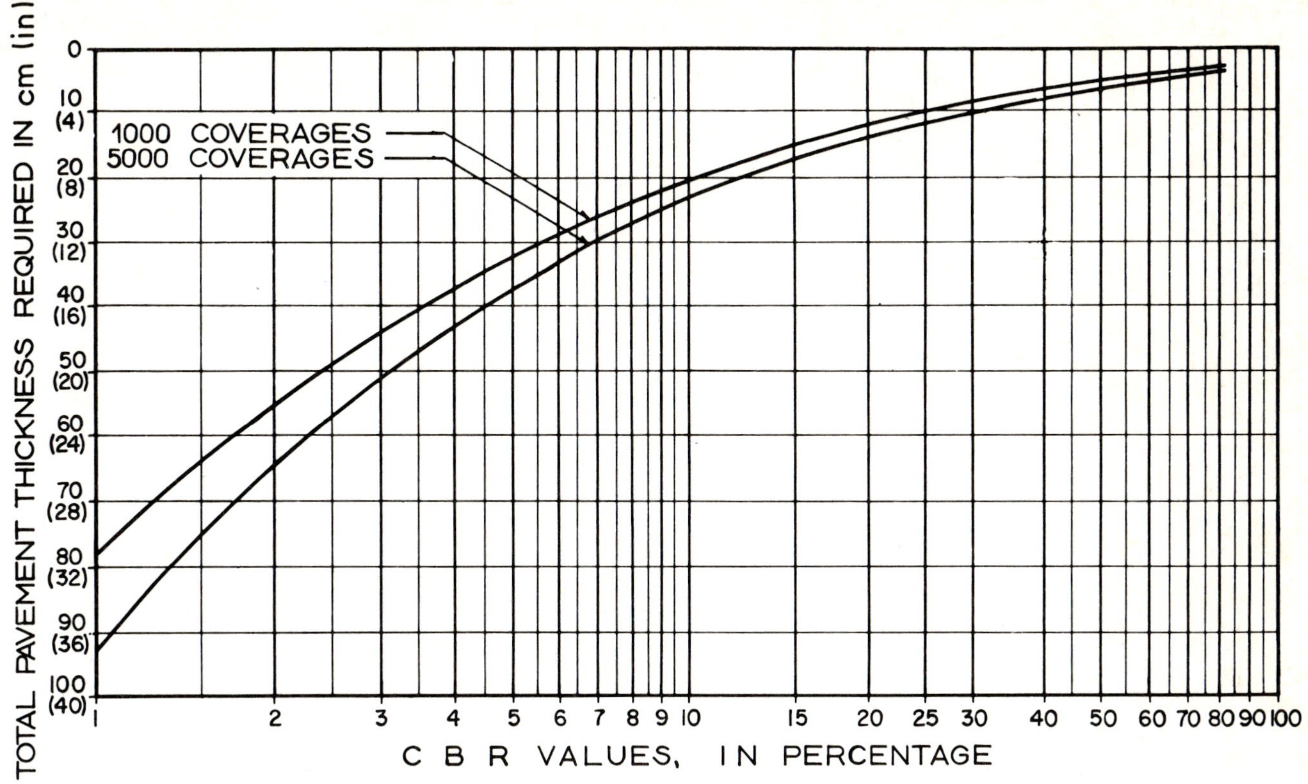

Fig. 9-18 Flexible-pavement design curves as a function of the *CBR* for the DC-3 Aircraft

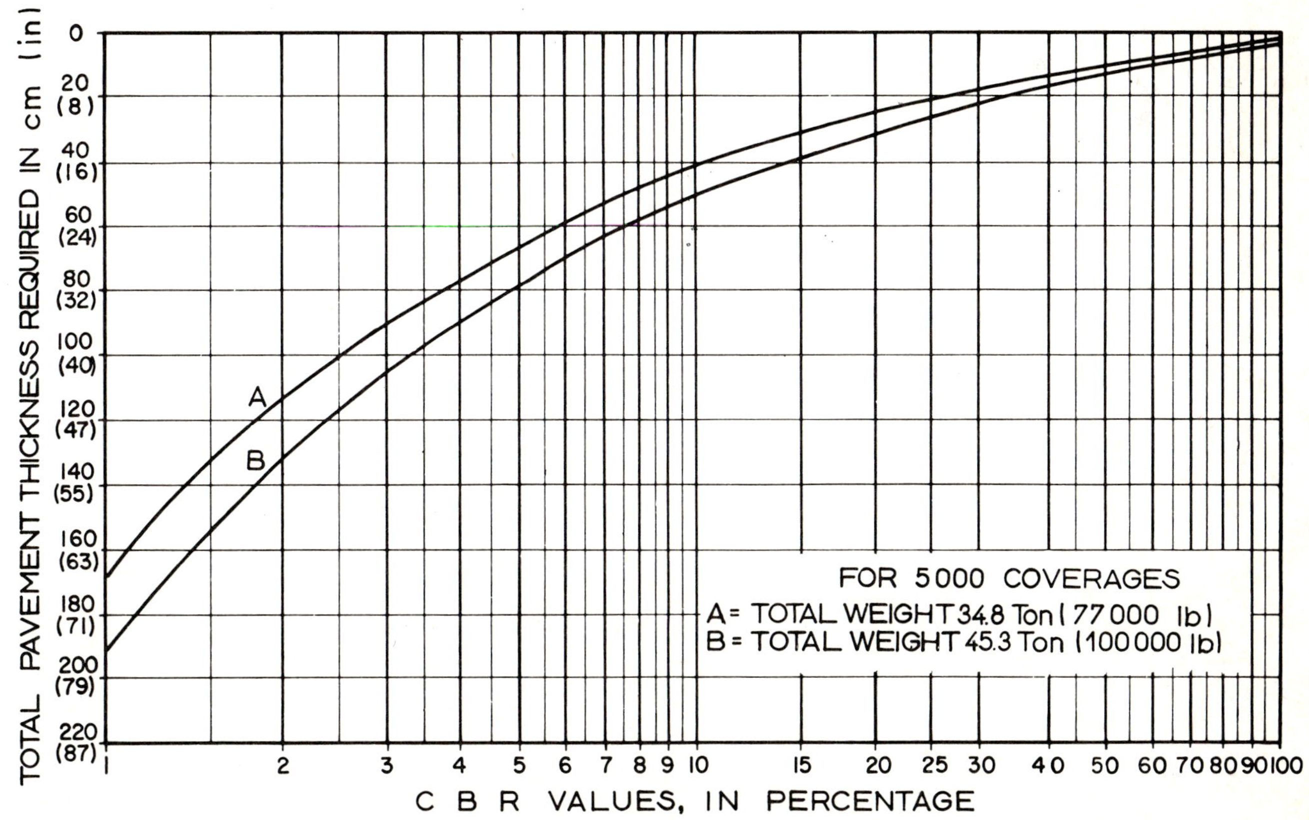

Fig. 9-19 Flexible-pavement design curves as a function of the *CBR* for the DC-6 Aircraft

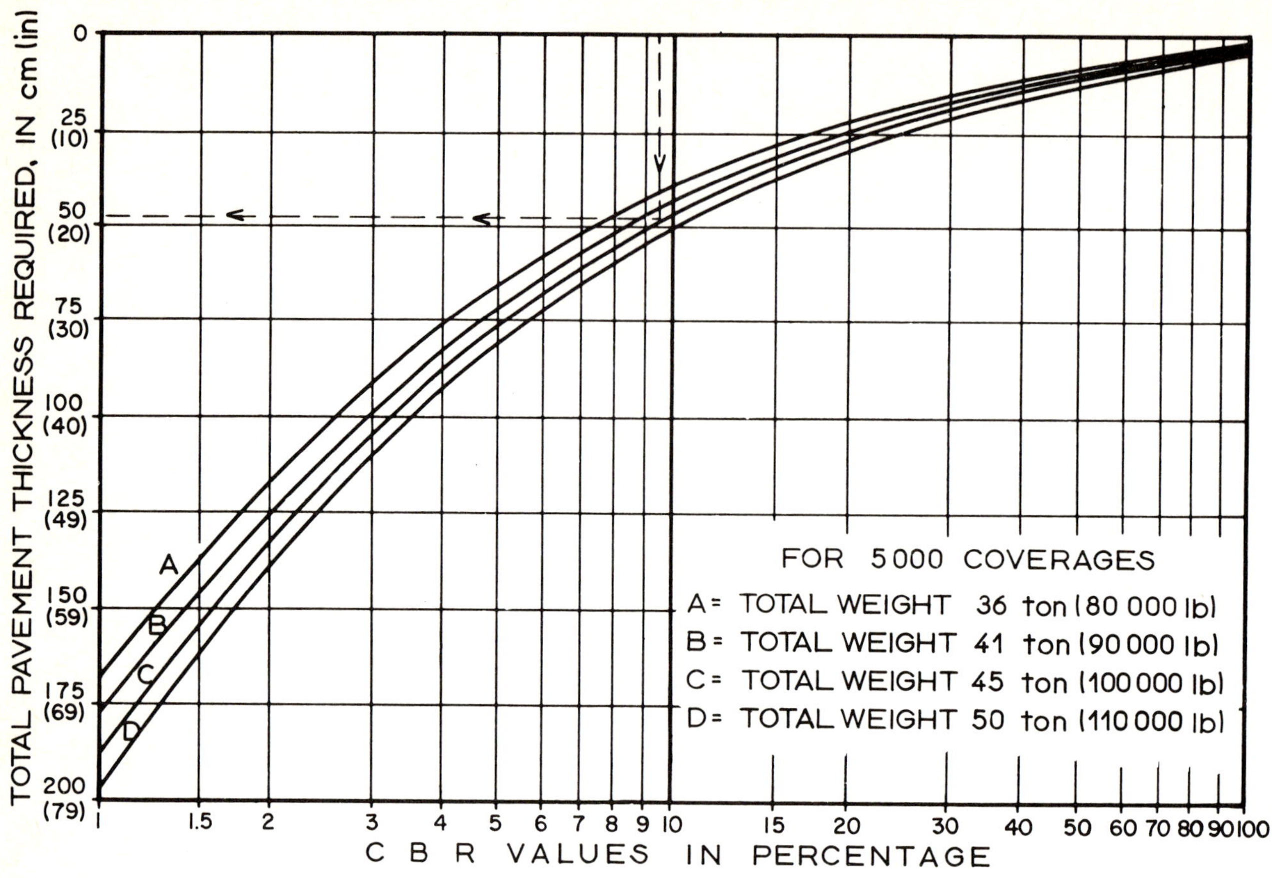

Fig. 9-20 Flexible-pavement design curves as a function of the *CBR* for the DC-9 Aircraft

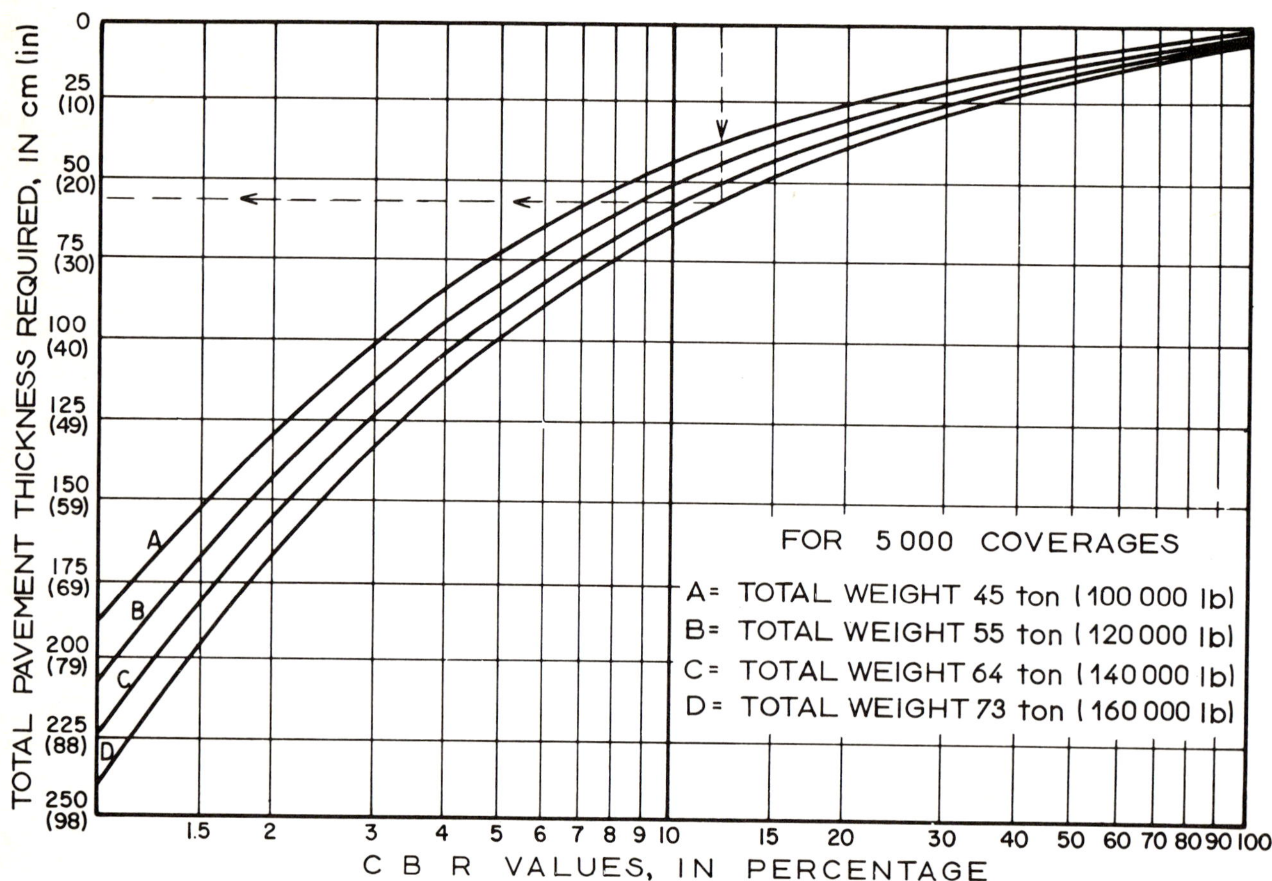

Fig. 9-21 Flexible-pavement design curves as a function of the *CBR* for the Boeing 727 Aircraft

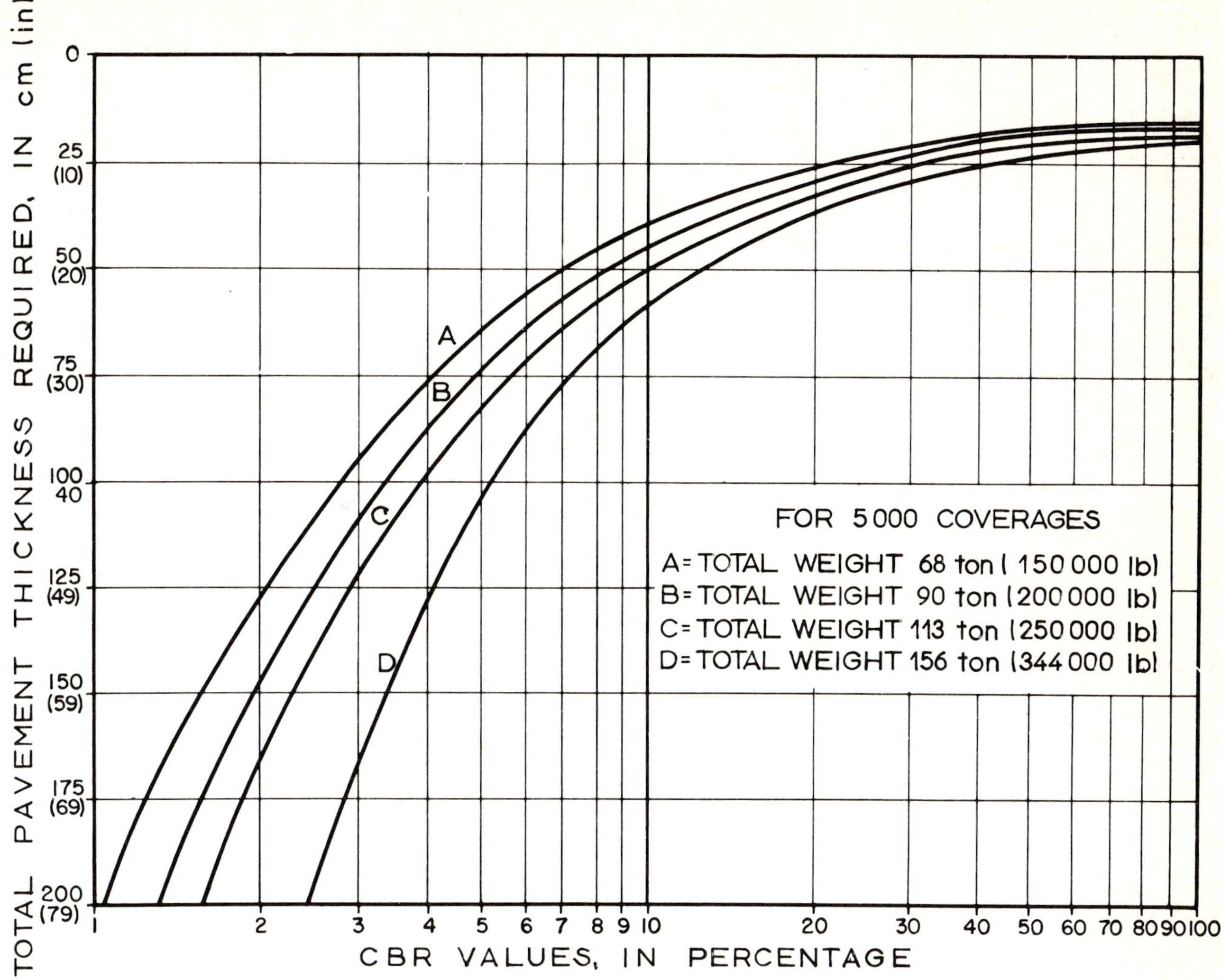

Fig. 9-22 Flexible-pavement design curves as a function of the *CBR* for the DC-8-63 Aircraft

specific previously plotted graph exists. However, the design engineer will find it more convenient to use the graphs included in this chapter.

Design graphs are used as follows. Once the *CBR* for the upper layer of fill has been determined, the corresponding graph will give the total thickness of all paving materials required above the fill in order to achieve satisfactory performance. A question immediately arises which is not directly found by the method: what material should be used to provide the pavement thickness required. Although not stipulated by the design method, it is only logical that a better quality material should be used than the fill itself. The most suitable types of materials are indicated by the criteria that have been given earlier. A subgrade layer is added with the major condition that the thickness selected, plus the thickness of whatever materials are placed on top of it, must provide the total thickness required to cover the fill.

Subgrades are often tentatively designed with a thickness of about 50 cm (20 in) for runways and 30 cm (12 in) or thereabouts· for roads. The remaining thickness must be provided by layers of sub-base and base course. Following the same test procedures already described, it will now be possible to find the *CBR* for the subgrade material and with the same graph to find the total thickness of the remaining pavement layers required to cover the subgrade. The final subgrade thickness is then the difference between the total pavement thickness required above the fill and the total above the subgrade. With the *CBR* for the sub-base, a similar procedure is adopted, once again using the corresponding design graph, but this time obtaining the thickness necessary for the base course.

This is clearly a chain method, layer by layer progressing upward from the fill surface. It is common practice not to consider the thickness of the final pavement layer in these analyses, so that this automatically becomes an additional assurance of safety.

It can be seen that a design method like that described is not more correct than the experience and wise judgment of the engineer who applies it, and if used blindly may lead to either a highly conservative or an inadequate design. To give a concrete example, it is assumed that the fill requires a total overlying thickness of 1.20 m (4 ft). One engineer might solve the problem by using a 20 cm (8 in) layer of doubtful quality subgrade, with the remaining meter of a magnificent quality base course. Another engineer might use 1 m of poor-quality subgrade, giving the rest of the thickness (which would now be greater than 20 cm (8 in) with various different layers of graded materials. In this way very different designs can be developed, with strengths and deformabilities of successive layers that are completely different. The behavior of the pavements in each of the alternatives would be different.

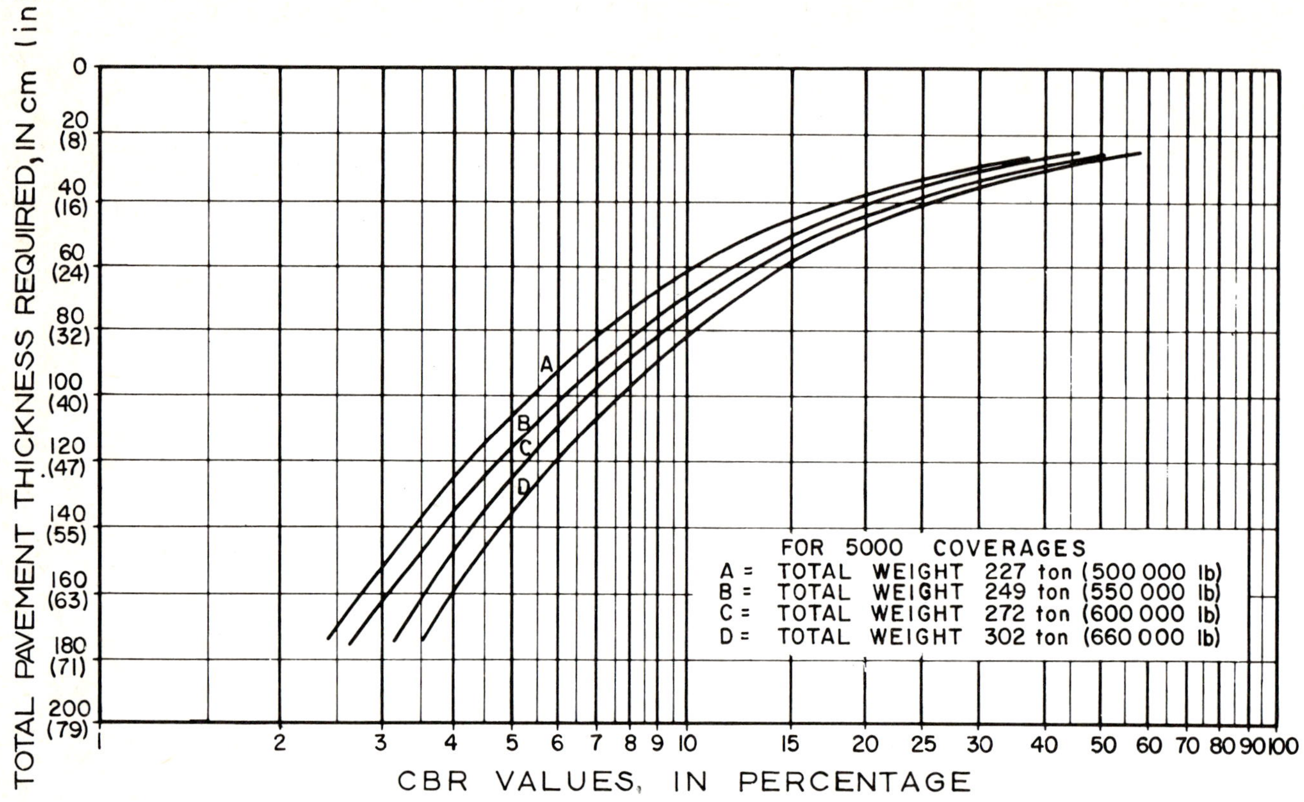

Fig. 9-23 Flexible-pavement design curves as a function of the *CBR* for the Boeing 747 Aircraft

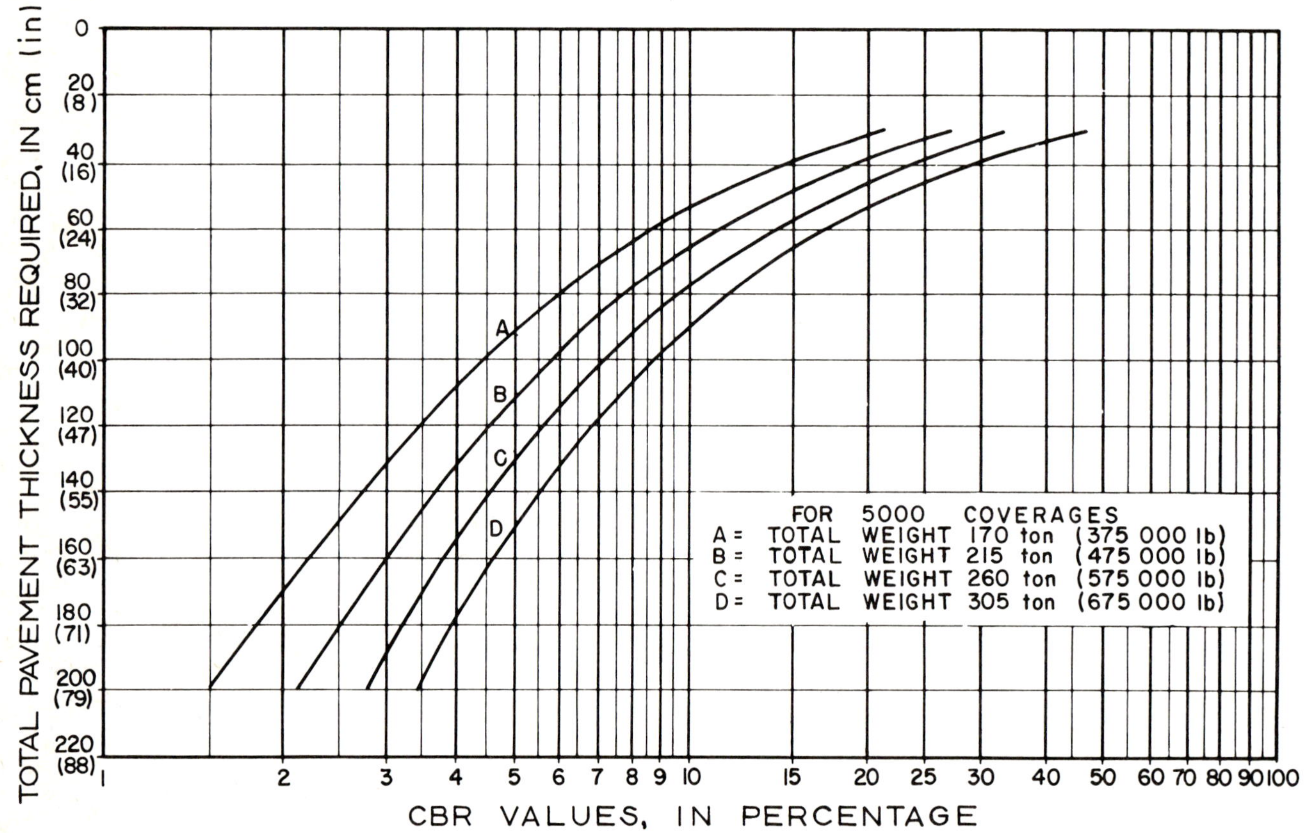

Fig. 9-24 Flexible-pavement design curves as a function of the *CBR* for the Boeing SST (supersonic aircraft)

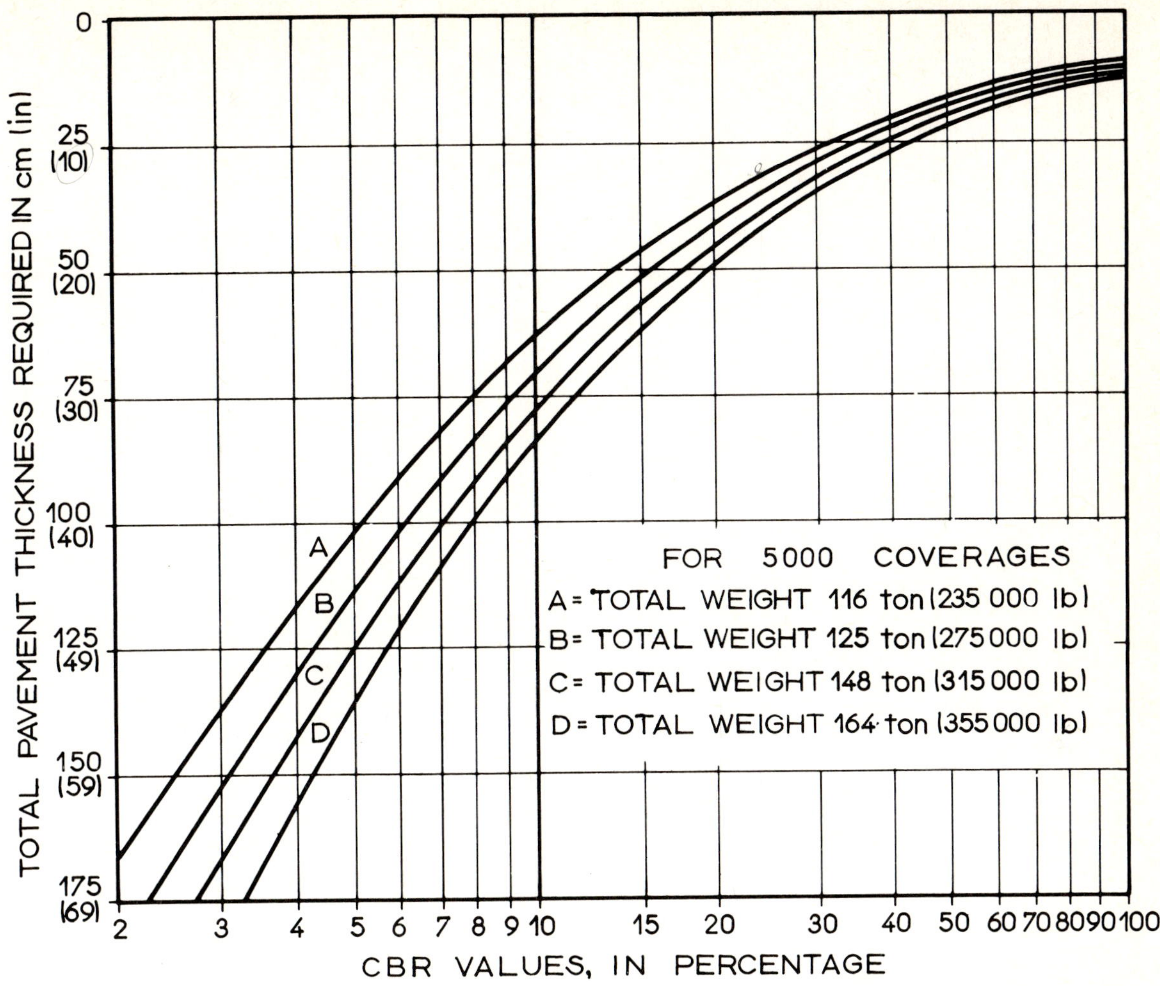

Fig. 9-25 Flexible-pavement design curves as a function of the *CBR* for the Concorde (supersonic aircraft)

The design method under discussion must therefore be handled by someone with ample experience in pavement performance and a deep understanding of materials. The method is of little value if there is no preconceived idea as to the way in which the available materials are to be used. It has been said that an engineer gifted with this understanding of materials will have limited use for design curves; this is probably true. The meaning of certain ideas stated earlier in this chapter can now be understood. The use of the *CBR* design method is intimately related to the experience of the engineers who apply it. A drastic change in the design method used by any organization might be dangerous, when engineers adopt a method in which they have no experience, because there is currently no method that is scientific and rational enough to be adopted without taking these extremely important subjective factors into account.

In relation to the above, many of the difficulties in conceiving the *CBR* method as being rational stem from the fact that the thickness of the overlying material is determined to be only a function of the characteristics of the material that is to be covered. (However representative the test may be for defining these characteristics). Obviously, the characteristics of the overlying material itself must be included in the thickness required.

As a final observation, we stress the danger of using empirical criteria based on a correlation between the results of a test and complex structural behavior in such a fast-changing field as pavement technology. Personal experience which is really the most important factor, is hard to extrapolate. For example, loads increase from one year to the next, both in runways and highways, so that the continuous use of the method implies a certain amount of experimental extrapolation which, initially, is not based on observed behavior.

.2 Application to Roads

Firstly we will consider the version of the method proposed by the Corps of Engineers, based on the test described in Appendix 9b. The thickness of the pavement can be obtained from the graphs in Fig. 9-26a, also supplied by the Corps of Engineers. The graphs provide the thicknesses for different *CBR* values and different wheel loads, selected with the equivalent load criteria already described. Furthermore, they give the thickness required for the pavement to tolerate 10^6 repetitions of the load. It is advisable to use the graphs in Fig. 9-26a in conjunction with McLeod's criteria for considering the type and combination of traffic [1,49]. In these criteria the thickness of the pavement varies linearly with the logarithm of the number of load repetitions. The conclusion is also reached that with 1/4 of the thickness required for 10^6 repetitions, the pavement will fail with just one repetition of the same load. In this method a load of 2.27 t (5,000 lb) is presumed to be the

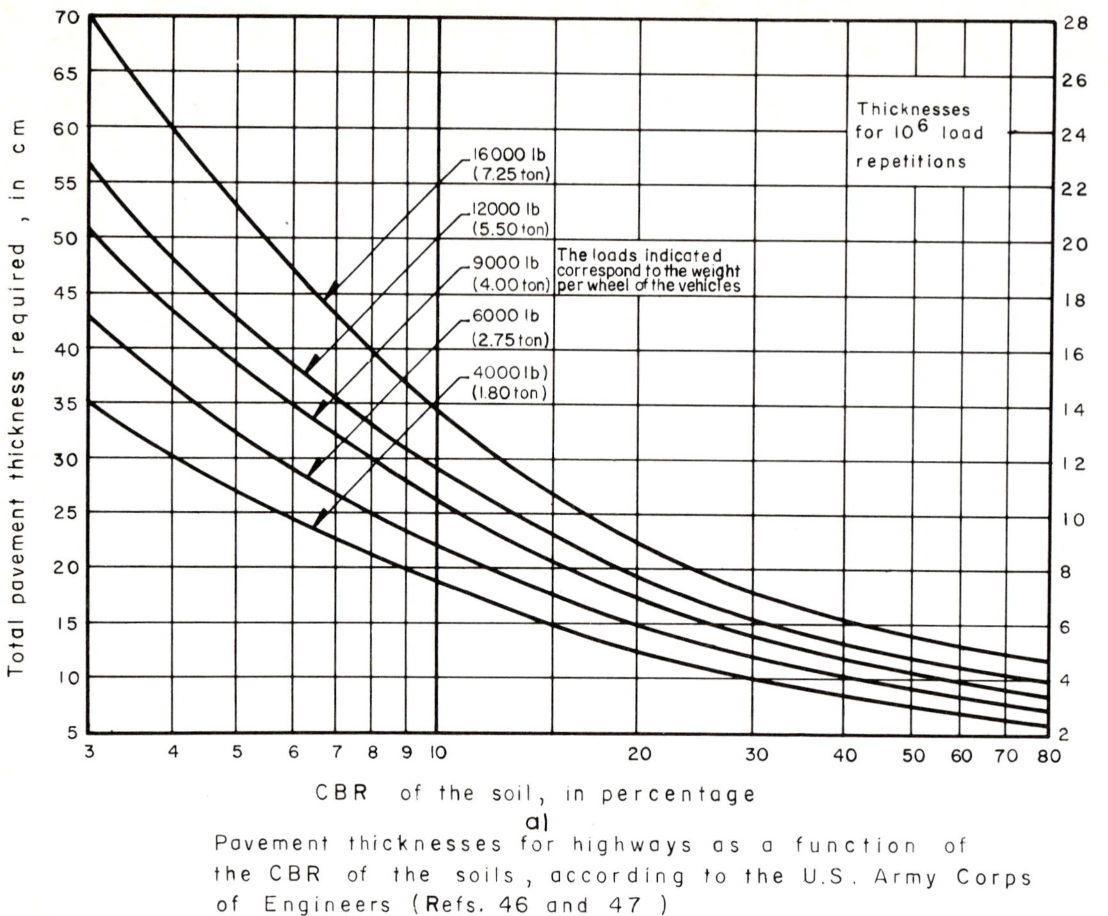

a)
Pavement thicknesses for highways as a function of the CBR of the soils, according to the U.S. Army Corps of Engineers (Refs. 46 and 47)

Fig. 9-26 Flexible-pavement design curves for highways as a function of the *CBR*

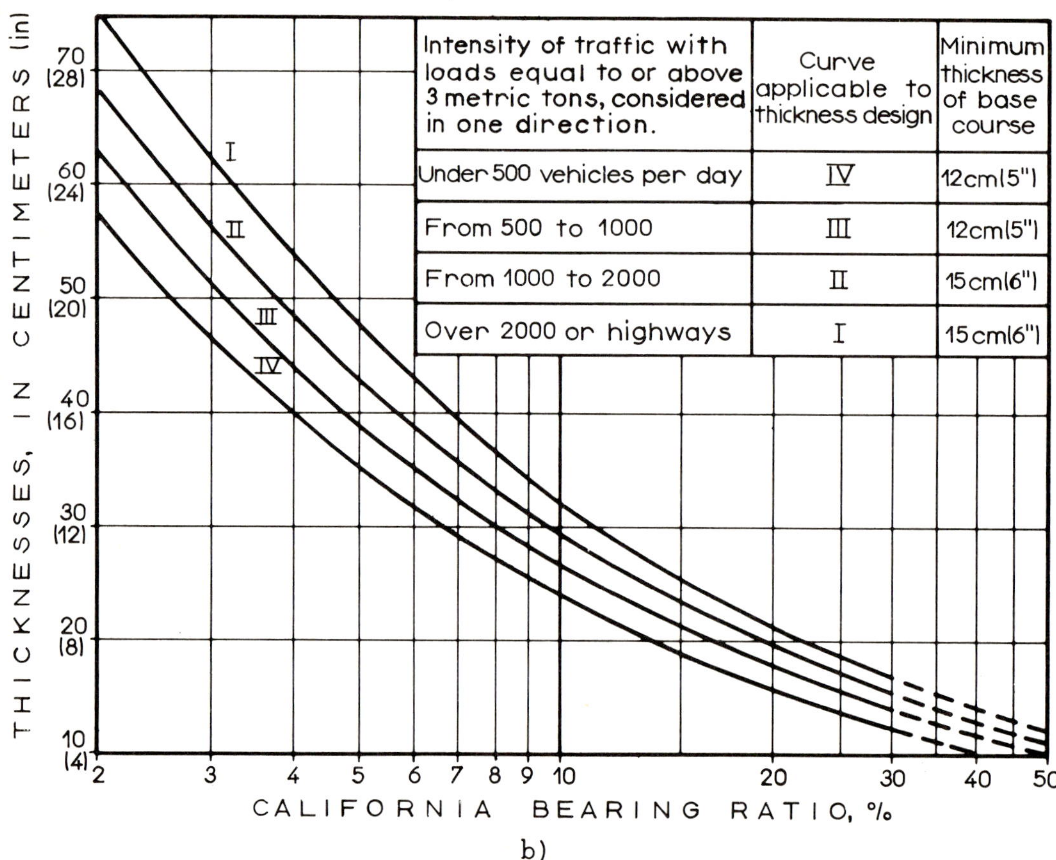

b)
Curves for calculating the minimum thickness of sub-base plus base course for flexible road pavements, as a function of the CBR of the subgrade, according to Mexican Ministry of Public Works practice (Ref. 15)

Fig. 9-26b Flexible-pavement design curves for highways as a function of the *CBR*

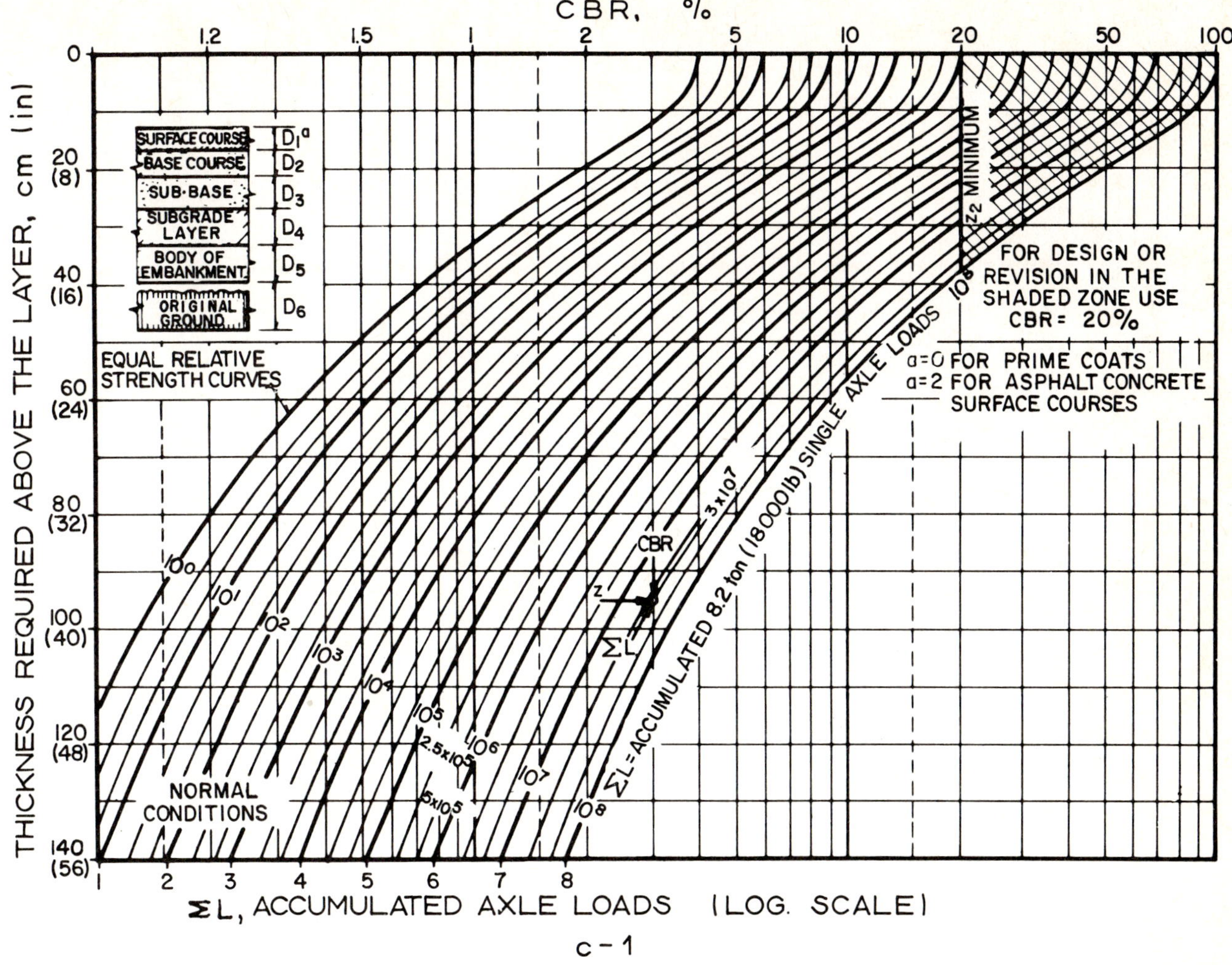

Fig. 9-26c.1 Flexible-pavement design curves for highways as a function of the *CBR*

equivalent wheel load to be used in the design. Varied traffic, of different types and weights, must then be reduced to the number of 2.27 t (5,000 lb) wheels that would produce the same effect. The system followed by the California Division of Highways for solving this problem was mentioned in §9.3. McLeod, however, developed a somewhat different solution as illustrated by Fig. 9-27.

First, using the curves in Fig. 9-26a and with the *CBR* for the soil under consideration (assumed to be 8% in the graph in Fig. 9-27), the total thicknesses can be found for the different wheel loads that are given. These are marked in Fig. 9-27 as ordinates on the abscissa corresponding to 10^6 repetitions. Then 1/4 of the thicknesses required to tolerate 10^6 repetitions are marked as ordinates for an abcissa of one repetition. It is recalled that these are regarded as the thicknesses at which the pavement will fail with the one load application. Between the ordinates thus plotted, straight lines can be drawn on the basis that the thickness of the pavement is a linear function of the logarithm of the number of repetitions. The set of straight lines appears in Fig. 9-27; each of them represents the variation in the thickness of pavement required with the varying numbers of repetitions for the load that is noted. In Fig. 9-27 a straight line appears for the load of 2.27 t (5,000 lb), in Fig. 9-26a there is no graph for this load: the values were obtained by interpolation.

Next, the point representing the design condition must be established (Point *A* in Fig. 9-27); this corresponds to one million repetitions of the 2.27 t (5,000 lb) load (the engineer may select another design condition, but if he does not use one million repetitions, the graphs in Fig. 9-26a will no longer be of use to him). A horizontal line is drawn through *A* which intersects, at different points, the straight lines corresponding to the different loads. The abscissa for each of these points gives the number of repetitions of the load that is indicated, which is equivalent to one million repetitions of an equivalent load of 2.27 t (5,000 lb). In Fig. 9-27, for example, it can be seen that 4,000 repetitions of the 5.5 t (12,000 lb) load are practically equivalent to the design conditions. By dividing one million by the number of repetitions of each of the loads (straight lines) in Fig. 9-27 the designer can find the same effects as the design condition. In this way the so-called equivalency factors are obtained for the different loads. For the 5.5 t (12,000 lb) load, for example, the equivalency factor is 1,000,000/4,000 = 250. The equivalency factors for each load are multiplied by the number of vehicles of that type that are expected to travel daily over the pavement being designed (the annual daily average). This product must be added to all the similar ones for the different loads. The total sum will give the daily number of repetitions of the 2.27 t (5,000 lb) wheel load, which, according to McLeod, are equivalent to the damaging effects of the traffic anticipated for the pavement.

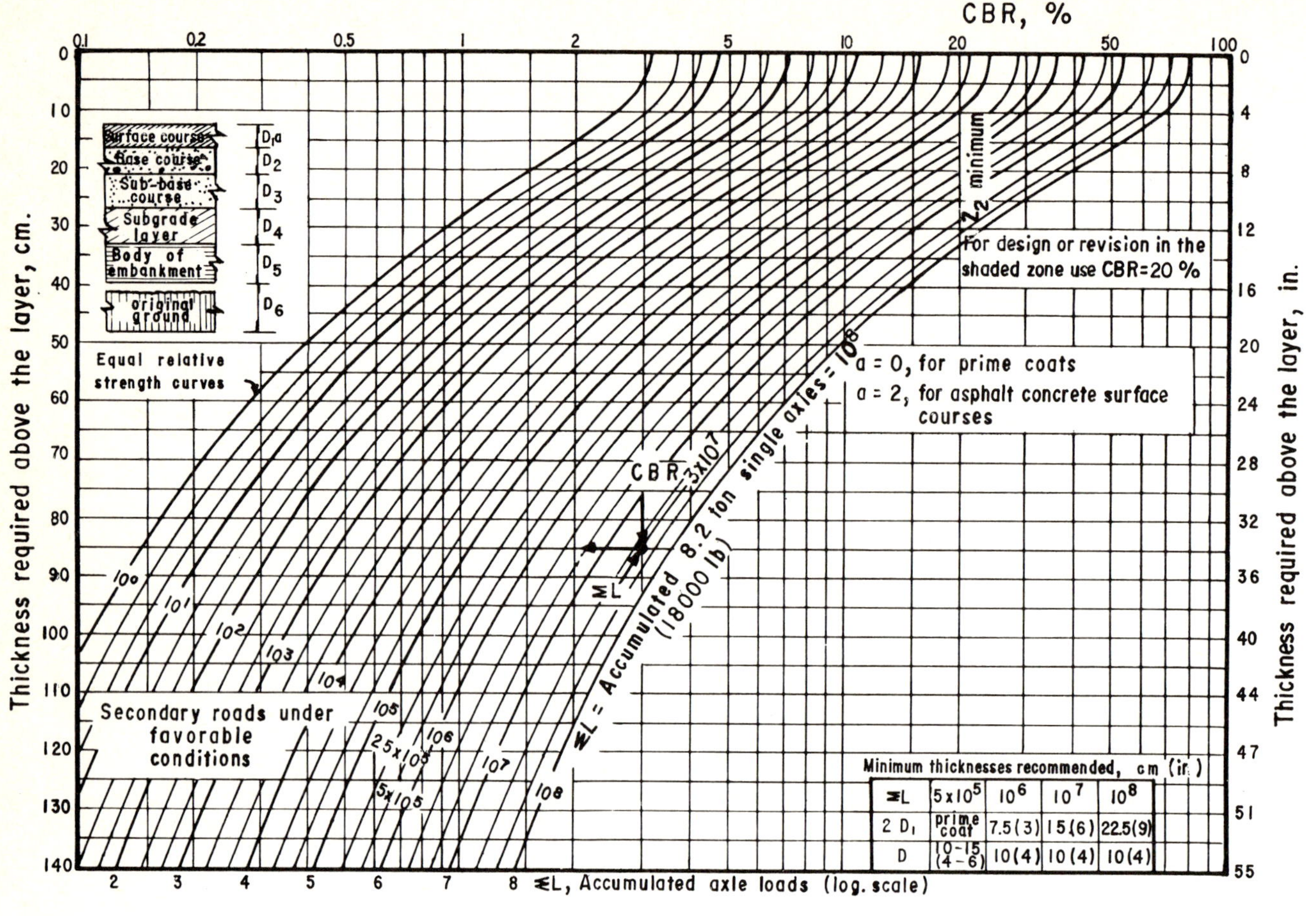

ROADS WITH LOW TRAFFIC VOLUMES

Trickness design graph for road pavements, by the Engineering Institute of the Mexican National Autonomous University (Ref. 79)

Fig. 9-26c.2 Flexible pavement design curves for highways as a function of the *CBR*

The above calculation must be extended to a reasonable service life, frequently 10 years, and some rate of traffic growth must also be considered for that period. In this way, a final grand total of equivalent load repetitions will be obtained for those 10 years. Entering this on the abscissa in Fig. 9-27, the design thickness can be read on the ordinate corresponding to the equivalent wheel load of 2.27 t (5000 lb).

The same general comments can be made with regard to the application of the *CBR* design method to roads as to those made for runways. The procedures are the same. There are the added problems (which have nothing to do with the method) that in roads it is hard to find out much about locally available construction materials, poorer quality materials are generally used, and there is not so much information available on the probable methods of compaction and subdrainage problems. The pronounced variations that are observed in *CBR* test results, regardless of quality and of the care taken by the operator, have been discussed earlier. It is important to find out how these differences influence the pavement thickness when the method is used by different organizations. The information contained in Fig. 9-28 [40] illustrates this point. This gives the results of thickness determinations by the design curves used by 11 of the North American states. The large differences that are obtained applying the *CBR* method with different graphs can be readily appreciated, and they have probably all been obtained with perfectly reasonable interpretations of the design method.

There is far more to the information in Fig. 9-28 than just the differences in the thicknesses obtained. Apart from thickness, these differences are affected by certain variations of the test that have been established by different organizations in their attempts at achieving an adequate representation of field conditions in the laboratory. The water contents of the specimen on which the test is conducted usually constitute a major difference, varying from one organization to another and ranging from moderate partial saturation to total saturation of the specimen as an extreme condition. This is not very realistic in many regions of low rainfall and hot sun. Important differences may also stem from the use of static compaction by some organizations (although by far the minority) for the preparation of their specimens, whereas most use dynamic compaction.

Regardless of this, however, by applying procedures they consider reasonable (and which undoubtedly are), responsible organizations utilize the *CBR* method for designs which are significantly different. These differences are greater than can be explained by economic factors or different opinions of safety. It is hard to regard a design method which leads to such very different final results, as the ultimate that can be developed by knowledge and experience of different engineers. In Mexico too, the *CBR* method is the design method most widely used for highway pavements. Fig. 9-26b shows pavement design graphs as a function of the *CBR*, currently used by the Mexican Ministry of Public Works [1].

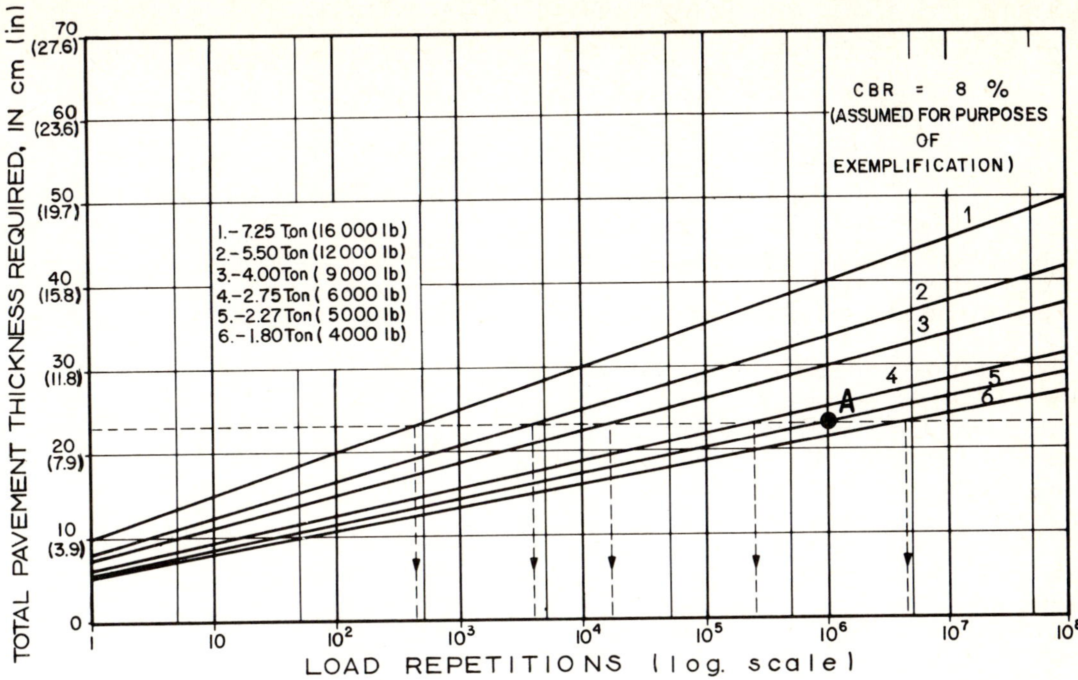

Fig. 9-27 McLEOD's method for homogenizing traffic for application of the *CBR* method to roads, according to the U.S. Army Corps of Engineers [1]

Figure 9-26c includes a design graph for highway pavements, which was developed by the work team at the Engineering Institute of the University of Mexico [79].

This graph is, in the opinion of the authors, a highly promising new approach to a traditional solution of the design problem. First, it deals with the number of load repetitions ($\sum L$) that will be tolerated by a certain thickness of pavement before it fails (failure being defined as the appearance of a permanent deflection of 2.5 cm (1 in) throughout 20% of the paved area). Second, the *CBR* concept is now handled more rationally. In the traditional methods, the *CBR* of each of the soils in a pavement is determined in the same manner, with little regard for the position of the layer of soil within the pavement structure. It is not easy to establish the *CBR* that will be required for the different soils according to their position within the fill or pavement; therefore this is done on the basis of empirical specifications. S. CORRO and his group of collaborators have different ideas. They acknowledge that the *CBR* should be distributed throughout the thickness of the firm section of the pavement in a way similar to the vertical normal stress distribution according to BOUSSINESQ's theory. Thus, there is a distribution curve for the required *CBR* similar to the typical BOUSSINESQ curve, higher near the

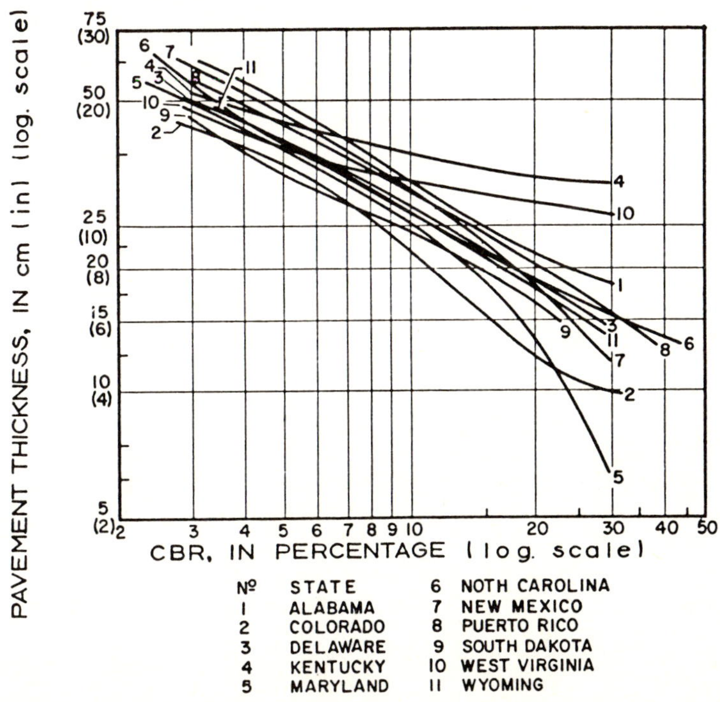

Fig. 9-28 Difference in the thickness of pavements using the *CBR* method, according to the different design curves used in several states of the U.S.A. [40]

surface and less at greater depths. Naturally this distribution cannot be continuous; the requirements of layered construction work will lead to stepped reductions in the required *CBR*, but the criterion of correspondence between the vertical stresses transmitted and the *CBR* required to support them is preserved.

Thus, the relationship between the thickness and the *CBR* is similar to a typical BOUSSINESQ curve (Fig. 9-26c) Several of these relations are drawn for the different load repetitions anticipated or design service lives, this service life being expressed as a number of load repetitions that must be tolerated without failure. The spacing between these service life curves on the thickness — *CBR* relations is experimental. Thus a graph such as Fig. 9-26c can be used during design or evaluation. For design purposes, an anticipated service life (Σ *L*) will be established for the pavement and with the *CBR* for one layer, the protective thickness that is required above it will be determined. For evaluation purposes, once the thickness and the *CBR* of a layer are known, it will be possible to estimate the service life that can be expected before a failure condition is reached. The most interesting application of the graph in the evaluation of existing pavements is that the different steps comprising the *CBR* — thickness curve can be plotted on it (each layer has a *CBR* and a certain thickness, thus forming a step) in order to determine the service life that corresponds to each. In this way the engineer can estimate not only the risk of failure and the moment at which it will occur, but also determine the layer of the entire pavement structure (including the fill) in which the failure can be expected to take place.

It is noted that with a design system like the above, any formal distinction between pavement layers, subgrade and fill disappears, and a single structural unit is handled, which is rational. The graph in Fig. 9-26c was obtained from ample experimental evidence, and is therefore reliable for analysis and design.

9.7.2 HVEEM'S Method

HVEEM and CARMANY [14,42,48], of the California Division of Highways, have developed a method for designing flexible pavement thicknesses, which is based on the laboratory tests that are described in Appendix 9c and that have been discussed previously. The method is similar to those including the *CBR* in which an experimental correlation must be established between the results of laboratory tests and the performance of the existing pavements. All the problems that make it hard to establish this correlation, and which were discussed for the *CBR* method, are also present in this method. However, HVEEM'S method is favored by the laboratory tests on which it is based which are more rational and provide a more complete representation of what really occurs in pavements than the *CBR* test. In addition, the method has been developed exclusively for roads. The design method has multiple requirements, in that the pavement that is ultimately accepted must satisfy conditions relating to the swell pressure, the pressure to exude water, the stability value obtained in a stabilometer, and the tensile strength in the upper layers of pavement measured with a cohesiometer.

The pavement thickness which overcomes the swell pressure of the subgrade as obtained in the laboratory, p_e, will be:

$$e_e = \frac{p_e}{m} \text{ (cm)}$$

$$e_e = 1728 \frac{p_e}{m} \text{ (in)} \qquad (9\text{-}7)$$

where: e_e is the thickness in cm (in) required to neutralize the swell pressure in the subgrade due to the weight of the overlying layers of pavement; p_e is the swell pressure in kg/cm^2 (lb/in^2), determined as in Appendix 9c; γ_m is the mean unit weight of the pavement structure, in kg/cm^3 (lb/ft^3).

Because the swell pressure in the subgrade depends on the water content of that layer and its variations, and because this water content is a function of the compacting water content, the California method requires the preparation of at least three specimens with three different moisture contents. Thus three pavement thicknesses are obtained for three swell pressures, one for each moisture specimen.

For the stability value obtained in HVEEM'S stabilometer, the corresponding thickness of pavements is obtained:

$$e_R = 0.098\ (TI)\ (100\text{-}R) \text{ (cm)}$$

$$e_R = 0.0032\ (TI)\ (100\text{-}R) \text{ (ft)} \qquad (9\text{-}8)$$

where: e_R is the thickness of pavement necessary in accordance with the strength of the soil, according to the stabilometer test, in cm (ft); *TI* is the traffic index, calculated with Eq. (9-1) and the procedures described in §9.3; and *R* is the HVEEM stability number, computed with Eq. (9-4) in §9.6.3.

To ensure that stresses that would generate unsuitable pore water pressures are not transmitted to the soil, the California method stipulates that for design, a value for *R* higher than that corresponding to an exudation pressure of 21 kg/cm^2 (300 psi) must not be used. Because the exudation pressure increases with increasing *R*, the exudation limit pressure, limits the design value of *R*.

A detailed description follows of the procedure for designing flexible pavement thicknesses using the California or HVEEM — CARMANY method.

Three specimens first are prepared in accordance with the kneading procedures described in Chapter 4 with three different water contents, so that two of them have an exudation pressure of less than 21 kg/cm^2 (300 psi), while the third one has a higher exudation pressure (or one specimen has a high exudation pressure and the other two a lower one). The exudation pressure of the three specimens must range between 7 and 56 kg/cm^2 (100 to 800 psi). However in those soils where very high swell pressures are anticipated, it may be necessary to work with an almost dry specimen in order to achieve sufficiently low swell pressures to satisfy the intersection of the expansion value curve and the *R* value curve that will be discussed later.

Next the exudation pressure values for each of the specimens is determined. The specimens will absorb free water until saturation is achieved in the swell pressure test as described in Appendix 9c. The same specimens are then placed in Hveem's Stabilometer, to obtain their *R* value. As a result of the above procedures, for each molding moisture content a specimen is obtained for which the exudation pressure, swell pressure and stability value are known.

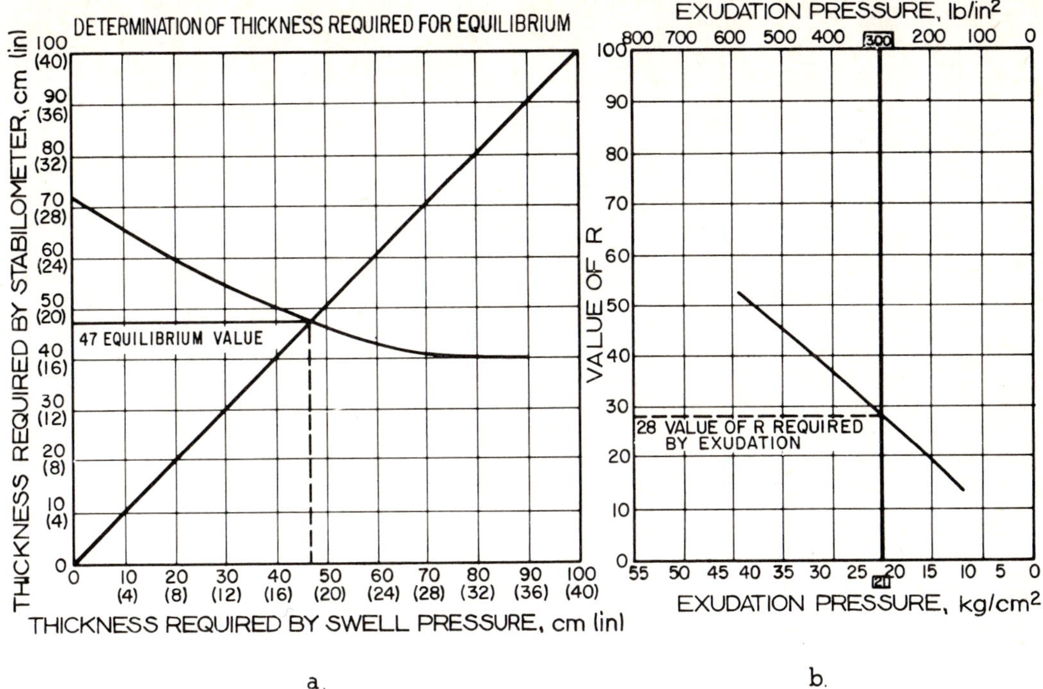

Fig. 9-29 Determination of the thickness of a pavement using HVEEM's method [42]

By applying Eq. (9-7) the thickness of the pavement can be computed to control expansion for each of the specimens. By applying Eq. (9-8) with the previously computed traffic index, it will be possible to find the thickness of pavement covering material required to satisfy the stability requirement.

Now graphs like the ones shown in Fig. 9-29 are plotted. Part *a* compares the required structural thickness with that required to control expansion. The intersection of this curve with a straight 45° line indicates the thickness that satisfies both conditions at once. In Part *b*, the graph depicts values of *R* for the corresponding exudation pressures. Bearing in mind that the maximum permissible exudation pressure is 21 kg/cm^2 (300 psi), it will be possible to obtain in this graph a corresponding maximum value of *R*. With this value of *R*, Eq. (9-8) and a known traffic index, another structural thickness will be obtained compatible with the exudation pressure. Figure 9-30 provides a nomograph which avoids the repeated use of Eq. (9-8) and is also useful for the calculation

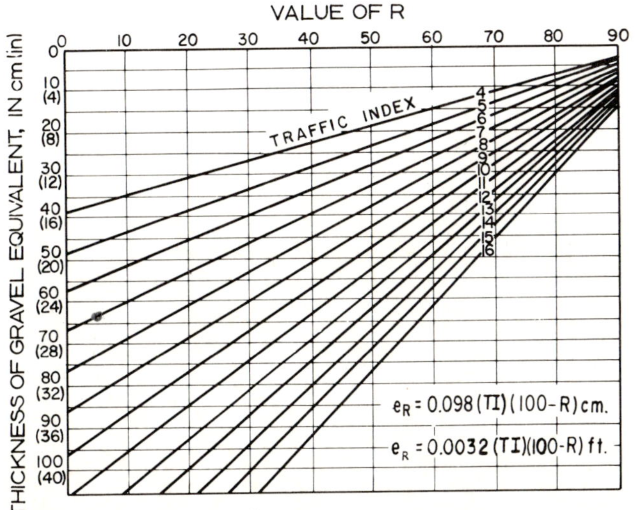

Fig. 9-30 Chart for calculation of Eq. (9-8) [42]

which gave the thickness required for structural stability. The thickness that satisfies both conditions obtained in Part *a* of Fig. 9-29 must be compared with the thickness determined by exudation pressure obtained using Part *b* of the same figure. The higher value will be the pavement structural thickness required above the soil under consideration.

Note that the method is applied similar to the *CBR* method. At first it is applied to specimens of fill material, which makes it possible to establish the thickness required above this material to ensure satisfactory performance. HVEEM referred to this required gravel pavement material thickness as *gravel equivalent*; the thickness of material with the characteristics of a common base course. If the method is repeated for the subgrade, preparing specimens with that material, it will be possible to find the thickness of gravel (or pavement) required to protect it. The same procedure can be followed upwards until the pavement structure is complete. The same comments apply as in the case of the *CBR* method, for there are many different solutions leading to very different pavements designs with different performances. The same incongruity exists as for the *CBR* in that the characteristics of the covering layer material are not taken into account when determining the thickness required to cover the deeper material.

On the basis of local experience in the State of California, HVEEM and his collaborators give thickness equivalencies for different layers commonly used for pavement construction including the *gravel* which they define as a *protective covering material*. Table 9-11 [42] gives the factors for the materials that are listed; these are numbers which, when multiplied by the thickness of the pavement layer that is proposed, give the thickness of the corresponding *gravel equivalent*.

Note that the gravel equivalent factors vary with the traffic index for the asphaltic surface courses, owing to the stiffening of asphalts with time, especially when traffic indexes are low. The three cement-treated base courses shown correspond to the different amounts of stabilizing material that are used in California engineering practice.

Table 9-11
Gravel equivalent factors for different flexible-pavement structures [42]

Type of material	Traffic index	Gravel equivalent factor
Asphalt concrete surface courses	5	2.5
	6	2.3
	7	2.2
	8	2.0
	9	1.9
	10	1.8
	11	1.7
	12	1.6
	13	1.6
	14	1.5
Asphalt stabilized base courses		1.2
Cement treated base courses A		1.7
Cement treated base courses B		1.5
Cement treated base courses C		1.2
Base courses of crushed granular material		1.1
Natural granular sub-bases and base courses		1.0

9.7.3 Criteria of the North American Asphalt Institute

The procedure proposed by the Asphalt Institute for designing flexible pavements [50] consists of determining the thickness of the pavement structure from a combination of a particular method of estimating the volume of traffic, a parameter representing the strength or deformability of the bearing material or fill (*CBR*, value of *R* or bearing value obtained in plate-bearing tests), a measure of the load spreading quality of the materials available and the predicted construction procedures. In this system, thickness equivalencies are recommended for different materials in accordance with their relative load spreading quality, thus providing a means of evaluating different possible combinations and finding the most appropriate, economical ones. These equivalencies are based on a comparison with the quality of asphalt-concrete layers used as a reference. The method refers to highways.

The traffic anticipated is characterized by the *Design Traffic Number* (*DTN*), which is the daily average of equivalent 8.2 ton (18,000 lb) loads on a single axle assembly, anticipated for the period of project design. This is normally established as 20 years by the Asphalt Institute. The mechanical properties of the fill material, subgrade layer, sub-base and base course are characterized by the usual tests in current pavement technology (*CBR*, HVEEM's stability value, or plate-bearing modulus or coefficient). The *CBR* test recommended by the Asphalt Institute is that proposed by the Corps of Engineers, described earlier in this chapter. If plate-bearing tests are used, preference is shown for MCLEOD's procedure, which has also been described.

Once the index value or values have been determined for the strength of the material and the *DTN* applicable to the particular case, the thickness of protective covering material is obtained from the nomographs in Figs. 9-31 and 9-32, established from experience by the Asphalt Institute.

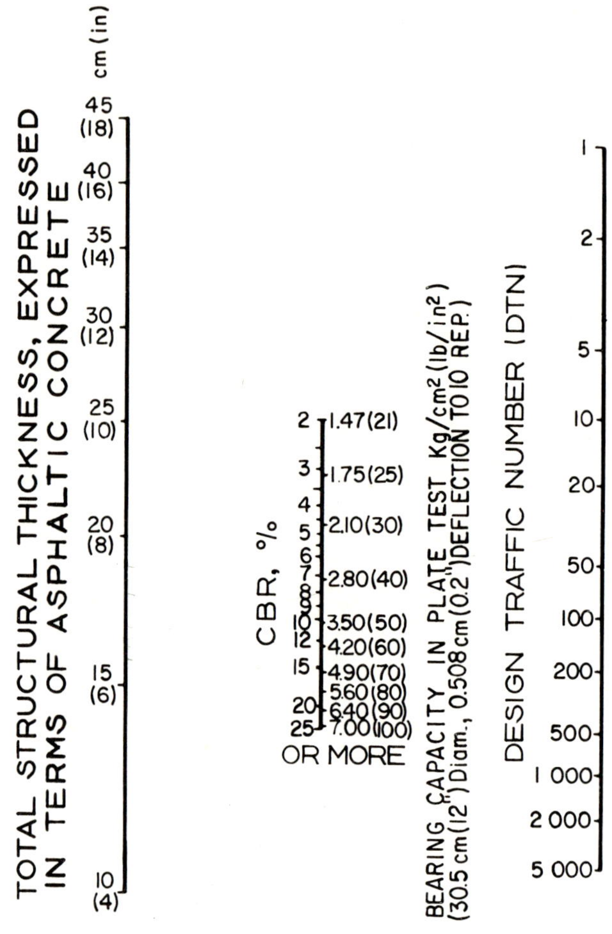

Fig. 9-31 Nomograph by the North American Asphalt Institute for determining the thickness of flexible pavements using the *CBR* or plate bearing tests [50]

In Fig. 9-31, the bearing capacity in the plate-bearing test refers to the pressure (in kg/cm^2) which when applied 10 times to a 30.5 cm (12 in) plate, produces a 0.508 cm (0.2 in) deflection (see Appendix 9a). The *CBR* that is indicated is that obtained with the Corps of Engineers method. As it can be seen, Fig. 9-31 enables thickness determinations to be made on the basis of either *CBR* or plate-bearing test results. The corresponding column of the nomograph was constructed using a laboratory correlation between the two methods of testing the soil's capacity, which may prove useful in other aspects of practical engineering.

The Asphalt Institute gives the thickness of protective covering or paving materials, that is required above a given soil in terms of a thickness of asphalt concrete. This leads to different pavement layer alternatives than using the normal gravel layers and applying the equivalency factors that are described in detail later. The design method is similar to the one that is used with the *CBR* or with HVEEM's stability number, so that nothing new has been added conceptually. Once again design is made with the aid of a nomograph, the major advantage being the experience of whoever plotted them.

The nomographs in Figs. 9-31 and 9-32 are for a design period of 20 years. However, the Asphalt Institute establishes procedures for other design periods. The Institute considers that the end of the service life has been reached when the serviceability index of the pavement reaches 2.5. At this time, pavement rehabilitation or reconstruction will be necessary. The serviceability index is a concept proposed by the AASHTO. It is based on the average of ratings from 1 (bad) to 5 (good) given to the pavement by a group of users, based on their judgement.

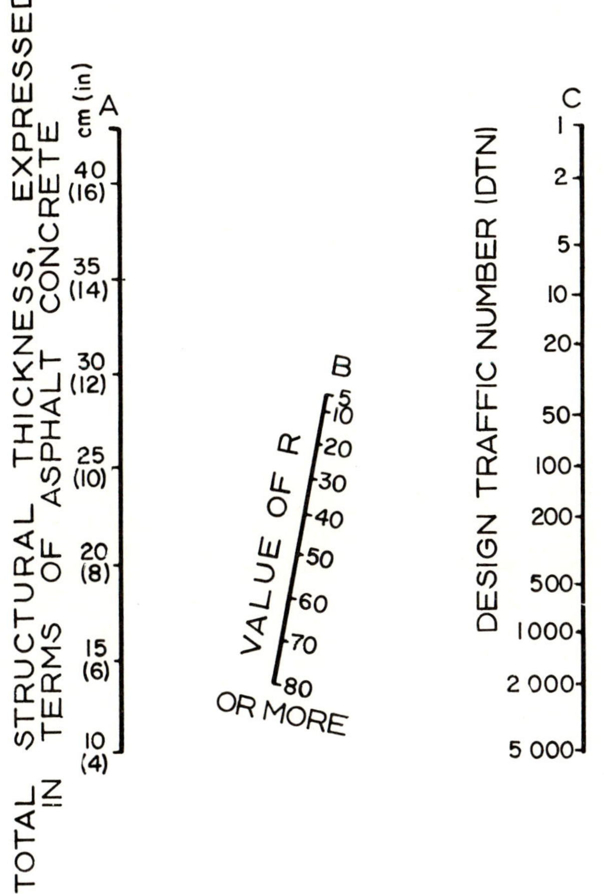

Fig. 9-32 Nomograph by the North American Asphalt Institute for determining the thickness of flexible pavements using HVEEM's stability value [50]

In order to determine the *Design Traffic Number* (*DTN*) first estimate the average number of vehicles for the first year the road is in service. This is based on previous traffic analyses, as well as economic and social studies. This number is known as the *Initial Traffic Number* (*ITN*). On the basis of readings from traffic counting devices and traffic classification for short periods, the percentage of heavy vehicles during this first year must also be estimated, as well as that part of the percentage in the design lane, for roads of more than two lanes. The Asphalt Institute suggests the distribution of heavy vehicles that should be considered in the design lane for different numbers of lanes (Table 9-12). These should not be used for conditions different from those prevailing in U.S.A., which are the ones for which they were established. On the basis of the data in Table 9-12 (or another similar table for different local conditions) it will be possible to estimate the daily average number of heavy vehicles in the design lane (in just one direction). From the previous analyses, it will also be possible to estimate the average weight of the heavy vehicles, considering the legal load limit established by the authorities for a single axle.

Table 9-12

Percentage of total two-directional heavy-vehicle traffic to be considered in the design lane

Total number of lanes	Percentage of trucks to be considered in design lane
2	50
4	45 (fluctuates between 35 and 48)
6 or more	40 (fluctuates between 25 and 48)

With all the above information, the Initial Traffic Number (*ITN*) is found, using the nomograph in Fig. 9-33 [50]. For this, the mean value of truck loads is entered on *D*. A line is

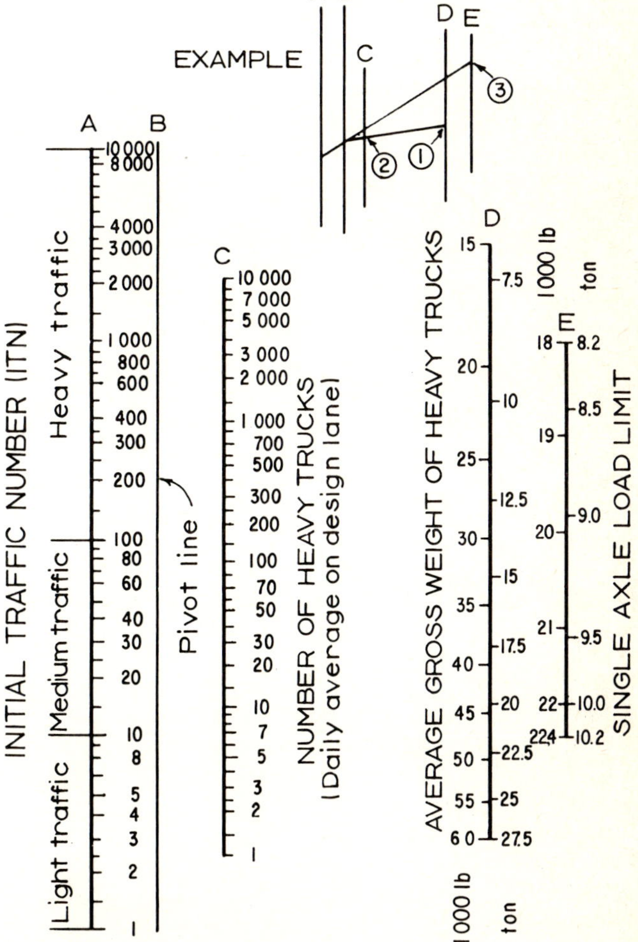

Fig. 9-33 Traffic analysis chart, Method of the North American Asphalt Institute [50]

drawn from this point to the number of heavy trucks in the design lane on *C*. This line is extended until it intersects *B*. Now the legal load limit for a single axle is established on *E*. A line is drawn from this point to the one previously found on *B* and is extended to *A*, where the *ITN* can finally be read.

For the service life (or design period) of the pavement under consideration, which is usually 20 years if the general recommendations of the Asphalt Institute are observed, the annual rate of traffic growth must be estimated. The Institute has made the corresponding planning and statistical analyses. With the design period and the rate of traffic growth, the correction factor that must be applied to the initial traffic number can be found in Table 9-13. The product of the *ITN* and the correction factor is the Design Traffic Number (*DTN*) which is required in Figs. 9-31 and 9-32. In Table, 9-13 the factor has been calculated in accordance with compound interest rules from the growth rate in per cent.

Table 9-13

Correction factors for the *ITN* in order to obtain the *DTN* [50]

Design period Years	Annual rate of traffic growth in percentage 0	2	4	6	8	10
1	0.05	0.05	0.05	0.05	0.05	0.05
2	0.10	0.10	0.10	0.10	0.10	0.10
4	0.20	0.21	0.21	0.22	0.22	0.23
6	0.30	0.32	0.33	0.35	0.37	0.39
8	0.40	0.43	0.46	0.50	0.53	0.57
10	0.50	0.55	0.60	0.66	0.72	0.80
12	0.60	0.67	0.75	0.84	0.95	1.07
14	0.70	0.80	0.92	1.05	1.21	1.40
16	0.80	0.93	1.09	1.28	1.52	1.80
18	0.90	1.07	1.28	1.55	1.87	2.28
20	1.00	1.21	1.49	1.84	2.29	2.86
25	1.25	1.60	2.08	2.74	3.66	4.92
30	1.50	2.03	2.80	3.95	5.66	8.22
35	1.75	2.50	3.68	5.57	8.62	13.55

The thickness of asphalt concrete must be converted to a more conventional pavement layering. This substitution can be done using an infinite variety of combinations, permitting the evaluation of a sufficient number of practical alternatives for a realistic design to be achieved. The alternatives consider the local availability of materials, their characteristics, the treatments required and the cost of handling and transportation. The Asphalt Institute proposes the equivalency factors in Table 9-14 for the most-used conventional layers. Some very common layer types, which were not subjected to experimental research, are not included in the Table. The *CBR* values that are given refer to the Corps of Engineers technique; those from other versions of the test cannot, therefore, be used.

The Asphalt Institute also specifies the minimum thicknesses of asphalt concrete that should be used in the surface course when asphaltic base courses are used. These thicknesses are given in Table 9-15 [50]. When common granular untreated base courses are used instead of asphaltic base courses, the minimum thickness of asphalt concrete proposed by the Asphalt Institute can be found in Fig. 9-34 [50]. Use of this figure may lead to surfacing thicknesses which are excessive

Table 9-14

Thickness equivalency factors between conventional layers and layers of asphalt concrete [50]

Conventional layers	Equivalency factor
Asphalt-sand base courses, plant mixes	1.3
Asphaltic base courses prepared with liquid or emulsified asphalts	1.4
High-quality granular base courses ($CBR > 100\%$)	2.0
Poor quality granular base courses ($CBR > 20\%$)	2.7

Table 9-15

Minimum thicknesses for asphalt concrete surface courses over asphaltic base courses [50]

Design traffic number (*DTN*)	Minimum thickness cm	in
Under 10 (light traffic)	5	2
Between 10 and 100 (medium traffic)	7	3
Over 100 (heavy traffic)	10	4

for some countries. They are seldom used, even in roads with high traffic volumes, where surface courses of even a prime coat, have given excellent results for roads with modest volumes of traffic.

Once the total thickness of asphaltic concrete required has been determined, the necessary thickness of surface course must be subtracted from it. The remaining thickness of asphalt concrete is converted to thicknesses of conventional layers, using equivalency factors.

9.7.4 Other Methods of Design

As has already been pointed out, other methods of pavement design of practical interest are described in [7]. Two methods will be briefly discussed here. These are promising efforts in the search for a theoretical justification of current ideas regarding design. They seek to establish an agreement between the theoretical research into pavement behavior and the results of experience based on the commonly-used empirical design methods.

Reference [51] presents a method which is being developed by the National Laboratory of Civil Engineering in Lisbon. The mechanical characteristics of a layer of pavement are expressed in terms of a strength factor, *F*, determined by the ratio:

$$F = \frac{p}{\frac{\delta}{d}} \quad (9\text{-}9)$$

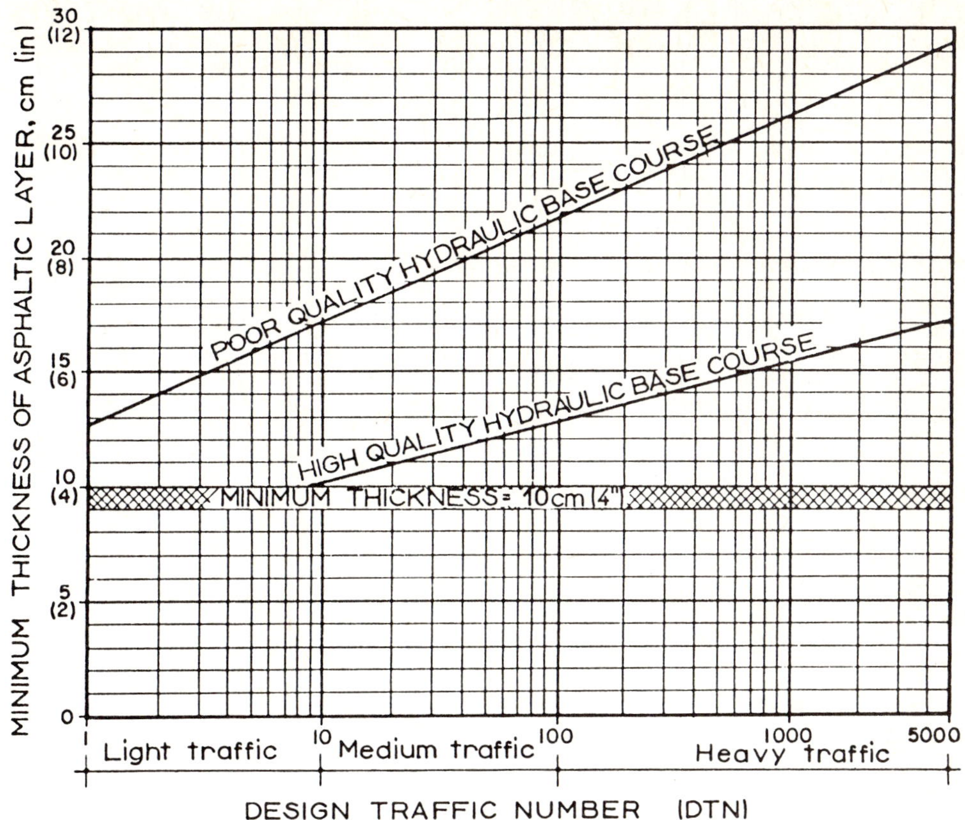

MINIMUM REQUIREMENTS FOR HYDRAULIC BASE COURSE MATERIALS

TYPE OF TEST	NORMS	
	LOW QUALITY	HIGH QUALITY
MINIMUM CBR	20	100
MINIMUM R VALUE	55	80
MAXIMUM LIQUID LIMIT	25	25
MAXIMUM PLASTICITY INDEX	6	NP
MINIMUM SAND EQUIVALENT	25	50
MAXIMUM PERCENTAGE OF MATERIAL PASSING Nº 200 SIEVE	12	7

Fig. 9-34 Minimum thicknesses of asphalt-concrete surface course over granular base courses, according to the North American Asphalt Institute [50]

where p is the pressure that is applied to a plate with a diameter d to produce a settlement δ. In this method, the strength factor that is necessary for the layer under consideration is correlated with the strength factor corresponding to the supporting soil. By conducting research on existing pavements, it will also be possible to correlate the values of the strength factor for a layer with the *CBR* for the same layer, to apply the previous experience in the use of the *CBR* method to this new method.

Reference [52] summarizes trends of Russian technology in the problem of flexible pavement design. A method based on the elasticity theory applied to multiple layer systems is proposed, with the utilization of material parameters obtained from dynamic tests. Two methods are suggested for determining layer thicknesses. One limits the maximum tolerable vertical deflection in a test with repeated dynamic loading that is maintained at a level that does not cause progressive failure or soil fatigue. The other method uses the concept of local shear failure and establishes as the design thickness the thickness that does not permit local shear failure in any of the pavement layers under design loads. The first method of analysis requires determination of the moduli of elasticity of the different layers under heavy loads, whereas the second method requires determination of conventional shear strength parameters. In both cases field information is used to prepare the final method of design. An emphasis on freezing phenomena is seen in this design method. Freezing is of critical importance for countries with a climate that is as harsh as that of the USSR and of less importance in more moderate climates. Reference [53] presents the evolution of the USSR design together with a series of nomographs, which greatly simplify its use. Experience in application of this method is also assessed and is reported as favorable.

Reference [54] presents a systematization of design methodology, based on the facilities currently available to organizations specializing in massive scale pavement construction. In it many of the variables that govern the behavior of a flexible pavement are taken into consideration.

9.8 Analysis of Existing Pavements with a View to Reconstruction or Strengthening

The need frequently arises to analyze the state of an existing pavement in order to decide whether it should be repaired (instead of replaced) and what the cost of repairs would be. Although this is a universal problem, it is particularly common in the transportation networks of developing countries, where there is rapid traffic growth, insufficient construction money and lack of suitable maintenance. Also responsible for the need to widen and reconstruct roads is the policy of gradual investment. This leads to *stage* construction and to pavement design which takes into consideration only conditions prevailing at the time of construction, with the result that the service life of the roads is relatively brief. Further investments are withheld until a stronger, wider pavement is justified by subsequent traffic growth. This policy of construction investment provides for greater availability of limited resources and permits attention to a greater number of projects, but it does make it necessary to strengthen and widen existing roads. In this section, methods of evaluating existing flexible pavements will be briefly reviewed, with special emphasis on some of the work methods in which applied soil mechanics plays a part.

Pavement rehabilitation problems may be enormously varied and range from the addition of *rejuvenating* surface treatment coats and simple overlays to total reconstruction. Also to be considered are the problems caused by the widening of existing road sections. Rehabilitation due to normal traffic growth is usually solved by the use of overlays, whereas reconstruction work is necessary for pavements showing incipient failure, either the appearance of excessive rutting or very high levels of elastic deflection that are detected with the instruments that are currently available and which will be discussed later in this book.

Establishing a criterion for rehabilitation is strictly a matter of reviewing the circumstances responsible for the unsatisfactory performance of the pavement. It is far more complicated than simply observing the appearance of superficial cracks. *Unsatisfactory* rarely is the result of a catastrophic slide. Rehabilitation may prove necessary in a pavement that is appropriately supporting very high volumes and loads of traffic, but for which maintenance costs are excessively high. The following are the principal criteria that are usually considered when determining the need for rehabilitation [55]:

1) Level of serviceability: This will vary with the type of road, and character of the traffic, particularly the speed, and with climate.
2) Structural condition: This concept refers to the capacity of the pavement to support current traffic loads and to continue to do so in the near future without progressive damage.
3) Surface conditions: The surface of a pavement (irregularities, waves, ruts, and cracks) is not necessarily associated with structural capacity, although a lack of structural capacity will rapidly influence the surface geometry of the pavement. Many defects in surface conditions can be easily corrected with methods which bring about no real improvement in structural conditions.
4) Safety: This is usually assessed on the basis of accident statistics.
5) Cost: This refers not only to the expenditure required for rehabilitation, but also to continuing maintenance and operational costs.

The serviceability index is frequently estimated on the basis of the opinion of a group of users, who travel along the road in normal conditions and qualify it in some way. Attempts have also been made to develop a rating by formulas based on statistical analysis of readings obtained using mechanical measuring devices on the pavement surface [55].

In addition to the level of serviceability, it is important to consider the conditions of the wearing surface such as cracks, permanent deformations and any other type of distress, which may not be attributed to structural insufficiency. Whenever the rehabilitation of a pavement is considered, a detailed survey should always be made of any areas of deterioration that can be observed in the wearing surface and their possible relation to drainage and subdrainage conditions, local topography, soils, cut or fill areas or the junction between cut and fill and similar features that can influence the general behavior of a pavement. Table 9-16 shows a schematic diagram used by the Mexican Ministry of Public Works for carrying out surveys of distress in flexible pavements.

The structural capacity of a pavement, is related to the deflection measured in that pavement. Deflections in a flexible pavement under a static load may be determined with equipment such as the Benkelman Beam or a Dehlen curvimeter. A deflectometer of the Dynaflect type enables deflections to be measured when the load that is applied to the pavement is moving or transient, [56] contains a description of these types of equipment, together with the procedure to follow with each of them.

The Benkelman Beam is illustrated schematically in Fig. 9-35 (Plate 9-13). A fixed arm D is placed on a level with the pavement, resting at three points (one point A and two points B). A moving arm D_1 is attached to the fixed arm by a rotating articulation at the point that is illustrated with point C

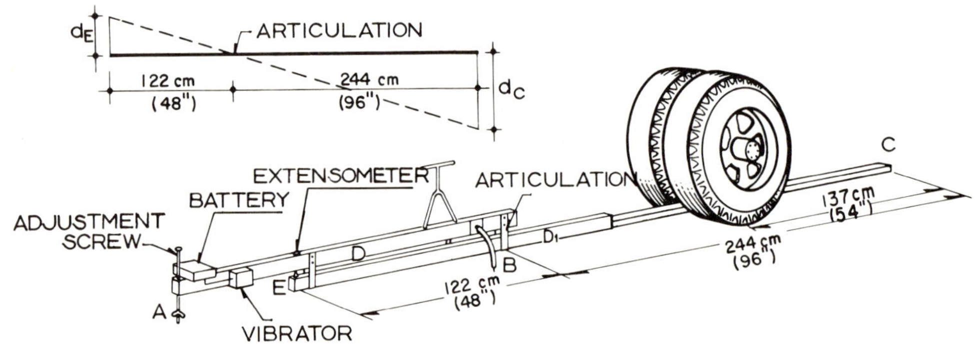

Fig. 9-35 Diagram of Benkelman beam for measuring deflection

Table 9-16

Table used by the Mexican Ministry of Public Works for carrying out surveys of distress in flexible pavements

Road or airport: ______________________

Observer: ______________________

Date: ______________________

Stretch from to

Scale	Type	Item
	Type of crack	Fissures
	Type of crack	Longitudinal
	Type of crack	Transversal
	Type of crack	Polyhedral (7.5cm)(3")approx.
	Type of crack	Polyhedral (15 cm)(6")approx.
	Type of crack	Map cracking (>30cm)(12")
	Type of crack	Reflection
	Type of crack	Smaller than 0.3175cm (1/8")
	Type of crack	Smaller than 0.635 cm (1/4")
0: None	Type of crack	Larger than 0.635 cm (1/4")
1: Minor		Local detachment
2: Moderate		General detachment
3: Major		Pronounced transversal deformation
4: Severe		Longitudinal deformation
-10	Very good	Warping
- 9		Subgrade settlement
- 8	A	Shallow patching
- 7	Good	Deep patching
- 6	B	Localized reconstruction
- 5	Moderate	Shallow roughness
- 4	C	Surface drainage
- 3	Poor	Subdrainage
- 2	D	General conditions
- 1	Very poor	General qualification
0		Work required

Observations (drainage) ______________________

Observations: ______________________

Plate 9-13 Benkelman beam operation

supported on the pavement. When the tires of a loaded truck are placed in such a way that point C of the moving arm is midway between them (note that this is not the position shown in the figure), this point will drop a certain amount owing to deflection caused in the pavement surface by the load on the tires. For this reason, D_1 will rotate around D, which was previously levelled (the dimensions of the beam are such that the position of D is not affected by the deflection caused by the tires' load) and the extensometer as is illustrated will thus record a reading. If the loaded tires are now withdrawn, point C will recover any elastic deflection it may have suffered and the extensometer will make a further reading in the same manner as before. With the two extensometer readings, it is possible to find how much point E has moved during the operation, and with the geometry of the beam the elastic recovery of C, on removal of the tires, will also be obtained, as illustrated in the diagram appearing in the same figure. Note that the recovery of C on removal of the load has been measured, and not the deflection that occurs when this load is applied. The different organizations that have made the method popular use different loads on the dual tire assembly.

Dynaflect is an electromechanical system which measures the changing deflection of a pavement surface when subjected to an oscillating (sine) load. The complicated details of the Dynaflect are beyond the scope of this book; it is transported on a trailer drawn by a vehicle in which the measuring instruments are placed (Plate 9-14). The meter works on the basis of a generator of dynamic force applied to the pavement (impact), the effects of which are recorded on a system of seismographs (geophones). One important advantage of the device is that it does not require any fixed reference point on the pavement surface where the measurements are taken. Another is that it operates automatically and is consequently free from operator errors; also it functions at relatively high trailer speeds. Figure 9-36 shows a set of deflection curves supplied by Dynaflect. The curves refer to readings from five geophones at different distances from the point of application of a dynamic load. The further the geophones are from the impact, the smaller the readings they give. The reading from the nearest geophone is usually taken as the calculation value, but by plotting the readings from all five, a graph is obtained with slope, changes in direction and changes in slope that can given an experienced interpreter a qualitative image of the rigidity of the pavement in the thickness affected by the dynamic loading. In this way, Dynaflect carries out a form of geophysical study of the affected thickness.

Plate 9-14 Dynaflect device for measuring deflection

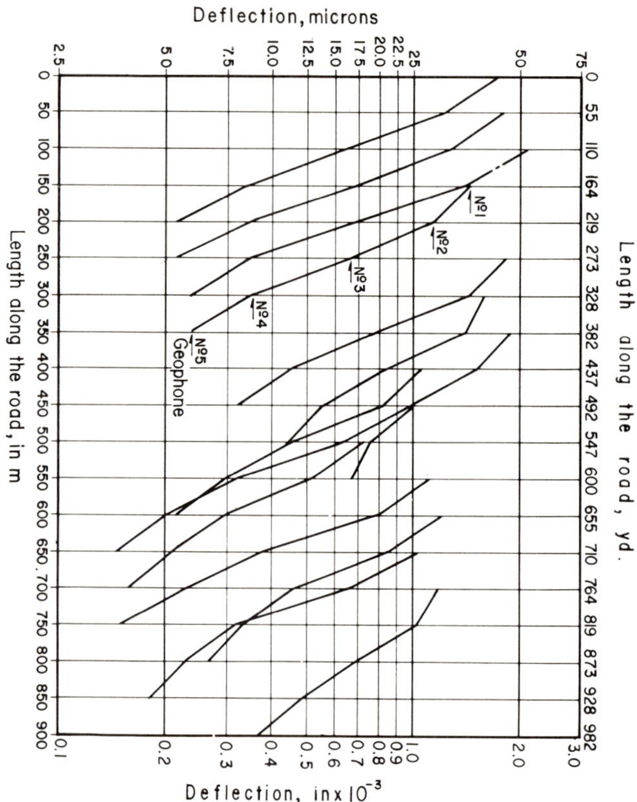

Fig. 9-36 Deflection curves determined with Dynaflect

Evaluation of the structural capacity of a pavement must include an analysis of the strength of the materials forming each of its layers, including the subgrade, and in some cases the fill. The final evaluation of the structural capacity must take into consideration the results obtained using both deflection and strength criteria. This is particularly important because the correlations existing between the deflection, thickness and quality of pavements, together with the traffic to which they are subjected, have been obtained by several different organizations under their own local conditions, and therefore an evaluation based only on these deflection might not be valid in all situations. The method of deflections integrates the combined effect of the pavement system layers but not their distribution depthwise. The inability of surface deflection to diagnose the depth of the weaker layers is their greatest limitation.

The choice of equipment for measuring deflections must be based on availability, cost and progress requirements. The cost of a Benkelman Beam is considerably lower than that of a Dynaflect, but the speed and efficiency of the Dynaflect are far greater.

Another means of evaluating the structural capacity of a pavement, which has been suggested by various organizations, such as the Canadian Department of Transportation [57] is to perform plate-bearing tests as described in Appendix 9a. This method has been frequently used by the Mexican Ministry of Public Works for evaluating the structural capacity of existing airport pavements, following the criteria also recommended by the Canadian Department of Transportation, which will be discussed in detail later in this chapter.

A procedure for evaluating the structural capacity of runways developed in England [58] consists of determining the so-called *load classification number* (*LCN*) by a plate-bearing test. This number represents the structural capacity of the pavement. The *LCN* is a system for classifying both an airport pavement and the aircraft that use it. In it the characteristics of the aircraft are compared with the structural capacity of the pavements. The *LCN* of the pavement is obtained from a field study which is based on plate-bearing tests. The *LCN* of the aircraft depends on the geometry and size of the wheels, tire pressure and characteristics of the pavement. A runway is regarded as suitable for the operation of an aircraft when the *LCN* of the pavement is greater than that of the aircraft. Both the *LCN* of the runway and the *LCN* of the aircraft are obtained using the parameters that were mentioned above and correlation charts of experimental origin [58]. The evaluation procedure provides guides to airport operation. Thus, for example, when the *LCN* of the design aircraft is from 1.1 to 1.25 times the *LCN* of the pavement, a further 3,000 operations, approximately, can be permitted with some degree of reliability, before undertaking a new evaluation. When the *LCN* of the aircraft is more than twice that of the pavement, the latter can only be used in case of emergency.

Once the foregoing conditions or criteria have been assessed (level of serviceability, surface conditions and structural capacity of the pavement), which can be regarded as the *parameters* of the problem, a decision can be made regarding the type of rehabilitation that is most suitable. It is here that those less tangible factors must be considered. They include the anticipated increase in traffic volume and vehicle loads to which the pavement will be subjected, the cost of rehabilitation work and its relation to the availability of funds for its execution, the service life anticipated or desired for the rehabilitation work and the cost of pavement maintenance. Another important factor, especially in roads which carry essential consumer goods or which encourage economic development, is the interruptions or delays in transport, caused by rehabilitation, and their influence on transportation costs for the users. Reference [55] contains some procedures for detailed analysis of these costs, which are beyond the scope of this book.

With all these considerations, some qualitative and others quantitative (but without which a generous proportion of engineering experience would lose their significance), the design engineer or group responsible for the study will determine the work required to rehabilitate the road or runway. Rehabilitation can range from the simple application of protective prime coats (or surface treatment) to construction of reinforcing overlays or the complete reconstruction of a pavement. One form of rehabilitation which is used by some organizations consists of gradually enhancing the structural capacity of a pavement, with *programmed* construction of overlays, to fit the increased volume and intensity of traffic loads. This procedure requires repeated evaluation of the general conditions of the pavement, so that the moment reinforcement is required can be determined without failure occurring. The success of this approach depends largely on the attention that is devoted to the maintenance and repair of pavements.

9.8.1 Thickness of Reinforcement Layers for Pavements Based on Deflection Readings

Once the engineer has analyzed all the factors described in the foregoing section and has decided that reinforcement of the pavement is the appropriate means of rehabilitation, the amount of this reinforcement must be determined, and the plans and specifications that are to govern construction must be established. The plans should include any drainage and/or subdrainage systems that may prove necessary, as well as shoulder repair and other tasks which ensure maximum pavement performance. It is good practice to design the reinforcement layer for the most critical structural conditions that may be encountered in the road. However, this criterion is not always the most appropriate one, especially if the availability of funds for rehabilitation work is considered. Furthermore, frequent variations in the thickness of the reinforcement layers in short sections of pavement, to try to meet different structural capacity requirements, may lead to construction procedures that are not practical. These might have adverse effects on cost. A careful balance between structural requirements and ease of construction lead a design which satisfies the need for reinforcement with maximum economy and practical construction conditions.

The reinforcement design procedures that are described below are applicable to flexible pavement, even if they include layers stabilized with materials such as asphalt, cement and lime. The pavement reinforcement is designed with an *overlay* of asphaltic concrete or a combination of asphalt concrete and layers of a granular material that can be stabilized or treated with the above-mentioned materials.

The methods presented here are based on the deflections measured on the surface of the pavement, by a Benkelman Beam with loads corresponding to a single axle with a dual tire assembly. If another device is used, for example a Dynaflect, correlations exist which enable the deflections obtained (Fig. 9-37) to be converted to deflections of the Benkelman type. These correlations must be used with caution, owing to the fact that they are subject to local conditions. The figure shows the correlation given by the manufacturer of Dynaflect (Dresser Atlas Co.), for Benkelman Beam tests made with a 6.8 ton (15,000 lb) axle load. This is the original experimental correlation. Also given is the correlation with the Benkelman Beam made with an 8.2 ton (18,000 lb) axle load. This graph was obtained by computation and not experiment, by establishing a coefficient of correlation between the Benkelman deflections for the two different loads and presuming that Dynaflect deflections also vary in the same proportion. Lastly the figure presents the experimental correlation, which was developed by the California Division of Highways, between the 6.8 ton (15,000 lb) axle load on the Benkelman Beam and the Dyna-

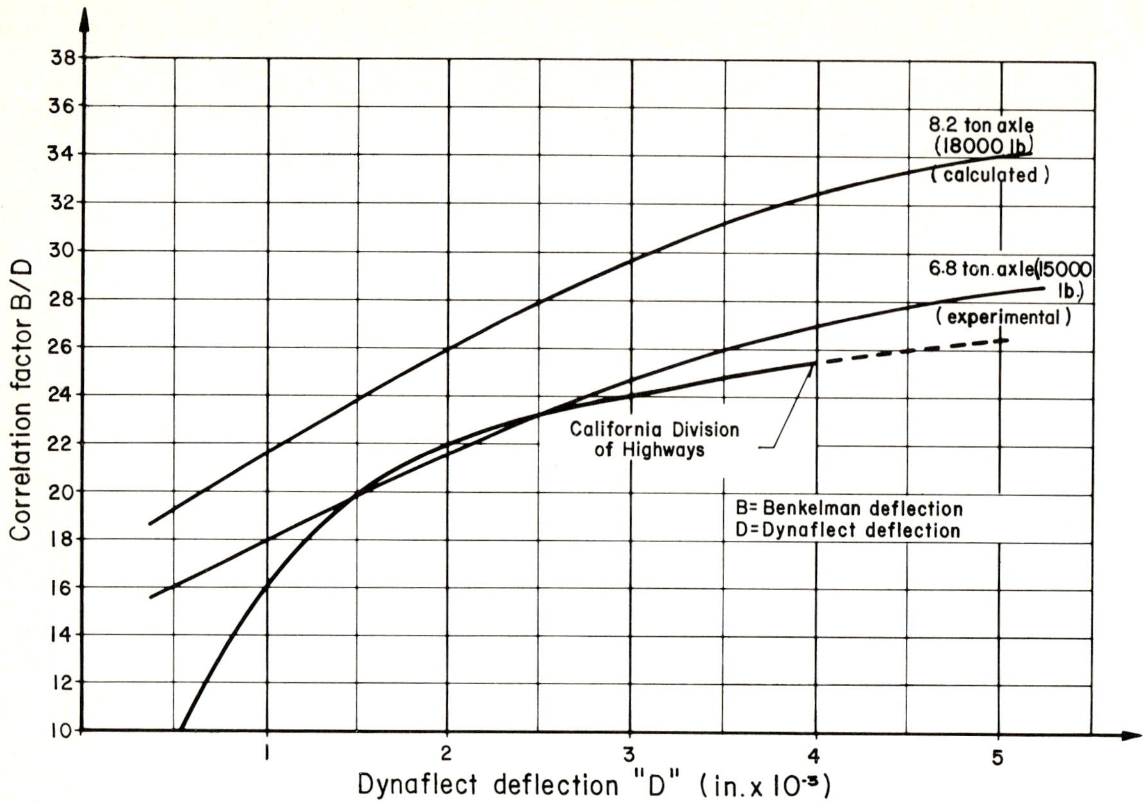

Fig. 9-37 Comparison of Benkelman Beam and Dynaflect operations. (The curve for the 15,000 lb axle supplied by the manufacturer of Dynaflect)

flect, after using both types of equipment for several years. The variations of the foregoing correlations demonstrate that any organization that is interested must obtain its own correlations that fits its pavement conditions.

.1 The California Method [56]

This method has been developed on the basis of observation of the performance of reinforced pavements. Its purpose is to establish a maximum limit for the deflection that can be permitted in a pavement structure, as a measure of its structural capacity. This limit is a function of the thickness of the asphaltic wearing course and the number of applications of a 2,270 kg (5,000 lb) wheel load that must be supported by the pavement. Figure 9-38 presents a graph which permits determination of the level of tolerable deflection in the surface of a pavement. The set of straight lines corresponds to the different thickness of the existing asphaltic layer (or to a single thickness of a cement-treated base course). The abscissa represents the number of repetitions of a 2,270 kg (5,000 lb) wheel load. It can be observed that a maximum limit of tolerable deflection that is defined on the ordinate axis is 40 thousandths of an inch. When there is no cement-treated layer in the pavement being evaluated, only the thickness of the surface course is considered for classification of the structure. When there is a cement-treated base course 15 cm (6 in) thick or more, Curve 7 is used. The above facts may appear surprising unless the relatively good-quality homogeneity that is usually observed in the pavement structures in the State of California is taken into consideration. If the method is applied blindly by other organizations there is a risk of applying the same curve to two very different types of pavements, although the thickness of the surface courses is the same. There may be significant differences in their structural behavior. The differences are greater the thinner the surface course.

When evaluating a pavement by the California method, the following steps are followed:

1. With the aid of the graph in Fig. 9-38 and the traffic analysis described earlier in this Chapter, the number of equivalent wheel loads of 2,270 kg (5,000 lb), yields the tolerable Benkelman deflection.
2. Next, the real deflections occurring in the pavement under analysis must be measured. For this purpose, a Benkelman Beam is used (if Dynaflect equipment is used the correlation in Fig. 9-37 will be adopted) with a loaded truck with 6,810 kg (15,000 lbs) on its rear axle (dual). A measurement interval of about 8 m (23.6 ft) is recommended by California, but can be varied slightly according to the conditions of the pavement under evaluation.
3. The value of pavement deflections must be determined in such a way that 20% of those obtained are greater and the remaining 80% correspondingly smaller. This is the statistically significant deflection and is represented by δ_{80}.
4. The value of δ_{80} is compared with the tolerable deflection obtained in Step *1*. On making this comparison, it should be remembered that the maximum tolerable deflection will be 0.101 cm (0.040 in). If δ_{80} is smaller than the tolerable deflection, the California method considers that no reinforcement other than a prime coat (surface treatment) is required by the pavement. If δ_{80} is greater than the tolerable deflection a percentage of reduction in the deflection that is measured must be determined in accordance with the following expression:

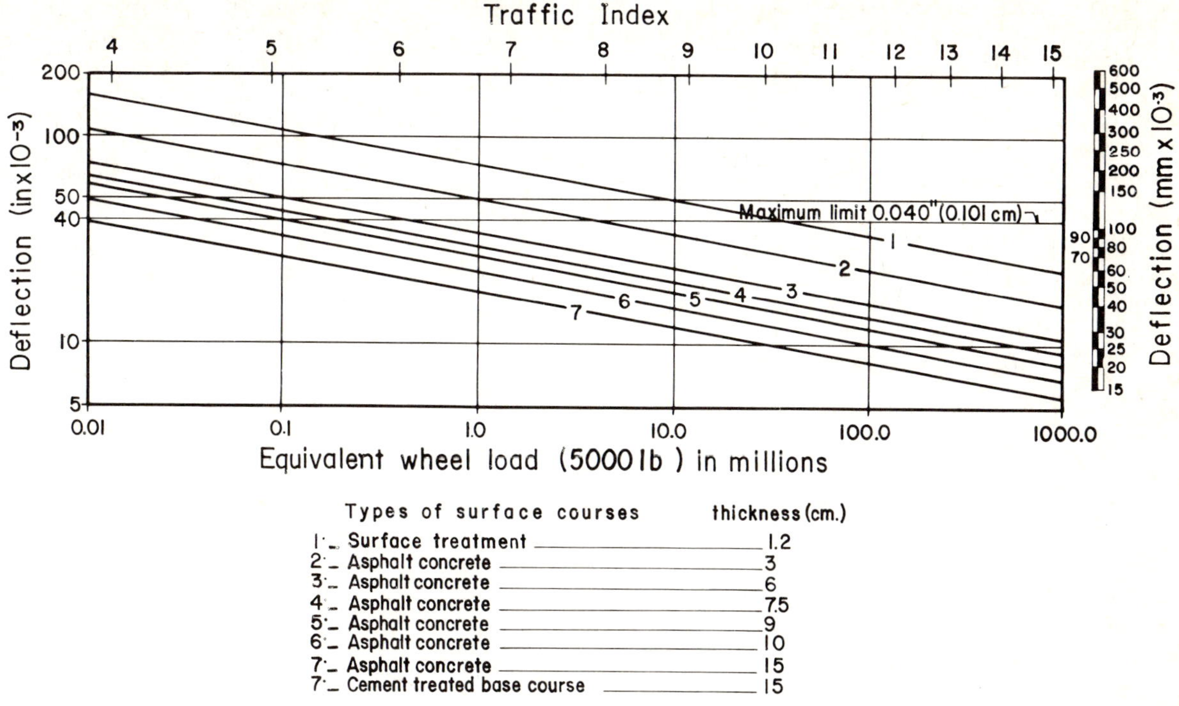

Fig. 9-38 Graph for the determination of tolerable deflection using the California Method [56]

$$R_\delta = \frac{\delta_{80} - \delta_{tol}}{\delta_{80}} 100 \qquad (9\text{-}10)$$

5. The value of R_δ is applied to the graph in Fig. 9-39 in order to obtain the thicknesses of gravel equivalent that is required to reinforce the pavement. Table 9-11 provides the relation between these thicknesses and those of the practical reinforcement layers appropriate for use in each case.

The California method advises revision of the reinforcement thicknesses obtained by following the above steps. With the thicknesses of gravel equivalent already obtained and converted to a thickness of asphaltic concrete surface course, the tolerable level or deflection must again be found (Fig. 9-38), so that it can be compared with the real deflection measured with the BENKELMAN Beam in the field. The level of tolerable deflection should now be higher than the real deflection; if this is not so, the entire sequence must be repeated using a different trial-and-error procedure, until a thickness of reinforcing gravel equivalent is obtained which, when converted to asphalt concrete with the coefficients in Table 9-11, will give a level of tolerable deflection equal to or greater than the field deflection. The reason for this check is that when the thickness of the surface course is changed, according to Fig. 9-38, the tolerance level of deflection of that pavement must also be changed.

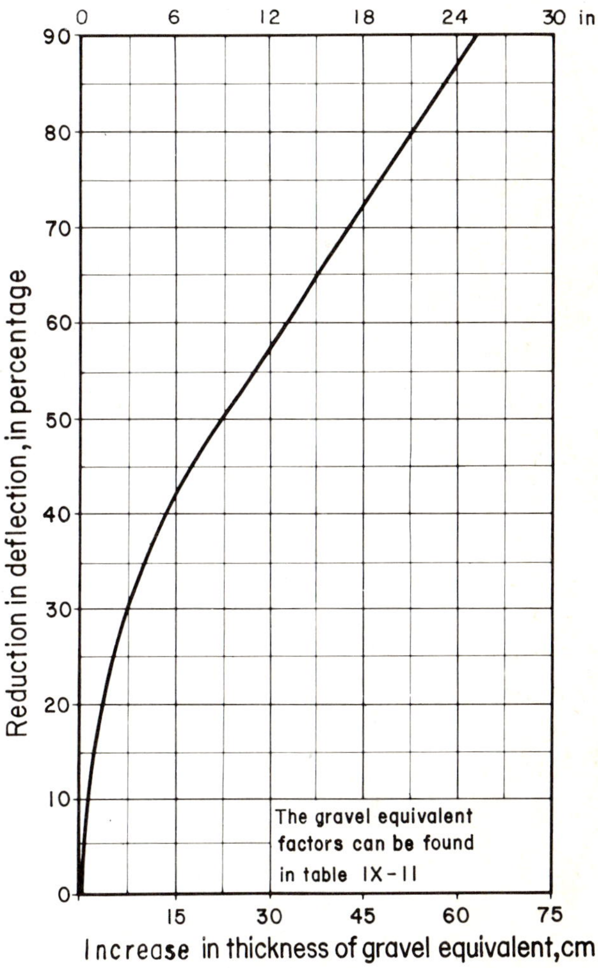

Fig. 9-39 Increase in the thickness of a pavement (gravel equivalent) as a function of the deflection reduction factor. California Method [56]

.2 Asphalt Institute Method [59]

This method is also based on establishing a limit of deflection of the pavement structure. This is a function of the number and intensity of load applications to which the pavement under evaluation is subjected.

The first step in application of the Asphalt Institute method is determining the design traffic number already discussed in this chapter. Next the deflections in the pavement being evaluated must be measured by the BENKELMAN Beam or an equivalent method. It is specified that the number of points studied must be not fewer than 15 per km (25 per mile) or 10 in each test section, which implies a stretch of road to which more or less uniform distress characteristics are attributed. The points studied must be distributed at random over the selected section. The arithmetical average of all the deflections obtained must be calculated, together with the standard deviation of the set of values corresponding to the section or stretch of road under study. The deflections are measured beneath a load of 4,100 kg (9,000 lbs) on a dual tire assembly (8,200 kg (18000 lb) per axle).

The characteristic deflection is defined by the following equation:

$$\delta_c = (\bar{x} + 2s)\ (f)\ (c) \qquad (9\text{-}11)$$

where: $\bar{x}$ is the arithmetical average of the individual values for deflection in the stretch of road under consideration; s the standard deviation of the same values in the same stretch of road; f an adjustment factor related to the temperature of the surface course; and c an adjustment factor which varies according to the time of year at which the measurements are taken ($c = 1$ for the period representing the most critical conditions of the pavement).

Figure 9-40 gives a graph for finding the value of the adjustment factor from the temperature of the surface course. The adjustment factor for the time of year must be obtained either by taking the readings during the period most critical for the pavement ($c = 1$) or else by taking a series of readings which cover different times of the year, and calculating for each the relation between these readings and the one corresponding to the critical period. This should not prove too difficult for an organization devoted to the construction and conservation of pavements on a large scale.

The characteristic deflection and the design traffic number are entered on the graph in Fig. 9-41 to find the thickness of asphalt concrete required to reinforce a pavement. Using the

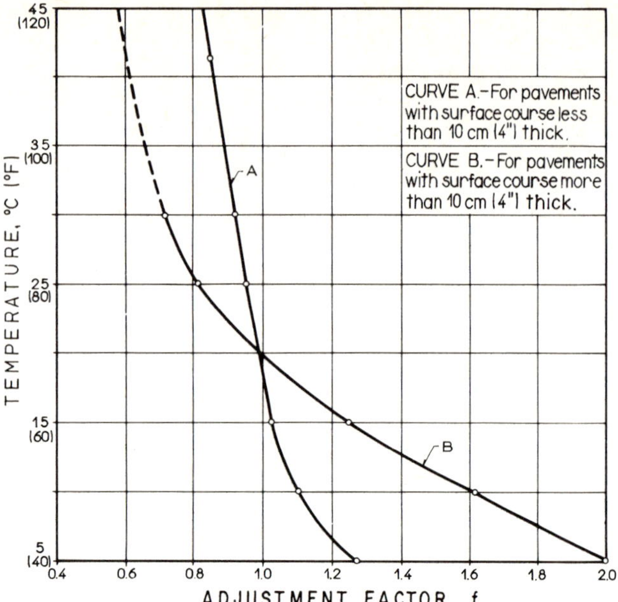

Fig. 9-40 Temperature correction factors in Benkelman Beam deflections (Method of Assessment by the North American Asphalt Institute)

equivalency coefficients between thicknesses of asphalt concrete and layer thicknesses of a different nature, which were given earlier in this chapter, different alternatives for the required reinforcement can be designed. However this last possibility is not explicitly mentioned in the original bibliographical sources proposing this method.

On the basis of its evaluation studies, the Asphalt Institute gives a guide for forecasting deterioration. By adopting it, it will be possible to estimate how long it may be before a pavement in a good condition requires reinforcement, based on the real characteristic deflection and the annual rate of traffic growth for the road under consideration.

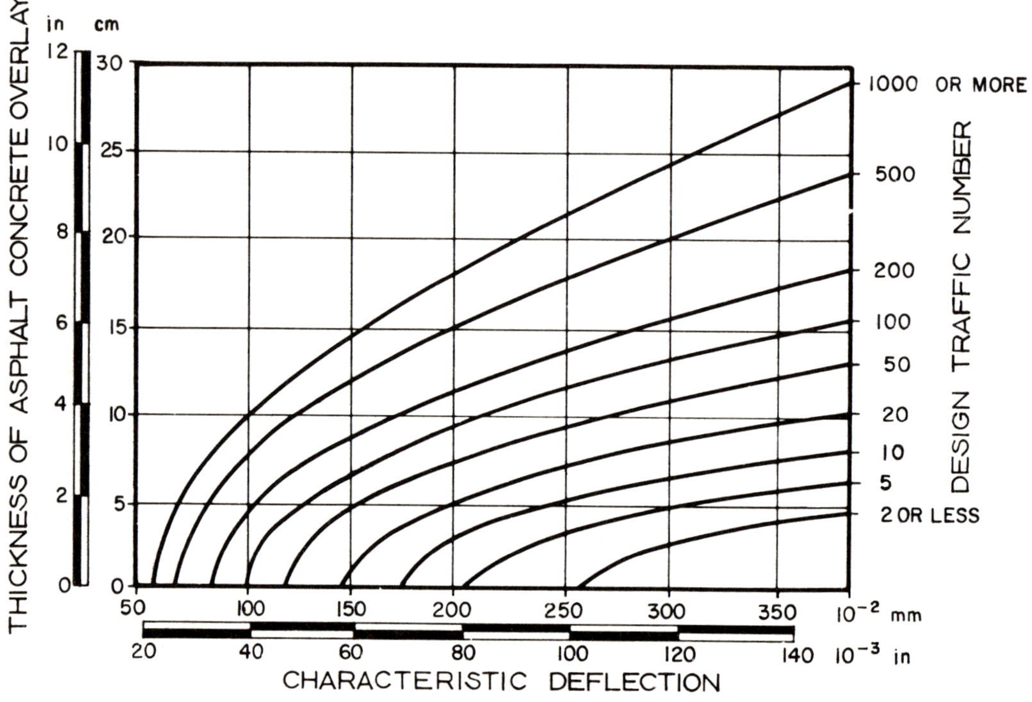

Fig. 9-41 Thicknesses of reinforcing overlay as a function of the characteristic deflection of the pavement, according to the North American Asphalt Institute

Figure 9-42 gives a graph which, with the current characteristic deflection of the road, makes it possible to obtain the highest design traffic number that may be achieved by the road if repairs are not required. This number is compared with the real design traffic number for the road in question. If this real *DTN* is smaller than the one found from the graph, the pavement requires no reinforcement at the present time. If both traffic numbers are equal, it is time for reinforcement, but if the pavement has a *DTN* higher than the one computed from the graph, reinforcement should have been carried out already.

If the *DTN* for the pavement is smaller than the one found from the graph in Fig. 9-42 and knowing the annual rate of traffic growth, it will be possible to estimate the time that must elapse before the pavement reaches a *DTN* equal to the one calculated from the graph. This permits estimating the moment when the pavement will have to be repaired. Table 9-13 enables the necessary calculations to be carried out.

.3 Canadian Method for Evaluating the State of Active Runways [57]

This evaluation method is valid only for runways, unlike the previous methods, which referred exclusively to highways. The departure point is the expression originally introduced by McLeod [7]:

$$e = K \log \frac{P}{S} \qquad (9\text{-}12)$$

where: e is the total thickness, in cm or in., of the pavement being evaluated, above the upper level of the subgrade layer; K a design constant, in cm or in., which depends on the diameter of the assumed circular loaded area; P the design wheel load, in kg (lb); and S the subgrade support, in kg (lb).

For a given aircraft, the equivalent load for its tire assembly is supplied by the manufacturer; consequently P will in future be regarded as a known factor in all problems.

Likewise, the value of e for a given pavement will be found from the corresponding exploratory soundings.

Figure 9-43 [57] depicts K, as a function of the diameter of the load-applying area, whether an aircraft tire or a test plate, for both methods of load application are used. If the load is applied by an aircraft, because the equivalent load and the tire pressure with which that aircraft operates are known, the contact area, assumed to be circular, can be obtained and the diameter can be computed. When carrying out a plate-bearing test, the diameter of the plate is known. Thus, in a pavement evaluation problem, Eq. (9-12) can be used with S as the only unknown factor.

If a plate-bearing test is carried out on the wearing course, P can be determined for 10 load repetitions producing a deflection of 1.27 cm (0.5 in). For this, the test sequence shown in Appendix 9a is followed. With this value, it is possible to obtain S, which is the total support of the pavement under consideration, separating it from Eq. (9-12). The plate-bearing tests may be replaced, sometimes economically, by taking readings with a Benkelman Beam using aircraft loads. In Canada, experimental correlations between the two test methods have been developed, which are reported to be sufficiently reliable for overlay design. Fig. 9-44 shows the correlation that is referred to, expressed in British units, which are the ones commonly used.

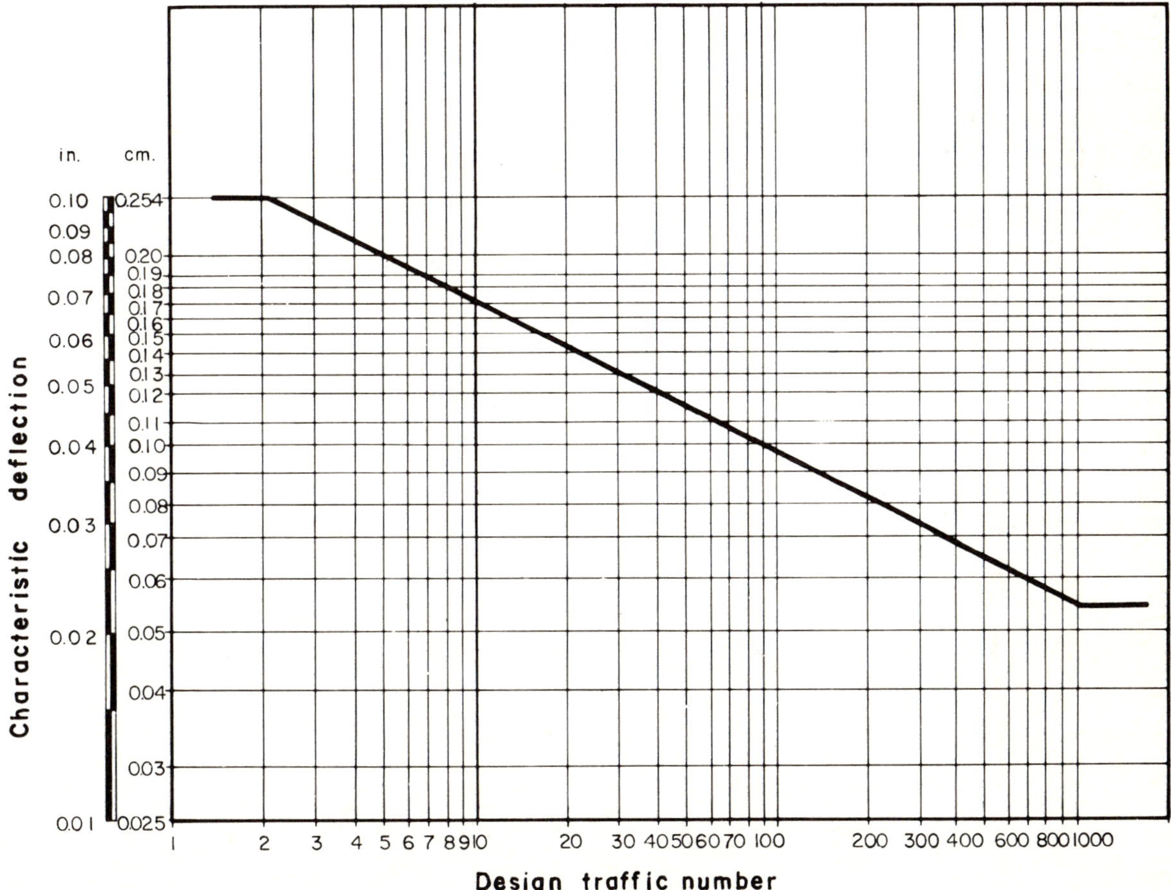

Fig. 9-42 Critical traffic number (as from which reinforcement is required) for a given characteristic deflection. North American Asphalt Institute [59]

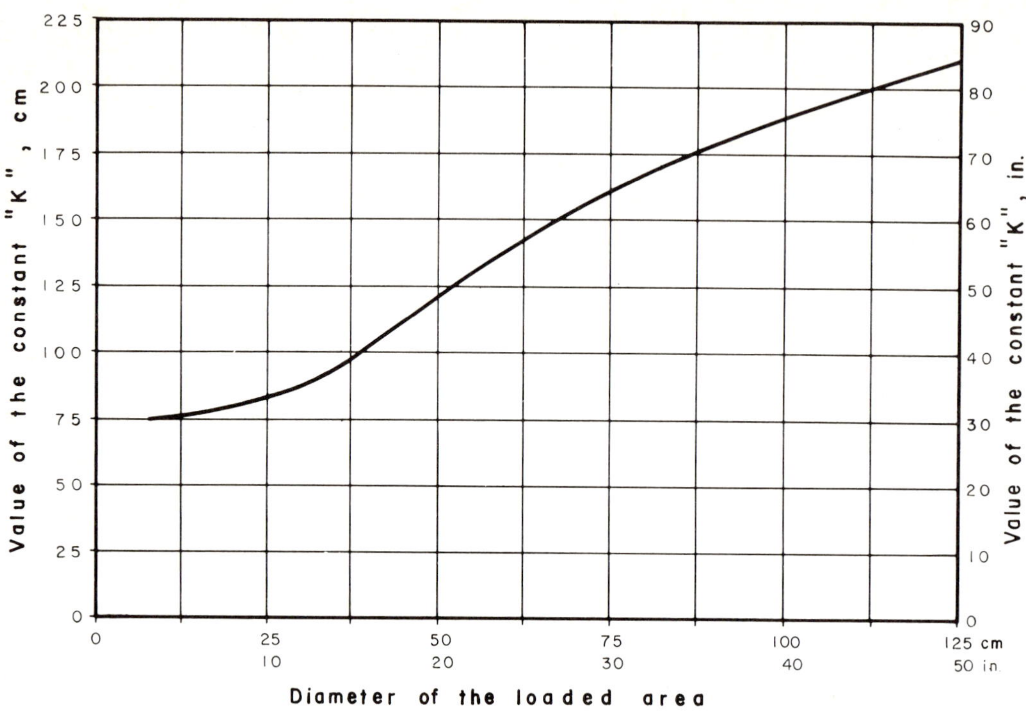

Fig. 9-43 Values of the design constant K as a function of the circular area to which loading is applied. Canadian Method of evaluation of runways [57]

Bearing in mind that there is a correlation between the readings taken with a BENKELMAN Beam and a Dynaflect (Fig. 9-37), it follows that either of these devices can be used to evaluate the total support that can be attributed to a pavement being evaluated.

In the Canadian method, the characteristic BENKELMAN Beam deflection is determined with the expression:

$$\delta_c = \bar{x} + 2s \tag{9-13}$$

This is similar to Eq. (9-11), but without any correction factors. $\bar{x}$ is the arithmetical average of the BENKELMAN Beam readings for the stretch of runway or taxiway under consideration, and s is the standard deviation of these readings.

If the critical operating aircraft and its equivalent design wheel are now considered, total support S can be computed again with Eq. (9-12). This is the support required by a pavement with the thickness under consideration, if its performance beneath the critical aircraft is to be adequate throughout the

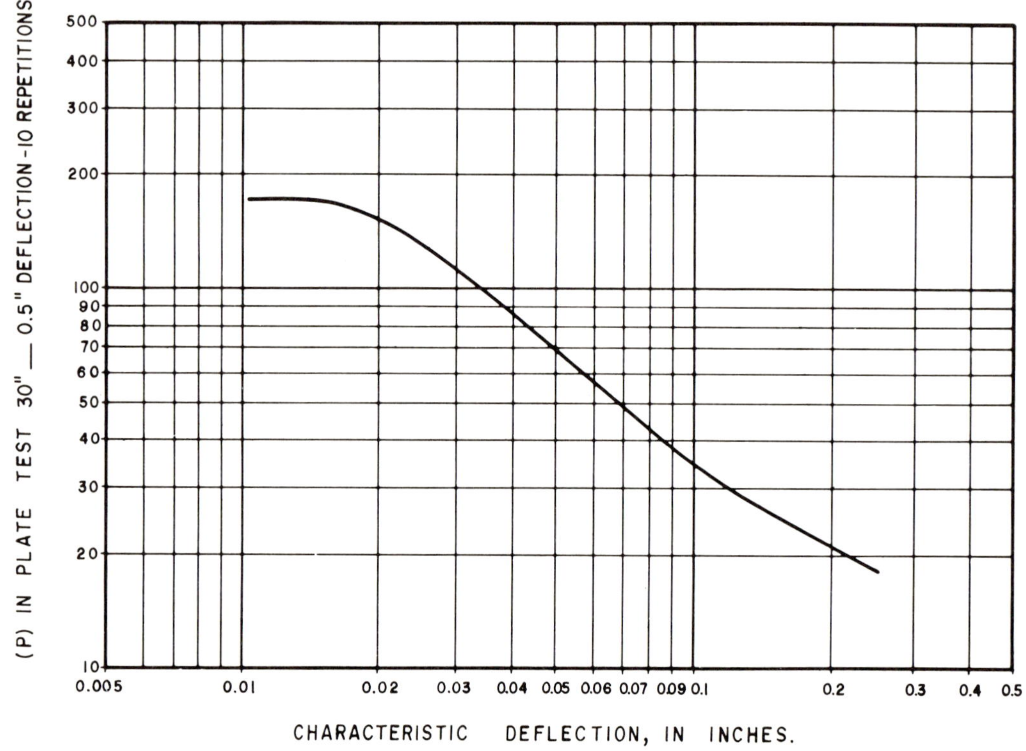

Fig. 9-44 Correlation between results of plate-bearing tests and measurements of deflection with a BENKELMAN beam [57]

number of operations predicted in the pavement design. This support value will be referred to as S_n. From this it is possible to compute the so-called surcharge factor, f_s.

$$f_s = \frac{S_n}{S} \tag{9-14}$$

Canadian experience has indicated the number of operations that can be supported by a runway from the time of study as a function of the surcharge factor. When the pavement has sustained this number of operations, it should be reinforced, unless a careful inspection at that time reveals that distress has not progressively developed in which case it will be possible to permit additional operations. The evaluation process is repeated as often as is wished, so long as no significant increase is noted in pavement distress. Table 9-17 [57] shows the correlation between the surcharge factor and the number of operations admissible before distress is likely, according to the Canadian experience.

Table 9-17

Correlation between the surcharge factor, f_s, and the number of operations permitted before reinforcement [57]

Surcharge factor, f_s	Number of operations between inspections
Less than 1.25	3,000
1.25 to 1.50	300
1.50 to 2.00	30
More than 2.0	Use only in emergency

Computation of the thickness of the reinforcement required in each case is regarded by the Canadian method as a normal design problem; that is, the method enables determining the moment when reinforcement will prove necessary. Once that moment has arrived, the thickness to be used is found by applying a pavement design procedure giving the necessary thicknesses for whatever loading condition may then prevail.

9.8.2 Determination of Thickness of Pavement Reinforcements Using Conventional Design Methods

These procedures simply apply a method of new pavement design to an existing pavement, considering current traffic conditions, and the usual predictions of increase plus the characteristics of the existing materials. In this way, it is possible to obtain the thickness that will be required by the pavement under present and future conditions. The present thickness of the pavement can be found by an exploration study; it is compared with that required and the difference represents the reinforcement required. The evaluations that are most likely to be carried out are on pavements that have caught the attention of the project engineers owing to their distressed state or the high level of their maintenance costs.

Once the thickness of the reinforcement has been estimated, there will be numerous possibilities for carrying out the work. Of these, the most appropriate is selected, bearing in mind the availability of materials, ease of construction, obstructions to established traffic and costs of the different alternatives.

9.9 Influence of Expansive Soils on Flexible Pavements

In many places clay soils, or those with a large clay content, have to be used for highway construction. Sometimes these soils have pronounced swell characteristics. They are referred to as active soils; it is their characteristic to undergo important changes in volume with variations in water content. These soils are sometimes incorporated in the body of embankments and occasionally in the subgrade layer (although good designs prohibit them in subgrades). In arid regions, active soils usually have very low water contents. Owing to hot sunshine and low humidity they often lose part of the water that is mixed into them during compaction. As a result, as time passes, their water content increases on account of their natural tendency to accumulate water by capillary rise or vapor movement under surfaces covered by pavements or as a result of the wetting that accompanies rainy seasons. The soils in such arid zones will always exhibit changes in water content due to climatic effects because of their high capillary potential. These changes have highly adverse effects on their volumetric stability and shear strength.

The presence of expansive soils in flexible pavement systems becomes increasingly important with the natural increase in traffic loads and volumes and corresponding increases in serviceability requirements from year to year. Currently not enough information is available regarding the properties of expansive soils and their effects. These are no reliable methods for identifying them in the field or in the laboratory, nor reliable construction procedures that enable them to be handled without greatly jeopardizing future performance. Reference [60] is an excellent summary of the current state of knowledge on the subject, and is the source of much of the information that is included in the pages that follow.

The principal effects that may be suffered by an expansive (or high volume change) soil in a flexible pavement are:

- Shrinkage due to drying
- Expansion due to wetting
- Development of swell pressures in the confined soils in which expansion is restricted
- Reduction in shear strength modulus of elasticity and bearing capacity and increase in compressibility as a consequence of expansion.

It is not unusual for several of these effects to occur simultaneously or consecutively. The distress caused in a flexible pavement by these effects is usually of one or more of the following:

- Rising or sinking of the surface course throughout a considerable length, which brings about unevenness wave-like irregularities, without cracking or other visible damage
- Longitudinal cracking
- Significant localized deformations, for example around culverts, and bridge abutments usually accompanied by cracking
- Generalized cracking (alligator) in the surface course, with a tendency towards disintegration.

Of the above types of distress, the first is the most frequent one. It can be measured with road roughness indicators. Some organizations are introducing specifications as to the maximum tolerable values [61]. Longitudinal cracking usually accompanies a rise or a sinking of the pavement surface.

Any attempt at considering the effects of expansive soils on the performance of a flexible pavement should require estimating moisture conditions and other significant properties prevailing at the time of construction, and of the changes in water content that are expected to occur during the service life of the pavement, as well as their influence on these properties. It is also useful to establish criteria for the classification of fine soils in the laboratory, or even better in the field, which will enable the presence of expansive soils to be detected reliably and simply. In this way the engineer will be duly prepared to deal with them.

In [62], SKEMPTON proposed a classification criterion based on clay activity (Eq. (1-23), Chapter 1), according to which soils are classified as shown in Table 9-18.

Table 9-18

Classification of fine soils in accordance with their tendency to expand, after SKEMPTON [61]

Activity	Soil category
Lower than 0.75	Inactive soil
0.75 to 1.25	Normal soil
Higher than 1.25	Active soil

According to the above classification, montmorillonites and bentonites (smecktites) are active, illites normal and kaolinites inactive. An attempt has been made to correlate SKEMPTON's concept of activity with the swelling potential of clays, but the correlations that have been found are not very reliable [60,63].

The U.S. Bureau of Reclamation classifies clays from the point of view of the intensity of their swelling potential [64]. To define this potential, the so-called *Degree of Expansion* is measured. This is the percentage of expansion of a sample of soil that is air-dried, then placed in a consolidometer, and submerged in water, subjected to a vertical pressure of 0.07 kg/cm^2 (1 lb/in^2). The swelling potential is defined in terms of several other characteristics of clay, besides the degree of expansion, the most important ones being the shrinkage limit, the plasticity index, the percentage of particles smaller than one micron and the free expansion. This last concept is defined by Eq. (9-15). A 10 cm^3 sample of air-dried soil, made up from the fraction of material passing No 40 sieve is introduced into a 100 cm^3 graduate filled with water. The new volume of the sample when it reaches the bottom of the graduate, is measured.

$$F.E. = \frac{V - V_0}{V_0} 100 \quad (9\text{-}15)$$

where: $F.E.$ is the free expansion of the soil, in percentage; V the volume of the sample after expansion, in cm^3 (in^3); and V_0 the volume of the sample before expansion = 10 cm^3 (0.61 in^3). A soil with a high swelling potential may have a free expansion higher than 100%.

Putting together all the foregoing factors, the U.S. Bureau of Reclamation classifies soils as in Table 9-19. Figure 9-45 is a graphical depiction of the data contained in Table 9-19. It classifies expansive soils as a function of the Plasticity Index and the Percentage of particles finer than one micron. In [64] correlations are shown for the Plasticity Index, shrinkage limit and the content of particles finer than one micron with the volume change of a specimen in the consolidometer, when submerged in water and confined with a vertical pressure of 0.07 kg/cm^2 (1 psi). The difference between these correlations is significant, so that it is hard to use them for predicting the behavior of a certain soil.

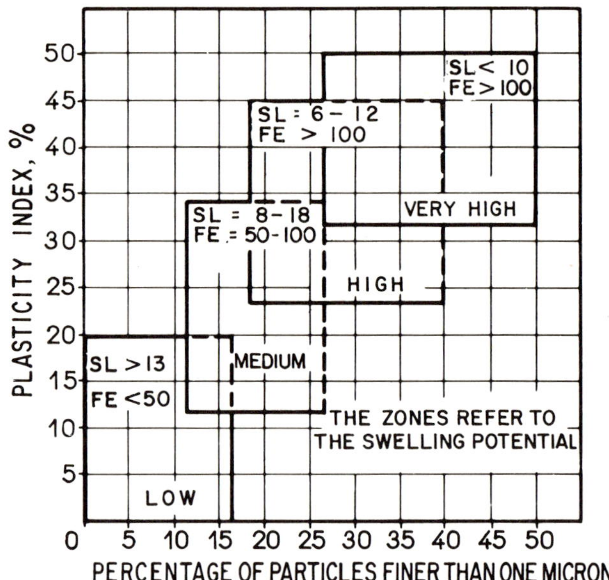

Fig. 9-45 Classification of expansive soils according to the U.S. Bureau of Reclamation [60,64]

Table 9-19

Classification of expansive soils according to U.S. Bureau of Reclamation: HOLTZ and GIBBS [60,64]

Swelling potential	Expansion in consolidometer under vertical pressure of 0.07 kg/cm^2 (1 lb/in^2)	Shrinkage limit	Plasticity index	Percentage of particles finer than 1 μm	Free expansion
	%	%	%	%	%
Very high	> 30	< 10	> 32	> 37	> 100
High	20-30	6-12	23-45	18-37	> 100
Medium	10-20	8-18	12-34	12-27	50-100
Low	< 10	> 13	< 20	< 17	< 50

McDowell [65], in his classification system, defines a percentage of volume change in the sample of soil subject to capillary absorption and to a chamber pressure of 0.07 kg/cm^2 (1 psi), using a triaxial apparatus for pavement design of the Texas Department of Highways. The time allowed for capillary absorption depends on the plasticity of the clay; the number of days is equal to the plasticity index, when this is higher than 15. It was found that the volume change for any initial soil condition can be correlated (to a certain extent) with the plasticity index, giving a criterion for the classification of clays (Fig. 9-46).

Seed and his collaborators [66] define the swelling potential as the percentage of vertical expansion of a compacted sample, with its optimum water content and maximum density (AASHTO standard test) when it is placed in a consolidometer and submerged in water, and confined by a vertical pressure of 0.07 kg/cm^2 (1 psi). They express the swelling potential with the following equation:

$$S.P. = KC^x \quad (9\text{-}16)$$

where: $S.P.$ is the swelling potential; C the percentage of particles smaller than 2 microns; x the number which depends on the type of clay; and K a factor which depends on the type of clay minerals.

For the tests reported by Seed and his collaborators, x was taken as 3.44. For the same conditions it was found that:

$$K = 3.6 \cdot 10^{-5} \cdot A^{2.44}$$

where A is the activity of the clay as defined by Skempton. Because A is related to the plasticity index and to the percentage of particles smaller than 2 microns, a direct relation can be established between the swelling potential and the plasticity index. This relationship is shown in Table 9-20. For classification purposes, Seed proposes the swelling potential values that are shown in Table 9-21. The chief disadvantage of the investigations carried out by Seed is that they were conducted on artificial soils, prepared in the laboratory, which gives rise to doubts concerning the representativity of the results.

Lambe [67] refers the characteristics of expansive soils to the so-called swell index, which he measures with a special apparatus [60] which he designed. The swell index is the swell pressure that is developed by a specimen of compacted clay after 2 hours in this apparatus.

Of all the aforementioned classification methods, the most convincing one is that proposed by the U.S. Bureau of Reclamation because it takes into consideration a greater number of factors. In all the current systems, however, correlations are used which are neither very reliable nor well verified.

There are constant improvements in the laboratory techniques for measuring the tendency of soils to expand and the swell pressures that are produced under different circumstances. For this purpose, several different systems are used, which are

Table 9-20

Correlation between the swelling potential and the plasticity index, I_p, after Seed et al. [60,66]

I_p (%)	Swelling pontential
10	0.4- 1.5
20	2.2- 3.8
30	5.7-12.2
40	11.8-25.0
50	20.1-42.6

Table 9-21

Classification of soils according to their swelling potential after Seed et al. [60,66]

Swelling characteristics of the soils	Swelling potential, %
Low	0- 1.5
Medium	1.5- 5.0
High	5.0-25.0
Very high	Over 25.0

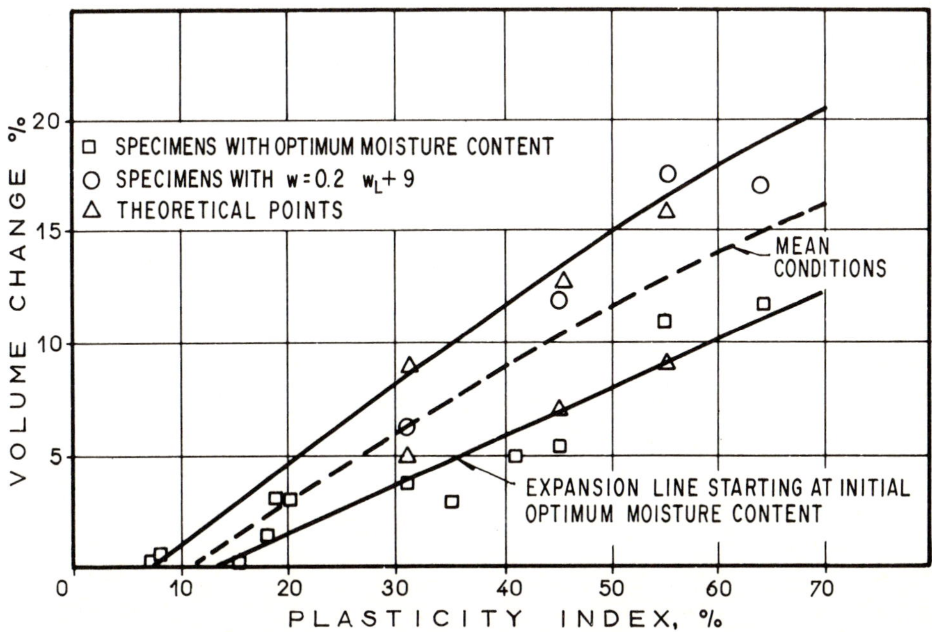

Fig. 9-46 Correlation between volumetric expansion and the plasticity index, according to McDowell [60,65]

described in detail in [60]. In some cases consolidometers are used, measuring either the pressure with which the soil expands or the vertical counterpressure that is required to impede expansion. In other cases, special devices are used in which a submerged sample pushes a piston as it expands, and the piston in turn applies pressure to a calibrated ring, bar or beam, so that by measuring the deformation of these elements the swell pressure for very restricted expansion can be found.

The conditions of specimen preparation vary from one technique to another, in search of maximum representativity. The counterpressures or confining stress imposed on the specimen sometimes reproduce the weight of a hypothetical overlying pavement.

Reference [68] includes a diagram of a typical device of this type, designed by SEED and his collaborators, and similar to HVEEM's expansiometer shown in Fig. 9-16. This is an example of such types of apparatus, Fig. 9-47 shows another version by PALIT [60] which illustrates the use of a calibrated ring.

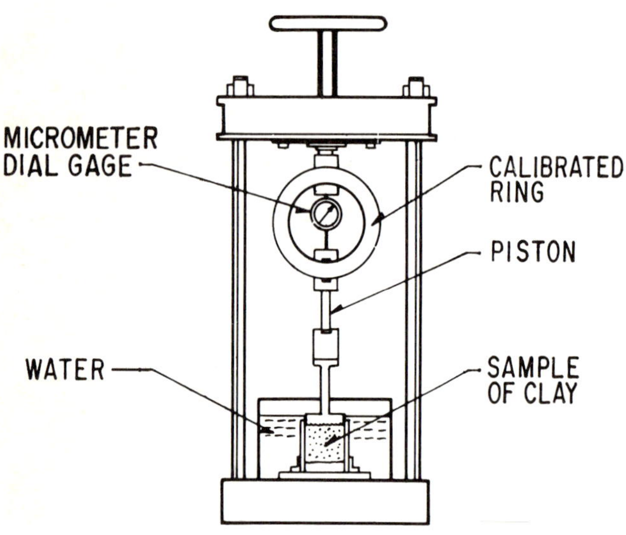

Fig. 9-47 Apparatus for measuring swell pressures [60]

The swell pressures that are obtained in the laboratory depend on the conditions and periods of saturation, and on the sequence with which loads are applied and expansion is permitted [69]. They also depend on the time that is allowed before a reading is taken, for it has been observed that a certain period of time must elapse for the total swelling potential to develop. Figure 9-48 shows the results of investigations by SEED and his collaborators [60,68], which shows how the swell pressure developing in the soil varies with time of inundation. In the figure, swell pressures are presented after one day and seven days as a function of the ultimate water content in the expansion test.

The period of time that elapses in the laboratory before swell pressure ceases to be generated and the equilibrium is reached depends on the nature of the clayey minerals and the size of the sample. It is longer for montmorillonites and minimal for kaolinitic clays.

It has been found that when the fabric of a natural soil is destroyed and the soil is afterwards compacted to the same original dry density and water content, its swelling potential increases [60]. This can be partially explained in terms of the energy that is supplied to the remolded clay during compaction and which is released when the clay is moistened during the expansion process. Whereas in its natural state, the soil had already released much of its energy in previous wetting and drying throughout its history. It has been observed that the swelling potential of compacted soils is greater in soils compacted with static methods than with dynamic methods. This fact, and the reasons for it, have been described in detail in Chapter 4 while discussing the structure acquired by soils subjected to different methods of compaction. The foregoing facts lead to the practical recommendation that expansive soils taken from borrow areas should be disturbed as little as possible, and should be compacted by kneading or tamping methods.

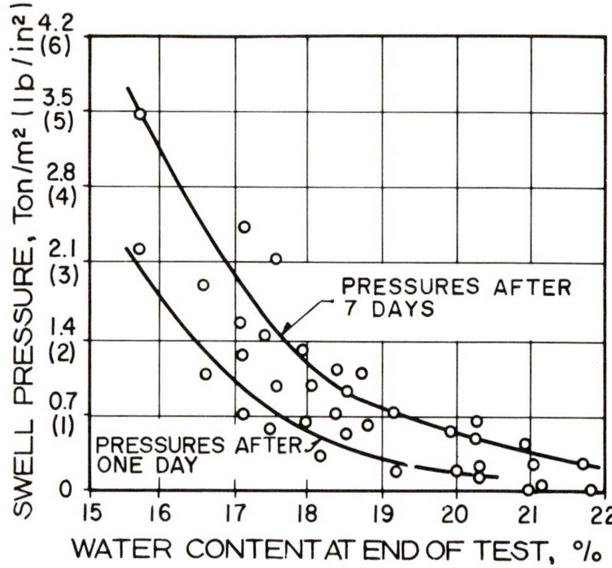

Fig. 9-48 Effect of elapsed time on the swell pressure [60,68]

The foregoing facts lead to the practical recommendation that expansive soils taken from borrow areas should be disturbed as little as possible, and should be compacted by kneading or tamping methods.

An important aspect of the problem is predicting the swelling potential in the field. Before a pavement is constructed, the clay in the active zone undergoes continuous changes in water content, dry density, and also in suction pressure. During the dry season, the suction pressure increases, likewise the dry density, as the water content drops. If, during this dry period, the surface of the soil is covered, contrary effects will occur. Evaporation is reduced, the water content will increase and the suction pressure and dry density will drop. This will be accompanied by expansion and heave in the surface of the ground.

After a few years, the central section of a pavement achieves an equilibrium in the distribution of its water content and of the suction pressure with depth. This is important, because knowledge of the time that has elapsed since construction of the pavement, and consequently the time required for equilibrium to be achieved, makes it easier to predict the ultimate heave in the soil surface.

Because the magnitude of the expansion depends on the composition of the clay, its structure, stress history, initial moisture content and dry density during construction of the pavement, as well as the pavement surcharge and the negative load of the water in contact with the clay, prediction of this expansion magnitude is not a simple task. Several methods have originated from attempts to achieve it, the most widely used ones being those by McDOWELL and JENNINGS.

In developing McDowell's method [65] undisturbed samples were obtained at different depths within fills. Their volume change for absorption by capillarity under various confining pressures were measured, using the Texas Highway triaxial test. The results are expressed in a series of master curves, Fig. 9-49, [65]. Each curve reflects the variation of volumetric expansion of a soil with different confining pressures, as given on the abcissa axis. The curve number is the percentage of volume expansion measured with a standard confining pressure of 0.07 kg/cm^2 (1 lb/in^2).

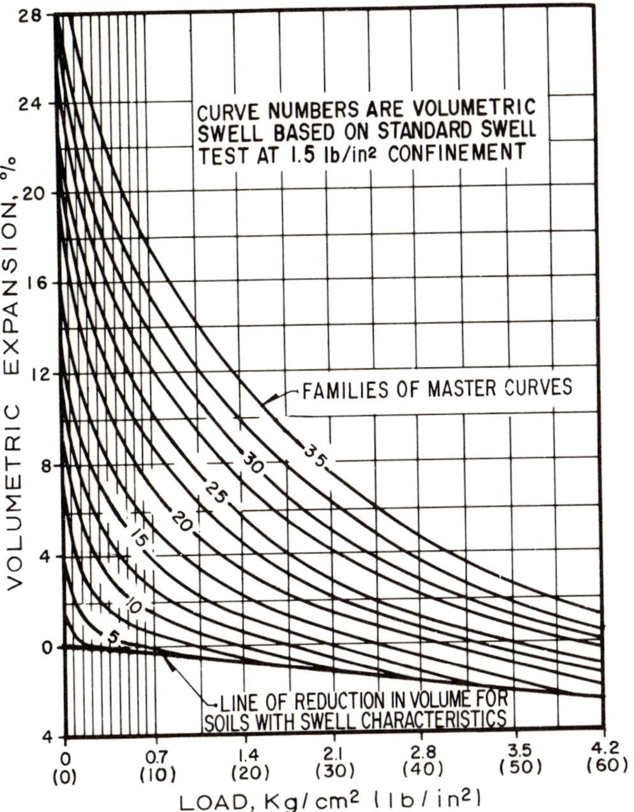

Fig. 9-49 Relation of loading to volume change in expansive clays

In using this curve an undisturbed sample of soil is tested by the Texas Highway procedure, THD-80 at a standard confining pressure of 0.07 kg/cm^2 (1 lb/in^2). Its volumetric expansion in percentage defines the expansion-pressure master curve to be used on Fig. 9-49. (The percentage sign is omitted from the master curve numbers). Once the master curve for that soil has been identified, the volumetric expansion for any vertical confining pressure can be read from the curve. In order to convert the volumetric expansion to linear, the volumetric expansion is multiplied by 1/3.

As a result of a study of various embankments in Texas, McDowell found that the low water content, w_i, existing in the embankments prior to pavement construction, is expressed by:

$$w_i = 0.2\, w_L + 9 \quad (9\text{-}17)$$

Clays with the foregoing initial conditions, subject to capillary absorption under a confining pressure of 0.07 kg/cm^2 (1 psi) in the laboratory, expand until they reach a final water content, w_f:

$$w_f = 0.47\, w_L + 2 \quad (9\text{-}18)$$

For typical inorganic clays the above final condition corresponds to a ratio of the water content over the plastic limit of 1.28 for a plasticity index of 30.

For the initial conditions given in Eq. (9-17), the percentage of volume change for capillary absorption under a confining pressure of 0.07 kg/cm^2 (1 psi) can be directly related to the plasticity index as follows:

$$\frac{\Delta V}{V}\ (\%) = 0.37\, I_p - 5 \quad (9\text{-}19)$$

For these conditions, if the plasticity index of the clay is constant with depth, using the set of master curves given by McDowell (Fig. 9-49), an expression can be developed for the surcharge pressure required to prevent expansion as a function of the plasticity index:

$$p_0 = 0.5\, I_p - 5 \quad (9\text{-}20)$$

where p_0 is the surcharge pressure, in t/m^2, required to prevent expansion. With the value of the plasticity index, the volumetric expansion is computed with Eq. (9-17), selecting the appropriate master curve in Fig. 9-49. The surcharge required to overcome expansion will be the abscissa of the point at which the master curve intersects the line for zero volumetric expansion. By integrating the expansions for the range of pressures corresponding to the depth of the active zone, it is possible to obtain the probable volumetric expansion for the conditions studied by McDowell. The heave is 1/3 the volumetric expansion. Table 9-22 presents surface heave as a function of the plasticity index of the clay profile, assuming uniform homogeneous stratum to an appreciable depth.

Table 9-22

Surface heave as a function of the plasticity index, I_p, [60]

I_p (%)	Surface heave	
	cm	in
10	0	0
20	1	0.40
30	4	1.57
40	7	2.76
50	13	5.11

Jennings [70] proposes another method for predicting the probable expansion of a layer of active soil. It is based on a correlation between the expansions measured in existing pavements and those predicted on the basis of a double consolidation test which is described later. It is assumed that the amount of expansion does not depend on the stress path followed by the soil. The test is made on undistrubed samples, from which two identical specimens are prepared for testing in the consolidometer; one with its natural water content and the other with the water content it achieves after it is allowed to expand when submerged in water while subjected to a small confining pressure. In both cases, the specimens are tested by applying the load in stages, as is usual. The two compressibility curves that are obtained in the two tests are superimposed, so that their virgin sections coincide (Fig. 9-50), by adjusting the submerged curve downwards.

Knowing the surcharge that will act on the soil at the level from which the sample was obtained, the voids ratio corresponding to the original soil consolidated under that surcharge (Point A) can be found on the curve corresponding to the natural water content. Next the equilibrium suction must be estimated on the submerged soil curve, adding this value to the surcharge pressure. In this manner, Point B can be obtained on the compressibility curve for the previously submerged sample (adjusted). JENNINGS uses the value of Δe (Fig. 9-50) or variation of the voids ratio in both cases to predict expansion. The method has given good results in South Africa, but it may not function equally well in other regions, because expansion depends on the stress path to which the soil is subjected.

Lastly, in [60] (which is recommended) the following sequence is proposed for predicting surface heave, in accordance with current knowledge on the subject:

a) Determine the depth of the active zone

b) Obtain undisturbed samples of the clay at regular intervals in the active zone

c) Estimate the suction pressure that is anticipated in the active zone

d) Conduct expansion tests on undisturbed samples in consolidometers, permitting the clay to expand in contact with free water. Each sample will be subjected to a surcharge with a vertical stress equal to the surcharge pressure, plus an additional load equal to the suction pressure anticipated for equilibrium conditions

e) Integrate the percentage of expansion obtained from swell tests with depth

It should be noted that for depths below the active zone, zero expansion is anticipated for samples loaded with a pressure equal to the surcharge plus the equilibrium pore tension that is expected.

When swell pressure is measured in the laboratory with a totally submerged sample (the usual way), and expansion of the soil is permitted, the pore pressure reaches zero when equilibrium is achieved at the end of the expansion process. This indicates that the swell pressure that will develop in the field, in places where expansive clays have reached equilibrium, will be lower than the ones obtained for the same soils in the laboratory, because the water content of clays in the field probably does not usually reach saturation. Thus, the field swell pressure can be expressed as:

$$p_c = p_l - ks \tag{9-21}$$

where: p_c is the field swell pressure, under partial saturation conditions; p_l the maximum swell pressure obtained in the laboratory under saturation conditions, when pore pressure in the specimen reaches zero; k an adjustment factor which must be studied, both to establish its value in each situation and to relate it to the different parameters which determine the behavior of the soil; and s is a suction value prevailing in the soil when in equilibrium, prior to loading. (In [60] an explanation will be found of the suction concept, which has already been dealt with by many soil mechanics papers).

In [71], the following ratio is established for p_l:

$$p_l = \frac{C}{C'} s \tag{9-22}$$

C and C' are two numbers which depend on physical characteristics of the soil, such as the voids ratio and the relative density of the solids, on how the suction phenomenon occurs in the soil and saturation conditions. In totally saturated soils $C = C'$, and consequently $p_l = s$. This result has been experimentally verified by WARKENTIN [72]. For partially saturated soils, $C < C'$, and p_l is a fraction of the initial suction value. Some observations have indicated that in relatively dry clays p_l may be about 10% of initial suction.

Current research does not yet permit general conclusions to be established regarding the value of k in Eq. (9-21). In compressible soils with a high degree of saturation, it seems reasonable to regard it as being very close to unity.

Taking into consideration the foregoing facts, it is possible to approximate the field swell pressure that can be anticipated for a certain situation, if the value of soil suction is known (note that in Eq. (9-21) p_c is computed on the basis of the suction and swell pressure measured in the laboratory). Coefficient k must be tentatively regarded as equal to one, until further information is available concerning its real nature and variability). Consequently, it is desirable to establish a correlation between suction, which is hard to evaluate in routine problems, and some simple soil properties. Correlations have been found [60,73] between the water content corresponding to various suction values and the plastic limit of Israeli clays. However correlations like these are questionable for use in other local conditions.

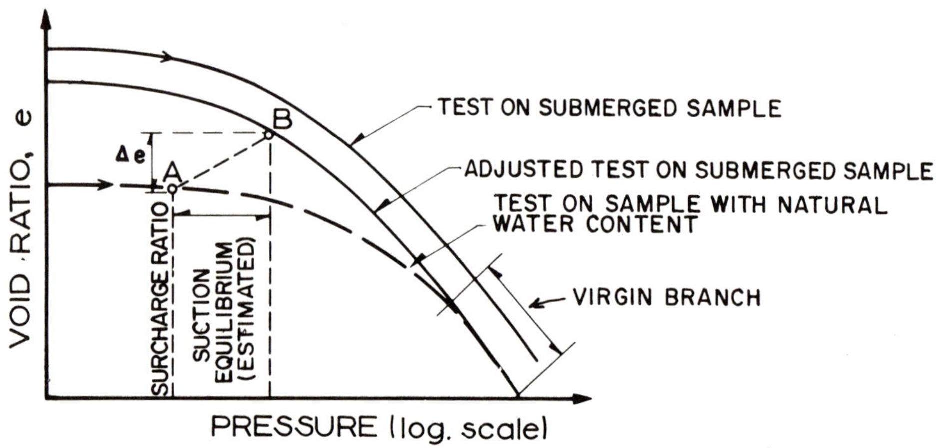

Fig. 9-50 Prediction of surface heave using the double oedometer test, according to JENNINGS [70]

Therefore this type of investigation should be repeated in each region, until a better knowledge of the phenomena permits a general theoretical treatment.

Figure 9-51 [74] shows the variation between the swell pressure measured in the laboratory and the initial dry density of the sample. Although this refers to a specific soil, it is representative of the usual relationship between the two

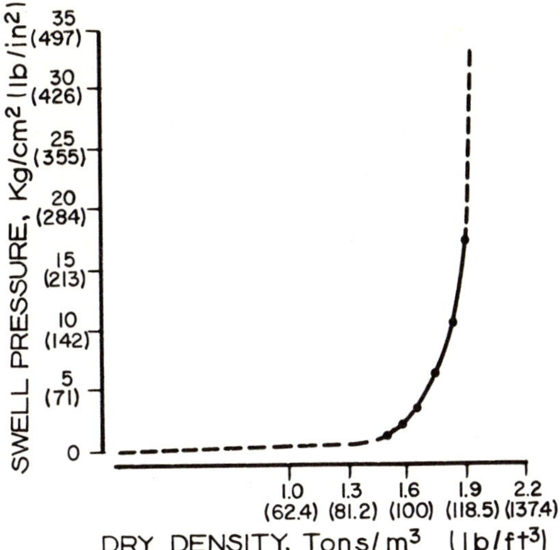

Fig. 9-51 Effect of the initial dry density of an expansive soil on swell pressure [74]

factors, and it conveniently illustrates the great practical importance of not compacting expansive soils beyond the levels where swell becomes significant. Expansive soils are a dramatic example of the dangers of regarding compaction as being better, the higher the dry density produced in the soil. Reference [75] includes a set of correlations between the swelling characteristics of soils and their common properties. Although data based on tests conducted on 270 samples of undisturbed natural clays obtained from very different places are presented, the correlations may not be valid for *any* clay from *any* location. However, in view of the limited current knowledge, they do provide useful information, even if they only serve as a guide for establishing the order of magnitude of the problem that may be posed, Figs. 9-52 and 9-53 give the two most useful correlations.

As has already been mentioned, longitudinal cracking in the pavement close to the shoulders is one of the most frequent types of distress in embankment sections of expansive soils. In Chapter 4 of Volume II when the problem of longitudinal cracking was discussed, reference was made to the mechanisms by which the volume changes associated with cyclic wetting and drying can produce this cracking. The maximum volume changes occur in the zones close to the shoulders, for in these zones there is less restriction to deformation, and the exposed shoulder provides the maximum opportunity for change in the water content. Figure 9-54 [76] shows the volume changes observed in a typical section; they were measured at different points at increasing distances from the line center.

At the beginning of this section the principal effects suffered by expansive soils were briefly mentioned, along with the types of distress that most frequently occur in pavements built on them. Expansion causes deterioration and very

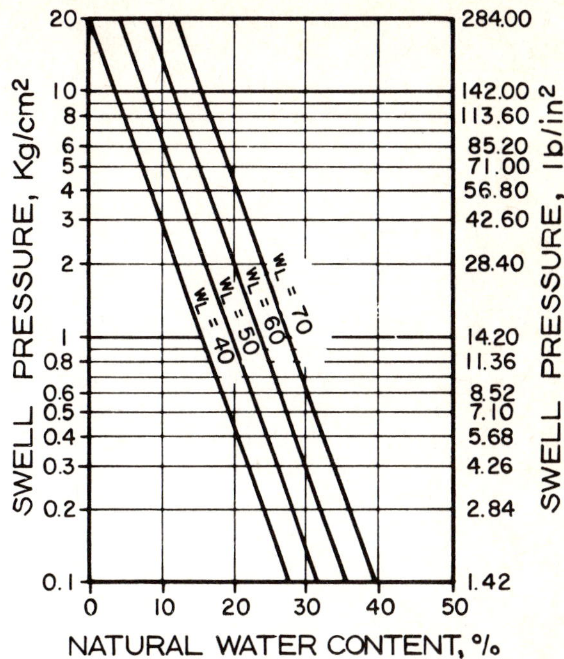

Fig. 9-52 Correlation between the swell pressure of a soil to its liquid limit and its natural water content [75]

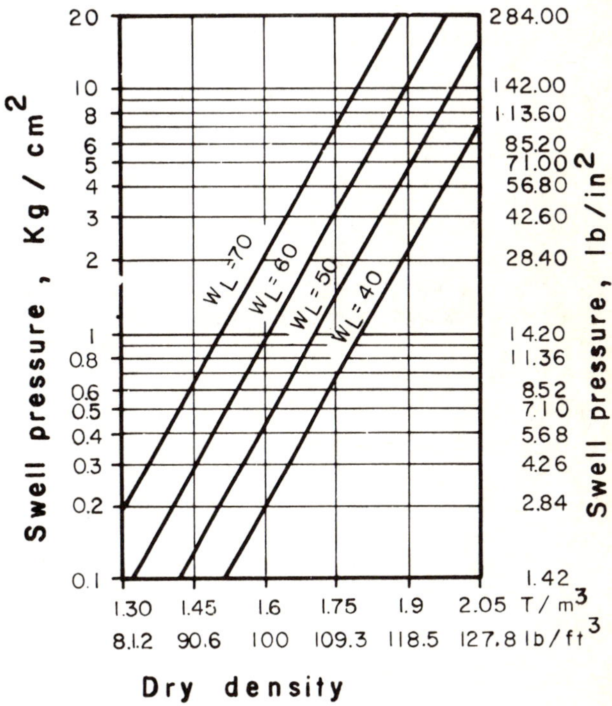

Fig. 9-53 Correlation between the swell pressure of a soil and its liquid limit and initial dry density [75]

appreciable reductions in the service life of these pavements. Figure 9-55 [75] illustrates the evolution of the Serviceability Index for pavements constructed in different sections of the same highway on expansive and non-expansive soils; the damage caused by the expansive soils can be readily appreciated.

When designing or constructing pavements on expansive soils, the chief aim is to avoid changes in water content, so that warping, distortion and cracking are reduced to a

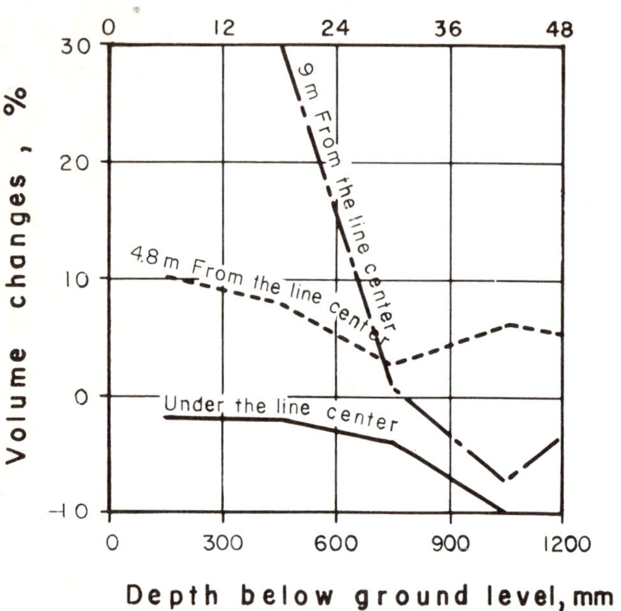

Fig. 9-54 Volume changes at several points in the cross-section of a road [76]

minimum. No method accomplishes this task with entire success, even though use of the appropriate method for a specific situation may lead to a significant reduction in distress. All the practical methods for avoiding the distress that is caused by expansion in a susceptible soil can be divided into three broad categories:

1) Replacement, or improvement by incorporating an inactive soil, of all or part of the active thickness (or of the layer of pavement that is susceptible to activity).
2) Neutralizing the previously evaluated swell pressure by placing a sufficiently heavy surcharge on the ground or the layer of pavement under consideration. The surcharge generally consists of a weight of earth or sub-base.
3) Reduction or control of the changes in water content in susceptible soils by drainage, subdrainage, utilization of impervious protective coats and other methods.

Some brief comments on these methods are given below:

Removal or improvement of soils: Replacement of expansive clay is an excellent solution from the mechanical point of view, but it is frequently impracticable, either due to cost or excavation difficulties. The material that is removed must be replaced by another, but inactive, one which will have to be duly transported, spread and compacted, all of which increases the cost of the solution. The incorporation of inactive materials in suitable proportions to reduce expansion to appropriate levels is an appropriate solution when the inactive materials are readily available, and when the problems are detected before construction commences. In Mexican engineering experience, there are many cases in which incorporation of 10 to 15% of another material has permitted the use of originally expansive soils in the body of embankments or even in the subgrade layer of many roads, with good subsequent performance. If the problem is detected in time by the geotechnical survey, a simple and rapid laboratory study will, in many cases, enable a mixture to be designed which will have good performance and will enable inexpensive soils to be used which would otherwise have been rejected.

In recent years, considerable experience has been gleaned in the reduction of soil expansion by additives which bring about a chemical reaction. Slaked lime has given good results in the treatment of layers that are not very thick (15 to 30 cm) (6 to 12 in). Solutions of this type based on soil stabilization, have the double advantage that they not only help solve expansion problems, but also enhance structural capacity and improve pavement design. It has also been seen that stabilized layers provide a useful barrier to wetting and drying. When conducting economic analyses involving stabilized layers, reductions in pavement thickness as a consequence of use should not be forgotten.

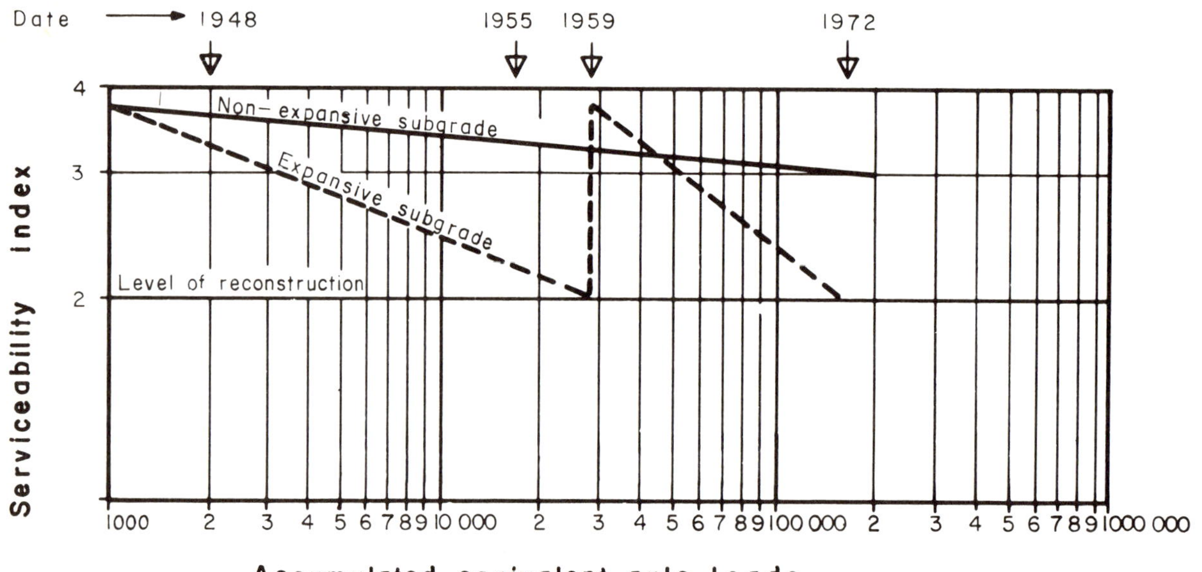

Fig. 9-55 Comparison of evolution of the serviceability index with time referring to parts of a highway built on expansive and non-expansive soils [76]

Surcharges: Neutralization of swell pressure by means of surcharges impedes expansion. The problem arises when expansive soils are located in the upper part of an embankment or in the subgrade. In cases like these, utilization of the surcharge criterion will result in very thick pavements, that will be expensive. It has already been mentioned that in HVEEM's design criterion, which is used by the California Division of Highways, this neutralizing approach has become a routine. When surcharges are used, it must be kept in mind that the shoulders of a road are subject to little loading and must also be protected, which implies the need for further large quantities of material that must be compacted.

Subdrainage and other methods: Reduction of changes in the water content of the active soil to a minimum gives magnificent results (Plate 9-15). It should be kept in mind that much of the serious distress in pavements constructed on expansive clays occurs as a consequence of changes in the water content in the embankment sides. Therefore subdrainage must be designed to give special protection in these zones. When designing drainage and subdrainage for these purposes, it is advisable to observe the following rules:

Plate 9-15 Pavement failure due to absence of or deficiency in subdrainage

1) The shoulders of the road should extend outwards significantly beyond what is geometrically convenient, perferably with a width comparable to the depth of active soil. Material for road widening must not be active.
2) Any surface drainage installations that cause an accumulation of water should be kept as far away from the pavement as possible. The same applies to wooded zones or barriers of vegetation consisting of trees or large shrubs.
3) Cut sections are more detrimental than embankment sections.
4) Any capillarity-breaking layer or one which drains away water infiltrating from the upper layers of pavement, or any impervious membrane that is included to protect the layer of active soil, should extend as far as the lateral ditch underdrains.
5) Lateral ditch underdrains may provide some water control to reduce these problems. The ditches must be filled with a relatively fine filter material, so that they will not only intercept lateral flow towards the inside of the section, but will also provide a source of moisture during the dry season.

It is helpful to construct the pavement during the time of year when the water content of the natural soils is close to the equilibrium value anticipated for the body of the embankments and the subgrade layer. The less the possibility of an expansive clay drying out, the smaller will be any future variations in the water content. Consequently any measures leading to these results are appropriate. During construction, the briefest possible period of time should be allowed to elapse between clearing and stripping the ground, spreading the material and compacting it, and the complete construction of the pavement, especially in dry hot weather. Volume changes can be effectively reduced by covering expansive clays with layers of granular material; the coarser this granular material, the better it preserves the menisci of capillary water in the upper part of the clay, thus restricting expansion.

When the clays are relatively dry, water should be added, preferably until values close to the plastic limit are reached. This procedure is difficult if it is attempted when spreading and compacting the material; it is usually difficult to increase the water content by more than 2% under these circumstances. The water content can be far more readily increased in the borrow area.

Delaying the final paving of a section constructed on expansive clays does not usually prove of great assistance in minimizing future changes in water content and is sometimes harmful. There is some disagreement as to the effectiveness of using *impervious* membranes to protect the sub-grade or the expansive material in an embankment. These membranes, which are made of asphalt, have been successfully used in Texas [77]. The use of capillarity-breaking layers may be another alternative for protecting the body of the embankment from important changes in water content. Although the amount of asphalt reported for constructing protective asphalt membranes is not very large (6 to 7 l/m^2 (1.3 to 1.5 gal/yd^2)), so that the solution is not prohibitively expensive. The capillarity-breaking layer may in the long term provide a good solution. However the hydraulic gradients due to the effects of suction in dry clays that are separated from wetter soils by an asphalt membrane, may be sufficient to encourage very large flows through the small cracks or fissures that gradually develop in this thin asphalt layer.

Reference [78] comments on the good results achieved by submerging subgrades that show a tendency to swell in sufficient water just before construction commences. This flooding brings about a large part of the possible swell so that after construction whatever volume changes may occur will be far smaller than if the method were not used. A technique must be used that will permit water to penetrate the expansive materials in a homogeneous manner. This has sometimes been achieved by boring mini-wells that are sufficiently deep, with a diameter of about 10 cm (4 in), which enable water to be incorporated into the natural ground or embankments. This method has been used in areas where high buildings are to be constructed; no experience is available in relation to highway construction. In any case, the intense wetting of expansive materials prior to construction should be regarded as an operation which improves their future behavior, at the risk of future shrinkage during dry weather if the soils become too wet.

9.10 Treatment of the Upper Part of Railroad Embankments

As is known, under the rails and cross-ties of a railroad, a layer of crushed sound material is placed which is referred to as ballast. Table 9-23 gives the usual thicknesses of ballast as a function of the annual tonnage carried by the railroad.

Table 9-23
Common thicknesses of ballast in railroads

Annual tonnage		Thickness of ballast	
$t \times 10^6$	$kips \times 10^6$	cm	in
Over 5	Over 11	40 or more	16 or more
2 to 5	4.4 to 11	30	12
Under 2	Under 4.4	20	8

A maximum particle size of 7.6 cm (3 in) is usually specified for ballast, but in our opinion it should be a little smaller, for example about 5.1 cm (2 in) or 6.4 cm (2.5 in). Ballast is usually prepared with less than 10% of the material finer than No 4 sieve, and the material passing that sieve should be controlled by its sand equivalent. The Mexican Ministry of Public Works demands a minimum sand equivalent of 80%. The minimum density of ballast material is usually controlled also, preference being given to a heavy material with mechanical properties of resistance to weathering and abrasion.

For many years, it was common practice to lay ballast directly on top of the railroad embankment. This often results in it becoming embedded in the embankment and contaminated by intrusion of fine material into the ballast voids, when the material of these embankments is fine (silts or clays). This practice, along with the insufficient compaction that was common in railroad embankments, often leads to stability problems, especially if it is recalled that ballast, by its very permeable nature, allows water to seep into the embankments, with the inevitable consequences of softening the soil.

This situation led to the once popular custom of improving the quality of the upper part of embankments as much as possible, initially by compacting it a little better, and subsequently by constructing a subgrade layer with the same functions and qualities as the subgrades for highway pavements. The construction of an appropriate subgrade led to a substantial improvement in railroad track performance. However, when the embankment or subgrade materials are fine, which is quite often the case, the problems remain of mutual embedment and loss of strength as a consequence of uncontrolled wetting.

For several years, it has been common railroad practice to construct between the subgrade layer of the embankment and the ballast, another layer, called sub-ballast. Its principal functions are:

— to enable a reduction in the thickness of ballast by providing additional structural strength with a layer of more economical material
— to provide the ballast with a supporting layer that is firmer than the subgrade, thus improving the structural strength of the entire structure
— to impede interpenetration of the ballast and the subgrade
— and to act as a drain both protecting the subgrade from any water that might reach it directly from the ballast and acting as a filter to prevent any very fine materials from rising and contaminating the ballast.

From the above it can be deduced that the functions of the layer of sub-ballast are similar to those of the sub-base in a highway. Also if the subgrade consists of materials that are reasonably resistant and stable in the presence of water, the layer of sub-ballast will not be necessary. It is only justified when the sub-grade is made up of fines: *CL, ML, OL, MH* or *CH* materials. The Mexican Ministry of Public Works specifies the same subgrade requirements for a railroad as for a highway, so that *OH* materials or *MH* or *CH* materials with a liquid limit exceeding 100% will not be included in this layer. In other words, when the subgrade is predominantly sandy, a sub-ballast layer will not be necessary.

The minimum thickness for the layer of sub-ballast is usually 30 cm (12 in), but reaches 40 cm (16 in) over weaker subgrades. Some organizations have adopted the requirement of constructing a layer of sub-ballast 20 to 30 cm thick over subgrades consisting of *GC, SP, SM* and *SC* soils, when these have questionable strength. Sub-ballasts are sometimes specified when the *CBR* of the subgrade built of these soils is lower than 20%. The requirements for sub-ballast quality frequently coincide with those for the sub-base of a highway. Normally natural materials requiring no crushing and/or sieving are preferred, but sometimes materials have to be used that do require these treatments.

APPENDIX 9A
PLATE BEARING TEST (McLeod) [1]

9a.1 General Observations

The test is used to measure the bearing value of soils at any depth, either in the natural ground, the embankment or any layer of a flexible pavement. The characteristics of the test depend on the design, as described below:

Highways:

30.5 cm (12 in) plate
0.508 cm (0.2 in) deflection limit
Ten load repetitions

Runways:
76.2 cm (30 in) plate
1.27 cm (0.5 in) deflection limit
Ten load repetitions

9a.2 Test Equipment

.1 Reaction System

This is usually provided by a truck with a minimum weight of 12 (26.4 kips) for highway tests and 20 (44 kips) for runway tests or a suitable heavy construction machine. A rigid structure must be available, that is capable of tolerating these thrusts. The bumper or framework of the vehicle or machine is usually suited to this purpose.

.2 Loading System

This consists of a hydraulic jack, with a calibrated gage to measure loads and other accessories and a set of circular plates each with a minimum thickness of 2.5 cm (1 in) and the following dimensions:

1. Highways: At least two plates will be used with diameters of 30.5 cm (12 in) and 15.24 cm (6 in).
2. Runways: At least four plates will be used, with diameters of 76.3 cm (30 in), 60.95 cm (24 in), 45.7 cm (18 in) and 30.5 cm (12 in). It is preferable to include another plate with a diameter of 15.24 cm (6 in).

.3 Deflection Measurement System

This consists of two or more micrometer dial gages with sensitivity of 0.01 mm (0.0004 in). If two are used, they should be placed opposite one another at 180°; if three are used, with an angle of 120° between them, and if four are used (which is preferable), they should be placed at right angles. The micrometers must be at a minimum distance of 2.5 cm (1 in) from the edge of the largest bearing plate. Chronometers must be used to time the test

9a.3 Test Procedure

1) The plate is centered beneath the hydraulic jack, placing it on a thin layer of fine sand or gypsum, so as to provide the bottom plate with a uniform foundation. The other plates are placed in such a way that they are concentric to this one, with decreasing diameters.
2) In order to adjust the systems of loading and control, a load producing a deflection no smaller than 0.25 mm (0.01 in) and no larger than 0.5 mm (0.02 in) is applied rapidly, and withdrawn quickly. Next, half of the previous load is applied, and the micrometers are set at zero in order to commence the test. This load is known as the adjustment or seating load.
3) Next a load producing a deflection of approximately 1 mm (0.04 in) is applied, and is sustained until the rate of deflection is of about 0.025 mm per minute (0.001 in/min), for three consecutive minutes. This load is then withdrawn and recovery is observed until a change of 0.025 mm per minute (0.001 in) is reached for an additional three minutes. The same load is applied and withdrawn in this manner six times, and all the micrometer readings are recorded.
4) The load is increased until a deflection of approximately 5.1 mm (0.2 in) is achieved, applying it and withdrawing it six times, as before.
5) Lastly, the load is increased until a deflection of approximately 10.1 mm (0.4 in) is achieved, following the same procedure.
6) In all cases, the final point of each stage is reached when a rate of deflection or recovery of 0.025 mm per minute (0.001 in/min) or lower is reached during three consecutive minutes.
7) The deflection for each load and for any time is determined by taking the arithmetical average of the readings from all the micrometer dial gages.
8) During the test, auxiliary data such as temperature, weather conditions, operator, time at which testing took place and any extraordinary conditions, should be recorded.

9a.4 Computations and Plotting of Graphs

1) For each load repetition the deflection is determined when the rate of deflection reaches 0.025 mm per minute (0.001 in/min).
2) The loads that are read on the jack gauge will be corrected using the calibration curve corresponding to the loading equipment used. This is obtained from recent calibration of the jack in the laboratory. To these corrected loads is added the adjustment load (corrected in the same manner) and the dead loads of the equipment, such as the weight of the plates, and of the hydraulic jack system. The sum of all these loads is the corrected total load. In each case, the corrected total load must be related to the corresponding deflection, obtained as indicated in *1*, tabulating the two variables for the six repetitions indicated for each case.
3) Using the above data, a correction must be determined for this deflection, plotting the corrected total loads against the deflection corresponding to the fifth load repetition (Fig. 9a-1).

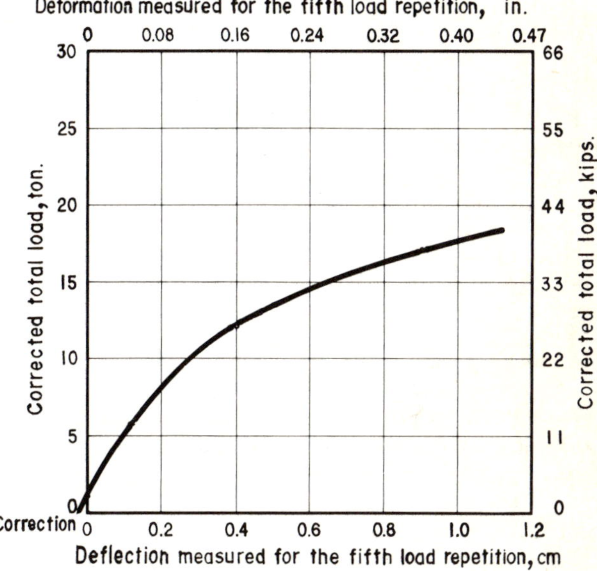

Fig. 9a-1 Determination of the correction of the initial point for the measurement of deflection

4) The correction for the deflection obtained from Fig. 9a-1 is added to each of the deflections measured, retabulating the values mentioned in *2*.
5) Graphs are plotted of the corrected deflection against the number of repetitions of each corrected total load (on a log. scale), extrapolating the straight lines that result up to ten repetitions. If a point falls outside the straight line, it is discarded (Fig. 9a-2).
6) Lastly, the corrected total loads are plotted against the deflections corresponding to ten load repetitions (Fig. 9a-3). From this graph, the bearing value of the soil (modulus of reaction) is computed, entering the pre-established deflection for each case, as was already indicated at the beginning of this supplement.
7) According to McLeod, if it is wished, the method can be extrapolated for a greater number of load repetitions.

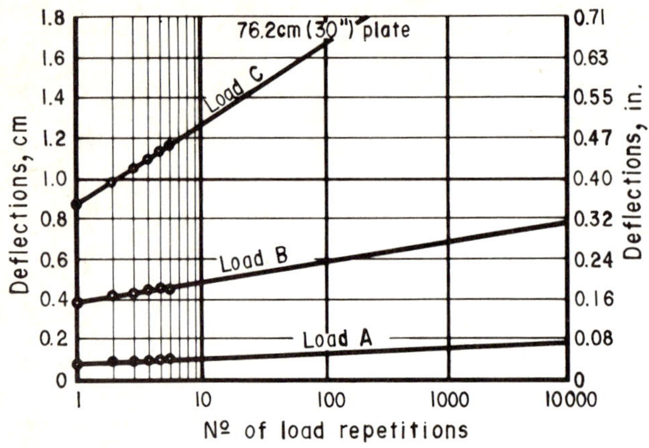

Fig. 9a-2 Influence on deflection of repeated loadings for each of the three load magnitudes

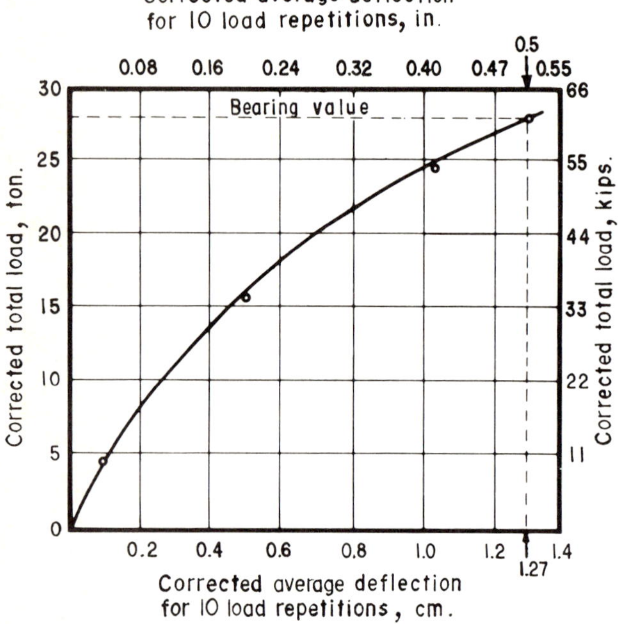

Fig. 9a-3 Corrected total load versus corrected deflection for 10 repeated loadings

APPENDIX 9B
CBR TEST (U.S. ARMY CORPS OF ENGINEERS) [7,40]

9b.1 General Observations

Experience has shown that even the smallest differences in the *CBR* test procedure can lead to enormous differences in test results. For this reason, test procedures must be followed in detail step by step, and even then difficulties may still arise. For materials such as coarse aggregates, the procedure is not entirely satisfactory. Several tests are necessary in order to determine a reasonable mean value. In some cases (in which the content of aggregates is so small that it does not affect the stability of the soil) the particles can be removed. This will avoid incongruities in test results. However, for most soils the methods presented here have proved satisfactory. The paragraphs that follow include the procedures and equipment recommended for tests conducted on remolded and compacted samples, undisturbed specimens and field determinations.

9b.2 Equipment

The equipment used for the preparation and testing of remolded specimens is as follows:

1) A cylindrical mold 15.2 cm (6 in) in diameter and 17.8 cm (7 in) high, equipped with a collar extension 5.1 cm (2 in) high and a plate. The base plate and collar can be attached to either end of the cylinder. When a group of molds is used, it is advisable to have an additional base plate, because two plates are required when the mold is inverted during preparation of the specimen.
2) A metal spacer disc 15 cm (5-15/16 in) in diameter and 6.3 cm (2.5 in) high, to act as a false bottom in the cylindrical mold during compaction.
3) A compactor similar to the one used in the AASHTO Modified compaction test weighing 4.54 kg (10 lbs) and with a diameter of 5.1 cm (2 in) on the striking face, and a drop of 45.8 cm (18 in).
4) Adjustable rod and perforated plate, tripod and micrometer with a sensitivity of 0.0025 cm (0.001 in) to measure soil expansion.
5) A ring weighing 2.27 kg (5 lb) and several specially designed ring weights, each weighing 2.27 kg (5 lb), suitable for application as a surcharge on the soil surface, during the saturation and penetration process.
6) Penetrating piston 4.9 cm (1.92 in) in diameter and approximately 10 cm (4 in) long.
7) Testing device or screw jack with its special frame, either of which can be used to introduce the piston into the specimen at a rate of 0.127 cm/min (0.05 in/min) and weigh the load with a sensitivity of 1 kg/cm^2 on the piston.
8) General laboratory equipment, such as mixing trays, spatulas, levelling instruments, scales, immersion tank, dishes for determining water content, oven, etc.

9b.3 Preparation of Remolded Test Specimens

This procedure is such that the *CBR* values are obtained from test specimens having the same density and water content as those anticipated for the field. Generally, for most materials, the most critical condition of the prototype is when it has absorbed the maximum quantity of water. For this reason and for the purpose of obtaining a conservative result, the design value adopted by the U.S. Army Corps of Engineers is the *CBR* obtained after the specimens have been soaked in water for four days. Throughout this time they are confined in a mold by a surcharge that exerts a pressure equal to that anticipated for the finished pavement. The procedure that is described has been formulated on the basis of numerous studies and should generally be followed.

1) The sample is dried until crumbly. Drying should be done in the open air or else using an oven, so long as the temperature of the sample does not exceed 60°C. Next the lumps are broken up, care being exercised not to crush the particles. Material larger than 1.9 cm (3/4 in) is removed, replacing it by an equal weight of material with sizes between the No 4 and 3/4 in sieves, mixing the sample thoroughly.
2) The method of compaction used is generally a dynamic PROCTOR type test.

The modifications made by the Corps of Engineers include changes in the weight of the tamper from 2.5 kg (5.5 lb) to 4.54 kg (10 lb), in the height of the compactor's fall from 30.5 cm (12 in) to 45.8 cm (18 in) and in the compaction of the tests specimens in the mold from three equal layers to five equal layers slightly smaller than 2.54 cm (1 in) each. Fifty five blows are applied to each layer. No material is re-used. During compaction, the mold is placed on a concrete floor or on a concrete pedestal. Wooden floors will change the result considerably.

A sufficient number of specimens is compacted with different water contents for the purpose of establishing the optimum water content and maximum density. If the compaction characteristics of the material are familiar to the operators it will be sufficient to compact four or five specimens with water contents within about two per cent of the optimum. These specimens are prepared using different compactive efforts, which are usually the PROCTOR standard effort, PROCTOR modified effort and an effort even smaller than the PROCTOR standard one. Specimens are thus obtained which, with different water contents, reach different dry densities, enabling a satisfactory study of variation of the *CBR* with water content and compactive effort which are the factors which affect it most. The height of fall of the compactor should be carefully controlled, along with the uniform distribution of the blows over the surface of the specimen.

The results are plotted on a water content versus density graph, a curve being drawn through the points thus obtained.

3) The mold with the collar extension is fixed to the base plate, and a spacer disc is inserted over this plate. A coarse filter paper or fine wire mesh is fitted on the upper surface of the disc.
4) Samples must be compacted for the *CBR* test following the procedure described in Step *2*, using the compactive efforts and water contents recommended in §9b.6. After compacting the sample, the collar is removed and the surface of the specimen is trimmed plane. A mesh or a coarse filter paper is placed on the upper surface and a perforated base plate is fixed to the upper part of the mold. The mold is inverted, removing the base plate and the spacer disc, and density is determined.
5) The adjustable rod is placed over the plate on the surface of the soil. A ring-shaped weight is applied on the plate to produce a load intensity equal to the weight anticipated for the finished pavement material, within ± 2.27 kg (5 lb), but never smaller than 4.54 kg (10 lb). The mold is then submerged with the weights, so that water has free access to the specimen from above and below. Initial measurements are taken. The specimen is soaked for 4 days and its expansion (or contraction) measured. A smaller period of immersion is permitted for pervious soils, if it is clear that the maximum water content has been reached. Finally, expansion is computed as a percentage of specimen height.
6) The water on the soil surface is then removed by tipping the mold, and the specimen is allowed to drain for fifteen minutes. Care should be exercised not to disturb the surface of the specimen during removal of free water. Both the perforated plate and the surcharge weights are withdrawn and the specimen is weighed, so that it is now ready for the penetration test.

9b.4 Penetration Test

Since the test procedure that is currently used is the same for all types of specimens, it will not be necessary to repeat it for each type of soil. The procedure described in the paragraphs that follow is also applicable to undisturbed field tests, once the test surface has been duly prepared.

1) A surcharge is applied to all the soils that is sufficient to produce a load intensity equal to that anticipated for the finished pavement (within ± 2.27 kg (5 lb)), but no smaller than 4.54 kg (10 lb). If the sample has been previously soaked, the surcharge should be equal to that applied during the saturation period. To minimize soil rising in the hole in the center of the surcharge weights, it is advisable to place a 2.27 kg (5 lb) surcharge disc with a circular perforation on the surface of the soil before introducing the piston and before applying the remaining weights.
2) The penetrating piston is introduced with an initial seating load of 4.54 kg (10 lb) and the gauges for measuring stress and deflection are set at zero. This initial load is indispensable for ensuring satisfactory settlement of the piston, and is considered as zero load when determining the pressure-penetration relationship.
3) Loading is applied to the penetrating piston so that the rate of application is approximately 0.127 cm/min (0.05 in/min). Load readings should be obtained for deformations of 0.063, 0.127, 0.190, 0.25, 0.51, 0.76, 1.02, 1.27 cm (0.025, 0.05, 0.075, 0.1, 0.2, 0.3, 0.4 and 0.5 in). In manually operated devices, it may be necessary to take load readings at smaller intervals so as to control the penetration speed.
4) The water content in the upper 2.5 cm (1 in) is determined, and in the case of laboratory tests, a mean water content for the entire thickness of the sample is also determined.
5) The pressure applied by the penetrometer is computed and the stress-penetration curve is plotted. To obtain real penetration pressures from the test data, the zero is adjusted to correct for surface irregularities affecting the initial shape of the curve.
6) Corrected pressure values are determined for 0.25 and 0.51 cm (0.1 in and 0.2 in) penetration, from which the *CBR* values are obtained by dividing these pressures by the standard ones of 70 and 105 kg/cm^2 (1,000 and 1,500 psi respectively). Each ratio is multiplied by 100 to obtain the percent relation. Generally the *CBR* is selected for 0.25 cm (0.1 in) penetration. If the *CBR* for 0.51 cm (0.2 in) penetration is greater than the previous one, the test must be repeated. If the test repetition gives similar results, the *CBR* for 0.51 cm (0.2 in) is used.

9b.5 Test Data and Results

The test data and results that should be given are as follows:

1) Compaction procedure (if different from that specified).
2) Compactive effort
3) Water content on compaction of the specimen
4) Density
5) Surcharge during saturation and penetration
6) Expansion of sample
7) Water content after saturation
8) Optimum water content and maximum density, determined by means of the AASHTO Modified compaction test, described in §9b.3
9) Pressure-penetration curve (Fig. 9b.1)

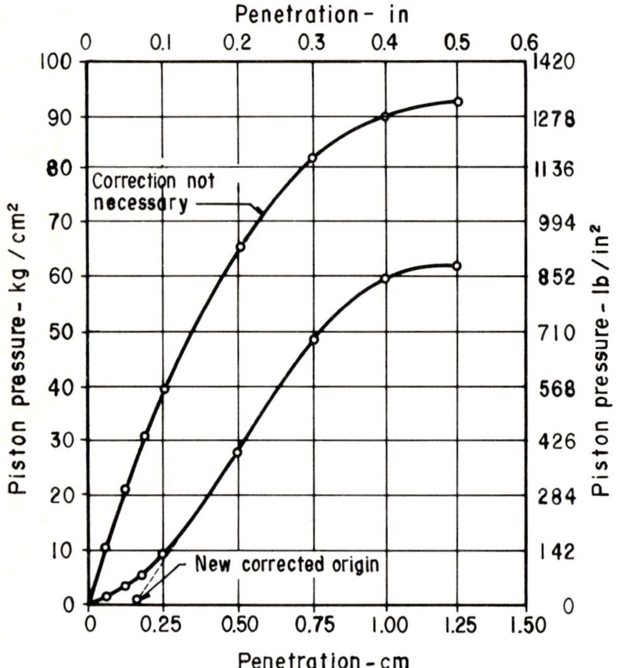

Fig. 9b-1 Pressure-penetration graph in a *CBR* test

9b.6 Preparation Procedure for Remolded Samples

When testing remolded specimens by the California method, all the subgrade layers and base courses have been grouped into three categories according to their behavior during saturation:

1) Cohesionless sands and gravels, 2) cohesive soils and 3) highly expansive soils. The first category generally includes *GW, GP, SW* and *SP* soils. The second category includes *GM, GC, SM, SC, ML, CL* and *OL* soils. Highly expansive soils generally include those classified as *MH, CH* and *OH*. The procedure for each of these categories is given separately.

Cohesionless sands and gravels: Generally speaking, cohesionless soils can be easily compacted using special rollers or by means of traffic, until they reach their maximum density specified by the AASHTO modified method. The test consists of 55 blows per layer, with a water content corresponding to saturation of the sample, so as to achieve maximum density. If saturation does not lower the *CBR* of a cohesionless sand or gravel, it can be omitted in any subsequent tests conducted on the same material.

Cohesive soils: The soils in this category are tested in order to obtain data to show their behavior with a range of water contents anticipated for field conditions. Compaction curves are plotted for 55, 25 and 10 blows per layer, each specimen being soaked and penetrated so as to obtain a complete family of curves that will show the relationship between density, water content and *CBR*. To help determine the validity of compaction data, maximum density versus compactive effort (work per unit volume) are plotted on a semi-logarithmic scale. The points thus obtained generally give a straight line.

Expansive soils: Test procedures for highly expansive soils are the same as described for cohesive soils. However, the objectives of the test program are not exactly the same. Tests on expansive soils are aimed at determining the water content and density that produce minimum expansion. The water content and density appropriate for this case are not necessarily the optimum values obtained from the AASHTO modified test. Generally, minimum expansion and maximum saturated *CBR* values correspond to a water content that is slightly higher than optimum. When highly expansive soils are tested, it may be necessary to prepare samples for a wider range of water contents and densities than those normally used, so that the relationship between the water content, density, expansion and *CBR* of a given soil can be established.

A careful study of test results obtained by an experienced engineer will permit the selection of the water content and density appropriate for field requirements. It should be observed that there is a possibility that the design thickness may depend on compaction requirements instead of the *CBR* in some cases.

9b.7 Procedure for Preparing Undisturbed Samples

Tests on undisturbed samples of existing soil are used during design when no compaction is required, and also to correlate field tests with the water content prevailing at the time of field testing with the results that would be obtained for these some soils using the design water content. For this last purpose, two tests must be conducted on the specimens, one with the design water content and another with the field water content, so as to determine the correlation necessary for interpretating field tests. In this case, the reduction in *CBR* that occurs during wetting should be applied as a correction factor to the field tests.

If it is wished to reduce specimen disturbance to a minimum, great care and considerable patience will be necessary. Undisturbed samples can be satisfactorily obtained using steel cylinders, flexible unfoldable galvanized metal covers or specially designed boxes. If no adequate lateral support is provided at the sides of the sample, variable non-representative *CBR* values will be obtained. For fine-graded materials, the use of molds and metal covers is satisfactory. The annular space around the sample (cut or carved from a pedestal) can be filled with paraffin or with a mixture of paraffin and 10% resin to provide support. For coarse (gravelly) soils the box method is recommended. The sample is covered with waxed paper or paraffin so as to avoid loss of moisture during transportation to the laboratory.

Saturation or penetration tests are carried out as previously explained, after removal of the paper or paraffin from the surface of the specimen in the case of molds or metal covers, or else after the surface of the samples in boxes has been levelled off with a thin layer of sand, when necessary. Calculations and test results are recorded as indicated earlier.

9b.8 Field Tests

Under certain conditions, the field test is a satisfactory test for determining the in situ *CBR* of a material. Basically the procedure for penetration is the same as that described in §9b.4.

The field test can be used in any of the following conditions:

1) When the field density and water content are such that the degree of saturation is 80% or higher.
2) When the material is coarse and cohesionless, so that it is not affected by changes in water content.
3) When the material has been at the same location for several years. In these cases the water content may fluctuate between narrow limits. The field test is regarded as providing a satisfactory indication of the bearing capacity.

APPENDIX 9C
HVEEM'S TESTS

9c.1 Introduction

F. N. HVEEM's method for designing pavement thicknesses is based on determination of the confined strength of the materials that will be used in the pavement, by means of the following laboratory tests.

9c.2 Determination of Swell Pressure and *R* Value

.1 General Observations

This test procedure is recommended for soils that are to be used in base courses, sub-bases or subgrades for a road structure.

.2 Test Equipment

1) Mechanical compactor and accessories.
2) Loading device with minimum capacity of 5 t (11 kips) and a sensitivity of 0.05 t.
3) Compaction molds with an inside diameter of 10.16 cm (4 in) and 12.70 cm (5 in) high.
4) Rubber disc 10.16 cm (4 in) in diameter.
5) Metal extension 10.1 cm (3-31/32 in) in diameter by 12.5 cm (5 in) high.
6) Exudation apparatus.
7) Perforated bronze and filter paper discs.
8) Swell pressure apparatus and accessories (Fig. 9-16).
9) Test ring having a strain of 0.001 cm per kg (0.00018 in/lb).
10) HVEEM's stabilometer and accessories (Fig. 9-15).
11) Hollow metal cylinder 10.16 cm (4 in) in diameter by 15.24 cm (6 in) high.
12) Micrometer dial gages, an ALLEN key, scales, trays, spoons and spatulas.

.3 Preparation of Soil Specimens

1) Soil is prepared for the test by separating from the coarse aggregates the fine particles adhering to them. Lumps of clay are broken up so that they will pass N° 4 sieve.
2) The grain-size distribution of the soil is adjusted when part of the material is retained on the 19.1 mm (3/4 in) sieve as follows. If 75% or more passes this sieve, the part of the sample passing the sieve is retained in the test specimen; if less than 75% of the sample passes the sieve, the fraction of the sample passing the 25.4 mm (1 in) sieve is used.
3) A quantity of material sufficient to form a compacted specimen 10.16 cm (4 in) in diameter by 6.35 cm (2.5 in) high is thoroughly mixed with about 1/2 to 2/3 of the amount of water required to saturate it. The sample is placed in a covered jar and is allowed to rest 12 hours. Then a quantity of water sufficient to saturate it is added.
4) The soil is compacted in the mold by means of a mechanical compactor as follows:

— The mold is placed in its holding device, inserting the rubber disc into the base of the mold.

— The compactor is started, adjusting air pressure to 1.05 kg/cm^2 (15 psi)

— The material is emptied into the mold in 20 equal parts, so that each part corresponds to one blow from the compactor. Once this operation is completed, 10 more blows are given to settle and level the material.

— The rubber disc is placed on the upper part of the specimen and 100 blows are applied using an air pressure that is calibrated to obtain a tamping pressure of 24.5 kg/cm^2 (350 psi).

— If free water appears around the base of the mold before 100 blows have been applied, compaction of the sample must immediately be stopped and the number of blows applied recorded.

5) The exudation pressure of the compacted specimen is determined in the following manner:

— The mold is removed from the mechanical compactor, the compacted surface of the specimen is levelled and smoothed off using a flat-ended bar.

— The perforated bronze disc is placed on top of the compacted specimen in the mold and over it is placed a piece of filter paper.

— The mold is inverted, so that the filter paper is at the bottom, and is placed on the contact plate of the exudation apparatus.

— The contact plate, with the mold containing the specimen, is placed on the compressor plate, centering it so that loading will be uniform. The metal extension is placed on top of the specimen, and the specimen is lowered to meet the contact plate. Next the head of the compressor is lowered to meet the metal extension.

— A load increment of 910 kg per minute (2000 lb/min) is applied to the specimen. The pressure at which five of the six outer lights of the exudation apparatus come on is recorded as the exudation pressure. However, if free water appears around the base of the mold and at least three lights come on, the load at that moment is recorded and exudation pressure computed. The light bulbs are connected individually by two conductor wires to the upper part of the sample. When water appears on the surface, contact is established between the terminals of the wires through the water, and the bulbs light up.

6) Guided by the results obtained for the previous specimen, three more are prepared with different water contents, so as to obtain a set of exudation pressures ranging between 7 kg/cm^2 (100 psi) and 56 kg/cm^2 (800 psi). These specimens are then subjected to the swell test and the stabilometer test.

.4 Swell Pressure Test

1) The specimen is allowed to rest for at least half an hour after conclusion of the exudation test.
2) The extensometer is placed on the upper bar of the swell pressure device.
3) The rod and the perforated disc are placed on the surface of the compacted specimen in the mold. The mold is placed in the swell pressure apparatus.
4) The rotating plate is turned until a strain of 0.002 cm (0.001 in) is generated in the specimen.
5) Approximately 200 cm^3 of water are poured on to the specimen in the mold and swell pressure is allowed to develop for 16 hours.
6) At the end of the saturation period, the strain on the steel bar is read to within ± 0.00025 cm (0.0001 in). When the displacement is in excess of 0.025 cm (0.01 in), the swell pressure apparatus should be recalibrated.
7) Swell pressure is determined using the following expression:

$$p_e = kd$$

where: p_e is the swell pressure shown by the soil in kg/cm^2 (lb/in^2); k the calibration constant of the steel bar (kg/cm^2 per 0.00025 cm) (lb/in^2 per 0.0001 in); d the strain in 0.00025 cm (0.0001 in) read on the extensometer.

.5 Stabilometer Test

1) After the swell test, water is eliminated from the mold containing the specimen, and with the help of the metal extension the sample is transferred to the stabilometer.
2) The metal extension is placed on top of the specimen, which is then positioned in the loading machine.
3) The head of the loading machine is lowered until it adjusts to fit the metal extension.
4) A horizontal pressure of 0.35 kg/cm^2 (5 psi) is applied to the specimen by means of the stabilometer pump.
5) A vertical load is applied at a velocity of the extension of 0.12 cm per minute (0.05 in/min).
6) The horizontal pressure is recorded when the vertical load is 910 kg (2000 lb); (11.2 kg/cm^2) (160 lb/in^2).
7) Loading is stopped at 910 kg (2,000 lb) and is immediately reduced to 455 kg (1,000 lb). Using the stabilometer pump, the horizontal pressure is reduced to 0.35 kg/cm^2 (5 psi). This leads to a further reduction in vertical load that should be disregarded.
8) The pump handle is turned until horizontal pressure increases from 0.35 kg/cm^2 (5 psi) to 7 kg/cm^2 (100 psi). Note is taken of the number of turns required for this purpose. This quantity is displacement D in turns of the specimen.
9) The R value (resistance) is computed using Eq. (9-4).

9c.3 Determination of the Cohesiometer Value

.1 General Observations

The cohesiometer test provides a measure of the tensile strength of a compacted asphaltic mixture, a cemented soil or any mixture of aggregates. Generally the test is not performed on poorly cemented materials, but cohesiometer values are assigned to them on the basis of experience.

.2 Test Equipment

1) Cohesiometer with complete equipment (Fig. 9-17)
2) Thermometer
3) Scales with a capacity of 10 kg (22 lb) (sensitivity of one gram).

.3 Test Procedure

1) Cohesiometer tests are conducted on the specimens previously used in the stabilometer.
2) The sample is put in an oven at 60°C for two hours.
3) The cohesiometer device is calibrated in such a way that shot flows into the receptacle at the end of the lever arm at a rate of 1,800 ± 20 grams per minute.
4) The heating unit of the cohesiometer chamber is adjusted to maintain a steady temperature of 60°C.
5) The system is secured with a safety catch. The specimen is taken out of the oven and is securely centered with the upper plates parallel to the upper face of the test tube. The temperature in the cohesiometer chamber is allowed to reach 60°C ± 1° before testing commences.
6) The safety catch is released and the shot is allowed to fall into the receptacle until the specimen in the test tube breaks as indicated by a sudden drop of the lever.
7) Should the specimen be flexible or ductile rather than crumbly, the fall of the shot is stopped when the end of the lever has dropped 12 mm (0.47 in) below the horizontal.
8) The shot in the receptacle is weighed to within ± 1 gram, and this weight is recorded.

.4 Calculation

The cohesiometer value is calculated using Eq. (9-5) which was given in the text of this chapter.

APPENDIX 9D CONSIDERATION OF SOIL EXPANSION IN DESIGN THICKNESSES

The Israel Highway Department [60] has proposed a method for taking soil expansion into account when using the *CBR* design method of the U.S. Army Corps of Engineers, which was referred to in the text of this chapter.

In this design method, the relationships between the water content, dry density, *CBR* of saturated specimens and soil expansion characteristics are determined. The Corps of Engineers method includes determination of the *CBR* of the expanded soil in their test conditions. In this way it becomes possible to draw a figure like Fig. 9d-1 for the soil tested [80].

This figure gives the usual compaction graphs corresponding to (1) the AASHTO modified test (2) the AASHTO standard test, (3) a test employing intermediate effort that is used by the Corps of Engineers, and (4) the dynamic test that is performed by the Mexican Ministry of Public Works. It also gives *CBR* curves, obtained by testing each saturated specimen that has been soaked in water for four days. The expansion of these same specimens has been measured by recording the difference in the heights of soil inside the mold. Equal expansion curves have also been plotted in the figure. Swell percentages correspond to the figures inside circles, while equal *CBR* values correspond to those in squares.

Under these conditions a suitable maximum tolerable expansion percentage can be determined. In this figure, it is 2%, which is the Israeli criterion. This criterion establishes as a boundary, a vertical line tangent to the 2% expansion line. To the left of this line no *CBR* is acceptable. A further two horizontal boundaries are established by considering the compaction specification for each case. In the figure it is assumed that compaction will be controlled by the Mexican Ministry of Public Works dynamic test, which demands in-situ compaction of between 100% and some other lower percentage, corresponding to dry densities of 1,760 and 1,697 kg/m^3 (110 and 106 lb/ft^3). Compaction specifications are often guided by the shape and layout of the equal expansion curves, because compaction percentages that are high in relation to a certain control curve might fall into zones where expansion is excessive.

In this manner 17% is defined in the figure as the minimum tolerable water content for soil compaction. Consideration of the maximum tolerable water content for soil compaction in the field provides the criterion for fixing the ultimate vertical boundary which limits the shaded zone in the figure. It can be readily seen that the higher the water content, the less will be the future expansion of the soil; however, if water contents

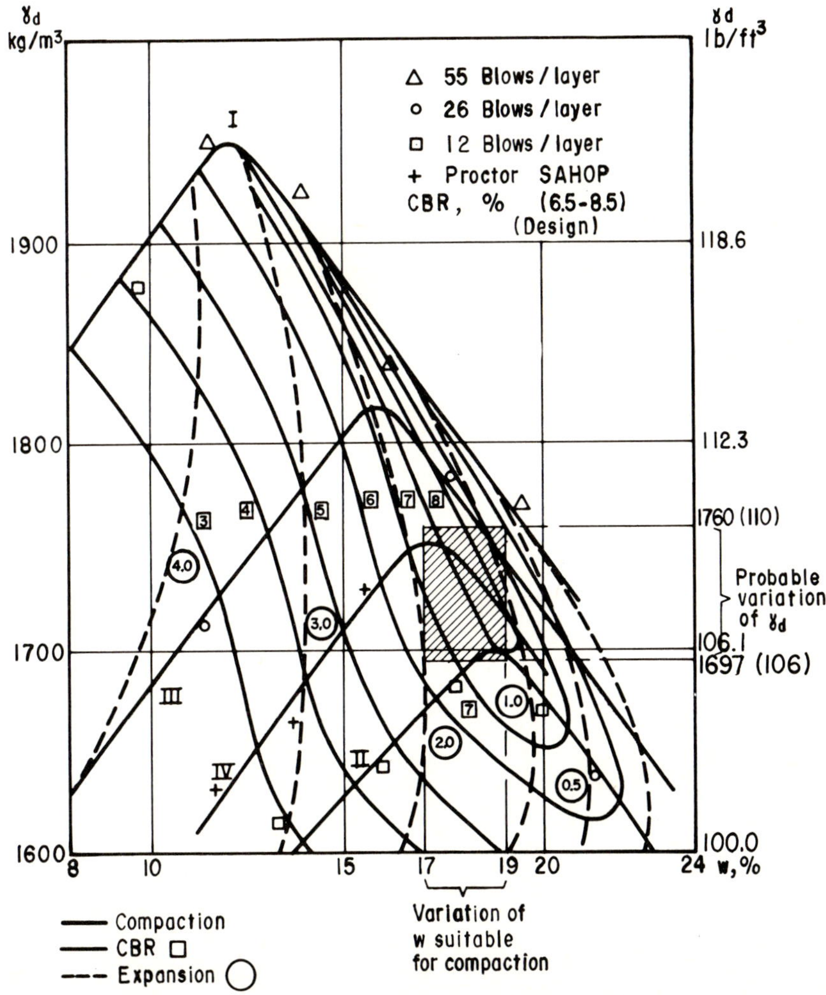

Fig. 9d-1 Handling of data for finding thicknesses of pavement, considering the expansion of the soils. Method of U.S. Corps of Engineers [80]

are too high zones with a low *CBR* will be invaded. If this situation is tolerated it will lead to uneconomical designs. In the case of this figure, 19% is regarded as the maximum tolerable water content for field compaction, in accordance with this criterion.

In this way the shaded zone in Fig. 9d-1 is determined and the entire range of *CBR* values within that zone (6.5 to 8.5) must be anticipated and taken into account in design. In the example a *CBR* of 7% was adopted for use in the design graphs.

The above data and criteria are applied using the normal curves appearing in the text of this Chapter, with the advantage that the limitation due to expansion is taken into account, so long as the conventional method of design that is employed allows for soil expansion in its laboratory testing.

APPENDIX 9E SOME PROBLEMS AND SOLUTIONS

9e.1 HVEEM's Method

Design the structural section of a flexible pavement, using the California Division of Highways method (HVEEM) for a road over ground consisting of soils with the following characteristics:

.1 Embankments

The embankment materials are mostly of volcanic origin and made up of inorganic silts with medium plasticity and low to high compressibility (*ML* and *MH*) and high resilience. Some mixtures of soils and small rock fragments with very variable properties are also present.

.2 Traffic Analysis

In order to conduct a traffic analysis for this road the results of vehicle counts performed by the corresponding authority during 1972 are used.

In HVEEM's (California) method, traffic is expressed in terms of the number of equivalent wheel loads of 5,000 lb (*EWL*) that are anticipated for the design life of the pavement. Once the *EWL* has been determined, the traffic index is computed using Eq. (9-1). In this method the effect of traffic action on the pavement is not considered.

The traffic analysis based on Fig. 9-10 is shown in Table 9e-1. The equivalency factors for each type of vehicle are obtained from Table 9-4.

In this problem an annual rate of growth of 7% is presumed which is constant for the different types of vehicles, and a design period of 10 years.

Calculation of the traffic growth factor for the 10-year period has not been included in detail in the text of this chapter, because it is not a conceptual aspect of the subject. The opportunity will be taken here to illustrate how it is obtained by traditional methods. In a subsequent problem, a simpler alternative method developed by the Engineering Institute of the Mexican National Autonomous University will be presented.

The formula for obtaining the traffic growth factor is:

$$F_p = \frac{1 + \dfrac{(ADT)_f}{(ADT)_i}}{2}$$

where *ADT* is the average daily traffic, which is the traffic that uses the road on an average day of the year. Indices *f* and *i* refer to the final and initial moments of the 10-year period. The above expression is not computed; we regard it as beyond the specific interests of this book.

Now the value of the Traffic Index (*TI*) will be computed using Eq. (9-1). For this the value of *EWL* is required, and

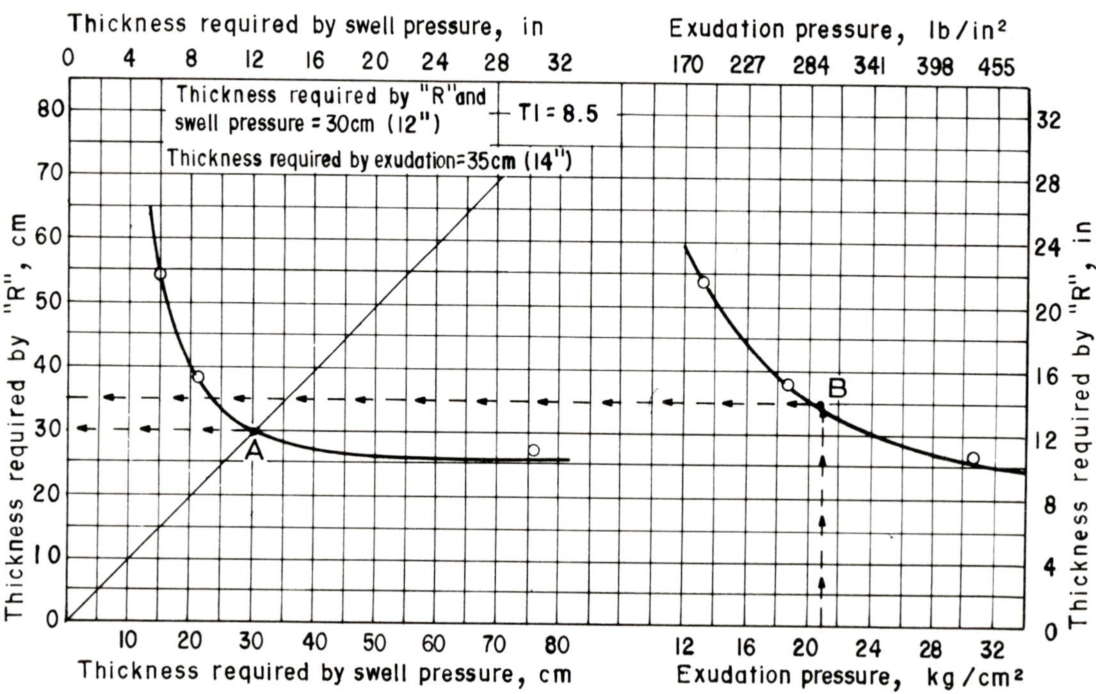

Fig. 9e-1 Study of pavement thicknesses (gravel equivalent)

this in turn must be evaluated in terms of the traffic growth factor (F_p) for the design period (in this case 10 years) and of the load equivalency factors for the different vehicles in relation to a single-axle load of 2,270 kg (5,000 lb) per wheel, which are given in Table 9-4.

The formula for calculation of *EWL* is:

$$EWL = p \sum EWL \cdot F_p$$

where p represents the 10-year period and F_p is computed with the expression that was given above. In the problem, assuming that the traffic increases at an interest rate of 7% per year:

$$(ADT)_f = (ADT)_i (1 + 0.07)^{10}$$

Giving the ratio:

$$\frac{(ADT)_f}{(ADT)_i} = 2$$

Therefore:

$$F_p = \frac{1 + 2}{2} = 1.5$$

and:

$$EWL = 10 \times 1.5 \times 433{,}630 = 6{,}504{,}450$$

Now the value of *TI* can be calculated using Eq. (9-1).

$$TI = 6.7 \left(\frac{EWL}{10^6}\right)^{0.119} \quad (9\text{-}1)$$

$$TI = 6.7\,(6.5)^{0.119} = 8.5$$

.3 Thickness Analysis of the Pavement Structure

The pavement thickness analysis was performed taking into consideration the laboratory results of R value (stability), swell pressure and exudation pressure tests, all of which were conducted on each of three specimens prepared with material from the road embankments. The specimens have different water contents.

The thickness of the structural section, for the R concept, was determined using Eq. (9-8), in which a traffic index of 8.5 and the stability value, Eq. (9-4), corresponding to each specimen tested were used. Thus three thicknesses are obtained for the R concept. Similarly the three thicknesses of pavement required to overcome the swell pressures that developed in the test specimens were determined by applying Eq. (9-7).

In the left-hand portion of Fig. 9e-1 thickness curves for R and for expansion are presented, Point A determining the thickness that satisfies expansion and stability requirements simultaneously, which in this case is 30 cm (12 in) of gravel equivalent.

The pavement thickness relating to exudation pressure was determined by means of the stability thickness-exudation pressure curve, considering a value of 21 kg/cm^2 (300 lb/in^2) for this exudation pressure, as was indicated in the text of this chapter. The total pavement thickness in *gravel equivalent* is found to be 35 cm (13.8 in) under these conditions. When comparing the two pavement thicknesses obtained, a structural section of 35 cm (13.8 in) of gravel equivalent will be used for design.

To calculate the thicknesses corresponding to the surface course, base course and sub-base, the following procedures are observed:

Surface course: For construction of the surface course, asphalt concrete will be used over a base course of crushed material with an R stability value of 82 found in the laboratory. With a traffic index $TI = 8.5$, using the chart in Fig. 9-30, this gives a thickness of 15 cm of gravel equivalent for the surface course. Using Table 9-11 a gravel factor of 1.95 is found for the asphalt concrete. The thickness of the asphalt concrete is 15/1.95 = 7.7 cm (3 in). A thickness of surface course of 7.5 cm (3 in) will be used in design.

Base course: for the base course, a good-quality crushed material with a gravel factor of 1.1 will be used (Table 9-11). The thickness of the base course in gravel equivalent is determined as follows:

The thickness of the base course is 35 − 15 = 20 cm (8 in) and the actual thickness of the base course = $\frac{20}{1.1}$ = 18.5 cm (7.3 in).

In summary, the pavement structure will be as follows: A wearing course of good-quality asphalt concrete 7.5 cm (3 in) thick over a hydraulic base course of crushed gravel 18.5 cm (7.3 in) thick.

9e.2 Asphalt Institute Method

Design the structural section of a flexible pavement using the Asphalt Institute Method. The physical characteristics of the natural soil are described in Problem No *1*, §9e.1.

.1 Traffic Analysis

The procedure consists of calculating the Design Traffic Number (*DTN*) which is obtained from the volume of mixed traffic converted to equivalent single-axle loads of 8.2 (18,000 lb), according to factors supplied by that method. The initial Average daily Traffic $(ADT)_i$ is 12,240 vehicles, according to Problem No *1* (Table 9e-1). The number of heavy trucks in the design lane is calculated using the following:

$$N = (ADT)_i \times \frac{A}{100} \times \frac{B}{100}$$

Where: A is the percentage of heavy trucks in the two directions and B the percentage of heavy trucks in the design lane.

In this case the heavy vehicles total 1,664 (Table 9e-1) and therefore: $A = 13.6\%$. Table 9-12 enables B to be computed. If the design is a four-lane highway, the result is: $B = 45\%$.

Therefore:

$$N = 12240 \times \frac{13.6}{100} \times \frac{45.0}{100} = 750$$

with this value the nomograph in Fig. 9-33 is then used to obtain the Initial Traffic Number (*ITN*). For this purpose, the mean weight of the heavy trucks is assumed to be 15 t (33 kips).

Table 9e-1
Traffic analysis based on Fig. 9-10, used in Problem 9e.1 based on HVEEM's method

Type of vehicle	Average daily traffic in two directions	Average daily traffic in one direction		Constant *EWL*		(*EWL*)
A	10,576	5,288		—		—
B	684	342	×	280	=	95,760
C2	734	367	×	280	=	102,760
C3	116	58	×	930	=	53,940
T2S2	28	14	×	1,320	=	18,480
T3S3	102	51	×	3,190	=	162,690
Total	12,240	6,120			$\sum CE =$	433,630

Using the nomograph, the result is:

$$ITN = 480 \text{ (corresponding to intense traffic)}$$

Design period = 10 years

Annual rate of growth = 7%

Adjustment factor for initial traffic = 0.69 (Table 9-13)

$$DTN = ITN \times 0.69 = 480 \times 0.69 = 331$$

.2 Structural Analysis of the Pavement

For designing the total thickness, an R (stability) value of 55 is found from laboratory tests for the subgrade material, with 331 for the Design Traffic Number. Using the nomograph in Fig. 9-32, thickness of 19 cm (7.5 in) (in asphalt concrete) is obtained.

A hydraulic base course made of high-quality crushed material will be employed. In the graph in Fig. 9-34, the minimum thickness of asphalt surface course required over a high-quality hydraulic base course is obtained. The minimum thickness of asphalt concrete is 14 cm (5.5 in) and therefore, the thickness of granular base course is 19 − 14 = 5 cm (in asphalt concrete) (7.5 − 5.5 = 2 in).

The equivalency factor corresponding to a high-quality granular base course is obtained using Table 9-14; it is 2.01. The actual thickness of the granular base course is: 5 × 2.01 = 10 cm (2 × 2.01 = 4 in).

In summary, the pavement structure will consist of an asphalt concrete wearing course 14 cm (5.5 in) thick over a hydraulic granular base course 10 cm (4 in) thick.

9e.3 N.A.U.M Method

Design of the structural section of a road with a flexible pavement using the procedure recommended by the Engineering Institute of the Mexican National Autonomous University. The physical characteristics are the same as those described in Problem No *1*, §9e.1.

.1 Traffic Survey

In this procedure the volume of actual mixed traffic is converted to equivalent single-axle loads of 8.2 t (18000 lb), by the appropriate application of the coefficients of traffic damage for typical vehicles (Fig. 9e-2). For this case, the following classification of two-directional traffic is considered, which has already been presented in Table 9e-1

$$A \begin{cases} A_p = 8{,}460 \\ A_c = 2{,}116 \end{cases}$$
$$B = 684$$
$$C_2 = 734$$
$$C_3 = 116$$
$$T_2S_2 = 28$$
$$T_3S_3 = 102$$
$$\overline{\quad 12{,}240}$$

$$ADT = 12{,}240, \text{ mixed traffic, in two directions}$$
$$A_c = 0.20A \text{ (Problem No } 1)$$
$$A_p = 0.80A \text{ (Problem No } 1)$$

The Engineering Institute of the Mexican National Autonomous University has based its work on the classification of traffic and the damage coefficients for different types of vehicles that can be obtained from AASHTO tests (Fig. 9-10), to develop its own chart of vehicle classification and coefficients of damage. On the basis of experiements conducted on the circular test road described in the text of this chapter, and the study of performance of experimental sections it studies at different points throughout the Mexican national highway network [15], the Engineering Institute has re-examined, utilizing Mexican experience, the evaluation of damage to pavements caused by different vehicles. The data show coefficients of damage for depths of 0, 15, 22.5 and 30 cm (0, 6, 9 and 12 in). Their chart, which is conceptually similar to the one in Fig. 9-10, appears in Fig. 9e-2 and is used to solve this problem. It was taken from [81].

Determining the number of equivalent 8.2 t (18000 lb) axle loads: Table 9e-2 presents the procedure for converting mixed traffic to the corresponding equivalent 8.2 t (18000 lb) single-axle loads, using the design lane (*ADT*). In this table it is considered that all the vehicles are loaded and travel in two directions.

The number of vehicles in the design lane (Column *3*) is obtained by multiplying the *ADT* in Column (*2*) by the coefficient of 40% distribution, which was selected as a function of the number of lanes (4 in this case), in accordance

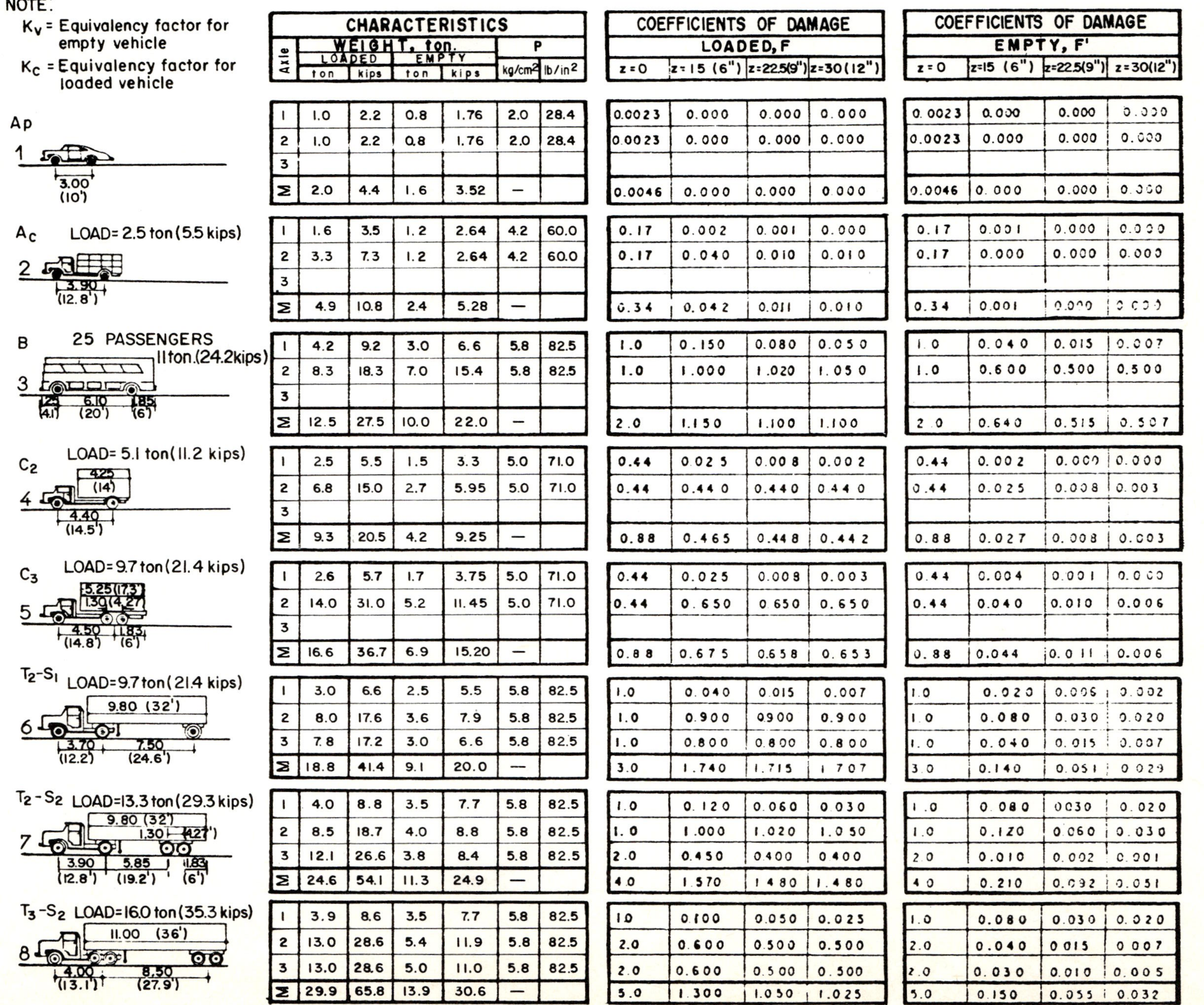

NOTE:
K_v = Equivalency factor for empty vehicle
K_c = Equivalency factor for loaded vehicle

Vehicle	Axle	Characteristics: Weight, ton — Loaded (ton)	Loaded (kips)	Empty (ton)	Empty (kips)	P (kg/cm²)	P (lb/in²)	Coefficients of Damage, Loaded, F: z=0	z=15 (6")	z=22.5 (9")	z=30 (12")	Coefficients of Damage, Empty, F': z=0	z=15 (6")	z=22.5 (9")	z=30 (12")
Ap 1 — 3.00 (10')	1	1.0	2.2	0.8	1.76	2.0	28.4	0.0023	0.000	0.000	0.000	0.0023	0.000	0.000	0.000
	2	1.0	2.2	0.8	1.76	2.0	28.4	0.0023	0.000	0.000	0.000	0.0023	0.000	0.000	0.000
	3														
	Σ	2.0	4.4	1.6	3.52	—		0.0046	0.000	0.000	0.000	0.0046	0.000	0.000	0.000
Ac 2 — LOAD= 2.5 ton (5.5 kips); 3.90 (12.8')	1	1.6	3.5	1.2	2.64	4.2	60.0	0.17	0.002	0.001	0.000	0.17	0.001	0.000	0.000
	2	3.3	7.3	1.2	2.64	4.2	60.0	0.17	0.040	0.010	0.010	0.17	0.000	0.000	0.000
	3														
	Σ	4.9	10.8	2.4	5.28	—		0.34	0.042	0.011	0.010	0.34	0.001	0.000	0.000
B 3 — 25 PASSENGERS 11 ton (24.2 kips); 1.25 (4.1'), 6.10 (20'), 1.85 (6')	1	4.2	9.2	3.0	6.6	5.8	82.5	1.0	0.150	0.080	0.050	1.0	0.040	0.015	0.007
	2	8.3	18.3	7.0	15.4	5.8	82.5	1.0	1.000	1.020	1.050	1.0	0.600	0.500	0.500
	3														
	Σ	12.5	27.5	10.0	22.0	—		2.0	1.150	1.100	1.100	2.0	0.640	0.515	0.507
C_2 4 — LOAD= 5.1 ton (11.2 kips); 4.25 (14); 4.40 (14.5')	1	2.5	5.5	1.5	3.3	5.0	71.0	0.44	0.025	0.008	0.002	0.44	0.002	0.000	0.000
	2	6.8	15.0	2.7	5.95	5.0	71.0	0.44	0.440	0.440	0.440	0.44	0.025	0.008	0.003
	3														
	Σ	9.3	20.5	4.2	9.25	—		0.88	0.465	0.448	0.442	0.88	0.027	0.008	0.003
C_3 5 — LOAD= 9.7 ton (21.4 kips); 5.25 (17.3'); 1.30 (4.27'); 4.50 (14.8'); 1.83 (6')	1	2.6	5.7	1.7	3.75	5.0	71.0	0.44	0.025	0.008	0.003	0.44	0.004	0.001	0.000
	2	14.0	31.0	5.2	11.45	5.0	71.0	0.44	0.650	0.650	0.650	0.44	0.040	0.010	0.006
	3														
	Σ	16.6	36.7	6.9	15.20	—		0.88	0.675	0.658	0.653	0.88	0.044	0.011	0.006
T_2-S_1 6 — LOAD= 9.7 ton (21.4 kips); 9.80 (32'); 3.70 (12.2'); 7.50 (24.6')	1	3.0	6.6	2.5	5.5	5.8	82.5	1.0	0.040	0.015	0.007	1.0	0.020	0.006	0.002
	2	8.0	17.6	3.6	7.9	5.8	82.5	1.0	0.900	0.900	0.900	1.0	0.080	0.030	0.020
	3	7.8	17.2	3.0	6.6	5.8	82.5	1.0	0.800	0.800	0.800	1.0	0.040	0.015	0.007
	Σ	18.8	41.4	9.1	20.0	—		3.0	1.740	1.715	1.707	3.0	0.140	0.051	0.029
T_2-S_2 7 — LOAD= 13.3 ton (29.3 kips); 9.80 (32'); 1.30 (4.27'); 3.90 (12.8'); 5.85 (19.2'); 1.83 (6')	1	4.0	8.8	3.5	7.7	5.8	82.5	1.0	0.120	0.060	0.030	1.0	0.080	0.030	0.020
	2	8.5	18.7	4.0	8.8	5.8	82.5	1.0	1.000	1.020	1.050	1.0	0.120	0.060	0.030
	3	12.1	26.6	3.8	8.4	5.8	82.5	2.0	0.450	0.400	0.400	2.0	0.010	0.002	0.001
	Σ	24.6	54.1	11.3	24.9	—		4.0	1.570	1.480	1.480	4.0	0.210	0.092	0.051
T_3-S_2 8 — LOAD= 16.0 ton (35.3 kips); 11.00 (36'); 4.00 (13.1'); 8.50 (27.9')	1	3.9	8.6	3.5	7.7	5.8	82.5	1.0	0.100	0.050	0.025	1.0	0.080	0.030	0.020
	2	13.0	28.6	5.4	11.9	5.8	82.5	2.0	0.600	0.500	0.500	2.0	0.040	0.015	0.007
	3	13.0	28.6	5.0	11.0	5.8	82.5	2.0	0.600	0.500	0.500	2.0	0.030	0.010	0.005
	Σ	29.9	65.8	13.9	30.6	—		5.0	1.300	1.050	1.025	5.0	0.150	0.055	0.032

Fig. 9e-2 Coefficients of traffic damage for typical vehicles

Table 9e-2

Procedure for converting mixed traffic to the equivalent 8.2 t (18000 lb) single axle loads using the design lane (*ADT*), §9e.3, N.A.U.M. Method

1 Type of vehicle	2 *ADT*, two directions	3 No. of vehicles design lane	4 Coefficient of design		5 No. of equivalent 8.2 t (18000 lb) (18000 lb) axles	
			$z = 0$	z = 15 cm (6 in)	$z = 0$	z = 15 cm (6 in)
Ap	8,460	3,384	0.005	0.000	16.9	0.0
Ac	2,116	846	0.34	0.042	287.6	35.5
B	684	274	2.00	1.15	548.0	315.1
C2	734	294	0.88	0.465	258.7	136.7
C3	116	46	0.88	0.675	40.5	31.1
T2S2	28	11	4.00	1.57	44.0	17.3
T3S3	102	41	5.00	1.30	205.0	53.3
Total	12,240	4,896			T_0 = 1400.7	T'_0 = 589.0

with the recommendations made by the Engineering Institute, Table 9e-3.

The number of equivalent axle loads (Column *5*) for each line is determined by multiplying the number of vehicles for the design lane (Column *3*) by the coefficient of equivalency for the corresponding damage (Column *4*). The sum of these partial results is obtained at the foot of Column *5* for two depth values z. Each of these totals represents the traffic equivalent to 8.2 t (18000 lb) single-axle loads referring to the design lane for an average day in 1972. The results are as follows:

$$T_0 = 1{,}400.7 \text{ equivalent axle loads } (z = 0)$$
$$T'_0 = 589.0 \text{ equivalent axle loads } [z = 15 \text{ cm (6 in)}]$$

Table 9e-3

Coefficient of distribution as a function of the number of lanes as used in the Engineering Institute of the Mexican Autonomous University, §9e.3, [81]

Number of lanes in both directions	Coefficient of distribution for the design lane
2	50
4	40-50
6 or more	30-40

Calculation of accumulated equivalent axle loads: The accumulated number of equivalent 8.2 t (18000 lb) axle loads for a service period of n years is computed using the following expression:

$$\sum L_n = C' \cdot T_0$$

where: $\sum L_n$ is the accumulated equivalent axle loads for n years of service, and rate of growth r in equivalent 8.2 t (1800 lb) axle loads; T_0 the average daily traffic during the first year of service for the design lane, in equivalent 8.2 t (18000 lb) axle loads; and C' the coefficient of accumulated equivalent axle loads for n years of service and an annual rate of growth, r, which can be obtained using the following equation:

$$C' = 365 \sum_{j=1}^{j=n} (1 + r)^{j-1}$$

The graphical solution to the above expression is given in Fig. 9e-3. Considering a design period of 10 years and an annual rate of growth of 7%, the coefficient of accumulated equivalent axle loads is determined using this graph. The following result is obtained: C' = 5,100, which when multiplied by T_0 and T'_0 gives the equivalent axle loads for depths z = 0 cm and z = 15 cm (6 in).

$$\sum L'_{10} = C'T'_0 = 5{,}100 \times 589 = 3{,}003{,}900 \text{ equivalent axle loads } [z = 15 \text{ cm (6 in)}].$$

$$\sum L_{10} = C'T_0 = 5{,}100 \times 1{,}400.7 = 7{,}143{,}570 \text{ equivalent axle loads } (z = 0).$$

.2 Structure Design of the Road

In accordance with the characteristics of the road under consideration, it is felt that the structural design should be based on the graph in Fig. 9e-4, which represents normal design conditions, this is the same as Fig. 9-26c described in the text of the chapter. It is repeated for assisting with the solution of this problem. Using the accumulated equivalent axle load data already obtained:

3,003,900 equivalent axle loads, z = 15 cm (6 in)
7,143,570 equivalent axle loads, z = 0

An equal relative strength curve is plotted on the graph in Fig. 9e-4. This is the curve that is represented by a dashed line.

The layer thicknesses of the road structure will now be determined. Here they include the subgrade layer, sub-base, base course and surface course. The equal relative strength curve and the strength properties of each of the materials

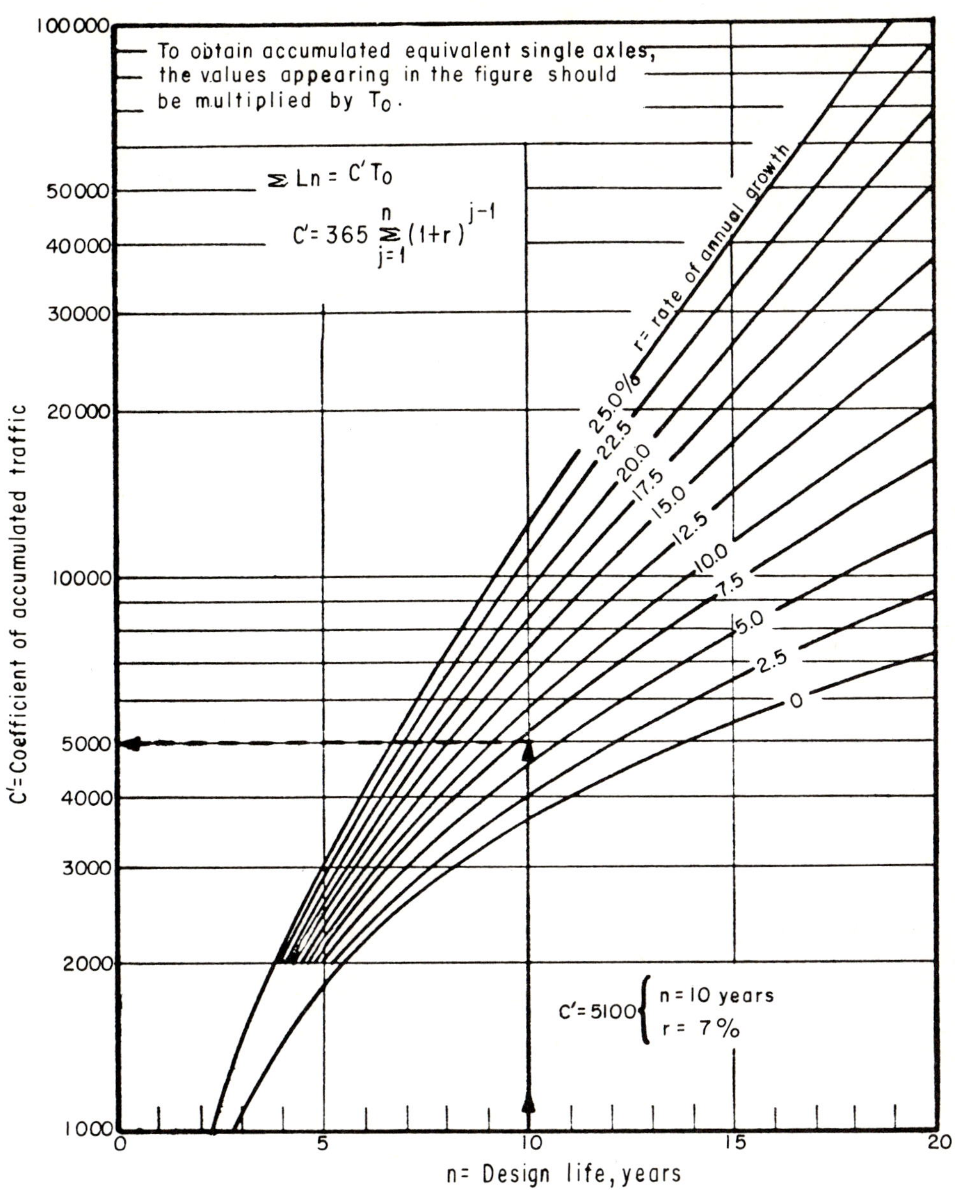

ΣL_n = Accumulated traffic after n years of service in equivalent 8.2 ton. (18000 lb.) axles.

C' = Coefficient of accumulated traffic for n years of service and an annual rate of growth, r %

T_0 = Average daily traffic per lane in the first year of service, in equivalent 8.2 ton. (18000 lb.) axles.

$T_0 = \Sigma N_i F_i + \Sigma N_i' F_i'$

N_i , N_i' = Average daily traffic per lane of type i vehicles (loaded or empty, respectively) during the first year of service.

F_i , F_i' = Coefficient of relative damage produced by each journey made by vehicle i (loaded or empty, respectively), in equivalent 8.2 ton. (18000 lb.) axles.

Fig. 9e-3 Graph for estimating accumulated equivalent axle loads

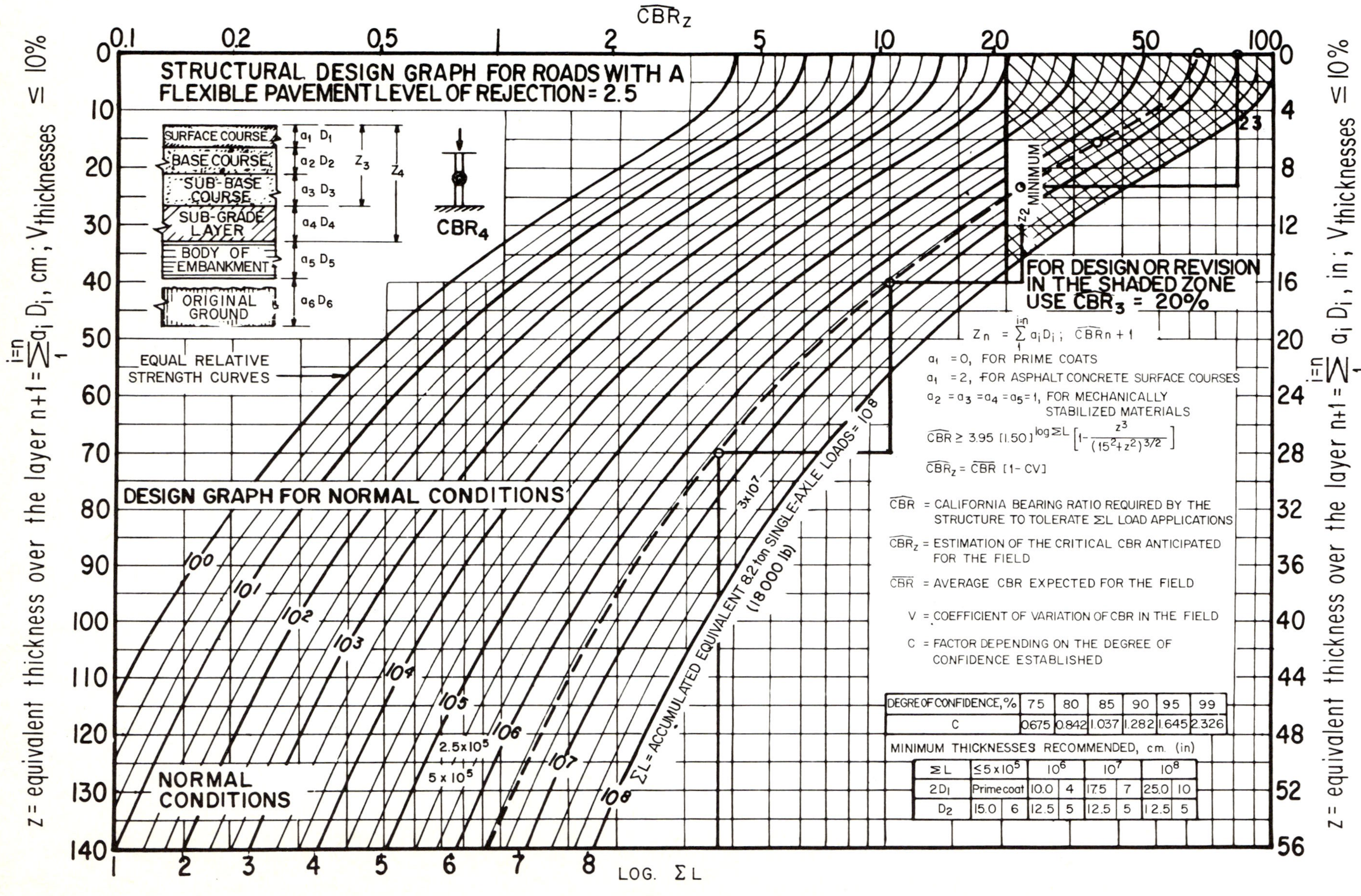

Fig. 9e-4 Graph of structural design for roads with a flexible pavement (Engineering Institute of the Mexican National Autonomous University)

that will make up the layers are required. To determine these strengths, physical tests are conducted, either in the laboratory or in the field, on materials from the borrow areas that will be used for the project construction. (For example, California Bearing Ratio, with the Corps of Engineers Method, triaxial tests, etc.)

Once the *CBR* values have been determined for the different materials, the critical design *CBR* corresponding to each layer is then computed using the following equation:

$$CBR_c = CBR\,(1 - CV)$$

where: CBR_c is the critical *CBR* value for design purposes; *CBR* the mean value for each material; *C* a factor that depends on the confidence limit (See Fig. 9e-4); and *V* is a coefficient of variability of test values. (A confidence limit of 90% is recommended for use with Fig. 9e-4).

The CBR_c values obtained for this case are as follows:

Natural soil	3.5 %
Subgrade	10%
Sub-base	21%
Base course	80%

With the strength data of the materials and the graph in Fig. 9e-4, the different layer thicknesses are obtained using the following procedure:

The total thickness of equivalent pavement materials that must be laid on the natural ground is determined by drawing a vertical line commencing at the point where the *CBR* is 3.5%, continuing until it intercepts the equal relative strength curve at a point, referred to as the critical point, which when plotted on the ordinate axis (z) gives a total thickness of 71 cm (28 in). The thickness of the subgrade layer is equal to the vertical distance between the critical points corresponding to the *CBR* values of 3.5% and 10%, determined in a similar manner. A thickness of 17 cm (6.7 in) was obtained for the sub-base layer. Both are shown in the figure.

The difference between the total thickness and the sum of the thicknesses of the subgrade and the sub-base is equal to the thickness required for the base course and surface course, in gravel equivalent, that is:

$$71 - (31 + 17) = 23 \text{ cm}$$
$$28 - (12.2 + 6.7) = 9.1 \text{ in}$$

which in accordance with the following structural thicknesses equation gives:

$$z_n = \sum_{1}^{n} a_i D_i$$

$$z = 23 \text{ cm } (9.1 \text{ in}) = a_1 D_1 + a_2 D_2$$

where: D_1 is the thickness of asphalt surface course, in cm (in); D_2 the thickness of base course, in cm (in); a_1 and a_2 are coefficients of equivalency of real thickness to gravel equivalent, $a_1 = 2$, for asphalt concrete and $a_2 = 1$, for hydraulic base courses.

The minimum thickness of asphalt surface course required is 5 cm (2 in), obtained using the table in Fig. 9e-4, as a function of the accumulated equivalent axle loads for a depth $z = 15$ cm (6 in).

From the above, the following equation is obtained:

$$a_1 D_1 + a_2 D_2 = 2 \times 5 + D_2 = 10 + D_2 = 23 \text{ cm}$$

$$a_1 D_1 + a_2 D_2 = 2 \times 2 + D_2 = 4 + D_2 = 9.1 \text{ in}$$

Therefore the thickness of the hydraulic base course, $D_2 = 13$ cm (5 .1 in).

The final road structure will thus be:

— Subgrade layer	31 cm (12.2 in)
— Sub-base layer	17 cm (6.7 in)
— Hydraulic base course	13 cm (5.1 in)
— Asphalt concrete surface course.	5 cm (2 in)

† The Problems in Appendix 9e are by courtesy of M.R. Carrizales.

REFERENCES

1. AGUIRRE, L. M., RICO, A., SÁNCHEZ, D. and SOSA, R., "Proyecto de espesores de pavimentos flexibles en carreteras y aeropistas," Communication to the Seminar on earthfills and pavements of the Mexican Ministry of Public Works, México, 1972.
2. ANON. *Especificaciones Generales de Construction,* Publications of the Mexican Ministry of Public Works, México, 1971, Second Edn., Eighth Part (Book 1).
3. PADILLA CORONA, E., *Estudio del comportamiento de pavimentos flexibles en tramos representativos de la red nacional,* Publication of the Institute of Engineering of the National Autonomous University of Mexico, sponsored by the Ministry of Public Works, México, 1972.
4. RATNARAJAH, A., "The Effect of Climatic Factors on Benkelman Beam Deflections in the Melbourne Area of Victoria, Australia," *III. International Conference on Structural Design of Asphalt Pavements,* London, 1972, Vol. I.
5. CORRO, S. and PRADO, G., "Análisis del comportamiento estructural de pavimentos flexibles en la pista circular. Experiment 1971-1972," Publication of the Institute of Engineering of the National Autonomous University of Mexico, México, 1972.
6. YODER, E. J., *Principles of Pavement Design,* John Wiley and Sons, 1967, Chap. 2.
7. JUÁREZ BADILLO, E. and RICO, A., *Mecánica de Suelos. Vol. II. Teoría y Aplicaciones de la Mecánica de Suelos,* Limusa: México, 1973, Chap. X.
8. HVEEM, F. N. and SHERMAN, G. B., "California Method for the Structural Design of Flexible Pavements," California Division of Highways, Materials and Research Deparment, Sacramento, Calif., 1958.
9. Reference [7], Chap. II.

10. Burmister, D. M., "The Theory of Stresses and Displacements in Layered Systems and Application to the Design of Airport Runways," *Proc. Highway Research Board,* 1943, Vol. 23.

11. Burmister, D. M., "The General Theory of Stresses and Displacements in Layered Soil Systems," *Journal of Applied Physics,* 1945, Vol. 16.

12. Burmister, D.M., "Evaluation of Pavement Systems of the WASHO Road Test by Layered Systems Methods," *Highway Research Board,* 1958, Bull. 177.

13. Yoder, E. J., *Principles of Pavement Design,* John Wiley and Sons, 1957, Chap. 4.

14. "Flexible Pavement Structural Section Design Guide for California Cities and Counties," County Engineers Association of California, Sacramento, Calif., 1968.

15. Corro, S., "Diseño de Pavimentos flexibles. Performance of experimental stretches," Publication of the Institute of Engineering of the UNAM, Sponsored by the Ministry of Public Works, México, 1970, No. 240.

16. Irick, P. E. and Hudson, W. R., "Guidelines for Satellite Studies of Pavement Performance," National Cooperative Highway Research Program, National Academy of Sciences, Washington, 1964, No. 2A.

17. Hveem, F. N., "Types and Causes of Failure in Highway Pavements," California Division of Highways, Materials and Research Department, Sacramento, Calif., 1958.

18. Chu, T. Y., Humphries, W. K. and Chen, S. N., "A Study of Subgrade Moisture Conditions in Connection with the Design of Flexible Pavement Structures," *III. International Conference on Structural Design of Asphalt Pavements,* London, 1972, Vol. I.

19. McCullough, B. F., "Distress Mechanism General," Highway Research Board, Structural Design of Asphalt Concrete Pavement Systems, Washington D. C., 1971, Special Report 126.

20. Moavenzadeh, F., "Damage and Distress in Highway Pavements," ibid.

21. Reference [7], Chap. I.

22. Porter, O. J., "The Preparation of Subgrades," *Proc. Highway Research Board,* 1938, Vol. 18.

23. U.S. Army Corps of Engineers, "Revised Method of Thickness Design for Flexible Highway Pavements at Military Installations," 1961, Tech. Report No. 3.

24. Hank, R. J. and Scrivner, F. H., "Some Numerical Solutions of Stresses in Two and Three Layered Systems," *Proc. Highway Research Board,* 1948, Vol. 28.

25. Fox, L., "Computation of Traffic Stresses in a Simple Road Structure," *Proc. II. ICSMFE,* Rotterdam, 1948.

26. Jones, A., "Tables of Stresses in Three-Layer Elastic Systems," Highway Research Board, 1962, Bull. 342.

27. Jeuffroy, G. and Bachelez, J., "Note on a Method of Analysis for Pavements," *Proc. Ann Arbor Conference,* Michigan University, Ann Arbor, Mich., 1962.

28. Sowers, G. F. and Vessic, A. B., "Vertical Stresses in Subgrades Beneath Statically Loaded Flexible Pavements," Highway Research Board, 1962, Bull. 342.

29. Ashton, J. E. and Moavenzadeh, F., "Analysis of Stresses and Displacements in a Three-Layered Viscoelastic System," *Procs. II. International Conference on Structural Design of Asphalt Pavements,* Michigan University, Ann Arbor, 1967.

30. Huang, Y. H., "Stresses and Displacements in Visco-elastic Layered Systems under Circular Loaded Areas," ibid.

31. Ishibara, K. and Kimura, T., "The Theory of Visco-elastic Two Layer Systems and the Conceptions of its Application to the Pavement Design," ibid.

32. Perloff, W. H. and Moavenzadeh, F., "Deflection of Visco-elastic Medium due to a Moving Load," ibid.

33. Jones, A., "The Calculation of Stress, Strain and Displacements in Layered Systems having Constant and Variable Elastic Parameters," ibid.

34. Hicks, R. G. and Monismith, C. L., "Prediction of the Resilient Response of Pavements Containing Granular Layers Using Non-Linear Elastic Theory," *Procs. III International Conference on Structural Design of Asphalt Pavements,* London, 1972.

35. Moavenzadeh, F. and Elliot, J. F., "A Stochastic Approach to Analysis and Design of Highway Pavements," ibid.

36. De Barros, S. T., "A Critical Review of Present Knowledge of the Problem of Rational Thickness Design of Flexible Pavements," Highway Research Board, 1965, Record No. 71.

37. Finn, F. N., Keshavan Nair and Monismith, C. L., "Applications of Theory in the Design of Asphalt Pavements," *Procs. III International Conference on Structural Design of Asphalt Pavements,* London, 1972.

38. Reference [6], Chap. 8.

39. The Asphalt Institute, *Soils Manual,* College Park, Maryland, 1963, MS 10.

40. Reference [6], Chap. 15.

41. Kansas State Highway Commission, 'Design of Flexible Pavements using the Triaxial Compressional Test," Highway Research Board, 1947, Bull. No. 8.

42. California Division of Highways, *Materials Manual. Testing and Control Procedures. Vol. I,* Materials and Research Department, Sacramento, Cal., 1964.

43. Texas Highway Department, *Manual of Testing Procedures. Test 117 E. Vol. I,* Austin, Texas, 1965.

44. Reference [6], Chap. 14.

45. "Development of C.B.R, Flexible Pavement Design Methods for Airfields," *Procs. ASCE,* 1950, Vol. 115.

46. U.S. Army Engineer Waterways Experiment Station, "Mathematical Expression of CBR Relations," *Technical Manual,* Vicksburg, Miss., 1956, No. 3441.

47. U.S. Army Engineer Waterways Experiment Station, "Developing a Set of CBR Design Curves," *Instruction Report* Vicksburg, Miss., 1969, No. 4.

48. Hveem, F. N. and Carmany, R. M., "The Factors Underlying a Rational Design of Pavements," *Proc. Highway Research Board,* 1948.

49. McLeod, N. W., "Flexible Pavements Thickness Requirements," *Proc. Assoc. of Asphalt Paving Technologists,* 1956, Vol. 25.

50. The Asphalt Institute, "Thickness Design, Full Depth Asphalt Pavement Structures for Highways and Streets," *Manual Series,* College Park, Maryland. 1969, No. 1.

51. Nascimento, U., Seguro, J. M., Da Costa, E. and Sequeira, P., "A Method of Designing Pavements for Roads and Airports," *Proc. V. ICSMFE,* Paris, 1961, Vol. II.

52. Ivanov, N., Krivisski, A., Tcherkassov, I., Babkov, V. and Biroulia, A., "Certains aspects de la mecánique des chaussées souples," ibid.

53. Krivisski, A., "Design of Flexible Pavements for Major Highways," *III. International Conference on Structural Design of Asphalt Pavements,* London, 1972, Vol. I.

54. Brown, S. F. and Pell, P. S., "A Fundamental Structural Design Procedure for Flexible Pavements," ibid.

55. Highway Research Board, "Pavement Rehabilitation. Materials and Techniques," NCHRP Synthesis 9, 1972.

56. California Division of Highways, *Materials Manual Testing and Control Procedures,* Test Method N° Calif. 356 A. Materials and Research Department, Sacramento, Cal., 1969.

57. Department of Transport, *Airport Development Engineering Planning and Construction Manual, Section I: Design and Evaluation of Flexible and Rigid Pavements,* Ottawa, 1969.

58. International Civil Aviation Organization, "Aerodromes, Air Routes and Ground Aids Division," *Aerodrome Manual,* Part IV, Document 7920 AN/865.

59. The Asphalt Institute, "Asphalt Overlays and Pavement Rehabilitation," *Manual Series*, College Park, Maryland, 1969, No. 17.

60. Kassiff, G., Livneh, M. and Wiseman, G., *Pavements on Expansive Clays,* Jerusalem Academic Press: Jerusalem, 1969.

61. Williams, A. A. B., "The Deformation of Roads Resulting from Moisture Changes in Expansive Soils in South Africa", in, *Moisture Equilibria and Moisture Changes in Soils Beneath Covered Areas,* Butterworths: Australia, 1965.

62. Skempton, A. W., "The Colloidal Activity of Clays," *Procs. of the III. ICSMFE,* Zurich, 1953, Vol. I.

63. Kassiff, G. and Holland, J. E., "The Expansive Properties of Dooen Clays as Applied to Buried Pipes," *Civ. Eng. Trans. of the Institution of Engineers,* Australia, 1966, Vol. C.E. 8.

64. Holtz, W. G. and Gibbs, H. J., "Engineering Properties of Expansive Clays," *Trans. A.S.C.E.,* 1957, Vol. 121.

65. McDowell, C., "Interrelationship of Load, Volume Change and Layer Thickness of Soils to the Behavior of Engineering Structures," *Proc. HRB,* 1956, Vol. 35.

66. Seed, H. B., Richard, W. I. and Raymond, L., "Prediction of Swelling Potential for Compacted Clays," *A.S.C.E., Journal of Soil Mechanics and Foundations* Division, 1962, Vol. SM3-88.

67. Lambe, T. W., "The Character and Identification of Expansive Soils," A Technical Studies Report, Federal Housing Administration (PHQ-701) Washington, D.C., 1960.

68. Seed, H. B., Mitchell, J. K. and Chan, C. K., Studies of Swell and Swell Pressure Characteristics of Compacted Clay, *Bulletin of H.R.B.,* Washington, D.C. 1962, No. 313.

69. Juárez Badillo, E. and Rico, A., *Mecánica de Suelos. Vol. III. Flujo de Agua en Suelos.* Limusa: México, 1969, Appendix IV.

70. Jennings, J. E., "The Theory and Practice of Construction on Partly Saturated Soils as Applied to South African Conditions," International Conference on Expansive Clay Soils, Texas, 1965.

71. Baker, R. and Kassiff, G., "Mathematical Analysis of Swell Pressure with Time for Partly Saturated Clays," *Canadian Geotechnical Journal,* 1968, Vol. 5, No. 4.

72. Warkentin, B. P., "Water Retention and Swelling Pressure of Clay Soils," *Canadian Journal of Soil Science,* Agricultural Institute of Canada, 1962, Vol. 42, No. 1.

73. Livneh, M., Kinsky, Y. and Zaslavsky, D., "The Relationships between the Suction Curves and the Consistency Limits," Technion Research and Development Foundation, Haifa, 1967. (Mentioned in Ref. [59]).

74. Chen, F. H., "The Basic Physical Property of Expansive Soils," III. International Conference on Expansive Soils, Haifa, Israel, 1973.

75. Vijayvergiya, V. N. and Ghazzaly, O. I., "Prediction of Swelling Potential for Natural Clays," ibid.

76. Merwe, C. P. van der and Ahronovitz, M., "The Behavior of Flexible Pavements on Expansive Soils in Rhodesia," ibid.

77. Harris, F. A., "Asphalt Membranes in Expressway Construction," Highway Research Board, Washington, D.C. 1963, Bull. No. 7.

78. Blight, G. E. and De Wet, J. A., "The Acceleration of Heave by Flooding," in, *Moisture Equilibria and Moisture Changes in Soils Beneath Covered Areas,* Butterworths: London, 1965.

79. Corro C., S. "Diseño Estructural de Carreteras con Pavimentos Flexibles," Design Graphs, Technical Publication N° 322 of the Institute of Engineering of the UNAM, (Sponsored by the Mexican Ministry of Public Works), México, October, 1973.

80. Sosa, R., "Pavimentos sobre Suelos Expansibles," VII. National Meeting on Soil Mechanics, Mexican Society of Soil Mechanics, Guadalajara, 1974, Vol. II.

CHAPTER 10

RIGID PAVEMENTS

10.1 Introduction

As already mentioned, the basic structural element of a rigid pavement is a concrete slab. This slab rests on a layer of selected material, referred to as the *sub-base*. When the subgrade is of sufficiently good quality, the concrete slab can be placed directly on top of it, thus avoiding the need for a special sub-base. As can be seen, the distinction between a sub-base, subgrade and even the upper part of a fill is in this case largely a question of nomenclature. The objective is to provide the concrete slab with a sufficiently uniform and firm foundation that it will at no point suffer from lack of support. How this is accomplished and what layers of soil are required depends on the quality of the materials that are used, the degrees of compaction employed and local conditions of climate and drainage.

The concretes that are used generally have a relatively high strength, between 200 and 400 kg/cm^2 (2840 and 5680 lb/in^2). The slabs may be of plain, reinforced or pre-stressed concrete. When plain or reinforced concrete is used, the slabs are similar in size, typically squares measuring 3 × 3 m or 5 × 5 m (10 × 10 ft or 16 × 16 ft), but there is a trend towards an increase in these dimensions. Pre-stressed concrete permits the use of continuous surfaces with a much larger area. This, plus the considerable economies in thickness that can be achieved in this way, lead to an increasing use of pre-stressed concrete.

The principal factors affecting slab thickness are loads that must be supported, vehicle tire pressures, the modulus of reaction of the bearing soil and the mechanical properties of the concrete used. In all design methods, it is commonly assumed that complete contact exists at all times between the slab and the bearing soil. This assumption requires evaluation of other factors in addition to the calculated support, because effects such as pumping or warping can be responsible for the partial disappearance of this assumed contact. These conditions have to be avoided, generally by employing rules of the art.

It has been shown [1] that the slab thickness that is required depends only to a small extent on the value of the modulus (or coefficient) of reaction of the supporting ground, (the measure of soil support that enters into the design of the concrete slab). Therefore most rigid pavement designs pay little attention to the quality of the supporting materials, sub-base, subgrade or embankment. They are allowed to vary between rather wide limits, with specifications based on empirical rules being established to prevent pumping or other harmful effects. Applied soil mechanics has, therefore, played only a minor role in the development of rigid pavement technology.

It is not intended in this chapter to make a detailed analysis of the methods for designing concrete slabs or of the structural considerations behind the rules for designing the different types of joints or the spacing between them. We feel that applied soil mechanics has little, if any, relation to these problems. Therefore we touch upon them only lightly. Some attention will be given to the desirable characteristics for the soils on which the slab rests, both the rules of the art that prevail, as well as application of certain general soil mechanics standards already discussed in previous chapters. In our opinion there has been insufficient research on this subject. This implies a lack of interest, by geotechnical engineers probably because the quality of the soils does not play a very dominant role. The structural support of wheel loads is supplied by the concrete slab, which is far stronger.

10.2 Properties of the Sub-Base and Bearing Soils

Interest in the quality of the support that is given to the concrete slabs dates back only to the second World War. Before that time, the slabs were placed directly on top of the fill or virgin ground, which was at most provided with a subgrade layer. There was little concern for whether the materials in contact with the concrete were sands, clays or silts. The increase in the numbers and weights of vehicles on roads, and the great increase in number and weight of aircraft at airports, made adequate support of the concrete slabs more important. As a result, roads carrying heavy traffic and the most runways must include an appropriate sub-base on a well compacted fill. This sub-base consists of one or more layers of granular material, often stabilized. Occasionally when the subgrade fully satisfies the characteristics that are desirable for the sub-base, construction of a special sub-base can be omitted.

The chief functions of the sub-base in a rigid pavement are as follows:

1) To provide the concrete slab with uniform support
2) To increase the bearing capacity of the supporting soils, which is why it is commonly used on fills and weaker subgrades
3) To minimize the consequences of any volume changes that may occur in the fill or subgrade soils.
4) To minimize the consequences of frost action in fill or subgrade soils.
5) To prevent pumping.
6) To act as a drain or an insulating layer

Given the comparative rigidity and high strength of concrete slabs, the stresses that are transmitted to the sub-base are small; therefore strength alone is not usually an important requirement. The slab performs adequately if its support is uniform and is consistent throughout the life of the pavement. A good support should include gradual transitions wherever there are abrupt changes in the bearing capacity of underlying ground, as is often the case between a cut and an embankment or between an embankment and firm ground.

When the width of traffic lanes is such that the wheels of heavy vehicles are likely to remain within the inner area of the slabs, the stress that will reach the sub-base is approximately 3 to 4% of the tire pressure acting on the surface of the slab [2]. This is likely with lanes over 3.50m (11.5 ft) wide. With lanes about 3 m (10 ft) wide, the outer tires of heavy vehicles travel very close to the outer edge of the slabs; and in these zones greater stresses are induced in the sub-base, possibly reaching about 0.5 kg/cm^2 (7.1 lb/in^2) [2]. The foregoing data, which are supplied by field tests, explain why the traffic stresses applied to the sub-base do not usually impose critical soil loads, even in the most adverse cases.

The functions listed above for the sub-base of a rigid pavement are usually satisfied by a well compacted, relatively coarse granular material with a fairly uniform grain-size distribution. When such materials are unavailable, stabilization not only greatly improves poor-quality materials, reducing pumping and susceptibility to volume change, but also improves the bearing surface, providing adequate strength and greater uniformity.

Volume changes in the supporting ground, caused by variations in water content, may lead to loss of uniform slab support, especially if swell-susceptible materials are compacted dry of optimum. The same problems may arise if materials that were compacted with an appropriate water content are allowed to over-dry as a result of evaporation before being covered by the pavement slab. Swell-susceptible materials with excessive water contents may also give rise to problems, for subsequent shrinkage on loss of the excess moisture may lead to local reductions in slab support, particularly along the pavement edges. The ideal compacting water content is probably a compromise between the field optimum value and 1 or 2 per cent higher, but the exact value must be established, taking into consideration the prevailing climatic and constructional conditions [3, 4].

Careful attention should be paid to swell susceptibility in fill materials too, for if they undergo severe volume changes, important deformation will occur in the surface of the sub-base, with the corresponding problems of loss of support, even when the subgrade and the sub-base are made of non-susceptible soils. Arid regions are especially dangerous for high volume change soils. After construction, the water content of most subgrades and sub-bases increase to near the plastic limit of the component soils [5], which is usually slightly above the field optimum. This helps explain why a compacted soil which will undergo minimum volume changes during the life of the pavement is recommended.

Compaction control tests that use high levels of specific energy, and consequently find lower values of optimum moisture, may cause trouble with expansive soils or when engineers believe that the optimum value for laboratory compaction represents the ideal water content for field compaction. The water content of soils compacted at low moistures will increase appreciably during the life of the pavement, with the inevitable swelling. The best approach is to refer the compacting water content to the optimum field value and not to that of a laboratory test (see Chapter 4). For expansive soils, if an appropriate value for the optimum field water content cannot be obtained, the optimum content for a standard AASHTO type compaction test will be a better standard than the optimum value for a test using higher specific energy, such as the AASHTO modified test. Figure 10-1 [5] illustrates the foregoing ideas, in accordance with various discussions throughout this book.

The upper part of the figure shows the index properties of the soil and the compaction curves obtained in AASHTO standard and modified tests. The three lower parts show the behavior of a subgrade composed of that soil with reference to expansion, absorption of water and *CBR*, obtained once the equilibrium condition has been reached under rigid pavement slabs. As has been discussed previously an indiscriminate increase in compaction may lead to adverse results, in this case in expansive soils (see Chapter 4 for further details).

When expansive fills or subgrades must be used because no other more suitable materials are available, one of the precautions discussed in Chapter 9 should be taken. The solution most often used in rigid pavement construction is probably to use a covering layer of a material that is not susceptible to volume changes. This has the added advantage of acting as a surcharge on the expansive materials.

Excellent results have also been obtained for eliminating swell susceptibility in fine soils by adding portland cement (stabilization). Soil stabilization methods will be included elsewhere in this book; consequently no further details will be added here about these much-used solutions.

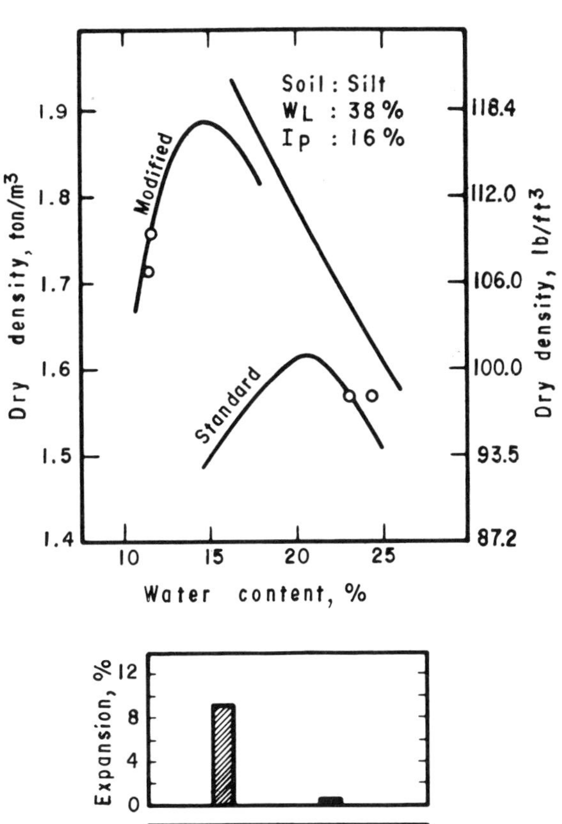

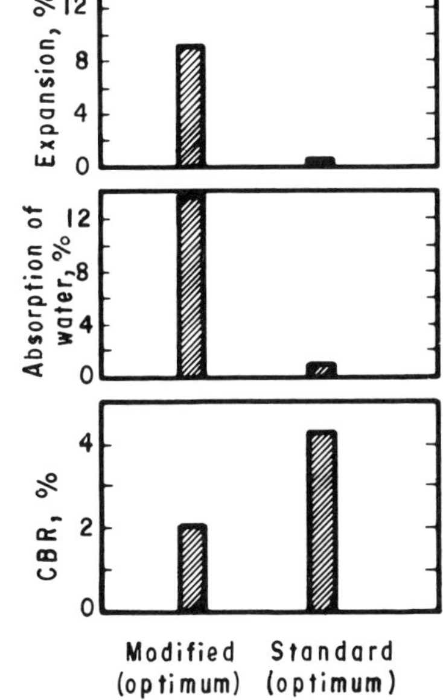

Fig. 10-1 Swelling, absorption and *CBR* characteristics of a soil compacted in AASHTO standard and modified tests, referred to maximum density [5]

In Mexico, frost action does not represent a practical problem of any significance, but in other regions it may be a critical condition. References [6-9] contain information on this subject, together with different soil classifications according to frost-susceptibility. From these it is deduced that the most frost resistant soils are clean gravels, coarse clean sands and mixtures of these soils. The most problem prone ones are clays, silts and some very fine sands. Subdrainage is an important method for protecting frost-susceptible areas [10].

Pumping is a special phenomenon of rigid pavements, that is highly undesirable; it occurs frequently unless special precautions are taken. When the axle load passes over a crack or joint in a slab, this slab deflects downward, transferring pressure to the supporting material beneath it, Plate 10-1. If the soil is very wet or saturated, most of this pressure will be transferred to the water, which will escape through the crack or joint. After the load has passed, the slab rises recovering from the deflection. This movement sucks water underneath the slab. The water is forced out with the next downward deflection under a wheel load. If the soil is erosive the water will emerge dirty, gradually creating a void beneath the slab. Pumping water in and out of the void aggravates the phenomenon. Furthermore, the consequent remolding that occurs turns the soil into a mud or suspension, further aggravating the phenomenon, making the pumping more serious. The process ends with pavement failure when the slab breaks under loading owing to lack of support. For pumping to develop, the material supporting the slab must be both resilient and very fine, especially of the *MH* or *CH* type; free water must also be present, or the soil very moist or saturated. The more numerous the load repetitions the worse the pumping [11, 12, 13]. For this reason, it occurs more frequently in roads than in runways. Note that in good design a *MH* or *CH* soil should not be placed directly beneath a pavement slab. The foregoing explanation should therefore also apply to the clayey fraction of the sub-base, especially if its percentage is high.

Plate 10-1 The pumping phenomenon, subsequently leading to destruction of the slabs.

In order of susceptibility to pumping, *CH* and *MH* soils are followed by *CL*, *ML* and *SM* soils. If the soils on which the slab rests are granular and have not been sufficiently compacted, pumping is combined with densification, with similarly destructive effects.

Experimental evidence indicates that pumping effects are most likely to appear when the volume of traffic in the stretch of road under consideration is over 300 to 400 vehicles per day [2].

For traffic volumes in excess of 1,000 heavy vehicles per day, the following requirements are recommended for the sub-base [2], in addition to it consisting of materials that are not susceptible to pumping.

— The maximum size of the particles must be not exceed ⅓ of the sub-base thickness

— The sub-base must not contain more than 15% of materials finer than N° 200 sieve

— The plasticity index of the material must be less than 6

— The liquid limit of the component material must be under 25%

To avoid densification, granular sub-bases should be carefully compacted to 100% or more of the AASHTO standard maximum.

Under rigid pavements, both uniform-graded sub-bases and relatively well-graded ones have been used. The grading requirements stipulated by the Mexican Ministry of Public Works are the same as specified in Chapter 9 for base courses for flexible pavements.

Among the well-graded mixtures, sands and silts with 17 to 20% material passing the N° 200 sieve have been widely used throughout the world. Good performance has been reported where there is little frost action (to which silty materials are highly susceptible). Mixtures of river gravels and sands have shown highly satisfactory behavior, with maximum particle sizes of 2.5 cm (1 in) and materials finer than the N° 200 sieve less than 10 to 15%. Sub-bases made of well-graded crushed materials have been used with maximum particle sizes of 3.8 cm (1-1/2 in) and as much as 25% finer than the N° 200 sieve. However, the fines are frequently limited to 10 to 15%. Results are reported as satisfactory, although costs may be high compared to those for untreated materials. The use of stabilized sub-bases, particularly cement-treated ones, is widespread throughout the world.

The uniform graded mixtures most commonly used include sands from dunes and similar deposits, with a maximum of 10% passing the N° 200 sieve. Mixtures of sand and gravel with a maximum particle size of 2.5 cm (1 in) have shown excellent behavior when the percentage finer than the N° 200 sieve is between 5 and 10%. River bank gravels are specified with a maximum particle size of 3.8 cm (1-1/2 in) and with 6 to 8% passing the N° 200 sieve. Mixtures of sand and shells have given excellent results under heavy traffic and severe environments in coastal areas.

Open-graded sub-bases may become contaminated by fine materials forced upward from the subgrade or from the fill. These eventually lead to serious deficiencies in performance. It is therefore advisable to design the sub-base gradation so the layer filters out the subgrade or fill. The U.S. Army Corps of Engineers [2] makes the following recommendations:

To ensure draining capacity: $D_{s15} \geqslant 5\ D_{sr15}$ (10-1)

To prevent infiltration: $D_{s15} \leqslant D_{sr85}$ (10-2)

$D_{s50} \leqslant 25 D_{sr50}$ (10-3)

The coefficient of uniformity of the sub-base should not be below 20.

In the above recommendations, D_s and D_{sr} refer to sub-base and subgrade sizes, respectively.

If the sub-base material is used as a covering for perforated pipes in underdrains, it should also satisfy the filter requirements described in Chapter 7.

The above specifications are often too rigid for routine use in roads, for they require materials that are hard to obtain under natural conditions and expensive to produce. In many cases where there is a very open-graded sub-base above a very fine fill or subgrade, sufficient protection will be provided by a thin filter layer of medium-graded material to prevent soil extrusion into the sub-base.

The shoulders of a section paved with concrete slabs deserve attention. If they are constructed with expensive materials, cracking will occur at the edges of the slabs, and the shoulder will be rapidly destroyed by traffic. It is therefore essential to use non-expansive materials in the pavement shoulders, and these materials must be compacted to about 95% of the AASHTO standard test.

When non-swell-susceptible materials are unavailable, soil-cement should be regarded as a suitable, but expensive, solution.

10.3 Rigid Pavement Design

Most concrete pavements are designed on the basis of the WESTERGAARD equations which use the value of the modulus (or coefficient) of subgrade reaction, k, obtained in a plate-bearing test conducted on that subgrade (or sub-base on the subgrade). It has become a custom in some organizations to include the effect of the sub-base that will subsequently be placed over the subgrade, by correcting the value of k. This will be explained in detail later. In this case it is preferable to measure k directly on the sub-base with a plate-bearing test, in the same manner as described for the subgrade.

Reference [14] describes the theoretical foundations of the equations that enable the determination of the concrete slab thickness that will be capable of supporting the different loads that are likely to act on it. Such an analysis is regarded as being beyond the scope of this book, since it does not involve soil mechanics theories. Also, the design engineer who has no special theoretical interest in the subject no longer requires knowledge either of the theories leading to these design formulas or of the formulas themselves. Graphs and charts are now available which give the thickness in terms of design variables. The Portland Cement Association (PCA) has developed very complete design graphs [5], some of which are reproduced here.

In design graphs, the characteristics of the concrete are described by the modulus of flexural strength, MR, which is expressed as a stress. This obtained experimentally by testing a standard beam. However it can be estimated on the basis of correlations with the value of f'_c, the resistance of the concrete to unconfined compression 28 days after placing. The correlation is not too reliable and is influenced by the type of cement that is used and particularly the nature of the aggregates.

The typical correlation is:

$$0.10\, f'_c \leqslant MR \leqslant 0.17\, f'_c \quad (10\text{-}4)$$

In Mexico, it appears appropriate [15] to use the relation:

$$MR = 0.12 f'_c \quad (10\text{-}5)$$

This value of MR corresponds to failure in flexure. The corresponding value which appears in the design graphs is the working one. The failure strength should be divided by a factor of safety between 1.75 and 2.

Design graphs are usually based on a modulus of elasticity of the concrete of 280,000 kg/cm^2 (4 × 10^6 psi) and a POISSON's ratio of 0.15. In the graphs for designing runways, it is assumed that the load acts some distance from the slab edge; this is a reasonable hypothesis if the slabs are connected with the appropriate joints. For highways, it is assumed that the load is applied to the joint between the slabs and the corners of the slabs are protected with devices for suitably transferring the loads to adjacent slabs.

The modulus of subgrade reaction, k, is defined in accordance with an expression that was presented in Chapter 9:

$$k = \frac{p}{\Delta} \quad (10\text{-}6)$$

where p is the pressure that is applied to the soil and Δ the corresponding deflection. In practice, the value of k is obtained in a plate-bearing test, handling information of the type that was shown schematically in Fig. 9-12, earlier. The plate-bearing test that is conducted on subgrades for rigid pavements is commonly of a more simple nature than those that have recently been developed for flexible pavements.

For both roads and runways, circular 76.2 cm (30 in) diameter plates are employed. The plate must be level at the beginning of the test and is placed on a thin layer of sand or plaster. Deflection of the plate during testing is measured with micrometer dial gages that must be supported by beams placed outside the deflected zone. As in the plate-bearing tests described for flexible pavements, it is also customary for rigid pavement tests to use several different plates with decreasing diameters to provide stiffness. Loads are applied using calibrated hydraulic jacks and are transferred by a heavy supporting machine, usually a ballasted truck.

The load increments applied to the plate should not exceed 10% of the maximum estimated load. Each increment is maintained until there is no further settlement. A rate of settlement of about 0.005 cm/min (0.002 in/min) is the usual limit that can be detected. Once the loading process is over, the load is released gradually, using decrements equal to the increments applied. In rigid pavement design, it is common practice (Corps of Engineers) to select the modulus of subgrade reaction on the basis of a pressure value of 0.7 kg/cm^2 (10 psi) [16]. The value of k is found with Eq. (10-6), using the pressure that is indicated and the total corresponding deflection obtained from the pressure-total deflection curve that was plotted from the test results.

It is not usually possible to conduct field tests under the most detrimental water conditions in the subgrade. What are assumed to be the service conditions, are not sufficiently severe. It is preferable to correct the value of k that is obtained on the subgrade just as it is, in order that a value can be reached that will be representative of anticipated future wet conditions. This is accomplished by conducting simple compression tests on a specimen compacted to represent the field moisture conditions in the subgrade at the time of testing, and on a second specimen compacted with the water content that is assumed to represent the worst anticipated future moisture (saturation moisture is usually employed). In both specimens deflection is determined under a load of 0.7 kg/cm^2 (10 psi). The value of the corrected modulus of reaction is obtained from:

$$k_c = \frac{d_1}{d_2} k \qquad (10\text{-}7)$$

where: k_c is the value of the modulus of subgrade reaction corresponding to future service conditions, represented by subjecting the specimen to saturation conditions or to some other less conservative condition that is presumed to be possible; d_1 is the total deflection shown by the specimen under a pressure of 0.7 kg/cm^2 (10 psi) with the material at the same moisture and density as the subgrade at the time of the plate-bearing test that was conducted in the field; d_2 is the total deflection under the same load, but the specimen is prepared with a water content that represents the subgrade's most detrimental future service conditions. As already stated, these conditions are often represented by saturating the material after compaction; and k is the modulus of subgrade reaction, obtained in the plate-bearing test that was carried out in the field.

The plate-bearing test is also used to find the modulus of subgrade reaction under an exisiting concrete pavement, for evaluation purposes. In this case the value of k is obtained by dividing the pressure applied to the subgrade (through a hole in the slab) by the characteristic deflection of the plate under the pressure (again the value of 0.7 kg/cm^2 (10 psi) is usually employed).

Lastly, the plate-bearing test has also been used in existing rigid pavements to evaluate the modulus of reaction of the pavement structure as a whole, applying the load to the plate on top of the concrete slab and using a set of micrometer dial gages to measure both the pavement and subgrade deflection that occurs. Here, the modulus of reaction is calculated as the quotient of the load applied divided by the total volume of deflection that occurred under that load.

Recently, in runways, plates larger than 76 cm (30 in) have been used, to reproduce better the loading conditions of the large landing gear of modern aircraft.

The Portland Cement Association recommends another criterion for evaluation of k for Eq. (10-6) [5]. According to this criterion, the value of k is obtained by dividing the pressure that produces a deflection of 0.13 cm (0.5 in) in the plate by the amount of that deflection.

In materials with a high modulus of reaction, the plate bends to a certain extent and the results of the test are therefore distorted. Figure 10-2 [16] enables a correction to be made for this bending.

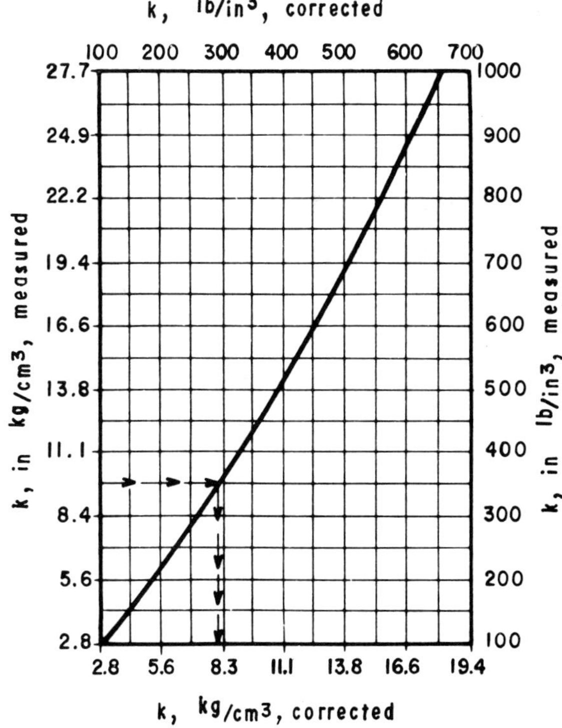

Fig. 10-2 Graph for correcting the value of k affected by bending of the plate [16]

As already mentioned, there is usually a sub-base over the subgrade of rigid pavements whose modulus of reaction has already been calculated. Its purpose is to improve slab support. The effect of the sub-base is taken into consideration by conducting plate-bearing tests on it. If this is not practical its effect can be estimated by correcting the value of k of the subgrade alone so as to obtain the final value to be used in the design graphs, Fig. 10-3 [5] estimates this correction for non-stabilized sub-bases, Fig. 10-4 gives the same information for stabilized sub-bases.

Sub-base thickness in rigid pavements is not determined by calculations, but has been established by tradition. It is not less than 10 cm (4 in); 15 cm (6 in) probably is an appropriate minimum.

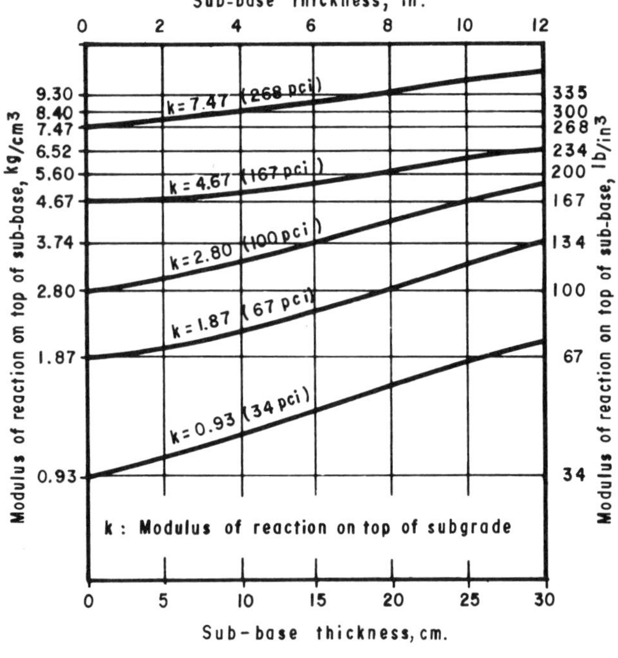

Fig. 10-3 Graph for estimating the value of k on top of sub-base, once it is known on top of subgrade; Non-stabilized sub-bases [5]

Fig. 10-4 Graph for estimating the value of k on top of sub-base, once it is known on top of subgrade; Cement-stabilized sub-bases [5]

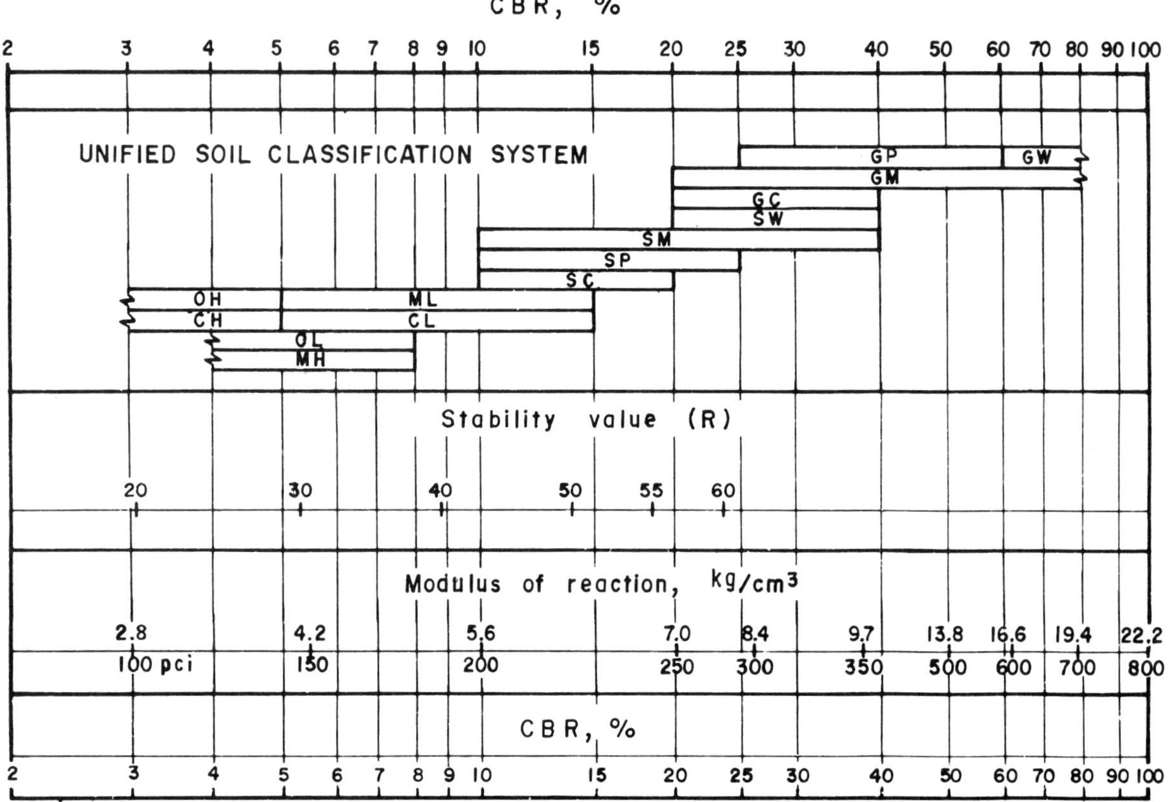

Fig. 10-5 Comparison between several different strength indices that can be used in sub-bases for rigid pavements [5]

If smaller dimensions are recommended in the design, the sub-base might turn out to be too thin at some point because of inevitable construction irregularities. Under nomal conditions, thicknesses of over 20 cm (8 in) are not usually employed. However, if either the fill or subgrade materials are susceptible to frost action or expansion, considerably thicker sub-bases may prove necessary. The information in Fig. 10-5 [5] is presented for comparison. It illustrates the relation between HVEEM's stability values (*R*), the *CBR* and the modulus of reaction of different soils as classified according to the Unified Soil Classification System. The information in Fig. 10-5 is a mere estimation and in no way should it be used for design. Like so much of the information in this book, its value is qualitative; its sole objective being guidance.

Figures 10-6 to 10-13 [17] present the design graphs for rigid runway pavements corresponding to typical heavy aircraft. In each case dimensions of the landing gear, tire pressure and other factors in design are specified. All the graphs function in the same way. With the flexural strength of the concrete *MR*, the modulus of reaction of the supporting soil and the weight of the wheel assembly of the aircraft, it is possible to obtain the concrete slab thickness by drawing a horizontal line for the value of *MR*, until it meets the straight loading curve. From the point thus obtained, another vertical line must be drawn to meet the curve for the corresponding value of the modulus of reaction. Through the new point thus established, a horizontal line is drawn which gives the slab thickness.

Attention is drawn to the relatively minor influence of the modulus of reaction of the bearing soil on the resulting slab thickness in these graphs. Reference [18] contains a fairly complete study of this influence and quantifies it for different circumstances.

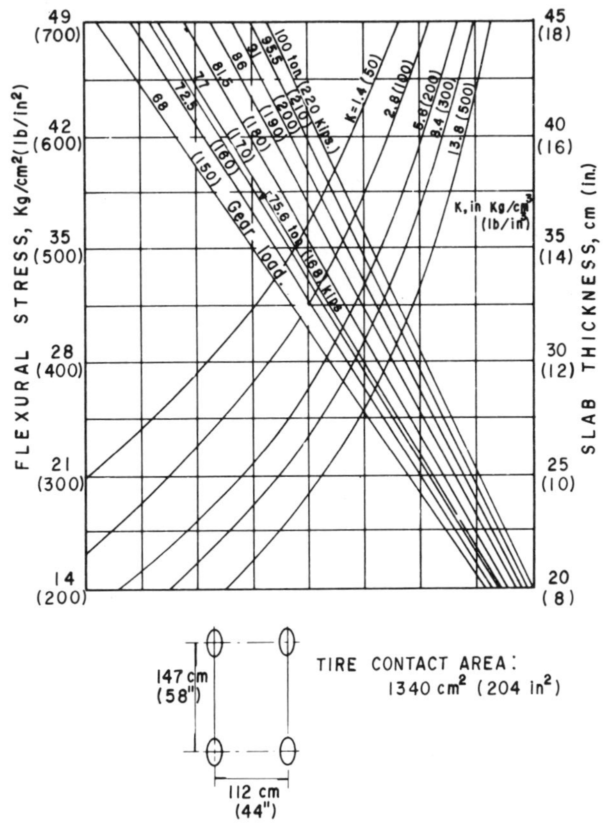

Fig. 10-7 Concrete slab thickness design graph; B-747 aircraft [17]

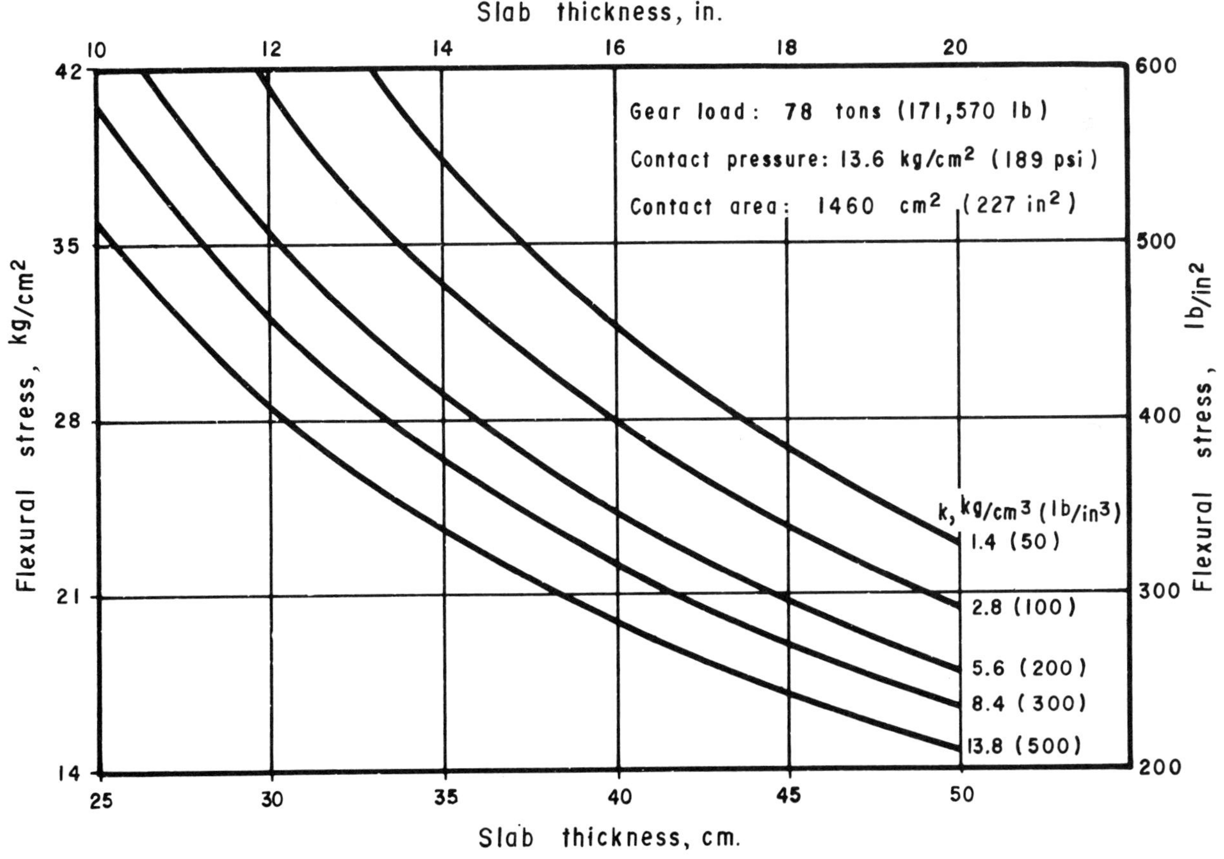

Fig. 10-6 Concrete slab thickness design graph; Concorde aircraft [17]

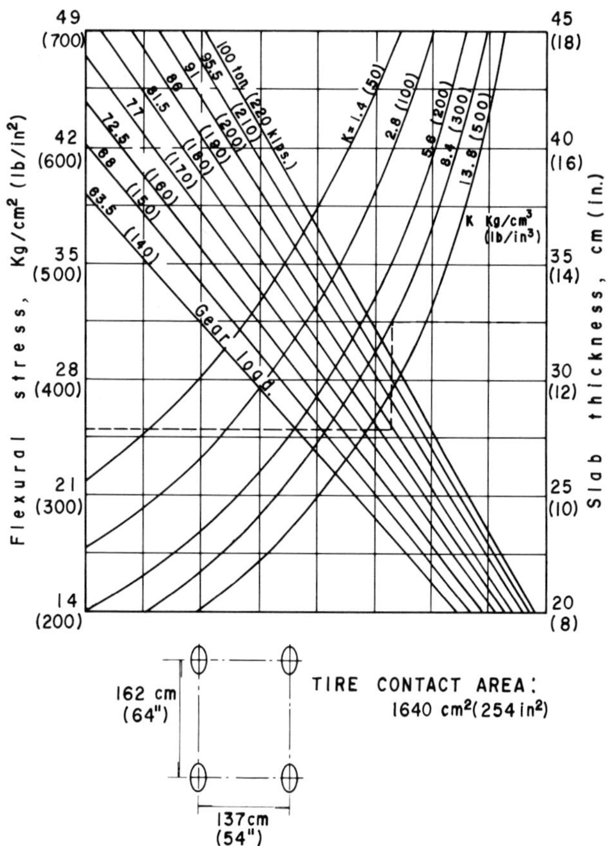

Fig. 10-8 Concrete slab thickness design graph; DC-10 aircraft [17]

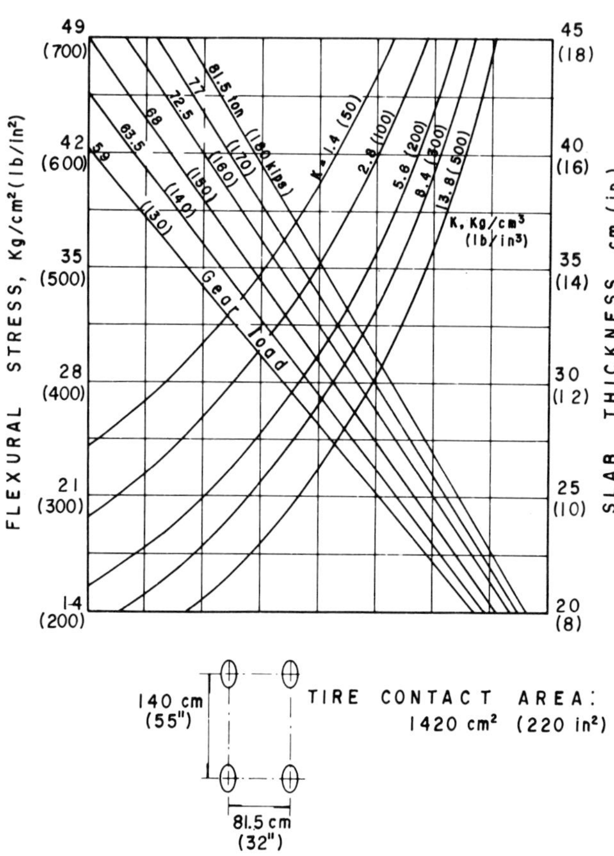

Fig. 10-10 Concrete slab thickness design graph; DC-8 aircraft (Series 62 and 63) [17]

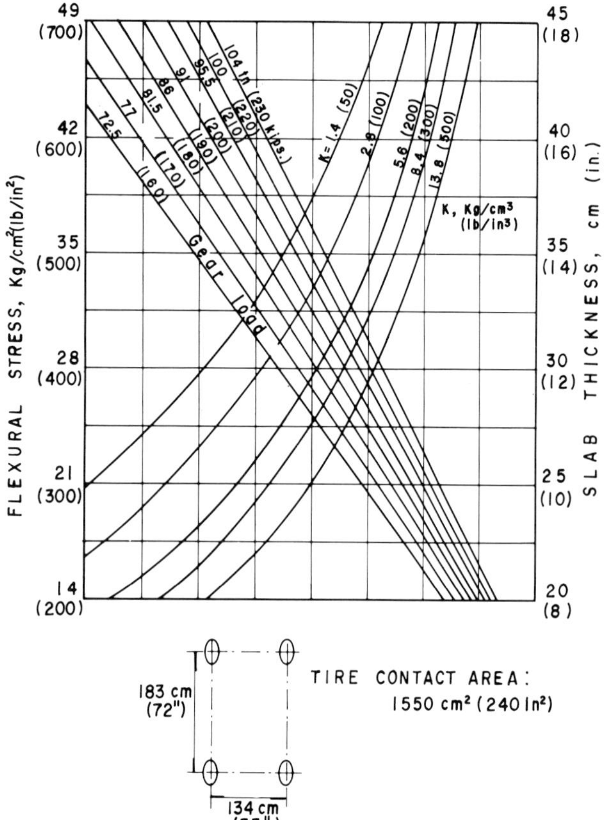

Fig. 10-9 Concrete slab thickness design graph; Lockheed L-500 aircraft [17]

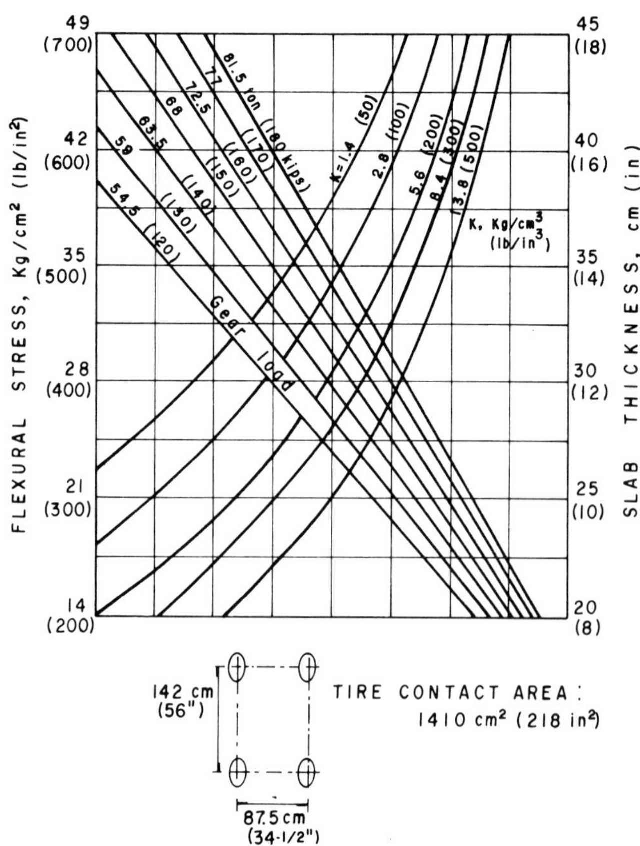

Fig. 10-11 Concrete slab thickness design graph; B-707 aircraft [17]

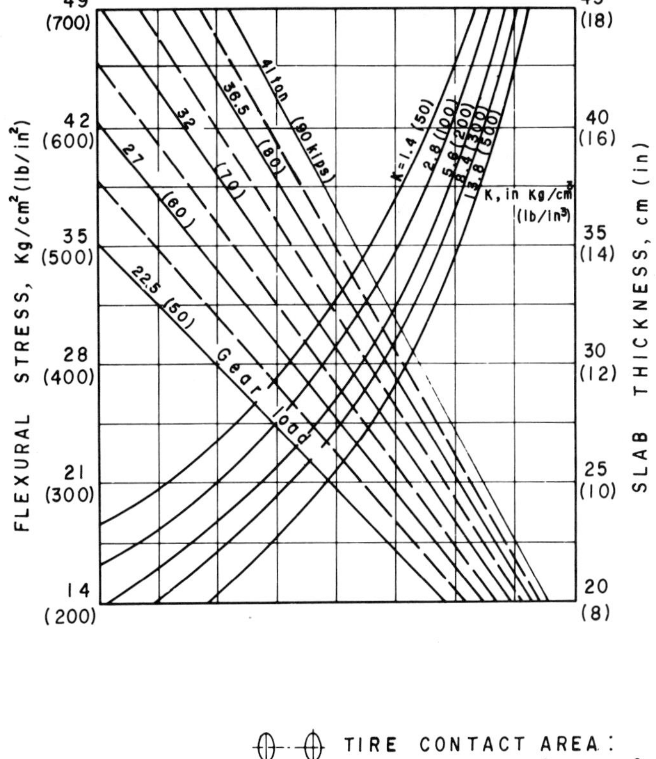

Fig. 10-12 Concrete slab thickness design graph; B-727 aircraft [17]

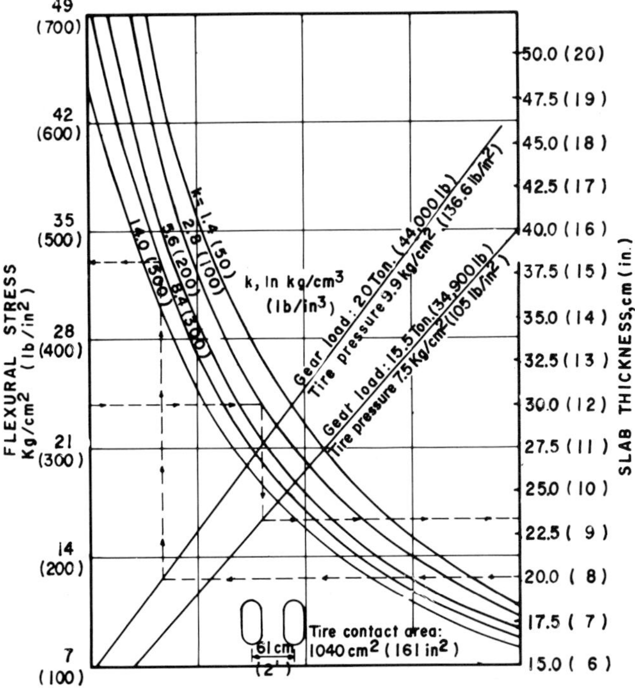

Fig. 10-13 Concrete slab thickness design graph; DC-9 aircraft [17]

For highways, rigid pavement thickness design procedures vary considerably according to the experience of the organization that recommends them. A brief description follows of the method proposed by the PCA, as presented in [2].

The first thing that must be done is to evaluate the so-called stress ratio:

$$R_r = \frac{MR\ (\text{acting})}{MR\ (\text{available})} \qquad (10\text{-}8)$$

To accomplish this, the value of the flexural stress that is produced in the slabs by wheel loads must be known, together with the flexural strength of the concrete on which the design is based.

Next, the number of load repetitions corresponding to the value of the stress ratio that is obtained must be established. The correlation between the two values is empirical and is given in Table 10-1, on the basis of the experience compiled by the PCA. In this Table it is assumed that a load on the slab that produces an *MR* value such that the stress ratio is less than 0.5, can be applied any number of times without inducing failure. If the stress ratio is 0.51, the corresponding load can be applied 400,000 times before failure of the slab. A load resulting in a stress ratio of 0.85 can be applied only 30 times before causing failure of the slab. As will be seen, a factor of safety is also recommended to be applied to the traffic load in order to obtain the loading value on which estimation of the acting *MR* is based. This factor of safety is 1.2 for important highways with large volumes of heavy traffic, 1.1 for highways or roads subjected to moderate volumes of heavy traffic, and 1.0 for highways and roads with little or no heavy traffic.

Table 10-1

Correlation between the stress ratio of a rigid highway pavement and the number of repetitions of the corresponding load it can tolerate without failure [2]

Stress ratio	Allowable repetitions	Stress ratio	Allowable repetitions
0.51	400,000	0.69	2,500
0.52	300,000	0.70	2,000
0.53	240,000	0.71	1,500
0.54	180,000	0.72	1,100
0.55	130,000	0.73	850
0.56	100,000	0.74	650
0.57	75,000	0.75	490
0.58	57,000	0.76	360
0.59	42,000	0.77	270
0.60	32,000	0.78	210
0.61	24,000	0.79	160
0.62	18,000	0.80	120
0.63	14,000	0.81	90
0.64	11,000	0.82	70
0.65	8,000	0.83	50
0.66	6,000	0.84	40
0.67	4,500	0.85	30
0.68	3,500		

Application of the PCA method requires knowledge of the distribution of traffic in terms of both single axles and tandem axles. The designer will have to know how many single-axle vehicles with loads of 12, 10, 8 and 6 t (26,400, 22,000, 17,600 and 13,200 lb) will travel over the pavement in question, just to give a random example. Similarly he will need to know what types of tandem axles will travel over the pavement and what the load on each one will be. The method depends, therefore, on a careful traffic analysis.

Also essential is knowledge of the modulus of subgrade reaction and its corrected value when there is a sub-base. An appropriate design value must also be found for the *MR* of the concrete that will be used, based on testing.

Once all this information has been obtained, the design sequence is illustrated by using the hypothetical data presented in Table 10-2 [2]. The first column presents the types of loads anticipated both for single axles and tandem axles, in accordance with the traffic analysis that has been carried out. The second column contains the same values multiplied by the factor of safety for loading, which was assumed to be 1.2.

Now it is time to use the design graphs in Figs 10-14 and 10-15. First Fig. X-14 is considered, which refers to single-axle loads. Using the different axle loads given in the table, and entering the graphs as shown by the dashed lines with arrows, it is possible to find the value of the acting *MR*, assuming a slab thickness (in this case 21.5 cm (8-1/2 in)) and knowing the corrected value of *k*, the modulus of reaction of the subgrade and sub-base (in Table 10-2, *k* is assumed to be 3.9 kg/cm^3 (140 pci)). The values of *MR* that are applied by traffic to the slab, are entered in the third column of the table.

In the illustrative example the design value for the *MR* of the concrete to be used in the slabs will be 49.6 kg/cm^2 (705 psi). With this value it is possible to enter the stress ratio (R_r) in the fourth column of Table 10-2. With the values of R_r it will be possible to determine the allowable repetitions of the load in question before failure is induced from Column *5* of Table 10-1. The corresponding values are entered in Column *5* of Table 10-2.

The repetitions that are actually anticipated for each load during the useful life of the pavement should be entered in Column *6* of Table 10-2. This information depends on the traffic estimation that has been made and the corresponding prediction of future growth. These estimates are assumed to have been made by one of the methods in use today, two of which were described in Chapter 9.

The numbers in Column *6* are divided by the corresponding ones in Column *5*, expressing the quotient as a percentage. The PCA refers to this as the *percentage of total pavement capacity used*. This expresses to what extent each of the loads will contribute to the ultimate failure of the pavement. Each will be fewer repetitions than those that would be required for just one of the loads to produce failure on its own. Column *7* tabulates these percentages, zero being the value assigned to those cases where the load is sufficiently low that it to be repeated infinitely without causing failure.

The sum of all the percentages in Column *7* is an indication of the total capacity of the pavement. In the case summarized in Table 10-2, the numbers totalled 69%, which is regarded as low. The ideal value for the sum in question is 100%. Even higher figures are permitted by the PCA method, so long as they do not exceed 125%. Naturally the acceptable percentage will depend on the importance of the road, future traffic growth and other factors that have been discussed.

For another design using the data given in Table 10-2, the same calculation sequence would be used, this time with a slab thickness smaller than the 21.5 cm (8-1/2 in) considered in this example. Different slab thicknesses would be tried, until one was found that would produce a suitable total percentage of utilization of the structural capacity of the concrete.

The trial design procedure that has been described here could be carried out with emphasis on other variables, for example

Table 10-2

Thickness design chart for rigid highway pavements after the Portland Cement Association [2]

Load		Load × F_s		*MR* (acting)		R_r	Allowable repetitions	Expected repetitions	Percentage of total capacity used
t	kips	t	kips	kg/cm^2	lb/in^2	—	—	—	—
Single axles									
13.6	30	16.3	36.0	26.0	369	0.52	300,000	3,100	1
12.7	28	15.2	33.4	25.0	355	0.51	400,000	3,100	1
11.8	26	14.2	31.2	23.3	331	< 0.50	No limit	—	0
10.9	24	13.1	28.8	—	—	—	,, ,,	—	0
10.0	22	12.0	26.4	—	—	—	,, ,,	—	0
Tandem axles									
24.5	54	29.4	64.8	29.3	416	0.59	42,000	3,100	7
23.6	52	28.3	62.3	28.2	402	0.57	75,000	3,100	4
22.6	50	27.2	60.0	27.4	390	0.55	130,000	30,360	23
21.8	48	26.1	57.4	26.6	378	0.54	180,000	30,360	17
20.8	46	25.0	54.0	25.6	364	0.52	300,000	48,140	16
20.0	44	24.0	52.8	24.5	348	< 0.50	No limit	—	0
19.0	42	22.8	50.2	—	—	—	,, ,,	—	0
18.1	40	21.8	48.0	—	—	—	,, ,,	—	0
									Σ : 69%

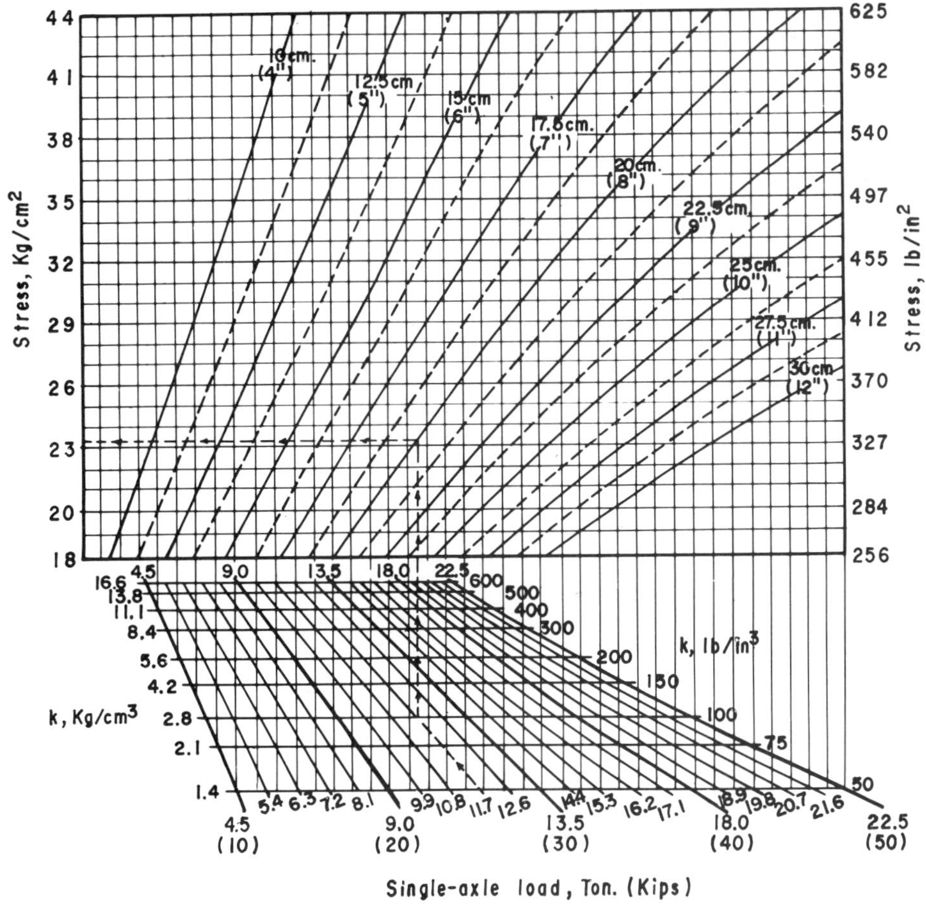

Fig. 10-14 Design graph for single axle loads. Rigid highway pavements [2]

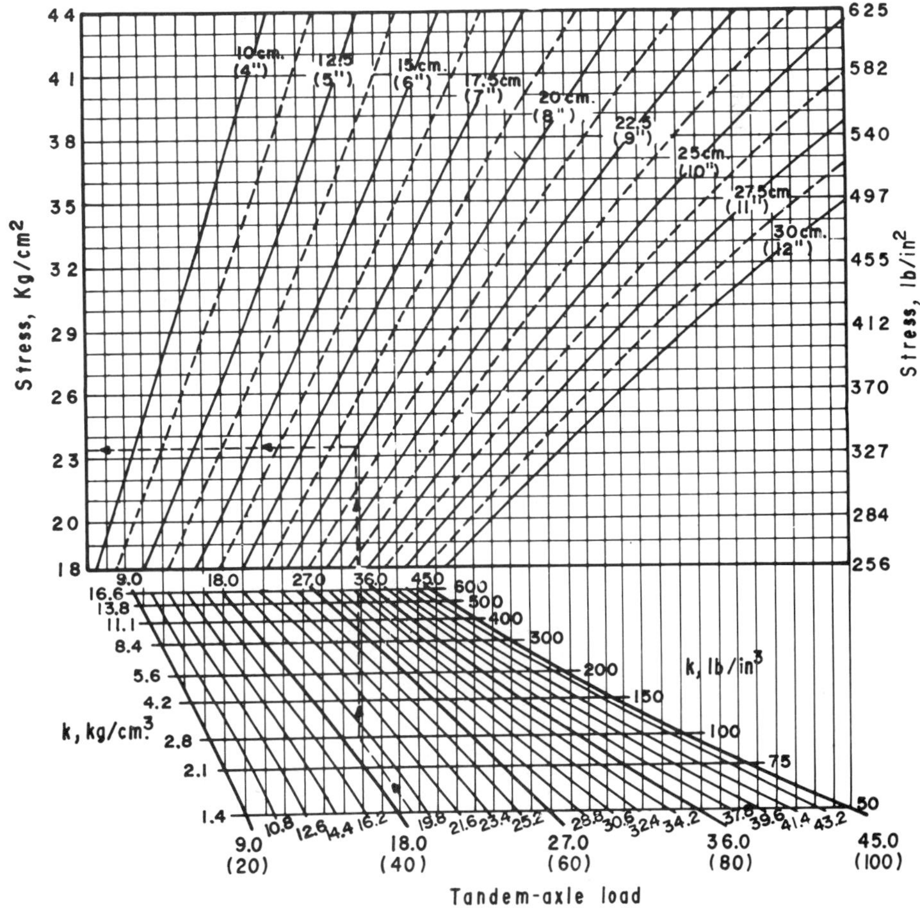

Fig. 10-15 Design graph for tandem axle loads. Rigid highway pavements [2]

reducing the design value of *MR*, which means using a weaker, and consequently less expensive, concrete.

The above method depends on a careful traffic analysis, which implies a prediction of the future growth. Designs will frequently have to be established for cases in which no traffic analysis has been made. It will then be necessary to resort to less sophisticated design procedures. Figure 10-16 [19] is a graph for determining the thickness of rigid highway pavements, only as a function of the heaviest load transferred by a single-axle vehicle with a dual wheel assembly. The graph is based on the design where the corners of the slabs are equipped with devices for suitably transferring the loads to adjacent slabs.

The graph in Fig. 10-16 uses the same values for the mechanical properties of the concrete and supporting soil as those in similar graphs. It is based on formulas by PICKETT [19] which are semi-empirical, based on data for field behavior. To use the graph, the load corresponding to the heaviest anticipated dual axle must be multiplied by a safety factor, 1.2 and the modulus of flexural strength of the concrete at failure must be divided by an appropriate safety factor, 2, in order to obtain a working value.

Although the thickness design method initially presented is based on a complete traffic analysis, it should not be regarded as lacking practical sense. The use of a concrete pavement in a highway for which traffic data are unknown is inconceivable. Concrete pavements are only economical for important roads with high loads and volumes of traffic; such traffic will obviously be carefully studied, before design is undertaken.

10.4 Joints

The joints used in rigid pavements can be grouped into four main categories:

— Contraction joints

— Expansion joints

— Construction joints

— Warping or hinged joints

Moreover, joints are usually described as longitudinal or transverse, according to their direction in relation to the runway or road.

The purpose of contraction joints is to relieve the tension stresses used by shrinkage of the concrete. The purpose of expansion joints is to enable the concrete slabs to expand towards one another during very hot weather without suffering damage. Construction joints are used during interruptions in pouring operations; their purpose is to ensure structural continuity. Lastly, the purpose of warping or hinged joints is to avoid any cracking along the central axis of pavements or along the junctions between the different rows of slabs, which would make the edges of the slab curl upwards when loaded in the center.

Joints are usually formed either by making a groove in the concrete, but maintaining continuity through the joint by placing concrete against concrete, by forming a groove which is filled with some appropriate material, or by establishing continuity across the groove with smooth steel bars (dowel bars) or corrugated steel bars (tie bars).

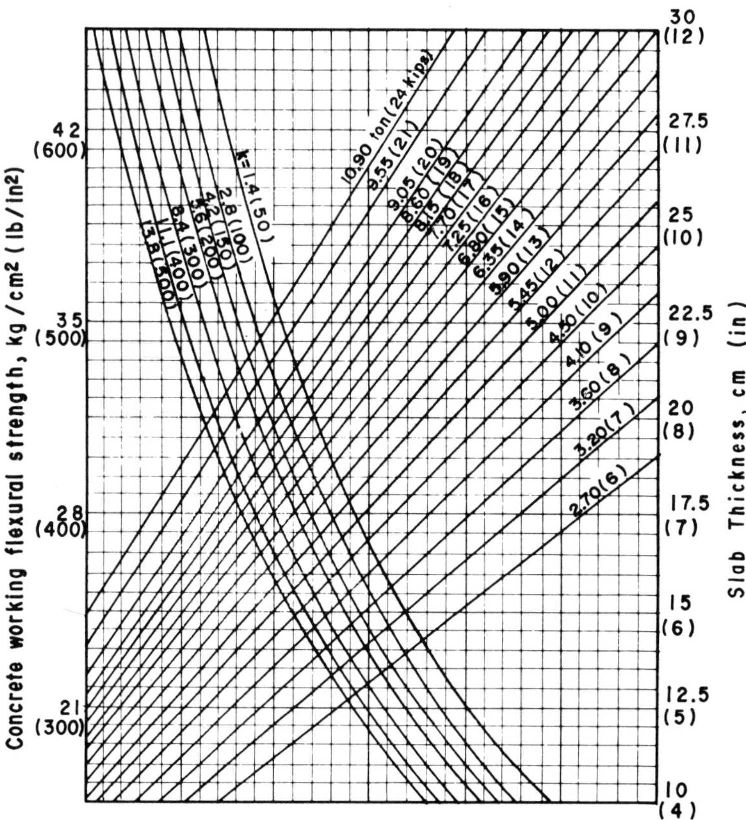

Fig. 10-16 Design chart by the PCA for finding the thickness of rigid road pavements subjected to loads from dual wheel assemblies

Figure 10-17 [20] shows the most common types of joints for highway pavements (Plates 10-2 to 10-4). The difference between contraction joints *a* and *b* is that in *a* the groove is incomplete, the hollow that forms is filled with a plastic sealing material and the crack will subsequently continue to develop of its own accord; whereas in joint *b*, two slabs are brought together, their eventual surface separation crack is sealed with the plastic material that fills the upper groove that is made. In all contraction joints a lubricated dowel bar is used so that the slabs can retract without generating tensile stresses that might cause cracking. Parts *c* and *d* show typical expansion joints, *c* has a dowel bar to prevent vertical differential movement with an expansion cap to enable relative horizontal movement, while *d* is a simple joint, in which the edge of the slab has been thickened

Plate 10-2 Preparations for the installation of joints in runways

Plate 10-3 Runway construction. Assembly and jointing stage

for added resistance to flexural cracking. Type *e* is a typical example of a construction joint, with a shear key and corrugated tie bar for both vertical and transverse load transfer. Lastly, *f* is a typical warping or hinged joint, also protected by a corrugated tie bar.

In highways, dowel and tie bars are recommended in expansion joints. They are usually omitted from contraction joints when the spacing between the joints is less than 6 m (20 ft), unless local service conditions are particularly severe, as in the case of intersections or points where pavements of different rigidities join. The matched joint (Fig 10-17e) is also used as a warping joint. Wherever such effects are feared, it will be advisable to equip the joints with tie bars.

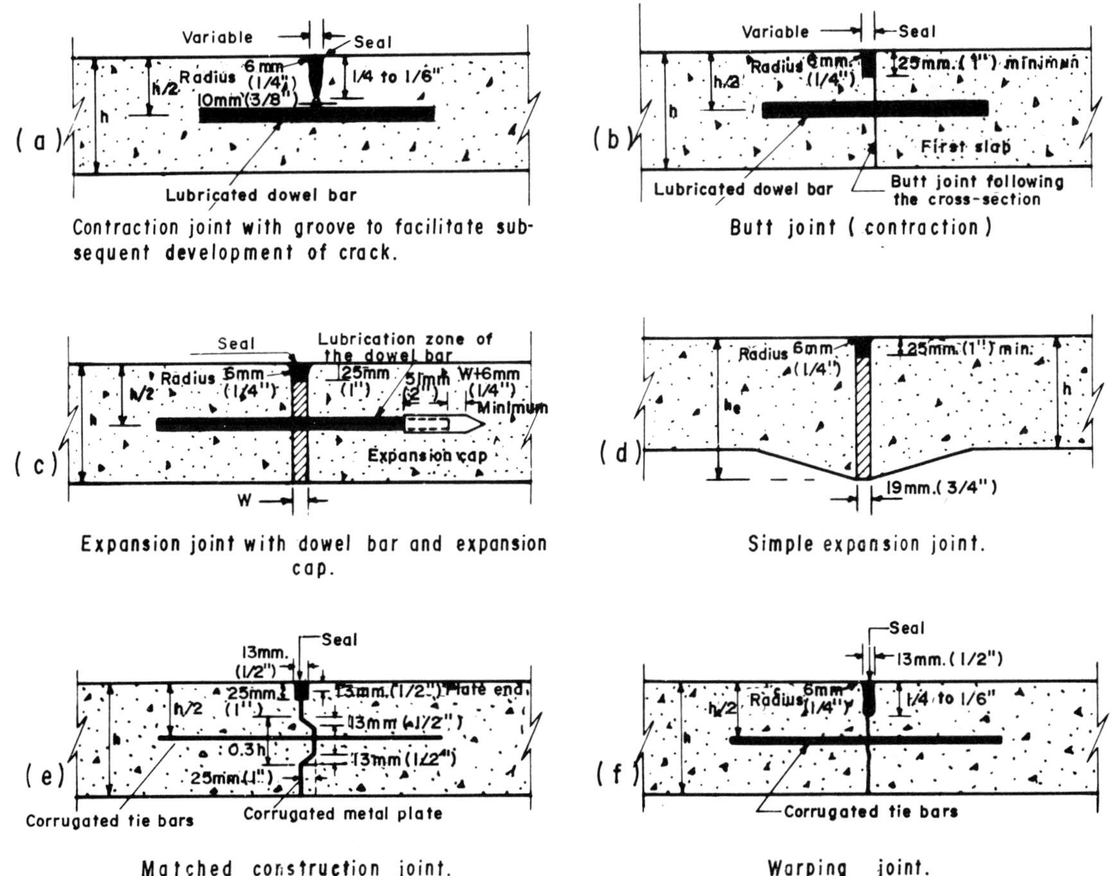

Fig. 10-17 Common joints in rigid pavements [20]

Plate 10-4 Runway construction. Assembly and jointing stage

Table 10-3 groups the minimum requirements recommended for installing dowel bars and load transfer devices in pavements of different thicknesses. The typical device is a smooth, round steel rod. The corrugated tie bar is not generally intended as a typical load transfer device. It is installed to resist tension stresses that may be generated by friction between the slab and the sub-base. Whenever these elements prove necessary the spacing recommendations in Table 10-3 [20] should also be observed.

In runways, the same types of joints as in roads are used (Fig. 10-17), but the importance of installing them carefully in the correct position is even greater. This usually means more extensive use of dowel bars and tie bars. The position of all these elements for a successful design is a complex matter far beyond the scope of this book. For the purpose of illustration, Fig. 10-18 [5] gives a design graph, proposed by the PCA, which enables determination of the minimum spacing between load transfer devices in commercial airport pavement slabs.

10.5 Failures Common in Rigid Pavements

Failure in rigid pavements is usually caused by one of two principal factors. The first involves defects in the slab and includes 1) deficiencies in the concrete due to the use of unsuitable materials and aggregates, break-up as a result of reaction of the aggregates with alkalis in the cement, deterioration from the salts used to protect the concrete from extreme cold in regions with severe climatic conditions, and 2) construction defects or structural insufficiency of the slab, such as the incorrect installation or insufficient number of load transfer devices, frictional restrictions imposed on slab movement by the sub-base, warping of the slabs or poor performance of contraction and expansion joints. The other main cause of failure in rigid pavements is the inadequate structural interaction of slab, sub-base, subgrade, fill and even the foundation soil. This category includes failures due to pumping, warping and breaking of corners or edges owing to insufficient slab support. Failures are often caused by a combination of factors, rather than just one, which only complicates their diagnosis and correction.

Unsuitable aggregates with poor durability cause cracks. They start as very closely spaced capillary cracks and develop in semicircles around joints or the edges of the slabs. The phenomenon is progressive and usually ends in total break-up of the slab.

Other common defects causing break-up of the concrete stem from errors in its manufacture: the use of a mixture that is too wet or aggregates containing too many fines, dirty aggregates or deterioration from the use of salts. During the curing period, concretes sometimes suffer excessive cracking due to contraction. Typical cracks of this kind are short and distributed at random over the slab surface, both in the longitudinal and the transverse directions. This cracking has nothing to do with traffic stresses. Traffic can, however, later enlarge them.

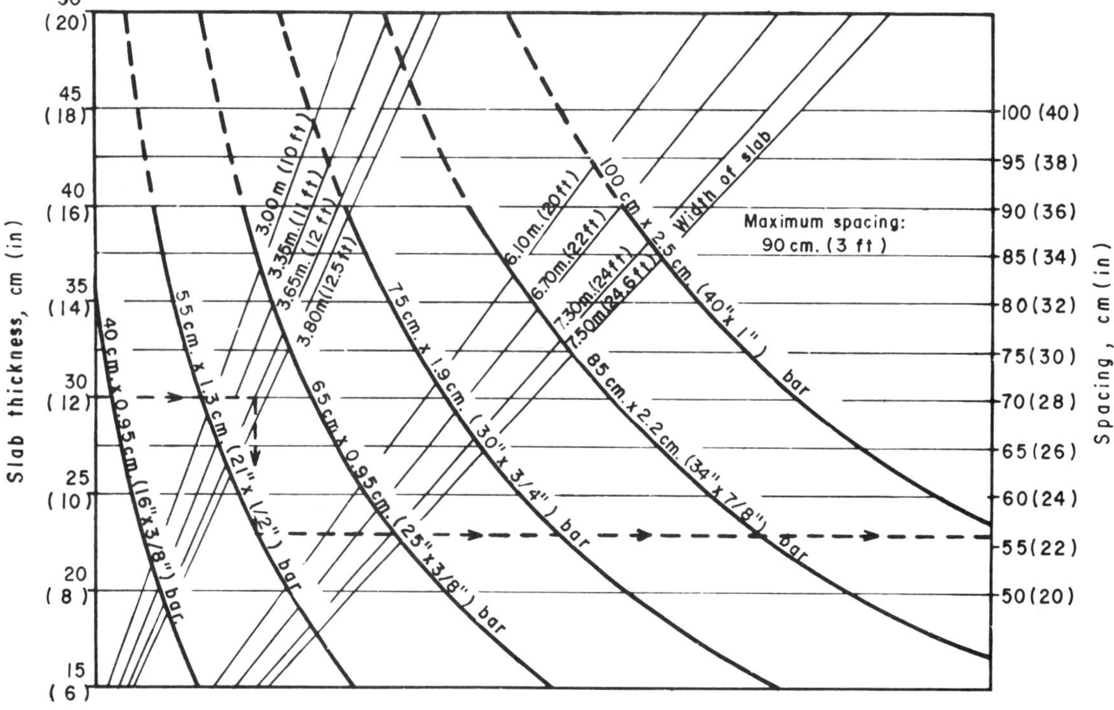

Fig. 10-18 Minimum spacing between load transfer devices in concrete slabs in runway pavements [5]

Table 10-3

Spacing recommended for load transfer devices between slabs in rigid highway pavements [20]†

Type and class of steel	Working stress		Pavement thickness		No. 4 bars Total length		No. 4 bars Spacing in cm (in) Width of strip						No. 5 bars Total length		No. 5 bars spacing in cm (in) Width of strip					
	kg/cm²	lb/in²	cm	in	cm	in	3m (10 ft)		3.3m (11ft)		3.6m (12ft)		cm	in	3m (10ft)		3.3m (11ft)		3.6m (12ft)	
			15	6			115	45.0	105	41.3	95	37.3			120	47.2	120	47.2	120	47.2
Structural			17.5	7			98	38.5	90	35.4	82	32.2			120	47.2	120	47.2	120	47.2
grade bar	1,500	21,300	20	8	50	20	85	33.4	77	30.3	70	27.5	60	23.6	120	47.2	120	47.2	112	44.0
or rod			22.5	9			75	29.5	70	27.5	62	24.4			120	47.2	107	42.0	100	39.3
			25	10			67	26.3	62	24.4	57	22.4			107	42.0	97	38.1	90	35.4
Medium grade			15	6			120	47.2	120	47.2	117	46.0			120	47.2	120	47.2	120	47.2
bar or rod			17.5	7			120	47.2	110	43.3	100	39.3			120	47.2	120	47.2	120	47.2
	1,900	27,000	20	8	60	24	105	41.3	95	37.3	87	34.2	68	26.7	120	47.2	120	47.2	120	47.2
			22.5	9			92	36.1	85	33.4	77	30.3			120	47.2	120	47.2	120	47.2
			25	10			85	33.4	77	30.2	70	27.5			120	47.2	120	47.2	120	47.2
High strength			15	6			120	47.2	120	47.2	120	47.2			120	47.2	120	47.2	120	47.2
section, bar			17.5	7			120	47.2	120	47.2	120	47.2			120	47.2	120	47.2	120	47.2
or rod	2,300	32,600	20	8	68	27	120	47.2	117	46.0	107	42.0	83	32.6	120	47.2	120	47.2	120	47.2
			22.5	9			115	45.0	105	41.3	95	37.3			120	47.2	120	47.2	120	47.2
			25	10			102	40.1	92	36.2	85	33.4			120	47.2	120	47.2	120	47.2

spacing between tie bars should not exceed 1.2m (4ft)

Cracks caused by defective performance of dowel-bars are almost always a result of poor lubrication of these bars, which inhibits the free movement for which they were designed. Excessive spacing between dowels is another cause of problems.

Another common defect caused by the poor performance of joints occurs when they are insufficient in number (too widely spaced) along an important stretch of pavement. The concrete creates its own contraction and expansion joints by cracking, but these cracks are spaced at irregular intervals, giving the pavement a distressed appearance. They do not always cause a genuine structural deficiency, because the cracks that form do not release shear stresses and in principle work like real joints. The performance of these natural cracks is seldom satisfactory in the long term, for they lack the appropriate plastic seal which keeps water out. Moreover the concrete within them gradually becomes detached and is ground into a powder with a self-abrasive action that enlarges the cracks. Water in the enlarged cracks sometimes leads to pumping phenomena. Naturally, insufficient slab thickness leads to structural cracking under traffic loads.

Of the phenomena influencing sub-base material, pumping, which has already been mentioned, is the most important. It leads to slab destruction, particularly in corner zones. In border slabs, when the shoulder material generates appreciable restriction to slab movement by friction, (which is commonly the case with sandy materials) cracks more or less parallel to the pavement edge develop.

Movements in the foundation ground or in large thicknesses of compressible fill cause cracking of the slabs when large differential settlements occur in a short time over short stretches of pavement. Experience indicates, however, that the long term flexibility of rigid pavements in the face of these problems is considerably greater than is usually believed.

Moreover, cracks due to long term settlement seldom lead to serious deficiencies in pavement performance, especially if they are appropriately sealed as they appear.

In runways, systematic structural damage is fairly common in wheel assembly channelization areas. This is a sign that the loads exceed the design value, either in their magnitude or number of repetitions; any runway presenting this defect should therefore be structurally strengthened.

10.6 Evaluation of Existing Pavements.

Until recent years, relatively little attention was given to the problem of evaluating the condition and performance of rigid pavements. Pavements were divided into two categories, satisfactory ones and those in need of repair, but there was no systematic methodology relating performance, level of serviceability, and programmed repairs. Moreover it was unusual for a design to include any performance objective. Carey and Irick [21] developed the Serviceability Index concept similar to that established for flexible pavements. It is based on somewhat arbitrary user-opinion ratings where the pavement is graded on a scale usually ranging from 0 to 5.

Much of the research involved in establishing the serviceability condition of a rigid pavement at a given moment is related to the surficial state of the constituent slabs. This type of evaluation, which can be carried out with specialized equipment, such as road roughness indicators, or profilometers [10, 22, 23] will not be described here, it is beyond the scope of this book.

Further research will be required to establish reliable correlations between the surficial state of pavements and subjective human interpretation of condition, so that equations can be developed that will enable pavement serviceability indices to be established on the basis of technically relevant factors. Furthermore, requirements for aircraft, operation and safety, related to the roughness of rigid runway pavements are needed.

It is currently assumed that if a set of serviceability index figures for a pavement, covering a reasonable period of time, are correlated with the history of traffic growth and climatic conditions, a clear indication is obtained of the conditions of that pavement from which a maintenance or reconstruction program can be readily formulated.

Evaluation of an existing rigid pavement should also include consideration of the structural condition of the whole support system: the slab, sub-base, subgrade, fill material and underlying ground. It is with reference to this second phase of the evaluation that some comments will be made in the paragraphs that follow.

It is of vital importance in these evaluation studies, which might be referred to as structural analyses, that the tests conducted do not destroy any part of the pavement structure. This can be accomplished by taking measurements on the pavement surface, that relate to the structural characteristics that are of interest. Currently there are several such methods, which can be grouped into three categories:

1) Methods in which the structural response of a pavement to a static load or to the application of a known slow-moving load is measured
2) Methods in which load repetitions are used
3) Me·hods in which measurements are taken with nuclear instruments, or some other form of radiation

In the methods belonging to the first category, deflection of the pavement surface is usually measured. The BENKELMAN Beam, already mentioned in relation to flexible pavements, has application in rigid pavements too. The aim is to establish the amount of deflection that indicates a need for strengthening or reconstructing the pavement. Plate-bearing tests have already been mentioned earlier in this chapter. When conducted on the pavement surface, they are another good example of evaluation tests of the first category.

Among the methods in which load repetitions are employed, the ones most often used are those where instruments are placed in selected test sections exposed to the action of real traffic. [24, 26] describe the measuring instruments required, together with some typical installations.

For the time being, the use of nuclear equipment is practically limited to the determination of water content and density of the concrete or underlying layers. Radar can measure pumping voids below pavements.

In the evaluation of rigid pavements, tests have been devised which are destructive in that they interrupt the structural continuity at a specific point in the pavement. They usually involve the extraction of concrete cores or soil sampling followed by testing with the usual methods.

If an evaluation study leads to the conclusion that a rigid pavement has a structural deficiency, it will have to be strengthened. Currently this is accomplished by adding a layer of asphalt concrete or Portland cement concrete.

10.7 Strengthening of Rigid Pavements

.1 Pavements strengthening by asphalt concrete

In [27] the following formula is proposed for calculating the thickness of asphaltic surfacing to be laid over the existing slabs once all the concrete that is seriously cracked or damaged has been removed and replaced and the slabs that are not so seriously damaged have been repaired.

$$e_r = 2.5\,(Fh_r - h_e) \tag{10-9}$$

where: e_r is the thickness of asphaltic surfacing required, in cm (in); h_r is the thickness of rigid pavement required in order to achieve successful pavement performance, in cm (in); h_e is the existing thickness of rigid pavement, in cm (in); and F is an empirical factor that varies from 0.7 to 1.0. It depends on the characteristics of the soil supporting the slabs, expressed by means of the modulus of subgrade reaction, for example. The value of 0.7 may correspond to moduli of reaction of about 11.1 kg/cm^3 (400 pci), while 1.0 should be taken for values of k of 1.4 kg/cm^3 (50 pci) or below.

It can be seen, therefore, that the only thing accomplished by Eq. (10-9) is that it expresses in equivalent flexible pavement thickness the strengthening contribution of the rigid pavement, which now becomes a part of a semi rigid, semi flexible system. The values that are given by Eq. (10-9) can be divided by 1.5 when the existing concrete pavement has a cement-treated sub-base.

.2 Pavement strengthening by Portland cement concrete

In [5] three cases are distinguished in the solution of this problem:

1) ***There is partial contact*** between the old concrete and the new: This condition is considered to exist when there is no separating element between the two slabs, so that the strengthening layer comes into direct contact with old slab. This is a favourable field condition, which usually ensures a good performance of both slabs. The formula for calculating the thickness of the strengthening layer is [28]

$$e_r = \sqrt[1.4]{h_r^{1.4} - C h_e^{1.4}} \tag{10-10}$$

where: e_r is the thickness of Portland cement concrete required to strengthen the pavement, in cm (in); h_r is the thickness of rigid pavement required over the sub-base and the existing bearing ground, to ensure good performance, in cm (in); h_e is the thickness of existing rigid pavement that is to be strengthened, in cm (in); and C is a coefficient which depends on the existing structural condition. It has a value of 1.0 when the general conditions of the existing pavement are good; 0.75 when there are cracks at the corners of the slabs, but no signs of progressive or recent failures; lastly, C has a value of 0.35 when the general conditions of the existing pavement are poor, with abundant cracks.

2) ***There is no reliable contact*** between the old concrete and the new. This is the case where some layer is placed between the two slabs to break continuity. This may be an asphaltic prime coat, a plastic sheet, a layer of granular material or a layer of asphalt concrete. This type of separator may be necessary when the surface of the existing pavement is extremely uneven or in very

poor condition, or when for some reason it is wished to raise the grade substantially. In this case [5] proposes the following:

$$e_r = \sqrt{h_r^2 - Ch_e^2} \quad (10\text{-}11)$$

where the letters have the same meaning as in Eq. (10-10) and C has the values given previously. Lack of a strong contact between the two slabs leads to greater thicknesses of strengthening material.

Where it is wished to raise the grade and a thick layer of granular material is placed on top of the existing pavement, it may prove necessary to evaluate the modulus of reaction of this layer in place above the old pavement and calculate the strengthening layer of concrete as though it were a new rigid pavement.

3.) ***There is complete contact*** between the old concrete and the new: This contact is achieved by means of careful preparation of the surface of the existing pavement and by placing a layer of effective binding material between the two pavements. This layer may consist of cement slurry and sand or epoxy resins [29]. This type of preparation is used particularly for strengthening runway pavements and results in genuine unity of action of the two slabs. In this case the thickness of the strengthening layer is given by the expression:

$$e_r = h_r - h_e \quad (10\text{-}12)$$

where the letters have the same meanings as before.

Complete contact is only recommended when the structural condition of the existing pavement is good, or when any structural defects that may have occurred have been repaired already.

In many pavement strengthening programs, a far more favourable condition can be achieved with interesting economic consequences if, before applying Eqs. (10-10) and (10-11), the value of C is increased by repairing the surface of the slabs or replacing the most damaged ones. This is an active alternative that should always be given due consideration.

The joints that are installed and the cracks in the existing pavement will affect the strengthening layer, unless precautions are taken to avoid it. This is particularly true in pavements having complete contact and partial contact between the strengthening layer and the pavement. In the case of joints, such effects are usually avoided by making the new ones coincide with the ones in the old pavement. When spacing between the joints in the existing pavement is insufficient and additional ones are required, it is advisable to install these new joints throughout the entire thickness of the new strengthened pavement (including the original concrete), especially in the cases of complete and partial contact. When no binding layer is placed between the old pavement and the new one, this precaution can be omitted, this is a practical advantage of strengthening layers without a reliable contact, that makes them attractive when the existing pavement is suffering the consequences of inadequate jointing. In this design there is considerable freedom to the placement of one type of joint in the strengthening layer compared to a different type of joint in the old pavement.

Far harder to avoid is the effect of cracks in the original pavement on the layer of strengthening pavement. The situation may be relieved by the correct sealing of these cracks with plastic materials or the use of strengthening layers with no binding materials, but the problem will undoubtedly not be eliminated. What must be taken into consideration, however, is that the later appearance of cracks in the layer of strengthening material does not necessarily signify a structural deficiency in the pavement, especially if the evolution of these cracks is carefully controlled.

APPENDIX 10A
WORKED PROBLEM

10a.1 Solved Problems — Rigid Pavement Design

Design a rigid pavement using the Portland Cement Association (PCA) method. The soil referred to in this problem is the same one described in Problem N° *1* of Appendix 9e.1. The general data are: 4-lane highway in mountainous region; Design Period = 20 years; Modulus of subgrade reaction k = 5.6 kg/cm³ (200 pci)

.1 Traffic Analysis

The volume of mixed traffic that is taken for designing the rigid pavement (DT) is determined using the following expression proposed by the PCA (in the Table 10-2 relevant traffic data were provided, but in this problem they will be calculated using the PCA method).

$$DT = \frac{100p}{100 + T_{ph}(j-1)} \times \frac{5000\,N}{KD}$$

where: DT is for mixed traffic, valid for design purposes; P is the number of passenger cars, including pick-ups, per lane per hour; N is the number of lanes in both directions; T_{ph} is the percentage of trucks during peak hours = 2/3 of the percentage of heavy vehicles in both directions; j is the number of passenger cars equivalent to one truck, which may be for: Mountainous ground = 4, Flat ground = 2; k is the hourly traffic volume for design (VHD) which is expressed as a percentage of the DT and may be: for highways with high volumes of traffic = 15%, for highways with medium volumes of traffic = 12%; D is the maximum traffic in one direction, in percentage, during peak hours, which varies from 50 to 75% and may be for highways with high volumes of traffic = 67%, and for highways with medium volumes of traffic = 60%.

Application of the above equation to the case under consideration leads to: Percentage of heavy vehicles $= \frac{1{,}664}{12{,}240} = 14\%$ (in two directions).

This is obtained by adding B, $C2$, $C3$, $T2\text{-}S2$ and $T3\text{-}S2$ of Table 9e-1 in Appendix 9e.1, and dividing the result by the total number of vehicles.

P = 1,200 vehicles

This value of P is obtained as follows: If the value of cars and pick-ups is obtained from a direct traffic count, this value will be used, but if, as is often the case, such information is not available, P is obtained from Table 10a-1.

Table 10a-1

Number of cars (including pick-ups) per lane per hour (according to the Portland Cement Association, PCA)

Type of road	Number of cars, P
Urban motorways	1500
Suburban motorways	1200
Motorways	1000
Roads with medium traffic volume	700–900
Roads with low traffic volume	500–700

In this problem P = 1,200 was assumed

N = 4 lanes

T_{ph} = 2/3 of the percentage of heavy vehicles = ⅔ × 14 = 9.4%.

j = 4

K = 12%

D = 60%

Thus, the design traffic will be:

$$DT = \frac{100 \times 1200}{100+9.4\,(4-1)} \times \frac{5000 \times 4}{12 \times 60} = 25800$$

(Two directions)

Therefore the number of heavy vehicles = 25,800 × 0.14 = 3,610 trucks per day, and the number of trucks in one direction will be 1,800 per day.

The hourly average volume of heavy vehicles in one direction (vph) is:

$$vph = \frac{25800}{2 \times 24} = 538$$

Using Table 10a-2, the percentage of trucks in the design lane is obtained. The resulting percentage of trucks in the design lane is 88%. The number of heavy trucks in the design lane for a period of 20 years is 1,800 trucks in one direction/day × 0.88 × 365 days/year × 20 years = 10,500,000 trucks during the design period.

The value of 10,500,000 heavy vehicles is distributed over the number of vehicles corresponding to each type of axle anticipated for the road under design. The aim is to reach an axle distribution like the one presented in Table 10-2, in the text of this chapter. This can be accomplished with data from direct traffic counts or by extrapolating regional information that is already available. In this problem, it is assumed that the axle loads exerted on the pavement will be the ones appearing in Column *1* of Table 10a-3. The distribution factors for each type of axle will be the ones in Column *3* of the same table. These are given for every thousand axles, that is to say for every 1,000 heavy vehicles travelling along the design lane in the 20-year period, which is a number equal to the coefficient corresponding to that axle. With this information, Table 10a-3 is self-explanatory. With this table it is possible to establish the number of load repetitions expected for each axle during the design period (20 years).

Table 10a-2

Percentage of trucks in the design lane for 4-lane highways

Hourly average of heavy vehicles in one direction (hundreds of trucks)	Percentage of heavy vehicles in the design lane, %
2	96
4	90
8	84
12	80
16	77
20	76
24	74
28	74
32	75
36	77

Table 10a-3

Distribution of heavy vehicles

Load per axle t	kips	Accumulated equivalent axle loads	Distribution factor for every 1000 axles	Anticipated load repetitions
Single axles				
13.6	30	10.500,000	0.30	3,150
12.7	28	10.500,000	0.50	5,250
11.8	26	10.500,000	56.0	588,000
10.9	24	10.500,000	60.0	630,000
10.0	22	10.500,000	78.0	819,000
Tandem axles				
24.5	54	10.500,000	0.30	3,150
23.6	52	10.500,000	0.50	5,250
22.6	50	10.500,000	1.50	15,750
21.8	48	10.500,000	10.0	105,000
20.8	46	10.500,000	5.7	60,000
20.0	44	10.500,000	3.0	31,500
19.0	42	10.500,000	3.5	36,750
18.1	40	10.500,000	4.0	42,000

Table 10a-4
Rigid pavement design problem — First trial

1 Load per axle		2 Load per axle + 20% of impact		3 Acting stress		4 Stress ratio	5 Permissible load repetitions	6 Anticipated load repetitions	7 Percentage of total capacity used
t	kips	t	kips	kg/cm^2	lb/in^2				
Single axles									
13.6	30	16.3	36.0	26	369	0.54	180,000	3,150	2
12.7	28	15.2	33.4	25	355	0.52	300,000	5,250	2
11.8	26	14.2	31.2	23.6	335	0.49	—	588,000	0
10.9	24	13.1	28.8	21.0	298			630,000	0
10.0	22	12.0	26.4	20.0	284	—	—	819,000	0
Tandem axles									
24.5	54	29.4	64.8	28.8	410	0.60	32,000	3,150	10
23.6	52	28.3	62.3	27.8	396	0.58	57,000	5,250	9
22.6	50	27.2	60.0	26.5	376	0.55	130,000	15,750	12
21.8	48	26.1	57.4	25.2	358	0.52	300,000	105,000	35
20.8	46	25.0	54.0	24.3	345	0.51	400,000	60,000	15
20.0	44	24.0	52.8	23.8	338	0.49	—	31,500	0
19.0	42	22.8	50.2	22.5	319	0.47	—	36,750	0
18.1	40	21.8	48.0	22.0	312	0.46	—	42,000	0
									$\sum = 85\%$

.2 Structural Analysis of the Pavement

Design data: Subgrade modulus $k = 5.6$ kg/cm^3 (200 pci); Factor of safety for load 1.2 (Highways with medium volumes of traffic); $MR = 48$ kg/cm^2 (680 psi) (design data); A hydraulic sub-base 10cm (4 in) thick will be laid over the subgrade. The modulus of reaction k of the sub-base $= 6.1$ kg/cm^3 (220 pci), estimated by means of the graph in Fig. 10-3.

The following presents in tabulated form design calculations for the rigid pavement, with a trial (first) thickness of 20 cm (8 in). Reference should be made to Table 10a-4 and also the explanation of Table 10-2 in the text.

The stress acting on the slab (Column *3*) in kg/cm^2 (psi) was determined using the design graphs in Figs. 10-14 (single axles) and 10-15 (tandem axles), for which the following elements were necessary (in the order used): load per axle in tons (single or tandem), modulus of reaction, k, of the sub-base and trial slab thickness. The stress ratio was obtained by dividing the acting or wheel load, stress by the modulus of rupture of the concrete.

The allowable load repetitions (Column *5*) were obtained using Table 10-1. The number of load repetitions anticipated (Column *6*) was obtained by studying an estimated distribution based on the general distribution described above the number of heavy vehicles in the design lane (10,500,000).

Lastly, the table illustrates that the percentage used of the capacity of this assumed pavement is 85%; this is a rather low value, although perhaps acceptable. Another trial could be conducted with a slab thickness of 18 or 19 cm (7.1 or 7.5 in).

$^+$ This problem is by courtesy of M.R. Carrizales.

REFERENCES

1. Yoder, E. J., *Principles of Pavement Design*, John Wiley and Sons, 1967, Chap. 18.
2. Portland Cement Association, "Thickness Design for Concrete Pavements, Subgrades, Sub-bases and Shoulder for Concrete Pavement", *P.C.A. Concrete Information,* Paving Bureau, Chicago, III. 1966.
3. McDowell, Ch., "The Relationship of Laboratory Testing to Design for Pavements and Structures on Expansive Soils", *Quarterly of the Colorado School of Mines*, 1959, Vol. 54, No. 4.
4. Parcher, J. B. and Liu, P. Ch., "Some Swelling Characteristics of Compacted Clays", *Journal of Soil Mechanics and Foundation Division. Proc. ASCE*, 1965, Vol.SM3-91.
5. Packard, R. G., "Design of Concrete Airport Pavements", *P.C.A. Engineering Bulletin*, Chicago, II., 1973.
6. Juárez Badillo, E. and Rico, A., *Mecánica de Suelos. Vol: II. Teoria y aplicaciones de la Mecánica de Suelos,* Limusa: México, 1973, Chap. I.
7. Terzaghi, K., "Permafrost", *Harvard Soil Mechanics Series*, Harvard University Cambridge, Mass., 1952, No. 37.
8. Highway Research Board, "Highway Pavement Design in Frost Areas", Symposium, Part I, *HRB*, 1959, Bull. No. 225.

9. Highway Research Board, "Highway Pavement Design in Frost Areas", *HRB*, 1963, Record No. 33.
10. Highway Research Board, "Rigid Pavement Design. State of-the-Art", *HRB*, 1968, Spec. Report No. 95.
11. Yoder, E. J., "Pumping of Highway and Airfield Pavements", *Proc. HRB*, 1957, Vol. 36.
12. Highway Research Board, "Performance of Granular Subbases under Concrete", *HRB*, 1958, Bull. No. 202.
13. Childs, L. D. and Kapernick, J. W., "Tests of Concrete Pavement on Gravel Sub-bases", *Proc. ASCE*, Highway Division, 1958, Vol. 84, HW3.
14. Reference [1], Chap. 3.
15. Juárez Badillo, E. and Rico, A., *Mecánica de Suelos. Vol: II. Teoria y Aplicaciones de la Mecánica de Suelos*, Limusa: México, 1973, Chap. X.
16. Reference [1], Chap. 8.
17. Packard, R. G., "Computer Program for Airport Pavement Design", *P.C.A. Special Report*, Chicago, Il., 1967.
18. Lewis, K. H. and Harr, M. E., "Analysis of Concrete Slabs on Ground Subjected to Warping and Moving Loads", *Highway Research Record N°291. Design, Evaluation and Performance of Pavement Systems* (19 Reports), Highway Research Board, Washington, D.C., 1969.
19. Portland Cement Association, "Concrete Pavement Design for Roads and Streets", *PCA Concrete Information*, Paving Bureau, Chicago, Il., 1951.
20. IMCYC, "Prática recomendada para el diseño de pavimentos de concreto", Trans. Mexican Cement and Concrete Institute, México, D.F., 1968, A. C. of the ACI Manual 325-58.
21. Carey, W. N. and Irick, P. E., "The Pavement Serviceability Performance Concept", *Highway Research Board* Washington, D.C., 1960, Bull No. 250.
22. Hveem, F. N., "Devices for Recording and Evaluating Pavement Roughness", *Highway Research Board*, Washington, D.C., 1960, Bull. No. 264.
23. Spangler, E. B. and Kelly, W. J. G. M. R., "Road Profilometer. A Method for Measuring Profile", *Highway Research Board,* Washington, D.C., 1955, Record No. 121.
24. Heukelom, W. and Klump. A. J. G., "Dynamic Testing as a Means of Controlling Pavements During and After Construction", *Proc. II. International Conference on Structural Design of Asphalt Pavements,* Michigan University, Ann Arbor, Mich., 1962.
25. Nijboer, L. W. and Metcalf, C. T., "Dynamic Testing at the AASHTO Road Test", ibid.
26. U.S. Army Corps of Engineers, "A Procedure for Determining Elastic Moduli of Soils by Field Vibratory Technics", U.S. Army Waterways Experiment Station, Vicksburg, Miss., 1963.
27. Reference [1], Chap. 20,
28. U.S. Army Corps of Engineers, "Rigid Airfield Pavements", *Engineering and Design Manual,* Dept. of the Army, 1958, EM 1110-45 303.
29. ACI, "Guide for the Use of Epoxy Compounds with Concrete", *ACI Journal,* 1962, Vol. 59, No. 9.

OTHER RELATED REFERENCES (CHAPTERS 9 & 10)

American Society of Civil Engineers, *Proc. 4th International Conference on Expansive Soils,* Denver, Colorado, June, 1980, 2 Vols.

Bickard, R. and Zwingelstein, R., "Dimensionnement des Couches de forme non traitees", Colloque International sur la Gestion des Ouvrager, Paris, 1981.

Blot, G., "Methode de Mesure de L'etat de surface D'une Chaussée. Presentation. Applications et Developpements Eventuels", Colloque International sur la Gestion des Ouvrager, Paris, 1981.

Carmichael III, R. F., Roberts, F. L., and Treybig, H. J., "A Procedure for Evaluating the Effects of Legal Load Limits on Pavement Costs", A paper prepared for presentation at the 58th Annual Meeting of the Transportation Research Board, Washington, D.C., 1979.

Dempsey, B. J. and Robnett, Q. L., "Influence of Precipitation, Joints, and Sealing on Pavement Drainage", A paper prepared for the 58th Annual Meeting of the Transportation Research Board, Washington, D.C., January, 1979.

Greer, W., "Pavement Design for a 3.5 Million Pound Vehicle", Law Engineering Testing Company, U.S.A., 1982.

Grenstein, J., "Pavement Evaluation and Upgrading of Low-Cost Roads", Berger (Lois) International Incorporated, U.S.A., 1982.

Hartgen, D. T., Shupon, J. J., Parrella, F. T., and Koeppel, K. W., "Visual Scales of Pavement Condition: Development, Validation, and Use", New York Department of Transportation, U.S.A., 1982.

Her Majesty's Stationary Office, "A guide to the structural design of bitumen-surfaced roads in tropical and subtropical countries", *Road Note 31,* 3rd. Edn. 1977.

Highway Research Board, "Design, Evaluation, and Performance of Pavement Systems", Highway Research Record 291, Washington, D.C., 1969.

Hveem, F. N., "Pavement Deflections and Fatigue Failures", Reprinted from Bulletin 114, Highway Research Board, Washington D.C.

Kennedy, C. K., Fevre, P., and Clarke, C. S., "Pavement deflection: equipment for measurement in the United Kingdom", Transport and Road Research Laboratory, Department of the Environment, Department of Transport, TRRL Laboratory Report 834.

Kennedy, T. W., Roberts, F. L., and Rauhut, J. B., "Distresses and Related Material Properties for Premium Pavements", A paper prepared for presentation at the 58th Annual Meeting of the Transportation Research Board, January, 1979.

Laboratoires des Ponts et Chaussées, "Bitumes et enrobés bitumineux", *Bulletin de Liaison. Special number V,* December, 1977.

Leech, D. and Powell, W. D., "Levels of compaction of dense coated macadam achieved during pavement construction", Transport and Road Research Laboratory, Department of the Environment, 1974, TRRL Laboratory Report 619.

LEECH, D. and SELVES, N. W., "Modified rolling to improve compaction of dense coated Macadam", Transport and Road Research Laboratory, Department of the Environment, 1976, TRRL Laboratory Report 724.

LENKE, L. R., "Suggested Improved Methodology for Relating Objective Profile Measurement with Subjective User Evaluation", New Mexico University, 1982.

LISTER, N. W. and POWELL, W. D., "The compaction of bituminous base and base-course materials and its relation to pavement performance", Transport and Road Research Laboratory, Department of the Environment, Department of Transport, 1977, TRRL Supplementary Report 260.

LUKANEN, E. O., "Evaluation of Full-Depth Asphalt Pavements", Investigation No. 195, Interim Report 1977, Minnesota Department of Transportation, Preprint for the 58th Annual Meeting of the Transportation Research Board, Washington, D.C., 1979.

McKENZIE, D. W., "Road Profile Evaluation for Compatible Pavement Evaluation", New York Department of Transportation, U.S.A., 1982.

MOULTON, L. K., and SEALS, R. K., "Determination of the In-Situ Permeability of Bases and Sub-bases", *Public Roads*, Vol. 43, No. 4, pp. 134-143, March, 1980.

NORMAN, P. J., SNOWDON, R. A., and JACOBS, J. C., "Pavement deflection measurements and their application to structural maintenance and overlay design", Transport and Road Research Laboratory, Department of the Environment, 1973, TRRL Laboratory Report 571.

PARSONS, A. W., "The rapid measurement of the moisture condition of earthwork material", Transport and Road Research Laboratory, Department of the Environment, 1976, TRRL Laboratory Report 750.

PAUTE, J. L., "Comportement des Sols Supports de Chaussées a L'appareil Triaxial a Chargements repetes", Colloque International sur la Gestion des Ouvrager, Paris, 1981.

PETERSON, D. E., "Evaluation of Pavement Maintenance Strategies", TRB National Cooperative Highway Research Program, Synthesis of Highway Practice 77, Washington, D.C., 1981.

PHANG, W. A., CHONG, G. J., "Ontario Flexible Pavement Distress Assessment for Use in Pavement Management", Ontario Ministry of Transportation and Communications, 1982.

POTTER, J. F., "Deformation of Road Pavements: correlation between elastic theory and measured behaviour of rolled asphalt road-bases", Transport and Road Research Laboratory, Department of the Environment, Department of Transport, 1977, TRRL Laboratory Report 784.

POWELL, W. D. and LEECH, D., "Rolling requirements to improve compaction of dense roadbase and basecourse Macadam", Transport and Road Research Laboratory, Department of the Environment, 1976, TRRL Laboratory Report 727.

"Recommendations on Methods to be Used for Testing Aggregates", Extracts from the Report presented at the XVI. World Road Congress by the Technical Committee on Testing of Road Materials, Vienna, Sept. 16-21, 1979.

SHAHIN, M. Y., DARTER, M.I., and KOHN, S. D., "Airfield Pavement Condition Evaluation and Determination of Rehabilitation Needs", Paper offered for presentation at the 58th Annual Meeting of the Transportation Research Board, Washington, D.C., 1979.

SPANGLER, E. B., "Inertial Profilometer Uses in the Pavement Management Process", Surface Dynamics Incorporated, U.S.A., 1982.

THROWER, E. N., "A deformation model for flexible road pavements", Transport and Road Research Laboratory, Department of the Environment, 1975, TRRL Supplementary Report 183 UC.

THROWER, E. N., "Pavement stresses and mechanical triaxial testing of pavement materials", Transport and Road Research Laboratory, Department of the Environment, 1974, TRRL Supplementary Report 100 UC.

Transportation Research Board, "An Introduction to Nondestructive Structural Evaluation of Pavements", Transportation Research Circular 189, Washington, D.C., 1978.

Transportation Research Board, "Concrete Pavement Construction and Joints and Loader-Truck Production Studies", Transportation Research Record 535, Washington, D.C., 1975.

Transportation Research Board, "Cost-Effectiveness of Transportation Services for Handicapped Persons", Highway Research Program Report 261, Washington, D.C., Sept. 1983.

Transportation Research Board, "Evaluation and Analysis of Flexible Pavements and Properties", Transportation Research Record 755, Commission on Sociotechnical Systems, National Research Council, National Academy of Sciences, Washington, D.C., 1980.

Transportation Research Board, "Layered Pavement Systems", Transportation Research Record 810, Washington, D.C., 1981.

Transportation Research Board, "Locating Voids Beneath Pavement Using Pulsed Electromagnetic Waves, "National Cooperative Highway Research. Program, Report 237, Washington, D.C., 1981.

Transportation Research Board, "Pavement Roughness and Skid Properties", Transportation Research Record 836, Washington, D.C., 1981.

Transportation Research Board, "Proceedings of the National Seminar on Asphalt Pavement Recycling", Transportation Research Record 780, Washington, D.C., 1980.

Transportation Research Board, "Wetlands and Roadside Management", Transportation Research Record 969, Washington, D.C. 1984.

TREYBIG, H. J., "Mechanistic Pavement Overlay Design", A paper prepared for presentation at the 58th Annual Meeting of the Transportation Research Board, Washington, D.C., January, 1979.

WEEKS, R., "Instrumentation for a fully automatic asphalt mixing plant", Transport and Road Research Laboratory, Department of the Environment, 1973, TRRL Report LR 578.

WITCZAK, M. W. "Design of Full Depth Asphalt Airfield Pavements", *Proc. Third International Conference on the Structural Design of Asphalt Pavements,* London, England, 1972.

CHAPTER 11

DRAINAGE STRUCTURES IN HIGHWAY PRACTICE

11.1 Introduction

As has been discussed in earlier chapters, highway drainage involves several different aspects, each of which must be given individual attention. Chapter 7 dealt with the methods based on theory and experience, which have gradually been introduced for controlling the infiltration of subsurface water into roads. *Subdrainage* was the term adopted for the techniques for controlling this water. *Drainage* refers to methods of controlling any *surface* flow affecting a road, regardless of whether the water initially falls on to the road itself or outside it.

In highways the most imposing drainage structures are bridges and culverts, which are the principal elements responsible for cross drainage. They enable large volumes of water, including streams and rivers, to cross the highway more or less perpendicular to it. Bridges are major, and culverts are minor drainage structures. The dividing line between the two is not clearly defined. In Mexico, engineers define a bridge as a structure with at least one span longer than 6m (20 ft). The term *culvert* is limited to structures with spans less than 6 m (20 ft) long, regardless of whether there are several of these spans, and an overall structure that exceeds this limit. Organizations in other parts of the world use similar, though not necessarily indentical criteria to distinguish between bridges and culverts.

The only aspect of bridges of interest in applied soil mechanics is their foundation problems, which were discussed in an earlier chapter. Also described (although only briefly) were the soil exploration methods appropriate for them (Chapter 3), with access embankments and some others which have been discussed earlier in this book. Therefore, it is not necessary to give further details concerning these most interesting structures.

Although culverts are similar to bridges from many points of view, there are two distinct differences which make them worthy of separate treatment.

First, there are far more culverts than bridges. Second, as individual structures they represent far less investment. Consequently a greater effort of study is dedicated to a bridge than to a culvert. Soil surveys and laboratory investigations are routinely carried out for bridges. Their foundations are often the object of highly sophisticated designs, using the most advanced techniques for point-bearing and friction piles, foundation cylinders and caissons. (No reference is made here to the highly important, sometimes elaborate hydrological and hydraulic studies which are an essential part of bridge design). On the other hand, the foundation studies that are conducted for each culvert are generally brief because the bearing capacity that is assumed for the ground is not very high (typically, capacities between 1.0 and 3.0 kg/cm^2 (2 to 6 kips/ft^2) are sufficient. Exploration is often restricted to a simple visual inspection of samples obtained from open wells or pits, using post-hole augers or light simple excavators that are cheap to operate.

The foundation and structural design are usually standardized, hydraulic studies that are conducted for bridge design as routine practice are not customarily made for culverts. Despite the above general criteria, the total investment represented by culverts for a given stretch of railroad or road is larger than that represented by bridge construction for the same stretch. In Mexico there are typically at least 3 or 4 culverts per kilometer (5 or 6 per mile), and the total investments for these culverts may be as much as 20% of the total cost of the road. In view of such large investments, and considering that the collapse of a culvert will cause a local but total interruption of traffic the authors wonder whether the attention which is traditionally given to these structures is sufficient. The experience of the authors is that culvert failures occur frequently. Even if failures only cause abnormal maintenance the ultimate costs are excessive. However, experience also indicates that most culverts do not fail as a consequence of foundation problems or from deficiencies in the application of soil mechanics. Instead, most failures are due to a lack of attention to the hydraulic capacity of the structure, including the transportation of solids, sedimentation and the protection of the structures against the onslaught of water. It therefore seems that the chief concern of the engineer in improving the methods of culvert design and construction should be more careful attention to all the hydraulic aspects of the problem. Owing to the numerous structures involved, it is impossible to carry out a detailed hydraulic study for each of them. Also, anyone attempting to conduct such studies will find it impossible because of the lack of essential statistical data concerning the behaviour of streams and rivers. Again the problem of insufficient data that presents itself in so many aspects of highway design and construction is faced, and it must be dealt with in the same manner as described for similar problems in other parts of this book. As a general principle, attempts at a detailed study of each case are futile. Instead, methods for obtaining regional information, at a reasonable cost will enable sufficient data to be obtained for a particular stretch or zone of highway. These general data used in conjunction with generous margins for what is not known are the bases for reasonable hydraulic sections, and economical designs that require limited maintenance. The geological surveys already discussed, and particularly photo-interpretation, provide invaluable assistance. Most useful of all will be regional hydrological studies, which are beyond the scope of this book. At a very low cost per kilometer of road, these studies have shown themselves capable of providing information of a general nature that is sufficient to protect against any important errors.

Although the most notorious and frequent problems in culvert performance are hydraulic ones, there are nevertheless others that relate to applied soil mechanics. Disregarding the obvious foundation problems, the remaining important problems of this type are scour, erosion or piping in access or approach embankments. Compaction of the layer of soil above and on the sides of the culvert is also important. Incorrect location of the culvert in relation to the stream or gully that is to be drained or to the embankment that protects it often originates or aggravates these phenomena.

Geotechnical studies for culverts have the same limitations as those already described for hydraulic problems. For example, since the recommendations for foundation design must be based on brief, simple studies, it is advisable for these to be entrusted to engineers with a solid technical background and experience in applied soil mechanics. We therefore recommend that the specialists in charge of the general geotechnical surveys be made responsible for these recommendations as well.

In addition to culverts and bridges in a railroad or highway, other lesser known structures are provided that are not included in this specialized field. These collect and eliminate surface water which would otherwise cause damage, and are usually referred to as *Complementary Drainage Structures*.

The following will be discussed in this chapter:

— Transverse grade

— Curbs

— Shoulders

— Chutes

— Drainage pipes

— Berms

— The appropriate use of vegetation

— Training walls

— Ditches

— Intercepting ditches

— Intercepting channels

Apart from the above structures or measures whose water control functions are well known, there are other structures for collecting, diverting and eliminating water, which are built in accordance with the specific needs of each situation, but which do not fit any general classification.

Complementary drainage structures should be used neither universally nor in a routine manner. They should be built only where required, otherwise money will be spent unnecessarily and results may even be detrimental.

A brief analysis follows of these structures, including the criteria for their location and construction. Unfortunately few hard facts are available on the subject, for although they are of great practical importance few have been carefully studied or seriously investigated. Their design and construction are still arts and are far from becoming scientific methodologies.

11.2 Transverse Grade

Cross-section grade or transverse slope is provided in highways and runways to direct any water falling on to them to flow towards the shoulders on either side. In straight sections of normal two-lane asphalt concrete highways, grade is commonly 2% from the center line of the road to the corresponding shoulder. In roads with a rigid pavement, grade may be a little flatter, for example about 1.5%. In curves, the slope is superimposed on the necessary superelevation. The latter quickly dominates, with no sudden changes in the transverse grade from the higher shoulder to the lower one. In curves, and in the transition from a straight alignment to a curved alignment, there is usually a portion of road where the development of suitable cross drainage grade is somewhat complicated. This problem must be solved individually in each case, but is aided by the existence of longitudinal grade.

In runways too, slope is provided from the center line towards the shoulders, and is generally 1.5%. In Mexico, 1.25% is now widely used.

In highways with more than two lanes, two typical situations are encountered. Either there is a relatively narrow central median or a very wide one, usually covered with grass. In the first case, the transverse slope is commonly from the median towards the two outside shoulders, but in the second case combined slopes are common practice. The surface slopes from the center line of each lane towards both shoulders, including the median, where a drainage channel is usually provided.

On curves surface water frequently ponds in the part of the shoulder that borders on the pavement on the highest side.

This occurs when the level of the pavement surface is somewhat higher than that of the shoulder. To avoid the formation of this ponding zone and consequent infiltration, the edge of the wearing surface should be beveled (Fig. 11-1).

When embankments are constructed on soft soils, the transverse grades tend to disappear with time, for greater settlement occurs in the centre of the cross-section than in the shoulders. Settlement calculations enable this difference to be estimated,

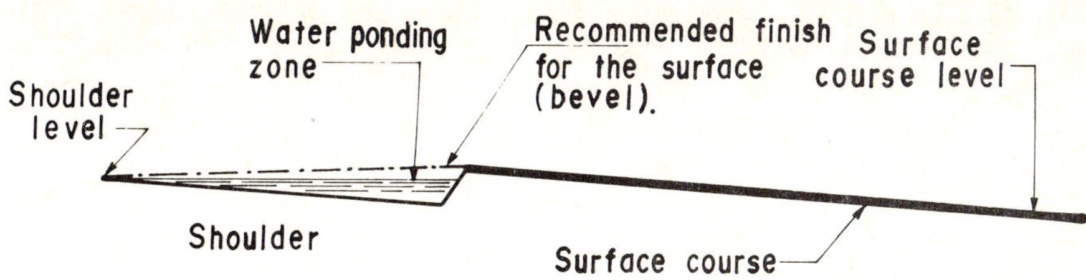

Fig. 11-1 Method of finishing off the surfacing to avoid the formation of puddles on the inside of curves

so that it can be allowed for in design. The initial tranverse grade is exaggerated in order to avoid, or at least minimize this problem. This has practical importance, because any re-levellings that might prove necessary in the future would demand the use of surfacing material, which is the most expensive pavement material.

In temporarily surfaced roads, transverse grade should be at least 4% so as to encourage rapid transverse runoff. In these minor roads there is a tendency for vehicles to travel along the central portion, which leads to the formation of ruts in the wearing surface. The resulting outward displacement of material causes very detrimental ponding if the transverse grade is not appreciable.

In highway with 4 or more lanes and a central median, a serious problem is usually posed by the accumulation on the median of all the water that collects on the highest part of the curves with superelevation. Sometimes, water has flooded the central median, overflowing on to the other part of the highway. A simple way to avoid this danger, is to provide small gaps in the median, so that the water can flow through them to the opposite side. The permanent solution to this problem is to construct an underground collector beneath the median, to which the water flows through storm channels that are provided in the median. The water that accumulates in the underground collector must then be suitably eliminated. This is an expensive solution because it will be necessary at every curve in the road. There is no cheap general solution to this problem. When the conformation of the surface of the curve is favourable, the underground collector could be replaced by a concrete box with a single inlet and corresponding outlet. This will seldom be feasible though, for most curves are wide. Another solution is to construct a small ditch in the gap in the median, with all due respect for the traffic hazard and the guards that would need to be imposed for such a depression adjacent to the high-speed lanes. The problem is a difficult one and its effects on traffic are extremely adverse. The solution best suited to the geometrical conditions in each specific case must be found.

Plate 11-1 Shoulder, curb and embankment with abundant vegetation

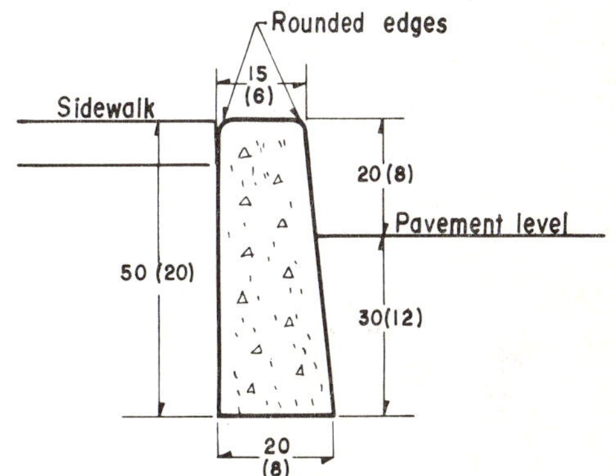

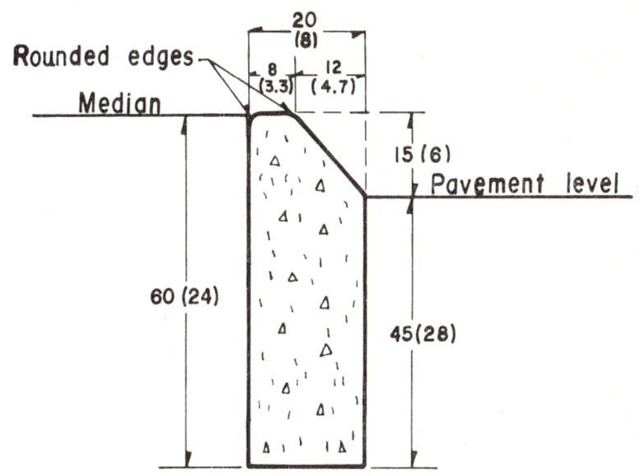

IN BOTH CASES :

1.- Dimensions in cm. (in.)

2.- Expansion joints of tar paper 0.3 cm (1/8") thick are placed at a maximum spacing of 6.0 m. (20 ft.)

3.- Concrete of $f'_c = 200\ kg/cm^2$ (3000 lb/in^2).

Fig. 11-2 Typical curbs [1]

11.3 Curbs

In urban zones, curbs are built along the edge of sidewalks to support them and prevent them from sliding onto the wearing surface. At the same time they help to protect sidewalks from traffic action. In highways, curbs are built for the same purpose, along sidewalks on bridges, toll gates and underpasses and on some medians built to separate traffic lanes in highways or at crossings and crossing islands (Plate 11-1)

Although drainage is not the chief function of curbs, they nevertheless collect surface flow from the wearing surface, conveying it towards specially provided outlets. (Some typical curb shapes are shown in Fig. 11-2, [1]). The trapezoidal shape is used to provide greater resistance to overturning: the same objective is met by the height of the cross-section, enabling embedment of an adequate portion of curb.

In some countries a buried curb is installed between the edge of the wearing surface and the shoulders. Its purpose is to protect the pavement by providing the edges with the confinement they would otherwise lack. If the curb is suitably painted, it will also provide clear indication of the edge of the road, and if the outboard part is corrugated it will warn the motorist who allows his vehicle to leave the wearing surface [1]. If this design is adopted, provision should be made for it in the construction of the lower layers of pavement to avoid any obstruction of lateral drainage of the pavement [2].

Curbs are usually made of concrete, but stone can be used if readily available and if labor is not expensive. The curing of concrete will always be a troublesome and sometimes expensive problem in curb construction, especially where water is scarce. As a rule, 6 sprinklings per day are required. Successful results have been obtained with the use of products which accelerate the curing process or which seal-in the moisture. For concrete curb construction, sliding wooden or steel molds are used. Steel is preferable because it is easier to handle and lasts longer, and gives a better finished curb. Concrete should always be vibrated.

Some engineers believe that curbs especially when fairly high, provide a psychological obstacle to motorists, causing channelizing which reduces effective lane width. To avoid this, they should not stand more than 15 to 20 cm (6 to 8 in) above the paved surface.

11.4 Dikes

Dikes are small ridges which are placed on the outer side of the shoulder in straight sections, on the edge opposite the cut in cut-and-fill sections, or on the inner side of bends in fill sections. They form a barrier to intercept and convey storm water to ditch outlets and chutes, thus protecting embankment slopes from the erosion and saturation that might otherwise be caused by water falling onto the surface of the road [1].

In Mexican practice, trapezoidal dikes of asphaltic concrete or Portland cement concrete are generally used (Fig. 11-3, [1]). In this figure, the Type *a* dike is securely anchored to the shoulder material thus protecting the alignment. Anchorage is not continuous, but intermittent. For example, small sections, 8 to 10 cm long every 6 m (3 to 4 in every 20 ft). The Type *b* dike is the one most often seen in Mexican highways.

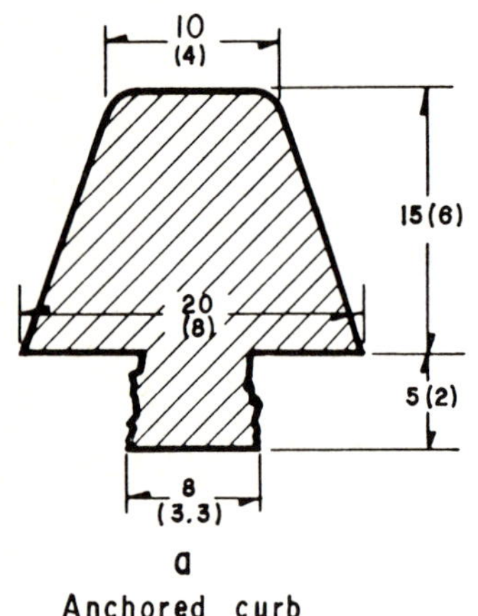

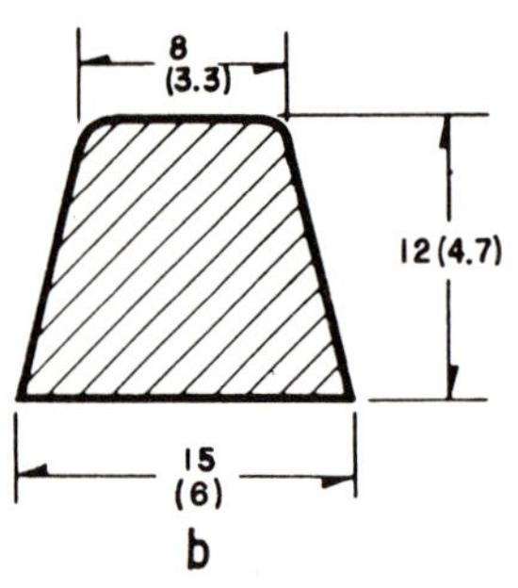

Anchored curb

Asphalt concrete curbs made with stony material with a maximum particle size of 3/4" and asphalt cement No. 6 in a proportion of approximately 100 kg in one cubic meter of stony material (6 lb of asphalt cement in one cubic foot of stony material)

All dimensions are in centimeters and in inches.

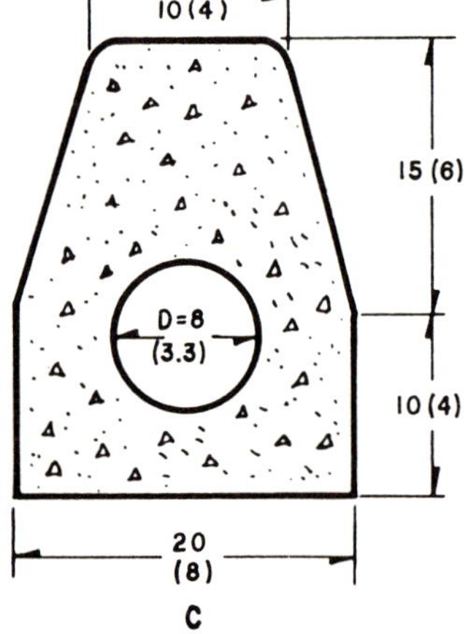

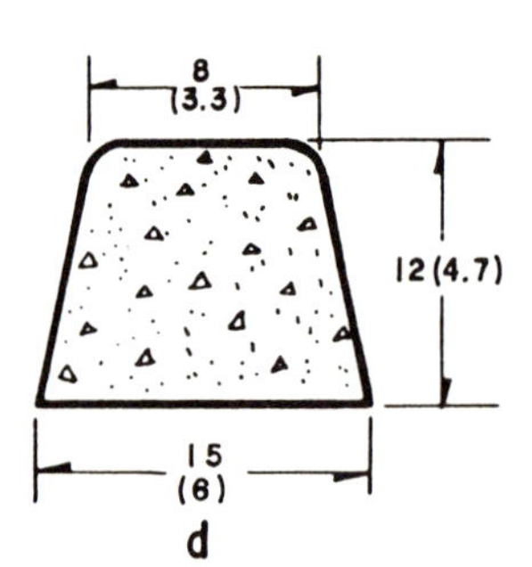

Portland cement concrete curbs, with $f'_c = 150\ kg/cm^2\ (2000 lb/in^2)$

Fig. 11-3 Types of curb commonly used in Mexican highway practice [1]

A dike must be high enough to hold back storm water, but not high enough to give a sensation of confinement to the motorist who finds himself obliged to park on the shoulder or even travel along it in case of emergency. Excessive dike height may also impede opening of parked vehicle doors. Dike height should not exceed 25 cm (10 in), but 12 to 15 cm (5 to 6 in) is adequate in the majority of cases.

Dikes are built preferably of asphaltic concrete or Portland cement concrete. The use of stone may be considered where plentiful, and where the employment of mass labor is wished. For the construction of asphaltic concrete or Portland cement concrete dikes, sliding metal or wooden molds are used. Special machines are available which are equipped with a mold of the structure into which the corresponding material is fed by a *worm*. Construction is faster and cheaper than with sliding molds. When machines are used, it is essential to maintain control of their operation speed, for it is on this that the structural consistency and successful finish of the curb depend. When asphaltic concrete is used temperature control is also very important. Very high temperatures result in crumbly, poorly cemented structures, while low temperatures lead to structures with low density and large voids owing to the use of an excessively viscous material. 130°C (266°F) is perhaps a suitable temperature under normal circumstances [1]. Table 11-1 [1] gives the gradation limits to be observed for the stony aggregate in asphaltic concrete dike construction. Portland cement concrete dikes require expansion joints; these are usually installed at intervals of 10 m (33 ft). Special care must also be taken with the concrete curing.

Table 11-1

Gradation requirements for stony materials used in asphalt concrete for curbs, according to Mexican engineering practice [1]

Sieve	% passing, by weight
3/4 in	100
1/2 in	100-85
3/8 in	100-75
No. 4	80-60
No. 8	60-45
No. 50	30-18
No. 200	15- 5

The anticipated discharge to be conveyed by a dike can be calculated as a function of the area drained (between chutes), the maximum rainfall per hour and its duration.

An increase in the longitudinal grade of a road leads to an increase in the flow velocity of the water that is confined by the dikes, with a consequent reduction in the requirement of hydraulic area; depth and width of the sheet of water. All these effects are desirable and support the criterion that a slight longitudinal grade is advisable in roads. These effects are also enhanced if the roughness coefficient of the shoulders is low; therefore, good surfacing is advisable in these zones. The depth and width of the water that is channelized by the dike are functions that are not greatly affected by longitudinal grade, therefore an adequate runoff can be ensured with only small grades. Reference [3] describes methods of hydraulic analysis for dikes, which are beyond the scope of this book.

In order to enable efficient maintenance of the system, careful attention must be paid to the connection of the dikes with the chutes or ditch outlets that ultimately eliminate the water from the surface of the road. For this purpose, a slight depression is recommended in the surface of the shoulder close to the chute entrance. In some European countries dikes are built in the shape of an L, the vertical portion being the dike proper and the horizontal portion, about 50 cm (20 in) wide, forming part of the shoulder. If this horizontal portion is tilted sufficiently, it may act like a ditch, channelizing runoff appropriately and establishing an adequate connection with chutes and ditch outlets. The use of a dike like this one is costly and is justified only for primary roads in exceptionally rainy zones, or when highly erosible materials are to be protected.

The dike is usually connected to the drainage chute by two bends, confining the depressed zone of the shoulder. The bend corresponding to the upstream part of the dike (with respect to the chute) is usually made wider than the downstream bend, to enable the water to flow more freely.

Like all complementary drainage structures, dikes should not be regarded as routine solutions to be used indiscriminately, but should be designed only where really necessary. A dike is an obstacle to rapid drainage in the transverse direction. Dikes should only be used, therefore, in places where water running off the top of an embankment is likely to cause damage to slopes of highly erosible material. Slope protection using vegetal cover is an important alternative to the construction of dikes, because the embankment materials that are well protected by plant species no longer erode. Another type of slope protection that reduces the need for dikes is that obtained in a natural manner by building very low embankments (less than 1.50 m (5 ft) high), where water cannot achieve erosive speeds. Lastly, other factors to consider are the intensity and duration of rainfall.

Dike maintenance is expensive. It is sometimes made unnecessary when slopes in time become covered with vegetation. In such cases dikes should be eliminated.

11.5 Drainage Chutes

Drainage chutes are channels which are connected to dikes and run transversally down road embankment slopes to convey any rainwater running off the shoulders to places far from the embankments where it can do no damage. They are structures with a very steep grade, and this is responsible for most of the dangers they involve (Plates 11-2 to 11-4).

In highways, they are installed in embankments on the fill side of cut-and-fill sections (usually at the entrance and exit), and on the inner sides of curves in embankment sections.

In straight alignments, they are installed at intervals of 60 to 100 m (197 to 328 ft) but the spacing varies with the longitudinal grade of the road and local rainfall. Figure 11-4 [1] gives the ground plan of a typical chute built of masonry, a longitudinal cross-section and a perspective showing the position of the chute in relation to the road.

The capacity of the chute inlet will depend on the difference between the total discharge along the curb and the depth of water in a section immediately before the inlet. IZZARD [3, 4] gives the following formula for calculating the length of the chute inlet, which takes into consideration the abrupt change in direction of the water at that point:

Plate 11-2 A masonry chute with metal bottom, functioning properly

Plate 11-3 Integration of a chute with the general drainage system

$$L_u = \frac{Q}{0.386\,(a + y)^{3/2}} \quad \text{(m);}$$

$$L_u = \frac{Q}{0.7\,(a + y)^{3/2}} \quad \text{(ft)} \qquad \text{(11-1)}$$

where: L_u is the length of the chute inlet, in m (ft); Q is the discharge that reaches the chute and must go down it, in m^3/sec (ft^3/sec); a is the difference between the level of the shoulder and the most depressed section from the chute inlet to the curb, in m (ft) (Typically it is about 0.06 m (2.4 in)); and y is the depth of water flowing along the shoulder, in a section close to the chute inlet, in m (ft).

Plate 11-4 Chute installed in the median of a two lane roadway

Considering how difficult it is for the entire volume of flow running along a curb to be successfully captured by a chute, in view of the abrupt turn the water makes, it is usually acknowledged that only 80 to 90% of the water is actually captured.

Despite the above formula, for simplicity all chute inlets are usually the same, with very similar discharge capacities. The different requirements are met by varying the spacing between chutes accordingly. If L is the design length of the chute inlet and I_u is the length required to intercept the entire discharge that reaches it, Fig. 11-5 [3] gives the portion of the total dis-

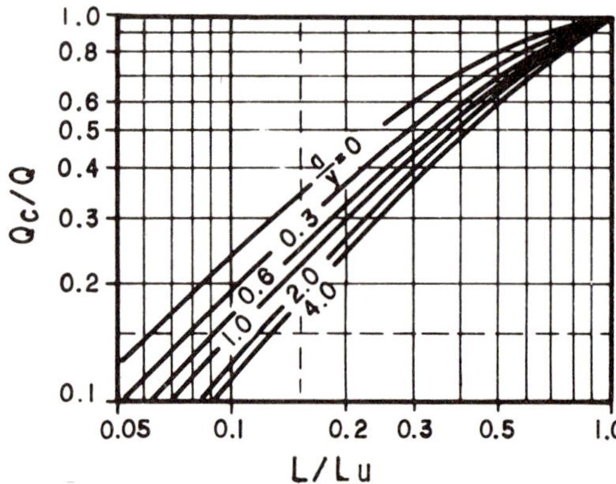

Fig. 11-5 Portion of the total discharge that can be captured by a chute inlet of length L [3]

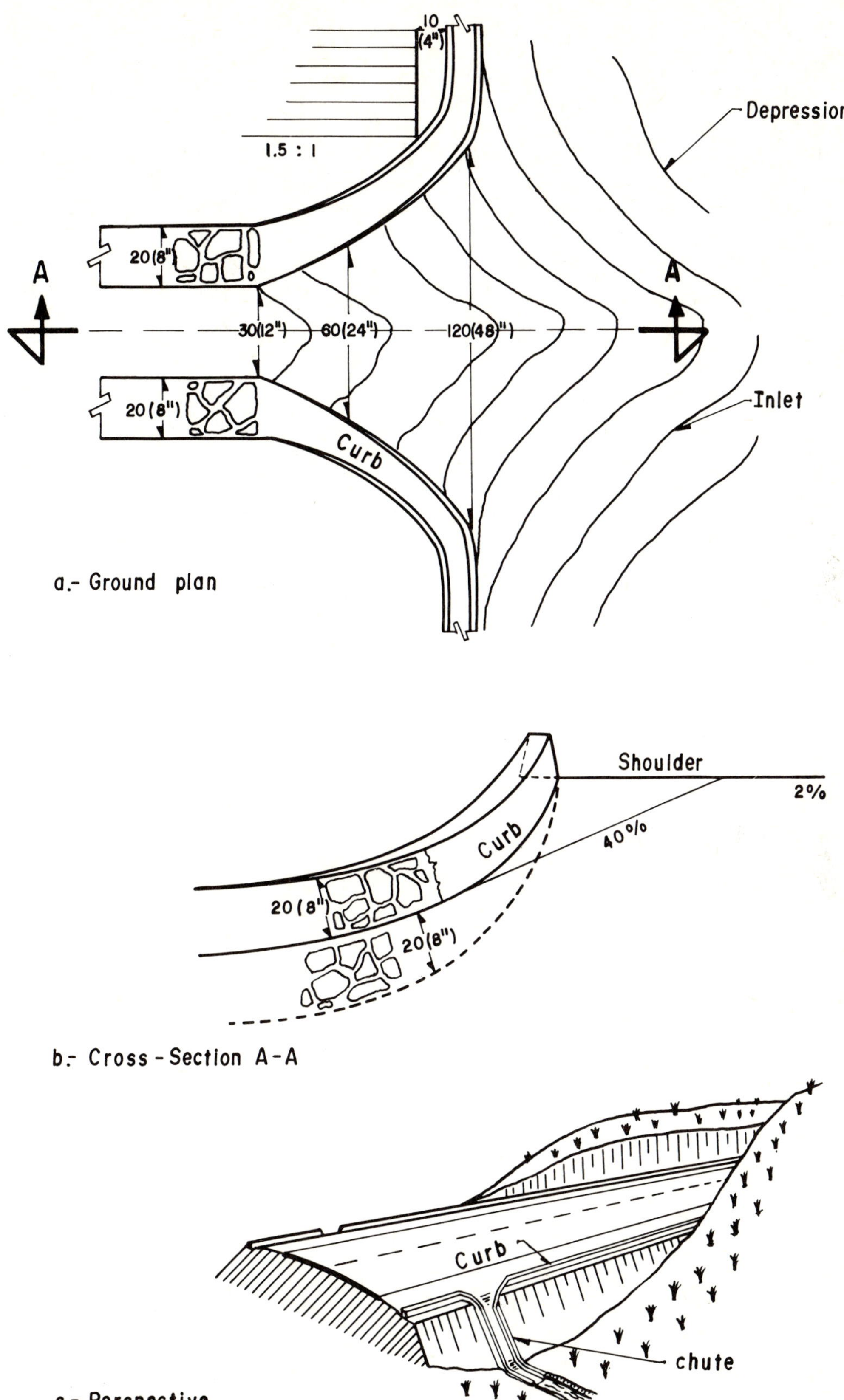

Fig. 11-4 Drainage chute

charge that can be captured by the inlet of length L. Q represents the total discharge reaching the inlet, and Q_c the discharge that is actually captured.

Figure 11-6 [3], which is complementary to Fig. 11-5, enables calculation of the length of inlet required (L_u) for the chute to capture the entire discharge Q that reaches it. In these figures, a and y have the meanings described in Eq. (11-1).

The chute proper is the paved channel that goes from the inlet at the top of the embankment to the toe of that embankment, or further still, to where the water is finally discharged so that it will cause no damage. It usually has a standard cross-section (Fig 11-4). The hydraulic design is based on the height at the chute edges, which depends on the discharge at the chute inlet, [3] gives the criteria for hydraulic design.

The high speeds at which the water goes down can cause erosion at the toe of the embankment, Plate 11-5. An energy dissipator is one solution. An alternative would be to prolong the chute, flaring the outlet to lessen erosive energy. Energy dissipation is a more serious problem if a pipe chute is used. An orthodox energy dissipation structure is too costly, so other methods have been developed to lessen the energy acquired by the water in its descent. A simple solution is to make the bottom surface of the chute very rough, thus provoking a strongly turbulent downrush entraining air. This often reduces the energy of the flow sufficiently. Very uneven masonry or steps built into the chute bottom also help dissipate energy. When the embankment is high and the discharge is considerable, the problem of erosion at the chute exit at the toe of the slope can cause dangerous head-cutting from downstream, resulting in destruction of the chute. In these cases, special care will have to be exercised by the engineer and he will be obliged to construct dissipation structures and discharge channels wherever they may prove necessary.

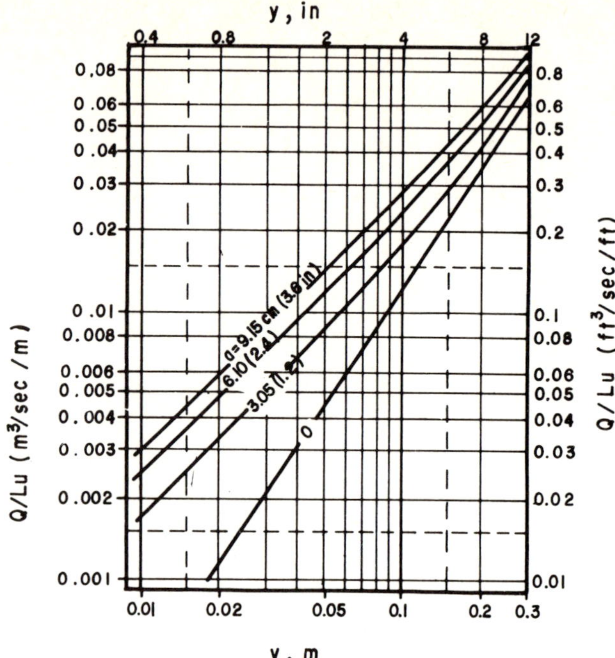

Fig. 11-6 Length of chute inlet required to capture the entire discharge reaching it [3]

The roughness of the chute bottom can be increased by partially sinking stones into the concrete, when this is the construction material used. Details of the hydraulic performance of these rough chute bottoms can also be seen in [3]. Also taken from this reference is Fig. 11-7, which gives the allowable outlet velocities for the rough-bottomed chute as a function of the embankment material, the natural ground at the point of discharge and the protection that is provided at the location.

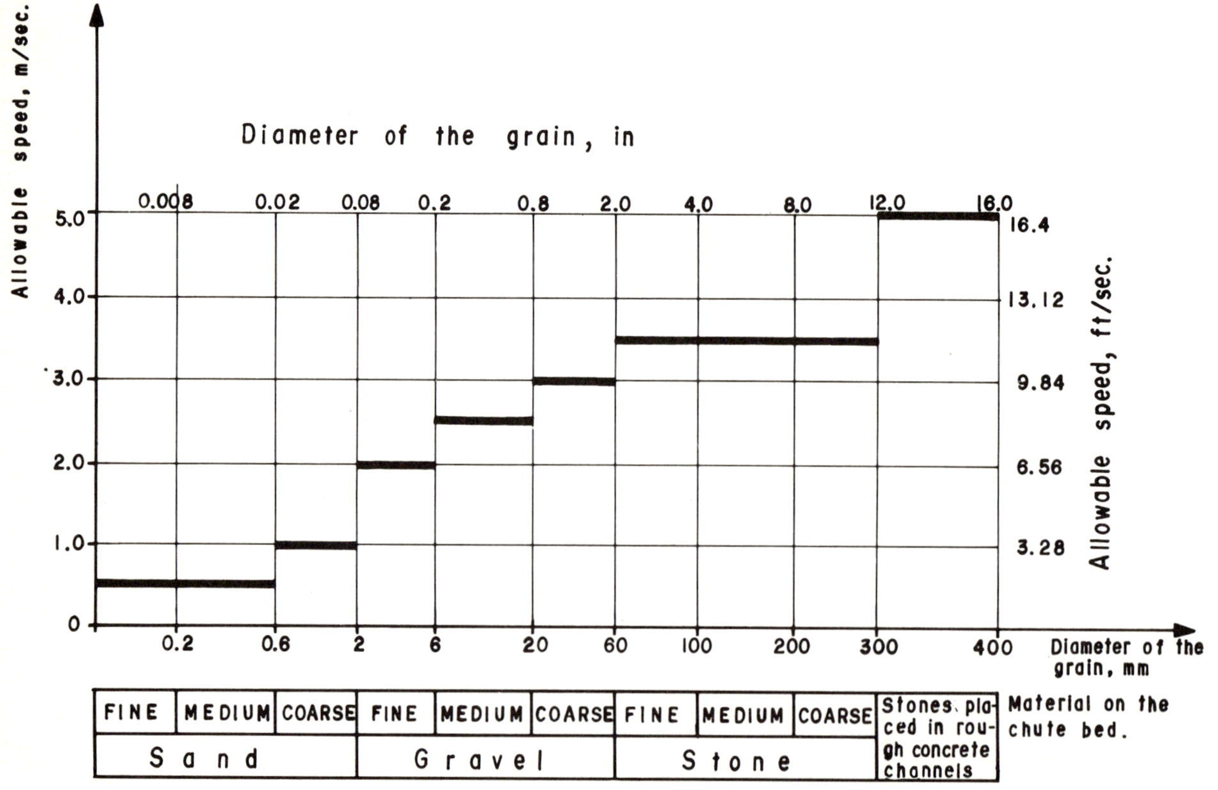

Fig. 11-7 Allowable outlet velocities for rough-bottomed chutes [3]

Plate 11-5 Masonry chute lacking adequate protection at the outlet

It is interesting to note that hydraulic calculations [1] show that the maximum velocity of water in rough-bottomed chutes is achieved at a very short distance from the chute entrance; consequently attention to erosion at the chute exit should concentrate on the materials at the point where discharge occurs.

An important aspect of the design of chutes is their stability within the body of the embankment (Plates 11-6, 11-7). For this reason they are usually embedded in the embankment in such a way that their lateral walls are at the same level as the adjoining slope. The installation of chutes directly on top of a slope is inappropriate because any overflow will follow the outside of the chute and undermine it.

Plate 11-6 Defective anchorage in metal chute

Plate 11-7 Chute destroyed by lack of anchorage and other protective systems

Chutes are often built of masonry with cement mortar in a proportion of 1 : 4. They can also be made of concrete, as has already been mentioned or of a half section of corrugated galvanized iron pipe with bolted joints. In this last case, at both the top and bottom ends of the pipe there must be a section of masonry or concrete chute. It is highly advisable for the pipe to be anchored with masonry tie plates at several intermediate points.

In very high embankments it may be advisable to install longitudinal channels in the surface of the slope in order to collect and eliminate water falling directly onto the slope. These discharge into chutes that carry the water down the slope (Plate 11-8).

Plate 11-8 System of chutes in a rural road. Note the successful integration with the topography

Chutes are also installed for eliminating water captured by roadside ditches and intercepting ditches, which will be discussed later. In this case there is a critical zone where both structures meet, for there is danger that the water will seep underneath the chute eroding and weakening the bedding material beneath it, with consequent failure [5]. To avoid this danger, the transition from ditch to chute should be wide and smooth, and the chute should have a cutoff wall at its entrance, to protect it against the effects of underseepage. This cutoff wall may be as deep as 50 cm (20 in) (Plates 11-9, 11-10).

The same comments apply to chutes as those made for other complementary drainage structures. They are not used in a routine manner, but only if really necessary when required to protect erosible fill materials that do not enjoy sufficient protection from other methods, such as vegetation.

In low specification roads, pipe culverts can often be seen to discharge above the toe of the embankment, above the level of the stream bed that originated them and onto the embankment body. As will be discussed later this practice cannot be recommended. Should such a situation arise, it is essential to install a chute or pipe of the type described in the following section of this chapter at the culvert exit. Such a chute could be far wider than the conventional ones that have been discussed previously. Its hydraulic capacity would have to be sufficient to eliminate the entire discharge from the culvert.

Chutes require high maintenance costs and careful observation, for they suffer distortions on account of all the movements that commonly occur in embankment slopes, even when the stability is adequate.

Plate 11-9 Chute functioning correctly

Plate 11-10 Chute protecting a culvert outlet

11.6 Drainage Pipes

These are structures with functions similar to that of chutes. They consist of a pipe resting on the surface of sloping ground or buried in it. The difference between drainage pipes and chutes is really a matter of nomenclature; many engineers regard drainage pipes as pipe chutes.

The pipes that have been used with the greatest success are made of steel, with joints that are capable of absorbing minor movements due to changes in temperature or settlement of the embankment or the ground in which the pipe is installed.

At locations where rainfall is slight or where flow velocity is not likely to be too high, portland cement concrete pipes can also be used. If the concrete is effectively protected against erosion, such pipes can be used for higher flow velocities. Vitrified clay bell-and-spigot pipes are also used. The minimum diameter of drainage pipes should be 45 cm (18 in), but larger diameters, of 60 cm (24 in) or more, are not unusual where the need to eliminate large discharges is anticipated. Small drainage pipes have the disadvantage that they are hard to inspect, which may sometimes make soundings necessary.

One of the most frequent applications of pipes is when a "thalweg" exists along the crest of a cut. The water that falls there cannot be allowed to run freely down the cut slope, because the large flow could erode the soil, neither can it be channelized into the ditch for the same reason. A drainage pipe provides the usual solution to this problem, conveying the water underneath the road surface to a place where it can do no damage.

11.7 Berms

Berms have already been discussed previously (Chapter 6), although predominantly in relation to embankment stability. Also discussed were the steps which, like berms, are used to enhance slope stability in cuts.

Some other functions of these berms or steps are to aid surface drainage, control flood waters and convey and eliminate water. It is these particular functions that will be discussed here.

The berms that are built in embankments for drainage usually have a rise to tread ratio of about 1:1 to 1:1.5 and small dimensions; that is, they are genuine steps. These ratios may be increased to 1:2 or 1:3 for berms built on natural ground to control the runoff that may threaten a road below, similar to terraces built in sloping farm land to provide protection against erosion [1]. The steps built to intercept water running down a cut face usually have a rise:tread ratio that depends on the steepness of the cut, and do not therefore go beyond 0.75:1 or 1:1.

The effect of a berm or step is to reduce the erosive power of water running down the face of an embankment, cut slopes or natural slopes. They are better able to channelize the water collected if they are provided with an adequate slope towards chutes, drainage pipes or other discharge structures. Otherwise this water will erode the slopes, carrying with it material which will cause silting in the ditches or will infiltrate or seep into the slope itself, with adverse effects on its stability.

Infiltration of water from berms or steps can be serious, especially for cut steps in pervious materials. The steps might aggravate the problems in their longitudinal direction where there are long stretches with little gradient. In these cases, either the steps must be protected or not used at all. Protection may consist of sloping the step slightly towards the inside of the cut, providing a ditch with a steep enough slope to eliminate rapidly collecting water, or carefully paving the tread of the steps, including the ditch with an impervious material. For the ditch, soil-cement, soil-asphalt or even concrete can be employed.

The materials most susceptible to seepage in steps are jointed or cracked rocks, especially if their bedding dips toward the road, and residual soils containing inherited cracks and similar structures in a weak formation. Other soils which are very susceptible on account of their make-up, include loess and many silty materials. The consequences of water infiltrating into steps of susceptible materials are so serious that wherever waterproofing is suspect it is better to avoid steps.

Small shrubs are sometimes planted on these steps. Eventually they provide effective protection against erosion of the slope but their roots can open up cracks in the step surface.

11.8 Vegetation

One of the most effective means of protecting embankment, cut slopes and natural ground slopes against the erosive action of surface water is to plant vegetation. This retards runoff, greatly reducing the energy of the water, and encourages an equilibrium condition in the water content of the soils (Plate 11-11).

Plate 11-11 The median of a highway protected by vegetation and showing a lined ditch

Whenever there is vegetation, the engineer should treat it with due respect. The systematic de-forestation and excesive removal and uprooting of plants in the right of way or in the influence zone of a road is one of the worst mistakes a highway engineer can make. Instead, he should encourage all aspects of vegetal protection. When no such vegetation exists, it can be planted intentionally. As has been discussed, the planting of vegetation must be entrusted to specialists who will select species that are suited to the region and which will grow with minimum care and attention.

Climbing plants or thick grasses are particularly useful on slopes, whereas shrubs give better results when used to form protective barriers on natural ground.

11.9 Training Walls

Training walls made of earth or occasionally masonry, are built to channelize waters. They are placed either in the natural ground adjacent to the road, so that the water will discharge into gorges or natural streams, or else at the entrances to culverts and bridges, enabling the water to flow smoothly through these structures. This second type of training wall is by far the most common one and should be included in a general hydrologic hydraulic study, which is beyond the objectives of this book. The training wall on natural ground should respond to a topographical need, generally connected with "thalwegs" and gullies which, if the walls did not exist, would discharge their water in such a way as to damage the road. With the training wall, this water is diverted towards a natural stream that will eliminate it without undue risk.

Typically, training walls are built with material from excavations. This excavation is usually more or less parallel to the wall, and should not be a deep trench. In the case of intercepting walls that are built upstream from a road (for example to take the water that is collected by a gully towards a gorge and for which a bridge will probably have been constructed) it is advisable for the excavation to be made upstream from the wall. The floor of the channel is sloped in such a way that any water that falls on to it will also be conveyed towards the natural stream. If this channel is deep and uniform, the training wall will be unnecessary. The problem will have been solved with an intercepting channel, which is an alternative solution to be considered. Construction of a training wall will have to be considered whenever the channel resulting from excavation is neither deep enough nor appropriately shaped, or when the borrow area is not immediately adjacent and parallel to the wall.

Training walls of earth are usually built with slopes of 2:1 or 3:1, with a height seldom over 2 m (6.5 ft) and a crest width of about 50 cm (20 in). In many regions they are built by hand, compacting the material that is used in an elementary fashion. Tampers are used for this purpose. If the water is expected to flow fairly rapidly down the upstream slope, it is necessary to protect this slope with stone or replace the training wall by a low masonry wall.

Before constructing the training wall, the ground underneath it must be cleared, but without unduly damaging or interfering with the adjacent vegetation. The vegetation and topsoil is stockpiled upstream, and afterwards placed on the training wall slope in order to encourage the growth of vegetation.

The training walls which channelize water towards culverts and drainage structures are generally better built than those described above, for they must withstand the onslaught of fast-flowing water. Some common practices include slope protection using rockfill, construction with good quality masonry, and even concrete deflector walls.

Often the embankment slopes of a road will act as training walls for channelizing flow towards drainage structures. These situations must be carefully evaluated, so that the corresponding methods of erosion protection can be planned, such as vegetation, rockfill and masonry or concrete walls, depending on the velocities and rates of flow anticipated.

Engineers in charge of highway drainage sometimes neglect the discharge of the water that is collected and channelized, downstream from the road. This discharge sometimes causes damage to farm land, pastureland and farmhouses. It should be remembered that the road, by interrupting the general drainage of a zone and concentrating the discharge of waters that may cross it at isolated points, may well cause hitherto non-existent hydraulic problems downstream. The anticipation and solution of these problems and the elimination of any flows that might cause damage to drainage structures, without affecting third parties, is an obligation of all engineers in charge of highway drainage. This obligation often leads to the construction of training walls and channels which will take the waters and discharge them where they will affect no one. The training walls built for this purpose are frequently long and large.

Also of importance are the training walls that have to be constructed because of a nearby river that is likely to cause erosion, a flood zone, or the perimeter channels that have to be made around airport runways over swampy ground, flood zones or lakes. In all these cases where high training walls have to be erected, on the stability of which the life of the road may depend, appropriate technological studies will be made, and a special design established on the basis of a detailed geotechnical survey. The methods of conducting such surveys are the same as those employed for the road they are to protect; likewise the methods of construction.

11.10 Ditches

Ditches are such widely used complementary drainage structures, that many engineers object to their inclusion in a list of *complementary* structures. Here the adjective does not, however, imply that they are seldom used, but merely qualifies a group of structures with common objectives.

Ditches are channels which are built at the edges of the traveled surface of a road, on the same side as a cut. In cut-and-fill sections, therefore, there is a ditch on just one side and in through-cuts, on both sides. The ditch is placed at the edge of the shoulder, in direct contact with the cut. Its position enables it to receive rainfall from the slope and from the area between the crest of the cut and any separate intercepting ditch, or from the natural ground upstream from the cut if there is no intercepting ditch. The ditch can also receive water falling on the pavement when the cross-section slope of the road is appropriate (See Plates 11-11 to 11-14).

Plate 11-12 Lined ditch showing inlet to a relief culvert

Plate 11-13 Avoidance of ditches, providing instead a shoulder protected by vegetation

Plate 11-14 Lined ditches in a formation where minor rockfalls are likely to obstruct them

The hydraulic capacity of a ditch is the chief factor that determines its ability to channelize and eliminate rapidly the water it collects. The volume of flow to be drained depends on the watershed area, the coefficient of runoff and the intensity of the rainfall during a time equal to that of concentration. The detailed hydraulic design [3, 6] is beyond the scope of this book. It is usually complicated by the lack of appropriate records of rainfall intensities, which have to be estimated from accounts of the local inhabitants or data from the nearest meteorological stations, either of which introduces uncertainty in the calculations that can be made.

The minimum longitudinal grade of a ditch is 0.5%. The velocity of the water is a compromise between the limits of deposition and erosion, both of which are undesirable. Table 11-2 [7] gives, as a guide, the maximum velocity of water flowing over the materials listed that does not cause erosion. Despite the high velocities indicated in the last line of Table 11-2, we recommend limiting the velocity in ditches to 3.00 m/sec (10ft/sec) when rip-rapped and 4.00 m/sec (13 ft/sec) with concrete lining [1].

The discharge that can be eliminated by a ditch is a function of its longitudinal grade, but it is doubtful whether it can exceed 0.5 m^3/sec (18 ft^3/sec) [1]. Higher values lead to overflow. As a guide, Table 11-3 [9] gives the discharges calculated for the ditch in Fig. 11-8 for different grades and rates of flow.

Ditches are usually made with a trapezoidal or triangular cross-section. In Mexico, triangular shapes are far more common (Fig 11-8). On the side next to the road the slope is a minimum of 3:1, preferably 4:1, and on the side next to the cut it closely

Table 11-2
Maximum values for non-erosive water velocities in ditches [7]

Material	Velocity m/sec	Velocity ft/sec
Fine sands and silts	0.40-0.60	1.3- 2.0
Sandy clay	0.50-0.75	1.6- 2.5
Clay	0.75-1.00	2.5- 3.3
Firm clay	1.00-1.50	3.3- 4.9
Silty gravel	1.00-1.50	3.3- 4.9
Fine gravel	1.50-2.00	4.9- 6.5
Soft shales	1.50-2.00	4.9- 6.5
Coarse gravel	2.00-3.50	6.5-11.5
Riprap	3.00-4.50	9.8-14.7
Unweathered rocks and concrete	4.50-7.50	14.7-24.6

Table 11-3
Discharge values for the triangular ditch shown in Fig. 11-8 for different grades of the road and water velocities [9]

Road grade %	Water Velocity m/sec	Water Velocity ft/sec	Discharge m^3/sec	Discharge ft^3/sec
1	0.63	2.06	0.11	3.88
2	0.89	2.92	0.15	5.28
3	1.09	3.57	0.19	6.70
4	1.26	4.14	0.22	7.75
5	1.41	4.63	0.24	8.45
6	1.54	5.06	0.27	9.52

follows the inclination of the cut. A sheet of water no larger than 30 cm (12 in) is anticipated.

The rectangular cross-section has been abandoned for traffic engineering reasons. The steep walled channel arouses a sense of danger in the motorist travelling close to it. For the same reason, the trapezoidal cross-section is less favored, except when the edge adjacent to the road is very flat. The wide triangular cross-section is the most appropriate one today and is easy to build. It is made with a motor grader, after the subgrade is placed. It is also the easiest one from the maintenance viewpoint. In railroads, some of the foregoing virtues of the triangular ditch disappear, resulting in the more frequent use of the other two cross-sections (although the triangular cross-section is also used).

When ditches are paved, masonry or Portland cement concrete are usually employed. In the first case, mortar in a proportion of 1:4 (90 kg of cement to every cubic meter or 150 lb to every cubic yard of masonry) is used. For concrete, either cast-in-place or pre-cast slabs can be used. The smoother finish of concrete makes it more efficient hydraulically than masonry rip-rap. Concrete also enables faster construction. The slabs that are used are generally about 1 m to 3 m (3 to 10 ft) long, with sealed joints to avoid leaks. It is common policy in many countries not to pave ditches at all for economy. Also the protection of ditches with vegetation is both simple and economical if velocities of flow are not very high, 1 or 1.5 m/sec (3 or 5 ft/sec) [8]. However, the hydraulic capacity of the ditch may be reduced by the corresponding increase in the coefficient of roughness.

Soil-cement and soil-asphalt have seldom been used in Mexico for paving ditches but they are used somewhat more frequently in some other countries. These materials are recommended with sandy soils that are likely to reach a high erosion resistance and permanency with relatively low contents of stabilizing material. Proportions of portland cement of about 5% to 8%, by weight, and of asphaltic cement of about 4% to 6%, also by weight, are common in most jobs. Before adopting a solution of this type, careful consideration should be given to all the construction difficulties it implies. These include mixing the stabilizer, transporting and placing procedures, and compacting the mixtures. A careful economic analysis will indicate if the use of stabilized products would be unwise in cases where at first sight they seemed highly beneficial. The durability of these paving materials is always briefer than that of concrete and masonry; therefore, they involve more serious maintenance problems. These paving materials are usually compacted with manual vibratory equipment.

Ditches have occasionally been used in embankment sections (Fig. 11-9). A curved alignment is illustrated with the corresponding superelevation. At the crest, a type of ditch is shown which is sometimes built to fulfill the function that in other cases would be served by curbs. This solution may be efficient from the hydraulic point of view in regions where rainfall is intense or for very wide highways. Moreover, it will be essential to pay careful attention to the traffic engineering aspects related with

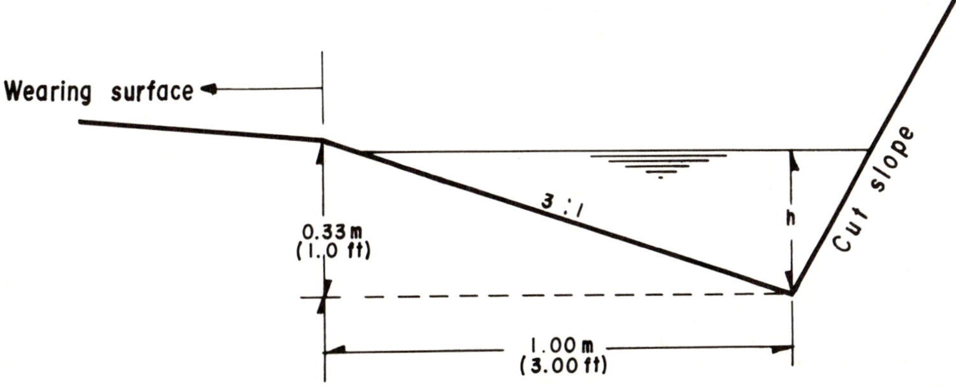

Fig. 11-8 Typical triangular ditch cross-section

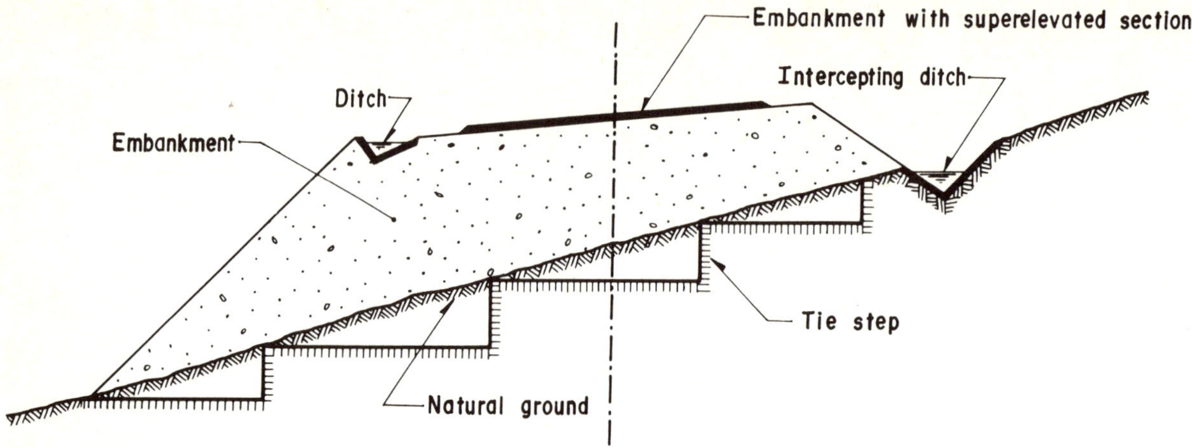

Fig. 11-9 Ditches in embankment cross-sections

this practice, which will require that the ditch be built on the other side of the shoulder. This leads to very wide highways; otherwise the traffic hazard might be considerable. This solution demands the construction of drainage chutes, although fewer may be required than in the conventional treatment using curbs, owing to the larger hydraulic area of the ditch. In some cases it appears that the construction of chutes could be avoided by prolonging the ditch with an appropriate slope until the water is discharged onto the natural ground. In this way the ditch would occupy different elevations along the embankment in relation to the surface, and would eventually lead to its toe. Should this approach be adopted, attention must be paid to the location of the ditch in the slope, which is a zone where the materials are poorly compacted and susceptible to movement, and thus very vulnerable to the action of water. Ditches that are installed either in the crest of the embankment or in its slope will have to be systematically paved with concrete, to prevent infiltration of water into the embankment.

Figure 11-9 also shows another type of ditch which is sometimes built in embankment sections. It is an intercepting ditch at the toe of the upstream slope. Its purpose is to prevent water from accumulating in that zone and possibly seeping in underneath the embankment, which would give rise to stability problems. These ditches too must be systematically paved with concrete to prevent water from infiltrating the soil. They are an expensive solution, but one that may prove wise in many cases, and indispensable in many places where failed zones are to be rehabilitated.

The relationship between the levels of the water in the ditch and the pavement is important. Because of the draining function of the base course, the surface of the water in the ditch must be kept below the bottom of the base course. It is best if the water is kept below the bottom of the sub-base too, to avoid wetting this layer if the ditch is not paved. Fig. 11-10a shows the ideal position of the ditch in relation to the pavement layers under these circumstances. If, however, the ditch is paved and water-

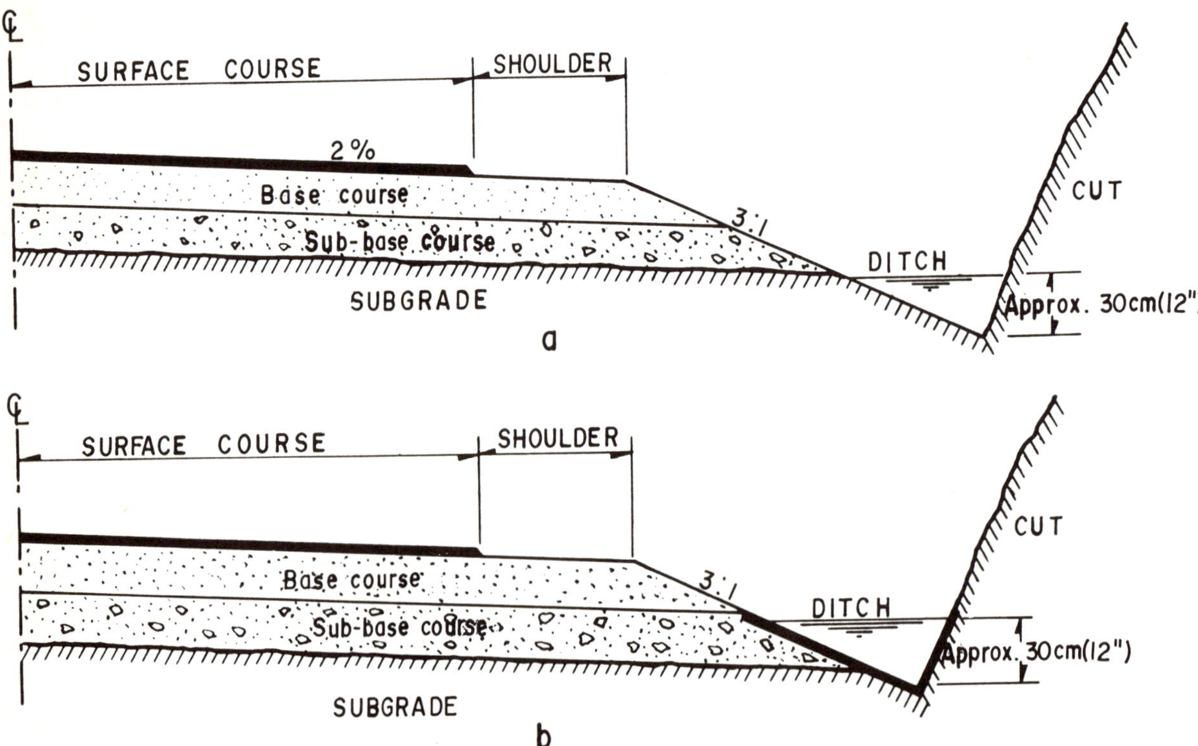

Fig. 11-10 Best position of the ditch in relation to the pavement. Cut section with no subdrainage

proofed, it will not have to be made so deep. It will be sufficient if the water in it stays below the level of the base course, for there will no longer be a danger of the water that is collected seeping into the sub-base. Fig. 11-10b illustrates this situation, in a ditch.

Because the thickness of the base course and sub-base together may be 40 cm (16 in), and often more, the layout in Fig. 11-10a may result in a deep ditch; the upper layers of pavement are thicker than the depth that would have to be excavated in order to achieve the necessary hydraulic capacity. Considering that the slope of the ditch on the side of the road will be at least 3:1, it is concluded that such a requirement will lead to an increase of one meter in the width of the roadway in cut-and-fill sections, and two meters in through-cuts, which will be expensive. In the case of the paved ditch (Fig. 11-10b), this requirement will lead to increases in road width of only about half of those just indicated, with a corresponding increase in construction costs. In this last case, a debatable point is whether it would be advisable to excavate the ditch to below the sub-base, so as to encourage its draining function.

The above requirements are frequently ignored on account of the costs involved. Some engineers prefer to ignore them because they feel it is not advisable for a base course or sub-base to end in a 3:1 slope without any confinement at the edge of the shoulder. This objection is reasonable, but not insuperable. It would simply give rise to the situation prevailing in all embankments and cut-and-fill sections. The decision is best limited to cost factors. This leads to the practice commonly seen in many countries, where a ditch, paved or unpaved, commences in the upper level of the base course, right at the edge of the shoulder. In this manner, the base course and sub-base are exposed to the invasion of water from the ditch, particularly when, as so often happens, the ditch is blocked by minor collapses or other causes. This practice leads to pavement failures which would otherwise not occur.

The authors feel it would be hard to establish any general rule regarding this problem. They believe that in each case the design should take into account the quantity of water to be eliminated, the duration of rainy seasons in the zone and the qualities of the materials in the pavement as a whole, from the foundation ground to the wearing surface.

Of course if the cut section has lateral subdrainage the problem does not arise, and all the layers extend to subdrain and discharge into it. Fig. 7-17, showed how the ditch could be installed above the subdrain, next to the shoulder, without any problems of drainage into the pavement layers.

If any of the pavement layers has been specifically designed as a draining layer or as a capillarity breaking layer (Chapter 7), this condition must be taken into account when considering the above problems. The draining layer must always have an outlet, and the capillarity breaking layer must never be allowed to become flooded; otherwise their respective functions will be cancelled.

In railroads, ditches are systematically installed so that the water in them remains below the lower level of ballast. With regard to the sub-ballast, the same comments apply as for the sub-base in highways. It is the practice for some railroad constructors to commence their ditches at the bottom of the sub-ballast with due respect for its draining function. To avoid seepage problems, railroad tunnels must always have ditches, built along the same principles. When the tunnel floor is of rock or is paved, the ditch is often a simple trench with vertical sides below the ballast. The tunnel floor can also be made to slope from both sides towards the center, where a perforated pipe is installed or a simple ditch dug. Railroad ditches are designed only by considerations relating to hydraulic capacity, because the psychological limitation of the driver does not have to be considered as in roads where ditch slopes have to be very flat. The slopes of these railroad ditches are therefore often very steep or even vertical, which alleviates the problem of excavation depth, without increasing the width in cut sections.

In developing areas, it is common practice to construct a road, giving it a temporary surfacing and opening it to traffic, paving it permanently only when traffic has developed sufficiently. This practice leads to the need for temporary ditches, paving them when necessary, with soil-cement, for example. Without this drainage damage may often occur, which converts the routine reshaping of the cut section for final paving work into an extremely expensive reconstruction (Plate 11-15).

Plate 11-15 Erosion in a cut due to absence of temporary ditches

Even if a road is to be paved immediately, it is good practice to make temporary ditches in cuts to make work easier. This practice has been made routine by some organizations. It cannot be justified, for not all cuts really need these temporary ditches.

When a road that was originally surfaced only temporarily is paved in a permanent manner, the error illustrated in Fig. 11-11 is not unusual. First, the permanent ditch has been built with lining from the shoulder of the road, to which the previous comments apply. However there is a more serious problem. The width of the temporary surfacing (Level I) and that of the permanently paved road (Level *II*) are usually the same. The new level is usually raised but the same dimensions are maintained in the permanent ditch as in the temporary ditch (dimensions m and d in the figure). The combination of these

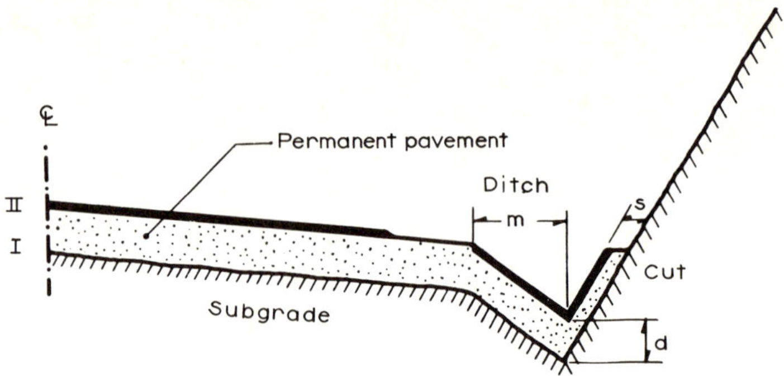

Fig. 11-11 A common defect when converting a temporary ditch into a permanent one

conditions produces a small step of width *s*. Sometimes, when the ditch is paved with concrete, the builder will extend this paving to cover the step completely. However, it is often left unpaved, permitting infiltration and consequently causing increased moisture in the pavement. Furthermore, the paved ditch becomes unstable and vulnerable to damage. It is better if the step is avoided, either by extending the permanent ditch as necessary, or by slightly increasing the width of the permanent pavement. If the new ditch is to be left unpaved, the normal practice will be to extend it as far as necessary, thus avoiding the formation of the step. In this way the engineer will no longer feel compelled to economize on paving concrete, which is undoubtedly the reason for the step.

When to pave ditches, which is usually costly, or when to save, is one of the most debatable aspects of complementary drainage. With the large number of decisive factors involved, it is seldom possible to establish any general rules. Usually, paving can be omitted when there is no danger either of erosion of the ditch bottom by the water that flows along it, or wetting of the upper layers of the pavement from eventual seepage from the ditch. Consequently it is not necessary to pave ditches cut in rock or in soils consisting of large fragments, or those ditches that will carry only light, occasional flows (either because the tributary area of the ditch is small, or because storms at the location will be very brief and sporadic).

Seepage of water from the ditches into the pavement is not serious when the cut bed, upper layers of pavement, subgrade and embankments are very pervious, or when pavement of a rock cut includes a very pervious and open base course, there is a good transverse grade in the bottom of the base course and a good longitudinal grade in the cut. It is seldom necessary to pave ditches in cuts with a very steep longitudinal grade, so long as the ditch bottom is not vulnerable to erosion.

Some of the requirements implied by the above conditions are contradictory. For example, some very pervious materials are highly erosible, so that the decision whether to pave a ditch or not should be based on consideration of many conflicting general and local factors. Although the decision is often problematical it is also one where only a small margin of error can be permitted. Incorrect decision may lead to serious consequences.

In recent years some engineers have advanced the opinion that ditches should be totally omitted in roads paved with asphaltic courses or concrete slabs. In such cases the paved surface is extended throughout the entire width of the shoulder to the toe of the cut slope. At that point a small curb is often constructed, which is usually no more than two centimeters high in order to provide a good connection with the cut slope. To hasten elimination of the water that collects in this zone, the slope of the cross-section in the shoulder is increased from the usual 1.5 or 2% to 4%. In the opinion of its supporters, this practice, in conjunction with an appropriate longitudinal grade, is sufficient to cause efficient drainage. This practice has economic and maintenance advantages. The costly and troublesome job of cleaning ditches is replaced by the far simpler task of cleaning the shoulder of the road, which is easily mechanized.

So far there is not sufficient experience to recommend this design generally, it is logical and appears very attractive for certain specific cases, such as roads with ditches where the cross-section is to be widened. Eliminating these ditches would be an attractive means of achieving this new width. In our opinion the use of ditches has been abused. It has become such a routine practice to install them in cut-and-fill sections on the cut side and in curved sections with superelevation on the embankment side, where there is little tendency for water to collect. A situation where it is better to omit ditches is in through-cuts, when one of the sides is very low and water collects on that low side as a result of transverse grade or superelevation. Instead of an expensive ditch, with paving, it may be far more cheaper and convenient to eliminate the low side of the cut by extending or daylighting the excavation. This is illustrated in Fig. 11-12.

At the end of their course, ditches discharge by chutes into culverts, gullies, and natural streams. As has been discussed, connection between ditches and chutes is a critical transition that should be carefully designed in each individual case.

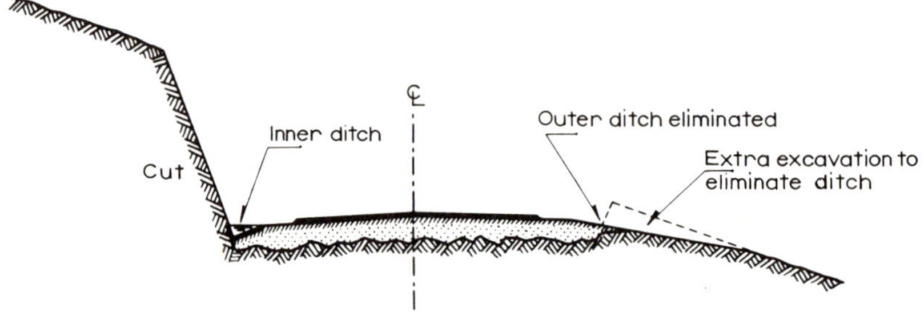

Fig. 11-12 Elimination of outer ditches in through cuts

11.11 Intercepting Ditches

These are the channels that are excavated in natural ground or formed with small training walls and located close to cut slopes on the upstream side, to intercept surface runoff from above, so as to avoid erosion of the slope and to avoid blocking the ditches and covering the pavement with the water and the material it carries with it (Fig. 11-13) (Plate 11-16).

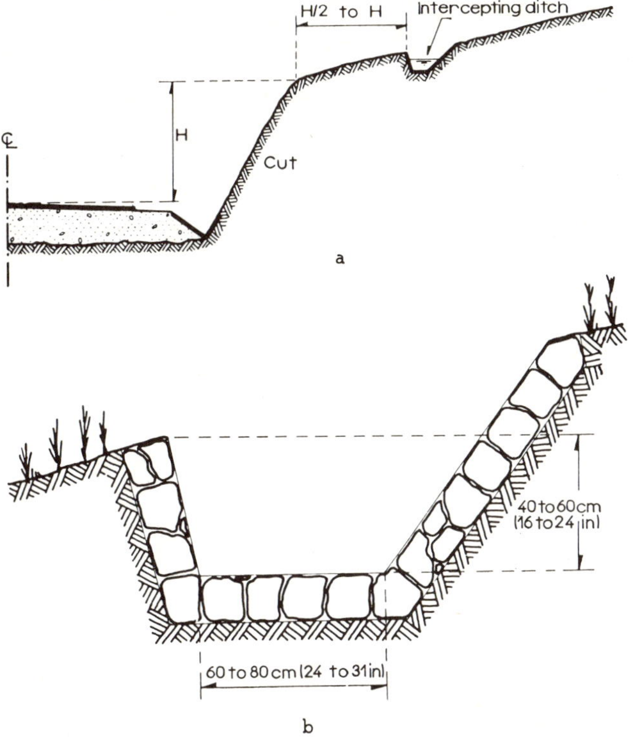

Fig. 11-13 Intercepting ditch at top of slope

The distance at which the intercepting ditch is built from the top of the cut is variable and depends on the height of the cut. Between the intercepting ditch and the cut there must be no area likely to cause significant uncontrolled runoff. At the same time the ditch must not be placed too close to the cut, so that the ground where it is installed will not be affected by any minor slides or settlement, or any compaction work that might subsequently prove necessary. In a cut of moderate height, the intercepting ditch is often installed at a distance from the top of the cut between the height of the cut and half of that value. In high cuts, the closest part of the intercepting ditch may be at an approximate distance of 8 to 10 m (26 to 33 ft) from the top of the cut.

Plate 11-16 A carefully waterproofed intercepting ditch at the top of a slope with a complementary step in the cut

The intercepting ditch should be parallel to the top of the cut. In this way the channel will have a longitudinal slope. If the hill in which the cut is made is very steep, however, a parallel ditch might have excessively steep grades. In this case, it should follow the hill surface profile with the ends of the ditch turned away from the road. These ends should cut the contour lines so that the channel will have a suitable slope.

The intercepting ditch must convey the water it collects to gullies or natural streams. Structures must be provided to carry the water across the road. In order to avoid excessive channel length it is customary to design the ends of the ditch with very steep grades, so that they will function like chutes.

The cross-section of the channel is determined by its hydraulic capacity, which in turn depends on the frequency and intensity of rainfall, the size of the drained area and the characteristics of that area that control surface runoff. References [3,6] give criteria for hydraulic design, first estimating the volume of flow anticipated and subsequently relating this information to the slope, to find a hydraulic section. This analysis is usually conducted for several different location alternatives, so that the different costs can be compared, and excavation needs established from the cross-section required for the channel. Owing to lack of hydrologic data leading to enormous uncertainties in hydraulic and hydrological analyses, intercepting ditch designs are usually standardized although a rational design should be preferred. It consists of a channel with a trapezoidal cross-section and a floor 60 to 80 cm (24 to 32 in) wide. The slopes are selected in accordance with the nature of the ground. Channel depth is usually from 40 to 60 cm (16 to 24 in). In unpaved intercepting ditches, the upstream slope must be flat so as to avoid erosion, but flattening becomes less imperative when ditches are paved. When a channel is dug, the intercepting ditches are excavated by hand or using lightweight machines such as ditchers, lightweight tractors, graders and tractor mounted backhoes. Excavated material should be placed on the downstream side of the ditch (at least 1m (3 ft) from it), or better, should be removed.

Sometimes intercepting ditches are made by building a small earth dike using material selected from some borrow pit or from an excavation on the upstream side of the dike. This dike should be placed on suitably cleared ground and consist of appropriate, well compacted materials.

It is common practice to make intercepting ditches directly in the natural ground, without paving them. Some comments will be made about this practice in the paragraphs that follow. The same materials are generally used for paving intercepting ditches as were described for ditches. For intercepting ditches, however, paving operations are complicated by hauling the materials to high locations. All the arguments for determining whether or not an intercepting ditch should be paved are similar to those discussed for ditches; likewise the considerations included in Table 11-2. Intercepting ditches present the special case of their very steep end portions, where paving is usually far more necessary.

The frequent inappropriate methods of paving intercepting ditches causes many engineers to wonder whether these complementary drainage structures should be used at all. To save money, highway engineers prefer not to pave them. Consequently there is a pervious section transporting water at the top of the cut. If the cut slope is made of a relatively pervious soil or a soil mixture that is susceptible to changes in moisture, this ditch will cause water to seep into the body of the cut slope, with the consequences that have been discussed previously. This is why it is not unusual, in highways and railroads where unpaved intercepting ditches have been used, to observe that the beginning of the failure surface at the top of the cut slope coincides with these ditches. The failure surface would possibly not have developed without the intercepting ditch. Not only does it add water, but it is also a discontinuity that concentrates stresses, promoting progressive failure.

Wherever an intercepting ditch is useful or necessary, either it should be paved or not installed, for the hazards of installing it under adverse conditions (possible total failure of the cut) by far outweigh the possible benefits (protection of the slope surface from erosion and the ditches or the top of the cut from uncontrolled surface runoff). A poor intercepting ditch often leads to large-scale collapse. If it is not installed where required, a zone will be created with poor stability, which can be readily detected and remedied by several methods, including the construction of a *good* intercepting ditch.

The above considerations for designing intercepting ditches is based on their routine use as complementary drainage structures where they are either useless or even detrimental, in spite of high costs.

The criterion for determining the need for intercepting ditches should be based on topography and the nature of the materials in the cuts, adjacent embankments and adjacent natural ground. The topography is largely responsible for the runoff that can be expected on a slope. For example, in hills that slope very steeply towards gullies on either side, most surface runoff finds the gullies in such slopes and flows parallel to the road and not towards it. In such cases, intercepting ditches will not be necessary, particularly if the ground is protected by vegetal cover or the surface is not very pervious. At other times, the catchment area at the crest of the cut will be very small, because of topography. Naturally the slope of this small area must be considered too.

The nature of the materials to be protected is decisive. Often intercepting ditches are installed where soils are highly resistant to erosion or very well protected. The authors have even seen intercepting ditches laboriously hewn into cuts made of sound rock where there were no special runoff problems, a complete waste of money.

To summarize, the advisability of constructing intercepting ditches must be considered first in cuts that are not protected by appropriate topography; those made in natural slopes and hillsides with a constant grade towards the road over extensive areas, where there is a significant rain catchment area. Second, they are installed in cuts of erosible materials likely to produce appreciable erosion of solids, such as silts, silty sands, clays and slope accumulations of mixtures of coarse soils and finer, varied filler materials. In all these cases the intercepting ditch *must* be paved; otherwise greater erosion and infiltration can occur than the ditch is intended to avoid. Occasionally in rock cuts intercepting ditches may be advisable to collect surface runoff that would otherwise discharge into the roadside ditches. Here, the need for paving may be necessary as in pervious soils. This is the case, for example, in jointed rock masses, where the joints are filled with water-erodible materials, and especially if blocks of rock with dip which falls to the road or stratified rock which dip towards the road.

Engineers in some countries, including Mexico, have adopted the practice in minor roads (where low cost is essential) of building a channel to act as an intercepting ditch parallel to the road in all zones where the ground slopes towards the road over extensive areas. This is particularly common practice in hillsides with few natural water courses in which flow could concentrate and culverts could be built. Sometimes these intercepting ditches are installed in ground with such a slight slope that it could be termed flat. In this case, they are installed close to the road, almost like roadside ditches at the upstream toe of low embankments, although it is not unusual to see them on both sides. To reduce cost, these ditches are not paved.

This practice deserves comment. In our opinion it cannot be adopted in such a routine manner as its most fervent supporters would wish.

First, the hydraulic cross-section of the ditch must be considered. If it is small, as is usual in normal intercepting ditches, the amount of runoff it can intercept must be established, and the necessary design calculations made before it is accepted. Sometimes the runoff if allowed to reach the road with its ditches and its temporary surfacing will cause no serious problems. At other times, however, the capacity of the ditch is insufficient to intercept even a small part of the anticipated runoff. Most of the water then falls onto the road, which makes the ditch superfluous Instead, water will be encouraged to seep into cuts and pavements, causing loss of stability. However, in minor roads cuts are not very high, and the stability problems caused by wetting are minimal. In those places where an appreciable amount of runoff that is likely to cause damage, can be seen on the side of the hill, a better practice is to replace the unpaved intercepting ditch by a well designed intercepting channel with a cross-section based on an appropriate hydraulic calculation. It is located well away from the road so that the lack of paving will not matter. These channels can be used in conjunction with training walls, with the advantage that both structures can be built by hand at low cost.

The authors do not understand the purpose of installing protective ditches at the toe of low slopes, in ground that is almost flat, on both sides of the road. It is more appropriate to intercept the waters up the slope, in the direction of increasing steepness. Furthermore, an unpaved ditch or trench at the toe of a low slope is likely to cause instability in the embankments and the rather modest pavement and generally poor conditions in that portion of the road. We prefer intercepting channels at a reasonable distance from the road, or training walls to divert the water towards natural watercourses. The comments made in the foregoing paragraphs about calculation of the hydraulic capacity of all these structures are valid here too. The water that appears downstream from the road should be channelized rather than intercepted, so that it will cause no damage to nearby farmhouses or agricultural land.

The maintenance of intercepting ditches deserves special consideration; it is always difficult because they are usually pretty inaccessible once the road is in service. As a consequence, they are seldom inspected, and their imperfections thus become masked. They frequently suffer very serious deteriorations which causes serious problems in the future. The authors have seen unpaved intercepting ditches of usual dimensions which, as a result of erosion, have become 3 to 4 m (10 to 13 ft) deep, enabling the water they collect to seep into the body of the cut with the worst possible consequences. At other times, this erosion leads to the disappearance of originally appropriate ditch grades, creating zones of stagnant water, which also encourage seepage into the soil. Even in paved intercepting ditches, a lack of maintenance will lead to infiltration through the cracks which will inevitably appear in the paving materials, particularly in soil-cement and soil-asphalt, but even in masonry and concrete, too.

Maintenance of intercepting ditches is usually by hand, because it is hard to maneuver any necessary equipment. This is another reason why maintenance is frequently neglected. To summarize, the experience of the authors is that maintenance of intercepting ditches is so badly neglected, and with such serious consequences, that engineers should take this circumstance seriously into consideration before recommending them in routine design. If there is not provision for regular maintenance, it will probably be better not to install ditches, especially if they are left unpaved.

Lastly, we stress that intercepting ditches very often have such steep slopes that paving them is essential. This is also necessary for the chutes that are usually constructed at the ends of these ditches.

11.12 Intercepting Channels

Intercepting channels are built to divert surface water that would otherwise run onto the wearing surface of a road, causing erosion or detrimental deposition. They are frequently constructed to intercept runoff from hillsides that slope towards the road, but also to convey water to culvert entrances or to control the flow that is discharged by these culverts. In the first case, the function of the intercepting channel is similar to that of an intercepting ditch, and many of the comments previously made about these structures apply. However, the expression *intercepting channels* is usually reserved for those that are built at relatively large distances from the road and which are not related to one specific cut, but which protect a fairly long stretch of road, regardless of its cross-section.

Intercepting channels are dug by hand or using lightweight equipment, such as ditchers, graders, small tractors or tractor-mounted excavators. The excavation material should be placed on the downstream side of the channel. Channel slopes will depend on the material that is excavated and the channel dimension. Slopes of 1:1 or 1-1/2 (H) to 1 (V) are common.

The dimensions of the channel should be selected on the basic of a hydraulic study, particularly where the flows to be handled are appreciable, [3] amd [6] give criteria for conducting such studies, which are otherwise beyond the scope of this book.

Since intercepting channels are usually made at a considerable distance from the road, they can often be left unpaved without any major hazards to the construction. This is not a rigid rule; the danger of permitting the infiltration which occurs through the unpaved channel bottom must be considered in each individual case. Protective measures can be installed wherever necessary. It is more frequent in intercepting channels than in intercepting ditches to encounter situations where omitting paving will not lead to serious consequences.

The material generally used for paving channels is masonry; concrete is employed only in the most important cases. The paved surface should be as smooth as possible, to reduce friction and thus increase the hydraulic efficiency.

The function of the channels that are built as complements to culverts is to improve the hydraulic performance of these structures. Situations where paving is either advisable or essential are more frequently encountered in such channels.

11.13 Geotechnical Considerations in Culvert Design

Wherever surface runoff discharges into a natural stream, either seasonal or perennial, a structure must be provided that will permit these waters to cross the road. Such structures are bridges and culverts. Culverts in Mexico are structures with spans less than 6 m (20 ft) Plates 11-17, 11-18. Depending on their hydraulic importance, culverts may consist of either one or several concrete pipes, a masonry arch over masonry or concrete walls, or reinforced concrete slabs on abutments of masonry or reinforced concrete. All these are what are known as rigid structures, because the deformations they may undergo under load are very small. Also available are flexible culverts of corrugated metal. These are often in the form of cylindrical pipes. However, other cross-sections, such as oval-shaped and elliptical ones, approp-

Plate 11-17 Typical culvert in a highway. Concrete slab construction

Plate 11-18 Construction of a concrete culvert

riate for handling larger volumes of flow than those discharged by pipes, or even for forming short tunnels and underpasses are becoming widespread. In these metal structures, deformations under earth pressures are appreciable, and this imposes notable geotechnical differences, which will be described in detail later.

When required by the volume of discharge or in some cases by the local topography, any of these structures can be duplicated, one next to the other, thus forming multiple culverts.

The hydraulic analysis is the fundamental problem of culverts, which is beyond the scope of this book. It is dealt with in detail in [3], one of the many papers on the subject. Here, discussion is limited to the geotechnical problems relating to culvert performance.

As a rule, a culvert somewhat reduces the area of the natural stream, causing the formation of a pond at the entrance and an increase in the velocity of flow, both inside the structure and at the outlet. The depth of the pond and the increase in flow velocity depend on the hydraulic design. If the pond is deep and lasting, it may cause internal erosion, piping or softening of the embankment. If it rises above the top of the embankment, it will lead to a catastrophic surface erosion failure. The earthwork is not capable of tolerating such an overtopping. Certainly construction of the necessary culvert will always be more economical than erosion repair. Culverts should have been designed so that their entrance will never be submerged. The above problems are therefore associated with insufficient maintenance, that allows obstructions in the hydraulic area of the culvert due to sediments, and collection of debris. This is one of the reasons why the periodical inspection and systematic cleaning of these structures is so vital, together with the use of special protective structures that prevent blockages.

The increase in the outlet velocity may lead to head-cutting erosion from downstream in the soils that receive the onslaught of the water. This encourages the use of cutoff walls, outlet aprons, diversion channels, and energy dissipators.

As far as possible, a culvert should follow the slope of the natural stream. Any abrupt change in the direction of the water at either end of the structure restricts flow and makes a larger cross-section necessary. However, there are occasions when the natural slope is too steep, and causes very high flow velocities inside the structure and at the exit, which cause erosion of the culvert, with the consequent danger of very serious leakages. In such cases, the locations shown in the profiles in Fig. 11-14 are often used. Part *a* of the figure shows the normal installation recommended for any arch culvert, box culvert, or one made of reinforced concrete slabs. Part *b* represents a case where the slope of the culvert is reduced by installing it inside an embankment. This type of installation can only be made with pipes, adding cutoff walls at the inlet and outlet chutes or drainage pipes. Very special care must be exercised to avoid any possible leakage in the culvert that could erode the embankment. Part *c* shows a typical relief culvert, of the type that is constructed at intermediate points in very long cuts, to relieve ditch flow or to drain the central median of a highway. Here too it is essential to observe all the necessary precautions, some of which are illustrated in the figure.

Often when the grade of the natural stream bed is excessively steep, the culvert is installed in the gully slope. This is frequently regarded as practical or advisable with grades of 15% or steeper. This location may prove satisfactory, but it may often have serious defects which may jeopardize not only the hydraulic efficiency of the culvert, but also the safety of the embankment.

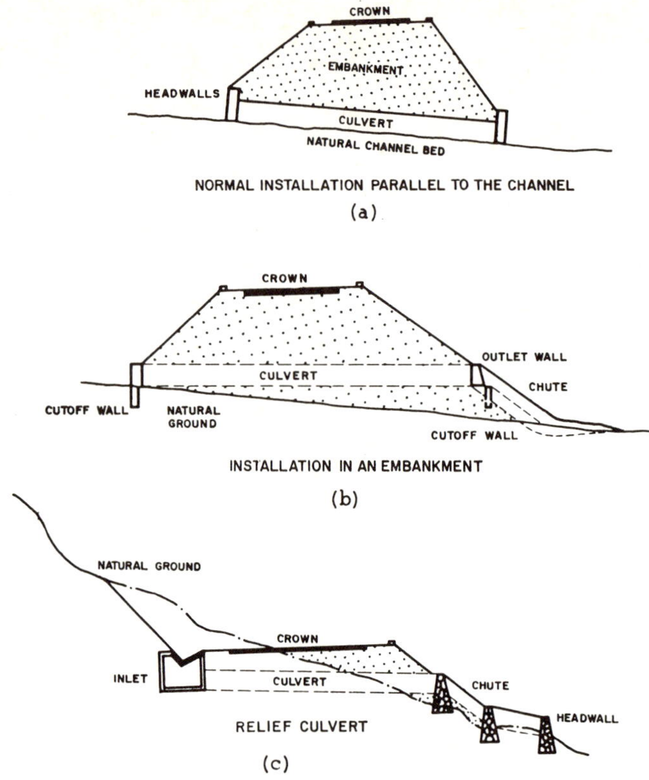

Fig. 11-14 Different culvert profiles

The entrance to the culvert must be located on a level with the natural stream at the toe of the embankment slope, so that all the water will be able to flow through it without any abrupt changes in direction or grade. To satisfy this requirement, training walls or some minor modification of the natural stream are often necessary. If the culvert entrance is above the toe of the embankment slope and above the level of the natural stream, an area is created where deposition and infiltration will occur. Flow should be discharged into the natural stream below. So as not to make the culvert too long, flow is usually conveyed from the culvert exit to the natural stream by a channel, pipe or other similar structure, controlling head-cutting from downstream. Fig. 11-15 shows a correct culvert installation and an incorrect one which is nevertheless frequently employed.

Another important aspect of culvert installation is alignment. It is natural and convenient to give the culvert the same alignment as the stream, so that the course of the stream will not be altered and thus avoiding erosion and ponding. If the natural stream is too skewed in relation to the road, this culvert alignment may prove too long. It may be best to cause the stream to cross the road in a more perpendicular direction. This requires a series of changes in the direction of flow, which is acceptable only if achieved by means of remedial channels which carry the water without causing erosive turbulence.

The geotechnical recommendations for culvert installation could be summarized as follows:

1) Whenever possible, culverts should be installed at the bottom of natural streams, wihout any abrupt changes in vertical or horizontal alignment.

2) When the alignment of the natural stream bed is not followed, culverts should be installed in a trench in firm soil.

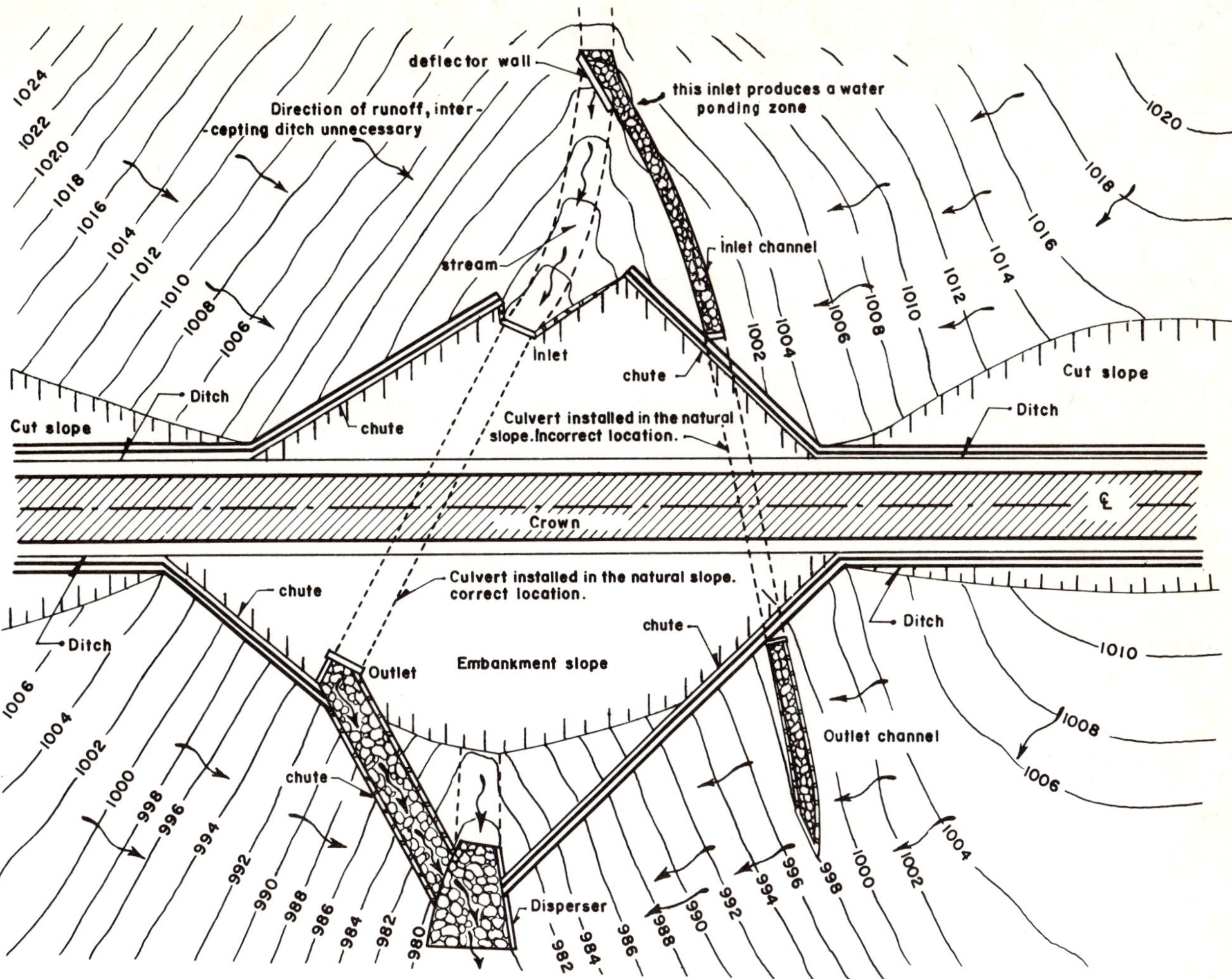

Fig. 11-15 Location of culverts in ravine slopes

3) For any location other than the bottom of a natural stream, a careful economic study should be conducted to establish that the cost of maintenance of the location chosen does not outweigh the saving in the cost of construction that is obtained with it.

4) When culverts do not follow the alignment of the natural stream, special care should be taken that their entrance and exit are suited to the flow, with no abrupt recesses or projections to encourage turbulence or erosion. The elimination of such obstacles usually saves cost as well as trouble.

5) The hydraulic grade within the culvert must be such that the rate of flow within it is equal to or greater than that of the natural stream for the corresponding stretch.

6) Any contractions in the flow of water should be avoided in culverts.

The boundary between the earthy fill material and the culvert proper is always a critical zone because compaction of bedding material is difficult; infiltration is consequently encouraged, which may cause piping or wetting of the fill material. For this reason, pavements overlaying culverts often exhibit distress and extra careful attention should always be given to this area during construction.

Culvert foundations are similar to those discussed for shallow foundations in Chapter 8. Pipes are not included in these considerations because the stresses they transfer to the natural ground are very low. As already mentioned, a major problem is the inadequate attention paid to culvert studies, owing to the large numbers of these structures and their low individual cost. As a result, recommendations for culvert foundations are usually based on soil inspection or superficial studies that are conducted by trained specialists in the application of the principles of soil mechanics to these problems. The principles followed by these specialists were listed in Chapter 8. The above considerations do not, of course, exclude the need to study all special cases.

When embankments are constructed on very soft compressible ground, the settlements they suffer are very detrimental to any drainage structures that are installed underneath them. These settlements can destroy conventional rigid structures, or deform flexible structures beyond tolerable limits. The foundation problem may sometimes be solved by installing the structure in the fill material, which is firmer than the foundation ground in this case. However, this solution is dependent on whether the hydraulic performance of the culvert is affected by the higher position, or a ponding zone is created under the culvert floor causing water to seep into the embankment. Generally the foregoing method is suitable in embankments built on swampy, flooded ground or in places where, owing to the softness of the natural ground, part of the embankment becomes embedded in it from the start. Of large culverts, concrete caissons or box culverts [10] are the structures which transmit the lowest levels of stress to the ground (considering cases where pipes are no longer able to solve the hydraulic problem). These are, moreover, the structures which tolerate best the settlement of the embankment over the compressible foundation ground, for although they may suffer cracks that will have to be caulked, their performance is not essentially jeopardized by that settlement. Because they transfer to the ground similar stresses to those transferred by the embankment itself, differential settlement problems, which would otherwise be very serious, are eliminated.

11.13.1 Flexible Culverts

Flexible culverts are those consisting of pipes or arches of corrugated steel, lined or unlined and installed in the ground under an embankment in one or more rows.

When designing these structures, it is essential to consider the influence of both dead loads and live loads. The former are due to the dead weight (total or partial) of the earth laid over the structure (bedding). Live loads are due to the weight of equipment and vehicles travelling over the structure before or after placement of its protective earth bedding. The impacts produced by moving loads, and in certain cases the vibrations transmitted by these loads, are also live loads. The effect of the live load diminishes as the thickness of the bedding increases, and as traffic speeds increase.

Besides the vertical effects of the loads considered, lateral and longitudinal pressures induced by the vertical loads also exist along the axis of the structure.

The minor settlement or deformation suffered by a flexible metal culvert relieves the vertical stress in the structure considerably better than a rigid culvert. This is due to the arching phenomenon [11], described in Chapter 5 and Chapter 14. This reduces the vertical earth pressure acting on the culvert arch to less than that which would normally be exerted by the thickness of bedding overlying it. The effect can be quantified approximately by referring to the theory described in the above mentioned chapters. From the structural point of view, it is usually specified that the arch of a flexible culvert must not settle more than 5% of its maximum vertical dimension. This limit is ample enough to include the deformation required for arching to occur, and the phenomenon therefore takes place above flexible metal culverts of the type used in engineering practice. The arching effect is more pronounced in sands than in clays. It is influenced by vibrations, which tend to reduce it, especially in sands. However, it should be recalled that a minimum thickness of bedding is required for the development of arching effects of any practical importance. These limits are also briefly discussed in [11].

If arching effects are assumed to be non-existent, the effects of the combination of live and dead load on a culvert are shown for two specific situations in Figs. 11-16 and 11-17, with reference to roads and railroads, respectively. In both cases it has been considered that the dead load, due to the bedding of earth, increases linearly with depth. The effect of the live load, (*H-20* for roads and COOPER *E-72* plus 50% impact for railroads, in these figures) obeys a hyperbolic law, varying with depth. The total load, which is the sum of both, is illustrated in the two figures.

Considering the arching effect, the two graphs give representative results for thin layers of bedding where the arching is not well developed. For thicker layers of bedding, the dead load is no longer a linear function of depth, but increases up to a limit, beyond which it remains uniform. In each graph it can be seen that there is a thickness of bedding for which the combined loads produce a minimum effect.

In order to tolerate loads adequately, the entire length of the culvert must rest on a homogeneous continuous soil. If the natural ground is not homogeneous, weak or compressible materials should be replaced by compacted material. A bedding should be laid beneath the structure, preferably of dense sand. In embankments built on compressible soils, the differential effect caused by greater settlement in the center than at the sides may call for an appropriate camber.

The strength and performance of any type of flexible drainage structure depends largely on the quality of the adjacent lateral fill and bedding material and the standards observed for placing it [12]. This fill should be unaffected by water as far as possible;

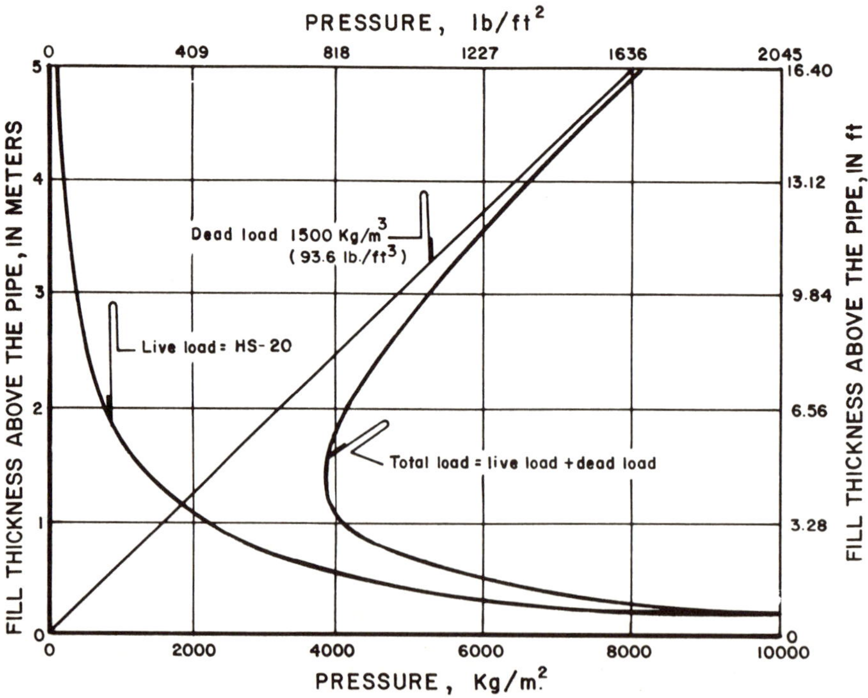

Fig. 11-16 Combination of dead and live loads on flexible drainage pipes in highways for the conditions indicated

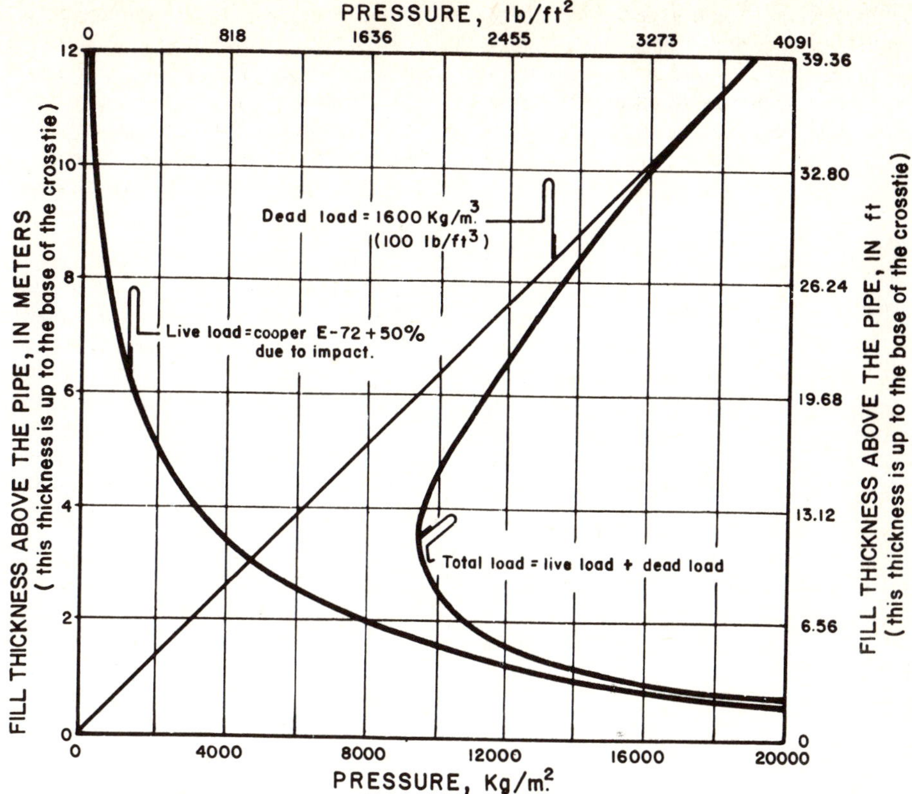

Fig. 11-17 Combination of dead and live loads on flexible drainage pipes in railroads for the conditions indicated

be non-susceptible to expansion, cracking and piping and be readily compacted. In Chapter 3, some guides were given for estimating the susceptibility of soils to these phenomena; they are valid in the case now under consideration.

To avoid distortion of the metal structure, bedding material must be extended from the center of the arch out towards the two sides simultaneously. Lateral fill material should be laid simultaneously on both sides to balance loading. Covering procedures are best commenced in the longitudinal direction, from the center of the pipe towards the two ends. When laying the fill, care must be taken that the layers are properly compacted. Neglect of compaction is the cause of numerous failures in flexible drainage structures. Compaction increases soil stability and as shear strength increases, there is a balancing of the lateral earth thrusts exerted by the fill against the structure. Compaction of the bedding increases the beneficial effects of arching, greatly reducing the vertical loads on the structure. Within an area with a width twice the diameter of the structure, there should be no material that has not been carefully compacted. Compaction can be done using manual equipment, or by machines so long as they do not cause any damage to the culvert. A frequent mistake during compaction of the bedding material is to allow heavy compaction equipment to travel over the structure before a sufficiently thick protective covering has been placed. This has been the cause of frequent failures.

If the layer of bedding above the culvert is not very thick, lateral thrusts may predominate, which tend to raise the crown. In this case, cross-sections of the arch type are recommended. These are wider and not so high. This tends to increase the amount of bedding and neutralize the detrimental effect mentioned.

Flexible culverts often suffer structural distress (excessive deformations, and settlement) during their life. These effects are usually caused by poor compaction of fill materials, which causes lateral earth thrusts that are greater than those considered in the design (which take into account the thicknesses of bedding and fill after careful compaction). The only solution to such problems is a radical one — removal of the loose fill, replacing it by another well compacted material.

When the ground beneath a flexible culvert is compressible and the culvert is installed in it lengthwise, the joints between the assembled metal plates will open, due to the greater settlement occurring under the center of the embankment than under the shoulders. To prevent this problem and keep water from seeping through the open joints, a corrugated steel expander ring is placed inside. The corrugations of this ring must coincide with those of the culvert plates. The ring expands from within, thus acting as a seal. Where the ring functions in harmony with the culvert wall, it can be regarded as a structural reinforcement. Should this inner ring be required to act as a seal, a layer of asphalt, neoprene or some other similar flexible material should be placed between it and the culvert structure.

It has often been observed that the maintenance of both flexible and rigid culverts and their auxiliary structures (headwalls, energy dissipators, channelizing structures, and discharge chutes) is sadly neglected. This leads to possible embankment damage and a briefer service life for the road and the culverts. Silting is particularly harmful. Efficient maintenance includes the installation of channelizing structures and all measures required to correct defects or omissions in construction and to improve hydraulic performance.

Piping erosion is recognized by irregularities, and hollows in the embankment water that appears on the surface, or damp spots and other signs of internal flow, especially in the downstream slope of the embankment. If the piping is not very advanced, the best remedy is the installation of a filter in the downstream embankment slope around the culvert. If, however, piping is advanced, it will be necessary to replace the affected soil and then install a filter, constructing galleries through the embankment for access to the damaged zone.

A common cause of piping is holes left unsealed inside the culvert. These holes may initially have been necessary for hauling and hoisting the culvert parts. They are particularly dangerous when the fill around the culvert is very susceptible to erosion (fine sands and non-plastic silts with $I_p<10$). Fill material, sucked into the holes by the flowing water, triggers a progressive erosion process that leads to the failure of the culvert and pavement due to lack of support. There are even cases where, instead of flowing through the culvert, water will cross the embankment in an erosion tunnel that forms *around* the conduit. The holes should be sealed during construction of the culvert.

In very clayey embankments, a prolonged drought may cause shrinkage cracks in the soil around the culvert, providing a natural inlet for water. When this is the case, all these cracks must be sealed by tunneling along them and introducing new, well compacted material. Protection of embankment slopes with vegetal cover contributes greatly to preventing crack problems.

Of the several different types of drainage structures currently in use in highway engineering, none should be regarded as the optimum solution to all problems. All have their advantages and disadvantages. A description follows here of the advantages and disadvantages of flexible metal culverts.

The chief advantages stem from the fact that flexible metal culverts are manufactured in accordance with strict standards, which minimizes serious defects of fabrication. Also, their strength is enormous compared to their weight. The inherent advantages of flexibility have already been mentioned. Metal pipes are structurally capable, even in soils with a very low bearing capacity, for they transfer very small pressures to the foundation ground. They are easy to install and handle and are available in a large variety of cross-sections, sizes and gauges of sheet iron, which provide numerous combinations for achieving the design best suited to each case.

The chief disadvantage of metal culverts is their high cost in comparison with masonry and even concrete structures, which usually prove cheaper in places where the foundation ground poses no special bearing capacity problems. Also, metal structures are damaged by corrosive waters as well as by sand and gravel in suspension in the water, unless the steel is protected by extremely costly covering processes. Concrete and masonry tolerate the erosion of fast-flowing waters better.

11.13.2 Rigid Culverts

The study of culverts made of rigid materials, such as reinforced concrete, commences with an analysis of the loads to which the structures will be subjected, for these loads have an especially important influence on design.

.1 Study of Dead Loads

For design, the two traditional loads must be considered: dead and live loads. Dead loads are caused by the earth above the rigid pipe. A simplistic analysis might consider this to be equal to the dead weight of the material above the pipe. Engineers assumed this to be correct for many years. Today, however, it is known that the force of the overlying soil may be greater or smaller than the dead weight, and only by a rare coincidence will it be equal to this weight. This is due to shear stresses that act between a prism of soil of the same width as the pipe diameter, situated above the pipe and extending as far as the surface of the ground or embankment, and the masses of soil on either side of the prism. The shear is mobilized by relative movement. The shear stresses act in an upward direction, when the prism under consideration tries to sink in relation to the adjacent masses and the force of the prism on the pipe is smaller than its dead weight. If, however, the adjacent masses subside in relation to the prism, the boundary shear stresses develop in a downward direction, adding to the dead weight of the prism and the force of this prism on the pipe will be greater than its dead weight.

For calculating dead loads, rigid pipe culverts are classified into four main groups, in accordance with the installation conditions which influence the magnitude and direction of the shear stresses mentioned above. These groups are shown in Fig. 11-18.

Pipes without any overlying embankment (Part *a* of the figure) are installed in narrow ditches below the level of the natural ground. Above them only the backfill replacing the soil excavated of the ditch is laid. Pipes with overlying embankments (Parts *b* and *c*) may or may not be placed in a ditch within the natural ground. Such a ditch is beneficial, because the looser the fill that is laid over the pipe, the smaller the vertical load. This loose fill material does not need to occupy the entire ditch, a 30 to 40 cm (12 to 16 in) layer is sufficient to achieve a beneficial arching or shear.

Part *d* of the figure shows an installation that successfully reduces the load acting on a pipe installed in the embankment. This is known as the imperfect ditch system. First the pipe is installed in the natural ground without any ditch. Then well compacted fill material is packed against the sides of the pipe to a distance of twice the pipe diameter and a height of about 40 cm (16 in) above its topmost point. Next a ditch equal in width to the pipe diameter is dug along the pipe to within about 10 cm (4 in) of the top of the pipe. The ditch is then filled with loose, compressible material, after which the embankment is built, compacting it in the conventional manner. In this system, the more compressible the fill material, the greater will be the reduction in the dead load acting on the pipe. MARSTON [13, 14] has suggested the addition of straw or dead leaves to the fill in the ditch in order to increase its compressibility. He has also developed a theory which permits evaluation of the dead load acting on the concrete pipe under the different installation conditions shown in Fig. 11-18.

The case of pipes in ditches (Fig. 11-18a) is analyzed, with the following nomenclature, with reference to Fig. 11-19 [15]

W_m = dead load acting on a horizontal plane touching the top of the rigid pipe.

γ_m = specific weight of the soil in the state in which it is found.

F = vertical load on the horizontal plane at level h.

D = external diameter of the rigid pipe.

B_t = width of the ditch at the level of the top of the pipe.

H = depth of the ditch to the horizontal plane touching the top of the pipe.

h = distance from the surface of the natural ground to a horizontal plane in the fill.

C_d = coefficient of loading.

$\varnothing$ = angle of internal friction of the fill material.

$\varnothing_t$ = angle of friction between the fill material and the ditch wall ($\varnothing_t \leqslant \varnothing$).

K = coefficient of earth pressure.

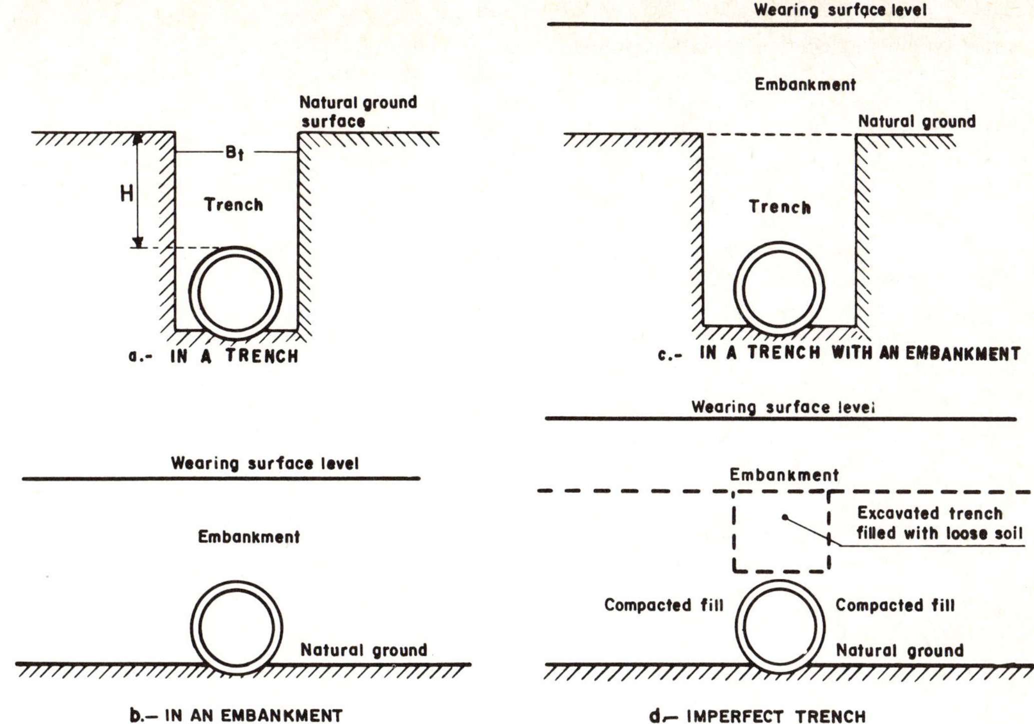

Fig. 11-18 Some different drainage systems and their installation

With reference to Fig. 11-19 and analyzing the equilibrium of the fill material at depth h, the following equation can be written for a unit portion of pipe:

$$F + \gamma_m B_t\, dh = F + dF = 2K \tan \varnothing_t \frac{F}{B_t}\, dh \quad (11\text{-}2)$$

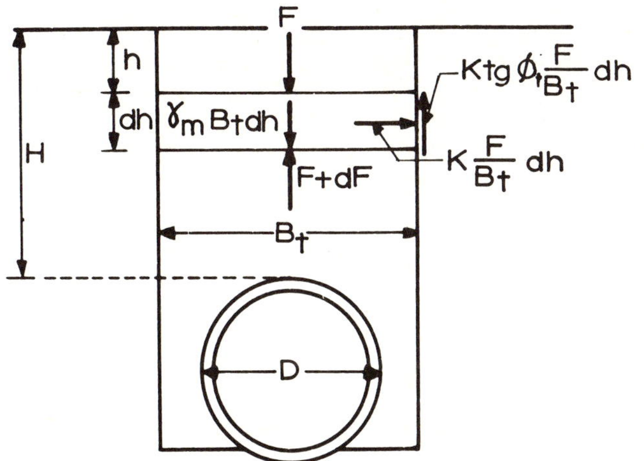

Fig. 11-19 Derivation of the formula giving the dead load on pipes in ditches

Note that since at least part of the fill is always placed in a loose state, it will tend to sink, and as a result the shear stresses in the ditch walls will act in an upward direction, which reduces the load on the pipe. In Eq. (11-2), MARSTON regarded K as being the coefficient of active earth pressure K_A. This is questionable because the walls of the ditch probably do not yield sufficiently under the thrust. From this point of view, it seems that the coefficient of earth pressure at rest (K_0) or some value between K_A and K_0 might be more reasonable. When evaluating the shear stresses in the ditch walls, MARSTON considered that they develop in unison with the peak shear strength at all points of the wall, and this is not very realistic either. Nevertheless, one consideration tends to compensate for the other, and specialists who usually apply MARSTON's formulas usually report satisfactory pipe behaviour, when the installation requirements are fully satisfied.

Equation (11-2) leads to a differential linear equation, the solution of which for the boundary condition $F = 0$ for $h = 0$ is:

$$F = \gamma_m B_t^2 \frac{1 - e^{-\frac{2KH \tan \varnothing_t}{B_t}}}{2K \tan \varnothing_t} \quad (11\text{-}3)$$

which, at a depth $h = H$ can be written thus:

$$W_m = C_d \gamma_m B_t^2 \quad (11\text{-}4)$$

where C_d is an adimensional loading factor equal to:

$$C_d = \frac{1 - e^{-\frac{2KH \tan \varnothing_t}{B_t}}}{2K \tan \varnothing_t} \quad (11\text{-}5)$$

In the above, e is the basis of the Napierian logarithms. Equation (11-14) permits calculation of the dead load on a unit of pipe length. Any homogeneous system of units can be used. C_d is a function of the product $K \tan \varnothing_t$ and the relation H/B_t. In the graphs in Fig. 11-20, it is given for different types of soils.

If the pipe is very rigid, as is generally the case with concrete pipes, it should be capable of supporting almost the entire load given by Eq. (11-4), for it will be far more rigid than the fill material that is placed on either side of it. If, however, the pipe is flexible and the soil on either side of it is suitably compacted, the rigidity of both may be similar. In order to consider the load that is supported by the pipe it will be necessary here to multiply the value given by Eq. (11-4) by the ratio D/B_t.

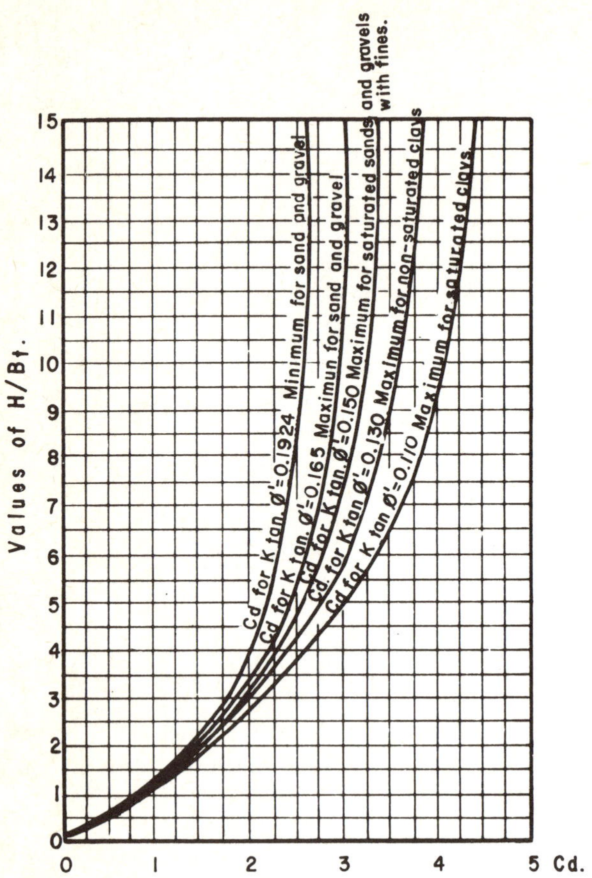

Fig. 11-20 Values of the coefficient of loading C_d

Often the walls of the ditch in which the pipe is installed are not vertical, but slightly sloping, which produces a variable B_t dimension. In this case, the width measured on the horizontal plane touching the top of the pipe is introduced into Eq. (11-4). This analysis will be valid only if the ditch slopes are steep. If they are flat the load acting on the pipe will have to be analyzed as in the case of the pipe installed in an embankment.

The case of pipes in embankments is illustrated in Part *b* of Fig. 11-18. Now it can be imagined that two vertical planes are drawn tangential to the pipe and extend to the surface of the embankment. The shear stresses that develop on these planes as a consequence of the movement of the internal prism in relation to the adjacent masses of soil will influence the load that will ultimately act on the pipe. If the internal prism tends to settle in relation to the adjacent masses, there will be a beneficial arching effect and the load acting on the pipe will be smaller than the weight of that internal prism. If, however, the adjacent masses tend to settle in relation to the prism, the load on the pipe will be greater than that corresponding to the weight of the column of soil above it. To quantify the dead load that will act in a specific case, movement of the horizontal plane touching the topmost point of the pipe is considered. The relative movement of points on that plane (the critical plane) at the top of the pipe and on the sides of it, is analyzed. Settlement of the critical plane at the sides of the pipe is equal to the displacement, S_g, that is suffered by the surface of the natural ground and between the natural ground the critical plane (S_m) Fig. 11-21. Thus, the settlement of the critical plane at the sides of the pipe will be $S_m + S_g$. The settlement that is suffered by the point on the critical plane above the top of the pipe is also found by adding two components. The first is the amount of settlement of the bottom of the pipe, S_f (as a rule $S_f > S_g$, for S_f includes the settlement of the natural ground, plus the embedment of the pipe within it) and the second is the structural deformation of the pipe in the vertical direction under the effect of the acting load, d_c. Thus, the amount dropped by the critical plane above the top of the pipe is $S_f + d_c$. The relative movement on the critical plane is equal to $(S_m + S_g) — (S_f + d_c)$.

r_a defines a settlement relation, as follows:

$$r_a = \frac{(S_m + S_g) — (S_f + d_c)}{S_m} \tag{11-6}$$

which expresses the relation between the relative movement on the critical plane and the reduction of the embankment at the sides of the pipe.

A positive settlement relation indicates that the adjacent soil prisms move more than the internal prism, and therefore the load on the pipe exceeds the weight of the prism above the pipe. Conversely, a negative settlement relation is a sign of favourable arching and a pipe load that is less than the weight of the prism above. The design relation, p, is defined as the thickness of the embankment between the natural ground and the critical plane divided by the width of the concrete pipe, D. Consequently, the thickness of the embankment at the sides of the pipe is expressed by pD.

In high embankments, the effect of arching on the pipe does not extend for the entire height, but dissipates as the elevation increases in relation to the pipe. If the embankment above the

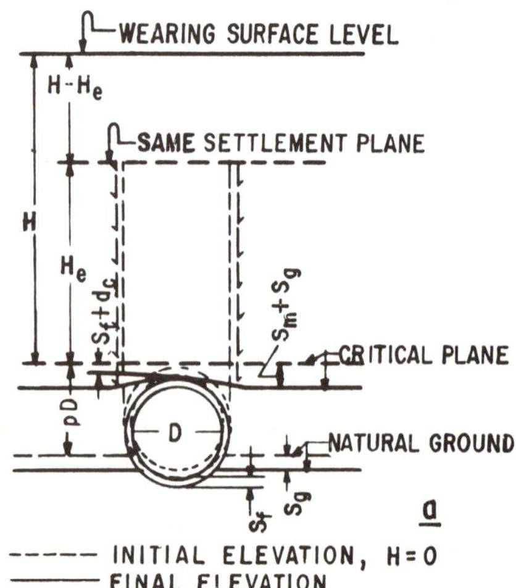

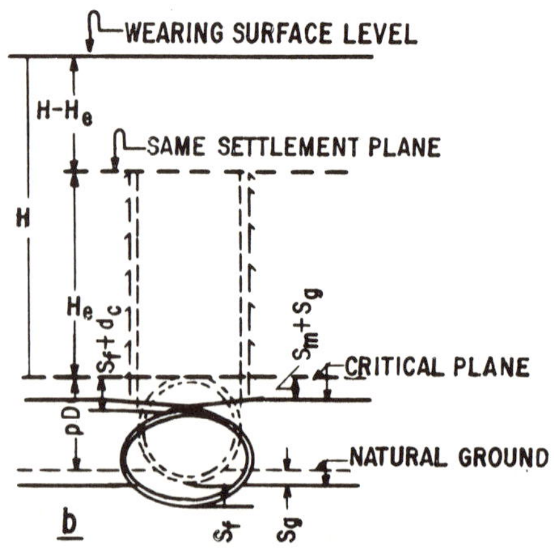

Fig. 11-21 Pipes in an embankment

pipe is high enough there is a height H_e, at which these effects are no longer perceptible. The horizontal plane at a height H_e above the pipe is referred to as the equal relative settlement plane, because it settles the same amount above the top of the pipe as it does to the sides of it. On the equal relative settlement plane, the previously mentioned shear stresses on the imaginary vertical planes tangential to the sides of the pipe, no longer exist.

The formula that is developed by MARSTON's loading theory for rigid pipes installed in embankments (Fig. 11-18b) is:

$$W_m = C_c \gamma_m D^2 \qquad (11\text{-}7)$$

where the letters are defined earlier in this section, and C_c is a loading coefficient given by the expressions:

$$C_c = \frac{e^{\pm 2K \tan \varnothing H/D} - 1}{\pm 2K \tan \varnothing}, \text{ for } H \leq H_e \qquad (11\text{-}8)$$

$$C_c = \frac{e^{\pm 2K \tan \varnothing H_e/D} - 1}{\pm 2K \tan \varnothing} + \left(\frac{H}{D} - \frac{H_e}{D}\right) \cdot e^{\pm 2K \tan \varnothing H_e/D}, \text{ for } H > H_e \qquad (11\text{-}9)$$

Plus signs should be used when the settlement relation is positive and minus signs when it is negative.

In the above formulas, H_e indicates the position of the equal relative settlement plane (Fig. 11-21). This can be evaluated using the expression:

$$\left[\frac{1}{2K \tan \varnothing} \pm \left(\frac{H}{D} - \frac{H_e}{D}\right) \pm \frac{r_a p}{3}\right] \frac{e^{\pm 2K \tan \varnothing H_e/D} - 1}{\pm 2K \tan \varnothing} \pm \frac{1}{2}\left(\frac{H_e}{D}\right)^2 \pm \frac{r_a p}{3}\left(\frac{H}{D} - \frac{H_e}{D}\right) e^{\pm 2K \tan \varnothing H_e/D} - \frac{1}{2K \tan \varnothing}\frac{H_e}{D} \mp \frac{H H_e}{D^2} = \pm \frac{r_a pH}{D} \qquad (11\text{-}10)$$

The upper signs must be used with a positive settlement relation and the lower signs with a negative relation.

Figure 11-22 is a graph which gives the value of C_c as a function of those of the H/D relation and of the product $r_a p$.

The graph makes application of the involved Eqs. (11-8) and (11-10) unnecessary. It provides directly the values of C_c that are required in order to apply Eq. (11-7). When $r_a p = 0$, two things may occur: $r_a = 0$, and the settlement of the critical plane is the same at the sides and top of the pipe, or else $p = 0$, and the pipe is installed in a ditch with a depth of one diameter. In both cases, constant C_c turns out to be equal to H/D, and the load on the pipe is identical to the weight of the embankment above it and

$$W_m = \frac{H}{D} \gamma_m D^2 = \gamma_m HD \qquad (11\text{-}11)$$

For negative values of the product $r_a p$, r_a is the negative one, since p is always positive, if it exists, and the load on the pipe is smaller than the weight of the overlying earth. The critical plane sinks deeper at the top of the pipe than at its sides. In this case, the value of C_c depends on that of $K \tan \varnothing$, Eqs. (11-8) and (11-9) and increases as the value of this product decreases, and it is therefore conservative to calculate it with

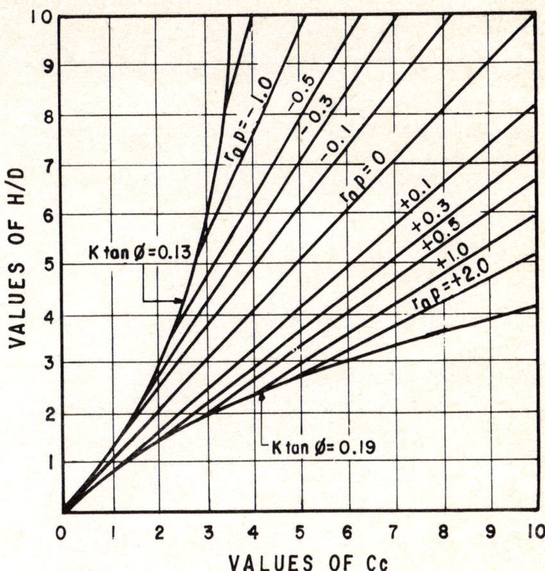

Fig. 11-22 Values of C_c

a low, although realistic, value of $K \tan \varnothing$. In the graph, the curves with negative r_a refer to $K \tan \varnothing = 0.13$, corresponding to a clay embankment. The curves corresponding to negative r_a extend from the graphical representation of Eq. (11-8), the thicker line. The intersection of the lines gives the value of H_e corresponding to each value of $r_a p$ starting from its ordinate. When r_a is positive, the product $r_a p$ is too. Different values of this product produce the lines to the right of the one which was drawn at 45° for $r_a p = 0$. In this case, the value of C_c increases with $K \tan \varnothing$; for conservative design calculate the lines with a high but realistic value of the product. In Fig. 11-22 the value used was $K \tan \varnothing = 0.19$, which corresponds to granular soils without fines.

These curves also extend from the line which represents Eq. (11-8) for positive r_a. Once again H_e can be estimated on the basis of the ordinates of the starting points. In order to apply these graphs, an evaluation cannot be made of the settlement relation r_a with which the pipe under design is to work. This setback is overcome by commencing with an assumed value of r_a based on the performance of existing culverts. Table 11-4 gives values recommended by experience for the relation under consideration [15].

For culverts in ditches, but with overlying embankments, (Fig. 11-18c), the dead load per meter of pipe can be estimated using Eq. (11-12) below, which refers to Fig. 11-23.

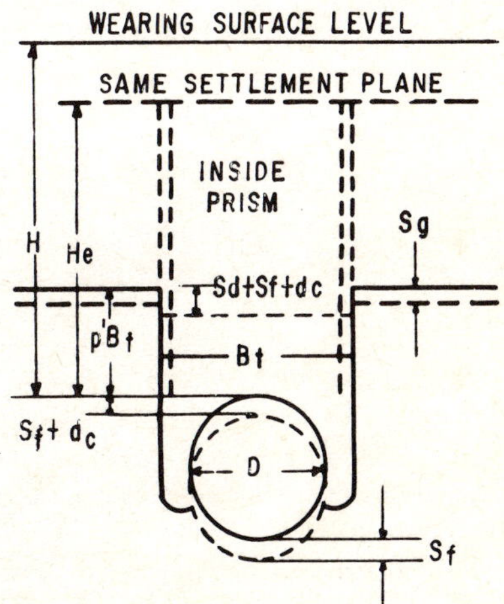

Fig. 11-23 Pipe in a ditch under an embankment

Table 11-4

Values of the settlement relation, r_a, for design purposes [15]

Prevailing conditions	r_a
Rigid pipe on rock or non-yielding soil	+ 1.0
Rigid pipe on compressible soil	0 to + 0.5
Rigid pipe on common soil	+ 0.5 to + 0.8

$$W_m = C_n \gamma_m B_t^2 \qquad (11\text{-}12)$$

where B_t is the width of the ditch and C_n is a loading coefficient that is obtained from the graphs in Fig. 11-24.

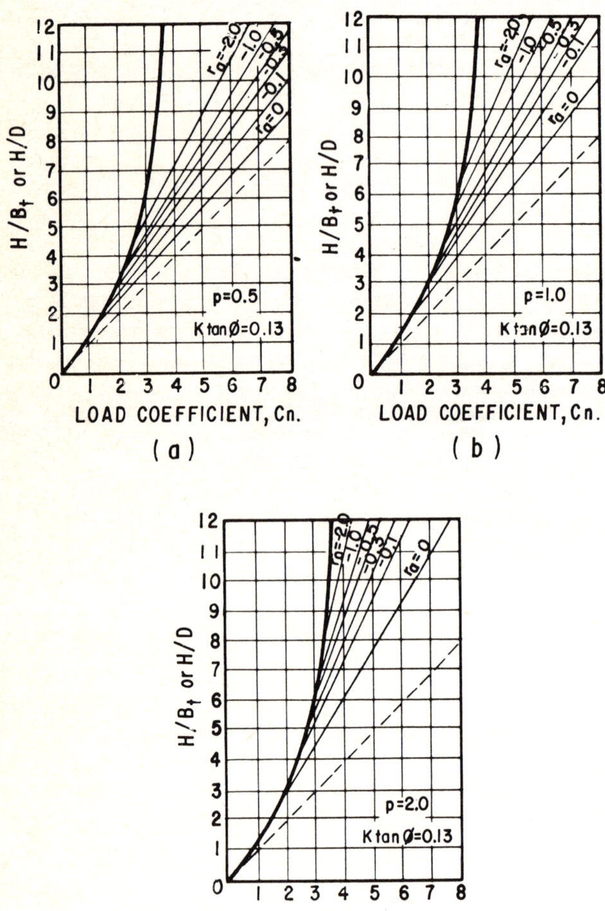

Fig. 11-24 Values of the coefficient of loading C_n

In these graphs, values of the design relation p (see Fig. 11-23) are used of 0.5, 1.0 and 2.0, respectively. For intermediate values of p, a linear interpolation of the values of C_n obtained can be made. The significance of the different curves shown in the graphs is similar to that already discussed for culverts installed underneath embankments. For calculation purposes, a value of $K \tan \varnothing = 0.13$ was used, which is conservative. Experience provides little information for establishing the settlement relation to use in the design, values between -0.3 and -0.5 being assumed adequate in this case.

For culverts that are installed in imperfect ditches (Fig. 11-18d), the equation to apply is Eq. (11-12), substituting D for B_t which is the width of the excavated ditch. Thus, for this case the expression will be:

$$W_m = C_n \gamma_m D^2 \qquad (11\text{-}13)$$

where C_n is obtained also from the graphs in Fig. 11-24, but using the H/D relation instead of H/B_t. The value of p is equal to the depth of the excavated trench, divided by D.

.2 Study of Live Loads

As already indicated, culverts also support live loads produced by the highway, railroad or aircraft which travel over them. The effects of the live load depend largely on the thickness of the earth bedding that is laid over the pipe, and are smaller, the thicker the layer of bedding.

All the experiments conducted up to the present time, both on pipes in ditches and those installed in the natural ground under an embankment, indicate that a static superficial load like that produced by a stationary wheel transmits stresses to the embankment soil that can be evaluated to an acceptable approximation by the BOUSSINESQ theory for a linearly elastic semi-infinite, homogeneous, isotropic medium. The loads applied to the culverts are, however, produced by moving vehicles. This is usually taken into consideration in the formulas that are used to calculate live load, introducing into them a factor higher than one, called the impact factor, to represent the dynamic effect. For rigid culverts under highway and runway embankments, HOLL [15] proposes the following expression for the calculation of live loads:

$$W_v = \frac{1}{L} w_o F_i P \qquad (11\text{-}14)$$

Where: W_v is the average live load acting on the culvert per unit of pipe length; L the length of a longitudinal section of the pipe, when each section measures one meter or less. If the culvert sections are longer, or a continuous pipe is employed, L must be taken as equal to 1 m (effective length); W_o is the influence factor of the superficial load; P_i the impact factor, usually between 1.5 and 2; and P is the wheel load, regarded as a concentrated load.

The influence factor of the superficial load, w_o, depends on the effective length, L, the diameter D (or width in the case of a concrete slab or box culvert) of the structure, the depth at which the top of the pipe is located from the surface of the embankment, H, and the position of the wheel load in relation to the area of the pipe that is projected on a horizontal plane across the top of the pipe. On introduction of the parameters: $m = L/H$; $n = D/H$ where m and n are interchangeable, the influence factor w_o can be calculated with the FADUM graph corresponding to a uniform load applied to a rectangular area [16]. This graph gives the influence factor for a point located on the vertical drawn through a corner of the rectangular area. Similarly, for this case the graph is applied when the wheel load, P, is located above a corner of the area of the pipe in which it is wished to calculate the load per unit of length (this area is the projection of the pipe on a horizontal plane across its top, as already stated). Although the FADUM curve is now apparently applied to a completely different case than that for which it was computed, the values of the influence factors are nonetheless valid, as indicated by HOLL [15]. If the wheel load is situated over the center of the rectangular area (position in which influence of the load is maximum), W_o will be obtained by multiplying by four the value obtained by considering one of the four equal parts into which the rectangular area can be divided, because the wheel load will now be in the corner of all four.

For the case of a rigid culvert installed under a railroad embankment, a different method must be used to calculate the live load exerted on the structure. It is assumed that the load exerted by the driving axles of the locomotive is evenly distributed over a rectangular area equal in length to the distance between the end driving axles and equal in width to the length of the railroad cross-ties or sleepers. The effect on the culvert of the load thus obtained can be evaluated by applying the same FADUM graph. Here, too, the load exerted by the locomative must be multiplied by an impact factor, generally estimated at 1.75, when the layer of fill above the culvert is less than two meters (six and a half feet) thick. This is reduced by 0.10 for every additional meter (three feet) of fill material, until the factor becomes unity.

The effect of loads, either live or dead, on culverts is reflected in stresses and strains on the culvert, but these are not the object of study at this point. A matter of considerable importance in relation to structural criteria is the increase in length and change in shape that are suffered by a pipe installed on compressible ground beneath high embankments. In these cases, a pipe with flexible joints is recommended, provided with an appropriate camber, rather than a continuous rigid pipe, in which deformation of the ground would lead to the development of prohibitive stresses or cracks.

11.13.3 Installation of Culverts in the Field

Both rigid and flexible culverts are designed to resist average earth pressures corresponding to a certain thickness of fill. However, as was seen in § 11.13.1 and 11.13.2, the relative movements that induce soil arching, may lead to appreciable variations in these anticipated average conditions. The engineer in charge of culvert installation in the field must, therefore, be fully aware of these problems, so that he will be able to interpret the specific conditions that may arise, and decide whether his structure will be required to support pressures greater or smaller than those corresponding to those in design handbooks or typical projects (which he must also be familiar with). He will then make any changes that may prove necessary in the installation, so that the structure will perform adequately.

.1 Unyielding Foundation Ground

Unyielding foundations of rock or hard, firm soils, cause stress concentrations and undue loads, if they are not treated properly. If the culvert is installed in a ditch, average pressure will always be somewhat relieved by the overlying soil backfill prism. There will be a greater height of fill material above the culvert than at the sides, and consequently the differential settlement suffered by this material will be at least a little more pronounced. This effect will be far more notable in flexible culverts, owing to their own tendency to distort under load.

If the culverts are in an embankment (Fig. 11-18), the foregoing beneficial effect is inverted. Here the height of the fill material at the sides of the culvert will also be somewhat greater than above it, and as a result the settlements at the sides of the culvert will be greater than those above it, which will cause the vertical load to rise above the average. Obviously if the culvert is flexible, its own settlement will neutralize (and possibly invert the above effect, actually relieving the pressure). If it is rigid this relief does not occur, and the condition is the severest one that may occur in a rigid culvert. In cases like this it will always be wise to dig a ditch or leave a portion of very loose bedding over the top of the culvert (imperfect ditch) which, as it deforms, will neutralize the increase in pressure. When the culvert is to be installed in an embankment, there is another alternative, which is usually the best — a layer of supporting bedding beneath it. The rock or hard soil throughout the entire width of the culvert must be excavated, to a minimum depth of 30 cm (12 in). Besides providing a uniform support, without any irregularities that might cause undesirable concentrations of pressure, as would occur with rock, this bedding helps to solve the pressure problem if it is not highly compacted and therefore able to settle somewhat beneath a high embankment. If, on the other hand, the embankment is very low, it will be advisable to compact the bedding material very carefully, for pressures will not be a problem, but settlements may well cause damage to the pavement.

.2 Yielding Foundation Ground

Peat formations, clayey soils or soils composed to mixtures of clay with other materials, in which moisture conditions are relatively high, provide a yielding foundation.

Usually, the ground beneath the culvert will settle by the same amount as at the sides, and vertical pressures will tend to become uniform and relieve themselves. As always, the relief will be more pronounced in flexible culverts. In order to standardize pressure conditions beneath and to the sides of the structure, it is advisable to construct a supporting bedding of granular material, which should extend to at least one diameter on each side and should have a minimum thickness of 20 cm (8 in).

Usually culverts are installed before embankment construction commences, so that surface drainage conditions in the zone will not be affected, even temporarily. However, when the soils under the embankments are soft, culvert installation is best postponed until all the major settlements expected have taken place, that is several months after construction of the embankments.

11.14 Mechanisms of Erosion by Water and Soil Resistance

Erosion, which is a process causing soils to disintegrate and be washed away, is of such great practical importance in the design, construction, and maintenance of roads, that it is appropriate to dedicate further attention to it in this chapter which deals with the complementary structures that are conceived and constructed largely for combating this phenomenon.

Researchers have paid little serious attention to this topic. Although some studies can be found on the subject of practical methods of combating erosion, the ultimate causes of the phenomenon are not well understood. Neither are the criteria for evaluating the erosion potential of a site nor are the methods for rational correction and prevention rationally interpreted.

The studies that have been conducted by the National Laboratory of Civil Engineering in Lisbon on the theory of erosion by water (which is the type of erosion that affects highways most), represent an important exception in an otherwise not very promising general panorama. Reference [18], which provides a framework for these comments, goes beyond the obvious effects of the phenomenon, with a view to establishing its generation and performance mechanisms, along with the forces developed by the soils in order to resist it. The paper refers to the erosion caused by falling rain and by surface runoff from the same source.

Table 11-5 [18] gives the different mechanisms by which rains can cause erosion. Erosion by rain has two principal causes: the impact of the raindrops on the soil and the washing away of soil by the water that runs along the surface of the ground.

The kinetic energy of drops of falling rain increases with the intensity of the precipitation, but the increment becomes gradually smaller as the intensity increases, so that the kinetic energy asymptotically approaches a limiting value, which appears to be the same for all heavy storms. The reason for this phenomenon is that a maximum stable size is reached by drops (5 or 6 mm) (0.2 or 0.24 in) in diameter. Although heavier precipitation produces larger drops they are unstable and break up as they fall [19]. There is also a minimum drop size for producing erosion. When rain is driven by wind and falls obliquely, its kinetic energy increases, for the new oblique arrival speed is higher than that of original vertical fall. This lends practical importance to the orientation of slopes in relation to wind.

Figure 11-25 shows a laminar runoff sheet, with a uniform thickness, over the surface of sloping ground. It can be seen that as it flows the water applies to the ground surface the following tangential stress:

$$\tau_a = \gamma_w h_w \sin\beta \cos\beta \qquad (11\text{-}15)$$

This expression implies that the water develops a shear stress with the soil below.

Table 11-5
Effects of rainfall on the erosion of soils [18]

Direct or indirect erosive action of rainfall	Active mechanism	Direct or indirect erosive effects
Impact of the raindrops	Disintegration	Erosion due to laminar runoff Erosion due to concentrated runoff (torrents)
Surface runoff	Disintegration Transportation	Differential erosion due to different resistance in different layers of ground
Infiltration	Perched water-table Raising of the water-table	Landslides Internal erosion, piping etc.
Wetting and drying	Expansion and contraction	Fissuring Loss of cohesion Seasonal flows

Since:

$$h_w \cos\beta = a \qquad (11\text{-}16)$$

therefore:

$$\tau_a = \gamma_w a \sin\beta \qquad (11\text{-}17)$$

can be regarded as the hydraulic gradient of the runoff.

When this shear stress reaches a critical value, characteristic of each type of ground, the particles start to disintegrate and erosion commences. This critical value is the erosive energy characteristic of each soil and slope. Assuming that the runoff consists of clear water, the depth of water and the rate of runoff will be related by the expression:

$$Q = a_w v \text{ (Unit strip of ground)} \qquad (11\text{-}18)$$

where v is the velocity of the runoff.

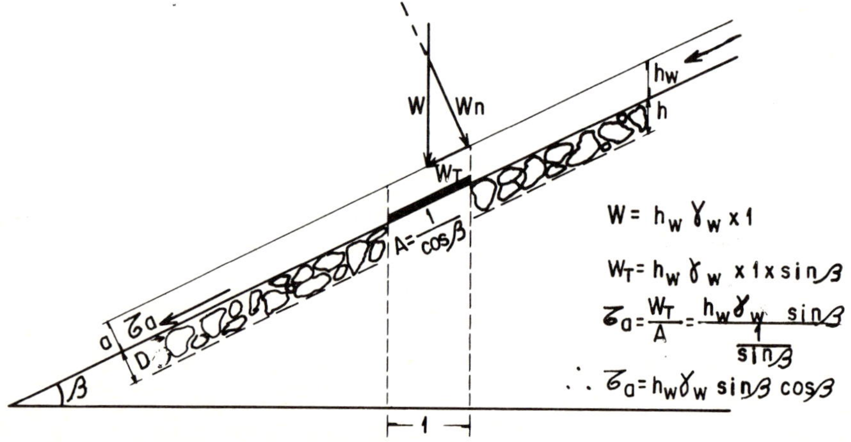

Fig. 11-25 Uniform sheet runoff on unlimited sloping ground

Moreover, the rate of flow existing at distance L from the top of the ground, can be related to the intensity of precipitation, (for example, in cm/min) by means of a coefficient of runoff, which expresses how much of the water that falls runs off and how much seeps into the ground, evaporates or is retained in any way:

$$Q = C\,I\,L \tag{11-19}$$

(After the time of concentration has elapsed).

Where C is the coefficient of runoff and I the intensity of precipitation. Comparison of Eqs. (11-18) and (11-19) leads to the following equation:

$$a_w v = C\,I\,L \tag{11-20}$$

If the runoff carries with it earthy solids in suspension, Eq. (11-17) will become:

$$\tau_a = (\gamma_s\, a_s + \gamma_w\, a_w) \sin \beta \tag{11-21}$$

where γ_s is the specific weight of the solids carried and a_s is the thickness of solids that can be considered. Concentration of the suspension, S, is defined thus:

$$S = \frac{\gamma_s\, a_s}{\gamma_w\, a_w} \tag{11-22}$$

as a result of which Eq. (11-21) becomes:

$$\tau_a = (1 + S)\, \gamma_w\, a_w \sin \beta \tag{11-23}$$

However, if Eq. (11-20) is relieved of a_w, it can be written as:

$$\tau_a = C\,I\,\gamma_w \frac{1+S}{v} L \sin \beta \tag{11-24}$$

In this manner, the value of the shear stress is expressed in terms of physical parameters that are familiar to the engineer. Equation (11-24) is only valid for a laminar runoff of uniform thickness.

If runoff tends to concentrate, forming small torrents as a consequence of irregularities in the ground, as usually happens, the runoff strip should no longer be regarded as having a unit width, as has been done up to now. Instead, the real width of the small torrent that forms must be taken into consideration. In this case, the height a_w and the velocity, v, of the runoff will also vary. When runoff becomes concentrated, it is easier to reach the limiting erosive stress for the storm. This is due to various factors, an important one being that any torrent that forms on sloping ground represents a steeper inclination for the particles on the surface, which will consequently suffer an increment in shear stress due to the component of weight parallel to the surface. The runoff rate also increases.

Another effect of rainwater is that it seeps into the ground and consequently modifies the regimen of subsurface waters. Figure 11-26 shows a slope with its water-table at Depth 2. For a distance ds covered by the water within the slope, the hydraulic gradient is produced both by the potential energy head and by pressure. Therefore (Fig. 11-26):

$$i = \frac{dz}{ds} + \frac{1}{\gamma_w} \frac{du}{ds} \tag{11-25}$$

If h_w is a height of water equivalent to pressure u, the following will be true:

$$i = \frac{dz}{ds} + \frac{dh_w}{ds} \tag{11-26}$$

If it is assumed that the slope is homogeneous and isotropic in permeability, the rainwater will penetrate vertically downwards, and the gradient corresponding to the energy head will therefore be one; ($ds = dz$) and:

$$i = 1 + \frac{dh_w}{dz} \tag{11-27}$$

If the water is at rest within the slope, $i = 0$, this equilibrium condition can consequently be expressed as:

$$dh_w = -\,dz \tag{11-28}$$

This condition is represented in Fig. 11-26b by the straight line ABC. Above the W.T., $h_w = \frac{u}{\gamma_w}$ will have to be negative for equilibrium to exist. Below this level, z is negative and h_w will be positive for equilibrium. Because the pressure head at the W. T. is zero by definition, the water above this level must have a negative pressure equivalent to its height above the W.T. (Chapter 7). Since the pressure developed by the water above the W.T. is due to capillary phenomena (Chapters 1 and 7) and depends on the capillary tension in the water, it follows that if the water is in equilibrium above the W.T. its

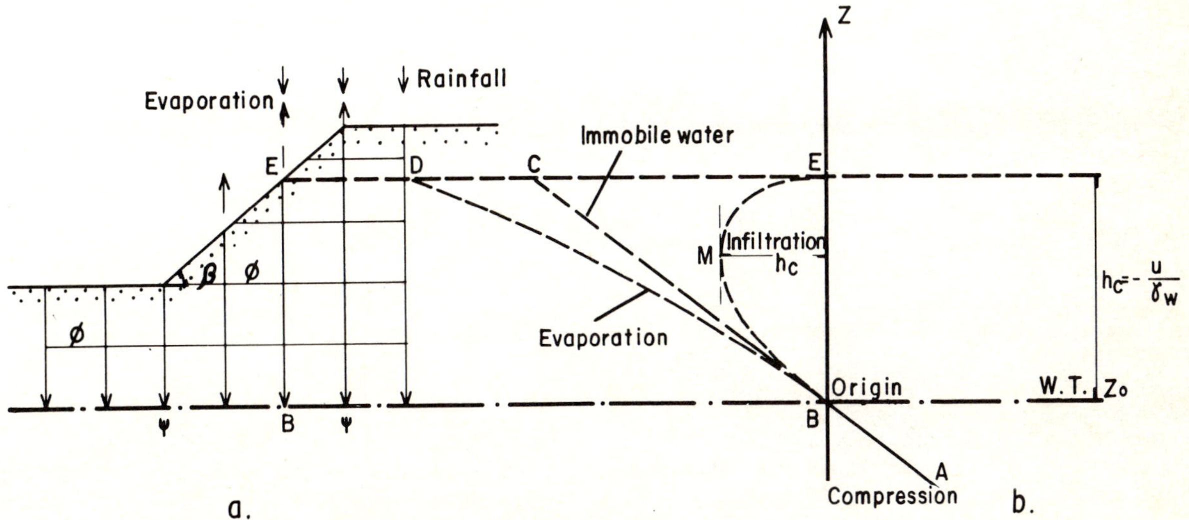

Fig. 11-26 Infiltration and evaporation mechanisms in a slope [18]

physical height above that level must be equal to the capillary height, h_c, of the soil. In other words:

$$-h_w = h_c = -\frac{u}{\gamma_v}$$

If the height of the water above the W.T. is different from the corresponding capillary height, there will be a hydraulic gradient of:

$$i = 1 - \frac{dh_c}{dz} \quad (11\text{-}29)$$

as is deduced from Eq. (11-27) and the foregoing considerations.

In [18] this gradient is expressed as:

$$i = 1 + i_c \quad (11\text{-}30)$$

where:

$$i_c = -\frac{dh_c}{dz} \quad (11\text{-}31)$$

This is the capillary or suction gradient. If the total i is positive, the water descends; if negative, it rises.

When it rains, the surface of the soil is moistened and often saturated. Therefore, the radii of the menisci in that zone will increase, capillary tension decreases, §7.5, and h_c decreases. Both in Eq. (11-29) and Eq. (11-31) this effect can be seen to feed the subsoil water by means of a downward flow. If the soil is saturated by the rain, $h_c = 0$ and this feeding process will be maximised.

Also in §7.5 it was seen that evaporation causes an increase in capillary tension, and consequently in h_c, which leads to upward flow.

Figure 11-26b reproduces the original pressure diagram ABC and its changes both for the evaporation and infiltration of rainwater. In the case of infiltration, the gravitational gradient of the water (which is one in homogeneous isotropic soil) reverses, changing appreciably, Eq. (11-29), from capillary tension to compression with a corresponding increase in the infiltration rate at the surface. The opposite occurs during evaporation.

It can be seen that the suction gradient i_c, Eq. (11-31), is equivalent to -1 at B (W.T) during infiltration. In order to travel from the original position at E, with an initial energy head of h_c, at point B the water will have covered a distance equal to h_c. This gradient is neutralized at an intermediate point M, where h_c has a maximum value. During rainfall (assuming that capillary tension is neutralized on the surface of the ground), the rain will be responsible for $h_c = 0$ at E. It will also be zero at B, because this point is located on the W.T. Although the capillary tension is different from zero between E and B, it has a maximum value, as shown by the curve BME, which shows the distribution of tension throughout the saturation zone.

This gradient is maximum on the surface of the ground.

The total hydraulic gradient of the flow, Eq. (11-30), has a maximum value on the surface of the ground, is more than 1 between E and M, under 1 between M and B, becoming zero at B at the water-table. These gradients determine the discharge rate ($v = ki$), which diminishes constantly from the surface, reaching zero at the water-table. This is responsible for accumulation of water in the saturation zone, above the water-table. The tensions in the water in that zone diminishing constantly, and a temporary perched water-table forms above the original water-table. Since dissipation of capillary tension within the slope due to infiltration occurs at a constant depth below the soil surface, the mass of perched water has a surface contour more or less parallel to the slope. This mass of water tends to flow downward due to gravitational effects, and surfaces near the toe of the slope. This effect increases the erosive tendencies of the water inside the slope and the parallel flow contributes to increase the shear stress that was expressed in Eq. (11-15). This increase is quantified in [18] giving rise to the following expression:

$$\frac{\Delta\tau_a}{\tau_a} = \frac{\sin^2\beta + n\cos^2\beta}{1-n}\,\frac{\gamma_w}{S_s} \quad (11\text{-}32)$$

n being the porosity of the soil, and S_s the relative density of the solids.

Figure 11-27 [18] shows the increase in acting shear stress in a slope for different inclinations and different values of n, which, as is seen in Eq. (11-32), determines the increase in erosive power on account of the mass of perched water flowing parallel to the slope. The figure shows that, for example in a slope with a 45° inclination in relation to the horizontal, the effect of the flow parallel to the surface increases the shear stresses on that surface by 37% for n = 30%, but by almost 80% for a porosity value of 60%.

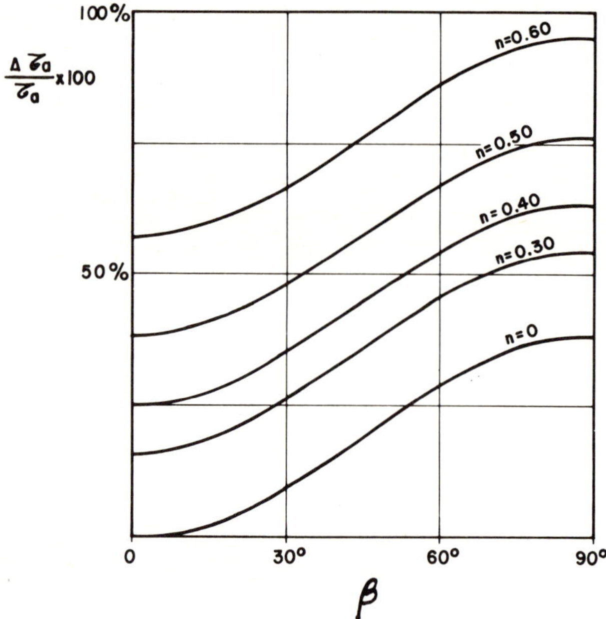

Fig. 11-27 Increase in the acting shear stress in a slope owing to its inclination and the porosity of the constituent soil [18]

Table 11-6 [18] summarizes the principal conclusions of the analysis of the erosion potential of rain, specifying the chief parameters that influence the phenomenon. These parameters include those describing the rainfall (its intensity and duration being by far the most important ones), the climate, the soil and the slope geometry.

In Reference [18] an analysis of the soil's resistance to erosion is also considered. The soil's resistance to the removal and washing away of its grains can be described by the traditional shear strength relationship:

Table 11-6
Main parameters affecting erosion due to rainfall [18]

Direct or indirect erosive actions	Parameters inherent in rain or climate	Parameters inherent in the soil or in slope geometry
Impact of the drops	Intensity of the rain (up to a limit) Wind velocity during storm	Orientation of the slope relative to winds
Surface runoff	Intensity and duration of rainfall	Inclination of slope Area of exposed surface of slope Number of gullies formed Coefficient of runoff Water velocity Concentration of solids carried
Infiltration	Duration of rainfall	Inclination of slope Porosity, permeability
Wetting and drying	Alternation of wet and dry seasons Intensity of solar action Rainfall	Balance of conditions for infiltration (protection, permeability, slope) and for evaporation (orientation to sun, protection etc.)

$$s = c + \overline{\sigma} \tan \varnothing$$

With reference to Fig. 11-25, the following is true:

$$\overline{\sigma} = \gamma'_m h \cos^2 \beta \qquad (11\text{-}33)$$

Where h corresponds to the size of the first row of grains (D, in the figure). The submerged or effective specific weight γ'_m is used if the inside of the slope is saturated by perched water. Bearing in mind the geometrical relation between h and D, Eq. (11-33) can also be written as:

$$\overline{\sigma} = \gamma'_m D \cos \beta \qquad (11\text{-}34)$$

and the resistance will be:

$$s = c + \gamma'_m D \cos \beta \tan \varnothing \qquad (11\text{-}35)$$

If flow is rapid enough to create turbulence, the second term of the second part of Eq. (11-35) will be somewhat reduced, which can be expressed by using a coefficient lower than one. However, no information is available about this.

Modifying Eq. (11-34) another decisive factor is the furrows and torrents that may have formed on the slope, with the result that the inclination of the particles on the surface will not be β, but somewhat higher.

For granular, cohesionless soils, most of the information concerning their resistance to erosion is based on stability studies conducted on soils at the bottom of nearly level channels [20]. These studies make it clear [18] that there is a well defined relation between the size of the soil grains and the erosive energy that moves them. In Fig. 11-28 [18], the shaded zone represents the relation, Eqs. (11-15) or (11-17), between the average diameter of the grains and the erosive energy. In our opinion this is an appropriate characterization that is accepted by many engineers.

In cohesionless materials with an average diameter of less than 5mm, the resistance to erosion is considerably more than in those of a larger size. There may be two main reasons for this. First, there appears to be a considerable increase in the critical value of the tolerable shear stress. Second, in these small particles the influence of particle arrangement and cementation is greater, despite the sediments in the water between the particles. Whatever the explanation, the greater resistance is an experimental fact, and it is correspondingly displayed in Fig. 11-28.

For cohesive soils, current information is far more uncertain and does little more than establish a critical velocity at which erosion does not occur, as was shown, for example, in Table 11-2. If these velocities are converted to the corresponding erosive energy, using Eqs. (11-18) and (11-17), with an estimated rate of flow, practical recommendations can be reached, similar to those included in Table 11-7. These refer to cohesive soils that are placed on the bottom of channels, corresponding to the Russian practice described in [18] and [20].

The data given in Table 11-7, show that a major factor is the influence of compaction on the resistance of cohesive soils to erosion. For similar soils, resistance may increase from 15 to 20 times as the soil passes from a loose state to a very well compacted one.

One of the effects that contributes most to erosion in cohesive soils is the expansion on the surface which occurs during wetting, this is not neutralized or restrained by any counterpressure, consequently it develops freely.

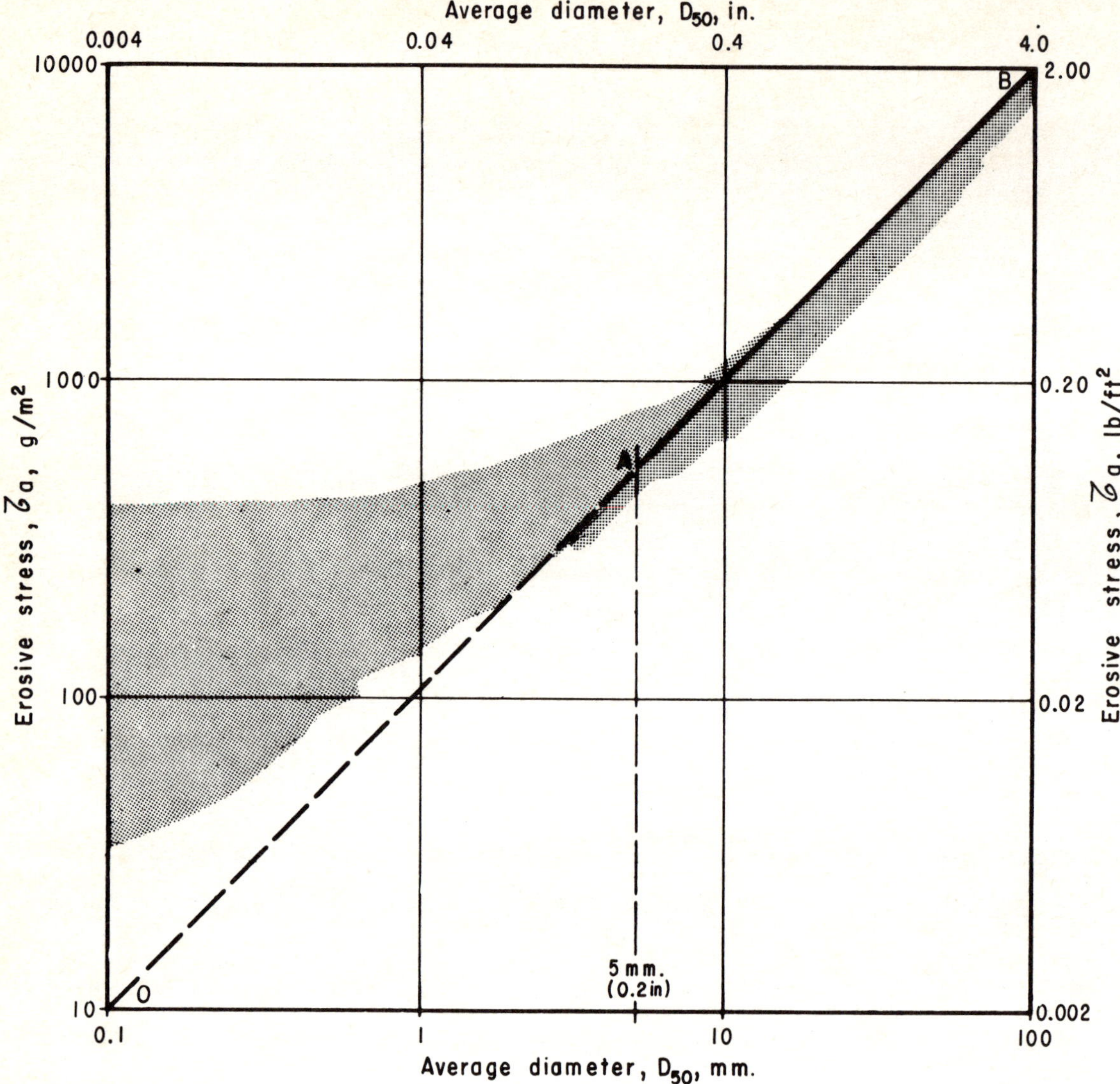

Fig. 11-28 Relation recommended by engineering practice between the average diameter of the particles and the allowable erosive energy (Channel bottom experiences) [18]

Table 11-7

Shear stresses causing erosion in channel bottoms in cohesive soils [18,20]

Material on the bottom	Consistency of the material							
	Loose		Poorly compacted		Compacted		Well compacted	
	g/m²	kips/ft²	g/m²	kips/ft²	g/m²	kips/ft²	g/m²	kips/ft²
Sandy clay	180	0.037	700	0.143	1470	0.300	2800	0.572
Very clayey soils	140	0.029	670	0.137	1370	0.280	2540	0.520
Pure clays	108	0.022	560	0.115	1260	0.258	2380	0.488
Slightly clayey soils	90	0.018	430	0.088	960	0.197	1540	0.315

Plate 11-19 Effect of erosion. Note the exposed chute

Plate 11-20 Erosion caused by infiltration in an incorrectly placed drainage pipe in the body of the embankment

11.15 Vegetative Cover, General Criteria

Of all the types of soil protection that have been mentioned (masonry, concrete, soil-cement, soil-asphalt, clays and vegetation), special attention is deserved by vegetation. Some of the other types possess their own technology which, although beyond the objectives of this book, is nevertheless familiar to many highway construction engineers. Such is the case for masonry and concrete. Other types of protection, such as soil-cement and soil-asphalt, will be described later in this book. However, although vegetation possesses an extremely complete technology, and man has developed many different sciences around it, it is usually outside the range of information familiar to highway engineers. We therefore feel that some general comments should be made here on the subject of the criteria for the utilization of vegetal protection, but without going into those sciences that are beyond the scope of this book. These observations are based largely on material taken from [21].

Vegetal cover serves two different purposes: the defence and protection of natural soils or slopes, and the visual pleasure of the road-user.

The chief benefits of vegetation in protection against erosion are,

1) Protects soil against the impact of raindrops
2) Reduces flow velocity by increasing the roughness of the ground surface.
3) Encourages infiltration through hollows produced by roots, and animals burrows

To meet these objectives, vegetal cover must be *thick*. It should consist of selected species, that will flourish under local conditions; consequently the assistance of a specialized botanist is always advisable in important problems. As already observed in Chapter 6, the species characteristic of the region offer some degree of assurance. These are, nevertheless, not entirely free from difficulties, because they are made to grow in conditions that may be different from those to which they are accustomed. For example, some plants that grow naturally at a certain location on flat ground may experience difficulty in growing on a slope. Also, the vegetation that is planted at a given location must suit the problem. For example, the species that are to cover a slope underlain by homogeneous soil must be different from those planted on a nearby slope consisting of

Plate 11-21 Effect of erosion in an embankment. Destruction of the curb due to absence of chutes

Plate 11-22 Situation that may eventually be reached if erosion is not controlled

blocks of rock, in which the soil requiring protection is only in the joints, or in a third slope of stratified soils where each stratum is of different fertility, infiltration rate and moisture retention.

Another important function already mentioned in Chapter 6 is the reduction of the water content of the upper layers of soil, by the transpiration of plants.

In canals, ditches, intercepting ditches and similar structures, vegetation also plays the important role of protecting the banks and edges from the action of running water, when the water velocities do not undermine the vegetation and the vegetation does not restrict flow. The importance of the improved appearance that is achieved with vegetation, has not been fully comprehended (or has not at least been adopted as a systematic line of action) by engineers in many countries.

The effect on the motorist is not always restricted to appearance, with all the psychological repercussions it may have, but to many other aspects, too. There is, for example, currently a growing interest in the study of vegetation as a means of combating noise, this is important for airports or urban roads and railroads.

When a program for planting vegetation in canals or on slopes is initiated, the first factor to take into consideration is that the ground to be protected is not a soil in the agricultural sense of the word. Neither its structure, its texture and mineralogy have the characteristics required to encourage plant life, nor does it contain the micro-organisms and organic matter of a so-called topsoil. Consequently, a covering of topsoil is almost always necessary. Topsoil removed for embankment; pavement and ditch construction must therefore be systematically conserved. The engineer who decides to plant vegetal cover at the close of embankment construction often finds that all the topsoil has been removed and disposed of as though it were waste material, by careless handling or to save the cost of stockpiling. It is a good practice to store the topsoil resulting from clearing and similar operations, so that better vegetal growth will be promoted at the end of construction work at very low costs.

This will usually compensate for the precautions taken. Topsoil should not be stored in very high piles, and should preferably be oriented in a north-south direction. It is advisable to protect it from sunshine and erosion by brush or even temporary grassing.

When topsoil is laid on a slope, it is essential to notice whether it is sufficiently retained by the slope inclination; otherwise expensive berms will have to be constructed. This is why it is often wise for a slope to be given a smaller inclination than that which is strictly required for stability. For the same reason, a slope that is to be protected by a vegetal cover should not have a smooth finish.

Any springs, weepholes, superficial streams should be diverted, for water will wash away recently laid topsoil. In these cases, surface drainage or subdrainage will be necessary, according to the nature of the problem.

When vegetal cover is to be employed on cut or embankment slopes, it is not enough to analyze the general conditions of the local climate, for these local conditions are altered by the slopes. Owing to the incidence of solar radiation, which increases the temperature of the soil, steepness, which encourages runoff and makes the soil drier during almost all the year, and exposure to wind, soils are exposed to generally worse-than-average conditions.

Vegetation can be planted by the direct seeding of appropriate topsoil. This can be done by hand or using mechanical or hydraulic methods. At other times, strips of turf or ready-planted soil are laid out like a mosaic or grid. This method is appropriate for grasses and herbaceous plants. Plants usually need to be watered several times before they settle in. Trees and shrubs, which are used particularly as a protection against aeolian erosion, sand invasion or as screens against noise, are usually planted once they have reached sufficient size, so that they will be able to provide immediate protection. While not fully established, they demand closer attention and more frequent watering.

Vegetation is usually one of the most economical methods of slope protection and the easiest one to maintain, especially when anticipated at the design stage, so that space is provided for storing topsoil and for plantings during road building operations.

REFERENCES

1. GÓMEZ CANTÚ, M. and URQUIJO, O., "Aspectos generales del Drenaje en la construcción, Construcción de Obras Complementarias", Communication to the Seminar on Drainage of the Mexican Ministry of Public Works, México, D.F., 1970.
2. ESCARIO, J. L. and ESCARIO, V., *Caminos.*, Publication of the Technical Superior School of Road Channel and Port Engineers, Madrid, 1960, Vol. I, Chap. V.
3. SOTELO, G., "Drenaje en Carreteras y Aeropuertos", Publication of the Institute of Engineering of the UNAM (Sponsored by the Mexican Ministry of Public Works), México, D.F., 1971.
4. LINSELY, R. K. and FRANZINI, J. B., *Water Resources Engineering*, McGraw Hill, 1964.
5. JUÁREZ BADILLO, E. and RICO, A., *Mecánica de Suelos. Vol. III. Flujo de Agua en Suelos*, Limusa: México, D.F., 1969, Chap. 6.
6. SPRINGALL, R., "Drenaje en cuencas pequeñas", Publication of the Institute of Engineering of the UNAM (Sponsored by the Mexican Ministry of Public Works). México, D.F., 1969.
7. RUBIO, J., OLIVERA, F. and PEÑA, F., "Drenaje en Caminos de Mano de Obra". Communication to the Seminar on Drainage of the Mexican Ministry of Public Works, México, D.F., 1973.
8. BRUCE, A. G. and CLARKESON, J., *Highway Design and Construction*, International Textbook, 1956, Chap. 6.
9. ETCHARREN, R. "Manual de Caminos vecinales," 6th Part, Co-edition of the Mexican Road Association and of Representations and Engineering Services, S.A. México, D.F., 1969.
10. DÍAZ DE COSSÍO, R. and SIESS, C. P., "Recommendations for protecting culverts with reinforced concrete", Technical Report to the U.S. Army Corps of Engineers, University of Illinois, Urbana, I11., 1959.

11. Juárez Badillo, E. and Rico, A., *Mecánica de Suelos, Vol. II. Teoría y Aplicaciones de la Mecánica de Suelos*, Limusa: México, 1973, Chap. 4.

12. Moreno Pecero, G. and Rico, A., "Instructivo de Campo para la construcción de alcantarillas flexibles", Publication of the Ministry of Public Works, México, D.F., 1964.

13. Marston, A. and Anderson, A. O., "The theory of Loads on Pipes in Ditches and Tests of Cement and Clay Drain Tile and Sewer Pipe", Iowa Engineering Experimental Station: Bulletin No 31, 1913.

14. Marston, A., "The Theory of External Loads on closed conduits in the Light of the Latest Experiments", Iowa Engineering Experimental Station: Bulletin No 96, 1930.

15. Spangler, M. G., *Foundation Engineering*, Edited by G.A. Leonards, McGraw Hill, 1962, Chap. 14.

16. Reference [11], Chap. 2.

17. Orozco, J., "Metodología para el anteproyecto de Caminos", Publication of the Geotechnical Department of the General Direction of Technical Services of the Mexican Ministry of Public Works, México, D.F., 1973.

18. Nascimento, U., "Teoría de erosao de Taludes", Publication of the Civil Engineering Laboratory (Course 142), Lisbon, 1973.

19. Hudson, N. N., "An Introduction to the Mechanics of Soil Erosion under conditions of sub-tropical Rainfall", *Trans. of Rhodesia Scientific Ass.*,Vol. XLIX, Part I, 1961.

20. Lane, E.W., "Progress Report on Studies on the Design of Stable Channels by the U.S. Bureau of Reclamation", *Proc. ASCE*, 1953.

21. Caldeira Cabral, "Revestimentos vegetais," Publication of the Civil Engineering Laboratory, Lisbon, 1973.

OTHER RELATED REFERENCES

Israelsen, C. E., Clyde, C. G., Fletcher, J. E., Israelsen, E. K., Haws, F. W., Packer, P. E., and Farmer, E.E., "Erosion Control During Highway Construction", Manual on Principles and Practices. Transportation Research Board, National Cooperative Highway Research Program, Report 221.

Leflaive, E., "Les Geotextiles: Pourquoi et Quel Avenir", Colloque International sur la Gestion des Ouvrager, Paris, 1981.

Transportation Research Board, "Erosion, Sedimentation, Flood Frequency, and Bridge Testing", Transportation Research Record 832, Washington, D.C. 1981.

Transportation Research Board, "Engineering Fabrics in Transportation Construction", Transportation Research Record 916, Washington, D.C., 1983.

CHAPTER 12

BORROW MATERIALS

12.1 Introduction

One of the principal costs of highway construction and maintenance is materials: rock, gravel, sand and other soils. Consequently their location and selection is one of the fundamental problems faced by the civil engineer in close collaboration with the geologist. Experience shows that if sufficient importance is given to these tasks, deposits of appropriate materials will usually be found close to the location where they are to be used, thus lowering haulage costs, which are usually the largest part of the overall costs. Sometimes, appropriate materials will be found in nearby zones which were previously thought to be available only in distant borrow areas. For these reasons, it is not surprising that the systematic search for these materials and their rational exploitation is gaining the attention of interested specialists.

The organizations dedicated to the design and construction of highways in every country have at their disposal an ever-increasing supply of detailed information about the availability of materials in each zone that is to be crossed by a highway. Unfortunately, this information is frequently mislaid once the project is completed. Therefore the engineers in charge of new construction are confronted with the problem of searching for suitable materials where others had found them previously. It appears to be truly urgent and important that all the information that is obtained daily about appropriate borrow materials, location, usable volumes, engineering qualities and methods of improvement be centralized. Accomplishment of this task on a nationwide level will enable designers as well as construction firms in the country to gain considerable savings in their search for borrow materials, and at the same time benefit from all the experience of those who have used the same borrow pit on a previous occasion, either for the same purposes or for similar ones. A task like this will be never-ending, but will quickly lead to highly satisfactory results.

For many years the location of borrow materials depended on ordinary methods of exploration, from the simple examination of the ground to the use of open pits, post-hole augers, hand drills and even drilling rigs. More recently geophysical studies, which have enormous potential, have gradually been added to the available techniques, saving a great deal of human time, effort and exploration costs.

In this chapter, emphasis will be placed on borrow materials, on the understanding that a great deal of what is said will be applicable to roadside borrow pits as well as to the materials that are obtained as a result of balancing cuts and fills across the centerline and along the centerline. Distinctions will have to be established between borrow materials consisting of rock and those consisting of soil. The boundary between the two materials is harder to establish as accurately in this case as in others. Rock presents many different degrees of alteration and hardness. Moreover, hard and soft may be interlayered with both well-indurated rock and poorly-indurated soils.

The first point in the classification of borrow materials is the evaluation of the qualities of the rocks or soils that are present. This is usually difficult to establish in a quantitative manner. Where rocks are concerned, there are two main points that deserve attention [1]. The first refers to the physical changes that may be undergone by the rock because of fragmentation during loosening, excavation, handling and placement. The second refers to the physico-chemical alteration (either hardening or softening) which may take place during the useful life of the project. Although the same factors must be considered for soils, they take on even greater importance in rocks. Soils have already suffered important physico-chemical transformations during the earlier decomposition of the parent rock. Rocks, especially sound ones that have been recently crushed or broken, have not been subjected previously to intense weathering processes; such weathering can have very damaging consequences.

Table 12-1 [1] provides a useful preliminary evaluation of the different types of rocks, in their characteristics as construction materials. A final diagnosis, however, depends on so many specific factors that it cannot possibly be made on the basis of the empirical information contained in this table alone.

For final evaluation, field and laboratory tests should be performed on the rocks that make up the borrow area under consideration. The best field test is to make an excavation similar to that which will afterwards be employed on a large scale, so as to find out objectively the nature of the material obtained. This test will have to be conducted on a sufficiently large scale to be realistic.

Table 12-1
Characteristics of some rocks as construction materials [1]

Rock	Excavation method	Fragmentation	Susceptibility to weathering
Granite Diorite	Explosives	Irregular fragments, depending on the blasting pattern	Likely to be resistant
Basalt	Explosives	Irregular fragments depending on joints and cracks	Likely to be resistant
Tuff	Excavators or explosives	Irregular fragments, often with excessive fines	Many forms deteriorate rapidly
Sandstone	Excavators or explosives	Slabby to irregular, depending on bedding	Depends on character of cementing
Conglomerate	Excavators or explosives	Excess fines, depending on cementing	Some deteriorate into silty sands
Siltstone Claystone shale	Excavating machines rippers	Small blocks to thin slabs and chips	Many slake or break down rapidly into clay: should be considered suspect unless tests prove otherwise
Limestones: massive	Explosives	Irregular fragments, sometimes slabby	Shaley seams deteriorate, otherwise resistant except to acids
Coquina Chalk	Excavating machines rippers	Porous fragments, excess fines common	Some porous forms soften on wetting others become partially cemented with alternate wetting and drying
Quarzite	Explosives	Irregular fragments, very angular	Likely to be resistant
Slate Schist	Explosives	Irregular or slabby fragments, depending on laminations	Some deteriorate on alternate wetting and drying
Gneiss	Explosives	Irregular fragments, sometimes elongated	Likely to be resistant
Mine and Industrial wastes	Excavating machines rippers	Depends on material, in most cases irregular	Most forms except igneous rock mine waste, should be considered susceptible to weathering or biochemical change until experience or tests prove otherwise

The possibility of deterioration of the excavated rock with time is far more difficult to establish. The most reliable indication is obtained by observing places where the rock has been exposed for a long period of time, such as old railroad and highway cuts or river banks.

The preliminary evaluation of soils is based largely on previous experience. The Unified System of Soil Classification helps in most cases, for alongside the classification of a specific group it gives typical behavior patterns. The detailed evaluation of borrow soils should be based on laboratory or large scale field tests.

12.2 Location of Borrow Materials

Few practical aspects of highway construction are so important and at the same time so elusive as those for developing standard criteria and techniques for locating borrow materials. It is so important that a design is not complete and ready for construction if it does not contain a complete and detailed list of the borrow pits from which the soils and rocks required for construction are to be taken. The expression *borrow pits* must be taken in the general sense, referring not only to the cuts from which an embankment or a cut-and-fill section are built by balancing cut-and-fill volumes but also to the materials in the natural ground in which a roadside borrow pit will be dug.

There is more to the location of borrow materials than the discovery of a place where there is a sufficient volume of exploitable soils or rocks that can be used in constructing a specific portion of highway (Plate 12-1). They must comply with the specifications for quality and the requirements of volume. The problem has many other implications. The borrow pits selected should be the *best* of all those available from several different points of view. First, the quality of the materials must be judged for the purpose they are to serve. Second, the pits must be accessible and suitable for efficient and economical excavation. Third, they must be as close as possible to the site, so as to minimize costly hauling materials over large distances. Fourth, the materials selected must lead to simple and economical construction procedures for processing (if needed) and laying and final placement in the structure, with minimum treatment. Fifth, the location of the borrow pits must be such that their exploitation will not lead to any complicated or lengthy legal problems and will not unduly affect the local

Plate 12-1 A typical river bank

inhabitants, giving rise to social injustices or adversely affect the environment. It is clear that in some practical cases many of these requirements will contradict one another. The engineer is confronted with the task of selecting the borrow pits that are best able to reconcile these contradictions.

These conditions are followed by others that involve the interrelations between materials available. For example, in the natural state there may be a difference in the engineering qualities of two materials that are considered for a certain purpose. However this difference may be neutralized or even reversed if the worst material is given suitable treatment, so that it is stabilized or if the design is modified so that the material that was not originally appropriate can now be used. This interrelation between construction materials and highway design is so essential that a design is useless if it does not consider the project as a whole entity, including the borrow materials that are available and the purpose for which they are intended.

The complicated evaluation process just described must commence with a simple location survey. By the end of this, the engineer should have a map showing all the possible purposes for which the available borrow materials may be useful in his project, probably after excluding others because of obvious deficiencies. From among all these borrow materials, the engineer will have to explore his options with regard to his projects needs.

Borrow areas can be located with the help of air photo-interpretation or direct ground surveys. The latter are in turn aided by photo-interpretation and geophysical prospecting.

In Chapter 3, photo-interpretation and geophysical prospecting methods were briefly discussed; consequently it is not necessary to give any further details about them at this point. Nevertheless, it is important to stress that photo-interpretation is unrivalled as a method of exploring large areas at a low cost, in a manner that may well be as accurate as ground surface reconnaissance, particularly if the organization that is looking for the materials employs geologists or geotechnical engineers who are well trained in the application of the method. This is one of the ways where applied geology may contribute more effectively to highway technology.

Whether used as the only method of detection or as a complement to photo-interpretation, a ground survey or investigation of the future borrow area is indispensable. It must determine not only the possibility of exploitation, but also the degree of difficulty involved, including problems that might be caused by surface or subsurface water, volumes of material available, and legal restrictions. The engineer conducting these preliminary studies must always consider local experience, from which he will learn many useful facts that might otherwise easily pass unnoticed (Plate 12-3).

Usually, borrow materials must be located for the embankments, subgrade, sub-base, base course and wearing course of highway pavements. In railroads, they must be located for the embankments, subgrade, sub-ballast and ballast. In runways the requirements are the same as for highways. Additional borrow materials may be required for making concrete, stone masonry or other special products. A single borrow pit may often meet several of these requirements, if the material it produces is subjected to different excavation methods and processing.

Borrow fill is generally plentiful and easily located in most regions of the world; almost all materials that can be economically excavated can be used for this purpose. The exceptions were described earlier, (*MH, CH* and *OH* soils with a liquid limit higher than 100%, and P_t soils). Problems may, however, arise on account of the poor quality of the materials due to a high moisture level in lacustrine plains, flood zones, delta deposits, large alluvial costal plains and other zones where very fine materials are abundant. In these cases, it is not unusual to have to look for borrow supplies outside these zones.

Borrow pits producing fill should not be established too far apart from one another, so that hauling distances are not excessive. In most practical cases the optimum spacing is that where an equilibrium is reached between the cost of hauling on the one hand, and that of clearing and preparing the borrow pit on the other. The resulting distances are not usually more than 5 km (3 mi) between borrow pits. Special cases may exist where the distances are far greater, especially in agricultural areas where land expropriation costs are very high.

Plate 12-2 Exploitation of gravel and sand in a river bed

Plate 12-3 Exploration using geophysical prospecting methods

Reference was made in Chapter 9 to the materials that can be used for subgrade layers and those that must be rejected, according to Mexican engineering practice. A further requirement which determines the selection of borrow materials is homogeneity over significant distances, thus minimizing variations in the structures and thicknesses of the overlying layers of pavement. The distance between borrow pits can in this case be as great as 10 km (6 mi).

Materials for the sub-base and base course, apart from having to meet the foregoing requirement, depend largely on the mechanical treatments to which they are subjected in order to bring their quality up to the required levels. This requires special equipment and complex processing plants, which are best not moved very far. For these reasons, the borrow pits are usually far more widely spaced; distances between them of about 50 km (30 mi) are not unusual. Borrow materials for the subgrade layer are usually found in broad, low hills in formations of very altered rock, in the silty-sandy zones of river deposits, in pyroclastic volcanic deposits, such as cineritic or tuffaceous cones, and in sandy layers in stratified formations. Materials for the sub-base and base course are usually found in river banks, rocky hillsides and cliffs, and high steep hills (Plate 12-2).

The materials for asphalt or Portland cement concretes are almost always obtained by crushing material from sound rock formations. Masonry is obtained from rock formations. Rubble stones can be found in fractured rocks or loose fragments on the ground surface.

12.3 Exploration and Sampling in Borrow Areas

Exploration of an area where it is intended to establish a borrow pit has the following objectives:

1) To determine the nature of the deposit, including its geology, history of previous excavation, relations with surface runoff, ground water, and possible mineral rights
2) Determine depth, thickness, extent and composition of the strata of soil or rock that are to be excavated
3) Analyse subsurface water, including the position of the water-table, its variations, and possible flow of ground water into excavations
4) Assess the properties of the soils and rocks, for the purposes intended as well as the purposes for which they had been used previously.

The investigation is divided into three stages:

1) Preliminary visual survey, and review of any previous data. A geological study however simple it may be, is essential.
2) Preliminary exploration, where by means of simple, rapid procedures, information can be obtained about the thickness and composition of the subsoil, the depth of the water-table and other data which will make it possible to determine whether the zone is suitable for a borrow pit with the characteristics required. These data are used to determine if further exploration is warranted, and, if so, what should be done.
3) Final exploration, where by means of soundings, pits, borings and laboratory tests, the engineering characteristics of the soils and rocks are determined in enough detail for design, as well as establishing the cost of materials.

Exploration for locating and evaluating borrow areas is usually by photo-interpretation, borings and geophysical prospecting. Although it is seldom necessary to explore to depths greater than 10 m (33 ft), preliminary and ultimate boring methods may differ little. The open pit or drilled well, and holes drilled by post-hole auger or helicoidal drills [2,3] are the methods most often employed for soils. The difference between the preliminary and the ultimate study is usually the number of borings. There are sufficient in the final investigation to corroborate the preliminary information, give a clear description of the different formations that exist, and to compute as precisely as necessary the volume of material that will be available. When carrying out final exploration studies in soils, it is important not to exclude the more sophisticated exploration methods or those capable of going to greater depths. For these methods drilling rigs will be used, with techniques of the type described in [2,3]. In rock formations, the usual practice is to pay careful attention to the results of the preliminary study for the extension of the borrow area and the volume of available material. The reason for this is that for exploration in rock, rotary core-drilling with drilling rigs must be employed, all of which is costly. Because of cost, core drilling is often omitted, except in important projects involving a considerable amount of rock. However this is often false economy. Inadequate data on rock borrows often causes gross errors in design and bad cost estimates.

In Chapter 3, reference was made to the value of photo-interpretation and geophysical prospecting techniques in soil studies. A good photo-interpretation study with limited field verification may cover the preliminary survey stage quickly, and may be sufficient for detecting possible borrow areas. Geophysical methods provide economical and rapid means of computing the volumes of soft and hard materials under consideration and distinguish between the different formations. Of these, the geoseismic method is by far the most common one.

Borrow materials are sampled so that the engineering characteristics that determine their use can be found in the laboratory. There is no rigid rule for establishing the number of borings required. Some organizations establish a specific number of borings based on the cubic meters of material to be exploited. However this takes into account neither the homogeneity, the heterogeneity of the formation nor other geological features. Consequently this is not a satisfactory exploration guide. Instead it is preferable to make enough borings to define the range of the characteristics of the borrow materials under consideration, considering the local geological conditions, and the faces that are to be exploited.

The sample size that is extracted will depend on the purpose for which the material is intended. For fill materials, grain-size distribution and plasticity limit tests are often made for soil identification, followed by compaction tests, and coefficient of volume change. All of these usually requires samples weighing between 50 and 100 kg (110 and 220 lb) minimum. These tests help to identify the qualities of the soil when compacted in an embankment.

For subgrade, sub-base and base materials, California Bearing Ratio or similar tests are performed in addition to the above, depending on the design method that is to be used. The whole range of tests that are conducted on asphaltic wearing course materials are not included here; they are beyond the scope of this book.

Most tests on paving materials are similar to those for fill materials; they are usually conducted with closer attention to detail and in greater numbers. For example, a grain-size distribution curve for fill material often does no more than determine the portions of gravel, sand and fines, whereas for paving materials the complete curve is necessary. Similarly, compaction tests and California Bearing Ratio tests are sometimes performed with greater attention to detail in the case of materials for the sub-grade and other layers of pavement than for other lower layers of fill.

With borrow materials for paving, one set of tests is frequently made during a preliminary study stage; other tests of a final nature are performed subsequently. The first set of test results will permit the selection of the most promising zones within the borrow area and help in establishing rational utilization alternatives between several different neighboring borrow areas.

Table 12-2 presents the types of tests that are carried out on different borrow materials, according to the purpose for which they are intended. The tests are divided into three types: (1) those for classification purposes; (2) those that establish the quality of the materials which, among other things, will make it possible to determine if the specifications are met; and (3) design tests. As was already observed previously these criteria may differ from one organization to another. The table has been drawn up on the assumption that the basic test for pavement design is the California Bearing Ratio. The choice of tests to determine the swell characteristics of the soils often varies between designers.

All the tests mentioned in Table 12-2 have already been discussed earlier in this book, with the exception of the *Sand Equivalent Test.* Originally developed by F. N. HVEEM of the California Division of Highways, this test has won enormous popularity. Although reference was made to this test in Chapter 9, it deserves a fuller description here, because its greatest value is to determine the quality of soils or crushed borrow materials. References [4-8], contain detailed information about the test, and Appendix 12a includes brief instructions as to how it should be conducted.

All the earthy materials that are used in embankments and pavements contain some fine particles. These fines and their activity are major factors in the mechanical behavior of the soil as a whole. The Sand Equivalent Test was developed by HVEEM to evaluate quantitatively the quantity and activity of the fines in the mixture of particles that constitute the borrow soil.

The test consists of introducing a pre-established amount of the fraction of soil passing N° 4 sieve into a standard test-tube, partially filled with a solution which, among other effects, encourages sedimentation of the fines. After being vigorously shaken to homogenize the suspension, the test tube is left to rest in a vertical position for 20 minutes. At the end of this time the sedimentation profile can be seen on the bottom. This usually consist of two readily distinguishable layers: a lower one which will contain practically all the particles of sand, and another overlying one made up of the particles of clay and fines that have settled out of suspension during those 20 mins, under the flocculating effect that is produced by the solution. The degree of flocculation depends on the concentration that is used. Thus, if the clay-fines are predominately montmorillonitic or bentonitic with high colloidal activity, the standard solution, when allowed to act for 20 minutes, will succeed in flocculating and depositing only a part of the clay, whereas if the clay-fines are kaolinitic with far lower colloidal activity, most or all the clay-fines will be deposited in the 20 minute period.

By studying the sedimentation profile, a volumetric index can be established for the respective portions of materials contained in the original soil; sands or clay-fines. As will be seen, moreover, the sedimentation profile also gives a qualitative idea of the *activity* of the clayey fraction.

Table 12-2

Laboratory tests that are conducted on soils extracted from borrow areas according to their utilization

I. Embankments
- a) Classification: Plasticity limits; Grain-size distribution
- b) Quality: Maximum unit weight; Sometimes, California Bearing Ratio
- c) Utilization: Natural moisture content

II. Subgrade layer
- a) Classification: Plasticity limits; Grain-size distribution
- b) Quality: Maximum unit weight; California Bearing Ratio; Swell; Sand equivalent
- c) Design: Determination of *CBR* (U.S. Army Corps of Engineers Method) or else: HVEEM tests, or else: Texas triaxial tests
- d) Utilization: Natural moisture content

III. Base course and sub-base
- a) Classification: Plasticity limits; Grain-size distribution
- b) Quality: Maximum unit weight; *CBR*; Sand equivalent; Swell
- c) Design: If it is wished to draw up a structural design by layers, the tests indicated for the subgrade layer must be performed
- d) Utilization: Natural moisture content

IV. Asphaltic wearing course
- a) Classification: Plasticity limits; Grain-size distribution
- b) Quality: Wearing and/or durability tests; Sand equivalent; Swell; Affinity with asphalt; Tests to determine the shape of the aggregates
- c) Design: MARSHALL test, or else: HVEEM tests; The optimum asphalt content can be determined by the C.K.E. method also

The solution is largely calcium chloride, which is a flocculating material. To it is added some glycerine, which has a stabilizing effect which makes the test more consistent when repeated by different operators, and formaldehyde, which sterilizes the solution so as to neutralize the possible growth of any micro-organisms in the original soil. The base of the solution is distilled or clear, ion-free, water. The amount of calcium chloride determines the flocculating power of the solution. With different amounts of calcium chloride, very different volumes of clay are deposited, and thus very different sand equivalents are obtained, because this parameter is determined by the respective thicknesses of sand and clay in the sedimentation profile.

Once the sedimentation profile has been obtained and note has been taken of the level of the top of the clay-fines layer, a standard weight tamper is introduced into the test tube and is allowed to sink into the clay. It stops when the layer of sand is reached. The height at which this tamper comes to a halt is measured; this is regarded as the upper boundary of the layer of sand. The tamper is necessary because during deposition there is no sharply defined boundary between the sand and the clay-fines. The Sand Equivalent is defined as:

$$S.E. = 100 \frac{\text{Reading corresponding to the upper boundary of the sand layer}}{\text{Reading corresponding to the upper boundary of the clay-fines layer}} \tag{12-1}$$

A sand equivalent of zero is found in a pure clay; the higher the sand equivalent, the greater is the proportion of sand in the material.

So far, the sand equivalent test does no more than establish a volumetric relation between the sand and the flocculated clay contents of the sample. If this were so, it would give no more information than a quick mechanical analysis using sieve Nos. 4 and 200, which is perhaps a simpler test to conduct. The value of the sand equivalent test is that it also gives an idea of the activity of the clayey fraction.

The first variable is the concentration of the flocculating solution that is used; on its flocculating power determines the amount of clay that is deposited during the test, and therefore variations in the sand equivalent. Hveem hypothesized that the nature of the fine fraction would influence the strength of the soil as a whole. He expressed this strength by means of the parameter R, obtained in the stabilometer discussed in Chapter 9.

Figure 12-1 [8] illustrates this effect: the influence of the activity of the fines on the R value given by the stabilometer. Note that R for 5% bentonite is similar to that of 21% kaolinite in the gravel studied. That is the level selected by Hveem for establishing the concentration of the standard solution. The standard solution to use in sand equivalent tests is such that it will give the same sand equivalent in a gravel with 5% bentonite as in the same gravel with 21% kaolinite.

A different concentration of calcium chloride gives different sand equivalent values. The concentration selected by Hveem has no special significance nor is it necessarily the most appropriate for establishing the best correlation between the value of the sand equivalent and the activity of the fines.

Once the concentration of the solution has been established and the test has been standardized, the activity of the fines can be seen in the results of the test in at least two ways. First, the quantity of clay that is deposited in the 20 minute period will

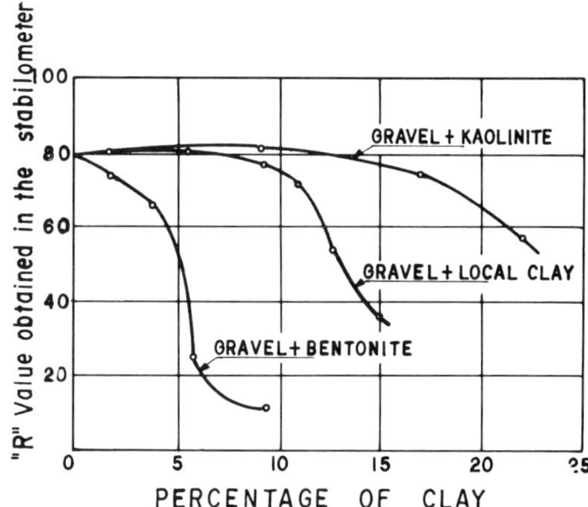

Fig. 12-1 Effect of clay on the R value [8]

vary according to the content and activity of the colloidal fraction of the clay; and second, the activity of the fines will affect the flocculent structure of the clay deposited in the standard solution. Closer or looser structures will be reflected in different heights reached by the clay, and fines and therefore different values of the sand equivalent, even for the same quantity of clay. Generally the looser the flocculent structure, the greater will be the activity of the fines and the smaller the sand equivalent obtained.

The test has not been studied in sufficient detail, to establish quantitative correlations between the result of the sand equivalent and the basic engineering properties, such as strength, compressibility, stress-strain relations, and permeability. Neither is it possible to say, whether such correlations can be established even in an approximate sense. What remains is to gauge the test in relation to the personal experience of field engineers; from this point of view the test reveals the relative effects of clay that is helpful in controlling the quality of borrow materials.

Figures 12-2 to 12-6, all taken from [8] show experimental correlations between the value of the sand equivalent and different soil properties or conditions. Figs. 12-2 and 12-4 should be interpreted with care, with respect to the abscissa. In

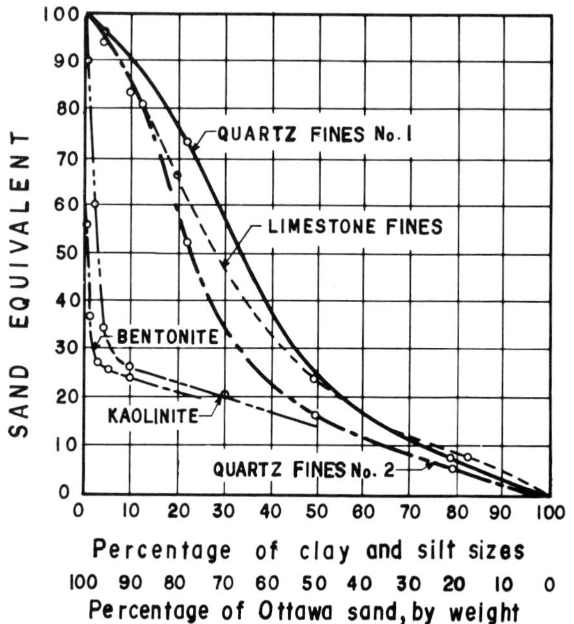

Fig. 12-2 Effect of various fine materials on the sand equivalent [8]

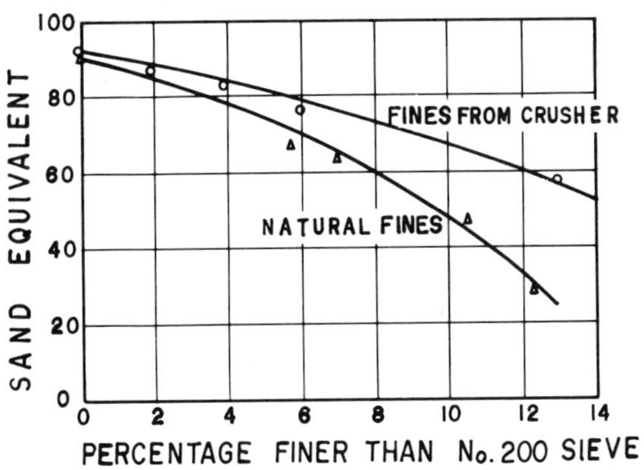

Fig. 12-3 Effect of dust on the sand equivalent of the fine aggregate in plant mix for wearing surface [8]

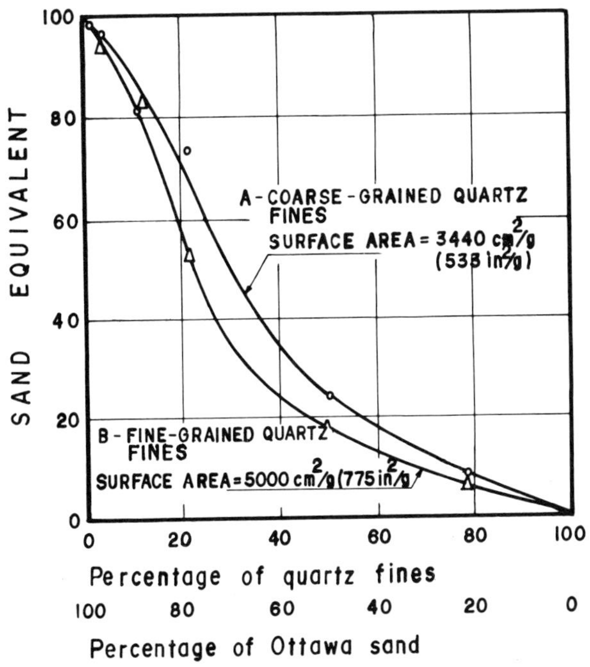

Fig. 12-4 Quartz dust and Ottawa sand. Effect of particle size on sand equivalent [8]

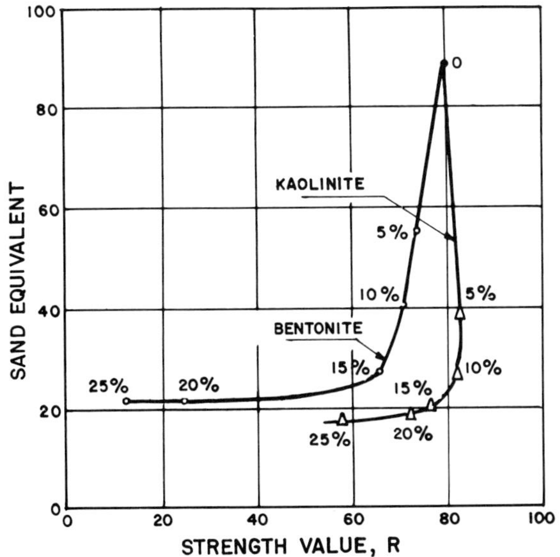

Fig. 12-5 Strength value in relation to sand equivalent in crushed rock with variable clay content [8].

RATIO OF COMPONENTS OF MIXTURE			
COMPONENTS IN LOWER Proportion	COMPONENTS IN HIGHER PROPORTION		
	SAND	SILT	CLAY
SAND	o	ф	ϴ
SILT	⦶	I	+
CLAY	ϴ	+	–

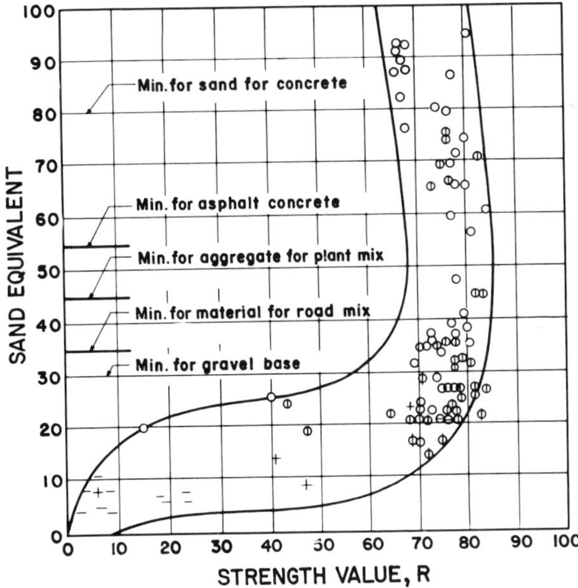

Fig. 12-6 Sand equivalent in relation to strength value [8]

Fig. 12-2, for example, either the ground rocks or the clays that are indicated are mixed with a certain percentage of Ottawa sand, so as to obtain the different graphs which describe how the sand equivalent varies with an increase in the percentages of rock fines or clay-fines in the sand matrix, Fig. 12-4 is similar. In Appendix 12a, the laboratory test is described in more practical detail.

The principal tests that are performed on rocks for borrow are those that determine how the rocks are fragmented and their susceptibility to weathering and abrasion. Table 12-3 contains a list of the index tests that are commonly carried out on rocks. In many practical cases some or even all of them may be omitted; the material is selected on the basis of observation of the borrow area and previous experience.

Table 12-3

The most common index tests for rocky materials in order to determine their engineering behavior [9]

Density of solids
Dry unit weight
Water content
Porosity
Weathering index
Permeability (water)
Permeability (air)
Alterability
Strength
Deformability

12.4 Borrow Materials [10,11]

12.4.1 Alteration of Rocks

The materials found by engineers at locations where exploitation is possible are soils or rocks that have formed after undergoing numerous changes by geologic evolution or revolution (Plate 12-4).

Plate 12-4 Exploitation of a large borrow pit in weathered rock

Probably the entire earth's crust was originally a viscous fluid which slowly hardened into igneous rocks. Weathering processes, encouraged by cracking and tectonic activities, gradually turned the original rock into residual materials or residual soils the *in-situ* products of decomposition, solution and disintegration. As has already been seen, many of these products subsequently may be transported by gravity, wind, water or ice, and redeposited in new locations and environments, becoming transported soils.

Soils that have been transported and deposited at a new location, continue to weather at this new location; some are retransported and redeposited in a new environment. Others harden by consolidation and cementation, forming sedimentary rocks. Marine organisms in both fresh and salt water, play a fundamental role in transforming these sediments produced by weathering into rocks. Sedimentary rocks are subject to the same distortion and fracturing produced by tectonics in igneous rocks. They are affected by environmental changes, which weather them, transforming them into new residual soils where erosion, transportation and redeposition continue.

Apart from being subjected to weathering and erosion, sedimentary rocks that become covered by accumulated sediments, are also subjected to increases in temperature, and new stress states. As a consequence, their minerals may change chemically, or be physically reoriented producing metamorphic rocks. These new rocks may resemble their parent rocks, but are sometimes more crystalline, denser and harder. Most metamorphics exhibit foliation: the orientation of minerals in layers. Metamorphic rocks are subjected to weathering wherever they are exposed to the effect of an external environment, and are altered into residual soils. These may subsequently suffer erosion, be transported, mixed and deposited in new sedimentary soils. Igneous rocks also can be metamorphosed by heat, pressure or shear stress, but the changes they suffer are usually less drastic than those that occur in sedimentary rocks. Lastly, metamorphic rocks can be retransformed into igneous rocks by heat, pressure and the addition of new minerals from fluid masses.

In Chapter 2, the principal rocks that are found by engineers in the earth's crust, together with the most common sediments that can be produced by these rocks, were described from a mineralogical viewpoint. Likewise, a brief description was given of the geological characteristics of rocks. Emphasis will not be placed on these aspects here, however some attention will be given to the dynamics of the alteration of rocks, to the early stages of residual and transported soils, to a description (although brief), of the formations that may be met by the engineer in his search for borrow materials, and the materials that most often appear in those formations. Table 12-4 describes the transformation of rocks into soils, the different types of soils and the influence of weathering and transportation factors.

When a rock that was not previously exposed is subjected to weathering, it adapts itself to the new environment that is imposed on it by changes in its physical and mineralogical makeup. The same thing happens when the rock is fragmented or crushed and placed in an engineering structure. For this reason, the engineer must regard all handling processes as a step toward future alteration. In most cases, the alteration processes that are triggered by the engineer occur so slowly that the useful lifetime of the structure represents an insignificant period. This is not always so, and cases of very rapid alteration are of great interest from a practical point of view. A typical example is the alteration of mudstones and shales into clays, especially in the presence of alternate wetting and drying.

The rocks that are used in different highway structures are generally subjected to compression, and often to abrasion and impact. All these are further causes of alteration. Among the different types of alteration, particle breakage leads to fundamental changes in paving materials in a brief period of time. Water and neutral pressures may also produce notorious effects during the useful lifetime of the highway. The engineer must, therefore, investigate the resistance to alteration bearing these facts in mind.

Certain uses in highway structures may impose very special conditions on the aggregates that are employed. Concrete technology, which is outside the scope of this book, offers numerous examples, but concrete is not the only material that imposes such conditions. The stony aggregate for asphalt mixtures must satisfy certain mechanical requirements; otherwise it may be rejected. Rocks containing a high percentage of siliceous minerals (acid rocks) are sometimes unsuitable for use in wearing courses, because they do not bind satisfactorily with the asphalt. Another cause of problems in asphalt mixtures are aggregates which alter rapidly into clays, as occurs with some basalts, despite the excellent quality usually achieved when this rock is first used in asphalt mixtures.

Clay minerals are the final product of the chemical alteration of rocks with their mineralogy related to that of the parent rock as well as the environment of weathering. For example, granites tend to form kaolinitic clays, whereas basalts, which are rich in ferromagnesian minerals, produce montmorillonitic clays. Chapter 6 described the weathering profile and its variations with the originating rock.

12.4.2 Treatments

Borrow materials that are to be used in embankments are not usually given any type of special treatment and are used just as they are. In this natural state they must meet the specifications for quality and construction. Special processing is usually unreasonably expensive, considering the large volumes. At the most, the largest particles will be removed, or two materials mixed together before compaction.

Table 12-4
Dynamics of the weathering of rocks and the formation of residual and transported soils

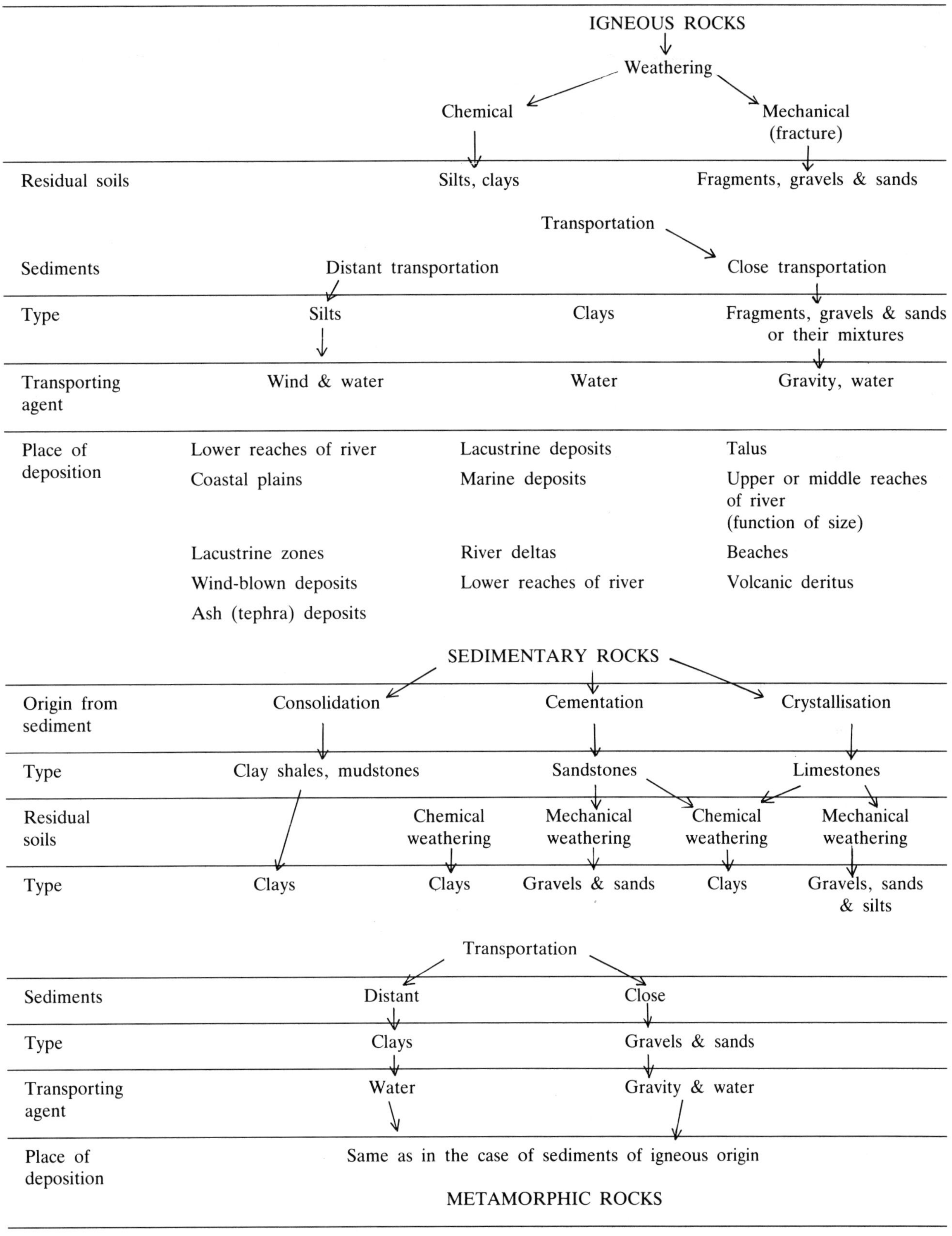

IGNEOUS ROCKS

Weathering

Chemical — Mechanical (fracture)

Residual soils	Silts, clays		Fragments, gravels & sands

Transportation

Sediments	Distant transportation		Close transportation
Type	Silts	Clays	Fragments, gravels & sands or their mixtures
Transporting agent	Wind & water	Water	Gravity, water
Place of deposition	Lower reaches of river	Lacustrine deposits	Talus
	Coastal plains	Marine deposits	Upper or middle reaches of river (function of size)
	Lacustrine zones	River deltas	Beaches
	Wind-blown deposits	Lower reaches of river	Volcanic deritus
	Ash (tephra) deposits		

SEDIMENTARY ROCKS

Origin from sediment	Consolidation		Cementation		Crystallisation
Type	Clay shales, mudstones		Sandstones		Limestones
Residual soils		Chemical weathering	Mechanical weathering	Chemical weathering	Mechanical weathering
Type	Clays	Clays	Gravels & sands	Clays	Gravels, sands & silts

Transportation

Sediments	Distant	Close
Type	Clays	Gravels & sands
Transporting agent	Water	Gravity & water
Place of deposition	Same as in the case of sediments of igneous origin	

METAMORPHIC ROCKS

Temperature, pressure etc. acting on the sediments. Similar residual and transported soils as from the other rocks (gravels, sands, silts or clays)

Materials for paving purposes are, however, usually subjected to different treatments to make them suitable for their functions. The most common treatments are:

Elimination of oversize particles: The purpose is to eliminate from borrow materials, those particles with a maximum size greater than that considered in the design (often about 7.5 cm (3in)). This sorting is often done by hand, or by a single coarse screen, known as a grizzley.

Breaking up: This operation is generally performed on hard borrow soils, partially weathered rock or materials with the consistency of poorly cemented agglomerates, tuffs or calcretes (caliches). It is often accomplished with machinery equipped with disk ploughs or with sheepsfoot rollers or smiliar rollers with pointed lugs.

Screening: This process is employed to achieve the appropriate grain-size distribution in a frictional material or to eliminate high percentages of particles larger than the maximum size required or smaller than the minimum size. The over-size or undersize particles are wasted or used in the embankment. Screens are employed to control gradation; washing is sometimes necessary to remove fines.

Screening facilities for eliminating large sizes are usually very simple. Normally the material is handled by gravity, the material that passes a particular screen being picked up in a truck. This method can cause segregation, and a material that is not uniformly mixed. When a satisfactory proportion of materials is required in different sizes, screening plants are required with a vibratory screen installed on two or three levels. The vibration rate is usually 1,200 cycles per minute. These plants are generally used in conjunction with crushing equipment.

Revolving screens are becoming increasingly popular, with concentric cylindrical screens. The material passes from one screen to another, following a different path from the center to the periphery of the system, depending on its size. This type of plant is sometimes more efficient in producing the specified proportions of particle sizes.

Crushing: Crushing is employed in order to produce base course and pavement aggregates from very coarse natural materials or rock fragments. It is also used to increase the angularity of rounded water worn aggregates. Crushing is normally divided into several different steps or stages, depending on the final product required: primary, secondary and tertiary crushing.

Crushing is usually done in elaborate plants (Plate 12-5) which include feeders, conveyor belts, screening plants, material elevators as well as jaw crushers, gyratory crushers, impact crushers, and rod mills for fine aggregates.

The relation between the sizes of the particle in the initial and final stages of the process is important, for it determines the type of equipment that must be used and the cost of the operation. Also important is the shape of the crushed particle, for on it depends largely the subsequent mechanical behavior. An equidimensional shape with sharp edges is the most desirable.

Washing: Washing is employed for materials contaminated with silt, clay or organic matter. It is frequently used in conjunction with crushing or screening.

There are several different washing systems; jetting during screening, washing tanks where the material is stirred with mechanical blades while water is sprayed onto it under pressure, and screw classifiers where an auger in a sloping water-filled trough causes fines to float away in suspension.

12.4.3 Types of Borrow Areas

The typical sources of material supplies are the roadside borrow pit, longitudinal and transverse cuts that balance fills and designated borrow pits. Some general information follows about typical natural deposits that make good borrow sources, whether they are a part of the cut on the right of way or designated pits (Plates 12-6 to 12-9).

Plate 12-5 View of a material crushing plant

Plate 12-6 Roadside borrow pit

Plate 12-7 Roadside borrow pit

River deposits are given the generic name of alluvial deposits. All along its course the water has the opportunity to erode very different materials; therefore alluvial deposits are made up of a wide variety of materials. Deposition obeys hydraulic laws that govern particle size. The capacity of the water to erode and transport sediments depends on the velocity and rate of discharge of the water. This determines the erosion in the upper reaches, where the river usually has steep gradients and consequently high velocities, that make the water capable of carrying very coarse particles of the size of cobbles, gravel and sand. Even boulders are moved by rolling downstream. In the middle reaches, river grades become flatter; the velocity of flow is reduced with a corresponding reduction in erosive power. Materials the size of gravel and sand are commonly deposited in middle river reaches, where materials can be found for borrow. In the final stage of its course, the river gradients become flat and the water loses a great deal of velocity. It eventually diverges or seeks an outlet to the sea, into a lake or into another larger river. In this stage, the erosive power is reduced even further, particularly close to the river's mouth. In this flatter zone the river deposits the finest materials; fine sand, silts and clays. If flow is very slow at the mouth of the river, a delta forms with a predominance of sand or silt.

To the foregoing regimen, which is largely dependent on flow velocity, is added the effect of the volume of flow on the erosive power. This controls the amount of particles in suspension rather than particle size.

Plate 12-8 Detail of a coarse alluvial deposit

In the upper part of the river, the volume of flow is small, the power to erode large quantities is limited. The volume of flow increases in the middle reaches, and particularly in the lower reaches. As a consequence of this effect, the capacity of the river to erode and transport large amounts of materials increases as it approaches the sea. The effect of the volume of flow is important in rivers that tend to undergo large discharges or flash floods, particularly because these high discharges are usually accompanied by abnormally high velocities, that erode volumes of coarser particles. These are deposited suddenly where the gradient flattens, in complex, mixed size braided deposits combined with erratic deposition.

The mineralogy of the sediments that can be found in a river bed depend largely on the nature of the formations through which the river passes. A typical example of this are the numerous rivers on the Pacific side of the Mexican Republic and many rivers in the South American Andes. In these places, the chains of high mountains are very close to the sea, leaving a very narrow coastal plain. The largest portions of the rivers

Plate 12-9 A typical alluvial deposit

(which are never long) cross very steep zones, which give the stream great erosive power. Also, owing to the heavy rains falling in these regions at certain times of the year as well as to the thawing of snow in the highest mountains, these rivers have enormous volumes of flow during certain months. As a result of all this, at these times of year the river rushes down towards the tiny coastal plain carrying large quantities of coarse particles, sand size and larger, which have been torn away from the rocky mountain formations. The velocity at which the river rushes into the plain and the enormous volume of flow results in large-scale flooding of the short flat flood plain near the sea, and the transport and deposition of gravels and sands in the flood plains and sands at the mouth of the river. In the rivers on the Pacific side of the Mexican Republic, this phenomenon is aggravated by the large formations of altered granite in the mountains through which the river passes, which provide enormous quantities of sand and gravel. In such areas, the engineer may find banks of sand and gravel in zones where other more tranquil rivers deposit only fine sediments.

To summarize, the deposits that can be found in river valleys, flood plains and alluvial terraces and fans are variable, not only in their mineralogical nature, but also in grain size, depending on the development of the stream, its hydraulic regimen and the formations that are crossed.

In areas where mountains or hills meet coastal plains, thick deposits are frequently found at the foot of the slopes. These are large gently sloping, undulating fan shaped formations or silty sands and gravels that have been deposited by rivers as they suddenly lose their velocity on entering the plain.

Lakes act as sediment basins for materials transported by the streams that discharge into them. When a river enters a lake, it deposits the coarser sediments it carries at the edge of the lake, the grain size depending on the previous regimen of the river. The deposit is in the form of a delta, where sands or silts can be found. The finest sediments are carried into the lake by the river water and are deposited in the deepest zones in uniform horizontal layers. During rainy seasons, most of the sediment deposited in the lake body is made up of silt sizes and coarse clays brought in by the river; the finer clays are deposited during periods of low flow when the velocities are small and lake waters are calmer. Because of this, lake deposits are usually stratified, consisting of fairly homogeneous, level, layers of fine materials, with strata of silty clays alternating with strata of very fine clays. The stagnancy of lacustrine zones usually encourages the growth and deposition of organic matter. It is not unusual for organic clays to be found in deep lake deposits, and peats in shallower parts. Siliceous skeletons of micro-organisms and calcareous shells are also common in lake deposits.

In arid regions, including Mexico, lacustrine deposits are often found in places where the original lake has disappeared a long time ago. Left behind may be thick deposits of soft silt and clay with layers of salt or other evaporated soluble matter.

In Mexico, as elsewhere, rivers are frequently found in desert and mountainous regions which do not discharge into any body of water. Instead they disappear, spreading out over a flat zone and forming an alluvial fan and playas. These rivers only very sporadically become torrents and do not carve themselves channels. When they suddenly lose the confinement they had throughout the mountainous part, they spread out with low velocities, unable to transport any materials. Practically all the sediments that are carried by the river are deposited in these fans and the playas below with only limited sorting; consequently these deposits are very heterogeneous, with abundant gravel, sands and silts.

Wind is another fundamental transportation element. It blows relatively coarse particles of soil along the top of the ground and suspends and transports silts and very fine sands. The distance they are carried depends on the size of the particles and the force of the wind. It varies from a few meters for coarse sands to many kilometers for very fine sands and silts.

A typical wind-blown deposit is loess. Loess soils are silts that originate from glacial deposits or desert floors, where they were initially picked up by wind. Primary loess is made up of particles of silt just as they were deposited by the wind, without undergoing any subsequent chemical alteration. Secondary loess has undergone some chemical alteration, generally by water. There is a large predominance of silt particles in loess soils, because sands are too coarse to be blown long distances through the air and clays are not so easily eroded by wind. The particles acquire an extremely loose honeycomb structure during deposition. Calcium carbonates and iron oxides and occasionally clays are deposited in their voids, providing a strength that is lost if these cementing agents become dissolved in water should the soil become saturated. For this reason, many engineers prefer to expose loess soils in vertical cuts, obtaining more stability than in sloping banks, which expose the freshly cut soil to rains and subsequent softening.

Loess soils provide good, abundant borrow materials for embankment construction. They pose, however, elastic rebound problems when used in the subgrade layer. They should not, therefore, be accepted for this purpose unless special tests show them to be sufficiently rigid. These materials may be greatly improved by compaction; however they are very sensitive to excess moisture. Since loess soils are deep deposits covering wide areas, other materials are seldom available in zones that are covered by them, and should be sought beyond the limits of the loess formation. Owing to their high porosity, there are few streams in loess areas and deposits of gravels or sands are unlikely.

Dunes are typical wind-blown formations of sand. Unfortunately the sand is too uniform and rounded for many purposes.

Glacial deposits are another possible source of construction materials, although in tropical regions such as Mexico they are few. They may be formed either by the moving ice or by melt waters. The ice deposits are very heterogeneous consisting of cobbles and gravel packed into a sandy clayey matrix. When caused by melt water, they resemble river deposits in reaches of high flow, although glaciers have an even greater capacity to transport coarse particles.

Residual soils are another common source of construction materials. Their nature varies greatly in accordance with that of the parent rock and the degree of alteration. Sedimentary rocks, except sandstones and conglomerates, produce very clayey soils. Igneous rocks produce either sandy or clayey soils, depending on how moist the alteration environment may be, and the mineralogy of the rock. Quartz-rich igneous rocks produce sandy silts, whereas those of a basic nature almost always produce clays.

Residual soils often contain a wide variety of particle sizes, and shapes since they have undergone no sorting process like those transported by water or wind. Depending on the predominating size and plasticity these residual soils may provide borrow materials for embankments or subgrades. For sub-grades they have to be sorted by hand or screened, eliminating fragments of partially weathered rock that are larger than 7.5 cm (3 in). From residual soils produced by quartz-rich or only slightly weathered rocks, gravelly sands can be obtained for sub-bases or base courses. However these may require cement or lime treatment, washing to eliminate excess fines, or partial crushing to eliminate the sizes that are too large.

An excellent source of paving materials is sound rock except shales and mudstones which can revert back to clays. Rock formations require fragmentation by blasting and then they must be totally crushed and screened for gradation. Some require treatment to improve some specific characteristic, such as their affinity with asphalt. During exploitation of these borrow materials, special care must be exercised to avoid slightly weathered zones or contamination from clay that fills

Plate 12-10 Preparing a bed of rock for blasting (drilling)

fractures or cracks. Washing is sometimes required to eliminate these undesirable materials.

A few comments will now be made on some special materials that have been frequently used in Mexican engineering practice.

Small shells: These are formations made up of calcareous residue of mollusc shells that are sometimes found in large volumes in seaside zones. Small shells generally show an advanced degree of weathering and are made up of thin layers which are fragments of the original shell. The grain-size distribution of the material is usually defective; their flat shape fractures easily and does not satisfy wearing requirements. However, acceptable results have been obtained with such materials in streets, roads, runways for light aircraft and even in asphaltic base courses where the method of mixing at the site is employed. The shells appear to interlock and form a strong layer that resists light traffic. The materials fracture under heavy loads.

Slags: These are porous, glassy, very hard materials, which when crushed produce limited fines. They have been used successfully as a base course in many places. Their excessive hardness is a disadvantage for crushing. Care must be taken to test their reaction to cement when they are used in concrete. Some swell; others accelerate steel corrosion.

Mining wastes: These are abundant in mining regions. They are stony materials which usually have uniform grain-size distributions as a consequence of the industrial process that produced them. According to the type of process, particle size may vary greatly, from the very fine sands to 5 or 7 cm (2 or 2-3/4 in).

12.5 Exploitation or Development of Borrow Materials

In order to excavate rock or soil, certain types of equipment are employed, the characteristics and uses of which have been established on the basis of previous construction experience. Selection of the appropriate equipment for a particular case will depend on three basic factors:

— Availability of equipment
— Type of material to be excavated
— Distance the material must be hauled

Once the type of equipment has been established, its size is a function of the volume of material to be moved, the time in which it must be accomplished and the space available for maneuvering.

In regions where industrial development is limited, the availability of equipment is a major factor. There are enormous varieties of machines which, when used in conjunction with one another in a well organized system, enable very efficient and economical borrow. However in many regions such specialized equipment is not available. In nations with limited foreign exchange the equipment has to be imported. Import duties offset the economy specialized equipment can provide. Thus, with the exception of the industrial nations, borrow materials usually have to be exploited using traditional types of equipment that serve several different purposes and can therefore be employed on a wide variety of projects. Manual labor, using a pick and shovel, can be very effective in the appropriate situations.

Table 12-5 presents the types of equipment frequently used by the highway engineer for excavating borrow materials. The usual means of transportation is also given, according to the type of material and the distance it must be hauled (Plates 12-11 to 12-16).

Plate 12-11 Typical bank of borrow material for use in fills

Plate 12-12 Tractor equipped with a ripper for breaking up earthy materials and weathered rocks

Plate 12-13 Tractor-drawn scraper

Figure 12-7 shows some of the operations usually required to prepare a borrow area for excavation. These include clearing the ground surface, and possibly loosening the material in order to simplify loading and hauling.

Table 12-5

Common equipment for exploiting and transporting borrow materials

Type of material	Cleaning and clearing (if necessary)	Preparation of borrow area	Excavation and loading		Transportation	
			Maximum size, m (ft)	Equipment	Distance, m (ft)	Equipment
ROCKS						
Sound rock (weathered on the surface)	Crawler tractor with tilting front blade	Drilling and blasting according to type of rock and maximum size to be obtained	$0.75 < x < 2.00$ $(2.5 < x < 6.5)$	Mechanical shovel	Under 150 (500)	Dumper truck
			$0.30 < x < 0.75$ $(1 < x < 2.5)$	Mechanical shovel or front loader	From 150 to 2500 (500 to 8200)	Dump car or truck
					From 2500 to 100,000 (8200 to 328,000)	Truck or trailer
			$0.075 < x < 0.30$ $(0.25 < x < 1)$	Mechanical shovel or front loader	Over 100,000 (328,000)	Railroad if available, truck or trailer
Weathered rock (very weathered on the surface)	Crawler or pneumatic tired tractor with inclinable front blade	Drilling and blasting, ripping and hand rock-block fragmentation	$0.30 < x < 0.75$ $(1 < x < 2.5)$	Mechanical shovel or front loader	Under 150 (500)	Dumper truck
					From 150 to 2500 (500 to 8200)	Dump car or truck
			$0.075 < x < 0.30$ $(0.25 < x < 1)$	Mechanical shovel or front loader	Over 2500 (8200)	Truck or trailer
Very weathered rock (soil and small fragments on the surface)	Crawler or pneumatic tired tractor with inclinable front blade. Scraper drawn by pneumatic tired tractor or self-powered scraper	Ripping and hand rock-block fragmentation or only ripping	$0.075 < x < 0.75$ $(0.25 < x < 2.5)$	Mechanical shovel or front loader	Under 150 (500)	Dumper truck
					From 150 to 2500 (500 to 8200)	Dump car or truck
					Over 2500 (8200)	Truck or trailer
		Ripping, scarification	$x < 0.075$ $(x < 0.25)$	Scraper	Under 150 (500)	Scraper drawn by crawler tractor or self-powered scraper
					From 150 to 2500 (500 to 8200)	Scraper drawn by pneumatic tired tractor or self-powered scraper
SOILS						
Alluvial Deposits	Crawler or pneumatic tired tractor with inclinable front blade	Ripping, scarification and hand fragmentation	$0.30 < x < 0.75$ $(1 < x < 2.5)$	Mechanical shovel or front loader	Under 150 (500)	Dumper truck
		Ripping, scarification	$0.075 < x < 0.30$ $(0.25 < x < 1)$		From 150 to 2500 (500 to 8200)	Dump car or truck
	Dredge	None	$x < 0.075$ Below the W.T. $(x < 0.25)$	Clamshell or dragline	Over 2500 (8200)	Truck or trailer
	Crawler or pneumatic tired tractor with inclinable front blade. Scraper drawn by crawler tractor or self-powered scraper	Ripping, scarification	$x < 0.075$ $(x < 0.25)$ Above W.T.	Scraper	Under 150 (500)	Scraper drawn by crawler tractor or self-powered scraper
					From 150 to 2500 (500 to 8200)	Scraper drawn up pneumatic tired tractor or self-powered scraper
Sands Silts and Clays	Crawler or pneumatic tired tractor with inclinable front blade	Ripping or scarification when dense, cemented or hard	$x < 0.005$ $(x < 0.016)$	Mechanical shovel	Under 150 (500)	Dumper truck
				Elevating grader	From 150 to 2500 (500 to 8200)	Dump car or truck
				Front end loader	Over 2500 (8200)	Truck or trailer
	Scraper drawn by crawler tractor or self-powered scraper	Ripping or scarification when dense, cemented or hard	$x < 0.005$ $(x < 0.016)$	Scraper	Under 150 (500)	Scraper drawn by crawler tractor or self-powered scraper
					From 150 to 250 (500 to 820)	Scraper drawn by pneumatic tired tractor or self-powered scraper
	Clamshell or dragline	None	$x < 0.005$ $(x < 0.016)$	Clamshell or dragline	Under 150 (500)	Truck
					From 150 to 2500 (500 to 8200)	Dump car or truck
	Marine dredge	None	Below the W.T.	Marine dredge	Carried hydraulically to sedimentation pond	

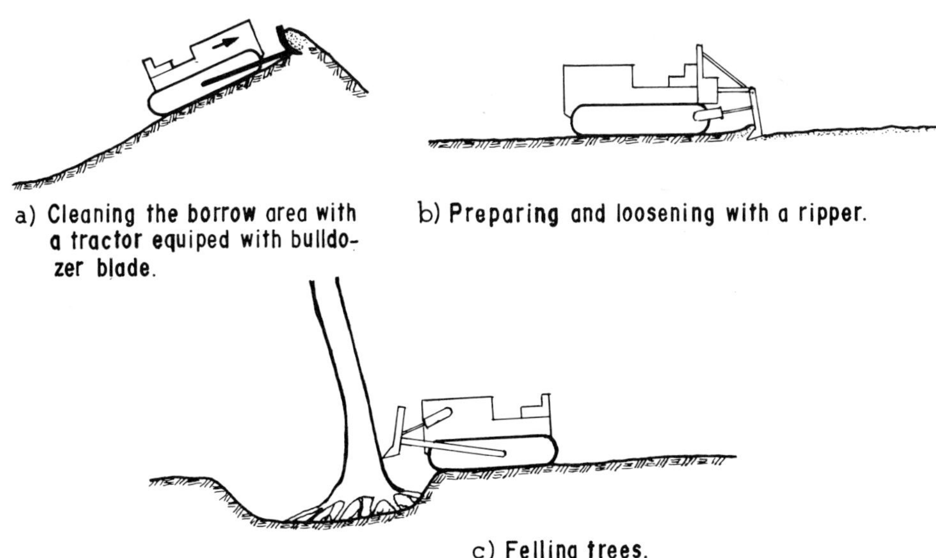

Fig. 12-7 Clearing and preparing a borrow area [12]

Figure 12-8 illustrates some situations where a mechanical shovel is employed. It is very frequently used (Table 12-5). The characteristics of the shovel proper are very variable, depending on the nature and relative position of the borrow pit. The ordinary bucket is used to cut rocky materials or soils, when located in steep faces or piles. The operation with the dragline is used when the material must be lifted, as occurs when it is located below the level of the equipment or when it is under water. The clam-shell is useful for selecting rock fragments from a mixture of abundant rock fragments and soils.

Figure 12-9 shows work with a front end loader, which is widely used in highway practice. Figures 12-10 and 12-11 show scrapers, which may be self-loading, or only serve for transportation purposes, being loaded by a different operation.

There is increasing use of heavy tractors equipped with rippers or ploughs for breaking up weak or layered materials to such a degree that they can then be removed by the tractor itself or by other types of equipment, thus avoiding drilling and blasting operations, which are always slower and more costly. The

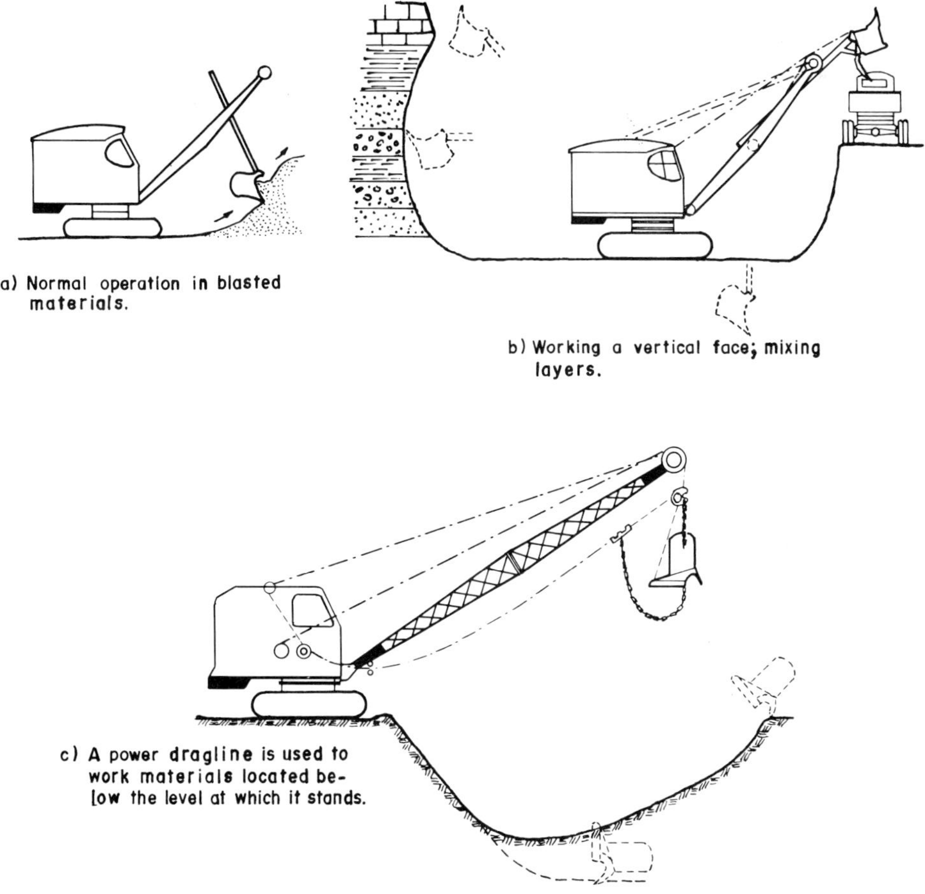

Fig. 12-8 Excavating borrow material with a mechanical shovel

Plate 12-14 Excavation of borrow material with a mechanical shovel

Plate 12-15 A large volcanic stone ("tezontle") quarry

tractor is also used as an excavator and bulldozer thanks to its front blade, but the excavation depth is usually restricted to no more than 50 cm (20 in). For these jobs the crawler tractor is usually employed. The pneumatic-tired tractor is more effective for towing equipment over short distances (between 150 m and 2,500 m) (164 and 2730 yd).

In heavy construction work, the use of self-driven and self-loading scrapers is becoming increasingly widespread when the nature of the material permits. They are very rapid and versatile types of equipment, both with regard to the materials they can handle and the distance over which hauling proves economical. Their self-loading capacity is frequently assisted by pushing them with a tractor, which helps to break up the material, during excavation. The motor scraper alone does the hauling. Non-self-driven scrapers are towed, generally by tractors with tires, and operate efficiently over short haulage distances.

Also winning increasing popularity for excavating borrow materials are front end loaders (sometimes termed mucking machines) with articulated arms, either on crawlers or tires. The former are more powerful and capable of working with larger rock fragments or in harder terrain, but the latter move faster and turn readily. Front end loaders have been used to haul materials over very short distances, of less than one hundred meters.

The mechanical shovel demands well defined faces and large volumes of material, so that it does not have to be frequently moved from one location to another. Most operate on crawlers, which enables them to adapt to most types of terrain, maintaining their stability even on fairly steep slopes. Mechanical shovels on tires are far more maneuverable, but are somewhat less stable.

Materials are usually hauled along highways by truck. Very short journeys and very long journeys are an exception. For short distances, industrial wagons drawn by pneumatic-tired

Plate 12-16 Storage of paving materials

Fig. 12-9 Front end loader excavating borrow material

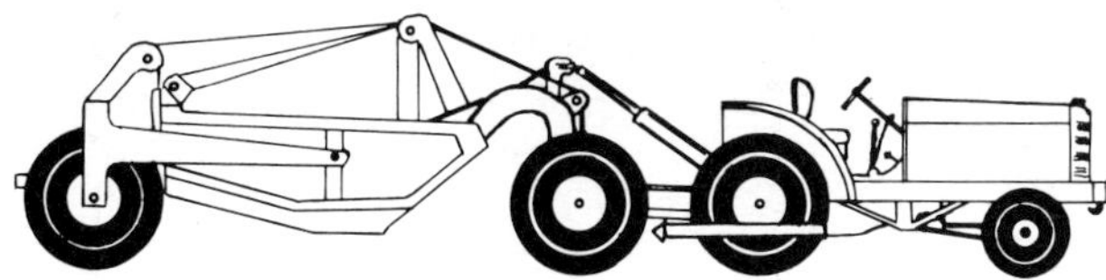

Fig. 12-10 A self-loading scraper

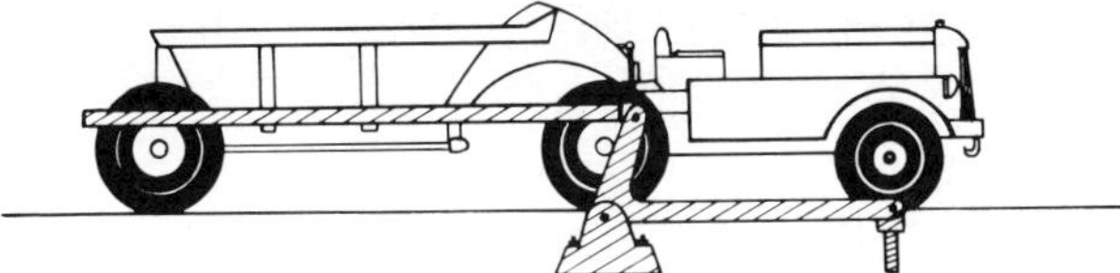

Fig. 12-11 Bottom dump trailer for transporting borrow material

tractors or other similar prime movers can be used, whereas for very long distances rail or barge transport is more economical.

In the excavation of borrow materials, it is essential to establish a suitable relation between the capacity of the equipment that excavates and removes the material and that of the hauling vehicles. In this way costly interferences or wasted time can be avoided. The carrying capacity of the hauling vehicles should preferably be a whole multiple of the capacity of the excavation or loading equipment.

A key operation in excavating rock is drilling and blasting. This will not be dealt with here, since it represents a technology which is beyond this text. Some specialized works, such as [13] may prove useful to those engineers who have a special interest in this topic.

There are special cases which frequently arise in the excavation of borrow materials, which deserve some special comments. In the same face in the borrow pit there are sometimes several strata of different materials, all of which can be used although their qualities are different. Excavation work for embankment materials should be conducted in such a way that the different materials will be well mixed, so that the final product will be as homogeneous as possible. Otherwise there will be layers with different properties distributed at random throughout the embankment, which produce different strengths and compressibilities.

In certain Karstic zones, of which the Yucatán Peninsula is a good example, there are large plains where the materials that can be used for embankments (more or less fine materials of limestone origin, referred to as *sascab* in Yucatán) are found below a crust of hard limestone rock, one to one and a half meters thick. In order to extract the underlying material, the protective covering must be broken. This requires drilling and blasting. In these areas roadside borrow pits are not feasible. Instead separate deep pits of borrow materials are developed, where by breaking a minimum surface area of hard rock a maximum volume of material can be obtained. This is preferable owing to the enormous expense of blasting. In these flat areas, embankments are seldom high, and the hard limestone would have to be broken up into very small pieces, which is usually not economical, because closely spaced blast holes are required. It is generally preferable to waste this rocky material, removing it in large slabs.

The lightweight embankments that are usually required over very soft, compressible soils generally impose restrictions on the use of borrow materials and encourage long distance hauling. Volcanic stone (*Tezontle*) which is porous foam-like basalt, scoria or pumice, is widely used in Mexico for these purposes, and therefore enjoys considerable confidence. Porous Volcanic stone banks often are contaminated by a sound basalt face, with a high unit weight, which must therefore be carefully avoided. At other times tezontle banks are very high and for safety reasons must be excavated from above or in benches. As a result, the materials roll considerably before falling into a heap on the ground, producing a high percentage of dust. This is wasted by selective excavation because it causes an excessive increase in the unit weight of the material. This problem is often ingeniously prevented by constructing flat ramps where the material does not roll so much, and which at the same time can be exploited from below without danger. Sometimes it is necessary to screen the material.

In many coastal plains, swampy zones or ancient lacustrine basins, there are often no materials on the surface that are suitable for embankments, and even less so for pavements. It has already been mentioned that in these cases hills should be located where the probability of finding better quality materials will be greater. If such hills do not exist, materials will have to be excavated from low mounds or terraces, where the soils are usually too moist. The moisture not only impedes their immediate utilization, but also hinders the operation of excavation equipment. Under such circumstances, it has been found useful to open several different faces in each bank, extracting from each a layer of material no more than 50 cm (20 in) thick, and alternating this procedure in the different faces. In this way, each face is left untouched for a few days, which gives it the opportunity to dry by evaporation, after which a new layer can be extracted.

Where the roadside borrow pit is employed in rainy zones, it has been found useful to commence excavation in the part furthest from the road, so that traffic will not be interrupted at intermediate points of the haulage trip.

Sometimes in very fine materials, it is advisable to program excavation operations in such a way that the necessary amount of water can be added in the borrow area so that they can subsequently be compacted in the embankment. Very fine soils, like high plasticity clays, have low permeabilities, and consequently water takes a considerable time to penetrate into them. Such time is practically impossible to accommodate in the embankment. Thus, it has been found expedient to irrigate the borrow pit soils or to remove the soil, place it in thin layers to which water is added by sprinkling. Once sufficient time for the material to absorb the water has elapsed, the soil is loaded and transported to its final destination, where it is immediately compacted so as to avoid loss of water by evaporation.

A similar situation occurs when very fine material in a borrow pit contains the right amount of water for compaction. The operations must be programmed in such a way that this water will not be lost, either in the pit or in the embankment. In such situations the material sometimes dries out because the pit was left open too long due to poor planning; when the soil was eventually laid in the embankment, clods formed that were so hard to break up and moisten that it was found preferable to waste all the material.

Borrow pits located in fluvial deposits must be excavated at the time of year when the river is at its lowest level; otherwise there is a risk of excavation becoming impossible during periods of heavy flooding and the progress of the entire project being interrupted. A further problem is the contamination by the fine soils in suspension which are transported by rivers during their flooding.

Some materials, such as weak tuffs and soft limestones, are so fragile that they undergo appreciable degradation during loading and hauling. All unnecessary handling, such as temporary stockpiling or transfers from one deposit to another should be avoided with such materials.

In rock quarries, where the stratum of usable material overlies another with unsuitable properties (for example a lava flow overlying a layer of clay) the face should be worked in such a way that the floor of unsuitable material is always covered by a layer of 30 to 50 cm (12 to 20 in) of fragments from the usable material, so as to avoid possible contamination.

Last, precautions are necessary in handling materials that are stored for future use. All materials made up of different sized particles tend to segregate when dropped from a conveyor or truck onto a slope below. To overcome this, the material should be taken from the bottom of the deposit when it is reloaded, so that all the particle sizes are duly mixed. The material should never be taken in horizontal layers from the top of the deposit.

12.6 Identification of the Degree of Alteration of Rocks

The problem of alteration or weathering of rocks should be considered, especially with a view of identifying the degree of alteration that prevails at the time of excavation and predicting if a rock will alter additionally during the useful lifetime of the project. This alteration or weathering might directly affect the strength of a mass, and can have an influence on slope stability, and on foundation capacity. Rapid alteration may lead to important changes in the permeability of the mass, which will be a concern for dam builders, and to a lesser extent for road builders. It will influence the strength, deformability and permeability of paving materials or ballast, which are of importance to highway engineers.

Table 12-6, which is taken from [14] provides a summary of the comments included in this section, and distinguishes the alteration of rocks after excavation and construction from those that remain undisturbed in the ground.

By the alteration of a rock, is implied any change it may undergo that may be of interest to the civil engineer. Weathering is a specific form of alteration, where the changes are caused by atmospheric agents and groundwater. Alteration is distinguished from erosion, which implies disintegration and loss of the material. In accordance with this conception [14], erosion is a particular mode of weathering (and weathering is a particular case of alteration). Table 12-7 [14] describes the various forms of alteration and weathering, distinguishing its agents and their specific effects. Table 12-8, also taken from [14] relates the different mechanisms that act on civil engineering materials with the final effects they produce, and the means by which the different mechanisms act.

Table 12-7
Weathering agents and their effects [14]

Processes	Agents	Effects
Weathering	Atmospheric	Changes in strength, deformability, color, texture etc.
Wear: Erosion Others	 Atmospheric Others	Changes in the surface geometry

Alteration usually debilitates the properties of rocks from the civil engineering point of view although it occasionally improves them. The major concern of the specialist is centered on those cases where alteration is detrimental. The limit reached by alteration is the total destruction of the interparticle and mineralogical bonds of the minerals of the rock [15].

Alteration mechanisms are essentially disintegration by fissuration or by loss of mineral strength and bonding, and dissolution by internal erosion or solution. As a result of these mechanisms, the rock loses weight and its porosity increases. Its capacity to absorb water increases and with it its tendency to expand. Loss of weight occurs only by loss of material, whereas absorption of water is influenced both by loss of material and disintegration. Expansion during absorption is influenced by disintegration, and is consequently used as an indication of that alteration mechanism.

The absorption of water that is shown by a specimen of rock, therefore, serves as an indication of the degree of alteration, although the specific mechanism of that alteration is not shown.

Table 12-6
Nature of problems caused by rock weathering in Civil Engineering [14]

Utilization of rock		Safety problems		Surface characteristics	Aesthetic problems
		Stability	Permeability		
The rock as a borrow material		X	X	X	X
The rock in-situ	Foundations	X	X	—	—
	Cuts	X	X	—	—
	Tunnels	X	X	X	—

Table 12-8
Mechanisms producing weathering [14]

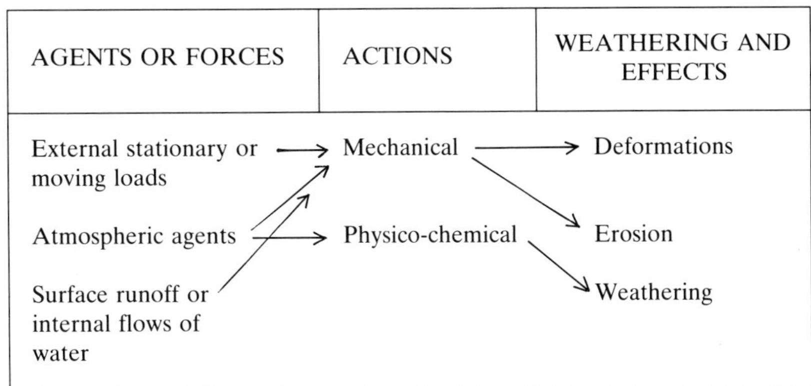

AGENTS OR FORCES	ACTIONS	WEATHERING AND EFFECTS
External stationary or moving loads	Mechanical	Deformations
Atmospheric agents	Physico-chemical	Erosion
Surface runoff or internal flows of water		Weathering

If it is wished to demonstrate the specific effect of disintegration, tests are conducted where both the absorption and expansion of the specimen are analyzed as water is added. If it is wished to discover the influence of the loss of material on alteration, it is necessary to investigate in the laboratory both the absorption of the water and changes in solid weight. Lastly, if it is wished to conduct a thorough study of the alteration mechanisms the absorption of water, the expansion of the specimen and the changes in its weight will have to be investigated [14,16,17].

Apart from these simple indications of alteration, there are more that are helpful, such as changes in density, strength, or permeability, all of which are affected by alteration.

Figure 6-17 relates the strength of a rock, in this case granite, with its absorption of water in the laboratory. Since absorption tests have been used for many years, there is already an appreciable amount of useful experience for correlating these two effects. The expansion of rock specimens has been far less frequently measured, so that it is correspondingly more difficult to establish reliable experimental correlations. Figure 12-12 gives some typical results that can be obtained with tests of expansion. The test is performed with total immersion of a cylindrical specimen, to the top of which an extensometer has been connected.

Reference [18] presents some techniques for developing tests to measure porosity and permeability with water and air, which have served as a basis for many laboratory techniques.

The properties of altered rocks that determine their utilization in civil engineering projects, such as their strength, deformability and permeability, are more meaningful indications of the degree of alteration than properties such as the absorption of water. It is more difficult to establish the engineering significance of absorption; it demands experimental correlations that are always subjective and unsure.

The alterability of a rock can be defined as the rate with which alteration processes evolve. Knowledge of this property is helpful for establishing criteria regarding the effects the alteration of the rock may have during the useful lifetime of a project. A rock's alterability is not a constant even with a constant

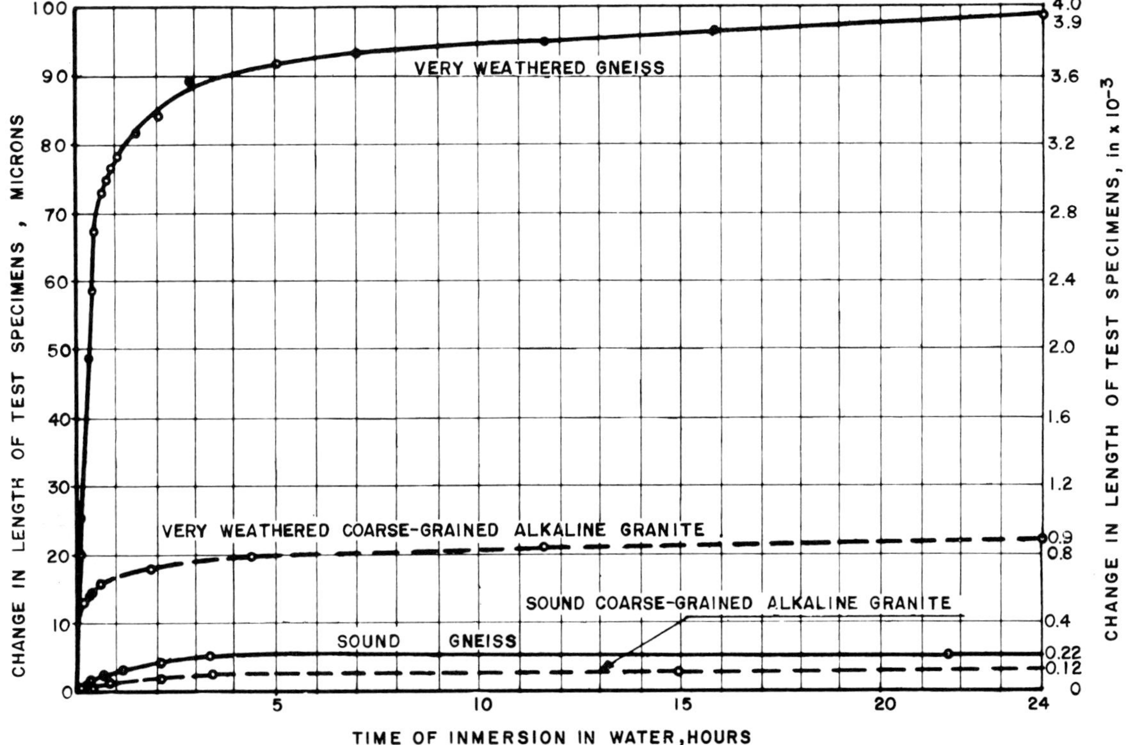

Fig. 12-12 Typical results of swell tests on rock samples [14]

environment because the factors that influence it change as the rock weathers. There is no analytical or laboratory technique to determine what the alterability of a rock will be in a specific situation. However criteria for testing can be established [14]:

— Determine the alteration conditions (largely environmental) to which the rock will be subjected.
— Among the alteration factors select the one which is most significant in the particular case studied. Define the initial state of the rock in relation to this parameter as well as the final state that would have to be reached for alteration of the rock to represent a significant danger to structural behavior.
— Subject a sample of the rock to an alteration process that is representative of that that will occur in the project, in order to establish the alterability of the material in relation to that process. In this way, it will be possible to estimate the time it will take for the rock alteration to become a danger to the project under consideration.

The above procedure has three main difficulties. The first is to discover the alteration conditions prevailing in the rock at the start of the project. The second is to establish a laboratory test that will reproduce the conditions to which the rock will be exposed in the future. These tests, although they can be reasonably conceived, would demand excessive performance time. It is questionable if the process can be accelerated realistically. The third difficulty is to specify what degree of future alteration will jeopardize the performance of the structure.

Attempts have been made to combat these difficulties by means of accelerated weathering and aging tests. These are correlated with observed behavior of similar rocks with similar degrees of alteration that have been experienced in other projects. The borrow areas that have previously provided material for structures are identified, so that the alteration of a single rock type used in different environments in different structures can be measured, after years of use. Ultimately these data will help to establish correlations with the behavior of similar materials that are to be used for the first time. Lastly, the engineer who hopes to find a reasonable solution to these problems should sieze every possible opportunity of observing the behavior of masses of altered rock and existing structures made of it, so that he will gain experience in estimating the potential degrees of alteration in different rocks.

Often, the most reasonable conclusion that can be drawn from all studies and observations is as follows:

If rock *A* has shown satisfactory behavior in a structure, and rock *B* is to be used in another structure with similar conditions, and if the alterability of *B* seems to be smaller than that of *A*, then *B* can be recommended for similar use.

APPENDIX 12A
SAND EQUIVALENCE TEST

12a.1 Objective

The objective of the tests is to serve as a quick field measurement of the presence or absence of fine materials of a clayey nature which might prove detrimental to coarse grained soils and stony aggregates.

12a.2 Apparatus

To conduct this test the following equipment is required (Fig. 12a-1, Plates 12a-1, 12a-2):

a) A transparent cylinder, graduated to measure volumes, with an interior diameter of 3.18 cm (1-1/4 in) and an approximate height of 43 cm (17 in), with graduations in tenths of a centimeter from the bottom up to a height of 38.1 cm (15 in).

b) An irrigator tube made of copper or brass, with an exterior diameter of 0.64 cm (1/4 in). One of the ends of the tube will be flattened, forming a wedge-shaped point. Close to the tip, through the side of the wedge, two lateral perforations will be made, using a N° 60 drill (diameter = 1 mm = 0.04 in).

c) A bottle with a capacity of 3.8 liters (one gallon), with siphon equipment, consisting of a stopper with two holes and a bent copper tube. The bottle will be placed 91.8 cm (3 ft) above the work bench.

d) A piece of 0.48 cm (3/16 in) rubber hose, with a clamp to close it. This tubing will be used to connect the irrigator to the siphon.

e) A tared tamper, consisting of a metal rod 46 cm (18 in) long with a cone-shaped foot with a diameter of 2.5 cm (1 in) at its lower end. This foot will be equipped with three small screws for centering it within the cylinder. A lid, which adjusts to the upper part of the cylinder and enables the rod to slide loosely through the center serves to center the upper part of the rod in the cylinder. A weight is added to the upper end of the rod so as to obtain a total weight of 1 kg (2.2 lb) for the device.

f) A beaker or sample container, with a capacity of 88 ml (3 fl ounces).

g) A wide-mouthed funnel for depositing the sample in the cylinder.

h) Stock solution consisting of:

Anhydrous calcium chloride	454 g (1 lb)
USP glycerine	2,050 g (4.5 lb)
Formaldehyde (volumetric solution at 40°C)	47 g (0.1 lb)

Dissolve the calcium chloride in 1.89 l (half a gallon) of water.

Cool and filter the solution through WHATMAN No 12 filter paper or equivalent. To the filtered solution, add the glycerine and the formaldehyde. Mix well and dilute to one gallon. Either distilled or demineralized tap water can be used. When there is any doubt about the quality of tap water, this should be

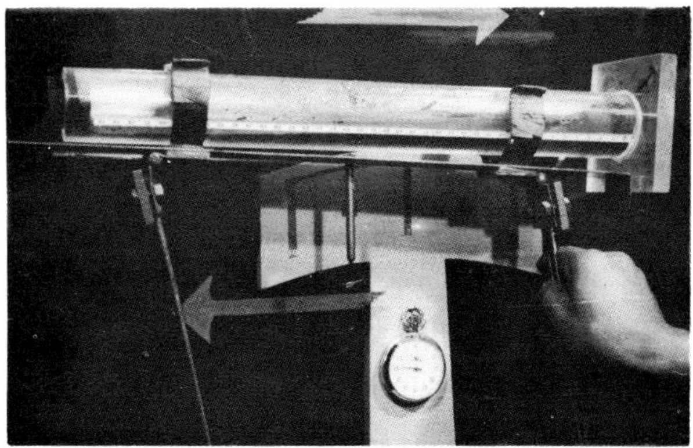

Plate 12a-1 Manual shaker used in the sand equivalent test

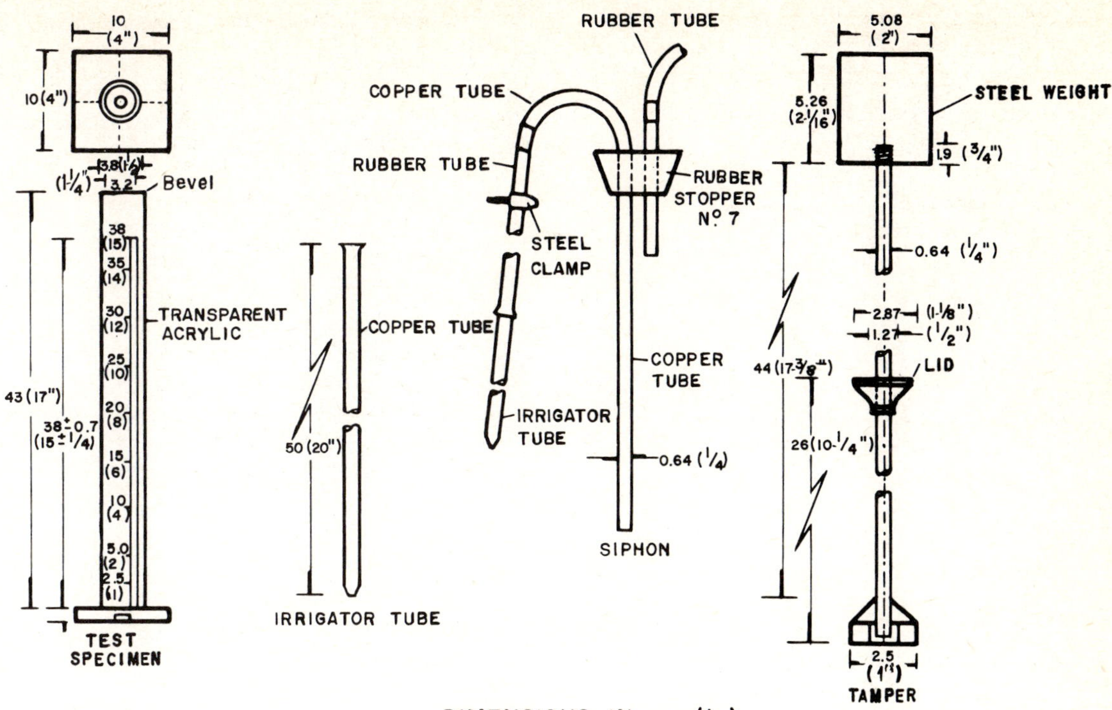

Fig. 12a-1 Sand equivalent test equipment

tested comparing the results of the sand equivalent value obtained from identical samples, using solutions made with the dubious water on the one hand and distilled water on the other hand.

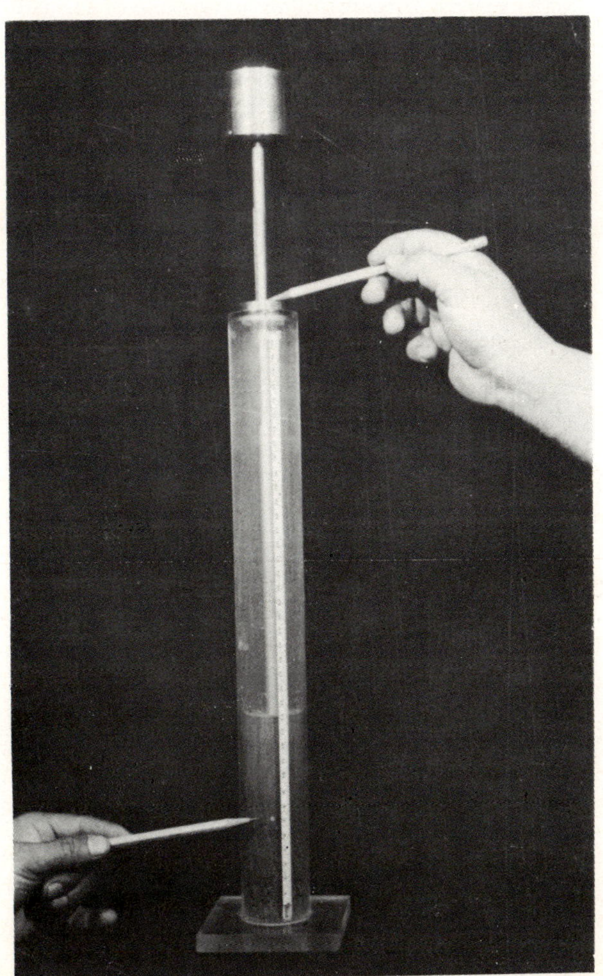

Plate 12a-2 Measuring the sand equivalent

i) Working solution

Dilute in 3.8 l (1 gallon) of tap water 88 ml of the stock solution. The 11.2 cm (4.4 in) mark on the graduated cylinder corresponds to the required 88 ml (3 fl ounces).

12a.3 Test Procedure

The test is conducted following these steps (Fig. 12a-2):

a) The material to be tested must be the portion of the sample passing a No 4 sieve. If, therefore, the sample contains coarse rock particles, it must be sieved through the No 4 sieve after breaking up the lumps of finer material. If the original sample is dry, water must be added to it before sieving. If the coarse aggregate has a covering that will not separate during the sieving operation, dry the coarse aggregate and rub between the hands, adding the resulting dust to the sieved material.

b) Start the siphon by blowing into the top of the bottle through a small tube with the clamp released. Once this has been done, the apparatus will be ready for use.

c) By means of the siphon introduce the working solution into the cylinder to a height of 10 cm (4 in).

d) Empty into the cylinder the contents of a beaker or container full of the prepared sample of soil. The capsule when full will contain approximately 110 g (0.25 lb) of loose material (as an average). Bang the bottom of the cylinder several times firmly against the palm of the hand, so that any air bubbles will escape, and also to speed up saturation of the sample. Allow the mixture to rest for 10 minutes.

e) When the ten minutes have elapsed, cover with a cylindrical stopper and shake hard in a longitudinal direction from side to side, holding it in a horizontal position. There should be 90 cycles in approximately 30 seconds, with a horizontal movement of 20.5 cm (8 in). A cycle consists of a complete oscillating movement. In order to shake the sample satisfactorily at this frequency, the operator must shake only with his forearms, relaxing body and shoulders.

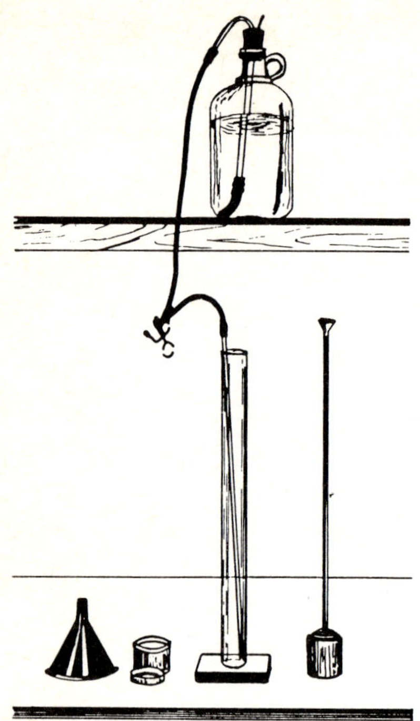

Fig. 12a-2 General view of sand equivalent test equipment

f) Remove the stopper and insert the irrigator tube. Rinse the sides downwards, and afterwards insert the tube until it reaches the bottom of the cylinder. Separate the clayey material from the sandy material, suspending it in the solution by means of a gentle downward movement of the irrigator tube, at the same time turning the cylinder slowly. When the level of the liquid reaches 38.1 cm (15 in), raise the irrigator tube gently without interrupting the flow, so that the level of the liquid is maintained approximately at 38.1 cm (15 in) while the tube is being withdrawn. Reduce the flow just before the tube is completely withdrawn and adjust the final liquid level to 38.1 cm (15 in). Leave the cylinder completely alone for 20 minutes. Any vibration or movement of the cylinder during this time will alter the normal settlement of the clay in suspension, provoking erroneous results.

g) Once the 20 minute period has elapsed, make a note of the upper level of the clay in suspension. Take the reading to the nearest 0.2 cm (0.1 in).

h) Introduce the weighted tamper slowly into the cylinder, until it rests on the sand below the clay. Turn the rod gently, without pushing it downwards, until one of the centering screws can be seen. Make a note of the level corresponding to the center of the screw.

12a.4 Calculations

Calculate the sand equivalent using the following formula:

$$\text{Sand equivalent} = \frac{\text{Reading corresponding to the upper boundary of the sand layer}}{\text{Reading corresponding to the upper boundary of the clay layer}} \times 100 \quad (12\text{-}1)$$

If the sand equivalent is lower than the value specified, perform two more tests with the same material and take the mean value of the three as the sand equivalent.

12a.5 Procedure for Dismantling the Apparatus

In order to empty the cylinder, cover it, turn it upside down and shake it up and down until the sand plug disintegrates. Empty immediately. Wash twice with water. Do not expose plastic cylinders to direct sunlight longer than necessary.

REFERENCES

1. Sowers, G. F., *Introductory Soil Mechanics and Foundations: Geotechnical Engineering,* Fourth Edition, Macmillan: New York. Collier Macmillan: London, 1970.
2. Juárez Badillo, E. and Rico, A., *Mecánica de Suelos. Vol. I: Fundamentos de la Mecánica de Suelos,* Limusa: México, 1972, Appendix.
3. Hvorslev, M. J., "Subsurface Exploration and Sampling of Soil for Civil Engineering Purposes", Waterway Experiment Station, Vicksburg, 1949.
4. California Division of Highways, *Materials Manual. Vol. I. Sand Equivalent Test 217-E,* Sacramento, Calif., 1963.
5. AASHTO, "Standard Method for Sand Equivalent Test", Designation: T-176-65, Washington, D.C., 1965.
6. Hveem, F. N., "Sand Equivalent Test", Materials and Research Department, California Division of Highways, Sacramento, Calif., 1952.
7. Hveem, F. N., "Degradation of Aggregates", 44th Annual Meeting of the American Association of State Highway Officials, San Francisco, Calif., 1958.
8. Hveem, F. N., "Sand Equivalent Test for Control of Materials During Construction", Procs. 32 Annual Meeting of the HRB, Washington, D.C., 1953.
9. Federal Commission of Electricity, *Manual de Diseño de Obras Civiles,* México, 1969, Vol. I, Section C.
10. As Reference [1].
11. Krynine, D. P. and Judd, W. R., *Principles of Engineering Geology and Geotechnics,* McGraw Hill: 1957, Chap. 3.
12. Nichols, H. L., *Moving the Earth,* Van Nostrand, 1955, Chap. 13.
13. Langefors, U. and Kihlström, B., in *The Modern Techniques of Rock Blasting,* Almqvist and Wiksell, Gebers Förlag AB: Stockholm, 1963.
14. Nascimento, U., "O Problema da Alterabilidade das Rochas em Engenharia Civil", National Laboratory of Civil Engineering. Publication N° 363, Lisbon, 1970.
15. Nascimento, U., Branco, F. and Castro, E., "Identification of Petrification in Soils", Proc. *VI. ICSMFE,* Montreal, 1965.
16. Knight, R. G., *Road Aggregates,* Edward Arnold: London, 1948.
17. Woods, K. B., *Highway Engineering Handbook,* McGraw Hill, 1960.
18. Ferran, I. and Thenoz, B., "L'alterabilité des Roches, ses Facteurs, sa Prévision," *Annales de L'Institute Tech. et des Trav.* Pub. N° 215, Series 78, Toulouse, 1965.

CHAPTER 13

FIELD INSTRUMENTATION

13.1 Introduction

In spite of the progress that has been made in Soil Mechanics in the last few decades, few important problems in the field of Applied Soil Mechanics can be resolved by theoretical solutions that are so satisfactory that the engineer feels free of anxiety for the subsequent behavior of the project and the accuracy of his predictions.

It has been said [1] that the differences between theory and reality are larger and more complex in the field of applied soil mechanics than in any other branch of civil engineering. This is due both to the complexities of soil as a foundation and as building material and to the limitations of time and cost that prevent the engineer from obtaining the levels of quantity and quality of engineering data that he needs, even in important problems. These two circumstances usually discourage any attitude of indifference or overconfidence in the behavior of existing structures or the geotechnical solutions that are adopted in a given case (Plate 13-1).

As a consequence engineers now insist on observing the behavior of existing structures and measuring their movements throughout their useful lifetime. Such observations, when properly conducted and interpreted, make it possible not only to determine the behavior of a structure and the evolution of its stability and service conditions, but also to verify its design, and theories that were used in developing that design. Thus the observation of prototypes not only provides information about a particular structure, but is also a valuable verification of theoretical concepts applied to real structures. Such verification is seldom possible for the engineer; its importance is evident to all but those having an unflexible, dogmatic view of engineering. Those who understand that engineering theories and conceptions are at best imperfect attempts at explaining the behavior of nature (which operates on a scale of complexity which has proved beyond the capacity of human comprehension) welcome such verifications.

However, the potential value of field observations and measurements, followed by the corresponding interpretation, is even greater than is indicated by the foregoing paragraphs.

Plate 13-1 View of a zone of the Tijuana-Ensenada highway, where serious instability problems were solved with the aid of field instrumentation. The photograph illustrates the relation between geological conditions and genuine engineering problems

New theoretical concepts or construction methods can be developed from analyses of the information that is acquired. Thus field observations are a valuable form of experimental research, capable of generating new ideas and contributing to the progress of soil mechanics.

In highways, the uncertainties which justify the need to observe existing structures in the field and to verify solutions adopted are fully satisfied. Nevertheless, field observation techniques have been employed far less extensively in highways than in other fields of applied soil mechanics, such as earth dams. This is because many highway engineers regard these techniques as being too sophisticated, costly, and unnecessary for building a satisfactory structure. Consequently, they oppose spending time and money on the observation of prototypes, overlooking that the amounts required are always insignificant fractions of the total cost of the road. Other specialists in different branches of engineering, such as dams, are better disposed towards the careful observation of prototypes, and incorporate instrumentation routinely in larger dams.

A second reason why observations of field behavior are conducted less frequently in highways than in other structures is that it is useless to carry out the extensive observation and measurement program for projects where no other comparable information is available regarding the geology and soil mechanics of the project. Those cases where highway engineers obtain large amounts of geotechnical information about a specific problem are relatively few. Thus the routine problems faced by the highway engineer are not sufficiently well defined to justify the observation of structural behavior in the field. This situation is clearly not incongruous, for it has already been said that the study of transportation routes (with the exception of airports), requires less detailed information than other structures. This is inevitable on account of their characteristics. With limited soil and rock data it would be difficult to interpret the observations made in routine cases.

The foregoing plus some doubtful designs, determine the norm of conduct that appears to be appropriate for structures of this type. Observation of the behavior of earth structures and solutions in the field should not be excluded from highway and railroad projects. The special nature of these structures will confine instrumentation to important situations, outside routine practice, because of the time and cost of studies they involve and particularly the consequences of failure. The use of field observation techniques must be maintained within sensible limits, omitting them where interpretation is impossible owing to a lack of general geotechnical information, nor engaging in repetitive measurements. The greatest risk of a very long term observation program that covers the useful life of a structure (or an important fraction of it), is the neglect of regular observations. The more structures that are to be observed, the greater is the risk.

Field measurements for verifying structural behavior are conducted using an ever-increasing variety of equipment and instruments. It is to this characteristic that the observation technique owes its term of Field Instrumentation. Earth structures are *instrumented* with a set of measuring devices that determine the evolution of the structure's behavior, including significant movements and the long term stability.

In highways, there are two problems that typically demand field instrumentation. The first includes the construction of embankments on soft compressible soils, and assessing settlements, time rates of settlement and changes in stability. Second are the problems connected with the stability of natural hillsides and slopes moving on what is suspected to be already existing slip surfaces. In this case, it is essential to determine how these movements occur in order to develop a successful system for correction.

Tunnels are another family of highway structures which must be frequently instrumented in order to determine the earth loads and deformations, which are always difficult to predict. Problems of earth thrust against walls and struts also can be minimized by instrumentation because there is such a wide uncertainty in their design loads.

In this chapter a brief description will be given of the instrumentation procedures that are currently in use, and the equipment available. The principal conclusions that can be drawn from an instrumentation and measurement program will be briefly discussed. In some cases reference will be made to important instrumentation jobs that have been conducted on different highway projects throughout Mexico under the direction of the Ministry of Public Works.

The final comment concerns operation. An instrumentation program, whatever its purposes may be, should be conceived and established within the general framework of the project, as part of the whole. It will probably even be closely related to the design; the results obtained from the measurement program during the early stages of construction often cause modifications in the remaining project designs. An instrumentation program that does not take the design into consideration will never lead to valuable results and may prove useless or impossible to carry out.

13.2 Instrumentation in Embankments on Soft Soils

The purpose of instrumentation in embankments (Plate 13-2) constructed on soft, compressible soils is to measure one or more of the following phenomena:

— Settlements
— The pore pressures and the changes in pore pressure with time under the embankment, in order to find out both the evolution of the consolidation in the natural ground, and the effect of pore pressures on the safety factor
— The horizontal displacements of the natural ground
— The vertical stresses exerted by the embankment on the natural ground and their distribution with depth
— The changes in the strength of the natural ground under the embankment load.

Some comments follow on each of these measurements.

Plate 13-2 View of an instrumented test embankment

13.2.1 Measurement of Settlements

.1 Leveling

The most obvious and simple method of measuring the settlements of an embankment is to place a series of fixed points distributed over its surface and level them periodically. When the embankment is paved with asphalt or concrete a series of nails may be used as the points to be leveled. In structures where the surface is made of earth, it is advisable to sink a small concrete block into the soil with a tube or other indicator in its center, which stands out slightly above the ground.

The most critical of all leveling operations is selecting the benchmark, which is not influenced by the movements of the embankment. Often this point must be situated at very great distances from the embankment that is to be measured, because the surface of the soft compressible soils often suffer important superficial movements at considerable distance from the embankment. Such movements are caused by pumping for the purpose of agricultural irrigation or water supply; so the benchmark must be placed beyond the zone of influence of these movements. Hills and rock outcrops in the neighborhood of the embankment that is to be measured may provide suitable reference points. At other locations, immobile structures can be used such as those supported on point-bearing piles that go below the compressible strata into truly firm foundations. At other times the benchmark may be obtained by driving a tube or heavy rod through the soft soils until it comes to rest on rocky or firm layers. In this case the tube is provided with an outer flexible jacket to absorb any negative skin friction that may be produced [2].

Once the benchmark has been selected, another one should be established directly on the surface of the natural ground, at a distance of about 100 m (300 ft) from the embankment that is to be measured. This second benchmark, which probably will move, provides a temporary base for leveling points situated on the embankment. It also makes it possible to detect any movements that are experienced by the foundation ground due to causes other than the embankment load. Constant comparison of the position of the secondary benchmark in relation to the fixed one will provide the correction that may be necessary in the vertical movements of the points on the embankment, on account of any local movement of the surface of the foundation ground.

Precision leveling techniques are necessary, using devices capable of detecting differences of elevation one kilometer apart to the nearest millimeter.

The most appropriate technique for establishing benchmarks on an embankment is a grid like one shown in Fig. 13-1. A wide spacing is used where only the surface profiles at the centerline, and pavement edges are required. If contours of settlement, such as those shown in Fig. 13-1, are needed, a closer spacing is required, depending on the compressible soil. The contours of Fig. 13-1 correspond to a test embankment constructed by the Mexican Ministry of Public Works for obtaining information about the behavior of a projected highway across Texcoco Lake near Mexico City. The characteristics of the soil were presented in Fig. 6-60 earlier.

The instrument of the test embankment was planned before construction. The benchmarks on the embankment are located between the base of the embankment and the natural ground, on 40 × 40 cm (16 × 16 in) concrete slabs in the center of which a tube was placed. The tube was lengthened as the height of the

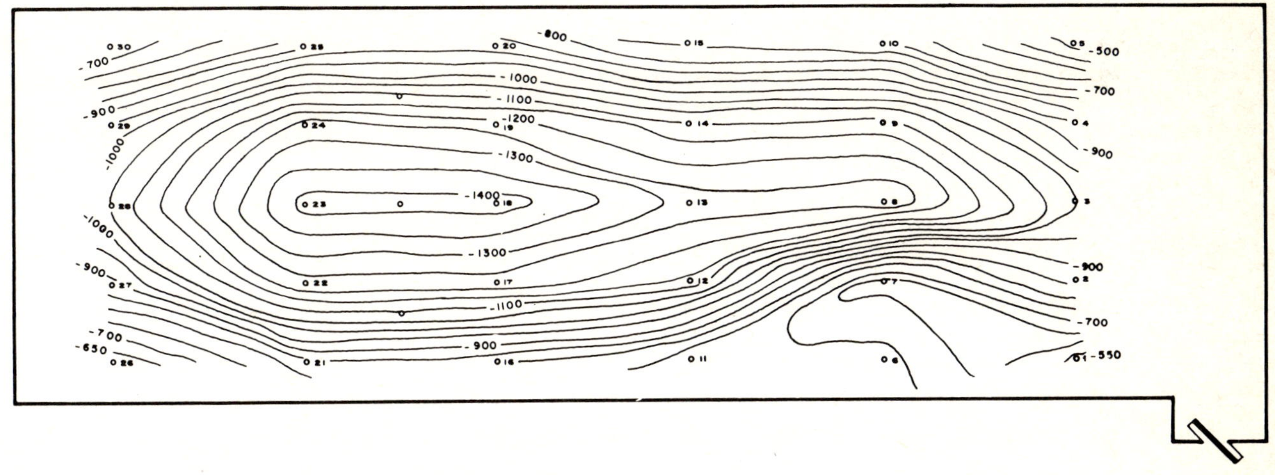

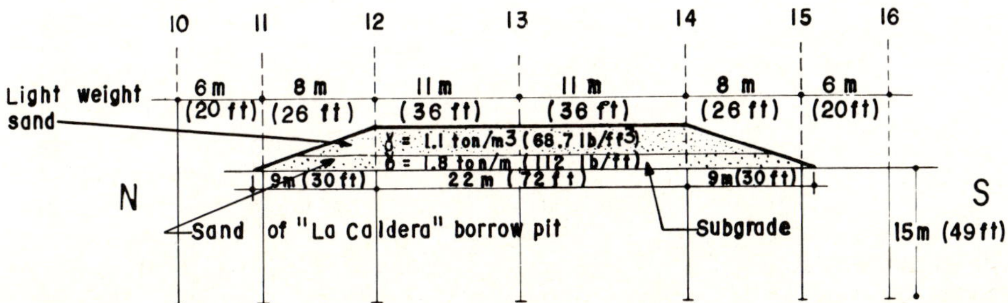

Fig. 13-1 Equal relative settlement curves in an embankment on soft soil. Test embankment on Texcoco Lake

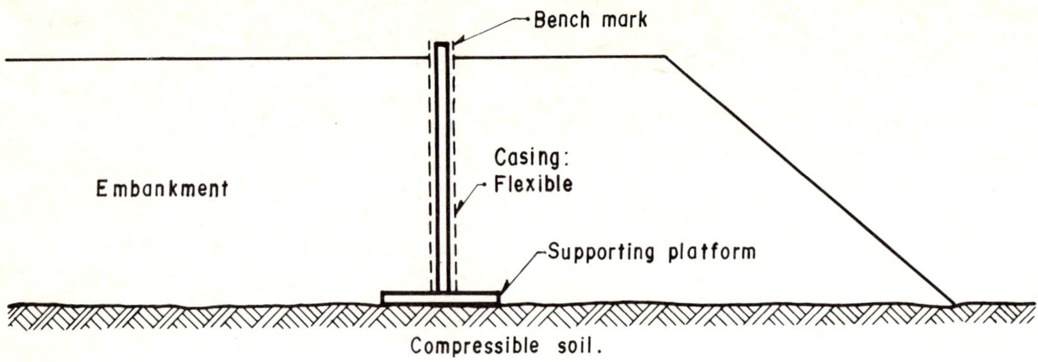

Fig. 13-2 A bench mark for controlling settlements by leveling

embankment increased. This tube can be braced for protection during the deformation processes. Fig. 13-2 is a schematic drawing of the benchmark that was used in this case.

The set-up in Fig. 13-2 has some advantages. For example, subsequent leveling will give a very precise image of the foundation settlement profile of the embankment, which is useful. The movements of the benchmarks closer to the surface of the embankment include the settlement within the embankment. If there are significant movements within the embankment and the embankment is higher than 4 or 5 m (13 to 16 ft), negative skin friction will develop in the rod or tube of the device and the concrete slab may settle more than the foundation on account of this surcharge. As a result the settlements measured may be greater than those that have actually taken place. Under such circumstances, the rod of the benchmark should be braced, and protected within an outer corrugated casing that can compress. The bracing can be rubber disks within the casing.

.2 The Settlement Measuring Torpedo

In this measurement system, a special tube is installed in a previously drilled hole. This is made up of sections joined together by external sliding couplings which permit telescopic play, so that the sections can gradually become closer to one another as the surrounding ground consolidates [3]. Thus, there is a variation in the apparent length of the system of tubes. These are installed in such a way as to include the entire thickness where settlement is to be measured. The distances between the small steps that are formed on the inside at the junction between one section of tubing and the corresponding coupling, will decrease as the soil settles.

A measuring instrument called a torpedo (Fig. 13-3) is introduced into the tube. It is provided with small extensible feet, which indicate the point an abrupt change occurs in the diameter of the tube whenever one of the small steps between tube and coupling is reached. In this way it is possible to measure from the surface the position of the steps in relation to their initial position, and deduce the settlements that have occurred. The tube sections are usually 1.5 to 3 m (5 to 10 ft) long.

This device has the advantage that is not only makes it possible to measure the settlement close the surface, but also at different depths within the soil. Settlement profiles are obtained like those that were shown in Fig. 6-44 earlier, (which correspond to the same Mexican test embankments already mentioned). By repeating readings periodically, the evolution of the settlements with time can be obtained at the different depths.

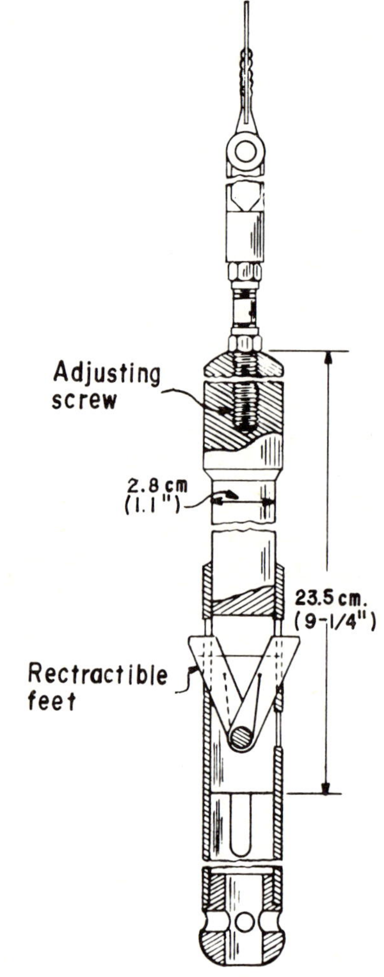

Fig. 13-3 Settlement torpedo [1]

.3 Pressure Cells

French technology [4] had developed a settlement measuring device which is shown schematically in Fig. 13-4. A plastic cell 9.5 cm (3.7 in) thick and 17 cm (6.7 in) in diameter is placed under the embankment, at the point where settlements are to be measured. The cell is partially filled with a liquid (generally water). At a distance that is beyond the influence of the settlements of the embankment, a fixed base is placed, and on it is installed a control panel with a device for applying pressure with nitrogen or carbon dioxide and a mercury gauge which controls the pressure of the liquid within the cell, transmitted

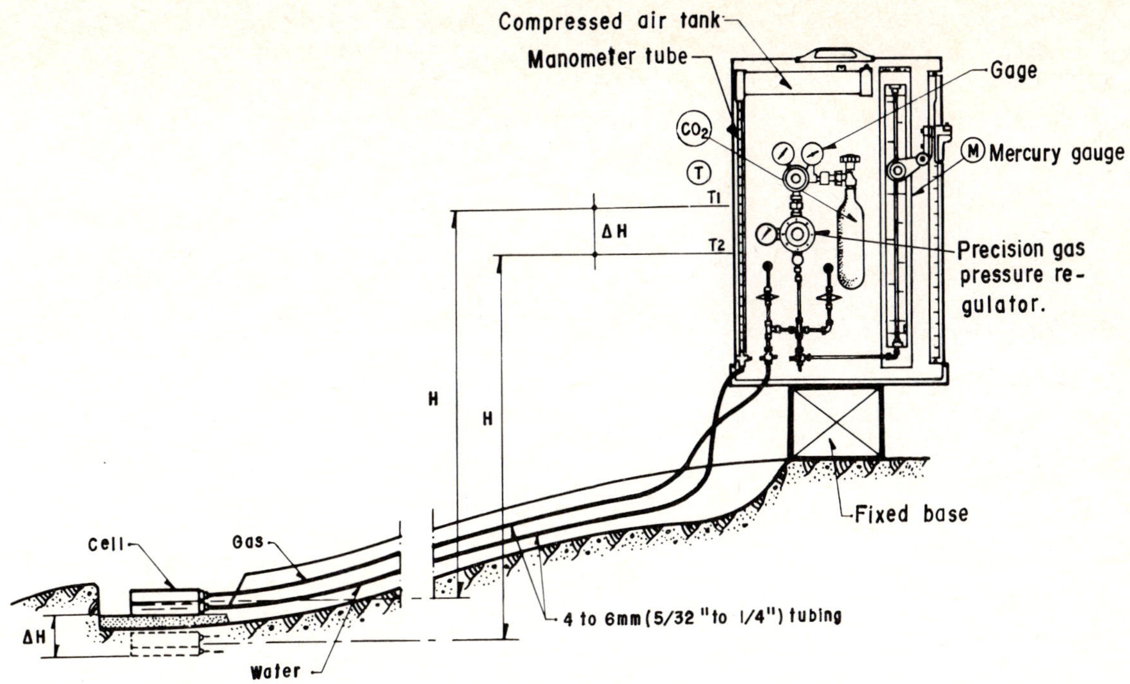

Fig. 13-4 French settlement gauge [4]

to it by means of the gas. Another tube runs from the cell to the control panel and is installed to one side of a vertical gauge, T, so that any pressure applied by the gas is transferred to the liquid in the cell and from there to the connecting tube between the cell and the vertical tube T, up to a certain height on the gauge. Under the circumstances, it is assumed that a pressure, p, is applied to the liquid in the cell, which rises to height T_1 on the gauge, T. After a time period has elapsed within the measurement program, the same pressure, p, will be applied to the liquid in the cell, but this cell will by now have settled by $\triangle H$. Accordingly, the liquid on gauge T will rise only as far as T_2, the difference between T_1 and T_2 being equivalent to the settlement of the cell.

As it can be seen, the device is ingenious and easy to handle, and in no way hinders construction equipment. The cells can be installed at any depth, so that settlement can be measured at any point. The precision of the device is to within about 0.5 cm (0.2 in), unless more precise reading of gauge T is possible.

The disadvantages of the device are the time one must wait for the liquid pressure to stabilize, especially when the control panel is a long way from the cell (20 min for 100 m) (328 ft), the isolation required by the cells when water is present, especially salt water, and the possible difficulties in finding a place to install the fixed base at a reasonable distance. The water in the cell, tube and manometer must be kept at uniform temperature.

Reference [5] mentions a similar device used in California, also based on the principle of establishing communication between a liquid in a receptacle inside the ground at the place where it is wished to measure settlement, and an arm of the tube, placed on a control panel outside the influence of the movements of the embankment. This device works on the principle of communicating vessels and is not activated by gas pressure except atmospheric. Therefore its use should be regarded as more restricted, although it may lead to satisfactory results in many cases.

A similar device which was originally proposed by TERZAGHI [6] has been used particularly in building settlements.

.4 Selection of Benchmarks and Their Number

Settlements in embankments on soft soils are generally measured under one of two conditions: either in an existing embankment with a view to finding out its behavior, or in a short test embankment or section, for the purpose of obtaining data for designing a far larger embankment.

In either case it is advisable to establish the points at which the settlements are to be studied in instrumented sections. The sections are usually more numerous in test embankments than in performance control problems. In either case the number of points depends on the importance of the project, the heterogeneity of the formations that are compressed and the degree of difficulty of the problem, from the point of view of soil mechanics.

In heterogeneous zones, where important differential settlements are anticipated, it will be wise to measure total settlement in sections not more than 50 m (164 ft) apart, but in the case of very homogeneous formations, the sections may be as far apart as 200 m (656 ft) or more. In test embankments, sections are usually selected at regular intervals throughout the embankment, often every 50 m (164 ft).

In discontinuous zones, it is helpful to install an instrumented section. Examples of these would be the access embankments to a bridge or overpass supported on point-bearing piles, or the abrupt termination of the compressible zone, or special zones such as abandoned river beds.

The settlement points should cover the entire cross-section of the embankment, to aid interpretation. For evaluating existing roads, each section usually has five settlement points, one on the axis, one on each shoulder and one at the toe of each embankment. In test embankments, the number of settlement points is far greater (see, for example, Fig. 13-1).

By periodical measurements, the evolution of settlements with time is determined. Readings need not be made at equal time intervals; they are more frequent at the beginning of the loading and become gradually more widely spaced at time goes by. One or two readings are usually taken daily during the construction period, to discover any instantaneous deformations as well as the beginning of the consolidation process. Subsequently, readings can be taken weekly during the first three months after embankment construction. The frequency is reduced until measurements are taken monthly during the first three years, and twice-yearly from then on. Naturally these are not rigid rules for the frequency of measurements; they should be adapted to suit each individual case.

13.2.2 Measurement of Lateral Movements in the Foundation Ground

These are usually measured for different reasons. First, part of settlements is due to lateral displacements in the compressible strata. This part is not taken into account by TERZAGHI'S consolidation theory, which only considers settlements by compressibility (by change of volume, but not by changes in shape due to the action of shear stresses). Second, failures in embankments on soft soils are preceded by continuing lateral displacements of the foundation ground below and adjacent to them. Thus, the lateral movement can be a precursor of embankment foundation failure.

.1 Surface Control

When embankments are built on soft soils, it is also appropriate to measure the horizontal movements of the ground, both on the surface and at the depths affected. There are no appreciable differences from the methodology for measuring settlements. Here too, one of the most important tasks is selecting the reference points or lines in zones not affected by the movements. Often in very extensive movements [7] the boundaries of the moving zone are not well-defined; therefore the benchmarks have to be selected conservatively. Measuring horizontal movements on the surface of the ground is simplified when it is wished to define only differential or relative movements between different points without having to establish their absolute movements.

.2 Inclinometers

Often it is not enough to establish horizontal displacements on only the ground surface. It is also necessary to determine how the soft foundation ground moves at different depths in the soil.

Almost all the instruments developed for these purposes are based on the same principle. A fairly flexible tube is introduced in the ground. Its initial shape (usually vertical) is distorted as horizontal displacements occur so that the distorted tube provides an image of the soil displacements that may have taken place. This image can be found by introducing an instrument that is sensitive to tilting into the tube. A. CASAGRANDE [8] describes one of the first large-scale studies of this type. Tubes with diameters of 5 cm (2 in) were placed in numerous wells distributed throughout the foundation ground at the foot of a large embankment that was being built across the Great Salt Lake. The purpose was to define the position of any slip surface that might eventually form. This was achieved by recovering the tubes after the sliding and observing the permanent deformation they had undergone.

In four test embankments that were built by the Mexican Ministry of Public Works on Texcoco Lake to determine the behavior of the foundation ground under the heavy load transmitted by highway embankments [9], some simple 5 and 7 cm (2 and 2-3/4 in) diameter tubes were used. A rigid bar of a suitable length was introduced into the tube; the depth at which deformation of the tube prevented the bar from passing is the depth of the shear surface. These very simple devices, often provide very useful information at a relatively low cost thus avoiding the use of more complicated instruments, which are costly, perhaps imported and not always available at the time they are needed.

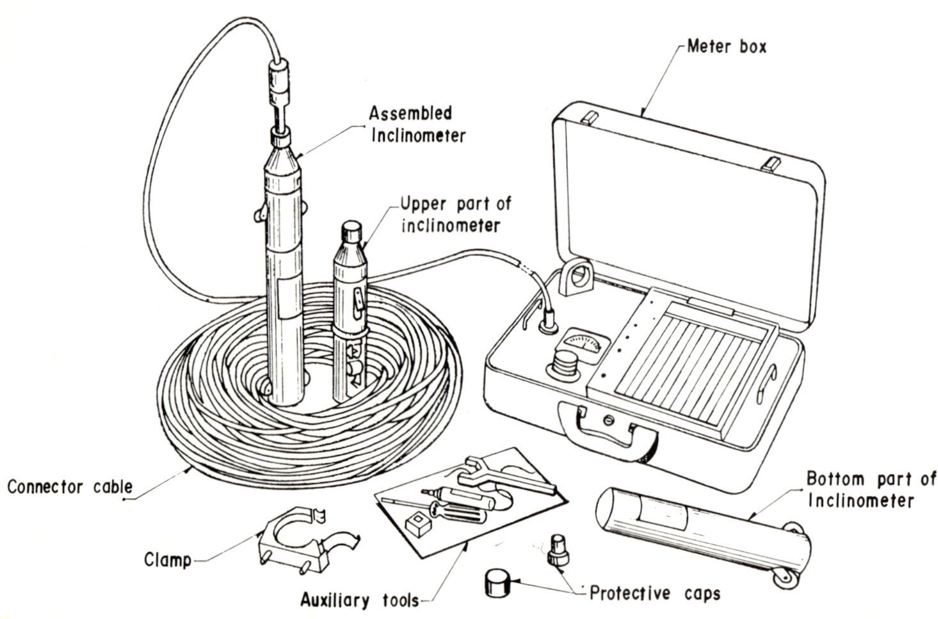

Fig. 13-5 Slope indicator instrument

Reference [1] mentions different types of rather more elaborate inclinometers (Plantema, WIEGMANN, a model from the Swedish Geotechnical Institute). The currently most widely used is that developed by WILSON [1,10] and later modified by PARSONS and WILSON in 1956. It is a precise, compact and lightweight device for measuring earth movements to a depth of 170 m (558 ft). In Fig. 13-5 the complete inclinometer is shown.

The apparatus consists of a sensing unit, a box with the necessary electrical controls and connector cable. A tube is installed in the ground, grooved on two planes at right angles to one another. The measuring device enters the tube guided on little wheels which run along the opposite grooves. It can detect any tilting suffered by the tube, from its originally installed position (usually as near vertical as possible). The tubes have a diameter of 8.1 cm (3.2 in) and are 0.22 cm (0.09 in) thick, in sections 1.5 or 3 m (5 or 10 ft) long. Couplings for joining tube sections are usually 15 cm (6 in) long. This tubing can be the same as that used in the settlement measuring torpedo referred to previously (which is also a design by S. D. WILSON). The tube for use with the torpedo (often one well serves both purposes) should be joined with 30 cm (12 in) telescopic couplings to permit action of the inclinometer (Plates 13-3, 13-4). The sensing unit (Fig. 13-6) has an internal circuit which is a WHEATSTONE bridge

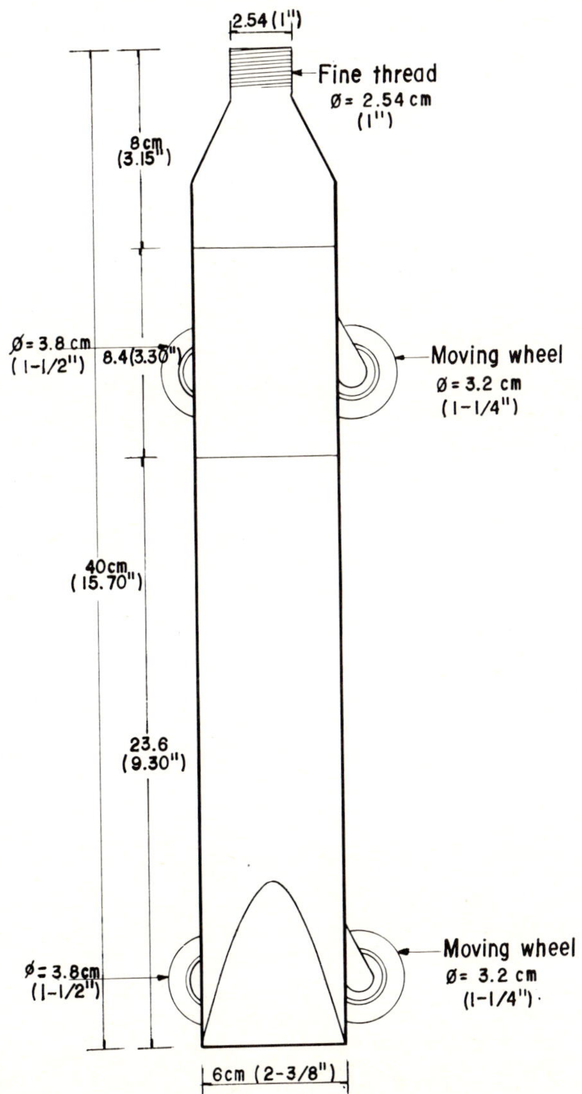

Fig. 13-6 Sketch showing assembled inclinometer of the WILSON type

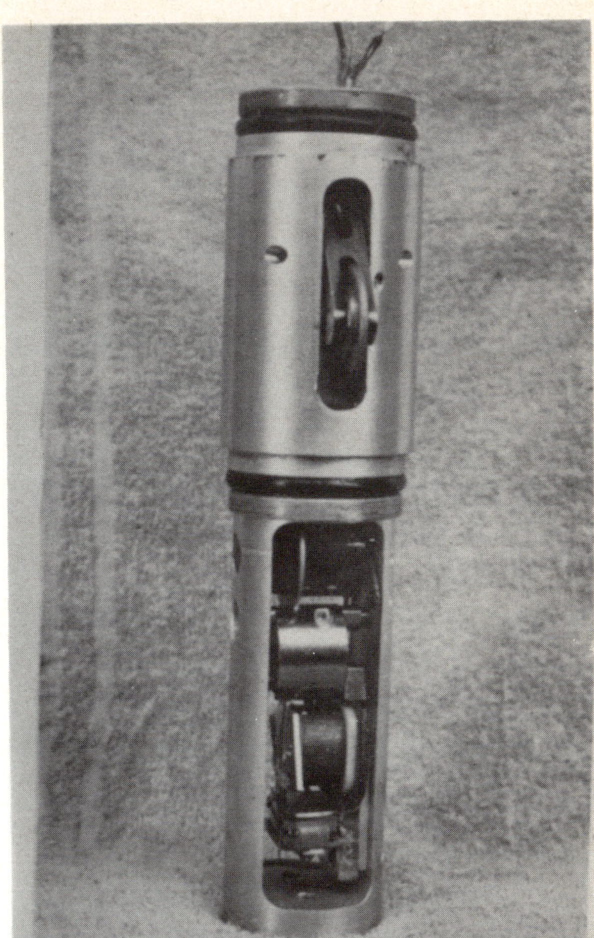

Plate 13-3 Wilson type inclinometer

Plate 13-4 Taking readings with an inclinometer

operated by a calibrated pendulum. When the inclinometer is vertical, the pendulum touches the center of a calibrated resistance, dividing it into two, the two arms or half of a WHEATSTONE bridge. The other half, plus a precision potentiometer, resistances and the necessary connections, are installed in the control box. The instrument is battery run. When the sensing unit is tilted by the tilted tube in which it is introduced, the pendulum remains vertical. The calibrated resistance changes in proportion to the tilt, and is measured by the control unit. Figure 13-7 shows a drawing of the circuits that are used in the inclinometer and the control box, which are connected by means of a cable. Figure 13-8 is a schematic drawing of the measuring unit, cut so as to show the interior.

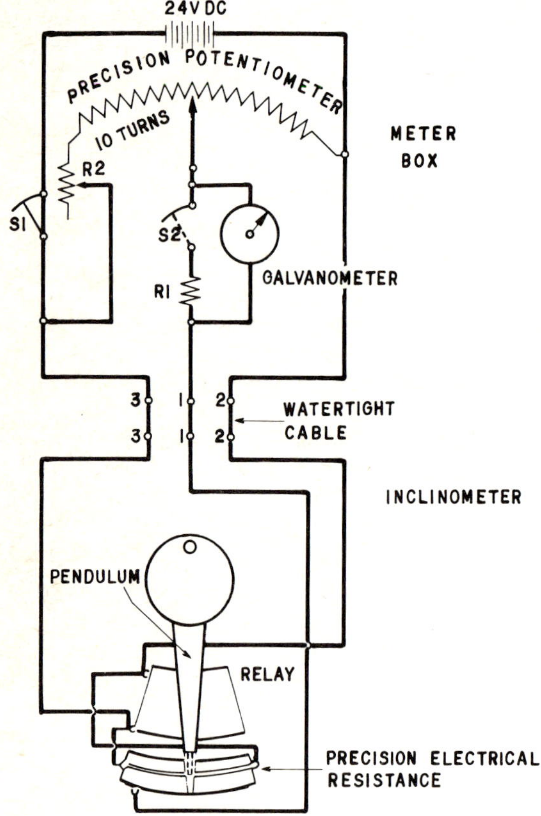

Fig. 13-7 Wiring diagram of the inclinometer connected with the meter box

The inclinometer is lowered down its tube with the aid of the equipment appearing in Fig. 13-9. As it is lowered, readings are obtained at regular intervals. By means of a previous simple calibration in the laboratory, the dials on the control box are made to give the tilt that corresponds to each electrical reading directly. Figure 13-10 shows how the special tube is deformed and the sensitive unit tilts when the system undergoes lateral displacements. The sensitivity of the instrument enables a minute of arc difference in tilt to be detected.

All inclinometer readings are usually taken in two orthogonal directions, using the groove layout that was mentioned. The purpose of this is to obtain the image of deformation in three dimensions. The planes determined by the opposed grooves should be oriented according to the estimated principal directions of the strain. Typically one set will be uphill-downhill, and the other set parallel to the slope.

The tube must be sufficiently flexible to follow the movements of thc ground closely and at the same time strong enough to

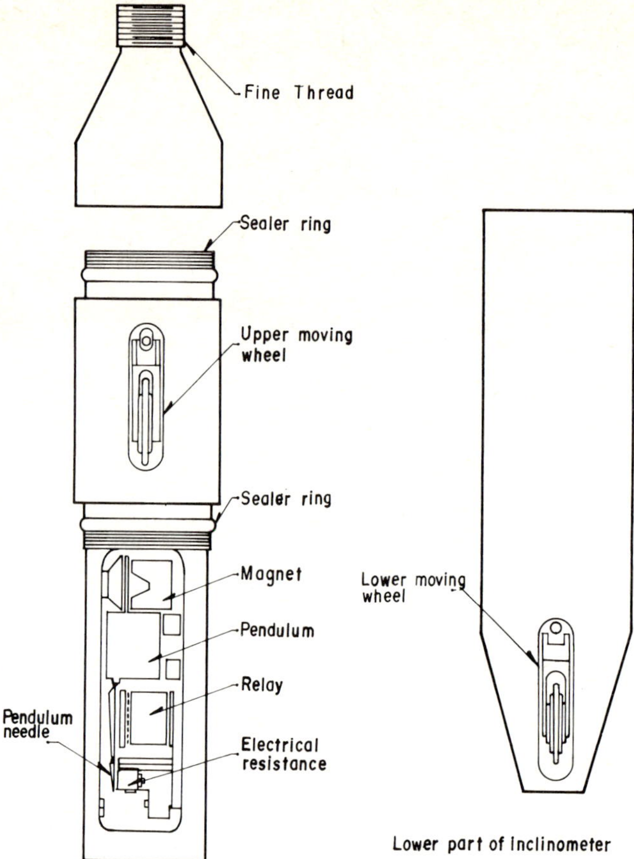

Fig. 13-8 Sensing unit of the WILSON inclinometer

tolerate installation maneuvers. The tube rigidity is a compromise in the design of the device, and some authors [4], have pointed out that the material used by WILSON in his commercial designs is too stiff. Mexican experience with regard to this is, however, satisfactory. By way of contrast, similar devices have been installed with square steel tubes.

Use of the apparatus is time consuming and costly. A computer is essential for processing and organizing the numerous measurements obtained. Excellent results can, however, be obtained. Figure 6-43, which was presented earlier, is an example of the information that can be obtained, which is both objective and clear.

The Swedish Geotechnical Institute has developed an apparatus with a pendulum similar to WILSON'S. However, the lower end of the pendulum, instead of modifying a resistance of an electrical circuit, is supported by an spring instrumented with systems of electrical strain gauges. When the pendulum is tilted, the spring bends changing the length of the metal filament of the electric strain gage. This alters its resistance, and the corresponding tilt is read by a control box at the ground surface in a manner similar to the WILSON inclinometer.

Geoconsult [20] has developed an inclinometer with a pendulum, which is based on the following principle (Fig. 13-11). The sensitive unit has two concentric cylinders, the inside one being able to turn within the outside one by an electrical motor which is controlled from the surface. The outside cylinder tilts following the deformations of a tube in the ground, similar to that employed for the other instruments. An electrodynamic balance containing a device for measuring intensities of current maintains the pendulum on the axis of the sensitive unit. The lower end of the pendulum is attached to a spring, so that the curvature of the

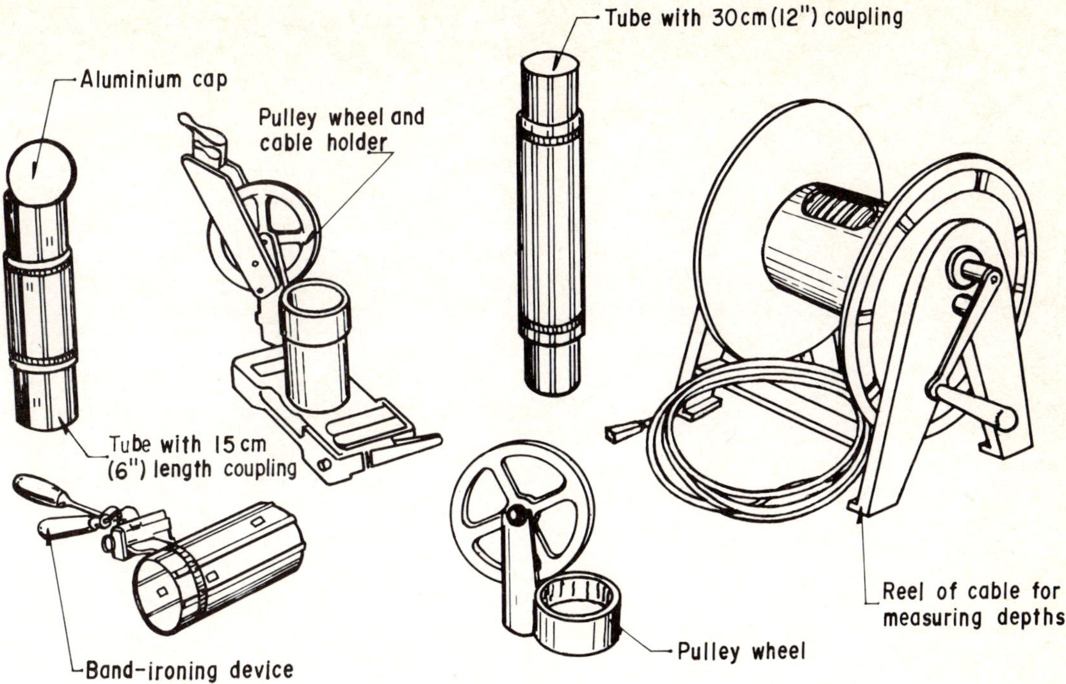

Fig. 13-9 Auxiliary equipment for lowering the inclinometer (Aluminium casing and assembly tools)

spring varies according to the amount the apparatus is tilted. The spring, balance and pendulum are parts of an electrical circuit, similar to that of the WILSON inclinometer, from which readings can be taken on the surface. The change in the curvature of the spring causes a change in the intensity of the circulating current, which is what is measured in this case. Furthermore, by rotating the inside cylinder of the sensitive unit the position can be found where the pendulum produces the minimum deviation from its original undeformed position. This makes it possible to measure inclinations or tilts in different directions with one position of the device. It also helps to minimize any minor warping suffered by the apparatus during tilting. This position of the pendulum at minimum tilt can be found, because to it corresponds the maximum intensity of current on the circuit in relation to any other position. By means of a simple previous calibration in the laboratory, it is possible to find the angle of inclination corresponding to each of the intensities of current measured.

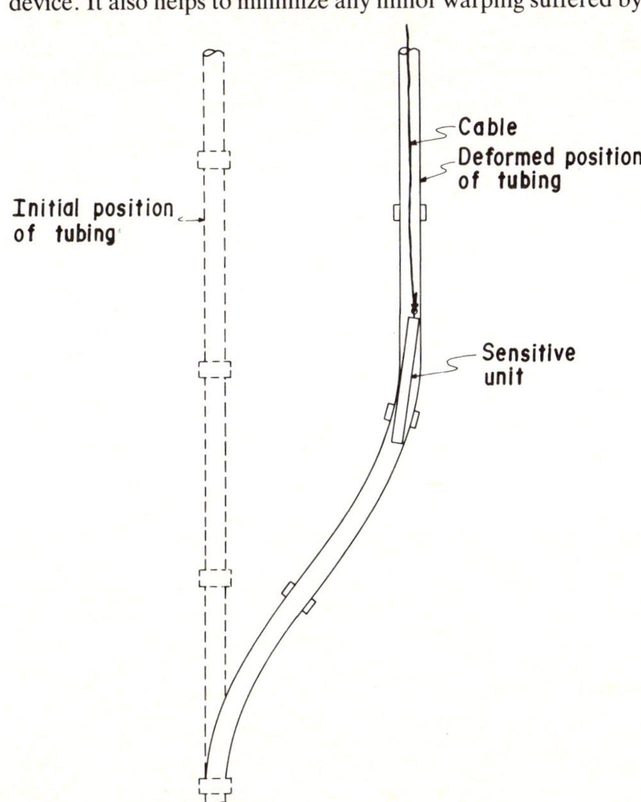

Fig. 13-10 Deformation of an inclinometer

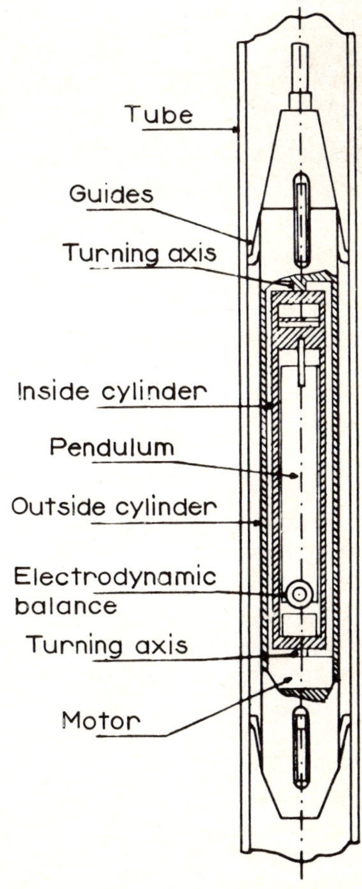

Fig. 13-11 Geoconsult inclinometer [20]

The Telemac inclinometer [20] is shown in Fig. 13-12. The sensitive unit is also a pendulum consisting of a flexible piece of metal from the lower part of which a weight is hung. On the upper part, the piece of flexible metal is firmly attached to the head of the sensitive unit. On this piece of flexible metal four vibrating wires are placed lengthwise at right angles on perpendicular planes. These are small metal wires that can be excited to vibrate by an electromagnet. If the tension of the wire varies from tilting of the pendulum, the frequency of the vibration changes. On a control panel on the surface of the ground, there is another identical wire, the lower end of which can be slightly moved with a micrometric screw, changing its tension. The circuit containing the control wire is connected with those containing each pair of opposed wires, so that the vibration frequencies of the wires can be compared.

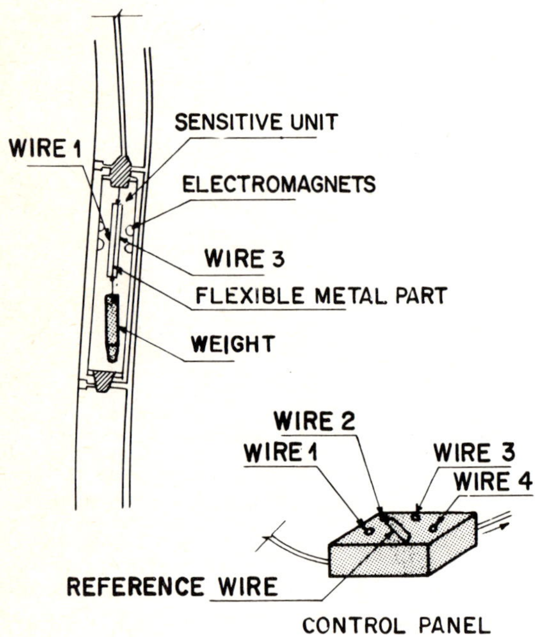

Fig. 13-12 Inclinometer with vibrating chords (Telemac) [20]

When the sensitive unit is tilted, the metal part containing the wires bends, so that the tension changes. A simple previous calibration in the laboratory makes it possible to find out the tilt corresponding to each position. In order to do this, the vibration frequency of the control wire must be equalled, varying its tension by a controllable amount by adjusting the micrometric screw.

By using the four chords, measurements can be taken on two perpendicular planes simultaneously. The two opposed wires on each plane (one with increasing tension and the other with decreasing tension) has the advantage that temperature effects are corrected automatically.

Installation of inclinometers [11] requires exacting care. First, it is essential to measure any drifting of the tube entrance, if it is free. If possible the tube should extend into a firm stratum, that can be regarded as immobile. In this case, the lower end of the tube is fixed and the position of the entrance can be by the displacement polygon found from integrating the tilt from the bottom up. But if the entire tube is embedded in soft deformable material, the entrance position will have to be measured by well controlled ground surveys. Then the tilt is integrated from the top down to find the position of its end. This surveying control must be very precisely done, comparable in precision to the reading of the tiltmeter or inclinometer apparatus, otherwise some of the significance of the readings may be lost. The control is established by triangulation from sufficiently distant points. The places in soft ground where such measurements are of interest are frequently subject to some type of local ground movements (for example, by pumping from deep water-bearing strata for irrigation). In such cases, the reference points distant from the control triangulation polygon can be located where they will take part in local movement, but sufficiently far so as not to be affected by the movement of the embankment. What it is of interest to measure is the lateral movement of the foundation ground, isolated from the local component.

At depths of about 20 m (66 ft) it is possible to limit errors to 2 mm (0.08 in) in the position of the lower end in inclinometers, whose entrance position is measured by ground surface surveys.

There are standards and precautions regarding the construction of the bore holes within which inclinometer tubes are placed in order to ensure contact between the tube and the surrounding ground. This contact is achieved by filling the space that may be left between the tube and the walls of the well with fine sand, vibrated in place to fill any voids.

A small amount of portland cement can be added to the sand to minimize any subsequent sand erosion or settlement. There should not be enough to make the sand rigid. Typically 2 to 4% will be sufficient if it is mixed well.

The tube should be installed in maximum movement zones. If several are installed to form a grid, unit strains can be measured and strain maps plotted, the closer the grid, the more accurate the maps. Control of the initial verticality of the tubes is important. Deviations of more than one or two degrees may restrict the use of inclinometers, because each different device has a different limit of tilt that can be measured.

Although the tube initially should be nearly vertical, the hole inclination or tilt is measured immediately after installation. The tube movement corresponds to the difference between the initial readings and each set of subsequent readings. When the tubes are longer than about 30 m (100 ft), twist in the tube can cause errors in the tilt direction. Special instruments are available to measure tube twist so the error can be corrected.

A satisfactory knowledge of the natural ground and its stratigraphy is essential for the interpretation of results and for establishing all installation details.

Inclinometer tubes can alter the hydraulic situation of the soil by providing a means of intercommunication between the water in several strata and levels, affecting the structure and especially the piezometers in the vicinity. When several watertables are suspected, the filling between the tube and the hole should be a mixture of sand, bentonite and cement to minimize water circulation. About 10% bentonite and 5% cement are typical. The mix should be impervious and firm enough to support the tube, but weak enough to move with the surrounding ground.

The inclinometer tube must be installed immediately after the well is bored. It can be done by introducing 4 coupled sections in one operation, using an appropriate tripod. The tube must be introduced in such a way that orientation of the grooves is as planned; minor adjustments can be made once it has been installed by making it turn slightly inside the well.

Reading frequency is very variable and depends on each particular project. Readings should be more frequent in the early stages of the measuring program, becoming less frequent as time goes by. A rigorous statistical control is maintained with the aid of specialized staff. At least two readings are taken at each point for each orientation of the apparatus for verification. The computer program that is developed to calculate displacements should not admit values with a deviation of more than 5%.

.3 Electrical Tape Detectors

There is a trend to measure lateral displacements with strain gauges that are increasingly smaller, slimmer and installed in tubes with a smaller diameter. This has led to the invention of electrical detector tape for locating shear surfaces along which the mass of soil is sliding.

The device consists of a plastic tape provided throughout with two conductor bands that are intermittently connected to known electrical resistances. The overall impression is that of a staircase fastened onto the strip of plastic. The entire device is covered with resins or similar waterproofing materials. The upper and lower ends of the tape are connected by cables to the control box at the ground surface. Readings are made of the total electrical resistance of the circuit. When shear occurs, the tape breaks and there is also, of course, a drastic change in the reading for the total resistance of the circuit. It is possible to discover how many of the resistances remain intact in the upper section of the tape, and how many in the lower section, thus locating the slip surface.

The manufacturers of these instruments indicate that they are also useful for detecting lateral displacements prior to any slide. For this they suggest introducing into the ground a plastic tube with four tapes placed at right angles (Fig. 13-13). When this tube deforms, the tapes will break in the tilting zones. The analysis of the data obtained will permit the reconstruction of the deformed tube. The authors lack experience in the use of detectors for these purposes. However, handling the instrument appears to be too delicate, and it is doubtful whether it is possible to measure lateral displacements, although the apparatus is precise enough to allow any break in the tape to be located to the nearest 15 or 20 cm (6 or 8 in) of depth.

The tapes are placed inside previously bored wells and it is recommended that they be embedded in weak concrete or weak cement grout.

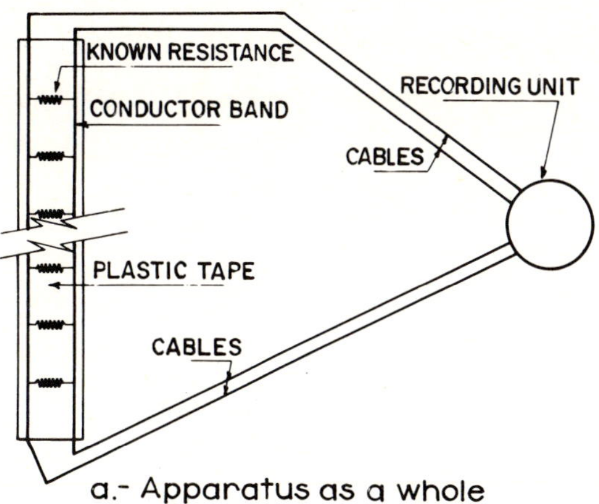

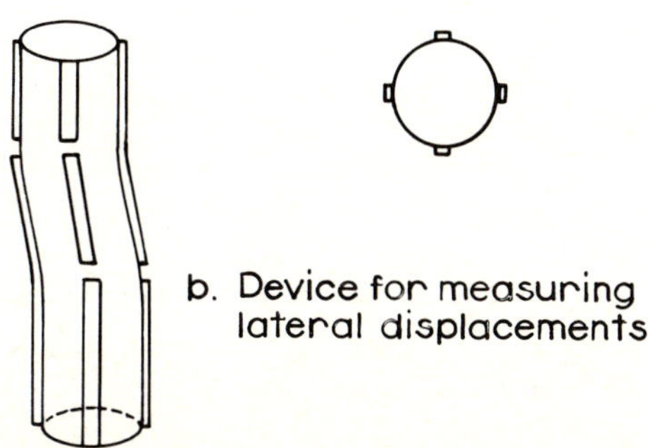

Fig. 13-13 Electrical tape detector

13.2.3 Measurement of Pore Pressure

In the construction of embankments on soft soils, it is desirable to measure the changes of the pore pressures in the subsoil water that exceed the hydrostatic pressure. This measurement has three objectives:

— To discover the hydraulic conditions inside the strata that make up the subsoil
— To indicate the changes in the degree of consolidation during the lifetime of the embankment. In Chapter 1 it was seen how the process of consolidation involves a transfer of pressure from the water saturating the soil to the solid structure of the soil. Initially, the entire load of the embankment will be supported by the water. Thus the neutral pressure immediately after placing the embankment equals the hydrostatic pressure plus that produced by the embankment weight. As the consolidation process progresses with the passage of time the excess pressure acquired by the water diminishes, with a corresponding increase in the effective pressure. Knowledge of the pressure in the water during the process, compared to the hydrostatic enables the stage of the consolidation to be established
— To determine the effective stress and thus the strength of the foundation ground underneath the embankment as consolidation progresses so as to determine the change of the safety factor (Chapter 6, Fig. 6-40)
— To check the performance of subdrainage structures or measurement system installed to control the flow of water towards excavations
— To measure pore pressures in slopes that could produce local sliding, before they can progress enough to cause a serious incident.

The devices for measuring pore pressure in the subsoil are known as piezometers. Under purely static conditions the pressure head at a point in the foundation ground is controlled by the position of the water-table. However, under natural circumstances this condition is not too frequent, except in very homogeneous, level soil masses. Furthermore, the load of any engineering structure changes the stress states, which usually is reflected in changes in the pore water pressure. The hydrostatic relations no longer represent the pore water pressure, and that part of soil strength that is related to pore water pressure.

Observations of the level of the water-table in small wells are often hard to interpret. Even under hydrostatic conditions, a coating of algae or sludge on the sounding walls formed during drilling, or by contact with air, is often enough to mask the presence of the water-table. An exploration well will gain water from all the strata that are intercepted in which the water-table is higher than that corresponding to the bottom of the well, and will lose water to all those strata where the water-table is lower. The significance of the height that is reached by the water in the well, is thus lost. If the water pressure in the soil adjacent to the observation well is not hydrostatic, the level of the water in the well will not always indicate the pore pressure changes with sufficient precision. For example in fine soils, which have low permeabilities, a long time will elapse before the volume of water necessary to raise (or lower) the height of water in the well can seep through the soil into the well to enable it to reflect a change in pore pressure.

In view of the above considerations, simple observation of the water surfaces in exploration wells is often not sufficient to enable conclusions to be drawn about pore pressure states, except in homogeneous previous soils. Such wells are inappropriate when the conditions of the subsoil are complex and the soils have low permeability such as when consolidation is taking place in a silt or clay soil.

A piezometer (Plates 13-5, 13-6) is a device which measures the pressure head of the water at the point where it is installed. All piezometers work by the principle of balancing the pore water pressure with counterpressure which is measured without much change in water volume in the system. There are various different types of piezometers, according to the type of counterpressure that is used.

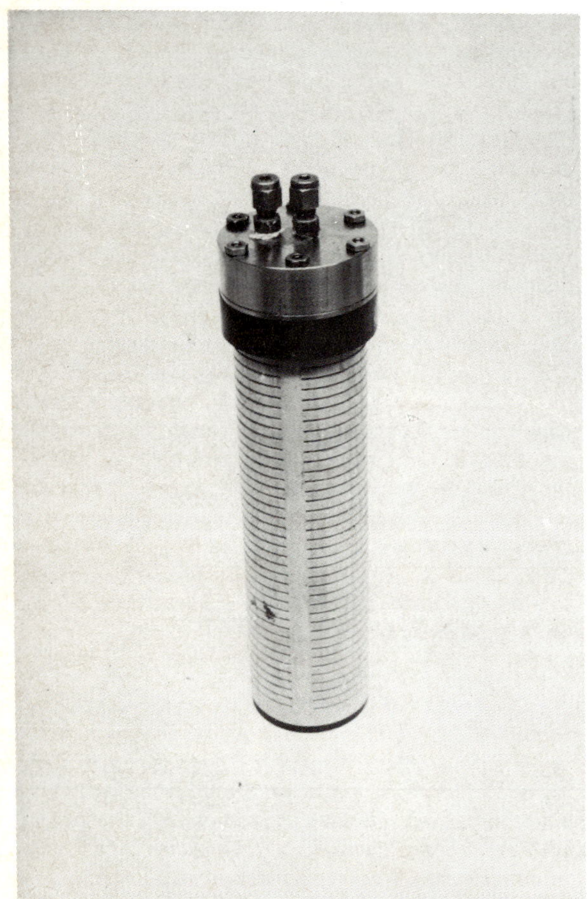

Plate 13-5 Pneumatic piezometer

Plate 13-6 Pneumatic piezometer

Figure 13-14 shows an early type of piezometer, known as the open piezometer, designed by A. CASAGRANDE [12,13]. In this apparatus, water enters the sensitive unit through the porous cell, filling it and establishing inside it the pressure prevailing in the subsoil. The water will rise through a small tube until it reaches a height where it produces a hydrostatic counterpressure which balances the pressure in the sensitive unit. The level of the water inside the outlet tube can be measured by electrical methods.

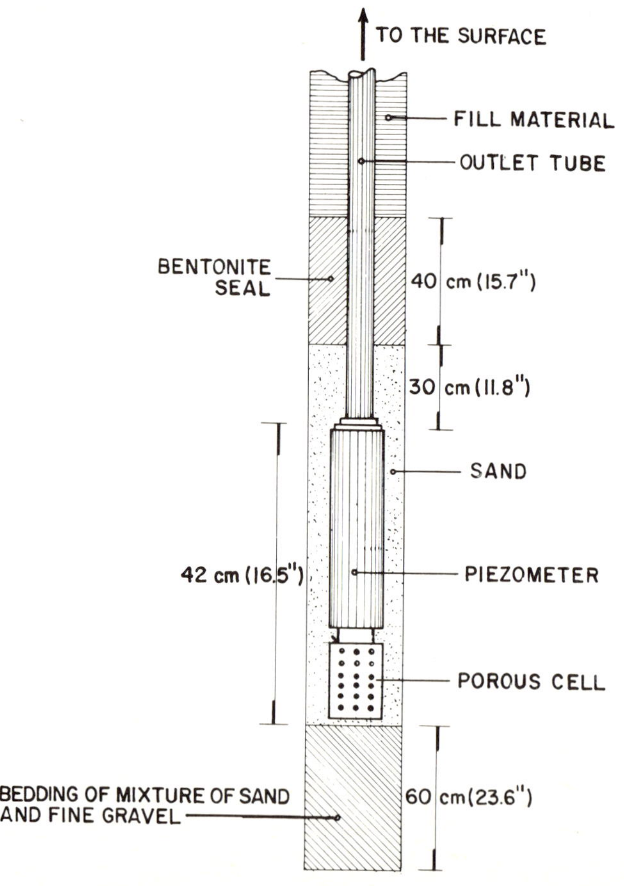

Fig 13-14 Open CASAGRANDE type piezometer

A probe that fits inside the outlet tube is fitted with two electrodes at the lower end, spaced 1 or 2 mm (0.04 or 0.08 in) apart. This is connected by an electric cable to an ohmmeter at the ground surface. When the electrodes touch the water the ohmmeter immediately jumps indicating a much lower resistance. The depth to the probe tip reflects the height of water in the outlet tube. A little grease between the electrodes prevents a capillary film between them and an uncertain meter change.

When pressure conditions are such that water runs out of the end of the piezometer tube, on the surface, pressures must be measured with a BOURDON gauge fitted over that end, following the sequence indicated in Ref. 14. Alternatively the tube can be extended upward to establish a hydrostatic pressure.

HVORSLEV [15] has indicated several serious disadvantages of the open piezometer. The most important one is the time that must elapse between any change in the pore pressure in the ground water and the response of the apparatus owing to the volume of water has to flow from the soil into the sensitive unit through the porous cell to establish internal equilibrium, and the corresponding change in the height of water in the outlet tube. The lower the soil permeability the slower the flow. Yet such low permeability soils are those in which piezometers are often installed. The time lag depends on the diameter of the outlet tube, the larger the tube, the greater the water volume and the greater the lag. Therefore it is usually thin, not larger than one or two centimeters (0.8 in). The dimensions and voids of the sensitive unit and the permeability of the soil also control the flow and the time lag. The sand filter around the sensitive unit greatly increases the efficiency of the inflow or outflow of the water, and this is one of the main reasons for installing it. Reference [1] gives data by HVORSLEV about the response time of open piezometers of different types. It can be seen that the phenomenon is not academic; frequently the response lag varies between many days and several months. The response time is also greatly influenced by anisotropic permeability.

Any bubbles of gas that may become lodged in the system may have a variety of effects. When they become lodged inside the sensitive unit or in the contact zone between the apparatus and the soil that surrounds it, they reduce permeability, hindering flow and increasing the response time. The change in volume of the gases as the pressure varies generally increases the response time of the apparatus. For these reasons the use of metal tubes is not recommended, for in them electrolytic reactions can generate gas. Plastic tubes, of the Saran type or similar, are universally used. Gas bubbles in small outlet tubes can raise the water level and produce false readings.

In order to eliminate the foregoing disadvantages and speed up the instrument's response to changes in pressure in the subsoil water, some piezometers have been developed which function with far smaller internal flow requirements and correspondingly far smaller time lags.

Figure 13-15 shows a model developed by French technology [4]. The apparatus consists of a sensitive unit with porous walls, with two tubes. One of them, marked *1* in the figure, is used when the piezometer is left open, when the ground in which it is installed is sufficiently pervious. In this case, the piezometer functions as was described above, and the pressure can be satisfactorily read either by determining the height of the water by means of an ohmmeter, probe or else by attaching a pressure gauge to the end. Tube number *2* connects the sensitive unit to a device for use in those low permeability soils where time lags would be large. Once the apparatus is installed, water will fill the sensitive unit and tube *2* as far as a pressure compensator,

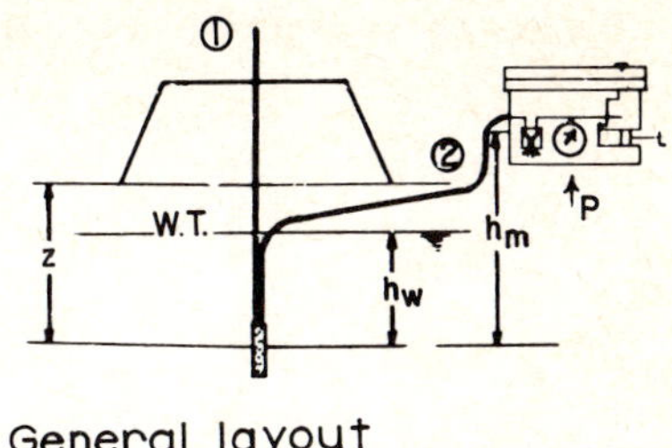

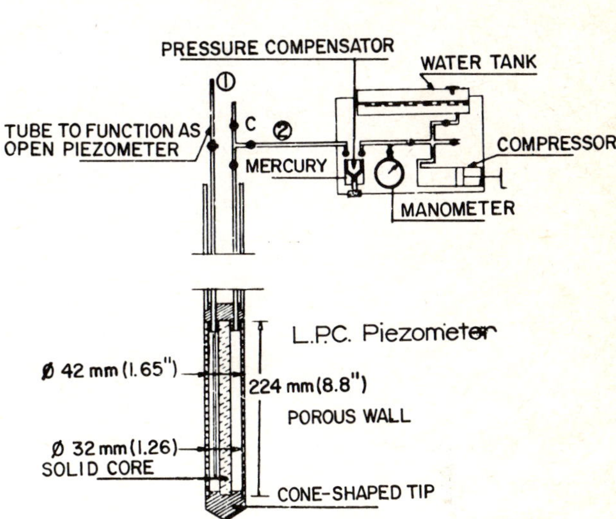

Fig. 13-15 L.P.C. piezometer [4]

which is a simple mercury gauge. Inside this gauge the same level of mercury is maintained in both legs of a *U*-tube with the aid of a compressor. The right-hand tube pressure balances that acting in the left-hand tube. Under these conditions, the apparatus gives its initial reading. The pressure of the compressor at that moment is read on the gauge. Any subsequent variation in subsoil pressure will produce unbalance in the mercury pressure compensator, which will be adjusted by means of the compressor, producing a new pressure which can be read on the gauge.

With reference to the upper part of the figure, if p is the gauge reading, h_m the difference between the heights in the sensitive unit and the mercury pressure compensator, and h_w the difference between the heights in the sensitive unit and the water-table, at a moment when the pressure compensator is in equilibrium, the total pressure of the water in the subsoil will be:

$$u = p + h_m \gamma_w \quad (13\text{-}1)$$

The hydrostatic pressure corresponding to the point where the sensitive unit is installed will be:

$$u_h = h_w \gamma_w \quad (13\text{-}2)$$

Consequently the pressure existing in the sensitive unit in excess of the hydrostatic value can be calculated with the following expression:

$$\Delta u = u - u_h = p + \gamma_w (h_m - h_w) \quad (13\text{-}3)$$

Experience of use of this apparatus shows that response times are less than three hours. It should be noted that the volume of water in the sensitive unit and in the outlet tube scarcely has to change for the instrument to respond. This converts the apparatus into a closed piezometer which operates with a near constant volume of water.

Apart from the advantage of its quick response, the apparatus is easy to handle, easy to install, precise and can be built with strong, cheap materials. On the other hand, it cannot measure very rapid changes in pressure, shorter than its response time, which makes it useless for measuring changes in pressure due to dynamic effects. It is also somewhat sensitive to changes in temperature.

The problem of delay in the response of piezometers due to the time required to mobilize the water which operates them is overcome by designing devices that operate with a nearly constant volume of water (closed piezometers). There are many types and designs. Reference [16] describes several of them, studying the response time in each case, [17] too, provides useful information about these problems.

Figure 13-16 shows a pneumatic piezometer which has frequently been used by Mexican and U.S. Engineers with success. The sensitive unit (Part *a* of the figure) has a porous cell, and porous stone into which the water on the outside penetrates, establishing its pressure state within both. As with all closed piezometer installations, the porous cell and porous stone are pre-saturated with deaerated water, because entrapped air will introduce a delay in the response of the device to water pressure change as well as measurement errors. The sensitive unit is installed in the same manner as in the open piezometer shown in Fig. 13-14.

The pressure exerted on the porous cell presses the Teflon membrane upwards. On the outside there is an air intake unit (Part *b* of the figure) with a tank which introduces compressed air to the sensitive unit through the plastic inlet tubing. The air intake device includes the tank (with its gauge) and a pressure regulator where this pressure is adjusted to a value close to that anticipated in the water that is to be measured. A precision gauge is installed immediately after the pressure regulator, so that the pressure with which the air will eventually reach the sensitive unit can be determined. Other gases such as nitrogen can be used instead of air to reduce chemical reactivity and entrapped moisture.

The injected gas will reach chamber A, which is a circular toroidal section (Part *a* of the figure) and presses against the membrane. When the force exerted by the gas pressure just exceeds that of the water pressure on the other side of the membrane, the membrane moves down slightly, allowing the gas to flow between the neoprene ring and the metal support. The gas exits through the outlet tube to the control panel (Part *b* of figure). The pressure of the air reaching the control panel is measured on a gage. This pressure is approximately equal to that of the water in the porous cell, though it must be somewhat greater in order to flow. So that the air pressure at the control panel represents the pressure of the water in the porous cell, a relief tap is provided on the control panel. When this tap is opened, the air pressure in excess of the minimum required to maintain general flow is dissipated. The reading on the control panel is made as the air pressure dissipates the moment the neoprene ring closes and impedes the flow air again. This method only gives the pressure of the water in the porous cell by means of a laboratory calibration curve previously plotted for the entire apparatus. The equilibrium of the teflon membrane does not indicate equality of the air and water pressures above and below it, since these two pressures are exerted over slightly

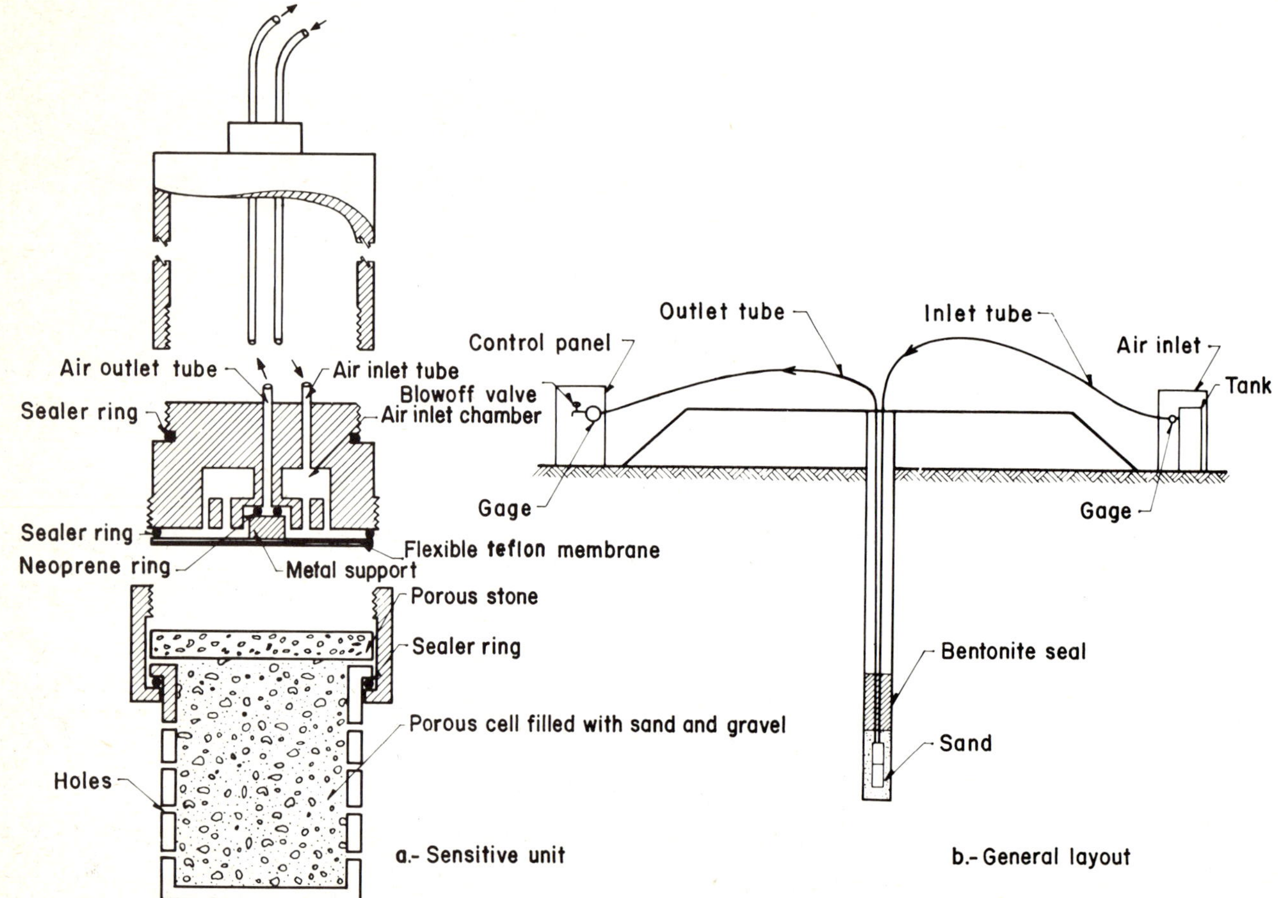

Fig. 13-16 Pneumatic type closed piezometer

different areas. The previously plotted calibration curve also accounts for other factors, such as the effect of the stiffness of the membrane and the dissipation of gas pressure in the tubes. The response time of these instruments is short being just a few hours for impervious soils, and a few minutes in pervious soils.

There are also electrical piezometers, two of which are illustrated in Fig. 13-17. The working principle of all these instruments are similar. There is a porous cell through which the water presses a flexible membrane upwards. Above the membrane is fixed the measuring device either a vibrating wire device or a system of electrical strain gauges that sense the force on the membrane.

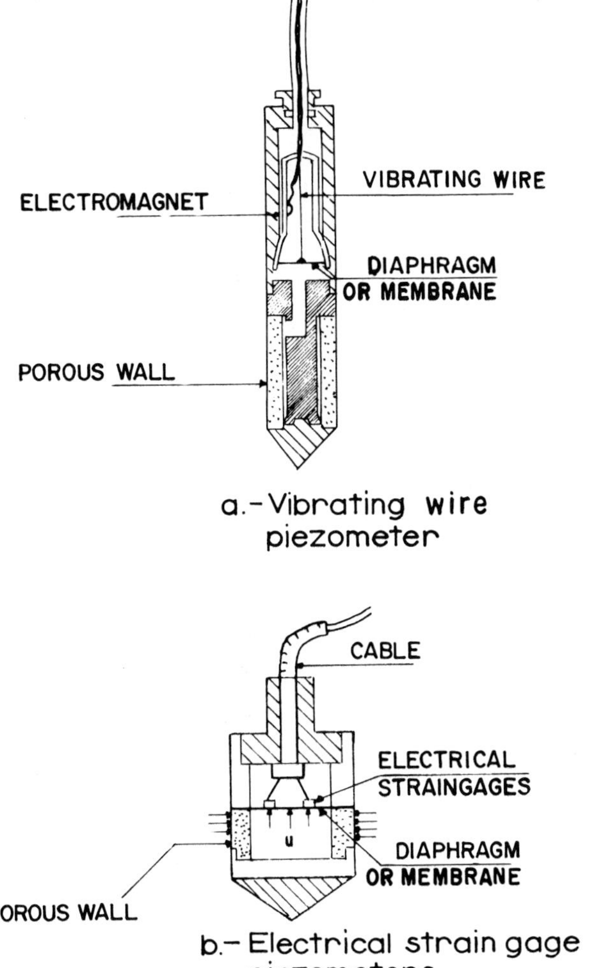

Fig. 13-17 Electrical piezometers

In the vibrating wire piezometer, one of these elements is provided inside the sensitive unit. The lower end of this wire is attached to the pressure measuring membrane or diaphragm. Initially the wire has a certain length and tension, so that when excited by an electromagnet it vibrates with the corresponding natural frequency. On a control panel on the surface of the ground there is another identical wire, the lower end of which can be moved slightly by adjusting micrometric screw. The circuits of both wires are connected so that their vibration frequencies can be compared, establishing the moment at which they are equal, which occurs in the initial position occupied by the apparatus.

When the membrane rises under the effect of the water pressure, the length and tension of the wire in the sensitive unit are reduced and accordingly, its natural vibration frequency. Consequently the micrometric screw on the wire on the control panel will have to be adjusted until the two frequencies are made equal. The movement of the control panel wire, as measured by the micrometer is converted into a corresponding pressure reading by means of a calibration conducted previously in the laboratory. The response of the apparatus is practically instantaneous. It is not greatly affected by problems deriving from the physico-chemical action of the water.

In the apparatus using systems of electrical strain gauges, small cells containing a metal filament with a resistance which varies with length, are fastened to the top of the pressure membrane. When the membrane deforms, the corresponding reading is taken. Currently, spiral electric resistance sensors are available which are highly satisfactory. Other European piezometer models are mentioned in [20].

The installation of a piezometer deserves as much attention as the design and construction of the piezometer itself. Inefficient sealing will spoil the performance of any apparatus. The same can be said of a poor filter. In deep piezometers installed in highly deformable soils, the tubing frequently brings about a self-driving phenomenon. The piezometer acts like a piston which generates a pressure at the tip, and gives false readings. Another source of error may be the change in position of the device as time goes by. These dangers must be avoided by isolating the apparatus and its connecting tubing from the movements of the ground around them.

Installation is closely related to stratigraphy. In interstratified clay and sand layers, special care must be taken to install the sensitive units of the instruments in whichever layer the pressure is to be measured. For monitoring consolidation it must be in the clay.

At least one piezometer should be installed in firm layers of subsoil at a depth where the normal pressure induced by the surcharge on the surface produces insignificant consolidation. This usually occurs when the stresses induced are only 5 or 10% of the surface pressure. It is best to install several piezometers on the same location, but at different depths.

In order to save drilling costs, several piezometers are sometimes placed in the same borehole, with impervious seals of bentonite between them. However, it is extremely difficult to make such multiple seals pressure tight. In our experience the cost of a separate hole for each piezometer is saved in fewer seal failures that make the pressure data meaningless.

An important danger to numerous piezometers is the corroding effect of impure water on their metal parts. Such corrosive water is only too common in the soft, compressible soils where piezometers have to be used. Isolation of susceptible metal parts is very difficult. The best way to combat this problem is to eliminate the use of metals, using instruments made entirely of plastics which are not affected by these phenomena.

Figure 13-18 is a graph showing the data that can be obtained from a piezometer installation. The instruments were pneumatic piezometers installed in the test embankments built by the Mexican Ministry of Public Works to obtain data for designing a highway across the Texcoco Lake near Mexico City. The embankment was 3 m (10 ft) high, plus one meter which corresponds to settlement of the embankment in the soft bottom of the lake. In one case the measurements of 3 piezometers installed in the same well are given, although for the sake of clarity they have been drawn separately. In the other case, the well contains two piezometers at different depths. The times at which the measurements are reported correspond to similar dates in the years that are indicated.

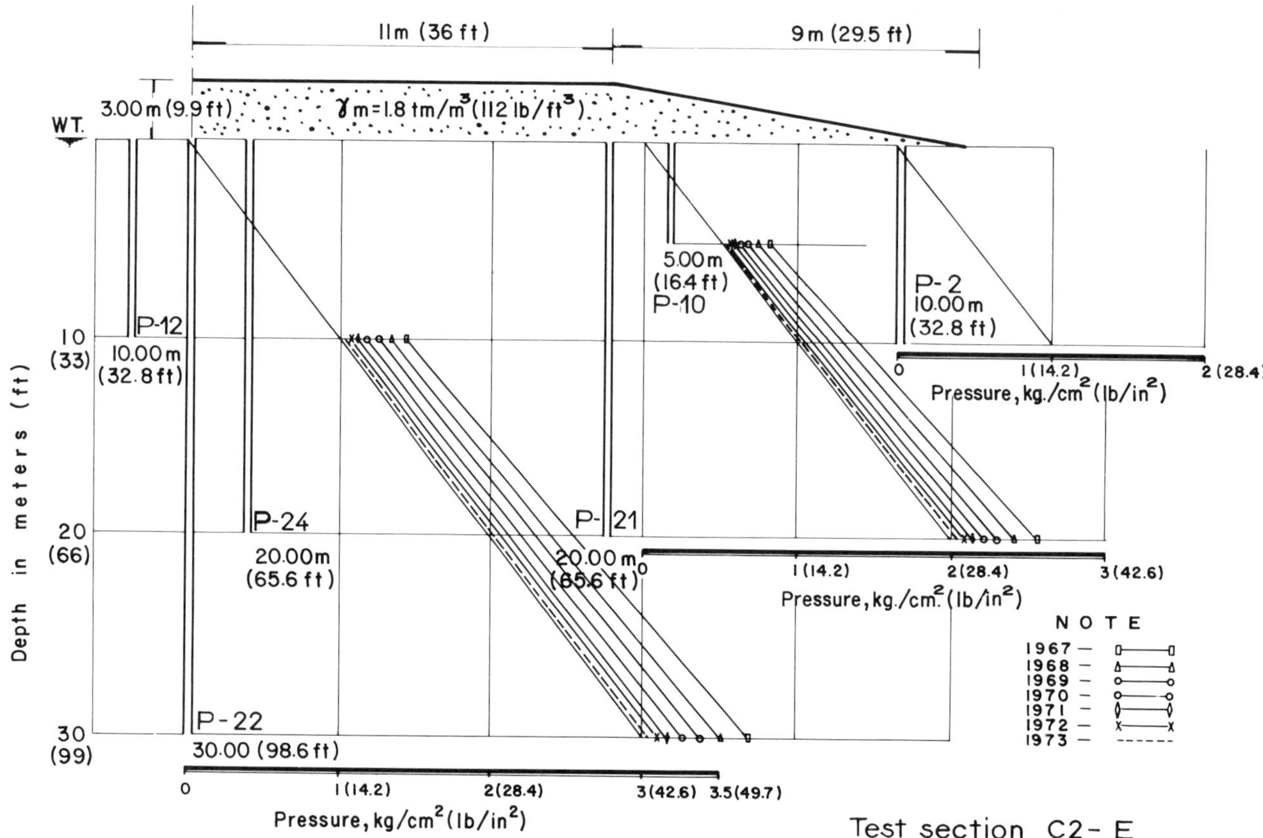

Fig. 13-18 Piezometric data under an embankment constructed on soft soils

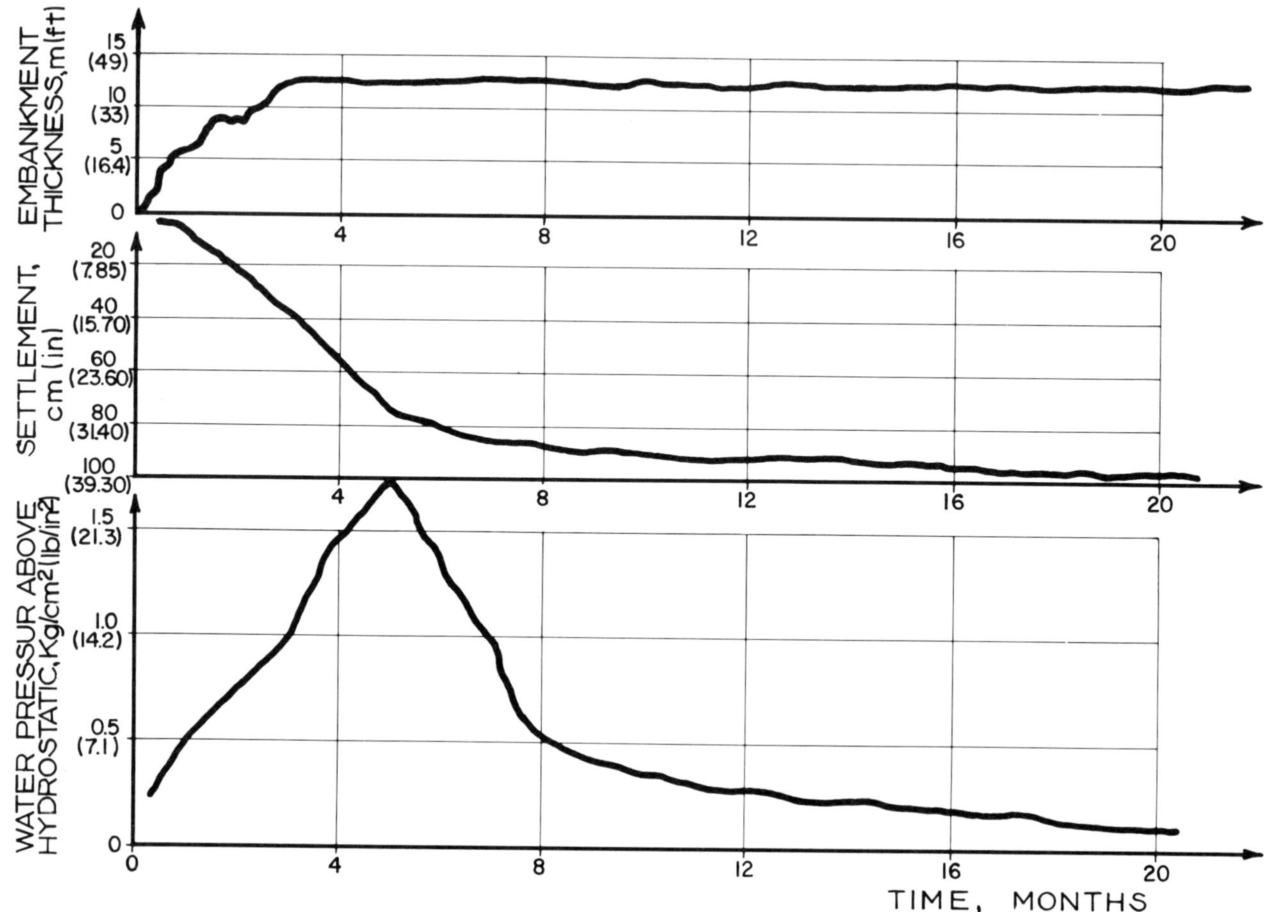

Fig. 13-19 Comparison of placement of the load, settlement and evolution of the piezometric readings in the access embankment to an underpass

Figure 13-19 shows schematically the data that can be obtained by comparing the evolution of settlement with that of pore water pressures. The readings were obtained under an access embankment 12.5 m (41 ft) high built on soft, compressible clayey subsoil. Settlement evolution shows that this is a case where many problems might be solved by constructing the embankment in advance, for practically all settlement took place in the first 5 or 6 months. Worth noting is the correspondence between the loading process and the rise in pore pressures, and between the settlement process and the drop in these pressures, always with a certain delay in the recovery of the pressures in relation to the changes in loading on the surface. Other examples of the use of piezometers and the information that can be obtained from them can be found in [18].

Selection of the piezometer to be used under embankments on soft soils depends largely on each particular case, but generally speaking the most suitable ones are those that have a quick response and are very resistant to the effects of saline and polluted waters.

13.2.4 Pressure transmitted by Embankments to the Foundation Ground

In stability analyses, it is usually assumed that the pressure transmitted by embankments to the natural ground is the product of the unit weight of the embankment material multiplied by the height of the structure. This is precise enough in most cases. However situations arise where it is necessary to find out the actual pressure that is applied. This may differ from the computed pressure, owing to uncertainty regarding the unit weight of the embankment, which varies with the heterogeneities in the material, and changes in the degree of compaction. Bearing in mind that many highway embankments on soft soils have to be designed with very low factors of safety (about 1.1 or 1.2), it is not hard to imagine cases where it is wise to obtain very exact values for the pressure transmitted.

The pressure varies from point to point, changing with time as the soft foundation settles non-uniformly. If the embankment is more rigid than the foundation, the pressure it exerts on the foundation becomes greater on the areas of less settlement and less on the areas of greater settlement pressure.

Where it is wished to verify the design hypothesis about pressure distribution with depth, or where it is wished to compare the settlements occurring at different depths with the normal vertical stresses that reach them, measurements should be made at different depths within the foundation as well as at the original ground surface.

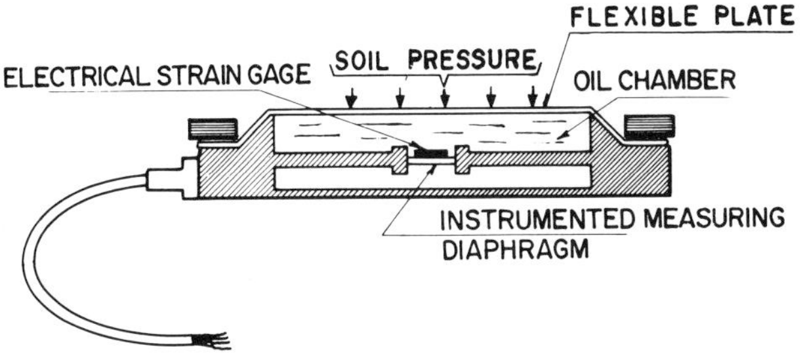

Fig. 13-20 Pressure cell

All vertical pressure meters are pressure cells which measure the total stresses applied to them. Most of the currently used meters are strain-gauge pressure cells or hydraulic pressure cells, similar to the piezometers.

The cell is a cylinder with a large diameter in comparison to its height, with a flexible plate in contact with the soil. Underneath it there is an oil-filled chamber, the purpose of which is to transmit the pressure from the flexible plate to the instrumented diaphragm, which is the sensing unit of the instrument. The sensing unit contains either the vibrating wire or electrical strain gauges as employed for electric piezometers. The pressures are read at the ground surface by similar electrical indicators as for the piezometers. These systems must be calibrated in the laboratory before installation (Plates 13-7 to 13-10).

Plate 13-7 Pressure cell (Interior View)

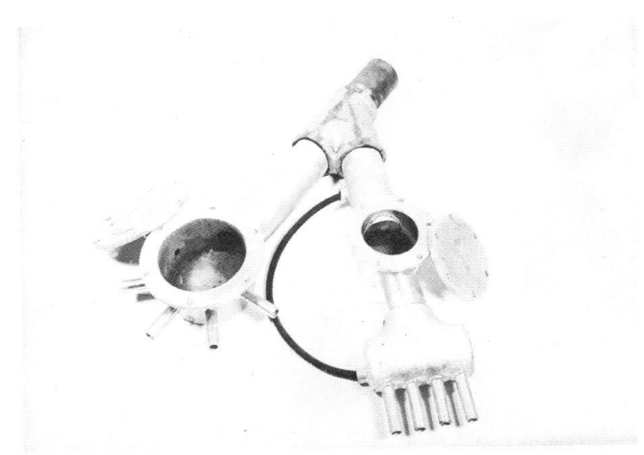

Plate 13-8 Pressure cells

Instruments like the one just described are used for any problems where it is wished to measure pressures, such as the determination of earth pressures on retaining walls, sheet piles, struts, etc. The cells are manufactured in diameters of 60 cm (24 in) or more, down to 5 mm (0.2 in) (with thicknesses of 2 mm (0.08 in)), which makes it possible to instrument small-scale laboratory models.

Like any device embedded in the soil, pressure cells alter the field of stresses of the soil mass in which they are installed. The ideal cell would be one with the same deformability conditions as the soil: Such cells are not available.

Hydraulic pressure cells are generally cheaper, stronger and sometimes more reliable. Some work at a constant volume and some use counterpressure. The constant volume systems [4] have a deformable cell filled with oil or water, which is buried in the soil. From it there is a tubing also filled with oil or water

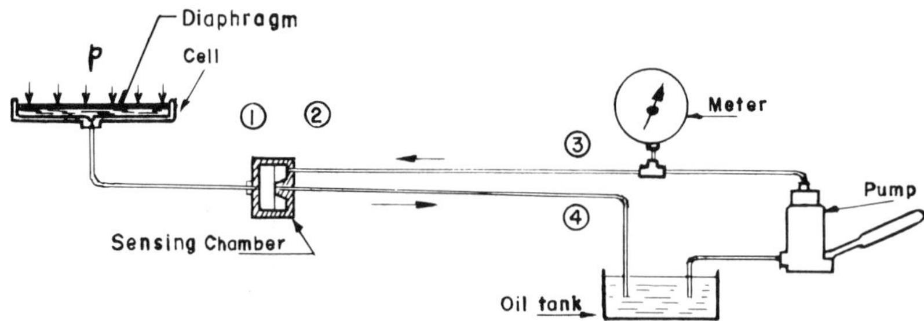

Fig. 13-21 Hydraulic pressure cell using counter pressure [4]

which is connected to a gauge. Direct readings of the pressure that is generated within the cell are taken from the gauge.

A piezometer that works on the principle of counterpressure is shown in Fig. 13-21 [4]. The device, which is of German design, by GLOTZL, consists of a cell, regulator chamber, hand-operated pump and control panel. The cell is similar to those already described. It is equipped with a flexible diaphragm through which the soil pressure is transmitted to oil or water inside the thin, wide cell.

Plate 13-9 An appropriate cell for measuring earth pressures on retaining walls.

The chamber communicates with the cell by a thin, rigid tube. It has two compartments separated by a second diaphragm. From the second compartment run two tubes. One is connected to a hand-operated pump and oil tank. The second tube returns oil to the tank.

Before making a measurement, the earth pressure, p, on the cell diaphragm is transmitted to the sensing chamber diaphragm. This closes the entrance to tube *4*. In order to make a measurement the oil pressure in tube *3* is increased to overcome the pressure on the cell side of the sensing diaphragm. The diaphragm moves outward, opening the entrance to tube *4* and allowing the oil to return. The oil pressure in tube *3* that just balances that pressure on the cell side of the sensing chamber is equal to the earth pressure. Like the piezometer cells, this hydraulic cell should be calibrated before installing it in the ground.

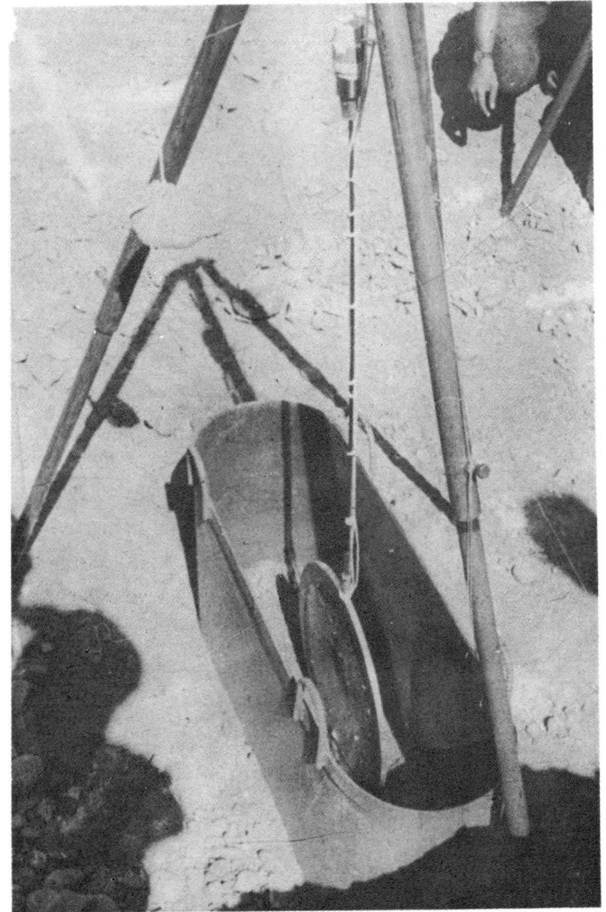

Plate 13-10 Installation of a pressure cell

13.3 Instrumentation of Embankments for Stability Studies

Measurements in existing embankments in order to study stability problems (other than those deriving from construction on soft, compressible soils), has become particularly important in dam construction, especially since the introduction of increasingly higher and more complex structures, e.g. [19]. The successes achieved in this field have led to the rapid development of even better instrumentation for the study of embankments and slope performance, which are equally applicable to embankments and natural hillsides in highway practice (Plate 13-11).

Plate 13-11 Exposed shear surface in an embankment

The instrumentation of embankments and slopes has several objectives:

1) To verify the behavior of the structures during construction in order to verify design hypotheses and the anticipated changes of the factor of safety. This objective is vital in high dams and may appear too sophisticated for highway embankments design. However, very high embankments are required by modern roads and railroads; still higher embankments will be required in the future. Their design will involve the same uncertainties as with dams. This objective will not be routine in highways, but used for unusual conditions.
2) To determine the performance of the structure through part or all of its useful lifetime. High rockfills or earthfills deform under their own weight in a manner which is far from being clearly understood. Similarly uncertain are the correlations between the structural behavior and the different construction methods now in use or those that might eventually be used. The virtues of each of these methods can only be satisfactorily demonstrated if sufficient data of real behavior are available.
3) To establish the kinematic conditions of slides already existing before the engineer commenced his activities, or slides caused by these activities, as may occur in cuts and natural slopes. Determining the shape of the shear surface, the nature, magnitude and seasonal variations of the movements, the changes in the relative position of the different earth or rock masses involved, according to the experience of the authors, are necessary for resolving problems like those presented in §6.1 of Chapter 6. The slides controlled on the Tijuana-Ensenada highway, which have been referred to at different places in this book are an example of the use of such data. In solving such problems, which are among the most difficult that are encountered in highway practice, the kinematic aspects are as important as those relating to strength which are traditionally considered in connection with slope stability problems. Field instrumentation is the only means available to the engineer for developing an accurate image of the slow movements that may be taking place; without this image any attempt at remedial work will be blind. References [21] and [22] serve as an example and justification of the foregoing statements.

The cost of a field instrumentation program for determining the behavior of slopes and natural hillsides should always justify itself in terms of potential saving of construction and maintenance costs of the specific project that is to be studied. This is not difficult in important problems. There is an additional advantage that cannot usually be expressed in cost savings on a specific project; the large amount of experience and knowledge that can be derived from instrumentation and which can be effectively used to save money on subsequent projects and similar situations. The authors are convinced that this advantage justifies reasonable amounts of field instrumentation in specific situations where direct cost savings cannot be demonstrated. References [3] and [24], which describe much of the experience that has been acquired from past instrumentation programs, illustrate this criterion adequately.

When a field instrumentation program is developed in embankments, cuts or natural slopes, information is usually sought on one or several of the following aspects [24]:

— Horizontal and vertical movements
— Stresses acting in the vertical or horizontal direction
— Pore pressures and their changes
— Seismic effects, including both the ground motion of the earthquake and the response of the earth structure
— Groundwater flow characteristics
— In situ measurement of mechanical properties, both in the embankment and the foundation ground

13.3.1 Surface Control

As in the case of embankments on soft soils, topographic observations must now be conducted on points on the surface of embankments, so that information about the direction and speed of the movements can be obtained. After completing several sets of readings, it will be possible to draw a topographical ground plan, in which the movement of each point can be represented by a vector. If there are sufficient points they give a picture of the surface movements and the speed with which they occur. The most difficult problem is usually to establish a fixed reference, beyond the influence of these movements, to which the movements of all the observation points can be referred. Large distances should be avoided, as they may induce errors. Reference [25] describes a high precision system used for controlling movements at points situated on the top of the Infiernillo Dam. In the landslides on the Tijuana-Ensenada Highway, for some of which ground plans were shown in Chapter 7, the observation points were distributed on axes crossing the slide zone. The two ends of each axis were beyond the moving zone. They defined a base line, which could be reconstructed in its original position each time observations were made. The movements manifest themselves by deviations of the different points in relation to the original base line. These can be determined by triangulation with the aid of any of the fixed points outside the moving line that may prove necessary (Plate 13-12). Reference [26] describes a case of surface control by means of triangulation, for the Netzahualcóyotl Dam in South-eastern Mexico.

Plate 13-12 Benchmark

It is often necessary to locate zones where tension or compression occurs. To do this some simple calibrated springs [24] have been developed, whose change in length can be measured. For more precise measurements of small movements such as creeps, these springs or wires can be installed in plastic tubes for protection, which are sometimes buried at shallow depths beneath the surface. The instrument used to record the movements is the same in these installations. It will be described in detail because it is used with slight variations in almost all devices for measuring horizontal displacements. An electrical potentiometer (Fig. 13-22) is an electrical resistance, generally laid out in a circle, on which an arm, A, can run which divides the initial resistance into two parts, R_1 and R_2. An axle, E, rotates when induced to do so by the tension it receives from a cable, C. This cable is always held taut by a calibrated spring, as shown in Fig. 13-22. Arm A is also part of the electrical circuit and receives current from a feeder cable. Resistances R_1 and R_2 are two arms of a WHEATSTONE bridge; the remainder is on a control panel on the ground surface.

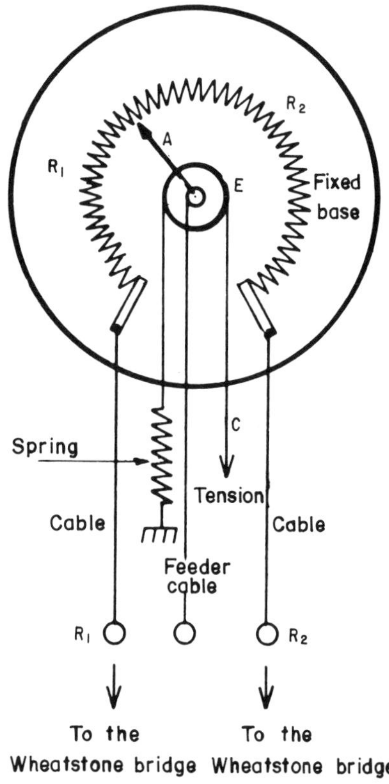

Fig. 13-22 Sketch showing a potentiometer used in mechanisms for measuring horizontal displacements

The operation is as follows. The wire for measuring displacement on the surface inside the plastic tubing, is fastened to an anchor plate that is embedded in the ground. When the plate moves horizontally, the initial tension of cable C will change, axle E will turn, moving arm A. The WHEATSTONE bridge will record a change, which by a previous laboratory calibration will indicate what displacement has occurred. When high precision is required with this device, it is usually necessary to compensate for the variations in the length of the wires by changes in temperature. This can be achieved by temperature sensing thermocouples on the installation to indicate the changes in length that are due to thermal expansion.

13.3.2 Vertical Settlements and Movements

These measurements can be made in two ways. Either the instruments are placed in such a way that vertical displacements can be measured at numerous points on the same surface, or else they are placed vertically so that the displacement of various points on the same vertical line can be measured, thus obtaining the settlement of strata or zones with a known thickness (Plate 13-13).

An instrument that is frequently used for installations of the second category is the settlement torpedo described in §13.2.2.

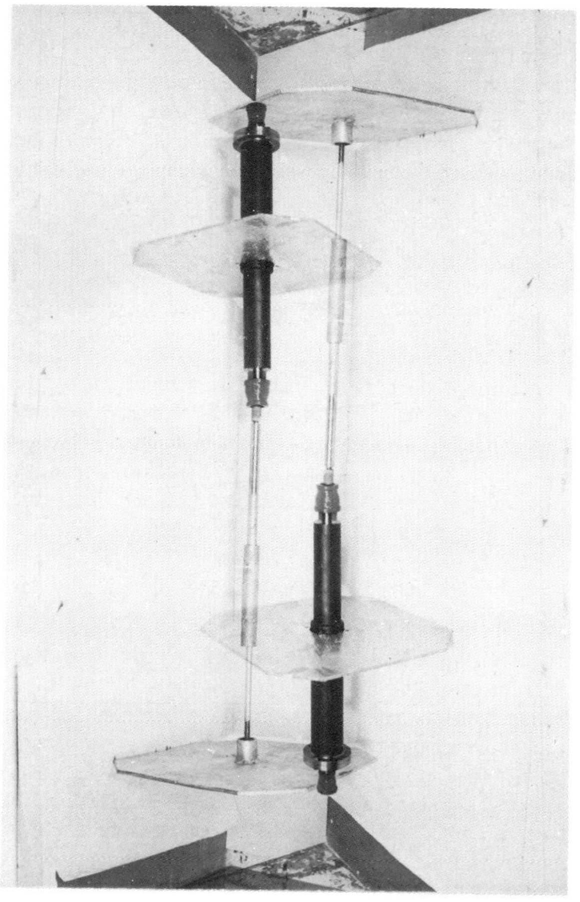

Plate 13-13 Device for measuring vertical movements

Reference [27] describes a similar instrument that has often been employed in dams. It consists of a series of telescopic tubes with cross-sections of 3.8 and 5.1 cm (1.5 and 2 in), which are installed alternately. They are anchored to the embankment material by cross arms at right angles to the tubes. A torpedo similar to the one described finds the depths of each change in diameter. These depths change with the soil or rockfill settlement.

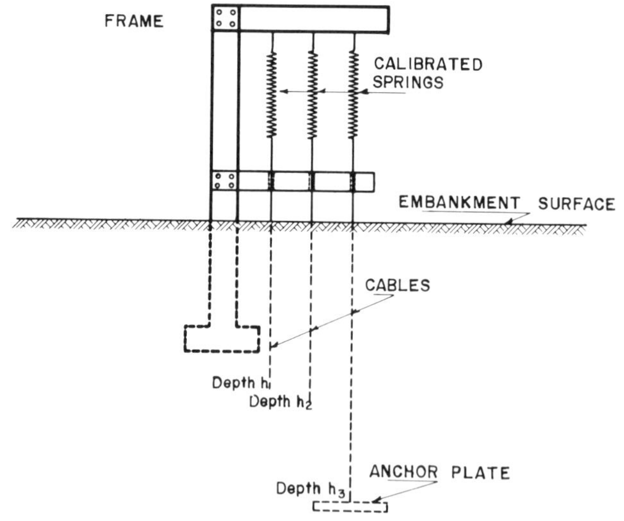

Fig. 13-23 Device for measuring relative settlements on a vertical plane inside an embankment [24]

A very elementary but effective system for measuring the relative settlement of several points in the embankment on a given vertical line is shown in Fig. 13-23 [24]. A metal frame is firmly driven into the ground surface. To it are joined a series of calibrated springs, which are connected to cables, at the lower end of which there is an anchor plate at different depths. With soil settlement, each plate moves, stretching the calibrated spring. On the cables and the arm of the frame, there are marks that permit measuring the amount the corresponding anchor plate has dropped. The cables are often placed in a common hole. If the vertical movements of the embankment surface are known, relative settlements within the embankment can be converted to elevation changes.

Figure 13-24 [24] shows another device for measuring vertical displacements at several points on different levels in the same well. Several anchors, two of which are shown in the drawing, are installed in a hole that is 8 to 10 cm (3 to 4 in) wide, either without any casing or with a very light casing in the case of the finest soils and pure sands. These anchors can be of any of the many existing types, but the drawing shows an expansive type which, after being introduced, increases its diameter and drives itself into the soil or carves its way into the rock, breaking any existing well bracing. The anchor is connected by a wire under constant tension to a potentiometer. A change in the level of the anchor can be interpreted from previous calibration, and from this the corresponding vertical displacement is found; [24,28,29] describe other similar devices that may have advantages in various types of soils.

Most devices that are installed measure the vertical movements of several points on a horizontal plane and are similar to those described in §13.2.1. Reference [24] describes in detail a variant that was installed in the Oroville Dam (U.S.A.).

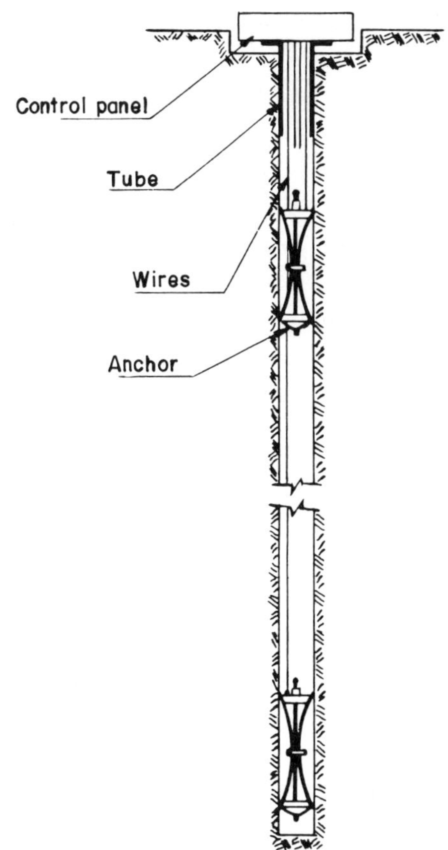

Fig. 13-24 Apparatus for measuring settlements in embankments [24]

13.3.3 Horizontal Movements

The inclinometers already described earlier in this chapter are the instruments most widely used for critical slope instrumentation, when horizontal displacements are to be measured. In the case of slopes, whether natural hillsides, cuts or embankments, the most frequent use of these instruments is for detecting the position of an ancient or recently formed slip surface and assessing the nature and magnitude of the movements that take place on it. Their value for determining a slip surface with moving masses was described in Fig. 7-37. This is an example of an increasingly popular use of field instrumentation. References [19,26,30,31] give examples of the use of inclinometers in different earth structures, mostly in hydraulic projects. References [21] and [51] describe a very intensive use of these instruments in important stability problems of natural slopes and large embankments in the construction of a highway, [22] describes another instrumentation program involving inclinometers in highway construction.

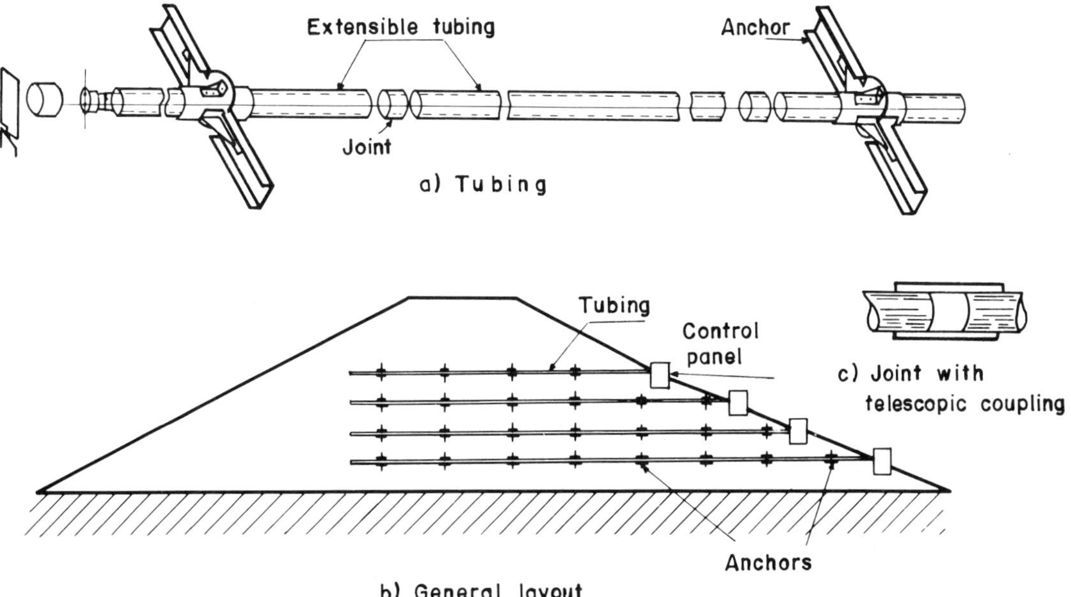

Fig. 13-25 Device for measuring horizontal movements in an embankment

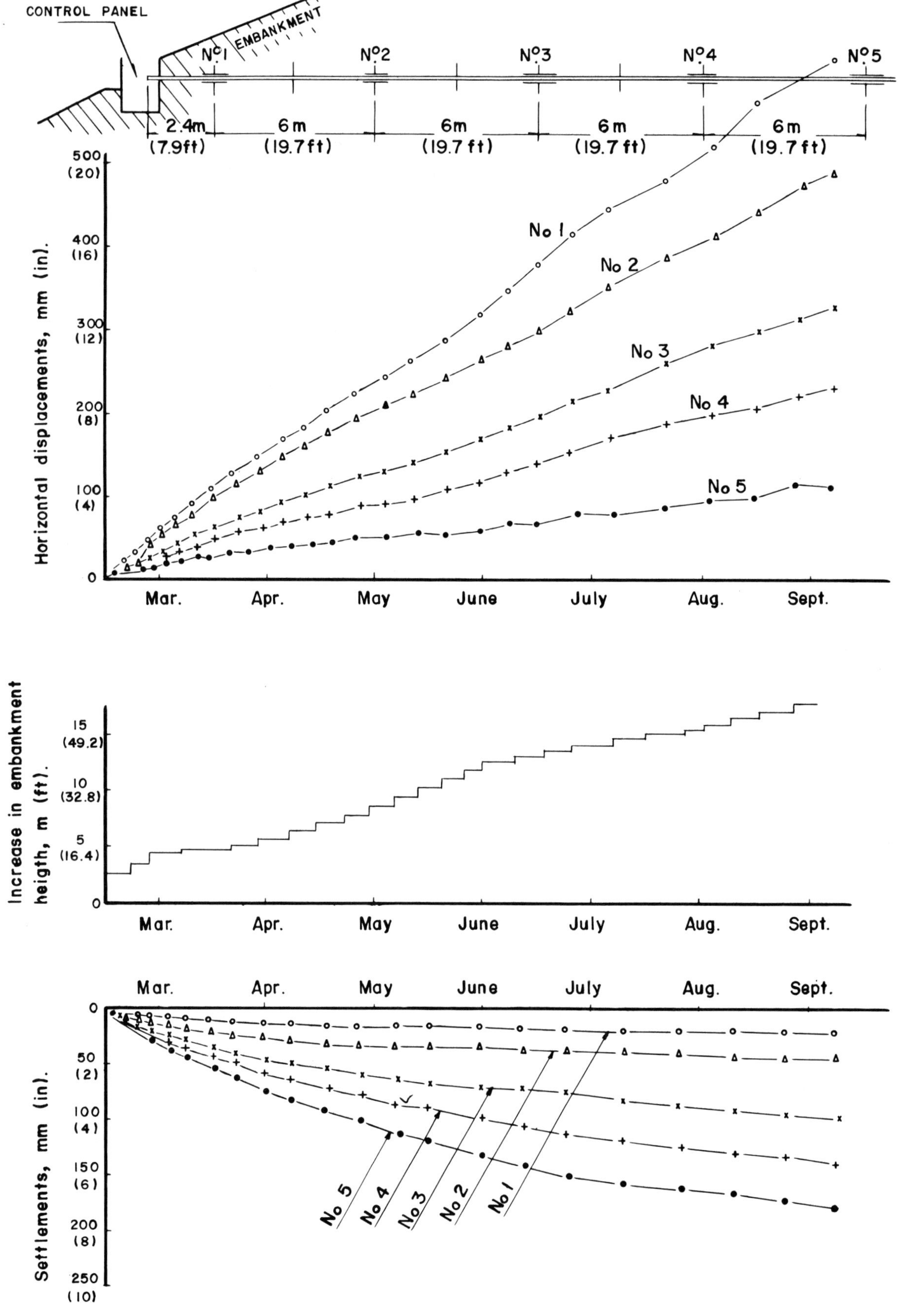

Fig. 13-26 Type of information that can be obtained with a device for measuring horizontal and vertical movements within an embankment

Figure 13-25 shows another type of device for measuring horizontal movements which has been used frequently in embankments. The device consists of a series of tubes (Part *a* of the figure), each equipped with extensions at right angles which serve to anchor it to the embankment material. Each section of tube reflects the movements of the embankment around it through its anchor, because they are joined with telescopic couplings that permit relative movement (Part *c* of the figure). All the tubes, connected together are placed in the desired position during construction of the embankment (Part *b* of the figure).

There are several different measurement systems. One is to install inside the tubing a cable at a constant tension, connected to an electrical potentiometer, using the system already described in previous pages for measuring settlement. The Japanese technique uses a measuring device similar to the WILSON inclinometer, which can be introduced by hand at any point in the tubing to detect the position of the telescopic couplings. Thanks to its tilt, the measuring unit can determine the deformation in the tubing at any instant in the life of the embankment. The relative elevation of the inclinometer can be detected by means of a gauge. For this purpose a chamber inside the device is partially filled with a liquid; the height of this liquid is measured by the gauge. From the changes, the elevation of points in the tube can be found, Fig. 13-26 shows the type of information that can be obtained from these devices.

Californian engineers [32] have developed a very simple instrument which enables the horizontal and vertical displacements within an embankment to be measured (Fig. 13-27). In a trench that is dug at the time of construction, a telescopic plastic tubing like the one shown in Part *a* of the figure is installed. Inside the tubing there are cables, each joined to rectangular metal anchors in such a way that each anchor and its cable does not interfere with the other cables attached to other anchors. All the cables meet in a meter box (Part *b* of the figure) which is installed on a concrete base in the outer part of the embankment. At the time of reading, the position of this base is found by survey methods.

Inside the meter box there is a scale which enables the position of marks on each cable to be measured. A weight holds each cable at a constant tension. Vertical movements can be measured on this device by installing a water-filled tube at each section and observing the level of this water. Greater precision in the measurement of horizontal movements could be achieved

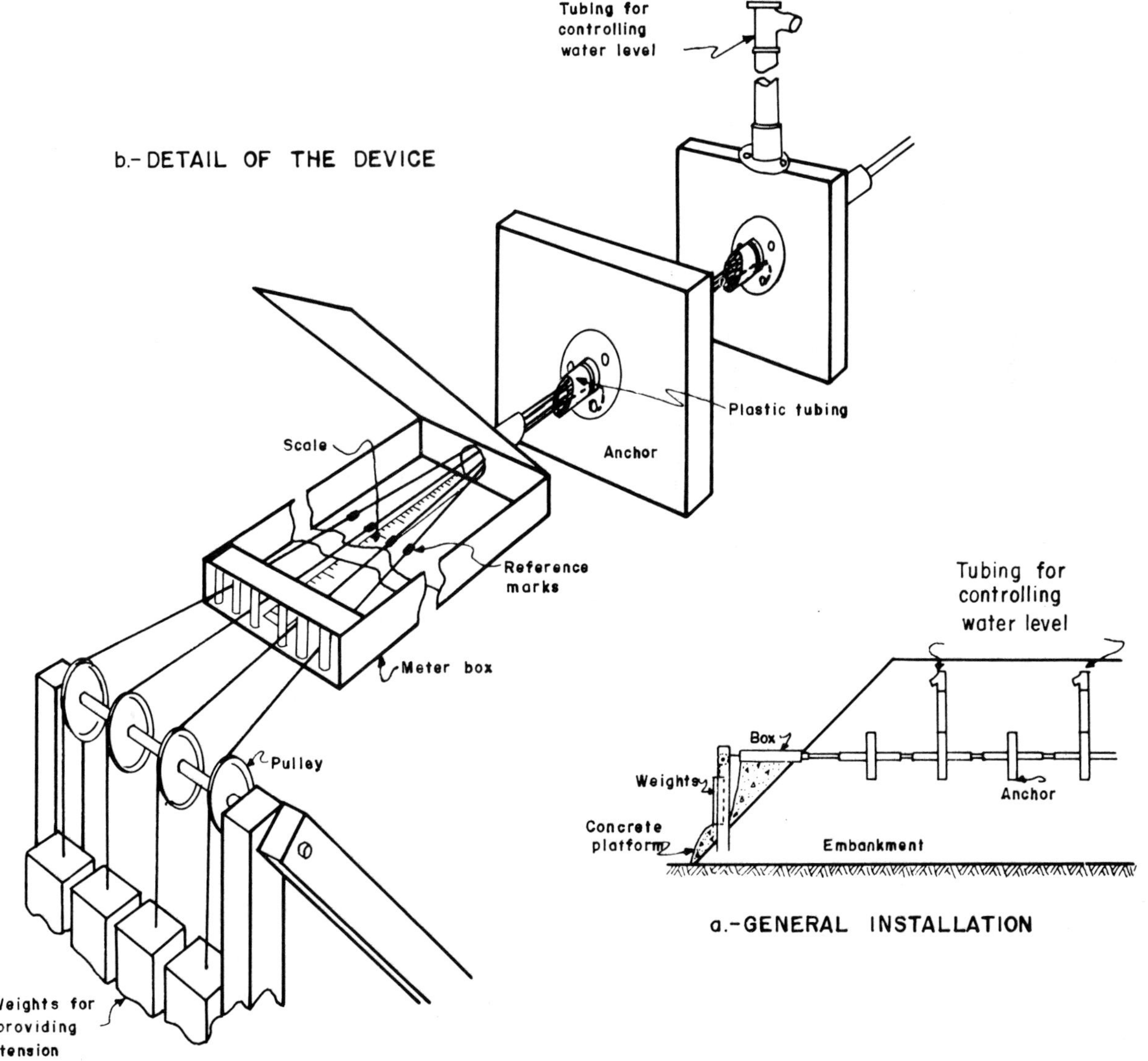

Fig. 13-27 Device for measuring vertical and horizontal movements in an embankment

using electrical potentiometers to measure the cable movement as previously described.

Many of the instruments that have been described for measurement of vertical movements can be employed to measure horizontal movements, by simply modifying their position inside the embankment (Plates 13-14).

Plate 13-14 Installation of horizontal strain gauges

Reference [25] describes a type of extensometer that can measure horizontal displacements in three directions of a plane simultaneously (Fig. 13-28). It was installed by MARSAL and his collaborators in the Infiernillo Dam (Plate 13-15).

The instrument body is placed in the plane on which displacements are to be measured. It is equipped with three legs made of metal or plastic telescopic tubing, with an anchor plate at the end of each, embedded in the embankment material. A vertical tube leads the necessary connections to a meter box on the surface of the embankment (Part *b* of the figure). Inside each of the three legs there is a cable joined to the anchor. The end of the cable is maintained at a constant tension by a coil spring. Any movement of the anchor is transmitted to the cable and to an axle located in the body of the device, which rotates a potentiometer arm of the type described in Fig. 13-22. As already described, the changes on the potentiometer are read on a WHEATSTONE bridge. The resistance change indicates the movement that has taken place, using a previous laboratory calibration. Determination of the movements in three directions is valuable when interpreting the readings theoretically. The anchors can be placed 3 to 4 m (10 to 13 ft) from the body of the apparatus.

One of the critical aspects of the performance of these instruments is their installation, which should be oriented in the directions in which the greatest movements are anticipated. This is particularly true close to the ends of the embankment, where the movements are most complex and it is hard to predict which will predominate. The instruments installed with different orientations permit detailed measurements regardless of direction.

Figure 13-29 depicts another type of strain gauge that measures the displacement of points on a plane normal to its axis. A plastic tube is installed in the ground in sections with telescopic couplings, so that the outlet of the tube on the embankment surface and its deepest end are totally immobilized. Inside the tube a taut wire is suspended, with a device to hold it in its initial position. This device may be a spring located on the surface of the ground. The tube is telescopic, so that it can absorb vertical movements, (which have to be measured by another different procedure). Each telescopic coupling con-

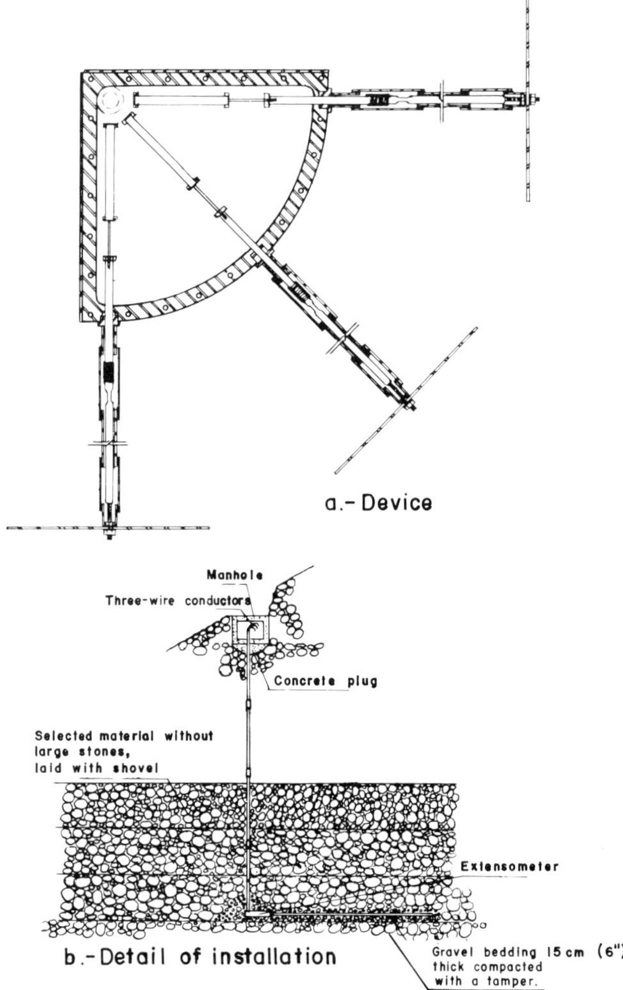

Fig. 13-28 Horizontal extensometer installed in the Infiernillo Dam [25]

stitutes a measuring unit. It contains a fixture with a fork at one end with an electrical resistance between its prongs, and a counterweight, W, at the other end (Part *c* of the figure). The counterweight maintains the electrical resistance in constant contact with the central wire, as the fixture tilts about an axle supported on each telescoping tubing section. Part *b* of the figure describes the measuring principle. When the tube section moves horizontally the resistance moves with it, shifting the point of contact with the vertical taught wire, and changing the relation of R_1 to R_2.

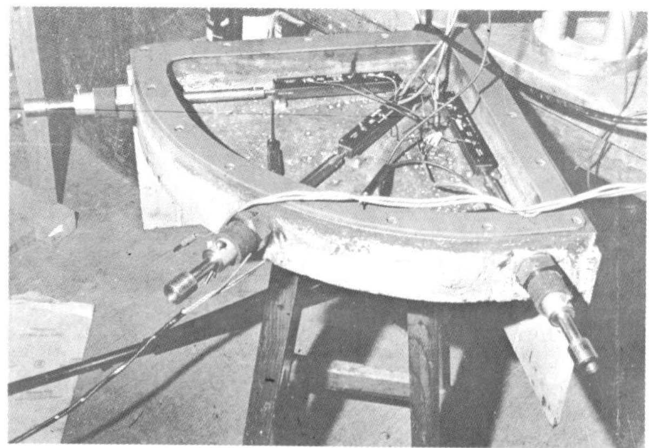

Plate 13-15 Horizontal extensometer used in the Infiernillo Dam

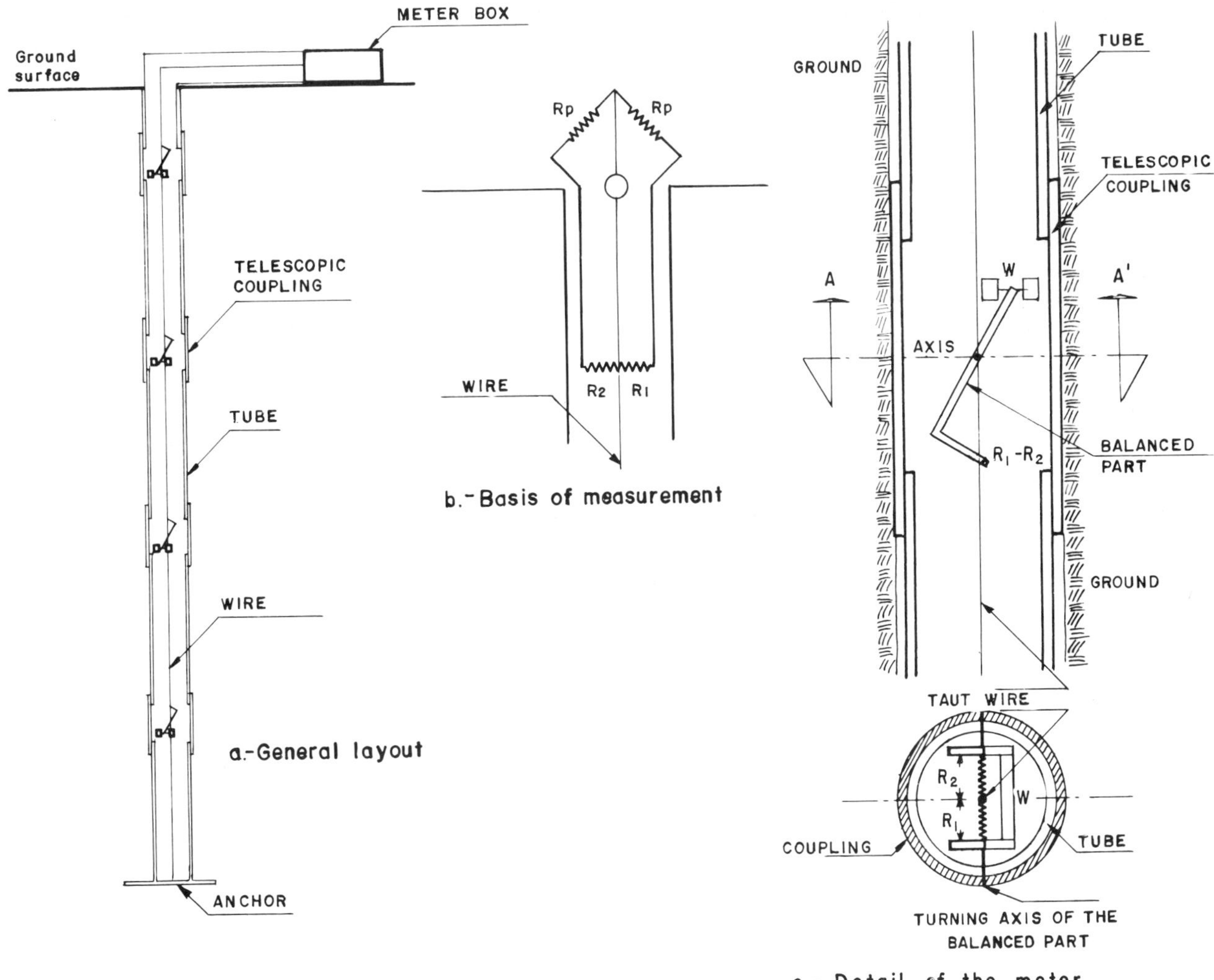

Fig. 13-29 Vertical strain gauge

On the ground surface, there is a WHEATSTONE bridge which contains two of its resistances. The other two (R_1 and R_2) are provided by the resistance contained in the measuring unit, which is divided into two sections by the central wire. Thus, with a previous simple electrical calibration in the laboratory it is possible to find from the surface the lateral displacement of the tubing at the level of the measuring unit under consideration. Displacement of the tube causes a relative displacement of the electrical resistance in relation to the taut central wire, which modifies the values of R_1 and R_2 and makes it possible to take a reading on the bridge.

By installing several measuring units, it is possible to obtain an image of the deformation of the tube as time passes. This image is similar to the one that would be found with an inclinometer. The instrument measures only very small displacements, for as soon as deformation is great enough, the weighted part of the measuring units will come into contact with the walls of the tube and the instrument becomes useless; however, measurements are very accurate. It is only possible to measure displacements in the direction in which the internal resistance is installed ($R_1 - R_2$). The instrument should be oriented with the resistances in the direction of the displacement, if it is known. If it is not, two units can be installed with their resistances at right angles so as to obtain the components of the displacements.

In large highway embankments, the usefulness of a device like the one just described is restricted, for the magnitude of the displacements that are usually of interest in such embankments are greater than those the instrument is capable of measuring. These devices are more useful in tunnel excavation problems, when it is wished to measure the small displacements that take place in a mass of soil or rock as a consequence of excavation work. Devices of this type were installed for this purpose during construction of the Angostura Dam in Mexico. This device can also be installed with its tubing in a horizontal position, thus providing a very sensitive instrument for measuring small vertical displacements.

Also worthy of attention is the strain gauge developed by German engineers (IDEL strain gauge). A plastic tube is placed horizontally in the embankment, in the direction in which it is wished to measure displacements. This tube is telescopic and is anchored at intervals to the material around it. Other installations utilize a smooth tube with metal rings or anchors outside. The initial position of these anchors is carefully observed when the device is installed. Any horizontal displacement of the soil modifies the relative position of the anchors or rings accordingly. Their new position is determined by the change in magnetic field produced by the ring on a radio-wave generating special sounding rod, a *metal detector*. The moving anchors are the only elements made of metal in the entire apparatus.

13.3.4 Measurement of Porewater Pressures

As in the case of embankments built on soft soils, porewater pressures are measured with piezometers in all instrumentation problems connected with the stability analysis of natural hillsides and slopes. The types of instruments and their working principles are similar to those described previously in this chapter, but their use in the problems now under analysis presents certain peculiarities which deserve a comment.

First, these piezometers are less frequently exposed to the effects of polluted or saline waters than the equipment installed in the soft soils which are commonly found in zones where there are stagnant water, swamps, lagoons or salt marshes. Consequently, instruments with metal parts can be used, and less attention will have to be paid to corrosion problems.

Their use in large embankments often endangers the measuring unit, especially the outlet tubing, to puncture or strangulation caused by stones in the soil or shear from soil movements. As with all piezometers, those installed in embankments and hillsides may trap bubbles of air that block the tubes or chambers inside the recording unit or increase the height of the water column.

With these instruments there is also the problem of the response time that was referred to for instruments installed in soft soils, although the permeability of the soils that normally exist in embankments and hillsides is often sufficient that the lag in response to new pressure states is less serious than in soft clays. Another two factors combine to make the response of piezometers installed in embankments and hillsides more rapid than those installed in soft, compressible soils. First, it is usually possible in embankments to place around the instrument a wide, firm layer of sand, which contains a large volume of easily mobilized water. Second, in large highway embankments under normal conditions pore water pressures change very slowly as time passes. Readings are taken at relatively wide intervals, after enough time for the external conditions to establish themselves inside the instrument. Embankments or hillsides in the process of failure and suffering important displacements are exceptions; here information corresponding to short periods of time is essential.

The need to instrument large highway embankments, often involves installing piezometers in partially saturated soils, where the air in the voids is under great pressure. It becomes necessary to differentiate between the part of the piezometric reading measuring air and that referring to water pressure. The problem is usually solved using porous ceramic walls in the measuring unit, which allow the air to pass freely, thus reducing its pressure [33].

When the hillsides or slopes that are instrumented are liable to suffer displacements, these must be taken into careful consideration when installing the piezometers so as to avoid breakage or strangulation of the measuring tubes.

Open piezometers are frequently used because they are economical and strong and easy to install and read when the problem of time lag is of no importance. Interpretation of readings proves difficult in partially saturated soils. In embankments liable to suffer settlements, piezometers are best installed inside telescopic struts, and in weak materials, tubing can be made of metal and as strong as is wished. Pneumatic piezometers have the advantages already discussed. They require minimum volumes of water, they can be easily drained, are easy to operate, and are small and easy to install.

As a final comment, when installing piezometers in large embankments for determining the changes of stability, it must be remembered that the installation will have to last a long time. Consequently safe, reliable equipment must be selected and installed in such a way that precautions are taken against all predictable adverse circumstances.

13.4 Installation Problems

There are many problems that are characteristic of all engineering instrument installations, which are worth mention and comment. Instruments must very often remain buried in the soil for a long time, often below the water-table or subject to its fluctuations. This is a severe condition and restricts or often excludes any possibility of repair or replacement. Electric equipment is prone to insulation leakage and failure. Often the most important and interesting changes take place very slowly and are masked by secondary effects, such as variations in temperature and barometric pressure and fluctuations in the water-table. This leads to serious interpretation problems, demanding good judgment so that erroneous or discordant readings can be rejected and attention given only to essential data and corrections made for secondary influences. Most measurements are relative between two points. In order to establish absolute movements, reliable fixed references are essential, at some distance from the area of movement. In many instrumentation problems related to soil and rock mechanics, knowledge of the behavior of structures during construction is essential. This means that measuring instruments have to be installed which interfere with the freedom of movement of men and equipment. This usually causes friction, opposition to measurement programs, and possibly even the deterioration or deliberate damage to measuring equipment.

As a consequence of the above considerations, it can be said that the equipment and instruments for a field instrumentation program must:

- — Be sturdy, strong and easy to handle.
- — Be simple, with the smallest possible number of movable parts, and preferably not electrically run.
- — Be easy to repair.
- — Be as accessible as possible yet protected from vandalism.
- — Provide data that are easy to obtain and interpret.

Many instrumentation programs eventually require the assistance of a computer, which leads to significant expenses over a long period of time.

A well-designed program for a test installation should:

- — Take into consideration the purpose or objective of the tests.
- — Define if the test is essential to the project, only advisable (and to what degree), complementary, or only of value to future projects.
- — Determine the possibility of including the program of installation and tests in the construction program.
- — Consider the time required and available for acquiring or manufacturing the equipment, checking, overhauling, calibrating and installing it, and for acquiring and constructing any auxiliary devices that may prove necessary.
- — Consider the time and cost of making readings.
- — Assess the time in which preliminary and ultimate conclusions can be obtained, comparing it with the need for information that may have been recommended, so as to

determine whether the information obtained is in agreement with the specific requirements of the case.

— Assess the risks to which men and equipment will be subjected, programming adequate protective measures, and convenient access for reading and servicing equipment.

— Conduct an economic analysis to determine whether the cost of instrumentation will affect the project under consideration directly or whether it can be shared out between several different projects, taking into consideration the quantifiable benefits which will be directly or indirectly due to the instrumentation program.

Execution of the test program must also take into account the following specific points. The soil or rock mechanics specialist, in conjunction with the instrumentation expert, must draw up one or more mental images of the behavior of the structure under study and of the probable evolution of the tests as time passes, allowing for the possibility of gradually modifying these images as appears appropriate on the basis of the actual information obtained. The ultimate form of the report must be prepared, including graphs and auxiliary treatments. It will be advisable to take readings at more frequent intervals than would at first appear necessary, in case the performance of the structure proves different from the one anticipated. It is advisable to take into consideration all phenomena that interfere with or might interfere with measurements throughout the measurement period. The probability of phenomena occurring which do not directly interfere with the test, but which may influence it, should be considered. By anticipating such phenomena, it is possible to distinguish between effects and interferences. This differentiation between relevant information and other unforseen information that may come with it is one of the principal goals of a good instrumentation program. Due consideration should be given to the possible loss of data owing to instrument failure or man power failure and to other problems such as mislaid information, mistakes in recording or identifying data, and faulty electrical or pipe connections. The means of obtaining general information relevant to the test must be studied, so that data from different instruments and teams can be correctly correlated. This is not a one-solution problem. An attempt should also be made at handling comparable data simultaneously. The possibility should be considered of achieving simultaneity of data by interpolation or extrapolation of other data that are not strictly simultaneous. Constant inspection of equipment and installations is essential, providing maintenance and conducting calibrations and repairs. The time required to accomplish all this should always be considered in the total cost of the entire instrumentation program. Readings should be taken at more frequent intervals when an important change is expected or has already occurred in loading and hydraulic or environmental conditions; also when an earthquake or torrential rains have occurred or a slide is expected.

13.5 Other Problems of Interest

13.5.1 Earth Pressure and Retaining Walls

A conclusion from Chapter 5 was the need to measure the magnitude of earth pressures exerted by soils on retaining walls. Only in this way will it be possible to compare and verify the different theories that are available to the designer and acquire a reasonable knowledge of the pressures exerted by different types of soils against the different types of retaining walls.

Almost all the investigation work that is carried out on earth pressure is conducted using groups of pressure cells which are installed between fill and wall. Many of the cells described earlier in this chapter can be used for this purpose. However, a few further comments on this topic are justified.

Almost all the pressure cells that have been used in problems of earth thrust are of three types. An example of the first type is the GOLDBECK cell, Fig. 13-30. [1,34]. The pressure acts against a piston which can deflect inward, bending a diaphragm. When this occurs, an electrical contact is established and a circuit is closed. The instant the circuit is closed can be found by any simple ohmmeter connected to the lead wires. A device enables compressed air to be injected into a chamber inside the cell at the pressure necessary counterbalancing the earth pressure, breaking the electrical contact and interrupting the flow of current. At this instant the air pressure just equals the earth pressure. The GOLDBECK cell was one of the first pressure measuring devices to be developed. It has several practical disadvantages. The most important is that the piston must move significantly, changing the stress field. Most cells of this type become useless after a few years, either because of condensation of moisture in the air chamber or deterioration of the electrical contacts.

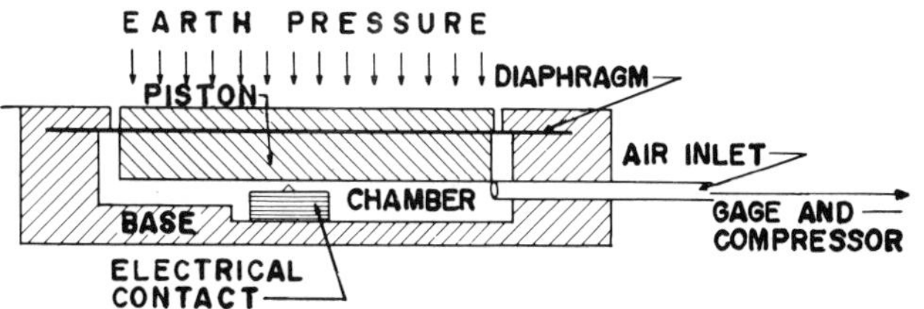

Fig. 13-30 GOLDBECK cell [1]

The CARLSON cell [1,35,36] is shown schematically in Fig. 13-31. The earth pressure acts through a diaphragm on a narrow chamber filled with mercury. The mercury presses on the diaphragm, modifying the length of a flexible rod fitted with electrical strain gauges. The change in the electrical resistance can be measured on a bridge located above ground in a manner similar to that described in other parts of this chapter. Previous calibration of the instrument is required.

The U.S. Waterways Experiment Station, has developed a cell similar to the CARLSON cell, also widely used in the U.S.A. with the electrical gauges located directly above the diaphragm, two in the tension zone and two in the compression zone.

In the CARLSON cell, it has been possible to reduce the effects of temperature to almost imperceptible levels. This is achieved by using extremely thin layers of mercury (two or three hundredths of a centimeter thick). It is a highly sensitive device, however it is resistant to normal handling. In the W.E.S. cell, oil is used instead of mercury. The most delicate feature of this type of cell is the connection between the wires of the electrical gauges and the diaphragm, and the most frequent possibility of failure of the device is that the strain gauge may suffer creep. Because of creep some specialists regard the long-term behavior of the CARLSON cell as safer than that of the W.E.S. cell.

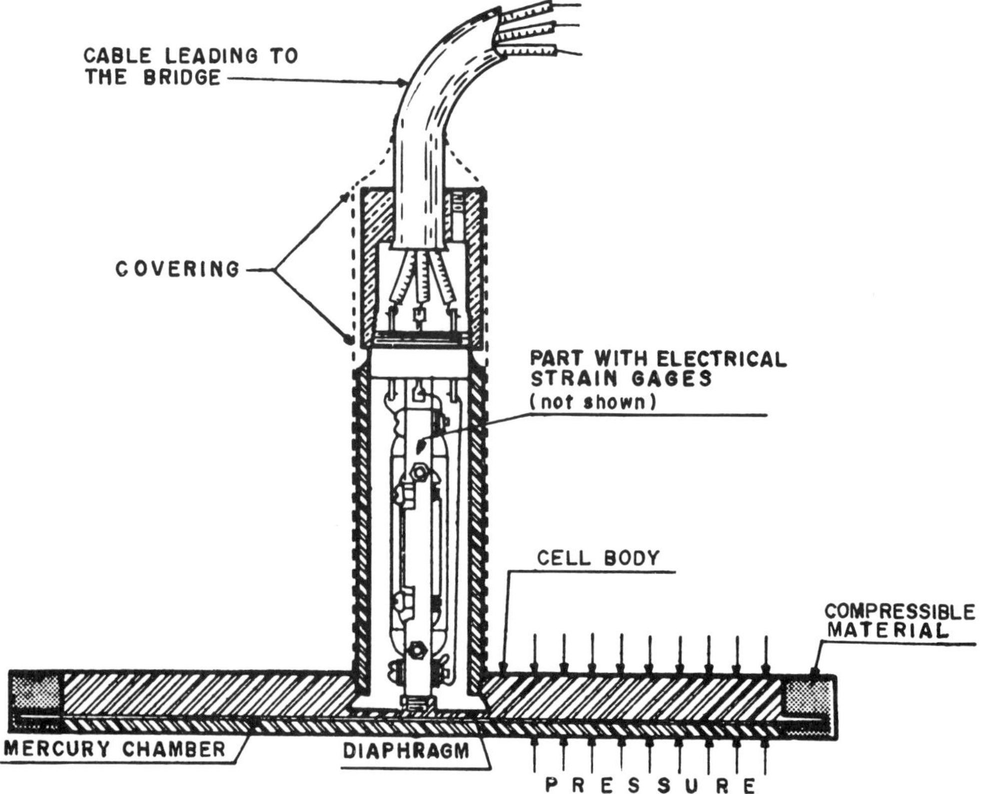

Fig. 13-31 CARLSON cell [1,35]

However, the W.E.S. cell is less liable to be affected by changes in the resistance of the connecting cables.

Apart from the above three cells, it has already been mentioned that all the cell types described in §13.2.4 of this chapter can be used to measure earth pressure. These same cells embedded in different pavement layers are used to measure the stresses transmitted by moving loads.

The compaction of the fill or embankment material around these cells is critical to their reliable operation. This is an operation that must be done by hand, achieving the same density as that in the rest of the conventionally compacted embankment. If the ground around the instrument is not as dense as the surrounding fill, lower pressures will be read than those at similar points without the cell; the opposite is true if compaction is excessive. Another cause of problems is the difference between the compressibility of the cell and that of the earth around it. Many such devices break down because the connecting cables break as a result of movements in the fill. This can be prevented by laying the cables in wavy lines that can absorb embankment movements.

13.5.2 Tunnels

The design and construction of tunnels through soils and rocks poses many problems deserving special instrumentation. The principal problems are [1]:

— Magnitude and distribution of the earth pressure on the tunnel.
— Loads that are exerted on struts and temporary coverings.
— Movements of soil or rock in the neighborhood of the tunnel during construction.
— Movements of soil or soft, broken rock at points that are relatively distant from the tunnel, as a consequence of excavation.
— Movements on the surface of the ground above the tunnel.

Most instrumentation programs for tunnels are restricted to the construction period; only a few are extended to measure the long term behavior of the tunnel or adjoining structures. References [37-40] are descriptions of highly successful instrumentation programs, [41] lists some more recent examples.

When a tunnel is constructed through soils, material tends to move towards the excavation, causing displacements in the adjacent ground. These displacements may cause damage to nearby buildings or structures. Once the tunnel is actually in operation, minor movements continue. In all cases, it is hard to predict and interpret these movements on the basis of existing theories, which are often not capable of taking into account all the geological heterogeneities and the complexities of the ground. Therefore, to determine the ground movements, measurement of behavior in the field is necessary.

Instrumentation programs for tunnels usually pursue at least one of the following objectives:

— Measurement of earth or rock pressures, already acting and as influenced by the tunneling.
— Measurement of strains or distortions in struts and coverings.
— Measurement of ground movements influenced by the tunnel.
— Measurement of the pore pressures in that ground.
— Measurement of movements of the tunnel as a whole.

The instruments for measuring displacements in the ground or pore water pressures are similar to many of those described earlier in this chapter. The devices with wires subjected to constant tension are especially suitable for installation in tunnels. These will be further discussed in Chapter 14, on Tunnels, but although the geotechnical aspects of tunnels will not be dealt with here, a description of some of the instruments most widely used is appropriate.

Figure 13-32 shows how instruments can be installed to measure the stress conditions prevailing on the exposed surface of a rock or hard soil through which a tunnel is dug. These stresses are not the ones that prevailed inside the mass before excavation. An extensometer is anchored at two points, movement between them can be measured. With other similar devices, the extensometer is replaced by an electrical strain gauge whose length change is reflected in the electrical resistance which can be measured as previously described.

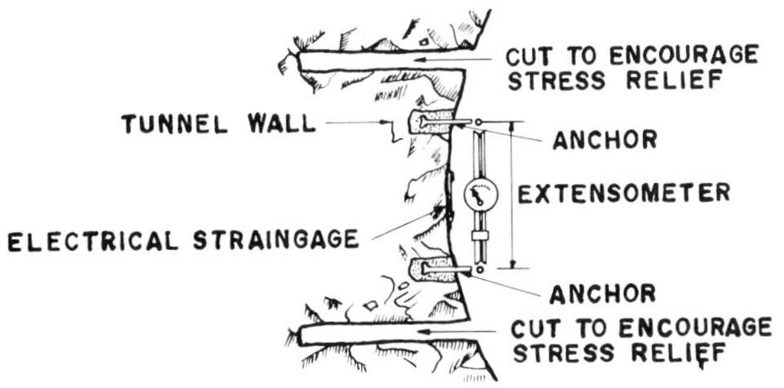

Fig. 13-32 Device for measuring pressures on the exposed surface of a tunnel [48]

Devices like these enable processes of stress relaxation to be controlled around the excavation. To aid this control, cuts are made in zones close to the device, where the soil or rock can expand, relieving itself of its pressures. If these devices are placed around the grooves made in the material, sufficient information can be obtained to determine the principal stresses and their evolution.

Figure 13-33 shows a device that may include either extensometers or electrical strain gauges (both of which are shown in the figure). It permits analysis of stress relaxation about a bore hole. This hole should be small enough that the stress relaxation that takes place around it will not exceed one third of the original value for the stresses. This is the limit beyond which the state of stress would no longer be elastic. Thus the values of the elasticity moduli that are obtained from the measurements and calculations can retain a reasonable physical significance.

Figure 13-34 [48] shows a third method for measuring the stress states near the surface of a tunnel excavation in soft rocks or firm soils. The principle of restoring stresses is employed here.

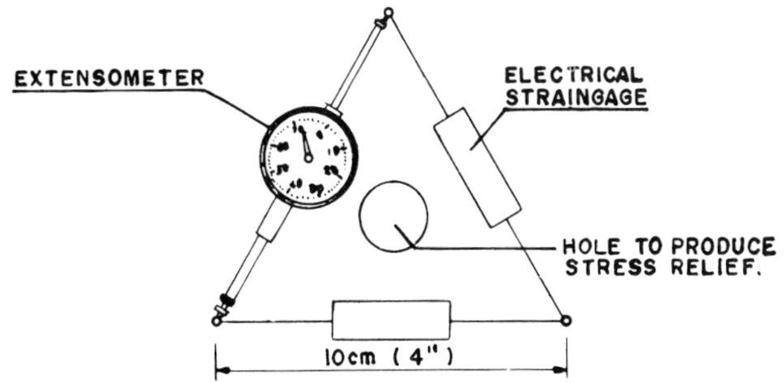

Fig. 13-33 Device for measuring stress relaxation about a hole [48]

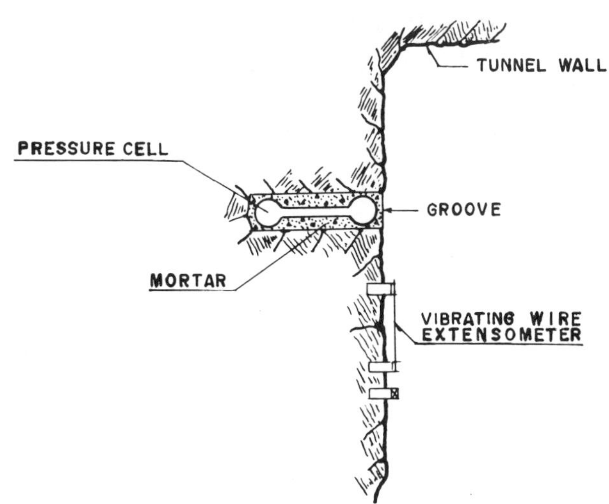

Fig. 13-34 Measurement of pressures on the exposed surface of a tunnel by the method of restoring stresses [48]

First a wire at constant tension is placed on the surface of the material to measure its vibration frequency. This type of device for measuring lengths operates on the principle of the vibrating wire which was discussed earlier in this chapter. Next, a groove is cut in the excavation face, which produces a relaxation of the initial stress. Then, a cell that generates pressure is introduced into the groove (for example, a Freyssinet or flat jack) and is firmly embedded with cement mortar. The cell is then expanded until the original stress state is restored, which is detected because the vibrating wire returns to its initial length and vibration frequency.

Instruments are also used for measuring stresses inside the material through which the tunnel is excavated at points located inside the mass, away from the exposed excavation surface. Stress relaxation measurements are used in some devices [48], with instruments similar to those previously described, but adapted so that they can be introduced into holes with small diameters (7 to 10 cm (2-3/4 to 4 in)). In another system a hole with a diameter of 18 to 20 cm (7 to 8 in) is bored to the point where it is wished to measure the existing stresses. At the bottom of this hole an instrument is placed with a series of strain gauges laid out in crown formation. Stress relaxation is produced by subsequently boring a smaller hole 4 to 5 cm (1-5/8 to 2 in) in diameter, starting from the bottom of the previous one and with a common axis. This relaxation, leads to a variation in the readings of the previously installed strain gauges. During the final stage, a cylindrical jack is introduced into the smaller hole and is expanded so that the stress required to return the strain gauges to their initial position can be measured. This represents the rocks stress before drilling the small hole.

Lastly, various indirect methods have been attempted for measuring the stresses inside the mass of soil about the tunnel. Habib [49] and others have correlated the velocity of sound wave propagation produced in the soil or rock with the acting stresses. However, investigations have shown that small changes in the speed of sound may correspond to changes in stress of many hundreds of kilograms per square centimeter. Therefore, the method is not sufficiently precise for anything more than finding the order of magnitude of stresses.

The purpose of most instrumentation programs that are conducted in tunnels is to measure earth and rock pressures on struts and supports or the stresses acting on different parts of them. The later implies measuring strain in wood or steel parts. These measurements are often taken in pilot tunnels to obtain useful information for designing the full size strut parts. The measurements are taken by making reference points on the strut, either in a pilot tunnel or on the real one, and observing the relative movements, either by surveying methods or preferably by some mechanical device [48]. The relative movements can be interpreted to show absolute movement by relating at least one point with a fixed reference outside the zone of movement.

The forces acting on the strut parts can be measured with longitudinal extensometers of the type described elsewhere in this chapter, positioned, for example, as is seen in Fig. 13-35 (radial layout). Star arrangements or Delta arrangements are also common.

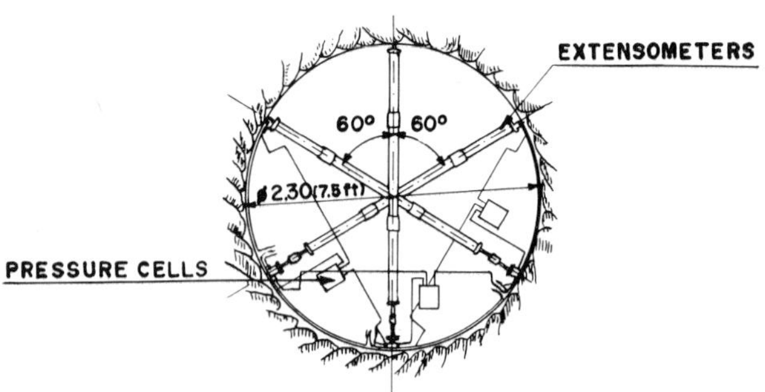

Fig. 13-35 Radial layout of instruments for measuring displacements [48]

The earth pressures can be measured using cells, some of which have been mentioned. Figure 13-36 shows the cell installation that was used in the Chicago subway to measure the pressures exerted by a soft clay on a permanent lining of concrete [50].

An important fact that has been proved both by the measurements taken in the Chicago subway and in tunnels beneath Moscow and Leningrad [48], is that the stresses that develop in the struts are frequently greatly affected by circumstances that have nothing to do with the overall pressure exerted by the soil.

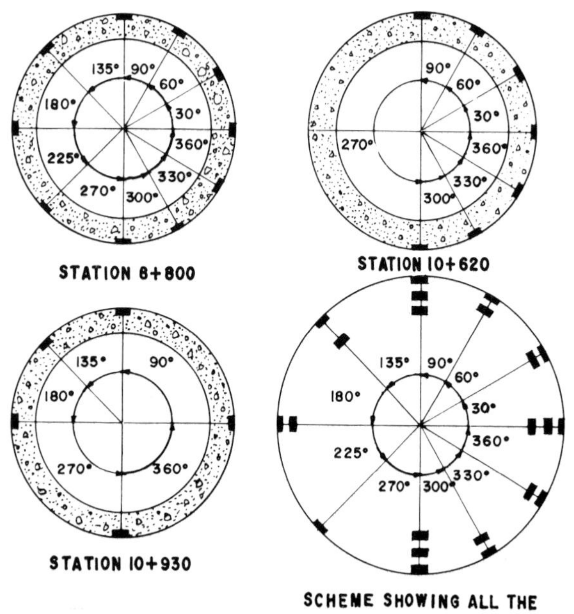

Fig. 13-36 Location of pressure cells in a circular tunnel [48]

Some of the most influential circumstances may be a support system that distorts from its original shape (for example, elliptical shapes in circular tunnels), incorrect grouting between the lining and soil or rock, inappropriate fill in areas of over excavation and either excessively rigid or very weak supports.

13.6 Instrumentation in Laboratory Investigations

Laboratory research, whether in material tests or models, offers extensive opportunities for the utilization of sophisticated instrumentation. This is a field where there is a great deal yet to be investigated. Reference [42] is a summary of several aspects of these problems, and gives a detailed description of various instrumentation techniques that depend on the scale of the tests performed.

The equipment employed in laboratories for measuring displacements should generally be of high precision, because of the relatively small movements that have to be detected in small scale models or laboratory test specimens. For this reason, great care must be taken with the disturbing effect caused by the measuring instrument.

Most of the instruments employed for measuring displacements in soils are mechanical or electrical [43]. The mechanical ones (see, for example [44]) usually consist of a very thin metal rod, 1.5 mm (0.06 in) approximately. This is installed in a slightly larger tube, that supports the road but which excludes soil particles between the two. At the lower end of the rod and protruding from the tube there is an extension which anchors the rod to the soil. Rod and strut have their upper ends supported in such a way that the end of the rod activates a micrometer outside the soil mass. The micrometer measures the displacements of the anchor in the soil at the lower end. Depending on how the instrument is installed, horizontal or vertical movements can be measured.

Electrical displacement gauges almost always operate by determining the change in the distance between two small discs placed close to one another in the soil. In one model [45], each disc has an iron armature which penetrates a solenoid coil with an alternating current imposed on it. Any movement of one disc in relation to the other makes the iron armature penetrate its coil to a greater or lesser degree, and causes a corresponding change in the electrical impedance of the coil-armature system. By previous calibration it is possible to find the relative displacement that corresponds to each change in electrical impedance.

Reference [46] describes another similar device that consists of two small discs each with an electrical coil. The disks are placed in the soil separated by a nominal distance. They are connected by wires to gauges on the surface. The change in their spacing is reflected by a change in the electrical inductance between the coils. Very small pressure cells, which are highly appropriate for laboratory work, has already been described. Almost all the instruments of this type use electrical resistance strain gauges, which make the creation of tiny instruments possible.

Reference [47] describes an investigation where instrumentation was extensively used for measuring stresses and strains. It was conducted on a model pier for a large bridge, to verify the safety of its projected foundation.

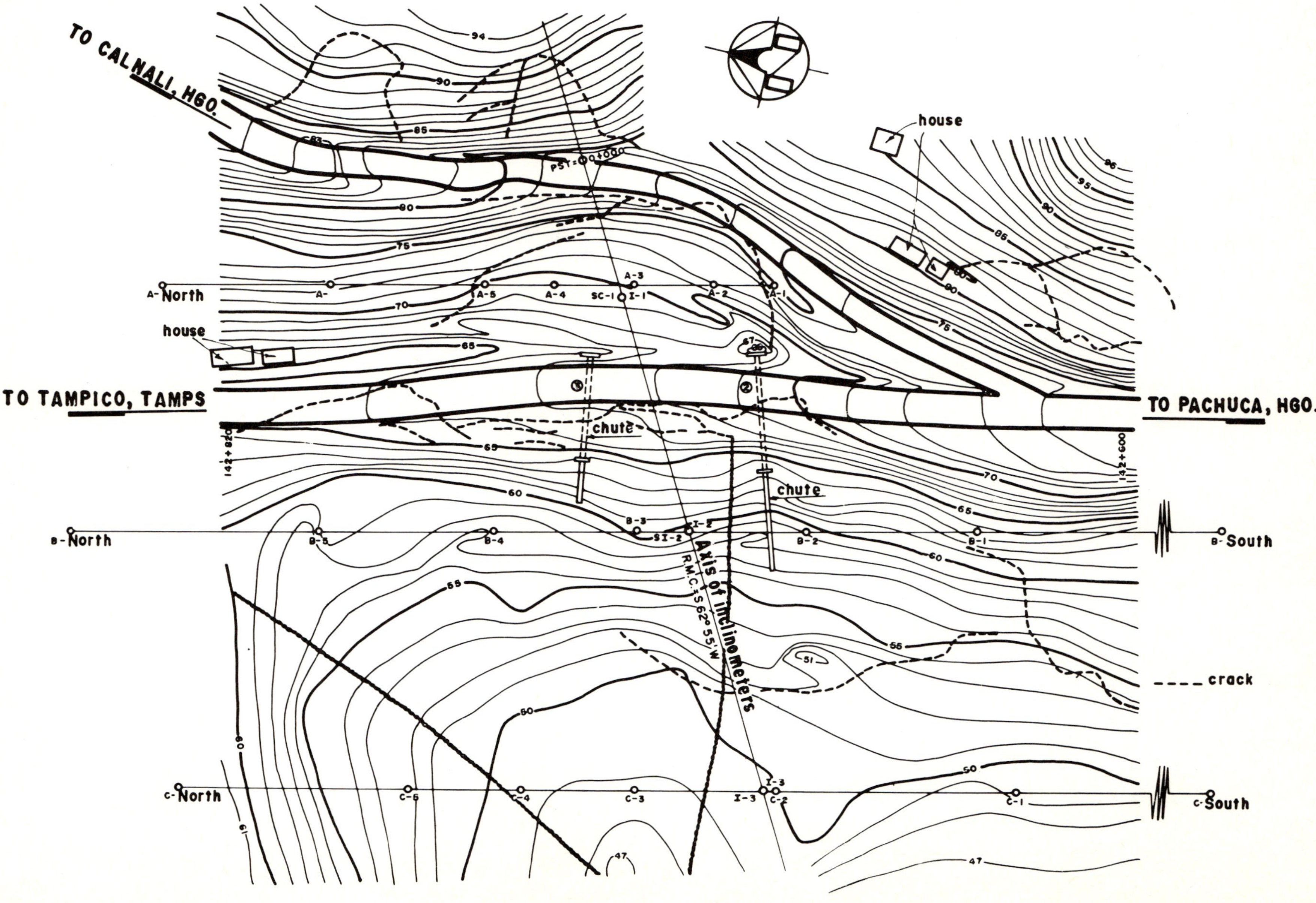

Fig. 13a-1 Topographic map or plan of the slide zone

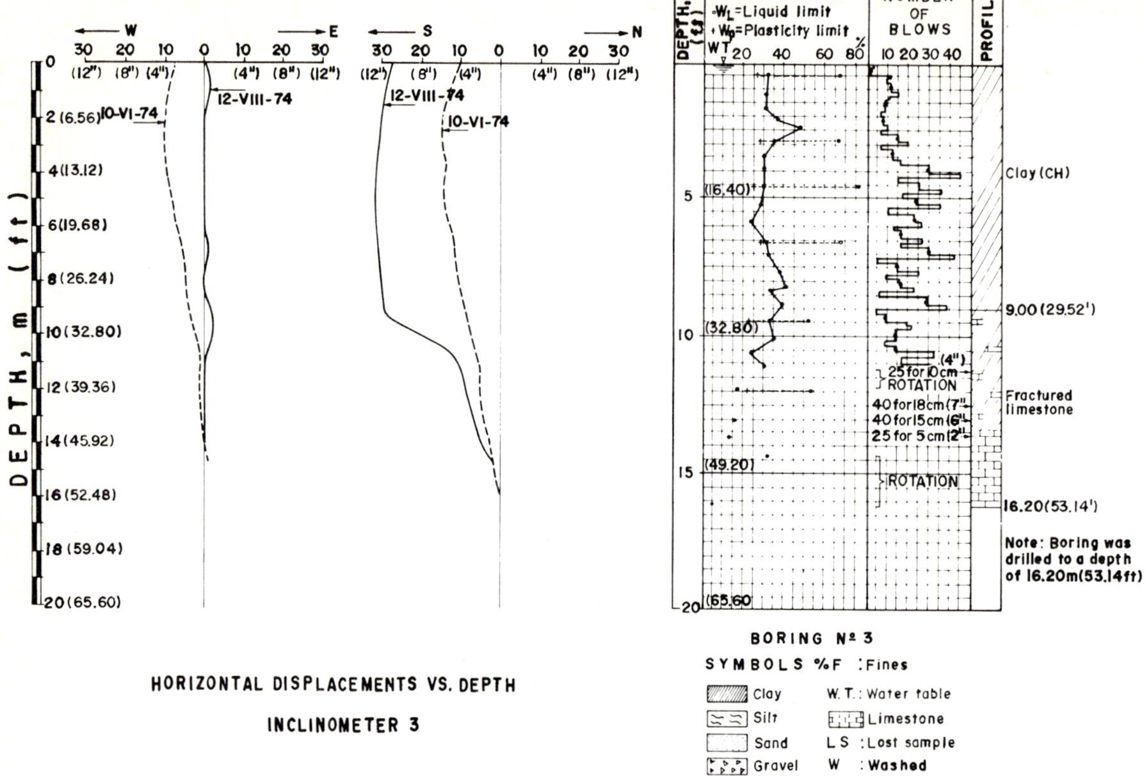

Fig. 13a-2 Soil properties and inclinometer data

APPENDIX 13A
A PRACTICAL CASE

For the purpose of stabilizing a natural hillside (Fig. 13a.1), with slides which were affecting the operation of a road as well as finding the kinematic mechanism of these movements, the horizontal and vertical displacements of the ground were measured. Three lines of surface control points and three Slope-Indicators were installed, on an axis which coincided approximately with the axis of symmetry of the slide zone. In this way the shape and depth of the slip surface could be found.

The lines of control points were selected once a detailed topographical survey of the zone had been made. Some points were selected outside the slide zone, at each end of it. These provided a base line; the movements of other points on it were compared to the original position.

The information obtained from the borings made for the slope indicators, along with the results of the readings taken, are shown in Fig. 13a.2.

The data obtained were used to draw up the soils profile shown in Fig. 13a.3, where the slip surface on which the movements take place can be seen.

The tubing of the slope indicators permitted determination of the position of the water-table, making it possible to establish its influence on the movements.

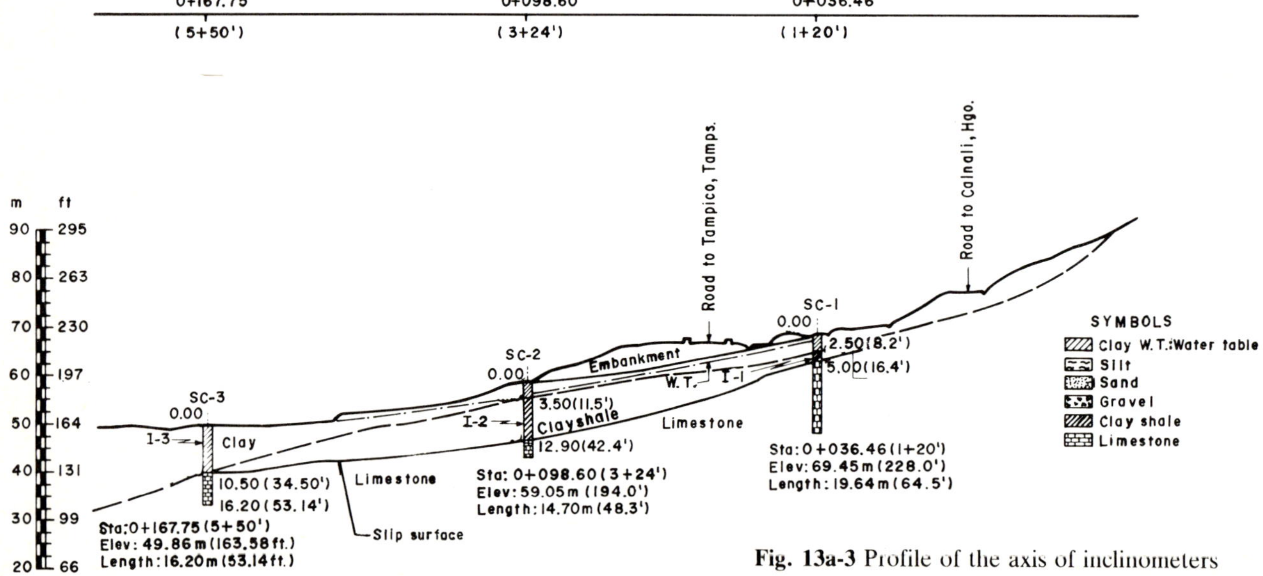

Fig. 13a-3 Profile of the axis of inclinometers

REFERENCES

1. Shannon, W. L., Wilson, S. D. and Meese, R. H., "Field Problems: Field Measurements", Chapter 13 of *Foundation Engineering,* G. A. Leonards Ed., McGraw Hill, 1962.
2. Vargas, M., "Building Settlement Observations in Sao Paulo", *Proc. of the II. ICSMFE,* Rotterdam, 1948, Vol. IV.
3. Rico, A., Moreno, G. and Hernández, R., "Instrumentación de campo en carreteras", *VIII. National Congress of Civil Engineering,* México, 1969.
4. Laboratoire des Ponts et Chaussées, *Etude de Remblais sur Sols compressibles,* Dunod: París, 1971, Chap. 6.
5. California Division of Highways, "Notes on Procedures, Testing Methods and Use of Materials for Highway Purposes", Materials and Research Department, Sacramento, California, 1968.
6. Terzaghi, K., "Settlement of Structures in Europe and Methods of Observations", *Trans. ASCE.,* Vol. 103, 1938.
7. Berbower, R. F., "Subsidence Problem in the Long Beach Harbor District", *Proc. ASCE,* Vol. 85 N° WW2, 1959.
8. Casagrande, A., "An Unsolved Problem of Embankment Stability on Soft Ground", *Procs. of the I. Panamerican Conference on SMFE.* México, 1959.
9. Rico, A., Moreno, G. and García, G., "Test Embankments on Texcoco Lake", *Procs. of the VII. ICSMFE,* México, 1969.
10. Parsons, J. O. and Wilson, S. D., "Safe Loads on Dog – Leg Piles", *Trans. ASCE,* Vol. 121, 1956.
11. Rico, A., Contribution to Discussion on Session 3, *Proc. III. Panamerican Conference on SMFE,* Caracas, Venezuela, 1967, Vol. III.
12. Casagrande, A., "Soil Mechanics in the Design and Construction of the Logan Airport", *Journal, Boston Society of Civil Engineers,* Vol. 36. No 2, 1949.
13. Casagrande, A., "Piezometers for Pore Pressure Measurements in Clay", Chair notes, Harvard University, 1958, Mentioned in Ref. [1].
14. Juárez Badillo, E. and Rico, A., *Mecánica de Suelos. Vol. I. Fundamentos de la Mecánica de Suelos,* Limusa: México, 1973, Appendix.
15. Hvorslev, M. J., "Time Lag and Soil Permeability in Ground Water Observations", Waterways Experiment Station Bull. No 36, Vicksburg, Miss., 1951.
16. Brooker, E. W. and Lindberg, D. A., "Field Measurement of Pore Pressure in High Plasticity Soils", *Procs. International Research Conference on Expansive Clay Soils,* University of Texas A. & M., 1965.
17. Penman, A. D. M., *A Study of the Response Time of Various Types of Piezometers. Pore Pressure and Suction in Soils,* Butterworths: London, 1961.
18. Highway Research Board, H. R. Record No 463 (6 reports), National Academy of Sciences. Washington, D. C., 1973.
19. Marsal, R. J. and Ramírez de Arellano, L., "Performance of El Infiernillo Dam. Symposium on Stability and Performance of Slopes and Embankments", ASCE. Soil Mechanics and Foundations Division, Berkeley. Cal., 1966.
20. Laboratoire des Ponts et Chaussées, "Remblais sur sols compressibles", Special Bulletin T, Paris, 1973.
21. Rico, A., Springall, G. and Springall, J., "Deslizamientos en la autopista Tijuana-Ensenada", *Proc. VII. International Conference on Soil Mechanics and Foundation Engineering,* México, 1969.
22. De Fries, C. K. and Pardo, E., "Grandes rellenos fundados sobre suelo residual", *Proc. IV. Panamerican Conference on SMFE,* Sn. Juan, Puerto Rico, 1971, Vol. II.
23. Casagrande, A., "Evaluation of Embankment Performance", Conference at A.S.C.E. Soil Mechanics and Foundations Division: Stability and Performance of Slopes and Embankments, Berkeley. Cal., 1966.
24. Wilson, S. D., "Investigation of Embankment Performance", ibid.
25. Marsal, R. J. and Ramírez de Arellano, L., "Presa del Infiernillo. Observaciones durante el periodo de construcción y el primer llenado", Publication of the Federal Commission of Electricity, México, 1965.
26. Gamboa, J. and Benassini, A., "Behavior of Netzahualcoyotl Dam during Construction", *A.S.C.E. Journal of Soil Mechanics and Foundations Division,* Vol. 93. SM4, 1967.
27. U.S. Bureau of Reclamation, Earth Manual, Denver, Col., 1960.
28. Bjerrum, L., Kenney, T. C. and Kjaernsli, B., "Measuring Instruments for Strutted Excavations", *A.S.C.E. Journal of the Soil Mechanics and Foundations Division,* Vol. 91 SM1., 1965.
29. Rouse, G. C., Richardson, J. T. and Misterek, D. J., "Measurements of Rock Deformations in Foundations for Mass Concrete Dams", *Instruments and Apparatus for Soil and Rock Mechanics,* A.S.T.M., 1965.
30. Sainz Ortiz, I., "Zumpango Test Embankment", *A.S.C.E. Journal of the Soil Mechanics and Foundations Division,* Vol. 93 SM4, 1967.
31. Kaufman, R. L. and Weaver, F. L., "Stability of Atchafalaya Levees", *A.S.C.E. Journal of the Soil Mechanics and Foundations Division,* Vol. 93 SM4, 1967.
32. Howe, D. R. and Leech, L. R., "Movement within Large Fills. Materials and Research Department. California Division of Highways", Provisional report prepared under the direction of T. Smith, Sacramento, Cal., 1966.
33. Bishop, A. W., Kennard, M. F. and Penman, A. D. M., "Pore Pressure Observations at Selset Dam", *Proc. Conference on Pore Pressure and Suction in Soils,* Butterworths: London, 1961.
34. Golbeck, A. T. and Smith, E. B., "An Apparatus for Determining Soil Pressures", *Proc. A.S.T.M.,* Vol. 16 No 2, 1916.
35. Hanna, T. H., *Foundation Instrumentation,* Trans Tech Publications Clausthal-Zellerfeld FRG, 1973, Chap. 4.

36. Carlson, R. W. and Pirtz, D., "Development of a Device for the Direct Measurement of Compressive Stress", *Journal American Concrete Institute,* 1951.

37. Terzaghi, K., "A Liner Plate Tunnel on the Chicago Subway", *Trans. A.S.C.E.,* Vol. 108, 1943.

38. Peck, R. B., "The Measurement of Earth Pressures on the Chicago Subway", *Bull. A.S.T.M.,* August, 1941.

39. Peck, R. B., "Earth Pressure Measurements in Open Cuts Chicago Subway", *Trans. A.S.C.E.,* Vol. 108, 1943.

40. Housel, W. S., "Earth Pressure on Tunnels", *Trans. A.S.C.E.,* Vol. 108, 1943.

41. Reference [35], Chap. 6.

42. Roscoe, K. H., "The Influence of Strains in Soil Mechanics", Rankine Lecture, *Geotechnique*, Vol. 20 No 2, 1970.

43. Reference [35], Chap. 6.

44. Carr, R. W. and Hanna, T. H., "Sand Movement Measurements near Anchor Plates", *Procs. A.S.C.E.,* Vol. 97 SM5, 1971.

45. Eggestad, A., "Deformation Measurements below a Model Footing on the Surface of Dry Sand", *Procs. European Conference of Soil Mechanics and Foundation Engineering,* Wiesbaden, 1963, Vol. I.

46. Morgan, J. R. Contribution to Discussion on Session 4, *Procs. Roscoe Memorial Symposium,* Cambridge University, 1971, Mentioned in [43].

47. Rico, A., Sosa, R., Quintero, M., Aztegui, E. and Rangel, M., "Bridge "Mariano García Sela" (Metlac), Physical model of the foundation of pier No 2", Journal Engineering, Faculty of Engineering, UNAM. 1970.

48. Széchy, K., *The Art of Tunneling,* Akademiai Kiadó: Budapest, 1967, Chap. 3.

49. Habib, P., "Determination du module d'elasticité des roches en place", Annales de l'Institut B.T.P., September 1950.

50. Terzaghi, K., "Shield Tunnels of the Chicago Subway", *Journal of the Boston Society of Civ. Eng.,* 1942.

51. Rico, A., Springall, G., Springall, J., Moreno, G. and Mendoza, J. A., "Landslides on the Tijuaja-Ensenada Highway", Final Report, Publications of the Mexican Ministry of Public Works (Spanish and English edition) México, 1975.

OTHER RELATED REFERENCES

Baguelin, F., Jézéquel, J. F. and Shields, D. H., *The Pressuremeter and Foundation Engineering,* Trans Tech Publications, Clausthal-Zellerfeld, FRG, 1978.

Hanna, T. H., *Field Instrumentation in Geotechnical Engineering,* Trans Tech Publications, Clausthal-Zellerfeld, FRG, 1985.

Javor, T., "General Recommendations for the Vibrating-wire Measuring Method and its Equipment", Rilem Recommendations TBS.-5, Czechoslovakia, 1984.

Pignaud, M., and Chaput, D., "Coulis de Scellements D'appareiles de Mesures dans les Sols", Colloque International sur la Gestion des Ouvrager, Paris, 1981.

CHAPTER 14

TUNNELLING

14.1 Introduction

The purpose of this chapter is to describe concepts that serve as an introduction to the design and construction of tunnels. Three aspects in particular will be dealt with 1) the evaluation of the pressures acting on temporary or permanent tunnel linings, 2) the measurement and interpretation of the deformations in the structure and the adjacent ground, and 3) the construction methods most commonly employed, briefly considering their influence on the general behavior of the structure and the development of any pressures and deformations that may take place. As in the cases already discussed in problems of earth thrust on retaining structures (Chapter 5) the construction procedure that is employed in tunnels is largely responsible for the pressures and behavior that are likely to occur.

This chapter pays special attention to the difficulties already so often referred to elsewhere in this book, which inevitably arise when soil mechanics is separated from rock mechanics and applied geology. Tunnel construction is fundamentally linked with the last two disciplines. Rare are the tunnels that are driven through soils alone but these few cases may pose some of the most difficult problems in the art of tunnel construction. Moveover, severely fractured rocks or ones that are only moderately broken whose joints and fractures are filled with soil, exhibit behaviors which cannot be readily divorced from soil mechanics. Tunnels are perhaps the structures where separation of the three geotechnical disciplines may prove hardest. The chapter should be approached bearing this in mind.

Throughout this book the emphasis is on soils; however, rocks are not excluded. Soft rocks behave in many ways like hard clays and fractured rocks such as angular gravels. Therefore the applicability of the ideas presented here to these in-between materials is discussed. Tunnelling in hard continuous rock is not discussed in detail. The problems in rock are more of construction than design but many of the principles discussed in this chapter do apply to rock tunnels.

The use of tunnels in highway technology is most unevenly distributed. They are a familiar solution to railroad designers and builders throughout the world because of the limitations imposed by railroad grades. In roads their use depends to a certain extent on the personal preference of the designers and the customs and traditions prevailing in the region. In the road networks of some nations tunnels are common structures. One even wonders if they are not too frequently employed. Examples are many European countries and Japan. In the road networks of many other nations, however, tunnels are rare despite mountainous topography.

Whether or not a tunnel should be built at a specific location on a road depends foremost on the topography, on whether by tunnelling through an obstacle appreciable road length can be saved, observing appropriate requirements of curvature and slope. In railroads, the operation costs produced by the demands of curvature and slope are of paramount importance; consequently railroad designers are accustomed to considering the tunnels in their analyses of alternatives, often adopting it as an ultimate solution.

In roads, slope and curvature requirements are far more flexible. In many regions where traffic intensities are relatively low, longer steeper roads are preferred to the higher costs of construction, illumination and ventilation of tunnels. Due consideration of these criteria will result in a higher proportion of railroad tunnels than road tunnels in a well conceived transportation network. However it is also probable that these structures will often prove an economical and appropriate solution in roads.

The construction of a tunnel involves many hazards and uncertainties, despite the progress that has been made in tunnelling techniques in recent years, according to many experienced highway engineers. In tunnels more than in any other types of highway structure, situations arise which cannot be readily predicted by exploration and preliminary studies, leading to unexpected extra work, time and money, which upset construction programs and cause economic, social and political difficulties. Naturally the greater the extent of the exploration and preliminary studies that are conducted, the smaller the risks. All this influences the highway constructors in many countries to avoid tunnels. However, railroad builders in the same countries may be of the opinion that tunnels can be built quite safely by observing very reasonable programming procedures. It is not unusual among railroad and highway designers for there to exist a friendly but lively controversy about the virtues of tunnels, their construction hazards, and the scope of their application.

The authors are of the opinion that current techniques have made tunnel construction as safe and successful as any other large-scale engineering project. They believe that the decision to construct a tunnel should in most cases be a matter of economics and availability of equipment, and should be determined by the usual comparisons of costs of construction, operation and maintenance. The tunnel shoud be systematically considered among the variety of solutions available and should be constructed whenever it proves to be the most economical and safest: the appropriate design.

In roads, the art of tunnelling can be regarded from a broader range of applications. Often the tunnel provides an appropriate alternative to the construction of culverts, drainage and diversion structures. In Chapter 7 of this book reference was made to tunnels as excellent subdrainage structures.

The analysis of earth pressures on tunnel linings, the objective of this chapter, employs a methodology that is based on theoretical studies, but modified and influenced by intuition and experience. The design of such linings is both an art and a science [1].

There are theoretical solutions which permit the calculation of stresses and strains in tunnels through idealized materials, but the designer must adapt these calculations to real materials, with properties that are not necessarily identical to those of the ideal materials considered. The properties of the real material can seldom be reliably determined; moreover they may vary greatly over short distances. The assumed concepts in theoretical solutions are therefore often not very satisfactory. Consequently both theoretical and experimental investigations into the behavior of tunnels must continue until better methods of analysis can be found. It is the opinion of specialists that the combination of theoretical studies and empirical knowledge resulting from the observation and measurement of the performance of existing tunnels provide the most promising means of achieving a substantial improvement in the methods of design which are today available and of developing new and better methods in the future.

A tunnel lining is often thought of as a relatively rigid element in contact with the surrounding less rigid soil material. This is a limited view, for there are some cases where other types of supporting systems such as shotcrete and anchor bolts are sufficient to solve the problem, and still others where the material in which the tunnel is driven is capable of supporting itself.

Tunnel linings or support systems may be temporary or permanent. The distinction between the two is not often very significant in relation to soil mechanics; the chief difference between them is the construction procedures and durability of the materials they require. A tunnel lining study consists of two steps. First, it includes an evaluation of what is wanted of the lining throughout the changing conditions to which it may be subjected during its useful lifetime, and second it includes an investigation to determine how each of these requirements can be achieved. A lining may perform different functions at different times. The direction of the heaviest loads and the distribution of the stresses and strains around the tunnel change from the initial stages of excavation to the final equilibrium stage.

The principal requirements of a permanent lining are related to strength, stability, durability, control of water, and deformability during the service life. In a temporary lining, the chief requirement is usually the immediate retention of the excavated material. This requirement depends largely on the nature of the soil and ground water, the excavation methods employed and the procedures used to construct the support itself. Both temporary and permanent linings are greatly influenced by environmental factors.

The first requirement for a satisfactory tunnel [2] is that it may be constructed safely, so that it will continue to fulfill its functions on its own or with the help of a lining. The second requirement is that construction must not cause damage to neighbouring structures. Often in urban areas where the construction of tunnels is desirable to divert traffic from surface streets, nearby structures may be influenced by tunnelling work. A third condition that must be fulfilled by a satisfactory tunnel is that it must have minimum maintenance throughout its service life and remain relatively unaffected by the influences to which it may be subjected. The most important of these is earth pressure; however other hazards such as accidents or fires in the tunnel must be considered.

14.2 General Concepts of Tunnel Behavior

The excavation of a tunnel alters the stress conditions in the original soil or rock, which initially is a mass in equilibrium within a field of gravity. Tunnelling produces changes in a continuous manner or by stages, until a condition is eventually reached in the now discontinuous mass in which relatively unchanging state of ultimate, but changed, equilibrium is reached. This final condition includes the establishment of new hydraulic conditions in the soil or rock and the end of the deformations and changes in stress produced by the excavation.

When a tunnel is excavated, a zone with changing stresses is created. There is generally an increment in vertical pressures in front of the excavation which moves forward with it. At the excavation face, the stress states are complex and three-dimensional. They tend to become two-dimensional as the tunnel face moves ahead, and the opening is left behind. The changing stress states caused by the excavation cannot take place without deformation occurring in the soil or rock. When there are linings, these deform also. The resulting deformations evolve with time, which add new variables to the process. Excavation also brings about changes in neighbouring pore water pressures. Since the tunnel is eventually under atmospheric pressure ground water above it tends to flow into it. Consequently, the change of ground water pressures constitutes another important variable.

In soils with a fairly low permeability, adaptation of the pore water pressures to the new stress states is by no means instantaneous, thus adding another time-dependent mechanism. The effect of pore pressure changes and influences the pressures in the solid structure of the soil. The increase in effective pressures as the pore pressures gradually dissipate leads to changing shear stresses and strains in the medium. The whole picture is complicated even further by viscoplastic effects, such as creep.

The construction of a tunnel not only changes the stress states inside the soil or rock, but often it changes the medium itself. Blasting usually reduces the strength of rocks and hard soils around the cavity. Other excavation methods such as shield tunnelling cause remolding of the neighboring soils. Only in tunnels that can be excavated using hand tools and where no support is required can disturbance of the cavity materials be considered really small.

Most tunnels in soils have to be supported at some stage during construction. Often the support is required to ensure immediate stability. Sometimes, even before excavation commences the soils involved have to be improved. Supports are generally rigid, made of wood or steel. However in the early stages other more flexible types of support are sometimes employed even when a rigid support is constructed shortly afterwards. It is also common practice to keep construction of a support or lining system some distance behind the advancing excavation face. In this case a partial stress relief occurs in the unsupported part of the excavation as well as movements in the ground which occur before the lining is installed. Sometimes a support is built which subsequently expands and presses against the tunnel walls. This process induces an increment in the stress states both in the lining and in the ground. These new stresses produce corresponding deformations.

Even when there are no apparent hazards of collapse, caving in or closing of the tunnel cavity, lining may be required to maintain the deformations in the excavation within tolerable limits. Very large deformations may cause excessive distortion of the soil or rock above, causes subsidence of the ground surface and damage to neighbouring structures. Moreover, large distortions often produce undesirable reductions in the shear strength of sensitive soils. Thus, the timely installation of a support may impede the eventual development of heavy earth pressures. Expansive materials that creep or rocks with joints filled with clayey soils cannot achieve a satisfactory ultimate equilibrium condition if they are not supported, even if initially they show no sign of instability.

It is impossible and also inadvisable to totally prevent deformations of the cavity. Some deformation is necessary to achieve a favourable pressure distribution. The engineer determines the amount of minimum movement that can be advantageously tolerated and at what point deformation will start to be adverse. Deformation is often controlled by wedges or blocking in the lining. The loads carried by a support system or lining depend on the condition of the soil at the time of support installation. If the soil has reached an ultimate equilibrium condition before the lining is installed, this lining may receive no subsequent thrusts, but if the lining is installed before ultimate equilibrium is achieved, it becomes a new boundary condition to the state or pre-existing stresses and strains, so that these states will evolve in a different manner than if the support had not been installed.

For a time at least, excavation reduces the radial normal stresses on the soil or rock in the tunnel wall to that of atmospheric pressure. This causes deformations in the walls such that the stresses that finally act on any lining that is installed will in no way resemble the ones that originally existed in the soil or rock.

Tunnelling causes radical changes in the pore water pressures of the subsoil. These may be temporary or permanent, depending on the relative permeability of the soil and the lining. Construction of a tunnel below the ground water-table generally lowers the neighbouring water-table. This causes an increase in the effective stresses in the soil mass and changes in the weights of this mass, resulting in settlements that are usually irreversible. If the tunnel lining is impervious or precautions are taken to add water by inflow wells to the groundwater, the water-table will usually recover after a time. If the contrary is true, the tunnel can be a permanent drain. Restoring the ground water level demands linings that will support hydrostatic pressures, whereas allowing the tunnel to remain as a drain demands that all precautions be taken to ensure that the tunnel will continue to function efficiently as a drain, without this interfering with its principal function as a transportation route.

A design method that uses to best advantage the capacity of the soil or rock to support itself will usually be economical and safe. An appropriate choice of lining systems and installation procedures may minimize earth loads and make the soil or rock bear the greater part of its pressures unaided. It is a fundamental rule that a lining should be regarded as a restriction which is installed to help the soil or rock support itself. In order to do this, the control of movements is very important, because large movements may loosen and weaken the material, making it lose strength and the ability to support itself. Allowing the appropriate degree of displacement to develop is an essential aspect of the design of tunnel linings. How the movements are controlled depends largely on the properties of the soil or rock and the nature of the lining. For certain specific conditions there exists an optimum balance between flexibility and rigidity.

14.3 Earth and Rock Pressure in Tunnels

Under natural conditions, soils and rocks are subject to their own self-weight and to the weight of the materials and any structures above. As a consequence, stresses and strains develop between the individual material particles. So long as a rock or hard soil is confined, the interparticle displacements corresponding to the acting stresses will be small. The stresses are stored in the material, sometimes reaching very high values. If the material thus stressed is able to move, by excavating within it displacements will take place. If the stresses are high enough there may be explosions, in which fragments of the rock or hard soil are violently ejected. If the residual stresses have not gone beyond the elastic limit of the material, the displacements will be of an elastic nature.

Any excavations that are made in a mass create an empty space towards which displacements occur. The weights of the overlying materials are a load distributed over the roof of the resulting cavity. The strength of the soil or rock, scarcely mobilized before excavation as a consequence of impeded deformation that prevailed, now supports the unbalanced load.

In order to keep the cavity open, auxiliary retaining elements are usually necessary. The pressure received by these elements from the soil or rock acting against them is the load, or pressure for which the support is designed.

Determining these pressures for design is one of the most difficult problems confronted by a geotechnical engineer. This is not only due to establishing the initial or residual stress in the virgin soil or rock, but also because the stress state is modified around the open cavity. This modification depends on many factors that are hard to evaluate, such as the rigidity and strength of the soil or rock, the size of the opening, the excavation method used, the shape and rigidity of the supports that are employed and the time the cavity is left unsupported.

The history of residual stresses in a rocky mass or a firm soil is reflected in the cracks, fissures or other defects that may exist. However it is usually extremely hard to establish which of these stresses still act. Sedimentary soils are deposited in a more or less uninterrupted and uniform manner, acquiring a stratification profile that is usually little disturbed by external influences. In such soils residual stresses are not generally very high and in simplified pressure analyses it is common practice to consider only the vertical stresses due to the weight of overlying horizontal strata. In residual soils, on the other hand, the situation more closely resembles that prevailing in rocky masses. The heritage from the original rock manifests itself not only in cracks, fissures and fractures, but also in residual stresses that are greater than those produced by the weight of the materials above. The same occurs in formations of fragmented or fissured rock, which can be described as either soils or rocks. Thus basic differences, that are unfortunately not clearly defined, must be established between formations of sedimentary soils and those of residual soils or other materials that might be regarded as having mixed behavior. It is probable that in formations of residual soil or in formations of fragmented or highly fissured rock (with the fissures filled with soil) it would be reasonable to estimate pressures with the methods of rock mechanics, whereas in formations of sedimentary soils simpler patterns can be used, which consider only the vertical stress of the soil weight and the horizontal stress caused by the vertical stress.

Current experience [3] demonstrates that stress states in rock do not establish themselves with the relation between horizontal and vertical stresses that is commonly found in soils. There may be an initial hydrostatic state and on this are superimposed tectonic stresses, often acting at right angles or in the opposite direction to the gravitational field. The random orientation of joints in rocks demonstrates that a vertical major principal stress is not as common in rocks as it is usually considered in soils.

In transported soils, especially clayey deposits, it is observed that lateral pressures do not develop in accordance with what might be expected from the corresponding deformations. Instead there is a gradual change from the initial condition, in which the horizontal stress is a fraction of the vertical stress, until an ultimate condition is reached which is more nearly hydrostatic.

On the basis of TERZAGHI's works [4-6] it has become customary to express the pressures of rock or firm soil, exerted on supports after the excavation of a tunnel, as the weight of a mass of a certain height above the tunnel. This is the mass that would be the first to fall were no supports installed. Deformations in the support system cause a subsequent arching of the mass above the roof and reduction of the pressure. TERZAGHI distinguishes the case of pressures exerted by soft plastic soils where subsequent deformation of the support produces no stress reduction. The magnitude of pressures in soft soils does not depend greatly on when the support is installed or what its characteristics are, although pressure distribution is certainly affected by these factors. Pressure from rocks and firm soils, however, is affected by when the support is installed, because the deformations that follow excavation vary greatly with time.

The reasons why pressures develop on the supports are as follows [1, 3]:

— Loosening of the mass of rock or firm soil
— The weight of overlying mass
— Tectonic or other residual forces
— Expansion of the soil or rock in which the tunnel is excavated.

These mechanisms can be categorized by three types of pressures in tunnel roofs:
— Loosening pressure
— Pressure due to the tunnel depth and rock weight
— Swelling pressure.

These pressures may develop individually or together. The predominate type of pressure that develops in a given case depends largely on the nature of the material above and around the tunnel. Three cases are distinguished:

— Sound massive rocks
— Soft or weathered rocks and hard residual soils, some fractured rocks
— Soft residual or transported soils, closely fractured rocks.

14.3.1 Loosening Pressure

When the mass of rock or firm soil over the tunnel roof is loosened due to excavation and the weight of the overlying load, the behavior of the mass may resemble that of a mass of granular soil in a silo when a door is opened in the bottom of the silo. This is the pressure condition considered by TERZAGHI in his analyses of thrusts in tunnels [6, 7]; the rock loosening pressure is represented by the mechanisms proposed by him.

TERZAGHI's conception of the mechanism of these pressures includes predominantly soil arching. This effect may be visualized as follows. An extensive mass of soil resting on a rigid horizontal surface is assumed. For some reason part of the horizontal surface subsides a little; the soil above that part also tends to subside. The downward movement of this part of the soil in relation to the rest, which has remained immobile because it is firmly supported, is opposed by the shear strength that develops between the subsiding soil and the stationary remainder. This shear strength partially supports the subsiding mass in its original position, thus reducing the pressure of the soil on the part of the bearing surface that subsided. As a result, there is an increase in the pressure exerted by the stationary soil on the immobile parts of the horizontal bearing surface.

Consequently, pressures are transferred from the part of the surface which subsides to the stationary supports. This effect recalls the manner in which a structural arch works; hence the name *arching effect*. The practical consequence of the arching effect on supporting elements where deformation is restricted is a concentration of pressure, and where subsidence occurs, is a reduction in the pressure. Thus the pressure diagram is modified.

The experiment that will now be described enables the arching effect to be visualized [7]. Subsequent interpretation leads to the classic formulation of the phenomenon made by TERZAGHI [7, 9].

A balance is placed on a table. Above one of the balance pans a vertical glass or lucite cylinder is placed in such a way that it does not touch the pan, being held by an independent support which stands on the table. On the other balance pan a receptacle with water fitted with a drain cock is placed. The water extracted is collected in a graduated test tube. On the scalepan underneath the glass cylinder a counterweight is placed to balance the weight of the receptacle on the other pan when it no longer contains any water. Fig. 14-1 illustrates the arrangement of this test apparatus.

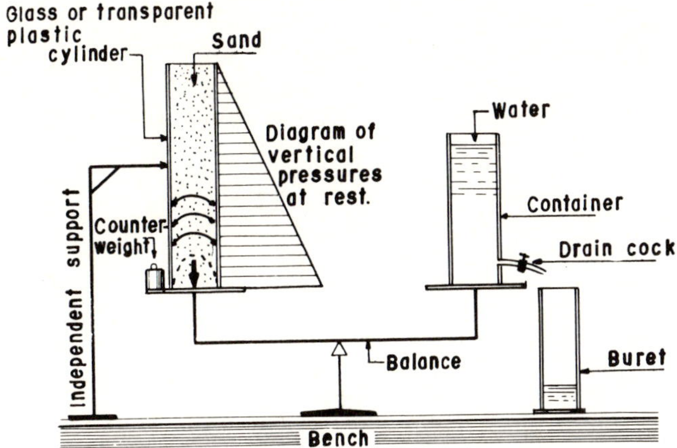

Fig. 14-1 Experiment to illustrate the effect of "arching" or shear transfer of forces in granular soils

Once the cylinder has been placed very close to the balance (but without touching it when the scales are absolutely still), it is filled with sand which is dropped through the top. The sand thus rests directly on top of the scalepan. At the same time, water is placed in the counterweighted receptacle on the other scalepan, so that the weight of the water is equal to that of the sand on the first pan. Under these conditions the pans are left alone, ensuring, of course, that they are balanced. If the drain cock of the water containing receptacle is now opened, allowing the water to flow into the test tube, it will be observed that the scales do not become unlevelled, although the weight of water that is lost may be considerable.

When only a small fraction of the original water is left in the receptacle, it will be noted that the scales become unlevelled, and sand spills from the cylinder through the space underneath it caused by the movement of the scales.

A suggestive interpretation of the experiment is to assume that what occurs in the cylinder is that when the scalepan tends to drop any yield under the sand, the sand starts to undergo arching, transmitting its weight to the walls of the cylinder by friction. This effect reduces the load of the sand on the balance. As water continues to drain from the receptacle on the other pan, the first pan (under the sand) will continue to subside by an imperceptible amount, but sufficient to enable a greater development of the arching effect in the lower zone of the sand. The upper part of the sand is supported on the arches (or in this case it would be better to say *vaults* formed in the lower granular mass. Unbalance of the balance occurs when the weight of the water is practically equal to the weight of the sand contained in the semi-ellipsoid of revolution, indicated in the figure by a broken line. This mass of sand has no other possible means of support. Once the equilibrium is broken, this volume of sand yields and runs out permitting the collapse of the arches or vaults, with the result that all the sand spills. This arching effect is also often referred to as *silo action*, since it occurs in grain storage silos, when grain is withdrawn.

The arching theories that have been most carefully studied refer to two specific problems. The first considers a stratum of sand of infinite extension, but finite thickness, resting on an infinite base of which a narrow section of infinite length subsides; in other words, a problem of plane strain is analysed. The second problem considers the case of a vertical supporting element which turns about its upper end, causing arching in the mass of fill. Both problems are illustrated schematically in Fig. 14-2 (Parts *a* and *c*).

TERZAGHI distinguishes three types of arching theories in the treatment of the first of the two problems.

1) Theories in which the equilibrium of the sand located immediately above the subsiding zone is considered, without investigating whether the results are compatible with the equilibrium of the sand outside the zone.

2) Theories based on the hypothesis that the entire mass of sand above the subsiding boundary is in a critical equilibrium condition. This hypothesis is not compatible with available experimental data.

3) Theories in which it is assumed that the vertical sections *ad* and *bc* (Fig. 14-2a) passing through the ends of the subsiding belt are slip surfaces and that the pressure on the subsiding boundary is equal to the difference between the total weight of the sand above that boundary and the frictional strength developed along the slip surfaces. The real slip surfaces are *ae* and *bf*, which are curves, as indicated by experimental data, with a greater separation on the surface than the width of the subsiding zone. Consequently the friction along the assumed vertical surfaces cannot be fully developed for these surfaces are not the correct slip surfaces. This fact leads to an error tending towards unreliability.

Calculations based on the three theories lead to different results. The arching phenomenon has not been measured often enough in full size structures to permit assessment of

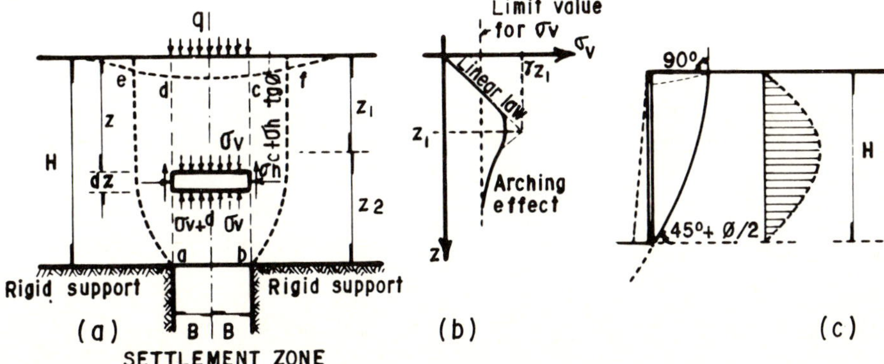

Fig. 14-2 The two problems most frequently involving "arching"

the validity of any of them. The simplest group to analyze is the third, from which the following theory has been used in tunnel analyses.

It assumes that the strength of the soil is given by COULOMB's law:

$$s = c + \sigma \tan \varnothing$$

Initially a surcharge q is assumed to act on the surface of the ground. Figure 14-2a shows a prismatic element of soil with a thickness dz situated at depth z. The vertical stress on the upper face is called σ_v and the horizontal stress on the lateral faces is assumed to be:

$$\sigma_h = K\sigma_v \tag{14-1}$$

where K is a coefficient of earth pressure.

Considering the vertical equilibrium of the elemental prism, the following equation is obtained:

$$2B\gamma dz = 2B(\sigma_v + d\sigma_v) - 2B\sigma_v + 2cdz + 2K\sigma_v \tan \varnothing \, dz \tag{14-2}$$

Simplifying we obtain:

$$\frac{d\sigma_v}{dz} + \sigma_v \frac{K}{B} \tan \varnothing = \gamma - \frac{c}{B} \tag{14-3}$$

which is a linear differential equation of the first order and first degree. By solving it, the following expression is obtained:

$$\sigma_v = e^{-\int Pdz}\left[Qe^{-\int Pdz}\, dz + C\right]$$

where:

$$P = \frac{K}{B} \tan \varnothing \quad \text{and} \quad Q = \gamma - \frac{c}{B}$$

Consequently:

$$\sigma_v = e^{-z(K/B)\tan \varnothing}\left[\left(\gamma - \frac{c}{B}\right)\int e^{z(K/B)\tan \varnothing}\, dz + C\right]$$

and:

$$\sigma_v = \frac{B\left(\gamma - \frac{c}{B}\right)}{K \tan \varnothing} + Ce^{-z(K/B)\tan \varnothing} \tag{14-4}$$

Taking into account the problem under consideration, the following boundary condition can be written:

$$\sigma_v = q \quad \text{if} \quad z = 0 \tag{14-5}$$

Applying this condition to Eq.(14-4):

$$\sigma_v = \frac{B\left(\gamma - \frac{c}{B}\right)}{K \tan\varnothing}\left[1 - e^{-K \tan \varnothing (z/B)}\right] + qe^{-K \tan \varnothing (z/B)} \tag{14-6}$$

where e is the base of natural logarithms. If the material that makes up the stratum under study is purely *frictional* ($c = 0$), the above equation reduces to:

$$\sigma_v = \frac{B\gamma}{K \tan \varnothing}\left(1 - e^{-K \tan \varnothing (z/B)} + qe^{-K \tan \varnothing (z/B)}\right) \tag{14-7}$$

If surcharge q is zero, Eq.(14-7) reduces to:

$$\sigma_v = \frac{B\gamma}{K \tan \varnothing}\left(1 - e^{-K \tan \varnothing (z/B)}\right) \tag{14-8}$$

When z becomes great, approaching ∞, the value of σ_v for a stratum of clean sand, without any surcharge, becomes

$$\sigma_v = \frac{B\gamma}{K \tan \varnothing} \tag{14-9}$$

a constant. It can be seen, therefore, that in this case the vertical pressure in the sand no longer increases in proportion to depth. Instead its graph becomes curved, approaching asymptotically the value found by Eq.(14-9). Thus, according to the theory described, the pressure that acts on the subsiding boundary is smaller than would be deduced from the depth of this boundary. Observing Eq.(14-9) and considering for example $\varnothing = 30°$ and $K = 1$:

$$\sigma_v = 2B\gamma \tag{14-10}$$

This indicates that for these values the pressure that is exerted on the subsiding zone is limited to that of a column of sand of height $2B$: the width of the subsiding zone. It is important to observe in Eq.(14-9) that the value of the vertical pressure σ_v is proportional to the width of the subsiding zone, $2B$.

On the other hand, however, data from the experimental observation of sands [9] have shown that the value of K increases from 1, very close to the center of the subsiding boundary, to 1.5 at an elevation $2B$ above this point. At elevations higher than $5B$ approximately, it appears that the boundary subsidence no longer influences the stress state of the sand. These experimental facts impose the hypothesis that the shear strength of the sand is mobilized only in the lower zone of thickness z_2 of the slip surfaces ad and bc. With this hypothesis, the upper part of the mass of sand acts on z_2 simply as a surcharge q and the pressure on the subsiding boundary must therefore be calculated using Eq.(14-7).

If z (Fig. 14-2a) is the depth throughout which there are no shear stresses in the vertical slip surfaces, the following will be true: $q = \gamma z_1$. Consequently, for this value of q and for $z = z_2$ the depth at which the shear strength of the sand is mobilized, Eq.(14-7) becomes:

$$\sigma_v = \frac{B\gamma}{K \tan \varnothing}\left[1 - e^{-K \tan \varnothing (z_2/B)}\right] + \gamma z_1 e^{-K \tan \varnothing (z_2/B)} \tag{14-11}$$

when z_2 approaches ∞ the value of σ_v becomes the same constant value as in Eq.(14-9).

Consequently, when part of the lower boundary of a very thick mass of sand subsides, the pressure on this subsiding zone is not equal to that corresponding to the entire height of the sand above it, but reaches a lower value given by Eq. (14-9), regardless of depth.

For example, if $\varnothing = 40°$, $K = 1$, $z_l = 4B$, the pressure of the sand increases with depth according to a hydrostatic law to a depth of $z_1 = 4B$. Below this value, the pressure is described by Eq. (14-11). It diminishes as the depth increases, approaching asymptotically the value in Eq. (14-9). The theory indicates that at a depth of more than $8B$, the influence of the weight of the sand on the thickness z_1 is already negligible, for at this depth the value of σ_v is already sufficiently close to the final constant value. It can also be said that at an elevation of over $4B$ or $6B$ above the center of the subsiding zone, the pressure on this subsiding zone is no longer influenced by the state of stresses prevailing in the upper layers of sand.

The transition from the fully mobilized shear strength in the lower part of slip surfaces *ad* and *bc* to the zero value in the upper part of these hypothetical slip surfaces is probably gradual. Consequently the change in the vertical normal stress with depth will also be gradual, not reaching γz_1. Below z_1 it drops until it approaches that given by Eq. (14-9). In Fig. 14-2b the solid line represents the real variation of σ_v, verified with measurements, whereas the broken lines represent the theoretical sudden change at depth z_1.

Returning to the initial example of arching, recent experiments show that when the opening in the rigid horizontal surface on which the sand mass rests suddenly subsides, the pressure in the subsiding zone falls suddenly to its minimum value and arching is maximum, but if subsidence becomes greater, the pressure increases gradually, although the value corresponding to the weight of the column of overlying soil (geostatic pressure) is not reached. This phenomenon is shown in Fig. 14-3, which illustrates KÜNCZL's experiences [3].

Subsidence in an opening in the rigid supporting surface of a soil mass produces arching similar to that caused by the subsidence of a tunnel roof under the weight of soil above it. If the roof is left unsupported (which is equivalent to allowing part of the support to subside, the pressure above it will be at least partially regenerated. Wedges of soil or rock detach themselves and fall, becoming larger as time goes by (Fig. 14-4) [3]. The value of the angle α which develops, depends on the cohesion of the soil, increasing with it. The maximum height affected above the roof before final equilibrium is restored will be:

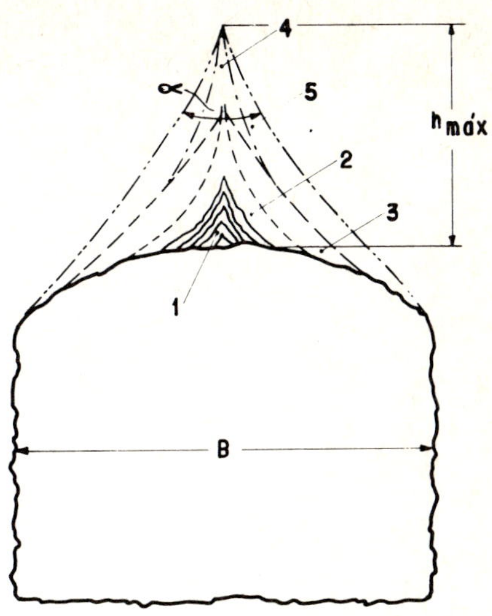

Fig. 14-4 Upbreaking or stoping process of wedges above roof of tunnel [3]

$$h_{\text{max}} = \frac{B_t}{2\tan\dfrac{\alpha}{2}} \qquad (14\text{-}12)$$

The magnitude of the pressure is almost proportional to the area of the wedge in the detachment process, or equivalent to the square of the cavity opening.

The weight of the loosened detached wedge is responsible for what has been referred to previously as loosening pressure. From the foregoing observations regarding the evolution of the detached wedge with time and the development of arching in accordance with the levels of strain that are permitted in the roof, it becomes clear how loosening pressure will depend largely on the timeliness with which the support is installed, the nature of this suport and the care exercised during construction. For example, excessive blasting may lead to exaggerated pressures due to loosening of the tunnel roof.

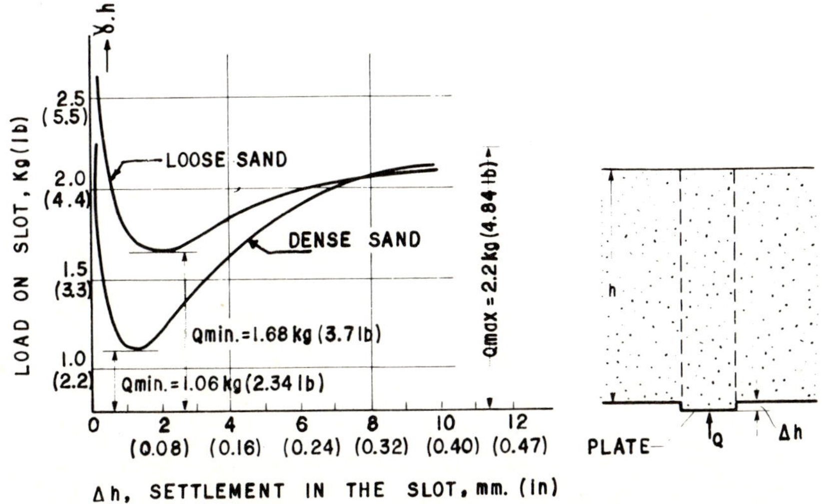

Fig. 14-3 Pressure on a slot, "Silo" effect or vertical "arching" [3]

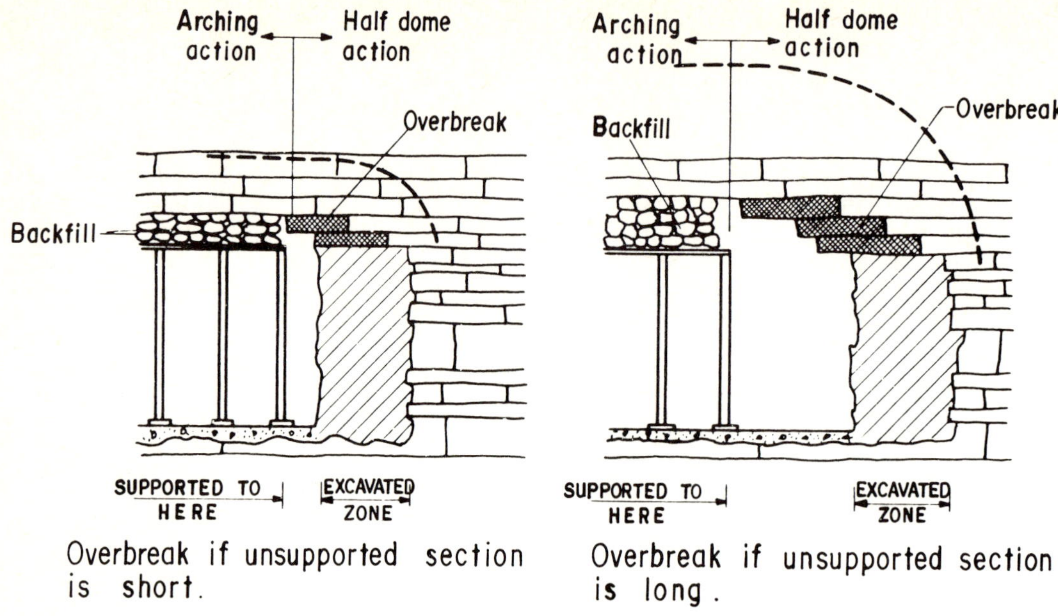

Fig. 14-5 Increase of overbreak as a function of the length of unsupported section [3]

The height of the loosened zone increases with the length of the cavity parallel to the tunnel axis that is left unsupported. Figure 14-5 shows how these effects vary as a function of the length of cavity that is left unsupported at the excavation face. In Part *a* the unsupported section is relatively short. In the supported section arching develops, whereas at the excavation face three-dimensional arching occurs. This is referred to as the half-dome effect. In Part *b* of the figure a far longer portion has been left unsupported. The unsupported part, loosened as a result of excavation, is far wider, and the half-dome as a result of excavation, is far wider, and the half-dome effect includes far larger masses, so that the weights of the loosened zone are greater.

When a shallow excavation is conducted carelessly, the loosened zone may extend upward to the surface, with cracks forming around it. In this event, loosening pressure in the roof may become equal to the total weight of the overlying mass (geostatic pressure).

The arching effects already mentioned, first described by Terzaghi, become more pronounced the more frictional and less cohesive the soil or the fissured or fragmented rock. This does not mean that these effects do not occur to some extent in cohesive soils. In these soils the subsequent increase in pressure and ultimate equilibrium occurs more slowly; the reduction in pressure due to arching will be more pronounced, the greater the cohesion.

14.3.2 Squeezing (Extreme Mountain Pressure)

This condition develops when the secondary pressures that come from the tunnel superimposed on the primary stresses existing in the mass prior to excavation, exceed the strength of the rock or soil surrounding the tunnel. This occurs not only in the roof but also in the sides and even in the floor of the tunnel. The condition may be caused by superposition of stresses, in which the original stresses in the soil or rock combine with those that are produced as a consequence of excavation. However it may also be a natural condition existing in the soil or rock prior to excavation when there exists a latent plastic state, in which the plastic flow of the rock or very firm soil has been impeded by confinement from neighbouring masses. Karman's experiences have shown that under triaxial compression even the hardest materials may become plasticized. All depends on the magnitude of the stresses that are applied.

In the case of a mountain through which a tunnel is driven, the latent plastic condition of the material depends on the magnitude of vertical pressures being sufficiently large (usually associated with appreciable overburden depths) and lateral expansions being sufficiently impeded. Lateral tectonic forces can augment the condition. It does not manifest itself until lateral confinement disappears and deformation can take place, which occurs when the tunnel is excavated.

When there are squeezing pressures, movements start in the tunnel walls; the pressure only makes its presence felt when the engineer attempts to impede these movements. If these movements are allowed to develop freely, they often gradually diminish of their own accord, and a zone will be created around the cavity which will be devoid of these pressures. The effect of this plasticized region around the cavity is to produce a zone of relaxed or relieved stresses. In this zone the material is at its shear strength, a state of plastic flow.

Squeezing or extreme mountain pressure is largely a manifestation of geostatic pressures, which depends on overburden weight and geological structure related to tectonic disturbances. Many authors [3] regard this pressure as a primary pressure, like the initial stress states prevailing prior to excavation of the tunnel. Other authors regard it as a secondary pressure, since it manifests itself when the tangential pressures acting in the tunnel walls reach a critical value, exceeding the unconfined strength of the material.

The effect of these pressures depends on the nature of the rock or soil. In clays, clay shales or clayey slates, flow occurs around the entire perimeter of the cavity. As a consequence stresses are relieved and redistributed, the plastic zone increases in size and equilibrium may eventually be reached. On the other hand, in weaker materials with very high stresses, the flow may continue until the cavity closes. When a support system is installed, the plastic zone is restricted and stability may be achieved, so long as the structural system is capable of resisting the heavy pressures that are produced as a consequence of the restriction. This depends on the flexibility of the support and the extent to which the plastic zone may have developed before it is installed.

In softer rocks or soils, development of the protective zone around the tunnel opening is slow and is characterized by plastic flow from the original perimeter inwards. In harder materials, however, this phenomenon takes the form of successive fracturing. The plastic zone develops more extensively the higher the stresses involved and the softer the rock.

As already mentioned, the squeezing pressure exerted on a support depends on the flexibility of the support and the nature of the soil or rock. If minor movement is sufficient for the plastic zone to develop fully and pressures to be relieved, it is good design to install a support that will be able to yield the necessary amount. For these cases, MOHR [3] has recommended that a layer of resilient material be laid between the support and the tunnel to permit the necessary movement. At other times, the movements in the walls and roof of the tunnel required for the plastic zone to develop is so great that they cannot be allowed to develop completely. In this case, the pressures exerted increase in proportion to the effort of the movement that is prevented. These pressures must be resisted by sufficiently strong supports, spaced as necessary. In soils and soft rocks, the development of the protective zone is slow; consequently, the initial pressures may increase for several months after the support has been installed. Squeezing pressure may reach thousands of tons per square meter, making any support system irrelevant. For these cases of intense pressure, immediate installation of a rigid support system can cause failure. If the permanent lining has to be built before the deformation necessary for a substantial and definitive relief of pressure can develop, a space should be left between the lining and the rock or soil face. It is filled with a porous yielding material such as slag.

From the foregoing definition it becomes clear that squeezing pressure can develop with a small depth of cover, if the strength of the material is low. For example [3], in weak clay shales and slates, squeezing pressures have been measured for overburden pressures of 200 t/m^2 (40 kips/ft^2). In lignite, pressures of 60 t/m^2 (12 kips/ft^2) have proved sufficient.

Whereas in the case of loosening pressures the best procedure is usually excavate rapidly and construct a support system against the excavation face as soon as possible, where squeezing pressures develop no rigid rule can be given, regardless of the nature of the soil or rock. In hard rock masses, squeezing pressures often are accompanied by explosive or *popping* rock [7]. This does not effect the ultimate stability to any appreciable degree. In such cases, it is advisable to use a rigid support system, installing it immediately after excavation, to protect workers and equipment. In soft materials, it is not advisable to construct a support system immediately, for this would not allow time for the development of protective plastic zones. Moreover, a support system cannot be considered which will resist all the squeezing pressure that may develop. The wisest solution in many situations is to build a temporary support system which can be gradually replaced as squeezing of the material permits. The criteria for dealing with those cases where minor movement is sufficient to relieve pressure have already been discussed.

Figure 14-6 [3] shows experience obtained from a tunnel in clayey soil, where a space was left between the support system and the walls and filled with ash. Attention is drawn to the way in which the radial normal stress diminishes as the radius of the circular cavity diminishes, and also to how quickly the pressure on the ash increases.

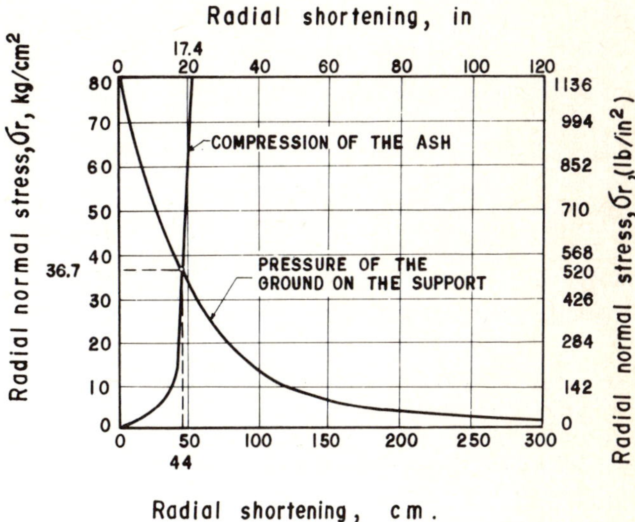

Fig. 14-6 Reduction in pressures as a function of deformation of a deformable layer between tunnel wall and support [3]

14.3.3 Swelling Pressure

Under certain conditions swelling pressure has been found in clays, clay shales, altered slates and in other rocks interstratified with clay. The explanation to this expansion was offered by TERZAGHI [6] and is based on non-uniform stress relaxations caused by migration of pore water from the soil or rock in more highly stressed zones to less stressed zones, which expand accordingly. As a result of excavation, all zones on the perimeter of the cavity are less stressed zones, especially those on the working face where it has not yet been possible to install a support system. Also critical are the tunnel floor and the lower parts of the walls, unless a complete perimetral support system is employed. Under the circumstances, there will be migration of water from the pores in the mass to the perimeter of the cavity, with the corresponding expansion.

At the present time, it is not possible to predict swelling pressures. Measurements have shown that they can be very intense (tens of kilograms per square centimeter) and increase over a long period of time (weeks or months). The greater the deformation that is permitted, the more will be the relief of pressure on any support. Thus the procedure for minimizing the pressure consists of permitting deformation of the soil or rock to a certain limit and then building a support system capable of resisting any subsequent pressures. The correct moment for construction and evaluation of the amount of pressure that may still develop require great experience and engineering intuition.

It is often very hard to distinguish swelling pressures from squeezing pressure. The problem is made even more difficult by the fact that expansive soils and rocks usually have very low strain moduli where the development of squeezing pressures can be expected even with thin overburden. High lateral pressures or pressures in the roof very close to tunnel portals in relatively stable formations are often a sign of swelling pressures. A more precise identification of swelling pressures is only possible by investigating the physical properties of the soils or rocks very carefully.

Swelling pressures are often a consequence of chemical transformations of soils or rocks by an increasing water content. A typical case is the transformation of anhydrite into gypsum.

From the foregoing observations it becomes clear that the magnitude of the swelling pressures that develop in any specific soil or rock depends largely on the nature of the support system that is employed and the amount of expansion that this system must absorb.

14.4 Theoretical Approach to Evaluation of Pressures on Tunnels

The ground in which a tunnel is driven can be regarded either as a continuum or a discontinuum. A continuum is a material in which the mean of the micro-mechanical properties, such as stress or density, can be used to determine the mechanical behavior of the material on an engineering level [1]. When the ground is regarded as a continuum, the behavior of a tunnel may be approximated by analyses based on the theories of continuum mechanics, principally elasticity and plasticity theories. In the paragraphs that follow, a brief description will be given of the principles with which these disciplines approach the problem of tunnels. At first sight they lead to conclusions that are somewhat discouraging because continuum mechanics demands characteristics of homogeneity and uniformity which are present in soils and rocks. However, a brief review of these methods is essential for the evaluation of pressures and the rational design of tunnel support systems.

14.4.1 Elastic Analysis

The application of elasticity theory to rocks and soils has been attempted by establishing a relationship between stresses and strains. This relationship may be established at the time of initial loading (E_0, the slope of the stress-strain curve for initial loading) or in subsequent cycles when an apparent strain hardening of the material has taken place (E, slope of the stress-strain curve in a subsequent loading cycle).

Another elastic constant is the familiar Poisson's ratio, μ (or sometimes ν). In soils and rocks it is somewhat variable, changing with the stress field and with cycles of loading.

The theory of elasticity enables a fairly simple characterization of the stress states about a circular cavity driven in the elastic continuum. Other cavity shapes lead to mathematical equations, which make it either difficult or impossible to obtain direct solutions. However the solution for the circular cavity, even in these cases, can be a reasonable approximation or guide. This can be obtained by considering a plane strain condition in an unsupported cavity, for various combinations of horizontal and vertical loads. These are shown in Fig. 14-7 [1, 10], which illustrates the nomenclature usually employed in the analysis of elasticity problems. The ultimate stress state around the tunnel can be given once the cavity has been opened (KIRSCH'S solution, [1, 10, 11]:

$$\sigma_r = \frac{1}{2}\sigma_z\left[(1 + K_0)\left(1 - \frac{a^2}{r^2}\right) + (1 - K_0)\left(1 + 3\frac{a^4}{r^4} - 4\frac{a^2}{r^2}\right)\cos 2\theta\right] \tag{14-13a}$$

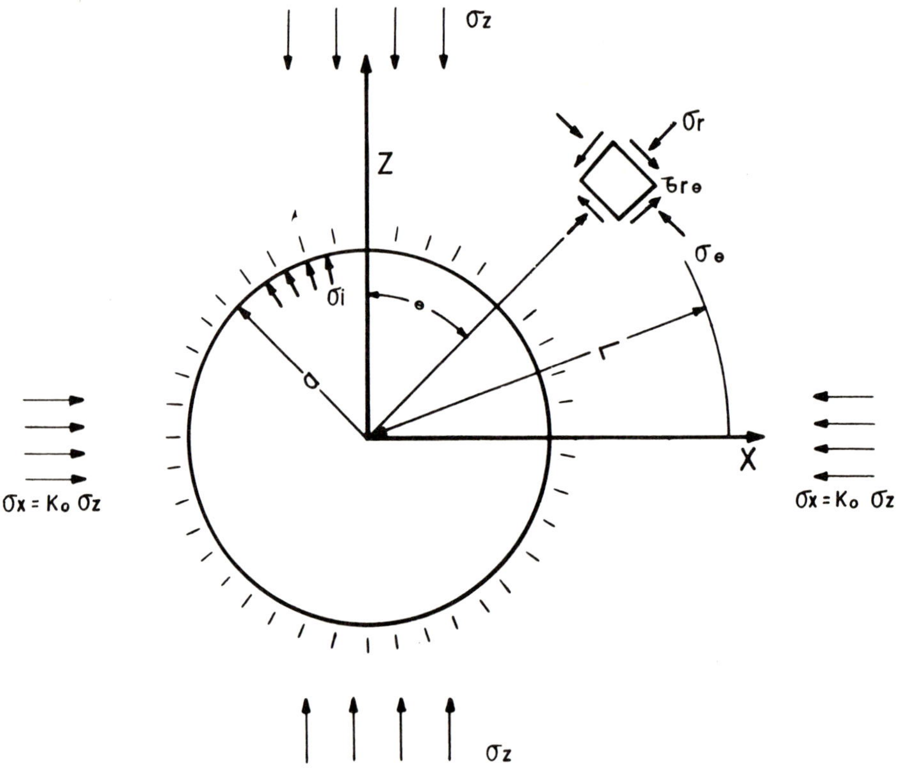

Fig. 14-7 Nomenclature corresponding to KIRSCH'S solution for a cylindrical hole in a isotropic homogeneous elastic medium [1, 10]

$$\sigma_\theta = \frac{1}{2}\sigma_z\left[(1 + K_0)\left(1 + \frac{a^2}{r^2}\right) - (1 - K_0)\left(1 + 3\frac{a^4}{r^4}\right)\cos 2\theta\right] \quad (14\text{-}13b)$$

$$\tau_{r\theta} = -\frac{1}{2}\sigma_z(1 - K_0)\left(1 - 3\frac{a^4}{r^4} + 2\frac{a^2}{r^2}\right)\sin 2\theta \quad (14\text{-}13c)$$

K_0 is the coefficient of at-rest earth pressure, as defined in Chapter 5.

In the tunnel walls, the stresses can be obtained by simplifying the foregoing equations for the case $r = a$.

$$\sigma_r = 0 = \tau_{r\theta}$$
$$\sigma_\theta = \sigma_z\left[(1 + K_0) - 2(1 - K_0)\cos 2\theta\right] \quad (14\text{-}14)$$

At the crown $\theta = 0$, Fig. 14-7:

$$\sigma_r = 0 = \tau_{r\theta}$$
$$\sigma_\theta = \sigma_z(3K_0 - 1) \quad (14\text{-}15)$$

It can be seen that $\sigma\theta$ is zero for $K_0 = \frac{1}{3}$. If $K_0 < \frac{1}{3}$, according to this analysis, fissures will open at the crown.

Reference [1] also gives equations which permit calculation of the deformations about the tunnel as a consequence of the previous stress states. At distance r from the center of the unsupported circular tunnel, with radius a, the inward radial displacement produced by excavation of the tunnel, according to the theory of elasticity is:

$$u = \sigma_z\frac{1 + \mu}{E}\frac{a^2}{r} \quad (14\text{-}16)$$

In the tunnel wall, where $r = a$, it is:

$$u = \sigma_z\frac{1 + \mu}{E}a \quad (14\text{-}17)$$

Tangential displacements (around the tunnel ring) are zero.

If inside the tunnel there is a pressure σ_i against the roof, Eqs.(14-16) or (14-17) give the displacements it produces, since by superposition σ_z can be replaced by $-\sigma_i$.

An average radial displacement around the entire circumference of the tunnel is:

$$u_m = \frac{1}{2}(1 + K_0)\sigma_z a\frac{1 + \mu}{E} \quad (14\text{-}18)$$

Figure 14-8 depicts the stress distribution about the circular cavity represented by KIRSCH's equations, Eq.(14-13), corresponding to $\sigma_z = \sigma_x$; $K_0 = 1$.

In practice, elastic solutions can very seldom be used for design because neither rocks nor soils are homogeneous, isotropic, linearly elastic materials. The precision that can be expected of elastic solutions diminishes when the material in which the tunnel is driven is so weak that the stresses around the perimeter approach failure. The solutions become completely inacceptable in many soils or where rock is weakened by blasting. CORDING [12] has nevertheless stated that elastic solutions may prove useful for calculating the near-elastic displacements that occur immediately after a cavity is opened in rock or firm soils.

14.4.2 Plastic and Elasto-Plastic Analysis

Tunnels are structures for which plastic analyses offer very promising fields of application. Indeed, there are numerous cases of tunnels where the pressure in the soil or rock is such that failure occurs locally forming a restricted plastic zone with constrained deformation around the cavity. Calculation for design of a support system for such cases would benefit greatly if the plastic equilibrium of the cavity could be established. The mathematical analyses based on plasticity theory are more complicated than those of elasticity theory. Therefore even more simplifying hypotheses have to be made. In plastic analyses it is usual to assume $\sigma_z = \sigma_x$, $K_0 = 1$. The weight of the materials extracted from the cavity is disregarded and it is assumed that the normal stress in the direction of the tunnel axis is a principal stress. TRESCA's plastic flow conditions are used [13, 14] for purely cohesive materials ($\varnothing = 0$) or COULOMB's already familiar failure criteria for materials with $c \neq 0$ and $\varnothing \neq 0$.

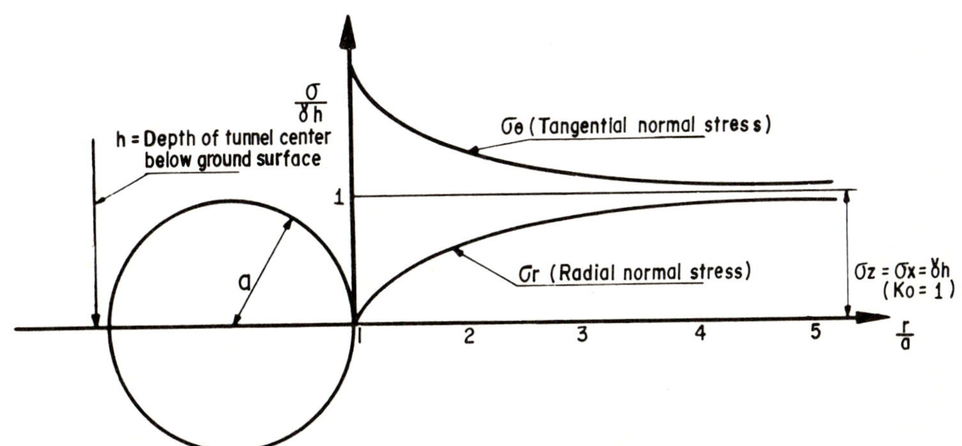

Fig. 14-8 Stress distribution around a circular tunnel in an isotropic homogeneous elastic medium, Poisson's ratio = 0.5, after KIRSCH [1]

.1 Plastic Analyses in Cohesive Materials ($\varnothing = 0$)

Plastic flow develops when:

$$\sigma_1 - \sigma_3 \geqslant 2c_u \qquad (14\text{-}19)$$

where c_u is the shear strength obtained in an unconsolidated-undrained triaxial test, or as half of the simple unconfined compression strength (q_u). While the difference between the principal stresses is less than $2c_u$, the material remains more or less elastic and plastic flow does not occur.

The field of the stresses acting on the tunnel is made up of a vertical pressure, σ_z, a horizontal pressure, $\sigma_x = \sigma_z$ ($K_0 = 1$, by hypothesis) and an internal pressure σ_i applied from inside the tunnel along the whole perimeter of the circular cavity.

If $\sigma_z - \sigma_i \leqslant c_u$, no plastic zone will form about the tunnel, but if $\sigma_z - \sigma_i > c_u$, one will develop which will extend to a distance R, beyond the center of the tunnel (Fig. 14-9).

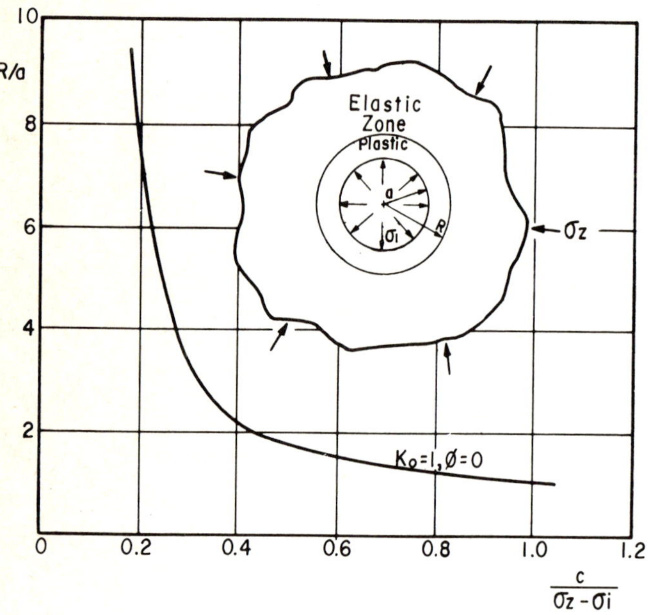

Fig. 14-9 Extension of the plastic zone around a circular tunnel, purely cohesive material [1]

According to theory:

$$R = a e^{\frac{\sigma_z - \sigma_i}{2c_u} - \frac{1}{2}} \qquad (14\text{-}20)$$

In the plastic zone, for $a \leqslant r \leqslant R$, the stresses are (see Fig. 14-7 for nomenclature):

$$\sigma_r = \sigma_i + 2c_u \ln \frac{r}{a}$$

$$\sigma_\theta = \sigma_r + 2c_u$$

$$\sigma_y = \frac{1}{2}(\sigma_r + \sigma_\theta)$$

σ_y is the normal stress acting in the direction corresponding to the tunnel axis. The plastic zone is assumed to maintain constant volume despite stress changes (μ – 1/2).

The shear stress $\tau_{r\theta}$ is zero at all points, by symmetry. For deduction of Eq.(14-20) for the case $\sigma_i = 0$, see [15].

On the boundary between elastic and plastic zones:

$$\sigma_R = \sigma_z - c_u \qquad (14\text{-}21)$$

If there is no pressure acting inside the tunnel, $\sigma_i = 0$. In this case the radius of the plastic zone from Eq.(14-20) is:

$$R = a e^{\frac{1}{2}\left(\frac{\sigma_z}{c_u} - 1\right)} \qquad (14\text{-}22)$$

Displacements on the boundary between the plastic and elastic zones towards the inside of the tunnel are:

$$u_R = (\sigma_z - \sigma_R)\frac{1+\mu}{E}R = c_u \frac{1+\mu}{E}R \qquad (14\text{-}23)$$

Contrarily, if there is a pressure σ_i inside the tunnel, which will be the case if there is a support system exerting a uniform pressure on the whole perimeter of the cavity, the radial displacement of the walls of that cavity towards the center will be:

$$\frac{u_a}{a} = 1 - \sqrt{\frac{1}{1+A}} \qquad (14\text{-}24)$$

$$A = 2c_u \frac{1+\mu}{E} e^{\frac{\sigma_z - \sigma_i}{c_u} - 1} \qquad (14\text{-}25)$$

.2 Plastic Analyses in Cohesive Frictional Materials

The plastic flow criterion that customarily used these cases is the Mohr-Coulomb, which can be written:

$$\sigma_1 = \sigma_3 \frac{1 + \sin \varnothing}{1 - \sin \varnothing} + \frac{2c \cos \varnothing}{1 - \sin \varnothing} \qquad (14\text{-}26)$$

The value for the cohesion c should be interpreted in its broadest sense. It is the ordinate at the origin of the strength envelope plotted on $\tau - \sigma$ coordinates. It is obtained from a set of triaxial tests (of whatever type is most representative for the problem under consideration). The strength envelope usually can be approximated by a straight line, within the interval of normal stresses used.

Here too it is usually assumed that $\sigma_z = \sigma_x$ ($K_0 = 1$) and that the volume of the material remains constant ($\mu = 1/2$).

The theory indicates that for:

$$\sigma_z \leq \frac{\sigma_i + c \cos \varnothing}{1 - \sin \varnothing} \qquad (14\text{-}27)$$

no plastic zone develops, and the material remains in the elastic state. For higher values of σ_z an annular plastic zone develops with a radius:

$$R = a\left[(1 - \sin \varnothing)\frac{\sigma_z + c \cot \varnothing}{\sigma_i + c \cot \varnothing}\right]^{\frac{1-\sin \varnothing}{2 \sin \varnothing}} \tag{14-28}$$

Within the plastic zone ($a \leqslant r \leqslant R$), the stresses are:

$$\sigma_r = -c \cot \varnothing + (\sigma_i + c \cot \varnothing)\left(\frac{r}{a}\right)^{\frac{2 \sin \varnothing}{1-\sin \varnothing}}$$

$$\sigma_\theta = -c \cot \varnothing + (\sigma_i + c \cot \varnothing)\frac{1 + \sin \varnothing}{1 - \sin \varnothing}\left(\frac{r}{a}\right)^{\frac{2 \sin \varnothing}{1-\sin \varnothing}}$$

$$\sigma_y = \frac{1}{2}(\sigma_r + \sigma_\theta) \tag{14-29}$$

On the boundary between plastic and elastic zones, the stresses are:

$$\begin{aligned} \sigma_r &= \sigma_z (1 - \sin \varnothing) - c \cos \varnothing \text{ (radial)} \\ \sigma_\theta &= \sigma_z (1 + \sin \varnothing) + c \cos \varnothing \text{ (tangential)} \end{aligned} \tag{14-30}$$

In the elastic zone ($r \geqslant R$) the elastic state prevails.

Figure 14-10 [1] gives the radius of the plastic zone for different cases of σ_z, σ_i, c and values of the angle $\varnothing$. Figure 14-11 [1] depicts different stress distributions about a circular tunnel. Stresses are expressed as radial and tangential in their variation with distance to the center of the cavity. In the figure details are given of the specific c and $\varnothing$ on which the example illustrated is based. For derivation of Equations (14-27) to (14-30) see [16].

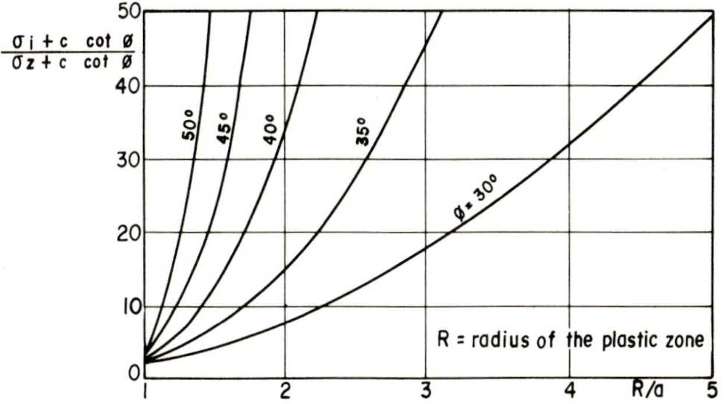

Fig. 14-10 Radius of the plastic zone as a function of the values of c and $\varnothing$ for different combinations of values of σ_i and σ_z [1]

In the foregoing set of expressions it can be seen that if $\sigma_i = c = 0$, the radius of the plastic zone becomes infinite. This means that the cavity will close of its own accord. These equations remain valid only so long as the soil maintains its strength during the period of plastic flow. As already stated, this does not usually occur because flow usually leads to a structural breakdown and a reduction in strength. As a result a process similar to progressive failure is triggered. Occasional attempts have been made to allow for this effect by selecting reduced values of the strength parameters of the material for calculation purposes [1].

Reference [17] describes analyses which permit determination of the plastic zone and stress distributions in materials which obey plastic flow laws different from the MOHR-COULOMB one.

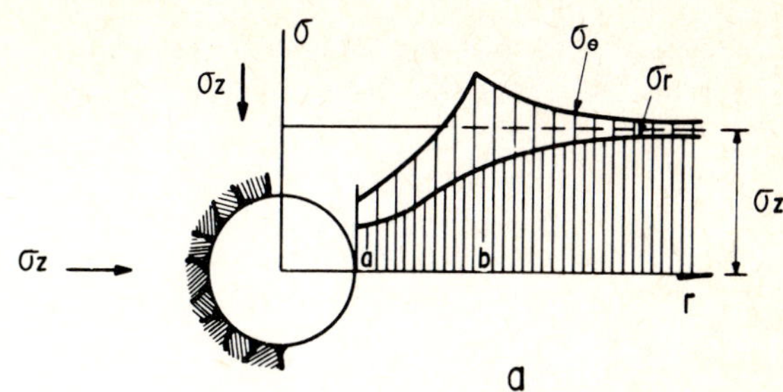

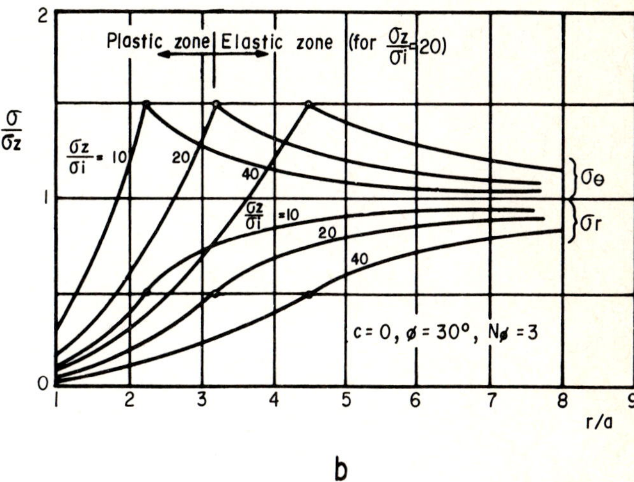

Fig. 14-11 Stress distribution around a cylindrical gallery, a particular illustrative case [1, 10]

There are no theoretical studies [1] which consider a condition other than $K_0 = 1$ or in which the influence of the overburden depth is included.

According to the theory of plasticity, radial displacement towards the inside of the tunnel on the boundary between the plastic and elastic zones can be described thus:

$$u_R = (\sigma_z - \sigma_R)\frac{1 + \mu}{E} R \tag{14-31}$$

where σ_r and R are given by Eqs.(14-30) and (14-28). As previously, it is assumed that the volume of the plastic zones remains constant.

14.4.3 Analyses Based on Continuum Mechanics.

As was mentioned, these analyses are based on hypotheses relating to the material and by others of an operational type. It would be unusual for such assumptions to be an accurate depiction of a real soil or rock in which a tunnel is driven. They will more closely approximate the behavior in massive, uniform materials, such as solid rock masses and some deposits of clay or sand. However, it should always be borne in mind that the uniformity and apparent homogeneity of a formation do not necessarily signify that these qualities are verified on the level of continuum mechanics. Further, the hypotheses concerning

the stress-strain relationships of materials imposed by the theories usually cause disagreements between theory and reality.
It is hard to believe that these theoretical analyses can be used in practice as any more than an initial calculation stage, destined to provide a qualitative view of the problem. In order to formulate criteria and acquire a better understanding of the problem it may therefore be useful to repeat the theoretical analysis many times, assuming different properties of the material, different configurations of the problem and different stress conditions. Thus it will be possible to obtain an idea of the relative importance of the different factors.

Before accepting any analysis based on continuum mechanics, the results must be proved reasonable in the field or laboratory. At the present time not enough measurements are available. They should cover wide ranges of geological conditions and types of support systems and linings. Eventually they may lead to the improvement or establishment of new empirical or semi-empirical methods of design.

14.4.4 Visco-Elastic Analyses

Attempts have been made to describe the behavior of soils around tunnels on the basis of visco-elastic models, with the assumption that an inelastic behavior exists in the material which varies with time [18]. The model used to represent the behavior of the material is that of MAXWELL-KELVIN for the distortional component and the theory of elasticity for the volumetric component. The results of such analyses enable the initial and long-term horizontal and vertical pressures to be obtained. Such analyses conclude that the horizontal pressure increases with time, reaching values close to those for the vertical pressure found in plastic analyses.

14.5 Empirical Methods for the Calculation of Pressures on Tunnels

The lack of agreement between the theoretical approaches of continuum mechanics for the evaluation of pressures acting in tunnels before and after excavation and before and after installation of a support system or lining, and the limited measurements of tunnel behavior has led to empirical or semi-empirical rules. These have been developed from experience to determine or estimate these pressures by simple methods which use the information that is usually available.

At the present time these empirical methods are required to help determine the interaction between soil and support, which depends on many slight differences in the properties of the soil as well as the characteristics of shape and rigidity of the support. These cannot be handled by theories of applied mathematics but are of paramount importance for the determination of pressure conditions.

It is not the authors' intention to imply that the empirical methods that will be described are perfect. It has already been stated that few measurements have been taken in tunnels in soils, so that the total available experience is incomplete. Moreover *blind experience* that does not measure the behavior of the materials employed in a specific project, under the specific conditions of that project, usually leads engineers to interpretations and conclusions that are very different from those that would have been reached had more been known about the real mechanisms prevailing in the actual project. This *blind* experience often has the further disadvantage that, under the guise of objectivity, it is sometimes not questioned, analyzed or criticized by those having access to it. Empirical methods for the calculation of pressures in tunnels are undoubtedly affected by these limitations in the quality and accuracy of the experience obtained. They must be compared with the entire range of real conditions, adjusting them to agree with new levels of information that may become available. Only by keeping his mind open to new possibilities will the engineer be able to use these methods for an important project. He should feel obliged to measure the specific conditions he may encounter so that he can modify and adjust his initial design conceptions as required.

The ultimate purpose of empirical methods is estimation of external loads on tunnel supports. This is also the most difficult task. These loads are often selected regardless of strain, although in some cases criteria exist for establishing adaptations to these strain conditions.

The soil behavior mechanisms that are employed in empirical methods are simple. For example, the load may be expressed as a function of the weight of the soil and of the angle of internal friction (which are alone not sufficient to describe total soil behavior).

The first empirical or semi-empirical methods for evaluating pressures in tunnels are mentioned in [19] and [20]. These refer to the works by SPANGLER for evaluating pressures on flexible culverts using MARSTON's Theory (Chapter 9 and [21]). These are based on early contributions by TERZAGHI, later completed in [6] and will be described in detail further on.

A description follows of the most widely used methods of evaluating vertical pressures and loads on supports. Later, the calculation of lateral and acting pressures on the tunnel floor will be discussed.

14.5.1 TERZAGHI's Method

TERZAGHI's tunnel loadings [6] are based on his concepts of arching in soils, which were presented in §14.3.1. He defines the vertical rock or soil load on the tunnel as the weight of material which would tend to fall from the roof were it not supported. The rock load values presented are classified in accordance with the nature of the rock or soil; however a full definition is not given of the boundaries between the different categories.

Figure 14-12 illustrates the nomenclature employed in presentating the method, H_p is the rock or soil zone acting on the tunnel roof. If the value of the rock load is greater than zero and the tunnel is unsupported, the material that drops down on the roof tends to penetrate the tunnel gradually. As a result the shape of the roof becomes irregular.

The roof load depends on the nature of the rock and particularly the details of structure such as cracking and the degree of weathering. If the rock is sound or moderately cracked, the tunnel roof may support itself, or require only a relatively light support system. Whereas if cracking or alteration is very considerable, the pressure on the support may become an earth thrust. Often many different conditions prevail along the axis of a tunnel, and the engineer must always be prepared to modify any preconceived design criteria in view of the conditions he discovers in the actual project.

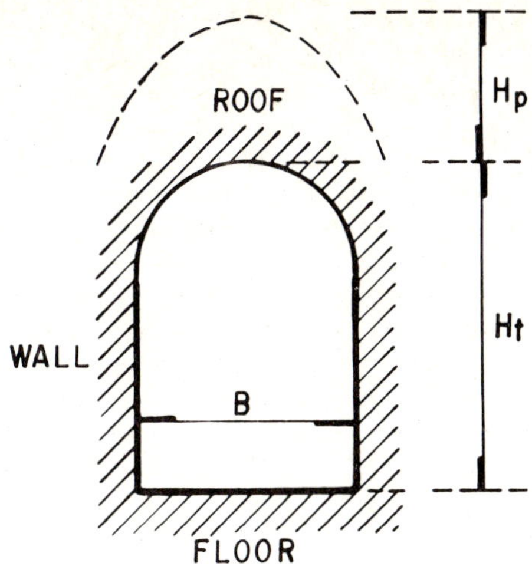

Fig. 14-12 Cross-section of a tunnel

The load acting on the supports depends to a certain extent on the stress state existing in the mass of rock before the tunnel is bored. The relationship between the vertical pressure exerted by the rock and the horizontal pressure acting in that same section depends chiefly on the geological history of the rock. It may vary enormously. Generally the vertical pressure is greater in undisturbed rock masses. In a folded rock the horizontal pressure depends on whether the horizontal forces that caused the folding have disappeared or not. If they have not, the horizontal pressure may be greater than the vertical, limited only by the compressive strength of the rock. Measuring the stress state inside a mass of rock requires special techniques beyond this treatise. Consequently the existence of strong horizontal pressures is only deduced from external signs, such as the appearance of explosive (popping) rock fairly near the ground surface.

.1 Tunnels in Solid Intact Rocks or Firm Soils

Theory has demonstrated that in solid rock the modification of the overall stress state of the mass imposed by the tunnel is rapidly neutralized as the depth of the tunnel increases. For depths in the order of one diameter the effect of excavation is already very small.

As already stated, in the walls of an unsupported tunnel the radial stress (acting in a directional normal to the wall) is zero and the circumferential stress in the direction of the tangent is approximately double that which existed prior to excavation of the tunnel. An element of the tunnel wall is subject to a stress state to a certain extent similar to that of a specimen of rock that is tested under simple unconfined compression. Failure occurs when the circumferential stress becomes equal to the unconfined compressive strength of the rock. Very great circumferential stresses, which correspond to the weight of rock above the tunnel may be compatible to the equilibrium load of more than a thousand meters. If the rock is sound and strong the tunnel in solid rock may not require any support system.

A problem does, however, frequently arise in tunnels in solid strong rock which makes it necessary to provide sufficient support to ensure protection of the workers during the construction period. This problem is usually referred to as explosive or *popping* rock. Often splinters of rock suddenly break away from the roof and walls of tunnels in solid rock, and are projected at tremendous speed, with the consequent danger to the workers. The phenomenon occurs when rock in the walls or roof of a tunnel is subject to states of intense elastic strain or when strong squeezing pressures exist. This may be due to the permanent nature of horizontal pressures, left by non-dissipated tectonic folding processes or other causes that have not yet been clearly established. It is most prevalent in surfaces that are flat or have a very long radius. Figure 14-13 shows the formation of an explosive slab. It can be seen that it is analogous to buckling in a heavily loaded column. The remedy against explosive rock is to provide the walls and roof of the tunnel with an element which will exert a force against them that will neutralize their tendency to expand. The pressure necessary for this purpose is small and any support that is capable of tolerating about 2 t/m^2 will be sufficient to meet this objective. The best remedy is a smooth tunnel wall that curves continuously at the minimum radius.

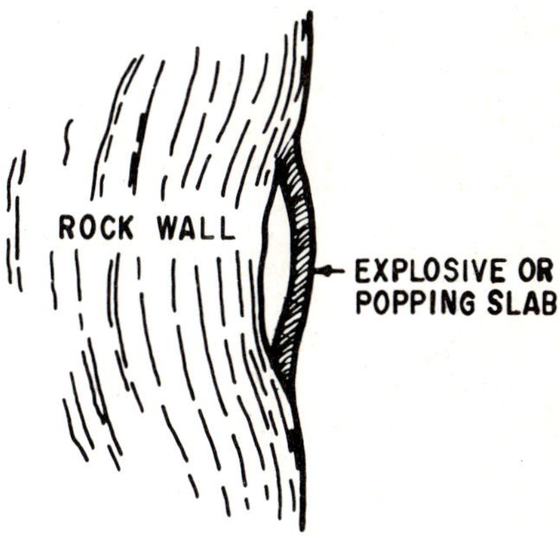

Fig. 14-13 Explosive or popping slab

At times, if the phenomenon of explosive rock adopts very large proportions, overbreak occurs in the tunnel walls and roof behind the support. In this event the support should be designed to tolerate the maximum thrust corresponding to this type of rock. In any case, the support must be tightly wedged or continuously acting against the tunnel walls.

.2 Tunnels in Stratified Rock

Stratified rock poses the problem that it breaks easily along its stratification planes and the joints that are transverse to the stratification or bedding. When stratification is horizontal in these rocks, the effect known as *bridge action* occurs. The rock acts like a slab, supporting itself without any need for bracing so long as the tensile strength of the under side of the slab is greater than the bending stresses (Fig. 14-14). If tension

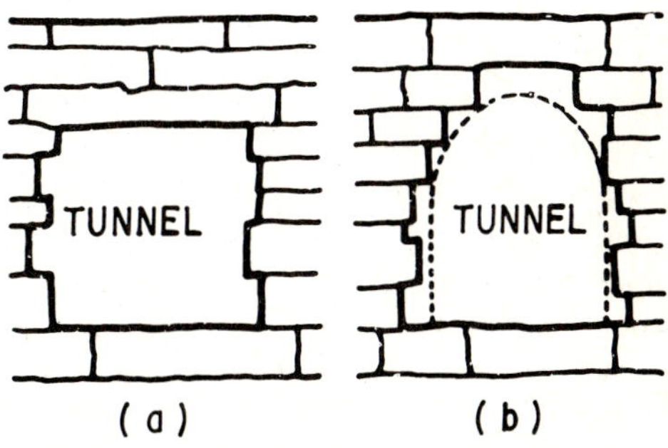

Fig. 14-14 Bridge action in stratified rock a) with widely spaced transverse joints b) with closely spaced transverse joints

Plate 14-1 Problems at the entrance portal to a tunnel in jointed rock

stresses are greater than the strength of the rock slabs, the tunnel roof will crack and will then require a support system to support the broken slabs (Plate 14-1).

The effect of blasting at the tunnel face during driving leads to an overbreak which depends on the distance between the joints in the rock, the quantity and power of the explosives, the drill hole spacing and the distance between the supported section and the unsupported working face (the round length). Even in cases where total development of the overbreak is permitted by not supporting the tunnel face and roof, seldom does the cavity that forms above the roof of the face as a result of collapse exceed the value $0.5B$, where B is the width of the tunnel. This only occurs in the case of very jointed rock. Thus in practice it would not be reasonable for the rock load on the support to exceed that value, which constitutes an upper limit for design purposes. If the support system at the open face of the tunnel is constructed rapidly and the space between the support and the vault caused by blasting operations is duly wedged with rock fragments or supported by rock bolts, rock loads of less than $0.5B$ can be achieved.

If the stratification planes of the rock are vertical, the amount of overbreak depends largely on the distance between the unsupported working face and the place where the support system commences. Here the masses of rock are sustained by friction on their stratification planes. The roof of the support system only has to tolerate the difference between the rock weight and that friction. In many formations the situation is more favorable than might appear at first sight; the load seldom exceeds the weight of the rock that is loosened by blasting operations. A rock load value of about $0.25B$ (B being the width of the tunnel) seems to ensure satisfactory roof support conditions. However if the cracks are open due to flexure or weathering so the friction is limited, the load can be equivalent to several diameters. If stratification planes are inclined in relation to the tunnel axis, thrusts are exerted not only against the tunnel roof, but also against the wall that is intercepted by stratification. Figure 14-15 shows the procedure proposed by TERZAGHI for calculating these thrusts. Wedge *aed* pushes wall *ac* of the support and tries to penetrate the tunnel. The magnitude of this thrust per unit of tunnel length can be computed assuming that along *de* there is no adhesion between rock and rock and that along *ce* rupture has also occurred, so that the mass *cefg* thrusts against the tunnel roof.

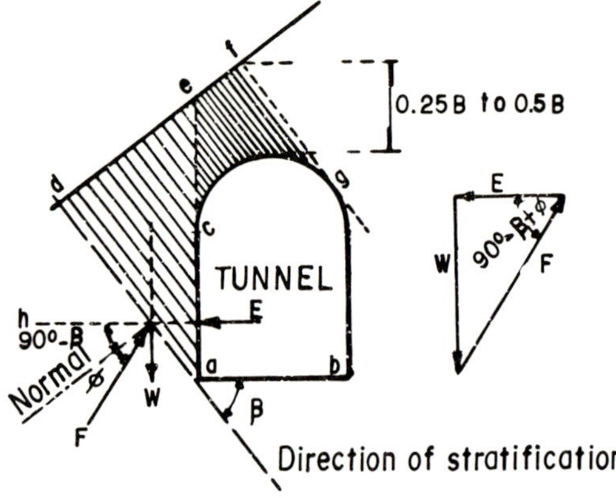

Fig. 14-15 Calculation of thrusts in stratified rock in sloping planes of weakness

Wedge *ade*, therefore, is in equilibrium under its weight W, reaction F along *ad* and thrust E on the wall. Since the magnitude and direction of W and the direction of F and E are known ($\varnothing$ is the apparent angle of internal friction along the stratification planes of the rock under consideration) the corresponding triangle of forces can be drawn and the value of E obtained. The value of angle $\varnothing$ depends not only on the nature of the rock, but also on the water pressure that may exist in the stratification planes of that rock. Experience has shown that in rock masses containing clay in their stratification surfaces, $\varnothing$ may be as low as 15°, whereas it will be in the order of 25° to 35° if the rock is clean. The magnitude of the rock load that is exerted by wedge *cefg* on the tunnel roof may vary from $0.5B$ for gently dipping stratification to $0.25B$ for cases where the stratification is very steep.

.3 Tunnels in Fissured Rock

Fissures frequently occur in a direction parallel to the ground surface. In these rocks, problems of overbreak and support are very similar to the ones dealt with in the case of stratified rocks. If fissures occur at random, failure to install a support system generally leads to arching, especially above the tunnel roof. Frequently, however, owing to the irregular nature of the fissure surface, the friction and bond between the rock plays an important role. Consequently the thrust on the tunnel walls is often small and similarly small in the roof, corresponding to a rock load equivalent to a height of one quarter of the tunnel width.

When this type of rock is subject to pronounced elastic strain conditions such as from high lateral tectonic stresses, the problem of explosive rock is also posed, which should be avoided as was described previously.

.4 Tunnels in Crushed Rock

Into this category fall a large variety of formations, from very coarse broken rock to rock that is so crushed that its behavior is really that of an angular gravel or angular sand (Plate 14-2).

Plate 14-2 Tunnel in crushed rock

In these rocks, the phenomenon known as arching is typical, and indicates the capacity of the rock located above a tunnel roof to transmit the pressure due to its own weight to the masses on either side. The arching effect is the same as that previously discussed for sands. It occurs as a consequence of stress relief in the tunnel roof, Fig. 14-16 depicts a rock mass affected by this phenomenon.

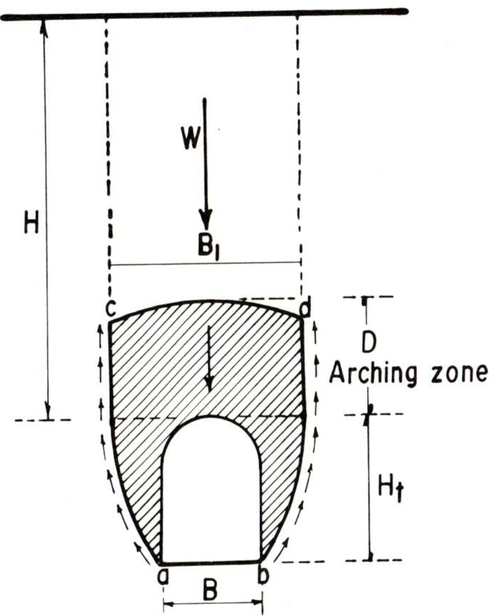

Fig. 14-16 Arching above a tunnel cavity in soil or closely fractured rock

To determine the load acting on the tunnel roof, taking the arching effect into consideration, theories like that discussed earlier or results of laboratory tests performed on sands can be analyzed. These tests, which are fairly representative of the behavior of sands or crushed rocks above the water-table, lead to conclusions of practical interest. Figure 14-16 shows a rock mass affected by arching. The weight of this mass tends to penetrate the tunnel so long as no appropriate support is installed. However much of the load is transferred to the adjoining rock masses by the friction that develops on surfaces *ac* and *bd*. Attention is drawn to the width of the arching zone, B_1, which is greater than the width of the tunnel. It is also observed that the thickness D of the arching zone is approximately equal to $1.5B$. Above this height, stresses in the rock mass remain practically unaltered by the tunnel excavation. The fractured rock in the tunnel roof has only to yield slightly for the load on the support system to decrease, so that it is even smaller than the weight of the arching zone, $\gamma B_1 D$. Thus, a minimum H_p value is obtained. If the deformation of the crown of the tunnel arch continues to increase beyond the point of minimum load, the rock load increases again. It approaches a maximum equivalent to H_p as strain increases, which is nevertheless smaller than equivalent to D. Typically, for reasons that are hard to establish, the ultimate rock load is somewhere between H_{pmin} and H_{pmax}.

Once the support system has been installed and wedged against the tunnel roof, the rock load increases with time, at a gradually decreasing rate, until an ultimate value is reached. According to TERZAGHI, this is when:

$$H_{P_{ULT}} = 1.15 \, H_P$$

where H_p is equivalent to the rock load originally acting on the support. This value is reached regardless of the depth to which the tunnel is driven below the ground surface.

The initial value of H_p depends on B_1 and, according to TERZAGHI, is:

$$H_p = CB_1 \qquad (14\text{-}32)$$

where C is a constant which depends on the compactness of the rock and the distance the tunnel roof may have subsided before the support system was installed. If the rock is totally crushed to the extent that its appearance is that of a sand, B_1 reaches the value: $B + H_t$. The height of the mass producing rock load on the tunnel roof, H_p, can be estimated, according to Eq. (14-32), with the values in Table 14-1 obtained from model tests in dry sands.

The mean pressure on the tunnel walls can be estimated by applying theories of earth pressure in sands, with the equation:

$$p_h = 0.3\,\gamma\,(0.5H_t + H_p) \qquad (14\text{-}33)$$

where γ is the specific weight of the mass of totally crushed rock and all the other symbols have the meaning already defined.

As already stated, these values for rock load and horizontal pressure increase with time by approximately 15%, and this increment must be taken into account in design.

Experience has shown that in real tunnels the loads are usually far closer to the minimum than to the maximum given in Table 14-1. This suggests that the deformation of the tunnel roof, which takes place during excavation, is sufficient to cause total development of arching in the rock mass.

Table 14-1
Roof loads in crushed rock

Totally crushed rock, equivalent to sand	H_p	Yield of tunnel roof
Dense	Min: $0.27\,(B + H_t)$ Max: $0.60\,(B + H_t)$	$0.01\,(B + H_t)$ $0.15\,(B + H_t)$ or more
Loose	Min: $0.47\,(B + H_t)$ Max: $0.60\,(B + H_t)$	$0.02\,(B + H_t)$ $0.15\,(B + H_t)$ or more

From the foregoing considerations it becomes clear that in crushed rock it is advisable to construct the support system immediately, wedging it correctly.

If the tunnel is driven below the water-table, model tests have shown that the arching phenomenon is not interfered with by the flow that occurs towards the tunnel. The tunnel acts like an underground drain, but owing to seepage forces the rock load becomes practically double. Upward flow affects appreciably the bearing capacity at the base of the struts in the support system of the tunnel walls.

.5 Tunnels in Broken Rock

The term *broken* includes rock with closely spaced joints, cracks and fissures, so that it behaves almost as independent blocks between which there is little interaction. The joints between blocks may be narrow or wide, and may or may not be filled with finer materials. The mechanical behavior of these formations is similar to that of coarse-grained dense cohesionless sands and gravels. If the joints between the blocks are distributed at random, pressures frequently occur not only in the tunnel roof, but in the walls too (Plate 14-3).

Plate 14-3 Tunnel in fragmented rock

The rock load in these formations is determined by laws similar to those that govern arching in sands. Thus, load H_p on the roof of a tunnel excavated at a considerable depth is independent of that depth and is proportional to the sum of $B + H_t$.

Experience indicates that these rocks do not adapt immediately to the new stress state caused by excavation of the tunnel. Immediately after blasting, some blocks in the working face zone fall into the tunnel, causing arching at that face. A dome of unstable blocks tends to form between the face and the supported section of the tunnel. Under the circumstances, the working face supports itself for a limited time, known as the *stand up* time. At the end of this period blocks again fall, forming another larger dome of unstable rock. If the cavity remains unsupported, the effect is progressive. The fall of a quantity of rock leads to instability in a larger dome-shaped mass which in turn subsequently falls. The time the unstable mass supports itself depends on the shape and size of its blocks, the width of the joints, the matrix between the cracks and the distance between the working face and the supported section of tunnel. The time that elapses between blasting and the falling of the first dome of unstable rock is known as the stand-up time, t_p. This period is attributed both to the viscous strength of the matrix that fills the joints and to the progressive sliding of the surfaces between the blocks.

Even if an appropriate support system is constructed and carefully wedged against the rock during the stand up period, the rock load on the tunnel roof tends to increase with time for two reasons. The first is that as the working face advances the three-dimensional dome effect is replaced by the two-dimensional arching effect, which is less efficient. The second is that the wedging of the support against the rock does not completely stop the settlement of the crown under the new stress conditions caused by excavation. These continuing movements bring about a gradual increase in the rock load, which continues until the blocks have achieved their ultimate position. The total increase of the rock load and the time that elapses before it reaches its constant value depend to a large degree on how carefully the support is wedged against the rock. If this operation is adequately conducted, this time does not usually exceed a week. Paradoxically, if the space between the support and the rock is not properly filled and the support is not appropriately wedged, the initial rock load may be even smaller than that which occurs when these operations are conducted satisfactorily. However the load increases over several months and its ultimate value is eventually far higher than that reached with appropriate fill and wedging.

The stand up time increases rapidly when the spacing between struts is reduced. The minimum distance that should be left between the working face and the supported section can be somewhat greater than that required during blasting. This is usually in the order of 6/10 of the width, B, of the tunnel, but varies with the type of rock and very rarely exceeds 5 or 6 m (16 or 20 ft). However, if the duration of drilling and blasting is longer than the stand up time, the support system must be installed close to the working face.

The stand up time influences the planning of blasting, scaling, mucking and support operations. If this period is only a little longer than that required to ventilate the working face after blasting collapses will be inevitable at the face. The greater the difference between these two times, the greater will be the time for constructing the support system, and consequently collapses will be proportionally fewer. Where the stand up time equals the time required for ventilation of the tunnel and installation of a support system in the exposed face, there will be few collapses if operations are appropriately conducted.

There is no well defined boundary between the crushed rock discussed in the last section and the broken rock now under consideration. Consequently, the equivalent height for rock load for broken rock may vary from $0.25B$, (moderately jointed rock) also to the highest possible values for crushed rock. Arbitrarily, two types of broken rock can be distinguished when estimating rock load: moderately broken rock or severely broken rock. On the basis of measurements taken in railroad tunnels through the Alps, some representatives values of H_p have been reached in moderately and severely broken rock. In tunnels with water, through moderately broken rock, H_p may initially be zero, subsequently increasing to several meters. If the rock is severely broken, the initial value of H_p may be higher. Table 14-2 summarizes these experiences.

In dry tunnels the values of H_p may be smaller than in tunnels where water is present. However, in tunnel design it is always advisable to consider the most critical condition, for the permanent absence of water is very hard to ensure.

Table 14-2
Roof loads in cracked rock

Type of rock	Rock load height H_p	
	Initial	Ultimate
Moderately cracked	0	0.25 B to 0.35 $(B + H_t)$
Very cracked	0 to 0.6 $(B + H_t)$	$0.35(B + H_t)$ to $1.10(B + H_t)$

The fact that the joints between blocks of rock are occupied by clay may be very important at times when the tunnel is dry, for dry clay acts like a cementing agent on account of its shear strength. When the tunnel becomes wet, however, this shear strength dissipates rapidly, and so it would not be wise to rely on it, except in very special cases. Consequently, it is advisable to use the values given in Table 14-2, whatever the appearance of the rock during construction.

.6 Tunnels in Altered Rock and Clay

As already indicated, chemical alteration transforms the most rocks, including igneous rocks and metamorphics such as gneisses, schists and slates, into clays. Alteration may affect the entire mass of rock or may only occur close to fissures, cracks and joints. The mechanical and hydraulic properties of an altered rock are radically different from those of the original rock and resemble silty sands, sandy silts and sandy silty clays.

When a tunnel is driven in altered rock, arching occurs which is similar to that discussed for broken and crushed rocks: the rock load is far smaller than the pressure corresponding to the weight of all the material overlying the excavation. Nevertheless, in altered rocks the arching is modified by phenomena which do not occur in other types of rock.

In altered rock or clay, the stand up time is far longer than in sands and crushed or broken rocks. Because of this, excavation of the tunnel face by stages rarely proves necessary. On the other hand, however, the increase that takes place with time in the variation of their shear strength with normal pressure, and unaltered rocks.

The most significant properties of the badly weathered rocks in tunnels are 1) their tendency to swell when relieved of loads, 2) the variation of their shear strength with normal pressure and 3) the speed with which they react to changes in stress.

When the weathered rock is relieved of pressures, it tends to swell and in the presence of water it may swell significantly depending on the new minerals present. These may be clay minerals of the montmorillonite or smecktite group, and similar minerals such as vermiculite and chlorite. Sufficient emphasis has already been placed on this phenomenon in other parts of this book. When a tunnel is driven in these materials, the clays and similar minerals in the zones close to the edge of the cavity undergo a reduction in confining pressures and consequently expand, taking water from the materials further from the tunnel. This causes a reduction in the shear strength of the weathered rock next to the tunnel walls. Some engineers believe that the moisture generally prevailing in tunnels softens the clay in the roof and walls. This is rarely true; usually a sample of clay that is taken from the wall and left inside the tunnel in contact with the tunnel air dries out appreciably in just a few days.

When a tunnel in clay is not suitably supported, material flows slowly from the walls, floor and roof and tends to close up the cavity. This is referred to as a plastic flow. During this process, and due to the expansion that occurs simultaneously, the shear strength of the clay is reduced to a minimum, where it remains practically constant. This final strength is referred to as *ultimate or residual cohesion*. It is clear that the time required for expansion and loss of strength depends on the initial permeability of the minerals and their water adsorption properties in general. For a given tunnel and a given depth, swell velocity increases rapidly with the dimensions of the unsupported section of the tunnel. Consequently expansion problems can usually be avoided by installing the support system close to the working face.

When excavation work advances ahead of the support system, the three-dimensional dome action that occurs at the working face is replaced by the two-dimensional arching, which is naturally less effective. As a result, expansion tends to increase, especially in the floor and walls of the tunnel. Flow of material towards the cavity is accompanied by strain which lengthens a clay element in the radial direction and shortens it in the circumferential direction. This strain mobilizes the internal friction and apparent cohesion of the material, so that the moment the clay starts to flow into the cavity, the neighbouring materials on the boundaries of the tunnel start to function like a cylindrical arch around the entire cavity. This firm material on the perimeter of the cavity is referred to as a resistant cylinder and is extremely valuable for supporting pressure from the weathered rock or clay located furthest from the excavation.

As soon as the support system has been installed and suitably wedged, plastic flow ceases, although the clay may not have adapted to the new stress state resulting from excavation. Consequently the flow tendency may not have been fully neutralized. The result is that pressure against the supports often increases, although at a decreasing rate. The period of time over which this pressure increase takes place can vary from a few weeks to many months.

The swell capacity of clays depends largely on the pressure at which they have been consolidated. In preconsolidated clays, swell capacity is great, the rate of swell and plastic flow is slow and the pressure increase on tunnel supports is appreciable and slow. If the tunnel is near the ground surface, the ultimate value of pressure on the support may exceed the overburden pressure.

Hard clays are frequently very cracked. They crumble readily when they flow under pressure in the walls of a tunnel, because this flow causes a reduction in the length of any soil element in the circumferential direction. As a result of these effects, clays often fall from the tunnel roof and the stand-up time or bridge action period in these materials is generally restricted by the crumbling effect just referred to.

In soft clays the stand-up time or bridge action period lacks significance, for these materials flow from the start.

All the foregoing mechanisms may be present in rocks containing a sufficient quantity of clay. This clay may be a product of decomposition of the rock proper or have some other origin.

The rock may be crushed and jointed or even mechanically intact, but its properties as far as flow or swell capacity are concerned are nevertheless determined by those of the clay it contains.

The few tests that have been conducted on rocks with plastic flow but little or no swell capacity indicate that the height of rock load H_p is proportional to $(B + H_t)$, but with a coefficient of proportion that is higher than in the case of very cracked rock. H_p increases for several weeks from the moment excavation commences and also increases with the depth of the tunnel below the ground surface. The highest pressures reported by TERZAGHI in tunnels at depths of one or two hundred meters indicated that the corresponding H_p value increased from an initial value of 1.10 $(B + H_t)$ to a final value of 2.1 $(B + H_t)$. However this can increase to about 4.50 $(B + H_t)$ during the months following excavation. Experience also shows that pressure on the walls is in the order of one third of that on the roof, and the pressure on the floor is about half that on the roof.

In expansive rocks the concepts for preconsolidated clays are applicable. The stand up time or bridge action period depends particularly on the rate of expansion and the spacing between any fissures that may exist in the rock. The initial rock load is almost exclusively due to wedging; it increases over a long period of time, sometimes several months, until very high pressures are reached.

In expansive rocks failure of a strut or rib is accompanied by an almost instantaneous but localized pressure relief. The intact struts adjacent to the failed strut are usually able to prevent a catastrophic failure for a few days. Pressure increases again when a new strut replaces the original one, but its ultimate load is now lower than that reached previously. Strut or rib failure usually increases the load on the adjacent struts. If they cannot support the increase, they also fail. The result can be a chain reaction failure of several struts in succession, and a roof fall. When the supports are not circular, the increase in water contant and reduction in shear strength occurring when the rock close to the cavity expands may cause the struts to penetrate the tunnel floor, triggering a general collapse of the support system. For this reason circular supports should be regarded as indispensable.

Little reliable information is available for evaluating loads in expansive rocks. In tunnels close to the ground surface, the rock load may be considerably greater than that corresponding to overburden depth. In deep tunnels pressures of 10 kg/cm^2 (142 lb/in^2) have sometimes been observed, and in exceptional cases as high as 20 kg/cm^2 (284 lb/in^2). This last value is roughly equivalent to a cover depth of 80 m (260 ft) of rock above the tunnel roof. These pressures indicate that even in expansive rocks the arching effect is important. Because expansion leads to a reduction in the pressures exerted by the soil or rock so long as there are no restrictions, some engineers recommend a space between the support and the excavation; 10 to 15 cm (4 to 6 in) is often satisfactory.

The following procedure was recommended by TERZAGHI for constructing support systems. Circular steel ribs capable of tolerating the swell pressure of the rock are installed. As a consequence, the rock flows around these ribs, overcoming the strength of the relatively weak struts that are placed between them. Once these struts have yielded, they are withdrawn, the expanded material is cut away and the elements in between are reconstructed. In this way, the pressure is gradually controlled without having to replace the entire support system and without having to make it excessively strong.

A very important aspect is recognition of the swell capacity of the rock before commencing excavation. For this purpose TERZAGHI recommends that specimens of fresh rock be taken and submerged in water and their increase in volume measured. An increment of less than 2% indicates that the rock is not expansive, in the sense that has been dealt with here. This point is important, not only for evaluating rock load, but also for deciding how carefully the support system must be wedged. Indeed it was seen that in all the aforementioned types of rock careful wedging of the support reduces not only the period of pressure increase, but also the ultimate value of this pressure. However, in clearly expansive rocks reference has already been made to the advisability of leaving a space between the strut and the walls of the excavation, in order to reduce the ultimate value of the pressure on the support. Thus, a correct evaluation of the swell capacity of the rock determines the construction procedures that should be adopted.

TERZAGHI'S recommendations are summarized in Table 14-3 [6, 7], and also include those relating to the type of support system that is required in each case.

The limits in the load height H_p column refer to the condition of the rock or soil. There are also limits according to the period over which the load is measured, assuming that the tunnel is temporarily unsupported. For this purpose, the initial load can be regarded as zero in sound intact, sound stratified or moderately fissured or cracked formations.

In very broken rocks, the value of the intial load is equivalent to 0.6 $(B + H_t)$. In gravels and sands it varies between 0.54 $(B + H_t)$ and 1.2 $(B + H_t)$. If the sand or gravel are in a loose state, the minimum value for the initial load will be 0.94 $(B + H_t)$ and the recommended minimum value for the ultimate load will be 1.1 $(B + H_t)$. In all the other cases referred to in Table 14-3, it is advisable to regard the ultimate load as acting from the time of first support.

In sound, moderately stratified, fissured or broken rocks (or broken soils), the loading values given in Table 14-3 can be reduced by half when the tunnel is above the water-table.

14.5.2 PROTODYAKONOV'S Method [3]

This evaluation method also is essentially based on arching theories and has been developed for granular materials, although its use has been extended to include rocks and other types of soils with satisfactory results reported by Russian engineers, who have employed the method most often. For granular materials, PROTODYAKONOV assumed that above the cavity an arch develops that can be regarded as having three hinges, the equilibrium of which can be evaluated along the line AOB if there are only compression stresses and no bending stresses (Fig. 14-17).

.1 Granular Materials

It is considered that excavation of the tunnel produces arching. The equilibrium of the mass below the arch that forms is maintained by the stresses that develop along line AOB (Fig. 14-17). The arch is assumed to be a parabola.

The forces acting in any section DO of the arch are:

a) The resulting force, T, of the reactions that act from the right side above point 0.

Table 14-3

Rock or soil loads on tunnels [6,7] (The values in column H_p refer to the ultimate load occurring on an unsupported tunnel)

Conditions of rock or soil	Load height, H_p	Remarks
Unweathered sound intact rock	Zero	Light support if there is "popping" or explosive rock
Unweathered sound stratified rock	0 to 0.5 B	When necessary, light support
Moderately fissured rock	0 to 0.25 B	Light support if there is popping or explosive rock
Moderately fragmented rock	0.25 B to 0.35 $(B + H_t)$	Support in the ceiling, seldom in the walls and never in the floor
Very fragmented rock	0.35 $(B + H_t)$ to 1.10 $(B + H_t)$	Support in the ceiling and walls
Crushed but unweathered rock	1.10 $(B + H_t)$	Circular support elements recommended
Gravel and sand	0.62 $(B + H_t)$ to 1.4 $(B + H_t)$	Circular support elements recommended
Rock that deforms plastically at low stresses	1.10 $(B + H_t)$ to 2.10 $(B + H_t)$	Circular support elements advisable
Rock that deforms plastically at high stresses	2.10 $(B + H_t)$ to 4.50 $(B + H_t)$	Circular support elements advisable
Expansive rock	Up to 70 m (230 ft) independent of value of $(B + H_t)$	Circular support elements indispensable

b) $p = \sigma_z x$, the result of the vertical pressures acting on the section of arch under consideration.

c) The tangential reaction R' at point D, caused by the actions of the section of arch to the left of D.

Taking moments of these forces in relation to point D, the following equation is obtained:

$$M_D = -Ty + \sigma_z \frac{x^2}{2} = 0 \qquad (14\text{-}34)$$

whence:

$$\sigma_z \frac{x^2}{2} = Ty \qquad (14\text{-}35)$$

Assuming one hinge at A and others at 0 and B, reaction R acting at A is resolved into horizontal and vertical components. The vertical reaction supports the arch and the horizontal reaction tends to open it. Any displacement must be resisted by the friction forces developing on plane AB, subject to vertical pressures. From this, the Horizontal Reaction by Friction, H_1 can be expressed as:

$$H_1 = Vf \qquad (14\text{-}36)$$

where H_1 represents the restriction to displacement of point A by friction.

$$V = \sigma_z \frac{B}{2} \qquad (14\text{-}37)$$

and $f = \tan \varnothing$ for granular materials, where $\varnothing$ is the angle of internal friction of the material.

The horizontal reaction H should include both the effect of H_1 and that of the lateral pressure σ_x, so that the following expression is finally developed:

$$H = \sigma_z \frac{B}{2} f - \sigma_x h \qquad (14\text{-}38)$$

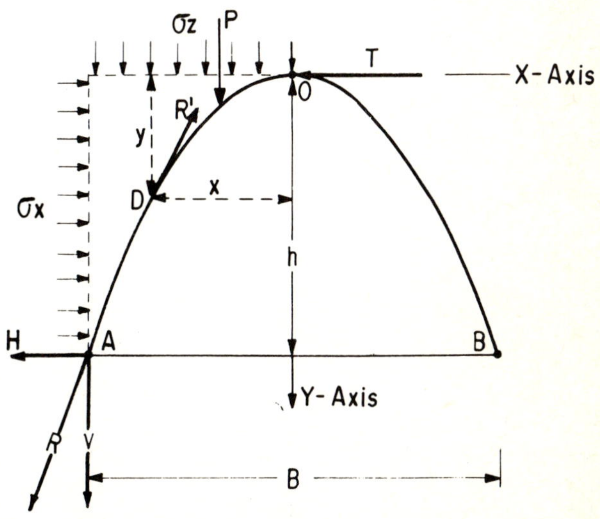

Fig. 14-17 Explanatory diagram of PROTODYAKONOV's method [3]

Equation (14-35) is that of a parabola passing through D. If this equation is applied at point A, where $x = B/2$, $y = h$ and $T = H$ and also taking into consideration Eq.(14-38), the following expression will be obtained:

$$\sigma_z \frac{B^2}{8} = \left[\sigma_z \frac{B}{2} f - \sigma_x h\right] h \tag{14-39}$$

whence:

$$\sigma_x = \sigma_z \frac{B}{2} \frac{4hf - B}{4h^2} \tag{14-40}$$

The height of the natural arch that forms is associated with the maximum stress σ_x that can be developed by the material. This provides a mathematical means of finding h. It consists of deriving σ_x in relation to h and separating this value from the derived one equalled to zero:

$$\frac{d\sigma_x}{dh} = \sigma_z \frac{B}{2} \frac{B/2 - fh}{h^3} = 0 \tag{14-41}$$

Separating h:

$$h = \frac{B}{2f} \tag{14-42}$$

This gives the height of the natural arch that develops in the material above the cavity, under stress conditions that are compatible with equilibrium.

Substituting this value of h in Eq.(14-40):

$$\sigma_x = \frac{\sigma_z}{2} f^2 \tag{14-43}$$

This value is further included in Eq.(14-38), in order to obtain:

$$H = T = \sigma_z \frac{B}{2} f - \sigma_z \frac{B}{4} f = \sigma_z \frac{B}{4} f \tag{14-44}$$

The above can then be combined in the equation for the parabola passing through D, Eq.(14-35) to obtain:

$$\sigma_z \frac{x^2}{2} = Ty = \sigma_z \frac{B}{4} fy$$

whence:

$$y = \frac{2x^2}{Bf} \tag{14-45}$$

The load acting on the tunnel is the weight of the material below that parabola. The load of the material that is left above and to the sides of the parabola is transmitted by arching to the outside of the tunnel, without exerting pressure on the supports. The area under the parabola has the following value:

$$A = \frac{2}{3} Bh \tag{14-46}$$

Consequently, the load received by the support per unit of tunnel length will be:

$$C = \frac{2}{3} \gamma_m Bh \tag{14-47}$$

γ_m being the unit weight of the covering material. Bearing in mind the value found for h, Eq.(14-42), the following is true with $f = \tan \varnothing$ as in the granular materials under consideration:

$$C = \frac{1}{3} \gamma_m \frac{B^2}{f} \tag{14-48}$$

The above expression permits calculation of the load that must be carried by the supports per unit of tunnel length as a function of elements that are either known or can be reasonably evaluated. If load C is regarded as acting on an area B, the vertical pressure to be supported is:

$$\sigma_v = \frac{1}{3} \gamma_m \frac{B}{f} \tag{14-49}$$

The formula reached by PROTODYAKONOV for the pressure on the support is similar to the one obtained by TERZAGHI in his arching analyses leading to Eq.(14-9).

.2 Extension to Cohesive Materials and Materials with $c \neq 0$ and $\varnothing = 0$

PROTODYAKONOV's method has been verified by numerous experiences which have shown that (with the exception of thinner covering depths) the pressures observed do not depend on the depth at which the tunnel is located.

These experiences have made it possible to extend the application of this method, which was originally established for granular materials. Its extension to cohesive soils has been accomplished by adding a coefficient of relation between the horizontal reaction due to friction and the vertical reaction, different from the value tan $\varnothing$, (which was used previously in granular materials).

In purely cohesive soils, this coefficient is referred to as the strength factor and is expressed [3] as:

$$f = \frac{c}{q_u} \tag{14-50}$$

Thus, in the general case of soils that are both cohesive and frictional, the strength factor will be expressed as follows:

$$f = \frac{c}{q_u} + \tan \varnothing \tag{14-51}$$

In rocks, PROTODYAKONOV's method has been used with a strength factor equal to:

$$f = \frac{q_{uc}}{100} \tag{14-52}$$

where q_{uc} is the strength of the rock, in an unconfined test performed on a cubic specimen. The empirical value 100 corresponds to strength in kg/cm^2. Table 14-4 [3] gives a series of empirical values for the strength factor for various different soils and rocks.

Table 14-4
Empirical values for the strength factor, f, in Protodyakonov's method [3]

Degree of strength	Type of rock or soil	γ_m		q_{uc}		Strength factor, f
		kg/m^3	lb/ft^3	kg/cm^2	lb/in^2	
Very high	Massive granites, quarzites or unweathered basalts and in general unweathered hard rocks of exceptionally high strength	2800–3000	175–187	2000	28400	20
Very high	Nearly massive granites, porphyries, siliceous slates, well indurated sandstones and unweathered limestones	2600–2700	162–168	1500	21300	15
High	Jointed granites and similar hard but cracked formations. Fairly well indurated sandstones and limestones. High strength conglomerates. Strong ironstones	2500–2600	156–162	1000	14200	10
High	Limestones in general. Weathered granites. Ironstones. Relatively solid sandstones. Marbles	2500	156	800	11360	8
Moderately High	Sandstones	2400	150	600	8520	6
Moderately High	Slates	2300	144	500	7100	5
Medium	Clay shales. Poorly indurated limestones and sandstones. Fractured or variably cemented conglomerates	2400–2800	150–175	400	5680	4
Medium	Clay shales. Clayey slates. Marls	2400–2600	150–162	300	4260	3
Moderately Low	Soft clay shales. Closely fractured limestones. Gypsums. Blocky sandstone. Poorly cemented gravels	2200–2600	137–162	200–150	2140–2130	2–1.5
Moderately Low	Gravels. Fragmented clay shales and slates. Hard slope deposits. Hard clays	2000	125	—	—	1.5
Low	Dense clay, clayey soils	1700–2000	106–125	—	—	1.0
Low	Loess. Sand and gravel formations. Sandy-clayey or silty-clayey soils	1700–1900	106–118	—	—	0.8
Soils	Soils with fibrous organic matter. Peats. Wet sands	1600–1800	100–112	—	—	0.6
Granular Soils	Sands and gravels	1400–1600	87–100	—	—	0.5
Plastic Soils	Soft silts and clays	—	—	—	—	0.3

According to Soviet practice, which includes the construction on numerous tunnels in soils and rocks, PROTODYAKONOV'S method is verified for depths ranging from $B/2 \tan \varnothing$ to $B/\tan \varnothing$.

The two main objections that have been made to the method can be summarized as follows:

1) The height of the loading arch, Eq.(14-42), is a linear function of the width of the tunnel. Some engineers believe that this relation is too simple to take into consideration all the factors of importance.
2) Table 14-4, which gives values of the strength factor employed by this method for a wide range of cases, is not definitive, so that the choice of this factor fluctuates between very wide limits.

In addition to all the foregoing considerations some further criteria have been proposed by Soviet engineers [3]:

1) When the tunnel is built in loose sedimentary rock, with $\varnothing < 40°$ and a depth of cover of less than $2.5B$, the support must be designed to carry the geostatic pressure corresponding to the entire depth of cover.
2) In stratified soils, only the properties of the stratum immediately above the tunnel roof should be taken into consideration. The higher overlying strata are only of interest when calculating vertical normal pressures (σ_z).
3) When a tunnel is built in clays below the water-table, the effect of expansion of the clay should in some way be taken into consideration. This can be done by taking an additional weight equal to that of a column of soil from the tunnel roof to the water-table.

14.5.3 BIERBÄUMER'S Method [3]

This method was developed during the construction of large tunnels through the Alps. According to this theory, the load that acts on the tunnel corresponds to the weight of the material that is included inside a parabola of height $h = \alpha H$ (Fig. 14-18).

The mathematical approach to the theory is intended to establish the value of α to be used in each case. Figure 14-19 [3] gives force diagrams on the basis of which the value of α can be

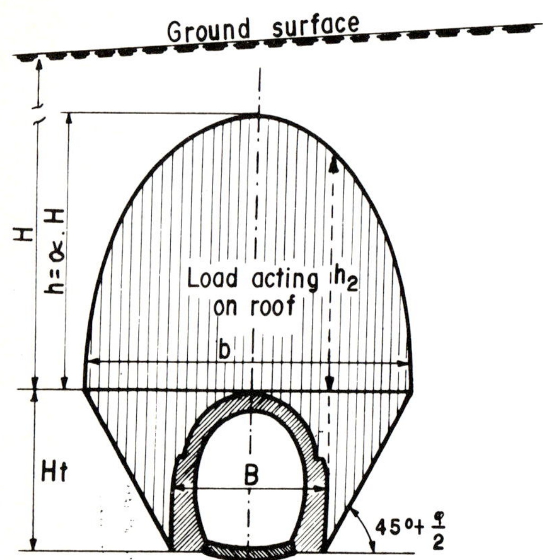

Fig. 14-18 Schematic drawing of the acting load in BIERBÄUMER'S theory [3]

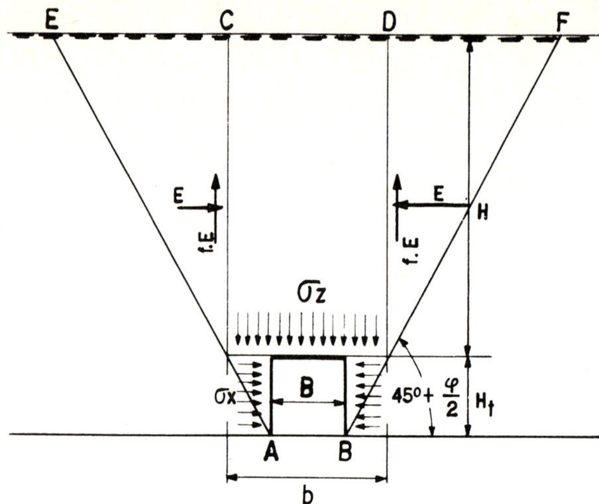

Fig. 14-19 Operational force diagram from BIERBÄUMER'S theory [3]

found. Dimension b is the width of the load parabola in the section corresponding to the tunnel roof, which is determined by drawing two lines at $45° + \varnothing/2$ outward from points A and B. The weight of the mass that gravitates on the tunnel roof is counteracted by the two friction forces that develop along the vertical planes through points C and D and also by the friction produced by the pressures σ_x that are exerted laterally against the tunnel walls. The direction taken by the analysis implies that the material in which the tunnel is driven is of a frictional nature.

After a mathematical procedure, which is omitted as it does not add to the conceptual outlook, BIERBÄUMER arrives at the following expression for α:

$$\alpha = 1 - \frac{\tan \varnothing \tan^2 (45° - \varnothing/2)\, H}{B + 2H_t \tan (45° - \varnothing/2)} \tag{14-53}$$

Table 14-5 gives a series of values of α for the conditions that are quoted with reference to single and double-track railroad tunnels.

For every thin depths of cover, $\alpha = 1$ should always be taken For thick depths of cover, whenever $H > 5(2H_t \tan (45° - \varnothing/2))$ the coefficient α becomes independent of the depth of cover and becomes the constant value:

$$\alpha = \tan^4 \left(45° - \frac{\varnothing}{2}\right) \tag{14-54}$$

14.5.4 Other Methods for Calculating Pressures on Tunnels

Reference [3] which is very complete, discusses several theories for estimating the vertical pressure that is exerted on tunnel roofs. Although some of these theories are very attractive from the mechanical viewpoint, none of them have been satisfactorily verified by experimental work. Some of the most reasonable ones are FENNER'S, ESZTÓ'S, KOMMERELL'S and BALLA'S [22].

Table 14-5

Values of the coefficient, α in BIERBÄUMERS theory [3]
Single track railroad tunnel, B = 7 m (23 ft); H_t = 8 m (26 ft)

H_m (ft) \ Ø		15°	20°	25°	30°	35°	40°	45°
20	(65)	0.80	0.79	0.78	0.77	0.76	0.74	0.72
30	(98)	0.70	0.69	0.67	0.65	0.63	0.61	0.58
40	(131)	0.60	0.58	0.56	0.54	0.51	0.48	0.44
50	(164)	0.50	0.48	0.44	0.42	0.38	0.34	0.30
75	(246)	0.42	0.38	0.32	0.26	0.21	0.17	0.12
100	(328)	0.36	0.32	0.26	0.20	0.15	0.12	0.09
125	(410)	0.35	0.28	0.22	0.17	0.12	0.09	0.07
150	(492)	0.35	0.24	0.19	0.14	0.10	0.08	0.06
175	(574)	0.35	0.24	0.17	0.12	0.08	0.06	0.04
200	(656)	0.35	0.24	0.17	0.11	0.07	0.05	0.04
> 200	(> 656)	0.35	0.24	0.17	0.11	0.07	0.05	0.03

Twin track railroad tunnel, B = 10 m (33 ft); H_t = 8 m (26 ft)

H_m (ft) \ φ		15°	20°	25°	30°	35°	40°	45°
20	(65)	0.86	0.84	0.84	0.83	0.83	0.83	0.83
30	(98)	0.79	0.76	0.76	0.73	0.73	0.73	0.73
40	(131)	0.72	0.68	0.66	0.64	0.64	0.63	0.63
50	(164)	0.65	0.60	0.58	0.55	0.54	0.53	0.53
75	(246)	0.48	0.42	0.37	0.33	0.31	0.29	0.29
100	(328)	0.39	0.36	0.29	0.24	0.18	0.15	0.11
125	(410)	0.35	0.30	0.24	0.19	0.14	0.11	0.08
150	(492)	0.35	0.28	0.20	0.16	0.11	0.09	0.06
175	(574)	0.35	0.24	0.18	0.13	0.09	0.07	0.05
200	(656)	0.35	0.24	0.17	0.12	0.08	0.06	0.04
> 200	(> 656)	0.35	0.24	0.17	0.11	0.07	0.05	0.03

14.5.5 Evaluation of Some Vertical Pressure Theories

Reference [18] contains an analysis of interest to engineers confronted by the practical need to estimate the vertical pressures exerted on tunnel roof supports. By applying the various available theories, results will be obtained that are sometimes very different. The question arises, therefore, of which is the most reasonable criterion to adopt for selection of a theory. Theory alone provides little guidance. None of the theories proposed in previous pages can meet all the requirements of a valid analysis in the field of continuum mechanics. All contain elements that depend on engineering *art*, introducing effects supported by ample experimental evidence, such as arching, which do not fit in with applied continuum mechanics. It appears impossible to choose a theory that is better than the others. In view of this situation, an evaluation like that of [18] may be of help. TERZAGHI'S, PROTODYAKONOV'S and BIERBÄUMER'S methods, among others, are compared with the results of the application of a visco-elastic model in which the vertical pressure is assumed to be the total geostatic pressure. To this condition is added a very large ratio of vertical pressure to horizontal pressure, which results in a tunnel that is subject to very great pressures. This does not necessarily require a very sturdy support system, because the high ratio of σ_z to σ_x causes reduced bending moments. A condition like this is probably similar to the long-term condition actually existing in many real tunnels.

Having designed the same tunnel using the different methods, PROTODYAKONOV'S led to a support system which was 7.5% more expensive than that corresponding to the visco-elastic model, 5.4% more expensive than TERZAGHI'S method and 0.6% more expensive than BIERBÄUMER'S method.

An assessment of the safety conditions of a supported tunnel established that PROTODYAKONOV'S method led to a design that was 25% safer than the visco-elastic model; TERZAGHI'S to a design that was 18% safer, and BIERBÄUMER's to a design 2% safer. From this point of view and under the conditions discussed in [18], PROTODYAKONOV's would probably have been the most appropriate method, for it led to an increase in safety of 25% with a cost increment of only 7.5%. Obviously these figures represent only one case. They might prove very different under circumstances and design conditions different from those assumed. The analysis is quoted as an example of a possible methodology for discerning which might be the most suitable theory for design.

Table 14-6 [3] serves a purely illustrative purpose by giving pressures observed in existing tunnels. These values may serve to establish criteria for preliminary design purposes or for qualifying the results obtained from a certain calculation.

14.5.6 Methods for Calculating Lateral Pressures

The magnitude of lateral pressures may be no less significant than that of vertical pressures for the design of a tunnel support system. When discussing squeezing or *genuine mountain* pressures, it was mentioned that tunnel walls may be the critical element in relation to this stress. Moreover this is not the only case where lateral pressure may play a role that is just as important as the vertical pressure. The magnitude of lateral pressures depends more on the deformations of the support system than that of the pressures in the roof; therefore their magnitude is influenced, as with no other type of pressure, by the strength and rigidity of the support (Plate 14-4).

Residual stresses, which are a product of the previous geological history of the material in which the tunnel is excavated, also affect lateral pressures far more than vertical pressures.

Lateral pressure in soils is usually established on the basis of the same ideas that are developed in the classic earth pressure theories against walls. The lateral (horizontal) pressure is equal

Plate 14-4 Distortion of tunnel walls due to lateral pressure

Table 14-6
Observed rock pressures in tunnels [3]

Type of material	Roof pressures, σ_z				Lateral pressure, $\sigma_x = \frac{1}{2}\sigma_z$ or $\frac{1}{3}\sigma_z$				Bottom pressure		Support		Remarks
	Initial condition		Ultimate condition		Initial condition		Ultimate condition				Construction procedure	Degree of stress that it supports	
	t/m²	lb/in²	t/m²	lb/in²	t/m²	lb/in²	t/m²	lb/in²	t/m²	lb/in²			
Unweathered or slightly fragmented rock	0	0	8–12	11–17	—	—	—	—	—	—	Light skeleton lagging	None	Slight loosening pressure
Very cracked rock. Soft rock with thin covering	10	14	30–35	42–50	—	—	3	4	4–6	6–9	Solid skeleton lagging	Low	Loosening pressure tending to increase with time
Fragmented rock. Moderately cemented conglomerates. Agglomerates	15–25	21–35	30–40	42–57	5–10	7–14	5–15	7–21	10	14	Very sturdy skeleton lagging	Moderate	Takes time for equilibrium conditions to be reached
Closely fractured rock, saturated. Thick covering	25–35	35–50	40–60	57–85	10	14	10	14	15	21	Very tight, strong lagging. Circular support element is advisable	Considerable	Stabilization of pressure conditions very difficult
Fragments, gravels and other soils. Very soft rock under heavy pressures. Very thick coverings	40–60	57–85	100–150	142–213	20	28	15	21	30	42	Circular support element	Hard to control expansion pressures which may reach rupture point	Stabilization of pressures only possible after developing large displacements, sometimes lasting months or even years

to the vertical pressure multiplied by a coefficient, the value of which depends on the horizontal displacement that is permitted. Thus, lateral pressure is a function of the depth of cover above the tunnel. Consequently, at very great depths lateral pressure determined in this way may prove even greater than vertical pressure, which is reduced by arching effects. At the same time, at these great depths, there may also be an appreciable increase in the frictional strength of the soil that can be considered in the horizontal direction.

Some lateral pressures observed in tunnels have been included in Table 14-6. In this table it can be seen how lateral pressure normally ranges between one quarter and one third of vertical pressure; however the table does not include any case with heavy lateral stresses of tectonic origin, which alter the above relations completely. Reference [3] also quotes the case of tunnels through the Hoover Dam (U.S.A.), where lateral pressures were observed up to three times as great as those acting in the roof.

According to TERZAGHI, horizontal pressure can be estimated using the Equation:

$$p_h = 0.3\gamma_m (0.5H_t + H_p) \tag{14-55}$$

where H_t is the height of the tunnel and H_p is the height of cover material above the roof considered for calculation purposes. In granular soils or broken rock, TERZAGHI proposes the expression:

$$p_h = \gamma_m \frac{H}{N_\varnothing} \tag{14-56}$$

where H is the total distance from the tunnel roof to the surface of the natural ground above it and $N_\varnothing$ has the usual meaning in earth pressure theories (Chapter 5). This is equivalent to considering an active earth pressure state in the tunnel walls.

According to Soviet practice [3], considering a parabolic pressure for the vertical load, as in PROTODYAKONOV's or BIERBÄUMER's method, horizontal pressure can be computed considering a linear distribution between a value acting at roof level (P_{h1}), and another value acting at floor level (P_{h2}), as shown in Fig. 14-18. These two values will be:

$$p_{h1} = \gamma_m \frac{h_2}{N_\varnothing} - 2c \frac{1}{\sqrt{N_\varnothing}} \tag{14-57}$$

$$p_{h2} = \gamma_m \frac{h_2 + H_t}{N_\varnothing} - 2c \frac{1}{\sqrt{N_\varnothing}} \tag{14-58}$$

with h_2 having the meaning shown in Fig. 14-18, and other terms from the familiar equations corresponding to the active state. $N_\varnothing$ has the usual meaning of $\tan^2 (45° + \varnothing/2)$.

The meaning of both expressions within the familiar context of earth thrust theories is clear. Also to be noted is that the foregoing expressions are valid for frictional, cohesive materials. PROTODYAKONOV has proposed an expression for the total thrust acting in the tunnel wall as:

$$E = \gamma_m H_t \frac{1}{N_\varnothing} \left\{ \frac{2}{3 \tan \varnothing} \left[B + H_t \frac{1}{\sqrt{N_\varnothing}} \right] + \frac{H_t}{2} \right\} \tag{14-59}$$

In the relevant literature, the point at which this thrust is applied is not indicated. It would seem reasonable to locate it at half of the tunnel height.

During construction of large diameter Alpine tunnels in Switzerland, BONNARD measured lateral pressures which are rather more related with the total depth of cover above the tunnel than with the height reduced by arching consideration. BONNARD thus proposed the following expression:

$$p_h = \gamma_m H \frac{1}{N_\varnothing} - 2c \frac{1}{\sqrt{N_\varnothing}} \tag{14-60}$$

H is again the total depth of cover above the tunnel.

14.5.7 Calculating Pressures in the Tunnel Floor

Pressures in the floor are usually reactions to pressures in the roof, but owing to the effect of absorption of the neighboring masses of soil or rock, floor pressures are usually much lower than the corresponding roof pressures. TERZAGHI gave empirical evidence which verifies an estimation of floor pressure of about half of roof pressure.

Floor pressures are also greatly influenced by the construction procedures adopted. The *squeezing* or *genuine mountain* pressures and swell pressures usually play a leading role in converting floor pressure into an important problem in tunnels. In practice, therefore, serious problems in tunnel floors are associated with loose sands and, very particularly, saturated clays. The floor is an unloaded zone, whereas there may be important compression stresses at the sides, which may trigger bottom failure in the form of floor heave. In susceptible soils or rocks this heave may also be brought about by swelling associated with water absorption of the minerals.

The development of floor pressures is closely related to the appearance of important lateral pressures, as is shown in Fig. 14-20 [3], which depicts measurements taken during construction of a tunnel in Europe.

It can be seen how there was practically no heave in the bottom of the pilot excavation until excavation of the vault commenced, and how expansion increased rapidly when the vertical walls of the drift were built. The moment there was an increase

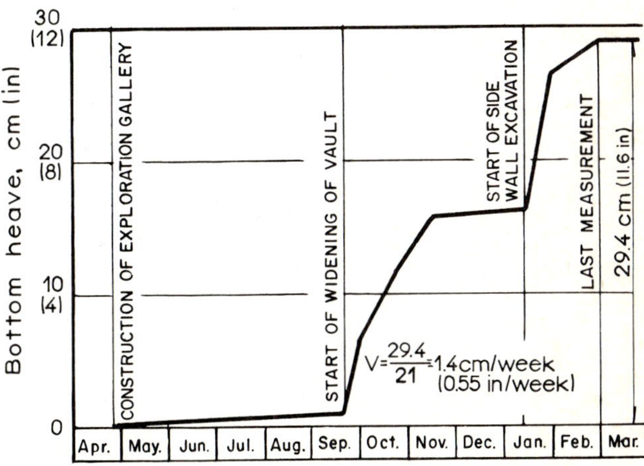

Fig. 14-20 Pressures on a tunnel floor showing its relation to lateral pressures [3]

in the width of the drift and a greater vertical load was transferred laterally, causing greater horizontal pressures, the heave increased dramatically.

The pressures in tunnel floors rarely lead to serious problems except for cases of squeezing pressures or swelling pressures. Computation of the pressures generated by the combined effect of vertical and horizontal pressures is complicated and is not regarded as necessary in this book, since the phenomenon is seldom really important. Reference [3] gives various procedures for conducting these analyses, where necessary.

14.5.8 Influence of Time on the Development of Pressures

As already mentioned, the pressure that is eventually received by a tunnel support may have several different origins. The foregoing methods can only quantify that corresponding to the weight of the materials involved, whether or not it is partially neutralized by arching. Moreover the pressures that develop will depend largely on the time that elapses before the tunnel support system is installed.

Excavation and construction are conducted in systematically repeated cycles. Drilling, loading, blasting, ventilation, mucking, installation of the temporary support system and construction of the permanent support system are usually the most significant stages. During excavation (drilling, loading, and blasting), the best protected sections are those that are close to the excavated face, for here the material is supported on three sides, instead of the usual two in the rest of the tunnel. At the excavated face a half dome forms, which is capable of standing longer unsupported than an arch. The time that the half dome can be left unsupported before fissures appear or expansions start to cause collapses or reductions in the section is known as the *stand up time* (bridge action period) and depends on the nature of the material, the width of the cavity and the distance that is left between the excavated face and the support system.

The operation sequence for construction should be such that the time the half dome at the face is left unsupported is shorter than the stand up time. Figure 14-21 [3] shows the relationship that can be expected between the time elapsing before the temporary support system is installed and the development of either excessive deflection or pressure in the tunnel roof. It is assumed that a construction cycle inside the tunnel will take a little longer than the stand up time, which is not an unusual situation. It is noted how the load of the material (H_p) increases at the face and in the cavity and how the ultimate values depend on the care that may have been exercised during driving operations. Curve c_1 refers to a careful driving operation, where no loose blocks are left unwedged, whereas c_2 refers to a less careful operation. Heights $H_{p\ max}$ and $H'_{p\ max}$ are the respective loads of material on the tunnel roof, corresponding to the two curves.

Figure 14-22 [3] shows a case where the stand up time is compared with the construction sequence, assuming that this includes blasting.

Considerable overbreak is caused if the stand up time is shorter than the ventilation period required for access to the face after blasting. The same thing occurs if the stand up time is only slightly longer than the ventilation period. It is advisable for the stand up time to last at least until mucking operations are fairly well advanced. The ideal situation is when the stand up time lasts until erection of the temporary support system commences. It may be advisable to reduce the advance or blasting round, Length l (Fig 14-21) when it is wished to prolong the stand up time in relation to the time required to commence construction of the temporary support system. In fact, this is one of the most obvious solutions for adapting excavation operations to suit the nature of the materials and the width of the cavity. Alternatively, temporary support by shotcrete and rock bolting the roof can precede the mucking. This is the reason this method is replacing all others for support.

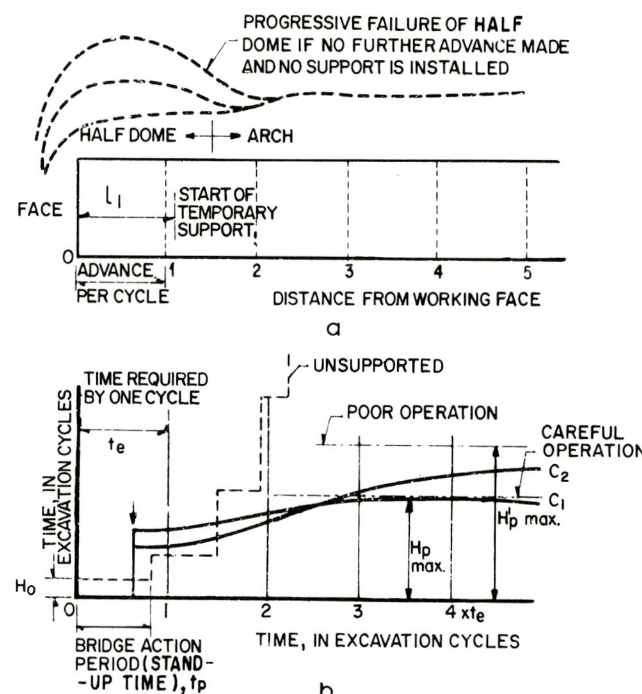

Fig. 14-21 Relation between the time before installation of supports and the development of pressure in the tunnel roof [3]

In wide tunnels, the stand up time or bridge action period is far shorter than in narrow tunnels. After a section has been excavated, the load of the material in the half dome that gradually forms always increases with time, but, as is shown by Fig. 14-21, this tendency may vary considerably, depending on the care that is taken when driving the tunnel. A careful operation that minimizes crack or joints opening and overbreak leads to smaller load increments than a careless operation, and the ultimate equilibrium condition is reached far more rapidly (Curves c_1 and c_2 in Fig. 14-21).

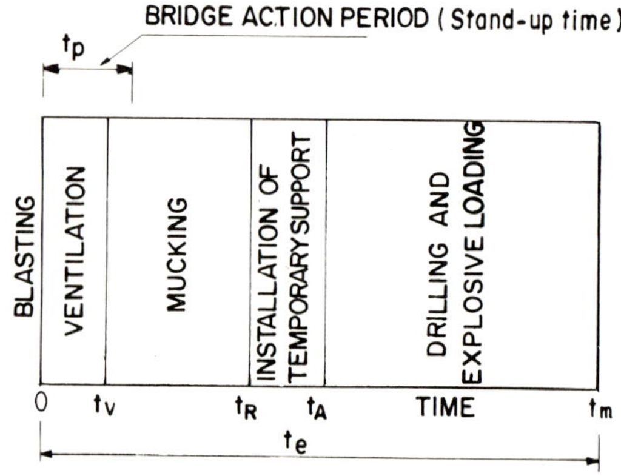

Fig. 14-22 Comparison of the bridge action period (stand-up time), t_p, with the construction operation sequence [3]

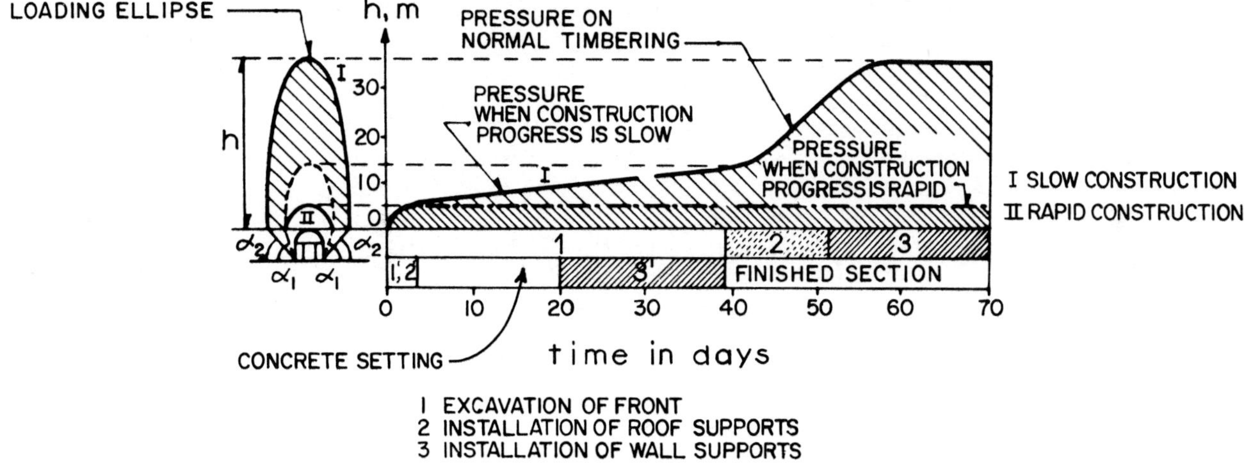

Fig. 14-23 Effect of construction time on the development of pressure in the tunnel roof [3]

Figure 14-23 [3] illustrates the effect of construction time on the development of pressure in the tunnel roof. Numbers *1*, *2* and *3* respectively represent the excavation at the working face, the ultimate conformation of the vault, and the excavation required to form the side walls. The figure depicts both slow tunnelling and rapid tunnelling. It clearly illustrates the difference both in the magnitude of the pressures that develop and in the mass of material that gravitates on the roof. In the case presented, which was reported by Bendel for a Belgian tunnel, the width of the load influence zone was defined with angles $\alpha_1 = 45° + \varnothing/2$ (with the horizontal) for the rapid construction procedure and with angles $\alpha_2 = 45° - \varnothing/2$ for the slow procedure. The figure refers to a pressure calculation procedure that has not been studied in this chapter, in which the subsiding zone above the roof is regarded as elliptical.

In cohensionless soils, the stand up time (bridge action period) is produced by capillary tension. Because it is small, and changes quickly, the temporary support system as close as possible to the excavated face. Special excavation methods, such as supported drifts or shields, frequently have to be employed in these cases. The stand up time is generally far longer in weathered rocks than in broken rocks. Many residual soils have longer stand up time than deposited soils because of their significant strength with no effective confinement. (True cohesion). Figure 14-24 shows estimates based on experience and experiments of Terzaghi on the pressure that is exerted by different types of rock or soil. The pressures that would appear were the roof left unsupported are indicated by broken lines.

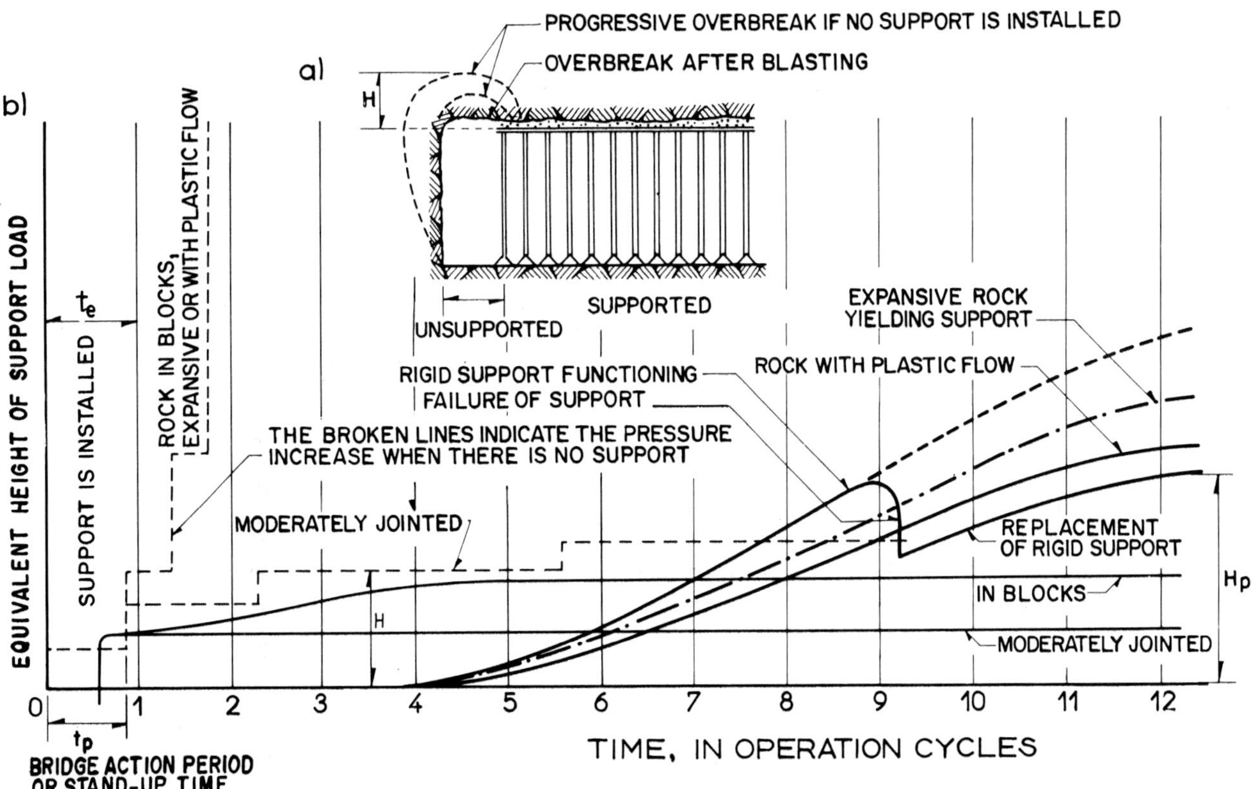

Fig. 14-24 Variation in roof pressure with time, according to Terzaghi

In sound rocks, the stand up time is practically unlimited and in moderately jointed rocks where the joints are tight the equilibrium condition is reached as excavation is completed. An interesting case is that of expansive rock, whose pressure curve, when unsupported is shown with a dot-and-dash line, whereas the pressure measured on a rigid support is shown with a solid line. Attention is drawn to the abrupt drop in pressure when the support failed and the subsequent recovery when it was replaced. Figure 14-24 is drawn assuming a stand up time (t_p) equal to 80% of the building operation cycle (t_e), which is taken as 1 on the horizontal time axis.

14.5.9 Hewett and Johannesson Method for Calculation of Pressures in Circular Tunnels in Soils.

Figure 14-25 [19] gives a diagram of vertical, lateral and bottom pressures proposed by Hewett and Johannesson in 1922 for circular tunnels in soils. The respective pressures that appear in the figure are as follows:

1) The weight of the upper half of the tunnel lining.
2) The weight of the soil inside the area marked *2*.
3) A uniform, vertical upward load which balances the combined effect of loads *1* and *2*.
4) The weight of the subsiding earth above the crown of the tunnel, which can be estimated with one of the methods already mentioned.
5) A uniform upward reaction which balances the effect of load *4*.
6) Horizontal pressure due to the weight of water that may exist from the crown of the tunnel upwards.
7) Horizontal pressure due to the water that may exist between the crown and the lowest point of the tunnel.
8) Horizontal pressure due to the soil above the crown of the tunnel, considered equal to the product of the unit weight (submerged) of the soil multiplied by the height H_p that causes vertical pressure and by a coefficient of earth thrust, K. It is usually assumed that K may vary between the values corresponding to the active state and the passive state:

$$\frac{1}{N_\varnothing} \leqslant K \leqslant N_\varnothing \qquad (14\text{-}61)$$

In practice the following values are used, based on experience.

$$K = \frac{1}{3} \text{ to } 1 \qquad (14\text{-}62)$$

9) Horizontal pressure due to the earth existing between the crown and the lowest point of the tunnel. At any point, this pressure is equal to the product of the unit weight of the soil (submerged, if this is the case) multiplied by the vertical distance between the crown of the tunnel and the point at which the pressure is measured and by the coefficient of earth thrust, K.

A serious disadvantage of the use of the diagrams in Fig. 14-25 is the very wide range of the value of K and the absence of reliable criteria for establishing it within the interval between active and passive earth pressure coefficient limits. These diagrams are presented for historical reasons, for this was one of the first attempts made at evaluating soil pressures in tunnels.

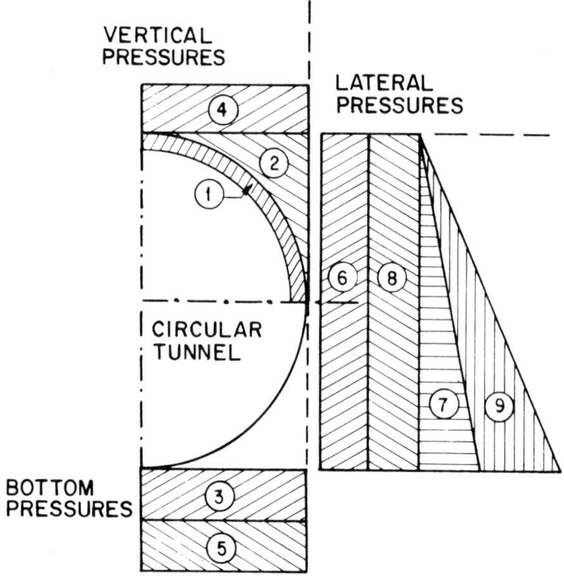

Fig. 14-25 Pressures acting on a circular tunnel in soils, according to Hewett and Johannesson [19]

14.5.10 Some Recent Experience in Calculating Pressures in Tunnels

.1 Method of Calculating Pressures used in the Design of the San Francisco Subway, U.S.A. [1]

A description follows of the criteria used in estimate radial pressure during construction of the San Francisco subway, which was built mainly through granular materials under water, but with the occasional appearance of formations of soft clay. The cross-section of the cavity is circular.

Criteria were established on the basis of building a tunnel lining that would not be very rigid in relation to the surrounding environment, so that both elements would move in the same way. Accordingly the design recommendations include both the effects of earth weight and stresses in the lining due to distortion. Figure 14-26 [1] shows the criteria adopted for sands and Fig. 14-27 those adopted for soft clays.

The recommendations refer to the radial pressure, σ_r, that acts uniformily on the entire perimeter of the cavity. Arching was considered only in granular soils and for overburden exceeding 8 m (26 ft). For depths exceeding 11 m (36 ft) it was estimated that arching reduces the effective overburden thickness to half.

For sandy soils three conditions are presented, depending on the overburden thickness (Fig. 14-26):

Condition a: For overburden thicknesses of 11 m (36 ft) to 23 m (75 ft), with the water-table at a height z above the crown of the tunnel, the radial pressure is:

$$\sigma_r = \left(11 + \frac{x}{2}\right) \gamma + (z + D)\, \gamma_w \left[\text{t/m}^2\right]$$

$$\sigma_r = \left(36 + \frac{x}{2}\right) \gamma + (z + D)\, \gamma_w \left[\text{kips/ft}^2\right] \qquad (14\text{-}63)$$

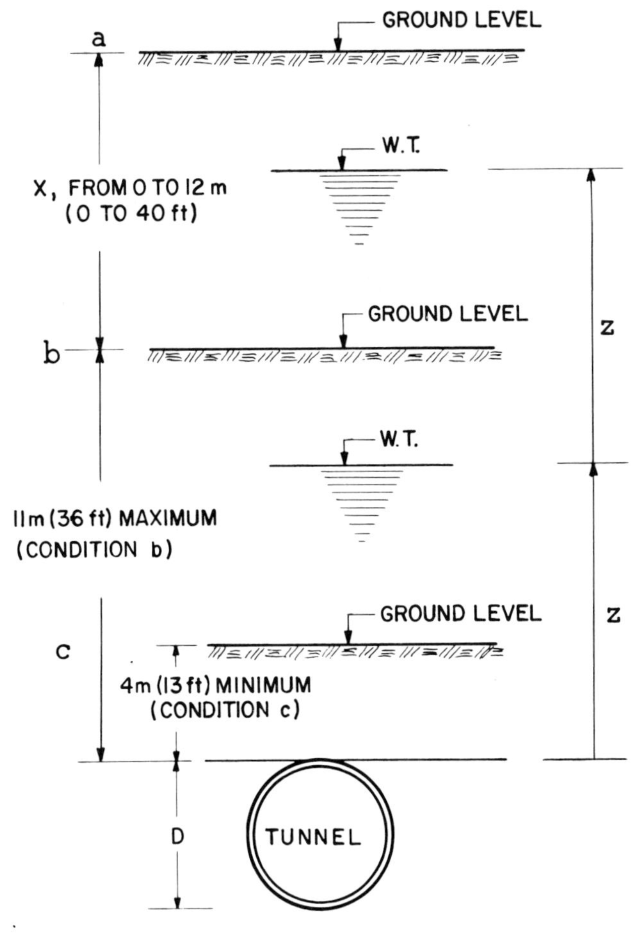

Fig. 14-26 Criteria for estimating radial pressures in sandy soils. San Francisco Subway, U.S.A. [1]

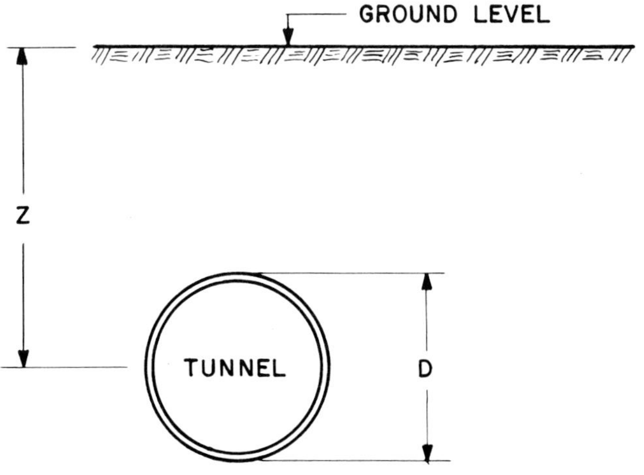

Fig. 14-27 Criteria for estimating radial pressures in normally consolidated soft clays. San Francisco Subway, U.S.A. [1]

γ should be regarded as submerged when the material is in such condition. Also to be taken into account are the effects of bending of the lining, considering that the vertical diameter of the tunnel may vary by + 1.6 cm (0.63 in). In Eq.(14-63), γ and γ_w are in t/m^3 (kips/ft^3), z and D in m (ft) and σ_r is in t/m^2 (kips/ft^2).

Condition b: For an overburden depth of between 4 and 11 m (13 and 36 ft) and a water-table at a height z above the key of the tunnel the radial pressure is:

$$\sigma_r = 11\gamma + (z + D)\,\gamma_w \left[\text{t/m}^2\right]$$

$$\sigma_r = 36\,\gamma + (z + D)\,\gamma_w \left[\text{kips/ft}^2\right] \qquad (14\text{-}64)$$

γ should be considered submerged when the material is in such condition. Also to be taken into account are the effects of bending of the lining, considering that the vertical diameter of the tunnel may vary by 1.27 cm (0.5 in). In Eq.(16-64), γ and γ_w are also in t/m^3 (kips/ft^2), z and D in m (ft) and σ_r in t/m^2 (kips/ft^2).

Condition c: For an overburden depth of between 0 and 4 m (13 ft) the radial pressure is:

$$\sigma_r = \left(4 + \frac{D}{2}\right)\gamma + (z + D)\,\gamma_w \left[\text{t/m}^2\right]$$

$$\sigma_r = \left(13 + \frac{D}{2}\right)\gamma + (z + D)\,\gamma_w \left[\text{kips/ft}^2\right] \qquad (14\text{-}65)$$

The bending effects that will have to be taken into account are the same as those in Condition *b*. For construction in the soft clay zone, no arching effect was considered, and the radial pressure was proposed as simply equal to γ_z, using γ'_m when the soil is below the water-table. In addition, the effects of bending resulting for the lining should be considered, acknowledging that the horizontal diameter of the tunnel may suffer a reduction of − 2.2 cm (0.86 in).

.2 Calculation of Pressures in Recently Constructed Tunnels in the Valley of Mexico

In the Valley of Mexico several large tunnels have been recently built or are in the process of construction through both the soft clay formation typical of the zone and peripheral formations of a sandy or alluvial fan type.

The general criterion for evaluating pressures has been based on the concept discussed in §14.7 of this chapter; if the lining is sufficiently flexible, the stresses are redistributed and the ground exerts a uniform radial pressure on the circular tunnel. For design purposes it was generally considered that this uniform radial pressure is equal to that produced by the total weight of the overburden thickness at tunnel level, which is conservative. This criterion was applied regardless of the type of soil.

.3 Controlled Displacement Support — the NATM Method

A system of tunnel support has been developed that maximizes the ability of the soil or rock to sustain itself. This is done by controlling the displacement in the tunnel walls and roof by combinations of shotcrete, mesh-reinforced shotcrete, rock bolts, and lightweight ribs. The system is known as the New Austrian Tunnelling Method (*NATM*) because such Austrian experts as Rabcewicz, Lauffer, Mueller and Golser developed the methodology of correlating theoretical displacement estimates with observed roof and wall movements for different levels of support. However the support methods are really neither new nor Austrian. They have evolved through worldwide experience with rock bolts and shotcrete in tunnels since the 1950's. They are replacing the older systems of liner plates or ribs and blocking, permitting faster tunnelling at lower cost in a wide variety of geological environments.

Simplistic pictures of the stress distribution around a tunnel were seen in Figs. 14-7 and 14-8. Although these applied to elastic conditions, they illustrate the manner with which the stresses vary in the vicinity of the tunnel wall, so long as very large strains or failure do not disrupt the continuity of the mass.

The relation of the tangential and radial stresses in the tunnel wall to strength as depicted in a MOHR diagram for a relatively strong and a relatively weak material are shown in Fig. 14-28a and 14-28b respectively. For each material there are three MOHR envelopes. The highest envelope represents very short-term, the middle, intermediate term and the lowest, long-term loading. For the stronger material with a major principal stress of $\sigma_T = \sigma_I = 2\gamma z$ and $\sigma_r = \sigma_3 = 0$, the circle of stresses for the surface of the tunnel is below the envelopes and the material does not fail. For the weaker material and a major principal stress, $\sigma_T = 2\gamma z$, the minor principal stress cannot drop to zero (unless the short term envelope is sufficiently high). In order to provide safety for intermediate and long term loads a minor principal stress, σ_r, must be applied to the tunnel wall. This is the function of the support system. If the MOHR envelope is very steep close to the vertical or τ axis, as is typical of fractured rock or dense angular cohensionless soil, only a small minor principal stress, σ_r, may be necessary to prevent failure; for a clay, a relatively large minor principal stress may be necessary.

The way in which the minor principal stress can be introduced depends on the stand-up time (without support) which is related to the very short-term material strength. If that very-short-term strength is not sufficient to enable the material to stand a short time with no support (stand up time = 0) then continuous support by forepoling, compressed air or a shield may be necessary. If the very short-term strength is large enough to provide some stand-up time, then it is possible to introduce support that provides sufficient minor principal stress to control strains and prevent immediate failure.

The continuing tangential stress produces an inward deflection of the tunnel wall, depending on the (1) length of time since excavation, (2) the short, intermediate and long term strength and rigidity of the material, and (3) particularly on the radial stress, $\sigma_r = \sigma_3$. If a radial stress could be introduced equal to the stress in the material before tunnelling, there would be no deflection. However, this is a very large stress. As the wall and

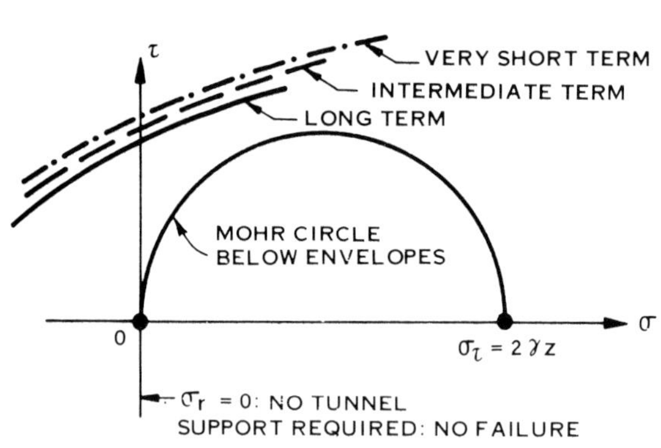

a. MOHR ENVELOPE, σ_r AND σ_t IN TUNNEL WALL AND ROOF OF STRONG MATERIAL

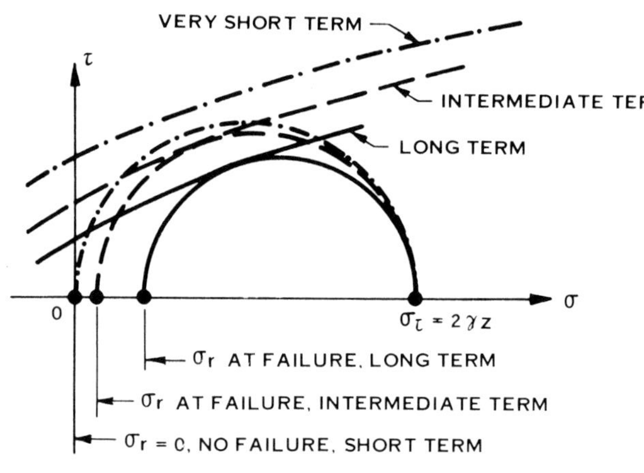

b. MOHR ENVELOPE, σ_r AND σ_t IN TUNNEL WALL AND ROOF OF WEAK MATERIAL

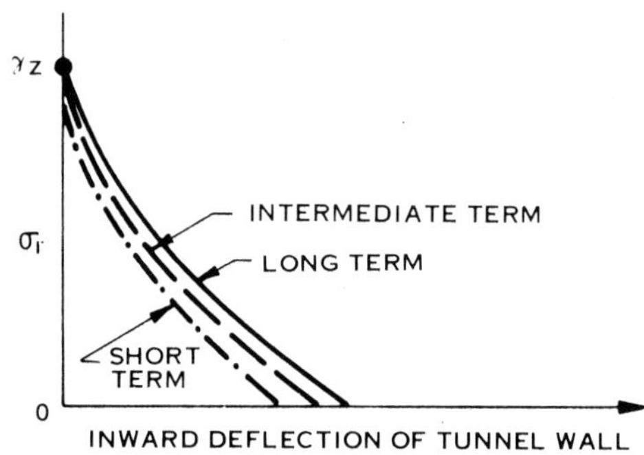

c. RADIAL STRESS AS A FUNCTION OF TUNNEL WALL DEFLECTION IN STRONG MATERIAL

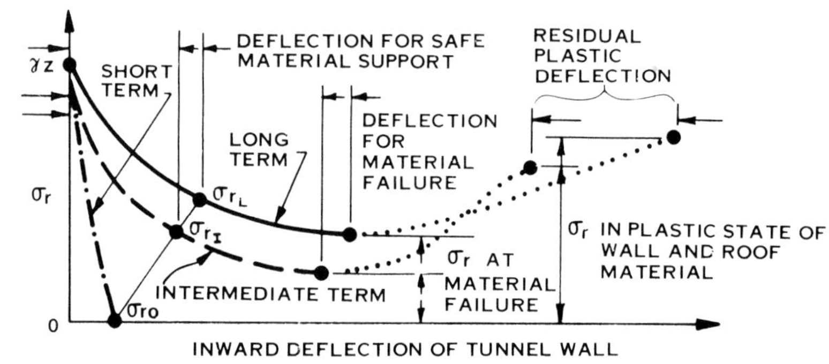

d. RADIAL STRESS AS A FUNCTION OF TUNNEL WALL OR ROOF DEFLECTION IN WEAK MATERIAL

STRESS, STRENGTH, DEFLECTION AND TUNNEL SUPPORT

Fig. 14-28 New Austrian Tunnelling Method

roof displace or deflect inward, the radial stress decreases correspondingly, as shown in Fig. 14-28 parts *c* and *d*, for a strong and a weak material respectively. For a strong material σ_r can be zero indefinitely; the wall will deflect inward without material failure.

For a weaker material with short term loading the radial stress can drop to zero, σ_{ro}, within the stand up time. This is illustrated by the lowest, dash-dot curve of Fig. 14-28d. However as time elapses, the material weakens, first to its intermediate and finally to its long term rigidities and strengths. This causes further deflection. By installing a support that develops a stress σ_{rI}, the deflection can be stopped before it reaches the limit for material failure. This must be done within the stand up time. This can be seen on the middle dashed curve of Fig. 14-28d. The added deflection that ensues during support installation and as the support deflects under load and develops resistance, σ_{rI}, is shown by the line from σ_{ro} to σ_{rI}. Long term deflection-stress is shown by the upper, solid line. With the passage of time the support stress σ_{rI} increases to σ_{rL}, accompanied by more deflection. The successive support stresses σ_{rI} and σ_{rL} must be great enough that the deflection does not reach that required for material failure.

Once failure occurs, the material breaks up, losing its continuity; consequently its strength drops. The radial stress increases with further deflection beyond that required for material failure. Ultimately, plastic equilibrium is reached at the residual strength of the material. This is equivalent to the plastic equilibrium described by TERZAGHI.

The objective of the *NATM* is to mobilize the strength of the material immediately surrounding the opening (the tunnel ring or tunnel arch) to support the unbalanced load in so far as possible. The minimum theoretical support stress occurs at the instant of material failure as illustrated in Fig. 14-28d. Because any deflection beyond failure causes disruption of the tunnel arch, loss of strength and a rapid increase in stress, it is unsafe to permit deflections to approach those at failure. The essence of the *NATM* is to allow enough deflection to reduce the support required as much as possible but consistent with maintaining safety against material failure and arch disruption.

The *NATM* requires placing the support within the available stand up time. It is placed in such intimate contact with the material and is sufficiently rigid that the inward deflection of the wall with the passage of time causes a corresponding increase in the radial stress generated by the support. If the support is sufficiently rigid, the radial stress it generates will equal the radial stress required to stop further displacement at a safe value: less than the displacement that allows failure, Fig. 14-28d. If the support is too rigid, the support stress will be high. If it is too flexible, the deflection will be excessive, failure may occur, and then the support required will be even higher.

Often the initial and smallest level of *NATM* support is 5 to 10 cm of dry shotcrete. It can be placed in an hour or two and develops significant strength in two to three hours. It is intimately in contact with the exposed material and once hard begins to develop resistance as the wall or roof displaces inward. Depending on the ultimate rigidity and support stress needed, additional layers of shotcrete, reinforced by steel mesh can be added until as much as 30 cm or reinforced shotcrete are placed. Dowels or tensioned rock bolts are often added after the first shotcrete layer to enhance the minor principal stress. Tensioning the bolts helps to minimize displacement, providing more rigid support.

If the material is cemented soil or rock, with widely spaced cracks, a pattern of dowels or rock bolts without shotcrete may be sufficient. Dowels (untensioned grouted bolts) develop their stress as the wall displaces inward. Tensioned bolts are preloaded, and thus develop support stress with little or no wall displacement. Dowels and tensioned bolts have the additional function of providing tensile strength and structural continuity in fractured materials. They maintain the physical shape of the tunnel ring or arch and enhance the strength of the material of which it is composed.

Application of the *NATM* requires estimating the displacement-support stress character of the material as well as the displacement that can produce failure. It requires a prediction of the stand up time and a knowledge of the fracture patterns, and other material features that influence the need for support.

The initial support that is applied is based on experience. The displacement of the roof and walls is measured and compared with the displacement-support prediction. If the displacement is greater than anticipated, or is increasing with time, more support is installed (more layers of shotcrete, more bolts) until satisfactory equilibrium is reached. Based on observed deflection in one section of tunnel, the support applied in the next increment of the same tunnel material is increased, or decreased. Displacement measurement and adjustment of the degree of support are essential to successful use of the method. Thus it requires good integration of design predictions with performance observations, adjusting the design as the work progresses.

14.6 Geological Considerations in Design and Construction of Tunnels in Soils

The geological formation through which a tunnel is driven is largely responsible for the construction and performance problems that will be met. Much of what was seen previously in relation to the evaluation of the pressures acting in tunnels was presented in a manner that did not take into account the geological situation that might prevail at a specific location. If this led to the idea that the formulas and evaluations reached are independent of the specific geological conditions prevailing at a given location, this would be one of the most dangerous and unreal conclusions that could have been reached. No method of pressure evaluation is powerful enough to cover all the possibilities of local geological accidents and the influences these might have on the pressures that are produced in a tunnel and on its performance. Evaluation methods and construction methods should be regarded as standards that are reasonable only under *normal* conditions. Local geological features and peculiarities often impose special conditions which will make the situation radically different, even though there may be no change in the geometry of the tunnel or the constitution of the materials through which it is driven.

The geological aspects of soils which may affect tunnel design and performance have not been thoroughly investigated. Far more careful investigations have been made in relation to tunnels in rocks. Also lacking is a clear distinction between the behavior or residual and deposited soils, which is fundamental. Residual soils present all the problems deriving from their inherited structures, which inevitably make their behavior resemble that of many rocks. On the other hand, rocks, when very broken or so weathered that the mineral bonds are destroyed exhibit behavior similar to those of soils. The same is true of closely jointed rocks or rocks whose joints are filled with soils. Consequently, so long as specific information is not produced as a result of investigation work, it will be reasonable to compare many aspects of the behavior of residual soils to those of fissured and partially weathered rocks. Transported deposited soils, which are typically stratified, will exhibit a different behavior, far less closely involved with geological structural defects.

The identification and classification of geological defects should be carried out in successive stages. First comes a general stage [23] such as the identification of a geological fault. Second there is a diagnosis stage for the recognition of problems associated with that defect and determination of the most important parameters that govern them. An example of this would be a study to compare the spacing between joints in rocks or a residual soil with the width of a tunnel excavation. A third stage applies the identification and classification of geological features or defects to aspects of design and construction. An example of work carried out during this stage would be a study of the effects of blasting or other loosening techniques to that system of features or defects. Characteristics of materials in relation to the tunnelling problems that may be confronted are found in [1] which mentions other bibliographical sources.

From the point of view of tunnel construction, soils are strictly discontinuous because of their very constitution and the discontinuities and weakness planes they contain. However in many cases they can be dealt with under a general hypothesis of continuity. This is implicitly accepted in most pressure evaluation methods.

The validity of the conclusions that are reached in any analysis depend on whether or not the material can be considered a continuum. If the material is indeed a continuum, the stresses will propagate in a uniform symmetrical manner; deformations will be symmetrical too. Discontinuous media, on the other hand, are usually not uniform and their properties depend on location and direction. Whether the material through which a tunnel is driven behaves more as a continuum or as a discontinuum is largely a matter of scale. If the proportions of its structural irregularities are small in relation to the size of the masses affected by the tunnel the behavior will be similar to that of a continuum. Most clays behave as cohesive continuous media, and most sands and gravels as cohesionless continuous media of a frictional nature. Residual soils often behave as discontinuous, depending on the spacing between their joints and other cracks. Here, the nature, geometry and scale of their discontinuities are of paramount importance. However, when the inter-joint spacing is very large compared to the width of the cavity, the residual soil again behaves like a continuum.

When referring to the inter-joint spacing, it is essential to consider which spacings are significant for they usually vary greatly. Often the smaller spacings are more important than the average spacing. For a material to be regarded as a continuum, the usual ratio for the spacing between joints to the width of the cavity is smaller than 1:50.

The characteristics that determine the behavior of a residual soil are:

— The spacing, regularity, orientation (direction) and nature of the discontinuities

— The properties of the solid soil mass between discontinuities

— The degree of alteration or fracturing

The spacing between joints and irregularities that can be seen inside the cavity are only a distorted image of reality. It depends on how the curved faces of the cavity intercept the more nearly plane systems of joints and fractures. When there are several fracture systems, the average spacing between fractures is usually between 2/3 and 1/3 of the maximum spacing in each of the systems. References [24-28] contain studies and classifications of the fracture systems in rocks and soils, together with conclusions about the engineering behavior of the materials in which they are found.

Groundwater plays an important role in determining behavior. For example, a fine sand is frictional if it is dry, cohesive if it is moist and has the behavior of a fluid if it is below the water-table. The effect of water is particularly important in large bodies of cohensionless materials.

In clays, whether residual or transported, the most significant parameters are usually the shear strength and modulus of elasticity and their relation to roof pressure. There is a very close connection between the foregoing relationship and the stability of the tunnel. When the soil is weak or has a low modulus, the cavity is likely to be unstable and should be supported. For increasing strength and modulus, the cavity diameter may be reduced by plastic flow of the clay. Beyond some strength it becomes very stable, although some very hard dry clays may have expansion problems. High strengths usually are a sign of preconsolidation and consequently may herald possible problems of expansion, fissuration, and even cracking.

In continuous materials, most behavior does not depend much on the width of the cavity. However strains are roughly proportional to it. In discontinuous materials, on the other hand, behavior usually varies greatly with the width of the cavity. The stand up time (bridge action period) is one of the factors that is dependent on the continuity or discontinuity of the materials and on the relationship with the width of the cavity in the latter case. According to Lauffert [1], the reduction in the stand up time in discontinuous materials is proportional to the first or second power of the length of tunnel that is left unsupported, depending on the quality of the material and the ratio of the inter-joint spacing to the width of the cavity.

The concept of roof load, which was discussed in pressure evaluation methods, is based on the concept that a certain height of material must be carried by the support. This is a gravitational concept. In cohesionless continuous masses, the concepts of arching and redistribution of loads are directly based on the force of gravity, but if the materials are cohesive and continuous, the effect of gravity should be considered in a somewhat more indirect manner, depending on the relationship existing between vertical earth pressure and shear strength. In expansive materials, gravitational effects may lose all significance. From the foregoing considerations, it is deduced that methods for evaluating roof load have more meaning in cohesionless continua, and only as an extrapolation of variable reliability can they be extended to include cohesive continua.

It becomes clear from the foregoing discussion that the pressures produced by discontinuous masses are controlled by joints and fissures and the spacing between them as well as by direct gravitational effects. This is another case where application of the pressure evaluation methods discussed here has variable reliability and should only be used considering the structuration problems of the material. Methods like Terzaghi's or Protodyakonov's include these effects in their recommendations and adjustment factors.

14.7 The Performance of Support Systems in Tunnels Through Soils

In order to comprehend some of the concepts that will be discussed later in this section, some theoretical considerations will first be presented. A mass of soil with a horizontal surface is assumed. At depth z, the vertical normal stress will be γz and the horizontal stress $K_o \gamma z$ before the tunnel is excavated. If a circular support (a tunnel lining) is immediately installed inside the mass [1], the stress state in the mass at the time of installation does not differ from that existing previously. If the circular support is flexible, but capable of supporting the compressive stresses produced by the earth around it.

After a time, the circular (flexible) ring can only be in equilibrium if the radial normal stresses affecting it are equal in all directions. In order to fulfill this condition, the horizontal normal stress produced by the earth must increase (if $K_o < 1$), whereas the vertical normal stresses must diminish. This situation can only be reached by means of deformation of the support system by the necessary amount at all points and in the required direction. The circle becomes slightly elliptical.

If the lining ring tolerates the uniform radial stresses that are produced, a satisfactory equilibrium condition will be achieved, so long as the amount of strain required does not exceed tolerable limits of deformation of the ring as well as any nonstructural limits. The amount of deformation that the ring undergoes before it reaches its equilibrium condition depends on the characteristics of the soil around it and their variation with time, the ring stiffness, the dimensions of the tunnel and the depth at which it is driven.

If the lining were initially elliptical, with its smaller axis equal to K_o times the larger (vertical) axis, this shape would bring about a nearly uniform distribution of the radial stress about the tunnel, even if no deformation occurs immediately after installation.

The concepts presented here are summarized in Fig. 14-29 [1]. Part *a* of the figure shows the stress distribution on a circular tunnel immediately after installation of the support, assuming that this is instantaneous. Part *b* shows the ultimate nearly uniform distribution of the radial stress that is reached after the tunnel deforms as shown. Lastly, Part *c* shows the initially elliptical cross-section referred to.

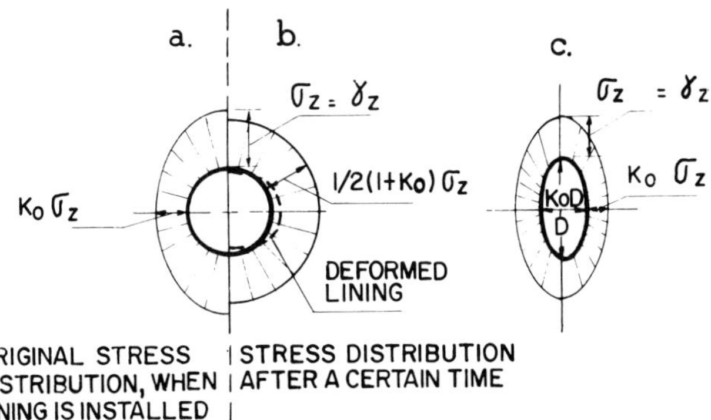

Fig. 14-29 Distribution of radial stresses about flexible tunnel linings or supports [1], in normally consolidated material

The favorable distribution that can be reached by the radial stress about a circular tunnel is achieved at the expense of the change of shear stresses in the surrounding mass produced by the soil deformation. If these shear stresses remain essentially constant after redistribution has taken place and their level is appropriate, the mass of soil will not suffer any subsequent distortions. However if this is not so, the mass of soil will eventually act on the support system and will cause further displacements in it, sometimes over a very long period of time. Thus, the strength and rigidity of the excavated material, the deformability of the support system and the interrelation between them are of fundamental importance.

If the support were infinitely rigid, the pressures developing immediately after its installation would be the same as in the case of the flexible support. Subsequently it would not deform and the non-uniform pressure distribution would continue indefinitely. Bending moments will develop in the support system in accordance with stress states that are no longer uniformly radial. This is a basic disadvantage of rigid supports, which apart from being designed to withstand radial stresses, must also be capable of supporting bending moments. Once again, an elliptical support with its smaller axis K_o times the larger (vertical) one would undergo practically zero bending moments, even if it were rigid. This illustrates the great influence of the shape of the cavity and the initial stress distribution in the case of rigid support systems.

14.7.1 Stability at the Working Face

This is important, especially in tunnels driven in soft ground, for it influences the construction methods to be used and the uniformity and magnitude of the loads on the supports [1].

.1 Tunnels in Clays

In clays, the stability of the working face depends on their undrained strength. The long-term strength, in terms of effective stresses will only be of importance if the face is left exposed and unsupported for a long period of time.

In principle, it is possible to analyze the short-term stability conditions of the working face of a tunnel in clay, although in practice many factors may have an influence which are not taken into account by theory. In [29] the equilibrium of a mass of clay about a narrow horizontal cut made in a wall that supports it is analyzed.

Assuming a cylindrical slip surface like the one shown in Fig. 14-30 failure of the clay occurs when the total vertical pressure corresponding to the center of the horizontal cut (which obviously represents the excavated face of a tunnel) exceeds 6.28 times the value of c_u, which is the undrained strength of the clay. If the cut has a circular aperture, this ratio would increase to 7.5. If the width of the cut is appreciable in comparison with the depth at which it was located, the critical value of the ratio will be [1]:

$$\left(\frac{\sigma_z}{c_u}\right)_{crit} = \frac{2\dfrac{z}{B} + \pi - 1}{1 + \dfrac{B}{6z}} \qquad (14\text{-}66)$$

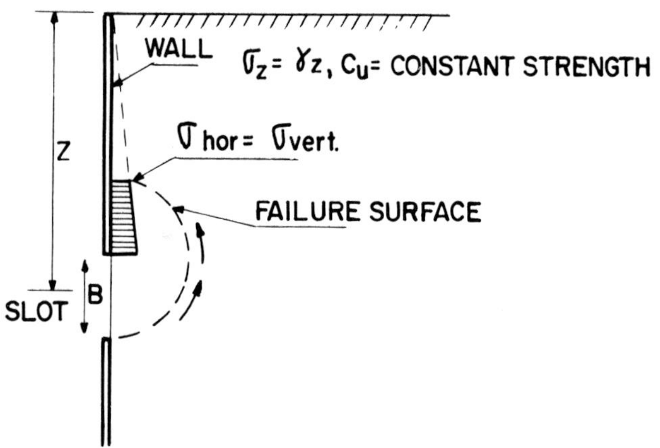

Fig. 14-30 Failure mechanism about a horizontal slot in clay [1, 29]

σ_z/c_u is called the *surcharge factor* or *ratio*.

When the strength of the clay increases with depth, failure occurs in the face when $\sigma_z > 6c_u$ (for $z/B < 4$ or 5).

In tunnels, the situation of the working face is not identical to that presented in Fig. 14-30. However, there is experimental evidence that the critical values of σ_z/c_u mentioned above represent the general order of magnitude [31-32]. The exposure time of the working face, that is the time that it is left unsupported, is very important. At the time of excavation negative pore pressures are usually induced in the clay, so that as time goes by there is a dissipation of these tensions and a reduction in clay strength.

Hard fissured clays are very sensitive to the deformation processes that accompany stress redistributions, and consequently in such soils signs of instability may appear in the face for values of $\sigma_z/c_u < 6$.

Clay attempts to flow plastically towards the cavity at the working face of the tunnel. The degree of flow depends on the construction procedure, the tunnelling speed, the stiffness of the clay and the ratio σ_z/c_u [33]. If σ_z/c_u is smaller than 2 or 3, movements usually are small and of a somewhat elastic nature. For values beyond this limit, large elasto-plastic deformations occur. It is often observed that the clay ceases to move towards the cavity when tunnelling operations are brought to a halt, unless the face is on the point of failure [1]. Consequently, movements occurring when construction is stopped will always be a sign of danger.

It appears, therefore, that the situation of the working face will be more favorable the more uniform, smooth and well finished the excavated face in clay. This is one of the reasons why shield tunnelling has become so popular.

When a cohesive material is pervious enough for pore pressure to dissipate during exposure or when the material is exposed long enough for this to occur, the stability of the face is determined by the acting effective stresses and by the shear strength expressed in terms of effective stresses rather than by the undrained strength. This can be either a favorable or unfavorable condition; the result in each specific case depends on the nature of the clay. In other words, the evolution of the stability condition with time obeys no hard-and-fast rules, but depends on the material and the circumstances in each case.

.2 Tunnels in Friction Soils

Sands, gravels and totally fractured rocks are continuous materials of a frictional nature. Silts and clayey or silty soils are sometimes hard to classify; their behavior will depend on the nature and proportion of their fine fraction. The criteria included in CASAGRANDE's Plasticity Chart (Chapter 2) are useful for establishing a reasonable classification.

The essential element for determining the stability conditions of the excavated face in frictional soils is the groundwater condition. Above the water-table, these materials when dry lack bearing capacity, when their slope is steeper than the natural equilibrium slope (Chapter 6), but on the other hand they are very stable if the face is allowed to adopt an inclination equal to Ø. Sometimes sands develop apparent cohesion by capillarity, which can be used to advantage so long as the support is installed before it dissipates. Using conventional excavation methods, the absence of cohesion restricts the size of the cavity, as well as the time that the material can be left unsupported. Usually frictional soils above the water-table are the hardest ones to excavate. Methods resorted to have been tunnelling in sections of limited size (drifts), carefully supporting each section before excavating the next, a shield which supports the soil immediately after excavation, and other complicated and costly methods.

When granular soils are below the water-table, the stability of the construction face of a tunnel depends particularly on the existence of some cohesion, capable of counteracting the seepage forces generated by the seepage of groundwater into the tunnel, which acts like a drain at atmospheric pressure. When there is no cohesion, the face must be supported. Probably the most widely used method is the compressed air method, despite its high cost and the dangers and discomfort it causes to the workers. Also employed, of course, are methods of lowering the water-table under the tunnel floor, for example by pumping.

.3 Tunnels With Working Faces in Two or More Types of Soils

The working face of large tunnels is often in two or more different soils. Currently the problem is dealt with by superposition, analyzing each type of soil individually.

Difficult formations include those with strata of clean sand and gravel overlying other far less pervious ones, with the water-table in the frictional strata. In such cases it is very difficult (but not impossible) to drain the granular soils. Furthermore it is essential to support them from the early stages of tunnelling, for example by using compressed air. In difficult cases, cement grouting or freezing has been resorted to produce cohesion in frictional soils.

Also difficult are formations where frictional strata are confined above and below by impervious cohesive strata. In these cases, pressure of the groundwater in the frictional soils may be high. When an attempt is made to lower the water-table by pumping, the pressure in these confined water-bearing layers cannot always be readily lowered. Thus, large quantities of water are left which produce appreciable horizontal flows towards the tunnel. These cases require extremely careful tunnelling procedures, with the soil always well supported by carefully controlled shields, or air pressure.

Often tunnelling causes a drop in the water levels, with the cavity acting as a drain. These water levels recover after the tunnel has been lined, which increases the loads on it. A remedy is to foresee this situation and to design the structure to function like a permanent drain. This, however, can cause subsidence above the tunnel due to consolidation produced by the increased effective stress.

14.7.2 Stability and Deformation in Unsupported Walls

Much of what has been said about the tunnel face is valid for the walls as well. However, a few special comments are felt worth-while [1].

.1 Unsupported Walls in Clay

It has been seen that in clays, which behave as purely cohesive materials, a plastic zone can develop about any circular cavity that is made. If the material is considered to be elastic and $K_0 = 1$, then the tangential stress in the tunnel will be $\sigma_\theta = 2\sigma_z$, Eq.(14-15). A plastic zone is produced when σ_θ is greater than the unconfined compressive strength of the clay, $q_u = 2c_u$ (Chapter 1). The radius of the plastic zone depends on the ratio σ_z/c_u, Eq.(14-20). This zone disappears for $\sigma_z/c_u \leq 1$.

If $K_0 \neq 1$, the tangential stresses vary at the different points of the perimeter of the cavity and can be computed using Eq.(14-15).

In order to describe the behavior of the unsupported walls of a tunnel driven in a continuum, a description of the surcharge factor (R_s) is useful. This is the ratio of the maximum tangential stress that can be computed theoretically in the tunnel wall, using the theory of elasticity, to the simple compressive strength of the material in which the tunnel is driven. If $K_0 = 1$, the surcharge factor will be:

$$R_{s1} = \frac{\sigma_z}{c_u} \qquad (14\text{-}67)$$

Thus, the plastic zone will develop if $R_s > 1$, and the extent of this zone will also depend on the value of R_s.

If there is a pressure σ_i acting inside the tunnel against the walls, the value of R_s is adjusted as follows:

$$R_s = \frac{\sigma_{\theta max} - 2\sigma_i}{q_u} \qquad (14\text{-}68)$$

As already indicated, real materials do not behave elastically; consequently both the stresses and strains that occur are different from those than can be computed with the elasticity theory. This is particularly true if K_0 is very different from 1. Real distributions of the tangential stress about a tunnel are more uniform than the theoretical ones [1]. Consequently, the critical surcharge factors calculated elastically may be exceeded without plastic zones developing, when K_0 is fairly different from one.

In the case of the working face of a tunnel in fat clay, it was seen in §14.7.1 that the critical surcharge factor is approximately $R_{s1} = 6$. A higher value is a sign of impending failure. The same limit is usually accepted as valid for tunnel walls in clay. For shallow tunnels, the critical value of R_{s1} is smaller.

Construction procedures affect stability conditions appreciably. When the shield tunnelling method is used, very large shear stresses are transmitted to the ground, which produce structural disturbance and modifies the stress states and the strength in a manner that is impossible to predict.

The displacements that occur in tunnel walls in clay depend largely on the extent of the plastic zone which forms about them. If this zone is large, the displacements will be also; whereas if it is small, they will likewise tend to be small. Close to the face, the displacements of the walls are greatly influenced by the stability of the face itself. If this is reasonably good, the displacements of the walls are usually about 50% of those that may occur away from the face. Experience indicates that the influence of the face is felt at a distance of about one or two tunnel diameters from it [30, 34, 35].

.2 Unsupported Walls in Frictional Soils

By comparing the data in Figs. 14-9 and 14-10 [1], it can be seen that the radius of the plastic zone that can be expected about the tunnel is far smaller in frictional materials than in clays. Figure 14-10 shows that even very low values for cohesion that may be exhibited by the frictional material have an important influence on the radius of the plastic zone that forms.

In sandy soils, the pore pressures developing close to the walls of the tunnel as a result of excavation dissipate almost immediately. Since the plastic zone that may develop is narrow, it follows that this zone should be analyzed in terms of effective stresses. However, such an analysis is practically impossible because the pore pressure varies greatly with time and with the distance to the wall of the cavity and it is nearly impossible to determine it accurately at any point.

A rough but perhaps reasonable analysis can be conducted [1] using Eqs. (14-13) and employing the total stresses, σ_z, and the effective strength parameters $\varnothing$ and c (if this last parameter is significant). This is equivalent to assuming that pore pressure becomes zero immediately following excavation of the tunnel walls. The same Reference [1], however, indicates that the validity of this analysis has not been verified. Another method for analyzing the stability of the unsupported walls and roof of a tunnel driven in sands was proposed by BALLA [22].

In frictional materials, the instability of the walls does not manifest itself in large displacements or plastic flows. Instead, above the water-table, the sand or gravel will break apart in fragments held together by capillarity. Below it may flow under the thrust of seepage forces that have not been dissipated. Signs of instability are less obvious and occur more erratically than in clays and the deformations caused are more irregular. If, on the other hand, the frictional material is basically stable, the pressures it will exert on a support that is installed will be relatively small and the displacements it undergoes will be small too.

Sometimes the strength of the sand in the walls diminishes with time. Drying, for example, dissipates the apparent cohesion of the material. In this event, the material may start to break apart and flow like dry sand in an hour glass. It is common that the time the frictional material is able to support itself is roughly in inverse proportion to the width of the tunnel opening. However, there is no reliable means or evaluating this time in terms of the strength or other relevant characteristics of the material.

.3 Unsupported Walls in Very Hard Soils or in Partially Altered Rocks

If the soils or altered rocks are continuous, an analysis based on elastic considerations will lead to reasonable results, §14.4, assuming that the value of K_0 can be estimated. With these materials the usual risk is the fracturing they may suffer under large stresses, which weaken them enormously.

14.7.3 Radial Strains and Loads on Tunnel Supports

The design of a tunnel support is divided into three well differentiated stages. First, it must be capable of tolerating the circumferential forces that may develop. Second, it must be capable of supporting the bending moments that may occur on planes normal to the tunnel axis. And lastly, it must be capable of absorbing any local irregularity that may occur in loading or strain.

The basis for establishing the radial stresses the support should resist is illustrated in Fig. 14-31 [1], which shows the radial stresses as a function of the average radial displacements towards the center of the cavity.

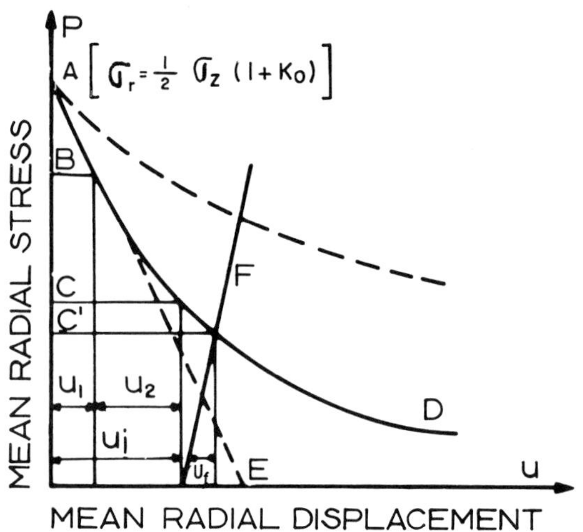

Fig. 14-31 Relation between mean radial stresses on a circular support and the corresponding radial displacements [1]

As an initial approximation, the average radial stress (if there is no radial displacement) could be regarded as the mean value of the horizontal and vertical pressures existing in the center of the tunnel prior to construction. The vertical normal pressure is $\sigma_z = \gamma z$ and the horizontal normal pressure will be $\sigma_x = K_0\,\sigma_z$. Consequently, assuming this approximation,the average radial stress, if there is no radial displacement, will be:

$$\sigma_r = \frac{1}{2}\sigma_z(1 + K_0) \qquad (14\text{-}69)$$

In the figure this stress value is represented by point A. Should the radius of the cavity diminish uniformly, the average radial stress acting on the support would diminish too, as represented by the solid line in the figure. The shape of this curve depends on the stress-strain behavior of the soil and construction time. The broken line AE would represent an elastic response of the soil. A curve like AD is called the *reaction curve* of the soil.

The reaction curve cannot be calculated for a given case, but an idea of its shape can be obtained by means of field observations combined with theoretical reasoning. In the following considerations, the shape of the curve will be regarded as having been approximated adequately. The value u_1 represents the displacement that the section of tunnel undergoes from the moment construction commences until the final design shape is achieved. If, as soon as the section is excavated, a circumferential support is installed that is capable of avoiding any subsequent displacement, the pressure exerted by the roof and the walls will no longer be that indicated by point A, but by B. It is somewhat smaller because the soil moves during installation. Normally, even after installation of the support a radial displacement will occur (u_2 in the figure), with the result that the radial stress will be reduced to the value indicated by point C. This displacement is due to the support not making perfect contact with the soil.

In addition, the support will deflect and this deflection will have to be added to the radial displacements, the effect of which can be estimated by curve F, corresponding to the reaction of the support. This represents the relationship between the radial load that is carried by the support structure and its corresponding radial displacement.

The ultimate load on the support will be given by the intersection of the two reaction curves: for the soil and the support. The point C' will be the ultimate stress and u_f the total displacement that can be expected.

If the tunnel is constructed with an internal air pressure, σ_i, the relationship between the radial stresses and the radial strains will be displaced downwards to a distance σ_i, above the stress axis, so that point A will be transformed into A', as is shown in Fig. 14-32.

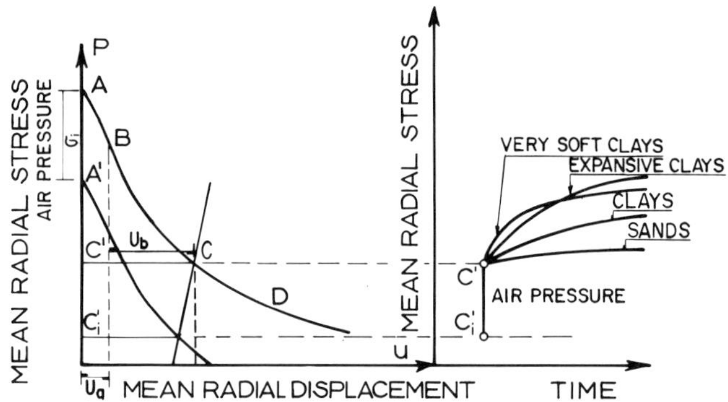

Fig. 14-32 Relation between radial normal stresses and radial displacements in a tunnel excavated using the compressed air method [1], increase in pressure on support as time goes by [2]

It can be seen how in principle only a support that can tolerate a radial stress C'_i is required in this case, although this value can seldom be used in practice and the support will always have to be designed for C′, for there is a constant risk that the air pressure may for some reason cease.

In Fig. 14-32, u_a represents the average radial displacement at a point in the tunnel close to the working face [2]. If when u_a develops a support were installed that would impede any subsequent movement, the pressure on it would no longer be A, but the ordinate of point B. If from u_a a further radial displacement of u_b is permitted, pressure on the support would be reduced to the ordinate of C. A support capable of impeding all radial displacements greater than $u_a + u_b$ should be designed for this last pressure.

Part b of Fig. 14-32 illustrates the change in pressure that may take place with time. Several different soils are illustrated. Experience indicates that the pressure increases with the logarithm of time at a decreasing rate. However it does so in a very different manner in the different types of soils. This subsequent time dependent increment should be considered in addition to the design loads that are reached in accordance with the reasoning shown in Part a of Fig. 14-32. All the foregoing estimations can lead to a reasonable evaluation of the mean radial stress for which a particular support should be designed. The change that occurs in this stress with time, of Fig. 14-32b causes an elevation in the soil's reaction curve like that shown by the broken line in Fig. 14-31. In almost any case, a radial normal stress smaller than σ_r can be expected, possibly of the order of the one given by Eq. (14-69). Only expansive clays may exceed this value.

When the support is in contact with the soil around it, the load it receives and displacements it suffers depend on the interaction of soil and support. If contact is not perfect, there may still be deformation in the soil with only a partial restriction from the support, but if contact is firm, any further displacement and the load that is carried will depend largely on the relative rigidity of soil and its support.

The compressibility of a support subjected to radial loading is expressed as:

$$S_a = p \frac{a}{u} \tag{14-70}$$

where the ratio u/a represents the relative reduction in the radius of the tunnel when the support is subjected to peripheral pressure p. If t is the thickness of the support and a the external diameter, the compressibility, S_a, for a circular support can be computed as:

$$S_a = E_a \frac{t}{a} \tag{14-71}$$

E_a being the modulus of elasticity of the support material. Or, for a relatively thick support:

$$S_a = \frac{E_a t}{a - t/2} \tag{14-72}$$

The rigidity of the soil is harder to evaluate, and in materials that are not perfectly elastic it depends on the displacement it undergoes. An upper limit could be obtained from an elastic analysis. If S_m refers to the rigidity of the material and E_m to the modulus of elasticity of that material, taking Eq.(14-17) into account, for the tunnel wall the following equation will be true:

$$\frac{u}{a} = \frac{p}{S_m} = p \frac{1 + \mu}{E_m} \tag{14-73}$$

where p refers to the pressure around the entire perimeter of the support. The rigidity of the soil has been defined accordingly as:

$$S_m = \frac{E_m}{1 + \mu} \tag{14-74}$$

If the average displacement of the tunnel wall, u_m, is considered the rigidity of the material could also be obtained through $u_{m/a} = p/S_m$ that is:

$$u_m = a \frac{p}{S_m}$$

Setting this equal to Eq.(14-18) we obtain:

$$S_m = \frac{2E_m}{(1 + K_0)(1 + \mu)} \tag{14-75}$$

If the behavior of the soil is approximately elastic, its interaction with the support could be described as is shown in Fig. 14-33 [1].

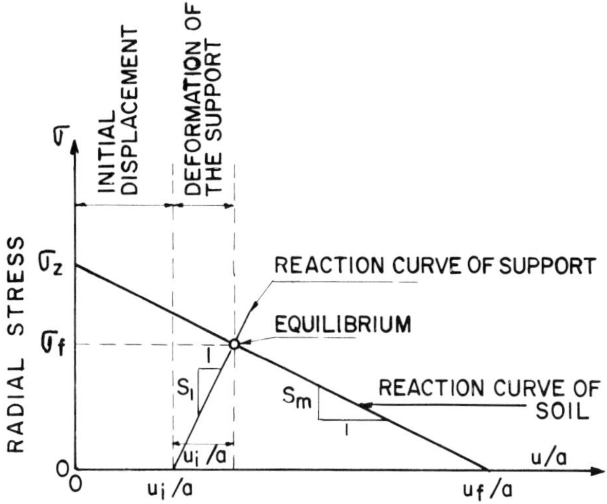

Fig. 14-33 Soil-support interaction curves (elastic materials) [1]

If it is wished to reduce the displacement to zero at every point of the circular support, the maximum pressure that this support should exert radially on the surrounding ground is σ_z, which is the total vertical pressure exerted by the soil. (Equation (14-69) made it possible to calculate the mean radial stress that must be balanced with zero displacement, but in Fig. 14-29 it can be seen that a maximum equal to σ_z may occur). On the other hand, if the development of all possible strains were permitted before the soil ceased to move of its own accord, an ultimate radial displacement would occur, which will be called u_f, which could be calculated under elastic hypotheses, using formulas already discussed in the corresponding sections. As is seen in Fig. 14-33, these two points define the soil's reaction curve (see Fig. 14-31), if it behaves elastically. This curve has a slope S_m. Indeed, u_f can be calculated using Eq.(14-17) so that:

$$\frac{u_f}{a} = \sigma_z \frac{1 + \mu}{E_m}$$

which is the value on the abscissa in Fig. 14-33. The corresponding ordinate has a value of σ_z, so that the slope of the straight curve will be $E_m/(1 + \mu)$, which is the value of S_m according to Eq.(14-74).

The support will be constructed after a certain displacement u_i, which takes place as soon as the section of tunnel is excavated. If displacements continue to occur, loads will be transmitted to the support, following the reaction curve of that support, with a slope S_a (Fig. 14-33). Where the two straight reaction curves intercept one another, the pressure on the support will be sufficient to maintain the displacement of the ground at a value not exceeding $u_i + u_a$ and the support will have deformed by u_a.

This method of evaluation is of limited practical use, for soils are not elastic at the levels of strain occurring in tunnels. The method is useful as a working guide, but in order to apply it to a real case, a far more satisfactory knowledge of the reaction curves of the soil and the support is necessary. These curves depend on the properties of the soil and construction procedures employed, and must be plotted on the basis of information obtained from instrumentation programs.

14.7.4 Distortion of Tunnel Supports

As was seen at the beginning of this section, when a support is flexible it always undergoes some distortion between the initial and final installation stages. This distortion is the result of the changes that take place in the horizontal and vertical diameters. It was also seen how in these cases the bending moments that are brought about in the support will probably be of little importance. The design requirement is that the change in the diameters be kept within tolerable limits, which depends on the function of the tunnel, the nature of the soil and the degree in which the shape of the tunnel corresponds to the equilibrium corresponding to the initial stress state.

All supports have a certain amount of rigidity; and when as a result of this the changes in diameters are smaller than those corresponding to an ideally flexible support of the same shape, bending moments develop in the support. These may be reduced if part of the soil deformation is allowed to take place before installation of the support. On the other hand, the support will develop no bending moments if it can deform by the amount corresponding to the changes in diameter that would be suffered by an ideally flexible support without too much effort. In Fig. 14-29, it can be seen how the conservative design is the one where it will deform with the necessary increase in the horizontal diameter (circular tunnel).

The rigidity of the support does not greatly influence the amount of distortion that takes place. In soft clays, typical values for distortion range from 0.3 to 0.7% (defined as the percentage of shortening or lengthening of the radius in relation to the original radius). In stiff clays distortion usually ranges from 0 to 0.3%. Reference [1] includes data about distortion observed in several tunnels, which may serve as a norm for establishing the distortion that can be expected in similar tunnels.

14.7.5 General Design Recommendations

Current experience makes it quite clear that all the design methods which assume that tunnel supports are affected by pressures corresponding to active or passive states, or even ones corresponding to earth pressure at rest, are rarely valid. Furthermore, design methods which assume that a support is subjected to the effect of a fixed loading system, regardless of construction procedures or of what may happen in the soil, should be rejected. Considerations such as whether a temporary plus a permanent support system, or just a permanent one will be employed are fundamental. The following design stages are proposed in [1]:

1) Make a reasonable estimate of the loads or compressive circumferential pressures that will develop on the support
2) Make an estimate of the distortions that will be experienced by the support
3) Consider the possibility that buckling may occur in some parts of the support
4) Consider any external circumstance that may be of importance in the case under consideration.

The compression stresses on the support are related to the mean radial pressure in the following expression:

$$\sigma_c = p \frac{a}{t} \tag{14-76}$$

where: σ_c is the compression stress on the support; p is the mean radial pressure suffered by the support; a is the external radius of the support; and t is the thickness of the support.

In clays, p depends on σ_z and on the surcharge factor already discussed. In soft or medium clays, the surcharge factor may be very high and the value of p may remain close to $\frac{1}{2}(1 + K_0)\,\sigma_z$.

Since K_0 is usually close to one, it follows that p may be as high as σ_z. In stiff preconsolidated clays, K_0 may be higher than one, but the surcharge factor tends to be far smaller. The two effects neutralize one another, so that $p = \sigma_z$ again represents an acceptable upper limit.

In frictional continuous materials, there is usually little load increment with time, and the design pressure can be far smaller than σ_z. The radial pressure can be estimated on the basis of the ideas expressed in §14.7.3. A danger in this case is that cavities or voids may be left between the soil and support, causing large increases in the pressure on the support when they subsequently collapse. Arching is the main reason why the pressures applied by these soils are relatively low. In weak soils the reaction curves for the soil and the support usually intercept one another at about σ_z.

.1 Comments About Different Support Systems

Again referring to [1] some comments will now be made about the influence of certain characteristics of the different support systems.

Flexibility of the supports: Both theory and experience indicate that flexibility is a quality that should be sought rather than avoided. Although it is true that a rigid support resists distortion, it is also true that the reduction in distortion that is achieved is not very appreciable, even though the required strength is greatly increased. The rigid support has to tolerate far heavier loads and often rigidity works against stability.

The highest costs are in resisting the bending moment due to non-uniform pressures. It has already been stated how reasonable distortion of the support causes the pressures to be more uniform and prevents bending moments. This leads to far lighter and more economical sections, better able to withstand pressures without failing [36].

In accordance with the foregoing considerations, criteria concerning the types of materials that will be suitable for support construction can be established. For those installations where distortions are acceptable on account of the nature of the soil, supports must have a high compressive strength, so as to be capable of resisting the equal radial stresses, and including the vertical pressure corresponding to the depth of cover. They should not be too rigid, so that they will be able to deform enough to avoid the significant bending moments. Lastly, they should be made of a lightweight materials that can be easily handled. Unfortunately, a support that meets all these requirements may not fit shields and other drilling devices currently in use. This calls for an improvement in the design of the total tunnelling system, making the support more independent of the excavation system [1].

Need for temporary supports: The need for temporary supports before the final support system is installed depends on the settlement and loss of material that may arise, the type of soil and the cost of the design under consideration.

In the case of clays, experience indicates [1] that, except for expansive materials, the rapid construction of a temporary support eliminates many of the risks deriving from the development of high pressures and large displacements above the permanent support system.

Once the temporary support has been constructed, appropriate criteria for erecting a good permanent support system can be obtained from it. For example, the pressures that will act on it can be very reasonably estimated using the criteria given in this same section based on the displacements observed in the temporary support.

If the ultimate support system is built as excavation progresses, it will have to support far greater pressures and distortions than those it would undergo if a temporary support were constructed first. However should this procedure be adopted, a relatively flexible support is advisable, bearing in mind that the distortions required to reach the pressure state corresponding to a ultimate equilibrium are usually small.

All studies indicate that the evolution of the pressure state and the changes in shape that are experienced by the temporary support, occur to the advantage of the requirements to be met by the permanent support system. The ideal solution is to construct the permanent support once the temporary support has reached the equilibrium condition. This can seldom be accomplished in view of the usual urgency of construction. Whenever possible, an attempt should be made to restore groundwater pressure conditions before installing the support system. Furthermore, since this restoration of pressure usually causes little distortion and seldom large pressure increments, it does not represent a serious problem even in the permanent support system.

The suitability of the decision to postpone construction of the support system until pressures have been appreciably relieved depends on the nature of the support. If it is very rigid, the longer installation is postponed, the more economical will it be, but if the support is flexible the advantages of postponing construction are smaller.

In expansive clays considerable economies can be made by constructing support and lining systems after a substantial part of the expansion has taken place. However this may not do much to relieve subsequent distortions; the pressures are relieved even though there is no certainty as to when. This procedure should be handled with care because if excessive swelling is permitted in clay, the material will be structurally weakened and become very unstable.

It is usually beneficial to prestress supports and linings, for this leads to a better contact with the ground and more favorable stress distributions about the tunnel.

.2 Permeability of Supports

An impervious tunnel below the water-table must resist not only earth pressure (which are considered here in terms of effective stresses), but also hydrostatic pressures. On the other hand, a perfectly pervious tunnel will have to support only effective soil pressure, although the flow of water that is induced towards the tunnel is accompanied by seepage forces in the soil which will have to be considered. These can be estimated in a flow net. As a consequence, in cohesive soils the conditions of pervious and impervious tunnels are usually similar, whereas in frictional soils a pervious tunnel is usually subject to smaller loads.

Water may pose construction problems by penetrating into the tunnel. This has been combatted by grouting the pervious zones around the tunnel, which serves the dual purpose of reducing the volume of flow entering the tunnel and stabilizing potentially unstable zones.

It should be understood that so-called impervious linings are often more pervious than clayey soils. Consequently, a tunnel below the water-table is always a drain to a greater or lesser extent, and minimum precautions should be taken with respect to this.

In many cases, the slow seepage of water into the tunnel and the draining effect of the cavity are sufficient to cause a permanent and very appreciable reduction in the water pressures around it. It has also been observed that the distribution of the water pressure around an impervious tunnel lining below the water-table is far from regular, chiefly on account of soil stratification effects.

It is normal practice to design supposedly impervious tunnel linings considering smaller water pressures than the hydrostatic one corresponding to the position of the tunnel in relation to the water-table, 1) where soils are fissured, 2) have pervious interstratifications, or 3) where the water is unable to concentrate around the tunnel in a greater proportion than it can be drained.

.3 Effect of Irregularities in the Straight Section

Here reference is made to the possible effect of changes in the straight section of the tunnel on pressures on the supports and the behavior of these supports. These changes may be caused by very close parallel tunnels, intersection with tunnels or other drifts, and the effect of branch lines, stations, sidings, and crossings. If the soil or rock behaves elastically, the effects of neighbouring tunnels can be judged by applying the principle of superposition of stresses and strains. In elasto-plastic analyses, the problem is no longer simple, but the superposition of theoretical solutions may still provide guidance. When dealing with materials that do not come under the hypotheses of continuum mechanics the problem becomes even more complex.

Earlier in this chapter it was seen how supports and linings are designed to resist peripheral pressures which are usually smaller than the weights of overlying earth as a consequence of radial displacements. Consideration will now be given to the effects that can be expected from different common irregularities [37].

Pressures in adjacent tunnels or ones crossing at different levels: In these cases, it can generally be assumed that the pressures on the support of one of the tunnels will not exceed those corresponding to the weight of the overlying earth, as an upper design limit. General design criteria are the same as for a single tunnel.

Pressures at tunnel intersections: In this case, very complex interaction phenomena take place, [38] includes a partial study of these phenomena for the case of 90° crossings. It is concluded that stresses increase by 60% at the corner of the intersection in elastic materials; they correspond to those for a single tunnel at distances of one diameter from the intersection.

Before constructing an intersection, the support system in the first tunnel should have reached ultimate equilibrium with the soil or rock around it. Then, in order to make the intersection, most of the support in the zone has to be removed, which has two effects:

- The pressures previously resisted by the support that is removed must be transferred to the parts that have not been removed, by arching.
- The support is weakened, so that even after completion of the intersection, it cannot resist the pressures it did initially.

At the present time, with these problems still so much in need of investigation, it is usually recommended that the supports at intersections be designed in such a way that they will transfer the radial load to the other sections of the support, adjacent to the intersection zone [37]. The pressure in the sections of the tunnels adjacent to the intersection should be regarded as double that which would be obtained for a single tunnel. This excess dissipates over a distance of one tunnel diameter, commencing at the intersection, and the dissipation is approximately linear.

In practice, it is hard to adapt the design of the support to suit the foregoing linear relation. Consequently for a distance of one tunnel diameter the supports are usually constructed with the requirements of the section at the intersection. If the diameters of the two tunnels are different, the larger one should be considered for designing the supports in the four openings.

Distortions in adjacent tunnels or ones crossing at different levels: When the diameter of the two tunnels is nearly same (that is, when one is not larger than two thirds of the other), the distortion occurring in one as a result of construction of the other depends on the diameters of both, the spacing between them and the order in which they are excavated. The standard distortion of a single tunnel is defined by the ratio $\triangle a/a$, a being the radius of the first tunnel (assuming that the second tunnel does not exist). It can be estimated on the basis of the ideas expressed earlier in this chapter or by distortions observed in similar tunnels, as in [1]. Figure 14-34a [37] gives the distortions that can be expected for each tunnel on account of the presence of the other adjacent one. The Roman numerals indicate the order in which excavation took place. The figure refers to tunnels that are close to one another. When they are more than four radii apart (4 a), the added distortion caused by their interaction is zero and the distortion in each tunnel will be that corresponding to a single tunnel. In in-between cases, linear interpolation can be used.

The recommendations for the distortion induced when a tunnel of small diameter is close to one of far greater diameter (less than two thirds of the larger) are given in Fig. 14-34b. In this case, the distortion of each tunnel can be estimated as though it were a single tunnel, but if the two are close to each other, construction of the small one can be expected to induce localized areas of greater deformation around the larger one. This deformation is estimated as double the one that would be brought about by distortion if the large tunnel were on its own.

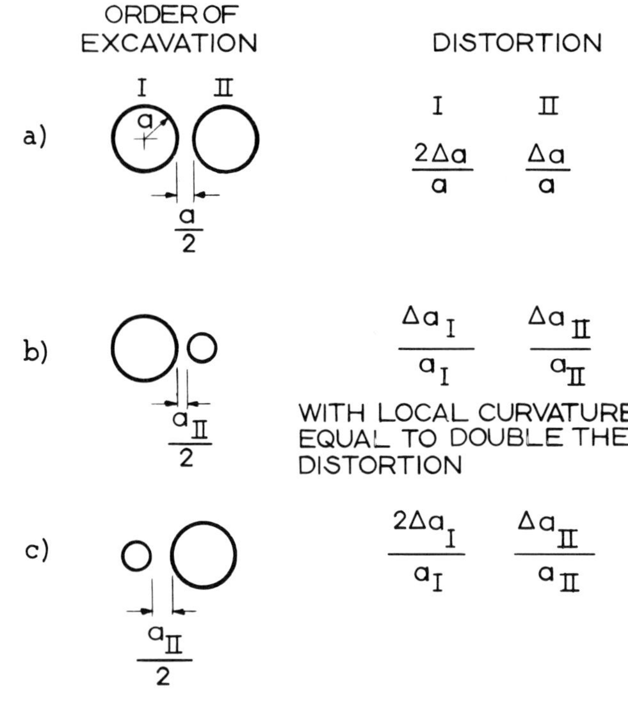

Fig. 14-34 Distortions due to the presence of another adjacent tunnel [37]

Lastly, Part *c* of Fig. 14-34 illustrates the case where a tunnel of larger diameter is built close to another existing one of smaller diameter (less than two thirds of the larger one).

Distortions at tunnel intersections: It can be assumed [37] that distortions of the support sections next to the intersection zone will be double those for a single tunnel. This is reduced linearly to that of the single tunnel at a distance of one diameter from the intersection.

Tunnel widening: Widening of a tunnel for any reason demands a change in the support or lining system. This justifies a very careful study, since it implies far more serious pressure and displacement conditions.

Very different construction procedures may be required. A tunnel that was originally excavated full face, when widened may have to be excavated in sections or drifts. These are supported individually, withdrawing the material between these zones in a subsequent operation.

However, from the point of view of pressures and displacements, a widening process can be regarded as a single tunnel of greater diameter and be dealt with accordingly. It is emphasized that it is normal for support requirements to be far greater because of soil or rock disturbance even in cases where there is relatively little change in the diameter.

14.8 Settlements Related to the Construction of Tunnels in Soils

When tunnels are built in open country, loss of material inside the cavity or movement of surface soil as a consequence of excavation are only of importance if they affect the stability of the supports, but in built-up, heavily populated areas, movements on the surface affect neighbouring structures. Therefore evaluating these possible movements before tunnelling and after they occur becomes necessary. Prediction of the settlements associated with tunnelling usually requires the evaluation of two different factors: first, the soil losses that take place during construction. These are the volumes of material that are withdrawn from the cavity over and above the volume of the cavity proper; second, the shape of the depressed zone that is caused by the tunnel on the ground surface. These two aspects will be dealt with individually.

14.8.1 Soil Loss

As has been seen, excavation of a tunnel causes horizontal and vertical movements in the soil about it, all of which under normal circumstances leads to a depression in the ground surface. Also, whether the tunnel is supported or not, leakages, soil flows, and radial displacements will occur in the material towards the axis of the cavity. The shape of the depression is not easily related to the maximum settlement that may occur on the surface. The volume of the depression is equal to the volume of soil lost in the cavity, plus the changes in volume of the mass of soil over and around the tunnel. For the time being, however, it is enough to assume that the volume of the depression in the ground surface is equal to the volume of soil lost in the cavity [37]. In other words, volume changes in the subsiding soil mass are disregarded.

Experience has shown that, except in certain expansive clays, the movements necessary for the total strength of the soil to be mobilized often occur around tunnels. This justifies the criterion according to which supports and linings are designed to maintain soil displacements at the lowest practical levels, which plays an important role in reducing soil loss and accordingly the volume of the depression and maximum settlements. Even with these criteria, however, soil loss is often considerable and the depression in the ground surface is pronounced in many cases [2].

The magnitude of settlements depends largely on the nature of the soil, and the construction precautions adopted. In a clay with plastic flow, for example, soil loss will tend to be considerable, but can be avoided to a large extent with a good support system. In a dense silt, on the other hand, the amount of soil lost is usually small; however, if the material is allowed to run and flow during construction, settlements on the surface may be appreciable.

Soil loss also depends largely on the method of construction, particularly related to the soil and groundwater condition. At the present time it is not possible to specify what portion of the loss of soil can be attributed to a given construction procedure and how much this loss can be blamed on carelessness during the construction sequence. Unfortunately the information available about soil loss is scarce and poorly organized, but useful nonetheless.

The depression that forms in the surface of the ground along the axis of the tunnel is usually symmetrical to it, unless the loss of soil is caused by very localized mechanisms, such as a large flow of sand into the cavity. If soil conditions remain uniform throughout a long section of the tunnel axis, the shape of the depression in the surface remains about the same and the maximum settlement in it (δ_{max}) does not change significantly from point to point. Average maximum settlement refers to the average of all the δ_{max} values, this quantity being represented by δ'_{max}. The value of δ'_{max} is an excellent reference for assessing the effects of excavation of a tunnel on adjacent zones. Lastly, there is the value of the largest settlement that can be found in all sections of the depression (δ''_{max}).

.1 Cohesionless Granular Soils

In these materials the roof, walls and working face of the tunnel must be fully supported, unless cohesion is provided by grouting. If operations are conducted carefully and there are no flows of sand into the cavity, soil loss can be maintained at very low levels, and the depression in the ground surface can be almost completely avoided. However if sand flows into the tunnel, which may happen, especially in loose soils, then very large depressions may form.

Settlement prediction in these materials is therefore very uncertain for it depends primarily on the detailed construction procedure. It is also influenced by the density of the sand and how dry it is. If the material has apparent cohesion by capillarity (tunnels above the water-table), problems may be considerably less important. In tunnels below the water-table water flow can erode the sand into the tunnel. The best solution often is to lower the water-table until this apparent cohesion is created and there is little or no water seepage into the tunnel. Even with these techniques, loose sands may densify and their volume reduce owing to changes in the stress state which take place as a consequence of excavation or the development of pore pressures which neutralize capillary tensions.

Granular soils are usually dewatered by wells driven in the working face of the tunnel or beside the tunnel from the ground surface. All the precautions listed in Chapter 7 of this book must be taken. Dewatering usually involves the problem that the sandy formation is not strictly uniform and there are some zones of finer sands and others of coarser sands, so that the efficiency of drains and the drainage times are variable. Drainage must be designed considering not only the possibility of sand flows, but also of piping and internal erosion phenomena.

The use of internal air pressure in the tunnel affects soil loss phenomena only slightly, but may help neutralize the seepage gradients towards the cavity in tunnels situated below the water-table. In tunnels above the water-table, the soil may dry out as a result of air escaping through it; any beneficial apparent cohesion developed by capillarity is eliminated.

Cement grouting in granular soils can be an important means of reducing soil loss. When grouting is not accompanied by drainage to lower the water-table, however, a serious danger may arise. Heavy water pressures may be exerted on small areas of sand which the grouting material has not reached and cause serious sand flows into the tunnel.

The specific values for the soil loss that can be expected in a tunnel driven in sand in a particular case must be taken from previous experience. References [37] and [39] give data from observations from several recently constructed tunnels, which may serve as a guide.

When parallel tunnels are constructed, soil loss in the second tunnel may not be very different from that which occurred in the one constructed first. If the tunnels are close together the depressed areas add. The depression becomes wider and somewhat deeper.

.2 Cohesive Frictional Soils

Into this category fall many types of soils, from clayey sands and sandy clays to plastic silts; also the many residual soils including loess and some calcareous clays. All these materials usually have a stress-strain relationship that is almost linear until the stress required to break the bond between the particles is reached. Also they frequently exhibit joints, cracks or inherited structural defects, which encourage slides and flows. They usually require some support from the initial excavation stage.

The soil loss and settlements associated with these materials are usually small [2]. These values can be reduced still further if during the stand up time the support is carefully wedged against the soil. The shield tunnelling procedure with liner plates is often employed for this purpose. There is also the danger of material sliding towards the unsupported cavity, and piping. Both phenomena may have very serious consequences here, for these materials are usually highly sensitive to seepage pressures. In such types of soils, groundwater control is essential, using wells or similar techniques for lowering the water-table and even internal air pressure in the tunnel.

.3 Stiff Non-Expansive Clays

The behavior of these materials from the point of view of soil loss is usually highly favorable, unless they have a well developed secondary fissuration structure. Moreover they are not very sensitive to flowing water. Soil loss usually takes place through the fairly light temporary support system, consisting of discontinuous, often weak members, which is often employed in these materials due to plastic flow towards the cavity. English technology recommends the use of shield tunnelling procedures in stiff clays and reports negligible soil loss [2].

.4 Saturated Soft Clays

Experimental evidence indicates that excavation of a tunnel in soft clays causes a structural alteration of the soil around it, which may extend to several diameters from the face and tunnel wall. Owing to this alteration, soil losses tend to be significant because of a scarcely perceptible but continuous plastic flow towards the center of the cavity. A corresponding depression forms in the ground surface. Most movements usually occur during excavation, mucking and installation of the support system, and diminish once the supports are in place. Consequently it is not unusual for the movements to escape the attention of the engineers in charge of the project.

An important effect in these tunnels is that the usual settlements due to soil loss may be followed by others caused by consolidation of the remolded zone around the cavity, under the weight of the overlying soil, a phenomenon which may last a long time.

Shield tunnelling is the procedure most often adopted in soft clays, and systems have been developed which tend to produce a minimum alteration of the material around the cavity.

In Sections 14.7.1.1 and 14.7.2.1, it was seen that the stability conditions in a tunnel in clay are a function of the surcharge relationship σ_z/c_u. When there is internal air pressure equal to σ_i, the relationship can be written [1] thus:

$$R_s = \frac{\sigma_z - \sigma_i}{c_u} \tag{14-77}$$

The theoretical loss of soil that will occur in a tunnel driven in clay can be written [37,39] as:

$$V_p = A \tag{14-78}$$

for $R_s \geqslant 1$, this is:

$$A = 3\frac{c_u}{E}e^{R_s-1} \tag{14-79}$$

for $R_s \leqslant 1$:

$$V_p = 3R_s\frac{c_u}{E} \tag{14-80}$$

Figure 14-35 [37] shows two examples of the theoretical variation in the volume of soil loss (V_p) as a function of the relationship R_s, Eq. (14-77), for several values of the c_u/E ratio for clay. It also shows losses observed in tunnels with certain specific conditions that are indicated. For a given construction procedure, the loss of soil varies with the different values of R_s over a fairly narrow range. The figure also illustrates the beneficial effect of internal air pressure.

It can be noted that the soil losses observed are far smaller than those given for the two theoretical curves that are plotted. This may be attributed to the restraining effect of the shield, in

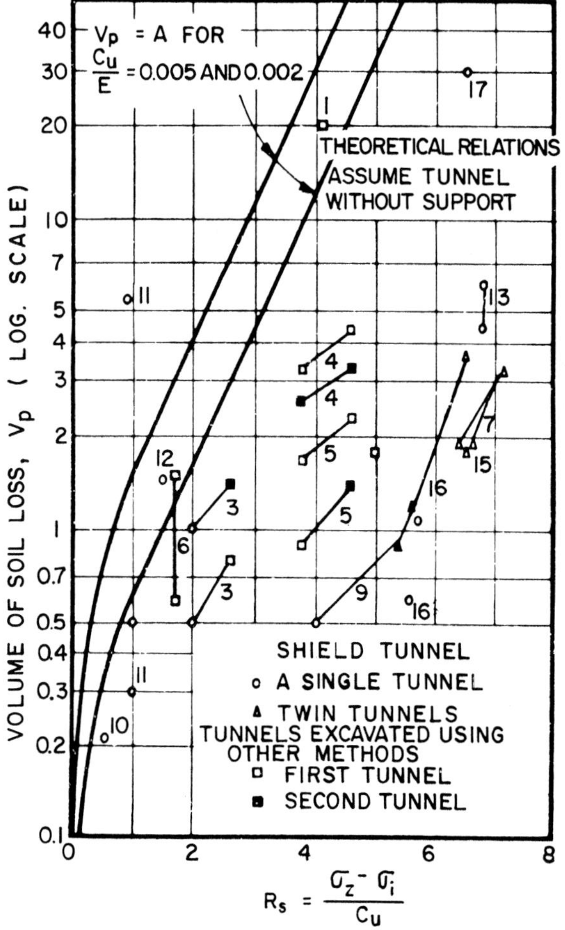

Fig. 14-35 Soil losses for tunnels in clay [37]

the case of tunnels constructed by this method. However when tunnels are driven by other methods or using hand tools, it is far harder to give a reasonable explanation of the low soil loss values, for in these cases it generally depends on many construction details that are hard to evaluate. Cases *2*, *3* and *5* in Fig. 14-35 are tunnels driven in Chicago (U.S.A.) clays, practically with the same methods, whereas *4* is a tunnel built at the same location, but using procedures that were acknowledged as being less efficient.

In double tunnels, the situation is different from that of single tunnels. Observations suggest that when twin tunnels are driven in clay using the shield method, the soil losses in the second tunnel are similar to those that occurred in the first. For tunnels excavated using hand tools, however, there may be appreciable difference [37]. See, for example, case *3* in Fig. 14-35, showing that the volume of soil loss in the second tunnel was in the order of double the volume that occurred in the first. In cases *4* and *5* of the same figure, the two tunnels had a common central dividing wall and this apparently made soil loss in the second tunnel even smaller than in the first.

14.8.2 Depression in the Ground Surface

In this section, some indications will be given about the procedures for estimating the shape of the depression which occurs in the ground surface as a consequence of tunnelling, and for evaluating the maximum settlement in that depression.

The profile of the depression resembles a normal frequency distribution curve (Gauss curve) which is shown in Fig. 14-36, together with one of its geometrical relations which will be of interest here. In order to use these relations to best advantage, in practical cases the shape of the depression is usually recognized as such.

The volume of the depression equal to the area below the Gauss curve is:

$$V_d = 2.5i\,\delta_{max} \qquad (14\text{-}81)$$

where δ_{max} is the maximum settlement of the section, and i is the standard deviation corresponding to the curve (Fig. 14-36) which is the abscissa of the inflection point of the curve.

The value of i can be estimated from the relationship [37,39]:

$$\frac{i}{a} = K\left(\frac{z_0}{2a}\right)^{0.8} \qquad (14\text{-}82)$$

where a is the radius of the circular tunnel or half of the width of the cavity should it have a different shape. z_o is the depth of the tunnel axis and K is a coefficient that can be regarded as equal to unity.

.1 Cohesionless Granular Soils

In these soils either expansion or shrinkage may occur as a result of the changes in stresses induced by excavation, depending on whether the sand is in a loose or dense state. Consequently, the volume of the depression is not always the same as that of the soil loss in the cavity. The value of i may be different from that given by Eq. (14-82), deduced under the hypothesis that settlement in the ground surface is not influenced by volume changes in the soil mass.

The value of the parameter i, which defines the width of the depression, depends on construction details, as well as the position of the water-table. Figure 14-37 [37] contains information about the width parameter of the depression in several modern tunnels in sands in different parts of the world.

Figure 14-38 [2] shows the shape of the depression above a pair of twin tunnels in dense sand above the water-table. The depression curves are very wide, an effect that can be attributed to the dryness of the sand and its complete lack of cohesion. The distance between the two tunnels was sufficient for excavation of the second not to have pronounced effects on the first.

The depression curve for a tunnel in dense sand below the water-table is shown in Part *a* of Fig. 14-39. Part *b* shows the final result of the evolution of this depression when a second tunnel was built next to it.

The influence of a second tunnel on the depression caused by the first depends largely on the distance between the two as well as the distance compared to their depth. If they are very close to each other compared to their depth, construction of the first tunnel loosens the ground above the site that will be occupied by the second, and the settlements above this one will increase.

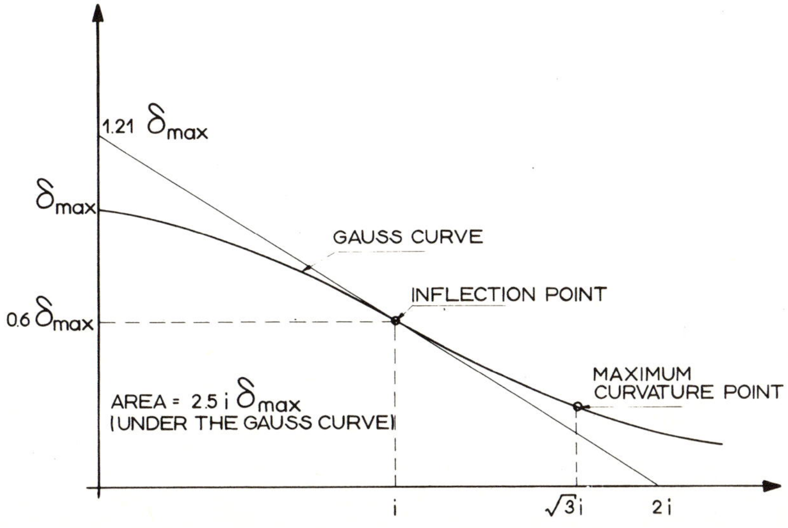

Fig. 14-36 Gauss curve and some of its geometric relations

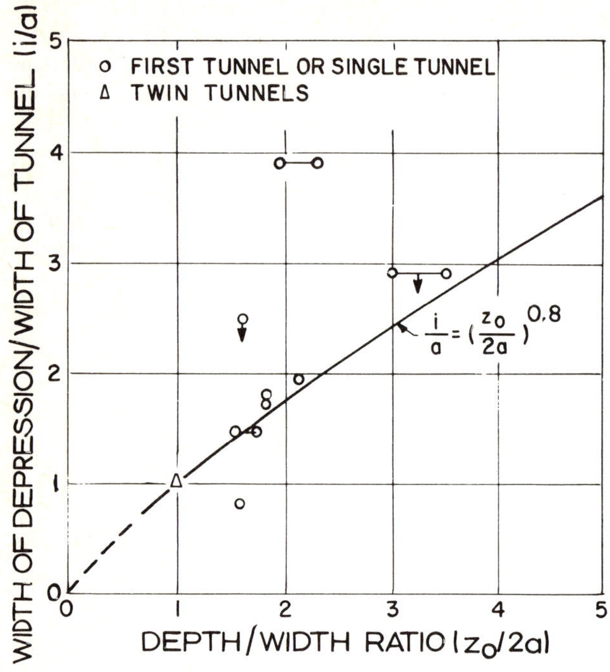

Fig. 14-37 Width of the depression in the ground surface above tunnels in sand [37]

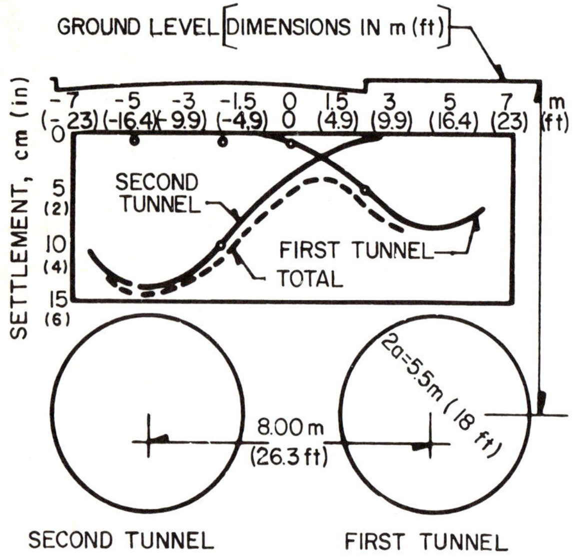

Fig. 14-38 Depression in the ground surface above twin tunnels in dense sand above the W.T. [2]

.2 Cohesive Frictional Soils

Figure 14-40 [2] shows the distribution of settlements over tunnels constructed as part of the same underground transportation system in the city of San Francisco, U.S.A.

The figure corresponds to construction of a single tunnel. Construction of a second tunnel, with an approximate distance of 10m (33 ft) from center to center, caused practically no subsequent effect. In these soils, both the settlements and the movements occurring inside the mass are generally very small, when the groundwater is carefully controlled.

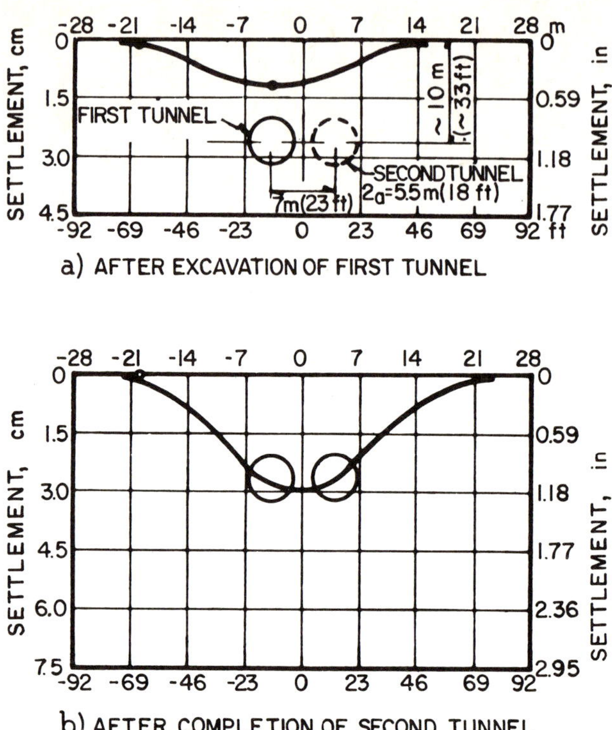

Fig. 14-39 Depression in the ground surface above twin tunnels in dense sand below the W.T. [2]

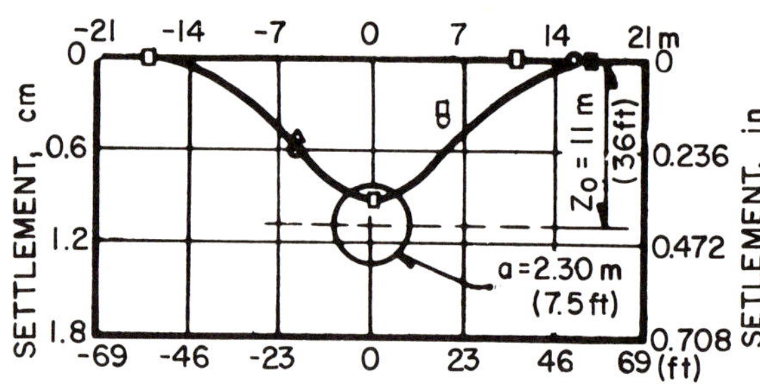

Fig. 14-40 Depression in the ground surface above a tunnel in dense clayey, well-drained sand [2]

.3 Stiff Non-Expansive Clays

The depression in the surface of these soils, corresponding to the minimum soil losses that have already been mentioned, is usually small where the value of δ_{max} is concerned and not very extensive. Figure 14-41 [2] is an illustrative example corresponding to a tunnel excavated using hand tools and with wooden supports.

The use of steel skeletons wedged against the walls and roof as a temporary support prior to construction of a concrete lining, usually greatly reduces the settlement in stiff clays. The danger of clays with a tendency towards plastic flow can be very effectively combatted using suitably wedged circular steel supports. It has been said [2] that the problem of settlement is practically non-existent in tunnels where a relationship $\sigma_z/c_u < 4$ coexists with a careful, well-organized, appropriate construction technique.

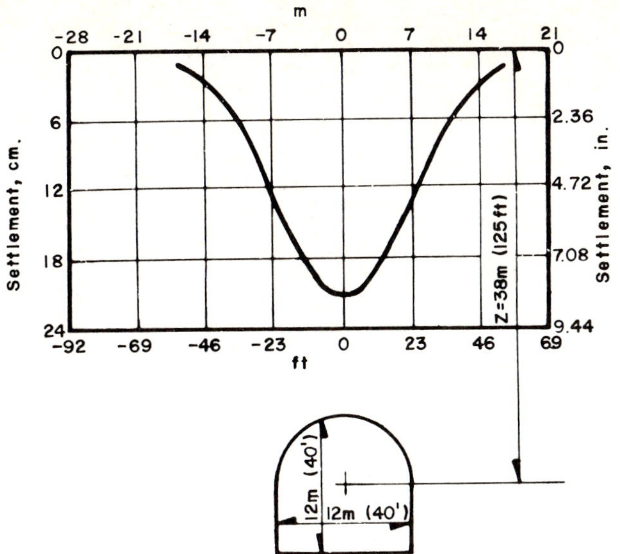

Fig. 14-41 Depression in the ground surface above a large railroad tunnel in stiff clay [2]

.4 Saturated Soft Clays

Figure 14-42 [37] shows the variation in the width of the depression in the ground surface (expressed in terms of the value $2i$) as a function of the relationship between the depth of the tunnel axis and its radius ($z_o/2a$) for a series of tunnels driven in soft clay. Equation (14-82) appears in the same figure represented by a solid line. In order to use Fig. 14-42 in extrapolations for design evaluation of twin tunnels it is recommended that the value of a be taken as half of the total width of the two tunnels.

By observing the figure, the excellent approximation given by Eq. (14-82) will immediately be appreciated. This should not be too surprising, since displacement of the soft clay around the tunnel occurs almost without any drainage, that is without any volume change, which is the hypothesis under which the equation was deduced.

The shape of the depression above tunnels in soft clay is shown in Figs. 14-43 and 14-44 [2]. This shape is fairly consistent, regardless of whether the shield driving method is employed or some other method. In these cases, a second tunnel frequently causes important effects, similar to those in the first tunnel, with a depression tending somewhat asymmetrically towards the second one.

It is to be noted that the magnitude of the settlements caused in these soils is large, even if careful construction precautions are observed. The rate of settlement is usually fairly slow, which is an advantage.

Reference has already been made to the additional problem in soft clays of settlements due to consolidation after excavation. These settlements may be particularly important in tunnels built with the shield method, owing to the remolding that is brought about by this method.

.5 Determination of the Shape of the Depression

The shape of the depression that forms above a future tunnel can be established with reasonable approximation on the basis of information from real cases and the idea that this shape corresponds to a normal frequency distribution curve.

Figure 14-45 [2] summarizes the information given in many of the previous figures in this section, and other similar ones which are becoming available in the literature. In it is plotted

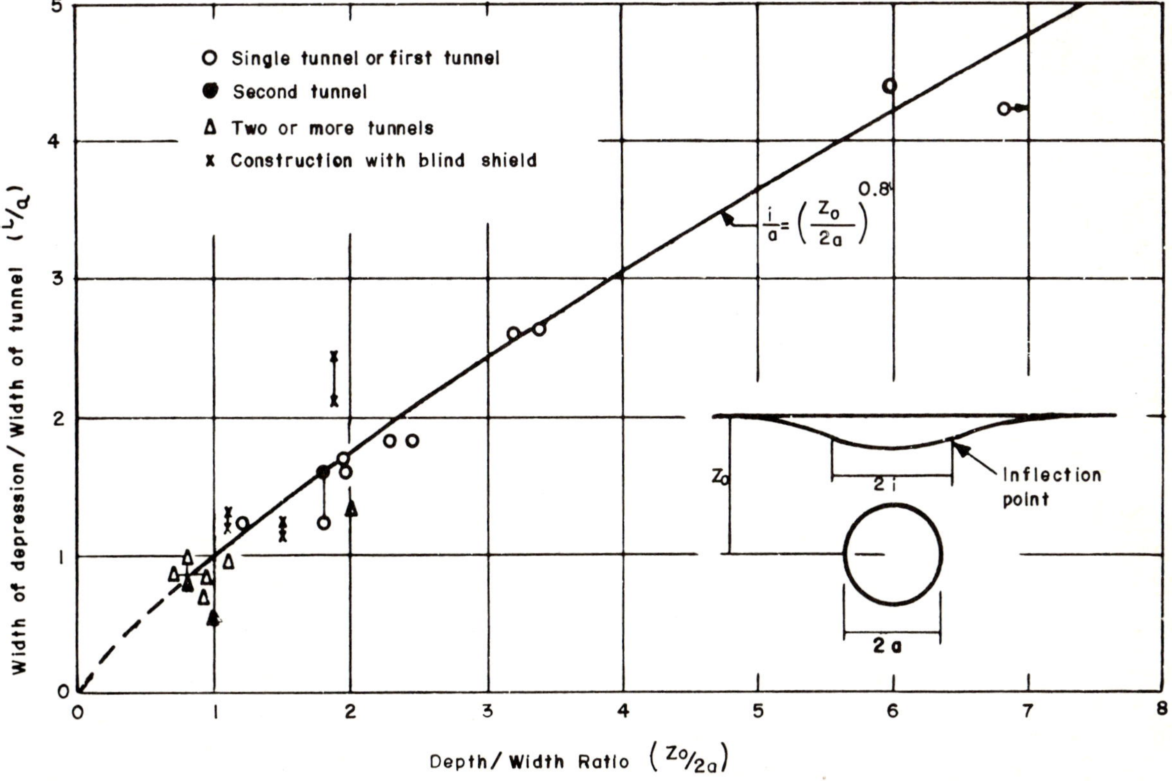

Fig. 14-42 Width of the depression in the ground surface above tunnels in soft clay [37]

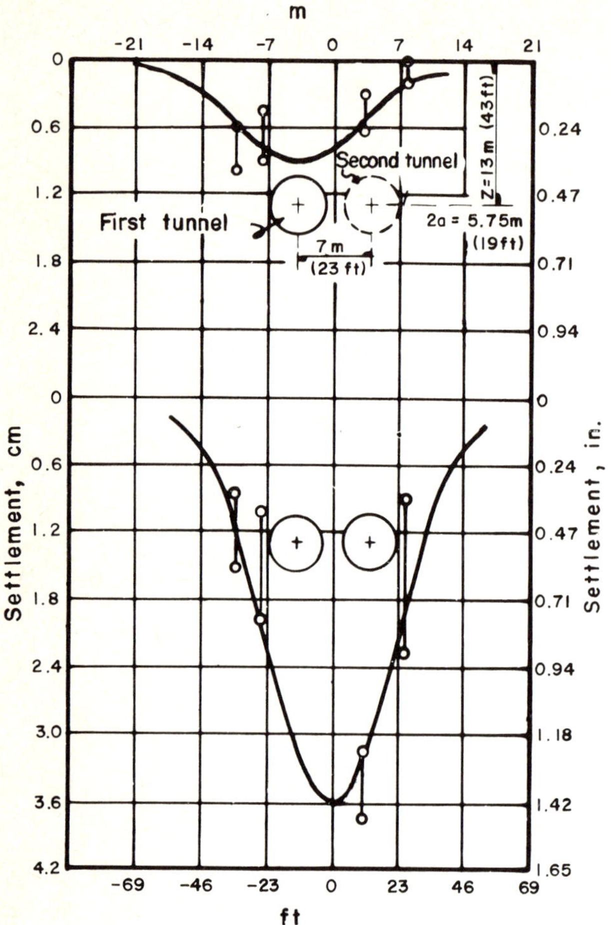

Fig. 14-43 Depression in the ground surface above twin tunnels in soft clay [2]

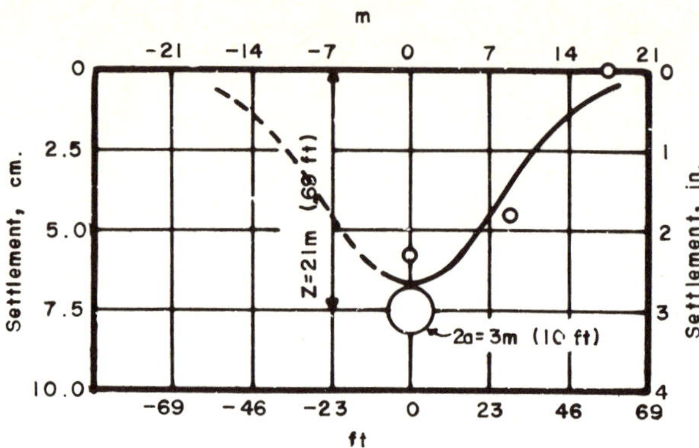

Fig. 14-44 Depression in the ground surface above a culvert in soft clay [2]

the relationship *i/a* against the relationship *z/2a*. Here, *a* is the radius of the tunnel, *z* the depth of the tunnel axis and *i* the abscissa of the inflection point of the Gauss curve. The quantity $2a'$ is the distance from the center of one tunnel to the center of a second adjacent tunnel, when there are two.

In the figure, an attempt has been made to separate the results according to the types of soil, despite the uncertainties implied by the scarce information available as yet. This serves to estimate the shape of the depression in the ground surface above a future tunnel.

.6 Settlement Control

Tunnels in clay: Naturally, as already mentioned, soil loss can be reduced by the rapid installation of supports at the working face. The theoretical curves in Fig. 14-35 were obtained under the hypothesis that the only support in the tunnel was that provided by an internal air pressure. An increase in this air pressure brings about a reduction in soil loss which is always beneficial.

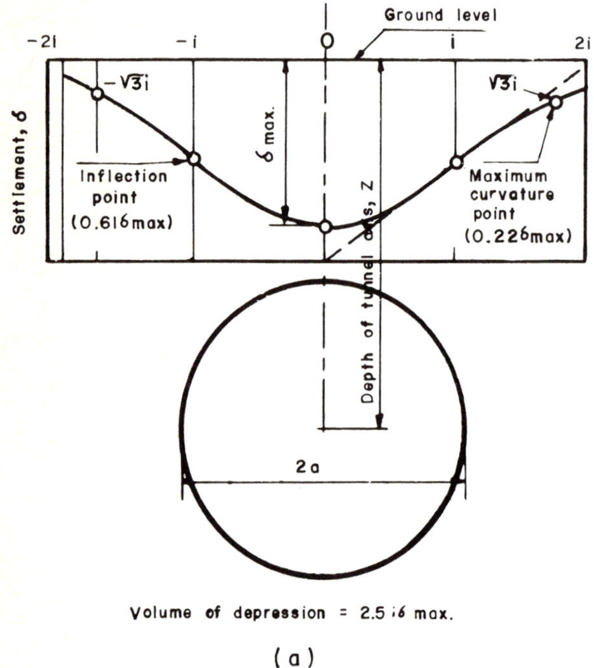

(a)

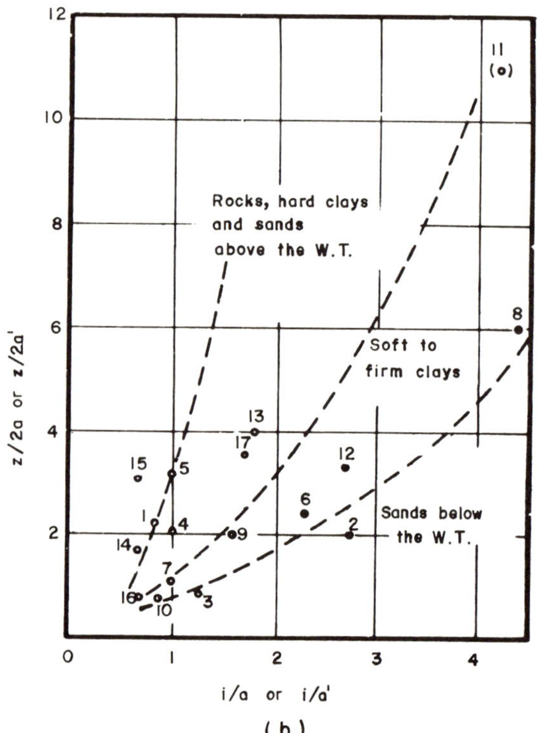

(b)

Fig. 14-45 Relation between the width of the surface depression (*i/a*) and the depth of the cavity (*-z/a*) for tunnels in various different materials [2]

In principle, the use of compressed air to provide support at the working face is appropriate, but [40] describes at least one case where a wedged concrete lining constructed immediately after the shield passed did not lead to satisfactory results.

For relationships where $R_s < 2.5$ and when the clay is not very fissured, settlements can be minimized by using prefabricated liner plates, which are wedged against the soil after the shield has passed. It should not, however, be forgotten that in several parts of this chapter emphasis has been placed on the fact that the immediate construction of any wedged support can lead to an appreciable increase in the pressures exerted by the soil against that support.

In Fig. 14-35 [37], it can be seen that theoretically for an unsupported tunnel with $R_s = 3$ there would be a soil loss of the order of 7% of the excavated volume. Also theoretically, if it were wished to reduce this loss to 2%, $R_s = 1.7$ would be required. From this it would be possible to estimate the large increment in internal air pressure that would be required to achieve this reduction. The above considerations indicate how difficult and costly it may prove to try to reduce soil loss, which is why these reductions should be attempted only in zones where settlement is liable to cause very considerable damage and where the operation can be controlled carefully.

Tunnels in sand: In sands an attempt can be made to reduce soil loss by many procedures, the efficiency of which varies from one case to another. It will always be advisable to reduce to a minimum the localized suction that is produced at the working face as a consequence of the action of mining tools, even when shield tunnelling methods are used.

Below the water-table the use of compressed air usually gives very good results, but if the pressure is considerable, it may contribute to the withdrawal of water from the sand, which leads to collapses. These hazards can be lessened by sprinkling the exposed sand face. Sometimes clay coverings have been used on the exposed area of the working face, which may be effective for impeding sand flows in the cavity that forms above that face. Subdrainage is an excellent technique for impeding sand flows. Cement grouting has shown rather disappointing results at the excavated face, but has proved more effective when used to stabilize tunnel walls or floors.

.7 Summary of Calculation Sequences

With the information given in this chapter, principally taken from [2] and [37], the following sequence can be adopted for estimating the settlements that are liable to take place over a future tunnel.

Tunnels in clay:

1) Estimate the volume of soil loss by means of Fig. 14-34
2) Use Fig. 14-42 or Fig. 14-45 to estimate the width of the depression in the ground surface as a function of the parameter i
3) Assuming that the volume of soil loss is equal to the volume of the depression in the ground surface, use Eq. (14-81) to calculate maximum settlement
4) Make a suitable adjustment if it is likely that the soil mass above the tunnel may suffer internal volume changes. These are usually negligible, however
5) If twin tunnels are to be constructed, make a further adjustment in view of the presence of the second tunnel; this is usually negligible, too

Tunnels in sand:

1) Estimate the volume of soil loss, which should be done on the basis of data obtained from existing tunnels similar to the one projected [37]. Soil losses reported in the literature generally range between 1 and 3% of the total excavated volume
2) Use Fig. 14-37 or Fig. 14-45 to estimate the width and shape of the depression in the ground surface
3) Using Eq. (14-81) it is now possible to compute δ_{max} under the same hypothesis used for clays
4) Make a suitable adjustment in view of the volume changes that may be suffered by the granular soil above the tunnel as a consequence of excavation. These are usually appreciable here
5) If twin tunnels are to be constructed, the settlement above the second tunnel may be similar to the one that occurred above the first. If the granular soil contains no fines, settlement above the second tunnel may be even greater than the one which occurred above the first

.8 Depressions in Ground Surface above Mexico City Metro Tunnels

Reference [41] gives some interesting information about the settlement measured above a section of shield tunnel constructed to house the Mexico City Metro in ancient alluvial fan deposits, as well as four inverted siphons located in clayey lake deposits in the same city [42, 43] which were constructed to give continuity to four large collectors which were intercepted by the Metro. In the case of the four siphons, the surcharge relationships σ_z/c_u were as given in Table 14-7.

When driving the Metro tunnel through ancient alluvial fans, a shield with a diameter of 9.15 m (30 ft) was used with a visor or overhang and open front. Liner plates were installed immediately. In the four siphon tunnels, a shield with a diameter of 2.95 m (9.7 ft) was used with an open front that was tilted 25° to form a visor. In the Metro tunnel, liner plates were used as a permanent lining; so that it can be said that no temporary support system was employed. In the siphons, the temporary

Table 14-7
Values of the surcharge factor, R_s, in tunnels in Mexico City [41]

Inverted Siphon	Surcharge relation	
	With c_u obtained from direct shear tests	With c_u obtained from unconfined compression test
"2nd of April"	6.2	3.4
"Manuel Gonzalez" (Two twin tunnels)	6.1	4.3
"Obero Mundial"	6.1	4.3

support system consisted of metal liner plates; the lining was ring-shaped and made of portland cement concrete. In both cases the space between the perimeter of the lining and the soil was grouted using a grouting pressure equal to the weight of the soil at the level of the tunnel axis.

In all cases pumping was carried out ahead of the shield to prevent seepage forces from directing themselves towards the working face and encouraging soil loss.

Figure 14-46 shows the greatest settlements observed. In the *2 de Abril* siphon, the maximum settlement reported did not exceed 9 cm (3-1/2 in).

In the siphons, the depression is generally very slight until the shield approaches the benchmark. Beyond the benchmark, the depression developed rapidly at a rate of 1 to 2 cm (0.4 to 1.8 in) per day. This rate of settlement was sustained for several days, after which it dropped to values in the order of 10% of the previous ones. These continued no more than one month, then came to a total halt.

In the Metro tunnel the settlements measured were smaller, chiefly because of the density of the soil and effectiveness of the expansion induced by the liner plates. A slow rate of settlement is also observed ahead of the shield under the benchmark; this rate doubled or tripled in the next two or three days after the shield passed. Not more than a week later settlements gradually reduced until eventually they ceased completely.

When locating the values of parameter i on the graph in Fig. 14-45, it was observed that the points corresponding to the siphons fall into the zone regarded as corresponding to sands below the water-table. Figure 14-47 [41] shows the position of these points.

In the case of the *Obrero Mundial* and *Manuel González II* siphons driven in soft clay, the boundary indicated in [2] between soft clays and sands below the water-table might be con-

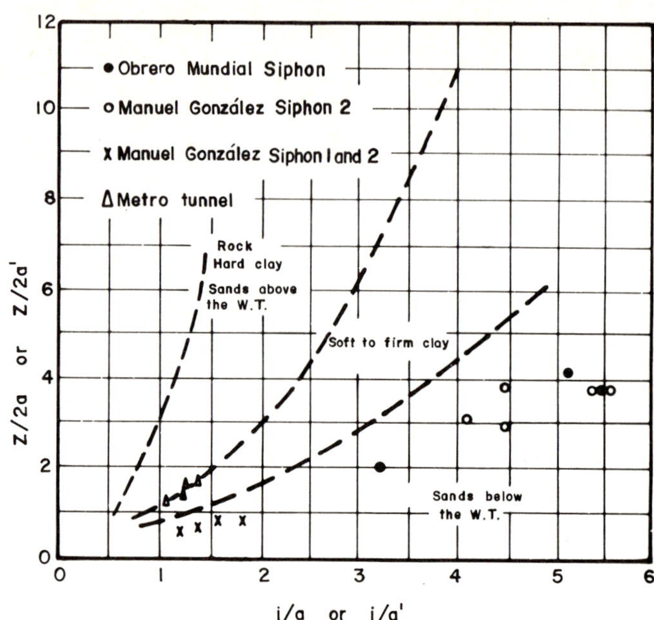

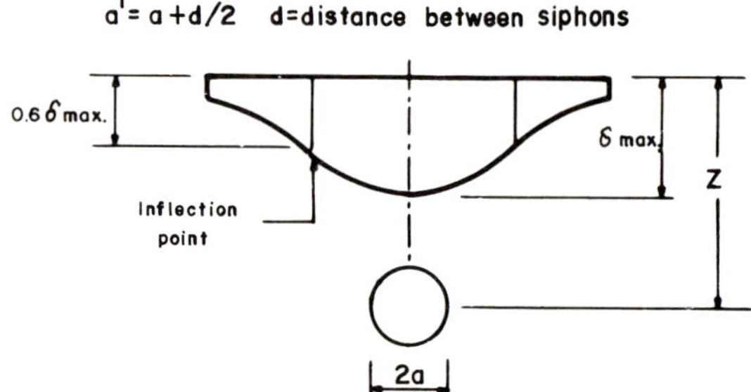

Fig. 14-47 Location of Mexico City tunnels on the PECK zoning chart [2, 41]

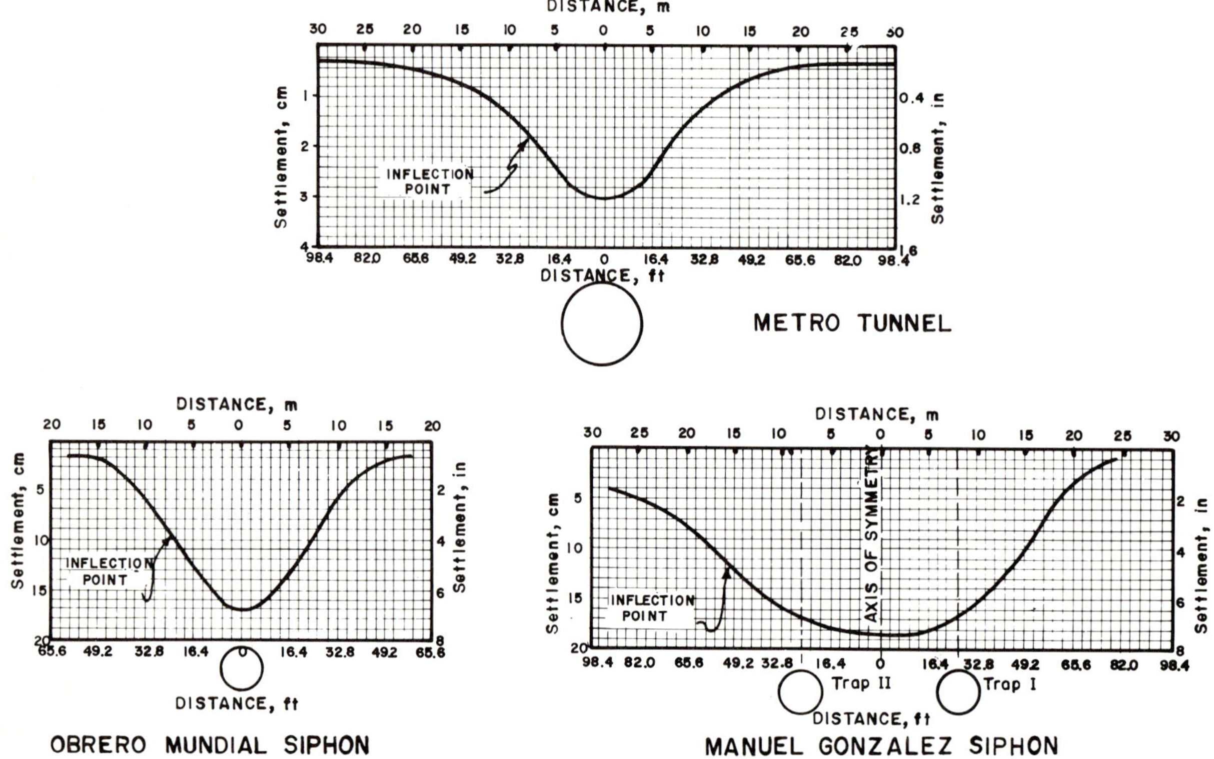

Fig. 14-46 Depression in the ground surface and settlements in Mexico City tunnels [41]

sidered to be slightly lower in the Mexico City clays. The points corresponding to the *Manuel González I* and *II* siphon and the Metro tunnel are located in a zone which appears to be conflictive on the PECK zoning chart, but can be regarded as reasonable.

The total volumes of the depression in the ground surface in all cases agreed reasonably with the estimation that can be made on the basis of PECK's ideas contained in [2, 37], which were commented on earlier in this chapter.

On the basis of all the measurements taken, Reference [41] tentatively presents curves for predicting the magnitude of the settlements and the shape of the depression in the ground surface above similar tunnels that may be constructed in Mexico Valley clay at some future date. This important information appears in Fig. 14-48.

It can be seen that the data are fairly dispersive for estimating settlement, although the tentative curve is plotted in a conservative manner. For the shape of the depression, the available information is reliable and consistent.

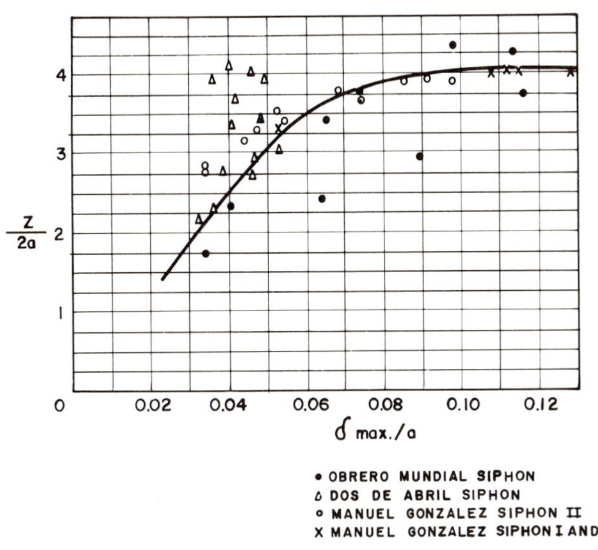

a) Estimate of maximum settlement

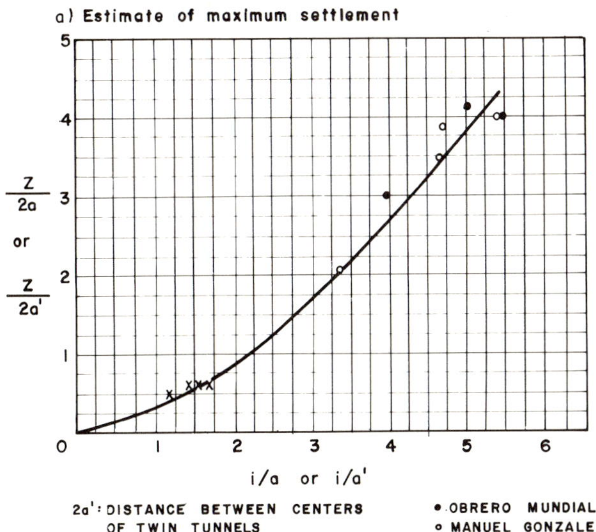

b) Estimate of shape of surface depression

Fig. 14-48 Tentative curves for estimating the maximum settlement and the shape of the surface depression above shield tunnels in Mexico City clay [41]

In [41] a fact which has been greatly emphasized in this chapter, is how a good building procedure can greatly reduce settlement hazards above tunnels in very treacherous soils. Also described in detail is what is referred to as *damage to neighbouring structures* which was remarkably insignificant, despite all the tunnels being built in densely populated urban zones.

14.9 Some Aspects of Shallow Tunnelling

There are numerous methods that can be employed for tunnelling in soils. However, discussion of them is beyond the scope and objectives of this book; the sole aim of this section is to make a few comments of a very general nature, with special emphasis on the shield-driving method, which is frequently employed nowadays and which becomes ever more widespread with the development of better equipment.

Tunnelling in soils is not a task that most engineers take pleasure in doing very frequently. The uncertainties involved are numerous at all stages of the design and construction. Consequently a tunnel usually holds many surprises which are almost always disagreeable during the construction period, and which result in a waste of time and money. Moreover, tunnelling in soils most often takes place in urban areas, where surprises are particularly troublesome, especially when involving settlement of the ground surface and damage to neighbouring buildings.

In view of all this, the engineer often prefers to open cuts, the walls and bottom of which are supported with suitable bracing, currently available in many different types. Eventually the open cut is covered with a roof which functions in the same way as the original ground surface. This is the method that has been adopted for the construction of many underground railroads when the depth of the open excavation required did not exceed 10 m (33 ft), which is the maximum economical depth for most applications of this method (Plate 14-5). In urban areas, this

Plate 14-5 Construction of an open section of the Mexico City subway (Courtesy of Solum, SA)

Plate 14-6 Bracing the lateral bearing walls in a section of the Mexico City subway (Courtesy of Solum, SA)

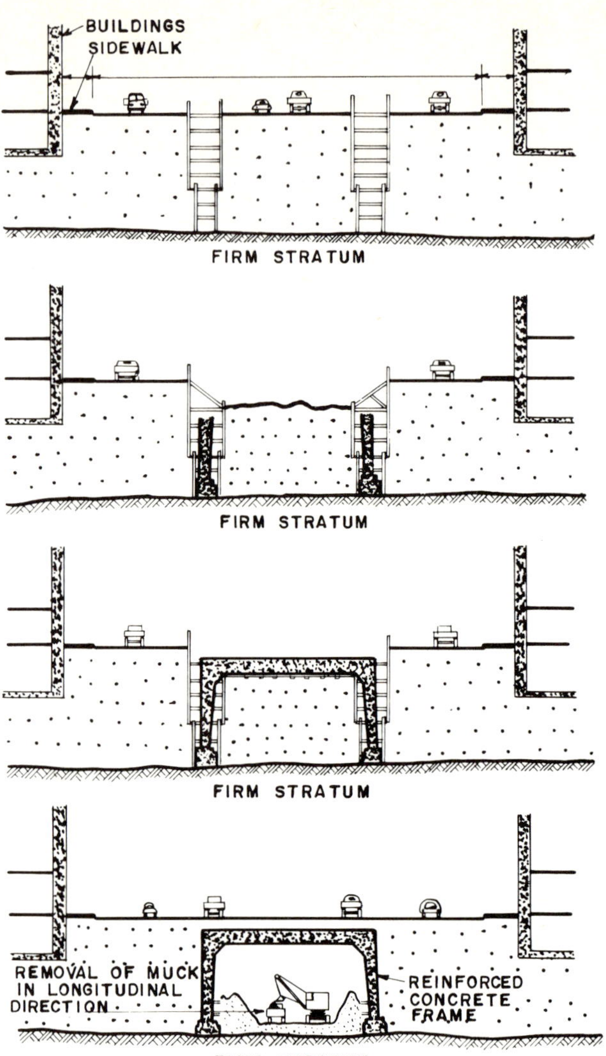

Fig. 14-49 Cut-and-cover method of excavation, with minimum effects on the ground surface [3]

procedure usually interferes with traffic (Plate 14-6) damages the aesthetic appearance of cities and damages trade and the normal exchange of goods and services. Therefore many tunnelling methods have been developed which combine the underground excavation technique and the cut and cover method, with a view to interfering as little as possible with normal conditions of life on the surface. Figure 14-49 [3], for example, shows a construction sequence along these lines. Figure 14-50 [3] shows the method of construction using slurry displaced concrete walls (*Milanese walls* or *Icos screens*) which has been adopted for large sections of the Mexico City Metro. This procedure reduces to a minimum the inconvenience that is caused by an excavation job to the normal life of a city.

Another practical method for building shallow tunnels is to install pre-cast concrete caissons in a previous excavation open to the sky. This method is particularly popular for tunnelling under rivers, and in this case the pre-cast segments are usually floated into position and then sunk.

For underground excavation without affecting the surface, shield tunnelling is one of the most widely used methods. Currently shields come in many different models with numerous variations in their operation methods, but the schematic drawing in Fig. 14-51 [3] covers the essential principles for all cases. The perimeter of the shield is slightly larger than and in the shape of the tunnel that is to be built. In the case in the figure, the front of the shield is equipped with a cutting edge that is capable of penetrating a soft soil, forced ahead by a set of jacks which rest on the lining segments that have already been installed. The back of the shield extends into a tail, which makes it possible to carry out the necessary maneuvers with speed, so that the cavity can be lined as the shield advances. In soils that are not so soft, which do not allow the shield to advance under simple pressure but which can squeeze, cave-in, or erode into the cavity, the entire front of the shield is fitted with a metal bulkhead which holds back the soil ahead of it. Openings make it possible to pass appropriate drilling and excavating tools, and remove the muck that is excavated (Plates 14-7, 14-8).

The operation cycle, which is also shown in Fig. 14-51, consists of:

1) Excavation and temporary support (sometimes with removable bulkheads or hydraulic fingers) of the working face to a suitable depth.
2) Advancement of the shield, the jacks reacting on the lining segments already installed.
3) Installation of another segment of lining at the tail of the shield, which enables the shield to continue advancing by the same procedure.
4) Grouting the gap between the lining and the soil.

The advantages of shield tunnelling are that:

1) The tunnel section can be built in its full size
2) Support is provided at all times and in all directions in the ground through which the tunnel is driven

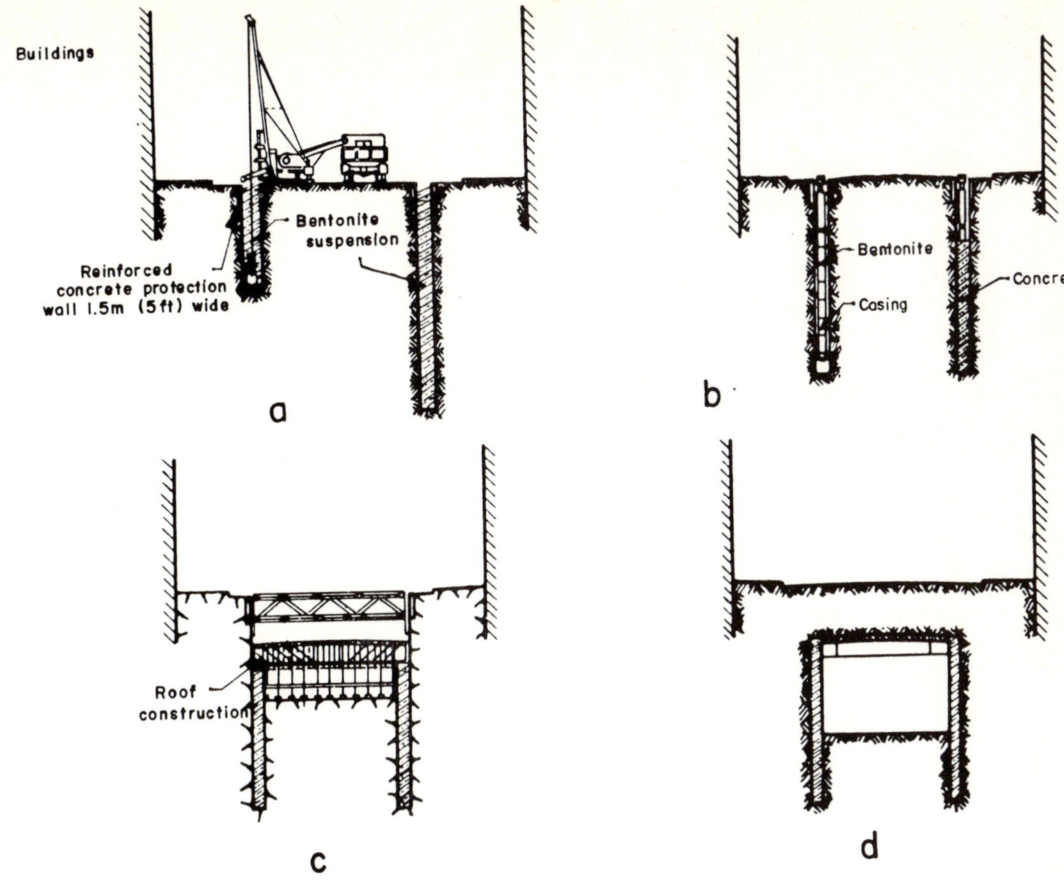

Fig. 14-50 Excavation method using ICOS screens or Milanese walls [3]

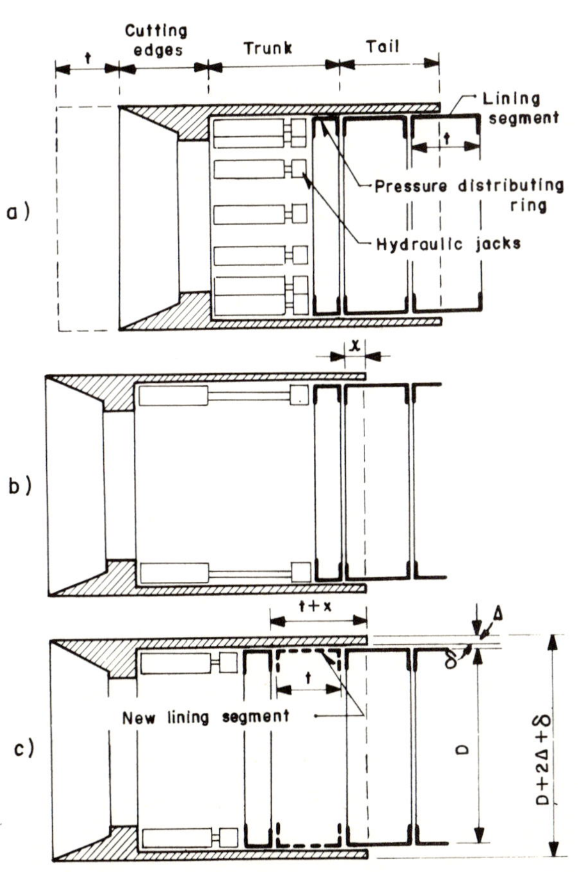

Fig. 14-51 Schematic drawing showing principle of shield tunnelling [3]

3) The lining can be installed immediately, avoiding temporary supports and giving very little opportunity for excessive deformation of the adjacent ground
4) Construction progresses very fast
5) Excessive deformation of the ground is avoided and soil loss is satisfactorily controlled, all of which tends to reduce settlements in the ground surface.

Plate 14-7 Shield tunnelling in an underground section of the Mexico City subway. Note the lining of the sections already excavated, along with the front-loading mucking machine and conveyor belt to remove muck (Courtesy of Solum, SA)

Plate 14-8 Shield tunnelling in an underground section of the Mexico City subway. Operations for installing liner plates (Courtesy of Solum, SA)

Advancement of the shield, driven by hydraulic jacks which react against previously installed lining segments, is accomplished by overcoming the following resistances [3,44], which assume the shield advances under pressure, through a sufficiently soft soil):

W_1 = Friction between the outer surface of the shield jacket and the ground

W_2 = Friction between the tail of the shield and the lining segments installed beneath it

W_3 = The passive strength (or horizontal bearing) of the soil in front of the shield exerted against the cutting edges

The Soviet literature referred to in [3] provides equations for evaluating the three resistances. Accordingly (see Fig. 14-52):

$$W_1 = \left[\sigma_x\left(1 + \frac{\sigma_x}{\sigma_z}\right) 2LD + G_p\right] f_1 \quad (14\text{-}83)$$

where:

σ_z, is the vertical pressure exerted by the soil on the shield skin

σ_x, is the horizontal pressure exerted by the soil on the shield skin

L, is the length of the shield

D, is the diameter of the shield

G_p, is the weight of the shield

f_1, is the coefficient of friction between the shield skin and the soil.

The second resistance is equivalent to:

$$W_2 = f_2 Q_t \quad (14\text{-}84)$$

where:

Q_t, is the weight of the lining segments installed beneath the tail of the shield.

f_2, is the coefficient of friction between the tail of the shield and the segments of lining beneath it.

Lastly, the third resistance is equivalent to:

$$W_3 = \pi D_k \delta \sigma_z K_p \quad (14\text{-}85)$$

where:

D_k, is the diameter of the shield measured at the center of-the cutting δ, edges is the thickness of the cutting edge

K_p, is the coefficient of passive earth pressure

There are a wide variety of shields, but the circular ones usually give the best results, especially if high pressures are exerted by the soil. Construction practice recommends length to diameter ratios of about 0.75. The longer the shield, the easier it will be to keep it aligned, but the more difficult it will be for it to make any turn that may prove necessary. Shields that are too long demand very high pressures from the hydraulic jacks.

During tunnelling operations, there is usually a tendency for shields to turn about their longitudinal axis, owing to oblique stratifications, asymmetrical external pressures or because excavation is not carried out in a uniform manner. It can be fitted with stabilizer fins to avoid this effect.

The first soil support behind the shield and below the tail is usually provided by steel or concrete liner plates. This system permits a very rapid and economical operation, so long as the liner plates are capable of supporting radial pressure from the ground as well as longitudinal loads from the jacks. By way of illustration, Fig. 14-53 [3] shows the liner plate arrangement used in the Moscow subway.

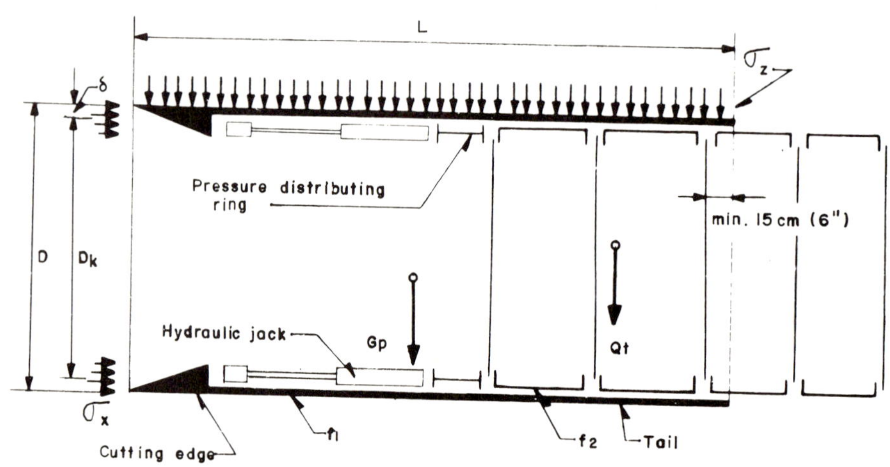

Fig. 14-52 Forces acting against the advancing shield [3]

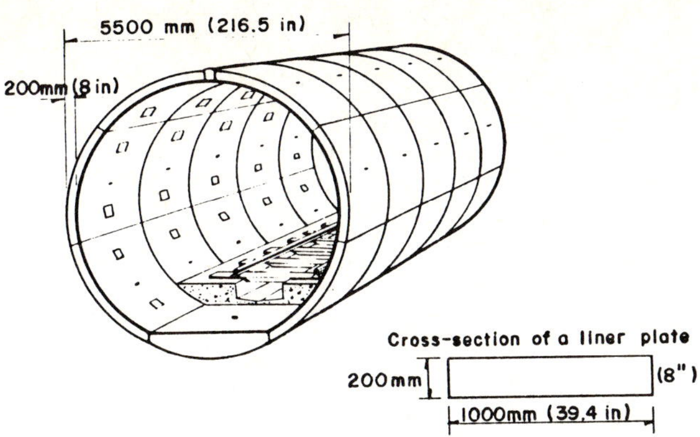

Fig. 14-53 Liner plate arrangement in the Moscow subway [3]

Often liner plates are also the permanent lining system. In such cases the permeability of the joints becomes critical. However, current technology has developed several means of combatting this [3].

The excavation operation involves the entire cross-section of the tunnel. Serious problems may arise in stabilizing the working face, because large areas are involved. Mining tools are very varied. The pneumatic hammer is very common, as are rotary cutting tools. In larger shields, tracked excavators and backhoes are employed. It is good practice to fill the annular space left around the support with fine gravel or stiff grout when the tail is withdrawn. This is equal to the thickness of the tail plus any over-excavation. This practice is appropriate to avoid settlements. Subsequently this fine gravel is sealed with cement grouting to establish a homogeneous condition around the tunnel and to prevent water flow. At times gravel need not be employed; however grouting of the empty spaces behind the shield is always advisable to reduce pressures and settlements.

Despite all the precautions, the passing of the shield is usually perceptible on the ground surface because it causes certain disturbances in the excavated soil. Figure 14-54 [2] shows a typical situation.

Part *a* of the figure shows three successive positions adopted by a line of originally vertical supports which was about 6 m (20 ft) from the center of a circular tunnel of the same diameter driven through soft clays during construction of the San Francisco Subway. As the shield approaches, the clay moves slightly towards the outside of the tunnel, but when the tail has gone by, it returns rapidly towards the tunnel. Part *b* of the figure shows the settlement curves for three benchmarks on the surface. Attention is drawn to how the rate of settlement increases greatly when the shield approaches the benchmark, maintaining this high rate of settlement even after the device has passed. Settlements stabilize when the shield has moved on sufficiently far, which in this case proved to be about 15 m (50 ft).

Fig. 14-55 [41] shows the surface settlements during excavation of the three siphons and the section of Mexico City Metro to which reference has already been made in previous pages. The same tendency is perceived, with maximum settlement rates immediately before and after the shield passes the benchmark.

As has been seen in different sections of this chapter, with shield tunnelling methods (as well as others) it is often necessary to lower groundwater levels. Reference [45] describes the procedure used for this during construction of the Tacubaya tunnel of the Mexico City Metro, which crosses alluvial fans made up of sands and silts with clay lenses, layers of gravels and heterogeneously distributed boulders. The water pressures were lowered using deep pumped-wells 45 cm (18 in) in diameter, located every 20 m (65 ft) on both sides of the tunnel and drilled to 9 m (30 ft) below the future tunnel floor. The wells were equipped with submerged electric pumps. Pumping started 60 m (200 ft) ahead of the working face, commencing the operation 10 days before the shield was due to pass and continuing for 30 days afterwards.

Reference [3] contains a very comprehensive, detailed description of a wide variety of shields and numerous different construction procedures. It describes the construction operations conducted at the same time during tunnelling. This reference is particularly recommended to engineers who are confronted with the construction of tunnels in soils.

14.10 Compressed Air Tunnelling

Shields are commonly used when a tunnel is to be driven through fairly well consolidated, saturated alluvial deposits made of gravels and sands with deposits or lenses of plastic silts or clays, which usually have a low bearing capacity. When deposits of this type are well below the water-table, the clayey strata often have a very low shear strength, causing stability problems at the working face. Lenses of loose sand, even under relatively low water pressures, tend to pipe and flow towards the working face.

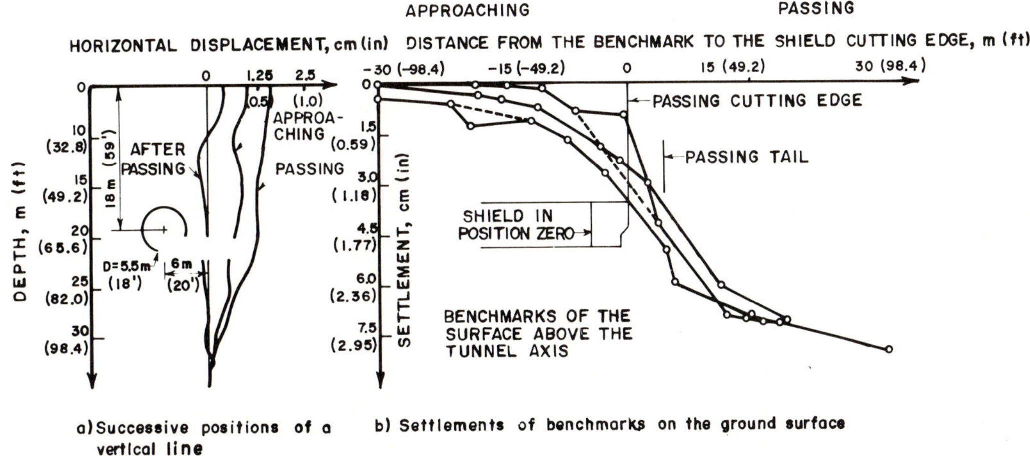

Fig. 14-54 Effect of the passing shield on soft clay. San Francisco subway [2]

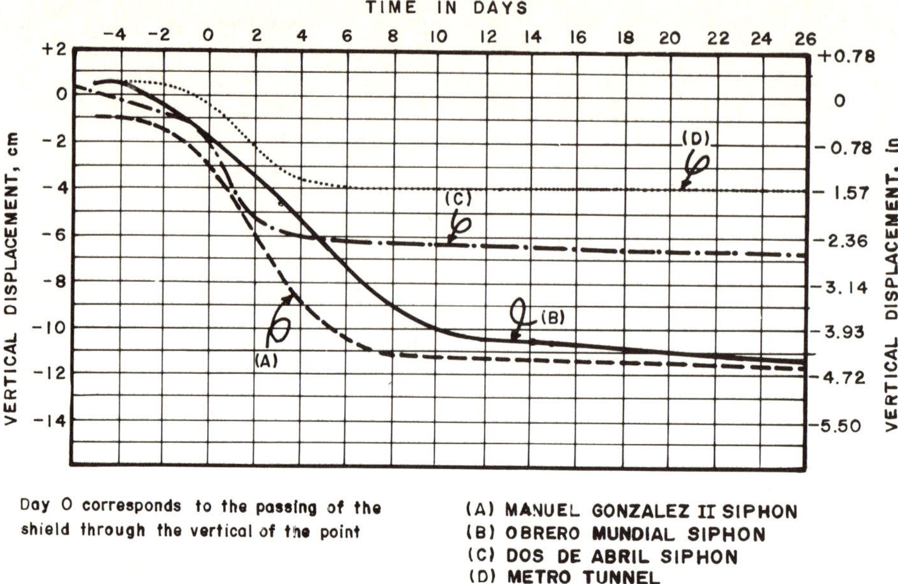

Fig. 14-55 Shield tunnelling with a double chamber of compressed air [3]

As has been pointed out several times in this chapter, there is a need in such cases to lower the water pressures at excavation level. This can be attempted by two methods: first, using normal subdrainage procedures, as discussed, in the course of which complex construction problems may arise (see, for example §14.7.1.2 and .3); and second, impeding the discharge of water into the tunnel by increasing the air pressure inside the tunnel. This creates a counterpressure from the inside outwards which will maintain the water in the soil pores partially or totally. Owing to its cost and construction complications, this second method is employed where pumping, electro-osmosis and other methods of subdrainage are unreliable. Consequently, the compressed air method is most often employed in very fine sands, sandy silts, clayey sands or the more special cases discussed earlier in this chapter which present difficulties in subdrainage.

Compressed air has been used in underground tunnelling for many centuries and the system is being more and more used of late. A working chamber is installed close to the working face, extending the shortest practical distance towards the section of tunnel that has already been excavated. Installation of the chamber is greatly assisted by the shield tunnelling method which employs a strong, reasonably waterproof lining. Once grouting has been completed the lining rapidly acquires the capacity to bear the entire soil load. Thus, at a very short distance from the face (10 to 20 diameters) it is possible to install an air-tight bulkhead, which will separate the completed tunnel section from the working area, where the working face is located. It is into this last zone that the compressed air is injected and communication through the bulkhead is achieved by means of air-tight compartments: the air-lock. Sometimes the working area is divided into several compartments each excavated separately, for the face may require far more stabilization than the zone where the support system is being installed below the tail of the shield. The pressure in the second more distant chamber in this case is often about half of that at the working face. The two chambers must be separated by an airtight bulkhead and communication between them must also be by means of airlocks, going through chambers and airlocks twice which makes the system expensive, slow and complicated. This limits the double chamber system, the advantages of which are smaller air pressure gradients across the airlocks. Figure 14-56 [3] shows such a double chamber device.

The air under pressure is injected into the working chamber or chambers and must subsequently be maintained in them. The magnitude depends on the hydraulic head that has to be neutralized, the characteristics of the soil and the size of the tunnel [46]. If the ground has some cohesion, the pressure

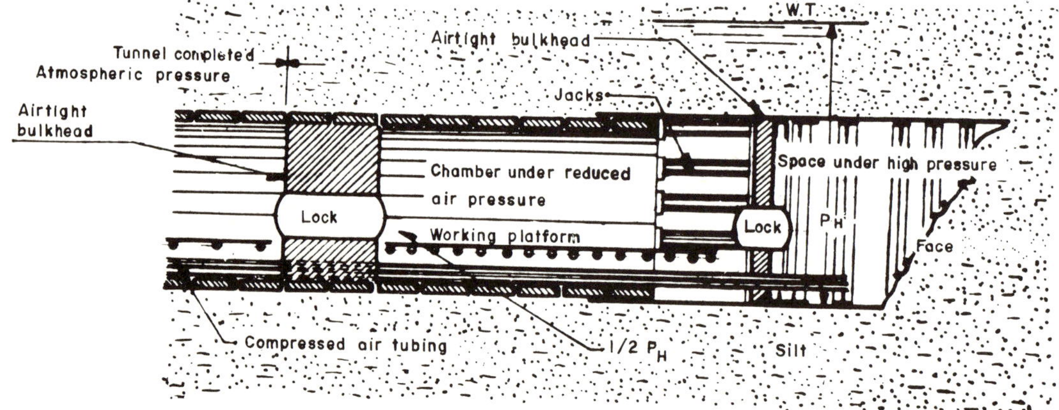

Fig. 14-56 Time-settlement curves for surface benchmarks. Excavations in the Mexico City subway

required to keep the tunnel dry is approximately equal to the water pressure prevailing at the level of the tunnel floor. If the ground is very pervious, this pressure cannot usually be maintained for fear of air escaping through the soil at the crown, where the air pressure exceeds the water pressure. In this case, the usual solution is to employ an air pressure equal to the one that will balance that of the water on the crown of the tunnel, or at most in the middle section of the tunnel. In such cases the entire cavity or the lower half (depending on the alternative used) will not keep dry by air pressure. In loose cohesionless soils, through which the shield can be easily driven, air pressure may be lower than the water pressure on the crown of the tunnel, so long as no work is conducted ahead of the shield. If it is necessary to work in front of the shield, the pressure of the air must be a combination of the water and soil pressures.

Apart from the obvious problems of the cost and delay of operation cycles, the principal problem in compressed air tunnelling is the safety and decompression of the men working inside the chamber. Careful consideration should always be given to this. Owing to the position of the tunnel, the uniform pressure of the air inside the chamber is resisted by external pressures (water and ground) that are smaller in the roof and greater in the floor or invert, with a difference that depends on the diameter of the shield. This difference is marked m in Fig. 14–57 [3]. If it is wished to maintain the entire tunnel dry, air pressure p_H in the working chamber must be equal to the external pressure corresponding to that of the tunnel floor. In this case there is an unbalanced pressure acting outwards above the invert which becomes equal to m in the roof. This in balance must be counteracted by the resistance to air flow that is developed in the strata of soil between the roof of the tunnel and the external water-table. Otherwise a blow-out will occur with the corresponding drop in pressure inside the chamber and inflow of water, accompanied by a serious risk to the men in the chamber.

The risk of a blow-out is greater the larger the diameter of the tunnel. The resistance of the ground to air flow should always be investigated and any predictable blow out should be compared with the capacity of the compressor system to supply air to the chamber and maintain sufficient pressure inside it, in spite of the leak. If this balance is not favorable, the working chamber must be made more airtight. Sometimes layers of clay or grouting have been used for this purpose. If this quality cannot be ensured, compressed air cannot be used to fully balance the water pressure.

The resistance offered by the soil to blow-out determines the maximum air pressure. Experience demonstrates that to keep the working chamber dry it is usually sufficient to provide an air pressure that is somewhat smaller than the water pressure that is to be neutralized. Also affecting this difference are the losses of hydraulic head which occur in all cases. The air pressure required to keep the chamber dry is usually 75% to 80% of the value that theoretically would be estimated by considering the hydraulic heads that must be balanced [3].

There are, however, exceptional cases which require air pressures even higher than the prevailing hydraulic head. One of these would be when a pervious sandy stratum is strongly confined along its upper boundary by a very airtight clay stratum. In this case, the air that leaves the working area through the face is trapped under the clay, causing an increase in water pressure in the water-bearing stratum.

The length of the working area with compressed air should be as small as possible to reduce air leakage, but sufficient space must be provided for grouting and other work requirements, including hauling muck.

It has already been discussed in this chapter that internal air pressure not only permits dry construction but also influences the pressures that are exerted on the tunnel, often leading to more favorable stresses. Sometimes the effect has been attributed to simple mechanisms such as air pressure expelling water from the soil voids around the tunnel, creating capillary tension and bringing about a corresponding increase in shear strength. It has also been said [47] that air pressure helps support clayey soils. Moreover, it appears that these effects are far more noticeable in the working face than in the walls and roof and help to increase stand up time (bridge action period). The effect of drying about the tunnel may cause cracking in some clays and many silts, eventually leading to highly unstable situations. Consequently, it is essential to balance the pressure that is used very carefully. In tunnels with a large diameter serious problems arise on account of the large difference in the pressures that is required to balance the water pressure between roof and floor.

In tunnels in coarse clean sand, air penetrates the material readily and produces a dome of dried material ahead of and above the front of the shield. This dome is the material which must be carried by a temporary support or stabilized with grouting. However, if the sand is fine there will be a beneficial effect of apparent cohesion due to capillary tension inside the relatively dry dome, which will help make it self-supporting. This effect does not exist in coarse sands or cohesionless

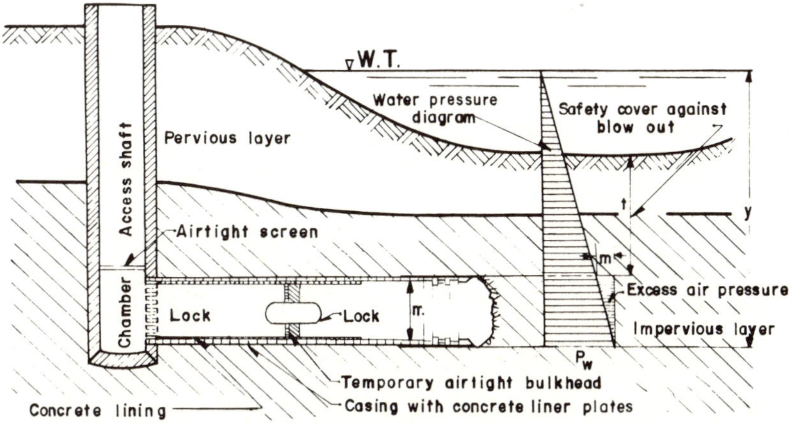

Fig. 14-57 Pressure conditions in a shield tunnel driven under compressed air [3]

gravels. There is the added problem that air penetrates these soils very readily and consequently the volume of the dome that forms is large. Because the air leaks into the sand or gravel it is difficult to maintain a constant air pressure in the working chamber in these coarse soils. In these cases the remedy has frequently been to apply a layer of clay or plastic silt to the exposed areas of gravel or coarse cohesionless sand and at the same time making the support or lining as close as possible to the working face.

Reference [3] gives an expression after SCHENCK and WAGNER for computing the time required for the desiccation of a soil stratum of thickness H to take place when subjected to an air pressure:

$$t = \frac{1}{2} \frac{H^2}{v_t (\sigma_i - h_c)} \tag{14-86}$$

where:

t, is the time required for desiccation of the soil stratum

H, is the desiccated thickness

v_f, is the rate of water seepage in that soil

h_c, is the capillary height in the same soil.

Figure 14-58 [46] is a schematic drawing showing the performance of personnel and muck air-locks between the working chamber with compressed air and the rest of the tunnel at atmospheric pressure. The reduction in air pressure for the workmen, termed decompression requires from several minutes to several hours. The time increases with increasing air pressure and also with the length of time the workman has experienced the maximum air pressure. For example for half hour of work at the maximum permissible pressure, 4 hours of decompression is required. Thus the cost of useful labor is increased 8 times with the maximum air pressure.

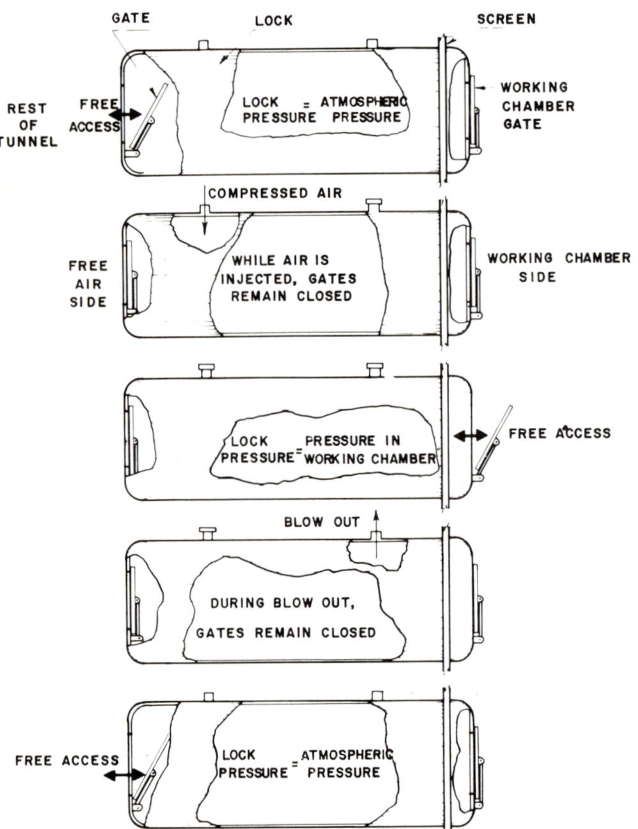

Fig. 14-58 Schematic drawing of the communication between the working chamber and the rest of the tunnel [46]

APPENDIX 14A WORKED PROBLEMS

14a.1 Problem Data

In order to construct a tunnel, exploration work, sampling and laboratory tests were performed which permitted completion of the geological profile along the axis that is shown in Fig. 14a.1.

The center part of the valley is made up of a superficial stratum of soft clay, with a simple shear strength of 1 kg/cm^2 (14.2 lb/in^2). The value for cohesion in a quick test was 0.5 kg/cm^2 (7.1 lb/in^2). The submerged unit weight is in the order of 1.7 t/m^3 (106 lb/ft^3) and the modulus of elasticity 50 kg/cm^2 (710 lb/in^2). At the toe of the mountains, a formation of loose dry sand was located with an angle of friction of 32°, with zero cohesion and a unit weight of about 1.5 t/m^3 (94 lb/ft^3).

The tunnel will have a circular cross-section with a radius of 4 m (13 ft) and will be located at an average depth of 20 m (65 ft), both in the clay and the sand. In all cases the water-table was located underneath the grade of the tunnel.

To compute pressures, the elastic, plastic methods after TERZAGHI, PROTODYAKONOV and BIERBÄUMER will be employed. A typical section in clay and a typical section in sand will be analyzed.

14a.2 Elastic and Plastic Analysis

.1 Section AA' (in clay)

Using KIRSCH's equations, Eq.(14-13) and assuming a value of 0.8 for the coefficient of earth thrust K, in-situ the stress state on the perimeter of the excavation ($a = r$) is given by Eq.(14-14) and in this case will be:

$$\sigma_r = \tau_{r\theta} = 0 \; ; \; \sigma_\theta = \sigma_z (1.8 - 0.4 \cos 2\theta)$$

The maximum circumferential stress will occur when:

$$\frac{d\sigma_\theta}{d\theta} = 0 = 0.4 \sin 2\theta$$

whence:

$$\sin 2\theta = 0$$

thus:

for minimum stress: $\theta = 0°, 180°$
for maximum stress: $\theta = 90°, 270°$

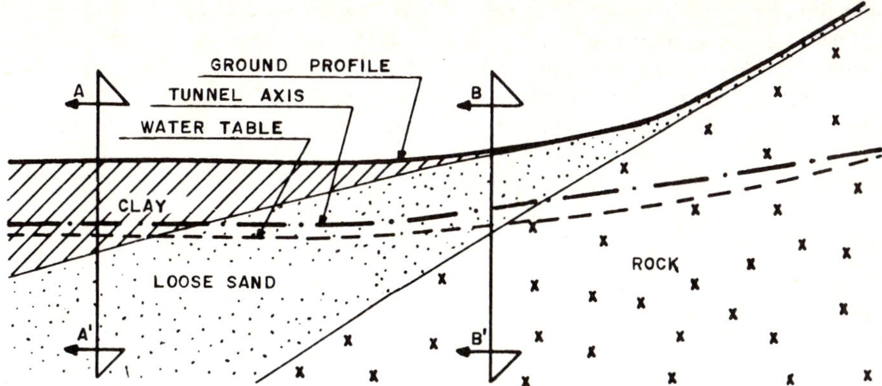

Fig. 14a-1 Geologic profile for worked examples

Thus, the maximum stress will occur half way up the side walls and will generally be:

$$\sigma_{\theta max} = \sigma_z \left[(1 + K_0) + 2(1 - K_0)\right] = \sigma_z(3 - K_0)$$

In order for tension stresses to appear in a given tunnel, it is essential for $K_0 = 3$, which implies that the horizontal stresses must be three times as large as the vertical stresses, which can only occur if very strong residual tectonic stresses exist.

In this particular case, the maximum circumferential stress is:

$$\sigma_\theta = (3 - K_0)\,\gamma z = (3 - 0.8)\ 1.5 \times 20 = 66 \text{ t/m}^2$$
$$= (3 - 0.8)\ 0.094 \times 65 = 13.4 \text{ kips/ft}^2$$

Since the compressive strength is far smaller, $q_u < \sigma_\theta$ $10 \ll 66$ t/m^2 ($2 \ll 13.4$ kips/ft^2) the soil around the tunnel will not resist this stress and a support will have to be installed. In this case elastic analysis is no longer possible because the soil will shear. Therefore plastic analysis will have to be resorted to so long as the plasticization condition is met, Eq.(14-19):

$$\sigma_\theta - \sigma_r \geqslant 2c_u$$

i.e.:

$$66 - 0 \geqslant 2 \times 5 \qquad 13.4 - 0 \geqslant 2 \times 1$$
$$66 \geqslant 10 \text{ t/m}^2 \qquad 13.4 \geqslant 2 \text{ kips/ft}^2$$

For the purpose of calculation, it is considered that $K_0 = 1$; in other words there is a state of hydrostatic stresses in the soil mass before tunnelling commences.

In order to prevent a plastic zone, the support must exert a thrust against the walls of the excavation such that:

$$\sigma_\theta - \sigma_i = 2c_u \qquad \sigma_i = \sigma_\theta - 2c_u$$

thus:

$$\sigma_i = 66 - 10 = 56 \text{ t/m}^2 \qquad \sigma_i = 13.4 - 2 = 11.4 \text{ kips/ft}^2$$

If a plastic zone of 10 m (33 ft) develops around the tunnel under impeded strain, then:

$$\frac{R}{a} = \frac{10}{4} = 2.5 \qquad \frac{R}{a} = \frac{33}{13} = 2.5$$

where R is the radius of the plastic zone.

Entering this value in the graph in Fig. 14-9, the following equation is obtained:

$$\frac{2c_u}{\sigma_\theta - \sigma_i} = 0.35 \quad ; \quad \sigma_i = \sigma_\theta - \frac{2c_u}{0.35}$$

thus:

$$\sigma_i = 66 - \frac{10}{0.35} = 66 - 28.6 = 37.4 \text{ t/m}^2$$
$$= 13.4 - \frac{2}{0.35} = 13.4 - 5.7 = 7.7 \text{ kips/ft}^2$$

which is the value of the radial pressure with which the support must be designed. The design must assure that the support can deform enough for the plastic zone to develop.

.2 Section *BB'* (in sand)

Since sand has no cohesion, it will require support in all cases. The maximum circumferential stress occurs in the side walls of the tunnel. Assuming a value $K_0 = 0.4$, it will be:

$$\sigma_\theta = 1.5 \times 20\ (3 - 0.4) = 78 \text{ t/m}^2$$
$$\sigma_\theta = 0.094 \times 65\ (3 - 0.4) = 15.9 \text{ kips/ft}^2$$

Applying the flow criterion represented by Eq.(14-26), the following expression is obtained:

$$\sigma_\theta \leqslant \sigma_r \frac{1 + \sin \varnothing}{1 - \sin \varnothing} + \frac{2c \cos \varnothing}{1 - \sin \varnothing} \leqslant 0$$

which indicates the need for the support. In order to prevent a plastic zone from developing in accordance with Eq.(14-27):

$$\sigma_\theta \leqslant \frac{\sigma_i + c \cos \varnothing}{1 - \sin \varnothing} = \frac{\sigma_i}{1 - \sin \varnothing}$$

whence:

$$\sigma_i \geqslant \sigma_\theta\ (1 - \sin \varnothing) = 78\ (1 - 0.53) = 78 \times 0.47 = 36.6 \text{ t/m}^2$$
$$= 15.9\ (1 - 0.53) = 15.9 \times 0.47$$
$$= 7.5 \text{ kips/ft}^2$$

if in this case a plastic zone 20 m (65 ft) in diameter is accepted, according to equation (14-28):

$$R = a\left[(1 - \sin \varnothing)\frac{\sigma_\theta + c \cot \varnothing}{\sigma_i + c \cot \varnothing}\right]^{\frac{1-\sin\varnothing}{2 \sin\varnothing}}$$

$$20 = 4\,(1 - \sin 32°)\frac{78 + 0}{\sigma_i + 0}\ \frac{1 - \sin 32°}{2 \sin 32°}$$

$$20 = 4\left(0.47\ \frac{78}{\sigma_i}\right)\frac{0.47}{2 \times 0.53}$$

$$20 = 4\left(\frac{36.6}{\sigma_i}\right)^{0.443}$$

$$\therefore \sigma_i = \left(\frac{36.6^{0.443}}{5}\right)\frac{1}{0.443}$$

$$= 0.98 \text{ t/m}^2$$

$$65 = 13\,(1 - \sin 32°)\frac{15.9 + 0}{\sigma_i + 0}\ \frac{1 - \sin 32°}{2 \sin 32°}$$

$$65 = 13\left(0.47\,\frac{15.9}{\sigma_i}\right)\frac{0.47}{2 \times 0.53}$$

$$65 = 13\left(\frac{7.5}{\sigma_i}\right)^{0.443}$$

$$\therefore \sigma_i = \left(\frac{7.5^{0.443}}{5}\right)\frac{1}{0.443}$$

$$= 0.2 \text{ kips/ft}^2$$

In sands it is evident that by allowing the development of a plastic zone the strength of the sand is mobilized and the pressure on the support diminishes rapidly, whereas in the case of clays the reduction in pressure on the support is far less.

14a.3 TERZAGHI's Method

.1 Section AA' (in clay)

For the case of soft cohesive soils, TERZAGHI recommends that it be considered that above the tunnel roof there is a mass of soil of height H_p, the value of which lies between:

$$1.1\,(B + H_t) < H_p < 2.1\,(B + H_t)$$

If for the case under consideration:

$$B = 2r = 2 \times 4 = 8 \text{ m } [2 \times 13 = 26 \text{ ft}]$$
$$H_t = 2r = 8 \text{ m (26 ft)}$$

Applying these values to the foregoing inequality:

$$1.1\,(8 + 8) < H_p < 2.1\,(8 + 8)$$
$$[1.1\,(26 + 26) < H_p < 2.1\,(26 + 26)]$$

that is:

$$17.6 < H_p < 33.6$$
$$[57 < H_p < 110]$$

However, in this case the maximum depth of cover above the tunnel roof is:

$$H_{p\,max} = 20 - 4 = 16 \text{ m}$$
$$= 65 - 13 = 52 \text{ ft}$$

The unit weight of the clay is 1.7 t/m³ (0.106 kips/ft³); therefore, the pressure on the roof of the support will be:

$$\sigma_y = 16 \times 17 = 27.2 \text{ t/m}^2$$
$$= 52 \times 0.106 = 5.5 \text{ kips/ft}^2$$

.2 Section BB' (in sand)

$$\gamma = 1.5 \text{ t/m}^3 = 0.094 \text{ kips/ft}^3$$

In this case TERZAGHI recommends consideration of the following expression:

$$0.62\,(B + H_t) < H_p < 1.4\,(B + H_t)$$

whence:

$$9.92 < H_p < 22.4$$
$$32.4 < H_p < 73.3$$

Taking into account the overburden thickness the following is computed

$$9.92 < H_p < 16$$
$$32.4 < H_p < 52$$

Consequently σ_y will be

$$14.88 < \sigma_y < 24 \text{ t/m}^2$$
$$3.0 < \sigma_y < 4.9 \text{ kips/ft}^2$$

The highest value obtained is that for design:

$$\sigma_y = 24 \text{ t/m}^2 = 4.9 \text{ kips/ft}^2$$

14a.4 PROTODYAKONOV's Method

.1 Section AA' (in clay)

In accordance with the method proposed by PROTODYAKONOV, the vertical pressure on the roof of the support is, Eq.(14-49):

$$\sigma_r = \frac{1}{3}\gamma_m \frac{B}{f}$$

where: γ_m = 1.7 t/m³ (0.106 kips/ft³) and B = 8 m (26 ft). The value of f, the strength factor, can be obtained from Eq.(14-50), where:

$$f = \frac{c}{q_u} = \frac{5}{10} = 0.5 \quad \left[f = \frac{c}{q_u} = \frac{1}{2} = 0.5\right]$$

Alternately the value proposed in Table 14-4 can be taken:

$$f = 0.3$$

and consequently:

$$\sigma_v = \frac{1}{3}\,(1.7)\,\frac{8}{0.3} = 15.1 \text{ t/m}^2$$

$$\sigma_v = \frac{1}{3}\,(0.106)\,\frac{26}{0.3} = 3.07 \text{ kips/ft}^2$$

.2 Section *BB*′ (in sands)

In this case $f = \tan \varnothing$ and with $\varnothing = 32°$,

$$f = 0.625.$$

This is applied to Eq.(14-49):

$$\sigma_v = \frac{1}{3}\,(1.5)\,\frac{8}{0.625} = 6.4\ \text{t/m}^2$$

$$\sigma_v = \frac{1}{3}\,(0.094)\,\frac{26}{0.625} = 1.3\ \text{kips/ft}^2$$

14a.5 Bierbäumer's Method

.1 Section *AA*′ (in clay)

The overburden depth in this method is: $h = \alpha H$ where α is a coefficient given by Table 14-5. The appropriate value is: $\alpha = 0.85$. There is a limitation that must be satisfied:

$$H < 5\left[2H_t \tan(45° - \varnothing/2)\right];\ 16 < 80$$

When this is met, the value $\alpha = 1.0$ must be taken.

Consequently:

$$h = 1 \times H = 16\ \text{m (52 ft)}$$

and:

$$\sigma_v = \gamma h = 1.7 \times 16 = 27.2\ \text{t/m}^2$$
$$= 0.106 \times 52 = 5.5\ \text{kips/ft}^2$$

which is the same as obtained by Terzaghi's method.

.2 Section *BB*′ (in sands)

For sands, Bierbäumer's method proposes:

$$\alpha = 1 - \frac{\tan \varnothing \tan^2(45° - \varnothing/2)\,H}{B + 2H_t \tan(45° - \varnothing/2)}$$

that is:

$$\alpha = 1 - \frac{\tan 32° \tan^2(45° - 16°)\,16}{8 + 16 \tan(45° - 16°)}$$

$$\alpha = 1 - \frac{0.625 \cdot 0.307 \cdot 16}{8 + 16 \cdot 0.554}$$

$$\alpha = 1 - \frac{0.192 \cdot 16}{8 + 8.86} = 1 - \frac{3.07}{16.86} = 1 - 0.182 = 0.818$$

$$\alpha = 1 - \frac{\tan 32° \tan^2(45° - 16°)\,52}{26 + 52 \tan(45° - 16°)}$$

$$\alpha = 1 - \frac{0.625 \cdot 0.307 \cdot 52}{26 + 52 \cdot 0.554}$$

$$\alpha = 1 - \frac{0.192 \cdot 52}{26 + 28.81} = 1 - \frac{9.98}{54.81} = 1 - 0.182$$
$$= 0.818$$

The overburden depth will be

$$h = \alpha H = 0.818 \times 16 = 13.1\ \text{m}$$
$$= 0.818 \times 52 = 43.0\ \text{ft}$$

and the vertical pressure will be

$$\sigma_v = 13.1 \times 1.5 = 19.7\ \text{t/m}^2$$
$$\sigma_v = 43.0 \times 0.094 = 4.0\ \text{kips/ft}^2$$

14a.6 Method Employed in the San Francisco Tunnel

.1 Section *AA*′ (in clay)

The vertical pressure is regarded as equal to the pressure exerted by the overburden:

$$\sigma_r = \gamma z = 1.7 \times 16 = 27.2\ \text{t/m}^2$$
$$= 0.106 \times 52 = 5.5\ \text{kips/ft}^2$$

This is the same as recommended by Terzaghi and Bierbäumer.

.2 Section *BB*′ (in sand)

This corresponds to a assumed in the San Francisco Subway; consequently:

$$\sigma_r = \left(11 + \frac{x}{2}\right)\gamma + (z + D)\,\gamma_w\ \left[\text{t/m}^2\right]$$

$$\sigma_r = \left(36 + \frac{x}{2}\right)\gamma + (z + D)\,\gamma_w\ \left[\text{kips/ft}^2\right]$$

where:

σ_r = uniform radial pressure on the tunnel
x = distance in excess of 11 m (36 ft) between the crown of the tunnel and the natural ground surface = 16 − 11 = 5 m (52 − 36 = 16 ft).
γ = unit weight of the material, in t/m³ (kips/ft³) = 1.5 t/m³ (0.094 kips/ft³).
z = height of the water-table above the roof of the tunnel, in meters (feet) = 0 m (0 ft).
D = diameter of the tunnel.
γ_w = unit weight of water = 1 t/m³ (0.0624 kips/ft³).

In the case under analysis it seems reasonable to eliminate the second part of the foregoing equation (water-table under tunnel floor):

$$\sigma_r = \left(11 + \frac{x}{2}\right)\gamma = \left(11 + \frac{5}{2}\right)1.5 = 13.5 \times 1.5$$
$$= 20.25\ \text{t/m}^2$$

$$\sigma_r = \left(36 + \frac{x}{2}\right)\gamma = \left(36 + \frac{16}{2}\right)0.094 = 44 \times 0.094$$
$$= 4.14\ \text{kips/ft}^2$$

14a.7 Settlement Calculation

.1 Section *AA*′ (in clay)

Assuming that compressed air is not employed and taking into account the circumferential stress obtained from the plastic analysis, the value of the factor R_s on application of Eq.(14-77) is:

$$R_s = \frac{\sigma_z - \sigma_i}{c_u} = \frac{34.8}{5} = 6.95 \geqslant 1$$

Consequently to calculate the volume of soil loss, Eq.(14-79) should be used, obtaining:

$$V_p = \frac{3c_u}{E} e^{R_s - 1} = \frac{3.5}{500} e^{5.95} = 11.85$$

From Eq.(14-82), assuming that $K = 1$, the value of i is:

$$i = ak \left(\frac{z_0}{2a}\right)^{0.8} = 4\left(\frac{20}{8}\right)^{0.8} = 8.32 \text{ m (27.3 ft)}$$

From Eq.(14-81), equalizing the volume of the depression to the volume of soil loss:

$$V_d = V_p = 2.5 i \delta_{max}$$

$$\delta_{max} = \frac{V_P}{2.5i} = \frac{11.85}{2.5 \times 8.32} = 0.57 \text{ m (1.86 ft)}$$

which is the maximum settlement which will occur in the ground surface right along the tunnel axis.

.2 Section *BB'* (in sands)

Supposing that compressed air is not employed and considering that the volume of soil loss of tunnels driven in sand above the water-table varies by 1% to 3%, it is appropriate to estimate a volume of soil loss equivalent to 2% of the excavated volume:

$$V_p = 0.02\,\pi r^2 = 0.02\pi\, 4^2 = 1.0 \text{ m}^3/\text{m}$$
$$= 0.02\pi\, 13^2 = 10.7 \text{ ft}^3/\text{ft}$$

In this case, using Fig. 14-42, the following value is obtained for i:

$$i = 2.1 \times 4 = 8.4 \text{ m (27.5 ft)}$$

similar to that obtained on application of Eq.(14-82).

From Eq.(14-81), equalizing the volume of the depression to the volume of soil loss, the following expressions are obtained:

$$V_D = V_p = 2.5 i \delta_{max}$$

$$\delta_{max} = \frac{1.0}{2.5 \times 8.4} = 0.0475 \text{ m (0.156 ft)}$$

which is the value for maximum settlement, which in this case also will take place immediately above the tunnel.

REFERENCES

1. Deere, D. U., Peck, R. B., Monsees, J.E. and Schmidt, R., "Design of Tunnel Liners and Support Systems", Final Report. Department of Civil Engineering, University of Illinois, Urbana, III, 1969.
2. Peck, R. B., "Deep Excavations and Tunneling in Soft Ground", *Proc. VII. ICSMFE*, Mexico, 1969, State-of the Art Volume.
3. Szechy, K., *The Art of Tunneling*, Akademiai Kiadó. Budapest, 1967, Chap 3.
4. Terzaghi, K., "A Liner Plate Tunnel on the Chicago Subway", *Trans. A.S.C.E.*, Vol. 108, 1943.
5. Terzaghi, K., "Shield Tunnels of the Chicago Subway", *Journal of the Boston Society of Civil Engineering*, 1942.
6. Terzaghi, K., "Load on Tunnel Supports", Chapter 4 of *Rock Tunneling with Steel Supports*, R.V. Proctor and T.L. White Eds., The Commerical Shearing and Stamping Co, 1956.
7. Juárez-Badillo, E. and Rico, A., *Mecánica de Suelos. Vol. II: Teoría y Aplicaciones de la Mecánica de Suelos*, Limusa: México, 1973, Chap. IV.
8. Terzaghi, K., *Theoretical Soil Mechanics*, John Wiley, 1956, Chap. 5.
9. Terzaghi, K., "Stress Distribution in Dry and in Saturated Sand Above the Yielding Trap-door", *Procs. I ICSMFE*, Cambridge, Mass. 1936, Vol.I.
10. Habib, P., "Características Fundamentales de las Rocas", Journal "Ingeniería", Faculty of Engineering UNAM, México, July, 1963.
11. Timoshenko, S. and Goodier, J.N., *Theory of Elasticity*, McGraw Hill, 1951.
12. Cording, E.J., "The Stability During Construction of Three Large Under Ground Openings in Rock", Tech. Report 1-813, U.S. Army Engineer Waterway Experiment Station, Vicksburg, 1968.

13. Hill, R., *The Mathematical Theory of Plasticity*, Clarendon Press: Oxford, 1950.

14. Prager, W. and Hodge, P. G., *Theory of Perfectly Plastic Solids*, John Wiley, 1961.

15. Obert, L. and Duvall, W. I., *Rock Mechanics and the Design of Structures in Rock*, John Wiley, 1967.

16. Sirieys, P. M., "Champs de contrainte autour des tunnels circulaires en elastoplasticité", *Rock Mechanics and Engineering Geology*, Vol. II No 1, 1964.

17. Hobbs, D. W., "A study of the behavior of a Broken Rock under Triaxial Compression, and its Application to Mine Roadways", *Intern. Jour. of Rock Mechanics and Mining Sciences*, Vol. V No 3, 1966.

18. Escalante, E., "Estudio comparativo de Teorías usadas para el Diseño del Revestimiento de Túneles Profundos", Thesis, Faculty of Engineering, UNAM, Mexico, 1971.

19. Hewett, B.H.M. and Johannesson, S., *Shield and Compressed Air Tunneling*, McGraw Hill, 1922.

20. Spangler, M. G., *Soil Engineering*, International Textbook Co, 1960.

21. Terzaghi, K., *Theoretical Soil Mechanics*, John Wiley, 1943, Chaps. 10 and 17.

22. Balla, A., "Rock Pressure Determined from Shearing Resistance", *Procs. International Conference on Soil Mechanics*, Budapest, 1963.

23. Coates, D.F., "Classification of Rock for Rock Mechanics", *International Journal on Rock Mechanics and Mining Sciences*, Vol. 1 No 3, 1964.

24. John, K. W., "An Approach to Rock Mechanics", *Journal A.S.C.E.*, No 88 SM 4, 1962.

25. Deere, D. U., "Technical Description of Rock Cores for Engineering Purposes", *Felsmechanik und Ingeneurgeologie*, Vol. I No 1, 1963.

26. Deere, D. U., Hendron, A. J., Patton, F. D. and Cording, E. J., "Design of Surface and Near-surface Construction in Rock", *Procs. VIII. Symposium on Rock Mechanics.*, The American Institute of Mining, Metallurgical and Petroleum Engineers: New York, 1967.

27. Coates, D. F., "Rock Mechanics Principles", Department of Mines and Technical Surveys Branch, Monography 874, Ottawa, 1965.

28. Scott, J. H. and Carroll, R. D., "Surface and Underground Geophysical Studies at the Straight Creek Tunnel Site Colorado", Highway Research Record No 185, HRB, 1967.

29. Broms, B. B. and Bennermark, H., "Stability of Clay in Vertical Openings", *Journal A.S.C.E*, Vol. 93 SM1, 1967.

30. Ward, W. H. and Thomas, H. S. H., "The Development of Earth Loading and Deformation in Tunnel Linings in London Clay", *Proc. VI. International Conference on Soil Mechanics and Foundation Engineering*, Montreal, 1965, Vol. II.

31. Eden, W. J. and Bozozuk, M., "Earth Pressure in Ottawa Out-fall Sewer Tunnel", XXI Canadian Soil Mechanics Conference, Winnipeg, 1968.

32. De Beer E. E. and Buttiens, E., "Construction de Reservoirs pour Hydrocarbures Liquefiés dans L'argile de Boom a Anvers", *Travaux*, September and October, 1966.

33. Cravioto, J. M. and Villarreal, A., "Experiencia reciente en la Construcción de Túneles y Lumbreras en la Ciudad de México", Publication of Estrella, S.A. Building Co: México, 1969.

34. Abel, J. F., "Tunnel Mechanics", *Quarterly Colorado School of Mines*, Vol. 62 No 2, Denver, 1967.

35. Burke, H. H., "Garrison Dam Tunnel Test Section Investigations", *Journal A.S.C.E*, Vol. 93 SM4, 1957.

36. Kérisel, J., "Poussée des Terres sur les Ouvrages et Tunnels", *General Report IV. ICSMFE*, London, 1957.

37. Peck, R. B., Deere, D. U., Monsees, J. E., Parker, H. W. and Schmidt, B., "Some Design Considerations in the Selection of Underground Support Systems", Final Report, Department of Civil Engineering, University of Illinois, Urbana, III, 1969.

38. Riley, W. F., "Stresses at Tunnel Intersections", *Journal A.S.C.E*, Vol. 90 EM2, 1964.

39. Schmidt, B., "Settlements and Ground Movements Associated with Tunneling in Soil", Thesis, University of Illinois, Urbana, III, 1969.

40. Moretto, O., Panel Discussion, VII. ICSMFE, Session IV, México, D. F., 1969.

41. Tinajero, J. and Vieitez, L., "Asentamientos en la Vecindad de Túneles con Escudo", *Proc. IV. Panamerican Conference on SMFE*, Sn. Juan, Puerto Rico, 1971, Vol. II.

42. Juárez-Badillo, E. and Rico, A., *Mecánica de Suelos, Vol. II: Teoría y Aplicaciones de la Mecánica de Suelos*, Limusa: México, 1973, Chap. 12.

43. Marsal, R. J. and Mazari, M., "El Subsuelo de la Ciudad de México," Publication of the Institute of Engineering, UNAM México, 1969.

44. Farjeat, E., "Excavación con Escudos", Solum, S.A. Special Report, México, 1973.

45. Paniagua, W. and Tinajero, J., "Excavaciones con Escudo para el Metro de la Ciudad de México", Publication of Solum, S.A. México, 1971.

46. Ayestarán, L., "Aire Comprimido en Túneles", Internal Publication of Solum, S.A. México, 1974.

47. Richardson, H. W. and Mayo, R. S., *Practical Tunnel Driving*, McGraw Hill, 1941.

OTHER RELATED REFERENCES

American Institute of Mining, Metallurgical, and Petroleum Engineers, *Proceedings of the 1981 Rapid Excavation and Tunneling Conference*, San Francisco, California, May, 1981.

Bhelke, S. E., Subba Rao, S. R., and More, D. M., "Severe Problems Encountered in Construction of Tail Race Tunnel of Penche Hydroelectric Project", *1st Latin-American Conference on Underground Construction*, Caracas, Venezuala, 1984.

FUKUSHIMA, K., "Excavation and Supporting of a Large Inclined Shaft in Heavily Shattered Rock", *1st Latin American Congress of Underground Construction,* Caracas, Venezuela, 1984.

MORENO TALLÓN, E., *Las clasificaciones geomecánicas de las rocas, aplicadas a las obras subterráneas*, Estudios y Proyec tos Técnicos Industriales: Madrid, 1980.

MORENO TALLÓN, E., "Aplicaciones de las medidas de Convergencia al diseño y comprobación de sostenimientos en túneles", *Proc. Symposium on the Underground Industrial Use*, Madrid, April, 1981, Vol.I.

MORENO TALLÓN, E., Utilización y comparación de métodos declasificación de rocas en excavaciones subterráneas, ibid.

MORENO TALLÓN, E., Comportamiento del bulonado en función de la calidad de la roca, ibid.

NEWMARK, N. M., "Design of Tunnel Linings", *Proc. 13th Symposium on Rock Mechanics*, Urbana, III, 1971.

SCHAEFER, M., and ROISIN, V., "Calcul D'une Section de Tunnel Circulaire dans des Terrains non Homogenes", *1st. Latin American Congress of Underground Construction*, Caracas, Venezuela, 1984.

University of Illinois, "Shotcrete Practice in Underground Construction", PB-248 765, Prepared for Department of Transportation, August, 1975.

CHAPTER 15

SPECIAL TOPICS

15.1 Introduction

This Chapter will deal with some topics of a special nature which have not been discussed in previous chapters dedicated to specific subjects. These special topics complete the information or techniques that are covered in various different chapters and cannot therefore be included in any particular one of them. In view of this, it is evident that the hitherto uninterrupted practice of devoting each chapter of this book to one specific subject must now be abandoned. The special topics treated have thus no particular relation to one another.

15.2 Frost Action in Soils

.1 General Comments

It is a known fact that at freezing point water solidifies and increases in volume. Both the temperature at which it freezes and the coefficient of expansion depend on the pressure that prevails. Under atmospheric pressure, water freezes at 0°C (32°F), whereas under a pressure of 600 atmospheres (8520 lb/in^2) water freezes at −5°C (23°F) and under 1,100 atmospheres (15620 lb/in^2) at −10°C (14°F). The respective coefficients of expansion are 0.09 for 1 atmosphere (14.2 lb/in^2), 0.102 for 600 (8520 lb/in^2) and 0.112 for 1,100 atmospheres (15620 lb/in^2).

When water freezes in masses of clean gravel or sand, it increases in volume, but not necessarily by 10% of the initial volume of the voids, as suggested by the coefficients of expansion mentioned, since water may drain off during freezing. If water is homogeneously distributed throughout the soil, as is usually the case, freezing affects the entire mass, and there will be no formation of layers or isolated lenses of ice. These will, however, form when previously existing masses of free water freeze in-situ.

In many relatively fine soils, such as saturated silts or saturated silty sands, the effect of freezing depends largely on the rate at which the temperature drops [1]. Rapid cooling causes water to freeze on the spot, but if the drop in temperature is gradual, most of the water collects forming small layers of ice parallel to the surface that is exposed to the cold. The result is alternate layers of frozen soil and thin layers of ice.

Under natural conditions, in silty soils that are exposed to a pronounced drop in temperature, layers of ice several centimeters thick will form. The formation of layers of clean ice indicates migration of water from the voids towards the center of freezing. This water may come from the voids in the soil that is freezing or else be absorbed from a water-bearing layer, located underneath the frozen zone, Fig. 15-1 [1] shows the different possibilities for a specimen of fine soil.

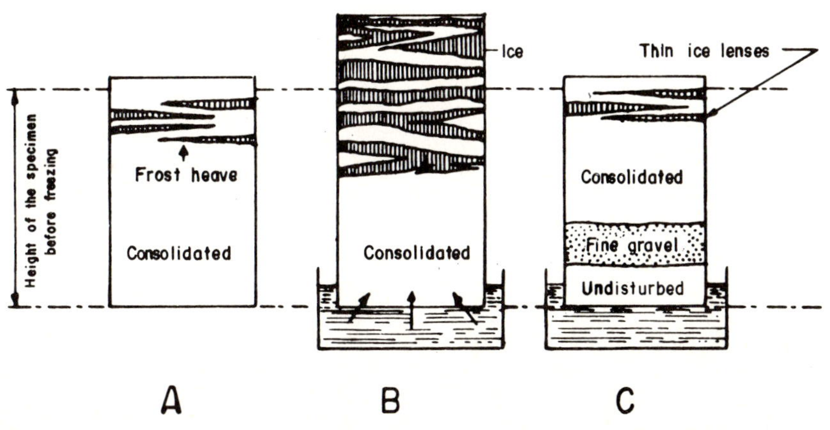

Fig. 15-1 Cases of ice formation in fine soils, after TERZAGHI [1]

Specimen A rests on a solid, impervious base, whereas the lower part of B and C is submerged in water. In the three cases, the temperature of the upper faces is maintained below the freezing point of water. In A, the water from which the thin layers of ice form comes from the lower part of the specimen, whereas in B the water comes from the underlying source passing through the lower unfrozen part. TERZAGHI [2] referred to specimen A as a closed system because the total water content of the soil mass does not vary. By contrast, B is an open system. C, although it might appear to be an open system, is in fact closed, owing to the effect of the presence of the layer of fine gravel, which breaks the capillary rise from the water source into the freezing zone.

In system A, the water from which the ice lenses form rises from the lower part of the specimen, which induces consolidation in this zone similar to the one that is produced when water rises by capillarity towards an evaporation surface. The process continues until the water content of the lower part of the specimen is reduced to the shrinkage limit, so long as the porewater temperature is low enough. The limit of the total volume increment in a closed system will therefore correspond to the volume increment undergone by the water due to freezing. Typically this fluctuates between 3% and 5% of the total volume of the soil, depending on the compressibility of the unfrozen zone.

In open systems (system B), ice lenses initially start to form on account of water rising from the lower levels of the soil; consequently the lower part of the soil consolidates from the start. However, as the process progresses, the amount of water that is drawn from the source of free water gradually increases until the flows originating in the lower part of the sample and in the free source become equal, and from that moment on the water content in the lower part of the sample remains constant.

Observation in regions where very low temperatures prevail for long periods of time shows that when a natural soil works as an open system ice lenses several meters thick may form in it. An open system can be converted into a closed system by simply introducing a layer of clean gravel similar to the one shown in specimen C in Fig. 15-1 between the freezing surface and the water-table. As water can no longer rise by capillarity through the layer of coarse soil, from this layer upwards the soil behaves like a closed system.

It has been found that ice lenses do not develop unless, in addition to the necessary climatic conditions, there is a certain minimum percentage of fine particles in the soil. Also important is the degree of uniformity of the particles and the type of stratification. The quantitative effect of each of these factors has not yet been fully established.

Generally speaking, a soil is said to be frost-susceptible when appreciable lenses of pure ice can develop in it, increasing the average water content in the frozen zone.

.2 Frost Effects

When water freezes in the voids of a soil under a moderate pressure, it acts like a wedge which separates the solid particles and increases the volume of the voids. If the soil is non-frost-susceptible, like gravels and sands, or if it behaves like a closed system, this volume increment has an upper limit of about 10% of the initial volume of the voids, which is why in a formation with a horizontal surface, the elevation of that surface cannot be greater than:

$$h = 0.1nH \tag{15-1}$$

where n is the porosity of the soil and H the thickness in which frost effects are felt. However, in an open system consisting of frost-susceptible soil, frost heave may be far greater than the limit indicated in Fig. 15-1. The pressure exerted by the frozen soil on expanding is hard to measure with any degree of precision, but it is appreciable. Theoretically it may reach extraordinary values, greatly exceeding the usual superimposed structural loads. Thus, structures that are erected on the soil will be lifted with it.

During the spring thaw, the frozen zone melts, which takes several weeks and is accompanied by settlement of the subsoil. The mode of settlement depends on whether or not ice lenses have formed during the freezing period. In non-frost-susceptible soils, the maximum possible settlement will also be given by Eq.(15-1). In frost-susceptible soils, in which pure ice crystals have formed, settlement due to thawing includes not only the volume of the ice, but also the structural collapse of the cavities where the crystals were lodged, which may prove to be an important effect. In addition the soil between the lenses becomes very soft and loads punch through them.

The large area loading of a pavement system will squeeze free water up into the base course and down into the consolidated unfrozen zone, causing irregular settlements that are almost as great as the total thickness of the frozen lenses. The limit of settlement can be the entire thickness of the frozen zone if there are concentrated loads at the surface. They punch through the thawed ice and the softened soil between the lenses. Structures suffering these settlements usually meet with serious difficulties, which are heightened by the large differential settlements. These effects usually cause serious damage to roads and runways.

In slopes and hillsides, frost action causes soil particles to move viscously towards the toe of the slope. If the material is non-frost-susceptible, during freezing the boundary of the slope displaces perpendicularly to its initial position, and during thawing each of its points descends vertically, with a resulting net displacement towards the toe of the slope. If soils are susceptible, especially if they are silty, most particle displacement occurs during the subsequent melting of the ice lenses that developed during the freezing period, and is in the form of viscous flow parallel to the slope surface. This phenomenon is known as solifluction.

Freezing of the free water in the soil behind a retaining wall causes an increase in pressure against it, which is far greater in frost-susceptible soils. Repetitions of this pressure increment may eventually cause collapse of the wall. In concrete structures shear failure may occur in the section between the wall and the foundation slab.

The thickness of the ice lenses that form in frost-susceptible soils depends on many factors, including the degree of susceptibility of the soil, favorable drainage conditions (both with regard to the absorption and release of water), the intensity of the cold and especially its duration.

Solutions that have been adopted to avoid frost damage in the upper layers of soil can be grouped into three different categories:

1) Replacement of frost-susceptible soils by other non-susceptible ones, to below the depth of penetration of the 0°C freezing temperature.

2) Adequate drainage, to lower the water-table to a greater depth than the maximum height of capillary rise into the soils freezing zones.
3) Conversion of a pre-existing open system into a closed one. This is achieved by placing at the approximate depth of frost a layer of coarse, non-capillary material. Subsequently the excavation is filled with the original material.

These solutions have been applied principally in roads and runways.

In addition to the volume changes mentioned, thawing lowers the shear strength of soils within the previously frozen zone and consequently reduces their load-carrying capacity. Much of the water in the frozen lenses is forced by the weight of the soil and that of the pavement above into the soil between the ice lenses and into the consolidated soil below, raising the soils water content and increasing its pore pressure. These soils become soft, and in some cases semi-liquid. Under concentrated wheel loads these soft soils are sometimes extruded through the pavement in small muddy geysers. The pavement ruptures in the form of irregular holes are termed *potholes*. The formation of the softened soil and pavement damage is termed *spring break-up*.

.3 Classification of Soils for Frost Susceptibility

According to A. Casagrande [3] a soil is usually non-frost-susceptible if less than 3% of its particles are smaller than 0.02 mm (0.0008 in). A material will start to show signs of susceptibility between 3% and 10% of this content, depending on the grain size distribution.

Frost-susceptible soils can be classified as shown in Table 15-1, which is widely used by specialists throughout the world and is based on classic studies by Terzaghi and Casagrande. In this table, the soils are grouped by order of increasing susceptibility.

The most dangerous soils from the point of view of frost action are those combining fine grain-sizes with moderately high permeability. Clays interstratified with numerous thin layers of sand are therefore extremely frost-susceptible, as are silts and silty sands followed by low plasticity clays.

In general terms, F_4 soils are not recommended when intense frost action is feared, and are particularly inadvisable in roads and runways.

Table 15-1
Classification of soils according to frost susceptibility

Group	Type of soil
F_1	Gravels with 3 to 20% of the particles smaller than 0.02 mm (0.0008 in)
F_2	Sands with 3 to 15% of the particles smaller than 0.02 mm (0.0008 in)
F_3–*a*	Gravels with more than 20% of the particles smaller than 0.02 mm (0.0008 in)
F_3–*b*	Sands (without silt sizes) with more than 15% of the particles smaller than 0.02 mm (0.0008 in)
F_3–*c*	Clays (except finely stratified) with $I_p > 12$
F_4–*a*	All inorganic silts, including sandy silts
F_4–*b*	Fine silty clays with more than 15% of the particles smaller than 0.02 mm (0.0008 in)
F_4–*c*	Clays with $I_p < 12$
F_4–*d*	Finely stratified clays

.4 Freezing Index

The depth of the frozen zone of a soil depends on how much and how long the air temperature is below 0°C (32°F). In order to take both factors into account in the depth of frost penetration the concept of the *Freezing Index* (I_c) has been established.

For the purposes of the following considerations, by a number of degree/days (°C-days) will be understood the difference between the mean daily temperature at a given location and the freezing point of water. If temperature is expressed in degrees centigrade, the freezing point of water is 0°C and the number of degree/days coincides with that measured by the mean daily temperature at the location. 10°C/day, for example, may correspond to a temperature of 10°C lasting one day or a temperature of 1°C lasting 10 days.

If for a given winter an accumulative graph is plotted for degree-days versus time in days, a curve like that in Fig. 15-2 is obtained. In this graph, the Freezing Index can be calculated as the number of degree-days between the maximum and minimum points on the curve. Therefore the Freezing Index is related to a specific winter. The Normal Freezing Index is the mean value of the freezing indices for a specific location considered over a long period of time, usually 10 years or more.

Frost penetration is related to the Freezing Index. Fig. 15-3 [4] shows one such relationship obtained for U.S.A. conditions by the U.S. Corps of Engineers. Clearly relationships like this one can be used only as an approximate guide in other regions subject to different climatic cycles, because the depth depends also on the location of the water-table, the height of capillary rise, and the thermal conductivity of the soil-pavement system.

The principal application of this concept is in roads and runways design, where the degree-day curves are plotted to find the minimum thickness of non-frost-susceptible materials which are required to protect the soil located beneath the subgrade from frost action. It is a common practice to give these protective thicknesses in terms of the Normal Freezing Index for the regions under consideration, the thickest layers corresponding of course to the highest indices.

.5 Frost Effects in Pavements

Generally speaking frost conditions are not severe in Mexico and other Spanish-speaking countries which do not suffer the climatic extremes that must exist for frost penetration to reach

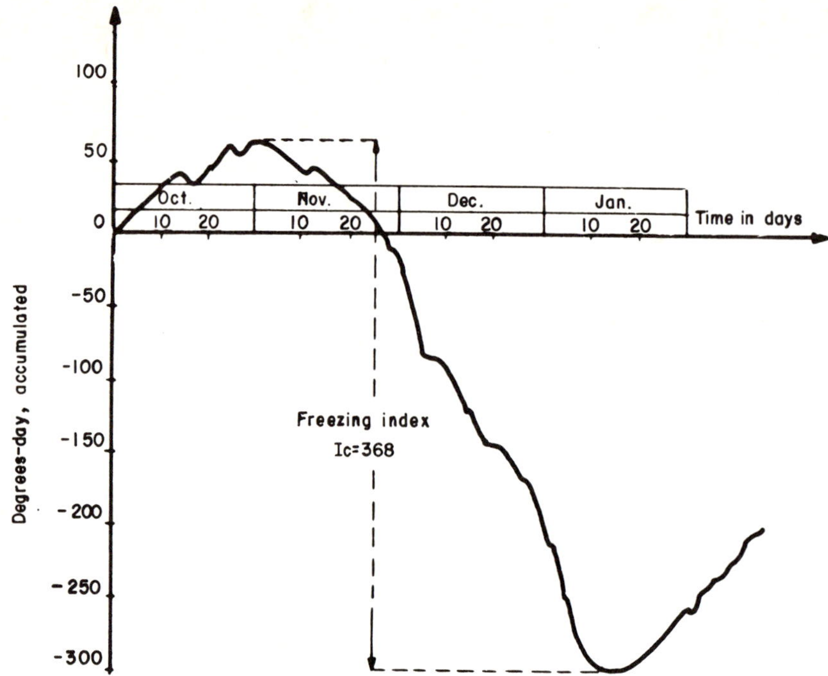

Fig. 15-2 Determination of the freezing index [1]

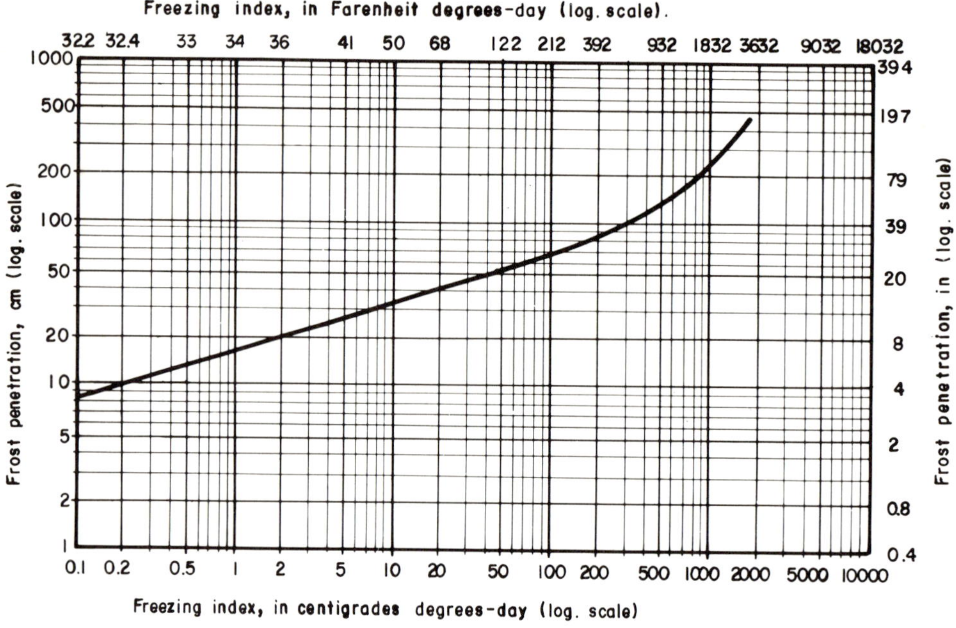

Fig. 15-3 Frost penetration as a function of the freezing index in a well-drained, non-susceptible granular material, on the basis of information from [4,5]

appreciable depths, last a long time and produce adverse effects during the thaw. This is a practical rule for problems involving shallow foundations and other similar ones in Mexico. Only in the central northern zone of Mexico does frost affect layers of soil several centimeters thick, but this is always of little significance in problems of this type.

On the other hand it is worthwhile for engineers of other regions to make some comments on the influence of frost phenomena on the strength and behavior of pavements. These structures are highly susceptible to the effects of severe climatic action involving low temperatures. Thaw phenomena, in particular, cause a drastic reduction in the strength of pavement systems affected by freezing. Figure 15-4 [4] is just one of numerous testimonies to this statement.

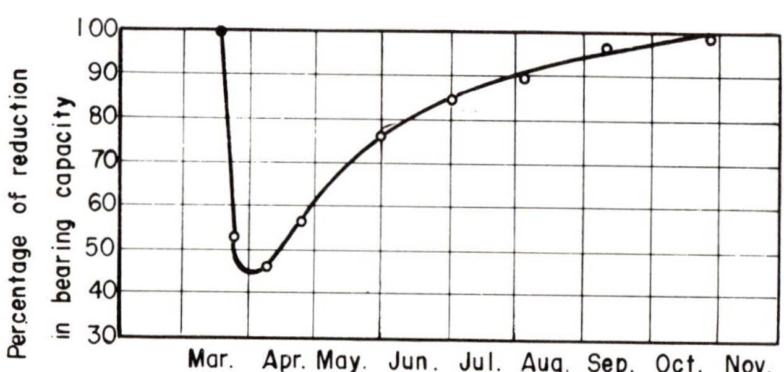

Fig. 15-4 Seasonal variation of the load-carrying capacity of a pavement exposed to different freezing conditions [4]

The figure shows in detail the reduction in the bearing capacity of a pavement subgrade (measured, for example, with plate-loading tests) when subjected to severe frost action, expressed as a percentage of the initial value. The dramatic effects of the spring thaw are illustrated and how they continue long after the actual thaw process has taken place. In the case in the figure, the bearing capacity of the subgrade dropped to 45% of the initial value. In rigid pavements, pumping in soils that are susceptible to this phenomenon and even cracking of slabs are usually very closely related to the spring thaw.

Design Recommendations for Runways: The criteria for the design of runway pavements in zones where appreciable frost penetration is anticipated usually include two aspects. First, it is customary to use adequate thicknesses of non-susceptible materials, so that frost penetration in susceptible underlying strata will be of little importance, and second, when the foregoing condition is not met or only partially, pavements are designed taking into consideration the reduced bearing capacity of the subgrade and the upper layers of fill, due to thaw conditions.

In all F_4 soils (Table 15-1), it is recommended to use layers of non-frost-susceptible material that will restrict frost penetration in the subgrade and upper part of the fill. Designs are usually based on the coldest year of a period of 10, or on the Average Freezing Index for the three coldest years out of 30. Penetration depth can be determined accurately enough on the basis of the data in Fig. 15-3. With this thickness of base course, obviously freezing would not penetrate the subgrade. The U.S. Army Corps of Engineers [5], however, considers that it is too conservative to use this thickness of base course made up of a costly granular material. It is an insulator which reduces the frost penetration estimated from Fig. 15-3. A certain penetration is tolerable in the susceptible materials that may exist beneath the granular base course. Figure 15-5 [5] permits determination of the admissible depth of frost pene-

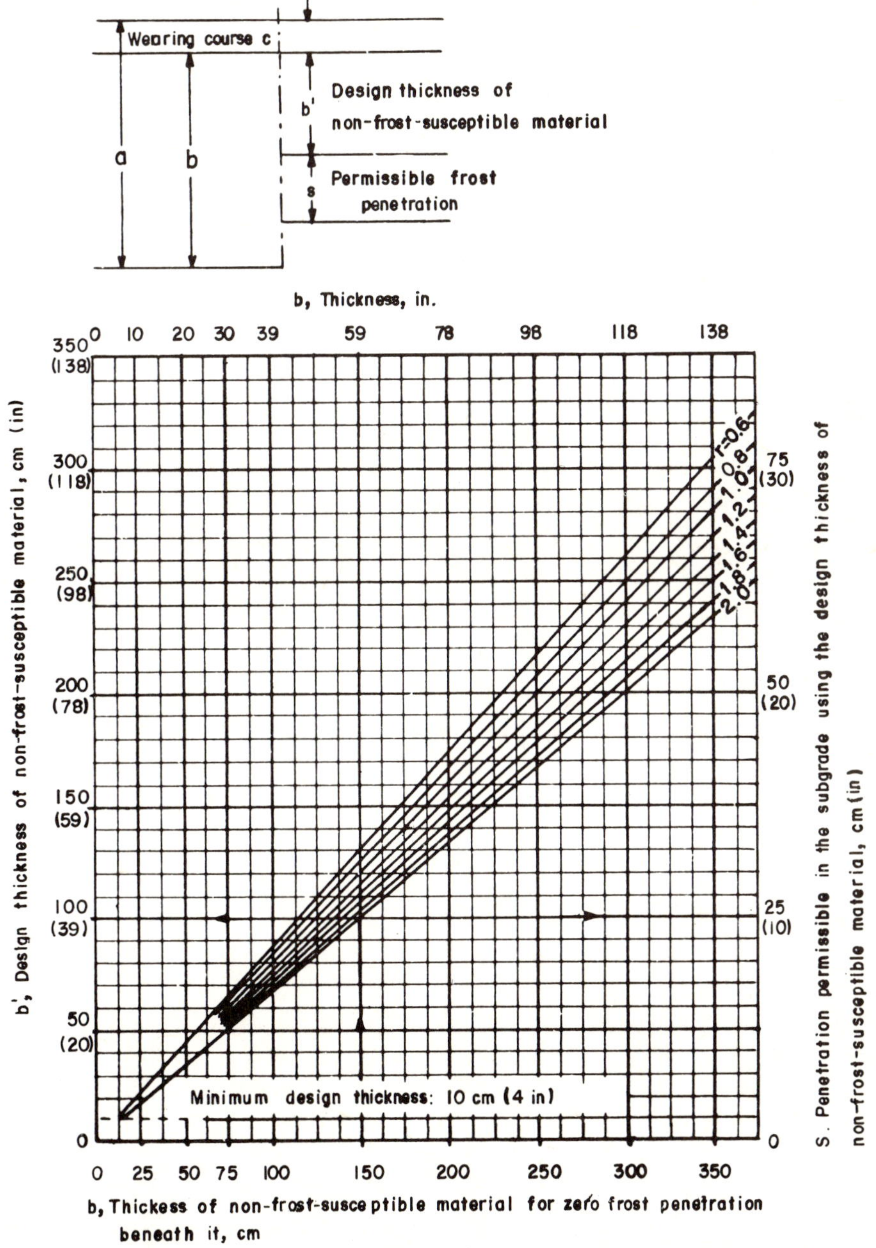

Fig. 15-5 Graph by the U.S. Corps of Engineers for the determination of permissible frost penetration in runways under a given thickness of non-frost-susceptible materials and the corrected thickness of non-frost-susceptible base course [5]

tration into non-frost-susceptible materials (which can be the surface plus base course plus the sub-base, depending on the materials of which this sub-base is composed). The figure functions as follows. If c is the thickness of the wearing course and a is the total depth of frost penetration the thickness of a non-susceptible base course b should correspond to a–c for no frost action. This value for b should be entered on the abscissa in the graph of Fig. 15-5, in accordance with the corresponding r value, r being the relationship between the water content of the subgrade and the water content of the granular base course. Using this graph, it will be possible to obtain a corrected thickness of granular base course, which is adequate according to the U.S. Corps of Engineers, and a permissible depth of frost penetration in the subgrade material.

What is meant by a non-frost-susceptible granular base course in the foregoing description should be clearly comprehended. The expression includes the entire thickness of non-frost-susceptible granular materials that are laid, which may include only the base course proper (which should always be non-frost-susceptible) or the base course plus the sub-base and upper part of the subgrade if the quality of these layers is reliable too. In this case all the deeper layers can be frost susceptible.

To prevent contamination of the granular base course by the sub-grade during thawing, the U.S. Army Corps of Engineers recommends that the lower 10 cm (4 in) of that base course be designed as a filter. If the combined thickness of wearing course and base course obtained by the foregoing analyses proves to be over 1.8 m (6 ft), the Corps of Engineers recommends a special study to determine whether a rigid pavement would not be more economical than a flexible one.

The Corps of Engineers also permits a design which considers a reduced structural capacity in the subgrade during thawing in accordance with the second of the two general design criteria mentioned above, but only if the subgrade materials belong to categories F_1, F_2 or F_3 (Table 15-1). Figure 15-6 [5] gives the combined pavement thicknesses (wearing course and base course) that are recommended for different total aircraft loads with double tandem landing gear of the type used by commercial aircraft (for example, DC-8, Boeing 707) weighing up to 80 t. When designing pavements, it is recommendable to compare these two criteria and choose the one that results in the greater thickness of non-frost-susceptible material.

It is emphasized that the criteria and graphs given here refer to countries where frost action is far more severe than in Mexico. This is evident from the thicknesses of non-susceptible material that are handled. It should also be noted that the basis of the foregoing recommendations is empirical, and consequently their blind use would not be justified in any country other than the U.S.A., where they originated.

Design Recommendations for Roads: Figure 15-7 [6] gives some specifications after the U.S. Army Corps of Engineers that are a typical example of recommendations that appear in literature

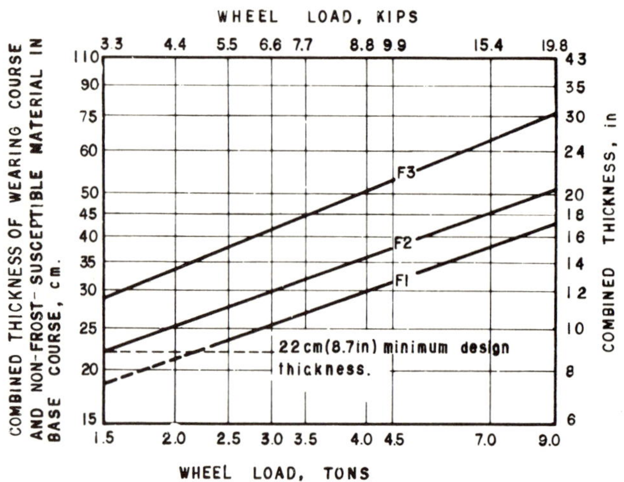

Fig. 15-7 Design curves for frost penetration in roads, according to the U.S. Corps of Engineers [6]

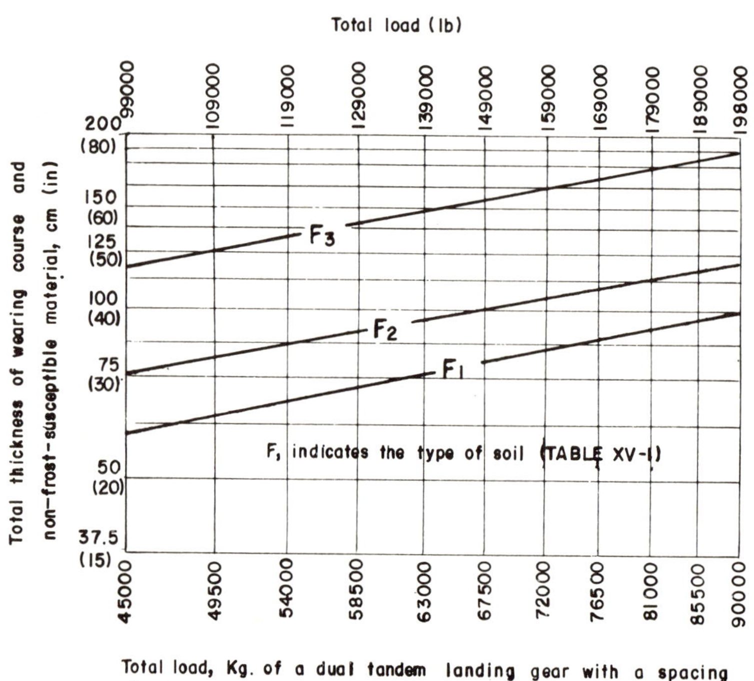

Fig. 15-6 Combined thickness of wearing course and layer or layers of non-frost-susceptible materials recommended for runways, beneath various total aircraft loads [5]

for taking into account the influence of frost on the thickness of flexible road pavements.

The figure gives the combined thickness of wearing course and layer or layers of non-frost-susceptible material (depending on whether or not the sub-base is made of susceptible material) which should be employed, as a function of the wheel load of the design vehicle and the nature of the subgrade materials, according to the classification given in Table 15-1. Note that F_4 soils are not recommended for use in the subgrade layer when there is a likelihood of frost. The Corps of Engineers recommends a fully protective design similar to the one described for runways (Figs. 15-3 and 15-5) if the risk of frost is severe.

15.3 Undermining by Scour

Scour is a natural phenomenon which affects principally river and stream beds, but which is not restricted to these, since material may be removed from the bottom or banks of any stream or mass of water in motion, as may be the case of a coastal current, a swamp or even a canal. Scour is of interest to the highway engineer on account of the frequent need to span streams, chiefly rivers and swamps with bridges, the supports of which are normally located in mid-stream. From this point of view, the highway engineer is chiefly concerned with the three types of scour that affect his bridges most:

1) General scour, which is the suspension of materials in the river-bed as a result of flooding. This type of scour is independent of the presence of any bridge and takes place before the crossing is constructed. Any bridge support must on principle be founded below the depth of general scour.
2) Local scour, which occurs in the neighbourhood of bridge piers located underwater is a consequence of the diversion of flow paths and turbulence caused by the pier itself. If this phenomenon progresses sufficiently, reaching depths below the pier footing, total collapse of the pier will occur (Plates 15-1, 15-2).
3) Bank scour, due to narrowing of the channel is caused by invasion by the access embankments to the crossing structure. This narrowing causes a reduction in the hydraulic area of the channel, with a corresponding increase in the velocity and erosive power of the water.

In streams any of the above three types of erosion, and other possible varieties such as scour on river banks in bends are a matter of equilibrium between the contribution of solid material brought by the water in suspension to a certain hydraulic section and the capacity of the river to remove material from that section. When a river is in flood, there is an increase in the velocity of the water and consequently its erosive capacity. The possibility of material being eroded from the riverbed depends on the relationship between the flow velocity and that required to erode and carry away existing material. Flow velocity depends on the hydraulic characteristics of the river and the intensity of the flood, whereas the velocity required to erode the material away, depends on the characteristics of the material in the riverbed, the depth of the water and the amount of solids already in suspension.

Plate 15-1 Bridge failure due to scour

Plate 15-2 Effect of scour on the shallow foundation of a bridge pier

Until recently the characteristics of riverbed material considered were the average diameter of the particles of frictional soils, and the dry unit weight of cohesive soils, which are not the only relevant properties for representing the resistance of a soil to the erosive action of water. This is one of the reasons why attempts to establish a mathematical model of the scour phenomenon have proved so unsatisfactory and why the predictions based on these models are so often in serious disagreement with real observations.

Reference [7] gives the theories and formulas that have been used. It includes both the results that have been reached by U.S. engineers as well as by Soviet engineers, who have dedicated considerable systematic effort to these problems. In this reference the reader will find some specific formulas for computing the three types of scour mentioned above, including discussion of these formulas and some attempts to compare their respective effectiveness and their most appropriate fields of application. Reference [8] includes some of the expressions used in U.S. practice, together with a general discussion of the problem of scour.

The models and formulas included in [7] will not be repeated here, but some comments on the philosophy of anti-scour design are appropriate, for this phenomenon is a frequent cause of bridge failure.

Attention should be given to probability of survival of a structure and the reinforcing that is justified considering increasingly scour conditions. In the past it was often heard that a bridge should be designed on the basis of the flood that occurs once every 50 years. Figure 15-8 [8] shows how arbitrary this number is. The figure represents an analysis in which the risk of failure of a bridge is quantified in terms of the cost of its probable loss compared with the amount that must be spent to protect against different probabilities of failure. Curve *1* rep-

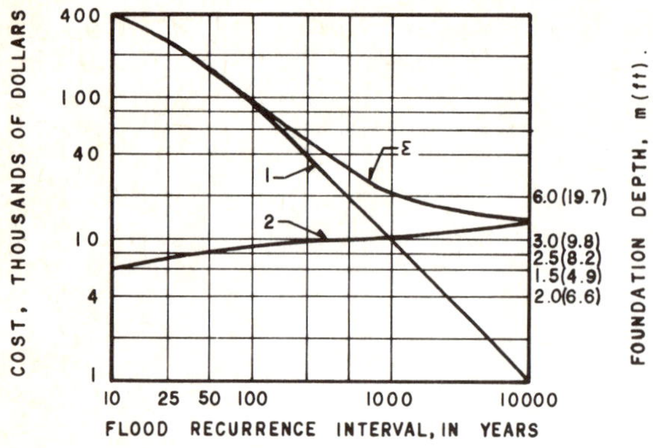

Fig. 15-8 Analysis of scour hazard for a bridge pier [8]

resents the product of the cost of the bridge superstructure multiplied by the annual probability of failure for increasing flood danger (the probability is smaller the greater the interval of recurrence considered). Curve *2* represents the increase in the cost of the bridge substructure for growing scour depths, given by calculations considering increasing probability of flood scour. Curve Σ gives the sum of the ordinates of the previous two curves for increasing flood danger.

It should be noted that the total cost of the bridge affected by the probability of total failure continues to decrease for floods with an interval of recurrence of 10,000 years, a period which seems exaggerated for the design of a bridge with a useful life of 50 to 100 years. This means that the *sensible* depth for a bridge foundation cannot be limited simply by an analysis of the probability of risk, similar to the type often conducted for other engineering projects. The acceptance of a design flood less than the maximum theoretical presumes a probability of failure. Therefore a design depth for the foundations must be a matter for careful decision. A bridge is seldom designed in a routine manner (except for very special cases) for catastrophic conditions like the ones implied by occurrences of 1 in 1000 years or less frequently.

Currently most engineers design their pier footings below the depth of general scour. By calculating the depth of local scour, a limit is obtained below which piers should be placed, using common sense to establish the depth of pier footings. (Deep foundations will be used for levels that cannot be reached economically with shallow foundations). These limits vary according to the different bearing elevations. It is advisable to calculate several different levels, so as to be able to perceive the influence of elevation on scour depth. All these calculations, together with an analysis of the conditions of the soils profile in the riverbed, will assist the engineer in a wise choice of depth. A careful depth usually corresponds to intervals of flood recurrence for design purposes ranging from 100 years for small structures and 1000 years or more for large ones.

In addition to the information given in [7] reference should also be made to scour aggravated by the access embankments to bridges, or embankments of great length across swamps. Methods of protecting against undermining the toe of access embankments and bridge piers on underwater footings are discussed in [9-13]. These are based on model studies for the protection of embankments and piers. Figure 15-9 summarizes the typical recommendations for embankment protection. Part *a* of the figure shows a typical scour profile at the toe of an unprotected embankment. Part *b* shows the influence and interaction of the embankment and the neighbouring piers, also without protection. Both drawings are based on model studies and help visualize the scour phenomenon. Part *c* shows a means of protecting both the access embankment and the adjacent pier with riprap (coarse rock fill). References [7] and [14] give further information about these types of protection, which are widely used. Part *d* shows a protection with sheet-piling, which is a costly method and must be designed on the basis of a careful estimation of scour depths. Part *e* refers to

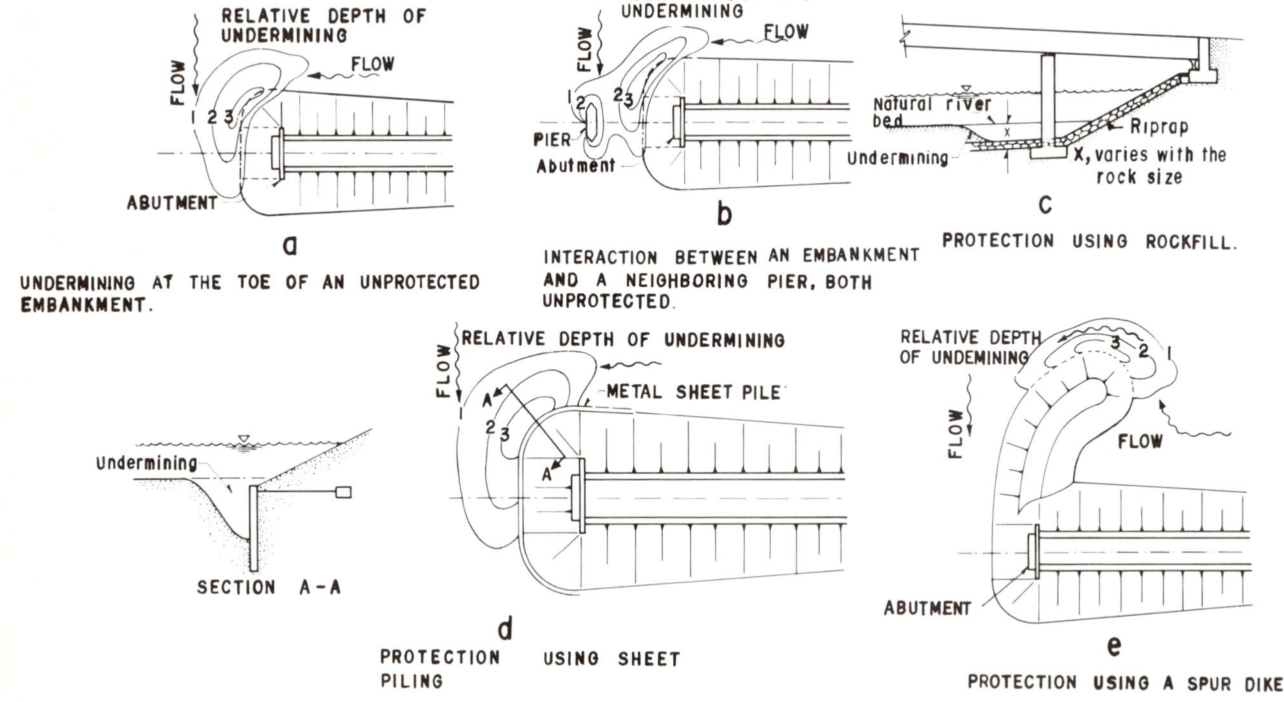

Fig. 15-9 Different types of protection against scour in approach embankments and bridge piers [8]

protection with a spur dike, which is a very reliable method according to Mexican experience (Plates 15-3, 15-4).

Protection of river banks against scour and undermining is essential when roads pass very close to erosion zones. Riprap is probably the most frequently used means of protection, but spur dikes are also very widespread.

Plate 15-3 Very thorough protection against general scour

Plate 15-4 Scour. Remedial work on a pier of the Cocoraqui bridge on the Mexico City — Nogales road.

The greatest lateral displacements usually taken place on bends, due to the effect of centrifugal forces causing a super-elevation of the water on the outside, which in turn produces a current on the bottom from the outside to the inside. When this current is added to the normal river current, a helicoidal current results which carries the riverbed materials downstream and towards the inside bank. The result is erosion from the outside of the bend towards the inside downstream, which produces deeper channels and sedimentation on the inside bank. The higher velocity on the outside of the bend plus the rotation of the mass of water in the bend, leads to a greater erosive power on the outside of the bend. This usually forms steeper banks on the outside of the bend. The ultimate result of erosion here is progressive collapse and small bank-slides.

Bank protection with rockfill inhibits the progress of the erosive cycle, because a material that cannot be eroded away is placed at the edge and on the bottom of the channel. A filter is usually necessary between the riprap and the natural ground in the bank, to prevent the finer soil from slowly eroding through the riprap. The advantage of riprap or rockfill protection, along with that of any other bank protection (walls, sheetpiling, etc.) is that there is little reduction in the hydraulic area of the channel, and the natural configuration of the bank is retained. The main disadvantages are the need for very careful maintenance and the cost, especially if there are no convenient stone quarries in the neighbourhood.

In areas where there is no rock coarse enough for riprap, wire baskets filled with gravel or coarse aggregate, termed *gabions* can be stacked to form bank protection walls (Plate 15-5) (Gabions can also be stacked up like large stones to make gravity retaining walls). Where there is no stone, cloth bags filled with about 9 parts sand to 1 part portland cement can be used as riprap. The proportion of cement is increased if there is severe erosion or if the water chemistry is unfavorable to cement.

Plate 15-5 Bank protection with gabions

Mattresses or blankets of various types are used for bank erosion protection. The oldest form, employed by the ancient Chinese consists of woven sticks of wood or bamboo, as large as 10 m square. These are lashed together as needed. Concrete slabs, linked by chains or cables have been similarly used. A variant is a double walled blanket of porous textile which is *inflated* by pumping it full of sand cement mortar. These have the advantage of easy installation in almost any size, and when filled they conform to the river bank shape.

Spur dikes are structures that lean against or are embedded in the riverbank and extend into the channel. Their function is to divert high velocity flow so that it can no longer reach the bank. Within these zones of relatively still water, deposits may even accumulate. Spur dikes reduce the hydraulic area of the channel and are expensive. Although their design is largely intuitive they often give excellent results. They are easy to repair and the erosion or loss of one dike has little effect on the rest of the system. The reduction in hydraulic area, which increases the average water velocity is sometimes unfavorable, but it can be an advantage if the river is navigable, by producing greater depths so long as the spur dikes do not affect the shipping channel.

The following factors affect spur dike design [15]:

— River Geometry: The radii of the bends and the length of the tangents compared with the stable width of the river.

— Length of dikes
— Spacing between dikes
— Slope of the crest
— Angle of orientation in relation to the river bank
— Permeability of the spur dike (controlled by the materials of which it is made).
— Scour on bends and local scour at the end of the spur dike

When locating a spur dike, the first step is to draw a line that is parallel to the axis of the river, which represents the extent of future spur dikes. This line virtually defines a new bank, and consequently a new river width, which will determine the stability of the new channel and its topographical characteristics.

When a riverbed is made up of sands and silts and it is wished to protect a uniform bend, all spur dikes should be the same length and angle of orientation, and the spacing between them should be constant. In a set of spur dikes, the first three upstream ones should vary in length, the first one being no longer than the depth of the channel, the length of the next two increasing progressively until the fourth dike is the same length as all the remaining ones.

The working length of spur dikes is usually somewhere between the average depth of the river and one quarter of the average width of the riverbed. Spur dikes are not usually anchored to the river bank.

To calculate the spacing between dikes, the angle α that is formed by the spur dike with the downstream bank is considered, along with the theoretical widening of the stream (β) when passing the end of the dike. For straight river banks, the spacing is usually between 5 and 6 times the working length when the dikes are perpendicular to the bank. Spacing criteria for spur dikes with α angles other than 90°, also for straight river banks, are given in Fig. 15-10 [15].

For curved river banks, the spacing of the spur dikes is between 2.5 and 4 times their working length, (L_T) if the curve is regular. If the curve is irregular, the length and spacing should be established on the basis that the line of flow deflected by a spur dike meets the river bank at the origin of the following dike, but not before. This same criterion can be employed to select the spacing between 2.5 and 4 for regular curves. When spur dikes are embedded in the riverbank, slightly greater spacings may be adopted, but never exceeding 8 L_T and 6 L_T spacings for straight and curved river banks respectively.

Spur dikes should be built with their tops sloping down towards the center of the river, so that the end that is in the water should be at a maximum height of 50 cm (20 in) above the present channel bed. In this way local scour around the end of the dike is greatly reduced and material is saved. It has also been seen that deposition tendencies are encouraged in the space between spur dikes.

Spur dikes can point downstream, upstream or be perpendicular to the bank. The first is the most common orientation. In straight sections, spur dikes usually form an angle of about 70° with the bank (generally pointing in the downstream direction). In curved sections, the angles are reduced and may be as small as 30° on very irregular bends. Generally speaking, the smaller the value of the angle, the smaller the spacing between the dikes. The experience depicted in [15] indicates that in straight river banks an angle of orientation of about 70° is preferable to a perpendicular orientation. On bends, angles less than 40° are seldom economical and direct bank protection should be considered.

Spur dikes can be built with a wide variety of materials, ranging from mattresses using poles, tree trunks and rock fragments, to prefabricated concrete slabs and gabions. Sheetpiles and rockfills are probably the most common solutions. In sheet-piling methods, compartments are often built which are filled with stone or gravel. Wooden piles are driven into the ground with planks or small poles between them. Large plastic bags filled with sand or sand-cement are useful where scour is pronounced and wood or rock is not available. If a spur dike is to remain permanently within the main channel, it should be impervious. However if its function is only to reduce the velocity of flow in a certain zone where it is wished to encourage deposition, it should be pervious.

Local undermining at the end of spur dikes may be appreciable. In order to prevent this it is advisable to cover the part of the riverbed on which the dike is to rest with a layer of stone at least 30 cm (12 in) thick. As already mentioned, it has been observed that local scour is greatly reduced if spur dikes are built with an appreciable longitudinal slope.

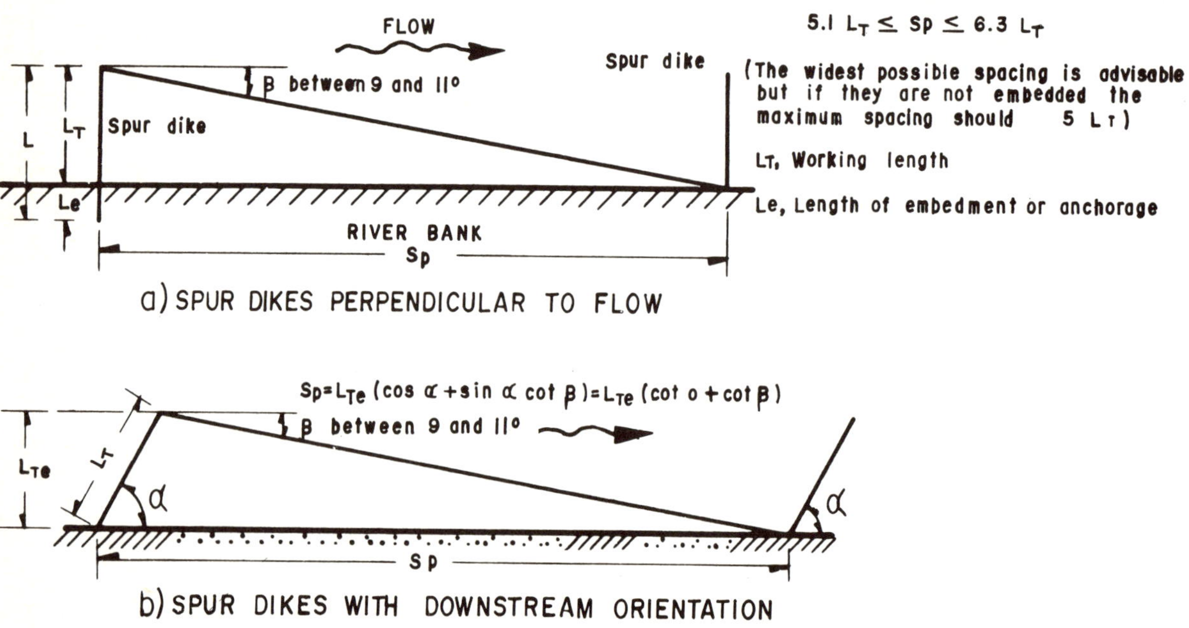

Fig. 15-10 Location of spur dikes on straight banks (Ref. 15)

15.4 Vibroflotation

Vibroflotation is the process by which sands and sandy soils are compacted at depth to improve their load bearing capacity, so that they will be able to support shallow foundations.

In typical processes, a gigantic vibrator cylinder (about 40 cm (16 in) in diameter by 1.80 m (6 ft) long) equipped with jets is suspended from a crane. The action of the jets, combined with the weight of the equipment and vibration enables the cylinder to be driven to depths of 15 m (50 ft) or even more. Once the cylinder has been driven into the ground the flow of jet water is reduced. The vibrator inside it applies horizontal vibrations to the soil, densifying a cylindrical mass of soil 2 to 3 m in diameter. At the same time the equipment is slowly withdrawn. Sand is added to the annular space around the vibrator. The hole below the vibrator, and any depression that forms at the ground surface (Plate 15-6).

Plate 15-6 Vibroflotation in constructing the fishing port of Alvarado, Veracruz (Courtesy of CIESA)

During the slow withdrawal process the pressure against the vibrator increases as the sand is compacted; the variation in the energy required of the vibrator motor indicates the degree of densification and the duration of vibration. The most appropriate vibration frequency should be, and is, usually determined for each specific case by performance tests, if the device is capable of a frequency change (Plate 15-7).

With the foregoing procedures the soil is compacted in cylindrical columns 2.50 to 3.00 m (8 to 10 ft) in diameter. The method is only applicable, of course, in free-draining granular materials, where water is not an obstacle to particle rearrangement.

The vibroflotation technique was invented by STEUERMAN, who described it for the first time in Russian literature and applied it in different foundation jobs for buildings in Germany [15]. Much of the experience in the use of the method is

Plate 15-7 Vibroflotation: Effect of the vibrator. (Courtesy of CIESA)

described in [17]. Mexican and U.S. engineers have used vibroflotation in many important projects.

The degree of compaction that is achieved with the method is greatest in the central 1 m of the compacted column and diminishes with radial distance. As already mentioned, in clean sands and gravels, the radius of significant compaction can be as much as 1.50 m (5 ft). It is occasionally more but is less than 1 m (3.3 ft) in sands with more than 20% fines. Experience indicates that the most appropriate soils are clean coarse to medium sands. However small amounts of coarse particles improves the transmission of the vibrations [18]; consequently it has become a fairly widespread practice to mix into the fill sand some gravels. Figure 15-11 [19] indicates the range of grain-size distributions within which the vibroflotation method is effective. Experience has also shown that when vibroflotation is applied appropriately, relative densities of 70% can be achieved.

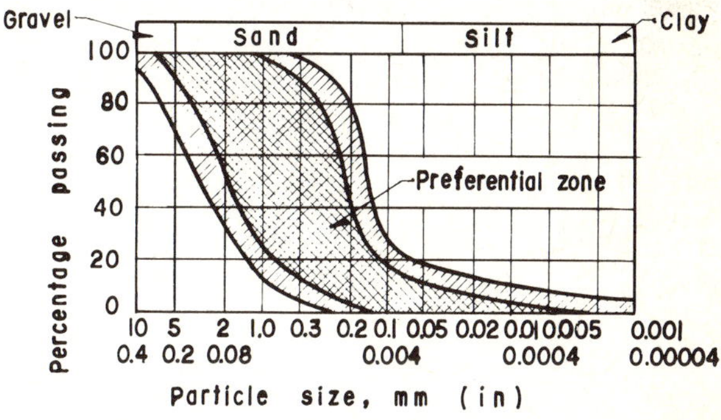

Fig. 15-11 Grain-size distribution range suitable for vibroflotation [19]

The spacing between the points where the vibrator is applied is not uniform. This spacing depends on the fines content of the sand, and on the grain-size distribution. Reference [20] contains a method for determining the spacing required between application points in order to obtain a specified minimum relative density. For guidance, the information contained in Fig. 15-12 [21] is included. In this case the method was applied to a low to medium density sand, inter-stratified with thick layers of clay. The grain-size distribution of the sand included 100% passing No 4 sieve and 15% passing No 200 sieve (as a maximum), and a very uniform grain-size distribution. Attention is drawn in the figure to the small spacings used, and the

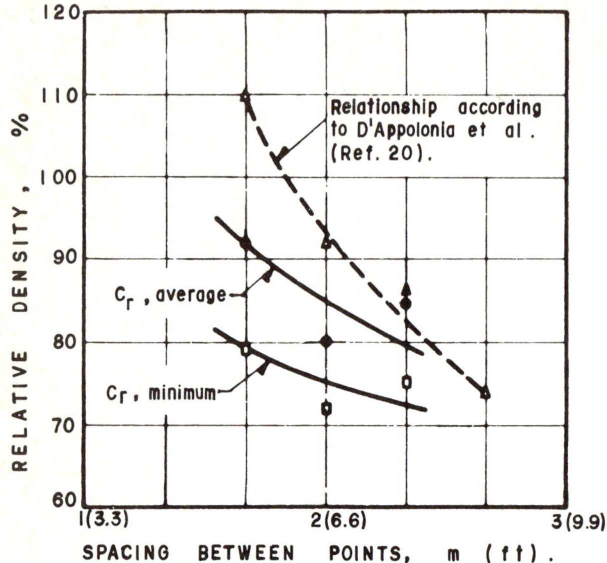

Fig. 15-12 Relationship between the relative density and the spacing between the points at which vibroflotation is applied [21]

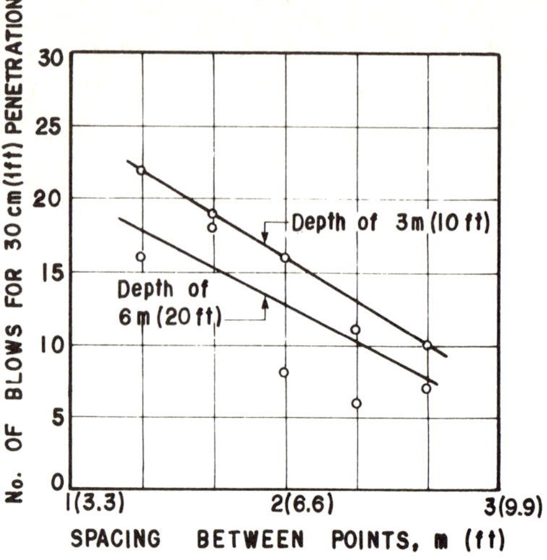

Fig. 15-13 Relationship between the spacing between vibrator application points and the standard penetration resistance of sands [21]

influence on density. In this case the method proposed by D'Appolonia et al. [20] led to the prediction of relative densities which were much greater than the ones really obtained.

Reference [21] also includes a comparative study of the compaction achieved using the vibroflotation procedure and piles. Vibroflotation was by far the most successful for the soils dealt with in the investigation. For the same project, Fig. 15-13 [21] gives an interesting correlation between the spacing between vibration points and the resistances obtained in a standard penetration test. Also worth noting is the variation in the resistance with depth.

The vibroflotation procedure introduces enough water into the soil that the level of the water-table has little effect on its applicability. The soils become nearly saturated from the jetting. On the other hand, clayey strata in the soil profile are important because they impede drainage and they may be mixed with the sand and thus resist compaction, as the vibrator passes through them [22].

In recent years, a completely new and different use has been found for vibroflotation techniques [19]. This is the improvement of foundation conditions in soft clayey soils or organic deposits [23,24] by stone columns. The vibrator is driven into the ground, forming a vertical well. The hole is then filled with gravel or crushed stone as the device is slowly withdrawn towards the surface. The fill material is compacted by the action of the vibrator. The softer the ground, the greater will be the diameter of the gravel column that forms. The diameter of these columns is usually between 80 cm (2.6 ft) and 1.30 m (4.3 ft). The columns are laid out in a triangular arrangement, with spacings ranging from 1.50 to 3.0 m (5 to 10 ft). After construction of the columns, a layer of granular material about 1.0 m (3.3 ft) thick is laid over the top of the columns and helps both to distribute the load that may be produced by an embankment, for example, and to drain the foundation ground.

As is explained in [24], initially the columns act like piles, but as the load above them increases, they tend to deform laterally, causing a passive thrust in the surrounding ground. The draining function speeds up consolidation and the increase in the shear strength of the clay.

15.5 Preloading Systems for Improving the Behavior of Soft Soils Under Embankments

15.5.1 Introduction

Preloading is the process by which an external pressure is applied to a soil prior to the erection or completion of a structure on that soil. The purpose of preloading is to bring about a consolidation in the soil, predominately primary but sometimes secondary. The preload is maintained sufficiently long that an adequate, predetermined amount of settlement will already have taken place, and a similarly predetermined shear strength increase will have developed. The soil is improved sufficiently that after the preload is removed the structure will have enhanced bearing capacity and reduced settlement. The method is helpful when the foundation ground is clayey, soft and compressible. Preloading was discussed at length in Chapter 3. The ultimate objective is usually to achieve under the preload the same settlement that was predicted for the future structure, but in the shortest time corresponding to the heaviest load. Although this technique has been employed for several centuries, its rational use subject to analyses can be said to have commenced in the 1940's [25].

Almost all compressible soils can be improved by preloading [26]. The method has been applied to organic and inorganic silts, any type of clay, peats, ash, and even waste fills. In many successful applications, the soils have had water contents ranging from 20% to 1,000%. In recent years there has been increasing use of preloading with peats. The method was not in the past regarded as appropriate because engineers assumed that little could be achieved in the necessarily limited application time because peats undergo large amounts of long term and continuing settlement. An obvious condition for application of this method is that the ground must tolerate the preload without failing.

An essential prerequisite for application of preloading techniques is a good knowledge of the soil profile and the properties of the different component strata. Use of the method, particularly in its applications to roads, demands an accurate preliminary assessment of the amount of settlement that will take place from the structure as well as the evolution of that settlement with time. When application of this method proves justified, it is not unusual for the savings that can be obtained with it to far outweigh the expenses involved by the preliminary study.

Current preloading methods are extremely varied. The principal ones are as follows [26]:

- Weight of earth fill.
- Load of water in tanks or a membrane-lined lagoon constructed expressly for this purpose [27].
- Lowering of the water-table [28], including the use of vacuum stabilization techniques [29].
- Vacuum stabilization techniques alone [30], although to the authors' knowledge this method has not yet been used in real practice.

Of the above methods by far the most common ones are the transmission of a load exerted by earth fill or by water. The methods based on lowering of the water-table are feasible when there is a pervious layer above soft soils and when the water-table in that stratum is located at a sufficient height above the soft ground.

The most important practical condition for the use of preloading is sufficient time, sometimes several years, before final construction. This can be analyzed by comparing the total project program with consolidation times. Not to be forgotten in the analysis is the time required for soil exploration and for preload construction.

When subsoil conditions are very poor and the loads that must be supported are small and fairly uniform, preloading usually will be the most appropriate of all the methods considered (Plate 15-8).

Plate 15-8 Constructing a test embankment to determine the conditions under which a highway will have to be built on soft ground. Note the cracking in this ground

15.5.2 Influence of Profile Characteristics and Properties of Soil Strata

Thin layers of free-draining sand or silt within the part of the soil profile that is affected by the preload can accelerate the rate of consolidation greatly. If the surcharge to be used is small and narrow (as for a highway embankment) even the thinnest free-draining strata will have a pronounced effect on the rate of consolidation. However a danger lies in the possible development of high pore pressures in the layers of sand or silt, which will lower the shear strength of these layers considerably [26]. These pressures can be controlled by relief-wells.

If the required surcharge is large and wide, the effect of sandy strata on the acceleration of the consolidation process is less important, at least in the central zones of the surcharge layer. The relief of pore pressures is now restricted principally to the parts of the layers that are under or outside the edge of the surcharge. As time passes, excess pore pressures and loss of soil strength can develop beyond the limit of the preloaded area.

Under very extensive surcharges, the water-table adjacent to the preload area rises, partly neutralizing the effects of high pore pressures in sand seams. The mechanism by which the water-table rises under large areas of loaded ground has already been discussed in other parts of this book. This effect is particularly notable when the preload is hydraulic fill, which is saturated. In such cases the pore water pressure dissipates slowly, restricting the time rate of consolidation. This can be relieved by providing subdrainage inside the surcharge, generally by perforated pipes or vertical drains.

The development of pore pressures in sandy layers under surcharges may have very serious consequences in reducing the shear strength of these strata, which leads to translational slides. When no surface drainage is provided the pore pressures accompanying the preload can cause the water-table to rise into the surcharge.

The coefficient of consolidation, C_v (an important calculation element when evaluating preloads) when obtained in the laboratory from relatively undisturbed samples is usually fairly high for pressures below the preconsolidation load (see Chapter 1) but diminishes appreciably for pressures approaching it. At higher loads, it generally increases again [32]. If the samples employed in the laboratory are more disturbed, the value of C_v is fairly low for pressures below the preconsolidation load, but increases with increasing load. For pressures greater than the preconsolidation load the values approach those for undisturbed samples [26]. For remolded clays, even at the highest pressures, the value of C_v is below that obtained in undisturbed samples of the same clay. The values of the coefficient of compressibility, a_v (see Chapter 1) are not so susceptible to the degree of disturbance of the samples, but are nevertheless affected by it [26]. When preloads are used, the pressures applied to the ground are not usually very high, which indicates the range of consolidation stresses to be studied.

Figure 15-14 [33] shows the effect of a temporary surcharge that exceeds the final load on settlements due to primary consolidation. It can be seen how the load corresponding to $H + \Delta H$ (height of earth) produces in time t the same settlement as height H would produce over a considerably longer period.

The coefficient of secondary consolidation, C_α, depends on the effective pressure sustained by the soil. For pressures below the preconsolidation load it is usually very small, but it increases when the pressure on the clay approaches this limit. It remains practically constant or diminishes slightly for higher pressures [26]. The coefficient of secondary consolidation, C_α, is equal to the deformation corresponding to a cycle of the logarithmic time scale on a secondary consolidation curve (see Chapter 1), Fig. 15-15 [26] shows the typical variation of C_α in relation to the effective pressure acting on the clay.

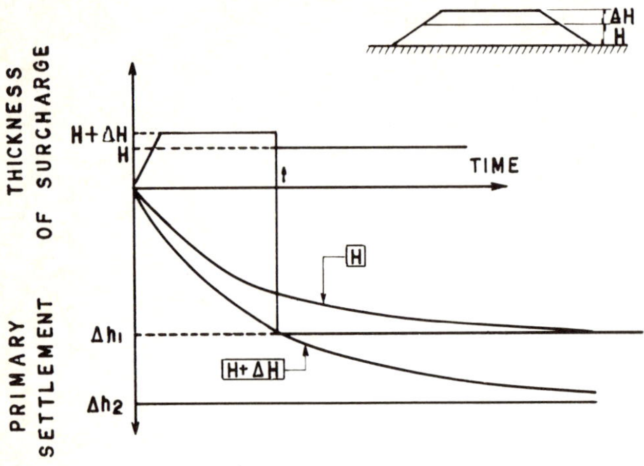

Fig. 15-14 Variation in the primary settlements on placement of a surcharge [33]

Laboratory tests suggest that continuing settlements due to secondary consolidation, can be minimized by preloading the soil with a surcharge that exceeds the final load, and then removing the surcharge. Such tests indicate that preloading has two effects. Figure 15-16 [26] shows a primary and secondary consolidation curve for a sample of clay subjected to a pressure similar to that which will act on the soil in the field. It also illustrates what happens when different portions of that pressure are withdrawn. It can be seen that for some time after partial removal of the pressure no settlement due to secondary consolidation occurs. There may even be some expansion not shown in the figure. Subsequently, secondary consolidation does appear (broken lines in the figure), although at a slower rate than originally. The time required for secondary consolidation to appear is longer the greater the amount of unloading that has been performed on the sample. The rate of subsequent settlement decreases with the amount of unloading indicating that the ultimate C_∞ is smaller the greater the amount of unloading that has taken place. Figure 15-17 [26] indicates by how much the value of C_∞ can be expected to drop in terms of the percentage by which the original stress is reduced by the unloading, based on both field and laboratory data.

The implications of the foregoing information in practical surcharge operations are obvious. It can be seen that the greater the excess or surcharge preload applied, the smaller

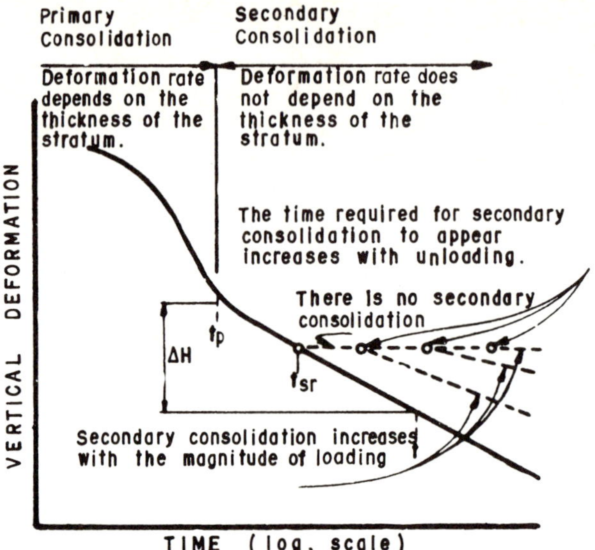

Fig. 15-16 Effect of partial unloading on the consolidation of a clay sample [26]

and slower will be the settlements corresponding to secondary consolidation once the surcharge load is withdrawn, leaving only the load corresponding to the height of the final embankment. Apart from other factors that are not considered here, the limit to the preload that is employed will be the maximum load that can be applied to the soil without causing failure.

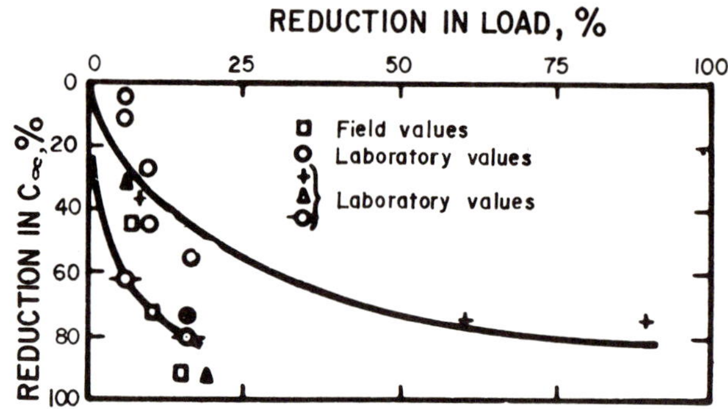

Fig. 15-17 Relationship between the coefficient of secondary consolidation and unloading in a clay [26]

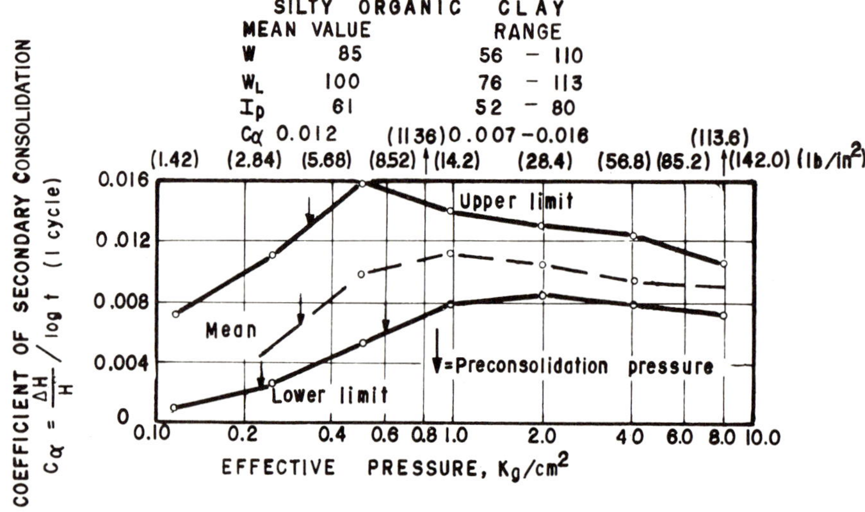

Fig. 15-15 Variation of the coefficient of secondary consolidation with effective pressure on the clay sample [26]

References [34,35] describe examples which illustrate the advantages of appreciable surcharges.

Some attempts have been made to correlate the soil properties that determine the design of a preload with the readily obtainable index soil properties. Naturally these correlations should be used with a judicious skepticism. They may be slightly more reliable when they have been verified in similar projects involving similar soils in the same region. Figure 15-18 [26] shows a correlation between the values for the coefficient of secondary consolidation, C_α, and the water content and void ratio of the soil. These correlations are based on laboratory tests using one-dimensional consolidation. They should not be employed even for preliminary design studies if the soil is likely to undergo lateral deformations. The coefficient of secondary consolidation for peats and similar materials with very high void ratios does not increase so rapidly at void ratios exceeding 5, as can be seen in Fig. 15-18. This helps eliminate the impression that secondary consolidation makes preloading less favorable in peats. Reference [26] includes other correlations that may be useful in practical cases.

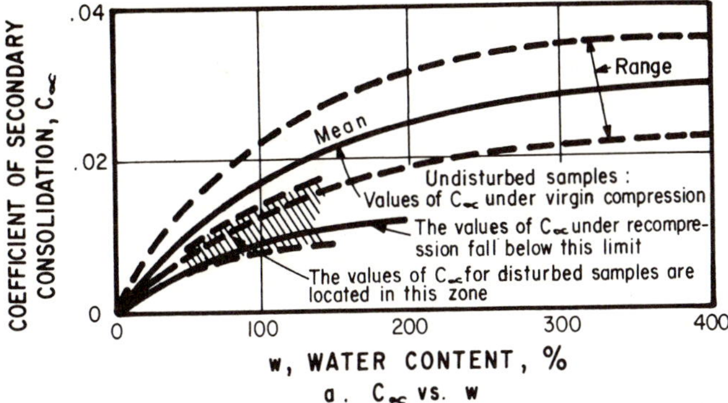

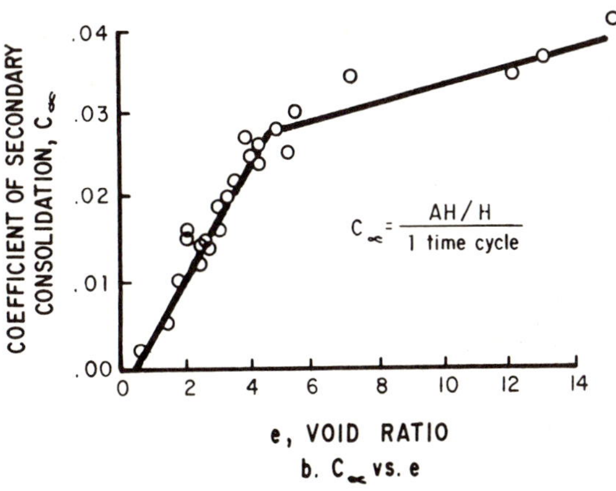

Fig. 15-18 Correlation between the coefficient of secondary consolidation of a clay and its *w* and *e* [26]

15.5.3 Some Guidelines for Preload Design

When studying surcharges it is advisable to consider primary and secondary consolidation as parts of the same continuous process. Moreover, this is in harmony with some current theoretical views on the consolidation of soils. However, the division into primary and secondary consolidation (separate, although partially concurrent phenomena) continues to be useful in design.

.1 Primary Settlement Control

The effect of a temporary surcharge on the settlement computed using TERZAGHI's theory (primary consolidation) has already been discussed in relation to Fig. 15-14. Such a graph is a basis for design decisions. Final design will be strongly influenced by the time available for surcharge loading. Often adequate structural performance can be achieved without the surcharge producing the total settlement that will be caused by the ultimate embankment. Frequently it is sufficient to produce a substantial portion of that settlement, knowing that the remainder will be absorbed by the structure when in operation.

On withdrawal of the surcharge before all excess pore pressure has been dissipated (average percentage of consolidation less than 100 per cent) the upper and lower parts of the stratum will be consolidated to an above average degree, whereas the central parts will be consolidated to a below average degree. This is due to the closeness of the draining boundaries as shown in detail in Fig. 15-19, From [26]. It also follows from the theory of primary consolidation in Chapter 1. The figure illustrates the pore pressures in the stratum when the degree of primary consolidation is of 50%. If the stratum is drained through the two pervious boundaries, it can be seen that the two upper and lower portions, of thickness 0.44 H, have a degree of consolidation above 50%, whereas the center part of the stratum, of thickness 1.12 H, has a lower degree (H is half the stratum thickness). If settlement calculations are made for average degree of consolidation $U = 50\%$, the upper and lower parts will be pre-consolidated, whereas the central part will have a degree of consolidation below the one computed. Thus, the upper and lower parts will tend to expand, whereas the central part will undergo further consolidation if the soil is unloaded to half the preload (which is almost equivalent to the load that would ultimately produce the same average consolidation; the two effects tend to compensate for one another).

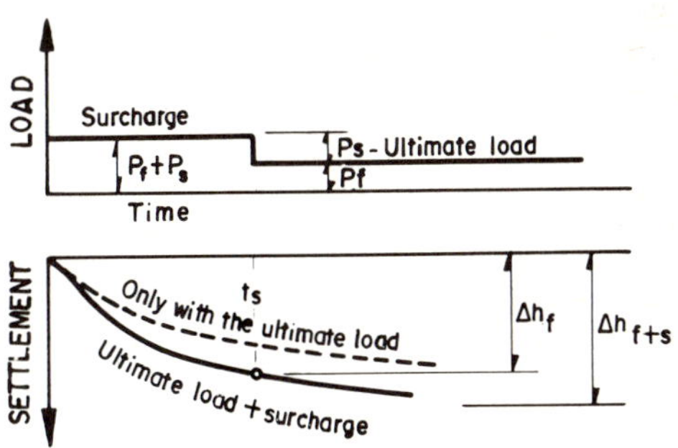

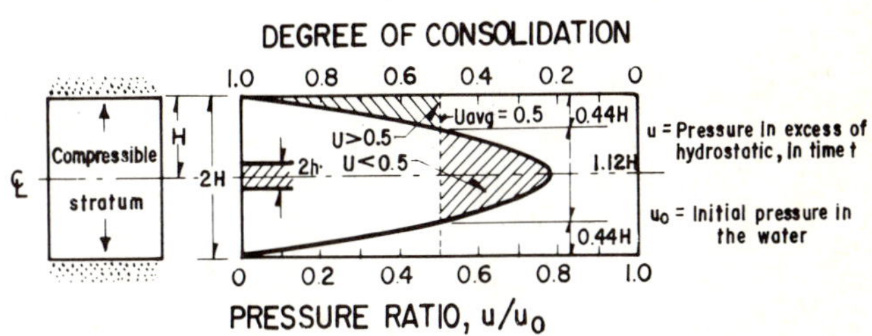

Fig. 15-19 Effect of a surcharge on the primary consolidation of a stratum drained top and bottom [26]

However settlement exceeds expansion, because the coefficients of compressibility are higher for consolidation than for expansion. Thus routine calculations predict smaller primary consolidation after removal of the surcharge than those measured. This can be avoided if the degree of consolidation at the removal of the surcharge corresponding to the central part of the stratum is used for calculation. Some engineers use an intermediate value between that of the central part and the mean value of U for the stratum.

A description follows of the procedure for evaluating surcharges on normally consolidated soils, using the degrees of consolidation corresponding to the center of the compressible layer.

With reference to Part *a* of Fig. 15-19 if the surcharge causes a certain settlement in the ground, the degree of consolidation of the center of the compressible layer (U_{f+s}) at the time of withdrawal of the surcharge compared with the settlement under the ultimate structural load is

$$U_{f+s}\,\Delta h_{f+s} = \Delta h_f \qquad (15\text{-}2)$$

where Δh_{f+s} and Δh_f are the respective final primary settlements corresponding to the ultimate load plus the surcharge, and to the ultimate load alone. These settlements are for the center of the stratum and correspond to those that would occur if that center had a thickness of $2h$.

Consequently:

$$U_{f+s} = \frac{\Delta h_f}{\Delta h_{f+s}} \qquad (15\text{-}3)$$

which is the degree of primary consolidation that must be acquired by the center of the stratum if no further settlements are to take place after withdrawal of the surcharge.

Figure 15-19b illustrates this condition. If the soil is originally loaded with a pressure $p_f + p_s$ (assuming $p_s = p_f$ in this case) when the average degree of consolidation of a stratum of thickness $2H$ is 50%, the degree of consolidation of the center of the stratum will be 22%. If at this moment 78% of the total load $p_s + p_f$ is withdrawn, the soil at the center will undergo no further settlement due to the remaining load. Any part of the stratum other than the center will expand. The expansion is greatest close to the boundaries of the stratum. The method of analysis that is proposed [26] considers these expansions are negligible. This is not a reasonable procedure for it requires the withdrawal of 78% of the total load instead of the 50% which should be withdrawn in order to leave the ultimate load p_f. In this procedure the surcharge must left longer, until U_{f+s} in the center of the stratum is 50 per cent and on withdrawal of half of the total load, no further settlement will take place.

Figure 15-20 [26] relates the degree of primary consolidation that must be obtained at the center of a compressible stratum with the surcharge that is applied expressed by the pressure ratio p_s/p_f (see Fig. 15-19a), and with the pressure ratio p_f/p_o in which p_f is the pressure that is transmitted to the center of the compressible stratum by the structure and p_o is the initial pressure at that level. The figure refers only to normally consolidated clays.

Knowing the load of the embankment that is eventually to be erected, and knowing the value of p_o in the compressible soil, it will be possible to make an initial estimate of an appropriate surcharge intensity, p_s. Figure 15-20 [26] gives the degree of primary consolidation, U_{f+s} required in the central plane of the compressible stratum so that no primary settlement will take place when the surcharge is removed.

This value for the degree of consolidation can now be entered in the chart in Fig. 15-21 [26]. Using the curve that represents the ratios for the central plane of the compressible stratum ($z/H = 1$) it will be possible to find the average time factor for the stratum, corresponding to the degree of consolidation in the center of the stratum.

The chart also gives many other curves for different z/H values. That corresponding to the average ratios for the entire compressible stratum is especially notable.

With the average time factor, T, obtained in Fig. 15-21, it will be possible to calculate consolidation time, using the expression:

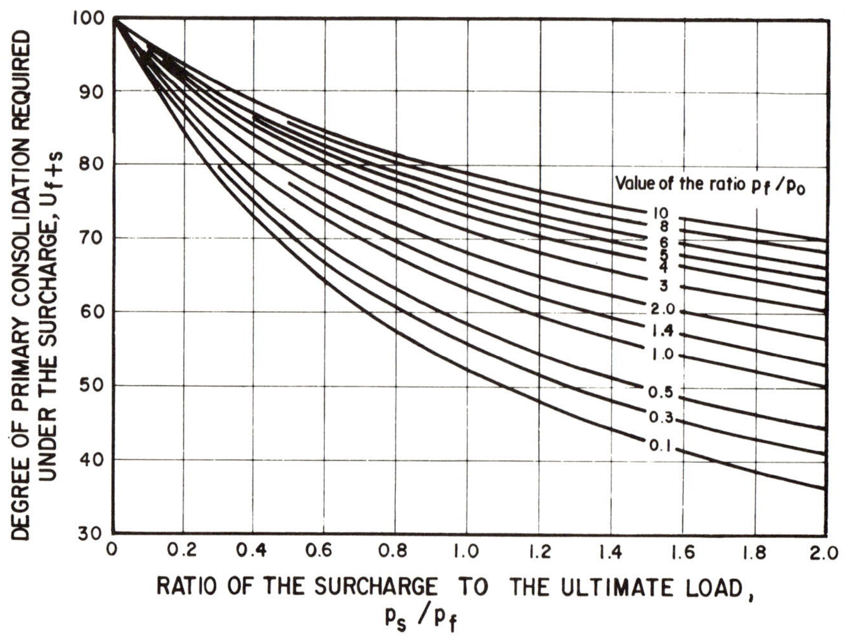

Fig. 15-20 Degree of primary consolidation produced at the center of a compressible stratum (normally consolidated) under different surcharges [26]

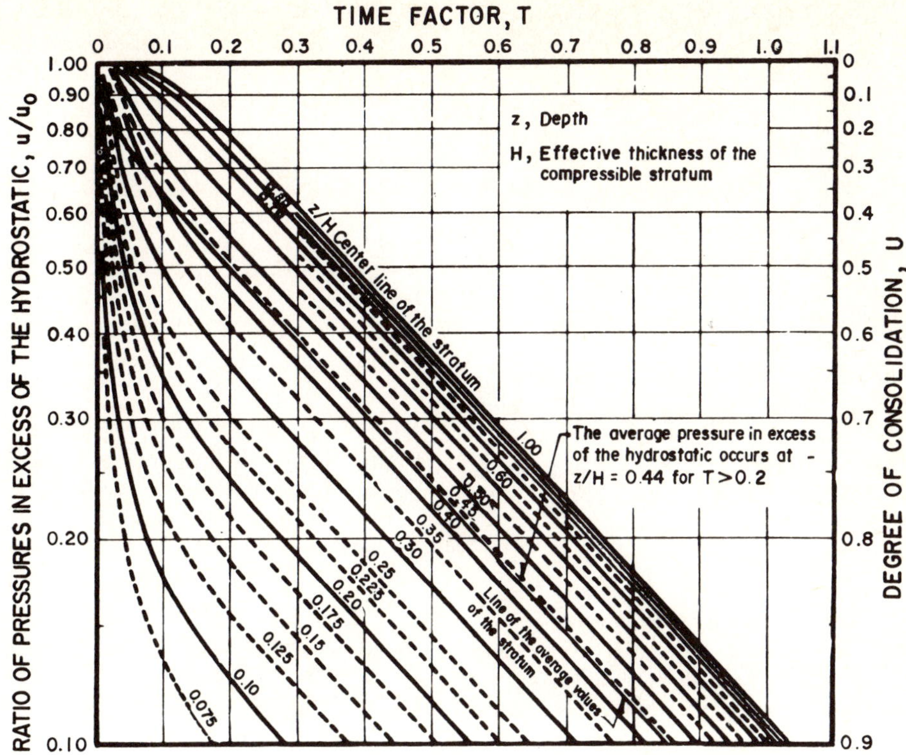

Fig. 15-21 Relationship between the time factor (T) in a stratum of thickness $2H$ and the degree of consolidation of its central plane [26]

$$t_s = T\frac{H^2}{C_v} \qquad (15\text{-}4)$$

where: t_s is the time after which the surcharge is withdrawn, which was estimated at an earlier stage of the procedure. It is the time in which the soil consolidated under pressure $p_s + p_f$. H is the effective thickness of the compressible layer: a stratum with pervious strata above and below has a thickness of $2\,H$. C_v is the coefficient of consolidation of the compressible stratum, described in Chapter 1.

Equation (15-4) permits the evaluation of the time the arbitrarily selected surcharge should be allowed to act, such that on withdrawal the compressible layer will undergo no subsequent settlement due to primary consolidation. This time is compared with that available according to the construction program. If the time available is inadequate, the procedure should be repeated using a different surcharge, until a surcharge is found which will meet the proposed objectives in a reasonable time. If such a surcharge load cannot be found, then the surcharge method is not suitable for the problem. The surcharge value that is selected will also be limited by the bearing capacity of the natural ground.

If it is not wished to introduce the refinement of working with the degree of primary consolidation in the center of the compressible layer (for those cases where the average degrees of consolidation for the entire thickness, $2H$, are felt to be sufficient) a simplified procedure for computing the necessary surcharge is illustrated in Fig. 15-14 or Part *a* of Fig. 15-19. First calculate the two time-settlement curves (under p_f and under $p_s + p_f$), considering the average values for the entire compressible layer. On the curve that includes the surcharge find the time corresponding to the total settlement under the permanent load. This will be the surcharge time, t_s, which will now have to be considered.

So far the purpose of the surcharge has been to eliminate primary settlement. If it were sufficient to eliminate only a part of it, then the procedure is similar to the previous one (remembering that the charts were drawn up with a view of eliminating all settlement).

.2 Secondary Settlement Control

For most soils settlement due to secondary consolidation is smaller and far slower than that due to primary consolidation. Consequently once the primary is controlled, it is less necessary to control the second. However, for some soils, settlement due to secondary consolidation plays an appreciable part. In micaceous and in highly organic soils, post-construction secondary can exceed the primary.

In most soils, the rate of secondary settlement decreases with time, obeying a logarithmic law (see Chapter 1). The magnitude of secondary consolidation is proportional to the thickness of the compressible layer (or at least to the thickness of that layer at the time when primary consolidation has virtually stopped).

The secondary settlement that will take place after primary consolidation has terminated, can be computed using the expression:

$$\Delta H_\alpha = C_\alpha H_p \log \frac{t}{t_p} \qquad (15\text{-}5)$$

where: ΔH_α is the settlement attributed to secondary consolidation. C_α is the coefficient of secondary consolidation. H_p is the total thickness of the compressible stratum the instant primary consolidation can be regarded as having terminated. t_p is the time in which it is considered that primary consolidation terminated. t is the time at which settlement by secondary consolidation is computed.

Earlier in this section, the coefficient C_α was defined. From this definition a method can be devised for its evaluation in the laboratory. The standard consolidation test is performed on samples from under the central part of wide embankments, as was also discussed previously. The concepts of secondary consolidation, which are essential for the correct comprehension of what follows, can be consulted in Chapter 1 of this book or in [36] and [37].

As already indicated, for computation purposes, primary and secondary consolidation will be considered as separate and successive processes, although this does not agree with the modern concepts of total consolidation; Fig. 15-22 [26,36] summarizes TAYLOR's approach to secondary consolidation. According to TAYLOR, secondary consolidation under various different constant effective pressures develops in accordance with the lines shown in the figure, these last values on a logarithmic scale, [39] explains the origin of the constant time curves shown.

TAYLOR drew the lines corresponding to the different times parallel to that for primary consolidation, whereas today these lines would probably not be drawn in a parallel direction. However this detail is irrelevant conceptually. The figure shows that when the soil is subjected to a permanent load and a surcharge, ultimate settlement takes place which is equivalent to total primary settlement under a permanent load, plus part of the settlement that the soil would undergo due to secondary consolidation due to the permanent load alone. Point A of the figure represents total primary settlement under the permanent load, and point B total primary settlement under the permanent load and the surcharge. The difference in the void ratios between points A and B corresponds to the secondary settlement under the permanent load alone, which has occurred as a result of the surcharge.

In this way, the surcharge reduces the secondary consolidation that will take place under the permanent load once that surcharge is removed. This could have been expected from the information in Figs. 15-16 and 15-17. Reference [26] includes an explanation based on the variations that occur during the consolidation processes in the principal horizontal and vertical stresses. The moment the surcharge is withdrawn, the soil which was originally normally consolidated became preconsolidated (point C in Fig. 15-22). This assumes that secondary consolidation depends primarily on the magnitude of the acting shear stresses, and that the overconsolidation achieved with the surcharge causes a reduction in the secondary consolidation that would occur under the permanent load, once it is left to act on its own. The mechanism in preconsolidated soils is conceptually the same, but when the surcharge is withdrawn the expansion is negligible.

Figure 15-23 [26] permits application of the foregoing concepts to the design of surcharges to partially compensate settlements due to secondary consolidation. Once again use will be made of the degree of consolidation at the center of the stratum of thickness $2H$, considering a small element of thickness $2h$. It is the settlement of this element, which is assumed to be representative of the center, that will be considered.

Assuming Δh_{sr} to be the settlement that takes place in the center of the compressible stratum on removal of the surcharge, and t_{sr} the time of this removal, the following equation is obtained:

$$\Delta h_{sr} = \Delta h_f + \Delta h_{sc} \tag{15-6}$$

where:

Δh_f is the total primary settlement under the permanent load, and Δh_{sc} is the amount of secondary settlement that would occur under the permanent load, that takes place during the surcharge action period.

The following equation must also be true:

$$\Delta h_{sr} = U_{f+s}\,\Delta h_{f+s} \tag{15-7}$$

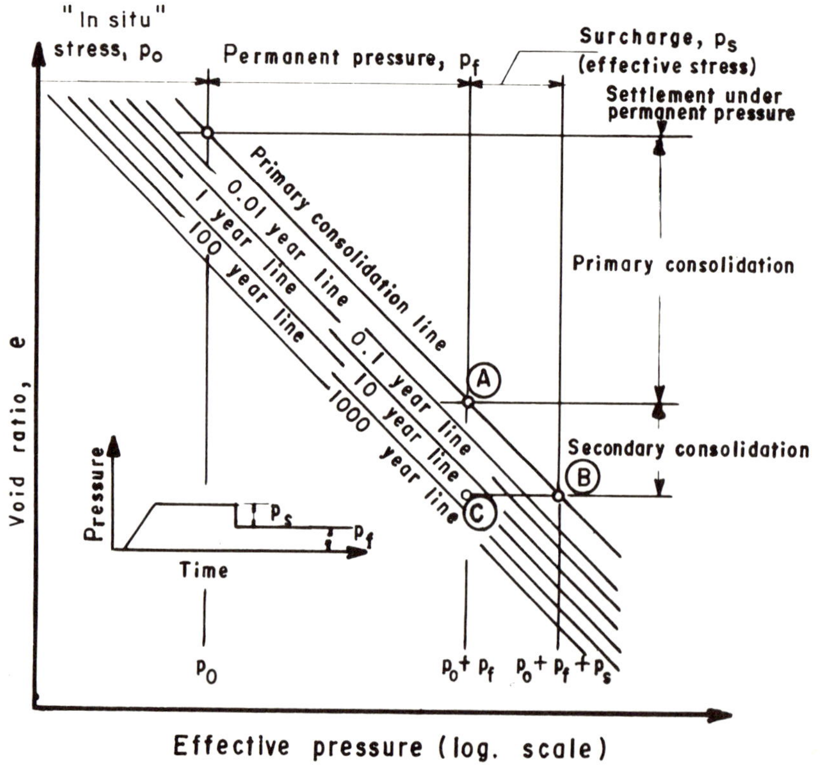

Fig. 15-22 Use of surcharges to reduce secondary consolidation in normally consolidated soils [26,36]

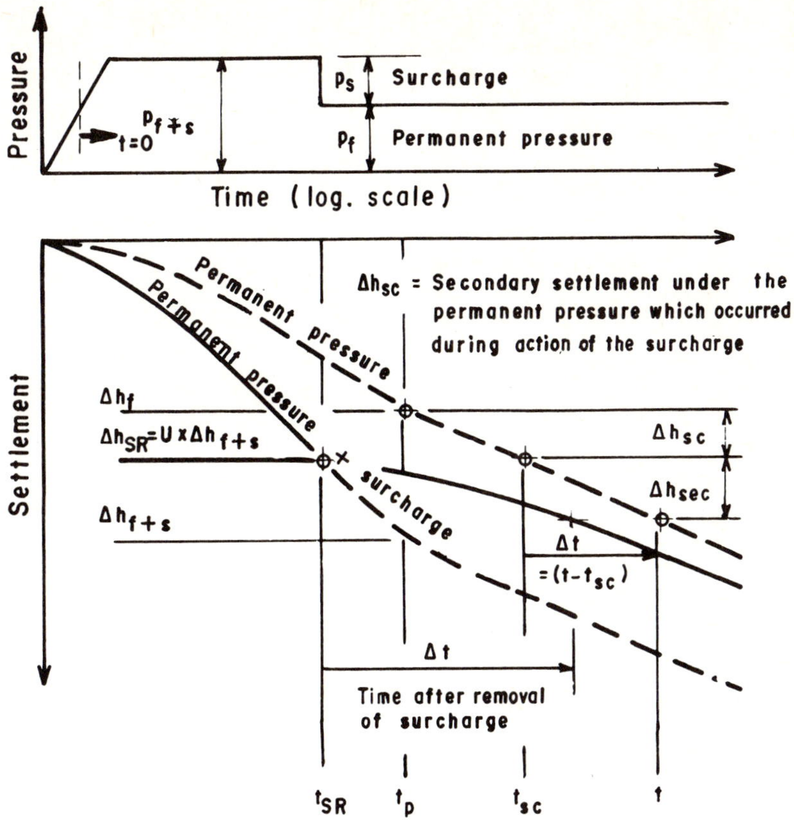

Fig. 15-23 Partial compensation of secondary consolidation using a temporary surcharge [26]

where:

U_{f+s} is the degree of consolidation of the center of the compressible layer on removal of the surcharge, and Δh_{f+s} is the total primary settlement of the soil under the permanent load and the surcharge.

Thus, the degree of consolidation at the center of the compressible layer on removal of the surcharge can be expressed as:

$$U_{f+s} = \frac{1}{\Delta h_{f+s}} (\Delta h_f + \Delta h_{sc}) \tag{15-8}$$

or

$$U_{f+s} = \frac{1}{\Delta h_{f+s}} \left[\Delta h_f + C_\alpha (H - \Delta h_f) \log \frac{t_{sc}}{t_p} \right] \tag{15-9}$$

In order to understand the above, Equation (15-5) should be reviewed. The values of Δh_f and Δh_{f+s} can be computed thus [40] (See also Chapter 1):

$$\begin{aligned} \Delta h_f &= (m_v)_f \, \Delta p_f \, h \\ \Delta h_{f+s} &= (m_v)_{f+s} \, \Delta p_{f+s} h \end{aligned} \tag{15-10}$$

where m_v is the coefficient of volume change of the compressible soil. This is somewhat different when the soil is subjected to a permanent load (p_f), or the permanent load and a surcharge (p_{f+s}). However in most cases, especially if the surcharge is not too great in relation to the permanent load, approximately it can be said:

$$(m_v)_f = (m_v)_{f+s} = m_v \tag{15-11}$$

In the following, Eq.(15-11) is regarded as valid. If there is a significant difference in m_v, either the average should be used or a step by step evaluation.

Taking into consideration Eq.(15-10), Eq.(15-9) can be written:

$$U_{f+s} = \frac{1}{m_v \, \Delta p_{f+s} \, h} \left[m_v \, \Delta p_f \, h + C_\alpha (h - m_v \, \Delta p_f \, h) \log \frac{t_{sc}}{t_p} \right]$$

from which the following equation is obtained:

$$U_{f+s} = \frac{\Delta p_f}{\Delta p_{f+s}} \left[1 + C_\alpha \left(\frac{1}{m_v \, \Delta p_f} - 1 \right) \log \frac{t_{sc}}{t_p} \right] \tag{15-12}$$

Equation (15-12) expresses the degree of consolidation that must be acquired by the center of the compressible layer in order to achieve total primary settlement under the permanent load plus the desired amount of secondary settlement.

All the terms in Eq.(15-12) can be found from laboratory consolidation tests or estimated from index properties based on empirical correlations. Δp_f and Δp_{f+s} can be obtained by applying the Boussinesq or similar stress distribution theories to compute the pressure that will be transmitted to the center of the compressible layer by the permanent load, and by the permanent load together with the surcharge. C_α is the coefficient of secondary consolidation computed as the slope of the secondary consolidation curve (strain corresponding to one cycle of the logarithmic time scale) for the stress range involved. The term

m_v can be evaluated by a consolidation test [40]. The term t_{sc} is usually regarded as equal to the useful life of the structure. Thus, Eq.(15-12) gives the degree of consolidation required under the surcharge, so that the permanent load will suffer no significant secondary settlement during its useful life. The term t_p is the time when it can be considered that the entire primary settlement has occurred under the permanent load.

The value for the degree of total consolidation required is given by Eq.(15-12), which now includes the secondary effect. This is entered in the graph in Fig. 15-21, using the line corresponding to the center of the stratum to obtain a value for the time factor, T. This, as in the case of the prevention of primary settlement discussed previously, will provide the surcharge time, t_{sr}, that should be left once the surcharge has been selected, in order to achieve the U_{f+s} computed in the center of the compressible layer. If this surcharge application time does not fit project requirements, another computation will have to be made, either using a heavier surcharge or a lower value for t_{sc}.

If a lower t_{sc} value is adopted, Eq.(15-5) will permit evaluation of the amount of secondary settlement that is avoided under the new conditions. This is compared with the total secondary settlement that can be expected during the useful life of the permanent load in order to determine the effectiveness of the surcharge.

The heavier the surcharge employed and the longer the application period, the smaller will be the secondary settlements occurring on its removal. However there is a limit for the surcharge which, when exceeded, will lead to *expansions* of the compressible soil after removal.

15.6 Collapsing Soils

15.6.1 Introduction

Some soils of low density, particularly in arid regions where desiccation prevails undergo a rapid loss of soil volume upon wetting. This leads to appreciable subsidence in the ground surface, accompanied by a rapid drop in strength and an internal structural collapse, all of which takes place when the soil absorbs large quantities of water.

Literature in the English language terms the rapid process of volume loss as *collapse* and the soil that undergoes it as *collapsing* [41]. Until recently engineers took little interest in these phenomena because large-scale engineering activity was unusual in arid zones. Currently, with important projects for carrying water to these zones, the problem has now become a social and economic one.

The phenomenon can occur in a wide variety of soils, although many have rounded grains. The median grain-size is often that of silts, but larger particles, including sands and gravels, are not unusual. A small clay content is also a common feature of collapsing soils. Collapse problems can occur in a wide variety of soil formations. Many are in loess and other clayey and silty aeolian deposits, but others occur in alluvial and residual formations and even in man-made fills. All have two characteristics in common [41]: first, a loose structure, as shown by a fairly high void ratio, and second a water content below the one corresponding to saturation.

Collapse is closely related to the water content of these soils; if the degree of saturation is low, susceptibility is more pronounced. This is why the problems appear in zones where desiccation is appreciable [42]. When part of the soil in these arid zones is covered with an impermeable stratum, the water content increases in the covered area, because evaporation is impeded. This increase in the water content is a frequent cause of collapse. However, dry climate and intense surface evaporation are not indispensable conditions, and cases of collapse have been observed where dry soils in humid zones have become wet, and their water contents approached saturation.

Figure 15-24 [41] shows a collapse which occurred during a laboratory consolidation test and illustrates the scale on which the void ratio changed. A sample of fine sand with 10% montmorillonite in a specimen 2.5 cm (1 in) thick was loaded to 8 kg/cm^2 (114 lb/in^2). The sample was then unloaded to 0.53 kg/cm^2 (7.5 lb/in^2) and was subsequently loaded again to just over 2 kg/cm^2 (28.4 lb/in^2). All these operations were conducted at the natural very low water content of the sample. The sample was then allowed to absorb all the water it could through its upper and lower faces. The figure shows the abrupt decrease in void ratio that occurred as a consequence of the absorption of water. After collapse the test was continued, loading and unloading the specimen with its new increased water content

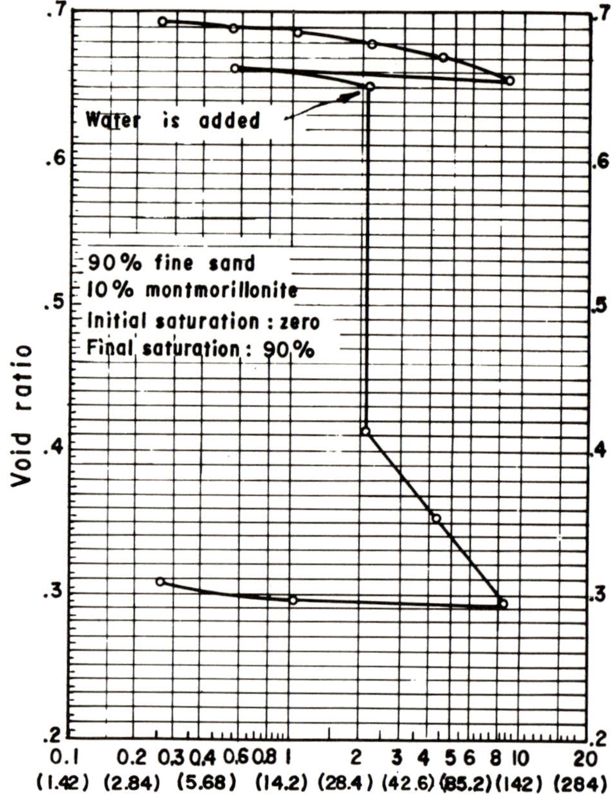

Fig. 15-24 Soil collapse in a consolidation test [41]

Analysis of a test like that reported in Fig. 15-24 indicates that settlements due to collapse may be large. However large void ratio changes are not necessarily so, for there are a wide variety of different collapsible soils.

Reference [43] includes a study of consolidation tests conducted on numerous specimens of a loess, at its natural water content and also after wetting. The data show that the compressibilities were far greater after wetting, and that the void ratio for a soil with its natural water content decreases suddenly on the addition of water.

15.6.2 Causes of Collapse

Most of the cases of collapse investigated in engineering practice have found soils with a honeycomb structure and somewhat rounded particles cemented together in some way. In all cases, the cementing agent was subject to weakening or dissolving in water. The grains fall into the voids, when the cementation between them weakens or disappears.

The shear strength between the grains can be expressed by COULOMB's law, ∅ being the angle of internal friction and c a final resultant of all the effects of interparticle interlocking and cementing.

In many instances, the bonding or cohesion between the particles is at least partially due to the capillary tension of the water between them. When the water content of the soil is below the shrinkage limit approximately, the water forms menisci between the soil particles as a consequence of the water-air boundary in the voids [44]. As a consequence, compression develops between the grains of soil: an increase in effective stress. An increase in the water content of the soil causes an immediate dissipation of the capillary tension in the water, with a corresponding reduction in the effective stresses between the grains. There is a loss of shear strength, accompanied by a rapid loss of volume, if the soil has a high void ratio, [45] includes the effect of air pressure in the voids as a factor in the collapse.

These capillary forces plus gravity explain much of the collapse behavior of silts, sands and gravelly sands. However they have a considerably smaller influence on very fine materials, in which microscopic physico-chemical forces such as osmosis and VAN DER WAALS, are important. Soils are mixtures, in which some part is played by many kinds of forces (which explains the wide variety of soil behaviors observed).

Collapsing soils usually have a wide variety of particle shapes and sizes, with sands, large quantities of silt and often a certain amount of clay. When the coarser grains are surrounded by particles of clay, the origin and stress history of the clayey fraction are important. The clay may be residual or transported. When the residual clays dry out, the strength of the soil as a whole may be very high. However it is rapidly and appreciably reduced when the clayey fraction absorbs water and the clay crystals are hydrated and separate.

Figure 15-25 [41] shows a typical arrangement of residual clays serving to bind coarser grains together. The clay particles adopt a parallel or orientated arrangement and their grains separate when the soil absorbs water, greatly reducing the strength of the soil.

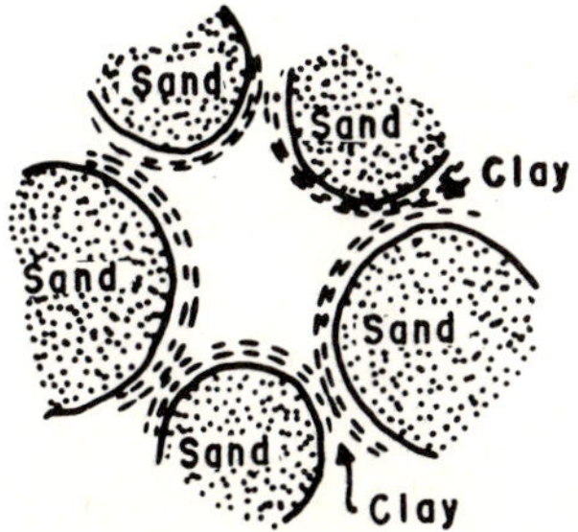

Fig. 15-25 Diagram showing arrangement of residual clays binding sand particles together [41]

In transported clays that have been deposited in a continuous medium, such as water, a situation may arise like that shown in Fig. 15-26 [41] in which particles of flocculated clay fill the space between the coarser grains. This is a result of the movement of water towards the voids between the coarser particles, as a consequence of drying. Under these conditions, large capillary tensions develop in the clay.

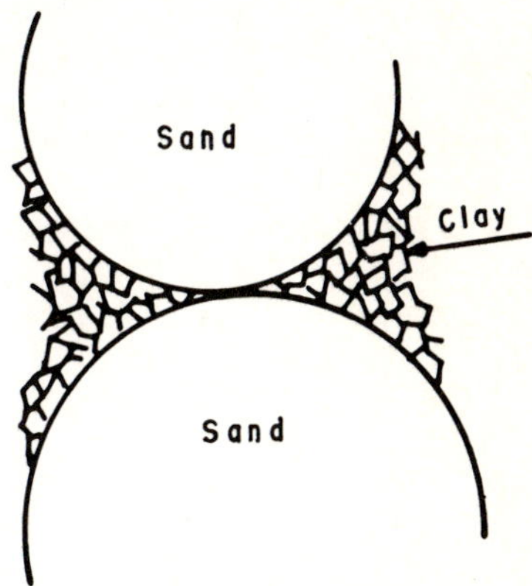

Fig. 15-26 Diagram showing arrangement of transported clays binding sand particles together [41]

When a structure like that in Fig. 15-26 absorbs water, the capillary tensions are relieved and the concentration of ions in the fluid phase is greatly reduced. Reference [46] shows that the reduction in the concentration of ions in the liquid phase becomes permanent on the appearance of repulsive forces between the particles of clay and a pronounced tendency for the flocculated structure of the clay to evolve towards a more dispersed arrangement. As a consequence, there is a reduction in the cementing of the clay between the coarser particles, and a reduction in the shear strength of the soil as a whole.

The grain structure or fabric of the deposited clay that is shown in Fig. 15-26 is not the only possible one. In some environments the clay may form flocculated structures in isolated individual lumps, which act like independent silt sized grains in producing capillary tension which binds the larger particles of sand together. It is also possible that under appropriate conditions the clayey fraction of the soil forms intensely flocculated structures occupying the entire space between the grains of sand. The grains lose their direct contact in the arrangement shown in Fig. 15-27 [41].

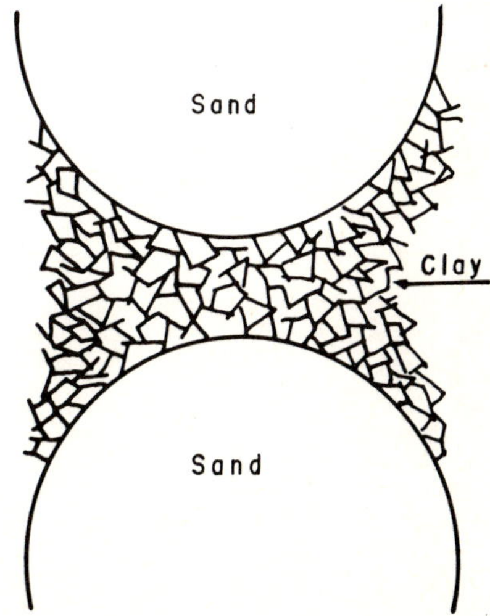

Fig. 15-27 Another possible arrangement of the clayey fraction in a soil [41]

When such an arrangement is subjected to intense drying, it becomes strong and almost incompressible under small loads. However if the load is great the clay structure may suddenly collapse without a change in water content it becoming dispersed with considerable volumetric deformation.

Capillary forces or clayey materials are not the only agents capable of binding particles of silt or sand together. Other natural cementing agents such as iron oxides and calcium carbonate are particularly important in tropical regions.

15.6.3 Identification and Use of Collapsing Soils

The identification of collapsing soils and a fairly accurate prediction of the collapse they are liable to suffer is essential in geotechnical engineering. It is also necessary to determine the evolution of collapse with time.

The identification of collapsing soils is based on tests that have been developed by geotechnical engineers to meet different objectives. Depending on the degree of care with which it is wished to conduct indentification, very simple or complicated, time-consuming tests can be employed.

For rapid identification, the liquid limit of soils has been employed [41,47,48]. If, under natural circumstances, the void ratio of a given soil is higher than that at its liquid limit, on absorbing water the soil will lose strength. Before saturation is achieved, the soil will undergo considerable structural collapse accompanied by an appreciable reduction in volume.

The slopes of lines such as those shown in Fig. 15-28 [41] are useful indications of the activity of the clayey fraction of soils and the risk of collapse. The figure refers to data obtained in the San Joaquín Valley, in California, U.S.A., where collapsing soils are frequently found and have already caused many serious problems.

The soils represented by the most horizontal lines (*F* and *G*) are the ones that run the greatest risk of serious collapse. Soils like *E*, with a steeper slope, expand when they absorb water. The disadvantage of a graph like Fig. 15-28 is that it does not include the water content or the degree of saturation at which collapse may take place. There are soils which collapse long before reaching 100% saturation, whereas there are others which retain their strength at degrees of saturation very close to 100%. Nor should it be forgotten that there are soils in which the risk of collapse is related only to increases in the external load.

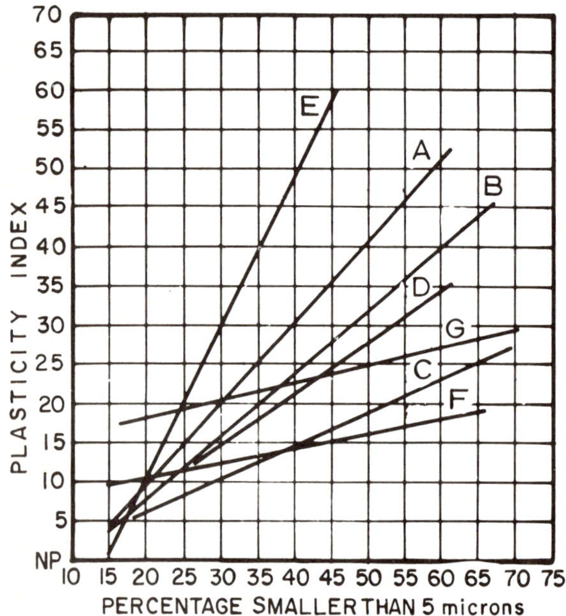

Fig. 15-28 Clay activity as a measurement of the susceptibility of soils to collapse [41]

For a more definitive identification of collapsing soils, consolidation tests have been used. JENNINGS [49] proposes a simple procedure. Two specimens of the same soil are tested in consolidometers, one with the natural water content and the other saturated. By comparing the two compressibility curves, the amount of additional settlement that is caused by saturation is obtained. The results for a specific case are shown in Fig. 15-29 [41,50].

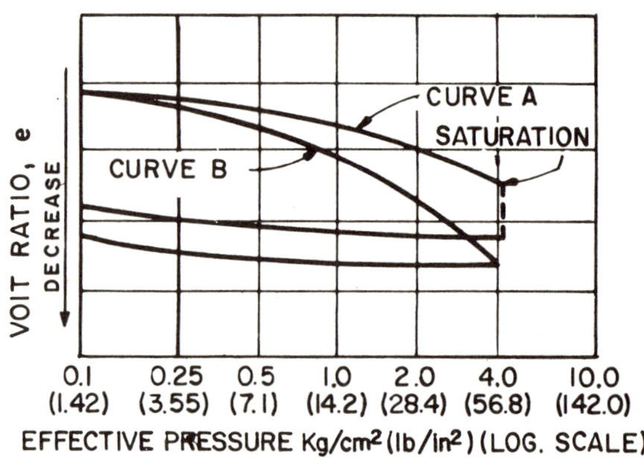

Fig. 15-29 Double consolidation test on two samples of the same soil; one saturated and the other at its natural water content [41,50]

Curve *A* corresponds to a sample of a fine clayey sand with 86% sand and 14% montmorillonite, which at its natural water content was loaded to an effective pressure of 4 kg/cm^2 (57 lb/in^2). The sample was then saturated. Curve *B* was plotted from a presaturated sample of the same soil, also loaded to 4 kg/cm^2 (57 lb/in^2). The effect of saturation under load in the case of curve *A* is evident. In curve *B* smaller void ratios are reached than in *A*; however, presaturation prevents collapse. Tests where a curve like *A*, after saturation of the sample, reaches about the same final void ratio as curve *B*, corresponding to the pre-saturated sample, can be a reasonable means of identifying collapsing soils.

Figure 15-30 [41,50] shows another test procedure for evaluating the risk of collapse. Two samples of the same soil (the same fine clayey sand with 14% montmorillonite, already mentioned) were tested in consolidometers, one sample with a water content of 3% and another with 6%, almost the same. In the first sample (broken line) saturation and collapse were induced at a pressure of approximately 3 kg/cm^2 (43 lb/in^2), while in the second sample (solid line) saturation and collapse were induced at an approximate pressure of 0.6 kg/cm^2 (8.5 lb/in^2).

It can be seen that under a heavier external load a collapse of larger proportions will occur. A test like this one may serve to estimate the collapse that is anticipated under the pressure to which the soil will be subjected in the field. It will also be seen how in this case the two samples achieve compressibility curves that eventually coincide. As expected, the sample with the lower initial water content also suffered a greater collapse. However in the test in Fig. 15-30 there are two reasons for the greater collapse occurring in Sample *1*: the lower initial water

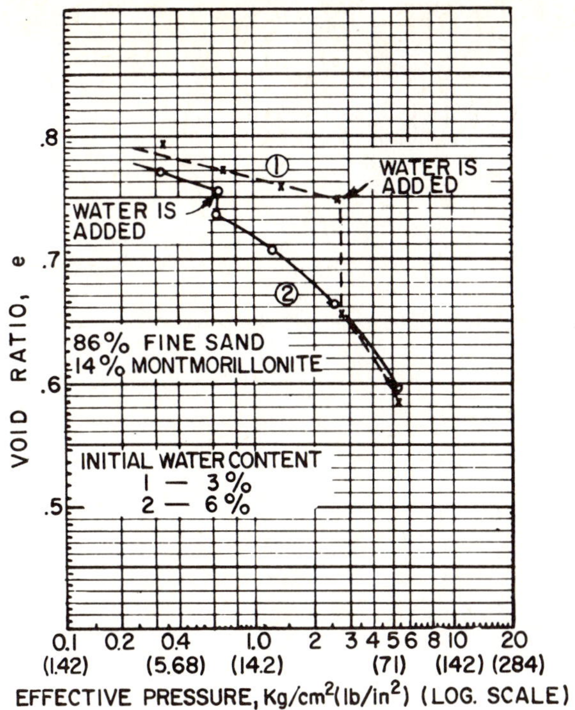

Fig. 15-30 Effect of the external load and the initial water content on the collapse of a susceptible soil [41,50]

content and the greater effective pressure acting at the time of saturation. Both higher load and lower initial moisture lead to a greater collapse in the same soil.

The best test that can be conducted on a soil that is suspected of being prone to collapse is in the field, in which the water content of the soil is increased under the real load. This is an expensive and slow test; therefore seldom performed.

Future tests based on the microphysical investigation of soils, using the electronic microscope, and x-ray diffraction are promising but so far have produced no practical criteria.

Lastly, for identifying and preventing of collapse problems in the field, the reader should study the practical information in [41]. According to this work, the water content of all the soils studied under maximum collapse conditions ranged from 13% to 39%. However, this might not apply to all soils and therefore should be considered only as an example of the lower water contents associated with collapse.

The problems of collapsing soils are more closely related to the construction of shallow foundations or hydraulic projects than to roads. There are cases where their presence might affect roads. High embankments constructed on collapsing soils might suffer sudden settlements from sudden intense rains with extremely serious consequences. Almost all the methods of preventing these problems are related to improvement of the natural ground. In extreme cases pre-flooding the soil and even vibroflotation, have sometimes been used.

15.7 Anchorages in Soils

15.7.1 Introduction

Anchorage techniques have made spectacular progress in the last 40 years. Originally conceived for reasonably intact rocks, their use has extended to broken and weathered formations and also to soils. Since their introduction, the daring, innovative applications of the different types of anchorages has increased enormously; it is not unusual for an anchor to be subjected to far greater forces than would have been considered possible a decade ago.

This growing confidence in anchorage techniques and the improvement in their reliability and capacity is not a consequence of progress in the theoretical field, nor is it the result of systematic and careful experimental work [51]. Anchorage theories are still limited, and anchor design is largely empirical. Test programs have been limited to obtain valid general hypotheses than can form the basis of fully rational design. Anchorage techniques have developed largely in tunnelling and rock slope support. Until a few years ago, almost the only types of tunnel supports were wooden or steel frames, the construction of which caused considerable obstruction to construction and consumed a great deal of precious time. In many cases, anchors provide a highly efficient and rapidly installed alternative.

Almost all the anchors that have been developed can be classified in two main categories. Expanding anchors are usually bars which are introduced into a pre-bored hole and are fitted with an expander at their embedded end, designed to increase the cross-section like a miniature footing once the anchor has been correctly installed. It thus offers a very strong resistance to extraction if sufficiently embedded. There are many different types of expanders. Figure 15-31 [52] shows one which is a satisfactory representation of many others.

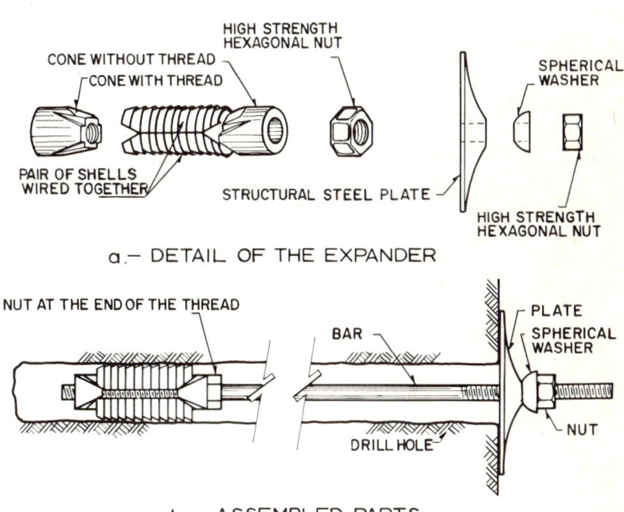

Fig. 15-31 Tension anchors [52]

In the other type of anchors, known as friction anchors, a hole is drilled in the material to be anchored, introducing into it the anchorage bar. Against this bar is installed a tube of sufficient diameter to permit cement grouting material to be injected through it, so that the bar is embedded in this grouting material and through it makes contact with the soil or rock. The bar may be fitted with an expander, like the one used with tension anchors deformations like reinforcing steel or a hook system which enables it to adhere better to the grouting material, Fig. 15-32 [52,53,54] shows some different types of friction anchors.

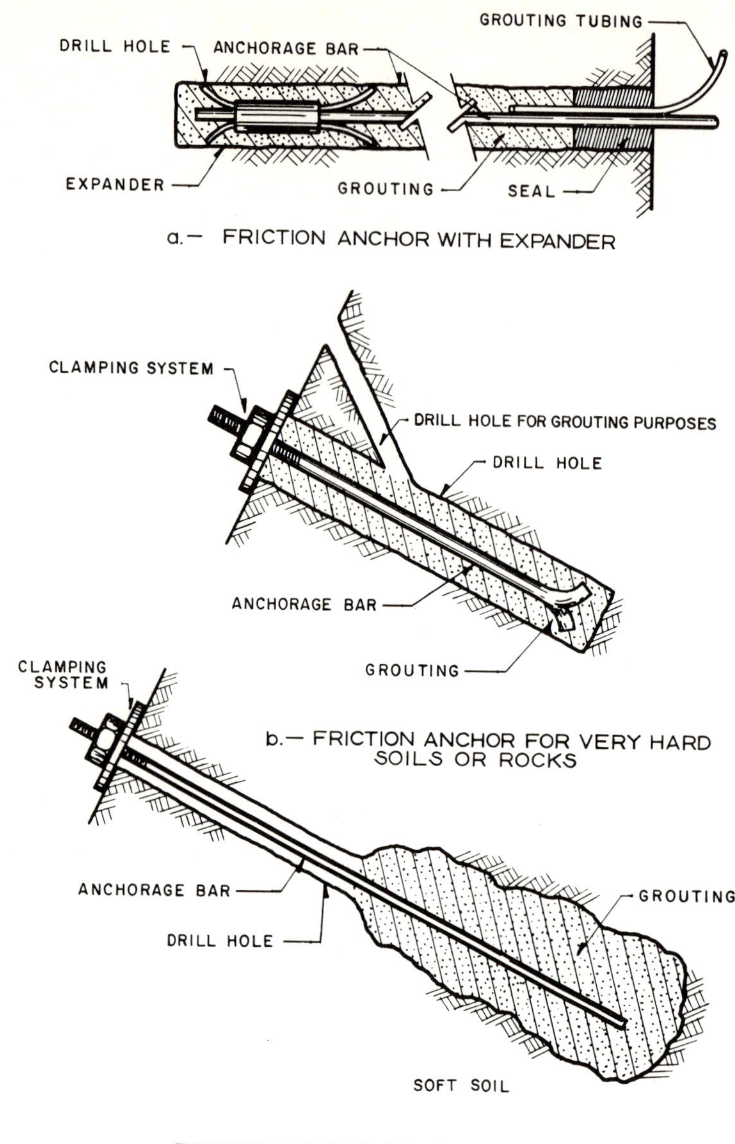

Fig. 15-32 Friction anchors [52,53,54]

15.7.2 Sequences for Computing Anchorages

Because there is no complete theory available for anchorages, nor any mathematical model to fully represent the anchor-soil or anchor-rock compatibility no reliable computation methods can be recommended. The design of an anchorage system is largely a matter of wise judgment exercised by an experienced engineer. There are, however, certain guides that can be recommended, without pretending that they can take the place of ample previous experience.

In [51] Habib discusses the lateral surface of the anchorage system that many rely on as the source of strength. Thus, an anchor is analyzed like a friction pile (Chapter 8) multiplying the lateral contact area of the anchor by the cohesion or adhesion of purely cohesive soils or by a coefficient of friction ($f = K\gamma z \tan \varnothing$) for cohesionless soils. In the latter case a value between 0.5 and 1 is often used for K. This criterion applies to friction anchors. It is far more complex to establish the strength of tension anchors, where everything depends on the bearing that is achieved between the expander and the soil and the local strength of the soil in the anchorage zone. It seems at present that this type of anchor can be reasonably designed only on the basis of experience, representative scale tests, or the bearing capacity of deeply embedded foundations. If the bearing of the expander is sufficiently great, an upper limit for the load that can be carried by the anchor is the stresses corresponding to the elastic limit of the anchorage bar.

Reference [54] recommends that two types of analysis be considered. In one a failure is considered where the anchor and its mortar jacket pull out of the drill hole owing to insufficient lateral friction. In this case, adhesion values of about 2 to 3 kg/cm^2 (28 to 43 lb/in^2) are recommended. The other failure criterion refers to the joint action of the bar and soil or fractured or soft rock. It considers a cone whose surface forms with the bar an angle of 45° $-$ $\varnothing/2$ (see Fig. 15-33).

Of these two analyses, the one should be selected where the capacity of the anchor is smaller; however the cone type failure occurs when the anchor is short and the material weak.

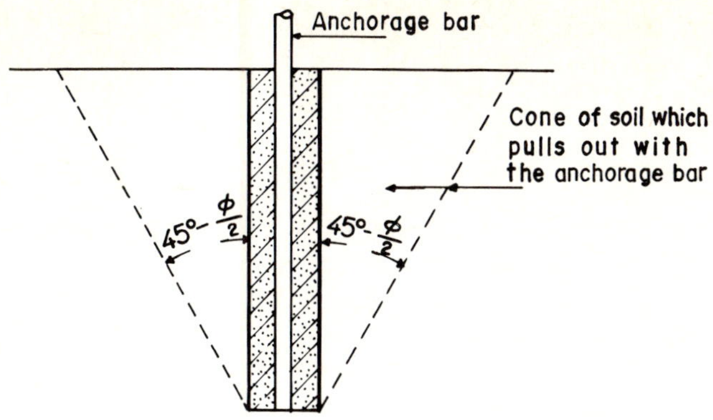

Fig. 15-33 Criterion for determining the capacity of a friction anchor which fails, tearing the soil [54]

In all the foregoing analyses a limit value should be considered which is the resistance of the anchorage bar to extraction from the grouting material in which it is embedded. This is governed by the bond between the bar and the grouting material. If the bar has deformations or expanders and hooks, the extraction load may increase greatly, to the extent of making the analysis irrelevant, if the grouting material is of good quality. These analyses refer basically to friction anchors.

Reference [52] expresses the criterion that if the mortars, resins or cement used for grouting purposes in friction anchors in rock are correctly employed and the length of embedment sufficient, the extraction load will be determined by the strength of the anchorage bar. It has been confirmed both by information from specialized literature and by tests reported by the author of [52] that the working capacity of anchors, usually established in such a way that the stress on the anchorage does not exceed two thirds of the elastic limit of the bar, can be achieved without difficulty if suitable materials are used for grouting and this process is carefully executed.

Reference [52] also gives some general rules with regard to the length of anchors and the spacing between them, relating these dimensions to those of the opening in the case of underground projects. It suggests that the length of anchors should be between 1/3 and 1/2 of the maximum width of the opening; similar values are recommended for the spacing between anchors. The diameters most frequently used for anchorage bars are from 2 to 3 cm (0.8 to 1.2 in).

For designing anchorages to support tunnel roofs in rocks or hard soils, a method of calculation advanced by Roguinsky [52] is sometimes used. He considers that above the tunnel an arch forms in which the soil or rock works under compression. It is this arch which carries the pressures that are induced in the tunnel roof and walls, Fig. 15-34 [52].

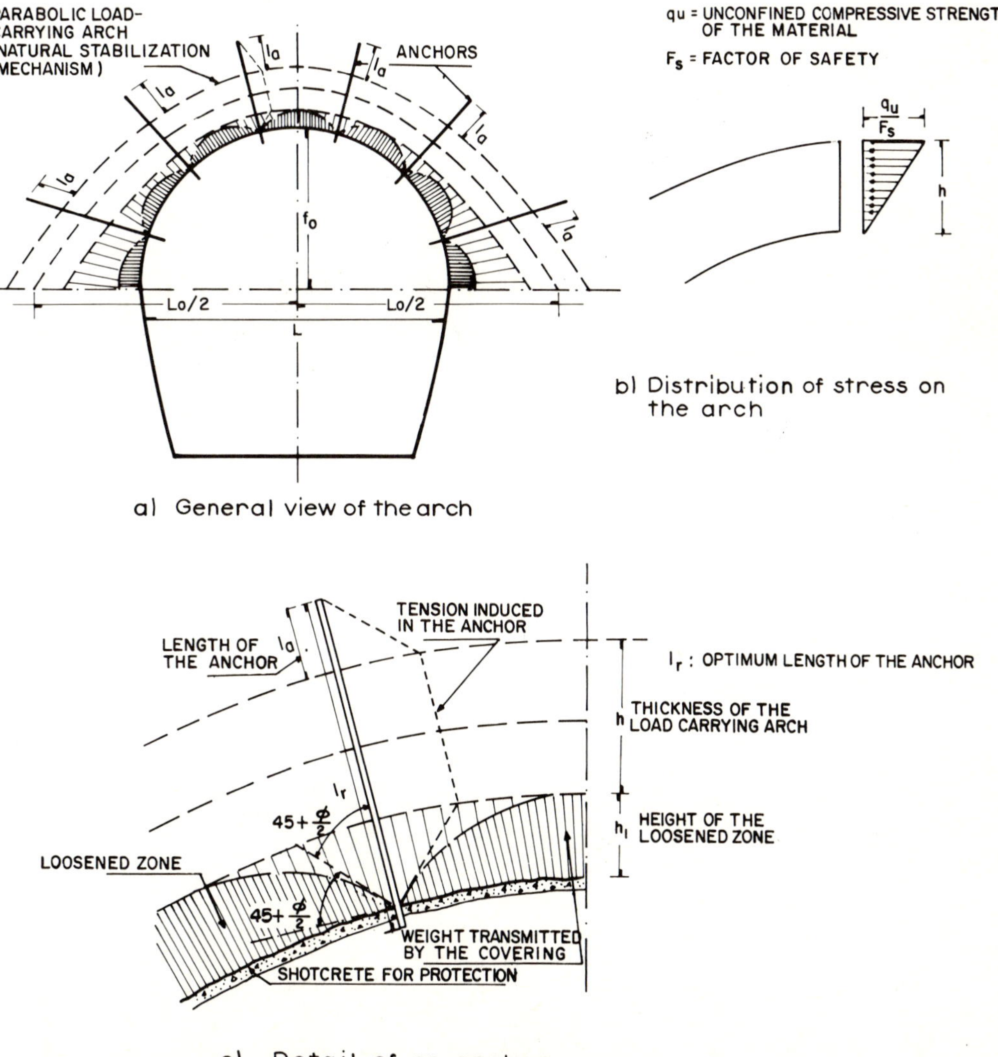

Fig. 15-34 Hypothesis for designing the lengths of friction anchors placed in weathered rhyolites to dense sand [52]

The material underneath the arch becomes loosened and lacks strength, and has to be supported by the anchorage. The shape of the load-carrying arch depends on the excavation and on the shear strength of the material in which it forms. The reaction force at the supports must not be more inclined than the angle of friction of the material. With this criterion it is possible to determine the level reached by the arch, and once this position has been determined its thickness can be computed by means of an equilibrium equation between the vertical pressure acting on the arch and the unconfined compressive strength of the material in which the arch forms:

$$\frac{P_v L_0^2}{8f_0} = \frac{q_u}{F.S.} \frac{h}{2} \qquad (15\text{-}13)$$

Reference [55] mentions an expression which permits calculation of the tension of an anchorage bar installed in clay, assuming an annular failure mechanism in which a cone of clay comes away with the bar when it fails:

$$T = \pi d L \Upsilon h \tan \varnothing \left(1 + \frac{h}{d} \tan \varnothing\right) \qquad (15\text{-}14)$$

where: P_v is the vertical pressure acting on the load-carrying arch. L_0 is the width of the arch. f_0 is the rise of the arch. $F.S.$ is the factor of safety that is employed for calculation.

The above equation considers a linear stress distribution in the arch, as is shown in Part *b* of the figure. Now the weight of the loosened rock below the arch can be found, and suitable supports can be designed in accordance with this value. These supports may be the anchors themselves, although the principal function of these elements is to assist in the successful performance of the tunnel roof as a whole.

where: d is the diameter of the anchorage bar. L is the length of the anchorage bar. h is the thickness of soil above the anchorage bar, computable for purposes of calculation of the normal pressure on the bar.

In the above formula the submerged unit weight should be used when appropriate, and the angle of strength of the soil in terms of effective stress. The equation includes an effect of precompression of the soil under grouting pressure.

For the same failure conditions as before, the following expression [55] can again be used:

$$T = \pi d L (c + \overline{\sigma} \tan \varnothing) \qquad (15\text{-}15)$$

where the letters have the same meaning as already seen, the c and $\varnothing$ are the strength parameters of the clay in terms of effective stress.

15.7.3 Anchorage Strength: Field Tests

Limited information is available about field tests on anchorages [51], and consequently it is very hard to draw any general conclusions. Reference [52] presents some interesting results, which are given in Figs. 15-35 and 15-36. In both cases they are the results of extraction tests performed on cement-grouted anchors installed in poor quality rocks (so weathered and broken that they could be considered residual soils). The graphs in Fig. 15-35 correspond to anchorages in an andesite with pockets of loose sand and soft clayey silt. Those in Fig. 15-36 refer to a rhyolite, almost weathered to silty sand.

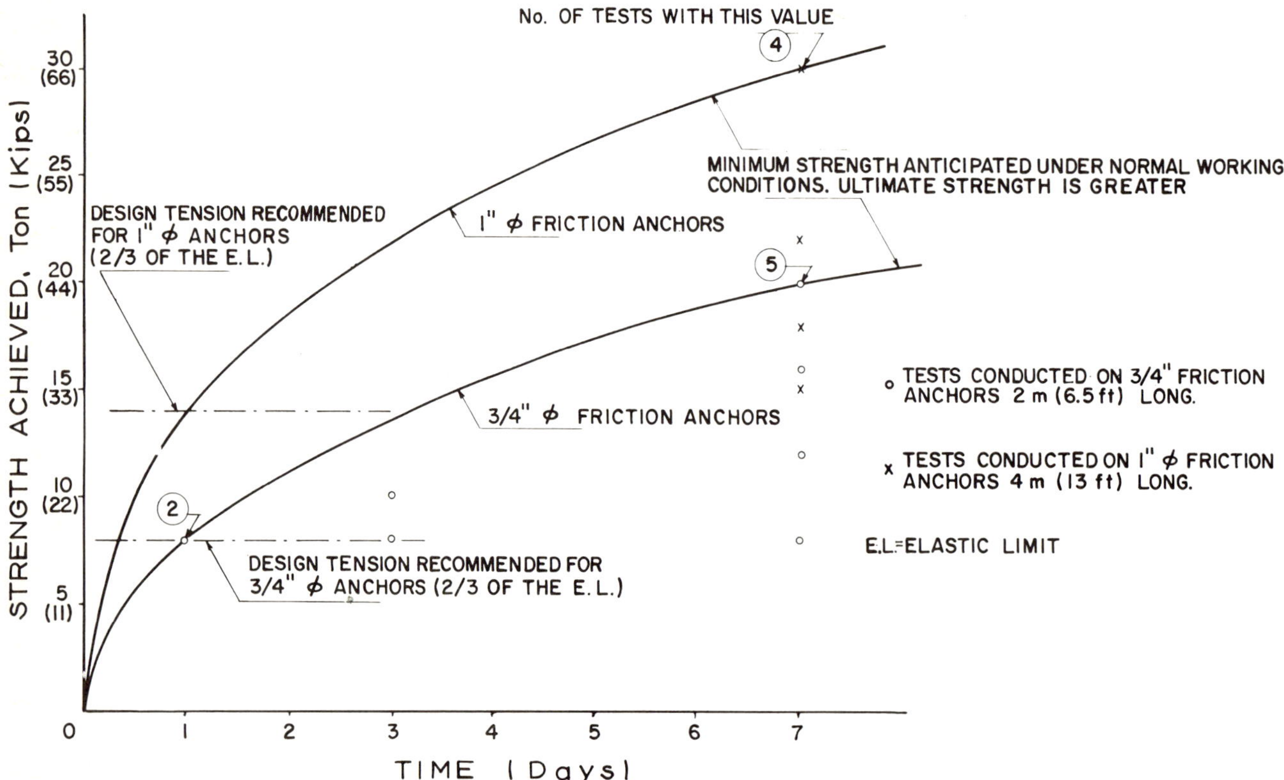

Fig. 15-35 Results of extraction tests on anchors installed in fractured andesite with pockets of sand and clayey silt [52]

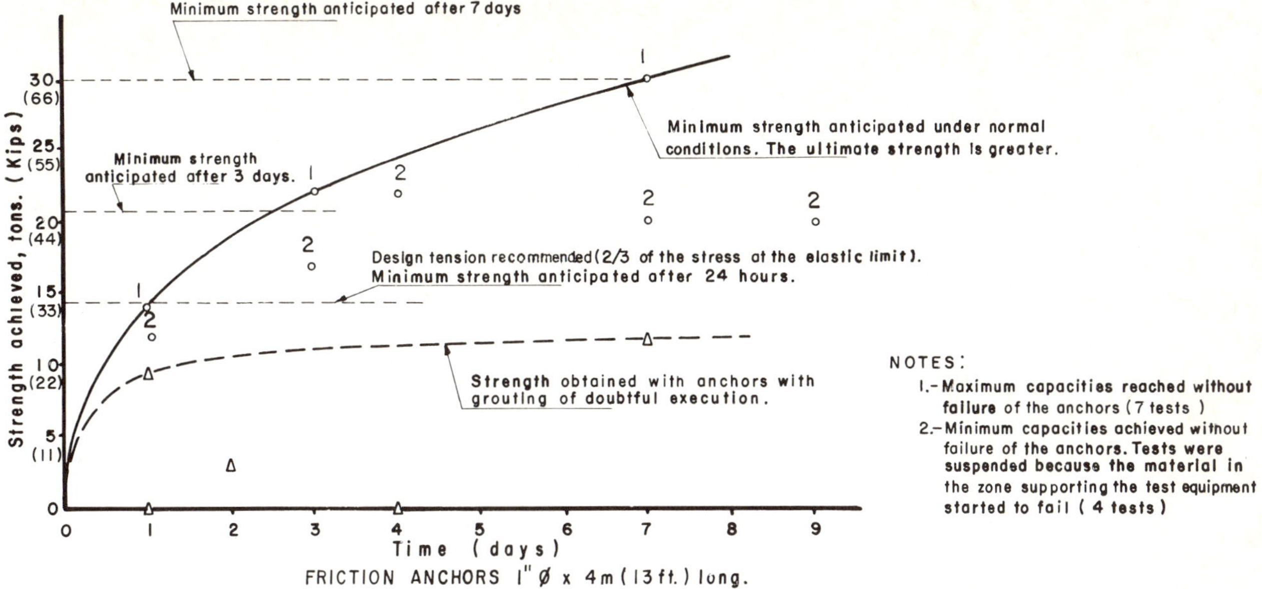

Fig. 15-36 Results of extraction tests on anchors installed in weathered rhyolite to dense sand [52]

The results of the extraction tests indicate that when anchors are installed in accordance with a good construction practice and the anchor is long enough, extraction loads are determined by the ultimate strength of the bar (Plate 15-9). However, when unsuitable materials are used, grouting is defective or the anchor is too short the extraction load is reduced considerably. In well made anchors, failures that are not a result of rupture of the bar are due to lack of adhesion between the bar and the grouting material. A very important fact is that in many of the tests reported there was no failure between the anchor and the soil. However in other tests the limit of capacity is the shear between the grout and the soil or rock.

Plate 15-9 Extraction test on an anchor in a tunnel (Courtesy of Geosistemas, S. A.).

Reference [52] lists a series of tests and measurements conducted to determine anchor tension. Measurements are given for a volcanic tuff weathered to a very dense silty sand, but cemented in places, with an unconfined compressive strength of about 40 kg/cm^2 (568 lb/in^2). The work was carried out inside the Central Emission tunnel in the Valley of Mexico. The tension in the anchors was measured with strain gauges, Fig. 15-37 [52] shows the arrangement of the anchors, and Figs. 15-38 and 15-39, from the same reference, the principal results.

The tension recorded was partically constant throughout much of the length of the anchor, but fell to zero at the end which is embedded in the soil. The information in Fig. 15-38, shows the tension is relieved shortly after installation of the anchor, but subsequently rises in all tests of a similar type.

Reference [55] describes a set of tests that were performed on friction anchors installed in clays, Fig. 15-40 refers to one such test, in which a gradually increasing tension was applied to a cement-grouted anchor bar, and the respective movements of the anchor head were measured.

Reference [56] describes a type of anchor based on the ideas of L. Ménard. It is constructed by introducing a membrane into a pre-bored hole in the soil. This membrane subsequently expands, and the entire cavity that is produced is filled with cement mortar by hydraulic pressure. The tension is developed by a set of 6 wires which are sunk into the mortar. In the reference, numerous field tests are presented that depict the behavior of these anchors. Two cases are described where performance tests were conducted. In the first, the anchors were installed in a sandy soil resulting from the decomposition of granite, and in the second in a stratified soil with lenses of clay, sandy clay and thin layers of fine sand. Anchor lengths ranged from 6 to 13 m (20 to 43 ft), Fig. 15-41 [56] gives a summary of the most relevant information obtained from these field tests.

Part *a* of the figure shows displacement of the anchor head versus the extraction load that is applied, comparing the anchor described with a cement mortar pile of similar geometry. In this case both the anchor and the pile were 6 m (20 ft) long. Part *b* shows the distribution of axial force throughout the anchor, obtained from strain gauges attached to the wires, which permitted calculation of the axial stresses at different points, and the axial forces. Part *c* gives the displacement of the heads of two anchors with a constant extraction load as a function of time. It is noted that even for very high loads, the *creep* undergone by the anchors is negligible after the initial deformation. Lastly, Part *d* shows the stress-displacement curves at different depths, in the soil in contact with the anchors.

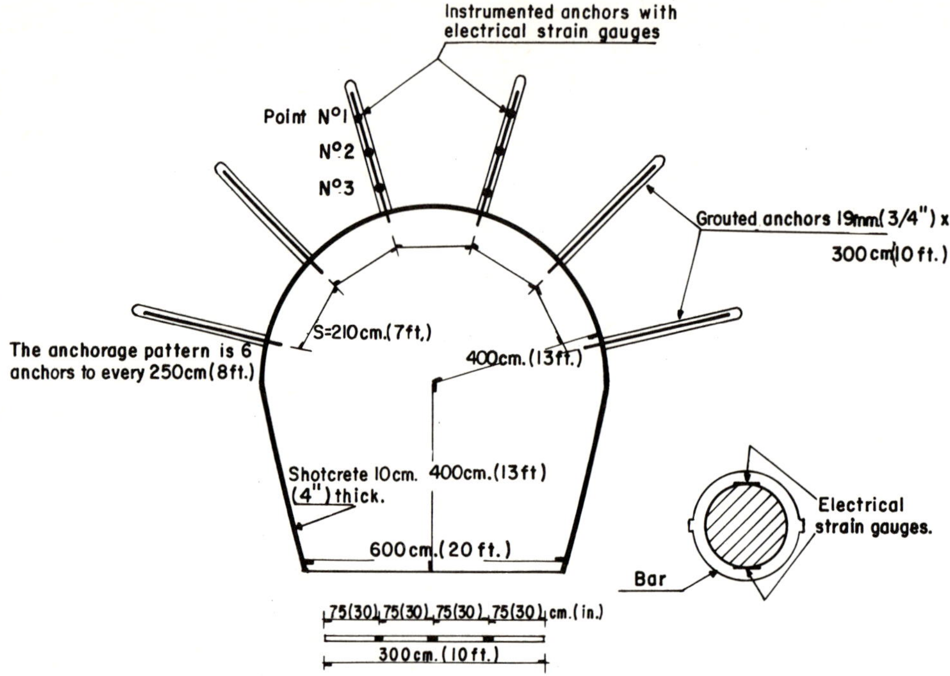

Fig. 15-37 Position of instrumented anchors [52]

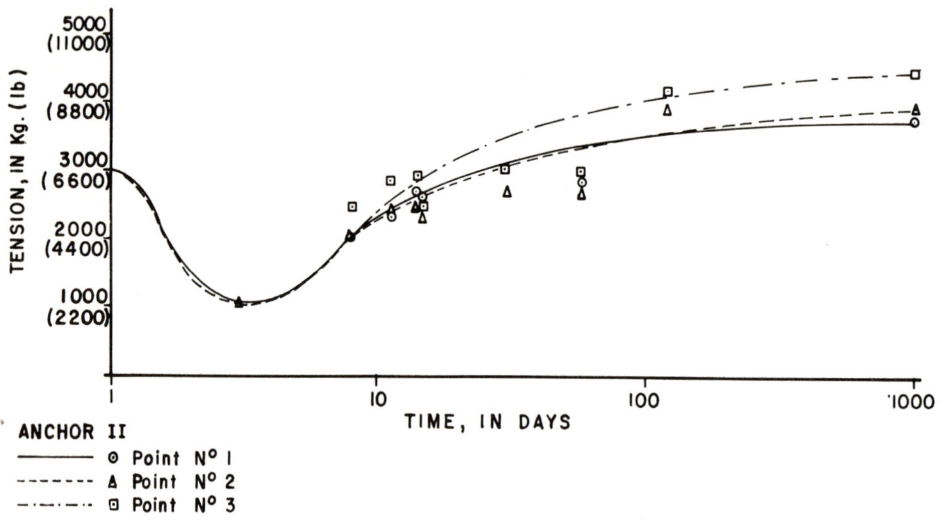

Fig. 15-38 Variation of the tension induced in relation to time [52]

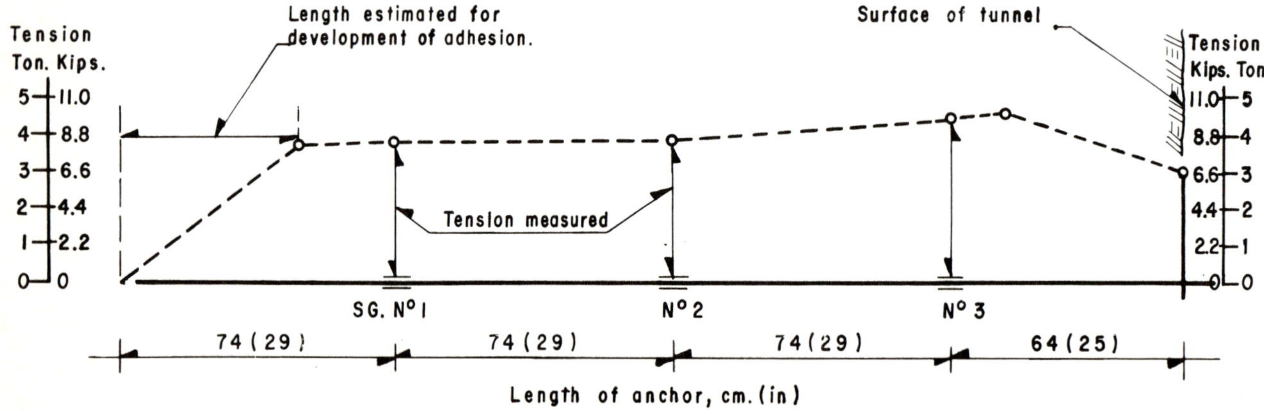

Fig. 15-39 Distribution of the tension induced throughout the anchor [52]

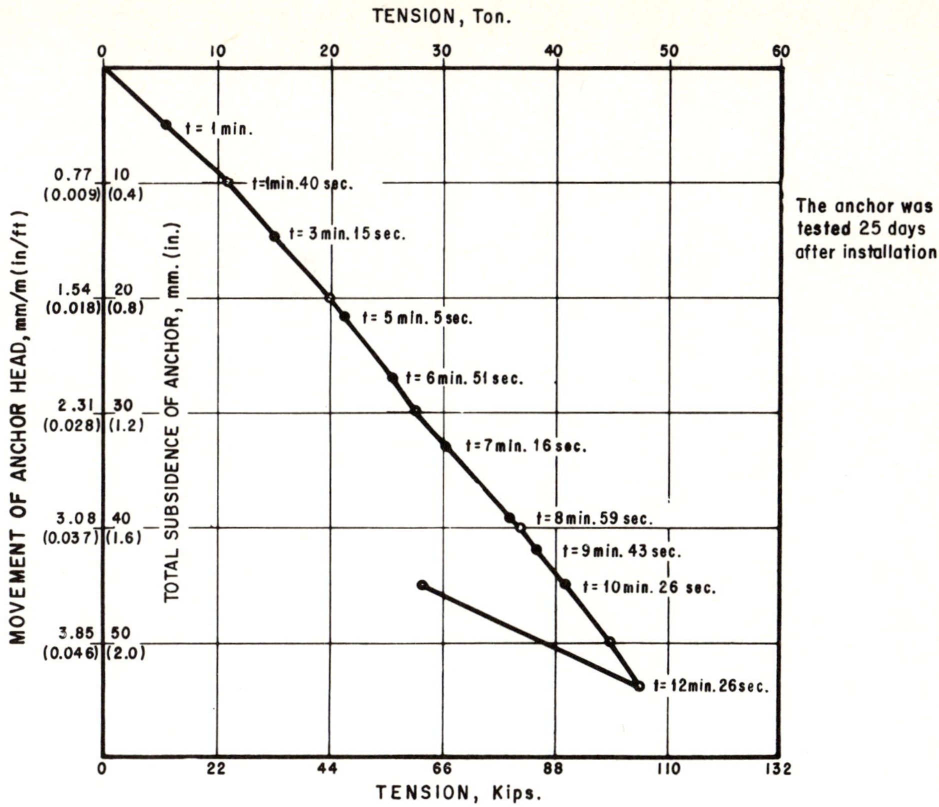

Fig. 15-40 Behavior of a friction anchor in clay measured in an extraction test [55]

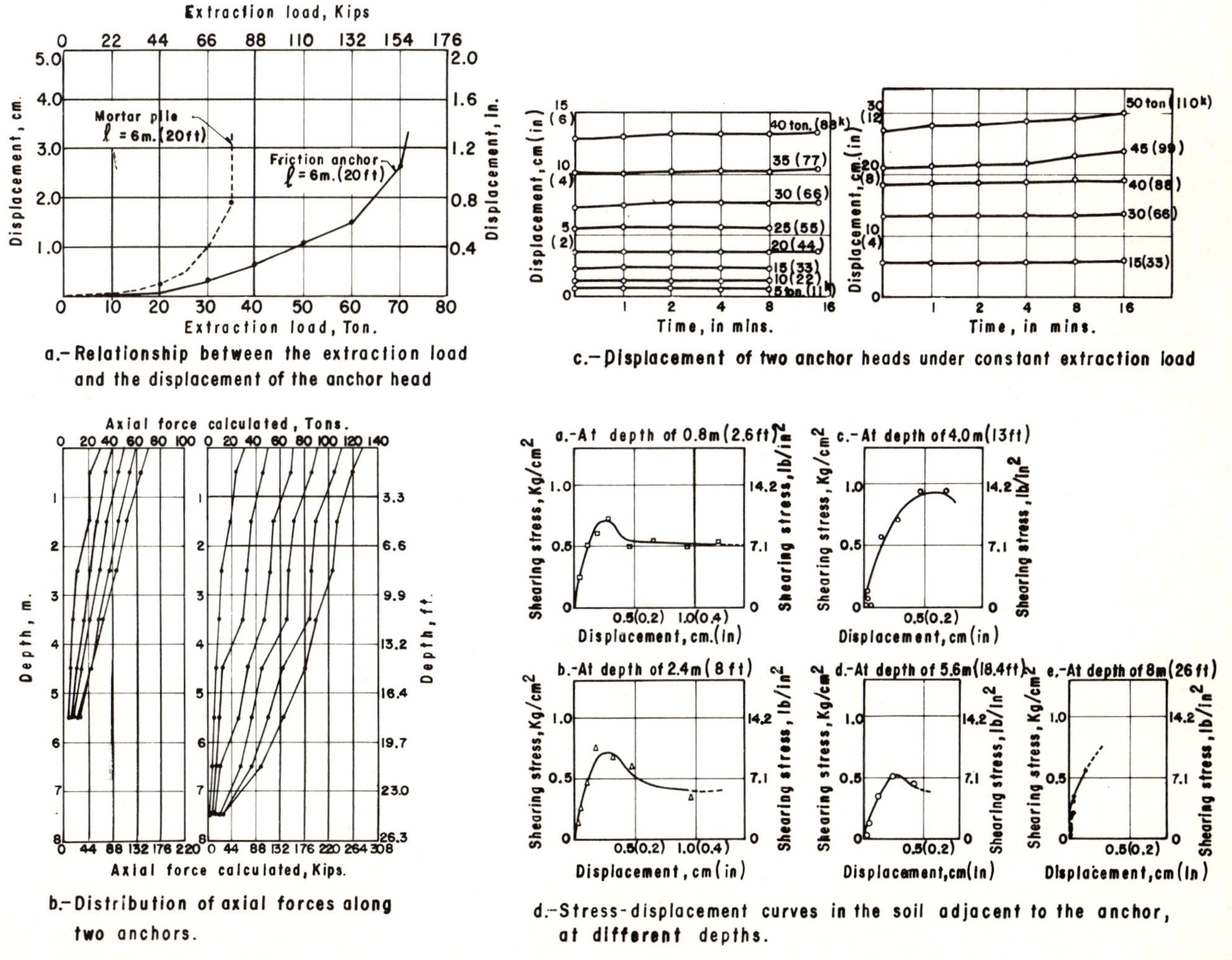

a.- Relationship between the extraction load and the displacement of the anchor head

c.- Displacement of two anchor heads under constant extraction load

b.- Distribution of axial forces along two anchors.

d.- Stress-displacement curves in the soil adjacent to the anchor, at different depths.

Fig. 15-41 Result of field tests on a friction anchor [56]

APPENDIX 15A SOLVED PROBLEMS

15a.1 Scour Problem

A bridge is to be built over a river with the subsoil stratigraphy shown in Fig. 15a.1. Characteristics of the bridge are also indicated in the figure.

The hydraulic data at the crossing site are as follows:

Q_d = design volume of flow = 1,370 m^3/sec (48500 ft^3/sec)
v = velocity = 2.3 m/sec (7.5 ft/sec)
T_r = interval of flood recurrence = 50 years

In Fig. 15a.1 the elevation of the river surface is indicated in accordance with flood data. The axis of the bridge is perpendicular to the direction of flow. It is wished to compute the depth of general scour in the riverbed and the depth of local scour around the bridge piers.

.1 General Scour

To determine general scour, the Lischtvan-Levediev criterion, [7] will be used.

Stratum of fine, slightly silty, low-density sand, d_m = 0.7 mm (0.028 in)

for cohesionless soils:

$$H_s = \left(\frac{\alpha H_0^{1.667}}{0.68\, \beta d_m^{0.28}} \right)^{\frac{1}{1+X}} \tag{15a-1}$$

where:

H_s = depth of general scour, in m
H_0 = depth before erosion, in m

$$\alpha = \frac{Q_d}{H_m^{1.667}\, B_e \mu} \tag{15a-2}$$

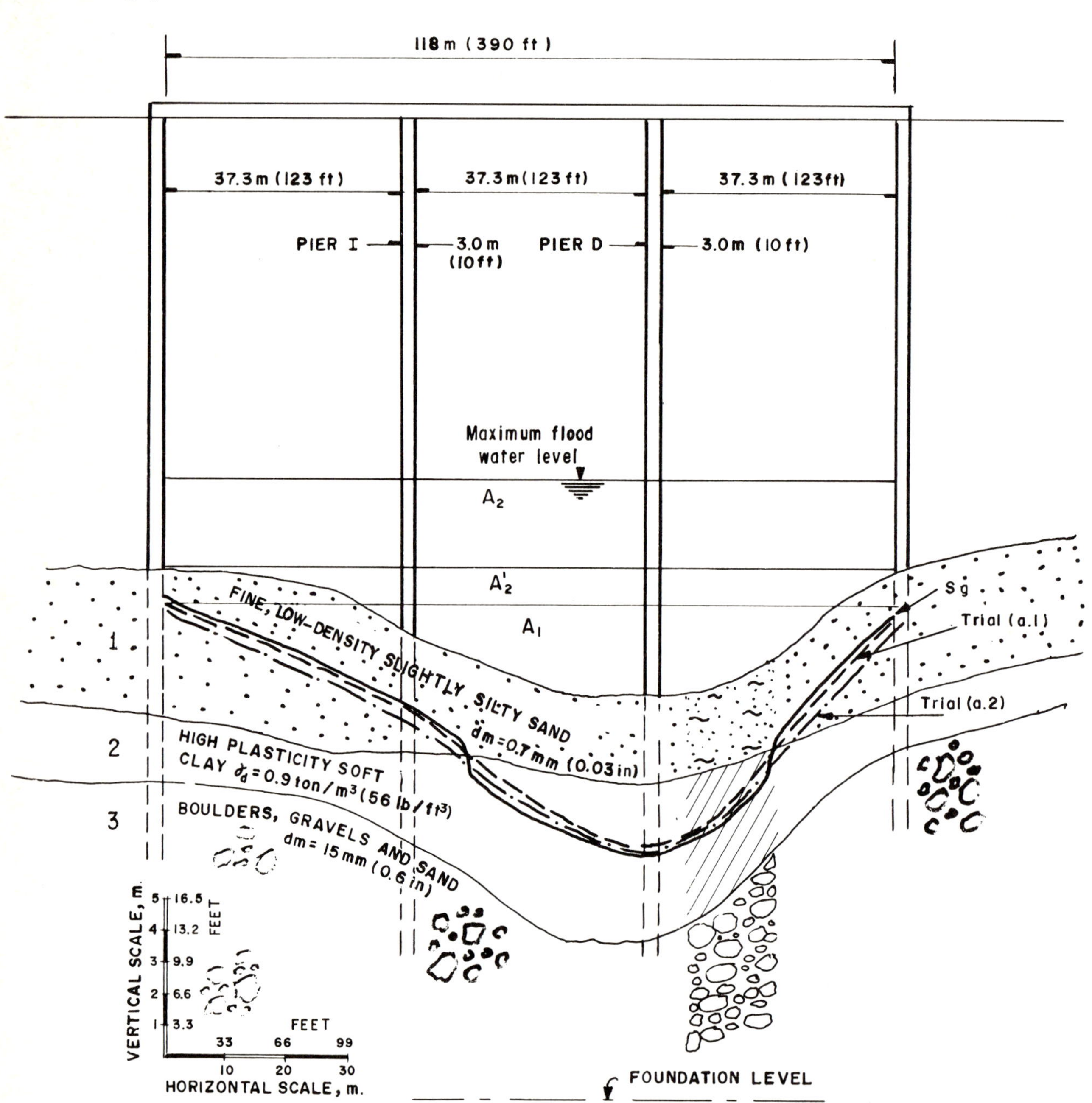

Fig. 15a-1 Profile in the crossing zone

With the water level data, a hydraulic area $A = 595.7\ m^2$ (6400 ft^2) is obtained.

B_e = effective width = 112 m (368 ft) (the effective width is obtained by subtracting the width of the piers from the total width, when the angle of incidence of the stream in relation to the axis of the pier is zero).

H_m = average depth = $\frac{A}{B_e}$ = 5.32 m (17.5 ft)

μ = coefficient of contraction = 0.98 (Table A-3.4 in [7]) with a space of 37 m (122 ft) between two piers and v = 2.3 m/sec (7.55 ft/sec).

Applying Eq.(15a-2) the following expressions are obtained:

$$\alpha = \frac{1370}{5.32^{1.667} \times 112 \times 0.98} = 0.769$$

β = 0.97 (Table A-3.2 in [7]) for T_r = 50 years

d_m = 0.7 mm (0.028 in), obtained from results of gradation tests

$\frac{1}{1+X}$ = 0.71 (Table A-3.3, in [7]) with d_m = 0.7 mm (0.028 in).

Using Eq.(15a-1)

$$H_s = \left(\frac{0.769\ H_0^{1.667}}{0.68 \times 0.97 \times 0.7^{0.28}}\right)^{0.71} = (1.288\ H_0^{1.667})^{0.71} \quad (15a\text{-}3)$$

In order to determine the probable scour profile along the riverbed, values for H_0 corresponding to different points are calculated from Eq.(15a-3) and given in Table 15a-1.

The depth of general scour, S_g, is obtained by subtracting H_0 from H_s, Table 15a-1. The scour profile is shown in *trial (a.1)* in Fig. 15a-1.

It will be observed that erosion affects the clay stratum, and it will therefore be necessary to calculate the scour in this stratum.

Stratum of high plasticity, soft clay, γ_d = 0.9 t/m^3

Cohesive soils (using Eqs. from [7])

$$H_s = \left(\frac{\alpha H_0^{1.667}}{0.60\ \beta \gamma_d^{1.18}}\right)^{\frac{1}{1+X}} \quad (15a\text{-}4)$$

$$\alpha = \frac{Q_d}{H_m^{1.667}\ \beta_e \mu}$$

γ_d = 0.9 t/m^3 (56 lb/ft^3)

$\frac{1}{1+X}$ = 0.68 (Table A-3.3 in [7]) with γ_d = 0.9 t/m^3 (56 lb/ft^3).

Applying Eq.(15a-4):

$$H_s = \left(\frac{0.769\ H_0^{1.667}}{0.60 \times 0.97 \times 0.9^{1.18}}\right)^{0.68} = (1.496\ H_0^{1.667})^{0.68}$$

Using different H_0 values we obtain the values for H_s and S_g given in Table 15a-2.

In Fig. 15a-1, these results are given in *trial (a.2)*. The general scour profile S_g is shown in Fig. 15a-1.

Table 15a-1
Calculation of H_s and S_g from H_0 using Eq. 15a-3, §15a.1.1

H_0		H_s		S_g	
m	ft	m	ft	m	ft
2.8	9.2	4.05	13.3	1.25	4.1
4.8	15.8	7.66	25.2	2.86	9.4
6.8	22.4	11.57	37.9	4.77	15.6
3.0	9.9	4.39	14.4	1.39	4.6

Table 15a-2
Calculation of H_s and S_g from H_0 using Eq. 15a-4, §15a.1.1

H_0		H_s		S_g	
m	ft	m	ft	m	ft
2.8	9.2	4.22	13.9	1.42	4.7
4.8	15.8	7.78	25.5	2.98	9.8
6.8	22.4	11.55	38.0	4.75	15.6
3.0	9.9	4.57	15.0	1.57	5.2

.2 Local Scour

For computation it is assumed that local scour around the piers occurs *after* general scour. Under these circumstances, local scour will be computed in the deposit of high plasticity clay using YAROSLAVTZIEV's method [7].

Local scour s_0 is given by:

$$s_0 = K_f\, K_v\, (e + K_H) \frac{v'^2}{g} - 30\, d \quad (15a\text{-}5)$$

where:

s_0 = scour depth, in m.

K_f = coefficient which depends generally on the shape of the nose of the pier and the angle of incidence between the flow and the axis of this flow = 12.4. (Fig. A-III, [7] rectangular pier, angle of incidence = 0).

K_v = coefficient defined by the expression:

$$\log K_v = -\ 0.28 \sqrt[3]{v'^2/gb_1} = 0.73$$

(Fig. A-III.8, [7] with b = 3 m (9.9 ft), v' = 1.82 m/sec (6 ft/sec), g = 9.81 m/sec^2 (32 ft/sec^2) v'^2/gb_1 = 0.1126).

v' = average velocity of flow upstream of the pier, after general erosion has taken place, in m/sec.

To obtain the velocity v' corresponding to the increased hydraulic area A', this is calculated in the crossing profile, and v' will be:

$$v' = \frac{Q_d}{A'} \qquad (15a\text{-}6)$$

Applying Eq.(15a-6) with $A' = 753.8\ m^2$ (8120 ft^2), $v' = 1.82$ m/sec (6 ft/sec) is obtained.

b_1 = Projection of the section of the pier on a plane perpendicular to flow. When the angle of incidence is 0, b_1 is equal to the width b of the pier; $b_1 = b = 3$ m (9.9 ft).

e = correction factor, the value of which depends on the place where the piers are installed = 0.6 (piers constructed in the main channel).

K_H = coefficient that takes into account the depth of the channel, defined by the expression:

$\log K_H = 0.17 - 0.35\ H/b_1 = 0.08$
(Fig. A-III.9 in [7]; $H/b_1 = 3.867$)

H = depth of the channel in front of the pier = 11.60 m (38 ft).

d = equivalent diameter for cohesive soils = 0.01 m (0.033 ft) (Table A-III.8 of [7]) with γ_d = 0.9 t/m^3 (56 lb/ft^3).

Substituting values in Eq.(15a-5)

$$S_o = 12.4 \times 0.73\ (0.6 + 0.08)\ \frac{1.82^2}{9.81} - 30 \times 0.01$$

$$S_o = 2.4 \text{ m } (7.9 \text{ ft})$$

The local scour around the bridge piers will be of about 2.4 m (7.9 ft).

(This problem is by courtesy of AGUSTÍN DEMÉNEGHI COLINA)

15a.2 Preload Problem

An embankment is to be constructed with a height of 2.5 m (8.2 ft). The first 6.0 m (19.7 ft) of the foundation subsoil is a high plasticity, soft, compressible clay. Underneath there is an indefinite thickness of a medium fine sand. Determine the influence of a preload one meter thick and another two meters thick on the future behavior of the embankment. The geometric characteristics, along with the mechanical properties of the compressible soil in the foundation ground and those of the body of the embankment, are as follows (see Fig. 15a-2):

Mechanical properties of Stratum *1*:

c_u = Apparent cohesion obtained in an unconfined compression test = 2 t/m^2 (412 lb/ft^2).

e_0 = Initial void ratio = 6.0

w = Natural water content = 300%

a_v = Coefficient of compressibility, obtained from the compressibility curve = 0.14 m^2/t (0.68 ft^2/kip).

k = Coefficient of permeability = 10^{-7} cm/sec (2×10^{-7} ft/min) = 10^{-9} m/sec.

m_v = Modulus of volumetric compressibility = $a_v/1 + e_0$ = 0.020 m^2/t (0.1 ft^2/kip).

.1 Calculation of the settlement in the 2.5 m (8.2 ft) embankment under surcharge 1.0 m (3.3 ft) thick, considering 100% primary consolidation.

1) The stressed zone, *pressure bulb* will be determined and the vertical stress acting on the plane located through the center of the compressible stratum is found using Fig. 3-3 (Chapter 3) see Table 15a-3.

   For this case: a = 5.0 m (16.4 ft), c = 3.5 m (11.5 ft), $p = 1.7 \times 2.5 = 4.25\ t/m^2$ ($p = 0.106 \times 8.2 = 0.87\ kips/ft^2$)

2) The settlement due to primary consolidation is calculated

   For z = 3.0 m (9.9 ft), $\sigma_z = 4.08\ t/m^2$ (840 lb/ft^2)

   Since m_v is constant for the entire stratum:

   $$\Delta H_f = m_v\,\Delta p' H$$

   A value of $\Delta p'$, representative of the 6.0 m (19.7 ft) is:

   $$\Delta p'_m = 4.04\ t/m^2\ (830\ lb/ft^2)$$

   H = thickness of the compressible stratum = 6.0 m (19.7 ft) consequently:

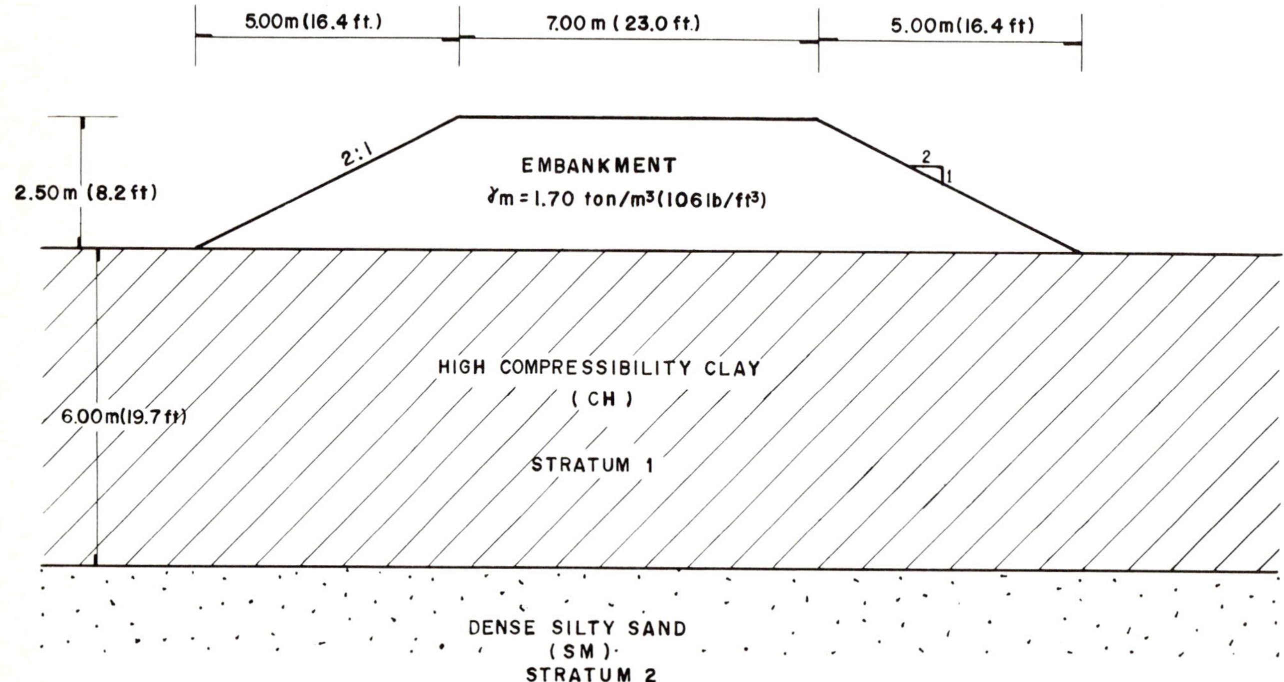

Fig. 15a-2 Profile of the section under analysis

Table 15a-3
Calculation of the vertical stress acting on plane, §15a.2.1

z (m)	a/z	c/z	I	$\sigma_z = 2\,I.p$
1	5.0	3.5	0.5	4.25
2	2.5	1.75	0.49	4.17
3	1.67	1.17	0.48	4.08
4	1.25	0.88	0.45	3.83
5	1.00	0.70	0.43	3.65
6	0.83	0.58	0.39	3.32

$\Delta H_f = 0.020 \times 4.04 \times 6.0 = 0.49$ m
$(\Delta H_f = 0.1 \times 0.83 \times 19.7 = 1.62$ ft)

3) The time in which the different percentages of primary consolidation take place is computed:

The following Equation is employed:

$$t = \frac{a_v \gamma_w H^2 T}{k(1+e_0)} \tag{1-48}$$

where γ_w = unit weight of water = 1.0 t/m³ (62.4 lb/ft³), and
T = time factor for the different percentages of consolidation. Note that the compressible layer is assumed to be drained through its upper and lower boundaries (effective thickness: 3.0 m (9.9 ft)).
t = 2083 T (days). With the aid of Table 1-1 of Chapter 1, Table 15a-4 can be computed.

Column Δh_f was obtained by multiplying the total settlement ΔH_f by the different degrees of consolidation, U%.

Similarly, after calculating the total settlement under the embankment and the surcharges 1 and 2 m (3.3 and 6.6 ft) thick (which is done next) the two following columns of Table 15a-4 are calculated by the same procedure.

4) Calculation of the settlement that will be experienced by the compressible stratum for a surcharge with an additional thickness of 1.0 m (3.3 ft).

It is assumed that for the surcharges γ_m = 1.7 t/m³ (106 lb/ft³).

$$p = (2.5 + 1.0)\,1.7 = 5.95 \text{ t/m}^2$$
$$p = (8.2 + 3.3)\,106 = 1220 \text{ lb/ft}^2$$

In view of the homogeneity of the bearing layer, settlements are computed on the basis of the data for the center of that stratum (z = 3.0 m (9.9 ft); I = 0.96. Table 15a-3)

$$\sigma_z = 0.96 \times 5.95 = 5.71 \text{ t/m}^2 \ (1180 \text{ lb/ft}^2) = \Delta p'$$

$$\therefore \Delta H_{f+s} = m_v\,\Delta p'\,H = 0.020 \times 5.71 \times 6 = 0.69 \text{ m}$$
$$= 0.1 \times 1.18 \times 19.7 = 2.27 \text{ ft}$$

The time and the different percentages of primary consolidation are given in Table 15a-4.

In Fig. 15a-3, the time evolution of primary settlements is shown for settlements ΔH_f and ΔH_{f+s} (surcharge = 1 m (3.3 ft)).

Also, the degree of consolidation in the center of the compressible stratum, U_{f+s}, on removal of the surcharge must fulfil the following condition:

$$U_{f+s}\,\Delta h_{f+s} = \Delta h_f \tag{15-2}$$

$$\therefore U_{f+s} = \frac{\Delta h_f}{\Delta h_{f+s}} \tag{15-3}$$

Substituting values:

$$U_{f+s} = \frac{0.49}{0.69} = 0.71$$

$$\therefore U_{f+s} = 71\%$$

For a U_{f+s} of 71%, Δh_{f+s} = 0.48 m (1.57 ft) and t = 844 days (Table 15a-4)

It is considered that 150 days (5 months) is sufficient time for the surcharge to meet the requirements of the work program.

Table 15a-4
Calculation of settlement, §15a.2.1

U (%)	T	t, days	Δh_f		Δh_{f+s}		ΔH_{f+s}	
			m	ft	m	ft	m	ft
0	0	0	0	0	0	0	0	0
10	0.008	17	0.05	0.17	0.07	0.23	0.09	0.30
20	0.031	65	0.10	0.33	0.14	0.46	0.18	0.59
30	0.071	148	0.15	0.50	0.21	0.69	0.26	0.85
40	0.126	262	0.20	0.65	0.28	0.92	0.35	1.15
50	0.197	410	0.25	0.82	0.35	1.15	0.44	1.44
60	0.287	598	0.29	0.95	0.41	1.34	0.52	1.71
70	0.405	844	0.34	1.12	0.48	1.58	0.61	2.00
80	0.565	1176	0.39	1.28	0.55	1.80	0.70	2.30
90	0.848	1765	0.44	1.44	0.62	2.04	0.79	2.60
95	1.127	2343	0.47	1.54	0.66	2.17	0.84	2.76
100	∞	∞	—	—	—	—	—	—
					S = 1.0 m (3.3 ft)		S = 2.0 m (6.6 ft)	

For t = 150 days, the percentage of primary consolidation that would occur for Δh_f can also be computed with the data in Table 15a-4. If t = 150 days, Δh_{f+s} is 21 cm (8.3 in) and corresponds approximately to 40% primary consolidation without any surcharge, which would require 262 days.

.2 If the surcharge is 2.0 m (6.6 ft) thick

$$p = (2.5 + 2.0)\ 1.7 = 7.65 \text{ t/m}^2$$
$$p = (8.2 + 6.6)\ 106 = 1575 \text{ lb/ft}^2$$

At a depth of 3.0 m (9.9 ft)

$$\sigma_z = 0.96 \times 7.65 = 7.35 \text{ t/m}^2\ (1520 \text{ lb/ft}^2) = \Delta p' \quad \text{(Table 15a-3)}$$

$$\therefore h_{f+s} = m_v\ \Delta p'\ H = 0.020 \times 7.35 \times 6 = 0.88 \text{ m}$$
$$= 0.1 \times 1.52 \times 19.7 = 2.90 \text{ ft}$$

The time and the different percentages in which primary consolidation takes place are given in Table 15a-4.

In Fig. 15a-3, the time rate of primary settlement is shown, for settlements ΔH_f and ΔH_{f+s}, for s = 2.0 m (6.6 ft). For this case:

$$U_{f+s}\ h_{f+s} = \Delta h_f$$

$$\therefore U_{f+s} = \frac{\Delta h_f}{\Delta h_{f+s}} = \frac{0.49}{0.88} = 0.55$$

Consequently, U_{f+s} = 55% in the center of the stratum.

According to Table 15a-4, this degree of consolidation corresponds to the following equation:

$$\Delta h_{f+s} = 0.48 \text{ m (1.57 ft) in } t = 504 \text{ days}$$

Now it is possible to analyse what would happen if the 2.0 m (6.6 ft) surcharge were applied for 150 days (5 months).

If t = 150 days, the percentage of primary consolidation that will occur for Δh_f can be computed with the data in Table 15a-4. If t = 150 days, Δh_{f+s} is 26 cm (10.2 in) and corresponds to approximately 50% primary consolidation without any surcharge, which would require 410 days.

(This problem is by courtesy of GABRIEL GARCÍA ALTAMIRANO of the Mexican Ministry of Public Works)

15a.3 Anchor Problem

At the toe of the embankment in Fig. 15a-4 it is wished to leave a space for future use. For this purpose the soil is retained by sheet piling or a prefabricated wall, the upper part of which will be restrained by anchors. The design of these anchors is the object of the present problem.

A surcharge of 4 t/m (2.7 kips/ft) is placed on the embankment, in the position indicated. The characteristics of the embankment material, which is a silty sand, are as follows:

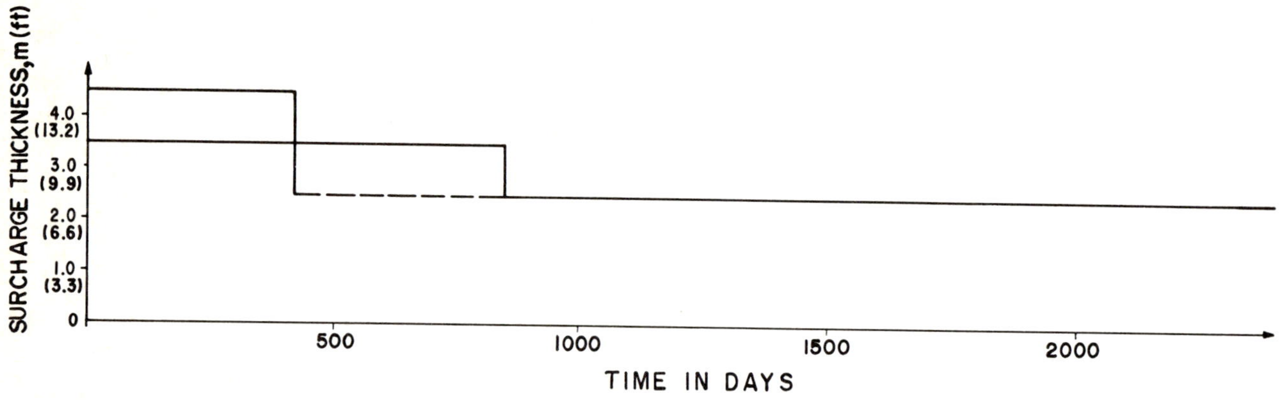

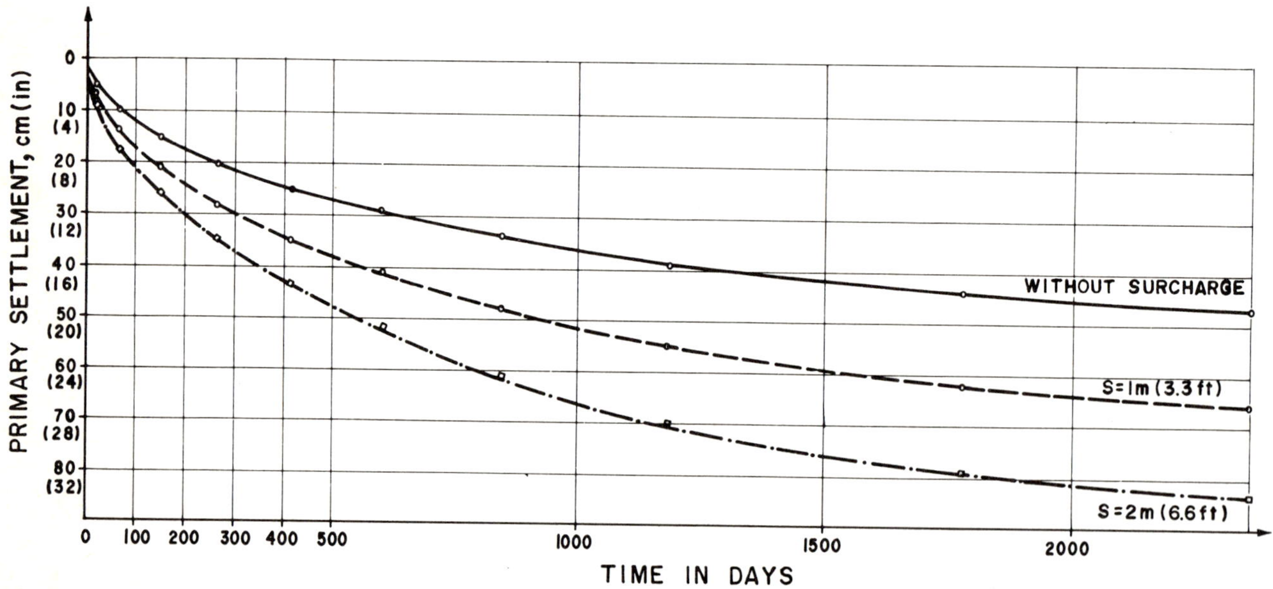

Fig. 15a-3 Evolution of settlement with time

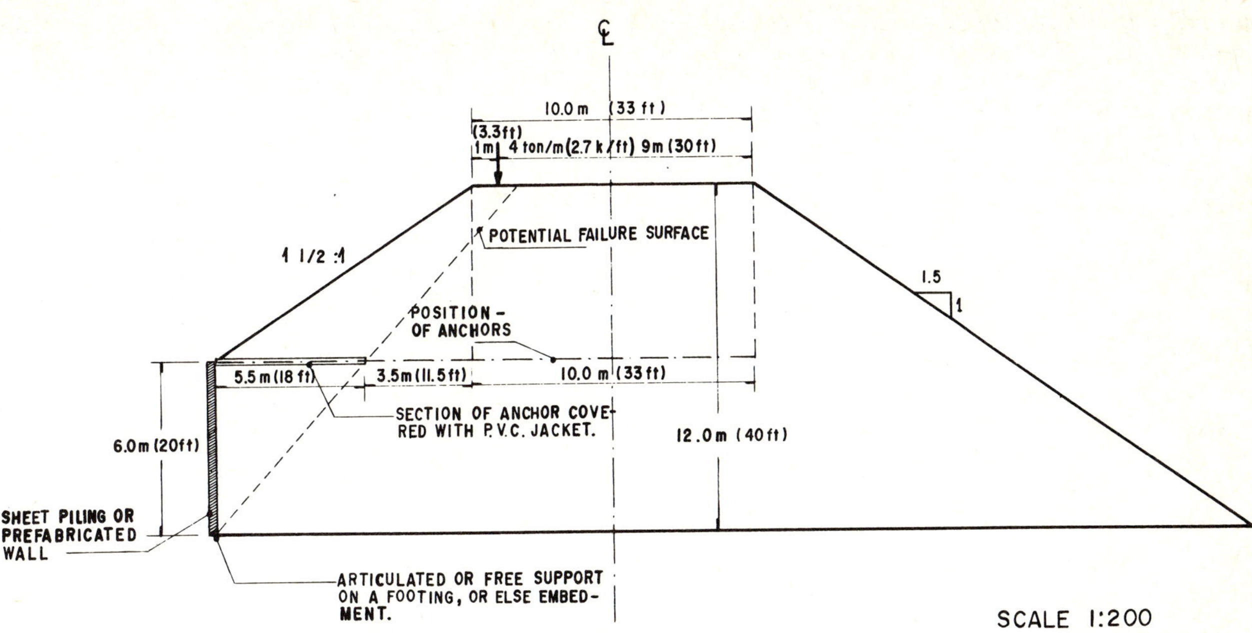

Fig. 15a-4 Explanatory diagram

$$
\begin{aligned}
c &= 1\ \text{t/m}^2\ (0.206\ \text{kips/ft}^2) \\
\varnothing &= 30^\circ \\
d &= 1800\ \text{kg/m}^3\ (112\ \text{lb/ft}^3) \\
m &= 2000\ \text{kg/m}^3\ (125\ \text{lb/ft}^3)
\end{aligned}
$$

The aim is to design the anchors necessary to restrain the prefabricated wall or sheet piling.

Solution:

The first step will be to compute the thrust that the embankment exerts on the retaining structure installed. In order to do this, COULOMB's method (Chapter 5) will be used, as applied to a cohesive frictional soil. Fig. 15a-5 is a diagram of the critical sliding wedge, which was found using a trial and error procedure, according to methods described in Chapter 5 of this book.

The weight of the critical mass is:

$$
\begin{aligned}
W &= \text{Area} \times \gamma_m \times 1\text{m} \\
W &= 39\ \text{m}^2 \times 2\ \text{t/m}^3 \times 1\ \text{m} = 78\ \text{t} \\
&= 416\ \text{ft}^2 \times 0.125\ \text{kips/ft}^3 \times 3.3\ \text{ft} = 172\ \text{kips} \\
C &= cl = 16.2\ \text{m} \times 1\ \text{t/m}^2 \times 1\ \text{m} = 16.2\ \text{t} \\
&= 53\ \text{ft} \times 0.206\ \text{kips/ft}^2 \times 3.3\ \text{ft} = 36\ \text{kips} \\
C' &= Hc = 6\ \text{m} \times 1\ \text{t/m}^2 \times 1\ \text{m} = 6\ \text{t} \\
&= 19.7\ \text{ft} \times 0.206\ \text{kips/ft}^2 \times 3.3\ \text{m} = 13\ \text{kips}
\end{aligned}
$$

The surcharge force on a meter of wall length is 4 t (8.8 kips).

If $\varnothing = 30^\circ$, $\delta = \varnothing/2$ is assumed, considering that the retaining structure is not very rough. Part *b* of Fig. 15a-5 shows the force polygon that is computed, plus the resultant of normal and friction forces along the slip surface (F), plus the thrust E against the wall. In this force polygon $E = 10$ t (22 kips). Part *c* of the figure shows calculation of the horizontal thrust component (E_h) and the distribution of this force at the upper and lower ends of the retaining structure. It is assumed that the thrust is applied at a height of 2 m (6.6 ft) measured from the natural ground surface.

$$
\begin{aligned}
E_h &= E \cos 15^\circ = 9.4\ \text{t}\ (20.6\ \text{kips}) \\
T &= 9.4 \times \frac{2}{6} = 3.13\ \text{t}\ (6.9\ \text{kips})
\end{aligned}
$$

This force of 3.13 t (6.9 kips) must be carried by the anchors for every linear meter of wall. With a factor of safety of 3, the anchors will be designed for 9.39 t/m (6.3 kips/ft).

Equation (15-15) is now employed:

$$T = \pi d L\,(c + \bar{\sigma} \tan \varnothing) \tag{15-15}$$

In Part *d* of Fig. 15a-5 the distribution of vertical pressure on the anchor is shown, as obtained from Fig. 15a-4. Note that the resistance corresponding to the length of anchor inside the critical wedge is disregarded, because it is enclosed in a jacket, which leaves 3.50 m (11.5 ft) at a variable pressure and 10 m (32.8 ft) at a constant vertical pressure.

Applying the Equation:

$$T = 3.5\,\pi d\,(1 + 9 \tan 30^\circ) + 10\,\pi d\,(1 + 10.8 \tan 30^\circ)$$

Consequently:

$$T = 21.6\,\pi d + 72\,\pi d = 93.6\,\pi d$$

Since T is 9.39, then:

$$
\begin{aligned}
9.39 &= 93.6\,\pi d \\
d &= \frac{9.39}{93.6\,\pi} = 0.0318\ \text{m}\ (0.104\ \text{ft})
\end{aligned}
$$

The bar diameter must be 3.18 cm (1.25 in) if anchors are installed every meter.

(This problem is by courtesy of RUBÉN REYES REYES, EUGENIO RAMÍREZ RODRÍGUEZ and ALFONSO JASSO DE LEÓN)

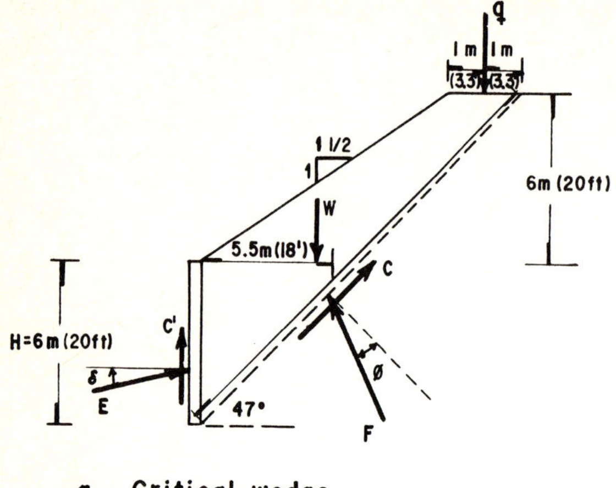

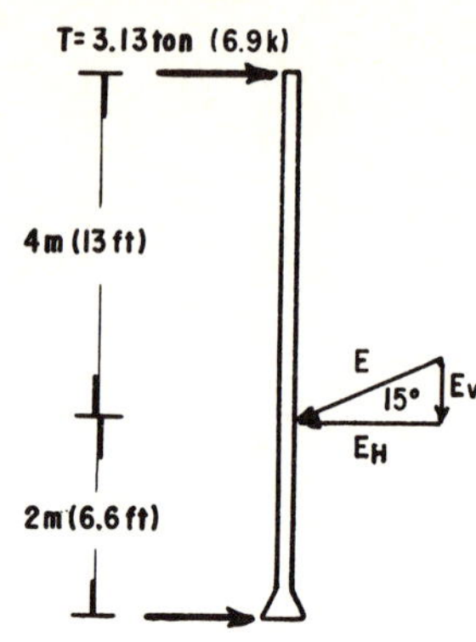

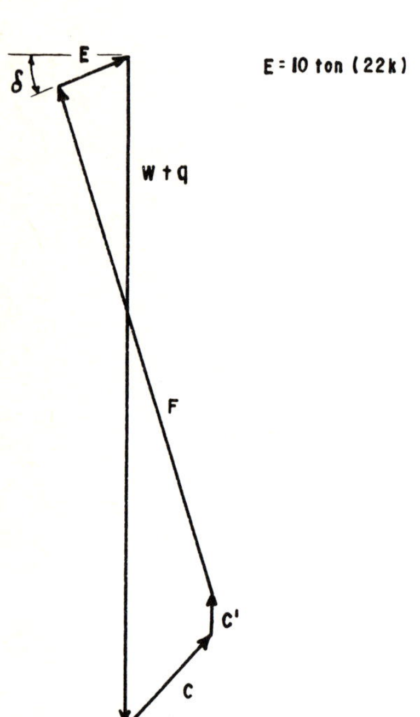

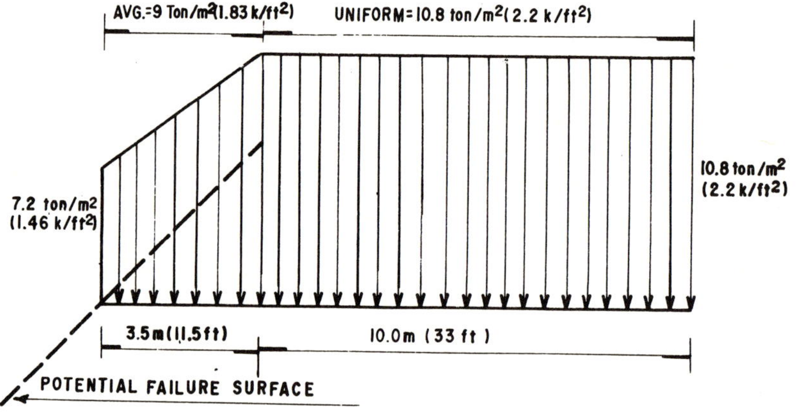

NOTE:
One meter (one foot) in length of wall or sheet piling was considered.

Fig. 15a-5 Operational diagram

REFERENCES

1. JUÁREZ-BADILLO, E. and RICO, A., *Mecánica de Suelos. Vol. II: Teoría y Aplicaciones de la Mecánica de Suelos,* Limusa: México, 1973, Chap. 1.
2. TERZAGHI, K., *Permafrost,* Harvard Soil Mechanics Series No 37, Harvard University, Cambridge, Mass., 1952.
3. CASAGRANDE, A., Unpublished chair notes, 1948.
4. YODER, E. J., *Principles of Pavement Design,* John Wiley, 1967, Chap. 5.
5. Reference [4], Chap. 14.
6. Reference [4], Chap. 15.
7. JUÁREZ BADILLO, E. and RICO, A., "Undermining by Scour," *Mecánica de Suelos. Vol. III: Flujo de Agua en Suelos,* Limusa: México, 1974, Appendix III.
8. Highway Research Board, "Scour at Bridge Waterways," NCHRP Synthesis 5, National Research Council, Washington, D.C., 1970.
9. LAURSEN, E. M., "Scour at Bridge Crossings", Iowa H.R.B., Bull. No 8., University of Iowa, 1958.
10. LAURSEN, E. M. and TOCH. A., "Scour around Bridge Piers and Abutments", Iowa H.R.B., Bull No 4, University of Iowa, 1956.
11. KARAKI, S. S., "Hydraulic Model Study of Spur Dikes for Highway Bridge Openings", Report CER 59 SSK 36, Colorado State University, Fort Collins, 1959.
12. LEVI, E. and LUNA H., "Dispositifs pour Reduire l'affouillement au Pied des Piles de Ponts", *IX. AIRH Conference,* Dubrovnik, 1961.
13. MAZA, J. A. and SÁNCHEZ-BRIBIESCA, J. L., "Socavación y Protección al pie de Pilas de Puente", *II. Latinamerican Conference on Hydraulics,* Caracas, Venezuela, 1966.
14. SEARCY, J. K., "Use of Riprap for Bank Protection", Hydraulic Engineering Circ. No 11, Bureau of Public Roads, 1967.
15. MAZA, J. A., "Diseño de Espigones", *Proc. VI. Latinamerican Conference on Hydraulics,* Colombia, 1974.
16. STEUERMAN, S., "A New Soil Compacting Device", *Engineering News Record,* July, 1939.
17. D'APPOLONIA, E., "Loose Sands: Their Compaction by Vibroflotation", *A.S.T.M. Special Technical Publication No 156,* 1953.
18. CAQUOT, A. and KERISEL, J., *Traité de Mécanique des Sols,* Gauthiers Villars: 1966.
19. MITCHELL, J. K., "In Place Treatment of Foundation Soils. Placement and Improvement of Soil to Support Structures", A.S.C.E. Special Meeting, Cambridge, Mass., 1968.
20. D'APPOLONIA, E., MILLER, C. E. and WARE, T. M., "Sand Compaction by Vibroflotation", *Trans. A.S.C.E.,* Vol. 20, 1955.
21. BASORE, C. E. and BOITANO, J. D., "Sand Densification by Piles and Vibroflotation. Placement and Improvement of Soil to Support Structures", Reference [18], ibid.
22. WEBB, D. L. and HALL, R. I., "Effects of Vibroflotation on clayey sands. Placement and Improvement of Soil to Support Structures", Reference [18], ibid.
23. WATT, A. J., DE BOER, B. B. and GREENWOOD, D. A., "Loading Tests on Structures Founded on Soft Cohesive Soil Strengthened by Compacted Granular Columns", *Proc. III. Asian Regional Conference on SMFE,* September 1967.
24. LUCE, A. F., "The Strengthening in Depth of Weak Soils by Vibration at Tolka Quay", The Institution of Civil Engineers of Ireland, 1968.
25. U.S. Army Engineer Waterways Experiment Station, "Review of Soils and Foundation Design and Field Observations, Morganza Floodway Control Structure", Technical Memorandum No 3-384, Vicksburg, Miss, 1954.
26. JOHNSON, S. J., "Precompression for Improving Foundation Soils. Placement and Improvement of Soil to Support Structures", Reference [18], ibid.
27. DARRAGH, R. D., "Controlled Water Tests to Preload Tank Foundations", *Journal of Soil Mechanics and Foundations Division. A.S.C.E.,* Vol. 90 SM5, 1964.
28. SCOTT, R. F., "Oil Tank Sites Preloaded by Wellpoint System", *Canadian Engineering Journal,* August, 1959.
29. HALTON, G. R., LOUGHNEY, R. W. and WINTER, E., "Vacuum Stabilization of Subsoil beneath Runway Extension at Philadelphia International Airport", *Procs. VI. ICSMFE,* Montreal, 1965, Vol. II.
30. KJELLMAN, W., "Consolidation of Clay Soil by Means of Atmospheric Pressure", Procs. Conference on Soil Stabilization, M.I.T, 1952.
31. YORK, D. L., "An Example of Changes in Ground Water Levels During Consolidation", A.S.C.E. Annual Meeting, New York, 1967.
32. PENMAN, A.O.M. and WATSON, G. H., "The improvement of a Tank Foundation by the Weight of its own Test Load", *Procs. VI. ICSMFE,* Montreal, 1965, Vol. II.
33. DREYFUS, G., *Etude des Remblais sur sols compressibles,* Laboratoire des Ponts et Chaussés, Dunod: Paris, 1971.
34. SIMMONS, N.E., "Consolidation Investigation on Undisturbed Fornebu Clay", Norwegian Geotechnical Institute, Publication No 62, 1965.
35. KAPP, M. S. et al., "Construction on Marshland Deposits: Treatment and Results", Highway Research Board Record No 133, 1966.
36. TAYLOR, D. W., "Research on Consolidation of Clays", Department of Civil and Sanitary Engineering, M.I.T. Serial 82, 1942.
37. BJERRUM, L., "Engineering Geology of Normally-Consolidated Marine Clays as Related to the Settlement of Buildings", VII. Rankine Lecture, Geotechnique, Vol. 17, 1967.
38. BROOKER, E. W. and IRELAND, H. O., "Earth Pressures at Rest Related to Stress History", *Canadian Geotechnical Journal,* Vol. II No 1, 1965.
39. JUÁREZ-BADILLO, E. and RICO, A., *Mecánica de Suelos. Vol. I: Fundamentos de la Mecánica de Suelos,* Limusa: México, 1972, Annexe 12-b.
40. Ibid. Chap. 10.

41. DUDLEY, J. H., "Review of Collapsing Soils", *Procs. A.S.C.E. Soil Mechanics and Foundations Division,* SM3, 1970.

42. JENNINGS, J. E., "The Theory and Practice of Construction on Partly Saturated Soils as Applied to South African Conditions, Engineering Effects of Moisture Changes in Soils", Procs. International Research and Engineering Conference on Expansive Clay Soils, University of Texas, A & M, 1965.

43. GIBBS, H. J. and BARA, J. P., "Predicting Surface Subsidence from Basic Soil Tests", Soils Engineering Report No EM 658, U.S. Dept. of Interior, Bureau of Reclamation, Denver, Co. 1962.

44. Reference [39], Chap. 8.

45. BLIGHT, G. E., "Effective Stress on Unsaturated Soils", *Journal A.S.C.E. Soil Mechanics and Foundations Division,* Vol. 93 SM2, 1967.

46. Reference [39], Chap. II, Supplement II-a.

47. GIBBS H. J., Properties which Divide Loose and Dense Uncemented Soils, Earth Lab. Report No EM-608, Bureau of Reclamation, U.S. Department of Interior, Denver, CO. 1961.

48. GIBBS, H. J. and BARA, J. P., Stability Problems of Collapsing Soils, Journal A.S.C.E. Soil Mechanics and Foundations Division, Vol. 93 SM4, 1967.

49. JENNINGS, J. E. and KNIGHT, K., The Additional Settlement of Foundations Due to a Collapse of Structures of Sandy Soils on Wetting, Procs. IV. ICSMFE, London, 1957, Vol. III.

50. MOORE, H. E., "The Engineering Properties of Silty Soils, Snake River Canyon, State of Washington", U.S. Army Corps of Engineers, Walla Walla District, 1967.

51. HABIB, P., "Les Ancorages, netament en Terrain meuble", VII. International Conference on Soil Mechanics and Foundation Engineering, (México, D.F.) Special session No 15. Publication of the Société de Diffusion des Techniques du Batiment et des Travaux Publiques, Paris, 1969.

52. BELLO, A. "Sistemas de Soporte en Excavaciones Subterráneas mediante anclas de fricción." *Proc. III. International Conference on Rock Mechanics,* Denver, Co., 1974.

53. JORGE, G. R. and LE TIRANT I.R.P., "Réinjectable spécial pour Terrains meubles, Karstiques on a faibles Caractéristiques Géotechniques", VII. International Conference on Soil Mechanics and Foundation Engineering, Special Session No 15, México, 1969.

54. SOWERS, G. F., *Introductory Soil Mechanics and Foundations: Geotechnical Engineering,* Fourth Edition. MacMillan: New York, Collier MacMillan: London, 1979, Chap. 11.

55. COSTA NUNES, J., "Anchorage Tests in Clays for the Construction of the Sao Paulo Subway", VII. International Conference on Soil Mechanics and Foundation Engineering, Special Session No 15, México, 1969.

56. MORI, H. and ADACHI, K., "Anchorage by an Inflated Cylinder on Soft Ground", VII. International Conference on Soil Mechanics and Foundation Engineering, Special Session No 15, México, 1969.

OTHER RELATED REFERENCES

Transportation Research Board, "Frost Action and Risk Assessment in Soil Mechanics," Transportation Research, Record 809, Washington, D.C., 1981.

Transportation Research Board, "Frost Action on Transportation Facilities," Transportation Research Record 918, Washington, D.C., 1983.

CHAPTER 16

SOIL STABILIZATION

16.1 Introduction

Often the geotechnical engineer will find the soils he is obliged to use for a particular purpose at a specific location in some way unsuitable. He can choose between three alternatives [1]:

— Acceptance of the material just as he finds it, but considering its poorer qualities realistically for the design in question.
— Elimination or avoidance of the unsatisfactory material replacing it by another with suitable properties
— Modification of the properties of the existing material, so that it will be able to meet higher requirements

The last alternative leads to soil stabilization techniques. Strictly speaking there are numerous different procedures for improving soils in order to make them more suitable for some specific use. This is referred to as stabilization. The following list includes the most common procedures:

— Stabilization by mechanical means, of which the best known is compaction, (although soil mixing is also very frequently used)
— Stabilization by drainage, which has already been discussed in sufficient detail
— Stabilization by electrical methods, of which electro-osmosis [2] and the use of electrometalic piles are probably the best known
— Stabilization using heat and calcination, also discussed previously in Chapter 6 of this book
— Stabilization by chemical or biochemical methods, generally by the addition of stabilizing agents, such as cement, lime and asphalt

On account of the great variability in the composition of soils, each method can only be applied to a limited number of soil types. Sometimes this variability affects just a few meters of highway while at other times it affects several kilometers. However, in order to apply a particular method economically it should be adapted to several different types of soils, sometimes with significant variations. The use of the *optimum* procedure for each soil type is usually out of the question.

From the start it must be acknowledged that stabilization is not always advisable, and is not equally beneficial in all cases. Consequently, the set of soil properties it is wished to improve and the relationship between what will be achieved by improving them and the effort and money that will have to be invested, must always be kept very much in mind. Only by weighing up these factors carefully will it be possible to use soil stabilization appropriately.

The soil properties most often requiring improvement by stabilization are:

— Volumetric stability
— Strength
— Permeability
— Compressibility
— Durability (resistance to change of any of the above)

Often it is possible to employ treatments which will simultaneously improve several of these properties, but one must also be prepared for contradictory evolutions, where the improvement of one property causes the deterioration of another. Stabilization should not be regarded only as a remedial measure. Some of the best applications of these techniques are rather more preventive measures against adverse conditions that are likely to develop subsequently. A description follows of some of the soil properties that can most readily be improved by stabilization.

.1 Volumetric Stability

This expression refers to soils that swell and shrink as a result of changes in their water content related to seasonal rainfall variations or engineering activities. Stabilization is usually employed as an alternative treatment for these soils, replacing loads and pervious layers or adding water, which are the lines of action most frequently taken and which have been discussed elsewhere in this book. The objective is to transform the swelling clay either into a stiff, or non-moisture sensitive mass, with its particles bound together strongly enough to resist internal swelling and shrinking pressures. This is achieved by chemical or thermal treatments, which will be described later in this chapter. Experience, largely governed by cost, has shown that chemical treatments are particularly useful for clays within 3 m (10 ft) of the ground surface, whereas thermal treatments have been applied to clays at greater depths.

Often when treating superficial layers of potentially expanding and shrinking clay, only a limited thickness of the upper part of the layer is stabilized to minimize cost. This is sufficient so long as the swelling pressure produced by the untreated thickness is correctly balanced.

.2 Strength

Various different stabilization methods have proved useful for improving the strength of many soils. However, before discussing this in greater detail, attention is drawn to the fact that they all lose much of their effectiveness with large organic contents. This is unfortunate, because many of the most serious strength problems occur in organic soils. Table 16-1 [1] shows the influence of the content of organic matter on the stabilization of otherwise similar soils.

Compaction is a type of mechanical stabilization, one of the chief objectives of which is to increase soil strength. In Chapter 4 and other parts of this book it was shown how the use of greater compaction intensities does not always lead to higher strength values, particularly if strength is to be maintained within reasonable limits for long periods of time. The following are some of the methods of stabilization most frequently used to increase soil strength:

— Compaction
— Vibroflotation
— Preloading
— Drainage
— Mechanical stabilization with mixtures of other soils
— Chemical stabilization with cement, lime or liquid additives.

With the exception of the last two methods, all have been discussed elsewhere in this book. Stabilization with heat has also been employed, but far less often, because of the high cost.

.3 Permeability

It is not usually very difficult to substantially modify the permeability of soil formations by methods such as compaction and grouting. In clayey materials, deflocculants (for example, polyphosphates) may also result in a significant reduction in permeability. However these chemicals also reduce the resistance of the soil to internal erosion: piping. The use of flocculants (often hydrated lime or gypsum) raises permeability. Currently some wax-like substances are available which, when added to the soil in the form of an emulsion, may greatly reduce its permeability. However the use of these substances must be carefully evaluated, for it is not unusual for them to seriously reduce the shear strength of some soils.

In general terms, not counting mechanical stabilization, stabilization methods for modifying soil permeability are usually independent from those employed to modify volumetric stability or strength.

.4 Compressibility

Compaction is a routine stabilization method which brings about an important decrease in soil compressibility. Some information on the subject was given in Chapter 4. However, compaction is not the only means of stabilization that influences compressibility; most stabilization methods mentioned earlier in this chapter reduce it.

.5 Durability

Durability involves resistance to weathering, erosion, traffic abrasion and future changes in other properties. Consequently in roads durability problems are usually related to soils that are close to the surface. These problems may affect natural and stabilized soils alike. The poorest behavior of stabilized soils is either a consequence of inadequate design (such as an incorrect choice of stabilizing agent, or a serious error in its use) or poor construction [1,3].

Table 16-1

Effect of organic matter on stabilization results [1]

Type of soil	Depth		Content of organic matter	7-day unconfined compressive strength (Specimens compacted to 95% AASHTO Standard)					
	m (ft)		%	kg/cm^2 (lb/in^2)					
—	—		—	Unstabilized		With 10% cement		With 10% lime	
Humus	0.45	1.5	2.65	3.80	54	15.50	220	1.90	27
Humus	1.60	5.2	0.22	3.80	54	36.00	512	47.00	668
Organic clay	0.10	0.3	13.70	1.05	15	1.83	26	2.25	32
Organic clay	0.60	2.0	2.50	6.30	90	20.00	284	1.83	26
Organic clay	0.10	0.3	11.70	3.15	45	7.00	100	5.60	80
Organic clay	0.45	1.5	2.00	5.00	71	20.00	284	16.20	230
Organic clay	0.10	0.3	10.30	3.90	55	4.20	60	4.90	70
Organic clay	0.80	2.6	2.40	5.00	71	41.00	582	26.80	381
Topsoil	0.10	0.3	3.10	3.90	55	30.00	426	11.20	159
Topsoil	0.45	1.5	1.10	5.00	71	42.00	596	22.50	320

In current engineering practice, there is no field or laboratory criteria to enable a reliable determination of the durability of a stabilized soil. For this reason the property of durability is currently one of the hardest to evaluate quantitatively.

From the brief discussion so far, a major problem is the multiplicity of soil stabilization objectives. We emphasize that stabilization objectives may frequently be contradictory; moreover a stabilization technique that improves one soil may be harmful to another. A large number of stabilization methods have been developed by modern technology; some successful, others technically successful but uneconomical. The many uncertainties involved in relation to the short-term and particularly the long-term behavior of stabilized materials plus the contradictory technical and economic feasibility provide numerous doubts. The result is an irrational preference for one method or another. It is not hard to see why in one country some techniques have become construction routine, whereas others are never used. In another country, perhaps just as technically advanced, these last methods are regarded as trustworthy and some of the former methods are rejected.

Some stabilization methods are gradually becoming generally accepted. Leaving aside mechanical stabilization techniques, which are accepted everywhere, chemical stabilization with cement, lime and asphalt are appearing more and more often in highway construction throughout the world, especially in pavement technology.

In addition some of the stabilization methods mentioned previously are also extensively used in highway construction, such as drainage and preloading, or in other fields, such as electrical stabilization of foundations.

16.2 Soil Identification for Stabilization Purposes

The Unified Soil Classification System (Chapter 2) is almost universally employed for the grouping of soils for highway construction. However, the system either disregards, or does not sufficiently emphasis, certain soil characteristics which hold a special importance in stabilization problems, especially chemical stabilization. These characteristics are mineralogical composition, permeability, the influence of the local environment, such as climate or vegetation, and the previous geologic history. These are especially important in fine grained soils. It is for this reason that when confronting soil stabilization problems, it is usually necessary to augment the classification of the Unified System with some additional information about these features. Some authors have even drawn up complete soil classification systems specifically for stabilization. References [1,4,5] mention some such systems. Regardless of whether a better definition of stabilization needs and responses is achieved with a specially conceived classification system, its introduction establishes an element of confusion and nulifies the advantages of using one soil classification system as a general reference. The authors are not in favor of using a different classification system for each specific application, no matter how complete the system may be and how useful it may prove. We prefer a system for general use, even though we are aware that it may prove somewhat incomplete when employed in certain specific fields. In such cases we feel that the best solution is to add the relevant information to the general-purpose classification system.

The following pages present some ideas which may help the highway engineer to recognize the composition of the soils he encounters in his projects, and to augment the information he may obtain from the Unified Classification System. Table 16-2 [1] suggests some conclusions that can be drawn from the profile analysis that is obtained.

Table 16-2
Qualitative information concerning mineralogical composition based on observation of the soil profile [1]

Observed profile	Possible dominant minerals
Mottled clays, with red, orange or white coloration	Kaolinites
Mottled clays with yellow, orange or grey coloration	Montmorillonites
Black or dark grey clays	Montmorillonites
Brown or reddish-brown clays	Illites or also with some montmorillonites
White or light grey clays	Kaolinites and bauxites
Micaceous soils	Micas
Small, readily separable crystals	Gypsums
Soft, disseminated nodules, acid soluble	Carbonates
Hard, reddish-brown nodules	Iron minerals, laterites
Numerous cracks, wide deep and 5-6 cm (2 in) or less apart	Illites rich in calcium or montmorillonites
— cracks 30 cm (12 in) or more apart	Illites
Crumbly, open textured soil with appreciable clay content	Usually carbonates and kaolin. Never montmorillonite, seldom illite
Crumbly, open textured black soils with appreciable clay content	Organic soils, peats
Crumbly, open textured soils with low clay content	Carbonates. Sands and silts
Soils of rough appearance when surface exposed to weathering	Montmorillonite. Salinity
Thin horizons of whitish soils less than 60 cm (24 in) below the surface	Above whitish horizons: fine silts; below: unstable clays. Perched water may exist on the whitish horizons

Table 16-3, also taken from [1] gives some indications of stabilization alternatives for different common materials.

Lastly, Table 16-4 [1] indicates the typical response of some important soil minerals to different stabilization methods.

The saturation or non-saturation of soils has important implications in their stabilization. Saturated fine soils sometimes can be satisfactorily treated with lime, but may be very inappropriate for treatment with asphalt or emulsions. The asphalt may not penetrate into the voids and usually does not adhere to the grains, whereas emulsions suffer uncontrolled ruptures. The strengths acquired as a result of cement stabilization are usually considerably lower in saturated fine grained soils than in the same soils when only partially saturated. In non-saturated fine soils, there may be a highly favorable response to lime or cement, but the homogeneous incorporation of water as well as the stabilizing agent may be so difficult that stabilization becomes impossible.

16.3 Mechanical Stabilization

This section discusses stabilization by mixing soils. It will not refer to the most common and routine type of mechanical treatment, compaction, which has already been dealt with in Chapter 4.

Stabilization by mixing goes back to the 1920's [6], and was probably first quantified in the U.S.A. Some of the famous pioneers in Soil Mechanics, such as TERZAGHI, CASAGRANDE and HOGENTOGLER, were involved with the development of these techniques.

When soil mixtures are designed for the purpose of achieving specific properties, the grain-size distribution is usually the most relevant requirement in the coarse fraction, and plasticity in the fine fraction.

Table 16-3
Some typical problems and stabilisation possibilities for various common soils [1]

Type of soil	Problems and usual stabilization
Sandy soils	When the grain size distribution is uniform, stabilization by the addition of other soils may be appropriate. The characteristics of clean sands may improve with cement or asphalt.
Silty soils with some clay	Generally speaking, the only economical treatment is compaction.
Silty soils with little or no clay	There are no economical treatments. Their use on exposed surfaces should be avoided because of the dust produced on drying.
Cracked clayey soils	Respond to stabilization with lime.
Open-textured clayey soils without cracks	Respond very well to compaction.
Soft clays	Respond to stabilization with lime.

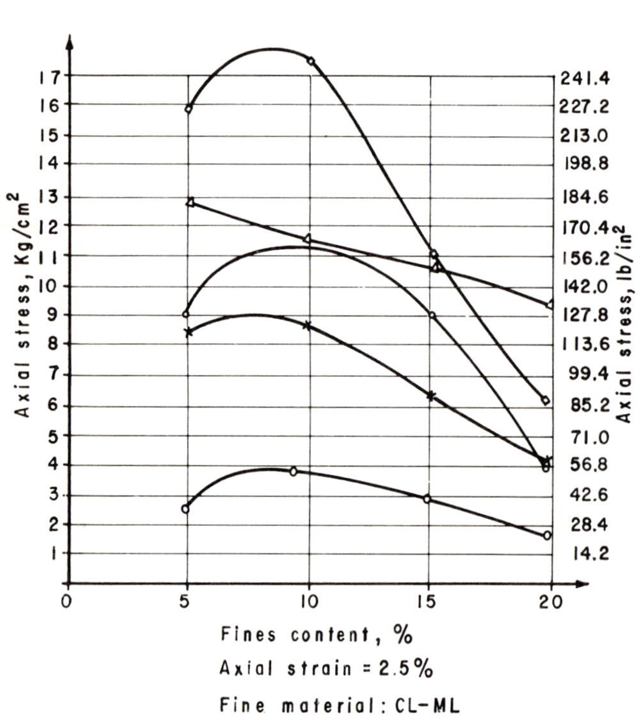

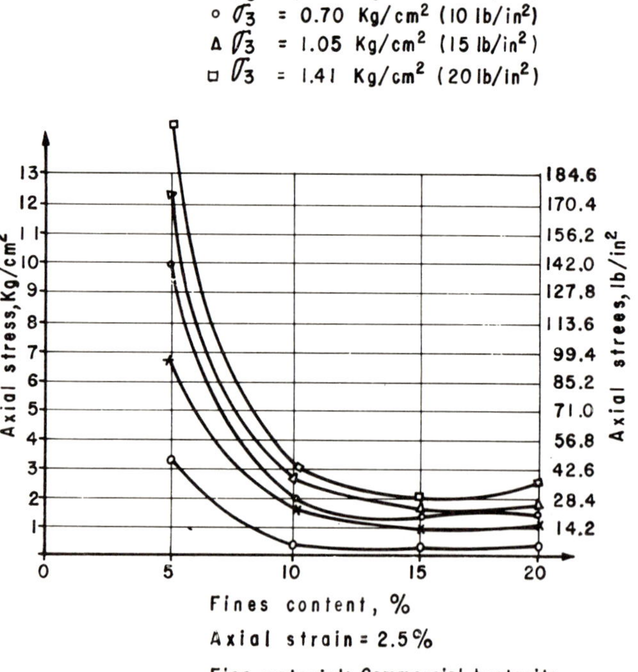

Fig. 16-1 Effect of the addition of fines on the strength of a crushed stone [7]

Table 16-4
Response of some typical minerals to different stabilization methods [1]

Typical mineral or soil component	Stabilization recommended	Objective
Organic matter	Mechanical stabilization	Other methods are ineffective
Sands	Mixture with fine, non-plastic materials	Mechanical stability
	Cement	To increase the strength
	Asphalt	To impart cohesion
Silts	Do not respond to methods of stabilization in use	
Allophanes	Lime or mixtures of lime and gypsum	To increase the strength
Kaolin	Sand	Mechanical stability
	Cement	To increase short term strength
	Lime	To improve workability and impart long term strength
Illite	Cement	To increase short term strength
	Lime	To improve workability and impart long term strength
Montmorillonite	Lime	To improve workability and impart short term strength
Chlorite	Cement	As yet there is no conclusive experience of the effects

The maximum size of the particles in the mixture is important, because particles that are too large are hard to work with and result in very rough road surfaces. An excessive proportion of coarse sizes leads to mixtures with a marked tendency to segregate. Large amounts of fine materials (smaller than No 40 sieve) makes it hard to achieve satisfactory strength and rigidity and may lead to road surfaces that are too smooth and muddy when wet, and dusty when dry.

Reference [7] contains a laboratory study conducted to assess the effect of the addition of fines (material finer than No 200 sieve) to the crushed rocks that are usually best as base course materials. Some comments of a purely qualitative nature were made about this addition of fines in Chapter 9. In the investigation under consideration varying percentages ranging from 5 to 20% of (1) a relatively inert fine material (*CL-ML*), (2) a kaolinite and (3) a commercial bentonite were added to a crushed basalt with a maximum particle size of 3.8 cm (1-1/2 in). A study was made of the effects of these fines on the mixtures obtained with regard to the triaxial strength, stress-strain ratio and *CBR*.

Figure 16-1 shows the effect of the *CL-ML* fines on the maximum strength obtained in different specimens prepared by dynamic compaction (energy corresponding to the compaction method recommended by the Texas Highway Department, U.S.A.) [8] and subsequently allowed to absorb water. The maximum strengths measured in the Texas triaxial chamber mentioned in Chapter 9, and those corresponding to a 2.5% unit strain of the sample are given. The curves correspond to mixtures with *CL-ML* and with bentonite. The effect of the plasticity of the bentonite fines can be seen immediately, although the smaller influence of the less plastic fines is notable.

Figure 16-2 [7] refers to the stress-strain behavior of mixtures of the same materials, where the specimens were allowed to absorb water freely. The curves that are shown were obtained with a confinement pressure of 0.7 kg/cm^2 (10 lb/in^2) in the triaxial chamber.

The influence of the activity of the fines on the deformability of the mixture is great. The more inert fines have a moderate effect on the stress-strain behavior of the mixtures, but this situation changes drastically when the fines are bentonite.

Figure 16-3 [7] depicts the strength envelopes of the mixtures with different percentages of *CL-ML* fines and also commercial bentonite. In both cases after the specimens were allowed to absorb water freely.

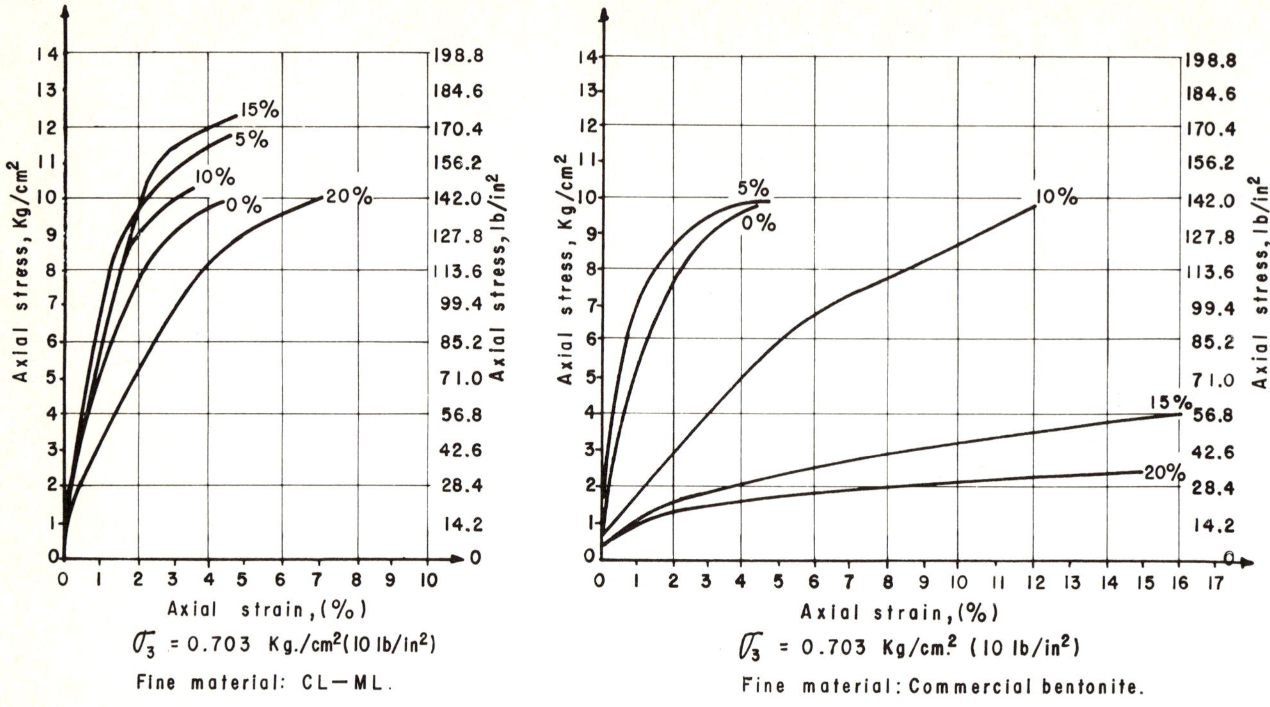

Fig. 16-2 Effect of the addition of fines on the stress-strain relation of a crushed stone [7]

Lastly, Fig. 16-4 [7] gives the variation in the *CBR* of the different mixtures with the fines content that is used in them. In this case, the specimens were prepared using the compactive effort corresponding to the AASHTO standard test, and were subsequently saturated.

The results of research such as that in [7] corroborate the experience of many specialists, who say they object to using indiscriminate or arbitrary quantities of fines in soil mixtures for stabilization. Satisfactory behavior cannot be expected of a mixture containing more than 8 to 10% fines smaller than the No 200 sieve. Moreover this limit may be acceptable only if the fines are relatively inert. This is very hard to control in highway engineering in view of the procedures that may be employed to measure the activity of a bank material. If the fines gradually become more active, fines contents of about 10% of the mixture would be detrimental based on this research. Reference [7] warns the engineer and constructor against the use of fines in base course mixtures, pointing out how variable any initial percentage that is accepted may prove in practice. It discusses the effects of contamination by very small particles extruded up from lower pavement layers as well as particle separation and breakage by the abrasive action of traffic as the principal causes of the increase in the initial percentage of fines with time.

Figure 16-5 [9] gives information concerning the optimum fines content required by different coarse soils in order to achieve the maximum dry unit weight (in this case, by the AASHTO

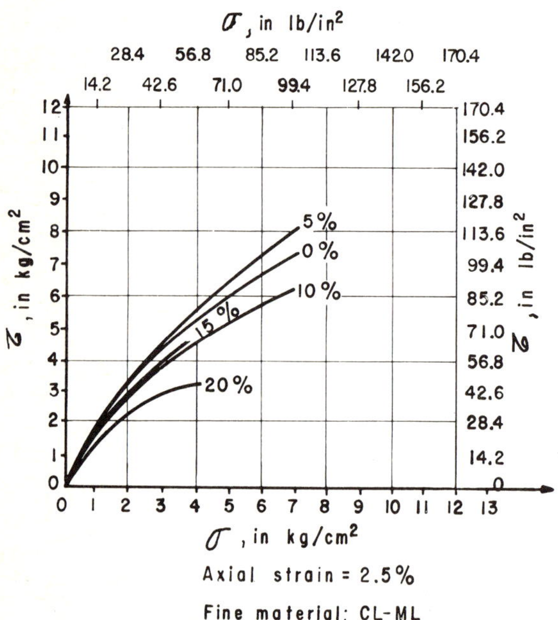

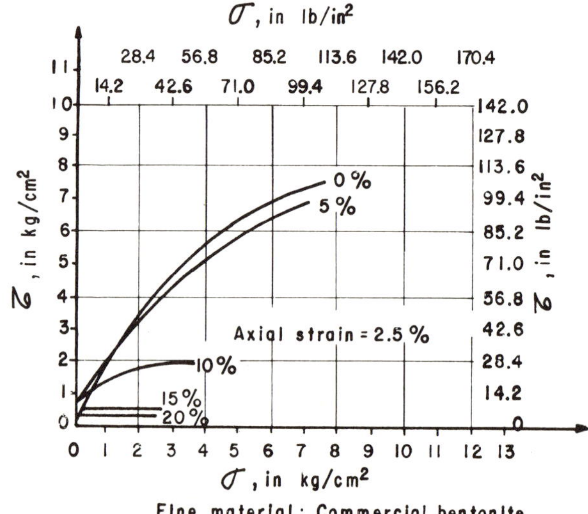

Fig. 16-3 Strength envelopes for a crushed stone mixed with different percentages of finc material [7]

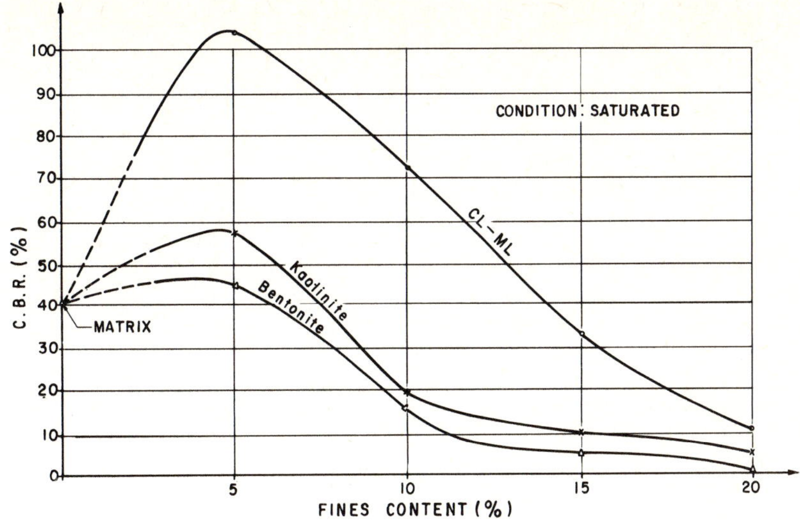

Fig. 16-4 Effect of the percentage of fines on the *CBR* of different mixtures [7]

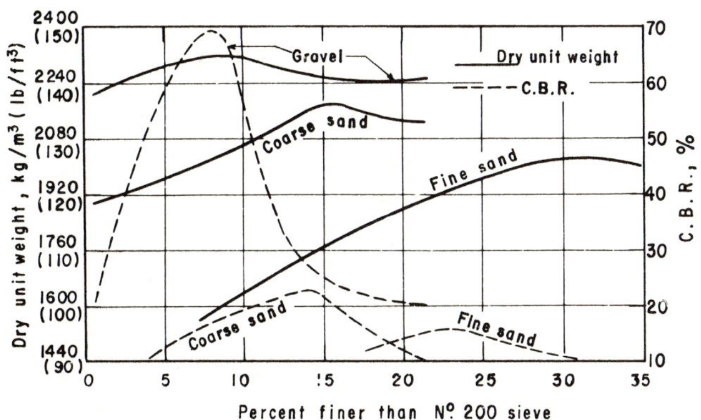

Fig. 16-5 Optimum *CBR* and compacted dry unit weight as a function of the fines content of sand and gravel [9]

standard test) and the highest *CBR*. It can be seen how the optimum fines content tends to increase the finer the soil, optimum percentages of about 25% resulting in the highest *CBR*.

The information given in Fig. 16-5 is familiar to highway engineers and those who direct project control laboratories. However, the *optimum* concept that is shown in this figure and in other similar ones should be carefully considered. The *optimum* fines content required to produce the highest unit weight or even the highest *CBR* is by no means the same optimum that is required in order to achieve the best strength and rigidity of the soil in a layer of pavement or an embankment. Reference [7] shows that the fines contents which in Fig. 16-5 produce the highest unit weight or the highest *CBR* result in mixtures which behave either somewhat unreliably or else totally inappropriately, depending on the activity of the fines. Also [7] which includes a study of the *CBR* of such mixtures, shows similar results for two different mixtures, one containing a low activity kaolin and the other a highly active bentonite. Although the effects of these two different fines on the strength and deformability of the soil are radically different, the data show the results of the *CBR* tests are only slightly dependent on the nature of the fines that are used. This means that the test is unduly insensitive to this fundamental proportion.

The results of Fig. 16-5 show only a limited range of the physical condition of the soil mixtures: they are more dense or less readily penetrated if the spaces between the coarse particles are filled with fines (but without greatly exceeding this value). However these properties are not characteristic of the strength or deformability under varied stress conditions, changing effects of water, or in the long term.

From the foregoing observations it is deduced that the most critical aspect of stabilization with mixtures of other soils is the criterion that is adopted for assessing the properties of the new soil. If, as is not uncommon, the purpose of stabilization is to achieve a higher unit weight, the risk is run of adding excessively large quantities of fines (Fig. 16-5). If the purpose of stabilization is to achieve a higher *CBR* it will be easy to make the same mistake, apart from the fact that the test does not discriminate sufficiently the quality of the fines that are used. Control of the quality of the fines by measuring their plasticity is not sufficient, perhaps because of the very rapid changes in plasticity that take place in the borrow pits. By the time the test can be made (1/2 day minimum) the soil can have been excavated, placed and compacted. The value of a soil mixture should be judged by the basic soil properties, which are generally the triaxial strength, the stress-strain ratio, in some cases, permeability, or compressibility.

There are several different methods for preparing mixtures of two soils in order to produce a third soil with a grain-size distribution that will ensure the desirable properties, previously established by a laboratory study or by specifications. The procedure is described as follows:

A soil is divided into various fractions by grain size, the percentage of each being determined. It is wished to modify one or more of these percentages by the addition of another soil of a known grain-size distribution.

If A, B, C . . . are the percentages of soils *1, 2, 3* . . . passing a certain sieve which are to be combined to form a single soil, and if a, b, c . . . are the percentages of these soils *1, 2, 3* . . . to be included in the combination, the percentage of the mixture that will pass the sieve in question will be given by the equation:

$$p = aA + bB + cC + \dots \qquad (16\text{-}1)$$

where the percentage passing is, and will be further, expressed as the decimal volume fraction, i.e. 70% = 0.7.

Such mixtures involve several problems. Sometimes it will be necessary to determine the complete grain-size distribution curve for the mixture; at other times it will be sufficient to establish one or more appropriate percentages of certain specific sizes. An example of the latter case would be modification of the fine fraction of a soil in order to improve plasticity or permeability.

Assuming a mixture of just two soils, Eq. (16-1) becomes:

$$p = aA + bB \qquad (16\text{-}2)$$

Clearly: $a + b = 100\%$ or 1.0; consequently:

$$a = 1 - b$$

Substituting in Eq. (16-2):

$$p = (1 - b)\,A + bB = A - Ab + bB$$
$$p - A = b(B - A)$$

Consequently:

$$b = \frac{p - A}{B - A} \qquad (16\text{-}3)$$

Similarly:

$$a = \frac{p - B}{A - B} \qquad (16\text{-}4)$$

Equations (16-3) and (16-4) give the percentages of soils *1* and *2* to be combined for percentage p of the mixture to pass the sieve specified; p must be selected by the design engineer bearing in mind practical requirements.

To illustrate the foregoing mechanism, an example is taken from [4] Table 16-5 gives data for Soils *1* and *2*, which are to be mixed to make a material that will meet the specifications indicated. Figure 16-6 shows the grain-size distribution curves for materials *1* and *2*, as well as the specified mix.

As is pointed out in [5], approximate solutions can be reached by trial and error procedures, in the laboratory. These trial and error procedures are simplified by estimating the fractions of soils *1* and *2* which can be employed most readily. For example, in Fig. 16-6, predominately coarse soil, *1*, is used to provide the coarse sizes, while the fine grained soil, *2*, will provide the fines.

By way of illustration, the percentage of the mixture that must pass No 8 sieve will be found, typically the arithmetical average of the limits indicated. Consequently:

$$p = \frac{35 + 50}{2} = 42.5\%$$

Thus, applying Eq.(16-3), the following expression is obtained:

$$b = \frac{p - A}{B - A} = \frac{42.5 - 3.2}{82 - 3.2} = \frac{39.3}{78.8} = 50\%$$

By using 50% of each component soil to prepare the mixture, the requirement for the percentage of the mix passing the No 8 sieve will be met; however, as Fig. 16-6 shows, not with regard to all the other sieve sizes. The first trial procedure can, however, be based on this criterion and will lead to Table 16-6 [4].

The mixture obtained was within the desired specifications, although the material passing No 200 sieve came rather close to the limit and the coarser particles slightly exceeded the desired range. Should this limit be exceeded, a second trial procedure would be necessary, increasing slightly the percentage of material *1*, which contains no fines, and reducing accordingly the percentage of material *2*, which is the one that contributes to the fine fraction. However it will seldom be possible to mix two natural soils in proper proportions to fit a desired gradation. The resulting grain-size distribution for the mixture in Table 16-6 is also shown in Fig. 16-6.

Table 16-5
Problem illustrating mixture of two soils to obtain required grading [4]
Data of the problem

Soils \ Sieve	3/4 in	1/2 in	3/8 in	No. 4	No. 8	No. 30	No. 50	No. 100	No. 200
1	100	90	59	16	3.2	1.1	0	0	0
2	100	100	100	96	82	51	36	21	9.2
Specification	100	80–100	70–90	50–70	35–50	18–29	13–23	8–16	4–10

Table 16-6
First trial solution

	3/4 in	1/2 in	3/8 in	No. 4	No. 8	No. 30	No. 50	No. 100	No. 200
0.5 Soil 1	50.0	45.0	29.5	8.0	1.6	0.6	0	0	0
0.5 Soil 2	50.0	50.0	50.0	48.0	41.0	25.5	18.0	10.5	4.6
Mixture	100.0	95.0	79.5	56.0	42.6	26.1	18.0	10.5	4.6
Specification desired	100.0	80–100	70–90	50–70	35–50	18–29	13–23	8–16	4–10

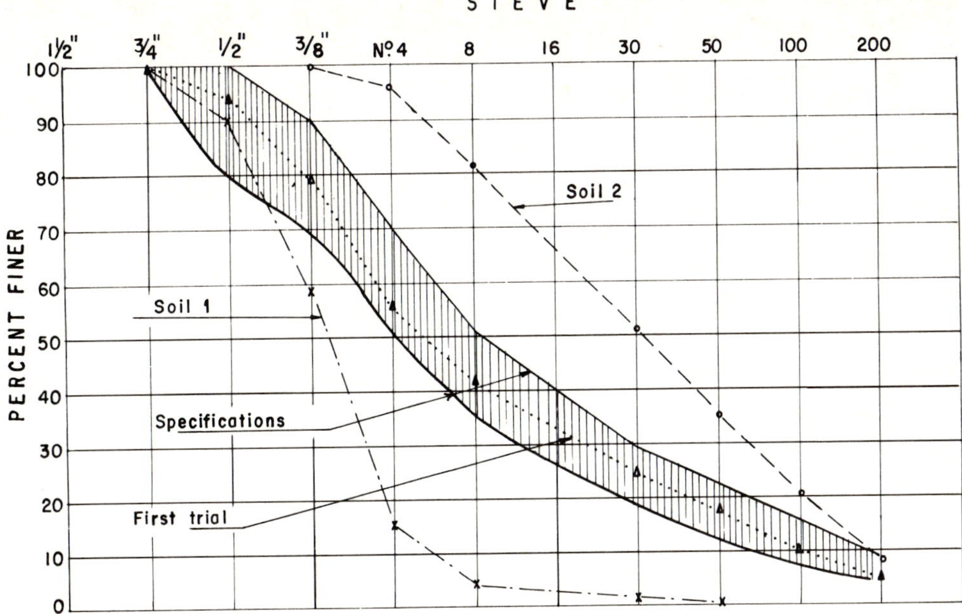

Fig. 16-6 Grain-size distribution curves for Soils *1* and *2* of the resulting mixture [4]

The foregoing method is also suitable to a graphical solution, which is particularly useful for those cases where it is difficult to visualize which of the component soils should contribute most to the formation of different fractions of the mixture. This occurs primarily when the grain-size distribution curves of the component materials intersect one another.

To commence the graphical method, a square is drawn (Fig. 16-7) with grain-size percentage scales on the sides of the square as shown. On the vertical scale on the right, points are plotted in accordance with the grain-size distribution of material *1*, plotting the corresponding sieve size against each percentage. The same is done for material *2* on the vertical scale on the left. Joining these points by straight lines, as in the figure, a straight line will be obtained for each sieve size. On

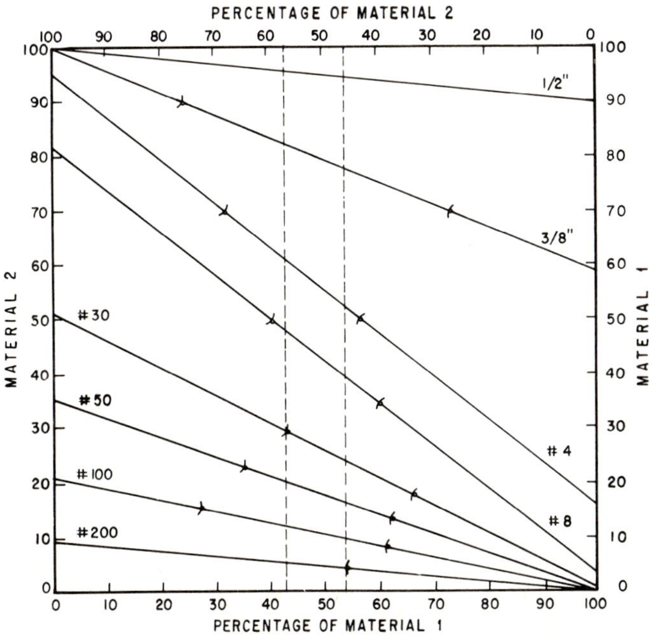

Fig. 16-7 Graphical method for designing mixtures of two materials [4]

each of these straight lines, on the basis of the vertical percentage scales through which they pass, the limits of the specification to which the mixture is to be subjected are marked (see Table 16-5 for interpretation of the limits indicated in Fig. 16-7).

The figure is constructed in such a way that any of the points indicated on one of the sloping lines, which represents a certain percentage of the mixture of the size corresponding to the line, makes it possible to read on the upper and lower horizontal scales the respective percentages of materials *1* and *2* that should be added to the mixture in order to produce the same percentage as that indicated, in the corresponding size.

Thus, if two vertical lines (see broken lines in the figure) are drawn through the closest of all the points that correspond to specification limits for the mixture, a zone is formed within the limits of which the percentages employed of soils *1* and *2* will automatically lead to a mixture within all the specifications indicated. As previously stated, mixing two soils will not in general always produce the desired gradation.

Other methods exist which enable more than two components to be handled. They are not included here because they add nothing to the conceptual panorama, [4,5] and [6] describe some of these procedures.

The mixing may in itself cause problems, especially when its components are fines. The use of special machinery may simplify work appreciably. Cohesive soils usually require pulverization before mixing operations. Disc plows, harrows, etc. can be used for this purpose.

A use of soil mixtures which is not so common, but nevertheless of great potential, is the one which tends to modify the mineralogical composition and the nature of the interchangeable mineral ions. The compressibility of montmorillonite is ten times greater than that of kaolin, and the liquid limit of a sodium montmorillonite may be five times greater than that of a ferrous montmorillonite [10]. The proportions of fines of one type or another may be modified, achieving very appreciable changes in some specific property. Typical examples of this technique are the addition of fine sands to soils that have a high percentage of particles susceptible to elastic rebound, or the addition of percentages of bentonite to reduce the permeability of soils. Reference [10] mentions a case in which the permeability of a silty sand was lowered from 10^{-4} to 10^{-9} cm/sec. (2×10^{-4} to 2×10^{-9} ft/min) by the addition of 10% bentonite to the original soil.

16.4 Soil Stabilization with Cement

Soil stabilization with cement is currently one of the most widely used methods. It commenced in 1917, when AMIES patented an initial procedure for improving soils by mixing them with variable proportions of Portland cement [11]. Since then, the use of soil-cement, which is the name that has been adopted for this mixture, has extended throughout the whole world, and is becoming increasingly popular in highway engineering, and particularly with pavement design and construction.

The chemical phenomena that take place between soil and cement, when both are mixed with the appropriate amount of water, are not yet fully comprehended. They are largely due to the reactions of cement with siliceous soil components, that bind gravels, sands and silts into synthetic, weak rock-like materials. This is the fundamental effect in predominately coarse soils. Moreover, the calcium hydrate that forms as a result of contact of the cement with water, releases calcium ions, which greedily take water from between the clay laminae. The result of this process is a reduction in the porosity and plasticity of the clayey soil, together with an increase in its strength and durability.

The beneficial soil-cement reaction is strongly impeded or neutralized when the soil contains organic matter. The organic acids capture the calcium ions that are released by the initial reaction of the cement, thus hindering the binding action of the cement in coarse soils and the stabilization of the laminar particles in clays. For this reason, most specifications limit a soil's content of organic matter to 1 or 2%, by weight, if it is to be stabilized with cement.

The presence of sulphates, usually sodium, calcium, or magnesium is also detrimental. The sulphate ion weakens cement. The less soluble sulphates deprive the binders of the moisture they require to function.

The effect of cement on clayey soils is more complicated and not so readily comprehended as in coarser soils. First, a primary effect takes place, in which hydration of the cement produces hydrated calcium silicates and aluminates, calcium hydroxide and calcium ions, which raise the concentration of electrolytes in the pore water, increasing its *pH*. A secondary process follows in two stages. In the first, an interchange takes place between the calcium ions and others absorbed by the clay minerals, which typically leads to flocculation of the clay. During the second stage, pozzolanic chemical reactions take place between the lime and the elements that compose the clay crystals. The siliceous and alumina components react with the calcium compounds to form cementing agents. The final result of this reaction is the transformation of a clayey structure, that was originally flocculated and vaporous, into a hard aggregate. Also in this second stage, the calcium hydroxide that is gradually consumed may be replaced by calcium compounds released during the initial cement hydration process.

The calcium on the surfaces exposed to the air absorbs carbon dioxide, becoming calcium carbonate with the corresponding effect of further cementation between the clay particles.

Montmorillonitic clays react most with cement, followed by illites and kaolins. However, the increase in the strength of the soil as a result of cement stabilization is not proportional to the potential reaction; instead, the strength gain is inversely proportional to the reactivity of the clay.

The reactivity of clays increases with the degree of pulverization. In montmorillonites and illites, pulverization is encouraged by the addition of small percentages of lime to the cement. Subsequent addition of the cement becomes easier.

Almost all types of cement can be successfully used for soil stabilization, but those normally employed are those with normal setting and strength characteristics. To counteract the effects of organic matter, high-strength cements are recommended, and when mixing is performed at low temperatures quick-setting high early strength cements or those containing calcium chloride as an additive may be appropriate [5].

Cement particles generally range from 0.5 to 100 microns, equivalent size with an average size of about 20 microns. Hydration of the largest particles will probably never take place, and particles measuring 10 microns may take months to become hydrated. Therefore the finer ground cements are recommended, which also produce higher strengths. However fine ground cements are more expensive, and the choice depends therefore on weighing added costs against possible benefits. In many soil cement designs the coarsest particles of common cements are replaced by an inert fill, which reduces the consumption of water and avoids high hydration temperatures and cracking. However, despite these refinements, which are becoming increasingly popular, most cement stabilization is with common *Type I* portland cement.

The mechanical properties (strength, etc.) of any soil that does not contain excessive amounts of organic matter can be improved by the addition of cement. The only restriction is that it may be hard to achieve an adequate cement mixture, especially in soft, moist clays. Coarse sands and gravels are also difficult to mix; however they seldom require stabilization because their properties are already satisfactory enough. Moreover they may suffer serious cracking problems if treated.

Many engineers maintain that a cement-treated soil should not have particles larger than 8 cm (3.14 in) or one third of the thickness of the treated layer. It is also commonly specified that not more than 50% of the soil should pass No 200 sieve, its liquid limit should not exceed 50% and its plasticity index should not exceed 18% [1]; otherwise uneconomical amounts of cement may be needed.

Some high plasticity clays which exceed these limits have been successfully treated with cement after a previous treatment with 2 or 3% of the same cement or hydrated lime. The pre-treatment improves the workability of the soil and lowers its plasticity. The curing time for this pre-treatment is typically 2 to 3 days.

Sometimes small amounts of about 0.5 or 1% cement have been added to crushed materials to improve their workability and compactibility. As already mentioned, many builders also try to improve these qualities by mixing in fine-grained soils, but the results of this procedure are uncertain or unsuitable. In the case of crushed materials containing plastic fines smaller than the No 200 sieve, good results have been obtained not only by washing, but also by mixing the materials with percentages of about 2 to 3% cement.

The only requirement for the water that is used for stabilization is that it must not possess a high content of organic matter, sulphates, or other salts. The amount of water that is used depends on compaction rather than stabilization requirements.

Cement is the most expensive component of the mixture and the amount required determines the economic feasibility of a stabilization process. In addition the properties of the resulting mixture also largely depend on the amount of cement that is employed.

During the early years of soil-cement stabilization, proportions were determined on the basis of a durability test [1]. Specimens are prepared with different amounts of cement, cured for seven days and then subjected to twelve freeze-and-thaw or wetting-and-drying cycles. The loss in weight of these specimens after any loose material had been brushed away is taken as a measure of the durability that can be expected. Also measured in these tests is the volume increase of the specimens due to expansion. Selection of the proportion of cement is based on previous experience and the designer's personal judgment of the acceptable volume change and weight loss. English technology also developed a strength requirement which is less exposed to the unreliability of a purely subjective opinion. The specimens were subjected to an unconfined compression test. A minimum strength of 17.5 kg/cm^2 (250 psi) is specified, after a 7 day curing period of 25°C (77° F). The water content of the samples is maintained constant during curing.

A strength of 17.5 kg/cm^2 (250 lb/in^2) is achieved in coarse soils, such as gravels, with extremely low cement contents, and with the usual contents strengths of about 100 kg/cm^2 (1420 lb/in^2) can be obtained. Here the result is a mixture with the consistency of a plain concrete, which, although very strong, is also highly susceptible to cracking. The large cracks that form in the treated material are often reflected in any overlying layers. As a consequence of these phenomena, it is now the practice to regard a layer of cement-treated soil as somewhat flexible, made up of interlocked blocks of soil-cement. To achieve this flexibility with interlocking, unconfined compressive strengths not exceeding about 55 kg/cm^2 (800 psi) are suggested for soil-cement mixtures based on experience. Higher strength values result in a rigid layer behavior. It is even proposed that cement-stabilized layers be opened to traffic at the earliest opportunity, thus encouraging limited break-up, which will promote the flexible behavior of the pavement [1].

It is now customary to proportion soil-cement, mixtures to produce a minimum and maximum strength, generally measured by an unconfined compression test. The durability tests already mentioned are also employed to provide fuller information. Recent attempts have also been made to introduce more sophisticated strength tests, such as the Texas triaxial test previously described.

Attempts to relate soil-cement proportions to *CBR* tests have not led to conclusive results. The *CBR* values obtained for soil-cement mixture, and especially those made up of coarse soils, are so high that they cannot be interpreted.

In conclusion, no reliable criterion is available currently for determining the appropriate cement content for a given soil. The criteria that are employed by engineers designing soil cement mixtures are strongly subject to personal experience and interpretation. For many engineers a visual examination of the specimens, as well as the added cost compared with the benefits continue to play an important part in the final judgment. The behavior of soil-cement under repeated loading including the effects of accumulative deformation have not been sufficiently studied.

Current tests involve three fundamental aspects of soil-cement mixtures [12]:

- The amount of cement required to produce the desired soil characteristics
- The amount of water that must be added for compaction (and hydration)
- The unit weight to which the mixture must be compacted, according to the requirements of the layer in which it is to be used

In the laboratory, selection of the initial percentages to be employed in the different test specimens usually poses a problem. Table 16-7 [12] provides some guidance. Table 16-8 [12] permits a closer estimation for soils of a frictional nature. It is based on the grain-size percentages together with the unit weight the mixtures reaches after dynamic compaction, often in an AASHTO-type test. With the aid of the table it is then possible to adjust the cement contents with a view to conducting durability or strength tests.

Table 16-9 [12] is equivalent to 16-8, but refers to clayey, plastic soils. This table includes the concept of the soil group index (*IG*) which is defined as:

$$IG = 0.2a + 0.005ac + 0.01bd \qquad (16\text{-}5)$$

where:

a = refers to the percentage of material passing No 200 sieve; a = 0 if this percentage is lower than

Table 16-7

Percentage of cement to be tested initially in different types of soil [12]

Type of soil	Percentage by weight of cement usually required by the finished layer	Percentage by weight of cement to be used initially in compaction tests	Percentage by weight of cement to be used initially in durability tests
GW, GP, GM, SW	3–8	5–6	3–7
SC, GC	5–9	7	5–9
SP, SM	7–11	9	7–11
ML	7–12	10	8–12
CL, OL, MH	8–13	10	8–12
CH	9–15	12	10–14
OH, P_t	10–16	13	11–15

Table 16-8
Percentage of cement to be mixed with specimens for durability tests [12] Frictional soils

Material retained on No. 4 sieve	Material smaller than 0.5 mm (0.02 in)	Tentative percentage by weight of cement — for the unit weights indicated in Kg/m^3 (lb/ft^3)					
—	—	1680 to 1750 (105-110)	1750 to 1830 (110–115)	1830 to 1910 (115–120)	1910 to 1990 (120–125)	1990 to 2070 (125–130)	2070 to more (130 or more)
%	%	%	%	%	%	%	%
0–14	0–19	10	9	8	7	6	5
	20–39	9	8	7	7	5	5
	40–50	11	10	9	8	6	5
15–29	0–19	10	9	8	6	5	5
	20–39	9	8	7	6	6	5
	40–50	12	10	9	8	7	6
30–45	0–19	10	8	7	6	5	5
	20–39	11	9	8	7	6	5
	40–50	12	11	10	9	8	6

35%, and a = 40 if it is higher than 75%. For percentages between 35 and 75, a is a whole number between 1 and 40 selected proportionally. (a = 1 for 35% and a = 40 for 75%).

b = also refers to the percentage of material passing No 200 sieve; b = 0 if this percentage is lower than 15%, and b = 40 if it is higher than 55%. For percentages between 15 and 55, b is a whole number between 1 and 40 selected accordingly (b = 1 for 15% and b = 40 for 55%).

c = refers to the liquid limit of the fraction of soil smaller than No 40 sieve. If $W_L < 40\%$, then $c = 0$. If $W_L > 60\%$, then c = 60. For percentages between these limits c will be a whole number between 0 and 60 selected proportionally.

d = refers to the plasticity index of the soil. If $I_p < 10\%$, $d = 0$. If $I_p > 30\%$, $d = 30$. For percentages between these limits d will be a whole number between 0 and 30, selected proportionally.

Tables 16-7, 16-8 and 16-9 are based on laboratory experience and should be regarded only as a guide to the selection of percentages. In no way should they be interpreted as a substitute for the laboratory work that should always be undertaken in each specific case. References [12,13,14] also give full details of all test procedures, both for compaction tests and durability tests with wetting and drying cycles. Such details are, however, felt unnecessary in this book.

Table 16-10 [1] is a summary of the properties usually required of soil-cement mixtures for highway purposes. Like so many similar tables, Table 16-10 cannot be regarded as a totally reliable recommendation to be blindly trusted by the designer, but simply as a guide to assist him in his laboratory studies or in the exercise of his selective capacity, which he must ultimately employ in each case.

16.4.1 Mechanical Properties of Soil-Cement Mixtures

In the paragraphs that follow, some mechanical properties of soil-cement mixtures will be mentioned, and their variations with certain characteristics of practical interest will be briefly analysed. Reference will also be made to a thickness design method for soil-cement pavement layers.

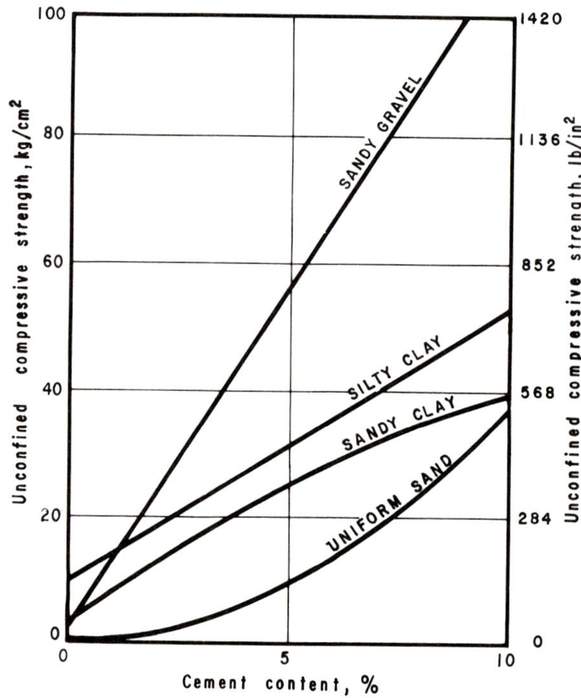

Fig. 16-8 Variation in the unconfined compressive strength of cement-treated specimens of various typical soils [1]

Table 16-9

Percentage of cement to be mixed with specimens for durability tests [12] Plastic soils

Index of soil category	Material between 0.05 mm and 0.005 mm (0.002 in — 0.0002 in)	Total percentage, by weight, of cement — for the indicated unit weights in Kg/m^3 (lb/ft^3)						
—	—	1140 to 1510 (70–95)	1510 to 1590 (95–100)	1590 to 1670 (100–105)	1670 to 1750 (105–110)	1750 to 1830 (110–115)	1830 to 1910 (115–120)	1910 or more (over 120)
—	%	%	%	%	%	%	%	%
0–3	0–19	12	11	10	8	8	7	7
	19–39	12	11	10	9	8	8	7
	39–59	13	12	11	9	9	8	8
	60 or more	—	—	—	—	—	—	—
4–7	0–19	13	12	11	9	8	7	7
	19–39	13	12	11	10	9	8	8
	39–59	14	13	12	10	10	9	8
	60 or more	15	14	12	11	10	9	9
8–11	0–19	14	13	11	10	9	8	8
	19–39	15	14	11	10	9	9	9
	39–59	16	14	12	11	10	10	9
	60 or more	17	15	13	11	10	10	10
12–15	0–19	15	14	13	12	11	9	9
	19–39	16	15	13	12	11	10	10
	39–59	17	16	14	12	12	11	10
	60 or more	18	16	14	13	12	11	11
16–20	0–19	17	16	14	13	12	11	10
	19–39	18	17	15	14	13	11	11
	39–59	19	18	15	14	14	12	12
	60 or more	20	19	16	15	14	13	12

.1 Unconfined Compressive Strength

Figure 16-8 [1] shows the variation of the unconfined compressive strength of soil-cement mixtures with the cement content, using different soils. In most soils, strength increases almost linearly with the cement content, but the slope of the curves does vary greatly from one soil to another.

Figure 16-9 [14] shows the effect of the content of material passing the No 4 sieve on the unconfined compressive strength. Well-graded materials with a range of particles between No 4 sieve and fines, require less cement than uniform-graded soils. The figure refers to three specimens with very similar grain-size distributions, predominantly coarse (maximum size = 1 in, and with less than 25% passing No 40 sieve). They are natural, uncrushed materials, with a similar mineralogical composition. It is observed that material *B*, which is the most continuously graded, without predominance of fine or coarse particles, exhibits the highest strength value.

Figure 16-10 [14] shows the effect of the finest fraction of the soil on the unconfined compressive strength, expressed as a ratio of the percentage passing the No 200 sieve and that passing the No 30 sieve. The specimens tested are all sands or sandy soils, the characteristics of which are shown in Table 16-11 [14]. Several specimens of the same natural soil were tested with variable fine fractions. Results are shown for five different natural soils.

All the specimens were prepared with 3% cement. It can be seen that to obtain minimum strength values of 15 kg/cm^2 (213 lb/in^2) with this content of cement, the ratio of the percentage passing the No 200 sieve to that passing the No 30 sieve must be no smaller than 0.15. It was also observed that this minimum strength value demanded a unit weight of at least 1,900 kg/m^3 (118 lb/ft^3) in all the sandy soils tested. Figure 16-10 illustrates that the fines content encourages development of the strength of soil-cement mixtures, so long as the component soil is cohesionless. However, the percentage of fines in natural soils can seldom be modified, because of high cost.

Table 16-10

Properties commonly demanded of soil-cement mixtures [1]

Type of Layer	Unconfined compressive strength (1)		*CBR* (2)	Expansion	Weight loss in wetting-and-drying test (3)
—	kg/cm^2	lb/in^2	%	%	%
Sub-bases. Fill material for ditches	3.5–10.5	50–150	20–80	< 2	< 7
Sub-bases, or bases for very light traffic (4)	7–14	100–200	50–150	< 2	< 10
Bases for heavy traffic	14–56	200–800	200–600	< 2	< 14
Embankment protection against erosion and action of water	> 56	> 800	600	< 2	< 14

(1) After a 7-day curing period at constant water content. The strength of similar specimens submerged in water should not be more than 20% lower

(2) After a 4-day period of immersion in water

(3) The durability test is only worthwhile if water can penetrate the mixture.

(4) In this case the strength may be lower in well-drained zones.

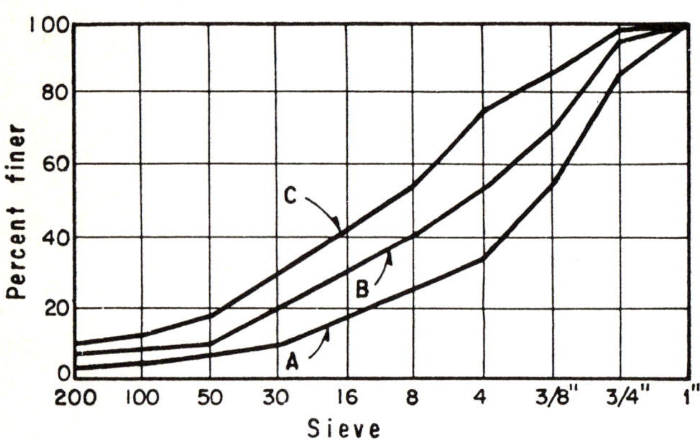

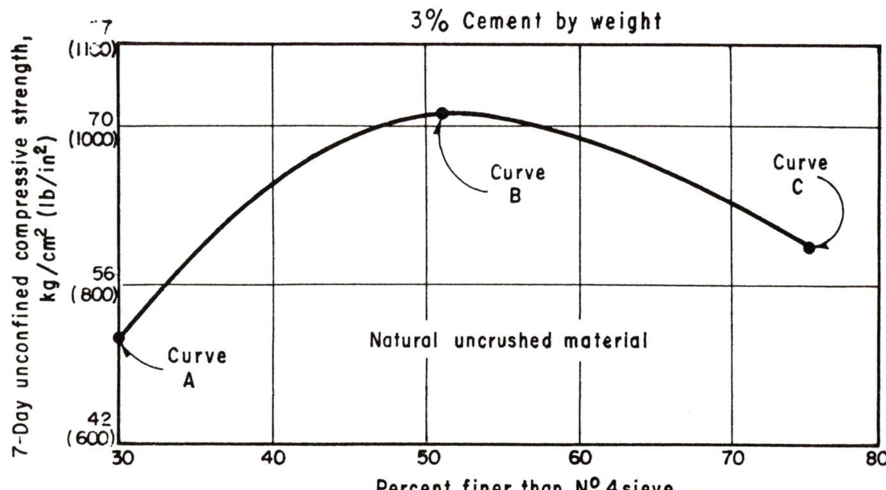

Fig. 16-9 Effect of the percentage of material smaller than No 4 sieve on the unconfined compressive strength [14]

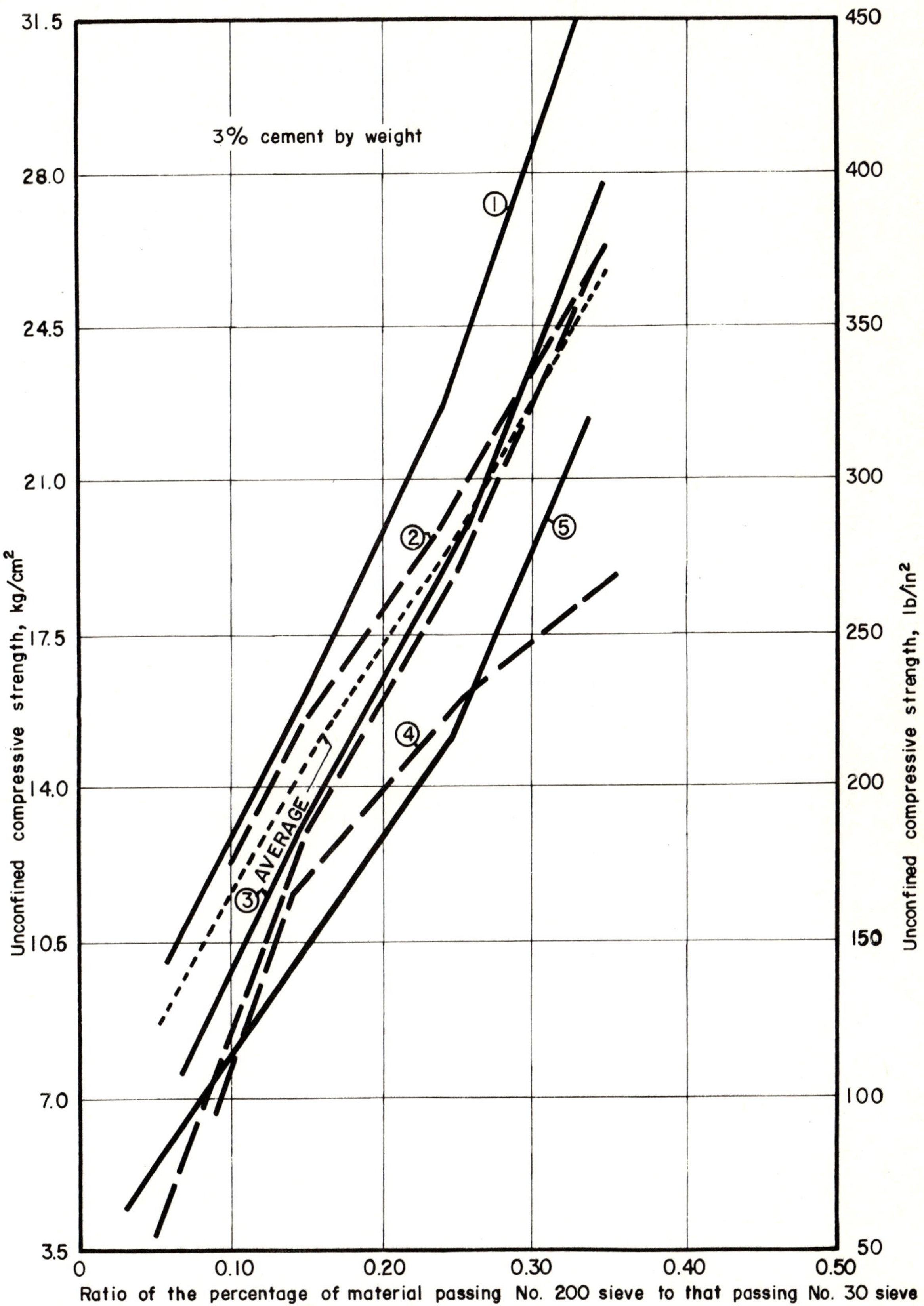

Fig. 16-10 Effect of the fine fraction of frictional soils on the unconfined compressive strength [14]

Table 16-11
Characteristics of the materials mentioned in Fig. 16-10

Material	Description	Percentage passing sieve No. 200	Unit weight after compaction Kg/m^3	lb/ft^3
1	Sandy soil	3 8 12 19	1860 1930 1980 2080	116 120 124 130
2	Sandy soil	9 13.5 23 33	1850 1900 1950 2000	116 119 122 125
3	Sandy soil	7 15 27 35	1680 1730 1980 2060	105 108 111 128
4	Sandy soil	9 14 24 34	1800 1850 1910 1940	112 115 119 121
5	Sandy gravel	1 6 11 16	1815 1930 1990 2080	114 120 124 130
6 (Not included in Fig. 16-10)	Sand	2 7 12.5 19	1850 1960 2030 2090	115 122 127 131

Figure 16-11 [14] shows the effect of the unit weight achieved with compaction on the unconfined compressive strength obtained. The same natural soils are employed as those listed in Table 16-11. Note that the cleanest sands and sandy gravels do not acquire high strength values in relation to the other soils tested. This is probably due to the relatively low proportion of cement used. As already mentioned, with higher percentages of cement clean cohesionless soils become genuine plain concretes; Fig. 16-12 [1] gives similar information, but for a clay stabilized with 10% cement.

Figure 16-13 [14] shows the effect of adding different percentages of small balls of hard, active clay to well-graded, clean materials. As might be expected, the material suffers a notable reduction in strength. (The reduction in durability is even worse).

Figure 16-14 [14] shows the effect of delaying compaction once the mixtures have been prepared. This effect is clearly a reduction in strength. Bearing in mind these results and the realities imposed by practical engineering, the specifications in almost all countries prohibit a lapse of more than two hours from the moment soil-cement is mixed at a given location to the end of the compaction process. Reference [1] points out that a considerably higher proportion of strength is lost than that indicated in the figure when cement contents are greater.

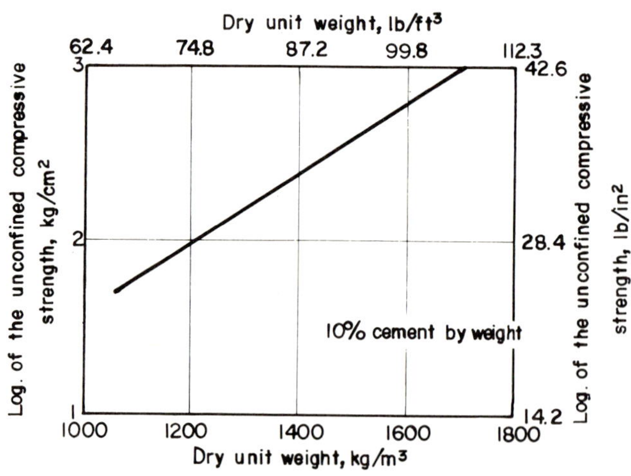

Fig. 16-12 Effect of the density achieved by compaction on the unconfined compressive strength of a cement-treated clay [1]

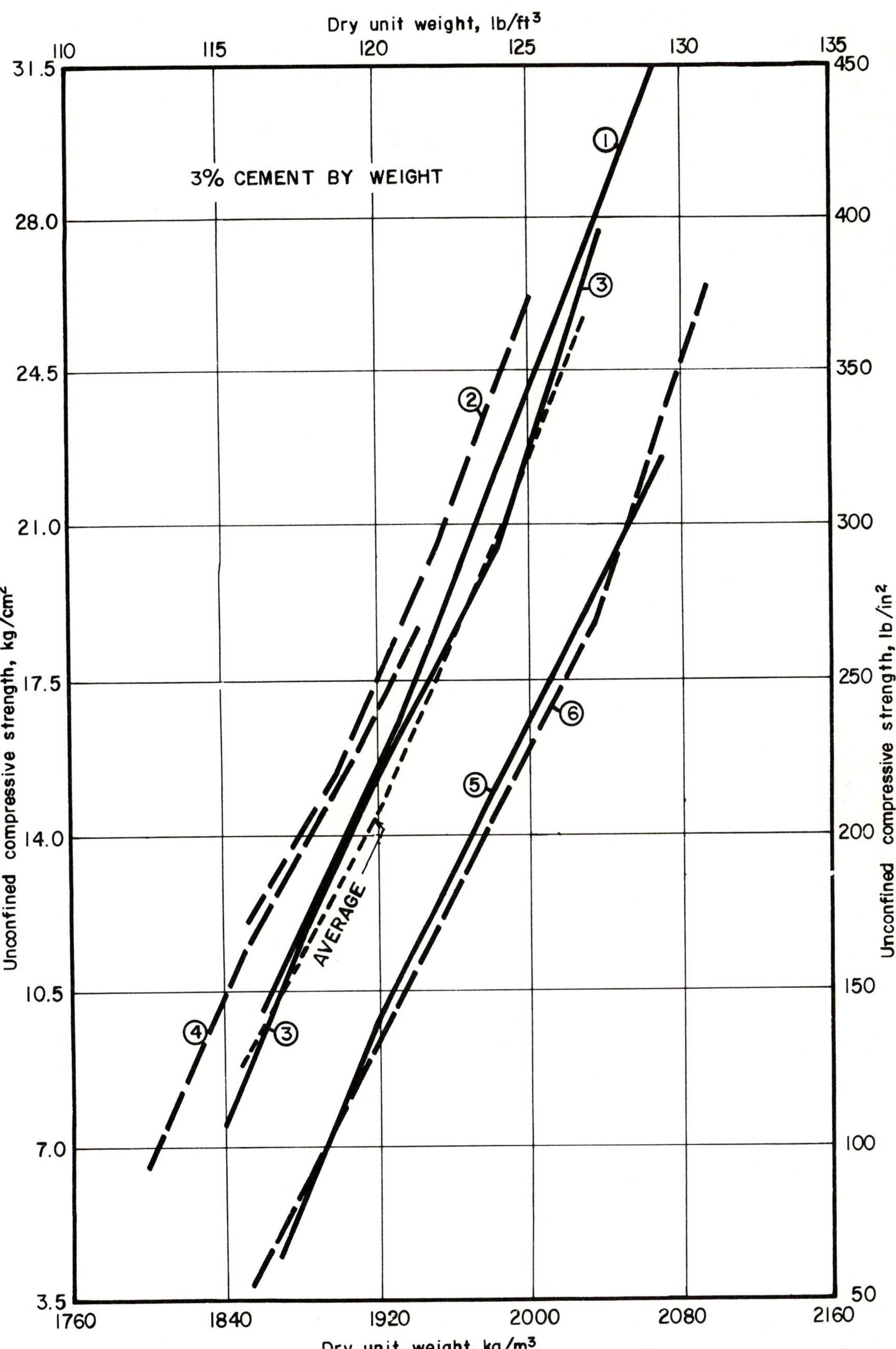

Fig. 16-11 Effect of compaction on the unconfined compressive strength of sandy soils [14]

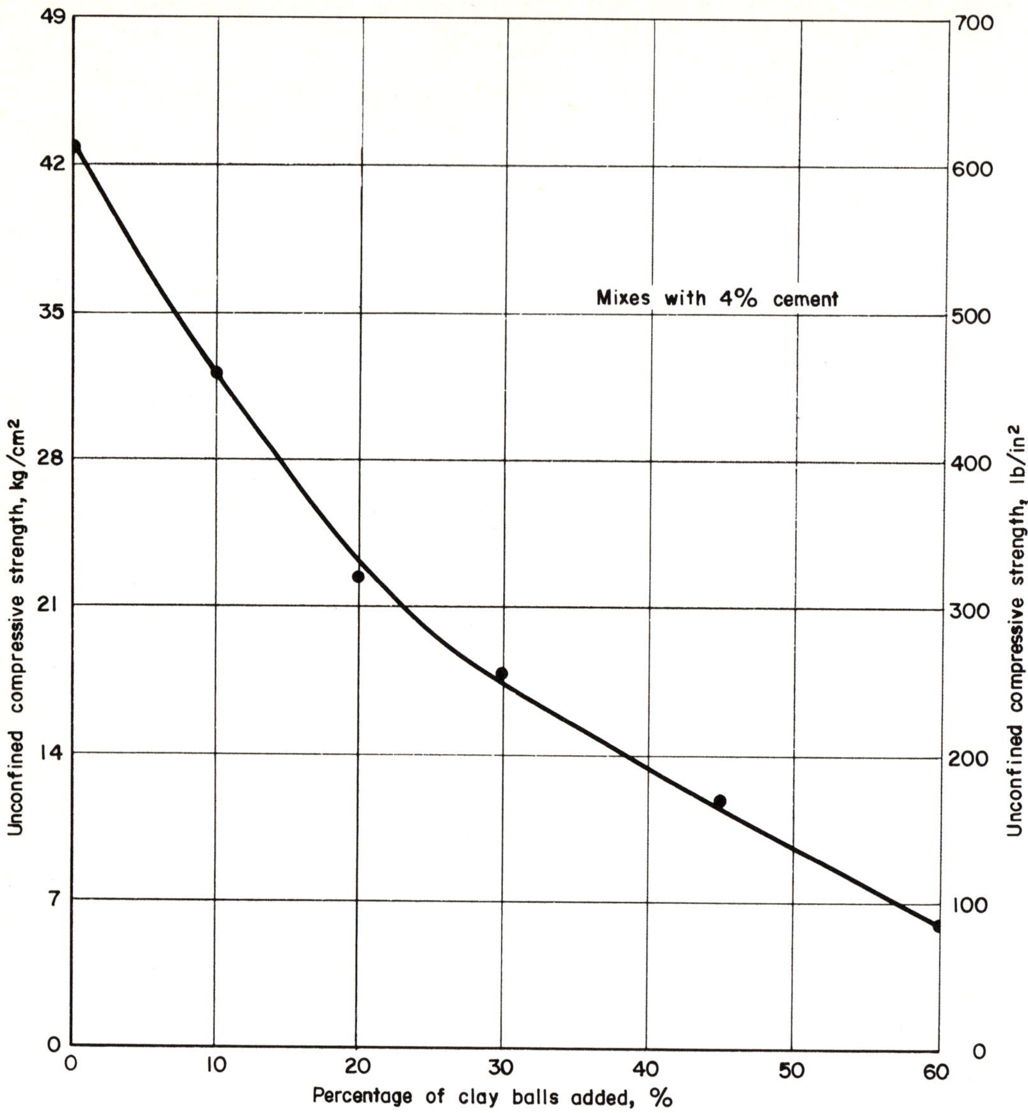

Fig. 16-13 Effect of adding small balls of hardened clay on the strength of well-graded frictional materials [14]

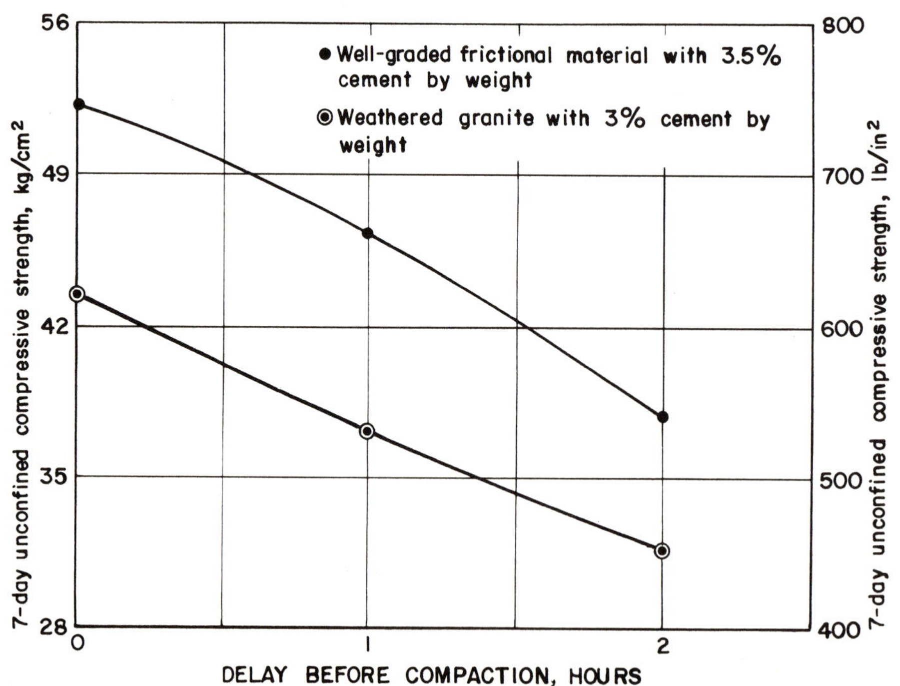

Fig. 16-14 Effect of delayed compaction on material strength [14]

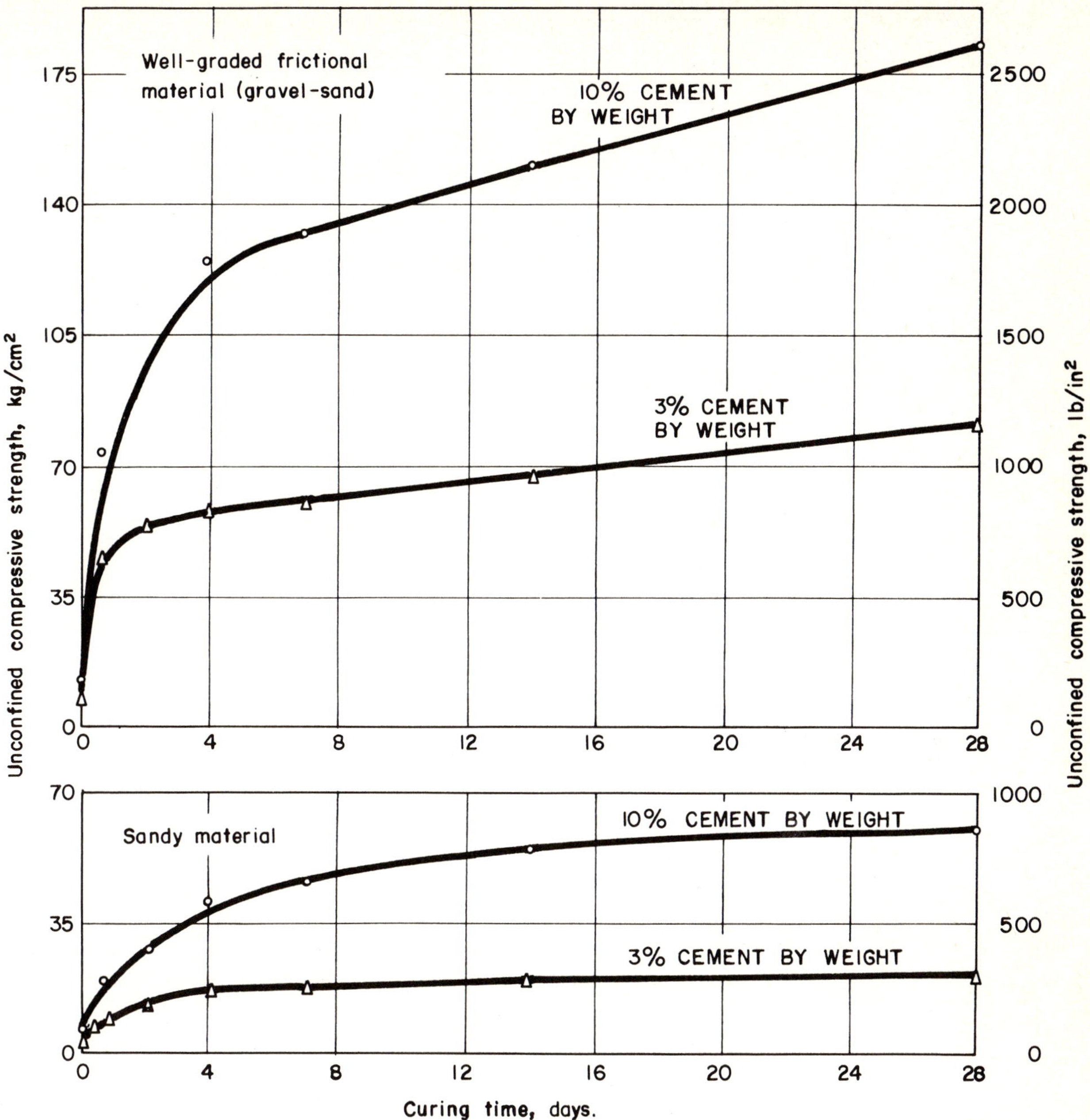

Fig. 16-15 Influence of curing time on the unconfined compressive strength of various mixtures [14]

Figure 16-15 [14] expresses the influence of curing time on the strength obtained. The strength increases when the curing time to which the mixture is subjected after compaction increases, although with time this increment is by no means constant, similar to genuine concrete.

Figure 16-16 [14] provides the same information, but with reference to different types of soils. Again it is noted that clean sand does not prove to be a very appropriate material for stabilization with low cement contents. In Fig. 16-15 a sand achieves high strength values with 10% cement, but both Fig. 16-14 and Fig. 16-16 show that similar sands with cement contents as high as 5% achieve far lower strength values. This is of course due to the high proportion of voids that is characteristic of coarse soils. These have to be filled with a large amount of cement if the soil particles are to be adequately bound together. For this reason the cement content required by sands and gravels can be drastically reduced by adding a fine, inert material to partially fill the voids.

Figure 16-17 [14] shows the effect of curing temperature and curing time, on the unconfined compressive strength of soil-cement specimens. In agreement with traditional experience in the field of concrete, the higher the curing temperature, the greater the strength acquired by a layer of soil-cement. Thus higher strength values can be obtained when layers of soil-cement are cured during hot weather.

Reference [14] which presents much of the early California experience on the subject now under discussion, is also the source of the information in Fig. 16-18. Here the effect is studied of adding to the soil-cement mixtures indicated different percentages of the additives shown. It can be seen that the effect of these additives is not very pronounced in silty materials but may be a little more so in mixtures of gravel and sand. However, it can be concluded that the additives tested did not lead to any substantial improvement in the strengths obtained.

At one time it was thought that it would be helpful to add small amounts of asphalt emulsion to soil-cement mixtures. To a

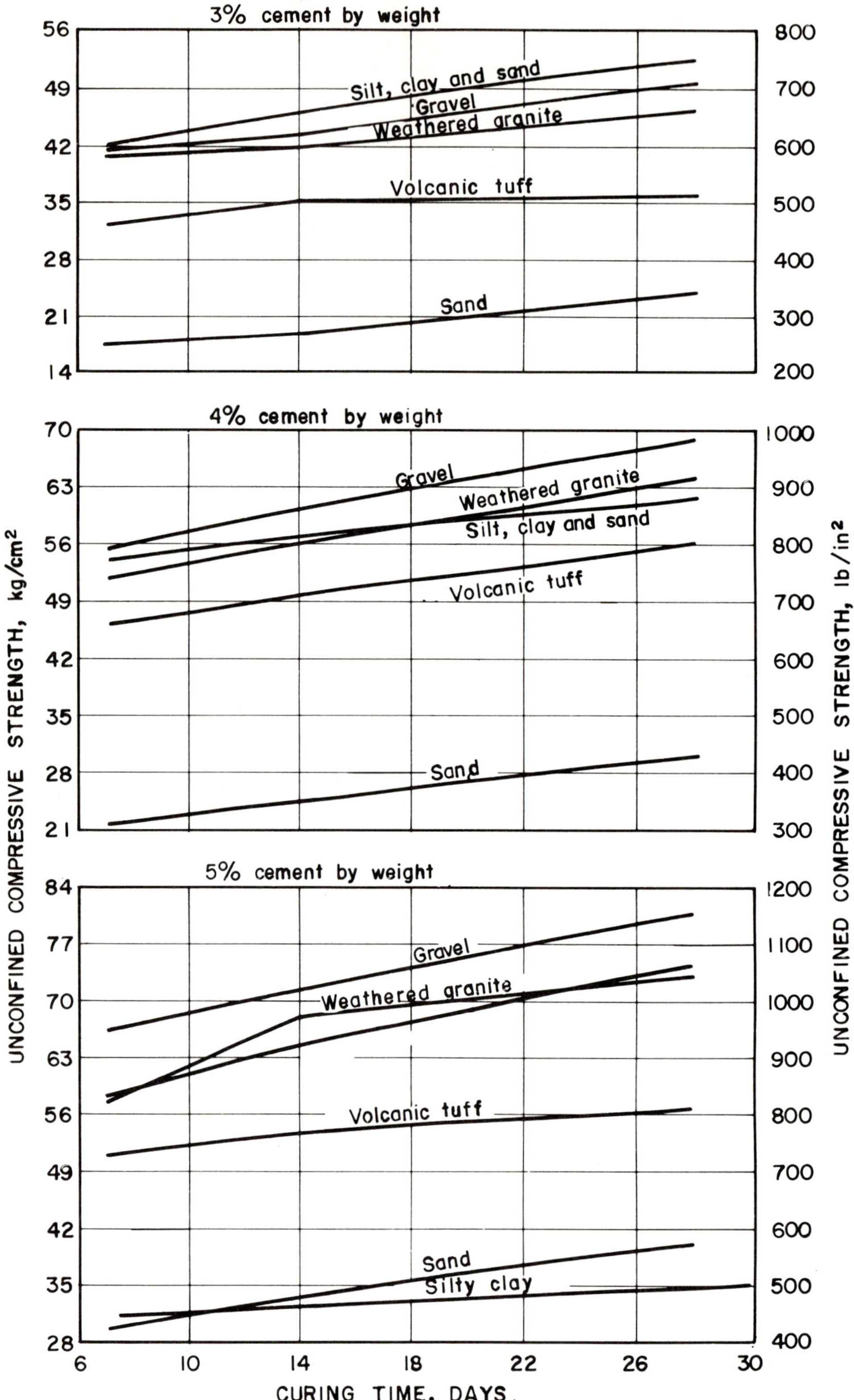

Fig. 16-16 Influence of curing time on the unconfined compressive strength of various materials [14]

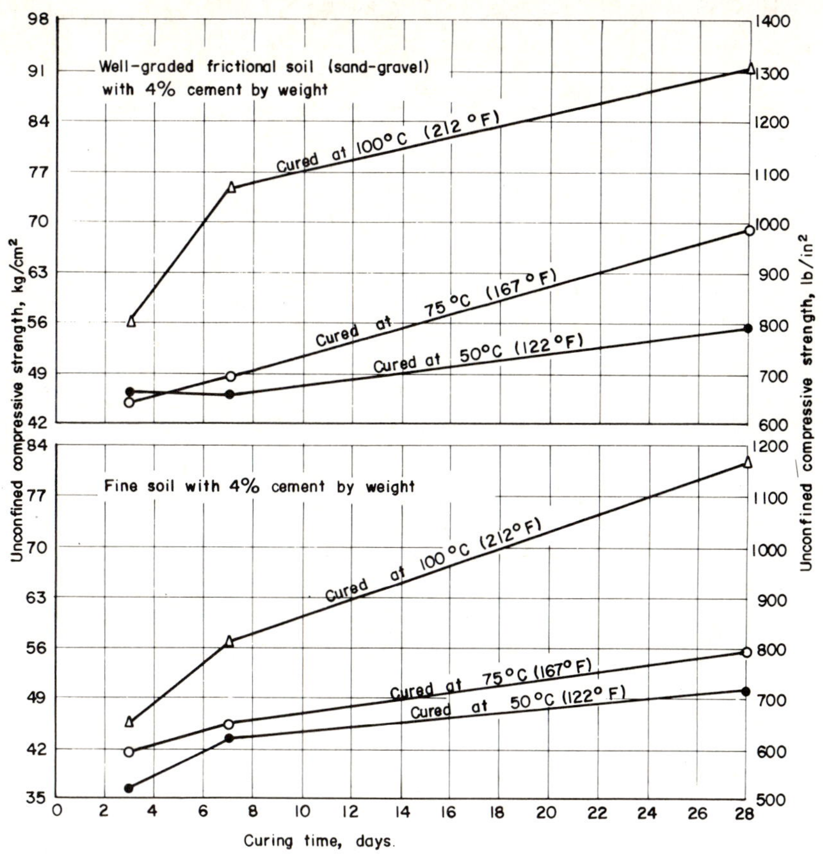

Fig. 16-17 Influence of temperature and curing time on unconfined compressive strength [14]

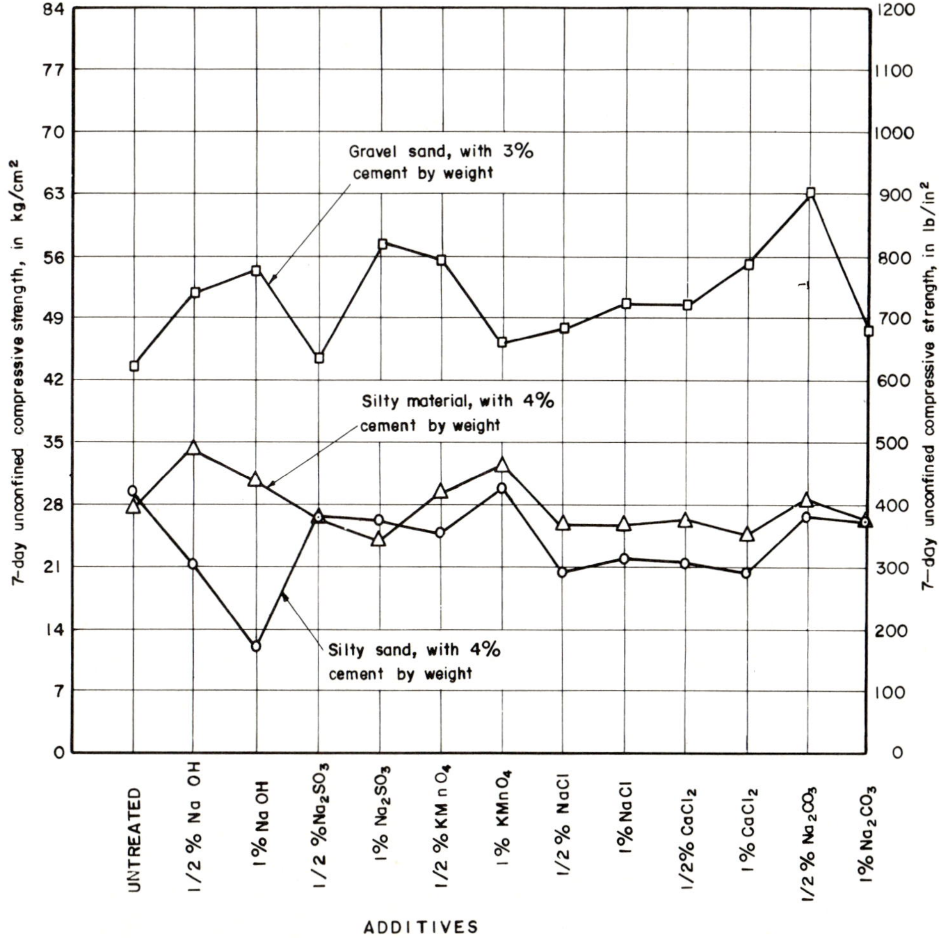

Fig. 16-18 Influence of various chemical additives on the unconfined compressive strength [14]

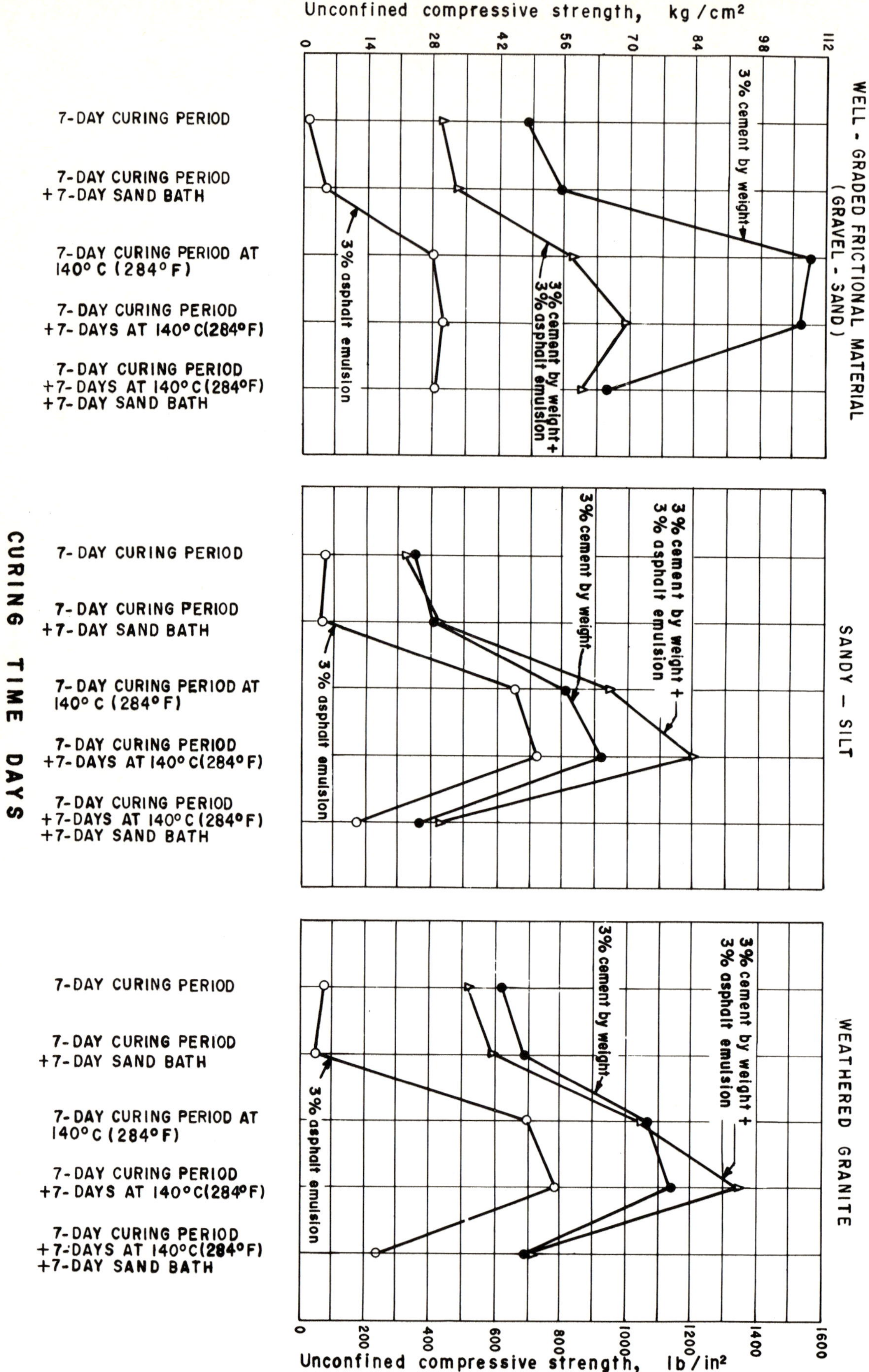

Fig. 16-19 Effect of unconfined compression when different percentages of asphalt emulsion are added to specimens of soil-cement [4]

certain extent this is what is done when during reconstruction work, an old wearing course is scarified and mixed with an old base and is relaid, adding cement. This is a useful technique for renewing old pavement layers, Fig. 16-19 [14] is a study of these aspects, including the effect of the addition of asphalt, of curing time and curing temperature. It can be seen that the addition of asphalt does not essentially improve the strength of the mixtures and increases the cost of the layer. Therefore, the addition of asphalt is not a good routine practice. On the other hand, in the case of the reconstruction work referred to, the addition of cement to old wearing courses and bases, (to which new material is often added to complete the required thickness) enables a far more homogeneous strength to be achieved.

Figure 16-20 [1] shows the reduction in unconfined compressive strength that is obtained when the soil-cement mixture is saturated. The results refer to a low plasticity clay, in one case compacted and tested at its optimum water content, and in the other case submerged in water after compaction and before testing.

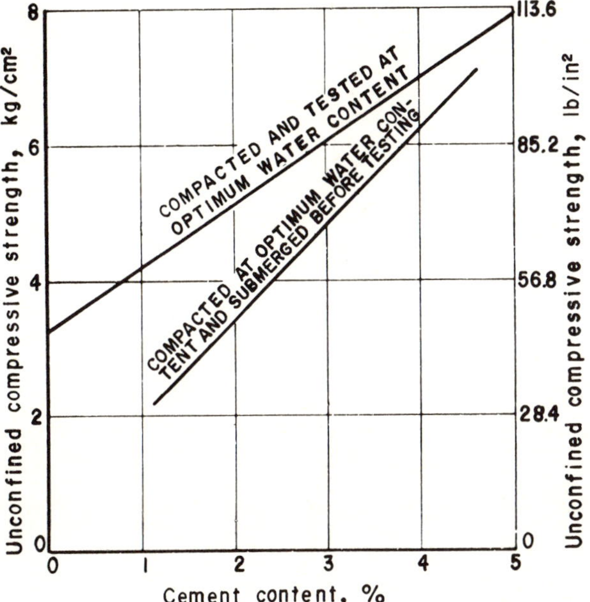

Fig. 16-20 Effect of a reduction in the unconfined compressive strength on saturation of the soil-cement specimen before testing [1]

Lastly, [15] describes the effect of storage time in a moist room on the unconfined compressive strength of nine different soils. The specimens were protected against free water until the day of testing, but were submerged in water one hour before the test, Fig. 16-21 illustrates the results obtained.

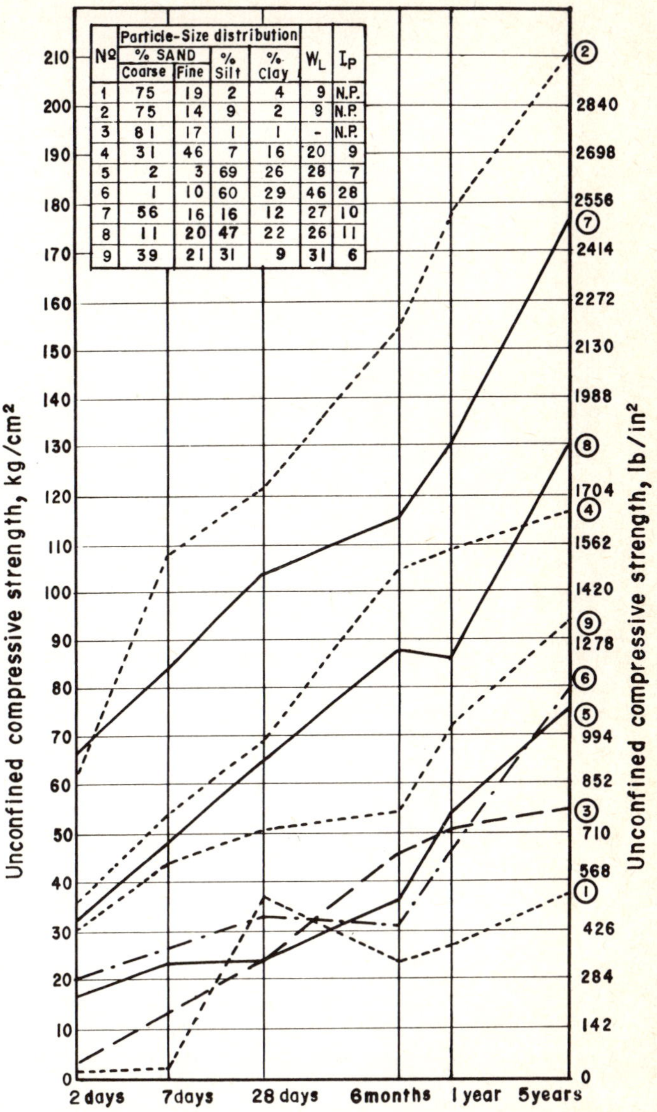

Nº	Particle-Size distribution % SAND Coarse	Fine	% Silt	% Clay	W_L	I_P
1	75	19	2	4	9	N.P.
2	75	14	9	2	9	N.P.
3	81	17	1	1	-	N.P.
4	31	46	7	16	20	9
5	2	3	69	26	28	7
6	1	10	60	29	46	28
7	56	16	16	12	27	10
8	11	20	47	22	26	11
9	39	21	31	9	31	6

Fig. 16-21 Effect of storage time of soil-cement specimens in a moist room on their unconfined compressive strength [15]

.2 Other Properties

Other studies conducted on specimens of soil-cement show a non-linear stress-strain behavior for loads above 60% of the maximum, and relatively linear for lower values. Fig. 16.22 [1] shows the typical behavior under repeated loading, which indicates considerable unrecovered strain between one loading cycle and the next.

Another characteristic of cement-stabilized clays is their tendency to undergo accumulative strain (creep) under constant loading. This is illustrated in Fig. 16-23. Unfortunately [1], from which Fig. 16-23 was taken, does not specify the degree of loading under which the experiment was performed.

For some time it was thought that creep in cement-stabilized clays under constant tension stresses was one of the main reasons for deformation and cracking of the layers. It was later

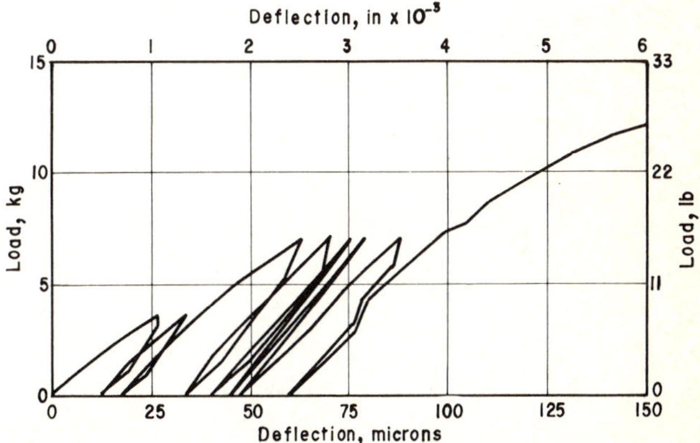

Fig. 16-22 Effect of repeated loading on a specimen of clay treated with 10% cement [1]

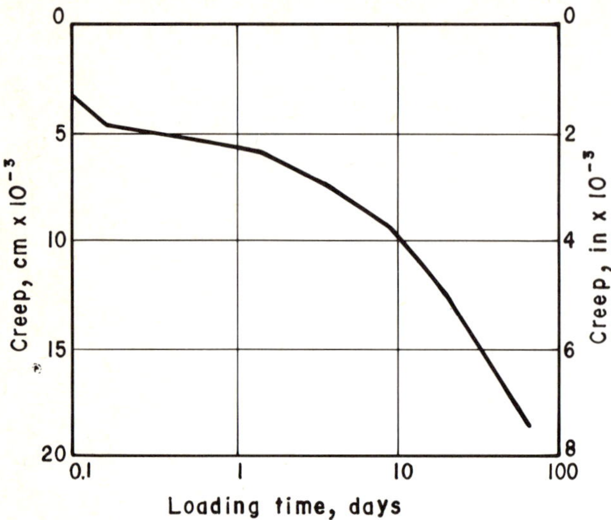

Fig. 16-23 Effect of cumulative deformation under constant loading in specimens of clay stabilized with 10% cement [1]

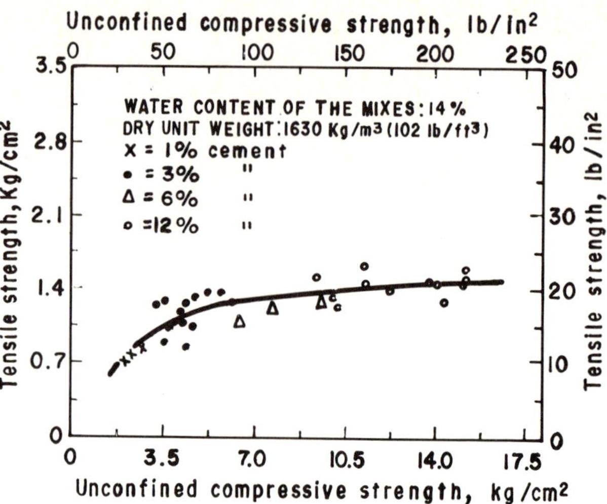

Fig. 16-24 Relation between the unconfined compressive strength and the tensile strength of cement-treated silt specimens [20]

shown, however, that the shrinkage of cement stabilized clays at constant water contents may be a more important cause of cracking and deformation.

Reference [20] describes an investigation for obtaining information about the tensile strength of cement-treated soils. In it is proposed an expression for predicting this strength as a function of the cement content of the soil the curing time for the conditions of the soils employed and both tensile and unconfined compressive strength. The curing time and the cement content are factors which control the tensile strength as well as the unconfined compressive strength.

Reference [20] also gives a correlation between the tensile and unconfined compressive strengths, which is only valid for the specific test conditions adopted. This correlation is given in Fig. 16-24. The tensile strength was measured in a uniaxial tension test performed on cylindrical specimens, and the compressive strength by the traditional methods. Further investigations into the tensile strength of stabilized soils can be found in [21-25].

Recently interest has also been taken in the properties of cement-treated soils under dynamic loading [26]. Table 16-12 shows the variation in the static and dynamic properties of sands and clays treated with different percentages of cement (by weight). It shows the static properties in relation to a modulus of elasticity and the dynamic properties in relation to a dynamic modulus of shear strength.

In soil-cement investigations, special interest has been taken in pavement fatigue phenomena caused by the effects of repeated

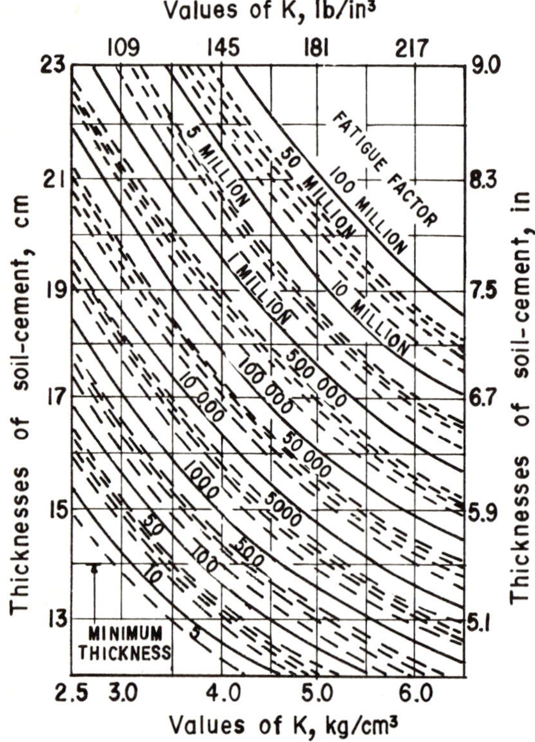

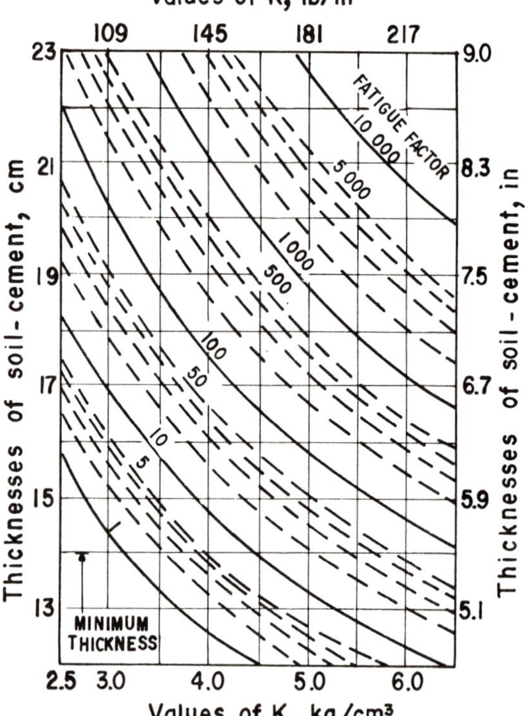

Fig. 16-25 Thickness design graphs for soil-cement layers [27,28]

Table 16-12
Effect of the cement content on the static and dynamic properties of mixtures [26]

Soil	Cement content	Modulus of elasticity		Dynamic modulus of shear strength	
—	%	kg/cm^2	lb/in^2	kg/cm^2	lb/in^2
Sand	0	45	638	950	13500
	2	144	2040	1930	27400
	4	165	2340	2700	38350
	6	480	6820	4750	67350
Clay	0	55	780	1260	17860
	2	56	794	1300	18460
	4	150	2130	1830	26000
	6	600	8520	4900	69500

loading. The number of load repetitions that lead to failure has been related to the radius of curvature of the part of the treated layer that bends under a loaded tire (or under a loading plate, in a test of this type) [27,28]. In these studies it has been seen that the effect of the modulus of subgrade reaction (K) is small and that the cement content has a smaller influence than might initially be expected on the results of repeated loading tests. The magnitude of the repeated load in relation to the maximum load is of importance, as is the nature of the soil in the mixture. The soils were divided into two categories:

— Granular soils, with less than 35% finer than the No 200 sieve (*SC, SP, SM, GC* soils, in general).
— Fine soils, with more than 35% finer than the No 200 sieve (*ML, CL, OL, MH, CH, OH* and P_t soils in general).

As a result of these studies, equations have been derived which limit the radius of curvature of the part of the layer bent by the load, after application of an established number of load repetitions. By way of an example, the following equations [27] are mentioned:

granular:

$$R = \frac{0.3h^{3/2}}{h - 1.2} R_c N^{0.025} \qquad (16\text{-}6)$$

fine material:

$$R = \frac{0.3h^{3/2}}{h - 1.2} R_c N^{0.050} \qquad (16\text{-}7)$$

where:

R = permissible radius of curvature in cm for a number of load repetitions N.
R_c = critical radius of curvature in cm causing failure of the treated layer with N load repetitions, which depends on the degree of loading employed and the characteristics of the mixture.
h = thickness of the treated layer, in cm
N = number of load repetitions, applied either by a loaded tire or a plate representing traffic conditions.

16.4.2 Construction Methods

Construction of a soil-cement layer is performed in the following stages:

— Scarification, pulverization and pre-wetting the soil, when necessary.
— Proportioning and spreading the cement
— Application of water
— Mixing of materials
— Compaction
— Finishing
— Curing

Coarse, frictional materials do not usually require scarification, but plastic soils usually do. Lumps larger than 3 cm are not accepted at the time of mixing with the cement; it is usually specified that 80% of the lumps must pass No 4 sieve [12]. A maximum size of 5 cm is often established for individual soil particles. Pre-wetting may help scarification and pulverization, often reducing operation times.

Sandy soils can be mixed with cement in almost any condition, but in clayey soils a below-optimum water content is usually required. Proportioning of the mixture can be done by mechanical means,by spacing of cement bags or in a central plant. Mechanical spreading is done with a cement spreader attached to a truck. Self-driven mixers are also available. The cement is proportioned either as a percentage of the volume or weight of the soil. Reference [12] contains some useful diagrams which assist with field calculations in proportioning operations.

When no sprinkler is available, the cement can be poured directly from bags, spreading it as evenly as possible and afterwards harrowing it.

The batcher plant is the best system for mixing the cement with soil and water. With this system an extra 2% above the optimum water content is often added in dry climates to compensate for loss during handling. Without a plant it is not so easy to add water accurately; similarly any other method of mixing the three basic components is not so reliable. There are mixing devices, which take the soil covered with cement from central ridges or winrows and mix it several successive times, at the same time adding water, which is provided by tank trucks. Three passes of the mixer are usual, water being added during the second pass.

Multiple mixers make mixing easier and more efficient. They lift the material, pulverize it and mix it with a first set of rotors. Water is added during the passing of a second set of rotors, and a third set stir all the components and spread the material, leaving it ready for initial compaction.

To summarize, it can be said that there are numerous methods of spreading and mixing soil-cement, from the central ridge or winrow process where it is spread with motor graders, to the use of different mixers which make one or several passes, to the central mixing plants.

The better the soil-cement mixture, the more uniform the material and the higher the strength obtained. The quality of the mix achieved is not a linear function of the mixing energy [16], but depends on the field procedure employed. The field strength values are rarely equalled in the laboratory for the same proportions. This is generally attributed to deficient mixing. Homogeneity of the mixture is also essential. Mixing time is an important fact in the achievement of both efficient mixing and homogeneity. Strength increases greatly with mixing time, but this increase is not linear; for a given method the advantages of mixing time become less with more mixing. The use of too much time may encourage segregation of the soil particles, which reduces the homogencity of the mixture. Also, hydration of the cement is taking place, so that an unduly long mixing operation may break the newly created bonds between the cement and soil particles.

In [1] the *Mixing Efficiency* is defined as the relation between the strength of the material when mixed in the field and compacted in the laboratory and the strength of the material when both mixed and compacted in the laboratory. In actual practice an efficiency above 80% is unusual; normal values are about 60%.

The compaction of soil-cement follows the same principles as that of soils, described in Chapter 4. Sheepsfoot compacters are the most widely used for initial compaction of cohesive soils whereas rubber tired or vibratory rollers are more effective for granular soils. As is normal in compaction, pneumatic rollers with overlapping wheel paths are gradually gaining popularity for almost all soil types, although they are most frequently used for relatively granular soils without a very high fines content. Drum rollers are used chiefly for finishing-off operations.

The water content of a compacted layer of soil-cement must be maintained until the cement is adequately hydrated. For this purpose the compacted layer is often covered with a material that will prevent loss of moisture. Asphalt, plastics, waterproof paper, straw or moistened fabrics are the materials most often employed. If laying of these materials is delayed, the surface of the soil-cement layer should be maintained moist by light, continuous sprinklings.

It is important to verify the degree of mixing and the homogeneity of the mixture obtained: whether or not the design proportions have been observed. This is a problem of quality control.

Reference [17] mentions a procedure that has been developed for measuring the *Uniformity Index* of the mixture as a function of a random cement distribution and the homogeneity achieved. The strength of a number of specimens subjected to statistical sampling is evaluated, and although the procedure is more reliable than a mere analysis control of a sample taken for each specific unit of area (which is not very reliable but widely used method) it is so complicated that it is of little value for routine jobs.

Reference [16] suggests a method which consists of adding to the cement a small amount of radioactive material, which can then be traced in the layer using a suitable measuring device. The precision depends on how homogeneously the radioactive material is mixed with the cement. Mixing efficiency, as defined previously, has also been used as a measure of the results obtained in the field.

Reference [18] describes in great detail a test which has been successfully used in different parts of the world, including Mexico, to determine in an accurate and simple manner the cement content of samples obtained from the treated layer. This test, which has been named the *titration test*, originated in California [19].

The test specimen is a mixture of fragments of soil-cement taken from the tested layer. It is pulverized and suspended in water. After a period of rest; the coarse fraction of the soil is deposited and a suspension of fine soil and cement will remain in the water. There are two different versions of the test: the acid-alkali version established for cases where there is no reaction between the soil particles and hydrochloric acid, and the constant neutralization method to be used in those cases where such a reaction does exist, such as in calcareous soils.

In the acid-alkali test, hydrochloric acid and phenolphthalein are added to the suspension of soil-cement and water. Drops of sodium hydroxide are then added, measuring the volume necessary for the mixture to acquire a characteristic red color. With this volume, the cement content of the sample can be directly measured, using a previous calibration curve on which the volumes of sodium hydroxide required to produce the reaction in a carefully prepared set of laboratory specimens have been measured, mixing a similar soil with different percentages of cement.

In the constant neutralization test, the suspension of soil-cement and water is mixed only with phenolphthalein, subsequently adding drops of hydrochloric acid. The quantity of this acid required to produce a red coloration reveals the cement content of the specimen, using a similar calibration curve.

Apart from giving full details of the test, [18,19] describe a laboratory method that has been established to determine which version of the test should be used for different soils. By using the titration test, the cement content of a large number of specimens can be easily measured. If a representative sampling is performed, it will be possible to enhance the quality of the final product obtained.

Table 16-13 [1] gives the typical by weight percentages of cement that are usually required with different types of soil in treated pavement layers.

Table 16-13

Some typical cement contents for various types of soils in pavement technology [1]

Type of soil	Percentage cement by weight
Crushed rock	0.5–2 1)
Well-graded sandy-clayey gravels	2–4
Well-graded sands	2–4
Uniform sands	4–6 2)
Sandy clay	4–6
Silty clay	6–8
Clay	8–12
Very active clay	12–15 3)
Organic soils	10–15 4)

1) Cement is used chiefly to improve workability, to reduce the sensitivity of the soil to the compaction water content and to avoid layer deformations under heavy construction equipment.

2) Compaction becomes very difficult, and there is risk of segregation of the cement.

3) Mixing may be very difficult and a pretreatment with lime is usually of great assistance.

4) Pretreatment with lime or 2% calcium chloride may be of considerable assistance in minimizing inability of cement to set.

Thickness Design

Current thickness design methods for soil-cement layers pay special attention to fatigue, which has already been mentioned. For this, knowledge of traffic distribution is essential; including the single and double axle loads of different weights likely to occur during the design period.

The fatigue that is caused by each individual axle passing over the layer is evaluated by means of the fatigue factor, a parameter which expresses the overall traffic effect during the design life of the pavement. Table 16-14 [27,28] shows the *Basic Fatigue Factors*, which represent the equivalencies of the effects produced by different axle loads in relation to those caused by a single 8.2 t axle (18000 lb) or a double axle of 13.6 t (30,000 lbs).

The Basic Fatigue Factors in the Table are multiplied by the number of axles of each type (in thousands), according to the prevailing traffic distribution, and the total value of the sum of these products gives the design value for fatigue.

Table 16-14

Basic fatigue factors for the thickness design of soil-cement pavements [27,28]

Axle load		Soil-cement with granular matrix	Soil-cement with fine matrix
t	Kips	—	—
Single axle			
13.6	30	12500000	3530
12.7	28	1270000	1130
11.8	26	113000	337
10.9	24	8650	93
10.0	22	544	23.3
9.1	20	27	5.2
8.2	18	1	1.0
7.3	16	.12	0.16
6.4	14	.012	0.02
5.4	12	.001	0.002
Tandem axle			
22.7	50	12500000	3530
21.8	48	3210000	1790
20.9	46	792000	890
20.0	44	186000	431
19.1	42	41400	203
18.2	40	8650	93
17.3	38	1690	41.1
16.3	36	305	17.5
15.4	34	50.4	7.1
14.5	32	7.5	2.7
13.6	30	1.0	1.0
12.7	28	0.12	0.34
11.8	26	0.012	0.11
10.9	24	0.0010	0.03
10.0	22	—	0.008
9.1	20	—	0.002

With this value it is now possible to employ the graphs in Fig. 16-25 [27,28], in which the thickness of the cement-treated layer can be obtained for both granular and fine soils, as a function of the modulus of subgrade reaction (Chapter 9).

16.5 Soil Stabilization With Lime

Soil stabilization with lime seems to be the most ancient method of improving soils. There is evidence that the Appine Way, the access route to ancient Rome, was constructed using these techniques [11]. In general terms, stabilization techniques with hydrated lime are pretty similar to those with cement, but there are two differences which should be emphasized [1]. First, lime has a range of effectiveness which includes more clayey materials than cement, but which is less effective in more granular materials of a frictional nature. Second, the use of lime stabilization as a pre-treatment is becoming more and more widespread. This is especially significant for it is not essential for all the usual stabilization requirements to be met in these cases.

For stabilization purposes lime is usually employed hydrated, in the form of calcium hydroxides. Calcium carbonates have little stabilizing properties. Quicklime or calcium oxide is often used for pre-treating moist soils. The basic effect of lime appears to be the formation of calcium silicates due to a chemical reaction between lime and clay minerals, resulting in cementing compounds.

Lime is generally prepared by heating calcium carbonates, usually in the form of natural limestones, until they lose their carbon dioxide and become calcium oxides. The resulting material is quicklime, which is very unstable and eager for water. This makes it very difficult to handle and store, which is why it is usually hydrated immediately.

In order to produce a stabilizing lime, it is not essential to start with pure limestones, and a certain amount of impurities including magnesium carbonate and clay minerals, are tolerated. Table 16-15 [1] shows the properties usually required of the raw material to be used for obtaining a stabilizing lime.

Two types of chemical reaction may take place between a lime and a soil. The first is instantaneous and involves the capture of numerous calcium ions by the soil particles. This depresses their *double layer* (Chapter 1) because of the increase in the concentration of bivalent cations in the water. At the same time another effect takes place which tends to expand the double layer owing to the high *pH* of lime. This second reaction takes place over long periods of time and is in fact the true cementing reaction. Although knowledge of this reaction is by no means complete, it is attributed to an interrelation between

Table 16-15

Requirements that must be met by limestones and natural calcium carbonates in order to provide stabilizing lime [1]

Property	Quicklime (CaO)	Hydrated lime ($Ca(OH)_2$)
Calcium magnesium oxides	Not less than 92%	Not less than 95%
Carbon dioxide In the oven Out of the oven	 No more than 3% No more than 5%	 No more than 5% No more than 7%
Fineness	—	Not more than 12% retained on No. 180 sieve

the calcium ions of lime and the aluminic and siliceous components of soils. The reaction can be enhanced by adding volcanic or fly ash, rich in silica, to the soil. The formation of calcium silicates is attributed chiefly to kaolinites and that of calcium aluminates to montmorillonites. The cementing effect is largely due to the formation of calcium silicates and depends on the type of soil involved. From this point of view, lime stabilization is very different from cement stabilization.

Reference [29] presents some data about the effect of the addition of different amounts of various sodium silicates on the strength of the soil-lime obtained. Tests were conducted on a silty material and the results are shown in Table 16-16.

Table 16-16

Effect of the use of sodium silicates as an additive on the strength of lime-stabilized soils after curing for 7-days and one day of immersion. Silty soils [29]

Lime	Type of silicate	Quantity of silicate	Unconfined compressive strength of the mixture	
%	—	%	kg/cm^2	lb/in^2
2	$Na_3HSiO_4.5H_2O$	0	8.9	126
2	$Na_3HSiO_4.5H_2O$	2	8.4	119
2	$Na_3HSiO_4.5H_2O$	4	7.0	100
4	$Na_3HSiO_4.5H_2O$	0	9.0	128
4	$Na_3HSiO_4.5H_2O$	2	14.4	204
4	$Na_3HSiO_4.5H_2O$	4	16.2	230
6	$Na_3HSiO_4.5H_2O$	0	7.4	105
6	$Na_3HSiO_4.5H_2O$	2	16.9	240
6	$Na_3HSiO_4.5H_2O$	4	20.0	284
6	$Na_3HSiO_4.5H_2O$	5.3	16.7	238
6	$Na_2SiO_3.9H_2O$	8	16.1	229
6	Na_2SiO_3	3.4	15.7	223
6	$Na_2SiO_3.5H_2O$	6.0	16.8	238
6	Na_4SiO_4	3.8	11.7	166

It is observed that for very low lime contents the influence of the additive is not appreciable. With somewhat greater contents of hydrated lime, significant advantages can be obtained by using small amounts of an additive which is very easy to handle and mix.

Lime has little effect on highly organic soils or soils without clay. Its maximum effect is on clayey gravels, where it may result in mixtures that are sometimes even stronger than those that would be obtained with cement. It has been most frequently used with plastic clays, which become more workable and easy to compact. This is why it is often employed as a pretreatment, prior to a cement stabilization, (apart from the many cases where it is used as an actual stabilizer). The effect of lime is more dramatic in montmorillonites than in kaolin clays. With the former, a far more spectacular increase in strength and reduction in plasticity are achieved. Lime also provides clays with volumetric stability in the presence of changing water.

In lime stabilization the use of acid waters should be avoided. Sea water has frequently been used to compact soil-lime, but should be avoided where an asphaltic sealer coat is to be used on top of the treated layer; crystallization of the salts will detach the sealer coat. The amount of water to be employed is determined by compaction procedures. However if quicklime is used extra quantities may prove necessary to hydrate the lime in soils with less than 50% natural water content [1].

16.5.1 Properties of Soil-Lime Mixtures

A brief review follows of the most significant soil-lime properties and the factors that influence them.

.1 Plasticity

Lime greatly reduces the plasticity index of high plasticity soils such as montmorillonites and bentonites, but has little influence on the plasticity index of medium plasticity soils. It may even increase the plasticity index of low plasticity fine soils.

.2 Strength

Figure 16-26 [1] shows the variation in the unconfined compressive strength of various soils with the lime content. The general increase in strength up to lime contents of about 8% by weight can be readily appreciated. Beyond this limit strength is often relatively unaffected by an increase in the proportion of lime, except in the case of the more clayey materials where strength may continue to increase for lime contents of more than 10% [11]. In this aspect lime differs from cement, the strength of which continues to increase for contents as high as 20%.

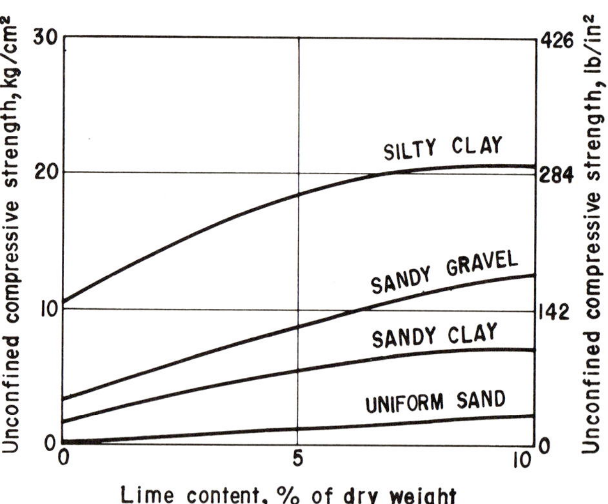

Fig. 16-26 Effect of the lime content on the unconfined compressive strength of various types of soil cured for seven days with hydrated lime [1]

Figure 16-27 [1] shows the effect of the age of the soil-lime mixture on the unconfined compressive strength of different soils. As in the case of cement, strength increases with time.

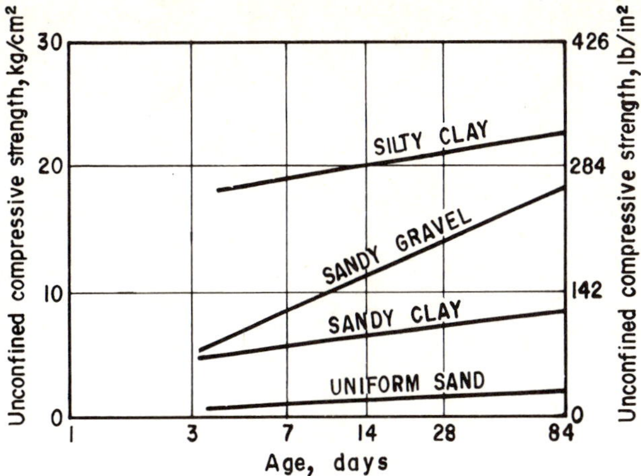

Fig. 16-27 Effect of the age of a mixture of 5% lime, by weight, with different types of soils [1]

Figure 16-28 [11] shows the influence of curing time on the unconfined compressive strength of soil-lime specimens. In all cases a silt was stabilized with 10% hydrated lime by weight, and the curves that are presented correspond to different additives, with the lime. The control curve contains none of these additives. The additives were used in proportions of 1 to 2% of the dry soil.

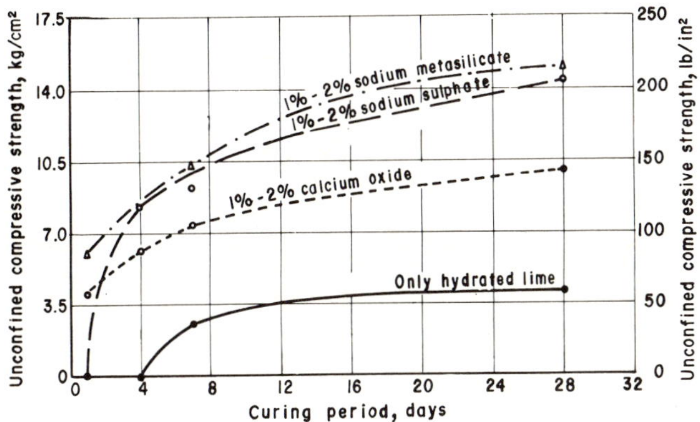

Fig. 16-28 Effect of curing time on the unconfined compressive strength obtained for specimens. Silt treated with 10% hydrated lime and the proportions of additive that are shown [11]

Figure 16-29 [1] shows the change in the strength of soil-cement and soil-lime mixtures during the first hours after stabilization. In this case, strength refers to vane tests conducted in the laboratory. By way of reference, the figure gives the strength of the unstabilized soil, which was a fat clay. It can be observed how the effect of the lime is initially a little greater, but how the cement has a greater effect once hydration commences.

The effect of organic matter on the strength can be seen in Fig. 16-30 [30]. The figure shows the variation in the unconfined compressive strength of a silty clay with variable contents of organic matter. The specimens were prepared with 12% hydrated lime, by weight, and with the optimum water content. In one case they were cured for 7 days and in the other 28.

Figure 16-31 [30] presents another interesting study on the effect of organic matter. In Part *a* of the figure quick strength envelopes are shown for specimens of a high plasticity clay, totally devoid of organic matter and with variable contents of

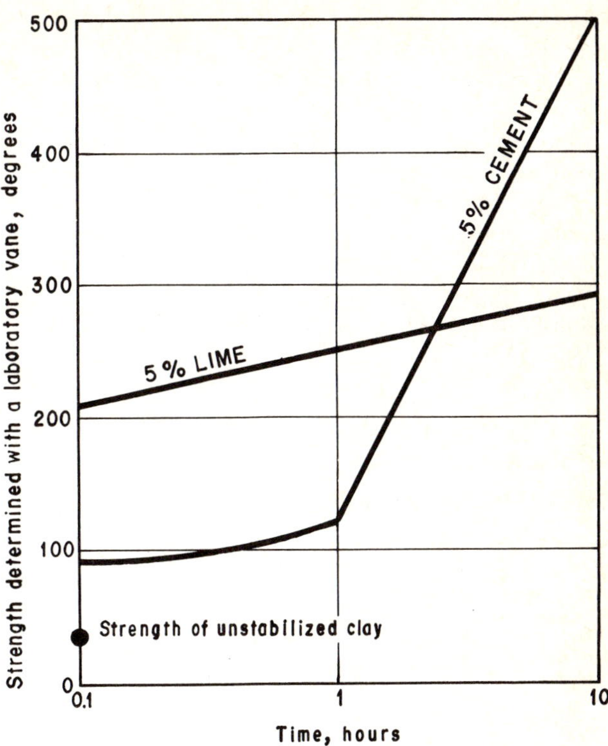

Fig. 16-29 Change in the strength of a stabilized clay during the first few hours after stabilization [1]

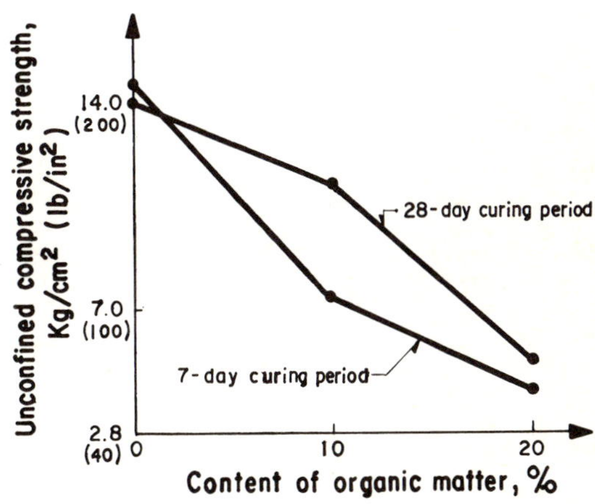

Fig. 16-30 Effect of the percentage of organic matter on the unconfined compressive strength of a silty clay treated with 12% hydrated lime [30]

hydrated lime. This curve, corresponding to the soil alone, is for comparison. The curing time was 28 days in all cases. Part *b* shows similar envelopes for a silty but far less active clay, with 20% organic matter. The specimens were also cured for 28 days and were compacted with the optimum water content corresponding to the same dynamic test that was employed to prepare the specimens in Part *a*. It can be seen that with organic matter a soil that is initially stronger will exhibit lower strengths than a high compressibility clay. It can also be seen how, under comparable conditions, the addition of lime causes an increase in the apparent cohesion of the material by cementation. Although Part *a* is an exception, in which the *CH* material with 4% lime shows greater apparent cohesion than the same material with 12% lime which cannot be easily explained.

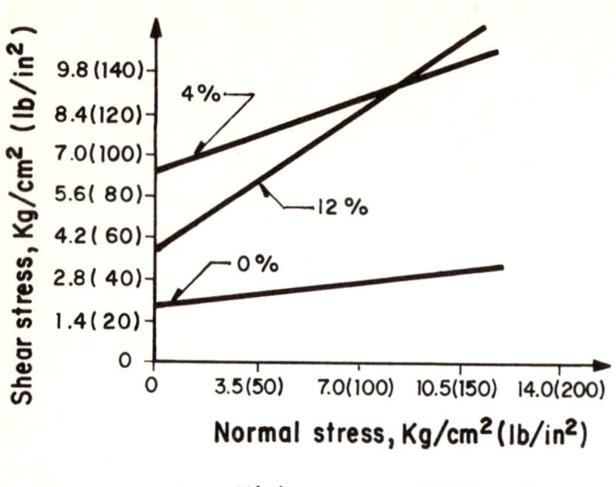

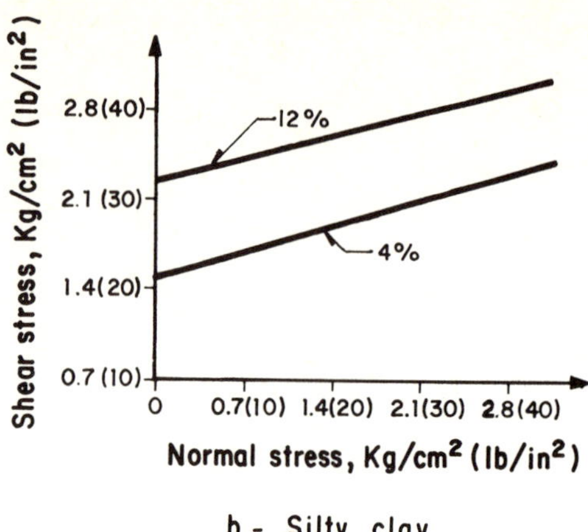

Fig. 16-31 Quick strength envelopes for a highly compressible clay without organic matter, and one with low plasticity and with 20% organic matter, both stabilized with lime [30]

.3 Other Properties

Figure 16-32 [1] shows the effect of the time that is allowed to elapse between the preparation and compaction of stabilized mixtures on the dry unit weight. A comparison is made between a mixture of active clay with 10% cement and another of the same clay with 10% lime. It can be seen that the effect of this time has far less importance in the case of the lime on account of the differences already discussed between the initial chemical reactions of the two stabilizers. From the information given in the figure, a rule can be deduced, according to which cement mixtures must be compacted immediately after preparation, while lime mixtures can be handled less rapidly, which leads to a more convenient and flexible construction process. However, it has been observed that after a few hours, lime mixtures require more water for compaction. Thus if too much time is allowed to elapse, the extra water that becomes necessary may lead to further costs.

Table 16-17 [1] shows the influence of lime on soil pulverization and handling. Data for a clay with $W_L = 73\%$, $W_p = 27\%$ and I_p of 46 are given. After treatment with 5% lime, the clay exhibited the same liquid limit, but its plastic limit increased to 40% and its I_p dropped to 33. The table reports the percentages of different sized lumps for both cases.

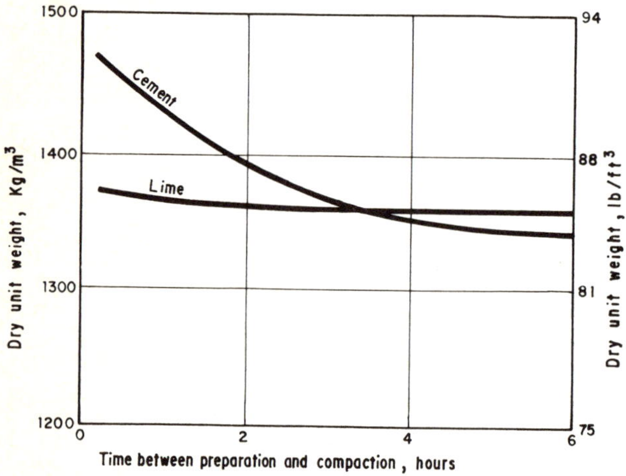

Fig. 16-32 Effect of the time between the preparation and compaction of mixtures of an active clay with 10% lime and cement, by weight, on the unit weight obtained [1]

Table 16-17
Effect of lime on the workability of a clay [1]

Sieve	Percentage of soil lumps passing sieve	
	Untreated clay	Treated with 5% lime
2 in	80	83
1 in	40	50
1/2 in	17	28
1/4 in	7	19
1/8 in	4	15

Soil-lime construction methods are similar to those for soil-cement, as are the laboratory tests that are currently employed to determine the quality of the soils obtained. A good guide for preparing mixtures for laboratory study purposes is to add 1% lime for every 10% of the fine fraction of the soil.

16.5.2 Construction Procedures

Table 16-18 [1] gives the typical lime contents for different types of soils. Again this information must be regarded as a guide, not as a substitute for the corresponding laboratory determinations.

The steps to follow for stabilizing pavement layers with lime are:

— Scarify the bearing material
— Pulverize the soil
— Sprinkle lime from bags or using a mechanical sprinkler
— Mix (by hand, with mixing equipment or machines)
— Add water, if necessary, to give the mixture its optimum compaction water content.
— Compact
— Shape surface
— Minimum curing period of 5 days
— Lay a protective surface

Thickness design

The most common practice in pavement technology is for a soil-lime layer to be given the same thickness that would be

Table 16-18

Usual contents of hydrated lime in different soils (percentage by weight of dry soil) [1]

Type of soil	Pretreatments	Stabilization
Crushed rock[1)]	2–4	Not recommended
Well-graded clayey gravels	1–3	3
Sands[2)]	Not recommended	Not recommended
Sandy clay	Not recommended	5
Silty clay	1–3	2–4
Plastic clay	1–3	3–8
Highly plastic clay	1–3	3–8
Organic soils	Not recommended	Not recommended

1) Only recommended if there are plastic fines

2) Lime is helpful for asphalt stabilization, because it improves adhesion. For loess soils quicklime is recommended

required by the soil alone. When *CBR* design methods are employed, specimens are prepared with different lime contents and different water contents, obtaining the *CBR* for each until the minimum design value is reached. This criterion should, however, be adjusted according to the strength of the soil-lime and other properties that may be considered of interest.

16.6 Stabilization With Asphalt

The improvement of soil properties by the addition of asphalt and asphalt products is a much used and often effective technique. Three types of product are employed for this purpose [11]:

- Bituminous products, which are anhydrous systems of hydrocarbons totally soluble in carbon bisulphide.
- Asphalt products derived from the distillation and refinement of oil (or natural asphalts, more rarely).
- Residual products of the destructive distillation of organic materials, such as coal, certain oils, lignites, peats and wood (tars).

The nature of these products is not very well defined and the detailed analysis of their respective compositions is beyond the objectives of this book (and probably beyond the interests of most civil engineers).

Asphaltic and bituminous products are normally too viscous to mix directly into soils. Because of this, they have to be heated, emulsified in water (emulsions) or diluted with a volatile solvent, such as gasoline.

Emulsions and cutback asphalts are the products most widely used for soil stabilization, but heated or diluted tars are also used. The usual types of cutback products are slow- and medium-curing, but for sands, rapid curing products have also been successfully used.

Table 16-19 [1] gives the most usual characteristics of some types of bituminous products used for stabilizations. The flowing oil most widely used is diesel fuel with a viscosity of between 0.03 and 0.07 degrees Stokes at 50°C (122°F). The solvents usually employed in diluted stabilizers are gasoline, kerosene and a mixture of kerosene and diesel fuel, which result in fast, medium and slow-curing cutbacks respectively. Table 16-20 [1] indicates some requirements frequently made of cutback asphalts.

Medium and slow-curing emulsified asphalts are used. Emulsions are very fine suspensions of asphalt particles in water and the asphalt binds with the soil when the suspension sets. The moment that this setting takes place determines the effectiveness of the asphalt-soil bond. If curing (demulsification) takes place rapidly, penetration will be inadequate. For this reason quick-setting emulsions are avoided. Anionic emulsions should have an asphalt content of at least 55%. The emulsifier and the stabilizing agents should leave a maximum residue of 0.2%, by weight, retained on the No 100 sieve after dilution in an equal volume of distilled water and a rest period of 10 minutes. The viscosity of these suspensions should be between 4 and 25 degrees Engler. The maximum temperature for field use should be 49°C (120°F).

The specifications for cationic emulsions are not so well known. A minimum asphalt content of 58% is usually required and a viscosity of between 3 and 24 degrees Engler.

Practically all types of soil respond to asphalt stabilization, including highly compressible and active clays, but the best results are obtained with sands and sandy gravels, to which asphalt gives cohesion and impermeability. The grain-size distribution of the soils is not critical, but there are usually certain requirements such as the following:

- Maximum particle size should be less than one third of the thickness of the compacted layer.
- More than 50% of the material should be smaller than No 4 sieve (or sometimes finer than the 3/16 in sieve).
- 35% of the material should be finer than No 40 sieve
- The amount retained on No 200 sieve should be from 10 to 50%
- The liquid limit of the fine fraction should be below 40.
- The plasticity index of the fine fraction should be below 18.

In very clean sands there may be problems of adhesion between the asphalt and the siliceous materials, which leads to detachment (stripping) of the stabilizing material and loss of its beneficial effects. Moist soils may have the disadvantage that when more liquid is added to them during the stabilization process, they become so soft that it is very hard to compact them. In the case of very clean sands, with no more than 3% moist material passing the No 200 sieve, pre-stabilization with 1 or 2% lime may successfully improve the adhesion between the asphalt and the particles of sand.

Any type of fresh water is acceptable for compaction moisture both for stabilization and for the subsequent compaction process. A heavy concentration of salts and organic matter is not recommended, for these are highly detrimental to the adhesion between the soil and the asphalt.

The stabilizing effects of asphalt are due to two mechanisms. The first is a bond that is established between the particles of soil through the asphalt. This provides the whole with *cohesion*. The second is protection of the soil against softening and swelling by water. The first mechanism is especially important in granular soils, whereas the second is rather more useful in cohesive soils.

Table 16-19
Common requirements for asphalt materials [1]

Asphaltic cements for preparing cutbacks							
Property	Units	Penetration 65		Penetration 90		Penetration 200	
		Min	Max	Min	Max	Min	Max
Relative density		0.99	—	0.98	—	0.97	—
Ignition point	°C °F	225 437	— —	220 428	— —	205 401	— —
Softening point	°C °F	46 115	58 146	42 108	53 127	33 91	44 111
Penetration at 25°C (77°F)		60	70	85	100	180	210
Temperature of use	°C °F	— —	— —	160 320	182 360	155 312	182 360

Fluxed natural asphalts							
Property	Units	Penetration 65		Penetration 90		Penetration 200	
		Min	Max	Min	Max	Min	Max
Relative density		1.18	1.26	1.16	1.24	1.1	1.20
Ignition point	°C °F	175 347	— —	175 347	— —	175 347	— —
Softening point	°C °F	42 108	55 131	40 104	51 124	33 91	46 115
Penetration at 25°C (77°F)		60	70	80	100	175	225
Solubility in carbon tetrachloride	%	68	—	70	—	74	—
Ash content	%	—	25	—	23	—	21
Temperature of use	°C °F	— —	— —	160 320	182 360	155 312	182 360

Coal tar							
Property	Units	Very heavy		Heavy		Light	
		Min	Max	Min	Max	Min	Max
Relative density		1.07	—	1.05	—	1.04	—
Viscosity (Stokes) at 50°C (122°F)	deg	5	8	1	2	0.25	0.50
Solubility in carbon disulphide	%	95	90	96	91	97	92
Water content	%	—	0.5	—	2	—	5
Temperature of use	°C °F	71 160	85 185	54 129	68 154	35 95	45 113

Petroleum tar							
Property	Units	For impregnation		For binding		For presealing	
		Min	Max	Min	Max	Min	Max
Relative density		1.09	—	1.05	—	1.05	—
Viscosity (Stokes) at 50°C (122°F)	deg	31	50	1	2	0.10	0.25
Solubility in carbon disulphide	%	—	92	—	95	—	95
Water content	%	—	0.5	—	2	—	5
Temperature of use	°C °F	88 190	93 200	57 135	71 160	18 65	35 95

Table 16-20
Properties most frequently required of cutback asphalts for stabilization [1]

Medium curing cutbacks							
Property	Units	Grade 0		Grade 1		Grade 2	
		Min	Max	Min	Max	Min	Max
Viscosity (Stokes) at 50°C (122°F)	deg	0.36	0.71	1.10	2.20	3.60	7.10
Ignition point	°C °F	38 100	— —	38 100	— —	49 120	— —
Penetration in the residue at 25°C (77°F)		100	250	100	250	100	250
Parts of kerosene (vol) to 100 parts asphaltic cement with penetration 90	%	78	78	51	51	37	37
Slow curing cutbacks							
Property	Units	Light		Medium		Heavy	
		Min	Max	Min	Max	Min	Max
Viscosity (Stokes) at 50°C (122°F)	deg	0.36	0.71	0.70	1.40	1.10	2.20
Proportion of cement with penetration 90		100	—	100	—	100	—
Proportion of diesel oil		60	—	45	—	30	—
Proportion of kerosene		50	—	38	—	25	—

16.6.1 Properties of Asphalt-Stabilized Soils

The following paragraphs discuss some points of interest in relation to certain important properties of asphalt-stabilized soils, along with the factors which influence their variation.

.1 Dry Unit Weight

Figure 16-33 [1] shows the variation in the dry unit weight of mixtures with the percentage of asphalt. The information refers to stabilization with cutback asphalts. It can be seen in Part *a* of the figure how asphalt causes a reduction in the maximum dry unit weight. This reduction is seldom critical, because stabilization produces an improvement in the mechanical properties which more than compensates for it. Part *b* of the figure shows that the addition of asphalt and the liquids that accompany it reduce the need for water required for compaction, which is an advantage at dry locations.

Cutback asphalts contain volatile solvents which have an important effect on the behavior of the mixture obtained. The higher the asphalt content, the smaller the loss in strength due to saturation compared to the untreated soil. On the other hand, when the proportion of solvents is excessive, the strength of the mixture, as well as its dry unit weight, diminish. There are optimum contents of solvents. Should these be exceeded, the risk would be run of reducing strength to undesirably low values. With emulsions and cutback asphalts water and solvents are added to the liquid fraction, which may cause an inappropriate increase in the water content of the mixture for compaction, Fig. 16-34 [11] summarizes this information.

Figure 16-35 [1] shows typical laboratory compaction curves (AASHTO modified tests) for various types of fine soils stabilized with the percentages that are indicated of medium-curing cutback asphalts. The *water content* employed for the mixture corresponds to the total content for the liquid phase.

.2 Strength

Figure 16-34 [11] included some basic information about the influence of the content of solvents in cutback asphalts on the unconfined compressive strength of the mixtures obtained. The higher the actual asphalt content, the better the behavior of the mixture obtained, within practical limits. However, this

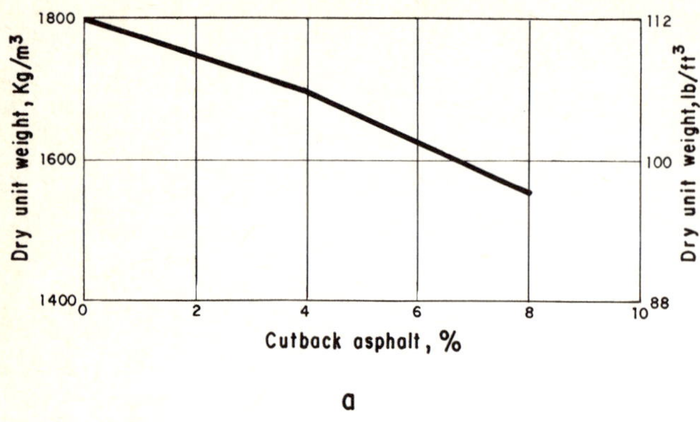

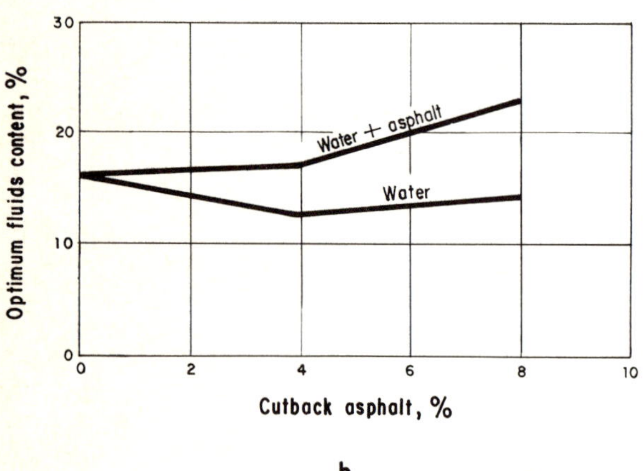

Fig. 16-33 Variation in the dry unit weight and the optimum water content of specimens stabilized with cutback asphalt [1]

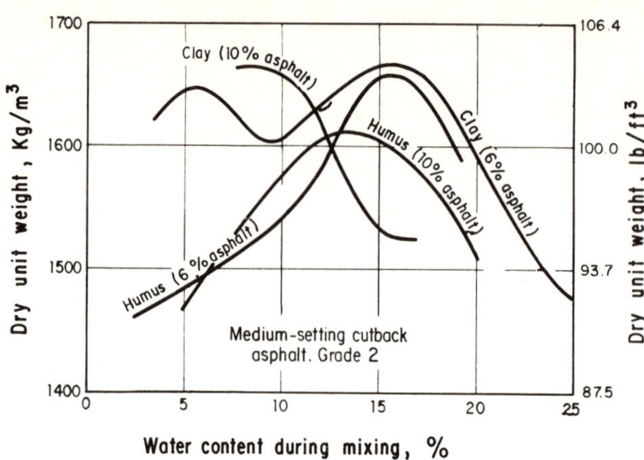

Fig. 16-35 Effect of asphalt stabilization on the compaction of specimens of various soils [1]

improvement does not necessarily refer to the same properties. In fine soils, an increase in the asphalt content does not influence the unconfined compressive strength, as is illustrated by Fig. 16-36 [11], in which the strength of a clayey silt, measured immediately after curing of the mixture, remains practically constant for highly variable asphalt contents. The mixture was prepared with asphalt emulsion. On the other hand, in more granular soils of a frictional nature, strength does increase with an increase in the asphalt content. However if this increase is excessive, strength drops again. Also, in fine soils the asphalt content does have an appreciable influence on the behavior of the mixture in the presence of water, as shown by Fig. 16-36. On re-wetting, the strength of the clayey silt and asphalt mixture increases with the asphalt content within the practical limits. This does not, however, mean that the asphalt content must be indiscriminately increased in fine soils. Beyond an optimum

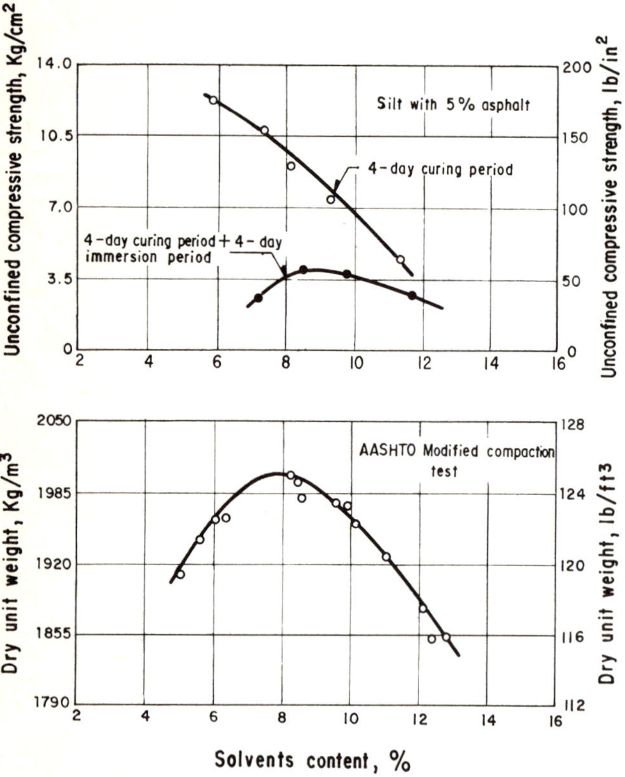

Fig. 16-34 Effect of the percentage of solvents on the dry unit weight and unconfined compressive strength of asphalt-stabilized specimens [11]

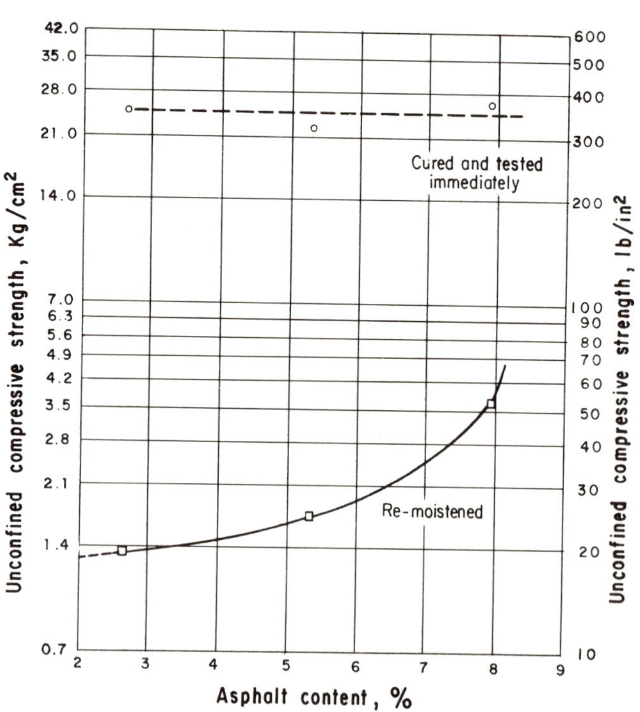

Fig. 16-36 Effect of the percentage of asphalt on the unconfined compressive strength of specimens of clayey silt stabilized with emulsion [11]

value the solvents in cutback asphalts or the water in emulsions cause an excessive increase in the liquid phase of the mixture, and the soil consequently becomes too plastic and loses strength. In granular or frictional soils, asphalt does little to improve the stability in the presence of water.

Figure 16-37 [11,31] shows the effect of mixing time on the unconfined compressive strength of the soil-asphalt mixture. Initial strengths are presented for the mixtures immediately after curing and after re-wetting. The soil used was a clayey silt, which was mixed with 5% cutback asphalt. Data are also given for another mixture to which 2% phosphorus pentoxide was added, with notable effects on the strength in all cases.

The curing time has an important influence on the strength of the mixtures. The longer this period and the higher the curing temperature, the greater the loss of solvents when cut-backs are used. The longer the re-wetting period the greater the amount of water absorbed by the soil-asphalt mixture. The strength of a soil-asphalt is roughly inversely proportional to the content of solvents at the time of testing, so that the more solvents lost, the greater the strength. Figure 16-38 [11,31] gives information based on tests on a large number of mixtures, prepared by very different methods and with very different soils. It illustrates the general effects of asphaltic stabilization.

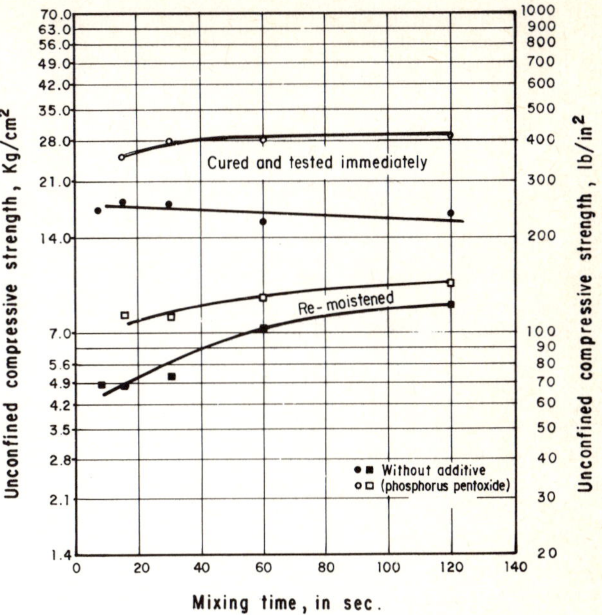

Fig. 16-37 Effect of mixing time on the unconfined compressive strength of specimens of clayey silt stabilized with 5% cutback asphalt, with or without an additive [11,31]

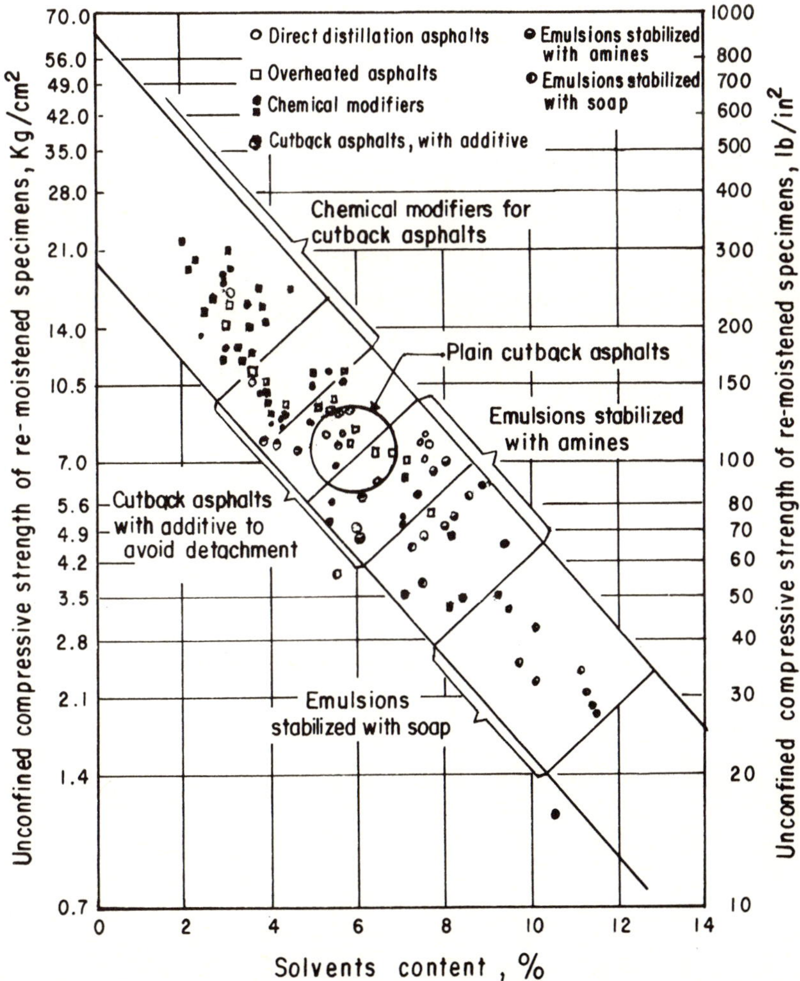

Fig. 16-38 Effect of the percentage of solvents on the unconfined compressive strength of specimens of various soils stabilized with asphalt products, after a rehydration period [11,31]

.3 California Bearing Ratio

When proportioning mixtures for asphalt-stabilized pavement layers, an extension of the *CBR* design method is frequently employed. As occurs with the unconfined compressive strength in fine soils, the *CBR* of the mixture usually increases with the asphalt content to an optimum amount, after which it drops again. Fairly high asphalt contents are employed in order to achieve mixtures that will be capable of tolerating the repeated action of water. The asphalt content is selected which will lead to a satisfactory strength and adequate stability in the presence of water. This stability is measured in the laboratory by means of *CBR* tests which are performed on pre-saturated specimens.

Figure 16-39 [1] gives the variation in the *CBR* with the asphalt content of one mixture tested after a normal curing period, and another tested after a period of immersion in water. The material is a calcareous sand-gravel, stabilized with a medium-curing cutback asphalt, [1] does not, however, mention the duration of the immersion period.

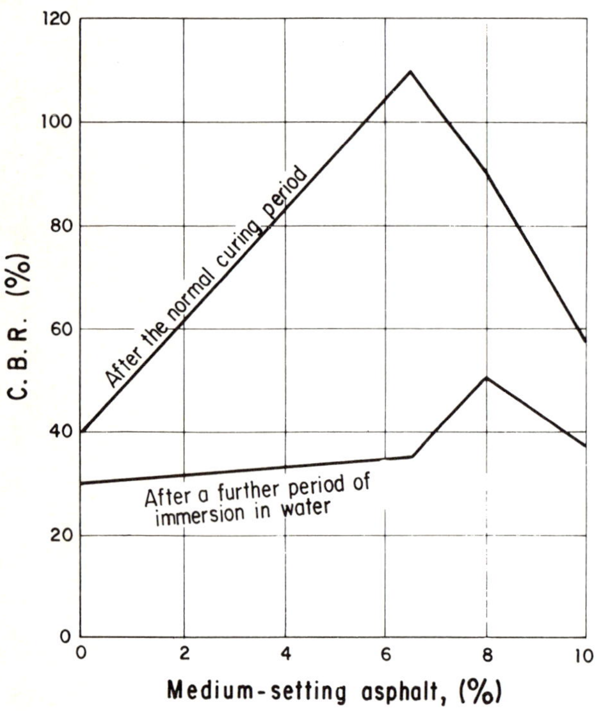

Fig. 16-39 Variation of the *CBR* with the asphalt content in specimens of sand-gravel tested after the normal curing period and after immersion in water [1]

There seems to be a definite relation between the *CBR* of asphalt-stabilized soils and temperature, [1] mentions some South African experiences which are quite conclusive. On the basis of these, a correlation is made between the *CBR* and the temperature of specimens of sand-asphalt. The reader is warned against the indiscriminate use of these local correlations, the validity of which is always questionable when applied to a different environment or material. However, they are a useful guide, and it is with this intention that the correlation in Fig. 16-40 [1] is presented.

According to these South African investigations, the unconfined compressive strength, or the triaxial compression strength on the same sand-asphalt mixtures is also influenced by temperature. An increase in temperature brings about a decrease in strength, which is logical considering how asphalt softens when heated.

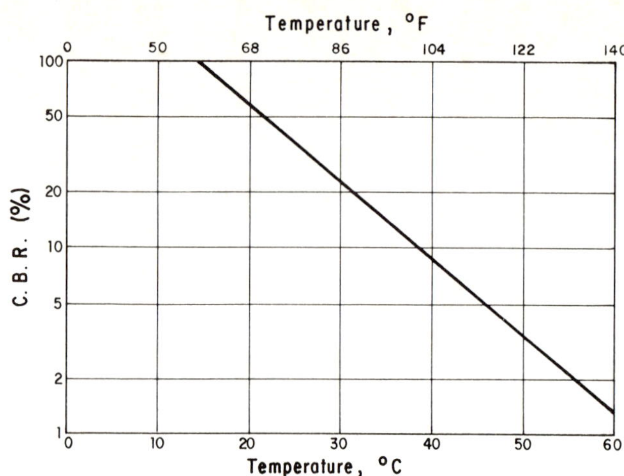

Fig. 16-40 Influence of temperature in the *CBR* test conducted on specimens of asphalt-stabilized sand [1]

.4 Stability in the Presence of Water

This is the principal characteristic that is sought when stabilizing fine cohesive soils with asphalt. Figure 16-36 [11] has shown how the stability of mixtures of fine soils and asphalt in the presence of water increases greatly with the asphalt content. Nevertheless it is advisable to draw the reader's attention to the information that is given in [32].

This describes the stabilization of a sandy clay with cutback asphalts. As would be expected in a predominantly cohesive soil, the unconfined compressive strength proved to be somewhat independent of the asphalt content, although the sand content of the samples causes some response to this factor. The unconfined compressive strength increases by 40% for the optimum asphalt content, compared to that of the original natural sandy clay. Moreover, when the asphalt content increased, the stability of the mixture in the presence of water improved until finally the solvents accompanying the cutback asphalt caused an excessive increase in the liquid phase of the mixture, making it unduly plastic and weak. Clearly this limit should not be exceeded, for the stability that is gained is outweighed by the strength that is lost. In this particular investigation, asphalt contents of more than 8 to 10% were found to be undesirable for these reasons. However, insofar as stability in the presence of water is concerned (which is measured by the degree of absorption of the samples during different immersion periods) it is observed that very high asphalt contents continue to provide satisfactory values. Figure 16-41 [32] shows the amount of water absorbed by two specimens during immersion periods of 1 and 28 days as a function of the asphalt content that was added to the sandy clay employed for the investigation. The asphalt used was a medium-setting cutback.

The mixing process that is employed has an appreciable influence on the stability of mixtures of fine soils and asphalt in the presence of water, Fig. 16-42 [32] shows the variation in the amount of water absorbed by specimens of sandy clay stabilized with 3% heavy oil for different mixing times and different immersion periods. It can be seen that when the mixing time is prolonged, the ultimate stability of the mixture in the presence of water is greatly reduced. This is attributed to grain breakage and to the subsequent diffusion of the asphalt over larger areas of soil, in films that are thinner and consequently more susceptible to penetration by water.

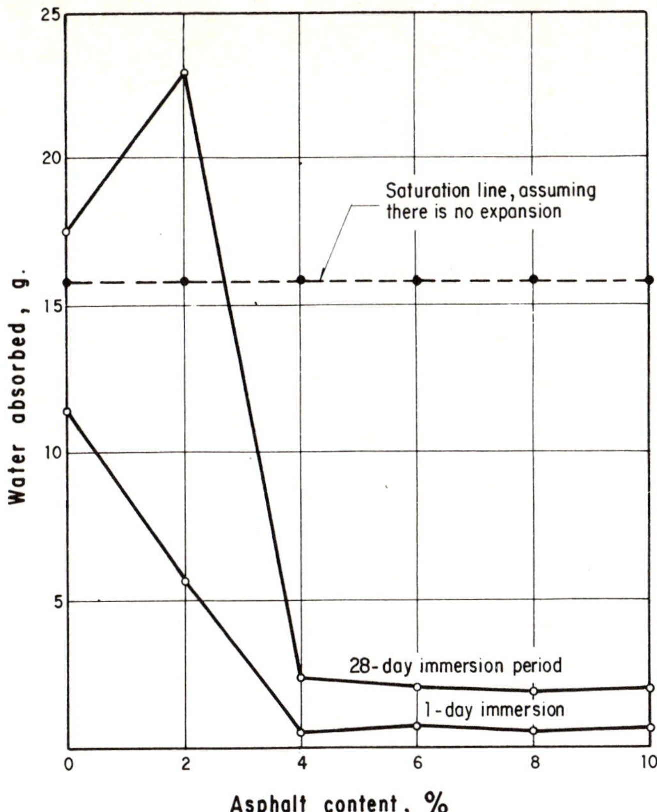

Fig. 16-41 Variation in water absorption with the asphalt content. Specimens of sandy clay stabilized with cutback asphalt [32]

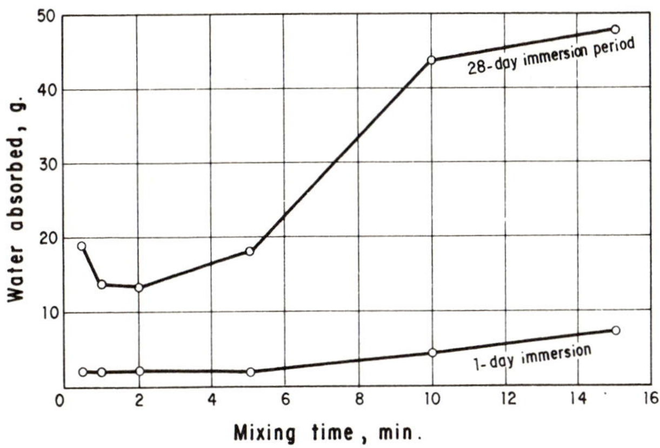

Fig. 16-42 Variation in water absorption with mixing time. Specimens of sandy clay with 3% heavy oil [32]

16.6.2 Construction Methods

From the foregoing observations it can be deduced that soil-asphalt mixtures are utilized for different purposes. In cohesionless soils, the objective may be to find the maximum strength which can be achieved using the optimum asphalt content. Another purpose may be to increase soil stability in the presence of water. This is frequent in cohesive soils or in soils of a frictional nature containing a large proportion of fines. In this case the asphalt contents can be higher, care being exercised that they do not exceed the limit at which the strength of the mixture may be affected. The use of asphalt to stabilize sands is currently so frequent that many authors regard it as a special topic. The type of laboratory test to conduct in each case is shown in previous pages. The designer is free to adopt the method he considers appropriate, since there is no well-established methodology for the design of soil-asphalt mixtures. Reference [33] provides a useful summary of the methods employed by some of the leading organizations in this field.

Selection of the type and amount of asphalt should be determined by laboratory tests that compare strength and stability characteristics. In highway engineering technology, tests are often conducted to find the *CBR* values in specimens that have been cured and subjected to periods of immersion in water. In these cases *CBR* values of at least 80% are frequently required for pavement layers.

Figure 16-43 [11,34] shows the percentages of asphalt that are suitable for different stabilizations. These percentages are related to the fraction of soil finer than the No 200 sieve.

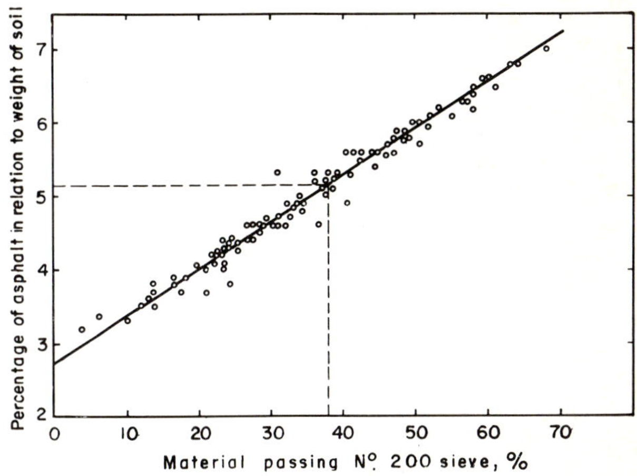

Fig. 16-43 Chart for determination of the percentage of asphalt for practical stabilization purposes [11,34]

The construction sequence of a layer of soil-asphalt is as follows:

— Pulverize the soil
— Add the necessary water for adequate mixing
— Add the asphalt product and mix with the soil
— Aerate to achieve an appropriate content of volatile solvents for compaction
— Compact
— Finish off
— Ventilate and cure

The mixing and the reduction of volatile solvents are usually the most critical stages of the above sequence. In clayey soils in particular, a high content of solvents may be helpful during mixing. Moreover, the solvents content required to encourage satisfactory compaction is usually higher than that producing adequate stability. Therefore, soil-asphalt mixtures normally require a period of aeration or ventilation after compaction. The more cohesive the soil, the more essential the ventilation period becomes.

When emulsions are employed, which are extremely appropriate for stabilizing fine soils, their liquid fraction should be evaporated after compaction, which implies the use of sufficiently thin layers.

Thickness Design

The *CBR* method is frequently used to design soil-asphalt layers. When this is so, the thickness assigned to the soil-asphalt layer is usually the same as the design thickness which would be adopted for the soil were it left unstabilized, which is very conservative. It is also common and more logical to establish some empirical equivalency between the thickness of the stabilized layer and a layer of gravel-equivalent, as was shown in Chapter 9.

16.7 Other Stabilization Methods

Among other soil stabilization methods which have not been specifically dealt with in other parts of this book, there are both chemical and physical ones. A few will be briefly described. They are far less frequently used than cement, lime and asphalt or the mechanical stabilization that is achieved by mixing soils.

16.7.1 Chemical Stabilizers

Chemical stabilizers can be inorganic or organic. The former are usually less restricted by patents and proprietary limitations. These can be subdivided into acid, neutral and alkaline stabilizers. Acid and alkaline stabilizers act by chemically attacking the soil components, particularly clay minerals, resulting in new cementing compounds. Neutral stabilizers alter the physical soil properties, such as unit weight.

A brief review follows of chemical stabilizers, based on [1,11,32].

.1 Phosphoric Acid and Phosphates

Stabilization with these products has been largely experimental. Further progress is unlikely since it has been found that the percentages required of these products are similar to those of cement or lime, but their price is several times higher.

The chief advantage of these stabilizers is the strength they produce in soils containing chlorides by comparison with the erratic results produced by cement and lime [35]. It has also been said that phosphoric acid has a highly beneficial effect on the dry unit weight of the mixture that is achieved. The foregoing advantages seem to indicate that these stabilizers will be of special value when mixed with young soils of a volcanic origin.

Phosphoric stabilizers are effective only with acid soils, and have no effect on alkaline soils, silts or sands. This is because the phosphoric reacts with the clay minerals to capture aluminum and precipitate a hydrated aluminum phosphate. Handling of these stabilizers is difficult because they are acids although they are not otherwise toxic substances.

.2 Sodium Chloride

Sodium chloride has often been employed as a rather short-lived stabilizer or to prevent dust on the soil surface.

Sodium chloride is effective in all soils, although far less so in those containing organic matter. Its effect is due to the colloidal reactions it produces and the alteration of the pore water characteristics. It normally acts as a flocculant, thus aiding compaction.

A rather special but promising use of common salt is the reduction it causes in the permeability of many clays, which makes it useful for treating expansive clays. Salt also increases soil strength [36].

The chief disadvantage of salt treatment is that the material is very soluble and is therefore easily washed away. This is why it is considered a short-lived stabilizer.

.3 Calcium Sulphates (Gypsum) and Calcium Chloride

The effect of these substances, their advantages and disadvantages, are very similar to those of sodium chloride. Their influence on compaction is usually far less pronounced. They sometimes cause an increase in permeability, which renders them far more susceptible to being washed away. Gypsum has frequently been employed as an additive to accelerate setting in soil-cement mixtures, but caution should be exercised here because of fluorescence problems.

.4 Sodium Hydroxide (Caustic Soda)

As yet the effects of caustic soda as a stabilizer are not well known. However, [1] mentions some highly positive results obtained in India where small amounts of this stabilizer were added to lateritic soils. Some recent laboratory investigations corroborate this benefit. Although the cost of caustic soda is high, it can be used in very small percentages.

The chief disadvantage of caustic soda is its caustic nature, which makes handling dangerous. Moreover it absorbs carbon dioxide and carbonates very rapidly when exposed to air thus losing its power.

The chief advantages are its easy application and its positive effect on soil compaction. The best results are obtained with kaolin clays of a lateritic nature and kaolin clays rich in aluminum. Figure 16-44 [1] shows the unconfined compressive strength of a kaolin stabilized with 10% sodium hydroxide.

In montmorillonites the effect of sodium hydroxide is small or zero.

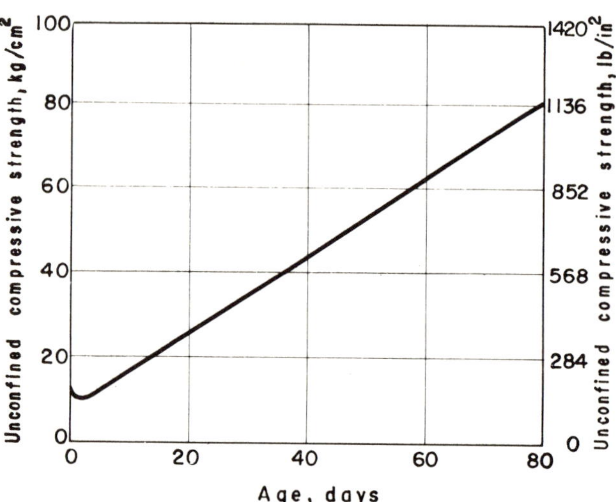

Fig. 16-44 Evolution of the strength of a kaolin stabilized with 10% caustic soda [1]

.5 Aluminum Salts

Reference [11] mentions an investigation in which a loess is stabilized with aluminum salts. This indicates that these substances have a very useful effect on soil stability or resistance to water, although this is not apparent from the strength of the mixtures, Fig. 16-45 shows both characteristics. For low water contents, the stabilizer has a negative effect on the strength, but when the water content of the soil increases, the samples containing a higher percentage of stabilizer lose proportionally far less strength than those that have not been stabilized or those with lower percentages of stabilizer.

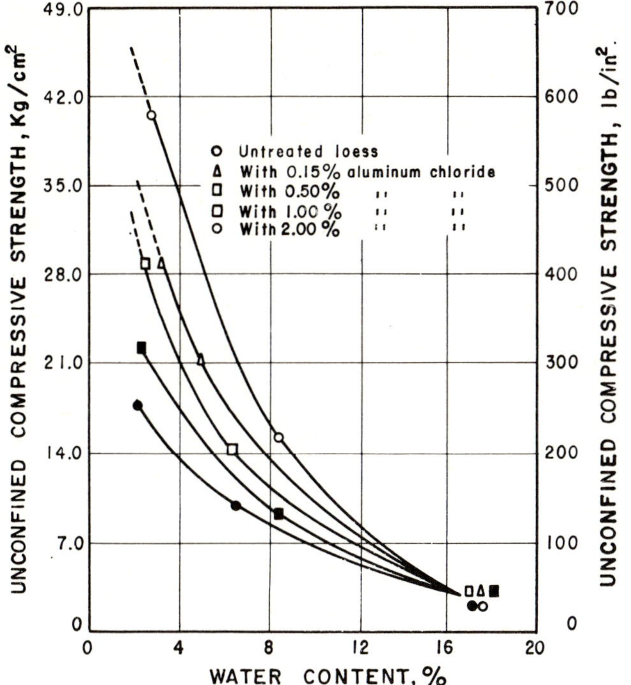

Fig. 16-45 Effect of treatment with aluminum salts on the stability of a loess in relation to the water content [11]

.6 Resins and Polymers

These are very long chains of molecules resulting from the linking of organic compounds, which are referred to as monomers. Resins are a form of natural polymer. Polymers can be added to soils in two different ways. First, monomers are added plus catalizing system which subsequently produces polymerization and the linking of particles to form a semi-solid. Second, the polymer is added in the form of a solid, solution or emulsion, which bonds the soil particles mechanically, similar to asphalts.

Vinsol is the resin most often used as a stabilizer, particularly in Europe. The results that have been achieved with it are variable, but encouraging. Its effect is restricted to the stabilization of mixtures by increasing their resistance to water [32]. This is the reaction generally attributed to all resins and polymers, but WINTERKORN has described some products which also tend to increase strength [37].

The amount of resins and polymers employed is normally between 1% and 2%. Most can only be used with acid soils. An important disadvantage of some of these products is their limited life due to bacterial degradation.

The possibility of mixing resins or polymers in strips and duly putting the soil together has been little investigated. Polymers may be cationic, anionic and non-ionic. The cationic ones have positive charges which create very strong electrical contacts with the negative charges of clay particles or fine siliceous sands. This mechanism may lead to an increase in soil strength. Unfortunately this effect brings about an appreciable increase in the resistance of the mixture to compaction. Moreover, these products are difficult to mix because of the small proportion of them that is used. Finally their costs are very high.

Anionic polymers have the same electrical charge as clay minerals, and consequently they tend to reduce the strength of the soils to which they are added. Consequently they favor compaction. Their effectiveness is very variable, according to the type of soil, and their use is probably seldom justified because common salt, which is far cheaper, is often just as effective or even more so.

Non-ionic polymers create important hydrogen bonds between the particles of clay, their *OH* groups associating with the oxygen in the clay particles.

16.7.2 Physical Stabilization

Many of the most widely used physical or mechanical stabilization techniques have been described in other parts of this book. Compaction, vibroflotation and cement grouting are among them. Others, such as electro-osmosis [38,39,40], although well known in Mexico, Europe, and North America have been merely touched on, since they are as yet seldom employed in highway engineering.

Special attention is drawn here to thermal stabilization techniques, either by heating or by cooling.

.1 Thermal Stabilization by Heating

Heat transforms any clay into a hard solid. At low temperatures the strength gain is by water loss, and is reversible. At a high enough temperature the process produces a mineral change which is irreversible and the strength acquired cannot be lost even by prolonged immersion. This occurs with temperatures of about 900°C (1650°F) which are inconveniently high for large-scale stabilizations. For practical purposes, it is sufficient to heat the clay until rehydration of the clay minerals becomes impossible, which occurs at a temperature between 200 and 400°C (390 and 750°F) [41-44]. Below 200°C (390°F), the effect on strength may be very considerable, but not irreversible.

Figure 16-46 [1,42] illustrates the changes in the water content and plasticity index after rewetting as a function of temperature. The variation in the plasticity index is expressed for two montmorillonites and two kaolins.

Heat is applied to the soil by means of a direct flame on the surface or by circulating heated gas. The direct flame method was first used in Rumania [45]. It is usually applied by drilling two sloping intercommunicating holes in the soil, combustion being caused at the point at which they meet. Fuel is introduced into the first hole and the hot combustion gases are allowed to escape through the second. The second method was developed by Soviet engineers [1] and consists of making a single hole in the soil, at the bottom of which the combustion chamber is placed, the temperature of which is controlled with excess pressure (in relation to atmospheric pressure). The soil is heated by hot compressed air passing through its pores. The Russian method is more effective but more complicated. It is limited to soils with sufficient air permeability that the hot gasses can diffuse through them.

In both methods, the range of influence of a heating point is not much more than one or two meters.

Heating is particularly useful for reducing the swell potential of clayey soils.

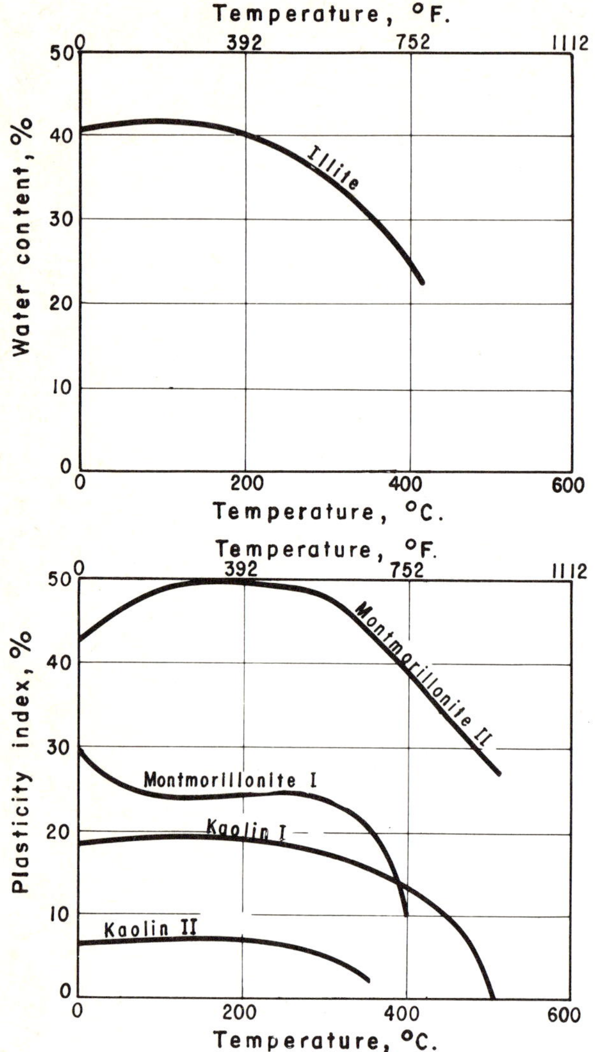

Fig. 16-46 Evaluation of the water content and the plasticity index after rehydration, as a function of temperature in the soils that are indicated [1,42]

.2 Thermal Stabilization by Cooling

Cooling above 0°C (32°F) may reduce the strength of fine soils as a consequence of increased interparticle repulsion, causing movement of the pore water due to the effect of thermal gradient. This leads to changes in soil behavior that are not easy to control. Consequently, all stabilization methods by cooling involve freezing the pore water and creating a hard, strong mass.

In sandy soils, water freezes at temperatures of about 0°C (32°F), but in clayey soils considerably lower temperatures may be required.

Freezing techniques have been used particularly for the construction of deep foundations in clay for very high buildings and for stabilizing soft silty soils so that tunnels can be driven. Recently Mexican engineers have had some extremely successful experiences in this field. In the Soviet Union, freezing techniques are used in ports, tunnels, mines and subway construction [46].

Freezing is achieved by sending cooling substances through piping networks installed in the subsoil.

16.8 Some Observations on the Behavior of Stabilized Pavements

Little has been written on the behavior of cement- or asphalt-stabilized layers in pavements constructed under different climatic conditions and subject to varying volumes of traffic.

Soil-cement has been studied more than any other mixture, perhaps because it is thought that it is the one with properties most different from those of the original soil [1]. Reference has already been made to the differences of opinion still existing between those engineers in favor of permitting total hardening of stabilized pavement layers before exposing them to traffic action and those who prefer break-up of the layers before full strength is achieved, so as to encourage a behavior more like that of flexible pavements. Japanese engineers appear to favor this last technique.

Generally, cement is used when it is wished to achieve a firm layer, of uniform strength, capable of providing a good homogeneous non compressible support, as would be the case of the sub-base of a rigid pavement or the base or sub-base of a flexible pavement subject to very heavy traffic action. At other times a firm layer is required in the structural section of the road or runway, capable of developing tension stresses which will protect a weak subgrade or earthfill against heavy traffic loads.

The effect of a thin layer of soil-cement as a protection against water seeping through from the surface to deeper layers of susceptible soil has not been very thoroughly investigated.

The literature on lime-treated soils is less specific about the evaluation of relative behaviors. Recently considerable interest has been aroused in a comparison of the results for a variety of lime-stabilized soils, especially different types of clays [47]. Otherwise the effects reported as having been achieved in most cases are improvements in the plasticity, the compaction characteristics and the volumetric stability of fine soils. A secondary benefit is the increase in strength, although this may be significant in relatively coarse granular soils.

The observed benefits of bituminous stabilizations are not as well documented as the benefits of cement. The objectives have been mentioned; to achieve strength or stability (resistance) of the mixture in the presence of water.

APPENDIX 16A PROBLEMS

16a.1 Soil Stabilization: Mixture of 3 Soils

Three aggregates, α, β and γ with the grain-size distributions shown in Fig. 16a-1 are to be mixed. In this way it should be possible to obtain a grain-size distribution within the specifications shown in the same figure.

The grain-size distributions for the aggregates and the specifications are those shown in Table 16a-1. The problem will be analysed employing the graphical solutions using triangular and rectangular diagram methods. As can be seen in Fig. 16a-1, simple analytical solution is not practical in this case, owing to the way in which the grain-size distributions overlap.

Table 16a-1
Data for soil stabilization problem mixture of 3 soils

AGGREGATE	PERCENTAGE PASSING SIEVE											1
	1½ in	1 in	¾ in	½ in	⅜ in	No. 4	No. 8	No. 16	No. 30	No. 50	No. 100	No. 200
Specified	100	88–100	80–100	70–100	61–90	45–69	34–50	26–38	18–29	12–22	7–16	4–10
α	100.0	100.0	91.5	71.0	58.0	30.0	10.0	2.0	—	—	—	—
β	100.0	100.0	100.0	89.0	80.0	67.5	57.0	47.5	40.0	33.5	28.5	24.0
γ	100.0	100.0	100.0	100.0	100.0	100.0	62.0	33.0	18.0	4.5	—	—

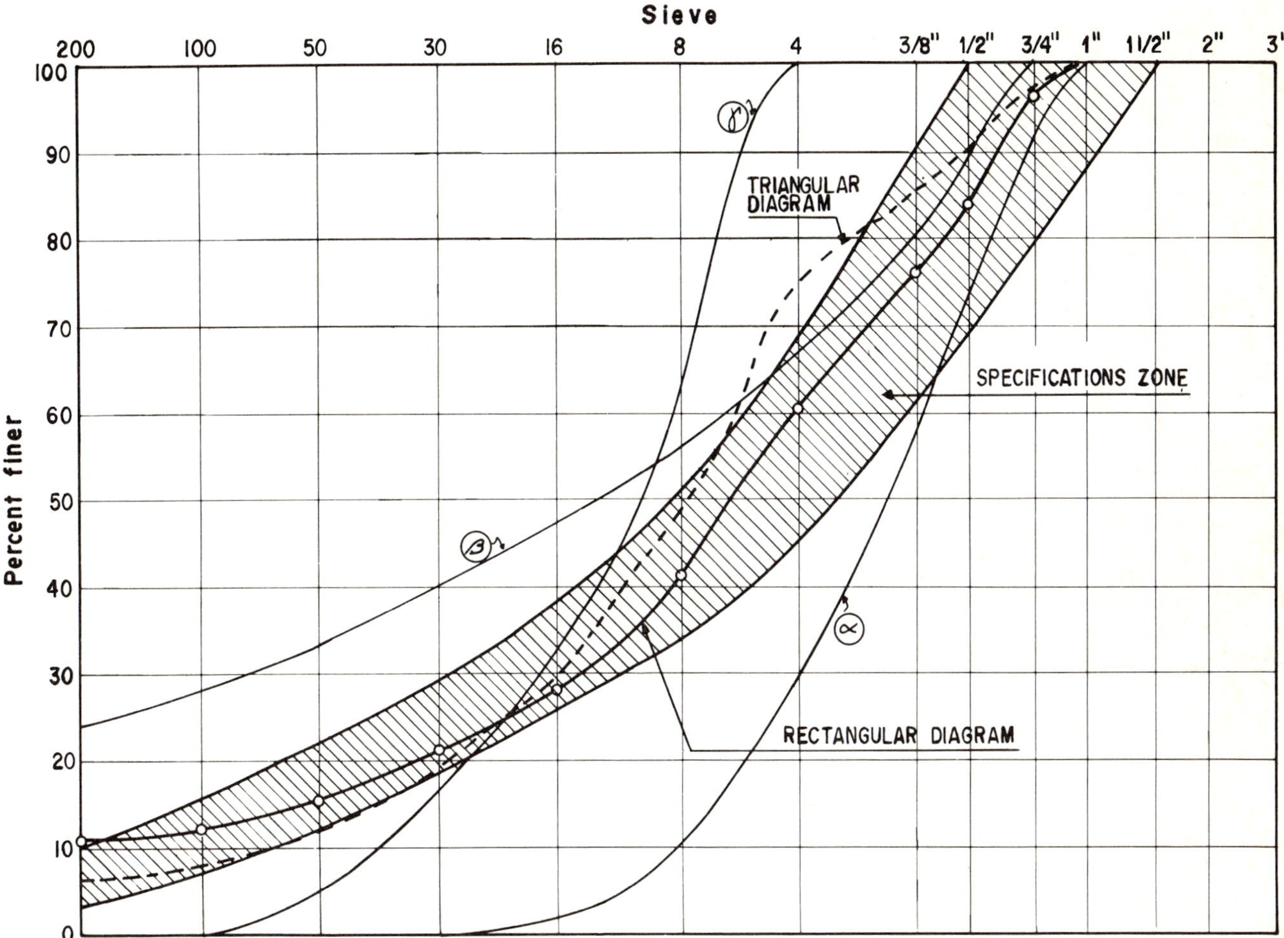

Fig. 16a-1 Grain-size distribution curves

.1 Triangular Diagram Method

This method, which is not described in detail in the text, is basically as follows:

An equilateral triangle is drawn, with each of its sides divided from 0 to 100, as is shown in Fig. 16a-2. On one of these scales is marked the percentage of soils passing No 200 sieve; on another the percentage passing No 4 sieve but retained on No 200 sieve, and on the third scale the percentage retained on No 4 sieve.

The sloping position of the numbers is worth observing, for in this way the readings will be plotted and evaluated.

From the grain-size distributions for the aggregates are determined the percentages corresponding to the aforementioned scales. The same is done with the specifications, so that a table like that shown in Fig. 16a-2 is obtained. Each of the aggregates is represented in the triangular diagram by a single point, and in turn each point represents a possible soil.

Using two of the coordinates the position of each aggregate is plotted on the triangular diagram, the third coordinate serving as a check.

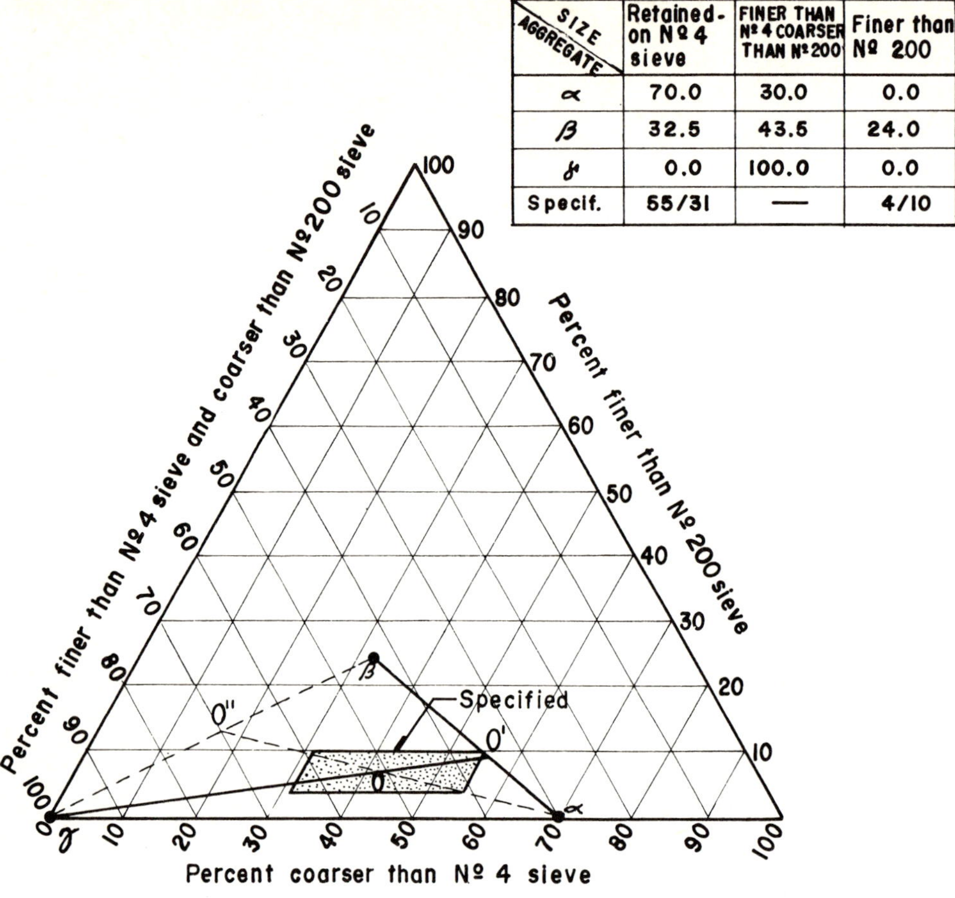

SIZE / AGGREGATE	Retained on Nº 4 sieve	FINER THAN Nº 4 COARSER THAN Nº 200	Finer than Nº 200
α	70.0	30.0	0.0
β	32.5	43.5	24.0
γ	0.0	100.0	0.0
Specif.	55/31	—	4/10

Fig. 16a-2 Triangular diagram

The specification zone is plotted on the triangular diagram. For this it is sufficient to plot the two grain sizes (that retained on No 4 sieve and that passing No 200 sieve, parallel to the respective scales.

Any point inside the triangle formed by the three soils would represent a mixture of the three different grain sizes. Consequently, for any mixture of the three aggregates to remain within the specifications, this triangle and the specification zone must overlap.

Thus, any point within the overlapping zone will represent an acceptable mixture of grain sizes, the optimum mixture being the one that is located at the center of gravity of this overlap (Point *0*).

The proportions of the three soils can now be obtained, as follows:

Draw a line from the point corresponding to soil γ to the center of gravity *0*, as far as *0'*, above the line joining the points or soils α and β. The relation of the segments *0–0'*, to γ*0'*, gives the proportion of soil γ required in order to achieve the grain-size characteristics corresponding to point *0*. Similarly the proportion between segments $\beta 0'$ and $\beta\alpha$ would give the proportion of β of the total of $(\alpha + \beta)$ (Note $\alpha + \beta = 1 - \gamma$) that should be included in the mixture. Similarly, the complement of one of the sum of the proportions of soils α and γ employed will give the proportion of soil β that should be used. Note that this last proportion could also be calculated by multiplying the ratio of segments $0'\alpha/\alpha\beta$ by the complement of one of the proportion of γ that was included.

In the case of Fig. 16a-2, the lengths of the segments measured were:

Percentage of

$$\gamma = \frac{00'}{0'\gamma} = \frac{14}{14 + 46} = \frac{14}{60} = 0.234$$

Percentage of

$$(\alpha + \beta) = 1.000 - 0.234 = 0.766$$

Percentage of

$$\beta = 0.766 \frac{12}{12 + 21} = 0.766 \frac{12}{33}$$

Percentage of

$$\beta = 0.278$$

Percentage of

$$\alpha = 0.766 \frac{21}{33} = 0.488$$

The percentages of the aggregates are:

A = 23.4%
B = 27.8%
C = 48.8%

100.0%

In accordance with the above percentages, the resulting grain-size distribution can be calculated as indicated in Table 16a-2.

The solid lines are those used to solve the problem in Fig. 16a-2. The intercepting line was drawn along the greater diagonal. The grain-size distribution curve obtained for the mixture of the three soils is represented by a broken line in Fig. 16a-1. In this case the curve for the percentage corresponding to No 4 sieve is slightly outside the specification zone.

Table 16a-2

Soil stabilization problem: mixture of 3 soils triangular diagram method

AGGREGATE	PERCENTAGE PASSING SIEVE											2
	1½ in	1 in	¾ in	½ in	⅜ in	No. 4	No. 8	No. 16	No. 30	No. 50	No. 100	No. 200
0.234 α	23.4	23.4	21.4	16.6	13.6	7.0	2.3	0.5	—	—	—	—
0.278 β	27.8	27.8	27.8	24.8	22.2	18.8	15.9	13.2	11.1	9.3	7.9	6.7
0.488 γ	48.8	48.8	48.8	48.8	48.8	48.8	30.2	16.1	8.8	2.2	—	—
Total	100.0	100.0	98.0	90.2	84.6	74.6	48.4	29.8	19.9	11.5	7.9	6.7
Specified	100	88/100	80/100	70/100	61/90	45/69	34/50	26/38	18/29	12/22	7/16	4/10

.2 Rectangular Diagram Method

Now the rectangular diagram method, which was described in the text with reference to soils, will be applied. For this purpose, the graphs in Fig. 16a-3 are plotted. First, aggregates α and β are combined, and to the mixture thus obtained is subsequently added aggregate γ.

To mix materials α and β the rules indicated in this Chapter for the case of two aggregates must be followed. In this case the zone within which all the conditions of the problem are satisfied is extremely narrow. The line through the middle of this zone was regarded as the optimum mixture, the values of which are shown with a dash-dot line in the diagram on the lower right-hand side of Fig. 16a-3. From this point the mixture of soils α and β is treated like a new soil which must be mixed with γ. In this method two aggregates are mixed each time, repeating the operation as many times as necessary.

From the figure the following results can be obtained:

Aggregate $\gamma = 18\%$
Aggregate $\alpha + \beta = 82\%$
Aggregate $\beta = 0.55 \times 82 = 45\%$
Aggregate $\alpha = 0.45 \times 82 = 37\%$

Coefficients 0.55 and 0.45 were obtained from the right-hand graph in Fig. 16a-3 and represent the respective percentages of aggregates β and α included in the mixture $\alpha + \beta$.

The total percentages required are therefore:

$$A = 37\%$$
$$B = 45\%$$
$$C = 18\%$$

In accordance with the percentages obtained, the resulting grain-size distribution could be calculated as shown in Table 16a-3.

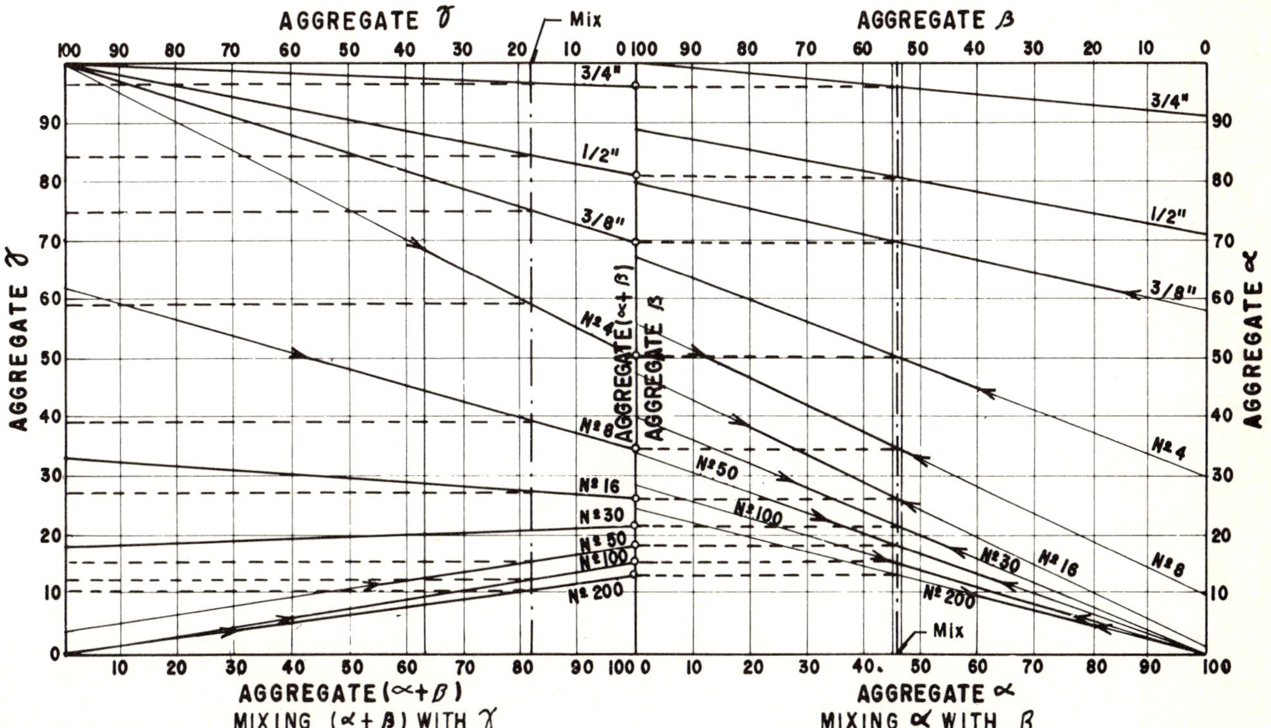

Fig. 16a-3 Rectangular diagram

Table 16a-3

Soil stabilization problem: mixture of 3 soils rectangular diagram method

AGGREGATE	PERCENTAGE PASSING SIEVE											3
	1½ in	1 in	¾ in	½ in	⅜ in	No. 4	No. 8	No. 16	No. 30	No. 50	No. 100	No. 200
0.370 α	37.0	37.0	33.8	26.2	21.4	11.1	3.7	0.7	—	—	—	—
0.450 β	45.0	45.0	45.0	40.0	36.0	30.3	26.6	21.4	18.0	15.0	12.8	10.8
0.180 γ	18.0	18.0	18.0	18.0	18.0	18.0	10.2	5.9	3.2	0.8	—	—
Total	100.0	100.0	96.8	84.2	75.4	59.4	40.5	28.0	21.2	15.8	12.8	10.8
Specified	100	88/100	80/100	70/100	61/90	45/69	34/50	26/38	18/29	12/22	7/16	4/10

Note that the total grain-size distribution is the same as the one that would have been achieved by simply obtaining projections of the intersections of the center line with the lines corresponding to each aggregate, above the ordinate on the extreme left of the graph in Fig. 16a-3.

This grain-size distribution, already plotted in Fig. 16a-1, is notably different from the one obtained with the triangular diagram. The procedure could be repeated inverting the order in which the aggregates are employed; that is to say, first mixing aggregate Υ with α and subsequently adding β.

In this case, the result indicated in Fig. 16a-4 would be obtained.

Here the respective percentages will be:

Aggregate $\beta = 38\%$
Aggregate $\alpha + \Upsilon = 62\%$
Aggregate $\alpha = 0.62 \times 49 = 30.4\%$
Aggregate $\Upsilon = 0.62 \times 51 = 31.6\%$

Resulting in:

$A = 30.4\%$
$B = 38\%$
$C = 31.6\%$

Observing the percentages obtained in this second trial and comparing them with those obtained in the first trial, it is concluded that the order in which the aggregates are handled in the procedure has an important influence on the grain-size distribution of the resulting mixture.

Attention is however drawn to the fact that the center or best fit lines were selected rather at random, since they are not usually very clearly defined.

In conclusion, it can be said that the grain-size distribution obtained may be quite variable, for analysis of Fig. 16a-1 shows that there are several possible answers. The strength values achieved for specimens prepared with the different proportions obtained, along with the relative cost of these mixes, will be decisive factors in the selection of the most appropriate grain-size distribution.

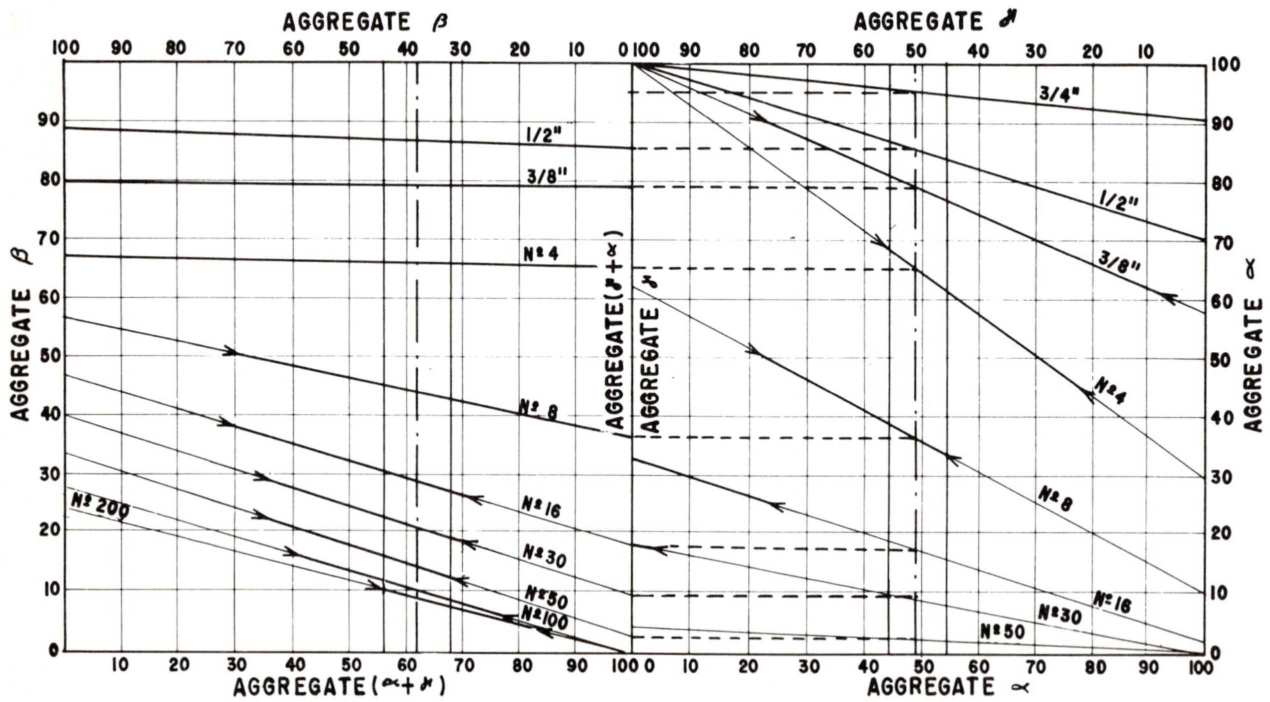

Fig. 16a-4 Another alternative using the rectangular diagram method

16a.2 Soil Stabilization: Addition of Lime

It is wished to estimate the layer thicknesses required for a flexible road pavement, using HVEEM's method.

Traffic indices of 7.9 and 11 will be considered.

The material at the location is a high compressibility clay, the geotechnical characteristics of which are as follows:

$$w_L = 62.7\%,\ w_p = 21.6\%,\ I_p = 41.1\%,\ S_s = 2.60$$

According to a mineralogical analysis conducted on the clay, it was found that the predominant clayey mineral is sodium montmorillonite.

For financial reasons, it is wished to employ this material for the subgrade, either compacting it with an appropriate water content or adding a suitable stabilizing agent.

Solution:

Since the clay contains a predominant amount of sodium montmorillonite, the stabilizing agent chosen was lime. For different lime contents the following variation in the soil classification was obtained:

Unstabilized clay $w_L = 62.7\%$; $I_p = 41.4\%$
Clay with 2% lime $w_L = 51\%$; $I_p = 31\%$
Clay with 4% lime $w_L = 44\%$; $I_p = 23\%$
Clay with 6% lime $w_L = 43\%$; $I_p = 19\%$

On the basis of the above figures, the percentage corresponding to 4% was chosen for the HVEEM's tests on the stabilized clay.

The results obtained in these tests are reported in Table 16a-4.

According to HVEEM's design method, it is necessary to determine the thickness corresponding to the stabilometer value for each compaction water content. It is also necessary to determine the thickness required to balance the swell pressure that develops on saturation of the compacted specimens with different water contents. For this last purpose, the material that is placed on top of the subgrade layer will be assumed to have unit weight of 2000 kg/m^3 = 0.002 kg/cm^3 (125 lb/ft^3).

Calculations follow for the unstabilized clay. To determine the thicknesses for R, the graph in Fig. 9-30 (Table 16a-5) was used.

The required thickness to prevent expansion is obtained by dividing the swell pressure (Table 16a-4) by the unit weight of the overlying material, which was assumed to be 2000 kg/m^3 (125 lb/ft^3).

The calculations corresponding to the lime-stabilized clay are given in Table 16a-6.

In Fig. 16a-5 the thicknesses obtained are presented graphically. As indicated in HVEEM's method, the final pavement thickness will be that corresponding to the intersection of the curve drawn through the points corresponding to the thicknesses determined for the different compaction water contents with a line drawn from the origin of the coordinates and forming an angle of 45° with the horizontal.

In accordance with this figure, the required thicknesses will be:

For the unstabilized clay		For the clay with 4% lime	
TI = 11	t = 88 cm (35 in)	TI = *11*	t = 28 cm (11 in)
TI = 9	t = 80 cm (31 in)	TI = *9*	t = 24 cm (9.5 in)
TI = 7	t = 72 cm (28 in)	TI = *7*	t = 20 cm (8 in)

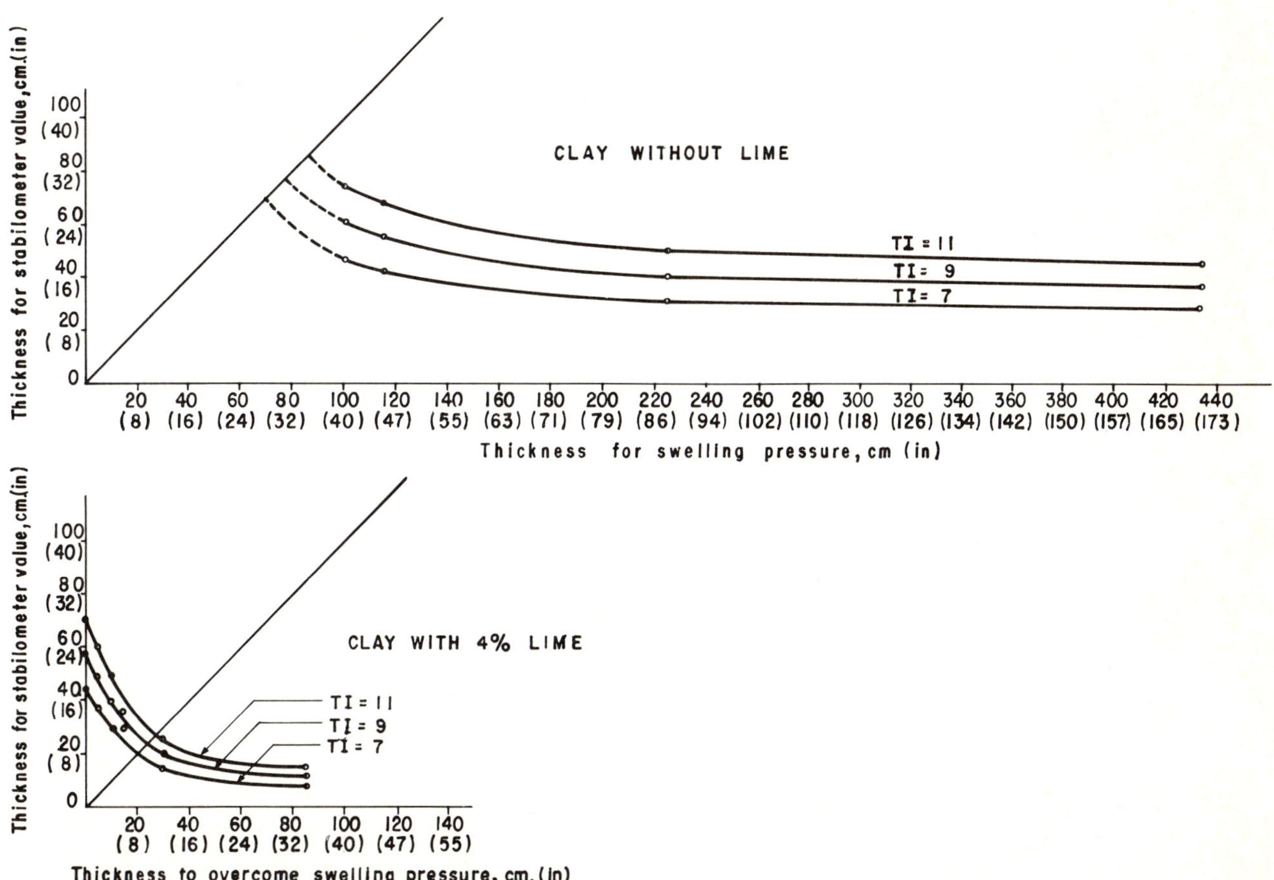

Fig. 16a-5 Thicknesses of gravel equivalent obtained for the pavement with and without lime, for different traffic indices

Table 16a-4

Results of Hveem tests on stabilized and unstabilized clay soil stabilization problem

Parameter	Units	Clay without lime addition							Clay with 4% lime addition						
Compaction moisture	%	21	22	23	24	25	26	27	21	22	23	24	25	26	27
Stabilometer value	*R*	60	55	45	38	32	27	24	87	77	67	55	45	35	28
Swelling pressure	kg/cm^2	0.87	0.45	0.29	0.23	0.20	0.19	0.18	0.17	0.06	0.03	0.02	0.01	0.00	0.00
	lb/in^2	12.40	6.40	4.10	3.30	2.80	2.70	2.60	2.40	0.85	0.43	0.28	0.14	0.00	0.00
Cohesion value	g/in^2	—	0.96	0.85	0.75	0.67	0.62	0.57	0.56	0.34	0.16	0	0	0	0
Unit weight	kg/m^3	1710	1680	1648	1616	1580	—	—	1626	1590	1552	1512	1480	—	—
	lb/ft^3	107	105	103	101	99	—	—	102	99	97	94	92	—	—

Table 16a-5

Soil stabilization problem: Addition of lime Calculations for unstabilized clay

Traffic Index	*w* %	*R*	Thickness due to *R*		Swelling pressure		Thickness due to Swelling pressure	
			cm	in	kg/cm^2	lb/in^2	cm	in
7	21	60	27.4	10.8	0.87	12.4	435	171.0
	22	55	30.5	12.0	0.45	6.4	225	88.5
	23	45	38.2	15.0	0.29	4.1	145	57.0
	24	38	42.0	16.5	0.23	3.3	115	45.2
	25	32	45.7	18.0	0.20	2.8	100	39.3
	26	27	49.5	19.5	0.19	2.7	95	37.3
	27	24	51.7	20.3	0.18	2.6	90	35.4
9	21	60	35.0	13.8	0.87	12.4	435	171.0
	22	55	39.6	15.6	0.45	6.4	225	88.5
	23	45	48.8	19.2	0.29	4.1	145	57.0
	24	38	55.0	21.6	0.23	3.3	115	45.2
	25	32	59.0	23.2	0.20	2.8	100	39.3
	26	27	64.0	25.2	0.19	2.7	95	37.3
	27	24	67.0	26.4	0.18	2.6	90	35.4
11	21	60	42.7	16.8	0.87	12.4	435	171.0
	22	55	48.8	19.2	0.45	6.4	225	88.5
	23	45	59.5	23.4	0.29	4.1	145	57.0
	24	38	67.1	26.4	0.23	3.3	115	45.2
	25	32	73.2	28.8	0.20	2.8	100	39.3
	26	27	78.0	30.6	0.19	2.7	95	37.3
	27	24	82.0	32.2	0.18	2.6	90	35.4

The design method also recommends consideration of the pressure under which the specimens exude water. Since the clay in question was highly impervious, this aspect was not taken into account.

According to the thicknesses obtained, it is found that the thickness of gravel equivalent required above the clay in question is substantially smaller when lime is added to the clay.

(The two problems in this Appendix are by courtesy of CARLOS FERNÁNDEZ LOAIZA)

Table 16a-6

Soil stabilization problem: Addition of lime : Calculations for clay stabilized with lime

Traffic Index	w %	R	Thickness due to R		Swelling pressure		Thickness due to Swelling pressure	
			cm	in	kg/cm^2	lb/in^2	cm	in
7	21	87	9.2	3.6	0.17	2.40	85	33.4
	22	77	15.2	6.0	0.06	0.85	30	11.8
	23	67	22.8	9.0	0.03	0.43	15	5.9
	24	55	30.5	12.0	0.02	0.28	10	3.9
	25	45	37.5	14.8	0.01	0.14	5	2.0
	26	35	44.3	17.4	0.00	0.00	0	0
	27	28	49.0	19.2	0.00	0.00	0	0
9	21	87	11.6	4.6	0.17	2.40	85	33.4
	22	77	20.4	8.0	0.06	0.85	30	11.8
	23	67	29.0	11.4	0.03	0.43	15	5.9
	24	55	39.7	15.6	0.02	0.28	10	3.9
	25	45	49.0	19.2	0.01	0.14	5	2.0
	26	35	57.0	22.4	0.00	0.00	0	0
	27	28	63.5	25.0	0.00	0.00	0	0
11	21	87	14.7	5.8	0.17	2.40	85	33.4
	22	77	15.2	6.0	0.06	0.85	30	11.8
	23	67	36.0	14.2	0.03	0.43	15	5.9
	24	55	49.0	19.2	0.02	0.28	10	3.9
	25	45	59.7	23.4	0.01	0.14	5	2.0
	26	35	70.5	27.7	0.00	0.00	0	0
	27	28	80.5	31.6	0.00	0.00	0	0

REFERENCES

1. Ingles, O. G. and Metcalf, J. B., *Soil Stabilization: Principles and Practice,* Butterworths: Sydney, 1972.
2. Juárez-Badillo, E. and Rico, A., *Mecánica de Suelos. Vol. III: Flujo de agua en suelos,* Limusa: México, 1974, Appendix II.
3. Smith, K. W. G., "Some Problems of Salts in Semi-arid Soils for Stabilization with Cement", *Procs. I. Australian Conference on Road Construction,* Camberra, 1962, Vol. I.
4. Fernández Loaiza, C., *Mejoramiento y Estabilización de Suelos,* Limusa: México, 1982.
5. Torrente M. and Sagüés L., *Estabilización de Suelos. Suelo Cemento,* Editores Técnicos Asociados, S.A: Barcelona, Spain, 1968.
6. Zalazar, L. M., "Estabilización de Suelos", Notes for the Possible Publication of a book, communicated personally to the authors, Buenos Aires, 1974.
7. Rico, A. and Orozco, J., "El Efecto de la Incorporación de Finos en el Comportamiento de Materiales para Base", *V. Panamerican Conference on SMFE,* Buenos Aires, 1975.
8. Texas Highway Department, Materials and Test Division, *Test Method Tex-114-E* (Revised in April, 1970), Austin, Tex., 1970.
9. Yoder, E. J., *Principles of Pavement Design,* John Wiley, 1967, Chap. II.
10. Lambe, T. W. and Whitman, R. V., *Soil Mechanics,* John Wiley, 1969.
11. Lambe, T. W., "Soil Stabilization", Chap. 4 of *Foundation Engineering,* G.A. Leonards Ed., Mc Graw-Hill, 1962.
12. Avitia, R., "Suelo-Cemento", Publication of the Mexican Cement and Concrete Institute (IMCYC) México, 1971.
13. Norling, L. T., "Standard Laboratory Tests for Soil-Cement. Their Development, Purpose and History of Use", Portland Cement Association, Chicago, Ill, 1963.
14. Hveem, F. N. and Zube, E., "California Mix Design for Cement Treated Bases", California Division of Highways, Materials and Research Department, Sacramento, Cal., 1963.
15. López Soto, V., "Estabilización de Suelos con Cemento en la Construcción de Carreteras", Thesis, Faculty of Engineering. U.N.A.M. México, 1967.

16. Baker, C. N., "Strength of Soil-Cement as a Function of Degree of Mixing", Highway Research Board Meeting, 1954.

17. Michaels, A. S. and Puzinauskas, V., "Evaluating Performance Characteristics of Mechanical Mixing Processes", *Chemical Engineering Progr,* Vol. 50, 1954.

18. Rodríguez Elizalde, M., "Determinación por Trituración del Contenido de Cemento Portland en Bases Estabilizadas", Thesis Faculty of Engineering, U.N.A.M. México, 1970.

19. California Division of Highways, "Determination of Cement or Lime Content in Treated Aggregate by the Titration Method", Materials and Research Department, Test Method No 338 D, Sacramento, Cal., 1968.

20. Chang Wang, M. and Huston, M. T., "Direct Tensile Stress and Strain of a Cement Stabilized Soil", Highway Research Board Record No 379, H.R.B. Washington, D.C., 1972.

21. Thompson, M. R., "Split Tensile Strength of Lime-Stabilized Soil", H.R.B. Record No 92, Highway Research Board, Washington, D.C., 1965.

22. Kennedy, T. W. and Hudson, W. R., "Application of Indirect Tensile Test to Stabilized Materials", H.R.B. Record No 235, Highway Research Board, Washington, D.C., 1968.

23. Moore, R. K., Kennedy, T. W. and Hudson, W. R., "Factors Affecting the Tensile Properties of Cement-Treated Materials", H.R.B. Record No 315, Highway Research Board, Washington, D.C., 1970.

24. Moore, R. K., Kennedy, T. W. and Kozuh, J. A., "Tensile Properties for the Design of Lime Treated Materials", H.R.B. Record No 351, Highway Research Board, Washington, D.C., 1971.

25. Kennedy, T. W., Moore, R. K. and Anagnos, J. N., "Estimations of Indirect Tensile Strengths for Cement Treated Materials", H.R.B. Record No 351, Highway Research Board, Washington, D.C., 1971.

26. Yung-Chiech Chiang and Yong S. Chae., "Dynamic Properties of Cement Treated Soils", H.R.B. Record No 379, Highway Research Board, Washington, D.C., 1972.

27. Rico, A. and Quintero, M., "El Suelo-Cemento en Caminos de bajo Costo", Journal of the Mexican Cement and Concrete Institute, México, 1974.

28. Portland Cement Association, "Thickness Design for Soil-Cement Pavements", Engineering Bulletin, 1970.

29. Hurley, C. H. and Thornburn, T. H., "Sodium Silicate Stabilization of Soils: A Review of the Literature", H.R.B. Bull. No 381, Highway Research Board, Washington, D.C., 1972.

30. Arman, A. and Munfakh, G. A., "Lime Stabilization Organic Soils", H.R.B. Bull. No 381, Highway Research Board, Washington, D.C., 1972.

31. Michaels, A. S. and Puzinauskas, V., "Chemical Additives as Aids of Asphalt Stabilization of Fine Grained Soils", H.R.B. Bull. No 129, Highway Research Board, Washington, D.C., 1956.

32. Road Research Laboratory, *Soil Mechanics for Road Engineers,* Her Majesty's Stationery Office, London, 1961.

33. Katti, R. K., Davidson, D. T. and Sheeler, J. B., "Water in Cutback Asphalt Stabilization of Soils", H.R.B. Bull. No 241, Highway Research Board, Washington, D.C., 1960.

34. Johnson, J. C., "The Place of Asphalt Stabilization in the Expanded Highway Program", National Convention of American Road Builders., Chicago, Ill., 1957.

35. Demirel, T. and Davidson, D. T., "Reactions of Phosphoric Acid with Clay Minerals", H.R.B. Bulletin No 318, Highway Research Board, Washington, D.C., 1961.

36. Lambe, T. W., "The Engineering Behaviour of Compacted Clay", *Procs. A.S.C.E.,* 1958.

37. Winterkorn, H. F., "Desarrollo del Método Anilina Furfural para la Estabilización de las Playas", Sixth National Meeting on Asphalt, Buenos Aires, 1952.

38. Juárez Badillo, F. and Rico, A., *Mecánica de Suelos, Vol. III: Flujo de Agua en Suelos,* Limusa: México, 1974, Appendix II.

39. Casagrande, L., "La Electrósmosis y Fenómenos Conexos", Journal *"Ingeniería"* México, April, 1962.

40. Tamez, E. and Flamand, C., *La Electrósmosis Aplicada a la Construcción,* Solum, S.A: México, 1963.

41. Aylmore, L. A. G., Quirk, J. P. and Sills, I. D., "Effects of Heating on the Swelling of Clay Minerals", H.R.B. Special Report No 103, Highway Research Board, Washington, D.C., 1969.

42. Chandrasekharan, E. C., Boominathan, S., Sadayan, E. and Setty, K. R. N., "Influence of Heat Treatment on the Pulverization and Stabilization Characteristics of Typical Tropical Soils", H.R.B. Special Report No 103, Highway Research Board, Washington, D.C., 1969.

43. Radhakrishnan, N., Katti, R. K. and Hussain, M., "Studies on the Thermal Stabilization of Black Cotton Soils", *Procs. III. Asian Regional Conference of Soil Mechanics and Foundation Engineering,* Haifa, Israel, 1967.

44. Wöhlbier, H. and Henning, D., "Effect of Preliminary Heat Treatment on the Shear Strength of Kaolinite Clay", H.R.B. Special Report No 103, Highway Research Board, Washington, D.C., 1969.

45. Beles, A. A. and Stanculescu, I. T., "Improving the Stability of Earth Masses", *Geotechnique,* Vol. 8, 1958.

46. Hakimov, H., *Theory and Practice of Artificial Freezing of Soils,* Publication of the USSR Academy of Sciences. 1957 (Mentioned in [11]).

47. Le Roux, A. and Riviere, A., "Traitement des Sols Argileux par la Chaux", Bulletin de Liaison des Laboratoires Routiers, No 40 Paris, 1969.

OTHER RELATED REFERENCES

Transportation Research Board, "Soil Stabilization 1982", Transportation Research Record 839, Washington, D.C., 1982.

Puiatti, D., Puig, J., and Schaeffner, M., "Traitement des Sols ã la Chaux Aerienne et aux Ciments, Methodologie des Etudes de Laboratoire", Colloque International sur la Gestion des Ouvrager, Paris, 1981.

CHAPTER 17

QUALITY CONTROL

17.1 Introduction

Clearly even the most careful design and the most ambitious and costly construction are not sufficient to ensure the creation of a useful, economical and durable engineering structure. Between the design and the structure there are steps to be taken and criteria that must be met if satisfactory results are to be achieved. A simplistic criterion might express this as doing things *well*, but this is not enough. A series of actions well done, each carefully conceived and executed, sometimes combine to an inadequate process.

A successful road, railroad or runway is a balance of a very large number of previous actions. It is not sufficient for each of these actions to be *well-done* to ensure the success of the project as a whole. On the contrary, in some cases success smiles upon processes in which many links in the chain of actions have been neglected, although other essential ones have been well taken care of. It is the joining of the links that must be fully comprehended. A realistic knowledge of the significance of each and its influence is the secret of a successful control process.

Perfect control of every single step leads to a rigid perfectionism, incompatible with the realities of heavy construction work. Determination of the vital processes and the exercise of a reasonable scientific vigilance over them is the secret of successful control.

The degree of perfection or care with which each action is carried out should fit the nature of the action and the consequences of sub-standard quality. In some cases carelessness or improvisation will be technically permissible, so long as the quality of the whole, so far as its intended purpose, is satisfactory.

For example it would be useless and expensive to insist on a sidewalk quality finish on the upper surface of a concrete structure that will be buried forever beneath the ground.

Quality control in engineering practice has currently become a complex science. It is a new field in its own right, with its own specific methodology and criteria. As such, its details are beyond the scope of this book. But at the same time, in the case of transportation structures, applied soil mechanics plays a very important part as a supporting discipline. Since many of the processes that are subjected to control involve soil mechanics, this discipline must provide the criteria for deciding what is of primary importance and what is only secondary. From these considerations the field tests or laboratory tests on which control decisions must be based and the limits and tolerances of construction must be developed.

Some engineers, even in highly responsible positions, are disinclined to understand the role played by geotechnical sciences in the design, construction and maintenance of roads. Consequently their major interest is to assure a reduction in first cost or execution time (although these are often imaginary). If these skeptics would only consider the composition of a current quality control laboratory for highway purposes, they would see that it is basically a laboratory for soil mechanics or rock mechanics (the latter, which is a relatively new and rapidly expanding discipline, is not yet so well known in many places, which is a serious limitation for the control units). This is equivalent to recognition that the success or failure of such projects is controlled by these two disciplines.

An important feature of a good control program is the previous determination of the degree of quality required. The simplest approach to this determination is to ask three fundamental questions [1]:

— What is desired?
— How can the activity that will achieve this desire be organized and programmed?
— How can it be determined whether what was desired has been fulfilled?

It seems illogical that most large organizations specializing in design and construction employ uniform control principles for all their projects, when the foregoing considerations indicate that different control principles and goals be established for different projects with different characteristics, risks and importance.

The above three questions are interrelated. What is needed could be established in a *closed system*, where the requirements are specified by the design, and the final results achieved serve as a guide for future projects. This course of action is inefficient, does not make the most of many possibilities of improvement and exposes important projects to defects that are difficult or impossible to correct. Instead, a system is required that can be *fed back*, so that requirements can be continually compared with partial achievements. The system for evaluating these partial achievements is in turn fed back by the requirements. At the same time, it is necessary to review project requirements constantly in the light of partial achievements that gradually appear feasible.

Furthermore, the first two questions are concerned with both the design and the contract. When establishing the design philosophy, the engineer must understand that the construction cannot be classified simply as good or bad, rejectable or acceptable. There will always be degrees of quality compared to optimum conditions. Possible variations in the construction materials and techniques should be considered as well as necessary (not arbitrary) with tolerance limits for almost all activities. These tolerance limits must be clearly specified in the contract. Only with this flexible framework will it be possible to establish realistically the engineer's aspirations and requirements.

The third question demands a system of inspection, sampling and tests which will permit an evaluation of construction realities, including project trends and variations. There are four basic requirements for designing this program [2]. First, it must be based on realistic aspirations, otherwise it will only lead to confusion. Second, it must be based on tests of relevant technical significance, for only these will provide adequate information about the true conditions of the project. Third, the condition that must be satisfied is that the inspection system must emphasize the fundamental aspects of the behavior of the structure and not secondary ones. Fourth, interpretation of the program must be clear and non-controversial, objective and scientific, minimizing subjective judgment when possible. when possible.

The quality control program should not focus on construction work alone, but also must look at the future maintenance. The organization in charge of control must continually improve its control methods, bearing in mind both construction costs and maintenance requirements.

The wording of the construction specifications often, without meaning to, determine many of the control objectives, many of the programs leading to the achievement of these objectives, and many of the methods that decide whether or not what was desired has been achieved. In other words, construction specifications have a major impact on the three fundamental questions which were asked above, which are the basis of control philosophy.

Unfortunately the attitude of some of the staff of large road-designing organizations towards following their specifications is not always healthy. There is a pronounced tendency to idealize the specifications in use like some religious dogma regarding them as beyond all criticism. What is expressed by the specifications cannot be questioned, and anyone who modifies them in any way is accused either of disrespect for the accepted techniques or a cheater.

The authors are aware of any large design organization's need to utilize standard or previous technical job specifications. This is the most efficient means of handling clearly and reasonably all the contractural, regulatory, and owner-engineer-contractor relations as well as providing uniformity of style and quality. However an exaggerated respect for any set of previous specifications will lead to narrow-mindedness and lack of flexibility in the design and construction techniques employed. Organizations that create an aura of holiness about their technical standards develop a strong feeling of opposition towards any change in these standards; their technique becomes fossilized as a result.

A set of specifications is the result of the team work of a few men chosen for their knowledge and experience. This team should be capable of excellent work, producing reasonable standards well suited to current considerations. However, each of these men must remember that the final recommendation produced by the team will be applied to a project, the characteristics and circumstances of which he does not fully know. This consideration forces him to be cautious. As a result it is not unusual for the blind observance of technical norms pre-established on an international or national level to lead to ultra conservative, outmoded projects, that are not optimal economically. The engineer who is not prepared to accept even slight modifications in the standards and specifications of his organization is implicitly acknowledging that the criteria produced by a group of possibly distinguished men many years before are more valuable than those of the current team. Such a narrow-minded attitude is not fair to this man's colleagues, and clearly sacrifices a great deal of their ability to make decisions that are adjusted to the current technology as well as the conditions of the specific project they are handling. A responsive progressive organization can handle these contradictions in a logical manner. Specifications should first be treated as the legal framework for technical activity, and second as the ultimate reference or instructions for this technical activity. They are valid so long as no restrictions, variations or minor adjustments are indicated. Each new large-scale design, therefore, should include its own additional specifications, which should be as audacious and original as needed and take into consideration data corresponding to the most recent experience and knowledge available to the profession.

In any quality control program, the technical specifications adopted must be competent, in that they must ensure the essential standards of job quality. They must be adjusted to suit the social and economic needs of the region as well as taking into consideration its topographical, environmental and traffic conditions. This is why the blind adoption of technical standards produced by other countries, however advanced they may seem from a strictly technological viewpoint, often leads to inappropriate policies. Specifications must also be very realistic, adjusted to what must be achieved in view of the needs of a specific project and to what can be achieved in view of the technological level (specialized labor adequacy of field laboratories, construction equipment and money flow) of the country that is to build and use them.

Specifications should also ensure that materials of acceptable quality are not rejected. This explains why errors in policy are sometimes made in regions that observe the standards that have been developed elsewhere. Usually those regions whose specifications are adopted blindly by others are more advanced both technically and economically. Consequently their roads, railroads and runways carry traffic volumes that are exceptional or unknown in some other region that adopts the specifications. As a result this region will reject many economical materials and techniques which could be adequately employed for its lower traffic levels and which may fit its construction technology better. The region that is economically less developed will very soon discover the inadequacy of the specifications it has adopted for its own particular purposes. This will lead to systematic violations, inevitably resulting in confusion or corruption. This is the price that will always have to be paid for the adoption of unrealistic specifications.

Another basic requirement of a set of specifications is that it must contain appropriate tolerance limits, which should be based on a thorough knowledge of the factors contributing to variations in the different construction qualities [3]. The consequences of exceeding such limits should be duly evaluated. It may be of assistance to draw up a system of classifying how critical any eventual discrepancies and defects may be. Reference [3] proposes the following classification:

Critical defect: One which could render the construction very dangerous if not remedied.

Important defect: One which may have a serious effect on performance, particularly durability.

Defect of little importance: One which may not affect performance very seriously.

Contractual defect: Transgression of the contract that will not have important consequences, other than legal.

In the case of construction or products that are a mixture of others, the specifications must permit ready recognition of the component responsible for the principal characteristics exhibited by the sample.

Another important aspect of any quality control program is the set of laboratory tests, which provides the methodological and technical basis of the evaluation program. Laboratory tests for control purposes must satisfy certain requirements. They must:

— Verify essential characteristics
— Be simple and readily standardized, and repeatable
— Be rapid
— Be easy to interpret
— Use economical equipment that is easy to adjust and calibrate and simple to use by technicians with limited training.

Only in this way will it be possible to obtain reliable results in field laboratories, where control must be carried out, without interfering with or impeding construction programs. In field laboratories it is unusual to find personnel and equipment of outstanding quality, and consequently the laboratory requirements for quality control must be particularly realistic, otherwise a great deal of dubious information will be generated. The need for speed is essential and requires no further discussion.

Another component of a quality control program is the method adopted to process the volumes of information from field and laboratory tests and from the reports of those who interpret them. This information must be processed for future use and circulated to all the levels of the organization that may be concerned. It is of exceptional value in the formation of institutional experience and the planning of future maintenance or reconstruction jobs.

The foregoing objectives demand the development of integrated systems for storing information, the availability of such information, continuing or periodical data analyses and diffusion mechanisms. Unless such systems operate correctly, it will be difficult to refer to *institutional experience*, even in organizations where certain personnel have adequate personal experience. The introduction of a system of this type is one of the heaviest responsibilities of the directors of any important road building organization. Inevitably any decisions relating to quality control must be based on schemes in which cost-efficiency relationships play an important part.

It is a common mistake in quality control programs to make the tests and evaluate them *after* execution of the project that is to be controlled. The quality control specialist has no alternative but to accept the defective structure or to reject it. This always gives rise to problems of time and money. A better system is to divide control into two clearly differentiated steps [4]:

— Control and inspection of materials to ensure the satisfaction of design requirements. Ideally this work should be carried out by the constructor, obligated by a contract, but under the engineer's inspection.
— Acceptance of the materials and the partially completed work associated with them by the engineer who represents the contracting firm and the design engineer representing the owner.

Control, inspection and acceptance criteria will be established in accordance with the general specifications of the owner and designer, giving due consideration to the standards of the constructor. In many countries the differing roles of the owner, engineer and builder may change the emphasis given by each. A logical solution is for the contractor to be responsible for quality control and the owner or engineer responsible for assuring the control is properly carried out and the work accepted. However, in many countries the owner or engineer frequently assumes full responsibility for control. This is not always an ideal solution, for it leads to excessive conflict between two groups which should be collaborators (constructor and owner and engineer party). It relieves the constructor of a sense of responsibility for the technical quality and technical direction of his work, tending to make him a mere construction manager.

In support of this viewpoint, it is regarded as desirable for the constructor to have his own laboratories and control methods.

The contracting party does not usually have much legal influence over the manner in which the contractor carries out his work, the equipment he uses or his administration unless it is obvious his methods will cause permanent harm. Consequently, the control by the owner or engineer can lead to numerous contradictions. The acceptance or rejection of the final achievement can depend on a whole series of operations that are carried out by the constructor, over which the owner engineer has only limited influence. Thus it is easy to see why the owner or engineer sometimes demands certain procedures which the contractor is not in a position to achieve, given his system of work and the equipment he employs. The logical alternative to these situations would clearly be for the constructor to be made responsible for the quality of his work, the owner and engineer being responsible for verification and final acceptance.

In many regions, quality control is still based on what might be referred to as specified index figures. For example, the quality of a compaction process is evaluated in relation to a fixed index figure, typically the percentage of compaction. For example, the work is considered satisfactory if 95% compaction or more is achieved in relation to a given test. Control is carried out by obtaining samples by different procedures, all of which are standardized, as will be seen. On testing each of the samples, no percentage of compaction must be below 95%. This system of measuring the quality of what has been achieved does not take the human factor into account. Any activity that is performed by men is subject to highly complex laws of variation, which are either impossible to define or too complex to be quantitatively detailed. Variability is the result of the heterogeneity of the materials and methods employed, and other circumstantial or environmental factors.

As a consequence of the foregoing ideas, if a specific value is to be observed, such as the 95% compaction mentioned above, a systematic attempt must be made to achieve a considerably higher average value on the job. Only in this way will values of 95% or higher be obtained considering the inevitable variations imposed by man and nature. Therefore, in order to achieve 95% systematically and not run significant risk of rejection, a considerably higher index figure must be sought on the job. This can lead to unnecessary expenses, if 95% *average* is assumed to be the degree of compaction that is appropriate and previously selected by the designer. Also, the systematic attempt to achieve a higher index figure than that selected for design purposes, merely to meet an artificial requirement imposed by control, may lead to serious technical deficiencies, such as over-compaction in the case described here and possibly excessive swelling.

If the construction engineer decides to avoid further expense and strives to achieve exactly 95% compaction, he can be sure that about half of the soil samples that are analysed for him by the quality control specialist will exhibit degrees of compaction below the 95% specified. Such a control problem will be met by anyone who bases quality on strict minimum index figures that must be achieved, whereas average was meant.

Considerable emphasis is placed here on the foregoing ideas, which are well known to all those concerned with highway design and construction, for they are the reason why good quality control is based on statistical concepts.

Before bringing these brief comments on quality control philosophy to a close, attention is drawn to another two important aspects. The first is that a quality control program must be conceived at the design stage, so that both the project and the construction staffs will give it careful consideration and be fully aware of its needs. When this procedure is not followed, control is confronted by numerous problems as it comes into conflict with the time limits of the program. In the same way, requirements to implement control (budget personnel, transportation, equipment, and laboratories) must be clearly provided for in the administration of the project.

Lastly, it would seem appropriate for quality control to function independently of construction and to a lesser extent design authorities. Only thus will it be possible to achieve the freedom of action and independence of opinion that are required by the objective criticism that is an essential part of control activity. If quality control is subordinated to the construction authority, it will be difficult if not impossible for the director of a large construction firm to have an objective and unbiased knowledge of construction activity and its defects, and possible means of remedying them. If control is directly dependent on design, it will be hard to establish to what extent defects in construction can be attributed to deficiencies in design.

Some common facts which cannot be easily disregarded by any large building firm should also be considered here. Some conflict between the design staff and the construction staff is inevitable. The two groups have somewhat different goals, for while the design group is after quality and may lapse into perfectionism, the construction group is bent on the rapid completion of programs and may lapse into short cuts and excessive haste. The maintenance team tends to be antagonistic toward both groups, for it inherits the errors or deficiencies of both. These different points of view do not have to end in personal confiicts but unfortunately they sometimes do. They are merely a logical, inevitable and probably not entirely unfavorable consequence of the respective responsibilities of the different teams. The quality control team must relate to all these groups, without becoming attached to any one of them, so that it will be in a position to exercise an independent judgment, often acting as referee and giving guidance to those responsible for directing operations.

From the foregoing considerations, it can be seen how important it is for the control team to have ample experience in human relations as well as in technical affairs. The worst error the control team can make is to change its position of a go-between, informant and judge of the success of the other teams to that of a critic, or even worse, that of a group seeking preeminence over the others. At many project meetings the constructor handles his own opinions; similarly the quality control man must handle his, including a whole set of control laboratory data. These data must be correlated with both the design data and experience on previous projects. If this correlation is missing there will be errors and conflicts.

The safest rule of conduct for a control group that aspires to long-term success is a team spirit and a consciousness of serving the community.

To summarize all the foregoing considerations, we have shown that quality control must have the following characteristics:

1) It must be capable of distinguishing significant flaws and deficiencies, separating the essential characteristics of the project from the secondary ones. This will lead to a flexible and diversified control, adapted to suit the needs of the project.
2) It must be capable of differentiating between the flaws and deficiencies inherent in project construction, and those due to sampling or laboratory tests.
3) It must be capable of an opportune vigilance over the materials to be used, so as to ensure the adequate behavior of those that are selected for a specific purpose. Ideally this aspect of control should be handled by the contractor, but verified independently. Quality control must also be capable of establishing clear, reliable standards for the acceptance or rejection of different stages of the project which should be left to the owner or his engineer.
4) It must be based on standards that are not only in line with the legal and contractual aspects of the project, but are also rapid, so that control will not materially interfere with the normal rate of construction.
5) It must be based on competent and realistic specifications, adapted to the materials available, the environment and needs of the project and its technical environment.
6) It must be based on objective, rapid and simple sampling techniques and laboratory tests. At the same time, these must be easy to interpret and form part of a scientific plan which will eliminate as far as possible any decisions based on personal appreciations.
7) It must be considered in the design, so that its interferences and needs will be duly programmed and will cause no unexpected delays.

8) It must represent criteria that are totally independent of those of the design and construction engineers.
9) It must be entrusted to capable staff with a desire to serve.

17.2 Fundamental Concepts of Statistical Quality Control Methods

In this section an attempt will be made to give the fundamental theoretical concepts of the statistical control of construction quality for road engineering purposes.

As already mentioned, all data obtained from repeated observations or from laboratory or field tests on similar materials are subject to variations. Table 17-1 [5], which refers to the unconfined compressive strength of specimens of a rock, is used to illustrate the characteristic layout and presentation of such variations, which typically occur in the data obtained from repeated observations of any particular parameter.

The first method for obtaining a general value for such a set of data that is representative but unique, is to find the arithmetic mean by dividing the sum of all the strength values by the number of specimens tested. However, further consideration of the problem will show that the simple arithmetic mean is not sufficient, for it does not indicate the range of values in the numbers that generate the mean obtained, or the frequency with which each value occurs.

A data table like Table 17-1 is often presented in the form of a histogram, like that appearing in Fig. 17-1. The histogram is constructed by presenting as ordinates a number of data points which are located within equal intervals of variation. The range in each interval are the abscissas. In the figure, the values for the compressive strength of the rock were grouped within intervals of 20 kg/cm^2 (284 lb/in^2). There were 23 specimens with a strength between 251 and 270 kg/cm^2 (3560 and 3840 lb/in^2). The arithmetic mean of all the values in Table 17-1 is 247 kg/cm^2 (3500 lb/in^2).

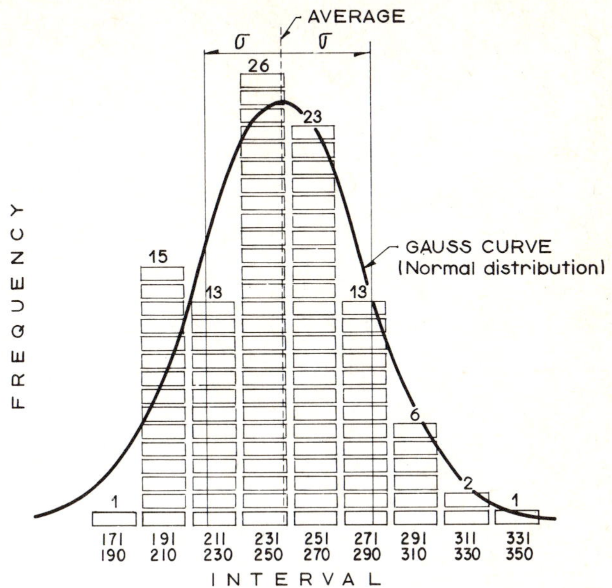

Fig. 17-1 Histograms of the data in Table 17-1 [5]

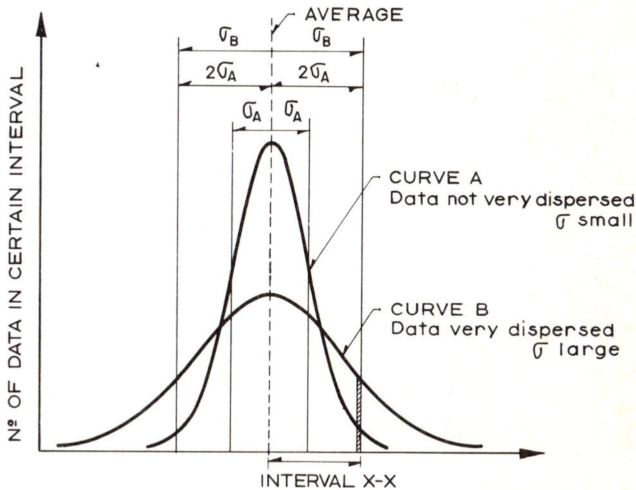

Fig. 17-2 Normal distribution curves [5]

Both experience and theory indicate that if the number of data handled is sufficiently large and the interval of variation selected sufficiently small and the factors causing variation occur at random, the histogram will approach a continuous data distribution curve. Many of the distributions that are of interest from the engineering viewpoint, with particular reference to quality control problems, are of the type known as normal or Gaussian distribution, shown superimposed on the histogram in Fig. 17-1. For the purposes of the following considerations, it will be assumed that all the data distributions that are handled are normal or Gaussian.

Figure 17-2 [5] shows two normal distribution curves, one high and narrow and the other low and wide. If both refer to the same number of data, the areas under them will be equal. Clearly the data on the high curve are closer to the mean, whereas on the lower curve they are scattered more widely about the mean.

If these curves have been obtained by measuring a certain magnitude of the same property of the same material by means of laboratory tests, then it can be said that method *A* (the high curve) depicts more consistent results than method *B* (the low curve).

A fundamental need in practical applications is to be able to measure the dispersion of data about the mean. A rough idea of this is given by the difference between the highest value and the lowest, but this does not measure distribution directly, which is fundamental. The standard deviation of the normal distribution curve, σ, is defined thus:

$$\sigma = \sqrt{\frac{\Sigma (x - \bar{x})^2}{n}} \qquad (17\text{-}1)$$

where x represents any value given and $\bar{x}$ is the mean of all the values; $x - \bar{x}$ will therefore be the deviation of any given value in relation to the mean. In this expression the square of the deviations is considered so as to eliminate the influence of the sign, which in some cases may be plus and in others minus. By dividing the sum of all the deviations by their number, what could be regarded as an average dispersion is obtained. The value σ^2 is termed the distribution variance and the units of the standard deviation are the same as those of the original data.

In the case of the data in Table 17-1, the standard deviation is σ = 32.7 kg/cm^2 (465 lb/in^2).

Table 17-1
Unconfined compressive strength values obtained for 100 samples of a rock [5]

Specimen	Strength		Specimen	Strength	
	kg/cm^2	lb/in^2		kg/cm^2	lb/in^2
1	247	3500	51	236	3350
2	249	3540	52	236	3350
3	241	3420	53	211	3000
4	197	2800	54	261	3700
5	252	3580	55	243	3450
6	252	3580	56	243	3450
7	241	3420	57	249	3540
8	197	2800	58	251	3560
9	304	4310	59	261	3700
10	276	3920	60	247	3500
11	249	3540	61	233	3310
12	322	4570	62	249	3540
13	348	4950	63	249	3540
14	241	3420	64	267	3790
15	249	3540	65	211	3000
16	194	2760	66	238	3390
17	236	3350	67	253	3600
18	233	3310	68	241	3420
19	208	2950	69	246	3500
20	231	3280	70	246	3500
21	261	3700	71	253	3600
22	304	4310	72	211	3000
23	288	4090	73	217	3080
24	308	4370	74	213	3020
25	281	3990	75	224	3180
26	265	3760	76	204	2900
27	279	3960	77	208	2950
28	314	4460	78	203	2880
29	308	4370	79	208	2950
30	293	4160	80	198	2810
31	283	4100	81	277	3940
32	239	3400	82	253	3600
33	246	3500	83	253	3600
34	288	4090	84	251	3560
35	300	4260	85	224	3180
36	286	4060	86	268	3800
37	281	3990	87	271	3850
38	288	4090	88	216	3060
39	277	3930	89	216	3060
40	268	3800	90	251	3560
41	267	3790	91	203	2880
42	257	3650	92	229	3250
43	267	3790	93	217	3080
44	227	3220	94	227	3220
45	236	3350	95	193	2740
46	257	3650	96	204	2900
47	273	3880	97	193	2740
48	268	3800	98	204	2900
49	257	3650	99	187	2660
50	270	3840	100	193	2740

An important characteristic of the normal distribution curve is that, regardless of its form, if the value of the standard deviation is plotted on both sides of the mean, a partial area is obtained between them which represents a fixed percentage of the data of the sample under observation (68.2%) slightly more than two thirds of all data. Similarly, if the value 2σ is plotted on both sides of the mean, a partial area is obtained which represents 95% of the population of the sample under consideration. A value of 99.7% of the sample data thus being obtained if 3σ is plotted on both sides of the mean, Fig. 17-3 [6] illustrates this.

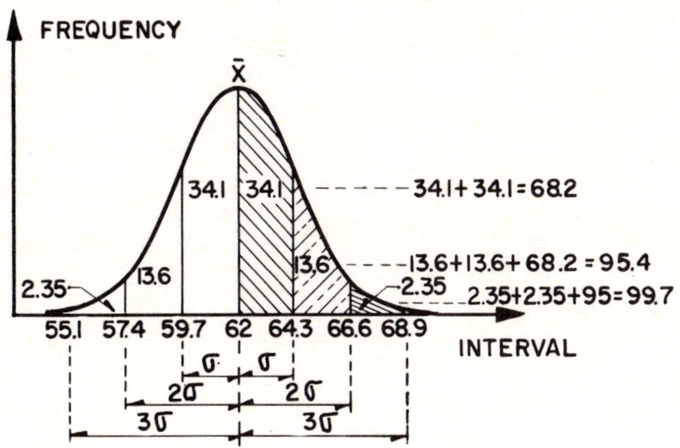

Fig. 17-3 Percentages of the area under the normal distribution curve, corresponding to different multiples of σ [6]

Referring back to Fig. 17-2, it is seen that the smaller the standard deviation the smaller the dispersion of data. For example, for a curve like *A*, a certain interval $x - \bar{x}$ may be included in the central portion extending $\pm 2\ \sigma$, about the mean $(\bar{x})$. This means that the deviation of 95% of the values about the mean is less than $x - \bar{x}$. In curve *B* this same interval may occur within the central portion of extension only $\pm\ \sigma$, which indicates that in distribution *B* only 68.2% of the data vary about the mean by less than $x - \bar{x}$. Thus, the smaller the standard deviation the smaller the dispersion of data.

In view of the characteristic just described, the standard deviation is clearly a measure of the dispersion of data about the mean. The larger the standard deviation (σ) the greater the interval containing the same percentage of data. For instance, in Fig. 17-2 the standard deviation of curve *A* is far smaller than that of *B*, so that to give a concrete example if the curves referred to two sets of results obtained from the same test of the same material conducted in two different laboratories, it could be said that laboratory *A* is far more consistent than laboratory *B*.

In practical applications, the standard deviation is often compared with the mean value of all the data, for with reference to Table 17-1 and Fig. 17-1, a standard deviation of 20 kg/cm^2 (284 lb/in^2) compared with an average rock strength of 150 kg/cm^2 (2130 lb/in^2) is not the same as compared with the average strength of another of 400 kg/cm^2 (5680 lb/in^2). This sequence of ideas leads to definition of the coefficient of variation:

$$V = \frac{\sigma}{\bar{x}} \tag{17-2}$$

where the symbols have the meaning seen previously. The coefficient of variation is non-dimensional and is usually expressed as a percentage.

Lastly, also frequently referred to is the variance, σ^2, of the data distribution. This concept has the advantage of the constancy of its sign, which always permits an arithmetical sum, whereas the standard deviation may occur on either side of the mean and must be handled algebraically.

When comparing data distributions with specification limits for these data, which is a very frequent situation, three different cases may arise, Fig. 17-4, [6]:

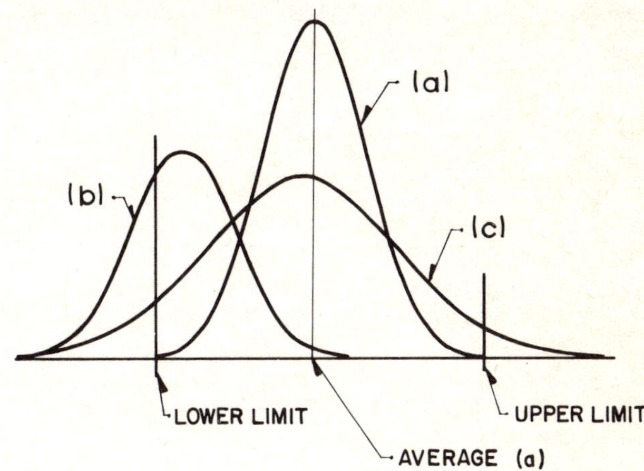

Fig. 17-4 Some interesting positions of a data distribution curve in relation to specification limits [6]

a) A small dispersion occurs (curve *a*), with most of the data within the specified limits. This indicates that the specifications adopted are realistic and data are being obtained with well controlled processes. However, the fact that all the data are within the limits might indicate that the sampling systems that are employed suffer from some consistent bias and do not provide all the types of samples.

b) A relatively small dispersion occurs with the mean very close to the lower specification limit (curve *b*). This may indicate either that construction procedure is inadequate, requiring improvement, or that the specifications are not realistic.

c) A large dispersion is obtained which makes it improbable that most of the data will fall within the specified limits most of the time. This situation indicates that the control of the quality of construction should be improved so as to reduce the dispersion obtained (or possible that the tolerance limits are not realistic and must be widened).

A fourth type of curve sometimes arises: one with two or more peaks. This signifies that these are non-random variables so that the curve is the superposition of two or more curves, each representing a true random variation. A common example is the percent compaction distribution which includes two different soils but only one reference was used in testing. A second example is when two different layer thicknesses were used for the same soil and same equipment.

The above possibilities must always be kept in mind when controlling the quality of materials or construction procedures (such as compaction, for example). The mental plan that is obtained on analysing them can be used to establish criteria with regard to two aspects:

— To establish the reliability of a given material, process, or test method, in relation to the requirements established by the specifications.
— To compare the requirements established by the specifications with the variability of other similar operations.

This method of analysis enables appropriate relationships to be visualized between the tolerance limits for actual operations and the limits established by the specifications, and provides reasonable methods for recognizing areas that may require more detailed study in order to determine whether it is necessary to improve the control, the methods of administration, or to change the specifications in use.

To ensure a sound statistical quality control program, the error inherent in the actual program must first be found and evaluated. For a certain confidence level, this error is given by the following expression [7,8]:

$$E_m = \frac{t\sigma}{\sqrt{n}} \tag{17-3}$$

where E_m is the error inherent in the control program, σ and n have the meanings already discussed, and t is a factor which expresses the confidence level at which it is desired to work with the data distribution available. With reference to Fig. 17-3, t would be *1* for a confidence level of 68.2%, *2* for a confidence level of 95.5%, or *3* for a confidence level of 99.7%. Naturally in-between values could be used, which appear in Table 17-11. In practice, the value of t is established according to the judgment of whoever is to use the control program.

It can be observed that E_m diminishes when the number of available data (n) increases. In reality E_m is the error that must inevitably be expected in evaluating a problem. It does not depend on the human element, but is a random variable.

The value of σ corresponding to a whole control program is made up of the values of σ corresponding to each individual operation. There will be operations producing data such as the materials, sampling, laboratory tests, and compilations. The total value of σ can be obtained from the expression:

$$\sigma = \sqrt{\sigma_1^2 + \sigma_2^2 + \sigma_3^2} \tag{17-4}$$

If x is the variable representing the data that are obtained and $\bar{x}$ is the mean value of these data, in many applications of statistics to quality control, definition of a new term will be necessary [9]:

$$z = \frac{x - \bar{x}}{\sigma} \tag{17-5}$$

From the above equation it is deduced that:

$$x = \bar{x} + \sigma z \tag{17-6}$$

which shows that this z is simply a new, transformed variable that is defined by Eq.(17-6). However, Eq.(17-5) leads to some useful rules. Let it be assumed that the mean compressive strength for a set of rock samples is 240 kg/cm^2 (3410 lb/in^2), and that the data are distributed in such a way that there is a standard deviation of 24 kg/cm^2 (341 lb/in^2). It is also assumed that design requires a minimum strength of 210 kg/cm^2 (3000 lb/in^2). The percentage of samples that can be expected to have a strength of 210 kg/cm^2 (3000 lb/in^2) or less must now be found. Applying Eq.(17-5) it can be seen that:

$$z = \frac{x - \bar{x}}{\sigma} = \frac{210 - 240}{24} = -1.2$$

$$\left(\frac{3000 - 3410}{341} = -1.2\right)$$

It is seen that 210 kg/cm^2 (3000 lb/in^2) corresponds to a deviation of 1.2σ to the left of the mean. In the tables of areas under normal distribution curves corresponding to different abscissas as a function of σ, which can be found in the usual statistical texts, it can be seen that for z = 1.2, 12% of the data remain outside the interval $x - \bar{x}$. Referring specifically to the example handled thus far, 12% of the specimens could be expected to have a strength below 210 kg/cm^2 (3000 lb/in^2).

These tables of areas under the normal distribution curve, Table 17-11, unlike the curves depicted in Fig. 17-3, generally appear as a function of z and not of x in statistical texts.

If 12% of samples with strength values equal to or lower than 210 kg/cm^2 (3000 lb/in^2) is considered dangerous, there are two alternative courses of action. Either the strength can be increased, or else the standard deviation of the data reduced. This second course of action is of little value in this particular example, but might be useful if the source of the data were a production process, like mixing concrete or soil compaction, in which measures could indeed be taken to reduce the variability of the results achieved. Full details of the foregoing criteria, and comparisons between the predicted results and those actually obtained can be found in [10], in a study relating to specimens of concrete. In all cases the low strength percentages computed with Eq.(17-5) and those actually obtained in the laboratory were very similar.

For a data population with a mean value, $\bar{x}'$, and a standard deviation, σ', if random samples of any size n are taken, the sample means ($\bar{x}$), the standard deviation of which will be σ, form a frequency distribution which does not coincide with that of the population. The $\bar{x}$ distribution will have a mean value ($\bar{\bar{x}}$) and a standard deviation ($\sigma\bar{x}$) of its own. The value of $\bar{\bar{x}}$ tends to be close to the population mean, $\bar{x}'$. The dispersion of the frequency distribution for the values of $\bar{x}$ depends not only on dispersion of the population, but also on sample size, n, so that the higher the value of n, the smaller the dispersion of the values of $\bar{x}$.

In the long term, as n, which is the number of items in each sample, increases, the mean value of $\bar{x}$ will tend to be the same as $\bar{x}'$ and the standard deviation of the values of $\bar{x}$ ($\sigma\bar{x}$) will be $\sigma'/\sqrt{n}$, σ' being the standard deviation of the population. For example, if n = 4, the standard deviation of the frequency distribution of the values of $\bar{x}$ tends to be half of the standard deviation of the population, but if n = 16, the standard deviation of the values of $\bar{x}$ will be only one quarter of that of the population.

Often $\sigma\bar{x}$ is referred to as the standard error of the mean of the $\bar{x}$ values.

Regardless of whether the distribution curve for the population is normal or not, it is true that the $\sigma\bar{x}$ expected is $\sigma'/\sqrt{n}$ and the $\bar{x}$ expected is $\bar{x}'$. If the universe is normal, statistics show that the frequency distribution expected for the values of $\bar{x}$ is also normal. But even on the basis of rectangular or triangular universes, the distribution of the values of $\bar{x}$ of the samples is also approximately normal.

The foregoing statements are only valid if the value of n is high ($n > 30$). If n is small, they are not entirely correct, but can be regarded as such because the error is not of great practical significance.

17.3 Sampling for Statistical Control Programs

A valid sampling operation is an essential requirement for drawing up a reasonable quality control program. It must take into consideration three essential factors [2]. First, it must cover only the requirements of the control program and no others. A sampling operation that goes any further will often be far more expensive than necessary. Second, sampling must take into consideration the homogeneity of the item being sampled. Those materials or operations with a natural tendency towards variability or dispersion must be sampled more extensively than homogeneous ones. For example more samples will have to be obtained of a natural subgrade material than of a plant-crushed base course material. Third, sampling must take into consideration the importance of the item to be sampled in relation to the project as a whole, and the technical and economic repercussions of its acceptance or rejection.

In a construction program, sampling operations are usually conducted in two stages. First, the material must be divided into a certain number of similarly sized lots, each representative of the material as a whole. In highway engineering practice, this initial division is often done considering similar sections of road or similar borrow areas. Next, each of these lots is sampled, generally in a laboratory, to obtain the specimens that will be subjected to analysis. The size of the lots depends largely on the value of the materials, and their constitution depends on the property it is wished to measure. For example, when sampling low-cost earth materials for construction purposes, the different stock piles that are accumulated can be regarded as the initial samples, sometimes consisting of thousands of cubic meters each. In more expensive materials, such as cement-stabilized soils, the initial sample is usually far smaller. The size of the initial samples may also be larger when the sampled material is homogeneous. The important thing is that each of the selected samples must be representative of the construction material as a whole. In compaction jobs, the initial sample would be a representative section of road. Reference [11] recommends the selection of no less than 50 such sections for the control of a given project.

Determining the number of initial samples to be taken for test purposes also depends on the homogeneity of the item tested, the cost of sampling and the representativity that can be attributed to each sample. In the application of statistical control methods it is a good practice to select samples to reflect a range of results that will differ from the mean by $\pm 3\sigma$, σ being the standard deviation of the distribution of these results. This implies working with a confidence level of 99.7% (see Fig. 17-3). Clearly the greater the number of samples, the larger the range of results obtained, but the mean will provide a better description of the true situation of the population being sampled.

The sampling methods currently in use are not always reasonable, and it is not hard to see how fundamental errors are made, as can be illustrated by a simple example. Imagine a certain product which leaves a plant in trucks, each of which carries 100 units. Assume also that 10 of these units are defective. A sampling criterion that is often used today would be the following. An inspector stops each truck and takes a random sample from it (one unit). If the inspection is favorable the truck moves on, but if the sample is defective the truck is rejected. Logically nine trucks will pass the inspection and the tenth will fail, although all the trucks are clearly in the same condition. A sampling criterion like this does not satisfy the basic requirement of accepting what should be accepted and rejecting only what should be rejected.

Although the foregoing example is elementary and a little far-fetched, the concepts it involves are valid. This is equivalent to sampling a truck load of aggregate or concrete or asphalt or a large volume of borrow material by taking a small amount of the material in a vessel and analysing it; similarly the quality of n cubic meters of material is judged according to the quality of just one cubic meter.

A sampling criterion that is often employed, and which may be reasonable if correctly used, is the so-called acceptance sampling [12]. According to this criterion, an item is termed defective when it does not meet certain pre-established specifications relating to one or more quality characteristics. A plan is drawn up as a function of three numbers. N is the number of items existing in the original lot from which the test sample will be taken, n is the number of items taken from the lot, which will represent the test sample, and c is the so-called acceptance number of the sample, which is the maximum number of defective items permitted (the number below a certain specification limit). Thus, sampling is automatically associated with the rejection criterion, for more than c defective items in the sample will result in the rejection of the original lot or sample. Most current quality control processes employ sampling procedures based on this criterion. When discrete items are sampled, a common practice is to establish a lot made up of 50 to 100, select 10% of the items at random (5 to 10) and evaluate this sample to establishing a number c. Often this number is zero, expressing the illusion that if the sample under study is perfect, the lot will be perfect, and the total material (population) to which both refer will also be perfect. In compaction jobs, application of this criterion is not so clear, but is nevertheless used. Usually a specific sampling section of a limited length is established every few kilometers or meters, and it is agreed that the results for this section will represent the lot or the entire section (entire population in statistical terms). The degree of compaction is measured; c is frequently (but not necessarily) zero. If c is defined as zero, no degree of compaction below this value is accepted.

With such a criterion a relationship will always be found between the percentage of the lots analyzed that will be accepted (which is usually referred to as the probability of acceptance, P), and the percentage of defective items in each lot (p). It is assumed that lots are made up of N items, of which only n are taken to be studied, with an acceptance number c. It is also assumed that the lots consist of 50 items, of which 5 are analyzed with an acceptance number $c = 0$. If 4% (on average) of the items in each lot are assumed to be defective, in each lot of 50 there are on average two unsatisfactory items. What expresses the relationship between the probability of acceptance (percentage of lots approved) and the percentage of defective items in each lot is the fact that when sampling 5 of the 50 items, one of the four unsatisfactory ones will not necessarily be obtained, so that there is a certain probability that the lot will pass as perfect. The foregoing relationship is known as the operating characteristic curve of the process. This shows for each defective portion of the lot (p) the probability of acceptance of the lot (P) if the sampling plan is followed, Fig. 17-5 shows the operating characteristic curve corresponding to $N = 50$, $n = 5$, $c = 0$.

It is easy to see how the different points that form the curve are obtained. Assume, for example, that $p = 4\%$ is the average percentage of defective items in each lot of 50 items ($N = 50$).

Assume also that from each lot 5 items are taken for testing ($n = 5$), and lastly let it be assumed that the population being studied is made up of 1000 lots of 50 items, that is 50000 items. The study will be conducted using the criterion $c = 0$. In other words, it is sufficient for one item of the sample of 5 to be defective for the corresponding lot to be rejected.

If in the lot of 50 there are 2 unsatisfactory items (4%), there will be 48 satisfactory ones and the probability of selecting a satisfactory item to make up the sample for study purposes will be 48/50. This operation must be repeated 5 times for the lot to be accepted. In this way the probability of acceptance will be $(48/50)^5$, or 80%, for 1000 lots, which corresponds to approximately 4% on the abscissas in Fig. 17-5.

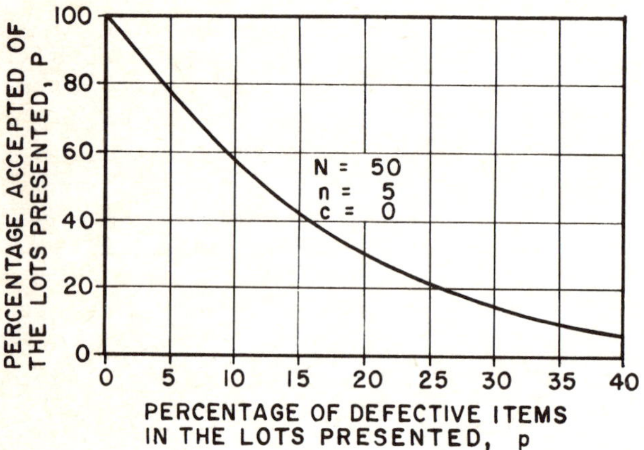

Fig. 17-5 Operating characteristic curve [12]

A sampling plan like that just described is seldom questioned in geotechnical quality control methodology, but it should be. This criterion gives the illusion that a perfect sample represents a perfect lot and a perfect mass (population) which is clearly not true since some defective items are mixed at random with the satisfactory ones in the lots. It has been assumed that the average percentage of unsatisfactory items per lot is 4, but this means that one lot may have 0% unsatisfactory items and another 6%.

This criterion also implies a further hypothesis: that the protection provided by a sampling system is constant if the ratio of sample size to lot size is also constant.

Figure 17-6 [12] illustrates how inexact this hypothesis is. It compares four operating curves corresponding to lots of 50, 100, 200 and 1000 items into which a population of 50000 items has been devided. In all cases the sample for study purposes is 10% of the lot (n = 5, 10, 20 and 100 items, selected at random).

The differences in the quality assurance provided by this sampling plan are impressive. It can be seen that the lots containing 4% defective items will be accepted 80% of the times when a sample of 10% of a lot of 50 is employed; 65% of the times when a lot of 100 items is employed, and less than 2% of the times when the sample is 10% of 1000 items in each lot. It is hard to fully rely on a sampling criterion that leads to such variations simply because of size, especially if it is considered that the realities of geotechnical engineering processes constantly impose drastic changes in sample size owing to problems of availability and cost.

Another means of interpreting the curves in Fig. 17-6 is illustrated in the following question. What will be the quality of the

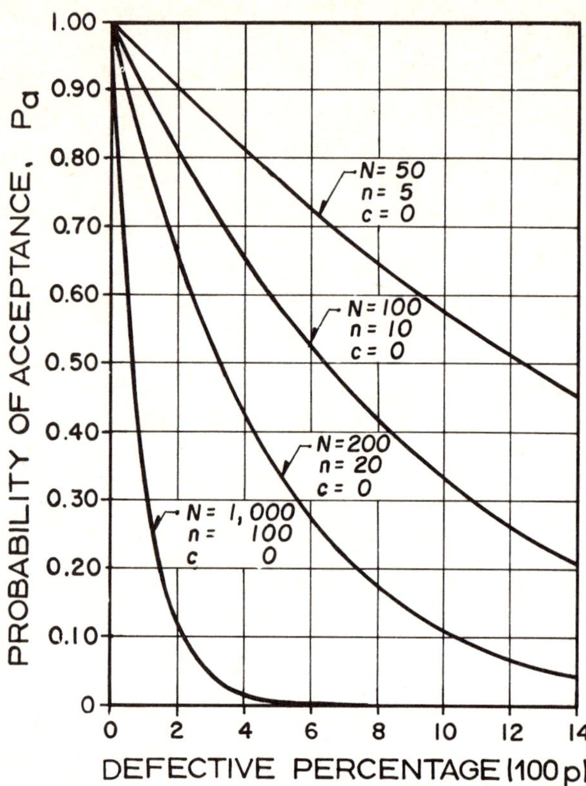

Fig. 17-6 Comparison between operating characteristic curves for four versions of a sampling plan with samples of 10% of the lot [12]

lot that will pass 50% of the times in each version of the sampling plan? In the figure it is seen that in lots consisting of 50 items, one with 12% of defective items will be accepted 50% of the time, but in lots of 100 items, then one with 6% of defective items will be accepted half of the times. This percentage is 3 for a lot consisting of 200 items and 0.65 for lots of 1000. Once again the lack of consistency of the sampling plan is clear.

The absolute size of the random sample taken is far more important than its size in relation to the size of the lot, as can be seen in Fig. 17-6. It is emphasized by Fig. 17-7 [12]. This figure presents four versions of the sampling plan now under consideration (N = 50, 100, 200 and 1000), but taking a sample of n = 20 items in every case.

First, attention is drawn to the four curves that are now very close to one another; the enormous spread previously seen has disappeared. The foregoing leads to a practical criterion for judging or planning a sampling procedure. The simple approach just discussed can be applied in those cases where the number of items in the sample to be evaluated is constant in all cases. It leads to an unprotected control operation when the criterion of a fixed fraction of the items of the sampled lots is adopted. In highway practice there are certain situations where it is easy to work with a fixed number of samples, as might be the case of crushing plants, asphalt plants or even many compaction jobs. However there are other situations where, owing to the availability of items or cost, samples with different populations have to be employed. It follows, therefore, that in the first case simple acceptance sampling may lead to reasonable control conditions, but in the second case the sampling operation has to be planned along different lines.

The acceptance number (c) does not have to be zero. From Figs. 17-6 and 17-7 it is clear that a perfect sample does not

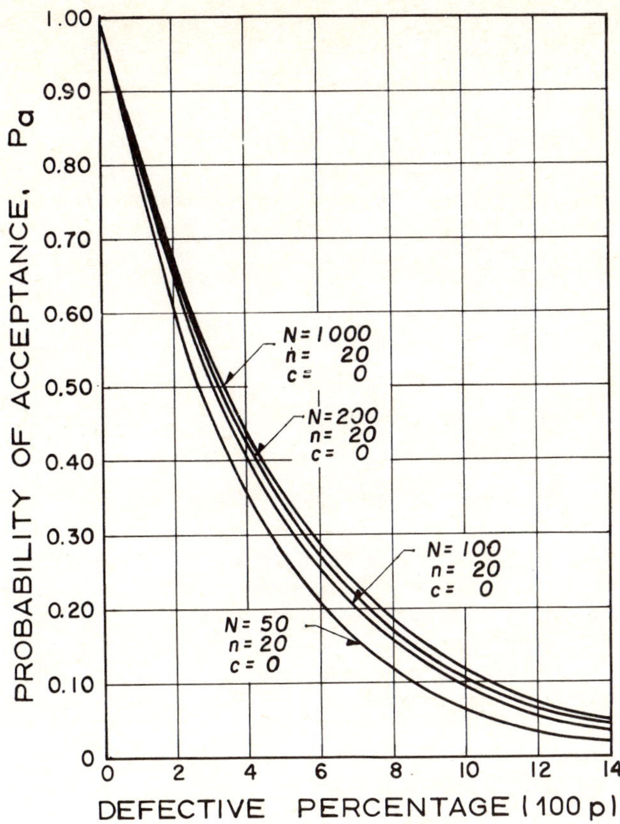

Fig. 17-7 Comparison between operating characteristic curves for four versions of sampling plan with samples of twenty items [12]

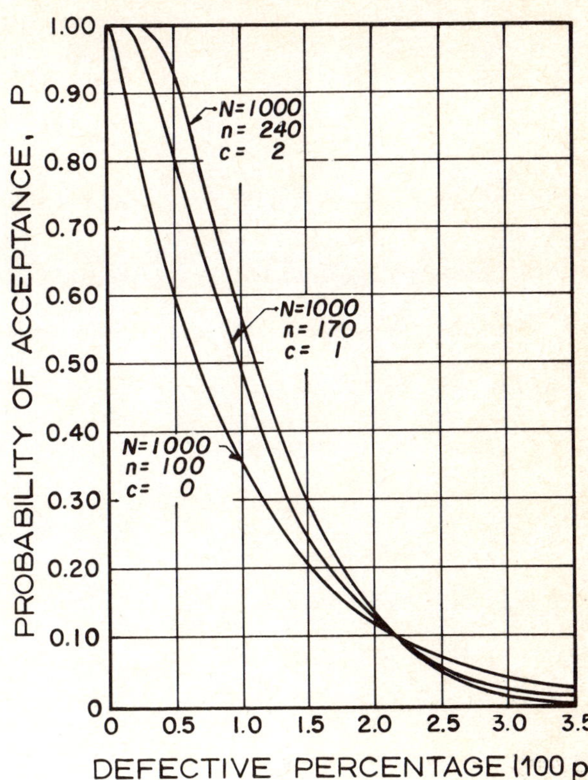

Fig. 17-8 Operating characteristic curves for three versions of acceptance sampling with P = 10% for a lot containing 22% defective items [12]

ensure a perfect lot. This conclusion shows the weakness of the common objections against permitting defective items in a lot. The protection desired against the acceptance of lots with defective items must consider larger sample sizes, since these provide better possibilities of discriminating between satisfactory and unsatisfactory lots.

Figure 17-8 [12] shows three versions of sampling. The first is for N = 1000, n = 100 and c = 0; the second for N = 1000, n = 170 and c = 1, and the third for N = 1000, n = 240 and c = 2. It is immediately noted that all the curves give equal protection against acceptance of a lot with 2.2% defective items. Versions c = 1 and c = 2 give somewhat better protection against rejection of satisfactory lots. Some further considerations on traditional sampling methods can be found in [12,13].

The above sampling plans are known as simple sampling, since the criterion for the acceptance or rejection of a representative lot of a given population is based on the analysis of a sample from that lot. Another approach used in industrial quality control problems, is so-called double sampling. This has the possibility of postponing the decision of acceptance or rejection of the lot until a second sample has been analysed. Usually in double sampling a highly satisfactory initial sample leads to immediate acceptance of the lot and a highly unsatisfactory initial sample to rejection. If neither of these two extremes occurs, the decision is based on the evaluation of both the first and second samples. An example of a double sampling plan might include the following numbers: N = 1000, n = 36, c_1 = 0, n_2 = 59 and c_2 = 3. It can be interpreted as follows.

— Inspect a first sample with 36 items from a lot with a population of 1000.
— Accept the lot if the first sample contains zero defective items.
— Reject the lot if the first sample contains more than 3 defective items.
— Inspect a second sample with 59 items if the first sample contains 1, 2 or 3 defective items.
— Accept the lot if the joint sample with 95 items (36 + 59) contains 3 defective items or less.
— Reject the lot if the joint sample contains more than 3 defective items.

Figure 17-9 [12] shows 3 operating characteristic curves relating to evaluation of the foregoing sampling plan. The result of the sampling process will be one of the following:

— Accept the lot after inspection of the first sample, if no defective item is found.
— Reject the lot if the first sample obtained contains more than 3 defective items.
— Accept the lot after inspection of the second sample if 3 defective items or less are obtained out of the joint total of 95, for the two samples.
— Reject the lot if more than 3 defective items are found in the joint sample.

Curve A (N = 1000, n = 36, c = 0) in Fig. 17-9 corresponds to the probabilities of accepting the lot after inspection of the first sample, for different percentages of defective items. Curve C (N = 1000, n = 36, c = 3) represents the probability that the lot will not be rejected after inspection of the first sample, in which case a second sample must be obtained. These two curves can be plotted in the manner discussed for the case of simple sampling. For any given percentage of defective items, the difference in the ordinates between curves A and C corresponds to the probability that a second sample will be required.

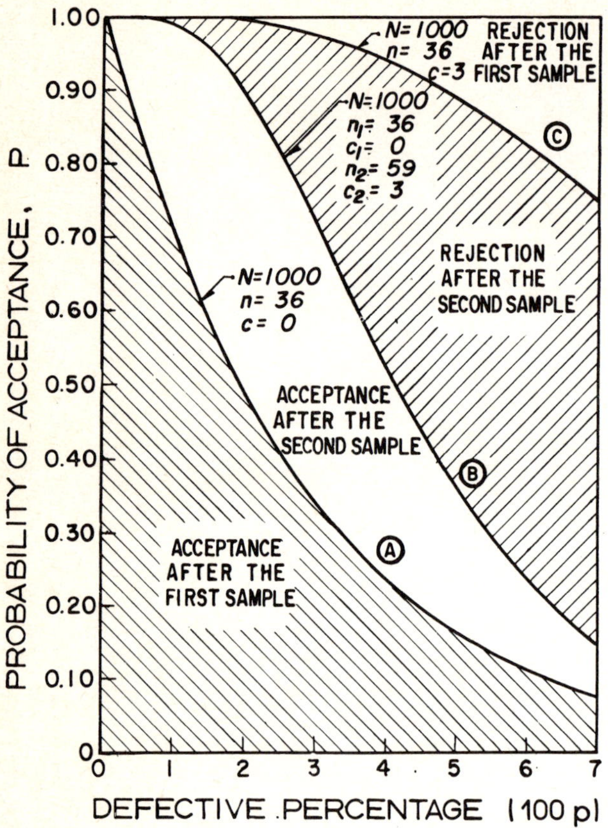

Fig. 17-9 Operating characteristic curves in a double sampling plan [12]

Curve *B* shows the behavior of the double sampling plan. To determine the ordinates of the points of this curve (the abscissas are determined by the percentage of defective items) it is necessary to calculate the probability of the lot being accepted after obtaining a second sample, which in the case that is described may occur in any of the following manners:

— Zero defective items in the first sample
— One defective item in the first sample and zero, one or two defective items in the second sample.
— Two defective items in the first sample and zero or one defective item in the second sample.
— Three defective items in the first sample and none in the second sample.

The probability of acceptance of the lot proves to be the sum of the probabilities of each of the four above-mentioned events occurring separately. This calculation belongs to the domain of probability analysis and is beyond the scope of this chapter, which must remain conceptual. It is recalled that quality control is a special branch of road technology, which has its own highly complex methodology, for which the references should be consulted.

Curve *B* in Fig. 17-9 is the result of the above-mentioned calculation and divides the space between curves *A* and *C* as indicated in the figure. References [7, 8, 12, 14] give details of the calculations required for obtaining curve *B*.

There are also multiple sampling plans, in which the decision of acceptance or rejection is based on the analysis of more than two samples [12].

There is, however, one fact that does not seem to be sufficiently acknowledged by traditional sampling methods, even those that go beyond simply taking a fixed percentage of the items of the lot and make use of rational improvements such as those discussed earlier. This is the variability of the results of any tests to which the samples analysed may be subjected. This variability is due both to problems deriving from the test procedure and others deriving from the material or the sampling processes. All these factors may be random variables, and the values resulting from any set of tests similarly vary randomly. The entire sampling and testing process must, therefore, be treated as a statistical process. Samples must be obtained truly at random, observing the rules that have been developed by mathematical statistics for this purpose. Tests conducted on the basis of what an inspector may consider satisfactory or unsatisfactory samples or ones indicative of the average situation, cannot be regarded as random sample tests. Only samples obtained following statistical rules can give a true indication of the quality of the materials or processes that are studied.

In quality control processes connected with highway practice, the result of a certain activity or the quality of a material must often be compared with previously specified limits on the basis of just a few test results. When statistical control methods are not employed, a requirement of an absolutist nature is established, demanding that all the values found must satisfy the specified limits. Some important disadvantages of this attitude have already been pointed out. The alternative is to establish acceptance criteria that will acknowledge that the values obtained in tests carried out on random samples will vary. If an absolute requirement is to hold any significance, it demands a large number of control tests. For example, if it is established that 95% of all the tested values must meet a certain specification and if 20 samples of 100 times are taken, which is a very high number, it will be sufficient for just one of the 20 tests to fail for the entire lot to be rejected. Statistical sampling, however, provides reasonable acceptance criteria in cases like this with no more than 4 or 5 tests.

References [15-17] describe some of the earliest attempts made at applying statistical sampling techniques to road construction projects. In [18] some sampling plans for the control of asphalt concrete hot mix are discussed, [19] analyses many of the specifications normally adopted for the control of plant manufactured asphalt concrete, and indicates how ambiguous and unpractical many of them are.

A *statistical* sampling plan must have the following characteristics [20]:

— It includes an objective sample selection procedure, based on the use of a table of probability statistics.

— It includes a clear procedure for the quantitative measurement of the characteristics of the sample and the standard error of this measurement. If the results of tests of the sample are used as a basis for decision, the rules that govern this decision are also included. An acceptance or rejection sampling plan indicates very clearly the levels on which these activities will take place.

A random number table is a strictly random selection of numbers with a pre-established total of numbers. Table 17-2 is an example with two-figure numbers. The numbers can be selected by placing the nine digits and the zero in a vessel and drawing them at random one by one, returning the number drawn immediately and entering each of the pairs of numbers in the table.

Once the table has been completed, it may function as though it consists of a greater number of figures. For example, Table

Table 17-2

A table of random numbers [12]

10	09	73	25	33	76	52	01	35	86	34	67	35	48	76	80	95	90	91	17	39	29	27	49	45
37	54	20	48	05	64	89	47	42	96	24	80	52	40	37	20	63	61	04	02	00	82	29	16	65
08	42	26	89	53	19	64	50	93	03	23	20	90	25	60	15	95	33	47	64	35	08	03	36	06
99	01	90	25	29	09	37	67	07	15	38	31	13	11	65	88	67	67	43	97	04	43	62	76	59
12	80	79	99	70	80	15	73	61	47	04	03	23	66	53	98	95	11	68	77	12	17	17	68	33
66	06	57	47	17	34	07	27	68	50	36	69	73	61	70	65	81	33	98	85	11	19	92	91	70
31	06	01	08	05	45	57	18	24	06	35	30	34	26	14	86	79	90	74	39	23	40	30	97	32
85	26	97	76	02	02	05	16	56	92	68	66	57	48	18	73	05	38	52	47	18	62	38	85	79
63	57	33	21	35	05	32	54	70	48	90	55	35	75	48	28	46	82	87	09	83	49	12	56	24
73	79	64	57	53	03	52	96	47	78	35	80	83	42	82	60	93	52	03	44	35	27	38	84	35
98	52	01	77	67	14	90	56	86	07	22	10	94	05	58	60	97	09	34	33	50	50	07	39	98
11	80	50	54	31	39	80	82	77	32	50	72	56	82	48	29	40	52	42	01	52	77	56	78	51
83	45	29	96	34	06	28	89	80	83	13	74	67	00	78	18	47	54	06	10	68	71	17	78	17
88	68	54	02	00	86	50	75	84	01	36	76	66	79	51	90	36	47	64	93	29	60	91	10	62
99	59	46	73	48	87	51	76	49	69	91	82	60	89	28	93	78	56	13	68	23	47	83	41	13
65	48	11	76	74	17	46	85	09	50	58	04	77	69	74	73	03	95	71	86	40	21	81	65	44
80	12	43	56	35	17	72	70	80	15	45	31	82	23	74	21	11	57	82	53	14	38	55	37	63
74	35	09	98	17	77	40	27	72	14	43	23	60	02	10	45	52	16	42	37	96	28	60	26	55
69	91	62	68	03	66	25	22	91	48	36	93	68	72	03	76	62	11	39	90	94	40	05	64	18
09	89	32	05	05	14	22	56	85	14	46	42	75	67	88	96	29	77	88	22	54	38	21	45	98
91	49	91	45	23	68	47	92	76	86	46	16	28	35	54	94	75	08	99	23	37	08	92	00	48
80	33	69	45	98	26	94	03	68	58	70	29	73	41	35	53	14	03	33	40	42	05	08	23	41
44	10	48	19	49	85	15	74	79	54	32	97	92	65	75	57	60	04	08	81	22	22	20	64	13
12	55	07	37	42	11	10	00	20	40	12	86	07	46	97	96	64	48	94	39	28	70	72	58	15
63	60	64	93	29	16	50	53	44	84	40	21	95	25	63	43	65	17	70	82	07	20	73	17	90
61	19	69	04	46	26	45	74	77	74	51	92	43	37	29	65	39	45	95	93	42	58	26	05	27
15	47	44	52	66	95	27	07	99	53	59	36	78	38	48	82	39	61	01	18	33	21	15	94	66
94	55	72	85	73	67	89	75	43	87	54	62	24	44	31	91	19	04	25	92	92	92	74	59	73
42	48	11	62	13	97	34	40	87	21	16	86	84	87	67	03	07	11	20	59	25	70	14	66	70
23	52	37	83	17	73	20	88	98	37	68	93	59	14	16	26	25	22	96	63	05	52	28	25	62
04	49	35	24	94	75	24	63	38	24	45	86	25	10	25	61	96	27	93	35	65	33	71	24	72
00	54	99	76	54	64	05	18	81	59	96	11	96	38	96	54	69	28	23	91	23	28	72	95	29
35	96	31	53	07	26	89	80	93	54	33	35	13	54	62	77	97	45	00	24	90	10	33	93	33
59	80	80	83	91	45	42	72	68	42	83	60	94	97	00	13	02	12	48	92	78	56	52	01	06
46	05	88	52	36	01	39	00	22	86	77	28	14	40	77	93	91	08	36	47	70	61	74	29	41
32	17	90	05	97	87	37	92	52	41	05	56	70	70	07	86	74	31	71	57	85	39	41	18	38
69	23	46	14	06	20	11	74	52	04	15	95	66	00	00	18	74	39	24	23	97	11	89	63	38
19	56	54	14	30	01	75	87	53	79	40	41	02	15	85	66	67	43	68	06	84	96	28	52	07
45	15	51	49	38	19	47	60	72	46	43	66	79	45	43	59	04	79	00	33	20	82	66	95	41
94	86	43	19	94	36	16	81	08	51	34	88	88	15	53	01	54	03	54	56	05	01	45	11	76
98	08	62	48	26	45	24	02	84	04	44	99	90	88	96	39	09	47	34	07	35	44	13	18	80
33	18	51	62	32	41	94	15	09	49	89	43	54	85	81	88	69	54	19	94	37	54	87	30	43
80	95	10	04	06	96	38	27	07	74	20	15	12	33	87	25	01	62	52	98	94	62	46	11	71
79	75	24	91	40	71	96	12	82	96	69	86	10	25	91	74	85	22	05	39	00	38	75	95	79
18	63	33	25	37	98	14	50	65	71	31	01	02	46	74	05	45	56	14	27	77	93	89	19	36
74	02	94	39	02	77	55	73	22	70	97	79	01	71	19	52	52	75	80	21	80	81	45	17	48
54	17	84	56	11	80	99	33	71	43	05	33	51	29	69	56	12	71	92	55	36	04	09	03	24
11	66	44	98	83	52	07	98	48	27	59	38	17	15	39	09	97	33	34	40	88	46	12	33	56
48	32	47	79	28	31	24	96	47	10	02	29	53	68	70	32	30	75	75	46	15	02	00	99	94
69	07	49	41	38	87	63	79	19	76	35	58	40	44	01	10	51	82	16	15	01	84	87	69	38

17-2 can be made to function as though it were made up of 4-figure numbers by simply considering two adjacent columns in each reading. In highway engineering practice sampling is frequently referred to the mileage or station of a section of road to indicate the place where a sample will be taken (for example, km 105 + 286 (65 mi + 1470ft) or (65 mi + Sta 14 + 70)). This sampling order can be indicated for a certain section by running over the table from the beginning and selecting all the random numbers that appear within that section. Sampling points are selected after it has been decided how many samples are to be taken from each section.

Assume, for instance, that in the section between Kms 125 + 250 and 142 + 300 (77 mi + Sta 44 + 35 and 88 mi + Sta. 23 + 23), it is wished to designate five sampling points for compaction control, selecting them at random. The three-column table will be used, since six figures are to be handled.

According to the table, the sampling points would be: 128 + 079 (79 + 3168), 125 + 507 (78 + 0016), 140 + 620 (87 + 2090), 131 + 165 (81 + 2743) and 135 + 462 (84 + 1004). Of course, in a random sampling plan these sampling points will not be equidistant from one another, neither does their location follow any of the usual laws for other types of plans. Reference [21] gives another sequence for obtaining random samples that can be employed when very systematic sampling is desired. It has been used in paving jobs in particular.

The procedure is based on use of Table 17-3 [21], which is another example of a random number table. To determine sampling points, the following steps must be followed.

— Determine the average distance at which it is wished to take samples for analysis. Thus, if the section is 5500 m (18000 ft) long and an average distance of 500 m (1640 ft) is desired, the number of samples required will be 11.

Table 17-3

A random number table for locating sampling points vertically and horizontally [21]

Col. No. 1			Col. No. 2			Col. No. 3			Col. No. 4			Col. No. 5			Col. No. 6			Col. No. 7		
A	B	C	A	B	C	A	B	C	A	B	C	A	B	C	A	B	C	A	B	C
15	.033	.576	05	.048	.879	21	.013	.220	18	.089	.716	17	.024	.863	30	0.30	.901	12	0.29	.386
21	.101	.300	17	.074	.156	30	.036	.853	10	.102	.330	24	.060	.032	21	.096	.198	18	.112	.284
23	.129	.916	18	.102	.191	10	.052	.746	14	.111	.925	26	.074	.639	10	.100	.161	20	.114	.843
30	.163	.434	06	.105	.257	25	.061	.954	28	.127	.840	07	.167	.512	29	.133	.388	03	.121	.656
24	.177	.397	28	.179	.447	29	.062	.507	24	.132	.271	28	.194	.776	24	.138	.062	13	.178	.640
11	.202	.271	26	.187	.844	13	.037	.887	19	.285	.899	03	.219	.166	20	.168	.564	22	.209	.421
16	.204	.012	04	.183	.482	24	.105	.849	01	.326	.037	29	.264	.284	22	.232	.953	16	.221	.311
08	.208	.418	02	.208	.577	07	.139	.159	30	.334	.938	11	.282	.262	14	.259	.217	29	.235	.356
19	.211	.798	03	.214	.402	01	.175	.641	22	.405	.295	14	.379	.994	01	.275	.195	28	.264	.941
29	.233	.070	07	.245	.080	33	.196	.873	05	.421	.282	13	.394	.405	06	.277	.475	11	.287	.199
07	.260	.073	15	.248	.831	26	.240	.981	13	.451	.212	06	.410	.157	02	.296	.497	02	.336	.992
17	.262	.308	29	.261	.087	14	.255	.374	02	.461	.023	15	.438	.700	26	.311	.144	15	.393	.488
25	.271	.180	30	.302	.883	06	.310	.043	06	.487	.539	22	.453	.635	05	.351	.141	19	.437	.655
06	.302	.672	21	.318	.088	11	.316	.653	08	.497	.396	21	.472	.824	17	.370	.811	24	.466	.773
01	.409	.406	11	.376	.936	13	.324	.585	25	.503	.893	05	.488	.118	09	.388	.484	14	.531	.014
13	.507	.693	14	.430	.814	12	.351	.275	15	.594	.603	01	.525	.222	04	.410	.073	09	.562	.678
02	.575	.654	27	.438	.676	20	.371	.535	27	.620	.894	12	.561	.980	25	.471	.530	06	.601	.675
13	.591	.318	03	.467	.205	08	.409	.495	21	.629	.841	08	.652	.508	13	.486	.779	10	.612	.859
20	.610	.321	09	.474	.138	16	.445	.740	17	.691	.580	18	.668	.271	15	.515	.867	26	.673	.112
12	.631	.597	10	.492	.474	03	.494	.929	09	.708	.689	30	.736	.634	23	.567	.798	23	.738	.770
27	.651	.281	13	.499	.892	27	.543	.387	07	.709	.012	02	.763	.253	11	.618	.502	21	.753	.614
04	.661	.953	19	.511	.520	17	.625	.171	11	.714	.049	23	.804	.140	28	.636	.148	30	.758	.851
22	.692	.089	23	.591	.770	02	.699	.073	23	.720	.695	25	.828	.425	27	.650	.741	27	.765	.563
05	.779	.346	20	.604	.730	19	.702	.934	03	.748	.413	10	.843	.627	16	.711	.508	07	.780	.534
09	.787	.173	24	.654	.330	22	.816	.802	20	.781	.603	16	.858	.849	19	.778	.812	04	.818	.187
10	.818	.837	12	.728	.523	04	.838	.166	26	.830	.384	04	.903	.327	07	.804	.675	17	.837	.353
14	.895	.631	16	.753	.344	15	.904	.116	04	.843	.002	09	.912	.382	08	.806	.952	05	.854	.818
26	.912	.376	01	.806	.134	28	.969	.742	12	.884	.582	27	.935	.162	18	.841	.414	01	.867	.133
28	.920	.163	22	.878	.884	09	.974	.046	29	.926	.700	20	.970	.582	12	.918	.114	08	.915	.538
03	.945	.140	25	.939	.162	05	.977	.494	16	.951	.601	19	.975	.327	03	.992	.399	25	.975	.584

Col. No. 8			Col. No. 9			Col. No. 10			Col. No. 11			Col. No. 12			Col. No. 13			Col. No. 14		
A	B	C	A	B	C	A	B	C	A	B	C	A	B	C	A	B	C	A	B	C
09	.042	.071	14	.061	.935	26	.038	.023	27	.074	.779	16	.073	.987	03	.033	.091	26	.035	.175
17	.141	.411	02	.065	.097	30	.066	.371	06	.084	.396	23	.078	.056	07	.047	.391	17	.089	.363
02	.143	.221	03	.094	.228	27	.073	.876	24	.098	.524	17	.096	.076	28	.064	.113	10	.149	.681
05	.162	.899	16	.122	.945	09	.095	.568	10	.133	.919	04	.153	.163	12	.066	.360	28	.238	.075
03	.285	.016	18	.158	.430	05	.180	.741	15	.187	.079	10	.254	.834	26	.076	.552	13	.244	.767
28	.291	0.34	25	.193	.469	12	.200	.851	17	.227	.767	06	.234	.623	30	.087	.101	24	.262	.366
08	.369	.557	24	.224	.572	13	.259	.327	20	.236	.571	12	.305	.616	02	.127	.187	08	.264	.651
01	.436	.386	10	.225	.223	21	.264	.681	01	.245	.988	25	.319	.901	06	.144	.068	18	.285	.311
20	.450	.289	09	.253	.838	17	.283	.645	04	.317	.291	01	.320	.212	25	.202	.674	02	.340	.131
18	.455	.789	20	.290	.120	23	.363	.063	29	.350	.911	08	.416	.372	01	.247	.025	29	.353	.478
23	.438	.715	01	.297	.242	20	.364	.366	26	.380	.104	13	.432	.556	23	.253	.323	06	.359	.270
14	.496	.276	11	.337	.760	16	.395	.363	28	.425	.864	02	.489	.827	24	.320	.651	20	.387	.248
15	.503	.342	19	.389	.064	02	.423	.540	22	.487	.526	29	.503	.787	10	.328	.365	14	.392	.694
04	.515	.693	13	.411	.474	08	.432	.736	05	.552	.511	15	.518	.717	27	.338	.412	03	.408	.077
16	.532	.112	20	.447	.893	10	.476	.468	14	.564	.357	28	.524	.998	13	.356	.991	27	.440	.280
22	.557	.357	22	.478	.321	03	.508	.774	11	.572	.306	03	.542	.352	16	.401	.792	22	.461	.830
11	.559	.620	29	.481	.993	01	.601	.417	21	.594	.197	19	.585	.462	17	.423	.117	16	.527	.003
12	.650	.216	27	.562	.403	22	.687	.917	09	.607	.524	05	.695	.111	21	.481	.838	30	.531	.486
21	.672	.320	04	.566	.179	29	.697	.862	19	.650	.572	07	.733	.838	08	.560	.401	25	.678	.360
13	.709	.273	08	.603	.753	11	.701	.605	18	.664	.101	11	.744	.948	19	.564	.190	21	.725	.014
07	.745	.687	15	.632	.927	07	.728	.498	25	.674	.428	18	.793	.748	05	.571	.054	05	.797	.595
30	.780	.285	06	.707	.107	14	.745	.679	02	.697	.674	27	.802	.967	18	.587	.584	15	.801	.927
19	.845	.097	28	.737	.161	24	.819	.444	03	.767	.928	21	.826	.487	15	.604	.145	12	.836	.294
26	.846	.366	17	.846	.130	15	.840	.823	16	.809	.529	24	.835	.832	11	.641	.298	04	.854	.982
29	.861	.307	07	.874	.491	25	.863	.568	30	.838	.294	26	.855	.142	22	.672	.156	11	.884	.928

Table 17-3 (contnd.)

A random number table for locating sampling points vertically and horizontally [21]

Col. No. 8			Col. No. 9			Col. No. 10			Col. No. 11			Col. No. 12			Col. No. 13			Col. No. 14		
A	B	C	A	B	C	A	B	C	A	B	C	A	B	C	A	B	C	A	B	C
25	.906	.874	05	.890	.828	06	.878	.215	13	.845	.470	14	.861	.462	20	.674	.887	19	.386	.832
24	.919	.809	23	.931	.659	18	.930	.601	08	.855	.524	20	.374	.625	14	.752	.881	07	.929	.932
10	.919	.809	26	.960	.365	04	.954	.827	07	.867	.718	30	.929	.056	09	.774	.560	09	.932	.206
06	.961	.504	21	.978	.194	28	.963	.004	12	.881	.722	09	.935	.582	29	.921	.752	01	.970	.692
27	.969	.811	12	.982	.183	19	.983	.020	23	.937	.872	22	.947	.797	04	.959	.099	23	.973	.082

Col. No. 15			Col. No. 16			Col. No. 17			Col. No. 18			Col. No. 19			Col. No. 20			Col. No. 21		
A	B	C	A	B	C	A	B	C	A	B	C	A	B	C	A	B	C	A	B	C
15	.023	.979	19	.062	.588	13	.045	.004	25	.027	.290	12	.052	.075	20	.030	.881	01	.010	.946
11	.118	.465	25	.080	.218	18	.086	.878	06	.057	.571	30	.075	.493	12	.034	.291	10	.014	.939
07	.134	.172	09	.131	.295	26	.126	.990	26	.059	.026	28	.120	.341	22	.043	.823	07	.032	.346
01	.139	.230	18	.136	.381	12	.128	.661	07	.105	.176	27	.145	.689	28	.143	.073	06	.093	.180
16	.145	.122	05	.147	.864	30	.146	.337	18	.107	.358	02	.209	.957	03	.150	.937	15	.151	.012
20	.165	.520	12	.158	.365	05	.169	.470	22	.128	.827	26	.272	.818	04	.154	.867	16	.185	.455
06	.185	.481	28	.214	.184	21	.244	.433	23	.156	.440	22	.299	.317	19	.158	.359	07	.227	.277
09	.211	.316	14	.215	.757	23	.270	.849	15	.171	.157	18	.306	.475	29	.304	.615	02	.304	.400
14	.248	.348	13	.224	.846	25	.274	.407	08	.220	.097	20	.311	.653	06	.369	.633	30	.316	.074
25	.249	.890	15	.227	.809	10	.290	.925	20	.252	.066	15	.348	.156	18	.390	.536	18	.328	.799
13	.252	.577	11	.280	.890	01	.323	.490	04	.268	.576	16	.381	.710	17	.403	.392	20	.352	.288
30	.273	.088	01	.331	.925	24	.352	.291	14	.275	.302	01	.411	.607	23	.404	.182	26	.371	.216
18	.277	.689	10	.399	.992	15	.361	.155	11	.297	.589	13	.417	.715	01	.415	.457	19	.448	.754
22	.372	.958	30	.417	.787	29	.374	.832	01	.358	.305	21	.472	.484	07	.437	.696	13	.487	.598
10	.461	.075	08	.439	.921	08	.432	.139	09	.412	.089	04	.478	.885	24	.446	.546	12	.546	.640
28	.519	.536	20	.472	.484	04	.467	.266	16	.429	.834	25	.479	.080	26	.435	.768	24	.550	.038
17	.520	.090	24	.498	.712	22	.508	.880	10	.491	.200	11	.566	.104	15	.511	.313	03	.604	.780
03	.523	.519	04	.516	.396	27	.632	.191	28	.542	.306	10	.576	.659	10	.517	.290	22	.621	.930
26	.573	.502	03	.548	.688	16	.661	.836	12	.563	.091	29	.665	.397	30	.556	.853	21	.629	.154
19	.634	.206	23	.597	.508	19	.675	.629	02	.593	.321	19	.739	.298	25	.561	.837	11	.634	.908
24	.635	.810	21	.681	.114	14	.680	.890	30	.692	.198	14	.749	.759	09	.574	.599	05	.696	.459
21	.679	.841	02	.739	.298	28	.714	.508	19	.705	.445	08	.756	.919	13	.613	.762	23	.710	.078
27	.712	.366	29	.792	.038	06	.719	.441	24	.709	.717	07	.793	.183	11	.698	.783	29	.726	.585
05	.780	.497	22	.829	.324	09	.735	.040	13	.820	.739	23	.834	.647	14	.715	.179	17	.749	.916
23	.861	.106	17	.834	.647	17	.741	.906	05	.848	.866	06	.837	.978	16	.770	.128	04	.802	.186
12	.865	.377	16	.909	.608	11	.747	.205	27	.867	.633	03	.849	.964	08	.815	.385	14	.835	.319
29	.882	.635	06	.914	.420	20	.850	.047	03	.883	.333	24	.851	.109	05	.872	.490	08	.870	.546
08	.902	.020	27	.958	.856	02	.859	.356	17	.900	.443	05	.859	.935	21	.885	.999	28	.871	.539
04	.951	.482	26	.981	.976	07	.870	.612	21	.914	.483	17	.863	.220	02	.958	.177	25	.971	.369
02	.977	.172	07	.983	.624	08	.916	.463	29	.950	.750	09	.863	.147	27	.961	.980	27	.984	.252

Col. No. 22			Col. No. 23			Col. No. 24			Col. No. 25			Col. No. 26			Col. No. 27			Col. No. 28		
A	B	C	A	B	C	A	B	C	A	B	C	A	B	C	A	B	C	A	B	C
12	.051	.032	26	.051	.187	08	.015	.521	02	.039	.006	16	.026	.102	21	.050	.952	29	.042	.039
11	.068	.980	03	.053	.256	16	.068	.994	16	.061	.599	01	.033	.886	17	.085	.403	07	.105	.293
17	.089	.309	29	.100	.159	11	.118	.400	26	.068	.054	04	.088	.686	10	.141	.624	25	.115	.420
01	.091	.371	13	.102	.465	21	.124	.565	11	.073	.812	22	.090	.602	05	.154	.157	09	.126	.612
10	.100	.709	24	.110	.316	18	.153	.158	07	.123	.649	13	.114	.614	06	.164	.841	10	.205	.144
30	.121	.744	18	.114	.300	17	.190	.159	05	.126	.658	20	.136	.576	07	.197	.013	03	.210	.054
02	.166	.056	11	.123	.208	26	.192	.676	14	.161	.189	05	.138	.228	16	.215	.363	23	.234	.533
23	.179	.529	09	.138	.182	01	.237	.030	18	.166	.040	10	.216	.565	08	.222	.520	13	.266	.799
21	.187	.051	06	.194	.115	12	.283	.077	28	.243	.171	02	.233	.610	13	.269	.477	20	.305	.603
22	.205	.543	22	.234	.480	03	.236	.318	06	.255	.117	07	.278	.357	02	.288	.012	05	.372	.223
28	.230	.688	20	.274	.107	10	.317	.734	15	.261	.928	30	.405	.273	25	.333	.633	26	.385	.111
19	.243	.001	21	.331	.292	05	.337	.844	10	.301	.811	06	.421	.807	28	.348	.710	30	.422	.315
27	.267	.990	08	.346	.085	25	.441	.336	24	.363	.025	12	.426	.583	20	.362	.961	17	.453	.783
15	.283	.440	27	.382	.979	27	.469	.786	22	.378	.792	08	.471	.708	14	.511	.989	02	.460	.916
16	.352	.089	07	.387	.865	24	.473	.237	27	.379	.959	18	.473	.738	26	.540	.903	27	.461	.841
03	.377	.648	28	.411	.776	20	.475	.761	19	.420	.557	19	.510	.207	27	.587	.643	14	.483	.095
06	.397	.769	16	.444	.999	06	.557	.001	21	.467	.943	03	.512	.329	12	.603	.745	12	.507	.375
09	.409	.428	04	.515	.993	07	.610	.238	17	.494	.225	15	.640	.329	29	.619	.895	28	.509	.748
14	.465	.406	17	.518	.827	09	.617	.041	09	.620	.081	09	.665	.354	23	.623	.333	21	.583	.804
13	.499	.651	05	.539	.620	13	.641	.648	30	.623	.106	14	.680	.884	22	.624	.076	22	.587	.993
04	.539	.972	02	.623	.271	22	.664	.291	03	.625	.777	26	.703	.622	18	.670	.904	16	.689	.339
18	.560	.747	30	.637	.374	19	.717	.232	08	.651	.790	29	.739	.394	11	.711	.253	06	.727	.298
26	.575	.892	14	.714	.364	02	.776	.504	12	.715	.599	25	.759	.386	01	.790	.392	04	.731	.814
29	.756	.712	15	.730	.107	02	.775	.504	23	.782	.093	24	.803	.602	04	.813	.611	08	.807	.983
20	.760	.920	19	.771	.552	29	.777	.548	20	.810	.371	27	.842	.491	19	.843	.732	15	.833	.757

Table 17-3 (contnd.)

A random number table for locating sampling points vertically and horizontally [21]

Col. No. 22			Col. No. 23			Col. No. 24			Col. No. 25			Col. No. 26			Col. No. 27			Col. No. 28		
A	B	C	A	B	C	A	B	C	A	B	C	A	B	C	A	B	C	A	B	C
05	.847	.925	23	.780	.662	14	.823	.223	01	.841	.726	21	.870	.435	03	.844	.511	19	.896	.464
25	.872	.891	10	.924	.888	23	.848	.264	29	.862	.009	28	.906	.367	30	.858	.299	18	.916	.384
24	.874	.135	12	.929	.204	30	.892	.817	25	.891	.873	23	.948	.367	09	.929	.199	01	.948	.610
08	.911	.215	01	.937	.714	28	.943	.190	04	.917	.264	11	.956	.142	24	.931	.263	11	.976	.799
07	.946	.065	25	.974	.398	15	.975	.962	13	.958	.990	17	.993	.989	15	.939	.947	24	.978	.633

To select the corresponding column of the random number table, cards numbered from 1 to 28 must be placed in a vessel and one of them drawn at random. This number of cards is arbitrary and is a function of the number of columns that are available in the random number table. In this case there are 28 columns.

Once one of the columns has been selected by the foregoing procedure, all the numbers smaller than or equal to the number of samples required (which was determined in the first step), are located in the corresponding sub-column *A*. Assuming, for example, that Column N° 20 has been selected at random and the number of samples required will be 11. In Sub-Column *B* will be found the factor by which the length of the section must be multiplied in order to determine the distance to the origin of all the sampling points. In this particular case Table 17-4 gives the relevant values.

In order to determine the location in the direction of the horizontal axis of the road, the total width of the section is multiplied by the decimal coefficient of sub-column *C* of the line corresponding to the number of each of the samples. From this product is subtracted the half-width of the section. If this difference proves positive, the sampling point will be located to the right of the line center of the road, and if the difference proves negative, it will be located to the left. In this example, a section width of 12 m (40 ft) will be considered (Table 17-5); Fig. 17-10 [12] then illustrates the sampling plan for the example described.

Table 17-4

Determination of the position of sampling points according to the longitudinal axis

Sample No.	Factor	Distance from the origin
01	0.415	5500 × 0.415 = 2 + 282.5
02	0.958	5500 × 0.958 = 5 + 169.0
03	0.150	5500 × 0.150 = 0 + 825.0
04	0.154	5500 × 0.154 = 0 + 847.0
05	0.872	5500 × 0.872 = 4 + 796.0
06	0.369	5500 × 0.369 = 2 + 029.5
07	0.437	5500 × 0.437 = 2 + 403.5
08	0.815	5500 × 0.815 = 4 + 482.5
09	0.574	5500 × 0.574 = 3 + 157.0
10	0.517	5500 × 0.517 = 2 + 843.5
11	0.698	5500 × 0.698 = 3 + 839.0

The foregoing procedure could be used in the frequent case of determining a representative value of the strength parameters of subgrades for pavement design. Another application is the preparation of a sampling plan to control the compaction of the different layers of soil that make up the structural section of a road.

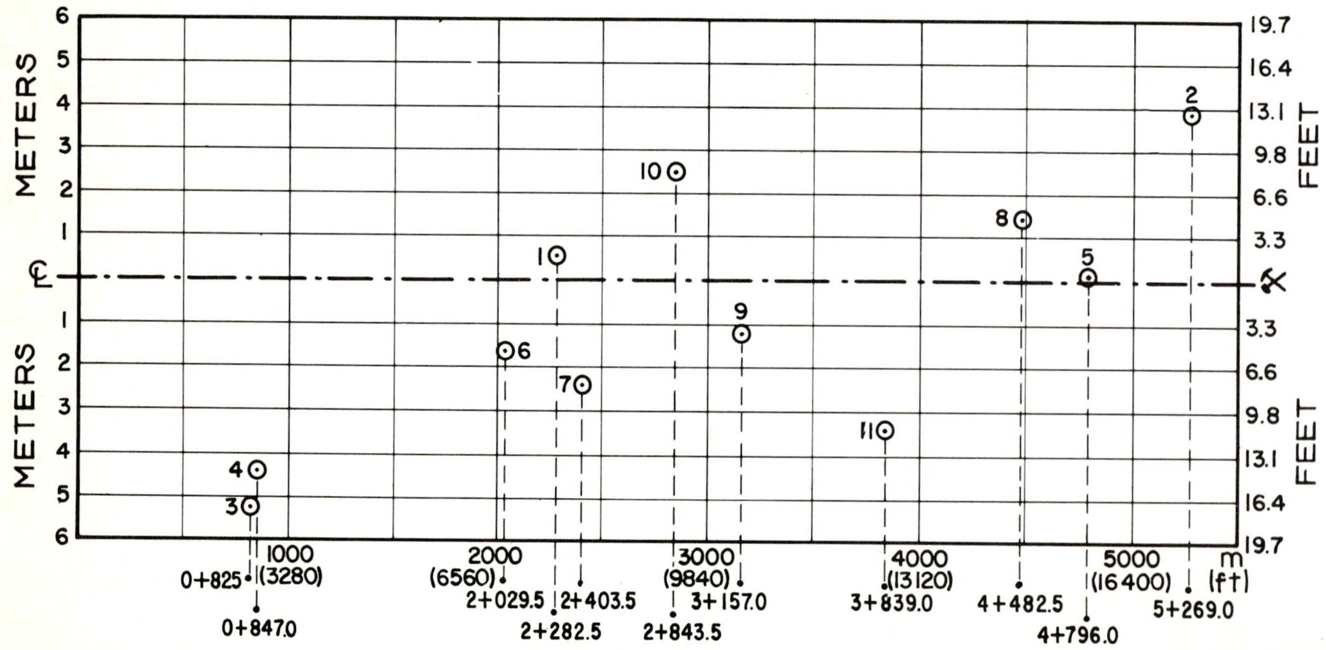

Fig. 17-10 Plotting of sampling points obtained from Table 17-2 [21]

Table 17-5
Determination of the position of sampling points in the cross-section

Sample No.	Factor	Product	Distance from centreline
01	0.457	12 × 0.457 = 5.5	0.5 Left
02	0.177	12 × 0.177 = 2.1	3.9 Left
03	0.937	12 × 0.937 = 11.2	5.2 Right
04	0.867	12 × 0.867 = 10.4	4.4 Right
05	0.490	12 × 0.490 = 5.9	0.1 Left
06	0.633	12 × 0.633 = 7.6	1.6 Right
07	0.696	12 × 0.696 = 8.4	2.4 Right
08	0.385	12 × 0.385 = 4.6	1.4 Left
09	0.599	12 × 0.599 = 7.2	1.2 Right
10	0.290	12 × 0.290 = 3.5	2.5 Left
11	0.783	12 × 0.783 = 9.4	3.4 Right

The enormous advantage of random sampling is that every item in a population stands the same chance of being drawn and tested. Any other sampling procedure is subject to the particular bias that may be adopted by the inspector, which may make the chances of selection different for every item in the population. Otherwise, the basic principles discussed elsewhere in this section apply to random sampling. For example, the larger the sample tested, the stronger the probability of detecting what is defective or what should be rejected.

17.4 Statistical Methods of Quality Control

17.4.1 Control Graphs

.1 Introduction

In a section of road there are 20 sampling points for compaction control (maximum dry density), and at each, 5 horizontally distributed tests. Table 17-6 summarizes the results of all the readings taken. With this data the two graphs that are shown in Fig. 17-11 are plotted. Part *a* gives the individually plotted readings for each sample. Also indicated is the nominal or expected value for the maximum dry density and the upper and lower tolerance limits that are assumed acceptable for the problem under consideration. As will be seen, these limits are not arbitrary, but are provided by the laws of statistics for a given production process. They are frequently established arbitrarily, but this is a vicious practice which breaks statistical laws and the control process is no longer a truly statistical process in such cases. In Part *b* the averages of the 5 tests for each of the 20 samples have been plotted.

Table 17-6
Maximum dry density readings taken at 20 compaction control points (in kg/m^3)

Sample No.	Value at each point on the cross-section					Mean ($\bar{x}$)	Standard deviation σ	Range R
1	1800	1750	1700	1650	1600	1700	70.7	200
2	1550	1550	1700	1600	1500	1580	67.8	200
3	1500	1500	1600	1500	1600	1540	49.0	100
4	1600	1650	1650	1600	1750	1650	54.8	150
5	1600	1700	1850	1850	1750	1750	94.9	250
6	1600	1600	1550	1650	1650	1610	37.4	100
7	1650	1650	1800	1600	1550	1650	83.7	250
8	1150	1650	1800	1750	1800	1630	243.5	650
9	2150	1800	1750	1200	1550	1690	312.1	950
10	1800	1750	1800	2050	2050	1890	131.9	300
11	1700	1900	1750	1700	1900	1790	91.7	200
12	1800	1900	1950	1950	2000	1920	67.8	200
13	1800	2000	1750	1300	1650	1700	230.2	700
14	1800	1750	1850	1700	1650	1750	70.7	200
15	1500	1850	1650	1700	1750	1690	115.7	350
16	1400	1550	1650	1650	1650	1580	124.9	250
17	1650	1700	1700	1650	1750	1650	83.7	250
18	1350	1400	1450	1350	1500	1410	58.3	150
19	1750	1800	1450	1350	1600	1590	171.5	450
20	1650	1750	1750	1950	1800	1780	69.3	300

$\bar{\bar{x}} = \dfrac{\Sigma \bar{x}}{n} = 1677.5$ kg/m^3 (104.5 lb/ft^3) $\qquad \bar{\sigma} = 111.48$ kg/m^3 (6.93 lb/ft^3)

$\bar{R} = \dfrac{\Sigma R}{n} = 312.5$ kg/m^3 (19.5 lb/ft^3) $\qquad \sigma\bar{x} = 115.15$ kg/m^3 (7.2 lb/ft^3)

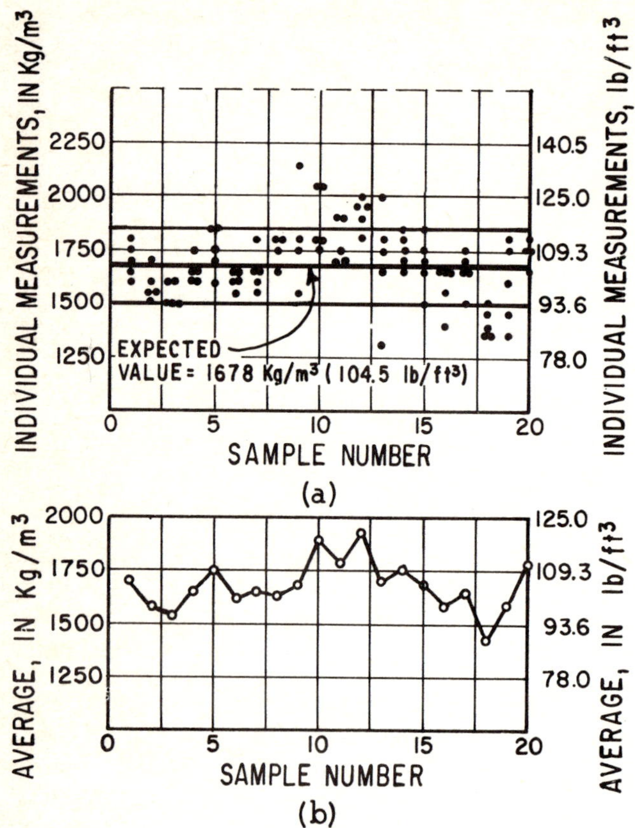

Fig. 17-11 Graphs of results of individual tests and their averages for a process

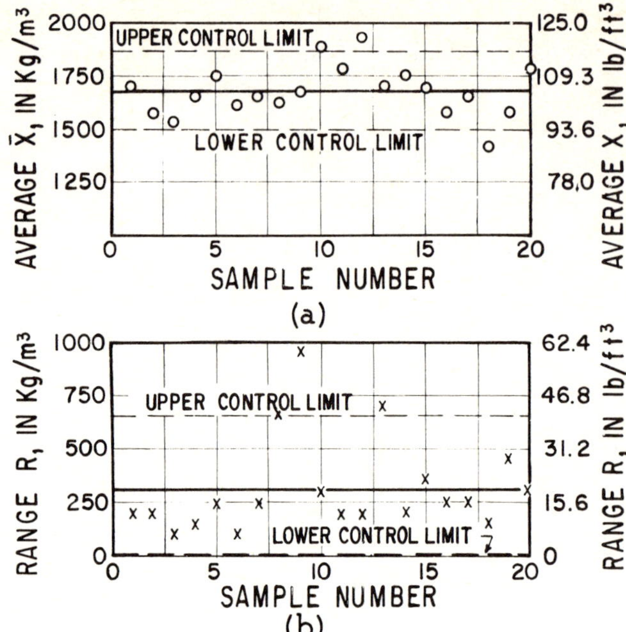

Fig. 17-12 Control graphs of a compaction process
a) Graph of averages
b) Graph of ranges

Both graphs show certain tendencies in the results obtained and indicate whether or not these results go beyond the specified limits. However, these graphs have limited use in a genuine quality control process.

Figure 17-12 shows two more useful control graphs which can be plotted using the same data in Table 17-6. In Part *a* the mean $\bar{x}$ values have been plotted. It is the same graph as Fig. 17-11b, but with the upper and lower acceptance limits added and without the line joining the points that are plotted. In a process subject to statistical control, the position of the upper and lower limits in the $\bar{x}$ graph is not completely arbitrary. The resulting population is subjected to a statistical control and the position of the upper and lower limits corresponding to that population can be calculated by the statistical methods themselves, or what amounts to the same thing, to a certain process for producing data or readings correspond upper and lower acceptance or rejection limits, with reference to the variations due to randomness or inherent in the production process. Thus, if a large number of data plotted on the control charts go beyond statistically selected limits, it is likely that their variation is due to other factors than those inherent in the process (and consequently inevitable). These non-inherent factors can and must be corrected. This is the fundamental information that can be provided by a control graph. A statistically drawn up control graph permits differentiation between the variables that are likely to be inevitable in data and those that can be controlled. Thus, by the mere presence of these last variations, it is possible to indicate when a particular process gets out of control and has to be modified or adjusted. Control graphs also indicate how many samples and which ones depict variations that must be corrected.

Part *b* of Fig. 17-12 shows a control graph plotted according to the range of each sample. In [12] formulas are given for calculation of the control limits that are established in these graphs.

.2 Mean $\bar{x}$ Control Graphs

The calculation of control limits can be based on different parameters:

Based on the average of the ranges: The formulas to apply are:

$$LS = \bar{\bar{x}} + A_2 \bar{R}$$
$$LI = \bar{\bar{x}} - A_2 \bar{R} \qquad (17\text{-}7)$$

where *LS* and *LI* are the upper and lower control limits, $\bar{R}$ is the average of the ranges obtained for each sample analysed, $\bar{\bar{x}}$ is the average of the samples analysed, and *A* is a coefficient which can be computed using Table 17-7.

With reference to the example presented in Table 17-6, $\bar{\bar{x}}$ is the average of all the $\bar{x}$ values, obtained by dividing by 20 the sum of all them, and $\bar{R}$ is the average of all the *R* values, calculated in the same way. In this case:

$$\bar{\bar{x}} = 1677.5 \text{ kg/m}^3 \text{ (105 lb/ft}^3\text{)}$$
$$\bar{R} = 312.5 \text{ kg/m}^3 \text{ (20 lb/ft}^3\text{)}$$

For $n = 5$, Table 17-7 gives a value $A_2 = 0.58$. With all these data the following is true:

$$LS = 1858.9 \text{ kg/m}^3 \text{ (116 lb/ft}^3\text{)}$$
$$LI = 1496.25 \text{ kg/m}^3 \text{ (93.6 lb/ft}^3\text{)}$$

The two limits are equidistant from the mean, one 181.3 above it and the other 181.3 below it.

These limits are obtained assuming that whatever the distribution of the original universe, all the other distributions that are handled are normal (This is only approximately true, as was

Table 17-7

Factors for determining control limits on the basis of R for $\bar{x}$ and R graphs [12]

No. of observations in the subgroup	Factor for the $\bar{x}$ graph	Factors for the R graph	
		Lower control limit	Upper control limit
n	A_2	D_3	D_4
2	1.88	0	3.27
3	1.02	0	2.57
4	0.73	0	2.28
5	0.58	0	2.11
6	0.48	0	2.00
7	0.42	0.08	1.92
8	0.37	0.14	1.86
9	0.34	0.18	1.82
10	0.31	0.22	1.78
11	0.29	0.26	1.74
12	0.27	0.28	1.72
13	0.25	0.31	1.69
14	0.24	0.33	1.67
15	0.22	0.35	1.65
16	0.21	0.36	1.64
17	0.20	0.38	1.62
18	0.19	0.39	1.61
19	0.19	0.40	1.60
20	0.18	0.41	1.59

mentioned, unless the original distribution of the universe is also normal). Also considered in all the normal distributions are levels of acceptance of $\bar{x} \pm 3\ \sigma$, which correspond to 99.7% of the area under the Gaussian curve (Fig. 17-3).

Based on the average of the standard deviations: For this technique, [12] proposes the following formulas:

$$LS = \bar{\bar{x}} + A_1\,\bar{\sigma}$$
$$LI = \bar{\bar{x}} - A_1\,\bar{\sigma} \qquad (17\text{-}8)$$

where all the code letters have the meanings already indicated, $\bar{\sigma}$ is the average of the standard deviations of the samples that are handled, and A_1 is a factor that can be obtained from Table 17-8 for different sample sizes (n).

In this particular case (Table 17-6), $n = 5$ and $A_1 = 1.6$. The mean value of σ is 111.48 (7). Consequently:

$$LS = 1855.9 \text{ kg/m}^3 \ (115.8 \text{ lb/ft}^3)$$
$$LI = 1499.2 \text{ kg/m}^3 \ (93.8 \text{ lb/ft}^3)$$

The limits are above and below the mean ($\bar{\bar{x}}$), at a distance of 178.4 from it.

Based on the mean ($\bar{x}'$) and the standard deviation (σ') of the original universe: In this case the formulae are:

$$LS = \bar{x}' + A\sigma'$$
$$LI = \bar{x}' - A\sigma' \qquad (17\text{-}9)$$

Application of the above formulas requires estimation of ($\bar{x}'$) and σ', but it has already been seen that if the sample is large enough:

$$\bar{\bar{x}} = \bar{x}' \text{ and } \sigma_{\bar{x}} = \sigma'/\sqrt{n}$$

In practice n is frequently small, which makes a slight adjustment of the foregoing calculations necessary. It is sufficient to consider $\bar{\bar{x}} = \bar{x}'$ and:

$$\sigma' = \frac{\sigma}{c_2} \qquad (17\text{-}10)$$

or else:

$$\sigma' = \frac{\bar{R}}{d_2} \qquad (17\text{-}11)$$

The coefficient c_2 permits evaluation of σ' as a function of $\bar{\sigma}$, which is easier to obtain than $\sigma_{\bar{x}}$. The factors c_2 and d_2 can be obtained from Table 17-9. The coefficient A in Eq.(17-9) can be obtained from Table 17-10. Application of Eq.(17-9) to the example under consideration here is left to the reader.

Equations (17-7), (17-8) and (17-9) lead to similar results, and their use is a matter of preference or convenience.

.3 Range (R) Control Graphs

In this case the control limits, according to [12] are given by the expressions:

$$LS = D_4\,\bar{R}$$
$$LI = D_3\,\bar{R} \qquad (17\text{-}12)$$

The values of factors D_3 and D_4 can be obtained from Table 17-7.

In the example in Table 17-6, it is recalled that:

$$R = 312.5 \text{ kg/m}^3 \ (20 \text{ lb/ft}^3)$$

In Table 17-7 it is seen that for $n = 5$, $D_3 = 0$ and $D_4 = 2.11$. Therefore, applying Eq.(17-12) the following limits are achieved:

$$LS = 659.38 \text{ kg/m}^3 \ (41.2 \text{ lb/ft}^3)$$
$$LI = 0$$

These limits are the ones that are shown in Part b of Fig. 17-12. The ranges that are within the control limits obtained and indicated in this figure correspond to inevitable variations inherent in whatever process is being conducted. If these variations exceed the tolerance limits imposed, or are not realistic, either the limits should be modified, or else the process itself should be adjusted or replaced by another that will be capable of producing the deviations desired. The deviations that go beyond the control limits obtained may not be inevitable, and the engineer can attempt to improve the application of his process in order to reduce them to the limits indicated by the control graph.

Reference [12] gives another alternative for calculating the control limits in the range control graph as a function of the mean and the standard deviation of the original universe ($\bar{x}$, σ'). For this purpose the following formulas are used:

Table 17-8

Factors for determination of control limits for $\bar{x}$ and σ graphs on the basis of $\bar{\sigma}$ [12]

No. of observations in subgroup	Factor for the $\bar{x}$ graph	Factors for the σ graph	
		Lower control limit	Upper control limit
n	A_1	B_3	B_4
2	3.76	0	3.27
3	2.39	0	2.57
4	1.88	0	2.27
5	1.60	0	2.09
6	1.41	0.03	1.97
7	1.28	0.12	1.88
8	1.17	0.19	1.81
9	1.09	0.24	1.76
10	1.03	0.28	1.72
11	0.97	0.32	1.68
12	0.93	0.35	1.65
13	0.88	0.38	1.62
14	0.85	0.41	1.59
15	0.82	0.43	1.57
16	0.79	0.45	1.55
17	0.76	0.47	1.53
18	0.74	0.48	1.52
19	0.72	0.50	1.50
20	0.70	0.51	1.49
21	0.68	0.52	1.48
22	0.66	0.53	1.47
23	0.65	0.54	1.46
24	0.63	0.55	1.45
25	0.62	0.56	1.44
30	0.56	0.60	1.40
35	0.52	0.63	1.37
40	0.48	0.66	1.34
45	0.45	0.68	1.32
50	0.43	0.70	1.30
55	0.41	0.71	1.29
60	0.39	0.72	1.28
65	0.38	0.73	1.27
70	0.36	0.74	1.26
75	0.35	0.75	1.25
80	0.34	0.76	1.24
85	0.33	0.77	1.23
90	0.32	0.77	1.23
95	0.31	0.78	1.22
100	0.30	0.79	1.21

$$
\begin{aligned}
LS &= D_2\,\sigma' \\
LI &= D_1\,\sigma'
\end{aligned}
\qquad (17\text{-}13)
$$

where σ' is obtained as previously indicated, and D_1 and D_2 can be obtained from Table 17-10.

.4 Graphs for Controlling Standard Deviations, σ

In this case the control limits, according to [12] are given by the following expressions:

$$
\begin{aligned}
LS &= B_4\,\bar{\sigma} \\
LI &= B_3\,\bar{\sigma}
\end{aligned}
\qquad (17\text{-}14)
$$

Table 17-9
Factors for estimating σ' on the basis of $\bar{R}$ or $\bar{\sigma}$ [12]

No. of observations in the subgroup	Estimation factor based on $\bar{R}$	Estimation factor based on $\bar{\sigma}$
n	$d_2 = \bar{R}/\sigma'$	$C_2 = \bar{\sigma}/\sigma'$
2	1.128	0.5642
3	1.693	0.7236
4	2.059	0.7979
5	2.326	0.8407
6	2.534	0.8686
7	2.704	0.8882
8	2.847	0.9027
9	2.970	0.9139
10	3.078	0.9227
11	3.173	0.9300
12	3.258	0.9359
13	3.336	0.9410
14	3.407	0.9453
15	3.472	0.9490
16	3.532	0.9523
17	3.588	0.9551
18	3.640	0.9576
19	3.689	0.9599
20	3.735	0.9619
21	3.778	0.9638
22	3.819	0.9655
23	3.858	0.9670
24	3.895	0.9684
25	3.931	0.9696
30	4.086	0.9748
35	4.213	0.9784
40	4.322	0.9811
45	4.415	0.9832
50	4.498	0.9849
55	4.572	0.9863
60	4.639	0.9874
65	4.699	0.9884
70	4.755	0.9892
75	4.806	0.9900
80	4.854	0.9906
85	4.898	0.9912
90	4.939	0.9916
95	4.978	0.9921
100	5.015	0.9925

where $\bar{\sigma}$ is obtained as already indicated using the data in Table 17-6 ($\bar{\sigma} = \Sigma\sigma/n$), and the factors B_3 and B_4 are provided by Table 17-8.

In Fig. 17-12 the graph for controlling the standard deviations of the data in Table 17-6 has not been plotted. This task is left to the reader. Considering that $\bar{\sigma}$ = 111.48 = (7) the control limits to obtain in this case are:

$$LS = 2.09 \times 111.48 = 232.99$$
$$(LS = 2.09 \times 7 = 14.63)$$
$$LI = 0$$

Reference [12] also provides formulas for calculation of these control limits in terms of the values $\bar{x}'$ and σ', which are:

$$LS = B_2\,\sigma' \qquad LI = B_1\,\sigma' \qquad (17\text{-}15)$$

Coefficient B_1 and B_2 are obtained from Table 17-10.

The foregoing equations for determining the control limits for the different graphs are not deduced in detail here, but they can be consulted in the already frequently mentioned [12,22,23] or in many texts on mathematical statistics including applications to quality control. Moreover, this deduction would be very simple on the basis of the concepts presented in Section 17.2 of this chapter. The factors obtained from tables and handled here are nothing more than a simple interrelation between statistical parameters already referred to.

.5 Observations Concerning the Use of Control Graphs

When dealing with a specific construction process or investigation the variation with which a laboratory or group of laboratories run a certain test, it is always possible and indeed easy to draw up a table of values like Table 17-6. It is equally easy to calculate the means, standard deviations and ranges of these values. A prerequisite is that they must be the product of a proper sampling operation, either based on the use of operating characteristic curves or a random sampling plan.

Once a table like 17-6 has been obtained, it will be equally practical and simple to plot the control graphs of the means, the ranges and the standard deviations of the data, and to calculate the control limits for these graphs. In this way, the engineer will be in a position to determine whether the values he is handling have reasonable variations or dispersions or if, on the contrary, there are some that can be reduced.

By comparing tolerance and control limits, the engineer will be better able to judge how realistic his tolerance limits are. If his tolerance limits prove narrower than the control limits, the solution will be to adopt a method of higher precision unless, as might often be the case in highway practice, he should realise that his tolerance limits, which were probably arbitrarily selected or based on experience, could be extended as far as the control limits of the process.

It is also worth noting that the methodology that has been given for the calculation of control limits includes a level of acceptance $\bar{x} \pm 3\sigma$, which represents a very unbending attitude. In a properly planned quality control process, the same requirement must not be observed for the construction of all types of roads or all the operations that may be involved. Anyone with a knowledge of elementary statistics would be perfectly capable of modifying the foregoing formulas, making the level of acceptance less demanding, for example $\bar{x} \pm 2\sigma$ (95%) or even $\bar{x} \pm \sigma$ (68%). Selection of a particular criterion depends not only on the importance of the project, but also on the risk of failure, the cost of the operation and other considerations. For example, if a relatively undemanding criterion has been adopted for the quality of the materials used for an unpretentious road, so as to avoid long hauling then a far more demanding criterion will be necessary for drainage problems. A satisfactory quality

Table 17-10

Factors for determining control limits of 3σ for $\overline{x}$, R and σ graphs on the basis of σ' [12]

No. of observations in subgroup	Factor for the $\overline{x}$ graph	Factors for the R graph		Factors for the σ graph	
		Lower control limit	Upper control limit	Lower control limit	Upper control limit
n	A	D_1	D_2	B_1	B_2
2	2.12	0	3.69	0	1.84
3	1.73	0	4.36	0	1.86
4	1.50	0	4.70	0	1.81
5	1.34	0	4.92	0	1.76
6	1.22	0	5.08	0.03	1.71
7	1.13	0.20	5.20	0.10	1.67
8	1.06	0.39	5.31	0.17	1.64
9	1.00	0.55	5.39	0.22	1.61
10	0.95	0.69	5.47	0.26	1.58
11	0.90	0.81	5.53	0.30	1.56
12	0.87	0.92	5.59	0.33	1.54
13	0.83	1.03	5.65	0.36	1.52
14	0.80	1.12	5.69	0.38	1.51
15	0.77	1.21	5.74	0.41	1.49
16	0.75	1.28	5.78	0.43	1.48
17	0.73	1.36	5.82	0.44	1.47
18	0.71	1.43	5.85	0.46	1.45
19	0.69	1.49	5.89	0.48	1.44
20	0.67	1.55	5.92	0.49	1.43
21	0.65			0.50	1.42
22	0.64			0.52	1.41
23	0.63			0.53	1.41
24	0.61			0.54	1.40
25	0.60			0.55	1.39
30	0.55			0.59	1.36
35	0.51			0.62	1.33
40	0.47			0.65	1.31
45	0.45			0.67	1.30
50	0.42			0.68	1.28
55	0.40			0.70	1.27
60	0.39			0.71	1.26
65	0.37			0.72	1.25
70	0.36			0.74	1.24
75	0.35			0.75	1.23
80	0.34			0.75	1.23
85	0.33			0.76	1.22
90	0.32			0.77	1.22
95	0.31			0.77	1.21
100	0.30			0.78	1.20

control will depend on due consideration of all these criteria and will be based on control graphs and good human team work.

The routine use of control graphs during any engineering process shows the engineer whether his process is still *under control*. This will be true so long as the values are within the control limits. Beyond these limits, the engineer will know that the process is out of control, requiring adjustment, improvement or modification.

Control graphs, therefore, indicate when a process should be reviewed, but not where. They indicate that something is wrong but do not say what, although engineers who are very familiar with them appear to develop a certain ability to detect the causes of the problems that lead to loss of control.

Also, the strict function of quality control may perhaps not have to go much further than control charts. Once a flaw is detected in the quality that is being obtained, it will be up to the different members of the control team to investigate the origin of the deficiency and establish the necessary measures for correcting it.

17.4.2 Statistical Estimation

.1 Methods for Estimating a Population Mean

A rational procedure for confronting quality control problems, to produce useful results, is as follows:

A population is made up of different values of the parameter it is wished to control. This population will have a mean ($\bar{x}'$) and a standard deviation σ'. In practice, there are two cases which may arise: in the first, the value of σ' is known and in the second it is not. Using one of the procedures already analysed, for example the random one, samples are drawn from the population. The mean and the standard deviation ($\bar{x}$ and σ) can be easily calculated using the methods already described in this chapter.

The step is establishing the confidence interval for the population mean once the desired confidence level has been selected. This statement deserves an explanation. Section 17.2 of this Chapter, Eq.(17-3), showed what is represented by the confidence level in a statistical estimation. It was also seen that its value (which determines the t factor) is selected or established by whoever is in control of the process. Clearly, the higher the value of t the greater the probability of the population mean being within a pre-established interval; consequently it can be said that for lower values of t stricter controls are necessary. It is worth mentioning that with a reduction in the value of t there is likewise a reduction in the inherent error of the statistical operation, Eq.(17-3). The confidence interval is the area around the *exact* value of the population mean within which its fluctuation is tolerated.

Thus, the initial statistical estimation to which reference is made is expressed as follows: What is the probability of the population mean always being located within certain limits, selected symmetrically to the mean, under the frequency distribution curve? Or else, what are the variability limits symmetrical to the mean for this mean to be situated between them with a pre-established probability? Basically the two questions are similar.

In the paragraphs that follow the methods for answering these two questions will be analysed. This is often referred to as conducing the initial statistical estimation of the population mean, distinguishing the case in which the standard deviation of the population is known from the one where it is not.

The standard deviation of the population is known: In road engineering problems this case is fairly common, for σ' can often be quite reliably estimated, even though its exact value may not be known.

Earlier it was seen that in a set of representative samples (as in Table 17-6), the mean of the averages of each sample can be said to be $\bar{\bar{x}}$.

$$\bar{\bar{x}} = 1677.5\ (105) = \bar{x}'$$

It was also seen that:

$$\sigma_{\bar{x}} = \frac{\sigma'}{N} \qquad (17\text{-}16)$$

where N is the number of items in each sample and $\sigma_{\bar{x}}$ is the standard deviation of the sample means, which in Table 17-6 proves to be:

$$\sigma_{\bar{x}} = \sqrt{\frac{\Sigma(\bar{\bar{x}} - \bar{x})^2}{n}} = 115.15\ (7.2)$$

Note that n refers to the number of lines in Table 17-6 (the number of samples that are handled) whereas N is the number of items in each of these samples. In the case of Table 17-6, $N = 5$. This distinction must always be kept in mind to avoid confusion in previous and subsequent stages.

References [8,9,23,24] or [25] together with [12] represent a good basic bibliography for people interested in the theoretical aspects mentioned in this chapter. They show that the confidence interval for the population mean can be expressed thus:

$$\bar{x} \pm t\frac{\sigma'}{n} \qquad (17\text{-}17)$$

where $\bar{x}$ is the mean of each sample (lines in Table 17-6), t is the factor that defines the confidence level adopted, σ' is the standard deviation of the population, which is assumed to be known, and N is the number of items in each sample.

Taking into consideration Eq.(17-16), it is deduced that the confidence interval for the population mean can also be expressed as:

$$\bar{x} \pm t\sigma_{\bar{x}} = 150\ (9.36) \qquad (17\text{-}18)$$

The value of t, as already stated, can be tabulated once and for all under the hypothesis that the distribution of $\bar{x}$ is normal. Table 17-11 [24] is an example.

As an example, the limits within which the population mean is expected to remain can be calculated from the data obtained for Sample No. 5 in Table 17-6 (5th line). A confidence level of 95% is desired. Using Eq.(17-18), the confidence interval would be:

Table 17-11

Values of t for different confidence levels [24]

Confidence level (%)	t
99.7	3.00
99.0	2.58
98.0	2.33
96.0	2.05
95.5	2.00
95.0	1.96
90.0	1.64
80.0	1.28
68.2	1.00
50.0	0.67

$1750 \pm 1.96 \times 115.15 = 1750 \pm 225.96$
$(109 \pm 1.96 \times 7.2 \quad = \quad 109 \pm 14.11)$

This signifies that on the basis of the data for Sample No. 5 in Table 17-6 it can be stated that there is a 95% probability that the population mean will be located within the foregoing interval. Fluctuation within this interval is inherent in the randomness of the process.

The foregoing ideas could also be used to solve the following problem:

Using the data for Sample No. 5 in Table 17-6, estimate the probability of the population mean being located within the interval:

1750 ± 150 (kg/m^3)
109 ± 9.4 (lb/ft^3)

From Eq.(17-18) it is deduced that:

$$t\sigma_{\bar{x}} = \pm 150 \ (\pm 9.4)$$

Consequently, since it has already been calculated that $\sigma_{\bar{x}} = 115.15$ (7.2), then:

$$t = \frac{150}{115.15} = \frac{9.4}{7.2} = \pm 1.3$$

The values of t corresponding to all confidence levels are given in Table 17-12 [24]. This table includes Table 17-11, but is more complete, going beyond the most usual values in practical calculations. It refers to the areas that occur under the normal distribution curve between the value of the mean and any value of t. Double this value will lead to the probability that part of the distribution will remain within the interval $0 \pm t$ (see comments on Fig. 17-3).

Table 17-12
Areas under the Normal Curve between 0 and t [24]

O t

t	0	1	2	3	4	5	6	7	8	9
0.0	.0000	.0040	.0080	.0120	.0160	.0199	.0239	.0279	.0319	.0359
0.1	.0398	.0438	.0478	.0517	.0557	.0596	.0636	.0675	.0714	.0754
0.2	.0793	.0832	.0871	.0910	.0948	.0987	.1026	.1064	.1103	.1141
0.3	.1179	.1217	.1255	.1293	.1331	.1368	.1406	.1443	.1480	.1517
0.4	.1554	.1591	.1628	.1664	.1700	.1736	.1772	.1808	.1844	.1879
0.5	.1915	.1950	.1985	.2019	.2054	.2088	.2123	.2157	.2190	.2224
0.6	.2258	.2291	.2324	.2357	.2389	.2422	.2454	.2486	.2518	.2549
0.7	.2580	.2612	.2642	.2673	.2704	.2734	.2764	.2794	.2823	.2852
0.8	.2881	.2910	.2939	.2967	.2996	.3023	.3051	.3078	.3106	.3133
0.9	.3159	.3189	.3212	.3238	.3264	.3289	.3315	.3340	.3365	.3389
1.0	.3413	.3438	.3461	.3485	.3508	.3531	.3554	.3577	.3599	.3621
1.1	.3643	.3665	.3686	.3708	.3729	.3749	.3770	.3790	.3810	.3830
1.2	.3849	.3869	.3888	.3907	.3925	.3944	.3962	.3980	.3997	.4015
1.3	.4032	.4049	.4066	.4082	.4099	.4115	.4131	.4147	.4162	.4177
1.4	.4192	.4207	.4222	.4236	.4251	.4265	.4279	.4292	.4306	.4319
1.5	.4332	.4345	.4357	.4370	.4382	.4394	.4406	.4418	.4429	.4441
1.6	.4452	.4463	.4474	.4484	.4495	.4505	.4515	.4525	.4535	.4545
1.7	.4554	.4564	.4573	.4582	.4591	.4599	.4608	.4616	.4625	.4633
1.8	.4641	.4649	.4656	.4664	.4671	.4678	.4686	.4693	.4699	.4706
1.9	.4713	.4719	.4726	.4732	.4738	.4744	.4750	.4756	.4761	.4767
2.0	.4772	.4778	.4783	.4788	.4793	.4798	.4803	.4803	.4812	.4817
2.1	.4821	.4826	.4830	.4834	.4838	.4842	.4846	.4850	.4854	.4857
2.2	.4861	.4864	.4868	.4871	.4875	.4878	.4881	.4884	.4887	.4890
2.3	.4893	.4896	.4898	.4901	.4904	.4906	.4909	.4911	.4913	.4916
2.4	.4918	.4920	.4922	.4925	.4927	.4929	.4931	.4932	.4934	.4936
2.5	.4938	.4940	.4941	.4943	.4945	.4946	.4948	.4949	.4951	.4952
2.6	.4953	.4955	.4956	.4957	.4959	.4960	.4961	.4962	.4963	.4964
2.7	.4965	.4966	.4967	.4968	.4969	.4970	.4971	.4972	.4973	.4974
2.8	.4974	.4975	.4976	.4977	.4977	.4978	.4979	.4979	.4980	.4981
2.9	.4981	.4982	.4982	.4983	.4984	.4984	.4985	.4985	.4986	.4986
3.0	.4987	.4987	.4987	.4988	.4988	.4989	.4989	.4989	.4990	.4990
3.1	.4990	.4991	.4991	.4991	.4992	.4992	.4992	.4992	.4993	.4993
3.2	.4993	.4993	.4994	.4994	.4994	.4994	.4994	.4995	.4995	.4995
3.3	.4995	.4995	.4995	4996	.4996	.4996	.4996	.4996	.4996	.4997
3.4	.4997	.4997	.4997	.4997	.4997	.4997	.4997	.4997	.4997	.4998
3.5	.4998	.4998	.4998	.4998	.4998	.4998	.4998	.4998	.4998	.4998
3.6	.4998	.4998	.4999	.4999	.4999	.4999	.4999	.4999	.4999	.4999
3.7	.4999	.4999	.4999	.4999	.4999	.4999	.4999	.4999	.4999	.4999
3.8	.4999	.4999	.4999	.4999	.4999	.4999	.4999	.4999	.4999	.4999
3.9	.5000	.5000	.5000	.5000	.5000	.5000	.5000	.5000	.5000	.5000

In Table 17-12 it is seen that for $t = \pm 1.3$ the probability is 80.64%, which is lower than the 95% in the previous example, because the interval of fluctuation that was selected here is smaller. Figure 17-13 shows the intervals of fluctuation of the population mean in the previous two examples.

When the standard deviation of the population, σ', is unknown, it can be estimated with a reasonable degree of reliability on the basis of the standard deviation of the distribution of the sample means ($\sigma_{\bar{x}}$) which will be known. Therefore most of the practical estimation problems fall into the category of those in which σ' is known. It is not, however, impossible to come up against some practical problem in which the value of σ' is unknown, in which case the problem should be handled as in the following.

The standard deviation of the population is unknown: In some practical problems the standard deviation of the population is not known and the closest approximation that can be obtained for its value is σ, which is the standard deviation of the sample. In this case the procedure for establishing the confidence interval for the population mean, once the confidence level at which it is wished to work has been established, is similar to that in the previous case; however, the distribution of the samples is no longer normal, but of the type known in statistical sciences as the Student distribution. As a result the problem has to be solved with a Student distribution table. The formula that gives the confidence interval is:

$$\bar{x} \pm t \frac{\sigma}{\sqrt{N}} \tag{17-19}$$

where $\bar{x}$ is the sample mean, N the size of the sample and t the factor corresponding to the confidence level adopted, calculated for $N = 1$, as is shown in the texts on statistics referred to previously. Lastly, σ is the standard deviation of the sample.

It can be intuitively seen that if the estimation of σ' is based on a small sample, the result obtained will be less reliable than if a larger sample is used.

Table 17-13 [24] shows the Student distribution.

Let us imagine, for example, a sample taken from the unconfined compressive strengths, which could be determined for a certain soil in a laboratory. Values are given in t/m^2 (lb/in^2). the mean of which is $\bar{x}$ = 15.53 t/m^2 (3.2 kips/ft^2) and the standard deviation σ = 1.98 t/m^2 (0.4 kips/ft^2).

A confidence level of 95% is acceptable and it is wished to find the confidence interval of the population mean for this confidence level. This interval will be $\bar{x} \pm t\sigma/\sqrt{N}$. Substituting:

$$15.53 \pm 2.31 \frac{1.98}{\sqrt{9}} = 15.53 \pm 1.53$$

$$\left(3.2 \pm 2.31 \frac{0.4}{\sqrt{9}} = 3.2 \pm 0.3\right)$$

Thus, in the population under analysis there is a 95% probability that the mean will be within the interval ± 1.53 (± 0.3) about the sample mean.

The value 2.31 in the foregoing calculations was obtained from Line 8 of Table 17-13 (corresponding to $N - 1$ in this case, since $N = 9$, in column $t_{0.975}$). The reason for this is that for $t_{0.975}$ there would be a tail of 0.025 under the distribution on the right (see diagram in Table). Since the Student distribution is symmetrical, there will be another tail also of 0.025 on the left-hand side, with a total of 0.05 outside the interval $\pm t_{0.975}$, so that to the factor $t_{0.975}$ corresponds an area of 95% under the curve, as desired. It is easy to see that the value of t with which the column of Table 17-13 that should be read is located, is given by the expression:

$$\frac{100 + \text{confidence level}}{2} \tag{17-20}$$

Another possible calculation might be as follows. For the same sample, what probability is there that the population mean will be located within the interval:

15.53 ± 0.92? (3.2 ± 0.19?)

Now the confidence interval would have to be set forth as follows:

$$15.53 \pm t \frac{1.98}{3} \left(3.2 \pm t \frac{0.4}{3}\right)$$

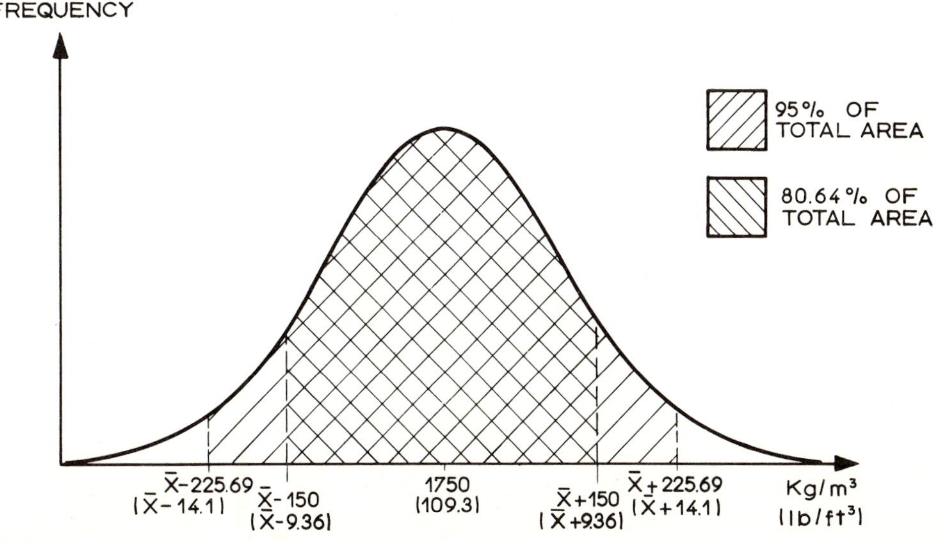

Fig. 17-13 Graph of confidence intervals for a population mean

Table 17-13

Values of t in the Student distribution [24]

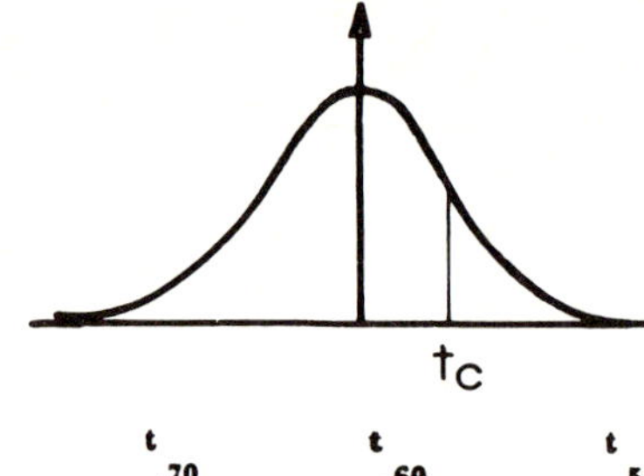

N-1	$t_{.995}$	$t_{.99}$	$t_{.975}$	$t_{.95}$	$t_{.90}$	$t_{.80}$	$t_{.75}$	$t_{.70}$	$t_{.60}$	$t_{.55}$
1	63.66	31.82	12.71	6.31	3.07	1.376	1.000	.727	.325	1.58
2	9.92	6.96	4.30	2.92	1.89	1.061	.816	.617	.289	.142
3	5.84	4.54	3.18	2.35	1.64	.978	.765	.584	.275	.138
4	4.60	3.75	2.78	2.13	1.53	.941	.741	.569	.271	.134
5	4.04	3.36	2.58	2.02	1.48	.920	.727	.560	.267	.132
6	3.71	3.14	2.45	1.94	1.44	.906	.718	.553	.265	.131
7	3.50	3.00	2.36	1.91	1.43	.896	.711	.549	.263	.130
8	3.36	2.90	2.31	1.86	1.40	.889	.706	.546	.262	.130
9	3.25	2.82	2.26	1.83	1.38	.883	.703	.543	.261	.129
10	3.17	2.76	2.23	1.81	1.37	.879	.700	.542	.260	.129
11	3.11	2.72	2.20	1.80	1.36	.876	.697	.540	.260	.129
12	3.06	2.68	2.18	1.78	1.36	.873	.695	.539	.259	.128
13	3.01	2.65	2.16	1.77	1.36	.871	.694	.538	.259	.128
14	2.98	2.62	2.14	1.76	1.34	.868	.693	.537	.258	.128
15	2.95	2.61	2.13	1.75	1.34	.866	.691	.536	.258	.128
16	2.92	2.58	2.12	1.75	1.34	.865	690	.535	.258	.128
17	2.90	2.57	2.11	1.74	1.33	.863	.689	.534	.257	.128
18	2.88	2.55	2.10	1.73	1.33	.862	.688	.534	.257	.128
19	2.87	2.54	2.09	1.73	1.33	.861	.688	.533	.257	.127
20	2.84	2.53	2.09	1.72	1.32	.860	.687	.533	.257	.127
21	2.83	2.52	2.08	1.72	1.32	.859	.686	.532	.256	.127
22	2.82	2.51	2.07	1.72	1.32	.858	.686	.532	.256	.127
23	2.81	2.50	2.07	1.71	1.32	.858	.685	.532	.256	.127
24	2.80	2.49	2.06	1.71	1.32	.857	.685	.531	.256	.127
25	2.79	2.48	2.06	1.71	1.32	.856	.684	.531	.256	.127
26	2.78	2.48	2.05	1.71	1.32	.856	.684	.531	.256	.127
27	2.77	2.47	2.05	1.71	1.31	.855	.683	.531	.256	.127
28	2.76	2.47	2.05	1.70	1.31	.855	.683	.530	.256	.127
29	2.76	2.46	2.04	1.70	1.31	.854	.683	.530	.256	.127
30	2.75	2.46	2.04	1.70	1.30	.853	.683	.530	.256	.127
40	2.70	2.43	2.02	1.68	1.30	.851	.681	.529	.255	.126
60	2.66	2.39	2.00	1.67	1.30	.848	.679	.528	.254	.126
120	2.62	2.36	1.98	1.66	1.29	.845	.677	.526	.254	.126
∞	2.58	2.33	1.96	1.645	1.28	.842	.674	.524	.253	.126

Consequently:

$$t\,\frac{1.98}{3} = 0.92 \left(t\,\frac{0.4}{3} = 0.19 \right)$$

Therefore:

$$t = \frac{3 \times 0.92}{1.98} = 1.39 \left(t = \frac{3 \times 0.19}{0.4} = 1.39 \right)$$

In Table 17-13 it is seen that for $N - 1 = 8$, which is the case under consideration here, $t_{90} = 1.40$, which is close enough to 1.39, t_{90} leaves a tail of 10% on either side under the curve, that is it corresponds to a confidence level of 80%, which is the answer to the question.

.2 Hypothesis Tests for the Population Mean

Standard deviation of the population (σ') is known: As has been seen, in the case of a large number of samples (as in Table 17-6, for example), it is possible to work on the hypothesis that the standard deviation of the population (σ') is known, since it can be estimated on the basis of $\sigma_{\bar{x}}$, which can always be evaluated.

Strictly, a similar result is obtained when working with a single very large sample. If the sample was made up of as many items as the original population, clearly $\sigma = \sigma'$. If the sample is large, σ may be a close approximation to σ', and it is possible to work on the hypothesis σ', which is known, estimating it on the basis of σ. In practice, this second possibility of finding σ' is usually accepted so long as $N \geqslant 30$.

To transform a simple statistical estimation into a quality control program, it becomes necessary to introduce the concept of hypothesis tests, also referred to as decision-making rules.

The foregoing establishes a distinction between the concept of statistical estimation and that of existence of a control based on this estimation. It has already been said that a statistical estimation simply makes it possible to establish what is the probability that in a given process a certain concept (in the foregoing analyses, the mean) will be located between given limits. In order to answer this question, statistical methods automatically take into account the nature and variability of the process under study. When a process is subject to control, not only must the limits within which a certain parameter varies under a certain confidence level be indicated, but also what this variation signifies within the process in question, indicating whether the variation that is observed is included in the inherent error of the process (inevitable according to the laws of statistics) or exceeds it with a deviation that, according to these laws, can be at least partially avoided. Eventually, the engineer must seek out its cause and correct it, which is no longer a problem of quality control.

Decision-making rules are the ingredient required to convert a statistical estimation into a control rule.

A hypothesis test or decision-making rule is any procedure that enables the decision to be based on statistical studies on samples from any process. The validity of a decision is determined by examining two possible courses of action. The first, known as the null hypothesis, consists of establishing the equality of two concepts ($c_1 = c_2$). The second known as the alternative hypothesis, consists of one of the following possibilities:

$$c_1 > c_2$$
$$c_1 < c_2$$
$$c_1 \neq c_2$$

A type *I* error is said to be made when a hypothesis is rejected which should have been accepted. When a hypothesis is accepted which should have been rejected, a type *II* error is said to be made.

When testing a hypothesis, there is always some risk of making a type *I* error. The maximum risk that is accepted of making a type *I* error is known as the *level of significance* of the test. Thus, if during a hypothesis test a level of significance of 5% is adopted, this implies acceptance of 5 possibilities out of 100 of rejecting the hypothesis, and it should therefore be accepted. The complement of 100 of the level of significance is known as the *confidence level.* In the foregoing example, there would be a confidence level of 95% of accepting the hypothesis that should have been accepted. Strictly, the confidence level just defined coincides with the confidence level discussed in previous pages.

Let it now be assumed that it is wished to test the hypothesis that a population mean ($\overline{x}'$) is equal to the value c_1, against the alternative hypothesis that this mean is equal to the value c_2, where $c_2 > c_1$. Assuming that $\overline{x}$ (the mean of the samples that are available) has a normal distribution, Fig. 17-14 [24] is a graph showing the relationship between the Type *I* and Type *II* errors corresponding to this case. The figure shows two Gaussian curves that would be obtained if the distribution of the values of $\overline{x}$ was normal and their mean ($\overline{x} = \overline{x}'$) was c_1, and similarly for c_2. Clearly the two curves should be identical, since they represent the same hypothetical distribution. They will be displaced by only $c_2 - c_1$.

The decision-making rule for accepting or rejecting a given hypothesis would be as follows:

If the hypothesis $\overline{x}' = c_1$ is acceptable, the curve on the left in Fig. 17-14 would represent the true distribution of the values of $\overline{x}$. This curve extends from $-\infty$ to $+\infty$ on the $\overline{x}$ axis, which means that any value of $\overline{x}$ that is obtained from a real sample could be the abscissa of a point on that curve, and any discriminatory criterion for finding if the hypothesis being tested is satisfied or not is impossible. In other words, any value of $\overline{x}$ obtained from a sample can be assigned to any distribution.

Assuming now that a parameter k is adopted on the $\overline{x}$ axis between c_1 and c_2, the following criterion could be established as reasonable. If the $\overline{x}$ value for a sample taken from the population is smaller than k, the hypothesis that c_1 is the distribution mean could be considered acceptable. On the other hand, if the sample mean proves to be greater than k, the hypothesis could be considered rejectable. In this way a type *I* error is made in all the cases where $\overline{x}$ is higher than k, for the test hypothesis ($\overline{x}' = c_1$) is rejected despite the sample mean moving under points on the normal curve on the left. According to the ideas previously expounded, area α in Fig. 17-14 automatically indicates the probability of a type *I* error if the previous decision-making criterion is adopted.

If $\overline{x} > k$, the hypothesis tested should be rejected and the alternative hypothesis ($x' = c_2$) should be accepted. In this case the normal curve on the right in Fig. 17-14 would be regarded as truly representing the distribution of the values of $\overline{x}$. This alternative hypothesis will be accepted so long as $\overline{x} > k$, but if $\overline{x} < k$ a hypothesis will have been accepted which should have been rejected; in other words, a type *II* error will have been made. Consequently, area β in Fig. 17-14 represents the risk of making a type *II* error when the decision sequence just described is followed.

It can be observed in Fig. 17-14 that an increase in the value of k leads to a reduction in area α, and as a consequence there is a reduction in the risk of a type *I* error, but area β increases with a heavier risk of a type *II* error. In [26] and many other texts on the subject, the following principle is established for all hypothesis tests for quality control purposes: of all the tests having the same risk of a type *I* error, the one should be chosen with the smallest risk of a type *II* error.

The following sequence is proposed for hypothesis tests [27,28]:

1) Establish the test or null hypothesis and the alternative hypothesis. The criterion for this should be based on careful handling of any previous information available on the problem in question.
2) Select the level of significance at which it is wished to work.
3) Select the type of distribution that will be considered for the sample means ($\overline{x}$). σ being known, as in this particular case,

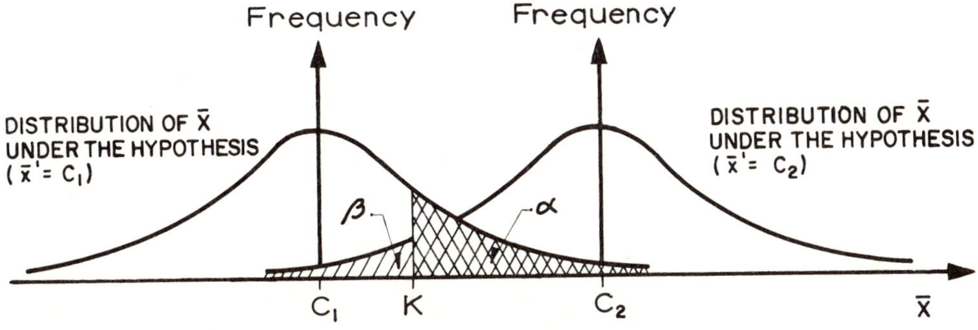

Fig. 17-14 Probabilities of Error Types *I* and *II* in hypothesis tests [24]

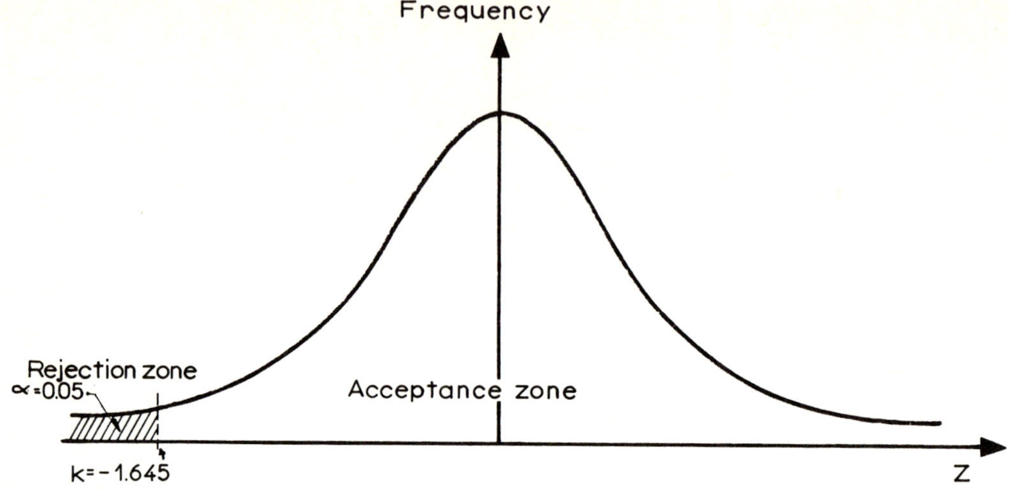

Fig. 17-15 Determination of the critical zone by means of the α value in a one-sided hypothesis test

it will be reasonable to regard the distribution of the values of $\bar{x}$ as normal.

4) Find the critical region, establishing the value of parameter k. This must be done with specific reference to the problem in hand.

5) Determine the acceptance zone and the rejection zone. The rejection zone is area α in Fig. 17-14.

6) Select a random sample consisting of N items. Determine parameter k_0 on the basis of the distribution assumed in step 3, which will serve as a basis for comparison with the value of k selected previously, for acceptance or rejection of the test hypothesis.

By way of an example, let us consider the sample represented by line 5 in Table 17-6. The sample mean is 1750 kg/m^3 (110 lb/ft^3) ($\bar{x}$). The standard deviation of the population (σ') will be assumed to be known and equal to $\sigma_{\bar{x}}\sqrt{N}$. The test or null hypothesis will consist of the following statement: The population mean ($\bar{x}'$) is 1677.5 kg/m^3 (105 lb/ft^3). The following alternative hypothesis will be adopted:

1) The population mean ($\bar{x}'$) is smaller than 1677.5 kg/m^3 (105 lb/ft^3).

2) It is wished to work with a confidence level of 95%. Consequently the level of significance of the hypothesis test will be $0.05 = \alpha$.

3) The distribution of the sample mean ($\bar{x}$) will be considered normal.

4) To select k the following argument will be followed, based on Fig. 17-15. Since the distribution of the values of $\bar{x}$ is normal, the corresponding Gaussian curve can be plotted. The value of α is 0.05, which means that area α must include 5% of the possibilities of $\bar{x}$.

   Equation (17-5) showed a convenient mechanism for changing the variable, which is frequently adopted in statistical calculations because the tables of areas under the normal curve available in published literature are usually expressed in terms of the new variable z, where $z = (x - \bar{x})/\sigma$.

   In the above it was wished to change variable x, with mean $\bar{x}$ and standard deviation σ, to z. Accordingly, variable $\bar{x}$, with mean $\bar{x}'$ and standard deviation $\sigma'/\sqrt{N}$ will now be changed to z, with the following result:

$$z = \frac{\bar{x} - \bar{x}'}{\sigma'/\sqrt{N}} \tag{17-21}$$

Using the *normalized* variable in Table 17-12, for a confidence level of 95% (level of significance: 0.05) it is possible to find the value of abscissa z for which 0.05 of the total area is on the left. In this case, the hypothesis test consists only of judging whether the population mean is equal to or smaller than 1677.5 kg/m^3 (105 lb/ft^3). In other words, it is only of interest to analyse an area α to the left of the normal curve and equal to the level of significance. The semi-area to the right of the normal curve is 0.5, but to the left there will be an area equal to only 0.45 of the total, α being the remaining 0.05. In Table 17-12 it is seen that for an area 0.45 the value of $z = t$ proves to be $-$ 1.645 (the minus sign is because it is to the left of the mean) which is obtained by interpolating between the areas 0.4495 and 0.4505 which are in the Table. Thus, in the case under analysis, $z = -1.645$. This is the value of k that must be selected in this case.

5) The acceptance zone will be the entire area under the normal curve to the right of k, and the rejection zone will be the area under the normal curve to the left of k.

6) The value of k_0 will be the particular value of z corresponding to the data for the sample obtained; in other words, it will be obtained by applying Eq.(17-21).

$$z = k_0 = \frac{\bar{x} - \bar{x}'}{\sigma'/\sqrt{N}}$$

$$= \frac{1750.0 - 1677.5}{115.15} \left(\frac{110 - 105}{7.2}\right)$$

It should be recalled (Table 17-6) that:

$$\frac{\sigma'}{\sqrt{N}} = \sigma_{\bar{x}} = 115.15 \text{ (7.2 lb/ft}^3\text{)}$$

Consequently:

$$z = k_0 = \frac{72.50}{115.15} = 0.63 \left(\frac{4.5}{7.2} = 0.63\right)$$

The plus sign in the result obtained immediately denotes that k_0 is located in the acceptance zone in Fig. 17-15. That is to say, it can be stated with 95% confidence that the population mean corresponding to the sample studied = 1677.5 kg/m^3 (105 lb/ft^3).

The potential of a calculation like that just described in a practical quality control problem is obvious. Let it be assumed that the compaction of a large section of road is to be controlled, and the measurements given in Table 17-6 correspond to a sub-section of that road. With these values it will be possible to find $\sigma_{\bar{x}}$. With this information and the previous calculation, it will be possible to estimate the compaction mean obtained in all the other sub-sections of the road, using a far simpler, quicker and more economic sampling process at the confidence level that is desired. In order to extend the statistical analysis conducted on just one section of road to cover the remaining sub-sections, the characteristics of the materials, compaction equipment and compaction procedures must remain the same. If in a particular sub-section it is found that during a hypothesis test using the same confidence level the population mean suddenly leaves the acceptance zone, it can be said that the compaction process has got out of control for some reason which will have to be investigated using the normal methods.

The previous example is known in statistics as the one-sided test and occurs when the alternative hypothesis expresses only a condition of *greater than* or *smaller than*. Other hypothesis tests may, however, occur like that in the following example: a two-sided test.

1) The following test or null hypothesis will be accepted: the population mean (Table 17-6) = 1677.5 kg/m^3 (105 lb/ft^3). The alternative hypothesis will be: the population mean ≠ 1677.5 kg/m^3 (105 lb/ft^3).
2) The confidence level at which it is wished to work is 95% (level of significance: $\alpha = 0.05$). Line 20 in Table 17-6, with a mean of 1780 kg/m^3 (112 lb/ft^3), is a sample of the population.
3) A normal distribution will be accepted for the sample mean ($\bar{x}$).
4) The value of k will be found in Table 17-12, considering that the value of z can now be to the right or left of the mean. Since the risk of the population mean being on one side or the other of the pre-established value is the same, the level of significance will be divided into two areas that are symmetrical to the normal distribution mean (Fig. 17-16). There will be two areas of 0.025, one on each side. In Table 17-12 it is seen that for an area 0.475 of half of the normal curve (that is 0.5 − 0.025), $t = k \pm 1.96$ is obtained.
5) The acceptance zone will be the entire area under the normal curve between the two diagonally lined areas in Fig. 17-16, and accordingly the rejection zone will be these two hatched areas.
6) The value of k_0 will be:

$$z = k_0 = \frac{\bar{x} - \bar{x}'}{\sigma'/\sqrt{N}} = \frac{1780.0 - 1677.5}{115.15}$$

$$= \frac{102.5}{115.15} = 0.89$$

$$= \frac{111.4 - 105.0}{7.2} = \frac{6.4}{7.2} = 0.89$$

As can be seen, the hypothesis proposed must be accepted. It is acceptable with 95% confidence, that is to say, 95 times out of 100 its acceptance will lead to no error, but 5 times out of 100 a type *I* error will be made. With reference to practical application, the same comments apply as previously.

Case where the standard deviation of the population (σ') is unknown.

In this case, estimation of the population mean should be based on the sample mean, but otherwise the statistical working methods are similar to those employed in the previous case. The distribution of the sample means is no longer normal, but of the *Student* type (Table 17-13).

The normalized variable equivalent to z, above, is:

$$t = \frac{\bar{x} - \bar{x}'}{\sigma/\sqrt{N}} \tag{17-22}$$

Using this normalized variable t and Table 17-13, it is possible to find the concrete value of t that corresponds to the level of significance with which the hypothesis test is performed (this is the value of k mentioned previously, which determines the rejection zone). The value of k to compare for test purposes will also be obtained by introducing into Eq.(17-22) the values corresponding to the available sample.

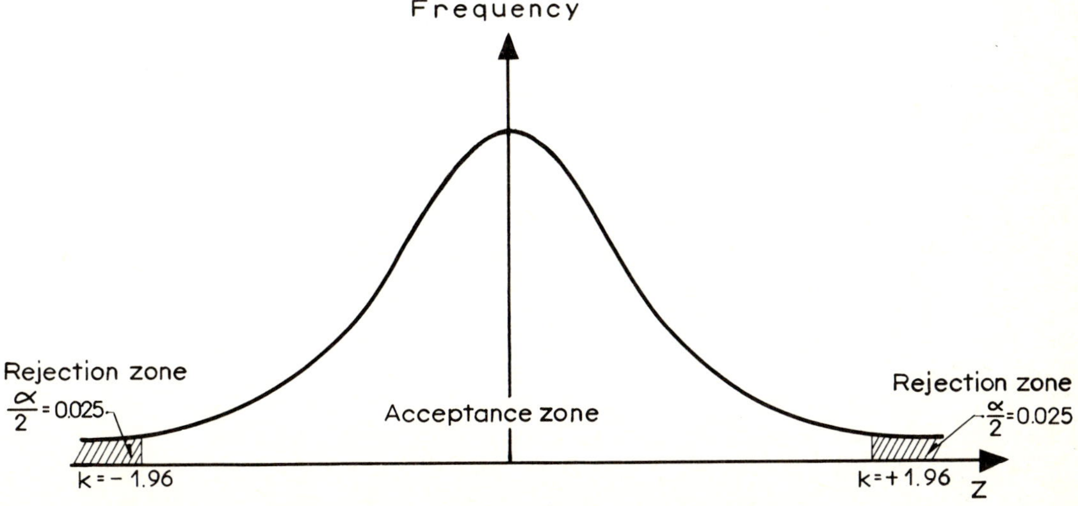

Fig. 17-16 Determination of the critical zone by means of the α value in a two-sided hypothesis test

By way of an example and considering as a sample the values of q_u given in Table 17-14 ($\bar{x}$ = 15.53 t/m^2 (3.16 Kips/ft^2); σ = 1.98 t/m^2 (0.4 Kips/ft^2) a hypothesis test will be conducted using the following null hypothesis: *the population mean = 16 t/m^2 (3.26 Kips/ft^2)* and the following alternative hypothesis: *the population mean < 16 t/m^2 (3.26 Kips/ft^2).*

A level of significance of 0.05 is desired, which is equivalent to a confidence level of 95%. In Table 17-13 it is seen that for a one-sided test like the one proposed, in Line 8 for $N - 1 = 8$, $t_{0.95} = \pm 1.86$ is obtained.

The value of k_0 will be:

$$t = k_0 = \frac{15.53 - 16.00}{1.98/\sqrt{9}} = \frac{-3 \times 0.47}{1.98} = -0.71$$

$$t = k_0 = \frac{3.16 - 3.26}{0.4/\sqrt{9}} = \frac{-3 \times 0.10}{0.4} = -0.71$$

-0.71 is greater than -1.86 (since this is a *smaller than* test the value of k to the left of the mean is taken, that is -1.86). From this it is deduced that the hypothesis should be accepted or, what comes to the same thing, the population mean is 16.0 t/m^2 (3.26 Kips/ft^2) with a 95% probability that there will be no type *I* error.

The above is an example of a one-sided test. Now a two-sided test will be described. Test hypothesis: The population mean of which the values in Table 17-14 are a sample, is 25.5 t/m^2 (5.2 Kips/ft^2) The alternative hypothesis will be that the population mean is not equal to 25.5 t/m^2 (5.2 Kips/ft^2).

Table 17-14
Values of q_u for a soil, t/m^2 (Kips/ft^2)

15.96 (3.26)	12.32 (2.52)	17.28 (3.52)
15.63 (3.19)	12.40 (2.53)	16.96 (3.46)
17.60 (3.59)	17.64 (3.60)	14.56 (2.96)

It is wished to work with a level of significance of 0.1 (confidence level = 90%) Table 17-13 is consulted, but bearing in mind that in a two-sided test with a confidence level of 90%, an area of 0.05 must be left on *either* side of the Student distribution, and since the Table gives values of t corresponding to areas on only one side of the distribution, the coefficient will be found in column $t_{0.95}$. Optionally Eq.(17-20) can be used to determine the index of t under which the coefficient will be found on the line for $N - 1$. In this case for $N - 1 = 8$ and $t_{0.95}$, $k = 1.86$ will be obtained.

The value of k_0 can now be found using the expression:

$$k_0 = \frac{15.53 - 25.50}{1.98/\sqrt{9}} = -\frac{3 \times 9.97}{1.98} = -15.1$$

$$k_0 = \frac{3.16 - 5.20}{0.4/\sqrt{9}} = -\frac{3 \times 2.04}{0.4} = -15.1$$

Clearly -15.1 falls outside the acceptance interval, which is limited by the values ± 1.86, so that the test hypothesis must be rejected and 25.5 does not represent the population mean for the confidence level established.

A test like this, regardless that the values used are well outside the acceptance interval, might be used to calibrate laboratory equipment or to find out at what point a process involving constant determination of the compressive strength of a material over a certain period of time gets out of control. In any case, a series of preliminary data are obtained which could be appropriately taken as the mean and the standard deviation of the sample. On a certain day the test equipment will give a series of values which are suspiciously different from the usual mean of the material tested. A hypothesis test like the one just described may tell the engineer whether this deviation is inherent in the process or whether it represents an effect that is out of control because the equipment requires adjustment or because there has been an error in testing, for example due to a new operator (no change in the material is considered).

.3 Methods for Estimating the Standard Deviation of a Population

As in the case of the mean, statistical estimation consists of determining the confidence interval for the standard deviation of the population after the confidence level to be adopted has been established. Texts on statistics show that the confidence interval of the standard deviation depends rather more on the variance concept (σ^2), which was mentioned in §17.2. They also show that the distribution of the variance of a normal distribution is not normal, but of the type known as the χ^2 distribution (Table 17-15). The equation limiting the confidence interval of the variance in this case is as follows:

$$\frac{N\sigma^2}{\chi_c^2} < \sigma'^2 < \frac{N\sigma^2}{\chi_{c'}^2} \qquad (17\text{-}23)$$

where: N is the number of items in the sample representing the population. σ is the standard deviation of that sample. The square of this value is the corresponding variance. σ' is the standard deviation of the population, the value of which is being estimated. χ_c^2 and $\chi_{c'}^2$ are the limits of the distribution variable now in use. Working with a confidence level of 90%, for example, both values can be obtained in Table 17-15. The area under the curve is as always equal to one. The limits χ_c^2 and $\chi_{c'}^2$ must be such that they divide the area into three portions, a central portion with a value equal to the confidence level selected and two lateral portions, respectively equal to half of the complement of one of the confidence level. If the confidence level is 90%, as stated, the values should be sought in the line for $N - 1$, columns $\chi_{0.05}^2$ and $\chi_{0.95}^2$.

By way of an example, let it be assumed that the sample is the one given by the values in Table 17-14, with a mean of 15.53 t/m^2 (3.16 Kips/ft^2) and a standard deviation of 1.98 t/m^2 (0.4 Kips/ft^2). The question that might be asked is within what interval of values will the standard deviation of the population be found with a 90% probability (level of confidence) on the basis of the standard deviation of the sample.

For this case $N = 9$, $\sigma^2 = 1.98^2 = 3.92$, $\chi_c^2 = 15.5$, $\chi_{c'}^2 = 2.73$. Therefore:

$$\frac{9 \times 3.92}{15.5} < \sigma'^2 < \frac{9 \times 3.92}{2.73}$$

the following interval being obtained:

Table 17-15

Values of χ^2 in the distribution of the same name

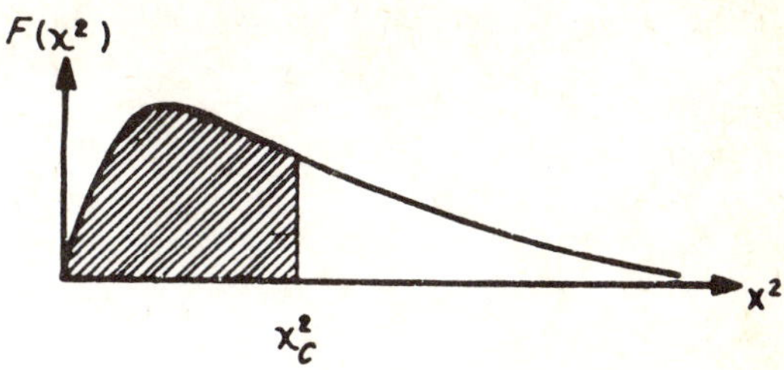

N-1	$\chi^2_{.995}$	$\chi^2_{.99}$	$\chi^2_{.975}$	$\chi^2_{.95}$	$\chi^2_{.90}$	$\chi^2_{.75}$	$\chi^2_{.50}$	$\chi^2_{.25}$	$\chi^2_{.10}$	$\chi^2_{.05}$	$\chi^2_{.025}$	$\chi^2_{.01}$	$\chi^2_{.005}$
1	7.88	6.63	5.02	3.84	2.71	1.32	.455	.102	.016	.0039	.0010	.0002	.0000
2	10.6	9.21	7.38	5.99	4.61	2.77	1.39	.575	.211	.103	.0506	.0201	.0100
3	12.8	11.3	9.35	7.81	6.25	4.11	2.37	1.21	.584	.352	.216	.115	.072
4	14.9	13.3	11.1	9.49	7.76	5.39	3.36	1.92	1.06	.711	.483	.297	.207
5	16.7	15.2	12.8	11.15	9.2	6.63	4.35	2.67	1.61	1.15	.831	.554	.413
6	18.5	16.8	14.4	12.6	10.6	7.84	5.35	3.45	2.20	1.64	1.24	.872	.676
7	20.3	18.5	16.0	14.1	12.0	9.04	6.35	4.25	2.83	2.18	1.69	1.24	.989
8	22.0	20.1	17.5	15.5	13.4	10.2	7.34	5.07	3.49	2.73	2.18	1.65	1.34
9	23.6	21.7	19.0	16.9	14.7	11.4	8.34	5.90	4.17	3.33	2.70	2.09	1.73
10	25.2	23.2	20.5	18.3	16.0	12.5	9.34	6.74	4.87	3.94	3.25	2.56	2.16
11	26.8	24.7	21.9	19.7	17.3	13.7	10.35	7.57	5.58	4.57	3.82	3.05	2.60
12	28.3	26.2	23.2	21.0	18.5	14.8	11.3	8.44	6.30	5.23	4.40	3.57	3.07
13	29.8	27.7	24.7	22.4	19.8	16.0	12.3	9.30	7.04	5.89	5.01	4.11	3.57
14	31.3	29.1	26.1	23.7	21.1	17.2	13.3	10.2	7.79	6.57	5.63	4.66	4.07
15	32.7	30.6	27.5	25.1	22.3	18.2	14.3	11.0	8.55	7.26	6.25	5.22	4.60
16	34.3	32.0	28.8	26.3	23.5	19.4	15.3	11.9	9.31	7.96	6.91	5.81	5.14
17	35.7	33.4	30.2	27.6	24.8	20.5	16.3	12.8	10.1	8.67	7.56	6.41	5.70
18	37.2	34.8	31.5	28.9	26.0	21.6	17.3	13.7	10.9	9.39	8.23	7.01	6.26
19	38.6	36.2	32.9	30.1	27.2	22.7	18.3	14.6	11.73	10.1	8.91	7.63	6.84
20	40.0	37.6	34.2	31.4	28.45	23.8	19.3	15.5	12.4	10.9	9.59	8.26	7.43
21	41.4	38.8	35.6	32.7	29.6	24.9	20.3	16.3	13.2	11.6	10.3	8.90	8.02
22	42.8	40.3	36.8	33 9	30.8	26.0	21.3	17.2	14.0	12.3	11.0	9.54	8.64
23	44.2	41.6	38.1	35.2	32.0	27.1	22.3	18.1	14.8	13.1	11.7	10.2	9.26
24	45.6	43.0	39.4	36.4	33.2	28.2	23.3	19.0	15.7	13.8	12.4	10.9	9.89
25	46.9	44.3	40.6	37.7	34.4	29.03	24.3	19.9	16.5	14.5	13.15	11.5	10.5
26	48.3	45.6	41.9	38.9	35.6	30.4	25.3	20.8	17.3	15.4	13.8	12.2	11.2
27	49.6	47.0	43.2	40.1	36.7	31.5	26.3	21.7	18.1	16.2	14.6	12.9	11.8
28	51.0	48.3	44.5	41.3	37.9	32.6	27.3	22.7	18.9	16.9	15.3	13.6	12.5
29	52.3	49.6	45.7	42.5	39.1	33.7	28.3	23.6	19.8	17.7	16.0	14.3	13.1
30	53.7	50.9	47.0	43.8	40.3	34.8	29.3	24.5	20.6	18.5	16.8	15.0	13.8
40	66.8	63.7	59.3	55.8	51.8	45.7	39.3	33.7	29.1	26.5	24.4	22.2	20.7
50	79.5	76.2	71.4	67.5	63.2	56.3	49.3	43.0	37.7	34.8	32.4	29.7	28.0
60	92.0	88.4	83.3	79.1	74.4	67.0	59.3	52.3	46.5	43.2	40.5	37 5	35.5
70	104.2	100.4	95.0	90.5	85.5	77.6	69.3	61.7	55.3	51.7	48.8	45.4	43.3
80	116.3	112.3	106.6	101.9	96.6	88.1	79.3	71.1	64.3	60.4	57.2	53.5	51.2
90	128.3	124.1	118.1	113.1	107.6	98.6	89.3	80.6	73.3	69.1	65.6	61.8	59.2
100	140.2	135.8	129.6	124.3	118.5	109.1	99.3	90.12	82.4	77.9	74.2	70.1	67.3

$$2.28 < \sigma'^2 < 12.92$$

Consequently the standard deviation of the population will be within the interval: $1.51 < \sigma' < 3.59$ $(0.3 < \sigma' < 0.73)$.

This means that on the basis of the data for the sample handled it can be said with 90% confidence that the standard deviation of the population to which the sample belongs is located between the limits indicated. If the confidence level is raised, the interval obtained will increase accordingly.

Equation (17-23) also lends itself to the solution of another aspect of the problem, which consists of finding what the probability is of the standard deviation of the population being located between certain pre-established limits. By way of an example, the probability will be calculated of the standard deviation in question being situated between the limits 2.0 and 3.0. Expression (17-23) can now be written thus:

$$4.0 = \frac{N\sigma^2}{\chi^2_c} < \sigma'^2 < \frac{N\sigma^2}{\chi^2_{c'}} = 9.0$$

Consequently:

$$\chi^2_c = \frac{9 \times 3.92}{4} = 8.82 \text{ and}$$

$$\chi^2_{c'} = \frac{9 \times 3.92}{9.0} = 3.92$$

In Table 17-15 it is seen that these limits correspond to $(N - 1 = 8)$ $\chi^2_{0.65}$ and to $\chi^2_{0.15}$. This means that on one side of the χ^2 distribution an area of 0.35 is left and on the other side another area of 0.15, which means that in all the confidence interval corresponds to an area of 1.0 − (0.35 + 0.15) = 0.50. Thus, there is a 50% probability that the standard deviation of the population will be located between the pre-established limits 2.0 and 3.0. This situation is represented in Fig. 17-17, where the characteristic shape of the χ^2 distribution can also be appreciated. It is worth observing that a χ^2 distribution becomes more and more like a normal distribution as N increases.

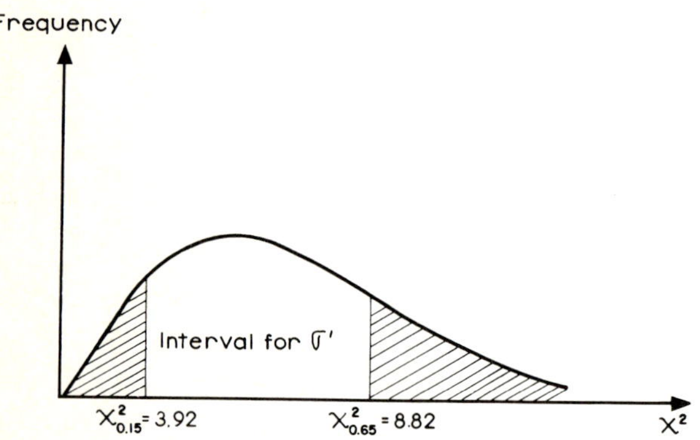

Fig. 17-17 Calculation of the probability of keeping within a pre-established interval

.4 Hypothesis Tests for the Standard Deviation of a Population

The mechanism of these hypothesis tests is the same as that described for the population mean. The distribution to apply here is systematically χ^2, which is the theoretical distribution of variance. In the case under analysis, one- or two-sided tests can also be employed. Considering either of the two sides, it is deduced from Eq.(17-23) that:

$$\chi^2 = \frac{N\sigma^2}{\sigma'^2}$$

which will be the formula to consider in this case in order to establish the value of k_0.

By way of illustration and using the data in Table 17-6, the following test or null hypothesis will be made: the standard deviation of the population is 258 kg/m³ (16 lb/ft³). The alternative hypothesis is: the standard deviation of the population is smaller than 258 kg/m³ (16 lb/ft³). Line *13* in Table 17-6 is regarded as the sample under analysis. It is wished to work with a confidence level of 95% (level of significance = 0.05).

The value of k is established using Table 17-15. Since the test is one-sided, and it is wished to separate any values of the deviation that are smaller than the one proposed, area α will now be located to the left in the χ^2 distribution, limited by the value $\chi^2_{0.05}$ which in the Table proves to be $(N-1=4)$; = 0.711. The value of k_0 will be obtained by applying:

$$\chi^2 = k_0 = \frac{5 \times 230.2^2}{258^2}\left(\frac{5 \times 14.4^2}{16^2}\right)$$

In Table 17-6 (Line *13*), it is seen that the standard deviation of the sample σ is 230.2 kg/m³ (14.4 lb/ft³). The value σ' = 258 kg/m³ (16 lb/ft³) is the test or null hypothesis. The result is: k_0 = 3.98, 3.98 > 0.711, and consequently falls in the acceptance zone, which indicates that the hypothesis should be accepted with 95% confidence and with a 5% risk of a type *I* error. Fig. 17-18 shows the situation that has been reached.

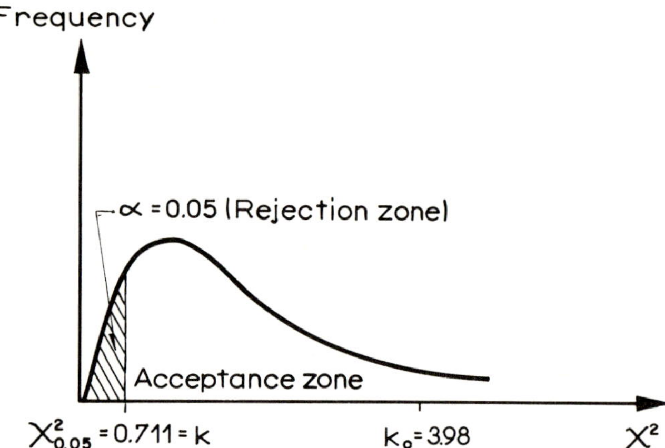

Fig. 17-18 One-sided hypothesis test for controlling variability

An example of a two-sided hypothesis test would be: Null hypothesis: the standard deviation of the population = 258 kg/m³ (16 lb/ft³). Alternative hypothesis: the standard deviation of the population ≠ 258 kg/m³ (16 lb/ft³). Now, if the same level of significance is adopted as previously, under the χ_2 distribution there will be rejection zones of 0.025 both on the left and the right. The corresponding values of k are obtained from Table 17-15, and are:

$$\chi^2_{0.025} = 0.483 \qquad \chi^2_{0.975} = 11.1$$

The value of k_0 would be the same as the one calculated in the previous example, that is 3.98. It is observed that 3.98 is within the interval of acceptance, and consequently the test or null hypothesis is acceptable with 95% probability. Fig. 17-19 illustrates this example.

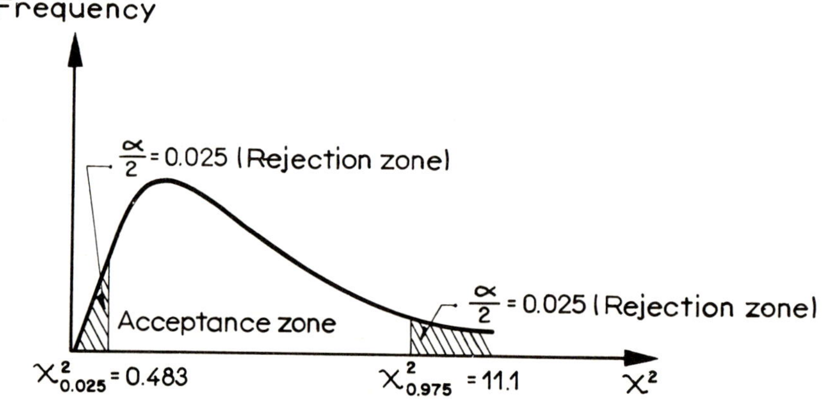

Fig. 17-19 Two-sided hypothesis test for controlling variability

Hypothesis tests for variance or for standard deviation, which is automatically involved too, have a practical application in problems in which the variability of a process is controlled. If a certain property is being measured, its standard deviation is known during the measurement period. If for any reason in a new sample a value for standard deviation appears which is different from the usual value, a test like this will indicate whether the change corresponds to characteristics due to the randomness of the process or whether it is due to factors that are not connected with the process, which can and should be corrected.

.5 Comparison of Two Means

It is often convenient to establish whether or not there are significant differences between the means of two samples for one and the same property. Generally the two samples belong to the same population, but the methods that are described here could be used to compare the means of two different populations. The purpose of these analyses may often be to determine whether the two samples really belong to the same population, perhaps so as to decide whether the process used to produce one of them is different from the process used to produce the other. For instance, it may be wished to discover whether one laboratory functions better than another when conducting a certain test, or whether in the same laboratory a certain detail added to the methodology of a test is or is not significant.

Comparison techniques include the performance of a hypothesis test, in which the null hypothesis is:

$$\bar{x}_A = \bar{x}_B$$

In other words, the equality of population means *A* and *B* is established. The alternative hypothesis may adopt any of the following three forms:

$$\bar{x}_A \neq \bar{x}_B \,;\, \bar{x}_A > \bar{x}_B \,;\, \bar{x}_A < \bar{x}_B.$$

The test in itself is conducted in exactly the same way as the others already studied, rejecting the null hypothesis when the statistical value of the test falls outside the acceptance zone. The first of the three alternative hypotheses leads to a two-sided test, while the other two lead to one-sided tests.

Since the distribution of the samples is affected by the standard deviation of the population, in order to conduct hypothesis tests it is necessary to have knowledge of this last value or to find it by some procedure. There are three possibilities [9].

- σ'_A and σ'_B are known
- The magnitudes of σ'_A and σ'_B are unknown, but it is known or assumed that they are equal.
- The magnitudes of σ'_A and σ'_B are unknown and it is not known if they are equal or different.

Of the foregoing possibilities, the first is unusual in practice, while the second is by far the most frequent one. The third possibility is not unusual and should be observed whenever there is good reason for thinking that σ'_A and σ'_B are indeed different. There are certain tests [29] which may help establish whether the standard deviations σ'_A and σ'_B are equal or different.

The procedures for conducting hypothesis tests in each of the above three cases will now be analysed.

σ'_A and σ'_B are known: In this case the statistical process can be dealt with like a compound process, with the variable $x_A - x_B$ playing the part that is usually assigned to x. The normalized variable can be obtained following the rule given by Eq.(17-5), although expressed as in Eq.(17-21) for the central interest of the analysis is a comparison of the mean values of the variable and not the variable itself.

The standard deviation of the compound process should be calculated with Eq.(17-4). Bearing in mind all the foregoing considerations, the normalized variable will be:

$$z = \frac{(\bar{x}_A - \bar{x}_B) - (\bar{x}'_A - \bar{x}'_B)}{\sqrt{\dfrac{\sigma'^2_A}{N_A} + \dfrac{\sigma'^2_B}{N_B}}} \qquad (17\text{-}24)$$

This variable will have a normal distribution. It will now be necessary to establish the confidence level at which it is wished to work, obtaining from Table 17-12 the value of z corresponding to this level k. This value should be compared with the one obtained from Eq.(17-24) calculated for the specific conditions of the problem, k_0.

By way of example, let it be assumed that two sections of a road are subjected to sampling for compaction purposes. Both samples could be similar to Table 17-6. Let it be assumed also that the sampling values proved to be:

$$\bar{x}_A = 1722.3 \text{ kg/m}^3 \ (107.6 \text{ lb/ft}^3)$$
$$N_A = 100$$
$$\sigma'_A = 110.4 \text{ kg/m}^3 \ (6.9 \text{ lb/ft}^3)$$

$$\bar{x}_B = 1689.5 \text{ kg/m}^3 \ (105.7 \text{ lb/ft}^3)$$
$$N_B = 81$$
$$\sigma'_B = 112.6 \text{ kg/m}^3 \ (7.9 \text{ lb/ft}^3)$$

It is wished to find out if there are any significant statistical differences between the mean values of the two populations. The confidence level for the conclusion is to be 95%. In other words, it is wished to find out with 95% confidence whether the compaction work conducted on the two sections of road is equivalent, assuming that the same materials have been used.

The following hypothesis test will be conducted:

Null hypothesis: $\bar{x}'_A = \bar{x}'_B$

Alternative hypothesis: $\bar{x}'_A \neq \bar{x}'_B$

The value of k can be obtained from Table 17-12. Since this is a two-sided test, the value of the area equal to 0.475 must be found, $t = \pm 1.96$ being obtained. Note that Table 17-11, which also includes 95%, could have been used. The value of k_0 can be obtained from Eq.(17-24):

$$Z = k_0 = \frac{1722.3 - 1689.5}{\sqrt{\dfrac{110.4^2}{100} + \dfrac{112.6^2}{81}}}$$

$$= \frac{38.2}{\sqrt{121.88 + 156.53}} = \frac{32.8}{16.7} = 1.96$$

By chance, the value of k_0 proved to be equal to the upper acceptance limit (1.96). Consequently the hypothesis that the compaction work carried out on the two sections is equivalent can be accepted.

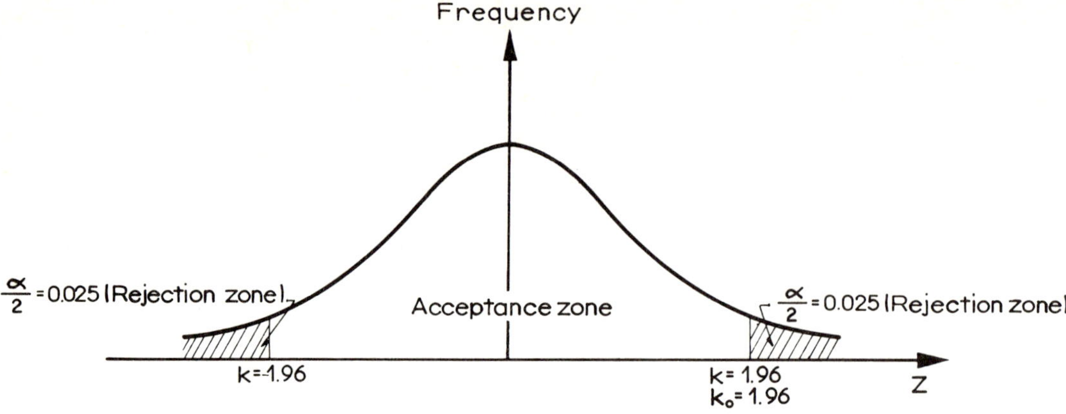

Fig. 17-20 Determination of acceptance and rejection zones for a hypothesis test for comparing two population means

A few comments are necessary at this point. First, the method proposed is a rational aid in the solution of many of the problems that usually arise in road building projects. It will undoubtedly be better to base decisions on a criterion like the one just described rather than on personal judgment. Second, it is clear from the example that statistical control methods are far superior to observational ones. Assuming, for example, that the degree of compaction of population A is 100%, the maximum dry density obtained in the control test is 1722.3 kg/m^3 (107.6 lb/ft^3). In this case, the value 1689.5 kg/m^3 (105.6 lb/ft^3) obtained as the mean of sample B will represent 98% of the degree of compaction. If the degree of compaction specified for the project is 100%, section B will be rejected by a strict (or conscientious) inspector. However, the section does not deserve to be rejected. Of course in the foregoing example the values are very close and the relevant discussion might prove a little academic. However, conceptually similar situations often arise with values that are rather more widely dispersed. The important point is that judgment based on personal opinion, however well it may be supported by sound previous experience, cannot hope to distinguish at a glance between the errors inherent in a random process, and those which are a result of defective work and which can be corrected.

σ'_A and σ'_B are unknown, but it is known or assumed that they are equal: In this case too the statistical process should be handled like a compound process, bearing in mind that the distribution of the sample means is not normal, but of the *Student* type already mentioned. The expression for the normalized variable is that given in Eq.(17-22) and in this particular case can be written thus:

$$t = \frac{(\bar{x}_A - \bar{x}_B) - (\bar{x}'_A - \bar{x}'_B)}{s\,(1/N_A + 1/N_B)^{1/2}} \qquad (17\text{-}25)$$

A discussion of those cases where use of the Student distribution leads to mathematically exact or only approximate solutions can be found in [30].

In the above expression:

$$s = \left(\frac{N_A\sigma_A^2 + N_B\sigma_B^2}{N_A + N_B - 2}\right)^{1/2} \qquad (17\text{-}26)$$

The mechanism of the hypothesis test here is similar to that of the others already described. The confidence level is selected and is used to obtain the corresponding values of t on line $N_A + N_B - 2$ in Table 17-13, taking into consideration whether the test is one or two-sided.

Using Eq.(17-25), a value must then be calculated for k_0, comparing it with k to determine whether it falls within the acceptance or rejection zone.

As an example, let it be assumed that a certain laboratory gives the results shown in Table 17-14 ($N_A = 9$, $\sigma_A = 1.98$ t/m^2 (0.4 $Kips/ft^2$); $\bar{x}_A = 15.53$ t/m^2 (3.16 $kips/ft^2$). Another laboratory gives results for a similar sample, measuring $\bar{x}_B = 17.0$ t/m^2 (3.47 $kips/ft^2$), which has a standard deviation of $\sigma_B = 1.82$ t/m^2 (0.37 $kips/ft^2$) and is made up of 12 items ($N_B = 12$).

It is wished to find out if the results of the two laboratories are statistically in agreement or whether there is any significant difference between them. A confidence level of 90% is desired.

The null hypothesis to be adopted will be that where $\bar{x}'_A = \bar{x}'_B$, and the alternative hypothesis $\bar{x}'_A < \bar{x}'_B$, with which a one-sided test is defined.

In Table 17-13 it can be seen that $k = t_{90} = -1.33$ (calculated on line $N_A + N_B - 2 = 19$).

$$s = \left(\frac{15.53 \times 1.98^2 + 17.0 \times 1.82^2}{19}\right)^{1/2}$$

$$= \left(\frac{60.88 + 56.27}{19}\right)^{1/2} = (6.16)^{1/2} = 2.48$$

With this value of s, Eq.(17-25) will be applied:

$$t = k_0 = \frac{15.53 - 17.00}{2.48\left(\frac{1}{9} + \frac{1}{12}\right)^{1/2}} = -\frac{1.47}{\frac{2.65}{6}\,2.48}$$

$$= -\frac{8.82}{6.57} = -1.34$$

Since -1.34 is smaller than -1.33, the null hypothesis must be rejected, concluding that the differences obtained between the results of the two laboratories are due to causes which go beyond the inherent error of the test and which must therefore be reviewed. Figure 17-21 illustrates this situation.

σ'_A and σ'_B are unknown and it is not known whether they are equal or different: The procedure here is again similar to that in the previous cases, the Student distribution being employed for the sample means and the following expressions:

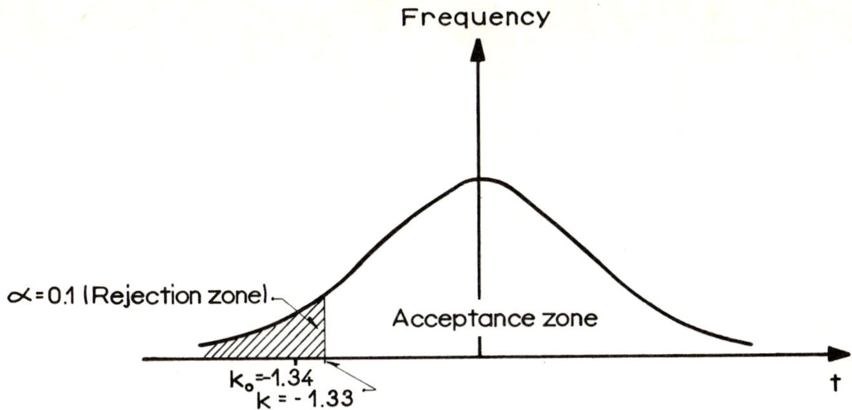

Fig. 17-21 Determination of acceptance and rejection zones for a hypothesis test for comparing two population means, using the Student t distribution

$$t = \frac{(\bar{x}_A - \bar{x}_B) - (\bar{x}'_A - \bar{x}'_B)}{(\sigma_A^2/N_A + \sigma_B^2/N_B)^{1/2}} \tag{17-27}$$

The line to be consulted in Table 17-13 in order to find the value of k, is given by expression:

$$r = \frac{1}{\dfrac{c^2}{N_A - 1} + \dfrac{(1-c)^2}{N_B - 1}} \tag{17-28}$$

where:

$$c = \frac{\sigma_A^2/N_A}{\sigma_A^2/N_A + \sigma_B^2/N_B} \tag{17-29}$$

17.4.3 Use of Statistical Control Methods

In Civil Engineering statistical control methods have not been as intensively and systematically employed as in many other industrial operations. This is a serious restriction in heavy construction techniques. With particular reference to road engineering, statistical concepts are seldom used for quality control or for risk evaluation problems, which are an important and little used aspect of this activity.

References [31-38] give the reader an idea of the degree of application of these methods in the techniques of certain countries. Reference [39] gives a very full description of how many of the methods described earlier in this Chapter can be applied to the control of concrete structures, but it also provides some general application criteria, which can easily be extended to road engineering practice.

It is a commonplace remark that statistical quality control methods lead to a rather idealistic methodology, beyond the true scope of the average engineer. This statement is debatable from more than one point of view. First, statistical quality control is more economical than the more traditional methods, since it demands less sampling and less laboratory work. Moreover, control work can be clearly and rapidly interpreted, whereas the more conventional methods, in which conclusions are based on data accumulated from numerous tests, have the disadvantage that nobody can find sufficient time to reach a rational interpretation of such a large amount of information.

Second, as the reader will probably have concluded by now, statistical control methods offer a logical analysis, which it would be very hard to achieve by simply accumulating observations. From this general outline, the reader will probably also have reached the conclusion that these methods are practical, economical and easy to apply, especially when control work is entrusted to specialized teams.

Control charts may perhaps be the most promising approach to control problems for routine work, using additional inference and approach analyses of hypothesis tests for the analysis of alternatives. In all events, there is a wide range of possibilities for organizing these methods, which makes them appropriate for road engineering purposes, for they can be very easily adapted to different project requirements.

An important advantage of a statistical quality control study is the opportunity it provides of analysing the risk of failure at different stages of a project and of acquiring an objective idea of what interval of tolerance should be permitted in different aspects of that project.

Figure 17-22 [2] shows two distributions for the same event, although considered from two rather different angles. The unbroken curve could be referred to as the resisting event and represents the response of the structural system to an external cause or stress. The broken line represents the distribution

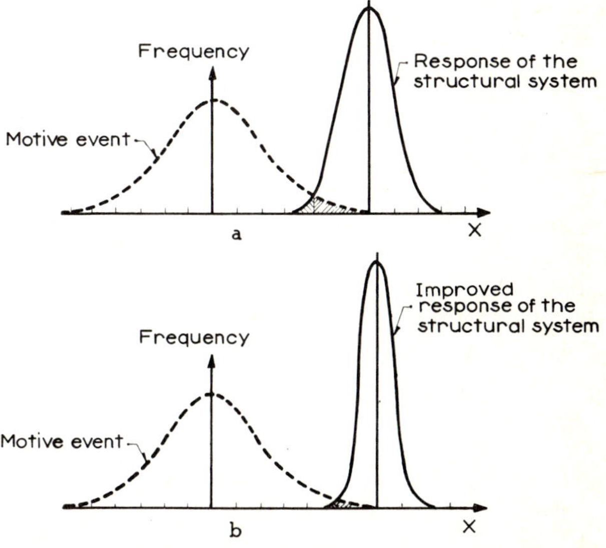

Fig. 17-22 Statistical interaction of a major event and the structural response [2]

corresponding to the motive event, or to the factor acting on the structure which may lead to failure. It is typical for the structural response to be to the right, for the engineer will always try to see that the values corresponding to the response of the structure for any factor likely to cause failure are higher than those with which this factor threatens the structure. The horizontal distance between the two statistical distributions will give a statistical view of the safety. In Fig. 17-22 it is acknowledged that any factor likely to cause failure will be a variable with a random distribution, and the structural response to this will have the same random characteristics. For example, the driving moment that threatens the stability of a slope is not a fixed pre-established constant, nor is the resisting moment which protects the structure. Both depend on a series of factors such as climate, flow conditions, vegetation, maintenance work, etc., some of which are random variables.

For a specific pre-selected confidence level, the risk of failure appears when the statistical distribution of the driving moment intersects the distribution that is anticipated for the structural response. The point at which both distributions intersect one another represents equilibrium in the occurrence of the critical and non-critical states. The area under the intersection zone (lined) represents the upper limit of the risk of failure for the factor under analysis. In this case, *failure* represents failure of the structural response to reach the level of confidence that has been established. *Failure* does not imply structural collapse here, and if the risk of failure under the two systems (that is, under the two distribution curves) is maintained smaller than the risk of failure anticipated in the design, the critical state is statistically not likely to lead to any structural problem. If the impulse is high, it may be that the structural response is not adequate to tolerate the impulse under analysis within the interval of confidence proposed.

In this last event, measures must be taken to reduce the area under both distributions (Part *b* of Fig. 17-22). This can be achieved by measures that increase the mean for the response distribution or by reducing the coefficient of variation, or both. It is likely that the engineer has limited influence over the driving event and its distribution.

The above activities raise costs and increase inspection work, (all of which must be compared with a less careful construction sequence based on a more conservative design). Either will transfer the response distribution towards the right; so the course of action selected will be that which result in lower cost and greater flexibility.

The foregoing considerations seldom lead to the precise evaluation of a real road construction project, possibly owing to laziness or inertia, rather than the true difficulties of the problem, but whatever the case this discussion is useful for indicating courses of action and formulating adequate criteria.

Another important aspect that will be shown by a statistical quality control program is the true relationship existing between the coefficients of variation of the different materials involved. Figure 17-23 [2] illustrates the type of information that can be obtained, and gives the correlation for the variations in the percentage of compaction and the unconfined compressive strength of a subgrade in a real situation which is fairly representative.

It is observed that relatively small changes in the percentage of compaction lead to very important changes in the strength of the subgrade. (Note the linear extrapolation of the relation, which is probably not realistic).

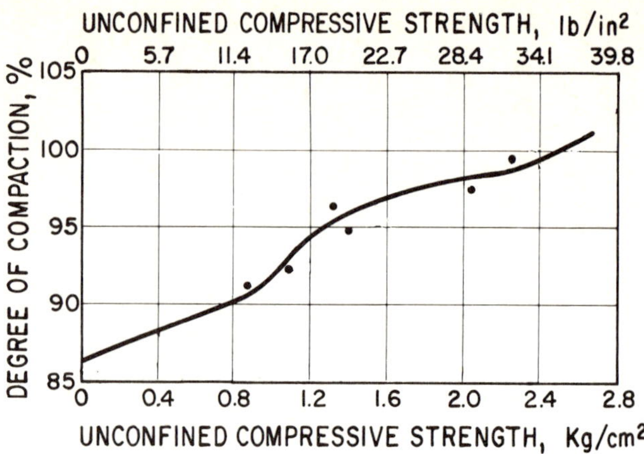

Fig. 17-23 Correlation between statistical variations in the degree of compaction and the unconfined compressive strength of a subgrade in an actual project [2]

This does not, of course, mean that over-compaction of the pavement layer is recommended. No layer should be compacted beyond what may be permanent, in view of the action of water, or beyond what is tolerated by expandibility and resilience of the soil. At any rate there is little value in providing strength that is greater than needed by the design. What is shown by the figure is the type of relationship between the two variables, so that criteria can be duly established for each individual case.

17.5 Further Comments On Compaction Control

Currently compaction control in the field is based on measurement of the dry density of the material handled, generally using the percentage of compaction concept (Chapter 4), or else on controlling the specific energy applied to the soil. Control sequences combining the two methods are not unusual.

Control based on density demands close inspection, sampling and laboratory tests to determine the density of samples of the compacted material obtained in the field. These tests permit determination of the total density of the soil mass, γ_m. For compaction purposes it is usually the dry density, γ_d, that is required. The relationship between the two concepts is given by the expression:

$$\gamma_d = \frac{\gamma_m}{1 + w} \tag{17-30}$$

so that determination of γ_d on the basis of γ_m demands determination of the water content of the soil (w). This represents a practical difficulty for compaction control methods based on density, since determination of w by the conventional soil mechanics methods demands an oven-drying process, the standard procedure for which requires from 18 to 24 hours. Because compaction jobs are conducted at a quick pace and control operations have to be rapid and run on the spot, the oven-drying period frequently prevents making control data available when needed. It is not unusual for a rejection order from the control engineer for a compacted layer to reach the construction engineer after several other layers have already been laid and compacted on top of it. In this way, the concern of control engineers and technicians when supervising compaction work with the density concept, extends not only to the density but also to the water content of the soil. This concern leads to the development of rapid compaction control methods, which will be discussed later in this section.

For the conventional control procedure based on the percentage of compaction, the sequence already described in Chapter 4 is followed. A laboratory test is pre-selected for control purposes. Based on experience it is presumed that this test is representative of the field compaction process. In principle the laboratory control test should also have served as a design norm by providing a representative material on which to measure strength, compressibility and other significant properties for the design in question. This control test has a maximum dry density and an optimum water content. Experience has shown that if the same density is obtained in the field, then the significant soil properties usually will also remain similar. As was shown in detail in Chapter 4 this is not satisfied in every case. However currently it is frequently employed as a working base.

Thus, in the field, a certain percentage of the maximum dry density of the control test is demanded based on the design requirements and this is the degree of compaction the constructor is obliged to produce.

When this criterion is applied literally, frequent readings have to be taken of the dry density obtained for the compacted material, with the disadvantage that w has to be determined, as already mentioned. This work sequence is widely employed today, but the need to determine w promptly is a continual cause of problems.

On account of the above, it has become a constant preoccupation in recent years to reduce to more practical amounts the times required by control tests. Thus, numerous rapid procedures for determining γ_m and w have been created, of which the more important ones will be discussed later in this chapter. Some empirical methods based on local experience have also been developed, which have made it possible to control compaction using the density method practically with only very limited need for laboratory tests. One of these methods will be described in the following section.

In the case of control by measuring the compactive effort transmitted to the soil during the field process, the entire responsibility for definition of the minimum and maximum compaction requirements rests on the field engineer in charge. Measurement of the effort is based on favorable data from a test embankment, which includes the number of passes of the available equipment, thickness of the layer, compaction water content and compactive effort. It is then assumed that the same compaction procedure when applied to the same soils at the job location will produce the same favorable results. In this case independent density tests are conducted to verify the results, but not at the same frequency as when density is specified. This method has been fairly widely used to control the compaction of coarse materials, on which conventional compaction tests cannot be performed in the laboratory on account of their size.

Whatever the case, a control process must include not only inspection and sampling criteria or criteria for interpreting test results, but also criteria governing the tests. Much of the information that follows will deal with this last aspect.

17.5.1 Compaction Control Based On Density

.1 Traditional Control Using Degree of Compaction

Table 17-16 indicates the principal tests that are currently used for measuring the density of a compacted material at the job location [40,41].

In type *I* tests a hole is bored in the compacted material, care being exercised not to lose any of the material that is extracted, for its wet weight and water content must then be measured. To obtain the dry density, it is necessary to know the volume of the hole. This is done by filling it with sand, water or oil. It will be possible to find out what part of an initial total weight was introduced into the hole, afterwards dividing this value by the density of the substance in order to obtain the desired volume.

When the substance used to fill the hole is sand, the sand cone device is frequently employed. This device is illustrated in Fig. 17-24, Plates 17-1, 17-2. It consists of a container filled with sand which is connected by means of a valve to a metal cone, the volume of which is known. The cone is placed over the boring in the embankment, the value is opened and the sand falls, filling the hole and the cone. The operation continues until the sand ceases to flow. At this point, the valve is closed and the entire device is withdrawn (deposit and cone together) and is weighed, comparing this weight with the initial weight of the deposit when full. The difference between the two weights is the weight of the sand which filled the hole plus that of the sand which filled the cone of known volume, and with this it is possible to calculate the weight of the sand which was required to fill the hole. Since the density of this sand is already known, the volume of the hole can therefore be deduced.

The greatest disadvantage of this test lies in its destructive nature and the time it takes to perform. It is also rather sensitive to handling during performance and to the previous calibrations that have to be carried out. This density of the sand that is employed as a reference influences the test results. It varies with moisture, vibration and even minute changes in gradation.

Table 17-16
Field tests to determine the dry density of compacted soils

Type I	Type II	Type III
Destructive		Non-destructive
Disturbed sample	Undisturbed sample	
1) Replacement with sand	1) Cubic sample (cut)	1) Nuclear device
2) Replacement with water: gravity or pressurized	2) Sampler	2) Seismic test
3) Replacement with oil: gravity or pressurized		3) Ultrasonic methods

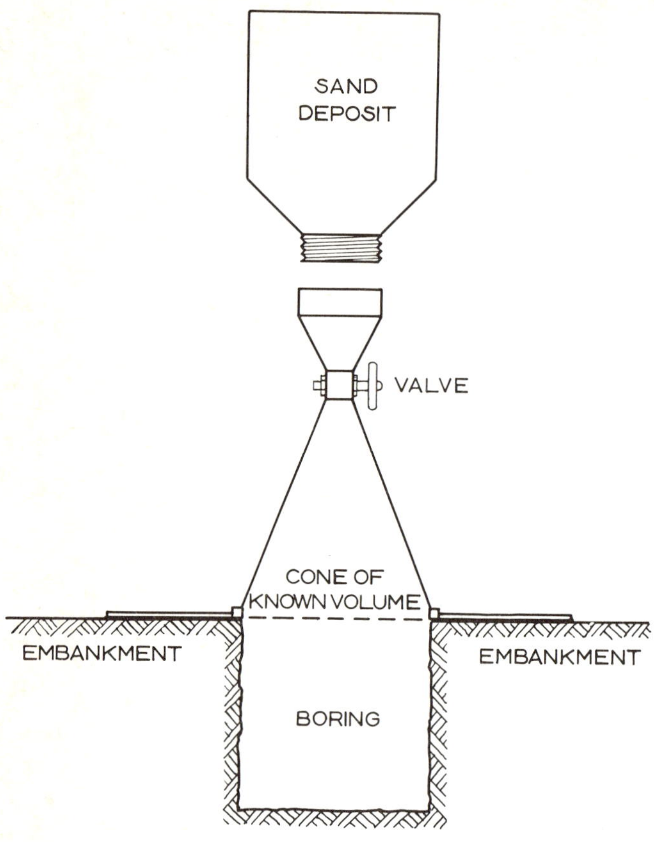

Fig. 17-24 Sand cone device

Plate 17-1 Sand cone device for measuring in-situ density

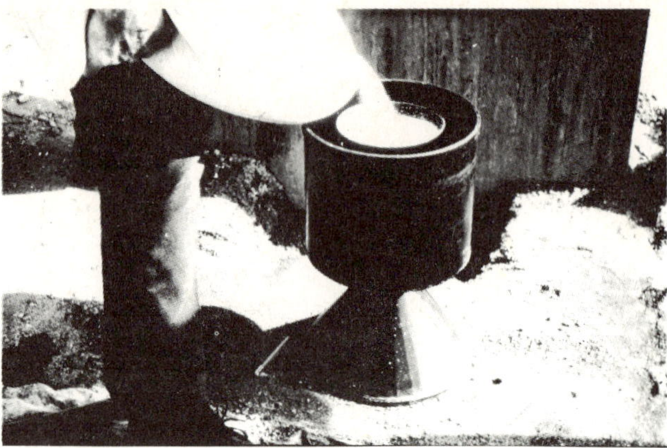

Plate 17-2 The sand cone device in position

It is good practice to calibrate the reference sand in the device before each test. This can be done easily by determining the weight of sand required to fill the volume of the cone. In general terms the test gives more exact results the larger the hole that is bored in the embankment.

In tests where the volume of soil extracted is replaced by water or oil, the hole must be lined with a closely fitting rubber or plastic membrane. This is hard to accomplish when the compacted material contains large particles which make the outer surface of the hole irregular. Sometimes a special device is used with a hand pump, which forces the water or oil from a graduated vessel into a balloon made of impervious material which is placed inside the boring. The pressure from the pump encourages a more satisfactory contact between the walls of the hole and the plastic or rubber membrane; the graduated receptacle automatically indicates the quantity of liquid that was used to fill the hole. Whatever the case, the larger the sampling hole the smaller will be the error due to the presence of large particles.

Determining the in-situ density (γ_m) using cut cubic or cylindrical samples is felt to deserve no further comments. In fine soils, the method is often quick and effective, but in soils containing coarse particles or those of a sandy nature obtaining the sample may become very complicated.

The use of samplers has also led to successful results in cohesive soils on account of speed. Some Mexican-designed models can be seen in [42]. Samplers usually consist of a split barrel, with a very thin inner jacket to house the sample, and they are driven under pressure or using light hammer blows. To simplify calculations, the weight and volume of the tube that collects the sample are usually known.

Paying particular attention to time-saving during control operations, a considerable effort has been made in the last few years to develop nuclear radiation methods for measuring density and water content. This is not the place to deal with the theoretical aspects of the performance of nuclear devices. For purely illustrative purposes, let it be said that all such devices for measuring density have a radioactive source (usually radium, cobalt 60 or cesium 137) which generally emits gamma rays that penetrate the soil and collide with the electrons from the external orbits of the atoms of that soil, rebounding with a reduced energy. On their return, these rays are captured by a detector. The energy lost in the collisions increases the probability of the rays being absorbed before reaching the detector. If a soil has a higher γ_m the gamma rays will collide more often with solid particles than in another less dense soil. Thus there is a correlation between the detector readings and the density of the soil mass (Plate 17-3).

Plate 17-3 Nuclear device for measuring water content

In devices for measuring water content (Plate 17-4) the radioactive source also emits high-power rapid neutrons (generally the source is a mixture of radium and beryllium) which gradually lose their energy when they collide with the nuclei of heavy atoms in the soil or with the nuclei of hydrogen atoms. Since hydrogen atoms are far lighter, their mass being comparable to that of rapid neutrons, the latter will lose far more energy when they collide with hydrogen atoms than when they collide with far heavier atoms. (They will lose almost half of their energy in each collision with a hydrogen atom). A receiver picks up and records the slow neutrons, the number of which depends on the number of hydrogen atoms intercepted. Consequently another correlation can be established, this time with the water content. (However if the soil contains other hydrogen atoms, including those attached to soil minerals, the reading can be wrong.

Plate 17-4 Nuclear device for measuring density

Nuclear devices are available in the form of meters which take readings on the surface, or sounding rods capable of taking depth readings. The advantage is that both moisture and density can be determined in a few minutes.

The disadvantage of traditional nuclear equipment [43,44] is that both the absorption of radiation and the slowing down of the rapid neutrons depended on the physico-chemical composition of the soil under analysis. The different physico-chemical structures of different soils produce different absorption, even for the same density, and the presence of hydrogen atoms *in the soil* itself, not in the groundwater, or the presence of other light atoms alter the correlation used to find the water content. As a consequence of all this, when some correlation curves supplied by the manufacturer are used, the density indicated may have differences of 300 or 400 kg/m^3. Moreover it is hard to measure the water content to within ± 3 or 4%. This disadvantage requires plotting a calibration curve for the *specific* material in which readings were to be taken. Thus some organizations restrict the use of nuclear measuring devices to those cases where very large volumes of a single material were compacted. This is often the case in earth dam technology, but seldom in road engineering, with the exception, perhaps, of airports.

Improvements have been made in the design of nuclear measuring devices, with the object of alleviating their disadvantages and achieving all-purpose devices not dependent on the results of the specific materials analysed. Although there have been definite improvements in nuclear methods over the last few years, the experience of North American engineers continues to be variable. Modern measuring devices have been tested on specially prepared groups of soils, the density or water content of which are determined with great care, and the results obtained for both properties using the calibration curve supplied by the manufacture (which is assumed to be of universal value) are too erratic to be tolerable for practical purposes. This is particularly true of the water content, for very small differences in this quality (1 or 2%) can cause substantial variations in the dry density results in the properties of compacted soils.

Modern nuclear devices are better than those available just a few years ago [45,46], and they will probably improve. Consequently the use of nuclear equipment for compaction control should not be overlooked by control engineers. Usually the devices can be used with relative versatility and without excessive changes in the calibration curve. It is likely that future equipment will prove very useful for solving the serious problem of the timely availability of results. A problem which remains unsolved is the one caused by the iron content of tested soils, this continues to have a substantial effect on the results obtained with nuclear equipment. Another problem is the air gap between the measuring device and the soil.

The principal advantage of nuclear devices is their operation speed, which reduces the calculation of γ_m and w from several hours to just a few minutes, which is compatible with job requirements. An important disadvantage [41] is the high price of these devices, but in the authors' opinion this is not in fact a serious disadvantage because this cost is far outweighed by the advantage of a timely control.

One last comment follows on the advantages and disadvantages of nuclear devices. When comparing the precision obtained with these devices with that obtained with the traditional laboratory methods [47,48,49] the latter are frequently considered to be *exact*, particularly when the comparison is made by engineers who are not accustomed to using statistical control methods. Such a comparison makes no sense. Indeed, some tables are presented in this chapter which show results obtained in actual projects for the density of compacted soils. These tables show the variations which are observed, regardless of the test method, much of which are undoubtedly due to the inherent inexactitude of the laboratory tests themselves. These variations have, moreover, been frequently reported in published literature.

The tests conducted with nuclear measuring devices are nondestructive. This represents a further advantage, for the relatively frequent sampling holes required by sampling control methods may cause weak spots in pavements, unless they are carefully filled and recompacted.

Seismic methods [50], which are also non-destructive, have recently been the object of considerable attention, although no efficient field procedure based on them has yet been developed. Investigations are still at the experimental stage.

The application of the principles of attenuating and transmitting ultrasonic waves to obtain correlations with the soil structure, result in indirect density measurements [51,52,53]. The devices transmit energy to the ground in the form of vibrations which are generally produced by an oscillator and amplified in a power amplifier [54]. A receiver, usually made up of several units, records the vibrations after they have penetrated a certain volume of soil and transmits their energy to a preamplifier which in turn sends them to an oscilloscope where they are recorded and measured, Fig. 17-25. [54] is a diagram of the installation. Such devices are also promising although they are still experimental. So far it appears that some correlation exists between the characteristics of wave transmission and soil density, but results still show appreciable scatter, which is incompatible with a general practical application. The method seems to be more promising in fine soils than in coarse soils.

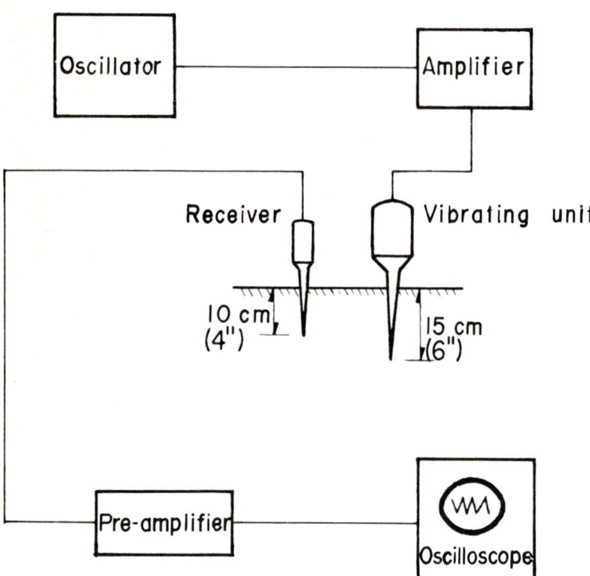

Fig. 17-25 Diagram of a sonic device for estimating the density of compacted soils [54]

Reference [54] also mentions some other methods that are being developed for measuring the density of compacted materials. Of special interest are those where electro-magnetic waves (generally X-rays) are transmitted to the soil and picked up by special detectors, low-frequency vibrations, and methods based on spectroscopic techniques.

.2 Control Using the Degree of Compaction and Typical Curves

Before commencing a description of the empirical control method referred to here, it is worth stressing the randomness of the density of compacted materials in the field.

Figure 17-26 [31] shows the densities obtained for three road embankment compaction processes, carried out by US engineers. The care taken with construction could be described as satisfactory, but not extraordinary. The processes can therefore be regarded as representative of a good normal construction practice.

In every case the compaction specified was 100% of the control test, in terms of dry density. It can be seen how densities varied from 84% to 112% of the intended value, which, in the specific case of the example quoted, is equivalent to obtaining dry densities from 1430 to 1910 kg/m^3 (89 to 119 lb/ft^3). The example shows the appreciable variations that are actually obtained when it is intended to compact an embankment to a minimum density. The hypothesis of normal distribution of the frequencies of the density values is reasonably satisfied in the three cases shown.

Figure 17-27 [31] shows a set of laboratory compaction curves obtained for materials extracted from a 7 km (4.35 mi) section of one of the above three embankments (No 3). This figure indicates the relatively large variation that is obtained for a 7 km (4.35 mi) section made up of materials that were presumed to be homogeneous.

However expert a control engineer may be, he will experience great difficulty in selecting the appropriate compaction curve to represent the material he has sampled from an embankment during a sampling operation in conjunction with the sand cone device, for example. It is clear, therefore, that the results this inspector would obtain would be highly erratic if he tried to compare all the samples taken from the 7 km (4.35 mi) section with a single control test.

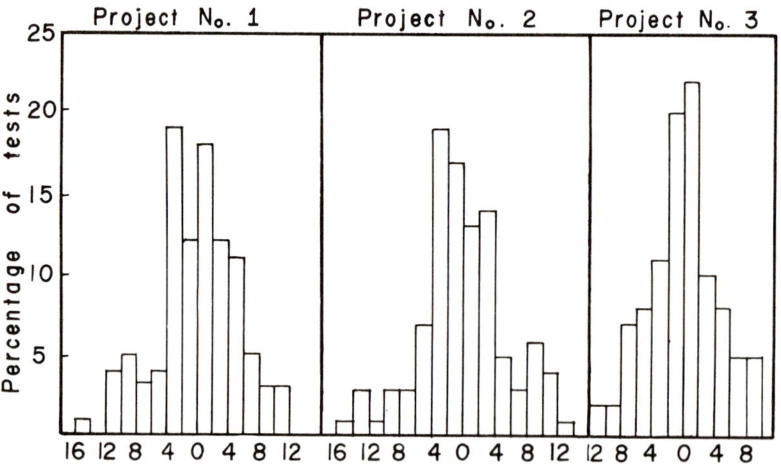

Fig. 17-26 Histograms of compaction results obtained on three construction projects [31]

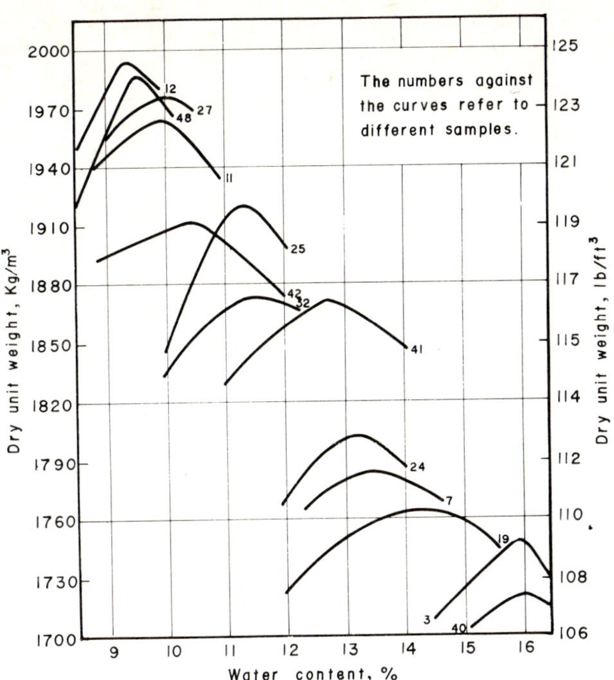

Fig. 17-27 Set of laboratory compaction curves for a real highway section [31]

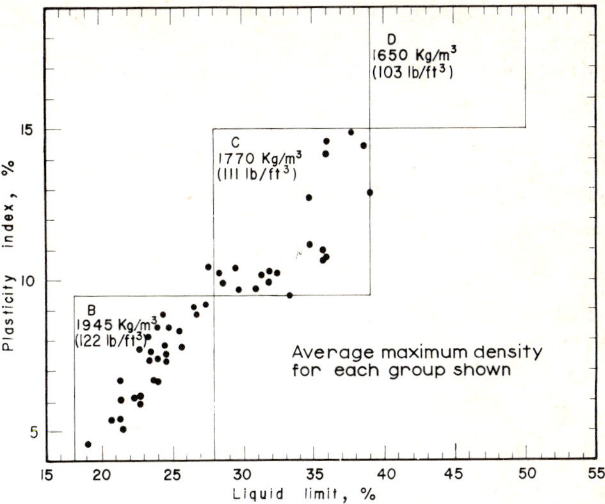

Fig. 17-28 Classification of compacted soils according to their index properties [31]

The engineer has several alternatives for classifying the material sampled in relation to the other materials present in the section, as indicated in Fig. 17-27. In this way his selection of the appropriate laboratory curve will be more rational and he will be able to relate the density obtained with the maximum value on that curve, thus reaching a more realistic value for the percentage of compaction represented by the sample.

The first alternative is to conduct numerous tests so as to classify the materials to be used in the section before construction commences, subsequently conducting the same tests on the sample obtained, so as to locate it within the section as a whole, correlating the compaction data for the sample with those for the soils showing the same index properties or grain size distribution. This technique is illustrated in Fig. 17-28 [31], which is a plasticity chart which groups all the materials studied in one of the three projects to which reference is made here. In the squares the maximum densities are recorded, based on subjecting the soils to a standard control test. The values indicated are the mean for the soils in each square.

In order to use this technique the control engineer carries out numerous classification tests on all the materials before construction commences, in order to achieve an acceptable correlation. This is a disadvantage of the method, particularly bearing in mind the frequent changes from one borrow pit to another and the variation in the properties of soils from the same pit, which are common in road engineering.

Another alternative for the engineer is to generate typical regional curves [31,40,41,55].

For a particular region in which the geological and soil formations have been identified and for which there is a list of suitable borrow pits for road building purposes, a set of laboratory compaction curves will be drawn up relating the total density or bulk density of the soil with the water content during testing. This will cover all the soil types in that region, or at least the most representative ones.

Figures 17-29 [31] and 17-30 [41] show two sets of curves, such as utilised in some US states, one corresponding to Indiana soils and the other to Ohio soils. The latter, in particular, have been adopted by some other states as a working basis, adding to them certain soils that are characteristic of their regions. This is the case, for example, of the curves developed by South Dakota and Louisiana.

It is assumed that a set of curves is available for a certain region in which a road is being built. The engineer can obtain his samples from different points in the compacted embankments, measuring the bulk density and the water content of each sample. Then, for each sample, he will be able to locate a point on his control curves and subsequently draw a curve to resemble the others. In this way he will be able to estimate the maximum bulk density he can expect in that sample, assuming that the compactive effort has been the same. This maximum bulk density can then be converted to maximum dry density, either by applying Eq.(17-30) or by interpolating on sight the maximum dry densities of the neighboring curves (which are indicated in the tables included in the typical curve figures). Once the engineer has found the maximum dry density that can be expected, he can then compare it with that exhibited by the sample (which he obtained previously by converting the bulk density into dry density using Eq.(17-30) and the water content measured for the sample). In this way he obtains the percentage of compaction of his material. This simple method may be more realistic than a comparison of his samples with a single laboratory control curve, disregarding the variation in the materials in his section, such as in Fig. 17-27. Such variations occur even in those cases where the material laid is apparently homogeneous.

The foregoing procedure is a refinement which is perfectly reasonable, and which may avoid many arguments between constructors and quality control engineers, or between government representatives and contracting firms. The method has a fundamental pre-requisite, a control curve chart, that is not impossible to obtain. It does, however, require a series of density studies for various materials and borrow pits, which can only be of use to constructors or state and government agencies.

Note that this procedure uses the degree of compaction concept; if it is employed with dry densities, there is the added disadvantage of the need to calculate the water content of the

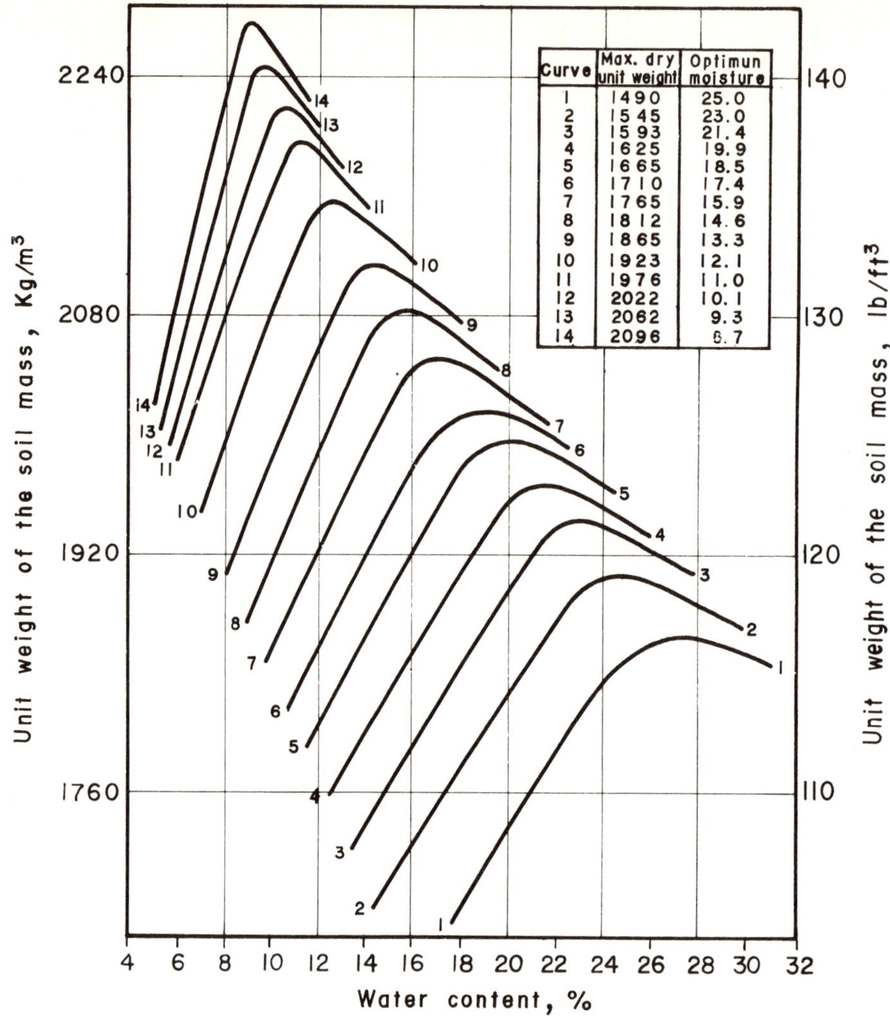

Fig. 17-29 Typical compaction curves for soils in the State of Indiana [3]

samples. Nuclear methods and all the quick methods that will be discussed in the next section may be of benefit to this method of work, as they are to the traditional method presented previously.

17.5.2 Quick Methods for Measuring Density

The nuclear methods that were presented earlier in this chapter are quick tests for compaction control, for they permit calculation of density in very brief times. At this point, however, the authors add the description of some other methods, all of which are designed to reduce the time required for control operations, making them more practical.

.1 HILF's Method [42,56,58]

The method described here was proposed by J.W. HILF and is intended to speed up the traditional field compaction control procedures.

The chief advantage of HILF's method is that the percentage of compaction achieved can be found in approximately one hour. This is possible because the method does not require knowledge of the water content of the sample obtained for control purposes. HILF also suggests a method for measuring the water content in the field, which, although not quite exact, is nevertheless adequate. It is performed directly on the material that is to be controlled, which means that any heterogeneities at the location are taken into account. The same advantage is true of the method proposed for compaction control.

A compaction test is made on a sample soil from an embankment which does not contain particles larger than No 4 sieve. The material must be protected against evaporation, so that its water content will not vary. It is compacted with some of the standard procedures at its field water content, w_f. The wet density of this sample will be γ_{mc}. The wet density of the sample as extracted from the embankment is γ_{mf}. This is easy to obtain by simply dividing the weight by the volume of the sample extracted. The densities γ_{mc} and γ_{mf} are for the same material with the same water content, but they are seldom equal, since the compactive effort and the method of applying this effort is different in the two cases.

Recalling Eq.(17-30) this leads to:

$$\gamma_m = \gamma_d (1 + w)$$

Therefore:

$$\frac{\gamma_{mf}}{\gamma_{mc}} = \frac{\gamma_{df}(1 + w_f)}{\gamma_{dc}(1 + w_f)} = \frac{\gamma_{df}}{\gamma_{dc}} = C \qquad (17\text{-}31)$$

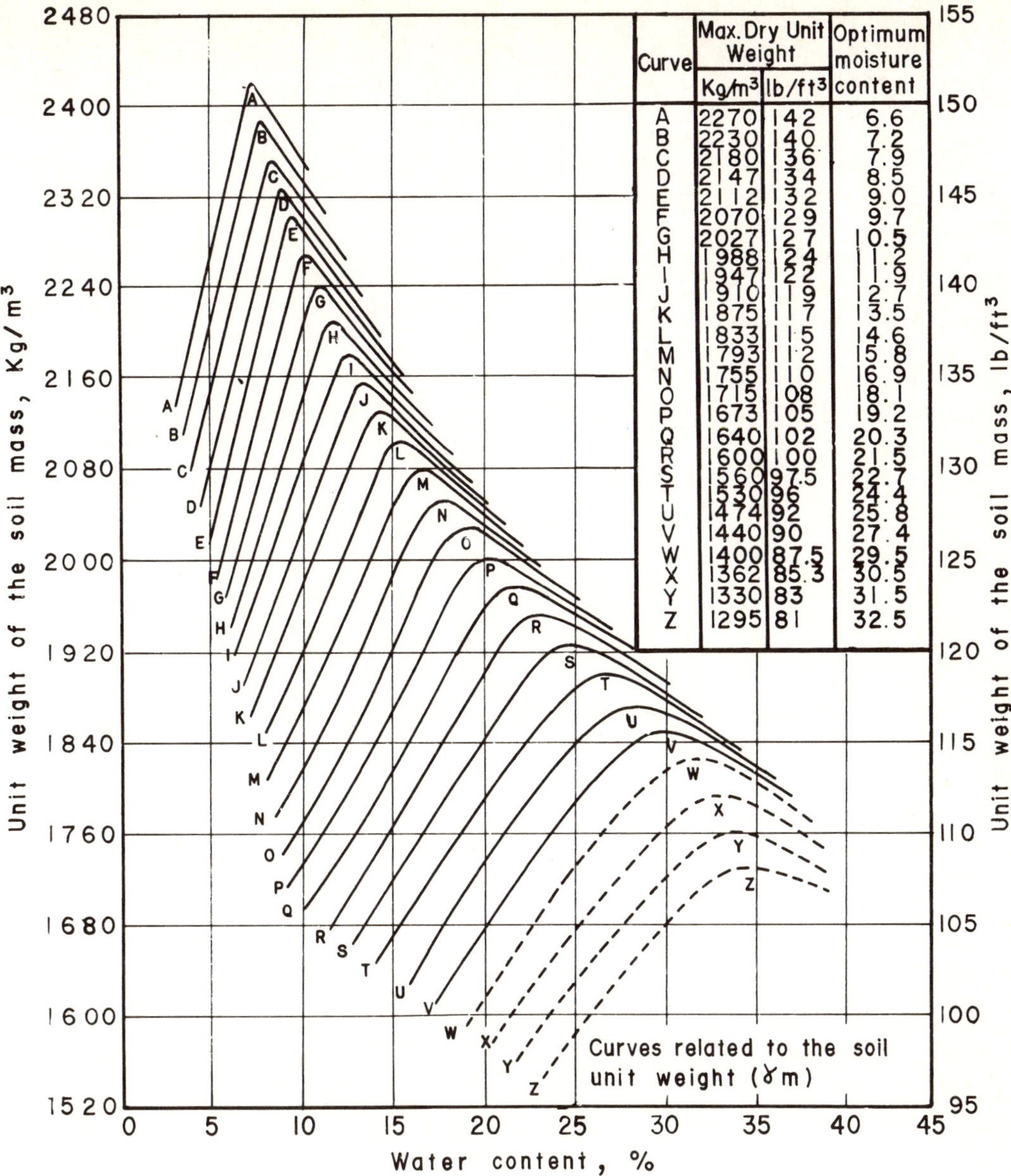

Curve	Max. Dry Unit Weight Kg/m³	Max. Dry Unit Weight lb/ft³	Optimum moisture content
A	2270	142	6.6
B	2230	140	7.2
C	2180	136	7.9
D	2147	134	8.5
E	2112	132	9.0
F	2070	129	9.7
G	2027	127	10.5
H	1988	124	11.2
I	1947	122	11.9
J	1910	119	12.7
K	1875	117	13.5
L	1833	115	14.6
M	1793	112	15.8
N	1755	110	16.9
O	1715	108	18.1
P	1673	105	19.2
Q	1640	102	20.3
R	1600	100	21.5
S	1560	97.5	22.7
T	1530	96	24.4
U	1474	92	25.8
V	1440	90	27.4
W	1400	87.5	29.5
X	1362	85.3	30.5
Y	1330	83	31.5
Z	1295	81	32.5

Fig. 17-30 Typical compaction curves for soils in the State of Ohio [41]

where w_f is the water content of the sample obtained in the field and γ_{df} and γ_{dc} are the dry field and test densities, respectively.

The ratio of the dry field density γ_{df} to the maximum dry density obtained in the laboratory with the optimum test moisture, w_0 (usually different from w_f), on which compaction control is based, can be obtained on the basis of the wet densities in a manner similar to the *C* ratio in Eq.(17-31). For this it will be necessary to evaluate the expression $\gamma_{dm}(1 + w_f)$ where γ_{dm} is the maximum dry density of the material obtainable in the laboratory. Consequently the foregoing expression represents the wet density of the material to which, once dry and as well compacted as possible, w_f is added. The wet density that truly corresponds to the maximum density, γ_{dm} is of course given by the expression $\gamma_{dm}(1 + w_0)$. This wet density can be converted to $\gamma_{dm}(1 + w_f)$ by simply dividing by $1 + w_0/1 + w_f$, which can in turn be written thus:

$$\frac{1 + w_0}{1 + w_f} = \frac{1 + w_0 + w_f - w_f}{1 + w_f} = 1 + \frac{w_0 - w_f}{1 + w_f} \tag{17-32}$$

Likewise, the wet density at any water content, w, $[\gamma_d(1 + w)]$, can be converted to the wet density corresponding to the field water content, w_f, $[\gamma_d(1 + w_f)]$, by dividing by the expression:

$$1 + \frac{w - w_f}{1 + w_f} \tag{17-33}$$

In other words

$$z = \frac{w - w_f}{1 + w_f} \tag{17-34}$$

When $w_f = 0$, $z = w$; in other words it is the water content as a percentage of the dry density of the soil. However, when $w_f > 0$, multiplying and dividing by the dry weight of the sample, W_s, the following expression is obtained:

$$z = \frac{wW_s - w_f W_s}{W_s(1 + w_f)} \tag{17-35}$$

The numerator in Eq.(17-35) is the weight of the water acquired by the sample with w in relation to its field water content, w_f. The denominator represents the weight of the mass of soil with its field water content, w_f. Consequently when $w_f \neq 0$, z is no longer a water content, but an increment in the weight of the water in the soil over and above the field water content, expressed as a percentage of the weight of the soil mass at the field water content.

If a fixed amount of water is added to a specimen of wet soil of any weight at the field water content, and is mixed until the moisture is evenly distributed throughout, z remains constant regardless of the amount of soil that has been compacted in the mold, because z does not depend on W_m in Eq.(17-35).

Consequently, the wet density resulting from the addition of water to the field sample can be converted to the wet density based on the field water content (the one it would have for the field water content) by simply dividing the wet weight obtained by $1 + z$, thus:

$$\gamma_d \frac{1 + w}{1 + z} = \frac{\gamma_d (1 + w)}{1 + \dfrac{w - w_f}{1 + w_f}} = \gamma_d (1 + w_f) \tag{17-36}$$

If the values of $\gamma_d (1 + w_f)$ are plotted as ordinate, and the corresponding values of z as abscissa, a curve like the lower one in Fig. 17-31 is obtained.

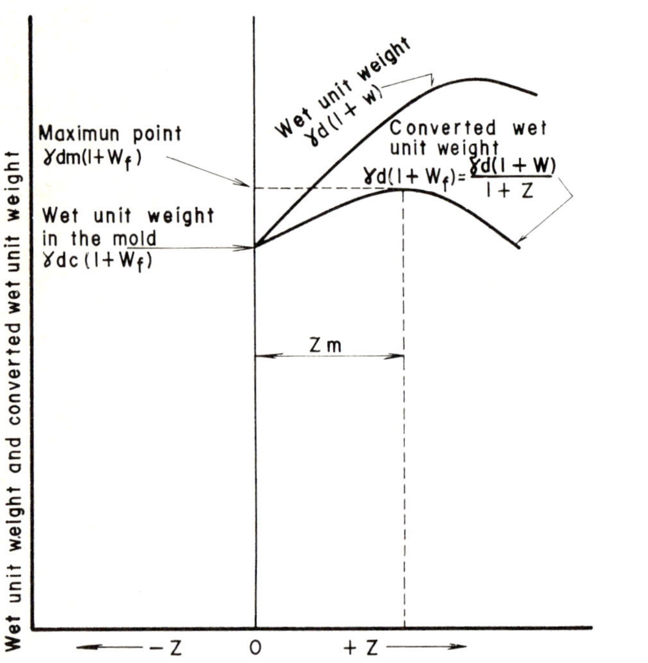

Fig. 17-31 Wet density curves versus z values

The values of $\gamma_d (1 + w_f)$ are usually referred to as converted wet densities (converted to the field water content, w_f). The wet densities before conversion are $\gamma_d (1 + w)$; if their values are plotted as ordinate, with the corresponding z values as abscissa, the upper curve in Fig. 17-31 is obtained.

Attention is now drawn to the lower curve in this figure. This curve represents the values of $\gamma_d (1 + w_f)$ plotted against the values of z. The maximum point on this converted water contents curve must be $\gamma_{dm} (1 + w_f)$ since w_f is constant and γ_d is the only variable. Consequently, the exact degree of compaction, G, can now be obtained using the expression:

$$G = \frac{\gamma_{df} (1 + w_f)}{\gamma_{dm} (1 + w_f)} = \frac{\gamma_{df}}{\gamma_{dm}} \tag{17-37}$$

The converted wet densities curve can be obtained by plotting the ordinates:

$$\gamma_{dc} (1 + w_f) ; \frac{\gamma_{d2} (1 + w_2)}{1 + z_2} ; \frac{\gamma_{d3} (1 + w_3)}{1 + z_3} \text{ etc.}$$

against the corresponding values of z. Also, $z = (w - w_f)/(1 + w_f)$ can be either negative or positive. In practice, the minimum number of points required is three, but if more points are plotted, it will be possible to determine the maximum value of the curve more exactly.

To summarize, the procedure for determination of the degree of compaction using HILF's method, is as follows:

First a sample of material is taken from the embankment that is being controlled, exercising care that it loses no moisture. Its wet weight and its volume are determined, after which it is possible to determine the wet unit weight of the sample $\gamma_{mf} = \gamma_{df} (1 + w_f)$. Next the sample is remolded and compacted, using whatever compaction test is selected, with the same field water content, w_f, thus obtaining the value of $\gamma_{mc} = \gamma_{df} (1 + w_f)$. This is the ordinate at the origin of the graph of converted wet weights versus values of z. A fixed amount of water is then added to the sample, it being decided that the value of z should be about two. Consequently, the necessary amount of water will be calculated as 2% of the wet weight of the sample under analysis. The sample thus moistened is compacted and $\gamma_{m2} = \gamma_{d2} (1 + w_2)$ is found, which enables determination of the converted wet weight $\gamma_{d2} (1 + w_2)/(1 + z_2)$. If the value thus obtained is higher than the previous one (ordinate at the origin), more water is added to the sample in a similar proportion, repeating the process until the converted wet densities start to fall, so that the curve will permit determination of the maximum ordinate. Should the second point be smaller than $\gamma_{dc} (1 + w_f)$ it will be located in the descending section of the curve, beyond the maximum converted wet weight, and consequently it will be necessary to plot the curve towards the dry side. In order to do this, the sample will be allowed to dry evenly and the weight of water lost will be determined. This when divided by the initial wet weight, will give the corresponding negative value of z. In this way, it is possible to obtain the points required to plot the entire curve on which the maximum ordinate required to calculate the degree of compaction can be easily seen. The method is practical only when the embankment has been compacted dry of optimum.

.2 The Use of Penetrometers [42,54]

In the methods for measuring density with penetrometers, this concept is correlated with the resistance of the compacted material to penetration by a pin, which is driven by a hammer dropped from a certain height. Theoretically there is no reasonable means of reliably correlating penetration and density. What is attempted is to establish a purely empirical relationship, based on measurements.

The origin of these techniques goes back to work by PROCTOR, in 1933, who designed the first soil penetrometer. Towards the mid-1950's, HOUSEL [42] carried out a considerable amount of research on the correlation referred to. Some more recent works are described in [57].

Figure 17-32 [54] is a diagram of a penetrometer which is representative of the various models in use. It is a bullet-ended high-strength steel bar, with an approximate diameter of 1.5 cm (0.6 in). The hammer head may weigh from 2.25 to 4.50 kg (5 to 10 lb) in almost all models, and the height of fall is also variable, typically about 50 cm (20 in).

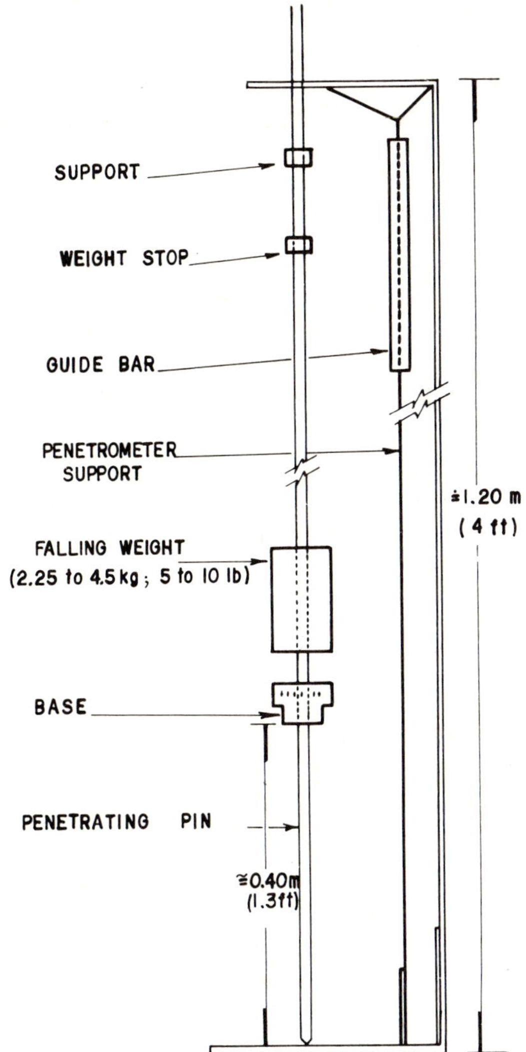

Fig. 17-32 Diagram of a penetrometer for measuring density [54]

Penetration is usually correlated with the dry density of the soil, which presents no difficulties since it takes place in the laboratory, without the pressures of practical compaction jobs.

It has sometimes been attempted to correlate on a large scale penetrations by a device like the one just described with the dry density of a large number of soils, working on a regional scale and using typical penetration curves similar to those described earlier for compaction processes. In the authors' opinion such correlations are inaccurate at best and are sometimes misleading.

Figure 17-33 shows the results obtained from a very complete study to evaluate the correlation, carried out by the authors of [54]. All the soils were compacted using a metal mold with a volume of 0.028 m^3 (one cubic foot), 20 cm (8 in) high. The water contents ranged from 0 to 20%, and different methods of compaction were employed, from the manual dynamic PROCTOR procedure to the vibratory method. The highest densities were obtained for dynamic PROCTOR compaction.

Once the specimens had been compacted, the penetrometer tests were performed, using a 2.27 kg (5 lb) ram for soils and a height of fall of 30.5 cm (1 ft), and for the crushed rock a 4.54 kg (10 lb) ram with a height of fall of 45.7 cm (18 in). The penetrometer was applied to the compacted samples, until a penetration of approximately 5 cm (2 in) was achieved. This was regarded as the initial penetration, after which the blows required to achieve a total penetration of 15 cm (6 in) (a further 10 cm = 4 in) were recorded, which was recorded as the final penetration. The resistance to penetration was calculated as the number of blows required to penetrate the final 10 cm (4 in), divided by 10 to obtain the blows per centimeter of penetration. The penetrometer was applied several times to each sample and the average of these penetration values was recorded as being the penetration of the sample.

As can be seen in Fig. 17-33, the correlation is not equally good in all the soils; very significant differences were obtained in some cases. It can also be seen, as was bound to happen, that the water content of the compacted materials has an extraordinary effect on the resistance to penetration. Thus the density-penetration correlation is not unique; when the water content changes, the correlation continues to exist, but the curve is different. As a result, every time it is wished to apply the method to a field problem a previous calibration will be necessary to determine the different curves corresponding to possible different water contents. This will not mean much more work than that required by the traditional control methods, since several different water contents are also necessary for the purposes of a compaction test. It does require a water content test in the field: the most time consuming task. Figure 17-34 [54] is an example of the type of correlation graph that might be obtained, and reflects the sensitivity of the materials' resistance to penetration to changes in the water content.

To summarize, it can be said that compaction control methods using penetrometers may be useful when a great deal of laboratory work is carried out previously to define the correlation curves corresponding to the different materials that are to be used in the project, but only when field moisture tests are conducted. This work and previous knowledge of the materials will prove difficult in many cases of road engineering. However, if good correlation curves are achieved, in which variations occur within the tolerance limits, the use of the penetrometer may simplify routine control work appreciably.

17.5.3 Quick Methods for Measuring Water Content

.1 HILF's method

Within the framework of his general rapid control method, HILF [42,56,58] also developed a procedure for obtaining the water content, which is described here with reference to Fig. 17-31.

Determining the maximum point on the converted wet weight versus values of z curve indicates whether the soil has the optimum moisture corresponding to the laboratory test that is being used (the special case where the maximum ordinate of the curve falls at a point of abscissa $z = 0$) or, more commonly,

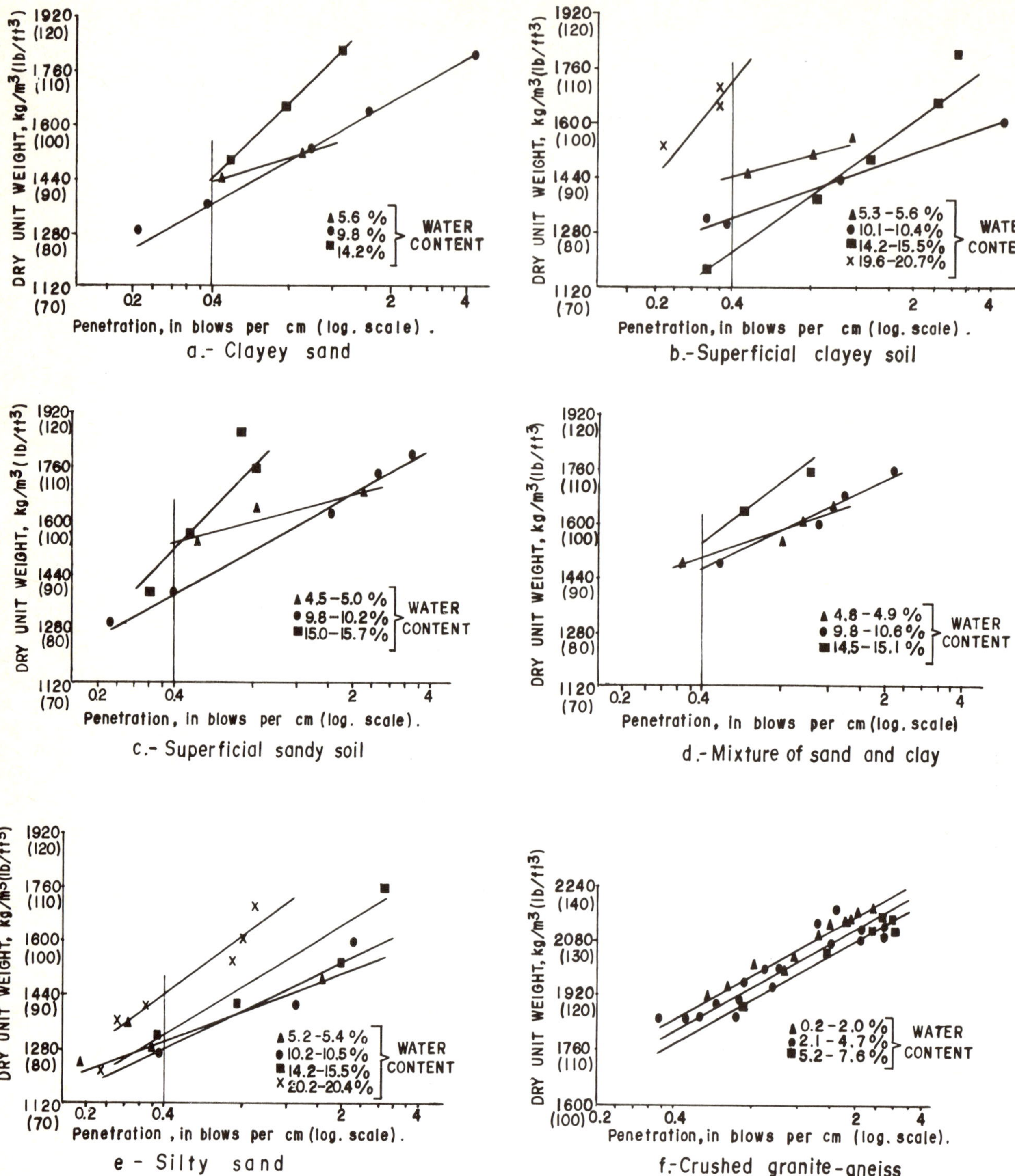

Fig. 17-33 Correlation between the penetration resistance of various soils and their dry density [54]

whether the moisture is above or below the optimum. However, the exact magnitude of the difference between the optimum water content and the field moisture is unknown. Hilf also determines this difference approximately, although not as precisely as in the case of compaction control. Nevertheless the error is small enough for the method is often acceptable for control purposes. the procedure proposed by Hilf is as follows:

From Eq.(17-34) it can be deduced that:

$$w_0 - w_f = z_m(1 + w_f) \tag{17-38}$$

where z_m represents the abscissa of the maximum ordinate on the converted wet weight versus values of z curve. If $z_m = 0$, $w_0 - w_f = 0$, and consequently the field water content will be the optimum. For values of $z_m \neq 0$, this difference is calculated with w_f, Eq.(17-38). Relating the first and third terms of Eq.(17-36):

$$1 + w_f = \frac{1 + w}{1 + z}$$

If $z = z_m$ for $w = w_{0'}$ then:

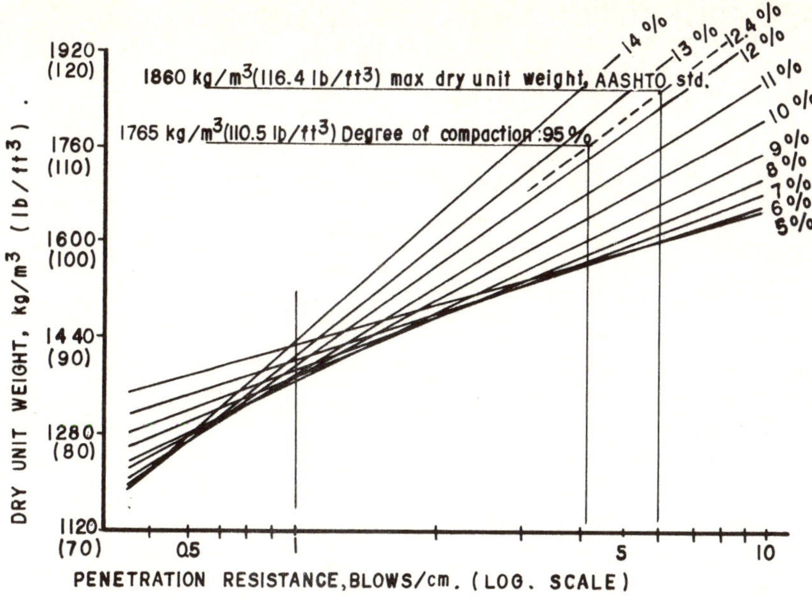

Fig. 17-34 Correlation between the penetration of a clayey sand and its dry density, taking into consideration changes in the water content [54]

$$1 + w_f = \frac{1 + w_0}{1 + z_m} \tag{17-39}$$

Introducing the value of $1 + w_f$ into Eq.(17-38):

$$w_0 - w_f = \frac{z_m}{1 + z_m}(1 + w_0) \tag{17-40}$$

In other words, the difference $w_0 - w_f$ now depends on w_0, which is not known either, if HILF's method is applied. For this either w_0 or w_f must be estimated in order to obtain the magnitude of $w_0 - w_f$. HILF holds that the error in estimating w_f or w_0 is greatly reduced when the difference between w_0 and w_f is calculated thus, and that this error is acceptable for control purposes.

So as not to have to estimate w_f or w_0 in each laboratory compaction test, a set of curves can be prepared to estimate w_0 at the maximum point of the converted wet weight curve. In other words for the same compaction method selected in the laboratory for field control, a relationship exists between the optimum moisture and the wet density of the sample with that optimum moisture. Such a curve, plotted for eighty soils compacted with the standard test of the US Bureau of Reclamation, is shown in Fig. 17-35.

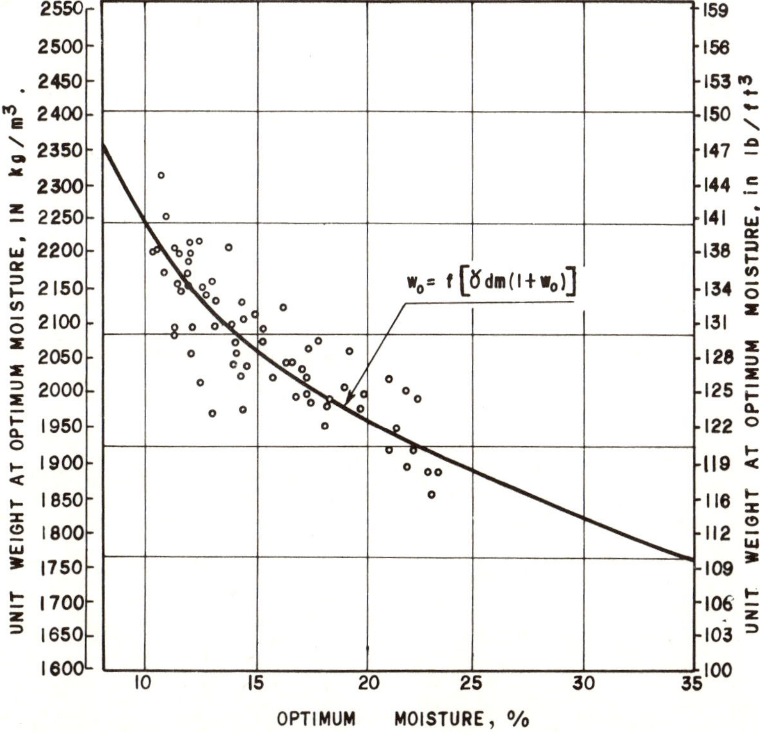

Fig. 17-35 Curves for wet density versus optimum water content in HILF's method [58]

For other compaction tests similar curves can be prepared. Then, since the maximum point of the converted wet weight curve is the wet weight of the sample with the optimum moisture, once this value is known it will be possible to estimate w_0 on the basis of the curve in Fig. 17-31 using a graph like the one shown in Fig. 17-35. Using Eq.(17-40) it is now possible to find the value of $w_0 - w_f$, which will indicate the difference between the moisture of the compacted material and the optimum laboratory moisture.

It is worth mentioning that the moisture control proposed by HILF compares the field values for the water content of the material with the optimum laboratory value. This will only be practical when field compaction conditions are reasonably represented in the laboratory tests. Only when the optimum field water content is so close to that obtained in the laboratory test will HILF's method have full significance. In many cases, however, the optimum field value will be very different from the laboratory optimum, and it will have to be determined in test embankments reproducing job conditions exactly. In this case, HILF's method is no longer applicable, unless the curve in Fig. 17-35 is obtained from test embankments constructed with the soils that are actually involved, and using the specific equipment that is to be employed on the project.

.2 The Volumetric Method [42]

When a weight of wet soil is introduced into a volume of water, this volume will increase by ΔV, which represents the volume of solids and that of the water in the soil introduced, so long as all the air in the soil has been eliminated. If the specific gravity of the solids (see Chapter 1) and the temperature of the mixture of soil and water are known, there will be sufficient data to determine the water content of the original soil, by the following analysis:

$$\Delta V = V_w + V_s = \frac{W_w}{\gamma_0} + \frac{W_s}{\gamma_s} \qquad (17\text{-}41)$$

but:

$$W_w = wW_s$$

and

$$W_s = \frac{W_m}{1 + w}$$

All the code letters have the usual meanings specified in Chapter 1.

Substituting the above values in Eq.(17-41) the following expression is obtained:

$$\Delta V = \frac{wW_s}{\gamma_0} + \frac{W_m}{(1 + w)\,\gamma_s} \qquad (17\text{-}42)$$

Whence:

$$\Delta V\,(1 + w)\,\gamma_s = \frac{wW_m}{\gamma_0}\,\gamma_s + W_m \qquad (17\text{-}43)$$

which can be written thus:

$$w\left(\Delta V - \frac{W_m}{\gamma_0}\right)\gamma_s = W_m - \Delta V\gamma_s \qquad (17\text{-}44)$$

which in turn leads to:

$$w = \frac{W_m - \Delta V\gamma_s}{\left(\Delta V - \frac{W_m}{\gamma_0}\right)\gamma_s} \qquad (17\text{-}45)$$

It is recalled that by definition:

$$S_s = \frac{\gamma_s}{\gamma_0}$$

which when substituted in the above expression leads to the following equation:

$$w = \frac{W_m - \Delta V s_s\,\gamma_0}{\Delta V s_s\,\gamma_0 - W_m\,s_s} \qquad (17\text{-}46)$$

This last expression permits calculation of the water content of the original soil in terms of known quantities and of s_s which must be determined in order to be able to apply the method. This limits the method. Since the test to determine specific gravity is complicated and long (it also includes oven-drying), use of Eq.(17-46) will only be worthwhile in those cases where a large number of compaction control operations must be carried out on a soil, the s_s of which does not change significantly.

A further disadvantage of the volumetric method is that all the air trapped in the original soil must be expelled. This is done by breaking up the soil before the procedure starts and shaking the sample vigorously once it is in the water. However these operations will not ensure total expulsion of the air in clayey soils, especially *CH* ones.

With the exception of the case already mentioned, Mexican engineers [42] are satisfied with the precision of the method.

.3 The Alcohol Method

There are currently two versions of this method which will be dealt with separately.

Method based on changes in the density of alcohol: This method is after G.J. BOUYOUCOS [42] and is based on alcohol being a liquid capable of forming an intimate solution with water. This occurs when a wet soil is broken up completely and mixed with a measured volume of alcohol, the density of which is known. The amount of water originally present in the soil, on mixing with the alcohol, will modify the density of the alcohol by an amount which will depend on the original density of the alcohol and the amount of water that is added. Using a fixed volume of alcohol with pre-established characteristics, it is possible, by a previous laboratory calibration, to measure the density of the water-alcohol mixture, which is obtained by filtering the mixture of soil, water and alcohol. Using a fixed wet weight for the soil that is broken up in the alcohol, calibration curves can be plotted so that the water content of the soil sample is obtained from the density of the water-alcohol mixture.

The density of the water-alcohol mixture, obtained by filtration, is measured with a conventional densimeter. In the laboratory it is found that the final density of a mixture of water and alcohol depends on the temperature at which the two liquids are mixed. Consequently this should be considered when plotting the calibration curves in the laboratory. These curves are shown in Figs. 17-36 and 17-37 [42], the first of which refers to conventional embankment materials and the second to soils

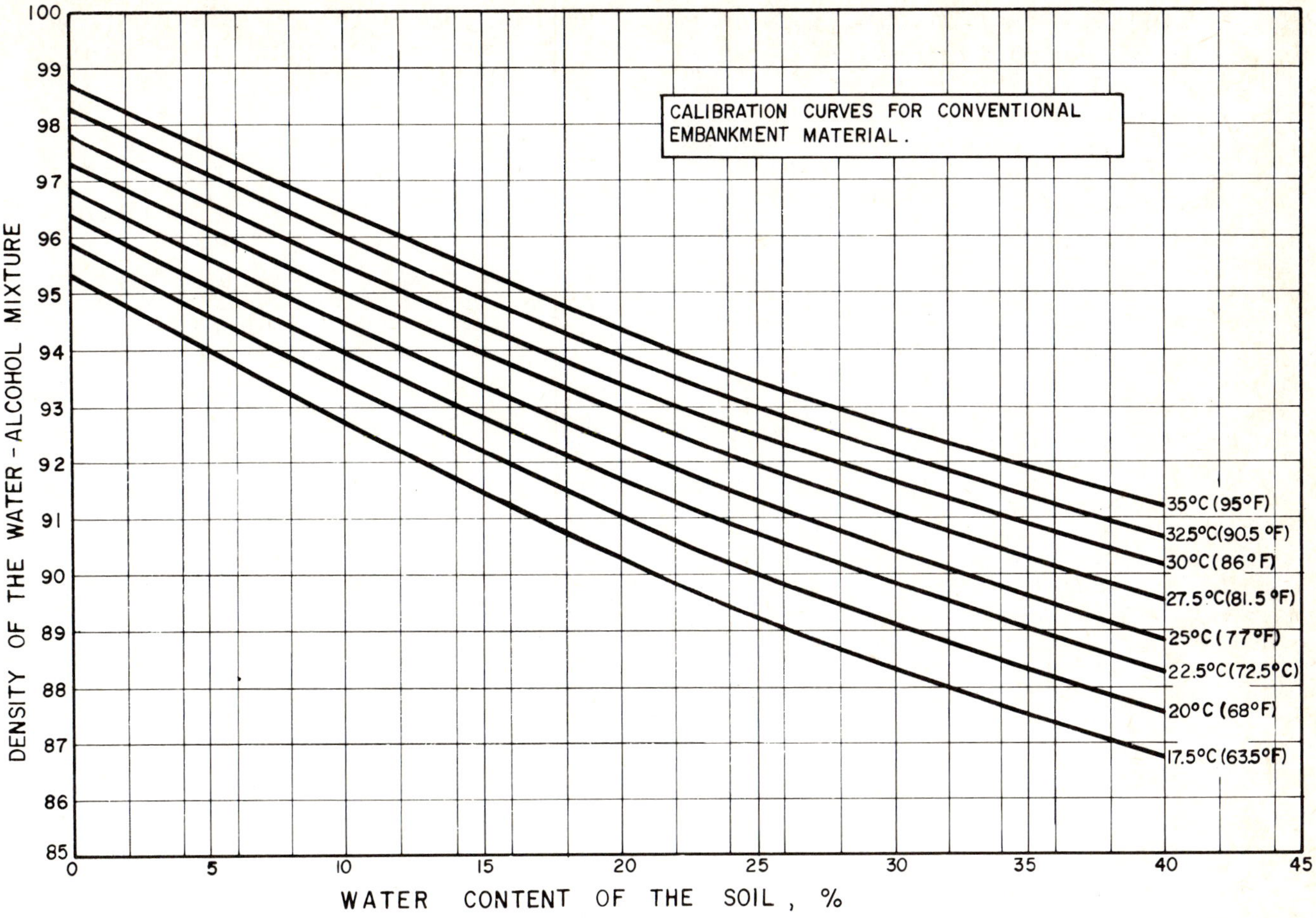

CALIBRATION CURVES FOR CONVENTIONAL EMBANKMENT MATERIAL.

DENSITY OF THE WATER - ALCOHOL MIXTURE

100 99 98 97 96 95 94 93 92 91 90 89 88 87 86 85

0 5 10 15 20 25 30 35 40 45

WATER CONTENT OF THE SOIL, %

35°C (95°F)
32.5°C (90.5 °F)
30°C (86 °F)
27.5°C (81.5 °F)
25°C (77 °F)
22.5°C (72.5°C)
20°C (68°F)
17.5°C (63.5°F)

Fig. 17-36 Alcohol-soil-moisture calibration curves for conventional embankment material

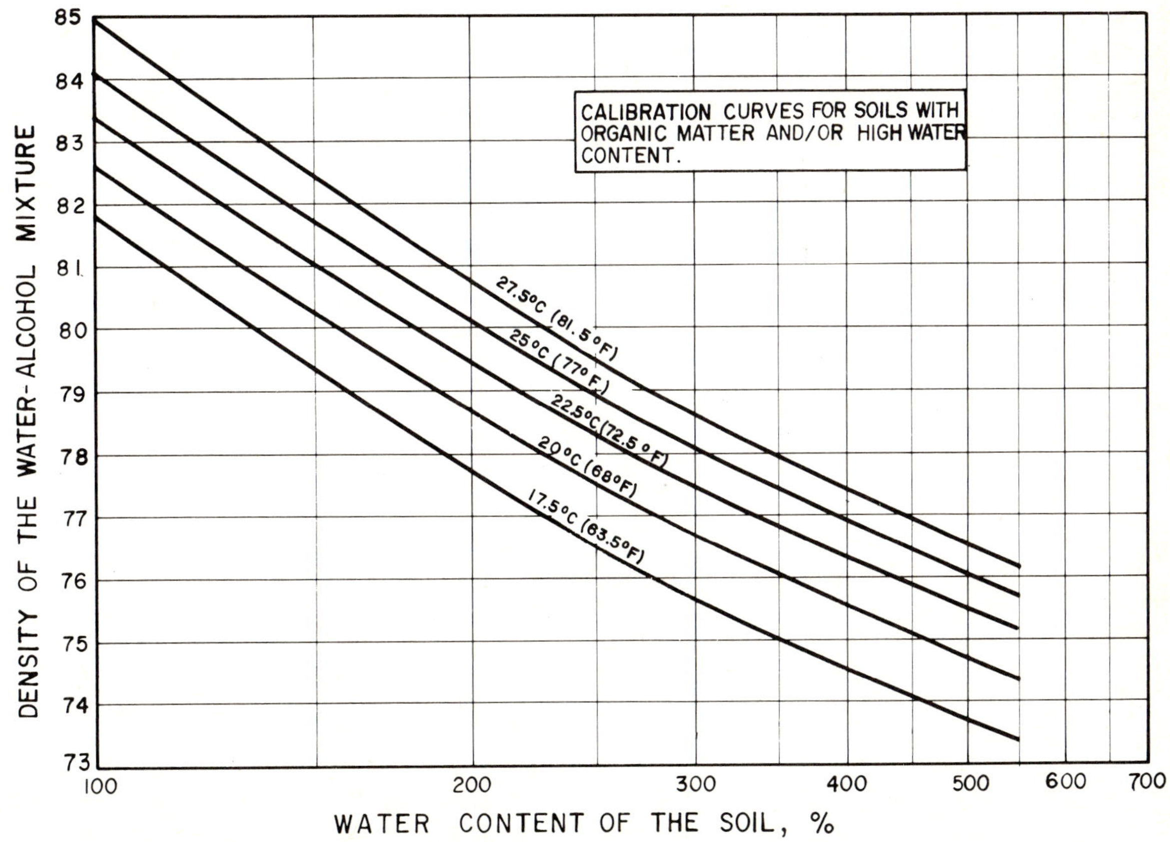

Fig. 17-37 Alcohol-soil-moisture calibration curves for soils with organic matter and/or a high water content

with a high content of organic matter and/or high content of water, examples of which should be hard to find in conventional road building practice.

Mexican engineers have employed this method with 100 g (3.5 oz.) wet weight of soil and 250 cm^3 (15 in^3) of alcohol. In studies to compare this method with the conventional one for finding the water content of a soil, the discrepancy observed in the value obtained by the quick method has been ± 3%. Often the discrepancies are smaller. As a rule, 5 to 10 minutes are allowed for the water and alcohol to mix completely. The error is probably the filtering process to separate the mineral phase of the soil from the water-alcohol mixture. This process is very slow and unreliable when the soil is clayey. It is very hard to achieve complete disintegration of the clayey soil in the alcohol, at least with the methods available to the field operator. In highly organic or clayey soils, mixing has been made far easier by the use of mechanical mixers, but this is seldom a routine field procedure.

Alcohol combustion method: This method is described in detail in [54,59]. It consists of breaking up the mixture of wet soil in a pre-established quantity of alcohol, which is subsequently burned until total combustion is achieved. The residue assumed to be left by this process is the mineral phase of the soil, now dry. The difference between the wet weight or bulk weight and the dry weight permits calculation of the water content in the already familiar manner.

In order to achieve effective elimination of the water, the mixing and burning are sometimes repeated two or three times. The usual quantities used are 100 g (3.5 oz) of wet soil, to which a minimum of about 30 g (1 oz) of alcohol are added, although this amount can be increased to double and even more in clayey soils.

Figure 17-38 [54] shows the results of a study conducted on numerous soil samples with clay contents ranging from 4 to 71%, for the purpose of analysing the effect of subjecting the soil to one, two or three alcohol burnings. In all cases the water contents obtained with the alcohol method are compared with those obtained with the traditional standardized method including the oven-drying process. In these graphs a line at 45° indicates the position of the points where the two methods give the same value for the water content. It can be observed how the errors decrease as the number of combustion processes is increased. It is also clear that the greater the water content of the soil, the more times the alcohol burning must be repeated.

This method cannot be employed in organic soils, soils containing water of hydration and in all those where combustion causes the disappearance of anything other than free water.

.4 Intensive Drying Methods

Under this heading are included many methods which seek a rapid, severe drying of the soil by exposing it to a direct flame, an infra-red lamp or microwave radiation. Most of these methods have the common disadvantage that they expose the soil to temperatures above the 110°C (230°F) indicated as the maximum value in the standardized method, which leads to the vaporization or combustion of substances other than free water and consequently false test results. Organic matter and water of hydration are examples of this.

In Table 17-17, the authors of [42] present the results of a statistical investigation conducted on 945 soils, divided up as shown. The errors found for each type of soil are analysed for different values of the water content, comparing in each case the standardized conventional oven-drying method at 110°C (230°F) with the method of drying by direct exposure to a very hot flame. Plus and minus signs indicate whether the number should be added to or subtracted from the water content obtained by exposure to the flame, in order to reach the value that was obtained by the standardized method.

Errors increase with increasing water content of the soil, and there is a marked tendency towards greater errors as the plasticity of the material increases.

In view of the influence of the water content on the properties of compacted soils, the error obtained with this method is excessive. Consequently the method is not recommended.

Reference [42] describes in detail an investigation on soil drying with an infra-red lamp. In this case errors of only ± 0.5% are obtained in relation to the conventional method. The method, therefore, is reasonably reliable. However, since the equipment requires significant electric current, the method is best suited for laboratory use, but less practical for routine control operations at the job location.

.5 Measurement of Pressure of Calcium Carbide Gases

This method [60] consists of placing a measured quantity of calcium carbide (typically 20 g) inside a hermetic chamber, subsequently adding a similar quantity of the moist soil. The measuring device sometimes includes a rotation system that

Table 17-17

Errors in the determination of the water content with the method of direct exposure to a flame as compared to standard oven drying [42]

Type of soil	S	ML	MH	CL	CH	CL — ML	CH — MH
No. of Samples	81	157	144	130	140	130	163
Water content							
0 — 10%	±2.0	+1.0	±2.0	+1.0	±2.5	±2.0	±2.0
10 — 20%	±2.5	+1.5	±2.5	+1.5	±3.0	±2.5	±2.5
20 — 30%	±3.0	±2.0	±3.0	±2.0	±3.5	±3.0	±3.0
30 — 40%	±3.0	±2.5	±4.0	±2.5	±4.0	±3.0	±3.5

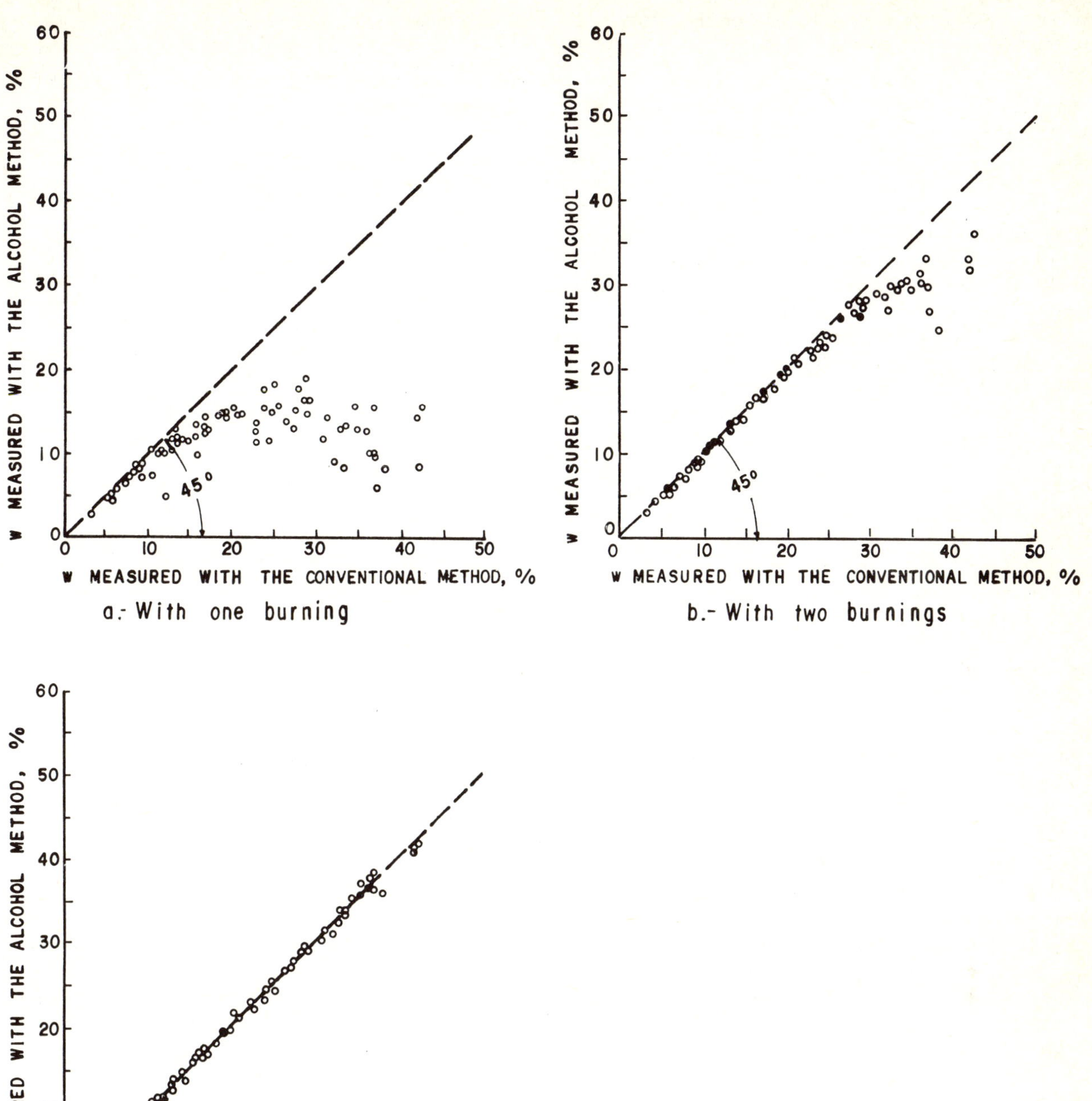

Fig. 17-38 Correlation between the alcohol combustion method and the conventional method for rapid determination of the water content

enables the two materials to mix thoroughly. Calcium carbide gasifies immediately on contact with water: the higher the water content of the soil, the more intense will be the gasification process that takes place in the chamber. The chamber is connected to a gauge which measures the pressure that is generated. This is empirically correlated with soil water content by previous calibration of the device in the laboratory. One of the drawbacks of this method is the small size of the soil sample, which leads to problems of representativity.

Reference [54] describes a comparative study between the results of this method and the conventional standard method, conducted on 35 soil samples, all with clay contents of over 70%. The water contents of the samples were extremely varied, Fig. 17-39 shows the correlation obtained in 183 comparisons of the 35 soils with different water contents. It is reported that 90% of the comparisons were within ± 2% of the standard water content; also that the errors or discrepancies found were not related to the clay content of the soil, although the precision of the method tends to diminish at greater water contents. However, the authors have found that the error often increases with the plasticity index.

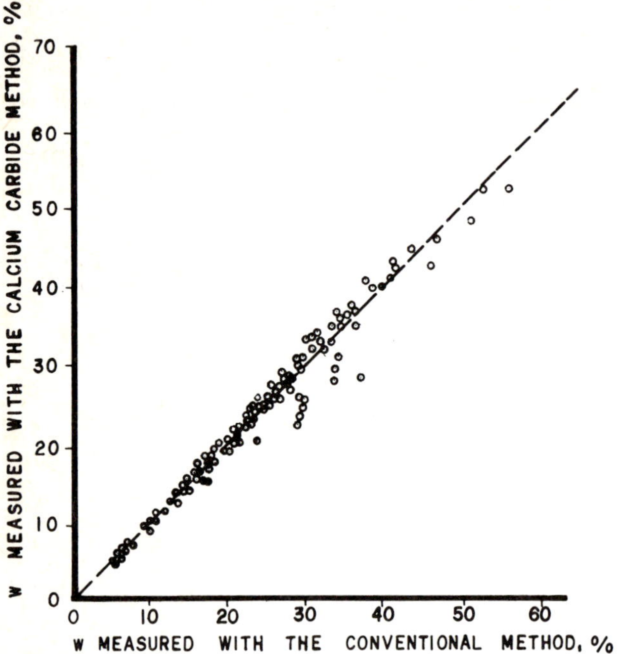

Fig. 17-39 Correlation between the calcium carbide method and the conventional method for rapid determination of the water content

17.5.4 Number of Samples in Compaction Control

The specifications of many large road design and construction organizations throughout the world usually establish a minimum number of compaction tests that must be performed on each section for control to be convincing. However, the criterion for establishing this minimum number is usually somewhat arbitrary and often governed only by consideration of time and money. As already indicated elsewhere in this chapter, statistical criteria permit the selection of samples and their number in accordance with more reasonable concepts than appearance alone. General sampling criteria have been discussed in considerable detail in this chapter, but it would seem appropriate at this point to emphasize some specific aspects of compaction control, with particular attention to the important problem of the number of samples required.

When establishing the size of a sampling lot for compaction control both statistical and economic considerations are taken into account. From the statistical viewpoint a homogeneous lot is essential, so that with a reasonable number of samples it will be possible to predict the properties of the universe that is to be controlled. From the economic viewpoint, the answer must be known as soon as possible; whether the lot produced is or is not acceptable. If it is rejected, the lot should be as small as practical so as to cause the least possible inconvenience.

For control purposes the size of a lot consisting of work produced is established by estimating the time required for sampling, laboratory tests, analysis of the information obtained, and reaching a decision of acceptance or rejection. The contractor's work will be delayed (or allowed to continue at risk) long enough for the control inspector to sample, conduct tests and reach a decision as to the quality of the work performed. This leads to the following [54]:

$$N = V_p (\Sigma t_R + t_D) \qquad (17\text{-}47)$$

where: N is the size of the lot subject to control, for example, in m/lot, V_p is the rate of progress of the work, for example in m/day, Σt_R is the total time employed for sampling and laboratory tests. (For all the samples taken from the control lot), and t_D is the time employed to analyse test results and reach a decision of acceptance or rejection.

Eq.(17-47) is used to establish the size of the control lot. For example, it may be decided to control sections several kilometers long. It is stressed that the lot should not be very large, because of the risk of rejection. On the other hand, if it is too small, control operations will be more expensive and the inspector's job will be rendered materially impossible because of insufficient time. Equation (17-47) attempts to strike a compromise between the two extremes.

The next question to determine is how many samples shall be taken from the control lot, [61] proposes the following formula for calculation of this number [54]:

$$n = \left(\frac{t\sigma'}{E_m}\right)^2 \qquad (17\text{-}48)$$

where: t is the confidence level at which it is wished to work, σ' is the standard deviation of the population, that is in this case the standard deviation of the compaction measurements for the entire control lot, and E_m is the inherent error of the compaction process which, in this case, is handled as an acceptable error and is established accordingly. Equation (17-48) is the same as Eq.(17-3).

Once the number of samples has been calculated, the inspector, who knows how long he needs to test each one, will be able to calculate the total time, Σt_R. Once the production rate is known (the amount advanced by the contractor per unit of time), and t_D has been estimated, Eq.(17-47) will tell him the value of N, in units of length per control lot. In other words, he will find out when in how many meters advanced by the contractor he must carry out a control cycle.

17.6 Geotechnical Surveys for Highways

This is the program under which Mexican engineering practice embraces the field and laboratory studies, visits and site inspections, analyses and calculations that lead to the recommendations that establish the geotechnical standards that must be observed in highway design and construction.

A geotechnical survey must provide the group in charge of highway design with all relevant information concerning the foundation soils and rocks, types of materials available and the utilization of available materials, indicating their probable future behavior and the treatments that will be required by all the soils and rocks that are to be employed, including the appropriate construction procedures to be adopted.

Emphasis has already been placed on the essentially simple and statistical nature of the studies, sampling and tests that are conducted to provide a basis for a geotechnical evaluation. This is a condition imposed by the highway (and runways for secondary airports) as a service to the general public. Low cost must always be kept in mind and determines the style and scope of the survey.

Geotechnical information must be presented in a simple, clear and methodical manner, translating the characteristics of the formations existing in the field and all pertinent data into numerical values for design computations and concise recommendations, which can be readily comprehended by the remaining members of the design group, even if they are not specialists in geotechnical disciplines.

A geotechnical survey can be divided into two stages. During the first stage, exploration work is conducted, data collected and laboratory tests performed. During the second stage, all available information is compiled and analyzed, and characterized concrete recommendations are drawn up, and the corresponding report is written.

17.6.1 Physiographical and Lithological Zoning

In order to simplify and organize field work, it is advisable to divide the route of the future highway into zones with similar characteristics. This is accomplished on the basis of physiography, vegetation and land use taking into consideration morphological characteristics. The zones are further divided into subzones on the basis of rock type, including major soil classes. Each of these subzones is then described in detail. Because each will present more or less homogeneous characteristics, each will have uniform classification and recommendations.

The description of each subzone must be done vertically, classifying each of its layers or strata. For this it is usually necessary to carry out soundings, take samples, conduct manual field tests as well as laboratory tests, especially in the case of soils. In the case of rocks, outcrops will have to be studied, establishing their lithology category and their structure.

For the initial zoning, the questionnaire appearing in Fig. 17-40 is completed. A questionnaire will have to be completed for each of the zones thus the geological and geomorphological knowledge of the engineer conducting the study will come into play. The availability of a photo-geological map of the region is of the greatest utility in this evaluation. The assistance of a geological engineer or engineering geologist is usually advisable at this stage of the survey. Figure 17-41 presents a table for more detailed analyses. In Part *1* the type of ground is largely classified in accordance with the magnitude of the earthworks that will be necessary in order to build the highway. In other words, the classification is based on the topographical characteristics of the area.

Morphological changes correspond to changes in the constituent materials. A morphological unit may be made up of different materials or a single type of material with different structural characteristics. In Part *2* of the figure, this is considered in detail. The zone under consideration is divided into a series of subzones in accordance with lithological characteristics. Observations include the degree of fracturing, degree of alteration and similar pertinent information. In Part *3* of Fig. 17-41 the most important factor to establish is the origin of the soils and, if possible, the type of accumulations in which they are found (alluvial, alluvial fan, fluvial terrace, swamp, salt marsh, lacustrine deposit, slope deposit, etc).

Table 17-18 contains a suggested, but not complete, list of the principal geotechnical problems that may be encountered in a

Table 17-18

Estimation of special geotechnical problems in surveys

Detail the problems relating to:

1) Lacustrine zones
2) Unstable hillsides
3) The poor quality of building materials
4) Heavily eroded zones.
5) Head-cutting of gullies from downstream
6) Failures
7) Instability of cliffs
8) Swampy zones
9) Flood zones
10) Adverse stratification or fracturing
11) Running water
12) High water table
13) Other problems

ROAD: ______________________________

____

STRETCH: ______________________________

____

SUB-ZONE: ______________________________

____

FROM STA ________ TO STA. ________

ORIGIN: ______________________________

____

DATE ______________

The purpose of this initial survey will be to zone the portion of road to be studied

Fig. 17-40 Questionnaire for initial geotechnical survey for necessity of further study and subdivision into zones

1. SUPERFICIAL OR MORPHOLOGICAL TOPOGRAPHICAL CHARACTERISTICS

(Put a cross in the box corresponding to the type of ground between the stations on the left.)

LOCATION		TYPE OF GROUND				
				HILLS		
From sta.	To sta.	Steep	Mountainous	Steep	Gentle	Flat
+	+					
+	+					
+	+					
+	+					
+	+					

2 GENERAL LITHOLOGICAL DESCRIPTION

LOCATION		TYPE OF ROCKS						
		IGNEOUS			SEDIMENTARY		METAMORPHIC	
			EXTRUSIVE					
From sta.	To sta.	INSTRUSIVE	LAVATIC	PYROCLASTIC	STRATIFIED	NON STRATIFIED	FOLIATED	NON FOLIATED

3 GENERAL DESCRIPTION OF SOILS

LOCATION		TYPES OF SOILS				
		RESIDUAL		TRANSPORTED		
From sta.	To sta.	FRICTIONAL	COHESIVE	WATER	WIND	GRAVITY

Fig. 17-41 Questionnaire for the physiographical zoning of highway routes

particular zone. Detection of these problems is very important for analyzing location alternatives, which is a major stage in highway design. Moreover, in the final design stages, each of these problems must be given special attention, considering possible solutions, along with their possible alternatives and cost, so that the most appropriate one can be selected. Detailed studies are very often required before sufficient information can be obtained about these special problems. Particularly detailed studies will be necessary for lacustrine or swampy zones, which are major causes of the stability and settlement problems in embankments on soft soils; unstable hillsides that may require very special methods of slope design and construction, and natural slopes showing signs of instability. In order to find out the nature of potential movements and future tendencies of a failed zone it may be necessary to establish lengthy (and costly) programs of field readings. Flood zones adjacent to large rivers usually demand large sections of protected embankments, numerous bridges and other drainage structures. In all these cases, very careful consideration must be given to the possible alternative of changing the location of the future highway so as to avoid these special problems.

17.6.2 Soil Data for Calculation of Mass Diagram

The correct calculation of a mass diagram, which is required for determining the excavation and fill sequence, utilization of available materials and cost of a project, depends largely on geotechnical considerations and topographic data that can be given to those in charge of the geometrical design of the highway.

Each location alternative that is studied must have its corresponding brief soil profile, plus detailed directions concerning the use of materials and the treatments to which they should be subjected. Figures 17-42 to 17-46 present a method of systematizing the relevant information.

.1 Table Giving Data for Calculation of Mass Diagram

The classification that appears in the third column of Fig. 17-42 refers to stony materials and soils, and was described in detail in Chapter 2. Besides providing the relevant group symbol, a very brief description of the materials should be added. For example, the typical classification of a fluvial deposit would be:

> Clean gravel, uniform, coarse, very hard, rounded, light grey, with 20% sand and 30% small fragments, dense (*GP–Fc*).

A fine residual soil might be described as:

> Clay, slightly sandy, moderate plasticity, reddish, not very moist and very firm, fissured throughout the 2 m (6.5 ft) thickness that is studied, with 4% rounded gravel and some isolated small rock fragments, with roots in the upper 30 cm (12 in) (*CL*).

The criteria for establishing the above classifications are developed from the information that was given in Chapter 2.

The column which appears under the heading *Probable treatment* refers to the mechanical processing that is recommended for each of the materials found, when placed in the embankment. The most frequent treatments are compaction for soils, ramming or pounding with a tractor or similar equipment (which is still used for very coarse materials) or dumping

TABLE FOR CALCULATING THE MASS DIAGRAM

ROAD ____________________
STRETCH ____________________
SUB-ZONE ____________________
ORIGIN ____________________ DATE ____________

STATION		STRATUM		CLASSIFICATION	PROBABLE TREATMENT	COEFFICIENT OF VOLUME CHANGE				CLASIFICATION FOR ESTIMATE	CUT		OBSERVATIONS
FROM	TO	Nº	THICKNESS						TRACTOR RAMMING	A B C	MAXIMUM HEIGHT	SLOPE	

Fig. 17-42 Table for calculating mass diagram in geotechnical surveys

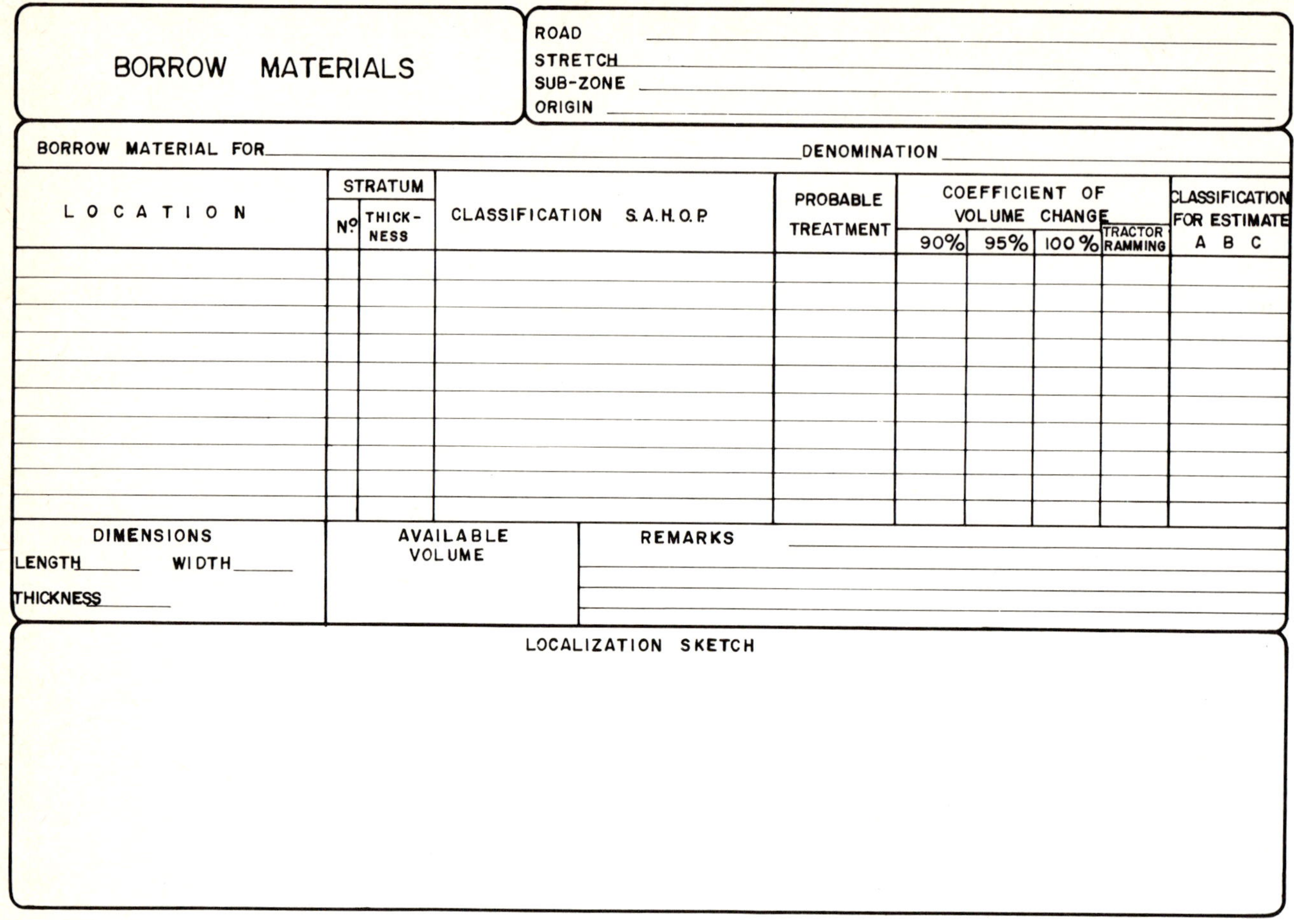

BORROW MATERIALS

ROAD ______
STRETCH ______
SUB-ZONE ______
ORIGIN ______

BORROW MATERIAL FOR ______ DENOMINATION ______

LOCATION	STRATUM		CLASSIFICATION S.A.H.O.P.	PROBABLE TREATMENT	COEFFICIENT OF VOLUME CHANGE				CLASSIFICATION FOR ESTIMATE
	No.	THICK-NESS			90%	95%	100%	TRACTOR RAMMING	A B C

DIMENSIONS	AVAILABLE VOLUME	REMARKS
LENGTH ____ WIDTH ____ THICKNESS ____		

LOCALIZATION SKETCH

Fig. 17-43 Table for detailing borrow materials in geotechnical surveys

rocky material to fill the first few meters at the bottom of gorges.

Ramming consists of driving a tractor over the coarse material spread in layers. It was discussed in Chapter 4. This is not the ideal treatment for the construction of large rockfills, but in regions where sophisticated machines are scarce it is still used to accommodate rock fragments in small, low embankments. The procedure is used only for very coarse materials, where the normal compaction procedures involving conventional equipment are impossible.

The coefficients of volume change of the materials to be used in the construction of embankments, help predict the cut-fill balance. The dry unit weight of a material in the field will never be the same as that of the original material, once extracted and placed in the embankment. On excavation, the volume of the material usually increases, decreasing again when compacted at its ultimate location, this decrease depending on the degree of compaction that is obtained. The coefficient of volume change is a number which expresses the relation between the dry unit weight of a material in its natural state and the dry unit weight when the material is compacted. It is best expressed as:

$$C_{vv} = \frac{\gamma_{dn}/\gamma_{dmax}}{G_c} \qquad (17\text{-}49)$$

where:

γ_{dn}, is the dry unit weight of the soil in its natural state in the field.

$\gamma_{d\,max}$, is the maximum dry unit weight for that soil by the compaction control test that is employed.

G_c, is the degree of compaction that is specified for the case under consideration, as defined in Chapter 4.

The coefficient of volume change permits determination of the volumes of the materials that are to be taken from borrow areas, so as to reach the volume that is required in the embankments. It is a vital piece of information for determining the real cost of a given project, yet it is often only estimated instead of being based on real data. Equation (17-49) cannot be used for rock fragments because these materials, on account of the size of their particles, cannot be subjected to ordinary compaction tests. In these materials, therefore, the coefficient of volume change has to be estimated or measured from test excavations and fills. By way of illustration, Table 17-19 gives some typical coefficients of volume change for different materials. These coefficients should be calculated from soil tests when possible, for their influence on the earthworks planning and economic evaluation in a particular project is vitally important. Obtaining the value most appropriate for each specific situation is usually worth the cost of testing.

The classification for cost estimation purposes appearing in the next column of the mass diagram calculation table (Fig. 17-42) answers a practical need of the organizations dedicated to the design and construction of highways on a large scale. With the assistance of contracting firms, the purpose is to classify the materials that are to be moved, to establish a price for each type of material, determine the degree of difficulty of the operations involved, and the equipment and methods that must be employed for this purpose, Mexican practice distinguishes

Table 17-19

Typical volume change coefficients for different materials depending on compaction

Type of material	Compacted 90%	Compacted 95%	Compacted 100%	Bulldozer or excavator impact (ramming)	Excavated (dumped)
SANDS					
Loose	0.87	0.82	0.78		1.00
Moderately dense	0.96	0.91	0.86		1.10
Dense	1.03	0.98	0.93		1.20
Very dense	1.11	1.05	1.00		1.28
NON-PLASTIC SILTS					
Very loose	0.82	0.78	0.74		1.06
Loose	0.91	0.86	0.82		1.17
Moderately dense	0.99	0.94	0.89		1.27
Dense	1.06	1.00	0.95		1.36
Very dense	1.11	1.05	1.00		1.43
CLAYS AND PLASTIC SILTS					
Very soft	0.78	0.74	0.70		1.08
Soft	0.87	0.82	0.78		1.20
Medium	0.95	0.90	0.85		1.30
Firm	1.01	0.96	0.91		1.40
Very firm	1.08	1.02	0.97		1.49
Hard	1.14	1.08	1.02		1.57
ROCKS					
Well weathered. Rocks showing very advanced physical and chemical alteration; poorly cemented, with appreciable soil-filled cracks; they crumble easily. A bulldozer or tractor-excavator can be used to obtain small fragments, gravels sands and clays.				1.00	1.10
Moderately weathered. Rocks showing moderately advanced physical and chemical alteration; moderately cemented; fractured. A plow, ripper or light explosive charge is required to obtain small and medium sized fragments, gravels and sands.				1.07	1.25
Slightly weathered. Rocks showing little physical and chemical alteration; well cemented; slightly fractured. Large explosive charges are required to obtain small, medium-sized and large fragments and gravels.				1.15	1.50
Unweathered. Rocks showing no physical or chemical alteration; slightly fissured or not at all; well cemented; dense. Large explosive charges are required to obtain large and medium-sized fragments.				1.25	1.75

between three types of materials: *A*, which can be readily excavated, for example with pick and shovel; *B*, which presents greater difficulties, but does not require the use of explosives, and *C*, which has to be extracted by means of blasting. It is common practice in Mexico to describe a material by means of three numbers, always totaling 100, which represent respectively the percentages of *A*, *B* and *C* material that make up the total to be removed. The final price for one cubic meter of excavated material is determined by these percentages and by the price previously indicated by the contracting firm for excavation of the same unit of volume in each of the three categories of material considered. In Mexico, an excavation classification system such as this serves no further purpose than as a guide to the cost calculations that are made by the organization that designs and compares alternatives. The Ministry of Public Works has the practice of public competition for pricing, with a single price per cubic meter of material placed in the embankment and treated as directed in the design regardless of classification.

This procedure avoids many legal problems or differences of opinion, for a payment classification system like that described above, that establishes different prices for different materials, involves a great deal of subjective interpretation and may result in greatly divergent conclusions from different specialists, all well-intentioned.

One of the most important parts of a geotechnical survey for a highway is the recommendations that are included for the slopes required by cuts and embankments. Emphasis has already been placed elsewhere in this book on the practical need for most of such recommendations to be based on brief low cost studies, and on how essential it is, therefore, that they be made by experienced specialists, who will be capable of using the limited information resulting from these studies to best advantage. When this recommendation is considered as part of the vital information that must be found in a geotechnical survey, it is easy to understand why these studies must be systematically entrusted to groups of experienced specialists, and how profitable it must be for any organization in charge of these projects to have such a team of specialists available. The recommendations for slope inclination are even more important than calculation of a mass diagram.

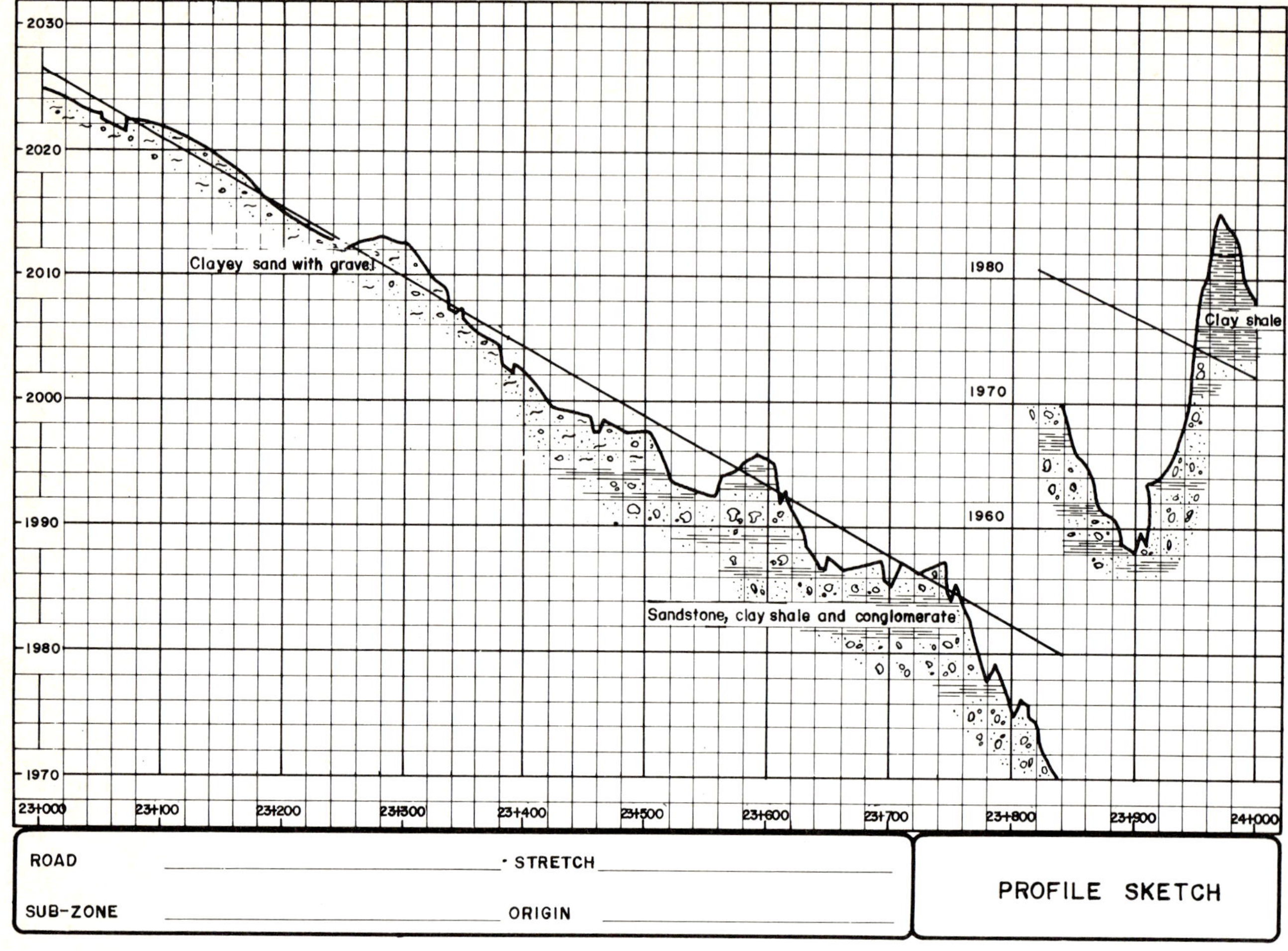

Fig. 17-44 Profile sketch for geotechnical survey (Elevation)

In addition to the mass diagram calculation table, it will be of value to indicate how the different materials found in the field can be used in embankments. Reference has already been made elsewhere in this book that the correct use of a material is dependent on its position within the embankment; for example, the stress levels in the same material will vary from one location to another, owing to the dead weight of the embankment itself and the effect of traffic as well as different conditions imposed by surface water and subsurface water. Thus, a material which in one location may function satisfactorily may be the cause of catastrophic failures if placed in some other location. Also, the treatments required by a material may vary depending on its position within the embankment. This is one of the most vital pieces of information provided by a geotechnical survey, and also one of those demanding greater knowledge and care of whoever determines it.

As a further complement to the mass diagram, it is useful to indicate the places where excavated steps will be built between the embankments and the foundation ground, where clearing or similar procedures will have to be carried out, plus other operations involving earthworks and leading to variations in the cost of the project.

.2 Borrow Material Diagrams

Figure 17-43 gives the data of all the borrow pits that are used for the construction of a highway.

The materials for building embankments are usually obtained from three different sources. The material obtained from excavating a cut is used to construct a neighbouring fill. This procedure is usually known as balancing cuts and fills. It is economical because it reduces volumes of waste and uses all suitable excavation materials. In many cases the balance that is achieved is not total, resulting either in insufficient material or waste, depending on whether or not the volumes of fill are in excess of those of the cut. Procedure is limited by the quality of the materials that are obtained on excavating cuts and that required for the embankments.

The second source of building materials is the so-called roadside borrow pit. Here the material required is taken from excavations that are parallel to the road and immediately adjacent to it, usually within the right of way. This procedure permits a reduction in material haul which represents an important part of total construction costs. The method is limited by the quality of the materials existing in the foundation ground. In flat, agricultural, flood susceptible or swampy zones, this may leave much to be desired. Also, the ditches made from excavation close to the road may cause serious humidity problems in the embankments, when they become filled with rainwater, and are difficult to drain, in flat ground where logically the roadside borrow pit is more often used. At other times, owing to the narrowness of the right of way (so as to avoid expenditure on the acquirement or expropriation of land) very deep ditches have to be dug, resulting in intensification of drainage and ponding problems. A roadside borrow pit is

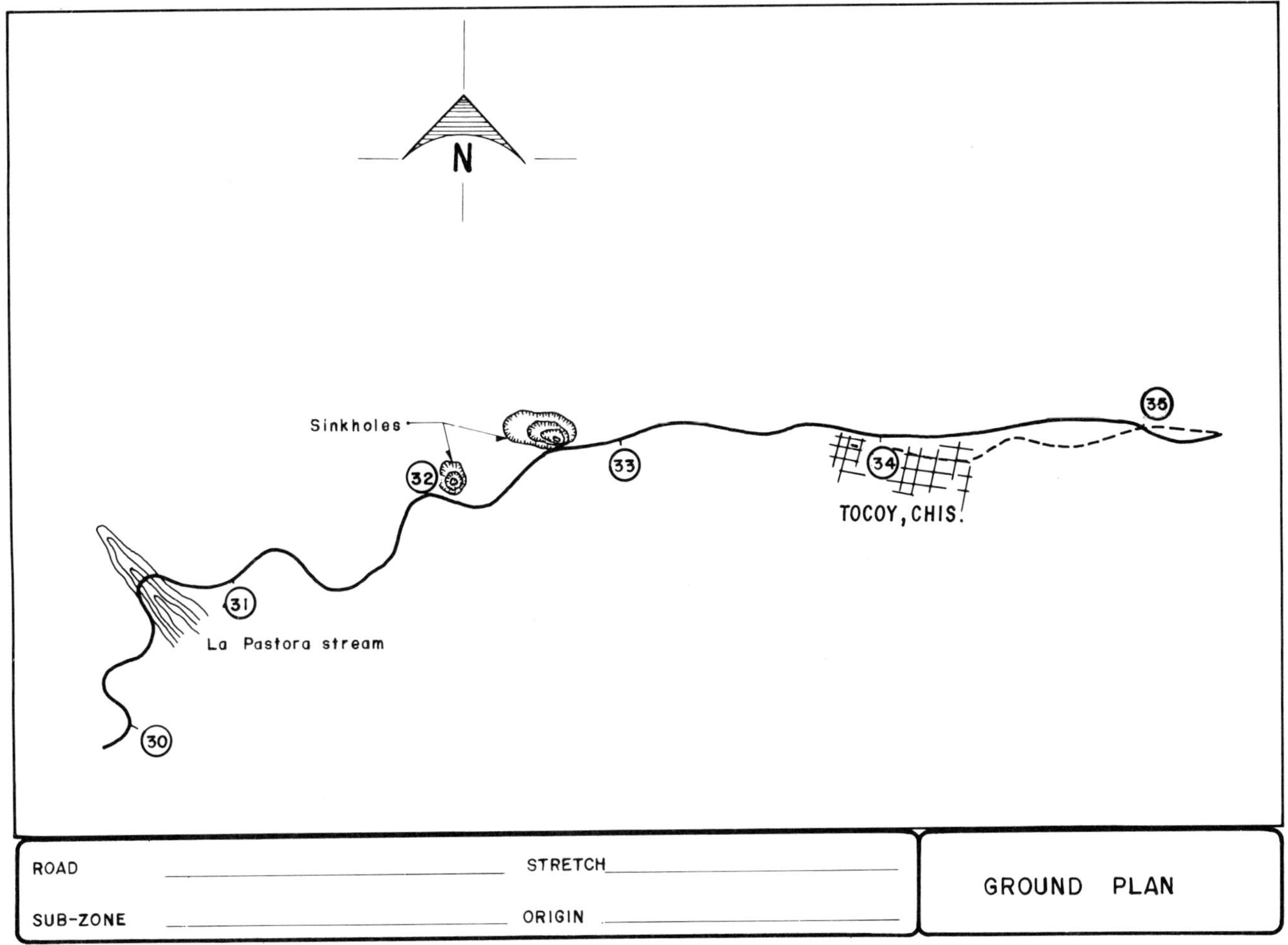

Fig. 17-45 Profile sketch for geotechnical survey (Plan)

not recommended by traffic engineers, for they are afraid of its adverse psychological influence on the motorist, leading to serious accidents. In view of all the foregoing factors, the roadside borrow pit should be used only when the materials it provides and the resulting ditches are appropriate, easy to drain, located at a reasonable distance from the road. Reference must be made here to another effect of the roadside borrow pit, which is rarely considered by the engineers who use it. The ditches produced by excavation, when well drained, are similar to a drain that lowers the water-table in neighbouring land. This land is often agricultural, and in quite a wide strip parallel to the excavation the originally moist condition favorable to farming is lost, replaced by arid land which is useless for crop-growing. If the losses due to unforeseen drainage were taken into account in the costs analyses of the engineer (which is rare), it will be seen that in some cases the roadside borrow pit is not so economically advantageous as it might at first appear.

The third method for obtaining earth or rock materials in highway practice is the location of a natural deposit or formation, consisting of a material with appropriate characteristics, which is excavated massively and is then hauled to the road and used. These are true borrow pits, about which more details will be found elsewhere in this book.

The balancing of cuts and fills, the roadside borrow pit and the use of banks should be described in the geotechnical studies. The first two methods must be given careful consideration once the stratigraphy and properties of the foundation ground adjacent to the road are known, plus the characteristics of the hills in which cuts will be made that are likely to produce suitable material for building embankments. Because this last case demands a knowledge of the subsoil at far greater depths than those usually reached by the exploration that is employed in highway geotechnical surveys, geophysical prospecting methods (Chapter 3) will provide a valuable complement. They not only give information about the rippability of soils and rocks, with a view to determining methods of excavation and costs, they also serve to estimate the quality of the materials and their eventual utilization in the partial or total construction of neighbouring embankments.

Banks or quarry sites of materials must be the object of special research, which will be described in greater detail elsewhere. For each one a table such as the one entitled *Borrow materials* (Fig. 17-43) will have to be completed. Most of the data provided by this table have been discussed, but in addition, good information will have to be supplied regarding utilization, the shape of the bank, the side where excavation should commence, the volume that can be utilized, the location and the treatments required according to the purpose for which the materials are intended.

The subgrade layer and the materials for the sub-base, base course and wearing course of flexible pavements, the sub-ballast and ballast for railroads, and the materials for making concrete usually come from specially located banks or quarries. The materials for the lower layers of embankments are often obtained by balancing cuts and fills or from roadside borrow

ROAD:		STRETCH		COMPLEMENTARY STRUCTURES
SUB-ZONE		ORIGIN		

COMPLEMENTARY DRAINAGE STRUCTURES

FROM STATION TO STATION	LINING OF DITCHES			CONSTRUCTION OF INTERCEPTING DITCHES			UNDERDRAIN			ROADSIDE CURBS			CHUTES LENGTH	OBSERVATIONS
	LEFT	RIGHT	LENGTH	LEFT	RIGHT	LENGTH	LEFT	RIGHT	LENGTH	LEFT	RIGHT	LENGTH		
TOTALS														

Fig. 17-46 Table for detailing necessity for complementary drainage structures in geotechnical survey

pits. Materials from special banks are becoming more and more widely used, especially in runways or stretches of highway or railroad where better quality materials are sought.

.3 Sketch of Soil Profile

This is the graph shown in Fig. 17-44. It contains a sketch of the soil profile in each of the zones or subzones that have been determined throughout the entire length of the future highway. This sketch should give all the information compiled during field observation and exploration work, along with complementary geophysical data when necessary.

.4 Ground Plan

In order to provide a graphical representation of the projected highway location, along with the principal topographical and geological features and those relating to population, a ground plan of the location is drawn to scale as in Fig. 17-45.

.5 Complementary Drainage Structures

As was already stated at the beginning of this chapter, in view of their characteristics and number, the recommendation of complementary drainage structures will depend on the knowledge and experience of specialists rather than on the results of detailed studies. For this there is no specific methodology that is trustworthy. Exception is made of culverts, which although drainage structures, do not fall into the category of those referred to as *complementary* in this book. Indeed, specific methodologies have been developed both for determining the location of culverts in relation to existing streams and runoff, and for their hydraulic characteristics.

Experience indicates that the specialists best able to establish specific recommendations concerning complementary drainage structures (in the sense employed in this chapter) are those in charge of the geotechnical surveys for highways, although it is advisable for them to consult and compare opinions frequently with the engineers responsible for the hydrological studies and conceptual design of major drainage structures (bridges) and minor drainage structures (culverts). The reason for this, as was discussed earlier in this chapter, is that so-called complementary structures are basically associated with cut and embankment protection, erosion control in soils, and the loss of stability in soils and rocks, all of which are dealt with by the specialist responsible for geotechnical surveys.

For this reason it is customary in Mexico for the geotechnical survey to include recommendations as to where and how to construct these complementary drainage structures. Such information is provided in Fig. 17-46, which will include details of where to build ditches, intercepting ditches, curbs and chutes, and how to build them, especially concerning paving them with watertight materials.

Subdrainage by means of a perforated pipe installed in a roadside ditch filled with filter material, has lately become so widespread and so useful that the corresponding recommendations must also be included in the geotechnical survey. Other more sophisticated subdrainage structures (Chapter 7) should be discussed in the geotechnical survey, even if the details of their design are in fact the object of a special survey.

APPENDIX 17A

WORKED PROBLEM

A quality control program has been drawn up for an embankment construction job utilizing the same soil and compaction procedure. In order to verify §17.3, a random sampling of the densities obtained was undertaken with the results presented in Table 17a-1. The purpose of sampling was to check 30 sampling points in each 10 km (6 mi) stretch. Table 17a-1 corresponds to the first 10 km (6 mi). With the foregoing data, the following facts are obtained:

$$\bar{x} = 1468 \text{ kg/m}^3 \ (92 \text{ lb/ft}^3)$$
$$\sigma = 136 \text{ kg/m}^3 \ (8.5 \text{ lb/ft}^3)$$

1) The stretch under control has a total length of 72.0 km (45 mi), and 185 observations will be necessary over and above the ones already indicated. Assuming that the construction materials, equipment and procedures remain constant throughout the entire stretch and that the distribution of the unit weights is normal, how many observations can be anticipated between 1420 kg/m^3 and 1520 kg/m^3 (89 and 95 lb/ft^3)?

   If the unit weights are distributed normally, a diagram like that in Fig. 17a-1 will be obtained. The probability of the observations being located between 1420 and 1520 kg/m^3 (89 and 95 lb/ft^3) is numerically equal to the hatched area in the figure.

Table 17a-1
Quality control data for embankment construction : random sampling of dry unit weight

Sampling point		Dry unit weight	
km	mi	kg/m^3	lb/ft^3
0 + 320	0 + 0350	1380	86
0 + 975	0 + 1065	1460	91
1 + 105	0 + 1205	1680	105
1 + 217	0 + 1330	1460	91
1 + 820	1 + 0230	1610	101
1 + 963	1 + 0387	1640	102
1 + 978	1 + 0403	1580	99
2 + 203	1 + 0650	1260	79
2 + 285	1 + 0738	1730	108
2 + 715	1 + 1210	1450	90
2 + 917	1 + 1430	1500	94
2 + 995	1 + 1520	1400	87
3 + 415	2 + 0215	1380	86
3 + 970	2 + 0822	1420	89
3 + 993	2 + 0848	1350	84
4 + 515	2 + 1420	1320	82
4 + 917	3 + 0098	1470	92
5 + 030	3 + 0223	1760	110
5 + 145	3 + 0348	1470	92
5 + 670	3 + 0923	1420	89
6 + 013	3 + 1300	1440	90
6 + 620	4 + 0202	1360	85
6 + 715	4 + 0306	1630	102
6 + 982	4 + 0597	1350	84
7 + 315	4 + 0962	1500	94
7 + 717	4 + 1403	1250	78
7 + 814	4 + 1507	1480	92
8 + 015	4 + 1725	1190	74
8 + 917	5 + 0953	1530	96
9 + 030	5 + 1076	1560	98

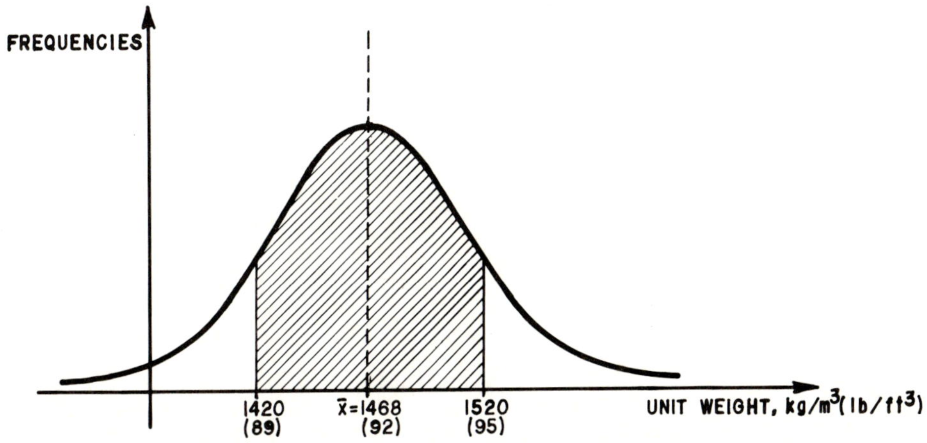

Fig. 17a-1 Distribution of observation frequencies

It can be determined with the aid of Table 17-12 and Eq.(17-5). Using the change of variable:

$$z = \frac{x - \bar{x}}{\sigma}$$

$$z_1 = \frac{1420 - 1468}{136} = \frac{-48}{136} = -0.35$$

$$= \left[\frac{89 - 92}{8.5} = -0.35\right]$$

$$z_2 = \frac{1520 - 1468}{136} = \frac{52}{136} = 0.38$$

$$= \left[\frac{95 - 92}{8.5} = 0.38\right] \qquad (17\text{-}5)$$

With the aid of Table 17-12 it is found that the area is equal to 0.2848, thus the probability of the observations being located within the interval considered is 28.48%. If 185 observations are conducted, the number of them within the interval is calculated as 0.2848 × 185 = 52.

2) Using the same procedures as in a), determine the interval within which 90% of the observations could be located symmetrically in relation to the mean. The values of x_1 and x_2 in the conditions in Fig. 17a-2 must now be determined. The values of z with which the conditions would be obtained are (from Table 17-12): $z_1 = -1.65$, $z_2 = 1.65$. From Eq.(17-6): $x = \bar{x} + z\sigma$. Therefore:

$$x_1 = 1468 - 1.65 \times 136 = 1468 - 224 = 1244$$
$$(92 - 1.65 \times 8.5 = 78)$$

$$x_2 = 1468 + 1.65 \times 136 = 1468 + 224 = 1692$$
$$(92 + 1.65 \times 8.5 = 106)$$

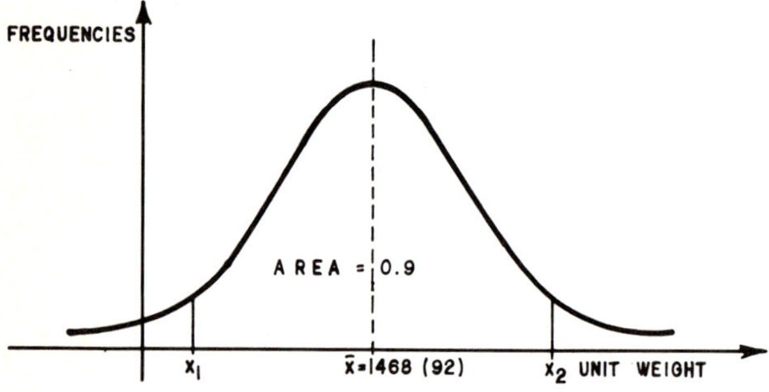

Fig. 17a-2 Calculation of the interval corresponding to 90% of the observations

It is to be expected (with 90% probability) that 90% of the observations beyond km 10 + 000 (6 mi) will be within the limits obtained above. Should it be concluded that the results obtained are outside the tolerance limits required by the design, the control inspector must take action to improve the compaction process.

3) Let the maximum density obtained in the corresponding laboratory test be 1400 kg/m^3 (87 lb/ft^3) and let a degree of compaction of 95% be specified in the design with a lower tolerance limit of 2% (93%). It is wished to find what percentage of control points beyond the first stretch will be rejected in view of the results for this controlled stretch. According to the above data, the minimum acceptable unit weight will be 1302 kg/m^3 (81 lb/ft^3). The percentage of rejected points anticipated is equal to the hatched area in Fig. 17a-3, which can be obtained from Eq.(17-5) and Table 17-12.

$$Z = \frac{1302 - 1468}{136} = \frac{-166}{136} = -1.22$$

$$\left[Z = \frac{81.6 - 92}{8.5} = \frac{-10.4}{8.5} = -1.22\right]$$

According to Table 17-12, the area proves equal to 0.3888 and the percentage of points rejected is 12.

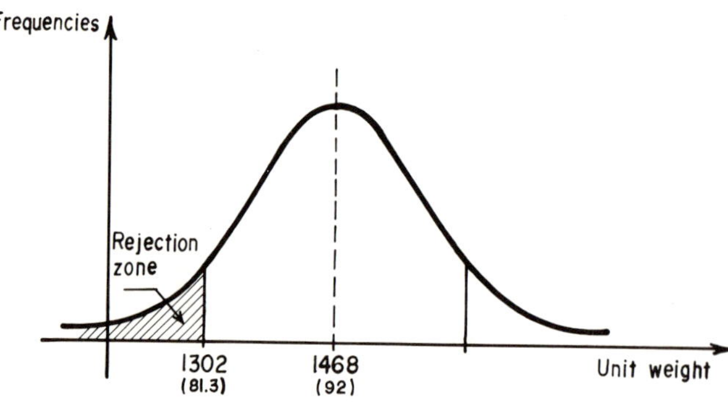

Fig. 17a-3 Calculation of the percentage of points under the lower limit selected

4) Let's assume now that the sampling stage is over, the local stretch has been subdivided into 7 groups (samples) and the mean control chart has been drawn up taking into consideration the following results for the mean and the standard deviation of the sample in the corresponding stretch (see Table 17a-2). The control limits for the mean control chart are obtained thus:

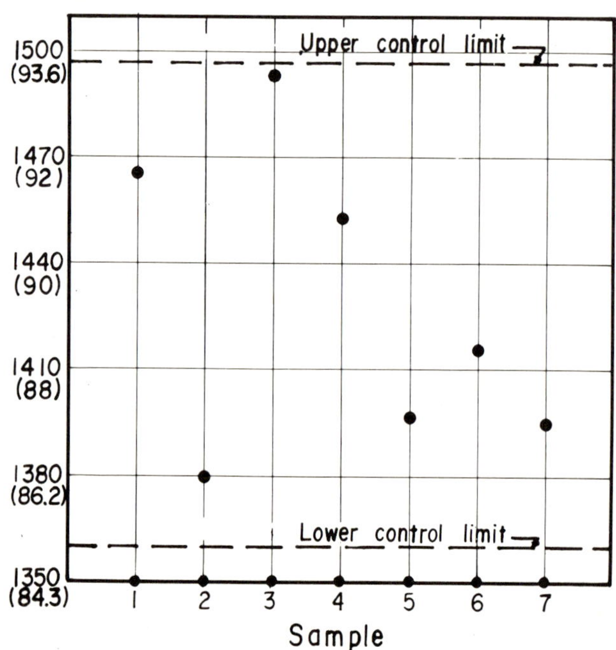

Fig. 17a-4 Measurement control chart

Table 17a-2
Mean and standard deviation data for stretch of highway in Table 17a-1, divided into 7 stretches

Stretch	Length		Mean dry unit wt., $\bar{x}$		Standard deviation, σ	
	km	mi	kg/m^3	lb/ft^3	kg/m^3	lb/ft^3
1	10		1468	91.5	136	8.48
2	10	6.2	1380	86.0	112	6.98
3	10	6.2	1495	93.2	131	8.16
4	10	6.2	1453	90.6	110	6.86
5	10	6.2	1397	87.1	121	7.54
6	10	6.2	1415	88.2	134	8.35
7	12	7.5	1395	86.9	108	6.73

$$\bar{\bar{x}} = \frac{\Sigma \bar{x}}{7} = \frac{10003}{7} = 1429$$

$$\left[= \frac{624}{7} = 89 \right]$$

$$\bar{\sigma} = \frac{\Sigma \sigma}{7} = \frac{852}{7} = 121.7$$

$$\left[= \frac{53.2}{7} = 7.6 \right]$$

Applying Eq.(17-8)

$$LS = \bar{\bar{x}} + A_1 \sigma$$
$$LI = \bar{\bar{x}} - A_1 \sigma$$

The coefficient A_1 is obtained from Table 17-8, taking into consideration that in all cases the number of items in the sample is 30. Thus, $A_1 = 0.56$.

$$LS = 1429 + 0.56\,(121.7) = 1497.15$$

$$\left[= 89 + 0.56\,(7.6) = 93.25 \right]$$

$$LI = 1429 - 0.56\,(121.7) = 1360.85$$

$$\left[= 89 - 0.56\,(7.6) = 84.75 \right]$$

It can be observed that in accordance with the control charts criterion the 7 sections (samples) are within the control limits, Fig. 17a-4. In other words, all the observations can be attributed to inherent variations of the compaction process.

(This problem is by courtesy of MIGUEL DE J. QUINTERO NÁRES)

REFERENCES

1. MC MAHON, T.F., THURMUL, F. and HALSTEAD, J. W., "Quality Assurance in Highway Construction, Part 1: Introduction and Concepts," *Public Roads*, Vol. 35 No 6, 1969.
2. YANG, N. C., *Design of Functional Pavements*, McGraw Hill, 1972, Chap. 4.
3. Highway Research Board, "Quality Assurance and Acceptance Procedures," Special Report 188, Washington, D. C., 1971.
4. ABDUN-NUR, E. A., "Design of Practical Specifications for Concrete Quality Control," Publication of the Mexican Cement and Concrete Institute, México, 1970.
5. DÍAZ DE COSSÍO, R., CASILLAS, J. and ROBLES, F., "Concreto Reforzado," Mexican Cement and Concrete Institute, Strength and Quality Control Indices, *Journal IMCYC*, No 11. México, 1964.
6. Highway Research Board, "Evaluation of Construction Control Procedures" (Interim Report), Report 34: Chapter 2, Washington, D. C., 1967.

7. Cramer, H., *Elements of Probability Theory,* 1960, Chap. 16.

8. Alexander, H. W., *Elements of Mathematical Statistics,* John Wiley and Sons, 1961, Chap. 4.

9. McLauglin, J. F. and Hanna, S. J., "Evaluation of Data," Civil Engineering Reprint No CE 222, Purdue University, Lafayette, Ind., 1966.

10. Wagner, W. K., "Effect of Sampling and Job Curing Procedures on Compressive Strength of Concrete," *Materials Research and Standards A. S. T. M.*, Vol. 3 No 8, 1963.

11. Sherman, G. B., Watkins, R. O. and Prysock, R. H., "A Statistical Analysis of Embankment Compaction," California Division of Highways, Materials and Research Department, Report No M. R. 631133-3, Sacramento, Cal., 1967.

12. Grant, E. L., *Statistical Quality Control*, 1974.

13. Simon, J. and Leflaive, E., "Nombre et Dimension des Echantillons: Incidence sur L'interpretation du Controle Symposium sur le Controle de la Qualité des Ouvrages Routiers," Organisation de cooperation et de Development Economiques, Aix en Provence, France, 1970.

14. Spiegel, M. R., *Statistics Theory and Problems*, McGraw Hill, 1961.

15. Hveem, F. N., "California's Experience with the Record Sampling Program," Procs. AASHO, Washington, D. C., 1962.

16. Cristiansen, R. B., "Record Sampling from a State's Viewpoint," Procs. WASHO., 1962.

17. Little, L. W., "The Record Sampling Program," Procs. WASHO., 1961.

18. Zuehlke, G. H., "Problems Relating to Control of Hot Bituminous Mixtures," Procs. AASHO, Washington, D. C., 1962.

19. Keyser, J. H. and Wade, P. F., "A study of Variations in the Testing and Production of Bituminous Mixtures," Highway Research Board, Record No 24, Washington, D. C., 1963.

20. Shook, J. F., "Significance of Test Results Obtained from Random Samples," Special Technical Publications No 362, A. S. T. M., 1963.

21. The Asphalt Institute, "Thickness Design Full Depth Asphalt Pavements Structures for Highways and Streets," Manual Series No 1, College Park, Md., 1969.

22. Kreyszie, E., *Introducción a la Estadística Matemática*, Limusa: México, 1973, Chap. 14.

23. Benjamin, J. R. and Cornell, C. A., *Probability Statistics and Decision for Civil Engineers*, McGraw Hill, 1970, Chap. 4.

24. Rascón, O. and Villarreal, A., *Introducción a probabilidades y Estadística*, Publication of the Institute of Engineering of the Faculty of Engineering. UNAM. México, D. F., 1972, Chap. 3.

25. Rascón, O., *Introducción a la Teoría de Probabilidades* (programmed text), Publication of the National Autonomous University. México, 1971.

26. Moreno, B. A. and Jauffred, M. F., *Elementos de Probabilidad y Estadística,* Representations and Services of Engineering S. A. México, 1969, Chap. 6.

27. Hicks, C. R., *Fundamental Concepts in the Design of Experiments*, Holt, Reinhart and Winston: New York, 1964.

28. Guenther, W. C., *Concepts of Statistical Inference*, McGraw Hill, 1968, Chap. 4.

29. Ostle, B., *Statistics in Research*, Iowa State University Press, Ames, Iowa, 1963.

30. Duncan, A. J., "Quality Control and Industrial Statistics," Richard D. Irwin, Inc. Homewood, Ill., 1959.

31. Yoder, E. J. and Williamson, T. G., "Techniques for Compaction Control," Purdue University Engineering Reprints, CE 249, Purdue University, Lafayette, Ind., 1969.

32. Mathews, D. H. and Hardman, R., "Le Risque Dans les Specifications, le Controle et la Reception des Materiaux destinés a la Construction Routier," *Procs. of the Simposium sur Controle de la Qualité des Ouvrages Routiers*, Centre D'etudes Techniques de l'equipement, Aix en Provence, 1970.

33. Schuhbauer, A., "Controle Statistique de la Qualité dans la Construction des Chaussées (German)," ibid.

34. Halstead, W. J. and McMahon, T. F., "Garantie de la Qualité en Construction Routiere aux Etats Unis d'Ameriqué," ibid.

35. Van de Fliert, C., Brouwers, J. A. C., Span, H. J. J. H. and Wester, K., "Controle de la Qualité des Chaussées aux Pays Bas," ibid.

36. Hondermarcq, H. and Doyen, A., "Le Controle des Travaux Routiers en Belgique," ibid.

37. Matsuno, S., "Situation actuelle de Controle par Sondage dans la Construction des Chaussés souplés au Japon," ibid.

38. Bonitzer, J., "Problemes specifiques dans le Controle de la Qualité des Travaux Routiers en France," ibid.

39. Manton Hall, A. W., "The Operating Characteristics of A. S. CA2. 1963, 1968 and 1973 versions," First Australian Conference on Engineering Materials, The University of New South Wales, 1974.

40. Johnson, A. W. and Sallberg, J. R., "Factors that Influence Field Compaction of Soils.," H. R. B. Bulletin No 272, Washington, D. C., 1960.

41. Highway Research Board, "Construction of Embankments," Synthesis of Highway Practice No 88, Washington D.C., 1970.

42. Saborío, J. and Zárate, M., "Métodos de Control Rápido de la Compactación y la Humedad en Terraplenes," Publication of the Mexican Ministry of Public Works, México, 1964.

43. Belcher, D. F., Cuykendall, T. R. and Sack, H. S., "The Measurements of Soil Moisture and Density by Neutron and Gamma Ray Scattering," U. S. Civil Aeronautics Admin. Tech. Dev., Report No 127, 1950.

44. Carlton, P. F., Belcher, D. J., Cuykendall, T. R. and Sack, H. S., "Modifications and Tests of Radioactive Probes for Measuring Soil Moisture and Density," U. S. Civil Aeronautics Admin. Tech. Dev., Report No 194, 1953.

45. Pocock, B. W., Smith, L. W., Schwartje, W. H. and Hanna, R. E., "The Michigan Nuclear Combination Density and Moisture Surface Gage," Michigan State Highway Department, Office of Testing and Research, Report No 316., 1959.

46. Carey, W. N. and Reynolds, J. F., "Some Refinements in Measurements of Surface Density by Gamma Ray Absorption," H. R. B. Special Report No 38, 1959.

47. Carey, W. N., Shook, J. F. and Reynolds, J. F., "Evaluation of Nuclear Moisture and Density Testing Equipment," A. S. T. M. Annual Meeting, 1960.

48. Brown, W. R., "Nuclear Testing Correlated and Applied to Compaction Control in Colorado," H. R .B. Bulletin No 360, 1962.

49. Worona, V. and Gunderman, W. G., "Field Evaluation of Nuclear Gages Used in Compaction Control of Embankments.," H. R. B. Record No 66, 1965.

50. Weaver, R. J. and Rebull, P. M., "Determination of Embankment Density by the Seismic Method," New York State Department of Transportation, 1966.

51. Barkan, D. D., *Dynamics of Bases and Foundations*, McGraw Hill, 1960.

52. Ferry, J. D., *Viscoelastic Properties of Polymers*, John Wiley and Sons, 1961.

53. McMaster, R. C. et al., "Development of Sonic and Ultrasonic Power Devices for Applications in Highway Engineering," First Annual Report, Engineering Experiment Station, Report No 221.1, Ohio State University, 1963.

54. Antrim, J. D., Brown, F. B., Busching, H. W., Chisman, J. A., Moore, J. H., Rostron, J. P. and Schwartz, A. E., "Rapid Test Methods for Field Control of Highway Construction," H. R. B. Report No 103, Washington, D. C., 1970.

55. Woods, K. B. and Litehiser, R. R., "Soil Mechanics Applied to Highway Engineering in Ohio," Ohio University Experiment Station, Bulletin No 99, Ohio State University, 1938.

56. Hilf, J. M., "Un Método rápido de Control en la Construcción de Terraplenes con Suelos Cohesivos," Proceedings of the Meeting of the A. S. T. M. D-18 Committee and the Mexican Society of Soil Mechanics, México, D. F., 1957.

57. Gioiosa, T. E., "A Rapid Method for Determining the In Place Density of a Pavement Base Course by a Drop Hammer Penetrometer," Thesis presented to the Department of Civil Engineering Clemson University, U. S. A., 1965.

58. Juárez-Badillo, E. and Rico, A., *Mecánica de Suelos. Vol. I; Fundamentos de la Mecánica de Suelos*, Limusa Ed., México, D. F., 1972, Chap. 13.

59. Tyner, H. L., "An Evaluation of Conventional and Nuclear Methods for Determining in Situ Density and Moisture Content of Soils," Thesis presented to the Department of Civil Engineering, Clemso University, U. S. A., 1966.

60. Blystone, J. R., Pelzner, A. and Steffens, G. P., "Moisture Content Determination by the Calcium Carbide Gas Pressure Method," *Public Roads*, Vol. 3 No 8, 1961.

61. Miller — Warden Associates, "Effects of Different Methods of Stockpiling Aggregates," NCHRP Project 10-3 (Interim Report), 1964.

Subject Index